List of the Elements with Their Atomic Symbols and Atomic Weights

W9-BEI-585

Name	Symbol	Atomic Number	Atomic Weight	Name	Symbol	Atomic Number	Atomic Weight
Actinium	Ac	89	227.028	Mendelevium	Md	101	(258)
Aluminum	Al	13	26.9815	Mercury	Hg	80	200.59
Americium	Am	95	(243)	Molybdenum	Mo	42	95.94
Antimony	Sb	51	121.76	Neodymium	Nd	60	144.24
Argon	Ar	18	39.948	Neon	Ne	10	20.1797
Arsenic	As	33	74.9216	Neptunium	Np	93	237.048
Astatine	At	85	(210)	Nickel	Ni	28	58.693
Barium	Ba	56	137.327	Niobium	Nb	41	92.9064
Berkelium	Bk	97	(247)	Nitrogen	N	7	14.0067
Beryllium	Be	4	9.01218	Nobelium	No	102	(259)
Bismuth	Bi	83	208.980	Osmium	Os	76	190.23
Bohrium	Bh	107	(264)	Oxygen	O	8	15.9994
Boron	B	5	10.811	Palladium	Pd	46	106.42
Bromine	Br	35	79.904	Phosphorus	P	15	30.9738
Cadmium	Cd	48	112.411	Platinum	Pt	78	195.08
Calcium	Ca	20	40.078	Plutonium	Pu	94	(244)
Californium	Cf	98	(251)	Polonium	Po	84	(209)
Carbon	C	6	12.011	Potassium	K	19	39.0983
Cerium	Ce	58	140.115	Praseodymium	Pr	59	140.908
Cesium	Cs	55	132.905	Promethium	Pm	61	(145)
Chlorine	Cl	17	35.4527	Protactinium	Pa	91	231.036
Chromium	Cr	24	51.9961	Radium	Ra	88	226.025
Cobalt	Co	27	58.9332	Radon	Rn	86	(222)
Copper	Cu	29	63.546	Rhenium	Re	75	186.207
Curium	Cm	96	(247)	Rhodium	Rh	45	102.906
Darmstadtium	Ds	110	(271)	Roentgenium	Rg	111	(272)
Dubnium	Db	105	(262)	Rubidium	Rb	37	85.4678
Dysprosium	Dy	66	162.50	Ruthenium	Ru	44	101.07
Einsteinium	Es	99	(252)	Rutherfordium	Rf	104	(261)
Erbium	Er	68	167.26	Samarium	Sm	62	150.36
Europium	Eu	63	151.965	Scandium	Sc	21	44.9559
Fermium	Fm	100	(257)	Seaborgium	Sg	106	(266)
Fluorine	F	9	18.9984	Selenium	Se	34	78.96
Francium	Fr	87	(223)	Silicon	Si	14	28.0855
Gadolinium	Gd	64	157.25	Silver	Ag	47	107.868
Gallium	Ga	31	69.723	Sodium	Na	11	22.9898
Germanium	Ge	32	72.61	Strontium	Sr	38	87.62
Gold	Au	79	196.967	Sulfur	S	16	32.066
Hafnium	Hf	72	178.49	Tantalum	Ta	73	180.948
Hassium	Hs	108	(269)	Technetium	Tc	43	(98)
Helium	He	2	4.00260	Tellurium	Te	52	127.60
Holmium	Ho	67	164.930	Terbium	Tb	65	158.925
Hydrogen	H	1	1.00794	Thallium	Tl	81	204.383
Indium	In	49	114.818	Thorium	Th	90	232.038
Iodine	I	53	126.904	Thulium	Tm	69	168.934
Iridium	Ir	77	192.22	Tin	Sn	50	118.710
Iron	Fe	26	55.847	Titanium	Ti	22	47.88
Krypton	Kr	36	83.80	Tungsten	W	74	183.84
Lanthanum	La	57	138.906	Uranium	U	92	238.029
Lawrencium	Lr	103	(260)	Vanadium	V	23	50.9415
Lead	Pb	82	207.2	Xenon	Xe	54	131.29
Lithium	Li	3	6.941	Ytterbium	Yb	70	173.04
Lutetium	Lu	71	174.967	Yttrium	Y	39	88.9059
Magnesium	Mg	12	24.3050	Zinc	Zn	30	65.39
Manganese	Mn	25	54.9381	Zirconium	Zr	40	91.224
Meitnerium	Mt	109	(268)				

Improve Your Understanding!

REGISTER NOW for

Mastering CHEMISTRY™

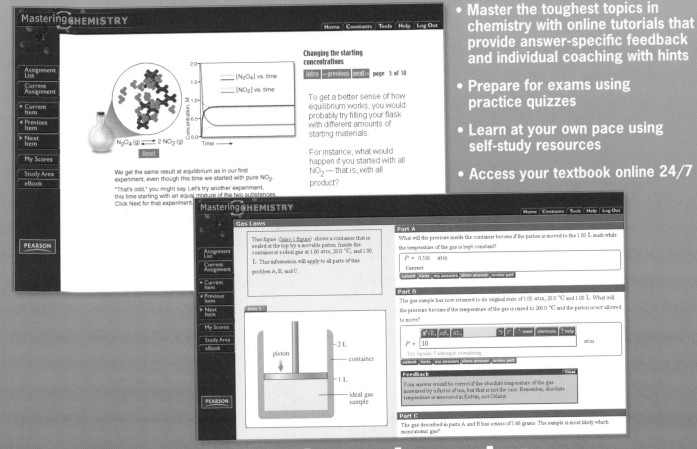

- **Master the toughest topics in chemistry with online tutorials that provide answer-specific feedback and individual coaching with hints**

- **Prepare for exams using practice quizzes**

- **Learn at your own pace using self-study resources**

- **Access your textbook online 24/7**

www.masteringchemistry.com

STUDENTS

To Register Using the Student Access Kit

Your textbook may have been packaged with a **MasteringChemistry** Student Access Kit. This kit contains your access code to this valuable website.

1. Go to www.masteringchemistry.com.
2. Click "New Students" under Register.
3. Select "Yes, I have an access code."
4. Follow the on-screen instructions to create your Login Name and Password.

To Purchase Access Online

If your textbook was not packaged with a **MasteringChemistry** Student Access Kit, you can purchase access online using a major credit card or PayPal.

1. Go to www.masteringchemistry.com.
2. Click "New Students" under Register.
3. Select "No, I need to purchase access online now."
4. Identify your book cover.
5. Follow the on-screen instructions to create your Login Name and Password.

INSTRUCTORS

To Request Access Online

1. Go to www.masteringchemistry.com.
2. Click "New Instructors" under Register.
3. Follow the instructions to request an instructor access code (if you don't have one yet). Then, register with this code to create your Login Name and Password.

Please contact your sales representative for more information.

TECHNICAL SUPPORT

For registration queries:
http://247pearsoned.custhelp.com

For all other queries:
www.masteringchemistry.com/support

If you receive a Course ID from your instructor: Enter this ID either when you first log in or after clicking "Join Course."

Fundamentals
of General, Organic,
and Biological Chemistry

SIXTH EDITION

John McMurry
Cornell University

Mary Castellion
Norwalk, Connecticut

David S. Ballantine
Northern Illinois University

Carl A. Hoeger
University of California, San Diego

Virginia E. Peterson
University of Missouri, Columbia

Prentice Hall

New York Boston San Francisco
London Toronto Sydney Tokyo Singapore Madrid
Mexico City Munich Paris Cape Town Hong Kong Montreal

Library of Congress Cataloging-in-Publication Data

Fundamentals of general, organic, and biological chemistry/John McMurry... [et al.]. —6th ed.
 p. cm.
 Rev. ed. of: Fundamentals of general, organic, and biological chemistry/John McMurry,
Mary E. Castellion, David S. Ballantine. 5th ed. ©2007.
 Includes index.
 ISBN 0-13-605450-1
 1. Chemistry—Textbooks. I. McMurry, John.
 QD31.3.M355 2010
 540—dc22

2008054670

Editor in Chief, Science: Nicole Folchetti
Acquisitions Editor: Dawn Giovanniello
Assistant Editor: Laurie Varites
Editor in Chief, Development: Ray Mullaney
Development Editor: Irene Nunes
Editorial Assistant: Lisa Tarabokjia
Marketing Manager: Elizabeth Averbeck
Marketing Assistant: Keri Parcells
Managing Editor, Chemistry and Geosciences:
 Gina M. Cheselka
Project Manager: Wendy Perez

Senior Operations Supervisor: Alan Fischer
Composition/Full Service: Macmillan Publishing
 Solutions
Production Editor, Full Service: Robert Walters
Art Editor: Connie Long
Art Studio: Precision Graphics
Art Director, Interior and Cover: Suzanne Behnke
Designer, Interior and Cover: Michael Fruhbeis
Photo Researcher: Eric Schrader
Cover Photo: David Fleetham/Alamy
Other image credits appear in the backmatter

© 2010, 2007, 2003, 1999, 1996, 1992 Pearson Education, Inc.
Pearson Prentice Hall
Pearson Education, Inc.
Upper Saddle River, NJ 07458

All rights reserved. No part of this book may be reproduced, in any form or by any means, without permission in writing from the publisher.

Pearson Prentice Hall™ is a trademark of Pearson Education, Inc.

Printed in the United States of America.

10 9 8 7 6 5 4 3 2

Prentice Hall
is an imprint of

www.pearsonhighered.com

ISBN-10: 0-13-605450-1
ISBN-13: 978-0-13-605450-4

About the Authors

John McMurry, educated at Harvard and Columbia, has taught approximately 17,000 students in general and organic chemistry over a 30-year period. A Professor of Chemistry at Cornell University since 1980, Dr. McMurry previously spent 13 years on the faculty at the University of California at Santa Cruz. He has received numerous awards, including the Alfred P. Sloan Fellowship (1969–71), the National Institute of Health Career Development Award (1975–80), the Alexander von Humboldt Senior Scientist Award (1986–87), and the Max Planck Research Award (1991).

David S. Ballantine received his B.S. in Chemistry in 1977 from the College of William and Mary in Williamsburg, VA, and his Ph.D. in Chemistry in 1983 from the University of Maryland at College Park. After several years as a researcher at the Naval Research Labs in Washington, DC, he joined the faculty in the Department of Chemistry and Biochemistry of Northern Illinois University, where he has been a professor for the past twenty years. He was awarded the Excellence in Undergraduate Teaching Award in 1998 and was recently named the departmental Director of Undergraduate Studies. In addition, he is the faculty advisor to the NIU Chemistry Club, an American Chemical Society Student Affiliate program.

Carl A. Hoeger received his B.S. in Chemistry from San Diego State University and his Ph.D. in Organic Chemistry from the University of Wisconsin, Madison in 1983. After a postdoctoral stint at the University of California, Riverside, he joined the Peptide Biology Laboratory at the Salk Institute in 1985 where he ran the NIH Peptide Facility while doing basic research in the development of peptide agonists and antagonists. During this time he also taught general, organic, and biochemistry at San Diego City College, Palomar College, and Miramar College. He joined the teaching faculty at University of Califiornia, San Diego in 1998. Dr. Hoeger has been teaching chemistry to undergraduates for over 20 years, where he continues to explore the use of technology in the classroom. In 2004 he won the Paul and Barbara Saltman Distinguished Teaching Award from UCSD. He is currently the General Chemistry coordinator at UCSD, where he is also responsible for the training and guidance of over 100 teaching assistants in the Chemistry and Biochemistry departments.

Virginia E. Peterson received her B.S. in Chemistry in 1967 from the University of Washington in Seattle, and her Ph.D. in Biochemistry in 1980 from the University of Maryland at College Park. Between her undergraduate and graduate years she worked in lipid, diabetes, and heart disease research at Stanford University. Following her Ph.D. she took a position in the Biochemistry Department at the University of Missouri in Columbia and is now an Associate Professor. Currently she is the Director of Undergraduate Advising for the department and teaches both senior capstone classes and biochemistry classes for nonscience majors. Awards include both the college level and the university-wide Excellence in Teaching Award and, in 2006, the University's Outstanding Advisor Award and the State of Missouri Outstanding University Advisor Award. Dr. Peterson believes in public service and in 2003 received the Silver Beaver Award for service from the Boy Scouts of America.

Brief Contents

Contents

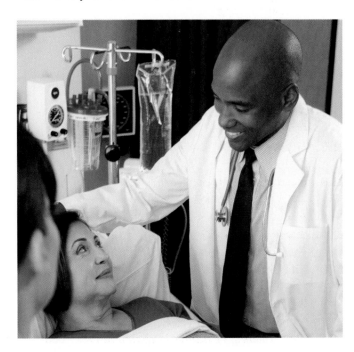

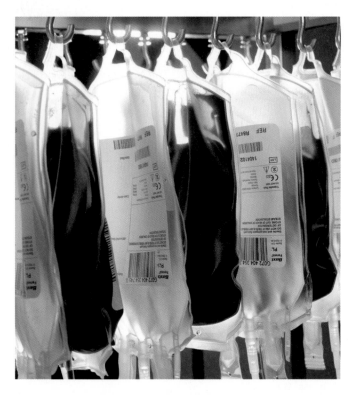

Applications

Preface

This textbook is primarily designed to provide students in the allied health sciences with an appropriate background in chemistry and biochemistry. But it also provides a general context for many of the chemical concepts so that students in other disciplines will gain a better appreciation of the importance of chemistry in everyday life. The coverage in this sixth edition includes sufficient breadth and depth to ensure adequate context and to provide students with opportunities to expand their knowledge.

To teach chemistry all the way from "What is an atom?" to "How do we get energy from glucose?" is a challenge. Throughout our general chemistry and organic chemistry coverage, the focus is on concepts fundamental to the chemistry of living things and everyday life. In our biochemistry coverage we strive to meet the further challenge of providing a context for the application of those concepts in biological systems. Our goal is to provide enough detail for thorough understanding while avoiding so much detail that students are overwhelmed. Many practical and relevant examples are included to illustrate the concepts and enhance student learning.

The material covered is ample for a two-term introduction to general, organic, and biological chemistry. While the general and early organic chapters contain concepts that are fundamental to understanding the material in biochemistry, the later chapters can be covered individually and in an order that can be adjusted to meet the needs of the students and the duration of the course.

The writing style is clear and concise and punctuated with practical and familiar examples from students' personal experience. Art work, diagrams, and molecular models are used extensively to provide graphical illustration of concepts to enhance student understanding. Since the true test of knowledge is the ability to apply that knowledge appropriately, we include numerous worked examples that incorporate consistent problem-solving strategies.

Regardless of their career paths, all students will be citizens in an increasingly technological society. When they recognize the principles of chemistry at work not just in their careers but in their daily lives, they are prepared to make informed decisions on scientific issues based on a firm understanding of the underlying concepts.

Organization

GENERAL CHEMISTRY: CHAPTERS 1–11 The introduction to elements, atoms, the periodic table, and the quantitative nature of chemistry (Chapters 1–3) is followed by chapters that individually highlight the nature of ionic and molecular compounds (Chapters 4 and 5). The next two chapters discuss chemical reactions and their stoichiometry, energies, rates, and equilibria (Chapters 6 and 7). Topics relevant to the chemistry of life follow: Gases, Liquids, and Solids (Chapter 8); Solutions (Chapter 9); and Acids and Bases (Chapter 10). Nuclear Chemistry (Chapter 11) closes the general chemistry sequence.

ORGANIC CHEMISTRY: CHAPTERS 12–17 These chapters concisely focus on what students must know in order to understand biochemistry. The introduction to hydrocarbons (Chapters 12 and 13) includes the basics of nomenclature, which is thereafter kept to a minimum. Discussion of functional groups with single bonds to oxygen, sulfur, or a halogen (Chapter 14) is followed by a short chapter on amines, which are so important to the chemistry of living things and drugs (Chapter 15). After introducing aldehydes and ketones (Chapter 16), the chemistry of carboxylic acids and their derivatives (including amides) is covered (Chapter 17), with a focus on similarities among the derivatives.

BIOLOGICAL CHEMISTRY: CHAPTERS 18–29 Rather than proceed through the complexities of protein, carbohydrate, lipid, and nucleic acid structure before getting to the roles of these compounds in the body, structure and function are integrated in this text. Protein structure (Chapter 18) is followed by enzyme and coenzyme chemistry (Chapter 19). After that we cover the function of hormones and neurotransmitters, and the action of drugs (Chapter 20). With enzymes introduced, the central pathways and themes of biochemical energy production can be described (Chapter 21). If the time you have available to cover biochemistry is limited, stop with Chapter 21 and your students will have an excellent preparation in the essentials of metabolism. The following chapters cover carbohydrate chemistry (Chapters 22 and 23), then lipid chemistry (Chapters 24 and 25). Next we discuss nucleic acids and protein synthesis (Chapter 26) and genomics (Chapter 27). The last two chapters cover protein and amino acid metabolism (Chapter 28) and provide an overview of the chemistry of body fluids (Chapter 29).

Changes to This Edition

COVERAGE OF GENERAL CHEMISTRY

Once again, there is a major emphasis in this edition on problem-solving strategies. This is reflected in expanded solutions in the Worked Example problems and the addition of more Key Concept Problems that focus on conceptual understanding. The most significant change in the Worked Example problems is the addition of a Ballpark Estimate at the beginning of many problems. The Ballpark Estimate provides an opportunity for students to evaluate the relationships involved in the problem and allows them to use an intuitive approach to arrive at a first approximation of the final answer. The ability to think through a problem before attempting a mathematical solution is a skill that will be particularly useful on exams, or when solving "real world" problems.

Other specific changes to chapters are provided below:

Chapter 1

- The Scientific Method is introduced in the text and reinforced in Applications presented in the chapter.

Chapter 3

- Discussion of the critical experiments of Thomson, Millikan, and Rutherford are included in the Application "Are Atoms Real" to provide historical perspective on the development of our understanding of atomic structure.
- Electron dot structures are introduced in Chapter 3 to emphasize the importance of the valence shell electronic configurations with respect to chemical behavior of the elements.

Chapter 4

- Electron dot structures are used to reinforce the role of valence shell electronic configurations in explaining periodic behavior and the formation of ions.

Chapter 5

- The two methods for drawing Lewis dot structures (the "general" method and the streamlined method for molecules containing C, N, O, X, and H) are discussed back-to-back to highlight the underlying principle of the octet rule common to both methods.

Chapter 6

- The concept of limiting reagents is incorporated in Section 6.7 in the discussion of reaction stoichiometry and percent yields.

Chapter 7

- The discussion of free energy and entropy in Section 7.4 has been revised to help students develop a more intuitive understanding of the role of entropy in spontaneous processes.
- Section 7.8 includes more discussion of how the equilibrium constant is calculated and what it tells us about the extent of reaction.

Chapter 8

- Sections 8.3–8.10 include more emphasis on use of the kinetic molecular theory to understand the behavior of gases described by the gas laws.
- Section 8.15 includes more discussion on the energetics of phase changes to help students understand the difference between heat transfer associated with a temperature change and heat transfer associated with the phase change of a substance.

Chapter 9

- Discussion of equivalents in Section 9.10 has been revised to emphasize the relationship between ionic charge and equivalents of ionic compounds.
- Discussion of osmotic pressure (Section 9.12) now includes the osmotic pressure equation and emphasizes the similarity with the ideal gas law.

Chapter 10

- Both the algebraic and logarithmic forms of K_w are presented in Section 10.8 to give students another approach to solving pH problems.
- The discussion of buffer systems now introduces the Henderson-Hasselbalch equation. This relationship makes it easier to identify the factors that affect the pH of a buffer system and is particularly useful in biochemical applications in later chapters.
- Discussion of common acid-base reactions has been moved back in the chapter to provide a more logical segue into titrations in Section 10.15.

Chapter 11

- Treatment of half-life in Section 11.5 now includes a generic equation to allow students to determine the fraction of isotope remaining after an integral or non-integral number of half-lives, which is more consistent with "real world" applications.
- The Applications in this chapter have been expanded to include discussion of new technologies such as Boron Neutron-Capture Therapy (BNCT), or to clear up misconceptions about current methods such as MRI.

COVERAGE OF ORGANIC CHEMISTRY

A major emphasis in this edition was placed on making the fundamental reactions organic molecules undergo much clearer to the reader, with particular vision toward those reactions encountered again in biochemical transformations. Also new to this edition is the expanded use and evaluation of line-angle structure for organic molecules, which are so important when discussing biomolecules. Most of the Applications have been updated to reflect current understanding and research.

Other specific changes to chapters are provided below:

Chapter 12

- This chapter has been significantly rewritten to provide the student with a stronger foundation for the organic chemistry chapters that follow.
- A clearer description of what a functional group is, as well as a more systematic approach to drawing alkane isomers have been made.

- The topic of how to draw and interpret line structures for organic molecules has been added, along with worked examples of such.
- The discussion of conformations has been expanded.

Chapter 13

- A more general discussion of cis and trans isomers has been added.
- The discussion of organic reaction types, particularly rearrangement reactions, have been simplified.

Chapter 14

- The topic of oxidation in organic molecules has been clarified.

Chapter 15

- The role of NO in human biology has been updated to reflect current research.

Chapter 16

- A more detailed discussion of what is meant by toxic or poisonous has been added.

Chapter 17

- A discussion of ibuprofen has been added.

COVERAGE OF BIOLOGICAL CHEMISTRY

New topics, such as the use of anabolic steroids in sports, have been added to many of these chapters to highlight the relevance of biochemistry in modern society. In this text, nutrition is not treated as a separate subject but is integrated with the discussion of each type of biomolecule.

Chapter 18

- The discussion of sickle cell anemia has been expanded and the role of an amino acid substitution on hemoglobin structure clarified.
- The Application *Prions—Proteins That Cause Disease* has been updated to reflect current research.

Chapter 19

- Incorporated the information about lead poisoning into the discussion of enzyme inhibition.

Chapter 20

- The discussion of anabolic steroids has been updated.
- The discussion of drugs and their interaction with the neurotransmitter acetylcholine has been expanded.

Chapter 21

- The discussion of ATP energy production has been revised.

Chapter 22

- An explanation of the chair conformation of glucose has been included to enhance understanding of the shape of cyclic sugars.
- The Application *Chirality and Drugs* has been updated.
- The Application *Cell Surface Carbohydrates and Blood Type* has been revised.

Chapter 23

- The explanation of substrate level phosphorylation has been expanded for clarity.
- The emerging medical condition referred to as Metabolic Syndrome has been added to the text discussion of diabetes.
- The Application *Diagnosis and Monitoring of Diabetes* has been updated to include metabolic syndrome.
- Section 23.11 now contains an expanded discussion of gluconeogenesis.
- The discussion of polysaccharides has been updated.

Chapter 24

- The description of the cell membrane has been expanded.
- A discussion of some inhibitors of Cox 1 and Cox 2 enzymes, important in inflammation, has been added.

Chapter 25

- The discussion of triacylglycerol synthesis has been expanded.
- The discussion of ketone body formation has been expanded.
- A thorough explanation of the biosynthesis of fatty acids has been added.

Chapter 26

- The Application *Viruses and AIDS* has been updated.
- Information about the 1918 influenza pandemic was included in the Application *"Bird Flu": The Next Epidemic?*

Chapter 27

- A discussion of the problems associated with using recombinant DNA for commercial protein manufacture has been added.
- In Section 27.5, new bioethical issues are pointed out to reflect modern concerns.
- The discussion of recombinant DNA and polymerase chain reactions has been moved to this chapter from Chapter 26.

Focus on Learning

WORKED EXAMPLES   Most Worked Examples, both quantitative and not quantitative, include an Analysis section that precedes the Solution. The Analysis lays out the approach to solving a problem of the given type. When appropriate, a "Ballpark Estimate" gives students an overview of the relationships needed to solve the problem, and provides an intuitive approach to arrive at a rough estimate of the answer. The Solution presents the worked-out example using the strategy laid out in the Analysis and, in many cases, includes expanded discussion to enhance student understanding. The use of the two-column format introduced in the fifth edition for quantitative problems has been applied to more Worked Examples throughout the text. Following the Solution there is a Ballpark Check that compares the calculated answer to the Ballpark Estimate, when appropriate, and verifies that the answer makes chemical and physical sense.

KEY CONCEPT PROBLEMS are integrated throughout the chapters to focus attention on the use of essential concepts, as do the ***Understanding Key Concepts*** problems at the end of each chapter. Understanding Key Concepts problems are designed to test students' mastery of the core principles developed in the chapter. Students thus have an opportunity to ask "Did I get it?" before they proceed. Most of these Key Concept Problems use graphics or molecular-level art to illustrate the core principles and will be particularly useful to visual learners.

PROBLEMS   The problems within the chapters, for which brief answers are given in an appendix, cover every skill and topic to be understood. One or more problems, many of which are *new* to this edition, follow each Worked Example and others stand alone at the ends of sections.

MORE COLOR-KEYED, LABELED EQUATIONS   It is entirely too easy to skip looking at a chemical equation while reading. We have extensively used color to call attention to the aspects of chemical equations and structures under discussion, a continuing feature of this book that has been judged very helpful.

MOLECULAR MODELS   Additional computer-generated molecular models have been introduced, including the use of *electrostatic-potential maps for molecular models*.

KEY WORDS   Every key term is boldfaced on its first use, fully defined in the margin adjacent to that use, and listed at the end of the chapter. These are the terms students must understand to get on with the subject at hand. Definitions of all Key Words are collected in the Glossary.

END-OF-CHAPTER SUMMARIES   Here, the answers to the questions posed at the beginning of the chapter provide a summary of what is covered in that chapter. Where appropriate, the types of chemical reactions in a chapter are also summarized.

Focus on Relevancy

Chemistry is often considered to be a difficult and tedious subject. But when students make a connection between a concept in class and an application in their daily lives the chemistry comes alive, and they get excited about the subject. The applications in this book strive to capture student interest and emphasize the relevance of the scientific concepts. The use of relevant applications makes the concepts more accessible and increases understanding.

- **Applications** are both integrated into the discussions in the text and set off from the text in Application boxes. Each boxed application provides sufficient information for reasonable understanding and, in many cases, extends the concepts discussed in the text in new ways. The boxes end with a cross-reference to end-of-chapter problems that can be assigned by the instructor. Some well-received Applications from previous editions that have been retained include *Breathing and Oxygen Transport, Buffers in the Body, Prions, Protein Analysis by Electrophoresis, The Biochemistry of Running,* and *DNA Fingerprinting.*
- **New Applications in this edition** include *Aspirin—A Case Study, Temperature-Sensitive Materials, Anemia—A Limiting Reagent Problem, GERD: Too Much Acid or Not Enough,* and *It's a Ribozyme!*

FOCUS ON MAKING CONNECTIONS AMONG GENERAL, ORGANIC, AND BIOLOGICAL CHEMISTRY   This can be a difficult course to teach. Much of what students are interested in lies in the last part of the course, but the material they need to understand the biochemistry is found in the first two-thirds. It is easy to lose sight of the connections among general, organic, and biological chemistry so we use a feature, **Concepts to Review**, to call attention to these connections. From Chapter 4 on, the Concepts to Review section at the beginning of the chapter lists topics covered in earlier chapters that form the basis for what is discussed in the current chapter.

We have also retained the successful concept link icons and Looking Ahead notes.

- **Concept link icons** ⊂⊃ are used extensively to indicate places where previously covered material is relevant to the discussion at hand. These links provide for cross-references and also serve to highlight important chemical themes as they are revisited.

- **Looking Ahead notes** call attention to connections between just-covered material and discussions in forthcoming chapters. These notes are designed to illustrate to the students why what they are learning will be useful in what lies ahead.

Making It Easier to Teach: Supplements for Instructors

MasteringChemistry™ **(www.masteringchemistry.com)** MasteringChemistry is the first adaptive-learning online homework system. It provides selected end-of-chapter problems from the text, as well as hundreds of tutorials with automatic grading, immediate answer-specific feedback, and simpler questions on request. Based on extensive research of precise concepts students struggle with, MasteringChemistry uniquely responds to your immediate needs, thereby optimizing your study time.

Instructor Resource Manual (0-32-161241-8) Features lecture outlines with presentation suggestions, teaching tips, suggested in-class demonstrations, and topics for classroom discussion.

Test Item File (0-32-161514-X) Updated to reflect the revisions in this text and contains questions in a bank of more than 2,000 multiple-choice questions.

Transparency Pack (0-32-161513-1) More than 225 full-color transparencies chosen from the text put principles into visual perspective and save you time while you are preparing for your lectures.

Instructor Resource Center on CD/DVD (0-32-161242-6) This CD/DVD provides an intergrated collection of resources designed to help you make efficient and effective use of your time. This CD/DVD features most art from the text, including figures and tables in PDF format for high-resolution printing, as well as four pre-built PowerPoint™ presentations. The first presentation contains the images/figures/tables embedded within the PowerPoint slides, while the second includes a complete modifiable lecture outline. The final two presentations contain worked "in chapter" sample exercises and questions to be used with Classroom Response Systems. This CD/DVD also contains movies and animations, as well as the TestGen version of the Test Item File, which allows you to create and tailor exams to your needs.

BlackBoard® **and WebCT**®—Practice and assessment materials are available upon request in these course management platforms.

Making It Easier to Learn: Supplements for Students

Study Guide and Full Solutions Manual (0-32-161238-8) and **Study Guide and Selected Solutions Manual (0-32-161239-6)**, both by Susan McMurry. The selected version provides solutions only to those problems that have a short answer in the text's Selected Answer Appendix (problems numbered in blue in the text). Both versions explain in detail how the answers to the in-text and end-of-chapter problems are obtained. They also contain chapter summaries, study hints, and self-tests for each chapter.

For the Laboratory

Exploring Chemistry: Laboratory Experiments in General, Organic and Biological Chemistry, 2nd Edition (0-13-047714-1) by Julie R. Peller of Indiana University. Written specifically to accompany Fundamentals of General, Organic and Biological Chemistry, this manual contains 34 fresh and accessible experiments specifically for GOB students.

Catalyst: The Prentice Hall Custom Laboratory Program for Chemistry. This program allows you to custom-build a chemistry lab manual that matches your content needs and course organization. You can either write your own labs using the Lab Authoring Kit tool, or select from the hundreds of labs available at www. prenhall.com/catalyst. This program also allows you to add your own course notes, syllabi, or other materials.

Acknowledgments

From conception to completion, the development of a modern textbook requires both a focused attention on the goals and the coordinated efforts of a diverse team. We have been most fortunate to have had the services of many talented and dedicated individuals whose efforts have contributed greatly to the overall quality of this text.

First and foremost, we are grateful to Kent Porter Hamann who, as senior editor of this text through many past revisions, provided exemplary leadership and encouragement to the team in the early stages of this project. Very special appreciation goes to Ray Mullaney, editor in chief of book development, who mentored the new team members and managed to coordinate the many and varied details. Irene Nunes, our developmental editor, worked closely with the authors to ensure accuracy and consistency. We also are grateful for the services of Wendy Perez, project manager; Laurie Varites, assistant editor; Lia Tarabokjia, and Jill Traut and Robert Walters, production project managers. Finally, special thanks also to Susan McMurry and Margaret Trombley, whose efforts on the Solutions Manuals and MasteringChemistry tutorial software, respectively, have added value to the overall package.

Finally, many instructors and students who have used the fifth edition have provided valuable insights and feedback and improved the accuracy of the current edition. We gratefully acknowledge the following reviewers for their contributions to the sixth edition:

Sheikh Ahmed, *West Virgina University*
Stanley Bajue, *CUNY-Medgar Evers College*
Daniel Bender, *Sacramento City College*
Dianne A. Bennett, *Sacramento City College*
Alfredo Castro, *Felician College*
Gezahegn Chaka, *Louisiana State University, Alexandria*
Michael Columbia, *Indiana University-Purdue University, Fort Wayne*
Rajeev B. Dabke, *Columbus State University*
Danae R. Quirk-Dorr, *Minnesota State University, Mankato*
Pamela S. Doyle, *Essex County College*
Marie E. Dunstan, *York College of Pennsylvania*

Karen L. Ericson, *Indiana University-Purdue University, Fort Wayne*
Charles P. Gibson, *University of Wisconsin, Oshkosh*
Clifford Gottlieb, *Shasta College*
Mildred V. Hall, *Clark State Community College*
Meg Hausman, *University of Southern Maine*
Ronald Hirko, *South Dakota State University*
L. Jaye Hopkins, *Spokane Community College*
Margaret Isbell, *Sacramento City College*
James T. Johnson, *Sinclair Community College*
Margaret G. Kimble, *Indiana University-Purdue University Fort Wayne*

Grace Lasker, *Lake Washington Technical College*
Ashley Mahoney, *Bethel University*
Matthew G. Marmorino, *Indiana University, South Bend*
Diann Marten, *South Central College, Mankato*
Barbara D. Mowery, *York College of Pennsylvania*
Tracey Arnold Murray, *Capital University*
Andrew M. Napper, *Shawnee State University*
Lisa Nichols, *Butte Community College*
Glenn S. Nomura, *Georgia Perimeter College*

Douglas E. Raynie, *South Dakota State University*
Paul D. Root, *Henry Ford Community College*
Victor V. Ryzhov, *Northern Illinois University*
Karen Sanchez, *Florida Community College, Jacksonville-South*
Mir Shamsuddin, *Loyola University, Chicago*
Jeanne A. Stuckey, *University of Michigan*
John Sullivan, *Highland Community College*
Deborah E. Swain, *North Carolina Central University*
Susan T. Thomas, *University of Texas, San Antonio*
Yakov Woldman, *Valdosta State University*

The authors are committed to maintaining the highest quality and accuracy and look forward to comments from students and instructors regarding any aspect of this text and supporting materials. Questions or comments should be directed to the lead co-author.

David S. Ballantine
dballant@niu.edu

Concise, Accessible, and Unique

QUANTITATIVE AND CONCEPTUAL

Worked Examples
These examples have been modified to emphasize both problem-solving strategies and conceptual understanding.

Analysis
Most Worked Examples include an Analysis section that precedes the solution. The Analysis lays out the approach to solving a problem of the given type.

NEW! Ballpark Estimates
Ballpark Estimates help students arrive at a rough estimate of the final answer based on an intuitive approach to the problem.

Solution
The Solution shows students how to apply the appropriate problem-solving strategy and guides them through the steps to follow in obtaining the answer.

Ballpark Check
Many of the Worked Examples culminate with a Ballpark Check that helps students quickly check whether the answer they have calculated in numerical Worked Examples is reasonable.

WORKED EXAMPLE | **8.4** Using Boyle's Law: Finding Volume at a Given Pressure

In a typical automobile engine, the fuel/air mixture in a cylinder is compressed from 1.0 atm to 9.5 atm. If the uncompressed volume of the cylinder is 750 mL, what is the volume when fully compressed?

ANALYSIS This is a Boyle's law problem because the volume and pressure in the cylinder change but the amount of gas and the temperature remain constant. According to Boyle's law, the pressure of the gas times its volume is constant:

$$P_1V_1 = P_2V_2$$

Knowing three of the four variables in this equation, we can solve for the unknown.

◄ A cut-away diagram of internal combustion engine shows movement of pistons during expansion and compression cycles.

BALLPARK ESTIMATE Since the pressure *increases* approximately tenfold (from 1.0 atm to 9.5 atm), the volume must *decrease* to approximately one-tenth, from 750 mL to about 75 mL.

SOLUTION

STEP 1: **Identify known information.** Of the four variables in Boyle's law, we know P_1, V_1, and P_2.

$P_1 = 1.0$ atm
$V_1 = 750$ mL
$P_2 = 9.5$ atm

STEP 2: **Identify answer and units.**

$V_2 = ??$ mL

STEP 3: **Identify equation.** In this case, we simply substitute the known variables into Boyle's law and rearrange to isolate the unknown.

$$P_1V_1 = P_2V_2 \implies V_2 = \frac{P_1V_1}{P_2}$$

STEP 4: **Solve. Substitute the known information into the equation.** Make sure units cancel so that the answer is given in the units of the unknown variable.

$$V_2 = \frac{P_1V_1}{P_2} = \frac{(1.0 \text{ atm})(750 \text{ mL})}{(9.5 \text{ atm})} = 79 \text{ mL}$$

BALLPARK CHECK Our estimate was 75 mL.

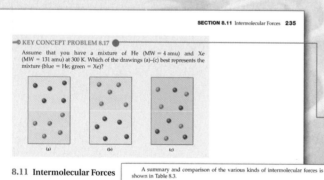

SECTION 8.11 Intermolecular Forces **235**

→ KEY CONCEPT PROBLEM 8.17

Assume that you have a mixture of He (MW = 4 amu) and Xe (MW = 131 amu) at 300 K. Which of the drawings (a)–(c) best represents the mixture (blue = He; green = Xe)?

(a)   (b)   (c)

8.11 Intermolecular Forces

What determines whether a substance is a gas... perature? Why does rubbing alcohol evapor... Why do molecular compounds have lower m... To answer these and a great many other such... nature of **intermolecular forces**—the forces tha... than within an individual molecule.

In gases, the intermolecular forces are neg... pendently of one another. In liquids and solid... strong enough to hold the molecules in close c... the intermolecular forces in a substance, the m... ecules, and the higher the melting and boiling...

There are three major types of intermolecu... sion, and *hydrogen bonding*. We will discuss ea...

Dipole–Dipole Forces

Recall from Sections 5.8 and 5.9 that many m... and may therefore have a net molecular polari... the positive and negative ends of different m... by what is called a **dipole–dipole force**. (Figure...

Dipole–dipole forces are weak, with stren... pared to the 70–100 kcal/mol typically found for the strength of a covalent bond. Nevertheless, the effects of dipole–dipole forces are important, as can be seen by looking at the difference in boiling points between polar and nonpolar molecules. Butane, for instance, is a nonpolar molecule with a molecular weight of 58 amu and a boiling point of −0.5 °C, whereas acetone has the same molecular weight yet boils 57 °C higher because it is polar. (Recall from Section 5.8 how molecular polarities can be visualized using electrostatic potential maps. (⬭, p. 000)

A summary and comparison of the various kinds of intermolecular forces is shown in Table 8.3.

TABLE 8.3 A Comparison of Intermolecular Forces

FORCE	STRENGTH	CHARACTERISTICS
Dipole–dipole	Weak (1 kcal/mol)	Occurs between polar molecules
London dispersion	Weak (0.5–2.5 kcal/mol)	Occurs between all molecules; strength depends on size
Hydrogen bond	Moderate (2–10 kcal/mol)	Occurs between molecules with O—H, N—H, and F—H bonds

⬭ Looking Ahead

Dipole–dipole forces, London dispersion forces, and hydrogen bonds are traditionally called "intermolecular forces" because of their influence on the properties of molecular compounds. But these same forces can also operate between different parts of a very large molecule. In this context, they are often referred to as "noncovalent interactions." In later chapters, we will see how noncovalent interactions determine the shapes of biologically important molecules such as proteins and nucleic acids.

polar molecules of similar size.

Dipole–dipole force The attractive force between positive and negative ends of polar molecules.

(a)   (b)

Key Concept Problems
These problems are integrated within the chapter, appearing at the end of a Worked Example or after the discussion of an important concept. These problems immediately focus students' attention on essential concepts and help them to test their understanding.

Looking Ahead
Looking Ahead Notes provide students with a preview of how the material being presented connects to the discussion in forthcoming chapters.

Concept Links
These links indicate where concepts in the text build on material from earlier chapters. This chain link icon provides a quick visual reminder that new material being discussed relates to a concept introduced previously.

This best-selling text showcases the hallmark strengths of the McMurry author team—a concise writing style, a unique approach to explaining the quantitative aspects of chemistry, and deep insight into chemical principles. The elements on each page—text, worked examples, figures, molecular structures and models, and the various learning aids described below—were designed to work together to help students develop an understanding to the quantitative and conceptual side of chemistry.

APPLICATION ▶ Greenhouse Gases and Global Warming

The mantle of gases surrounding the earth is far from the uniform mixture you might expect, consisting of layers that vary in composition and properties at different altitudes. The ability of the gases in these layers to absorb radiation is responsible for life on earth as we know it.

The *stratosphere*—the layer extending from about 12 km up to 50 km altitude—contains the ozone layer that is responsible for absorbing harmful UV radiation. The *troposphere* is the layer extending from the surface up to about 12 km altitude. It should not surprise you to learn that the troposphere is the layer most easily disturbed by human activities and that this layer has the greatest impact on the earth's surface conditions. Among those impacts, a process called the *greenhouse effect* is much in the news today.

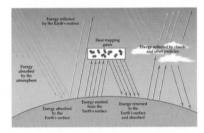

The greenhouse effect refers to the warming that occurs in the troposphere as gases absorb radiant energy. Much of the radiant energy reaching the earth's surface from the sun is reflected back into space, but some is absorbed by atmospheric gases—particularly those referred to as—*greenhouse gases* (GHGs)—water vapor, carbon dioxide, and methane. This absorbed radiation warms the atmosphere and acts to maintain a relatively stable temperature of 15 °C (59 °F) at the earth's surface. Without the greenhouse effect, the average surface temperature would be about −18 °C (0 °F)—a temperature so low that Earth would be frozen and unable to sustain life.

The basis for concern about the greenhouse effect is the fear that human activities over the past century have disturbed the earth's delicate thermal balance. Should increasing amounts of radiation be absorbed, increased atmospheric heating will result, and global temperatures will continue to rise.

Measurements show that the concentration of atmospheric CO_2 has been rising in the last 150 years, largely because of the increased use of fossil fuels, from an estimated 290 parts per million (ppm) in 1850 to current levels approaching 380 ppm. The increase in CO_2 levels correlates with a concurrent increase in average global temperatures, with the 11 years between 1995 and 2006 ranking among the 12-highest since recording of global temperatures began in 1850. The latest report of the Intergovernmental Panel on Climate Change published in November 2007 concluded that "Warming of the climate system is unequivocal, as is now evident from observations of increases in global average air and ocean temperatures, widespread melting of snow and ice and rising global average sea level. . . . Continued GHG emissions at or above current rates would cause further warming and induce many changes in the global climate system during the 21st century that would *very likely* be larger than those observed during the 20th century."

See Additional Problems 8.101 and 8.102 at the end of the chapter.

▲ **Concentrations of atmospheric [CO₂] and average temperatures have increased d[...] 150 years because of increased fo[...] serious changes in earth's climate s[...]**

© Crown copyright 2006 data provided by t[...]

Applications

These boxed Application essays connect the chemical concepts within each chapter to topics that are drawn from everyday life, clinical practice, health and nutrition, ecology, biotechnology, and chemical research.

Molecular Models

These computer-generated molecular models promote students' understanding of chemistry on a molecular level with electrostatic potential maps.

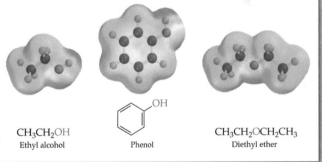

CH_3CH_2OH
Ethyl alcohol

OH
Phenol

$CH_3CH_2OCH_2CH_3$
Diethyl ether

6.5 Mole Relationships and Chemical Equations

In a typical recipe, the amount of ingredients needed are specified using a variety of units: The amount of flour, for example, is usually specified in cups, whereas the amount of salt or vanilla flavoring might be indicated in teaspoons. In chemical reactions, the appropriate unit to specify the relationship between reactants and products is the mole.

The coefficients in a balanced chemical equation tell how many *molecules*, and thus how many *moles*, of each reactant are needed and how many molecules, and thus moles, of each product are formed. You can then use molar mass to calculate reactant and product masses. If, for example, you saw the following balanced equation for the industrial synthesis of ammonia, you would know that 3 mol of H_2 (3 mol × 2.0 g/mol = 6.0 g) are required for reaction with 1 mol of N_2 (28.0 g) to yield 2 mol of NH_3 (2 mol × 17.0 g/mol = 34.0 g).

This number of moles of hydrogen . . . . . . reacts with this number of moles of nitrogen . . . to yield this number of moles of ammonia.

$$3\,H_2 + 1\,N_2 \longrightarrow 2\,NH_3$$

The coefficients can be put in the form of *mole ratios*, which act as conversion factors when setting up factor-label calculations. In the ammonia synthesis, for example, the mole ratio of H_2 to N_2 is 3:1, the mole ratio of H_2 to NH_3 is 3:2, and the mole ratio of N_2 to NH_3 is 1:2:

$$\frac{3\ \text{mol}\ H_2}{1\ \text{mol}\ N_2} \quad \frac{3\ \text{mol}\ H_2}{2\ \text{mol}\ NH_3} \quad \frac{1\ \text{mol}\ N_2}{2\ \text{mol}\ NH_3}$$

Worked Example 6.9 shows how to set up and use mole ratios.

Color and Labeling Equations

Color has been extensively used to call attention to the aspects of chemical equations and structures under discussion. The number of explanatory labels set in balloons that focus on the important details in chemical equations and structures have also been increased.

A Balanced Approach to General, Organic, and Biological Chemistry

UNDERSTANDING CORE PRINCIPLES

Summary: Revisiting the Chapter Goals

The chapter summary revisits the chapter goals that open the chapter. Each of the questions posed at the start of the chapter is answered by a summary of the essential information needed to attain the corresponding goal.

Key Words

All of the chapter's boldface terms are listed in alphabetical order and are cross-referenced to the page where it appears in the text.

PROBLEM 5.25

A white crystalline solid has a melting point of 128 °C. It is soluble in water, but the resulting solution does not conduct electricity. Is the substance ionic or molecular? Explain.

PROBLEM 5.26

Aluminum chloride ($AlCl_3$) has a melting point of 190 °C, whereas aluminum oxide (Al_2O_3) has a melting point of 2070 °C. Explain.

KEY WORDS

Binary compound, *p. 90*

Bond angle, *p. 91*

Bond length, *p. 92*

Condensed structure, *p. 102*

Coordinate covalent bond, *p. 103*

Covalent bond, *p. 105*

Double bond, *p. 109*

Electronegativity, *p. 110*

Lewis structure, *p. 110*

Lone pair, *p. 114*

Molecular compound, *p. 120*

Molecular formula, *p. 120*

Molecule, *p. 122*

Polar covalent bond, *p. 125*

Regular tetrahedron, *p. 126*

Single bond, *p. 130*

Structural formula, *p. 131*

Triple bond, *p. 132*

Valence-shell electron-pair repulsion (VSEPR) model, *p. 132*

SUMMARY: REVISITING THE CHAPTER GOALS

1. **What is a covalent bond?** A *covalent bond* is formed by the sharing of electrons between atoms rather than by the complete transfer of electrons from one atom to another. Atoms that share two electrons are joined by a *single bond* (such as C—C), atoms that share four electrons are joined by a *double bond* (such as C=C), and atoms that share six electrons are joined by a *triple bond* (such as C≡C). The group of atoms held together by covalent bonds is called a *molecule*.

   Electron sharing typically occurs when a singly occupied valence orbital on one atom *overlaps* a singly occupied valence orbital on another atom. The two electrons occupy both overlapping orbitals and belong to both atoms, thereby bonding the atoms together. Alternatively, electron sharing can occur when a filled orbital containing an unshared, *lone pair* of electrons on one atom overlaps a vacant orbital on another atom to form a *coordinate covalent bond*.

2. **How does the octet rule apply to covalent bond formation?** Depending on the number of valence electrons, different atoms form different numbers of covalent bonds. In general, an atom shares enough electrons to reach a noble gas configuration. Hydrogen, for instance, forms one covalent bond because it needs to share one more electron to achieve the helium configuration ($1s^2$). Carbon and other group 4A elements form four covalent bonds because they need to share four more electrons to reach an octet. In the same way, nitrogen and other group 5A elements form three covalent bonds, oxygen and other group 6A elements form two covalent bonds, and halogens (group 7A elements) form one covalent bond.

3. **How are molecular compounds represented?** Formulas such as H_2O, NH_3, and CH_4, which show the numbers and kinds of atoms in a molecule, are called *molecular formulas*. More useful are *Lewis structures*, which show how atoms are connected in molecules. Covalent bonds are indicated as lines between atoms, and valence electron lone pairs are shown as dots. Lewis structures are drawn by counting the total number of valence electrons in a molecule or polyatomic ion and then placing shared pairs (bonding) and lone pairs (nonbonding) so that all electrons are accounted for.

4. **What is the influence of valence-shell electrons on molecular shape?** Molecules have specific shapes that depend on the number of electron charge clouds (bonds and lone pairs) surrounding the various atoms. These shapes can often be predicted using the *valence-shell electron-pair repulsion* (*VSEPR*) model. Atoms with two electron charge clouds adopt linear geometry, atoms with three charge clouds adopt planar triangular geometry, and atoms with four charge clouds adopt tetrahedral geometry.

5. **When are bonds and molecules polar?** Bonds between atoms are *polar covalent* if the bonding electrons are not shared equally between the atoms. The ability of an atom to attract electrons in a covalent bond is the atom's *electronegativity* and is highest for reactive nonmetal elements on the upper right of the periodic table and lowest for metals on the lower left. Comparing electronegativities allows a prediction of whether a given bond is covalent, polar covalent, or ionic. Just as individual bonds can be polar, entire molecules can be polar if electrons are attracted more strongly to one part of the molecule than to another. Molecular polarity is due to the sum of all individual bond polarities and lone-pair contributions in the molecule.

6. **What are the major differences between ionic and molecular compounds?** *Molecular compounds* can be gases, liquids, or low-melting solids. They usually have lower melting points and boiling points than ionic compounds, many are water insoluble, and they do not conduct electricity when melted or dissolved.

Demanding, yet logical, this text sets itself apart by requiring students to master problem-solving strategies and conceptual understanding before moving on to the next concept.

Understanding Key Concepts

These problems at the end of each chapter allow students to test their mastery of the core principles developed in the chapter. Students have an opportunity to ask "Did I get it?" before they proceed.

UNDERSTANDING KEY CONCEPTS

5.27 Which of the drawings shown here is more likely to represent an ionic compound and which a covalent compound?

(a)

(b)

(c)

(d)

5.28 If yellow spheres represent sulfur atoms and red spheres represent oxygen atoms, which of the following drawings depicts a collection of sulfur dioxide molecules?

(a)

(b)

(c)

(d)

5.29 What is the geometry around the central atom in the following molecular models? (There are no "hidden" atoms; all atoms in each model are visible.)

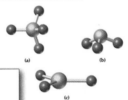

(a)

(b)

(c)

5.31 The ball-and-stick molecular model shown here is a representation of acetaminophen, the active ingredient in such over-the-counter headache remedies as Tylenol. The lines indicate only the connections between atoms, not whether the bonds are single, double, or triple (red = O, gray = C, blue = N, ivory = H).
(a) What is the molecular formula of acetaminophen?
(b) Indicate the positions of the multiple bonds in acetaminophen.
(c) What is the geometry around each carbon and each nitrogen?

Acetaminophen

5.32 The atom-to-atom connections in vitamin C (ascorbic acid) are as shown here. Convert this skeletal drawing to a Lewis electron-dot structure for vitamin C by showing the positions of any multiple bonds and lone pairs of electrons.

Vitamin C

5.33 The ball-and-stick molecular model shown here is a representation of thalidomide, a drug that causes terrible birth defects when taken by expectant mothers but has been approved for treating leprosy. The lines indicate only the connections between atoms, not whether the bonds are single, double, or triple (red = O, gray = C, blue = N, ivory = H).
(a) What is the molecular formula of thalidomide?
(b) Indicate the positions of the multiple bonds in thalidomide.

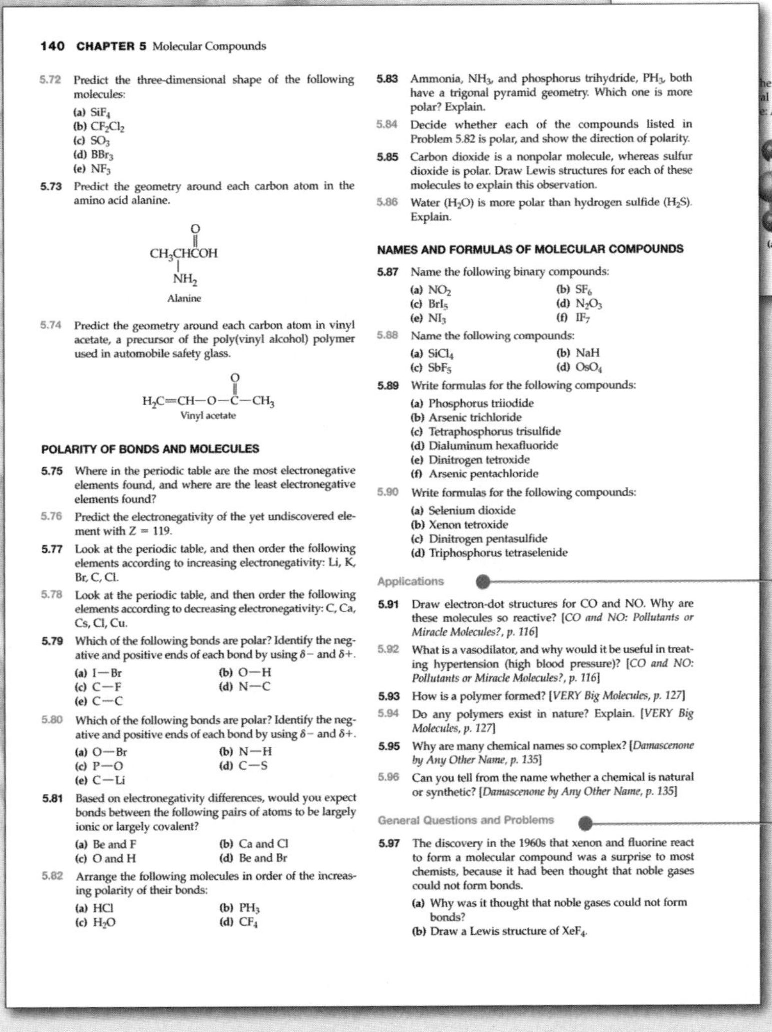

the following molecular models have a tetrahedral atom, and one does not. Which is the odd e: All peripheral atoms are visible.)

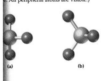

(a)

(b)

---

140 CHAPTER 5 Molecular Compounds

5.72 Predict the three-dimensional shape of the following molecules:
(a) SiF_4
(b) CF_2Cl_2
(c) SO_3
(d) BBr_3
(e) NF_3

5.73 Predict the geometry around each carbon atom in the amino acid alanine.

Alanine

5.74 Predict the geometry around each carbon atom in vinyl acetate, a precursor of the poly(vinyl alcohol) polymer used in automobile safety glass.

Vinyl acetate

POLARITY OF BONDS AND MOLECULES

5.75 Where in the periodic table are the most electronegative elements found, and where are the least electronegative elements found?

5.76 Predict the electronegativity of the yet undiscovered element with Z = 119.

5.77 Look at the periodic table, and then order the following elements according to increasing electronegativity: Li, K, Br, C, Cl.

5.78 Look at the periodic table, and then order the following elements according to decreasing electronegativity: C, Ca, Cs, Cl, Cu.

5.79 Which of the following bonds are polar? Identify the negative and positive ends of each bond by using $\delta-$ and $\delta+$.
(a) I—Br
(b) O—H
(c) C—F
(d) N—C
(e) C—C

5.80 Which of the following bonds are polar? Identify the negative and positive ends of each bond by using $\delta-$ and $\delta+$.
(a) O—Br
(b) N—H
(c) P—O
(d) C—S
(e) C—Li

5.81 Based on electronegativity differences, would you expect bonds between the following pairs of atoms to be largely ionic or largely covalent?
(a) Be and F
(b) Ca and Cl
(c) O and H
(d) Be and Br

5.82 Arrange the following molecules in order of the increasing polarity of their bonds:
(a) HCl
(b) PH_3
(c) H_2O
(d) CF_4

5.83 Ammonia, NH_3, and phosphorus trihydride, PH_3, both have a trigonal pyramid geometry. Which one is more polar? Explain.

5.84 Decide whether each of the compounds listed in Problem 5.82 is polar, and show the direction of polarity.

5.85 Carbon dioxide is a nonpolar molecule, whereas sulfur dioxide is polar. Draw Lewis structures for each of these molecules to explain this observation.

5.86 Water (H_2O) is more polar than hydrogen sulfide (H_2S). Explain.

NAMES AND FORMULAS OF MOLECULAR COMPOUNDS

5.87 Name the following binary compounds:
(a) NO_2
(b) SF_6
(c) BrI_5
(d) N_2O_3
(e) NI_3
(f) IF_7

5.88 Name the following compounds:
(a) $SiCl_4$
(b) NaH
(c) SbF_5
(d) OsO_4

5.89 Write formulas for the following compounds:
(a) Phosphorus triiodide
(b) Arsenic trichloride
(c) Tetraphosphorus trisulfide
(d) Dialuminum hexafluoride
(e) Dinitrogen tetroxide
(f) Arsenic pentachloride

5.90 Write formulas for the following compounds:
(a) Selenium dioxide
(b) Xenon dioxide
(c) Dinitrogen pentasulfide
(d) Triphosphorus tetraselenide

Applications

5.91 Draw electron-dot structures for CO and NO. Why are these molecules so reactive? [*CO and NO: Pollutants or Miracle Molecules?, p. 116*]

5.92 What is a vasodilator, and why would it be useful in treating hypertension (high blood pressure)? [*CO and NO: Pollutants or Miracle Molecules?, p. 116*]

5.93 How is a polymer formed? [*VERY Big Molecules, p. 127*]

5.94 Do any polymers exist in nature? Explain. [*VERY Big Molecules, p. 127*]

5.95 Why are many chemical names so complex? [*Damascenone by Any Other Name, p. 135*]

5.96 Can you tell from the name whether a chemical is natural or synthetic? [*Damascenone by Any Other Name, p. 135*]

General Questions and Problems

5.97 The discovery in the 1960s that xenon and fluorine react to form a molecular compound was a surprise to most chemists, because it had been thought that noble gases could not form bonds.
(a) Why was it thought that noble gases could not form bonds?
(b) Draw a Lewis structure of XeF_4.

Application Problems

Each boxed application essay throughout the text ends with a cross-reference to end-of-chapter problems. These problems help students test their understanding of the material and, more importantly, help students see the connection between chemistry and the world around them.

General Questions and Problems

These problems are cumulative, pulling together topics from various parts of the chapter and previous chapters. These help students synthesize the material just learned while helping them review topics from previous chapters.

Three Steps to Make Learning
Part of the Grade

MasteringChemistry™ emulates the instructor's office-hour environment, coaching students on problem-solving techniques by asking questions that reveal gaps in understanding, giving them the power to answer questions on their own. It tutors them individually— with feedback specific to their errors, offering optional simpler steps.

1. Submit an answer and receive immediate, error-specific feedback.

MasteringChemistry is the only system to provide instantaneous feedback specific to the most-common wrong answers, accumulated over eight years of capturing and researching student errors.

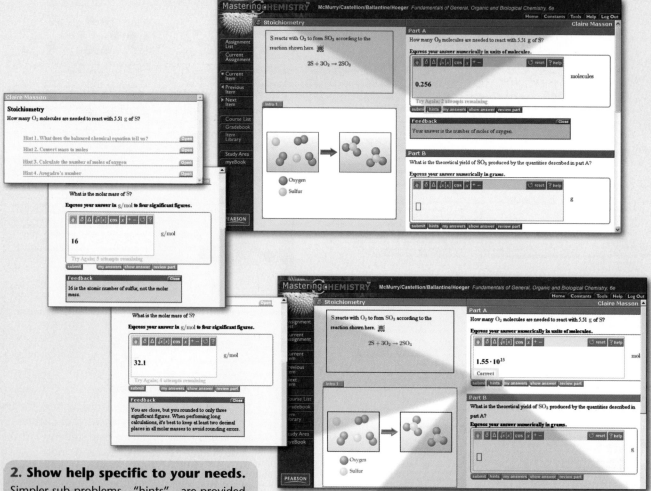

2. Show help specific to your needs.

Simpler sub-problems—"hints"—are provided upon request. From these, students can pick and choose only the help they need. These hints are built on data of key steps and concepts that students are found to struggle with nationally.

3. Partial credit means motivation for method.

MasteringChemistry is uniquely able to provide partial credit for the student's method (based on the simpler subproblems requested and errors made). Credit is at the heart of student motivation and so MasteringChemistry encourages students to focus on their method as well as their final answer.

Mastering CHEMISTRY™

www.masteringchemistry.com

Incorporate Dynamic Homework into Your Course with Automatic Grading

Homework Assignments of Ideal Difficulty and Duration

MasteringChemistry™ is unique in providing instructors with national data on difficulty and duration of every problem and tutorial. This allows you to quickly build homework assignments uniquely tailored to the ability of your students and goals of your course. Alternately, you can customize or choose pre-built weekly assignments to your needs.

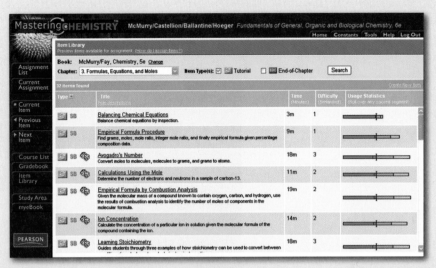

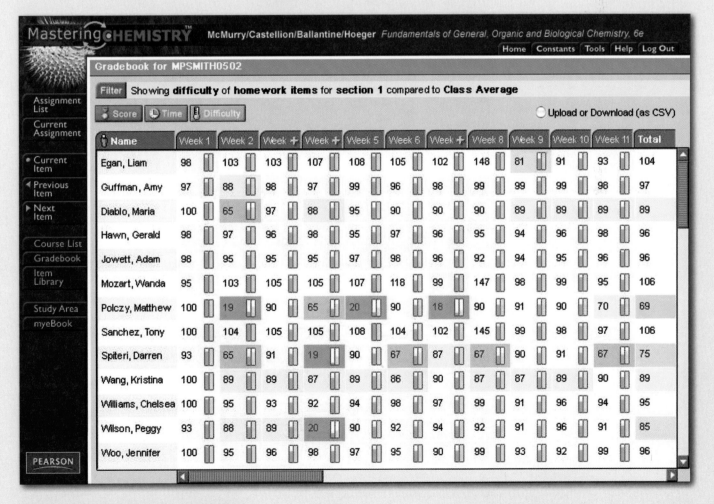

Unmatched Gradebook Capability

By capturing more detailed work of every student than any other homework system, MasteringChemistry provides the most powerful gradebook and diagnostics available. Spot students in trouble at a glance with the color-coded gradebook, effortlessly identify the most difficult problem (and step within that problem) in your last assignment, or critique the detailed work of any one student who needs more help. Compare results on any problem and any step with a previous class, or the national average.

CHAPTER 1

Matter and Life

▲ The variety of exotic colors and aromas featured in this marketplace is due to the chemical and physical properties of chemicals.

CHAPTER GOALS

We will begin in this chapter by looking at the following topics:

1. What is matter?

THE GOAL: Be able to discuss the properties of matter and describe the three states of matter.

2. How is matter classified?

THE GOAL: Be able to distinguish between mixtures and pure substances, and between elements and compounds.

3. What kinds of properties does matter have?

THE GOAL: Be able to distinguish between chemical and physical properties.

4. How are chemical elements represented?

THE GOAL: Be able to name and give the symbols of common elements.

L ook around you. Everything you see, touch, taste, and smell is made of chemicals. Many of these chemicals—those in rocks, trees, and your own body—occur naturally, but others are synthetic. Many of the plastics, fibers, and medicines that are so important a part of modern life do not occur in nature but have been created in the chemical laboratory.

Just as everything you see is made of chemicals, many of the natural changes you see taking place around you are the result of *chemical reactions*—the change of one chemical into another. The crackling fire of a log burning in the fireplace, the color change of a leaf in the fall, and the changes that a human body undergoes as it grows and ages are all the results of chemical reactions. To understand these and all other natural processes, you must have a basic understanding of chemistry.

As you might expect, the chemistry of living organisms is complex and it is not possible to jump right into it. Thus, the general plan of this book is to increase gradually in complexity, beginning in the first eleven chapters with a grounding in the scientific fundamentals that govern all of chemistry. In the next six chapters we look at the nature of the carbon-containing substances, or *organic chemicals*, that compose all living things. In the final twelve chapters we apply what we have learned to biological chemistry.

1.1 Chemistry: The Central Science

Chemistry is often referred to as "the central science" because it is crucial to all other sciences. In fact, as more and more is learned, the historical dividing lines between chemistry, biology, and physics are fading and current research is becoming more interdisciplinary. Figure 1.1 diagrams the relationship of chemistry and biological chemistry to some other fields of scientific study. Whatever the discipline in which you are most interested, the study of chemistry builds the necessary foundation.

Chemistry is the study of matter—its nature, properties, and transformations. **Matter**, in turn, is a catchall word used to describe anything physically real— anything you can see, touch, taste, or smell. In more scientific terms, matter is anything that has mass and volume. As with our knowledge of all the other sciences, our knowledge of chemistry has developed by application of a process called the **scientific method**. Starting with observations of the physical world, we form hypotheses to explain what we have observed. These hypotheses can then be tested by experiments to improve our understanding.

How might we describe different kinds of matter more specifically? Any characteristic that can be used to describe or identify something is called a **property**; size, color, and temperature are all familiar examples. Less familiar properties include *chemical composition*, which describes what matter is made of, and *chemical reactivity*, which describes how matter behaves. Rather than focus on the properties themselves, however, it is often more useful to think about *changes* in properties. Changes are of two types: *physical* and *chemical*. A **physical change** is one that does not alter the chemical makeup of a substance, whereas a **chemical change** is one that *does* alter a substance's chemical makeup. The melting of solid ice to give liquid water, for instance, is a physical change because the water changes only in form but not in chemical makeup. The rusting of an iron bicycle left in the rain, however, is

Chemistry The study of the nature, properties, and transformations of matter.

Matter The physical material that makes up the universe; anything that has mass and occupies space.

Scientific Method The systematic process of observation, hypothesis, and experimentation used to expand and refine a body of knowledge.

Property A characteristic useful for identifying a substance or object.

Physical change A change that does not affect the chemical makeup of a substance or object.

Chemical change A change in the chemical makeup of a substance.

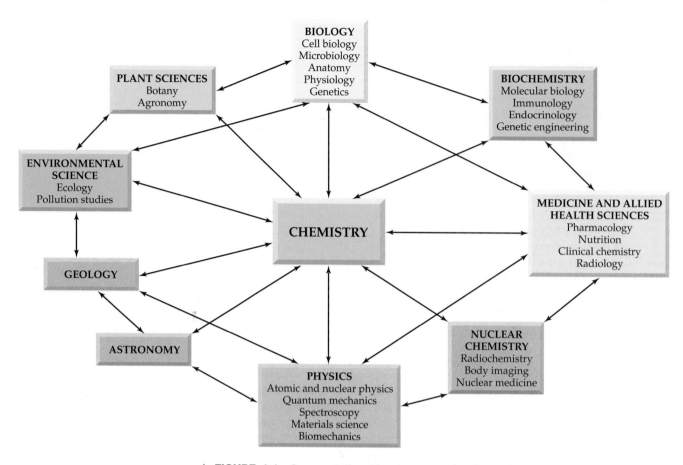

▲ **FIGURE 1.1** **Some relationships between chemistry, the central science, and other scientific and health-related disciplines.**

▲ **FIGURE 1.2** Samples of the pure substances water, sugar, and baking soda.

a chemical change because iron combines with oxygen and moisture from the air to give a new substance, rust.

Figure 1.2 shows several familiar substances—water, table sugar (sucrose), and baking soda (sodium bicarbonate)—and Table 1.1 lists their composition and some

TABLE 1.1   Some Properties of Water, Sugar, and Baking Soda

WATER	SUGAR (SUCROSE)	BAKING SODA (SODIUM BICARBONATE)
Physical properties		
Colorless liquid	White crystals	White powder
Odorless	Odorless	Odorless
Melting point: 0 °C	Begins to decompose at 160 °C, turning black and giving off water.	Decomposes at 270 °C, Giving off water and carbon dioxide.
Boiling point: 100 °C	—	—
Chemical properties		
Composition:*	Composition:*	Composition:*
11.2% hydrogen	6.4% hydrogen	27.4% sodium
88.8% oxygen	42.1% carbon	1.2% hydrogen
	51.5% oxygen	14.3% carbon
		57.1% oxygen
Does not burn.	Burns in air.	Does not burn.

Compositions are given by mass percent.

properties. Note in Table 1.1 that the changes occurring when sugar and baking soda are heated are chemical changes because new substances are produced.

PROBLEM 1.1

Which of the following are made of chemicals? Which might consist of "natural" chemicals and which of "synthetic" chemicals?

(a) Apple juice   **(b)** Laundry bleach

(c) Glass   **(d)** Coffee beans

PROBLEM 1.2

Identify each of the following as a physical change or a chemical change:

(a) A metal surface being ground   **(b)** Fruit ripening

(c) Wood burning   **(d)** A rain puddle evaporating

▲ Burning of potassium in water is an example of a chemical change.

1.2 States of Matter

Matter exists in three forms: solid, liquid, and gas. A **solid** has a definite volume and a definite shape that does not change regardless of the container in which it is placed; for example, a wooden block, marbles, or a cube of ice. A **liquid**, by contrast, has a definite volume but an indefinite shape. The volume of a liquid, such as water, does not change when it is poured into a different container, but its shape does. A **gas** is different still, having neither a definite volume nor a definite shape. A gas expands to fill the volume and take the shape of any container it is placed in, such as the helium in a balloon, or steam formed by boiling water (Figure 1.3).

Solid A substance that has a definite shape and volume.

Liquid A substance that has a definite volume but assumes the shape of its container.

Gas A substance that has neither a definite volume nor a definite shape.

(a) Ice: A solid has a definite volume and a definite shape independent of its container.

(b) Water: A liquid has a definite volume but a variable shape that depends on its container.

(c) Steam: A gas has both variable volume and shape that depend on its container.

◀ **FIGURE 1.3 The three states of matter—solid, liquid, and gas.**

Many substances, such as water, can exist in all three phases, or **states of matter**—the solid state, the liquid state, and the gaseous state—depending on the temperature. The conversion of a substance from one state to another is known as a **change of state** and is a common occurrence. The melting of a solid, the freezing or boiling of a liquid, and the condensing of a gas to a liquid are familiar to everyone.

State of matter The physical state of a substance as a solid, liquid, or gas.

Change of state The conversion of a substance from one state to another—for example, from liquid to gas.

WORKED EXAMPLE 1.1

Formaldehyde is a disinfectant, a preservative, and a raw material for plastics manufacture. Its melting point is −92 °C, and its boiling point is −19.5 °C. Is formaldehyde a gas, a liquid, or a solid at room temperature (25 °C)? (The symbol °C means degrees Celsius.)

ANALYSIS The state of matter of any substance depends on its temperature. How do the melting point and boiling point of formaldehyde compare with room temperature?

SOLUTION
At room temperature (25 °C), the formaldehyde temperature is above the boiling point and so the formaldehyde is a gas.

PROBLEM 1.3

Acetic acid, which gives the sour taste to vinegar, has a melting point of 16.7 °C and a boiling point of 118 °C. Does a bottle of acetic acid contain a solid or a liquid on a chilly morning with the window open and the laboratory at 10 °C?

Pure substance A substance that has a uniform chemical composition throughout.

Mixture A blend of two or more substances, each of which retains its chemical identity.

Element A fundamental substance that cannot be broken down chemically into any simpler substance.

Chemical compound A pure substance that can be broken down into simpler substances by chemical reactions.

Reactant A starting substance that undergoes change during a chemical reaction.

Product A substance formed as the result of a chemical reaction.

Chemical reaction A process in which the identity and composition of one or more substances are changed.

▲ Individually, sugar and water are pure substances. Honey, however, is a mixture composed mostly of sugar and water.

1.3 Classification of Matter

The first question a chemist asks about an unknown substance is whether it is a pure substance or a mixture. Every sample of matter is one or the other. Water and sugar alone are pure substances, but stirring some sugar into a glass of water creates a *mixture*.

What is the difference between a pure substance and a mixture? One difference is that a **pure substance** is uniform in its chemical composition and its properties all the way down to the microscopic level. Every sample of water, sugar, or baking soda, regardless of source, has the composition and properties listed in Table 1.1. A **mixture**, however, can vary in both composition and properties, depending on how it is made (⊂⊃, Section 9.1). The amount of sugar dissolved in a glass of water will determine the sweetness, boiling point, and other properties of the mixture. Note that you often cannot distinguish between a pure substance and a mixture just by looking. The sugar–water mixture *looks* just like pure water but differs on a molecular level.

Another difference between a pure substance and a mixture is that the components of a mixture can be separated without changing their chemical identities. Water can be separated from a sugar–water mixture, for example, by boiling the mixture to drive off the steam and then condensing the steam to recover the pure water. Pure sugar is left behind in the container.

Pure substances are themselves classified into two groups: those that can undergo a chemical breakdown to yield simpler substances and those that cannot. A pure substance that cannot be broken down chemically into simpler substances is called an **element** (⊂⊃, Section 1.5). Examples include hydrogen, oxygen, aluminum, gold, and sulfur. At the time this book was printed, 117 elements had been identified, and all the millions of other substances in the universe are derived from them.

Any pure material that *can* be broken down into simpler substances by a chemical change is called a **chemical compound**. The term *compound* implies "more than one" (think "compound fracture"). A chemical compound, therefore, is formed by combining two or more elements to make a new substance. Water, sugar, baking soda, and millions of other substances are examples. Water, for example, can be chemically changed by passing an electric current through it to produce hydrogen and oxygen. In writing this chemical change, the **reactant** (water) is written on the left, the **products** (hydrogen and oxygen) are written on the right, and an arrow connects the two parts to indicate a chemical change, or **chemical reaction**. The conditions necessary to bring about the reaction are written above and below the arrow.

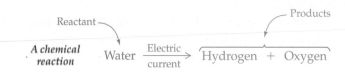

The classification of matter into mixtures, pure compounds, and elements is summarized in Figure 1.4.

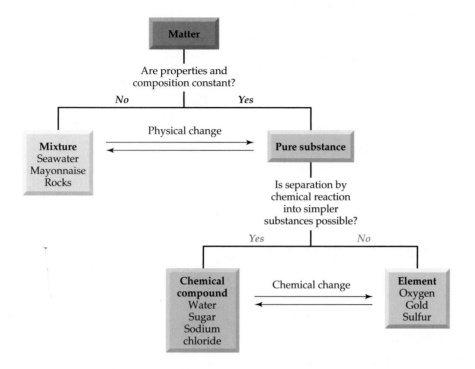

◀ **FIGURE 1.4** **A scheme for the classification of matter.**

WORKED EXAMPLE **1.2** Classifying Matter

Classify each of the following as a mixture or a pure substance:

   (a) Vanilla ice cream   **(b)** Sugar

ANALYSIS Refer to the definitions of pure substances and mixtures. Is the substance composed of more than one kind of matter?

SOLUTION

   (a) Vanilla ice cream is composed of more than one substance—cream, sugar, and vanilla flavoring. This is a mixture.

   (b) Sugar is composed of only one kind of matter—pure sugar. This is a pure substance.

PROBLEM 1.4

Classify each of the following as a mixture or a pure substance:

(a) Concrete   **(b)** The helium in a balloon

(c) A lead weight   **(d)** Wood

PROBLEM 1.5

Classify each of the following as a physical change or a chemical change:

(a) Dissolving sugar in water.

(b) Producing carbon dioxide gas and solid lime by heating limestone

(c) Frying an egg

▲ **The reactants.** The flat dish contains pieces of nickel, an element that is a typical, lustrous metal. The bottle contains hydrochloric acid, a solution of the chemical compound hydrogen chloride in water. These reactants are about to be combined in the test tube.

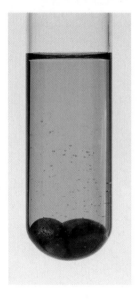

▲ **The reaction** As the chemical reaction occurs, the colorless solution turns green when water-insoluble nickel metal slowly changes into the water-soluble chemical compound nickel chloride. Gas bubbles of the element hydrogen are produced and rise slowly through the green solution.

▶ **The product** Hydrogen gas can be collected as it bubbles from the solution. Removal of water from the solution leaves behind the other product, a solid green chemical compound known as nickel chloride.

━○ KEY CONCEPT PROBLEM 1.6

Identify the process illustrated in the figure below as a chemical change or a physical change. Explain.

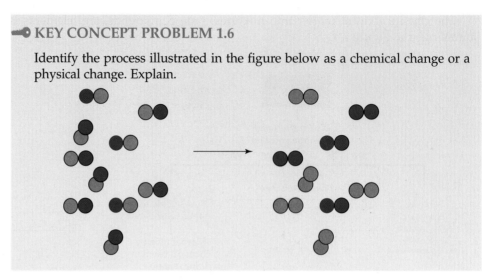

1.4 An Example of a Chemical Reaction

If we take a quick look at an example of a chemical reaction, we can reinforce some of the ideas discussed in the previous section. The element *nickel* is a hard, shiny metal, and the compound *hydrogen chloride* is a colorless gas that dissolves in water to give a solution called *hydrochloric acid*. When pieces of nickel are added to hydrochloric acid in a test tube, the nickel is slowly eaten away, the colorless solution turns green, and a gas bubbles out of the test tube. The change in color, the dissolving of the nickel, and the appearance of gas bubbles are indications that a chemical reaction is taking place.

Overall, the reaction of nickel with hydrochloric acid can be either written in words or represented in a shorthand notation using symbols, as shown below in brackets. We will explain the meaning of these symbols in the next section.

Reactants ⌐   Products ⌐

$$\overbrace{\text{Nickel} + \text{Hydrochloric acid}} \longrightarrow \overbrace{\text{Nickel (II) chloride} + \text{Hydrogen}}$$

$$[\text{Ni} + 2\,\text{HCl} \longrightarrow \text{NiCl}_2 + \text{H}_2]$$

1.5 Chemical Elements and Symbols

As of the date this book was printed, 117 chemical elements had been identified. Some are certainly familiar to you—oxygen, helium, iron, aluminum, copper, and gold, for example—but many others are probably unfamiliar—rhenium, niobium, thulium, and promethium. Rather than write out the full names of elements, chemists use a shorthand notation in which elements are referred to by one- or two-letter symbols. The names and symbols of some common elements are listed in Table 1.2, and a complete alphabetical list is given inside the front cover of this book.

Note that all two-letter symbols have only their first letter capitalized, whereas the second letter is always lowercase. The symbols of most common elements are the first one or two letters of the elements' commonly used names, such as

APPLICATION ▶ Aspirin—A Case Study

Acetylsalicylic acid, more commonly known as aspirin, is perhaps the first true wonder drug. It is used as an analgesic to reduce fevers and to relieve headaches and body pains. It possesses anticoagulant properties, which in low doses can help prevent heart attacks and minimize the damage caused by strokes. But how was it discovered, and how does it work? The "discovery" of aspirin is a combination of serendipity and the scientific method.

The origins of aspirin can be traced back to the ancient Greek physician Hippocrates in 400 B.C., who prescribed the bark and leaves of the willow tree to relieve pain and fever. In 1828 scientists isolated from willow bark a bitter-tasting yellow extract, called salicin, that was identified as the active ingredient responsible for the observed medical effects. Salicin could be easily converted to salicylic acid (SA), which by the late 1800s was being mass-produced and marketed. SA had an unpleasant taste, however, and often caused stomach irritation and indigestion.

The discovery of acetylsalicylic acid (ASA) has often been attributed to Felix Hoffman, a chemist working for the Bayer pharmaceutical labs, but the first synthesis of ASA was reported by a French chemist, Charles Gerhardt, in 1853. Nevertheless, Hoffman obtained a patent for ASA in 1900, and Bayer marketed the new drug, now called aspirin, in water-soluble tablets.

How does aspirin work? In 1971 the British pharmacologist John Vane discovered that aspirin suppresses the body's production of prostaglandins (⬤, Section 24.9), which are

▲ **Hippocrates.** The ancient Greek physician, prescribed a precursor of aspirin found in willowbark to relieve pain.

responsible for the pain and swelling that accompany inflammation. This discovery of this mechanism led to the development of new analgesic drugs.

Research continues to explore aspirin's potential for preventing colon cancer, cancer of the esophagus, and other diseases.

See Additional Problem 1.52 at the end of the chapter.

H (hydrogen) and Al (aluminum). Pay special attention, however, to the elements grouped in the last column to the right in Table 1.2. The symbols for these elements are derived from their original Latin names, such as Na for sodium, once known as *natrium*. The only way to learn these symbols is to memorize them; fortunately they are few in number.

Only 90 of the elements occur naturally; the remaining elements have been produced artificially by chemists and physicists. Each element has its own distinctive

TABLE 1.2 Names and Symbols for Some Common Elements

ELEMENTS WITH SYMBOLS BASED ON MODERN NAMES						ELEMENTS WITH SYMBOLS BASED ON LATIN NAMES	
Al	Aluminum	**Co**	Cobalt	**N**	Nitrogen	**Cu**	Copper (*cuprum*)
Ar	Argon	**F**	Fluorine	**O**	Oxygen	**Au**	Gold (*aurum*)
Ba	Barium	**He**	Helium	**P**	Phosphorus	**Fe**	Iron (*ferrum*)
Bi	Bismuth	**H**	Hydrogen	**Pt**	Platinum	**Pb**	Lead (*plumbum*)
B	Boron	**I**	Iodine	**Rn**	Radon	**Hg**	Mercury (*hydrargyrum*)
Br	Bromine	**Li**	Lithium	**Si**	Silicon	**K**	Potassium (*kalium*)
Ca	Calcium	**Mg**	Magnesium	**S**	Sulfur	**Ag**	Silver (*argentum*)
C	Carbon	**Mn**	Manganese	**Ti**	Titanium	**Na**	Sodium (*natrium*)
Cl	Chlorine	**Ni**	Nickel	**Zn**	Zinc	**Sn**	Tin (*stannum*)

TABLE 1.3 Elemental Composition of the Earth's Crust and the Human Body*

EARTH'S CRUST		HUMAN BODY	
Oxygen	46.1%	Oxygen	61%
Silicon	28.2%	Carbon	23%
Aluminum	8.2%	Hydrogen	10%
Iron	5.6%	Nitrogen	2.6%
Calcium	4.1%	Calcium	1.4%
Sodium	2.4%	Phosphorus	1.1%
Magnesium	2.3%	Sulfur	0.20%
Potassium	2.1%	Potassium	0.20%
Titanium	0.57%	Sodium	0.14%
Hydrogen	0.14%	Chlorine	0.12%

Mass percent values are given.

Chemical formula A notation for a chemical compound using element symbols and subscripts to show how many atoms of each element are present.

properties, and just about all of the first 95 elements have been put to use in some way that takes advantage of those properties. As indicated in Table 1.3, which shows the approximate elemental composition of the earth's crust and the human body, the naturally occurring elements are not equally abundant. Oxygen and silicon together account for nearly 75% of the mass in the earth's crust; oxygen, carbon, and hydrogen account for nearly all the mass of a human body.

Just as elements combine to form chemical compounds, symbols are combined to produce **chemical formulas**, which show by subscripts how many *atoms* (the smallest fundamental units) of different elements are in a given chemical compound. For example, the formula H_2O represents water, which contains 2 hydrogen atoms combined with 1 oxygen atom. Similarly, the formula CH_4 represents methane (natural gas), and the formula $C_{12}H_{22}O_{11}$ represents table sugar (sucrose). When no subscript is given for an element, as for carbon in the formula CH_4, a subscript of "1" is understood.

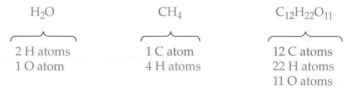

H_2O CH_4 $C_{12}H_{22}O_{11}$

2 H atoms	1 C atom	12 C atoms
1 O atom	4 H atoms	22 H atoms
		11 O atoms

PROBLEM 1.7

Look at the alphabetical list inside the front cover of this book and find the symbols for the following elements:

(a) Sodium, a major component in table salt

(b) Titanium, used in artificial hip and knee replacement joints

(c) Strontium, used to produce brilliant red colors in fireworks

(d) Yttrium, used in many lasers

(e) Fluorine, added to municipal water supplies to strengthen tooth enamel

(f) Hydrogen, used in fuel cells

PROBLEM 1.8

What elements do the following symbols represent?

(a) U   **(b)** Ca   **(c)** Nd

(d) K   **(e)** W   **(f)** Sn

PROBLEM 1.9

Identify the elements represented in each of the following chemical formulas, and tell the number of atoms of each element:

(a) NH_3 (ammonia)

(b) $NaHCO_3$ (sodium bicarbonate)

(c) C_8H_{18} (octane, a component of gasoline)

(d) $C_6H_8O_6$ (vitamin C)

1.6 Elements and the Periodic Table

The symbols of the known elements are normally presented in a tabular format called the **periodic table**, as shown in Figure 1.5 and the inside front cover of this book. We will have much more to say about the periodic table and how it is numbered in Sections 3.4–3.8 but will note for now that it is the most important organizing principle in chemistry. An enormous amount of information is embedded in the periodic table, information that gives chemists the ability to explain known chemical behavior of elements and to predict new behavior. The elements can be roughly divided into three groups: *metals*, *nonmetals*, and *metalloids* (sometimes called *semimetals*).

Periodic table A tabular format listing all known elements.

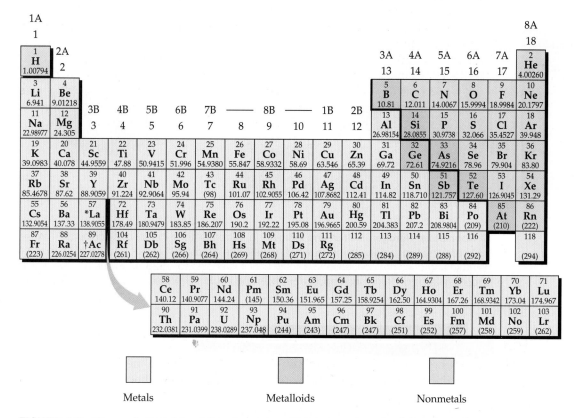

Metals   Metalloids   Nonmetals

FIGURE 1.5 The periodic table of the elements. Metals appear on the left, nonmetals on the right, and metalloids in a zigzag band between metals and nonmetals. The numbering system is explained in Section 3.4.

Ninety of the currently known elements are metals—aluminum, gold, copper, and zinc, for example. **Metals** are solid at room temperature (except for mercury), usually have a lustrous appearance when freshly cut, are good conductors of heat and electricity, and are malleable rather than brittle. That is, a metal can be pounded into a different shape rather than shattering when struck. Note that metals occur on the left side of the periodic table.

Metal A malleable element with a lustrous appearance that is a good conductor of heat and electricity.

(a)

(b)

(c)

▲ **Metals: Gold, zinc, and copper.** (a) Known for its beauty, gold is very unreactive and is used primarily in jewelry and in electronic components. (b) Zinc, an essential trace element in our diets, has industrial uses ranging from the manufacture of brass, to roofing materials, to batteries. (c) Copper is widely used in electrical wiring, in water pipes, and in coins.

Nonmetal An element that is a poor conductor of heat and electricity.

Seventeen elements are **nonmetals**. All are poor conductors of heat and electricity, eleven are gases at room temperature, five are brittle solids, and one is a liquid. Oxygen and nitrogen, for example, are gases present in air; sulfur is a solid found in large underground deposits. Bromine is the only liquid nonmetal. Note that nonmetals occur on the right side of the periodic table.

(a)

(b)

(c)

▲ **Nonmetals: Nitrogen, sulfur, and iodine.** (a) Nitrogen, (b) sulfur, and (c) iodine are essential to all living things. Pure nitrogen, which constitutes almost 80% of air, is a gas at room temperature and does not condense to a liquid until it is cooled to −328 °F. Sulfur, a yellow solid, is found in large underground deposits in Texas and Louisiana. Iodine is a dark violet, crystalline solid first isolated from seaweed.

Metalloid An element whose properties are intermediate between those of a metal and a nonmetal.

Only seven elements are **metalloids**, so-named because their properties are intermediate between those of metals and nonmetals. Boron, silicon, and arsenic are examples. Pure silicon has a lustrous or shiny surface, like a metal. But it is brittle, like a nonmetal, and its electrical conductivity lies between that of metals and nonmetals. Note that metalloids occur in a zigzag band between metals on the left and nonmetals on the right side of the periodic table.

Those elements essential for human life are listed in Table 1.4. In addition to the well-known elements carbon, hydrogen, oxygen, and nitrogen, less familiar elements such as molybdenum and selenium are also important.

⊂⊃ Looking Ahead

The elements listed in Table 1.4 are not present in our bodies in their free forms, of course. Instead, they are combined into many thousands of different chemical compounds. We will talk about some compounds formed by metals in Chapter 4 and compounds formed by nonmetals in Chapter 5.

TABLE 1.4　Elements Essential for Human Life*

ELEMENT	SYMBOL	FUNCTION
Carbon	C	These four elements are present in all living organisms
Hydrogen	H	
Oxygen	O	
Nitrogen	N	
Arsenic	As	May affect cell growth and heart function
Boron	B	Aids in the use of Ca, P, and Mg
Calcium*	Ca	Necessary for growth of teeth and bones
Chlorine*	Cl	Necessary for maintaining salt balance in body fluids
Chromium	Cr	Aids in carbohydrate metabolism
Cobalt	Co	Component of vitamin B_{12}
Copper	Cu	Necessary to maintain blood chemistry
Fluorine	F	Aids in the development of teeth and bones
Iodine	I	Necessary for thyroid function
Iron	Fe	Necessary for oxygen-carrying ability of blood
Magnesium*	Mg	Necessary for bones, teeth, and muscle and nerve action
Manganese	Mn	Necessary for carbohydrate metabolism and bone formation
Molybdenum	Mo	Component of enzymes necessary for metabolism
Nickel	Ni	Aids in the use of Fe and Cu
Phosphorus*	P	Necessary for growth of bones and teeth; present in DNA/RNA
Potassium*	K	Component of body fluids; necessary for nerve action
Selenium	Se	Aids vitamin E action and fat metabolism
Silicon	Si	Helps form connective tissue and bone
Sodium*	Na	Component of body fluids; necessary for nerve and muscle action
Sulfur*	S	Component of proteins; necessary for blood clotting
Zinc	Zn	Necessary for growth, healing, and overall health

C, H, O, and N are present in all foods. Other elements listed vary in their distribution in different foods. Those marked with an asterisk are macronutrients, essential in the diet at more than 100 mg/day; the rest, other than C, H, O, and N, are micronutrients, essential at 15 mg or less per day.

(a)

(b)

◀ **Metalloids: Boron and silicon.**
(a) Boron is a strong, hard metalloid used in making the composite materials found in military aircraft. (b) Silicon is well known for its use in making computer chips.

PROBLEM 1.10

Look at the periodic table inside the front cover, locate the following elements and identify them as metals or nonmetals:

(a) Cr, chromium   **(b)** K, potassium

(c) S, sulfur   **(d)** Rn, radon

PROBLEM 1.11

The seven metalloids are boron (B), silicon (Si), germanium (Ge), arsenic (As), antimony (Sb), tellurium (Te), and astatine (At). Locate them in the periodic table, and tell where they appear with respect to metals and nonmetals.

APPLICATION ▶   Mercury and Mercury Poisoning

Mercury, the only metallic element that is liquid at room temperature, has fascinated people for millennia. Egyptian kings were buried in their pyramids along with containers of mercury, alchemists during the Middle Ages used mercury to dissolve gold, and Spanish galleons carried loads of mercury to the New World in the 1600s for use in gold and silver mining. Even its symbol, Hg, from the Latin *hydrargyrum,* meaning "liquid silver," hints at mercury's uniqueness.

Much of the recent interest in mercury has concerned its toxicity, but there are some surprises. For example, the mercury compound Hg_2Cl_2 (called *calomel*) is nontoxic and has a long history of medical use as a laxative, yet it is also used as a fungicide and rat poison. Dental amalgam, a solid alloy of approximately 50% elemental mercury, 35% silver, 13% tin, 1% copper, and trace amounts of zinc, has been used by dentists for many years to fill tooth cavities. Yet exposure to elemental mercury *vapor* for long periods leads to mood swings, headaches, tremors, and loss of hair and teeth. The widespread use of mercuric nitrate, a mercury compound to make the felt used in hats, exposed many hatters of the eighteenth and nineteenth centuries to toxic levels of mercury. The eccentric behavior displayed by hatters suffering from mercury poisoning led to the phrase "mad as a hatter"—the Mad Hatter from Lewis Carroll's *Alice in Wonderland* is a parody of this stereotype.

Why is mercury toxic in some forms but not in others? It turns out that the toxicity of mercury and its compounds is related to solubility. Only soluble mercury compounds are toxic, because they can be transported through the bloodstream to all parts of the body where they react with different enzymes and interfere with various biological processes. Elemental mercury and insoluble mercury compounds become toxic only when converted into soluble compounds, reactions that are extremely slow in the body. Calomel, for example, is an insoluble mercury compound that passes

▲ The Mad Hatter's erratic behavior was a parody of the symptoms commonly associated with mercury poisoning.

through the body long before it is converted into any soluble compounds. Mercury alloys were considered safe for dental use because mercury does not evaporate readily from the alloys and it neither reacts with nor dissolves in saliva. Mercury vapor, however, remains in the lungs when breathed, until it is slowly converted into soluble compounds.

Of particular concern with regard to mercury toxicity is the environmental danger posed by pollution from both natural and industrial sources. Microorganisms present in lakes and streams are able to convert many mercury-containing wastes into a soluble and highly toxic compound called *methylmercury*. Methylmercury is concentrated to high levels in fish, particularly in shark and swordfish, which are then hazardous when eaten. Although the commercial fishing catch is now monitored carefully, more than 50 deaths from eating contaminated fish were recorded in Minimata, Japan, during the 1950s before the cause of the problem was realized.

See Additional Problem 1.53 at the end of the chapter.

SUMMARY: REVISITING THE CHAPTER GOALS

1. **What is matter?** *Matter* is anything that has mass and occupies volume—that is, anything physically real. Matter can be classified by its physical state as *solid*, *liquid*, or *gas*. A solid has a definite volume and shape, a liquid has a definite volume but indefinite shape, and a gas has neither a definite volume nor shape.

2. **How is matter classified?** A substance can be characterized as being either *pure* or a *mixture*. A pure substance is uniform in its composition and properties, but a mixture can vary in both composition and properties, depending on how it was made. Every pure substance is either an *element* or a *chemical compound*. Elements are fundamental substances that cannot be chemically changed into anything simpler. A chemical compound, by contrast, can be broken down by chemical change into simpler substances.

3. **What kinds of properties does matter have?** A *property* is any characteristic that can be used to describe or identify something. A *physical property* can be seen or measured without changing the chemical identity of the substance, (that is, color, melting point). A *chemical property* can only be seen or measured when the substance undergoes a *chemical change*, such as a chemical reaction.

4. **How are chemical elements represented?** Elements are represented by one- or two-letter symbols, such as H for hydrogen, Ca for calcium, Al for aluminum, and so on. Most symbols are the first one or two letters of the element name, but some symbols are derived from Latin names—Na (sodium), for example. All the known elements are commonly organized into a form called the *periodic table*. Most elements are *metals*, 17 are *nonmetals*, and 7 are *metalloids*.

KEY WORDS

Change of state, *p. 5*

Chemical change, *p. 3*

Chemical compound, *p. 6*

Chemical formula, *p. 10*

Chemical reaction, *p. 6*

Chemistry, *p. 3*

Element, *p. 6*

Gas, *p. 5*

Liquid, *p. 5*

Matter, *p. 3*

Metal, *p. 11*

Metalloid, *p. 12*

Mixture, *p. 6*

Nonmetal, *p. 12*

Periodic table, *p. 11*

Physical change, *p. 3*

Pure substance, *p. 6*

Product, *p. 6*

Property, *p. 3*

Reactant, *p. 6*

Scientific Method, *p. 3*

Solid, *p. 5*

State of matter, *p. 5*

UNDERSTANDING KEY CONCEPTS

The problems in this section are intended as a bridge between the Chapter Summary and the Additional Problems that follow. Primarily visual in nature, they are designed to help you test your grasp of the chapter's most important principles before attempting to solve quantitative problems. Answers to all Key Concept problems are at the end of the book following the appendixes.

1.12 Six of the elements at the far right of the periodic table are gases at room temperature. Identify them using the periodic table inside the front cover of this book.

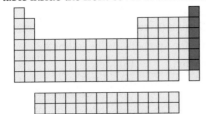

1.13 The so-called "coinage metals" are located near the middle of the periodic table. Identify them using the periodic table inside the front cover of this book.

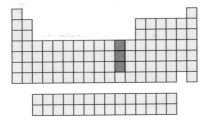

1.14 Identify the three elements indicated on the following periodic table and tell which is a metal, which is a nonmetal, and which is a metalloid.

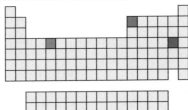

1.15 The radioactive element indicated on the following periodic table is used in smoke detectors. Identify it, and tell whether it is a metal, a nonmetal, or a metalloid.

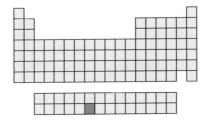

ADDITIONAL PROBLEMS

These exercises are divided into sections by topic. Each section begins with review and conceptual questions, followed by numerical problems of varying levels of difficulty. The problems are presented in pairs, with each even-numbered problem followed by an odd-numbered one requiring similar skills. The final section consists of unpaired General Questions and Problems that draw on various parts of the chapter and, in future chapters, may even require the use of concepts from previous chapters. Answers to all even-numbered problems are given at the end of the book following the appendixes.

CHEMISTRY AND THE PROPERTIES OF MATTER

1.16 What is chemistry?

1.17 Identify the following chemicals as natural or synthetic.

(a) The nylon used in stockings
(b) The substances that give roses their fragrance
(c) The yeast used in bread dough

1.18 Which of the following is a physical change and which is a chemical change?

(a) Boiling water
(b) Decomposing water by passing an electric current through it
(c) Dissolving sugar in water
(d) Exploding of potassium metal when placed in water
(e) Breaking of glass

1.19 Which of the following is a physical change and which is a chemical change?

(a) Steam condensing
(b) Milk souring
(c) Ignition of matches
(d) Breaking of a dinner plate
(e) Nickel sticking to a magnet
(f) Exploding of nitroglycerin

STATES AND CLASSIFICATION OF MATTER

1.20 Name and describe the three states of matter.

1.21 Name two changes of state, and describe what causes each to occur.

1.22 Sulfur dioxide is a compound produced when sulfur burns in air. It has a melting point of $-72.7\ °C$ and a boiling point of $-10\ °C$. In what state does it exist at room temperature $(25\ °C)$? (The symbol °C means degrees Celsius.)

1.23 Menthol, a chemical compound obtained from peppermint or other mint oils, melts at $45\ °C$ and boils at $212\ °C$. In what state is it found at:

(a) Room temperature $(25\ °C)$?　(b) $60\ °C$?　(c) $260\ °C$?

1.24 Classify each of the following as a mixture or a pure substance:

(a) Pea soup　　　　　(b) Seawater
(c) The contents of a propane tank
(d) Urine　　　　　　(e) Lead
(f) A multivitamin tablet

1.25 Classify each of the following as a mixture or a pure substance. If it is a pure substance, classify it as an element or a compound:

(a) Blood　　　　　　(b) Silicon
(c) Dishwashing liquid　(d) Toothpaste
(e) Gold　　　　　　(f) Gaseous ammonia

1.26 Classify each of the following as an element, a compound, or a mixture:

(a) Aluminum foil　(b) Table salt　(c) Water
(d) Air　　　　　　(e) A banana　(f) Notebook paper

1.27 Which of these terms, (i) mixture, (ii) solid, (iii) liquid, (iv) gas, (v) chemical element, (vi) chemical compound, applies to the following substances at room temperature?

(a) Gasoline　　　(b) Iodine　　　(c) Water
(d) Air　　　　　(e) Sodium bicarbonate

1.28 Hydrogen peroxide, often used in solutions to cleanse cuts and scrapes, breaks down to yield water and oxygen:

$$\text{Hydrogen peroxide} \longrightarrow \text{Water} + \text{Oxygen}$$

(a) Identify the reactants and products.
(b) Which of the substances are chemical compounds, and which are elements?

1.29 When sodium metal is placed in water, the following change occurs:

$$\text{Sodium} + \text{Water} \longrightarrow \text{Hydrogen} + \text{Sodium hydroxide}$$

(a) Identify the reactants and products.
(b) Which of the substances are elements, and which are chemical compounds?

ELEMENTS AND THEIR SYMBOLS

1.30 How many elements are presently known? About how many occur naturally?

1.31 Where in the periodic table are the metallic elements found? The nonmetallic elements? The metalloid elements?

1.32 Describe the general properties of metals, nonmetals, and metalloids.

1.33 What is the most abundant element in the earth's crust? In the human body? List the name and symbol for each.

1.34 What are the symbols for the following elements?

(a) Gadolinium (used in color TV screens)
(b) Germanium (used in semiconductors)
(c) Technetium (used in biomedical imaging)
(d) Arsenic (used in pesticides)
(e) Cadmium (used in rechargeable batteries)

1.35 What are the symbols for the following elements?

(a) Tungsten　　(b) Mercury　　(c) Boron
(d) Gold　　　　(e) Silicon　　(f) Argon
(g) Silver　　　(h) Magnesium

1.36 Give the names corresponding to the following symbols:

(a) N　　　　　(b) K　　　　(c) Cl
(d) Ca　　　　(e) P　　　　(f) Mn

1.37 Give the names corresponding to the following symbols:

(a) Te　　　　(b) Re　　　(c) Be
(d) Cr　　　　(e) Pu　　　(f) Mn

1.38 The symbol CO stands for carbon monoxide, a chemical compound, but the symbol Co stands for cobalt, an element. Explain how you can tell them apart.

1.39 Explain why the symbol for sulfur is S, but the symbol for sodium is Na.

1.40 What is wrong with the following statements? Correct them.

(a) The symbol for bromine is BR.
(b) The symbol for manganese is Mg.
(c) The symbol for carbon is Ca.
(d) The symbol for potassium is Po.

1.41 What is wrong with the following statements? Correct them.

(a) Carbon dioxide has the formula CO2.
(b) Carbon dioxide has the formula Co_2.
(c) Table salt, NaCl, is composed of nitrogen and chlorine.

1.42 What is wrong with the following statements? Correct them.

(a) "Fool's gold" is a mixture of iron and sulfur with the formula IrS_2.
(b) "Laughing gas" is a compound of nitrogen and oxygen with the formula NiO.

1.43 What is wrong with the following statements? Correct them.

(a) Soldering compound is an alloy of tin and lead with the formula TiPb.
(b) White gold is an alloy of gold and nickel with the formula GdNI.

1.44 Name the elements combined in the chemical compounds represented by the following formulas:

(a) $MgSO_4$ **(b)** $FeBr_2$ **(c)** CoP
(d) AsH_3 **(e)** $CaCr_2O_7$

1.45 How many atoms of what elements are represented by the following formulas?

(a) Propane (LP gas), C_3H_8
(b) Sulfuric acid, H_2SO_4
(c) Aspirin, $C_9H_8O_4$
(d) Rubbing alcohol, C_3H_8O

1.46 The amino acid glycine has the formula $C_2H_5NO_2$. What elements are present in glycine? What is the total number of atoms represented by the formula?

1.47 Benzyl salicylate, a chemical compound sometimes used as a sunscreen, has the formula $C_{14}H_{12}O_3$. What is the total number of atoms represented by the formula? How many are carbon?

1.48 What is the formula for ibuprofen: 13 carbons, 18 hydrogens, and 2 oxygens.

1.49 What is the formula for penicillin V: 16 carbons, 18 hydrogens, 2 nitrogens, 5 oxygens, and 1 sulfur.

1.50 Which of the following two elements is a metal, and which is a nonmetal?

(a) Osmium, a hard, shiny, very dense solid that conducts electricity
(b) Xenon, a colorless, odorless gas

1.51 Which of the following elements is likely to be a metal and which a metalloid?

(a) Tantalum, a hard, shiny solid that conducts electricity
(b) Germanium, a brittle, gray solid that conducts electricity poorly

Applications

1.52 The active ingredient in aspirin, acetylsalicylic acid (ASA), has the formula $C_9H_8O_4$ and melts at 140 °C.

Identify the elements and how many atoms of each are present in ASA. Is it a solid or a liquid at room temperature? [*Aspirin—A Case Study, p. 9*]

1.53 Why is Hg_2Cl_2 harmless when swallowed, yet elemental mercury is toxic when breathed? [*Mercury and Mercury Poisoning, p. 14*]

General Questions and Problems

1.54 Distinguish between the following:

(a) Physical changes and chemical changes
(b) Melting point and boiling point
(c) Elements, compounds, and mixtures
(d) Chemical symbols and chemical formulas
(e) Reactants and products
(f) Metals and nonmetals

1.55 Are the following statements true or false? If false, explain why.

(a) The combination of sodium and chlorine to produce sodium chloride is a chemical reaction.
(b) The addition of heat to solid sodium chloride until it melts is a chemical reaction.
(c) The formula for a chemical compound that contains lead and oxygen is LiO.
(d) By stirring together salt and pepper we can create a new chemical compound to be used for seasoning food.
(e) Heating sugar to make caramel is a physical change.

1.56 Which of the following are chemical compounds, and which are elements?

(a) H_2O_2 **(b)** Al   **(c)** CO
(d) N_2 **(e)** $NaHCO_3$

1.57 A white solid with a melting point of 730 °C is melted. When electricity is passed through the resultant liquid, a brown gas and a molten metal are produced. Neither the metal nor the gas can be broken down into anything simpler by chemical means. Classify each—the white solid, the molten metal, and the brown gas—as a mixture, a compound, or an element.

1.58 As a clear red liquid sits at room temperature, evaporation occurs until only a red solid remains. Was the original liquid an element, a compound, or a mixture?

1.59 Describe how you could physically separate a mixture of iron filings, table salt, and white sand.

1.60 Small amounts of the following elements in our diets are essential for good health. What is the chemical symbol for each?

(a) Iron   **(b)** Copper   **(c)** Cobalt
(d) Molybdenum   **(e)** Chromium   **(f)** Fluorine
(g) Sulfur

1.61 Obtain a bottle of multivitamins from your local drug store or pharmacy and identify as many of the essential elements from Table 1.4 as you can. How many of them are listed as elements, and how many are included as part of chemical compounds?

1.62 Consider the recently discovered or as yet undiscovered elements with atomic numbers 115, 117, and 119. Is element 115 likely to be a metal or a nonmetal? What about element 117? Element 119?

1.63 How high would the atomic number of a new element need to be to appear under uranium in the periodic table?

CHAPTER 2

Measurements in Chemistry

▲ The success of many endeavors, such as NASA's Deep Impact comet probe, depends on both the accuracy and the precision of measurements.

CONTENTS

CHAPTER GOALS

In this chapter we will deal with the following questions about measurement:

1. How are measurements made, and what units are used?

THE GOAL: Be able to name and use the metric and SI units of measure for mass, length, volume, and temperature.

2. How good are the reported measurements?

THE GOAL: Be able to interpret the number of significant figures in a measurement and round off numbers in calculations involving measurements.

3. How are large and small numbers best represented?

THE GOAL: Be able to interpret prefixes for units of measure and express numbers in scientific notation.

4. How can a quantity be converted from one unit of measure to another?

THE GOAL: Be able to convert quantities from one unit to another using conversion factors.

5. What techniques are used to solve problems?

THE GOAL: Be able to analyze a problem, use the factor-label method to solve the problem, and check the result to ensure that it makes sense chemically and physically.

6. What are temperature, specific heat, density, and specific gravity?

THE GOAL: Be able to define these quantities and use them in calculations.

How often do you make an observation or perform an action that involves measurement? Whenever you check the temperature outside, make a pitcher of lemonade, or add detergent to the laundry, you are making a measurement. Likewise, whenever you use chemical compounds—whether you are adding salt to your cooking, putting antifreeze in your car radiator, or choosing a dosage of medicine—quantities of substances must be measured. A mistake in measuring the quantities could ruin the dinner, damage the engine, or harm the patient.

2.1 Physical Quantities

Height, volume, temperature, and other physical properties that can be measured are called **physical quantities** and are described by both a number and a **unit** of defined size:

Number　　Unit

61.2 kilograms

Physical quantity A physical property that can be measured.

Unit A defined quantity used as a standard of measurement.

The number alone is not much good without a unit. If you asked how much blood an accident victim had lost, the answer "three" would not tell you much. Three drops? Three milliliters? Three pints? Three liters? (By the way, an adult human has only 5–6 liters of blood.)

Any physical quantity can be measured in many different units. For example, a person's height might be measured in inches, feet, yards, centimeters, or many other units. To avoid confusion, scientists from around the world have agreed on a system of standard units, called by the French name *Système International d'Unites* (International System of Units), abbreviated *SI*. **SI units** for some common physical quantities are given in Table 2.1. Mass is measured in *kilograms* (kg), length is measured in *meters* (m), volume is measured in *cubic meters* (m³), temperature is measured in *kelvins* (K), and time is measured in *seconds* (s, not sec).

SI units Units of measurement defined by the International System of Units.

SI units are closely related to the more familiar *metric units* used in all industrialized nations of the world except the United States. If you compare the SI and metric units shown in Table 2.1, you will find that the basic metric unit of mass is the *gram* (g) rather than the kilogram (1 g = 1/1000 kg), the metric unit of volume is the *liter* (L) rather than the cubic meter (1 L = 1/1000 m³), and the metric unit of temperature is the *Celsius degree* (°C) rather than the kelvin. The meter is the unit of

TABLE 2.1 Some SI and Metric Units and Their Equivalents

QUANTITY	SI UNIT (SYMBOL)	METRIC UNIT (SYMBOL)	EQUIVALENTS
Mass	Kilogram (kg)	Gram (g)	1 kg = 1000 g = 2.205 lb
Length	Meter (m)	Meter (m)	1 m = 3.280 ft
Volume	Cubic meter (m³)	Liter (L)	1 m³ = 1000 L = 264.2 gal
Temperature	Kelvin (K)	Celsius degree (°C)	See Section 2.9
Time	Second (s)	Second (s)	—

length and the second is the unit of time in both systems. Although SI units are now preferred in scientific research, metric units are still used in some fields. You will probably find yourself working with both.

In addition to the units listed in Table 2.1, many other widely used units are derived from them. For instance, units of *meters per second* (m/s) are often used for *speed*—the distance covered in a given time. Similarly, units of *grams per cubic centimeter* (g/cm^3) are often used for *density*—the mass of substance in a given volume. We will see other such derived units in future chapters.

One problem with any system of measurement is that the sizes of the units often turn out to be inconveniently large or small for the problem at hand. A biologist describing the diameter of a red blood cell (0.000 006 m) would find the meter to be an inconveniently large unit, but an astronomer measuring the average distance from the earth to the sun (150,000,000,000 m) would find the meter to be inconveniently small. For this reason, metric and SI units can be modified by prefixes to refer to either smaller or larger quantities. For instance, the SI unit for mass—the kilogram—differs by the prefix *kilo-* from the metric unit gram. *Kilo-* indicates that a kilogram is 1000 times as large as a gram:

$$1 \text{ kg} = (1000)(1 \text{ g}) = 1000 \text{ g}$$

Small quantities of active ingredients in medications are often reported in *milligrams* (mg). The prefix *milli-* shows that the unit gram has been divided by 1000, which is the same as multiplying by 0.001:

$$1 \text{ mg} = \left(\frac{1}{1000}\right)(1 \text{ g}) = (0.001)(1 \text{ g}) = 0.001 \text{ g}$$

A list of prefixes is given in Table 2.2, with the most common ones displayed in color. Note that the exponents are multiples of 3 for *mega-* (10^6), *kilo-* (10^3), *milli-* (10^{-3}), *micro-* (10^{-6}), *nano-* (10^{-9}), and *pico-* (10^{-12}). (The use of exponents is reviewed in Section 2.5.) The prefixes *centi-*, meaning 1/100, and *deci-*, meaning 1/10, indicate exponents that are not multiples of 3. *Centi-* is seen most often in the length unit *centimeter* (1 cm = 0.01 m), and *deci-* is used most often in clinical chemistry, where the concentrations of blood components are given in milligrams per deciliter (1 dL = 0.1 L). These prefixes allow us to compare the magnitudes of different numbers by noting how the prefixes modify a common unit. For example,

$$1 \text{ meter} = 10 \text{ dm} = 100 \text{ cm} = 1000 \text{ mm} = 1,000,000 \ \mu m$$

Such comparisons will be useful when we start performing calculations involving units in Section 2.7. Note also in Table 2.2 that numbers having five or more digits to the right of the decimal point are shown with thin spaces every three digits for convenience—0.000 001, for example. This manner of writing numbers is becoming more common and will be used throughout this book.

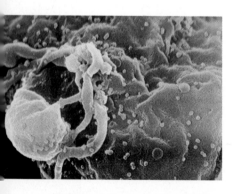

▲ The HIV-1 virus particles (in green) budding from the surface of a lymphocyte have an approximate diameter of 0.000 000 120 m.

TABLE 2.2   Some Prefixes for Multiples of Metric and SI Units

PREFIX	SYMBOL	BASE UNIT MULTIPLIED BY*	EXAMPLE
mega	M	$1{,}000{,}000 = 10^6$	1 megameter (Mm) = 10^6 m
kilo	k	$1000 = 10^3$	1 kilogram (kg) = 10^3 g
hecto	h	$100 = 10^2$	1 hectogram (hg) = 100 g
deka	da	$10 = 10^1$	1 dekaliter (daL) = 10 L
deci	d	$0.1 = 10^{-1}$	1 deciliter (dL) = 0.1 L
centi	c	$0.01 = 10^{-2}$	1 centimeter (cm) = 0.01 m
milli	m	$0.001 = 10^{-3}$	1 milligram (mg) = 0.001 g
micro	μ	$0.000\ 001 = 10^{-6}$	1 micrometer (μm) = 10^{-6} m
nano	n	$0.000\ 000\ 001 = 10^{-9}$	1 nanogram (ng) = 10^{-9} g
pico	p	$0.000\ 000\ 000\ 001 = 10^{-12}$	1 picogram (pg) = 10^{-12} g
femto	f	$0.000\ 000\ 000\ 000\ 001 = 10^{-15}$	1 femtogram = 10^{-15} g

*The scientific notation method of writing large and small numbers (for example, 10^6 for 1,000,000) is explained in Section 2.5.

PROBLEM 2.1

Give the full name of the following units:

(a) dL   (b) mg   (c) ns   (d) km   (e) μg

PROBLEM 2.2

Write the symbol for the following units:

(a) liter   (b) kilogram   (c) nanometer   (d) megameter

PROBLEM 2.3

Express the following quantities in terms of the basic unit (for example, 1 mL = 0.001 L):

(a) 1 nm   (b) 1 dg   (c) 1 km   (d) 1 μs   (e) 1 ng

(handwritten margin notes) use balance to measure ↓ mass – measure of the content ← weight = measure of mass times gravity ↑ use scale to measure

2.2 Measuring Mass

The terms *mass* and *weight*, though often used interchangeably, really have quite different meanings. **Mass** is a measure of the amount of matter in an object, whereas **weight** is a measure of the gravitational pull that the earth, moon, or other large body exerts on an object. Clearly, the amount of matter in an object does not depend on location. Whether you are standing on the earth or standing on the moon, the mass of your body is the same. On the other hand, the weight of an object *does* depend on location. Your weight on earth might be 140 lb, but it would only be 23 lb on the moon because the pull of gravity there is only about one-sixth as great.

At the same location, two objects with identical masses have identical weights; that is, gravity pulls equally on both. Thus, the *mass* of an object can be determined by comparing the *weight* of the object to the weight of a known reference standard. Much of the confusion between mass and weight is simply due to a language problem: We speak of "weighing" when we really mean that we are measuring mass by

Mass A measure of the amount of matter in an object.

Weight A measure of the gravitational force that the earth or other large body exerts on an object.

(a)   (b)

▶ **FIGURE 2.1** (a) The single-pan balance has a sliding counterweight that is adjusted until the weight of the object on the pan is just balanced. (b) A modern electronic balance.

▲ This pile of 400 pennies has a mass of about 1 kg.

comparing two weights. Figure 2.1 shows two types of balances used for measuring mass in the laboratory.

One kilogram, the SI unit for mass, is equal to 2.205 lb—too large a quantity for many purposes in chemistry and medicine. Thus, smaller units of mass such as the gram, milligram (mg), and microgram (μg), are more commonly used. Table 2.3 shows the relationships between metric and common units for mass.

TABLE 2.3   Units of Mass

UNIT	EQUIVALENT	UNIT	EQUIVALENT
1 kilogram (kg)	= 1000 grams = 2.205 pounds	1 ton	= 2000 pounds = 907.03 kilograms
1 gram (g)	= 0.001 kilogram = 1000 milligrams = 0.035 27 ounce	1 pound (lb)	= 16 ounces = 0.454 kilogram = 454 grams
1 milligram (mg)	= 0.001 gram = 1000 micrograms	1 ounce (oz)	= 0.028 35 kilogram = 28.35 grams
1 microgram (μg)	= 0.000 001 gram = 0.001 milligram		= 28,350 milligrams

2.3 Measuring Length and Volume

The meter is the standard measure of length, or distance, in both the SI and metric systems. One meter is 39.37 inches (about 10% longer than a yard), a length that is much too large for most measurements in chemistry and medicine. Other, more commonly used measures of length are the *centimeter* (cm; 1/100 m) and the *millimeter* (mm; 1/1000 m). One centimeter is a bit less than half an inch— 0.3937 inch to be exact. A millimeter, in turn, is 0.03937 inch, or about the thickness of a dime. Table 2.4 lists the relationships of these units.

Volume is the amount of space occupied by an object. The SI unit for volume— the cubic meter, m^3—is so large that the liter (1 L = 0.001 m^3 = 1 dm^3) is much more commonly used in chemistry and medicine. One liter has the volume of a cube 10 cm (1 dm) on edge and is a bit larger than one U.S. quart. Each liter is further divided into 1000 *milliliters* (mL), with 1 mL being the size of a cube 1 cm on edge, or 1 cm^3. In fact, the milliliter is often called a *cubic centimeter* (cm^3, or cc) in medical work. Figure 2.2 shows the divisions of a cubic meter, and Table 2.5 shows the relationships among units of volume.

TABLE 2.4   Units of Length

UNIT	EQUIVALENT	UNIT	EQUIVALENT
1 kilometer (km)	= 1000 meters = 0.6214 mile	1 mile (mi)	= 1.609 kilometers = 1609 meters
1 meter (m)	= 100 centimeters = 1000 millimeters = 1.0936 yards = 39.37 inches	1 yard (yd)	= 0.9144 meter = 91.44 centimeters
		1 foot (ft)	= 0.3048 meter = 30.48 centimeters
1 centimeter (cm)	= 0.01 meter = 10 millimeters = 0.3937 inch	1 inch (in.)	= 2.54 centimeters = 25.4 millimeters
1 millimeter (mm)	= 0.001 meter = 0.1 centimeter		

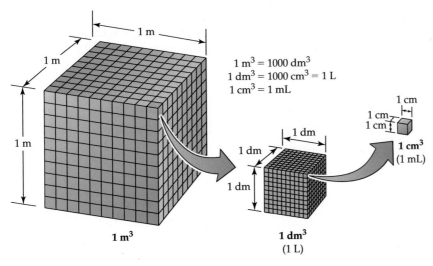

$$1 \text{ m}^3 = 1000 \text{ dm}^3$$
$$1 \text{ dm}^3 = 1000 \text{ cm}^3 = 1 \text{ L}$$
$$1 \text{ cm}^3 = 1 \text{ mL}$$

▲ **FIGURE 2.2** A cubic meter is the volume of a cube 1 m on edge. Each cubic meter contains 1000 cubic decimeters (liters), and each cubic decimeter contains 1000 cubic centimeters (milliliters). Thus, there are 1000 mL in a liter and 1000 L in a cubic meter.

TABLE 2.5   Units of Volume

UNIT	EQUIVALENT
1 cubic meter (m³)	= 1000 liters = 264.2 gallons
1 liter (L)	= 0.001 cubic meter = 1000 milliliters = 1.057 quarts
1 deciliter (dL)	= 0.1 liter = 100 milliliters
1 milliliter (mL)	= 0.001 liter = 1000 microliters
1 microliter (μL)	= 0.001 milliliter
1 gallon (gal)	= 3.7854 liters
1 quart (qt)	= 0.9464 liter = 946.4 milliliters
1 fluid ounce (fl oz)	= 29.57 milliliters

2.4 Measurement and Significant Figures

How much does a tennis ball weigh? If you put a tennis ball on an ordinary bathroom scale, the scale would probably register 0 lb (or 0 kg if you have a metric scale). If you placed the same tennis ball on a common laboratory balance, however, you might get a reading of 54.07 g. Trying again by placing the ball on an expensive analytical balance like those found in clinical and research laboratories, you might find a weight of 54.071 38 g. Clearly, the precision of your answer depends on the equipment used for the measurement.

Every experimental measurement, no matter how precise, has a degree of uncertainty to it because there is always a limit to the number of digits that can be determined. An analytical balance, for example, might reach its limit in measuring mass to the fifth decimal place, and weighing the tennis ball several times might produce slightly different readings, such as 54.071 39 g, 54.071 38 g, and 54.071 37 g. Also, different people making the same measurement might come up with slightly different answers. How, for instance, would you record the volume of the liquid shown in Figure 2.3? It is clear that the volume of liquid lies between 17.0 and 18.0 mL, but the exact value of the last digit must be estimated.

To indicate the precision of a measurement, the value recorded should use all the digits known with certainty, plus one additional estimated digit that is usually considered uncertain by plus or minus 1 (written as ±1). The total number of digits used to express such a measurement is called the number of **significant figures**. Thus, the quantity 54.07 g has four significant figures (5, 4, 0, and 7), and the quantity 54.071 38 g has seven significant figures. *Remember*: All but one of the significant figures are known with certainty; the last significant figure is only an estimate accurate to ±1.

▲ The tennis ball weighs 54.07 g on this common laboratory balance, which is capable of determining mass to about 0.01 g.

Significant figures The number of meaningful digits used to express a value.

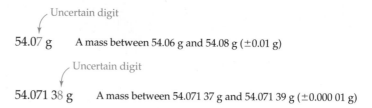

Uncertain digit

54.07 g   A mass between 54.06 g and 54.08 g (±0.01 g)

Uncertain digit

54.071 38 g   A mass between 54.071 37 g and 54.071 39 g (±0.000 01 g)

Deciding the number of significant figures in a given measurement is usually simple, but can be troublesome when zeros are involved. Depending on the circumstances, a zero might be significant or might be just a space-filler to locate the decimal point. For example, how many significant figures does each of the following measurements have?

94.072 g	Five significant figures (9, 4, 0, 7, 2)
0.0834 cm	Three significant figures (8, 3, 4)
0.029 07 mL	Four significant figures (2, 9, 0, 7)
138.200 m	Six significant figures (1, 3, 8, 2, 0, 0)
23,000 kg	*Anywhere* from two (2, 3) to five (2, 3, 0, 0, 0) significant figures

The following rules are helpful for determining the number of significant figures when zeros are present:

RULE 1. Zeros in the middle of a number are like any other digit; they are always significant. Thus, 94.072 g has five significant figures.

RULE 2. Zeros at the beginning of a number are not significant; they act only to locate the decimal point. Thus, 0.0834 cm has three significant figures, and 0.029 07 mL has four.

RULE 3. Zeros at the end of a number and *after* the decimal point are significant. It is assumed that these zeros would not be shown unless they were significant. Thus, 138.200 m has six significant figures. If the value were known to only four significant figures, we would write 138.2 m.

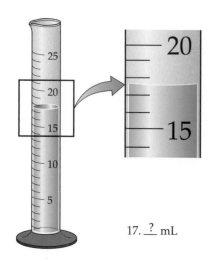

▲ **FIGURE 2.3** What is the volume of liquid in this graduated cylinder?

17. _?_ mL

RULE 4. Zeros at the end of a number and *before* an implied decimal point may or may not be significant. We cannot tell whether they are part of the measurement or whether they act only to locate the unwritten but implied decimal point. Thus, 23,000 kg may have two, three, four, or five significant figures. Adding a decimal point at the end would indicate that all five numbers are significant.

Often, however, a little common sense is useful. A temperature reading of 20 °C probably has two significant figures rather than one, because one significant figure would imply a temperature anywhere from 10 °C to 30 °C and would be of little use. Similarly, a volume given as 300 mL probably has three significant figures. On the other hand, a figure of 150,000,000 km for the distance between the earth and the sun has only two or three significant figures because the distance is variable. We will see a better way to deal with this problem in the next section.

One final point about significant figures: Some numbers, such as those obtained when counting objects and those that are part of a definition, are *exact* and effectively have an unlimited number of significant figures. Thus, a class might have *exactly* 32 students (not 31.9, 32.0, or 32.1), and 1 foot is defined to have *exactly* 12 inches.

▲ How many CDs are in this stack? (Answer: *exactly* 29.)

WORKED EXAMPLE **2.1** Significant Figures of Measurements

How many significant figures do the following measurements have?

    (a) 2730.78 m   **(b)** 0.0076 mL   **(c)** 3400 kg   **(d)** 3400.0 m^2

ANALYSIS All nonzero numbers are significant; the number of significant figures will then depend on the status of the zeros in each case. (Hint: which rule applies in each case?)

SOLUTION

    (a) Six (rule 1)   **(b)** Two (rule 2)
    (c) Two, three, or four (rule 4)   **(d)** Five (rule 3)

PROBLEM 2.4

How many significant figures do the following measurements have?

(a) 3.45 m   **(b)** 0.1400 kg

(c) 10.003 L   **(d)** 35 cents

━○ KEY CONCEPT PROBLEM 2.5

What is the temperature reading on the following Celsius thermometer? How many significant figures do you have in your answer?

2.5 Scientific Notation

Scientific notation A number expressed as the product of a number between 1 and 10, times the number 10 raised to a power.

Rather than write very large or very small numbers in their entirety, it is more convenient to express them using *scientific notation*. A number is written in **scientific notation** as the product of a number between 1 and 10, times the number 10 raised to a power. Thus, 215 is written in scientific notation as 2.15×10^2:

$$215 = 2.15 \times 100 = 2.15 \times (10 \times 10) = 2.15 \times 10^2$$

Notice that in this case, where the number is *larger* than 1, the decimal point has been moved *to the left* until it follows the first digit. The exponent on the 10 tells how many places we had to move the decimal point to position it just after the first digit:

$$215. = 2.15 \times 10^2$$

Decimal point is moved two places to the left, so exponent is 2.

To express a number *smaller* than 1 in scientific notation, we have to move the decimal point *to the right* until it follows the first digit. The number of places moved is the negative exponent of 10. For example, the number 0.002 15 can be rewritten as 2.15×10^{-3}:

$$0.002\ 15 = 2.15 \times \frac{1}{1000} = 2.15 \times \frac{1}{10 \times 10 \times 10} = 2.15 \times \frac{1}{10^3} = 2.15 \times 10^{-3}$$

$$0.002\ 15 = 2.15 \times 10^{-3}$$

Decimal point is moved three places to the right, so exponent is −3.

To convert a number written in scientific notation to standard notation, the process is reversed. For a number with a *positive* exponent, the decimal point is moved to the *right* a number of places equal to the exponent:

$$3.7962 \times 10^4 = 37,962$$

Positive exponent of 4, so decimal point is moved to the right four places.

For a number with a *negative* exponent, the decimal point is moved to the *left* a number of places equal to the exponent:

$$1.56 \times 10^{-8} = 0.000\ 000\ 015\ 6$$

Negative exponent of −8, so decimal point is moved to the left eight places.

Scientific notation is particularly helpful for indicating how many significant figures are present in a number that has zeros at the end but to the left of a decimal point. If we read, for instance, that the distance from the earth to the sun is 150,000,000 km, we do not really know how many significant figures are indicated. Some of the zeros might be significant, or they might merely act to locate the decimal point. Using scientific notation, however, we can indicate how many of the zeros are significant. Rewriting 150,000,000 as 1.5×10^8 indicates two significant figures, whereas writing it as 1.500×10^8 indicates four significant figures.

Scientific notation is not ordinarily used for numbers that are easily written, such as 10 or 175, although it is sometimes helpful in doing arithmetic. Rules for doing arithmetic with numbers written in scientific notation are reviewed in Appendix A.

WORKED EXAMPLE **2.2** Significant Figures and Scientific Notation

There are 1,760,000,000,000,000,000,000 molecules of sucrose (table sugar) in 1 g. Use scientific notation to express this number with four significant figures.

ANALYSIS Because the number is larger than 1, the exponent will be positive. You will have to move the decimal point 21 places to the left.

SOLUTION
The first four digits—1, 7, 6, and 0—are significant, meaning that only the first of the 19 zeros is significant. Because we have to move the decimal point 21 places to the left to put it after the first significant digit, the answer is 1.760×10^{21}.

▲ How many molecules are in this 1 g pile of table sugar?

WORKED EXAMPLE **2.3** Scientific Notation

The rhinovirus responsible for the common cold has a diameter of 20 nm, or 0.000 000 020 m (See Application on p. 28). Express this number in scientific notation.

ANALYSIS The number is smaller than 1, and so the exponent will be negative. You will have to move the decimal point eight places to the right.

SOLUTION
There are only two significant figures, because zeros at the beginning of a number are not significant. We have to move the decimal point 8 places to the right to place it after the first digit, so the answer is 2.0×10^{-8} m.

WORKED EXAMPLE **2.4** Scientific Notation and Unit Conversions

A clinical laboratory found that a blood sample contained 0.0026 g of phosphorus and 0.000 101 g of iron.

(a) Give these quantities in scientific notation.
(b) Give these quantities in the units normally used to report them—milligrams for phosphorus and micrograms for iron.

ANALYSIS Is the number larger or smaller than 1? How many places do you have to move the decimal point?

SOLUTION

(a) 0.0026 g phosphorus $= 2.6 \times 10^{-3}$ g phosphorus

$0.000\ 101$ g iron $= 1.01 \times 10^{-4}$ g iron

(b) We know that 1 mg $= 1 \times 10^{-3}$ g, where the exponent is -3. Expressing the amount of phosphorus in milligrams is straightforward because the amount in grams (2.6×10^{-3} g) already has an exponent of -3. Thus, 2.6×10^{-3} g $= 2.6$ mg of phosphorus.

$$(2.6 \times 10^{-3}\ \cancel{g})\left(\frac{1\ \text{mg}}{1 \times 10^{-3}\ \cancel{g}}\right) = 2.6\ \text{mg}$$

We know that 1 μg $= 1 \times 10^{-6}$ g, where the exponent is -6. Expressing the amount of iron in micrograms thus requires that we restate the amount in grams so that the exponent is -6. We can do this by moving the decimal point six places to the right:

$$0.000\ 101 \text{ g iron} = 101 \times 10^{-6} \text{ g iron} = 101\ \mu\text{g iron}$$

APPLICATION ▶ Powers of 10

I t is not easy to grasp the enormous differences in size represented by powers of 10 (scientific notation). A sodium atom has a diameter of 388 pm (3.88×10^{-10} m), a typical rhinovirus may have a diameter of 20 nm (2.0×10^{-8} m), and a single bacteria cell may be 3 μm long (3×10^{-6} m). Even though none of these objects can be seen with the naked eye, scientists know that the size of an atom relative to the size of a bacterium would be like comparing the size of a bowling ball with that of a baseball stadium. So how do scientists know the relative sizes of these items if we can not see them?

To study the miniature world of bacteria and viruses requires special tools. Optical light microscopes, for example, can magnify an object up to 1500 (1.5×10^3) times,

enabling us to study objects as small as 0.2 μm, such as bacteria. To study smaller objects, such as viruses, requires even higher magnification. Electron microscopy, which uses a beam of electrons instead of light, can achieve magnifications of 1×10^5 for scanning electron microscopy (SEM), and the ability to distinguish features on the order of 0.1 nm in length are possible with scanning tunneling microscopes (STM).

The change in magnification needed to distinguish features from 1 mm to 1 nm may not seem great, but it represents a million-fold (10^6) increase—enough to open a whole new world. Powers of 10 are powerful indeed.

See Additional Problems 2.78 and 2.79 at the end of the chapter.

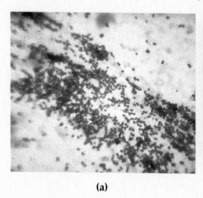 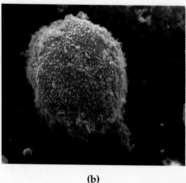 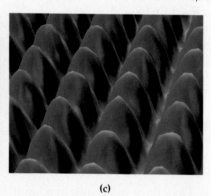

(a)   (b)   (c)

▲ **(a)** A light microscope image of bacteria; **(b)** a color-enhanced SEM image showing a T-lymphocyte white blood cell (in orange) with HIV particles (in blue) budding on the surface of the cell membrane; **(c)** an STM image of a "blue nickel" surface showing the orientation of individual metal atoms.

PROBLEM 2.6

Convert the following values to scientific notation:

(a) 0.058 g   **(b)** 46,792 m   **(c)** 0.006 072 cm   **(d)** 345.3 kg

PROBLEM 2.7

Convert the following values from scientific notation to standard notation:

(a) 4.885×10^4 mg   **(b)** 8.3×10^{-6} m   **(c)** 4.00×10^{-2} m

PROBLEM 2.8

Rewrite the following numbers in scientific notation as indicated:

(a) 630,000 with five significant figures
(b) 1300 with three significant figures
(c) 794,200,000,000 with four significant figures

PROBLEM 2.9

Ordinary table salt, or sodium chloride, is made up of small particles called *ions*, which we will discuss in Chapter 4. If the distance between a sodium ion and a chloride ion is 0.000 000 000 278 m, what is the distance in scientific notation? How many picometers is this?

2.6 Rounding Off Numbers

It often happens, particularly when doing arithmetic on a pocket calculator, that a quantity appears to have more significant figures than are really justified. For example, you might calculate the gas mileage of your car by finding that it takes 11.70 gallons of gasoline to drive 278 miles:

$$\text{Mileage} = \frac{\text{Miles}}{\text{Gallons}} = \frac{278 \text{ mi}}{11.70 \text{ gal}} = 23.760\ 684 \text{ mi/gal (mpg)}$$

Although the answer on a pocket calculator has eight digits, your measurement is really not as precise as it appears. In fact, as we will see below, your answer is good to only three significant figures and should be **rounded off** to 23.8 mi/gal.

How do you decide how many digits to keep? The full answer to this question is a bit complex and involves a mathematical treatment called *error analysis*, but for many purposes, a simplified procedure using just two rules is sufficient:

Rounding off A procedure used for deleting nonsignificant figures.

RULE 1. In carrying out a multiplication or division, the answer cannot have more significant figures than either of the original numbers. This is just a common-sense rule if you think about it. After all, if you do not know the number of miles you drove to better than three significant figures (278 could mean 277, 278, or 279), you certainly cannot calculate your mileage to more than the same number of significant figures.

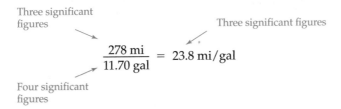

▲ Calculators often display more digits than are justified by the precision of the data.

RULE 2. In carrying out an addition or subtraction, the answer cannot have more digits after the decimal point than either of the original numbers. For example, if you have 3.18 L of water and you add 0.013 15 L more, you now have 3.19 L. Again, this rule is just common sense. If you do not know the volume you started with past the second decimal place (it could be 3.17, 3.18, or 3.19), you cannot know the total of the combined volumes past the same decimal place.

Volume of water at start ⟶ 3.18? ?? L ⟵ Two digits after decimal point
Volume of water added ⟶ + 0.013 15 L ⟵ Five digits after decimal point
Total volume of water ⟶ 3.19? ?? L ⟵ Two digits after decimal point

If a calculation has several steps, it is generally best to round off at the end after all the steps have been carried out, keeping the number of significant figures determined by the least precise number in your calculations. Once you decide how many digits to retain for your answer, the rules for rounding off numbers are straightforward:

RULE 1. If the first digit you remove is 4 or less, drop it and all following digits. Thus, 2.4271 becomes 2.4 when rounded off to two significant figures because the first of the dropped digits (a 2) is 4 or less.

RULE 2. If the first digit you remove is 5 or greater, round the number up by adding a 1 to the digit to the left of the one you drop. Thus, 4.5832 becomes 4.6 when rounded off to two significant figures because the first of the dropped digits (an 8) is 5 or greater.

WORKED EXAMPLE   **2.5** Significant Figures and Calculations:
Addition/Subtraction

Suppose that you weigh 124 lb before dinner. How much will you weigh after dinner if you eat 1.884 lb of food?

ANALYSIS When performing addition or subtraction, the number of significant figures you report in the final answer is determined by the number of digits in the least precise number in the calculation.

SOLUTION
Your after-dinner weight is found by adding your original weight to the weight of the food consumed:

$$\begin{array}{r} 124 \quad \text{lb} \\ \underline{1.884\ \text{lb}} \\ 125.884\ \text{lb} \quad \text{(Unrounded)} \end{array}$$

Because the value of your original weight has no significant figures after the decimal point, your after-dinner weight also must have no significant figures after the decimal point. Thus, 125.884 lb must be rounded off to 126 lb.

WORKED EXAMPLE   **2.6** Significant Figures and Calculations:
Multiplication/Division

To make currant jelly, 13.75 cups of sugar was added to 18 cups of currant juice. How much sugar was added per cup of juice?

ANALYSIS For calculations involving multiplication or division, the final answer cannot have more significant figures than either of the original numbers.

SOLUTION
The quantity of sugar must be divided by the quantity of juice:

$$\frac{13.75 \text{ cups sugar}}{18 \text{ cups juice}} = 0.763\,888\,89 \frac{\text{cup sugar}}{\text{cup juice}} \text{(Unrounded)}$$

The number of significant figures in the answer is limited to two by the quantity 18 cups in the calculation and must be rounded to 0.76 cup of sugar per cup of juice.

PROBLEM 2.10

Round off the following quantities to the indicated number of significant figures:

(a) 2.304 g (three significant figures)
(b) 188.3784 mL (five significant figures)
(c) 0.008 87 L (one significant figure)
(d) 1.000 39 kg (four significant figures)

PROBLEM 2.11

Carry out the following calculations, rounding each result to the correct number of significant figures:

(a) 4.87 mL + 46.0 mL   (b) 3.4 × 0.023 g
(c) 19.333 m − 7.4 m   (d) 55 mg − 4.671 mg + 0.894 mg
(e) 62,911 ÷ 611

2.7 Problem Solving: Converting a Quantity from One Unit to Another

Many activities in the laboratory and in medicine—measuring, weighing, preparing solutions, and so forth—require converting a quantity from one unit to another. For example: "These pills contain 1.3 grains of aspirin, but I need 200 mg. Is one pill enough?" Converting between units is not mysterious; we all do it every day. If you run 9 laps around a 400 meter track, for instance, you have to convert between the distance unit "lap" and the distance unit "meter" to find that you have run 3600 m (9 laps times 400 m/lap). If you want to find how many miles that is, you have to convert again to find that 3600 m = 2.237 mi.

The simplest way to carry out calculations involving different units is to use the **factor-label method**. In this method, a quantity in one unit is converted into an equivalent quantity in a different unit by using a **conversion factor** that expresses the relationship between units:

$$\text{Starting quantity} \times \text{Conversion factor} = \text{Equivalent quantity}$$

As an example, we said in Section 2.3 that 1 km = 0.6214 mi. Writing this relationship as a fraction restates it in the form of a conversion factor, either kilometers per mile or miles per kilometer.

Since 1 km = 0.6214 mi, then:

Conversion factors between kilometers and miles
$$\frac{1 \text{ km}}{0.6214 \text{ mi}} = 1 \quad \text{or} \quad \frac{0.6214 \text{ mi}}{1 \text{ km}} = 1$$

Note that this and all other conversion factors are numerically equal to 1 because the value of the quantity above the division line (the numerator) is equal in value to the quantity below the division line (the denominator). Thus, multiplying by a conversion factor is equivalent to multiplying by 1 and so does not change the value of the quantity being multiplied:

These two quantities are the same.
$$\frac{1 \text{ km}}{0.6214 \text{ mi}}$$
or
These two quantities are the same.
$$\frac{0.6214 \text{ mi}}{1 \text{ km}}$$

The key to the factor-label method of problem solving is that units are treated like numbers and can thus be multiplied and divided (though not added or subtracted) just as numbers can. When solving a problem, the idea is to set up an equation so that all unwanted units cancel, leaving only the desired units. Usually, it is best to start by writing what you know and then manipulating that known quantity. For example, if you know there are 26.22 mi in a marathon and want to find how many kilometers that is, you could write the distance in miles and multiply by the conversion factor in kilometers per mile. The unit "mi" cancels because it appears both above and below the division line, leaving "km" as the only remaining unit.

$$26.22 \text{ mi} \times \frac{1 \text{ km}}{0.6214 \text{ mi}} = 42.20 \text{ km}$$

Starting quantity   Conversion factor   Equivalent quantity

The factor-label method gives the right answer only if the equation is set up so that the unwanted unit (or units) cancel. If the equation is set up in any other way, the units will not cancel and you will not get the right answer. Thus, if you selected the incorrect conversion factor (miles per kilometer) for the above problem,

▲ These runners have to convert from laps to meters to find out how far they have run.

Factor-label method A problem-solving procedure in which equations are set up so that unwanted units cancel and only the desired units remain.

Conversion factor An expression of the numerical relationship between two units.

you would end up with an incorrect answer expressed in meaningless units:

$$\text{Incorrect } 26.22 \text{ mi} \times \frac{0.6214 \text{ mi}}{1 \text{ km}} = 16.29 \frac{\text{mi}^2}{\text{km}} \text{ Incorrect}$$

WORKED EXAMPLE **2.7** Factor Labels: Unit Conversions

Write conversion factors for the following pairs of units (use Tables 2.3–2.5):

(a) Deciliters and milliliters

(b) Pounds and grams

ANALYSIS Start with the appropriate equivalency relationship and rearrange to form conversion factors.

SOLUTION

(a) Since 1 dL = 0.1 L and 1 mL = 0.001L, then

$$1 \text{ dL} = (0.1 \text{ L}) \left(\frac{1 \text{ mL}}{0.001 \text{L}} \right) = 100 \text{ mL. The conversion factors are}$$

$$\frac{1 \text{ dL}}{100 \text{ mL}} \quad \text{and} \quad \frac{100 \text{ mL}}{1 \text{ dL}}$$

(b) $\dfrac{1 \text{ lb}}{454 \text{ g}}$ and $\dfrac{454 \text{ g}}{1 \text{ lb}}$

WORKED EXAMPLE **2.8** Factor Labels: Unit Conversions

(a) Convert 0.75 lb to grams.

(b) Convert 0.50 qt to deciliters.

ANALYSIS Start with conversion factors and set up equations so that units cancel appropriately.

SOLUTION

(a) Select the conversion factor from Worked Example 2.7(b) so that the "lb" units cancel and "g" remains:

$$0.75 \text{ lb} \times \frac{454 \text{ g}}{1 \text{ lb}} = 340 \text{ g}$$

(b) In this, as in many problems, it is convenient to use more than one conversion factor. As long as the unwanted units cancel correctly, two or more conversion factors can be strung together in the same calculation. In this case, we can convert first between quarts and milliliters, and then between milliliters and deciliters:

$$0.50 \text{ qt} \times \frac{946.4 \text{ mL}}{1 \text{ qt}} \times \frac{1 \text{ dL}}{100 \text{ mL}} = 4.7 \text{ dL}$$

PROBLEM 2.12

Write conversion factors for the following pairs of units:

(a) liters and milliliters   **(b)** grams and ounces

(c) liters and quarts

PROBLEM 2.13

Carry out the following conversions:

(a) 16.0 oz = ? g   **(b)** 2500 mL = ? L   **(c)** 99.0 L = ? qt

PROBLEM 2.14

Convert 0.840 qt to milliliters in a single calculation using more than one conversion factor.

PROBLEM 2.15

One international nautical mile is defined as exactly 6076.1155 ft, and a speed of 1 knot is defined as one international nautical mile per hour. What is the speed in meters per second of a boat traveling at a speed of 14.3 knots?

2.8 Problem Solving: Estimating Answers

The main drawback to using the factor-label method is that it is possible to get an answer without really understanding what you are doing. It is therefore best when solving a problem to first think through a rough estimate, or *ballpark estimate*, as a check on your work. If your ballpark estimate is not close to the final calculated solution, there is a misunderstanding somewhere and you should think the problem through again. If, for example, you came up with the answer 5.3 cm^3 when calculating the volume of a human cell, you should realize that such an answer could not possibly be right. Cells are too tiny to be distinguished with the naked eye, but a volume of 5.3 cm^3 is about the size of a walnut. The Worked Examples at the end of this section show how to estimate the answers to simple unit-conversion problems.

The factor-label method and the use of ballpark estimates are techniques that will help you solve problems of many kinds, not just unit conversions. Problems sometimes seem complicated, but you can usually sort out the complications by analyzing the problem properly:

STEP 1: Identify the information given, including units.

STEP 2: Identify the information needed in the answer, including units.

STEP 3: Find the relationship(s) between the known information and unknown answer, and plan a series of steps, including conversion factors, for getting from one to the other.

STEP 4: Solve the problem.

BALLPARK CHECK: Make a ballpark estimate at the beginning and check it against your final answer to be sure the value and the units of your calculated answer are reasonable.

Worked Examples 2.9, 2.10 and 2.11 illustrate how to use the analysis steps and ballpark checks as an aid in dosage calculations.

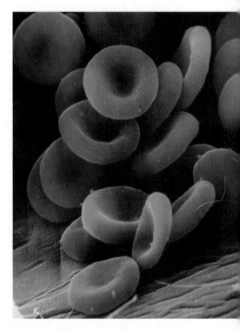

▲ What is the volume of a red blood cell?

| **WORKED EXAMPLE** | **2.9** Factor Labels: Unit Conversions |

A child is 21.5 in. long at birth. How long is this in centimeters?

ANALYSIS This problem calls for converting from inches to centimeters, so we will need to know how many cm are in an inch, and how to use this information as a conversion factor.

BALLPARK ESTIMATE It takes about 2.5 cm to make 1 in., and so it should take 2.5 times as many centimeters to make a distance equal to approximately 20 in., or about 20 in. × 2.5 = 50 cm.

SOLUTION

STEP 1: **Identify given information.**	Length = 21.5 in.
STEP 2: **Identify answer and units.**	Length = ?? cm
STEP 3: **Identify conversion factor.**	$1 \text{ in.} = 2.54 \text{ cm} \rightarrow \dfrac{2.54 \text{ cm}}{1 \text{ in.}}$
STEP 4: **Solve** Multiply the known length (in inches) by the conversion factor so that units cancel, providing the answer (in cm).	$21.5 \text{ in.} \times \dfrac{2.54 \text{ cm}}{1 \text{ in.}} = 54.6 \text{ cm}$ (Rounded off from 54.61)

BALLPARK CHECK How does this value compare with the ballpark estimate we made at the beginning? Are the final units correct?

WORKED EXAMPLE **2.10** Factor Labels: Concentration to Mass

A patient requires injection of 0.012 g of a pain killer available as a 15 mg/mL solution. How many milliliters should be administered?

ANALYSIS Knowing the amount of pain killer in 1 mL allows us to use the concentration as a conversion factor to determine the volume of solution that would contain the desired amount.

BALLPARK ESTIMATE One milliliter contains 15 mg of the pain killer, or 0.015 g. Since only 0.012 g is needed, a little less than 1.0 mL should be administered.

▲ How many milliliters should be injected?

SOLUTION

STEP 1: **Identify known information.**	Dosage = 0.012 g Concentration = 15 mg/mL
STEP 2: **Identify answer and units.**	Volume to administer = ?? mL
STEP 3: **Identify conversion factors.** Two conversion factors are needed. First, g must be converted to mg. Once we have the mass in mg, then we can calculate mL using the conversion factor of mL/mg.	$1 \text{ mg} = .001 \text{ g} \Rightarrow \dfrac{1 \text{ mg}}{0.001 \text{ g}}$ $15 \text{ mg/mL} \Rightarrow \dfrac{1 \text{ mL}}{15 \text{ mg}}$
STEP 4: **Solve.** Starting from the desired dosage, we use the conversion factors to cancel units, obtaining the final answer in mL.	$(0.012 \text{ g})\left(\dfrac{1 \text{ mg}}{0.001 \text{ g}}\right)\left(\dfrac{1 \text{ mL}}{15 \text{ mg}}\right) = 0.80 \text{ mL}$

BALLPARK CHECK Consistent with our initial estimate of a little less than 1 mL.

WORKED EXAMPLE **2.11** Factor Labels: Multiple Conversion Calculations

Administration of digitalis to control atrial fibrillation in heart patients must be carefully regulated because even a modest overdose can be fatal. To take differences between patients into account, dosages are sometimes prescribed in micrograms per kilogram of body weight (μg/kg). Thus, two people may differ greatly in weight, but both will receive the proper dosage. At a dosage of 20 μg/kg body weight, how many milligrams of digitalis should a 160 lb patient receive?

ANALYSIS Knowing the patient's body weight (in kg) and the recommended dosage (in μg/kg), we can calculate the appropriate amount of digitalis.

BALLPARK ESTIMATE Since a kilogram is roughly equal to 2 lb, a 160 lb patient has a mass of about 80 kg. At a dosage of 20 μg/kg, an 80 kg patient should receive 80 × 20 μg, or about 1600 μg of digitalis, or 1.6 mg.

SOLUTION

STEP 1: Identify known information.

STEP 2: Identify answer and units.

STEP 3: Identify conversion factors. Two conversions are needed. First, convert the patient's weight in pounds to weight in kg. The correct dose can then be determined based on μg digitalis/kg of body weight. Finally the dosage in μg is converted to mg.

STEP 4: Solve. Use the known information and the conversion factors so that units cancel, obtaining the answer in mg.

BALLPARK CHECK Close to our estimate of 1.6 mg.

Patient weight = 160 lb
Prescribed dosage = 20 μg digitalis/kg body weight
Delivered dosage = ?? mg digitalis

$$1 \text{ kg} = 2.205 \text{ lb} \rightarrow \frac{1 \text{ kg}}{2.205 \text{ lb}}$$

$$1 \text{ mg} = (0.001 \text{ g})\left(\frac{1 \mu g}{10^{-6} \text{ g}}\right) = 1000 \ \mu g$$

$$160 \text{ lb} \times \frac{1 \text{ kg}}{2.205 \text{ lb}} \times \frac{20 \ \mu g \text{ digitalis}}{1 \text{ kg}} \times \frac{1 \text{ mg}}{1000 \ \mu g}$$

$$= 1.5 \text{ mg digitalis (Rounded off)}$$

PROBLEM 2.16

(a) How many kilograms does a 7.5 lb infant weigh?

(b) How many milliliters are in a 4.0 fl oz bottle of cough medicine?

PROBLEM 2.17

Calculate the dosage in milligrams per kilogram body weight for a 135 lb adult who takes two aspirin tablets containing 0.324 g of aspirin each. Calculate the dosage for a 40 lb child who also takes two aspirin tablets.

2.9 Measuring Temperature

Temperature, the measure of how hot or cold an object is, is commonly reported either in Fahrenheit (°F) or Celsius (°C) units. The SI unit for reporting temperature, however, is the *kelvin* (K). (Note that we say only "kelvin," not "kelvin degree.")

The kelvin and the Celsius degree are the same size—both are 1/100 of the interval between the freezing point of water and the boiling point of water at atmospheric pressure. Thus, a change in temperature of 1 °C is equal to a change of 1 K. The only difference between the Kelvin and Celsius temperature scales is that they have different zero points. The Celsius scale assigns a value of 0 °C to the freezing point of water, but the Kelvin scale assigns a value of 0 K to the coldest possible temperature, sometimes called *absolute zero*, which is equal to −273.15 °C. Thus, 0 K = −273.15 °C, and +273.15 K = 0 °C. For example, a warm spring day with a temperature of 25 °C has a Kelvin temperature of 25 + 273.15 = 298 K (for most purposes, rounding off to 273 is sufficient):

$$\text{Temperature in K} = \text{Temperature in °C} + 273.15$$

$$\text{Temperature in °C} = \text{Temperature in K} - 273.15$$

For practical applications in medicine and clinical chemistry, the Fahrenheit and Celsius scales are used almost exclusively. The Fahrenheit scale defines the freezing point of water as 32 °F and the boiling point of water as 212 °F, whereas 0 °C and 100 °C are the freezing and boiling points of water on the Celsius scale. Thus, it takes 180 Fahrenheit degrees to cover the same range encompassed by only 100 Celsius degrees, and a Celsius degree is therefore exactly 180/100 = 9/5 = 1.8

Temperature The measure of how hot or cold an object is.

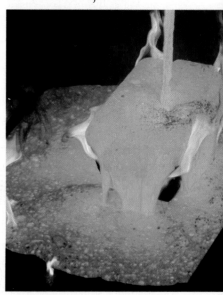

▲ Gold metal is a liquid at temperatures above 1064.4 °C?

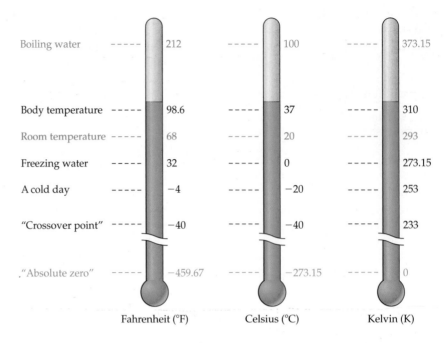

	Fahrenheit (°F)	Celsius (°C)	Kelvin (K)
Boiling water	212	100	373.15
Body temperature	98.6	37	310
Room temperature	68	20	293
Freezing water	32	0	273.15
A cold day	−4	−20	253
"Crossover point"	−40	−40	233
"Absolute zero"	−459.67	−273.15	0

▶ **FIGURE 2.4 A comparison of the Fahrenheit, Celsius, and Kelvin temperature scales.** One Fahrenheit degree is 5/9 the size of a kelvin or a Celsius degree.

times as large as a Fahrenheit degree. In other words, a change in temperature of 1.0 °C is equal to a change of 1.8 °F. Figure 2.4 gives a comparison of all three scales.

Converting between the Fahrenheit and Celsius scales is similar to converting between different units of length or volume, but is a bit more complex because two corrections need to be made—one to adjust for the difference in degree size and one to adjust for the different zero points. The size correction is made by using the relationship 1 °C = (9/5) °F and 1 °F = (5/9) °C. The zero-point correction is made by

APPLICATION ▶ ## Temperature–Sensitive Materials

Wouldn't it be nice to be able to tell if the baby's formula bottle is too hot? Or if the package of chicken you are buying for dinner has been stored appropriately? Temperature-sensitive materials are being used in these and other applications.

A class of materials called thermochromic materials change color as their temperature increases and they change from the liquid phase to a semicrystalline ordered state. These "liquid crystals" can be incorporated into plastics or paints, and can be used to monitor the temperature of the products or packages in which they are incorporated. For example, some meat packaging now includes a temperature strip that darkens when the meat is stored above a certain temperature, making the meat unsafe to eat. Hospitals and other medical facilities now routinely use temperature strips that, when placed under the tongue or applied to the forehead, change color to indicate the patient's body temperature. There are even clothes that change color based on the air temperature.

Can't decide what color bathing suit to wear to the beach? Pick one that will change color to fit your mood!

See Additional Problems 2.80 and 2.81 at the end of the chapter.

remembering that the freezing point is higher by 32 on the Fahrenheit scale than on the Celsius scale. Thus, if you want to convert from Celsius to Fahrenheit, you do a size adjustment (multiply °C by 9/5) and then a zero-point adjustment (add 32); if you want to convert from Fahrenheit to Celsius, you find out how many Fahrenheit degrees there are above freezing (by subtracting 32) and then do a size adjustment (multiply °C by 5/9). The following formulas show the conversion methods:

Celsius to Fahrenheit:

$$°F = \left(\frac{9\ °F}{5\ °C} \times °C \right) + 32\ °F$$

Fahrenheit to Celsius:

$$°C = \frac{5\ °C}{9\ °F} \times (°F - 32\ °F)$$

WORKED EXAMPLE **2.12** Temperature Conversions: Fahrenheit to Celsius

A body temperature above 107 °F can be fatal. What does 107 °F correspond to on the Celsius scale?

ANALYSIS Using the temperature (in °F) and the appropriate temperature conversion equation we can convert from the Fahrenheit scale to the Celsius scale.

BALLPARK ESTIMATE Note in Figure 2.4 that normal body temperature is 98.6 °F or 37 °C. A temperature of 107 °F is approximately 8 °F above normal; since 1 °C is nearly 2 °F, then 8 °F is about 4 °C. Thus, the 107 °F body temperature is 41 °C.

SOLUTION

STEP 1: **Identify known information.**

STEP 2: **Identify answer and units.**

STEP 3: **Identify conversion factors.**
We can convert from °F to °C using this equation.

STEP 4: **Solve.** Substitute the known temperature (in °F) into the equation.

Temperature = 107 °F

Temperature = ?? °C

$$°C = \frac{5\ °C}{9\ °F} \times (°F - 32\ °F)$$

$$°C = \frac{5\ °C}{9\ °F} \times (107\ °F - 32\ °F) = 42\ °C^*$$

BALLPARK CHECK Close to our estimate of 41 °C.
(Rounded off from 41.666 667 °C)

*It is worth noting that the 5/9 conversion factor in the equation is an exact conversion, and so does not impact the number of significant figures in the final answer.

PROBLEM 2.18

The highest land temperature ever recorded was 136 °F in Al Aziziyah, Libya, on September 13, 1922. What is this temperature on the Celsius scale?

PROBLEM 2.19

The use of mercury thermometers is limited by the fact that mercury freezes at −38.9 °C. To what temperature does this correspond on the Fahrenheit scale? On the Kelvin scale?

2.10 Energy and Heat

All chemical reactions are accompanied by a change in **energy**, which is defined in scientific terms as *the capacity to do work or supply heat* (Figure 2.5). Detailed discussion of the various kinds of energy will be included in Chapter 7, but for now we will look at the various units used to describe energy, and the way heat energy can be gained or lost by matter.

Energy is measured in SI units by the unit *joule* (J; pronounced "jool"), but the metric unit *calorie* (cal) is still widely used in medicine. One calorie is the amount of heat necessary to raise the temperature of 1 g of water by 1 °C. A *kilocalorie* (kcal), often called a *large calorie (Cal)* or *food calorie* by nutritionists, equals 1000 cal:

$$1000 \text{ cal} = 1 \text{ kcal} \qquad 1000 \text{ J} = 1 \text{ kJ}$$
$$1 \text{ cal} = 4.184 \text{ J} \qquad 1 \text{ kcal} = 4.184 \text{ kJ}$$

Not all substances have their temperatures raised to the same extent when equal amounts of heat energy are added. One calorie raises the temperature of 1 g of water by 1 °C but raises the temperature of 1 g of iron by 10 °C. The amount of heat needed to raise the temperature of 1 g of a substance by 1 °C is called the **specific heat** of the substance. It is measured in units of cal/(g·°C).

$$\text{Specific heat} = \frac{\text{Calories}}{\text{Grams} \times °\text{C}}$$

Specific heats vary greatly from one substance to another, as shown in Table 2.6. The specific heat of water, 1.00 cal/(g·°C), is higher than that of most other substances, which means that a large transfer of heat is required to change the temperature of a given amount of water by a given number of degrees. One consequence is that the human body, which is about 60% water, is able to withstand changing outside conditions.

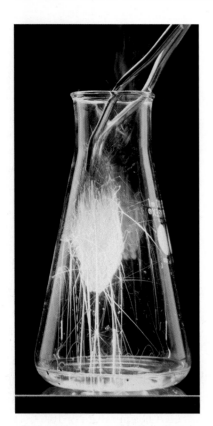

▲ **FIGURE 2.5 The reaction of aluminum with bromine releases energy in the form of heat.** When the reaction is complete, the products undergo no further change.

Energy The capacity to do work or supply heat.

Specific heat The amount of heat that will raise the temperature of 1 g of a substance by 1 °C.

TABLE 2.6 Specific Heats of Some Common Substances

SUBSTANCE	SPECIFIC HEAT [cal/(g · °C)]
Ethanol	0.59
Gold	0.031
Iron	0.106
Mercury	0.033
Sodium	0.293
Water	1.00

Knowing the mass and specific heat of a substance makes it possible to calculate how much heat must be added or removed to accomplish a given temperature change, as shown in Worked Example 2.13.

$$\text{Heat (cal)} = \text{Mass (g)} \times \text{Temperature change (°C)} \times \text{Specific heat}\left(\frac{\text{cal}}{\text{g} \cdot °\text{C}}\right)$$

WORKED EXAMPLE **2.13** Specific Heat: Mass, Temperature, and Energy

Taking a bath might use about 95 kg of water. How much energy (in calories) is needed to heat the water from a cold 15 °C to a warm 40 °C?

ANALYSIS From the amount of water being heated (95 kg) and the amount of the temperature change (40 °C − 15 °C = 25 °C), the total amount of energy needed can be calculated by using specific heat [1.00 cal/(g·°C)] as a conversion factor.

BALLPARK ESTIMATE The water is being heated 25 °C (from 15 °C to 40 °C), and it therefore takes 25 cal to heat each gram. The tub contains nearly 100,000 g (95 kg is 95,000 g), and so it takes about 25 × 100,000 cal, or 2,500,000 cal, to heat all the water in the tub.

SOLUTION

STEP 1: Identify known information.

STEP 2: Identify answer and units.

STEP 3: Identify conversion factors. The amount of energy (in cal) can be calculated using the specific heat of water (cal/g · °C), and will depend on both the mass of water (in g) to be heated and the total temperature change (in °C). In order for the units in specific heat to cancel correctly, the mass of water must first be converted from kg to g.

STEP 4: Solve. Starting with the known information, use the conversion factors to cancel unwanted units.

Mass of water = 95 kg

Temperature change = 40 °C − 15 °C = 25 °C

Heat = ?? cal

$$\text{Specific heat} = \frac{1.0 \text{ cal}}{\text{g} \cdot {}^{\circ}\text{C}}$$

$$1 \text{ kg} = 1000 \text{ g} \rightarrow \frac{1000 \text{ g}}{1 \text{ kg}}$$

$$95 \text{ kg} \times \frac{1000 \text{ g}}{\text{kg}} \times \frac{1.00 \text{ cal}}{\text{g} \cdot {}^{\circ}\text{C}} \times 25 \, {}^{\circ}\text{C} = 2{,}400{,}000 \text{ cal}$$

$$= 2.4 \times 10^6 \text{ cal}$$

BALLPARK CHECK Close to our estimate of 2.5 × 10⁶ cal.

PROBLEM 2.20

Assuming that Coca-Cola has the same specific heat as water, how much energy in calories is removed when 350 g of Coke (about the contents of one 12 oz can) is cooled from room temperature (25 °C) to refrigerator temperature (3 °C)?

PROBLEM 2.21

What is the specific heat of aluminum if it takes 161 cal to raise the temperature of a 75 g aluminum bar by 10.0 °C?

Density The physical property that relates the mass of an object to its volume; mass per unit volume.

2.11 Density

One further physical quantity that we will take up in this chapter is **density**, which relates the mass of an object to its volume. Density is usually expressed in units of grams per cubic centimeter (g/cm³) for solids and grams per milliliter (g/mL) for liquids. Thus, if we know the density of a substance, we know both the mass of a given volume and the volume of a given mass. The densities of some common materials are listed in Table 2.7.

$$\text{Density} = \frac{\text{Mass (g)}}{\text{Volume (mL or cm}^3)}$$

Most substances change their volume by expanding or contracting when heated or cooled, and densities are therefore temperature-dependent. For example, at 3.98 °C a 1 mL container holds exactly 1.0000 g of water (density = 1.0000 g/mL). As the temperature rises, however, the volume occupied by the water expands so that only 0.9584 g fits in the 1 mL container at 100 °C (density = 0.9584 g/mL). When reporting a density, the temperature must also be specified.

Although most substances contract when cooled and expand when heated, water behaves differently. Water contracts when cooled from 100 °C to 3.98 °C, but

▲ Which has the greater mass, the pillow or the brass cylinder? In fact, they have similar masses, but the brass cylinder has a higher *density* because of its smaller volume.

TABLE 2.7   Densities of Some Common Materials at 25 °C

SUBSTANCE	DENSITY*	SUBSTANCE	DENSITY*
Gases		Solids	
Helium	0.000 194	Ice (0 °C)	0.917
Air	0.001 185	Gold	19.3
Liquids		Human fat	0.94
Water (3.98 °C)	1.0000	Cork	0.22–0.26
Urine	1.003–1.030	Table sugar	1.59
Blood plasma	1.027	Balsa wood	0.12
		Earth	5.54

Densities are in g/cm³ for solids and g/mL for liquids and gases.

▲ Ice floats because its density is less than that of water.

below this temperature it begins to *expand* again. The density of liquid water is at its maximum of 1.0000 g/mL at 3.98 °C but decreases to 0.999 87 g/mL at 0 °C. When freezing occurs, the density drops still further to a value of 0.917 g/cm³ for ice at 0 °C. Since a less dense substance will float on top of a more dense fluid, ice and any other substance with a density less than that of water will float. Conversely, any substance with a density greater than that of water will sink.

Knowing the density of a liquid is useful because it is often easier to measure a liquid's volume rather than its mass. Suppose, for example, that you need 1.50 g of ethanol. Rather than use a dropper to weigh out exactly the right amount, it would be much easier to look up the density of ethanol (0.7893 g/mL at 20 °C) and measure the correct volume (1.90 mL) with a syringe or graduated cylinder. Thus, density acts as a conversion factor between mass (g) and volume (mL).

$$1.50 \text{ g ethanol} \times \frac{1 \text{ mL ethanol}}{0.7893 \text{ g ethanol}} = 1.90 \text{ mL ethanol}$$

WORKED EXAMPLE   **2.14** Density: Mass-to-Volume Conversion.

What volume of isopropyl alcohol (rubbing alcohol) would you use if you needed 25.0 g? The density of isopropyl alcohol is 0.7855 g/mL at 20 °C.

ANALYSIS The known information is the mass of isopropyl alcohol needed (25.0 g). The density (0.7855 g/mL) acts as a conversion factor between mass and the unknown volume of isopropyl alcohol.

BALLPARK ESTIMATE Because 1 mL of isopropyl alcohol contains only 0.7885 g of the alcohol, obtaining 1 g of alcohol would require almost 20% more than 1 mL, or about 1.2 mL. Therefore, a volume of about 25 × 1.2 mL = 30 mL is needed to obtain 25 g of alcohol.

SOLUTION

STEP 1: **Identify known information.**

Mass of rubbing alcohol = 25.0 g

Density of rubbing alcohol = 0.7855 g/mL

STEP 2: **Identify answer and units.**

Volume of rubbing alcohol = ?? mL

STEP 3: **Identify conversion factors.** Starting with the mass of isopropyl alcohol (in g), the corresponding volume (in mL) can be calculated using density (g/mL) as the conversion factor.

Density = g/mL → 1/density = mL/g

STEP 4: **Solve.** Starting with the known information, set up the equation with conversion factors so that unwanted units cancel.

$$25.0 \text{ g alcohol} \times \frac{1 \text{ mL alcohol}}{0.7855 \text{ g alcohol}} = 31.8 \text{ mL alcohol}$$

BALLPARK CHECK Our estimate was 30 mL.

APPLICATION ▶ Obesity and Body Fat

According to the U.S. Center for Disease Control, the U.S. population is suffering from a fat epidemic. Over the last 25 years, the percentage of adults (20 years or older) identified as obese increased from 15% in the late 1970s to nearly 33% in 2004. Even children and adolescents are gaining too much weight: the number of overweight children in all age groups increased by nearly a factor of 3, with the biggest increase seen among teenagers (from 5% to 17.4%). How do we define obesity, however, and how is it measured?

Obesity is defined by reference to *body mass index* (BMI), which is equal to a person's mass in kilograms divided by the square of his or her height in meters. Alternatively, a person's weight in pounds divided by the square of her or his height in inches is multiplied by 703 to give the BMI. For instance, someone 5 ft 7 in. (67 inches; 1.70 m) tall weighing 147 lb (66.7 kg) has a BMI of 23:

$$BMI = \frac{\text{weight (kg)}}{[\text{height (m)}]^2}, \quad \text{or} \quad \frac{\text{weight (lb)}}{[\text{height (in.)}]^2} \times 703$$

A BMI of 25 or above is considered overweight, and a BMI of 30 or above is obese. By these standards, approximately 61% of the U.S. population is overweight.

Health professionals are concerned by the rapid rise in obesity in the United States because of the link between BMI and health problems. Many reports have documented the correlation between health and BMI, including a recent study on more than one million adults. The lowest death risk from any cause, including cancer and heart disease, is associated with a BMI between 22 and 24. Risk increases steadily as BMI increases, more than doubling for a BMI above 29.

Fat is not the enemy, however, because having some body fat is not just good, it is necessary! The layer of adipose tissue (body fat) lying just beneath our skin acts as a shock absorber and as a thermal insulator to maintain body temperature. It also serves as a long-term energy storehouse (see Section 25.6). A typical adult body contains about 50% muscles and other cellular material, 24% blood and other fluids, 7% bone, and 19% body fat. Overweight sedentary individuals

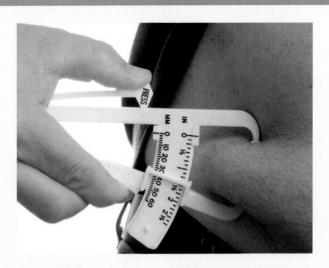

▲ A person's percentage body fat can be estimated by measuring the thickness of the fat layer under the skin.

have a higher fat percentage, whereas some world-class athletes have as little as 3% body fat. The problem occurs when the percentage of body fat is excessive.

An individual's percentage of body fat is most easily measured by the skinfold thickness method. The skin at several locations on the arm, shoulder, and waist is pinched, and the thickness of the fat layer beneath the skin is measured with calipers. Comparing the measured results to those in a standard table gives an estimation of percentage body fat.

As an alternative to skinfold measurement, a more accurate assessment of body fat can be made by underwater immersion. The person's underwater body weight is less than her weight on land because water gives the body buoyancy. The higher the percentage of body fat, the more buoyant the person and the greater the difference between land weight and underwater body weight. Checking the observed buoyancy on a standard table then gives an estimation of body fat.

See Additional Problems 2.82 and 2.83 at the end of the chapter.

Weight (lb)

Height	110	115	120	125	130	135	140	145	150	155	160	165	170	175	180	185	190	195	200
5'0"	21	22	23	24	25	26	27	28	29	30	31	32	33	34	35	36	37	38	39
5'2"	20	21	22	23	24	25	26	27	27	28	29	30	31	32	33	34	35	36	37
5'4"	19	20	21	21	22	23	24	25	26	27	27	28	29	30	31	32	33	33	34
5'6"	18	19	19	20	21	22	23	23	24	25	26	27	27	28	29	30	31	31	32
5'8"	17	17	18	19	20	21	21	22	23	24	24	25	26	27	27	28	29	30	30
5'10"	16	17	17	18	19	19	20	21	22	22	23	24	24	25	26	27	27	28	29
6'0"	15	16	16	17	18	18	19	20	20	21	22	22	23	24	24	25	26	26	27
6'2"	14	15	15	16	17	17	18	19	19	20	21	21	22	22	23	24	24	25	26
6'4"	13	14	15	15	16	16	17	18	18	19	19	20	21	21	22	23	23	24	24

Body Mass Index (numbers in boxes)

PROBLEM 2.22

Which of the solids whose densities are given in Table 2.7 will float on water, and which will sink?

PROBLEM 2.23

Chloroform, once used as an anesthetic agent, has a density of 1.474 g/mL. What volume would you use if you needed 12.37 g?

PROBLEM 2.24

A glass stopper that has a mass of 16.8 g has a volume of 7.60 cm³. What is the density of the glass?

2.12 Specific Gravity

Specific gravity The density of a substance divided by the density of water at the same temperature.

For many purposes, ranging from winemaking to medicine, it is more convenient to use *specific gravity* than density. The **specific gravity** (sp gr) of a substance (usually a liquid) is simply the density of the substance divided by the density of water at the same temperature. Because all units cancel, specific gravity is unitless:

$$\text{Specific gravity} = \frac{\text{Density of substance (g/mL)}}{\text{Density of water at the same temperature (g/mL)}}$$

At normal temperatures, the density of water is very close to 1 g/mL. Thus, the specific gravity of a substance is numerically equal to its density and is used in the same way.

The specific gravity of a liquid can be measured using an instrument called a *hydrometer*, which consists of a weighted bulb on the end of a calibrated glass tube, as shown in Figure 2.6. The depth to which the hydrometer sinks when placed in a fluid indicates the fluid's specific gravity: The lower the bulb sinks, the lower the specific gravity of the fluid.

Water that contains dissolved substances can have a specific gravity either higher or lower than 1.00. In winemaking, for instance, the amount of fermentation taking place is gauged by observing the change in specific gravity on going from grape juice, which contains 20% dissolved sugar and has a specific gravity of 1.082, to dry wine, which contains 12% alcohol and has a specific gravity of 0.984. (Pure alcohol has a specific gravity of 0.789.)

In medicine, a hydrometer called a *urinometer* is used to indicate the amount of solids dissolved in urine. Although the specific gravity of normal urine is about 1.003–1.030, conditions such as diabetes mellitus or a high fever cause an abnormally high urine specific gravity, indicating either excessive elimination of solids or decreased elimination of water. Abnormally low specific gravity is found in individuals using diuretics—drugs that increase water elimination.

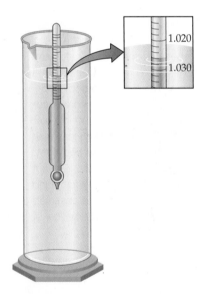

▲ **FIGURE 2.6 A hydrometer for measuring specific gravity.** The instrument has a weighted bulb at the end of a calibrated glass tube. The depth to which the hydrometer sinks in a liquid indicates the liquid's specific gravity.

PROBLEM 2.25

The sulfuric acid solution in an automobile battery typically has a specific gravity of about 1.27. Is battery acid more dense or less dense than pure water?

▲ The amount of fermentation that has taken place in the wine can be measured with a hydrometer.

SUMMARY: REVISITING THE CHAPTER GOALS

1. **How are measurements made, and what units are used?** A property that can be measured is called a *physical quantity* and is described by both a number and a label, or *unit*. The preferred units are either those of the International System of Units (*SI units*) or the *metric system*. Mass, the amount of matter an object contains, is measured in *kilograms* (kg) or *grams* (g). Length is measured in *meters* (m). Volume is measured in *cubic meters* (m³) in the SI system and in *liters* (L) or *milliliters* (mL) in the metric system. Temperature is measured in *kelvins* (K) in the SI system and in *degrees Celsius* (°C) in the metric system.

2. **How good are the reported measurements?** When measuring physical quantities or using them in calculations, it is important to indicate the exactness of the measurement by *rounding off* the final answer using the correct number of *significant figures*. All but one of the significant figures in a number is known with certainty; the final digit is estimated to ±1.

3. **How are large and small numbers best represented?** Measurements of small and large quantities are usually written in *scientific notation* as the product of a number between 1 and 10, times a power of 10. Numbers greater than 10 have a positive exponent, and numbers less than 1 have a negative exponent. For example, $3562 = 3.562 \times 10^3$, and $0.003\,91 = 3.91 \times 10^{-3}$.

4. **How can a quantity be converted from one unit of measure to another?** A measurement in one unit can be converted to another unit by multiplying by a *conversion factor* that expresses the exact relationship between the units.

5. **What techniques are used to solve problems?** Problems are best solved by applying the *factor-label method*, in which units can be multiplied and divided just as numbers can. The idea is to set up an equation so that all unwanted units cancel, leaving only the desired units. Usually it is best to start by identifying the known and needed information, then

KEY WORDS

Conversion factor, *p. 31*

Density, *p. 39*

Energy, *p. 38*

Factor-label method, *p. 31*

Mass, *p. 21*

Physical quantity, *p. 19*

Rounding off, *p. 29*

Scientific notation, *p. 26*

SI units, *p. 19*

Significant figures, *p. 24*

Specific gravity, *p. 42*

Specific heat, *p. 38*

Temperature, *p. 35*

Unit, *p. 19*

Weight, *p. 21*

decide how to convert the known information to the answer, and finally check to make sure the answer is reasonable both chemically and physically.

6. What are temperature, specific heat, density, and specific gravity? *Temperature* is a measure of how hot or cold an object is. The *specific heat* of a substance is the amount of heat necessary to raise the temperature of 1 g of the substance by 1 °C. Water has an unusually high specific heat, which helps our bodies to maintain an even temperature. *Density*, the physical property that relates mass to volume, is expressed in units of grams per milliliter (g/mL) for a liquid or grams per cubic centimeter (g/cm³) for a solid. The *specific gravity* of a liquid is the density of the liquid divided by the density of water at the same temperature. Because the density of water is approximately 1 g/mL, specific gravity and density have the same numerical value.

UNDERSTANDING KEY CONCEPTS

2.26 How many milliliters of water does the graduated cylinder in (a) contain, and how tall in centimeters is the paper clip in (b)? How many significant figures do you have in each answer?

(a)   (b)

2.27 Using a metric ruler, measure the following objects:
  (a) Length of your calculator
  (b) Width of a page in this text book
  (c) Height and diameter of a 12 ounce can

2.28 **(a)** What is the specific gravity of the following solution?
  (b) How many significant figures does your answer have?
  (c) Is the solution more dense or less dense than water?

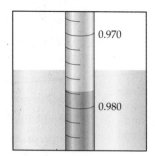

2.29 Assume that you have two graduated cylinders, one with a capacity of 5 mL (a) and the other with a capacity of 50 mL (b). Draw a line in each showing how much liquid you would add if you needed to measure 2.64 mL of water. Which cylinder do you think is more precise? Explain.

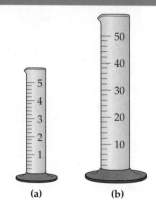

(a)   (b)

2.30 State the length of the pencil depicted in the accompanying figure in both inches and cm.

2.31 Assume that you are delivering a solution sample from a pipette. Figures (a) and (b) show the volume level before and after dispensing the sample, respectively. State the liquid level (in mL) before and after dispensing the sample, and calculate the volume of the sample.

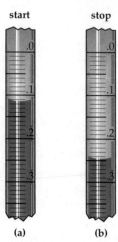

(a)   (b)

2.32 **(a)** Pour the contents of a 12 ounce can into a measuring cup. Record the volume in both fluid ounces and in cups. Which measurement is more precise?

(b) Pour the contents of a 12 ounce can into a 400 mL beaker. What is the volume in mL?

2.33 Assume that identical hydrometers are placed in ethanol (sp gr 0.7893) and in chloroform (sp gr 1.4832). In which liquid will the hydrometer float higher? Explain.

ADDITIONAL PROBLEMS

DEFINITIONS AND UNITS

2.34 What is the difference between a physical quantity and a number?

2.35 What is the difference between mass and weight?

2.36 What are the units used in the SI system to measure mass, volume, length, and temperature?

2.37 What are the units used in the metric system to measure mass, volume, length, and temperature?

2.38 What is the difference between a cubic decimeter (SI) and a liter (metric)?

2.39 What is the difference between a kelvin (SI) and a Celsius degree (metric)?

2.40 Give the full name of the following units:

(a) cL **(b)** dm **(c)** mm
(d) nL **(e)** mg **(f)** m^3
(g) cc

2.41 Write the symbol for the following units:

(a) nanogram **(b)** centimeter **(c)** microlliter
(d) micrometer **(e)** milligram

2.42 How many picograms are in 1 mg? In 35 ng?

2.43 How many microliters are in 1 L? In 20 mL?

SCIENTIFIC NOTATION AND SIGNIFICANT FIGURES

2.44 Express the following numbers in scientific notation:

(a) 9457 **(b)** 0.000 07
(c) 20,000,000,000 (four significant figures)
(d) 0.012 345 **(e)** 652.38

2.45 Convert the following numbers from scientific notation to standard notation:

(a) 5.28×10^3 **(b)** 8.205×10^{-2}
(c) 1.84×10^{-5} **(d)** 6.37×10^4

2.46 How many significant figures does each of the following numbers have?

(a) 237,401 **(b)** 0.300 **(c)** 3.01
(d) 244.4 **(e)** 50,000 **(f)** 660

2.47 How many significant figures are there in each of the following quantities?

(a) Distance from New York City to Wellington, New Zealand, 14,397 km

(b) Average body temperature of a crocodile, 25.6 °C

(c) Melting point of gold, 1064 °C

(d) Diameter of an influenza virus, 0.000 01 mm

(e) Radius of a phosphorus atom, 0.110 nm

2.48 The diameter of the earth at the equator is 7926.381 mi.

(a) Round off the earth's diameter to four significant figures, to two significant figures, and to six significant figures.

(b) Express the earth's diameter in scientific notation.

2.49 Round off each of the numbers in Problem 2.46 to two significant figures, and express them in scientific notation.

2.50 Carry out the following calculations, express each answer to the correct number of significant figures, and include units in the answers.

(a) 9.02 g + 3.1 g
(b) 88.80 cm + 7.391 cm
(c) 362 mL − 99.5 mL
(d) 12.4 mg + 6.378 mg + 2.089 mg

2.51 Carry out the following calculations, express the answers to the correct numbers of significant figures, and include units in the answers.

(a) $5,280 \dfrac{ft}{mi} \times 6.2 \ mi$

(b) 4.5 m × 3.25 m

(c) $2.50 \ g \div 8.3 \dfrac{g}{cm^3}$

(d) 4.70 cm × 6.8 cm × 2.54 cm

UNIT CONVERSIONS AND PROBLEM SOLVING

2.52 Carry out the following conversions:

(a) 3.614 mg to centigrams
(b) 12.0 kL to megaliters
(c) 14.4 μm to millimeters
(d) 6.03×10^{-6} cg to nanograms
(e) 174.5 mL to deciliters
(f) 1.5×10^{-2} km to centimeters

2.53 Carry out the following conversions. Consult Tables 2.3–2.5 as needed.

(a) 56.4 mi to kilometers and to megameters
(b) 2.0 L to to quarts and to fluid ounces
(c) 7 ft 2.0 in. to centimeters and to meters
(d) 1.35 lb to kilograms and to decigrams

2.54 Express the following quantities in more convenient units by using SI unit prefixes:

(a) 9.78×10^4 g **(b)** 1.33×10^{-4} L
(c) 0.000 000 000 46 g **(d)** 2.99×10^8 cm

2.55 Which SI unit prefix corresponds to each of the following multipliers?

(a) 10^3 **(b)** 10^{-3}
(c) 10^6 **(d)** 10^{-6}

2.56 The speed limit in Canada is 100 km/h.

(a) How many miles per hour is this?
(b) How many feet per second?

2.57 The muzzle velocity of a projectile fired from a 9 mm handgun is 1200 ft/s.

(a) How many miles per hour is this?
(b) How many meters per second?

2.58 The diameter of a red blood cell is 6×10^{-6} m. How many red blood cells are needed to make a line 1 in. long?

2.59 The Sears Tower in Chicago has an approximate floor area of 418,000 m^2. How many square feet of floor is this?

2.60 A normal value for blood cholesterol is 200 mg/dL of blood. If a normal adult has a total blood volume of 5 L, how much total cholesterol is present?

2.61 One bottle of aspirin holds 50 tablets, each containing 250 mg of aspirin, and sells for $1.95. Another bottle holds 100 tablets, each containing 200 mg of aspirin, and sells for $3.75. For each bottle, calculate the value in milligrams of aspirin per dollar. Which bottle is the better bargain?

2.62 The white blood cell concentration in normal blood is approximately 12,000 cells/mm^3 of blood. How many white blood cells does a normal adult with 5 L of blood have? Express the answer in scientific notation.

2.63 The recommended daily dose of calcium for an 18-year-old male is 1200 mg. If 1.0 cup of whole milk contains 290 mg of calcium and milk is his only calcium source, how much milk should an 18-year-old male drink each day?

ENERGY, HEAT, AND TEMPERATURE

2.64 What is the normal temperature of the human body (98.6 °F) in degrees Celsius? In kelvins?

2.65 The boiling point of liquid nitrogen, used in the removal of warts and in other surgical applications, is −195.8 °C. What is this temperature in kelvins and in degrees Fahrenheit?

2.66 Diethyl ether, a substance once used as a general anesthetic, has a specific heat of 0.895 cal/(g·°C). How many calories and how many kilocalories of heat are needed to raise the temperature of 30.0 g of diethyl ether from 10.0 °C to 30.0 °C?

2.67 Copper has a specific heat of 0.092 cal/(g·°C). When 52.7 cal of heat is added to a piece of copper, the temperature increases from 22.4 °C to 38.6 °C. What is the mass of the piece of copper?

2.68 Calculate the specific heat of copper if it takes 23 cal to heat a 5.0 g sample from 25 °C to 75 °C.

2.69 The specific heat of fat is 0.45 cal/(g·°C), and the density of fat is 0.94 g/cm^3. How much energy (in calories) is needed to heat 10 cm^3 of fat from room temperature (25 °C) to its melting point (35 °C)?

2.70 A 150 g sample of mercury and a 150 g sample of iron are at an initial temperature of 25.0 °C. If 250 cal of heat is applied to each sample, what is the final temperature of each? (See Table 2.6.)

2.71 When 100 cal of heat is applied to a 125 g sample, the temperature increases by 28 °C. Calculate the specific heat of the sample and compare your answer to the values in Table 2.6. What is the identity of the sample?

DENSITY AND SPECIFIC GRAVITY

2.72 Aspirin has a density of 1.40 g/cm^3. What is the volume in cubic centimeters of a tablet weighing 250 mg?

2.73 Gaseous hydrogen has a density of 0.0899 g/L at 0 °C. How many liters would you need if you wanted 1.0078 g of hydrogen?

2.74 What is the density of lead (in g/cm^3) if a rectangular bar measuring 0.500 cm in height, 1.55 cm in width, and 25.00 cm in length has a mass of 220.9 g?

2.75 What is the density of lithium metal (in g/cm^3) if a cube measuring 0.82 cm $\times$ 1.45 cm $\times$ 1.25 cm has a mass of 0.794 g?

2.76 What is the density of isopropyl alcohol (rubbing alcohol) in grams per milliliter if a 5.000 mL sample has a mass of 3.928 g at room temperature? What is the specific gravity of isopropyl alcohol?

2.77 Ethylene glycol, commonly used as automobile antifreeze, has a specific gravity of 1.1088 at room temperature. What is the volume of 1.00 kg of ethylene glycol? What is the volume of 2.00 lb of ethylene glycol?

Applications

2.78 The typical rhinovirus has a diameter of 20 nm, or 2.0×10^{-8} m

  (a) What is the length in centimeters?
  (b) How many rhinoviruses would need to be laid end to end to make a chain 1 in. long? [*Powers of 10, p. 28*]

2.79 Blood cells have a mean cell volume of 90 fL, or 9.0×10^{-14} L

  (a) Convert this volume to cubic centimeters.
  (b) The formula for the volume of a sphere is $V = 4\pi r^2/3$. Assuming a spherical shape, calculate the mean diameter of a blood cell in centimeters. [*Powers of 10, p. 28*]

2.80 A thermochromic plastic chip included in a shipping container for beef undergoes an irreversible color change if the storage temperature exceeds 28 °F. What is this temperature on the Celsius and Kelvin scales? [*Temperature-Sensitive Materials, p. 36*]

2.81 A temperature-sensitive bath toy undergoes several color changes in the temperature range from 37 °C to 47 °C. What is the corresponding temperature range on the Fahrenheit scale? [*Temperature-Sensitive Materials, p. 36*]

2.82 Calculate the BMI for an individual who is:
  (a) 5 ft 1 in. tall and weighs 155 lb
  (b) 5 ft 11 in. tall and weigh 170 lb
  (c) 6 ft 3 in. tall and weigh 195 lb

  Which of these individuals is likely to have increased health risks? [*Obesity and Body Fat, p. 41*]

2.83 Liposuction is a technique for removing fat deposits from various areas of the body. How many liters of fat would have to be removed to result in a 5.0 lb weight loss? The density of human fat is 0.94 g/mL. [*Measuring Body Fat, p.41*]

General Questions and Problems

2.84 Refer to the pencil in Problem 2.30. Using the equivalent values in Table 2.4 as conversion factors, convert the length measured in inches to centimeters. Compare the calculated length in centimeters to the length in centimeters measured using the metric ruler. How do the two values compare? Explain any differences.

2.85 Gemstones are weighed in carats, where 1 carat = 200 mg exactly. What is the mass in grams of the Hope diamond, the world's largest blue diamond, at 44.4 carats?

2.86 If you were cooking in an oven calibrated in Celsius degrees, what temperature would you use if the recipe called for 350 °F?

2.87 What dosage in grams per kilogram of body weight does a 130 lb woman receive if she takes two 250 mg tablets of penicillin? How many 125 mg tablets should a 40 lb child take to receive the same dosage?

2.88 A clinical report gave the following data from a blood analysis: iron, 39 mg/dL; calcium, 8.3 mg/dL; cholesterol, 224 mg/dL. Express each of these quantities in grams per deciliter, writing the answers in scientific notation.

2.89 The density of air at room temperature is 1.3 g/L. What is the mass of the air **(a)** in grams and **(b)** in pounds in a room that is 4.0 m long, 3.0 m wide, and 2.5 m high?

2.90 Approximately 75 mL of blood is pumped by a normal human heart at each beat. Assuming an average pulse of 72 beats per minute, how many milliliters of blood are pumped in one day?

2.91 A doctor has ordered that a patient be given 15 g of glucose, which is available in a concentration of 50.00 g glucose/1000.0 mL of solution. What volume of solution should be given to the patient?

2.92 Reconsider the volume of the sample dispensed by pipette in Problem 2.31. Assuming that the solution in the pipette has a density of 0.963 g/mL, calculate the mass of solution dispensed in the problem to the correct number of significant figures.

2.93 Today thermometers containing mercury are used less frequently than in the past because of concerns regarding the toxicity of mercury and because of its relatively high melting point (−39 °C), which means that mercury thermometers cannot be used in very cold environments because the mercury is a solid under such conditions. Alcohol thermometers, however, can be used over a temperature range from −115 °C (the melting point of alcohol) to 78.5 °C (the boiling point of alcohol).

(a) What is the effective temperature range of the alcohol thermometer in °F?

(b) The densities of alcohol and mercury are 0.79 g/mL and 13.6 g/mL, respectively. If the volume of liquid in a typical laboratory thermometer is 1.0 mL, what mass of alcohol is contained in the thermometer? What mass of mercury?

2.94 In a typical person, the level of glucose (also known as blood sugar) is about 85 mg/100 mL of blood. If an average body contains about 11 pints of blood, how many grams and how many pounds of glucose are present in the blood?

2.95 A patient is receiving 3000 mL/day of a solution that contains 5 g of dextrose (glucose) per 100 mL of solution. If glucose provides 4 kcal/g of energy, how many kilocalories per day is the patient receiving from the glucose?

2.96 A rough guide to fluid requirements based on body weight is 100 mL/kg for the first 10 kg of body weight, 50 mL/kg for the next 10 kg, and 20 mL/kg for weight over 20 kg. What volume of fluid per day is needed by a 55 kg woman? Give the answer with two significant figures.

2.97 Chloral hydrate, a sedative and sleep-inducing drug, is available as a solution labeled 10.0 gr/fluidram. What volume in milliliters should be administered to a patient who is meant to receive 7.5 gr per dose? (1 gr = 64.8 mg; 1 fluidram = 3.72 mL)

2.98 When 1.0 tablespoon of butter is burned or used by our body, it releases 100 kcal (100 food calories) of energy. If we could use all the energy provided, how many tablespoons of butter would have to be burned to raise the temperature of 3.00 L of water from 18.0 °C to 90.0 °C?

2.99 An archeologist finds a 1.62 kg goblet that she believes to be made of pure gold. When 1350 cal of heat is added to the goblet, its temperature increases by 7.8 °C. Calculate the specific heat of the goblet. Is it made of gold? Explain.

2.100 In another test, the archeologist in Problem 2.99 determines that the volume of the goblet is 205 mL. Calculate the density of the goblet and compare it with the density of gold (19.3 g/mL), lead (11.4 g/mL), and iron (7.86 g/mL). What is the goblet probably made of?

2.101 The density of sulfuric acid, H_2SO_4, is 1.83 g/mL. What volume of sulfuric acid is needed to obtain 98.0 g of H_2SO_4?

2.102 Sulfuric acid (Problem 2.101) is produced in larger amounts than any other chemical: 2.01×10^{11} lb worldwide in 2004. What is the volume of this amount in liters?

2.103 The caliber of a gun is expressed by measuring the diameter of the gun barrel in hundredths of an inch. A "22" rifle, for example, has a barrel diameter of 0.22 in. What is the barrel diameter of a .22 rifle in millimeters?

2.104 Amounts of substances dissolved in solution are often expressed as mass per unit volume. For example, normal human blood has a cholesterol concentration of about 200 mg of cholesterol/100 mL of blood. Express this concentration in the following units:

(a) milligrams per liter   **(b)** micrograms per liter
(c) grams per liter   **(d)** nanograms per microliter

2.105 The element gallium (Ga) has the second largest liquid range of any element, melting at 29.8 °C and boiling at 2204 °C at atmospheric pressure.

(a) Is gallium a metal, a nonmetal, or a metalloid?
(b) What is the density of gallium in grams per cubic centimeter at 25 °C if a 1.00 in. cube has a mass of 0.2133 lb?

2.106 A sample of water at 293.2 K was heated for 8 min and 25 s, and the heating was carried out so that the temperature increased at a constant rate of 3.0 °F/min. What is the final temperature of the water in degrees Celsius?

2.107 At a certain point, the Celsius and Fahrenheit scales "cross," and at this point the numerical value of the Celsius temperature is the same as the numerical value of the Fahrenheit temperature. At what temperature does this crossover occur?

2.108 Imagine that you place a piece of cork measuring 1.30 cm × 5.50 cm × 3.00 cm in a pan of water and that on top of the cork you place a small cube of lead measuring 1.15 cm on each edge. The density of cork is 0.235 g/cm³, and the density of lead is 11.35 g/cm³. Will the combination of cork plus lead float or sink?

CHAPTER 3

Atoms and the Periodic Table

▲ The mineral formations of the Giant's Causeway in Northern Ireland are an example of periodicity— the presence of regularly repeating patterns—found throughout nature.

CONTENTS

CHAPTER GOALS

We will take up the following questions about atoms and atomic theory:

1. What is the modern theory of atomic structure?

THE GOAL: Be able to explain the major assumptions of atomic theory.

2. How do atoms of different elements differ?

THE GOAL: Be able to explain the composition of different atoms according to the number of protons, neutrons, and electrons they contain.

3. What are isotopes, and what is atomic weight?

THE GOAL: Be able to explain what isotopes are and how they affect an element's atomic weight.

4. How is the periodic table arranged?

THE GOAL: Be able to describe how elements are arranged in the periodic table, name the subdivisions of the periodic table, and relate the position of an element in the periodic table to its electronic structure.

5. How are electrons arranged in atoms?

THE GOAL: Be able to explain how electrons are distributed in shells and subshells around the nucleus of an atom, how valence electrons can be represented as electron-dot symbols, and how the electron configurations can help explain the chemical properties of the elements.

C hemistry must be studied on two levels. In the past two chapters we have dealt with chemistry on the large-scale, or *macroscopic*, level, looking at the properties and transformations of matter that we can see and measure. Now we are ready to look at the sub-microscopic, or atomic level, studying the behavior and properties of individual atoms. Although scientists have long been convinced of their existence, only within the past twenty years have powerful new instruments made it possible to see individual atoms themselves. In this chapter we will look at modern atomic theory and at how the structure of atoms influences macroscopic properties.

3.1 Atomic Theory

Take a piece of aluminum foil and cut it in two. Then take one of the pieces and cut *it* in two, and so on. Assuming that you have extremely small scissors and extraordinary dexterity, how long can you keep dividing the foil? Is there a limit, or is matter infinitely divisible into ever smaller and smaller pieces? Historically, this argument can be traced as far back as the ancient Greek philosophers. Aristotle believed that matter could be divided infinitely, while Democritus argued (correctly) that there is a limit. The smallest and simplest bit that aluminum (or any other element) can be divided into and still be identifiable as aluminum is called an **atom**, a word derived from the Greek *atomos*, meaning "indivisible."

Chemistry is founded on four fundamental assumptions about atoms and matter, which together make up modern **atomic theory**:

- All matter is composed of atoms.
- The atoms of a given element differ from the atoms of all other elements.
- Chemical compounds consist of atoms combined in specific ratios. That is, only whole atoms can combine—one A atom with one B atom, or one A atom with two B atoms, and so on. The enormous diversity in the substances we see around us is based on the vast number of ways that atoms can combine with one another.
- Chemical reactions change only the way that atoms are combined in compounds. The atoms themselves are unchanged.

Atoms are extremely small, ranging from about 7.4×10^{-11} m in diameter for a hydrogen atom to 5.24×10^{-10} m for a cesium atom. In mass, atoms vary from 1.67×10^{-24} g for hydrogen to 3.95×10^{-22} g for uranium, one of the heaviest naturally occurring atoms. It is difficult to appreciate just how small atoms are, although it might help if you realize that a fine pencil line is about 3 *million* atoms across, and that even the smallest speck of dust contains about 10^{16} atoms.

Atom The smallest and simplest particle of an element.

Atomic theory A set of assumptions proposed by the English scientist John Dalton to explain the chemical behavior of matter.

▲ How small can a piece of aluminum foil be cut?

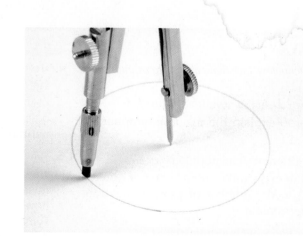

► Atoms are so small that the radius of this circle is about three million atoms wide.

Subatomic particles Three kinds of fundamental particles from which atoms are made: protons, neutrons, and electrons.

Proton A positively charged subatomic particle.

Neutron An electrically neutral subatomic particle.

Electron A negatively charged subatomic particle.

Atomic mass unit (amu) A convenient unit for describing the mass of an atom; 1 amu = $\frac{1}{12}$ the mass of a carbon-12 atom.

Nucleus The dense, central core of an atom that contains protons and neutrons.

Atoms are composed of tiny **subatomic particles** called *protons, neutrons,* and *electrons*. A **proton** has a mass of $1.672\ 622 \times 10^{-24}$ g and carries a positive (+) electrical charge; a **neutron** has a mass similar to that of a proton ($1.674\ 927 \times 10^{-24}$ g) but is electrically neutral; and an **electron** has a mass that is only 1/1836 that of a proton ($9.109\ 328 \times 10^{-28}$ g) and carries a negative (−) electrical charge. In fact, electrons are so much lighter than protons and neutrons that their mass is usually ignored. Table 3.1 compares the properties of the three fundamental subatomic particles.

TABLE 3.1 A Comparison of Subatomic Particles

NAME	SYMBOL	MASS		CHARGE (CHARGE UNITS)
		(GRAMS)	(AMU)	
Proton	p	$1.672\ 622 \times 10^{-24}$	1.007 276	+1
Neutron	n	$1.674\ 927 \times 10^{-24}$	1.008 665	0
Electron	e^-	$9.109\ 328 \times 10^{-28}$	$5.485\ 799 \times 10^{-4}$	−1

The masses of atoms and their constituent subatomic particles are so small when measured in grams that it is more convenient to express them on a *relative mass scale*. That is, one atom is assigned a mass, and all others are measured relative to it. The process is like deciding that a golf ball (46.0 g) will be assigned a mass of 1. A baseball (149 g), which is 149/46.0 = 3.24 times heavier than a golf ball, would then have a mass of about 3.24, a volleyball (270 g) would have a mass of 270/46.0 = 5.87, and so on.

The basis for the relative atomic mass scale is an atom of carbon that contains 6 protons and 6 neutrons. Such an atom is assigned a mass of exactly 12 **atomic mass units** (amu; also called a *dalton* in honor of the English scientist John Dalton, who proposed most of atomic theory as we know it), where 1 amu = $1.660\ 539 \times 10^{-24}$ g. Thus, for all practical purposes, both a proton and a neutron have a mass of 1 amu (Table 3.1). Hydrogen atoms are only about one-twelfth as heavy as carbon atoms and have a mass close to 1 amu, magnesium atoms are about twice as heavy as carbon atoms and have a mass close to 24 amu, and so forth.

Subatomic particles are not distributed at random throughout an atom. Rather, the protons and neutrons are packed closely together in a dense core called the **nucleus**. Surrounding the nucleus, the electrons move about rapidly through a large, mostly empty volume of space (Figure 3.1). Measurements show that the diameter of a nucleus is only about 10^{-15} m, whereas that of the atom itself is about 10^{-10} m. For comparison, if an atom were the size of a large domed stadium, the nucleus would be approximately the size of a small pea in the center of the playing field.

▲ The relative size of a nucleus in an atom is the same as that of a pea in the middle of this stadium.

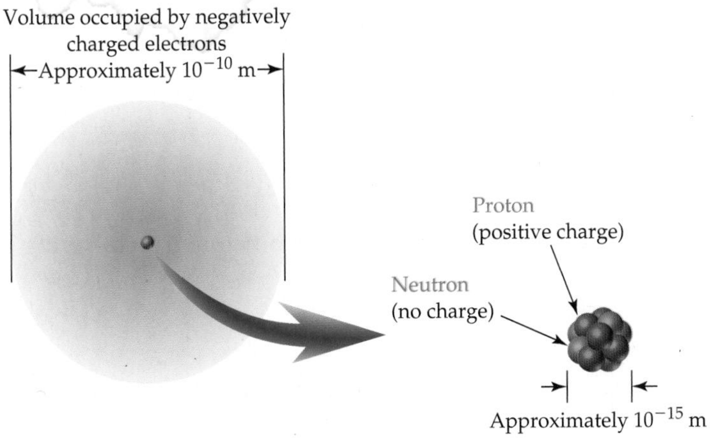

Volume occupied by negatively
charged electrons
←Approximately 10^{-10} m→

Proton
(positive charge)

Neutron
(no charge)

Approximately 10^{-15} m

◀ **FIGURE 3.1 The structure of
an atom.** Protons and neutrons are
packed together in the nucleus,
whereas electrons move about in the
large surrounding volume. Virtually all
the mass of an atom is concentrated
in the nucleus.

The structure of the atom is determined by an interplay of different attractive
and repulsive forces. Because unlike charges attract one another, the negatively
charged electrons are held near the positively charged nucleus. But because like
charges repel one another, the negatively charged electrons try to get as far away
from one another as possible, accounting for the relatively large volume they occupy.
The positively charged protons in the nucleus also repel one another, but are nevertheless held together by a unique attraction called the *nuclear strong force*, which is
beyond the scope of this text.

Electrons repel
one another

Protons repel
one another

Protons and electrons
attract one another

WORKED EXAMPLE **3.1** Atomic Mass Units: Gram-to-Atom Conversions

How many atoms are in a small piece of aluminum foil with a mass of 0.100 g? The mass of an atom of aluminum
is 27.0 amu.

ANALYSIS We know the sample mass in grams and the mass of one atom in atomic mass units. To find the number
of atoms in the sample, two conversions are needed, the first between grams and atomic mass units and the second
between atomic mass units and the number of atoms. The conversion factor between atomic mass units and grams
is 1 amu = $1.660\,539 \times 10^{-24}$ g.

BALLPARK ESTIMATE An atom of aluminum has a mass of 27.0 amu; since 1 amu $\sim 10^{-24}$ g, the mass of a single
aluminum atom is very small ($\approx 10^{-23}$ g). A very *large* number of atoms, therefore (10^{22} ?), is needed to obtain a mass
of 0.100 g.

SOLUTION

STEP 1: Identify known information.

Mass of aluminum foil = 0.100 g

1 Al atom = 27.0 amu

STEP 2: Identify unknown answer and units.

Number of Al atoms = ?

STEP 3: Identify needed conversion factors.
Knowing the mass of foil (in g) and the mass of
individual atoms (in amu) we need to convert from
amu/atom to g/atom.

1 amu = $1.660\,539 \times 10^{-24}$ g

$$\rightarrow \frac{1\ \text{amu}}{1.660\,539 \times 10^{-24}\ \text{g}}$$

STEP 4: Solve. Set up an equation using known
information and conversion factors so that unwanted
units cancel.

$$(0.100\ \text{g})\left(\frac{1\ \text{amu}}{1.660\,539 \times 10^{-24}\ \text{g}}\right)\left(\frac{1\ \text{Al atom}}{27.0\ \text{amu}}\right)$$

$$= 2.23 \times 10^{21}\ \text{Al atoms}$$

BALLPARK CHECK: Our estimate was 10^{22}, which is
within a factor of 10.

PROBLEM 3.1

What is the mass in atomic mass units of a nitrogen atom weighing 2.33×10^{-23} g?

PROBLEM 3.2

What is the mass in grams of 150×10^{12} iron atoms, each having a mass of 56 amu?

PROBLEM 3.3

How many atoms are in each of the following?

(a) 1.0 g of hydrogen atoms, each of mass 1.0 amu
(b) 12.0 g of carbon atoms, each of mass 12.0 amu
(c) 23.0 g of sodium atoms, each of mass 23.0 amu

PROBLEM 3.4

What pattern do you see in your answers to Problem 3.3? (We will return to this very important pattern in Chapter 6.)

APPLICATION ▶ Are Atoms Real?

Chemistry rests on the premise that matter is composed of the tiny particles we call atoms. Every chemical reaction and every physical law that governs the behavior of matter are explained by chemists in terms of atomic theory. But how do we know that atoms are real and not just an imaginary concept? And how do we know the structure of the atom? Our understanding of atomic structure is another example of the scientific method at work.

Dalton's atomic theory was originally published in 1808, but many prominent scientists dismissed it. Over the next century, however, several unrelated experiments provided insight into the nature of matter and the structure of the atom. Nineteenth-century investigations into electricity, for example, demonstrated that matter was composed of charged particles—rubbing a glass rod with a silk cloth would generate "static electricity," the same phenomenon that shocks you when you walk across a carpet and then touch a metal surface. It was also known that passing electricity through certain substances, such as water, decomposed the compounds into their constituent elements.

In 1897, J. J. Thomson used magnetic fields to deflect a beam of particles emitted from a cathode ray tube, demonstrating that these particles had a negative charge and were 1000 times lighter than the lightest charged particles found in aqueous solution (H^+). (⊂◯⊃ Section 6.10 and Chapter 10) This result implied that atoms were *not* the smallest particles of matter but could be divided into even smaller particles. In 1909, Robert Millikan studied the behavior of oil droplets in a

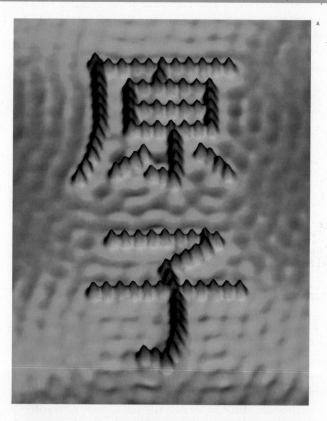

▲ Colored scanning tunneling micrograph of the surface of graphite, a form of the element carbon. The hexagonal pattern is related to an identical arrangement of the carbon atoms.

chamber containing electrically charged plates. When Millikan bombarded the atmosphere in the chamber with x-rays, the droplets acquired a negative charge. Millikan observed that by varying the strength of the electric field associated with the charged plates he could alter the rate of descent of the droplets or even suspend them in midair! From these results he was able to calculate the charge associated with the electron (16×10^{-19} coulombs). Finally, in 1910 Ernest Rutherford bombarded a gold foil with positively charged "alpha" particles emitted from radium during radioactive decay. The majority of these particles passed straight through the foil, but a small fraction of them were deflected, and a few even bounced backward. From these results Rutherford deduced that an atom consists mostly of empty space (occupied by the negatively charged electrons) and that most of the mass and all of the positive charge are contained in a relatively small, dense region that he called the nucleus.

We can now actually "see" and manipulate individual atoms through the use of a device called a *scanning tunneling microscope*, or STM (⬤⬤⬤ see Application on p. 28). With the STM, invented in 1981 by a research team at the IBM Corporation, magnifications of up to ten million have been achieved, allowing chemists to look directly at atoms. The accompanying photograph shows a computer-enhanced representation of carbon atoms in graphite that have been deposited on a copper surface.

Most present uses of the STM involve studies of surface chemistry, such as the events accompanying the corrosion of metals and the ordering of large molecules in polymers. Work is also under way using the STM to determine the structures of complex biological molecules.

See Additional Problems 3.90 and 3.91 at the end of the chapter.

3.2 Elements and Atomic Number

Atoms of different elements differ from one another according to how many protons they contain, a value called the element's **atomic number (Z)**. Thus, if we know the number of protons in an atom, we can identify the element. Any atom with 6 protons, for example, is a carbon atom because carbon has $Z = 6$.

Atoms are neutral overall and have no net charge because the number of positively charged protons in an atom is the same as the number of negatively charged electrons. Thus, the atomic number also equals the number of electrons in every atom of a given element. Hydrogen, $Z = 1$, has only 1 proton and 1 electron; carbon, $Z = 6$, has 6 protons and 6 electrons; sodium, $Z = 11$, has 11 protons and 11 electrons; and so on up to the newly discovered element with $Z = 118$. In a periodic table, elements are listed in order of increasing atomic number, beginning at the upper left and ending at the lower right.

Atomic number (Z) The number of protons in atoms of a given element; the number of electrons in atoms of a given element

The sum of the protons and neutrons in an atom is called the atom's **mass number (A)**. Hydrogen atoms with 1 proton and no neutrons have mass number 1; carbon atoms with 6 protons and 6 neutrons have mass number 12; sodium atoms with 11 protons and 12 neutrons have mass number 23; and so on. Except for hydrogen, atoms generally contain at least as many neutrons as protons, and frequently contain more. There is no simple way to predict how many neutrons a given atom will have.

Mass number (A) The total number of protons and neutrons in an atom

WORKED EXAMPLE **3.2** Atomic Structure: Protons, Neutrons, and Electrons

Phosphorus has atomic number $Z = 15$. How many protons, electrons, and neutrons are there in phosphorus atoms, which have mass number $A = 31$?

ANALYSIS The atomic number gives the number of protons, which is the same as the number of electrons, and the mass number gives the total number of protons plus neutrons.

SOLUTION

Phosphorus atoms, with Z = 15, have 15 protons and 15 electrons. To find the number of neutrons, subtract the atomic number from the mass number:

Mass number — — Atomic number
(sum of protons and neutrons) (number of protons)

$$31 - 15 = 16 \text{ neutrons}$$

WORKED EXAMPLE **3.3** Atomic Structure: Atomic Number and Atomic Mass

An atom contains 28 protons and has $A = 60$. Give the number of electrons and neutrons in the atom, and identify the element.

ANALYSIS The number of protons and the number of electrons are the same and are equal to the atomic number Z, 28 in this case. Subtracting the number of protons (28) from the total number of protons plus neutrons (60) gives the number of neutrons.

SOLUTION

The atom has 28 electrons and $60 - 28 = 32$ neutrons. Looking at the list of elements inside the front cover shows that the element with atomic number 28 is nickel (Ni).

PROBLEM 3.5

Use the list inside the front cover to identify elements with the following atomic numbers:

(a) Z = 75

(b) Z = 20

(c) Z = 52

PROBLEM 3.6

The cobalt used in cancer treatments has $Z = 27$ and $A = 60$. How many protons, neutrons, and electrons are in these cobalt atoms?

PROBLEM 3.7

A certain atom has $A = 98$ and contains 55 neutrons. Identify the element.

3.3 Isotopes and Atomic Weight

All atoms of a given element have the same number of protons—the atomic number Z characteristic of that element—but different atoms of an element can have different numbers of neutrons and therefore different mass numbers. Atoms with identical atomic numbers but different mass numbers are called **isotopes**. Hydrogen, for example, has three isotopes. The most abundant hydrogen isotope, called *protium*, has no neutrons and thus has a mass number of 1. A second hydrogen isotope, called *deuterium*, has one neutron and a mass number of 2; and a third isotope, called *tritium*, has two neutrons and a mass number of 3. Tritium is unstable and does not occur naturally in significant amounts, although it can be made in nuclear reactors (⬤◯⬤, see Section 11.11).

Isotopes Atoms with identical atomic numbers but different mass numbers.

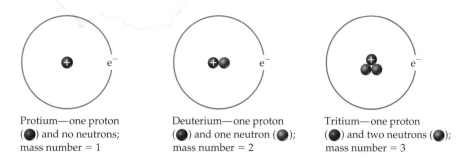

Protium—one proton
(●) and no neutrons;
mass number = 1

Deuterium—one proton
(●) and one neutron (●);
mass number = 2

Tritium—one proton
(●) and two neutrons (●);
mass number = 3

A specific isotope is represented by showing its mass number (A) as a super-script and its atomic number (Z) as a subscript in front of the atomic symbol, for example, $_Z^AX$, where X represents the symbol for the element. Thus, protium is $_1^1H$, deuterium is $_1^2H$, and tritium is $_1^3H$.

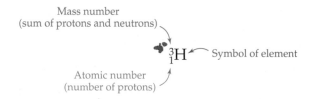

Mass number
(sum of protons and neutrons)

$_1^3H$ ←— Symbol of element

Atomic number
(number of protons)

Unlike the three isotopes of hydrogen, the isotopes of most elements do not have distinctive names. Instead, the mass number of the isotope is given after the name of the element. The $_{92}^{235}U$ isotope used in nuclear reactors, for example, is usually referred to as uranium-235, or U-235.

Most naturally occurring elements are mixtures of isotopes. In a large sample of naturally occurring hydrogen atoms, for example, 99.985% have mass number $A = 1$ (protium) and 0.015% have mass number $A = 2$ (deuterium). It is therefore useful to know the *average* mass of the atoms in a large sample, a value called the element's **atomic weight**. For hydrogen, the atomic weight is 1.008 amu. Atomic weights for all elements are given inside the front cover of this book.

To calculate the atomic weight of an element, the individual masses of the naturally occurring isotopes and the percentage of each must be known. The atomic weight can then be calculated as the sum of the masses of the individual isotopes for that element, or

$$\text{Atomic weight} = \Sigma(\text{isotopic abundance}) \times (\text{isotopic mass})$$

where the Greek symbol Σ indicates the mathematical summing of terms.

Chlorine, for example, occurs on earth as a mixture of 75.77% Cl-35 atoms (mass = 34.97 amu) and 24.23% Cl-37 atoms (mass = 36.97 amu). The atomic weight is found by calculating the percentage of the mass contributed by each isotope. For chlorine, the calculation is done in the following way (to four significant figures), giving an atomic weight of 35.45 amu:

Contribution from ^{35}Cl: (0.7577)(34.97 amu) = 26.4968 amu

Contribution from ^{37}Cl: (0.2423)(36.97 amu) = $\underline{\text{8.9578 amu}}$

Atomic weight = 35.4546 = 35.45 amu
(rounded to four significant figures)

The final number of significant figures in this case (four) was determined by the atomic masses. Note that the final rounding to four significant figures was not done until *after* the final answer was obtained.

Atomic weight The weighted average mass of an element's atoms.

WORKED EXAMPLE 3.4 Average Atomic Mass: Weighted-Average Calculation

Gallium is a metal with a very low melting point—it will melt in the palm of your hand. It has two naturally occurring isotopes: 60.4% is Ga-69, (mass = 68.9257 amu), and 39.6% is Ga-71, (mass = 70.9248 amu). Calculate the average atomic weight for gallium.

ANALYSIS We can calculate the average atomic mass for the element by summing up the contributions from each of the naturally occurring isotopes.

BALLPARK ESTIMATE The masses of the two naturally occurring isotopes of gallium differ by 2 amu (68.9 and 70.9 amu). Since slightly more than half of the Ga atoms are the lighter isotope (Ga-69), the average mass will be slightly less than halfway between the two isotopic masses; estimate = 69.8 amu.

SOLUTION

STEP 1: Identify known information.	Ga-69 (60.4% at 68.9257 amu)
	Ga-71 (39.6% at 70.9248 amu)
STEP 2: Identify the unknown answer and units.	Average atomic weight for Ga (in amu) = ?
STEP 3: Identify conversion factors or equations. This equation calculates the average atomic weight as a weighted average of all naturally occurring isotopes.	Atomic weight = Σ(isotopic abundance) $\times$ (isotopic mass)
STEP 4: Solve. Substitute known information and solve.	Atomic weight = (0.604) $\times$ (68.9257 amu) = 41.6311 amu
	+ (0.396) $\times$ (70.9248 amu) = 28.0862 amu
	Atomic weight = 69.7 amu
	(3 significant figures)

BALLPARK CHECK: Our estimate (69.8 amu) is close!

WORKED EXAMPLE 3.5 Identifying Isotopes from Atomic Mass and Atomic Number

Identify element X in the symbol $^{194}_{78}X$ and give its atomic number, mass number, number of protons, number of electrons, and number of neutrons.

ANALYSIS The identity of the atom corresponds to the atomic number—78.

SOLUTION

Element X has Z = 78, which shows that it is platinum. (Look inside the front cover for the list of elements.) The isotope $^{194}_{78}Pt$ has a mass number of 194, and we can subtract the atomic number from the mass number to get the number of neutrons. This platinum isotope therefore has 78 protons, 78 electrons, and 194 − 78 = 116 neutrons.

PROBLEM 3.8

Potassium (K) has two naturally occurring isotopes: K-39 (93.12%; mass = 38.9637 amu), and K-41 (6.88%; 40.9618 amu). Calculate the atomic weight for potassium. How does your answer compare with the atomic mass given in the list inside the front cover of this book?

PROBLEM 3.9

Bromine, an element present in compounds used as sanitizers and fumigants (for example, ethylene bromide), has two naturally occurring isotopes, with mass numbers 79 and 81. Write the symbols for both, including their atomic numbers and mass numbers.

PROBLEM 3.10

Complete the following isotope symbols:

(a) $^{11}_{5}?$ (b) $^{56}_{?}Fe$ (c) $^{37}_{17}?$

3.4 The Periodic Table

Ten elements have been known since the beginning of recorded history: antimony (Sb), carbon (C), copper (Cu), gold (Au), iron (Fe), lead (Pb), mercury (Hg), silver (Ag), sulfur (S), and tin (Sn). It is worth noting that the symbols for many of these elements are derived from their Latin names, a reminder that they have been known since the time when Latin was the language used for all scholarly work. The first "new" element to be found in several thousand years was arsenic (As), discovered in about 1250. In fact, only 24 elements were known up to the time of the American Revolution in 1776.

As the pace of discovery quickened in the late 1700s and early 1800s, chemists began to look for similarities among elements that might make it possible to draw general conclusions. Particularly important was Johann Döbereiner's observation in 1829 that there were several *triads*, or groups of three elements, that appeared to have similar chemical and physical properties. For example, lithium, sodium, and potassium were all known to be silvery metals that react violently with water; chlorine, bromine, and iodine were all known to be colored nonmetals with pungent odors.

Numerous attempts were made in the mid-1800s to account for the similarities among groups of elements, but the great breakthrough came in 1869 when the Russian chemist Dmitri Mendeleev organized the elements into a forerunner of the modern periodic table, introduced previously in Section 1.6 and shown again in Figure 3.2. The table has boxes for each element that give the symbol, atomic number, and atomic weight of the element:

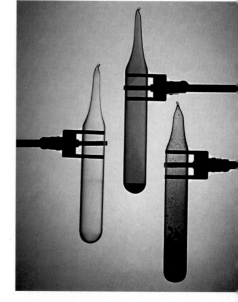

▲ Samples of chlorine, bromine, and iodine, one of Döbereiner's triads of elements with similar chemical properties.

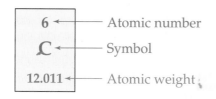

Beginning at the upper left corner of the periodic table, elements are arranged by increasing atomic number into seven horizontal rows, called **periods**, and 18 vertical columns, called **groups**. When organized in this way, *the elements in a given group have similar chemical properties*. Lithium, sodium, potassium, and the other elements in group 1A behave similarly. Chlorine, bromine, iodine, and the other elements in group 7A behave similarly, and so on throughout the table.

Note that different periods (rows) contain different numbers of elements. The first period contains only 2 elements, hydrogen and helium; the second and third periods each contain 8 elements; the fourth and fifth periods each contain 18; the sixth period contains 32; and the seventh period (incomplete as yet) contains 31 elements. Note also that the 14 elements following lanthanum (the *lanthanides*) and the 14 following actinium (the *actinides*) are pulled out and shown below the others.

Groups are numbered in two ways, both shown in Figure 3.2. The two large groups on the far left and the six on the far right are called the **main group elements** and are numbered 1A through 8A. The ten smaller groups in the middle of the table are called the **transition metal elements** and are numbered 1B through 8B. Alternatively, all 18 groups are numbered sequentially from 1 to 18. The 14 groups shown separately at the bottom of the table are called the **inner transition metal elements** and are not numbered.

Period One of the 7 horizontal rows of elements in the periodic table.

Group One of the 18 vertical columns of elements in the periodic table.

Main group element An element in one of the two groups on the left or the six groups on the right of the periodic table.

Transition metal element An element in one of the 10 smaller groups near the middle of the periodic table.

Inner transition metal element An element in one of the 14 groups shown separately at the bottom of the periodic table.

Main groups

Period	1A 1	2A 2					Transition metal groups								3A 13	4A 14	5A 15	6A 16	7A 17	8A 18

Main groups

	1A 1	2A 2												3A 13	4A 14	5A 15	6A 16	7A 17	8A 18
1	1 H 1.00794																		2 He 4.00260
2	3 Li 6.941	4 Be 9.01218	3B 3	4B 4	5B 5	6B 6	7B 7	8B 8	9	10	1B 11	2B 12	5 B 10.81	6 C 12.011	7 N 14.0067	8 O 15.9994	9 F 18.9984	10 Ne 20.1797
3	11 Na 22.98977	12 Mg 24.305											13 Al 26.98154	14 Si 28.0855	15 P 30.9738	16 S 32.066	17 Cl 35.4527	18 Ar 39.948
4	19 K 39.0983	20 Ca 40.078	21 Sc 44.9559	22 Ti 47.88	23 V 50.9415	24 Cr 51.996	25 Mn 54.9380	26 Fe 55.847	27 Co 58.9332	28 Ni 58.69	29 Cu 63.546	30 Zn 65.39	31 Ga 69.72	32 Ge 72.61	33 As 74.9216	34 Se 78.96	35 Br 79.904	36 Kr 83.80
5	37 Rb 85.4678	38 Sr 87.62	39 Y 88.9059	40 Zr 91.224	41 Nb 92.9064	42 Mo 95.94	43 Tc (98)	44 Ru 101.07	45 Rh 102.9055	46 Pd 106.42	47 Ag 107.8682	48 Cd 112.41	49 In 114.82	50 Sn 118.710	51 Sb 121.757	52 Te 127.60	53 I 126.9045	54 Xe 131.29
6	55 Cs 132.9054	56 Ba 137.33	57 *La 138.9055	72 Hf 178.49	73 Ta 180.9479	74 W 183.85	75 Re 186.207	76 Os 190.2	77 Ir 192.22	78 Pt 195.08	79 Au 196.9665	80 Hg 200.59	81 Tl 204.383	82 Pb 207.2	83 Bi 208.9804	84 Po (209)	85 At (210)	86 Rn (222)
7	87 Fr (223)	88 Ra 226.0254	89 †Ac 227.0278	104 Rf (261)	105 Db (262)	106 Sg (266)	107 Bh (264)	108 Hs (269)	109 Mt (268)	110 Ds (271)	111 Rg (272)	112 (285)	113 (284)	114 (289)	115 (288)	116 (292)		118 (294)

Lanthanides	58 Ce 140.12	59 Pr 140.9077	60 Nd 144.24	61 Pm (145)	62 Sm 150.36	63 Eu 151.965	64 Gd 157.25	65 Tb 158.9254	66 Dy 162.50	67 Ho 164.9304	68 Er 167.26	69 Tm 168.9342	70 Yb 173.04	71 Lu 174.967
Actinides	90 Th 232.0381	91 Pa 231.0399	92 U 238.0289	93 Np 237.048	94 Pu (244)	95 Am (243)	96 Cm (247)	97 Bk (247)	98 Cf (251)	99 Es (252)	100 Fm (257)	101 Md (258)	102 No (259)	103 Lr (262)

☐ Metals ■ Metalloids ☐ Nonmetals

▲ **FIGURE 3.2** **The periodic table of the elements.** Each element is identified by a one- or two-letter symbol and is characterized by an *atomic number*. The table begins with hydrogen (H, atomic number 1) in the upper left-hand corner and continues to the yet unnamed element with atomic number 118. The 14 elements following lanthanum (La, atomic number 57) and the 14 elements following actinium (Ac, atomic number 89) are pulled out and shown below the others.

Elements are organized into 18 vertical columns, or *groups*, and 7 horizontal rows, or *periods*. The two groups on the left and the six on the right are the *main groups*; the ten in the middle are the *transition metal groups*. The 14 elements following lanthanum are the *lanthanides*, and the 14 elements following actinium are the *actinides*; together these are known as the *inner transition metals*. Two systems for numbering the groups are explained in the text.

Those elements (except hydrogen) on the left-hand side of the zigzag line running from boron (B) to astatine (At) are *metals*, those elements to the right of the line are *nonmetals*, and most elements abutting the line are *metalloids*.

PROBLEM 3.11

Locate aluminum in the periodic table, and give its group number and period number.

PROBLEM 3.12

Identify the group 1B element in period 5 and the group 2A element in period 4.

PROBLEM 3.13

There are five elements in group 5A of the periodic table. Identify them, and give the period of each.

▲ Sodium, an alkali metal, reacts violently with water to yield hydrogen gas and an alkaline (basic) solution.

3.5 Some Characteristics of Different Groups

To see why the periodic table has the name it does, look at the graph of atomic radius versus atomic number in Figure 3.3. The graph shows an obvious *periodicity*—a repeating, rise-and-fall pattern. Beginning on the left with atomic number 1 (hydrogen), the sizes of the atoms increase to a maximum at atomic number 3 (lithium), then decrease to a minimum, then increase again to a maximum at atomic number 11 (sodium), then decrease, and so on. It turns out that the maxima occur for atoms of group 1A elements—Li, Na, K, Rb, Cs, and Fr—and the minima occur for atoms of the group 7A elements.

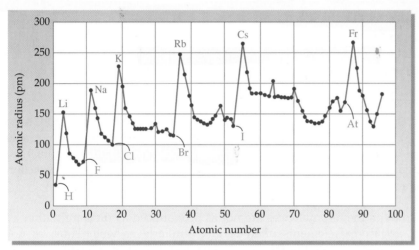

▲ **FIGURE 3.3   A graph of atomic radius in picometers (pm) versus atomic number shows a periodic rise-and-fall pattern.**   The maxima occur for atoms of the group 1A elements (Li, Na, K, Rb, Cs, Fr); the minima occur for atoms of the group 7A elements. Accurate data are not available for the group 8A elements.

There is nothing unique about the periodicity of atomic radii shown in Figure 3.3. The melting points of the first 100 elements, for example, exhibit similar periodic behavior, as shown in Figure 3.4. Many other physical and chemical properties can be plotted in a similar way with similar results. In fact, the various elements in a given group of the periodic table usually show remarkable similarities

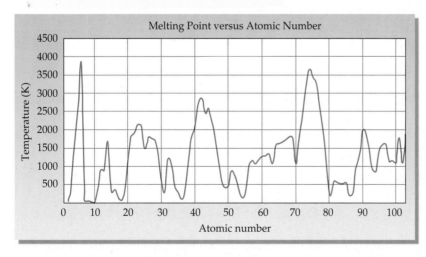

▲ **FIGURE 3.4   A graph of melting point versus atomic number shows periodic properties similar to the trend in Figure 3.3.**   While the maxima and minima are not as sharp as in Figure 3.3, the change in melting points of the elements still shows the same periodic trend.

Alkali metal An element in group 1A of the periodic table.

Alkaline earth metal An element in group 2A of the periodic table.

Halogen An element in group 7A of the periodic table.

Noble gas An element in group 8A of the periodic table.

▲ Magnesium, an alkaline earth metal, burns in air. Magnesium alloys are often used in welding rods.

▲ Chlorine, a halogen, is a toxic, corrosive green gas.

▲ Neon and other noble gases are used in neon lights, signs, and works of art.

in many of their chemical and physical properties. Look at the following four groups, for example:

- **Group 1A—Alkali metals:** Lithium (Li), sodium (Na), potassium (K), rubidium (Rb), cesium (Cs), and francium (Fr) are shiny, soft metals with low melting points. All react rapidly (often violently) with water to form products that are highly alkaline, or basic—hence the name *alkali metals*. Because of their high reactivity, the alkali metals are never found in nature in the pure state but only in combination with other elements.

- **Group 2A—Alkaline earth metals:** Beryllium (Be), magnesium (Mg), calcium (Ca), strontium (Sr), barium (Ba), and radium (Ra) are also lustrous, silvery metals, but are less reactive than their neighbors in group 1A. Like the alkali metals, the alkaline earths are never found in nature in the pure state.

- **Group 7A—Halogens:** Fluorine (F), chlorine (Cl), bromine (Br), iodine (I), and astatine (At) are colorful and corrosive nonmetals. All are found in nature only in combination with other elements, such as with sodium in table salt (sodium chloride, NaCl). In fact, the group name *halogen* is taken from the Greek word *hals*, meaning salt.

- **Group 8A—Noble gases:** Helium (He), neon (Ne), argon (Ar), krypton (Kr), xenon (Xe), and radon (Rn) are colorless gases. The elements in this group were labeled the "noble" gases because of their lack of chemical reactivity—helium, neon, and argon don't combine with any other elements, whereas krypton and xenon combine with very few.

Although the resemblances are not as pronounced as they are within a single group, *neighboring* elements often behave similarly as well. Thus, as noted in Section 1.6 and indicated in Figure 3.2, the periodic table can be divided into three major classes of elements—*metals*, *nonmetals*, and *metalloids* (metal-like). Metals, the largest category of elements, are found on the left side of the periodic table, bounded on the right by a zigzag line running from boron (B) at the top to astatine (At) at the bottom. Nonmetals are found on the right side of the periodic table, and seven of the elements adjacent to the zigzag boundary between metals and nonmetals are metalloids.

▭ Looking Ahead

Carbon, the element on which life is based, is a group 4A nonmetal near the top right of the periodic table. Clustered near carbon are other elements often found in living organisms, including oxygen, nitrogen, phosphorus, and sulfur. We will look at the subject of *organic chemistry*—the chemistry of carbon compounds—in Chapters 12–17, and move on to *biochemistry*—the chemistry of living things—in Chapters 18–29.

PROBLEM 3.14

Identify the following elements as metals, nonmetals, or metalloids:

(a) Ti   **(b)** Te   **(c)** Se

(d) Sc   **(e)** At   **(f)** Ar

PROBLEM 3.15

Locate **(a)** krypton, **(b)** strontium, **(c)** nitrogen, and **(d)** cobalt in the periodic table. Indicate which categories apply to each: (i) metal, (ii) nonmetal, (iii) transition element, (iv) main group element, (v) noble gas.

APPLICATION ▶ The Origin of Chemical Elements

Astronomers believe that the universe began some 15 billion years ago in an extraordinary moment they call the "big bang." Initially, the temperature must have been inconceivably high, but after 1 second, it had dropped to about 10^{10} K and subatomic particles began to form: protons, neutrons, and electrons. After 3 minutes, the temperature had dropped to 10^9 K, and protons began fusing with neutrons to form helium nuclei, ^4_2He.

Matter remained in this form for many millions of years until the expanding universe had cooled to about 10,000 K and electrons were then able to bind to protons and to helium nuclei, forming stable hydrogen and helium atoms.

The attractive force of gravity acting on regions of higher-than-average density slowly produced massive local concentrations of matter and ultimately formed billions of galaxies, each with many billions of stars. As the gas clouds of hydrogen and helium condensed under gravitational attraction and stars formed, their temperatures reached 10^7 K, and their densities reached 100 g/cm^3. Protons and neutrons again fused to yield helium nuclei, generating vast amounts of heat and light.

Most of these early stars probably burned out after a few billion years, but a few were so massive that, as their nuclear fuel diminished, gravitational attraction caused a rapid contraction leading to still higher core temperatures and higher densities—up to 5×10^8 K and 5×10^5 g/cm^3. Under such extreme conditions, larger nuclei were formed, including carbon, oxygen, silicon, magnesium, and iron. Ultimately, the

▲ The stars in the Milky Way galaxy condensed from gas clouds under gravitational attraction.

stars underwent a gravitational collapse resulting in the synthesis of still heavier elements and an explosion visible throughout the universe as a *supernova*.

Matter from exploding supernovas was blown throughout the galaxy, forming a new generation of stars and planets. Our own sun and solar system formed about 4.5 billion years ago from matter released by former supernovas. Except for hydrogen and helium, all the atoms in our bodies and our entire solar system were created more than five billion years ago in exploding stars. We and our world are made from the ashes of dying stars.

See Additional Problems 3.92 and 3.93 at the end of this chapter.

◄● KEY CONCEPT PROBLEM 3.16

Identify the elements shown in red and in blue on the following periodic table. For each, tell its group number, its period number, and whether it is a metal, nonmetal, or metalloid.

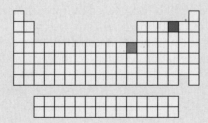

3.6 Electronic Structure of Atoms

Why does the periodic table have the shape it does, with periods of different length? Why are periodic variations observed in atomic radii and in so many other characteristics of the elements? And why do elements in a given group of the periodic table show similar chemical behavior? These questions occupied the thoughts of chemists for more than 50 years after Mendeleev, and it was not until well into the 1920s that the answers were established. Today, we know that *the properties of the elements are determined by the arrangement of electrons in their atoms.*

Our current understanding of the electronic structure of atoms is based on the now accepted *quantum mechanical model*, developed by Austrian physicist Erwin Schrödinger in 1926. Although a detailed discussion of the model is beyond the scope of this text, one of the fundamental assumptions of the model is that electrons have both particle-like and wave-like properties, and that the behavior of electrons can be described using a mathematical equation called a wave function. One consequence of this assumption is that electrons are not perfectly free to move about in an atom. Instead, each electron is restricted to moving about in only a certain region of space within the atom, depending on the energy level of the electron. Different electrons have different amounts of energy and thus occupy different regions within the atom. Furthermore, the energies of electrons are *quantized*, or restricted to having only certain values.

To understand the idea of quantization, think about the difference between stairs and a ramp. A ramp is *not* quantized because it changes height continuously. Stairs, by contrast, *are* quantized because they change height only by a fixed amount. You can climb one stair or two stairs, but you cannot climb 1.5 stairs. In the same way, the energy values available to electrons in an atom change only in steps rather than continuously.

The wave functions derived from the quantum mechanical model also provide important information about the location of electrons in an atom. Just as a person can be found by giving his or her address within a state, an electron can be found by giving its "address" within an atom. Furthermore, just as a person's address is composed of several successively narrower categories—city, street, and house number— an electron's address is also composed of successively narrower categories—*shell*, *subshell*, and *orbital*, which are defined by the quantum mechanical model.

The electrons in an atom are grouped around the nucleus into **shells**, roughly like the layers in an onion, according to the energy of the electrons. The farther a shell is from the nucleus, the larger it is, the more electrons it can hold, and the higher the energies of those electrons. The first shell (the one nearest the nucleus) can hold only 2 electrons, the second shell can hold 8, the third shell can hold 18, and the fourth shell can hold 32 electrons.

Shell number:	1	2	3	4
Electron capacity:	2	8	18	32

Within shells, electrons are further grouped into **subshells** of four different types, identified in order of increasing energy by the letters *s*, *p*, *d*, and *f*. The first shell has only an *s* subshell; the second shell has an *s* and a *p* subshell; the third shell has an *s*, a *p*, and a *d* subshell; and the fourth shell has an *s*, a *p*, a *d*, and an *f* subshell. Of the four types, we will be concerned mainly with *s* and *p* subshells because most of the elements found in living organisms use only these. A specific subshell is symbolized by writing the number of the shell, followed by the letter for the subshell. For example, the designation 3*p* refers to the *p* subshell in the third shell.

Shell number:	1	2	3	4
Subshell designation:	s	s , p	s , p , d	s , p , d , f

Note that the number of subshells in a given shell is equal to the shell number. For example, shell number 3 has 3 subshells.

Finally, within each subshell, electrons are grouped into **orbitals**, regions of space within an atom where the specific electrons are most likely to be found. There are different numbers of orbitals within the different kinds of subshells. A given *s* subshell has only 1 orbital, a *p* subshell has 3 orbitals, a *d* subshell has 5 orbitals, and an *f* subshell has 7 orbitals. Each orbital can hold only two electrons, which differ

▲ Stairs are *quantized* because they change height in discrete amounts. A ramp, by contrast, is not quantized because it changes height continuously.

Shell (electron) A grouping of electrons in an atom according to energy.

Subshell (electron) A grouping of electrons in a shell according to the shape of the region of space they occupy.

Orbital A region of space within an atom where an electron in a given subshell can be found.

in a property known as *spin*. If one electron in an orbital has a clockwise spin, the other electron in the same orbital must have a counterclockwise spin.

Shell number:	1	2	3	4
Subshell designation:	*s*	*s* , *p*	*s* , *p* , *d*	*s* , *p* , *d* , *f*
Number of orbitals:	1	1 , 3	1 , 3 , 5	1 , 3 , 5 , 7

Different orbitals have different shapes and orientations, which are described by the quantum mechanical model. Orbitals in *s* subshells are spherical regions centered about the nucleus, whereas orbitals in *p* subshells are roughly dumbbell-shaped regions (Figure 3.5). As shown in Figure 3.5(b), the three *p* orbitals in a given subshell are oriented at right angles to one another.

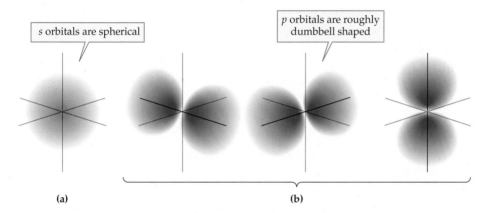

s orbitals are spherical

p orbitals are roughly dumbbell shaped

(a)   (b)

▲ **FIGURE 3.5 The shapes of *s* and *p* orbitals.** (a) The *s* orbitals and (b) the *p* orbitals. The three *p* orbitals in a given subshell are oriented at right angles to one another. Each orbital can hold only two electrons.

The overall electron distribution within an atom is summarized in Table 3.2 and in the following list:

- The first shell holds only 2 electrons. The 2 electrons have different spins and are in a single 1*s* orbital.
- The second shell holds 8 electrons. Two are in a 2*s* orbital, and 6 are in the three different 2*p* orbitals (two per 2*p* orbital).
- The third shell holds 18 electrons. Two are in a 3*s* orbital, 6 are in three 3*p* orbitals, and 10 are in five 3*d* orbitals.
- The fourth shell holds 32 electrons. Two are in a 4*s* orbital, 6 are in three 4*p* orbitals, 10 are in five 4*d* orbitals, and 14 are in seven 4*f* orbitals.

TABLE 3.2 Electron Distribution in Atoms

SHELL NUMBER:	1	2	3	4
Subshell designation:	*s*	*s* , *p*	*s* , *p* , *d*	*s* , *p* , *d* , *f*
Number of orbitals:	1	1 , 3	1 , 3 , 5	1 , 3 , 5 , 7
Number of electrons:	2	2 , 6	2 , 6 , 10	2 , 6 , 10 , 14
Total electron capacity:	2	8	18	32

WORKED EXAMPLE 3.6 Atomic Structure: Electron Shells

How many electrons are present in an atom that has its first and second shells filled and has 4 electrons in its third shell? Name the element.

ANALYSIS The number of electrons in the atom is calculated by adding the total electrons in each shell. We can identify the element from the number of protons in the nucleus, which is equal to the number of electrons in the atom.

SOLUTION
The first shell of an atom holds 2 electrons in its 1s orbital, and the second shell holds 8 electrons (2 in a 2s orbital and 6 in three 2p orbitals). Thus, the atom has a total of $2 + 8 + 4 = 14$ electrons and must be silicon (Si).

PROBLEM 3.17

What is the maximum number of electrons that can occupy the following subshells?

(a) 3p subshell

(b) 2s subshell

(c) 2p subshell

PROBLEM 3.18

How many electrons are present in an atom in which the 1s, 2s, and 2p subshells are filled? Name the element.

PROBLEM 3.19

How many electrons are present in an atom in which the first and second shells and the 3s subshell are filled? Name the element.

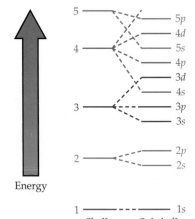

▲ Following the energy level diagram for the electronic orbitals, the orbital filling order can be predicted as indicated in this figure.

Electron configuration The specific arrangement of electrons in an atom's shells and subshells.

3.7 Electron Configurations

The exact arrangement of electrons in an atom's shells and subshells is called the atom's **electron configuration** and can be predicted by applying three rules:

RULE 1. **Electrons occupy the lowest-energy orbitals available, beginning with 1s and continuing in the order shown in Figure 3.6.** Within each shell, the orbital energies increase in the order s, p, d, f. The overall ordering is complicated, however, by the fact that some "crossover" of energies occurs between orbitals in different shells above the 3p level. The 4s orbital is lower in energy than the 3d orbitals, for example, and is therefore filled first.

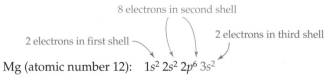

8 electrons in second shell

2 electrons in first shell

2 electrons in third shell

Mg (atomic number 12): $1s^2\ 2s^2\ 2p^6\ 3s^2$

RULE 2. **Each orbital can hold only two electrons, which must be of opposite spin.**

RULE 3. **Two or more orbitals with the same energy—the three p orbitals or the five d orbitals in a given shell, for example—are each half filled by one electron before any one orbital is completely filled by addition of the second electron.**

▲ **FIGURE 3.6 Order of orbital energy levels.** Above the 3p level, there is some crossover of energies among orbitals in different shells.

TABLE 3.3 Electron Configurations of the First 20 Elements

	ELEMENT	ATOMIC NUMBER	ELECTRON CONFIGURATION
H	Hydrogen	1	$1s^1$
He	Helium	2	$1s^2$
Li	Lithium	3	$1s^2\,2s^1$
Be	Beryllium	4	$1s^2\,2s^2$
B	Boron	5	$1s^2\,2s^2\,2p^1$
C	Carbon	6	$1s^2\,2s^2\,2p^2$
N	Nitrogen	7	$1s^2\,2s^2\,2p^3$
O	Oxygen	8	$1s^2\,2s^2\,2p^4$
F	Fluorine	9	$1s^2\,2s^2\,2p^5$
Ne	Neon	10	$1s^2\,2s^2\,2p^6$
Na	Sodium	11	$1s^2\,2s^2\,2p^6\,3s^1$
Mg	Magnesium	12	$1s^2\,2s^2\,2p^6\,3s^2$
Al	Aluminum	13	$1s^2\,2s^2\,2p^6\,3s^2\,3p^1$
Si	Silicon	14	$1s^2\,2s^2\,2p^6\,3s^2\,3p^2$
P	Phosphorus	15	$1s^2\,2s^2\,2p^6\,3s^2\,3p^3$
S	Sulfur	16	$1s^2\,2s^2\,2p^6\,3s^2\,3p^4$
Cl	Chlorine	17	$1s^2\,2s^2\,2p^6\,3s^2\,3p^5$
Ar	Argon	18	$1s^2\,2s^2\,2p^6\,3s^2\,3p^6$
K	Potassium	19	$1s^2\,2s^2\,2p^6\,3s^2\,3p^6\,4s^1$
Ca	Calcium	20	$1s^2\,2s^2\,2p^6\,3s^2\,3p^6\,4s^2$

Electron configurations of the first 20 elements are shown in Table 3.3. Notice that the number of electrons in each subshell is indicated by a superscript. For example, the notation $1s^2\,2s^2\,2p^6\,3s^2$ for magnesium means that magnesium atoms have 2 electrons in the first shell, 8 electrons in the second shell, and 2 electrons in the third shell.

As you read through the following electron configurations, check the atomic number and the location of each element in the periodic table (Figure 3.2). See if you can detect the relationship between electron configuration and position in the table.

- **Hydrogen ($Z = 1$):** The single electron in a hydrogen atom is in the lowest-energy, $1s$ level. The configuration can be represented in either of two ways:

$$\textbf{H}\quad 1s^1 \quad\text{or}\quad \frac{\uparrow}{1s^1}$$

In the written representation, the superscript in the notation $1s^1$ means that the $1s$ orbital is occupied by one electron. In the graphic representation, the $1s$ orbital is indicated by a line and the single electron in this orbital is shown by an up arrow ($\uparrow$). A single electron in an orbital is often referred to as being *unpaired*.

- **Helium ($Z = 2$):** The two electrons in helium are both in the lowest-energy, $1s$ orbital, and their spins are *paired*, as represented by up and down arrows ($\uparrow\downarrow$):

$$\textbf{He}\quad 1s^2 \quad\text{or}\quad \frac{\uparrow\downarrow}{1s^2}$$

- **Lithium ($Z = 3$):** With the first shell full, the second shell begins to fill. The third electron goes into the $2s$ orbital:

$$\textbf{Li}\quad 1s^2\,2s^1 \quad\text{or}\quad \frac{\uparrow\downarrow}{1s^2}\ \frac{\uparrow}{2s^1}$$

Because [He] has the configuration of a filled $1s^2$ orbital, it is sometimes substituted for the $1s^2$ orbital in depictions of electron pairing. Using this alternative shorthand notation, the electron configuration for Li is written [He] $2s^1$.

- **Beryllium ($Z = 4$):** An electron next pairs up to fill the $2s$ orbital:

$$\textbf{Be} \quad 1s^2\,2s^2 \quad \text{or} \quad \underset{1s^2}{\uparrow\downarrow} \; \underset{2s^2}{\uparrow\downarrow} \quad \text{or} \quad [\text{He}]\,2s^2$$

- **Boron ($Z = 5$), Carbon ($Z = 6$), Nitrogen ($Z = 7$):** The next three electrons enter the three $2p$ orbitals, one at a time. Note that representing the configurations with lines and arrows gives more information than the alternative written notation because the filling and pairing of electrons in individual orbitals within the p subshell is shown.

$$\textbf{B} \quad 1s^2\,2s^2\,2p^1 \quad \text{or} \quad \underset{1s^2}{\uparrow\downarrow} \; \underset{2s^2}{\uparrow\downarrow} \; \underset{2p^1}{\underbrace{\uparrow\;\_\;\_}} \quad \text{or} \quad [\text{He}]\,2s^2\,2p^1$$

$$\textbf{C} \quad 1s^2\,2s^2\,2p^2 \quad \text{or} \quad \underset{1s^2}{\uparrow\downarrow} \; \underset{2s^2}{\uparrow\downarrow} \; \underset{2p^2}{\underbrace{\uparrow\;\uparrow\;\_}} \quad \text{or} \quad [\text{He}]\,2s^2\,2p^2$$

$$\textbf{N} \quad 1s^2\,2s^2\,2p^3 \quad \text{or} \quad \underset{1s^2}{\uparrow\downarrow} \; \underset{2s^2}{\uparrow\downarrow} \; \underset{2p^3}{\underbrace{\uparrow\;\uparrow\;\uparrow}} \quad \text{or} \quad [\text{He}]\,2s^2\,2p^3$$

- **Oxygen ($Z = 8$), Fluorine ($Z = 9$), Neon ($Z = 10$):** Electrons now pair up one by one to fill the three $2p$ orbitals and fully occupy the second shell:

$$\textbf{O} \quad 1s^2\,2s^2\,2p^4 \quad \text{or} \quad \underset{1s^2}{\uparrow\downarrow} \; \underset{2s^2}{\uparrow\downarrow} \; \underset{2p^4}{\underbrace{\uparrow\downarrow\;\uparrow\;\uparrow}} \quad \text{or} \quad [\text{He}]\,2s^2\,2p^4$$

$$\textbf{F} \quad 1s^2\,2s^2\,2p^5 \quad \text{or} \quad \underset{1s^2}{\uparrow\downarrow} \; \underset{2s^2}{\uparrow\downarrow} \; \underset{2p^5}{\underbrace{\uparrow\downarrow\;\uparrow\downarrow\;\uparrow}} \quad \text{or} \quad [\text{He}]\,2s^2\,2p^5$$

$$\textbf{Ne} \quad 1s^2\,2s^2\,2p^6 \quad \text{or} \quad \underset{1s^2}{\uparrow\downarrow} \; \underset{2s^2}{\uparrow\downarrow} \; \underset{2p^6}{\underbrace{\uparrow\downarrow\;\uparrow\downarrow\;\uparrow\downarrow}}$$

At this point we may use the shorthand notation [Ne] to represent the electron configuration for a completely filled set of orbitals in the second shell.

- **Sodium to Calcium ($Z = 11 - 20$):** The pattern seen for lithium through neon is seen again for sodium ($Z = 11$) through argon ($Z = 18$) as the $3s$ and $3p$ subshells fill up. For elements having a third filled shell, we may use [Ar] to represent a completely filled third shell. After argon, however, the first crossover in subshell energies occurs. As indicated in Figure 3.6, the $4s$ subshell is lower in energy than the $3d$ subshell and is filled first. Potassium ($Z = 19$) and calcium ($Z = 20$) therefore have the following electron configurations:

$$\textbf{K} \quad 1s^2\,2s^2\,2p^6\,3s^2\,3p^6\,4s^1 \text{ or } [\text{Ar}]\,4s^1 \quad \textbf{Ca} \quad 1s^2\,2s^2\,2p^6\,3s^2\,3p^6\,4s^2 \text{ or } [\text{Ar}]\,4s^2$$

WORKED EXAMPLE **3.7** Atomic Structure: Electron Configurations

Show how the electron configuration of magnesium can be assigned.

ANALYSIS Magnesium, $Z = 12$, has 12 electrons to be placed in specific orbitals. Assignments are made by putting 2 electrons in each orbital, according to the order shown in Figure 3.6.

- The first 2 electrons are placed in the 1s orbital (1s^2).
- The next 2 electrons are placed in the 2s orbital (2s^2).
- The next 6 electrons are placed in the three available 2p orbitals (2p^6).
- The remaining 2 electrons are both put in the 3s orbital (3s^2).

SOLUTION

Magnesium has the configuration 1s^2 2s^2 2p^6 3s^2 or [Ne] 3s^2.

WORKED EXAMPLE   **3.8** Electron Configurations: Orbital-Filling Diagrams

Write the electron configuration of phosphorus, $Z = 15$, using up and down arrows to show how the electrons in each orbital are paired.

ANALYSIS Phosphorus has 15 electrons, which occupy orbitals according to the order shown in Figure 3.6.

- The first 2 are paired and fill the first shell (1s^2).
- The next 8 fill the second shell (2s^2 2p^6). All electrons are paired.
- The remaining 5 electrons enter the third shell, where 2 fill the 3s orbital (3s^2) and 3 occupy the 3p subshell, one in each of the three p orbitals.

SOLUTION

$$\text{P} \quad \underset{1s^2}{\uparrow\downarrow} \quad \underset{2s^2}{\uparrow\downarrow} \quad \underset{2p^6}{\underbrace{\uparrow\downarrow \ \uparrow\downarrow \ \uparrow\downarrow}} \quad \underset{3s^2}{\uparrow\downarrow} \quad \underset{3p^3}{\underbrace{\uparrow \ \uparrow \ \uparrow}}$$

PROBLEM 3.20

Write electron configurations for the following elements. (You can check your answers in Table 3.3.)

(a) C   (b) P   (c) Cl   (d) K

PROBLEM 3.21

Write electron configurations for the elements with atomic numbers 14 and 36.

PROBLEM 3.22

For an atom containing 33 electrons, identify the incompletely filled subshell, and show the paired and/or unpaired electrons in this subshell using up and down arrows.

◉ KEY CONCEPT PROBLEM 3.23

Identify the atom with the following orbital-filling diagram.

$$1s^2 \ 2s^2 \ 2p^6 \ 3s^2 \ 3p^6 \quad \underset{4s}{\updownarrow} \quad \underset{3d}{\updownarrow \ \ \updownarrow \ \ \updownarrow \ \ \updownarrow \ \ \updownarrow} \quad \underset{4p}{\uparrow \ \ \underline{\ \ } \ \ \underline{\ \ }}$$

3.8 Electron Configurations and the Periodic Table

How is an atom's electron configuration related to its chemical behavior, and why do elements with similar behavior occur in the same group of the periodic table? As shown in Figure 3.7, the periodic table can be divided into four regions, or *blocks*, of elements according to the electron shells and subshells occupied by *the subshell filled last*.

s-**Block element** A main group element that results from the filling of an *s* orbital.

- The main group 1A and 2A elements on the left side of the table (plus He) are called the *s*-**block elements** because an *s* subshell is filled last in these elements.

p-**Block element** A main group element that results from the filling of *p* orbitals.

- The main group 3A–8A elements on the right side of the table (except He) are the *p*-**block elements** because a *p* subshell is filled last in these elements.

d-**Block element** A transition metal element that results from the filling of *d* orbitals.

- The transition metals in the middle of the table are the *d*-**block elements** because a *d* subshell is filled last in these elements.

f-**Block element** An inner transition metal element that results from the filling of *f* orbitals.

- The inner transition metals detached at the bottom of the table are the *f*-**block elements** because an *f* subshell is filled last in these elements.

Thinking of the periodic table as outlined in Figure 3.7 provides a simple way to remember the order of orbital filling shown previously in Figure 3.6. Beginning at the top left corner of the periodic table, the first row contains only two elements (H and He) because only two electrons are required to fill the *s* orbital in the first shell, $1s^2$. The second row begins with two *s*-block elements (Li and Be) and continues with six *p*-block elements (B through Ne), so electrons fill the next available *s* orbital (2*s*) and then the first available *p* orbitals (2*p*). The third row is similar to the second row, so the 3*s* and 3*p* orbitals are filled next. The fourth row again starts with two *s*-block elements (K and Ca) but is then followed by ten *d*-block elements (Sc through Zn) and six *p*-block elements (Ga through Kr). Thus, the order of orbital filling is 4*s* followed by the first available *d* orbitals (3*d*) followed by 4*p*. Continuing through successive rows of the periodic table gives the entire filling order, identical to that shown in Figure 3.6.

$$1s \rightarrow 2s \rightarrow 2p \rightarrow 3s \rightarrow 3p \rightarrow 4s \rightarrow 3d \rightarrow 4p \rightarrow 5s \rightarrow$$
$$4d \rightarrow 5p \rightarrow 6s \rightarrow 4f \rightarrow 5d \rightarrow 6p \rightarrow 7s \rightarrow 5f \rightarrow 6d$$

But why do the elements in a given group of the periodic table have similar properties? The answer emerges when you look at Table 3.4, which gives electron

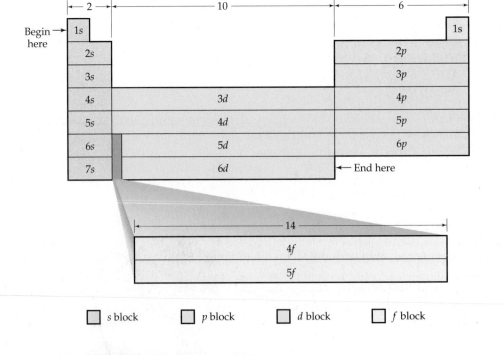

► **FIGURE 3.7 The blocks of elements in the periodic table correspond to filling the different types of subshells.** Beginning at the top left and going across successive rows of the periodic table provides a method for remembering the order of orbital filling: $1s \rightarrow 2s \rightarrow 2p \rightarrow 3s \rightarrow 3p \rightarrow 4s \rightarrow 3d \rightarrow 4p$, and so on.

TABLE 3.4   Valence-Shell Electron Configurations for Group 1A, 2A, 7A, and 8A Elements

GROUP	ELEMENT	ATOMIC NUMBER	VALENCE-SHELL ELECTRON CONFIGURATION
1A	Li (lithium)	3	$2s^1$
	Na (sodium)	11	$3s^1$
	K (potassium)	19	$4s^1$
	Rb (rubidium)	37	$5s^1$
	Cs (cesium)	55	$6s^1$
2A	Be (beryllium)	4	$2s^2$
	Mg (magnesium)	12	$3s^2$
	Ca (calcium)	20	$4s^2$
	Sr (strontium)	38	$5s^2$
	Ba (barium)	56	$6s^2$
7A	F (fluorine)	9	$2s^2\ 2p^5$
	Cl (chlorine)	17	$3s^2\ 3p^5$
	Br (bromine)	35	$4s^2\ 4p^5$
	I (iodine)	53	$5s^2\ 5p^5$
8A	He (helium)	2	$1s^2$
	Ne (neon)	10	$2s^2\ 2p^6$
	Ar (argon)	18	$3s^2\ 3p^6$
	Kr (krypton)	36	$4s^2\ 4p^6$
	Xe (xenon)	54	$5s^2\ 5p^6$

configurations for elements in the main groups 1A, 2A, 7A, and 8A. Focusing only on the electrons in the outermost shell, or **valence shell**, *elements in the same group of the periodic table have similar electron configurations in their valence shells.* The group 1A elements, for example, all have one **valence electron**, ns^1 (where n represents the number of the valence shell: $n = 2$ for Li; $n = 3$ for Na; $n = 4$ for K; and so on). The group 2A elements have two valence electrons (ns^2); the group 7A elements have seven valence electrons ($ns^2\ np^5$); and the group 8A elements (except He) have eight valence electrons ($ns^2\ np^6$). You might also notice that the group numbers from 1A through 8A give the numbers of valence electrons for the elements in each main group.

What is true for the main group elements is also true for the other groups in the periodic table: Atoms within a given group have the same number of valence electrons and have similar electron configurations. *Because the valence electrons are the most loosely held, they are the most important in determining an element's properties.* Similar electron configurations thus explain why the elements in a given group of the periodic table have similar chemical behavior.

Valence shell   The outermost electron shell of an atom.

Valence electron   An electron in the valence shell of an atom.

⬤⬤⬤ Looking Ahead

We have seen that elements in a given group have similar chemical behavior because they have similar valence electron configurations, and that many chemical properties exhibit periodic trends across the periodic table. The *chemical* behavior of nearly all the elements can be predicted based on their position in the periodic table, and will be examined in more detail in Chapters 4–6. Similarly, the *nuclear* behavior of the different isotopes of a given element is related to the configuration of the nucleus (that is, the number of neutrons and protons), and will be examined in Chapter 11.

WORKED EXAMPLE 3.9 Electron Configurations: Valence Electrons

Write the electron configuration for the following elements, using both the complete and the shorthand notations. Indicate which electrons are the valence electrons.

(a) Na   (b) Cl   (c) Zr

ANALYSIS Locate the row and the block in which each of the elements is found in Figure 3.7. The location can be used to determine the complete electron configuration and to identify the valence electrons.

SOLUTION

(a) Na (sodium) is located in the third row, and in the first column of the s-block. Therefore, all orbitals up to the 3s are completely filled, and there is one electron in the 3s orbital.

$$\textbf{Na:}\ 1s^2\ 2s^2\ 2p^6\ \underline{3s^1}\quad \text{or}\quad [\text{Ne}]\ \underline{3s^1}\quad \text{(valence electrons are underlined)}$$

(b) Cl (chlorine) is located in the third row, and in the fifth column of the p-block.

$$\textbf{Cl:}\ 1s^2\ 2s^2\ 2p^6\ \underline{3s^2\ 3p^5}\quad \text{or}\quad [\text{Ne}]\ \underline{3s^2\ 3p^5}$$

(c) Zr (zirconium) is located in the fifth row, and in the second column of the d-block. All orbitals up to the 4d are completely filled, and there are 2 electrons in the 4d orbitals. Note that the 4d orbitals are filled after the 5s orbitals in both Figures 3.6 and 3.7.

$$\textbf{Zr:}\ 1s^2\ 2s^2\ 2p^6\ 3s^1\ 3p^6\ 4s^2\ 3d^{10}\ 4p^6\ \underline{5s^2\ 4d^2}\quad \text{or}\quad [\text{Kr}]\ \underline{5s^2\ 4d^2}$$

WORKED EXAMPLE 3.10 Electron Configurations: Valence-Shell Configurations

Using n to represent the number of the valence shell, write a general valence-shell configuration for the elements in group 6A.

ANALYSIS The elements in group 6A have 6 valence electrons. In each element, the first two of these electrons are in the valence s subshell, giving ns^2, and the next four electrons are in the valence p subshell, giving np^4.

SOLUTION
For group 6A, the general valence-shell configuration is $ns^2\ np^4$.

WORKED EXAMPLE 3.11 Electron Configurations: Inner Shells versus Valence Shell

How many electrons are in a tin atom? Give the number of electrons in each shell. How many valence electrons are there in a tin atom? Write the valence-shell configuration for tin.

ANALYSIS The total number of electrons will be the same as the atomic number for tin (Z = 50). The number of valence electrons will equal the number of electrons in the valence shell.

SOLUTION

Checking the periodic table shows that tin has atomic number 50 and is in group 4A. The number of electrons in each shell is

Shell number:	1	2	3	4	5
Number of electrons:	2	8	18	18	4

As expected from the group number, tin has 4 valence electrons. They are in the $5s$ and $5p$ subshells and have the configuration $5s^2\,5p^2$.

PROBLEM 3.24

Write the electron configuration for the following elements, using both the complete and the shorthand notations. Indicate which electrons are the valence electrons.

(a) F   **(b)** Al   **(c)** As

PROBLEM 3.25

Identify the group in which all elements have the valence-shell configuration ns^2.

PROBLEM 3.26

For chlorine, identify the group number, give the number of electrons in each occupied shell, and write its valence-shell configuration.

◦ KEY CONCEPT PROBLEM 3.27

Identify the group number, and write the general valence-shell configuration (for example, ns^1 for group 1A elements) for the elements indicated in red in the following periodic table.

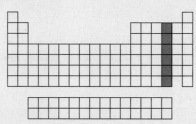

3.9 Electron-Dot Symbols

Valence electrons play such an important role in the behavior of atoms that it is useful to have a method for including them with atomic symbols. In an **electron-dot symbol**, dots are placed around the atomic symbol to indicate the number of valence electrons present. A group 1A atom, such as sodium, has a single dot; a group 2A atom, such as magnesium, has two dots; a group 3A atom, such as boron, has three dots; and so on.

Table 3.5 gives electron-dot symbols for atoms of the first few elements in each main group. As shown, the dots are distributed around the four sides of the element symbol, singly at first until each of the four sides has one dot. As more electron dots

Electron-dot symbol An atomic symbol with dots placed around it to indicate the number of valence electrons.

TABLE 3.5 Electron-Dot Symbols for Some Main Group Elements

1A	2A	3A	4A	5A	6A	7A	NOBLE GASES
H·							He:
Li·	·Be·	·B·	·C·	·N:	·O:	·F:	:Ne:
Na·	·Mg·	·Al·	·Si·	·P:	·S:	·Cl:	:Ar:
K·	·Ca·	·Ga·	·Ge·	·As:	·Se:	·Br:	:Kr:

are added they will form pairs, with no more than two dots on a side. Note that helium differs from other noble gases in having only two valence electrons rather than eight. Nevertheless, helium is considered a member of group 8A because its properties resemble those of the other noble gases and because its highest occupied subshell is filled ($1s^2$).

WORKED EXAMPLE 3.12 Electron Configurations: Electron-Dot Symbols

Write the electron-dot symbol for any element X in group 5A.

ANALYSIS The group number, 5A, indicates 5 valence electrons. The first four are distributed singly around the four sides of the element symbol, and any additional are placed to form electron pairs.

SOLUTION

·X: (5 electrons)

PROBLEM 3.28

Write the electron-dot symbol for any element X in group 3A.

PROBLEM 3.29

Write electron-dot symbols for radon, lead, xenon, and radium.

APPLICATION ▶   Atoms and Light

What we see as *light* is really a wave of energy moving through space. The shorter the length of the wave (the *wavelength*), the higher the energy; the longer the wavelength, the lower the energy.

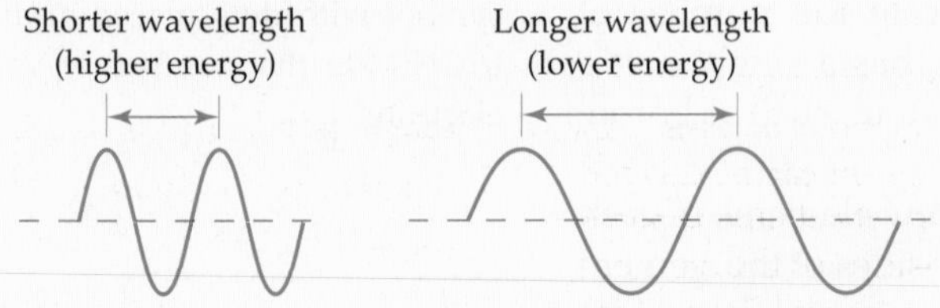

Shorter wavelength (higher energy)   Longer wavelength (lower energy)

Visible light has wavelengths in the range 400–800 nm, but that is just one small part of the overall *electromagnetic spectrum*, shown in the accompanying figure. Although we cannot see the other wavelengths of electromagnetic energy, we use them for many purposes and their names may be familiar to you: gamma rays, X rays, ultraviolet (UV) rays, infrared (IR) rays, microwaves, and radio waves, for example.

What happens when a beam of electromagnetic energy collides with an atom? Remember that electrons are located in orbitals based on their energy levels. An atom with its electrons in their usual, lowest-energy locations is said to be in its *ground state*. If the amount of electromagnetic energy is just right, an electron can be kicked up from its usual energy level to a higher one. Energy from an electrical discharge or in the form of heat can have the same effect. With one of its electrons promoted to a higher energy, an atom is said to be *excited*. The excited state does not last long, though, because the electron quickly drops back to its more stable, ground-state energy level, releasing its extra energy in the process. If the released energy falls in the range of visible light, we can see the result. Many practical applications, from neon lights to fireworks, are the result of this phenomenon.

In "neon" lights, noble gas atoms are excited by an electric discharge, giving rise to a variety of colors that depend on the gas—red from neon, white from krypton, and blue from argon. Similarly, mercury or sodium atoms excited by electrical energy are responsible for the intense bluish or yellowish light, respectively, provided by some street lamps. In the same manner, metal atoms excited by heat are responsible for the spectacular colors of fireworks—red from strontium, green from barium, and blue from copper, for example.

The concentration of certain metals in body fluids is measured by sensitive instruments relying on the same principle of electron excitation that we see in fireworks. These instruments determine the intensity of the flame color produced by lithium (red), sodium (yellow), and potassium (violet),

▲ The brilliant colors of fireworks are due to the release of energy from excited atoms as electrons fall from higher to lower energy levels.

yielding the concentrations of these metals given in most clinical lab reports. Calcium, magnesium, copper, and zinc concentrations are also found by measuring the energies of excited atoms.

See Additional Problems 3.94 and 3.95 at the end of the chapter.

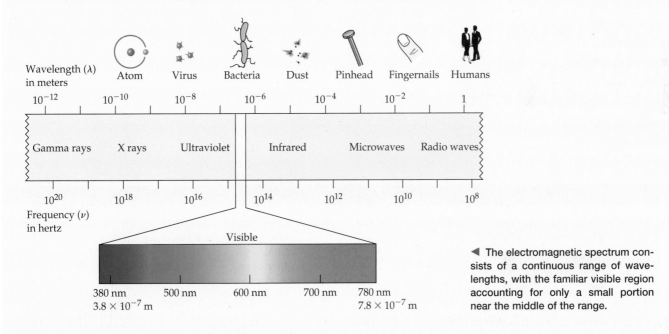

◄ The electromagnetic spectrum consists of a continuous range of wavelengths, with the familiar visible region accounting for only a small portion near the middle of the range.

SUMMARY: REVISITING THE CHAPTER GOALS

1. What is the modern theory of atomic structure? All matter is composed of *atoms*. An atom is the smallest and simplest unit into which a sample of an element can be divided while maintaining the properties of the element. Atoms are made up of subatomic particles called *protons, neutrons,* and *electrons*. Protons have a positive electrical charge, neutrons are electrically neutral, and electrons have a negative electrical charge. The protons and neutrons in an atom are present in a dense, positively charged central region called

KEY WORDS

Alkali metal, *p. 60*

Alkaline earth metal, *p. 60*

Atom, *p. 49*

Atomic mass unit (amu), *p. 50*

the *nucleus*. Electrons are situated a relatively large distance away from the nucleus, leaving most of the atom as empty space.

2. **How do atoms of different elements differ?** Elements differ according to the number of protons their atoms contain, a value called the element's *atomic number* (Z). All atoms of a given element have the same number of protons and an equal number of electrons. The number of neutrons in an atom is not predictable but is generally as great or greater than the number of protons. The total number of protons plus neutrons in an atom is called the atom's *mass number* (A).

3. **What are isotopes, and what is atomic weight?** Atoms with identical numbers of protons and electrons but different numbers of neutrons are called *isotopes*. The atomic weight of an element is the weighted average mass of atoms of the element's naturally occurring isotopes measured in *atomic mass units* (amu).

4. **How is the periodic table arranged?** Elements are organized into the *periodic table*, consisting of 7 rows, or *periods*, and 18 columns, or *groups*. The two groups on the left side of the table and the six groups on the right are called the *main group elements*. The ten groups in the middle are the *transition metal groups*, and the 14 groups pulled out and displayed below the main part of the table are called the *inner transition metal groups*. Within a given group in the table, elements have the same number of valence electrons in their valence shell and similar electron configurations.

5. **How are electrons arranged in atoms?** The electrons surrounding an atom are grouped into layers, or *shells*. Within each shell, electrons are grouped into *subshells*, and within each subshell into *orbitals*—regions of space in which electrons are most likely to be found. The *s* orbitals are spherical, and the *p* orbitals are dumbbell-shaped.

Each shell can hold a specific number of electrons. The first shell can hold 2 electrons in an *s* orbital ($1s^2$); the second shell can hold 8 electrons in one *s* and three *p* orbitals ($2s^2\ 2p^6$); the third shell can hold 18 electrons in one *s*, three *p*, and five *d* orbitals ($3s^2\ 3p^6\ 3d^{10}$); and so on. The *electron configuration* of an element is predicted by assigning the element's electrons into orbitals, beginning with the lowest-energy orbital. The electrons in the outermost shell, or *valence shell*, can be represented using electron-dot symbols.

UNDERSTANDING KEY CONCEPTS

3.30 Which of the following drawings represents an *s* orbital, and which represents a *p* orbital?

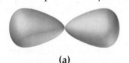

(a)   (b)

3.31 Where on the following outline of a periodic table do the indicated elements or groups of elements appear?

  (a) Alkali metals   **(b)** Halogens
  (c) Alkaline earth metals **(d)** Transition metals
  (e) Hydrogen   **(f)** Helium   **(g)** Metalloids

3.32 Is the element marked in red on the following periodic table likely to be a gas, a liquid, or a solid? What is the atomic number of the element in blue? Name at least one other element that is likely to be chemically similar to the element in green.

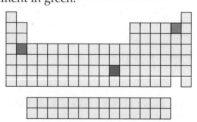

3.33 Use the blank periodic table at the top of the next page to show where elements matching the following descriptions appear.

(a) Elements with the valence-shell electron configuration $ns^2\, np^5$

(b) An element whose third shell contains two p electrons

(c) Elements with a completely filled valence shell

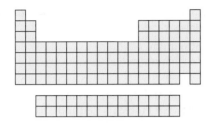

3.34 What atom has the following orbital-filling diagram?

$1s^2\, 2s^2\, 2p^6\, 3s^2\, 3p^6$ ⇅   ⇅ ⇅ ⇅ ⇅ ⇅   ⇅ ↓ ↓

                               $4s$ $3d$ $4p$

3.35 Use the orbital-filling diagram below to show the electron configuration for As:

$1s^2\, 2s^2\, 2p^6\, 3s^2\, 3p^6$ —   — — — — —   — — —

                               $4s$ $3d$ $4p$

ADDITIONAL PROBLEMS

ATOMIC THEORY AND THE COMPOSITION OF ATOMS

3.36 What four fundamental assumptions about atoms and matter make up modern atomic theory?

3.37 How do atoms of different elements differ?

3.38 Find the mass in grams of one atom of the following elements:

(a) Bi, atomic weight 208.9804 amu
(b) Xe, atomic weight 131.29 amu
(c) He, atomic weight 4.0026 amu

3.39 Find the mass in atomic mass units of the following:

(a) 1 O atom, with a mass of 2.66×10^{-23} g
(b) 1 Br atom, with a mass of 1.31×10^{-22} g

3.40 What is the mass in grams of 6.022×10^{23} N atoms of mass 14.01 amu?

3.41 What is the mass in grams of 6.022×10^{23} O atoms of mass 16.00 amu?

3.42 How many O atoms of mass 15.99 amu are in 15.99 g of oxygen?

3.43 How many C atoms of mass 12.00 amu are in 12.00 g of carbon?

3.44 What are the names of the three subatomic particles? What are their approximate masses in atomic mass units, and what electrical charge does each have?

3.45 Where within an atom are the three types of subatomic particles located?

3.46 Identify the following atoms:

(a) Contains 19 protons
(b) Contains 50 protons
(c) Has $Z = 30$

3.47 Identify the following atoms:

(a) Contains 15 electrons
(b) Contains 41 protons
(c) Has $Z = 27$

3.48 Give the number of neutrons in each naturally occurring isotope of argon: argon-36, argon-38, argon-40.

3.49 Give the number of protons, neutrons, and electrons in the following isotopes:

(a) Al-27   **(b)** $^{28}_{14}\text{Si}$
(c) B-11   **(d)** $^{115}_{47}\text{Ag}$

3.50 Which of the following symbols represent isotopes of the same element?

(a) $^{19}_{9}\text{X}$ **(b)** $^{19}_{10}\text{X}$
(c) $^{21}_{9}\text{X}$ **(d)** $^{21}_{12}\text{X}$

3.51 Complete the following isotope symbols:

(a) $^{206}_{?}\text{Po}$ **(b)** $^{224}_{88}?$
(c) $^{197}_{79}?$ **(d)** $^{84}_{?}\text{Kr}$

3.52 Name the isotope represented by each symbol in Problem 3.50.

3.53 Give the number of neutrons in each isotope listed in Problem 3.51.

3.54 Write the symbols for the following isotopes:

(a) Its atoms contain 6 protons and 8 neutrons.
(b) Its atoms have mass number 39 and contain 19 protons.
(c) Its atoms have mass number 20 and contain 10 electrons.

3.55 Write the symbols for the following isotopes:

(a) Its atoms contain 9 electrons and 10 neutrons.
(b) Its atoms have $A = 79$ and $Z = 35$.
(c) Its atoms have $A = 51$ and contain 23 electrons.

3.56 There are three naturally occurring isotopes of carbon, with mass numbers of 12, 13, and 14. How many neutrons does each have? Write the symbol for each isotope, indicating its atomic number and mass number.

3.57 The isotope of iodine with mass number 131 is often used in medicine as a radioactive tracer. Write the symbol for this isotope, indicating both mass number and atomic number.

3.58 Naturally occurring copper is a mixture of 69.17% ^{63}Cu with a mass of 62.93 amu and 30.83% ^{65}Cu with a mass of 64.93 amu. What is the atomic weight of copper?

3.59 Naturally occurring lithium is a mixture of 92.58% ^{7}Li with a mass of 7.016 amu and 7.42% ^{6}Li with a mass of 6.015 amu. What is the atomic weight of lithium?

THE PERIODIC TABLE

3.60 Why does the third period in the periodic table contain eight elements?

3.61 Why does the fourth period in the periodic table contain 18 elements?

3.62 Americium, atomic number 95, is used in household smoke detectors. What is the symbol for americium? Is americium a metal, a nonmetal, or a metalloid?

3.63 Antimony, Z = 51, is alloyed with lead for use in automobile batteries. What is the symbol for antimony? Is antimony a metal, a nonmetal, or a metalloid?

3.64 Answer the following questions for the elements from scandium through zinc:

(a) Are they metals or nonmetals?
(b) To what general class of elements do they belong?
(c) What subshell is being filled by electrons in these elements?

3.65 Answer the following questions for the elements from cerium through lutetium:

(a) Are they metals or nonmetals?
(b) To what general class of elements do they belong?
(c) What subshell is being filled by electrons in these elements?

3.66 For (a) rubidium (b) tungsten, (c) germanium, and (d) krypton, which of the following terms apply? (i) metal, (ii) nonmetal, (iii) metalloid (iv) transition element, (v) main group element, (vi) noble gas, (vii) alkali metal, (viii) alkaline earth metal

3.67 For (a) calcium, (b) palladium, (c) carbon, and (d) radon, which of the following terms apply? (i) metal, (ii) nonmetal, (iii) metalloid (iv) transition element, (v) main group element, (vi) noble gas, (vii) alkali metal, (viii) alkaline earth metal

3.68 Name an element in the periodic table that you would expect to be chemically similar to sulfur.

3.69 Name an element in the periodic table that you would expect to be chemically similar to magnesium

3.70 What elements in addition to lithium make up the alkali metal family?

3.71 What elements in addition to fluorine make up the halogen family?

ELECTRON CONFIGURATIONS

3.72 What is the maximum number of electrons that can go into an orbital?

3.73 What are the shapes and locations within an atom of s and p orbitals?

3.74 What is the maximum number of electrons that can go into the first shell? The second shell? The third shell?

3.75 What is the total number of orbitals in the third shell? The fourth shell?

3.76 How many subshells are there in the third shell? The fourth shell? The fifth shell?

3.77 How many orbitals would you expect to find in the last subshell of the fifth shell? How many electrons would you need to fill this subshell?

3.78 How many electrons are present in an atom with its $1s$, $2s$, and $2p$ subshells filled? What is this element?

3.79 How many electrons are present in an atom with its $1s$, $2s$, $2p$, and $3s$ subshells filled and with two electrons in the $3p$ subshell? What is this element?

3.80 Use arrows to show electron pairing in the valence p subshell of:

(a) Sulfur
(b) Bromine
(c) Silicon

3.81 Use arrows to show electron pairing in the $5s$ and $4d$ orbitals of:

(a) Strontium
(b) Technetium
(c) Palladium

3.82 Determine the number of unpaired electrons for each of the atoms in Problems 3.80 and 3.81.

3.83 Without looking back in the text, write the electron configurations for the following:

(a) Calcium, Z = 20
(b) Sulfur, Z = 16
(c) Fluorine, Z = 9
(d) Cadmium, Z = 48

3.84 How many electrons does the element with Z = 12 have in its valence shell? Write the electron-dot symbol for this element.

3.85 How many valence electrons do group 4A elements have? Explain. Write a generic electron-dot symbol for elements in this group.

3.86 Identify the valence subshell occupied by electrons in beryllium and arsenic atoms.

3.87 What group in the periodic table has the valence-shell configuration $ns^2\,np^3$?

3.88 Give the number of valence electrons and draw electron-dot symbols for atoms of the following elements:

(a) Kr
(b) C
(c) Ca
(d) K
(e) B
(f) Cl

3.89 Using n for the number of the valence shell, write a general valence-shell configuration for the elements in group 7A and in group 1A.

Applications

3.90 What is the advantage of using a scanning tunneling microscope rather than a normal light microscope? [*Are Atoms Real? p. 52*]

3.91 Before Rutherford's experiments, atomic structure was represented using a *"plum pudding" model* proposed by Thomson, in which negatively charged electrons were embedded in diffuse, positively charged matter like raisins in plum pudding. How might Rutherford's experimental results have been different if this model were correct? [*Are Atoms Real? p. 52*]

3.92 What are the first two elements that are made in stars? [*The Origin of Chemical Elements, p. 61*]

3.93 How are elements heavier than iron made? [*The Origin of Chemical Elements, p. 61*]

3.94 Which type of electromagnetic energy in the following pairs is of higher energy? [*Atoms and Light, p. 72*]

(a) Infrared, ultraviolet
(b) Gamma waves, microwaves
(c) Visible light, X rays

3.95 Why do you suppose ultraviolet rays from the sun are more damaging to the skin than visible light? [*Atoms and Light*, p. 72]

General Questions and Problems

3.96 What elements in addition to helium make up the noble gas family?

3.97 Hydrogen is placed in group 1A on many periodic charts, even though it is not an alkali metal. On other periodic charts, however, hydrogen is included with group 7A even though it is not a halogen. Explain. (Hint: draw electron-dot symbols for H and for the 1A and 7A elements.)

3.98 Tellurium ($Z = 52$) has a *lower* atomic number than iodine ($Z = 53$), yet it has a *higher* atomic weight (127.60 amu for Te versus 126.90 amu for I). How is this possible?

3.99 What is the atomic number of the yet undiscovered element directly below francium (Fr) in the periodic table?

3.100 Give the number of electrons in each shell for lead.

3.101 Identify the highest-energy occupied subshell in atoms of the following elements:

(a) Argon
(b) Magnesium
(c) Technetium
(d) Iron

3.102 What is the atomic weight of naturally occurring bromine, which contains 50.69% Br-79 of mass 78.92 amu and 49.31% Br-81 of mass 80.91 amu?

3.103 Naturally occurring magnesium consists of three isotopes: 78.99% ^{24}Mg with a mass of 23.99 amu, 10.00% ^{25}Mg with a mass of 24.99 amu, and 11.01% ^{26}Mg with a mass of 25.98 amu. Calculate the atomic weight of magnesium.

3.104 If you had one atom of hydrogen and one atom of carbon, which would weigh more? Explain.

3.105 If you had a pile of 10^{23} hydrogen atoms and another pile of 10^{23} carbon atoms, which of the two piles would weigh more? (See Problem 3.104.)

3.106 If your pile of hydrogen atoms in Problem 3.105 weighed about 1 gram, how much would your pile of carbon atoms weigh?

3.107 Based on your answer to Problem 3.106, how much would you expect a pile of 10^{23} sodium atoms to weigh?

3.108 An unidentified element is found to have an electron configuration by shell of 2 8 18 8 2. To what group and period does this element belong? Is the element a metal or a nonmetal? How many protons does an atom of the element have? What is the name of the element? Write its electron-dot symbol.

3.109 Germanium, atomic number 32, is used in building semiconductors for microelectronic devices. If germanium, has an electron configuration by shell of 2 8 18 4, in what orbital are the valence electrons?

3.110 Tin, atomic number 50, is directly beneath germanium (Problem 3.109) in the periodic table. What electron configuration by shell would you expect tin to have? Is tin a metal or a nonmetal?

3.111 A blood sample is found to contain 8.6 mg/dL of Ca. How many atoms of Ca are present in 8.6 mg? The atomic weight of Ca is 40.08 amu.

3.112 What is wrong with the following electron configurations?

(a) Ni $1s^2\ 2s^2\ 2p^6\ 3s^2\ 3p^6\ 3d^{10}$
(b) N $1s^2\ 2p^5$
(c) Si $1s^2\ 2s^2\ 2p\ \underline{\uparrow\downarrow}\ \underline{}\ \underline{}$
(d) Mg $1s^2\ 2s^2\ 2p^6\ 3s\ \underline{\uparrow\uparrow}$

3.113 Not all elements follow exactly the electron-filling order described in Figure 3.7. Atoms of which elements are represented by the following electron configurations? (Hint: count the total number of electrons!)

(a) $1s^2\ 2s^2\ 2p^6\ 3s^2\ 3p^6\ 3d^5\ 4s^1$
(b) $1s^2\ 2s^2\ 2p^6\ 3s^2\ 3p^6\ 3d^{10}\ 4s^1$
(c) $1s^2\ 2s^2\ 2p^6\ 3s^2\ 3p^6\ 3d^{10}\ 4s^2\ 4p^6\ 4d^5\ 5s^1$
(d) $1s^2\ 2s^2\ 2p^6\ 3s^2\ 3p^6\ 3d^{10}\ 4s^2\ 4p^6\ 4d^{10}\ 5s^1$

3.114 What similarities do you see in the electron configurations for the atoms in Problem 3.113? How might these similarities explain their anomalous electron configurations?

3.115 What orbital is filled last in the yet undiscovered element 117?

CHAPTER 4

Ionic Compounds

CONCEPTS TO REVIEW

The Periodic Table
(Sections 3.4 and 3.5)

Electron Configurations
(Sections 3.7 and 3.8)

▲ Stalagmites and stalactites, such as these in a cave in the Nangu Stone Forest in China, are composed of the ionic compounds calcium carbonate, $CaCO_3$, and magnesium carbonate, $MgCO_3$.

CONTENTS

CHAPTER GOALS

We will answer the following questions in this chapter:

1. **What is an ion, what is an ionic bond, and what are the general characteristics of ionic compounds?**

   THE GOAL: Be able to describe ions and ionic bonds, and give the general properties of compounds that contain ionic bonds.

2. **What is the octet rule, and how does it apply to ions?**

   THE GOAL: Be able to state the octet rule and use it to predict the electron configurations of ions of main group elements.

3. **What is the relationship between an element's position in the periodic table and the formation of its ion?**

   THE GOAL: Be able to predict what ions are likely to be formed by atoms of a given element.

4. **What determines the chemical formula of an ionic compound?**

   THE GOAL: Be able to write formulas for ionic compounds, given the identities of the ions.

5. **How are ionic compounds named?**

   THE GOAL: Be able to name an ionic compound from its formula or give the formula of a compound from its name.

6. **What are acids and bases?**

   THE GOAL: Be able to recognize common acids and bases.

There are more than 19 million known chemical compounds, ranging in size from small *diatomic* (two-atom) substances like carbon monoxide, CO, to deoxyribonucleic acid (DNA), which can contain several *billion* atoms linked together in a precise way. Clearly, there must be some force that holds atoms together in compounds; otherwise, the atoms would simply drift apart and no compounds could exist. The forces that hold atoms together are called *chemical bonds* and are of two major types: *ionic bonds* and *covalent bonds*. In this chapter, we look at ionic bonds and at the substances formed by them. In the next chapter, we will look at covalent bonds.

All chemical bonds result from the electrical attraction between opposite charges—between positively charged nuclei and negatively charged electrons. As a result, the way in which different elements form bonds is related to their different electron configurations.

4.1 Ions

A general rule noted by early chemists is that metals, on the left side of the periodic table, tend to form compounds with nonmetals, on the right side of the table. The alkali metals of group 1A, for instance, react with the halogens of group 7A to form a variety of compounds. Sodium chloride (table salt), formed by the reaction of sodium with chlorine, is a familiar example. The names and chemical formulas of some other compounds containing elements from groups 1A and 7A include:

Potassium iodide, KI	Added to table salt to provide iodide ion needed by the thyroid gland
Sodium fluoride, NaF	Added to many municipal water supplies to provide fluoride ion for the prevention of tooth decay
Sodium iodide, NaI	Used in laboratory scintillation counters to detect radiation (See Section 11.8)

Both the compositions and the properties of these alkali metal–halogen compounds are similar. For instance, the two elements always combine in a 1:1 ratio: one alkali metal atom for every halogen atom. Each compound has a high melting point (all are over 500 °C); each is a stable, white, crystalline solid; and each is soluble in water. Furthermore, the water solution of each compound conducts electricity, a property that gives a clue to the kind of chemical bond holding the atoms together.

Electricity can flow only through a medium containing charged particles that are free to move. The electrical conductivity of metals, for example, results from the

▲ A solution of sodium chloride in water conducts electricity, allowing the bulb to light.

movement of negatively charged electrons through the metal. But what charged particles might be present in the water solutions of alkali metal–halogen compounds? To answer this question, think about the composition of atoms. Atoms are electrically neutral because they contain equal numbers of protons and electrons. By gaining or losing one or more electrons, however, an atom can be converted into a charged particle called an **ion.**

Ion An electrically charged atom or group of atoms.

Cation A positively charged ion.

The *loss* of one or more electrons from a neutral atom gives a *positively* charged ion called a **cation** (**cat**-ion). As we saw in Section 3.8, sodium and other alkali metal atoms have a single electron in their valence shell and an electron configuration symbolized as ns^1, where n represents the shell number. By losing this electron, an alkali metal is converted to a cation.

A sodium *atom*, Na — A sodium *cation*, Na^+

Anion A negatively charged ion.

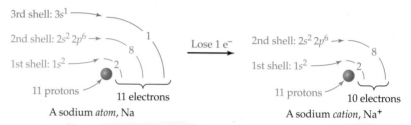

Conversely, the *gain* of one or more electrons by a neutral atom gives a *negatively* charged ion called an **anion** (**an**-ion). Chlorine and other halogen atoms have $ns^2\,np^5$ valence electrons and can easily gain an additional electron to fill out their valence subshell, thereby forming anions.

A chlorine *atom*, Cl — A chlorine *anion*, Cl^-

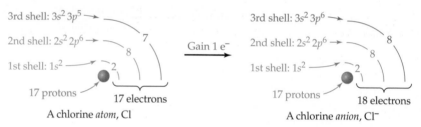

The symbol for a cation is written by adding the positive charge as a superscript to the symbol for the element; an anion symbol is written by adding the negative charge as a superscript. If one electron is lost or gained, the charge is +1 or −1 but the number 1 is omitted in the notation, as in Na^+ and Cl^-. If two or more electrons are lost or gained, however, the charge is ±2 or greater and the number *is* used, as in Ca^{2+} and N^{3-}.

PROBLEM 4.1

Magnesium atoms lose two electrons when they react. Write the symbol of the ion that results. Is it a cation or an anion?

PROBLEM 4.2

Sulfur atoms gain two electrons when they react. Write the symbol of the ion that results. Is it a cation or an anion?

KEY CONCEPT PROBLEM 4.3

Write the symbol for the ion depicted here. Is it a cation or an anion?

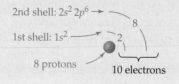

4.2 Periodic Properties and Ion Formation

The ease with which an atom loses an electron to form a positively charged cation is measured by a property called the atom's **ionization energy**, defined as the energy required to remove one electron from a single atom in the gaseous state. Conversely, the ease with which an atom *gains* an electron to form a negatively charged anion is measured by a property called **electron affinity**, defined as the energy released on adding an electron to a single atom in the gaseous state.

Ionization energy The energy required to remove one electron from a single atom in the gaseous state.

Electron affinity The energy released on adding an electron to a single atom in the gaseous state.

Ionization energy
(energy is added)   Atom + Energy $\longrightarrow$ Cation + Electron

Electron affinity
(energy is released)   Atom + Electron $\longrightarrow$ Anion + Energy

The relative magnitudes of ionization energies and electron affinities for elements in the first four rows of the periodic table are shown in Figure 4.1. Because ionization energy measures the amount of energy that must be *added* to pull an electron away from a neutral atom, the small values shown in Figure 4.1 for alkali metals (Li, Na, K) and other elements on the left side of the periodic table mean that these elements lose an electron easily. Conversely, the large values shown for halogens (F, Cl, Br) and noble gases (He, Ne, Ar, Kr) on the right side of the periodic table mean that these elements do not lose an electron easily. Electron affinities, however, measure the amount of energy *released* when an atom gains an electron. Although electron affinities are small compared to ionization energies, the halogens nevertheless have the largest values and therefore gain an electron most easily, whereas metals have the smallest values and do not gain an electron easily:

Alkali metal {
Small ionization energy—-electron easily lost
Small electron affinity—-electron not easily gained
Net result: Cation formation is favored.
}

Halogen {
Large ionization energy—-electron not easily lost
Large electron affinity—-electron easily gained
Net result: Anion formation is favored.
}

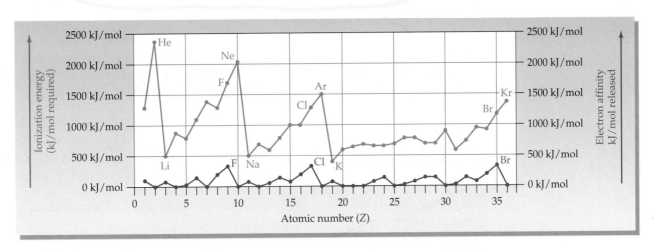

▲ **FIGURE 4.1   Relative ionization energies (red) and electron affinities (blue) for elements in the first four rows of the periodic table.**   Those elements having a value of zero for electron affinity do not accept an electron. Note that the alkali metals (Li, Na, K) have the lowest ionization energies and lose an electron most easily, whereas the halogens (F, Cl, Br) have the highest electron affinities and gain an electron most easily. The noble gases (He, Ne, Ar, Kr) neither gain nor lose an electron easily.

You might also note in Figure 4.1 that main group elements near the *middle* of the periodic table—boron (Z = 5), carbon (Z = 6, group 4A), and nitrogen (Z = 7, group 5A)—neither lose nor gain electrons easily and thus do not form ions easily. In the next chapter, we will see that these elements tend not to form ionic bonds but form covalent bonds instead.

Because alkali metals such as sodium tend to lose an electron, and halogens such as chlorine tend to gain an electron, these two elements (sodium and chlorine) will react with each other by transfer of an electron from the metal to the halogen (Figure 4.2). The product that results—sodium chloride (NaCl)—is electrically neutral because the positive charge of each Na^+ ion is balanced by the negative charge of each Cl^- ion.

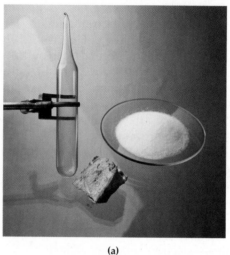

▶ **FIGURE 4.2** (a) Chlorine is a toxic green gas, sodium is a reactive metal, and sodium chloride is a harmless white solid. (b) Sodium metal burns with an intense yellow flame when immersed in chlorine gas, yielding white sodium chloride "smoke."

(a)

(b)

WORKED EXAMPLE **4.1** Periodic Trends: Ionization Energy

Look at the periodic trends in Figure 4.1, and predict where the ionization energy of rubidium is likely to fall on the chart.

ANALYSIS Identify the group number of rubidium (group 1A), and find where other members of the group appear in Figure 4.1.

SOLUTION
Rubidium (Rb) is the alkali metal below potassium (K) in the periodic table. Since the alkali metals Li, Na, and K all have ionization energies near the bottom of the chart, the ionization energy of rubidium is probably similar.

WORKED EXAMPLE **4.2** Periodic Trends: Formation of Anions and Cations

Which element is likely to lose an electron more easily, Mg or S?

ANALYSIS Identify the group numbers of the elements, and find where members of those groups appear in Figure 4.1.

SOLUTION
Magnesium, a group 2A element on the left side of the periodic table, has a relatively low ionization energy, and loses an electron easily. Sulfur, a group 6A element on the right side of the table, has a higher ionization energy, and loses an electron less easily.

PROBLEM 4.4

Look at the periodic trends in Figure 4.1, and predict approximately where the ionization energy of xenon is likely to fall.

PROBLEM 4.5

Which element in the following pairs is likely to lose an electron more easily?

(a) Be or B   **(b)** Ca or Co   **(c)** Sc or Se

PROBLEM 4.6

Which element in the following pairs is likely to gain an electron more easily?

(a) H or He   **(b)** S or Si   **(c)** Cr or Mn

4.3 Ionic Bonds

When sodium reacts with chlorine, the product is sodium chloride, a compound completely unlike either of the elements from which it is formed. Sodium is a soft, silvery metal that reacts violently with water, and chlorine is a corrosive, poisonous, green gas (Figure 4.2a). When chemically combined, however, they produce our familiar table salt containing Na^+ ions and Cl^- ions. Because opposite electrical charges attract each other, the positive Na^+ ion and negative Cl^- ion are said to be held together by an **ionic bond**.

When a vast number of sodium atoms transfer electrons to an equally vast number of chlorine atoms, a visible crystal of sodium chloride results. In this crystal, equal numbers of Na^+ and Cl^- ions are packed together in a regular arrangement. Each positively charged Na^+ ion is surrounded by six negatively charged Cl^- ions, and each Cl^- ion is surrounded by six Na^+ ions (Figure 4.3). This packing arrangement allows each ion to be stabilized by the attraction of unlike charges on its six nearest-neighbor ions, while being as far as possible from ions of like charge.

Because of the three-dimensional arrangement of ions in a sodium chloride crystal, we cannot speak of specific ionic bonds between specific pairs of ions. Rather, there are many ions attracted by ionic bonds to their nearest neighbors. We therefore speak of the whole NaCl crystal as being an **ionic solid** and of such compounds as being **ionic compounds**. The same is true of all compounds composed of ions.

Ionic bond The electrical attractions between ions of opposite charge in a crystal.

Ionic solid A crystalline solid held together by ionic bonds.

Ionic compound A compound that contains ionic bonds.

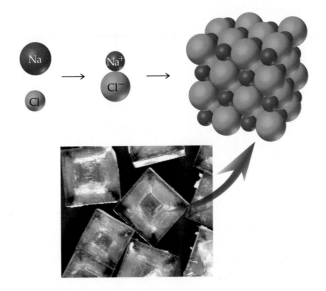

◀ **FIGURE 4.3 The arrangement of Na^+ and Cl^- ions in a sodium chloride crystal.** Each positively charged Na^+ ion is surrounded by six negatively charged Cl^- ions, and each Cl^- ion is surrounded by six Na^+ ions. The crystal is held together by ionic bonds—the attraction between oppositely charged ions.

▲ The melting point of sodium chloride is 801 °C.

4.4 Some Properties of Ionic Compounds

Like sodium chloride, ionic compounds are usually crystalline solids. Different ions vary in size and charge, however, and therefore are packed together in crystals in different ways. The ions in each compound settle into a pattern that efficiently fills space and maximizes ionic bonding.

Because the ions in an ionic solid are held rigidly in place by attraction to their neighbors, they cannot move about. Once an ionic solid is dissolved in water, however, the ions can move freely, thereby accounting for the electrical conductivity of these compounds in solution.

The high melting points and boiling points observed for ionic compounds are also accounted for by ionic bonding. The attractive force between oppositely charged particles is extremely strong, and the ions need to gain a large amount of energy by being heated to high temperatures for them to loosen their grip on one another. Sodium chloride, for example, melts at 801 °C and boils at 1413 °C; potassium iodide melts at 681 °C and boils at 1330 °C.

Despite the strength of ionic bonds, ionic solids shatter if struck sharply. A blow disrupts the orderly arrangement of cations and anions, and the electrical repulsion between ions of like charge that have been pushed together helps to split apart the crystal.

Ionic compounds dissolve in water if the attraction between water and the ions overcomes the attraction of the ions for one another. Compounds like sodium chloride are very soluble and can be dissolved to make solutions of high concentration. Do not be misled, however, by the ease with which sodium chloride and other familiar ionic compounds dissolve in water. Many other ionic compounds are not water-soluble, because water is unable to overcome the ionic forces in many crystals.

4.5 Ions and the Octet Rule

We have seen that alkali metal atoms have a single valence-shell electron, ns^1. The electron-dot symbol $X\cdot$ is consistent with this valence electron configuration. Halogens, having seven valence electrons, ns^2np^5, can be represented using $:\ddot{X}\cdot$ as the electron-dot symbol. Noble gases can be represented as $:\ddot{X}:$, since they have eight valence electrons, ns^2np^6. Both the alkali metals and the halogens are extremely reactive, undergoing many chemical reactions and forming many compounds. The noble gases, however, are quite different. They are the least reactive of all elements.

Now look at sodium chloride and similar ionic compounds. When sodium or any other alkali metal reacts with chlorine or any other halogen, the metal transfers an electron from its valence shell to the valence shell of the halogen. Sodium thereby changes its valence-shell electron configuration from $2s^22p^63s^1$ in the atom to $2s^22p^6(3s^0)$ in the Na^+ ion, and chlorine changes from $3s^23p^5$ in the atom to $3s^23p^6$ in the Cl^- ion. *In so doing, both sodium and chlorine gain noble gas electron configurations, with 8 valence electrons.* The Na^+ ion has 8 electrons in the $n = 2$ shell, matching the electron configuration of neon. The Cl^- ion has 8 electrons in the $n = 3$ shell, matching the electron configuration of argon.

$$Na \quad + \quad Cl \quad \longrightarrow \quad Na^+ \quad + \quad Cl^-$$

$$1s^2\,2s^2\,2p^6\,3s^1 \qquad 1s^2\,2s^2\,2p^6\,3s^2\,3p^5 \qquad \underbrace{1s^2\,2s^2\,2p^63s^0}_{\text{Neon configuration}} \quad \underbrace{1s^2\,2s^2\,2p^6\,3s^2\,3p^6}_{\text{Argon configuration}}$$

$$Na\cdot \quad + \quad \cdot\ddot{\underset{\cdot\cdot}{Cl}}: \quad \longrightarrow \quad Na \quad + \quad :\ddot{\underset{\cdot\cdot}{Cl}}:$$

Evidently there is something special about having 8 valence electrons (filled s and p subshells) that leads to stability and lack of chemical reactivity. In fact,

APPLICATION ▶ Minerals and Gems

I f you are wearing a sapphire, ruby, emerald, or zircon ring, you have a crystalline ionic compound on your finger. The gem came from the earth's crust, the source of most of our chemical raw materials. These gemstones are *minerals*, which to a geologist means that they are naturally occurring, crystalline chemical compounds. (To a nutritionist, by contrast, a "mineral" is one of the metal ions essential to human health.)

Sapphire and ruby are forms of the mineral *corundum*, or aluminum oxide (Al_2O_3). The blue of sapphire is due to traces of iron and titanium ions also present in the crystal, and the red of ruby is due to traces of chromium ions. Many minerals are *silicates*, meaning that their anions contain silicon and

oxygen combined in a variety of ways. Zircon is zirconium silicate ($ZrSiO_4$), and emerald is a form of the mineral beryl, $Be_3Al_2(Si_6O_{18})$, which is composed of Al^{3+}, Be^{2+}, and silicate rings.

Many minerals not sufficiently beautiful for use in jewelry are valuable as sources for the exotic metals so essential to our industrial civilization. Extraction of the metals from these minerals requires energy and a series of chemical reactions to convert the metal ions into free elements—the reverse of ion formation from atoms.

See Additional Problem 4.82 at the end of the chapter.

(a)

(b)

(c)

▲ Many minerals are silicates, which are compounds made up of polyatomic anions of silicon and oxygen. Their colors are determined largely by the cations they contain. (a) A crystal of tourmaline, a complex silicate of boron, aluminum, and other elements. (b) Dioptase from Namibia and (c) epidote from Pakistan.

observations of a great many chemical compounds have shown that main group elements frequently combine in such a way that each winds up with 8 valence electrons, a so-called *electron octet*. This conclusion is summarized in a statement called the **octet rule**:

Octet rule   Main group elements tend to undergo reactions that leave them with 8 valence electrons.

Put another way, main group *metals* tend to lose electrons when they react so that they attain an electron configuration like that of the noble gas just *before* them in the periodic table, and reactive main group *nonmetals* tend to gain electrons when they react so that they attain an electron configuration like that of the noble gas just *after* them in the periodic table. In both cases, the product ions have filled *s* and *p* subshells in their valence electron shell.

WORKED EXAMPLE **4.3** Electron Configurations: Octet Rule for Cations

Write the electron configuration of magnesium ($Z = 12$). Show how many electrons a magnesium atom must lose to form an ion with a valence octet, and write the configuration of the ion. Explain the reason for the ion's charge, and write the ion's symbol.

ANALYSIS Write the electron configuration of magnesium as described in Section 3.7 and count the number of electrons in the valence shell.

SOLUTION

Magnesium has the electron configuration $1s^2 2s^2 2p^6 3s^2$. Since the second shell contains an octet of electrons ($2s^2 2p^6$) whereas the third shell is only partially filled ($3s^2$), magnesium can achieve a valence-shell octet by losing the two electrons in the $3s$ subshell. The result is formation of a doubly charged cation, Mg^{2+} with the neon configuration:

$$Mg^{2+} \quad 1s^2 2s^2 2p^6 \quad \text{(Neon configuration)(or [Ne])}$$

A neutral magnesium atom has 12 protons and 12 electrons. With the loss of 2 electrons, there is an excess of 2 protons, accounting for the +2 charge of the ion, Mg^{2+}.

WORKED EXAMPLE **4.4** Electron Configurations: Octet Rule for Anions

How many electrons must a nitrogen atom, $Z = 7$, gain to attain a noble gas configuration? Write the electron-dot and ion symbols for the ion formed.

ANALYSIS Write the electron configuration of nitrogen, and see how many more electrons are needed to reach a noble gas configuration.

SOLUTION

Nitrogen, a group 5A element, has the electron configuration $1s^2 2s^2 2p^3$. The second shell contains 5 electrons ($2s^2 2p^3$) and needs 3 more to reach an octet. The result is formation of a triply charged anion, N^{3-}, with 8 valence electrons, matching the neon configuration:

$$N^{3-} \quad 1s^2 2s^2 2p^6 \quad \text{(Neon configuration)} \quad :\ddot{N}:^{3-}$$

PROBLEM 4.7

Write the electron configuration of potassium, $Z = 19$, and show how a potassium atom can attain a noble gas configuration.

PROBLEM 4.8

How many electrons must an aluminum atom, $Z = 13$, lose to attain a noble gas configuration? Write the symbol for the ion formed.

⊸ KEY CONCEPT PROBLEM 4.9

Which atom in the reaction depicted here gains electrons, and which loses electrons? Draw the electron-dot symbols for the resulting ions.

$$X: + \cdot \ddot{Y} \cdot \rightarrow ?$$

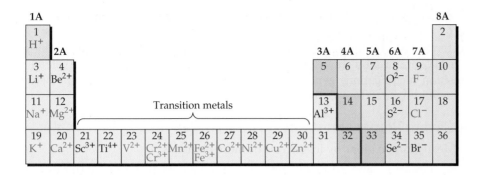

4.6 Ions of Some Common Elements

The periodic table is the key to understanding and remembering which elements form ions and which do not. As shown in Figure 4.4, atoms of elements in the same group tend to form ions of the same charge. The metals of groups 1A and 2A, for example, form only +1 and +2 ions, respectively. The ions of these elements all have noble gas configurations as a result of electron loss from their valence s subshells. (Note in the following equations that the electrons being lost are shown as products.)

Group 1A:　$M \cdot \rightarrow M^+ + e^-$
　　　　　　$(M = Li, Na, K, Rb, or Cs)$

Group 2A:　$M: \rightarrow M^{2+} + 2e^-$
　　　　　　$(M = Be, Mg, Ca, Sr, Ba, or Ra)$

Four of these ions, Na^+, K^+, Mg^{2+}, and Ca^{2+}, are present in body fluids, where they play extremely important roles in biochemical processes.

The only group 3A element commonly encountered in ionic compounds is aluminum, which forms Al^{3+} by loss of three electrons from its valence s and p subshells. Aluminum is not thought to be an essential element in the human diet, although it is known to be present in some organisms.

The first three elements in groups 4A (C, Si, Ge) and 5A (N, P, As) do not ordinarily form cations or anions, because either too much energy is required to remove an electron or not enough energy is released by adding an electron to make the process energetically favorable. The bonding of these elements is largely covalent and will be described in the next chapter. Carbon, in particular, is the key element on which life is based. Together with hydrogen, nitrogen, phosphorus, and oxygen, carbon is present in all the essential biological compounds that we will be describing throughout the latter half of this book.

The group 6A elements oxygen and sulfur form large numbers of compounds, some of which are ionic and some of which are covalent. Their ions have noble gas configurations, achieved by gaining two electrons:

Group 6A:　$\cdot \ddot{\text{O}} \cdot + 2\,e^- \longrightarrow :\ddot{\text{O}}:^{2-}$

　　　　　　$\cdot \ddot{\text{S}} \cdot + 2\,e^- \longrightarrow :\ddot{\text{S}}:^{2-}$

The halogens are present in many compounds as −1 ions, formed by gaining one electron:

Group 7A:　$\cdot \ddot{\text{X}}: + e^- \longrightarrow :\ddot{\text{X}}:^-$
　　　　　　$(X = F, Cl, Br, I)$

Transition metals lose electrons to form cations, some of which are present in the human body. The charges of transition metal cations are not as predictable as those of main group elements, however, because many transition metal atoms can lose one or more d electrons in addition to losing valence s electrons. For example, iron ($\ldots 3s^2 3p^6 3d^6 4s^2$) forms Fe^{2+} by losing two electrons from the $4s$ subshell and also forms Fe^{3+} by losing an additional electron from the $3d$ subshell. Looking at the iron configuration shows why the octet rule is limited to main group elements: Transition metal cations generally do not have noble gas configurations because they would have to lose *all* their d electrons.

Important Points about Ion Formation and the Periodic Table:

- Metals form cations by losing one or more electrons.
  - Group 1A and 2A metals form +1 and +2 ions, respectively (for example, Li^+ and Mg^{2+}) to achieve a noble gas configuration.
  - Transition metals can form cations of more than one charge (for example, Fe^{2+} and Fe^{3+}) by losing a combination of valence-shell s electrons and inner-shell d electrons.
- Reactive nonmetals form anions by gaining one or more electrons to achieve a noble gas configuration.
  - Group 6A nonmetals oxygen and sulfur form the anions O^{2-} and S^{2-}.
  - Group 7A elements (the halogens) form -1 ions; for example, F^- and Cl^-.
- Group 8A elements (the noble gases) are unreactive.
- Ionic charges of main group elements can be predicted using the group number and the octet rule.
  - For 1A and 2A metals, cation charge = group number
  - For nonmetals in groups 5A, 6A, and 7A, anion charge = 8 − (group number)

WORKED EXAMPLE **4.5** Formation of Ions: Gain/Loss of Valence Electrons

Which of the following ions is likely to form?

    (a) S^{3-} **(b)** Si^{2+} **(c)** Sr^{2+}

ANALYSIS Count the number of valence electrons in each ion. For main group elements, only ions with a valence octet of electrons are likely to form.

SOLUTION

    (a) Sulfur is in group 6A, has 6 valence electrons, and needs only two more to reach an octet. Gaining two electrons gives an S^{2-} ion with a noble gas configuration, but gaining three electrons does not. The S^{3-} ion is therefore unlikely to form.

    (b) Silicon is a nonmetal in group 4A. Like carbon, it does not form ions because it would have to gain or lose too many (4) electrons to reach a noble gas electron configuration. The Si^{2+} ion does not have an octet and will not form.

    (c) Strontium is a metal in group 2A, has only 2 outer-shell electrons, and can lose both to reach a noble gas configuration. The Sr^{2+} ion has an octet and therefore forms easily.

PROBLEM 4.10

Is molybdenum more likely to form a cation or an anion? Why?

PROBLEM 4.11

Which of the following elements can form more than one cation?

    (a) Strontium   **(b)** Chromium   **(c)** Bromine

PROBLEM 4.12

Write symbols, both with and without electron dots, for the ions formed by the following processes:

(a) Gain of two electrons by selenium

(b) Loss of two electrons by barium

(c) Gain of one electron by bromine

APPLICATION ▶ Salt

If you are like most people, you feel a little guilty about reaching for the salt shaker at mealtime. The notion that high salt intake and high blood pressure go hand in hand is surely among the most highly publicized pieces of nutritional lore ever to appear.

Salt has not always been held in such disrepute. Historically, salt has been prized since the earliest recorded times as a seasoning and a food preservative. Words and phrases in many languages reflect the importance of salt as a life-giving and life-sustaining substance. We refer to a kind and generous person as "the salt of the earth," for instance, and we speak of being "worth one's salt." In Roman times, soldiers were paid in salt; the English word "salary" is derived from the Latin word for paying salt wages (*salarium*).

Salt is perhaps the easiest of all minerals to obtain and purify. The simplest method, used for thousands of years throughout the world in coastal climates where sunshine is abundant and rainfall is scarce, is to evaporate seawater. Though the exact amount varies depending on the source, seawater contains an average of about 3.5% by mass of dissolved substances, most of which is sodium chloride. It has been estimated that evaporation of all the world's oceans would yield approximately *4.5 million cubic miles* of NaCl.

Only about 10% of current world salt production comes from evaporation of seawater. Most salt is obtained by mining the vast deposits of *halite*, or *rock salt*, formed by evaporation of ancient inland seas. These salt beds vary in thickness up to hundreds of meters and vary in depth from a few meters to thousands of meters below the earth's surface. Salt mining has gone on for at least 3400 years, and the Wieliczka mine in Galicia, Poland, has been worked continuously from A.D. 1000 to the present.

Now, back to the dinner table. What about the link between dietary salt intake and high blood pressure? There is no doubt that most people in industrialized nations have a relatively high salt intake, and also that high blood pressure among industrialized populations is on the rise. How closely, though, are the two observations related?

In a study called the DASH-Sodium study published in 2001, a strong correlation was found between a change in salt intake and a change in blood pressure. When volunteers cut back their salt intake from 8.3 g per day—roughly what Americans typically consume—to 3.8 g per day, significant

▲ In many areas of the world salt is still harvested by evaporation of ocean or tidal waters.

drops in blood pressure were found. The largest reduction in blood pressure was seen in people already diagnosed with hypertension, but subjects with normal blood pressure also lowered their readings by several percent.

What should an individual do? The best answer, as in so many things, is to use moderation and common sense. People with hypertension should make a strong effort to lower their sodium intake; others might be well advised to choose unsalted snacks, use less salt in preparing food, and read nutrition labels for sodium content.

See Additional Problems 4.83 and 4.84 at the end of this chapter.

4.7 Naming Ions

Main group metal cations in groups 1A, 2A, and 3A are named by identifying the metal, followed by the word "ion," as in the following examples:

$$K^+ \qquad Mg^{2+} \qquad Al^{3+}$$

Potassium ion   Magnesium ion   Aluminum ion

It is sometimes a little confusing to use the same name for both a metal and its ion, and you may occasionally have to stop and think about what is meant. For example, it is common practice in nutrition and health-related fields to talk about sodium or potassium in the bloodstream. Because both sodium and potassium *metals* react violently with water, however, they cannot possibly be present in blood. The references are to dissolved sodium and potassium *ions*.

For transition metals, such as iron or chromium, which can form more than one type of cation, a method is needed to differentiate these ions. Two systems are used. The first is an old system that gives the ion with the smaller charge the word ending -*ous* and the ion with the larger charge the ending -*ic*.

The second is a newer system in which the charge on the ion is given as a Roman numeral in parentheses right after the metal name. For example:

$$Cr^{2+} \qquad\qquad Cr^{3+}$$

Old name: Chrom*ous* ion   Chrom*ic* ion

New name: Chromium(II) ion   Chromium(III) ion

We will generally emphasize the new system in this book, but it is important to understand both because the old system is often found on labels of commercially supplied chemicals. The small differences between the names in either system illustrate the importance of reading a name very carefully before using a chemical. There are significant differences between compounds consisting of the same two elements but having different charges on the cation. In treating iron-deficiency anemia, for example, iron(II) compounds are preferable because the body absorbs them considerably better than iron(III) compounds.

The names of some common transition metal cations are listed in Table 4.1. Notice that the old names of the copper, iron, and tin ions are derived from their Latin names (*cuprum*, *ferrum*, and *stannum*).

Anions are named by replacing the ending of the element name with -*ide*, followed by the word "ion" (Table 4.2). For example, the anion formed by fluor*ine* is the fluor*ide* ion, and the anion formed by sul*fur* is the sul*fide* ion.

TABLE 4.1   Names of Some Transition Metal Cations

ELEMENT	SYMBOL	OLD NAME	NEW NAME
Chromium	Cr^{2+}	Chromous	Chromium(II)
	Cr^{3+}	Chromic	Chromium(III)
Copper	Cu^+	Cuprous	Copper(I)
	Cu^{2+}	Cupric	Copper(II)
Iron	Fe^{2+}	Ferrous	Iron(II)
	Fe^{3+}	Ferric	Iron(III)
Mercury	$*Hg_2^{2+}$	Mercurous	Mercury(I)
	Hg^{2+}	Mercuric	Mercury(II)
Tin	Sn^{2+}	Stannous	Tin(II)
	Sn^{4+}	Stannic	Tin(IV)

*This cation is composed of two mercury atoms, each of which has an average charge of +1.

TABLE 4.2 Names of Some Common Anions

ELEMENT	SYMBOL	NAME
Bromine	Br^-	Bromide ion
Chlorine	Cl^-	Chloride ion
Fluorine	F^-	Fluoride ion
Iodine	I^-	Iodide ion
Oxygen	O^{2-}	Oxide ion
Sulfur	S^{2-}	Sulfide ion

PROBLEM 4.13

Name the following ions:

(a) Cu^{2+} **(b)** F^- **(c)** Mg^{2+} **(d)** S^{2-}

PROBLEM 4.14

Write the symbols for the following ions:

(a) Silver(I) ion   **(b)** Iron(II) ion   **(c)** Cuprous ion   **(d)** Telluride ion

PROBLEM 4.15

Ringer's solution, which is used intravenously to adjust ion concentrations in body fluids, contains the ions of sodium, potassium, calcium, and chlorine. Give the names and symbols of these ions.

4.8 Polyatomic Ions

Ions that are composed of more than one atom are called **polyatomic ions**. Most polyatomic ions contain oxygen and another element, and their chemical formulas show by subscripts how many of each type of atom are combined. Sulfate ion, for example, is composed of one sulfur atom and four oxygen atoms, and has a -2 charge: $SO_4{}^{2-}$. The atoms in a polyatomic ion are held together by covalent bonds, of the sort discussed in the next chapter, and the entire group of atoms acts as a single unit. A polyatomic ion is charged because it contains a total number of electrons different from the total number of protons in the combined atoms.

The most common polyatomic ions are listed in Table 4.3. Note that the ammonium ion, NH_4^+, and the hydronium ion, H_3O^+, are the only cations; all the others

Polyatomic ion An ion that is composed of more than one atom.

▲ The brilliant blue mineral chalcanthite consists of the hydrated ionic compound copper sulfate, $CuSO_4$ $5H_2O$.

TABLE 4.3 Some Common Polyatomic Ions

NAME	FORMULA	NAME	FORMULA
Hydronium ion	H_3O^+	Nitrate ion	NO_3^-
Ammonium ion	NH_4^+	Nitrite ion	NO_2^-
Acetate ion	$CH_3CO_2^-$	Oxalate ion	$C_2O_4{}^{2-}$
Carbonate ion	$CO_3{}^{2-}$	Permanganate ion	MnO_4^-
Hydrogen carbonate ion (bicarbonate ion)	HCO_3^-	Phosphate ion	$PO_4{}^{3-}$
		Hydrogen phosphate ion	$HPO_4{}^{2-}$
Chromate ion	$CrO_4{}^{2-}$	Dihydrogen phosphate ion	$H_2PO_4^-$
Dichromate ion	$Cr_2O_7{}^{2-}$	Sulfate ion	$SO_4{}^{2-}$
Cyanide ion	CN^-	Hydrogen sulfate ion (bisulfate ion)	HSO_4^-
Hydroxide ion	OH^-		
Hypochlorite ion	OCl^-	Sulfite ion	$SO_3{}^{2-}$

Chlorate ClO_3^- Chlorite ClO_2^- Perchorlate ClO_4^-

APPLICATION ▶ Biologically Important Ions

The human body requires many different ions for proper functioning. Several of these ions, such as Ca^{2+}, Mg^{2+}, and HPO_4^{2-}, are used as structural materials in bones and teeth in addition to having other essential functions. Although 99% of Ca^{2+} is contained in bones and teeth, small amounts in body fluids play a vital role in transmission of nerve impulses. Other ions, including essential transition metal ions such as Fe^{2+}, are required for specific chemical reactions in the body. And still others, such as K^+, Na^+, and Cl^-, are present in fluids throughout the body.

Solutions containing ions must have overall neutrality, and several polyatomic anions, especially HCO_3^- and HPO_4^{2-}, are present in body fluids where they help balance the cation charges. Some of the most important ions and their functions are shown in the accompanying table.

See Additional Problems 4.85, 4.86, and 4.87 at the end of the chapter.

Some Biologically Important Ions

ION	LOCATION	FUNCTION	DIETARY SOURCE
Ca^{2+}	Outside cell; 99% of Ca^{2+} is in bones and teeth as $Ca_3(PO_4)_2$ and $CaCO_3$	Bone and tooth structure; necessary for blood clotting, muscle contraction, and transmission of nerve impulses	Milk, whole grains, leafy vegetables
Fe^{2+}	Blood hemoglobin	Transports oxygen from lungs to cells	Liver, red meat, leafy green vegetables
K^+	Fluids inside cells	Maintain ion concentrations in cells; regulate insulin release and heartbeat	Milk, oranges, bananas, meat
Na^+	Fluids outside cells	Protect against fluid loss; necessary for muscle contraction and transmission of nerve impulses	Table salt, seafood
Mg^{2+}	Fluids inside cells; bone	Present in many enzymes; needed for energy generation and muscle contraction	Leafy green plants, seafood, nuts
Cl^-	Fluids outside cells; gastric juice	Maintain fluid balance in cells; help transfer CO_2 from blood to lungs	Table salt, seafood
HCO_3^-	Fluids outside cells	Control acid–base balance in blood	By-product of food metabolism
HPO_4^{2-}	Fluids inside cells; bones and teeth	Control acid–base balance in cells	Fish, poultry, milk

are anions. These ions are encountered so frequently in chemistry, biology, and medicine that there is no alternative but to memorize their names and formulas. Fortunately, there are only a few of them.

Note in Table 4.3 that several pairs of ions— CO_3^{2-} and HCO_3^-, for example— are related by the presence or absence of a hydrogen ion, H^+. In such instances, the ion with the hydrogen is sometimes named using the prefix *bi-*. Thus, CO_3^{2-} is the carbonate ion and HCO_3^- is the bicarbonate ion; similarly, SO_4^{2-} is the sulfate ion and HSO_4^- is the bisulfate ion.

PROBLEM 4.16

Name the following ions:

(a) NO_3^- **(b)** CN^- **(c)** OH^- **(d)** HPO_4^{2-}

PROBLEM 4.17

Write the formulas of the following ions:

(a) Bicarbonate ion   (b) Ammonium ion

(c) Phosphate ion   (d) Permanganate ion

4.9 Formulas of Ionic Compounds

Since all chemical compounds are neutral, it is relatively easy to figure out the formulas of ionic compounds. Once the ions are identified, all we need to do is decide how many ions of each type give a total charge of zero. Thus, the chemical formula of an ionic compound tells the ratio of anions and cations.

If the ions have the same charge, only one of each ion is needed:

$$K^+ \quad \text{and} \quad F^- \quad \text{form} \quad KF$$
$$Ca^{2+} \quad \text{and} \quad O^{2-} \quad \text{form} \quad CaO$$

This makes sense when we look at how many electrons must be gained or lost by each atom in order to satisfy the octet rule:

$$K\cdot + \cdot\ddot{\underset{..}{F}}: \rightarrow K^+ + :\ddot{\underset{..}{F}}:^-$$
$$Ca: + \cdot\ddot{\underset{..}{O}}\cdot \rightarrow Ca^{2+} + :\ddot{\underset{..}{O}}:^{2-}$$

If the ions have different charges, however, unequal numbers of anions and cations must combine in order to have a net charge of zero. When potassium and oxygen combine, for example, it takes two K^+ ions to balance the -2 charge of the O^{2-} ion. Put another way, it takes two K atoms to provide the two electrons needed in order to complete the octet for the O atom:

$$2K\cdot + \cdot\ddot{\underset{..}{O}}\cdot \rightarrow 2K^+ + :\ddot{\underset{..}{O}}:^{2-}$$
$$2\,K^+ \quad \text{and} \quad O^{2-} \quad \text{form} \quad K_2O$$

The situation is reversed when a Ca^{2+} ion reacts with a Cl^- ion. One Ca atom can provide two electrons; each Cl atom requires only one electron to achieve a complete octet. Thus, there is one Ca^{2+} cation for every two Cl^- anions:

$$Ca: + 2\cdot\ddot{\underset{..}{Cl}}: \rightarrow Ca^{2+} + 2\,:\ddot{\underset{..}{Cl}}:^-$$
$$Ca^{2+} \quad \text{and} \quad 2\,Cl^- \quad \text{form} \quad CaCl_2$$

It sometimes helps when writing the formulas for an ionic compound to remember that, when the two ions have different charges, the number of one ion is equal to the charge on the other ion. In magnesium phosphate, for example, the charge on the magnesium ion is $+2$ and the charge on the polyatomic phosphate ion is -3. Thus, there must be 3 magnesium ions with a total charge of $3 \times (+2) = +6$ and 2 phosphate ions with a total charge of $2 \times (-3) = -6$ for overall neutrality:

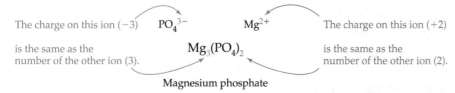

The charge on this ion (-3) $PO_4{}^{3-}$ Mg^{2+} The charge on this ion ($+2$)

is the same as the $Mg_3(PO_4)_2$ is the same as the
number of the other ion (3).   number of the other ion (2).

Magnesium phosphate

The formula of an ionic compound shows the lowest possible ratio of atoms in the compound and is thus known as a *simplest formula*. Because there is no such

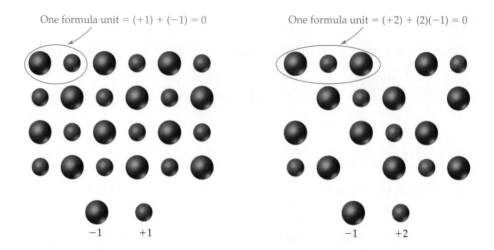

One formula unit = $(+1) + (-1) = 0$

One formula unit = $(+2) + 2(-1) = 0$

-1 $+1$ -1 $+2$

▶ **FIGURE 4.5 Formula units of ionic compounds.** The sum of charges on the ions in a formula unit equals zero.

Formula unit The formula that identifies the smallest neutral unit of an ionic compound.

thing as a single neutral *particle* of an ionic compound, however, we use the term **formula unit** to identify the smallest possible neutral *unit* (Figure 4.5). For NaCl, the formula unit is one Na^+ ion and one Cl^- ion; for K_2SO_4, the formula unit is two K^+ ions and one SO_4^{2-} ion; for CaF_2, the formula unit is one Ca^{2+} ion and two F^- ions; and so on.

Once the numbers and kinds of ions in a compound are known, the formula is written using the following rules:

- List the cation first and the anion second; for example, NaCl rather than ClNa.
- Do not write the charges of the ions; for example, KF rather than K^+F^-.
- Use parentheses around a polyatomic ion formula if it has a subscript; for example, $Al_2(SO_4)_3$ rather than Al_2SO_{43}.

WORKED EXAMPLE   **4.6** Ionic Compounds: Writing Formulas

Write the formula for the compound formed by calcium ions and nitrate ions.

ANALYSIS Knowing the formula and charges on the cation and anion (Table 4.4), we determine how many of each are needed to yield a neutral formula for the ionic compound.

SOLUTION
The two ions are Ca^{2+} and NO_3^-. Two nitrate ions, each with a -1 charge, will balance the $+2$ charge of the calcium ion.

Ca^{2+} Charge = $1 \times (+2) = +2$

$2\,NO_3^-$ Charge = $2 \times (-1) = -2$

Since there are two nitrate ions, the nitrate formula must be enclosed in parentheses:

$Ca(NO_3)_2$ Calcium nitrate

PROBLEM 4.18

Write the formulas for the ionic compounds that silver(I) forms with each of the following:

(a) Iodide ion

(b) Oxide ion

(c) Phosphate ion

PROBLEM 4.19

Write the formulas for the ionic compounds that sulfate ion forms with the following:

(a) Sodium ion   (b) Iron(II) ion   (c) Chromium(III) ion

PROBLEM 4.20

The ionic compound containing ammonium ion and carbonate ion gives off the odor of ammonia, a property put to use in smelling salts for reviving someone who has fainted. Write the formula for this compound.

PROBLEM 4.21

An *astringent* is a compound that causes proteins in blood, sweat, and other body fluids to coagulate, a property put to use in deodorants. Two safe and effective astringents are the ionic compounds of aluminum with sulfate ion and with acetate ion. Write the formulas of both.

KEY CONCEPT PROBLEM 4.22

Three ionic compounds are represented on this periodic table—red cation with red anion, blue cation with blue anion, and green cation with green anion. Give a likely formula for each compound.

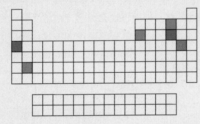

KEY CONCEPT PROBLEM 4.23

The ionic compound calcium nitride is represented here. What is the formula for calcium nitride, and what are the charges on the calcium and nitride ions?

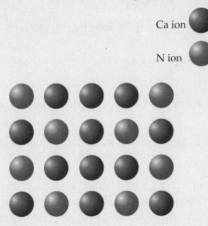

4.10 Naming Ionic Compounds

Just as in writing formulas for ionic compounds, these compounds are named by citing first the cation and then the anion, with a space between words. There are two kinds of ionic compounds, and the rules for naming them are slightly different.

Type I: Ionic compounds containing cations of main group elements (1A, 2A, aluminum). Since the charges on these cations do not vary, we do not need to specify the charge on the cation as discussed in Section 4.7.

Type II: Ionic compounds containing metals that can exhibit more than one charge. Since some metals, including the transition metals, often form more than one ion, we need to specify the charge on the cation in these compounds. Either the old (-ous, -ic) or the new (Roman numerals) system described in Section 4.7 can be used. Thus, $FeCl_2$ is called iron(II) chloride (or ferrous chloride), and $FeCl_3$ is called iron(III) chloride (or ferric chloride). Note that we do *not* name these compounds iron *di*chloride or iron *tri*chloride—once the charge on the metal is known, the number of anions needed to yield a neutral compound is also known and does not need to be included as part of the compound name. Ions of elements that form only one type of ion, such as Na^+ and Ca^{2+}, do not need Roman numerals. Table 4.4 lists some common ionic compounds and their uses.

⊂⊃ Looking Ahead

Because the formula unit for an ionic compound must be neutral, we can unambiguously write the formula from the name of the compound, and vice versa. As we shall see in Chapter 5, covalent bonding between atoms can produce a much greater variety of compounds. The rules for naming covalent compounds must be able to accommodate multiple combinations of elements (for example, CO and CO_2).   ⊂⊃

TABLE 4.4 Some Common Ionic Compounds and Their Applications

CHEMICAL NAME (COMMON NAME)	FORMULA	APPLICATIONS
Ammonium carbonate	$(NH_4)_2CO_3$	Smelling salts
Calcium hydroxide (hydrated lime)	$Ca(OH)_2$	Mortar, plaster, whitewash
Calcium oxide (lime)	CaO	Lawn treatment, industrial chemical
Lithium carbonate ("lithium")	Li_2CO_3	Treatment of manic depression
Magnesium hydroxide (milk of magnesia)	$Mg(OH)_2$	Antacid
Magnesium sulfate (Epsom salts)	$MgSO_4$	Laxative, anticonvulsant
Potassium permanganate	$KMnO_4$	Antiseptic, disinfectant*
Potassium nitrate (saltpeter)	KNO_3	Fireworks, matches, and desensitizer for teeth
Silver nitrate	$AgNO_3$	Antiseptic, germicide
Sodium bicarbonate (baking soda)	$NaHCO_3$	Baking powder, antacid, mouthwash, deodorizer
Sodium hypochlorite	NaOCl	Disinfectant; active ingredient in household bleach
Zinc oxide	ZnO	Skin protection, in calamine lotion

An antiseptic kills or inhibits growth of harmful microorganisms on the skin or in the body; a disinfectant kills harmful microorganisms but is generally not used on living tissue.

WORKED EXAMPLE   **4.7** Ionic Compounds: Formulas Involving Polyatomic Ions

Magnesium carbonate is used as an ingredient in Bufferin tablets. Write its formula.

ANALYSIS Since magnesium is a main group metal, we can determine its ionic compound formula by identifying the charges and formulas for the anion and the cation, remembering that the overall formula must be neutral.

SOLUTION
Look at the cation and the anion parts of the name separately. Magnesium, a group 2A element, forms the doubly positive Mg^{2+} cation; carbonate anion is doubly negative, CO_3^{2-}. Because the charges on the anion and cation are equal, a formula of $MgCO_3$ will be neutral.

WORKED EXAMPLE **4.8** Ionic Compounds: Formulas and Ionic Charges

Sodium and calcium both form a wide variety of ionic compounds. Write formulas for the following compounds:

(a) Sodium bromide and calcium bromide

(b) Sodium sulfide and calcium sulfide

(c) Sodium phosphate and calcium phosphate

ANALYSIS Using the formulas and charges for the cations and the anions (from Tables 4.2 and 4.3) we determine how many of each cation and anion are needed to yield a formula that is neutral.

SOLUTION

(a) Cations = Na^+ and Ca^{2+}; anion = Br^-: $NaBr$ and $CaBr_2$

(b) Cations = Na^+ and Ca^{2+}; anion = S^{2-}: Na_2S and CaS

(c) Cations = Na^+ and Ca^{2+}; anion = PO_4^{3-}: Na_3PO_4 and $Ca_3(PO_4)_2$

WORKED EXAMPLE **4.9** Naming Ionic Compounds

Name the following compounds, using Roman numerals to indicate the charges on the cations where necessary:

(a) KF (b) $MgCl_2$ (c) $AuCl_3$ (d) Fe_2O_3

ANALYSIS For main group metals, the charge is determined from the group number and no Roman numerals are necessary. For transition metals, the charge on the metal can be determined from the total charge(s) on the anion(s).

SOLUTION

(a) Potassium fluoride. No Roman numeral is necessary because a group 1A metal forms only one cation.

(b) Magnesium chloride. No Roman numeral is necessary because magnesium (group 2A) forms only Mg^{2+}.

(c) Gold(III) chloride. The three Cl^- ions require a +3 charge on the gold for a neutral formula. Since gold is a transition metal that can form other ions, the Roman numeral is necessary to specify the +3 charge.

(d) Iron(III) oxide. Because the three oxide anions (O^{2-}) have a total negative charge of −6, the two iron cations must have a total charge of +6. Thus, each is Fe^{3+}, and the charge on each is indicated by the Roman numeral (III).

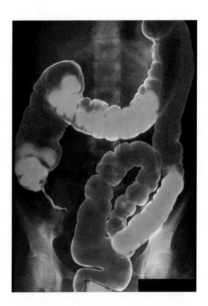

PROBLEM 4.24

Barium sulfate is an ionic compound swallowed by patients before having an X-ray of their gastrointestinal tract. Write its formula.

PROBLEM 4.25

The compound Ag_2S is responsible for much of the tarnish found on silverware. Name this compound, and give the charge on the silver ion.

▲ A barium sulfate "cocktail" is given to patients prior to an X-ray to help make the gastrointestinal tract visible on the film.

PROBLEM 4.26

Name the following compounds:

(a) SnO_2 (b) $Ca(CN)_2$ (c) Na_2CO_3

(d) Cu_2SO_4 (e) $Ba(OH)_2$ (f) $Fe(NO_3)_2$

PROBLEM 4.27

Write formulas for the following compounds:

(a) Lithium phosphate

(b) Copper(II) carbonate

(c) Aluminum sulfite

(d) Cuprous fluoride

(e) Ferric sulfate

(f) Ammonium chloride

KEY CONCEPT PROBLEM 4.28

The ionic compound formed between chromium and oxygen is shown here. Name the compound and write its formula.

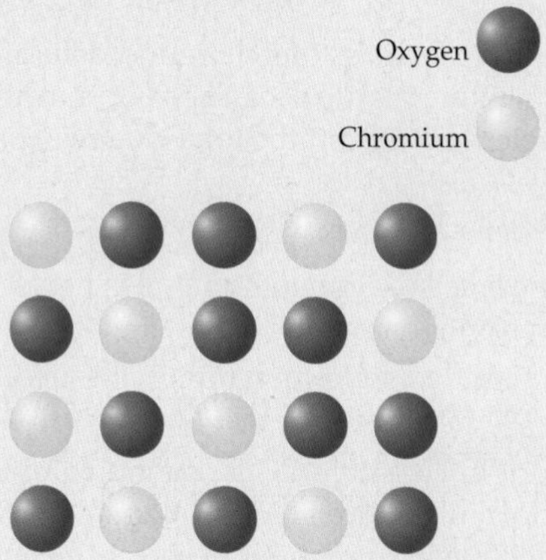

4.11 H^+ and OH^- Ions: An Introduction to Acids and Bases

Two of the most important ions we will be discussing in the remainder of this book are the hydrogen cation (H^+) and the hydroxide anion (OH^-). Since a hydrogen *atom* contains one proton and one electron, a hydrogen *cation* is simply a proton. When an acid dissolves in water, the proton typically attaches to a molecule of water to form the hydronium ion (H_3O^+), but chemists routinely use the H^+ and H_3O^+ ions interchangeably. A hydroxide anion, by contrast, is a polyatomic ion in which an oxygen atom is covalently bonded to a hydrogen atom. Although much of Chapter 10 is devoted to the chemistry of H^+ and OH^- ions, it is worth taking a preliminary look now.

The importance of the H^+ cation and the OH^- anion is that they are fundamental to the concepts of *acids* and *bases*. In fact, one definition of an **acid** is a substance that provides H^+ ions when dissolved in water, and one definition of a **base** is a substance that provides OH^- ions when dissolved in water.

Acid   A substance that provides H^+ ions in water; for example, HCl, HNO_3, H_2SO_4, H_3PO_4

Base   A substance that provides OH^- ions in water; for example, $NaOH$, KOH, $Ba(OH)_2$

Hydrochloric acid (HCl), nitric acid (HNO_3), sulfuric acid (H_2SO_4), and phosphoric acid (H_3PO_4) are among the most common acids. When any of these substances is dissolved in water, H^+ ions are formed along with the corresponding anion (Table 4.5).

TABLE 4.5   Some Common Acids and Ions Derived from Them

ACIDS		IONS	
Acetic acid	CH_3COOH	Acetate ion	*CH_3COO^-
Carbonic acid	H_2CO_3	Hydrogen carbonate ion (bicarbonate ion)	HCO_3^-
		Carbonate ion	CO_3^{2-}
Hydrochloric acid	HCl	Chloride ion	Cl^-
Nitric acid	HNO_3	Nitrate ion	NO_3^-
Nitrous acid	HNO_2	Nitrite ion	NO_2^-
Phosphoric acid	H_3PO_4	Dihydrogen phosphate ion	$H_2PO_4^-$
		Hydrogen phosphate ion	HPO_4^{2-}
		Phosphate ion	PO_4^{3-}
Sulfuric acid	H_2SO_4	Hydrogen sulfate ion	HSO_4^-
		Sulfate ion	SO_4^{2-}

*Sometimes written $C_2H_3O_2^-$ or as $CH_3CO_2^-$.

Different acids can provide different numbers of H^+ ions per acid molecule. Hydrochloric acid, for instance, provides one H^+ ion per acid molecule; sulfuric acid can provide two H^+ ions per acid molecule; and phosphoric acid can provide three H^+ ions per acid molecule.

Sodium hydroxide ($NaOH$; also known as *lye* or *caustic soda*), potassium hydroxide (KOH; also known as *caustic potash*), and barium hydroxide $[Ba(OH)_2]$ are examples of bases. When any of these compounds dissolves in water, OH^- anions go into solution along with the corresponding metal cation. Sodium hydroxide and potassium hydroxide provide one OH^- ion per formula unit; barium hydroxide provides two OH^- ions per formula unit, as indicated by its formula, $Ba(OH)_2$.

PROBLEM 4.29

Which of the following compounds are acids, and which are bases? Explain.

(a) HF

(b) $Ca(OH)_2$

(c) LiOH

(d) HCN

KEY CONCEPT PROBLEM 4.30

One of these pictures represents a solution of HCl, and one represents a solution of H_2SO_4. Which is which?

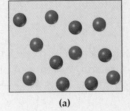
(a)

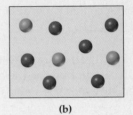
(b)

APPLICATION ▶ Osteoporosis

Bone consists primarily of two components, one mineral and one organic. About 70% of bone is the ionic compound *hydroxyapatite*, $Ca_{10}(PO_4)_6(OH)_2$, called the *trabecular*, or spongy, bone. This mineral component is intermingled in a complex matrix with about 30% by mass of fibers of the protein *collagen*, called the *cortical*, or compact, bone. Hydroxyapatite gives bone its hardness and strength, whereas collagen fibers add flexibility and resistance to breaking.

Total bone mass in the body increases from birth until reaching a maximum in the mid 30s. By the early 40s, however, an age-related decline in bone mass begins to occur in both sexes. Should this thinning of bones become too great and the bones become too porous and brittle, a clinical condition called *osteoporosis* can result. Osteoporosis is, in fact, the most common of all bone diseases, affecting approximately 25 million people in the United States. Approximately 1.5 million bone fractures each year are caused by osteoporosis at an estimated health-care cost of $14 billion.

Although both sexes are affected by osteoporosis, the condition is particularly common in postmenopausal women, who undergo cortical bone loss at a rate of 2–3% per year over and above that of the normal age-related loss. The cumulative lifetime bone loss, in fact, may approach 40–50% in women versus 20–30% in men. It has been estimated that half of all women over age 50 will have an osteoporosis-related bone fracture at some point in their life. Other risk factors in addition to sex include being thin, being sedentary, having a family history of osteoporosis, smoking, and having a diet low in calcium.

No cure yet exists for osteoporosis, but treatment for its prevention and management includes estrogen replacement therapy for postmenopausal women as well as several approved medications called *bisphosphonates* to prevent further bone loss. Calcium supplements are also recommended, as is appropriate weight-bearing exercise. In addition, treatment with sodium fluoride is under active investigation and shows considerable promise. Fluoride ion reacts with hydroxyapatite to give *fluorapatite*, in which OH^- ions are replaced by F^-, increasing both bone strength and density.

$$Ca_{10}(PO_4)_6(OH)_2 + 2\ F^- \longrightarrow Ca_{10}(PO_4)_6F_2$$
Hydroxyapatite　　　　　　　　　　Fluorapatite

See Additional Problems 4.88 and 4.89 at the end of the chapter.

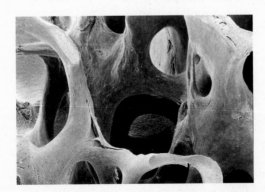 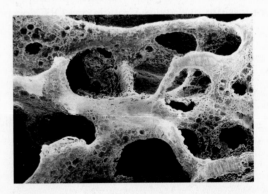

▲ Normal bone is strong and dense; a bone affected by osteoporosis is weak and spongy in appearance.

SUMMARY: REVISITING THE CHAPTER GOALS

1. **What is an ion, what is an ionic bond, and what are the general characteristics of ionic compounds?** Atoms are converted into *cations* by the loss of one or more electrons and into *anions* by the gain of one or more electrons. Ionic compounds are composed of cations and anions held together by *ionic bonds*, which result from the attraction between opposite electrical charges. Ionic compounds conduct electricity when dissolved in water, and they are generally crystalline solids with high melting points and high boiling points.

2. **What is the octet rule, and how does it apply to ions?** A valence-shell electron configuration of 8 electrons in filled *s* and *p* subshells leads to stability and lack of reactivity, as typified by the noble gases in group 8A. According to the *octet rule*, atoms of main group elements tend to form ions in which they have gained or lost the appropriate number of electrons to reach a noble gas configuration.

3. **What is the relationship between an element's position in the periodic table and the formation of its ion?** Periodic variations in *ionization energy*, the amount of energy that must be supplied to remove an electron from an atom, show that metals lose electrons more easily than nonmetals. As a result, metals usually form cations. Similar periodic variations in *electron affinity*, the amount of energy released on adding an electron to an atom, show that reactive nonmetals gain electrons more easily than metals. As a result, reactive nonmetals usually form anions. The ionic charge can be predicted from the group number and the octet rule. For main group metals, the charge on the cation is equal to the group number. For nonmetals, the charge on the anion is equal to 8 − (group number).

4. **What determines the chemical formula of an ionic compound?** Ionic compounds contain appropriate numbers of anions and cations to maintain overall neutrality, thereby providing a means of determining their chemical formulas.

5. **How are ionic compounds named?** Cations have the same name as the metal from which they are derived. Monatomic anions have the name ending *-ide*. For metals that form more than one ion, a Roman numeral equal to the charge on the ion is added to the name of the cation. Alternatively, the ending *-ous* is added to the name of the cation with the lesser charge and the ending *-ic* is added to the name of the cation with the greater charge. To name an ionic compound, the cation name is given first, with the charge of the metal ion indicated if necessary, and the anion name is given second.

6. **What are acids and bases?** The hydrogen ion (H^+) and the hydroxide ion (OH^-) are among the most important ions in chemistry because they are fundamental to the idea of acids and bases. According to one common definition, an *acid* is a substance that yields H^+ ions when dissolved in water, and a *base* is a substance that yields OH^- ions when dissolved in water.

KEY WORDS

Acid, *p. 99*

Anion, *p. 80*

Base, *p. 99*

Cation, *p. 80*

Electron affinity, *p. 81*

Formula unit, *p. 94*

Ion, *p. 80*

Ionic bond, *p. 83*

Ionic compound, *p. 83*

Ionic solid, *p. 83*

Ionization energy, *p. 81*

Octet rule, *p. 85*

Polyatomic ion, *p. 91*

UNDERSTANDING KEY CONCEPTS

4.31 Where on the blank outline of the periodic table are the following elements found?

(a) Elements that commonly form only one type of cation
(b) Elements that commonly form anions
(c) Elements that can form more than one type of cation
(d) Elements that do not readily form either anions or cations

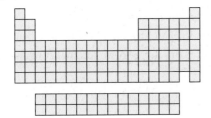

4.32 Where on the blank outline of the periodic table are the following elements found?

(a) Elements that commonly form +2 ions
(b) Elements that commonly form −2 ions
(c) An element that forms a +3 ion

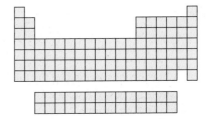

4.33 Which of these drawings represents a Ca atom, which an Na^+ ion, and which an O^{2-} ion?

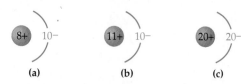

(a) (b) (c)

4.34 One of these drawings represents an Na atom, and one represents an Na^+ ion. Tell which is which, and explain why there is a difference in size.

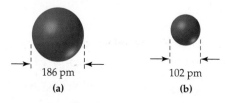

186 pm 102 pm
(a) (b)

4.35 One of these drawings represents a Cl atom, and one represents a Cl^- ion. Tell which is which, and explain why there is a difference in size.

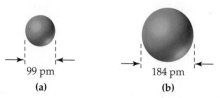

99 pm 184 pm
(a) (b)

4.36 Three ionic compounds are represented in this outline of the periodic table—red cation with red anion, blue cation with blue anion, and green cation with green anion. Give a likely formula for each compound, and tell the name of each.

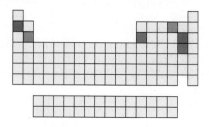

4.37 Each of these drawings (a)–(d) represents one of the following ionic compounds: $PbBr_2$, ZnS, CrF_3, Al_2O_3. Which is which?

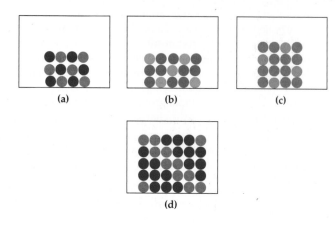

(a) (b) (c)

(d)

ADDITIONAL PROBLEMS

IONS AND IONIC BONDING

4.38 Write electron-dot symbols for the following atoms:
(a) Beryllium
(b) Neon
(c) Strontium
(d) Aluminum

4.39 Write electron-dot symbols for the following atoms:
(a) Nitrogen
(b) Selenium
(c) Iodine
(d) Strontium

4.40 Write equations for loss or gain of electrons by atoms that result in formation of the following ions from the corresponding atoms:
(a) Ca^{2+}
(b) Au^+
(c) F^-
(d) Cr^{3+}

4.41 Write symbols for the ions formed by the following:
(a) Gain of 3 electrons by phosphorus
(b) Loss of 1 electron by lithium
(c) Loss of 2 electrons by cobalt
(d) Loss of 3 electrons by thallium

4.42 Tell whether each statement about ions is true or false:
(a) A cation is formed by addition of one or more electrons to an atom.
(b) Group 4A elements tend to lose 4 electrons to yield ions with a +4 charge.
(c) Group 4A elements tend to gain 4 electrons to yield ions with a −4 charge.
(d) The individual atoms in a polyatomic ion are held together by covalent bonds.

4.43 Tell whether each statement about ionic solids is true or false:
(a) Ions are randomly arranged in ionic solids.
(b) All ions are the same size in ionic solids.
(c) Ionic solids can often be shattered by a sharp blow.
(d) Ionic solids have low boiling points.

IONS AND THE OCTET RULE

4.44 What is the *octet rule*?

4.45 Why do H and He not obey the octet rule?

4.46 What is the charge of an ion that contains 34 protons and 36 electrons?

4.47 What is the charge of an ion that contains 20 protons and 18 electrons?

4.48 Identify the element X in the following ions, and tell which noble gas has the same electron configuration.

(a) X^{2+}, a cation with 36 electrons
(b) X^-, an anion with 36 electrons

4.49 Element Z forms an ion Z^{3+}, which contains 31 protons. What is the identity of Z, and how many electrons does Z^{3+} have?

4.50 Write the electron configuration for the following ions:

(a) Rb^+
(b) Br^-
(c) S^{2-}
(d) Ba^{2+}
(e) Al^{3+}

4.51 The following ions have a noble gas configuration. Identify the noble gas.

(a) Ca^{2+}
(b) Li^+
(c) O^{2-}
(d) F^-
(e) Mg^{2+}

PERIODIC PROPERTIES AND ION FORMATION

4.52 Looking only at the periodic table, tell which member of each pair of atoms has the larger ionization energy and thus loses an electron less easily:

(a) Li and O
(b) Li and Cs
(c) K and Zn
(d) Mg and N

4.53 Looking only at the periodic table, tell which member of each pair of atoms has the larger electron affinity and thus gains an electron more easily:

(a) Li and S
(b) Ba and I
(c) Ca and Br

4.54 Which of the following ions are likely to form? Explain.

(a) Li^{2+}
(b) K^-
(c) Mn^{3+}
(d) Zn^{4+}
(e) Ne^+

4.55 Which of the following elements are likely to form more than one cation?

(a) Magnesium
(b) Silicon
(c) Manganese

4.56 Write the electron configurations of Cr^{2+} and Cr^{3+}.

4.57 Write the electron configurations of Co, Co^{2+}, and Co^{3+}.

4.58 Would you expect the ionization energy of Li^+ to be less than, greater than, or the same as the ionization energy of Li? Explain.

4.59 (a) Write equations for the loss of an electron by a K atom and the gain of an electron by a K^+ ion.
(b) What is the relationship between the equations?
(c) What is the relationship between the ionization energy of a K atom and the electron affinity of a K^+ ion?

SYMBOLS, FORMULAS, AND NAMES FOR IONS

4.60 Name the following ions:

(a) S^{2-}
(b) Sn^{2+}
(c) Sr^{2+}
(d) Mg^{2+}
(e) Au^+

4.61 Name the following ions in both the old and the new systems:

(a) Cu^{2+}
(b) Fe^{2+}
(c) Hg_2^{2+}

4.62 Write symbols for the following ions:

(a) Selenide ion
(b) Oxide ion
(c) Silver(I) ion

4.63 Write symbols for the following ions:

(a) Ferric ion
(b) Cobalt(II) ion
(c) Lead (IV) ion

4.64 Write formulas for the following ions:

(a) Hydroxide ion
(b) Bisulfate ion
(c) Acetate ion
(d) Permanganate ion
(e) Hypochlorite ion
(f) Nitrate ion
(g) Carbonate ion
(h) Dichromate ion

4.65 Name the following ions:

(a) SO_3^{2-}
(b) CN^-
(c) H_3O^+
(d) $PO_4{}^{3-}$

NAMES AND FORMULAS FOR IONIC COMPOUNDS

4.66 Write formulas for the compounds formed by the sulfate ion with the following cations:

(a) Aluminum
(b) Silver(I)
(c) Zinc
(d) Barium

4.67 Write formulas for the compounds formed by the carbonate ion with the following cations:

(a) Strontium
(b) Fe(III)
(c) Ammonium
(d) Sn(IV)

4.68 Write the formula for the following substances:

(a) Sodium bicarbonate (baking soda)
(b) Potassium nitrate (a backache remedy)
(c) Calcium carbonate (an antacid)
(d) Ammonium nitrate (first aid cold packs)

4.69 Write the formula for the following compounds:

(a) Calcium hypochlorite, used as a swimming pool disinfectant

(b) Copper(II) sulfate, used to kill algae in swimming pools

(c) Sodium phosphate, used in detergents to enhance cleaning action

4.70 Complete the table by writing in the formula of the compound formed by each pair of ions:

	S^{2-}	Cl^-	PO_4^{3-}	CO_3^{2-}
Copper(II)	CuS			
Ca^{2+}				
NH_4^+				
Ferric ion				

4.71 Complete the table by writing in the formula of the compound formed by each pair of ions:

	O^{2-}	HSO_4^-	HPO_4^{2-}	$C_2O_4^{2-}$
Na^+	Na_2O			
Zn^{2+}				
NH_4^+				
Ferrous ion				

4.72 Write the name of each compound in the table for Problem 4.70.

4.73 Write the name of each compound in the table for Problem 4.71.

4.74 Name the following substances:
(a) $MgCO_3$
(b) $Ca(CH_3CO_2)_2$
(c) $AgCN$
(d) $Na_2Cr_2O_7$

4.75 Name the following substances:
(a) $Fe(OH)_2$
(b) $KMnO_4$
(c) Na_2CrO_4
(d) $Ba_3(PO_4)_2$

4.76 Which of the following formulas is most likely to be correct for calcium phosphate?
(a) Ca_2PO_4
(b) $CaPO_4$
(c) $Ca_2(PO_4)_3$
(d) $Ca_3(PO_4)_2$

4.77 Fill in the missing information to give the correct formula for each compound:
(a) $Na_?SO_4$
(b) $Ba_?(PO_4)_?$
(c) $Ga_?(SO_4)_?$

ACIDS AND BASES

4.78 What is the difference between an acid and a base?

4.79 Identify the following substances as either an acid or a base:
(a) H_2CO_3
(b) HCN

(c) $Mg(OH)_2$
(d) KOH

4.80 Write equations to show how the substances listed in Problem 4.79 give ions when dissolved in water.

4.81 Name the anions that result when the acids in Problem 4.79 are dissolved in water.

Applications

4.82 What is the difference between a geologist's and a nutritionist's definition of a mineral? [*Minerals and Gems, p. 85*]

4.83 How is salt obtained? [*Salt, p. 89*]

4.84 What is the effect of normal dietary salt on most people? [*Salt, p. 89*]

4.85 Where are most of the calcium ions found in the body? [*Biologically Important Ions, p. 92*]

4.86 Excess sodium ion is considered hazardous, but a certain amount is necessary for normal body functions. What is the purpose of sodium in the body? [*Biologically Important Ions, p. 92*]

4.87 Before a person is allowed to donate blood, a drop of the blood is tested to be sure that it contains a sufficient amount of iron (men, 41 $\mu g/dL$; women, 38 $\mu g/dL$). Why is this required? [*Biologically Important Ions, p. 92*]

4.88 Name each ion in hydroxyapatite, $Ca_{10}(PO_4)_6(OH)_2$; give its charge; and show that the formula represents a neutral compound. [*Osteoporosis, p. 100*]

4.89 Sodium fluoride reacts with hydroxyapatite to give fluorapatite. What is the formula of fluorapatite? [*Osteoporosis, p. 100*]

General Questions and Problems

4.90 Explain why the hydride ion, H^-, has a noble gas configuration.

4.91 The H^- ion (Problem 4.90) is stable, but the Li^- ion is not. Explain.

4.92 Many compounds containing a metal and a nonmetal are not ionic, yet they are named using the Roman numeral system for ionic compounds described in Section 4.7. Write the chemical formulas for the following such compounds.
(a) Chromium(VI) oxide
(b) Vanadium(V) chloride
(c) Manganese(IV) oxide
(d) Molybdenum(IV) sulfide

4.93 The arsenate ion has the formula AsO_4^{3-}. Write the formula of lead(II) arsenate, used as an insecticide.

4.94 One commercially available calcium supplement contains calcium gluconate, a compound that is also used as an anticaking agent in instant coffee.
(a) If this compound contains one calcium ion for every two gluconate ions, what is the charge on a gluconate ion?
(b) What is the ratio of iron ions to gluconate ions in iron(III) gluconate, a commercial iron supplement?

4.95 The names given for the following compounds are incorrect. Write the correct name for each compound.

(a) Cu_3PO_4, copper(III) phosphate
(b) Na_2SO_4, sodium sulfide
(c) MnO_2, manganese(II) oxide
(d) $AuCl_3$, gold chloride
(e) $Pb(CO_3)_2$, lead(II) acetate
(f) Ni_2S_3, nickel(II) sulfide

4.96 The formulas given for the following compounds are incorrect. Write the correct formula for each compound.

(a) Cobalt(II) cyanide, $CoCN_2$
(b) Uranium(VI) oxide, UO_6
(c) Tin(II) sulfate, $Ti(SO_4)_2$
(d) Manganese(IV) oxide; MnO_4
(e) Potassium phosphate, K_2PO_4
(f) Calcium phosphide, CaP
(g) Lithium bisulfate, $Li(SO_4)_2$
(h) Aluminum hydroxide; $Al_2(OH)_3$

4.97 How many protons, electrons, and neutrons are in each of these following ions?

(a) $^{16}O^{2-}$
(b) $^{89}Y^{3+}$
(c) $^{133}Cs^+$
(d) $^{81}Br^-$

4.98 Element X reacts with element Y to give a product containing X^{3+} ions and Y^{2-} ions.

(a) Is element X likely to be a metal or a nonmetal?
(b) Is element Y likely to be a metal or a nonmetal?
(c) What is the formula of the product?
(d) What groups of the periodic table are elements X and Y likely to be in?

4.99 Identify each of the ions having the following charges and electron configurations:

(a) X^{3+}; $[Ar]\ 4s^0 3d^5$
(b) X^{2+}; $[Ar]\ 4s^0 3d^8$
(c) X^{6+}; $[Ar]\ 4s^0 3d^0$

CHAPTER 5

Molecular Compounds

▲ In living systems, as in this print by the Dutch artist M. C. Escher (1898–1972), the shapes of the component parts determine their contribution to the whole.

CONTENTS

CHAPTER GOALS

In this chapter, we will explore the following questions about molecules and covalent bonds:

1. What is a covalent bond?

THE GOAL: Be able to describe the nature of covalent bonds and how they are formed.

2. How does the octet rule apply to covalent bond formation?

THE GOAL: Be able to use the octet rule to predict the numbers of covalent bonds formed by common main group elements.

3. How are molecular compounds represented?

THE GOAL: Be able to interpret molecular formulas and draw Lewis structures for molecules.

4. What is the influence of valence-shell electrons on molecular shape?

THE GOAL: Be able to use Lewis structures to predict molecular geometry.

5. When are bonds and molecules polar?

THE GOAL: Be able to use electronegativity and molecular geometry to predict bond and molecular polarity.

6. What are the major differences between ionic and molecular compounds?

THE GOAL: Be able to compare the structures, compositions, and properties of ionic and molecular compounds.

We saw in the preceding chapter that ionic compounds are crystalline solids composed of positively and negatively charged ions. Not all substances, however, are ionic. In fact, with the exception of table salt (NaCl), baking soda ($NaHCO_3$), lime for the garden (CaO), and a few others, most of the compounds around us are *not* crystalline, brittle, high-melting ionic solids. We are much more likely to encounter gases (like those in air), liquids (such as water), low-melting solids (such as butter), and flexible solids like plastics. All these materials are composed of *molecules* rather than ions, all contain *covalent* bonds rather than ionic bonds, and all consist primarily of nonmetal atoms rather than metals.

5.1 Covalent Bonds

How do we describe the bonding in carbon dioxide, water, polyethylene, and the many millions of nonionic compounds that make up our bodies and much of the world around us? Simply put, the bonds in such compounds are formed by the *sharing* of electrons between atoms (in contrast to ionic bonds, which involve the complete transfer of electrons from one atom to another). The bond formed when atoms share electrons is called a **covalent bond**, and the group of atoms held together by covalent bonds is called a **molecule**. A single molecule of water, for example, contains two hydrogen atoms and one oxygen atom covalently bonded to one another. We might visualize a water molecule using a space-filling model as shown here:

Covalent bond A bond formed by sharing electrons between atoms.

Molecule A group of atoms held together by covalent bonds.

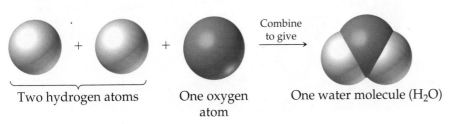

Two hydrogen atoms + One oxygen atom →(Combine to give) One water molecule (H_2O)

Recall that according to the *octet rule* (Section 4.5), main group elements tend to undergo reactions that leave them with completed outer subshells with eight valence electrons (or two for hydrogen), so that they have a noble gas electron configuration. (⊂▭⊃, p. 85) Although metals and reactive nonmetals can achieve an electron octet by gaining or losing an appropriate number of electrons to form ions, the nonmetals can also achieve an electron octet by *sharing* an appropriate number of electrons in covalent bonds.

107

As an example of how covalent bond formation occurs, let us look first at the bond between two hydrogen atoms in a hydrogen molecule, H_2. Recall that a hydrogen *atom* consists of a positively charged nucleus and a single, negatively charged $1s$ valence electron, which we represent as $H \cdot$ using the electron-dot symbols. When two hydrogen atoms come together, electrical interactions occur. Some of these interactions are repulsive—the two positively charged nuclei repel each other and the two negatively charged electrons repel each other. Other interactions, however, are attractive—each nucleus attracts both electrons and each electron attracts both nuclei (Figure 5.1). Because the attractive forces are stronger than the repulsive forces, a covalent bond is formed and the hydrogen atoms stay together.

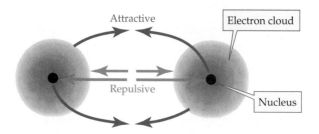

Attractive

Electron cloud

Repulsive

Nucleus

▲ **FIGURE 5.1** **A covalent H—H bond is the net result of attractive and repulsive forces.** The nucleus–electron attractions (blue arrows) are greater than the nucleus–nucleus and electron–electron repulsions (red arrows), resulting in a net attractive force that holds the atoms together to form an H_2 molecule.

In essence, the electrons act as a kind of "glue" to bind the two nuclei together into an H_2 molecule. Both nuclei are simultaneously attracted to the same electrons and are therefore held together, much as two tug-of-war teams pulling on the same rope are held together.

▶ The two teams are joined together because both are holding onto the same rope. In a similar way, two atoms are bonded together when both hold onto the same electrons.

Covalent bond formation in the H—H molecule can be visualized by imagining that the spherical $1s$ orbitals from the two individual atoms blend together and *overlap* to give an egg-shaped region in the H_2 molecule. The two electrons in the H—H covalent bond occupy the central region between the nuclei, giving both atoms a share in two valence electrons and the $1s^2$ electron configuration of the noble gas helium. For simplicity, the shared pair of electrons in a covalent bond is often represented as a line between atoms. Thus, the symbols H—H, H:H, and H_2 all represent a hydrogen molecule.

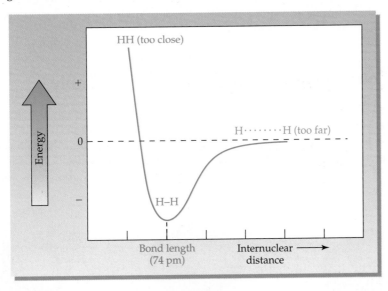

Two hydrogen atoms   A hydrogen molecule

As you might imagine, the magnitudes of the various attractive and repulsive forces between nuclei and electrons in a covalent bond depend on how close the atoms are to each other. If the atoms are too far apart, the attractive forces are small and no bond exists. If the atoms are too close, the repulsive interaction between nuclei is so strong that it pushes the atoms apart. Thus, there is an optimum point where net attractive forces are maximized and where the H_2 molecule is most stable. This optimum distance between nuclei is called the **bond length** and is 74 pm $(7.4 \times 10^{-11}$ m) in the H_2 molecule. On a graph of energy versus internuclear distance, the bond length corresponds to the minimum-energy, most stable arrangement (Figure 5.2).

Bond length The optimum distance between nuclei in a covalent bond.

◀ **FIGURE 5.2 A graph of potential energy versus internuclear distance for the H_2 molecule.** If the hydrogen atoms are too far apart, attractions are weak and no bonding occurs. If the atoms are too close, strong repulsions occur. When the atoms are optimally separated, the energy is at a minimum. The distance between nuclei at this minimum-energy point is called the *bond length*.

As another example of covalent bond formation, look at the chlorine molecule, Cl_2. An individual chlorine atom has seven valence electrons and the valence-shell electron configuration $3s^2 3p^5$. Using the electron-dot symbols for the valence electrons, each Cl atom can be represented as $:\ddot{C}l\cdot$. The 3s orbital and two of the three 3p orbitals are filled by two electrons each, but the third 3p orbital holds only one electron. When two chlorine atoms approach each other, the unpaired 3p electrons are shared by both atoms in a covalent bond. Each chlorine atom in the resultant Cl_2 molecule now "owns" six outer-shell electrons and "shares" two more, giving each a valence-shell octet like that of the noble gas argon. We can represent the formation of a covalent bond between chlorine atoms as

$$:\ddot{C}l\cdot + \cdot\ddot{C}l: \longrightarrow :\ddot{C}l:\ddot{C}l:$$

Such bond formation can also be pictured as the overlap of the 3p orbitals containing the single electrons, with resultant formation of a region of high electron density between the nuclei:

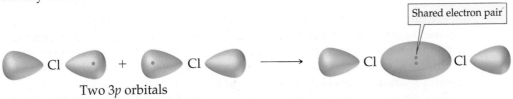

Two 3p orbitals

In addition to H_2 and Cl_2, five other elements always exist as *diatomic* (two-atom) molecules (Figure 5.3): Nitrogen (N_2) and oxygen (O_2) are colorless, odorless, nontoxic gases

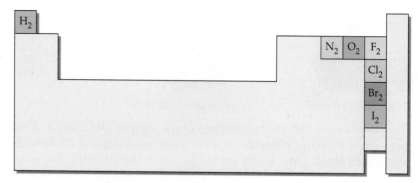

▶ **FIGURE 5.3** Diatomic elements in the periodic table.

present in air; fluorine (F_2) is a pale yellow, highly reactive gas; bromine (Br_2) is a dark red, toxic liquid; and iodine (I_2) is a violet, crystalline solid.

PROBLEM 5.1

Draw the iodine molecule using electron-dot symbols, and indicate the shared electron pair. What noble gas configuration do the iodine atoms have in an I_2 molecule?

5.2 Covalent Bonds and the Periodic Table

Molecular compound A compound that consists of molecules rather than ions.

Covalent bonds can form between unlike atoms as well as between like atoms, making possible a vast number of **molecular compounds**. Water molecules, for example, consist of two hydrogen atoms joined by covalent bonds to an oxygen atom, H_2O; ammonia molecules consist of three hydrogen atoms covalently bonded to a nitrogen atom, NH_3; and methane molecules consist of four hydrogen atoms covalently bonded to a carbon atom, CH_4.

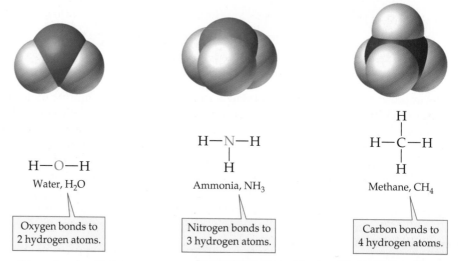

H—O—H
Water, H_2O

H—N—H
|
H
Ammonia, NH_3

H
|
H—C—H
|
H
Methane, CH_4

Oxygen bonds to 2 hydrogen atoms.

Nitrogen bonds to 3 hydrogen atoms.

Carbon bonds to 4 hydrogen atoms.

Note that in all these examples, each atom shares enough electrons to achieve a noble gas configuration: two electrons for hydrogen, and octets for oxygen, nitrogen, and carbon. Hydrogen, with one valence electron (H·), needs one more electron to achieve a noble gas configuration (that of helium, $1s^2$) and thus forms one covalent bond. Oxygen, with six valence electrons (·Ö·), needs two more electrons to have an octet; this happens when oxygen forms two covalent bonds. Nitrogen, with five valence electrons (·N̈·), needs three more electrons to achieve an octet and thus forms three covalent bonds. Carbon, with four valence electrons (·C̈·), needs four more electrons and thus forms four covalent bonds. Figure 5.4 summarizes the number of covalent bonds typically formed by common main group elements.

The octet rule is a useful guideline, but it has numerous exceptions. Boron, for example, has only three valence electrons it can share (·B̈·) and thus forms

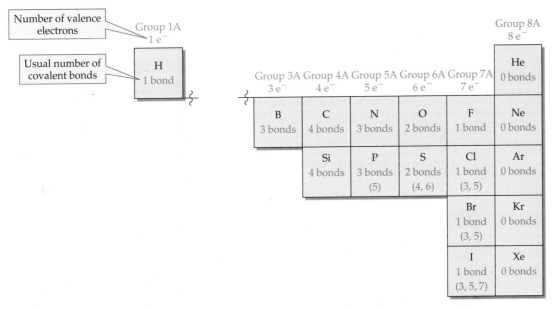

▲ **FIGURE 5.4** **Numbers of covalent bonds typically formed by main group elements to achieve octet configurations.** For P, S, Cl, and other elements in the third period and below, the number of covalent bonds may vary. Numbers in parentheses indicate other possible numbers of bonds that result in exceptions to the octet rule as explained in the text.

compounds in which it has only 3 covalent bonds and 6 electrons, such as BF_3. Exceptions to the octet rule are also seen with elements in the third row and below in the periodic table because these elements have vacant d orbitals that can be used for bonding. Phosphorus, for example, sometimes forms 5 covalent bonds (using 10 bonding electrons); sulfur sometimes forms 4 or 6 covalent bonds (using 8 and 12 bonding electrons, respectively); and chlorine, bromine, and iodine sometimes form 3, 5, or 7 covalent bonds. Phosphorus and sulfur, for example, form molecules such as PCl_5, SF_4, and SF_6.

BF_3
Boron trifluoride
(6 valence electrons on B)

PCl_5
Phosphorus pentachloride
(10 valence electrons on P)

SF_6
Sulfur hexafluoride
(12 valence electrons on S)

WORKED EXAMPLE **5.1** Molecular Compounds: Octet Rule and Covalent Bonds

Look at Figure 5.4 and tell whether the following molecules are likely to exist.

(a) Br—C—Br with Br above and CBr_3 below

(b) I—Cl, ICl

(c) H—F—H with H above and H below, FH_4

(d) H—S—H, H_2S

ANALYSIS Count the number of covalent bonds formed by each element and see if the numbers correspond to those shown in Figure 5.4.

SOLUTION

  (a) No. Carbon needs four covalent bonds but has only three in CBr_3.
  (b) Yes. Both iodine and chlorine have one covalent bond in ICl.
  (c) No. Fluorine only needs one covalent bond to achieve an octet. It cannot form more than one covalent bond because it is in the second period and does not have valence d orbitals to use for bonding.
  (d) Yes. Sulfur, which is in group 6A like oxygen, often forms two covalent bonds.

WORKED EXAMPLE 5.2 Molecular Compounds: Electron-Dot Symbols

Using electron-dot symbols, show the reaction between a hydrogen atom and a fluorine atom.

ANALYSIS The electron-dot symbols show the valence electrons for the hydrogen and fluorine atoms. A covalent bond is formed by the sharing of unpaired valence electrons between the two atoms.

SOLUTION
Draw the electron-dot symbols for the H and F atoms, showing the covalent bond as a shared electron pair.

$$H\cdot + \cdot\ddot{F}: \longrightarrow H:\ddot{F}:$$

WORKED EXAMPLE 5.3 Molecular Compounds: Predicting Number of Bonds

What are likely formulas for the following molecules?

  (a) $SiH_2Cl_?$ (b) $HBr_?$ (c) $PBr_?$

ANALYSIS The numbers of covalent bonds formed by each element should be those shown in Figure 5.4.

SOLUTION

  (a) Silicon typically forms 4 bonds: SiH_2Cl_2
  (b) Hydrogen forms only 1 bond: HBr
  (c) Phosphorus typically forms 3 bonds: PBr_3

PROBLEM 5.2

How many covalent bonds are formed by each atom in the following molecules? Draw molecules using the electron-dot symbols and lines to show the covalent bonds.

  (a) PH_3 (b) H_2Se (c) HCl   (d) SiF_4

PROBLEM 5.3

Lead forms both ionic and molecular compounds. Using Figure 5.4 and the periodic table, predict whether a molecular compound containing lead and chlorine is more likely to be $PbCl_4$ or $PbCl_5$.

PROBLEM 5.4

What are likely formulas for the following molecules?

(a) $CH_2Cl_?$ **(b)** $BH_?$ **(c)** $NI_?$ **(d)** $SiCl_?$

5.3 Multiple Covalent Bonds

The bonding in some molecules cannot be explained by the sharing of only two electrons between atoms. For example, the carbon and oxygen atoms in carbon dioxide (CO_2) and the nitrogen atoms in the N_2 molecule cannot have electron octets if only two electrons are shared:

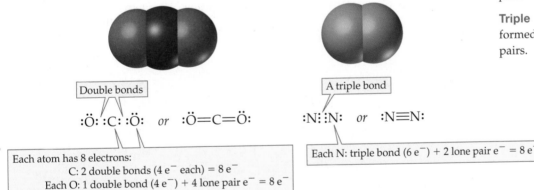

UNSTABLE—Carbon has only 6 electrons; each oxygen has only 7.

UNSTABLE—Each nitrogen has only 6 electrons.

The only way the atoms in CO_2 and N_2 can have outer-shell electron octets is by sharing *more* than two electrons, resulting in the formation of *multiple* covalent bonds. Only if the carbon atom shares four electrons with each oxygen atom do all atoms in CO_2 have electron octets, and only if the two nitrogen atoms share six electrons do both have electron octets. A bond formed by sharing two electrons (one pair) is a **single bond**, a bond formed by sharing four electrons (two pairs) is a **double bond**, and a bond formed by sharing six electrons (three pairs) is a **triple bond**. Just as a single bond is represented by a single line between atoms, a double bond is represented by two lines between atoms and a triple bond by three lines:

Single bond A covalent bond formed by sharing one electron pair.

Double bond A covalent bond formed by sharing two electron pairs.

Triple bond A covalent bond formed by sharing three electron pairs.

Double bonds

$:\ddot{O}::C::\ddot{O}:$ *or* $:\ddot{O}=C=\ddot{O}:$

A triple bond

$:N::N:$ *or* $:N\equiv N:$

Each atom has 8 electrons:
C: 2 double bonds (4 e^- each) = 8 e^-
Each O: 1 double bond (4 e^-) + 4 lone pair e^- = 8 e^-

Each N: triple bond (6 e^-) + 2 lone pair e^- = 8 e^-

Carbon, nitrogen, and oxygen are the elements most often present in multiple bonds. Carbon and nitrogen form both double and triple bonds; oxygen forms double bonds. Multiple covalent bonding is particularly common in *organic* molecules, which consist predominantly of the element carbon. For example, ethylene, a simple compound used commercially to induce ripening in fruit, has the formula C_2H_4. The only way for the two carbon atoms to have octets is for them to share four electrons in a carbon—carbon double bond:

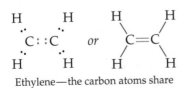

Ethylene—the carbon atoms share 4 electrons in a double bond.

▲ Acetylene is frequently used for welding metal because it burns with such a hot flame.

As another example, acetylene, the gas used in welding, has the formula C_2H_2. To achieve octets, the two acetylene carbons share six electrons in a carbon—carbon triple bond:

H : C⫶⫶C : H   *or*   H—C≡C—H

Acetylene—the carbon atoms share
6 electrons in a triple bond.

Note that in compounds with multiple bonds like ethylene and acetylene, each carbon atom still forms a total of 4 covalent bonds.

WORKED EXAMPLE 5.4 Molecular Compounds: Multiple Bonds

The compound 1-butene contains a multiple bond. In the following representation, however, only the connections between atoms are shown; the multiple bond is not specifically indicated. Identify the position of the multiple bond.

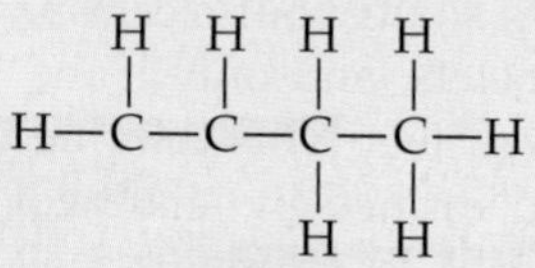

$$
\begin{array}{cccc}
\text{H} & \text{H} & \text{H} & \text{H} \\
| & | & | & | \\
\text{H}-\text{C}-\text{C}-\text{C}-\text{C}-\text{H} \\
& | & | & \\
& \text{H} & \text{H} &
\end{array}
$$

1-Butene

ANALYSIS Look for two adjacent atoms that appear to have fewer than the typical number of covalent bonds, and connect those atoms by a double or triple bond. Refer to Figure 5.4 to see how many bonds will typically be formed by hydrogen and carbon atoms.

SOLUTION

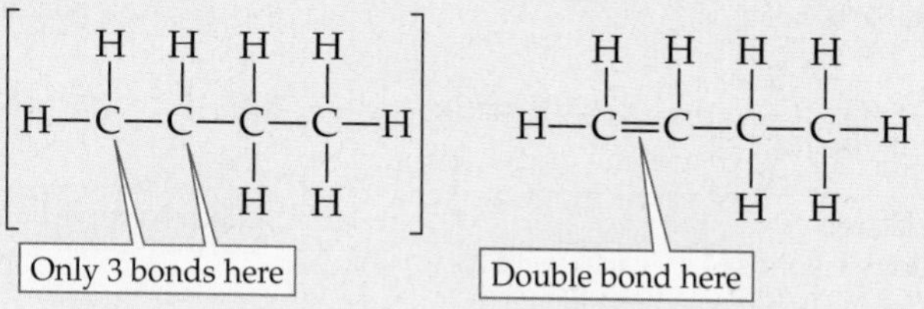

$$
\left[
\begin{array}{cccc}
\text{H} & \text{H} & \text{H} & \text{H} \\
| & | & | & | \\
\text{H}-\text{C}-\text{C}-\text{C}-\text{C}-\text{H} \\
& | & | & \\
& \text{H} & \text{H} &
\end{array}
\right]
\qquad
\begin{array}{cccc}
\text{H} & \text{H} & \text{H} & \text{H} \\
| & | & | & | \\
\text{H}-\text{C}=\text{C}-\text{C}-\text{C}-\text{H} \\
& & | & | \\
& & \text{H} & \text{H}
\end{array}
$$

Only 3 bonds here   Double bond here

WORKED EXAMPLE 5.5 Multiple Bonds: Electron-Dot and Line Structures

Draw the oxygen molecule (a) using the electron-dot symbols, and (b) using lines rather than dots to indicate covalent bonds.

ANALYSIS Each oxygen atom has six valence electrons, and will tend to form two covalent bonds to reach an octet. Thus, each oxygen will need to share four electrons to form a double bond.

SOLUTION

:Ö⫶⫶Ö: or   :Ö=Ö:

PROBLEM 5.5

Acetic acid, the organic constituent of vinegar, can be drawn using electron-dot symbols as shown below. How many outer-shell electrons are associated with each atom? Draw the structure using lines rather than dots to indicate covalent bonds.

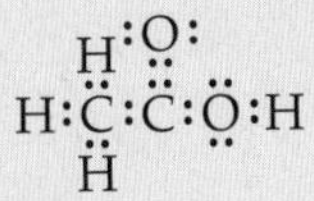

$$H:\overset{\displaystyle \ddot{O}:}{\underset{\displaystyle H}{C}}:\overset{\displaystyle }{C}:\ddot{O}:H$$

PROBLEM 5.6

Identify the position of the double bond in methyl ethyl ketone, a common industrial solvent with the following connections among atoms.

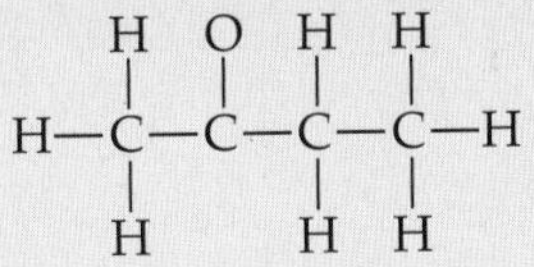

$$\begin{array}{c} \quad\ H \ \ \ O \ \ H \ \ H \\ \quad\ | \ \ \ \ | \ \ \ | \ \ \ | \\ H-C-C-C-C-H \\ \quad\ | \ \ \ \ \ \ \ \ | \ \ \ | \\ \quad\ H \ \ \ \ \ \ \ H \ \ H \end{array}$$

Methyl ethyl ketone

5.4 Coordinate Covalent Bonds

In the covalent bonds we have seen thus far, the shared electrons have come from different atoms. That is, the bonds result from the overlap of two singly occupied valence orbitals, one from each atom. Sometimes, though, a bond is formed by the overlap of a filled orbital on one atom with a vacant orbital on another atom so that both electrons come from the *same* atom. The bond that results in this case is called a **coordinate covalent bond**.

Coordinate covalent bond The covalent bond that forms when both electrons are donated by the same atom.

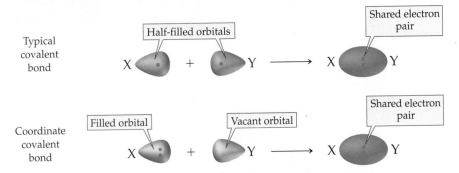

The ammonium ion, NH_4^+, is an example of a species with a coordinate covalent bond. When ammonia reacts in water solution with a hydrogen ion, H^+, the nitrogen atom donates two electrons from a filled valence orbital to form a coordinate covalent bond to the hydrogen ion, which has a vacant $1s$ orbital.

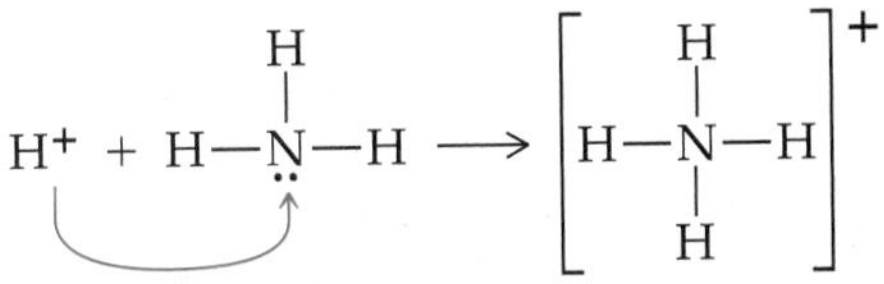

$$H^+ + H-\underset{\displaystyle \underset{\displaystyle H}{|}}{\overset{\displaystyle \overset{\displaystyle H}{|}}{\underset{..}{N}}}-H \longrightarrow \left[H-\underset{\displaystyle \underset{\displaystyle H}{|}}{\overset{\displaystyle \overset{\displaystyle H}{|}}{N}}-H \right]^+$$

Once formed, a coordinate covalent bond contains two shared electrons and is no different from any other covalent bond. All four covalent bonds in NH_4^+ are identical, for example. Note, however, that formation of a coordinate covalent bond results in unusual bonding patterns, such as an N atom with four covalent bonds rather than the usual three.

APPLICATION ▶ CO and NO: Pollutants or Miracle Molecules?

Carbon monoxide (CO) is a killer; everyone knows that. It is to blame for an estimated 3500 accidental deaths and suicides each year in the United States and is the number one cause of all deaths by poisoning. Nitric oxide (NO) is formed in combustion engines and reacts with oxygen to form nitrogen dioxide (NO_2), the reddish-brown gas associated with urban smog. What most people do not know, however, is that our bodies cannot function *without* these molecules. A startling discovery made in 1992 showed that CO and NO are key chemical messengers in the body, used by cells to regulate critical metabolic processes.

The toxicity of CO in moderate concentration is due to its ability to bind to hemoglobin molecules in the blood, thereby preventing the hemoglobin from carrying oxygen to tissues. The high reactivity of NO leads to the formation of compounds that are toxic irritants. At the same time, though, low concentrations of CO and NO are produced in cells throughout the body. Both CO and NO are highly soluble in water and can diffuse from one cell to another, where they stimulate production of a substance called *guanylyl cyclase*. Guanylyl cyclase, in turn, controls the production of another substance called *cyclic GMP*, which regulates many cellular functions.

Levels of CO production are particularly high in certain regions of the brain, including those associated with long-term memory. Evidence from experiments with rat brains suggests that a special kind of cell in the brain's hippocampus is signaled by transfer of a molecular messenger from a neighboring cell. The receiving cell responds back to the signaling cell by releasing CO, which causes still more messenger molecules to be sent. After several rounds of this back-and-forth communication, the receiving cell undergoes some sort of change that serves as a memory. When CO production is blocked, long-term memories are no longer stored, and those memories that previously existed are erased. When CO production is stimulated, however, memories are again laid down.

NO controls a seemingly limitless range of functions in the body. The immune system uses NO to fight infections and tumors. It also is used to transmit messages between nerve cells and is associated with the processes involved in learning and memory, sleeping, and depression. Its most advertised role, however, is as a *vasodilator*, a substance that allows blood vessels to relax and dilate. This discovery led to

▲ Los Angeles at sunset. Carbon monoxide is a major component of photochemical smog, but it also functions as an essential chemical messenger in our bodies.

the development of a new class of drugs that stimulate production of enzymes called nitric oxide synthases (NOS). These drugs can be used to treat conditions from erectile dysfunction (Viagra) to hypertension. Given the importance of NO in the fields of neuroscience, physiology, and immunology, it is not surprising that it was named "Molecule of the Year" in 1992.

See Additional Problems 5.91 and 5.92 at the end of the chapter.

▭ Looking Ahead

An entire class of substances is based on the ability of transition metals to form coordinate covalent bonds with nonmetals. Called *coordination compounds*, many of these substances have important roles in living organisms. For example, toxic metals can be removed from the bloodstream by forming water-soluble coordination compounds. As another example, we will see in Chapter 19 that essential metal ions are held in enzyme molecules by coordinate covalent bonds. ▭

5.5 Molecular Formulas and Lewis Structures

Formulas such as H_2O, NH_3, and CH_4, which show the numbers and kinds of atoms in one molecule of a compound, are called **molecular formulas**. Though important, molecular formulas are limited in their use because they do not provide information about how the atoms in a given molecule are connected.

Much more useful are **structural formulas**, which use lines to show how atoms are connected, and **Lewis structures**, which show both the connections among atoms and the placement of unshared valence electrons. In a water molecule, for instance, the oxygen atom shares two electron pairs in covalent bonds with two hydrogen atoms and has two other pairs of valence electrons that are not shared in bonds. Such unshared pairs of valence electrons are called **lone pairs**. In an ammonia molecule, three electron pairs are used in bonding and there is one lone pair. In methane, all four electron pairs are bonding.

Molecular formula A formula that shows the numbers and kinds of atoms in one molecule of a compound.

Structural formula A molecular representation that shows the connections among atoms by using lines to represent covalent bonds.

Lewis structure A molecular representation that shows both the connections among atoms and the locations of lone-pair valence electrons.

Lone pair A pair of electrons that is not used for bonding.

Lewis structures

Electron lone pairs

$$H-\ddot{O}-H$$
Water

$$H-\ddot{N}-H$$
Ammonia

$$H-\underset{|}{\overset{|}{C}}-H$$
Methane

▲ Ammonia is used as a fertilizer to supply nitrogen to growing plants.

Note how a molecular formula differs from an ionic formula, described previously in Section 4.10. A *molecular* formula gives the number of atoms that are combined in one molecule of a compound, whereas an *ionic* formula gives only a ratio of ions (Figure 5.5). The formula C_2H_4 for ethylene, for example, says that every ethylene molecule consists of two carbon atoms and four hydrogen atoms. The formula NaCl for sodium chloride, however, says only that there are equal numbers of Na^+ and Cl^- ions in the crystal; the formula says nothing about how the ions interact with one another.

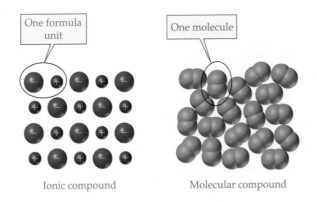

One formula unit

One molecule

Ionic compound   Molecular compound

◄ **FIGURE 5.5 The distinction between ionic and molecular compounds.** In ionic compounds, the smallest particle is an ion. In molecular compounds, the smallest particle is a molecule.

5.6 Drawing Lewis Structures

To draw a Lewis structure, you first need to know the connections among atoms. Sometimes the connections are obvious. Water, for example, can only be H—O—H because only oxygen can be in the middle and form two covalent bonds. Other times, you will have to be told how the atoms are connected.

Two approaches are used for drawing Lewis structures once the connections are known. The first is particularly useful for organic molecules like those found in living organisms because common bonding patterns are followed by the atoms. The second approach is a more general, stepwise procedure that works for all molecules.

Lewis Structures for Molecules Containing C, N, O, X (Halogen), and H

As summarized in Figure 5.4, carbon, nitrogen, oxygen, halogen, and hydrogen atoms usually maintain consistent bonding patterns:

- H forms one covalent bond.
- C forms four covalent bonds and often bonds to other carbon atoms.
- N forms three covalent bonds and has one lone pair of electrons.
- O forms two covalent bonds and has two lone pairs of electrons.
- Halogens (X = F, Cl, Br, I) form one covalent bond and have three lone pairs of electrons.

Carbon	Nitrogen	Oxygen	Halogen	Hydrogen
4 bonds	3 bonds	2 bonds	1 bond	1 bond

Relying on these common bonding patterns simplifies writing Lewis structures. In ethane (C_2H_6), a constituent of natural gas, for example, three of the four covalent bonds of each carbon atom are used in bonds to hydrogen, and the fourth is a carbon—carbon bond. There is no other arrangement in which all eight atoms can have their usual bonding patterns. In acetaldehyde (C_2H_4O), a substance used in manufacturing perfumes, dyes, and plastics, one carbon has three bonds to hydrogen, while the other has one bond to hydrogen and a double bond to oxygen.

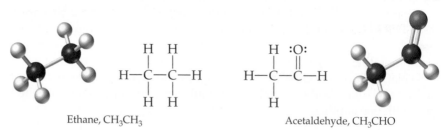

Ethane, CH_3CH_3

Acetaldehyde, CH_3CHO

Condensed structure A molecular representation in which bonds are understood by the order in which they are written rather than specifically shown.

Because Lewis structures are awkward for larger organic molecules, ethane is more frequently written as a **condensed structure** in which the bonds are not specifically shown. In its condensed form, ethane is CH_3CH_3, meaning that each carbon atom has three hydrogen atoms bonded to it (CH_3) and the two CH_3 units are bonded to each other. In the same way, acetaldehyde can be written as CH_3CHO. Note that neither the lone-pair electrons nor the C=O double bond in acetaldehyde are shown explicitly. You will get a lot more practice with such condensed structures in later chapters.

Many of the computer-generated pictures we will be using from now on will be *ball-and-stick models* rather than the space-filling models used previously. Space-filling models are more realistic, but ball-and-stick models do a better job of showing connections and molecular geometry. All models, regardless of type, use a consistent color code in which C is dark gray or black, H is white or ivory, O is red, N is blue, S is yellow, P is dark blue, F is light green, Cl is greenish yellow, Br is brownish red, and I is purple.

Space-filling

Ball-and-stick

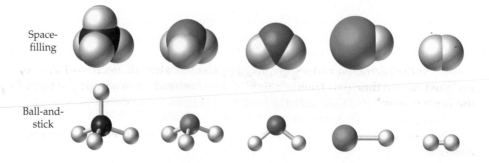

A General Method for Drawing Lewis Structures

A Lewis structure can be drawn for any molecule or polyatomic ion by following a five-step procedure. Take PCl_3, for example, a substance in which three chlorine atoms surround the central phosphorus atom.

STEP 1: Find the total number of valence electrons of all atoms in the molecule or ion. For a polyatomic ion, add one electron for each negative charge or subtract one for each positive charge. In PCl_3, for example, phosphorus (group 5A) has 5 valence electrons and chlorine (group 7A) has 7 valence electrons, giving a total of 26:

$$P + (3 \times Cl) = PCl_3$$
$$5e^- + (3 \times 7e^-) = 26e^-$$

In OH^-, the total is 8 electrons (6 from oxygen, 1 from hydrogen, plus 1 for the negative charge). In NH_4^+, the total is 8 (5 from nitrogen, 1 from each of 4 hydrogens, minus 1 for the positive charge).

STEP 2: Draw a line between each pair of connected atoms to represent the two electrons in a covalent bond. Remember that elements in the second row of the periodic table form the number of bonds discussed earlier in this section, whereas elements in the third row or lower can use more than 8 electrons and form more than the "usual" number of bonds (Figure 5.4). A particularly common pattern is that an atom in the third row or lower occurs as the central atom in a cluster. In PCl_3, for example, the phosphorus atom is in the center with the three chlorine atoms bonded to it:

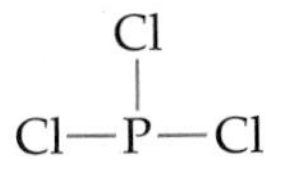

$$
\begin{array}{c}
\text{Cl} \\
| \\
\text{Cl}-\text{P}-\text{Cl}
\end{array}
$$

STEP 3: Add lone pairs so that each atom connected to the central atom (except H) gets an octet. In PCl_3, each Cl atom needs three lone pairs:

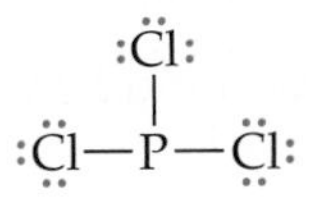

$$
\begin{array}{c}
:\ddot{\text{Cl}}: \\
| \\
:\ddot{\text{Cl}}-\text{P}-\ddot{\text{Cl}}:
\end{array}
$$

STEP 4: Place all remaining electrons in lone pairs on the central atom. In PCl_3, we have used 24 of the 26 available electrons—6 in three single bonds and 18 in the three lone pairs on each chlorine atom. This leaves 2 electrons for one lone pair on phosphorus:

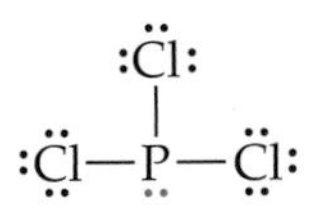

$$
\begin{array}{c}
:\ddot{\text{Cl}}: \\
| \\
:\ddot{\text{Cl}}-\ddot{\text{P}}-\ddot{\text{Cl}}:
\end{array}
$$

STEP 5: If the central atom does not yet have an octet after all electrons have been assigned, take a lone pair from a neighboring atom and form a multiple bond to the central atom. In PCl_3, each atom has an octet, all 26 available electrons have been used, and the Lewis structure is finished.

Worked Examples 5.6–5.8 shows how to deal with cases where this fifth step is needed.

WORKED EXAMPLE **5.6** Multiple Bonds: Electron Dots and Valence Electrons

Draw a Lewis structure for the toxic gas hydrogen cyanide, HCN. The atoms are connected in the order shown in the preceding sentence.

ANALYSIS Follow the procedure outlined in the text.

SOLUTION

STEP 1: Find the total number of valence electrons:

$$H = 1, C = 4, N = 5 \quad \text{Total number of valence electrons} = 10$$

STEP 2: Draw a line between each pair of connected atoms to represent bonding electron pairs:

$$\text{H—C—N} \quad \text{two bonds} = 4 \text{ electrons; 6 electrons remaining}$$

STEP 3: Add lone pairs so that each atom (except H) has a complete octet:

$$\text{H—C—N̈:}$$

STEP 4: All valence electrons have been used, and so step 4 is not needed. H and N have filled valence shells, but C does not.

STEP 5: If central atom (C in this case) does not yet have an octet, use lone pairs from a neighboring atom (N) to form multiple bonds. This results in a triple bond between the C and N atoms, as shown n the electron dot and ball-and stick representations below:

$$\text{H—C≡N:}$$

We can check the structure by noting that all 10 valence electrons have been used (in four covalent bonds and one lone pair) and that each atom has the expected number of bonds (one bond for H, three for N, and four for C).

WORKED EXAMPLE **5.7** Lewis Structures: Location of Multiple Bonds

Draw a Lewis structure for vinyl chloride, C_2H_3Cl, a substance used in making poly(vinyl chloride), or PVC, plastic.

ANALYSIS Since H and Cl form only one bond each, the carbon atoms must be bonded to each other, with the remaining atoms bonded to the carbons. With only four atoms available to bond with them, the carbon atoms cannot have four covalent bonds each unless they are joined by a double bond.

SOLUTION

STEP 1: The total number of valence electrons is 18; 4 from each of the two C atoms, 1 from each of the three H atoms, and 7 from the Cl atom.

STEP 2: Place the two C atoms in the center and divide the other four other atoms between them:

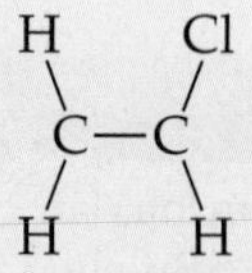

The five bonds account for 10 valence electrons, with 8 remaining.

STEP 3: Place 6 of the remaining valence electrons around the Cl atom so that it has a complete octet, and the remaining 2 valence electrons on one of the C atoms (either C, it does not matter):

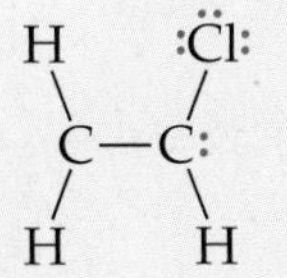

$$\begin{array}{ccc} H & & :\ddot{C}l: \\ \backslash & & | \\ & C-C: \\ / & & \backslash \\ H & & H \end{array}$$

When all the valence electrons are distributed, the C atoms still do not have a complete octet; they each need 4 bonds but have only 3.

STEP 5: The lone pair of electrons on the C atom can be used to form a double bond between the C atoms, giving each a total of four bonds (8 electrons). Placement of the double bond yields the Lewis structure and ball-an-stick model for vinyl chloride shown below:

$$\begin{array}{ccc} H & & :\ddot{C}l: \\ \backslash & & | \\ & C=C \\ / & & \backslash \\ H & & H \end{array}$$

All 18 valence electrons are accounted for in six covalent bonds and three lone pairs, and each atom has the expected number of bonds.

WORKED EXAMPLE **5.8** Lewis Structures: Octet Rule and Multiple Bonds

Draw a Lewis structure for sulfur dioxide, SO_2. The connections are O—S—O.

ANALYSIS Follow the procedure outlined in the text.

SOLUTION

STEP 1: The total number of valence electrons is 18, 6 from each atom:

$$S + (2 \times O) = SO_2$$
$$6e^- + (2 \times 6e^-) = 18e^-$$

STEP 2: O—S—O Two covalent bonds use 4 valence electrons.

STEP 3: $:\ddot{O}-S-\ddot{O}:$ Adding three lone pairs to each oxygen to give each an octet uses 12 additional valence electrons.

STEP 4: $:\ddot{O}-\underset{..}{\overset{..}{S}}-\ddot{O}:$ The remaining 2 valence electrons are placed on sulfur, but sulfur still does not have an octet.

STEP 5: Moving one lone pair from a neighboring oxygen to form a double bond with the central sulfur gives sulfur an octet (it does not matter on which side the S=O bond is written):

$$:\ddot{O}-\underset{..}{S}=\ddot{O}:$$

PROBLEM 5.7

Methylamine, CH_5N, is responsible for the characteristic odor of decaying fish. Draw a Lewis structure of methylamine.

PROBLEM 5.8

Add lone pairs where appropriate to the following structures:

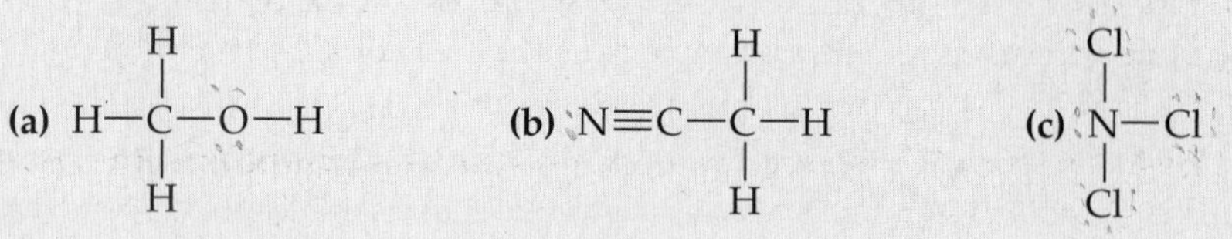

(a)
H
|
H—C—O—H
|
H

(b) $:N\equiv C—C—H$ with H above and H below the second C

(c)
Cl
|
N—Cl
|
Cl

PROBLEM 5.9

Draw Lewis structures for the following:

(a) Phosgene, $COCl_2$, a poisonous gas
(b) Hypochlorite ion, OCl^-, present in many swimming pool chemicals
(c) Hydrogen peroxide, H_2O_2
(d) Sulfur dichloride, SCl_2

PROBLEM 5.10

Draw a Lewis structure for nitric acid, HNO_3. The nitrogen atom is in the center, and the hydrogen atom is bonded to an oxygen atom.

KEY CONCEPT PROBLEM 5.11

The molecular model shown here is a representation of methyl methacrylate, a starting material used to prepare Lucite plastic. Only the connections between atoms are shown; multiple bonds are not indicated.

(a) What is the molecular formula of methyl methacrylate?
(b) Indicate the positions of the multiple bonds and lone pairs in methyl methacrylate.

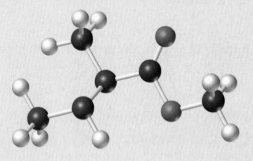

5.7 The Shapes of Molecules

Look back at the computer-generated drawings of molecules in the preceding section and you will find that the molecules are shown with specific shapes. Acetylene is *linear*, water is *bent*, ammonia is *pyramid-shaped*, methane is *tetrahedral*, and ethylene is flat, or *planar*. What determines such shapes? Why, for example, are the three atoms in water connected at an angle of 104.5° rather than in a straight line? Like so many properties, molecular shapes are related to the numbers and locations of the valence electrons around atoms.

Molecular shapes can be predicted by noting how many bonds and electron pairs surround individual atoms and applying what is called the **valence-shell electron-pair repulsion (VSEPR) model**. The basic idea of the VSEPR model is that the constantly moving valence electrons in bonds and lone pairs make up negatively

Valence-shell electron-pair repulsion (VSEPR) model A method for predicting molecular shape by noting how many electron charge clouds surround atoms and assuming that the clouds orient as far away from one another as possible.

charged clouds of electrons, which electrically repel one another. The clouds therefore tend to keep as far apart as possible, causing molecules to assume specific shapes. There are three steps to applying the VSEPR model:

STEP 1: Draw a Lewis structure of the molecule, and identify the atom whose geometry is of interest. In a simple molecule like PCl_3 or CO_2, this is usually the central atom.

STEP 2: Count the number of electron charge clouds surrounding the atom of interest. The number of charge clouds is simply the total number of lone pairs plus connections to other atoms. It does not matter whether a connection is a single bond or a multiple bond because we are interested only in the *number* of charge clouds, not in how many electrons each cloud contains. The carbon atom in carbon dioxide, for instance, has two double bonds to oxygen (O=C=O), and thus has two charge clouds.

STEP 3: Predict molecular shape by assuming that the charge clouds orient in space so that they are as far away from one another as possible. How they achieve this favorable orientation depends on their number, as summarized in Table 5.1.

TABLE 5.1 Molecular Geometry Around Atoms with 2, 3, and 4 Charge Clouds

NUMBER OF BONDS	NUMBER OF LONE PAIRS	TOTAL NUMBER OF CHARGE CLOUDS	MOLECULAR GEOMETRY		EXAMPLE
2	0	2		Linear	O=C=O
3	0	3		Planar triangular	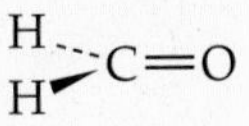 H, H > C=O
2	1			Bent	 O, O > S :
4	0	4	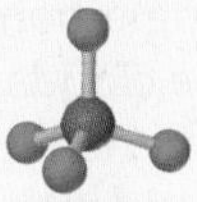	Tetrahedral	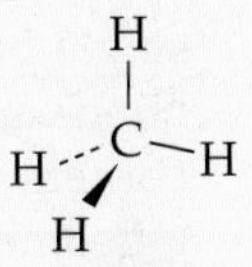 H—C—H with H, H
3	1		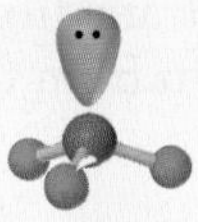	Pyramidal	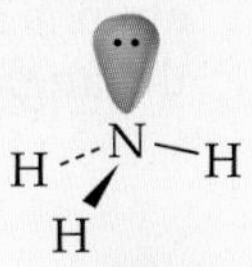 H—N—H with H
2	2		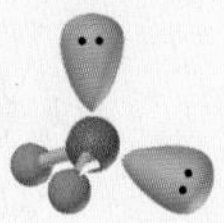	Bent	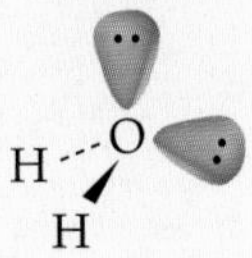 H—O with H

If there are only two charge clouds, as occurs on the central atom of CO_2 (two double bonds) and HCN (one single bond and one triple bond), the clouds are farthest apart when they point in opposite directions. Thus, both HCN and CO_2 are linear molecules, with **bond angles** of 180°:

$$\overset{180°}{\overgroup{H-C\equiv N{:}}}$$

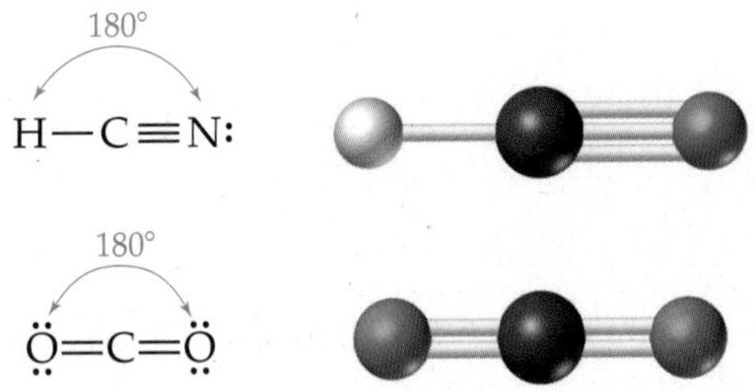

These molecules are linear, with bond angles of 180°.

$$\overset{180°}{\overgroup{\ddot{O}=C=\ddot{O}}}$$

Bond angle The angle formed by three adjacent atoms in a molecule.

When there are three charge clouds, as occurs on the central atom in formaldehyde (two single bonds and one double bond) and SO_2 (one single bond, one double bond, and one lone pair), the clouds will be farthest apart if they lie in a plane and point to the corners of an equilateral triangle. Thus, a formaldehyde molecule is planar triangular, with all bond angles near 120°. In the same way, an SO_2 molecule has a planar triangular arrangement of its three electron clouds, but one point of the triangle is occupied by a lone pair. The relationship of the three atoms themselves is therefore bent rather than linear, with an O—S—O bond angle of approximately 120°:

A formaldehyde molecule is planar triangular, with bond angles of roughly 120°. Note: solid wedges and dashed lines indicate bonds projecting out from and into the plane of the page, respectively.

An SO_2 molecule is bent, with a bond angle of approximately 120°.

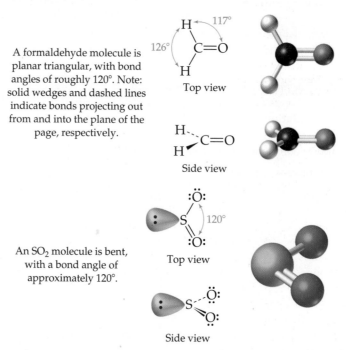

Top view

Side view

Top view

Side view

When there are four charge clouds, as occurs on the central atom in CH_4 (four single bonds), NH_3 (three single bonds and one lone pair), and H_2O (two single bonds and two lone pairs), the clouds can be farthest apart when they extend to the corners of a *regular tetrahedron*. As illustrated in Figure 5.6, a **regular tetrahedron** is a geometric solid whose four identical faces are equilateral triangles. The central atom is at the center of the tetrahedron, the charge clouds point to the corners, and the angle between lines drawn from the center to any two corners is 109.5°.

Regular tetrahedron A geometric figure with four identical triangular faces.

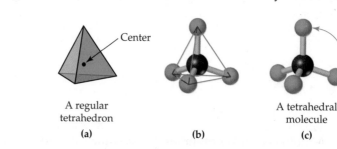

A regular tetrahedron		A tetrahedral molecule
(a)	(b)	(c)

▲ **FIGURE 5.6** **The tetrahedral geometry of an atom surrounded by four charge clouds.** The atom is located at the center of the regular tetrahedron, and the four charge clouds point toward the corners. The bond angle between the center and any two corners is 109.5°.

▲ The tetrahedral arrangement of atoms at the corners of a trigonal pyramid is clearly seen in this model.

Because valence-shell electron octets are so common, a great many molecules have geometries based on the tetrahedron. In methane (CH_4), for example, the carbon atom has tetrahedral geometry with H—C—H bond angles of exactly 109.5°. In ammonia (NH_3), the nitrogen atom has a tetrahedral arrangement of its four charge clouds, but one corner of the tetrahedron is occupied by a lone pair, resulting in an overall pyramidal shape for the molecule. Similarly, water, which has two corners of the tetrahedron occupied by lone pairs, has an overall bent shape.

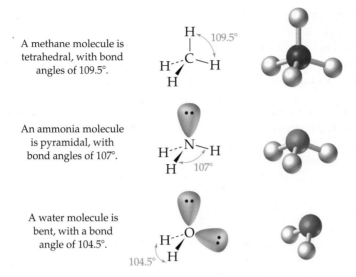

A methane molecule is tetrahedral, with bond angles of 109.5°.

An ammonia molecule is pyramidal, with bond angles of 107°.

A water molecule is bent, with a bond angle of 104.5°.

Note how the three-dimensional shapes of molecules like methane, ammonia, and water are shown. Solid lines are assumed to be in the plane of the paper; a dashed line recedes behind the plane of the paper away from the viewer; and a dark wedged line protrudes out of the paper toward the viewer. This standard method for showing three-dimensionality will be used throughout the rest of the book. Note also that the H—N—H bond angle in ammonia (107°) and the H—O—H bond angle in water (104.5°) are close to, but not exactly equal to, the ideal 109.5° tetrahedral value. The angles are diminished somewhat from their ideal value because the lone-pair charge clouds repel other electron clouds strongly and compress the rest of the molecule.

The geometry around atoms in larger molecules also derives from the shapes shown in Table 5.1. For example, each of the two carbon atoms in ethylene (H_2C=CH_2) has three charge clouds, giving rise to planar triangular geometry. It turns out that the molecule as a whole is also planar, with H—C—C and H—C—H bond angles of approximately 120°:

The ethylene molecule is planar, with bond angles of 120°.

Top view

Side view

Carbon atoms bonded to four other atoms are each at the center of a tetrahedron, as shown here for ethane, H_3C—CH_3:

The ethane molecule has tetrahedral carbon atoms, with bond angles of 109.5°.

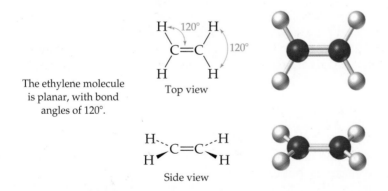

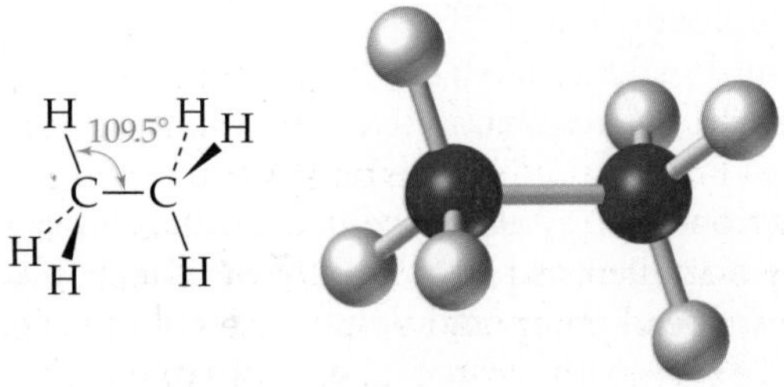

> **WORKED EXAMPLE** **5.9** Lewis Structures: Molecular Shape

What shape do you expect for the hydronium ion, H_3O^+?

ANALYSIS Draw the Lewis structure for the molecular ion and count the number of charge clouds around the central oxygen atom; imagine the clouds orienting as far away from one another as possible.

SOLUTION
The Lewis structure for the hydronium ion shows that the oxygen atom has four charge clouds (three single bonds and one lone pair). The hydronium ion is therefore pyramidal with bond angles of approximately 109.5°:

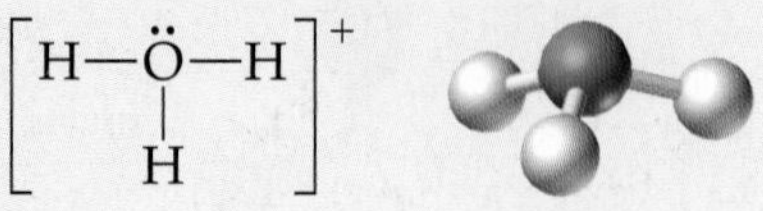

> **WORKED EXAMPLE** **5.10** Lewis Structures: Charge Cloud Geometry

Predict the geometry around each of the carbon atoms in an acetaldehyde molecule, CH_3CHO.

ANALYSIS Draw the Lewis structure and identify the number of charge clouds around each of the central carbon atoms.

SOLUTION
The Lewis structure of acetaldehyde shows that the CH_3 carbon has four charge clouds (four single bonds) and the CHO carbon atom has three charge clouds (two single bonds, one double bond). Table 5.1 indicates that the CH_3 carbon is tetrahedral, but the CHO carbon is planar triangular.

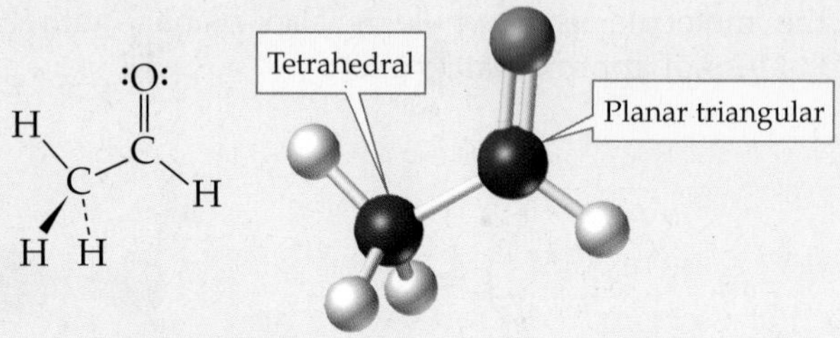

PROBLEM 5.12

Boron typically only forms three covalent bonds because it only has three valence electrons, but can form coordinate covalent bonds. Draw the Lewis structure for BF_4^- and predict the molecular shape of the ion.

PROBLEM 5.13

Predict shapes for the organic molecules chloroform, $CHCl_3$, and dichloroethylene, $Cl_2C{=}CH_2$.

PROBLEM 5.14

Electron-pair repulsion influences the shapes of polyatomic *ions* in the same way it influences neutral molecules. Draw electron-dot symbols and predict the shape of the ammonium ion, NH_4^+, and the sulfate ion, SO_4^{2-}.

APPLICATION ▶ VERY Big Molecules

How big can a molecule be? The answer is very, *very* big. The really big molecules in our bodies and in many items we buy are all *polymers*. Like a string of beads, a polymer is formed of many repeating units connected in a long chain. Each "bead" in the chain comes from a simple molecule that has formed chemical bonds at both ends, linking it to other molecules. The repeating units can be the same:

–a–a–a–a–a–a–a–a–a–a–a–a–

or they can be different. If different, they can be connected in an ordered pattern:

–a–b–a–b–a–b–a–b–a–b–a–b–

or in a random pattern:

–a–b–b–a–b–a–a–a–b–a–b–b–

Furthermore, the polymer chains can have branches, and the branches can have either the same repeating unit as the main chain or a different one:

```
        a
        |
        a–a–a–a–
        |
        a
        |
        a
        |
        a–a–a–a–
        |
        a
        |
        a
        |
        a–a–a–a–
        |
        a
```

or

```
        a
        |
        a–b–b–b–
        |
        a
        |
        a
        |
        a
        |
        a–b–b–b–
        |
        a
        |
        a
        |
        a–b–b–b–
        |
        a
```

Still other possible variations include complex, three-dimensional networks of "cross-linked" chains. The rubber used in tires, for example, contains polymer chains connected by cross-linking atoms of sulfur to impart greater rigidity.

We all use synthetic polymers every day—we usually call them "plastics." Common synthetic polymers are made by connecting up to several hundred thousand smaller molecules together, producing giant polymer molecules with masses up to several million atomic mass units. Polyethylene, for example, is made by combining as many as 50,000 ethylene molecules (H_2C═CH_2) to give a polymer with repeating —CH_2CH_2— units:

Many H_2C═CH_2 ⟶ —$CH_2CH_2CH_2CH_2CH_2CH_2$—

Ethylene   Polyethlene

The product is used in such items as chairs, toys, drain pipes, milk bottles, and packaging films.

Nature began to exploit the extraordinary variety of polymer properties long before humans did. In fact, despite great progress in recent years, there is still much to be learned about the polymers in living things. Carbohydrates and proteins are polymers, as are the giant molecules of deoxyribonucleic acid (DNA) that govern the reproduction of viruses, bacteria, plants, and all living creatures. Nature's polymer molecules, though, are larger and more complex than any that chemists have yet created. We will see the structures of these natural polymers in later chapters.

▲ The protective gear used by these firefighters is composed of advanced materials based on polymers.

See Additional Problems 5.93 and 5.94 at the end of the chapter.

PROBLEM 5.15

Hydrogen selenide (H_2Se) resembles hydrogen sulfide (H_2S) in that both compounds have terrible odors and are poisonous. What are their shapes?

KEY CONCEPT PROBLEM 5.16

Draw a structure corresponding to the molecular model of the amino acid methionine shown here and describe the geometry around the indicated atoms. (Remember the color key discussed in Section 5.6: black = carbon; white = hydrogen; red = oxygen; blue = nitrogen; yellow = sulfur.)

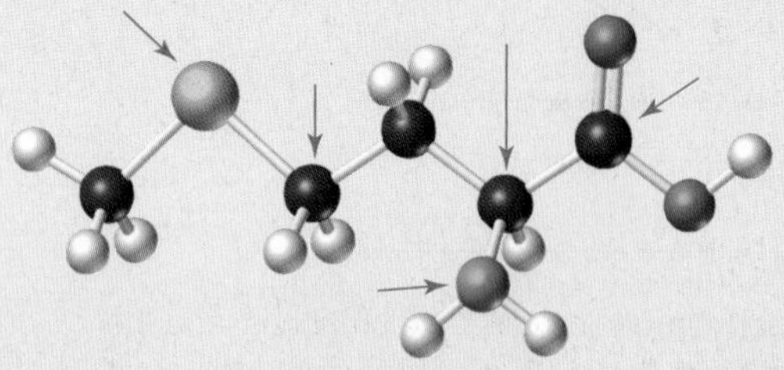

Methionine

5.8 Polar Covalent Bonds and Electronegativity

Polar covalent bond A bond in which the electrons are attracted more strongly by one atom than by the other.

Electrons in a covalent bond occupy the region between the bonded atoms. If the atoms are identical, as in H_2 and Cl_2, the electrons are attracted equally to both atoms and are shared equally. If the atoms are *not* identical, however, as in HCl, the bonding electrons may be attracted more strongly by one atom than by the other and may thus be shared unequally. Such bonds are said to be **polar covalent bonds**. In hydrogen chloride, for example, electrons spend more time near the chlorine atom than near the hydrogen atom. Although the molecule as a whole is neutral, the chlorine is more negative than the hydrogen, resulting in *partial* charges on the atoms. These partial charges are represented by placing a $\delta-$ (Greek lowercase *delta*) on the more negative atom and a $\delta +$ on the more positive atom.

A particularly helpful way of visualizing this unequal distribution of bonding electrons is to look at what is called an *electrostatic potential map*, which uses color to portray the calculated electron distribution in a molecule. In HCl, for example, the electron-poor hydrogen is blue, and the electron-rich chlorine is reddish-yellow:

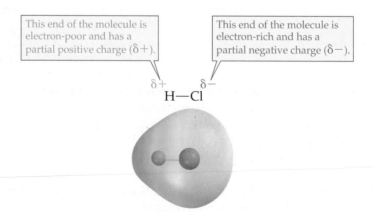

This end of the molecule is electron-poor and has a partial positive charge ($\delta+$).

This end of the molecule is electron-rich and has a partial negative charge ($\delta-$).

$$\overset{\delta+}{H}\!-\!\overset{\delta-}{Cl}$$

The ability of an atom to attract electrons in a covalent bond is called the atom's **electronegativity**. Fluorine, the most electronegative element, is assigned a value of 4, and less electronegative atoms are assigned lower values, as shown in Figure 5.7. Metallic elements on the left side of the periodic table attract electrons only weakly and have lower electronegativities, whereas the halogens and other reactive non-metal elements on the upper right side of the table attract electrons strongly and have higher electronegativities. Note also in Figure 5.7 that electronegativity generally decreases going down the periodic table within a group.

Electronegativity The ability of an atom to attract electrons in a covalent bond.

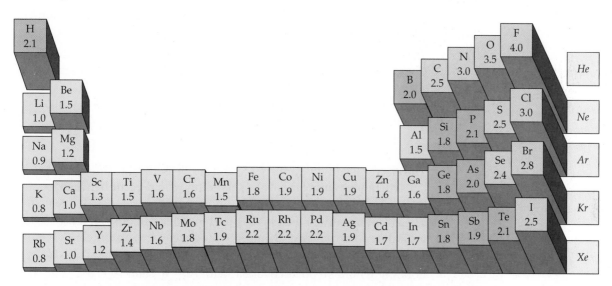

▲ FIGURE 5.7 Electronegativities of several main-group and transition-metal elements. Reactive nonmetals at the top right of the periodic table are most electronegative, and metals at the lower left are least electronegative. The noble gases are not assigned values.

Comparing the electronegativities of bonded atoms makes it possible to compare the polarities of bonds and to predict the occurrence of ionic bonding. Both oxygen (electronegativity 3.5) and nitrogen (3.0), for instance, are more electronegative than carbon (2.5). As a result, both C—O and C—N bonds are polar, with carbon at the positive end. The larger difference in electronegativity values shows that the C—O bond is the more polar of the two:

Less polar
$$\delta^+ C \longrightarrow N^{\delta-}$$
Electronegativity
difference:
3.0 − 2.5 = 0.5

More polar
$$\delta^+ C \longrightarrow O^{\delta-}$$
Electronegativity
difference:
3.5 − 2.5 = 1.0

As a rule of thumb, electronegativity differences of less than 0.5 result in non-polar covalent bonds, differences up to 1.9 indicate increasingly polar covalent bonds, and differences of 2 or more indicate substantially ionic bonds. The electronegativity differences show, for example, that the bond between carbon and fluorine is highly polar covalent, the bond between sodium and chlorine is largely ionic, and the bond between rubidium and fluorine is almost completely ionic:

$$\delta^+ C \longrightarrow F^{\delta-} \quad Na^+Cl^- \quad Rb^+F^-$$
Electronegativity
difference: 1.5   2.1   3.2

E.N difference		Type of bond
0 — 0.4	~	Covalent
0.5 — 1.9	~	Polar covalent
2.0 and above	~	Ionic

Note, though, that there is no sharp dividing line between covalent and ionic bonds; most bonds fall somewhere between two extremes.

Cⅅ **Looking Ahead**

The values given in Figure 5.7 indicate that carbon and hydrogen have similar electronegativities. As a result, C—H bonds are nonpolar. We will see in Chapters 12–25 how this fact helps explain the properties of organic and biological compounds, all of which have carbon and hydrogen as their principal constituents. ⅅ

WORKED EXAMPLE **5.11** Electronegativity: Ionic, Nonpolar, and Polar Covalent Bonds

Predict whether each of the bonds between the following atoms would be ionic, polar covalent, or nonpolar covalent. If polar covalent, which atom would carry the partial positive and negative charges?

(a) C and Br (b) Li and Cl (c) N and H (d) Si and I

ANALYSIS Compare the electronegativity values for the atoms and classify the nature of the bonding based on the electronegativity difference.

SOLUTION

(a) The electronegativity for C is 2.5, and for Br is 2.8; the difference is 0.3, indicating nonpolar covalent bonding would occur between these atoms.

(b) The electronegativity for Li is 1.0, and for Cl is 3.0; the difference is 2.0, indicating ionic bonding would occur between these atoms.

(c) The electronegativity for N is 3.0, and for H is 2.5; the difference is 0.5. Bonding would be polar covalent, with N = δ^-, and H = δ^+.

(d) The electronegativity for Si is 1.8, and for I is 2.5; the difference is 0.7. Bonding would be polar covalent, with I = δ^-, and Si = δ^+.

PROBLEM 5.17

Arrange the elements commonly bonded to carbon in organic compounds, H, N, O, P, and S, in order of increasing electronegativity.

PROBLEM 5.18

Use electronegativity differences to classify bonds between the following pairs of atoms as ionic, nonpolar covalent, or polar covalent:

(a) I and Cl (b) Li and O (c) Br and Br (d) P and Br

PROBLEM 5.19

Use the symbols $\delta+$ and $\delta-$ to identify the location of the partial charges on the polar covalent bonds formed between the following:

(a) Fluorine and sulfur

(b) Phosphorus and oxygen

(c) Arsenic and chlorine

5.9 Polar Molecules

Just as individual bonds can be polar, entire *molecules* can be polar if electrons are attracted more strongly to one part of the molecule than to another. Molecular polarity is due to the sum of all individual bond polarities and lone-pair contributions in the molecule and is often represented by an arrow pointing in the direction that electrons are displaced. The arrow is pointed at the negative end and is crossed at the positive end to resemble a plus sign, $(\delta+) \leftrightarrow (\delta-)$.

Molecular polarity depends on the shape of the molecule as well as the presence of polar covalent bonds and lone pairs. In water, for example, electrons are displaced away from the less electronegative hydrogen atoms toward the more electronegative oxygen atom so that the net polarity points between the two O—H bonds. In chloromethane, CH_3Cl, electrons are attracted from the carbon/hydrogen part of the molecule toward the electronegative chlorine atom so that the net polarity points along the C—Cl bond. Electrostatic potential maps show these polarities clearly, with electron-poor regions in blue and electron-rich regions in red.

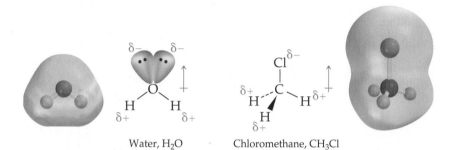

Water, H_2O Chloromethane, CH_3Cl

Furthermore, just because a molecule has polar covalent bonds does not mean that the molecule is necessarily polar overall. Carbon dioxide (CO_2) and tetra-chloromethane (CCl_4) molecules, for instance, have no net polarity because their symmetrical shapes cause the individual C=O and C—Cl bond polarities to cancel.

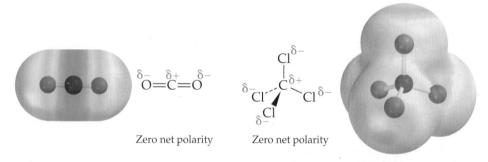

Zero net polarity   Zero net polarity

Polarity has a dramatic effect on the physical properties of molecules, particularly on melting points, boiling points, and solubilities. We will see numerous examples of such effects in subsequent chapters.

WORKED EXAMPLE   **5.12** Electronegativity: Polar Bonds and Polar Molecules

Look at the structures of (a) hydrogen cyanide (HCN) and (b) vinyl chloride (H_2C=CHCl) described in Worked Examples 5.6 and 5.7, decide whether the molecules are polar, and show the direction of net polarity in each.

ANALYSIS Draw a Lewis structure for each molecule to find its shape, and identify any polar bonds using the electronegativity values in Figure 5.7. Then decide on net polarity by adding the individual contributions.

SOLUTION

(a) The carbon atom in hydrogen cyanide has two charge clouds, making HCN a linear molecule. The C—H bond is relatively nonpolar, but the C≡N bonding electrons are pulled toward the electronegative nitrogen atom. In addition, a lone pair protrudes from nitrogen. Thus, the molecule has a net polarity:

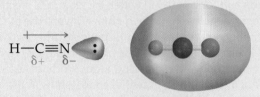

(b) Vinyl chloride, like ethylene, is a planar molecule. The C—H and C=C bonds are nonpolar, but the C—Cl bonding electrons are displaced toward the electronegative chlorine. Thus, the molecule has a net polarity:

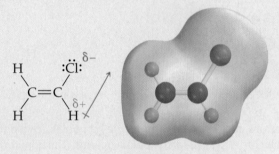

PROBLEM 5.20

Look at the molecular shape of formaldehyde (CH_2O) described on page 124, decide whether the molecule is polar, and show the direction of net polarity.

PROBLEM 5.21

Draw a Lewis structure for dimethyl ether (CH_3OCH_3), predict its shape, and tell whether the molecule is polar.

◆○ KEY CONCEPT PROBLEM 5.22

From this electrostatic potential map of methyllithium, identify the direction of net polarity in the molecule. Explain this polarity based on electronegativity values.

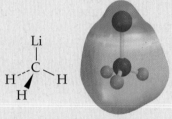

Methyllithium

5.10 Naming Binary Molecular Compounds

Binary compound A compound formed by combination of two different elements.

When two different elements combine, they form what is called a **binary compound**. The formulas of binary molecular compounds are usually written with the less electronegative element first. Thus, metals are always written before nonmetals, and a nonmetal farther left on the periodic table generally comes before a nonmetal farther right. For example,

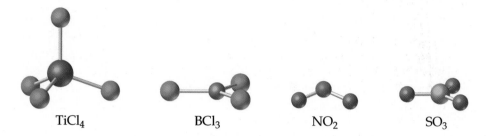

As we learned in Section 4.10, the formulas of ionic compounds indicate the number of anions and cations necessary for a neutral formula unit, which depends on the charge on each of the ions. With molecular compounds, however, many combinations of atoms are possible since nonmetals are capable of forming multiple covalent bonds. When naming binary molecular compounds, therefore, we must identify exactly how many atoms of each element are included in the molecular formula. The names of binary molecular compounds are assigned in two steps, using the prefixes listed in Table 5.2 to indicate the number of atoms of each element combined.

TABLE 5.2   Numerical Prefixes Used in Chemical Names

NUMBER	PREFIX
1	mono-
2	di-
3	tri-
4	tetra-
5	penta-
6	hexa-
7	hepta-
8	octa-
9	nona-
10	deca-

STEP 1: Name the first element in the formula, using a prefix if needed to indicate the number of atoms.

STEP 2: Name the second element in the formula, using an *-ide* ending as for anions (Section 4.8), along with a prefix if needed.

The prefix *mono-*, meaning one, is omitted except where needed to distinguish between two different compounds with the same elements. For example, the two oxides of carbon are named carbon *mon*oxide for CO and carbon *di*oxide for CO_2. (Note that we say *mon*oxide rather than *mono*oxide.) Some other examples are

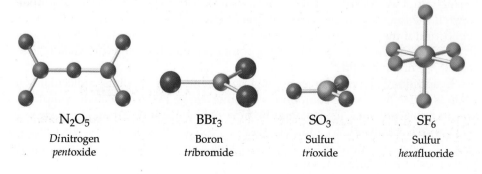

N_2O_5 — *Di*nitrogen *pent*oxide

BBr_3 — Boron *tri*bromide

SO_3 — Sulfur *tri*oxide

SF_6 — Sulfur *hexa*fluoride

WORKED EXAMPLE **5.13** Naming Molecular Compounds

Name the following compounds:

(a) N_2O_3 (b) $GeCl_4$ (c) PCl_5

SOLUTION

(a) Dinitrogen trioxide
(b) Germanium tetrachloride
(c) Phosphorus pentachloride

WORKED EXAMPLE **5.14** Writing Formulas for Molecular Compounds

Write molecular formulas for the following compounds:

(a) Nitrogen triiodide
(b) Silicon tetrachloride
(c) Carbon disulfide

SOLUTION

(a) NI_3 (b) $SiCl_4$ (c) CS_2

PROBLEM 5.23

Name the following compounds:

(a) S_2Cl_2 (b) ICl (c) ICl_3

PROBLEM 5.24

Write formulas for the following compounds:

(a) Selenium tetrafluoride
(b) Diphosphorus pentoxide
(c) Bromine trifluoride

5.11 Characteristics of Molecular Compounds

We saw in Section 4.4 that ionic compounds have high melting and boiling points because the attractive forces between oppositely charged ions are so strong that the ions are held tightly together. *Molecules*, however, are neutral, so there is no strong electrical attraction between different molecules to hold them together. There are, however, several weaker forces between molecules, called *intermolecular forces*, that we will look at in more detail in Chapter 8.

When intermolecular forces are very weak, molecules of a substance are so weakly attracted to one another that the substance is a gas at ordinary temperatures. If the forces are somewhat stronger, the molecules are pulled together into a liquid; and if the forces are still stronger, the substance becomes a molecular solid. Even so, the melting points and boiling points of molecular solids are usually lower than those of ionic solids.

In addition to having lower melting points and boiling points, molecular compounds differ from ionic compounds in other ways. Most molecular compounds are insoluble in water, for instance, because they have little attraction to strongly polar water molecules. In addition, they do not conduct electricity when melted because they have no charged particles. Table 5.3 provides a comparison of the properties of ionic and molecular compounds.

APPLICATION ▶ Damascenone by Any Other Name Would Smell as Sweet

What's in a name? According to Shakespeare's *Romeo and Juliet*, a rose by any other name would smell as sweet. Chemical names, however, often provoke less favorable responses: "It's unpronounceable." "It's too complicated." "It must be something bad."

Regarding pronunciation, chemical names are usually pronounced using every possible syllable. *Phenylpropanolamine*, for instance, a substance used in over-the-counter decongestants, is spoken with seven syllables: phen-yl-pro-pa-**nol**-a-mine.

Regarding complexity, the reason is obvious once you realize that there are more than 19 *million* known chemical compounds: The full name of a chemical compound has to include enough information to tell chemists the composition and structure of the compound. It is as if every person on earth had to have his or her own unique name that described height, hair color, and other identifying characteristics.

But does it really follow that a chemical with a really complicated name must be bad? These days, it seems that a different chemical gets into the news every week, often in a story describing some threat to health or the environment. The unfortunate result is that people sometimes conclude that everything with a chemical name is unnatural and dangerous. Neither is true, though. Acetaldehyde, for instance, is present naturally in most tart, ripe fruits and is often added in small amounts to artificial flavorings. When *pure*, however, acetaldehyde is also a flammable gas that is toxic and explosive in high concentrations.

Similar comparisons of desirable and harmful properties can be made for almost all chemicals, including water, sugar, and salt. The properties of a substance and the conditions surrounding its use must be evaluated before judgments are made. And damascenone, by the way, is the chemical largely responsible for the wonderful odor of roses.

▲ The scent of these roses contains the following: β-damascenone, β-ionone, citronellol, geraniol, nerol, eugenol, methyl eugenol, β-phenylethyl, alcohol, farnesol, linalool, terpineol, rose oxide, carvone, and many other natural substances.

See Additional Problems 5.95 and 5.96 at the end of the chapter.

TABLE 5.3 A Comparison of Ionic and Molecular Compounds

IONIC COMPOUNDS	MOLECULAR COMPOUNDS
Smallest components are ions (e.g., Na^+, Cl^-)	Smallest components are molecules (e.g., CO_2, H_2O)
Usually composed of metals combined with nonmetals	Usually composed of nonmetals with nonmetals
Crystalline solids	Gases, liquids, or low-melting solids
High melting points (e.g., NaCl = 801 °C)	Low melting points (H_2O = 0.0 °C)
High boiling points (above 700 °C) (e.g., NaCl = 1413 °C)	Low boiling points (e.g. H_2O = 100 °C; CH_3CH_2OH = 76 °C)
Conduct electricity when molten or dissolved in water	Do not conduct electricity
Many are water-soluble	Relatively few are water-soluble
Not soluble in organic liquids	Many are soluble in organic liquids

PROBLEM 5.25

A white crystalline solid has a melting point of 128 °C. It is soluble in water, but the resulting solution does not conduct electricity. Is the substance ionic or molecular? Explain.

PROBLEM 5.26

Aluminum chloride ($AlCl_3$) has a melting point of 190 °C, whereas aluminum oxide (Al_2O_3) has a melting point of 2070 °C. Explain.

SUMMARY: REVISITING THE CHAPTER GOALS

1. **What is a covalent bond?** A *covalent bond* is formed by the sharing of electrons between atoms rather than by the complete transfer of electrons from one atom to another. Atoms that share two electrons are joined by a *single bond* (such as C—C), atoms that share four electrons are joined by a *double bond* (such as C=C), and atoms that share six electrons are joined by a *triple bond* (such as C≡C). The group of atoms held together by covalent bonds is called a *molecule*.

   Electron sharing typically occurs when a singly occupied valence orbital on one atom *overlaps* a singly occupied valence orbital on another atom. The two electrons occupy both overlapping orbitals and belong to both atoms, thereby bonding the atoms together. Alternatively, electron sharing can occur when a filled orbital containing an unshared, *lone pair* of electrons on one atom overlaps a vacant orbital on another atom to form a *coordinate covalent bond*.

2. **How does the octet rule apply to covalent bond formation?** Depending on the number of valence electrons, different atoms form different numbers of covalent bonds. In general, an atom shares enough electrons to reach a noble gas configuration. Hydrogen, for instance, forms one covalent bond because it needs to share one more electron to achieve the helium configuration ($1s^2$). Carbon and other group 4A elements form four covalent bonds because they need to share four more electrons to reach an octet. In the same way, nitrogen and other group 5A elements form three covalent bonds, oxygen and other group 6A elements form two covalent bonds, and halogens (group 7A elements) form one covalent bond.

3. **How are molecular compounds represented?** Formulas such as H_2O, NH_3, and CH_4, which show the numbers and kinds of atoms in a molecule, are called *molecular formulas*. More useful are *Lewis structures*, which show how atoms are connected in molecules. Covalent bonds are indicated as lines between atoms, and valence electron lone pairs are shown as dots. Lewis structures are drawn by counting the total number of valence electrons in a molecule or polyatomic ion and then placing shared pairs (bonding) and lone pairs (nonbonding) so that all electrons are accounted for.

4. **What is the influence of valence-shell electrons on molecular shape?** Molecules have specific shapes that depend on the number of electron charge clouds (bonds and lone pairs) surrounding the various atoms. These shapes can often be predicted using the *valence-shell electron-pair repulsion (VSEPR)* model. Atoms with two electron charge clouds adopt linear geometry, atoms with three charge clouds adopt planar triangular geometry, and atoms with four charge clouds adopt tetrahedral geometry.

5. **When are bonds and molecules polar?** Bonds between atoms are *polar covalent* if the bonding electrons are not shared equally between the atoms. The ability of an atom to attract electrons in a covalent bond is the atom's *electronegativity* and is highest for reactive nonmetal elements on the upper right of the periodic table and lowest for metals on the lower left. Comparing electronegativities allows a prediction of whether a given bond is covalent, polar covalent, or ionic. Just as individual bonds can be polar, entire molecules can be polar if electrons are attracted more strongly to one part of the molecule than to another. Molecular polarity is due to the sum of all individual bond polarities and lone-pair contributions in the molecule.

6. **What are the major differences between ionic and molecular compounds?** *Molecular compounds* can be gases, liquids, or low-melting solids. They usually have lower melting points and boiling points than ionic compounds, many are water insoluble, and they do not conduct electricity when melted or dissolved.

UNDERSTANDING KEY CONCEPTS

5.27 Which of the drawings shown here is more likely to represent an ionic compound and which a covalent compound?

(a)   (b)

5.28 If yellow spheres represent sulfur atoms and red spheres represent oxygen atoms, which of the following drawings depicts a collection of sulfur dioxide molecules?

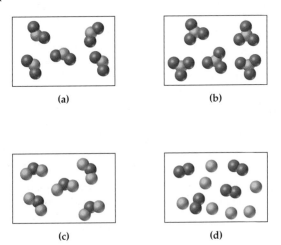

(a)   (b)

(c)   (d)

5.29 What is the geometry around the central atom in the following molecular models? (There are no "hidden" atoms; all atoms in each model are visible.)

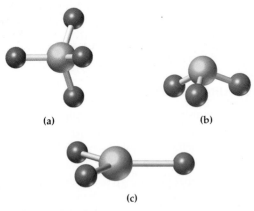

(a)   (b)

(c)

5.30 Three of the following molecular models have a tetrahedral central atom, and one does not. Which is the odd one?

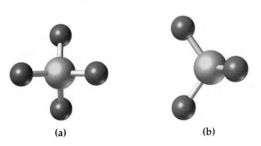

(a)   (b)

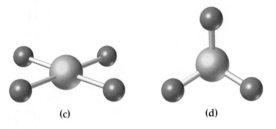

(c)   (d)

5.31 The ball-and-stick molecular model shown here is a representation of acetaminophen, the active ingredient in such over-the-counter headache remedies as Tylenol. The lines indicate only the connections between atoms, not whether the bonds are single, double, or triple (red = O, gray = C, blue = N, ivory = H).

(a) What is the molecular formula of acetaminophen?

(b) Indicate the positions of the multiple bonds in acetaminophen.

(c) What is the geometry around each carbon and each nitrogen?

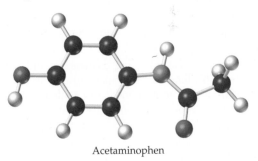

Acetaminophen

5.32 The atom-to-atom connections in vitamin C (ascorbic acid) are as shown here. Convert this skeletal drawing to a Lewis electron-dot structure for vitamin C by showing the positions of any multiple bonds and lone pairs of electrons.

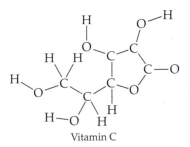

Vitamin C

5.33 The ball-and-stick molecular model shown here is a representation of thalidomide, a drug that causes terrible birth defects when taken by expectant mothers but has been approved for treating leprosy. The lines indicate only the connections between atoms, not whether the bonds are single, double, or triple (red = O, gray = C, blue = N, ivory = H).

(a) What is the molecular formula of thalidomide?

(b) Indicate the positions of the multiple bonds in thalidomide.

(c) What is the geometry around each carbon and each nitrogen?

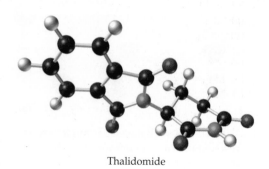

Thalidomide

5.34 Show the position of any electron lone pairs in this structure of acetamide, and indicate the electron-rich and electron-poor regions.

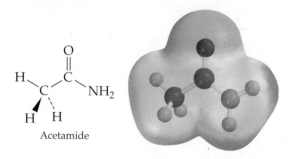

Acetamide

ADDITIONAL PROBLEMS

COVALENT BONDS

5.35 What is a covalent bond, and how does it differ from an ionic bond?

5.36 What is a coordinate covalent bond and how does it differ from a covalent bond?

5.37 Which of the following elements would you expect to form (i) diatomic molecules, (ii) mainly covalent bonds, (iii) mainly ionic bonds, (iv) both covalent and ionic bonds? (More than one answer may apply.)

(a) Oxygen
(b) Potassium
(c) Phosphorus
(d) Iodine
(e) Hydrogen
(f) Cesium

5.38 Identify the bonds formed between the following pairs of atoms as either covalent or ionic.

(a) Aluminum and bromine
(b) Carbon and fluorine
(c) Cesium and iodine
(d) Zinc and fluorine
(e) Lithium and chlorine

5.39 Write electron-dot symbols to show the number of covalent bonds in the molecules that are formed by reactions between the atoms in Problem 5.38.

5.40 Look up tellurium ($Z = 52$) in the periodic table and predict how many covalent bonds it is likely to form. Explain.

5.41 Germanium (atomic number 32) is an element used in the manufacture of transistors. Judging from its position in the periodic table, how many covalent bonds does it usually form?

5.42 Which of the following contains a coordinate covalent bond?

(a) $PbCl_2$ **(b)** $Cu(NH_3)_4{}^{2+}$ **(c)** $NH_4{}^+$

5.43 Which of the following contain a coordinate covalent bond?

(a) H_2O **(b)** $BF_4{}^-$ **(c)** H_3O^+

5.44 Tin forms both an ionic compound and a covalent compound with chlorine. The ionic compound is $SnCl_2$. Is the covalent compound more likely to be $SnCl_3$, $SnCl_4$, or $SnCl_5$? Explain.

5.45 A compound of gallium with chlorine has a melting point of 77 °C and a boiling point of 201 °C. Is the compound ionic or covalent? What is a likely formula?

5.46 Nitrous oxide, N_2O, has the following structure. Which bond in N_2O is a coordinate covalent bond?

$$:N\equiv N-\ddot{\underset{\cdot\cdot}{O}}:$$

Nitrous oxide

5.47 Thionyl chloride, $SOCl_2$, has the following structure. Which bond in $SOCl_2$ is a coordinate covalent bond?

Thionyl chloride

STRUCTURAL FORMULAS

5.48 Distinguish between the following:

(a) A molecular formula and a structural formula
(b) A structural formula and a condensed structure
(c) A lone pair and a shared pair of electrons

5.49 Assume that you are given samples of two white, crystalline compounds, one of them ionic and one covalent. Describe how you might tell which is which.

5.50 Give the total number of valence electrons in the following molecules:

(a) N_2 **(b)** CO
(c) CH_3CH_2CHO **(d)** OF_2

5.51 Add lone pairs where appropriate to the following structures:

(a) $C\equiv O$ **(b)** CH_3SH

(c) 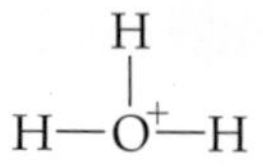 $H-\overset{+}{O}-H$ with H above, **(d)** 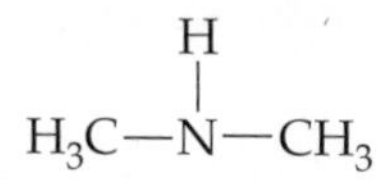 $H_3C-N-CH_3$ with H above

5.52 If a research paper appeared reporting the structure of a new molecule with formula C_2H_8, most chemists would be highly skeptical. Why?

5.53 Which of these possible structural formulas for $C_3H_6O_2$ is correct? Explain.

(a)

(b)

(c)

5.54 Convert the following Lewis structures into structural formulas in which lines replace the bonding electrons. Include the lone pairs.

(a) $H:\ddot{O}:\ddot{N}::\ddot{O}:$ (b) $H:\ddot{C}:C:::N:$ with H below   (c) $H:\ddot{F}:$

5.55 Convert the following Lewis structure for the nitrate ion into a line structure that includes the lone pairs. Why does the nitrate ion have a −1 charge?

5.56 Convert the following structural formulas into condensed structures.

(a)

(b)

(c)

5.57 Expand the following condensed structures into the correct structural formulas.

(a) $CH_3COCH_2CH_3$
(b) CH_3CH_2COOH
(c) $CH_3CH_2OCH_3$

5.58 Acetic acid is the major organic constituent of vinegar. Convert the following structural formula of acetic acid into a condensed structure.

DRAWING LEWIS STRUCTURES

5.59 Draw a Lewis structure for the following molecules:
(a) SiF_4
(b) $AlCl_3$
(c) CH_2O
(d) SO_2
(e) BBr_3
(f) NF_3

5.60 Draw a Lewis structure for the following molecules:
(a) Nitrous acid, HNO_2 (H is bonded to an O atom)
(b) Ozone, O_3 43416
(c) Acetaldehyde, CH_3CHO

5.61 Ethanol, or "grain alcohol," has the formula C_2H_6O and contains an O—H bond. Propose a structure for ethanol that is consistent with common bonding patterns.

5.62 Dimethyl ether has the same molecular formula as ethanol (Problem 5.61) but very different properties. Propose a structure for dimethyl ether in which the oxygen is bonded to two carbons.

5.63 Hydrazine, a substance used to make rocket fuel, has the formula N_2H_4. Propose a structure for hydrazine.

5.64 Tetrachloroethylene, C_2Cl_4, is used commercially as a dry cleaning solvent. Propose a structure for tetrachloroethylene based on the common bonding patterns expected in organic molecules. What kind of carbon–carbon bond is present?

5.65 Draw a Lewis structure for carbon disulfide, CS_2, a foul-smelling liquid used as a solvent for fats. What kind of carbon–sulfur bonds are present?

5.66 Draw a Lewis structure for hydroxylamine, NH_2OH.

5.67 The nitrate ion, NO_3^-, contains a double bond. Draw a Lewis structure for the ion and show why it has a negative charge.

5.68 Draw a Lewis structure for the following polyatomic ions:
(a) Formate, HCO_2^-
(b) Carbonate, CO_3^{2-}
(c) Sulfite, SO_3^{2-}
(d) Thiocyanate, SCN^-
(e) Phosphate, PO_4^{3-}
(f) Chlorite, ClO_2^- (chloride is the central atom)

MOLECULAR GEOMETRY

5.69 Predict the geometry and bond angles around atom A for molecules with the general formulas AB_3 and AB_2E, where B represents another atom and E represents an electron pair.

5.70 Predict the geometry and bond angles around atom A for molecules with the general formulas AB_4, AB_3E, and AB_2E_2, where B represents another atom and E represents an electron pair.

5.71 Sketch the three-dimensional shape of the following molecules:
(a) Chloroform, $CHCl_3$
(b) Hydrogen sulfide, H_2S
(c) Ozone, O_3
(d) Nitrogen triiodide, NI_3
(e) Chlorous acid, $HClO_2$

5.72 Predict the three-dimensional shape of the following molecules:

(a) SiF₄
(b) CF₂Cl₂
(c) SO₃
(d) BBr₃
(e) NF₃

5.73 Predict the geometry around each carbon atom in the amino acid alanine.

$$
\underset{\text{Alanine}}{\overset{\displaystyle \overset{O}{\underset{\|}{}}}{CH_3CHCOH}}
$$

$$
\underset{NH_2}{|}
$$

5.74 Predict the geometry around each carbon atom in vinyl acetate, a precursor of the poly(vinyl alcohol) polymer used in automobile safety glass.

$$
H_2C=CH-O-\overset{\overset{\displaystyle O}{\|}}{C}-CH_3
$$
Vinyl acetate

POLARITY OF BONDS AND MOLECULES

5.75 Where in the periodic table are the most electronegative elements found, and where are the least electronegative elements found?

5.76 Predict the electronegativity of the yet undiscovered element with $Z = 119$.

5.77 Look at the periodic table, and then order the following elements according to increasing electronegativity: Li, K, Br, C, Cl.

5.78 Look at the periodic table, and then order the following elements according to decreasing electronegativity: C, Ca, Cs, Cl, Cu.

5.79 Which of the following bonds are polar? Identify the negative and positive ends of each bond by using $\delta-$ and $\delta+$.

(a) I—Br
(b) O—H
(c) C—F
(d) N—C
(e) C—C

5.80 Which of the following bonds are polar? Identify the negative and positive ends of each bond by using $\delta-$ and $\delta+$.

(a) O—Br
(b) N—H
(c) P—O
(d) C—S
(e) C—Li

5.81 Based on electronegativity differences, would you expect bonds between the following pairs of atoms to be largely ionic or largely covalent?

(a) Be and F
(b) Ca and Cl
(c) O and H
(d) Be and Br

5.82 Arrange the following molecules in order of the increasing polarity of their bonds:

(a) HCl
(b) PH₃
(c) H₂O
(d) CF₄

5.83 Ammonia, NH₃, and phosphorus trihydride, PH₃, both have a trigonal pyramid geometry. Which one is more polar? Explain.

5.84 Decide whether each of the compounds listed in Problem 5.82 is polar, and show the direction of polarity.

5.85 Carbon dioxide is a nonpolar molecule, whereas sulfur dioxide is polar. Draw Lewis structures for each of these molecules to explain this observation.

5.86 Water (H₂O) is more polar than hydrogen sulfide (H₂S). Explain.

NAMES AND FORMULAS OF MOLECULAR COMPOUNDS

5.87 Name the following binary compounds:

(a) NO₂
(b) SF₆
(c) BrI₅
(d) N₂O₃
(e) NI₃
(f) IF₇

5.88 Name the following compounds:

(a) SiCl₄
(b) NaH
(c) SbF₅
(d) OsO₄

5.89 Write formulas for the following compounds:

(a) Phosphorus triiodide
(b) Arsenic trichloride
(c) Tetraphosphorus trisulfide
(d) Dialuminum hexafluoride
(e) Dinitrogen tetroxide
(f) Arsenic pentachloride

5.90 Write formulas for the following compounds:

(a) Selenium dioxide
(b) Xenon tetroxide
(c) Dinitrogen pentasulfide
(d) Triphosphorus tetraselenide

Applications

5.91 Draw electron-dot structures for CO and NO. Why are these molecules so reactive? [*CO and NO: Pollutants or Miracle Molecules?, p. 116*]

5.92 What is a vasodilator, and why would it be useful in treating hypertension (high blood pressure)? [*CO and NO: Pollutants or Miracle Molecules?, p. 116*]

5.93 How is a polymer formed? [*VERY Big Molecules, p. 127*]

5.94 Do any polymers exist in nature? Explain. [*VERY Big Molecules, p. 127*]

5.95 Why are many chemical names so complex? [*Damascenone by Any Other Name, p. 135*]

5.96 Can you tell from the name whether a chemical is natural or synthetic? [*Damascenone by Any Other Name, p. 135*]

General Questions and Problems

5.97 The discovery in the 1960s that xenon and fluorine react to form a molecular compound was a surprise to most chemists, because it had been thought that noble gases could not form bonds.

(a) Why was it thought that noble gases could not form bonds?
(b) Draw a Lewis structure of XeF₄.

5.98 Acetone, a common solvent used in some nail polish removers, has the molecular formula C_3H_6O and contains a carbon—oxygen double bond.

    (a) Propose two Lewis structures for acetone.

    (b) What is the geometry around the carbon atoms in each of the structures?

    (c) Which of the bonds in each structure are polar?

5.99 Draw the structural formulas for two compounds having the molecular formula C_2H_4O. What is the molecular geometry around the carbon atoms in each of these molecules? Would these molecules be polar or nonpolar? (Hint: there is one double bond.)

5.100 The following formulas are unlikely to be correct. What is wrong with each?

    (a) CCl_3 **(b)** N_2H_5

    (c) H_3S **(d)** C_2OS

5.101 Which of the compounds (a) through (d) contain one or more of the following: (i) ionic bonds, (ii) covalent bonds, (iii) coordinate covalent bonds?

    (a) $BaCl_2$ **(b)** $Ca(NO_3)_2$

    (c) BCl_4^- **(d)** $TiBr_4$

5.102 The phosphonium ion, PH_4^+, is formed by reaction of phosphine, PH_3, with an acid.

    (a) Draw the Lewis structure of the phosphonium ion.

    (b) Predict its molecular geometry.

    (c) Describe how a fourth hydrogen can be added to PH_3.

    (d) Explain why the ion has a +1 charge.

5.103 Compare the trend in electronegativity seen in Figure 5.7 (p. 129) with the trend in electron affinity shown in Figure 4.1 (p. 81). What similarities do you see? What differences? Explain.

5.104 Name the following compounds. Be sure to determine whether the compound is ionic or covalent so that you use the proper rules.

    (a) $CaCl_2$ **(b)** $TeCl_2$ **(c)** BF_3

    (d) $MgSO_4$ **(e)** K_2O **(f)** FeF_3

    (g) PF_3

5.105 Titanium forms both molecular and ionic compounds with nonmetals, as, for example, $TiBr_4$ and TiO_2. One of these compounds has a melting point of 39 °C, and the other has a melting point of 1825 °C. Which is ionic and which is molecular? Explain your answer in terms of electronegativities of the atoms involved in each compound.

5.106 Draw a Lewis structure for chloral hydrate, known in detective novels as "knockout drops." Indicate all lone pairs.

$$\begin{array}{ccc} Cl & O\text{—}H & \\ | & | & \\ Cl\text{—}C\text{—}C\text{—}O\text{—}H & & \text{Chloral hydrate} \\ | & | & \\ Cl & H & \end{array}$$

5.107 The dichromate ion, $Cr_2O_7^{2-}$, has neither Cr—Cr nor O—O bonds. Write a Lewis structure.

5.108 Oxalic acid, $H_2C_2O_4$, is a poisonous substance found in uncooked spinach leaves. If oxalic acid has a C—C single bond, draw its Lewis structure.

5.109 Identify the elements in the fourth row of the periodic table that form the following compounds.

    (a) $\ddot{O}{=}\ddot{X}{=}\ddot{O}$ **(b)** $:\!\ddot{F}\!-\!\ddot{X}\!-\!\ddot{F}\!:$

5.110 Write Lewis structures for molecules with the following connections, showing the positions of any multiple bonds and lone pairs of electrons.

    (a)
$$\begin{array}{ccccc} & O & & H & \\ & \| & & | & \\ Cl\text{—}C\text{—}O\text{—}C\text{—}H & & & \\ & & & | & \\ & & & H & \end{array}$$

    (b)
$$\begin{array}{ccc} & H & \\ & | & \\ H\text{—}C\text{—}C\text{—}C\text{—}H & \\ & | & \\ & H & \end{array}$$

CHAPTER 6

Chemical Reactions: Classification and Mass Relationships

CONCEPTS TO REVIEW

Problem Solving: Converting a Quantity from One Unit to Another

(Section 2.7)

Periodic Properties and Ion Formation

(Section 4.2)

H^+ and OH^- Ions: An Introduction to Acids and Bases

(Section 4.11)

▲ The A320 Airbus shown on the assembly line in Toulouse, France. Commercial air transport relies on oxidation-reduction reactions, from the production of lightweight materials used for construction to the generation of energy required for flight.

CONTENTS

CHAPTER GOALS

Among the questions we will answer are the following:

1. How are chemical reactions written?

THE GOAL: Given the identities of reactants and products, be able to write a balanced chemical equation or net ionic equation.

2. What is the mole, and why is it useful in chemistry?

THE GOAL: Be able to explain the meaning and uses of the mole and Avogadro's number.

3. How are molar quantities and mass quantities related?

THE GOAL: Be able to convert between molar and mass quantities of an element or compound.

4. What are the limiting reagent, theoretical yield, and percent yield of a reaction?

THE GOAL: Be able to take the amount of product actually formed in a reaction, calculate the amount that could form theoretically, and express the results as a percent yield.

5. How are chemical reactions of ionic compounds classified?

THE GOAL: Be able to recognize precipitation, acid–base neutralization, and redox reactions.

6. What are oxidation numbers, and how are they used?

THE GOAL: Be able to assign oxidation numbers to atoms in compounds and identify the substances oxidized and reduced in a given reaction.

A log burns in the fireplace, an oyster makes a pearl, a seed grows into a plant—these and almost all other changes you see taking place around you are the result of *chemical reactions*. The study of how and why chemical reactions happen is a major part of chemistry, providing information that is both fascinating and practical. In this chapter, we will begin to look at chemical reactions, starting with a discussion of how to represent them in writing. Next we will describe the mass relationships among substances involved in chemical reactions, and then introduce a few easily recognized classes of chemical reactions.

6.1 Chemical Equations

One way to view chemical reactions is to think of them as "recipes." Like recipes, all the "ingredients" in a chemical equation and their relative amounts are given, as well as the amount of product that would be obtained. Take, for example, a recipe for making s'mores, a concoction of chocolate, marshmallows, and graham crackers, which could be written as

Graham crackers + Roasted marshmallows + Chocolate bars ⟶ S'mores

This recipe, however, is simply a list of ingredients and gives no indication of the relative amounts of each ingredient, or how many s'mores we would obtain. A more detailed recipe would be

2 Graham crackers + 1 Roasted marshmallow + $\frac{1}{4}$ Chocolate bar ⟶ 1 S'more

In this case, the relative amounts of each ingredient are given, as well as the amount of the final product.

Let us extend this analogy to a typical chemical reaction. When sodium bicarbonate is heated in the range 50–100 °C, sodium carbonate, water, and carbon dioxide are produced. In words, we might write the reaction as

Sodium bicarbonate $\xrightarrow{\text{Heat}}$ Sodium carbonate + Water + Carbon dioxide

Just as in the recipe, the starting materials and final products are listed. Replacing the chemical names with formulas converts the word description of this reaction into a **chemical equation**:

$$2 \underbrace{NaHCO_3}_{\text{Reactant}} \xrightarrow{\text{Heat}} \underbrace{Na_2CO_3 + H_2O + CO_2}_{\text{Products}}$$

Chemical equation An expression in which symbols and formulas are used to represent a chemical reaction.

Reactant A substance that undergoes change in a chemical reaction and is written on the left side of the reaction arrow in a chemical equation.

Product A substance that is formed in a chemical reaction and is written on the right side of the reaction arrow in a chemical equation.

Balanced equation A chemical equation in which the numbers and kinds of atoms are the same on both sides of the reaction arrow.

Coefficient A number placed in front of a formula to balance a chemical equation.

Look at how this equation is written. The **reactants** are written on the left, the **products** are written on the right, and an arrow is placed between them to indicate a chemical change. Conditions necessary for the reaction to occur—heat in this particular instance—are often specified above the arrow.

Why is the number 2 placed before $NaHCO_3$ in the equation? The 2 is necessary because of a fundamental law of nature called the *law of conservation of mass*:

Law of conservation of mass   Matter is neither created nor destroyed in chemical reactions.

The bonds between atoms in the reactants are rearranged to form new compounds in chemical reactions, but none of the atoms disappear and no new ones are formed. As a consequence, chemical equations must be **balanced**, meaning that *the numbers and kinds of atoms must be the same on both sides of the reaction arrow*.

The numbers placed in front of formulas to balance equations are called **coefficients**, and they multiply all the atoms in a formula. Thus, the symbol "$2 NaHCO_3$" indicates two units of sodium bicarbonate, which contain 2 Na atoms, 2 H atoms, 2 C atoms, and 6 O atoms ($2 \times 3 = 6$, the coefficient times the subscript for O). Count the numbers of atoms on the right side of the equation to convince yourself that it is indeed balanced.

The substances that take part in chemical reactions may be solids, liquids, or gases, or they may be dissolved in a solvent. Ionic compounds, in particular, frequently undergo reaction in *aqueous solution*— that is, dissolved in water. Sometimes this information is added to an equation by placing the appropriate symbols after the formulas:

$$\underset{\text{Solid}}{(s)} \qquad \underset{\text{Liquid}}{(l)} \qquad \underset{\text{Gas}}{(g)} \qquad \underset{\text{Aqueous solution}}{(aq)}$$

Thus, the decomposition of solid sodium bicarbonate can be written as

$$2 NaHCO_3(s) \xrightarrow{\text{Heat}} Na_2CO_3(s) + H_2O(l) + CO_2(g)$$

WORKED EXAMPLE   **6.1** Balancing Chemical Reactions

Interpret in words the following equation for the reaction used in extracting lead metal from its ores. Show that the equation is balanced.

$$2 PbS(s) + 3 O_2(g) \longrightarrow 2 PbO(s) + 2 SO_2(g)$$

SOLUTION
The equation can be read as, "Solid lead(II) sulfide plus gaseous oxygen yields solid lead(II) oxide plus gaseous sulfur dioxide."

To show that the equation is balanced, count the atoms of each element on each side of the arrow:

On the left:	2 Pb	2 S	$(3 \times 2) O = 6 O$
On the right:	2 Pb	2 S	$2 O + (2 \times 2) O = 6 O$

From 2 PbO   From 2 SO_2

The numbers of atoms of each element are the same in the reactants and products, so the equation is balanced.

PROBLEM 6.1

Interpret the following equations in words:

(a) $CoCl_2(s) + 2 HF(g) \longrightarrow CoF_2(s) + 2 HCl(g)$
(b) $Pb(NO_3)_2(aq) + 2 KI(aq) \longrightarrow PbI_2(s) + 2 KNO_3(aq)$

PROBLEM 6.2

Which of the following equations are balanced?

(a) $HCl + KOH \longrightarrow H_2O + KCl$

(b) $CH_4 + Cl_2 \longrightarrow CH_2Cl_2 + HCl$

(c) $H_2O + MgO \longrightarrow Mg(OH)_2$

(d) $Al(OH)_3 + H_3PO_4 \longrightarrow AlPO_4 + 2 H_2O$

6.2 Balancing Chemical Equations

Just as a recipe indicates the appropriate amounts of each ingredient needed to make a given product, a balanced chemical equation indicates the appropriate amounts of reactants needed to generate a given amount of product. Balancing chemical equations can often be done using a mixture of common sense and trial-and-error. There are four steps:

STEP 1: Write an unbalanced equation, using the correct formulas for all reactants and products. For example, hydrogen and oxygen must be written as H_2 and O_2, rather than as H and O, since we know that both elements exist as diatomic molecules. Remember that *the subscripts in chemical formulas cannot be changed in balancing an equation because doing so would change the identity of the substances in the reaction.*

STEP 2: Add appropriate coefficients to balance the numbers of atoms of each element. It helps to begin with elements that appear in only one formula on each side of the equation, which usually means leaving oxygen and hydrogen until last. For example, in the reaction of sulfuric acid with sodium hydroxide to give sodium sulfate and water, we might balance first for sodium. We could do this by adding a coefficient of 2 for NaOH:

$$H_2SO_4 + NaOH \longrightarrow Na_2SO_4 + H_2O \quad \text{(Unbalanced)}$$
$$H_2SO_4 + 2\,NaOH \longrightarrow Na_2SO_4 + H_2O \quad \text{(Balanced for Na)}$$

Add this coefficient . . .   . . . to balance these 2 Na.

If a polyatomic ion appears on both sides of an equation, it is treated as a single unit. For example, the sulfate ion (SO_4^{2-}) in our example is balanced because there is one on the left and one on the right:

$$H_2SO_4 + 2\,NaOH \longrightarrow Na_2SO_4 + H_2O \quad \text{(Balanced for Na and sulfate)}$$

One sulfate here . . .   . . . and one here.

At this point, the equation can be balanced for H and O by adding a coefficient of 2 for H_2O

$$H_2SO_4 + 2\,NaOH \longrightarrow Na_2SO_4 + 2\,H_2O \quad \text{(Completely balanced)}$$

4 H and 2 O here.   4 H and 2 O here.

STEP 3: Check the equation to make sure the numbers and kinds of atoms on both sides of the equation are the same.

STEP 4: Make sure the coefficients are reduced to their lowest whole-number values. For example, the equation

$$2 H_2SO_4 + 4 NaOH \longrightarrow 2 Na_2SO_4 + 4 H_2O$$

is balanced but can be simplified by dividing all coefficients by 2:

$$H_2SO_4 + 2\,NaOH \longrightarrow Na_2SO_4 + 2\,H_2O$$

WORKED EXAMPLE **6.2** Balancing Chemical Equations

Write a balanced chemical equation for the Haber process, an important industrial reaction in which elemental nitrogen and hydrogen combine to form ammonia according to the following unbalanced reaction:

$$N_2(g) + H_2(g) \longrightarrow NH_3(g)$$

SOLUTION

STEP 1: Write an unbalanced equation, using the correct formulas for all reactants and products. The unbalanced equation is provided above. By examination, we see that only two elements, N and H, need to be balanced. Both these elements exist in nature as diatomic gases, as indicated on the reactant side of the unbalanced equation.

STEP 2: Add appropriate coefficients to balance the numbers of atoms of each element. Remember that the subscript 2 in N_2 and H_2 indicates that these are diatomic molecules (that is, two N atoms or two H atoms per molecule). Since there are two nitrogen atoms on the left, we must add a coefficient of 2 in front of the NH_3 on the right side of the equation to balance the equation with respect to N:

$$N_2(g) + H_2(g) \longrightarrow 2\,NH_3(g)$$

Now we see that there are two H atoms on the left, but six H atoms on the right. We can balance the equation with respect to hydrogen by adding a coefficient of 3 in front of the $H_2(g)$ on the left side:

$$N_2(g) + 3\,H_2(g) \longrightarrow 2\,NH_3(g)$$

STEP 3: Check the equation to make sure the numbers and kinds of atoms on both sides of the equation are the same.

On the left: $(1 \times 2)\,N = 2\,N$ $(3 \times 2)\,H = 6\,H$

On the right: $(2 \times 1)\,N = 2\,N$ $(2 \times 3)\,H = 6\,H$

STEP 4: Make sure the coefficients are reduced to their lowest whole-number values. In this case, the coefficients already represent the lowest whole-number ratios.

WORKED EXAMPLE **6.3** Balancing Chemical Equations

Natural gas (methane, CH_4) burns in oxygen to yield water and carbon dioxide (CO_2). Write a balanced equation for the reaction.

SOLUTION

STEP 1: Write the unbalanced equation, using correct formulas for all substances:

$$CH_4 + O_2 \longrightarrow CO_2 + H_2O \text{ (Unbalanced)}$$

STEP 2: Since carbon appears in one formula on each side of the arrow, let us begin with that element. In fact, there is only one carbon atom in each formula, so the equation is already balanced for that element. Next, note that there are four hydrogen atoms on the left (in CH_4) and only two on the right (in H_2O). Placing a coefficient of 2 before H_2O gives the same number of hydrogen atoms on both sides:

$$CH_4 + O_2 \longrightarrow CO_2 + 2\,H_2O \text{ (Balanced for C and H)}$$

Finally, look at the number of oxygen atoms. There are two on the left (in O_2) but four on the right (two in CO_2 and one in each H_2O). If we place a 2 before the O_2, the number of oxygen atoms will be the same on both sides, but the numbers of other elements will not change:

$$CH_4 + 2\,O_2 \longrightarrow CO_2 + 2\,H_2O \text{ (Balanced for C, H, and O)}$$

STEP 3: Check to be sure the numbers of atoms on both sides are the same.

On the left: 1 C   4 H $(2 \times 2)\,O = 4\,O$

On the right: 1 C $(2 \times 2)\,H = 4\,H$ $2\,O + 2\,O = 4\,O$

                                  From CO_2 From 2 H_2O

STEP 4: Make sure the coefficients are reduced to their lowest whole-number values. In this case, the answer is already correct.

WORKED EXAMPLE   **6.4** Balancing Chemical Equations

Sodium chlorate ($NaClO_3$) decomposes when heated to yield sodium chloride and oxygen, a reaction used to provide oxygen for the emergency breathing masks in airliners. Write a balanced equation for this reaction.

▲ The oxygen in emergency breathing masks comes from heating sodium chlorate.

SOLUTION

STEP 1: The unbalanced equation is

$$NaClO_3 \longrightarrow NaCl + O_2$$

STEP 2: Both the Na and the Cl are already balanced, with only one atom of each on the left and right sides of the equation. There are three O atoms on the left, but only two on the right. The O atoms can be balanced by placing a coefficient of 1½ in front of O_2 on the right side of the equation:

$$NaClO_3 \longrightarrow NaCl + 1^{1}/_{2}\,O_2$$

STEP 3: Checking to make sure the same number of atoms of each type occurs on both sides of the equation, we see one atom of Na and Cl on both sides, and three O atoms on both sides.

STEP 4: In this case, obtaining all coefficients in their smallest whole-number values requires that we multiply all coefficients by 2 to obtain

$$2\,NaClO_3 \longrightarrow 2\,NaCl + 3\,O_2$$

Checking gives

On the left: 2 Na 2 Cl $(2 \times 3)\,O = 6\,O$

On the right: 2 Na 2 Cl $(3 \times 2)\,O = 6\,O$

PROBLEM 6.3

Ozone (O_3) is formed in the earth's upper atmosphere by the action of solar radiation on oxygen molecules (O_2). Write a balanced equation for the formation of ozone from oxygen.

PROBLEM 6.4

Balance the following equations:

(a) $Ca(OH)_2 + HCl \longrightarrow CaCl_2 + H_2O$

(b) $Al + O_2 \longrightarrow Al_2O_3$

(c) $CH_3CH_3 + O_2 \longrightarrow CO_2 + H_2O$

(d) $AgNO_3 + MgCl_2 \longrightarrow AgCl + Mg(NO_3)_2$

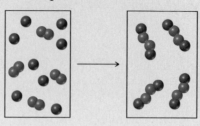

KEY CONCEPT PROBLEM 6.5

The following diagram represents the reaction of A (red spheres) with B_2 (blue spheres). Write a balanced equation for the reaction.

6.3 Avogadro's Number and the Mole

The balanced chemical equation indicates what is happening at the molecular level. Now let us imagine a laboratory experiment: the reaction of ethylene (C_2H_4) with hydrogen chloride (HCl) to prepare ethyl chloride (C_2H_5Cl), a colorless, low-boiling liquid used by doctors and athletic trainers as a spray-on anesthetic. The reaction is represented as

$$C_2H_4(g) + HCl(g) \rightarrow C_2H_5Cl(g)$$

In this reaction, one molecule of ethylene reacts with one molecule of hydrogen chloride to produce one molecule of ethyl chloride.

How, though, can you be sure you have a 1 to 1 ratio of reactant molecules in your reaction flask? Since it is impossible to hand-count the number of molecules correctly, you must weigh them instead. (This is a common method for dealing with all kinds of small objects: Nails, nuts, and grains of rice are all weighed rather than counted.) But the weighing approach leads to another problem. How many molecules are there in one gram of ethylene, hydrogen chloride, or any other substance? The answer depends on the identity of the substance because different molecules have different masses.

To determine how many molecules of a given substance are in a certain mass, it is helpful to define a quantity called *molecular weight*. Just as the *atomic* weight of an element is the average mass of the element's *atoms* (Section 3.3), the **molecular weight (MW)** of a molecule is the average mass of a substance's *molecules*. (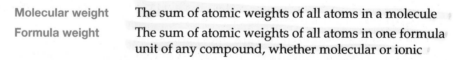 p. 55) Numerically, a substance's molecular weight (or **formula weight** for an ionic compound) is equal to the sum of the atomic weights for all the atoms in the molecule or formula unit.

▲ Ethyl chloride is often used as a spray-on anesthetic for athletic injuries.

Molecular weight	The sum of atomic weights of all atoms in a molecule
Formula weight	The sum of atomic weights of all atoms in one formula unit of any compound, whether molecular or ionic

For example, the molecular weight of ethylene (C_2H_4) is 28.0 amu, the molecular weight of HCl is 36.5 amu, and the molecular weight of ethyl chloride (C_2H_5Cl) is 64.5 amu. (The actual values are known more precisely but are rounded off here for convenience.)

For ethylene, C_2H_4:

Atomic weight of 2 C = 2 × 12.0 amu = 24.0 amu

Atomic weight of 4 H = 4 × 1.0 amu = 4.0 amu

MW of C_2H_4 = 28.0 amu

For hydrogen chloride, HCl:

$$\begin{array}{rl}
\text{Atomic weight of H} &= \ \ 1.0 \text{ amu} \\
\underline{\text{Atomic weight of Cl}} &= \underline{35.5 \text{ amu}} \\
\text{MW of HCl} &= 36.5 \text{ amu}
\end{array}$$

For ethyl chloride, C_2H_5Cl:

$$\begin{array}{rl}
\text{Atomic weight of 2 C} = 2 \times 12.0 \text{ amu} &= 24.0 \text{ amu} \\
\text{Atomic weight of 5 H} = 5 \times 1.0 \text{ amu} &= \ \ 5.0 \text{ amu} \\
\underline{\text{Atomic weight of Cl}} &\underline{= 35.5 \text{ amu}} \\
\text{MW of } C_2H_5Cl &= 64.5 \text{ amu}
\end{array}$$

How are molecular weights used? Since the mass ratio of *one* ethylene molecule to *one* HCl molecule is 28.0 to 36.5, the mass ratio of *any* given number of ethylene molecules to the same number of HCl molecules is also 28.0 to 36.5. In other words, a 28.0 to 36.5 *mass* ratio of ethylene and HCl always guarantees a 1 to 1 *number* ratio. *Samples of different substances always contain the same number of molecules or formula units whenever their mass ratio is the same as their molecular or formula weight ratio* (Figure 6.1).

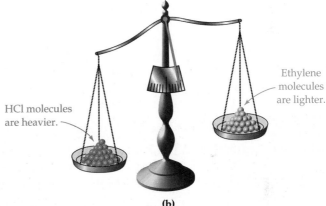

HCl molecules are heavier.

Ethylene molecules are lighter.

(a)   (b)

▲ **FIGURE 6.1** (a) Because one gumdrop weighs more than one jellybean, you cannot get equal numbers by taking equal weights. The same is true for atoms or molecules of different substances. (b) Equal numbers of ethylene and HCL molecules always have a mass ratio equal to the ratio of their molecular weights, 28.0 to 36.5.

A particularly convenient way to use this mass/number relationship for molecules is to measure amounts in grams that are numerically equal to molecular weights. If, for instance, you were to carry out your experiment with 28.0 g of ethylene and 36.5 g of HCl, you could be certain that you would have a 1 to 1 ratio of reactant molecules.

◀ These samples of water, sulfur, table sugar, mercury, and copper each contain 1 mol. Do they all weigh the same?

When referring to the vast numbers of molecules or formula units that take part in a visible chemical reaction, it is convenient to use a counting unit called a **mole**, abbreviated *mol*. One mole of any substance is the amount whose mass in grams—its *molar mass*—is numerically equal to its molecular or formula weight. One mole of ethylene has a mass of 28.0 g, one mole of HCl has a mass of 36.5 g, and one mole of ethyl chloride has a mass of 64.5 g.

> **Mole**   The amount of a substance whose mass in grams is numerically equal to its molecular or formula weight.

Just how many molecules are there in a mole? Experiments show that one mole of any substance contains 6.022×10^{23} formula units, a value called **Avogadro's number** (abbreviated N_A) after the Italian scientist who first recognized the importance of the mass/number relationship in molecules. Avogadro's number of formula units of any substance—that is, one mole—has a mass in grams numerically equal to the molecular weight of the substance.

Avogadro's number (N_A) The number of units in 1 mol of anything; 6.022×10^{23}.

$$1 \text{ mol HCl} = 6.022 \times 10^{23} \text{ HCl molecules} = 36.5 \text{ g HCl}$$

$$1 \text{ mol } C_2H_4 = 6.022 \times 10^{23} \text{ } C_2H_4 \text{ molecules} = 28.0 \text{ g } C_2H_4$$

$$1 \text{ mol } C_2H_5Cl = 6.022 \times 10^{23} \text{ } C_2H_5Cl \text{ molecules} = 64.5 \text{ g } C_2H_5Cl$$

How big is Avogadro's number? Our minds cannot really conceive of the magnitude of a number like 6.022×10^{23}, but the following comparisons will give you a sense of the scale:

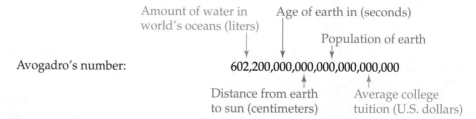

Avogadro's number: 602,200,000,000,000,000,000,000

| Amount of water in world's oceans (liters) | Age of earth in (seconds) | Population of earth |
| Distance from earth to sun (centimeters) | Average college tuition (U.S. dollars) |

> **WORKED EXAMPLE**   **6.5** Molar Mass and Avogadro's Number: Number of Molecules
>
> Pseudoephedrine hydrochloride ($C_{10}H_{16}ClNO$) is a nasal decongestant commonly found in cold medication. (a) What is the molar mass of pseudoephedrine hydrochloride? (b) How many molecules of pseudoephedrine hydrochloride are in a tablet that contains 30.0 mg of this decongestant ?
>
> **ANALYSIS** We are given a mass and need to convert to a number of molecules. This is most easily accomplished by using the molecular weight of pseudoephedrine hydrochloride calculated in part (a) as the conversion factor from mass to moles and realizing that this mass (in grams) contains Avogadro's number of molecules (6.022×10^{23}). ·
>
> **BALLPARK ESTIMATE** The formula for pseudoephedrine contains ten carbon atoms (each one of atomic weight 12.0 amu), so the molecular weight is greater than 120 amu, probably near 200 amu. Thus the molar mass should be near 200 g/mol. The mass of 30 mg of pseudoepinephrine HCl is less than the mass of 1 mol of this compound by a factor of roughly 10^4 (0.03 g versus 200 g), which means the number of molecules should also be smaller by a factor of 10^4 (on the order of 10^{19} in the tablet versus 10^{23} in 1 mol).
>
> **SOLUTION**
>
> (a) The molecular weight of pseudoephedrine is found by summing the atomic weights of all atoms in the molecule:
>
> | | | |
> |---|---|---|
> | Atomic Weight of 10 atoms of C: | 10×12.011 amu | $= 120.11$ amu |
> | 16 atoms of H: | 16×1.00794 amu $=$ | 16.127 amu |
> | 1 atom of Cl: | 1×35.4527 amu $=$ | 35.4527 amu |
> | 1 atom of N: | 1×14.0067 amu $=$ | 14.0067 amu |
> | 1 atom of O: | 1×15.9994 amu $=$ | 15.9994 amu |
>
> MW of $C_{10}H_{16}ClNO$ $= 201.6958$ amu $\rightarrow 201.70$ g/mol
>
> Remember that atomic mass in amu converts directly to molecular weight in g/mol. Also, following the rules for significant figures from Section 2.6, our final answer is rounded to the second decimal place.

(b) Since this problem involves unit conversions, we can use the step-wise solution introduced in Chapter 2.

STEP 1: Identify known information. We are given the mass of pseudoephedrine hydrochloride (in mg).

30.0 mg pseudoephedrine hydrochloride

STEP 2: Identify answer and units. We are looking for the number of molecules of pseudoephedrine hydrochloride in a 30 mg tablet.

?? = molecules

STEP 3: Identify conversion factors. Since the molecular weight of pseudoephedrine hydrochloride is 201.70 amu, 201.70 g contains 6.022×10^{23} molecules. We can use this ratio as a conversion factor to convert from mass to molecules. We will also need to convert from mg to g.

$$\frac{6.022 \times 10^{23} \text{ molecules}}{201.70 \text{ g}}$$

$$\frac{.001 \text{ g}}{1 \text{ mg}}$$

STEP 4: Solve. Set up an equation so that unwanted units cancel.

$$(30.0 \text{ mg pseudoephedrine hydrochloride}) \times \left(\frac{.001 \text{ g}}{1 \text{ mg}}\right) \times$$

$$\left(\frac{6.022 \times 10^{23} \text{ molecules}}{201.70 \text{ g}}\right)$$

$$= 8.96 \times 10^{19} \text{ molecules of pseudoephedrine hydrochloride}$$

BALLPARK CHECK: Our estimate for the number of molecules was on the order of 10^{19}, which is consistent with the calculated answer.

WORKED EXAMPLE **6.6** Avogadro's Number: Atom to Mass Conversions

A tiny pencil mark just visible to the naked eye contains about 3×10^{17} atoms of carbon. What is the mass of this pencil mark in grams?

ANALYSIS We are given a number of atoms and need to convert to mass. The conversion factor can be obtained by realizing that the atomic weight of carbon in grams contains Avogadro's number of atoms (6.022×10^{23}).

BALLPARK ESTIMATE Since we are given a number of atoms that is six orders of magnitude less than Avogadro's number, we should get a corresponding mass that is six orders of magnitude less than the molar mass of carbon, which means a mass for the pencil mark of about 10^{-6} g.

SOLUTION

STEP 1: Identify known information. We know the number of carbon atoms in the pencil mark.

3×10^{17} atoms of carbon

STEP 2: Identify answer and units.

Mass of carbon = ?? g

STEP 3: Identify conversion factors. The atomic weight of carbon is 12.01 amu, so 12.01 g of carbon contains 6.022×10^{23} atoms.

$$\frac{12.01 \text{ g carbon}}{6.022 \times 10^{23} \text{ atoms}}$$

STEP 4: Solve. Set up an equation using the conversion factors so that unwanted units cancel.

$$(3 \times 10^{17} \text{ atoms})\left(\frac{12.01 \text{ g carbon}}{6.022 \times 10^{23} \text{ atoms}}\right) = 6 \times 10^{-6} \text{ g carbon}$$

BALLPARK CHECK: The answer is of the same magnitude as our estimate and makes physical sense.

PROBLEM 6.6

Calculate the molecular weight of the following substances:

(a) Ibuprofen, $C_{13}H_{18}O_2$
(b) Phenobarbital, $C_{12}H_{12}N_2O_3$

PROBLEM 6.7

How many molecules of ascorbic acid (vitamin C, $C_6H_8O_6$) are in a 500 mg tablet?

PROBLEM 6.8

What is the mass in grams of 5.0×10^{20} molecules of aspirin ($C_9H_8O_4$)?

KEY CONCEPT PROBLEM 6.9

What is the molecular weight of cytosine, a component of DNA (deoxyribonucleic acid)? (Gray = C, blue = N, red = O, ivory = H.)

Cytosine

6.4 Gram–Mole Conversions

To ensure that we have the correct molecule to molecule (or mole to mole) relationship between reactants as specified by the balanced chemical equation, we can take advantage of the constant mass ratio between reactants, as indicated previously. The mass in grams of 1 mol of any substance (that is, Avogadro's number of molecules or formula units) is called the **molar mass** of the substance.

Molar mass The mass in grams of 1 mol of a substance, numerically equal to molecular weight.

> Molar mass = Mass of 1 mol of substance
> = Mass of 6.022×10^{23} molecules (formula units) of substance
> = Molecular (formula) weight of substance in grams

In effect, molar mass serves as a conversion factor between numbers of moles and mass. If you know how many moles you have, you can calculate their mass; if you know the mass of a sample, you can calculate the number of moles. Suppose, for example, we need to know how much 0.25 mol of water weighs. The molecular weight of H_2O is $(2 \times 1.0 \text{ amu}) + 16.0 \text{ amu} = 18.0 \text{ amu}$, so the molar mass of water is 18.0 g. Thus, the conversion factor between moles of water and mass of water is 18.0 g/mol:

Molar mass used as conversion factor

$$0.25 \text{ mol } H_2O \times \frac{18.0 \text{ g } H_2O}{1 \text{ mol } H_2O} = 4.5 \text{ g } H_2O$$

APPLICATION ▶ Did Ben Franklin Have Avogadro's Number? A Ballpark Calculation

t length being at Clapham, where there is on the common a large pond . . . I fetched out a cruet of oil and dropped a little of it on the water. I saw it spread itself with surprising swiftness upon the surface. The oil, though not more than a teaspoonful, produced an instant calm over a space several yards square which spread amazingly and extended itself gradually . . . making all that quarter of the pond, perhaps half an acre, as smooth as a looking glass. *Excerpt from a letter of Benjamin Franklin to William Brownrigg, 1773.*

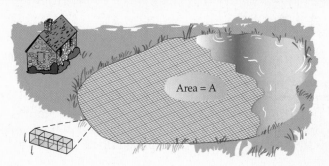

▲ What did these two have in common? [Benjamin Franklin (left), Amedeo Avogadro (right)]

Benjamin Franklin, author and renowned statesman, was also an inventor and a scientist. Every schoolchild knows of Franklin's experiment with a kite and a key, demonstrating that lightning is electricity. Less well known is that his measurement of the extent to which oil spreads on water makes possible a simple estimate of molecular size and Avogadro's number.

The calculation goes like this: Avogadro's number is the number of molecules in 1 mol of any substance. So if we can estimate both the number of molecules and the number of moles in Franklin's teaspoon of oil, we can calculate Avogadro's number. Let us start by calculating the number of molecules in the oil.

1. The volume (V) of oil Franklin used was 1 tsp = 4.9 cm^3, and the area (A) covered by the oil was 1/2 acre = 2.0 × 10^7 cm^2. We will assume that the oil molecules are tiny cubes that pack closely together and form a layer only one molecule thick. As shown in the accompanying figure, the volume of the oil is equal to the surface area of the layer times the length (l) of the side of one molecule: $V = A \times l$. Rearranging this equation to find the length then gives us an estimate of molecular size:

$$l = \frac{V}{A} = \frac{4.9 \text{ cm}^3}{2.0 \times 10^7 \text{ cm}^2} = 2.5 \times 10^{-7} \text{ cm}$$

2. The area of the oil layer is the area of the side of one molecule (l^2) times the number of molecules (N) of oil: $A = l^2 \times N$. Rearranging this equation gives us the number of molecules:

$$N = \frac{A}{l^2} = \frac{2 \times 10^7 \text{ cm}^2}{(2.5 \times 10^{-7} \text{ cm})^2} = 3.2 \times 10^{20} \text{ molecules}$$

3. To calculate the number of moles, we first need to know the mass (M) of the oil. This could have been determined by weighing the oil, but Franklin neglected to do so. Let us therefore estimate the mass by multiplying the volume (V) of the oil by the density (D) of a typical oil, 0.95 g/cm^3. (Since oil floats on water, it is not surprising that the density of oil is a bit less than the density of water, which is 1.00 g/cm^3.)

$$M = V \times D = 4.9 \text{ cm}^3 \times 0.95 \frac{g}{\text{cm}^3} = 4.7 \text{ g}$$

4. We now have to make one final assumption about the molecular weight of the oil before we complete the calculation. Assuming that a typical oil has MW = 200, then the mass of 1 mol of oil is 200 g. Dividing the mass of the oil (M) by the mass of 1 mol gives the number of moles of oil:

$$\text{Moles of oil} = \frac{4.7 \text{ g}}{200 \text{ g/mol}} = 0.024 \text{ mol}$$

5. Finally, the number of molecules per mole—Avogadro's number—can be obtained:

$$\text{Avogadro's number} = \frac{3.2 \times 10^{20} \text{ molecules}}{0.024 \text{ mol}} = 1.3 \times 10^{22}$$

The calculation is not very accurate, of course, but Ben was not really intending for us to calculate Avogadro's number when he made a rough estimate of how much his oil spread out. Nevertheless, the result is not too bad for such a simple experiment.

See Additional Problem 6.93 at the end of the chapter.

Alternatively, suppose we need to know how many moles of water are in 27 g of water. The conversion factor is 1 mol/18.0 g:

Molar mass used as conversion factor

$$27 \text{ g } H_2O \times \frac{1 \text{ mol } H_2O}{18.0 \text{ g } H_2O} = 1.5 \text{ mol } H_2O$$

Worked Examples 6.7 and 6.8 give more practice in gram–mole conversions.

WORKED EXAMPLE **6.7** Molar Mass: Mole to Gram Conversion

The nonprescription pain relievers Advil and Nuprin contain ibuprofen ($C_{13}H_{18}O_2$), whose molecular weight is 206.3 amu (Problem 6.6a). If all the tablets in a bottle of pain reliever together contain 0.082 mol of ibuprofen, what is the number of grams of ibuprofen in the bottle?

ANALYSIS We are given a number of moles and asked to find the mass. Molar mass is the conversion factor between the two.

BALLPARK ESTIMATE Since 1 mol of ibuprofen has a mass of about 200 g, 0.08 mol has a mass of about $0.08 \times 200 \text{ g} = 16 \text{ g}$.

SOLUTION

STEP 1: Identify known information.	0.082 mol ibuprofen in bottle
STEP 2: Identify answer and units.	mass ibuprofen in bottle = ?? g
STEP 3: Identify conversion factor. We use the molecular weight of ibuprofen to convert from moles to grams.	1 mol ibuprofen = 206.3 g $$\frac{206.3 \text{ g ibuprofen}}{1 \text{ mol ibuprofen}}$$
STEP 4: Solve. Set up an equation using the known information and conversion factor so that unwanted units cancel.	$$0.082 \text{ mol } C_{13}H_{18}O_2 \times \frac{206.3 \text{ g ibuprofen}}{1 \text{ mol ibuprofen}} = 17 \text{ g } C_{13}H_{18}O_2$$
	BALLPARK CHECK: The calculated answer is consistent with our estimate of 16 g.

WORKED EXAMPLE **6.8** Molar Mass: Gram to Mole Conversion

The maximum dose of sodium hydrogen phosphate (Na_2HPO_4, MW = 142.0 g/mol) that should be taken in one day for use as a laxative is 3.8 g. How many moles of sodium hydrogen phosphate, how many moles of Na^+ ions, and how many total moles of ions are in this dose?

ANALYSIS Molar mass is the conversion factor between mass and number of moles. The chemical formula Na_2HPO_4 shows that each formula unit contains 2 Na^+ ions and 1 HPO_4^{2-} ion.

BALLPARK ESTIMATE The maximum dose is about two orders of magnitude smaller than the molecular weight (approximately 4 g compared to 142 g). Thus, the number of moles of sodium hydrogen phosphate in 3.8 g should be about two orders of magnitude less than one mole. The number of moles of $NaHPO_4$ and total moles of ions, then, should be on the order of 10^{-2}.

SOLUTION

STEP 1: Identify known information. We are given the mass and molecular weight of Na_2HPO_4.	3.8 g Na_2HPO_4; MW = 142.0 amu
STEP 2: Identify answer and units. We need to find the number of moles of Na_2HPO_4, and the total number of moles of ions.	Moles of Na_2HPO_4 = ?? mol Moles of Na^+ ions = ?? mol Total moles of ions = ?? mol

STEP 3: Identify conversion factor. We can use the molecular weight of Na_2HPO_4 to convert from grams to moles.

$$\frac{1 \text{ mol Na}_2\text{HPO}_4}{142.0 \text{ g Na}_2\text{HPO}_4}$$

STEP 4: Solve. We use the known information and conversion factor to obtain moles of Na_2HPO_4; since 1 mol of Na_2HPO_4 contains 2 mol of Na^+ ions and 1 mol of HPO_4^{2-} ions, we multiply these values by the number of moles in the sample.

$$3.8 \text{ g Na}_2\text{HPO}_4 \times \frac{1 \text{ mol Na}_2\text{HPO}_4}{142.0 \text{ g Na}_2\text{HPO}_4} = 0.027 \text{ mol Na}_2\text{HPO}_4$$

$$\frac{2 \text{ mol Na}^+}{1 \text{ mol Na}_2\text{HPO}_4} \times 0.027 \text{ mol Na}_2\text{HPO}_4 = 0.054 \text{ mol Na}^+$$

$$\frac{3 \text{ mol ions}}{1 \text{ mol Na}_2\text{HPO}_4} \times 0.027 \text{ mol Na}_2\text{HPO}_4 = 0.081 \text{ mol ions}$$

BALLPARK CHECK: The calculated answers (0.027 mol Na_2HPO_4, 0.081 mol ions) are on the order of 10^{-2}, consistent with our estimate.

PROBLEM 6.10

How many moles of ethyl alcohol, C_2H_6O, are in a 10.0 g sample? How many grams are in a 0.10 mol sample of ethyl alcohol?

PROBLEM 6.11

12.01×8

Which weighs more, 5.00 g or 0.0225 mol of acetaminophen ($C_8H_9NO_2$)?

6.5 Mole Relationships and Chemical Equations

In a typical recipe, the amount of ingredients needed are specified using a variety of units: The amount of flour, for example, is usually specified in cups, whereas the amount of salt or vanilla flavoring might be indicated in teaspoons. In chemical reactions, the appropriate unit to specify the relationship between reactants and products is the mole.

The coefficients in a balanced chemical equation tell how many *molecules*, and thus how many *moles*, of each reactant are needed and how many molecules, and thus moles, of each product are formed. You can then use molar mass to calculate reactant and product masses. If, for example, you saw the following balanced equation for the industrial synthesis of ammonia, you would know that 3 mol of H_2 (3 mol $\times$ 2.0 g/mol = 6.0 g) are required for reaction with 1 mol of N_2 (28.0 g) to yield 2 mol of NH_3 (2 mol $\times$ 17.0 g/mol = 34.0 g).

This number of moles of hydrogen . . .   . . . reacts with this number of moles of nitrogen . . .   to yield this number of moles of ammonia.

$$3 \text{ H}_2 + 1 \text{ N}_2 \longrightarrow 2 \text{ NH}_3$$

The coefficients can be put in the form of *mole ratios*, which act as conversion factors when setting up factor-label calculations. In the ammonia synthesis, for example, the mole ratio of H_2 to N_2 is 3:1, the mole ratio of H_2 to NH_3 is 3:2, and the mole ratio of N_2 to NH_3 is 1:2:

$$\frac{3 \text{ mol H}_2}{1 \text{ mol N}_2} \quad \frac{3 \text{ mol H}_2}{2 \text{ mol NH}_3} \quad \frac{1 \text{ mol N}_2}{2 \text{ mol NH}_3}$$

Worked Example 6.9 shows how to set up and use mole ratios.

WORKED EXAMPLE   **6.9** Balanced Chemical Equations: Mole Ratios

Rusting involves the reaction of iron with oxygen to form iron(III) oxide, Fe_2O_3:

$$4\,Fe(s) + 3\,O_2(g) \longrightarrow 2\,Fe_2O_3(s)$$

(a) What are the mole ratios of the product to each reactant and of the reactants to each other?

(b) How many moles of iron(III) oxide are formed by the complete oxidation of 6.2 mol of iron?

ANALYSIS AND SOLUTION

(a) The coefficients of a balanced equation represent the mole ratios:

$$\frac{2\text{ mol }Fe_2O_3}{4\text{ mol }Fe} \quad \frac{2\text{ mol }Fe_2O_3}{3\text{ mol }O_2} \quad \frac{4\text{ mol }Fe}{3\text{ mol }O_2}$$

(b) To find how many moles of Fe_2O_3 are formed, write down the known information—6.2 mol of iron—and select the mole ratio that allows the quantities to cancel, leaving the desired quantity:

$$6.2\ \cancel{\text{mol Fe}} \times \frac{2\text{ mol }Fe_2O_3}{4\ \cancel{\text{mol Fe}}} = 3.1\text{ mol }Fe_2O_3$$

Note that mole ratios are exact numbers and therefore do not limit the number of significant figures in the result of a calculation.

PROBLEM 6.12

(a) Balance the following equation, and tell how many moles of nickel will react with 9.81 mol of hydrochloric acid.

$$Ni(s) + HCl(aq) \longrightarrow NiCl_2(aq) + H_2(g)$$

(b) How many moles of $NiCl_2$ can be formed in the reaction of 6.00 mol of Ni and 12.0 mol of HCl?

PROBLEM 6.13

Plants convert carbon dioxide and water to glucose ($C_6H_{12}O_6$) and oxygen in the process of photosynthesis. Write a balanced equation for this reaction, and determine how many moles of CO_2 are required to produce 15.0 mol of glucose.

6.6 Mass Relationships and Chemical Equations

It is important to remember that the coefficients in a balanced chemical equation represent molecule to molecule (or mole to mole) relationships between reactants and products. Mole ratios make it possible to calculate the molar amounts of reactants and products, but actual amounts of substances used in the laboratory are weighed out in grams. Regardless of what units we use to specify the amount of reactants and/or products (mass, volume, number of molecules, and so on), the reaction always takes place on a mole to mole basis. Thus, we need to be able to carry out three kinds of conversions when doing chemical arithmetic:

- **Mole to mole conversions** are carried out using *mole ratios* as conversion factors. Worked Example 6.9 at the end of the preceding section is an example of this kind of calculation.

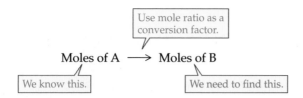

- **Mole to mass and mass to mole conversions** are carried out using *molar mass* as a conversion factor. Worked Examples 6.7 and 6.8 at the end of Section 6.4 are examples of this kind of calculation.

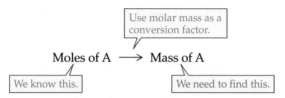

- **Mass to mass conversions** are frequently needed but cannot be carried out directly. If you know the mass of substance A and need to find the mass of substance B, you must first convert the mass of A into moles of A, then carry out a mole to mole conversion to find moles of B, and then convert moles of B into the mass of B (Figure 6.2).

Overall, there are four steps for determining mass relationships among reactants and products:

STEP 1: Write the balanced chemical equation.

STEP 2: Choose molar masses and mole ratios to convert the known information into the needed information.

STEP 3: Set up the factor-label expression and calculate the answer.

STEP 4: Check the answer against the ballpark estimate you made before you began your calculations.

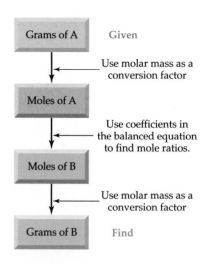

▲ **FIGURE 6.2 A summary of conversions between moles and grams for substances in a chemical reaction.** The numbers of moles tell how many molecules of each substance are needed, as given by the coefficients in the balanced equation; the numbers of grams tell what mass of each substance is needed.

WORKED EXAMPLE **6.10** Mole Ratios: Mole to Mass Conversions

In the atmosphere, nitrogen dioxide reacts with water to produce NO and nitric acid, which contributes to pollution by acid rain:

$$3\ NO_2(g)\ +\ H_2O(l)\ \longrightarrow\ 2\ HNO_3(aq)\ +\ NO(g)$$

How many grams of HNO_3 are produced for every 1.0 mol of NO_2 that reacts? The molecular weight of HNO_3 is 63.0 amu.

ANALYSIS We are given the number of moles of a reactant and are asked to find the mass of a product. Problems of this sort always require working in moles and then converting to mass, as outlined in Figure 6.2.

BALLPARK ESTIMATE The molar mass of nitric acid is approximately 60 g/mol, and the coefficients in the balanced equation say that 2 mol of HNO_3 are formed for each 3 mol of NO_2 that undergo reaction. Thus, 1 mol of NO_2 should give about 2/3 mol HNO_3, or 2/3 mol × 60 g/mol = 40 g.

SOLUTION

STEP 1: Write balanced equation.

$$3\ NO_2(g)\ +\ H_2O(l)\ \longrightarrow\ 2\ HNO_3(aq)\ +\ NO(g)$$

STEP 2: Identify conversion factors. We need a mole to mole conversion to find the number of moles of product, and then a mole to mass conversion to find the mass of product. For the first conversion we use the mole ratio of NO_2 to HNO_3 as a conversion factor, and for the mole to mass calculation we use the molar mass of HNO_3 (63.0 g) as a conversion factor.

$$\frac{2\ mol\ HNO_3}{3\ mol\ NO_2}$$

$$\frac{63.0\ g\ HNO_3}{1\ mol\ HNO_3}$$

STEP 3: **Set up factor labels.** Identify appropriate mole-ratio factor labels to convert moles NO_2 to moles HNO_3, and moles HNO_3 to grams.

$$1.0 \text{ mol } NO_2 \times \frac{2 \text{ mol } HNO_3}{3 \text{ mol } NO_2} \times \frac{63.0 \text{ g } HNO_3}{1 \text{ mol } HNO_3} = 42 \text{ g } HNO_3$$

STEP 4: BALLPARK CHECK

Our estimate was 40 g!

WORKED EXAMPLE **6.11** Mole Ratios: Mass to Mole/Mole to Mass Conversions

The following reaction produced 0.022 g of calcium oxalate (CaC_2O_4). What mass of calcium chloride was used as reactant? (The molar mass of CaC_2O_4 is 128.1 g, and the molar mass of $CaCl_2$ is 111.0 g.)

$$CaCl_2(aq) + Na_2C_2O_4(aq) \longrightarrow CaC_2O_4(s) + 2 NaCl(aq)$$

ANALYSIS Both the known information and that to be found are masses, so this is a mass to mass conversion problem. The mass of CaC_2O_4 is first converted into moles, a mole ratio is used to find moles of $CaCl_2$, and the number of moles of $CaCl_2$ is converted into mass.

BALLPARK ESTIMATE The balanced equation says that 1 mol of CaC_2O_4 is formed for each mole of $CaCl_2$ that reacts. Because the formula weights of the two substances are similar, it should take about 0.02 g of $CaCl_2$ to form 0.02 g of CaC_2O_4.

SOLUTION

STEP 1: **Write the balanced equation.**

$$CaCl_2(aq) + Na_2C_2O_4(aq) \longrightarrow CaC_2O_4(s) + 2 NaCl(aq)$$

STEP 2: **Identify conversion factors.** Convert the mass of CaC_2O_4 into moles, use a mole ratio to find moles of $CaCl_2$, and convert the number of moles of $CaCl_2$ to mass. We will need three conversion factors.

mass CaC_2O_4 to moles: $\dfrac{1 \text{ mol } CaC_2O_4}{128.1 \text{ g}}$

moles Ca C_2O_4 to moles $CaCl_2$: $\dfrac{1 \text{ mol } CaCl_2}{1 \text{ mol } CaC_2O_4}$

moles $CaCl_2$ to mass: $\dfrac{111.0 \text{ } CaCl_2}{1 \text{ mol } CaCl_2}$

STEP 3: **Set up factor-labels.** We will need to perform gram to mole and mole to mole conversions to get from grams CaC_2O_4 to grams $CaCl_2$.

$$0.022 \text{ g } CaC_2O_4 \times \frac{1 \text{ mol } CaC_2O_4}{128.1 \text{ g } CaC_2O_4} \times$$

$$\frac{1 \text{ mol } CaCl_2}{1 \text{ mol } CaC_2O_4} \times \frac{111.0 \text{ g } CaCl_2}{1 \text{ mol } CaCl_2} = 0.019 \text{ g } CaCl_2$$

STEP 4: BALLPARK CHECK

The calculated answer (0.019 g) is consistent with our estimate (0.02 g).

▲ The floral pattern on this glass was created by *etching*, a process based on the reaction of HF with glass.

PROBLEM 6.14

Hydrogen fluoride is one of the few substances that react with glass (which is made of silicon dioxide, SiO_2).

$$4 HF(g) + SiO_2(s) \longrightarrow SiF_4(g) + 2 H_2O(l)$$

(a) How many moles of HF will react completely with 9.90 mol of SiO_2?

(b) What mass of water (in grams) is produced by the reaction of 23.0 g of SiO_2?

PROBLEM 6.15

The tungsten metal used for filaments in light bulbs is made by reaction of tungsten trioxide with hydrogen:

$$WO_3(s) + 3 H_2(g) \longrightarrow W(s) + 3 H_2O(g)$$

How many grams of tungsten trioxide and how many grams of hydrogen must you start with to prepare 5.00 g of tungsten? (For WO_3, MW = 231.8 amu.)

6.7 Limiting Reagent and Percent Yield

All the calculations we have done in the last several sections have assumed that 100% of the reactants are converted to products. Only rarely is this the case in practice, though. Let us return to the recipe for s'mores presented previously:

$$2 \text{ Graham crackers} + 1 \text{ Roasted marshmallow} + \tfrac{1}{4} \text{ Chocolate bar} \longrightarrow 1 \text{ S'more}$$

When you check your supplies, you find that you have 20 graham crackers, 8 marshmallows, and 3 chocolate bars. How many s'mores can you make? (Answer = 8!) You have enough graham crackers and chocolate bars to make more, but you will run out of marshmallows after you have made eight s'mores. In a similar way, when running a chemical reaction we don't always have the exact amounts of reagents to allow all of them to react completely. The reactant that is exhausted first in such a reaction is called the **limiting reagent.** The amount of product you obtain if the limiting reagent is completely consumed is called the **theoretical yield** of the reaction.

Suppose that, while you are making s'mores, one of your eight marshmallows gets burned to a crisp. If this happens, the actual number of s'mores produced will be less than what you predicted based on the amount of starting materials. Similarly, chemical reactions do not always yield the exact amount of product predicted by the initial amount of reactants. More frequently, a majority of the reactant molecules behave as written but other processes, called *side reactions*, also occur. In addition, some of the product may be lost in handling. As a result, the amount of product actually formed—the reaction's **actual yield**—is somewhat less than the theoretical yield. The amount of product actually obtained in a reaction is usually expressed as a **percent yield**:

$$\text{Percent yield} = \frac{\text{Actual yield}}{\text{Theoretical yield}} \times 100\%$$

Limiting reagent The reactant that runs out first.

Theoretical yield The amount of product formed assuming complete reaction of the limiting reagent.

Actual yield The amount of product actually formed in a reaction.

Percent yield The percent of the theoretical yield actually obtained from a chemical reaction.

A reaction's actual yield is found by weighing the amount of product obtained. The theoretical yield is found by using the amount of limiting reagent in a mass to mass calculation like those illustrated in the preceding section (see Worked Example 6.11). Worked Examples 6.12–6.14 involve limiting reagent, percent yield, actual yield, and theoretical yield calculations.

WORKED EXAMPLE **6.12** Percent Yield

The combustion of acetylene gas (C_2H_2) produces carbon dioxide and water as indicated in the following reaction:

$$2 \, C_2H_2(g) + 5 \, O_2(g) \longrightarrow 4 \, CO_2(g) + 2 \, H_2O(g)$$

When 26.0 g of acetylene is burned in sufficient oxygen for complete reaction, the theoretical yield of CO_2 is 88.0 g. Calculate the percent yield for this reaction if the actual yield is only 72.4 g CO_2.

ANALYSIS The percent yield is calculated by dividing the actual yield by the theoretical yield and multiplying by 100.

BALLPARK ESTIMATE The theoretical yield (88.0 g) is close to 100 g. The actual yield (72.4 g) is about 15 g less than the theoretical yield. The actual yield is thus about 15% less than the theoretical yield, so the percent yield is about 85%.

SOLUTION

$$\text{Percent yield} = \frac{\text{Actual yield}}{\text{Theoretical yield}} \times 100 = \frac{72.4 \text{ g } CO_2}{88.0 \text{ g } CO_2} \times 100 = 82.3\%$$

BALLPARK CHECK The calculated percent yield agrees very well with our estimate of 85%.

WORKED EXAMPLE 6.13 Mass to Mole Conversions: Limiting Reagent and Theoretical Yield

The element boron is produced commercially by the reaction of boric oxide with magnesium at high temperature:

$$B_2O_3(l) + 3 Mg(s) \longrightarrow 2 B(s) + 3 MgO(s)$$

What is the theoretical yield of boron when 2350 g of boric oxide is reacted with 3580 g of magnesium? The molar masses of boric oxide and magnesium are 69.6 g/mol and 24.3 g/mol, respectively.

ANALYSIS To calculate theoretical yield, we first have to identify the limiting reagent. The theoretical yield in grams is then calculated from the amount of limiting reagent used in the reaction. The calculation involves the mass to mole and mole to mass conversions discussed in the preceding section.

SOLUTION

STEP 1: Identify known information. We have the masses and molar masses of the reagents.

2350 g B_2O_3, molar mass 69.6 g/mol
3580 g Mg, molar mass 24.3 g/mol

STEP 2: Identify answer and units. We are solving for the theoretical yield of boron.

Theoretical mass of B = ?? g

STEP 3: Identify conversion factors. We can use the molar masses to convert from masses to moles of reactants (B_2O_3, Mg). From moles of reactants, we can use mole ratios from the balanced chemical equation to find the number of moles of B produced. B_2O_3 is the limiting reagent, since complete reaction of this reagent yields less product (B).

$$(2350 \text{ g } B_2O_3) \times \frac{1 \text{ mol } B_2O_3}{69.6 \text{ g } B_2O_3} = 33.8 \text{ mol } B_2O_3$$

$$(3580 \text{ g Mg}) \times \frac{1 \text{ mol Mg}}{24.3 \text{ g Mg}} = 147 \text{ mol Mg}$$

$$33.8 \text{ mol } B_2O_3 \times \frac{2 \text{ mol B}}{1 \text{ mol } B_2O_3} = 67.6 \text{ mol B*}$$

$$147 \text{ mol Mg} \times \frac{2 \text{ mol B}}{3 \text{ mol Mg}} = 98.0 \text{ mol B}$$

(*limiting reagent!)

STEP 4: Solve. Once the limiting reagent has been identified (B_2O_3), the theoretical amount of B that should be formed can be calculated using a mole to mass conversion.

$$67.6 \text{ mol B} \times \frac{10.8 \text{ g B}}{1 \text{ mol B}} = 730 \text{ g B}$$

WORKED EXAMPLE 6.14 Mass to Mole Conversion: Percent Yield

The reaction of ethylene with water to give ethyl alcohol (CH_3CH_2OH) occurs in 78.5% actual yield. How many grams of ethyl alcohol are formed by reaction of 25.0 g of ethylene? (For ethylene, MW = 28.0 amu; for ethyl alcohol, MW = 46.0 amu.)

$$H_2C=CH_2 + H_2O \longrightarrow CH_3CH_2OH$$

ANALYSIS Treat this as a typical mass relationship problem to find the amount of ethyl alcohol that can theoretically be formed from 25.0 g of ethylene, and then multiply the answer by 78.5% to find the amount actually formed.

BALLPARK ESTIMATE The 25.0 g of ethylene is a bit less than 1 mol; since the percent yield is about 78%, a bit less than 0.78 mol of ethyl alcohol will form—perhaps about 3/4 mol, or 3/4 × 46 g = 34 g.

SOLUTION
The theoretical yield of ethyl alcohol is

$$25.0 \text{ g ethylene} \times \frac{1 \text{ mol ethylene}}{28.0 \text{ g ethylene}} \times \frac{1 \text{ mol ethyl alc.}}{1 \text{ mol ethylene}} \times \frac{46.0 \text{ g ethyl alc.}}{1 \text{ mol ethyl alc.}} = 41.1 \text{ g ethyl alcohol}$$

and so the actual yield is

$$41.1 \text{ g ethyl alc.} \times 0.785 = 32.3 \text{ g ethyl alcohol}$$

BALLPARK CHECK The calculated result (32.3 g) is close to our estimate (34 g).

PROBLEM 6.16

What is the theoretical yield of ethyl chloride in the reaction of 19.4 g of ethylene with 50 g of hydrogen chloride? What is the percent yield if 25.5 g of ethyl chloride is actually formed? (For ethylene, MW = 28.0 amu; for hydrogen chloride, MW = 36.5 amu; for ethyl chloride, MW = 64.5 amu.)

$$H_2C{=}CH_2 + HCl \longrightarrow CH_3CH_2Cl$$

PROBLEM 6.17

The reaction of ethylene oxide with water to give ethylene glycol (automobile antifreeze) occurs in 96.0% actual yield. How many grams of ethylene glycol are formed by reaction of 35.0 g of ethylene oxide? (For ethylene oxide, MW = 44.0 amu; for ethylene glycol, MW = 62.0 amu.)

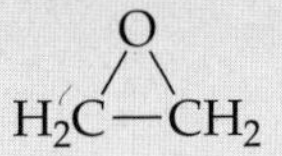

$$\underset{\text{Ethylene oxide}}{\overset{\displaystyle O}{\underset{\displaystyle H_2C{-}CH_2}{\bigwedge}}} + H_2O \longrightarrow \underset{\text{Ethylene glycol}}{HOCH_2CH_2OH}$$

⚷ KEY CONCEPT PROBLEM 6.18

Identify the limiting reagent in the reaction mixture shown below (red = A_2, blue = B_2). The balanced reaction is

$$A_2 + 2 B_2 \longrightarrow 2 AB_2$$

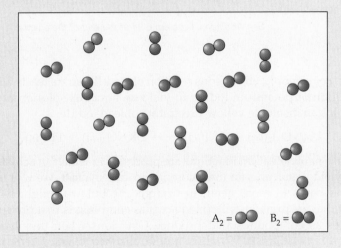

$A_2 =$ ⬤⬤ $B_2 =$ ⬤⬤

6.8   Classes of Chemical Reactions

One of the best ways to understand any subject is to look for patterns that help us categorize large amounts of information. When learning about chemical reactions, for instance, it is helpful to group the reactions of ionic compounds into three general classes: *precipitation reactions, acid–base neutralization reactions*, and *oxidation–reduction reactions*. This is not the only possible way of categorizing reactions nor does the list include all possibilities, but it is useful nonetheless. Let us look briefly at examples of each of these three reaction classes before studying them in more detail in subsequent sections.

- **Precipitation reactions** are processes in which an insoluble solid called a **precipitate** forms when reactants are combined in aqueous solution. Most precipitations take place when the anions and cations of two ionic compounds change

Precipitate   An insoluble solid that forms in solution during a chemical reaction.

APPLICATION ▶ Anemia – A Limiting Reagent Problem?

Anemia is the most commonly diagnosed blood disorder, with symptoms typically including lethargy, fatigue, poor concentration, and sensitivity to cold. While anemia has many causes, including genetic factors, the most common cause is insufficient dietary intake or absorption of iron.

Hemoglobin (abbreviated Hb), the iron-containing protein found in red blood cells, is responsible for oxygen transport throughout the body (⊂⊃, p. 266) Low iron levels in the body result in decreased production and incorporation of Hb in red blood cells. In addition, blood loss due to injury or to menstruation in women increases the body's demand for iron in order to replace lost Hb. In the United States, nearly 20% of women of child-bearing age suffer from iron-deficiency anemia compared to only 2% of adult men.

The recommended minimum daily iron intake is 8 mg for adult men and 18 mg for pre-menopausal women. One way to ensure sufficient iron intake is a well-balanced diet that includes iron-fortified grains and cereals, red meat, egg yolks, leafy green vegetables, tomatoes, and raisins. Vegetarians should pay extra attention to their diet since the iron in fruits and vegetables is not as readily absorbed by the body as the iron in meat, poultry, and fish. Vitamin supplements containing folic acid and either ferrous sulfate or ferrous glutonate can decrease iron deficiencies, and vitamin C increases the absorption of iron by the body.

▲ Can cooking in cast iron pots decrease anemia?

However, the simplest way to increase dietary iron may be to use cast iron cookware. Studies have demonstrated that the iron content of many foods increases when cooked in an iron pot. Other studies involving Ethiopian children showed that those who ate food cooked in iron cookware were less likely to suffer from iron-deficiency anemia than their playmates who ate similar foods prepared in aluminum cookware.

See Additional Problem 6.94 at the end of the chapter.

Salt An ionic compound formed from reaction of an acid with a base.

Oxidation–reduction (redox) reaction A reaction in which electrons are transferred from one atom to another.

▲ Reaction of aqueous $Pb(NO_3)_2$ with aqueous KI gives a yellow precipitate of PbI_2.

partners. For example, an aqueous solution of lead(II) nitrate reacts with an aqueous solution of potassium iodide to yield an aqueous solution of potassium nitrate plus an insoluble yellow precipitate of lead iodide:

$$Pb(NO_3)_2(aq) + 2\ KI(aq) \longrightarrow 2\ KNO_3(aq) + PbI_2(s)$$

- **Acid–base neutralization reactions** are processes in which an acid reacts with a base to yield water plus an ionic compound called a **salt**. We will look at both acids and bases in more detail in Chapter 10, but you might recall for the moment that we previously defined acids as compounds that produce H^+ ions and bases as compounds that produce OH^- ions when dissolved in water (Section 4.11). (⊂⊃, p. 99) Thus, a neutralization reaction removes H^+ and OH^- ions from solution and yields neutral H_2O. The reaction between hydrochloric acid and sodium hydroxide is a typical example:

$$HCl(aq) + NaOH(aq) \longrightarrow H_2O(l) + NaCl(aq)$$

Note that in this reaction, the "salt" produced is sodium chloride, or common table salt. In a general sense, however, *any* ionic compound produced in an acid–base reaction is also called a salt.

- **Oxidation–reduction reactions**, or **redox reactions**, are processes in which one or more electrons are transferred between reaction partners (atoms, molecules, or ions). As a result of this transfer, the number of electrons assigned to individual atoms in the various reactants change. When metallic magnesium reacts with iodine vapor, for instance, a magnesium atom gives an electron to each of two iodine atoms, forming an Mg^{2+} ion and two I^- ions. The charge on the magnesium changes from 0 to +2, and the charge on each iodine changes from 0 to −1:

$$Mg(s) + I_2(g) \longrightarrow MgI_2(s)$$

WORKED EXAMPLE **6.15** Classifying Chemical Reactions

Classify the following processes as a precipitation, acid–base neutralization, or redox reaction.

(a) $Ca(OH)_2(aq) + 2\,HBr(aq) \longrightarrow 2\,H_2O(l) + CaBr_2(aq)$
(b) $Pb(ClO_4)_2(aq) + 2\,NaCl(aq) \longrightarrow PbCl_2(s) + 2\,NaClO_4(aq)$
(c) $2\,AgNO_3(aq) + Cu(s) \longrightarrow 2\,Ag(s) + Cu(NO_3)_2(aq)$

ANALYSIS One way to identify the class of reaction is to examine the products that form and match them with the descriptions for the types of reactions provided in this section. By process of elimination, we can readily identify the appropriate reaction classification.

SOLUTION

(a) The products of this reaction are water and an ionic compound, or salt ($CaBr_2$). This is consistent with the description of an acid–base neutralization reaction.

(b) This reaction involves two aqueous reactants, $Pb(ClO_4)_2$ and NaCl, which combine to form a solid product, $PbCl_2$. This is consistent with a precipitation reaction.

(c) The products of this reaction are a solid, Ag(s), and an aqueous ionic compound, $Cu(NO_3)_2$. This does not match the description of a neutralization reaction, which would form *water* and an ionic compound. One of the products *is* a solid, but the reactants are not both aqueous compound; one of the reactants is *also* a solid (Cu). Therefore, this reaction would not be classified as a precipitation reaction. By process of elimination, then, it must be a redox reaction.

PROBLEM 6.19

Classify each of the following processes as a precipitation, acid–base neutralization, or redox reaction.

(a) $AgNO_3(aq) + KCl(aq) \longrightarrow AgCl(s) + KNO_3(aq)$
(b) $2\,Al(s) + 3\,Br_2(l) \longrightarrow 2\,AlBr_3(s)$
(c) $Ca(OH)_2(aq) + 2\,HNO_3(aq) \longrightarrow 2\,H_2O(l) + Ca(NO_3)_2(aq)$

6.9 Precipitation Reactions and Solubility Guidelines

Now let us look at precipitation reactions in more detail. To predict whether a precipitation reaction will occur on mixing aqueous solutions of two ionic compounds, you must know the **solubilities** of the potential products—how much of each compound will dissolve in a given amount of solvent at a given temperature. If a substance has a low solubility in water, then it is likely to precipitate from an aqueous solution. If a substance has a high solubility in water, then no precipitate will form.

Solubility is a complex matter, and it is not always possible to make correct predictions. As a rule of thumb, though, the following solubility guidelines for ionic compounds are useful.

Solubility The amount of a compound that will dissolve in a given amount of solvent at a given temperature.

General Rules on Solubility

RULE 1. **A compound is probably soluble if it contains one of the following *cations*:**
- Group 1A cation: $Li^+, Na^+, K^+, Rb^+, Cs^+$
- Ammonium ion: NH_4^+

▲ Azurite (blue) and malachite (green) are mineral forms of the insoluble copper (II) carbonate.

RULE 2. A compound is probably soluble if it contains one of the following *anions*:

- Halide: Cl^-, Br^-, I^- *except Ag^+, Hg_2^{2+}, and Pb^{2+} compounds*
- Nitrate (NO_3^-), perchlorate (ClO_4^-), acetate ($CH_3CO_2^-$), sulfate (SO_4^{2-}) *except Ba^{2+}, Hg_2^{2+}, and Pb^{2+} sulfates*

If a compound does *not* contain at least one of the ions listed above, it is probably *not* soluble. Thus, Na_2CO_3 is soluble because it contains a group 1A cation, and $CaCl_2$ is soluble because it contains a halide anion. The compound $CaCO_3$, however, is *insoluble* because it contains none of the ions listed above. The guidelines are given in a different form in Table 6.1.

TABLE 6.1 General Solubility Guidelines for Ionic Compounds in Water

SOLUBLE	EXCEPTIONS
Ammonium compounds (NH_4^+)	None
Lithium compounds (Li^+)	None
Sodium compounds (Na^+)	None
Potassium compounds (K^+)	None
Nitrates (NO_3^-)	None
Perchlorates (ClO_4^-)	None
Acetates ($CH_3CO_2^-$)	None
Chlorides (Cl^-)	
Bromides (Br^-)	Ag^+, Hg_2^{2+}, and Pb^{2+} compounds
Iodides (I^-)	
Sulfates (SO_4^{2-})	Ba^{2+}, Hg_2^{2+}, and Pb^{2+} compounds

Let us try a problem. What will happen if aqueous solutions of sodium nitrate ($NaNO_3$) and potassium sulfate (K_2SO_4) are mixed? To answer this question, look at the guidelines to find the solubilities of the two possible products, Na_2SO_4 and KNO_3. Because both have group 1A cations (Na^+ and K^+), both are water-soluble and no precipitation will occur. If aqueous solutions of silver nitrate ($AgNO_3$) and sodium carbonate (Na_2CO_3) are mixed, however, the guidelines predict that a precipitate of insoluble silver carbonate (Ag_2CO_3) will form.

$$2\,AgNO_3(aq) + Na_2CO_3(aq) \longrightarrow Ag_2CO_3(s) + 2\,NaNO_3(aq)$$

WORKED EXAMPLE | **6.16** Chemical Reactions: Solubility Rules

Will a precipitation reaction occur when aqueous solutions of $CdCl_2$ and $(NH_4)_2S$ are mixed?

SOLUTION

Identify the two potential products, and predict the solubility of each using the guidelines in the text. In this instance, $CdCl_2$ and $(NH_4)_2S$ might give CdS and NH_4Cl. Since the guidelines predict that CdS is insoluble, a precipitation reaction will occur:

$$CdCl_2(aq) + (NH_4)_2S(aq) \longrightarrow CdS(s) + 2\,NH_4Cl(aq)$$

PROBLEM 6.20

Predict the solubility of the following compounds:

(a) $CdCO_3$ **(b)** Na_2S **(c)** $PbSO_4$

(d) $(NH_4)_3PO_4$ **(e)** Hg_2Cl_2

APPLICATION ▶ Gout and Kidney Stones: Problems in Solubility

One of the major pathways in the body for the breakdown of the nucleic acids DNA and RNA is by conversion to a substance called *uric acid*, $C_5H_4N_4O_3$, so named because it was first isolated in 1776 from urine. Most people excrete about 0.5 g of uric acid every day in the form of sodium urate, the salt that results from an acid–base reaction of uric acid. Unfortunately, the amount of sodium urate that dissolves in water (or urine) is fairly low—only about 0.07 mg/mL at the normal body temperature of 37 °C. When too much sodium urate is produced or mechanisms for its elimination fail, its concentration in blood and urine rises, and the excess sometimes precipitates in the joints and kidneys.

Gout is a disorder of nucleic acid metabolism that primarily affects middle-aged men (only 5% of gout patients are women). It is characterized by an increased sodium urate concentration in blood, leading to the deposit of sodium urate crystals in soft tissue around the joints, particularly in the hands and at the base of the big toe. Deposits of the sharp, needle-like crystals cause an extremely painful inflammation that can lead ultimately to arthritis and even to bone destruction.

Just as increased sodium urate concentration in blood can lead to gout, increased concentration in urine can result in the formation of one kind of *kidney stones*, small crystals that precipitate in the kidney. Although often quite small, kidney stones cause excruciating pain when they pass through the ureter, the duct that carries urine from the kidney to the bladder. In some cases, complete blockage of the ureter occurs.

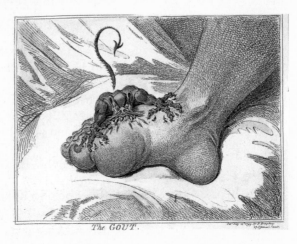
The GOUT.

▲ This old cartoon leaves little doubt about how painful gout can be.

Treatment of excessive sodium urate production involves both dietary modification and drug therapy. Foods such as liver, sardines, and asparagus should be avoided, and drugs such as allopurinol can be taken to lower production of sodium urate. Allopurinol functions by inhibiting the action of an enzyme called *xanthine oxidase*, thereby blocking a step in nucleic acid metabolism.

See Additional Problem 6.95 at the end of the chapter.

PROBLEM 6.21

Predict whether a precipitation reaction will occur in the following situations:

(a) $NiCl_2(aq) + (NH_4)_2S(aq) \longrightarrow$

(b) $AgNO_3(aq) + CaBr_2(aq) \longrightarrow$

6.10 Acids, Bases, and Neutralization Reactions

When acids and bases are mixed in the correct proportion, both acidic and basic properties disappear because of a **neutralization reaction**. The most common kind of neutralization reaction occurs between an acid (generalized as HA), and a metal hydroxide (generalized as MOH), to yield water and a salt. The H^+ ion from the acid combines with the OH^- ion from the base to give neutral H_2O, whereas the anion from the acid (A^-) combines with the cation from the base (M^+) to give the salt:

Neutralization reaction The reaction of an acid with a base.

$$\textit{A neutralization reaction:} \quad \underset{\text{Acid}}{HA(aq)} + \underset{\text{Base}}{MOH(aq)} \longrightarrow \underset{\text{Water}}{H_2O(l)} + \underset{\text{A salt}}{MA(aq)}$$

The reaction of hydrochloric acid with potassium hydroxide to produce potassium chloride is an example:

$$HCl(aq) + KOH(aq) \longrightarrow H_2O(l) + KCl(aq)$$

Another kind of neutralization reaction occurs between an acid and a carbonate (or bicarbonate) to yield water, a salt, and carbon dioxide. Hydrochloric acid reacts with potassium carbonate, for example, to give H_2O, KCl, and CO_2:

$$2\,HCl(aq) + K_2CO_3(aq) \longrightarrow H_2O(l) + 2\,KCl(aq) + CO_2(g)$$

The reaction occurs because the carbonate ion (CO_3^{2-}) reacts initially with H^+ to yield H_2CO_3, which is unstable and immediately decomposes to give CO_2 plus H_2O.

We will defer a more complete discussion of carbonates as bases until Chapter 10, but note for now that they yield OH^- ions when dissolved in water just as KOH and other bases do.

$$K_2CO_3(s) + H_2O(l) \xrightarrow{\text{Dissolve in water}} 2K^+(aq) + HCO_3^-(aq) + OH^-(aq)$$

⊂⊃ **Looking Ahead**

Acids and bases are enormously important in biological chemistry. We will see in Chapter 18, for instance, how acids and bases affect the structure and properties of proteins. ⊂⊃

WORKED EXAMPLE **6.17** Chemical Reactions: Acid–Base Neutralization

Write an equation for the neutralization reaction of aqueous HBr and aqueous $Ba(OH)_2$.

SOLUTION
The reaction of HBr with $Ba(OH)_2$ involves the combination of a proton (H^+) from the acid with OH^- from the base to yield water and a salt ($BaBr_2$).

$$2\,HBr(aq) + Ba(OH)_2(aq) \longrightarrow 2\,H_2O(l) + BaBr_2(aq)$$

PROBLEM 6.22

Write and balance equations for the following acid–base neutralization reactions:

(a) $CsOH(aq) + H_2SO_4(aq) \longrightarrow$ $(Cs)_2SO_4 + 2H_2O$

(b) $Ca(OH)_2(aq) + CH_3CO_2H(aq) \longrightarrow$

(c) $NaHCO_3(aq) + HBr(aq) \longrightarrow$

6.11 Redox Reactions

Oxidation-reduction (redox) reactions, the third and final category of reactions that we will discuss, are more complex than precipitation and neutralization reactions. Look, for instance, at the following examples and see if you can tell what they have in common. Copper metal reacts with aqueous silver nitrate to form silver metal and aqueous copper(II) nitrate; iron rusts in air to form iron(III) oxide; the zinc metal container on the outside of a battery reacts with manganese dioxide and ammonium chloride inside the battery to generate electricity and give aqueous zinc chloride plus manganese(III) oxide. Although these and many thousands of other reactions appear unrelated, all are examples of redox reactions.

$$Cu(s) + 2\,AgNO_3(aq) \longrightarrow 2\,Ag(s) + Cu(NO_3)_2(aq)$$

$$2\,Fe(s) + 3\,O_2(g) \longrightarrow Fe_2O_3(s)$$

$$Zn(s) + 2\,MnO_2(s) + 2\,NH_4Cl(s) \longrightarrow$$

$$ZnCl_2(aq) + Mn_2O_3(s) + 2\,NH_3(aq) + H_2O(l)$$

Historically, the word *oxidation* referred to the combination of an element with oxygen to yield an oxide, and the word *reduction* referred to the removal of oxygen from an oxide to yield the element. Today, though, the words have taken on a much broader meaning. An **oxidation** is now defined as the loss of one or more electrons by an atom, and a **reduction** is the gain of one or more electrons. Thus, an oxidation–reduction reaction, or redox reaction, is one in which *electrons are transferred from one atom to another.*

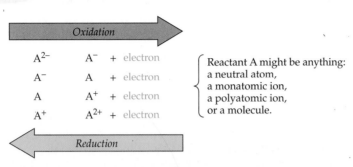

$$A^{2-} \quad A^- + \text{electron}$$
$$A^- \quad A + \text{electron}$$
$$A \quad A^+ + \text{electron}$$
$$A^+ \quad A^{2+} + \text{electron}$$

Reactant A might be anything: a neutral atom, a monatomic ion, a polyatomic ion, or a molecule.

Oxidation The loss of one or more electrons by an atom.

Reduction The gain of one or more electrons by an atom.

▲ The copper wire reacts with aqueous Ag^+ ion and becomes coated with metallic silver. At the same time, copper(II) ions go into solutions, producing the blue color.

Take the reaction of copper with aqueous Ag^+ as an example. Copper metal gives an electron to each of two Ag^+ ions, forming Cu^{2+} and silver metal. Copper is oxidized in the process, and Ag^+ is reduced. You can follow the transfer of the electrons by noting that the charge on the copper increases from 0 to +2 when it loses two electrons, whereas the charge on Ag^+ decreases from +1 to 0 when it gains an electron.

+2 electrons = reduced!

$$Cu(s) + 2\,Ag^+(aq) \longrightarrow Cu^{2+}(aq) + 2\,Ag(s)$$

0 charge   +1 charge   +2 charge   0 charge

−2 electrons = oxidized!

Similarly, in the reaction of aqueous iodide ion with bromine, iodide ion gives an electron to bromine, forming iodine and bromide ion. Iodide ion is oxidized as its charge increases from −1 to 0, and bromine is reduced as its charge decreases from 0 to −1.

+2 electrons = reduced!

$$2\,I^-(aq) + Br_2(aq) \longrightarrow I_2(aq) + 2\,Br^-(aq)$$

−1 charge   0 charge   0 charge   −1 charge

−2 electrons = oxidized!

As these examples show, oxidation and reduction always occur together. Whenever one substance loses an electron (is oxidized), another substance must gain that electron (be reduced). The substance that gives up an electron and causes the reduction—the copper atom in the reaction of Cu with Ag^+ and the iodide ion in the reaction of I^- with Br_2—is called a **reducing agent**. The substance that gains an electron and causes the oxidation—the silver ion in the reaction of Cu with Ag^+ and the bromine molecule in the reaction of I^- with Br_2—is called an **oxidizing agent**. The charge on the reducing agent increases during the reaction, and the charge on the oxidizing agent decreases.

Reducing agent A reactant that causes a reduction by giving up electrons to another reactant.

Oxidizing agent A reactant that causes an oxidation by taking electrons from another reactant.

Reducing agent   Loses one or more electrons
     Causes reduction
     Undergoes oxidation
     Becomes more positive (or less negative)

Oxidizing agent	Gains one or more electrons
	Causes oxidation
	Undergoes reduction
	Becomes more negative (or less positive)

Among the simplest of redox processes is the reaction of an element, usually a metal, with an aqueous cation to yield a different element and a different ion. Iron metal reacts with aqueous copper(II) ion, for example, to give iron(II) ion and copper metal. Similarly, magnesium metal reacts with aqueous acid to yield magnesium ion and hydrogen gas. In both cases, the reactant element (Fe or Mg) is oxidized, and the reactant ion (Cu^{2+} or H^+) is reduced.

$$Fe(s) + Cu^{2+}(aq) \longrightarrow Fe^{2+}(aq) + Cu(s)$$
$$Mg(s) + 2\,H^+(aq) \longrightarrow Mg^{2+}(aq) + H_2(g)$$

The reaction of a metal with water or aqueous acid (H^+) to release H_2 gas is a particularly important process. As you might expect based on the periodic properties discussed in Section 4.2, the alkali metals and alkaline earth metals (on the left side of the periodic table) are the most powerful reducing agents (electron donors), so powerful that they even react with pure water, in which the concentration of H^+ is very low. Ionization energy, which is a measure of how easily an element will lose an electron, tends to decrease as we move to the left and down in the periodic table. In contrast, metals toward the middle of the periodic table, such as iron and chromium, do not lose electrons as readily; they react only with aqueous acids but not with water. Those metals near the bottom right of the periodic table, such as platinum and gold, react with neither aqueous acid nor water. At the other extreme, the reactive nonmetals at the top right of the periodic table are extremely weak reducing agents and instead are powerful oxidizing agents (electron acceptors). This is, again, predictable based on the periodic property of electron affinity (⊂◻◻, Section 4.2) which becomes more energetically favored as we move up and to the right in the periodic table.

We can make a few generalizations about the redox behavior of metals and nonmetals.

1. In reactions involving metals and nonmetals, metals tend to lose electrons while nonmetals tend to gain electrons. The number of electrons lost or gained can often be predicted based on the position of the element in the periodic table. (⊂◻◻, Section 4.5)

2. In reactions involving nonmetals, the "more metallic" element (farther down and/or to the left in the periodic table) tends to lose electrons, and the "less metallic" element (up and/or to the right) tends to gain electrons.

    Redox reactions involve almost every element in the periodic table, and they occur in a vast number of processes throughout nature, biology, and industry. Here are just a few examples:

- **Corrosion** is the deterioration of a metal by oxidation, such as the rusting of iron in moist air. The economic consequences of rusting are enormous: It has been estimated that up to one-fourth of the iron produced in the United States is used to replace bridges, buildings, and other structures that have been destroyed by corrosion. (The raised dot in the formula $Fe_2O_3 \cdot H_2O$ for rust indicates that one water molecule is associated with each Fe_2O_3 in an undefined way.)

- **Combustion** is the burning of a fuel by rapid oxidation with oxygen in air. Gasoline, fuel oil, natural gas, wood, paper, and other organic substances of carbon and hydrogen are the most common fuels that burn in air. Even some metals, though, will burn in air. Magnesium and calcium are examples.

$$CH_4(g) + 2\,O_2(g) \longrightarrow CO_2(g) + 2\,H_2O(l)$$
Methane
(natural gas)
$$2\,Mg(s) + O_2(g) \longrightarrow 2\,MgO(s)$$

▲ Magnesium metal reacts with aqueous acid to give hydrogen gas and Mg^{2+} ion.

- **Respiration** is the process of breathing and using oxygen for the many biological redox reactions that provide the energy that living organisms need. We will see in Chapters 21–22 that energy is released from food molecules slowly and in complex, multistep pathways, but the overall result of respiration is similar to that of combustion reactions. For example, the simple sugar glucose ($C_6H_{12}O_6$) reacts with O_2 to give CO_2 and H_2O according to the following equation:

$$C_6H_{12}O_6 + 6\,O_2 \longrightarrow 6\,CO_2 + 6\,H_2O + \text{Energy}$$
Glucose
(a carbohydrate)

- **Bleaching** makes use of redox reactions to decolorize or lighten colored materials. Dark hair is bleached to turn it blond, clothes are bleached to remove stains, wood pulp is bleached to make white paper, and so on. The oxidizing agent used depends on the situation: hydrogen peroxide (H_2O_2) is used for hair, sodium hypochlorite (NaOCl) for clothes, and elemental chlorine for wood pulp, but the principle is always the same. In all cases, colored organic materials are destroyed by reaction with strong oxidizing agents.
- **Metallurgy,** the science of extracting and purifying metals from their ores, makes use of numerous redox processes. Worldwide, approximately 800 million tons of iron is produced each year by reduction of the mineral hematite, Fe_2O_3, with carbon monoxide.

$$Fe_2O_3(s) + 3\,CO(g) \longrightarrow 2\,Fe(s) + 3\,CO_2(g)$$

WORKED EXAMPLE | **6.18** Chemical Reactions: Redox Reactions

For the following reactions, indicate which atom is oxidized and which is reduced, based on the definitions provided in this section. Identify the oxidizing and reducing agents.

(a) $Cu(s) + Pt^{2+}(aq) \rightarrow Cu^{2+}(aq) + 2\,Pt(s)$
(b) $2\,Mg(s) + CO_2(g) \rightarrow 2\,MgO(s) + C(s)$

ANALYSIS The definitions for oxidation include a loss of electrons, an increase in charge, and a gain of oxygen atoms; reduction is defined as a gain of electrons, a decrease in charge, or a loss of oxygen atoms.

SOLUTION

(a) In this reaction, the charge on the Cu atom increases from 0 to 2+. This corresponds to a loss of 2 electrons. The Cu is therefore oxidized, and acts as the reducing agent. Conversely, each Pt^{2+} ion undergoes a decrease in charge from 2+ to 0, corresponding to a gain of 2 electrons per Pt^{2+} ion. The Pt^{2+} is reduced, and acts as the oxidizing agent.

(b) In this case, the gain or loss of oxygen atoms is the easiest way to identify which atoms are oxidized and reduced. The Mg atom is gaining oxygen to form MgO; therefore, the Mg is being oxidized and acts as the reducing agent. The C atom in CO_2 is losing oxygen. Therefore, the C atom is being reduced and acts as the oxidizing agent.

WORKED EXAMPLE | **6.19** Chemical Reactions: Identifying Oxidizing/ Reducing Agents

For the respiration and metallurgy examples discussed above, identify the atoms being oxidized and reduced, and label the oxidizing and reducing agents.

ANALYSIS Again, using the definitions of oxidation and reduction provided in this section, we can determine which atom(s) are gaining/losing electrons or gaining/losing oxygen atoms.

SOLUTION

$$\textit{Respiration:} \quad C_6H_{12}O_6 + 6\,O_2 \longrightarrow 6\,CO_2 + 6\,H_2O$$

Since the charge associated with the individual atoms is not evident, we will use the definition of oxidation/reduction as the gaining/losing of oxygen atoms. In this reaction, there is only one reactant besides oxygen ($C_6H_{12}O_6$), so we must determine *which* atom in the compound is changing. The ratio of carbon to oxygen in $C_6H_{12}O_6$ is 1:1, while the ratio in CO_2 is 1:2. Therefore, the C atoms are gaining oxygen and are oxidized; the C is the reducing agent. The O_2 is the oxidizing agent. Note that the ratio of hydrogen to oxygen in $C_6H_{12}O_6$ and in H_2O is 2:1. The H atoms are neither oxidized nor reduced.

$$\textit{Metallurgy:} \quad Fe_2O_3(s) + 3\,CO(g) \longrightarrow 2\,Fe(s) + 3\,CO_2(g)$$

The Fe_2O_3 is losing oxygen to form Fe(s); it is being reduced, and acts as the oxidizing agent. In contrast, the CO is gaining oxygen to form CO_2; it is being oxidized and acts as the reducing agent.

WORKED EXAMPLE **6.20** Chemical Reactions: Identifying Redox Reactions

For the following reactions, identify the atom(s) being oxidized and reduced:

(a) $2\,Al(s) + 3\,Cl_2(g) \longrightarrow 2\,AlCl_3(s)$

(b) $C(s) + 2\,Cl_2(g) \longrightarrow CCl_4(l)$

ANALYSIS Again, there is no obvious increase or decrease in charge to indicate a gain or loss of electrons. Also, the reactions do not involve a gain or loss of oxygen. We can, however, evaluate the reactions in terms of the typical behavior of metals and nonmetals in reactions.

SOLUTION

(a) In this case, we have the reaction of a metal (Al) with a nonmetal (Cl_2). Since metals tend to lose electrons and nonmetals tend to gain electrons, we can assume that the Al atom is oxidized (loss of electrons), and the Cl_2 is reduced (gains electrons).

(b) In this case, we have a reaction involving two nonmetals. The carbon is the more metallic element (farther to the left) and is more likely to lose electrons (oxidized). The less metallic element (Cl) will tend to gain electrons and be reduced.

PROBLEM 6.23

Identify the oxidized reactant, the reduced reactant, the oxidizing agent, and the reducing agent in the following reactions:

(a) $Fe(s) + Cu^{2+}(aq) \longrightarrow Fe^{2+}(aq) + Cu(s)$

(b) $Mg(s) + Cl_2(g) \longrightarrow MgCl_2(s)$

(c) $2\,Al(s) + Cr_2O_3(s) \longrightarrow 2\,Cr(s) + Al_2O_3(s)$

PROBLEM 6.24

Potassium, a silvery metal, reacts with bromine, a corrosive, reddish liquid, to yield potassium bromide, a white solid. Write the balanced equation, and identify the oxidizing and reducing agents.

APPLICATION ▶ Batteries

Imagine life without batteries: no cars (they do not start very easily without their batteries!), no heart pacemakers, no flashlights, no hearing aids, no portable computers, radios, cellular phones, or thousands of other things. Modern society cannot exist without batteries.

Although they come in many types and sizes, all batteries are based on redox reactions. In a typical redox reaction carried out in the laboratory—say, the reaction of zinc metal with Ag^+ to yield Zn^{2+} and silver metal—the reactants are simply mixed in a flask and electrons are transferred by direct contact between the reactants. In a battery, however, the two reactants are kept in separate compartments and the electrons are transferred through a wire running between them.

▲ Think of all the devices we use every day–laptop computers, cell phones, iPods—that depend on batteries.

The common household battery used for flashlights and radios is the *dry-cell*, developed in 1866. One reactant is a can of zinc metal, and the other is a paste of solid manganese dioxide. A graphite rod sticks into the MnO_2 paste to provide electrical contact, and a moist paste of ammonium chloride separates the two reactants. When the zinc can and the graphite rod are connected by a wire, zinc sends electrons flowing through the wire toward the MnO_2 in a redox reaction. The resultant electrical current can then be used to light a bulb or power a radio. The accompanying figure shows a cutaway view of a dry-cell battery.

$$Zn(s) + 2\ MnO_2(s) + 2\ NH_4Cl(s) \longrightarrow$$
$$ZnCl_2(aq) + Mn_2O_3(s) + 2\ NH_3(aq) + H_2O(l)$$

Closely related to the dry-cell battery is the familiar *alkaline* battery, in which the ammonium chloride paste is replaced by an alkaline, or basic, paste of NaOH or KOH. The alkaline battery has a longer life than the standard dry-cell battery because the zinc container corrodes less easily under basic conditions. The redox reaction is

$$Zn(s) + 2\ MnO_2(s) \longrightarrow ZnO(aq) + Mn_2O_3(s)$$

The batteries used in implanted medical devices such as pacemakers must be small, corrosion-resistant, reliable, and able to last up to 10 years. Nearly all pacemakers being implanted today—about 750,000 each year—use titanium-encased, lithium-iodine batteries, whose redox reaction is

$$2\ Li(s) + I_2(s) \longrightarrow 2\ LiI(aq)$$

See Additional Problem 6.96 at the end of the chapter.

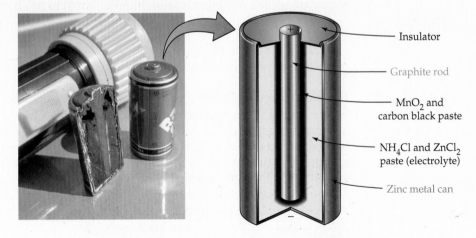

Insulator

Graphite rod

MnO_2 and carbon black paste

NH_4Cl and $ZnCl_2$ paste (electrolyte)

Zinc metal can

▲ A dry-cell battery. The cutaway view shows the two reactants that make up the redox reaction.

▲ Sulfur burns in air to yield SO_2. Is this a redox reaction?

Oxidation number A number that indicates whether an atom is neutral, electron-rich, or electron-poor.

6.12 Recognizing Redox Reactions

How can you tell when a redox reaction is taking place? When ions are involved, it is simply a matter of determining whether there is a change in the charges. For reactions involving metals and nonmetals, we can predict gain or loss of electrons as discussed previously. When molecular substances are involved, though, it is not as obvious. Is the combining of sulfur with oxygen a redox reaction? If so, which partner is the oxidizing agent and which is the reducing agent?

$$S(s) + O_2(g) \longrightarrow SO_2(g)$$

One way to evaluate this reaction is in terms of the oxygen gain by sulfur, indicating that S atoms are oxidized and O atoms are reduced. But can we also look at this reaction in terms of the gain or loss of electrons by the S and O atoms? Because oxygen is more electronegative than sulfur (Section 5.8), the oxygen atoms in SO_2 attract the electrons in the S—O bonds more strongly than sulfur does, giving the oxygen atoms a larger share of the electrons than sulfur. (⬤▢ , p. 129) By extending the ideas of oxidation and reduction to an increase or decrease in electron *sharing* instead of complete electron *transfer*, we can say that the sulfur atom is oxidized in its reaction with oxygen because it loses a share in some electrons, whereas the oxygen atoms are reduced because they gain a share in some electrons.

A formal system has been devised for keeping track of changes in electron sharing, and thus for determining whether atoms are oxidized or reduced in reactions. To each atom in a substance, we assign a value called an **oxidation number** (or *oxidation state*), which indicates whether the atom is neutral, electron-rich, or electron-poor. By comparing the oxidation number of an atom before and after reaction, we can tell whether the atom has gained or lost shares in electrons. Note that *oxidation numbers do not necessarily imply ionic charges*. They are simply a convenient device for keeping track of electrons in redox reactions.

The rules for assigning oxidation numbers are straightforward:

- **An atom in its elemental state has an oxidation number of 0.**

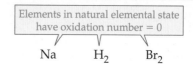

- **A monatomic ion has an oxidation number equal to its charge.**

- **In a molecular compound, an atom usually has the same oxidation number it would have if it were a monatomic ion.** Recall from Chapters 4 and 5 that the less electronegative elements (hydrogen and metals) on the left side of the periodic table tend to form cations, and the more electronegative elements (oxygen, nitrogen, and the halogens) near the top right of the periodic table tend to form anions. (⬤▢ , p. 88) Hydrogen and metals therefore have positive oxidation numbers in most compounds, whereas reactive nonmetals generally have negative oxidation numbers. Hydrogen is usually +1, oxygen is usually −2, nitrogen is usually −3, and halogens are usually −1:

$$\overset{+1}{H}-\overset{-1}{Cl} \qquad \overset{+1}{H}-\overset{-2}{O}-\overset{+1}{H} \qquad \overset{+1}{H}-\overset{-3}{N}-\overset{+1}{H}$$
$$\overset{\;}{H} \leftarrow +1$$

For compounds with more than one nonmetal element, such as SO_2, NO, or CO_2, the more electronegative element—oxygen in these examples—has a negative oxidation number and the less electronegative element has a positive oxidation number. Thus, in answer to the question posed at the beginning of this section, combining sulfur with oxygen to form SO_2 is a redox reaction because the oxidation number of sulfur increases from 0 to +4 and that of oxygen decreases from 0 to −2.

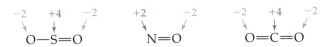

- **The sum of the oxidation numbers in a neutral compound is 0.** Using this rule, the oxidation number of any atom in a compound can be found if the oxidation numbers of the other atoms are known. In the SO_2 example just mentioned, each of the two O atoms has an oxidation number of −2, so the S atom must have an oxidation number of +4. In HNO_3, the H atom has an oxidation number of +1 and the strongly electronegative O atom has an oxidation number of −2, so the N atom must have an oxidation number of +5. In a polyatomic ion, the sum of the oxidation numbers equals the charge on the ion.

$$
\begin{array}{ccc}
+1 & -2 \ +5 & -2 \\
\searrow & \downarrow \downarrow & \swarrow \\
\mathrm{H-O-N=O} & & \text{Total} = 1 + 5 + 3(-2) = 0 \\
& \underset{-2}{\overset{|}{\mathrm{O}}} &
\end{array}
$$

Worked Examples 6.21 and 6.22 show further instances of assigning and using oxidation numbers.

WORKED EXAMPLE **6.21** Redox Reactions: Oxidation Numbers

What is the oxidation number of the titanium atom in $TiCl_4$? Name the compound using a Roman numeral (Section 4.11).

SOLUTION
Chlorine, a reactive nonmetal, is more electronegative than titanium and has an oxidation number of −1. Because there are four chlorine atoms in $TiCl_4$, the oxidation number of titanium must be +4. The compound is named titanium(IV) chloride. Note that the Roman numeral IV in the name of this molecular compound refers to the oxidation number +4 rather than to a true ionic charge.

WORKED EXAMPLE **6.22** Redox Reactions: Identifying Redox Reactions

Use oxidation numbers to show that the production of iron metal from its ore (Fe_2O_3) by reaction with charcoal (C) is a redox reaction. Which reactant has been oxidized, and which has been reduced? Which reactant is the oxidizing agent, and which is the reducing agent?

$$2\,Fe_2O_3(s) + 3\,C(s) \longrightarrow 4\,Fe(s) + 3\,CO_2(g)$$

SOLUTION
The idea is to assign oxidation numbers to both reactants and products, and see if there has been a change. In the production of iron from Fe_2O_3, the oxidation number of Fe changes from +3 to 0, and the oxidation number of C changes from 0 to +4. Iron has thus been reduced (decrease in oxidation number), and carbon has been oxidized (increase in oxidation number). Oxygen is neither oxidized nor reduced because its oxidation number does not change. Carbon is the reducing agent and Fe_2O_3 is the oxidizing agent.

$$
\begin{array}{ccccc}
+3 & -2 & 0 & 0 & +4 \ -2 \\
\searrow & \downarrow & \downarrow & \searrow & \downarrow \swarrow \\
2\,Fe_2O_3 & + & 3\,C & \longrightarrow & 4\,Fe + 3\,CO_2
\end{array}
$$

PROBLEM 6.25

What are the oxidation numbers of the metal atoms in the following compounds? Name each, using the oxidation number as a Roman numeral.

(a) VCl_3 **(b)** $SnCl_4$ **(c)** CrO_3 **(d)** $Cu(NO_3)_2$ **(e)** $NiSO_4$

PROBLEM 6.26

Assign an oxidation number to each atom in the reactants and products shown here to determine which of the following reactions are redox reactions:

(a) $Na_2S(aq) + NiCl_2(aq) \longrightarrow 2\,NaCl(aq) + NiS(s)$
(b) $2\,Na(s) + 2\,H_2O(l) \longrightarrow 2\,NaOH(aq) + H_2(g)$
(c) $C(s) + O_2(g) \longrightarrow CO_2(g)$
(d) $CuO(s) + 2\,HCl(aq) \longrightarrow CuCl_2(aq) + H_2O(l)$
(e) $2\,MnO_4^-(aq) + 5\,SO_2(g) + 2\,H_2O(l) \longrightarrow 2\,Mn^{2+}(aq) + 5\,SO_4^{2-}(aq) + 4\,H^+(aq)$

6.13 Net Ionic Equations

In the equations we have been writing up to this point, all the substances involved in reactions have been written using their full formulas. In the precipitation reaction of lead(II) nitrate with potassium iodide mentioned in Section 6.8, for example, only the parenthetical (*aq*) indicated that the reaction actually takes place in aqueous solution, and nowhere was it explicitly indicated that ions are involved:

$$Pb(NO_3)_2(aq) + 2\,KI(aq) \longrightarrow 2\,KNO_3(aq) + PbI_2(s)$$

In fact, lead(II) nitrate, potassium iodide, and potassium nitrate dissolve in water to yield solutions of ions. Thus, it is more accurate to write the reaction as an **ionic equation**, in which all the ions are explicitly shown:

Ionic equation An equation in which ions are explicitly shown.

An ionic equation: $Pb^{2+}(aq) + 2\,NO_3^-(aq) + 2\,K^+(aq) + 2\,I^-(aq) \longrightarrow$
$$2\,K^+(aq) + 2\,NO_3^-(aq) + PbI_2(s)$$

A look at this ionic equation shows that the NO_3^- and K^+ ions undergo no change during the reaction. They appear on both sides of the reaction arrow and act merely as **spectator ions**, that is, they are present but play no role. The actual reaction, when stripped to its essentials, can be described more simply by writing a **net ionic equation**, which includes only the ions that undergo change and ignores all spectator ions:

Spectator ion An ion that appears unchanged on both sides of a reaction arrow.

Net ionic equation An equation that does not include spectator ions.

Ionic equation: $Pb^{2+}(aq) + 2\,\cancel{NO_3^-}(aq) + 2\,\cancel{K^+}(aq) + 2\,I^-(aq) \longrightarrow$
$$2\,\cancel{K^+}(aq) + 2\,\cancel{NO_3^-}(aq) + PbI_2(s)$$

Net Ionic equation: $Pb^{2+}(aq) + 2\,I^-(aq) \longrightarrow PbI_2(s)$

Note that a net ionic equation, like all chemical equations, must be balanced both for atoms and for charge, with all coefficients reduced to their lowest whole numbers. Note also that all compounds that do *not* give ions in solution—all insoluble compounds and all molecular compounds—are represented by their full formulas.

We can apply the concept of ionic equations to acid–base neutralization reactions and redox reactions as well. Consider the neutralization reaction between KOH and HNO_3:

$$KOH(aq) + HNO_3(aq) \longrightarrow H_2O(l) + KNO_3(aq)$$

Since acids and bases are identified based on the ions they form when dissolved in aqueous solutions, we can write an ionic equation for this reaction:

Ionic equation: $K^+(aq) + OH^-(aq) + H^+(aq) + NO_3^-(aq) \longrightarrow$
$$H_2O(l) + K^+(aq) + NO_3^-(aq)$$

Eliminating the spectator ions (K^+ and NO_3^-) we obtain the net ionic equation for the neutralization reaction:

Net ionic equation: $\quad OH^-(aq) + H^+(aq) \longrightarrow H_2O(l)$

The net ionic equation confirms the basis of the acid–base neutralization; the OH^- from the base and the H^+ from the acid neutralize each other to form water.

Similarly, many redox reactions can be viewed in terms of ionic equations. Consider the reaction between $Cu(s)$ and $AgNO_3$ from Section 6.11:

$$Cu(s) + 2\,AgNO_3(aq) \longrightarrow 2\,Ag^+(aq) + Cu(NO_3)_2(aq)$$

The aqueous products and reactants can be written as dissolved ions:

Ionic equation: $\quad Cu(s) + 2\,Ag^+(aq) + 2\,NO_3^-(aq) \longrightarrow$
$$2\,Ag(s) + Cu^{2+}(aq) + 2\,NO_3^-(aq)$$

Again, eliminating the spectator ions (NO_3^-), we obtain the net ionic equation for this redox reaction:

Net ionic equation: $\quad Cu(s) + 2\,Ag^+(aq) \longrightarrow 2\,Ag(s) + Cu^{2+}(aq)$

It is now clear that the $Cu(s)$ loses two electrons and is oxidized, whereas each Ag^+ ion gains an electron and is reduced.

WORKED EXAMPLE　**6.23** Chemical Reactions: Net Ionic Reactions

Write balanced net ionic equations for the following reactions:

(a) $AgNO_3(aq) + ZnCl_2(aq) \longrightarrow$
(b) $HCl(aq) + Ca(OH)_2(aq) \longrightarrow$
(c) $6\,HCl(aq) + 2\,Al(s) \longrightarrow 2\,AlCl_3(aq) + 3\,H_2(g)$

SOLUTION

(a) The solubility guidelines discussed in Section 6.9 predict that a precipitate of insoluble AgCl forms when aqueous solutions of Ag^+ and Cl^- are mixed. Writing all the ions separately gives an ionic equation, and eliminating spectator ions Zn^{2+} and NO_3^- gives the net ionic equation.

Ionic equation: $2\,Ag^+(aq) + 2\,NO_3^-(aq) + Zn^{2+}(aq) + 2\,Cl^-(aq) \longrightarrow$
$$2\,AgCl(s) + Zn^{2+}(aq) + 2\,NO_3^-(aq)$$

Net ionic equation: $2\,Ag^+(aq) + 2\,Cl^-(aq) \longrightarrow 2\,AgCl(s)$

The coefficients can all be divided by 2 to give

Net ionic equation: $Ag^+(aq) + Cl^-(aq) \longrightarrow AgCl(s)$

A check shows that the equation is balanced for atoms and charge (zero on each side).

(b) Allowing the acid HCl to react with the base $Ca(OH)_2$ leads to a neutralization reaction. Writing the ions separately, and remembering to write a complete formula for water, gives an ionic equation. Then eliminating the spectator ions and dividing the coefficients by 2 gives the net ionic equation.

Ionic equation: $2\,H^+(aq) + \cancel{2\,Cl^-(aq)} + \cancel{Ca^{2+}(aq)} + 2\,OH^-(aq) \longrightarrow$
$2\,H_2O(l) + \cancel{Ca^{2+}(aq)} + \cancel{2\,Cl^-(aq)}$

Net ionic equation: $H^+(aq) + OH^-(aq) \longrightarrow H_2O(l)$

A check shows that atoms and charges are the same on both sides of the equation.

(c) The reaction of Al metal with acid (HCl) is a redox reaction. The Al is oxidized, since the oxidation number increases from $0 \rightarrow +3$, whereas the H in HCl is reduced from $+1 \rightarrow 0$. We write the ionic equation by showing the ions that are formed for each aqueous ionic species. Eliminating the spectator ions yields the net ionic equation.

Ionic equation: $6\,H^+(aq) + \cancel{6\,Cl^-(aq)} + 2\,Al(s) \longrightarrow$
$2\,Al^{3+}(aq) + \cancel{6\,Cl^-(aq)} + 3\,H_2(g)$

Net ionic equation: $6\,H^+(aq) + 2\,Al(s) \longrightarrow 2\,Al^{3+}(aq) + 3\,H_2(g)$

A check shows that atoms and charges are the same on both sides of the equation.

PROBLEM 6.27

Write net ionic equations for the following reactions:

(a) $Zn(s) + Pb(NO_3)_2(aq) \longrightarrow Zn(NO_3)_2(aq) + Pb(s)$
(b) $2\,KOH(aq) + H_2SO_4(aq) \longrightarrow K_2SO_4(aq) + 2\,H_2O(l)$
(c) $2\,FeCl_3(aq) + SnCl_2(aq) \longrightarrow 2\,FeCl_2(aq) + SnCl_4(aq)$

SUMMARY: REVISITING THE CHAPTER GOALS

1. How are chemical reactions written? Chemical equations must be *balanced;* that is, the numbers and kinds of atoms must be the same in both the reactants and the products. To balance an equation, *coefficients* are placed before formulas but the formulas themselves cannot be changed.

2. What is the mole, and why is it useful in chemistry? A *mole* refers to *Avogadro's number* (6.022×10^{23}) of formula units of a substance. One mole of any substance has a mass (a *molar mass*) equal to the molecular or formula weight of the substance in grams. Because equal numbers of moles contain equal numbers of formula units, molar masses act as conversion factors between numbers of molecules and masses in grams.

3. How are molar quantities and mass quantities related? The coefficients in a balanced chemical equation represent the numbers of moles of reactants and products in a reaction. Thus, the ratios of coefficients act as *mole ratios* that relate amounts of reactants and/or products. By using molar masses and mole ratios in factor-label calculations, unknown masses or molar amounts can be found from known masses or molar amounts.

4. What are the limiting reagent, theoretical yield, and percent yield of a reaction? The *limiting reagent* is the reactant that runs out first. The *theoretical yield* is the amount of product that would be formed based on the amount of the limiting reagent. The *actual yield* of a reaction is the amount of product obtained. The *percent yield* is the amount of product obtained divided by the amount theoretically possible and multiplied by 100%.

KEY WORDS

Actual yield, *p. 159*
Avogadro's number (N_A), *p. 150*
Balanced equation, *p. 144*
Chemical equation, *p. 143*
Coefficient, *p. 144*
Formula weight, *p. 148*
Ionic equation, *p. 174*
Law of conservation of mass, *p. 144*
Limiting reagent, *p. 159*
Molar mass, *p. 152*
Mole, *p. 150*
Molecular weight (MW), *p. 148*
Net ionic equation, *p. 174*
Neutralization reaction, *p. 165*

5. **How are chemical reactions of ionic compounds classified?** There are three common types of reactions of ionic compounds: *Precipitation reactions* are processes in which an insoluble solid called a *precipitate* is formed. Most precipitations take place when the anions and cations of two ionic compounds change partners. Solubility guidelines for ionic compounds are used to predict when precipitation will occur.

Acid–base neutralization reactions are processes in which an acid reacts with a base to yield water plus an ionic compound called a *salt*. Since acids produce H^+ ions and bases produce OH^- ions when dissolved in water, a neutralization reaction removes H^+ and OH^- ions from solution and yields neutral H_2O.

Oxidation–reduction (redox) reactions are processes in which one or more electrons are transferred between reaction partners. An *oxidation* is defined as the loss of one or more electrons by an atom, and a *reduction* is the gain of one or more electrons. An *oxidizing agent* causes the oxidation of another reactant by accepting electrons, and a *reducing agent* causes the reduction of another reactant by donating electrons.

6. **What are oxidation numbers, and how are they used?** *Oxidation numbers* are assigned to atoms in reactants and products to provide a measure of whether an atom is neutral, electron-rich, or electron-poor. By comparing the oxidation number of an atom before and after reaction, we can tell whether the atom has gained or lost shares in electrons and thus whether a redox reaction has occurred.

UNDERSTANDING KEY CONCEPTS

6.28   Assume that the mixture of substances in drawing (a) undergoes a reaction. Which of the drawings (b)–(d) represents a product mixture consistent with the law of conservation of mass?

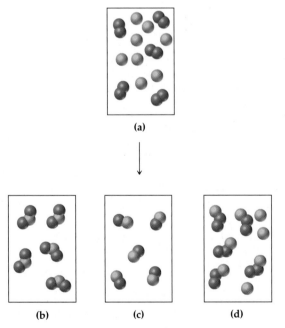

6.29   Reaction of A (green spheres) with B (blue spheres) is shown in the following diagram:

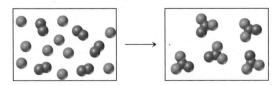

Which equation best describes the reaction?

(a) $A_2 + 2B \longrightarrow A_2B$ (b) $10A + 5B_2 \longrightarrow 5A_2B_2$
(c) $2A + B_2 \longrightarrow A_2B$ (d) $5A + 5B_2 \longrightarrow 5A_2B_2$

6.30   If blue spheres represent nitrogen atoms and red spheres represent oxygen atoms in the following diagrams, which box represents reactants and which represents products for the reaction $2NO(g) + O_2(g) \longrightarrow 2NO_2(g)$?

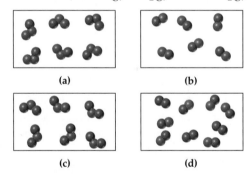

6.31   Methionine, an amino acid used by organisms to make proteins, can be represented by the following ball-and-stick molecular model. Write the formula for methionine, and give its molecular weight (red = O, gray = C, blue = N, yellow = S, ivory = H).

Methionine

6.32   The following diagram represents the reaction of A_2 (red spheres) with B_2 (blue spheres):

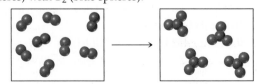

(a) Write a balanced equation for the reaction.

(b) How many moles of product can be made from 1.0 mol of A_2? From 1.0 mol of B_2?

6.33 Assume that an aqueous solution of a cation (represented as red spheres in the diagram) is allowed to mix with a solution of an anion (represented as yellow spheres). Three possible outcomes are represented by boxes (1)–(3):

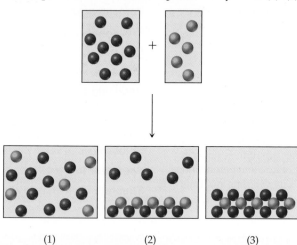

(1)   (2)   (3)

Which outcome corresponds to each of the following reactions?

(a) $2\,Na^+(aq) + CO_3{}^{2-}(aq) \longrightarrow$

(b) $Ba^{2+}(aq) + CrO_4{}^{2-}(aq) \longrightarrow$

(c) $2\,Ag^+(aq) + SO_3{}^{2-}(aq) \longrightarrow$

6.34 An aqueous solution of a cation (represented as blue spheres in the diagram) is allowed to mix with a solution of an anion (represented as green spheres) and the following result is obtained:

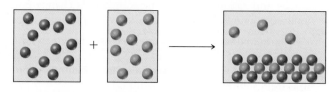

Which combinations of cation and anion, chosen from the following lists, are compatible with the observed results? Explain.

Cations: Na^+, Ca^{2+}, Ag^+, Ni^{2+}

Anions: Cl^-, $CO_3{}^{2-}$, $CrO_4{}^{2-}$, $NO_3{}^-$

6.35 The following drawing represents the reaction of ethylene oxide with water to give ethylene glycol, a compound used as automobile antifreeze. What mass in grams of ethylene oxide is needed to react with 9.0 g of water, and what mass in grams of ethylene glycol is formed?

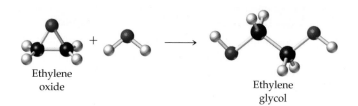

Ethylene oxide

Ethylene glycol

ADDITIONAL PROBLEMS

BALANCING CHEMICAL EQUATIONS

6.36 What is meant by the term "balanced equation"?

6.37 Why is it not possible to balance an equation by changing the subscript on a substance, say from H_2O to H_2O_2?

6.38 Write balanced equations for the following reactions:

(a) Gaseous sulfur dioxide reacts with water to form aqueous sulfurous acid (H_2SO_3).

(b) Liquid bromine reacts with solid potassium metal to form solid potassium bromide.

(c) Gaseous propane (C_3H_8) burns in oxygen to form gaseous carbon dioxide and water vapor.

6.39 Balance the following equation for the synthesis of hydrazine, N_2H_4, a substance used as rocket fuel.

$$NH_3(g) + Cl_2(g) \longrightarrow N_2H_4(l) + NH_4Cl(s)$$

6.40 Which of the following equations are balanced? Balance those that need it.

(a) $2\,C_2H_6(g) + 5\,O_2(g) \longrightarrow 2\,CO_2(g) + 6\,H_2O(l)$

(b) $3\,Ca(OH)_2(aq) + 2\,H_3PO_4(aq) \longrightarrow$
$$Ca_3(PO_4)_2(aq) + 6\,H_2O(l)$$

(c) $Mg(s) + O_2(g) \longrightarrow 2\,MgO(s)$

(d) $K(s) + H_2O(l) \longrightarrow KOH(aq) + H_2(g)$

6.41 Which of the following equations are balanced? Balance those that need it.

(a) $CaC_2 + 2\,H_2O \longrightarrow Ca(OH)_2 + C_2H_2$

(b) $C_2H_8N_2 + 2\,N_2O_4 \longrightarrow 2\,N_2 + 2\,CO_2 + 4\,H_2O$

(c) $3\,MgO + 2\,Fe \longrightarrow Fe_2O_3 + 3\,Mg$

(d) $N_2O \longrightarrow N_2 + O_2$

6.42 Balance the following equations:

(a) $Hg(NO_3)_2(aq) + LiI(aq) \longrightarrow LiNO_3(aq) + HgI_2(s)$

(b) $I_2(s) + Cl_2(g) \longrightarrow ICl_5(s)$

(c) $Al(s) + O_2(g) \longrightarrow Al_2O_3(s)$

(d) $CuSO_4(aq) + AgNO_3(aq) \longrightarrow$
$$Ag_2SO_4(s) + Cu(NO_3)_2(aq)$$

(e) $Mn(NO_3)_3(aq) + Na_2S(aq) \longrightarrow$
$$Mn_2S_3(s) + NaNO_3(aq)$$

(f) $NO_2(g) + O_2(g) \longrightarrow N_2O_5(g)$

(g) $P_4O_{10}(s) + H_2O(l) \longrightarrow H_3PO_4(aq)$

6.43 Write a balanced equation for the reaction of aqueous sodium carbonate (Na_2CO_3) with aqueous nitric acid (HNO_3) to yield CO_2, $NaNO_3$, and H_2O.

6.44 When organic compounds are burned, they react with oxygen to form CO_2 and H_2O. Write balanced equations for the combustion of the following:

(a) C_4H_{10} (butane, used in lighters)

(b) C_2H_6O (ethyl alcohol, used in gasohol and as race car fuel)

(c) C_8H_{18} (octane, a component of gasoline)

MOLAR MASSES AND MOLES

6.45 What is a mole of a substance? How many molecules are in 1 mol of a molecular compound?

6.46 What is the difference between molecular weight and formula weight? Between molecular weight and molar mass?

6.47 How many Na^+ ions are in a mole of Na_2SO_4? How many SO_4^{2-} ions?

6.48 How many moles of ions are in 1.75 mol of K_2SO_4?

6.49 How many calcium atoms are in 16.2 g of calcium?

6.50 What is the mass in grams of 2.68×10^{22} atoms of uranium?

6.51 Calculate the molar mass of each of the following compounds:

(a) Calcium carbonate, $CaCO_3$
(b) Urea, $CO(NH_2)_2$
(c) Ethylene glycol, $C_2H_6O_2$

6.52 How many moles of carbon atoms are there in 1 mol of each compound in Problem 6.51?

6.53 How many atoms of carbon and how many grams of carbon are there in 1 mol of each compound in Problem 6.51?

6.54 Caffeine has the formula $C_8H_{10}N_4O_2$. If an average cup of coffee contains approximately 125 mg of caffeine, how many moles of caffeine are in one cup?

6.55 How many moles of aspirin, $C_9H_8O_4$, are in a 500 mg tablet?

6.56 What is the molar mass of diazepam (Valium), $C_{16}H_{13}ClN_2O$?

6.57 Calculate the molar masses of the following substances:

(a) Aluminum sulfate, $Al_2(SO_4)_3$
(b) Sodium bicarbonate, $NaHCO_3$
(c) Diethyl ether, $(C_2H_5)_2O$
(d) Penicillin V, $C_{16}H_{18}N_2O_5S$

6.58 How many moles are present in a 4.50 g sample of each compound listed in Problem 6.57?

6.59 The recommended daily dietary intake of calcium for adult men and pre-menopausal women is 1000 mg/day. Calcium citrate , $Ca_3(C_6H_5O_7)_2$ (MW = 498.5 g/mol), is a common dietary supplement. What mass of calcium citrate would be needed to provide the recommended daily intake of calcium?

6.60 What is the mass in grams of 0.0015 mol of aspirin, $C_9H_8O_4$? How many aspirin molecules are there in this 0.0015 mol sample?

6.61 How many grams are present in a 0.075 mol sample of each compound listed in Problem 6.57?

6.62 The principal component of many kidney stones is calcium oxalate, CaC_2O_4. A kidney stone recovered from a typical patient contains 8.5×10^{20} formula units of calcium oxalate. How many moles of CaC_2O_4 are present in this kidney stone? What is the mass of the kidney stone in grams?

MOLE AND MASS RELATIONSHIPS FROM CHEMICAL EQUATIONS

6.63 At elevated temperatures in an automobile engine, N_2 and O_2 can react to yield NO, an important cause of air pollution.

(a) Write a balanced equation for the reaction.
(b) How many moles of N_2 are needed to react with 7.50 mol of O_2?
(c) How many moles of NO can be formed when 3.81 mol of N_2 reacts?

(d) How many moles of O_2 must react to produce 0.250 mol of NO?

6.64 Ethyl acetate reacts with H_2 in the presence of a catalyst to yield ethyl alcohol:

$$C_4H_8O_2(l) + H_2(g) \longrightarrow C_2H_6O(l)$$

(a) Write a balanced equation for the reaction.
(b) How many moles of ethyl alcohol are produced by reaction of 1.5 mol of ethyl acetate?
(c) How many grams of ethyl alcohol are produced by reaction of 1.5 mol of ethyl acetate with H_2?
(d) How many grams of ethyl alcohol are produced by reaction of 12.0 g of ethyl acetate with H_2?
(e) How many grams of H_2 are needed to react with 12.0 g of ethyl acetate?

6.65 The active ingredient in Milk of Magnesia (an antacid) is magnesium hydroxide, $Mg(OH)_2$. A typical dose (one tablespoon) contains 1.2 g of $Mg(OH)_2$. Calculate (a) the molar mass of magnesium hydroxide, and (b) the amount of magnesium hydroxide (in moles) in one teaspoon.

6.66 Ammonia, NH_3, is prepared for use as a fertilizer by reacting N_2 with H_2.

(a) Write a balanced equation for the reaction.
(b) How many moles of N_2 are needed for reaction to make 16.0 g of NH_3?
(c) How many grams of H_2 are needed to react with 75.0 g of N_2?

6.67 Hydrazine, N_2H_4, a substance used as rocket fuel, reacts with oxygen as follows:

$$N_2H_4(l) + O_2(g) \longrightarrow NO_2(g) + H_2O(g)$$

(a) Balance the equation.
(b) How many moles of oxygen are needed to react with 165 g of hydrazine?
(c) How many grams of oxygen are needed to react with 165 g of hydrazine?

6.68 One method for preparing pure iron from Fe_2O_3 is by reaction with carbon monoxide:

$$Fe_2O_3(s) + CO(g) \longrightarrow Fe(s) + CO_2(g)$$

(a) Balance the equation.
(b) How many grams of CO are needed to react with 3.02 g of Fe_2O_3?
(c) How many grams of CO are needed to react with 1.68 mol of Fe_2O_3?

6.69 Magnesium metal burns in oxygen to form magnesium oxide, MgO.

(a) Write a balanced equation for the reaction.
(b) How many grams of oxygen are needed to react with 25.0 g of Mg? How many grams of MgO will result?
(c) How many grams of Mg are needed to react with 25.0 g of O_2? How many grams of MgO will result?

6.70 Titanium metal is obtained from the mineral rutile, TiO_2. How many kilograms of rutile are needed to produce 95 kg of Ti?

6.71 In the preparation of iron from hematite (Problem 6.68) how many moles of carbon monoxide are needed to react completely with 105 kg of Fe_2O_3.

LIMITING REAGENT AND PERCENT YIELD

6.72 Once made by heating wood in the absence of air, methanol (CH_3OH) is now made by reacting carbon monoxide and hydrogen at high pressure:

$$CO(g) + 2 H_2(g) \longrightarrow CH_3OH(l)$$

(a) If 25.0 g of CO is reacted with 6.00 g of H_2, which is the limiting reagent?

(b) How many grams of CH_3OH can be made from 10.0 g of CO if it all reacts?

(c) If 9.55 g of CH_3OH is recovered when the amounts in part (b) are used, what is the percent yield?

6.73 In Problem 6.67 hydrazine reacted with oxygen according to the (unbalanced) equation:

$$N_2H_4(l) + O_2(g) \longrightarrow NO_2(g) + H_2O(g)$$

(a) If 75.0 kg of hydrazine are reacted with 75.0 kg of oxygen, which is the limiting reagent?

(b) How many kilograms of NO_2 are produced from the reaction of 75.0 kg of the limiting reagent?

(c) If 59.3 kg of NO_2 are obtained from the reaction in part (a), what is the percent yield?

6.74 Dichloromethane, CH_2Cl_2, the solvent used to decaffeinate coffee beans, is prepared by reaction of CH_4 with Cl_2.

(a) Write the balanced equation. (HCl is also formed.)

(b) How many grams of Cl_2 are needed to react with 50.0 g of CH_4?

(c) How many grams of dichloromethane are formed from 50.0 g of CH_4 if the percent yield for the reaction is 76%?

6.75 Cisplatin [$Pt(NH_3)_2Cl_2$], a compound used in cancer treatment, is prepared by reaction of ammonia with potassium tetrachloroplatinate:

$$K_2PtCl_4 + 2 NH_3 \longrightarrow 2 KCl + Pt(NH_3)_2Cl_2$$

(a) How many grams of NH_3 are needed to react with 55.8 g of K_2PtCl_4?

(b) How many grams of cisplatin are formed from 55.8 g of K_2PtCl_4 if the percent yield for the reaction is 95%?

TYPES OF CHEMICAL REACTIONS

6.76 Identify the following reactions as a precipitation, neutralization, or redox reaction:

(a) $Mg(s) + 2 HCl(aq) \longrightarrow MgCl_2(aq) + H_2(g)$

(b) $KOH(aq) + HNO_3(aq) \longrightarrow KNO_3(aq) + H_2O(l)$

(c) $Pb(NO_3)_2(aq) + 2 HBr(aq) \longrightarrow$ $PbBr_2(s) + 2 HNO_3(aq)$

(d) $Ca(OH)_2(aq) + 2 HCl(aq) \longrightarrow 2 H_2O(l) + CaCl_2(aq)$

6.77 Write balanced ionic equations and net ionic equations for the following reactions:

(a) Aqueous sulfuric acid is neutralized by aqueous potassium hydroxide.

(b) Aqueous magnesium hydroxide is neutralized by aqueous hydrochloric acid.

6.78 Write balanced ionic equations and net ionic equations for the following reactions:

(a) A precipitate of barium sulfate forms when aqueous solutions of barium nitrate and potassium sulfate are mixed.

(b) Zinc ion and hydrogen gas form when zinc metal reacts with aqueous sulfuric acid.

6.79 Which of the following substances are likely to be soluble in water?

(a) $ZnSO_4$ **(b)** $NiCO_3$

(c) $PbCl_2$ **(d)** $Ca_3(PO_4)_2$

6.80 Which of the following substances are likely to be soluble in water?

(a) Ag_2O **(b)** $Ba(NO_3)_2$

(c) $SnCO_3$ **(d)** Al_2S_3

6.81 Use the solubility guidelines in Section 6.9 to predict whether a precipitation reaction will occur when aqueous solutions of the following substances are mixed.

(a) $NaOH + HClO_4$ **(b)** $FeCl_2 + KOH$

(c) $(NH_4)_2SO_4 + NiCl_2$

6.82 Use the solubility guidelines in Section 6.9 to predict whether precipitation reactions will occur between the listed pairs of reactants. Write balanced equations for those reactions that should occur.

(a) NaBr and $Hg_2(NO_3)_2$ **(b)** $CuCl_2$ and K_2SO_4

(c) $LiNO_3$ and $Ca(CH_3CO_2)_2$ **(d)** $(NH_4)_2CO_3$ and $CaCl_2$

(e) KOH and $MnBr_2$ **(f)** Na_2S and $Al(NO_3)_3$

6.83 Write net ionic equations for the following reactions:

(a) $Mg(s) + CuCl_2(aq) \longrightarrow MgCl_2(aq) + Cu(s)$

(b) $2 KCl(aq) + Pb(NO_3)_2(aq) \longrightarrow$ $PbCl_2(s) + 2 KNO_3(aq)$

(c) $2 Cr(NO_3)_3(aq) + 3 Na_2S(aq) \longrightarrow$ $Cr_2S_3(s) + 6 NaNO_3(aq)$

6.84 Write net ionic equations for the following reactions:

(a) $2 AuCl_3(aq) + 3 Sn(s) \longrightarrow 3 SnCl_2(aq) + 2 Au(s)$

(b) $2 NaI(aq) + Br_2(l) \longrightarrow 2 NaBr(aq) + I_2(s)$

(c) $2 AgNO_3(aq) + Fe(s) \longrightarrow Fe(NO_3)_2(aq) + 2 Ag(s)$

REDOX REACTIONS AND OXIDATION NUMBERS

6.85 Where in the periodic table are the best reducing agents found? The best oxidizing agents?

6.86 Where in the periodic table are the most easily reduced elements found? The most easily oxidized?

6.87 In each of the following, tell whether the substance gains electrons or loses electrons in a redox reaction:

(a) An oxidizing agent

(b) A reducing agent

(c) A substance undergoing oxidation

(d) A substance undergoing reduction

6.88 For the following substances, tell whether the oxidation number increases or decreases in a redox reaction:

(a) An oxidizing agent

(b) A reducing agent

(c) A substance undergoing oxidation

(d) A substance undergoing reduction

6.89 Assign an oxidation number to each element in the following compounds or ions:

(a) N_2O_5 **(b)** SO_3^{2-}

(c) CH_2O **(d)** $HClO_3$

6.90 Assign an oxidation number to the metal in the following compounds:

(a) $CoCl_3$ **(b)** $FeSO_4$ **(c)** UO_3

(d) CuF_2 **(e)** TiO_2 **(f)** SnS

6.91 Which element is oxidized and which is reduced in the following reactions?

(a) $Si(s) + 2 Cl_2(g) \longrightarrow SiCl_4(l)$
(b) $Cl_2(g) + 2 NaBr(aq) \longrightarrow Br_2(aq) + 2 NaCl(aq)$
(c) $SbCl_3(s) + Cl_2(g) \longrightarrow SbCl_5(s)$

6.92 Which element is oxidized and which is reduced in the following reactions?

(a) $2 SO_2(g) + O_2(g) \longrightarrow 2 SO_3(g)$
(b) $2 Na(s) + Cl_2(g) \longrightarrow 2 NaCl(s)$
(c) $CuCl_2(aq) + Zn(s) \longrightarrow ZnCl_2(aq) + Cu(s)$
(d) $2 NaCl(aq) + F_2(g) \longrightarrow 2 NaF(aq) + Cl_2(g)$

Applications

6.93 What do you think are some of the errors involved in calculating Avogadro's number by spreading oil on a pond? [*Did Ben Franklin Have Avogadro's Number?*, p. 153]

6.94 Ferrous sulfate is one dietary supplement used to treat iron-deficiency anemia. What are the molecular formula and molecular weight of this compound? How many milligrams of iron are in 250 mg of ferrous sulfate? [*Anemia— A Limiting Reagent Problem?*, p. 162]

6.95 Sodium urate, the principal constituent of some kidney stones, has the formula $NaC_5H_3N_4O_3$. In aqueous solution, the solubility of sodium urate is only 0.067 g/L. How many moles of sodium urate dissolve in 1.00 L of water? [*Gout and Kidney Stones, p. 165*]

6.96 Identify the oxidizing and reducing agents in a typical dry-cell battery. [*Batteries, p. 171*]

General Questions and Problems

6.97 Zinc metal reacts with hydrochloric acid (HCl) according to the equation:

$$Zn(s) + 2 HCl(aq) \longrightarrow ZnCl_2(aq) + H_2(g)$$

(a) How many grams of hydrogen are produced if 15.0 g of zinc reacts?
(b) Is this a redox reaction? If so, tell what is reduced, what is oxidized, and reducing and oxidizing agents.

6.98 Lithium oxide is used aboard the space shuttle to remove water from the atmosphere according to the equation

$$Li_2O(s) + H_2O(g) \longrightarrow 2 LiOH(s)$$

(a) How many grams of Li_2O must be carried on board to remove 80.0 kg of water?
(b) Is this a redox reaction? Why or why not?

6.99 Balance the following equations.

(a) The thermite reaction, used in welding:

$$Al(s) + Fe_2O_3(s) \longrightarrow Al_2O_3(l) + Fe(l)$$

(b) The explosion of ammonium nitrate:

$$NH_4NO_3(s) \longrightarrow N_2(g) + O_2(g) + H_2O(g)$$

6.100 Batrachotoxin, $C_{31}H_{42}N_2O_6$, an active component of South American arrow poison, is so toxic that 0.05 μg can kill a person. How many molecules is this?

6.101 Look at the solubility guidelines in Section 6.9 and predict whether a precipitate forms when $CuCl_2(aq)$ and $Na_2CO_3(aq)$ are mixed. If so, write both the balanced equation and the net ionic equation for the process.

6.102 When table sugar (sucrose, $C_{12}H_{22}O_{11}$) is heated, it decomposes to form C and H_2O.

(a) Write a balanced equation for the process.
(b) How many grams of carbon are formed by the breakdown of 60.0 g of sucrose?
(c) How many grams of water are formed when 6.50 g of carbon are formed?

6.103 Although Cu is not sufficiently active to react with acids, it can be dissolved by concentrated nitric acid, which functions as an oxidizing agent according to the following equation:

$$Cu(s) + 4 HNO_3(aq) \longrightarrow$$
$$Cu(NO_3)_2(aq) + 2 NO_2(g) + 2 H_2O(l)$$

(a) Write the net ionic equation for this process.
(b) Is 35.0 g of HNO_3 sufficient to dissolve 5.00 g of copper?

6.104 The net ionic equation for the Breathalyzer test used to indicate alcohol concentration in the body is

$$16 H^+(aq) + 2 Cr_2O_7{}^{2-}(aq) + 3 C_2H_6O(aq) \longrightarrow$$
$$3 C_2H_4O_2(aq) + 4 Cr^{3+}(aq) + 11 H_2O(l)$$

(a) How many grams of $K_2Cr_2O_7$ must be used to consume 1.50 g of C_2H_6O?
(b) How many grams of $C_2H_4O_2$ can be produced from 80.0 g of C_2H_6O?

6.105 Ethyl alcohol is formed by enzyme action on sugars and starches during fermentation:

$$C_6H_{12}O_6 \longrightarrow 2 CO_2 + 2 C_2H_6O$$

If the density of ethyl alcohol is 0.789 g/mL, how many quarts can be produced by the fermentation of 100.0 lb of sugar?

6.106 Balance the following equations:

(a) $Al(OH)_3(aq) + HNO_3(aq) \longrightarrow$
$Al(NO_3)_3(aq) + H_2O(l)$
(b) $AgNO_3(aq) + FeCl_3(aq) \longrightarrow$
$AgCl(s) + Fe(NO_3)_3(aq)$
(c) $(NH_4)_2Cr_2O_7(s) \longrightarrow Cr_2O_3(s) + H_2O(g) + N_2(g)$
(d) $Mn_2(CO_3)_3(s) \longrightarrow Mn_2O_3(s) + CO_2(g)$

6.107 White phosphorus (P_4) is a highly reactive form of elemental phosphorus that reacts with oxygen to form a variety of molecular compounds, including diphosphorus pentoxide.

(a) Write the balanced chemical equation for this reaction.
(b) Calculate the oxidation number for P and O on both sides of the reaction, and identify the oxidizing and reducing agents.

6.108 The combustion of fossil fuels containing sulfur contributes to the phenomenon known as acid rain. The combustion process releases sulfur in the form of sulfur dioxide, which is converted to sulfuric acid in a process involving two reactions.

(a) In the first reaction, sulfur dioxide reacts with molecular oxygen to form sulfur trioxide. Write the balanced chemical equation for this reaction.
(b) In the second reaction, sulfur trioxide reacts with water in the atmosphere to form sulfuric acid. Write the balanced chemical equation for this reaction.
(c) Calculate the oxidation number for the S atom in each compound in these reactions.

CHAPTER 7

Chemical Reactions: Energy, Rates, and Equilibrium

CONCEPTS TO REVIEW

Energy and Heat
(Section 2.10)

Ionic Bonds
(Section 4.3)

Covalent Bonds
(Section 5.1)

Chemical Equations
(Section 6.1)

▲ Many spontaneous chemical reactions are accompanied by the release of energy, in some cases explosively.

CONTENTS

CHAPTER GOALS

In this chapter, we will look more closely at chemical reactions and answer the following questions:

1. What energy changes take place during reactions?

THE GOAL: Be able to explain the factors that influence energy changes in chemical reactions.

2. What is "free energy," and what is the criterion for spontaneity in chemistry?

THE GOAL: Be able to define enthalpy, entropy, and free-energy changes, and explain how the values of these quantities affect chemical reactions.

3. What determines the rate of a chemical reaction?

THE GOAL: Be able to explain activation energy and other factors that determine reaction rate.

4. What is chemical equilibrium?

THE GOAL: Be able to describe what occurs in a reaction at equilibrium and write the equilibrium equation for a given reaction.

5. What is Le Châtelier's principle?

THE GOAL: Be able to state Le Châtelier's principle and use it to predict the effect of changes in temperature, pressure, and concentration on reactions.

We have yet to answer many questions about reactions. Why, for instance, do reactions occur? Just because a balanced equation can be written does not mean it will take place. We can write a balanced equation for the reaction of gold with water, for example, but the reaction does not occur in practice, so your gold jewelry is safe in the shower.

Balanced, but does not occur
$$2 \text{ Au}(s) + 3 \text{ H}_2\text{O}(l) \longrightarrow \text{Au}_2\text{O}_3(s) + 3 \text{ H}_2(g)$$

To describe reactions more completely, several fundamental questions are commonly asked: Is energy released or absorbed when a reaction occurs? Is a given reaction fast or slow? Does a reaction continue until all reactants are converted to products or is there a point beyond which no additional product forms?

7.1 Energy and Chemical Bonds

There are two fundamental and interconvertible kinds of energy: *potential* and *kinetic*. **Potential energy** is stored energy. The water in a reservoir behind a dam, an automobile poised to coast downhill, and a coiled spring have potential energy waiting to be released. **Kinetic energy**, by contrast, is the energy of motion. When the water falls over the dam and turns a turbine, when the car rolls downhill, or when the spring uncoils and makes the hands on a clock move, the potential energy in each is converted to kinetic energy. Of course, once all the potential energy is converted, nothing further occurs. The water at the bottom of the dam, the car at the bottom of the hill, and the uncoiled spring no longer have potential energy and thus undergo no further change.

In chemical compounds, the attractive forces between ions (ionic bonds) or atoms (covalent bonds) are a form of potential energy. In many chemical reactions, this potential energy is often converted into **heat**—the kinetic energy of the moving particles that make up the compound. Because the reaction products have less potential energy than the reactants, we say that the products are *more stable* than the reactants. The term "stable" is used in chemistry to describe a substance that has little remaining potential energy and consequently little tendency to undergo further change. Whether a reaction occurs, and how much energy or heat is associated with the reaction, depends on the amount of potential energy contained in the reactants and products.

Potential energy Stored energy.

Kinetic energy The energy of an object in motion.

Heat A measure of the transfer of thermal energy.

7.2 Heat Changes during Chemical Reactions

Why does chlorine react so easily with many elements and compounds but nitrogen does not? What difference between Cl_2 molecules and N_2 molecules accounts for their different reactivities? The answer is that the nitrogen–nitrogen triple bond

Bond dissociation energy The amount of energy that must be supplied to break a bond and separate the atoms in an isolated gaseous molecule.

is much *stronger* than the chlorine–chlorine single bond and cannot be broken as easily in chemical reactions.

The strength of a covalent bond is measured by its **bond dissociation energy**, defined as the amount of energy that must be supplied to break the bond and separate the atoms in an isolated gaseous molecule. The triple bond in N_2, for example, has a bond dissociation energy of 226 kcal/mol, whereas the single bond in chlorine has a bond dissociation energy of only 58 kcal/mol:

$$:N:::N: \xrightarrow{\text{226 kcal/mol}} :\dot{N}\cdot \; + \; \cdot\dot{N}: \qquad N_2 \text{ bond dissociation energy } = 226 \text{ kcal/mol}$$

$$:\ddot{C}l:\ddot{C}l: \xrightarrow{\text{58 kcal/mol}} :\ddot{C}l\cdot \; + \; \cdot\ddot{C}l: \qquad Cl_2 \text{ bond dissociation energy } = 58 \text{ kcal/mol}$$

Endothermic A process or reaction that absorbs heat and has a positive ΔH.

Exothermic A process or reaction that releases heat and has a negative ΔH.

A chemical change like bond breaking that absorbs heat is said to be **endothermic**, from the Greek words *endon* (within) and *therme* (heat), meaning that *heat* is put *in*. The reverse of bond breaking is bond formation, a process that *releases* heat and is described as **exothermic**, from the Greek *exo* (outside), meaning that heat goes *out*. The amount of energy released in forming a bond is numerically the same as that absorbed in breaking it. When nitrogen atoms combine to give N_2, 226 kcal/mol of heat is released. Similarly, when Cl_2 molecules are pulled apart into atoms, 58 kcal/mol of heat is absorbed; when chlorine atoms combine to give Cl_2, 58 kcal/mol of heat is released.

$$:\dot{N}\cdot \; + \; \cdot\dot{N}: \longrightarrow :N:::N: \; + \; 226 \text{ kcal/mol heat released}$$

$$:\ddot{C}l\cdot \; + \; \cdot\ddot{C}l: \longrightarrow :\ddot{C}l:\ddot{C}l: \; + \; 58 \text{ kcal/mol heat released}$$

The same energy relationships that govern bond breaking and bond formation apply to every physical or chemical change. That is, the amount of heat transferred during a change in one direction is numerically equal to the amount of heat transferred during the change in the opposite direction. Only the *direction* of the heat transfer is different. This relationship reflects a fundamental law of nature called the *law of conservation of energy:*

Law of conservation of energy Energy can be neither created nor destroyed in any physical or chemical change.

If more energy could be released by an exothermic reaction than was consumed in its reverse, the law would be violated and we could "manufacture" energy out of nowhere by cycling back and forth between forward and reverse reactions—a clear impossibility.

In every chemical reaction, some bonds in the reactants are broken and new bonds are formed in the products. The difference between the energy absorbed in breaking bonds and the energy released in forming bonds is called the **heat of reaction** and is a quantity that we can measure. Heats of reaction that are measured when a reaction is held at constant pressure are represented by the abbreviation ΔH, where Δ (the Greek capital letter delta) is a general symbol used to indicate "a change in" and *H* is a quantity called **enthalpy**. Thus, the value of ΔH represents the **enthalpy change** that occurs during a reaction. The terms *enthalpy change* and *heat of reaction* are often used interchangeably, but we will generally use the latter term in this book.

Enthalpy (H) A measure of the amount of energy associated with substances involved in a reaction.

Heat of reaction ΔH = Energy of bonds formed in products minus
(Enthalpy change) energy of bonds broken in reactants

7.3 Exothermic and Endothermic Reactions

When the total strength of the bonds formed in the products is *greater* than the total strength of the bonds broken in the reactants, energy is released and a reaction is exothermic. All combustion reactions are exothermic; for example, burning 1 mol of methane releases 213 kcal of energy in the form of heat. The heat released in an exothermic reaction can be thought of as a reaction product, and ΔH is assigned a *negative* value because heat *leaves*.

An exothermic reaction—negative ΔH

Heat is a product.

$$CH_4(g) + 2 O_2(g) \longrightarrow CO_2(g) + 2 H_2O(l) + 213 \text{ kcal}$$

or

$$CH_4(g) + 2 O_2(g) \longrightarrow CO_2(g) + 2 H_2O(l) \qquad \Delta H = -213 \text{ kcal/mol}$$

Note that ΔH is given in units of kilocalories per mole, where "per mole" means the reaction of *molar amounts of products and reactants as represented by the coefficients of the balanced equation*. Thus, the value $\Delta H = -213$ kcal/mol refers to the amount of heat released when 1 mol (16.0 g) of methane reacts with O_2 to give 1 mol of CO_2 gas and 2 mol of liquid H_2O. If we were to double the amount of methane from 1 mol to 2 mol, the amount of heat released would also double.

The quantities of heat released in the combustion of several fuels, including natural gas (which is primarily methane), are compared in Table 7.1. The values are given in kilocalories per gram to make comparisons easier. You can see from the table why there is interest in the potential of hydrogen as a fuel.

TABLE 7.1   Energy Values of Some Common Fuels

FUEL	ENERGY VALUE (kcal/g)
Wood (pine)	4.3
Ethyl alcohol	7.1
Coal (anthracite)	7.4
Crude oil (Texas)	10.5
Gasoline	11.5
Natural gas	11.7
Hydrogen	34.0

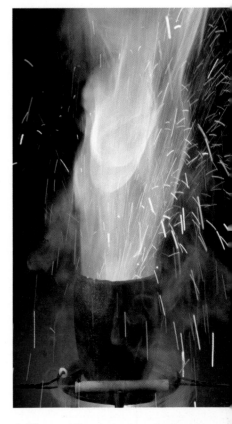

▲ The reaction between aluminum metal and iron(III) oxide, called the *thermite reaction*, is so strongly exothermic that it melts iron.

When the total energy of the bonds formed in the products is *less* than the total energy of the bonds broken in the reactants, energy is absorbed and a reaction is endothermic. The combination of nitrogen and oxygen to give nitrogen oxide (also known as nitric oxide), a gas present in automobile exhaust, is such a reaction. The heat added in an endothermic reaction is like a reactant, and ΔH is assigned a *positive* value because heat is *added*.

An endothermic reaction—positive ΔH

Heat is a reactant.

$$N_2(g) + O_2(g) + 43 \text{ kcal} \longrightarrow 2 NO(g)$$

or

$$N_2(g) + O_2(g) \longrightarrow 2 NO(g) \qquad \Delta H = +43 \text{ kcal/mol}$$

Important points about heat transfers and chemical reactions

- An exothermic reaction releases heat to the surroundings; ΔH is negative.
- An endothermic reaction absorbs heat from the surroundings; ΔH is positive.

- The reverse of an exothermic reaction is endothermic.
- The reverse of an endothermic reaction is exothermic.
- The amount of heat absorbed or released in the reverse of a reaction is equal to that released or absorbed in the forward reaction, but ΔH has the opposite sign.

Worked Examples 7.1–7.3 show how to calculate the amount of heat absorbed or released for reaction of a given amount of reactant. All that is needed is the balanced equation and its accompanying ΔH. Mole ratios and molar masses are used to convert between masses and moles of reactants or products, as discussed in Sections 6.5 and 6.6.

▲ Methane produced from rotting refuse is trapped and used for energy production to ease the energy crisis in California.

WORKED EXAMPLE   **7.1** Heat of Reaction

Methane undergoes combustion with O_2 according to the following equation:

$$CH_4(g) + 2\,O_2(g) \longrightarrow CO_2(g) + 2\,H_2O(l) \quad \Delta H = -213\frac{kcal}{mol\,CH_4}$$

How much heat is released during the combustion of 0.35 mol of methane?

ANALYSIS Since the value of ΔH for the reaction (213 kcal/mol) is negative, it indicates the amount of heat released when 1 mol of methane reacts with O_2. We need to find the amount of heat released when an amount other than 1 mol reacts, using appropriate factor-label calculations to convert from our known or given units to kilocalories.

BALLPARK ESTIMATE Since 213 kcal is released for each mole of methane that reacts, 0.35 mol of methane should release about one-third of 213 kcal, or about 70 kcal.

SOLUTION

To find the amount of heat released (in kilocalories) by combustion of 0.35 mol of methane, we use a conversion factor of kcal/mol:

$$0.35 \; \cancel{mol\,CH_4} \times \frac{-213 \; kcal}{1 \; \cancel{mol\,CH_4}} = -75 \; kcal$$

The negative sign indicates that the 75 kcal of heat is released.

BALLPARK CHECK The calculated answer is consistent with our estimate (70 kcal).

WORKED EXAMPLE   **7.2** Heat of Reaction: Mass to Mole Conversion

How much heat is released during the combustion of 7.50 g of methane (MW = 16.0 g/mol)?

$$CH_4(g) + 2\,O_2(g) \longrightarrow CO_2(g) + 2\,H_2O(l) \quad \Delta H = -213\frac{kcal}{mol\,CH_4}$$

ANALYSIS We can find the moles of methane involved in the reaction by using the molecular weight in a mass to mole conversion, and then use ΔH to find the heat released.

BALLPARK ESTIMATE Since 1 mol of methane (MW = 16.0 g/mol) has a mass of 16.0 g, 7.50 g of methane is a little less than 0.5 mol. Thus, less than half of 213 kcal, or about 100 kcal, is released from combustion of 7.50 g.

SOLUTION

Going from a given mass of methane to the amount of heat released in a reaction requires that we first find the number of moles of methane by including molar mass (in mol/g) in the calculation and then converting moles to kilocalories:

$$7.50 \text{ g } \cancel{CH_4} \times \frac{1 \text{ mol } \cancel{CH_4}}{16.0 \text{ g } \cancel{CH_4}} \times \frac{-213 \text{ kcal}}{1 \text{ mol } \cancel{CH_4}} = -99.8 \text{ kcal}$$

The negative sign indicates that the 99.8 kcal of heat is released.

BALLPARK CHECK Our estimate was −100 kcal!

WORKED EXAMPLE **7.3** Heat of Reaction: Mole Ratio Calculations

How much heat is released when 2.50 mol of O_2 reacts completely with methane?

$$CH_4(g) + 2\,O_2(g) \longrightarrow CO_2(g) + 2\,H_2O(l) \quad \Delta H = -213\frac{\text{kcal}}{\text{mol } CH_4}$$

ANALYSIS Since the ΔH for the reaction is based on the combustion of 1 mol of methane, we will need to perform a mole ratio calculation.

BALLPARK ESTIMATE The balanced equation shows that 213 kcal is released for each 2 mol of oxygen that reacts. Thus, 2.50 mol of oxygen should release a bit more than 213 kcal, perhaps about 250 kcal.

SOLUTION
To find the amount of heat released by combustion of 2.50 mol of oxygen, we include in our calculation a mole ratio based on the balanced chemical equation:

$$2.50 \text{ mol } \cancel{O_2} \times \frac{1 \text{ mol } \cancel{CH_4}}{2 \text{ mol } \cancel{O_2}} \times \frac{-213 \text{ kcal}}{1 \text{ mol } \cancel{CH_4}} = -266 \text{ kcal}$$

The negative sign indicates that the 266 kcal of heat is released.

BALLPARK CHECK The calculated answer is close to our estimate (250 kcal).

PROBLEM 7.1

In photosynthesis, green plants convert carbon dioxide and water into glucose ($C_6H_{12}O_6$) according to the following equation:

$$6\,CO_2(g) + 6\,H_2O(l) + 678 \text{ kcal} \longrightarrow C_6H_{12}O_6(aq) + 6\,O_2(g)$$

(a) Is the reaction exothermic or endothermic?
(b) What is the value of ΔH for the reaction?
(c) Write the equation for the reverse of the reaction, including heat as a reactant or product.

PROBLEM 7.2

The following equation shows the conversion of aluminum oxide (from the ore bauxite) to aluminum:

$$2\,Al_2O_3(s) \longrightarrow 4\,Al(s) + 3\,O_2(g) \quad \Delta H = +801 \text{ kcal/mol}$$

(a) Is the reaction exothermic or endothermic?
(b) How many kilocalories are required to produce 1.00 mol of aluminum?
(c) How many kilocalories are required to produce 10.0 g of aluminum?

▲ Events that lead to lower energy tend to occur spontaneously. Thus, water always flows *down* a waterfall, not up.

Spontaneous process A process or reaction that, once started, proceeds on its own without any external influence.

Entropy (S) A measure of the amount of molecular disorder in a system.

PROBLEM 7.3

How much heat is absorbed during production of 127 g of NO by the combination of nitrogen and oxygen?

$$N_2(g) + O_2(g) \longrightarrow 2\,NO(g) \quad \Delta H = +43 \text{ kcal/mol}$$

7.4 Why Do Chemical Reactions Occur? Free Energy

Events that lead to lower energy tend to occur spontaneously. Water falls downhill, for instance, releasing its stored (potential) energy and reaching a lower-energy, more stable position. Similarly, a wound-up spring uncoils when set free. Applying this lesson to chemistry, the obvious conclusion is that exothermic processes—those that release heat energy—should be spontaneous. A log burning in a fireplace is just one example of a spontaneous reaction that releases heat. At the same time, endothermic processes, which absorb heat energy, should not be spontaneous. Often, these conclusions are correct, but not always. Many, but not all, exothermic processes take place spontaneously, and many, but not all, endothermic processes are nonspontaneous.

Before exploring the situation further, it is important to understand what the word "spontaneous" means in chemistry, which is not quite the same as in everyday language. A **spontaneous process** is one that, once started, proceeds on its own without any external influence. The change does not necessarily happen quickly, like a spring suddenly uncoiling or a car coasting downhill. It can also happen slowly, like the gradual rusting away of an abandoned bicycle. A *nonspontaneous process*, by contrast, takes place only in the presence of a continuous external influence: Energy must be continually expended to rewind a spring or push a car uphill. The reverse of a spontaneous process is always nonspontaneous.

As an example of a process that takes place spontaneously yet absorbs heat, think about what happens when you take an ice cube out of the refrigerator. The ice spontaneously melts to give liquid water above 0 °C, even though it *absorbs* heat energy from the surroundings. What this and other spontaneous processes that absorb heat energy have in common is *an increase in molecular disorder, or randomness*. When the solid ice melts, the H_2O molecules are no longer locked in position but are now free to move around randomly in the liquid water.

The amount of disorder in a system is called the system's **entropy**, symbolized S and expressed in units of calories per mole-kelvin [cal/(mol · K)]. The greater the disorder, or randomness, of the particles in a substance or mixture, the larger the value of S (Figure 7.1). Gases have more disorder and therefore higher entropy than liquids because particles in the gas move around more freely than particles in the liquid. Similarly, liquids have higher entropy than solids. In chemical reactions, entropy increases when, for example, a gas is produced from a solid or when 2 mol of reactants split into 4 mol of products.

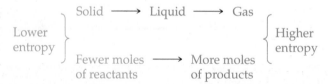

The entropy *change* for a process, ΔS, has a *positive* value if disorder increases because the process adds disorder to the system. The melting of ice to give water is an example. Conversely, ΔS has a *negative* value if the disorder of a system decreases. The freezing of water to give ice is an example.

It thus appears that two factors determine the spontaneity of a chemical or physical change: the release or absorption of heat, ΔH, and the increase or decrease in entropy, ΔS. *To decide whether a process is spontaneous, both the enthalpy change and*

APPLICATION ▶ Energy from Food

Any serious effort to lose weight usually leads to studying the caloric values of foods. Have you ever wondered how the numbers quoted on food labels are obtained?

Food is "burned" in the body to yield H_2O, CO_2, and energy, just as natural gas is burned in furnaces to yield the same products. In fact, the "caloric value" of a food is just the heat of reaction for complete combustion of the food (minus a small correction factor). The value is the same whether the food is burned in the body or in the laboratory. One gram of protein releases 4 kcal, 1 g of table sugar (a carbohydrate) releases 4 kcal, and 1 g of fat releases 9 kcal (see Table).

The caloric value of a food is usually given in "Calories" (note the capital C), where 1 Cal = 1000 cal = 1 kcal. To determine these values experimentally, a carefully dried and weighed food sample is placed together with oxygen into an instrument called a *calorimeter*, the food is ignited, the temperature change is measured, and the amount of heat given off is calculated from the temperature change. In the calorimeter, the heat from the food is released very quickly and the temperature rises dramatically. Clearly, though, something a bit different goes on when food is burned in the body, otherwise we would burst into flames after a meal!

It is a fundamental principle of chemistry that the total heat released or absorbed in going from reactants to products is the same, no matter how many reactions are involved.

Caloric Values of Some Foods

SUBSTANCE, SAMPLE SIZE	CALORIC VALUE (kcal)
Protein, 1 g	4
Carbohydrate, 1 g	4
Fat, 1 g	9
Alcohol, 1 g	7.1
Cola drink, 12 fl oz (369 g)	160
Apple, one medium (138 g)	80
Iceberg lettuce, 1 cup shredded (55 g)	5
White bread, 1 slice (25 g)	65
Hamburger patty, 3 oz (85 g)	245
Pizza, 1 slice (120 g)	290
Vanilla ice cream, 1 cup (133 g)	270

The body applies this principle by withdrawing energy from food a bit at a time in a long series of interconnected reactions rather than all at once in a single reaction. These and other reactions continually taking place in the body—called the body's *metabolism*—will be examined in later chapters.

See Additional Problems 7.70 and 7.71 at the end of the chapter.

▲ Eating this dessert gives your body 550 Calories. Burning the dessert in a calorimeter releases 550 kcal as heat.

▶ **FIGURE 7.1 Entropy and values of S.** The mixture on the right has more disorder and a higher entropy than the mixture on the left, and has a higher value of S. The value of the entropy change, ΔS, for converting the mixture on the left to that on the right is positive because entropy increases.

Stir

Entropy increases (positive ΔS)

the entropy change must be taken into account. We have already seen that a negative ΔH favors spontaneity, but what about ΔS? The answer is that an increase in molecular disorder (ΔS positive) favors spontaneity. A good analogy is the bedroom or office that seems to spontaneously become more messy over time (increase in disorder means increase in entropy, ΔS positive); to clean it up (a decrease in disorder, ΔS negative) requires an input of energy, a nonspontaneous process. Using our chemical example, the combustion of a log spontaneously converts large, complex molecules like lignin and cellulose (high molecular order, low entropy) into CO_2 and H_2O (smaller molecules with less molecular order and higher entropy). For this process, the level of disorder increases, and so ΔS is positive. The reverse process—turning CO_2 and H_2O back into cellulose—does occur in photosynthesis, but it requires a significant input of energy in the form of sunlight.

When enthalpy and entropy are both favorable (ΔH negative, ΔS positive), a process is spontaneous; when both are unfavorable, a process is nonspontaneous. Clearly, however, the two factors do not have to operate in the same direction. It is possible for a process to be *unfavored* by enthalpy (the process absorbs heat and so has a positive ΔH) and yet be *favored* by entropy (there is an increase in disorder and so ΔS is positive). The melting of an ice cube above 0 °C, for which $\Delta H = +1.44$ kcal/mol and $\Delta S = +5.26$ cal/(mol·K) is just such a process. To take both heat of reaction (ΔH) and change in disorder (ΔS) into account when determining the spontaneity of a process, a quantity called the **free-energy change (ΔG)** is needed:

Free-energy change

Heat of reaction   Temperature (in kelvins)   Entropy change

$$\Delta G = \Delta H - T\Delta S$$

Exergonic A spontaneous reaction or process that releases free energy and has a negative ΔG.

Endergonic A nonspontaneous reaction or process that absorbs free energy and has a positive ΔG.

The value of the free-energy change ΔG determines spontaneity. A negative value for ΔG means that free energy is released and the reaction or process is spontaneous. Such events are said to be **exergonic**. A positive value for ΔG means that free energy must be added and the process is nonspontaneous. Such events are said to be **endergonic**.

Spontaneous process ΔG is negative; free energy is released; process is exergonic.

Nonspontaneous process ΔG is positive; free energy is added; process is endergonic.

The equation for the free-energy change shows that spontaneity also depends on temperature (T). At low temperatures, the value of $T\Delta S$ is often small so that ΔH is the dominant factor. At a high enough temperature, however, the value of $T\Delta S$ can become larger than ΔH. Thus, an endothermic process that is nonspontaneous at low temperature can become spontaneous at a higher temperature. An example is the industrial synthesis of hydrogen by reaction of carbon with water:

$$C(s) + H_2O(l) \longrightarrow CO(g) + H_2(g) \qquad \begin{array}{ll} \Delta H = +31.3 \text{ kcal/mol} & \text{(Unfavorable)} \\ \Delta S = +32 \text{ cal/(mol} \cdot \text{K)} & \text{(Favorable)} \end{array}$$

The reaction has an unfavorable (positive) ΔH term but a favorable (positive) ΔS term because disorder increases when a solid and a liquid are converted into two gases. No reaction occurs if carbon and water are mixed together at 25 °C (298 K) because the unfavorable ΔH is larger than the favorable $T\Delta S$. Above about 700 °C (973 K), however, the favorable $T\Delta S$ becomes larger than the unfavorable ΔH, so the reaction becomes spontaneous.

Important points about spontaneity and free energy

- A spontaneous process, once begun, proceeds without any external assistance and is exergonic; that is, it has a negative value of ΔG.

- A nonspontaneous process requires continuous external influence and is endergonic; that is, it has a positive value of ΔG.

- The value of ΔG for the reverse of a reaction is numerically equal to the value of ΔG for the forward reaction, but has the opposite sign.

- Some nonspontaneous processes become spontaneous with a change in temperature.

⊂⊙⊃ Looking Ahead

In later chapters, we will see that a knowledge of free-energy changes is especially important for understanding how metabolic reactions work. Living organisms cannot raise their temperatures to convert nonspontaneous reactions into spontaneous reactions, so they must resort to other strategies, which we will explore in Chapter 21. ⊂⊙⊃

WORKED EXAMPLE **7.4** Entropy Change of Processes

Does entropy increase or decrease in the following processes?

(a) Smoke from a cigarette disperses throughout a room rather than remaining in a cloud over the smoker's head.

(b) Water boils, changing from liquid to vapor.

(c) A chemical reaction occurs: $3 H_2(g) + N_2(g) \longrightarrow 2 NH_3(g)$

ANALYSIS Entropy is a measure of molecular disorder. Entropy increases when the products are more disordered than the reactants; entropy decreases when the products are less disordered than the reactants.

SOLUTION

(a) Entropy increases because smoke particles are more disordered when they are randomly distributed in the larger volume.

(b) Entropy increases because H_2O molecules have more freedom and disorder in the gas phase than in the liquid phase.

(c) Entropy decreases because 4 mol of reactant gas particles become 2 mol of product gas particles, with a consequent decrease in freedom and disorder.

WORKED EXAMPLE **7.5** Spontaneity of Reactions: Enthalpy, Entropy, and Free Energy

The industrial method for synthesizing hydrogen by reaction of carbon with water has $\Delta H = +31.3$ kcal/mol and $\Delta S = +32$ cal/(mol · K). What is the value of ΔG for the reaction at 27 °C (300 K)? Is the reaction spontaneous or nonspontaneous at this temperature?

$$C(s) + H_2O(l) \longrightarrow CO(g) + H_2(g)$$

ANALYSIS The reaction is endothermic (ΔH positive) and does not favor spontaneity, whereas the ΔS indicates an increase in disorder (ΔS positive), which *does* favor spontaneity. Calculate ΔG to determine spontaneity.

BALLPARK ESTIMATE The unfavorable ΔH (+31.3 kcal/mol) is 1000 times greater than the favorable ΔS (+32 cal/mol · K), so the reaction will be spontaneous (ΔG negative) only when the temperature is high enough to make the $T\Delta S$ term in the equation for ΔG larger than the ΔH term. This happens at $T \geq 1000$ K. Since $T = 300$ K, expect ΔG to be positive and the reaction to be nonspontaneous.

SOLUTION
Use the free-energy equation to determine the value of ΔG at this temperature. (Remember that ΔS has units of *calories* per mole-kelvin, not kilocalories per mole-kelvin.)

$$\Delta G = \Delta H - T\,\Delta S$$
$$= +31.3\frac{\text{kcal}}{\text{mol}} - (300\text{ K})\left(+32\frac{\text{cal}}{\text{mol} \cdot \text{K}}\right)\left(\frac{1\text{ kcal}}{1000\text{ cal}}\right)$$
$$= +21.7\frac{\text{kcal}}{\text{mol}}$$

BALLPARK CHECK Because ΔG is positive, the reaction is nonspontaneous at 300 K, consistent with our estimate.

PROBLEM 7.4

Does entropy increase or decrease in the following processes?

(a) After raking your leaves into a neat pile, a breeze blows them all over your lawn.

(b) Gasoline fumes escape as the fuel is pumped into your car.

(c) $Mg(s) + Cl_2(g) \longrightarrow MgCl_2(s)$

PROBLEM 7.5

Lime (CaO) is prepared by the decomposition of limestone ($CaCO_3$).

$$CaCO_3(s) \longrightarrow CaO(s) + CO_2(g) \quad \Delta G = +31 \text{ kcal/mol at } 25 °C$$

(a) Does the reaction occur spontaneously at 25 °C?

(b) Does entropy increase or decrease in this reaction?

(c) Would you expect the reaction to be spontaneous at higher temperatures?

PROBLEM 7.6

The melting of solid ice to give liquid water has $\Delta H = 1.44$ kcal/mol and $\Delta S = +5.26$ cal/(mol · K). What is the value of ΔG for the melting process at the following temperatures? Is the melting spontaneous or nonspontaneous at these temperatures?

(a) −10 °C (263 K)

(b) 0 °C (273 K)

(c) +10 °C (283 K)

KEY CONCEPT PROBLEM 7.7

The following diagram portrays a reaction of the type A(s) ⟶ B(s) + C(g), where the different colored spheres represent different molecular structures. Assume that the reaction has $\Delta H = -23.5$ kcal/mol.

(a) What is the sign of ΔS for the reaction?

(b) Is the reaction likely to be spontaneous at all temperatures, nonspontaneous at all temperatures, or spontaneous at some but nonspontaneous at others?

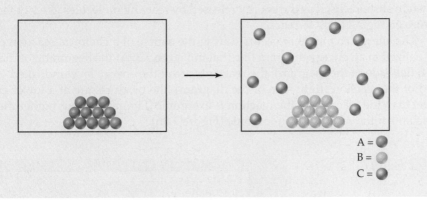

A =
B =
C =

7.5 How Do Chemical Reactions Occur? Reaction Rates

Just because a chemical reaction has a favorable free-energy change does not mean that it occurs rapidly. The value of ΔG tells us only whether a reaction *can* occur; it says nothing about how *fast* the reaction will occur or about the details of the molecular changes that take place during the reaction. It is now time to look into these other matters.

For a chemical reaction to occur, reactant particles must collide, some chemical bonds have to break, and new bonds have to form. Not all collisions lead to products, however. One requirement for a productive collision is that the colliding molecules must approach with the correct orientation so that the atoms about to form new bonds can connect. In the reaction of ozone (O_3) with nitric oxide (NO) to give oxygen (O_2) and nitrogen dioxide (NO_2), for example, the two reactants must collide so that the nitrogen atom of NO strikes a terminal oxygen atom of O_3 (Figure 7.2).

Another requirement for a reaction to occur is that the collision must take place with enough energy to break the appropriate bonds in the reactant. If the reactant particles are moving slowly, collisions might be too gentle to overcome the repulsion between electrons in the different reactants, and the particles will simply bounce apart. Only if the collisions are sufficiently energetic will a reaction ensue.

For this reason, many reactions with a favorable free-energy change do not occur at room temperature. To get such a reaction started, energy (heat) must be

▲ Matches are unreactive at room temperature but burst into flames when struck. The frictional heat produced on striking provides enough energy to start the combustion.

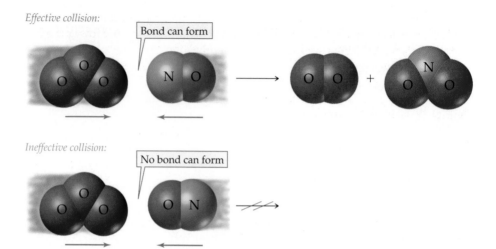

Effective collision:

Bond can form

Ineffective collision:

No bond can form

▶ **FIGURE 7.2 How do chemical reactions occur?** For a collision between NO and O_3 molecules to give O_2 and NO_2, the molecules must collide so that the correct atoms come into contact. No bond forms if the molecules collide with the wrong orientation.

added. The heat causes the reactant particles to move faster, thereby increasing both the frequency and the force of the collisions. We all know that matches burn, for instance, but we also know that they do not burst into flame until struck. The heat of friction provides enough energy for a few molecules to react. Once started, the reaction sustains itself as the energy released by reacting molecules gives other molecules enough energy to react.

The energy change that occurs during the course of a chemical reaction can be visualized in an energy diagram like that in Figure 7.3. At the beginning of the reaction (left side of the diagram), the reactants are at the energy level indicated. At the end of the reaction (right side of the diagram), the products are at a lower energy level than the reactants if the reaction is exergonic (Figure 7.3a) but higher than the reactants if the reaction is endergonic (Figure 7.3b).

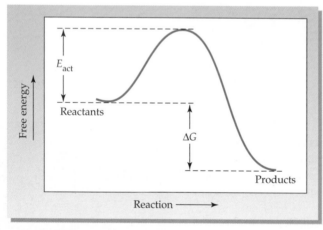

(a) An exergonic reaction

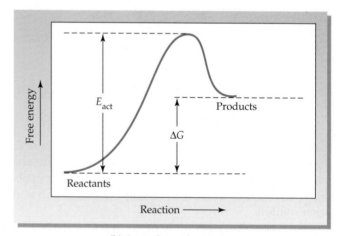

(b) An endergonic reaction

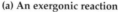

▲ **FIGURE 7.3 Reaction energy diagrams show energy changes during a chemical reaction.** A reaction begins on the left and proceeds to the right. (a) In an exergonic reaction, the product energy level is lower than that of reactants. (b) In an endergonic reaction, the situation is reversed. The height of the barrier between reactant and product energy levels is the activation energy, E_{act}. The difference between reactant and product energy levels is the free-energy change, ΔG.

Activation energy (E_{act}) The amount of energy necessary for reactants to surmount the energy barrier to reaction; determines reaction rate.

Reaction rate A measure of how rapidly a reaction occurs; determined by E_{act}.

Lying between the reactants and the products is an energy "barrier" that must be surmounted. The height of this barrier represents the amount of energy the colliding particles must have for productive collisions to occur, an amount called the **activation energy (E_{act})** of the reaction. The size of the activation energy determines the **reaction rate**, or how fast the reaction occurs. The lower the activation energy, the greater the number of productive collisions in a given amount of time, and the faster the reaction. Conversely, the higher the activation energy, the lower the number of productive collisions, and the slower the reaction.

Note that the size of the activation energy and the size of the free-energy change are unrelated. A reaction with a large E_{act} takes place very slowly even if it has a large negative ΔG. Every reaction is different; each has its own characteristic activation energy and free-energy change.

WORKED EXAMPLE **7.6** Energy of Reactions: Energy Diagrams

Draw an energy diagram for a reaction that is very fast but has a small negative free-energy change.

ANALYSIS A very fast reaction has a small E_{act}. A reaction with a small negative free-energy change is a favorable reaction with a small energy difference between starting materials and products.

SOLUTION

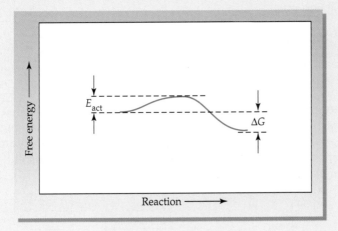

PROBLEM 7.8

Draw an energy diagram for a reaction that is very slow but highly favorable.

PROBLEM 7.9

Draw an energy diagram for a reaction that is slightly unfavorable.

7.6 Effects of Temperature, Concentration, and Catalysts on Reaction Rates

Several things can be done to help reactants over an activation energy barrier and thereby speed up a reaction. Let us look at some possibilities.

Temperature

One way to increase reaction rate is to add energy to the reactants by raising the temperature. With more energy in the system, the reactants move faster, so the frequency of collisions increases. Furthermore, the force with which collisions occur increases, making them more likely to overcome the activation barrier. As a rule of thumb, a 10 °C rise in temperature causes a reaction rate to double.

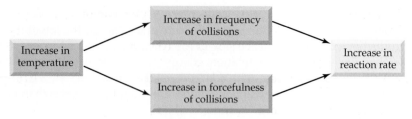

Concentration

Concentration A measure of the amount of a given substance in a mixture.

A second way to speed up a reaction is to increase the **concentrations** of the reactants. As the concentration increases reactants are crowded together, and collisions between reactant molecules become more frequent. As the frequency of collisions increases, reactions between molecules become more likely. Flammable materials burn more rapidly in pure oxygen than in air, for instance, because the concentration of O_2 molecules is higher (air is approximately 21% oxygen). Hospitals must therefore take extraordinary precautions to ensure that no flames are used near patients receiving oxygen. Although different reactions respond differently to concentration changes, doubling or tripling a reactant concentration often doubles or triples the reaction rate.

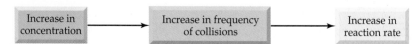

Catalysts

Catalyst A substance that speeds up the rate of a chemical reaction but is itself unchanged.

A third way to speed up a reaction is to add a **catalyst**—a substance that accelerates a chemical reaction but is itself unchanged in the process. For example, such metals as nickel, palladium, and platinum catalyze the addition of hydrogen to the carbon–carbon double bonds in vegetable oils to yield semisolid margarine. Without the metal catalyst, the reaction does not occur.

A catalyst does not affect the energy level of either reactants or products. Rather, it increases reaction rate either by letting a reaction take place through an alternative pathway with a lower energy barrier, or by orienting the reacting molecules appropriately. In a reaction energy diagram, the catalyzed reaction has a lower activation energy (Figure 7.4). It is worth noting that the free-energy change

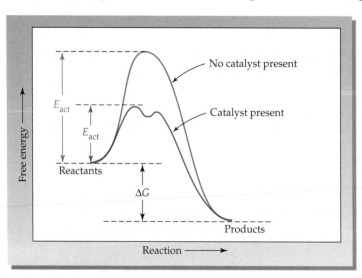

► **FIGURE 7.4 A reaction energy diagram for a reaction in the presence (green curve) and absence (red curve) of a catalyst.** The catalyzed reaction has a lower (E_{act}) because it uses an alternative pathway with a lower energy barrier. The free-energy change ΔG is unaffected by the presence of a catalyst.

for a reaction depends *only* on the difference in the energy levels of the reactants and products, and *not* on the pathway of the reaction. Therefore, a catalyzed reaction releases (or absorbs) the same amount of energy as an uncatalyzed reaction. It simply occurs more rapidly.

In addition to their widespread use in industry, we also rely on catalysts to reduce the air pollution created by exhaust from automobile engines. The catalytic converters in most automobiles are tubes packed with catalysts of two types (Figure 7.5). One catalyst accelerates the complete combustion of hydrocarbons and CO in the exhaust to give CO_2 and H_2O, and the other decomposes NO to N_2 and O_2.

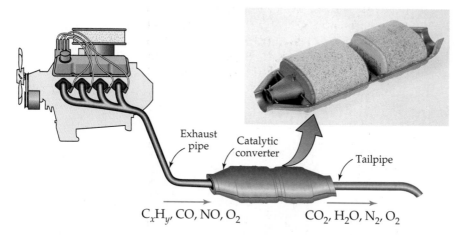

▶ **FIGURE 7.5 A catalytic converter.** The exhaust gases from an automobile pass through a two-stage catalytic converter. In one stage, carbon monoxide and unburned hydrocarbons are converted to CO_2 and H_2O. In the second stage, NO is converted to N_2 and O_2.

Table 7.2 summarizes the effects of changing conditions on reaction rates.

TABLE 7.2 Effects of Changes in Reaction Conditions on Reaction Rate

CHANGE	EFFECT
Concentration	Increase in reactant concentration increases rate.
	Decrease in reactant concentration decreases rate.
Temperature	Increase in temperature increases rate.
	Decrease in temperature decreases rate.
Catalyst added	Increases reaction rate.

Looking Ahead

The thousands of biochemical reactions continually taking place in our bodies are catalyzed by large protein molecules called *enzymes*, which promote reaction by controlling the orientation of the reacting molecules. Since almost every reaction is catalyzed by its own specific enzyme, the study of enzyme structure, activity, and control is a central part of biochemistry. We will look more closely at enzymes and how they work in Chapters 19 and 20.

PROBLEM 7.10

Ammonia is synthesized industrially by reaction of nitrogen and hydrogen in the presence of an iron catalyst according to the equation $3 H_2(g) + N_2(g) \longrightarrow 2 NH_3(g)$.

What effect will the following changes have on the reaction rate?

(a) The temperature is raised from 600 K to 700 K.

(b) The iron catalyst is removed.

(c) The concentration of H_2 gas is halved.

7.7 Reversible Reactions and Chemical Equilibrium

Many chemical reactions result in the virtually complete conversion of reactants into products. When sodium metal reacts with chlorine gas, for example, both are entirely consumed. The sodium chloride product is so much more stable than the reactants that, once started, the reaction keeps going until it is complete.

What happens, though, when the reactants and products are of approximately equal stability? This is the case, for example, in the reaction of acetic acid (the main organic constituent of vinegar) with ethyl alcohol to yield ethyl acetate, a solvent used in nail-polish remover and glue.

$$\underset{\text{Acetic acid}}{CH_3\overset{\overset{\displaystyle O}{\|}}{C}OH} + \underset{\text{Ethyl alcohol}}{HOCH_2CH_3} \underset{\text{Or this direction?}}{\overset{\text{This direction?}}{\rightleftharpoons}} \underset{\text{Ethyl acetate}}{CH_3\overset{\overset{\displaystyle O}{\|}}{C}OCH_2CH_3} + \underset{\text{Water}}{H_2O}$$

APPLICATION ▶ Regulation of Body Temperature

Maintaining normal body temperature is crucial. If the body's thermostat is unable to maintain a temperature of 37 °C, the rates of the many thousands of chemical reactions that take place constantly in the body will change accordingly, with potentially disastrous consequences.

If, for example, a skater fell through the ice of a frozen lake, *hypothermia* could soon result. Hypothermia is a dangerous state that occurs when the body is unable to generate enough heat to maintain normal temperature. All chemical reactions in the body slow down because of the lower temperature, energy production drops, and death can result. Slowing the body's reactions can also be used to advantage, however. During open-heart surgery, the heart is stopped and maintained at about 15 °C, while the body, which receives oxygenated blood from an external pump, is cooled to 25–32 °C.

Conversely, a marathon runner on a hot, humid day might become overheated, and *hyperthermia* could result. Hyperthermia, also called *heat stroke*, is an uncontrolled rise in temperature as the result of the body's inability to lose sufficient heat. Chemical reactions in the body are accelerated at higher temperatures, the heart struggles to pump blood faster to supply increased oxygen, and brain damage can result if the body temperature rises above 41 °C.

Body temperature is maintained both by the thyroid gland and by the hypothalamus region of the brain, which act together to regulate metabolic rate. When the body's environment changes, temperature receptors in the skin, spinal cord, and abdomen send signals to the hypothalamus, which contains both heat-sensitive and cold-sensitive neurons.

Stimulation of the heat-sensitive neurons on a hot day causes a variety of effects: Impulses are sent to stimulate the sweat glands, dilate the blood vessels of the skin, decrease muscular activity, and reduce metabolic rate. Sweating cools the body through evaporation; approximately 540 cal is removed by evaporation of 1.0 g of sweat. Dilated blood vessels cool the body by allowing more blood to flow close to

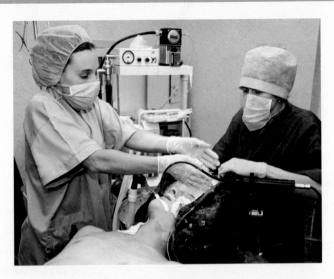

▲ The body is cooled to 25–32 °C by immersion in ice prior to open-heart surgery to slow down metabolism.

the surface of the skin where heat is removed by contact with air. Decreased muscular activity and a reduced metabolic rate cool the body by lowering internal heat production.

Stimulation of the cold-sensitive neurons on a cold day also causes a variety of effects: The hormone epinephrine is released to stimulate metabolic rate; peripheral blood vessels contract to decrease blood flow to the skin and prevent heat loss; and muscular contractions increase to produce more heat, resulting in shivering and "goosebumps."

One further comment: Drinking alcohol to warm up on a cold day actually has the opposite effect. Alcohol causes blood vessels to dilate, resulting in a warm feeling as blood flow to the skin increases. Although the warmth feels good temporarily, body temperature ultimately drops as heat is lost through the skin at an increased rate.

See Problems 7.72 and 7.73 at the end of the chapter.

Imagine the situation if you mix acetic acid and ethyl alcohol. The two begin to form ethyl acetate and water. But as soon as ethyl acetate and water form, they begin to go back to acetic acid and ethyl alcohol. Such a reaction, which easily goes in either direction, is said to be **reversible** and is indicated by a double arrow ($\rightleftharpoons$) in equations. The reaction read from left to right as written is referred to as the *forward reaction*, and the reaction from right to left is referred to as the *reverse reaction*.

Reversible reaction A reaction that can go in either direction, from products to reactants or reactants to products.

Now suppose you mix some ethyl acetate and water. The same thing occurs: As soon as small quantities of acetic acid and ethyl alcohol form, the reaction in the other direction begins to take place. No matter which pair of reactants is mixed together, both reactions occur until ultimately the concentrations of reactants and products reach constant values and undergo no further change. At this point, the reaction vessel contains all four substances—acetic acid, ethyl acetate, ethyl alcohol, and water—and the reaction is said to be in a state of **chemical equilibrium**.

Since the reactant and product concentrations undergo no further change once equilibrium is reached, you might conclude that the forward and reverse reactions have stopped. That is not the case, however. The forward reaction takes place rapidly at the beginning of the reaction but then slows down as reactant concentrations decrease. At the same time, the reverse reaction takes place slowly at the beginning but then speeds up as product concentrations increase (Figure 7.6). Ultimately, the forward and reverse rates become equal and change no further.

Chemical equilibrium A state in which the rates of forward and reverse reactions are the same.

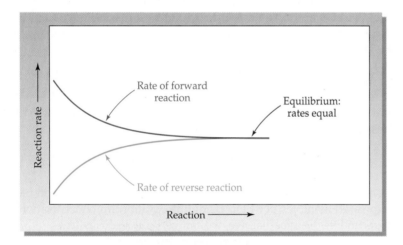

▲ **FIGURE 7.6 Reaction rates in an equilibrium reaction.** The forward rate is large initially but decreases as the concentrations of reactants drop. The reverse rate is small initially but increases as the concentrations of products increase. At equilibrium, the forward and reverse reaction rates are equal.

Chemical equilibrium is an active, dynamic condition. All substances present are continuously being made and unmade at the same rate, so their concentrations are constant at equilibrium. As an analogy, think of two floors of a building connected by up and down escalators. If the number of people moving up is the same as the number of people moving down, the numbers of people on each floor remain constant. *Individual people* are continuously changing from one floor to the other, but the *total populations* of the two floors are in equilibrium.

Note that it is not necessary for the concentrations of reactants and products at equilibrium to be equal (just as it is not necessary for the numbers of people on two floors connected by escalators to be equal). Equilibrium can be reached at any point between pure products and pure reactants. The extent to which the forward or reverse reaction is favored over the other is a characteristic property of a given reaction under given conditions.

▲ When the number of people moving up is the same as the number of people moving down, the number of people on each floor remains constant, and the two populations are in equilibrium.

7.8 Equilibrium Equations and Equilibrium Constants

Remember that the rate of a reaction depends on the number of collisions between molecules (Section 7.5), and that the number of collisions in turn depends on concentration, i.e., the number of molecules in a given volume (Section 7.6). For a reversible reaction, then, the rates of both the forward *and* the reverse reactions must depend on the concentration of reactants and products, respectively. When a reaction reaches equilibrium the rates of the forward and reverse reactions are equal, and the concentration of reactants and products remain constant. We can use this fact to obtain useful information about a reaction.

Let us look at the details of a specific equilibrium reaction. Suppose that you allow various mixtures of sulfur dioxide and oxygen to come to equilibrium with sulfur trioxide at a temperature of 727 °C and then measure the concentrations of all three gases in the mixtures.

$$2\,SO_2(g) + O_2(g) \rightleftharpoons 2\,SO_3(g)$$

In one experiment, we start with only 1.00 mol of SO_2 and 1.00 mol of O_2 in a 1.00 L container. In other words, the initial concentrations of reactants are 1.00 mol/L. When the reaction reaches equilibrium, we have 0.0620 mol/L of SO_2, 0.538 mol/L of O_2, and 0.938 mol/L of SO_3. In another experiment, we start with only 1.00 mol/L of SO_3. When this reaction reaches equilibrium, we have 0.150 mol/L of SO_2, 0.0751 mol/L of O_2, and 0.850 mol/L of SO_3. In both cases, we see that there is substantially more product than reactants when the reaction reaches equilibrium, regardless of the starting conditions. Is it possible to predict what the equilibrium conditions will be for any given reaction?

As it turns out, the answer is YES! No matter what the original concentrations, and no matter what concentrations remain at equilibrium, we find that a constant numerical value is obtained if the equilibrium concentrations are substituted into the expression

$$\frac{[SO_3]^2}{[SO_2]^2[O_2]} = 429 \quad \text{(at a constant temperature of 72\,°C)}$$

The square brackets in this expression indicate the concentration of each substance expressed as moles per liter. Using the equilibrium concentrations for each of the experiments described above, we can verify that this is true:

Experiment 1. $\quad \dfrac{[SO_3]^2}{[SO_2]^2[O_2]} = \dfrac{(0.938 \text{ mol/L})^2}{(0.0620 \text{ mol/L})^2(0.538 \text{ mol/L})} = 425$

Experiment 2. $\quad \dfrac{[SO_3]^2}{[SO_2]^2[O_2]} = \dfrac{(0.850 \text{ mol/L})^2}{(0.150 \text{ mol/L})^2(0.0751 \text{ mol/L})} = 428$

Within experimental error, the ratios of product and reactant concentrations at equilibrium yield the same result. Numerous experiments like those just described have led to a general equation that is valid for any reaction. Consider a general reversible reaction:

$$a\text{A} + b\text{B} + \cdots \rightleftharpoons m\text{M} + n\text{N} + \cdots$$

where A, B, ... are reactants; M, N, ... are products; and $a, b, \ldots, m, n, \ldots$ are coefficients in the balanced equation. At equilibrium, the composition of the reaction mixture obeys the following *equilibrium equation*, where K is the **equilibrium constant**.

Equilibrium equation $\quad K = \dfrac{[\text{M}]^m[\text{N}]^n \cdots}{[\text{A}]^a[\text{B}]^b \cdots}$ ⟵ Product concentrations

⟵ Reactant concentrations

Equilibrium constant

The equilibrium constant K is the number obtained by multiplying the equilibrium concentrations of the products and dividing by the equilibrium concentrations of the reactants, with the concentration of each substance raised to a power equal to its coefficient in the balanced equation. If we take another look at the reaction between sulfur dioxide and oxygen, we can now see how the equilibrium constant was obtained:

$$2\,SO_2\,(g)\ +\ O_2\,(g)\ \rightleftharpoons\ 2\,SO_3\,(g)$$

$$K = \frac{[SO_3]^2}{[SO_2]^2\,[O_2]}$$

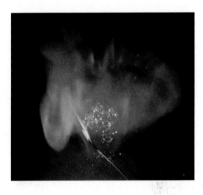

▲ The hydrogen gas in these soap bubbles reacts completely with oxygen when ignited to form water.

Note that if there is no coefficient for a reactant or product in the reaction equation it is assumed to be 1. The value of K varies with temperature—25 °C is assumed unless otherwise specified—and units are usually omitted.

The value of the equilibrium constant indicates the position of a reaction at equilibrium. If the forward reaction is favored, the product term $[M]^m[N]^n$ is larger than the reactant term $[A]^a[B]^b$, and the value of K is larger than 1. If instead the reverse reaction is favored, $[M]^m[N]^n$ is smaller than $[A]^a[B]^b$ at equilibrium, and the value of K is smaller than 1.

For a reaction such as the combination of hydrogen and oxygen to form water vapor, the equilibrium constant is enormous (3.1×10^{81}), showing how greatly the formation of water is favored. Equilibrium is effectively nonexistent for such reactions, and the reaction is described as *going to completion*.

On the other hand, the equilibrium constant is very small for a reaction such as the combination of nitrogen and oxygen at 25 °C to give NO (4.7×10^{-31}), showing what we know from observation—that N_2 and O_2 in the air do not combine noticeably at room temperature:

$$N_2(g) + O_2(g) \rightleftharpoons 2\,NO(g) \quad K = \frac{[NO]^2}{[N_2][O_2]} = 4.7 \times 10^{-31}$$

When K is close to 1, say between 10^3 and 10^{-3}, significant amounts of both reactants and products are present at equilibrium. An example is the reaction of acetic acid with ethyl alcohol to give ethyl acetate (Section 7.7). For this reaction, $K = 3.4$.

$$CH_3CO_2H + CH_3CH_2OH \rightleftharpoons CH_3CO_2CH_2CH_3 + H_2O$$

$$K = \frac{[CH_3CO_2CH_2CH_3][H_2O]}{[CH_3CO_2H][CH_3CH_2OH]} = 3.4$$

We can summarize the meaning of equilibrium constants in the following way:

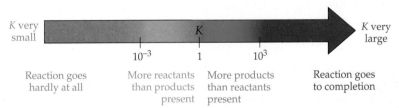

K very small			K very large
	10^{-3} $\qquad$ 1 $\qquad$ 10^3		
Reaction goes hardly at all	More reactants than products present	More products than reactants present	Reaction goes to completion

K much smaller than 0.001   Only reactants are present at equilibrium; essentially no reaction occurs.

K between 0.001 and 1   More reactants than products are present at equilibrium.

K between 1 and 1000	More products than reactants are present at equilibrium.
K much larger than 1000	Only products are present at equilibrium; reaction goes essentially to completion.

WORKED EXAMPLE   **7.7** Writing Equilibrium Equations

The first step in the industrial synthesis of hydrogen is the reaction of steam with methane to give carbon monoxide and hydrogen. Write the equilibrium equation for the reaction.

$$H_2O(g) + CH_4(g) \rightleftharpoons CO(g) + 3\,H_2(g)$$

ANALYSIS The equilibrium constant *K* is the number obtained by multiplying the equilibrium concentrations of the products (CO and H_2) and dividing by the equilibrium concentrations of the reactants (H_2O and CH_4), with the concentration of each substance raised to the power of its coefficient in the balanced equation.

SOLUTION

$$K = \frac{[CO][H_2]^3}{[H_2O][CH_4]}$$

WORKED EXAMPLE   **7.8** Equilibrium Equations: Calculating *K*

In the reaction of Cl_2 with PCl_3, the concentrations of reactants and products were determined experimentally at equilibrium and found to be 7.2 mol/L for PCl_3, 7.2 mol/L for Cl_2, and 0.050 mol/L for PCl_5.

$$PCl_3(g) + Cl_2(g) \rightleftharpoons PCl_5(g)$$

Write the equilibrium equation, and calculate the equilibrium constant for the reaction. Which reaction is favored, the forward one or the reverse one?

ANALYSIS All the coefficients in the balanced equation are 1, so the equilibrium constant equals the concentration of the product, PCl_5, divided by the product of the concentrations of the two reactants, PCl_3 and Cl_2. Insert the values given for each concentration, and calculate the value of *K*.

BALLPARK ESTIMATE At equilibrium, the concentration of the reactants (7.2 mol/L for each reactant) is higher than the concentration of the product (0.05 mol/L), so we expect a value of *K* less than 1.

SOLUTION

$$K = \frac{[PCl_5]}{[PCl_3][Cl_2]} = \frac{0.050\ \text{mol/L}}{(7.2\ \text{mol/L})(7.2\ \text{mol/L})} = 9.6 \times 10^{-4}$$

The value of *K* is less than 1, so the reverse reaction is favored. Note that units for K are omitted.

BALLPARK CHECK Our calculated value of *K* is just as we predicted: *K* < 1.

PROBLEM 7.11

Write equilibrium equations for the following reactions:

(a) $N_2O_4(g) \rightleftharpoons 2\,NO_2(g)$

(b) $CH_4(g) + Cl_2(g) \rightleftharpoons CH_3Cl(g) + HCl(g)$

(c) $2\,BrF_5(g) \rightleftharpoons Br_2(g) + 5\,F_2(g)$

PROBLEM 7.12

Do the following reactions favor reactants or products at equilibrium?

(a) Sucrose(aq) + $H_2O(l)$ $\rightleftharpoons$ Glucose(aq) + Fructose(aq) $K = 1.4 \times 10^5$
(b) $NH_3(aq)$ + $H_2O(l)$ $\rightleftharpoons$ $NH_4^+(aq)$ + $OH^-(aq)$ $K = 1.6 \times 10^{-5}$
(c) $Fe_2O_3(s)$ + $3\,CO(g)$ $\rightleftharpoons$ $2\,Fe(s)$ + $3\,CO_2(g)$ K(at 727°C) = 24.2

PROBLEM 7.13

For the reaction $H_2(g)$ + $I_2(g)$ $\rightleftharpoons$ $2\,HI(g)$, equilibrium concentrations at 25 °C are $[H_2]$ = 0.0510 mol/L, $[I_2]$ = 0.174 mol/L, and $[HI]$ = 0.507 mol/L. What is the value of K at 25 °C?

KEY CONCEPT PROBLEM 7.14

The following diagrams represent two similar reactions that have achieved equilibrium:

   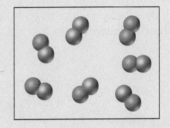

$A_2 + B_2 \longrightarrow 2\,AB$ $C_2 + D_2 \longrightarrow 2\,CD$

Which reaction has the larger equilibrium constant, and which has the smaller equilibrium constant?

7.9 Le Châtelier's Principle: The Effect of Changing Conditions on Equilibria

The effect of a change in reaction conditions on chemical equilibrium is predicted by a general rule called *Le Châtelier's principle:*

Le Châtelier's principle   When a stress is applied to a system at equilibrium, the equilibrium shifts to relieve the stress.

The word "stress" in this context means any change in concentration, pressure, volume, or temperature that disturbs the original equilibrium and causes the rates of the forward and reverse reactions to become temporarily unequal.

We saw in Section 7.6 that reaction rates are affected by changes in temperature and concentration, and by addition of a catalyst. But what about equilibria? Are they similarly affected? The answer is that changes in concentration, temperature, and pressure *do* affect equilibria, but that addition of a catalyst does not (except to reduce the time it takes to reach equilibrium). The change caused by a catalyst affects forward and reverse reactions equally so that equilibrium concentrations are the same in both the presence and the absence of the catalyst.

Effect of Changes in Concentration

Let us look at the effect of a concentration change by considering the reaction of CO with H_2 to form CH_3OH (methanol). Once equilibrium is reached, the concentrations of the reactants and product are constant, and the forward and reverse reaction rates are equal.

$$CO(g) + 2 H_2(g) \rightleftharpoons CH_3OH(g)$$

What happens if the concentration of CO is increased? To relieve the "stress" of added CO, according to Le Châtelier's principle, the extra CO must be used up. In other words, the rate of the forward reaction must increase to consume CO. Think of the CO added on the left as "pushing" the equilibrium to the right:

$$
\begin{array}{c}
[\text{CO} \longrightarrow] \\
CO(g) + 2 H_2(g) \rightleftharpoons CH_3OH(g)
\end{array}
$$

Of course, as soon as more CH_3OH forms, the reverse reaction also speeds up, and so some CH_3OH converts back to CO and H_2. Ultimately, the forward and reverse reaction rates adjust until they are again equal, and equilibrium is reestablished. At this new equilibrium state, the value of $[H_2]$ is lower because some of the H_2 reacted with the added CO and the value of $[CH_3OH]$ is higher because CH_3OH formed as the reaction was driven to the right by the addition of CO. The changes offset each other, however, so that the value of the equilibrium constant K remains constant.

$$CO(g) + 2 H_2(g) \rightleftharpoons CH_3OH(g)$$

If this increases . . .   . . . then this decreases . . .   . . . and this increases . . .

. . . but this remains constant. $K = \dfrac{[CH_3OH]}{[CO][H_2]^2}$

What happens if CH_3OH is added to the reaction at equilibrium? Some of the methanol reacts to yield CO and H_2, making the values of $[CO]$, $[H_2]$, and $[CH_3OH]$ higher when equilibrium is reestablished. As before, the value of K does not change.

If this increases . . .

$$CO(g) + 2 H_2(g) \rightleftharpoons CH_3OH(g)$$

. . . then this increases . . .   . . . and this increases . . .

. . . but this remains constant. $K = \dfrac{[CH_3OH]}{[CO][H_2]^2}$

Alternatively, we can view chemical equilibrium as a *balance* between the energy of the reactants (on the left) and the products (on the right). Adding more reactants tips the balance in favor of the reactants. In order to restore the balance, reactants must be converted to products, or the reaction must shift to the right. If, instead, we remove reactants, then the balance is too heavy on the products side and the reaction must shift left, generating more reactants to restore balance.

Finally, what happens if a reactant is continuously supplied or a product is continuously removed? Because the concentrations are continuously changing, equilibrium can never be reached. As a result, it is sometimes possible to force a reaction to produce large quantities of a desirable product even when the equilibrium constant is unfavorable. Take the reaction of acetic acid with ethanol to yield ethyl

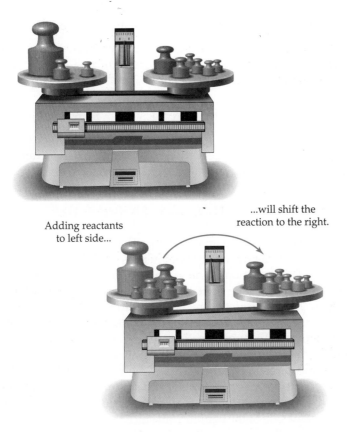

Adding reactants
to left side...

...will shift the
reaction to the right.

▲ Equilibrium represents a balance between the energy of reactants and products. Adding reactants (or products) to one side upsets the balance, and the reaction will proceed in a direction to restore the balance.

acetate, for example. As discussed in the preceding section, the equilibrium constant K for this reaction is 3.4, meaning that substantial amounts of reactants and products are both present at equilibrium. If, however, the ethyl acetate is removed as soon as it is formed, the production of more and more product is forced to occur, in accord with Le Châtelier's principle.

Continuously removing this product from the reaction forces more of it to be produced.

$$CH_3\overset{O}{\overset{\|}{C}}OH + CH_3CH_2OH \rightleftharpoons CH_3\overset{O}{\overset{\|}{C}}OCH_2CH_3 + H_2O$$
Acetic acid   Ethyl alcohol   Ethyl acetate

Metabolic reactions sometimes take advantage of this effect, with one reaction prevented from reaching equilibrium by the continuous consumption of its product in a further reaction.

Effect of Changes in Temperature and Pressure

We noted in Section 7.2 that the reverse of an exothermic reaction is always endothermic. Equilibrium reactions are therefore exothermic in one direction and endothermic in the other. Le Châtelier's principle predicts that an increase in temperature will cause an equilibrium to shift in favor of the endothermic reaction so the additional heat is absorbed. Conversely, a decrease in temperature will cause an equilibrium to shift in favor of the exothermic reaction so additional heat is released. In other words, you can think of heat as a reactant or product whose increase or decrease stresses an equilibrium just as a change in reactant or product concentration does.

Endothermic reaction (Heat is absorbed)	Favored by increase in temperature
Exothermic reaction (Heat is released)	Favored by decrease in temperature

In the exothermic reaction of N_2 with H_2 to form NH_3, for example, raising the temperature favors the reverse reaction, which absorbs the heat:

$$[\longleftarrow \text{Heat}]$$
$$N_2(g) + 3 H_2(g) \rightleftharpoons 2 NH_3(g) + \text{Heat}$$

We can also use the balance analogy to predict the effect of temperature on an equilibrium mixture; this time we think of heat as a reactant or product. Increasing the temperature of the reaction is the same as adding heat to the left side (for an endothermic reaction) or to the right side (for an exothermic reaction). The reaction then proceeds in the appropriate direction to restore "balance" to the system.

What about changing the pressure? Pressure influences an equilibrium only if one or more of the substances involved is a gas. As predicted by Le Châtelier's principle, decreasing the volume to increase the pressure in such a reaction shifts the equilibrium in the direction that decreases the number of molecules in the gas phase and thus decreases the pressure. For the ammonia synthesis, decreasing the volume *increases* the concentration of reactants and products, but has a greater effect on the reactant side of the equilibrium since there are more moles of gas phase reactants. Increasing the pressure, therefore, favors the forward reaction because 4 mol of gas is converted to 2 mol of gas.

$$[\text{Pressure} \longrightarrow]$$
$$\underbrace{N_2(g) + 3 H_2(g)}_{\text{4 mol of gas}} \rightleftharpoons \underbrace{2 NH_3(g)}_{\text{2 mol of gas}}$$

The effects of changing reaction conditions on equilibria are summarized in Table 7.3.

TABLE 7.3   Effects of Changes in Reaction Conditions on Equilibria

CHANGE	EFFECT
Concentration	Increase in reactant concentration or decrease in product concentration favors forward reaction.
	Increase in product concentration or decrease in reactant concentration favors reverse reaction.
Temperature	Increase in temperature favors endothermic reaction.
	Decrease in temperature favors exothermic reaction.
Pressure	Increase in pressure favors side with fewer moles of gas.
	Decrease in pressure favors side with more moles of gas.
Catalyst added	Equilibrium reached more quickly; value of K unchanged.

⊂⊃ Looking Ahead

In Chapter 21, we will see how Le Châtelier's principle is exploited to keep chemical "traffic" moving through the body's metabolic pathways. It often happens that one reaction in a series is prevented from reaching equilibrium because its product is continuously consumed in another reaction. ⊂⊃

APPLICATION ▶   Nitrogen Fixation

All plants and animals need nitrogen—it is present in all proteins and nucleic acids, and is the fourth most abundant element in the human body. Because the triple bond in the N_2 molecule is so strong, however, plants and animals cannot use the free element directly. It is up to nature and the fertilizer industry to convert N_2, which makes up 78% of the atmosphere, into usable nitrogen compounds in a process called *nitrogen fixation*. Plants use ammonia (NH_3), nitrates (NO_3^-), urea (H_2NCONH_2), and other simple, water-soluble compounds as their sources of nitrogen. Animals then get their nitrogen by eating plants or other plant-eating animals. All these processes contribute to the global *nitrogen cycle* (see figure).

Natural Nitrogen Fixation

Because of the very strong triple bond in the N_2 molecule, the rate of conversion of nitrogen to ammonia or nitrate is very slow under normal conditions. However, lightning in the atmosphere supplies sufficient energy to overcome the large activation energy and drive the endothermic reaction between N_2 and O_2 to form NO. Further reactions with O_2 or O_3 ultimately produce water-soluble nitrates that can be used by plants. Lightning produces small amounts of ammonia as well.

Nitrogen fixation also occurs as a result of bacterial activity in the soil. These bacteria are either free-living, such as *cyanobacteria*, or are associated symbiotically with plants, such as the *Rhizobium* found in the root nodules of legumes

(peas, beans, and clover). The biological fixation of nitrogen is represented by the following equation:

$$N_2 + 8\,H^+ + 8\,e^- + 16\,ATP \longrightarrow 2\,NH_3 + H_2 + 16\,ADP$$

where ATP and ADP are adenosine triphosphate and adenosine diphosphate respectively, in the reaction that serves as the major source of energy in many organisms (Chapter 21). This reaction is made possible by an enzyme complex called *nitrogenase*, which acts as a catalyst to lower the energy barrier.

Industrial Nitrogen Fixation

Most industrial nitrogen fixation results from fertilizer production utilizing the *Haber process*, the reaction between nitrogen and hydrogen to give ammonia:

The Haber process

$$N_2(g) + 3\,H_2(g) \longrightarrow 2\,NH_3(g)$$

This reaction is performed at high temperatures and pressures in the presence of an iron-based catalyst to increase the rate of reaction and shift the equilibrium in favor of ammonia. Since the reaction is reversible, an equilibrium would quickly be established between products and reactants; by removing ammonia as it forms, the reaction is continuously shifted to the right according to Le Châtelier's principle to maximize ammonia production.

See Additional Problems 7.74 and 7.75 at the end of the chapter.

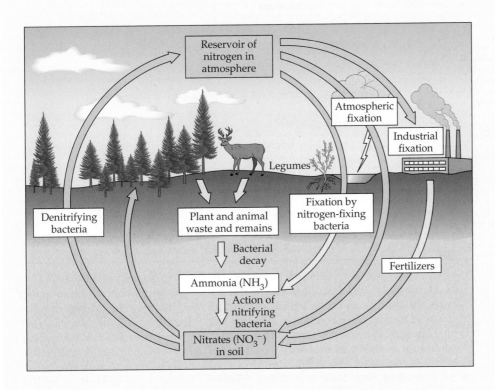

◀ The nitrogen cycle, showing how N_2 in the atmosphere is fixed, used by plants and animals, and then returned to the atmosphere.

WORKED EXAMPLE 7.9 Le Châtelier's Principle and Equilibrium Mixtures

Nitrogen reacts with oxygen to give NO:

$$N_2(g) + O_2(g) \rightleftharpoons 2\,NO(g) \quad \Delta H = +43\ \text{kcal/mol}$$

Explain the effects of the following changes on reactant and product concentrations:

(a) Increasing temperature

(b) Increasing the concentration of NO

(c) Adding a catalyst

SOLUTION

(a) The reaction is endothermic (positive ΔH), so increasing the temperature favors the forward reaction. The concentration of NO will be higher at equilibrium.

(b) Increasing the concentration of NO, a product, favors the reverse reaction. At equilibrium, the concentrations of both N_2 and O_2, as well as that of NO, will be higher.

(c) A catalyst accelerates the rate at which equilibrium is reached, but the concentrations at equilibrium do not change.

PROBLEM 7.15

Is the yield of SO_3 at equilibrium favored by a higher or lower pressure? By a higher or lower temperature?

$$2\,SO_2(g) + O_2(g) \rightleftharpoons 2\,SO_3(g) \quad \Delta H = -47\ \text{kcal/mol}$$

PROBLEM 7.16

What effect do the listed changes have on the position of the equilibrium in the reaction of carbon with hydrogen?

$$C(s) + 2\,H_2(g) \rightleftharpoons CH_4(g) \quad \Delta H = -18\ \text{kcal/mol}$$

(a) Increasing temperature

(b) Increasing pressure by decreasing volume

(c) Allowing CH_4 to escape continuously from the reaction vessel

KEY WORDS

Activation energy (E_{act}) p. 194

Bond dissociation energy, p. 184

Catalyst, p. 196

Chemical equilibrium, p. 199

Concentration, p. 196

SUMMARY: REVISITING THE CHAPTER GOALS

1. **What energy changes take place during reactions?** The strength of a covalent bond is measured by its *bond dissociation energy*, the amount of energy that must be supplied to break the bond in an isolated gaseous molecule. For any reaction, the heat released or absorbed by changes in bonding is called the *heat of reaction*, or *enthalpy change* (ΔH). If the total strength of the bonds formed in a reaction is greater than the total strength of the bonds broken, then heat is released (negative ΔH) and the reaction is said to be *exothermic*. If the total strength of the bonds formed in a reaction is less than the total strength of the bonds broken, then heat is absorbed (positive ΔH) and the reaction is said to be *endothermic*.

2. What is "free energy," and what is the criterion for spontaneity in chemistry? *Spontaneous reactions* are those that, once started, continue without external influence; nonspontaneous reactions require a continuous external influence. Spontaneity depends on two factors, the amount of heat absorbed or released in a reaction (ΔH) and the *entropy change* (ΔS), which measures the change in molecular disorder in a reaction. Spontaneous reactions are favored by a release of heat (negative ΔH) and an increase in disorder (positive ΔS). The *free-energy change* (ΔG) takes both factors into account, according to the equation $\Delta G = \Delta H - T\Delta S$. A negative value for ΔG indicates spontaneity, and a positive value for ΔG indicates nonspontaneity.

3. What determines the rate of a chemical reaction? A chemical reaction occurs when reactant particles collide with proper orientation and sufficient energy. The exact amount of collision energy necessary is called the *activation energy* (E_{act}). A high activation energy results in a slow reaction because few collisions occur with sufficient force, whereas a low activation energy results in a fast reaction. Reaction rates can be increased by raising the temperature, by raising the concentrations of reactants, or by adding a *catalyst*, which accelerates a reaction without itself undergoing any change.

4. What is chemical equilibrium? A reaction that can occur in either the forward or reverse direction is *reversible* and will ultimately reach a state of *chemical equilibrium*. At equilibrium, the forward and reverse reactions occur at the same rate, and the concentrations of reactants and products are constant. Every reversible reaction has a characteristic *equilibrium constant* (K), given by an *equilibrium equation*.

For the reaction: $aA + bB + \cdots \rightleftharpoons mM + nN + \cdots$

$$K = \frac{[M]^m[N]^n \cdots}{[A]^a[B]^b \cdots}$$

Product concentrations raised to powers equal to coefficients

Reactant concentrations raised to powers equal to coefficients

If K is larger than 1, the forward reaction is favored; if K is less than 1, the reverse reaction is favored.

5. What is Le Châtelier's principle? *Le Châtelier's principle* states that when a stress is applied to a system in equilibrium, the equilibrium shifts so that the stress is relieved. Applying this principle allows prediction of the effects of changes in temperature, pressure, and concentration.

UNDERSTANDING KEY CONCEPTS

7.17 What are the signs of ΔH, ΔS, and ΔG for the spontaneous conversion of a crystalline solid into a gas? Explain.

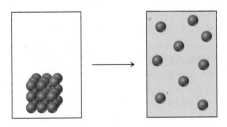

7.18 What are the signs of ΔH, ΔS, and ΔG for the spontaneous condensation of a vapor to a liquid? Explain.

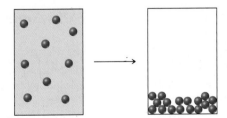

7.19 Consider the following spontaneous reaction of A_2 molecules (red) and B_2 molecules (blue):

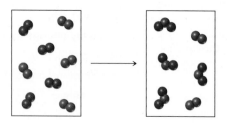

(a) Write a balanced equation for the reaction.
(b) What are the signs of ΔH, ΔS, and ΔG for the reaction? Explain.

7.20 Two curves are shown in the following energy diagram:

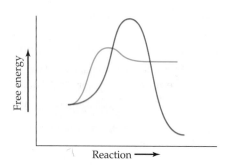

(a) Which curve represents the faster reaction, and which the slower?
(b) Which curve represents the spontaneous reaction, and which the nonspontaneous?

7.21 Two curves are shown in the following energy diagram. Which curve represents the catalyzed reaction, and which the uncatalyzed?

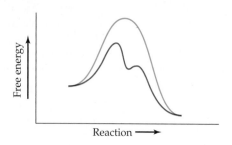

7.22 Draw energy diagrams for the following situations:
(a) A slow reaction with a large negative ΔG
(b) A fast reaction with a small positive ΔG

7.23 The following diagram portrays a reaction of the type $A(s) \longrightarrow B(g) + C(g)$, where the different colored spheres represent different molecular structures. Assume that the reaction has $\Delta H = +9.1$ kcal/mol.

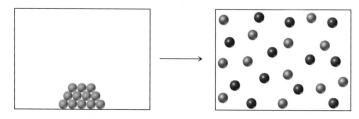

(a) What is the sign of ΔS for the reaction?
(b) Is the reaction likely to be spontaneous at all temperatures, nonspontaneous at all temperatures, or spontaneous at some but nonspontaneous at others?

ADDITIONAL PROBLEMS

ENTHALPY AND HEAT OF REACTION

7.24 Is the total enthalpy (H) of the reactants for an endothermic reaction greater than or less than the total enthalpy of the products?

7.25 What is meant by the term *heat of reaction*? What other name is a synonym for this term?

7.26 The vaporization of Br_2 from the liquid to the gas state requires 7.4 kcal/mol.
(a) What is the sign of ΔH for this process?
(b) How many kilocalories are needed to vaporize 5.8 mol of Br_2?
(c) How many kilocalories are needed to evaporate 82 g of Br_2?

7.27 Converting liquid water to solid ice releases 1.44 kcal/mol.
(a) What is the sign of ΔH for this process?
(b) How many kilocalories are released by freezing 2.5 mol of H_2O?
(c) How many kilocalories are released by freezing 32 g of H_2O?
(d) How many kilocalories are absorbed by melting 1 mol of ice?

7.28 Glucose, also known as "blood sugar," has the formula $C_6H_{12}O_6$.
(a) Write the equation for the combustion of glucose with O_2 to give CO_2 and H_2O.
(b) If 3.8 kcal is released by combustion of each gram of glucose, how many kilocalories are released by the combustion of 1.50 mol of glucose?
(c) What is the minimum amount of energy a plant must absorb to produce 15.0 g of glucose?

7.29 During the combustion of 5.00 g of octane, C_8H_{18}, 239.5 kcal is released.
(a) Write a balanced equation for the combustion reaction.
(b) What is the sign of ΔH for this reaction?
(c) How much energy is released by the combustion of 1.00 mol of C_8H_{18}?
(d) How many grams and how many moles of octane must be burned to release 450.0 kcal?
(e) How many kilocalories are released by the combustion of 17.0 g of C_8H_{18}?

ENTROPY AND FREE ENERGY

7.30 Which of the following processes results in an increase in entropy of the system?

(a) A drop of ink spreading out when it is placed in water
(b) Steam condensing into drops on windows
(c) Constructing a building from loose bricks

7.31 For each of the following processes, specify whether entropy increases or decreases. Explain each of your answers.

(a) Assembling a jigsaw puzzle
(b) $I_2(s) + 3 F_2(g) \longrightarrow 2 IF_3(g)$
(c) A precipitate forming when two solutions are mixed
(d) $C_6H_{12}O_6(aq) + 6 O_2(g) \longrightarrow 6 CO_2(g) + 6 H_2O(g)$
(e) $CaCO_3(s) \longrightarrow CaO(s) + CO_2(g)$
(f) $Pb(NO_3)_2(aq) + 2 NaCl(aq) \longrightarrow$
$PbCl_2(s) + 2 NaNO_3(aq)$

7.32 What is meant by a *spontaneous* process?

7.33 How is the sign of the free-energy change related to the spontaneity of a process?

7.34 What two factors affect the spontaneity of a reaction?

7.35 What is the difference between an exothermic reaction and an exergonic reaction?

7.36 Why are most spontaneous reactions exothermic?

7.37 Is it possible for a reaction to be nonspontaneous yet exothermic? Explain.

7.38 For the reaction

$$NaCl(s) \xrightarrow{\text{Water}} Na^+(aq) + Cl^-(aq), \quad \Delta H = +1.00 \text{ kcal/mol}$$

(a) Is this process endothermic or exothermic?
(b) Does entropy increase or decrease in this process?
(c) Table salt (NaCl) readily dissolves in water. Explain, based on your answers to parts (a) and (b).

7.39 For the reaction $2 Hg(l) + O_2(g) \longrightarrow 2 HgO(s)$, $\Delta H = -43 \text{ kcal/mol}$.

(a) Does entropy increase or decrease in this process? Explain.
(b) Under what conditions would you expect this process to be spontaneous?

7.40 The reaction of gaseous H_2 and liquid Br_2 to give gaseous HBr has $\Delta H = -17.4 \text{ kcal/mol}$ and $\Delta S = 27.2 \text{ cal/(mol} \cdot \text{K)}$.

(a) Write the balanced equation for this reaction.
(b) Does entropy increase or decrease in this process?
(c) Is this process spontaneous at all temperatures? Explain.
(d) What is the value of ΔG for the reaction at 300 K?

7.41 The following reaction is used in the industrial synthesis of PVC polymer:

$$Cl_2(g) + H_2C=CH_2(g) \longrightarrow$$
$$ClCH_2CH_2Cl(l) \quad \Delta H = -52 \text{ kcal/mol}$$

(a) Is ΔS positive or negative for this process?
(b) Is this process spontaneous at all temperatures? Explain.

RATES OF CHEMICAL REACTIONS

7.42 What is the activation energy of a reaction?

7.43 Which reaction is faster, one with $E_{act} = +10 \text{ kcal/mol}$ or one with $E_{act} = +5 \text{ kcal/mol}$? Explain.

7.44 Draw energy diagrams for exergonic reactions that meet the following descriptions:

(a) A slow reaction that has a small free-energy change
(b) A fast reaction that has a large free-energy change

7.45 Draw an energy diagram for a reaction whose products have the same free energies as its reactants. What is free-energy change in this case?

7.46 Give two reasons why increasing temperature increases the rate of a reaction.

7.47 Why does increasing concentration generally increase the rate of a reaction?

7.48 What is a catalyst, and what effect does it have on the activation energy of a reaction?

7.49 If a catalyst changes the activation energy of a forward reaction from 28.0 kcal/mol to 23.0 kcal/mol, what effect does it have on the reverse reaction?

7.50 For the reaction $C(s, \text{diamond}) \longrightarrow C(s, \text{graphite})$, $\Delta G = -0.693 \text{ kcal/mol}$ at 25 °C.

(a) According to this information, do diamonds spontaneously turn into graphite?
(b) In light of your answer to part (a), why can diamonds be kept unchanged for thousands of years?

7.51 The reaction between hydrogen gas and carbon to produce the gas known as ethylene is
$2 H_2(g) + 2 C(s) \longrightarrow H_2C=CH_2(g)$,
$\Delta G = +16.3 \text{ kcal/mol}$ at 25 °C.

(a) Is this reaction spontaneous at 25 °C?
(b) Would it be reasonable to try to develop a catalyst for the reaction run at 25 °C? Explain.

CHEMICAL EQUILIBRIA

7.52 What is meant by the term "chemical equilibrium"? Must amounts of reactants and products be equal at equilibrium?

7.53 Why do catalysts not alter the amounts of reactants and products present at equilibrium?

7.54 Write the equilibrium equations for the following reactions:

(a) $2 CO(g) + O_2(g) \rightleftharpoons 2 CO_2(g)$
(b) $C_2H_6(g) + 2 Cl_2(g) \rightleftharpoons C_2H_4Cl_2(g) + 2 HCl(g)$
(c) $HF(aq) + H_2O(l) \rightleftharpoons H_3O^+(aq) + F^-(aq)$
(d) $3 O_2(g) \rightleftharpoons 2 O_3(g)$

7.55 Write the equilibrium equations for the following reactions, and tell whether reactants or products are favored in each case.

(a) $S_2(g) + 2 H_2(g) \rightleftharpoons 2 H_2S(g) \ K = 2.8 \times 10^{-21}$
(b) $CO(g) + 2 H_2(g) \rightleftharpoons CH_3OH(g) \ K = 10.5$
(c) $Br_2(g) + Cl_2(g) \rightleftharpoons 2 BrCl(g) \ K = 58.0$
(d) $I_2(g) \rightleftharpoons 2I(g) \ K = 6.8 \times 10^{-3}$

7.56 For the reaction $N_2O_4(g) \rightleftharpoons 2 NO_2(g)$, the equilibrium concentrations at 25 °C are $[NO_2] = 0.0325 \text{ mol/L}$ and $[N_2O_4] = 0.147 \text{ mol/L}$. What is the value of K at 25 °C?

7.57 For the reaction $2 CO(g) + O_2(g) \rightleftharpoons 2 CO_2(g)$, the equilibrium concentrations at a certain temperature are $[CO_2] = 0.11 \text{ mol/L}$, $[O_2] = 0.015 \text{ mol/L}$, $[CO] = 0.025 \text{ mol/L}$.

What is the value of K at this temperature?

7.58 Use your answer from Problem 7.56 to calculate the following:

(a) $[N_2O_4]$ at equilibrium when $[NO_2] = 0.0250$ mol/L

(b) $[NO_2]$ at equilibrium when $[N_2O_4] = 0.0750$ mol/L

7.59 Use your answer from Problem 7.57 to calculate the following:

(a) $[O_2]$ at equilibrium when $[CO_2] = 0.18$ mol/L and $[CO] = 0.0200$ mol/L

(b) $[CO_2]$ at equilibrium when $[CO] = 0.080$ mol/L and $[O_2] = 0.520$ mol/L

7.60 Would you expect to find relatively more reactants or more products for the reaction in Problem 7.56 if the pressure is raised? Explain.

7.61 Would you expect to find relatively more reactants or more products for the reaction in Problem 7.57 if the pressure is lowered?

LE CHÂTELIER'S PRINCIPLE

7.62 Oxygen can be converted into ozone by the action of lightning or electric sparks:

$$3\, O_2(g) \rightleftharpoons 2\, O_3(g)$$

For this reaction, $\Delta H = +68$ kcal/mol and $K = 2.68 \times 10^{-29}$ at 25 °C.

(a) Is the reaction exothermic or endothermic?

(b) Are the reactants or the products favored at equilibrium?

(c) Explain the effect on the equilibrium of:

(1) Increasing pressure by decreasing volume

(2) Increasing the concentration of $O_2(g)$

(3) Increasing the concentration of $O_3(g)$

(4) Adding a catalyst

(5) Increasing the temperature

7.63 Hydrogen chloride can be made from the reaction of chlorine and hydrogen:

$$Cl_2(g) + H_2(g) \longrightarrow 2\, HCl(g)$$

For this reaction, $K = 26 \times 10^{33}$ and $\Delta H = -44$ kcal/mol at 25 °C.

(a) Is the reaction endothermic or exothermic?

(b) Are the reactants or the products favored at equilibrium?

(c) Explain the effect on the equilibrium of:

(1) Increasing pressure by decreasing volume

(2) Increasing the concentration of $HCl(g)$

(3) Decreasing the concentration of $Cl_2(g)$

(4) Increasing the concentration of $H_2(g)$

(5) Adding a catalyst

7.64 When the following equilibria are disturbed by increasing the pressure, does the concentration of reaction products increase, decrease, or remain the same?

(a) $2\, CO_2(g) \rightleftharpoons 2\, CO(g) + O_2(g)$

(b) $N_2(g) + O_2(g) \rightleftharpoons 2\, NO(g)$

(c) $Si(s) + 2\, Cl_2(g) \rightleftharpoons SiCl_4(g)$

7.65 For the following equilibria, use Le Châtelier's principle to predict the direction of the reaction when the pressure is increased by decreasing the volume of the equilibrium mixture.

(a) $C(s) + H_2O(g) \rightleftharpoons CO(g) + H_2(g)$

(b) $2\, H_2(g) + O_2(g) \rightleftharpoons 2\, H_2O(g)$

(c) $2\, Fe(s) + 3\, H_2O(g) \rightleftharpoons Fe_2O_3(s) + 3\, H_2(g)$

7.66 The reaction $CO(g) + H_2O(g) \rightleftharpoons CO_2(g) + H_2(g)$ has $\Delta H = -9.8$ kcal/mol. Does the amount of H_2 in an equilibrium mixture increase or decrease when the temperature is decreased?

7.67 The reaction $3\, O_2(g) \rightleftharpoons 2\, O_3(g)$ has $\Delta H = +68$ kcal/mol. Does the equilibrium constant for the reaction increase or decrease when the temperature increases?

7.68 The reaction $H_2(g) + I_2(g) \rightleftharpoons 2\, HI(g)$ has $\Delta H = -2.2$ kcal/mol. Will the equilibrium concentration of HI increase or decrease when:

(a) I_2 is added?

(b) H_2 is removed?

(c) A catalyst is added?

(d) The temperature is increased?

7.69 The reaction $Fe^{3+}(aq) + Cl^-(aq) \rightleftharpoons FeCl^{2+}(aq)$ is endothermic. How will the equilibrium concentration of $FeCl^{2+}$ change when:

(a) $Fe(NO_3)_3$ is added?

(b) Cl^- is precipitated by addition of $AgNO_3$?

(c) The temperature is increased?

(d) A catalyst is added?

Applications

7.70 Which provides more energy, 1 g of carbohydrate or 1 g of fat? [*Energy from Food, p. 189*]

7.71 How many Calories (that is, kilocalories) are in a 45.0 g serving of potato chips if we assume that they are essentially 50% carbohydrate, and 50% fats? [*Energy from Food, p. 189*]

7.72 Which body organs help to regulate body temperature? [*Regulation of Body Temperature, p. 198*]

7.73 What is the purpose of blood vessel dilation? [*Regulation of Body Temperature, p. 198*]

7.74 What does it mean to "fix" nitrogen, and what natural processes accomplish nitrogen fixation? [*Nitrogen Fixation, p. 207*]

7.75 The enthalpy change for the production of ammonia from its elements is $\Delta H = -22$ kcal/mol, yet this reaction does not readily occur at room temperature. Give two reasons why this is so. [*Nitrogen Fixation, p. 207*]

General Questions and Problems

7.76 For the unbalanced combustion reaction shown below, 1 mol of ethanol, C_2H_5OH, releases 327 kcal.

$$C_2H_5OH + O_2 \longrightarrow CO_2 + H_2O$$

(a) Write a balanced equation for the combustion reaction.

(b) What is the sign of ΔH for this reaction?

(c) How much heat (in kilocalories) is released from the combustion of 5.00 g of ethanol?

(d) How many grams of C_2H_5OH must be burned to raise the temperature of 500.0 mL of water from 20.0 °C to 100.0 °C? (The specific heat of water is 1.00 cal/g · °C. See Section 2.10)

(e) If the density of ethanol is 0.789 g/mL, calculate the combustion energy of ethanol in kilocalories/milliliter

7.77 For the production of ammonia from its elements, $\Delta H = -22\,\text{kcal/mol}$.

    (a) Is this process endothermic or exothermic?

    (b) How many kilocalories are involved in the production of 0.700 mol of NH_3?

7.78 Magnetite, an iron ore with formula Fe_3O_4, can be reduced by treatment with hydrogen to yield iron metal and water vapor.

    (a) Write the balanced equation.

    (b) This process requires 36 kcal for every 1.00 mol of Fe_3O_4 reduced. How much energy (in kilocalories) is required to produce 55 g of iron?

    (c) How many grams of hydrogen are needed to produce 75 g of iron?

    (d) This reaction has $K = 2.3 \times 10^{-18}$. Are the reactants or the products favored?

7.79 Hemoglobin (Hb) reacts reversibly with O_2 to form HbO_2, a substance that transfers oxygen to tissues:

$$\text{Hb}(aq) + O_2(aq) \rightleftharpoons \text{HbO}_2(aq)$$

Carbon monoxide (CO) is attracted to Hb 140 times more strongly than O_2 and establishes another equilibrium.

    (a) Explain, using Le Châtelier's principle, why inhalation of CO can cause weakening and eventual death.

    (b) Still another equilibrium is established when both O_2 and CO are present:

$$\text{Hb(CO)}(aq) + O_2(aq) \rightleftharpoons \text{HbO}_2(aq) + \text{CO}(aq)$$

Explain, using Le Châtelier's principle, why pure oxygen is often administered to victims of CO poisoning.

7.80 Many hospitals administer glucose intravenously to patients. If 3.8 kcal is provided by each gram of glucose, how many grams must be administered to maintain a person's normal basal metabolic needs of about 1700 kcal/day?

7.81 For the evaporation of water, $H_2O(l) \longrightarrow H_2O(g)$, at 100 °C, $\Delta H = +9.72\,\text{kcal/mol}$.

    (a) How many kilocalories are needed to vaporize 10.0 g of $H_2O(l)$?

    (b) How many kilocalories are released when 10.0 g of $H_2O(g)$ is condensed?

7.82 Ammonia reacts slowly in air to produce nitrogen monoxide and water vapor:

$$NH_3(g) + O_2(g) \rightleftharpoons NO(g) + H_2O(g) + \text{Heat}$$

    (a) Balance the equation.

    (b) Write the equilibrium equation.

    (c) Explain the effect on the equilibrium of:

        (1) Raising the pressure

        (2) Adding $NO(g)$

        (3) Decreasing the concentration of NH_3

        (4) Lowering the temperature

7.83 Methanol, CH_3OH, is used as race car fuel.

    (a) Write the balanced equation for the combustion of methanol.

    (b) $\Delta H = -174\,\text{kcal/mol}$ methanol for the process. How many kilocalories are released by burning 50.0 g of methanol?

7.84 Sketch an energy diagram for a system in which the forward reaction has $E_{act} = +25\,\text{kcal/mol}$ and the reverse reaction has $E_{act} = +35\,\text{kcal/mol}$.

    (a) Is the forward process endergonic or exergonic?

    (b) What is the value of ΔG for the reaction?

7.85 The thermite reaction (photograph, p. 185), in which aluminum metal reacts with iron(III) oxide to produce a spectacular display of sparks, is so exothermic that the product (iron) is in the molten state:

$$2\,Al(s) + Fe_2O_3(s) \longrightarrow 2\,Al_2O_3(s) + 2\,Fe(l)$$
$$\Delta H = -202.9\,\text{kcal/mol}$$

How much heat (in kilocalories) is released when 5.00 g of Al is used in the reaction?

7.86 How much heat (in kilocalories) is evolved or absorbed in the reaction of 1.00 g of Na with H_2O? Is the reaction exothermic or endothermic?

$$2\,Na(s) + 2\,H_2O(l) \longrightarrow 2\,NaOH(aq) + H_2(g)$$
$$\Delta H = -88.0\,\text{kcal/mol}$$

CHAPTER 8

Gases, Liquids, and Solids

▲ The three states of matter—solid (rock), liquid (lava, water), and gas (steam)—all come together as this lava stream flows into the ocean.

CONTENTS

CHAPTER GOALS

In this chapter, we will answer the following questions:

1. How do scientists explain the behavior of gases?

THE GOAL: Be able to state the assumptions of the kinetic–molecular theory and use these assumptions to explain the behavior of gases.

2. How do gases respond to changes in temperature, pressure, and volume?

THE GOAL: Be able to use Boyle's law, Charles's law, Gay-Lussac's law, and Avogadro's law to explain the effect on gases of a change in pressure, volume, or temperature.

3. What is the ideal gas law?

THE GOAL: Be able to use the ideal gas law to find the pressure, volume, temperature, or molar amount of a gas sample.

4. What is partial pressure?

THE GOAL: Be able to define partial pressure and use Dalton's law of partial pressures.

5. What are the major intermolecular forces, and how do they affect the states of matter?

THE GOAL: Be able to explain dipole–dipole forces, London dispersion forces, and hydrogen bonding, and recognize which of these forces affect a given molecule.

6. What are the various kinds of solids, and how do they differ?

THE GOAL: Be able to recognize the different kinds of solids and describe their characteristics.

7. What factors affect a change of state?

THE GOAL: Be able to apply the concepts of heat change, equilibrium, and vapor pressure to changes of state.

T he previous seven chapters dealt with matter at the atomic level. We have seen that all matter is composed of atoms, ions, or molecules; that these particles are in constant motion; that atoms combine to make compounds using chemical bonds; and that physical and chemical changes are accompanied by the release or absorption of energy. Now it is time to look at a different aspect of matter, concentrating not on the properties and small-scale behavior of individual atoms but on the properties and large-scale behavior of visible amounts of matter.

8.1 States of Matter and Their Changes

Matter exists in any of three phases, or *states*—solid, liquid, and gas. The state in which a compound exists under a given set of conditions depends on the relative strength of the attractive forces between particles compared to the kinetic energy of the particles. Kinetic energy (Section 7.1) is energy associated with motion, and is related to the temperature of the substance. In gases, the attractive forces between particles are very weak compared to their kinetic energy, so the particles move about freely, are far apart, and have almost no influence on one another. In liquids, the attractive forces between particles are stronger, pulling the particles close together but still allowing them considerable freedom to move about. In solids, the attractive forces are much stronger than the kinetic energy of the particles, so the atoms, molecules, or ions are held in a specific arrangement and can only wiggle around in place (Figure 8.1).

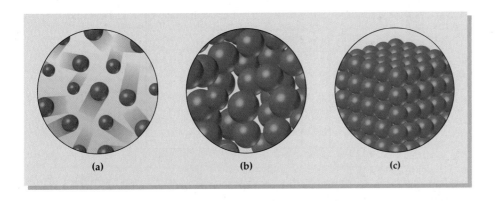

(a) (b) (c)

◀ **FIGURE 8.1 A molecular comparison of gases, liquids, and solids.** (a) In gases, the particles feel little attraction for one another and are free to move about randomly. (b) In liquids, the particles are held close together by attractive forces but are free to slide over one another. (c) In solids, the particles still move slightly, but are held in a specific arrangement with respect to one another.

Change of state The change of a substance from one state of matter (gas, liquid, or solid) to another.

The transformation of a substance from one state to another is called a *phase change*, or a **change of state**. Every change of state is reversible and, like all chemical and physical processes, is characterized by a free-energy change, ΔG. A change of state that is spontaneous in one direction (exergonic, negative ΔG) is nonspontaneous in the other direction (endergonic, positive ΔG). As always, the free-energy change ΔG has both an enthalpy term ΔH and a temperature-dependent entropy term ΔS, according to the equation $\Delta G = \Delta H - T\Delta S$. (You might want to reread Section 7.4 to brush up on these ideas.)(⬤▭, p. 188)

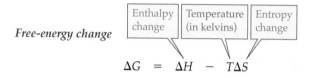

$$\Delta G = \Delta H - T\Delta S$$

The enthalpy change ΔH is a measure of the heat absorbed or released during a given change of state. In the melting of a solid to a liquid, for example, heat is absorbed and ΔH is positive (endothermic). In the reverse process—the freezing of a liquid to a solid—heat is released and ΔH is negative (exothermic). Look at the change between ice and water for instance:

Melting: $H_2O(s) \longrightarrow H_2O(l)$ $\Delta H = +1.44 \text{ kcal/mol}$

Freezing: $H_2O(l) \longrightarrow H_2O(s)$ $\Delta H = -1.44 \text{ kcal/mol}$

The entropy change ΔS is a measure of the change in molecular disorder or freedom that occurs during a process. In the melting of a solid to a liquid, for example, disorder increases because particles gain freedom of motion, so ΔS is positive. In the reverse process—the freezing of a liquid to a solid—disorder decreases as particles are locked into position, so ΔS is negative. Look at the change between ice and water:

Melting: $H_2O(s) \longrightarrow H_2O(l)$ $\Delta S = +5.26 \text{ cal/(mol·K)}$

Freezing: $H_2O(l) \longrightarrow H_2O(s)$ $\Delta S = -5.26 \text{ cal/(mol·K)}$

As with all processes that are unfavored by one term in the free-energy equation but favored by the other, the sign of ΔG depends on the temperature (Section 7.4). The melting of ice, for instance, is unfavored by a positive ΔH but favored by a positive ΔS. Thus, at a low temperature, the unfavorable ΔH is larger than the favorable

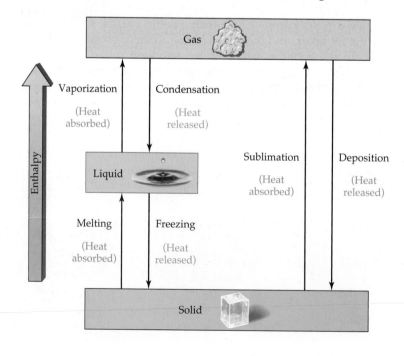

▶ **FIGURE 8.2 Changes of state.** The changes are endothermic from bottom to top and exothermic from top to bottom. Solid and liquid states are in equilibrium at the melting point; liquid and gas states are in equilibrium at the boiling point.

$T\Delta S$ so ΔG is positive and no melting occurs. At a higher temperature, however, $T\Delta S$ becomes larger than ΔH, so ΔG is negative and melting *does* occur. The exact temperature at which the changeover in behavior occurs is called the **melting point (mp)** and represents the temperature at which solid and liquid coexist in equilibrium. In the corresponding change from a liquid to a gas, the two states are in equilibrium at the **boiling point (bp)**.

Melting point (mp) The temperature at which solid and liquid are in equilibrium.

The names and enthalpy changes associated with the different changes of state are summarized in Figure 8.2. Note a solid can change directly to a gas without going through the liquid state—a process called *sublimation*. Dry ice (solid CO_2) at atmospheric pressure, for example, changes directly to a gas without melting.

Boiling point (bp) The temperature at which liquid and gas are in equilibrium.

WORKED EXAMPLE **8.1** Change of State: Enthalpy, Entropy, and Free Energy

The change of state from liquid to gas for chloroform, formerly used as an anesthetic, has $\Delta H = +6.98$ kcal/mol and a $\Delta S = +20.9$ cal/(mol·K).

(a) Is the change of state from liquid to gas favored or unfavored by ΔH? by ΔS?

(b) Is the change of state from liquid to gas favored or unfavored at 35 °C?

(c) Is this change of state spontaneous at 65 °C?

ANALYSIS A process will be favored if energy is released (ΔH = negative) and if there is a decrease in disorder (ΔS = positive). In cases in which one factor is favorable and the other is unfavorable, then we can calculate the free energy change to determine if the process is favored:

$$\Delta G = \Delta H - T\Delta S$$

When ΔG is negative, the process is favored.

SOLUTION

(a) The ΔH does NOT favor this change of state (ΔH = positive), but the ΔS does favor the process. Since the two factors are not in agreement, we must use the equation for free-energy change to determine if the process is favored at a given temperature.

(b) Substituting the values for ΔH and ΔS into the equation for free energy change we can determine if ΔG is positive or negative at 35 °C (308 K). Note that we must first convert degrees Celsius to kelvins and convert the ΔS from cal to kcal so the units can be added together.

$$\Delta G = \Delta H - T\Delta S = \left(\frac{6.98 \text{ kcal}}{\text{mol}}\right) - (308 \text{ K})\left(\frac{20.9 \text{ cal}}{\text{mol·K}}\right)\left(\frac{1 \text{ kcal}}{1000 \text{ cal}}\right)$$

$$= 6.98\frac{\text{kcal}}{\text{mol}} - 6.44\frac{\text{kcal}}{\text{mol}} = +0.54\frac{\text{kcal}}{\text{mol}}$$

Since the ΔG = positive, this change of state is not favored at 35 °C.

(c) Repeating the calculation using the equation for free-energy change at 65 °C (338 K):

$$\Delta G = \Delta H - T\Delta S = \left(\frac{6.98 \text{ kcal}}{\text{mol}}\right) - (338 \text{ K})\left(\frac{20.9 \text{ cal}}{\text{mol·K}}\right)\left(\frac{1 \text{ kcal}}{1000 \text{ cal}}\right)$$

$$= 6.98\frac{\text{kcal}}{\text{mol}} - 7.06\frac{\text{kcal}}{\text{mol}} = -0.08\frac{\text{kcal}}{\text{mol}}$$

Because ΔG is negative in this case, the change of state is favored at this temperature.

PROBLEM 8.1

The change of state from liquid H_2O to gaseous H_2O has $\Delta H = +9.72$ kcal/mol and $\Delta S = +26.1$ cal/(mol·K).

(a) Is the change from liquid to gaseous H_2O favored or unfavored by ΔH? By ΔS?

(b) What is the value of ΔG for the change from liquid to gaseous H_2O at 373 K?

(c) What are the values of ΔH and ΔS for the change from gaseous to liquid H_2O?

8.2 Gases and the Kinetic–Molecular Theory

Gases behave quite differently from liquids and solids. Gases, for instance, have low densities and are easily compressed to a smaller volume when placed under pressure, a property that allows them to be stored in large tanks. Liquids and solids, by contrast, are much more dense and much less compressible. Furthermore, gases undergo a far larger expansion or contraction when their temperature is changed than do liquids and solids.

The behavior of gases can be explained by a group of assumptions known as the **kinetic–molecular theory of gases**. We will see in the next several sections how the following assumptions account for the observable properties of gases:

Kinetic–molecular theory of gases A group of assumptions that explain the behavior of gases.

- **A gas consists of many particles, either atoms or molecules, moving about at random with no attractive forces between them.** Because of this random motion, different gases mix together quickly.

- **The amount of space occupied by the gas particles themselves is much smaller than the amount of space between particles.** Most of the volume taken up by gases is empty space, accounting for the ease of compression and low densities of gases.

- **The average kinetic energy of gas particles is proportional to the Kelvin temperature.** Thus, gas particles have more kinetic energy and move faster as the temperature increases. (In fact, gas particles move much faster than you might suspect. The average speed of a helium atom at room temperature and atmospheric pressure is approximately 1.36 km/s, or 3000 mi/hr, nearly that of a rifle bullet.)

- **Collisions of gas particles, either with other particles or with the wall of their container, are elastic; that is, the total kinetic energy of the particles is constant.** The pressure of a gas against the walls of its container is the result of collisions of the gas particles with the walls. The more collisions and the more forceful each collision, the higher the pressure.

A gas that obeys all the assumptions of the kinetic–molecular theory is called an **ideal gas**. In practice, though, there is no such thing as a perfectly ideal gas. All gases behave somewhat differently than predicted when, at very high pressures or very low temperatures, their particles get closer together and interactions between particles become significant. As a rule, however, most real gases display nearly ideal behavior under normal conditions.

Ideal gas A gas that obeys all the assumptions of the kinetic–molecular theory.

8.3 Pressure

We are all familiar with the effects of air pressure. When you fly in an airplane, the change in air pressure against your eardrums as the plane climbs or descends can cause a painful "popping." When you pump up a bicycle tire, you increase the pressure of air against the inside walls of the tire until the tire feels hard.

In scientific terms, **pressure (P)** is defined as a force (F) per unit area (A) pushing against a surface; that is, $P = F/A$. In the bicycle tire, for example, the pressure

Pressure (P) The force per unit area pushing against a surface.

you feel is the force of air molecules colliding with the inside walls of the tire. The units you probably use for tire pressure are pounds per square inch (psi), where 1 psi is equal to the pressure exerted by a 1 pound object resting on a 1 square inch surface.

We on earth are under pressure from the atmosphere, the blanket of air pressing down on us (Figure 8.3). Atmospheric pressure is not constant, however; it varies slightly from day to day depending on the weather, and it also varies with altitude. Due to gravitational forces, the density of air is greatest at the earth's surface and decreases with increasing altitude. As a result, air pressure is greatest at the surface: about 14.7 psi at sea level but only about 4.7 psi on the summit of Mt. Everest.

One of the most commonly used units of pressure is the *millimeter of mercury*, abbreviated *mmHg* and often called a *torr* (after the Italian physicist Evangelista Torricelli). This unusual unit dates back to the early 1600s when Torricelli made the first mercury *barometer*. As shown in Figure 8.4, a barometer consists of a long, thin tube that is sealed at one end, filled with mercury, and then inverted into a dish of mercury. Some mercury runs from the tube into the dish until the downward pressure of the mercury in the column is exactly balanced by the outside atmospheric pressure, which presses down on the mercury in the dish and pushes it up into the column. The height of the mercury column varies depending on the altitude and weather conditions, but standard atmospheric pressure at sea level is defined to be exactly 760 mm.

Gas pressure inside a container is often measured using an open-end *manometer*, a simple instrument similar in principle to the mercury barometer. As shown in Figure 8.5, an open-end manometer consists of a U-tube filled with mercury, with one end connected to a gas-filled container and the other end open to the atmosphere. The difference between the heights of the mercury levels in the two arms of the U-tube indicates the difference between the pressure of the gas in the container and the pressure of the atmosphere. If the gas pressure inside the container is less than atmospheric, the mercury level is higher in the arm connected to the container (Figure 8.5a). If the gas pressure inside the container is greater than atmospheric, the mercury level is higher in the arm open to the atmosphere (Figure 8.5b).

Pressure is given in the SI system (Section 2.1) by a unit named the *pascal* (Pa), where 1 Pa = 0.007500 mmHg (or 1 mmHg = 133.32 Pa). Measurements in pascals are becoming more common, and many clinical laboratories have made the

▲ **FIGURE 8.3 Atmospheric pressure.** A column of air weighing 14.7 lb presses down on each square inch of the earth's surface at sea level, resulting in what we call atmospheric pressure.

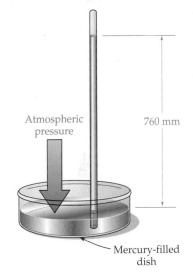

▲ **FIGURE 8.4 Measuring atmospheric pressure.** A mercury barometer measures atmospheric pressure by determining the height of a mercury column in a sealed glass tube. The downward pressure of the mercury in the column is exactly balanced by the outside atmospheric pressure, which presses down on the mercury in the dish and pushes it up into the column.

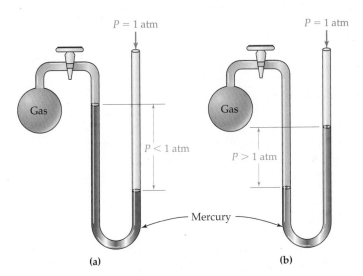

▲ **FIGURE 8.5 Open-end manometers for measuring pressure in a gas-filled bulb.** (a) When the pressure in the gas-filled container is lower than atmospheric, the mercury level is higher in the arm open to the container. (b) When the pressure in the container is higher than atmospheric, the mercury level is higher in the arm open to the atmosphere.

switchover. Higher pressures are often still given in *atmospheres* (atm), where 1 atm = 760 mmHg exactly.

$$\text{Pressure units: 1 atm} = 760 \text{ mmHg} = 14.7 \text{ psi} = 101{,}325 \text{ Pa}$$
$$1 \text{ mmHg} = 1 \text{ torr} = 133.32 \text{ Pa}$$

WORKED EXAMPLE 8.2 Unit Conversions (Pressure): psi, Atmospheres, and Pascals

A typical bicycle tire is inflated with air to a pressure of 55 psi. How many atmospheres is this? How many pascals?

ANALYSIS Using the starting pressure in psi, the pressure in atm and pascals can be calculated using the equivalent values in as conversion factors.

SOLUTION

STEP 1: **Identify known information.**	Pressure = 55 psi
STEP 2: **Identify answer and units.**	Pressure = ?? atm = ?? pascals
STEP 3: **Identify conversion factors.** Using equivalent values in appropriate units, we can obtain conversion factors to convert to atm and pascals.	$14.7 \text{ psi} = 1 \text{ atm} \rightarrow \dfrac{1 \text{ atm}}{14.7 \text{ psi}}$ $14.7 \text{ psi} = 101{,}325 \text{ Pa} \rightarrow \dfrac{101{,}325 \text{ Pa}}{14.7 \text{ psi}}$
STEP 4: **Solve.** Use the appropriate conversion factors to set up an equation in which unwanted units cancel.	$(55 \text{ psi}) \times \left(\dfrac{1 \text{ atm}}{14.7 \text{ psi}} \right) = 3.7 \text{ atm}$ $(55 \text{ psi}) \times \left(\dfrac{101{,}325 \text{ Pa}}{14.7 \text{ psi}} \right) = 3.8 \times 10^5 \text{ Pa}$

WORKED EXAMPLE 8.3 Unit Conversions (Pressure): mmHg to Atmospheres

The pressure in a closed flask is measured using a manometer. If the mercury level in the arm open to the sealed vessel is 23.6 cm higher than the level of mercury in the arm open to the atmosphere, what is the gas pressure (in atm) in the closed flask?

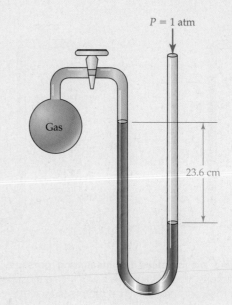

ANALYSIS Since the mercury level is higher in the arm open to the flask, the gas pressure in the flask is lower than atmospheric pressure (1 atm = 760 mmHg). We can convert the difference in the level of mercury in the two arms of the manometer from mmHg to atmospheres to determine the difference in pressure.

BALLPARK ESTIMATE The height difference (23.6 cm) is about one-third the height of a column of Hg that is equal to 1 atm (or 76 cm Hg). Therefore, the pressure in the flask should be about 0.33 atm lower than atmospheric pressure, or about 0.67 atm.

SOLUTION
Since the height difference is given in cm Hg, we must first convert to mmHg, and then to atm. The result is the difference in gas pressure between the flask and the open atmosphere (1 atm).

$$(23.6 \text{ cm Hg})\left(\frac{10 \text{ mmHg}}{\text{cm Hg}}\right)\left(\frac{1 \text{ atm}}{760 \text{ mmHg}}\right) = 0.311 \text{ atm}$$

The pressure in the flask is calculated by subtracting this difference from 1 atm:

$$1 \text{ atm} - 0.311 \text{ atm} = 0.689 \text{ atm}$$

BALLPARK CHECK: This result agrees well with our estimate, 0.67 atm.

PROBLEM 8.2

The air pressure outside a jet airliner flying at 35,000 ft is about 220 mmHg. How many atmospheres is this? How many pounds per square inch? How many pascals?

●○ KEY CONCEPT PROBLEM 8.3

What is the pressure of the gas inside the following manometer (in mmHg) if outside pressure is 750 mmHg?

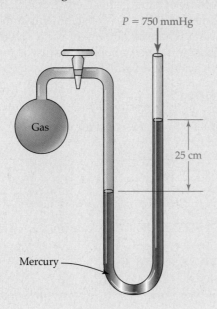

8.4 Boyle's Law: The Relation between Volume and Pressure

The physical behavior of all gases is much the same, regardless of identity. Helium and chlorine, for example, are completely different in their *chemical* behavior, but are very similar in many of their physical properties. Observations of many different gases by scientists in the 1700s led to the formulation of what are now called the **gas laws**, which make it possible to predict the influence of pressure (P), volume (V), temperature (T), and molar amount (n) on any gas or mixture of gases. We will begin by looking at *Boyle's law*, which describes the relation between volume and pressure.

Imagine that you have a sample of gas inside a cylinder that has a movable plunger at one end (Figure 8.6). What happens if you double the pressure on the gas by pushing the plunger down, while keeping the temperature constant? Since the gas particles are forced closer together, the volume of the sample decreases.

Gas laws A series of laws that predict the influence of pressure (P), volume (V), and temperature (T) on any gas or mixture of gases.

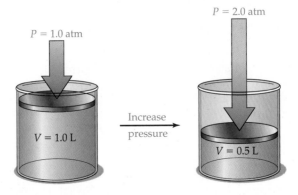

▶ **FIGURE 8.6 Boyle's law.** The volume of a gas decreases proportionately as its pressure increases. If the pressure of a gas sample is doubled, the volume is halved.

According to **Boyle's law**, the volume of a fixed amount of gas at a constant temperature is inversely proportional to its pressure, meaning that volume and pressure change in opposite directions. As pressure goes up, volume goes down; as pressure goes down, volume goes up (Figure 8.7). This observation is consistent with the kinetic–molecular theory. Since most of the volume occupied by gases is empty space, gases are easily compressed into smaller volumes. Since the average kinetic energy remains constant, the number of collisions must increase as the interior surface area of the container decreases, leading to an increase in pressure.

Boyle's law The volume of a gas is inversely proportional to its pressure for a fixed amount of gas at a constant temperature. That is, P times V is constant when the amount of gas n and the temperature T are kept constant. (The symbol $\propto$ means "is proportional to," and k denotes a constant value.)

$$\text{Volume } (V) \propto \frac{1}{\text{Pressure } (P)}$$

$$\text{or} \quad PV = k \quad \text{(A constant value)}$$

▶ **FIGURE 8.7 Boyle's law.** Pressure and volume are inversely related. Graph (a) demonstrates the decrease in volume as pressure increases, whereas graph (b) shows the linear relationship between V and $1/P$.

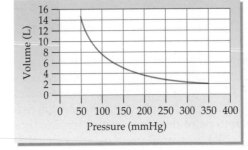

(a)

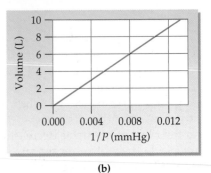

(b)

APPLICATION ▶ Blood Pressure

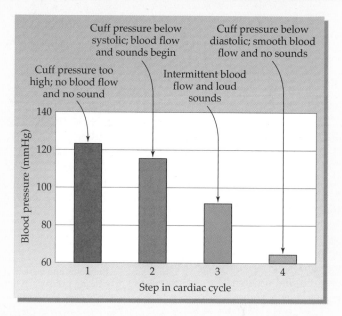

aving your blood pressure measured is a quick and easy way to get an indication of the state of your circulatory system. Although blood pressure varies with age, a normal adult male has a reading near 120/80 mmHg, and a normal adult female has a reading near 110/70 mmHg. Abnormally high values signal an increased risk of heart attack and stroke.

Pressure varies greatly in different types of blood vessels. Usually, though, measurements are carried out on arteries in the upper arm as the heart goes through a full cardiac cycle. *Systolic pressure* is the maximum pressure developed in the artery just after contraction, as the heart forces the maximum amount of blood into the artery. *Diastolic pressure* is the minimum pressure that occurs at the end of the heart cycle.

Blood pressure is most often measured by a *sphygmomanometer*, a device consisting of a squeeze bulb, a flexible cuff, and a mercury manometer. The cuff is placed around the upper arm over the brachial artery and inflated by the squeeze bulb to about 200 mmHg pressure, an amount great enough to squeeze the artery shut and prevent blood flow. Air is then slowly released from the cuff, and pressure drops. As cuff pressure reaches the systolic pressure, blood spurts through the artery, creating a turbulent tapping sound that can be heard through a stethoscope. The pressure registered on the manometer at the moment the first sounds are heard is the systolic blood pressure.

Sounds continue until the pressure in the cuff becomes low enough to allow diastolic blood flow. At this point, blood flow becomes smooth, no sounds are heard, and a diastolic

▲ The sequence of events during blood pressure measurement, including the sounds heard.

blood pressure reading is recorded on the manometer. Readings are usually recorded as systolic/diastolic, for example, 120/80. The accompanying Figure shows the sequence of events during measurement.

See Additional Problem 8.100 at the end of the chapter.

Because $P \times V$ is a constant value for a fixed amount of gas at a constant temperature, the starting pressure (P_1) times the starting volume (V_1) must equal the final pressure (P_2) times the final volume (V_2). Thus, Boyle's law can be used to find the final pressure or volume when the starting pressure or volume is changed.

Since $P_1 V_1 = k$ and $P_2 V_2 = k$

then $P_1 V_1 = P_2 V_2$

so $P_2 = \dfrac{P_1 V_1}{V_2}$ and $V_2 = \dfrac{P_1 V_1}{P_2}$

As an example of Boyle's law behavior, think about what happens every time you breathe. Between breaths, the pressure inside your lungs is equal to atmospheric pressure. When inhalation takes place, your diaphragm lowers and the rib cage expands, increasing the volume of the lungs and thereby decreasing the pressure inside them (Figure 8.8). Air must then move into the lungs to equalize their pressure with that of the atmosphere. When exhalation takes place, the diaphragm rises and the rib cage contracts decreasing the volume of the lungs and increasing pressure inside them. Now gases move out of the lungs until pressure is again equalized with the atmosphere.

▲ Air pressure in tires, typically reported in units of pounds per square inch (psi), is another example of Boyle's Law.

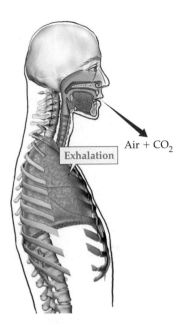

Inhalation

Air

Rest position

Exhalation

Air + CO_2

▶ **FIGURE 8.8 Boyle's law in breathing.** During inhalation, the diaphragm moves down and the rib cage moves up and out, thus increasing lung volume, decreasing pressure, and drawing in air. During exhalation, lung volume decreases, pressure increases, and air moves out.

Lung volume increases, causing pressure in lungs to *decrease*. Air flows *in*.

Lung volume decreases, causing pressure in lungs to *increase*. Air flows *out*.

WORKED EXAMPLE **8.4** Using Boyle's Law: Finding Volume at a Given Pressure

In a typical automobile engine, the fuel/air mixture in a cylinder is compressed from 1.0 atm to 9.5 atm. If the uncompressed volume of the cylinder is 750 mL, what is the volume when fully compressed?

ANALYSIS This is a Boyle's law problem because the volume and pressure in the cylinder change but the amount of gas and the temperature remain constant. According to Boyle's law, the pressure of the gas times its volume is constant:

$$P_1V_1 = P_2V_2$$

Knowing three of the four variables in this equation, we can solve for the unknown.

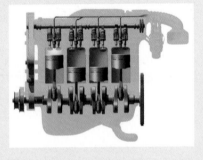

◀ A cut-away diagram of internal combustion engine shows movement of pistons during expansion and compression cycles.

BALLPARK ESTIMATE Since the pressure *increases* approximately tenfold (from 1.0 atm to 9.5 atm), the volume must *decrease* to approximately one-tenth, from 750 mL to about 75 mL.

SOLUTION

STEP 1: **Identify known information.** Of the four variables in Boyle's law, we know P_1, V_1, and P_2.

$P_1 = 1.0$ atm
$V_1 = 750$ mL
$P_2 = 9.5$ atm

STEP 2: **Identify answer and units.**

$V_2 = $?? mL

STEP 3: **Identify equation.** In this case, we simply substitute the known variables into Boyle's law and rearrange to isolate the unknown.

$$P_1V_1 = P_2V_2 \implies V_2 = \frac{P_1V_1}{P_2}$$

STEP 4: **Solve. Substitute the known information into the equation.** Make sure units cancel so that the answer is given in the units of the unknown variable.

$$V_2 = \frac{P_1V_1}{P_2} = \frac{(1.0 \text{ atm})(750 \text{ mL})}{(9.5 \text{ atm})} = 79 \text{ mL}$$

BALLPARK CHECK Our estimate was 75 mL.

PROBLEM 8.4

An oxygen cylinder used for breathing has a volume of 5.0 L at 90 atm pressure. What is the volume of the same amount of oxygen at the same temperature if the pressure is 1.0 atm? (Hint: Would you expect the volume of gas at this pressure to be greater or less than the volume at 90 atm?)

PROBLEM 8.5

A sample of hydrogen gas at 273 K has a volume of 3.2 L at 4.0 atm pressure. What is its pressure if its volume is changed to 10.0 L? To 0.20 L?

8.5 Charles's Law: The Relation between Volume and Temperature

Imagine that you again have a sample of gas inside a cylinder with a plunger at one end. What happens if you double the sample's Kelvin temperature while letting the plunger move freely to keep the pressure constant? The gas particles move with twice as much energy and collide twice as forcefully with the walls. To maintain a constant pressure, the volume of the gas in the cylinder must double (Figure 8.9).

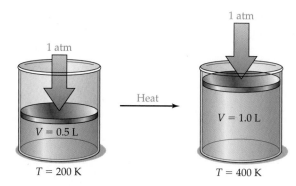

▲ The volume of the gas in the balloon increases as it is heated, causing a decrease in density and allowing the balloon to rise.

▲ **FIGURE 8.9 Charles's law.** The volume of a gas is directly proportionately to its Kelvin temperature at constant *n* and *P*. If the Kelvin temperature of the gas is doubled, its volume doubles.

According to **Charles's law**, the volume of a fixed amount of gas at constant pressure is directly proportional to its Kelvin temperature. Note the difference between *directly* proportional in Charles's law and *inversely* proportional in Boyle's law. Directly proportional quantities change in the same direction: As temperature goes up or down, volume also goes up or down (Figure 8.10).

Charles's law The volume of a gas is directly proportional to its Kelvin temperature for a fixed amount of gas at a constant pressure. That is, *V* divided by *T* is constant when *n* and *P* are held constant.

$$V \propto T \quad \text{(In kelvins)}$$

$$\text{or} \quad \frac{V}{T} = k \quad \text{(A constant value)}$$

$$\text{or} \quad \frac{V_1}{T_1} = \frac{V_2}{T_2}$$

This observation is consistent with the kinetic–molecular theory. As temperature increases, the average kinetic energy of the gas molecules increases, as does the

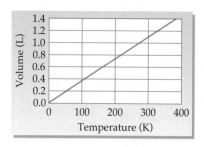

▲ **FIGURE 8.10 Charles's law.** Volume is directly proportional to the Kelvin temperature for a fixed amount of gas at a constant pressure. As the temperature goes up, the volume also goes up.

energy of molecular collisions with the interior surface of the container. The volume of the container must increase to maintain a constant pressure. As an example of Charles's law, think about what happens when a hot-air balloon is inflated. Heating causes the air inside to expand and fill the balloon. The air inside the balloon is less dense than the air outside the balloon, creating the buoyancy effect.

WORKED EXAMPLE   **8.5** Using Charles's Law: Finding Volume at a Given Temperature

An average adult inhales a volume of 0.50 L of air with each breath. If the air is warmed from room temperature (20 °C = 293 K) to body temperature (37 °C = 310 K) while in the lungs, what is the volume of the air exhaled?

ANALYSIS This is a Charles's law problem because the volume and temperature of the air change while the amount and pressure remain constant. Knowing three of the four variables, we can rearrange Charles's law to solve for the unknown.

BALLPARK ESTIMATE Charles's Law predicts an increase in volume directly proportional to the increase in temperature from 273 K to 310 K. The increase of less than 20 K represents a relatively small change compared to the initial temperature of 273 K. A 10% increase, for example, would be equal to a temperature change of 27 K; so a 20 K change would be less than 10%. We would therefore expect the volume to increase by less than 10%, from 0.50 L to a little less than 0.55 L.

SOLUTION

STEP 1: Identify known information. Of the four variables in Charles's law, we know T_1, V_1, and T_2.

$T_1 = 293$ K
$V_1 = 0.50$ L
$T_2 = 310$ K

STEP 2: Identify answer and units.

$V_2 = $?? L

STEP 3: Identify equation. Substitute the known variables into Charles's law and rearrange to isolate the unknown.

$$\frac{V_1}{T_1} = \frac{V_2}{T_2} \Rightarrow V_2 = \frac{V_1 T_2}{T_1}$$

STEP 4: Solve. Substitute the known information into Charles's law; check to make sure units cancel.

$$V_2 = \frac{V_1 T_2}{T_1} = \frac{(0.50 \text{ L})(310 \text{ K})}{293 \text{ K}} = 0.53 \text{ L}$$

BALLPARK CHECK This is consistent with our estimate!

PROBLEM 8.6

A sample of chlorine gas has a volume of 0.30 L at 273 K and 1 atm pressure. What is its volume at 250 K and 1 atm pressure? At 525 °C and 1 atm?

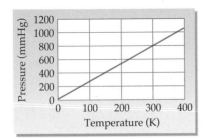

▲ **FIGURE 8.11 Gay-Lussac's law.** Pressure is directly proportional to the temperature in kelvins for a fixed amount of gas at a constant volume. As the temperature goes up, the pressure also goes up.

8.6 Gay-Lussac's Law: The Relation between Pressure and Temperature

Imagine next that you have a fixed amount of gas in a sealed container whose volume remains constant. What happens if you double the temperature (in kelvins)? The gas particles move with twice as much energy and collide with the walls of the container with twice as much force. Thus, the pressure in the container doubles. According to **Gay-Lussac's law**, the pressure of a fixed amount of gas at constant volume is directly proportional to its Kelvin temperature. As temperature goes up or down, pressure also goes up or down (Figure 8.11).

Gay-Lussac's law The pressure of a gas is directly proportional to its Kelvin temperature for a fixed amount of gas at a constant volume. That is, P divided by T is constant when n and V are held constant.

$$P \propto T \quad \text{(In kelvins)}$$

$$\text{or } \frac{P}{T} = k \quad \text{(A constant value)}$$

$$\text{or } \frac{P_1}{T_1} = \frac{P_2}{T_2}$$

▲ Because of the possibility of explosion (see warning label), aerosol containers cannot be incinerated, requiring disposal in landfills.

According to the kinetic–molecular theory, the kinetic energy of molecules is directly proportional to absolute temperature. As the average kinetic energy of the molecules increases, the energy of collisions with the interior surface of the container increases, causing an increase in pressure. As an example of Gay-Lussac's law, think of what happens when an aerosol can is thrown into an incinerator. As the can gets hotter, pressure builds up inside and the can explodes (hence the warning statement on aerosol cans).

WORKED EXAMPLE **8.6** Using Gay-Lussac's Law: Finding Pressure at a Given Temperature

What does the inside pressure become if an aerosol can with an initial pressure of 4.5 atm is heated in a fire from room temperature (20 °C) to 600 °C?

ANALYSIS This is a Gay-Lussac's law problem because the pressure and temperature of the gas inside the can change while its amount and volume remain constant. We know three of the four variables in the equation for Gay-Lussac's law, and can find the unknown by substitution and rearrangement.

BALLPARK ESTIMATE Gay-Lussac's law states that pressure is directly proportional to temperature. Since the Kelvin temperature increases approximately threefold (from about 300 K to about 900 K), we expect the pressure to also increase by approximately threefold, from 4.5 atm to about 14 atm.

SOLUTION

STEP 1: **Identify known information.** Of the four variables in Gay-Lussac's law, we know P_1, T_1 and T_2. (Note that T must be in kelvins.)

$P_1 = 4.5$ atm
$T_1 = 20 \text{ °C} = 293$ K
$T_2 = 600 \text{ °C} = 873$ K

STEP 2: **Identify answer and units.**

$P_2 = ?? $ atm

STEP 3: **Identify equation.** Substituting the known variables into Gay-Lussac's law, we rearrange to isolate the unknown.

$$\frac{P_1}{T_1} = \frac{P_2}{T_2} \implies P_2 = \frac{P_1 T_2}{T_1}$$

STEP 4: **Solve.** Substitute the known information into Gay-Lussac's law, check to make sure units cancel.

$$P_2 = \frac{P_1 T_2}{T_1} = \frac{(4.5 \text{ atm})(873 \text{ K})}{293 \text{ K}} = 13 \text{ atm}$$

BALLPARK CHECK Our estimate was 14 atm.

PROBLEM 8.7

Driving on a hot day causes tire temperature to rise. What is the pressure inside an automobile tire at 45 °C if the tire has a pressure of 30 psi at 15 °C? Assume that the volume and amount of air in the tire remain constant.

8.7 The Combined Gas Law

Since PV, V/T, and P/T all have constant values for a fixed amount of gas, these relationships can be merged into a **combined gas law**, which holds true whenever the amount of gas is fixed.

COMBINED GAS LAW $\quad \dfrac{PV}{T} = k \quad$ (A constant value)

$$\text{or} \quad \frac{P_1V_1}{T_1} = \frac{P_2V_2}{T_2}$$

If any five of the six quantities in this equation are known, the sixth quantity can be calculated. Furthermore, if any of the three variables T, P, or V is constant, that variable drops out of the equation, leaving behind Boyle's law, Charles's law, or Gay-Lussac's law. As a result, *the combined gas law is the only equation you need to remember for a fixed amount of gas*. Worked Example 8.7 gives a sample calculation.

Since $\qquad\qquad \dfrac{P_1V_1}{T_1} = \dfrac{P_2V_2}{T_2}$

At constant T: $\quad \dfrac{P_1V_1}{T} = \dfrac{P_2V_2}{T} \quad$ gives $\quad P_1V_1 = P_2V_2 \quad$ (Boyle's law)

At constant P: $\quad \dfrac{PV_1}{T_1} = \dfrac{PV_2}{T_2} \quad$ gives $\quad \dfrac{V_1}{T_1} = \dfrac{V_2}{T_2} \quad$ (Charles's law)

At constant V: $\quad \dfrac{P_1V}{T_1} = \dfrac{P_2V}{T_2} \quad$ gives $\quad \dfrac{P_1}{T_1} = \dfrac{P_2}{T_2} \quad$ (Gay-Lussac's law)

WORKED EXAMPLE **8.7** Using the Combined Gas Law: Finding Temperature

A 6.3 L sample of helium gas stored at 25 °C and 1.0 atm pressure is transferred to a 2.0 L tank and maintained at a pressure of 2.8 atm. What temperature is needed to maintain this pressure?

ANALYSIS This is a combined gas law problem because pressure, volume, and temperature change while the amount of helium remains constant. Of the six variables in this equation, we know P_1, V_1, T_1, P_2, and V_2, and we need to find T_2.

BALLPARK ESTIMATE Since the volume goes down by a little more than a factor of about 3 (from 6.3 L to 2.0 L) and the pressure goes up by a little less than a factor of about 3 (from 1.0 atm to 2.8 atm), the two changes roughly offset each other, and so the temperature should not change much. Since the volume-decrease factor (3.2) is slightly greater than the pressure-increase factor (2.8), the temperature will drop slightly ($T \propto V$).

SOLUTION

STEP 1: Identify known information. Of the six variables in combined-gas-law we know P_1, V_1, T_1, P_2, and V_2. (As always, T must be converted from Celsius degrees to kelvins.)

$P_1 = 1.0$ atm, $P_2 = 2.8$ atm
$V_1 = 6.3$ L, $V_2 = 2.0$ L
$T_1 = 25\ °C = 298$ K

STEP 2: Identify answer and units.

$T_2 = ??$ kelvin

STEP 3: Identify the equation. Substituting the known variables into the equation for the combined gas law and rearrange to isolate the unknown.

$$\frac{P_1V_1}{T_1} = \frac{P_2V_2}{T_2} \quad \Rightarrow \quad T_2 = \frac{P_2V_2T_1}{P_1V_1}$$

STEP 4: Solve. Solve the combined gas law equation for T_2; check to make sure units cancel.

$$T_2 = \frac{P_2V_2T_1}{P_1V_1} = \frac{(2.8\ \text{atm})(2.0\ \text{L})(298\ \text{K})}{(1.0\ \text{atm})(6.3\ \text{L})} = 260\ \text{K}\,(\Delta T = -38\ °C)$$

BALLPARK CHECK The relatively small decrease in temperature (38 °C, or 13% compared to the original temperature) is consistent with our prediction.

PROBLEM 8.8

A weather balloon is filled with helium to a volume of 275 L at 22 °C and 752 mmHg. The balloon ascends to an altitude where the pressure is 480 mmHg, and the temperature is −32 °C. What is the volume of the balloon at this altitude?

⚷ KEY CONCEPT PROBLEM 8.9

A balloon is filled under the initial conditions indicated below. If the pressure is then increased to 2 atm while the temperature is increased to 50 °C, which balloon on the right, (a) or (b), represents the new volume of the balloon?

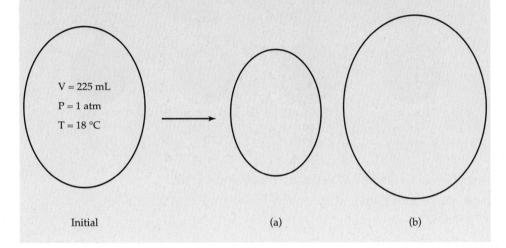

V = 225 mL
P = 1 atm
T = 18 °C

Initial   (a)   (b)

8.8 Avogadro's Law: The Relation between Volume and Molar Amount

Here we look at one final gas law, which takes changes in amount of gas into account. Imagine that you have two different volumes of a gas at the same temperature and pressure. How many moles does each sample contain? According to **Avogadro's law**, the volume of a gas is directly proportional to its molar amount at a constant pressure and temperature (Figure 8.12). A sample that contains twice the molar amount has twice the volume.

Avogadro's law The volume of a gas is directly proportional to its molar amount at a constant pressure and temperature. That is, V divided by n is constant when P and T are held constant.

Volume (V) $\propto$ Number of moles (n)

or $\dfrac{V}{n} = k$ (A constant value; the same for all gases)

or $\dfrac{V_1}{n_1} = \dfrac{V_2}{n_2}$

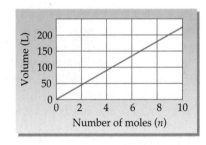

▲ **FIGURE 8.12 Avogadro's law.** Volume is directly proportional to the molar amount, n, at a constant temperature and pressure. As the number of moles goes up, the volume also goes up.

Because the particles in a gas are so tiny compared to the empty space surrounding them, there is no interaction among gas particles as proposed by the kinetic–molecular theory. As a result, the chemical identity of the particles does not matter and the value of the constant k in the equation $V/n = k$ is the same for all gases. It is therefore possible to compare the molar amounts of *any* two gases simply by comparing their volumes at the same temperature and pressure.

Notice that the *values* of temperature and pressure do not matter; it is only necessary that T and P be the same for both gases. To simplify comparisons of gas samples, however, it is convenient to define a set of conditions called **standard temperature and pressure (STP)**:

Standard temperature and pressure (STP)   0 °C (273.15 K); 1 atm (760 mmHg)

At standard temperature and pressure, 1 mol of any gas (6.02×10^{23} particles) has a volume of 22.4 L, a quantity called the **standard molar volume** (Figure 8.13).

Standard molar volume of any ideal gas at STP   22.4 L/mol

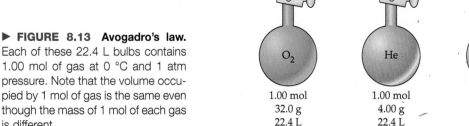

▶ **FIGURE 8.13 Avogadro's law.** Each of these 22.4 L bulbs contains 1.00 mol of gas at 0 °C and 1 atm pressure. Note that the volume occupied by 1 mol of gas is the same even though the mass of 1 mol of each gas is different.

O_2	He	F_2	Ar
1.00 mol	1.00 mol	1.00 mol	1.00 mol
32.0 g	4.00 g	38.0 g	39.9 g
22.4 L	22.4 L	22.4 L	22.4 L

WORKED EXAMPLE   **8.8** Using Avogadro's Law: Finding Moles in a Given Volume at STP

Use the standard molar volume of a gas at STP (22.4 L) to find how many moles of air at STP are in a room measuring 4.11 m wide by 5.36 m long by 2.58 m high.

ANALYSIS We first find the volume of the room and then use standard molar volume as a conversion factor to find the number of moles.

SOLUTION

STEP 1: Identify known information. We are given the room dimensions.

Length = 5.36 m
Width = 4.11 m
Height = 2.58 m

STEP 2: Identify answer and units.

Moles of air = ?? mol

STEP 3: Identify the equation. The volume of the room is the product of its three dimensions. Once we have the volume (in m³), we can convert to liters and use the molar volume at STP as a conversion factor to obtain moles of air.

$$\text{Volume} = (4.11 \text{ m})(5.36 \text{ m})(2.58 \text{ m}) = 56.8 \text{ m}^3$$

$$= 56.8 \text{ m}^3 \times \frac{1000 \text{ L}}{1 \text{ m}^3} = 5.68 \times 10^4 \text{ L}$$

$$1 \text{ mol} = 22.4 \text{ L} \rightarrow \frac{1 \text{ mol}}{22.4 \text{ L}}$$

STEP 4: Solve. Use the room volume and the molar volume at STP to set up an equation, making sure unwanted units cancel.

$$5.68 \times 10^4 \text{ L} \times \frac{1 \text{ mol}}{22.4 \text{ L}} = 2.54 \times 10^3 \text{ mol}$$

PROBLEM 8.10

How many moles of methane gas, CH_4, are in a 1.00×10^5 L storage tank at STP? How many grams of methane is this? How many grams of carbon dioxide gas could the same tank hold?

8.9 The Ideal Gas Law

The relationships among the four variables P, V, T, and n for gases can be combined into a single expression called the **ideal gas law**. If you know the values of any three of the four quantities, you can calculate the value of the fourth.

Ideal gas law $\quad \dfrac{PV}{nT} = R \quad$ (A constant value)

$\quad$ or $\quad PV = nRT$

The constant R in the ideal gas law (instead of the usual k) is called the **gas constant**. Its value depends on the units chosen for pressure, with the two most common values

Gas constant (R) The constant R in the ideal gas law, $PV = nRT$.

$$\text{For } P \text{ in atmospheres:} \quad R = 0.0821 \frac{\text{L} \cdot \text{atm}}{\text{mol} \cdot \text{K}}$$

$$\text{For } P \text{ in millimeters Hg:} \quad R = 62.4 \frac{\text{L} \cdot \text{mmHg}}{\text{mol} \cdot \text{K}}$$

In using the ideal gas law, it is important to choose the value of R having pressure units that are consistent with the problem and, if necessary, to convert volume into liters and temperature into kelvins.

Table 8.1 summarizes the various gas laws, and Worked Example 8.9 shows how to use the ideal gas law.

TABLE 8.1   A Summary of the Gas Laws

	GAS LAW	VARIABLES	CONSTANT
Boyle's law	$P_1V_1 = P_2V_2$	P, V	n, T
Charles's law	$V_1/T_1 = V_2/T_2$	V, T	n, P
Gay-Lussac's law	$P_1/T_1 = P_2/T_2$	P, T	n, V
Combined gas law	$P_1V_1/T_1 = P_2V_2/T_2$	P, V, T	n
Avogadro's law	$V_1/n_1 = V_2/n_2$	V, n	P, T
Ideal gas law	$PV = nRT$	P, V, T, n	R

WORKED EXAMPLE   **8.9** Using Ideal Gas Law: Finding Moles

How many moles of air are in the lungs of an average person with a total lung capacity of 3.8 L? Assume that the person is at 1.0 atm pressure and has a normal body temperature of 37 °C.

ANALYSIS This is an ideal gas law problem because it asks for a value of n when P, V, and T are known: $n = PV/RT$. The volume is given in the correct unit of liters, but temperature must be converted to kelvins.

SOLUTION

STEP 1: Identify known information.
We know three of the four variables in the ideal gas law.

$P = 1.0 \text{ atm}$

$V = 3.8 \text{ L}$

$T = 37 \,°\text{C} = 310 \text{ K}$

STEP 2: Identify answer and units.

Moles of air, $n = $?? mol

STEP 3: Identify the equation. Knowing three of the four variables in the ideal gas law, we can rearrange and solve for the unknown variable, n. (Note: because pressure is given in atm, we use the value of R that is expressed in atm:

$$R = 0.0821 \frac{L \cdot atm}{mol \cdot K}$$

$$PV = nRT \implies n = \frac{PV}{RT}$$

STEP 4: Solve. Substitute the known information and the appropriate value of R into the ideal gas law equation and solve for n.

$$n = \frac{PV}{RT} = \frac{(1.0 \text{ atm})(3.8 \text{ L})}{\left(0.0821 \dfrac{L \cdot atm}{mol \cdot K}\right)(310 \text{ K})} = 0.15 \text{ mol}$$

WORKED EXAMPLE 8.10 Using the Ideal Gas Law: Finding Pressure

Methane gas is sold in steel cylinders with a volume of 43.8 L containing 5.54 kg. What is the pressure in atmospheres inside the cylinder at a temperature of 20.0 °C (293.15 K)? The molar mass of methane (CH_4) is 16.0 g/mol.

ANALYSIS This is an ideal gas law problem because it asks for a value of P when V, T, and n are given. Although not provided directly, enough information is given so that we can calculate the value of n ($n = g/MW$).

SOLUTION

STEP 1: Identify known information. We know two of the four variables in the ideal gas law; V, T, and can calculate the third, n, from the information provided.

$V = 43.8$ L
$T = 37 \, °C = 310$ K

STEP 2: Identify answer and units.

Pressure, $P = ??$ atm

STEP 3: Identify equation. First, calculate the number of moles n of methane in the cylinder by using molar mass (16.0 g/mol) as a conversion factor. Then use the ideal gas law to calculate the pressure.

$$n = (5.54 \text{ kg methane})\left(\frac{1000 \text{ g}}{1 \text{ kg}}\right)\left(\frac{1 \text{ mol}}{16.0 \text{ g}}\right) = 346 \text{ mol methane}$$

$$PV = nRT \implies P = \frac{nRT}{V}$$

STEP 4: Solve. Substitute the known information and the appropriate value of R into the ideal gas law equation and solve for P.

$$P = \frac{nRT}{V} = \frac{(346 \text{ mol})\left(0.0821 \dfrac{L \cdot atm}{mol \cdot K}\right)(293 \text{ K})}{43.8 \text{ L}} = 190 \text{ atm}$$

PROBLEM 8.11

An aerosol spray can of deodorant with a volume of 350 mL contains 3.2 g of propane gas (C_3H_8) as propellant. What is the pressure in the can at 20 °C?

PROBLEM 8.12

A helium gas cylinder of the sort used to fill balloons has a volume of 180 L and a pressure of 2200 psi (150 atm) at 25 °C. How many moles of helium are in the tank? How many grams?

KEY CONCEPT PROBLEM 8.13

Show the approximate level of the movable piston in drawings (a) and (b) after the indicated changes have been made to the initial gas sample.

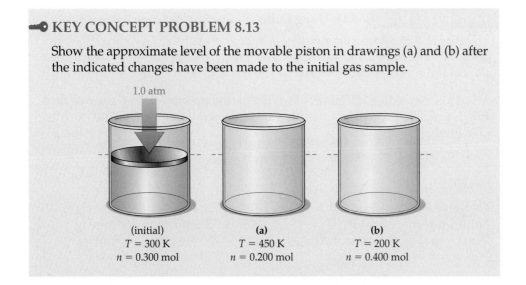

(initial)	(a)	(b)
$T = 300\ K$	$T = 450\ K$	$T = 200\ K$
$n = 0.300\ mol$	$n = 0.200\ mol$	$n = 0.400\ mol$

8.10 Partial Pressure and Dalton's Law

According to the kinetic–molecular theory, each particle in a gas acts independently of all others because there are no attractive forces between them and they are so far apart. To any individual particle, the chemical identity of its neighbors is irrelevant. Thus, *mixtures* of gases behave the same as pure gases and obey the same laws.

Dry air, for example, is a mixture of about 21% oxygen, 78% nitrogen, and 1% argon by volume, which means that 21% of atmospheric air pressure is caused by O_2 molecules, 78% by N_2 molecules, and 1% by Ar atoms. The contribution of each gas in a mixture to the total pressure of the mixture is called the **partial pressure**. of that gas. According to **Dalton's law**, the total pressure exerted by a gas mixture (P_{total}) is the sum of the partial pressures of the components in the mixture:

> **Partial pressure** The contribution of a given gas in a mixture to the total pressure.

Dalton's law $P_{total} = P_{gas\ 1} + P_{gas\ 2} + \cdots$

In dry air at a total air pressure of 760 mmHg, the partial pressure caused by the contribution of O_2 is 0.21×760 mmHg $= 160$ mmHg, the partial pressure of N_2 is 0.78×760 mmHg $= 593$ mmHg, and that of argon is 7 mmHg. *The partial pressure exerted by each gas in a mixture is the same pressure that the gas would exert if it were alone.* Put another way, the pressure exerted by each gas depends on the frequency of collisions of its molecules with the walls of the container. But this frequency does not change when other gases are present because the different molecules have no influence on one another.

To represent the partial pressure of a specific gas, we add the formula of the gas as a subscript to P, the symbol for pressure. You might see the partial pressure of oxygen represented as P_{O_2}, for instance. Moist air inside the lungs at 37 °C and atmospheric pressure has the following average composition at sea level. Note that P_{total} is equal to atmospheric pressure, 760 mmHg.

$$
\begin{aligned}
P_{total} &= P_{N_2} + P_{O_2} + P_{CO_2} + P_{H_2O} \\
&= 573\ \text{mmHg} + 100\ \text{mmHg} + 40\ \text{mmHg} + 47\ \text{mmHg} \\
&= 760\ \text{mmHg}
\end{aligned}
$$

The composition of air does not change appreciably with altitude, but the total pressure decreases rapidly. The partial pressure of oxygen in air therefore decreases with increasing altitude, and it is this change that leads to difficulty in breathing at high elevations.

WORKED EXAMPLE 8.11 Using Dalton's Law: Finding Partial Pressures

Humid air on a warm summer day is approximately 20% oxygen, 75% nitrogen, 4% water vapor, and 1% argon. What is the partial pressure of each component if the atmospheric pressure is 750 mmHg?

ANALYSIS According to Dalton's law, the partial pressure of any gas in a mixture is equal to the percent concentration of the gas times the total gas pressure (750 mmHg). In this case,

$$P_{total} = P_{O_2} + P_{N_2} + P_{H_2O} + P_{Ar}$$

SOLUTION

Oxygen partial pressure (P_{O_2}):	0.20×750 mmHg = 150 mmHg
Nitrogen partial pressure (P_{N_2}):	0.75×750 mmHg = 560 mmHg
Water vapor partial pressure (P_{H_2O}):	0.04×750 mmHg = 30 mmHg
Argon partial pressure (P_{Ar}):	0.01×750 mmHg = 8 mmHg

Total pressure = 748 mmHg → 750 mmHg (rounding to 2 significant figures!)

Note that the sum of the partial pressures must equal the total pressure (within rounding error).

▲ Because of the lack of sufficient oxygen at the top of Mt. Everest, most climbers rely on bottled oxygen supplies.

PROBLEM 8.14

Assuming a total pressure of 9.5 atm, what is the partial pressure of each component in the mixture of 98% helium and 2.0% oxygen breathed by deep-sea divers? How does the partial pressure of oxygen in diving gas compare with its partial pressure in normal air?

PROBLEM 8.15

Determine the percent composition of air in the lungs from the following composition in partial pressures: $P_{N_2} = 573$ mmHg, $P_{O_2} = 100$ mmHg, $P_{CO_2} = 40$ mmHg, $P_{H_2O} = 47$ mmHg; all at 37 °C and 1 atm pressure.

PROBLEM 8.16

The atmospheric pressure on the top of Mt. Everest, an altitude of 29,035 ft, is only 265 mmHg. What is the partial pressure of oxygen in the lungs at this altitude?

KEY CONCEPT PROBLEM 8.17

Assume that you have a mixture of He (MW = 4 amu) and Xe (MW = 131 amu) at 300 K. Which of the drawings (a)–(c) best represents the mixture (blue = He; green = Xe)?

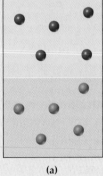

(a)

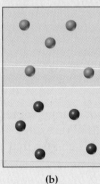

(b)

(c)

8.11 Intermolecular Forces

What determines whether a substance is a gas, a liquid, or a solid at a given temperature? Why does rubbing alcohol evaporate much more readily than water? Why do molecular compounds have lower melting points than ionic compounds? To answer these and a great many other such questions, we need to look into the nature of **intermolecular forces**—the forces that act *between different molecules* rather than within an individual molecule.

> **Intermolecular force** A force that acts between molecules and holds molecules close to one another.

In gases, the intermolecular forces are negligible, so the gas molecules act independently of one another. In liquids and solids, however, intermolecular forces are strong enough to hold the molecules in close contact. As a general rule, the stronger the intermolecular forces in a substance, the more difficult it is to separate the molecules, and the higher the melting and boiling points of the substance.

There are three major types of intermolecular forces: *dipole–dipole, London dispersion*, and *hydrogen bonding*. We will discuss each in turn.

Dipole–Dipole Forces

Recall from Sections 5.8 and 5.9 that many molecules contain polar covalent bonds and may therefore have a net molecular polarity. ($\bigcirc\!\!\!\bigcirc$, pp. 128, 130) In such cases, the positive and negative ends of different molecules are attracted to one another by what is called a **dipole–dipole force**. (Figure 8.14).

> **Dipole–dipole force** The attractive force between positive and negative ends of polar molecules.

▲ **FIGURE 8.14 Dipole–dipole forces.** The positive and negative ends of polar molecules are attracted to one another by dipole–dipole forces. As a result, polar molecules have higher boiling points than nonpolar molecules of similar size.

Dipole–dipole forces are weak, with strengths on the order of 1 kcal/mol compared to the 70–100 kcal/mol typically found for the strength of a covalent bond. Nevertheless, the effects of dipole–dipole forces are important, as can be seen by looking at the difference in boiling points between polar and nonpolar molecules. Butane, for instance, is a nonpolar molecule with a molecular weight of 58 amu and a boiling point of $-0.5\,°C$, whereas acetone has the same molecular weight yet boils $57\,°C$ higher because it is polar. (Recall from Section 5.8 how molecular polarities can be visualized using electrostatic potential maps. ($\bigcirc\!\!\!\bigcirc$, p. 128)

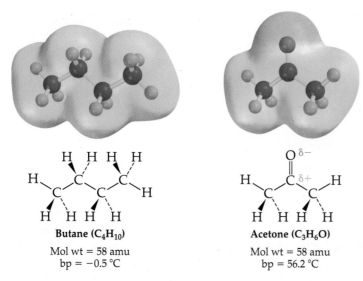

Butane (C_4H_{10})
Mol wt = 58 amu
bp = $-0.5\,°C$

Acetone (C_3H_6O)
Mol wt = 58 amu
bp = $56.2\,°C$

APPLICATION ▶ Greenhouse Gases and Global Warming

The mantle of gases surrounding the earth is far from the uniform mixture you might expect, consisting of layers that vary in composition and properties at different altitudes. The ability of the gases in these layers to absorb radiation is responsible for life on earth as we know it.

The *stratosphere*—the layer extending from about 12 km up to 50 km altitude—contains the ozone layer that is responsible for absorbing harmful UV radiation. The *troposphere* is the layer extending from the surface up to about 12 km altitude. It should not surprise you to learn that the troposphere is the layer most easily disturbed by human activities and that this layer has the greatest impact on the earth's surface conditions. Among those impacts, a process called the *greenhouse effect* is much in the news today.

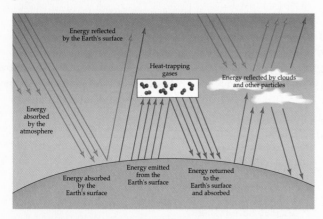

▲ Greenhouse gases (GHG) trap heat reflected from the earth's surface, resulting in the increase in surface temperatures known as global warming.

The greenhouse effect refers to the warming that occurs in the troposphere as gases absorb radiant energy. Much of the radiant energy reaching the earth's surface from the sun is reflected back into space, but some is absorbed by atmospheric gases—particularly those referred to as—*greenhouse gases* (GHGs)—water vapor, carbon dioxide, and methane. This absorbed radiation warms the atmosphere and acts to maintain a relatively stable temperature of 15 °C (59 °F) at the earth's surface. Without the greenhouse effect, the average surface temperature would be about −18 °C (0 °F)—a temperature so low that Earth would be frozen and unable to sustain life.

The basis for concern about the greenhouse effect is the fear that human activities over the past century have disturbed the earth's delicate thermal balance. Should increasing amounts of radiation be absorbed, increased atmospheric heating will result, and global temperatures will continue to rise.

Measurements show that the concentration of atmospheric CO_2 has been rising in the last 150 years, largely because of the increased use of fossil fuels, from an estimated 290 parts per million (ppm) in 1850 to current levels approaching 380 ppm. The increase in CO_2 levels correlates with a concurrent increase in average global temperatures, with the 11 years between 1995 and 2006 ranking among the 12-highest since recording of global temperatures began in 1850. The latest report of the Intergovernmental Panel on Climate Change published in November 2007 concluded that "Warming of the climate system is unequivocal, as is now evident from observations of increases in global average air and ocean temperatures, widespread melting of snow and ice and rising global average sea level. . . . Continued GHG emissions at or above current rates would cause further warming and induce many changes in the global climate system during the 21st century that would *very likely* be larger than those observed during the 20th century."

See Additional Problems 8.101 and 8.102 at the end of the chapter.

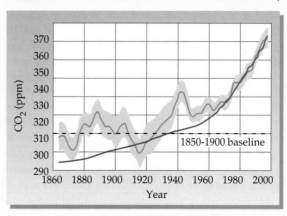

▲ Concentrations of atmospheric CO_2 and global average temperatures have increased dramatically in the last 150 years because of increased fossil fuel use, causing serious changes in earth's climate system.

© Crown copyright 2006 data provided by the Met office

London Dispersion Forces

London dispersion force The short-lived attractive force due to the constant motion of electrons within molecules.

Only polar molecules experience dipole–dipole forces, but all molecules, regardless of structure, experience *London dispersion forces*. **London dispersion forces** are caused by the constant motion of electrons within molecules. Take even a simple nonpolar molecule like Br_2, for example. Averaged over time, the distribution of electrons throughout the molecule is uniform, but at any given *instant* there may be more electrons at one end of the molecule than at the other (Figure 8.15). At that instant, the

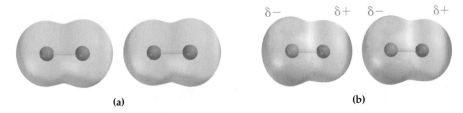

▲ **FIGURE 8.15** (a) Averaged over time, the electron distribution in a Br_2 molecule is symmetrical. (b) At any given instant, however, the electron distribution may be unsymmetrical, resulting in a temporary polarity that induces a complementary polarity in neighboring molecules.

molecule has a short-lived polarity. Electrons in neighboring molecules are attracted to the positive end of the polarized molecule, resulting in a polarization of the neighbor and creation of an attractive London dispersion force that holds the molecules together. As a result, Br_2 is a liquid at room temperature rather than a gas.

London dispersion forces are weak—in the range 0.5–2.5 kcal/mol—but they increase with molecular weight and amount of surface contact between molecules. The larger the molecular weight, the more electrons there are moving about and the greater the temporary polarization of a molecule. The larger the amount of surface contact, the greater the close interaction between different molecules.

The effect of surface contact on the magnitude of London dispersion forces can be seen by comparing a roughly spherical molecule with a flatter, more linear one having the same molecular weight. Both 2,2-dimethylpropane and pentane, for instance, have the same formula (C_5H_{12}), but the nearly spherical shape of 2,2-dimethylpropane allows for less surface contact with neighboring molecules than does the more linear shape of pentane (Figure 8.16). As a result, London dispersion forces are smaller for 2,2-dimethylpropane, molecules are held together less tightly, and the boiling point is correspondingly lower: 9.5 °C for 2,2-dimethylpropane versus 36 °C for pentane.

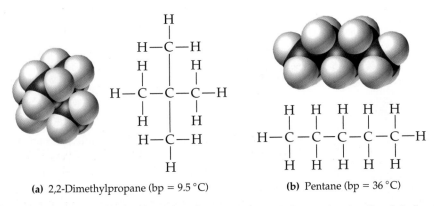

(a) 2,2-Dimethylpropane (bp = 9.5 °C)   **(b)** Pentane (bp = 36 °C)

▲ **FIGURE 8.16 London dispersion forces.** More compact molecules like 2,2-dimethylpropane have smaller surface areas, weaker London dispersion forces, and lower boiling points. By comparison, flatter, less compact molecules like pentane have larger surface areas, stronger London dispersion forces, and higher boiling points.

Hydrogen Bonds

In many ways, hydrogen bonding is responsible for life on earth. It causes water to be a liquid rather than a gas at ordinary temperatures, and it is the primary intermolecular force that holds huge biomolecules in the shapes needed to play their essential roles in biochemistry. Deoxyribonucleic acid (DNA) and keratin (Figure 8.17), for instance, are long molecular chains that form a coiled structure called an α-helix.

A **hydrogen bond** is an attractive interaction between an unshared electron pair on an electronegative O, N, or F atom and a positively polarized hydrogen atom

Hydrogen bond The attraction between a hydrogen atom bonded to an electronegative O, N, or F atom and another nearby electronegative O, N, or F atom.

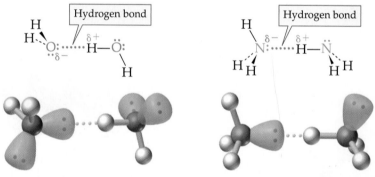

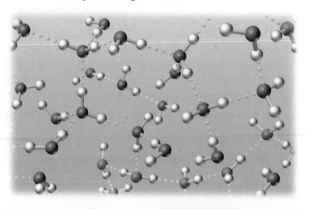

▶ **FIGURE 8.17** The α-helical structure of keratin results from hydrogen bonding along the amino acid backbone of the molecule.

bonded to another electronegative O, N, or F. For example, hydrogen bonds occur in both water and ammonia:

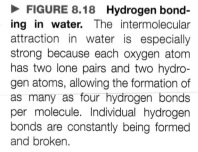

Hydrogen bonding is really just a special kind of dipole–dipole interaction. The O—H, N—H, and F—H bonds are highly polar, with a partial positive charge on the hydrogen and a partial negative charge on the electronegative atom. In addition, the hydrogen atom has no inner-shell electrons to act as a shield around its nucleus, and it is small so it can be approached closely. As a result, the dipole–dipole attractions involving positively polarized hydrogens are unusually strong, and hydrogen bonds result. Water, in particular, is able to form a vast three-dimensional network of hydrogen bonds because each H_2O molecule has two hydrogens and two electron pairs (Figure 8.18).

▶ **FIGURE 8.18 Hydrogen bonding in water.** The intermolecular attraction in water is especially strong because each oxygen atom has two lone pairs and two hydrogen atoms, allowing the formation of as many as four hydrogen bonds per molecule. Individual hydrogen bonds are constantly being formed and broken.

Hydrogen bonds can be quite strong, with energies up to 10 kcal/mol. To see the effect of hydrogen bonding, look at Table 8.2, which compares the boiling points of binary hydrogen compounds of second-row elements with their third-row counterparts. Because NH_3, H_2O, and HF molecules are held tightly together by hydrogen bonds, an unusually large amount of energy must be added to separate them in the boiling process. As a result, the boiling points of NH_3, H_2O, and HF are much higher than the boiling points of their second-row neighbor CH_4 and of related third-row compounds.

TABLE 8.2   Boiling Points for Binary Hydrogen Compounds of Some Second-row and Third-row Elements

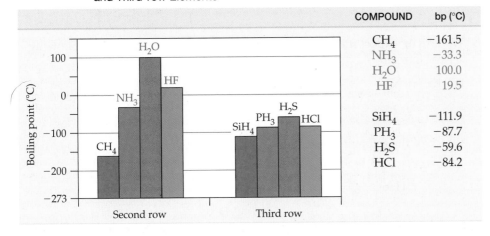

COMPOUND	bp (°C)
CH_4	−161.5
NH_3	−33.3
H_2O	100.0
HF	19.5
SiH_4	−111.9
PH_3	−87.7
H_2S	−59.6
HCl	−84.2

A summary and comparison of the various kinds of intermolecular forces is shown in Table 8.3.

TABLE 8.3   A Comparison of Intermolecular Forces

FORCE	STRENGTH	CHARACTERISTICS
Dipole–dipole	Weak (1 kcal/mol)	Occurs between polar molecules
London dispersion	Weak (0.5–2.5 kcal/mol)	Occurs between all molecules; strength depends on size
Hydrogen bond	Moderate (2–10 kcal/mol)	Occurs between molecules with O—H, N—H, and F—H bonds

Looking Ahead

Dipole–dipole forces, London dispersion forces, and hydrogen bonds are traditionally called "intermolecular forces" because of their influence on the properties of molecular compounds. But these same forces can also operate between different parts of a very large molecule. In this context, they are often referred to as "noncovalent interactions." In later chapters, we will see how noncovalent interactions determine the shapes of biologically important molecules such as proteins and nucleic acids.

WORKED EXAMPLE   **8.12** Identifying Intermolecular Forces: Polar versus Nonpolar

Identify the intermolecular forces that influence the properties of the following compounds:

   (a) Methane, CH_4 **(b)** HCl   **(c)** Acetic acid, CH_3CO_2H

ANALYSIS The intermolecular forces will depend on the molecular structure; what type of bonds are in the molecule (polar or non-polar) and how are they arranged.

SOLUTION

(a) Since methane contains only C—H bonds, it is a nonpolar molecule; it has only London dispersion forces.

(b) The H—Cl bond is polar, so this is a polar molecule; it has both dipole–dipole forces and London dispersion forces.

(c) Acetic acid is a polar molecule with an O—H bond. Thus, it has dipole–dipole forces, London dispersion forces, and hydrogen bonds.

PROBLEM 8.18

Would you expect the boiling points to increase or decrease in the following series? Explain.

(a) Kr, Ar, Ne

(b) Cl_2, Br_2, I_2

PROBLEM 8.19

Which of the following compounds form hydrogen bonds?

Methyl alcohol
(a)

Ethylene
(b)

Methylamine
(c)

PROBLEM 8.20

Identify the intermolecular forces (dipole–dipole, London dispersion, hydrogen bonding) that influence the properties of the following compounds:

(a) Ethane, CH_3CH_3

(b) Ethyl alcohol, CH_3CH_2OH

(c) Ethyl chloride, CH_3CH_2Cl

8.12 Liquids

Molecules are in constant motion in the liquid state, just as they are in gases. If a molecule happens to be near the surface of a liquid, and if it has enough energy, it can break free of the liquid and escape into the gas state, called **vapor**. In an open container, the now gaseous molecule will wander away from the liquid, and the process will continue until all the molecules escape from the container (Figure 8.19a). This, of course, is what happens during *evaporation*. We are all familiar with puddles of water evaporating after a rainstorm.

Vapor The gas molecules in equilibrium with a liquid.

▶ **FIGURE 8.19 The transfer of molecules between liquid and gas states.** (a) Molecules escape from an open container and drift away until the liquid has entirely evaporated. (b) Molecules in a closed container cannot escape. Instead, they reach an equilibrium in which the rates of molecules leaving the liquid and returning to the liquid are equal, and the concentration of molecules in the gas state is constant.

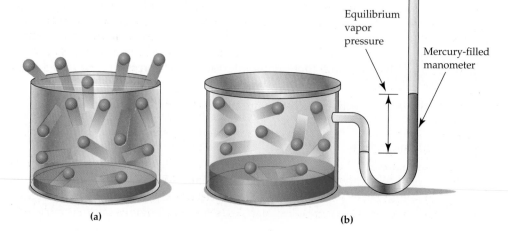

(a)

(b)

If the liquid is in a closed container, the situation is different because the gaseous molecules cannot escape. Thus, the random motion of the molecules occasionally brings them back into the liquid. After the concentration of molecules in the gas state has increased sufficiently, the number of molecules reentering the liquid becomes equal to the number escaping from the liquid (Figure 8.19b). At this point, a dynamic equilibrium exists, exactly as in a chemical reaction at equilibrium. Evaporation and condensation take place at the same rate, and the concentration of vapor in the container is constant as long as the temperature does not change.

Once molecules have escaped from the liquid into the gas state, they are subject to all the gas laws previously discussed. In a closed container at equilibrium, for example, the gas molecules make their own contribution to the total pressure of the gas above the liquid according to Dalton's law (Section 8.10). We call this contribution the **vapor pressure** of the liquid.

Vapor pressure depends on both temperature and the chemical identity of a liquid. As the temperature rises, molecules become more energetic and more likely to escape into the gas state. Thus, vapor pressure rises with increasing temperature until ultimately it becomes equal to the pressure of the atmosphere (Figure 8.20). At this point, bubbles of vapor form under the surface and force their way to the top, giving rise to the violent action observed during a vigorous boil. At an atmospheric pressure of exactly 760 mmHg, boiling occurs at what is called the **normal boiling point**.

▲ Because bromine is colored, it is possible to see its gaseous reddish vapor above the liquid.

Vapor pressure The partial pressure of gas molecules in equilibrium with a liquid.

Normal boiling point The boiling point at a pressure of exactly 1 atmosphere.

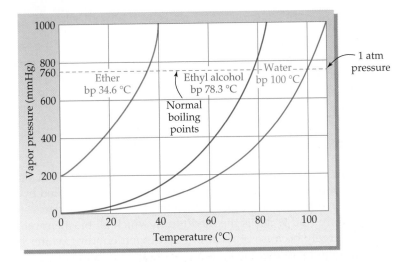

▲ **FIGURE 8.20** **A plot of the change of vapor pressure with temperature for ethyl ether, ethyl alcohol, and water.** At a liquid's boiling point, its vapor pressure is equal to atmospheric pressure. Commonly reported boiling points are those at 760 mmHg.

If atmospheric pressure is higher or lower than normal, the boiling point of a liquid changes accordingly. At high altitudes, for example, atmospheric pressure is lower than at sea level, and boiling points are also lower. On top of Mt. Everest (29,035 ft; 8850 m), atmospheric pressure is about 245 mmHg and the boiling temperature of water is only 71 °C. If the atmospheric pressure is higher than normal, the boiling point is also higher. This principle is used in strong vessels known as *autoclaves*, in which water at high pressure is heated to the temperatures needed for sterilizing medical and dental instruments (170 °C).

Many familiar properties of liquids can be explained by the intermolecular forces just discussed. We all know, for instance, that some liquids, such as water or gasoline, flow easily when poured, whereas others, such as motor oil or maple syrup, flow sluggishly.

The measure of a liquid's resistance to flow is called its *viscosity*. Not surprisingly, viscosity is related to the ease with which individual molecules move around in the liquid and thus to the intermolecular forces present. Substances such as gasoline, which have small, nonpolar molecules, experience only weak intermolecular

▲ Medical instruments are sterilized in this autoclave by heating them with water at high pressure.

▲ Surface tension allows a water strider to walk on water without penetrating the surface.

forces and have relatively low viscosities, whereas more polar substances such as glycerin [$C_3H_5(OH)_3$] experience stronger intermolecular forces and so have higher viscosities.

Another familiar property of liquids is *surface tension*, the resistance of a liquid to spread out and increase its surface area. The beading-up of water on a newly waxed car and the ability of a water strider to walk on water are both due to surface tension.

Surface tension is caused by the difference between the intermolecular forces experienced by molecules at the surface of the liquid and those experienced by molecules in the interior. Molecules in the interior of a liquid are surrounded and experience maximum intermolecular forces, whereas molecules at the surface have fewer neighbors and feel weaker forces. Surface molecules are therefore less stable, and the liquid acts to minimize their number by minimizing the surface area (Figure 8.21).

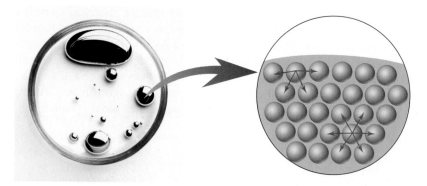

▲ **FIGURE 8.21** **Surface tension.** Surface tension is caused by the different forces experienced by molecules in the interior of a liquid and those on the surface. Molecules on the surface are less stable because they feel fewer attractive forces, so the liquid acts to minimize their number by minimizing surface area.

8.13 Water: A Unique Liquid

Ours is a world based on water. Water covers nearly 71% of the earth's surface, it accounts for 66% of the mass of an adult human body, and it is needed by all living things. The water in our blood forms the transport system that circulates substances throughout our body, and water is the medium in which all biochemical reactions are carried out. Largely because of its strong hydrogen bonding, water has many properties that are quite different from those of other compounds.

▲ The moderate year-round temperatures in San Francisco are due to the large heat capacity of the surrounding waters.

Water has the highest specific heat of any liquid (Section 2.10), giving it the capacity to absorb a large quantity of heat while changing only slightly in temperature. (⬭, p. 38) As a result, large lakes and other bodies of water tend to moderate the air temperature and climate of surrounding areas. Another consequence of the high specific heat of water is that the human body is better able to maintain a steady internal temperature under changing outside conditions.

In addition to a high specific heat, water has an unusually high *heat of vaporization* (540 cal/g), meaning that it carries away a large amount of heat when it evaporates. You can feel the effect of water evaporation on your wet skin when the wind blows. Even when comfortable, your body is still relying for cooling on the heat carried away from the skin and lungs by evaporating water. The heat generated by the chemical reactions of metabolism is carried by blood to the skin, where water moves through cell walls to the surface and evaporates. When metabolism, and therefore heat generation, speeds up, blood flow increases and capillaries dilate so that heat is brought to the surface faster.

Water is also unique in what happens as it changes from a liquid to a solid. Most substances are more dense as solids than as liquids because molecules are more closely packed in the solid than in the liquid. Water, however, is different. Liquid water has a maximum density of 1.000 g/mL at 3.98 °C but then becomes *less* dense as it cools. When it freezes, its density decreases still further to 0.917 g/mL.

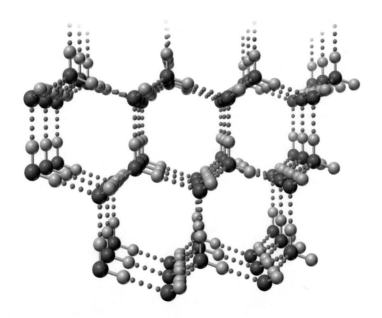

◀ **FIGURE 8.22 Ice.** Ice consists of individual H_2O molecules held rigidly together in an ordered manner by hydrogen bonds. The open cage-like crystal structure shows why ice is less dense than liquid water.

As water freezes, each molecule is locked into position by hydrogen bonding to four other water molecules (Figure 8.22). The resulting structure has more open space than does liquid water, accounting for its lower density. As a result, ice floats on liquid water, and lakes and rivers freeze from the top down. If the reverse were true, fish would be killed in winter as they became trapped in ice at the bottom.

8.14 Solids

A brief look around us reveals that most substances are solids rather than liquids or gases. It is also obvious that there are many different kinds of solids. Some, such as iron and aluminum, are hard and metallic; others, such as sugar and table salt, are crystalline and easily broken; and still others, such as rubber and many plastics, are soft and amorphous.

The most fundamental distinction between solids is that some are crystalline and some are amorphous. A **crystalline solid** is one whose particles—whether atoms, ions, or molecules—have an ordered arrangement extending over a long range. This order on the atomic level is also seen on the visible level, because crystalline solids usually have flat faces and distinct angles.

Crystalline solid A solid whose atoms, molecules, or ions are rigidly held in an ordered arrangement.

◀ Crystalline solids, such as the gypsum (left) and fluorite (right) shown here, have flat faces and distinct angles. These regular macroscopic features reflect a similarly ordered arrangement of particles at the atomic level.

▲ The sand on this beach is silica, SiO_2, a covalent network solid. Each grain of sand is essentially one large molecule.

Crystalline solids can be further categorized as ionic, molecular, covalent network, or metallic. *Ionic solids* are those like sodium chloride, whose constituent particles are ions. A crystal of sodium chloride is composed of alternating Na^+ and Cl^- ions ordered in a regular three-dimensional arrangement and held together by ionic bonds, as discussed in Section 4.3. (⊂◯⊃, p. 83) *Molecular solids* are those like sucrose or ice, whose constituent particles are molecules held together by the intermolecular forces discussed in Sections 8.11 and 8.12. A crystal of ice, for example, is composed of H_2O molecules held together in a regular way by hydrogen bonding (Figure 8.22). *Covalent network solids* are those like diamond (Figure 8.23) or quartz (SiO_2), whose atoms are linked together by covalent bonds into a giant three-dimensional array. In effect, a covalent network solid is one *very* large molecule.

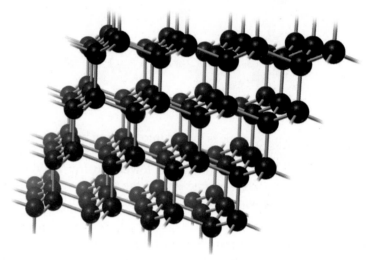

▲ **FIGURE 8.23** **Diamond.** Diamond is a covalent network solid—one very large molecule of carbon atoms linked by covalent bonds.

Metallic solids, such as silver or iron, can be viewed as vast three-dimensional arrays of metal cations immersed in a sea of electrons that are free to move about. This continuous electron sea acts both as a glue to hold the cations together and as a mobile carrier of charge to conduct electricity. Furthermore, the fact that bonding attractions extend uniformly in all directions explains why metals are malleable rather than brittle. When a metal crystal receives a sharp blow, no spatially oriented bonds are broken; instead, the electron sea simply adjusts to the new distribution of cations.

An **amorphous solid**, by contrast with a crystalline solid, is one whose constituent particles are randomly arranged and have no ordered long-range structure. Amorphous solids often result when liquids cool before they can achieve internal order, or when their molecules are large and tangled together, as happens in many polymers. Glass is an amorphous solid, as are tar, the gemstone opal, and some hard candies. Amorphous solids differ from crystalline solids by softening over a wide temperature range rather than having sharp melting points, and by shattering to give pieces with curved rather than planar faces.

A summary of the different types of solids and their characteristics is given in Table 8.4.

Amorphous solid A solid whose particles do not have an orderly arrangement.

8.15 Changes of State

What happens when a solid is heated? As more and more energy is added, molecules begin to stretch, bend, and vibrate more vigorously, and atoms or ions wiggle about with more energy. Finally, if enough energy is added and the motions become vigorous enough, particles start to break free from one another and the substance starts to melt. Addition of more heat continues the melting process until all particles have broken free and are in the liquid phase. The quantity of heat required to completely melt a substance once it reaches its melting point is called

TABLE 8.4   Types of Solids

SUBSTANCE	SMALLEST UNIT	INTERPARTICLE FORCES	PROPERTIES	EXAMPLES
Ionic solid	Ions	Attraction between positive and negative ions	Brittle and hard; high mp; crystalline	$NaCl$, KI, $Ca_3(PO_4)_2$
Molecular solid	Molecules	Intermolecular forces	Soft; low to moderate mp; crystalline	Ice, wax, frozen CO_2, all solid organic compounds
Covalent network	Atoms	Covalent bonds	Very hard; very high mp; crystalline	Diamond, quartz (SiO_2), tungsten carbide (WC)
Metal or alloy	Metal atoms	Metallic bonding (attraction between metal ions and surrounding mobile electrons)	Lustrous; soft (Na) to hard (Ti); high melting; crystalline	Elements (Fe, Cu, Sn, ...), bronze (CuSn alloy), amalgams (Hg + other metals)
Amorphous solid	Atoms, ions, or molecules (including polymer molecules)	Any of the above	Noncrystalline; no sharp mp; able to flow (may be very slow); curved edges when shattered	Glasses, tar, some plastics

its **heat of fusion**. After melting is complete, further addition of heat causes the temperature of the liquid to rise.

The change of a liquid into a vapor proceeds in the same way as the change of a solid into a liquid. When you first put a pan of water on the stove, all the added heat goes into raising the temperature of the water. Once the boiling point is reached, further absorbed heat goes into freeing molecules from their neighbors as they escape into the gas state. The quantity of heat needed to completely vaporize a liquid once it reaches its boiling point is called its **heat of vaporization**. A liquid with a low heat of vaporization, like rubbing alcohol (isopropyl alcohol), evaporates rapidly and is said to be *volatile*. If you spill a volatile liquid on your skin, you will feel a cooling effect as it evaporates because it is absorbing heat from your body.

It is important to know the difference between heat that is added or removed to change the *temperature* of a substance and heat that is added or removed to change the *phase* of a substance. Remember that temperature is a measure of the kinetic energy in a substance (⊂⊃ Section 7.1, p. 183). When a substance is above or below its phase change temperature (i.e., melting point or boiling point) adding or removing heat will simply change the kinetic energy and, hence, the temperature of the substance. The amount of heat needed to produce a given temperature change was presented previously (⊂⊃ Section 2.10), but is worth presenting again here:

$$\text{Heat (cal)} = \text{Mass (g)} \times \text{Temperature change (}^\circ\text{C)} \times \text{Specific heat}\left(\frac{\text{cal}}{\text{g} \times {}^\circ\text{C}}\right)$$

In contrast, when a substance is at its phase change temperature, heat that is added is being used to overcome the intermolecular forces holding particles in that phase. The temperature remains constant until *all* particles have been converted to the next phase. The energy needed to complete the phase change depends only on the amount of the substance, and the heat of fusion (for melting) or the heat of vaporization (for boiling).

$$\text{Heat (cal)} = \text{Mass (g)} \times \text{Heat of fusion}\left(\frac{\text{cal}}{\text{g}}\right)$$

$$\text{Heat (cal)} = \text{Mass (g)} \times \text{Heat of vaporization}\left(\frac{\text{cal}}{\text{g}}\right)$$

If the intermolecular forces are strong then large amounts of heat must be added to overcome these forces, and the heats of fusion and vaporization will be large. A list of heats of fusion and heats of vaporization for some common substances is given in Table 8.5. Butane, for example, has a small heat of vaporization since the

Heat of fusion The quantity of heat required to completely melt one gram of a substance once it has reached its melting point.

Heat of vaporization The quantity of heat needed to completely vaporize one gram of a liquid once it has reached its boiling point.

TABLE 8.5   Melting Points, Boiling Points, Heats of Fusion, and Heats of Vaporization of Some Common Substances

SUBSTANCE	MELTING POINT (°C)	BOILING POINT (°C)	HEAT OF FUSION (cal/g)	HEAT OF VAPORIZATION (cal/g)
Ammonia	−77.7	−33.4	84.0	327
Butane	−138.4	−0.5	19.2	92.5
Ether	−116	34.6	23.5	85.6
Ethyl alcohol	−117.3	78.5	26.1	200
Isopropyl alcohol	−89.5	82.4	21.4	159
Sodium	97.8	883	14.3	492
Water	0.0	100.0	79.7	540

APPLICATION ▶ Biomaterials for Joint Replacement

Freely movable joints in the body, such as those in the shoulder, knee, or hip, are formed by the meeting of two bones. The bony surfaces are not in direct contact, of course; rather, they are covered by cartilage for nearly frictionless motion and are surrounded by a fluid-containing capsule for lubrication. The hip, for instance, is a ball-and-socket joint, formed where the rounded upper end of the femur meets a cup-shaped part of the pelvic bone called the acetabulum.

Unfortunately, joints can wear out or fail, particularly when the cartilage is damaged by injury or diseased by degenerative arthritis. At some point, it may even be necessary to replace the failing joint—an estimated 500,000 joint-replacement surgeries are performed each year in the United States (170,000 hips, 325,000 knees). Although total joint replacement is not without problems, the lifetime of an artificial joint is nearly 20 years in 80% of the cases.

The first joint-replacement material, used in 1962, was stainless steel, but slow corrosion in the body led to its abandonment in favor of more resistant titanium or cobalt–chromium alloys. A typical modern hip-replacement joint consists of three parts: a polished metal ball to replace the head of the femur, a titanium alloy stem that is cemented into the shaft of the femur for stability, and a polyethylene cup to replace the hip socket.

Even with these materials, though, abrasion at the ball/cup contact can lead to joint wear. Over time, the constant repetitive movement of the polyethylene socket over the metal ball results in billions of microscopic polyethylene particles being sloughed off into the surrounding fluid. In addition, the cement holding the metal stem to the femur can slowly degrade, releasing other particles. The foreign particles are attacked by the body's immune system, resulting in the release of enzymes that cause the death of adjacent bone cells, loosening of the metal stem, and ultimate failure of the joint.

A potential solution to the problem involves the use of *biomaterials*—new materials that are created specifically for use in biological systems and do not provoke an immune response. It has been found, for instance, that the titanium

▲ This artificial hip joint is bonded to a thin layer of calcium phosphate to stimulate attachment of natural bone.

stem of the artificial joint can be bonded to an extremely thin layer of calcium phosphate, $Ca_3(PO_4)_2$, a close relative of hydroxyapatite $Ca_{10}(PO_4)_6(OH)_2$, the primary mineral constituent of bone (see the Chapter 4 Application, *Osteoporosis*). Natural bone then grows into the calcium phosphate, forming a strong natural bond to the stem, and making cement unnecessary. Other biomaterials are being designed to replace the polyethylene socket.

See Additional Problem 8.103 at the end of the chapter.

predominant intermolecular forces in butane (dispersion) are relatively weak. Water, on the other hand, has a particularly high heat of vaporization because of its unusually strong hydrogen bonding interactions. Thus, water evaporates more slowly than many other liquids, takes a long time to boil away, and absorbs more heat in the process. A so-called *heating curve*, which indicates the temperature and state changes as heat is added, is shown in Figure 8.24.

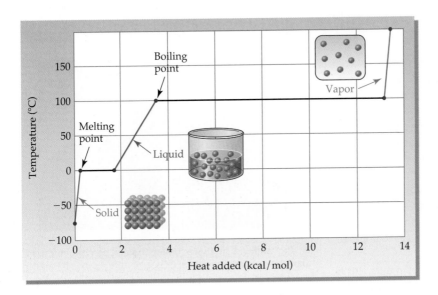

◀ **FIGURE 8.24 A heating curve for water, showing the temperature and state changes that occur when heat is added.** The horizontal lines at 0 °C and 100 °C represent the heat of fusion and heat of vaporization, respectively.

WORKED EXAMPLE   **8.13** Heat of Fusion: Calculating Total Heat of Melting

Naphthalene, an organic substance often used in mothballs, has a heat of fusion of 35.7 cal/g and a molar mass of 128.0 g/mol. How much heat in kilocalories is required to melt 0.300 mol of naphthalene?

ANALYSIS The heat of fusion tells how much heat is required to melt 1 g. To find the amount of heat needed to melt 0.300 mol, we need a mole to mass conversion.

BALLPARK ESTIMATE Naphthalene has a molar mass of 128.0 g/mol, so 0.300 mol has a mass of about one-third this amount, or about 40 g. Approximately 35 cal is required to melt 1 g, so we need about 40 times this amount of heat, or ($35 \times 40 = 1400$ cal = 1.4 kcal).

SOLUTION

STEP 1: Identify known information. We know heat of fusion (cal/g), and the number of moles of naphthalene.

Heat of fusion = 35.7 cal/g

Moles of naphthalene = 0.300 mol

STEP 2: Identify answer and units.

Heat = ?? cal

STEP 3: Identify conversion factors. First convert moles of naphthalene to grams using the molar mass (128 g/mol) as a conversion factor. Then use the heat of fusion as a conversion factor to calculate the total heat necessary to melt the mass of naphthalene.

$$(0.300 \text{ mol naphthalene})\left(\frac{128.0 \text{ g}}{1 \text{ mol}}\right) = 38.4 \text{ g naphthalene}$$

$$\text{Heat of fusion} = 35.7 \text{ cal/g}$$

STEP 4: Solve. Multiplying the mass of naphthalene by the heat of fusion then gives the answer.

$$(38.4 \text{ g naphthalene})\left(\frac{35.7 \text{ cal}}{1 \text{ g naphthalene}}\right) = 1370 \text{ cal} = 1.37 \text{ kcal}$$

BALLPARK CHECK The calculated result agrees with our estimate (1.4 kcal).

PROBLEM 8.21

How much heat in kilocalories is required to melt and boil 1.50 mol of isopropyl alcohol (rubbing alcohol; molar mass = 60.0 g/mol)? The heat of fusion and heat of vaporization of isopropyl alcohol are given in Table 8.5.

APPLICATION ▶ CO_2 as an Environmentally Friendly Solvent

How can CO_2 be a solvent? After all, carbon dioxide is a gas, not a liquid, at room temperature. Furthermore, CO_2 at atmospheric pressure does not become liquid even when cooled. When the temperature drops to −78 °C at 1 atm pressure, CO_2 goes directly from gas to solid (dry ice) without first becoming liquid. Only when the pressure is raised does liquid CO_2 exist. At a room temperature of 22.4 °C, a pressure of 60 atm is needed to force gaseous CO_2 molecules close enough together so they condense to a liquid. Even as a liquid, though, CO_2 is not a particularly good solvent. Only when it enters an unusual and rarely seen state of matter called the *supercritical state* does CO_2 become a remarkable solvent.

To understand the supercritical state of matter, think about the liquid and gas states at the molecular level. In the liquid state, molecules are packed closely together, and most of the available volume is taken up by the molecules themselves. In the gas state, molecules are far apart, and most of the available volume is empty space. In the supercritical state, however, the situation is intermediate between liquid and gas. There is *some* space between molecules, but not much. The molecules are too far apart to be truly a liquid, yet they are too close together to be truly a gas. Supercritical CO_2 exists when the pressure is above 72.8 atm and the temperature is above 31.2 °C. This pressure is high enough to force molecules close together and prevent them from expanding into the gas state. Above this temperature, however, the molecules have too much kinetic energy to condense into the liquid state.

Because open spaces already exist between CO_2 molecules, it is energetically easy for dissolved molecules to slip in, and supercritical CO_2 is therefore an extraordinarily good solvent. Among its many applications, supercritical CO_2 is used in the beverage and food processing industries to decaffeinate

▲ The caffeine in these coffee beans can be removed by extraction with supercritical CO_2.

coffee beans and to obtain spice extracts from vanilla, pepper, cloves, nutmeg, and other seeds. In the cosmetics and perfume industry, fragrant oils are extracted from flowers using supercritical CO_2. Perhaps the most important future application is the use of carbon dioxide for dry-cleaning clothes, thereby replacing environmentally harmful chlorinated solvents.

The use of supercritical CO_2 as a solvent has many benefits, including the fact that it is nontoxic and nonflammable. Most important, though, is that the technology is environmentally friendly. Industrial processes using CO_2 are designed as closed systems so that CO_2 is recaptured after use and continually recycled. No organic solvent vapors are released into the atmosphere and no toxic liquids seep into groundwater supplies, as can occur with current procedures using chlorinated organic solvents. The future looks bright for this new technology.

See Additional Problems 8.104 and 8.105 at the end of the chapter.

KEY WORDS

Amorphous solid, *p. 244*

Avogadro's law, *p. 229*

Boiling point (bp), *p. 217*

Boyle's law, *p. 222*

Change of state, *p. 216*

Charles's law, *p. 225*

SUMMARY: REVISITING THE CHAPTER GOALS

1. **How do scientists explain the behavior of gases?** According to the *kinetic–molecular theory of gases*, the physical behavior of gases can be explained by assuming that they consist of particles moving rapidly at random, separated from other particles by great distances, and colliding without loss of energy. Gas *pressure* is the result of molecular collisions with a surface.

2. **How do gases respond to changes in temperature, pressure, and volume?** *Boyle's law* says that the volume of a fixed amount of gas at constant temperature is inversely proportional to its pressure ($P_1V_1 = P_2V_2$). *Charles's law* says that the volume of a fixed amount of gas at constant pressure is directly proportional to its Kelvin temperature ($V_1/T_1 = V_2/T_2$). *Gay-Lussac's law* says that the pressure of a fixed amount of gas at

constant volume is directly proportional to its Kelvin temperature ($P_1/T_1 = P_2/T_2$). Boyle's law, Charles's law, and Gay-Lussac's law together give the *combined gas law* ($P_1V_1/T_1 = P_2V_2/T_2$), which applies to changing conditions for a fixed quantity of gas. *Avogadro's law* says that equal volumes of gases at the same temperature and pressure contain the same number of moles ($V_1/n_1 = V_2/n_2$).

3. **What is the ideal gas law?** The four gas laws together give the *ideal gas law*, $PV = nRT$, which relates the effects of temperature, pressure, volume, and molar amount. At 0 °C and 1 atm pressure, called *standard temperature and pressure (STP)*, 1 mol of any gas (6.02×10^{23} molecules) occupies a volume of 22.4 L.

4. **What is partial pressure?** The amount of pressure exerted by an individual gas in a mixture is called the *partial pressure* of the gas. According to *Dalton's law*, the total pressure exerted by the mixture is equal to the sum of the partial pressures of the individual gases.

5. **What are the major intermolecular forces, and how do they affect the states of matter?** There are three major types of *intermolecular forces*, which act to hold molecules near one another in solids and liquids. *Dipole–dipole forces* are the electrical attractions that occur between polar molecules. *London dispersion forces* occur between all molecules as a result of temporary molecular polarities due to unsymmetrical electron distribution. These forces increase in strength with molecular weight and with the surface area of molecules. *Hydrogen bonding*, the strongest of the three intermolecular forces, occurs between a hydrogen atom bonded to O, N, or F and a nearby O, N, or F atom.

6. **What are the various kinds of solids, and how do they differ?** Solids are either crystalline or amorphous. *Crystalline solids* are those whose constituent particles have an ordered arrangement; *amorphous solids* lack internal order and do not have sharp melting points. There are several kinds of crystalline solids: *Ionic solids* are those like sodium chloride, whose constituent particles are ions. *Molecular solids* are those like ice, whose constituent particles are molecules held together by intermolecular forces. *Covalent network solids* are those like diamond, whose atoms are linked together by covalent bonds into a giant three-dimensional array. *Metallic solids*, such as silver or iron, also consist of large arrays of atoms, but their crystals have metallic properties such as electrical conductivity.

7. **What factors affect a change of state?** When a solid is heated, particles begin to move around freely at the *melting point*, and the substance becomes liquid. The amount of heat necessary to melt a given amount of solid at its melting point is its *heat of fusion*. As a liquid is heated, molecules escape from the surface of a liquid until an equilibrium is reached between liquid and gas, resulting in a *vapor pressure* of the liquid. At a liquid's *boiling point*, its vapor pressure equals atmospheric pressure, and the entire liquid is converted into gas. The amount of heat necessary to vaporize a given amount of liquid at its boiling point is called its *heat of vaporization*.

UNDERSTANDING KEY CONCEPTS

8.22 Assume that you have a sample of gas in a cylinder with a movable piston, as shown in the following drawing:

Redraw the apparatus to show what the sample will look like after the following changes:

(a) The temperature is increased from 300 K to 450 K at constant pressure.

(b) The pressure is increased from 1 atm to 2 atm at constant temperature.

(c) The temperature is decreased from 300 K to 200 K and the pressure is decreased from 3 atm to 2 atm.

8.23 Assume that you have a sample of gas at 350 K in a sealed container, as represented in part (a). Which of the drawings (b)–(d) represents the gas after the temperature is lowered from 350 K to 150 K?

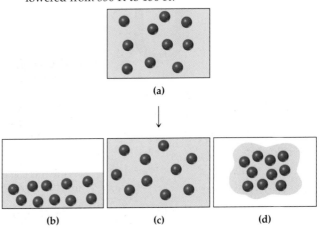

(a)

(b) (c) (d)

8.24 Assume that drawing (a) represents a sample of H_2O at 200 K. Which of the drawings (b)–(d) represents what the sample will look like when the temperature is raised to 300 K?

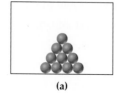
(a)

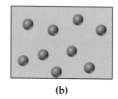
(b)

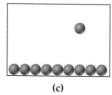
(c)

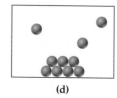
(d)

8.25 Three bulbs, two of which contain different gases and one of which is empty, are connected as shown in the following drawing:

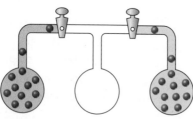

Redraw the apparatus to represent the gases after the stopcocks are opened and the system is allowed to come to equilibrium.

8.26 Redraw the following open-end manometer to show what it would look like when stopcock A is opened.

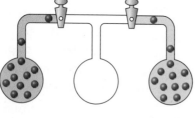

8.27 The following graph represents the heating curve of a hypothetical substance:

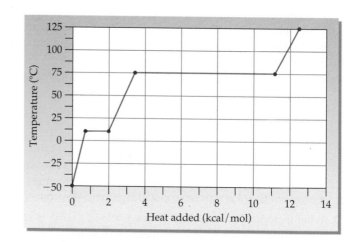

(a) What is the melting point of the substance?
(b) What is the boiling point of the substance?
(c) Approximately what is the heat of fusion for the substance in kcal/mol?
(d) Approximately what is the heat of vaporization for the substance in kcal/mol?

8.28 Show the approximate level of the movable piston in drawings (a)–(c) after the indicated changes have been made to the gas.

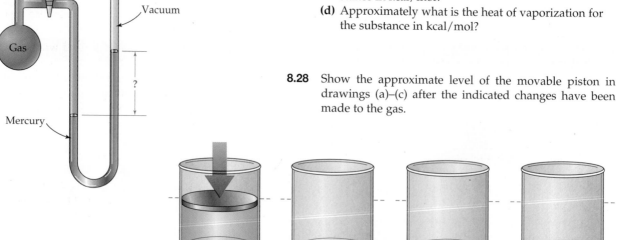

(initial)
$T = 25\,°C$
$n = 0.075$ mol
$P = 0.92$ atm

(a)
$T = 50\,°C$
$n = 0.075$ mol
$P = 0.92$ atm

(b)
$T = 175\,°C$
$n = 0.075$ mol
$P = 2.7$ atm

(c)
$T = 25\,°C$
$n = 0.22$ mol
$P = 2.7$ atm

8.29 What is the partial pressure of each gas—red, yellow, and green—if the total pressure inside the following container is 600 mmHg?

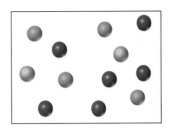

ADDITIONAL PROBLEMS

GASES AND PRESSURE

8.30 How is 1 atm of pressure defined?

8.31 List four common units for measuring pressure.

8.32 What are the four assumptions of the kinetic–molecular theory of gases?

8.33 How does the kinetic–molecular theory of gases explain gas pressure?

8.34 Convert the following values into mmHg:

(a) Standard pressure   (b) 25.3 psi
(c) 7.5 atm   (d) 28.0 in. Hg
(e) 41.8 Pa

8.35 Atmospheric pressure at the top of Mt. Whitney in California is 440 mmHg.

(a) How many atmospheres is this?
(b) How many pascals is this?

8.36 What is the pressure (in mmHg) inside a container of gas connected to a mercury-filled, open-end manometer of the sort shown in Figure 8.5 when the level in the arm connected to the container is 17.6 cm lower than the level in the arm open to the atmosphere, and the atmospheric pressure reading outside the apparatus is 754.3 mmHg?

8.37 What is the pressure (in atmospheres) inside a container of gas connected to a mercury-filled, open-end manometer of the sort shown in Figure 8.5 when the level in the arm connected to the container is 28.3 cm higher than the level in the arm open to the atmosphere, and the atmospheric pressure reading outside the apparatus is 1.021 atm?

BOYLE'S LAW

8.38 What is Boyle's law, and what variables must be kept constant for the law to hold?

8.39 Which assumption(s) of the kinetic–molecular theory explain the behavior of gases described by Boyle's Law? Explain your answer.

8.40 The pressure of gas in a 600.0 mL cylinder is 65.0 mmHg. To what volume must it be compressed to increase the pressure to 385 mmHg?

8.41 The volume of a balloon is 2.85 L at 1.00 atm. What pressure is required to compress the balloon to a volume of 1.70 L?

8.42 The use of chlorofluorocarbons (CFCs) as refrigerants and propellants in aerosol cans has been discontinued as a result of concerns about the ozone layer. If an aerosol can contained 350 mL of CFC gas at a pressure of 5.0 atm, what volume would this gas occupy at 1.0 atm?

8.43 A sample of neon gas used in a neon sign occupies a volume of 1.50 L at a pressure of 630 torr. Calculate the volume of the gas if the pressure in the glass tube of the sign is 793 torr.

CHARLES'S LAW

8.44 What is Charles's law, and what variables must be kept constant for the law to hold?

8.45 Which assumption(s) of the kinetic–molecular theory explain the behavior of gases described by Charles's Law? Explain your answer.

8.46 A hot-air balloon has a volume of 960 L at 18 °C. To what temperature must it be heated to raise its volume to 1200 L, assuming the pressure remains constant?

8.47 A hot-air balloon has a volume of 875 L. What is the original temperature of the balloon if its volume changes to 955 L when heated to 56 °C?

8.48 A gas sample has a volume of 185 mL at 38 °C. What is its volume at 97 °C?

8.49 A balloon has a volume of 43.0 L at 25 °C. What is its volume at −8 °C?

GAY-LUSSAC'S LAW

8.50 What is Gay-Lussac's law, and what variables must be kept constant for the law to hold?

8.51 Which assumption(s) of the kinetic–molecular theory explain the behavior of gases described by Gay-Lussac's Law? Explain your answer.

8.52 A glass laboratory flask is filled with gas at 25 °C and 0.95 atm pressure, sealed, and then heated to 117 °C. What is the pressure inside the flask?

8.53 An aerosol can has an internal pressure of 3.85 atm at 25 °C. What temperature is required to raise the pressure to 18.0 atm?

COMBINED GAS LAW

8.54 A gas has a volume of 2.84 L at 1.00 atm and 0 °C. At what temperature does it have a volume of 7.50 L at 520 mmHg?

8.55 A helium balloon has a volume of 3.50 L at 22.0 °C and 1.14 atm. What is its volume if the temperature is increased to 30.0 °C and the pressure is increased to 1.20 atm?

8.56 When H_2 gas was released by the reaction of HCl with Zn, the volume of H_2 collected was 75.4 mL at 23 °C and 748 mmHg. What is the volume of the H_2 at 0 °C and 1.00 atm pressure (STP)?

8.57 A compressed-air tank carried by scuba divers has a volume of 6.80 L and a pressure of 120 atm at 20 °C. What is the volume of air in the tank at 0 °C and 1.00 atm pressure (STP)?

8.58 What is the effect on the pressure of a gas if you simultaneously:

(a) Halve its volume and double its Kelvin temperature?
(b) Double its volume and halve its Kelvin temperature?

8.59 What is the effect on the volume of a gas if you simultaneously:

(a) Halve its pressure and double its Kelvin temperature?
(b) Double its pressure and double its Kelvin temperature?

8.60 A sample of oxygen produced in a laboratory experiment had a volume of 590 mL at a pressure of 775 mmHg and a temperature of 352 K. What is the volume of this sample at 25 °C and 800.0 mmHg pressure?

8.61 A small cylinder of helium gas used for filling balloons has a volume of 2.30 L and a pressure of 1850 atm at 25 °C. How many balloons can you fill if each one has a volume of 1.5 L and a pressure of 1.25 atm at 25 °C?

AVOGADRO'S LAW AND STANDARD MOLAR VOLUME

8.62 Explain Avogadro's law using the kinetic–molecular theory of gases.

8.63 What conditions are defined as standard temperature and pressure (STP)?

8.64 How many liters does 1 mol of gas occupy at STP?

8.65 How many moles of gas are in a volume of 48.6 L at STP?

8.66 Which sample contains more molecules: 1.0 L of O_2 at STP or 1.0 L of H_2 at STP? Which sample has the larger mass?

8.67 How many milliliters of Cl_2 gas must you have to obtain 0.20 g at STP?

8.68 What is the mass of CH_4 in a sample that occupies a volume of 16.5 L at STP?

8.69 Assume that you have 1.75 g of the deadly gas hydrogen cyanide, HCN. What is the volume of the gas at STP?

8.70 A typical room is 4.0 m long, 5.0 m wide, and 2.5 m high. What is the total mass of the oxygen in the room assuming that the gas in the room is at STP and that air contains 21% oxygen and 79% nitrogen?

8.71 What is the total mass of nitrogen in the room described in Problem 8.70?

IDEAL GAS LAW

8.72 What is the ideal gas law?

8.73 How does the ideal gas law differ from the combined gas law?

8.74 Which sample contains more molecules: 2.0 L of Cl_2 at STP, or 3.0 L of CH_4 at 300 K and 1.5 atm? Which sample weighs more?

8.75 Which sample contains more molecules: 2.0 L of CO_2 at 300 K and 500 mmHg, or 1.5 L of N_2 at 57 °C and 760 mmHg? Which sample weighs more?

8.76 If 2.3 mol of He has a volume of 0.15 L at 294 K, what is the pressure in atm?

8.77 If 3.5 mol of O_2 has a volume of 27.0 L at a pressure of 1.6 atm, what is its temperature in kelvins?

8.78 If 15.0 g of CO_2 gas has a volume of 0.30 L at 310 K, what is its pressure in mmHg?

8.79 If 20.0 g of N_2 gas has a volume of 4.00 L and a pressure of 6.0 atm, what is its temperature in degrees Celsius?

8.80 If 18.0 g of O_2 gas has a temperature of 350 K and a pressure of 550 mmHg, what is its volume?

8.81 How many moles of a gas will occupy a volume of 0.55 L at a temperature of 347 K and a pressure of 2.5 atm?

DALTON'S LAW AND PARTIAL PRESSURE

8.82 What is meant by *partial pressure*?

8.83 What is Dalton's law?

8.84 If the partial pressure of oxygen in air at 1.0 atm is 160 mmHg, what is the partial pressure on the summit of Mt. Whitney, where atmospheric pressure is 440 mmHg? Assume that the percent oxygen is the same.

8.85 Patients suffering from respiratory disorders are often treated in oxygen tents in which the atmosphere is enriched in oxygen. What is the partial pressure of O_2 in an oxygen tent consisting of 45% O_2 for an atmospheric pressure of 753 mmHg?

LIQUIDS AND INTERMOLECULAR FORCES

8.86 What is the vapor pressure of a liquid?

8.87 What is the value of a liquid's vapor pressure at its normal boiling point?

8.88 What is the effect of pressure on a liquid's boiling point?

8.89 What is a liquid's heat of vaporization?

8.90 What characteristic must a compound have to experience the following intermolecular forces?

(a) London dispersion forces
(b) Dipole–dipole forces
(c) Hydrogen bonding

8.91 In which of the following compounds are dipole–dipole attractions the most important intermolecular force?

(a) N_2 (b) HCN   (c) CCl_4
(d) $MgBr_2$ (e) CH_3Cl (f) CH_3CO_2H

8.92 Dimethyl ether (CH_3OCH_3) and ethanol (C_2H_5OH) have the same formula (C_2H_6O), but the boiling point of dimethyl ether is −25 °C while that of ethanol is 78 °C. Explain.

8.93 Iodine is a solid at room temperature (mp = 113.5 °C) while bromine is a liquid (mp = −7 °C). Explain in terms of intermolecular forces.

8.94 The heat of vaporization of water is 9.72 kcal/mol.

(a) How much heat (in kilocalories) is required to vaporize 3.00 mol of H_2O?
(b) How much heat (in kilocalories) is released when 320 g of steam condenses?

8.95 Patients with a high body temperature are often given "alcohol baths." The heat of vaporization of isopropyl alcohol (rubbing alcohol) is 159 cal/g. How much heat is removed from the skin by the evaporation of 190 g (about ½ cup) of isopropyl alcohol?

SOLIDS

8.96 What is the difference between an amorphous and a crystalline solid?

8.97 List three kinds of crystalline solids, and give an example of each.

8.98 The heat of fusion of acetic acid, the principal organic component of vinegar, is 45.9 cal/g. How much heat (in kilocalories) is required to melt 1.75 mol of solid acetic acid?

8.99 The heat of fusion of sodium metal is 630 cal/mol. How much heat (in kilocalories) is required to melt 262 g of sodium?

Applications

8.100 What is the difference between a systolic and a diastolic pressure reading? Is a blood pressure of 180/110 within the normal range? [*Blood Pressure, p. 223*]

8.101 What are the three most important greenhouse gases? [*Greenhouse Gases and Global Warming, p. 236*]

8.102 What evidence is there that global warming is occurring? [*Greenhouse Gases and Global Warming, p. 236*]

8.103 What is the mass ratio of calcium to phosphate in calcium phosphate, $Ca_3(PO_4)_2$? In hydroxyapatite, $Ca_{10}(PO_4)_6(OH)_2$? [*Biomaterials for Joint Replacement, p. 246*]

8.104 What is a supercritical fluid? [*CO_2 as an Environmentally Friendly Solvent, p. 248*]

8.105 What are the environmental advantages of using supercritical CO_2 in place of chlorinated organic solvents? [*CO_2 as an Environmentally Friendly Solvent, p. 248*]

General Questions and Problems

8.106 Use the kinetic–molecular theory to explain why gas pressure increases if the temperature is raised and the volume is kept constant.

8.107 Hydrogen and oxygen react according to the equation $2 H_2 + O_2 \longrightarrow 2 H_2O$. According to Avogadro's law, how many liters of hydrogen are required to react with 2.5 L of oxygen at STP?

8.108 If 3.0 L of hydrogen and 1.5 L of oxygen at STP react to yield water, how many moles of water are formed? What gas volume does the water have at a temperature of 100 °C and 1 atm pressure?

8.109 Approximately 240 mL/min of CO_2 is exhaled by an average adult at rest. Assuming a temperature of 37 °C and 1 atm pressure, how many moles of CO_2 is this?

8.110 How many grams of CO_2 are exhaled by an average resting adult in 24 hours? (See Problem 8.109.)

8.111 Imagine that you have two identical containers, one containing hydrogen at STP and the other containing oxygen at STP. How can you tell which is which without opening them?

8.112 When fully inflated, a hot-air balloon has a volume of 1.6×10^5 L at an average temperature of 375 K and 0.975 atm. Assuming that air has an average molar mass of 29 g/mol, what is the density of the air in the hot-air balloon? How does this compare with the density of air at STP?

8.113 A 10.0 g sample of an unknown gas occupies 14.7 L at a temperature of 25 °C and a pressure of 745 mmHg. What is the molar mass of the gas?

8.114 One mole of any gas has a volume of 22.4 L at STP. What are the molecular weights of the following gases, and what are their densities in grams per liter at STP?

(a) CH_4 (b) CO_2 (c) O_2

8.115 Gas pressure outside the space shuttle is approximately 1×10^{-14} mm Hg at a temperature of approximately 1 K. If the gas is almost entirely hydrogen atoms (H, not H_2), what volume of space is occupied by 1 mol of atoms? What is the density of H gas in atoms per liter?

8.116 Ethylene glycol, $C_2H_6O_2$, has one OH bonded to each carbon.

(a) Draw the Lewis dot structure of ethylene glycol.

(b) Draw the Lewis dot structure of chloroethane, C_2H_5Cl.

(c) Chloroethane has a slightly higher molar mass than ethylene glycol, but a much lower boiling point (3 °C versus 198 °C). Explain.

8.117 A rule of thumb for scuba diving is that the external pressure increases by 1 atm for every 10 m of depth. A diver using a compressed air tank is planning to descend to a depth of 25 m.

(a) What is the external pressure at this depth? (Remember that the pressure at sea level is 1 atm.)

(b) Assuming that the tank contains 20% oxygen and 80% nitrogen, what is the partial pressure of each gas in the diver's lungs at this depth?

8.118 The *Rankine* temperature scale used in engineering is to the Fahrenheit scale as the Kelvin scale is to the Celsius scale. That is, 1 Rankine degree is the same size as 1 Fahrenheit degree, and 0 °R = absolute zero .

(a) What temperature corresponds to the freezing point of water on the Rankine scale?

(b) What is the value of the gas constant R on the Rankine scale in (L·atm)/(°R·mol)?

8.119 Isooctane, C_8H_{18}, is the component of gasoline from which the term *octane rating* derives.

(a) Write a balanced equation for the combustion of isooctane to yield CO_2 and H_2O.

(b) Assuming that gasoline is 100% isooctane and that the density of isooctane is 0.792 g/mL, what mass of CO_2 (in kilograms) is produced each year by the annual U.S. gasoline consumption of 4.6×10^{10} L?

(c) What is the volume (in liters) of this CO_2 at STP?

CHAPTER 9

Solutions

▲ The giant sequoia relies on osmotic pressure—a colligative property of solutions—to transport water and nutrients from the roots to the treetops 300 ft up.

CONTENTS

CHAPTER GOALS

Among the questions we will answer are the following:

1. What are solutions, and what factors affect solubility?

THE GOAL: Be able to define the different kinds of mixtures and explain the influence on solubility of solvent and solute structure, temperature, and pressure.

2. How is the concentration of a solution expressed?

THE GOAL: Be able to define, use, and convert between the most common ways of expressing solution concentrations.

3. How are dilutions carried out?

THE GOAL: Be able to calculate the concentration of a solution prepared by dilution and explain how to make a desired dilution.

4. What is an electrolyte?

THE GOAL: Be able to recognize strong and weak electrolytes and nonelectrolytes, and express electrolyte concentrations.

5. How do solutions differ from pure solvents in their behavior?

THE GOAL: Be able to explain vapor pressure lowering, boiling point elevation, and freezing point depression for solutions.

6. What is osmosis?

THE GOAL: Be able to describe osmosis and some of its applications.

U p to this point, we have been concerned primarily with pure substances, both elements and compounds. In day-to-day life, however, most of the materials we come in contact with are mixtures. Air, for example, is a gaseous mixture of primarily oxygen and nitrogen; blood is a liquid mixture of many different components; and many rocks are solid mixtures of different minerals. In this chapter, we look closely at the characteristics and properties of mixtures, with particular attention to the uniform mixtures we call *solutions*.

9.1 Mixtures and Solutions

As we saw in Section 1.3, a *mixture* is an intimate combination of two or more substances, both of which retain their chemical identities. (⬤▭, p. 6) Mixtures can be classified as either *heterogeneous* or *homogeneous* as indicated in Figure 9.1, depending on their appearance. **Heterogeneous mixtures** are those in which the mixing is

Heterogeneous mixture A nonuniform mixture that has regions of different composition.

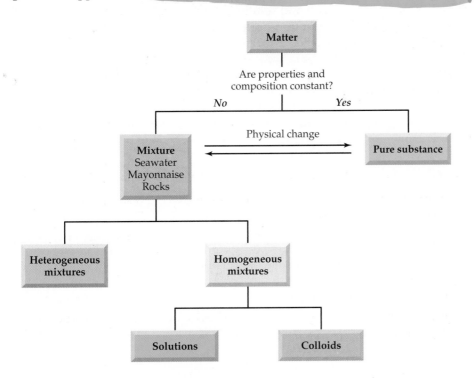

◀ **FIGURE 9.1 Classification of mixtures.** The components in heterogeneous mixtures are not uniformly mixed, and the composition varies with location. In homogeneous mixtures, the components are uniformly mixed at the molecular level.

255

Homogeneous mixture A uniform mixture that has the same composition throughout.

Solution A homogeneous mixture that contains particles the size of a typical ion or small molecule.

Colloid A homogeneous mixture that contains particles that range in diameter from 2 to 500 nm.

not uniform and which therefore have regions of different composition. Rocky Road ice cream, for example, is a heterogeneous mixture, with something different in every spoonful. Granite and many other rocks are also heterogeneous, having a grainy character due to the heterogeneous mixing of different minerals. **Homogeneous mixtures** are those in which the mixing *is* uniform and that therefore have the same composition throughout. Seawater, a homogeneous mixture of soluble ionic compounds in water, is an example.

Homogeneous mixtures can be further classified as either *solutions* or *colloids* according to the size of their particles. **Solutions**, the most important class of homogeneous mixtures, contain particles the size of a typical ion or small molecule—roughly 0.1–2 nm in diameter. **Colloids**, such as milk and fog, are also homogeneous in appearance but contain larger particles than solutions—in the range 2–500 nm diameter.

(a)

(b)

(c)

▲ (a) Wine is a solution of dissolved molecules, and (b) milk is a colloid with fine particles that do not separate out on standing. (c) An aerosol spray, by contrast, is a heterogeneous mixture of small particles visible to the naked eye.

Liquid solutions, colloids, and heterogeneous mixtures can be distinguished in several ways. For example, liquid solutions are transparent (although they may be colored). Colloids may appear transparent if the particle size is small, but they have a murky or opaque appearance if the particle size is larger. Neither solutions nor small-particle colloids separate on standing, and the particles in both are too small to be removed by filtration. Heterogeneous mixtures and large-particle colloids, also known as "suspensions," are murky and opaque and their particles will slowly settle on prolonged standing. House paint is an example.

Table 9.1 gives some examples of solutions, colloids, and heterogeneous mixtures. It is interesting to note that blood has characteristics of all three. About 45%

TABLE 9.1 Some Characteristics of Solutions, Colloids, and Heterogeneous Mixtures

TYPE OF MIXTURE	PARTICLE SIZE	EXAMPLES	CHARACTERISTICS
Solution	<2.0 nm	Air, seawater, gasoline, wine	Transparent to light; does not separate on standing; nonfilterable
Colloid	2.0–500 nm	Butter, milk, fog, pearl	Often murky or opaque to light; does not separate on standing; nonfilterable
Heterogeneous	>500 nm	Blood, paint, aerosol sprays	Murky or opaque to light; separates on standing; filterable

by volume of blood consists of suspended red and white cells, which settle slowly on standing; the remaining 55% is *plasma*, which contains ions in solution and colloidal protein molecules.

Although we usually think of solids dissolved in liquids when we talk about solutions, solutions actually occur in all three phases of matter (Table 9.2). Metal alloys like 14-karat gold (58% gold with silver and copper) and brass (10–40% zinc with copper), for instance, are solutions of one solid with another. For solutions in which a gas or solid is dissolved in a liquid, the dissolved substance is called the **solute** and the liquid is called the **solvent**. When one liquid is dissolved in another, the minor component is usually considered the solute and the major component is the solvent.

Solute A substance dissolved in a liquid.

Solvent The liquid in which another substance is dissolved.

TABLE 9.2 Some Different Types of Solutions

TYPE OF SOLUTION	EXAMPLE
Gas in gas	Air (O_2, N_2, Ar, and other gases)
Gas in liquid	Seltzer water (CO_2 in water)
Gas in solid	H_2 in palladium metal
Liquid in liquid	Gasoline (mixture of hydrocarbons)
Liquid in solid	Dental amalgam (mercury in silver)
Solid in liquid	Seawater (NaCl and other salts in water)
Solid in solid	Metal alloys such as 14-karat gold (Au, Ag, and Cu)

PROBLEM 9.1

Classify the following liquid mixtures as heterogeneous or homogeneous. Further classify each homogeneous mixture as a solution or colloid.

(a) Orange juice   **(b)** Apple juice

(c) Hand lotion   **(d)** Tea

9.2 The Solution Process

What determines whether a substance is soluble in a given liquid? Solubility depends primarily on the strength of the attractions between solute and solvent particles relative to the strengths of the attractions within the pure substances. Ethyl alcohol is soluble in water, for example, because hydrogen bonding (Section 8.11) is nearly as strong between water and ethyl alcohol molecules as it is between water molecules alone or ethyl alcohol molecules alone. (⬤▭⬤, p. 238)

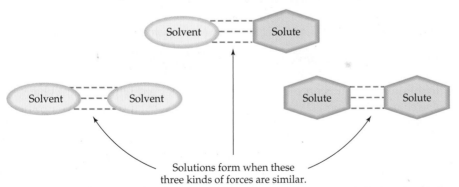

Solutions form when these three kinds of forces are similar.

A good rule of thumb for predicting solubility is that "like dissolves like," meaning that substances with similar intermolecular forces form solutions with one another, whereas substances with different intermolecular forces do not (Section 8.11). (⬤▭⬤, p. 235)

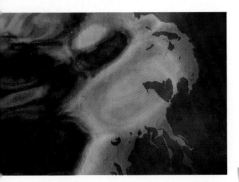

▲ Oil and water do not mix because they have different intermolecular forces, resulting in the formation of oil slicks.

Solvation The clustering of solvent molecules around a dissolved solute molecule or ion.

Polar solvents dissolve polar and ionic solutes; nonpolar solvents dissolve nonpolar solutes. Thus, a polar, hydrogen-bonding compound like water dissolves ethyl alcohol and sodium chloride, whereas a nonpolar organic compound like hexane (C_6H_{14}) dissolves other nonpolar organic compounds like fats and oils. Water and oil, however, do not dissolve one another, as summed up by the old saying, "Oil and water don't mix." The intermolecular forces between water molecules are so strong that after an oil–water mixture is shaken, the water layer re-forms, squeezing out the oil molecules.

Water solubility is not limited to ionic compounds and ethyl alcohol. Many polar organic substances, such as sugars, amino acids, and even some proteins, dissolve in water. In addition, small, moderately polar organic molecules such as chloroform ($CHCl_3$) are soluble in water to a limited extent. When mixed with water, a small amount of the organic compound dissolves, but the remainder forms a separate liquid layer. As the number of carbon atoms in organic molecules increases, though, water solubility decreases.

The process of dissolving an ionic solid in a polar liquid can be visualized as shown in Figure 9.2 for sodium chloride. When NaCl crystals are put in water, ions at the crystal surface come into contact with polar water molecules. Positively charged Na^+ ions are attracted to the negatively polarized oxygen of water, and negatively charged Cl^- ions are attracted to the positively polarized hydrogens. The combined forces of attraction between an ion and several water molecules pull the ion away from the crystal, exposing a fresh surface, until ultimately the crystal dissolves. Once in solution, Na^+ and Cl^- ions are completely surrounded by solvent molecules, a phenomenon called **solvation** (or, specifically for water, *hydration*). The water molecules form a loose shell around the ions, stabilizing them by electrical attraction.

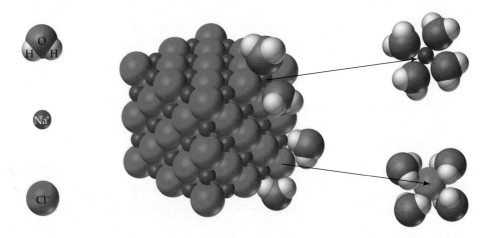

▲ **FIGURE 9.2 Dissolution of an NaCl crystal in water.** Polar water molecules surround the individual Na^+ and Cl^- ions at an exposed edge or corner, pulling them from the crystal surface into solution and surrounding them. Note how the negatively polarized oxygens of water molecules cluster around Na^+ ions and the positively polarized hydrogens cluster around Cl^- ions.

The dissolution of a solute in a solvent is a physical change since the solution components retain their chemical identities. Like all chemical and physical changes, the dissolution of a substance in a solvent has associated with it a heat change, or *enthalpy* change (Section 7.2). (⊂⊃, p. 184) Some substances dissolve exothermically, releasing heat and warming the resultant solution, whereas other substances dissolve endothermically, absorbing heat and cooling the resultant solution. Calcium chloride, for example, *releases* 19.4 kcal/mol of heat energy when it dissolves in water, but ammonium nitrate (NH_4NO_3) *absorbs* 6.1 kcal/mol of heat energy. Athletes and others take advantage of both situations when they use instant hot packs or cold packs to treat injuries. Both hot and cold packs consist of a pouch of water and a dry chemical, such as $CaCl_2$ or $MgSO_4$ for hot packs and NH_4NO_3 for cold packs. Squeezing the pack breaks the pouch and the solid dissolves, either raising or lowering the temperature.

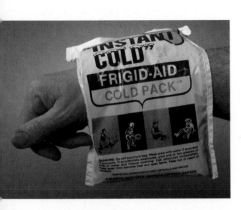

▲ Instant cold packs used to treat muscle strains and sprains often take advantage of the endothermic enthalpy of a solution of salts such as ammonium nitrate.

WORKED EXAMPLE **9.1** Formation of Solutions

Which of the following pairs of substances would you expect to form solutions?

(a) Carbon tetrachloride (CCl_4) and hexane (C_6H_{14}).

(b) Octane (C_8H_{18}) and methyl alcohol (CH_3OH).

ANALYSIS Identify the kinds of intermolecular forces in each substance (Section 8.11). Substances with similar intermolecular forces tend to form solutions.

SOLUTION

(a) Hexane contains only C—H and C—C bonds, which are nonpolar. Carbon tetrachloride contains polar C—Cl bonds, but they are distributed symmetrically in the tetrahedral molecule so that it too is nonpolar. The major intermolecular force for both compounds is London dispersion forces, so they will form a solution.

(b) Octane contains only C—H and C—C bonds and so is nonpolar; the major intermolecular force is dispersion. Methyl alcohol contains polar C—O and O—H bonds; it is polar and forms hydrogen bonds. The intermolecular forces for the two substances are so dissimilar that they do not form a solution.

PROBLEM 9.2

Which of the following pairs of substances would you expect to form solutions?

(a) CCl_4 and water

(b) Benzene (C_6H_6) and $MgSO_4$

(c) Hexane (C_6H_{14}) and heptane (C_7H_{16})

(d) Ethyl alcohol (C_2H_5OH) and heptanol ($C_7H_{15}OH$)

Hygroscopic Having the ability to pull water molecules from the surrounding atmosphere.

9.3 Solid Hydrates

Some ionic compounds attract water strongly enough to hold onto water molecules even when crystalline, forming what are called *solid hydrates*. For example, the plaster of Paris used to make decorative objects and casts for broken limbs is calcium sulfate hemihydrate, $CaSO_4 \cdot \frac{1}{2}H_2O$. The dot between $CaSO_4$ and $\frac{1}{2}H_2O$ in the formula indicates that for every two $CaSO_4$ formula units in the crystal there is also one water molecule present.

$$CaSO_4 \cdot \tfrac{1}{2}H_2O \quad \text{A solid hydrate}$$

After being ground up and mixed with water to make plaster, $CaSO_4 \cdot \frac{1}{2}H_2O$ gradually changes into the crystalline dihydrate $CaSO_4 \cdot 2\,H_2O$, known as *gypsum*. During the change, the plaster hardens and expands in volume, causing it to fill a mold or shape itself closely around a broken limb. Table 9.3 lists some other ionic compounds that are handled primarily as hydrates.

Still other ionic compounds attract water so strongly that they pull water vapor from humid air to become hydrated. Compounds that show this behavior, such as calcium chloride ($CaCl_2$), are called **hygroscopic** and are often used as drying agents. You might have noticed a small bag of a hygroscopic compound (probably silica gel, SiO_2) included in the packing material of a new MP3 player, camera, or other electronic device to keep humidity low during shipping.

▲ Plaster of Paris ($CaSO_4 \cdot \frac{1}{2}H_2O$) slowly turns into gypsum ($CaSO_4 \cdot 2\,H_2O$) when added to water. In so doing, the plaster hardens and expands, causing it to fill a mold.

TABLE 9.3   Some Common Solid Hydrates

FORMULA	NAME	USES
$AlCl_3 \cdot 6\,H_2O$	Aluminum chloride hexahydrate	Antiperspirant
$CaSO_4 \cdot 2\,H_2O$	Calcium sulfate dihydrate (gypsum)	Cements, wallboard molds
$CaSO_4 \cdot \frac{1}{2}H_2O$	Calcium sulfate hemihydrate (plaster of Paris)	Casts, molds
$CuSO_4 \cdot 5\,H_2O$	Copper(II) sulfate pentahydrate (blue vitriol)	Pesticide, germicide, topical fungicide
$MgSO_4 \cdot 7\,H_2O$	Magnesium sulfate heptahydrate (epsom salts)	Laxative, anticonvulsant
$Na_2B_4O_7 \cdot 10\,H_2O$	Sodium tetraborate decahydrate (borax)	Cleaning compounds, fireproofing agent
$Na_2S_2O_3 \cdot 5\,H_2O$	Sodium thiosulfate pentahydrate (hypo)	Photographic fixer

PROBLEM 9.3

Write the formula of sodium sulfate decahydrate, known as Glauber's salt and used as a laxative.

PROBLEM 9.4

What masses of Glauber's salt must be used to provide 1.00 mol of sodium sulfate?

9.4 Solubility

We saw in Section 9.2 that ethyl alcohol is soluble in water because hydrogen bonding is nearly as strong between water and ethyl alcohol molecules as it is between water molecules alone or ethyl alcohol molecules alone. So similar are the forces in this particular case, in fact, that the two liquids are **miscible**, or mutually soluble in all proportions. Ethyl alcohol will continue to dissolve in water no matter how much is added.

Most substances, however, reach a solubility limit beyond which no more will dissolve in solution. Imagine, for instance that you are asked to prepare a saline solution (aqueous NaCl). You might measure out some water, add solid NaCl, and stir the mixture. Dissolution occurs rapidly at first but then slows down as more and more NaCl is added. Eventually the dissolution stops because an equilibrium is reached when the numbers of Na^+ and Cl^- ions leaving a crystal and going into solution are equal to the numbers of ions returning from solution to the crystal. At this point, the solution is said to be **saturated**. A maximum of 35.8 g of NaCl will dissolve in 100 mL of water at 20 °C. Any amount above this limit simply sinks to the bottom of the container and sits there.

The equilibrium reached by a saturated solution is like the equilibrium reached by a reversible reaction (Section 7.7). (⬭ , p. 198) Both are dynamic situations in which no *apparent* change occurs because the rates of forward and backward processes are equal. Solute particles leave the solid surface and reenter the solid from solution at the same rate.

$$\text{Solid solute} \underset{\text{Crystallize}}{\overset{\text{Dissolve}}{\rightleftharpoons}} \text{Solution}$$

The maximum amount of a substance that will dissolve in a given amount of a solvent at a given temperature, usually expressed in grams per 100 mL (g/100 mL), is called the substance's **solubility**. Solubility is a characteristic property of a specific

Miscible   Mutually soluble in all proportions.

Saturated solution   A solution that contains the maximum amount of dissolved solute at equilibrium.

Solubility   The maximum amount of a substance that will dissolve in a given amount of solvent at a specified temperature.

solute–solvent combination, and different substances have greatly differing solubilities. Only 9.6 g of sodium hydrogen carbonate will dissolve in 100 mL of water at 20 °C, for instance, but 204 g of sucrose will dissolve under the same conditions.

9.5 The Effect of Temperature on Solubility

As anyone who has ever made tea or coffee knows, temperature often has a dramatic effect on solubility. The compounds in tea leaves or coffee beans, for instance, dissolve easily in hot water but not in cold water. The effect of temperature is different for every substance, however, and is usually unpredictable. As shown in Figure 9.3(a), the solubilities of most molecular and ionic solids increase with increasing temperature, but the solubilities of others (NaCl) are almost unchanged, and the solubilities of still others $[Ce_2(SO_4)_3]$ decrease with increasing temperature.

Solids that are more soluble at high temperature than at low temperature can sometimes form what are called **supersaturated solutions**, which contain even more solute than a saturated solution. Suppose, for instance, that a large amount of a substance is dissolved at a high temperature. As the solution cools, the solubility decreases and the excess solute should precipitate to maintain equilibrium. But if

Supersaturated solution A solution that contains more than the maximum amount of dissolved solute; a nonequilibrium situation.

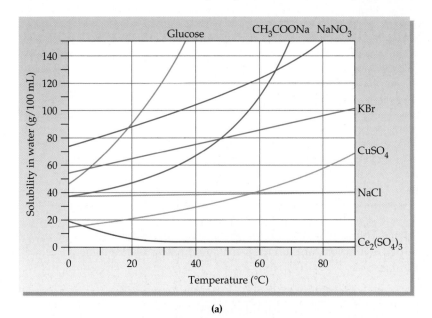

(a)

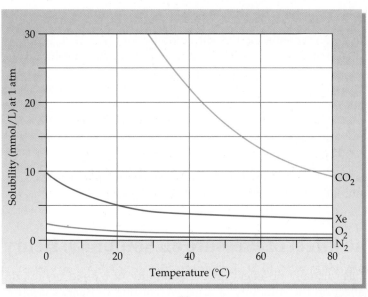

(b)

◄ **FIGURE 9.3 Solubilities of some (a) solids and (b) gases, in water as a function of temperature.** Most solid substances become more soluble as temperature rises (although the exact relationship is usually complex), while the solubility of gases decreases.

▲ **FIGURE 9.4 A supersaturated solution of sodium acetate in water.** When a tiny seed crystal is added, larger crystals rapidly grow and precipitate from the solution until equilibrium is reached.

the cooling is done very slowly, and if the container stands quietly, crystallization might not occur immediately and a supersaturated solution might result. Such a solution is unstable, however, and precipitation can occur dramatically when a tiny seed crystal is added to initiate crystallization (Figure 9.4).

Unlike solids, the influence of temperature on the solubility of gases *is* predictable: Addition of heat decreases the solubility of most gases, as seen in Figure 9.3(b) (helium is the only common exception). One result of this temperature-dependent decrease in gas solubility can sometimes be noted in a stream or lake near the outflow of warm water from an industrial operation. As water temperature increases, the concentration of dissolved oxygen in the water decreases, killing fish that cannot tolerate the lower oxygen levels.

WORKED EXAMPLE   **9.2** Solubility of Gases: Effect of Temperature

From the following graph of solubility versus temperature for O_2, estimate the concentration of dissolved oxygen in water at 25 °C and at 35 °C. By what percentage does the concentration of O_2 change?

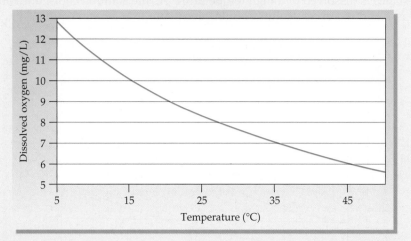

ANALYSIS The solubility of O_2 (on the *y*-axis) can be determined by finding the appropriate temperature (on the *x*-axis) and extrapolating. The percent change is calculated as

$$\frac{(\text{Solubility at 25 °C}) - (\text{Solubility at 35 °C})}{(\text{Solubility at 25 °C})} \times 100$$

SOLUTION

From the graph we estimate that the solubility of O_2 at 25 °C is approximately 8.3 mg/L and at 35 °C is 7.0 mg/L. The percent change in solubility is

$$\frac{8.3 - 7.0}{8.3} \times 100 = 16\%$$

PROBLEM 9.5

Look at the graph of solubility versus temperature in Figure 9.3, and estimate the solubility of KBr in water at 50 °C in g/100 mL.

9.6 The Effect of Pressure on Solubility: Henry's Law

Pressure has virtually no effect on the solubility of a solid or liquid, but it has a strong effect on the solubility of a gas. According to **Henry's law**, the solubility (or concentration) of a gas in a liquid is directly proportional to the partial pressure

(a) Equilibrium   **(b)** Pressure increase   **(c)** Equilibrium restored

◀ **FIGURE 9.5 Henry's law.** The solubility of a gas is directly proportional to its partial pressure. An increase in pressure causes more gas molecules to enter solution until equilibrium is restored between the dissolved and undissolved gas.

of the gas over the liquid. (Recall from Section 8.10 that each gas in a mixture exerts a partial pressure independent of other gases present (, p. 233). If the partial pressure of the gas doubles, solubility doubles; if the gas pressure is halved, solubility is halved (Figure 9.5).

> **Henry's law** The solubility (or concentration) of a gas is directly proportional to the partial pressure of the gas if the temperature is constant. That is, concentration (C) divided by pressure (P) is constant when T is constant,
>
> or $\dfrac{C}{P_{\text{gas}}} = k$ (At a constant temperature)

Henry's law can be explained using Le Châtelier's principle (Section 7.9), which states that when a system at equilibrium is placed under stress, the equilibrium shifts to relieve that stress. (, p. 203) In the case of a saturated solution of a gas in a liquid, an equilibrium exists whereby gas molecules enter and leave the solution at the same rate. When the system is stressed by increasing the pressure of the gas, more gas molecules go into solution to relieve that increase. Conversely, when the pressure of the gas is decreased, more gas molecules come out of solution to relieve the decrease.

$$\text{Gas} + \text{Solvent} \overset{[\text{Pressure} \longrightarrow]}{\rightleftharpoons} \text{Solution}$$

As an example of Henry's law in action, think about the fizzing that occurs when you open a bottle of soft drink or champagne. The bottle is sealed under greater than 1 atm of CO_2 pressure, causing some of the CO_2 to dissolve. When the bottle is opened, however, CO_2 pressure drops and gas comes fizzing out of solution.

Writing Henry's law in the form $P_{\text{gas}} = C/k$ shows that partial pressure can be used to express the concentration of a gas in a solution, a practice especially common in health-related sciences. Table 9.4 gives some typical values and illustrates the convenience of having the same unit for concentration of a gas in both air and blood.

TABLE 9.4   Partial Pressures and Normal Gas Concentrations in Body Fluids

	PARTIAL PRESSURE (mmHg)			
SAMPLE	P_{N_2}	P_{O_2}	P_{CO_2}	P_{H_2O}
Inspired air (dry)	597	159	0.3	3.7
Alveolar air (saturated)	573	100	40	47
Expired air (saturated)	569	116	28	47
Arterial blood	573	95	40	
Venous blood	573	40	45	
Peripheral tissues	573	40	45	

▲ The CO_2 gas dissolved under pressure comes out of solution when the bottle is opened and the pressure drops.

Compare the oxygen partial pressures in saturated alveolar air (air in the lungs) and in arterial blood, for instance. The values are almost the same because the gases dissolved in blood come to equilibrium with the same gases in the lungs.

If the partial pressure of a gas over a solution changes while the temperature is constant, the new solubility of the gas can be found easily. Because C/P is a constant value at constant temperature, Henry's law can be restated to show how one variable changes if the other changes:

$$\frac{C_1}{P_1} = \frac{C_2}{P_2} = k \quad \text{(Where } k \text{ is constant at a fixed temperature)}$$

Worked Example 9.3 gives an illustration of how to use this equation.

WORKED EXAMPLE **9.3** Solubility of Gases: Henry's Law

At a partial pressure of oxygen in the atmosphere of 159 mmHg, the solubility of oxygen in blood is 0.44 g/100 mL. What is the solubility of oxygen in blood at 11,000 ft, where the partial pressure of O_2 is 56 mmHg?

ANALYSIS According to Henry's law, the solubility of the gas divided by its pressure is constant:

$$\frac{C_1}{P_1} = \frac{C_2}{P_2}$$

Of the four variables in this equation, we know P_1, C_1, and P_2, and we need to find C_2.

BALLPARK ESTIMATE The pressure drops by a factor of about 3 (from 159 mmHg to 56 mmHg). Since the ratio of solubility to pressure is constant, the solubility must also drop by a factor of 3 (from 0.44 g/100 mL to about 0.15 g/100 mL).

SOLUTION

STEP 1: Identify known information. We have values for P_1, C_1, and P_2.

$P_1 = 159$ mmHg

$C_1 = 0.44$ g/100 mL

$P_2 = 56$ mmHg

STEP 2: Identify answer and units. We are looking for the solubility of O_2 (C_2) at a partial pressure P_2.

Solubility of O_2, $C_2 = $?? g/100 mL

STEP 3: Identify conversion factors or equations. In this case, we restate Henry's law to solve for C_2.

$$\frac{C_1}{P_1} = \frac{C_2}{P_2} \implies C_2 = \frac{C_1 P_2}{P_1}$$

STEP 4: Solve. Substitute the known values into the equation and calculate C_2.

$$C_2 = \frac{C_1 P_2}{P_1} = \frac{(0.44 \text{ g/100 mL})(56 \text{ mmHg})}{159 \text{ mmHg}} = 0.15 \text{ g/100 mL}$$

BALLPARK CHECK: The calculated answer matches our estimate.

PROBLEM 9.6

At 20 °C and a partial pressure of 760 mmHg, the solubility of CO_2 in water is 0.169 g/100 mL. What is the solubility of CO_2 at 2.5×10^4 mmHg?

PROBLEM 9.7

At a total atmospheric pressure of 1.00 atm, the partial pressure of CO_2 in air is approximately 4.0×10^{-4} atm. Using the data in Problem 9.6, what is the solubility of CO_2 in an open bottle of seltzer water at 20 °C?

9.7 Units of Concentration

Although we speak casually of a solution of, say, orange juice as either "dilute" or "concentrated," laboratory work usually requires an exact knowledge of a solution's concentration. As indicated in Table 9.5, there are several common methods for expressing concentration. The units differ, but all the methods describe how much solute is present in a given quantity of solution.

TABLE 9.5 Some Units for Expressing Concentration

CONCENTRATION MEASURE	SOLUTE MEASURE	SOLUTION MEASURE
Molarity, M	Moles	Volume (L)
Weight/volume percent, (w/v)%	Weight (g)	Volume (mL)
Volume/volume percent, (v/v)%	Volume*	Volume*
Parts per million, ppm	Parts*	10^6 parts*

Any units can be used as long as they are the same for both solute and solution.

Let us look at each of the four concentration measures listed in Table 9.5 individually, beginning with *molarity*.

Mole/Volume Concentration: Molarity

We saw in Chapter 6 that the various relationships between amounts of reactants and products in chemical reactions are calculated in *moles* (Sections 6.4–6.6). Thus, the most generally useful means of expressing concentration in the laboratory is **molarity (M)**, the number of moles of solute dissolved per liter of solution. For example, a solution made by dissolving 1.00 mol (58.5 g) of NaCl in enough water to give 1.00 L of solution has a concentration of 1.00 mol/L, or 1.00 M. The molarity of any solution is found by dividing the number of moles of solute by the number of liters of solution (solute + solvent):

$$\text{Molarity (M)} = \frac{\text{Moles of solute}}{\text{Liters of solution}}$$

Note that a solution of a given molarity is prepared by dissolving the solute in enough solvent to give a *final* solution volume of 1.00 L, not by dissolving it in an *initial* volume of 1.00 L. If an initial volume of 1.00 L were used, the final solution volume might be a bit larger than 1.00 L because of the additional volume of the solute. In practice, the appropriate amount of solute is weighed and placed in a *volumetric flask*, as shown in Figure 9.6. Enough solvent is then added to dissolve the solute, and further solvent is added until an accurately calibrated final volume is reached. The solution is then shaken until it is uniformly mixed.

(a)

(b)

(c)

◀ **FIGURE 9.6 Preparing a solution of known molarity.** (a) A measured number of moles of solute is placed in a volumetric flask. (b) Enough solvent is added to dissolve the solute by swirling. (c) Further solvent is carefully added until the calibration mark on the neck of the flask is reached, and the solution is shaken until uniform.

APPLICATION ▶ Breathing and Oxygen Transport

Like all other animals, humans need oxygen. When we breathe, the freshly inspired air travels through the bronchial passages and into the lungs. The oxygen then diffuses through the delicate walls of the approximately 150 million alveolar sacs of the lungs and into arterial blood, which transports it to all body tissues.

Only about 3% of the oxygen in blood is dissolved; the rest is chemically bound to *hemoglobin* molecules, large proteins with *heme* groups embedded in them. Each hemoglobin molecule contains four heme groups, and each heme group contains an iron atom that is able to bind one O_2 molecule. Thus, a single hemoglobin molecule can bind up to four molecules of oxygen. The entire system of oxygen transport and delivery in the body depends on the pickup and release of O_2 by hemoglobin (Hb) according to the following series of equilibria:

$$O_2(\text{lungs}) \rightleftharpoons O_2(\text{blood}) \quad (\text{Henry's law})$$

$$Hb + 4\,O_2(\text{blood}) \rightleftharpoons Hb(O_2)_4$$

$$Hb(O_2)_4 \rightleftharpoons Hb + 4\,O_2 \quad (\text{cell})$$

The delivery of oxygen depends on the concentration of O_2 in the various tissues, as measured by partial pressure (P_{O_2}, Table 9.4). The amount of oxygen carried by hemoglobin at any given value of P_{O_2} is usually expressed as a percent saturation and can be found from the curve shown in the accompanying figure. When $P_{O_2} = 100$ mmHg, the saturation in the lungs is 97.5%, meaning that each hemoglobin is carrying close to its maximum of four O_2 molecules. When $P_{O_2} = 26$ mmHg, however, the saturation drops to 50%.

So, how does the body ensure that enough oxygen is available to the various tissues? When large amounts of oxygen are needed—during a strenuous workout, for example—oxygen is released from hemoglobin to the hardworking, oxygen-starved muscle cells, where P_{O_2} is low. Increasing the

▲ At high altitudes, the partial pressure of oxygen in the air is too low to saturate hemoglobin sufficiently. Additional oxygen is therefore needed.

supply of oxygen to the blood (by breathing harder and faster) shifts all the equilibria toward the right, according to Le Châtelier's principle (Section 7.9), to supply the additional O_2 needed by the muscles.

What about people living at high altitudes? In Leadville, CO, for example, where the altitude is 10,156 ft, the P_{O_2} in the lungs is only about 68 mmHg. Hemoglobin is only 90% saturated with O_2 at this pressure, meaning that less oxygen is available for delivery to the tissues. The body responds by producing erythropoietin (EPO), a hormone (Chapter 20) that stimulates the bone marrow to produce more red blood cells and hemoglobin molecules. The increase in Hb provides more capacity for O_2 transport and drives the Hb + O_2 equilibria to the right.

World-class athletes use the mechanisms of increased oxygen transport associated with higher levels of hemoglobin to enhance their performance. High-altitude training centers have sprung up, with living and training regimens designed to increase blood EPO levels. Unfortunately, some athletes have also tried to "cheat" by using injections of EPO and synthetic analogs, and "blood doping" to boost performance. This has led the governing bodies of many sports federations, including the Olympic Committee, to start testing for such abuse.

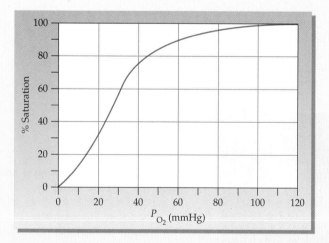

▲ An oxygen-carrying curve for hemoglobin. The percent saturation of the oxygen binding sites on hemoglobin depends on the partial pressure of oxygen (P_{O_2}).

See Additional Problem 9.92 at the end of the chapter.

Molarity can be used as a conversion factor to relate the volume of a solution to the number of moles of solute it contains. If we know the molarity and volume of a solution, we can calculate the number of moles of solute. If we know the number of moles of solute and the molarity of the solution, we can find the solution's volume.

$$\text{Molarity} = \frac{\text{Moles of solute}}{\text{Volume of solution (L)}}$$

$$\text{Moles of solute} = \text{Molarity} \times \text{Volume of solution}$$

$$\text{Volume of solution} = \frac{\text{Moles of solute}}{\text{Molarity}}$$

The flow diagram in Figure 9.7 shows how molarity is used in calculating the quantities of reactants or products in a chemical reaction, and Worked Examples 9.5 and 9.6 show how the calculations are done. Note that Problem 9.10 employs *millimolar* (mM) concentrations, which are useful in healthcare fields for expressing low concentrations such as are often found in body fluids (1 mM = 0.001 M).

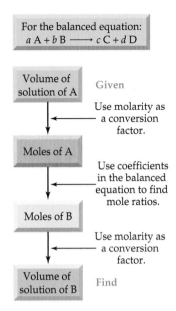

For the balanced equation:
$$a\,A + b\,B \longrightarrow c\,C + d\,D$$

Volume of solution of A — Given

Use molarity as a conversion factor.

Moles of A

Use coefficients in the balanced equation to find mole ratios.

Moles of B

Use molarity as a conversion factor.

Volume of solution of B — Find

▶ **FIGURE 9.7 Molarity and conversions.** A flow diagram summarizing the use of molarity for conversions between solution volume and moles to find quantities of reactants and products for chemical reactions in solution.

WORKED EXAMPLE **9.4** Solution Concentration: Molarity

What is the molarity of a solution made by dissolving 2.355 g of sulfuric acid (H_2SO_4) in water and diluting to a final volume of 50.0 mL? The molar mass of H_2SO_4 is 98.1 g/mol.

ANALYSIS Molarity is defined as moles of solute per liter of solution: M = mol/L. Thus, we must first find the number of moles of sulfuric acid by doing a mass to mole conversion, and then divide the number of moles by the volume of the solution.

BALLPARK ESTIMATE The molar mass of sulfuric acid is about 100 g/mol, so 2.355 g is roughly 0.025 mol. The volume of the solution is 50.0 mL, or 0.05 L, so we have about 0.025 mol of acid in 0.05 L of solution, which is a concentration of about 0.5 M.

SOLUTION

STEP 1: **Identify known information.** We know the mass of sulfuric acid and the final volume of solution.

Mass of H_2SO_4 = 2.355 g

Volume of solution = 50.0 mL

STEP 2: **Identify answer including units.** We need to find the molarity (M) in units of moles per liter.

STEP 3: **Identify conversion factors and equations.** We know both the amount of solute and the volume of solution, but first we must make two conversions: convert mass of H_2SO_4 to moles of H_2SO_4, using molar mass as a conversion factor, and convert volume from milliliters to liters:

$$\text{Molarity} = \frac{\text{Moles } H_2SO_4}{\text{Liters of solution}}$$

$$(2.355 \text{ g } H_2SO_4)\left(\frac{1 \text{ mol } H_2SO_4}{98.1 \text{ g } H_2SO_4}\right) = 0.0240 \text{ mol } H_2SO_4$$

$$(50.0 \text{ mL})\left(\frac{1 \text{ L}}{1000 \text{ mL}}\right) = 0.0500 \text{ L}$$

STEP 4: **Solve.** Substitute the moles of solute and volume of solution into the molarity expression.

$$\text{Molarity} = \frac{0.0240 \text{ mol } H_2SO_4}{0.0500 \text{ L}} = 0.480 \text{ M}$$

BALLPARK CHECK: The calculated answer is close to our estimate, which was 0.5 M.

WORKED EXAMPLE **9.5** Molarity as Conversion Factor: Molarity to Mass

A blood concentration of 0.065 M ethyl alcohol (EtOH) is sufficient to induce a coma. At this concentration, what is the total mass of alcohol (in grams) in an adult male whose total blood volume is 5.6 L? The molar mass of ethyl alcohol is 46.0 g/mol. (Refer to the flow diagram in Figure 9.7 to identify which conversions are needed.)

ANALYSIS We are given a molarity (0.065 M) and a volume (5.6 L), which allows us to calculate the number of moles of alcohol in the blood. A mole to mass conversion then gives the mass of alcohol.

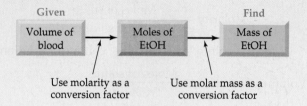

SOLUTION

$$(5.6 \text{ L blood})\left(\frac{0.065 \text{ mol EtOH}}{1 \text{ L blood}}\right) = 0.36 \text{ mol EtOH}$$

$$(0.36 \text{ mol EtOH})\left(\frac{46.0 \text{ g EtOH}}{1 \text{ mol EtOH}}\right) = 17 \text{ g EtOH}$$

WORKED EXAMPLE   **9.6** Molarity as Conversion Factor: Molarity to Volume

In our stomachs, gastric juice that is about 0.1 M in HCl aids in digestion. How many milliliters of gastric juice will react completely with an antacid tablet that contains 500 mg of magnesium hydroxide? The molar mass of $Mg(OH)_2$ is 58.3 g/mol, and the balanced equation is

$$2 \text{ HCl}(aq) + \text{Mg(OH)}_2(aq) \longrightarrow \text{MgCl}_2(aq) + 2 \text{ H}_2\text{O}(l)$$

ANALYSIS We are given the molarity of HCl and need to find the volume. We first convert the mass of $Mg(OH)_2$ to moles and then use the coefficients in the balanced equation to find the moles of HCl that will react. Once we have the moles of HCl and the molarity in moles per liter, we can find the volume. These conversions are summarized in the following flow diagram.

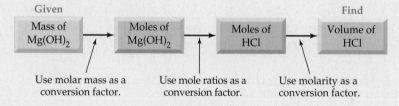

SOLUTION

$$[500 \text{ mg Mg(OH)}_2]\left(\frac{1 \text{ g}}{1000 \text{ mg}}\right)\left[\frac{1 \text{ mol Mg(OH)}_2}{58.3 \text{ g Mg(OH)}_2}\right] = 0.008\ 58 \text{ mol Mg(OH)}_2$$

$$[0.008\ 58 \text{ mol Mg(OH)}_2]\left[\frac{2 \text{ mol HCl}}{1 \text{ mol Mg(OH)}_2}\right]\left(\frac{1 \text{ L HCl}}{0.1 \text{ mol HCl}}\right) = 0.2 \text{ L } (200 \text{ mL})$$

PROBLEM 9.8

What is the molarity of a solution that contains 50.0 g of vitamin B_1 hydrochloride (molar mass = 337 g/mol) in 160 mL of solution?

PROBLEM 9.9

How many moles of solute are present in the following solutions?

(a) 175 mL of 0.35 M $NaNO_3$ **(b)** 480 mL of 1.4 M HNO_3

PROBLEM 9.10

The concentration of cholesterol ($C_{27}H_{46}O$) in blood is approximately 5.0 mM. How many grams of cholesterol are in 250 mL of blood?

PROBLEM 9.11

What mass (in grams) of calcium carbonate is needed to react completely with 65 mL of 0.12 M HCl according to the following equation?

$$2\,HCl(aq) + CaCO_3(aq) \longrightarrow CaCl_2(aq) + H_2O(l) + CO_2(g)$$

Weight/Volume Percent Concentration, (w/v)%

One of the most common methods for expressing percent concentration is to give the number of grams (weight) as a percentage of the number of milliliters (volume) of the final solution—called the **weight/volume percent concentration, (w/v)%**. Mathematically, (w/v)% concentration is found by taking the number of grams of solute per milliliter of solution and multiplying by 100%:

$$(w/v)\%\ \text{concentration} = \frac{\text{Mass of solute (g)}}{\text{Volume of solution (mL)}} \times 100\%$$

For example, if 15 g of glucose is dissolved in enough water to give 100 mL of solution, the glucose concentration is 15 g/100 mL or 15% (w/v):

$$\frac{15\ \text{g glucose}}{100\ \text{mL solution}} \times 100\% = 15\%\ (w/v)$$

To prepare 100 mL of a specific weight/volume solution, the weighed solute is dissolved in just enough solvent to give a final volume of 100 mL, not in an initial volume of 100 mL solvent. (If the solute is dissolved in 100 mL of *solvent*, the final volume of the *solution* will likely be a bit larger than 100 mL, since the volume of the solute is included.) In practice, solutions are prepared using a volumetric flask, as shown previously in Figure 9.5. Worked Example 9.7 illustrates how weight/volume percent concentration is found from a known mass and volume of solution.

WORKED EXAMPLE **9.7** Solution Concentration: Weight/Volume Percent

A solution of heparin sodium, an anticoagulant for blood, contains 1.8 g of heparin sodium dissolved to make a final volume of 15 mL of solution. What is the weight/volume percent concentration of this solution?

ANALYSIS Weight/volume percent concentration is defined as the mass of the solute in grams divided by the volume of solution in milliliters and multiplied by 100%.

BALLPARK ESTIMATE The mass of solute (1.8 g) is smaller than the volume of solvent (15 mL) by a little less than a factor of 10. The weight/volume percent should thus be a little greater than 10%.

SOLUTION

$$(w/v)\%\ \text{concentration} = \frac{1.8\ \text{g heparin sodium}}{15\ \text{mL}} \times 100\% = 12\%\ (w/v)$$

BALLPARK CHECK: The calculated (w/v)% is reasonably close to our original estimate of 10%.

> **WORKED EXAMPLE** 9.8 Weight/Volume Percent as Conversion Factor: Volume to Mass
>
> How many grams of NaCl are needed to prepare 250 mL of a 1.5% (w/v) saline solution?
>
> ANALYSIS We are given a concentration and a volume, and we need to find the mass of solute by rearranging the equation for (w/v)% concentration.
>
> BALLPARK ESTIMATE The desired (w/v)% value, 1.5%, is between 1 and 2%. For a volume of 250 mL, we would need 2.5 g of solute for a 1% (w/v) solution and 5.0 g of solute for a 2% solution. Thus, for our 1.5% solution, we need a mass midway between 2.5 and 5.0 g, or about 3.8 g.
>
> SOLUTION
>
> Since $(w/v)\% = \dfrac{\text{Mass of solute in g}}{\text{Volume of solution in mL}} \times 100\%$
>
> then $\text{Mass of solute in g} = \dfrac{(\text{Volume of solution in mL})[(w/v)]\%}{100\%}$
>
> $\qquad\qquad = \dfrac{(250)(1.5\%)}{100\%} = 3.75 \text{ g} = 3.8 \text{ g NaCl}$
>
> (2 significant figures)
>
> BALLPARK CHECK: The calculated answer matches our estimate.

> **WORKED EXAMPLE** 9.9 Weight/Volume Percent as Conversion Factor: Mass to Volume
>
> How many milliliters of a 0.75% (w/v) solution of the food preservative sodium benzoate are needed to obtain 45 mg?
>
> ANALYSIS We are given a concentration and a mass, and we need to find the volume of solution by rearranging the equation for (w/v)% concentration. Remember that 45 mg = 0.045 g.
>
> BALLPARK ESTIMATE A 0.75% (w/v) solution contains 0.75 g (750 mg) for every 100 mL of solution, so 10 mL contains 75 mg. To obtain 45 mg, we need a little more than half this volume, or a little more than 5 mL.
>
> SOLUTION
>
> Since $(w/v)\% = \dfrac{\text{Mass of solute in g}}{\text{Volume of solution in mL}} \times 100\%$
>
> then $\text{Volume of solution in mL} = \dfrac{(\text{Mass of solute in g})(100\%)}{(w/v)\%}$
>
> $\qquad\qquad = \dfrac{(0.045 \text{ g})(100\%)}{0.75\%} = 6.0 \text{ mL}$
>
> BALLPARK CHECK: The calculated answer is consistent with our estimate of a little more than 5 mL.

> **PROBLEM 9.12**
>
> In clinical lab reports, some concentrations are given in mg/dL. Convert a Ca^{2+} concentration of 8.6 mg/dL to weight/volume percent.

PROBLEM 9.13

What is the weight/volume percent concentration of a solution that contains 23 g of potassium iodide in 350 mL of aqueous solution?

PROBLEM 9.14

How many grams of solute are needed to prepare the following solutions?

(a) 125.0 mL of 16% (w/v) glucose ($C_6H_{12}O_6$)

(b) 65 mL of 1.8% (w/v) KCl

Volume/Volume Percent Concentration, (v/v)%

The concentration of a solution made by dissolving one liquid in another is often given by expressing the volume of solute as a percentage of the volume of final solution—the **volume/volume percent concentration, (v/v)%**. Mathematically, the volume of the solute (usually in milliliters) per milliliter of solution is multiplied by 100%:

$$\text{(v/v)\% concentration} = \frac{\text{Volume of solute (mL)}}{\text{Volume of solution (mL)}} \times 100\%$$

For example, if 10.0 mL of ethyl alcohol is dissolved in enough water to give 100.0 mL of solution, the ethyl alcohol concentration is (10.0 mL/100.0 mL) × 100% = 10.0% (v/v).

WORKED EXAMPLE **9.10** Volume Percent: Volume of Solution to Volume of Solute

How many milliliters of methyl alcohol are needed to prepare 75 mL of a 5.0% (v/v) solution?

ANALYSIS We are given a solution volume (75 mL) and a concentration [5.0% (v/v), meaning 5.0 mL solute/100 mL solution]. The concentration acts as a conversion factor for finding the amount of methyl alcohol needed.

BALLPARK ESTIMATE A 5% (v/v) solution contains 5 mL of solute in 100 mL of solution, so the amount of solute in 75 mL of solution must be about three-fourths of 5 mL, which means between 3 and 4 mL.

SOLUTION

$$(75 \text{ mL solution})\left(\frac{5.0 \text{ mL methyl alcohol}}{100 \text{ mL solution}}\right) = 3.8 \text{ mL methyl alcohol}$$

BALLPARK CHECK: The calculated answer is consistent with our estimate of between 3 and 4 mL.

PROBLEM 9.15

How would you use a 500.0 mL volumetric flask to prepare a 7.5% (v/v) solution of acetic acid in water?

PROBLEM 9.16

What volume of solute (in milliliters) is needed to prepare the following solutions?

(a) 100 mL of 22% (v/v) ethyl alcohol   **(b)** 150 mL of 12% (v/v) acetic acid

Parts per Million (ppm)

The concentration units weight/volume percent, (w/v)%, and volume/volume percent, (v/v)%, can also be defined as *parts per hundred*(pph) since 1% means one item per 100 items. When concentrations are very small, as often occurs in dealing with trace amounts of pollutants or contaminants, it is more convenient to use **parts per million (ppm)** or **parts per billion (ppb)**. The "parts" can be in any unit of either mass or volume as long as the units of both solute and solvent are the same:

$$\text{ppm} = \frac{\text{Mass of solute (g)}}{\text{Mass of solution (g)}} \times 10^6 \quad \text{or} \quad \frac{\text{Volume of solute (mL)}}{\text{Volume of solution (mL)}} \times 10^6$$

$$\text{ppb} = \frac{\text{Mass of solute (g)}}{\text{Mass of solution (g)}} \times 10^9 \quad \text{or} \quad \frac{\text{Volume of solute (mL)}}{\text{Volume of solution (mL)}} \times 10^9$$

To take an example, the maximum allowable concentration in air of the organic solvent benzene (C_6H_6) is currently set by government regulation at 1 ppm. A concentration of 1 ppm means that if you take a million "parts" of air in any unit—say, mL—then 1 of those parts is benzene vapor and the other 999,999 parts are other gases:

$$1 \text{ ppm} = \frac{1 \text{ mL}}{1,000,000 \text{ mL}} \times 10^6$$

Because the density of water is approximately 1.0 g/mL at room temperature, 1.0 L (or 1000 mL) of an aqueous solution weighs 1000 g. Therefore, when dealing with very dilute concentrations of solutes dissolved in water, ppm is equivalent to mg solute/L solution, and ppb is equivalent to μg solute/L solution. To demonstrate that these units are equivalent, the conversion from ppm to mg/L is as follows:

$$1 \text{ ppm} = \left(\frac{1 \text{ g solute}}{10^6 \text{ g solution}}\right)\left(\frac{1 \text{ mg solute}}{10^{-3} \text{ g solute}}\right)\left(\frac{10^3 \text{ g solution}}{1 \text{ L solution}}\right) = \frac{1 \text{ mg solute}}{1 \text{ L solution}}$$

WORKED EXAMPLE **9.11** ppm as Conversion Factor: Mass of Solution to Mass of Solute

The maximum allowable concentration of chloroform, $CHCl_3$, in drinking water is 100 ppb. What is the maximum amount (in grams) of chloroform allowed in a glass containing 400 g (400 mL) of water?

ANALYSIS We are given a solution amount (400 g) and a concentration (100 ppb). This concentration of 100 ppb means

$$100 \text{ ppb} = \frac{\text{Mass of solute (g)}}{\text{Mass of solution (g)}} \times 10^9$$

This equation can be rearranged to find the mass of solute.

BALLPARK ESTIMATE A concentration of 100 ppb means there are 100×10^{-9} g (1×10^{-7} g) of solute in 1 g of solution. In 400 g of solution, we should have 400 times this amount, or $400 \times 10^{-7} = 4 \times 10^{-5}$ g.

SOLUTION

$$\text{Mass of solute (g)} = \frac{\text{Mass of solution (g)}}{10^9} \times 100 \text{ ppb}$$

$$= \frac{400 \text{ g}}{10^9} \times 100 \text{ ppb} = 4 \times 10^{-5} \text{ g (or 0.04 mg)}$$

BALLPARK CHECK: The calculated answer matches our estimate.

PROBLEM 9.17

What is the concentration in ppm of sodium fluoride in tap water that has been fluoridated by the addition of 32 mg of NaF for every 20 kg of solution?

PROBLEM 9.18

The maximum amounts of lead and copper allowed in drinking water are 0.015 mg/kg for lead and 1.3 mg/kg for copper. Express these values in parts per million, and tell the maximum amount of each (in grams) allowed in 100 g of water.

9.8 Dilution

Many solutions, from orange juice to chemical reagents, are stored in high concentrations and then prepared for use by *dilution*—that is, by adding additional solvent to lower the concentration. For example, you might make up 1/2 gal of orange juice by adding water to a canned concentrate. In the same way, you might buy a medicine or chemical reagent in concentrated solution and dilute it before use.

The key fact to remember about dilution is that the amount of *solute* remains constant; only the *volume* is changed by adding more solvent. If, for example, the initial and final concentrations are given in molarity, then we know that the number of moles of solute is the same both before and after dilution, and can be determined by multiplying molarity times volume:

$$\text{Number of moles} = \text{Molarity (mol/L)} \times \text{Volume (L)}$$

Because the number of moles remains constant, we can set up the following equation, where M_1 and V_1 refer to the solution before dilution, and M_2 and V_2 refer to the solution after dilution:

$$\text{Moles of solute} = M_1V_1 = M_2V_2$$

This equation can be rewritten to solve for M_2, the concentration of the solution after dilution:

$$M_2 = M_1 \times \frac{V_1}{V_2} \quad \text{where} \quad \frac{V_1}{V_2} \quad \text{is a } dilution \ factor$$

The equation shows that the concentration after dilution (M_2) can be found by multiplying the initial concentration (M_1) by a **dilution factor**, which is simply the ratio of the initial and final solution volumes (V_1/V_2). If, for example, the solution volume *increases* by a factor of 5, from 10 mL to 50 mL, then the concentration must *decrease* to 1/5 its initial value because the dilution factor is 10 mL/50 mL, or 1/5. Worked Example 9.12 shows how to use this relationship for calculating dilutions.

The relationship between concentration and volume can also be used to find what volume of initial solution to start with to achieve a given dilution:

$$\text{Since} \quad M_1V_1 = M_2V_2$$
$$\text{then} \quad V_1 = V_2 \times \frac{M_2}{M_1}$$

In this case, V_1 is the initial volume that must be diluted to prepare a less concentrated solution with volume V_2. The initial volume is found by multiplying the final volume (V_2) by the ratio of the final and initial concentrations (M_2/M_1). For example, to decrease the concentration of a solution to 1/5 its initial value, the initial volume must be 1/5 the desired final volume. Worked Example 9.13 gives a sample calculation.

▲ Orange juice concentrate is diluted with water before drinking.

Dilution factor The ratio of the initial and final solution volumes (V_1/V_2).

Although the preceding equations and following examples deal with concentration units of molarity, it is worth noting that the dilution equation can be generalized to the other concentration units presented in this section, or

$$C_1V_1 = C_2V_2$$

WORKED EXAMPLE 9.12 Dilution of Solutions: Concentration

What is the final concentration if 75 mL of a 3.5 M glucose solution is diluted to a volume of 450 mL?

ANALYSIS The number of moles of solute is constant, so

$$M_1V_1 = M_2V_2$$

Of the four variables in this equation, we know the initial concentration M_1 (3.5 M), the initial volume V_1 (75 mL), and the final volume V_2 (450 mL), and we need to find the final concentration M_2.

BALLPARK ESTIMATE The volume increases by a factor of 6, from 75 mL to 450 mL, so the concentration must decrease by a factor of 6, from 3.5 M to about 0.6 M.

SOLUTION

Solving the above equation for M_2 and substituting in the known values gives

$$M_2 = \frac{M_1V_1}{V_2} = \frac{(3.5 \text{ M glucose})(75 \text{ mL})}{450 \text{ mL}} = 0.58 \text{ M glucose}$$

BALLPARK CHECK: The calculated answer is close to our estimate of 0.6 M.

WORKED EXAMPLE 9.13 Dilution of Solutions: Volume

Aqueous NaOH can be purchased at a concentration of 1.0 M. How would you use this concentrated solution to prepare 750 mL of 0.32 M NaOH?

ANALYSIS The number of moles of solute is constant, so

$$M_1V_1 = M_2V_2$$

Of the four variables in this equation, we know the initial concentration M_1 (1.0 M), the final volume V_2 (750 mL), and the final concentration M_2 (0.32 M), and we need to find the initial volume V_1.

BALLPARK ESTIMATE We want the solution concentration to decrease by a factor of about 3, from 1.0 M to 0.32 M, which means we need to dilute the 1.0 M solution by a factor of 3. This means the final volume must be about 3 times greater than the initial volume. Because our final volume is to be 750 mL, we must start with an initial volume of about 250 mL.

SOLUTION

Solving the above equation for V_1 and substituting in the known values gives

$$V_1 = \frac{V_2M_2}{M_1} = \frac{(750 \text{ mL})(0.32 \text{ M})}{1.0 \text{ M}} = 240 \text{ mL}$$

To prepare the desired solution, dilute 240 mL of 1.0 M NaOH with water to make a final volume of 750 mL.

BALLPARK CHECK: The calculated answer (240 mL) is reasonably close to our estimate of 250 mL.

PROBLEM 9.19

Hydrochloric acid is normally purchased at a concentration of 12.0 M. What is the final concentration if 100.0 mL of 12.0 M HCl is diluted to 500.0 mL?

PROBLEM 9.20

Aqueous ammonia is commercially available at a concentration of 16.0 M. How much of the concentrated solution would you use to prepare 500.0 mL of a 1.25 M solution?

PROBLEM 9.21

The Environmental Protection Agency has set the limit for arsenic in drinking water at 0.010 ppm. To what volume would you need to dilute 1.5 L of water containing 5.0 ppm arsenic to reach the acceptable limit?

9.9 Ions in Solution: Electrolytes

Look at Figure 9.8, which shows a light bulb connected to a power source through a circuit that is interrupted by two metal strips dipped into a beaker of liquid. When the strips are dipped into pure water, the bulb remains dark, but when they are dipped into an aqueous NaCl solution, the circuit is closed and the bulb lights. As mentioned previously in Section 4.1, this simple demonstration shows that ionic compounds in aqueous solution can conduct electricity. (⫘ , p. 79)

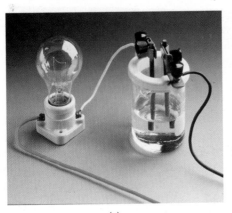
(a)

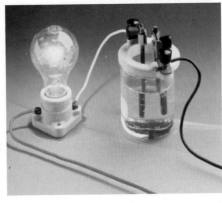
(b)

▲ **FIGURE 9.8 A simple demonstration shows that electricity can flow through a solution of ions.** (a) With pure water in the beaker, the circuit is incomplete, no electricity flows, and the bulb does not light. (b) With a concentrated NaCl solution in the beaker, the circuit is complete, electricity flows, and the light bulb glows.

Substances like NaCl that conduct an electric current when dissolved in water are called **electrolytes**. Conduction occurs because negatively charged Cl^- anions migrate through the solution toward the metal strip connected to the positive terminal of the power source, whereas positively charged Na^+ cations migrate toward the strip connected to the negative terminal. As you might expect, the ability of a solution to conduct electricity depends on the concentration of ions in solution. Distilled water contains virtually no ions and is nonconducting; ordinary tap water contains low concentrations of dissolved ions (mostly Na^+, K^+, Mg^{2+}, Ca^{2+}, and Cl^-) and is weakly conducting; and a concentrated solution of NaCl is strongly conducting.

Electrolyte A substance that produces ions and therefore conducts electricity when dissolved in water.

Strong electrolyte A substance that ionizes completely when dissolved in water.

Weak electrolyte A substance that is only partly ionized in water.

Nonelectrolyte A substance that does not produce ions when dissolved in water.

Ionic substances like NaCl that ionize completely when dissolved in water are called **strong electrolytes**, and molecular substances like acetic acid (CH_3CO_2H) that are only partially ionized are **weak electrolytes**. Molecular substances like glucose that do not produce ions when dissolved in water are **nonelectrolytes**.

Strong electrolyte; completely ionized

$$NaCl(s) \xrightarrow[\text{in water}]{\text{Dissolve}} Na^+(aq) + Cl^-(aq)$$

Weak electrolyte; partly ionized

$$CH_3CO_2H(l) \underset{\text{in water}}{\overset{\text{Dissolve}}{\rightleftharpoons}} CH_3CO_2^-(aq) + H^+(aq)$$

Nonelectrolyte; not ionized

$$Glucose(s) \underset{\text{in water}}{\overset{\text{Dissolve}}{\rightleftharpoons}} Glucose(aq)$$

9.10 Electrolytes in Body Fluids: Equivalents and Milliequivalents

What happens if NaCl and KBr are dissolved in the same solution? Because the cations (K^+ and Na^+) and anions (Cl^- and Br^-) are all mixed together and no reactions occur between them, an identical solution could just as well be made from KCl and NaBr. Thus, we can no longer speak of having an NaCl + KBr solution; we can only speak of having a solution with four different ions in it.

A similar situation exists for blood and other body fluids, which contain many different anions and cations. Since they are all mixed together, it is difficult to "assign" specific cations to specific anions or to talk about specific ionic compounds. Instead, we are interested only in individual ions and in the total numbers of positive and negative charges. To discuss such mixtures, we use a new term—*equivalents* of ions.

Equivalent For ions, the amount equal to 1 mol of charge.

Gram-equivalent For ions, the molar mass of the ion divided by the ionic charge.

For ions, one **equivalent (Eq)** is equal to the number of ions that carry 1 mol of charge. Of more practical use is the unit **gram-equivalent**, which is the mass of the ion that contains one mole of charge. It can be calculated simply as the molar mass of the ion divided by the absolute value of its charge.

$$\text{One gram-equivalent of ion} = \frac{\text{Molar mass of ion (g)}}{\text{Charge on ion}}$$

If the ion has a charge of +1 or −1, 1 gram-equivalent of the ion is simply the molar mass of the ion in grams. Thus, 1 gram-equivalent of Na^+ is 23 g, and 1 gram-equivalent of Cl^- is 35.5 g. If the ion has a charge of +2 or −2, however, 1 gram-equivalent is equal to the ion's formula weight in grams divided by 2. Thus, 1 gram-equivalent of Mg^{2+} is (24.3 g)/2 = 12.2 g, and 1 gram-equivalent of CO_3^{2-} is [12.0 g + (3 × 16.0 g)]/2 = 30.0 g. The gram-equivalent is a useful conversion factor when converting from volume of solution to mass of ions, as seen in Worked Example 9.14.

The number of equivalents of a given ion per liter of solution can be found by multiplying the molarity of the ion (moles per liter) by the charge on the ion. Because ion concentrations in body fluids are often low, clinical chemists find it more convenient to talk about *milliequivalents* of ions rather than equivalents. One **milliequivalent (mEq)** of an ion is 1/1000 of an equivalent. For example, the normal concentration of Na^+ in blood is 0.14 Eq/L, or 140 mEq/L.

$$1 \text{ mEq} = 0.001 \text{ Eq} \qquad 1 \text{ Eq} = 1000 \text{ mEq}$$

Note that the gram-equivalent for an ion can now be expressed as grams per equivalent or as mg per mEq.

Average concentrations of the major electrolytes in blood plasma are given in Table 9.6. As you might expect, the total milliequivalents of positively and negatively charged electrolytes must be equal to maintain electrical neutrality. Adding

TABLE 9.6   Concentrations of Major Electrolytes in Blood Plasma

CATION	CONCENTRATION (mEq/L)
Na^+	136–145
Ca^{2+}	4.5–6.0
K^+	3.6–5.0
Mg^{2+}	3

ANION	CONCENTRATION (mEq/L)
Cl^-	98–106
HCO_3^-	25–29
SO_4^{2-} and HPO_4^{2-}	2

the milliequivalents of positive and negative ions in Table 9.6, however, shows a higher concentration of positive ions than negative ions. The difference, called the *anion gap*, is made up by the presence of negatively charged proteins and the anions of organic acids.

WORKED EXAMPLE **9.14** Equivalents as Conversion Factors: Volume to Mass

The normal concentration of Ca^{2+} in blood is 5.0 mEq/L. How many milligrams of Ca^{2+} are in 1.00 L of blood?

ANALYSIS We are given a volume and a concentration in milliequivalents per liter, and we need to find an amount in milligrams. Thus, we need to calculate the gram-equivalent for Ca^{2+} and then use concentration as a conversion factor between volume and mass, as indicated in the following flow diagram:

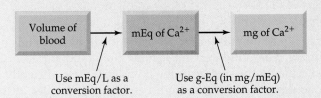

BALLPARK ESTIMATE The molar mass of calcium is 40.08 g/mol, and the calcium ion carries a charge of 2+. Thus, 1 g-Eq of Ca^{2+} equals about 20 g/Eq or 20 mg/mEq. This means that the 5.0 mEq of Ca^{2+} ions in 1.00 L of blood corresponds to a mass of 5.0 mEq Ca^{2+} × 20 mg/mEq = 100 mg Ca^{2+}.

SOLUTION

$$(1.00 \text{ L blood})\left(\frac{5.0 \text{ mEq Ca}^{2+}}{1.0 \text{ L blood}}\right)\left(\frac{20.04 \text{ mg Ca}^{2+}}{1 \text{ mEq Ca}^{2+}}\right) = 100 \text{ mg Ca}^{2+}$$

BALLPARK CHECK: The calculated answer (100 mg of Ca^{2+} in 1.00 L of blood) matches our estimate.

PROBLEM 9.22

How many grams are in 1 Eq of the following ions? How many grams in 1 mEq?

(a) K^+ (b) Br^- (c) Mg^{2+} (d) SO_4^{2-} (e) Al^{3+} (f) PO_4^{3-}

PROBLEM 9.23

Look at the data in Table 9.6, and calculate how many milligrams of Mg^{2+} are in 250 mL of blood.

9.11 Properties of Solutions

The properties of solutions are similar in many respects to those of pure solvents, but there are also some interesting and important differences. One such difference is that solutions have higher boiling points than the pure solvents; another is that solutions have lower freezing points. Pure water boils at 100.0 °C and freezes at 0.0 °C, for example, but a 1.0 M solution of NaCl in water boils at 101.0 °C and freezes at −3.7 °C.

Colligative property A property of a solution that depends only on the number of dissolved particles, not on their chemical identity.

The elevation of boiling point and the lowering of freezing point for a solution as compared with a pure solvent are examples of **colligative properties**—properties that depend on the *concentration* of a dissolved solute but not on its chemical identity. Other colligative properties are a lower vapor pressure for a solution compared with the pure solvent and *osmosis*, the migration of solvent molecules through a semipermeable membrane.

Colligative properties

- Vapor pressure is lower for a solution than for a pure solvent.
- Boiling point is higher for a solution than for a pure solvent.
- Freezing point is lower for a solution than for a pure solvent.
- Osmosis occurs when a solution is separated from a pure solvent by a semipermeable membrane.

Vapor Pressure Lowering in Solutions

We said in Section 8.12 that the vapor pressure of a liquid depends on the equilibrium between molecules entering and leaving the liquid surface. (⊂⊃, p. 241) Only those molecules at the surface of the liquid that are sufficiently energetic will evaporate. If, however, some of the liquid (solvent) molecules at the surface are replaced by other (solute) particles that do not evaporate, then the rate of evaporation of solvent molecules decreases and the vapor pressure of a solution is lower than that of the pure solvent (Figure 9.9). Note that the *identity* of the solute particles is irrelevant; only their concentration matters.

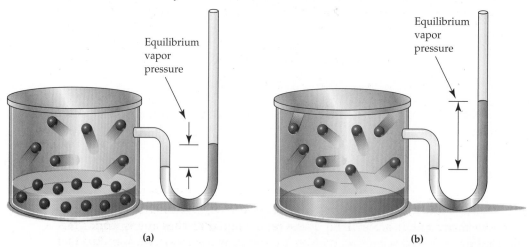

Equilibrium vapor pressure

Equilibrium vapor pressure

(a)

(b)

▲ **FIGURE 9.9 Vapor pressure lowering of solution.** (a) The vapor pressure of a solution is lower than (b) the vapor pressure of the pure solvent because fewer solvent molecules are able to escape from the surface of the solution.

APPLICATION ▶ Electrolytes, Fluid Replacement, and Sports Drinks

Athletes sweat. And the hotter the day, the more intense and longer lasting the activity, the more they sweat. Sweat loss during strenuous exercise on a hot day can amount to as much as 2 L/h, and the total sweat loss during a 24 h endurance run can exceed 16 L, or approximately 35 lb.

The composition of sweat is highly variable, not only with the individual but also with the time during the exercise and the athlete's overall conditioning. Typically, however, the Na^+ ion concentration in sweat is about 30–40 mEq/L, and that of K^+ ion is about 5–10 mEq/L. In addition, there are small amounts of other metal ions, such as Mg^{2+}, and there are sufficient Cl^- ions (35–50 mEq/L) to balance the positive charge of all these cations.

Obviously, all the water and dissolved electrolytes lost by an athlete through sweating must be replaced. Otherwise, dehydration, hyperthermia and heat stroke, dizziness, nausea, muscle cramps, impaired kidney function, and other difficulties ensue. As a rule of thumb, a sweat loss equal to 5% of body weight—about 3.5 L for a 150 lb person—is the maximum amount that can be safely allowed for a well-conditioned athlete.

Plain water works perfectly well to replace sweat lost during short bouts of activity up to a few hours in length, but

▲ Endurance athletes avoid serious medical problems by drinking sports drinks to replace water and electrolytes.

a carbohydrate–electrolyte beverage, or "sports drink," is much superior for rehydrating during and after longer activity in which substantial amounts of electrolytes have been lost. Some of the better known sports drinks are little more than overpriced sugar–water solutions, but others are carefully formulated and highly effective for fluid replacement. Nutritional research has shown that a serious sports drink should meet the following criteria. There are several dry-powder mixes on the market to choose from.

- The drink should contain 6–8% of soluble complex carbohydrates (about 15 g per 8 oz serving) and only a small amount of simple sugar for taste. The complex carbohydrates, which usually go by the name "maltodextrin," provide a slow release of glucose into the bloodstream. Not only does the glucose provide a steady source of energy, it also enhances the absorption of water from the stomach.

- The drink should contain electrolytes to replenish those lost in sweat. Concentrations of approximately 20 mEq/L for Na^+ ions, 10 mEq/L for K^+ ion, and 4 mEq/L for Mg^{2+} ions are recommended. These amounts correspond to about 100 mg sodium, 100 mg potassium, and 25 mg magnesium per 8 oz serving.

- The drink should be noncarbonated because carbonation can cause gastrointestinal upset during exercise, and it should not contain caffeine, which acts as a diuretic.

- The drink should taste good so the athlete will want to drink it. Thirst is a poor indicator of fluid requirements, and most people will drink less than needed unless a beverage is flavored.

In addition to complex carbohydrates, electrolytes, and flavorings, some sports drinks also contain vitamin A (as beta-carotene), vitamin C (ascorbic acid), and selenium, which act as antioxidants to protect cells from damage. Some drinks also contain the amino acid glutamine, which appears to lessen lactic acid buildup in muscles and thus helps muscles bounce back more quickly after an intense workout.

See Additional Problems 9.93 and 9.94 at the end of the chapter.

Boiling Point Elevation of Solutions

One consequence of the vapor pressure lowering for a solution is that the boiling point of the solution is higher than that of the pure solvent. Recall from Section 8.12 that boiling occurs when the vapor pressure of a liquid reaches atmospheric pressure. (⊂⊃, p. 241) But because the vapor pressure of a solution is lower than that of the pure solvent at a given temperature, the solution must be heated to a higher temperature for its vapor pressure to reach atmospheric pressure. Figure 9.10 shows a close-up plot of vapor pressure versus temperature for pure water and for

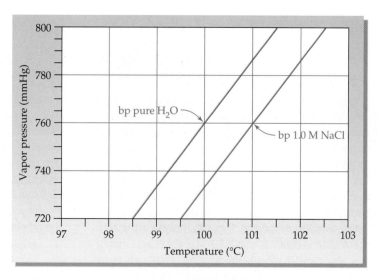

▲ **FIGURE 9.10   Vapor pressure and temperature.**   A close-up plot of vapor pressure versus temperature for pure water (red curve) and for a 1.0 M NaCl solution (green curve). Pure water boils at 100.0 °C, but the solution does not boil until 101.0 °C.

a 1.0 M NaCl solution. The vapor pressure of pure water reaches atmospheric pressure (760 mmHg) at 100.0 °C, but the vapor pressure of the NaCl solution does not reach the same point until 101.0 °C.

For each mole of solute particles added, regardless of chemical identity, the boiling point of 1 kg of water is raised by 0.51 °C, or

$$\Delta T_{boiling} = \left(0.51 \text{ °C} \frac{\text{kg water}}{\text{mol particles}} \right) \left(\frac{\text{mol particles}}{\text{kg water}} \right)$$

The addition of 1 mol of a molecular substance like glucose to 1 kg of water therefore raises the boiling point from 100.0 °C to 100.51 °C. The addition of 1 mol of NaCl per kilogram of water, however, raises the boiling point by 2 × 0.51 °C = 1.02 °C because the solution contains 2 mol of solute particles—Na^+ and Cl^- ions.

WORKED EXAMPLE   **9.15** Properties of Solutions: Boiling Point Elevation

What is the boiling point of a solution of 0.75 mol of KBr in 1.0 kg of water?

ANALYSIS The boiling point increases 0.51 °C for each mole of solute per kilogram of water. Since KBr is a strong electrolyte, there are 2 mol of ions (K^+ and Br^-) for every 1 mol of KBr that dissolves.

BALLPARK ESTIMATE The boiling point will increase about 0.5 °C for every 1 mol of ions in 1 kg of water. Since 0.75 mol of KBr produce 1.5 mol of ions, the boiling point should increase by (1.5 mol ions) × 0.5 °C/mol ions) = 0.75 °C

SOLUTION

$$\Delta T_{boiling} = \left(0.51 \text{ °C} \frac{\text{kg water}}{\text{mol ions}} \right) \left(\frac{2 \text{ mol ions}}{1 \text{ mol KBr}} \right) \left(\frac{0.75 \text{ mol KBr}}{1.0 \text{ kg water}} \right) = 0.77 \text{ °C}$$

The normal boiling point of pure water is 100 °C, so the boiling point of the solution increases to 100.77 °C.

BALLPARK CHECK: The 0.77 °C increase is consistent with our estimate of 0.75 °C.

PROBLEM 9.24

What is the boiling point of a solution of 0.67 mol of $MgCl_2$ in 0.50 kg of water?

PROBLEM 9.25

When 1.0 mol of HF is dissolved in 1.0 kg of water, the boiling point of the resulting solution is 100.5 °C. Is HF a strong or weak electrolyte? Explain.

⟜ KEY CONCEPT PROBLEM 9.26

The following diagram shows plots of vapor pressure versus temperature for a solvent and a solution.

(a) Which curve represents the pure solvent, and which the solution?

(b) What are the approximate boiling points of the pure solvent and the solution?

(c) What is the approximate concentration of the solution in mol/kg, if 1 mol of solute particles raises the boiling point of 1 kg of solvent by 3.63 °C?

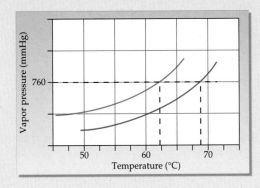

Freezing Point Depression of Solutions

Just as solutions have lower vapor pressure and consequently higher boiling points than pure solvents, they also have lower freezing points. Motorists in cold climates take advantage of this effect when they add "antifreeze" to the water in automobile cooling systems. Antifreeze is a nonvolatile solute, usually ethylene glycol ($HOCH_2CH_2OH$), that is added in sufficient concentration to lower the freezing point below the lowest expected outdoor temperature. In the same way, salt sprinkled on icy roads lowers the freezing point of ice below the road temperature and thus causes ice to melt.

Freezing point depression has much the same cause as vapor pressure lowering. Solute molecules are dispersed between solvent molecules throughout the solution, thereby making it more difficult for solvent molecules to come together and organize into ordered crystals.

For each mole of nonvolatile solute particles, the freezing point of 1 kg of water is lowered by 1.86 °C, or

$$\Delta T_{freezing} = \left(-1.86 \text{ °C} \frac{\text{kg water}}{\text{mol particles}}\right)\left(\frac{\text{mol particles}}{\text{kg water}}\right)$$

Thus, addition of 1 mol of antifreeze to 1 kg of water lowers the freezing point from 0.00 °C to −1.86 °C, and addition of 1 mol of NaCl (2 mol of particles) to 1 kg of water lowers the freezing point from 0.00 °C to −3.72 °C.

▲ A mixture of salt and ice is used to provide the low temperatures needed to make old-fashioned hand-cranked ice cream.

WORKED EXAMPLE **9.16** Properties of Solutions: Freezing Point Depression

The cells of a tomato contain mostly an aqueous solution of sugar and other substances. If a typical tomato freezes at –2.5 °C, what is the concentration of dissolved particles in the tomato cells (in moles of particles per kg of water)?

ANALYSIS The freezing point decreases by 1.86 °C for each mole of solute dissolved in 1 kg of water. We can use the decrease in freezing point (2.5 °C) to find the amount of solute per kg of water.

BALLPARK ESTIMATE The freezing point will decrease by about 1.9 °C for every 1 mol of solute particles in 1 kg of water. To lower the freezing point by 2.5 °C therefore requires about 1.5 mol of particles per kg of water.

SOLUTION

$$\Delta T_{\text{freezing}} = -2.5 \text{ °C}$$

$$= \left(-1.86 \text{ °C } \frac{\text{kg water}}{\text{mol solute particles}} \right) \left(\frac{?? \text{ mol solute particles}}{1.0 \text{ kg water}} \right)$$

We can rearrange this expression to

$$(-2.5 \text{ °C}) \left(\frac{1}{-1.86 \text{ °C}} \frac{\text{mol solute particles}}{\text{kg water}} \right) = 1.3 \frac{\text{mol solute particles}}{\text{kg water}}$$

BALLPARK CHECK: The calculated answer is relatively close to our estimate (1.5 mol/kg).

PROBLEM 9.27

What is the freezing point of a solution of 1.0 mol of glucose in 1.0 kg of water?

PROBLEM 9.28

When 0.5 mol of a certain ionic substance is dissolved in 1.0 kg of water, the freezing point of the resulting solution is −2.8 °C. How many ions does the substance give when it dissolves?

9.12 Osmosis and Osmotic Pressure

Osmosis The passage of solvent through a semipermeable membrane separating two solutions of different concentration.

Certain materials, including those that make up the membranes around living cells, are *semipermeable*. They allow water and other small molecules to pass through, but they block the passage of large solute molecules or ions. When a solution and a pure solvent, or two solutions of different concentration, are separated by a semipermeable membrane, solvent molecules pass through the membrane in a process called **osmosis**. Although the passage of solvent through the membrane takes place in both directions, passage from the pure solvent side to the solution side is favored and occurs more often. As a result, the amount of liquid on the pure solvent side decreases, the amount of liquid on the solution side increases, and the concentration of the solution decreases.

For the simplest explanation of osmosis, let us look at what happens on the molecular level. As shown in Figure 9.11, a solution inside a bulb is separated by a semipermeable membrane from pure solvent in the outer container. Solvent molecules in the outer container, because of their somewhat higher concentration, approach the membrane more frequently than do molecules in the bulb, thereby passing through more often and causing the liquid level in the attached tube to rise.

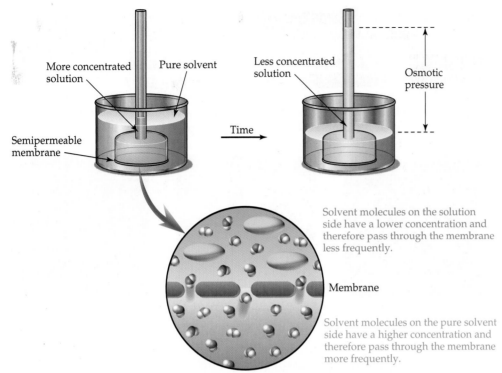

Solvent molecules on the solution side have a lower concentration and therefore pass through the membrane less frequently.

Membrane

Solvent molecules on the pure solvent side have a higher concentration and therefore pass through the membrane more frequently.

◀ **FIGURE 9.11 The phenomenon of osmosis.** A solution inside the bulb is separated from pure solvent in the outer container by a semipermeable membrane. Solvent molecules in the container have a slightly higher concentration than molecules in the bulb and therefore pass through the membrane more frequently. The liquid in the tube therefore rises until an equilibrium is reached. At equilibrium, the osmotic pressure exerted by the column of liquid in the tube is sufficient to prevent further net passage of solvent.

As the liquid in the tube rises, its increased weight creates an increased pressure that pushes solvent back through the membrane until the rates of forward and reverse passage become equal and the liquid level stops rising. The amount of pressure necessary to achieve this equilibrium is called the **osmotic pressure** (π) of the solution and can be determined from the expression

$$\pi = \left(\frac{n}{V}\right)RT$$

where n is the number of moles of particles in the solution, V is the solution volume, R is the gas constant (⊂⊃, p. 231), and T is the absolute temperature of the solution. Note the similarity between this equation for the osmotic pressure of a solution and the equation for the pressure of an ideal gas, $P = (n/V)RT$. In both cases, the pressure has units of atmospheres.

Osmotic pressures can be extremely high, even for relatively dilute solutions. The osmotic pressure of a 0.15 M NaCl solution at 25 °C, for example, is 7.3 atm, a value that supports a difference in water level of approximately 250 ft!

As with other colligative properties, the amount of osmotic pressure depends only on the concentration of solute particles, not on their identity. Thus, it is convenient to use a new unit, *osmolarity* (osmol), to describe the concentration of particles in solution. The **osmolarity** of a solution is equal to the number of moles of dissolved particles (ions or molecules) per liter of solution. A 0.2 M glucose solution, for instance, has an osmolarity of 0.2 osmol, but a 0.2 M solution of NaCl has an osmolarity of 0.4 osmol because it contains 0.2 mol of Na^+ ions and 0.2 mol of Cl^- ions.

Osmosis is particularly important in living organisms because the membranes around cells are semipermeable. The fluids both inside and outside cells must therefore have the same osmolarity to prevent buildup of osmotic pressure and consequent rupture of the cell membrane.

In blood, the plasma surrounding red blood cells has an osmolarity of approximately 0.30 osmol and is said to be **isotonic** with (that is, has the same osmolarity as) the cell contents. If the cells are removed from plasma and placed in 0.15 M NaCl (called *physiological saline solution*), they are unharmed because the osmolarity of the saline solution (0.30 osmol) is the same as that of plasma. If, however, red blood cells are placed in pure water or in any solution with an osmolarity much

Osmotic pressure The amount of external pressure applied to the more concentrated solution to halt the passage of solvent molecules across a semipermeable membrane.

Osmolarity (osmol) The sum of the molarities of all dissolved particles in a solution.

Isotonic Having the same osmolarity.

Hypotonic Having an osmolarity *less than* the surrounding blood plasma or cells.

Hypertonic Having an osmolarity *greater than* the surrounding blood plasma or cells.

lower than 0.30 osmol (a **hypotonic** solution), water passes through the membrane into the cell, causing the cell to swell up and burst, a process called *hemolysis*.

Finally, if red blood cells are placed in a solution having an osmolarity greater than the cell contents (a **hypertonic** solution), water passes out of the cells into the surrounding solution, causing the cells to shrivel, a process called *crenation*. Figure 9.12 shows red blood cells under all three conditions: isotonic, hypotonic, and hypertonic. Therefore, it is critical that any solution used intravenously be isotonic to prevent red blood cells from being destroyed.

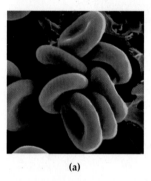 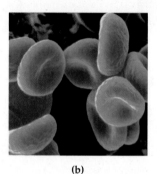 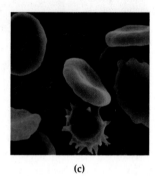

(a)   (b)   (c)

▶ **FIGURE 9.12** **Red blood cells.** In (a) an isotonic solution the blood cells are normal in appearance, but the cells in (b) a hypotonic solution are swollen because of water gain, and those in (c) a hypertonic solution are shriveled because of water loss.

WORKED EXAMPLE **9.17** Properties of Solutions: Osmolarity

The solution of glucose commonly used intravenously has a concentration of 5.0% (w/v) glucose. What is the osmolarity of this solution? The molecular weight of glucose is 180 amu.

ANALYSIS Since glucose is a molecular substance that does not give ions in solution, the osmolarity of the solution is the same as the molarity. Recall from Section 9.7 that a solution of 5.0% (w/v) glucose has a concentration of 5.0 g glucose per 100 mL of solution, which is equivalent to 50 g per liter of solution. Thus, finding the molar concentration of glucose requires a mass to mole conversion.

BALLPARK ESTIMATE One liter of solution contains 50 g of glucose (MW= 180 g/mol). Thus, 50 g of glucose is equal to a little more than 0.25 mol, so a solution concentration of 50 g/L is equal to about 0.25 osmol, or 0.25 M.

SOLUTION

STEP 1: Identify known information. We know the (w/v)% concentration of the glucose solution.

$$5.0\% \, (w/v) = \frac{5.0 \text{ g glucose}}{100 \text{ mL solution}} \times 100\%$$

STEP 2: Identify answer and units. We are looking for osmolarity, which in this case is equal to the molarity of the solution because glucose is a molecular substance and does not dissociate into ions.

$$\text{Osmolarity} = \text{Molarity} = \text{?? mol/liter}$$

STEP 3: Identify conversion factors. The (w/v)% concentration is defined as grams of solute per 100 mL of solution, and molarity is defined as moles of solute per liter of solution. We will need to convert from milliliters to liters and then use molar mass to convert grams of glucose to moles of glucose.

$$\frac{\text{g glucose}}{100 \text{ mL}} \times \frac{1000 \text{ mL}}{\text{L}} = \frac{\text{g glucose}}{\text{L}}$$

$$\frac{\text{g glucose}}{\text{L}} \times \frac{1 \text{ mol glucose}}{180 \text{ g glucose}} = \frac{\text{moles glucose}}{\text{L}}$$

STEP 4: Solve. Starting with the (w/v)% glucose concentration, we first find the number of grams of glucose in 1 L of solution and then convert to moles of glucose per liter.

$$\left(\frac{5.0 \text{ g glucose}}{100 \text{ mL solution}}\right)\left(\frac{1000 \text{ mL}}{1 \text{ L}}\right) = 50 \frac{\text{g glucose}}{\text{L solution}}$$

$$\left(\frac{50 \text{ g glucose}}{1 \text{ L}}\right)\left(\frac{1 \text{ mol}}{180 \text{ g}}\right) = 0.28 \text{ M glucose} = 0.28 \text{ osmol}$$

BALLPARK CHECK: The calculated osmolarity is reasonably close to our estimate of 0.25 osmol

WORKED EXAMPLE **9.18** Properties of Solutions: Osmolarity

What mass of NaCl is needed to make 1.50 L of a 0.300 osmol solution? The molar mass of NaCl is 58.44 g/mol.

ANALYSIS Since NaCl is an ionic substance that produces 2 mol of ions (Na^+, Cl^-) when it dissociates, the osmolarity of the solution is twice the molarity. From the volume and the osmolarity we can determine the moles of NaCl needed and then perform a mole to mass converstion.

SOLUTION

STEP 1: **Identify known information.** We know the volume and the osmolarity of the final NaCl solution.

$$V = 1.50 \text{ L}$$
$$0.300 \text{ osmol} = \left(\frac{0.300 \text{ mol ions}}{\text{L}}\right)$$

STEP 2: **Identify answer and units.** We are looking for the mass of NaCl.

$$\text{Mass of NaCl} = \text{ ?? g}$$

STEP 3: **Identify conversion factors.** Starting with osmolarity in the form (moles NaCl/L), we can use volume to determine the number of moles of solute. We can then use molar mass for the mole to mass conversion.

$$\left(\frac{\text{moles NaCl}}{\cancel{\text{L}}}\right) \times (\cancel{\text{L}}) = \text{moles NaCl}$$

$$(\cancel{\text{moles NaCl}}) \times \left(\frac{\text{g NaCl}}{\cancel{\text{mole NaCl}}}\right) = \text{g NaCl}$$

STEP 4: **Solve.** Use the appropriate conversions, remembering that NaCl produces two ions per formula unit, to find the mass of NaCl.

$$\left(\frac{0.300 \cancel{\text{ mol ions}}}{\cancel{\text{L}}}\right)\left(\frac{1 \text{ mol NaCl}}{2 \cancel{\text{ mol ions}}}\right)(1.50 \cancel{\text{ L}}) = 0.225 \text{ mol NaCl}$$

$$(0.225 \cancel{\text{ mol NaCl}})\left(\frac{58.44 \text{ g NaCl}}{\cancel{\text{mol NaCl}}}\right) = 13.1 \text{ g NaCl}$$

PROBLEM 9.29

What is the osmolarity of the following solutions?

(a) 0.35 M KBr   (b) 0.15 M glucose + 0.05 M K_2SO_4

PROBLEM 9.30

What is the osmolarity of a typical oral rehydration solution (ORS) for infants that contains 90 mEq/L Na^+, 20 mEq/L K^+, 110 mEq/L Cl^-, and 2.0% glucose? The molecular weight of glucose is 180 amu.

9.13 Dialysis

Dialysis is similar to osmosis, except that the pores in a dialysis membrane are larger than those in an osmotic membrane so that both solvent molecules and small solute particles can pass through, but large colloidal particles such as proteins cannot pass. (The exact dividing line between a "small" molecule and a "large" one is imprecise, and dialysis membranes with a variety of pore sizes are available.) Dialysis membranes include animal bladders, parchment, and cellophane.

Perhaps the most important medical use of dialysis is in artificial kidney machines, where *hemodialysis* is used to cleanse the blood of patients whose kidneys malfunction (Figure 9.13). Blood is diverted from the body and pumped through a long cellophane dialysis tube suspended in an isotonic solution formulated to contain many of the same components as blood plasma. These substances—glucose, NaCl, $NaHCO_3$, and KCl—have the same concentrations in the dialysis solution as they do in blood so that they have no net passage through the membrane.

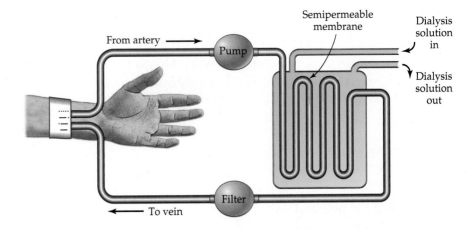

▶ **FIGURE 9.13 Operation of a hemodialysis unit used for purifying blood.** Blood is pumped from an artery through a coiled semipermeable membrane of cellophane. Small waste products pass through the membrane and are washed away by an isotonic dialysis solution.

APPLICATION ▶ Timed-Release Medications

There is much more in most medications than medicine. Even something as simple as a generic aspirin tablet contains a binder to keep it from crumbling, a filler to bring it to the right size and help it disintegrate in the stomach, and a lubricant to keep it from sticking to the manufacturing equipment. Timed-release medications are more complex still.

The widespread use of timed-release medication dates from the introduction of Contac decongestant in 1961. The original idea was simple: Tiny beads of medicine were encapsulated by coating them with varying thicknesses of a slow-dissolving polymer. Those beads with a thinner coat dissolve and release their medicine more rapidly; those with a thicker coat dissolve more slowly. Combining the right number of beads with the right thicknesses into a single capsule makes possible the gradual release of medication over a predictable time.

The technology of timed-release medications has become much more sophisticated in recent years, and the kinds of medications that can be delivered have become more numerous. Some medicines, for instance, either damage the stomach lining or are destroyed by the highly acidic environment in the stomach but can be delivered safely if given an *enteric coating*. The enteric coating is a polymeric material formulated so that it is stable in acid but reacts and is destroyed when it passes into the more basic environment of the intestines.

More recently, dermal patches have been developed to deliver drugs directly by diffusion through the skin. Patches are available to treat conditions from angina to motion sickness, as well as nicotine patches to help reduce cigarette cravings. One clever new device for timed release of medication through the skin uses the osmotic effect to force a drug from its reservoir. Useful only for drugs that do not dissolve in water, the device is divided into two compartments, one containing medication covered by a perforated membrane and the other containing a hygroscopic material (Section 9.3) covered by a semipermeable membrane. As moisture from the air diffuses through the membrane into the compartment with the hygroscopic material, the buildup of osmotic pressure squeezes the medication out of the other compartment through tiny holes.

▲ The small beads of medicine are coated with different thicknesses of a slow-dissolving polymer so that they dissolve and release medicine at different times.

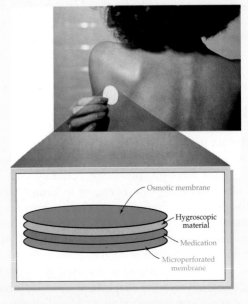

See Additional Problem 9.95 at the end of the chapter.

Small waste materials such as urea pass through the dialysis membrane from the blood to the solution side where they are washed away, but cells, proteins, and other important blood components are prevented by their size from passing through the membrane. In addition, the dialysis fluid concentration can be controlled so that imbalances in electrolytes are corrected. The wash solution is changed every 2 h, and a typical hemodialysis procedure lasts for 4–7 h.

As noted above, colloidal particles are too large to pass through a semipermeable membrane. Protein molecules, in particular, do not cross semipermeable membranes and thus play an essential role in determining the osmolarity of body fluids. The distribution of water and solutes across the capillary walls that separate blood plasma from the fluid surrounding cells is controlled by the balance between blood pressure and osmotic pressure. The pressure of blood inside the capillary tends to push water out of the plasma (filtration), but the osmotic pressure of colloidal protein molecules tends to draw water into the plasma (reabsorption). The balance between the two processes varies with location in the body. At the arterial end of a capillary, where blood pumped from the heart has a higher pressure, filtration is favored. At the venous end, where blood pressure is lower, reabsorption is favored, causing waste products from metabolism to enter the bloodstream.

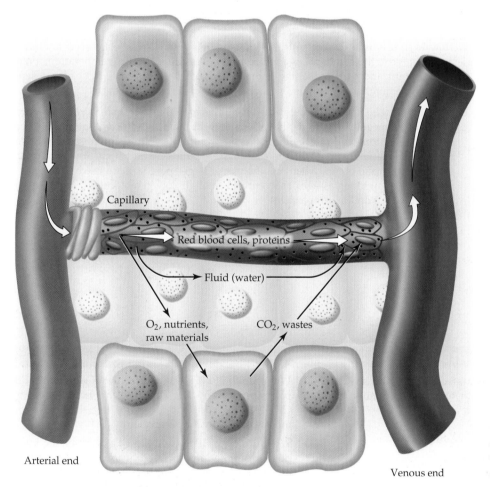

Capillary

Red blood cells, proteins

Fluid (water)

O_2, nutrients, raw materials

CO_2, wastes

Arterial end

Venous end

◀ The delivery of oxygen and nutrients to the cells and the removal of waste products are regulated by osmosis.

SUMMARY: REVISITING THE CHAPTER GOALS

1. What are solutions, and what factors affect solubility? Mixtures are classified as either *heterogeneous*, if the mixing is nonuniform, or *homogeneous*, if the mixing is uniform. *Solutions* are homogeneous mixtures that contain particles the size of ions and molecules (<2.0 nm diameter), whereas larger particles (2.0–500 nm diameter) are present in *colloids*.

KEY WORDS

Colligative property, *p. 278*

Colloid, *p. 256*

Dilution factor, *p. 273*

The maximum amount of one substance (the *solute*) that can be dissolved in another (the *solvent*) is called the substance's *solubility*. Substances tend to be mutually soluble when their intermolecular forces are similar. The solubility in water of a solid often increases with temperature, but the solubility of a gas decreases with temperature. Pressure significantly affects gas solubilities, which are directly proportional to their partial pressure over the solution (*Henry's law*).

2. **How is the concentration of a solution expressed?** The concentration of a solution can be expressed in several ways, including molarity, weight/weight percent composition, weight/volume percent composition, and parts per million. *Molarity*, which expresses concentration as the number of moles of solute per liter of solution, is the most useful method when calculating quantities of reactants or products for reactions in aqueous solution.

3. **How are dilutions carried out?** A dilution is carried out by adding more solvent to an existing solution. Only the amount of solvent changes; the amount of solute remains the same. Thus, the molarity times the volume of the dilute solution is equal to the molarity times the volume of the concentrated solution: $M_1 V_1 = M_2 V_2$.

4. **What is an electrolyte?** Substances that form ions when dissolved in water and whose water solutions therefore conduct an electric current are called *electrolytes*. Substances that ionize completely in water are *strong electrolytes*, those that ionize partially are *weak electrolytes*, and those that do not ionize are *nonelectrolytes*. Body fluids contain small amounts of many different electrolytes, whose concentrations are expressed as moles of ionic charge, or equivalents, per liter.

5. **How do solutions differ from pure solvents in their behavior?** In comparing a solution to a pure solvent, the solution has a lower vapor pressure at a given temperature, a higher boiling point, and a lower melting point. Called *colligative properties*, these effects depend only on the number of dissolved particles, not on their chemical identity.

6. **What is osmosis?** *Osmosis* occurs when solutions of different concentration are separated by a semipermeable membrane that allows solvent molecules to pass but blocks the passage of solute ions and molecules. Solvent flows from the more dilute side to the more concentrated side until sufficient *osmotic pressure* builds up and stops the flow. An effect similar to osmosis occurs when membranes of larger pore size are used. In *dialysis*, the membrane allows the passage of solvent and small dissolved molecules but prevents passage of proteins and larger particles.

UNDERSTANDING KEY CONCEPTS

9.31 Assume that two liquids are separated by a semipermeable membrane. Make a drawing that shows the situation after equilibrium is reached.

Before equilibrium

9.32 When 1 mol of HCl is added to 1 kg of water, the boiling point increases by 1.0 °C, but when 1 mol of acetic acid, CH_3CO_2H, is added to 1 kg of water, the boiling point increases by only 0.5 °C. Explain.

9.33 When 1 mol of HF is added to 1 kg of water, the freezing point decreases by 1.9 °C, but when 1 mol of HBr is added, the freezing point decreases by 3.7 °C. Which is more highly separated into ions, HF or HBr? Explain.

9.34 The graph at the top of page 289 shows the solubilities of two substances as a function of temperature. Which of the substances is a gas, and which is a liquid? Explain.

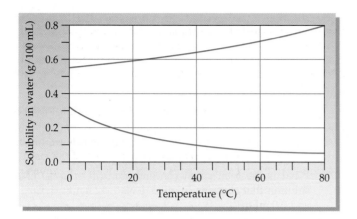

9.35 Assume that you have two full beakers, one containing pure water (blue) and the other containing an equal volume of a 10% (w/v) solution of glucose (green). Which of the drawings (a)–(c) best represents the two beakers after they have stood uncovered for several days and partial evaporation has occurred? Explain.

(a)   (b)   (c)

9.36 A beaker containing 150.0 mL of 0.1 M glucose is represented by (a). Which of the drawings (b)–(d) represents the solution that results when 50.0 mL is withdrawn from (a) and then diluted by a factor of 4?

(a)   (b)   (c)   (d)

9.37 The following diagram shows parts of the vapor pressure curves for a solvent and a solution. Which curve is which?

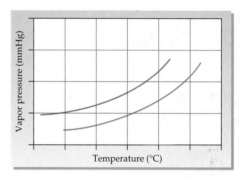

ADDITIONAL PROBLEMS

SOLUTIONS AND SOLUBILITY

9.38 What is the difference between a homogeneous mixture and a heterogeneous one?

9.39 How can you tell a solution from a colloid?

9.40 What characteristic of water allows it to dissolve ionic solids?

9.41 Why does water not dissolve motor oil?

9.42 Which of the following are solutions?

(a) Italian salad dressing
(b) Rubbing alcohol
(c) Algae in pond water
(d) Black coffee

9.43 Which of the following pairs of liquids are likely to be miscible?

(a) H_2SO_4 and H_2O
(b) C_8H_{18} and C_6H_6
(c) Na_3PO_4, and H_2O
(d) CS_2 and CCl_4

9.44 The solubility of NH_3 in water at an NH_3 pressure of 760.0 mmHg is 51.8 g/100 mL. What is the solubility of NH_3 if the partial pressure is reduced to 225.0 mmHg?

9.45 The solubility of CO_2 gas in water is 0.15 g/100 mL at a CO_2 pressure of 760 mmHg. What is the solubility of CO_2 in a soft drink (which is mainly water) that was bottled under a CO_2 pressure of 4.5 atm?

CONCENTRATION AND DILUTION OF SOLUTIONS

9.46 Is a solution highly concentrated if it is saturated? Is a solution saturated if it is highly concentrated?

9.47 How is weight/volume percent concentration defined?

9.48 How is molarity defined?

9.49 How is volume/volume percent concentration defined?

9.50 How would you prepare 750.0 mL of a 6.0% (v/v) ethyl alcohol solution?

9.51 A dilute solution of boric acid, H_3BO_3 is often used as an eyewash. How would you prepare 500.0 mL of a 0.50% (w/v) boric acid solution?

9.52 Describe how you would prepare 250 mL of a 0.10 M NaCl solution.

9.53 Describe how you would prepare 1.50 L of a 7.50% (w/v) $Mg(NO_3)_2$ solution.

9.54 What is the weight/volume percent concentration of the following solutions?

(a) 5.8 g KCl in 75 mL of solution
(b) 15 g sucrose in 380 mL of solution

9.55 The concentration of glucose in blood is approximately 90 mg/100 mL. What is the weight/volume percent concentration of glucose? What is the molarity of glucose?

9.56 How many grams of each substance are needed to prepare the following solutions?

(a) 50.0 mL of 8.0% (w/v) KCl
(b) 200.0 mL of 7.5% (w/v) acetic acid

9.57 Which of the following solutions is more concentrated?

(a) 0.50 M KCl or 5.0% (w/v) KCl
(b) 2.5% (w/v) $NaHSO_4$ or 0.025 M $NaHSO_4$

9.58 If you had only 23 g of KOH remaining in a bottle, how many milliliters of 10.0% (w/v) solution could you prepare? How many milliliters of 0.25 M solution?

9.59 Over-the-counter hydrogen peroxide solutions are 3% (w/v). What is this concentration in moles per liter?

9.60 The lethal dosage of potassium cyanide in rats is 10 mg KCN per kilogram of body weight. What is this concentration in parts per million?

9.61 The maximum concentration set by the U.S. Environmental Protection Agency for lead in drinking water is 15 ppb. What is this concentration in milligrams per liter? How many liters of water contaminated at this maximum level must you drink to consume 1.0 μg of lead?

9.62 What is the molarity of the following solutions?
(a) 12.5 g $NaHCO_3$ in 350.0 mL solution
(b) 45.0 g H_2SO_4 in 300.0 mL solution
(c) 30.0 g NaCl dissolved to make 500.0 mL solution

9.63 How many moles of solute are in the following solutions?
(a) 200 mL of 0.30 M acetic acid, CH_3CO_2H
(b) 1.50 L of 0.25 M NaOH
(c) 750 mL of 2.5 M nitric acid, HNO_3

9.64 How many milliliters of a 0.75 M HCl solution do you need to obtain 0.0040 mol of HCl?

9.65 Nalorphine, a relative of morphine, is used to combat withdrawal symptoms in heroin users. How many milliliters of a 0.40% (w/v) solution of nalorphine must be injected to obtain a dose of 1.5 mg?

9.66 A flask containing 450 mL of 0.50 M H_2SO_4 was accidentally knocked to the floor. How many grams of $NaHCO_3$ do you need to put on the spill to neutralize the acid according to the following equation?

$$H_2SO_4(aq) + 2\,NaHCO_3(aq) \longrightarrow$$
$$Na_2SO_4(aq) + 2\,H_2O(l) + 2\,CO_2(g)$$

9.67 How many milliliters of 0.0200 M $Na_2S_2O_3$ solution are needed to dissolve 0.450 g of AgBr?

$$AgBr(s) + 2\,Na_2S_2O_3(aq) \longrightarrow$$
$$Na_3Ag(S_2O_3)_2(aq) + NaBr(aq)$$

9.68 How much water must you add to 100.0 mL of orange juice concentrate if you want the final juice to be 20.0% of the strength of the original?

9.69 How much water would you add to 100.0 mL of 0.500 M NaOH if you wanted the final concentration to be 0.150 M?

9.70 An aqueous solution that contains 285 ppm of potassium nitrate (KNO_3) is being used to feed plants in a garden. What volume of this solution is needed to prepare 2.0 L of a solution that is 75 ppm in KNO_3?

9.71 What is the concentration of a NaCl solution, in (w/v)%, prepared by diluting 65 mL of a saturated solution, which has a concentration of 37 (w/v)%, to 480 mL?

9.72 Concentrated (12.0 M) hydrochloric acid is sold for household and industrial purposes under the name "muriatic acid." How many milliliters of 0.500 M HCl solution can be made from 25.0 mL of 12.0 M HCl solution?

9.73 Dilute solutions of $NaHCO_3$ are sometimes used in treating acid burns. How many milliliters of 0.100 M $NaHCO_3$ solution are needed to prepare 750.0 mL of 0.0500 M $NaHCO_3$ solution?

ELECTROLYTES

9.74 What is an electrolyte?

9.75 Give an example of a strong electrolyte and a nonelectrolyte.

9.76 What does it mean when we say that the concentration of Ca^{2+} in blood is 3.0 mEq/L?

9.77 What is the total anion concentration (in mEq/L) of a solution that contains 5.0 mEq/L Na^+, 12.0 mEq/L Ca^{2+}, and 2.0 mEq/L Li^+?

9.78 Kaochlor, a 10% (w/v) KCl solution, is an oral electrolyte supplement administered for potassium deficiency. How many milliequivalents of K^+ are in a 30 mL dose?

9.79 Calculate the gram-equivalent for each of the following ions:
(a) Ca^{2+} **(b)** K^+
(c) SO_4^{2-} **(d)** PO_4^{3-}

9.80 Look up the concentration of Cl^- ion in blood in Table 9.6. How many milliliters of blood would be needed to obtain 1.0 g of Cl^- ions?

9.81 Normal blood contains 3 mEq/L of Mg^{2+}. How many milligrams of Mg^{2+} are present in 150.0 mL of blood?

PROPERTIES OF SOLUTIONS

9.82 Which lowers the freezing point of 2.0 kg of water more, 0.20 mol NaOH or 0.20 mol $Ba(OH)_2$? Both compounds are strong electrolytes. Explain.

9.83 Which solution has the higher boiling point, 0.500 M glucose or 0.300 M KCl? Explain.

9.84 Methanol, CH_3OH, is sometimes used as an antifreeze for the water in automobile windshield washer fluids. How many grams of methanol must be added to 5.00 kg of water to lower its freezing point to −10.0 °C? For each mole of solute, the freezing point of 1 kg of water is lowered 1.86 °C.

9.85 Hard candy is prepared by dissolving pure sugar and flavoring in water and heating the solution to boiling. What is the boiling point of a solution produced by adding 650 g of cane sugar (molar mass 342.3 g/mol) to 1.5 kg of water? For each mole of nonvolatile solute, the boiling point of 1 kg of water is raised 0.51 °C.

OSMOSIS

9.86 Why do red blood cells swell up and burst when placed in pure water?

9.87 What does it mean when we say that a 0.15 M NaCl solution is isotonic with blood, whereas distilled water is hypotonic?

9.88 Which of the following solutions has the higher osmolarity?
(a) 0.25 M KBr or 0.20 M Na_2SO_4
(b) 0.30 M NaOH or 3.0% (w/v) NaOH

9.89 Which of the following solutions will give rise to a greater osmotic pressure at equilibrium: 5.00 g of NaCl in 350.0 mL water or 35.0 g of glucose in 400.0 mL water? For NaCl, MW = 58.5 amu; for glucose, MW = 180 amu.

9.90 A pickling solution is prepared by dissolving 270 g of NaCl in 3.8 L of water. Calculate the osmolarity of the solution.

9.91 An isotonic solution must be approximately 0.30 osmol. How much KCl is needed to prepare 175 mL of an isotonic solution?

Applications

9.92 How does the body increase oxygen availability at high altitude? [*Breathing and Oxygen Transport, p. 266*]

9.93 What are the major electrolytes in sweat, and what are their approximate concentrations in mEq/L? [*Electrolytes, Fluid Replacement, and Sports Drinks, p. 279*]

9.94 Why is a sports drink more effective than plain water for rehydration after extended exercise? [*Electrolytes, Fluid Replacement, and Sports Drinks, p. 279*]

9.95 How does an enteric coating on a medication work? [*Timed-Release Medications, p. 286*]

General Questions and Problems

9.96 Hyperbaric chambers, which provide high pressures (up to 6 atm) of either air or pure oxygen, are used to treat a variety of conditions, ranging from decompression sickness in deep-sea divers to carbon monoxide poisoning.

    (a) What is the partial pressure of O_2 (in millimeters of Hg) in a hyperbaric chamber pressurized to 5 atm with air that is 18% in O_2?

    (b) What is the solubility of O_2 (in grams per 100 mL) in the blood at this partial pressure? The solubility of O_2 is 2.1 g/100 mL for $P_{O_2} = 1$ atm.

9.97 Express the solubility of O_2 in Problem 9.96(b) in units of molarity.

9.98 Uric acid, the principal constituent of some kidney stones, has the formula $C_5H_4N_4O_3$. In aqueous solution, the solubility of uric acid is only 0.067 g/L. Express this concentration in (w/v)%, in parts per million, and in molarity.

9.99 Emergency treatment of cardiac arrest victims sometimes involves injection of a calcium chloride solution directly into the heart muscle. How many grams of $CaCl_2$ are administered in an injection of 5.0 mL of a 5.0% (w/v) solution? How many milliequivalents of Ca^{2+}?

9.100 Nitric acid, HNO_3, is available commercially at a concentration of 16 M. What volume would you use to prepare 750 mL of a 0.20 M solution?

9.101 One test for vitamin C (ascorbic acid, $C_6H_8O_6$) is based on the reaction of the vitamin with iodine:

$$C_6H_8O_6(aq) + I_2(aq) \longrightarrow C_6H_6O_6(aq) + 2\,HI(aq)$$

    (a) If 25.0 mL of a fruit juice requires 13.0 mL of 0.0100 M I_2 solution for reaction, what is the molarity of the ascorbic acid in the fruit juice?

    (b) The Food and Drug Administration recommends that 60 mg of ascorbic acid be consumed per day. How many milliliters of the fruit juice in part (a) must a person drink to obtain the recommended dosage?

9.102 *Ringer's solution*, used in the treatment of burns and wounds, is prepared by dissolving 8.6 g of NaCl, 0.30 g of KCl, and 0.33 g of $CaCl_2$ in water and diluting to a volume of 1.00 L. What is the molarity of each component?

9.103 What is the osmolarity of Ringer's solution (see Problem 9.102)? Is it hypotonic, isotonic, or hypertonic with blood plasma (0.30 osmol)?

9.104 The typical dosage of statin drugs for the treatment of high cholesterol is 10 mg. Assuming a total blood volume of 5.0 L, calculate the concentration of drug in the blood in units of (w/v)%.

9.105 Assuming the density of blood in healthy individuals is approximately 1.05 g/mL, report the concentration of drug in Problem 9.104 in units of ppm.

9.106 In many states, a person with a blood alcohol concentration of 0.080% (v/v) is considered legally drunk. What volume of total alcohol does this concentration represent, assuming a blood volume of 5.0 L?

9.107 Ammonia is very soluble in water (51.8 g/L at 20 °C and 760 mmHg).

    (a) Show how NH_3 can hydrogen bond to water.

    (b) What is the solubility of ammonia in water in moles per liter?

9.108 Cobalt(II) chloride, a blue solid, can absorb water from the air to form cobalt(II) chloride hexahydrate, a pink solid. The equilibrium is so sensitive to moisture in the air that $CoCl_2$ is used as a humidity indicator.

    (a) Write a balanced equation for the equilibrium. Be sure to include water as a reactant to produce the hexahydrate.

    (b) How many grams of water are released by the decomposition of 2.50 g of cobalt(II) chloride hexahydrate?

9.109 How many milliliters of 0.150 M $BaCl_2$ are needed to react completely with 35.0 mL of 0.200 M Na_2SO_4? How many grams of $BaSO_4$ will be formed?

9.110 Many compounds are only partially dissociated into ions in aqueous solution. Trichloroacetic acid (CCl_3CO_2H), for instance, is partially dissociated in water according to the equation

$$CCl_3CO_2H(aq) \rightleftharpoons H^+(aq) + CCl_3CO_2{}^-(aq)$$

For a solution prepared by dissolving 1.00 mol of trichloroacetic acid in 1.00 kg of water, 36.0% of the trichloroacetic acid dissociates to form H^+ and $CCl_3CO_2{}^-$ ions.

    (a) What is the total concentration of dissolved ions and molecules in 1 kg of water?

    (b) What is the freezing point of this solution? (The freezing point of 1 kg of water is lowered 1.86 °C for each mole of solute particles.)

CHAPTER 10

Acids and Bases

CONCEPTS TO REVIEW

Acids, Bases, and Neutralization Reactions
(Sections 4.11 and 6.10)

Reversible Reactions and Chemical Equilibrium
(Section 7.7)

Equilibrium Equations and Equilibrium Constants
(Section 7.8)

Units of Concentration; Molarity
(Section 9.7)

Ion Equivalents
(Section 9.10)

▲ Acids are found in many of the foods we eat, including tomatoes, peppers, and these citrus fruits.

CONTENTS

CHAPTER GOALS

We have already touched on the subject of acids and bases on several occasions, but the time has come for a more detailed study that will answer the following questions:

1. What are acids and bases?

THE GOAL: Be able to recognize acids and bases and write equations for common acid–base reactions.

2. What effect does the strength of acids and bases have on their reactions?

THE GOAL: Be able to interpret acid strength using acid dissociation constants K_a and predict the favored direction of acid–base equilibria.

3. What is the ion-product constant for water?

THE GOAL: Be able to write the equation for this constant and use it to find the concentration of H_3O^+ or OH^-.

4. What is the pH scale for measuring acidity?

THE GOAL: Be able to explain the pH scale and find pH from the H_3O^+ concentration.

5. What is a buffer?

THE GOAL: Be able to explain how a buffer maintains pH and how the bicarbonate buffer functions in the body.

6. How is the acid or base concentration of a solution determined?

THE GOAL: Be able to explain how a titration procedure works and use the results of a titration to calculate acid or base concentration in a solution.

A cids! The word evokes images of dangerous, corrosive liquids that eat away everything they touch. Although a few well-known substances such as sulfuric acid (H_2SO_4) do indeed fit this description, most acids are relatively harmless. In fact, many acids, such as ascorbic acid (vitamin C), are necessary for life.

10.1 Acids and Bases in Aqueous Solution

Let us take a moment to review what we said about acids and bases in Sections 4.11 and 6.10 before going on to a more systematic study:

- An acid is a substance that produces hydrogen ions, H^+, when dissolved in water.
- A base is a substance that produces hydroxide ions, OH^-, when dissolved in water.
- The neutralization reaction of an acid with a base yields water plus a *salt*, an ionic compound composed of the cation from the base and the anion from the acid.

The above definitions of acids and bases were proposed in 1887 by the Swedish chemist Svante Arrhenius and are useful for many purposes. The definitions are limited, however, because they refer only to reactions that take place in aqueous solution. (We will see shortly how the definitions can be broadened.) Another issue is that the H^+ ion is so reactive it does not exist in water. Instead, H^+ reacts with H_2O to give the **hydronium ion**, H_3O^+, as mentioned in Section 4.11. When gaseous HCl dissolves in water, for instance, H_3O^+ and Cl^- are formed. As described in Section 5.8, electrostatic potential maps show that the hydrogen of HCl is positively polarized and electron-poor (blue), whereas the oxygen of water is negatively polarized and electron-rich (red):

Hydronium ion The H_3O^+ ion, formed when an acid reacts with water.

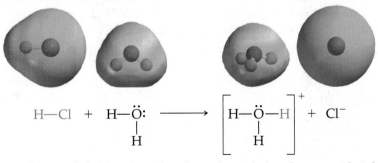

Thus, the Arrhenius definition is updated to acknowledge that an acid yields H_3O^+ in water rather than H^+. In practice, however, the notations H_3O^+ and $H^+(aq)$ are often used interchangeably.

The Arrhenius definition of a base is correct as far as it goes, but it is important to realize that the OH⁻ ions "produced" by the base can come from either of two sources. Metal hydroxides, such as NaOH, KOH, and Ba(OH)₂, are ionic compounds that already contain OH⁻ ions and merely release those ions when they dissolve in water. Ammonia, however, is not ionic and contains no OH⁻ ions in its structure. Nonetheless, ammonia is a base because it undergoes a reaction with water when it dissolves, producing NH_4^+ and OH⁻ ions:

This OH⁻ ion comes from NaOH.

$$NaOH(s) \xrightarrow[\text{in } H_2O]{\text{Dissolve}} Na^+(aq) + OH^-(aq)$$

This OH⁻ ion comes from H₂O.

$$H-\overset{..}{\underset{H}{N}}-H(g) + H_2O(l) \rightleftharpoons H-\overset{H}{\underset{H}{N^{\pm}}}-H(aq) + OH^-(aq)$$

The reaction of ammonia with water is a reversible process (Section 7.7) whose equilibrium strongly favors unreacted ammonia. (⬭, p. 198) Nevertheless, *some* OH⁻ ions are produced, so NH₃ is a base.

10.2 Some Common Acids and Bases

Acids and bases are present in a variety of foods and consumer products. Acids generally have a sour taste, and nearly every sour food contains an acid: Lemons, oranges, and grapefruit contain citric acid, for instance, and sour milk contains lactic acid. Bases are not so obvious in foods, but most of us have them stored under the kitchen or bathroom sink. Bases are present in many household cleaning agents, from perfumed toilet soap, to ammonia-based window cleaners, to the substance you put down the drain to dissolve hair, grease, and other materials that clog it.

Some of the most common acids and bases are listed below. It is a good idea at this point to learn their names and formulas, because we will refer to them often.

- **Sulfuric acid, H₂SO₄,** is probably the most important raw material in the chemical and pharmaceutical industries, and is manufactured in greater quantity worldwide than any other industrial chemical. Over 45 million tons are prepared in the United States annually for use in many hundreds of industrial processes, including the preparation of phosphate fertilizers. Its most common consumer use is as the acid found in automobile batteries. As anyone who has splashed battery acid on their skin or clothing knows, sulfuric acid is highly corrosive and can cause painful burns.

- **Hydrochloric acid, HCl,** or *muriatic acid* as it was historically known, has many industrial applications, including its use in metal cleaning and in the manufacture of high-fructose corn syrup. Aqueous HCl is also present as "stomach acid" in the digestive systems of most mammals.

- **Phosphoric acid, H₃PO₄,** is used in vast quantities in the manufacture of phosphate fertilizers. In addition, it is also used as an additive in foods and toothpastes. The tart taste of many soft drinks is due to the presence of phosphoric acid.

- **Nitric acid, HNO₃,** is a strong oxidizing agent that is used for many purposes, including the manufacture of ammonium nitrate fertilizer and military explosives. When spilled on the skin, it leaves a characteristic yellow coloration because of its reaction with skin proteins.

- **Acetic acid, CH₃CO₂H,** is the primary organic constituent of vinegar. It also occurs in all living cells and is used in many industrial processes such as the preparation of solvents, lacquers, and coatings.

▲ Hydrochloric acid, also known as muriatic acid, has many industrial applications.

- **Sodium hydroxide, NaOH,** also called *caustic soda* or *lye*, is the most commonly used of all bases. Industrially, it is used in the production of aluminum from its ore, in the production of glass, and in the manufacture of soap from animal fat. Concentrated solutions of NaOH can cause severe burns if allowed to sit on the skin for long. Drain cleaners often contain NaOH because it reacts with the fats and proteins found in grease and hair.

- **Calcium hydroxide, Ca(OH)₂,** or *slaked lime*, is made industrially by treating lime (CaO) with water. It has many applications, including its use in mortars and cements. An aqueous solution of Ca(OH)₂ is often called *limewater*.

- **Magnesium hydroxide, Mg(OH)₂,** or *milk of magnesia*, is an additive in foods, toothpaste, and many over-the-counter medications. Antacids such as Rolaids, Mylanta, and Maalox, for instance, all contain magnesium hydroxide.

- **Ammonia, NH₃,** is used primarily as a fertilizer, but it also has many other industrial applications including the manufacture of pharmaceuticals and explosives. A dilute solution of ammonia is frequently used around the house as a glass cleaner.

▲ Soap is manufactured by the reaction of vegetable oils and animal fats with the bases NaOH and KOH.

10.3 The Brønsted–Lowry Definition of Acids and Bases

The Arrhenius definition of acids and bases discussed in Section 10.1 applies only to reactions that take place in aqueous solution. A far more general definition was proposed in 1923 by the Danish chemist Johannes Brønsted and the English chemist Thomas Lowry. A **Brønsted–Lowry acid** is any substance that is able to give a hydrogen ion, H^+, to another molecule or ion. A hydrogen *atom* consists of a proton and an electron, so a hydrogen *ion*, H^+, is simply a proton. Thus, we often refer to acids as *proton donors*. The reaction need not occur in water, and a Brønsted–Lowry acid need not give appreciable concentrations of H_3O^+ ions in water.

Different acids can supply different numbers of H^+ ions, as we saw in Section 4.11. (p. 98) Acids with one proton to donate, such as HCl and HNO_3, are called *monoprotic acids*; H_2SO_4 is a *diprotic acid* because it has two protons to donate, and H_3PO_4 is a *triprotic acid* because it has three protons to donate. Notice that the acidic H atoms (that is, the H atoms that are donated as protons) are bonded to electronegative atoms, such as chlorine or oxygen.

Brønsted–Lowry acid A substance that can donate a hydrogen ion, H^+, to another molecule or ion.

H—Cl

Hydrochloric acid
(monoprotic)

$$\underset{\substack{| \\ O}}{\overset{\substack{O \\ ||}}{N}}-OH$$

Nitric acid
(monoprotic)

$$HO-\underset{\substack{|| \\ O}}{\overset{\substack{O \\ ||}}{S}}-OH$$

Sulfuric acid
(diprotic)

$$HO-\underset{\substack{| \\ OH}}{\overset{\substack{O \\ ||}}{P}}-OH$$

Phosphoric acid
(triprotic)

Acetic acid (CH_3CO_2H), an example of an organic acid, actually has a total of 4 hydrogens, but only the one bonded to the electronegative oxygen is positively polarized and therefore acidic. The 3 hydrogens bonded to carbon are not acidic. Most organic acids are similar in that they contain many hydrogen atoms, but only the one in the —CO_2H group (blue in the electrostatic potential map) is acidic:

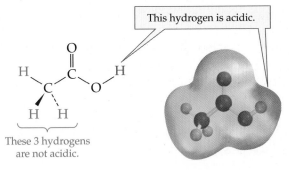

This hydrogen is acidic.

These 3 hydrogens are not acidic.

Brønsted–Lowry base A substance that can accept H^+ from an acid.

Whereas a Brønsted–Lowry acid is a substance that *donates* H^+ ions, a **Brønsted–Lowry base** is a substance that *accepts* H^+ from an acid. The reaction need not occur in water, and the Brønsted–Lowry base need not give appreciable concentrations of OH^- ions in water. Gaseous NH_3, for example, acts as a base to accept H^+ from gaseous HCl and yield the ionic solid $NH_4^+Cl^-$:

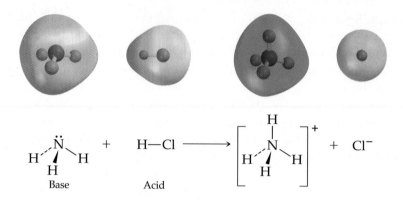

$$
\underset{\text{Base}}{H{-}\overset{\overset{\displaystyle \ddot{N}}{|}}{\underset{\displaystyle H}{\,}}H} \; + \; \underset{\text{Acid}}{H{-}Cl} \longrightarrow \left[H{-}\overset{\overset{\displaystyle H}{|}}{\underset{\displaystyle H}{N}}{-}H \right]^+ \; + \; Cl^-
$$

Putting the acid and base definitions together, *an acid–base reaction is one in which a proton is transferred.* The general reaction between proton-donor acids and proton-acceptor bases can be represented as

Electrons on base form bond with H^+ from acid.

$$B{:} \; + \; H{-}A \rightleftarrows B\overset{+}{-}H \; + \; A^-$$

$$B{:}^- \; + \; H{-}A \rightleftarrows B{-}H \; + \; A^-$$

where the abbreviation HA represents a Brønsted–Lowry acid and B: or B:⁻ represents a Brønsted–Lowry base. Notice in these acid–base reactions that both electrons in the product B—H bond come from the base, as indicated by the curved arrow flowing from the electron pair of the base to the hydrogen atom of the acid. Thus, the B—H bond that forms is a coordinate covalent bond (Section 5.4). (⊂⊙⊃, p. 115) In fact, a Brønsted–Lowry base *must* have such a lone pair of electrons; without them, it could not accept H^+ from an acid.

A base can be either neutral (B:) or negatively charged (B:⁻). If the base is neutral, then the product has a positive charge (BH^+) after H^+ has added. Ammonia is an example:

Adding an H^+ creates positive charge.

$$
\underset{\substack{\text{Ammonia}\\ \text{(a neutral base, B:)}}}{H{-}\overset{\overset{\displaystyle H}{|}}{\underset{\displaystyle H}{N}}{:}} \; + \; H{-}A \rightleftarrows \underset{\substack{\text{Ammonium ion}}}{H{-}\overset{\overset{\displaystyle H}{|}}{\underset{\displaystyle H}{\overset{+}{N}}}{-}H} \; + \; {:}A^-
$$

If the base is negatively charged, then the product is neutral (BH). Hydroxide ion is an example:

$$
\underset{\substack{\text{Hydroxide ion}\\ \text{(a negatively charged}\\ \text{base, B:}^-)}}{H{-}\ddot{O}{:}^-} \; + \; H{-}A \rightleftarrows \underset{\text{Water}}{H{-}\ddot{O}{-}H} \; + \; {:}A^-
$$

An important consequence of the Brønsted–Lowry definitions is that the *products* of an acid–base reaction can also behave as acids and bases. Many acid–base reactions are reversible (Section 7.8), although in some cases the equilibrium constant for the reaction is quite large. (⬤▭⬤, p. 200) For example, suppose we have as a forward reaction an acid HA donating a proton to a base B to produce A⁻. This product A⁻ is a base because it can act as a proton acceptor in the reverse reaction. At the same time, the product BH⁺ acts as an acid because it donates a proton in the reverse reaction:

Double arrow indicates reversible reaction.

$$\text{B:} + \text{H—A} \rightleftharpoons \text{:A}^- + \overset{+}{\text{B}}\text{—H}$$

Base   Acid   Base   Acid

Conjugate acid–base pair

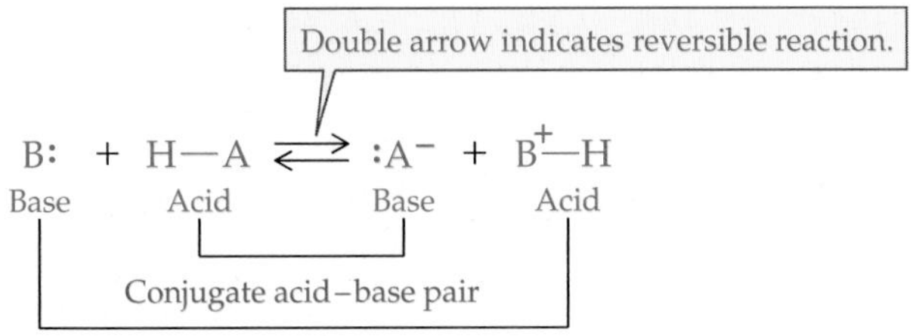

Pairs of chemical species such as B, BH⁺ and HA, A⁻ are called **conjugate acid–base pairs**. They are species that are found on opposite sides of a chemical reaction whose formulas differ by only one H⁺. Thus, the product anion A⁻ is the **conjugate base** of the reactant acid HA, and HA is the **conjugate acid** of the base A⁻. Similarly, the reactant B is the conjugate base of the product acid BH⁺, and BH⁺ is the conjugate acid of the base B. The number of protons in a conjugate acid–base pair is always one greater than the number of protons in the base of the pair. To give some examples, acetic acid and acetate ion, the hydronium ion and water, and the ammonium ion and ammonia all make conjugate acid–base pairs:

Conjugate acid–base pair Two substances whose formulas differ by only a hydrogen ion, H⁺.

Conjugate base The substance formed by loss of H⁺ from an acid.

Conjugate acid The substance formed by addition of H⁺ to a base.

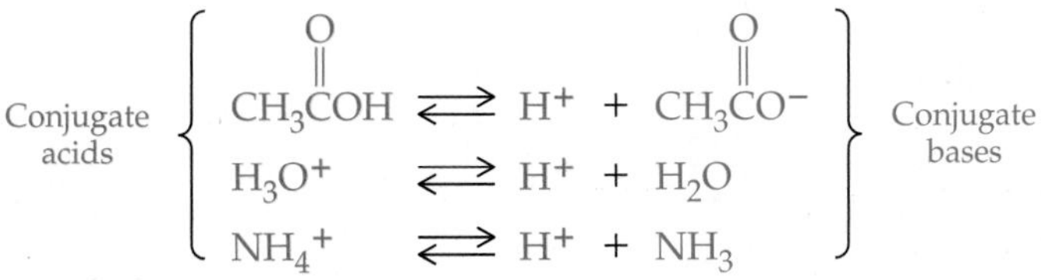

$$\text{Conjugate acids} \begin{cases} \text{CH}_3\overset{\text{O}}{\overset{\|}{\text{C}}}\text{OH} \rightleftharpoons \text{H}^+ + \text{CH}_3\overset{\text{O}}{\overset{\|}{\text{C}}}\text{O}^- \\ \text{H}_3\text{O}^+ \rightleftharpoons \text{H}^+ + \text{H}_2\text{O} \\ \text{NH}_4{}^+ \rightleftharpoons \text{H}^+ + \text{NH}_3 \end{cases} \text{Conjugate bases}$$

WORKED EXAMPLE **10.1** Acids and Bases: Identifying Brønsted–Lowry Acids and Bases

Identify each of the following as a Brønsted–Lowry acid or base:

(a) $PO_4{}^{3-}$

(b) $HClO_4$

(c) CN^-

ANALYSIS A Brønsted–Lowry acid must have a hydrogen that it can donate as H⁺, and a Brønsted–Lowry base must have an atom with a lone pair of electrons that can bond to H⁺. Typically, a Brønsted–Lowry base is an anion derived by loss of H⁺ from an acid.

SOLUTION

(a) The phosphate anion ($PO_4{}^{3-}$) is a Brønsted–Lowry base derived by loss of 3 H⁺ ions from phosphoric acid, H_3PO_4.

(b) Perchloric acid ($HClO_4$) is a Brønsted–Lowry acid because it can donate an H⁺ ion.

(c) The cyanide ion (CN^-) is a Brønsted–Lowry base derived by removal of an H⁺ ion from hydrogen cyanide, HCN.

WORKED EXAMPLE **10.2** Acids and Bases: Identifying Conjugate Acid–Base Pairs

Write formulas for

(a) The conjugate acid of the cyanide ion, CN^-

(b) The conjugate base of perchloric acid, $HClO_4$

ANALYSIS A conjugate acid is formed by adding H^+ to a base; a conjugate base is formed by removing H^+ from an acid.

SOLUTION

(a) HCN is the conjugate acid of CN^-.

(b) ClO_4^- is the conjugate base of $HClO_4$.

PROBLEM 10.1

Which of the following would you expect to be Brønsted–Lowry acids?

(a) HCO_2H (b) H_2S (c) $SnCl_2$

PROBLEM 10.2

Which of the following would you expect to be Brønsted–Lowry bases?

(a) SO_3^{2-} (b) Ag^+ (c) F^-

PROBLEM 10.3

Write formulas for:

(a) The conjugate acid of HS^- (b) The conjugate acid of PO_4^{3-}
(c) The conjugate base of H_2CO_3 (d) The conjugate base of NH_4^+

●○ KEY CONCEPT PROBLEM 10.4

For the reaction shown here, identify the Brønsted–Lowry acids, bases, and conjugate acid–base pairs.

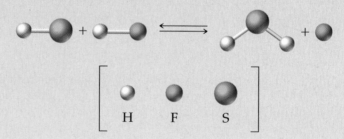

10.4 Water as Both an Acid and a Base

Water is neither an acid nor a base in the Arrhenius sense because it does not contain appreciable concentrations of either H_3O^+ or OH^-. In the Brønsted–Lowry sense, however, water can act as *both* an acid and a base. When in contact with a base, water reacts as a Brønsted–Lowry acid and *donates* a proton to the base. In its reaction with ammonia, for example, water donates H^+ to ammonia to form the ammonium ion:

$$NH_3 \ + \ H_2O \ \longrightarrow \ NH_4^+ \ + \ OH^-$$

Ammonia	Water	Ammonium ion	Hydroxide ion
(base)	(acid)	(acid)	(base)

When in contact with an acid, water reacts as a Brønsted–Lowry base and *accepts* H^+ from the acid. This, of course, is exactly what happens when an acid such as HCl dissolves in water, as discussed in Section 10.1.

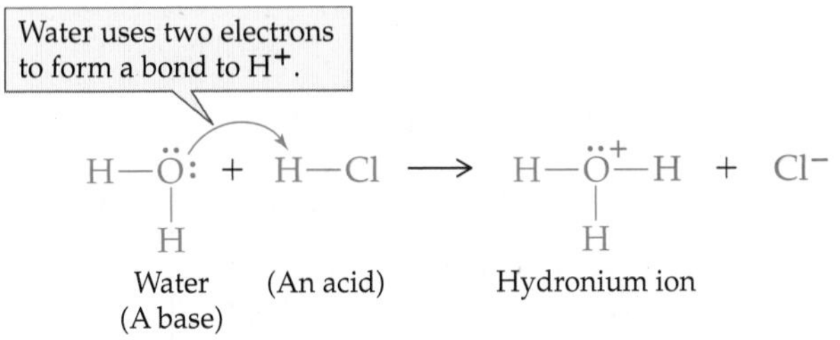

Substances like water, which can react as either an acid or a base depending on the circumstances, are said to be **amphoteric** (am-pho-**tare**-ic). When water acts as an acid, it donates H^+ and becomes OH^-; when it acts as a base, it accepts H^+ and becomes H_3O^+.

Amphoteric Describing a substance that can react as either an acid or a base.

PROBLEM 10.5

Is water an acid or a base in the following reactions?

(a) $H_3PO_4(aq) + H_2O(l) \longrightarrow H_2PO_4^-(aq) + H_3O^+(aq)$

(b) $F^-(aq) + H_2O(l) \longrightarrow HF(aq) + OH^-(aq)$

(c) $NH_4^+(aq) + H_2O(aq) \longrightarrow NH_3(aq) + H_3O^+(aq)$

10.5 Acid and Base Strength

Some acids and bases, such as sulfuric acid (H_2SO_4), hydrochloric acid (HCl), or sodium hydroxide (NaOH), are highly corrosive. They react readily and, in contact with skin, can cause serious burns. Other acids and bases are not nearly as reactive. Acetic acid (CH_3COOH, the major component in vinegar) and phosphoric acid (H_3PO_4) are found in many food products. Why are some acids and bases relatively "safe," while others must be handled with extreme caution? The answer lies in how easily they produce the active ions for an acid (H^+) or a base (OH^-).

As indicated in Table 10.1, acids differ in their ability to give up a proton. The six acids at the top of the table are **strong acids**, meaning that they give up a proton easily and are essentially 100% **dissociated**, or split apart into ions, in water. Those remaining are **weak acids**, meaning that they give up a proton with difficulty and are substantially less than 100% dissociated in water. In a similar way, the bases at the top of the table are **weak bases** because they have little affinity for a proton, and the bases at the bottom of the table are **strong bases** because they grab and hold a proton tightly.

Note that diprotic acids, such as sulfuric acid, undergo two stepwise dissociations in water. The first dissociation yields HSO_4^- and occurs to the extent of nearly 100%, so H_2SO_4 is a strong acid. The second dissociation yields SO_4^{2-} and takes place to a much lesser extent because separation of a positively charged H^+ from the negatively charged HSO_4^- anion is difficult. Thus, HSO_4^- is a weak acid:

$$H_2SO_4(l) + H_2O(l) \longrightarrow H_3O^+(aq) + HSO_4^-(aq)$$

$$HSO_4^-(aq) + H_2O(l) \rightleftharpoons H_3O^+(aq) + SO_4^{2-}(aq)$$

Strong acid An acid that gives up H^+ easily and is essentially 100% dissociated in water.

Dissociation The splitting apart of an acid in water to give H^+ and an anion.

Weak acid An acid that gives up H^+ with difficulty and is less than 100% dissociated in water.

Weak base A base that has only a slight affinity for H^+ and holds it weakly.

Strong base A base that has a high affinity for H^+ and holds it tightly.

TABLE 10.1 Relative Strengths of Acids and Conjugate Bases

		ACID		CONJUGATE BASE		
Increasing acid strength	Strong acids: 100% dissociated	Perchloric acid $HClO_4$	ClO_4^-	Perchlorate ion	Little or no reaction as bases	**Increasing base strength**
		Sulfuric acid H_2SO_4	HSO_4^-	Hydrogen sulfate ion		
		Hydriodic acid HI	I^-	Iodide ion		
		Hydrobromic acid HBr	Br^-	Bromide ion		
		Hydrochloric acid HCl	Cl^-	Chloride ion		
		Nitric acid HNO_3	NO_3^-	Nitrate ion		
		Hydronium ion H_3O^+	H_2O	**Water**		
	Weak acids	Hydrogen sulfate ion HSO_4^-	SO_4^{2-}	Sulfate ion	Very weak bases	
		Phosphoric acid H_3PO_4	$H_2PO_4^-$	Dihydrogen phosphate ion		
		Nitrous acid HNO_2	NO_2^-	Nitrite ion		
		Hydrofluoric acid HF	F^-	Fluoride ion		
		Acetic acid CH_3COOH	CH_3COO^-	Acetate ion		
	Very weak acids	Carbonic acid H_2CO_3	HCO_3^-	Bicarbonate ion	Weak bases	
		Dihydrogen phosphate ion $H_2PO_4^-$	HPO_4^{2-}	Hydrogen phosphate ion		
		Ammonium ion NH_4^+	NH_3	Ammonia		
		Hydrocyanic acid HCN	CN^-	Cyanide ion		
		Bicarbonate ion HCO_3^-	CO_3^{2-}	Carbonate ion		
		Hydrogen phosphate ion HPO_4^{2-}	PO_4^{3-}	Phosphate ion		
		Water H_2O	OH^-	**Hydroxide ion**	Strong base	

Perhaps the most striking feature of Table 10.1 is the inverse relationship between acid strength and base strength. *The stronger the acid, the weaker its conjugate base; the weaker the acid, the stronger its conjugate base.* HCl, for example, is a strong acid, so Cl^- is a very weak base. H_2O, however, is a very weak acid, so OH^- is a strong base.

Why is there an inverse relationship between acid strength and base strength? To answer this question, think about what it means for an acid or base to be strong or weak. A strong acid $H-A$ is one that readily gives up a proton, meaning that its conjugate base A^- has little affinity for the proton. But this is exactly the definition of a weak base—a substance that has little affinity for a proton. As a result, the reverse reaction occurs to a lesser extent, as indicated by the size of the forward and reverse arrows in the reaction:

> Larger arrow indicates forward reaction is stronger.

$$H-A + H_2O \rightleftarrows H_3O^+ + A^-$$

If this is a strong acid because it gives up a proton readily . . .

. . . then this is a weak base because it has little affinity for a proton.

In the same way, a weak acid is one that gives up a proton with difficulty, meaning that its conjugate base has a high affinity for the proton. But this is just the definition of a strong base—a substance that has a high affinity for the proton. The reverse reaction now occurs more readily.

$$H-A \; + \; H_2O \; \underset{\longleftarrow}{\overset{\longrightarrow}{\rightleftharpoons}} \; H_3O^+ \; + \; A^-$$

If this is a weak acid because it gives up a proton with difficulty . . .

Larger arrow indicates reverse reaction is stronger.

. . . then this is a strong base because it has a high affinity for a proton.

Knowing the relative strengths of different acids as shown in Table 10.1 makes it possible to predict the direction of proton-transfer reactions. *An acid–base proton-transfer equilibrium always favors reaction of the stronger acid with the stronger base, and formation of the weaker acid and base.* That is, the proton always leaves the stronger acid (whose weaker conjugate base cannot hold the proton) and always ends up in the weaker acid (whose stronger conjugate base holds the proton tightly). Put another way, in a contest for the proton, the stronger base always wins.

Stronger acid + Stronger base $\rightleftharpoons$ Weaker base + Weaker acid

To try out this rule, compare the reactions of acetic acid with water and with hydroxide ion. The idea is to write the equation, identify the acid on each side of the arrow, and then decide which acid is stronger and which is weaker. For example, the reaction of acetic acid with water to give acetate ion and hydronium ion is favored in the reverse direction, because acetic acid is a weaker acid than H_3O^+:

$$\underset{\text{Weaker acid}}{CH_3\overset{O}{\overset{\|}{C}}OH} \; + \; H_2O \; \rightleftharpoons \; CH_3\overset{O}{\overset{\|}{C}}O^- \; + \; \underset{\text{Stronger acid}}{H_3O^+} \qquad \text{Reverse reaction is favored.}$$

This base holds the proton less tightly . . .   . . . than this base does.

On the other hand, the reaction of acetic acid with hydroxide ion to give acetate ion and water is favored in the forward direction, because acetic acid is a stronger acid than H_2O:

$$\underset{\text{Stronger acid}}{CH_3\overset{O}{\overset{\|}{C}}OH} \; + \; OH^- \; \rightleftharpoons \; CH_3\overset{O}{\overset{\|}{C}}O^- \; + \; \underset{\text{Weaker acid}}{H_2O} \qquad \text{Forward reaction is favored.}$$

This base holds the proton more tightly . . .   . . . than this base does.

WORKED EXAMPLE   **10.3** Acid/Base Strength: Predicting Direction of H-transfer Reactions

Write a balanced equation for the proton-transfer reaction between phosphate ion (PO_4^{3-}) and water, and determine in which direction the equilibrium is favored.

ANALYSIS Look in Table 10.1 to see the relative acid and base strengths of the species involved in the reaction. The acid–base proton-transfer equilibrium will favor reaction of the stronger acid and formation of the weaker acid.

SOLUTION
Phosphate ion is the conjugate base of a weak acid (HPO_4^{2-}) and is therefore a relatively strong base. Table 10.1 shows that HPO_4^{2-} is a stronger acid than H_2O, and OH^- is a stronger base than PO_4^{3-}, so the reaction is favored in the reverse direction:

$$\underset{\text{Weaker base}}{PO_4^{3-}(aq)} \; + \; \underset{\text{Weaker acid}}{H_2O(l)} \; \rightleftharpoons \; \underset{\text{Stronger acid}}{HPO_4^{2-}(aq)} \; + \; \underset{\text{Stronger base}}{OH^-(aq)}$$

APPLICATION ▶ GERD—Too Much Acid or Not Enough?

Strong acids are very caustic substances that can dissolve even metals, and no one would think of ingesting them. However, the major component of the gastric juices secreted in the stomach is hydrochloric acid—a strong acid—and the acidic environment in the stomach is vital to good health and nutrition.

Stomach acid is essential for the digestion of proteins and for the absorption of certain micronutrients, such as calcium, magnesium, iron, and vitamin B_{12}. It also creates a sterile environment in the gut by killing yeast and bacteria that may be ingested. If these gastric juices leak up into the esophagus, the tube through which food and drink enter the stomach, they can cause the burning sensation in the chest or throat known as either heartburn or acid indigestion. Persistent irritation of the esophagus is known as gastro-esophageal reflux disease (GERD) and, if untreated, can lead to more serious health problems.

Hydrogen ions and chloride ions are secreted separately from the cytoplasm of cells lining the stomach and then combine to form HCl that is usually close to 0.10 M. The HCl is then released into the stomach cavity, where the concentration is diluted to about 0.01–0.001 M. Unlike the esophagus, the stomach is coated by a thick mucus layer that protects the stomach wall from damage by this caustic solution.

Those who suffer from acid indigestion can obtain relief using over-the-counter antacids, such as Tums or Rolaids (see Section 10.14, p. 320). Chronic conditions such as GERD, however, are often treated with prescription medications. GERD can be treated by two classes of drugs. Proton-pump inhibitors (PPI), such as Prevacid and Prilosec, prevent the production the H^+ ions in the parietal cells, while H2-receptor blockers (Tagamet, Zantac, and Pepcid) prevent the release of stomach acid into the lumen. Both drugs effectively decrease the production of stomach acid to ease the symptoms of GERD.

Ironically, GERD can also be caused by not having enough stomach acid—a condition known as *hypochlorhydria*. The valve that controls the release of stomach contents to the small intestine is triggered by acidity. If this valve fails to open because the stomach is not acidic enough, the contents of the stomach can be churned up into the esophagus.

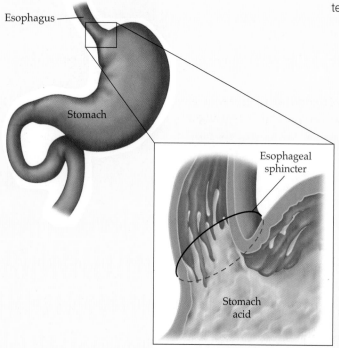

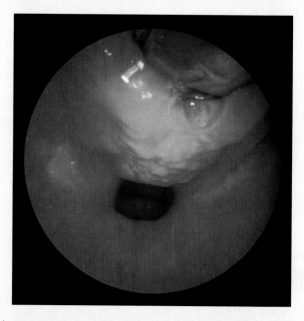

▲ If not treated, GERD can cause ulcers and scarring of esophageal tissue.

▲ The burning sensation and other symptoms associated with GERD are caused by the reflux of the acidic contents of the stomach into the esophagus.

See Additional Problem 10.96 at the end of the chapter.

PROBLEM 10.6

Use Table 10.1 to identify the stronger acid in the following pairs:

(a) H_2O or NH_4^+
(b) H_2SO_4 or CH_3CO_2H
(c) HCN or H_2CO_3

PROBLEM 10.7

Use Table 10.1 to identify the stronger base in the following pairs:

(a) F^- or Br^- **(b)** OH^- or HCO_3^-

PROBLEM 10.8

Write a balanced equation for the proton-transfer reaction between a hydrogen phosphate ion and a hydroxide ion. Identify each acid–base pair, and determine in which direction the equilibrium is favored.

KEY CONCEPT PROBLEM 10.9

From this electrostatic potential map of the amino acid alanine, identify the most acidic hydrogens in the molecule:

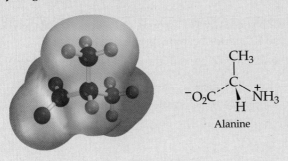

Alanine

10.6 Acid Dissociation Constants

The reaction of a weak acid with water, like any chemical equilibrium, can be described by an equilibrium equation (Section 7.8), where square brackets indicate the concentrations of the enclosed species in molarity (moles per liter). (⊂⊃, p. 200)

For the reaction $HA(aq) + H_2O(l) \rightleftharpoons H_3O^+(aq) + A^-(aq)$

We have $K = \dfrac{[H_3O^+][A^-]}{[HA][H_2O]}$

Because water is a solvent as well as a participant for the reaction, its concentration is essentially constant and has no effect on the equilibrium. Therefore, we usually put the equilibrium constant K and the water concentration $[H_2O]$ together to make a new constant called the **acid dissociation constant**, K_a. The acid dissociation constant is simply the hydronium ion concentration $[H_3O^+]$ times the conjugate base concentration $[A^-]$ divided by the undissociated acid concentration [HA]:

Acid dissociation constant $K_a = K[H_2O] = \dfrac{[H_3O^+][A^-]}{[HA]}$

For a strong acid, the H_3O^+ and A^- concentrations are much larger than the HA concentration, so K_a is very large. In fact, the K_a values for strong acids such as HCl are so large that it is difficult and not very useful to measure them. For a weak acid, however, the H_3O^+ and A^- concentrations are smaller than the HA concentration, so K_a is small. Table 10.2 gives K_a values for some common acids and illustrates several important points:

- Strong acids have K_a values much greater than 1 because dissociation is favored.
- Weak acids have K_a values much less than 1 because dissociation is not favored.

TABLE 10.2   Some Acid Dissociation Constants, K_a, at 25 °C

ACID	K_a	ACID	K_a
Hydrofluoric acid (HF)	3.5×10^{-4}	*Polyprotic acids*	
Hydrocyanic acid (HCN)	4.9×10^{-10}	Sulfuric acid	
Ammonium ion (NH_4^+)	5.6×10^{-10}	$\quad H_2SO_4$	Large
		$\quad HSO_4^-$	1.2×10^{-2}
Organic acids		Phosphoric acid	
Formic acid (HCOOH)	1.8×10^{-4}	$\quad H_3PO_4$	7.5×10^{-3}
Acetic acid (CH_3COOH)	1.8×10^{-5}	$\quad H_2PO_4^-$	6.2×10^{-8}
Propanoic acid	1.3×10^{-5}	$\quad HPO_4^{2-}$	2.2×10^{-13}
$\quad (CH_3CH_2COOH)$		Carbonic acid	
Ascorbic acid (vitamin C)	7.9×10^{-5}	$\quad H_2CO_3$	4.3×10^{-7}
		$\quad HCO_3^-$	5.6×10^{-11}

- Donation of each successive H^+ from a polyprotic acid is more difficult than the one before it, so K_a values become successively lower.
- Most organic acids, which contain the $-CO_2H$ group, have K_a values near 10^{-5}.

PROBLEM 10.10

Benzoic acid has $K_a = 6.5 \times 10^{-5}$, and citric acid has $K_a = 7.2 \times 10^{-4}$. Which of the two is the stronger acid?

10.7 Dissociation of Water

We saw previously that water is amphoteric – it can act as an acid when a base is present and as a base when an acid is present. What about when no other acids or bases are present, however? In this case, one water molecule acts as an acid while another water molecule acts as a base, reacting to form the hydronium and hydroxide ions:

$$H_2O(l) + H_2O(l) \rightleftharpoons H_3O^+(aq) + OH^-(aq)$$

Because each dissociation reaction yields one H_3O^+ ion and one OH^- ion, the concentrations of the two ions are identical. At 25 °C, the concentration of each is 1.00×10^{-7} M. We can write the equilibrium constant expression for the dissociation of water as

$$K = \frac{[H_3O^+][OH^-]}{[H_2O][H_2O]}$$

and $\quad K_a = K[H_2O] = \dfrac{[H_3O^+][OH^-]}{[H_2O]}$

where $\quad [H_3O^+] = [OH^-] = 1.00 \times 10^{-7}$ M $\quad$ (at 25 °C)

As both a reactant and a solvent, the concentration of water is essentially constant. We can therefore put the acid dissociation constant K_a and the water concentration $[H_2O]$ together to make a new constant called the **ion-product constant for water (K_w)**, which is simply the H_3O^+ concentration times the OH^- concentration. At 25 °C, $K_w = 1.00 \times 10^{-14}$.

Ion-product constant for water $\quad K_w = K_a[H_2O] = [H_3O^+][OH^-]$

$$= (1.00 \times 10^{-7})(1.00 \times 10^{-7})$$

$$= 1.00 \times 10^{-14} \quad \text{(at 25 °C)}$$

The importance of the equation $K_w = [H_3O^+][OH^-]$ is that it applies to all aqueous solutions, not just to pure water. Since the product of $[H_3O^+]$ times $[OH^-]$ is always constant for any solution, we can determine the concentration of one species if we know the concentration of the other. If an acid is present in solution, for instance, so that $[H_3O^+]$ is large, then $[OH^-]$ must be small. If a base is present in solution so that $[OH^-]$ is large, then $[H_3O^+]$ must be small. For example, for a 0.10 M HCl solution, we know that $[H_3O^+] = 0.10$ M because HCl is 100% dissociated. Thus, we can calculate that $[OH^-] = 1.0 \times 10^{-13}$ M:

$$\text{Since} \quad K_w \times [H_3O^+][OH^-] = 1.00 \times 10^{-14}$$

$$\text{we have} \quad [OH^-] = \frac{K_w}{[H_3O^+]} = \frac{1.00 \times 10^{-14}}{0.10} = 1.0 \times 10^{-13} \text{ M}$$

Similarly, for a 0.10 M NaOH solution, we know that $[OH^-] = 0.10$ M, so $[H_3O^+] = 1.0 \times 10^{-13}$ M:

$$[H_3O^+] = \frac{K_w}{[OH^-]} = \frac{1.00 \times 10^{-14}}{0.10} = 1.0 \times 10^{-13} \text{ M}$$

Solutions are identified as acidic, neutral, or basic (*alkaline*) according to the value of their H_3O^+ and OH^- concentrations:

Acidic solution: $\quad [H_3O^+] > 10^{-7}$ M $\quad$ and $\quad [OH^-] < 10^{-7}$ M

Neutral solution: $\quad [H_3O^+] = 10^{-7}$ M $\quad$ and $\quad [OH^-] = 10^{-7}$ M

Basic solution: $\quad [H_3O^+] < 10^{-7}$ M $\quad$ and $\quad [OH^-] > 10^{-7}$ M

WORKED EXAMPLE **10.4** Water Dissociation Constant: Using K_w to Calculate $[OH^-]$

Milk has an H_3O^+ concentration of 4.5×10^{-7} M. What is the value of $[OH^-]$? Is milk acidic, neutral, or basic?

ANALYSIS The OH^- concentration can be found by dividing K_w by $[H_3O^+]$. An acidic solution has $[H_3O^+] > 10^{-7}$ M, a neutral solution has $[H_3O^+] = 10^{-7}$ M, and a basic solution has $[H_3O^+] < 10^{-7}$ M.

BALLPARK ESTIMATE Since the H_3O^+ concentration is slightly *greater* than 10^{-7} M, the OH^- concentration must be slightly *less* than 10^{-7} M, on the order of 10^{-8}.

SOLUTION

$$[OH^-] = \frac{K_w}{[H_3O^+]} = \frac{1.00 \times 10^{-14}}{4.5 \times 10^{-7}} = 2.2 \times 10^{-8} \text{ M}$$

Milk is slightly acidic because its H_3O^+ concentration is slightly larger than 1×10^{-7} M.

BALLPARK CHECK The OH^- concentration is of the same order of magnitude as our estimate.

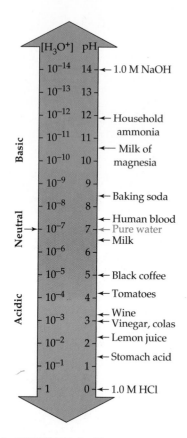

▲ **FIGURE 10.1 The pH scale and the pH values of some common substances.** A low pH corresponds to a strongly acidic solution, a high pH corresponds to a strongly basic solution, and a pH of 7 corresponds to a neutral solution.

p function The negative common logarithm of some variable, $pX = -\log(X)$.

pH A measure of the acid strength of a solution; the negative common logarithm of the H_3O^+ concentration.

▲ Adding only a teaspoonful of concentrated (6 M) hydrochloric acid lowers the pH of this pool from 7 to 6. Lowering the pH from 7 to 1 would take 400 gallons.

PROBLEM 10.11

Identify the following solutions as either acidic or basic. What is the value of $[OH^-]$ in each?

(a) Beer, $[H_3O^+] = 3.2 \times 10^{-5}$ M

(b) Household ammonia, $[H_3O^+] = 3.1 \times 10^{-12}$ M

10.8 Measuring Acidity in Aqueous Solution: pH

In many fields, from medicine to chemistry to winemaking, it is necessary to know the exact concentration of H_3O^+ or OH^- in a solution. If, for example, the H_3O^+ concentration in blood varies only slightly from a value of 4.0×10^{-8} M, death can result.

Although correct, it is nevertheless awkward to refer to low concentrations of H_3O^+ using molarity. If you were asked which concentration is higher, 9.0×10^{-8} M or 3.5×10^{-7} M, you would probably have to stop and think for a moment before answering. Fortunately, there is an easier way to express and compare H_3O^+ concentrations—the *pH scale*.

The pH of an aqueous solution is a number, usually between 0 and 14, that indicates the H_3O^+ concentration of the solution. A pH smaller than 7 corresponds to an acidic solution, a pH larger than 7 corresponds to a basic solution, and a pH of exactly 7 corresponds to a neutral solution. The pH scale and pH values of some common substances are shown in Figure 10.1

Mathematically, a **p function** is defined as the negative common logarithm of some variable. The **pH** of a solution, therefore, is the negative common logarithm of the H_3O^+ concentration:

$$pH = -\log[H^+] \quad (or[H_3O^+])$$

If you have studied logarithms, you may remember that the common logarithm of a number is the power to which 10 must be raised to equal the number. The pH definition can therefore be restated as

$$[H_3O^+] = 10^{-pH}$$

For example, in neutral water at 25 °C, where $[H_3O^+] = 1 \times 10^{-7}$ M, the pH is 7; in a strong acid solution where $[H_3O^+] = 1 \times 10^{-1}$ M, the pH is 1; and in a strong base solution where $[H_3O^+] = 1 \times 10^{-14}$ M, the pH is 14:

Acidic solution : $\quad pH < 7, \quad [H_3O^+] > 1 \times 10^{-7}$ M

Neutral solution : $\quad pH = 7, \quad [H_3O^+] = 1 \times 10^{-7}$ M

Basic solution : $\quad pH > 7, \quad [H_3O^+] < 1 \times 10^{-7}$ M

Keep in mind that the pH scale covers an enormous range of acidities because it is a *logarithmic* scale, which involves powers of 10 (Figure 10.2). A change of only 1 pH unit means a tenfold change in $[H_3O^+]$, a change of 2 pH units means a hundredfold change in $[H_3O^+]$, and a change of 12 pH units means a change of 10^{12} (a million) in $[H_3O^+]$.

To get a feel for the size of the quantities involved, think of a typical backyard swimming pool, which contains about 100,000 L of water. You would have to add only 0.10 mol of HCl (3.7 g) to lower the pH of the pool from 7.0 (neutral) to 6.0, but you would have to add 10,000 mol of HCl (370 kg!) to lower the pH of the pool from 7.0 to 1.0.

The logarithmic pH scale is a convenient way of reporting the relative acidity of solutions, but using logarithms can also be useful when calculating H_3O^+ and

OH⁻ concentrations. Remember that the equilibrium between H_3O^+ and OH^- in aqueous solutions is expressed by K_w, where

$$K_w = [H_3O^+][OH^-] = 1 \times 10^{-14} \quad \text{(at 25 °C)}$$

If we convert this equation to its negative logarithmic form, we obtain

$$-\log(K_w) = -\log(H_3O^+) - \log(OH^-)$$

$$-\log(1 \times 10^{-14}) = -\log(H_3O^+) - \log(OH^-)$$

$$or \quad 14.00 = pH + pOH$$

The logarithmic form of the K_w equation can simplify the calculation of solution pH from OH⁻ concentration, as demonstrated in Worked Example 10.7.

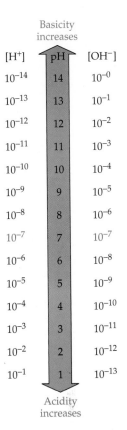

▲ **FIGURE 10.2 The relationship of the pH scale to H⁺ and OH⁻ concentrations.**

WORKED EXAMPLE **10.5** Measuring Acidity: Calculating pH from $[H_3O^+]$

The H_3O^+ concentration in coffee is about 1×10^{-5} M. What pH is this?

ANALYSIS The pH is the negative common logarithm of the H_3O^+ concentration: $pH = -\log[H_3O^+]$.

SOLUTION
Since the common logarithm of 1×10^{-5} M is −5, the pH is 5.0.

WORKED EXAMPLE **10.6** Measuring Acidity: Calculating $[H_3O^+]$ from pH

Lemon juice has a pH of about 2. What $[H_3O^+]$ is this?

ANALYSIS In this case, we are looking for the $[H_3O^+]$, where $[H_3O^+] = 10^{-pH}$.

SOLUTION
Since pH = 2.0, $[H_3O^+] = 10^{-2} = 1 \times 10^{-2}$ M.

WORKED EXAMPLE **10.7** Measuring Acidity: Using K_w to Calculate $[H_3O^+]$ and pH

A cleaning solution is found to have $[OH^-] = 1 \times 10^{-3}$ M. What is the pH?

ANALYSIS To find pH, we must first find the value of $[H_3O^+]$ by using the equation $[H_3O^+] = K_w/[OH^-]$. Alternatively, we can calculate the pOH of the solution and then use the logarithmic form of the K_w equation: pH = 14.00 − pOH.

SOLUTION
Rearranging the K_w equation, we have

$$[H_3O^+] = \frac{K_w}{[OH^-]} = \frac{1.00 \times 10^{-14}}{1 \times 10^{-3}} = 1 \times 10^{-11} \text{ M}$$

$$pH = -\log(1 \times 10^{-11}) = 11.0$$

Using the logarithmic form of the K_w equation, we have

$$pH = 14.0 - pOH = 14.0 - (-\log(OH^-))$$

$$pH = 14.0 - (-\log(1 \times 10^{-3}))$$

$$pH = 14.0 - 3.0 = 11.0$$

> **WORKED EXAMPLE** **10.8** Measuring Acidity: Calculating pH of Strong
> Acid Solutions
>
> What is the pH of a 0.01 M solution of HCl?
>
> **ANALYSIS** To find pH, we must first find the value of $[H_3O^+]$.
>
> **SOLUTION**
> Since HCl is a strong acid (Table 10.1), it is 100% dissociated, and the H_3O^+ concentration is the same as the HCl concentration: $[H_3O^+] = 0.01$ M, or 1×10^{-2} M, and pH = 2.0.

> **PROBLEM 10.12**
>
> Which solution has the higher H_3O^+ concentration, one with pH = 5 or one with pH = 9? Which has the higher OH^- concentration?
>
> **PROBLEM 10.13**
>
> Give the pH of solutions with the following concentrations:
>
> **(a)** $[H_3O^+] = 1 \times 10^{-5}$ M
>
> **(b)** $[OH^-] = 1 \times 10^{-9}$ M
>
> **PROBLEM 10.14**
>
> Give the hydronium ion concentrations of solutions with the following values of pH. Which of the solutions is most acidic? Which is most basic?
>
> **(a)** pH 13.0 **(b)** pH 3.0 **(c)** pH 8.0
>
> **PROBLEM 10.15**
>
> What is the pH of a 1×10^{-4} M solution of HNO_3?

10.9 Working with pH

Converting between pH and H_3O^+ concentration is easy when the pH is a whole number, but how do you find the H_3O^+ concentration of blood, which has a pH of 7.4, or the pH of a solution with $[H_3O^+] = 4.6 \times 10^{-3}$ M? Sometimes it is sufficient to make an estimate. The pH of blood (7.4) is between 7 and 8, so the H_3O^+ concentration of blood must be between 1×10^{-7} and 1×10^{-8} M. To be exact about finding pH values, though, requires a calculator.

Converting from pH to $[H_3O^+]$ requires finding the *antilogarithm* of the negative pH, which is done on many calculators with an "INV" key and a "log" key. Converting from $[H_3O^+]$ to pH requires finding the logarithm, which is commonly done with a "log" key and an "expo" or "EE" key for entering exponents of 10. Consult your calculator instructions if you are not sure how to use these keys. Remember that the sign of the number given by the calculator must be changed from minus to plus to get the pH.

The H_3O^+ concentration in blood with pH = 7.4 is

$$[H_3O^+] = \text{antilog}(-7.4) = 4 \times 10^{-8} \text{ M}$$

The pH of a solution with $[H_3O^+] = 4.6 \times 10^{-3}$ M is

$$pH = -\log(4.6 \times 10^{-3}) = -(-2.34) = 2.34$$

A note about significant figures: An antilogarithm contains the same number of digits that the original number has to the right of the decimal point. A logarithm contains the same number of digits to the right of the decimal point that the original number has

$$\text{antilog}(-7.4) = 4 \times 10^{-8} \qquad \log(4.6 \times 10^{-3}) = -2.34$$

| 1 digit after decimal point | 1 digit | 2 digits | 2 digits after decimal point |

WORKED EXAMPLE **10.9** Working with pH: Converting a pH to $[H_3O^+]$

Soft drinks usually have a pH of approximately 3.1. What is the $[H_3O^+]$ concentration in a soft drink?

ANALYSIS To convert from a pH value to an $[H_3O^+]$ concentration requires using the equation $[H_3O^+] = 10^{-pH}$, which requires finding an antilogarithm on a calculator.

BALLPARK ESTIMATE Because the pH is between 3.0 and 4.0, the $[H_3O^+]$ must be between 1×10^{-3} and 1×10^{-4}. A pH of 3.1 is very close to 3.0, so the $[H_3O^+]$ must be just slightly below 1×10^{-3} M.

SOLUTION
Entering the negative pH on a calculator (-3.1) and pressing the "INV" and "log" keys gives the answer 7.943×10^{-4}, which must be rounded off to 8×10^{-4} since the pH has only one digit to the right of the decimal point.

BALLPARK CHECK The calculated $[H_3O^+]$ of 8×10^{-4} M is between 1×10^{-3} M and 1×10^{-4} M and, as we estimated, just slightly below 1×10^{-3} M. (Remember, 8×10^{-4} is 0.8×10^{-3}.)

WORKED EXAMPLE **10.10** Working with pH: Calculating pH for Strong Acid Solutions

What is the pH of a 0.0045 M solution of $HClO_4$?

ANALYSIS Finding pH requires first finding $[H_3O^+]$ and then using the equation $pH = -\log[H_3O^+]$. Since $HClO_4$ is a strong acid (see Table 10.1), it is 100% dissociated, and so the H_3O^+ concentration is the same as the $HClO_4$ concentration.

BALLPARK ESTIMATE Because $[H^+] = 4.5 \times 10^{-3}$ M is close to midway between 1×10^{-2} M and 1×10^{-3} M, the pH must be close to the midway point between 2.0 and 3.0. (Unfortunately, because the logarithm scale is not linear, trying to estimate the midway point is not a simple process.)

SOLUTION
$[H_3O^+] = 0.0045$ M $= 4.5 \times 10^{-3}$ M. Taking the negative logarithm gives pH $= 2.35$.

BALLPARK CHECK The calculated pH is consistent with our estimate.

WORKED EXAMPLE **10.11** Working with pH: Calculating pH for Strong Base Solutions

What is the pH of a 0.0032 M solution of NaOH?

ANALYSIS Since NaOH is a strong base, the OH^- concentration is the same as the NaOH concentration. Starting with the OH^- concentration, finding pH requires either using the K_w equation to find $[H_3O^+]$ or calculating pOH and then using the logarithmic form of the K_w equation.

BALLPARK ESTIMATE Because $[OH^-] = 3.2 \times 10^{-3}$ M is close to midway between 1×10^{-2} M and 1×10^{-3} M, the pOH must be close to the midway point between 2.0 and 3.0. Subtracting the pOH from 14 would therefore yield a pH between 11 and 12.

SOLUTION

$$[OH^-] = 0.0032 \text{ M} = 3.2 \times 10^{-3} \text{ M}$$

$$[H_3O^+] = \frac{K_w}{(3.2 \times 10^{-3})} = 3.1 \times 10^{-12} \text{ M}$$

Taking the negative logarithm gives pH $= -\log(3.1 \times 10^{-12}) = 11.51$. Alternatively, we can calculate pOH and subtract from 14.00 using the logarithmic form of the K_w equation. For $[OH^-] = 0.0032$ M,

$$\text{pOH} = -\log(3.2 \times 10^{-3}) = 2.49$$

$$\text{pH} = 14.00 - 2.49 = 11.51$$

Since the given OH^- concentration included two significant figures, the final pH includes two significant figures beyond the decimal point.

BALLPARK CHECK The calculated pH is consistent with our estimate.

PROBLEM 10.16

Identify the following solutions as acidic or basic, estimate $[H_3O^+]$ values for each, and rank them in order of increasing acidity:

(a) Saliva, pH = 6.5   (b) Pancreatic juice, pH = 7.9

(c) Orange juice, pH = 3.7   (d) Wine, pH = 3.5

PROBLEM 10.17

Find the pH of the following solutions:

(a) Seawater with $[H_3O^+] = 5.3 \times 10^{-9}$ M

(b) A urine sample with $[H_3O^+] = 8.9 \times 10^{-6}$ M

PROBLEM 10.18

What is the pH of a 0.0025 M solution of HCl?

10.10 Laboratory Determination of Acidity

The pH of water is an important indicator of water quality in applications ranging from swimming pool and spa maintenance to municipal water treatment. There are several ways to measure the pH of a solution. The simplest but least accurate method is to use an **acid–base indicator**, a dye that changes color depending on

Acid–base indicator A dye that changes color depending on the pH of a solution.

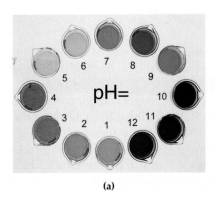

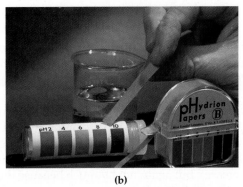

(a)   (b)

▲ **FIGURE 10.3 Finding pH.** (a) The color of universal indicator in solutions of known pH from 1 to 12. (b) Testing pH with a paper strip. Comparing the color of the strip with the code on the package gives the approximate pH.

the pH of the solution. For example, the well-known dye *litmus* is red below pH 4.8 but blue above pH 7.8 and the indicator *phenolphthalein* (fee-nol-**thay**-lean) is colorless below pH 8.2 but red above pH 10. To make pH determination particularly easy, test kits are available that contain a mixture of indicators known as *universal indicator* to give approximate pH measurements in the range 2–10 (Figure 10.3a). Also available are rolls of "pH paper," which make it possible to determine pH simply by putting a drop of solution on the paper and comparing the color that appears to the color on a calibration chart (Figure 10.3b).

A much more accurate way to determine pH uses an electronic pH meter like the one shown in Figure 10.4. Electrodes are dipped into the solution, and the pH is read from the meter.

▲ **FIGURE 10.4 Using a pH meter to obtain an accurate reading of pH.** Is milk of magnesia acidic or basic?

APPLICATION ▶ pH of Body Fluids

Each fluid in our bodies has a pH range suited to its function, as shown in the accompanying table. The stability of cell membranes, the shapes of huge protein molecules that must be folded in certain ways to function, and the activities of enzymes are all dependent on appropriate H_3O^+ concentrations.

pH of Body Fluids

FLUID	pH
Blood plasma	7.4
Interstitial fluid	7.4
Cytosol	7.0
Saliva	5.8–7.1
Gastric juice	1.6–1.8
Pancreatic juice	7.5–8.8
Intestinal juice	6.3–8.0
Urine	4.6–8.0
Sweat	4.0–6.8

Blood plasma and the interstitial fluid surrounding cells, which together comprise one-third of body water, have a slightly basic pH of 7.4. In fact, one of the functions of blood is to neutralize the acid by-products of cellular metabolism. The fluid within cells, called the *cytosol*, is slightly more acidic than the fluid outside, so a pH differential exists.

The strongly acidic gastric juice in the stomach has three important functions. First, gastric juice aids in the digestion of proteins by causing them to denature, or unfold. Second, it kills most of the bacteria we consume along with our food. Third, it converts the enzyme that breaks down proteins from an inactive form to the active form.

When the acidic mixture of partially digested food (*chyme*) leaves the stomach and enters the small intestine, it triggers secretion by the pancreas of an alkaline fluid containing bicarbonate ions, HCO_3^-. A principal function of this pancreatic juice and other intestinal fluids is to dilute and neutralize the hydrochloric acid carried along from the stomach.

Urine has a wide normal pH range, depending on the diet and recent activities. It is generally acidic, though, because one important function of urine is to eliminate a quantity of hydrogen ion equal to that produced by the body each day. Without this elimination, the body would soon be overwhelmed by acid.

See Additional Problem 10.97 at the end of the chapter.

10.11 Buffer Solutions

Much of the body's chemistry depends on maintaining the pH of blood and other fluids within narrow limits. This is accomplished through the use of **buffers**— combinations of substances that act together to prevent a drastic change in pH.

Buffer A combination of substances that act together to prevent a drastic change in pH; usually a weak acid and its conjugate base.

Most buffers are mixtures of a weak acid and a roughly equal concentration of its conjugate base—for example, a solution that contains 0.10 M acetic acid and 0.10 M acetate ion. If a small amount of OH^- is added to a buffer solution, the pH increases, but not by much because the acid component of the buffer neutralizes the added OH^-. If a small amount of H_3O^+ is added to a buffer solution, the pH decreases, but again not by much because the base component of the buffer neutralizes the added H_3O^+.

To see why buffer solutions work, look at the equation for the acid dissociation constant of an acid HA.

$$\text{For the reaction: } HA(aq) + H_2O(l) \rightleftharpoons A^-(aq) + H_3O^+(aq)$$

$$\text{we have} \quad K_a = \frac{[H_3O^+][A^-]}{[HA]}$$

Rearranging this equation shows that the value of $[H_3O^+]$, and thus the pH, depends on the ratio of the undissociated acid concentration to the conjugate base concentration, $[HA]/[A^-]$:

$$[H_3O^+] = K_a\frac{[HA]}{[A^-]}$$

In the case of the acetic acid–acetate ion buffer, for instance, we have

$$CH_3CO_2H(aq) + H_2O(l) \rightleftharpoons H_3O^+(aq) + CH_3CO_2^-(aq)$$
$$(0.10 \text{ M}) \hspace{6cm} (0.10 \text{ M})$$

$$\text{and} \quad [H_3O^+] = K_a\frac{[CH_3CO_2H]}{[CH_3CO_2^-]}$$

Initially, the pH of the 0.10 M acetic acid–0.10 M acetate ion buffer solution is 4.74. When acid is added, most is removed by reaction with $CH_3CO_2^-$. The equilibrium reaction shifts to the left, and as a result the concentration of CH_3CO_2H increases and the concentration of $CH_3CO_2^-$ decreases. As long as the changes in $[CH_3CO_2H]$ and $[CH_3CO_2^-]$ are relatively small, however, the ratio of $[CH_3CO_2H]$ to $[CH_3CO_2^-]$ changes only slightly, and there is little change in the pH.

When base is added to the buffer, most is removed by reaction with CH_3CO_2H. The equilibrium shifts to the right, and so the concentration of CH_3CO_2H decreases and the concentration of $CH_3CO_2^-$ increases. Here too, though, as long as the concentration changes are relatively small, there is little change in the pH.

The ability of a buffer solution to resist changes in pH when acid or base is added is illustrated in Figure 10.5. Addition of 0.010 mol of H_3O^+ to 1.0 L of pure water changes the pH from 7 to 2, and addition of 0.010 mol of OH^- changes the pH from 7 to 12. A similar addition of acid to 1.0 L of a 0.10 M acetic acid–0.10 M acetate ion buffer, however, changes the pH only from 4.74 to 4.68, and addition of base changes the pH only from 4.74 to 4.85.

As we did with K_w, we can convert the rearranged K_a equation to its logarithmic form to obtain

$$pH = pK_a - \log\left(\frac{[HA]}{[A^-]}\right)$$

$$\text{or} \quad pH = pK_a + \log\left(\frac{[A^-]}{[HA]}\right)$$

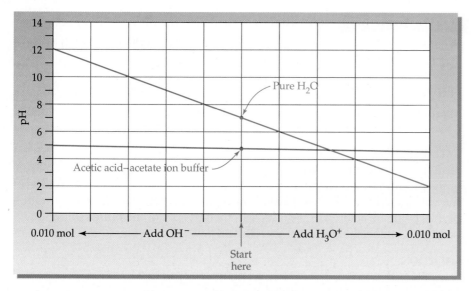

▲ **FIGURE 10.5 A comparison of the change in pH.** When 0.010 mol of acid and 0.010 mol of base are added to 1.0 L of pure water and to 1.0 L of a 0.10 M acetic acid–0.10 M acetate ion buffer, the pH of the water varies between 12 and 2, while the pH of the buffer varies only between 4.85 and 4.68.

This expression is known as the **Henderson–Hasselbalch equation** and is very useful in buffer applications, particularly in biology and biochemistry. Examination of the Henderson–Hasselbalch equation provides useful insights into how to prepare a buffer and into the factors that affect the pH of a buffer solution.

Henderson–Hasselbalch equation The logarithmic form of the K_a equation for a weak acid, used in applications involving buffer solutions.

The effective pH range of a buffer will depend on the pK_a of the acid HA and on the relative concentrations of HA and conjugate base A^-. In general, the most effective buffers meet the following conditions:

- The pK_a for the weak acid should be close to the desired pH of the buffer solution.
- The ratio of [HA] to $[A^-]$ should be close to 1, so that neither additional acid nor additional base changes the pH of the solution dramatically.
- The molar amounts of HA and A^- in the buffer should be approximately 10 times greater than the molar amounts of either acid or base you expect to add so that the ratio $[A^-]/[HA]$ does not undergo a large change.

WORKED EXAMPLE **10.12** Buffers: Selecting a Weak Acid for a Buffer Solution

Which of the organic acids in Table 10.2 would be the most appropriate for preparing a pH 4.15 buffer solution?

ANALYSIS The pH of the buffer solution depends on the pK_a of the weak acid. Remember that $pK_a = -\log(K_a)$.

SOLUTION
The K_a and pK_a values for the four organic acids in Table 10.2 are tabulated below. The ascorbic acid ($pK_a = 4.10$) will produce a buffer solution closest to the desired pH of 4.15.

ORGANIC ACID	K_a	pK_a
Formic acid (HCOOH)	1.8×10^{-4}	3.74
Acetic acid (CH_3COOH)	1.8×10^{-5}	4.74
Propanoic acid (CH_3CH_2COOH)	1.3×10^{-5}	4.89
Ascorbic acid (vitamin C)	7.9×10^{-5}	4.10

WORKED EXAMPLE **10.13** Buffers: Calculating the pH of a Buffer Solution

What is the pH of a buffer solution that contains 0.100 M HF and 0.120 M NaF? The K_a of HF is 3.5×10^{-4}, and so $pK_a = 3.46$.

ANALYSIS The Henderson–Hasselbalch equation can be used to calculate the pH of a buffer solution: $pH = pK_a + \log\left(\dfrac{[F^-]}{[HF]}\right)$.

BALLPARK ESTIMATE If the concentrations of F^- and HF were equal, the log term in our equation would be zero, and the pH of the solution would be equal to the pK_a for HF, which means pH = 3.46. However, since the concentration of the conjugate base ($[F^-] = 0.120$ M) is slightly higher than the concentration of the conjugate acid ($[HF] = 0.100$ M), then the pH of the buffer solution will be slightly higher (more basic) than the pK_a.

SOLUTION

$$pH = pK_a + \log\left(\frac{[F^-]}{[HF]}\right)$$

$$pH = 3.46 + \log\left(\frac{(0.120)}{(0.100)}\right) = 3.46 + 0.08 = 3.54$$

BALLPARK CHECK The calculated pH of 3.54 is consistent with the prediction that the final pH will be slightly higher than the pK_a of 3.46.

WORKED EXAMPLE **10.14** Buffers: Measuring the Effect of Added Base on pH

What is the pH of 1.00 L of the 0.100 M hydrofluoric acid–0.120 M fluoride ion buffer system described in Worked Example 10.13 after 0.020 mol of NaOH is added?

ANALYSIS Initially, the 0.100 M HF–0.120 M NaF buffer has pH = 3.54, as calculated in Worked Example 10.13. The added base will react with the acid as indicated in the neutralization reaction,

$$HF(aq) + OH^-(aq) \longrightarrow H_2O(l) + F^-(aq)$$

which means [HF] decreases and $[F^-]$ increases. With the pK_a and the concentrations of HF and F^- known, pH can be calculated using the Henderson–Hasselbalch equation.

BALLPARK ESTIMATE After the neutralization reaction, there is more conjugate base (F^-) and less conjugate acid (HF), and so we expect the pH to increase slightly from the initial value of 3.54.

SOLUTION
When 0.020 mol of NaOH is added to 1.00 L of the buffer, the HF concentration *decreases* from 0.100 M to 0.080 M as a result of an acid–base reaction. At the same time, the F^- concentration *increases* from 0.120 M to 0.140 M because additional F^- is produced by the neutralization. Using these new values gives

$$pH = 3.46 + \log\left(\frac{(0.140)}{(0.080)}\right) = 3.46 + 0.24 = 3.70$$

The addition of 0.020 mol of base causes the pH of the buffer to rise only from 3.54 to 3.70.

BALLPARK CHECK The final pH, 3.70, is slightly more basic than the initial pH of 3.54, consistent with our prediction.

PROBLEM 10.19

What is the pH of 1.00 L of the 0.100 M hydrofluoric acid–0.120 M fluoride ion buffer system described in Worked Example 10.13 after 0.020 mol of HNO_3 is added?

PROBLEM 10.20

The ammonia/ammonium buffer system is used to optimize polymerase chain reactions (PCR) used in DNA studies. The equilibrium for this buffer can be written as

$$NH_4^+(aq) \rightleftharpoons H^+(aq) + NH_3(aq)$$

Calculate the pH of a buffer that contains 0.050 M ammonium chloride and 0.080 M ammonia. The K_a of ammonium is 5.6×10^{-10}.

◄● KEY CONCEPT PROBLEM 10.21

A buffer solution is prepared using CN^- (from NaCN salt) and HCN in the amounts indicated. The K_a for HCN is 4.9×10^{-10}. Calculate the pH of the buffer solution.

= HCN   = CN^-

10.12 Buffers in the Body

The pH of body fluids is maintained by three major buffer systems. Two of these buffers, the carbonic acid–bicarbonate (H_2CO_3–HCO_3^-) system and the dihydrogen phosphate–hydrogen phosphate system, depend on weak acid–conjugate base interactions exactly like those of the acetate buffer system described in the preceding section:

$$H_2CO_3(aq) + H_2O(l) \rightleftharpoons HCO_3^-(aq) + H_3O^+(aq) \qquad pK_a = 6.37$$
$$H_2PO_4^-(aq) + H_2O(l) \rightleftharpoons HPO_4^{2-}(aq) + H_3O^+(aq) \qquad pK_a = 7.21$$

The third buffer system depends on the ability of proteins to act as either proton acceptors or proton donors at different pH values.

To illustrate the action of buffers in the body, take a look at the carbonic acid–bicarbonate system, the principal buffer in blood serum and other extracellular fluids. (The hydrogen phosphate system is the major buffer within cells.) Because carbonic acid is unstable and therefore in equilibrium with CO_2 and water, there is an extra step in the bicarbonate buffer mechanism:

$$CO_2(aq) + H_2O(l) \rightleftharpoons H_2CO_3(aq) \rightleftharpoons HCO_3^-(aq) + H_3O^+(aq)$$

As a result, the bicarbonate buffer system is intimately related to the elimination of CO_2, which is continuously produced in cells and transported to the lungs to be exhaled.

Because most CO_2 is present simply as the dissolved gas rather than as H_2CO_3, the acid dissociation constant for carbonic acid in blood can be written using $[CO_2]$:

$$K_a = \frac{[H_3O^+][HCO_3^-]}{[CO_2]}$$

which can be rearranged to

An increase in $[CO_2]$ raises $[H_3O^+]$ and lowers pH.

A decrease in $[CO_2]$ lowers $[H_3O^+]$ and raises pH.

$$[H_3O^+] = K_a\frac{[CO_2]}{[HCO_3^-]}$$

Converting this rearranged equation to the logarithmic form of the Henderson–Hasselbalch equation yields

$$pH = pK_a + \log\left(\frac{[HCO_3^-]}{[CO_2]}\right)$$

This rearranged equation shows that an increase in $[CO_2]$ makes the ratio of $[HCO_3^-]$ to $[CO_2]$ smaller, thereby decreasing the pH; that is, the blood becomes more acidic. Similarly, a decrease in $[CO_2]$ makes the ratio of $[HCO_3^-]$ to $[CO_2]$ larger, thereby increasing the pH; that is, the blood becomes more basic. At the normal blood pH of 7.4, the ratio $[HCO_3^-]/[CO_2]$ is about 20 to 1.

The relationships between the bicarbonate buffer system, the lungs, and the kidneys are shown in Figure 10.6. Under normal circumstances, the reactions shown in the figure are at equilibrium. Addition of excess acid (red arrows) causes formation of H_2CO_3 and results in lowering of H_3O^+ concentration. Removal of acid (blue arrows) causes formation of more H_3O^+ by dissociation of H_2CO_3. The maintenance of pH by this mechanism is supported by a reserve of bicarbonate ions in body fluids. Such a buffer can accommodate large additions of H_3O^+ before there is a significant change in the pH, a condition that meets the body's needs because excessive production of acid is a more common body condition than excessive loss of acid.

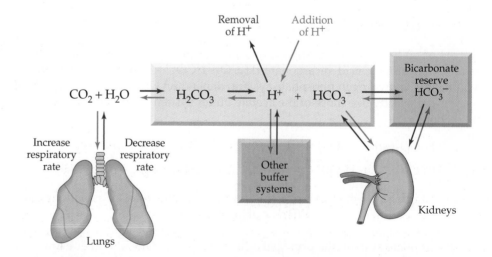

▶ **FIGURE 10.6 Relationships of the bicarbonate buffer system to the lungs and the kidneys.** The red and blue arrows show the responses to the stresses of increased or decreased respiratory rate and removal or addition of acid.

A change in the breathing rate provides a quick further adjustment in the bicarbonate buffer system. When the CO_2 concentration in the blood starts to rise, the breathing rate increases to remove CO_2, thereby decreasing the acid concentration (red arrows in Figure 10.6). When the CO_2 concentration in the blood starts to

APPLICATION ▶ Buffers in the Body: Acidosis and Alkalosis

A group of teenagers at a rock concert experience a collective fainting spell. A person taking high doses of aspirin for chronic pain appears disoriented and is having trouble breathing. An insulin-dependent diabetic patient complains of tiredness and stomach pains. An athlete who recently completed a highly strenuous workout suffers from muscle cramps and nausea. A patient on an HIV drug regimen experiences increasing weakness and numbness in the hands and feet. What do all these individuals have in common? They are all suffering from abnormal fluctuations in blood pH, resulting in conditions known as *acidosis* (pH < 7.35) or *alkalosis* (pH > 7.45).

The highly complex series of reactions and equilibria that take place throughout the body are very sensitive to pH—variations of even a few tenths of a pH unit can produce severe physiological symptoms. The carbonate–bicarbonate buffer system (Section 10.12) maintains the pH of blood serum at a fairly constant value of 7.4. The effective pH depends on the relative amounts of CO_2 and bicarbonate dissolved in the blood:

$$CO_2(aq) + H_2O(l)$$
$$\rightleftharpoons H_2CO_3(aq) \rightleftharpoons HCO_3^-(aq) + H_3O^+(aq)$$

Anything that significantly shifts the balance between dissolved CO_2 and HCO_3^- can raise or lower the pH. How does this happen, and how does the body compensate?

Respiratory acidosis can be caused by a decrease in respiration, which leads to a buildup of excess CO_2 in the blood and a corresponding decrease in pH. This could be caused by a blocked air passage due to inhaled food—removal of the blockage restores normal breathing and a return to the optimal pH. *Metabolic acidosis* results from an excess of other acids in the blood that reduce the bicarbonate concentration. High doses of aspirin (acetyl salicylic acid, Section 17.5), for example, increase the hydronium ion concentration and decrease the pH. Strenuous exercise generates excess lactate in the muscles, which is released into the bloodstream (Section 23.11). The liver converts lactate into glucose, which is the body's major source of energy; this process consumes bicarbonate ions, which decreases the pH. Some HIV drug therapies can damage cellular mitochondria (Section 21.3),

resulting in a buildup of lactic acid in the cells and blood stream. In the case of the diabetic patient, lack of insulin causes the body to start burning fat, which generates ketones and keto acids (Chapter 16), organic compounds that lower the blood pH.

The body attempts to correct acidosis by increasing the rate and depth of respiration—breathing faster "blows off" CO_2, shifting the CO_2–bicarbonate equilibrium to the left and lowering the pH. The net effect is rapid reversal of the acidosis. Although this may be sufficient for cases of respiratory acidosis, it provides only temporary relief for metabolic acidosis. A long-term solution depends on removal of excess acid by the kidneys, which can take several hours.

What about our teenage fans? In their excitement they have hyperventilated—their increased breathing rate has removed too much CO_2 from their blood and they are suffering from *respiratory alkalosis*. The body responds by "fainting" to decrease respiration and restore the CO_2 levels in the blood. When they regain consciousness, they will be ready to rock once again.

▲ Hyperventilation, the rapid breathing due to excitement or stress, removes CO_2 and increases blood pH resulting in respiratory alkalosis.

See Additional Problems 10.98 and 10.99 at the end of the chapter.

fall, the breathing rate decreases and acid concentration increases (blue arrows in Figure 10.6).

Additional backup to the bicarbonate buffer system is provided by the kidneys. Each day a quantity of acid equal to that produced in the body is excreted in the urine. In the process, the kidney returns HCO_3^- to the extracellular fluids, where it becomes part of the bicarbonate reserve.

Looking Ahead

In Chapter 29, we will see how the regulation of blood pH by the bicarbonate buffer system is particularly important in preventing *acidosis* and *alkalosis*.

10.13 Acid and Base Equivalents

We said in Section 9.10 that it is sometimes useful to think in terms of ion *equivalents* (Eq) and *gram-equivalents* (g-Eq) when we are primarily interested in an ion itself rather than the compound that produced the ion. (⬤◯◯, p. 276) For similar reasons, it can also be useful to consider acid or base equivalents and gram-equivalents.

When dealing with ions, the property of interest was the charge on the ion. Therefore, 1 Eq of an ion was defined as the number of ions that carry 1 mol of charge, and 1 g-Eq of any ion was defined as the molar mass of the ion divided by the ionic charge. For acids and bases, the property of interest is the number of H^+ ions (for an acid) or the number of OH^- ions (for a base) per formula unit. Thus, 1 **equivalent of acid** contains 1 mol of H^+ ions, and 1 g-Eq of an acid is the mass in grams that contains 1 mol of H^+ ions. Similarly, 1 **equivalent of base** contains 1 mol of OH^- ions, and 1 g-Eq of a base is the mass in grams that contains 1 mol of OH^- ions:

Equivalent of acid Amount of an acid that contains 1 mole of H^+ ions.

Equivalent of base Amount of base that contains 1 mole of OH^- ions.

$$\text{One gram-equivalent of acid} = \frac{\text{Molar mass of acid (g)}}{\text{Number of } H^+ \text{ ions per formula unit}}$$

$$\text{One gram-equivalent of base} = \frac{\text{Molar mass of base (g)}}{\text{Number of } OH^- \text{ ions per formula unit}}$$

Thus 1 g-Eq of the monoprotic acid HCl is

$$1 \text{ g-Eq HCl} = \frac{36.5 \text{ g}}{1 \text{ } H^+ \text{ per HCl}} = 36.5 \text{ g}$$

which is the molar mass of the acid, but one gram-equivalent of the diprotic acid H_2SO_4 is

$$1 \text{ g-Eq } H_2SO_4 = \frac{98.0 \text{ g}}{2 \text{ } H^+ \text{ per } H_2SO_4} = 49.0 \text{ g}$$

which is the molar mass divided by 2 because 1 mol of H_2SO_4 contains 2 mol of H^+.

$$\text{One equivalent of } H_2SO_4 = \frac{\text{Molar mass of } H_2SO_4}{2} = \frac{98.0 \text{ g}}{2} = 49.0 \text{ g}$$

Divide by 2 because H_2SO_4 is diprotic.

Using acid–base equivalents has two practical advantages: First, they are convenient when only the acidity or basicity of a solution is of interest rather than the identity of the acid or base. Second, they show quantities that are chemically equivalent in their properties; 36.5 g of HCl and 49.0 g of H_2SO_4 are chemically equivalent quantities because each reacts with 1 Eq of base. *One equivalent of any acid neutralizes one equivalent of any base.*

Because acid–base equivalents are so useful, clinical chemists sometimes express acid and base concentrations in *normality* rather than molarity. The **normality (N)** of an acid or base solution is defined as the number of equivalents (or milliequivalents) of acid or base per liter of solution. For example, a solution made by dissolving 1.0 g-Eq (49.0 g) of H_2SO_4 in water to give 1.0 L of solution has a concentration of 1.0 Eq/L, which is 1.0 N. Similarly, a solution that contains 0.010 Eq/L of acid is 0.010 N and has an acid concentration of 10 mEq/L:

$$\text{Normality (N)} = \frac{\text{Equivalents of acid or base}}{\text{Liters of solution}}$$

The values of molarity (M) and normality (N) are the same for monoprotic acids, such as HCl, but are not the same for diprotic or triprotic acids. A solution made by diluting 1.0 g-Eq (49.0 g = 0.50 mol) of the diprotic acid H_2SO_4 to a volume of 1.0 L has a *normality* of 1.0 N but a *molarity* of 0.50 M. For any acid or base, normality is always equal to molarity times the number of H^+ or OH^- ions produced per formula unit:

Normality of acid = (Molarity of acid) $\times$ (Number of H^+ ions produced per formula unit)

Normality of base = (Molarity of base) $\times$ (Number of OH^- ions produced per formula unit)

WORKED EXAMPLE   **10.15** Equivalents: Mass to Equivalent Conversion for Diprotic Acid

How many equivalents are in 3.1 g of the diprotic acid H_2S? The molar mass of H_2S is 34.0 g.

ANALYSIS The number of acid or base equivalents is calculated by doing a gram to mole conversion using molar mass as the conversion factor and then multiplying by the number of H^+ ions produced.

BALLPARK ESTIMATE The 3.1 g is a little less than 0.10 mol of H_2S. Since it is a diprotic acid, (two H^+ per mole), this represents a little less than 0.2 Eq of H_2S.

SOLUTION

$$(3.1 \ \text{g } H_2S)\left(\frac{1 \ \text{mol } H_2S}{34.0 \ \text{g } H_2S}\right)\left(\frac{2 \ \text{Eq } H_2S}{1 \ \text{mol } H_2S}\right) = 0.18 \ \text{Eq } H_2S$$

BALLPARK CHECK The calculated value of 0.18 is consistent with our prediction of a little less than 0.2 Eq of H_2S.

WORKED EXAMPLE   **10.16** Equivalents: Calculating Equivalent Concentrations

What is the normality of a solution made by diluting 6.5 g of H_2SO_4 to a volume of 200 mL? What is the concentration of this solution in milliequivalents per liter? The molar mass of H_2SO_4 is 98.0 g.

ANALYSIS Calculate how many equivalents of H_2SO_4 are in 6.5 g by using the molar mass of the acid as a conversion factor and then determine the normality of the acid.

SOLUTION

STEP 1: **Identify known information.** We know the molar mass of H_2SO_4, the mass of H_2SO_4 to be dissolved, and the final volume of solution.

MW of H_2SO_4 = 98.0 g/mol
Mass of H_2SO_4 = 6.5 g
Volume of solution = 200 mL

STEP 2: **Identify answer including units.** We need to calculate the normality of the final solution.

Normality = ?? (equiv./L)

STEP 3: **Identify conversion factors.** We will need to convert the mass of H_2SO_4 to moles, and then to equivalents of H_2SO_4. We will then need to convert volume from mL to L.

$$(6.5 \ \text{g } H_2SO_4)\left(\frac{1 \ \text{mol } H_2SO_4}{98.0 \ \text{g } H_2SO_4}\right)\left(\frac{2 \ \text{Eq } H_2SO_4}{1 \ \text{mol } H_2SO_4}\right)$$

$$= 0.132 \ \text{Eq } H_2SO_4 \text{ (don't round yet!)}$$

$$(200 \ \text{mL})\left(\frac{1 \ \text{L}}{1000 \ \text{mL}}\right) = 0.200 \ \text{L}$$

STEP 4: **Solve.** Dividing the number of equivalents by the volume yields the Normality.

$$\frac{0.132 \ \text{Eq } H_2SO_4}{0.200 \ \text{L}} = 0.66 \ \text{N}$$

The concentration of the sulfuric acid solution is 0.66 N, or 660 mEq/L.

PROBLEM 10.22

How many equivalents are in the following?

(a) 5.0 g HNO_3 (b) 12.5 g $Ca(OH)_2$ (c) 4.5 g H_3PO_4

PROBLEM 10.23

What are the normalities of the solutions if each sample in Problem 10.22 is dissolved in water and diluted to a volume of 300.0 mL?

10.14 Some Common Acid–Base Reactions

Among the most common of the many kinds of Brønsted–Lowry acid–base reactions are those of an acid with hydroxide ion, an acid with bicarbonate or carbonate ion, and an acid with ammonia or a related nitrogen-containing compound. Let us look briefly at each of the three types.

Reaction of Acids with Hydroxide Ion

One equivalent of an acid reacts with 1 Eq of a metal hydroxide to yield water and a salt in a neutralization reaction (Section 6.10): (, p. 165)

$$HCl(aq) + KOH(aq) \longrightarrow H_2O(l) + KCl(aq)$$

(An acid)   (A base)   (Water)   (A salt)

Such reactions are usually written with a single arrow because their equilibria lie far to the right and they have very large equilibrium constants ($K = 5 \times 10^{15}$; Section 7.8). The net ionic equation (Section 6.13) for all such reactions makes clear why acid-base equivalents are useful and why the properties of the acid and base disappear in neutralization reactions: The equivalent ions for the acid (H^+) and the base (OH^-) are used up in the formation of water.

$$H^+(aq) + OH^-(aq) \longrightarrow H_2O(l)$$

PROBLEM 10.24

Maalox, an over-the-counter antacid, contains aluminum hydroxide, $Al(OH)_3$, and magnesium hydroxide, $Mg(OH)_2$. Write balanced equations for the reaction of both with stomach acid (HCl).

Reaction of Acids with Bicarbonate and Carbonate Ion

Bicarbonate ion reacts with acid by accepting H^+ to yield carbonic acid, H_2CO_3. Similarly, carbonate ion accepts two protons in its reaction with acid. As mentioned on p. 315, though, that H_2CO_3 is unstable, rapidly decomposing to carbon dioxide gas and water:

$$H^+(aq) + HCO_3{}^-(aq) \longrightarrow [H_2CO_3(aq)] \longrightarrow H_2O(l) + CO_2(g)$$
$$2\,H^+(aq) + CO_3{}^{2-}(aq) \longrightarrow [H_2CO_3(aq)] \longrightarrow H_2O(l) + CO_2(g)$$

Most metal carbonates are insoluble in water—marble, for example, is almost pure calcium carbonate, $CaCO_3$—but they nevertheless react easily with aqueous acid. In fact, geologists often test for carbonate-bearing rocks by putting a few drops of aqueous HCl on the rock and watching to see if bubbles of CO_2 form (Figure 10.7). This reaction is also responsible for the damage to marble and

▲ **FIGURE 10.7 Marble.** Marble, which is primarily $CaCO_3$, releases bubbles of CO_2 when treated with hydrochloric acid.

limestone artwork caused by acid rain (See Application, p. 324). The most common application involving carbonates and acid, however, is the use of antacids that contain carbonates, such as Tums or Rolaids, to neutralize excess stomach acid.

▲ This limestone statue adorning the Rheims Cathedral in France has been severely eroded by acid rain.

PROBLEM 10.25

Write a balanced equation for each of the following reactions:

(a) $KHCO_3(aq) + H_2SO_4(aq) \longrightarrow$?

(b) $MgCO_3(aq) + HNO_3(aq) \longrightarrow$?

Reaction of Acids with Ammonia

Acids react with ammonia to yield ammonium salts, such as ammonium chloride, NH_4Cl, most of which are water-soluble:

$$NH_3(aq) + HCl(aq) \longrightarrow NH_4Cl(aq)$$

Living organisms contain a group of compounds called *amines*, which contain nitrogen atoms bonded to carbon. Amines react with acids just as ammonia does, yielding water-soluble salts. Methylamine, for example, an organic compound found in rotting fish, reacts with HCl:

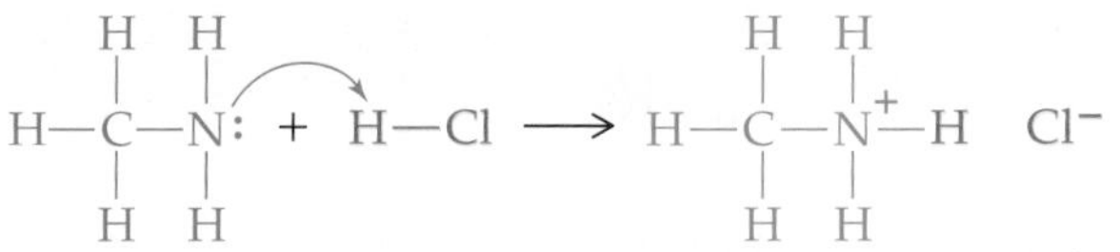

Methylamine   Methylammonium chloride

Looking Ahead

In Chapter 15, we will see that amines occur in all living organisms, both plant and animal, as well as in many pharmaceutical agents. Amines called amino acids form the building blocks from which proteins are made, as we will see in Chapter 18.

PROBLEM 10.26

What products would you expect from the reaction of ammonia and sulfuric acid in aqueous solution?

$$NH_3(aq) + H_2SO_4(aq) \longrightarrow ?$$

PROBLEM 10.27

Show how ethylamine ($C_2H_5NH_2$) reacts with hydrochloric acid to form an ethylammonium salt.

10.15 Titration

Determining the pH of a solution gives the solution's H_3O^+ concentration but not necessarily its total acid concentration. That is because the two are not the same thing. The H_3O^+ concentration gives only the amount of acid that has dissociated into ions, whereas total acid concentration gives the sum of dissociated plus undissociated acid. In a 0.10 M solution of acetic acid, for instance, the total acid concentration is 0.10 M, yet the H_3O^+ concentration is only 0.0013 M (pH = 2.89) because acetic acid is a weak acid that is only about 1% dissociated.

Titration A procedure for determining the total acid or base concentration of a solution.

The total acid or base concentration of a solution can be found by carrying out a **titration** procedure, as shown in Figure 10.8. Let us assume, for instance, that we want to find the acid concentration of an HCl solution. (We could equally well need to find the base concentration of an NaOH solution.) We begin by measuring out a known volume of the HCl solution and adding an acid–base indicator. Next, we fill a calibrated glass tube called a *buret* with an NaOH solution of known concentration, and we slowly add the NaOH to the HCl until neutralization is complete (the *end point*), identified by a color change in the indicator.

(a)   (b)

▲ **FIGURE 10.8** **Titration of an acid solution of unknown concentration with a base solution of known concentration.** (a) A measured volume of the acid solution is placed in the flask along with an indicator. (b) The base of known concentration is then added from a buret until the color change of the indicator shows that neutralization is complete (the *end point*).

Reading from the buret gives the volume of the NaOH solution that has reacted with the known volume of HCl. Knowing both the concentration and volume of the NaOH solution then allows us to calculate the molar amount of NaOH, and the coefficients in the balanced equation allow us to find the molar amount of HCl that has been neutralized. Dividing the molar amount of HCl by the volume of the HCl solution gives the concentration. The calculation thus involves mole–volume conversions just like those done in Section 9.7. (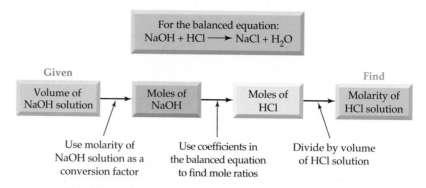, p. 265) Figure 10.9 shows a flow diagram of the strategy, and Worked Example 10.17 shows how to calculate total acid concentration.

For the balanced equation:

$$NaOH + HCl \longrightarrow NaCl + H_2O$$

Given
| Volume of NaOH solution | → | Moles of NaOH | → | Moles of HCl | → | Molarity of HCl solution |
Find

Use molarity of NaOH solution as a conversion factor

Use coefficients in the balanced equation to find mole ratios

Divide by volume of HCl solution

▲ **FIGURE 10.9** **A flow diagram for an acid–base titration.** This diagram summarizes the calculations needed to determine the concentration of an HCl solution by titration with an NaOH solution of known concentration. The steps are similar to those shown in Figure 9.7.

WORKED EXAMPLE **10.17** Titrations: Calculating Total Acid Concentration

When a 5.00 mL sample of household vinegar (dilute aqueous acetic acid) is titrated, 44.5 mL of 0.100 M NaOH solution is required to reach the end point. What is the acid concentration of the vinegar in moles per liter, equivalents per liter, and milliequivalents per liter? The neutralization reaction is

$$CH_3CO_2H(aq) + NaOH(aq) \longrightarrow CH_3CO_2{}^-Na^+(aq) + H_2O(l)$$

ANALYSIS To find the molarity of the vinegar, we need to know the number of moles of acetic acid dissolved in the 5.00 mL sample. Following a flow diagram similar to Figure 10.9, we use the volume and molarity of NaOH to find the number of moles. From the chemical equation, we use the mole ratio to find the number of moles of acid, and then divide by the volume of the acid solution. Because acetic acid is a monoprotic acid, the normality of the solution is numerically the same as its molarity.

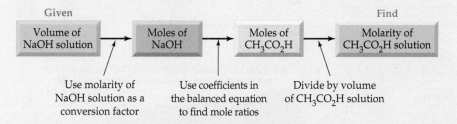

BALLPARK ESTIMATE The 5.00 mL of vinegar required nearly nine times as much NaOH solution (44.5 mL) for complete reaction. Since the neutralization stoichiometry is 1:1, the molarity of the acetic acid in the vinegar must be nine times greater than the molarity of NaOH, or 0.90 M.

SOLUTION
Substitute the known information and appropriate conversion factors into the flow diagram, and solve for the molarity of the acetic acid:

$$(44.5 \text{ mL NaOH})\left(\frac{0.100 \text{ mol NaOH}}{1000 \text{ mL}}\right)\left(\frac{1 \text{ mol CH}_3\text{CO}_2\text{H}}{1 \text{ mol NaOH}}\right)$$

$$\times \left(\frac{1}{0.005\ 00 \text{ L}}\right) = 0.890 \text{ M CH}_3\text{CO}_2\text{H}$$

$$= 0.890 \text{ N CH}_3\text{CO}_2\text{H}$$

Expressed in milliequivalents, this concentration is

$$\frac{0.890 \text{ Eq}}{\text{L}} \times \frac{1000 \text{ mEq}}{1 \text{ Eq}} = 890 \text{ mEq/L}$$

BALLPARK CHECK The calculated result (0.890 M) is very close to our estimate of 0.90 M.

PROBLEM 10.28

A titration is carried out to determine the concentration of the acid in an old bottle of aqueous HCl whose label has become unreadable. What is the HCl concentration if 58.4 mL of 0.250 M NaOH is required to titrate a 20.0 mL sample of the acid?

APPLICATION ▶ Acid Rain

As the water that evaporates from oceans and lakes condenses into raindrops, it dissolves small quantities of gases from the atmosphere. Under normal conditions, rain is slightly acidic, with a pH close to 5.6, because of atmospheric CO_2 that dissolves to form carbonic acid:

$$CO_2(aq) + H_2O(l) \rightleftharpoons$$
$$H_2CO_3(aq) \rightleftharpoons HCO_3^-(aq) + H_3O^+(aq)$$

In recent decades, however, the acidity of rainwater in many industrialized areas of the world has increased by a factor of over 100, to a pH between 3 and 3.5.

The primary cause of this so-called *acid rain* is industrial and automotive pollution. Each year, large power plants and smelters pour millions of tons of sulfur dioxide (SO_2) gas into the atmosphere, where some is oxidized by air to produce sulfur trioxide (SO_3). Sulfur oxides then dissolve in rain to form dilute sulfurous acid (H_2SO_3) and sulfuric acid (H_2SO_4):

$$SO_2(g) + H_2O(l) \longrightarrow H_2SO_3(aq)$$
$$SO_3(g) + H_2O(l) \longrightarrow H_2SO_4(aq)$$

Nitrogen oxides produced by the high-temperature reaction of N_2 with O_2 in coal-burning plants and in automobile engines further contribute to the problem. Nitrogen dioxide (NO_2) dissolves in water to form dilute nitric acid (HNO_3) and nitric oxide (NO):

$$3 NO_2(g) + H_2O(l) \longrightarrow 2 HNO_3(aq) + NO(g)$$

Oxides of both sulfur and nitrogen have always been present in the atmosphere, produced by such natural sources as volcanoes and lightning bolts, but their amounts have increased dramatically over the last century because of industrialization. The result is a notable decrease in the pH of rainwater in more densely populated regions, including Europe and the eastern United States.

Many processes in nature require such a fine pH balance that they are dramatically upset by the shift that has occurred in the pH of rain. Some watersheds contain soils that have high "buffering capacity" and so are able to neutralize acidic compounds in acid rain (Section 10.11). Other areas, such as the northeastern United States and eastern Canada, where soil buffering capacity is poor, have experienced negative ecological effects. Acid rain releases aluminum salts from soil, and the ions then wash into streams. The low pH and increased aluminum levels are so toxic to fish and other organisms that many lakes and streams in these areas are devoid of aquatic life. Massive tree die-offs have occurred throughout central and eastern Europe as acid rain has lowered the pH of the soil and has leached nutrients from leaves.

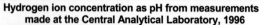

Hydrogen ion concentration as pH from measurements made at the Central Analytical Laboratory, 1996

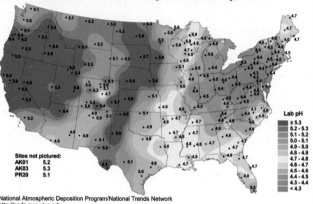

National Atmospheric Deposition Program/National Trends Network
http://nadp.sws.uiuc.edu

Hydrogen ion concentration as pH from measurements made at the Central Analytical Laboratory, 2006

National Atmospheric Deposition Program/National Trends Network
http://nadp.sws.uiuc.edu

▲ These maps compare the average pH of precipitation in the United States in 1996 and in 2006. During this period, total acid deposition in much of the eastern United States decreased substantially.

Fortunately, acidic emissions in the United States have been greatly reduced in recent years as a result of the Clean Air Act Amendments of 1990. Industrial emissions of SO_2 and nitrogen oxides decreased by over 40% from 1990 to 2007, resulting in a decrease in acid rain depositions, particularly in the eastern United States and Canada (see accompanying figure). While significant reductions have been realized, most environmental scientists agree that additional reductions in these pollutant emissions are necessary to ensure the recovery of affected lakes and streams.

See Additional Problems 10.100 and 10.101 at the end of the chapter.

PROBLEM 10.29

How many milliliters of 0.150 M NaOH are required to neutralize 50.0 mL of 0.200 M H_2SO_4? The balanced neutralization reaction is:

$$H_2SO_4(aq) + 2\,NaOH(aq) \longrightarrow Na_2SO_4(aq) + 2\,H_2O(l).$$

PROBLEM 10.30

A 21.5 mL sample of a KOH solution of unknown concentration requires 16.1 mL of 0.150 M H_2SO_4 solution to reach the end point in a titration. What is the molarity of the KOH solution?

10.16 Acidity and Basicity of Salt Solutions

It is tempting to think of all salt solutions as neutral; after all, they come from the neutralization reaction between an acid and a base. In fact, salt solutions can be neutral, acidic, or basic, depending on the ions present, because some ions react with water to produce H_3O^+ and some ions react with water to produce OH^-. To predict the acidity of a salt solution, it is convenient to classify salts according to the acid and base from which they are formed in a neutralization reaction. The classification and some examples are given in Table 10.3.

TABLE 10.3 Acidity and Basicity of Salt Solutions

ANION DERIVED FROM ACID THAT IS:	CATION DERIVED FROM BASE THAT IS:	SOLUTION	EXAMPLE
Strong	Weak	Acidic	NH_4Cl, NH_4NO_3
Weak	Strong	Basic	$NaHCO_3$, KCH_3CO_2
Strong	Strong	Neutral	$NaCl$, KBr, $Ca(NO_3)_2$
Weak	Weak	More information needed	

The general rule for predicting the acidity or basicity of a salt solution is that the stronger partner from which the salt is formed dominates. That is, a salt formed from a strong acid and a weak base yields an acidic solution because the strong acid dominates; a salt formed from a weak acid and a strong base yields a basic solution because the base dominates; and a salt formed from a strong acid and a strong base yields a neutral solution because neither acid nor base dominates. Here are some examples.

Salt of Strong Acid + Weak Base $\longrightarrow$ Acidic Solution

A salt such as NH_4Cl, which can be formed by reaction of a strong acid (HCl) with a weak base (NH_3), yields an acidic solution. The Cl^- ion does not react with water, but the NH_4^+ ion is a weak acid that gives H_3O^+ ions:

$$NH_4^+(aq) + H_2O(l) \rightleftharpoons NH_3(aq) + H_3O^+(aq)$$

Salt of Weak Acid + Strong Base $\longrightarrow$ Basic Solution

A salt such as sodium bicarbonate, which can be formed by reaction of a weak acid (H_2CO_3) with a strong base (NaOH), yields a basic solution. The Na^+ ion does not react with water, but the HCO_3^- ion is a weak base that gives OH^- ions:

$$HCO_3^-(aq) + H_2O(l) \rightleftharpoons H_2CO_3(aq) + OH^-(aq)$$

Salt of Strong Acid + Strong Base ⟶ Neutral Solution

A salt such as NaCl, which can be formed by reaction of a strong acid (HCl) with a strong base (NaOH), yields a neutral solution. Neither the Cl^- ion nor the Na^+ ion reacts with water.

Salt of Weak Acid + Weak Base

Both cation and anion in this type of salt react with water, so we cannot predict whether the resulting solution will be acidic or basic without quantitative information. The ion that reacts to the greater extent with water will govern the pH—it may be either the cation or the anion.

WORKED EXAMPLE **10.18** Acidity and Basicity of Salt Solutions

Predict whether the following salts produce an acidic, basic, or neutral solution:

    (a) $BaCl_2$ **(b)** NaCN   **(c)** NH_4NO_3

ANALYSIS Look in Table 10.1 (p. 300) to see the classification of acids and bases as strong or weak.

SOLUTION

    (a) $BaCl_2$ gives a neutral solution because it is formed from a strong acid (HCl) and a strong base $[Ba(OH)_2]$.

    (b) NaCN gives a basic solution because it is formed from a weak acid (HCN) and a strong base (NaOH).

    (c) NH_4NO_3 gives an acidic solution because it is formed from a strong acid (HNO_3) and a weak base (NH_3).

PROBLEM 10.31

Predict whether the following salts produce an acidic, basic, or neutral solution:

(a) K_2SO_4 **(b)** Na_2HPO_4 **(c)** MgF_2 **(d)** NH_4Br

KEY WORDS

Acid dissociation constant (K_a), *p. 303*

Acid–base indicator, *p. 310*

Amphoteric, *p. 299*

Brønsted–Lowry acid, *p. 295*

Brønsted–Lowry base, *p. 296*

Buffer, *p. 312*

Conjugate acid, *p. 297*

Conjugate acid–base pair, *p. 297*

Conjugate base, *p. 297*

SUMMARY: REVISITING THE CHAPTER GOALS

1. **What are acids and bases?** According to the *Brønsted–Lowry definition*, an acid is a substance that donates a hydrogen ion (a proton, H^+) and a base is a substance that accepts a hydrogen ion. Thus, the generalized reaction of an acid with a base involves the reversible transfer of a proton:

$$B: + H—A \rightleftharpoons A:^- + H—B^+$$

In aqueous solution, water acts as a base and accepts a proton from an acid to yield a *hydronium ion*, H_3O^+. Reaction of an acid with a metal hydroxide, such as KOH, yields water and a salt; reaction with bicarbonate ion (HCO_3^-) or carbonate ion (CO_3^{2-}) yields water, a salt, and CO_2 gas; and reaction with ammonia yields an ammonium salt.

2. **What effect does the strength of acids and bases have on their reactions?** Different acids and bases differ in their ability to give up or accept a proton. A *strong acid* gives up a proton easily and is 100% *dissociated* in aqueous solution; a *weak acid* gives up a proton with difficulty, is only slightly dissociated in water, and establishes an equilibrium between dissociated and undissociated forms. Similarly, a *strong base* accepts and holds a proton

readily, whereas a *weak base* has a low affinity for a proton and establishes an equilibrium in aqueous solution. The two substances that are related by the gain or loss of a proton are called a *conjugate acid–base pair*. The exact strength of an acid is defined by an *acid dissociation constant*, K_a:

For the reaction $\quad HA + H_2O \rightleftharpoons H_3O^+ + A^-$

we have $\quad\quad K_a = \dfrac{[H_3O^+][A^-]}{[HA]}$

A proton-transfer reaction always takes place in the direction that favors formation of the weaker acid.

3. **What is the ion-product constant for water?** Water is *amphoteric*; that is, it can act as either an acid or a base. Water also dissociates slightly into H_3O^+ ions and OH^- ions; the product of whose concentrations in any aqueous solution is the *ion-product constant for water*, $K_w = [H_3O^+][OH^-] = 1.00 \times 10^{-14}$ at 25 °C.

4. **What is the pH scale for measuring acidity?** The acidity or basicity of an aqueous solution is given by its *pH*, defined as the negative logarithm of the hydronium ion concentration, $[H_3O^+]$. A pH below 7 means an acidic solution; a pH equal to 7 means a neutral solution; and a pH above 7 means a basic solution.

5. **What is a buffer?** The pH of a solution can be controlled through the use of a *buffer* that acts to remove either added H^+ ions or added OH^- ions. Most buffer solutions consist of roughly equal amounts of a weak acid and its conjugate base. The bicarbonate buffer present in blood and the hydrogen phosphate buffer present in cells are particularly important examples.

6. **How is the acid or base concentration of a solution determined?** Acid (or base) concentrations are determined in the laboratory by *titration* of a solution of unknown concentration with a base (or acid) solution of known strength until an indicator signals that neutralization is complete.

UNDERSTANDING KEY CONCEPTS

10.32 Identify the Brønsted–Lowry acid and base in the following reactions:

$\bigcirc$ = H $\quad\quad$ ● = C $\quad\quad$ ● = O $\quad\quad$ ● = F

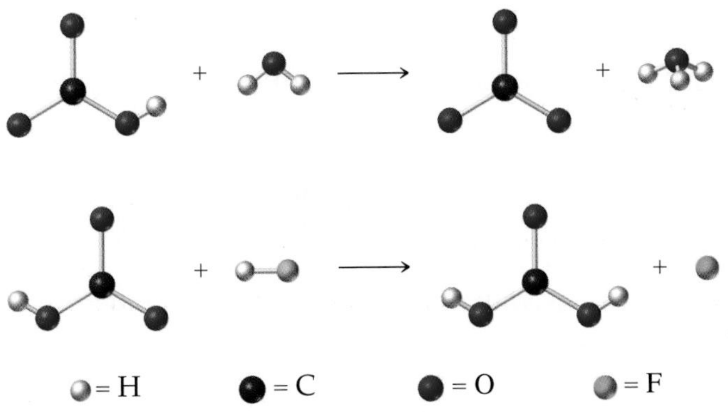

10.33 An aqueous solution of OH^-, represented as a blue sphere, is allowed to mix with a solution of an acid H_nA, represented as a red sphere. Three possible outcomes are depicted by boxes (1)–(3), where the green spheres represent A^{n-}, the anion of the acid:

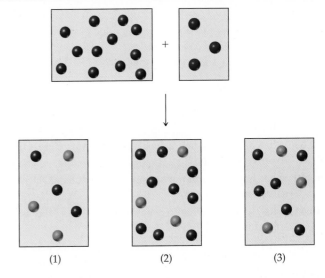

(1) $\quad\quad\quad$ (2) $\quad\quad\quad$ (3)

Which outcome corresponds to the following reactions?

(a) $HF + OH^- \longrightarrow H_2O + F^-$

(b) $H_2SO_3 + 2\,OH^- \longrightarrow 2\,H_2O + SO_3^{2-}$

(c) $H_3PO_4 + 3\,OH^- \longrightarrow 3\,H_2O + PO_4^{3-}$

10.34 Electrostatic potential maps of acetic acid (CH_3CO_2H) and ethyl alcohol (CH_3CH_2OH) are shown. Identify the most acidic hydrogen in each, and tell which of the two is likely to be the stronger acid.

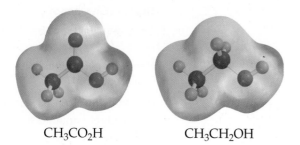

$$CH_3CO_2H \qquad CH_3CH_2OH$$

10.35 The following pictures represent aqueous acid solutions. Water molecules are not shown.

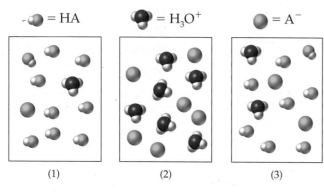

$$\bullet = HA \qquad \bullet = H_3O^+ \qquad \bullet = A^-$$

(1)   (2)   (3)

(a) Which picture represents the weakest acid?
(b) Which picture represents the strongest acid?
(c) Which picture represents the acid with the smallest value of K_a?

10.36 The following pictures represent aqueous solutions of a diprotic acid H_2A. Water molecules are not shown.

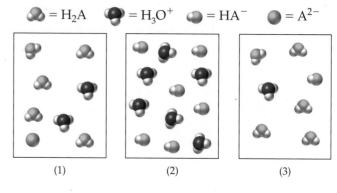

$$\bullet = H_2A \qquad \bullet = H_3O^+ \qquad \bullet = HA^- \qquad \bullet = A^{2-}$$

(1)   (2)   (3)

(a) Which picture represents a solution of a weak diprotic acid?
(b) Which picture represents an impossible situation?

10.37 Assume that the red spheres in the buret represent H^+ ions, the blue spheres in the flask represent OH^- ions, and you are carrying out a titration of the base with the acid. If the volumes in the buret and the flask are identical and the concentration of the acid in the buret is 1.00 M, what is the concentration of the base in the flask?

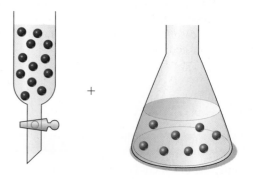

+

ADDITIONAL PROBLEMS

ACIDS AND BASES

10.38 What happens when a strong acid such as HBr is dissolved in water?

10.39 What happens when a weak acid such as CH_3CO_2H is dissolved in water?

10.40 What happens when a strong base such as KOH is dissolved in water?

10.41 What happens when a weak base such as NH_3 is dissolved in water?

10.42 What is the difference between a monoprotic acid and a diprotic acid? Give an example of each.

10.43 What is the difference between H^+ and H_3O^+?

10.44 Which of the following are strong acids? Look at Table 10.1 (p. 300) if necessary.

  (a) $HClO_4$ **(b)** H_2CO_3
  (c) H_3PO_4 **(d)** NH_4^+
  (e) HI   **(f)** $H_2PO_4^-$

10.45 Which of the following are weak bases? Look at Table 10.1 (p. 300) if necessary.

  (a) NH_3 **(b)** $Ca(OH)_2$
  (c) HPO_4^{2-} **(d)** $LiOH$
  (e) CN^- **(f)** NH_2^-

BRØNSTED–LOWRY ACIDS AND BASES

10.46 Identify the following substances as a Brønsted–Lowry base, a Brønsted–Lowry acid, or neither:

  (a) HCN   **(b)** $CH_3CO_2^-$
  (c) $AlCl_3$ **(d)** H_2CO_3
  (e) Mg^{2+} **(f)** $CH_3NH_3^+$

10.47 Label the Brønsted–Lowry acids and bases in the following equations, and tell which substances are conjugate acid–base pairs.

  (a) $CO_3^{2-}(aq) + HCl(aq) \longrightarrow HCO_3^-(aq) + Cl^-(aq)$
  (b) $H_3PO_4(aq) + NH_3(aq) \longrightarrow H_2PO_4^-(aq) + NH_4^+(aq)$

(c) $NH_4^+(aq) + CN^-(aq) \rightleftharpoons NH_3(aq) + HCN(aq)$
(d) $HBr(aq) + OH^-(aq) \longrightarrow H_2O(l) + Br^-(aq)$
(e) $H_2PO_4^-(aq) + N_2H_4(aq)$
$$\rightleftharpoons HPO_4^{2-}(aq) + N_2H_5^+(aq)$$

10.48 Write the formulas of the conjugate acids of the following Brønsted–Lowry bases:

(a) $ClCH_2CO_2^-$
(b) C_5H_5N
(c) SeO_4^{2-}
(d) $(CH_3)_3N$

10.49 Write the formulas of the conjugate bases of the following Brønsted–Lowry acids:

(a) HCN
(b) $(CH_3)_2NH_2^+$
(c) H_3PO_4
(d) $HSeO_3^-$

10.50 The hydrogen-containing anions of many polyprotic acids are amphoteric. Write equations for HCO_3^- and $H_2PO_4^-$ acting as bases with the strong acid HCl and as acids with the strong base NaOH.

10.51 Write balanced equations for proton-transfer reactions between the listed pairs. Indicate the conjugate pairs, and determine the favored direction for each equilibrium.

(a) HCl and PO_4^{3-}
(b) HCN and SO_4^{2-}
(c) $HClO_4$ and NO_2^-
(d) CH_3O^- and HF

10.52 Tums, a drugstore remedy for acid indigestion, contains $CaCO_3$. Write an equation for the reaction of Tums with gastric juice (HCl).

10.53 Write balanced equations for the following acid–base reactions:

(a) $LiOH + HNO_3 \longrightarrow$
(b) $BaCO_3 + HI \longrightarrow$
(c) $H_3PO_4 + KOH \longrightarrow$
(d) $Ca(HCO_3)_2 + HCl \longrightarrow$
(e) $Ba(OH_2) + H_2SO_4 \longrightarrow$

ACID AND BASE STRENGTH: K_a AND pH

10.54 How is K_a defined? Write the equation for K_a for the generalized acid HA.

10.55 Rearrange the equation you wrote in Problem 10.54 to solve for $[H_3O^+]$ in terms of K_a.

10.56 How is K_w defined, and what is its numerical value at 25 °C?

10.57 How is pH defined?

10.58 A solution of 0.10 M HCl has a pH = 1.00, whereas a solution of 0.10 M CH_3COOH has a pH = 2.88. Explain.

10.59 Calculate $[H_3O^+]$ for the 0.10 M CH_3COOH solution in Problem 10.58. What percent of the weak acid is dissociated?

10.60 Write the expressions for the acid dissociation constants for the three successive dissociations of phosphoric acid, H_3PO_4, in water.

10.61 Find K_a values in Table 10.2, and decide which acid in the following pairs is stronger:

(a) HCO_3H or HF
(b) HSO_4^- or HCN
(c) $H_3PO_4^-$ or HPO_4^{2-}
(d) $CH_3CH_2CO_2H$ or CH_3CO_2H

10.62 Which substance in the following pairs is the stronger base? Look at Table 10.1 if necessary.

(a) OH^- or PO_4^{3-}
(b) Br^- or NO_2^-
(c) NH_3 or OH^-
(d) CN^- or H_2O
(e) I^- or HPO_4^{2-}

10.63 Based on the K_a values in Table 10.1, rank the following solutions in order of increasing pH: 0.10 M HCOOH, 0.10 M HF, 0.10 M H_2CO_3, 0.10 M HSO_4^-, 0.10 M NH_4^+.

10.64 The electrode of a pH meter is placed in a sample of urine, and a reading of 7.9 is obtained. Is the sample acidic, basic, or neutral? What is the concentration of H_3O^+ in the urine sample?

10.65 A 0.10 M solution of the deadly poison hydrogen cyanide, HCN, has a pH of 5.2. Is HCN acidic or basic? Is it strong or weak?

10.66 Normal gastric juice has a pH of about 2. Assuming that gastric juice is primarily aqueous HCl, what is the HCl concentration?

10.67 Human spinal fluid has a pH of 7.4. Approximately what is the H_3O^+ concentration of spinal fluid?

10.68 What is the approximate pH of a 0.10 M solution of a strong monoprotic acid? Of a 0.10 M solution of a strong base, such as KOH?

10.69 Calculate the pOH of each solution in Problems 10.64–10.67.

10.70 Approximately what pH do the following H_3O^+ concentrations correspond to?

(a) Fresh egg white: $[H_3O^+] = 2.5 \times 10^{-8}$ M
(b) Apple cider: $[H_3O^+] = 5.0 \times 10^{-4}$ M
(c) Household ammonia: $[H_3O^+] = 2.3 \times 10^{-12}$ M

10.71 What are the OH^- concentration and pOH for each solution in Problem 10.70? Rank the solutions according to increasing acidity.

10.72 What are the H_3O^+ and OH^- concentrations of solutions that have the following pH values?

(a) pH 4
(b) pH 11
(c) pH 0
(d) pH 1.38
(e) pH 7.96

10.73 About 12% of the acid in a 0.10 M solution of a weak acid dissociates to form ions. What are the H_3O^+ and OH^- concentrations? What is the pH of the solution?

BUFFERS

10.74 What are the two components of a buffer system? How does a buffer work to hold pH nearly constant?

10.75 Which system would you expect to be a better buffer: $HNO_3 + Na^+NO_3^-$, or $CH_3CO_2H + CH_3CO_2^-Na^+$? Explain.

10.76 The pH of a buffer solution containing 0.10 M acetic acid and 0.10 M sodium acetate is 4.74.

   (a) Write the Henderson–Hasselbalch equation for this buffer.

   (b) Write the equations for reaction of this buffer with a small amount of HNO_3 and with a small amount of NaOH.

10.77 Which of the following buffer systems would you use if you wanted to prepare a solution having a pH of approximately 9.5?

   (a) 0.08 M $H_2PO_4^-$/0.12 M HPO_4^{2-}

   (b) 0.08 M NH_4^+/0.12 M NH_3

10.78 What is the pH of a buffer system that contains 0.200 M hydrocyanic acid (HCN) and 0.150 M sodium cyanide (NaCN)? The pK_a of hydrocyanic acid is 9.31.

10.79 What is the pH of 1.00 L of the 0.200 M hydrocyanic acid–0.150 M cyanide ion buffer system described in Problem 10.78 after 0.020 mol of HCl is added? After 0.020 mol of NaOH is added?

10.80 What is the pH of a buffer system that contains 0.15 M NH_4^+ and 0.10 M NH_3? The pK_a of NH_4^+ is 9.25.

10.81 How many moles of NaOH must be added to 1.00 L of the solution described in Problem 10.80 to increase the pH to 9.25?

CONCENTRATIONS OF ACID AND BASE SOLUTIONS

10.82 What does it mean when we talk about acid *equivalents* and base *equivalents*?

10.83 How is normality defined as a means of expressing acid or base concentration?

10.84 Calculate the gram-equivalent for each of the following acids and bases.

   (a) HNO_3 **(b)** H_3PO_4

   (c) KOH   **(d)** $Mg(OH)_2$

10.85 What mass of each of the acids and bases in Problem 10.84 is needed to prepare 500 mL of 0.15 N solution?

10.86 How many milliliters of 0.0050 N KOH are required to neutralize 25 mL of 0.0050 N H_2SO_4? To neutralize 25 mL of 0.0050 N HCl?

10.87 What is the normality of a 0.12 M H_2SO_4 solution? Of a 0.12 M H_3PO_4 solution?

10.88 How many equivalents of an acid or base are in the following?

   (a) 0.25 mol $Mg(OH)_2$

   (b) 2.5 g $Mg(OH)_2$

   (c) 15 g CH_3CO_2H

10.89 What mass of citric acid (triprotic, $C_6H_5O_7H_3$) contains 152 mEq of citric acid?

10.90 What are the molarity and the normality of a solution made by dissolving 5.0 g of $Ca(OH)_2$ in enough water to make 500.0 mL of solution?

10.91 What are the molarity and the normality of a solution made by dissolving 25 g of citric acid (triprotic, $C_6H_5O_7H_3$) in enough water to make 800 mL of solution?

10.92 Titration of a 12.0 mL solution of HCl requires 22.4 mL of 0.12 M NaOH. What is the molarity of the HCl solution?

10.93 What volume of 0.085 M HNO_3 is required to titrate 15.0 mL of 0.12 M $Ba(OH)_2$ solution?

10.94 Titration of a 10.0 mL solution of KOH requires 15.0 mL of 0.0250 M H_2SO_4 solution. What is the molarity of the KOH solution?

10.95 If 35.0 mL of a 0.100 N acid solution is needed to reach the end point in titration of 21.5 mL of a base solution, what is the normality of the base solution?

Applications

10.96 The concentration of HCl when released to the stomach cavity is diluted to between 0.01 and 0.001 M [*GERD— Too Much Acid or Not Enough? p. 302*]

   (a) What is the pH range in the stomach cavity?

   (b) Write a balanced equation for the neutralization of stomach acid by $NaHCO_3$.

   (c) How many grams of $NaHCO_3$ are required to neutralize 15.0 mL of a solution having a pH of 1.8?

10.97 Which body fluid is most acidic? Which is most basic? [*pH of Body Fluids, p. 311*]

10.98 Metabolic acidosis is often treated by administering bicarbonate intravenously. Explain how this treatment can increase blood serum pH. [*Buffers in the Body: Acidosis and Alkalosis, p. 317*]

10.99 The normal $[HCO_3^-]$/$[CO_2]$ ratio in the blood is about 20 to 1 at a pH = 7.4. What is this ratio at pH = 7.3 (acidosis) and at pH = 7.5 (alkalosis)? [*Buffers in the Body: Acidosis and Alkalosis, p. 317*]

10.100 Rain typically has a pH of about 5.6. What is the H_3O^+ concentration in rain? [*Acid Rain, p. 324*]

10.101 Acid rain with a pH as low as 1.5 has been recorded in West Virginia. [*Acid Rain, p. 324*]

   (a) What is the H_3O^+ concentration in this acid rain?

   (b) How many grams of HNO_3 must be dissolved to make 25 L of solution that has a pH of 1.5?

General Questions and Problems

10.102 Alka-Seltzer, a drugstore antacid, contains a mixture of $NaHCO_3$, aspirin, and citric acid, $C_6H_5O_7H_3$. Why does Alka-Seltzer foam and bubble when dissolved in water? Which ingredient is the antacid?

10.103 How many milliliters of 0.50 M NaOH solution are required to titrate 40.0 mL of a 0.10 M H_2SO_4 solution to an end point?

10.104 Which solution contains more acid, 50 mL of a 0.20 N HCl solution or 50 mL of a 0.20 N acetic acid solution? Which has a higher hydronium ion concentration? Which has a lower pH?

10.105 One of the buffer systems used to control the pH of blood involves the equilibrium between $H_2PO_4^-$ and HPO_4^{2-}. The pK_a for $H_2PO_4^-$ is 7.21.

    (a) Write the Henderson–Hasselbalch equation for this buffer system.

    (b) What HPO_4^{2-} to $H_2PO_4^-$ ratio is needed to maintain the optimum blood pH of 7.40?

10.106 A 0.15 M solution of HCl is used to titrate 30.0 mL of a $Ca(OH)_2$ solution of unknown concentration. If 140 mL of HCl is required, what is the concentration (in molarity) of the $Ca(OH)_2$ solution?

10.107 Which of the following combinations produces an effective buffer solution? Assuming equal concentrations of each acid and its conjugate base, calculate the pH of each buffer solution.

    (a) NaF and HF   **(b)** $HClO_4$ and $NaClO_4$

    (c) NH_4Cl and NH_3 **(d)** KBr and HBr

10.108 One method of analyzing ammonium salts is to treat them with NaOH and then heat the solution to remove the NH_3 gas formed.

$$NH_4^+(aq) + OH^-(aq) \longrightarrow NH_3(g) + H_2O(l)$$

    (a) Label the Brønsted–Lowry acid–base pairs.

    (b) If 2.86 L of NH_3 at 60 °C and 755 mmHg is produced by the reaction of NH_4Cl, how many grams of NH_4Cl were in the original sample?

10.109 One method of reducing acid rain is *scrubbing* the combustion products before they are emitted from power plant smoke stacks. The process involves addition of an aqueous suspension of lime (CaO) to the combustion chamber and stack, where the lime reacts with SO_2 to give calcium sulfite ($CaSO_3$).

    (a) Write the balanced chemical equation for this reaction.

    (b) How much lime is needed to remove 1 kg of SO_2?

10.110 Sodium oxide, Na_2O, reacts with water to give NaOH.

    (a) Write a balanced equation for the reaction.

    (b) What is the pH of the solution prepared by allowing 1.55 g of Na_2O to react with 500.0 mL of water? Assume that there is no volume change.

    (c) How many milliliters of 0.0100 M HCl are needed to neutralize the NaOH solution prepared in (b)?

CHAPTER 11

Nuclear Chemistry

CONCEPTS TO REVIEW

Atomic Theory
(Section 3.1)

Elements and Atomic Number
(Section 3.2)

Isotopes
(Section 3.3)

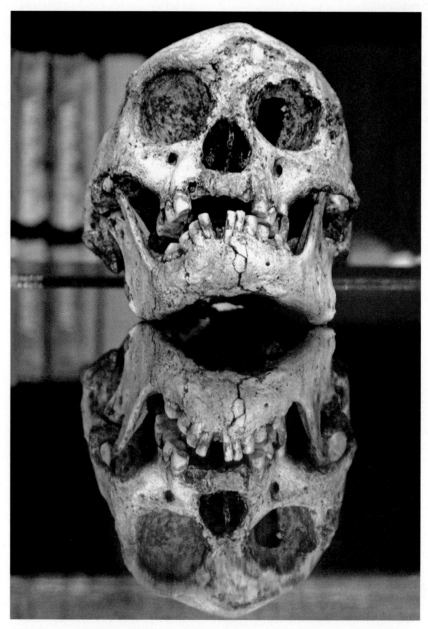

▲ The age of this skull, from a hominid discovered in 2004 in Indonesia, is estimated to be 18,000 years.

CONTENTS

CHAPTER GOALS

In this chapter we will answer the following questions about nuclear chemistry:

1. What is a nuclear reaction, and how are equations for nuclear reactions balanced?

THE GOAL: Be able to write and balance equations for nuclear reactions.

2. What are the different kinds of radioactivity?

THE GOAL: Be able to list the characteristics of three common kinds of radiation—α, β, and γ (alpha, beta, and gamma).

3. How are the rates of nuclear reactions expressed?

THE GOAL: Be able to explain half-life and calculate the quantity of a radioisotope remaining after a given number of half-lives.

4. What is ionizing radiation?

THE GOAL: Be able to describe the properties of the different types of ionizing radiation and their potential for harm to living tissue.

5. How is radioactivity measured?

THE GOAL: Be able to describe the common units for measuring radiation.

6. What is transmutation?

THE GOAL: Be able to explain nuclear bombardment and balance equations for nuclear bombardment reactions.

7. What are nuclear fission and nuclear fusion?

THE GOAL: Be able to explain nuclear fission and nuclear fusion.

I n all of the reactions we have discussed thus far, only the *bonds* between atoms have changed; the chemical identities of atoms themselves have remained unchanged. Anyone who reads the paper or watches television knows, however, that atoms *can* change, often resulting in the conversion of one element into another. Atomic weapons, nuclear energy, and radioactive radon gas in our homes are all topics of societal importance, and all involve *nuclear chemistry*—the study of the properties and reactions of atomic nuclei.

11.1 Nuclear Reactions

Recall from Section 3.2 that an atom is characterized by its *atomic number, Z*, and its *mass number, A*. (⬤⬤, p. 53) The atomic number, written below and to the left of the element symbol, gives the number of protons in the nucleus and identifies the element. The mass number, written above and to the left of the element symbol, gives the total number of **nucleons**, a general term for both protons (p) and neutrons (n). The most common isotope of carbon, for example, has 12 nucleons: 6 protons and 6 neutrons: $^{12}_{6}C$.

Nucleon A general term for both protons and neutrons.

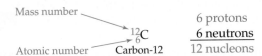

Mass number — Atomic number → $^{12}_{6}C$ Carbon-12

6 protons
6 neutrons
12 nucleons

Atoms with identical atomic numbers but different mass numbers are called *isotopes* (Section 3.3), and the nucleus of a specific isotope is called a **nuclide**. (⬤⬤, p. 54) Thirteen isotopes of carbon are known—two occur commonly (^{12}C and ^{13}C) and one (^{14}C) is produced in small amounts in the upper atmosphere by the action of neutrons from cosmic rays on ^{14}N. The remaining ten carbon isotopes have been produced artificially. Only the two commonly occurring isotopes are stable indefinitely. The others undergo spontaneous **nuclear reactions**, which change their nuclei. Carbon-14, for example, slowly decomposes to give nitrogen-14 plus an electron, a process we can write as

Nuclide The nucleus of a specific isotope of an element.

Nuclear reaction A reaction that changes an atomic nucleus, usually causing the change of one element into another.

$$^{14}_{6}C \longrightarrow {}^{14}_{7}N + {}^{0}_{-1}e$$

The electron is often written as $^{0}_{-1}e$, where the superscript 0 indicates that the mass of an electron is essentially zero when compared with that of a proton or neutron, and the subscript −1 indicates that the charge is −1. (The subscript in this instance is not a true atomic number.)

Nuclear reactions, such as the spontaneous decay of ^{14}C, are distinguished from chemical reactions in several ways:

- A *nuclear* reaction involves a change in an atom's nucleus, usually producing a different element. A *chemical* reaction, by contrast, involves only a change in distribution of the outer-shell electrons around the atom and never changes the nucleus itself or produces a different element.

- Different isotopes of an element have essentially the same behavior in chemical reactions but often have completely different behavior in nuclear reactions.

- The rate of a nuclear reaction is unaffected by a change in temperature or pressure or by the addition of a catalyst.

- The nuclear reaction of an atom is essentially the same whether it is in a chemical compound or in an uncombined, elemental form.

- The energy change accompanying a nuclear reaction can be up to several million times greater than that accompanying a chemical reaction. The nuclear transformation of 1.0 g of uranium-235 releases 3.4×10^8 kcal, for example, whereas the chemical combustion of 1.0 g of methane releases only 12 kcal.

11.2 The Discovery and Nature of Radioactivity

The discovery of *radioactivity* dates to the year 1896 when the French physicist Henri Becquerel made a remarkable observation. While investigating the nature of phosphorescence—the luminous glow of some minerals and other substances that remains when the light is suddenly turned off—Becquerel happened to place a sample of a uranium-containing mineral on top of a photographic plate that had been wrapped in black paper and put in a drawer to protect it from sunlight. On developing the plate, Becquerel was surprised to find a silhouette of the mineral. He concluded that the mineral was producing some kind of unknown radiation, which passed through the paper and exposed the photographic plate.

Marie Sklodowska Curie and her husband, Pierre, took up the challenge and began a series of investigations into this new phenomenon, which they termed **radioactivity**. They found that the source of the radioactivity was the element uranium (U) and that two previously unknown elements, which they named polonium (Po) and radium (Ra), were also radioactive. For these achievements, Becquerel and the Curies shared the 1903 Nobel Prize in physics.

Further work on radioactivity by the English scientist Ernest Rutherford established that there were at least two types of radiation, which he named *alpha* (α) and *beta* (β) after the first two letters of the Greek alphabet. Shortly thereafter, a third type of radiation was found and named for the third Greek letter, *gamma* (γ).

Subsequent studies showed that when the three kinds of radiation are passed between two plates with opposite electrical charges, each is affected differently. Alpha radiation bends toward the negative plate and must therefore have a positive charge. Beta radiation, by contrast, bends toward the positive plate and must have a negative charge, whereas gamma radiation does not bend toward either plate and has no charge (Figure 11.1).

> **Radioactivity** The spontaneous emission of radiation from a nucleus.

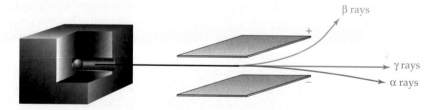

▲ **FIGURE 11.1   The effect of an electric field on α, β, and γ, radiation.**   The radioactive source in the shielded box emits radiation, which passes between the two electrically charged plates. Alpha radiation is deflected toward the negative plate, β radiation is deflected toward the positive plate, and γ radiation is not deflected.

Another difference among the three kinds of radiation soon became apparent when it was discovered that alpha and beta radiations are composed of small particles with a measurable mass, whereas **gamma (γ) radiation** consists of high-energy electromagnetic waves and has no mass (see the Application "Atoms and Light" in Chapter 3). (⬭ , p. 72) Rutherford was able to show that a **beta (β) particle** is an electron (e⁻) and that an **alpha (α) particle** is simply a helium nucleus, He^{2+}. (Recall that a helium *atom* consists of two protons, two neutrons, and two electrons. When the two electrons are removed, the remaining helium nucleus, or α particle, has only the two protons and two neutrons.)

Yet a third difference among the three kinds of radiation is their penetrating power. Because of their relatively large mass, α particles move slowly (up to about one-tenth the speed of light) and can be stopped by a few sheets of paper or by the top layer of skin. Beta particles, because they are much lighter, move at up to nine-tenths the speed of light and have about 100 times the penetrating power of α particles. A block of wood or heavy protective clothing is necessary to stop β radiation, which can otherwise penetrate the skin and cause burns and other damage. Gamma rays move at the speed of light (3.00×10^8 m/s) and have about 1000 times the penetrating power of α particles. A lead block several inches thick is needed to stop γ radiation, which can otherwise penetrate and damage the body's internal organs.

The characteristics of the three kinds of radiation are summarized in Table 11.1. Note that an α particle, even though it is an ion with a 2+ charge, is usually written using the symbol ^4_2He without the charge. A β particle is usually written $^0_{-1}e$, as noted previously.

Gamma (γ) radiation Radioactivity consisting of high-energy light waves.

Beta (β) particle An electron (e⁻), emitted as radiation.

Alpha (α) particle A helium nucleus (He^{2+}), emitted as α radiation.

TABLE 11.1 Characteristics of α, β, and γ Radiation

TYPE OF RADIATION	SYMBOL	CHARGE	COMPOSITION	MASS (AMU)	VELOCITY	RELATIVE PENETRATING POWER
Alpha	$\alpha, ^4_2He$	+2	Helium nucleus	4	Up to 10% speed of light	Low (1)
Beta	$\beta, ^0_{-1}e$	−1	Electron	1/1823	Up to 90% speed of light	Medium (100)
Gamma	$\gamma, ^0_0\gamma$	0	High-energy radiation	0	Speed of light (3.00×10^8 m/s)	High (1000)

11.3 Stable and Unstable Isotopes

Every element in the periodic table has at least one radioactive isotope, or **radioisotope**, and more than *3300* radioisotopes are known. Their radioactivity is the result of having unstable nuclei, although the exact causes of this instability are not fully understood. Radiation is emitted when an unstable radioactive nucleus, or **radionuclide**, spontaneously changes into a more stable one.

For elements in the first few rows of the periodic table, stability is associated with a roughly equal number of neutrons and protons (Figure 11.2). Hydrogen, for example, has stable ^1_1H (protium) and ^2_1H (deuterium) isotopes, but its ^3_1H isotope (tritium) is radioactive. As elements get heavier, the number of neutrons relative to protons in stable nuclei increases. Lead-208 ($^{208}_{82}Pb$), for example, the most abundant stable isotope of lead, has 126 neutrons and 82 protons in its nuclei. Nevertheless, of the 35 known isotopes of lead, only 3 are stable whereas 32 are radioactive. In fact, there are only 264 stable isotopes among all the elements. All isotopes of elements with atomic numbers higher than that of bismuth (83) are radioactive.

Most of the more than 3300 known radioisotopes have been made in high-energy particle accelerators by reactions that will be described in Section 11.10. Such isotopes are called *artificial radioisotopes* because they are not found in nature. All isotopes of the transuranium elements (those heavier than uranium) are artificial. The much smaller number of radioactive isotopes found in the earth's crust, such as $^{238}_{92}U$, are called *natural radioisotopes*.

Aside from their radioactivity, different radioisotopes of the same element have the same chemical properties as stable isotopes, which accounts for their great

Radioisotope A radioactive isotope.

Radionuclide The nucleus of a radioactive isotope.

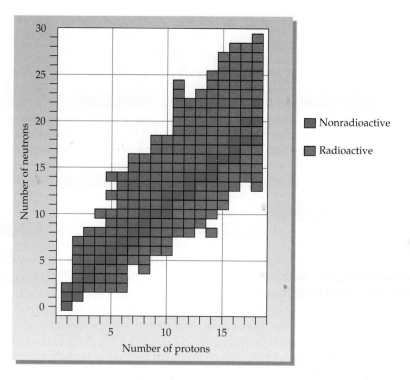

▲ **FIGURE 11.2** **A plot of the numbers of neutrons and protons for known isotopes of the first 18 elements.** Stable (nonradioactive) isotopes of these elements have equal or nearly equal numbers of neutrons and protons.

usefulness as *tracers*. A chemical compound tagged with a radioactive atom undergoes exactly the same reactions as its nonradioactive counterpart. The difference is that the tagged compound can be located with a radiation detector and its whereabouts determined, as discussed in the Application "Body Imaging" on page 352.

11.4 Nuclear Decay

Think for a minute about the consequences of α and β radiation. If radioactivity involves the spontaneous emission of a small particle from an unstable atomic nucleus, then the nucleus itself must undergo a change. With that understanding of radioactivity came the startling discovery that atoms of one element can change into atoms of another element, something that had previously been thought impossible. The spontaneous emission of a particle from an unstable nucleus is called **nuclear decay**, or *radioactive decay*, and the resulting change of one element into another is called **transmutation**.

Nuclear decay The spontaneous emission of a particle from an unstable nucleus.

Transmutation The change of one element into another.

> **Nuclear decay:** Radioactive element $\longrightarrow$ New element + Emitted particle

We now look at what happens to a nucleus when nuclear decay occurs.

Alpha Emission

When an atom of uranium-238 ($^{238}_{92}U$) emits an α particle, the nucleus loses two protons and two neutrons. Because the number of protons in the nucleus has now changed from 92 to 90, the *identity* of the atom has changed from uranium to thorium. Furthermore, since the total number of nucleons has decreased by 4, uranium-238 has become thorium-234 ($^{234}_{90}Th$) (Figure 11.3).

Note that the equation for a nuclear reaction is not balanced in the usual chemical sense because the kinds of atoms are not the same on both sides of the arrow. Instead, we say that a nuclear equation is balanced when the number of nucleons is the same on both sides of the equation and when the sums of the charges on the nuclei plus any ejected subatomic particles (protons or electrons) are the same on

$$^{238}_{92}\text{U} \longrightarrow \ ^{234}_{90}\text{Th} + \ ^{4}_{2}\text{He}$$

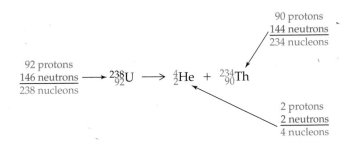

$^{238}_{92}\text{U}$	$^{234}_{90}\text{Th}$	$^{4}_{2}\text{He}$
92 protons	90 protons	2 protons
146 neutrons	144 neutrons	2 neutrons
238 total	234 total	4 total

◀ **FIGURE 11.3 Alpha emission.** Emission of an α particle from an atom of uranium-238 produces an atom of thorium-234.

both sides. In the decay of $^{238}_{92}\text{U}$ to give $^{4}_{2}\text{He}$ and $^{234}_{90}\text{Th}$, for example, there are 238 nucleons and 92 nuclear charges on both sides of the nuclear equation.

$$\underset{\underline{238 \text{ nucleons}}}{\overset{92 \text{ protons}}{146 \text{ neutrons}}} \longrightarrow \ ^{238}_{92}\text{U} \longrightarrow \ ^{4}_{2}\text{He} + \ ^{234}_{90}\text{Th}$$

90 protons
144 neutrons
234 nucleons

2 protons
2 neutrons
4 nucleons

WORKED EXAMPLE   11.1 Balancing Nuclear Reactions: Alpha Emission

Polonium-208 is one of the α emitters studied by Marie Curie. Write the equation for the α decay of polonium-208, and identify the element formed.

ANALYSIS Look up the atomic number of polonium (84) in the periodic table, and write the known part of the nuclear equation, using the standard symbol for polonium-208:

$$^{208}_{84}\text{Po} \longrightarrow \ ^{4}_{2}\text{He} + \ ?$$

Then calculate the mass number and atomic number of the product element, and write the final equation.

SOLUTION
The mass number of the product is $208 - 4 = 204$, and the atomic number is $84 - 2 = 82$. A look at the periodic table identifies the element with atomic number 82 as lead (Pb).

$$^{208}_{84}\text{Po} \longrightarrow \ ^{4}_{2}\text{He} + \ ^{204}_{82}\text{Pb}$$

Check your answer by making sure that the mass numbers and atomic numbers on the two sides of the equation are balanced:

Mass numbers: $208 = 4 + 204$ Atomic numbers: $84 = 2 + 82$

PROBLEM 11.1

High levels of radioactive radon-222 ($^{222}_{86}\text{Rn}$) have been found in many homes built on radium-containing rock, leading to the possibility of health hazards. What product results from α emission by radon-222?

PROBLEM 11.2

What isotope of radium (Ra) is converted into radon-222 by α emission?

Beta Emission

Whereas α emission leads to the loss of two protons and two neutrons from the nucleus, β emission involves the *decomposition* of a neutron to yield an electron and a proton. This process can be represented as

$$\,^{1}_{0}n \longrightarrow \,^{1}_{1}p + \,^{0}_{-1}e$$

where the electron ($\,^{0}_{-1}e$) is ejected as a β particle, and the proton is retained by the nucleus. Note that the electrons emitted during β radiation come from the *nucleus* and not from the occupied orbitals surrounding the nucleus.

The net result of β emission is that the atomic number of the atom increases by 1 because there is a new proton. The mass number of the atom remains the same, however, because a neutron has changed into a proton leaving the total number of nucleons unchanged. For example, iodine-131 ($\,^{131}_{53}I$), a radioisotope used in detecting thyroid problems, undergoes nuclear decay by β emission to yield xenon-131 ($\,^{131}_{54}Xe$):

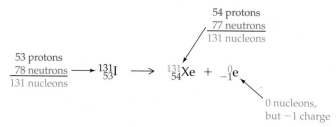

Note that the superscripts (mass numbers) are balanced in this equation because a β particle has a mass near zero, and the subscripts are balanced because a β particle has a charge of -1.

WORKED EXAMPLE **11.2** Balancing Nuclear Reactions: Beta Emission

Write a balanced nuclear equation for the β decay of chromium-55.

ANALYSIS Write the known part of the nuclear equation:

$$\,^{55}_{24}Cr \longrightarrow \,^{0}_{-1}e + ?$$

Then calculate the mass number and atomic number of the product element, and write the final equation.

SOLUTION

The mass number of the product stays at 55, and the atomic number increases by 1, $24 + 1 = 25$, so the product is manganese-55.

$$\,^{55}_{24}Cr \longrightarrow \,^{0}_{-1}e + \,^{55}_{25}Mn$$

Check your answer by making sure that the mass numbers and atomic numbers on the two sides of the equation are balanced:

Mass numbers: $55 = 0 + 55$ Atomic numbers: $24 = -1 + 25$

PROBLEM 11.3

Carbon-14, a β emitter, is a rare isotope used in dating archaeological artifacts. Write a nuclear equation for the decay of carbon-14.

PROBLEM 11.4

Write nuclear equations for β emission from the following radioisotopes:

(a) ${}^{3}_{1}\mathrm{H}$ **(b)** ${}^{210}_{82}\mathrm{Pb}$ **(c)** ${}^{20}_{9}\mathrm{F}$

Gamma Emission

Emission of γ rays, unlike the emission of α and β particles, causes no change in mass or atomic number because γ rays are simply high-energy electromagnetic waves. Although γ emission can occur alone, it usually accompanies α or β emission as a mechanism for the new nucleus that results from a transmutation to get rid of some extra energy.

Since γ emission affects neither mass number nor atomic number, it is often omitted from nuclear equations. Nevertheless, γ rays are of great importance. Their penetrating power makes them by far the most dangerous kind of external radiation for humans and also makes them useful in numerous medical applications. Cobalt-60, for example, is used in cancer therapy as a source of penetrating γ rays that kill cancerous tissue.

$${}^{60}_{27}\mathrm{Co} \longrightarrow {}^{60}_{28}\mathrm{Ni} + {}^{0}_{-1}\mathrm{e} + {}^{0}_{0}\gamma$$

Positron Emission

In addition to α, β, and γ radiation, there is another common type of radioactive decay process called *positron emission*, which involves the conversion of a proton in the nucleus into a neutron plus an ejected **positron**, ${}^{0}_{1}\mathrm{e}$ or β^{+}. A positron, which can be thought of as a "positive electron," has the same mass as an electron but a positive charge. This process can be represented as

Positron A "positive electron," which has the same mass as an electron but a positive charge.

$${}^{1}_{1}\mathrm{p} \longrightarrow {}^{1}_{0}\mathrm{n} + {}^{0}_{1}\mathrm{e}$$

The result of positron emission is a decrease in the atomic number of the product nucleus because a proton has changed into a neutron, but no change in the mass number. Potassium-40, for example, undergoes positron emission to yield argon-40, a nuclear reaction important in geology for dating rocks. Note once again that the sum of the two subscripts on the right of the nuclear equation ($18 + 1 = 19$) is equal to the subscript in the ${}^{40}_{19}\mathrm{K}$ nucleus on the left.

19 protons / 21 neutrons / 40 nucleons →

$${}^{40}_{19}\mathrm{K} \longrightarrow {}^{40}_{18}\mathrm{Ar} + {}^{0}_{1}\mathrm{e}$$

${}^{0}_{1}\mathrm{e}$: 0 nucleons but 1 charge

${}^{40}_{18}\mathrm{Ar}$: 18 protons / 22 neutrons / 40 nucleons

Electron Capture

Electron capture, symbolized E.C., is a process in which the nucleus captures an inner-shell electron from the surrounding electron cloud, thereby converting a proton into a neutron. The mass number of the product nucleus is unchanged, but the atomic number decreases by 1, just as in positron emission. The conversion of mercury-197 into gold-197 is an example:

Electron capture (E.C.) A process in which the nucleus captures an inner-shell electron from the surrounding electron cloud, thereby converting a proton into a neutron.

${}^{197}_{80}\mathrm{Hg}$: 80 protons / 117 neutrons / 197 nucleons

${}^{0}_{-1}\mathrm{e}$: Inner-shell electron

${}^{197}_{79}\mathrm{Au}$: 79 protons / 118 neutrons / 197 nucleons

$${}^{197}_{80}\mathrm{Hg} + {}^{0}_{-1}\mathrm{e} \longrightarrow {}^{197}_{79}\mathrm{Au}$$

Do not plan on using this reaction to get rich, however. Mercury-197 is not one of the naturally occurring isotopes of Hg and is typically produced by transmutation reactions as discussed in Section 11.10.

Characteristics of the five kinds of radioactive decay processes are summarized in Table 11.2.

TABLE 11.2   A Summary of Radioactive Decay Processes

PROCESS	SYMBOL	CHANGE IN ATOMIC NUMBER	CHANGE IN MASS NUMBER	CHANGE IN NUMBER OF NEUTRONS
α emission	^4_2He or α	-2	-4	-2
β emission	$^0_{-1}e$ or β^{-}*	$+1$	0	-1
γ emission	$^0_0\gamma$ or γ	0	0	0
Positron emission	^0_1e or β^{+}*	-1	0	$+1$
Electron capture	E.C.	-1	0	$+1$

*Superscripts are used to indicate the charge associated with the two forms of beta decay; β^-, or a beta particle, carries a -1 charge, while β^+, or a positron, carries a $+1$ charge.

WORKED EXAMPLE   **11.3** Balancing Nuclear Reactions: Electron Capture, Positron Emission

Write balanced nuclear equations for the following processes:

(a) Electron capture by polonium-204: $^{204}_{84}\text{Po} + ^0_{-1}e \longrightarrow$?

(b) Positron emission from xenon-118: $^{118}_{54}\text{Xe} \longrightarrow ^0_1e +$?

ANALYSIS The key to writing nuclear equations is to make sure that the number of nucleons is the same on both sides of the equation and that the number of charges is the same.

SOLUTION

(a) In electron capture, the mass number is unchanged and the atomic number decreases by 1, giving bismuth-204: $^{204}_{84}\text{Po} + ^0_{-1}e \longrightarrow ^{204}_{83}\text{Bi}$

Check your answer by making sure that the number of nucleons and the number of charges are the same on both sides of the equation:

Mass number: $204 + 0 = 204$ Atomic number: $84 + (-1) = 83$

(b) In positron emission, the mass number is unchanged and the atomic number decreases by 1, giving iodine-118: $^{118}_{54}\text{Xe} \longrightarrow ^0_1e + ^{118}_{53}\text{I}$.

CHECK! Mass number: $118 = 0 + 118$ Atomic number: $54 = 1 + 53$

PROBLEM 11.5

Write nuclear equations for positron emission from the following radioisotopes:

(a) $^{38}_{20}\text{Ca}$ **(b)** $^{118}_{54}\text{Xe}$ **(c)** $^{79}_{37}\text{Rb}$

PROBLEM 11.6

Write nuclear equations for electron capture by the following radioisotopes:

(a) $^{62}_{30}\text{Zn}$ **(b)** $^{110}_{50}\text{Sn}$ **(c)** $^{81}_{36}\text{Kr}$

KEY CONCEPT PROBLEM 11.7

The red arrow in this graph indicates the changes that occur in the nucleus of an atom during a nuclear reaction. Identify the isotopes involved as product and reactant, and name the type of decay process.

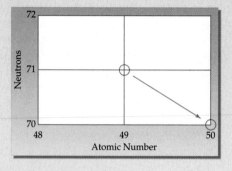

11.5 Radioactive Half-Life

The rate of radioactive decay varies greatly from one radioisotope to another. Some radioisotopes, such as uranium-238, decay at a barely perceptible rate over billions of years, but others, such as carbon-17, decay within thousandths of a second.

Rates of nuclear decay are measured in units of **half-life ($t_{1/2}$)**, defined as the amount of time required for one-half of a radioactive sample to decay. For example, the half-life of iodine-131 is 8.021 days. If today you have 1.000 g of $^{131}_{53}I$, then 8.021 days from now you will have only 50% of that amount (0.500 g) because one-half of the sample will have decayed into $^{131}_{54}Xe$. After 8.021 more days (16.063 days total), you will have only 25% (0.250 g) of your original $^{131}_{53}I$ sample; after another 8.021 days (24.084 days total), you will have only 12.5% (0.125 g); and so on. Each passage of a half-life causes the decay of one-half of whatever sample remains. The half-life is the same no matter what the size of the sample, the temperature, or any other external conditions. There is no known way to slow down, speed up, or otherwise change the characteristics of radioactive decay.

Half-life ($t_{1/2}$) The amount of time required for one-half of a radioactive sample to decay.

$$1.000 \text{ g } ^{131}_{53}I \xrightarrow[\text{days}]{8} \begin{array}{c} 0.500 \text{ g } ^{131}_{53}I \\ 0.500 \text{ g } ^{131}_{54}Xe \end{array} \xrightarrow[\text{days}]{8} \begin{array}{c} 0.250 \text{ g } ^{131}_{53}I \\ 0.750 \text{ g } ^{131}_{54}Xe \end{array} \xrightarrow[\text{days}]{8} \begin{array}{c} 0.125 \text{ g } ^{131}_{53}I \\ 0.875 \text{ g } ^{131}_{54}Xe \end{array} \longrightarrow$$

One half-life   Two half-lives   Three half-lives
                       (16 days total)   (24 days total)

The fraction of radioisotope remaining after the passage of each half-life is represented by the curve in Figure 11.4 and can be calculated as

$$\text{fraction remaining} = (0.5)^n$$

where n is the number of half-lives that have elapsed.

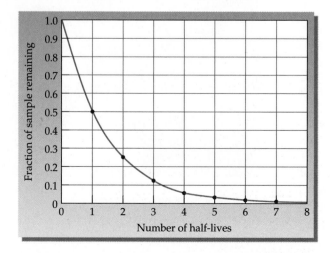

◀ **FIGURE 11.4 The decay of a radioactive nucleus over time.** All nuclear decays follow this curve, whether the half-lives are measured in years, days, minutes, or seconds. That is, the fraction of sample remaining after one half-life is 0.50, the fraction remaining after two half-lives is 0.25, the fraction remaining after three half-lives is 0.125, and so on.

APPLICATION ▶ Medical Uses of Radioactivity

The origins of nuclear medicine date from 1901 when the French physician Henri Danlos first used radium in the treatment of a tubercular skin lesion. Since that time, the use of radioactivity has become a crucial part of modern medical care, both diagnostic and therapeutic. Current nuclear techniques can be grouped into three classes: (1) in vivo procedures, (2) radiation therapy, and (3) imaging procedures. The first two are described here, and the third one is described on page 352 in the Application "Body Imaging."

In Vivo Procedures

In vivo studies—those that take place inside the body—are carried out to assess the functioning of a particular organ or body system. A *radiopharmaceutical* agent is administered, and its path in the body—whether absorbed, excreted, diluted, or concentrated—is determined by analysis of blood or urine samples.

Among the many in vivo procedures utilizing radioactive agents is a simple method for the determination of whole-blood volume by injecting a known quantity of red blood cells labeled with radioactive chromium-51. After a suitable interval to allow the labeled cells to be distributed evenly throughout the body, a blood sample is taken and blood volume is calculated by comparing the concentration of labeled cells in the blood with the quantity of labeled cells injected. This and similar procedures are known as *isotope dilution* and are described by

$$R_{sample} = R_{tracer}\left(\frac{W_{sample}}{W_{system} + W_{tracer}}\right)$$

where R_{sample} is the counting rate of the analyzed sample, R_{tracer} is the counting rate of the tracer added to the system, and W refers to either the mass or volume of the analyzed sample, added tracer, or total system as indicated.

Therapeutic Procedures

Therapeutic procedures—those in which radiation is purposely used as a weapon to kill diseased tissue—involve either external or internal sources of radiation. External radiation therapy for the treatment of cancer is often carried out with γ rays emanating from a cobalt-60 source. The highly radioactive source is shielded by a thick lead container and has a small opening directed toward the site of the tumor. By focusing the radiation beam on the tumor, the tumor receives the full exposure while exposure of surrounding parts of the body is minimized. Nevertheless, enough healthy tissue is affected so that most patients treated in this manner suffer the effects of radiation sickness.

Internal radiation therapy is a much more selective technique than external therapy. In the treatment of thyroid disease, for example, a radioactive substance such as iodine-131 is administered. This powerful β emitter is incorporated into the iodine-containing hormone thyroxine, which concentrates in the thyroid gland. Because β particles penetrate no farther than several millimeters, the localized ^{131}I produces

a high radiation dose that destroys only the surrounding diseased tissue. To treat some tumors, such as those in the female reproductive system, a radioactive source is placed physically close to the tumor for a specific amount of time.

Boron neutron-capture therapy (BNCT) is a relatively new technique in which boron-containing drugs are administered to a patient and concentrate in the tumor site. The tumor is then irradiated with a neutron beam from a nuclear reactor. The boron absorbs a neutron and undergoes transmutation to produce an alpha particle and a lithium nucleus. These highly energetic particles have very low penetrating power and can kill nearby tumor tissue while sparing the healthy surrounding tissue. Because one disadvantage of BNCT is the need for access to a nuclear reactor, this treatment is available only in limited locations.

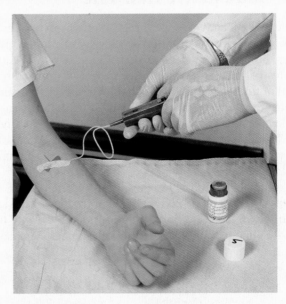

▲ A person's blood volume can be found by injecting a small amount of radioactive chromium-51 and measuring the dilution factor.

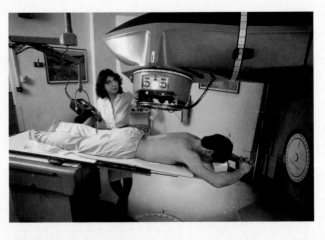

▲ T/B/W when new photo chosen.

See Additional Problems 11.70 and 11.71 at the end of the chapter.

The half-lives of some useful radioisotopes are given in Table 11.3. As you might expect, radioisotopes that are used internally for medical applications have fairly short half-lives so that they decay rapidly and do not remain in the body for prolonged periods.

TABLE 11.3   Half-Lives of Some Useful Radioisotopes

RADIOISOTOPE	SYMBOL	RADIATION	HALF-LIFE	USE
Tritium	$^{3}_{1}H$	β	12.33 years	Biochemical tracer
Carbon-14	$^{14}_{6}C$	β	5730 years	Archaeological dating
Sodium-24	$^{24}_{11}Na$	β	14.959 hours	Examining circulation
Phosphorus-32	$^{32}_{15}P$	β	14.262 days	Leukemia therapy
Potassium-40	$^{40}_{19}K$	β, β^{+}	1.277×10^{9} years	Geological dating
Cobalt-60	$^{60}_{27}Co$	β, γ	5.271 years	Cancer therapy
Arsenic-74	$^{74}_{33}As$	β^{+}	17.77 days	Locating brain tumors
Technetium-99m*	$^{99m}_{43}Tc$	γ	6.01 hours	Brain scans
Iodine-131	$^{131}_{53}I$	β	8.021 days	Thyroid therapy
Uranium-235	$^{235}_{92}U$	α, γ	7.038×10^{8} years	Nuclear reactors

*The m in technetium-99m stands for metastable, *meaning that the nucleus undergoes γ emission but does not change its mass number or atomic number.*

WORKED EXAMPLE   **11.4** Nuclear Reactions: Half-Life

Phosphorus-32, a radioisotope used in leukemia therapy, has a half-life of about 14 days. Approximately what percent of a sample remains after 8 weeks?

ANALYSIS Determine how many half-lives have elapsed. For an integral number of half-lives, we can multiply the starting amount (100%) by 1/2 for each half-life that has elapsed.

SOLUTION
Since one half-life of $^{32}_{15}P$ is 14 days (2 weeks), 8 weeks represents four half-lives. The fraction that remains after 8 weeks is thus

Four half-lives

Final Percentage $= 100\% \times (0.5)^{4} = 100\% \times \left(\frac{1}{2} \times \frac{1}{2} \times \frac{1}{2} \times \frac{1}{2}\right)$

$\qquad\qquad\qquad = 100\% \times \frac{1}{16} = 6.25\%$

WORKED EXAMPLE   **11.5** Nuclear Reactions: Half-Life

As noted on page 341 and in Table 11.3, iodine-131 has a half-life of about 8 days. Approximately what fraction of a sample remains after 20 days?

ANALYSIS Determine how many half-lives have elapsed. For a non-integral number (i.e., fraction) of half-lives, use the equation below to determine the fraction of radioisotope remaining.

$$\text{fraction remaining} = (0.5)^{n}$$

BALLPARK ESTIMATE Since the half-life of iodine-131 is 8 days, an elapsed time of 20 days is 2.5 half-lives. The fraction remaining should be between 0.25 (fraction remaining after two half-lives) and 0.125 (fraction remaining after three half-lives). Since the relationship between the number of half-lives and fraction remaining is not linear (see Figure 11.4), the fraction remaining will not be exactly halfway between these values but instead will be slightly closer to the lower fraction, say 0.17.

SOLUTION

$$\text{fraction remaining } = (0.5)^n = (0.5)^{2.5} = 0.177$$

BALLPARK CHECK: The fraction remaining is close to our estimate of 0.17.

PROBLEM 11.8

The half-life of carbon-14, an isotope used in archaeological dating, is 5730 years. What percentage of $^{14}_{6}C$ remains in a sample estimated to be 17,000 years old?

KEY CONCEPT PROBLEM 11.9

What is the half-life of the radionuclide that shows the following decay curve?

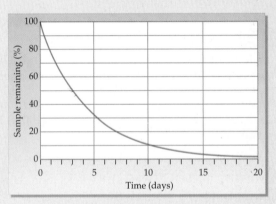

Decay series A sequential series of nuclear disintegrations leading from a heavy radioisotope to a nonradioactive product.

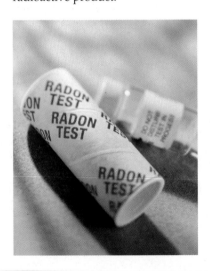

▲ Radon gas can be detected and measured by this home detection kit.

11.6 Radioactive Decay Series

When a radioactive isotope decays, nuclear change occurs and a different element is formed. Often, this newly formed nucleus is stable, but sometimes the product nucleus is itself radioactive and undergoes further decay. In fact, some radioactive nuclei undergo a whole **decay series** of nuclear disintegrations before they ultimately reach a nonradioactive product. This is particularly true for the isotopes of heavier elements. Uranium-238, for example, undergoes a series of 14 sequential nuclear reactions, ultimately stopping at lead-206 (Figure 11.5).

One of the intermediate radionuclides in the uranium-238 decay series is radium-226. Radium-226 has a half-life of 1600 years and undergoes α decay to produce radon-222, a gas. Rocks, soil, and building materials that contain radium are sources of radon-222, which can seep through cracks in basements and get into the air inside homes and other buildings. Radon itself is a gas that passes in and out of the lungs without being incorporated into body tissue. If, however, a radon-222 atom should happen to decay while in the lungs, the solid decay product polonium-218 results. Further decay of the ^{218}Po emits α particles, which can damage lung tissue.

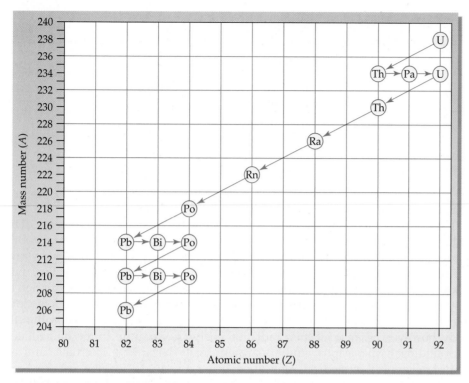

▲ **FIGURE 11.5** **The decay series from $^{238}_{92}U$ to $^{206}_{82}Pb$.** Each isotope except for the last is radioactive and undergoes nuclear decay. The long slanted arrows represent α emissions, and the short horizontal arrows represent β emissions.

11.7 Ionizing Radiation

High-energy radiation of all kinds is often grouped together under the name **ionizing radiation**. This includes not only α particles, β particles, and γ rays, but also *X rays* and *cosmic rays*. **X rays** are like γ rays; they have no mass and consist of high-energy electromagnetic radiation. The only difference between them is that the energy of X rays is somewhat less than that of γ rays (see the Application "Atoms and Light" in Chapter 3). **Cosmic rays** are not rays at all but are a mixture of high-energy particles that shower the earth from outer space. They consist primarily of protons, along with some α and β particles.

The interaction of any kind of ionizing radiation with a molecule knocks out an electron, converting the atom or molecule into an extremely reactive ion:

$$\text{Molecule} \xrightarrow[\text{radiation}]{\text{ionizing}} \text{Ion} + e^-$$

The reactive ion can react with other molecules nearby, creating still other fragments that can in turn cause further reactions. In this manner, a large dose of ionizing radiation can destroy the delicate balance of chemical reactions in living cells, ultimately causing the death of an organism.

A small dose of ionizing radiation may not cause visible symptoms but can nevertheless be dangerous if it strikes a cell nucleus and damages the genetic machinery inside. The resultant changes might lead to a genetic mutation, to cancer, or to cell death. The nuclei of rapidly dividing cells, such as those in bone marrow, the lymph system, the lining of the intestinal tract, or an embryo, are the most readily damaged. Because cancer cells are also rapidly dividing they are highly susceptible to the effects of ionizing radiation, which is why radiation therapy is an effective treatment for many types of cancer (see Application on p. 342). Some properties of ionizing radiation are summarized in Table 11.4.

Ionizing radiation A general name for high-energy radiation of all kinds.

X rays Electromagnetic radiation with an energy somewhat less than that of γ rays.

Cosmic rays A mixture of high-energy particles—primarily of protons and various atomic nuclei—that shower the earth from outer space.

TABLE 11.4 Some Properties of Ionizing Radiation

TYPE OF RADIATION	ENERGY RANGE*	PENETRATING DISTANCE IN WATER**
α	3–9 MeV	0.02–0.04 mm
β	0–3 MeV	0–4 mm
X	100 eV–10 keV	0.01–1 cm
γ	10 keV–10 MeV	1–20 cm

*The energies of subatomic particles are often measured in electron volts (eV): 1 eV = 6.703 × 10⁻¹⁹ cal.
**Distance at which one-half of the radiation is stopped.

The effects of ionizing radiation on the human body vary with the energy of the radiation, its distance from the body, the length of exposure, and the location of the source outside or inside the body. When coming from outside the body, γ rays and X rays are potentially more harmful than α and β particles because they pass through clothing and skin and into the body's cells. Alpha particles are stopped by clothing and skin, and β particles are stopped by wood or several layers of clothing. These types of radiation are much more dangerous when emitted within the body, however, because all their radiation energy is given up to the immediately surrounding tissue. Alpha emitters are especially hazardous internally and are almost never used in medical applications.

Health professionals who work with X rays or other kinds of ionizing radiation protect themselves by surrounding the source with a thick layer of lead or other dense material. Protection from radiation is also afforded by controlling the distance between the worker and the radiation source because radiation intensity (I) decreases with the square of the distance from the source. The intensities of radiation at two different distances, 1 and 2, are given by the equation

$$\frac{I_1}{I_2} = \frac{d_2{}^2}{d_1{}^2}$$

For example, suppose a source delivers 16 units of radiation at a distance of 1.0 m. Doubling the distance to 2.0 m decreases the radiation intensity to one-fourth:

$$\frac{16 \text{ units}}{I_2} = \frac{(2 \text{ m})^2}{(1 \text{ m})^2}$$

$$I_2 = 16 \text{ units} \times \frac{1 \text{ m}^2}{4 \text{ m}^2} = 4 \text{ units}$$

WORKED EXAMPLE   **11.6** Ionizing Radiation: Intensity versus Distance from the Source

If a radiation source gives 75 units of radiation at a distance of 2.4 m, at what distance does the source give 25 units of radiation?

ANALYSIS Radiation intensity (I) decreases with the square of the distance (d) from the source according to the equation

$$\frac{I_1}{I_2} = \frac{d_2{}^2}{d_1{}^2}$$

We know three of the four variables in this equation (I_1, I_2, and d_1), and we need to find d_2.

BALLPARK ESTIMATE In order to decrease the radiation intensity from 75 units to 25 units (a factor of 3), the distance must *increase* by a factor of $\sqrt{3} = 1.7$. Thus, the distance should increase from 2.4 m to about 4 m.

SOLUTION

STEP 1: Identify known information. We know three of the four variables.

$I_1 = 75$ units

$I_2 = 25$ units

$d_1 = 2.4$ m

STEP 2: Identify answer and units.

$d_2 = $??? m

STEP 3: Identify equation. Rearrange the equation relating intensity and distance to solve for d_2.

$$\frac{I_1}{I_2} = \frac{d_2{}^2}{d_1{}^2}$$

$$d_2{}^2 = \frac{I_1 d_1{}^2}{I_2} \quad \Rightarrow \quad d_2 = \sqrt{\frac{I_1 d_1{}^2}{I_2}}$$

STEP 4: Solve. Substitute in known values so that unwanted units cancel.

$$d_2 = \sqrt{\frac{(75 \text{ units})(2.4 \text{ m})^2}{(25 \text{ units})}} = 4.2 \text{ m}$$

BALLPARK CHECK The calculated result is consistent with our estimate of about 4 m.

PROBLEM 11.10

A β-emitting radiation source gives 250 units of radiation at a distance of 4.0 m. At what distance does the radiation drop to one-tenth its original value?

11.8 Detecting Radiation

Small amounts of naturally occurring radiation have always been present, but people have been aware of it only within the past 100 years. The problem is that radiation is invisible. We cannot see, hear, smell, touch, or taste radiation, no matter how high the dose. We can, however, detect radiation by taking advantage of its ionizing properties.

The simplest device for detecting exposure to radiation is the photographic film badge worn by people who routinely work with radioactive materials. The film is protected from exposure to light, but any other radiation striking the badge causes the film to fog (remember Becquerel's discovery). At regular intervals, the film is developed and compared with a standard to indicate the radiation exposure.

Perhaps the best-known method for detecting and measuring radiation is the *Geiger counter*, an argon-filled tube containing two electrodes (Figure 11.6). The inner walls of the tube are coated with an electrically conducting material and given a negative charge, and a wire in the center of the tube is given a positive charge. As radiation enters the tube through a thin window, it strikes and ionizes argon atoms, which briefly conduct a tiny electric current between the walls and the center electrode. The passage of the current is detected, amplified, and used to produce a clicking sound or to register on a meter. The more radiation that enters the tube, the more frequent the clicks. Geiger counters are useful for seeking out a radiation source in a large area and for gauging the intensity of emitted radiation.

The most versatile method for measuring radiation in the laboratory is the *scintillation counter*, a device in which a substance called a *phosphor* emits a flash of light when struck by radiation. The number of flashes are counted electronically and converted into an electrical signal.

▲ This photographic film badge is a common device for monitoring radiation exposure.

▲ Radiation is conveniently detected and measured using this scintillation counter, which electronically counts the flashes produced when radiation strikes a phosphor.

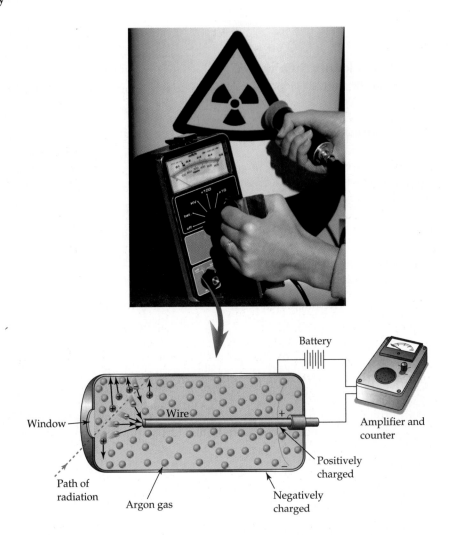

▶ **FIGURE 11.6 A Geiger counter for measuring radiation.** As radiation enters the tube through a thin window, it ionizes argon atoms and produces electrons that conduct a tiny electric current between the walls and the center electrode. The current flow then registers on the meter.

11.9 Measuring Radiation

Radiation intensity is expressed in different ways, depending on what characteristic of the radiation is measured (Table 11.5). Some units measure the number of nuclear decay events, while others measure exposure to radiation or the biological consequences of radiation.

TABLE 11.5 Common Units for Measuring Radiation

UNIT	QUANTITY MEASURED	DESCRIPTION
Curie (Ci)	Decay events	Amount of radiation equal to 3.7×10^{10} disintegrations per second
Roentgen (R)	Ionizing intensity	Amount of radiation producing 2.1×10^9 charges per cubic centimeter of dry air
Rad	Energy absorbed per gram of tissue	1 rad = 1 R
Rem	Tissue damage	Amount of radiation producing the same damage as 1 R of X rays
Sievert (Sv)	Tissue damage	1 Sv = 100 rem

The *curie* (Ci), the *millicurie* (mCi), and the *microcurie* (μCi) measure the number of radioactive disintegrations occurring each second in a sample. One curie is the decay rate of 1 g of radium, equal to 3.7×10^{10} disintegrations per second; 1 mCi = 0.001 Ci = 3.7×10^7 disintegrations per second; and 1 μCi = 0.000 001 Ci = 3.7×10^4 disintegrations per second.

APPLICATION ▶ Irradiated Food

The idea of irradiating food to kill harmful bacteria is not new; it goes back almost as far as the earliest studies on radiation. Not until the 1940s did serious work get under way, however, when U.S. Army scientists found that irradiation increased the shelf-life of ground beef. Nevertheless, widespread civilian use of the technique has been a long time in coming, spurred on in recent years by outbreaks of food poisoning that resulted in several deaths.

The principle of food irradiation is simple: Exposure of contaminated food to ionizing radiation—usually γ rays produced by cobalt-60 or cesium-137—destroys the genetic material of any bacteria or other organisms present, thereby killing them. Irradiation will not, however, kill viruses or prions (⬤▭⬤, p. 584), the cause of mad-cow disease. The food itself undergoes little if any change when irradiated and does not itself become radioactive. The only real argument against food irradiation, in fact, is that it is *too* effective. Knowing that irradiation will kill nearly all harmful organisms, a food processor might be tempted to cut back on normal sanitary practices!

Food irradiation has been implemented to a much greater extent in Europe than in the United States. The largest marketers of irradiated food are Belgium, France, and the Netherlands, which irradiate between 10,000 and 20,000 tons of food per year. One of the major concerns in the United States is the possible generation of *radiolytic products*, compounds formed in food by exposure to ionizing radiation. The U.S. Food and Drug Administration, after studying the matter extensively, has declared that food irradiation is safe and that it does not appreciably alter the vitamin or other nutritional content of food. Spices, fruits, pork, and vegetables were

approved for irradiation in 1986, followed by poultry in 1990 and red meat, particularly ground beef, in 1997. In 2000, approval was extended to whole eggs and sprouting seeds. Should the food industry adopt irradiation of meat as its standard practice, such occurrences as the 1993 Seattle outbreak of *E. coli* poisoning caused by undercooked hamburgers will become a thing of the past.

NON - IRRADIATED -   IRRADIATED - (0.2 M RAD)

STRAWBERRIES -
15 DAYS STORAGE 38°F (4°C)

▲ Irradiating food kills bacteria and extends shelf-life. The strawberries in these two containers were picked at the same time, but only the batch on the right was irradiated.

See Additional Problems 11.72 and 11.73 at the end of the chapter.

The dosage of a radioactive substance administered orally or intravenously is usually given in millicuries. To calculate the size of a dose, it is necessary to determine the decay rate of the isotope solution per milliliter. Because the emitter concentration is constantly decreasing as it decays, the activity must be measured immediately before administration. Suppose, for example, that a solution containing iodine-131 for a thyroid function study is found to have a decay rate of 0.020 mCi/mL and the dose administered is to be 0.050 mCi. The amount of the solution administered must be

$$\frac{0.05 \text{ mCi}}{\text{Dose}} \times \frac{1 \text{ mL }^{131}\text{I solution}}{0.020 \text{ mCi}} = 2.5 \text{ mL }^{131}\text{I solution/dose}$$

- **Roentgen** The *roentgen* (R) is a unit for measuring the ionizing intensity of γ or X radiation. In other words, the roentgen measures the capacity of the radiation for affecting matter. One roentgen is the amount of radiation that produces 2.1×10^9 units of charge in 1 cm^3 of dry air at atmospheric pressure. Each collision of ionizing radiation with an atom produces one ion, or one unit of charge.

- **Rad** The *rad* (radiation absorbed dose) is a unit for measuring the energy absorbed per gram of material exposed to a radiation source and is defined as the absorption of 1×10^{-5} J of energy per gram. The energy absorbed varies with the type of material irradiated and the type of radiation. For most purposes, though, the roentgen and the rad are so close that they can be considered identical when used for X rays and γ rays: 1 R = 1 rad.

- **Rem** The *rem* (roentgen equivalent for man) measures the amount of tissue damage caused by radiation, taking into account the differences in energy of different types of radiation. One rem is the amount of radiation that produces the same effect as 1 R of X rays.

  Rems are the preferred units for medical purposes because they measure equivalent doses of different kinds of radiation. For example, 1 rad of α radiation causes 20 times more tissue damage than 1 rad of γ rays, but 1 rem of α radiation and 1 rem of γ rays cause the same amount of damage. Thus, the rem takes both ionizing intensity and biological effect into account, whereas the rad deals only with intensity.

- **SI Units** In the SI system, the *becquerel* (Bq) is defined as one disintegration per second. The SI unit for energy absorbed is the *gray* (Gy; 1 Gy = 100 rad). For radiation dose, the SI unit is the *sievert* (Sv), which is equal to 100 rem.

The biological consequences of different radiation doses are given in Table 11.6. Although the effects seem fearful, the average radiation dose received annually by most people is only about 0.27 rem. About 80% of this *background radiation* comes from natural sources (rocks and cosmic rays); the remaining 20% comes from medical procedures such as X rays and from consumer products. The amount due to emissions from nuclear power plants and to fallout from testing of nuclear weapons in the 1950s is barely detectable.

TABLE 11.6 Biological Effects of Short-Term Radiation on Humans

DOSE (REM)	BIOLOGICAL EFFECTS
0–25	No detectable effects
25–100	Temporary decrease in white blood cell count
100–200	Nausea, vomiting, longer-term decrease in white blood cells
200–300	Vomiting, diarrhea, loss of appetite, listlessness
300–600	Vomiting, diarrhea, hemorrhaging, eventual death in some cases
Above 600	Eventual death in nearly all cases

▲ After the 1986 nuclear-reactor disaster at Chernobyl, security police in Ukraine limited access to the area within 30 km of the site. (©*Shirley Clive/Greenpeace International*)

PROBLEM 11.11

Radiation released during the 1986 Chernobyl nuclear power plant disaster is expected to increase the background radiation level worldwide by about 5 mrem. By how much will this increase the annual dose of the average person. Express your answer as a percentage.

PROBLEM 11.12

A solution of selenium-75, a radioisotope used in the diagnosis of pancreatic disease, is found just prior to administration to have an activity of 44 μCi/mL. How many milliliters should be administered intravenously for a dose of 175 μCi?

11.10 Artificial Transmutation

Artificial transmutation The change of one atom into another brought about by a nuclear bombardment reaction.

Very few of the approximately 3300 known radioisotopes occur naturally. Most are made from stable isotopes by **artificial transmutation**, the change of one atom into another brought about by nuclear bombardment reactions.

When an atom is bombarded with a high-energy particle, such as a proton, neutron, α particle, or even the nucleus of another element, an unstable nucleus is created in the collision. A nuclear change then occurs, and a different element is produced. For example, transmutation of ^{14}N to ^{14}C occurs in the upper atmosphere when neutrons produced by cosmic rays collide with atmospheric nitrogen. In the collision, a neutron dislodges a proton (^{1}H) from the nitrogen nucleus as the neutron and nucleus fuse together:

$$^{14}_{7}\text{N} + ^{1}_{0}\text{n} \longrightarrow ^{14}_{6}\text{C} + ^{1}_{1}\text{H}$$

Artificial transmutation can lead to the synthesis of entirely new elements never before seen on earth. In fact, all the *transuranium elements*—those elements with atomic numbers greater than 92—have been produced by bombardment reactions. For example, plutonium-241 (^{241}Pu) can be made by bombardment of uranium-238 with α particles:

$$^{238}_{92}\text{U} + ^{4}_{2}\text{He} \longrightarrow ^{241}_{94}\text{Pu} + ^{1}_{0}\text{n}$$

Plutonium-241 is itself radioactive, with a half-life of 14.35 years, decaying by β emission to yield americium-241, which in turn decays by α emission with a half-life of 432.2 years. (If the name *americium* sounds vaguely familiar, it is because this radioisotope is used in smoke detectors.)

$$^{241}_{94}\text{Pu} \longrightarrow ^{241}_{95}\text{Am} + ^{0}_{-1}\text{e}$$

Note that all the equations just given for artificial transmutations are balanced. The sum of the mass numbers and the sum of the charges are the same on both sides of each equation.

▲ Smoke detectors contain a small amount of americium-241. The α particles emitted by this radioisotope ionize the air within the detector, causing it to conduct a tiny electric current. When smoke enters the chamber, conductivity drops and an alarm is triggered.

WORKED EXAMPLE **11.7** Balancing Nuclear Reactions: Transmutation

Californium-246 is formed by bombardment of uranium-238 atoms. If four neutrons are also formed, what particle is used for the bombardment?

ANALYSIS First write an incomplete nuclear equation incorporating the known information:

$$^{238}_{92}\text{U} + ? \longrightarrow ^{246}_{98}\text{Cf} + 4\,^{1}_{0}\text{n}$$

Then find the numbers of nucleons and charges necessary to balance the equation. In this instance, there are 238 nucleons on the left and $246 + 4 = 250$ nucleons on the right, so the bombarding particle must have $250 - 238 = 12$ nucleons. Furthermore, there are 92 nuclear charges on the left and 98 on the right, so the bombarding particle must have $98 - 92 = 6$ protons.

SOLUTION
The missing particle is $^{12}_{6}$C.

$$^{238}_{92}\text{U} + ^{12}_{6}\text{C} \longrightarrow ^{246}_{98}\text{Cf} + 4\,^{1}_{0}\text{n}$$

PROBLEM 11.13

What isotope results from α decay of the americium-241 in smoke detectors?

PROBLEM 11.14

The element berkelium, first prepared at the University of California at Berkeley in 1949, is made by α bombardment of $^{241}_{95}$Am. Two neutrons are also produced during the reaction. What isotope of berkelium results from this transmutation? Write a balanced nuclear equation.

APPLICATION ▶ Body Imaging

We are all familiar with the appearance of a standard X-ray image, produced when X rays pass through the body and the intensity of the radiation that exits is recorded on film. X-ray imaging is, however, only one of a host of noninvasive imaging techniques that are now in common use.

Among the most widely used imaging techniques are those that give diagnostic information about the health of various parts of the body by analyzing the distribution pattern of a radioactively tagged substance in the body. A radiopharmaceutical agent that is known to concentrate in a specific organ or other body part is injected into the body, and its distribution pattern is monitored by an external radiation detector such as a γ-ray camera. Depending on the medical condition, a diseased part might concentrate more of the radiopharmaceutical than normal and thus show up on the film as a radioactive hot spot against a cold background. Alternatively, the diseased part might concentrate less of the radiopharmaceutical than normal and thus show up as a cold spot on a hot background.

Among the radioisotopes most widely used for diagnostic imaging is technetium-99m, whose short half-life of only six hours minimizes the patient's exposure to radioactivity. Bone scans using this nuclide, such as that shown in the accompanying photograph, are an important tool in the diagnosis of cancer and other conditions.

Several other techniques now used in medical diagnosis are made possible by *tomography*, a technique in which computer processing allows production of images through "slices" of the body. In X-ray tomography, commonly known as *CAT* or *CT* scanning (computerized tomography), the X-ray source and an array of detectors move rapidly in a circle around a patient's body, collecting up to 90,000 readings. CT scans can detect structural abnormalities such as tumors without the use of radioactive materials.

Combining tomography with radioisotope imaging gives cross-sectional views of regions that concentrate a radioactive substance. One such technique, *positron emission tomography* (PET), utilizes radioisotopes that emit positrons and ultimately yield γ rays. Oxygen-15, nitrogen-13, carbon-11, and

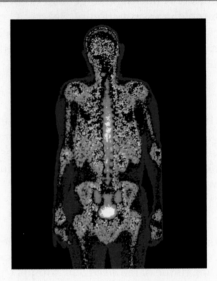

▲ A bone scan carried out with radioactive technetium-99*m*. Color has been added to help the visualization.

fluorine-18 are commonly used for PET because they can be readily incorporated into many physiologically active compounds. An ^{18}F-labeled glucose derivative, for instance, is useful for imaging brain regions that respond to various stimuli. The disadvantage of PET scans is that the necessary radioisotopes are so short-lived that they must be produced on-site immediately before use. The cost of PET is therefore high, because a hospital must install and maintain the necessary nuclear facility.

Magnetic resonance imaging (MRI) is a medical imaging technique that uses powerful magnetic and radio-frequency fields to interact with specific nuclei in the body (usually the nuclei of hydrogen atoms) to generate images in which the contrast between soft tissues is much better than that seen with CT. The original name for this technique was *nuclear* magnetic resonance imaging, but the *nuclear* was eliminated because in the public mind this word conjured up negative images of ionizing radiation. Ironically, MRI does not involve any nuclear radiation at all.

See Additional Problems 11.74 and 11.75 at the end of the chapter.

PROBLEM 11.15

Write a balanced nuclear equation for the reaction of argon-40 with a proton:

$$\ce{^{40}_{18}Ar} + \ce{^{1}_{1}H} \longrightarrow ? + \ce{^{1}_{0}n}$$

11.11 Nuclear Fission and Nuclear Fusion

In the preceding section, we saw that particle bombardment of various elements causes artificial transmutation and results in the formation of new, usually heavier elements. Under very special conditions with a very few isotopes, however, different kinds of nuclear events occur. Certain very heavy nuclei can split apart, and certain very light nuclei can fuse together. The two resultant processes—**nuclear fission** for the fragmenting of heavy nuclei and **nuclear fusion** for the joining together of light nuclei—have changed the world since their discovery in the late 1930s and early 1940s.

The huge amounts of energy that accompany these nuclear processes are the result of mass-to-energy conversions and are predicted by Einstein's equation

$$E = mc^2$$

where E = energy, m = mass change associated with the nuclear reaction, and c = the speed of light (3.0×10^8 m/s). Based on this relationship, a mass change as small as 1 μg results in a release of 2.15×10^4 kcal of energy!

Nuclear Fission

Uranium-235 is the only naturally occurring isotope that undergoes nuclear fission. When this isotope is bombarded by a stream of relatively slow-moving neutrons, its nucleus splits to give isotopes of other elements. The split can take place in more than 400 ways, and more than 800 different fission products have been identified. One of the more frequently occurring pathways generates barium-142 and krypton-91, along with two additional neutrons plus the one neutron that initiated the fission:

$$^1_0n + {}^{235}_{92}U \longrightarrow {}^{142}_{56}Ba + {}^{91}_{36}Kr + 3\,{}^1_0n$$

As indicated by the balanced nuclear equation above, *one* neutron is used to initiate fission of a ^{235}U nucleus, but *three* neutrons are released. Thus, a nuclear **chain reaction** can be started: 1 neutron initiates one fission that releases 3 neutrons. The 3 neutrons initiate three new fissions that release 9 neutrons. The 9 neutrons initiate nine fissions that release 27 neutrons, and so on at an ever faster pace (Figure 11.7). It is worth noting that the neutrons produced by fission reactions are highly energetic. They possess penetrating power greater than α and β particles, but less than γ rays. In a nuclear fission reactor, the neutrons must first be slowed down to allow them to react. If the sample size is small, many of the neutrons escape before initiating additional fission events, and the chain reaction stops. If a sufficient amount of ^{235}U is present, however—an amount called the **critical mass**—then the chain reaction becomes self-sustaining. Under high-pressure conditions that confine the ^{235}U to a small volume, the chain reaction occurs so rapidly that a nuclear explosion results. For ^{235}U, the critical mass is about 56 kg, although the amount can be reduced to approximately 15 kg by placing a coating of ^{238}U around the ^{235}U to reflect back some of the escaping neutrons.

An enormous quantity of heat is released during nuclear fission—the fission of just 1.0 g of uranium-235 produces 3.4×10^8 kcal, for instance. This heat can be used to convert water to steam, which can be harnessed to turn huge generators and produce electric power. The use of nuclear power is much more advanced in some countries than in others, with Lithuania and France leading the way by generating about 86% and 77%, respectively, of their electricity in nuclear plants. In the United States, only about 22% of the electricity is nuclear-generated.

Two major objections that have caused much public debate about nuclear power plants are safety and waste disposal. Although a nuclear explosion is not possible under the conditions that exist in a power plant, there is a serious potential radiation hazard should an accident rupture the containment vessel holding the nuclear fuel and release radioactive substances to the environment. Perhaps even more important is the problem posed by disposal of radioactive wastes from nuclear plants. Many of these wastes have such long half-lives that hundreds or even thousands of years must elapse before they will be safe for humans to approach. How to dispose of such hazardous materials safely is an unsolved problem.

Nuclear fission The fragmenting of heavy nuclei.

Nuclear fusion The joining together of light nuclei.

Chain reaction A reaction that, once started, is self-sustaining.

Critical mass The minimum amount of radioactive material needed to sustain a nuclear chain reaction.

▲ Energy from the fission of uranium-235 is used to produce steam and generate electricity in this nuclear power plant.

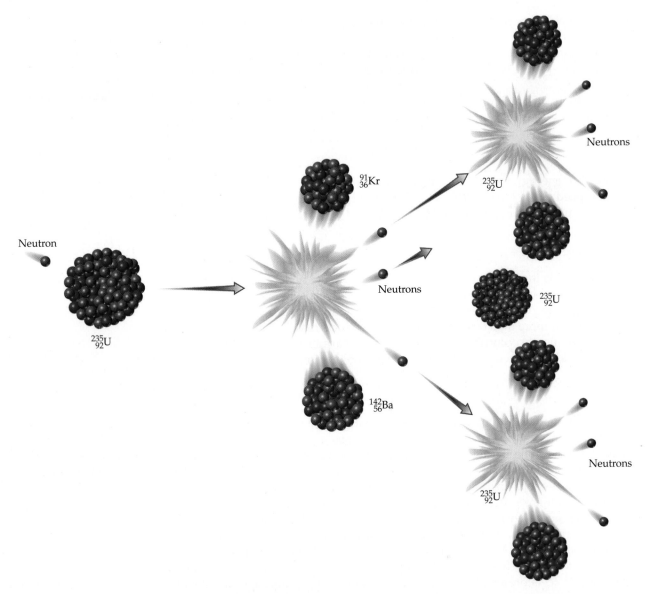

▲ **FIGURE 11.7** **A chain reaction.** Each fission event produces additional neutrons that induce more fissions. The rate of the process increases at each stage. Such chain reactions usually lead to the formation of many different fission products in addition to the two indicated.

PROBLEM 11.16

What other isotope besides tellurium-137 is produced by nuclear fission of uranium-235?

$$^{235}_{92}\text{U} + {}^{1}_{0}\text{n} \longrightarrow {}^{137}_{52}\text{Te} + 2\,{}^{1}_{0}\text{n} + ?$$

Nuclear Fusion

Just as heavy nuclei such as ^{235}U release energy when they undergo *fission*, very light nuclei such as the isotopes of hydrogen release enormous amounts of energy when they undergo *fusion*. In fact, it is just such a fusion reaction of hydrogen nuclei to produce helium that powers our sun and other stars. Among the processes thought to occur in the sun are those in the following sequence leading to helium-4:

$$^{1}_{1}\text{H} + {}^{2}_{1}\text{H} \longrightarrow {}^{3}_{2}\text{He}$$
$$^{3}_{2}\text{He} + {}^{3}_{2}\text{He} \longrightarrow {}^{4}_{2}\text{He} + 2\,{}^{1}_{1}\text{H}$$
$$^{3}_{2}\text{He} + {}^{1}_{1}\text{H} \longrightarrow {}^{4}_{2}\text{He} + {}^{0}_{1}\text{e}$$

APPLICATION ▶ Archaeological Radiocarbon Dating

Biblical scrolls are found in a cave near the Dead Sea. Are they authentic? A mummy is discovered in an Egyptian tomb. How old is it? The burned bones of a man are dug up near Lubbock, Texas. How long ago did humans live on the North American continent? Using a technique called *radiocarbon dating*, archaeologists can answer these and many other questions. (The Dead Sea Scrolls are 1900 years old and authentic, the mummy is 3100 years old, and the human remains found in Texas are 9900 years old.)

Radiocarbon dating depends on the slow and constant production of radioactive carbon-14 atoms in the upper atmosphere by bombardment of nitrogen atoms with neutrons from cosmic rays. Carbon-14 atoms combine with oxygen to yield $^{14}CO_2$, which slowly mixes with ordinary $^{12}CO_2$ and is then taken up by plants during photosynthesis. When these plants are eaten by animals, carbon-14 enters the food chain and is distributed evenly throughout all living organisms.

As long as a plant or animal is living, a dynamic equilibrium is established in which the organism excretes or exhales the same amount of ^{14}C that it takes in. As a result, the ratio of ^{14}C to ^{12}C in the living organism is the same as that in the atmosphere—about 1 part in 10^{12}. When the plant or animal dies, however, it no longer takes in more ^{14}C. Thus, the $^{14}C/^{12}C$ ratio in the organism slowly decreases as ^{14}C undergoes radioactive decay. At 5730 years (one ^{14}C half-life) after the death of the organism, the $^{14}C/^{12}C$ ratio has decreased by a factor of 2; at 11,460 years after death, the $^{14}C/^{12}C$ ratio has decreased by a factor of 4; and so on.

By measuring the amount of ^{14}C remaining in the traces of any once-living organism, archaeologists can determine how long ago the organism died. Human hair from well-preserved remains, charcoal or wood fragments from once-living trees, and cotton or linen from once-living plants are all useful sources for radiocarbon dating. The accuracy of the technique lessens as a sample gets older and the amount of ^{14}C it contains diminishes, but artifacts with an age of 1000–20,000 years can be dated with reasonable accuracy.

▲ Radiocarbon dating has determined that these charcoal paintings in the Lascaux cave in France are approximately 15,000 years old.

See Additional Problems 11.76 and 11.77 at the end of the chapter.

Under the conditions found in stars, where the temperature is on the order of 2×10^7 K and pressures approach 10^5 atmospheres, nuclei are stripped of all their electrons and have enough kinetic energy that nuclear fusion readily occurs. The energy of our sun, and all the stars, comes from thermonuclear fusion reactions in their core that fuse hydrogen and other light elements transmuting them into heavier elements. On earth, however, the necessary conditions for nuclear fusion are not easily created. For more than 50 years scientists have been trying to create the necessary conditions for fusion in laboratory reactors, including the Tokamak Fusion Test Reactor (TFTR) at Princeton, New Jersey, and the Joint European Torus (JET) at Culham, England. Recent advances in reactor design have raised hopes that a commercial fusion reactor will be realized within the next 20 years.

If the dream becomes reality, controlled nuclear fusion can provide the ultimate cheap, clean power source. The fuel is deuterium (^2H), available in the oceans in limitless amounts, and there are few radioactive by-products.

▲ A researcher stands inside the D3D Tokamak nuclear fusion reactor at the General Atomics facility in San Diego, California.

SUMMARY: REVISITING THE CHAPTER GOALS

1. **What is a nuclear reaction, and how are equations for nuclear reactions balanced?** A *nuclear reaction* is one that changes an atomic nucleus, causing the change of one element into another. Loss of an α particle leads to a new atom whose atomic number is 2 less than

KEY WORDS

Alpha (α) particle, *p. 335*
Artificial transmutation, *p. 350*

that of the starting atom. Loss of a β particle leads to an atom whose atomic number is 1 greater than that of the starting atom:

$$\alpha \text{ emission: } {}^{238}_{92}\text{U} \longrightarrow {}^{234}_{90}\text{Th} + {}^{4}_{2}\text{He}$$

$$\beta \text{ emission: } {}^{131}_{53}\text{I} \longrightarrow {}^{131}_{54}\text{Xe} + {}^{0}_{-1}\text{e}$$

A nuclear reaction is balanced when the sum of the *nucleons* (protons and neutrons) is the same on both sides of the reaction arrow and when the sum of the charges on the nuclei plus any ejected subatomic particles is the same.

2. What are the different kinds of radioactivity? *Radioactivity* is the spontaneous emission of radiation from the nucleus of an unstable atom. The three major kinds of radiation are called *alpha* (α), *beta* (β), and *gamma* (γ). Alpha radiation consists of helium nuclei, small particles containing two protons and two neutrons (${}^{4}_{2}\text{He}$); β radiation consists of electrons (${}^{0}_{-1}\text{e}$); and γ radiation consists of high-energy light waves. Every element in the periodic table has at least one radioactive isotope, or *radioisotope*.

3. How are the rates of nuclear reactions expressed? The rate of a nuclear reaction is expressed in units of *half-life* ($t_{1/2}$), where one half-life is the amount of time necessary for one half of the radioactive sample to decay.

4. What is ionizing radiation? High-energy radiation of all types—α particles, β particles, γ rays, and X rays—is called *ionizing radiation*. When any of these kinds of radiation strikes an atom, it dislodges an electron and gives a reactive ion that can be lethal to living cells. Gamma rays and X rays are the most penetrating and most harmful types of external radiation; α and β particles are the most dangerous types of internal radiation because of their high energy and the resulting damage to surrounding tissue.

5. How is radioactivity measured? Radiation intensity is expressed in different ways according to the property being measured. The *curie* (Ci) measures the number of radioactive disintegrations per second in a sample; the *roentgen* (R) measures the ionizing ability of radiation; the *rad* measures the amount of radiation energy absorbed per gram of tissue; and the *rem* measures the amount of tissue damage caused by radiation. Radiation effects become noticeable with a human exposure of 25 rem and become lethal at an exposure above 600 rem.

6. What is transmutation? *Transmutation* is the change of one element into another brought about by a nuclear reaction. Most known radioisotopes do not occur naturally but are made by bombardment of an atom with a high-energy particle. In the ensuing collision between particle and atom, a nuclear change occurs and a new element is produced by *artificial transmutation*.

7. What are nuclear fission and nuclear fusion? With a very few isotopes, including ${}^{235}_{92}\text{U}$, the nucleus is split apart by neutron bombardment to give smaller fragments. A large amount of energy is released during this *nuclear fission*, leading to use of the reaction for generating electric power. *Nuclear fusion* results when small nuclei such as those of tritium (${}^{3}_{1}\text{H}$) and deuterium (${}^{2}_{1}\text{H}$) combine to give a heavier nucleus.

UNDERSTANDING KEY CONCEPTS

11.17 Magnesium-28 decays by β emission to give aluminum-28. If yellow spheres represent ${}^{28}_{12}\text{Mg}$ atoms and blue spheres represent ${}^{28}_{13}\text{Al}$ atoms, how many half-lives have passed in the following sample?

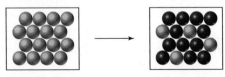

11.18 Write a balanced nuclear equation to represent the decay reaction described in Problem 11.17.

11.19 Refer to Figure 11.4 and then make a drawing similar to those in Problem 11.17 representing the decay of a sample of ${}^{28}_{12}\text{Mg}$ after approximately four half-lives have passed.

11.20 Write the symbol of the isotope represented by the following drawing. Blue spheres represent neutrons and red spheres represent protons.

11.21 Shown below is a portion of the decay series for plutonium-241 (${}^{241}_{94}\text{Pu}$). The series has two kinds of arrows: shorter arrows pointing right and longer arrows pointing left. Which arrow corresponds to an α emission, and which to a β emission? Explain.

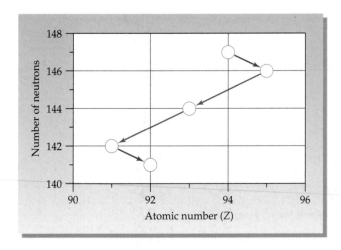

11.24 What is the half-life of the radionuclide that shows the following decay curve?

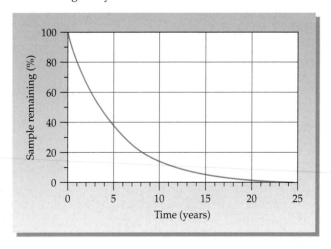

11.22 Identify and write the symbol for each nuclide in the decay series shown in Problem 11.21.

11.23 Identify the isotopes involved, and tell the type of decay process occurring in the following nuclear reaction:

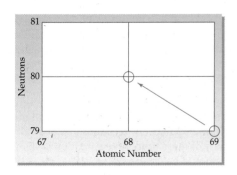

11.25 What is wrong with the following decay curve? Explain.

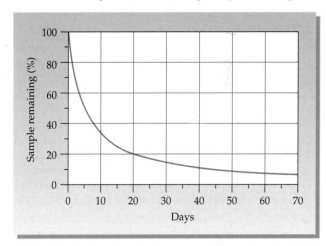

ADDITIONAL PROBLEMS

RADIOACTIVITY

11.26 What does it mean to say that a substance is radioactive?

11.27 Describe how α radiation, β radiation, γ radiation, positron emission, and electron capture differ.

11.28 List several ways in which a nuclear reaction differs from a chemical reaction.

11.29 What word is used to describe the change of one element into another?

11.30 What symbol is used for an α particle in a nuclear equation?

11.31 What symbol is used for a β particle in a nuclear equation? For a positron?

11.32 Which kind of radiation, α, β, or γ, has the highest penetrating power, and which has the lowest?

11.33 What happens when ionizing radiation strikes an atom in a chemical compound?

11.34 How does ionizing radiation lead to cell damage?

11.35 What are the main sources of background radiation?

11.36 How can a nucleus emit an electron during β decay when there are no electrons present in the nucleus to begin with?

11.37 What is the difference between an α particle and a helium atom?

NUCLEAR DECAY AND TRANSMUTATION

11.38 What does it mean to say that a nuclear equation is balanced?

11.39 What are transuranium elements, and how are they made?

11.40 What happens to the mass number and atomic number of an atom that emits an α particle? A β particle?

11.41 What happens to the mass number and atomic number of an atom that emits a γ ray? A positron?

11.42 How does nuclear fission differ from normal radioactive decay?

11.43 What characteristic of uranium-235 fission causes a chain reaction?

11.44 What products result from radioactive decay of the following β emitters?

(a) $^{35}_{16}S$ (b) $^{24}_{10}Ne$ (c) $^{90}_{38}Sr$

11.45 What products result from radioactive decay of the following α emitters?

    (a) $^{190}_{78}Pt$ **(b)** $^{208}_{87}Fr$ **(c)** $^{245}_{96}Cm$

11.46 Identify the starting radioisotopes needed to balance each of these nuclear reactions:

    (a) $? + ^4_2He \rightarrow ^{113}_{49}In$

    (b) $? + ^4_2He \rightarrow ^{13}_7N + ^1_0n$

11.47 Identify the product radioisotope needed to balance each of these nuclear reactions:

    (a) $^{140}_{55}Cs \rightarrow ? + ^0_{-1}e$

    (b) $^{248}_{96}Cm \rightarrow ? + ^4_2He$

11.48 Balance the following equations for the nuclear fission of $^{235}_{92}U$:

    (a) $^{235}_{92}U + ^1_0n \rightarrow ^{160}_{62}Sm + ^{72}_{30}Zn + ? \, ^1_0n$

    (b) $^{235}_{92}U + ^1_0n \rightarrow ^{87}_{35}Br + ? + 3 \, ^1_0n$

11.49 Complete and balance the following nuclear equations:

    (a) $^{126}_{50}Sn \rightarrow ^0_{-1}e + ?$

    (b) $^{210}_{88}Ra \rightarrow ^4_2He + ?$

    (c) $^{76}_{36}Kr + ^0_{-1}e \rightarrow ?$

11.50 For centuries, alchemists dreamed of turning base metals into gold. The dream finally became reality when it was shown that mercury-198 can be converted into gold-198 on bombardment by neutrons. What small particle is produced in addition to gold-198? Write a balanced nuclear equation for the reaction.

11.51 Cobalt-60 (half-life = 5.3 years) is used to irradiate food, to treat cancer, and to disinfect surgical equipment. It is produced by irradiation of cobalt-59 in a nuclear reactor. It decays to nickel-60. Write nuclear equations for the formation and decay reactions of cobalt-60.

11.52 Bismuth-212 attaches readily to monoclonal antibodies and is used in the treatment of various cancers. It is formed after the parent isotope undergoes a decay series consisting of four α decays and one β decay. What is the parent isotope for this decay series?

11.53 Meitnerium-266 ($^{266}_{109}Mt$) was prepared in 1982 by bombardment of bismuth-209 atoms with iron-58. What other product must also have been formed? Write a balanced nuclear equation for the transformation.

HALF-LIFE

11.54 What does it mean when we say that strontium-90, a waste product of nuclear power plants, has a half-life of 28.8 years?

11.55 What percentage of the original mass remains in a radioactive sample after two half-lives have passed? After three half-lives? After 3.5 half-lives?

11.56 Selenium-75, a β emitter with a half-life of 120 days, is used medically for pancreas scans. Approximately how much selenium-75 would remain from a 0.050 g sample that has been stored for one year?

11.57 Approximately how long would it take a sample of selenium-75 to lose 99% of its radioactivity? (See Problem 11.56.)

11.58 The half-life of mercury-197 is 64.1 hours. If a patient undergoing a kidney scan is given 5.0 ng of mercury-197, how much will remain after 7 days? After 30 days?

11.59 Gold-198, a β emitter used to treat leukemia, has a half-life of 2.695 days. The standard dosage is about 1.0 mCi/kg body weight.

    (a) What is the product of the β emission of gold-198?

    (b) How long does it take a 30.0 mCi sample of gold-198 to decay so that only 3.75 mCi remains?

    (c) How many millicuries are required in a single dosage administered to a 70.0 kg adult?

MEASURING RADIOACTIVITY

11.60 Describe how a Geiger counter works.

11.61 Describe how a film badge works.

11.62 Describe how a scintillation counter works.

11.63 Why are rems the preferred units for measuring the health effects of radiation?

11.64 Approximately what amount (in rems) of short-term exposure to radiation produces noticeable effects in humans?

11.65 Match each unit in the left column with the property being measured in the right column:

    1. curie   **(a)** Ionizing intensity of radiation

    2. rem   **(b)** Amount of tissue damage

    3. rad   **(c)** Number of disintegrations per second

    4. roentgen **(d)** Amount of radiation per gram of tissue

11.66 Technetium-99m is used for radioisotope-guided surgical biopsies of certain bone cancers. A patient must receive an injection of 28 mCi of technetium-99m 6–12 hours before surgery. If the activity of the solution is 15 mCi, what volume should be injected?

11.67 Sodium-24 is used to study the circulatory system and to treat chronic leukemia. It is administered in the form of saline (NaCl) solution, with a therapeutic dosage of 180 μCi/kg body weight. How many milliliters of a 6.5 mCi/mL solution are needed to treat a 68 kg adult?

11.68 A selenium-75 source is producing 300 rem at a distance of 2.0 m. What is its intensity at 25 m?

11.69 If a radiation source has an intensity of 650 rem at 1.0 m, what distance is needed to decrease the intensity of exposure to below 25 rem, the level at which no effects are detectable?

Applications

11.70 What are the three main classes of techniques used in nuclear medicine? Give an example of each. [*Medical Uses of Radioactivity, p. 342*]

11.71 A 2 mL solution containing 1.25 μCi/mL is injected into the blood stream of a patient. After dilution, a 1.00 mL sample is withdrawn and found to have an activity of 2.6×10^{-4} μCi. Calculate total blood volume. [*Medical Uses of Radioactivity, p. 342*]

11.72 What is the purpose of food irradiation, and how does it work? [*Irradiated Food, p. 349*]

11.73 What kind of radiation is used to treat food? [*Irradiated Food, p. 349*]

11.74 What are the advantages of CT and PET relative to conventional X rays? [*Body Imaging, p. 352*]

11.75 What advantages does MRI have over CT and PET imaging? [*Body Imaging, p. 352*]

11.76 Why is ^{14}C dating useful only for samples that contain material from objects that were once alive? [*Archaeological Radiocarbon Dating, p. 355*]

11.77 Some dried beans with a $^{14}C/^{12}C$ ratio one-eighth of the current value are found in an old cave. How old are the beans? [*Archaeological Radiocarbon Dating, p. 355*]

General Questions and Problems

11.78 Harmful chemical spills can often be cleaned up by treatment with another chemical. For example, a spill of H_2SO_4 might be neutralized by addition of $NaHCO_3$. Why is it that the harmful radioactive wastes from nuclear power plants cannot be cleaned up just as easily?

11.79 Why is a scintillation counter or Geiger counter more useful for determining the existence and source of a new radiation leak than a film badge?

11.80 Technetium-99m, used for brain scans and to monitor heart function, is formed by decay of molybdenum-99.

   (a) By what type of decay does ^{99}Mo produce ^{99m}Tc?
   (b) Molybdenum-99 is formed by neutron bombardment of a natural isotope. If one neutron is absorbed and there are no other by-products of this process, from what isotope is ^{99}Mo formed?

11.81 The half-life of technetium-99m (Problem 11.80) is 6.01 hours. If a sample with an initial activity of 15 μCi is injected into a patient, what is the activity in 24 hours, assuming that none of the sample is excreted?

11.82 Plutonium-238 is an α emitter used to power batteries for heart pacemakers.

   (a) Write the balanced nuclear equation for this emission.
   (b) Why is a pacemaker battery enclosed in a metal case before being inserted into the chest cavity?

11.83 Sodium-24, a beta-emitter used in diagnosing circulation problems, has a half-life of 15 hours.

   (a) Write the balanced nuclear equation for this emission.
   (b) What fraction of sodium-24 remains after 50 hours?

11.84 High levels of radioactive fallout after the 1986 accident at the Chernobyl nuclear power plant in what is now Ukraine resulted in numerous miscarriages and many instances of farm animals born with severe defects. Why are embryos and fetuses particularly susceptible to the effects of radiation?

11.85 One way to demonstrate the dose factor of ionizing radiation (penetrating distance × ionizing energy) is to think of radiation as cookies. Imagine that you have four cookies—an α cookie, a β cookie, a γ cookie, and a neutron cookie. Which one would you eat, which would you hold in your hand, which would you put in your pocket, and which would you throw away?

11.86 What are the main advantages of fusion relative to fission as an energy source? What are the drawbacks?

11.87 Write a balanced nuclear equation for

   (a) α emission of ^{162}Re and
   (b) β emission of ^{188}W.

11.88 Balance the following transmutation reactions:

   (a) $^{253}_{99}Es + ? \rightarrow ^{256}_{101}Md + ^{1}_{0}n$
   (b) $^{250}_{98}Cf + ^{11}_{5}B \rightarrow ? + 4\,^{1}_{0}n$

11.89 The most abundant isotope of uranium, ^{238}U, does not undergo fission. In a *breeder reactor*, however, a ^{238}U atom captures a neutron and emits two beta particles to make a fissionable isotope of plutonium, which can then be used as fuel in a nuclear reactor. Write the balanced nuclear equation.

11.90 Boron is used in *control rods* for nuclear reactors because it can absorb neutrons and emit α particles. Balance the equation

$$^{10}_{5}B + ^{1}_{0}n \longrightarrow ? + ^{4}_{2}He$$

11.91 Thorium-232 decays by a ten-step series, ultimately yielding lead-208. How many α particles and how many β particles are emitted?

11.92 Californium-246 is formed by bombardment of uranium-238 atoms. If four neutrons are formed as by-products, what particle is used for the bombardment?

CHAPTER 12

Introduction to Organic Chemistry: Alkanes

CONCEPTS TO REVIEW

Covalent Bonds
(Sections 5.1 and 5.2)

Multiple Covalent Bonds
(Section 5.3)

Drawing Lewis Structures
(Section 5.6)

VSEPR and Molecular Shapes
(Section 5.7)

Polar Covalent Bonds
(Section 5.8)

▲ The gasoline, kerosene, and other products of this petroleum refinery are primarily mixtures of simple organic compounds called alkanes.

CONTENTS

CHAPTER GOALS

In this and the next five chapters, we will look at the chemistry of organic compounds, beginning with answers to the following questions:

1. **What are the basic properties of organic compounds?**

   THE GOAL: Be able to identify organic compounds and the types of bonds contained in them.

2. **What are functional groups, and how are they used to classify organic molecules?**

   THE GOAL: Be able to classify organic molecules into families by functional group.

3. **What are isomers?**

   THE GOAL: Be able to recognize and draw constitutional isomers.

4. **How are organic molecules drawn?**

   THE GOAL: Be able to convert between structural formulas and condensed or line structures.

5. **What are alkanes and cycloalkanes, and how are they named?**

   THE GOAL: Be able to name an alkane or cycloalkane from its structure, or write the structure, given the name.

6. **What are the general properties and chemical reactions of alkanes?**

   THE GOAL: Be able to describe the physical properties of alkanes and the products formed in the combustion and halogenation reactions of alkanes.

As knowledge of chemistry slowly grew in the 1700s, mysterious differences were noted between compounds obtained from living sources and those obtained from minerals. It was found, for instance, that chemicals from living sources were often liquids or low-melting solids, whereas chemicals from mineral sources were usually high-melting solids. Furthermore, chemicals from living sources were generally more difficult to purify and work with than those from minerals. To express these differences, the term *organic chemistry* was introduced to mean the study of compounds from living organisms, and *inorganic chemistry* was used to refer to the study of compounds from minerals.

Today we know that there are no fundamental differences between organic and inorganic compounds: The same scientific principles are applicable to both. The only common characteristic of compounds from living sources is that they contain the element carbon as their primary component. Thus, organic chemistry is now defined as the study of carbon-based compounds.

Why is carbon special? The answer derives from its position in the periodic table. As a group 4A nonmetal, carbon atoms have the unique ability to form four strong covalent bonds. Also, unlike atoms of other elements, carbon atoms can readily form strong bonds with other carbon atoms to produce long chains and rings. As a result, only carbon is able to form such an immense array of compounds, from methane with one carbon atom to DNA with billions of carbons.

12.1 The Nature of Organic Molecules

Let us begin a study of **organic chemistry**—the chemistry of carbon compounds—by reviewing what we have seen in earlier chapters about the structures of organic molecules:

Organic chemistry The study of carbon compounds.

- **Carbon is tetravalent; it always forms four bonds** (Section 5.2). In methane, for example, carbon is connected to four hydrogen atoms:

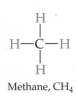

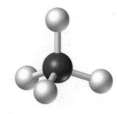

Methane, CH_4

- **Organic molecules have covalent bonds** (Section 5.2). In ethane, for example, the bonds result from the sharing of two electrons, either between two C atoms or a C and a H atom:

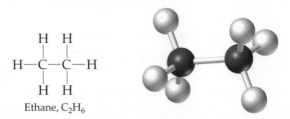

Ethane, C_2H_6

- **Organic molecules contain polar covalent bonds when carbon bonds to an electronegative element on the right side of the periodic table** (Section 5.8). In chloromethane, for example, the electronegative chlorine atom attracts electrons more strongly than carbon, resulting in polarization of the C—Cl bond so that carbon has a partial positive charge, $\delta+$, and chlorine has a partial negative charge, $\delta-$. In electrostatic potential maps (Section 5.8), the chlorine atom is therefore in the red region of the map and the carbon atom in the blue region:

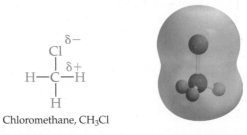

Chloromethane, CH_3Cl

- **Carbon forms multiple covalent bonds by sharing more than two electrons with a neighboring atom** (Section 5.3). In ethylene, for example, the two carbon atoms share four electrons in a double bond; in acetylene (also called ethyne), the two carbons share six electrons in a triple bond:

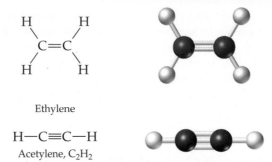

Ethylene

H—C≡C—H

Acetylene, C_2H_2

- **Organic molecules have specific three-dimensional shapes** (Section 5.7). When carbon is bonded to four atoms, as in methane, CH_4, the bonds are oriented toward the four corners of a regular tetrahedron with carbon in the center. Such three-dimensionality is commonly shown using normal lines for bonds in the plane of the page, dashed lines for bonds receding behind the page, and wedged lines for bonds coming out of the page:

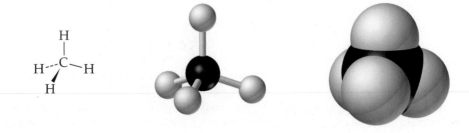

- **Organic molecules often contain hydrogen, nitrogen, and oxygen, in addition to carbon** (Section 5.6). Nitrogen can form single, double, and triple bonds to carbon; oxygen can form single and double bonds:

$$C-N \quad C-O \quad C-H$$
$$C=N \quad C=O$$
$$C\equiv N$$

Covalent bonding makes organic compounds quite different from the inorganic compounds we have been concentrating on up to this point. For example, inorganic compounds such as NaCl have high melting points and high boiling points because they consist of large numbers of oppositely charged ions held together by strong electrical attractions (Section 4.4). (⚭, p. 84) Organic compounds, by contrast, consist of atoms joined by covalent bonds into individual molecules. Because the molecules are attracted to one another only by weak non-ionic intermolecular forces (Section 8.11), organic compounds generally have lower melting and boiling points than inorganic salts. (⚭, p. 235) Because of their relatively low melting and boiling points, simple organic compounds are usually liquids at room temperature, and a few are gases.

Other important differences between organic and inorganic compounds include solubility and electrical conductivity. Whereas many inorganic compounds dissolve in water to yield solutions of ions that conduct electricity (Section 9.9), most organic compounds are insoluble in water, and almost all of those that are soluble do not conduct electricity. (⚭, p. 275) Only small polar organic molecules, such as glucose and ethyl alcohol, or large molecules with many polar groups, such as some proteins, dissolve in water. This lack of water solubility for organic compounds has important practical consequences, varying from the difficulty in removing greasy dirt and cleaning up environmental oil spills to drug delivery.

◀ Oil spills can be a serious environmental problem because oil is insoluble in water.

⚭ Looking Ahead

The interior of a living cell is largely a water solution that contains many hundreds of different compounds. In later chapters, we will see how cells use membranes composed of water-insoluble organic molecules to enclose their watery interiors and to regulate the flow of substances across the cell boundary.   ⚭

12.2 Families of Organic Molecules: Functional Groups

More than 18 *million* organic compounds are described in the scientific literature. Each of these 18 million compounds has unique physical properties, such as a melting point and a boiling point, each has unique chemical properties, and many of them have unique biological properties (both desired and undesired). How can we ever understand them all?

Chemists have learned through experience that organic compounds can be classified into families according to their structural features and that the chemical behavior of the members of a family is often predictable based on these specific grouping of atoms. Instead of 18 million compounds with seemingly random chemical reactivity, there are just a few general families of organic compounds whose chemistry falls into simple patterns.

The structural features that allow us to classify organic compounds into distinct chemical families are called **functional groups**. A functional group is an atom or group of atoms that has a characteristic physical and chemical behavior. Each functional group is always part of a larger molecule, and a molecule may have more than one class of functional group present, as we shall soon see. An important property of functional groups is that a given functional group *tends to undergo the same reactions in every molecule of which it is a part*. For example, the carbon–carbon double bond is a common functional group. Ethylene (C_2H_4), the simplest compound with a double bond, undergoes many chemical reactions similar to those of oleic acid ($C_{18}H_{34}O_2$), a much larger and more complex compound that also contains a double bond. Both, for example, react with hydrogen gas in the same manner, as shown in Figure 12.1. These identical reactions with hydrogen are typical: *The chemistry of an organic molecule is primarily determined by the functional groups it contains, not by its size and complexity*.

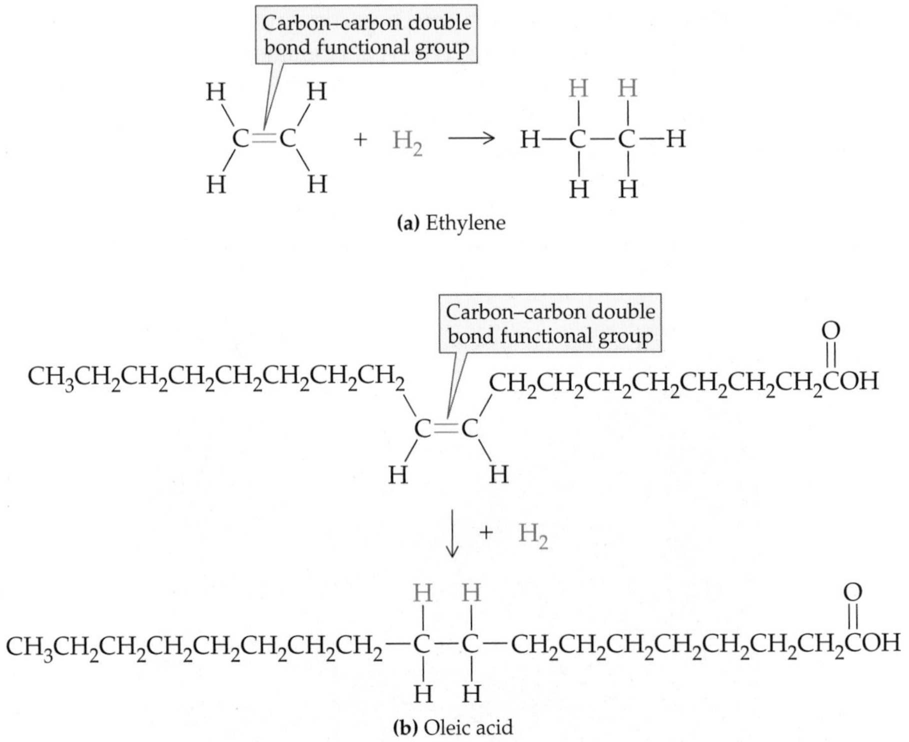

Functional group An atom or group of atoms within a molecule that has a characteristic physical and chemical behavior.

(a) Ethylene

(b) Oleic acid

▶ **FIGURE 12.1 The reactions of (a) ethylene and (b) oleic acid with hydrogen.** The carbon–carbon double-bond functional group adds 2 hydrogen atoms in both cases, regardless of the complexity of the rest of the molecule.

Table 12.1 lists some of the most important families of organic molecules and their distinctive functional groups. Compounds that contain a $C=C$ double bond, for instance, are in the *alkene* family; compounds that have an —OH group bound to a tetravalent carbon are in the *alcohol* family; and so on.

Much of the chemistry discussed in this and the next five chapters is the chemistry of the families listed in Table 12.1, so it is best to memorize the names and become familiar with their structures now. Note that they fall into three groups:

Hydrocarbon An organic compound that contains only carbon and hydrogen.

- The first four families in Table 12.1 are **hydrocarbons**, organic compounds that contain only carbon and hydrogen. *Alkanes* have only single bonds and contain no functional groups. As we will see, the absence of functional groups makes alkanes relatively unreactive. *Alkenes* contain a carbon–carbon double-bond functional group; *alkynes* contain a carbon–carbon triple-bond functional group; and *aromatic* compounds contain a six-membered ring of carbon atoms with three alternating double bonds.

- The next four families in Table 12.1 have functional groups that contain only single bonds and have a carbon atom bonded to an electronegative atom.

TABLE 12.1 Some Important Families of Organic Molecules

FAMILY NAME	FUNCTIONAL GROUP STRUCTURE*	SIMPLE EXAMPLE	NAME ENDING
Alkane	Contains only C—H and C—C single bonds	CH_3CH_3 Ethane	-ane
Alkene		$H_2C=CH_2$ Ethylene	-ene
Alkyne	—C≡C—	H—C≡C—H Acetylene (Ethyne)	-yne
Aromatic		Benzene	None
Alkyl halide	—C—X (X=F, Cl, Br, I)	CH_3—Cl Methyl chloride	None
Alcohol	—C—O—H	CH_3—OH Methyl alcohol (Methanol)	-ol
Ether	—C—O—C—	CH_3—O—CH_3 Dimethyl ether	None
Amine	—C—N	CH_3—NH_2 Methylamine	-amine
Aldehyde		Acetaldehyde (Ethanal)	-al
Ketone		Acetone	-one
Carboxylic acid		Acetic acid	-ic acid
Anhydride		Acetic anhydride	None
Ester		Methyl acetate	-ate
Amide		Acetamide	-amide

*The bonds whose connections are not specified are assumed to be attached to carbon or hydrogen atoms in the rest of the molecule.

Alkyl halides have a carbon–halogen bond; *alcohols* have a carbon–oxygen bond; *ethers* have two carbons bonded to the same oxygen; and *amines* have a carbon–nitrogen bond.

- The remaining families in Table 12.1 have functional groups that contain a carbon–oxygen double bond: *aldehydes, ketones, carboxylic acids, anhydrides, esters,* and *amides.*

WORKED EXAMPLE **12.1** Molecular Structures: Identifying Functional Groups

To which family of organic compounds do the following compounds belong? Explain.

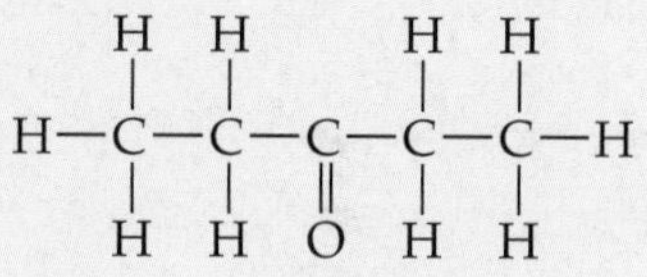

ANALYSIS Identify each functional group from the list in Table 12.1, and name the corresponding family to which the compound belongs.

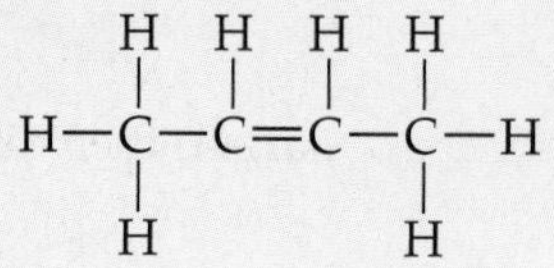

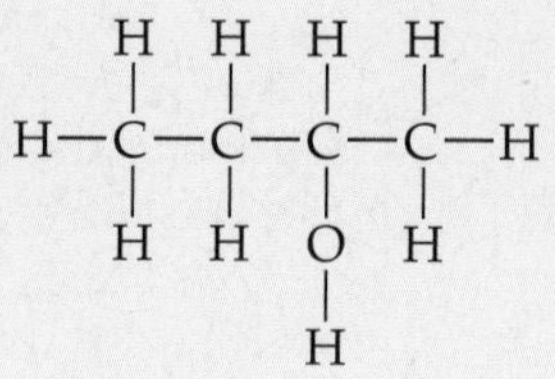

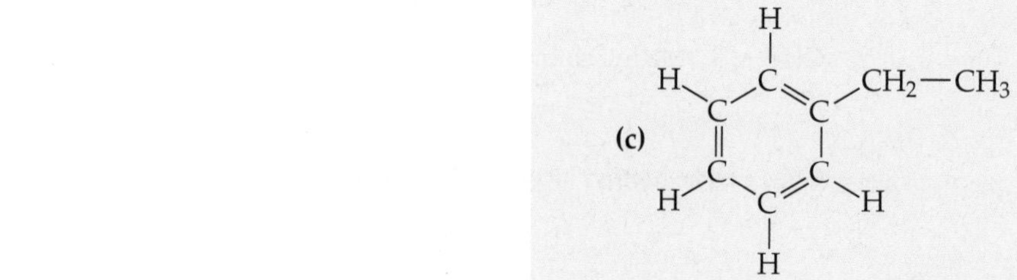

SOLUTION

- **(a)** This compound contains only carbon and hydrogen atoms, so it is a *hydrocarbon*. There is a carbon–carbon double bond, so it is an *alkene*.
- **(b)** The O—H group bonded to tetravalent carbon identifies this compound as an *alcohol*.
- **(c)** This compound also contains only carbon and hydrogen atoms, which identifies it as a *hydrocarbon*. The six-membered carbon ring with alternating double bonds also identifies this compound as an *aromatic* hydrocarbon compound.
- **(d)** The carbon–oxygen double bond in the middle of the carbon chain (as opposed to at either end of the chain) identifies this compound as a *ketone*.

WORKED EXAMPLE **12.2** Molecular Structures: Drawing Functional Groups

Given the family of organic compounds to which the compound belongs, propose structures for compounds having the following chemical formulas.

- **(a)** An amine having the formula C_2H_7N
- **(b)** An alkyne having the formula C_3H_4
- **(c)** An ether having the formula $C_4H_{10}O$

ANALYSIS Identify the functional group for each compound from Table 12.1. Once the atoms in this functional group are eliminated from the chemical formula, the remaining structure can be determined. (Remember that each carbon atom forms four bonds, nitrogen forms three bonds, oxygen forms two bonds, and hydrogen forms only one bond.)

SOLUTION

(a) Amines have a C—NH$_2$ group. Eliminating these atoms from the formula leaves 1 C atom and 5 H atoms. Since only the carbons are capable of forming more than one bond, the 2 C atoms must be bonded together. The remaining H atoms are then bonded to the carbons until each C has 4 bonds.

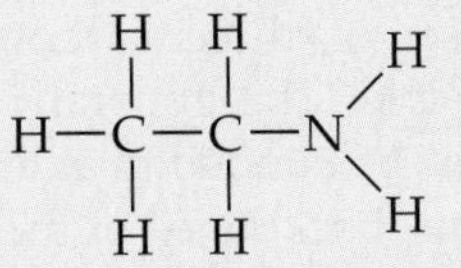

$$\text{H}-\underset{\underset{\text{H}}{|}}{\overset{\overset{\text{H}}{|}}{\text{C}}}-\underset{\underset{\text{H}}{|}}{\overset{\overset{\text{H}}{|}}{\text{C}}}-\text{N}\begin{matrix} \nearrow \text{H} \\ \searrow \text{H} \end{matrix}$$

(b) The alkynes contain a C≡C bond. This leaves 1 C atom and 4 H atoms. Attach this C to one of the carbons in the triple bond, and then distribute the H atoms until each carbon has a full complement of four bonds.

$$\text{H}-\underset{\underset{\text{H}}{|}}{\overset{\overset{\text{H}}{|}}{\text{C}}}-\text{C}{\equiv}\text{C}-\text{H}$$

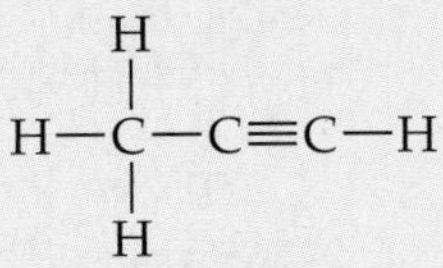

(c) The ethers contain a C—O—C group. Eliminating these atoms leaves 2 C atoms and 10 H atoms. The C atoms can be distributed on either end of the ether group, and the H atoms are then distributed until each carbon atom has a full complement of four bonds.

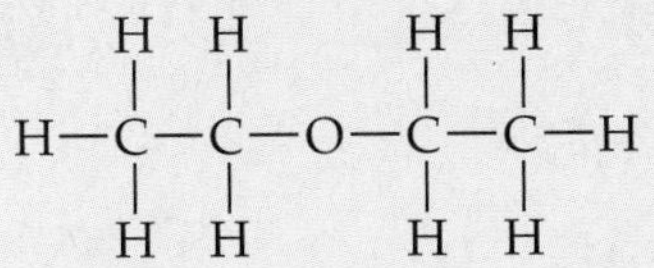

$$\text{H}-\underset{\underset{\text{H}}{|}}{\overset{\overset{\text{H}}{|}}{\text{C}}}-\underset{\underset{\text{H}}{|}}{\overset{\overset{\text{H}}{|}}{\text{C}}}-\text{O}-\underset{\underset{\text{H}}{|}}{\overset{\overset{\text{H}}{|}}{\text{C}}}-\underset{\underset{\text{H}}{|}}{\overset{\overset{\text{H}}{|}}{\text{C}}}-\text{H} \quad \text{or} \quad \text{H}-\underset{\underset{\text{H}}{|}}{\overset{\overset{\text{H}}{|}}{\text{C}}}-\underset{\underset{\text{H}}{|}}{\overset{\overset{\text{H}}{|}}{\text{C}}}-\underset{\underset{\text{H}}{|}}{\overset{\overset{\text{H}}{|}}{\text{C}}}-\text{O}-\underset{\underset{\text{H}}{|}}{\overset{\overset{\text{H}}{|}}{\text{C}}}-\text{H}$$

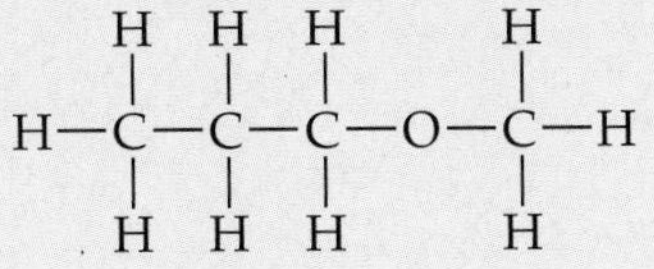

PROBLEM 12.1

Many organic compounds contain more than one functional group. Locate and identify the functional groups in (a) lactic acid, from sour milk; (b) methyl methacrylate, used in making Lucite and Plexiglas; and (c) phenylalanine, an amino acid found in proteins.

(a) 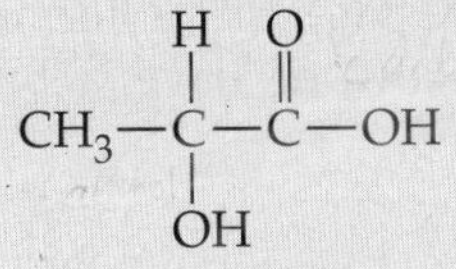 $\text{CH}_3-\underset{\underset{\text{OH}}{|}}{\overset{\overset{\text{H}}{|}}{\text{C}}}-\overset{\overset{\text{O}}{\|}}{\text{C}}-\text{OH}$

(b) $\text{CH}_2{=}\underset{\underset{\text{CH}_3}{|}}{\text{C}}-\overset{\overset{\text{O}}{\|}}{\text{C}}-\text{O}-\text{CH}_3$

(c) $\text{H}-\text{C}\underset{\underset{\text{H}}{|}}{\overset{\overset{\text{H}}{\diagdown}}{\text{C}{=}\text{C}}}\dots\text{C}-\text{CH}_2-\underset{\underset{\text{NH}_2}{|}}{\overset{\overset{\text{H}}{|}}{\text{C}}}-\overset{\overset{\text{O}}{\|}}{\text{C}}-\text{OH}$

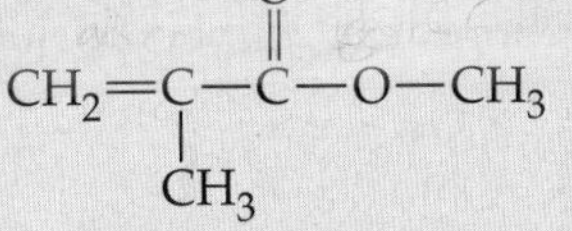

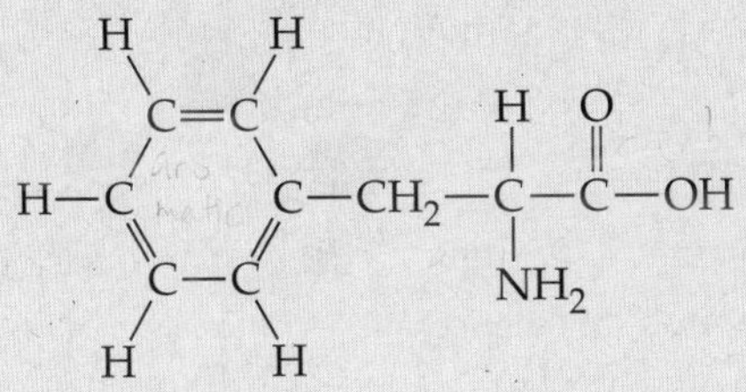

PROBLEM 12.2

Propose structures for molecules that fit the following descriptions:

(a) $\text{C}_2\text{H}_4\text{O}$ containing an aldehyde functional group

(b) $\text{C}_3\text{H}_6\text{O}_2$ containing a carboxylic acid functional group

12.3 The Structure of Organic Molecules: Alkanes and Their Isomers

Alkane A hydrocarbon that has only single bonds.

Hydrocarbons that contain only single bonds belong to the family of organic molecules called **alkanes**. Imagine how one carbon and four hydrogens can combine, and you will realize there is only one possibility: methane, CH_4. Now imagine how two carbons and six hydrogens can combine—only ethane, CH_3CH_3, is possible. Likewise with the combination of three carbons with eight hydrogens—only propane, $CH_3CH_2CH_3$, is possible. The general rule for *all* hydrocarbons except methane is that each carbon *must* be bonded to at least one other carbon. The carbon atoms bond together to form the "backbone" of the compound, with the hydrogens on the periphery. The general formula for alkanes is C_nH_{2n+2}, where n is the number of carbons in the compound.

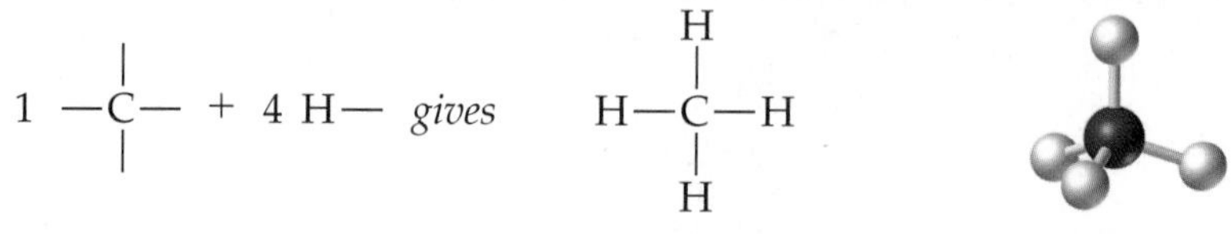

Methane

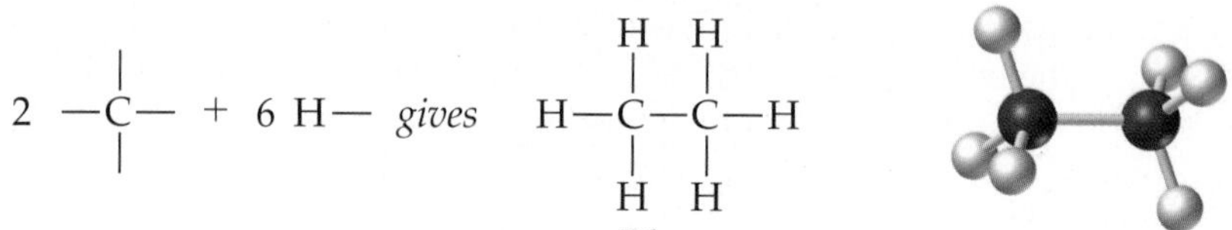

Ethane

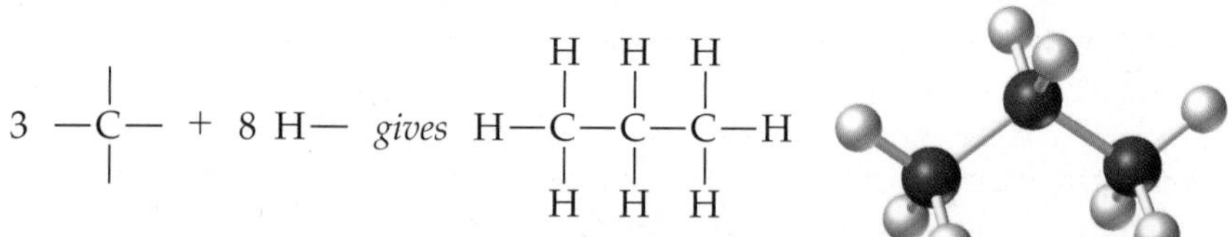

Propane

As larger numbers of carbons and hydrogens combine, the ability to form *isomers* arises. Compounds that have the same molecular formula but different structural formulas are called **isomers** of one another. For example, there are two ways in which molecules that have the formula C_4H_{10} can be formed: The four carbons can either be joined in a contiguous row or have a branched arrangement:

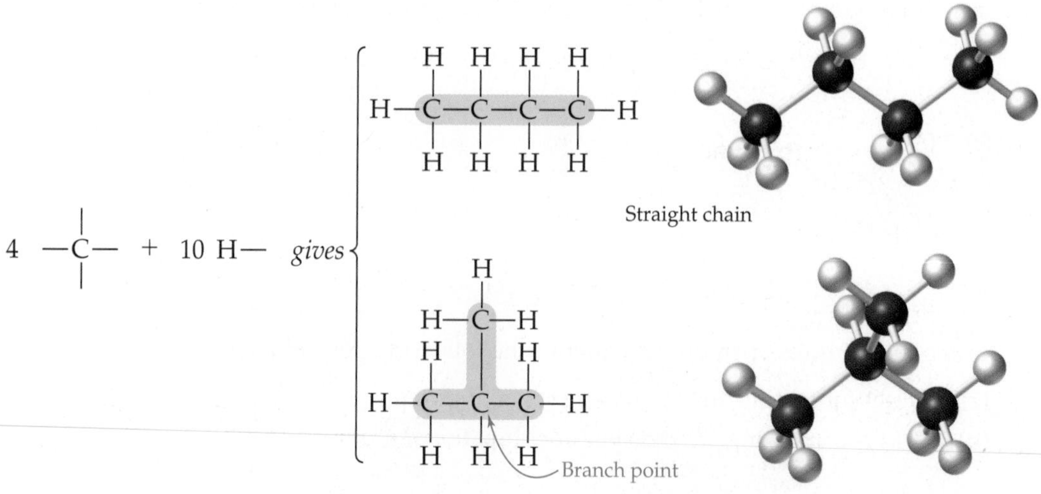

Isomers Compounds with the same molecular formula but different structures.

Straight chain

Branch point

Branched chain

The same is seen with the molecules that have the formula C_5H_{12}, for which three isomers are possible:

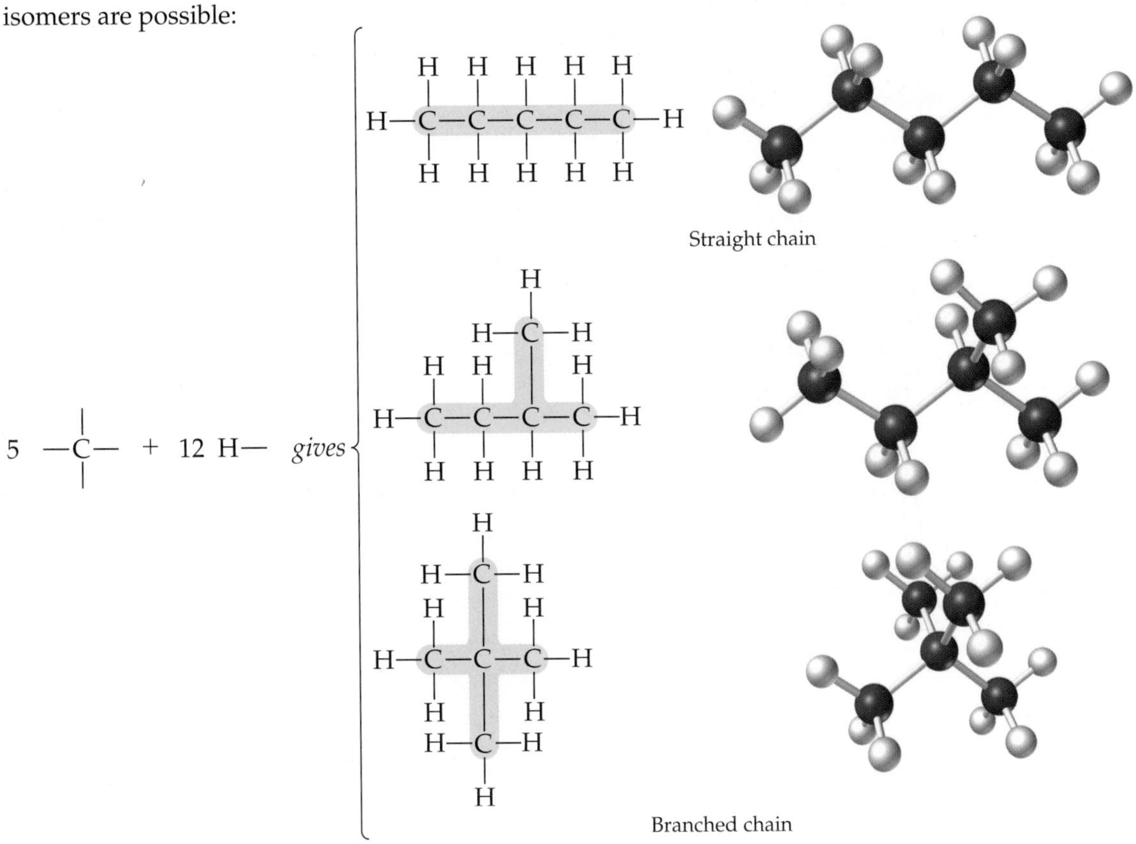

Straight chain

$$5 \ -\overset{|}{\underset{|}{C}}- \ + \ 12 \ H- \quad gives$$

Branched chain

Compounds with all their carbons connected in a continuous chain are called **straight-chain alkanes**; those with a branching connection of carbons are called **branched-chain alkanes**. Note that in a straight-chain alkane, you can draw a line through all the carbon atoms without lifting your pencil from the paper. In a branched-chain alkane, however, you must either lift your pencil from the paper or retrace your steps to draw a line through all the carbons.

The two isomers of C_4H_{10} and the three isomers of C_5H_{12} shown above are **constitutional isomers**—compounds with the same molecular formula but with different connections among their constituent atoms. Needless to say, the number of possible alkane isomers grows rapidly as the number of carbon atoms increases.

Constitutional isomers of a given molecular formula are chemically completely different from one another. They have different structures, different physical properties such as melting and boiling points, and potentially different physiological properties. For example, ethyl alcohol and dimethyl ether both have the formula C_2H_6O, but ethyl alcohol is a liquid with a boiling point of 78.5 °C and dimethyl ether is a gas with a boiling point of −23 °C (Table 12.2). Clearly, molecular formulas by

Straight-chain alkane An alkane that has all its carbons connected in a row.

Branched-chain alkane An alkane that has a branching connection of carbons.

Constitutional isomers Compounds with the same molecular formula but different connections among their atoms.

TABLE 12.2 Some Properties of Ethyl Alcohol and Dimethyl Ether

NAME AND MOLECULAR FORMULA	STRUCTURE	BOILING POINT	MELTING POINT	PHYSIOLOGICAL ACTIVITY
Ethyl alcohol C_2H_6O		78.5 °C	− 117.3 °C	Central-nervous-system depressant
Dimethyl ether C_2H_6O		− 23 °C	− 138.5 °C	Nontoxic; anesthetic at high concentration

themselves are not very useful in organic chemistry; a knowledge of structure is also necessary.

WORKED EXAMPLE **12.3** Molecular Structures: Drawing Isomers

Draw all isomers that have the formula C_6H_{14}.

ANALYSIS Knowing that all the carbons must be bonded together to form the molecule, find all possible arrangements of the 6 carbon atoms. Begin with the isomer that has all 6 carbons in a straight chain, then draw the isomer that has 5 carbons in a straight chain, using the remaining carbon to form a branch, then repeat for the isomer having 4 carbons in a straight chain and 2 carbons in branches. Once each carbon backbone is drawn, arrange the hydrogens around the carbons to complete the structure. (Remember that each carbon can only have *four* bonds.)

SOLUTION
The straight-chain isomer contains all 6 carbons bonded to form a chain with no branches. The branched isomers are drawn by starting with either a 5-carbon chain or a 4-carbon chain, and adding the extra carbons as branches in the middle of the chain. Hydrogens are added until each carbon has a full complement of 4 bonds.

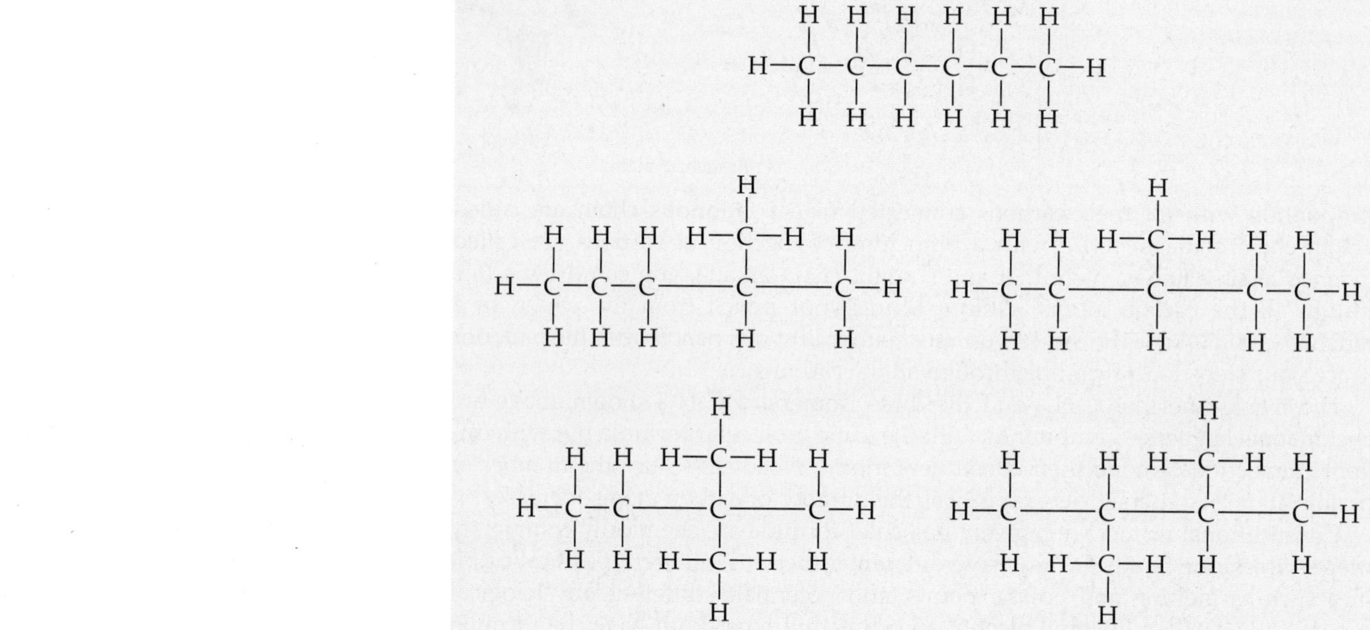

PROBLEM 12.3

Draw the straight-chain isomer with the formula C_7H_{16}.

PROBLEM 12.4

Draw any two of the seven branched-chain isomers with the formula C_7H_{16}.

12.4 Drawing Organic Structures

Condensed structure A short-hand way of drawing structures in which C—C and C—H bonds are understood rather than shown.

Drawing structural formulas that show every atom and every bond in a molecule is both time-consuming and awkward, even for relatively small molecules. Much easier is the use of **condensed structures** (Section 5.6), which are simpler but still show the essential information about which functional groups are present and how atoms are connected. ($\textbf{CO}$, p. 118) In condensed structures, C—C and C—H single bonds are not shown; rather, they are "understood." If a carbon atom has 3 hydrogens bonded to it, we write CH_3; if the carbon has 2 hydrogens bonded to it, we write CH_2; and so on. For example, the 4-carbon, straight-chain alkane (butane) and its branched-chain isomer (2-methylpropane) can be written as the following condensed structures:

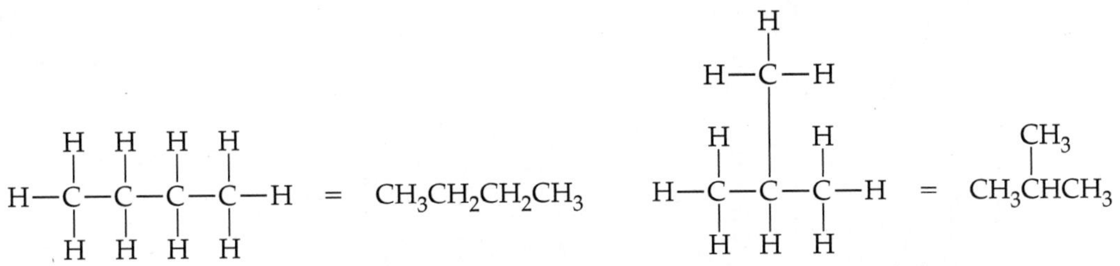

	Butane		2-Methylpropane	
	Structural formula	Condensed formula	Structural formula	Condensed formula

Note in these condensed structures for butane and 2-methylpropane that the horizontal bonds between carbons are not usually shown—the CH_3 and CH_2 units are simply placed next to one another—but that the vertical bond in 2-methyl-propane *is* shown for clarity.

Occasionally, as a further simplification, not all the CH_2 groups (called **methylenes**) are shown. Instead, CH_2 is shown once in parentheses, with a subscript to the right of the) indicating the number of methylene units strung together. For example, the 6-carbon straight-chain alkane (hexane) can be written as

Methylene Another name for a CH_2 unit.

$$CH_3CH_2CH_2CH_2CH_2CH_3 = CH_3(CH_2)_4CH_3$$

WORKED EXAMPLE **12.4** Molecular Structures: Writing Condensed Structures

Write condensed structures for the isomers from Worked Example 12.3.

ANALYSIS Eliminate all horizontal bonds, substituting reduced formula components (CH_3, CH_2, and so on) for each carbon in the compound. Show vertical bonds to branching carbons for clarity.

SOLUTION

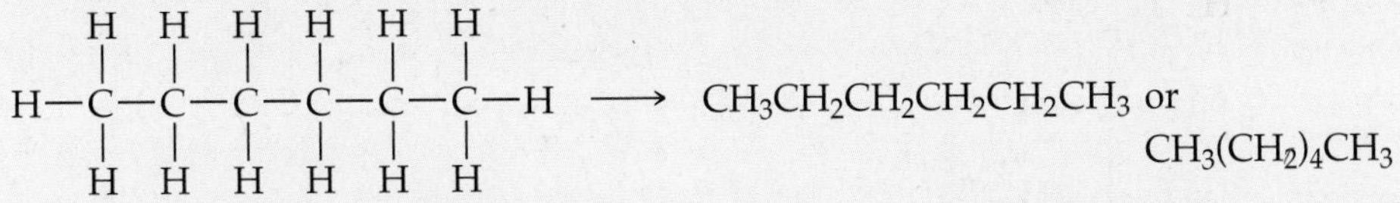

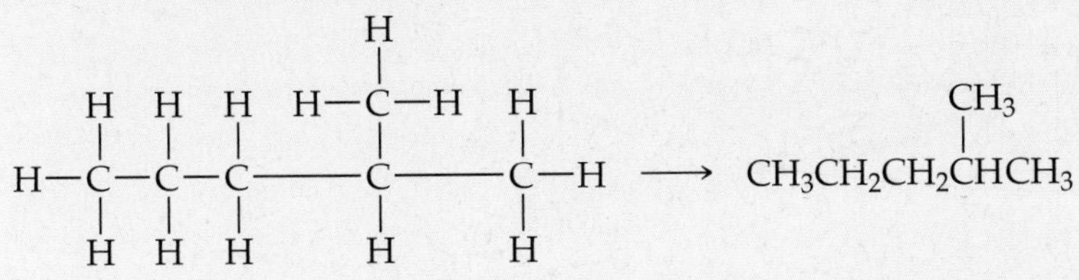

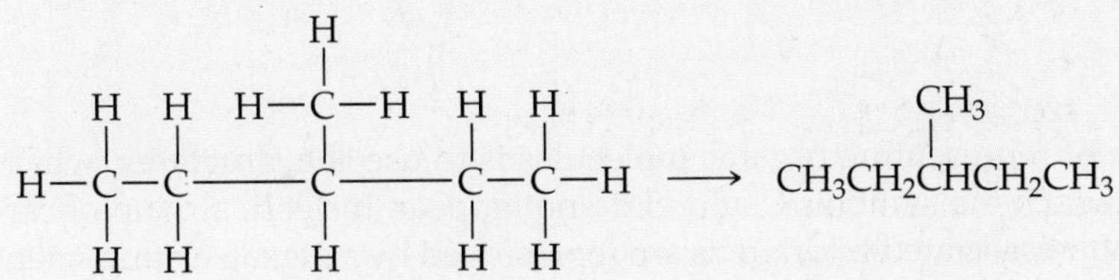

$$CH_3CH_2CH_2CH_2CH_2CH_3 \text{ or}$$
$$CH_3(CH_2)_4CH_3$$

$$CH_3CH_2CH_2\overset{\overset{\textstyle CH_3}{|}}{C}HCH_3$$

$$CH_3CH_2\overset{\overset{\textstyle CH_3}{|}}{C}HCH_2CH_3$$

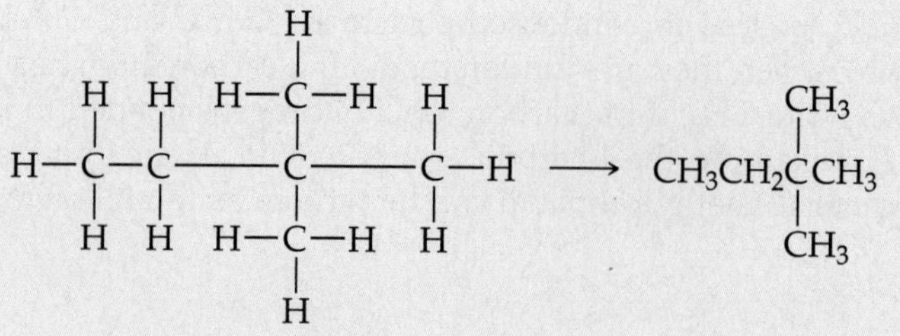

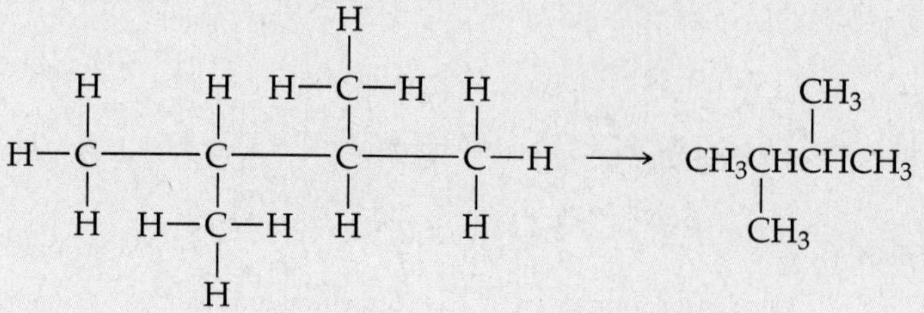

PROBLEM 12.5

Draw the following three isomers of C_5H_{12} as condensed structures:

(a) 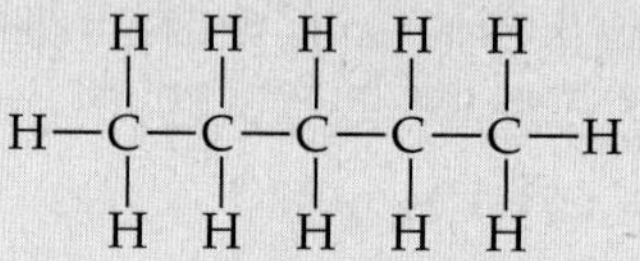 Pentane

(b) 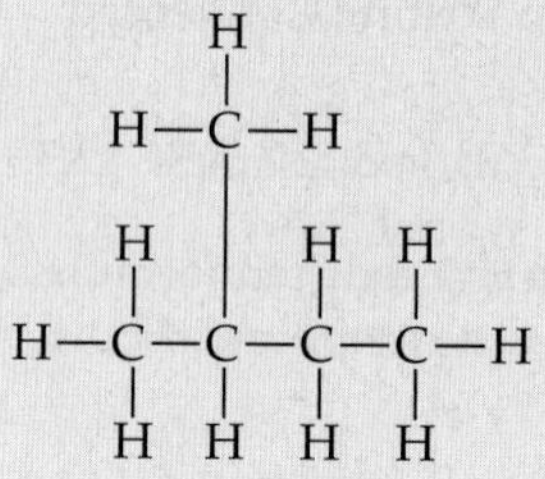 2-Methylbutane

(c) 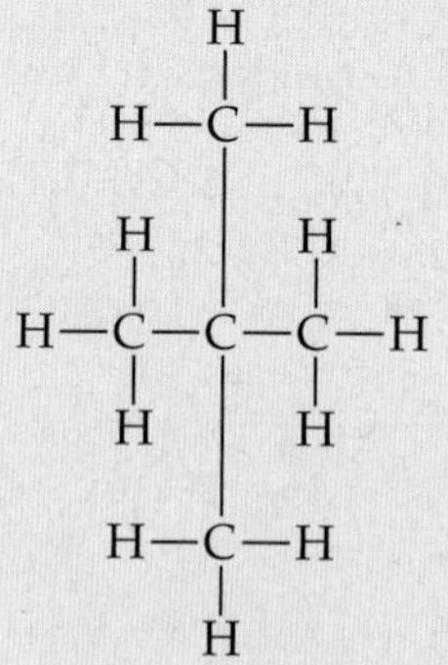 2,2-Dimethylpropane

Line structure A shorthand way of drawing structures in which carbon and hydrogen atoms are not shown. Instead, a carbon atom is understood to be wherever a line begins or ends and at every intersection of two lines, and hydrogens are understood to be wherever they are needed to have each carbon form four bonds.

Another way of representing organic molecules is to use **line structures**, which are structures in which the symbols C and H do not appear. Instead, a chain of carbon atoms and their associated hydrogens are represented by a zigzag arrangement of short lines, with any branches off the main chain represented by additional lines. The line structure for 2-methylbutane, for instance, is

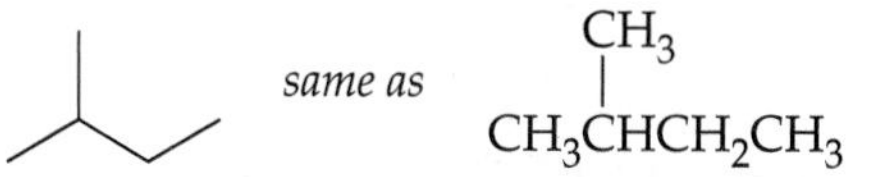

same as
CH3
CH3CHCH2CH3

Line structures are a simple and quick way to represent organic molecules without the clutter arising from showing all carbons and hydrogens present. Chemists, biologists, pharmacists, doctors, and nurses all use line structures to conveniently convey to one another very complex organic structures. Another advantage is that a line structure gives a more realistic depiction of the angles seen in a carbon chain.

Drawing a molecule in this way is simple, provided one follows these guidelines:

1. Each carbon-carbon bond is represented by a line.
2. Anywhere a line ends or begins, as well as any vertex where two lines meet, represents a carbon atom.
3. Any atom other than another carbon or a hydrogen attached to a carbon must be shown.
4. Since a neutral carbon atom forms four bonds, all bonds not shown for any carbon are understood to be the number of carbon-hydrogen bonds needed to have the carbon form four bonds.

CH3
CH2
CH3
CH3
CH
CH3
CH2
CH3
CH
CH3

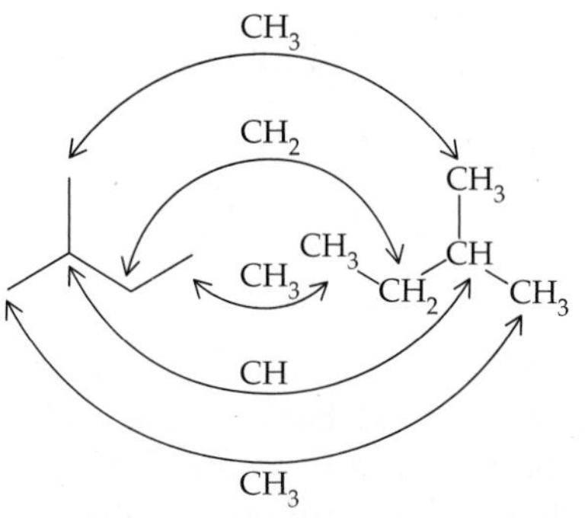

Converting line structures to structural formulas or to condensed structures is simply a matter of correctly interpreting each line ending and each intersection in a line structure. For example, the common pain reliever ibuprofen has the condensed and line structures

CH3
HC=CH
O
CH3CHCH2—C
C—CH—C—OH
CH—CH
CH3
OH
O

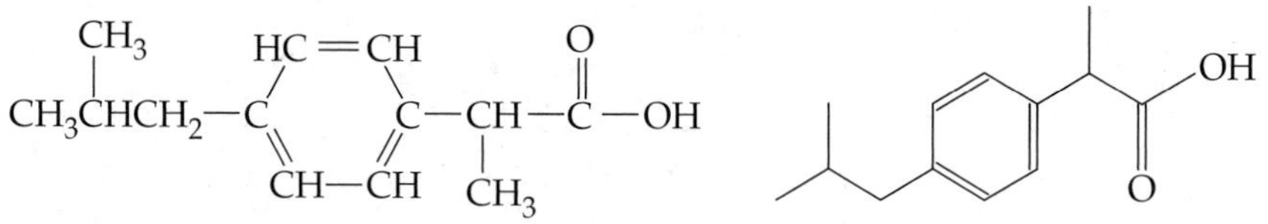

Finally, it is important to note that chemists and biochemists often use a mixture of structural formulas, condensed structures, and line structures to represent the molecules they study. As you progress through this textbook, you will see many complicated molecules represented in this way, so it is a good idea to get used to thinking interchangably in all three formats.

WORKED EXAMPLE **12.5** Molecular Structures: Converting Condensed Structures to Line Structures

Convert the following condensed structures to line structures:

(a)
$$CH_3CH_2CHCHCH_2CH_3$$
with CH_3 on the upper carbon and CH_3 on the lower carbon

(b)
$$CH_3CHCH-C\ CH_2CH_3$$
with OH and Cl above, and CH_3 and CH_3 below

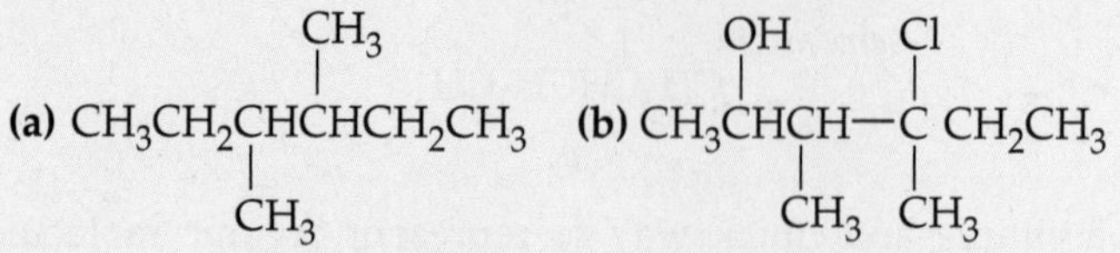

ANALYSIS Find the longest continuous chain of carbon atoms in the condensed structure. Begin the line structure by drawing a zigzag line in which the number of vertices plus line ends equals the number of carbon atoms in the chain. Show branches coming off the main chain by drawing vertical lines at the vertices as needed. Show all atoms that are not carbons or are not hydrogens attached to carbons.

SOLUTION

(a) Begin by drawing a zigzag line in which the total number of ends + vertices equals the number of carbons in the longest chain (here 6, with the carbons numbered for clarity):

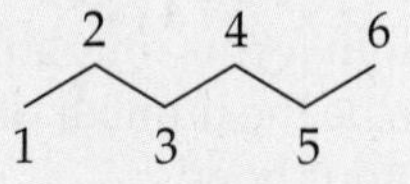

Looking at the condensed structure, you see CH_3 groups on carbons 3 and 4; these two methyl groups are represented by lines coming off those carbons in the line structure:

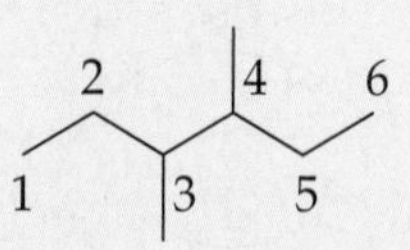

This is the complete line structure. Notice that the hydrogens are not shown, but understood. For example, carbon 4 has three bonds shown: one to carbon 3, one to carbon 5, and one to the branch methyl group; the fourth bond this carbon must have is understood to be to a hydrogen.

(b) Proceed as in (a), drawing a zigzag line for the longest chain of carbon atoms, which again contains 6 carbons. Next draw a line coming off each carbon bonded to a CH_3 group (carbons 3 and 4). Both the OH and the Cl groups must be shown to give the final structure:

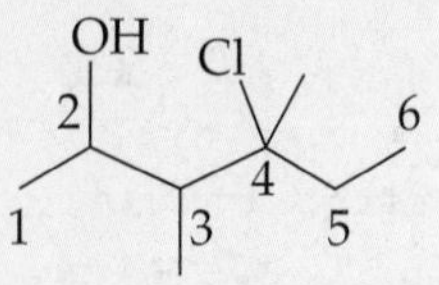

Note from this line structure than it does not matter in such a two-dimensional drawing what direction you show for a group that branches off the main chain, as long as it is attached to the correct carbon. This is true for condensed structures as well. Quite often the direction that a group is shown coming off a main chain of carbon atoms is chosen simply for aesthetic reasons.

WORKED EXAMPLE **12.6** Molecular Structures: Converting Line Structures to Condensed Structures

Convert the following line structures to condensed structures:

(a)

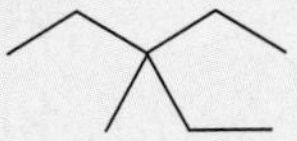

(b)

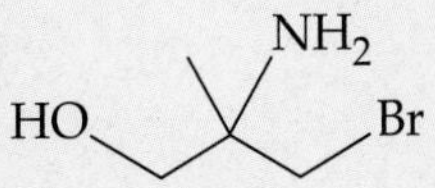

ANALYSIS Convert all vertices and line ends to carbons. Write in any noncarbon atoms and any hydrogens bonded to a noncarbon atom. Add hydrogens as needed so that each carbon has four groups attached. Remove lines connecting carbons except for branches.

SOLUTION

(a) Anywhere a line ends and anywhere two lines meet, write a C:

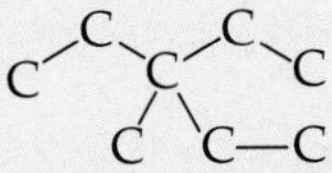

Because there are no atoms other than carbons and hydrogens in this molecule, the next step is to add hydrogens as needed to have four bonds for each carbon:

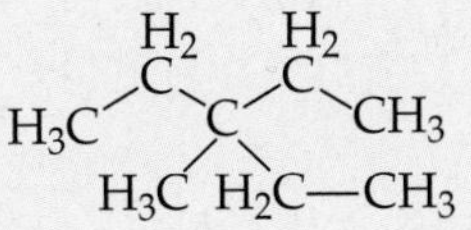

Finally, eliminate all lines except for branches to get the condensed structure:

$$CH_3CH_2 \overset{\overset{\displaystyle CH_3}{|}}{\underset{\underset{\displaystyle CH_2CH_3}{|}}{C}} CH_2CH_3$$

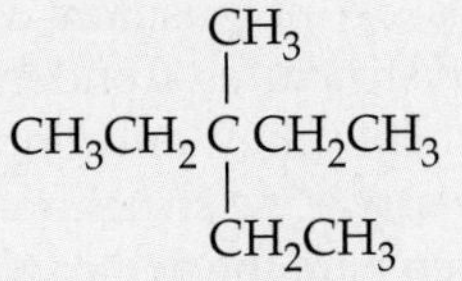

(b) Begin the condensed structure with a drawing showing a carbon at each line end and at each intersection of two lines:

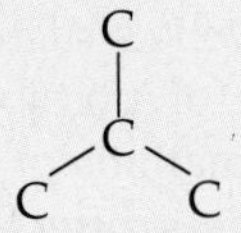

Next write in all the noncarbon atoms and the hydrogen bonded to the oxygen. Then add hydrogens so that each carbon forms four bonds:

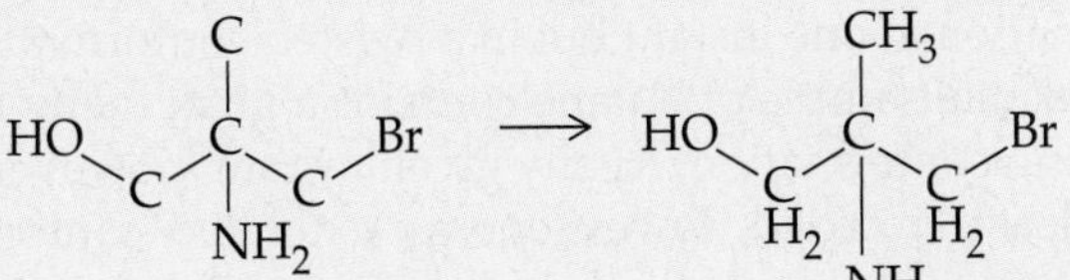

Eliminate all lines except for branches for the completed condensed structure:

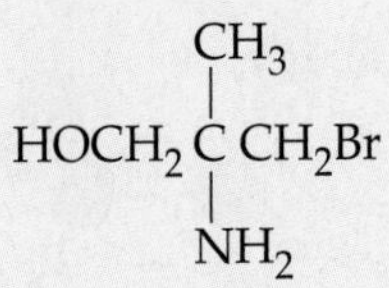

$$HOCH_2 \overset{\overset{\displaystyle CH_3}{|}}{\underset{\underset{\displaystyle NH_2}{|}}{C}} CH_2Br$$

PROBLEM 12.6

Convert the following condensed structures to line structures:

(a)

$$CH_3CH_2C \overset{\displaystyle CH_2CH_3}{\underset{\displaystyle CH_2OH}{|}}CH_2CH_2CH_3$$

(b)

$$CH_3CH\,\overset{\displaystyle CH_2CH_3}{\underset{\displaystyle CH_3CHCH_2CH_3}{|}}CH\;\overset{\displaystyle CH_3}{|}CH_2CHCH_3$$

(c)

$$CH_3C\overset{\displaystyle Br}{\underset{\displaystyle \overset{|}{CH_3}\;\overset{|}{CHCH_2CH_3}}{|}}CHCH_2CH_2CHCH_2CH_3\;\overset{\displaystyle Cl}{|}$$

PROBLEM 12.7

Convert the following line structures to condensed structures:

(a)   (b)

12.5 The Shapes of Organic Molecules

Every carbon atom in an alkane has its four bonds pointing toward the four corners of a tetrahedron, but chemists do not usually worry about three-dimensional shapes when writing condensed structures. Condensed structures do not imply any particular three-dimensional shape; they only indicate the connections between atoms without specifying geometry. Line structures do try to give some feeling for the shape of a molecule, but even here the ability to show three-dimensional shape is limited unless dashed and wedged lines are used for the bonds (page 125).

Butane, for example, has no one single shape because *rotation* takes place around carbon–carbon single bonds. The two parts of a molecule joined by a carbon–carbon single bond in a noncyclic structure are free to spin around the bond, giving rise to an infinite number of possible three-dimensional geometries, or **conformations**. The various conformations of a molecule such as butane are called **conformers** of one another. Conformers differ from one another as a result of rotation around carbon–carbon single bonds. Although the conformers of a given molecule have different three-dimensional shapes and different energies, the conformers cannot be separated from one another. A given butane molecule might be in its fully extended conformation at one instant but in a twisted conformation an instant later (Figure 12.2). An actual sample of butane contains a great many molecules that are constantly changing conformation. At any given instant, however, most of the molecules have the least crowded, lowest-energy extended conformation shown in Figure 12.2a. The same is true for all other alkanes: At any given instant, most molecules are in the least crowded conformation.

Conformation The specific three-dimensional arrangement of atoms in a molecule.

Conformer Molecular structures having identical connections between atoms; that is, they represent identical compounds.

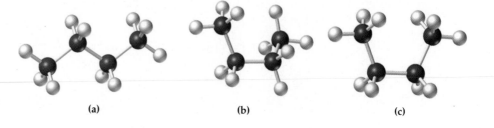

▶ **FIGURE 12.2 Some conformations of butane (there are many others as well).** The least crowded, extended conformation in (a) is the lowest-energy one.

(a)   (b)   (c)

As long as any two structures have identical connections between atoms, they are conformers of each other and represent the same compound, no matter how the structures are drawn. Sometimes you have to mentally rotate structures to see whether they are conformers or different molecules. To see that the following two structures represent conformers of the same compound rather than two isomers, picture one of them flipped right to left so that the red CH₃ groups are on the same side:

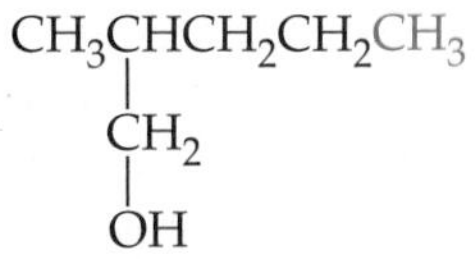

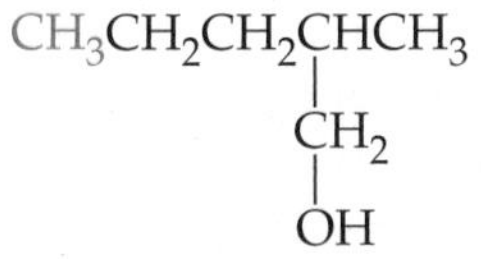

$$CH_3CHCH_2CH_2CH_3 \qquad CH_3CH_2CH_2CHCH_3$$
$$\quad\quad | \qquad\qquad\qquad\qquad\qquad |$$
$$\quad\quad CH_2 \qquad\qquad\qquad\qquad\quad CH_2$$
$$\quad\quad | \qquad\qquad\qquad\qquad\qquad |$$
$$\quad\quad OH \qquad\qquad\qquad\qquad\quad OH$$

Another way to determine whether two structures are conformers is to name each one using the IUPAC nomenclature rules (Section 12.6). If two structures have the same name, they are conformers of the same compound.

WORKED EXAMPLE   **12.7** Molecular Structures: Identifying Conformers

The following structures all have the formula C_7H_{16}. Which of them represent the same molecule?

(a) 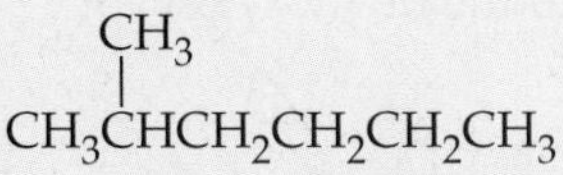
$$\begin{array}{c} CH_3 \\ | \\ CH_3CHCH_2CH_2CH_2CH_3 \end{array}$$

(b)
$$\begin{array}{c} CH_3 \\ | \\ CH_3CH_2CH_2CH_2CHCH_3 \end{array}$$

(c)
$$\begin{array}{c} CH_3 \\ | \\ CH_3CH_2CH_2CHCH_2CH_3 \end{array}$$

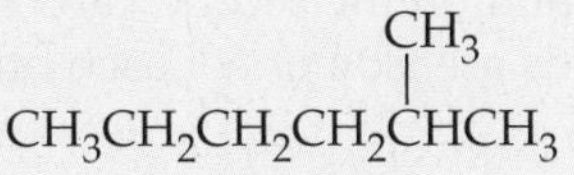

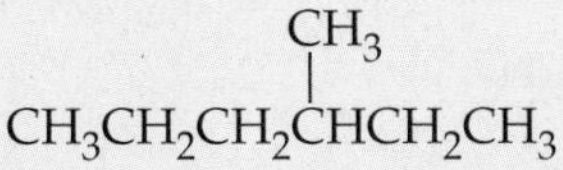

ANALYSIS Pay attention to the *connections* between atoms. Do not get confused by the apparent differences caused by writing a structure right to left versus left to right. Begin by identifying the longest chain of carbon atoms in the molecule.

SOLUTION
Molecule (a) has a straight chain of 6 carbons with a —CH₃ branch on the second carbon from the end. Molecule (b) also has a straight chain of 6 carbons with a —CH₃ branch on the second carbon from the end and is therefore identical to (a). That is, (a) and (b) are conformers of the same molecule. The only difference between (a) and (b) is that one is written "forward" and one is written "backward." Molecule (c), by contrast, has a straight chain of 6 carbons with a —CH₃ branch on the *third* carbon from the end and is therefore an isomer of (a) and (b).

WORKED EXAMPLE   **12.8** Molecular Structures: Identifying Conformers and Isomers

Are the following pairs of compounds the same (conformers), isomers, or unrelated?

(a) 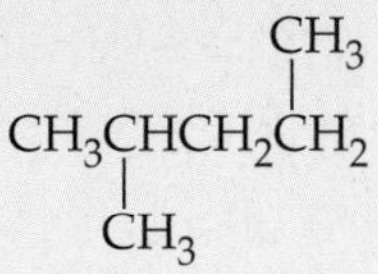 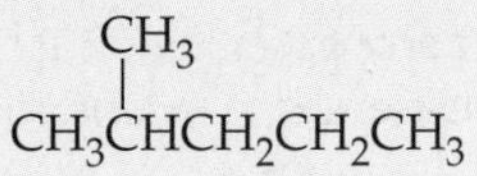
$$\begin{array}{c} CH_3 \\ | \\ CH_3CHCH_2CH_2 \\ | \\ CH_3 \end{array} \qquad \begin{array}{c} CH_3 \\ | \\ CH_3CHCH_2CH_2CH_3 \end{array}$$

(b) 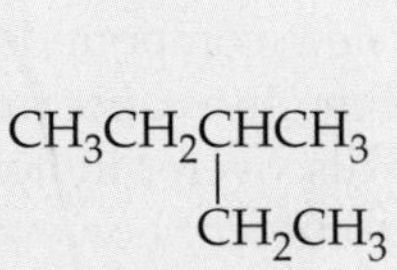 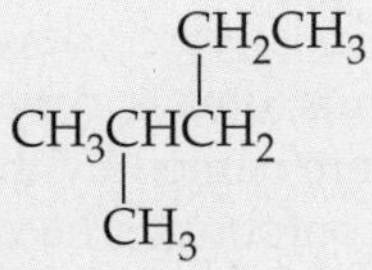
$$\begin{array}{c} CH_3CH_2CHCH_3 \\ | \\ CH_2CH_3 \end{array} \qquad \begin{array}{c} CH_2CH_3 \\ | \\ CH_3CHCH_2 \\ | \\ CH_3 \end{array}$$

(c)
$$CH_3CH_2OCH_3 \qquad \begin{array}{c} O \\ \| \\ CH_3CH_2CH \end{array}$$

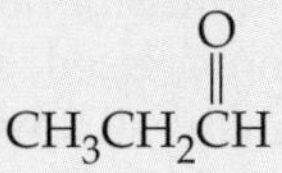

ANALYSIS First compare molecular formulas to see if the compounds are related, and then look at the structures to see if they are the same compound or isomers. Find the longest continuous carbon chain in each, and then compare the locations of the substituents connected to the longest chain.

SOLUTION

(a) Both compounds have the same molecular formula (C_6H_{14}) so they are related. Since the $-CH_3$ group is on the second carbon from the end of a 5-carbon chain in both cases, these structures represent the same compound and are conformers of each other.

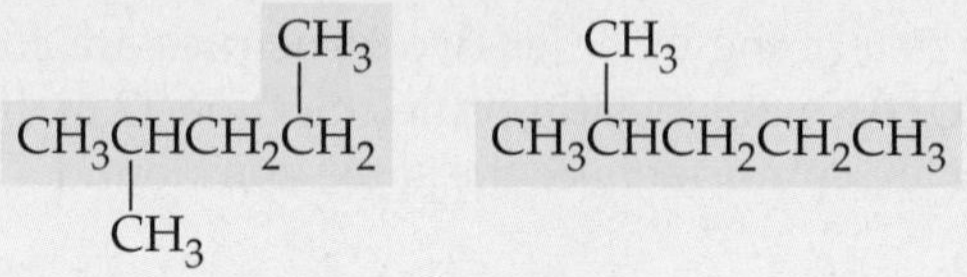

$$CH_3 \qquad\qquad CH_3$$
$$| \qquad\qquad\qquad |$$
$$CH_3CHCH_2CH_2 \qquad CH_3CHCH_2CH_2CH_3$$
$$|$$
$$CH_3$$

(b) Both compounds have the same molecular formula (C_6H_{14}) and the longest chain in each is 5 carbon atoms. A comparison shows, however, that the $-CH_3$ group is on the middle carbon atom in one structure and on the second carbon atom in the other. These compounds are isomers of each other.

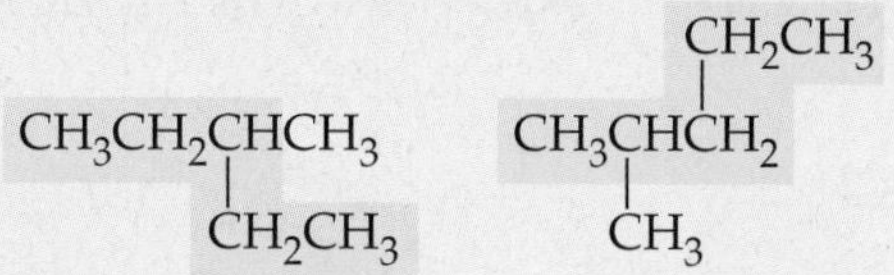

$$\qquad\qquad\qquad\qquad\qquad\qquad CH_2CH_3$$
$$\qquad\qquad\qquad\qquad\qquad\qquad |$$
$$CH_3CH_2CHCH_3 \qquad CH_3CHCH_2$$
$$\qquad\qquad |\qquad\qquad\qquad\qquad |$$
$$\qquad\quad CH_2CH_3 \qquad\qquad CH_3$$

(c) These compounds have different formulas (C_3H_8O and C_3H_6O), so they are unrelated; they are neither conformers nor isomers of each other.

PROBLEM 12.8

Which of the following structures are conformers?

$$CH_3 \quad CH_3 \qquad\qquad\qquad\qquad\qquad CH_3$$
$$| \qquad |\qquad\qquad\qquad\qquad\qquad\qquad\qquad |$$
(a) 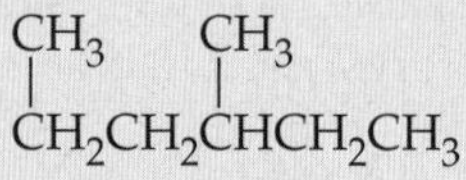 $CH_2CH_2CHCH_2CH_3$ \qquad (b) 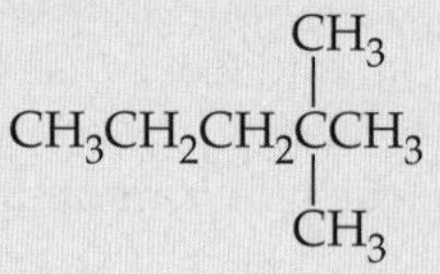 $CH_3CH_2CH_2CCH_3$
$$\qquad\qquad\qquad\qquad\qquad\qquad\qquad\qquad\qquad |$$
$$\qquad\qquad\qquad\qquad\qquad\qquad\qquad\qquad\quad CH_3$$

$$\qquad\qquad\qquad CH_3$$
$$\qquad\qquad\qquad |$$
(c) 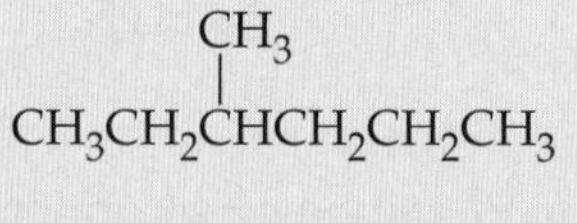 $CH_3CH_2CHCH_2CH_2CH_3$

PROBLEM 12.9

There are 18 isomers with the formula C_8H_{18}. Draw condensed structures for as many as you can and then convert those condensed structures to line structures.

12.6 Naming Alkanes

When relatively few pure organic chemicals were known, new compounds were named at the whim of their discoverer. Thus, urea is a crystalline substance first isolated from urine, and the barbiturates were named by their discoverer in honor of his friend Barbara. As more and more compounds became known, however, the need for a systematic method of naming compounds became apparent.

The system of naming (*nomenclature*) now used is one devised by the International Union of Pure and Applied Chemistry, IUPAC (pronounced **eye**-you-pack). In the IUPAC system for organic compounds, a chemical name has three

APPLICATION ▶ Displaying Molecular Shapes

Molecular shapes are critical to the proper functioning of biological molecules. The tiniest difference in shape can cause two compounds to behave differently or to have different physiological effects in the body. It is therefore critical that chemists be able both to determine molecular shapes with great precision and to visualize these shapes in useful ways.

Three-dimensional shapes of molecules can be determined by *X-ray crystallography*, a technique that allows us to "see" molecules in a crystal using X-ray waves rather than light waves. The molecular "picture" obtained by X-ray crystallography looks at first like a series of regularly spaced dark spots against a lighter background. After computerized manipulation of the data, however, recognizable molecules can be drawn. Relatively small molecules like morphine are usually displayed in either a ball-and-stick format (a), which emphasizes the connections among atoms, or in a space-filling format (b), which emphasizes the overall shape. Chemists find this useful in designing new organic molecules that have specific biological properties that make the molecules useful as therapeutic agents.

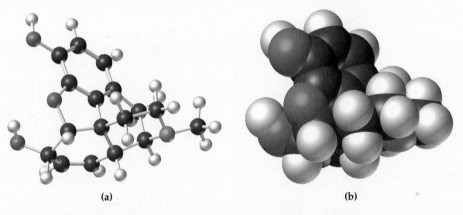

(a)　　　　　　　　　　　　　　(b)

▲ This computer-generated model of morphine is displayed in both (a) ball-and-stick format and (b) space-filling format.

Enormous biological molecules like enzymes and other proteins are best displayed on computer terminals where their structures can be enlarged, rotated, and otherwise manipulated for the best view. An immunoglobulin molecule, for instance, is so large that little detail can be seen in ball-and-stick or space-filling views. Nevertheless, a cleft inside the molecule is visible if the model is rotated in just the right way.

See Additional Problem 12.64 at the end of the chapter.

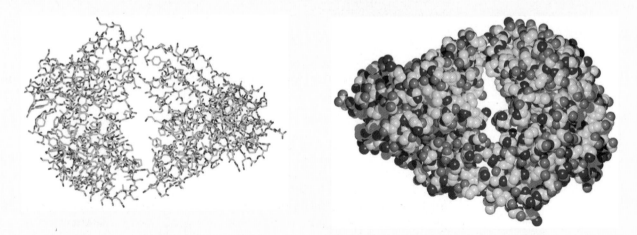

▲ Computer-generated models of an immunoglobulin, one of the antibodies in blood that protect us from harmful invaders such as bacteria and viruses.

Substituent An atom or group of atoms attached to a parent compound.

parts: *prefix, parent,* and *suffix.* The prefix specifies the location of functional groups and other **substituents** in the molecule; the parent tells how many carbon atoms are present in the longest continuous chain; and the suffix identifies what family the molecule belongs to:

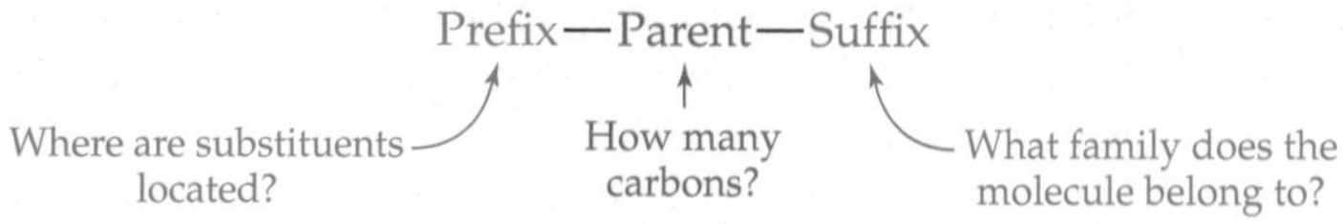

Prefix — Parent — Suffix

Where are substituents located? How many carbons? What family does the molecule belong to?

Straight-chain alkanes are named by counting the number of carbon atoms and adding the family suffix *-ane.* With the exception of the first four compounds— *meth*ane, *eth*ane, *prop*ane, and *but*ane—whose parent names have historical origins, the alkanes are named from Greek numbers according to the number of carbons present (Table 12.3). Thus, *pent*ane is the 5-carbon alkane, *hex*ane is the 6-carbon alkane, and so on. The first ten alkane names are so common that they should be memorized.

TABLE 12.3 Names of Straight-Chain Alkanes

NUMBER OF CARBONS	STRUCTURE	NAME
1	CH_4	*Meth*ane
2	CH_3CH_3	*Eth*ane
3	$CH_3CH_2CH_3$	*Prop*ane
4	$CH_3CH_2CH_2CH_3$	*But*ane
5	$CH_3CH_2CH_2CH_2CH_3$	*Pent*ane
6	$CH_3CH_2CH_2CH_2CH_2CH_3$	*Hex*ane
7	$CH_3CH_2CH_2CH_2CH_2CH_2CH_3$	*Hept*ane
8	$CH_3CH_2CH_2CH_2CH_2CH_2CH_2CH_3$	*Oct*ane
9	$CH_3CH_2CH_2CH_2CH_2CH_2CH_2CH_2CH_3$	*Non*ane
10	$CH_3CH_2CH_2CH_2CH_2CH_2CH_2CH_2CH_2CH_3$	*Dec*ane

Alkyl group The part of an alkane that remains when a hydrogen atom is removed.

Methyl group The $-CH_3$ alkyl group.

Ethyl group The $-CH_2CH_3$ alkyl group.

Substituents such as $-CH_3$ and $-CH_2CH_3$ that branch off the main chain are called **alkyl groups.** An alkyl group can be thought of as the part of an alkane that remains when 1 hydrogen atom is removed to create an available bonding site. For example, removal of a hydrogen from methane gives the **methyl group,** $-CH_3$, and removal of a hydrogen from ethane gives the **ethyl group,** $-CH_2CH_3$. Notice that these alkyl groups are named simply by replacing the *-ane* ending of the parent alkane with an *-yl* ending:

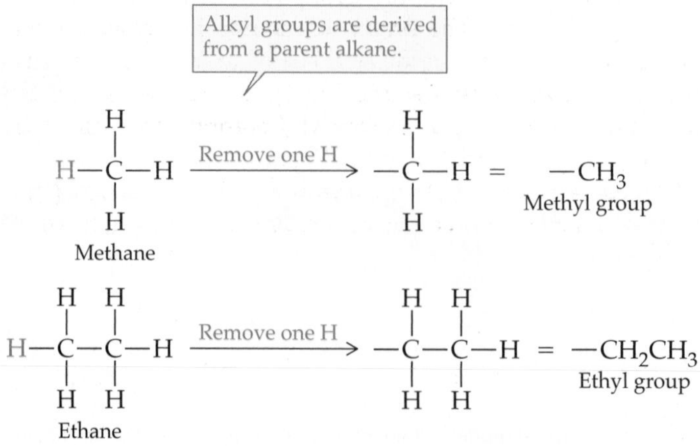

Alkyl groups are derived from a parent alkane.

H—C—H   Remove one H →   —C—H = $-CH_3$
Methane   Methyl group

H—C—C—H   Remove one H →   —C—C—H = $-CH_2CH_3$
Ethane   Ethyl group

Both methane and ethane have only one "kind" of hydrogen. It does not matter which of the 4 methane hydrogens is removed, so there is only one possible methyl group. Similarly, it does not matter which of the 6 equivalent ethane hydrogens is removed, so only one ethyl group is possible.

The situation is more complex for larger alkanes, which contain more than one kind of hydrogen. Propane, for example, has two different kinds of hydrogens. Removal of any one of the 6 hydrogens attached to an end carbon yields a straight-chain alkyl group called **propyl**, whereas removal of either one of the 2 hydrogens attached to the central carbon yields a branched-chain alkyl group called **isopropyl**:

Propyl group The straight-chain alkyl group $-CH_2CH_2CH_3$.

Isopropyl group The branched-chain alkyl group $-CH(CH_3)_2$.

$$-\underset{\underset{H}{|}}{\overset{\overset{H}{|}}{C}}-\underset{\underset{H}{|}}{\overset{\overset{H}{|}}{C}}-\underset{\underset{H}{|}}{\overset{\overset{H}{|}}{C}}-H \ = \ -CH_2CH_2CH_3$$

Propyl group (straight chain)

Remove H from end carbon

$$H-\underset{\underset{H}{|}}{\overset{\overset{H}{|}}{C}}-\underset{\underset{H}{|}}{\overset{\overset{H}{|}}{C}}-\underset{\underset{H}{|}}{\overset{\overset{H}{|}}{C}}-H$$

Propane   Remove H from inside carbon

$$H-\underset{\underset{H}{|}}{\overset{\overset{H}{|}}{C}}-\underset{\underset{H}{|}}{\overset{|}{C}}-\underset{\underset{H}{|}}{\overset{\overset{H}{|}}{C}}-H \ = \ CH_3CHCH_3 \quad \text{or} \quad (CH_3)_2CH-$$

Isopropyl group (branched chain)

It is important to realize that alkyl groups are not compounds but rather are simply partial structures that help us name compounds. The names of some common alkyl groups are listed in Table 12.4; you will want to memorize them.

TABLE 12.4 Some Common Alkyl Groups*

			CH_3
CH_3-	CH_3CH_2-	$CH_3CH_2CH_2-$	CH_3CH-
Methyl	Ethyl	*n*-Propyl	Isopropyl
		CH_3	
$CH_3CH_2CH_2CH_2-$	$CH_3CHCH_2CH_3$	CH_3CHCH_2-	CH_3CCH_3
n-Butyl	*sec*-Butyl	Isobutyl	CH_3
			tert-Butyl

*The red bond shows the connection to the rest of the molecule.

Notice that four butyl (four-carbon) groups are listed in Table 12.4: butyl, *sec*-butyl, isobutyl, and *tert*-butyl. The prefix *sec*- stands for *secondary*, and the prefix *tert*- stands for *tertiary*, referring to the number of other carbon atoms attached to the branch point. There are four possible substitution patterns, called *primary, secondary, tertiary,* and *quaternary*. A **primary (1°) carbon atom** has 1 other carbon attached to it (typically indicated as an —R group in the molecular structure), a **secondary (2°) carbon atom** has 2 other carbons attached, a **tertiary (3°) carbon atom** has 3 other carbons attached, and a **quaternary (4°) carbon atom** has 4 other carbons attached:

Primary (1°) carbon atom A carbon atom with 1 other carbon attached to it.

Secondary (2°) carbon atom A carbon atom with 2 other carbons attached to it.

Tertiary (3°) carbon atom A carbon atom with 3 other carbons attached to it.

Quaternary (4°) carbon atom A carbon atom with 4 other carbons attached to it.

$$R-\underset{\underset{H}{|}}{\overset{\overset{H}{|}}{C}}-H \qquad R-\underset{\underset{H}{|}}{\overset{\overset{R}{|}}{C}}-H \qquad R-\underset{\underset{R}{|}}{\overset{\overset{R}{|}}{C}}-H \qquad R-\underset{\underset{R}{|}}{\overset{\overset{R}{|}}{C}}-R$$

Primary carbon (1°) has one other carbon attached.

Secondary carbon (2°) has two other carbons attached.

Tertiary carbon (3°) has three other carbons attached.

Quaternary carbon (4°) has four other carbons attached.

*The symbol **R** is used here and in later chapters as a general abbreviation for any organic substituent.* You should think of it as representing the **R**est of the molecule, which we are not bothering to specify. The R is used to allow you to focus on a particular structural feature of a molecule without the "clutter" of the other atoms in the molecule detracting from it. The R might represent a methyl, ethyl, or propyl group, or any of a vast number of other possibilities. For example, the generalized formula R—OH for an alcohol might refer to an alcohol as simple as CH_3OH or CH_3CH_2OH or one as complicated as cholesterol.

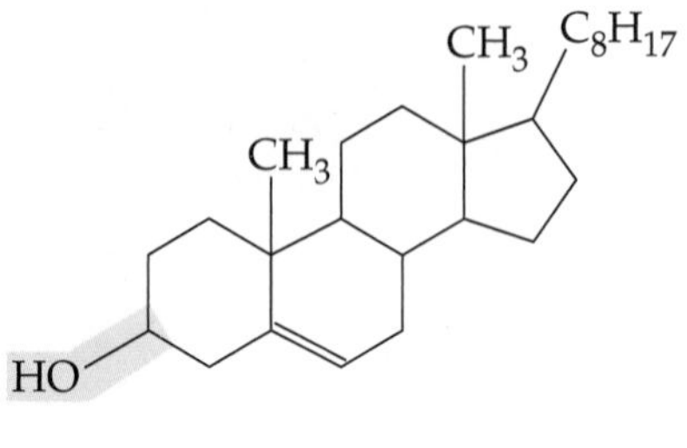

Branched-chain alkanes can be named by following four steps:

STEP 1: Name the main chain. Find the longest continuous chain of carbons, and name the chain according to the number of carbon atoms it contains. The longest chain may not be immediately obvious because it is not always written on one line; you may have to "turn corners" to find it.

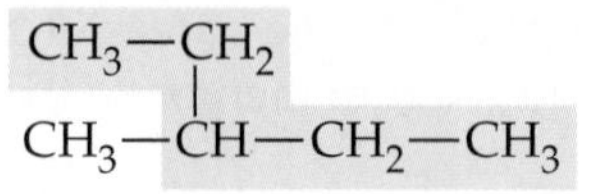

$$CH_3-CH_2$$
$$CH_3-CH-CH_2-CH_3$$

Name as a substituted pentane, not as a substituted butane, because the *longest* chain has five carbons.

STEP 2: Number the carbon atoms in the main chain, beginning at the end nearer the first branch point:

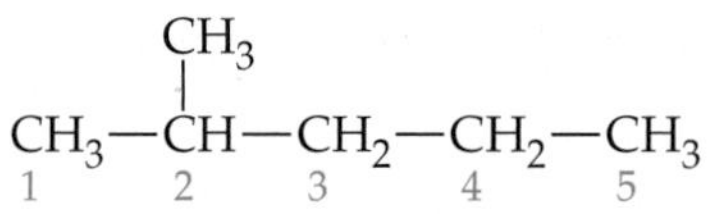

$$CH_3$$
$$CH_3-CH-CH_2-CH_2-CH_3$$
$$12345$$

The first (and only) branch occurs at C2 if we start numbering from the left, but would occur at C4 if we started from the right by mistake.

STEP 3: Identify the branching substituents, and number each according to its point of attachment to the main chain:

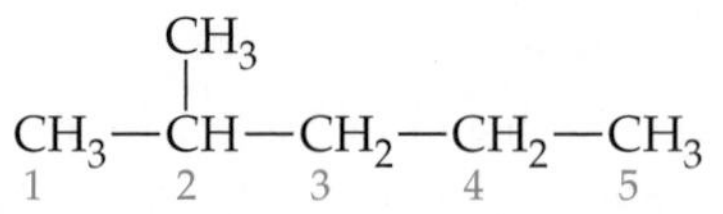

$$CH_3$$
$$CH_3-CH-CH_2-CH_2-CH_3$$
$$12345$$

The main chain is a pentane. There is one —CH_3 substituent group connected to C2 of the chain.

If there are two substituents on the same carbon, assign the same number to both. There must always be as many numbers in the name as there are substituents.

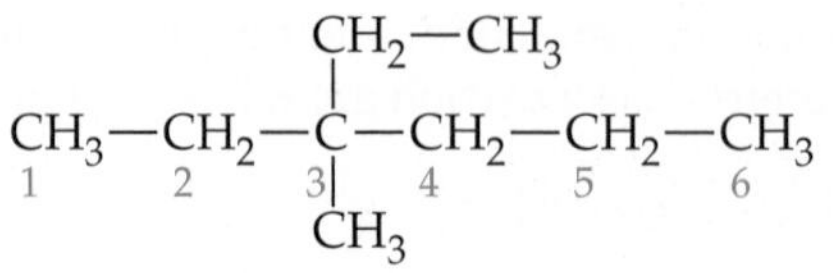

$$CH_2-CH_3$$
$$CH_3-CH_2-C-CH_2-CH_2-CH_3$$
$$123456$$
$$CH_3$$

The main chain is a hexane. There are two substituents, a —CH_3 and a —CH_2CH_3, both connected to C3 of the chain.

STEP 4: Write the name as a single word, using hyphens to separate the numbers from the different prefixes and commas to separate numbers if necessary. If two or more different substituent groups are present, cite them in alphabetical order. If two or more identical substituents are present, use one of the prefixes *di-, tri-, tetra-,* and so forth, but do not use these prefixes for alphabetizing purposes.

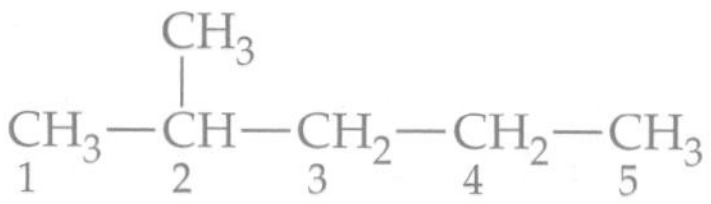

$$CH_3-\underset{\underset{1}{|}}{\overset{\overset{CH_3}{|}}{CH}}-\underset{3}{CH_2}-\underset{4}{CH_2}-\underset{5}{CH_3}$$

2-Methylpentane (a five-carbon main chain with a 2-methyl substituent)

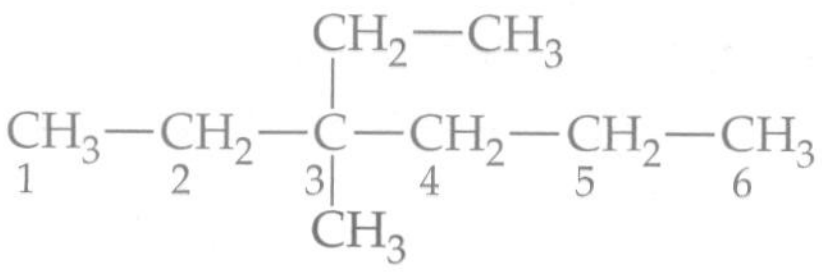

$$CH_3-\underset{1}{CH_2}-\underset{3}{\overset{\overset{CH_2-CH_3}{|}}{C}}-\underset{4}{CH_2}-\underset{5}{CH_2}-\underset{6}{CH_3}$$
$$\underset{}{\overset{}{|}}$$
$$CH_3$$

3-Ethyl-3-methylhexane (a six-carbon main chain with 3-ethyl and 3-methyl substituents cited alphabetically)

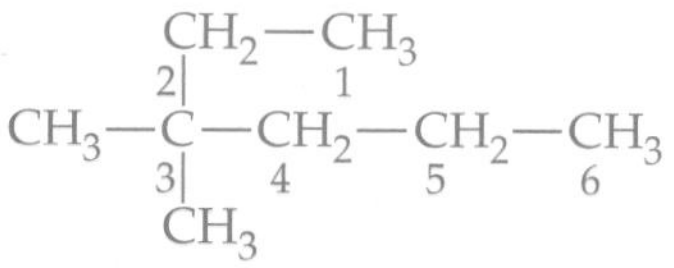

$$CH_3-\underset{3}{\overset{\overset{\overset{2|}{CH_2-CH_3}}{}}{C}}-\underset{4}{CH_2}-\underset{5}{CH_2}-\underset{6}{CH_3}$$
$$\overset{3|}{\underset{}{CH_3}}$$

3,3-Dimethylhexane (a six-carbon main chain with two 3-methyl substitutents)

WORKED EXAMPLE   **12.9** Naming Organic Compounds: Alkanes

What is the IUPAC name of the following alkane?

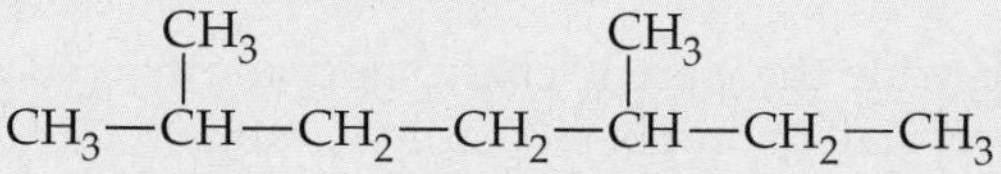

$$CH_3-\overset{\overset{CH_3}{|}}{CH}-CH_2-CH_2-\overset{\overset{CH_3}{|}}{CH}-CH_2-CH_3$$

ANALYSIS Follow the four steps outlined in the text.

SOLUTION

STEP 1: The longest continuous chain of carbon atoms is seven, so the main chain is a *hept*ane.

STEP 2: Number the main chain beginning at the end nearer the first branch:

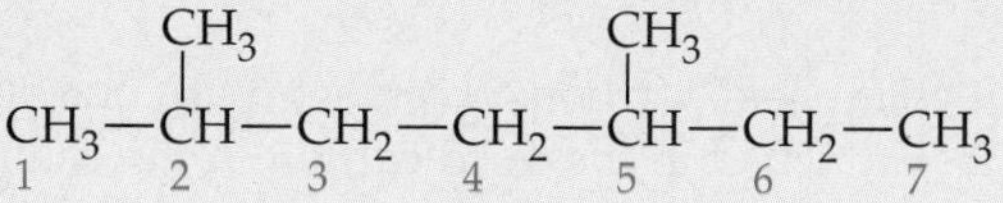

$$CH_3-\underset{2}{\overset{\overset{CH_3}{|}}{CH}}-\underset{3}{CH_2}-\underset{4}{CH_2}-\underset{5}{\overset{\overset{CH_3}{|}}{CH}}-\underset{6}{CH_2}-\underset{7}{CH_3}$$
$$\underset{1}{}$$

STEP 3: Identify and number the substituents (a 2-methyl and a 5-methyl in this case):

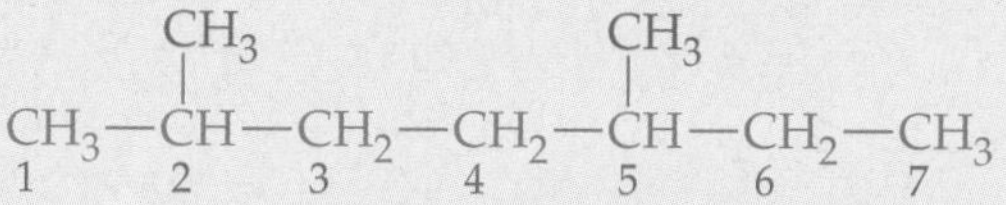

$$CH_3-\underset{2}{\overset{\overset{CH_3}{|}}{CH}}-\underset{3}{CH_2}-\underset{4}{CH_2}-\underset{5}{\overset{\overset{CH_3}{|}}{CH}}-\underset{6}{CH_2}-\underset{7}{CH_3}$$
$$\underset{1}{}$$

Substituents: 2-methyl, 5-methyl

STEP 4: Write the name as one word, using the prefix *di-* because there are two methyl groups. Separate the two numbers by a comma, and use a hyphen between the numbers and the word.

Name: 2, 5-Dimethylheptane

WORKED EXAMPLE   **12.10** Molecular Structure: Identifying 1°, 2°, 3°, and 4° Carbons

Identify each carbon atom in the following molecule as primary, secondary, tertiary, or quaternary.

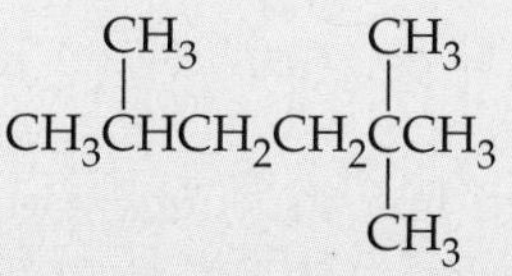

$$CH_3\overset{\overset{CH_3}{|}}{C}HCH_2CH_2\overset{\overset{CH_3}{|}}{C}CH_3$$
$$\underset{\underset{CH_3}{|}}{}$$

ANALYSIS Look at each carbon atom in the molecule, count the number of other carbon atoms attached, and make the assignment accordingly: primary (1 carbon attached); secondary (2 carbons attached); tertiary (3 carbons attached); quaternary (4 carbons attached).

SOLUTION

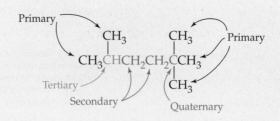

WORKED EXAMPLE **12.11** Molecular Structures: Drawing Condensed Structures from Names

Draw condensed structures corresponding to the following IUPAC names:

(a) 2,3-Dimethylpentane

(b) 3-Ethylheptane

ANALYSIS Starting with the parent chain, add the named alkyl substituent groups to the appropriately numbered carbon atom(s).

SOLUTION

(a) The parent chain has 5 carbons (*pent*ane), with two methyl groups (—CH_3) attached to the second and third carbon in the chain:

$$CH_3CH_3$$
$$|\quad|$$
$$CH_3CH\ CH\ CH_2CH_3$$
$$1\quad2\quad3\quad4\quad5$$

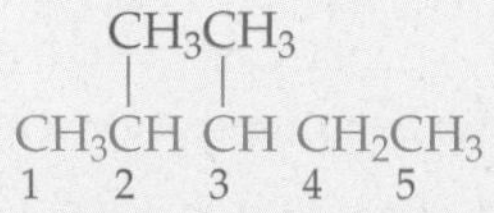

(b) The parent chain has 7 carbons (*hept*ane), with one ethyl group (—CH_2CH_3) attached to the third carbon in the chain:

$$CH_2CH_3$$
$$|$$
$$CH_3CH_2CH\ CH_2CH_2CH_2CH_3$$
$$1\quad2\quad3\quad4\quad5\quad6\quad7$$

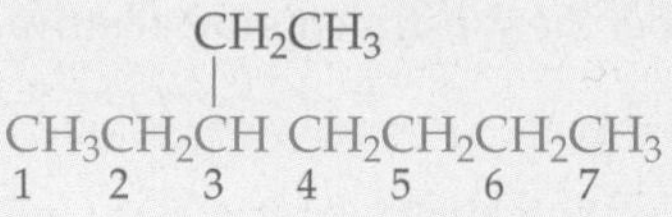

PROBLEM 12.10

What are the IUPAC names of the following alkanes?

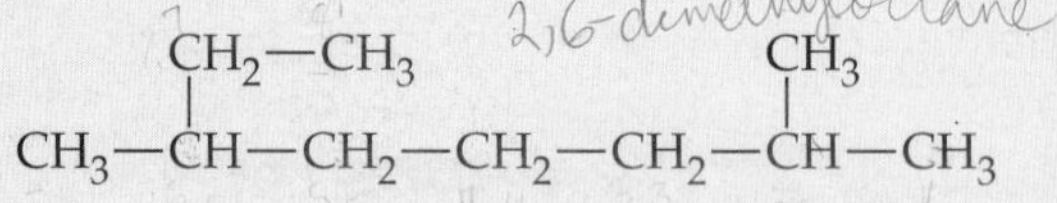

2,6-dimethyloctane

$$CH_2{-}CH_3 \qquad\qquad CH_3$$
$$|\qquad\qquad\qquad\qquad\quad |$$
(a) $CH_3{-}CH{-}CH_2{-}CH_2{-}CH_2{-}CH{-}CH_3$

$$CH_2{-}CH_3$$
$$|$$
(b) $CH_3{-}CH_2{-}CH_2{-}CH_2{-}C{-}CH_2{-}CH_3$
3,3-diethylheptane
$$CH_2{-}CH_3$$

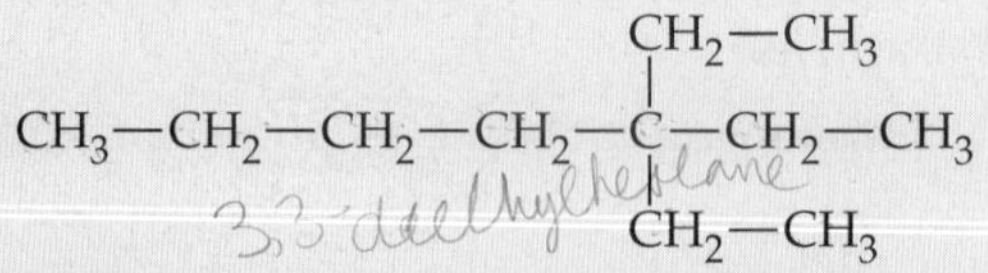

PROBLEM 12.11

Draw both condensed and line structures corresponding to the following IUPAC names and label each carbon as primary, secondary, tertiary, or quaternary:

(a) 3-Methylhexane   (b) 3,4-Dimethyloctane

(c) 2,2,4-Trimethylpentane

PROBLEM 12.12

Draw and name alkanes that meet the following descriptions:

(a) An alkane with a tertiary carbon atom

(b) An alkane that has both a tertiary and a quaternary carbon atom

KEY CONCEPT PROBLEM 12.13

What are the IUPAC names of the following alkanes?

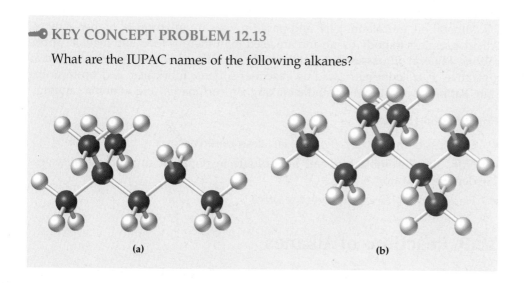

(a)   (b)

12.7 Properties of Alkanes

Alkanes contain only nonpolar C—C and C—H bonds, so the only intermolecular forces influencing them are weak London dispersion forces (Section 8.11). (🔗, p. 235) The effect of these forces is shown in the regularity with which the melting and boiling points of straight-chain alkanes increase with molecular size (Figure 12.3). The first four alkanes—methane, ethane, propane, and butane—are gases at room temperature and pressure. Alkanes with 5–15 carbon atoms are liquids; those with 16 or more carbon atoms are generally low-melting, waxy solids.

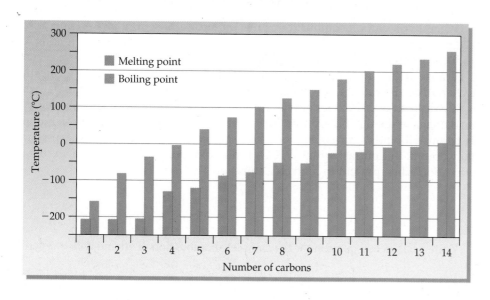

◀ **FIGURE 12.3** The boiling and melting points for the C_1–C_{14} straight-chain alkanes increase with molecular size.

In keeping with their low polarity, alkanes are insoluble in water but soluble in nonpolar organic solvents, including other alkanes (Section 9.2). (🔗, p. 257) Because alkanes are generally less dense than water, they float on its surface.

▲ The waxy coating that makes these apples so shiny is a mixture of higher-molecular-weight alkanes.

Low-molecular-weight alkanes are volatile and must be handled with care because their vapors are flammable. Mixtures of alkane vapors and air can explode when ignited by a single spark.

The physiological effects of alkanes are limited. Methane, ethane, and propane gases are nontoxic, and the danger of inhaling them lies in potential suffocation due to lack of oxygen. Breathing the vapor of higher alkanes in large concentrations can induce loss of consciousness. There is also a danger in breathing droplets of liquid alkanes because they dissolve nonpolar substances in lung tissue and cause pneumonia-like symptoms.

Mineral oil, petroleum jelly, and paraffin wax are mixtures of higher alkanes. All are harmless to body tissue and are used in numerous food and medical applications. Mineral oil passes through the body unchanged and is sometimes used as a laxative. Petroleum jelly (sold as Vaseline) softens, lubricates, and protects the skin. Paraffin wax is used in candle making, on surfboards, and in home canning.

Properties of Alkanes

- Odorless or mild odor; colorless; tasteless; nontoxic
- Nonpolar; insoluble in water but soluble in nonpolar organic solvents; less dense than water
- Flammable; otherwise not very reactive

12.8 Reactions of Alkanes

Alkanes do not react with acids, bases, or most other common laboratory *reagents* (a substance that causes a reaction to occur). Their only major reactions are with oxygen (combustion) and with halogens (halogenation).

Combustion

Combustion—A chemical reaction that produces a flame, usually because of burning with oxygen.

The reaction of an alkane with oxygen occurs during **combustion**, an oxidation reaction that commonly takes place in a controlled manner in an engine or furnace (Section 7.3). (⊂⊃, p. 184) Carbon dioxide and water are the products of complete combustion of any hydrocarbon, and a large amount of heat is released (ΔH is a negative number). Some examples were given in Table 7.1.

$$CH_4(g) + 2\ O_2(g) \longrightarrow CO_2(g) + 2\ H_2O(l) \qquad \Delta H = -213\ \text{kcal/mol}$$

When hydrocarbon combustion is incomplete because of faulty engine or furnace performance, carbon monoxide and carbon-containing soot are among the products. Carbon monoxide is a highly toxic and dangerous substance, especially so because it has no odor and can easily go undetected (See the Application "CO and NO: Pollutants or Miracle Molecules?" in Chapter 5). Breathing air that contains as little as 2% CO for only one hour can cause respiratory and nervous system damage or death. The supply of oxygen to the brain is cut off by carbon monoxide because it binds strongly to blood hemoglobin at the site where oxygen is normally bound. By contrast with CO, CO_2 is nontoxic and causes no harm, except by suffocation when present in high concentration.

> **PROBLEM 12.14**
>
> Write a balanced equation for the complete combustion of ethane with oxygen.

Halogenation

The second notable reaction of alkanes is *halogenation*, the replacement of an alkane hydrogen by a chlorine or bromine in a process initiated by heat or light. Only one

H at a time is replaced; however, if allowed to react for a long enough time, all H's will be replaced with halogens. Complete chlorination of methane, for example, yields carbon tetrachloride:

$$CH_4 + 4\,Cl_2 \xrightarrow{\text{Heat or light}} CCl_4 + 4\,HCl$$

Although the above equation for the reaction of methane with chlorine is balanced, it does not fully represent what actually happens. In fact, this reaction, like many organic reactions, yields a mixture of products:

$$CH_4 + Cl_2 \longrightarrow CH_3Cl + HCl$$

$$\downarrow^{Cl_2}$$
$$CH_2Cl_2 + HCl$$
$$\downarrow^{Cl_2}$$
$$CHCl_3 + HCl$$
$$\downarrow^{Cl_2}$$
$$CCl_4 + HCl$$

CH_3Cl, chloromethane
CH_2Cl_2, dichloromethane
$CHCl_3$, chloroform
CCl_4, carbon tetrachloride

When we write the equation for an organic reaction, our attention is usually focused on converting a particular reactant into a desired product; any minor by-products and inorganic compounds (such as the HCl formed in the chlorination of methane) are often of little interest and are ignored. Thus, it is not always necessary to balance the equation for an organic reaction as long as the reactant, the major product, and any necessary reagents and conditions are shown. A chemist who plans to convert methane into bromomethane might therefore write the equation as

$$CH_4 \xrightarrow[\text{Light, heat}]{Br_2} CH_3Br \qquad \text{Like many equations for organic reactions,}$$
$$\text{this equation is not balanced.}$$

In using this convention, it is customary to put reactants and reagents above the arrow and conditions, solvents, and catalysts below the arrow.

PROBLEM 12.15

Write the structures of all possible products with 1 and 2 chlorine atoms that form in the reaction of propane with Cl_2.

12.9 Cycloalkanes

The organic compounds described thus far have all been open-chain, or *acyclic*, alkanes. **Cycloalkanes**, which contain rings of carbon atoms, are also well known and are widespread throughout nature. To form a closed ring requires an additional C—C bond and the loss of 2 H atoms; the general formula for cycloalkanes, therefore, is C_nH_{2n}, which, as we will find in the next chapter, is the same as that for alkenes. Compounds of all ring sizes from 3 through 30 and beyond have been prepared in the laboratory. The two simplest cycloalkanes—cyclopropane and cyclobutane—contain 3 and 4 carbon atoms, respectively:

Cycloalkane An alkane that contains a ring of carbon atoms.

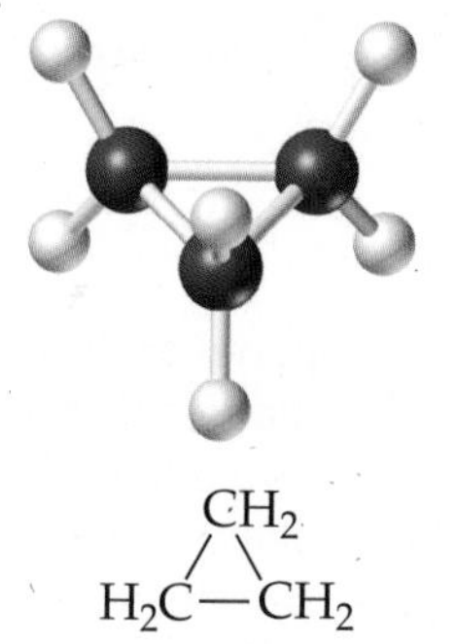

CH$_2$
H$_2$C—CH$_2$
Cyclopropane
(mp -128 °C, bp -33 °C)

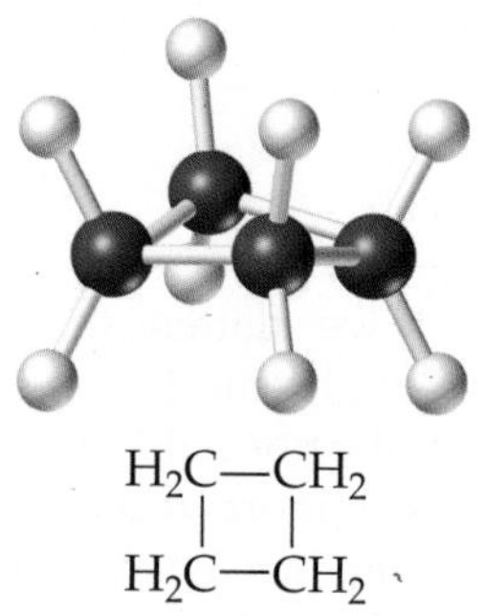

H$_2$C—CH$_2$
H$_2$C—CH$_2$
Cyclobutane
(mp -50 °C, bp -12 °C)

APPLICATION ▶ Petroleum

Many alkanes occur naturally throughout plants and animals, but natural gas and petroleum deposits provide the world's most abundant supply. *Natural gas* consists chiefly of methane, along with smaller amounts of ethane, propane, and butane. *Petroleum* is a complex liquid mixture of hydrocarbons that must be separated, or *refined*, into different *fractions* before it can be used. The vast majority of petroleum products are used as fuels and lubricating oils in the transportation industry, but petroleum-based products are also found in plastics, medicines, food items, and a host of other products.

▲ The crude petroleum from this off-shore well is an enormously complex mixture of alkanes and other organic molecules.

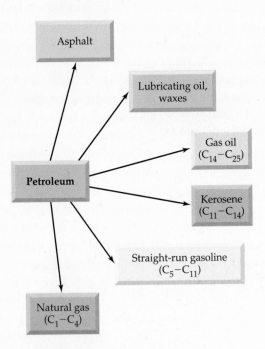

Petroleum has been used throughout history in such diverse applications as a component in mortar and for coating walls and boat hulls. Native Americans used it in magic and paints, and often sold it to pioneers for medicinal uses (as Seneca oil or Genesee oil). The modern petroleum industry began in 1859 in Oil Creek, Pennsylvania, when American oil pioneer E. L. Drake drilled the first oil-producing well.

Petroleum refining begins with distillation to separate the crude oil into three main fractions according to boiling points: straight-run gasoline (bp 30–200 °C), kerosene (bp 175–300 °C), and gas oil (bp 275–400 °C). The residue is then further distilled under reduced pressure to recover lubricating oils, waxes, and asphalt. But distillation of petroleum is just the beginning of the process for making gasoline. It has long been known that straight-chain alkanes burn less smoothly in engines than do branched-chain alkanes, as measured by a compound's *octane number*. Heptane, a straight-chain alkane and particularly poor fuel, is assigned an octane rating of 0, whereas 2,2,4-trimethylpentane, a branched-chain alkane commonly known as isooctane, is given a rating of 100. Straight-run gasoline, with its high percentage of unbranched alkanes, is thus a poor fuel. Petroleum chemists, however, have devised methods to remedy the problem. One of these methods, called *catalytic cracking*, involves taking the kerosene fraction ($C_{11}-C_{14}$) and "cracking" it into smaller C_3-C_5 molecules at high temperature. These small hydrocarbons are then catalytically recombined to yield C_7-C_{10} branched-chain molecules that are perfectly suited for use as automobile fuel.

In addition to the obvious fuel-based applications, petroleum is the source of paraffin wax (used in candy, candles, and polishes) and petroleum jelly (used in medical products and toiletries). Petroleum "feedstocks" are often converted into the basic chemical building blocks used to produce plastics, rubber, synthetic fibers, drugs, and detergents.

See Additional Problems 12.65 and 12.66 at the end of the chapter.

Note that if we flatten the rings in cyclopropane and cyclobutane, the C—C—C bond angles are 60° and 90°, respectively—values that are considerably compressed from the normal 109.5° tetrahedral value. As a result, these compounds are less stable and more reactive than other cycloalkanes. The five-membered (cyclopentane) ring has nearly ideal bond angles. So does the six-membered (cyclohexane) ring, which accomplishes this nearly ideal state by adopting a puckered, nonplanar shape called a *chair conformation*, further discussion of which, while important, is beyond the scope of this textbook. Both cyclopentane and cyclohexane rings are therefore stable, and many naturally occurring and biochemically active molecules contain such rings.

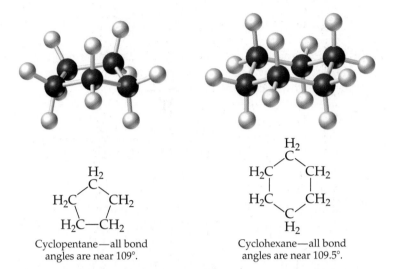

Cyclopentane—all bond angles are near 109°.

Cyclohexane—all bond angles are near 109.5°.

Cyclic and acyclic alkanes are similar in many of their properties. Cyclopropane and cyclobutane are gases, whereas larger cycloalkanes are liquids or solids. Like alkanes, cycloalkanes are nonpolar, insoluble in water, and flammable. Because of their cyclic structures, however, cycloalkane molecules are more rigid and less flexible than their open-chain counterparts. Rotation is not possible around the carbon–carbon bonds in cycloalkanes without breaking open the ring.

12.10 Drawing and Naming Cycloalkanes

Even condensed structures become awkward when we work with large molecules that contain rings. Thus, line structures are used almost exclusively in drawing cycloalkanes, with *polygons* used for the cyclic parts of the molecules. A triangle represents cyclopropane, a square represents cyclobutane, a pentagon represents cyclopentane, and so on.

Cyclopropane   Cyclobutane   Cyclopentane   Cyclohexane   Cycloheptane

Methylcyclohexane, for example, looks like this in a line structure:

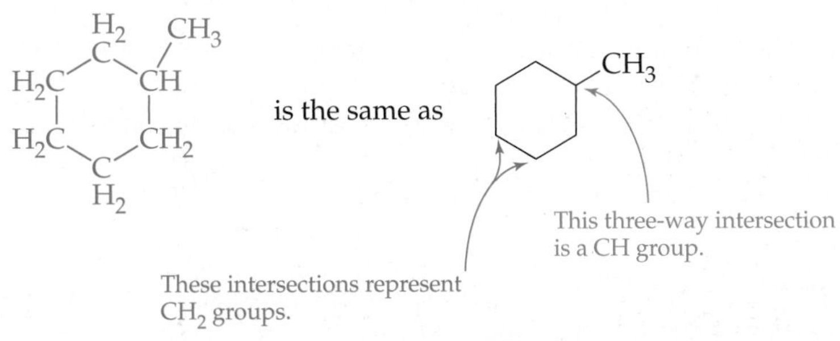

is the same as

These intersections represent CH_2 groups.

This three-way intersection is a CH group.

Cycloalkanes are named by a straightforward extension of the rules for naming open-chain alkanes. In most cases, only two steps are needed:

STEP 1: Use the cycloalkane name as the parent. That is, compounds are named as alkyl-substituted cycloalkanes rather than as cycloalkyl-substituted alkanes. If there is only one substituent on the ring, it is not even necessary to assign a number because all ring positions are identical.

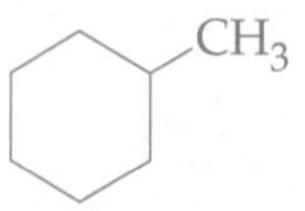

Parent compound: Cyclohexane
Name: Methylcyclohexane
(not cyclohexylmethane)

STEP 2: Identify and number the substituents. Start numbering at the group that has alphabetical priority, and proceed around the ring in the direction that gives the second substituent the lower possible number.

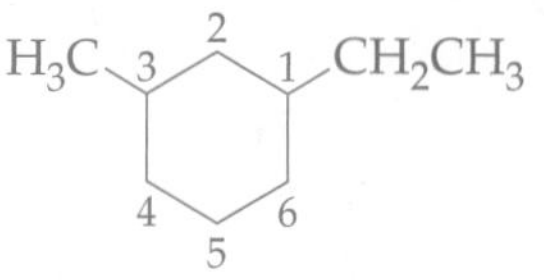

1-ethyl-3-methylcyclohexane
(not 1-ethyl-5-methylcyclohexane or
1-methyl-3-ethylcyclohexane or
1-methyl-5-ethylcyclohexane)

WORKED EXAMPLE 12.12 Naming Organic Compounds: Cycloalkanes

What is the IUPAC name of the following cycloalkane?

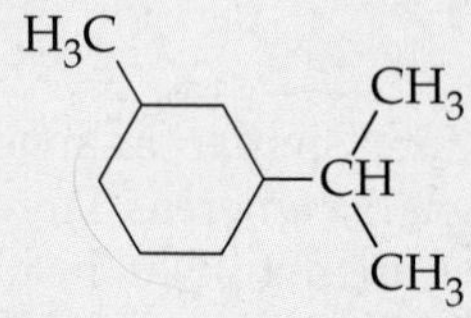

ANALYSIS First identify the parent cycloalkane, then add the positions and identity of any substituents.

SOLUTION

STEP 1: The parent cycloalkane contains 6 carbons (*hexane*); hence, *cyclohexane*.

STEP 2: There are two substituents; a *methyl* ($-CH_3$) and an *isopropyl* (CH_3CHCH_3). Alphabetically, the isopropyl group is given priority (number 1); the methyl group is then found on the third carbon in the ring.

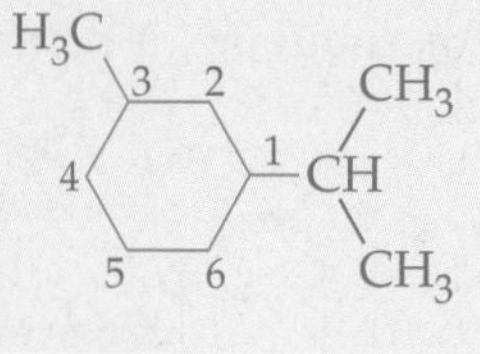

1-isopropyl-3-methylcyclohexane

WORKED EXAMPLE 12.13 Molecular Structures: Drawing Line Structures
for Cycloalkanes

Draw a line structure for 1,4-dimethylcyclohexane.

ANALYSIS This structure consists of a 6-carbon ring with two methyl groups attached at positions 1 and 4. Draw a hexagon to represent a cyclohexane ring, and attach a $-CH_3$ group at an arbitrary position that becomes the first carbon in the chain, designated as C1. Then count around the ring to the fourth carbon (C4), and attach another $-CH_3$ group.

SOLUTION

Note that the second methyl group is written here as H_3C- because it is attached on the left side of the ring.

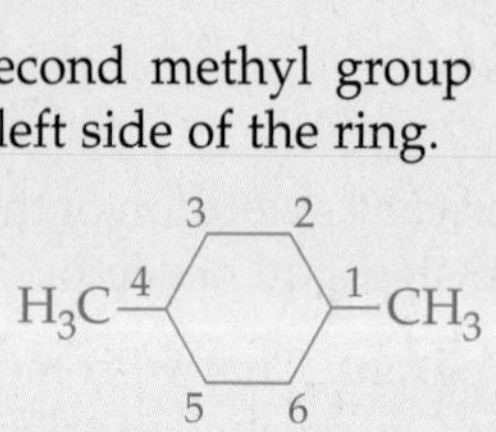

1,4-dimethylcyclohexane

PROBLEM 12.16

What are the IUPAC names of the following cycloalkanes?

(handwritten: 1-ethyl-3-isopropylcyclopentane)

(a) H₃C—⟨ring⟩—CH₂CH₃   **(b)** CH₃CH₂—⟨ring⟩—CH(CH₃)₂

(handwritten: 1-ethyl-4-methylcyclohexane)

PROBLEM 12.17

Draw line structures that represent the following IUPAC names:

(a) 1,1-Diethylcyclohexane   **(b)** 1,3,5-Trimethylcycloheptane

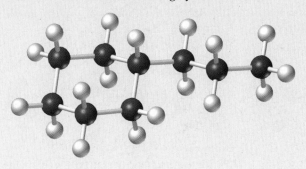

KEY CONCEPT PROBLEM 12.18

What is the IUPAC name of the following cycloalkane?

SUMMARY: REVISITING THE CHAPTER GOALS

1. **What are the basic properties of organic compounds?** Compounds made up primarily of carbon atoms are classified as organic. Many organic compounds contain carbon atoms that are joined in long chains by a combination of single (C—C), double (C=C), or triple (C≡C) bonds.

2. **What are functional groups, and how are they used to classify organic molecules?** Organic compounds can be classified into various families according to the functional groups they contain. A *functional group* is a part of a larger molecule and is composed of a group of atoms that has characteristic structure and chemical reactivity. A given functional group undergoes nearly the same chemical reactions in every molecule where it occurs. In this chapter we focused primarily on *alkanes*, hydrocarbon compounds that contain only single bonds between all C atoms.

3. **What are isomers?** *Isomers* are compounds that have the same formula but different structures. Isomers that differ in their connections among atoms are called *constitutional isomers*.

4. **How are organic molecules drawn?** Organic compounds can be represented by *structural formulas* in which all atoms and bonds are shown, by *condensed structures* in which not all bonds are drawn, or by *line structures* in which the carbon skeleton is represented by lines and the locations of C and H atoms are understood.

5. **What are alkanes and cycloalkanes, and how are they named?** Compounds that contain only carbon and hydrogen are called *hydrocarbons*, and hydrocarbons that have only single bonds are called *alkanes*. A *straight-chain alkane* has all its carbons connected in a row, a *branched-chain alkane* has a branching connection of atoms somewhere along its chain, and a *cycloalkane* has a ring of carbon atoms. Alkanes have the general formula C_nH_{2n+2}, whereas cycloalkanes have the formula C_nH_{2n}. Straight-chain alkanes are named by adding the family ending *-ane* to a parent; this tells how many carbon atoms are present. Branched-chain alkanes are named by using the longest continuous chain of carbon atoms for the parent and then identifying the *alkyl groups* present as branches off the main chain. The position of the substituent groups on the main chain are identified by numbering the carbons in the chain so that the substituents have the lowest number. Cycloalkanes are named by adding *cyclo-* as a prefix to the name of the alkane.

Quaternary (4°) carbon
atom, *p. 381*

Secondary (2°) carbon
atom, *p. 381*

Straight-chain alkane, *p. 369*

Substituent, *p. 380*

Tertiary (3°) carbon
atom, *p. 381*

6. What are the general properties and chemical reactions of alkanes? Alkanes are generally soluble only in nonpolar organic solvents, have weak intermolecular forces, and are nontoxic. Their principal chemical reactions are *combustion*, a reaction with oxygen that gives carbon dioxide and water, and *halogenation*, a reaction in which hydrogen atoms are replaced by chlorine or bromine.

SUMMARY OF REACTIONS

1. Combustion of an alkane with oxygen (Section 12.8):

$$CH_4 + 2\,O_2 \longrightarrow CO_2 + 2\,H_2O$$

2. Halogenation of an alkane to yield an alkyl halide (Section 12.8):

$$CH_4 + Cl_2 \longrightarrow CH_3Cl + HCl$$

UNDERSTANDING KEY CONCEPTS

12.19 How many hydrogen atoms are needed to complete the hydrocarbon formulas for the following carbon backbones?

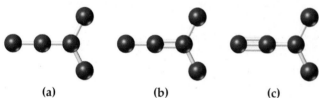

(a)   (b)   (c)

12.20 Convert the following models into condensed structures (gray = C; ivory = H; red = O):

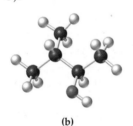

(a)   (b)

12.21 Convert the following models into line drawings (gray = C; ivory = H; red = O; blue = N):

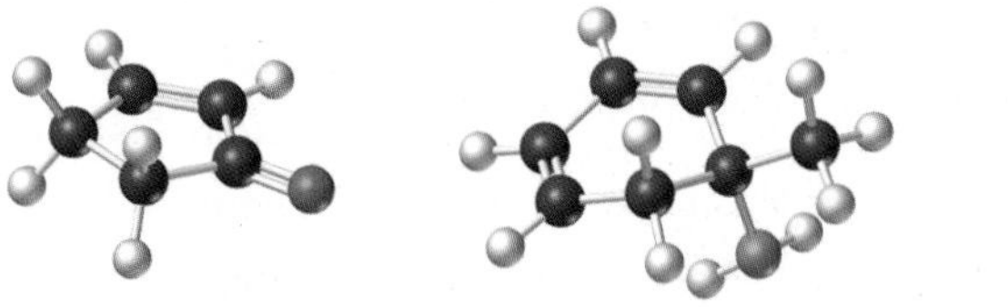

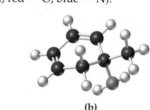

(a)   (b)

12.22 Identify the functional groups in the following compounds:

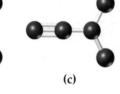

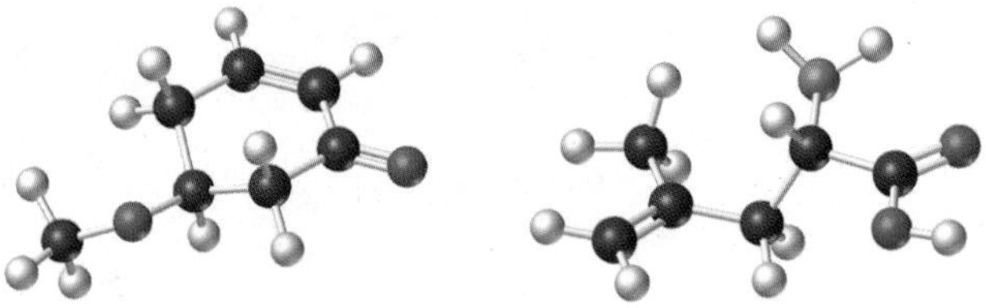

(a)   (b)

12.23 Give systematic names for the following alkanes:

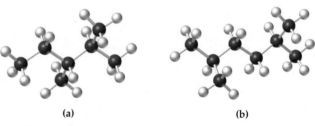

(a)   (b)

12.24 Give systematic names for the following cycloalkanes:

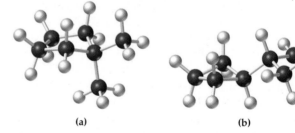

(a)   (b)

12.25 The following two compounds are isomers, even though both can be named 1,3-dimethylcyclopentane. What is the difference between them?

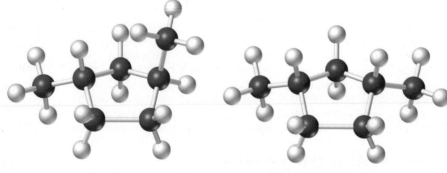

(a)   (b)

ADDITIONAL PROBLEMS

ORGANIC MOLECULES AND FUNCTIONAL GROUPS

12.26 What characteristics of carbon make possible the existence of so many different organic compounds?

12.27 What are functional groups, and why are they important?

12.28 Why are most organic compounds nonconducting and insoluble in water?

12.29 If you were given two unlabeled bottles, one containing hexane and one containing water, how could you tell them apart?

12.30 What is meant by the term *polar covalent bond*? Give an example of such a bond.

12.31 Give examples of compounds that are members of the following families:

(a) Alcohol   (b) Amine
(c) Carboxylic acid   (d) Ether

12.32 Locate and identify the functional groups in the following molecules ($-SO_2N-$ is a sulfanilamide):

(a)

Viagra

(b)

Aspirin

12.33 Identify the functional groups in the following molecules:

(a)

Vitamin A

(b)

Ambien

12.34 Propose structures for molecules that fit the following descriptions:

(a) A ketone with the formula $C_5H_{10}O$
(b) An ester with the formula $C_6H_{12}O_2$
(c) A compound with the formula $C_2H_5NO_2$ that is both an amine and a carboxylic acid

12.35 Propose structures for molecules that fit the following descriptions:

(a) An amide with the formula C_4H_9NO
(b) An aldehyde that has a ring of carbons, $C_6H_{10}O$
(c) An aromatic compound that is also an ether, $C_8H_{10}O$

ALKANES AND ISOMERS

12.36 What requirement must be met for two compounds to be isomers?

12.37 If one compound has the formula C_5H_{10} and another has the formula C_4H_{10}, are the two compounds isomers? Explain.

12.38 What is the difference between a secondary carbon and a tertiary carbon? Between a primary carbon and a quaternary carbon?

12.39 Why is it not possible for a compound to have a *quintary* carbon (five R groups attached to C)?

12.40 Give examples of compounds that meet the following descriptions:

(a) An alkane with two tertiary carbons
(b) A cycloalkane with only secondary carbons

12.41 Give examples of compounds that meet the following descriptions:

(a) A branched-chain alkane with only primary and quaternary carbons
(b) A cycloalkane with three substituents

12.42 There are three isomers with the formula C_3H_8O. Draw the condensed structure and the line structure for each isomer.

12.43 Write condensed structures for the following molecular formulas. (You may have to use rings and/or multiple bonds in some instances.)

(a) C_2H_7N
(b) C_4H_8 (Write the line structure as well.)
(c) C_2H_4O
(d) CH_2O_2 (Write the line structure as well.)

12.44 How many isomers can you write that fit the following descriptions?

(a) Alcohols with formula $C_4H_{10}O$
(b) Amines with formula C_3H_9N
(c) Ketones with formula $C_5H_{10}O$

12.45 How many isomers can you write that fit the following descriptions?

(a) Aldehydes with formula $C_5H_{10}O$
(b) Esters with formula $C_4H_8O_2$
(c) Carboxylic acids with formula $C_4H_8O_2$

12.46 Which of the following pairs of structures are identical, which are isomers, and which are unrelated?

(a) $CH_3CH_2CH_3$ and $\underset{\underset{CH_2CH_3}{|}}{CH_3}$

(b) $CH_3-\underset{\underset{H}{|}}{N}-CH_3$ and $CH_3CH_2-\underset{\underset{H}{|}}{N}-H$

(c) $CH_3CH_2CH_2$—O—CH_3 and

$CH_3CH_2CH_2$—$\overset{\overset{\displaystyle O}{\|}}{C}$—$CH_3$

(d) CH_3—$\overset{\overset{\displaystyle O}{\|}}{C}$—$CH_2CH_2CH(CH_3)_2$ and

CH_3CH_2—$\overset{\overset{\displaystyle O}{\|}}{C}$—$CH_2CH_2CH_2CH_3$

(e) CH_3CH=$CHCH_2CH_2$—O—H and

$CH_3CH_2\underset{\underset{\displaystyle CH_3}{|}}{CH}$—$\overset{\overset{\displaystyle O}{\|}}{C}$—H

12.47 Which structure(s) in each group represent the same compound, and which represent isomers?

(a)
```
     H  H  H  H
     |  |  |  |
  H—C—C—C—C—H
     |  |  |  |
     H  H  H  H
           |
        H—C—H          H  H  H
        |  |           |  |  |
     H  H        H—C—C—C—H
     |  |           |     |
  H—C—C—C—H         H     H
     |  |  |        H—C—H
     H  H  H           |
                       H
```

(b) $CH_3\underset{\underset{\displaystyle Br}{|}}{CH}\overset{\overset{\displaystyle CH_3}{|}}{C}HCH_3$ $CH_3\underset{\underset{\displaystyle Br}{|}}{CH}\overset{\overset{\displaystyle CH_3}{|}}{C}HCH_3$

$\underset{\underset{\displaystyle Br}{|}}{CH_2}\overset{\overset{\displaystyle CH_3}{|}}{C}HCH_2CH_3$

12.48 What is wrong with the following structures?

(a) CH_3=$CHCH_2CH_2OH$

(b) CH_3CH_2CH=$\overset{\overset{\displaystyle O}{\|}}{C}$—$CH_3$

(c) $CH_2CH_2CH_2C$≡$\overset{\overset{\displaystyle CH_3}{|}}{C}CH_3$

12.49 There are two things wrong with the following structure. What are they?

ALKANE NOMENCLATURE

12.50 What are the IUPAC names of the following alkanes?

(a) $CH_3CH_2CH_2CH_2\underset{\underset{\displaystyle CH_3}{|}}{C}H\overset{\overset{\displaystyle CH_2CH_3}{|}}{C}HCH_2CH_3$

(b) $CH_3CH_2CH_2\underset{\underset{\displaystyle CH_2CH_3}{|}}{C}H\overset{\overset{\displaystyle CH_3CHCH_3}{|}}{C}HCH_2CHCH_3$

(c) $CH_3\overset{\overset{\displaystyle CH_3}{|}}{C}CH_2CH_2CH_2\overset{\overset{\displaystyle CH_3}{|}}{C}HCH_3$
$\underset{\underset{\displaystyle CH_3}{|}}{}$

(d) $CH_3CH_2CH_2\overset{\overset{\displaystyle CH_2CH_2CH_2CH_3}{|}}{C}CH_3$
$\underset{\underset{\displaystyle CH_3CHCH_3}{|}}{}$

(e) $CH_3\overset{\overset{\displaystyle CH_3\ CH_3}{|\ \ |}}{C}CH_2\overset{}{C}CH_3$
$\underset{\underset{\displaystyle CH_3\ CH_3}{|\ \ |}}{}$

(f) $CH_3CH_2\overset{\overset{\displaystyle CH_3CH_2\ CH_3}{|\ \ \ \ \ |}}{C}CH_2\overset{}{C}H$
$\underset{\underset{\displaystyle CH_3CH_2\ CH_3}{|\ \ \ \ \ |}}{}$

(g) $CH_3(CH_2)_7\overset{\overset{\displaystyle CH_3}{|}}{C}$—$CH_3$
$\underset{\underset{\displaystyle CH_3}{|}}{}$

12.51 Give IUPAC names for the five isomers with the formula C_6H_{14}.

12.52 Write condensed structures for the following compounds:

(a) 4-*tert*-Butyl-3,3,5-trimethylheptane
(b) 2,4-Dimethylpentane
(c) 4,4-Diethyl-3-methyloctane
(d) 3-Isopropyl-2,3,6,7-tetramethylnonane
(e) 3-Isobutyl-1-isopropyl-5-methylcycloheptane
(f) 1,1,3-Trimethylcyclopentane

12.53 Draw structures that correspond to the following IUPAC names:

(a) 1,1-Dimethylcyclopropane
(b) 1,2,3,4-Tetramethylcyclopentane
(c) 4-*tert*-Butyl-1,1-dimethylcyclohexane
(d) Cycloheptane
(e) 1,3,5-Triisopropylcyclohexane
(f) 1,3,5,7-Tetramethylcyclooctane

12.54 Name the following cycloalkanes:

(a)

(b)

(c)

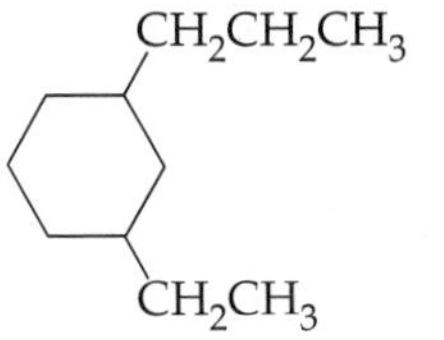

CH$_2$CH$_2$CH$_3$

CH$_2$CH$_3$

(d)

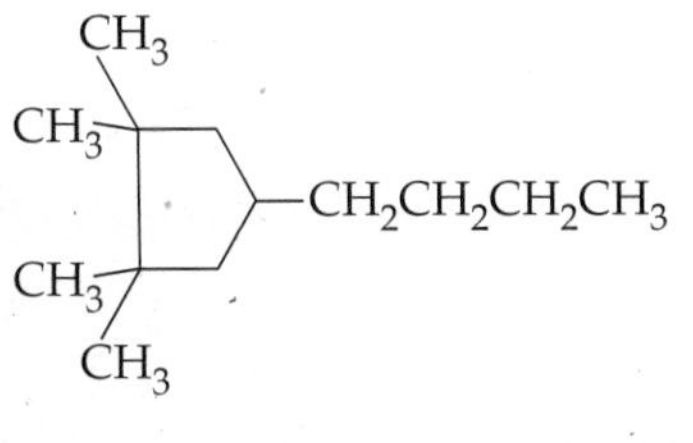

CH$_3$
CH$_3$
CH$_3$
CH$_3$
—CH$_2$CH$_2$CH$_2$CH$_3$

12.55 Name the following cycloalkanes:

(a)

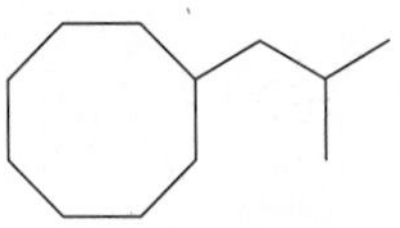

(b)

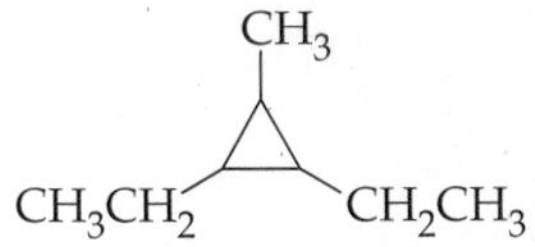

CH$_3$

CH$_3$CH$_2$ CH$_2$CH$_3$

(c)

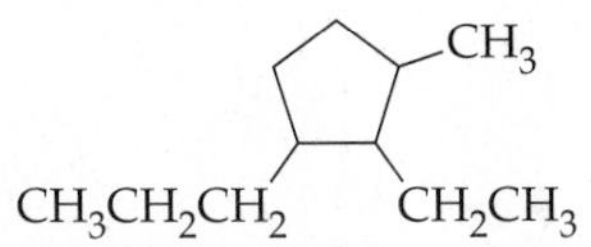

CH$_3$

CH$_3$CH$_2$CH$_2$ CH$_2$CH$_3$

12.56 The following names are incorrect. Tell what is wrong with each, and provide the correct names.

(a)

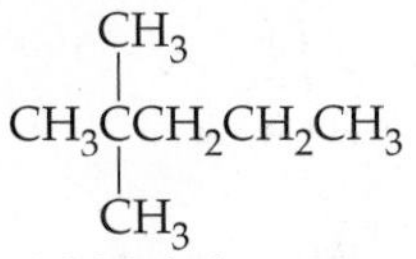

CH$_3$
CH$_3$CCH$_2$CH$_2$CH$_3$
CH$_3$

2,2-Methylpentane

(b)

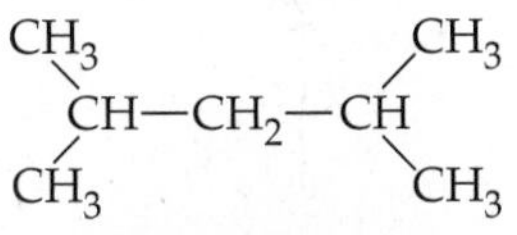

CH$_3$ CH$_3$
CH—CH$_2$—CH
CH$_3$ CH$_3$

1,1-Diisopropylmethane

(c)

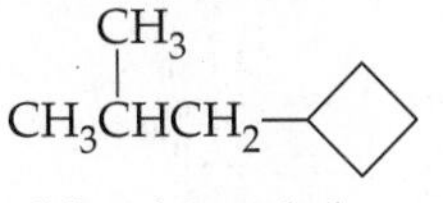

CH$_3$
CH$_3$CHCH$_2$—⟨◇⟩

1-Cyclobutyl-2-methylpropane

12.57 The following names are incorrect. Write the structural formula that agrees with the apparent name, and then write the correct name of the compound.

(a) 2-Ethylbutane
(b) 2-Isopropyl-2-methylpentane
(c) 5-Ethyl-1,1-methylcyclopentane
(d) 3-Ethyl-3,5,5-trimethylhexane
(e) 1,2-Dimethyl-4-ethylcyclohexane
(f) 2,4-Diethylpentane
(g) 5,5,6,6-Methyl-7,7-ethyldecane

12.58 Draw structures and give IUPAC names for the nine isomers of C$_7$H$_{16}$.

12.59 Draw the structural formulas and name all cyclic isomers with the formula C$_6$H$_{12}$.

REACTIONS OF ALKANES

12.60 Propane, commonly known as LP gas, burns in air to yield CO$_2$ and H$_2$O. Write a balanced equation for the reaction.

12.61 Write a balanced equation for the combustion of isooctane, C$_8$H$_{16}$, a component of gasoline.

12.62 Write the formulas of the three singly chlorinated isomers formed when 2,2-dimethylbutane reacts with Cl$_2$ in the presence of light.

12.63 Write the formulas of the seven doubly brominated isomers formed when 2,2-dimethylbutane reacts with Br$_2$ in the presence of light.

Applications

12.64 Why is it important to know the shape of a molecule? [*Displaying Molecular Shapes, p. 379*]

12.65 How does petroleum differ from natural gas? [*Petroleum, p. 388*]

12.66 What types of hydrocarbons burn most efficiently in an automobile engine? [*Petroleum, p. 388*]

General Questions and Problems

12.67 Identify the functional groups in the following molecules:

(a) Testosterone, a male sex hormone

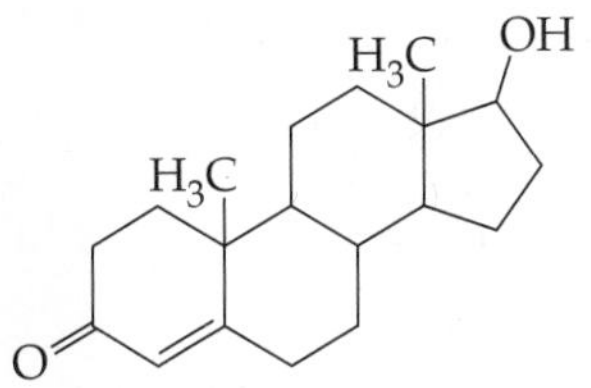

(b) Aspartame, an artificial sweetener

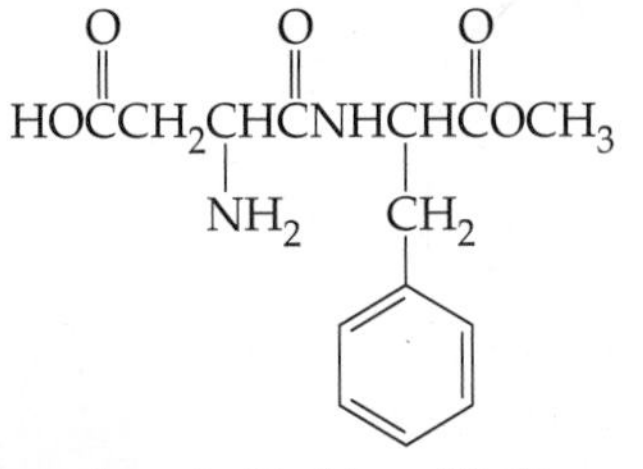

O   O   O
HOCCH$_2$CHCNHCHCOCH$_3$
NH$_2$ CH$_2$

12.68 Label each carbon in Problem 12.67 as primary, secondary, tertiary, or quaternary.

12.69 If someone reported the preparation of a compound with the formula C$_3$H$_9$ most chemists would be skeptical. Why?

12.70 Most lipsticks are about 70% castor oil and wax. Why is lipstick more easily removed with petroleum jelly than with water?

12.71 When cyclopentane is exposed to Br$_2$ in the presence of light, reaction occurs. Write the formulas of:

(a) All possible monobromination products
(b) All possible dibromination products

12.72 Which do you think has a higher boiling point, pentane or neopentane (2,2-dimethylpropane)? Why?

12.73 Propose structures for the following:

(a) An aldehyde, C$_4$H$_8$O
(b) An iodo-substituted alkene, C$_5$H$_9$I
(c) A cycloalkane, C$_7$H$_{14}$
(d) A diene (dialkene), C$_5$H$_8$

CHAPTER 13

Alkenes, Alkynes, and Aromatic Compounds

CONCEPTS TO REVIEW

VSEPR and Molecular Shapes
(Section 5.7)

The Shapes of Organic
Molecules
(Section 12.5)

Naming Alkanes
(Section 12.6)

▲ Flamingos owe their color to alkene pigments in their diet. Without these compounds, their feathers eventually turn white.

CONTENTS

CHAPTER GOALS

In this chapter, we will answer the following questions:

1. What are alkenes, alkynes, and aromatic compounds?

THE GOAL: Be able to recognize the functional groups in these three families of unsaturated organic compounds and give examples of each.

2. How are alkenes, alkynes, and aromatic compounds named?

THE GOAL: Be able to name an alkene, alkyne, or simple aromatic compound from its structure, or write the structure, given the name.

3. What are cis–trans isomers?

THE GOAL: Be able to identify cis–trans isomers of alkenes and predict their occurrence.

4. What are the categories of organic reactions?

THE GOAL: Be able to recognize and describe addition, elimination, substitution, and rearrangement reactions.

5. What are the typical reactions of alkenes, alkynes, and aromatic compounds?

THE GOAL: Be able to predict the products of reactions of alkenes, alkynes, and aromatic compounds.

6. How do organic reactions take place?

THE GOAL: Be able to show how addition reactions occur.

I n this and the remaining four chapters on organic chemistry, we examine some families of organic compounds whose functional groups give them characteristic properties. Compounds in the three families described in this chapter all contain carbon–carbon multiple bonds. *Alkenes*, such as ethylene, contain a double-bond functional group; *alkynes*, such as acetylene, contain a triple-bond functional group; and *aromatic compounds*, such as benzene, contain a six-membered ring of carbon atoms usually pictured as having three alternating double bonds with properties different from those of a typical alkene. All three functional groups are widespread in nature and are found in many biologically important molecules.

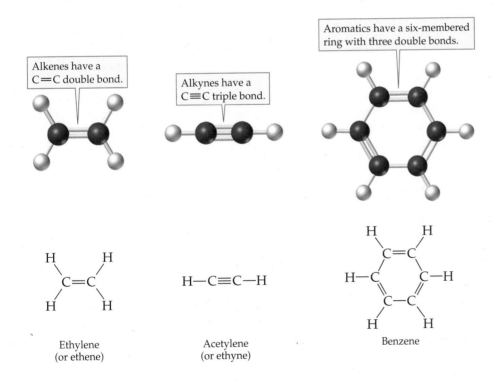

Alkenes have a C=C double bond.

Alkynes have a C≡C triple bond.

Aromatics have a six-membered ring with three double bonds.

Ethylene (or ethene)

Acetylene (or ethyne)

Benzene

13.1 Alkenes and Alkynes

Alkanes, introduced in Chapter 12, are often referred to as **saturated** because each carbon atom in an alkane forms four single bonds. Because this is the maximum number of bonds a carbon can form, no more atoms can be added to the alkane

Saturated A molecule whose carbon atoms bond to the maximum number of hydrogen atoms.

Unsaturated A molecule that contains a carbon–carbon multiple bond to which more hydrogen atoms can be added.

Alkene A hydrocarbon that contains a carbon–carbon double bond.

Alkyne A hydrocarbon that contains a carbon–carbon triple bond.

molecule—in other words, the molecule is saturated. Alkenes and alkynes, however, are said to be **unsaturated** because they contain carbon–carbon multiple bonds. Additional atoms can be added by converting these multiple bonds to single bonds. **Alkenes** are hydrocarbons that contain carbon–carbon double bonds, and **alkynes** are hydrocarbons that contain carbon–carbon triple bonds.

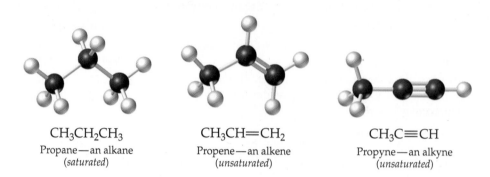

$CH_3CH_2CH_3$
Propane—an alkane
(*saturated*)

$CH_3CH{=}CH_2$
Propene—an alkene
(*unsaturated*)

$CH_3C{\equiv}CH$
Propyne—an alkyne
(*unsaturated*)

▲ Polyethylene, the most widely produced polymer in the world, is made from ethylene, a simple alkene.

Simple alkenes are made in vast quantities in the petroleum industry by thermal "cracking" of the alkanes that make up petroleum. Alkanes are rapidly heated to high temperatures (750–900 °C), which causes them to crack apart into reactive fragments that then reunite or rearrange into lower-molecular-weight molecules such as ethylene ($H_2C{=}CH_2$) and propene ($CH_3CH{=}CH_2$).

Most of the organic chemicals used in making drugs, explosives, paints, plastics, and pesticides are synthesized by routes that begin with alkenes. More ethylene is produced in the United States each year than any other organic chemical—more than 28 million tons in 2007, most of it used for making polyethylene. In fact, ethylene production worldwide was at a staggering 121 million tons in 2007, demonstrating just how important an industrial starting material it is.

Curiously, ethylene is also formed in the leaves, flowers, and roots of plants, where it acts as a hormone to control seedling growth, stimulate root formation, and regulate fruit ripening. The ability of ethylene to hasten ripening was discovered indirectly by citrus growers who at one time ripened green oranges in rooms heated with kerosene stoves. When the stoves were replaced with other heaters, the fruit, to their surprise, no longer ripened. Ethylene from incomplete combustion of the kerosene, not heat from the stoves, had caused the ripening. Today, fruit is intentionally exposed to ethylene to continue the ripening process during shipment to market.

13.2 Naming Alkenes and Alkynes

In the IUPAC system, alkenes and alkynes are named by a series of rules similar to those used for alkanes (Section 12.6). (⫘, p. 378) The parent names indicating the number of carbon atoms in the main chain are the same as those for alkanes, with the -*ene* suffix used in place of -*ane* for alkenes and the -*yne* suffix used for alkynes. The names of alkenes and alkynes also contain a number, called an *index number*, indicating the position of the multiple bond. The main chain in any unsaturated molecule is numbered so that the molecule's name has the lowest index number possible for that multiple bond. As we continue to study functional groups in later chapters we will see this lowest possible index number rule for functional groups again.

STEP 1: Name the parent compound. Find the longest chain containing the double or triple bond, and name the parent compound by adding the suffix -*ene* or -*yne* to the name for the main chain. If there is more than one double or triple bond, the number of multiple bonds is indicated using a numerical prefix (*diene* = two double bonds, *triene* = three double bonds, and so forth).

$CH_3CH_2CH_2CH{=}CH_2$ Name as a *pentene*—a five-carbon chain containing a double bond.

$CH_3CH_2CH_2C{\equiv}CCH_3$ Name as a *hexyne*—a six-carbon chain containing a triple bond.

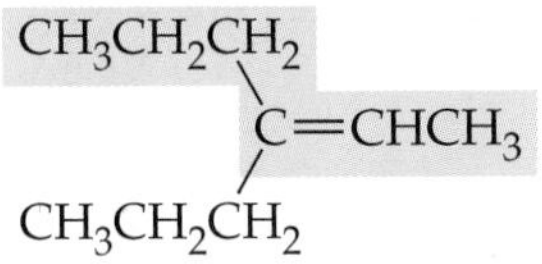

CH₃CH₂CH₂
       C=CHCH₃
CH₃CH₂CH₂

Name as a *hexene*—a six-carbon chain containing a double bond . . .

CH₃CH₂CH₂
       C=CHCH₃
CH₃CH₂CH₂

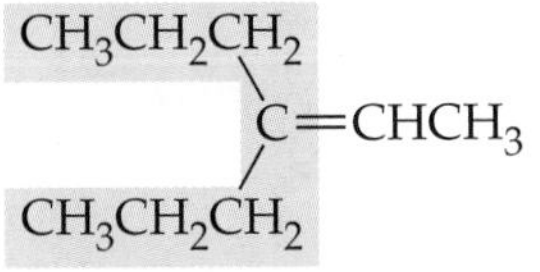

. . . *not* as a heptene, because the double bond is not contained in the seven-carbon chain.

STEP 2: Number the carbon atoms in the main chain, beginning at the end nearer the multiple bond. If the multiple bond is an equal distance from both ends, begin numbering at the end nearer the first branch point.

$$CH_3CH_2CH_2CH{=}CHCH_3$$
$$6\quad 5\quad 4\quad 3\quad\ 2\quad 1$$

Begin at this end because it's nearer the double bond.

$$CH_3CHCH{=}CHCH_2CH_3$$
with CH₃ on carbon 2
$$1\quad 2\quad 3\quad 4\quad 5\quad 6$$

Begin at this end because it's nearer the first branch point.

$$CH_3C{\equiv}CCH_2CH_2CHCH_3$$
with CH₃ on carbon 7
$$1\quad 2\quad 3\ 4\quad 5\quad 6\ 7\quad 8$$

Begin at this end because it's nearer the triple bond.

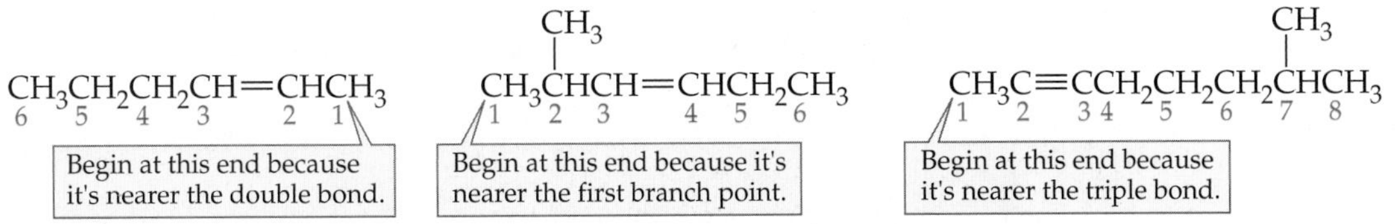

Cyclic alkenes are called **cycloalkenes** and are quite common. The double-bond carbon atoms in substituted cycloalkenes are numbered 1 and 2 so as to give the first substituent the lower number:

Cycloalkene A cyclic hydrocarbon that contains a double bond.

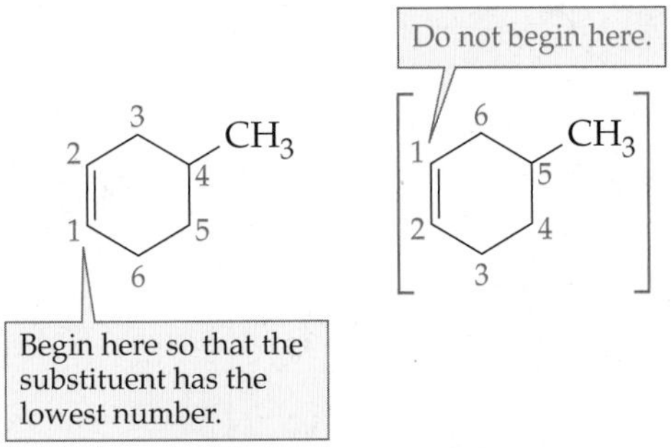

Do not begin here.

Name as a cyclohexene.

Begin here so that the substituent has the lowest number.

Cyclic alkynes are rare, and even those that are known are far too reactive to be readily available. For these reasons, we will spend no time discussing them.

STEP 3: Write the full name. Assign numbers to the branching substituents, and list the substituents alphabetically. Use commas to separate numbers and hyphens to separate words from numbers. Indicate the position of the multiple bond in the chain by giving the number of the *first* multiple-bonded carbon. If more than one double bond is present, identify the position of each and use the appropriate name ending (for example, 1,3-buta*diene* and 1,3,6-hepta*triene*).

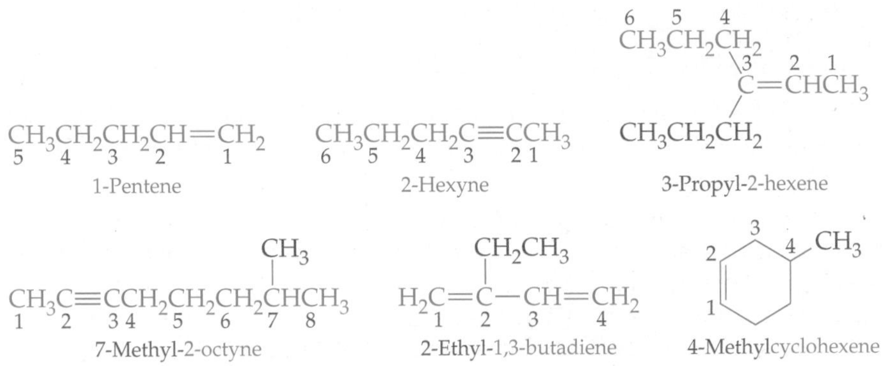

$$CH_3CH_2CH_2CH{=}CH_2$$
$$5\quad 4\quad 3\quad 2\quad 1$$
1-Pentene

$$CH_3CH_2CH_2C{\equiv}CCH_3$$
$$6\quad 5\quad 4\quad 3\quad 2\,1$$
2-Hexyne

CH₃CH₂CH₂ (6 5 4)
     3 2 1
     C=CHCH₃
CH₃CH₂CH₂
3-Propyl-2-hexene

$$CH_3C{\equiv}CCH_2CH_2CHCH_3$$
with CH₃ on carbon 7
$$1\quad 2\quad 3\ 4\quad 5\quad 6\ 7\quad 8$$
7-Methyl-2-octyne

$$H_2C{=}C{-}CH{=}CH_2$$
with CH₂CH₃ on carbon 2
$$1\quad 2\quad 3\quad 4$$
2-Ethyl-1,3-butadiene

4-Methylcyclohexene

For historical reasons, there are a few alkenes and alkynes whose names do not conform to the IUPAC rules. For instance, the 2-carbon alkene $H_2C\!=\!CH_2$ should properly be called *ethene*, but the name *ethylene* has been used for so long that it is now accepted by the IUPAC. Similarly, the 3-carbon alkene *propene* ($CH_3CH\!=\!CH_2$) is usually called *propylene*. The simplest alkyne, $HC\!\equiv\!CH$, should be known as *ethyne* but is almost always called *acetylene*.

WORKED EXAMPLE 13.1 Naming Organic Compounds: Alkenes

What is the IUPAC name of the following alkene?

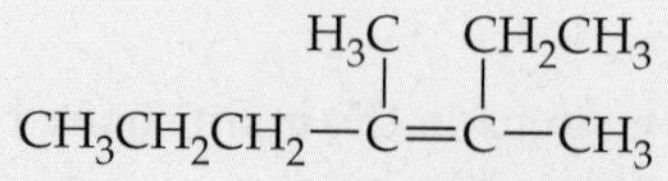

$$
\begin{array}{cc}
H_3C & CH_2CH_3 \\
| & | \\
CH_3CH_2CH_2\!-\!C\!=\!C\!-\!CH_3
\end{array}
$$

ANALYSIS Identify the parent compound as the longest continuous chain that contains the double bond. The location of the double bond and any substituents are identified by numbering the carbon chain from the end nearer the double bond.

SOLUTION

STEP 1: The longest continuous chain containing the double bond has 7 carbons—*heptene*. In this case, we have to turn a corner to find the longest chain:

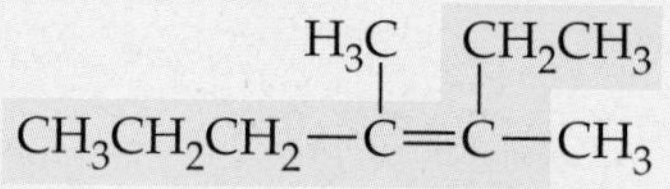

$$
\begin{array}{cc}
H_3C & CH_2CH_3 \\
| & | \\
CH_3CH_2CH_2\!-\!C\!=\!C\!-\!CH_3
\end{array}
$$
Name as a *heptene*.

STEP 2: Number the chain from the end nearer the double bond. The first double-bond carbon is C4 starting from the left end, but C3 starting from the right:

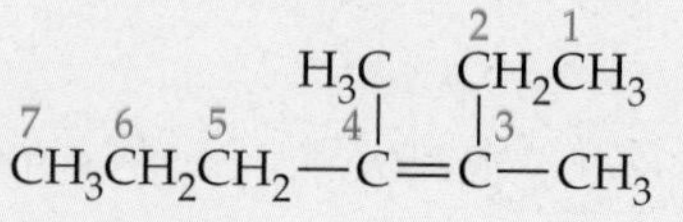

$$
\begin{array}{cc}
& {}^{2}\;\;{}^{1} \\
H_3C & CH_2CH_3 \\
{}^{4}| & |{}^{3} \\
{}^{7}\;\;{}^{6}\;\;{}^{5} \\
CH_3CH_2CH_2\!-\!C\!=\!C\!-\!CH_3
\end{array}
$$
Name as a substituted *3-heptene*.

STEP 3: Two methyl groups are attached at C3 and C4.

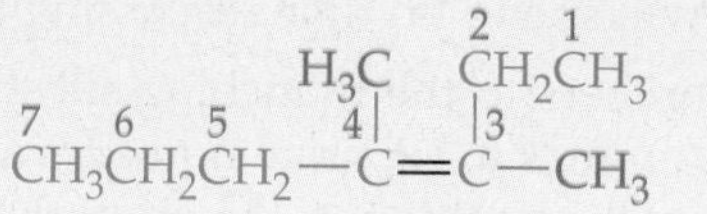

$$
\begin{array}{cc}
& {}^{2}\;\;{}^{1} \\
H_3C & CH_2CH_3 \\
{}^{4}| & |{}^{3} \\
{}^{7}\;\;{}^{6}\;\;{}^{5} \\
CH_3CH_2CH_2\!-\!C\!=\!C\!-\!CH_3
\end{array}
$$
Substituents: 3-methyl, 4-methyl

Name: 3,4-Dimethyl-3-heptene

WORKED EXAMPLE 13.2 Molecular Structures: Alkenes

Draw the structure of 3-ethyl-4-methyl-2-pentene.

ANALYSIS Identify the parent name (*pent*) and the location of the double bond and other substituents by numbering the carbons in the parent chain.

SOLUTION

STEP 1: The parent compound is a 5-carbon chain with the double bond between C2 and C3:

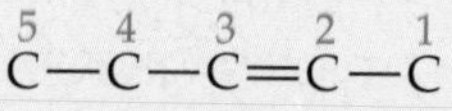

$$
\overset{5}{C}\!-\!\overset{4}{C}\!-\!\overset{3}{C}\!=\!\overset{2}{C}\!-\!\overset{1}{C}
$$
2-Pentene

STEP 2: Add the ethyl and methyl substituents on C3 and C4, and write in the additional hydrogen atoms so that each carbon atom has four bonds:

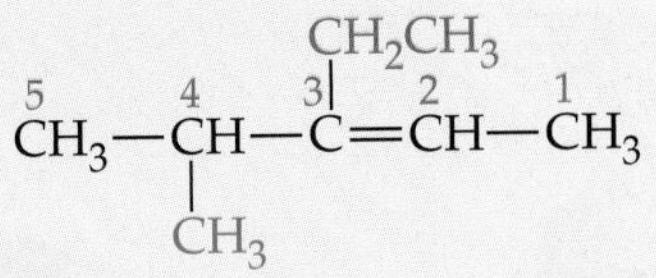

$$\overset{5}{CH_3}-\overset{4}{CH}-\overset{3}{\underset{|}{C}}=\overset{1}{CH}-CH_3 \qquad \text{3-Ethyl-4-methyl-2-pentene}$$

with CH_2CH_3 at position 3 and CH_3 below C4.

PROBLEM 13.1

What are the IUPAC names of the following compounds?

(a) 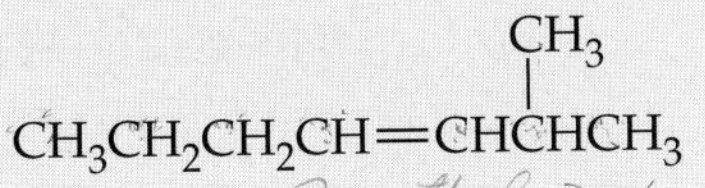 $CH_3CH_2CH_2CH{=}CHCHCH_3$ with CH_3 substituent

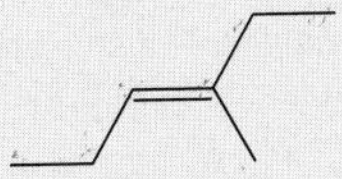

2-methyl-3-heptene

(b) 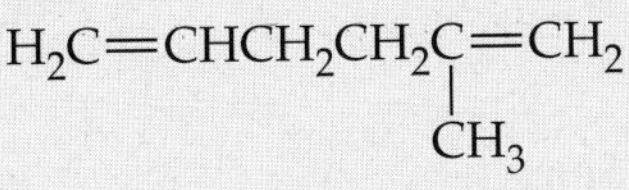 $H_2C{=}CHCH_2CH_2C{=}CH_2$ with CH_3 substituent

2-methyl-1,5-hexadiene

(c) 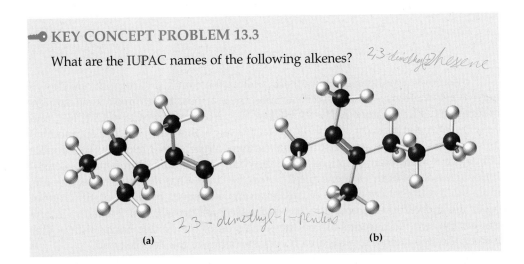 *3-methyl-3-hexene*

PROBLEM 13.2

Draw structures corresponding to the following IUPAC names:

(a) 3-Methyl-1-heptene

(b) 4,4-Dimethyl-2-pentyne

(c) 2-Methyl-3-hexene

(d) 2,2,3-Trimethyl-3-hexene

●━● KEY CONCEPT PROBLEM 13.3

What are the IUPAC names of the following alkenes? *2,3-dimethyl-2-hexene*

(a) *2,3-dimethyl-1-pentene*

(b)

13.3 The Structure of Alkenes: Cis–Trans Isomerism

Alkenes and alkynes differ from alkanes in shape because of their multiple bonds. Methane is tetrahedral, but ethylene is flat (planar) and acetylene is linear (straight), as predicted by the VSEPR model discussed in Section 5.7. (⊂⊃, p. 122)

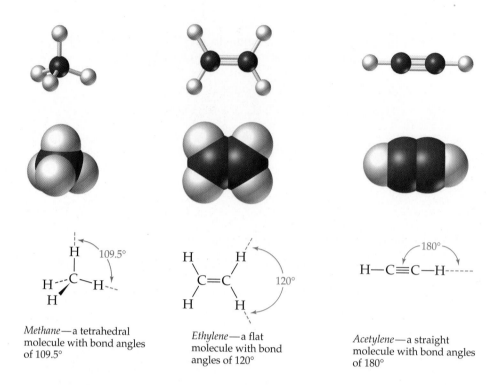

Methane—a tetrahedral molecule with bond angles of 109.5°

Ethylene—a flat molecule with bond angles of 120°

Acetylene—a straight molecule with bond angles of 180°

The two carbons and four attached atoms that make up the double-bond functional group lie in a plane. Unlike the situation in alkanes, where free rotation around the C—C single bond occurs (Section 12.5), there is no rotation around a double bond, making a new kind of isomerism possible for alkenes. (⊂⊃, p. 376) As a consequence of this rigid nature, alkenes possess *ends* and *sides*:

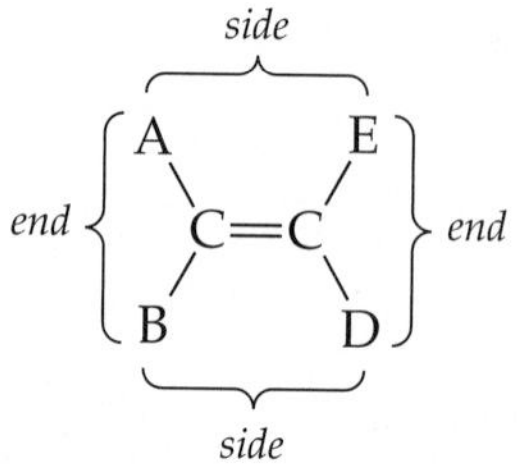

side

$$end \left\{ \begin{array}{c} A \quad\quad E \\ C{=}C \\ B \quad\quad D \end{array} \right\} end$$

side

To see this new kind of isomerism, look at the four C_4H_8 compounds shown here. When written as condensed structures, there appear to be three alkene isomers of formula C_4H_8: 1-butene [$CH_2{=}CHCH_2CH_3$], 2-butene [$CH_3CH{=}CHCH_3$], and 2-methylpropene [$(CH_3)_2CH{=}CH_2$]. In fact, though, there are *four*. The compounds 1-butene and 2-butene are constitutional isomers of each other because their double bonds occur at different positions along the chain, and 2-methylpropene is a constitutional isomer of both 1-butene and 2-butene because it has the same molecular formula but a different connection of carbon atoms. However, because rotation cannot occur around carbon–carbon double bonds, *there are two different 2-butenes*. In one isomer, the two CH_3 groups are on the same side of the double bond; in the other isomer, the two methyl groups are on opposite sides of the double bond.

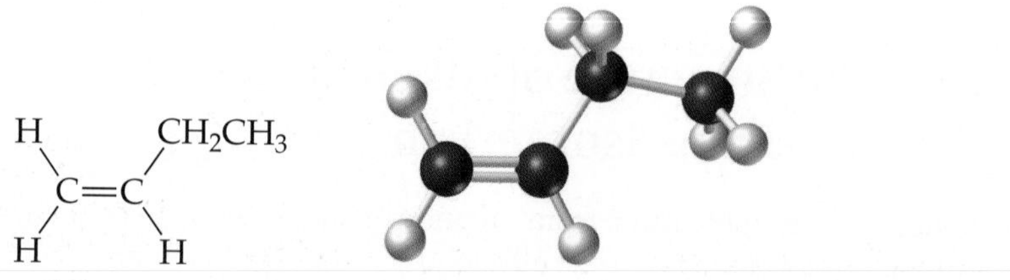

1-Butene

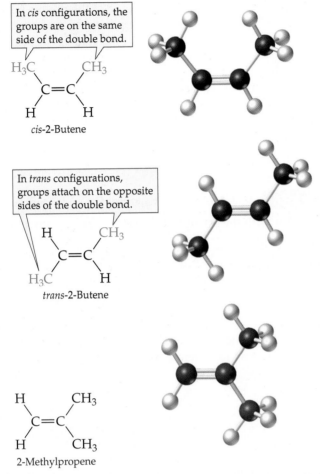

In *cis* configurations, the groups are on the same side of the double bond.

cis-2-Butene

In *trans* configurations, groups attach on the opposite sides of the double bond.

trans-2-Butene

2-Methylpropene

The two 2-butenes are called **cis–trans isomers**. They have the same formula and connections between atoms but have different three-dimensional structures because of the way that groups attach to different sides of the double bond. The isomer with its methyl groups on the same side of the double bond is named *cis*-2-butene, and the isomer with its methyl groups on opposite sides of the double bond is named *trans*-2-butene.

Cis–trans isomerism is possible whenever an alkene has two *different* substituent groups on each of its ends. (This means that in the above drawing illustrating the sides and ends of an alkene molecule, A ≠ B and D ≠ E.) If one of the carbons comprising the double bond is attached to two identical groups, cis–trans isomerism cannot exist. In 2-methyl-1-butene, for example, cis–trans isomerism is not possible because C1 is bonded to two identical groups (hydrogen atoms). To convince yourself of this, mentally flip either one of these two structures top to bottom; note that it becomes identical to the other structure:

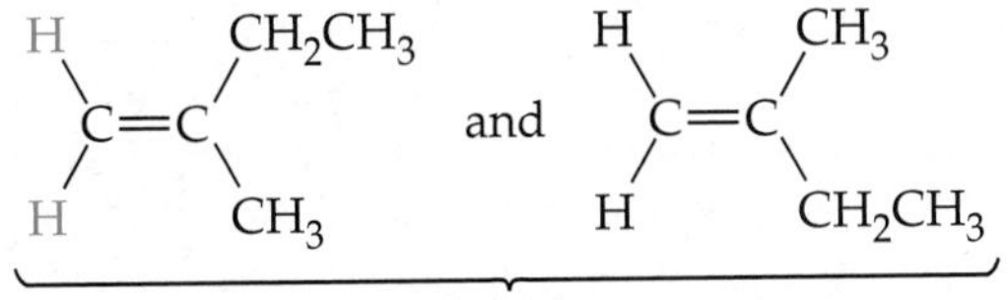

Cis–trans isomer Alkenes that have the same connections between atoms but differ in their three-dimensional structures because of the way that groups attach to different sides of the double bond.

and

These compounds are identical. Because the carbon left of the double bond has two H atoms attached, cis–trans isomerism is impossible.

2-Methyl-1-butene

In 2-pentene, however, the structures do not become identical when one of them is flipped, so cis–trans isomerism does occur:

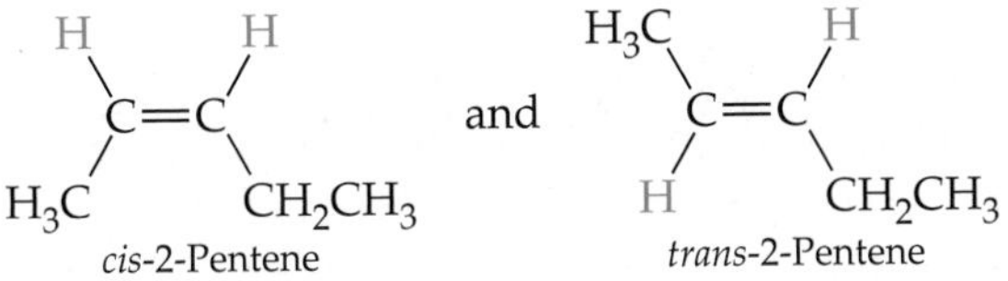

and

These compounds are not identical. Neither carbon of the double bond has two identical groups attached to it.

cis-2-Pentene

trans-2-Pentene

The two substituents that are on the same side of the double bond in an alkene are said to be cis to each other, and those on opposite sides of the double bond are said to be trans to each other. In our generic molecule above showing ends and sides, for example, A and E are cis to each other, B and D are cis to each other, B and E are trans to each other, A and D are trans to each other. Thus, in alkenes, the terms cis and trans are used in two ways: (i) as a *relative* term to indicate how various groups are attached to the double-bond carbons and (ii) in nomenclature as a way to indicate how the longest chain in the molecule goes in, through, and out of the double bond.

WORKED EXAMPLE   **13.3** Molecular Structure: Cis and Trans Isomers

Draw structures for both the cis and trans isomers of 2-hexene.

ANALYSIS First, draw a condensed structure of 2-hexene to see which groups are attached to the double-bond carbons:

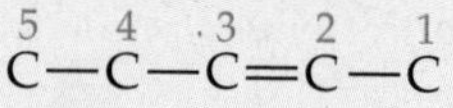

$$\overset{5}{C}-\overset{4}{C}-\overset{3}{C}=\overset{2}{C}-\overset{1}{C} \quad \text{2-Pentene}$$

Next, draw two double bonds. Choose one end of each double bond, and attach its groups in the same way to generate two identical partial structures:

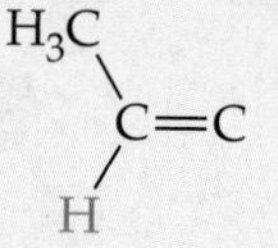

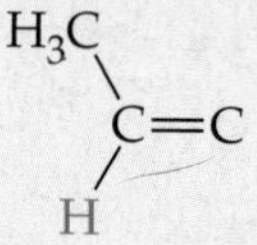

Finally, attach groups to the other end in the two possible ways.

SOLUTION

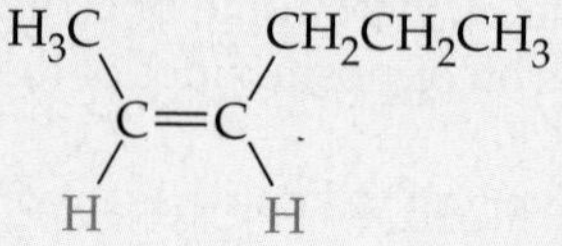

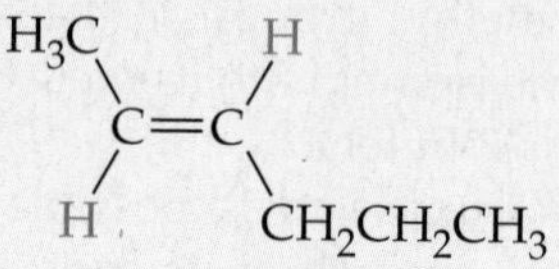

The structure with the 2 hydrogens on the same side of the double bond is the cis isomer, and that with the 2 hydrogens on opposite sides is the trans isomer.

PROBLEM 13.4

Which of the following substances exist as cis–trans isomers? Draw both isomers for those that do.

(a) 3-Heptene

(b) 2-Methyl-2-hexene

(c) 5-Methyl-2-hexene

PROBLEM 13.5

Draw the cis and trans isomers of 3,4-dimethyl-3-hexene.

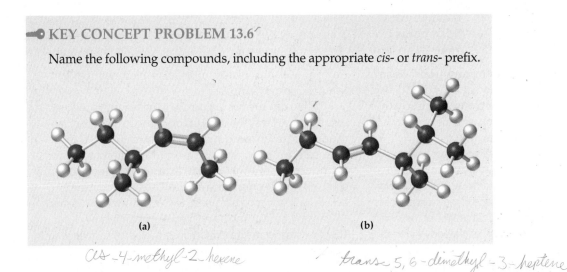

KEY CONCEPT PROBLEM 13.6

Name the following compounds, including the appropriate *cis*- or *trans*- prefix.

(a)   (b)

cis-4-methyl-2-hexene   *trans-5,6-dimethyl-3-heptene*

13.4 Properties of Alkenes and Alkynes

Alkenes and alkynes resemble alkanes in many respects. The bonds in alkenes and alkynes are nonpolar, and the physical properties of these compounds are influenced mainly by weak London dispersion forces (Section 8.11). (⬭ , p. 235) Alkenes and alkynes with 1–4 carbon atoms are gases, and boiling points increase with the size of the molecules.

Like alkanes, alkenes and alkynes are insoluble in water, soluble in nonpolar solvents, and less dense than water. They are flammable and nontoxic, although those that are gases present explosion hazards when mixed with air. Unlike alkanes, alkenes are quite reactive because of their double bonds. As we will see in the next section, alkenes undergo addition of various reagents to their double bonds to yield saturated products.

Properties of Alkenes and Alkynes

- Nonpolar; insoluble in water; soluble in nonpolar organic solvents; less dense than water
- Flammable; nontoxic
- Alkenes display cis–trans isomerism when each double-bond carbon atom has different substituents
- Chemically reactive at the multiple bond

13.5 Types of Organic Reactions

Before looking at the chemistry of alkenes we should first discuss some general reactivity patterns that make the task of organizing and categorizing organic reactions much simpler. Four particularly important kinds of organic reactions are discussed in this section: *additions, eliminations, substitutions,* and *rearrangements.*

- **Addition Reactions** Additions occur when two reactants add together to form a single product with no atoms "left over." We can generalize the process as

These two reactants   ... to give this
add together ...   single product.

$$A + B \longrightarrow C$$

The most common addition reactions encountered in organic chemistry are those in which a reagent adds across a carbon–carbon multiple bond to give a

Addition reaction A general reaction type in which a substance X—Y adds to the multiple bond of an unsaturated reactant to yield a saturated product that has only single bonds.

APPLICATION ▶ The Chemistry of Vision

Does eating carrots really improve your vision? Although carrots probably do not do much to help someone who is already on a proper diet, it is nevertheless true that the chemistry of carrots and the chemistry of vision are related. Both involve alkenes.

Carrots, peaches, sweet potatoes, and other yellow vegetables are rich in beta-carotene, a purple-orange alkene that provides our main dietary source of vitamin A. The conversion of beta-carotene to vitamin A takes place in the mucosal cells of the small intestine, where enzymes cut the molecule in half to yield an alcohol. Vitamin A in excess of the body's immediate needs is stored in the liver, from which it can be transported to the eye. In the eye, vitamin A is converted into a compound called *retinal*, which undergoes cis–trans isomerization of its C11–C12 double bond to produce 11-*cis*-retinal.

Reaction with the protein *opsin* then produces the light-sensitive substance *rhodopsin*.

The human eye has two kinds of light-sensitive cells, *rod cells* and *cone cells*. The three million or so rod cells are primarily responsible for seeing in dim light, whereas the 100 million cone cells are responsible for seeing in bright light and for the perception of bright colors. When light strikes the rod cells, cis–trans isomerization of the C11–C12 double bond occurs via a rearrangement reaction (Section 13.5, p. 405) and 11-*trans*-rhodopsin, also called metarhodopsin II, is produced. This cis–trans isomerization is accompanied by a change in molecular geometry, which in turn causes a nerve impulse to be sent to the brain where it is perceived as vision. Metarhodopsin II is then changed back to 11-*cis*-retinal for use in another vision cycle.

See Additional Problems 13.70 and 13.71 at the end of the chapter.

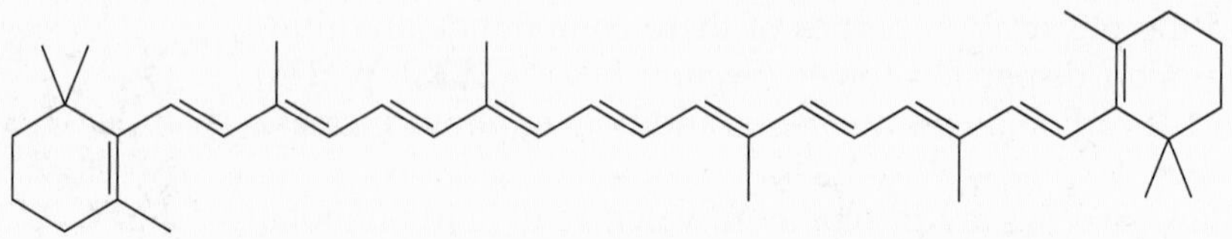

β-Carotene

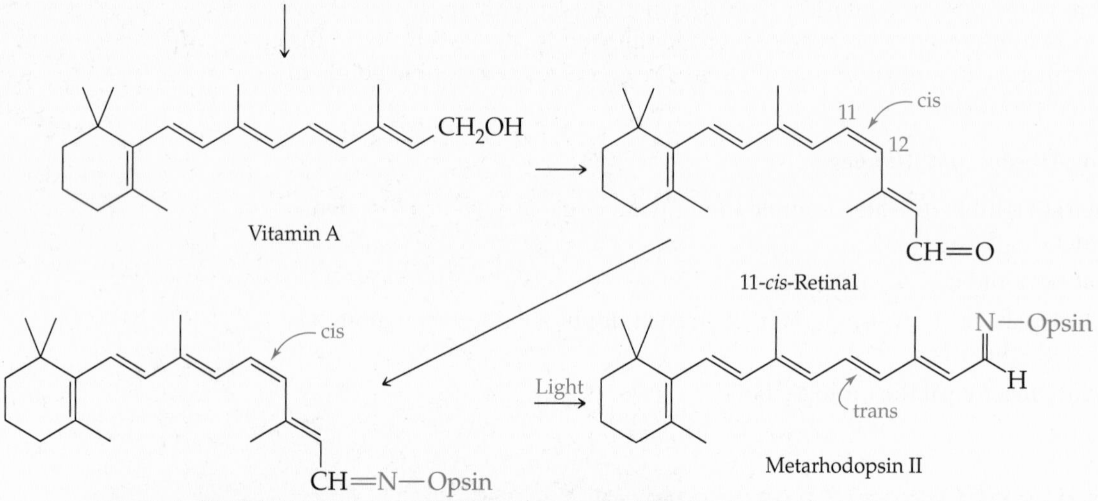

Vitamin A

11-*cis*-Retinal

Rhodopsin

Metarhodopsin II

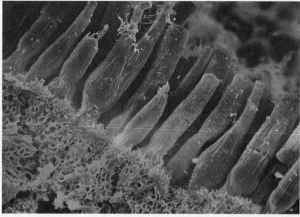

◀ Rod cells in the frog eye.

product that contains two (for alkenes) or four (for alkynes) new single bonds. This process can be generalized as

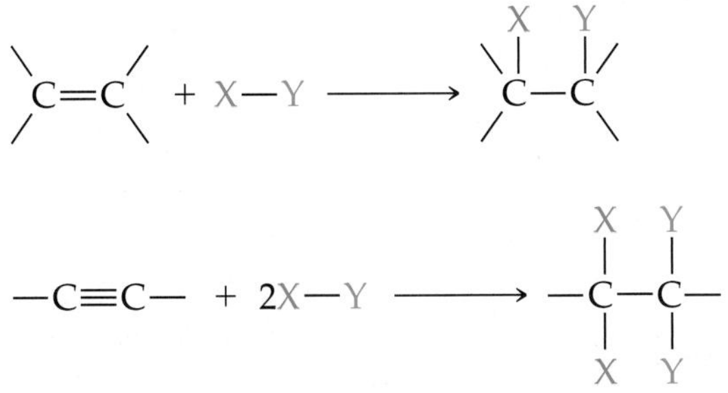

$$\text{C=C} + \text{X—Y} \longrightarrow \text{C—C with X, Y}$$

$$-\text{C}\equiv\text{C}- + 2\text{X—Y} \longrightarrow -\text{C—C-}$$

As an example of an addition reaction, we will soon see that alkenes, such as ethylene, react with H_2 to yield alkanes:

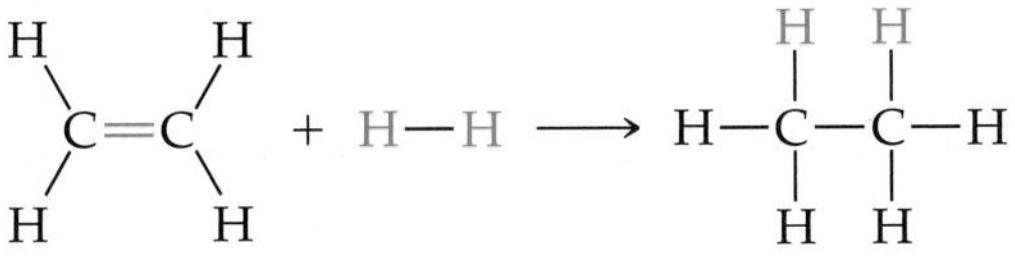

$$\text{Ethylene} + \text{H—H} \longrightarrow \text{Ethane}$$

- **Elimination Reactions** Eliminations are the opposite of addition reactions. Eliminations occur when a single reactant splits into two products, a process we can generalize as

This one reactant … $A \longrightarrow B + C$ … splits apart to give these two products.

In most cases, an elimination reaction converts the starting material to a product that has two fewer carbon–carbon single bonds and a carbon–carbon multiple bond in their place:

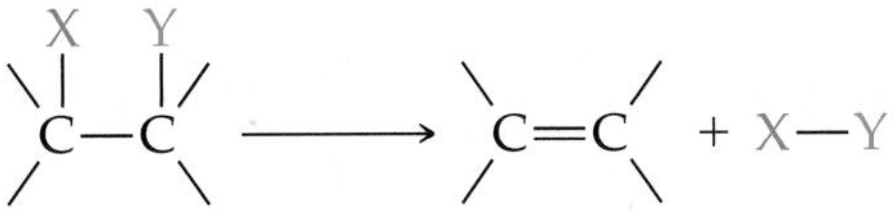

$$\text{C—C with X, Y} \longrightarrow \text{C=C} + \text{X—Y}$$

As an example of an elimination reaction, we will see in the next chapter that an alcohol, such as ethanol, splits apart into an alkene and water when treated with an acid catalyst:

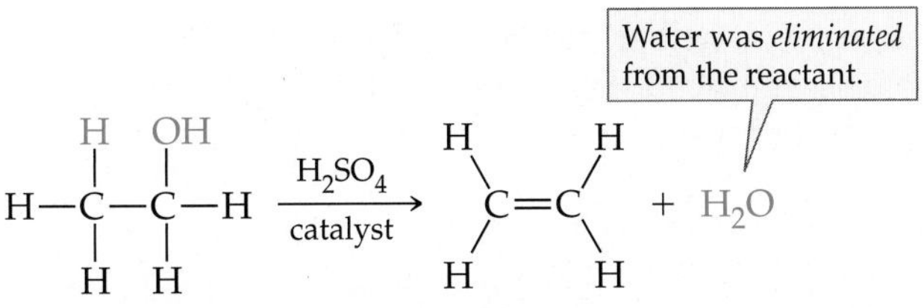

Water was *eliminated* from the reactant.

$$\text{Ethanol} \xrightarrow[\text{catalyst}]{H_2SO_4} \text{Ethylene} + H_2O$$

- **Substitution Reactions** Substitutions occur when two reactants exchange parts to give two new products, a process we can generalize as

These two reactants exchange parts … $AB + C \longrightarrow AC + B$ … to give these two products.

Elimination reaction A general reaction type in which a saturated reactant yields an unsaturated product by losing groups from two adjacent carbons.

Substitution reaction A general reaction type in which an atom or group of atoms in a molecule is replaced by another atom or group of atoms.

As an example of a substitution reaction, we saw in Section 12.8 that alkanes, such as methane, react with Cl_2 in the presence of ultraviolet light to yield alkyl chlorides. A—Cl group substitutes for the —H group of the alkane, and two new products result:

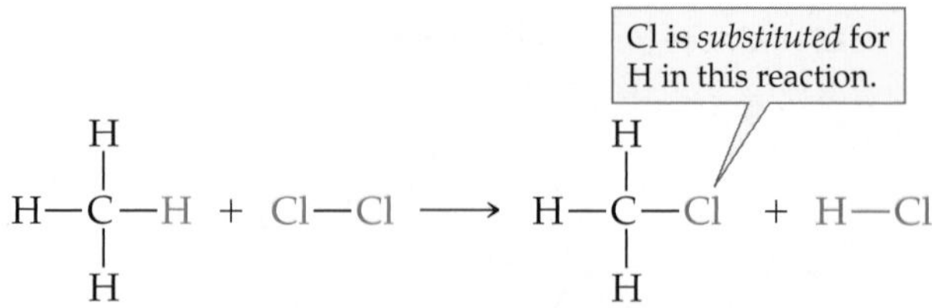

> Cl is *substituted* for H in this reaction.

Methane   Chloromethane

Rearrangement reaction A general reaction type in which a molecule undergoes bond reorganization to yield an isomer.

• **Rearrangement Reactions** Rearrangements are complicated processes that occur when a single reactant undergoes a reorganization of bonds and atoms to yield a single product that is an isomer of the reactant. A generalized example of one type of rearrangement seen in organic chemistry is:

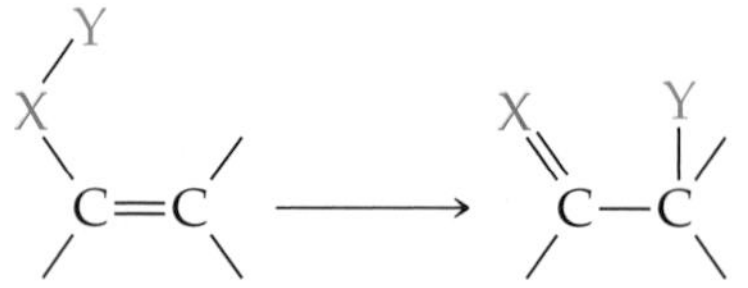

Rearrangement reactions are important in organic chemistry as well as biochemistry. Because of their complex nature, however, we will not discuss them in detail in this book. A simple example of a common rearrangement is the conversion of *cis*-2-butene into its isomer *trans*-2-butene by treatment with an acid catalyst:

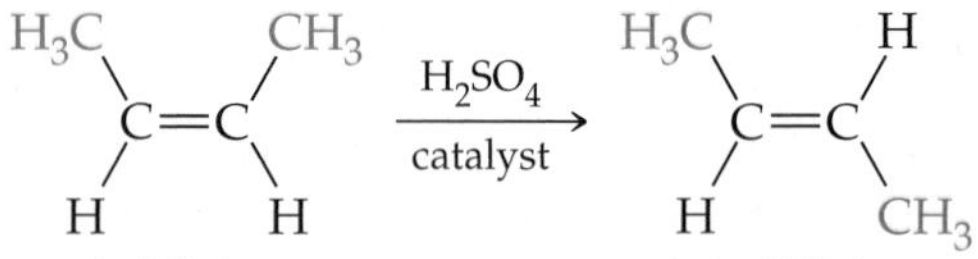

cis-2-Butene   *trans*-2-Butene

WORKED EXAMPLE **13.4** Identifying Reactions of Alkenes

Classify the following alkene reactions as addition, elimination, or substitution reactions:

(a) $CH_3CH{=}CH_2 + H_2 \longrightarrow CH_3CH_2CH_3$

(b) $CH_3CH_2CH_2OH \xrightarrow[\text{catalyst}]{H_2SO_4} CH_3CH{=}CH_2 + H_2O$

(c) $CH_3CH_2Cl + KOH \longrightarrow CH_3CH_2OH + KCl$

ANALYSIS Determine whether atoms have been added to the starting compound (addition), removed from the starting compound (elimination), or switched with another reactant (substitution).

SOLUTION

(a) Two H atoms have been *added* in place of the double bond, so this is an *addition* reaction.

(b) A water molecule (H_2O) has been formed by *removing* an H atom and an —OH group from adjacent C atoms, forming a double bond in the process, so this is an *elimination* reaction.

(c) The reactants (CH_3CH_2Cl and KOH) have *traded* the —OH and the —Cl substituent groups, so this is a *substitution* reaction.

PROBLEM 13.7

Classify the following reactions as an addition, elimination, substitution, or rearrangement:

(a) $CH_3Br + NaOH \longrightarrow CH_3OH + NaBr$

(b) $H_2C{=}CH_2 + HCl \longrightarrow CH_3CH_2Cl$

(c) $CH_3CH_2Br \longrightarrow H_2C{=}CH_2 + HBr$

13.6 Reactions of Alkenes and Alkynes

Most of the reactions of carbon–carbon multiple bonds are *addition reactions*. A generalized reagent we might write as X—Y adds to the multiple bond in the unsaturated reactant to yield a saturated product that has only single bonds:

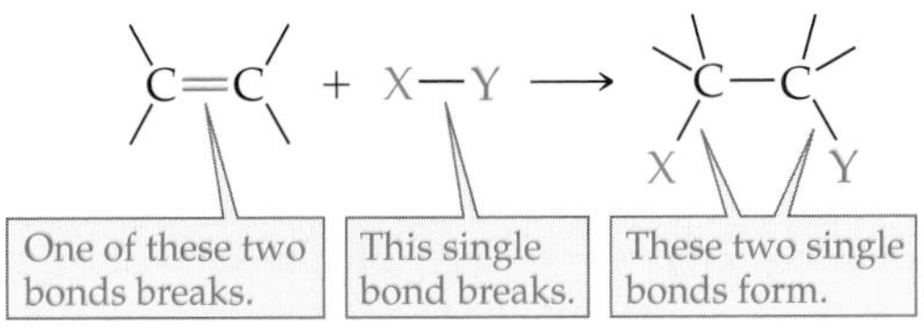

One of these two bonds breaks. | This single bond breaks. | These two single bonds form.

An addition reaction

Alkenes and alkynes react similarly in many ways, but we will look mainly at alkenes in this chapter because they are more commonly found in nature and industrially are used as precursors to other organic molecules. (Simple molecules that are prepared in large quantities for this purpose are known as *chemical feedstocks*.)

Addition of H₂ to Alkenes and Alkynes: Hydrogenation

Alkenes and alkynes react with hydrogen in the presence of a metal catalyst such as palladium to yield the corresponding alkane product:

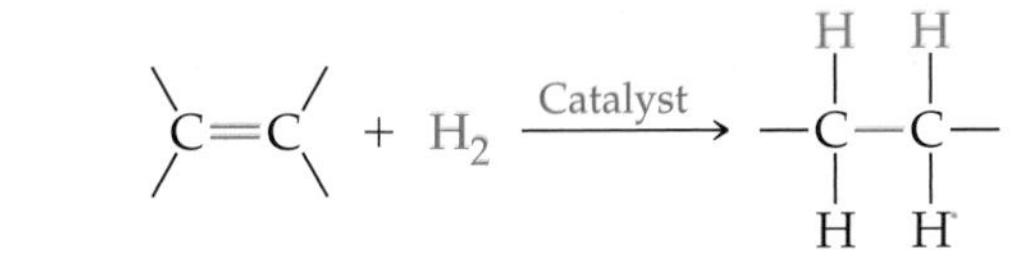

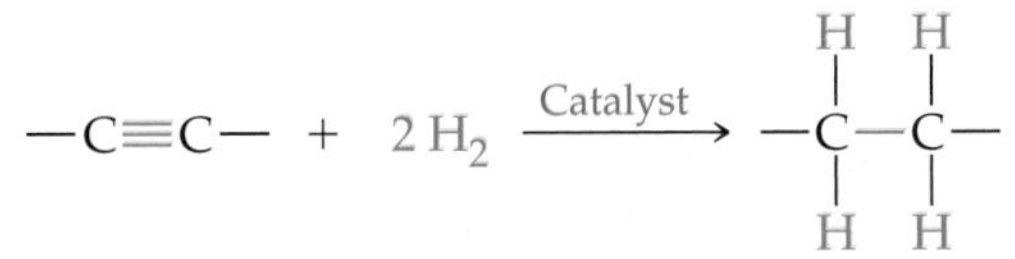

For example,

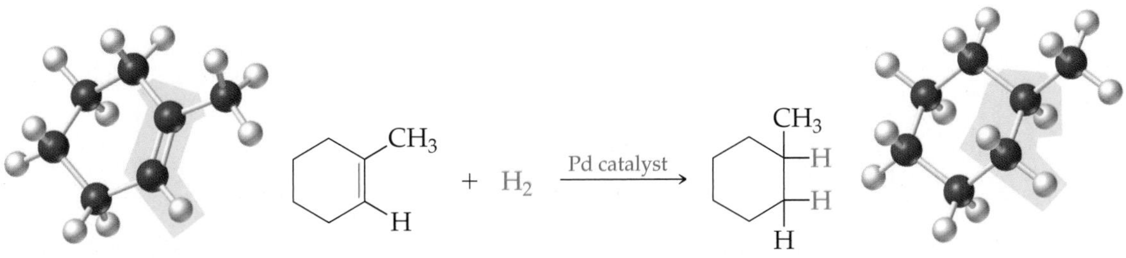

1-Methylcyclohexene

Methylcyclohexane (85% yield)

The addition of hydrogen to an alkene, a process called **hydrogenation**, is used commercially to convert unsaturated vegetable oils, which contain numerous double bonds, to the saturated fats used in margarine and cooking fats. This process has come under intense scrutiny in recent years because it creates *trans* fatty acids in the product (see Application "Butter and its Substitutes," Chapter 24, p. 774). We will see the structures of these fats and oils in Chapter 24.

Hydrogenation The addition of H_2 to a multiple bond to give a saturated product.

WORKED EXAMPLE   **13.5** Organic Reactions: Addition

What product would you obtain from the following reaction?

$$CH_3CH_2CH_2CH=CHCH_3 + H_2 \xrightarrow{Pd} ?$$

ANALYSIS Rewrite the reactant, showing a single bond and two partial bonds in place of the double bond:

$$CH_3CH_2CH_2CH{-}CHCH_3$$

Then, add a hydrogen to each carbon atom of the double bond, and rewrite the product in condensed form:

$$CH_3CH_2CH_2\underset{\underset{H}{|}}{CH}{-}\underset{\underset{H}{|}}{CH}CH_3 \quad \text{is the same as} \quad \underset{\text{Hexane}}{CH_3CH_2CH_2CH_2CH_2CH_3}$$

SOLUTION
The reaction is

$$CH_3CH_2CH_2CH=CHCH_3 + H_2 \xrightarrow{Pd} CH_3CH_2CH_2CH_2CH_2CH_3$$

PROBLEM 13.8

Write the structures of the products from the following hydrogenation reactions:

(a) $CH_3CH_2CH=CH_2 + H_2 \xrightarrow{Pd} ?$

(b) *cis*-2-Butene $+ H_2 \xrightarrow{Pd} ?$

(c) *trans*-2-Butene $+ H_2 \xrightarrow{Pd} ?$

(d) 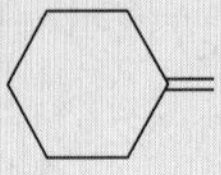 ⬡=$CH_2 + H_2 \xrightarrow{Pd} ?$

Addition of Cl₂ and Br₂ to Alkenes: Halogenation

Halogenation   The addition of Cl_2 or Br_2 to a multiple bond to give a dihalide product.

Alkenes react with the halogens Br_2 and Cl_2 to give 1,2-dihaloalkane addition products in a **halogenation** reaction:

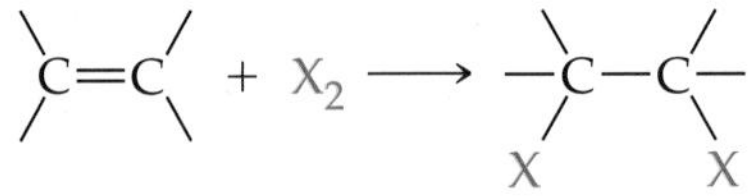

$$\underset{/}{\overset{\backslash}{C}}=\underset{\backslash}{\overset{/}{C}} + X_2 \longrightarrow \underset{\underset{X}{|}}{\overset{\backslash}{-}C}\underset{\underset{X}{|}}{\overset{/}{C-}} \qquad \begin{array}{l}\text{(A 1,2-dihaloalkane}\\ \text{where } X = Br \text{ or } Cl)\end{array}$$

For example,

$$\underset{H}{\overset{H}{\underset{|}{}}}C=C\underset{H}{\overset{H}{}} + Cl_2 \longrightarrow H{-}\underset{\underset{Cl}{|}}{C}{-}\underset{\underset{Cl}{|}}{C}{-}H$$

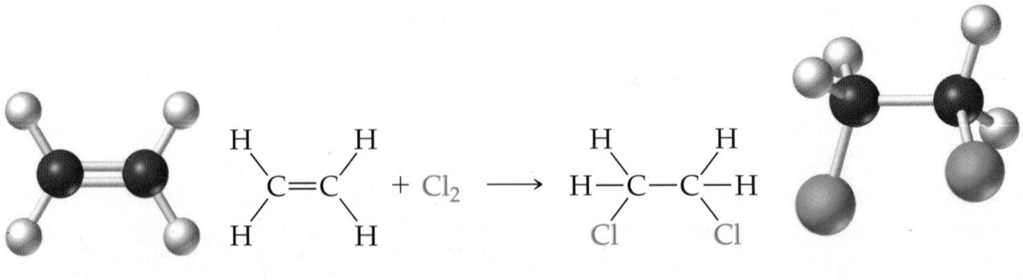

Ethylene   1, 2-Dichloroethane

This reaction, the first step in making poly(vinyl chloride) plastics, is used to manufacture nearly eight million tons of 1,2-dichloroethane each year in the United States.

Reaction with Br_2 also provides a convenient test for the presence in a molecule of a carbon–carbon double or triple bond (Figure 13.1). A few drops of a reddish-brown solution of Br_2 are added to a sample of an unknown compound. Immediate disappearance of the color as the bromine reacts to form a colorless dibromide reveals the presence of the multiple bond. This test can be used to determine the level of unsaturation of fats (Chapter 24). Although chlorine also adds to double bonds, it is not used to test for their presence because it is a gas at room temperature, harder to handle, and its color in solution is a very light yellow and difficult to see changes in.

(a)

(b)

◀ **FIGURE 13.1 Testing for unsaturation with bromine.** (a) No color change results when the bromine solution is added to hexane (C_6H_{14}). (b) Disappearance of the bromine color when it is added to 1-hexene (C_6H_{12}) indicates the presence of a double bond.

PROBLEM 13.9

What products would you expect from the following halogenation reactions?

(a) 2-Methylpropene + Br_2 ⟶ ?   (b) 1-Pentene + Cl_2 ⟶ ?

(c) $CH_3CH_2CH{=}\overset{\displaystyle CH_3}{\underset{\displaystyle CH_3}{C}CH_2CHCH_3}$ + Cl_2 ⟶ ?

Addition of HBr and HCl to Alkenes

Alkenes react with hydrogen bromide (HBr) to yield *alkyl bromides* (R—Br) and with hydrogen chloride (HCl) to yield *alkyl chlorides* (R—Cl), in what are called **hydrohalogenation** reactions:

Hydrohalogenation The addition of HCl or HBr to a multiple bond to give an alkyl halide product.

Hydrohalogenation:
Addition of HBr or HCl to a double bond.

$\underset{H}{\overset{HBr}{\longrightarrow}}$ —C—C— (An alkyl bromide)
        H   Br

$\overset{HCl}{\longrightarrow}$ —C—C— (An alkyl chloride)
        H   Cl

The addition of HBr to 2-methylpropene is an example:

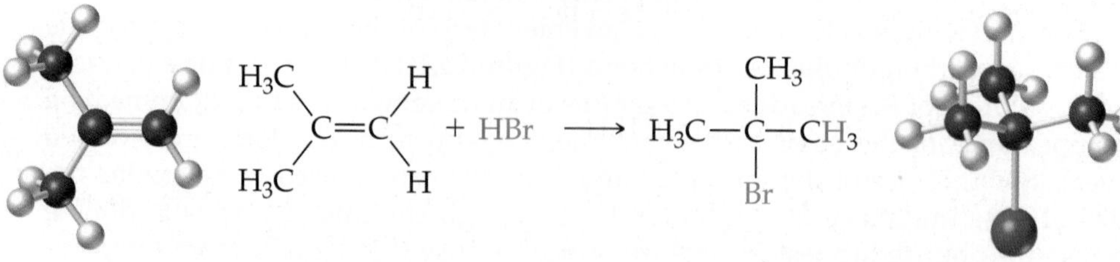

$$H_3C \backslash C=C / H \atop H_3C \diagup \quad \backslash H \quad + HBr \longrightarrow H_3C-\underset{Br}{\overset{CH_3}{\underset{|}{\overset{|}{C}}}}-CH_3$$

2-Methylpropene   2-Bromo-2-methylpropane

Look carefully at the above example. Only one of the two possible addition products is obtained. 2-Methylpropene *could* add HBr to give 1-bromo-2-methylpropane, but it does not; it gives only 2-bromo-2-methylpropane.

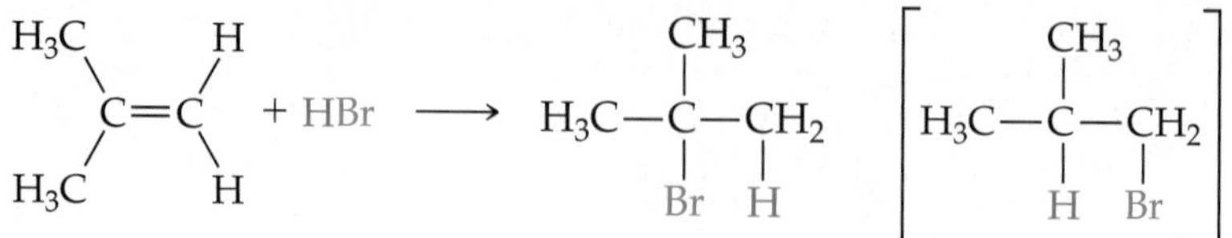

$$H_3C \backslash C=C / H \atop H_3C \diagup \quad \backslash H \quad + HBr \longrightarrow H_3C-\underset{Br \quad H}{\overset{CH_3}{\underset{|}{\overset{|}{C}}}}-CH_2 \qquad \left[H_3C-\underset{H \quad Br}{\overset{CH_3}{\underset{|}{\overset{|}{C}}}}-CH_2 \right]$$

2-Methylpropene   2-Bromo-2-methylpropane (Sole product)   1-Bromo-2-methylpropane (Not formed)

This result is typical of what happens when HBr and HCl add to an alkene in which one of the double-bond carbons has more hydrogens than the other (an unsymmetrically substituted alkene). The results of such additions can be predicted by **Markovnikov's rule**, formulated in 1869 by the Russian chemist Vladimir Markovnikov:

Markovnikov's rule   In the addition of HX to an alkene, the H attaches to the double-bond carbon that has the larger number of H atoms *directly* attached to it, and the X attaches to the carbon that has the smaller number of H atoms attached.

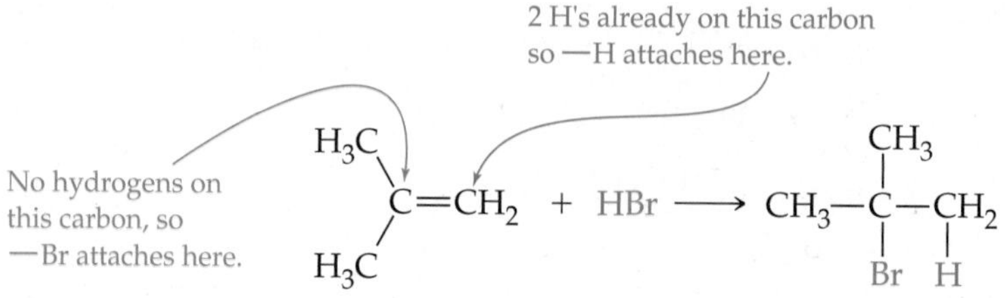

2 H's already on this carbon so —H attaches here.

No hydrogens on this carbon, so —Br attaches here.

$$H_3C \backslash C=CH_2 \atop H_3C \diagup \quad + HBr \longrightarrow CH_3-\underset{Br \quad H}{\overset{CH_3}{\underset{|}{\overset{|}{C}}}}-CH_2$$

The scientific reason behind Markovnikov's rule is a powerful and important principle in organic chemistry. Examining this reason, which has to do with the stability of intermediates known as *carbocations* that form during the reaction, is unfortunately beyond the scope of this text.

If an alkene has equal numbers of H atoms attached to the double-bond carbons, both possible products are formed in approximately equal amounts:

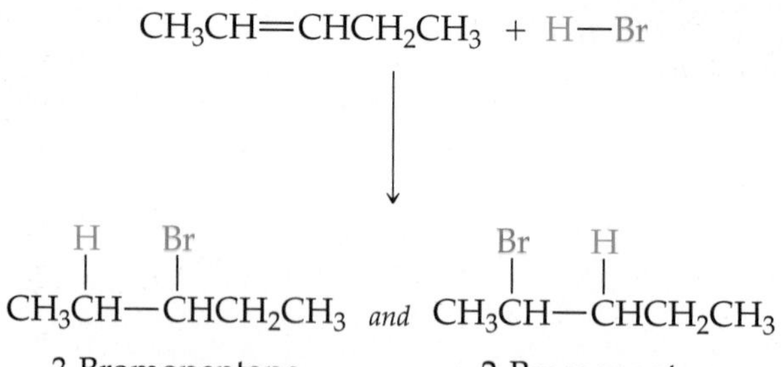

$$CH_3CH=CHCH_2CH_3 + H-Br$$

↓

$$CH_3\underset{\overset{|}{H}}{\overset{\overset{H}{|}}{C}}H-\underset{\overset{|}{Br}}{\overset{\overset{Br}{|}}{C}}HCH_2CH_3 \quad and \quad CH_3\underset{\overset{|}{Br}}{\overset{\overset{Br}{|}}{C}}H-\underset{\overset{|}{H}}{\overset{\overset{H}{|}}{C}}HCH_2CH_3$$

3-Bromopentane   2-Bromopentane

(1:1 ratio)

WORKED EXAMPLE **13.6** Organic Reactions: Markovnikov's Rule

What product do you expect from the following reaction?

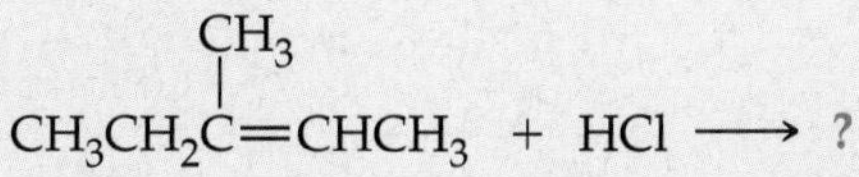

$$\underset{\text{CH}_3}{\overset{|}{\text{CH}_3\text{CH}_2\text{C}}}{=}\text{CHCH}_3 \ + \ \text{HCl} \ \longrightarrow \ ?$$

ANALYSIS The reaction of an alkene with HCl leads to formation of an alkyl chloride addition product according to Markovnikov's rule. To make a prediction, look at the starting alkene and count the number of hydrogens attached to each double-bond carbon. Then write the product by attaching H to the carbon with more hydrogens and attaching Cl to the carbon with fewer hydrogens.

SOLUTION

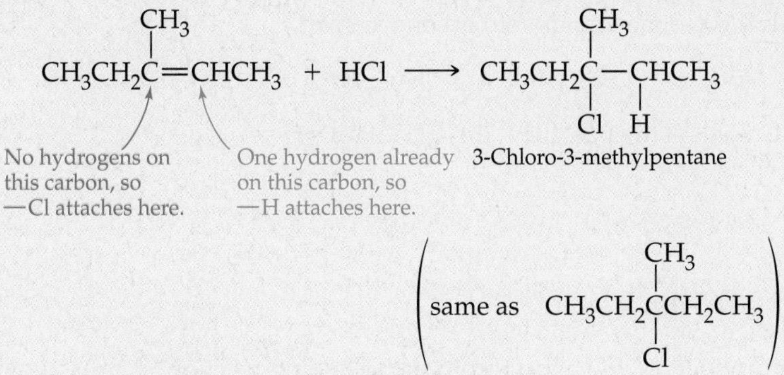

No hydrogens on this carbon, so —Cl attaches here.

One hydrogen already on this carbon, so —H attaches here.

3-Chloro-3-methylpentane

$$\left(\text{same as} \quad \underset{\text{Cl}}{\overset{\overset{\text{CH}_3}{|}}{\text{CH}_3\text{CH}_2\text{CCH}_2\text{CH}_3}} \right)$$

WORKED EXAMPLE **13.7** Organic Reactions: Markovnikov's Rule

From what alkene might 2-chloro-3-methylbutane be made?

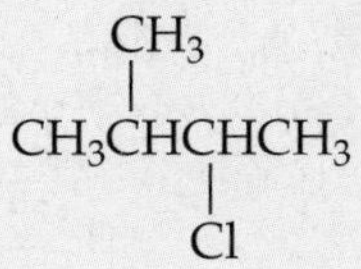

$$\underset{\text{Cl}}{\overset{\overset{\text{CH}_3}{|}}{\text{CH}_3\text{CHCHCH}_3}}$$

2-Chloro-3-methylbutane

ANALYSIS 2-Chloro-3-methylbutane is an alkyl chloride that might be made by addition of HCl to an alkene. To generate the possible alkene precursors, remove the —Cl group and an —H atom from adjacent carbons, and replace with a double bond:

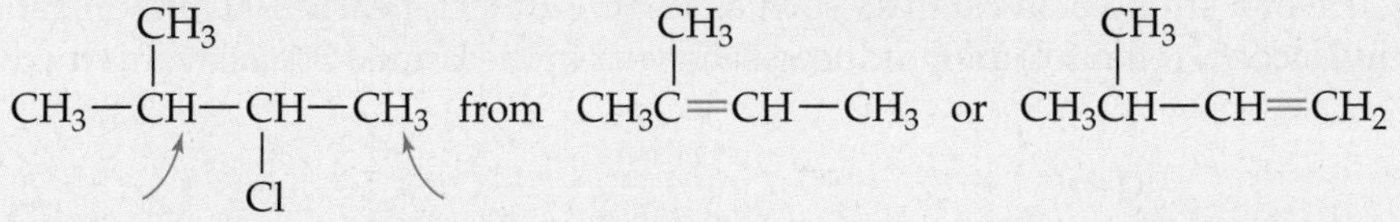

2-Methyl-2-butene   3-Methyl-1-butene

Look at the possible alkene addition reactions to see which is compatible with Markovnikov's rule. In this case, addition to 3-methyl-1-butene is compatible. Note that if HCl is added to 2-methyl-1-butene the Cl will be attached to the wrong carbon.

SOLUTION

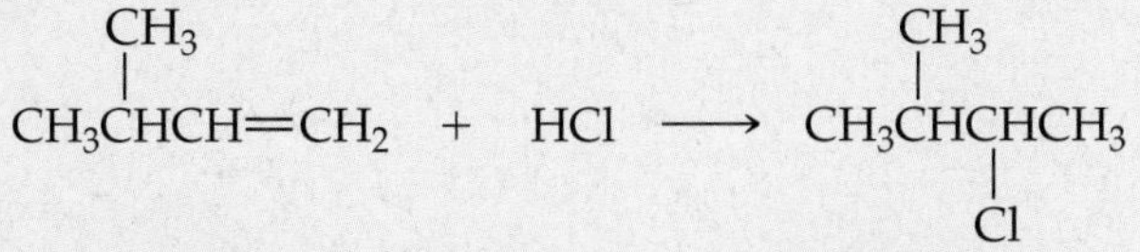

$$\underset{}{\overset{\overset{\text{CH}_3}{|}}{\text{CH}_3\text{CHCH}}}{=}\text{CH}_2 \ + \ \text{HCl} \ \longrightarrow \ \underset{\text{Cl}}{\overset{\overset{\text{CH}_3}{|}}{\text{CH}_3\text{CHCHCH}_3}}$$

3-Methyl-1-butene   2-Chloro-3-methylbutane

PROBLEM 13.10

What products do you expect from the following reactions?

(a) 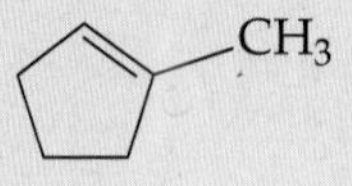 [cyclopentene with CH₃] + HCl ⟶ ?   (b) 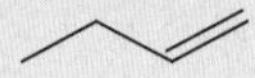 [propene] + HBr ⟶ ?

(c) CH₃C=CHCH₃ with CH₃ substituent + HCl ⟶ ?

PROBLEM 13.11

From what alkenes are the following alkyl halides likely to be made? (Careful, there may be more than one answer.)

(a) 3-Chloro-3-ethylpentane   (b) CH₃CHCCH₃ with H₃C, Br, and CH₃ substituents

KEY CONCEPT PROBLEM 13.12

What product do you expect from the following reaction?

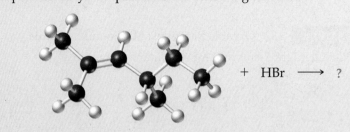 + HBr ⟶ ?

Addition of Water to Alkenes: Hydration

Hydration The addition of water to a multiple bond to give an alcohol product.

A water molecule (H—OH) can be considered to be one type of H—X, but an alkene will not react with pure water alone. If, however, a small amount of a strong acid catalyst, such as H_2SO_4, is added, an addition reaction takes place to yield an *alcohol* (R—OH). This **hydration** reaction occurs on treatment of the alkene with water in the presence of a strong acid catalyst, such as H_2SO_4. In fact, nearly 100 million gallons of ethyl alcohol (ethanol) are produced each year in the United States by this method.

$$C=C + H-O-H \xrightarrow[\text{catalyst}]{H_2SO_4} -\overset{|}{\underset{H}{C}}-\overset{|}{\underset{O-H}{C}}-$$

An alcohol

For example,

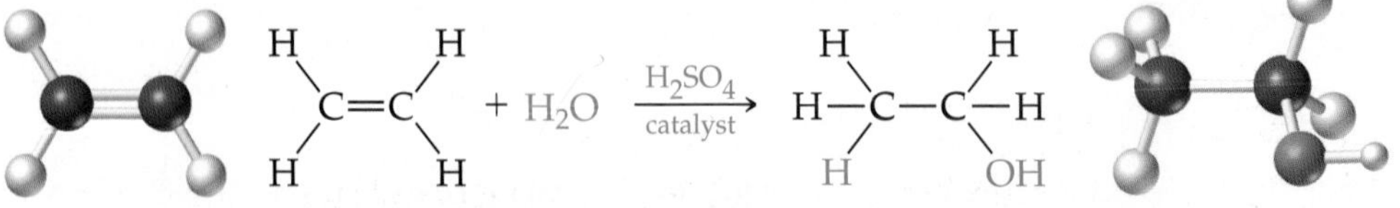

$$\underset{\text{Ethylene}}{\overset{H}{\underset{H}{C}}=\overset{H}{\underset{H}{C}}} + H_2O \xrightarrow[\text{catalyst}]{H_2SO_4} \underset{\text{Ethyl alcohol}}{H-\overset{H}{\underset{H}{C}}-\overset{H}{\underset{OH}{C}}-H}$$

As with the addition of HBr and HCl, we can use Markovnikov's rule to predict the product when water adds to an unsymmetrically substituted alkene.

Hydration of 2-methylpropene, for example, gives 2-methyl-2-propanol:

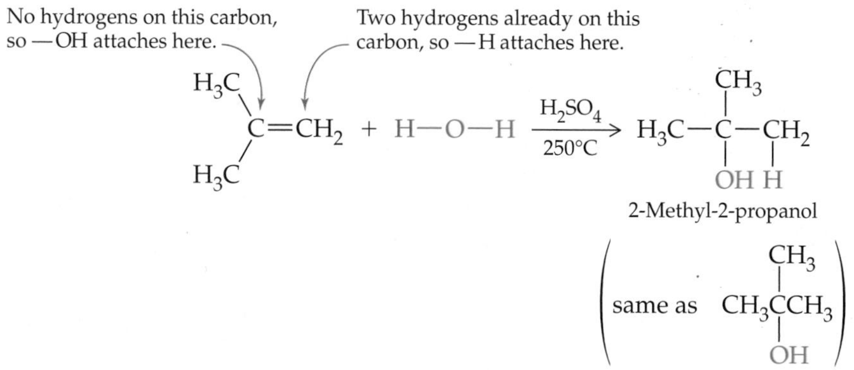

No hydrogens on this carbon, so —OH attaches here. Two hydrogens already on this carbon, so —H attaches here.

$$H_3C \backslash$$
$$C{=}CH_2 \;+\; H{-}O{-}H \xrightarrow[250°C]{H_2SO_4} H_3C{-}\underset{\underset{OH\;H}{|\;\;\;|}}{\overset{\overset{CH_3}{|}}{C}}{-}CH_2$$
$$H_3C /$$

2-Methyl-2-propanol

$$\left(\text{same as} \quad CH_3\underset{\underset{OH}{|}}{\overset{\overset{CH_3}{|}}{C}}CH_3 \right)$$

> **WORKED EXAMPLE**   **13.8** Reaction of Alkenes: Hydration
>
> What product(s) do you expect from the following hydration reaction?
>
> $$CH_3CH{=}CHCH_2CH_3 + H_2O \longrightarrow ?$$
>
> **ANALYSIS** Water is added to the double bond, with a H atom added to one carbon and an —OH group added to the other carbon of the double bond.
>
> **SOLUTION**
> Because this is *not* an unsymmetrically substituted alkene, we can add the —OH group to either carbon:
>
> $$CH_3{-}\underset{\underset{OH\;\;\;H}{|\;\;\;\;\;|}}{\overset{\overset{H\;\;\;\;\;H}{|\;\;\;\;\;|}}{C{-}\!-\!C}}{-}CH_3CH_3 \quad \text{or} \quad CH_3{-}\underset{\underset{H\;\;\;\;OH}{|\;\;\;\;\;|}}{\overset{\overset{H\;\;\;\;\;H}{|\;\;\;\;\;|}}{C{-}\!-\!C}}{-}CH_3CH_3$$
>
> 2-pentanol   3-pentanol

> **PROBLEM 13.13**
>
> What products do you expect from the following hydration reactions?
>
> **(a)** 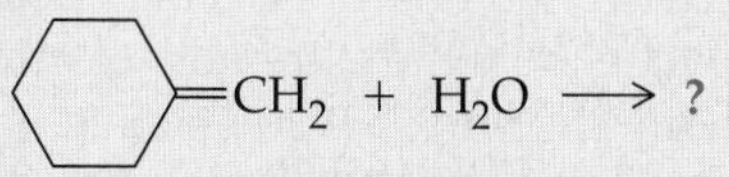 ⬡=CH₂ + H₂O ⟶ ?   **(b)** 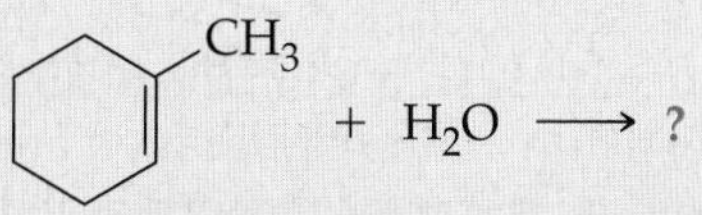 ⬡—CH₃ + H₂O ⟶ ?
>
> **(c)** $CH_3CH{=}CHCH_2CH_3 + H_2O \longrightarrow$? *(two possible products)*
>
> **PROBLEM 13.14**
>
> From what alkene reactant might 3-methyl-3-pentanol be made? 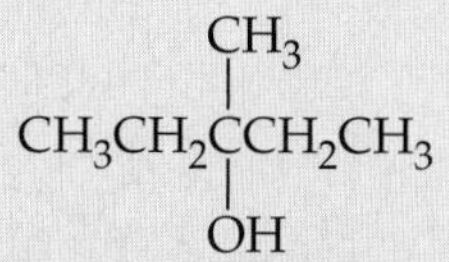
>
> $$CH_3CH_2\underset{\underset{OH}{|}}{\overset{\overset{CH_3}{|}}{C}}CH_2CH_3$$
> 3-Methyl-3-pentanol

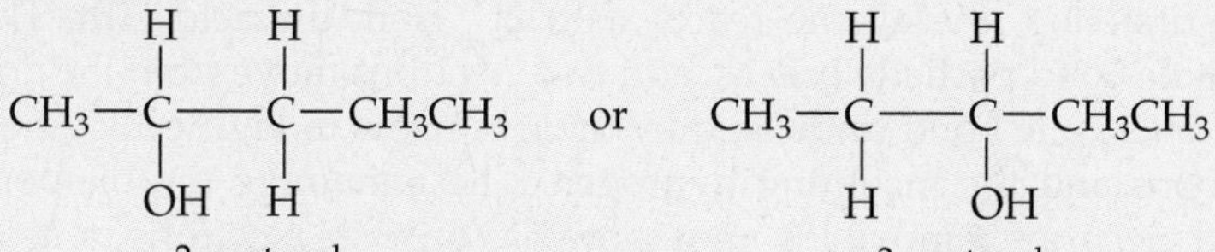

13.7 How Alkene Addition Reactions Occur

How do alkene addition reactions take place? Do two molecules, say ethylene and HBr, simply collide and immediately form a product molecule of bromoethane, or is the process more complex? Detailed studies show that alkene addition reactions take place in two distinct steps, as illustrated in Figure 13.2 for the addition of HBr to ethylene.

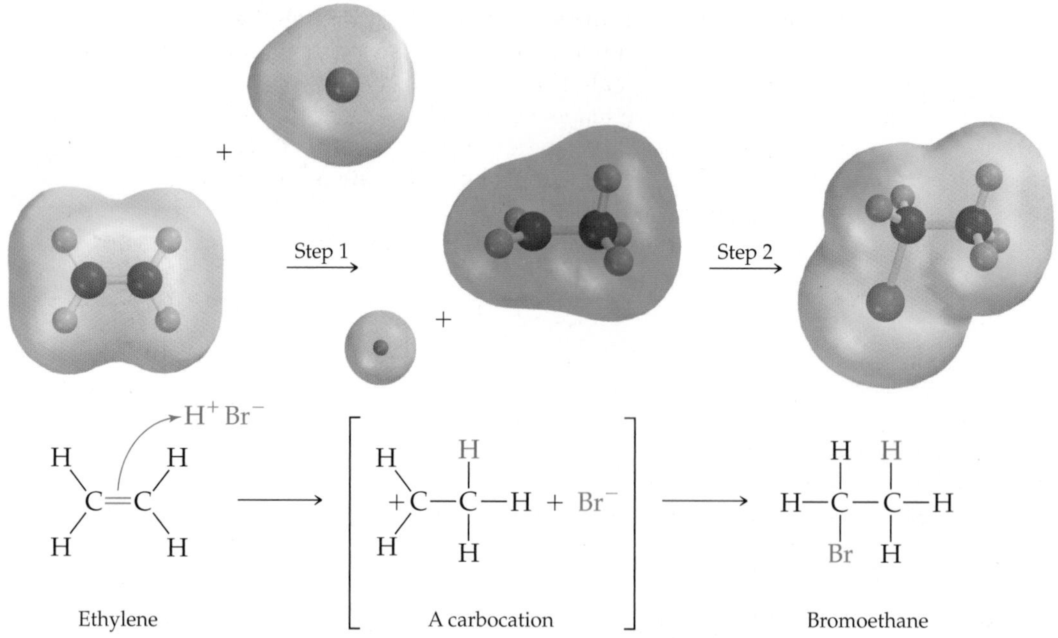

▲ **FIGURE 13.2** **The mechanism of the addition of HBr to an alkene.** The reaction takes place in two steps and involves a carbocation intermediate. In the first step, two electrons move from the C═C double bond to form a C─H bond. In the second step, Br⁻ uses two electrons to form a bond to the positively charged carbon.

In the first step, the alkene reacts with H^+ from the acid HBr. The carbon–carbon double bond partially breaks, and two electrons move from the double bond to form a new single bond (indicated by the red arrow in Figure 13.2) between one of the carbons and the incoming hydrogen. The remaining double-bond carbon, having had electrons removed from it, now has only six electrons in its outer shell and bears a positive charge. Unlike sodium ion (Na^+) and other metal cations, which are unreactive and easily isolated in salts like NaCl, carbons that possess a positive charge, or *carbocations*, are highly reactive. As soon as the carbocation is formed by reaction of an alkene with H^+, it immediately reacts with Br^- to form a neutral product. Note how electrostatic potential maps (Section 5.8) illustrate the electron-rich (red) nature of the ethylene double bond and the electron-poor (blue) nature of the H atom in HBr. (▭▭, p. 128). Also, note the extremely electron-poor (blue) nature of the carbocation and you readily see why it is so reactive!

Reaction mechanism A description of the individual steps by which old bonds are broken and new bonds are formed in a reaction.

A description of the individual steps by which old bonds are broken and new bonds are formed in a reaction is called a **reaction mechanism**. Although we will not examine many reaction mechanisms in this book, they are an important part of organic chemistry. Mechanisms allow chemists to classify thousands of seemingly unrelated organic reactions into only a few categories and help us to understand what is occurring during a reaction. Their study is essential to our ever-expanding ability to understand biochemistry and the physiological effects of drugs. If you continue your study of chemistry, you will see reaction mechanisms often.

PROBLEM 13.15

Remembering Markovnikov's rule (Section 13.6), draw the structure of the carbocation formed during the reaction of 2-methylpropene with HCl.

Polymer A large molecule formed by the repetitive bonding together of many smaller molecules.

Monomer A small molecule that is used to prepare a polymer.

13.8 Alkene Polymers

A **polymer** is a large molecule formed by the repetitive bonding together of many smaller molecules called **monomers**. As we will see in later chapters, biological polymers occur throughout nature. Cellulose and starch are polymers built from sugars, proteins are polymers built from amino acids, and the DNA that makes up our genetic heritage is a polymer built from nucleic acids. Although the basic idea is the same, synthetic polymers are much simpler than biopolymers because the starting monomer units are usually small, simple organic molecules.

Many simple alkenes (often called *vinyl monomers* because the partial structure $H_2C=CH-$ is known as a *vinyl group*) undergo *polymerization* reactions when treated with the proper catalyst. Ethylene yields polyethylene on polymerization, propylene yields polypropylene, and styrene yields polystyrene. The polymer product might have anywhere from a few hundred to a few thousand monomer units incorporated into a long, repeating chain.

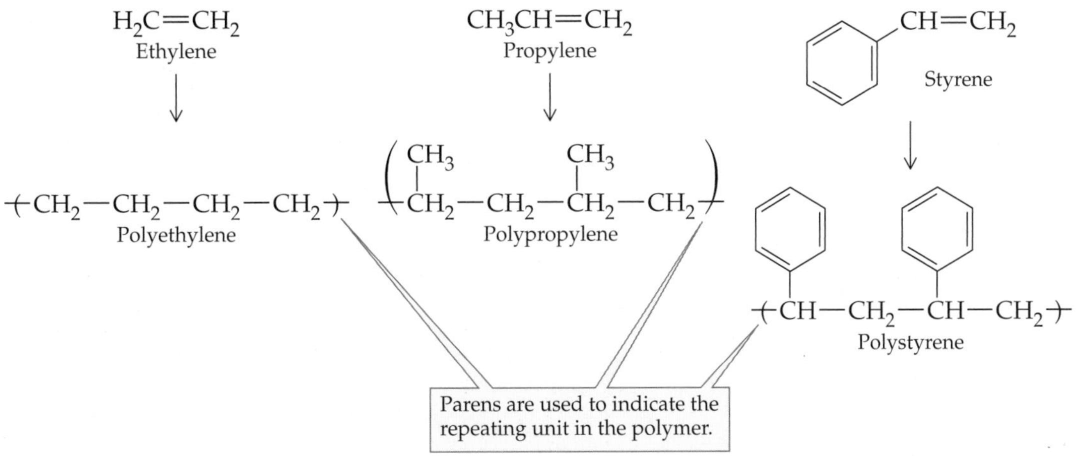

Parens are used to indicate the repeating unit in the polymer.

The fundamental reaction in the polymerization of an alkene monomer resembles the addition to a carbon–carbon double bond described in the preceding section. The reaction begins by addition of a species called an *initiator* to an alkene; this results in the breaking of one of the bonds making up the double bond. A reactive intermediate that contains an unpaired electron (known as a *radical*) is formed in this step, and it is this reactive intermediate that adds to a second alkene molecule. This produces another reactive intermediate, which adds to a third alkene molecule, and so on. Because the result is continuous addition of one monomer after another to the end of the growing polymer chain, polymers formed in this way are called *chain-growth polymers*. The basic repeating unit is enclosed in parentheses, and the subscript n indicates how many repeating units are in the polymer:

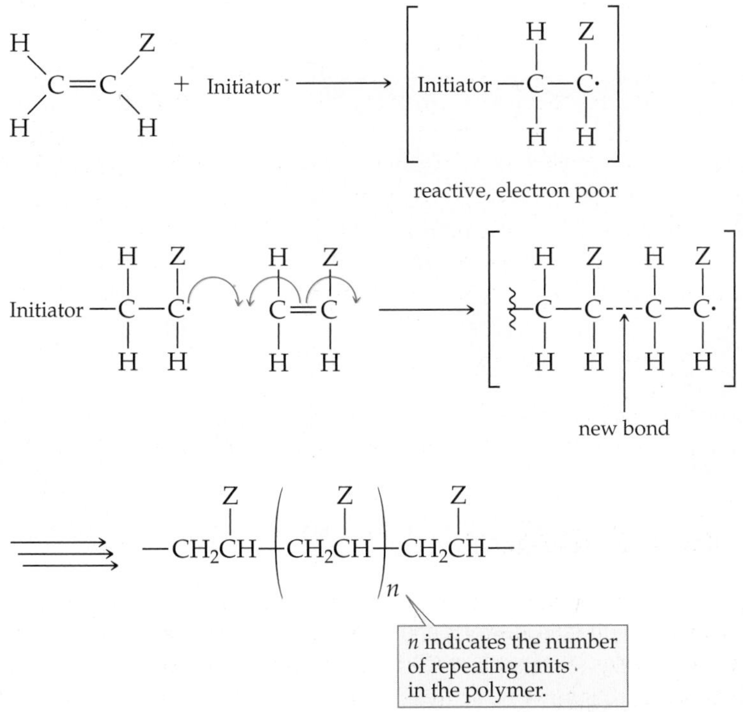

reactive, electron poor

new bond

n indicates the number of repeating units in the polymer.

TABLE 13.1 Some Alkene Polymers and Their Uses

MONOMER NAME	MONOMER STRUCTURE	POLYMER NAME	USES
Ethylene	$H_2C=CH_2$	Polyethylene	Packaging, bottles
Propylene	$H_2C=CH-CH_3$	Polypropylene	Bottles, rope, pails, medical tubing
Vinyl chloride	$H_2C=CH-Cl$	Poly(vinyl chloride)	Insulation, plastic pipe
Styrene	$H_2C=CH-\phenyl$	Polystyrene	Foams and molded plastics
Styrene and butadiene	$H_2C=CH-\phenyl$ and $H_2C=CHCH=CH_2$	Styrene-butadiene rubber (SBR)	Synthetic rubber for tires
Acrylonitrile	$H_2C=CH-C{\equiv}N$	Orlon, Acrilan	Fibers, outdoor carpeting
Methyl methacrylate	$H_2C=\overset{\displaystyle O \atop \displaystyle \|}{\underset{\displaystyle CH_3}{C}}COCH_3$	Plexiglas, Lucite	Windows, contact lenses, fiber optics
Tetrafluoroethylene	$F_2C=CF_2$	Teflon	Nonstick coatings, bearings, replacement heart valves and blood vessels

Variations in the substituent group Z attached to the double bond impart different properties to the product, as illustrated by the alkene polymers listed in Table 13.1. Polymer rigidity is controlled by addition of a small amount of a cross-linking agent, typically 1–2% of a dialkene, whose role is to covalently link two chains of monomer units together.

The properties of a polymer depend not only on the monomer but also on the average size of the huge molecules in a particular sample and on how extensively they cross-link and branch. The long molecules in straight-chain polyethylene pack

(a)   (b)

▲ (a) High-density polyethylene is strong and rigid enough to be used in many kinds of bottles and toys. (b) Low-density polyethylene is moisture-proof but flexible and is used in plastic bags and packaging.

closely together, giving a rigid material called *high-density polyethylene*, which is mainly used in bottles for products such as milk and motor oil. When polyethylene molecules contain many branches (due to the Z groups present), they cannot pack together as tightly and instead form a flexible material called *low-density polyethylene*, which is used mainly in packaging materials.

Polymer technology has come a long way since the development of synthetic rubber, nylon, Plexiglas, and Teflon. The use of polymers has changed the nature of activities from plumbing and carpentry to clothing and auto manufacture. In the healthcare fields, the use of inexpensive, disposable equipment is now common.

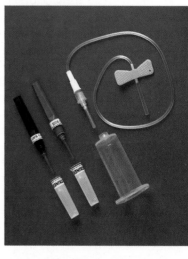

▲ These disposable polypropylene syringes are used once and then discarded.

WORKED EXAMPLE 13.9 Reactions of Alkenes: Polymerization

Write the structure of a segment of polystyrene, used in foams and molded plastics. The monomer is

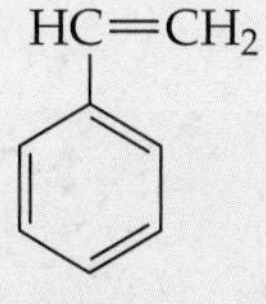

$$HC=CH_2$$

ANALYSIS The polymerization reaction resembles the addition of two monomer units to either end of the double bond.

SOLUTION
Draw three molecules of styrene with the double bonds aligned next to each other; then add the monomer units together with single bonds, eliminating the double bonds in the process:

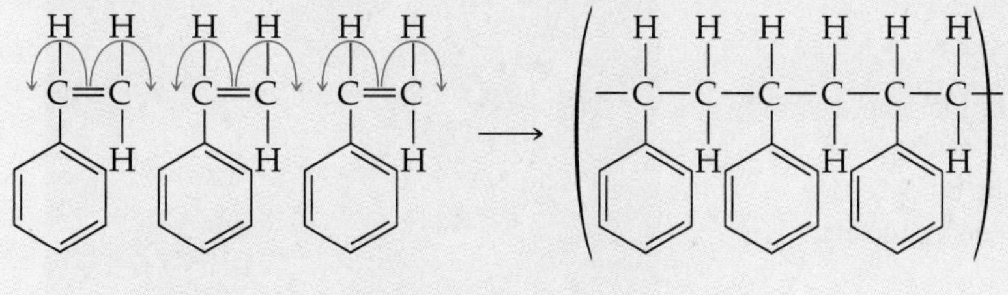

PROBLEM 13.16

Write the structure of a segment of poly(vinyl acetate), a polymer used for the springy soles in running shoes. The structure of the monomer is

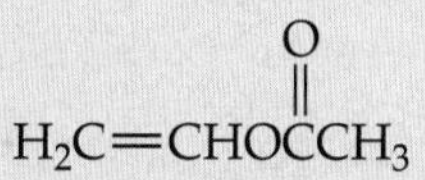

$$H_2C=CHOCCH_3 \quad \text{Vinyl acetate}$$

with a carbonyl (C=O) above the O.

PROBLEM 13.17

Write the structures of the monomers used to make the following polymers.

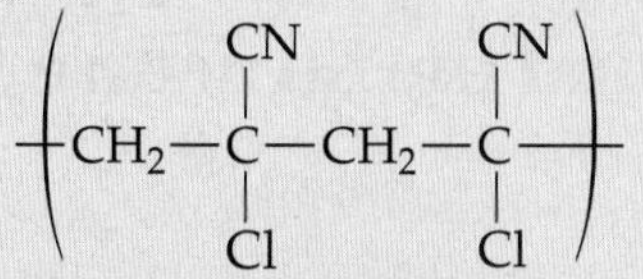

(a)
$$\left(CH_2-\underset{\underset{Cl}{\overset{CN}{|}}}{C}-CH_2-\underset{\underset{Cl}{\overset{CN}{|}}}{C}\right)$$

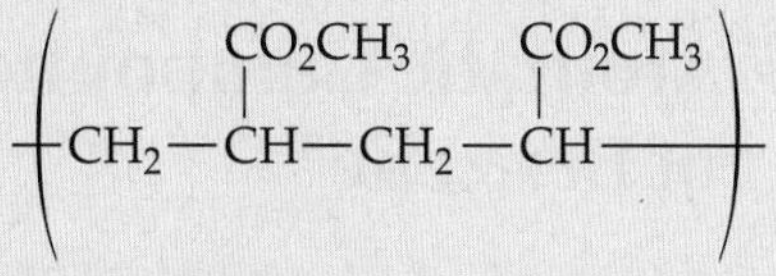

(b)
$$\left(CH_2-\underset{\overset{CO_2CH_3}{|}}{CH}-CH_2-\underset{\overset{CO_2CH_3}{|}}{CH}\right)$$

APPLICATION ▶ Polymer Applications—Currency

Polymers exhibit a wide range of properties, depending on both their chemical structure and how they are processed. What typically distinguishes polymeric materials from natural products, however, is their durability. Their resistance to degradation makes polymers attractive for applications including construction materials, storage containers, and food packaging. But did you ever think that the dollar bills in your wallet may someday be made from man-made polymers? This is already the case in many parts of the world, and the use of polymer-based currency is being investigated by more countries including the United States.

The first polymer-based banknotes were issued in 1988 in Australia. Since then, 17 other countries have adopted polymer-based currencies, including Brazil, Indonesia, Romania, Thailand, Vietnam, and Zambia. Made from biaxially oriented polypropylene (BOPP), polymer currency exhibits enhanced durability with lifetimes typically four times greater than paper currency of the same denomination. In addition, polymer banknotes can incorporate security features not available to paper currency. Many of the additional security features cannot be reproduced by photocopying or scanning, making counterfeiting more difficult.

The BOPP substrate is produced by simultaneously stretching the extruded polymer in two directions (called the transverse and machined directions), after which the film is heat-set to maintain the biaxial orientation. The BOPP substrate is then processed through the following steps:

- Opacifying—applying ink to each side of the note. In many cases, one area is intentionally left clear for the creation of a unique security feature called an optically variable device (OVD), a mark that changes colors when viewed at different angles.

- Sheeting—the substrate is cut into sheets prior to the printing process.

- Printing—the characteristic images and features are printed on the substrate using a variety of printing processes, including traditional offset, intaglio, and letterpress printing.

- Overcoating—a coating of protective varnish is added.

In addition to enhanced durability, the polymer currency is more resistant to folding and tearing, resists soiling, and is easier to machine process, which leads to greater handling efficiency. It is also 100% recyclable. Unlike paper currency,

▲ The first polymer banknote was released in 1988 to commemorate Australia's Bicentenary. Note the OVD in the upper right-hand corner.

▲ BOPP being produced using the "bubble" process, in which a thick-walled tube of extruded polypropylene is stretched in one direction (transverse) by applying air pressure, while simultaneously pulling down on the tube to stretch in the machined direction.

which is typically burned or placed in a landfill, the polymer currency can be shredded and reused.

See Additional Problem 13.72 at the end of the chapter.

13.9 Aromatic Compounds and the Structure of Benzene

In the early days of organic chemistry, the word *aromatic* was used to describe many fragrant substances from fruits, trees, and other natural sources. It was soon realized, however, that substances grouped as aromatic behave differently from most other organic compounds. Today, chemists use the term **aromatic** to refer to the class of compounds that contain benzene-like rings.

Benzene, the simplest aromatic compound, is a flat, symmetrical molecule with the molecular formula C_6H_6. It is often represented as cyclohexatriene, a six-membered carbon ring with three double bonds. Though useful, the problem with this representation is that it gives the wrong impression about benzene's chemical reactivity and bonding. Because benzene appears to have three double bonds, you might expect it to react with H_2, Br_2, HCl, and H_2O to give the same kinds of addition products that alkenes do. But this expectation is wrong. Benzene and other aromatic compounds are much less reactive than alkenes and do not normally undergo addition reactions.

Aromatic The class of compounds containing benzene-like rings.

▲ The odor of cherries is due to benzaldehyde, an aromatic compound.

Benzene's relative lack of chemical reactivity is a consequence of its structure. If you were to draw a six-membered ring with alternating single and double bonds, where would you place the double bonds? There are two equivalent possibilities (Figure 13.3b), neither of which is fully correct by itself. Experimental evidence shows that all six carbon–carbon bonds in benzene are identical, so a picture with three double bonds and three single bonds cannot be correct.

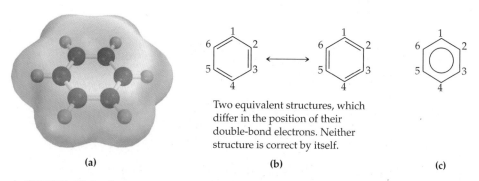

Two equivalent structures, which differ in the position of their double-bond electrons. Neither structure is correct by itself.

(a)   (b)   (c)

▲ **FIGURE 13.3** **Some representations of benzene.** (a) An electrostatic potential map shows the equivalency of the carbon–carbon bonds. Benzene is usually represented by the two equivalent structures in (b) or by the single structure in (c).

The properties of benzene are best explained by assuming that its true structure is an *average* of the two equivalent conventional structures. Rather than being held between specific pairs of atoms, the double-bond electrons are instead free to move over the entire ring. Each carbon–carbon bond is thus intermediate between a single bond and a double bond. The name **resonance** is given to this phenomenon where the true structure of a molecule is an average among two or more possible conventional structures, and a special double-headed arrow (⟷) is used to show the resonance relationship.

Because the real structure of benzene is intermediate between the two forms shown in Figure 13.3b, it is difficult to represent benzene with the standard conventions using lines for covalent bonds. Thus, we sometimes represent the double bonds as a circle inside the six-membered ring, as shown in Figure 13.3c. It is more common, though, to draw the ring with three double bonds, with the understanding that it is an aromatic ring with equivalent bonding all around. It is this convention that we use in this book.

Simple aromatic hydrocarbons like benzene are nonpolar, insoluble in water, volatile, and flammable. Unlike alkanes and alkenes, however, several aromatic

Resonance The phenomenon where the true structure of a molecule is an average among two or more conventional structures.

hydrocarbons are toxic. Benzene itself has been implicated as a cause of leukemia, and the dimethyl-substituted benzenes are central nervous system depressants.

Everything we have said about the structure and stability of the benzene ring also applies to the ring when it has substituents, such as in the bacteriocidal agent hexachlorophene and the flavoring ingredient vanillin:

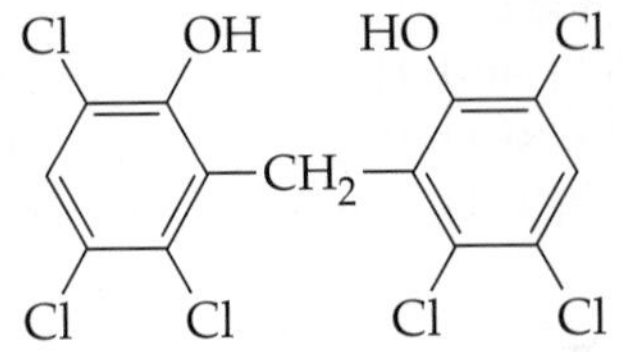

Hexachlorophene
(a germicide)

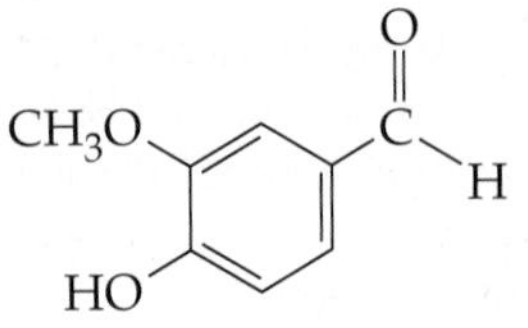

Vanillin
(vanilla flavoring)

The benzene ring is also present in many biomolecules and retains its characteristic properties in these compounds as well. In addition, aromaticity is not limited to rings that contain only carbon. For example, many compounds classified as aromatics have one or more nitrogen atoms in the ring. Pyridine, indole, and adenine are three examples:

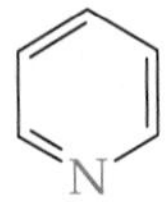

Pyridine 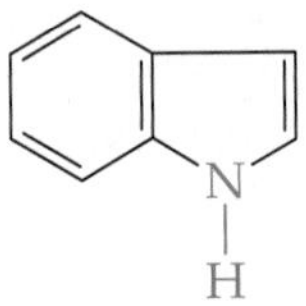   Indole 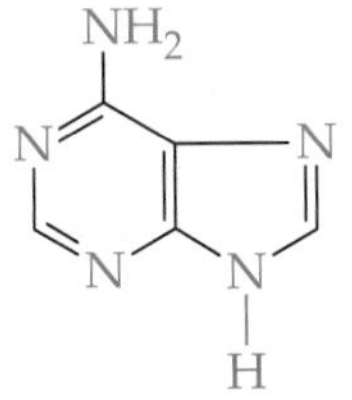   Adenine

These and all other compounds that contain a substituted benzene ring, or a similarly stable six-membered ring in which double-bond electrons are equally shared around the ring, are classified as aromatic compounds.

APPLICATION ▶ Polycyclic Aromatic Hydrocarbons and Cancer

The definition of the term *aromatic* can be extended beyond simple monocyclic (one-ring) compounds to include *polycyclic* aromatic compounds—substances that have two or more benzene-like rings joined together by a common bond. Naphthalene, familiar for its use in mothballs, is the simplest and best-known polycyclic aromatic compound.

In addition to naphthalene, there are many polycyclic aromatic compounds that are more complex. Benz[a]pyrene, for example, contains five benzene-like rings joined together; ordinary graphite (the "lead" in pencils) consists of enormous two-dimensional sheets of benzene-like rings stacked one on top of the other.

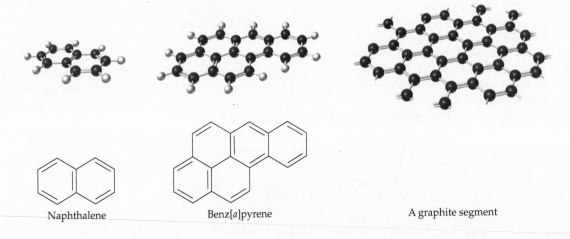

Naphthalene   Benz[a]pyrene   A graphite segment

Perhaps the most notorious polycyclic aromatic hydrocarbon is benz[a]pyrene, one of the carcinogenic (cancer-causing) substances found in chimney soot, cigarette smoke, and charcoal-broiled meat. Exposure to even a tiny amount is sufficient to induce a skin tumor in susceptible mice.

After benz[a]pyrene is taken into the body by eating or inhaling, the body attempts to rid itself of the foreign substance by converting it into a water-soluble metabolite called a *diol epoxide*, which can be excreted. Unfortunately, the diol epoxide metabolite reacts with and binds to cellular DNA, thereby altering the DNA and leading to mutations or cancer.

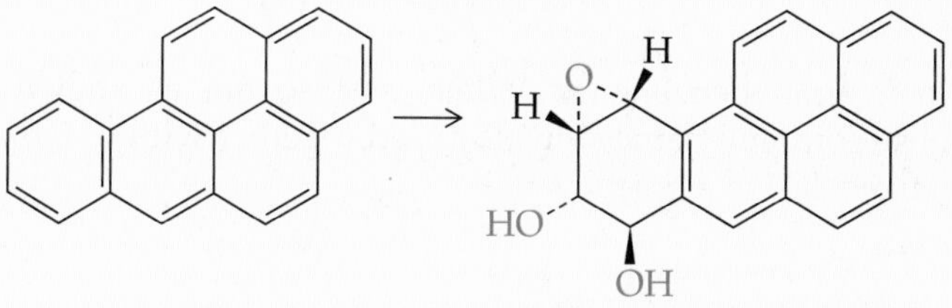

Benz[a]pyrene   A diol epoxide

Even benzene can cause certain types of cancer on prolonged exposure, so breathing the fumes of benzene and other volatile aromatic compounds in the laboratory should be avoided.

See Additional Problems 13.73 and 13.74 at the end of the chapter.

13.10 Naming Aromatic Compounds

Substituted benzenes are named using *-benzene* as the parent. Thus, C_6H_5Br is bromobenzene, $C_6H_5CH_2CH_3$ is ethylbenzene, and so on. No number is needed for monosubstituted benzenes because all the ring positions are identical.

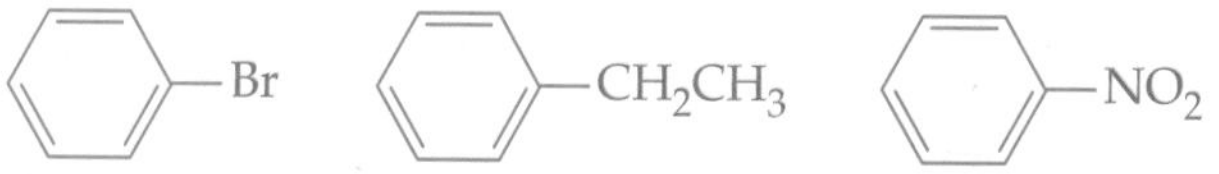

Bromobenzene   Ethylbenzene   Nitrobenzene

When a benzene has more than one substituent present the positions of those substituents are indicated by numbers, just as in naming cycloalkanes. Disubstituted benzenes are unique in that the relational descriptors *o-* (*ortho*), *m-* (*meta*), and *p-* (*para*) may be used in place of 1,2-, 1,3-, and 1,4-, respectively. The terms *ortho-*, *meta-* or *para-* (or their single letter equivalents) are then used as prefixes:

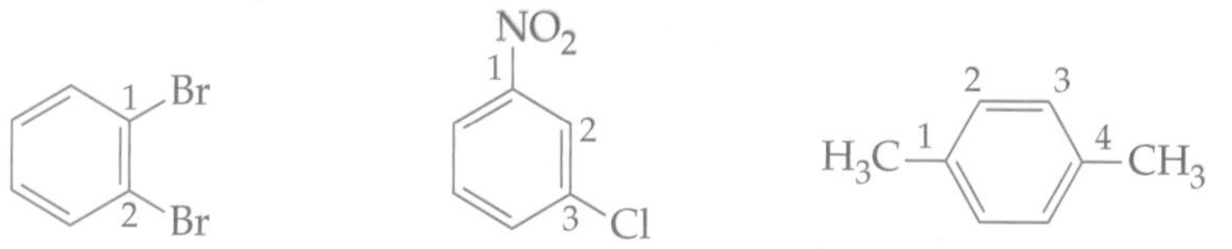

1,2-Dibromobenzene   3-Chloronitrobenzene   1,4-Dimethylbenzene
ortho-Dibromobenzene   *meta*-Chloronitrobenzene   *para*-Dimethylbenzene
o-Dibromobenzene   *m*-Chloronitrobenzene   *p*-Dimethylbenzene

While any one of these three nomenclature schemes are acceptable, we will almost exclusively use *o-*, *m-*, and *p-* in naming these compounds.

Many substituted aromatic compounds have common names in addition to their systematic names. For example, methylbenzene is familiarly known as *toluene,*

TABLE 13.2 Common Names of Some Aromatic Compounds

STRUCTURE	NAME	STRUCTURE	NAME
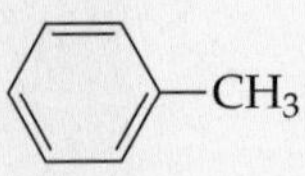 ⬡—CH$_3$	Toluene	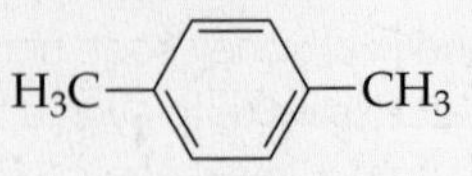 H$_3$C—⬡—CH$_3$	*para*-Xylene
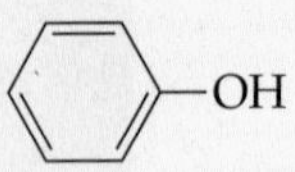 ⬡—OH	Phenol	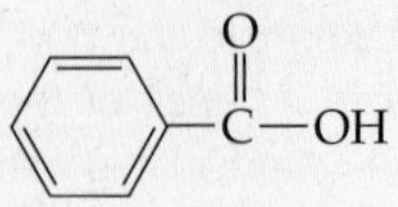 ⬡—C(=O)—OH	Benzoic acid
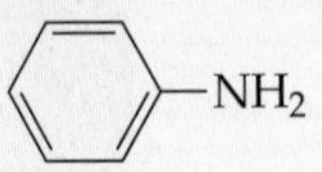 ⬡—NH$_2$	Aniline	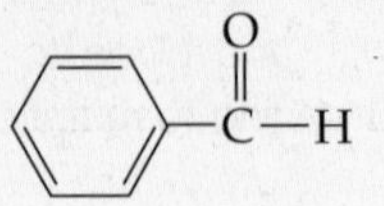 ⬡—C(=O)—H	Benzaldehyde

hydroxybenzene as *phenol*, aminobenzene as *aniline*, and so on, as shown in Table 13.2. Frequently, these common names are also used together with o- (*ortho*), m- (*meta*), or p- (*para*) prefixes. For example:

Cl—⬡—CH$_3$ HO—⬡—NO$_2$ ⬡(Br)—NH$_2$

p-Chlorotoluene 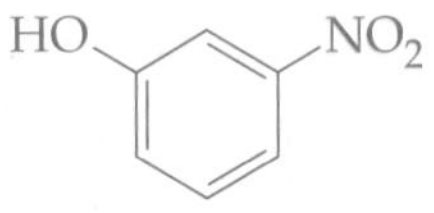   *m*-Nitrophenol 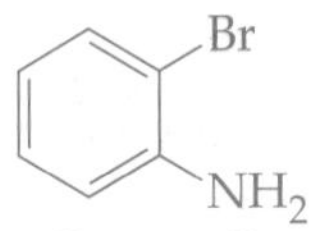   *o*-Bromoaniline

Occasionally, the benzene ring itself may be considered a substituent group attached to another parent compound. When this happens, the name **phenyl** (pronounced **fen**-nil and commonly abbreviated Ph—) is used for the C$_6$H$_5$—unit:

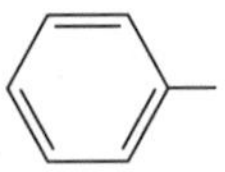

Phenyl The C$_6$H$_5$— group.

⬡—   ⬡—CH(CH$_2$CH$_3$)(CH$_2$CH$_2$CH$_2$CH$_3$)

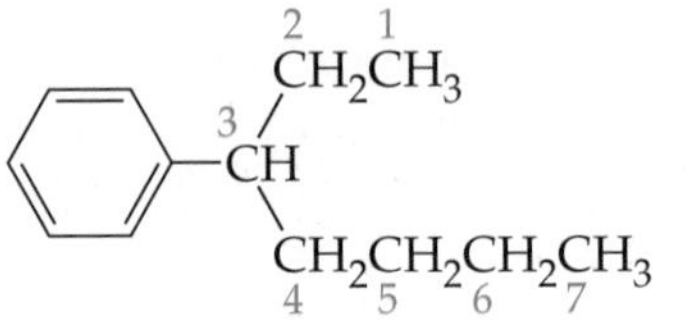

A phenyl group   3-Phenylheptane
C$_6$H$_5$—

WORKED EXAMPLE **13.10** Naming Organic Compounds: Aromatic Compounds

Name the following aromatic compound:

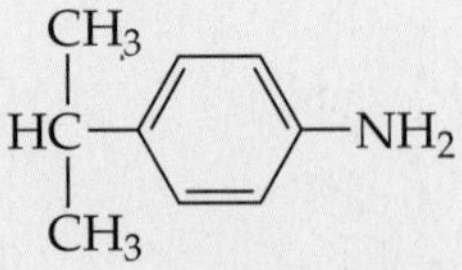

(CH$_3$)$_2$HC—⬡—NH$_2$

ANALYSIS First identify the parent organic compound, then identify the location of substituent groups on the benzene ring either by number, or by *ortho*, *meta*, or *para*.

SOLUTION
The parent compound is a benzene ring with an amine group (*aminobenzene*, which is commonly known as *aniline*). The substituent group is attached at the

C4 or para position relative to the amino group. The propyl group is attached to the benzene ring by the middle carbon, so it is *isopropyl*.

> The substituent group is at the para position.

$$CH_3 \quad \overset{3}{\underset{4}{\text{HC}}} \quad \overset{2}{\underset{}{\bigcirc}} \quad \overset{1}{\underset{6}{}} - NH_2$$
$$CH_3 \quad 5$$

> The propyl group is attached to the middle carbon, so it is isopropyl.

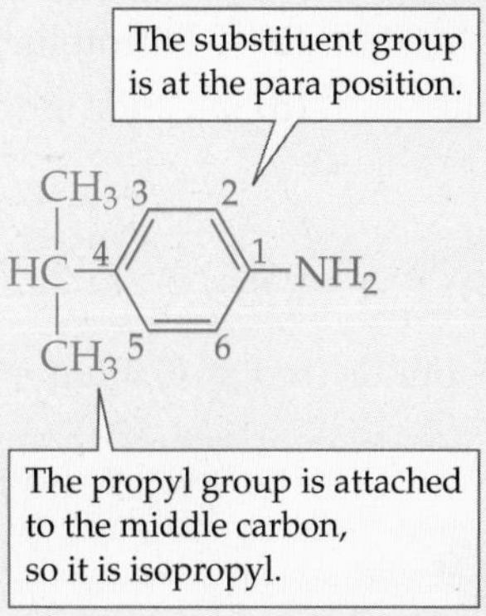

Name: paraisopropylaniline, or 4-isopropylaminobenzene

WORKED EXAMPLE **13.11** Molecular Structures: Aromatic Compounds

Draw the structure of *m*-chloroethylbenzene.

ANALYSIS *m*-Chloroethylbenzene has a benzene ring with two substituents, chloro and ethyl, in a meta relationship (that is, on C1 and C3).

SOLUTION

Since all carbons in the benzene ring are equivalent, draw a benzene ring and attach one of the substituents—for example, chloro—to any position:

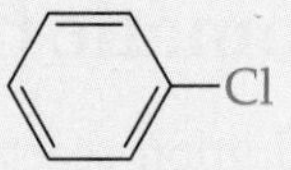

Now go to a meta position two carbons away from the chloro-substituted carbon, and attach the second (ethyl) substituent:

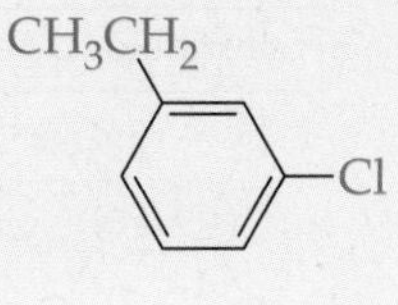

$$CH_3CH_2$$

—Cl   *m*-Chloroethylbenzene

PROBLEM 13.18

What are the IUPAC names for the following compounds?

(a) HO *m-ethylphenol*

—CH$_2$CH$_3$

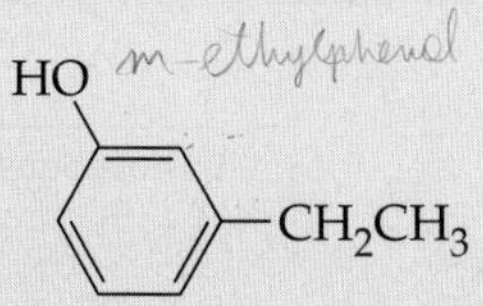

(b) Br *p-bromoaniline*

NH$_2$

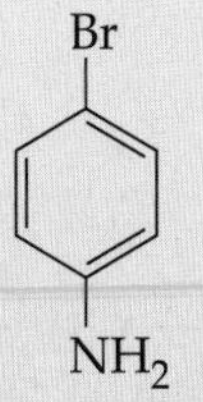

(c)

$$CH_3$$
—C—CH$_3$
$$CH_2CH_3$$

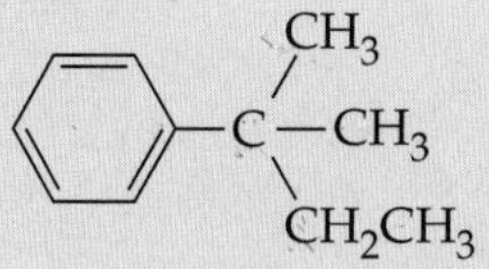

PROBLEM 13.19

Draw structures corresponding to the following names:

(a) *o*-Iodobromobenzene   **(b)** *o*-Nitrotoluene

(c) *m*-Diisopropylbenzene   **(d)** *p*-Chlorophenol

KEY CONCEPT PROBLEM 13.20

Name the following compounds (red = O, blue = N, brown = Br):

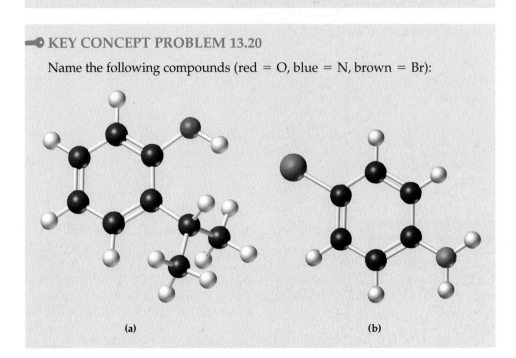

(a)   (b)

13.11 Reactions of Aromatic Compounds

Unlike alkenes, which undergo addition reactions, aromatic compounds usually undergo *substitution* reactions. That is, a group Y substitutes for one hydrogen atom on the aromatic ring without changing the ring itself. It does not matter which of the six ring hydrogens in benzene is replaced because all six are equivalent.

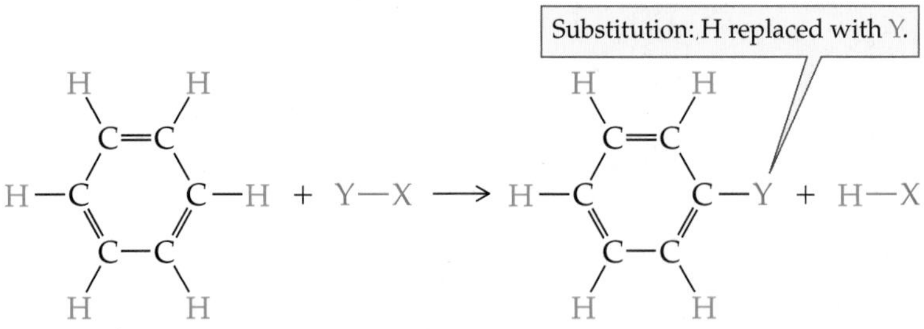

Substitution: H replaced with Y.

$$H\text{—}C \quad C\text{—}H + Y\text{—}X \longrightarrow H\text{—}C \quad C\text{—}Y + H\text{—}X$$

Nitration The substitution of a nitro group ($-NO_2$) for a hydrogen on an aromatic ring.

Nitration is the substitution of a *nitro group* ($-NO_2$) for one of the ring hydrogens. The reaction occurs when benzene reacts with nitric acid in the presence of sulfuric acid as catalyst:

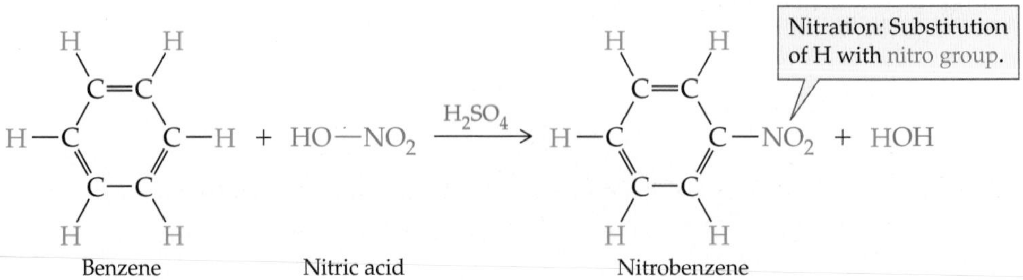

Nitration: Substitution of H with nitro group.

$$H\text{—}C \quad C\text{—}H + HO\text{—}NO_2 \xrightarrow{H_2SO_4} H\text{—}C \quad C\text{—}NO_2 + HOH$$

Benzene   Nitric acid   Nitrobenzene

Nitration of aromatic rings is a key step in the synthesis both of explosives like TNT (trinitrotoluene) and of many important pharmaceutical agents. Nitrobenzene itself is the industrial starting material for the preparation of aniline, which is used to make many of the brightly colored dyes in clothing.

Halogenation is the substitution of a halogen atom, usually bromine or chlorine, for one of the ring hydrogens. The reaction occurs when benzene reacts with Br_2 or Cl_2 in the presence of iron as catalyst:

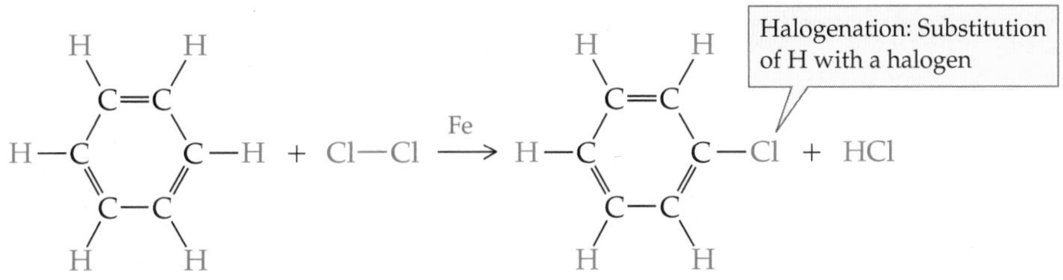

▲ Samples of the original aniline dyes mauveine (violet) and alizarin (red) synthesized by William Perkin in 1856 and 1869, respectively. Aniline itself is made from nitrobenzene.

Benzene + Chlorine →(Fe) Chlorobenzene + HCl

> Halogenation: Substitution of H with a halogen

Sulfonation is the substitution of a sulfonic acid group ($-SO_3H$) for one of the ring hydrogens. The reaction occurs when benzene reacts with concentrated sulfuric acid and SO_3:

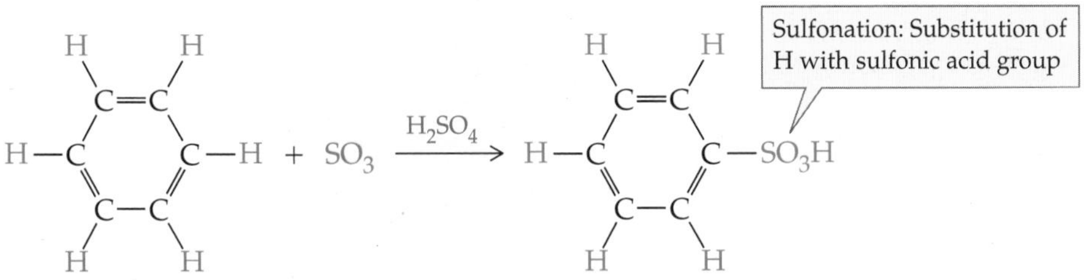

Benzene + SO_3 →(H_2SO_4) Benzenesulfonic acid

> Sulfonation: Substitution of H with sulfonic acid group

Halogenation The substitution of a halogen group ($-X$) for a hydrogen on an aromatic ring.

Sulfonation The substitution of a sulfonic acid group ($-SO_3H$) for a hydrogen on an aromatic ring.

Aromatic-ring sulfonation is a key step in the synthesis of such compounds as the sulfa-drug family of antibiotics:

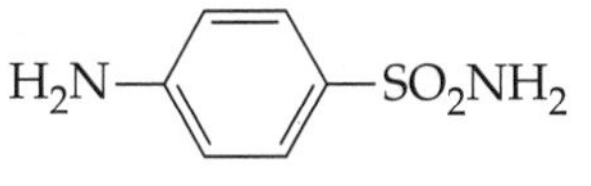

$$H_2N - \text{C}_6\text{H}_4 - SO_2NH_2$$

Sulfanilamide—a sulfa antibiotic

PROBLEM 13.21

Write the products from the reaction of the following reagents with *p*-xylene (*p*-dimethylbenzene).

(a) Br_2 and Fe

(b) HNO_3 and H_2SO_4

(c) SO_3 and H_2SO_4

PROBLEM 13.22

Reaction of Br_2 and Fe with toluene (methylbenzene) can lead to one or more of *three* substitution products. Show the structure of each.

APPLICATION ▶ Why We See Color

The purple dye mauve, the plant pigment cyanidin, and a host of other organic compounds are brightly colored. What do all these compounds have in common? If you look carefully at each structure, you will see that each has numerous alternating double and single bonds.

Mauve
(the first synthetic dye)

Cyanidin
(bluish-red color in flowers
and cranberries)

In describing benzene, we noted that the double-bond electrons are spread out, or *delocalized*, over the whole molecule. The same phenomenon occurs whenever there are many alternating double and single bonds in a molecule: The double-bond electrons form a delocalized region of electron density in the molecule that is capable of absorbing light. Organic compounds such as benzene, with small numbers of delocalized electrons, absorb in the ultraviolet region of the electromagnetic spectrum, which our eyes cannot detect. Compounds with longer stretches of alternating double and single bonds absorb in the visible region.

The color that we see is complementary to the color that is absorbed; that is, we see what is left of the white light after certain colors have been absorbed. For example, the plant pigment cyanidin absorbs greenish-yellow light and thus appears reddish-blue. The same principle applies to almost all colored organic compounds: They contain large regions of delocalized electrons and absorb some portion of the visible spectrum.

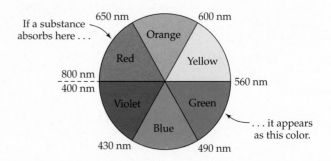

▲ Using an artist's color wheel, it is possible to determine the observed color of a substance by knowing the color of the light absorbed. Observed and absorbed colors are complementary. Thus, if a substance absorbs red light, it has a green color.

See Additional Problems 13.75 and 13.76 at the end of the chapter.

SUMMARY: REVISITING THE CHAPTER GOALS

1. **What are alkenes, alkynes, and aromatic compounds?** *Alkenes* are hydrocarbons that contain a carbon–carbon double bond, and *alkynes* are hydrocarbons that contain a carbon–carbon triple bond. *Aromatic compounds* contain six-membered, benzene-like rings and are usually written with three double bonds. In fact, however, there is equal bonding between neighboring carbon atoms in benzene rings because the double-bond electrons are symmetrically spread around the entire ring. All three families are said to be *unsaturated* because they have fewer hydrogens than corresponding alkanes.

2. **How are alkenes, alkynes, and aromatic compounds named?** Alkenes are named using the family ending *-ene*; alkynes use the family ending *-yne*. Disubstituted benzenes have the suffix *-benzene* as the parent name, and positions of the substituents are indicated with the prefixes *ortho-* (1,2 substitution), *meta-* (1,3 substitution), or *para-* (1,4 substitution).

3. **What are cis–trans isomers?** Cis–trans isomers are seen in disubstituted alkenes as a consequence of the lack of rotation around carbon–carbon double bonds. In the cis isomer, the two substituents are on the same side of the double bond; in the trans isomer, they are on opposite sides of the double bond.

4. **What are the categories of organic reactions?** The four categories of organic reactions are *addition, elimination, substitution,* and *rearrangement* reactions. Addition reactions occur when two reactants add together to form a single product with no atoms left over. Eliminations occur when a single reactant splits into two products. Substitution reactions occur when two reactants exchange parts to give two new products. Rearrangement reactions occur when a single reactant undergoes a reorganization of bonds and atoms to yield a single isomeric product.

5. **What are the typical reactions of alkenes, alkynes, and aromatic compounds?** Alkenes and alkynes undergo addition reactions to their multiple bonds. Addition of hydrogen to an alkene (*hydrogenation*) yields an alkane product; addition of Cl_2 or Br_2 (*halogenation*) yields a 1,2-dihaloalkane product; addition of HBr and HCl (*hydrohalogenation*) yields an alkyl halide product; and addition of water (*hydration*) yields an alcohol product. *Markovnikov's rule* predicts that in the addition of HX or H_2O to a double bond, the H becomes attached to the carbon with more Hs and the X or OH becomes attached to the carbon with fewer Hs.

   Aromatic compounds are unusually stable but can be made to undergo substitution reactions, in which one of the ring hydrogens is replaced by another group ($C_6H_6 \rightarrow C_6H_5Y$). Among these substitutions are *nitration* (substitution of $-NO_2$ for $-H$), *halogenation* (substitution of $-Br$ or $-Cl$ for $-H$), and *sulfonation* (substitution of $-SO_3H$ for $-H$).

6. **How do organic reactions take place?** A description of the individual steps by which old bonds are broken and new bonds are formed in a reaction is called a *reaction mechanism*. The addition reaction of an alkene with HX takes place in two steps. In the first step, the alkene uses two electrons to bond to H^+, giving a positively charged species called a *carbocation*. In the second step, the carbocation reacts with the halide ion to give the final product.

Hydrohalogenation, *p. 411*
Markovnikov's rule, *p. 412*
Monomer, *p. 416*
Nitration, *p. 426*
Phenyl, *p. 424*
Polymer, *p. 416*
Reaction mechanism, *p. 416*
Rearrangement reaction, *p. 408*
Resonance, *p. 421*
Saturated, *p. 397*
Substitution reaction, *p. 407*
Sulfonation, *p. 427*
Unsaturated, *p. 398*

SUMMARY OF REACTIONS

1. **Reactions of alkenes and alkynes** (Section 13.6)

   (a) Addition of H_2 to yield an alkane (hydrogenation):

An alkene:

An alkyne:

   (b) Addition of Cl_2 or Br_2 to yield a dihalide (halogenation):

   (c) Addition of HCl or HBr to yield an alkyl halide (hydrohalogenation):

   (d) Addition of H_2O to yield an alcohol (hydration):

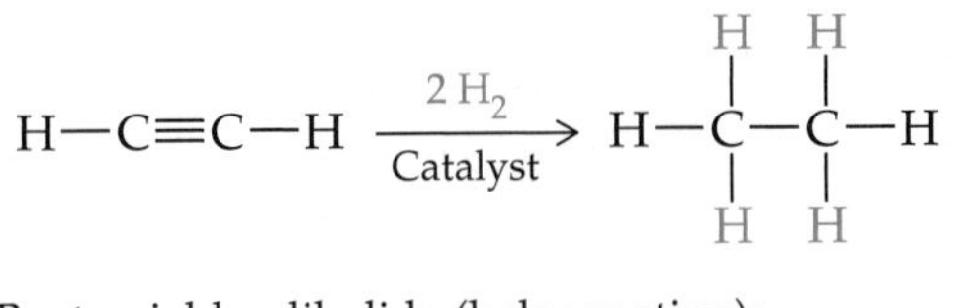

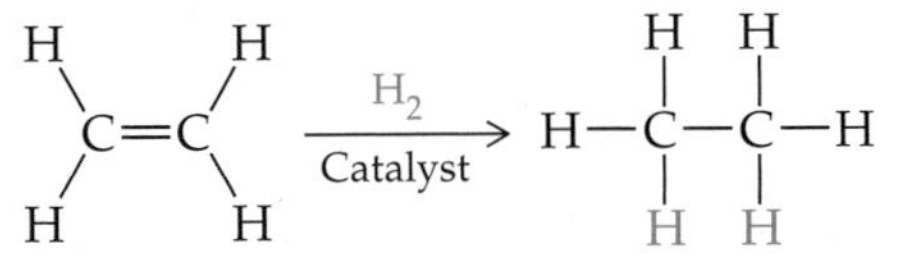

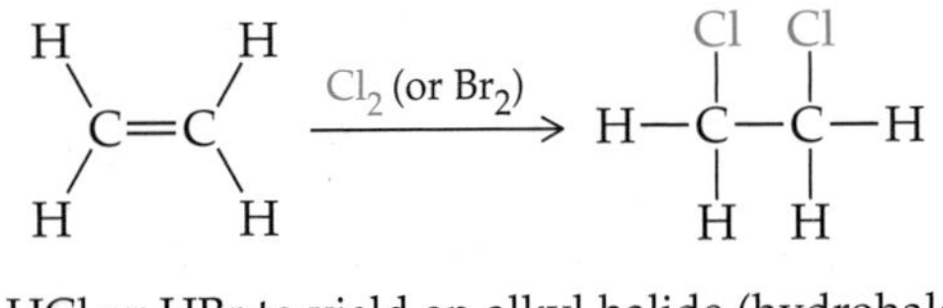

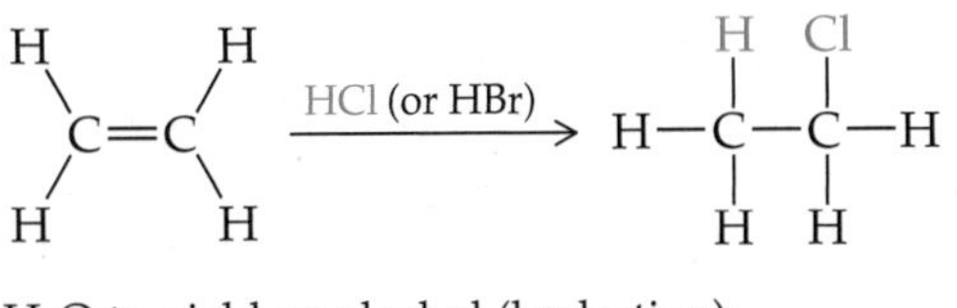

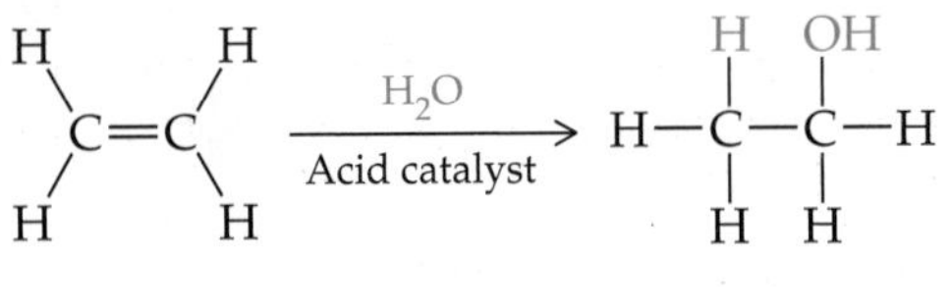

2. **Reactions of aromatic compounds** (Section 13.11)

(a) Substitution of an —NO_2 group to yield a nitrobenzene (nitration):

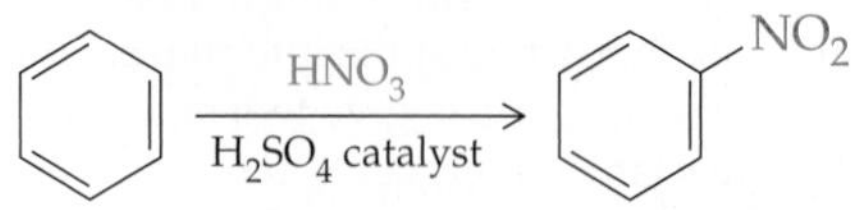

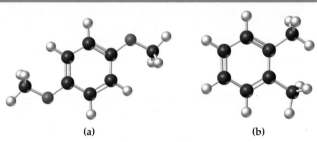

(b) Substitution of a Cl or Br atom to yield a halobenzene (halogenation):

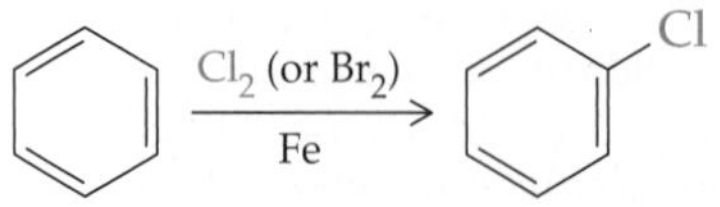

(c) Substitution of an —SO_3H group to yield a benzenesulfonic acid (sulfonation):

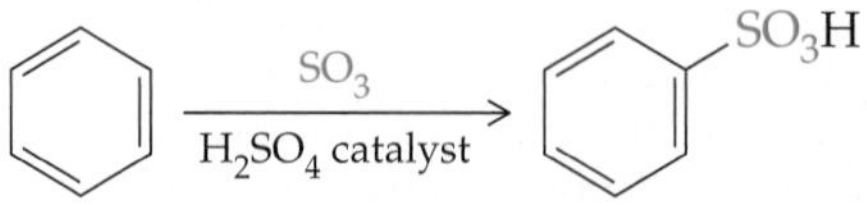

UNDERSTANDING KEY CONCEPTS

13.23 Name the following alkenes, and predict the products of their reaction with (1) HBr, (2) H_2O, and an acid catalyst.

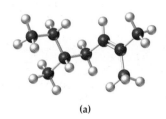

(a) 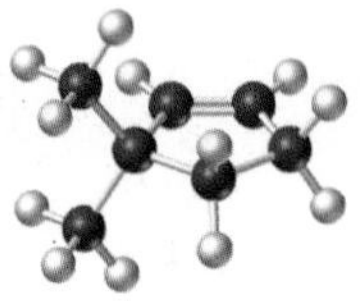   (b)

13.24 Name the following alkynes:

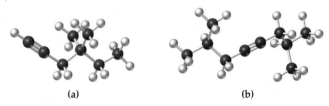

(a)   (b)

13.25 Give IUPAC names for the following substances (red = O, brown = Br):

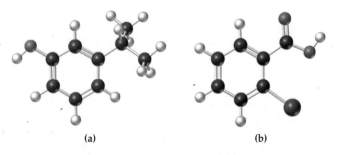

(a)   (b)

13.26 Draw the product from reaction of the following substances with (1) Br_2 and iron catalyst, (2) SO_3 and H_2SO_4 catalyst (red = O):

(a)   (b)

13.27 Alkynes undergo hydrogenation to give alkanes, just as alkenes do. Draw and name the products that would result from hydrogenation of the alkynes shown in Problem 13.24.

13.28 We saw in Section 13.9 that benzene can be represented by either of two resonance forms, which differ in the positions of the double bonds in the aromatic ring. Naphthalene, a polycyclic aromatic compound, can be represented by *three* forms with different double-bond positions. Draw all three structures, showing the double bonds in each (the following molecular model of naphthalene shows only the connections among atoms).

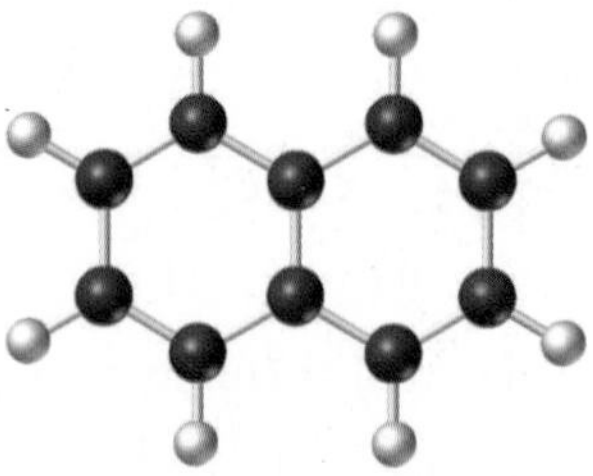

13.29 The following structure is that of a carbocation intermediate in the reaction of an alkene with HCl. Draw the structure of the alkene reactant.

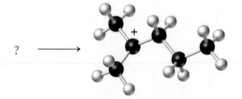

? $\longrightarrow$

ADDITIONAL PROBLEMS

NAMING ALKENES, ALKYNES, AND AROMATIC COMPOUNDS

13.30 Why are alkenes, alkynes, and aromatic compounds said to be unsaturated?

13.31 Not all compounds that smell nice are called "aromatic," and not all compounds called "aromatic" smell nice. Explain.

13.32 What family-name endings are used for alkenes, alkynes, and substituted benzenes?

13.33 What prefixes are used in naming the following?

(a) A 1,3-disubstituted benzene
(b) A 1,4-disubstituted benzene

13.34 Write structural formulas for compounds that meet the following descriptions:

(a) An alkene with 7 carbons
(b) An alkyne with 5 carbons
(c) A substituted aromatic hydrocarbon with a total of 8 carbons

13.35 Write structural formulas for compounds that meet the following descriptions:

(a) An alkene, C_6H_{12}, that cannot have cis–trans isomers and whose longest chain is 5 carbons long
(b) An aromatic alkene, $C_{10}H_{12}$, that has cis–trans isomers

13.36 What are the IUPAC names of the following compounds?

(a) $CH_3CH_2CH_2CH{=}CH_2$ (b) $CH_3CH\,C{\equiv}C\,CHCH_3$ with CH_3 groups on the two CH carbons

(c) $(CH_3)_2C{=}C(CH_3)_2$ (d) $CH_3CH{=}C{-}C{=}CH_2$ with CH_3 and CH_2CH_3 substituents

(e) cyclohexene with CH_3, CH_2CH_3, CH_3 substituents

(f) cyclobutane ring with CH_2CH_3, CH_3, CH(CH_3)

13.37 Give IUPAC names for the following aromatic compounds:

(a) CH_3-benzene-CH_3 with CH_3
(b) Br-benzene-NO_2
(c) HO-benzene-NO_2

13.38 Draw structures corresponding to the following IUPAC names:

(a) trans-3-Hexene (b) 2-Methyl-3-hexene
(c) 2-Methyl-1,3-butadiene (d) cis-3-Heptene
(e) m-Nitrotoluene (f) o-Chlorophenol
(g) m-Dipropylbenzene

13.39 Draw structures corresponding to the following names:

(a) Aniline (b) Phenol
(c) m-Xylene (d) Toluene
(e) Benzoic acid (f) p-Nitroaniline
(g) o-Chlorotoluene
(h) 3,3-Diethyl-6-methyl-4-nonene

13.40 Seven alkynes have the formula C_6H_{10}. Draw and name five of them.

13.41 Draw and name all aromatic compounds with the formula C_7H_7Br.

13.42 Excluding cis–trans isomers, five alkenes have the formula C_5H_{10}. Draw structures for as many as you can, and give their IUPAC names.

13.43 How many dienes (compounds with two double bonds) are there with the formula C_5H_8? Draw and name as many as you can. Ignore cis–trans isomers.

ALKENE CIS–TRANS ISOMERS

13.44 What requirement must be met for an alkene to show cis–trans isomerism?

13.45 Why do alkynes not show cis–trans isomerism?

13.46 Which alkene(s) in Problem 13.42 can exist as cis–trans isomers?

13.47 Which compound(s) in Problem 13.43 can exist as cis–trans isomers?

13.48 Draw structures of the following compounds, indicating the cis or trans geometry of the double bond if necessary:

(a) cis-3-Heptene
(b) cis-4-Methyl-2-pentene
(c) trans-2,5-Dimethyl-3-hexene

13.49 Draw structures of the double-bond isomers of the following compounds:

(a) Cl, Cl / H, CH_3 on C=C
(b) $CH_3CH_2CH_2$, OCH_3 / H, CH_3 on C=C

13.50 Which of the following pairs are isomers, and which are identical?

(a) H_3C, Br / H, Br on C=C and Br, H / Br, CH_3 on C=C

(b) CH_3CH_2, Cl / Cl, H on C=C and H, Cl / Cl, CH_2CH_3 on C=C

13.51 Draw the other cis–trans isomer for the following molecules:

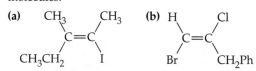

KINDS OF REACTIONS

13.52 What is the difference between a substitution reaction and an addition reaction?

13.53 Give an example of an addition reaction.

13.54 If 2-methyl-2-pentene were somehow converted into 1-hexene, what kind of reaction would that be?

13.55 If bromocyclohexane were somehow converted into cyclohexene, what kind of reaction would that be?

13.56 Identify the type of reaction for the following:

(a) [benzene with CH₃ group] $\xrightarrow[\text{Light}]{\text{Br}_2}$ [benzene with CH₂Br group] + HBr

(b) [cyclohexene with OH and CH₃] $\longrightarrow$ [cyclohexanone with CH₃ and H]

13.57 Identify the type of reaction for the following:

(a) $\underset{\displaystyle |}{CH_3}CHCH_2CH_2CH_2Br$ + NaCN $\longrightarrow$

$CH_3\underset{\displaystyle |}{C}HCH_2CH_2CH_2C{\equiv}N$ + NaBr (with CH₃)

(b) $2\ CH_3-\overset{\displaystyle O}{\overset{\|}{C}}-H \xrightarrow{\text{NaOH}} CH_3-\overset{O-H}{\underset{H}{\overset{|}{C}}}-CH_2-\overset{\displaystyle O}{\overset{\|}{C}}-H$

REACTIONS OF ALKENES AND ALKYNES

13.58 Write equations for the reaction of 1,2-dimethylcyclohexene with the following:

(a) H₂ and Pd catalyst   **(b)** Br₂
(c) HBr   **(d)** H₂O and H₂SO₄ catalyst

13.59 Write equations for the reaction of 1-methylcyclohexene with the reagents shown in Problem 13.58.

13.60 What alkene could you use to make the following products? Draw the structure of the alkene, and tell what other reagent is also required for the reaction to occur.

(a) $CH_3\overset{Cl}{\underset{}{\overset{|}{C}}}H\overset{Cl}{\underset{}{\overset{|}{C}}}H\overset{CH_3}{\underset{CH_3}{\overset{|}{C}}}CH_3$ **(b)** $CH_3CH_2CH_3$

(c) $CH_3\overset{Br}{\underset{}{\overset{|}{C}}}HCH_2CH_3$ **(d)** [cyclohexane with Cl]

(e) [cyclohexane with Cl and CH₂Cl]

13.61 2,2-Dibromo-3-methylpentane can be prepared by an addition reaction of excess HBr with an alkyne. Draw the structure of the alkyne, name it, and write the reaction.

13.62 Draw the carbocation formed as an intermediate when HCl adds to styrene (phenylethylene).

13.63 4-Methyl-1-pentyne reacts with HBr in a 1:1 molar ratio to yield an addition product, $C_6H_{11}Br$. Draw the structures

of two possible products. Assuming that Markovnikov's rule is followed, predict which of the two structures you drew is formed, and draw the carbocation involved as an intermediate.

13.64 Polyvinylpyrrolidone (PVP) is often used in hair sprays to hold hair in place. Draw a few units of the PVP polymer. The vinylpyrrolidone monomer unit has the structure

[structure: CH=CH₂ attached to N of pyrrolidinone ring with O]

13.65 Saran, used as a plastic wrap for foods, is a polymer with the following structure. What is the monomer unit of Saran?

[polymer chain structure with H, Cl substituents]

REACTIONS OF AROMATIC COMPOUNDS

13.66 Under ordinary conditions, benzene reacts with only one of the following reagents. Which of the four is it, and what is the structure of the product?

(a) H₂ and Pd catalyst
(b) Br₂ and iron catalyst
(c) HBr
(d) H₂O and H₂SO₄ catalyst

13.67 Write equations for the reaction of *p*-dichlorobenzene with the following:

(a) Br₂ and Fe catalyst
(b) HNO₃ and H₂SO₄ catalyst
(c) H₂SO₄ and SO₃
(d) Cl₂ and Fe catalyst

13.68 Aromatic compounds do not normally react with hydrogen in the presence of a palladium catalyst. If very high pressures (200 atm) and high temperatures are used, however, one aromatic molecule, toluene, adds three molecules of H₂ to give an addition product. What is a likely structure for the product?

13.69 The explosive trinitrotoluene, or TNT, is made by carrying out three successive nitration reactions on toluene. If these nitrations take place in the ortho and para positions relative to the methyl group, what is the structure of TNT?

Applications

13.70 What is the difference in the purpose of the rod cells and the cone cells in the eye? [*The Chemistry of Vision, p. 406*]

13.71 Describe the isomerization that occurs when light strikes the rhodopsin in the eye. [*The Chemistry of Vision, p. 406*]

13.72 The structure of polypropylene is similar to polyethylene, but has methyl groups (—CH₃) attached to the carbon-chain backbone. The methyl groups can be placed on the same side of the carbon chain (*isotactic*) or alternating on opposite sides of the backbone (*syndiotactic*). Draw structures for both forms of polypropylene (*Polymer Applications—Currency, p. 420*)

13.73 What is a polycyclic aromatic hydrocarbon? [*Polycyclic Aromatic Hydrocarbons and Cancer, p. 422*]

13.74 How does benz[*a*]pyrene cause cancer? [*Polycyclic Aromatic Hydrocarbons and Cancer, p . 422*]

13.75 Naphthalene is a white solid. Does it absorb light in the visible or in the ultraviolet range? [*Why We See Color, p. 428*]

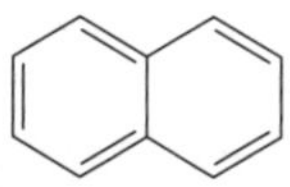

Naphthalene

13.76 Tetrabromofluorescein is a purple dye often used in lipsticks. If the dye is purple, what color does it absorb? [*Why We See Color, p. 428*]

General Questions and Problems

13.77 Why do you suppose small-ring cycloalkenes like cyclohexene do not exist as cis–trans isomers, whereas large ring cycloalkenes like cyclodecene *do* show isomerism?

13.78 Salicylic acid (*o*-hydroxybenzoic acid) is used as starting material to prepare aspirin. Draw the structure of salicylic acid.

13.79 "Superglue" is an alkene polymer made from the monomer unit

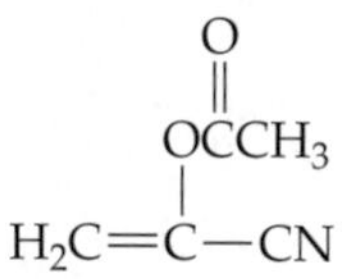

Draw a representative segment of the structure of superglue.

13.80 The following names are incorrect by IUPAC rules. Draw the structures represented by the following names, and write their correct names.

(a) 2-Methyl-4-hexene
(b) 1,3-Dimethyl-1-hexyne
(c) 2-Isopropyl-1-propene
(d) 1,4,6-Trinitrobenzene
(e) 1,2-Dimethyl-3-cyclohexene
(f) 3-Methyl-2,4-pentadiene

13.81 Assume that you have two unlabeled bottles, one with cyclohexane and one with cyclohexene. How could you tell them apart by carrying out chemical reactions?

13.82 Assume you have two unlabeled bottles, one with cyclohexene and one with benzene. How could you tell them apart by carrying out chemical reactions?

13.83 The compound *p*-dichlorobenzene has been used as an insecticide. Draw its structure.

13.84 Menthene, a compound found in mint plants, has the formula $C_{10}H_{18}$ and the IUPAC name 1-isopropyl-4-methyl-cyclohexene. What is the structure of menthene?

13.85 Cinnamaldehyde, the pleasant-smelling substance found in cinnamon oil, has the structure

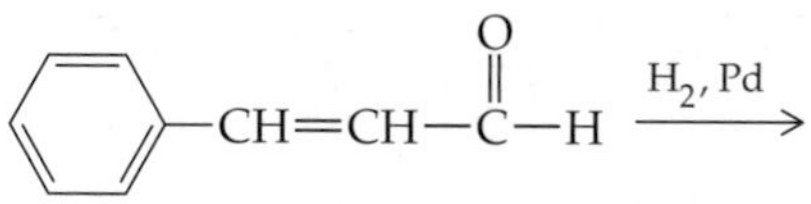

What product would you expect to obtain from reaction of cinnamaldehyde with hydrogen and a palladium catalyst?

13.86 Predict the products of the following reactions:

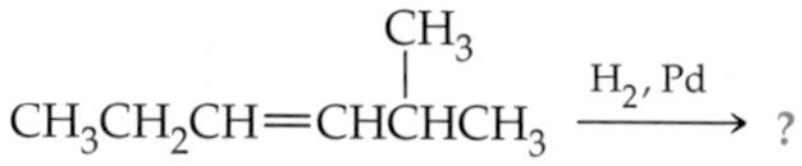

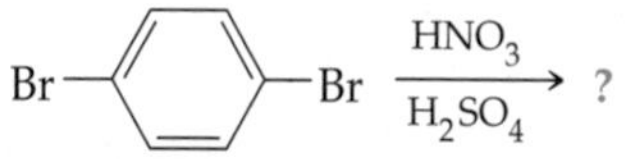

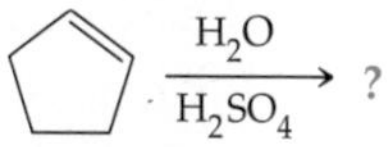

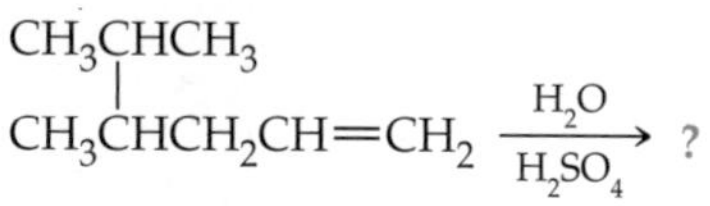

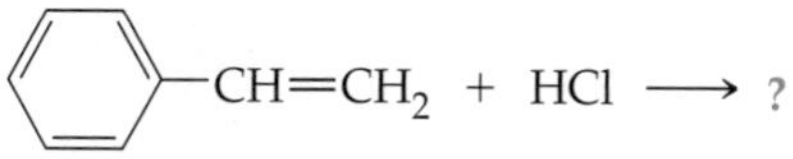

13.87 Two products are possible when 2-pentene is treated with HBr. Write the structures of the possible products, and explain why they are made in about equal amounts.

13.88 Benzene is a liquid at room temperature, but naphthalene, another aromatic compound (Problem 13.75), is a solid. Account for this difference in physical properties.

13.89 Ocimene, a compound isolated from the herb basil, has the IUPAC name 3,7-dimethyl-1,3-6-octatriene.

(a) Draw its structure.
(b) Draw the structure of the compound formed if enough HBr is added to react with all the double bonds in ocimene.

13.90 Describe how you could prepare the following compound from an alkene. Draw the formula of the alkene, name it, and list the inorganic reactants needed for the conversion.

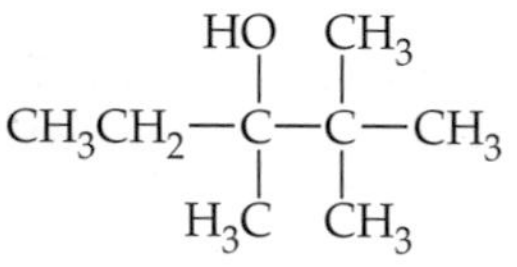

13.91 Which of the following compounds are capable of cis–trans isomerism?

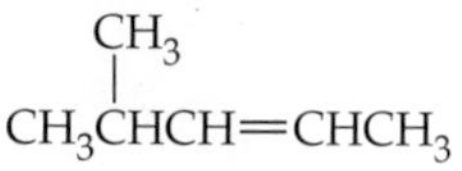

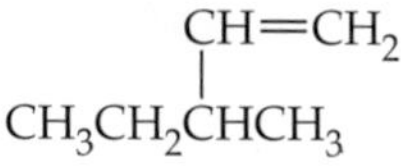

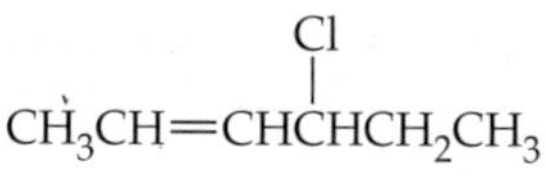

CHAPTER 14

Some Compounds with Oxygen, Sulfur, or a Halogen

CONCEPTS TO REVIEW

Oxidation and Reduction
(Section 6.11)

Hydrogen Bonds
(Section 8.11)

Acid Dissociation Constants
(Sections 10.6–10.7)

Naming Alkanes
(Section 12.6)

▲ Corn is a rich source of many important chemicals, most of which contain oxygen.

CONTENTS

CHAPTER GOALS

Questions we will answer in this chapter include the following:

1. What are the distinguishing features of alcohols, phenols, ethers, thiols, and alkyl halides?

THE GOAL: Be able to describe the structures and uses of compounds with these functional groups.

2. How are alcohols, phenols, ethers, thiols, and alkyl halides named?

THE GOAL: Be able to give systematic names for the simple members of these families and write their structures, given the names.

3. What are the general properties of alcohols, phenols, and ethers?

THE GOAL: Be able to describe such properties as polarity, hydrogen bonding, and water solubility.

4. Why are alcohols and phenols weak acids?

THE GOAL: Be able to explain why alcohols and phenols are acids.

5. What are the main chemical reactions of alcohols and thiols?

THE GOAL: Be able to describe and predict the products of the dehydration of alcohols and of the oxidation of alcohols and thiols.

The past two chapters dealt primarily with hydrocarbons, but most organic compounds contain other elements in addition to carbon and hydrogen. In this chapter, we will concentrate on functional groups that have single bonds to the electronegative atoms oxygen, sulfur, and the halogens. *Alcohols* and *phenols*, for example, are compounds that have an organic group bonded to an —OH group. In alcohols, the carbon to which the —OH is bound is tetravalent, and in phenols, the organic group is an aromatic ring. Alcohols are used widely in industry and are among the most abundant of all naturally occurring molecules. *Ethers* have two organic groups bonded to the same oxygen and are particularly useful as solvents. *Thiols*, which contain an organic portion bonded to an —SH group, are widespread in living organisms, and *alkyl halides*, which contain an organic group bonded to a halogen, are valuable in many industrial processes.

14.1 Alcohols, Phenols, and Ethers

An **alcohol** is a compound that has an —OH group (a *hydroxyl group*) bonded to a saturated, alkane-like carbon atom; a **phenol** has an —OH group bonded directly to an aromatic, benzene-like ring; and an **ether** has an oxygen atom bonded to two organic groups. Compounds in all three families can be thought of as organic relatives of water in which one or both of the H_2O hydrogens has been replaced by an organic substituent. For example,

Alcohol A compound that has an —OH group bonded to a saturated, alkane-like carbon atom, R—OH.

Phenol A compound that has an —OH group bonded directly to an aromatic, benzene-like ring, Ar—OH.

Ether A compound that has an oxygen atom bonded to two organic groups, R—O—R.

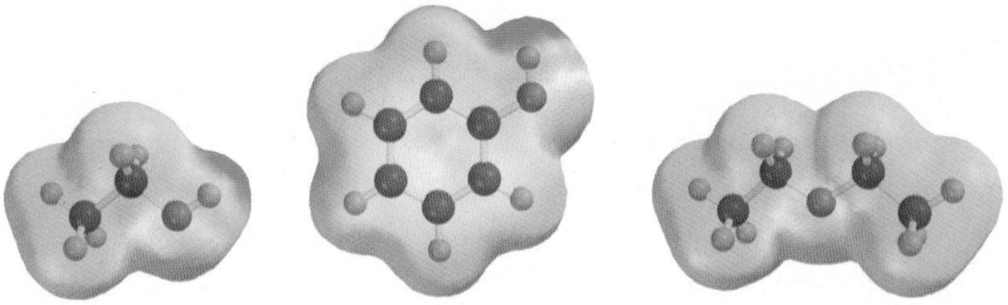

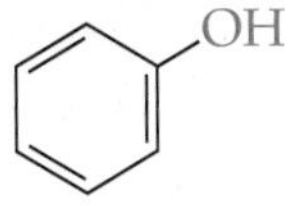

CH_3CH_2OH
Ethyl alcohol

Phenol

$CH_3CH_2OCH_2CH_3$
Diethyl ether

The structural similarity between alcohols and water also leads to similarities in many of their physical properties. For example, compare the boiling points of ethyl alcohol, dimethyl ether, propane, and water:

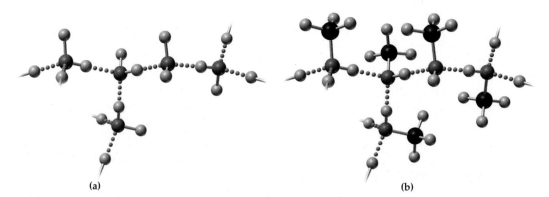

Ethyl alcohol, dimethyl ether, and propane have similar molecular weights, yet ethyl alcohol boils more than 100 °C higher than the other two. In fact, the boiling point of ethyl alcohol is close to that of water. Why should this be?

We said in Section 8.11 that the high boiling point of water is due to hydrogen bonding—the attraction between a lone pair of electrons on the electronegative oxygen in one molecule with the positively polarized —OH hydrogen on another molecule. (⬭, p. 235) This attraction holds molecules together and prevents their easy escape into the vapor phase. In a similar manner, hydrogen bonds form between alcohol (or phenol) molecules (Figure 14.1). Alkanes and ethers do not have hydroxyl groups, however, and cannot form hydrogen bonds. As a result, they have lower boiling points. Ethers, with the exception of their polarity, resemble alkanes in many of their properties.

▲ **FIGURE 14.1 The formation of hydrogen bonds in water (a) and in alcohols (b).** Because of the hydrogen bonds (shown in red), the easy escape of molecules into the vapor phase is prevented, resulting in high boiling points.

PROBLEM 14.1

Identify each of the following compounds as an alcohol, a phenol, or an ether:

(a) CH₃CH₂CHCH₃   (b) ⬡—OH   (c) ⬡—OH
        |
        OH

(d) ⬡—CH₂OH   (e) ⬡—OCH₃   (f) CH₃CHOCH₂CH₃
                                        |
                                        CH₃

PROBLEM 14.2

What is the difference between a hydroxyl group and hydroxide ion?

14.2 Some Common Alcohols

Simple alcohols are among the most commonly encountered of all organic chemicals. They are useful as solvents, antifreeze agents, and disinfectants, and they are involved in the metabolic processes of all living organisms.

Methyl Alcohol (CH_3OH, Methanol)

Methyl alcohol, the simplest member of the alcohol family, is commonly known as *wood alcohol* because it was once prepared by heating wood in the absence of air. Today it is made in large quantities by reaction of carbon monoxide with hydrogen. Methanol is used industrially as a starting material for preparing formaldehyde ($H_2C{=}O$); and methyl *tert*-butyl ether (MTBE), once an octane booster added to gasoline.

Methyl alcohol
(CH_3OH, methanol)

$$CO(g) + 2\,H_2(g) \xrightarrow[\text{High pressure, 250 °C}]{\text{Cu catalyst}} CH_3OH(l)$$

Methyl alcohol is colorless, miscible with water, and toxic to humans when ingested or inhaled. It causes blindness in low doses (about 15 mL for an adult) and death in larger amounts (100–250 mL).

Ethyl Alcohol (CH_3CH_2OH, Ethanol)

Ethyl alcohol is one of the oldest known pure organic chemicals; its production by fermentation of grain and sugar goes back many thousands of years. Sometimes called *grain alcohol*, ethyl alcohol is the "alcohol" present in all table wines (10–15%), beers (3–7%), and distilled liquors (35–90%). During fermentation, starches or complex sugars are broken down and ultimately converted to simple sugars ($C_6H_{12}O_6$), which are then converted to ethyl alcohol:

Ethyl alcohol
(CH_3CH_2OH, ethanol)

$$C_6H_{12}O_6 \xrightarrow{\text{Yeast enzymes}} 2\,CH_3CH_2OH + 2\,CO_2$$

The maximum alcohol concentration produced by fermentation alone is about 14% by volume, but higher alcohol concentrations can be produced by distillation of the fermentation product or by addition of a distilled product such as brandy. Alcohol for nonbeverage use is often *denatured* by addition of an unpleasant tasting and toxic substance such as methyl alcohol, camphor, or kerosene. The denatured alcohol that results is exempt from the tax applied to the sale of consumable alcohol.

Industrially, most ethyl alcohol is made by hydration of ethylene (Section 13.6). (⬭, p. 414) Distillation yields a 95% ethyl alcohol plus 5% water mixture, and subsequent removal of the water gives 100% ethyl alcohol, known as *absolute alcohol*. In some states, a blend of ethyl alcohol and gasoline called *gasohol* (or E85) is commercially available. Gasohol is a desirable fuel because it produces fewer air pollutants than gasoline.

▲ Distillation using homemade stills was a common practice during Prohibition (1920–1933).

Isopropyl Alcohol [$(CH_3)_2CHOH$]

Isopropyl alcohol, often called *rubbing alcohol*, is used as a 70% mixture with water for rubdowns and in astringents because it cools the skin through evaporation and causes pores to close. It is also used as a solvent for medicines, as a sterilant for instruments, and as a skin cleanser before drawing blood or giving injections. The "medicinal" odor we associate with doctors' offices is often that of isopropyl alcohol. Although not as toxic as methyl alcohol, isopropyl alcohol is much more toxic than ethyl alcohol.

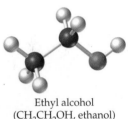

Isopropyl alcohol
($(CH_3)_2CHOH$)

Ethylene Glycol ($HOCH_2CH_2OH$)

Ethylene glycol, a *dialcohol* (meaning it has two —OH groups), is a colorless liquid that is miscible with water and insoluble in nonpolar solvents. Its major use today

Ethylene glycol
HOCH₂CH₂OH

is as an antifreeze or a coolant, or as a starting material for the manufacture of poly-ester films and fibers, such as Dacron. Ethylene glycol is slowly being phased out for use as an automobile antifreeze. In humans, ethylene glycol is a central nervous system depressant; a lethal dose for an adult is about 100 mL. It is also toxic to dogs and cats at doses of about 5 mL per kilogram of body weight. Because of the slight-ly sweet taste of ethylene glycol, accidental poisoning from antifreeze leaked from vehicles was a real concern for both parents of young children and pet owners. For this reason, modern antifreezes use propylene glycol (CH₃CH(OH)CH₂OH), which is tasteless and essentially nontoxic.

Glycerol (HOCH₂CH(OH)CH₂OH)

Like ethylene glycol, the *trialcohol* (three —OH groups in the molecule) glycerol is a colorless liquid that is miscible with water. Unlike ethylene glycol, it is not toxic, but it does have a sweet taste that makes it useful in candy and prepared foods. Often called *glycerin*, glycerol is also used in cosmetics and tobacco as a moisturizer, in plastics manufacture, in antifreeze and shock-absorber fluids, and as a solvent. In Chapter 24, we will see that the glycerol molecule also provides the structural backbone of animal fats and vegetable oils.

▲ A 50% (v/v) mixture of ethylene glycol in water is used as automobile antifreeze.

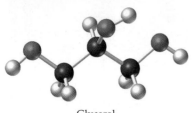

Glycerol
HOCH₂CH(OH)CH₂OH

14.3 Naming Alcohols

Common names of many alcohols containing one —OH group identify the alkyl group and then add the word *alcohol*. Thus, the two-carbon alcohol is ethyl alcohol, the three-carbon alcohol is propyl alcohol, and so on.

In the IUPAC system, alcohols are named using the *-ol* ending for the parent compound:

STEP 1: Name the parent compound. Find the longest chain that has the hydroxyl substituent attached, and name the chain by replacing the *-e* ending of the corresponding alkane with *-ol*:

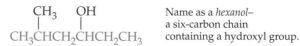

Name as a *hexanol*–
a six-carbon chain
containing a hydroxyl group.

If the compound is a cyclic alcohol, add the *-ol* ending to the name of the parent cycloalkane. For example,

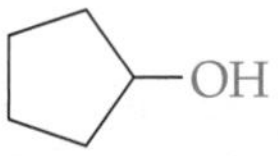

Cyclopentanol

STEP 2: Number the carbon atoms in the main chain. Begin at the end nearer the hydroxyl group, ignoring the location of other substituents:

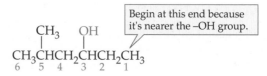

Begin at this end because it's nearer the –OH group.

In a cyclic alcohol, begin with the carbon that bears the —OH group and proceed in a direction that gives the other substituents the lowest possible numbers:

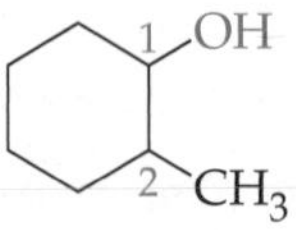

14.4 Properties of Alcohols

Alcohols are much more polar than hydrocarbons because of the electronegative oxygen atom that withdraws electrons from the neighboring atoms. Because of this polarity, hydrogen bonding (Section 8.11) occurs and has a strong influence on alcohol properties. (⊂⊃, p. 235)

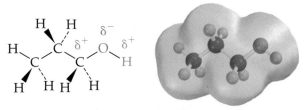

1-Propanol

Straight-chain alcohols with up to 12 carbon atoms are liquids, and each boils at a considerably higher temperature than the related alkane. Alcohols with a small organic part, such as methanol and ethanol, resemble water in their solubility behavior. Methanol and ethanol are miscible with water, with which they can form hydrogen bonds, and these two alcohols can dissolve small amounts of many ionic compounds. Nevertheless, both are also miscible with many organic solvents.

Alcohols with a larger organic part, such as 1-heptanol, are much more like alkanes and less like water. 1-Heptanol is nearly insoluble in water, for example, and cannot dissolve ionic compounds but does dissolve alkanes. The reason is that, in order for water and another liquid to be miscible, water molecules must be able to entirely surround a molecule of the other liquid; the larger the organic (or "greasy") portion of an alcohol molecule, the harder this is to accomplish:

$$CH_3-OH \qquad CH_3CH_2CH_2CH_2CH_2CH_2CH_2-OH$$

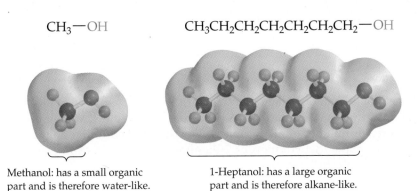

Methanol: has a small organic part and is therefore water-like.

1-Heptanol: has a large organic part and is therefore alkane-like.

Alcohols with two or more —OH groups can form more than one hydrogen bond. They are therefore higher boiling and more water-soluble than similar alcohols with one —OH group. Compare 1-butanol and 1,4-butanediol for example:

$$CH_3CH_2CH_2CH_2OH$$
1-Butanol

{ bp 117 °C, water solubility of 7 g/100 ml.

$$HOCH_2CH_2CH_2CH_2OH$$
1,4-Butanediol

{ Added – OH raises bp to 230 °C and gives miscibility with water

---

PROBLEM 14.6

Which of the following compounds has the highest boiling point?

(a) $CH_3CH_2CH_2OH$ **(b)** $CH_3CH_2OCH_3$ **(c)** $CH_3CH_2CH_3$

PROBLEM 14.7

Rank the following compounds according to their water solubility, most soluble first. Explain your ranking.

(a) $CH_3(CH_2)_{10}CH_2OH$ (b) 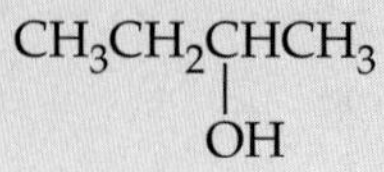 $CH_3CH_2CHCH_3$ (c) $CH_3CH_2OCH_3$
                                                      |
                                                      OH

14.5 Reactions of Alcohols

Dehydration

Dehydration The loss of water from an alcohol to yield an alkene.

Alcohols undergo loss of water (**dehydration**) on treatment with a strong acid catalyst. The —OH group is lost from one carbon, and an —H is lost from an adjacent carbon to yield an alkene product:

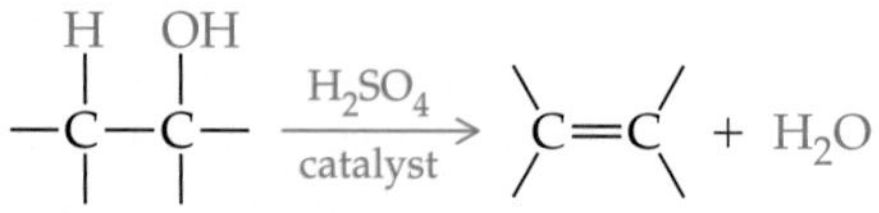

An alcohol → An alkene

For example,

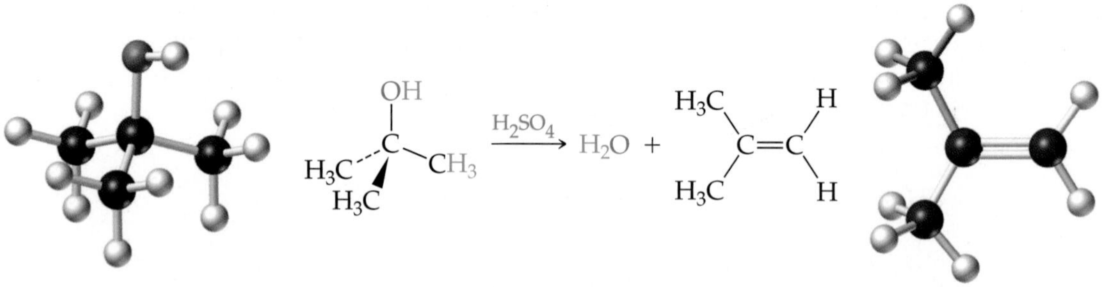

tert-Butyl alcohol   2-Methylpropene

When more than one alkene can result from dehydration of an alcohol, a mixture of products is usually formed. A good rule of thumb is that the major product has the greater number of alkyl groups directly attached to the double-bond carbons. For example, dehydration of 2-butanol leads to a mixture containing 80% 2-butene and only 20% 1-butene:

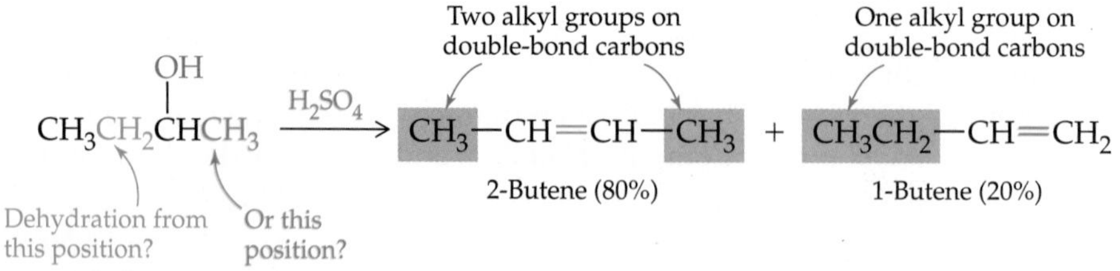

2-Butene (80%)   1-Butene (20%)

WORKED EXAMPLE   **14.2** Organic Reactions: Dehydration

What products would you expect from the following dehydration reaction? Which product will be major, and which minor?

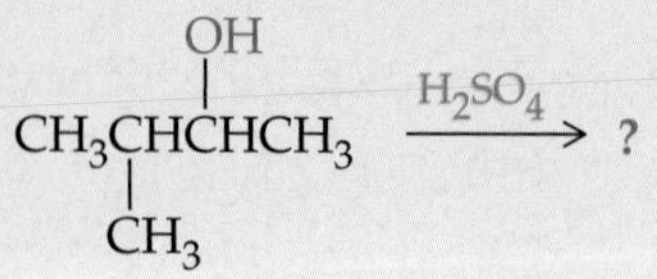

ANALYSIS Find the hydrogens on carbons next to the OH-bearing carbon, and rewrite the structure to emphasize these hydrogens:

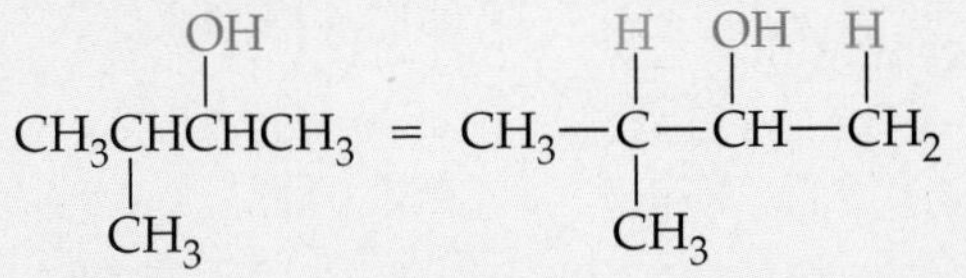

$$
\underset{\underset{CH_3}{|}}{CH_3CHCHCH_3} \quad = \quad \underset{\underset{CH_3}{|}}{CH_3-\overset{H}{\underset{|}{C}}-\overset{OH}{\underset{|}{CH}}-\overset{H}{\underset{|}{CH_2}}}
$$

Then, remove the possible combinations of —H and —OH, drawing a double bond where —H and —OH have come from:

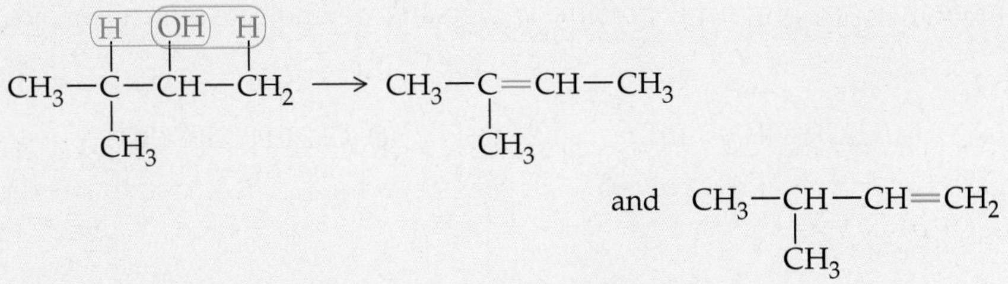

$$
\underset{\underset{CH_3}{|}}{CH_3-\overset{\boxed{H}}{\underset{|}{C}}-\overset{\boxed{OH}}{\underset{|}{CH}}-\overset{\boxed{H}}{\underset{|}{CH_2}}} \longrightarrow \underset{\underset{CH_3}{|}}{CH_3-C=CH-CH_3}
$$

and $\underset{\underset{CH_3}{|}}{CH_3-CH-CH=CH_2}$

Finally, determine which alkene has the larger number of alkyl substituents on its double-bond carbons and is therefore the major product.

SOLUTION

2-Methyl-2-butene
major product (three alkyl groups)

and

3-Methyl-1-butene
minor product (one alkyl group)

WORKED EXAMPLE **14.3** Organic Reactions: Dehydration

What alcohol(s) yield 4-methyl-2-hexene on dehydration?

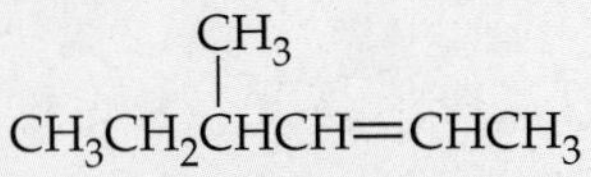

$$
\underset{4\text{-Methyl-2-hexene}}{\overset{\overset{CH_3}{|}}{CH_3CH_2CHCH=CHCH_3}}
$$

ANALYSIS The double bond in the alkene is formed by removing —H and —OH from adjacent carbons of the starting alcohol. This removal occurs in two possible ways, depending on which carbon is bonded to the —OH and to the —H.

SOLUTION

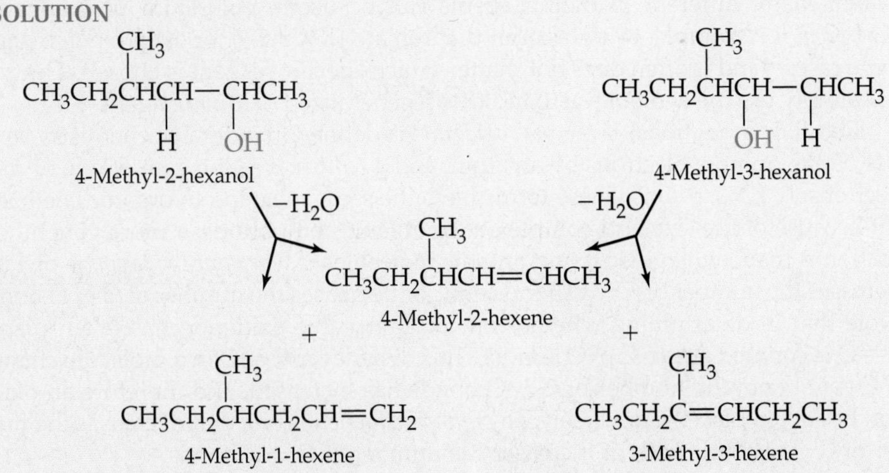

$$
\underset{4\text{-Methyl-2-hexanol}}{\overset{\overset{CH_3}{|}}{CH_3CH_2CHCH-CHCH_3}} \atop \underset{H \quad OH}{}
$$

$$
\underset{4\text{-Methyl-3-hexanol}}{\overset{\overset{CH_3}{|}}{CH_3CH_2CHCH-CHCH_3}} \atop \underset{OH \quad H}{}
$$

−H₂O

−H₂O

$$
\underset{4\text{-Methyl-2-hexene}}{\overset{\overset{CH_3}{|}}{CH_3CH_2CHCH=CHCH_3}}
$$

+

+

$$
\underset{4\text{-Methyl-1-hexene}}{\overset{\overset{CH_3}{|}}{CH_3CH_2CHCH_2CH=CH_2}}
$$

$$
\underset{3\text{-Methyl-3-hexene}}{\overset{\overset{CH_3}{|}}{CH_3CH_2C=CHCH_2CH_3}}
$$

Dehydration of 4-methyl-2-hexanol yields 4-methyl-2-hexene as the major product, along with 4-methyl-1-hexene. Dehydration of 4-methyl-3-hexanol also gives 4-methyl-2-hexene but as the minor product, along with 3-methyl-3-hexene as the major product.

PROBLEM 14.8

What alkenes might be formed by dehydration of the following alcohols? If more than one product is possible in a given case, indicate which is major.

$$\text{OH} \quad \text{CH}_3$$

(a) $CH_3CH_2CH_2OH$ (b) ⬡—OH (c) $CH_3\overset{|}{C}HCH_2\overset{|}{C}HCH_3$

PROBLEM 14.9

What alcohols yield the following alkenes on dehydration?

(a) $(CH_3)_2C\!=\!C(CH_3)_2$ (b) $CH_3CH_2CH\!=\!CH_2$

➤● KEY CONCEPT PROBLEM 14.10

What alkene(s) might be formed by dehydration of the following alcohol (red = O)?

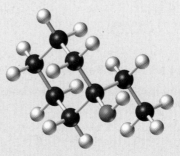

Oxidation

Carbonyl group The C=O functional group.

Primary and secondary alcohols are converted into *carbonyl*-containing compounds on treatment with an oxidizing agent. A **carbonyl group** (pronounced car-bo-**neel**) is a functional group that has a carbon atom joined to an oxygen atom by a double bond, C=O. Many different oxidizing agents can be used—potassium permanganate ($KMnO_4$), for example, or potassium dichromate ($K_2Cr_2O_7$), or even oxygen gas in some cases—and it often does not matter which specific reagent is chosen. Thus, we will simply use the symbol [O] to indicate a generalized oxidizing agent.

Recall from Section 6.11 that an *oxidation* is defined in inorganic chemistry as the loss of one or more electrons by an atom, and a *reduction* as the gain of one or more electrons. (◁◼▷, p. 167) These terms have the same meaning in organic chemistry, but because of the size and complexity of organic compounds, a more general distinction is made when discussing organic molecules. An *organic oxidation* is one that increases the number of C—O bonds and/or decreases the number of C—H bonds. (Note that in determining whether or not an organic oxidation has taken place a C=O is counted as *two* C—O bonds. Thus, whenever C—O in a molecule changes to C=O bond, the number of C—O bonds has increased, and therefore an oxidation has taken place.) Conversely, an *organic reduction* is one that decreases the number of C—O bonds and/or increases the number of C—H bonds.

In the oxidation of an alcohol, two hydrogen atoms are removed from the alcohol and converted into water during the reaction by the oxidizing agent [O]. One hydrogen comes from the —OH group, and the other hydrogen from the carbon atom bonded to the —OH group. In the process, a new C—O bond is formed and a C—H bond is broken:

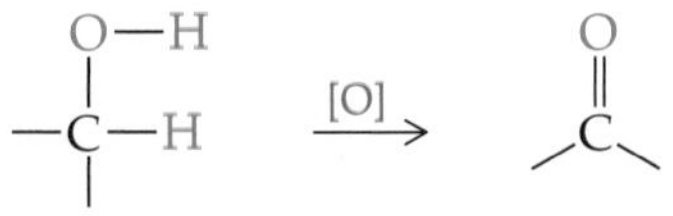

An alcohol   A carbonyl compound

Different kinds of carbonyl-containing products are formed, depending on the structure of the starting alcohol and on the reaction conditions. Primary alcohols (RCH_2OH) are converted either into *aldehydes* ($RCH=O$) if carefully controlled conditions are used, or into *carboxylic acids* (pronounced car-box-**ill**-ic) (RCO_2H) if an excess of oxidant is used:

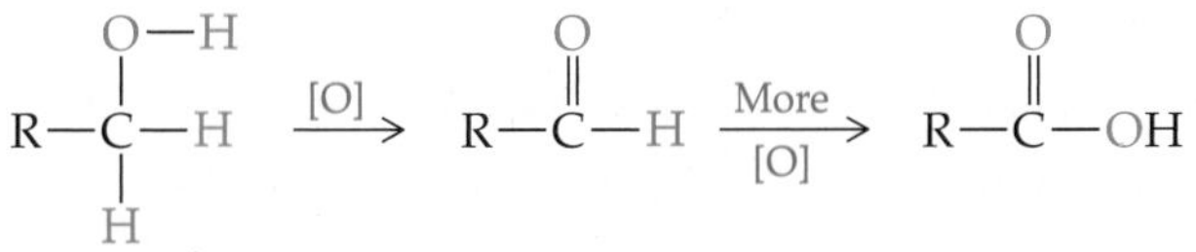

A primary alcohol   An aldehyde   A carboxylic acid

For example,

1-Butanol   Butanal   Butanoic acid

Secondary alcohols (R_2CHOH) are converted into *ketones* ($R_2C=O$) on treatment with oxidizing agents:

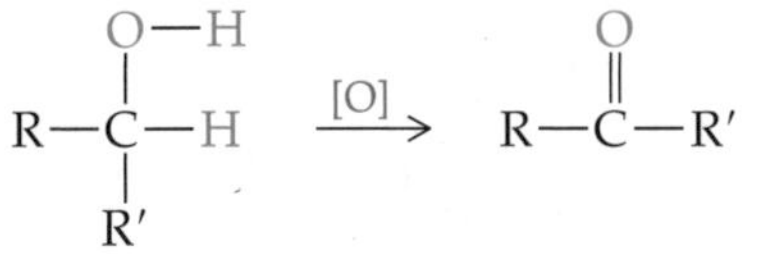

A secondary alcohol   A ketone

For example,

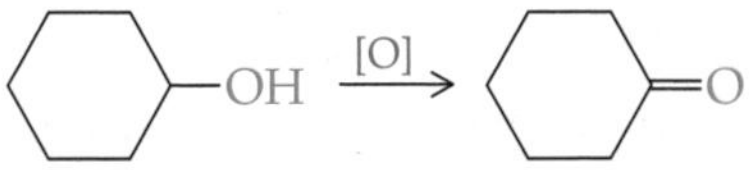

Cyclohexanol   Cyclohexanone

Tertiary alcohols do not normally react with oxidizing agents because they do not have a hydrogen on the carbon atom to which the —OH group is bonded:

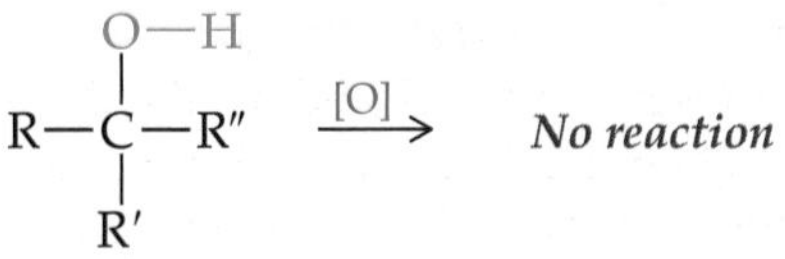

A tertiary alcohol

Looking Ahead

In Chapter 23, we will see that alcohol oxidations are critically important steps in many key biological processes. When lactic acid builds up in tired, overworked muscles, for example, the liver removes it by oxidizing it to pyruvic acid. Our bodies, of course, do not use $K_2Cr_2O_7$ or $KMnO_4$ for the oxidation; instead, they use specialized, highly selective enzymes to carry out this chemistry. Regardless of the details, though, the net chemical transformation is the same whether carried out in a laboratory flask or in a living cell.

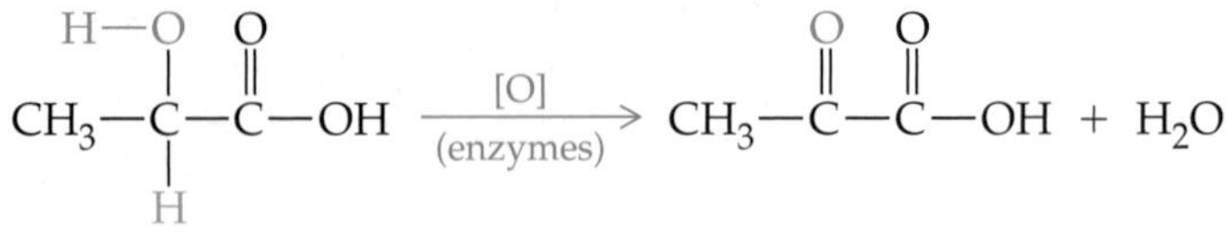

Lactic acid → Pyruvic acid

WORKED EXAMPLE 14.4 Organic Reactions: Oxidation

What is the product of the following oxidation reaction?

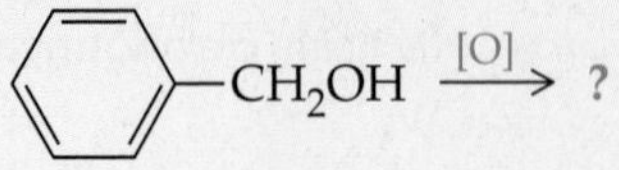

Benzyl alcohol

ANALYSIS The starting material is a primary alcohol, so it will be converted first to an aldehyde and then to a carboxylic acid. To find the structures of these products, first redraw the structure of the starting alcohol to identify the hydrogen atoms on the hydroxyl-bearing carbon:

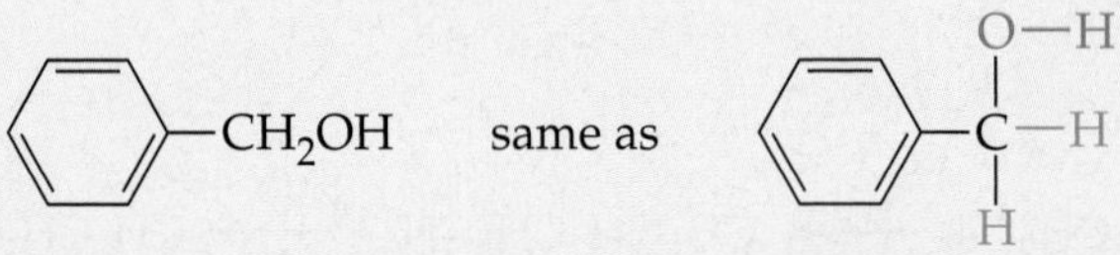

Next, remove 2 hydrogens, one from the —OH group and one from the hydroxyl-bearing carbon. In their place, make a C=O double bond. This is the aldehyde product that forms initially. Finally, convert the aldehyde to a carboxylic acid by replacing the hydrogen in the —CH=O group with an —OH group.

SOLUTION

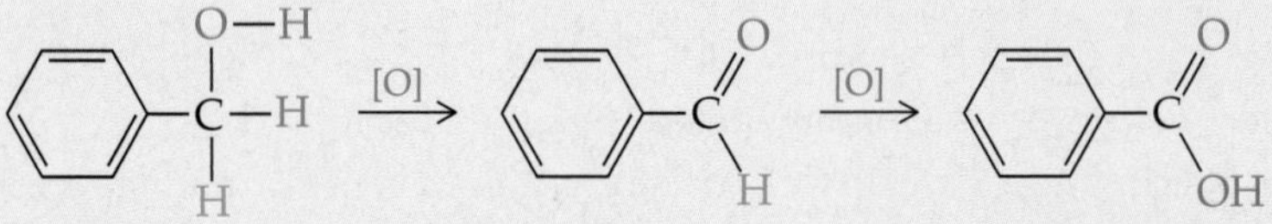

Aldehyde   Carboxylic acid

PROBLEM 14.11

What products would you expect from oxidation of the following alcohols?

(a) $CH_3CH_2CH_2OH$ (b) 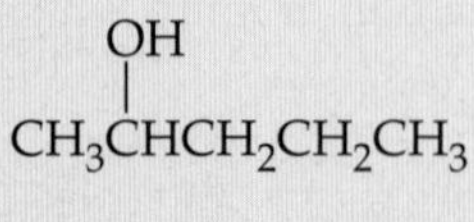 $CH_3\overset{\overset{\displaystyle OH}{|}}{C}HCH_2CH_2CH_3$ (c)

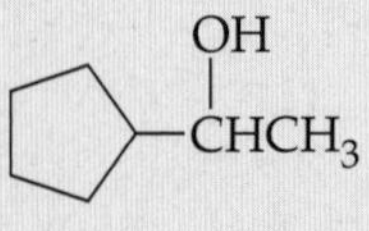

PROBLEM 14.12

From what alcohols might the following carbonyl-containing products have been made?

(a) 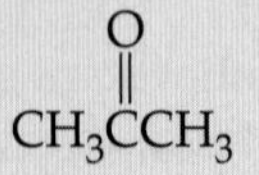 $CH_3\overset{\overset{\displaystyle O}{||}}{C}CH_3$ (b) 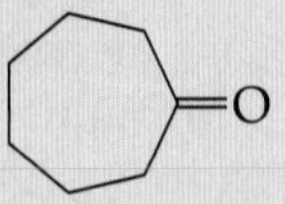   (c) 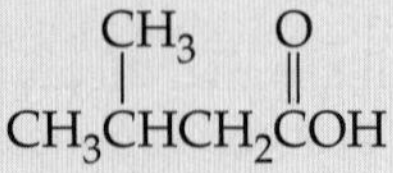 $CH_3\overset{\overset{\displaystyle CH_3}{|}}{C}HCH_2\overset{\overset{\displaystyle O}{||}}{C}OH$

◀─○ KEY CONCEPT PROBLEM 14.13

From what alcohols might the following carbonyl-containing products have been made (red = O, red-brown = Br)?

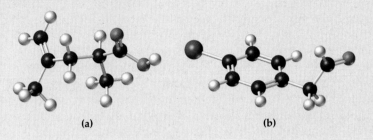

(a)   (b)

APPLICATION ▶   Ethyl Alcohol as a Drug and a Poison

Ethyl alcohol is classified for medical purposes as a central nervous system (CNS) depressant. Its direct effects (being "drunk") resemble the response to anesthetics and are quite predictable. At a blood alcohol concentration of 80–300 mg/dL, motor coordination and pain perception are affected, accompanied by loss of balance, slurred speech, and amnesia. At a concentration of 300–400 mg/dL, there may be nausea and loss of consciousness. Further increases in blood alcohol levels cause progressive loss of protective reflexes in stages like those of surgical anesthesia. Above 600 mg/dL of blood alcohol, spontaneous respiration and cardiovascular regulation are affected, ultimately resulting in death.

The passage of ethyl alcohol through the body begins with its absorption in the stomach and small intestine, followed by rapid distribution to all body fluids and organs. In the pituitary gland, alcohol inhibits the production of a hormone that regulates urine flow, causing increased urine production and dehydration. In the stomach, ethyl alcohol stimulates production of acid. Throughout the body, it causes blood vessels to dilate, resulting in flushing of the skin and a sensation of warmth as blood moves into capillaries beneath the surface. The result, though, is not a warming of the body but an increased loss of heat at the surface, making alcoholic beverages a poor choice in cold weather.

Ethyl alcohol metabolism in the liver is a two-step process: oxidation of the alcohol to acetaldehyde followed by oxidation of the aldehyde to acetic acid. One of the hydrogen atoms lost in the oxidation at each stage binds to the biochemical oxidizing agent NAD$^+$ (nicotinamide adenine dinucleotide, Chapter 23), and the other leaves as a hydrogen ion, H$^+$. When continuously present in the bodies of chronic alcoholics, alcohol and acetaldehyde are toxic, leading to devastating physical and metabolic deterioration. The liver usually suffers the worst damage because it is the major site of alcohol metabolism.

$$CH_3CH_2OH \xrightarrow[\substack{\text{Alcohol} \\ \text{dehydrogenase} \\ \text{enzyme}}]{NAD^+} CH_3\overset{\displaystyle O}{\overset{\|}{C}}H \xrightarrow[\substack{\text{Aldehyde} \\ \text{dehydrogenase} \\ \text{enzyme}}]{NAD^+} CH_3\overset{\displaystyle O}{\overset{\|}{C}}OH$$

The quick and uniform distribution of ethyl alcohol in body fluids, the ease with which it crosses lung membranes, and its ready oxidizability provide the basis for tests of blood alcohol concentration. The Breathalyzer test measures alcohol concentration in expired air by the color change that occurs when the bright yellow-orange oxidizing agent potassium dichromate (K$_2$Cr$_2$O$_7$) is reduced to blue-green chromium(III). The color change can be interpreted by instruments to give an accurate measure of alcohol concentration in the blood. As an alternative method, the Intoxilyzer test uses a beam of infrared light to measure blood alcohol levels. In many states, driving with a blood alcohol level above 0.10% (100 mg/dL) is illegal, and more than 50% of states have lowered the legal limit to 0.08%.

▲ The Breathalyzer test measures blood alcohol concentration.

See Additional Problems 14.54–14.57 at the end of the chapter.

14.6 Phenols

The word *phenol* is the name both of a specific compound (hydroxybenzene, C_6H_5OH) and of a family of compounds. Phenol itself, formerly called carbolic acid, is a medical antiseptic first used by Joseph Lister in 1867. Lister showed that the occurrence of postoperative infection dramatically decreased when phenol was used to cleanse the operating room and the patient's skin. Because phenol numbs the skin, it also became popular in topical drugs for pain and itching and in treating sore throats.

The medical use of phenol is now restricted because it can cause severe skin burns and has been found to be toxic, both by ingestion and by absorption through the skin. The once common use of phenol for treating diaper rash is especially hazardous because phenol is more readily absorbed through a rash. Only solutions containing less than 1.5% phenol or lozenges containing a maximum of 50 mg of phenol are now allowed in nonprescription drugs. Many mouthwashes and throat lozenges contain alkyl-substituted phenols such as thymol as active ingredients for pain relief.

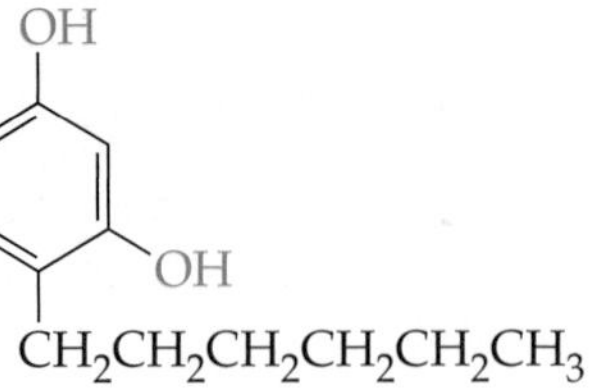

4-Hexylresorcinol
(a topical anesthetic)

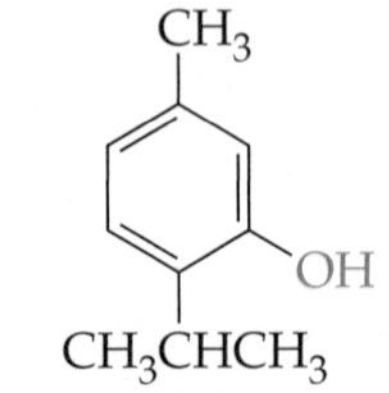

Thymol
(a topical anesthetic; occurs
naturally in the herb thyme)

Alkyl-substituted phenols such as the cresols (methylphenols) are common as *disinfectants* in hospitals and elsewhere. By contrast with an antiseptic, which safely kills microorganisms on living tissue, a disinfectant should be used only on inanimate objects. The germicidal properties of phenols can be partially explained by their ability to disrupt the permeability of cell walls of microorganisms.

Phenols are usually named with the ending *-phenol* rather than *-benzene* even though their structures include a benzene ring. For example,

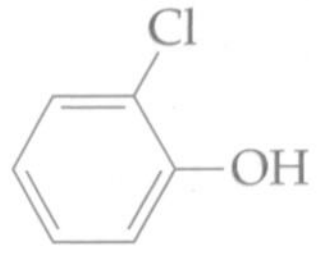

o-Chlorophenol 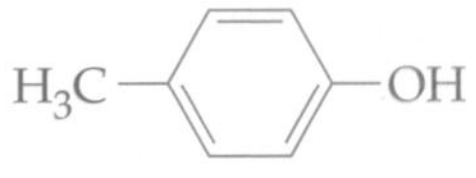   *p*-Methylphenol

The properties of phenols, like those of alcohols, are influenced by the presence of the electronegative oxygen atom and by hydrogen bonding. Most phenols are water-soluble to some degree and have higher melting and boiling points than similarly substituted alkylbenzenes. They are in general less soluble in water than alcohols are.

Biomolecules that contain a hydroxyl-substituted benzene ring include the amino acid tyrosine as well as many other compounds.

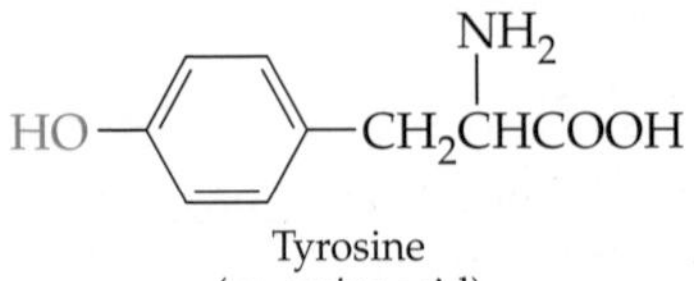

Tyrosine
(an *amino* acid)

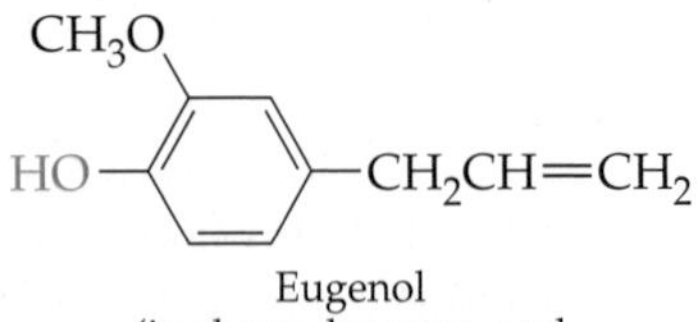

Eugenol
(in cloves, bananas, and
other fruits; used for
toothache pain)

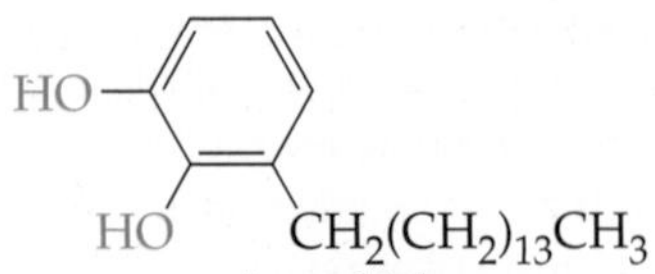

A urushiol
(skin irritant in
poison ivy)

▲ Careful! The urushiol in this poison ivy plant causes severe skin rash.

PROBLEM 14.14

Draw structures for the following:

(a) *m*-Nitrophenol   **(b)** *o*-Ethylphenol

PROBLEM 14.15

Name the following compounds:

(a) 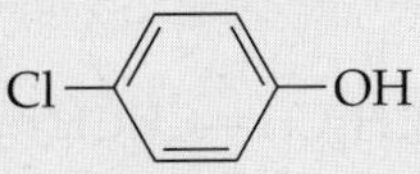 Cl—⬡—OH   **(b)** 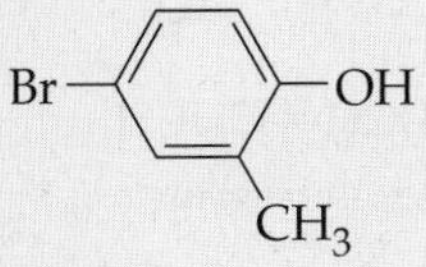 Br—⬡—OH
                                        |
                                       CH₃

14.7 Acidity of Alcohols and Phenols

Alcohols and phenols are very weakly acidic because of the positively polarized OH hydrogen. They dissociate slightly in aqueous solution and establish equilibria between neutral and anionic forms:

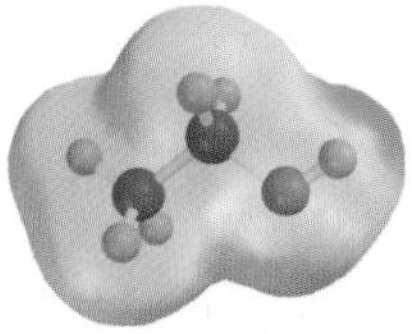

$$CH_3CH_2OH \; \underset{\text{water}}{\overset{\text{Dissolve in}}{\rightleftharpoons}} \; CH_3CH_2O^- + H_3O^+$$

An alcohol

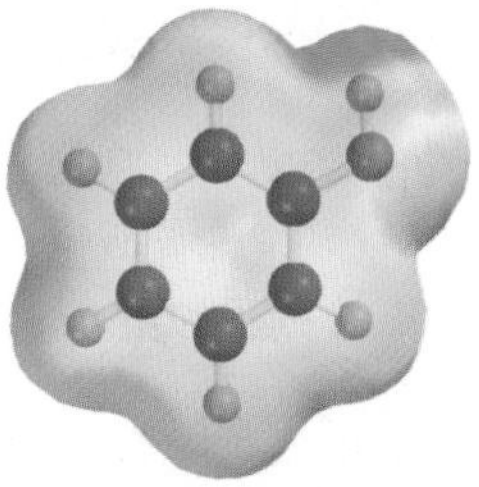

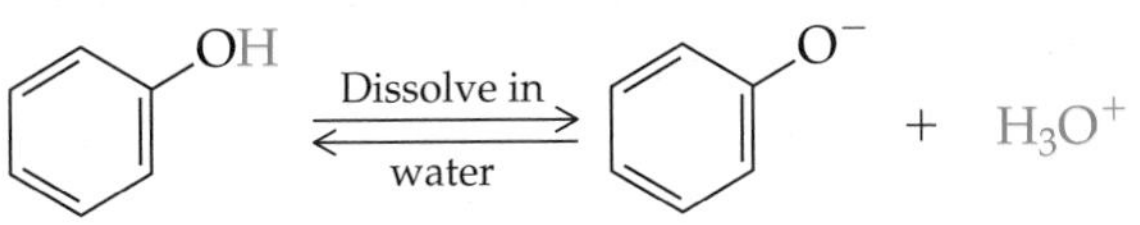

⬡—OH $\underset{\text{water}}{\overset{\text{Dissolve in}}{\rightleftharpoons}}$ ⬡—O⁻ $+ \; H_3O^+$

A phenol

Methanol and ethanol are about as acidic as water itself (Section 10.7), with K_a values near 10^{-15}. By comparison, acetic acid has a K_a of 10^{-5}. (⬭, p. 304) In fact, they are so little dissociated in water that their aqueous solutions are neutral (pH 7). Thus, an **alkoxide ion** (RO^-), or anion of an alcohol, is as strong a base as hydroxide ion, OH^-. An alkoxide ion is produced by reaction of an alkali metal with an alcohol, just as hydroxide ion is produced by reaction of an alkali metal with water. For example,

Alkoxide ion The anion resulting from deprotonation of an alcohol, RO^-.

$$2\,H_2O + 2\,Na \; \longrightarrow \; 2\,Na^+\,OH^- + H_2$$
$$2\,CH_3OH + 2\,Na \; \longrightarrow \; 2\,Na^{+\,-}OCH_3 + H_2$$

Methanol   Sodium methoxide

In contrast to alcohols, phenols are about 10,000 times more acidic than water. Phenol itself, for example, has $K_a = 1.0 \times 10^{-10}$. This acidic property means that phenols react with dilute aqueous sodium hydroxide to give a phenoxide ion. (Alcohols do not react in this way with sodium hydroxide.)

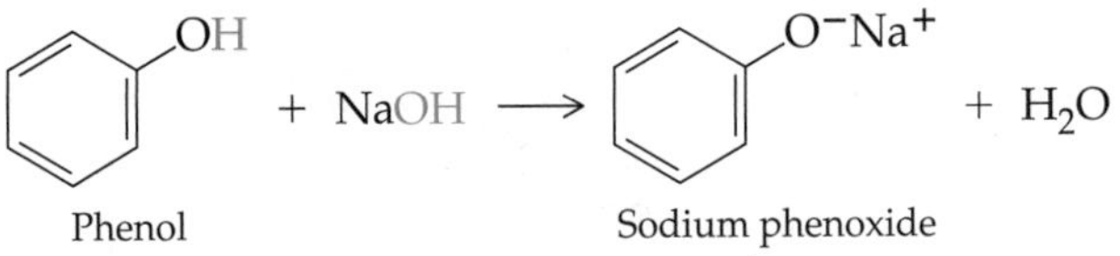

⬡—OH $+ \; NaOH \; \longrightarrow$ ⬡—O⁻Na⁺ $+ \; H_2O$

Phenol   Sodium phenoxide

APPLICATION ▶ Phenols as Antioxidants

If you occasionally read the labels on food packages, which you should do now that you are studying organic chemistry, the names butylated hydroxytoluene and butylated hydroxyanisole, or their abbreviations BHT and BHA, are probably familiar. You can find them on most cereal, cookie, and cracker boxes. Both compounds are substituted phenols.

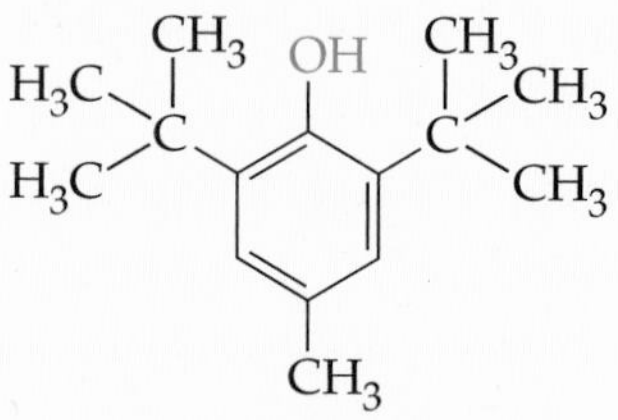

Butylated hydroxytoluene
(BHT)

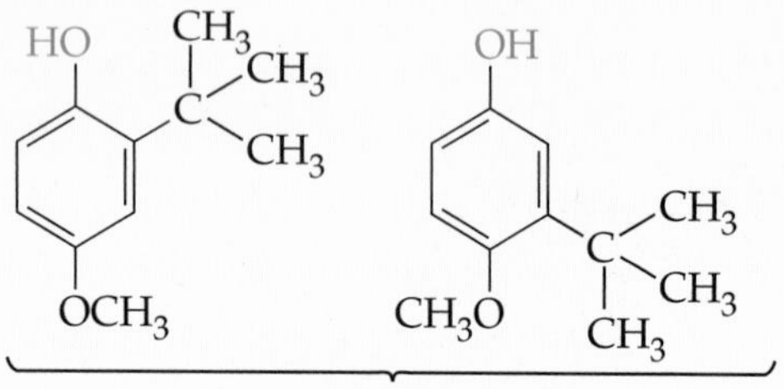

Butylated hydroxyanisole
(BHA)

Foods that contain unsaturated fats—those having carbon–carbon double bonds—become rancid when oxygen from the air reacts with their double bonds, producing ketones and other oxygen-containing substances with bad smells and tastes. The chemistry of oxidative rancidity is complex but involves the formation of reactive substances that contain unpaired electrons and are known as *free radicals*. Each free radical reacts further with O_2 in a *chain reaction* that ultimately leads to the destruction of a great many fat molecules:

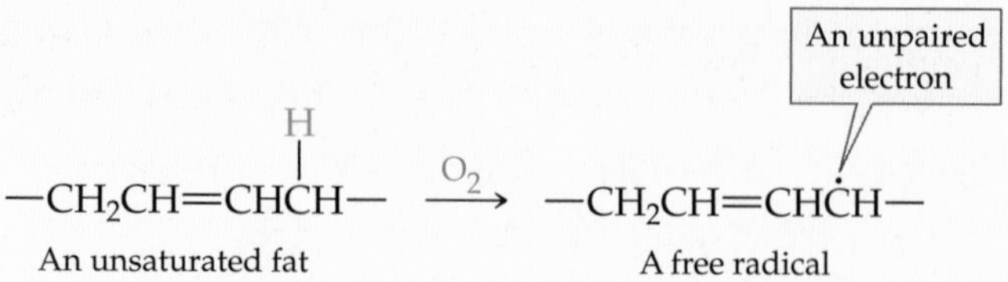

An unpaired electron

An unsaturated fat → A free radical

BHT and BHA prevent oxidation by donating a hydrogen atom from their —OH group to the free radical as soon as it forms, thereby converting the radical back to the starting fat and interrupting the destructive chain reaction. In the process, the BHA or BHT is converted into a stable and unreactive free radical, which causes no damage. Vitamin E, a natural antioxidant within the body, acts similarly.

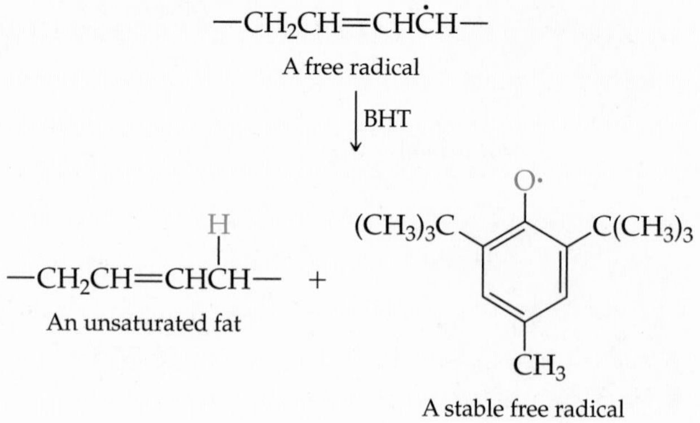

A free radical

BHT

An unsaturated fat +

A stable free radical

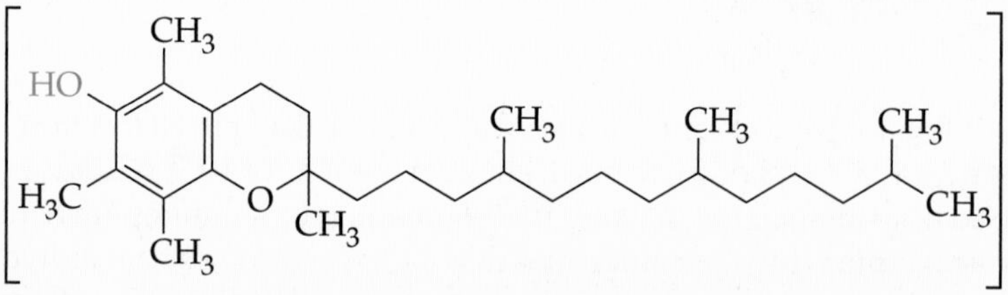

Vitamin E (a naturally occurring antioxidant)

The formation of free radicals is suspected of playing a role in both cancer and the normal aging of living tissue. Although there is no conclusive evidence, antioxidants may well be effective in slowing the progress of both conditions.

▲ Most processed foods contain the antioxidants BHT and BHA to preserve freshness and increase shelf life.

See Additional Problems 14.58 and 14.59 at the end of the chapter.

14.8 Ethers

Ethers—compounds with two organic groups bonded to the same oxygen atom—are named by identifying the two organic groups and adding the word *ether*. (The compound frequently referred to simply as "ether" is actually diethyl ether.)

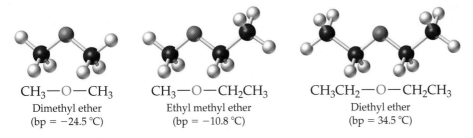

CH$_3$—O—CH$_3$
Dimethyl ether
(bp = −24.5 °C)

CH$_3$—O—CH$_2$CH$_3$
Ethyl methyl ether
(bp = −10.8 °C)

CH$_3$CH$_2$—O—CH$_2$CH$_3$
Diethyl ether
(bp = 34.5 °C)

Compounds that contain the oxygen atom in a ring are classified as cyclic ethers and are often referred to by their common names.

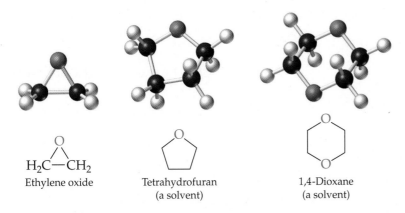

H$_2$C—CH$_2$
Ethylene oxide

Tetrahydrofuran
(a solvent)

1,4-Dioxane
(a solvent)

An —OR group is referred to as an **alkoxy group**; —OCH$_3$ is a *methoxy* group, —OCH$_2$CH$_3$ is an *ethoxy* group, and so on. These names are used when the ether functional group is present in a compound that also has other functional groups. For example,

Alkoxy group An —OR group.

CH$_3$CH$_2$OCH$_2$CH$_2$OH

2-Ethoxyethanol

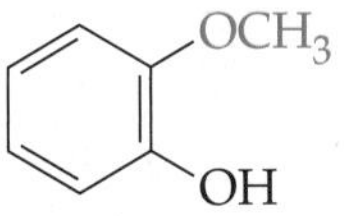

o-Methoxyphenol

Although they contain polar C—O bonds, ethers lack the —OH group of water and alcohols, and thus do not form hydrogen bonds to one another. Simple ethers are therefore higher boiling than comparable alkanes but much lower boiling than alcohols. The oxygen atom in ethers can hydrogen-bond with water, causing dimethyl ether to be water-soluble and diethyl ether to be partially miscible with water. As with alcohols, ethers with larger organic groups are often insoluble in water.

Ethers are alkane-like in many of their properties and do not react with most acids, bases, or other reagents. Ethers do, however, react readily with oxygen, and the simple ethers are highly flammable. On standing in air, many ethers form explosive *peroxides*, compounds that contain an O—O bond. Thus, ethers must be handled with care and stored in the absence of oxygen.

Diethyl ether, the best-known ether, is now used primarily as a solvent but was for many years a popular anesthetic. Its value as an inhalation anesthetic was discovered in the 1840s, and it was a mainstay of the operating room until the 1940s. Although it acts quickly and is very effective, ether is far from ideal as an anesthetic

▲ The maturation of this silkworm moth is controlled by a hormone that contains a three-membered ether ring.

because it has a long recovery time and it often induces nausea. Moreover, its effectiveness is strongly counterbalanced by its hazards. Diethyl ether is a highly volatile, flammable liquid whose vapor forms explosive mixtures with air.

Diethyl ether has now been replaced by safer, less flammable anesthetics such as enflurane and isoflurane (see the Application "Inhaled Anesthetics," p. 454). Both compounds were products of an intensive effort during the 1960s during which more than 400 halogenated ethers were synthesized in a search for improved anesthetics.

Enflurane   Isoflurane

Ethers are found throughout the plant and animal kingdoms. Some are present in plant oils and are used in perfumes; others have a variety of biological roles. Juvenile hormone, for example, is a cyclic ether that helps govern the growth of the silkworm moth. The three-membered ether ring (an *epoxide* ring) in the juvenile hormone is unusually reactive because of strained 60° bond angles.

Anethole—a flavoring agent
found in anise and fennel

Juvenile hormone—an insect hormone
found in the silkworm moth

| WORKED EXAMPLE | **14.5** Molecular Structures: Drawing Ethers and Alcohols |

Draw the structure for 3-methoxy-2-butanol.

ANALYSIS First identify the parent compound, then add numbered substituents to appropriate carbons in the parent chain.

SOLUTION
The parent compound is a 4-carbon chain with the —OH attached to C2.

$$\underset{4}{C}-\underset{3}{C}-\underset{2}{\overset{\overset{\displaystyle OH}{|}}{C}}-\underset{1}{C} \quad \text{2-butanol}$$

The 3-methoxy substituent indicates that a methoxy group (—OCH$_3$) is attached to C3.

$$\underset{4}{C}-\underset{3}{\underset{\underset{\displaystyle CH_3}{|}}{\overset{\displaystyle O}{C}}}-\underset{2}{\overset{\overset{\displaystyle OH}{|}}{C}}-\underset{1}{C} \quad \text{3-methoxy}$$

Finally, add hydrogens until each carbon atom has a total of four bonds.

$$\underset{4}{CH_3}-\underset{3}{\underset{\underset{\displaystyle CH_3}{|}}{\overset{\displaystyle O}{CH}}}-\underset{2}{\overset{\overset{\displaystyle OH}{|}}{CH}}-\underset{1}{CH_3} \quad \text{Name: 3-methoxy-2-butanol}$$

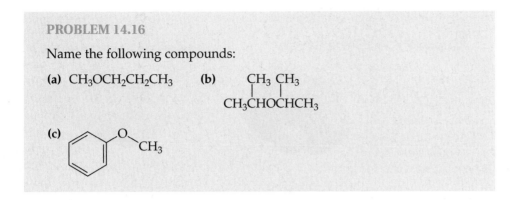

PROBLEM 14.16

Name the following compounds:

(a) $CH_3OCH_2CH_2CH_3$ **(b)**

$$\begin{array}{cc} CH_3 & CH_3 \\ | & | \\ CH_3CHOCHCH_3 \end{array}$$

(c)

14.9 Thiols and Disulfides

Sulfur is just below oxygen in group 6A of the periodic table, and many oxygen-containing compounds have sulfur analogs. For example, **thiols** (R—SH), also called *mercaptans*, are sulfur analogs of alcohols. The systematic name of a thiol is formed by adding *-thiol* to the parent hydrocarbon name. Otherwise, thiols are named in the same way as alcohols.

Thiol A compound that contains an —SH group, R—SH.

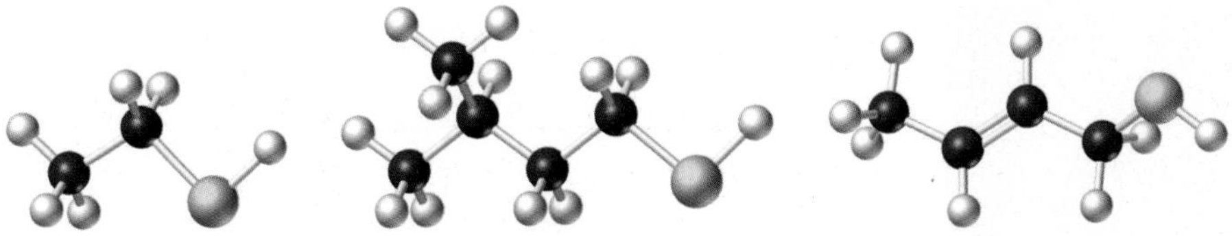

CH_3CH_2SH

Ethanethiol

$$\begin{array}{c} CH_3 \\ | \\ CH_3CHCH_2CH_2SH \end{array}$$

3-Methyl-1-butanethiol

$CH_3CH{=}CHCH_2SH$

2-Butene-1-thiol

▲ Skunks repel predators by releasing several thiols with appalling odors.

The most outstanding characteristic of thiols is their appalling odor. Skunk scent is caused by two of the simple thiols shown above, 3-methyl-1-butanethiol and 2-butene-1-thiol. Thiols are also in the air whenever garlic and onions are being sliced, or when there is a natural gas leak. Natural gas itself is odorless, but a low concentration of methanethiol (CH_3SH) is added as a safety measure to make leak detection easy.

Thiols react with mild oxidizing agents such as Br_2 in water to yield **disulfides**, RS—SR. Two thiols join together in this reaction, the hydrogen from each is lost, and a bond forms between the 2 sulfurs:

Disulfide A compound that contains a sulfur–sulfur bond, RS—SR.

$$\underset{\text{Two thiol molecules}}{RSH + HSR} \xrightarrow{[O]} \underset{\text{A disulfide}}{RSSR}$$

For example,

$$\underset{\text{Methanethiol}}{H_3C-S-H + H-S-CH_3} \xrightarrow{[O]} \underset{\text{Dimethyl disulfide}}{CH_3-S-S-CH_3 + H_2O}$$

The reverse reaction occurs when a disulfide is treated with a reducing agent, represented by [H]:

$$RSSR \xrightarrow{[H]} RSH + RSH$$

APPLICATION ▶ Inhaled Anesthetics

William Morton's demonstration in 1846 of ether-induced anesthesia during dental surgery ranks as one of the most important medical breakthroughs of all time. Before that date, all surgery had been carried out with the patient fully conscious. Use of chloroform ($CHCl_3$) as an anesthetic quickly followed Morton's work, popularized by Queen Victoria of England, who in 1853 gave birth to a child while anesthetized by chloroform.

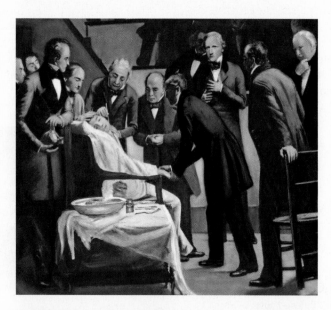

▲ William Morton performed the first public demonstration of ether as an anesthetic on October 16, 1846, at Massachusetts General Hospital.

Hundreds of substances have subsequently been shown to act as inhaled anesthetics. Halothane, enflurane, isoflurane, and methoxyflurane are at present the most commonly used agents in hospital operating rooms. All four are potent at relatively low doses, are nontoxic, and are nonflammable, an important safety feature.

Despite their importance, surprisingly little is known about how inhaled anesthetics work in the body. Remarkably, the potency of different inhaled anesthetics correlates well with their solubility in olive oil, leading many scientists to believe that anesthetics act by dissolving in the fatty membranes surrounding nerve cells. The resultant changes in the fluidity and shape of the membranes apparently decrease the ability of sodium ions to pass into the nerve cells, thereby blocking the firing of nerve impulses.

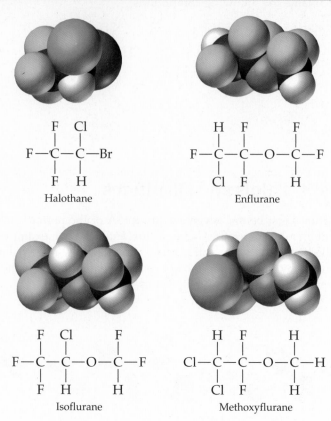

Depth of anesthesia is determined by the concentration of anesthetic agent that reaches the brain. Brain concentration, in turn, depends on the solubility and transport of the anesthetic agent in the bloodstream and on its partial pressure in inhaled air. Anesthetic potency is usually expressed as a *minimum alveolar concentration* (MAC), defined as the concentration of anesthetic in inhaled air that results in anesthesia in 50% of patients. As shown in the following Table, nitrous oxide, N_2O, is the least potent of the common anesthetics and methoxyflurane is the most potent agent; a partial pressure of only 1.2 mm Hg is sufficient to anesthetize 50% of patients.

Relative Potency of Inhaled Anesthetics

ANESTHETIC	MAC (%)	MAC (Partial pressure, mm Hg)
Nitrous oxide		>760
Enflurane	1.7	13
Isoflurane	1.4	11
Halothane	0.75	5.7
Methoxyflurane	0.16	1.2

See Additional Problems 14.60 and 14.61 at the end of the chapter.

Thiols are important biologically because they occur as a functional group in the amino acid cysteine, which is part of many proteins:

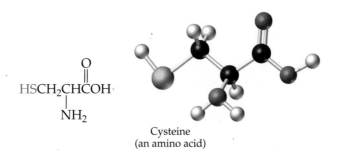

$$HSCH_2\underset{\underset{NH_2}{|}}{CH}\overset{\overset{O}{||}}{C}OH$$

Cysteine
(an amino acid)

The easy formation of S—S bonds between two cysteines helps pull large protein molecules into the shapes they need to function. Hair protein, for example, is unusually rich in —S—S— and —SH groups. When hair is "permed," some disulfide bonds are broken and others are then formed. As a result, the hair proteins are held in a different shape (Figure 14.2). The importance of the disulfide linkage will be discussed further in Section 18.8.

$$\xrightarrow{[O]}$$

SH   SH   S—S

▲ **FIGURE 14.2** **Chemistry can curl your hair.** A permanent wave results when disulfide bridges are formed between —SH groups in hair protein molecules.

PROBLEM 14.17

What disulfides would you obtain from oxidation of the following thiols?

(a) $CH_3CH_2CH_2SH$

(b) 3-Methyl-1-butan ethiol (skunk scent)

14.10 Halogen-Containing Compounds

The simplest halogen-containing compounds are the **alkyl halides**, RX, where R is an alkyl group and X is a halogen. Their common names consist of the name of the alkyl group followed by the halogen name with an *-ide* ending. The compound CH_3Br, for example, is commonly called *methyl bromide*.

Systematic names consider the halogen atom as a substituent on a parent alkane. The parent alkane is named in the usual way by selecting the longest continuous chain (Section 12.6) and numbering from the end nearer the first substituent, either alkyl or halogen. (⊂⊃, p. 378) The *halo*-substituent name is then given as a prefix, just as if it were an alkyl group. A few common halogenated compounds, such as chloroform ($CHCl_3$), are also known by nonsystematic names.

Alkyl halide A compound that has an alkyl group bonded to a halogen atom, R—X.

$$\underset{\underset{\text{1-Chloropropane}}{}}{\overset{3\quad 2\quad 1}{CH_3CH_2CH_2Cl}}$$

$$\underset{\underset{6\quad 5\quad 4\quad 3\quad 2\quad 1}{\text{2-Bromo-5-methylhexane}}}{CH_3\underset{\underset{CH_3}{|}}{CH}CH_2CH_2\underset{\underset{Br}{|}}{CH}CH_3}$$

$$\underset{\underset{\text{(Chloroform)}}{\text{Trichloromethane}}}{CHCl_3}$$

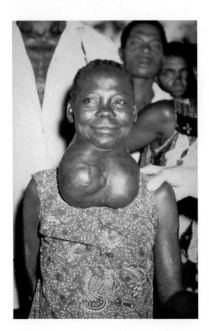

▲ Goiter, characterized by swelling of the thyroid gland, is caused by a dietary iodine deficiency.

Halogenated organic compounds have a variety of medical and industrial uses. Ethyl chloride is used as a topical anesthetic because it cools the skin through rapid evaporation; halothane is an important anesthetic. Chloroform was once employed as an anesthetic and as a solvent for cough syrups and other medicines, but is now considered too toxic for such uses. Bromotrifluoromethane, CF_3Br, is useful for extinguishing fires in aircraft and electronic equipment because it is nonflammable and nontoxic, and it evaporates without a trace.

Although a large number of halogen-containing organic compounds are found in nature, especially in marine organisms, few are significant in human biochemistry. One exception is thyroxine, an iodine-containing hormone secreted by the thyroid gland. A deficiency of iodine in the human diet leads to a low thyroxine level, which causes a swelling of the thyroid gland called *goiter*. To ensure adequate iodine in the diet of people who live far from an ocean, potassium iodide is sometimes added to table salt (to create the product we know as *iodized salt*).

Thyroid gland hormone; deficiency causes goiter

Thyroxine

Halogenated compounds are also used widely in industry and agriculture. Dichloromethane (CH_2Cl_2, methylene chloride), trichloromethane ($CHCl_3$, chloroform), and trichloroethylene ($Cl_2C{=}CHCl$) are used as solvents and degreasing agents, although their use is diminishing as less polluting alternatives become available. Because these substances are excellent solvents for the greases in skin, continued exposure often causes dermatitis.

The use of halogenated herbicides such as 2,4-D and fungicides such as Captan has resulted in vastly increased crop yields in recent decades, and the widespread application of chlorinated insecticides such as DDT is largely responsible for the progress made toward worldwide control of malaria and typhus. Despite their enormous benefits, chlorinated pesticides present problems because they persist in the environment and are not broken down rapidly. They remain in the fatty tissues of organisms and accumulate up the food chain as larger organisms consume smaller ones. Eventually, the concentration in some animals becomes high enough to cause harm. In an effort to maintain a balance between the value of halogenated pesticides and the harm they can do, the use of many has been restricted, and others have been banned altogether.

2,4-D   Captan   DDT

PROBLEM 14.18

Give systematic names for the following alkyl halides:

(a)

(b) $CH_3CH_2CHCH_2CHCH_2CH_3$ with CH_3 and Br substituents

APPLICATION ▶   Chlorofluorocarbons and the Ozone Hole

Newspaper stories about a "hole" in the ozone layer have appeared with regularity in recent years. What began as speculation about potential problems is now accepted as fact: Up to 65% of the stratospheric ozone over the South Pole disappears in the polar spring when the so-called *ozone hole* develops, before returning to near normal levels in the autumn. More recently, a similar though less dramatic decrease in polar ozone has been found over the North Pole.

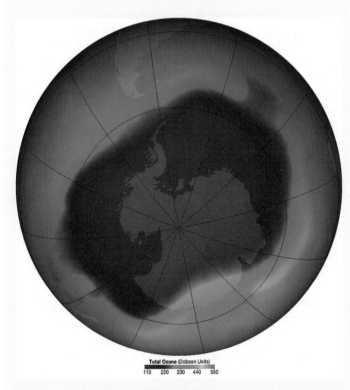

Total Ozone (Dobson Units)
110 220 330 440 550

▲ **False color image of the ozone hole over the Antarctic September 21–30, 2006, showing the largest ozone hole ever observed for more than one day, averaging 10.6 million square miles. The blue and purple are where there is the least ozone, and the greens, yellows, and reds are where there is more ozone.**

Although toxic to all life forms at high concentrations, ozone (O_3) is critically important in the upper atmosphere because it acts as a shield to protect the earth from intense solar radiation. If the global ozone layer were depleted, more solar radiation would reach the earth, causing an increase in the incidence of skin cancer and eye cataracts. Diminished growth of oceanic plankton and stunting of the growth of terrestrial crops are also predicted.

The principal cause of ozone depletion is the presence in the atmosphere of a group of alkyl halides called *chlorofluorocarbons*, or CFCs, familiar to many by the common name of *freon*. The chlorofluorocarbons are simple alkyl halides in which all the hydrogens of an alkane have been replaced by chlorine and fluorine. Fluorotrichloromethane (CCl_3F, Freon 11) and dichlorodifluoromethane (CCl_2F_2, Freon 12) are two of the most common CFCs in industrial use.

Because they are inexpensive, stable, nontoxic, nonflammable, and noncorrosive, CFCs are ideal as propellants in aerosol cans, as refrigerants, as solvents, and as fire extinguishing agents. In addition, they are used for blowing bubbles into foamed plastic insulation. Unfortunately, the stability that makes CFCs so useful also causes them to persist in the environment. It is calculated that a CFC molecule released at ground level takes an average of 15 years to get to the upper atmosphere and then can stay there for about a century before breaking up into its elements. Once in the upper atmosphere, the CFC molecule undergoes a complex series of reactions that ultimately result in ozone destruction, with one CFC molecule able to destroy up to 100,000 ozone molecules.

The process begins when ultraviolet (UV) light strikes a CFC molecule and a C—Cl bond breaks, producing a reactive chlorine atom, Cl. The chlorine atom then reacts with ozone to yield oxygen and ClO:

$$CFCl_3 \xrightarrow{\text{UV light}} CFCl_2 + Cl$$
$$Cl + O_3 \longrightarrow ClO + O_2$$

Recognition of the problem led the U.S. government in 1980 to ban the use of CFCs as aerosol propellants, although they are still used as refrigerants. Worldwide efforts to reduce CFC use began with an international agreement reached in 1987, and a total ban on the industrial production and atmospheric release (though not continued use) of CFCs took effect in 1996. Recent atmospheric data indicate that the level of CFCs in the stratosphere has leveled off, but the magnitude of the ozone hole over the Antarctic has not exhibited any sign of decreasing, reaching an all-time high in 2006. Although modest decreases have been observed in the years since 2006, it will be a decade or more before we can confidently say whether or not the ozone layer is recovering, and will be in the middle of the twenty-first century before the ozone hole over the South Pole disappears.

See Additional Problems 14.62 and 14.63 at the end of the chapter.

SUMMARY: REVISITING THE CHAPTER GOALS

1. **What are the distinguishing features of alcohols, phenols, ethers, thiols, and alkyl halides?** An *alcohol* has an —OH group (a *hydroxyl* group) bonded to a saturated, alkane-like carbon atom; a *phenol* has an —OH group bonded directly to an aromatic ring; and an *ether* has an oxygen atom bonded to two organic groups. *Thiols* are sulfur analogs of alcohols, R—SH. *Alkyl halides* contain a halogen atom bonded to an alkyl group, R—X.

    The —OH group is present in many biochemically active molecules. Phenols are notable for their use as disinfectants and antiseptics; ethers are used primarily as solvents. Thiols are found in proteins. Halogenated compounds are rare in human biochemistry but are widely used in industry as solvents and in agriculture as herbicides, fungicides, and insecticides.

2. **How are alcohols, phenols, ethers, thiols, and alkyl halides named?** Alcohols are named using the *-ol* ending, and phenols are named using the *-phenol* ending. Ethers are named by identifying the two organic groups attached to oxygen, followed by the word *ether*. Thiols use the name ending *-thiol*, and alkyl halides are named as halo-substituted alkanes.

3. **What are the general properties of alcohols, phenols, and ethers?** Both alcohols and phenols are like water in their ability to form hydrogen bonds. As the size of the organic part increases, alcohols become less soluble in water. Ethers do not hydrogen-bond and are more alkane-like in their properties.

4. **Why are alcohols and phenols weak acids?** Like water, alcohols and phenols are weak acids that can donate H^+ from their —OH group to a strong base. Alcohols are similar to water in acidity; phenols are more acidic than water and will react with aqueous NaOH.

5. **What are the main chemical reactions of alcohols and thiols?** Alcohols undergo loss of water (*dehydration*) to yield alkenes when treated with a strong acid, and they undergo *oxidation* to yield compounds that contain a *carbonyl group* (C=O). Primary alcohols (RCH_2OH) are oxidized to yield either aldehydes (RCHO) or carboxylic acids (RCO_2H), secondary alcohols (R_2CHOH) are oxidized to yield ketones (R_2C=O); and tertiary alcohols are not oxidized. Thiols react with mild oxidizing agents to yield *disulfides* (RSSR), a reaction of importance in protein chemistry. Disulfides can be reduced back to thiols.

SUMMARY OF REACTIONS

1. **Reactions of alcohols** (Section 14.5)

   (a) Loss of H_2O to yield an alkene (dehydration):

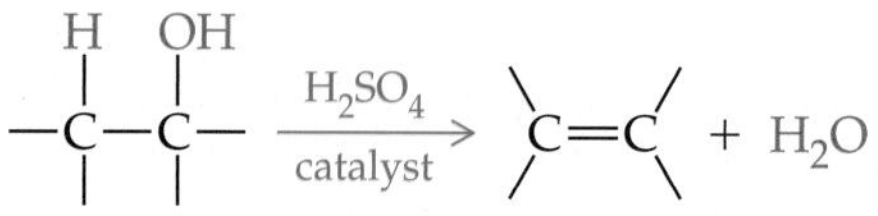

   An alcohol → An alkene + H_2O

   (b) Oxidation to yield a carbonyl compound:

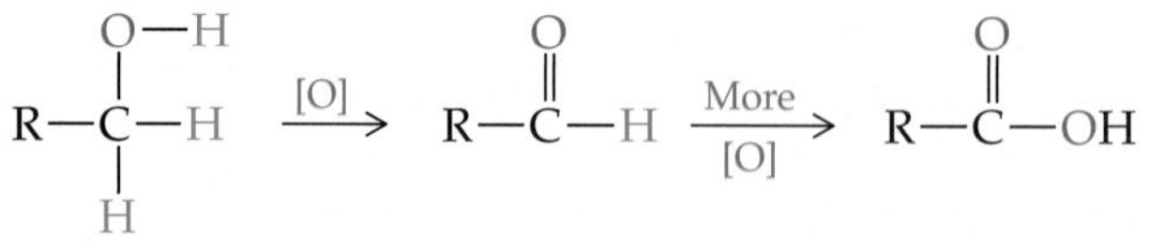

   A primary alcohol → An aldehyde → A carboxylic acid

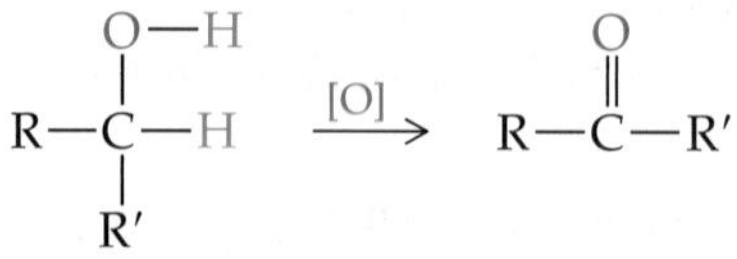

   A secondary alcohol → A ketone

2. **Reactions of thiols** (Section 14.9); oxidation to yield a disulfide:

   $$RSH + HSR \xrightarrow{[O]} RSSR$$

   Two thiol molecules → A disulfide

UNDERSTANDING KEY CONCEPTS

14.19 Give IUPAC names for the following compounds (gray = C, red = O, ivory = H):

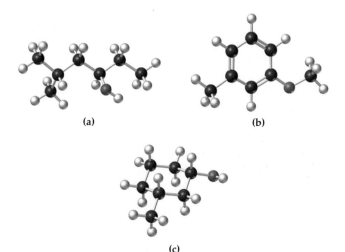

(a)   (b)

(c)

14.20 Predict the product of the following reaction:

$\xrightarrow{\text{H}_2\text{SO}_4}$

14.21 Predict the product of the following reaction:

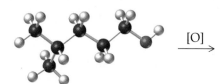

$\xrightarrow{[\text{O}]}$

14.22 Predict the product of the following reaction (gray = C, yellow = S, ivory = H):

$\xrightarrow{[\text{O}]}$

14.23 What alcohols might the following carbonyl compounds have been made from (reddish-brown = Br)?

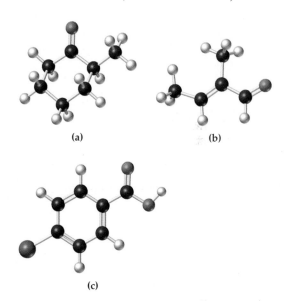

(a)   (b)

(c)

ADDITIONAL PROBLEMS

ALCOHOLS, ETHERS, AND PHENOLS

14.24 How do alcohols, ethers, and phenols differ structurally?

14.25 What is the structural difference between primary, secondary, and tertiary alcohols?

14.26 Why do alcohols have higher boiling points than ethers of the same molecular weight?

14.27 Which is the stronger acid, ethanol or phenol?

14.28 The steroidal compound prednisone is often used to treat poison ivy and poison oak inflammations. Identify the functional groups present in prednisone.

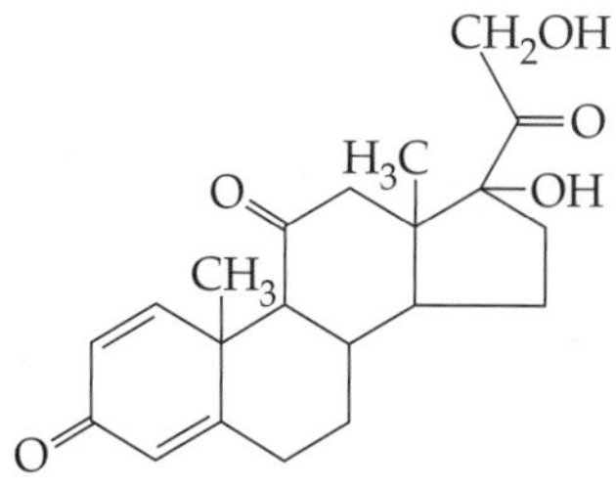

Prednisone

14.29 Vitamin E has the structure shown on page 450. Identify the functional group to which each oxygen belongs.

14.30 Give systematic names for the following alcohols:

(a) 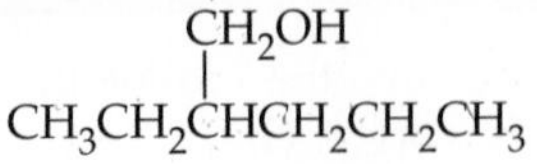
$$CH_3CH_2\overset{\overset{\displaystyle CH_2OH}{|}}{C}HCH_2CH_2CH_3$$

(b) $(CH_3)_2CHCH_2CH_2OH$

(c) 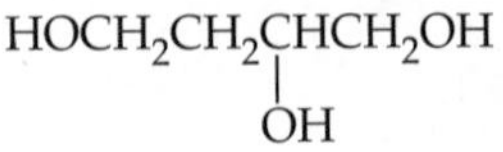
$$HOCH_2CH_2\overset{\overset{\displaystyle}{|}}{C}HCH_2OH$$
$$\qquad\qquad\quad OH$$

(d)

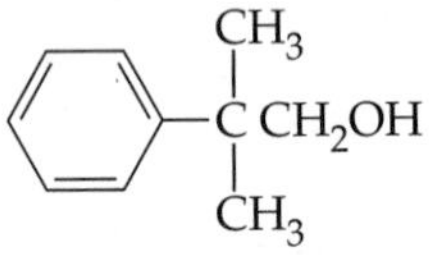

(e)

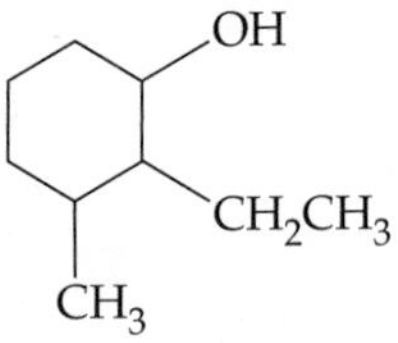

(f) 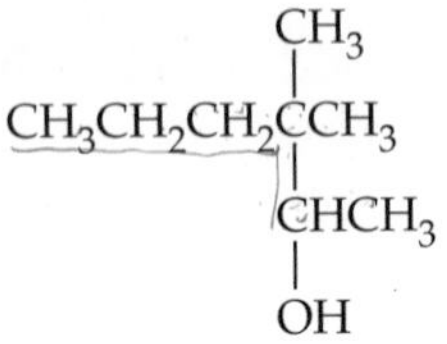
$$CH_3CH_2CH_2\overset{\overset{\displaystyle CH_3}{|}}{\underset{\underset{\displaystyle OH}{|}}{\underset{\displaystyle CHCH_3}{C}}}CH_3$$

14.31 Give systematic names for the following compounds:

(a)

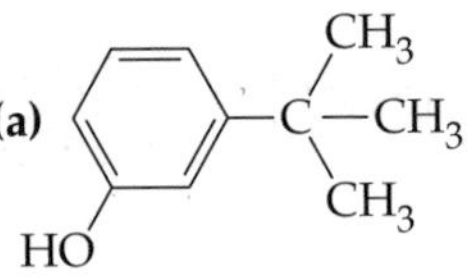

(b) 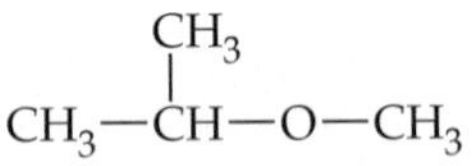
$$CH_3-\overset{\overset{\displaystyle CH_3}{|}}{C}H-O-CH_3$$

(c)

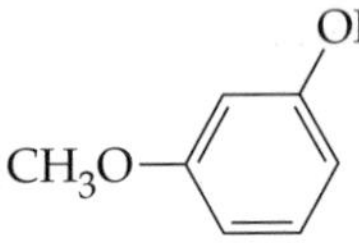

(d)

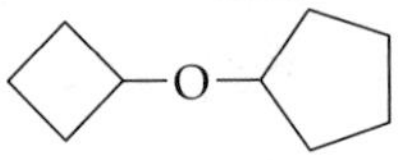

(e)

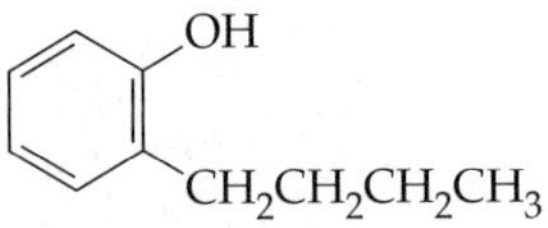

(f) $CH_3CH_2CH_2OCH_2CH_2CH_3$

14.32 Draw structures corresponding to the following names:

(a) 2,4-Dimethyl-3-heptanol
(b) 2,2-Dimethylcyclohexanol
(c) 5-Ethyl,5-methyl-1-heptanol
(d) 3-Ethyl-2-hexanol
(e) 2,3,7-Trimethylcyclooctanol
(f) 3,3-Diethyl-1,6-hexanediol

14.33 Draw structures corresponding to the following names:

(a) Isopropyl phenyl ether
(b) o-Dihydroxybenzene (catechol)
(c) p-Bromophenyl *tert*-butyl ether

(d) *m*-Nitrophenol
(e) 2,4-Diethoxy-3-ethylpentane
(f) 4-Methoxy-3-methyl-1-pentene

14.34 Identify each alcohol named in Problem 14.30 as primary, secondary, or tertiary.

14.35 Locate the alcohol functional groups in prednisone (Problem 14.28), and identify each as primary, secondary, or tertiary.

14.36 Arrange the following 6-carbon compounds in order of their expected boiling points, and explain your ranking:

(a) Hexane
(b) 1-Hexanol
(c) Dipropyl ether

14.37 Glucose is much more soluble in water than 1-hexanol even though both contain 6 carbons. Explain.

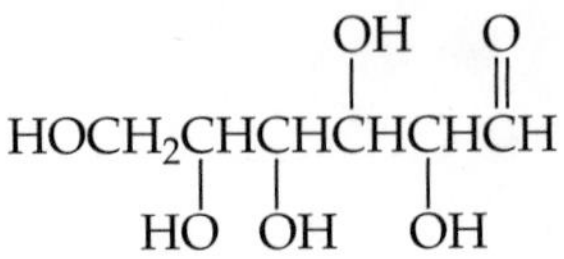

$$HOCH_2\overset{\overset{\displaystyle OH}{|}}{C}H\overset{}{C}H\overset{\overset{\displaystyle OH}{|}}{C}H\overset{}{C}H\overset{\overset{\displaystyle O}{\|}}{C}H \qquad \text{Glucose}$$
$$\qquad\quad HO\quad OH\quad OH$$

REACTIONS OF ALCOHOLS

14.38 What type of product is formed on oxidation of a secondary alcohol?

14.39 What structural feature prevents tertiary alcohols from undergoing oxidation reactions?

14.40 What product(s) can form on oxidation of a primary alcohol?

14.41 What type of product is formed on reaction of an alcohol with Na metal?

14.42 Assume that you have samples of the following two compounds, both with formula C_7H_8O. Both compounds dissolve in ether, but only one of the two dissolves in aqueous NaOH. How could you use this information to distinguish between them?

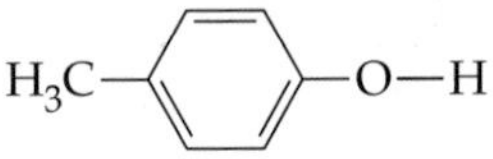

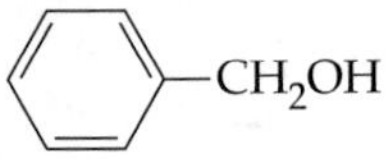

14.43 Assume that you have samples of the following two compounds, both with formula $C_7H_{14}O$. What simple chemical reaction will allow you to distinguish between them? Explain.

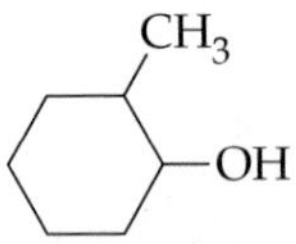

14.44 The following alkenes can be prepared by dehydration of either an appropriate alcohol or an appropriate diol. Show the structure of the alcohol in each case. If the alkene can arise from dehydration of more than one alcohol, show all possibilities.

(a)

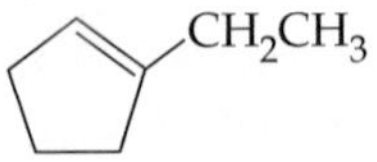

(b) 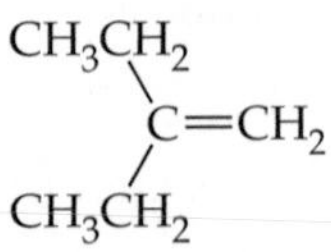
$$\overset{\displaystyle CH_3CH_2}{\underset{\displaystyle CH_3CH_2}{C}}=CH_2$$

(c) 3-Hexene

(d)

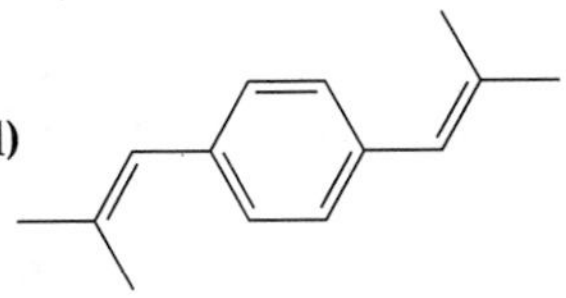

(e) 1,4-Pentadiene

(f)

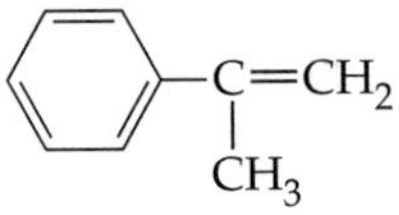

$C=CH_2$
CH_3

14.45 What alkenes might be formed by dehydration of the following alcohols? If more than one product is possible, indicate which you expect to be major.

(a)

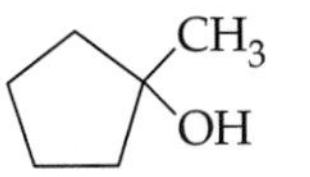

CH_3
OH

(b) 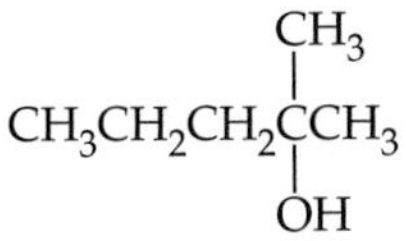 $CH_3CH_2CH_2CCH_3$
CH_3 (above)
OH (below)

(c) 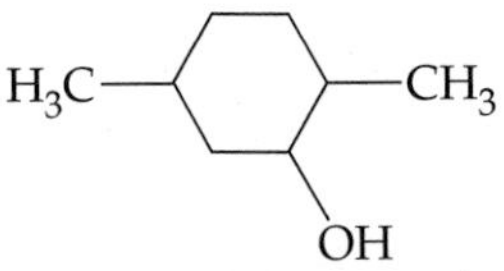 H_3C—⟨ring⟩—CH_3
OH

(d) 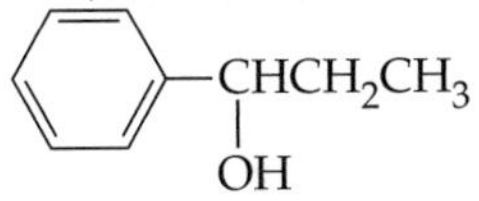 ⟨ring⟩—$CHCH_2CH_3$
OH

(e) 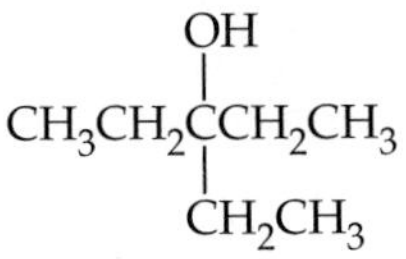 $CH_3CH_2CCH_2CH_3$
OH (above)
CH_2CH_3 (below)

14.46 What carbonyl-containing products would you obtain from oxidation of the following alcohols? If no reaction occurs, write "NR."

(a) 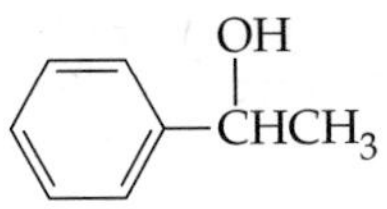 ⟨ring⟩—$CHCH_3$
OH

(b) 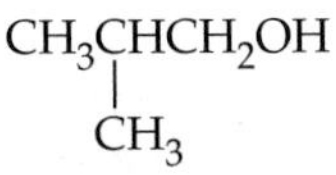 CH_3CHCH_2OH
CH_3

(c) 3-Methyl-3-pentanol

(d) 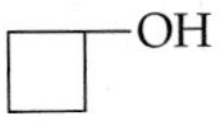 ⟨square⟩—OH

(e) 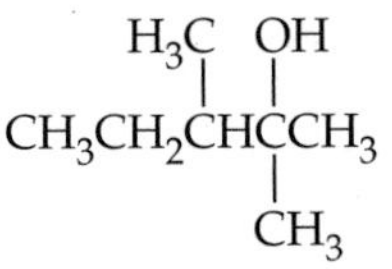 $CH_3CH_2CHCCH_3$
H_3C OH (above)
CH_3 (below)

(f) 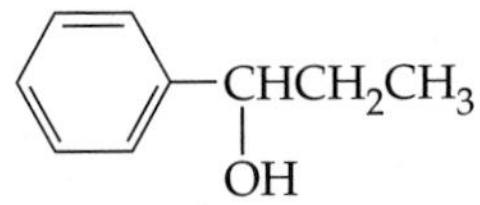 ⟨ring⟩—$CHCH_2CH_3$
OH

14.47 What alcohols would you oxidize to obtain the following carbonyl compounds?

(a)

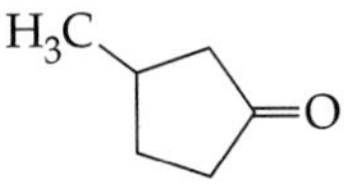

H_3C ⟨ring⟩ $=O$

(b)

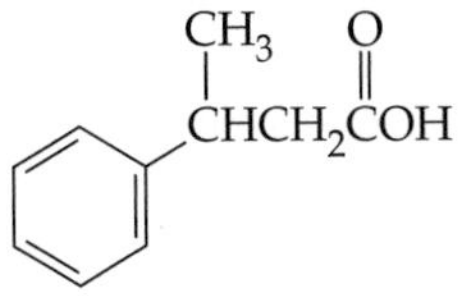

CH_3 O
$CHCH_2COH$

(c) 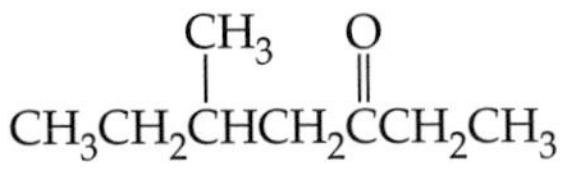 $CH_3CH_2CHCH_2CCH_2CH_3$
CH_3 (above left) O (above right)

THIOLS AND DISULFIDES

14.48 What is the most noticeable characteristic of thiols?

14.49 What is the structural relationship between a thiol and an alcohol?

14.50 The amino acid cysteine forms a disulfide when oxidized. What is the structure of the disulfide?

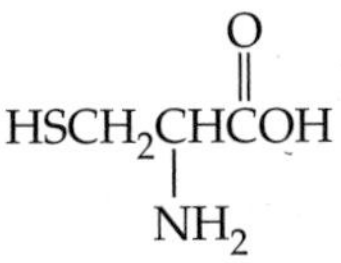

O
$HSCH_2CHCOH$ Cysteine
NH_2

14.51 Oxidation of a dithiol such as 1,5-hexanedithiol forms a cyclic disulfide. Draw the structure of the cyclic disulfide.

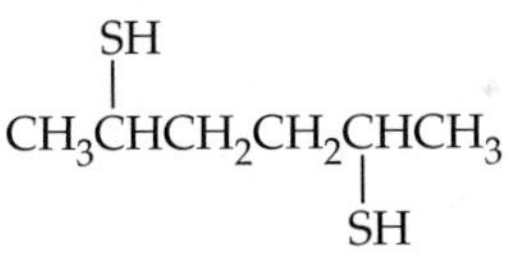

SH
$CH_3CHCH_2CH_2CHCH_3$
SH
2,5-Hexanedithiol

14.52 The boiling point of propanol is 97 °C, much higher than that of either ethanethiol (37 °C) or chloroethane (13 °C), even though all three compounds have similar molecular weights. Explain.

14.53 Propanol is very soluble in water, but ethanethiol and chloroethane are only slightly soluble. Explain.

Applications

14.54 Is ethanol a stimulant or a depressant? [*Ethyl Alcohol as a Drug and a Poison, p. 447*]

14.55 At what blood alcohol concentration does speech begin to be slurred? What is the approximate lethal concentration of ethyl alcohol in the blood? [*Ethyl Alcohol as a Drug and a Poison, p. 447*]

14.56 Cirrhosis of the liver is a common disease of alcoholics. Why is the liver particularly affected by alcohol consumption? [*Ethyl Alcohol as a Drug and a Poison, p. 447*]

14.57 Describe the basis of the Breathalyzer test for alcohol concentration. [*Ethyl Alcohol as a Drug and a Poison, p. 447*]

14.58 What is a free radical? [*Phenols as Antioxidants, p. 450*]

14.59 What vitamin appears to be a phenolic antioxidant? [*Phenols as Antioxidants, p. 450*]

14.60 What substance was used as the first general anesthetic? [*Inhaled Anesthetics, p. 454*]

14.61 How is "minimum alveolar concentration" for an anesthetic defined? [*Inhaled Anesthetics, p. 454*]

14.62 Ozone is considered to be an air pollutant at the earth's surface. Why is it beneficial in the upper atmosphere? [*Chlorofluorocarbons and the Ozone Hole, p. 457*]

14.63 Chlorofluorocarbons (CFCs) are still widely used as coolants in refrigerators and air-conditioners, but states have legislated methods for disposal of CFC-containing appliances. Why? [*Chlorofluorocarbons and the Ozone Hole, p. 457*]

General Questions and Problems

14.64 Name all ether and alcohol isomers with formula $C_4H_{10}O$, and write their structural formulas.

14.65 Thyroxine (Section 14.10) is synthesized in the body by reaction of thyronine with iodine. Write the reaction, and tell what kind of process is occurring.

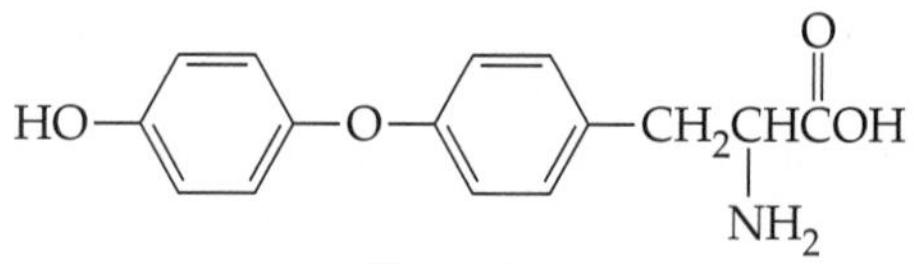

Thyronine

14.66 1-Propanol is freely soluble in water, 1-butanol is marginally soluble, and 1-hexanol is essentially insoluble. Explain.

14.67 Phenols undergo the same kind of substitution reactions that other aromatic compounds do (Section 13.11). Formulate the reaction of *p*-methylphenol with Br_2 to give a mixture of two substitution products.

14.68 What is the difference between an antiseptic and a disinfectant?

14.69 Write the formulas and IUPAC names for the following common alcohols:

(a) Rubbing alcohol
(b) Wood alcohol
(c) Grain alcohol
(d) Diol used as antifreeze

14.70 Name the following compounds:

(a) 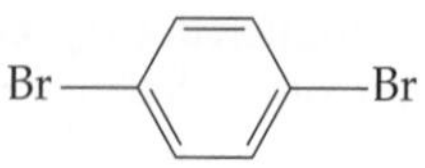 Br—⟨benzene ring⟩—Br

(b) 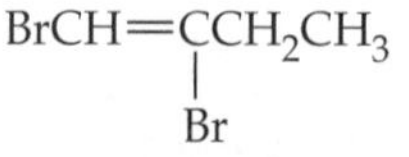 $BrCH=CCH_2CH_3$ with Br substituent

(c) 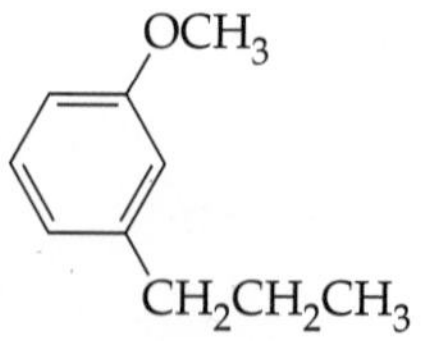 ⟨benzene ring with OCH_3 and $CH_2CH_2CH_3$⟩

(d) 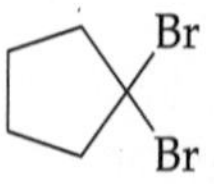 ⟨cyclopentane with two Br⟩

(e) 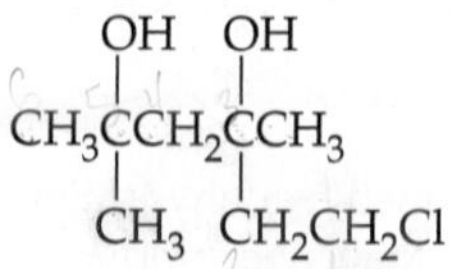 $CH_3CCH_2CCH_3$ with OH, OH, CH_3, CH_2CH_2Cl

(f) 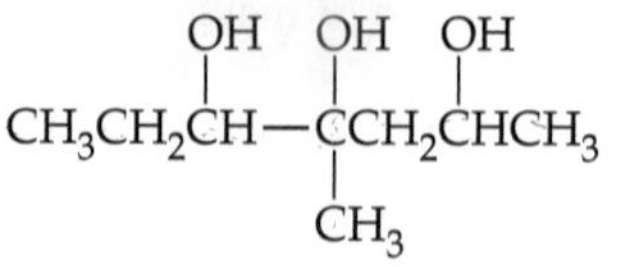 $CH_3CH_2CH—CCH_2CHCH_3$ with OH, OH, OH, CH_3

(g) 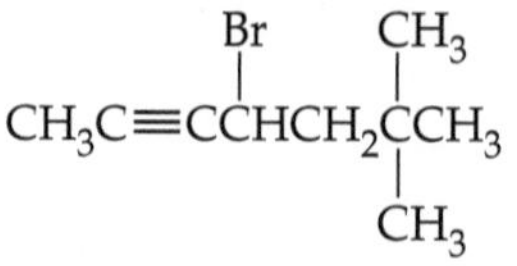 $CH_3C≡CCHCH_2CCH_3$ with Br, CH_3, CH_3

(h) 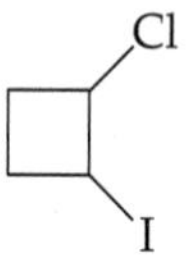 ⟨cyclobutane with Cl and I⟩

14.71 Complete the following reactions:

(a) 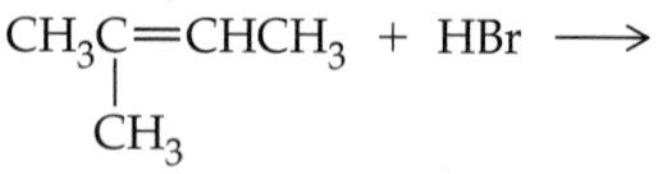 $CH_3C=CHCH_3 + HBr \longrightarrow$ with CH_3 substituent

(b) 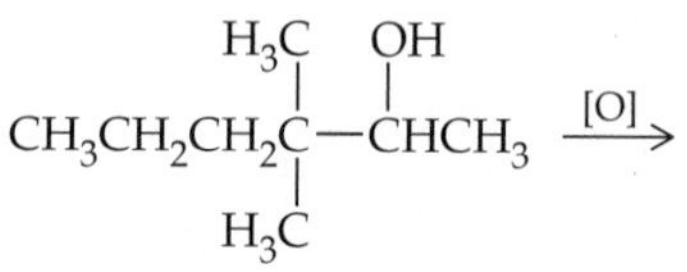 $CH_3CH_2CH_2C—CHCH_3 \xrightarrow{[O]}$ with H_3C, OH, H_3C

(c) 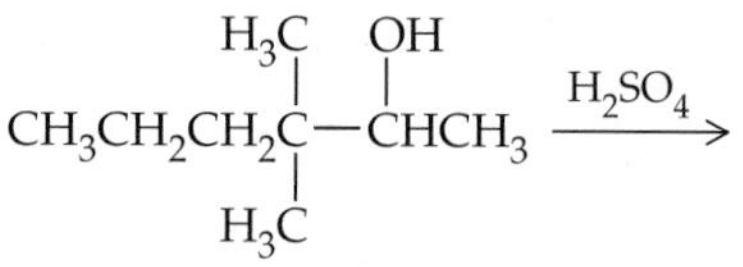 $CH_3CH_2CH_2C—CHCH_3 \xrightarrow{H_2SO_4}$ with H_3C, OH, H_3C

(d) 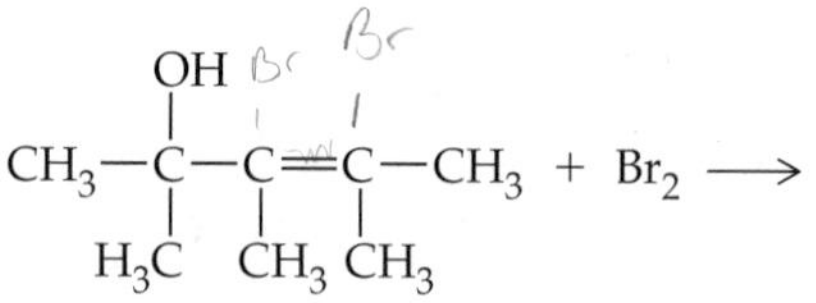 $CH_3—C—C=C—CH_3 + Br_2 \longrightarrow$ with OH, Br, H_3C, CH_3, CH_3

(e) $2(CH_3)_3CSH \xrightarrow{[O]}$

(f) 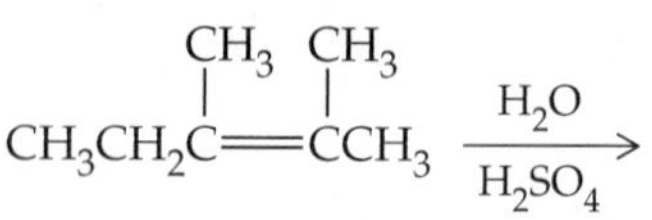 $CH_3CH_2C=CCH_3 \xrightarrow[H_2SO_4]{H_2O}$ with CH_3, CH_3

(g) 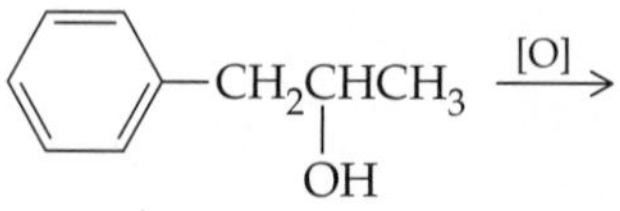 ⟨benzene ring⟩—$CH_2CHCH_3 \xrightarrow{[O]}$ with OH

14.72 The odor of roses is due to geraniol.

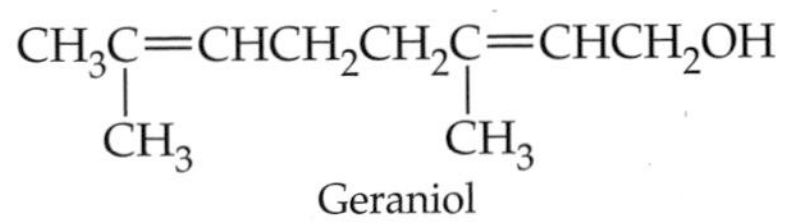

$$CH_3C{=}CHCH_2CH_2C{=}CHCH_2OH$$
$$\quad\,|\qquad\qquad\quad\;|$$
$$\quad CH_3\qquad\qquad\;\,CH_3$$

Geraniol

(a) What is the systematic name of geraniol?

(b) When geraniol is oxidized, the aldehyde citral, one of the compounds responsible for lemon scent, is formed. Write the structure of citral.

14.73 Concentrated ethanol solutions can be used to kill microorganisms. At low concentrations, however, such as in some wines, the microorganisms can survive and cause oxidation of the alcohol. What is the structure of the acid formed?

14.74 "Flaming" desserts, such as cherries jubilee, use the ethanol in brandy or other distilled spirits as the flame carrier. Write the equation for the combustion of ethanol.

14.75 We said in Chapter 13 that H_2SO_4 catalyzes the addition of water to alkenes to form alcohols. In this chapter, however, we saw that H_2SO_4 is also used to dehydrate alcohols to make alkenes. Why do you think that sulfuric acid can serve two purposes—aiding in both hydration and dehydration?

CHAPTER 15

Amines

Concepts to Review

Lewis structures
(Section 5.6)

Acids and bases
(Sections 6.10, 10.1, 10.2)

Organic molecules
(Sections 12.1, 12.2)

▲ The blooming of the titan arum plant, also known as the "*corpse flower*," is an extremely rare event, occurring to date only 12 times in the United States. The nickname derives from the plant's strong odor, which resembles the odor given off by rotting flesh. Tho odor is caused by the many amine compounds that exude from the blossoming plant.

CONTENTS

CHAPTER GOALS

In this chapter, we will consider the following questions:

1. What are the different types of amines?

THE GOAL: Be able to recognize primary, secondary, tertiary, and heterocyclic amines, as well as quaternary ammonium ions.

2. How are amines named?

THE GOAL: Be able to name simple amines and write their structures, given the names.

3. What are the general properties of amines?

THE GOAL: Be able to describe amine properties such as hydrogen bonding, solubility, boiling point, and basicity.

4. How do amines react with water and acids?

THE GOAL: Be able to predict the products of the acid–base reactions of amines and ammonium ions.

5. What are alkaloids?

THE GOAL: Be able to describe the sources of alkaloids, name some examples, and tell how their properties are typical of amines.

We have up to now looked at hydrocarbons and at organic molecules with single bonds joining carbon to oxygen, sulfur, or a halogen. We now turn our attention to amines, which have single bonds between carbon and nitrogen. The amines are present in so many kinds of essential biomolecules and important pharmaceutical agents that they are worthy of a chapter to themselves.

From a biochemical standpoint, many of the molecules that carry chemical messages (such as the neurotransmitters, Chapter 20) are relatively simple amines with extraordinary powers. Histamine, the compound that initiates hay fever and other allergic reactions, is an amine; you have experienced its power firsthand if you have ever had an insect bite. In addition, many of the drugs that have been developed to mimic or to control the activity of histamine—the antihistamines present in cold and allergy medications—are amines. The amino group (—NH$_2$) is important in the formation and stability of the proteins (⊂◯⊃, see Sections 18.7–18.8); heterocyclic amines play a crucial part in the function of DNA and RNA (Chapter 26). These are but a few examples of the roles played by amines.

15.1 Amines

Amines contain one or more organic groups bonded to nitrogen in compounds with the general formulas RNH_2, R_2NH, and R_3N. In the same way that alcohols and ethers can be thought of as organic derivatives of water, amines are organic derivatives of ammonia. In general, they are classified as *primary* (1°), *secondary* (2°), or *tertiary* (3°), according to how many organic groups are individually bound *directly* to the nitrogen atom. The organic groups (represented below by colored rectangles) may be large or small, they may be the same or different, or they may be connected to one another through a ring (Section 15.3). The ball-and-stick models shown here illustrate these amine classes. (All the R groups here are ethyl groups for this purpose.)

Amine A compound that has one or more organic groups bonded to nitrogen: primary, RNH_2; secondary, R_2NH; or tertiary, R_3N.

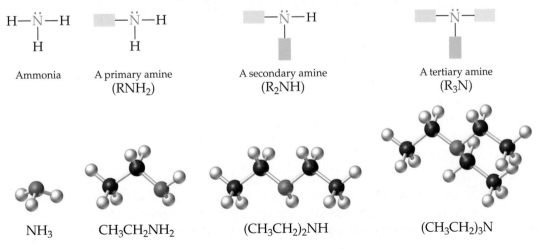

| Ammonia | A primary amine (RNH_2) | A secondary amine (R_2NH) | A tertiary amine (R_3N) |

NH_3 $CH_3CH_2NH_2$ $(CH_3CH_2)_2NH$ $(CH_3CH_2)_3N$

Quaternary ammonium ion A positive ion with four organic groups bonded to the nitrogen atom.

Note that each amine nitrogen atom has a lone pair of electrons. The lone pair, although not always shown, is always there and is responsible in large part for the chemistry of amines. When a fourth group bonds to the nitrogen through this lone pair, the product is a **quaternary ammonium ion**, which has a permanent positive charge and forms ionic compounds with anions [for example, $(CH_3CH_2)_4N^+ Cl^-$]:

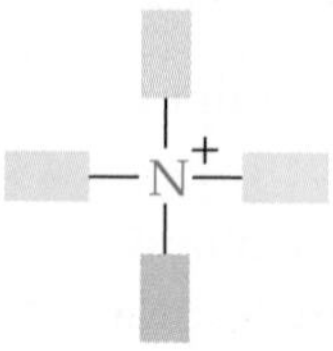

$$\begin{array}{c} \big| \\ -N^+ - \\ \big| \end{array}$$

A quaternary
ammonium ion
(R_4N^+)

The groups bonded to the amine nitrogen atom may be alkyl or aryl groups and may or may not contain other functional groups. For example:

CH_3NH_2

Methylamine
(a primary alkyl amine)

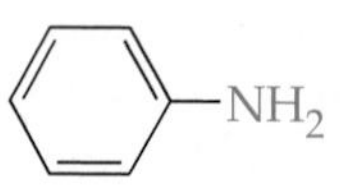

—NH_2

Aniline
(a primary
aromatic amine)

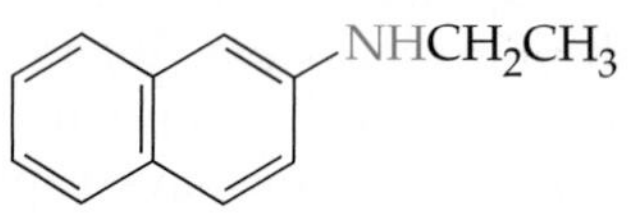

—$NHCH_2CH_3$

N-Ethylnaphthylamine
(a secondary aromatic amine)

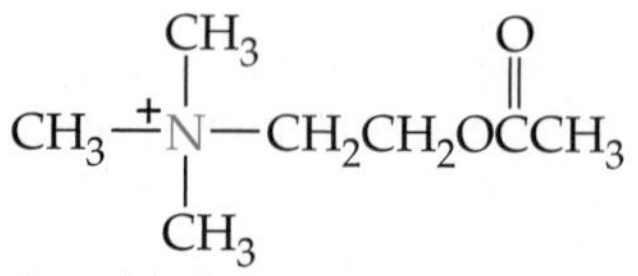

$$CH_3-\overset{+}{\underset{\underset{CH_3}{|}}{\overset{\overset{CH_3}{|}}{N}}}-CH_2CH_2O\overset{O}{\overset{\|}{C}}CH_3$$

Acetylcholine, a neurotransmitter
(a quaternary ammonium ion)

Primary alkyl amines (RNH_2) are named by identifying the alkyl group attached to nitrogen and adding the suffix *-amine* to the alkyl group name.

Some examples of naming primary amines

$CH_3CH_2-NH_2$

Ethylamine

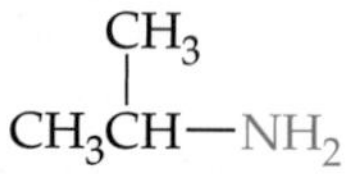

$$CH_3\overset{\overset{CH_3}{|}}{CH}-NH_2$$

Isopropylamine

—NH_2

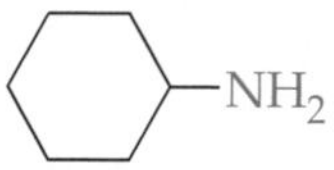

Cyclohexylamine

Simple, nonheterocyclic secondary (R_2NH) and tertiary (R_3N) amines (those possessing two or three identical groups on the nitrogen, respectively) are named by adding the appropriate prefix, *di-* or *tri-*, to the alkyl group name along with the suffix *-amine*.

Some examples of naming simple 2° and 3° amines

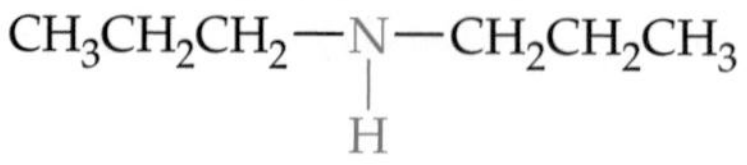

$$CH_3CH_2CH_2-\overset{\underset{\underset{H}{|}}{}}{N}-CH_2CH_2CH_3$$

Dipropylamine

$$CH_3CH_2-\overset{\underset{\underset{CH_2CH_3}{|}}{}}{N}-CH_2CH_3$$

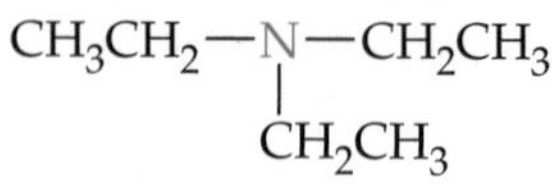

Triethylamine

When the R groups in secondary or tertiary amines are different, the compounds are named as *N*-substituted derivatives of a primary amine. The parent compound chosen as the primary amine is the one that contains the largest of the R groups; all other groups are considered to be *N*-substituents (*N* because they are attached directly to nitrogen). The following compounds, for example, are named as propylamines because the propyl group in each is the largest alkyl group:

Some examples of naming more complex 2° and 3° amines

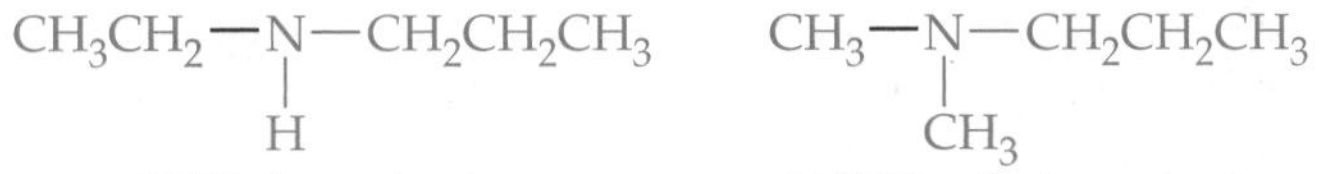

$$CH_3CH_2\text{—}\underset{\underset{H}{|}}{N}\text{—}CH_2CH_2CH_3 \qquad CH_3\text{—}\underset{\underset{CH_3}{|}}{N}\text{—}CH_2CH_2CH_3$$

N-Ethylpropylamine   *N,N*-Dimethylpropylamine

Heteroyclic amines (Section 15.3) are an important family of amines in which the nitrogen is part of the ring structure; the nomenclature of these compounds is too complicated to discuss here and will be briefly addressed as needed.

The —NH_2 functional group is an **amino group**, and when this group is a substituent, *amino-* is used as a prefix in the name of the compound. Aromatic amines are an exception to this rule and are primarily known by their historical, or common, names. The simplest aromatic amine is known by its common name aniline, and derivatives of it are named as anilines:

Amino group The —NH_2 functional group.

Aniline   —$NHCH_3$ $H_2NCH_2CH_2COOH$

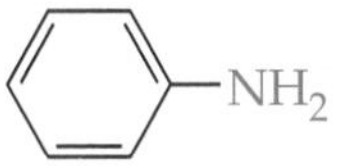

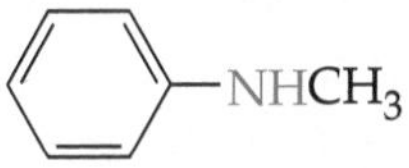

Aniline   *N*-Methylaniline   3-Aminopropanoic acid

The amino acids that make up all proteins have the general structure shown here—a primary amino group bonded to the carbon atom next to a *carboxylic acid group* (—COOH), with a side chain on the same carbon:

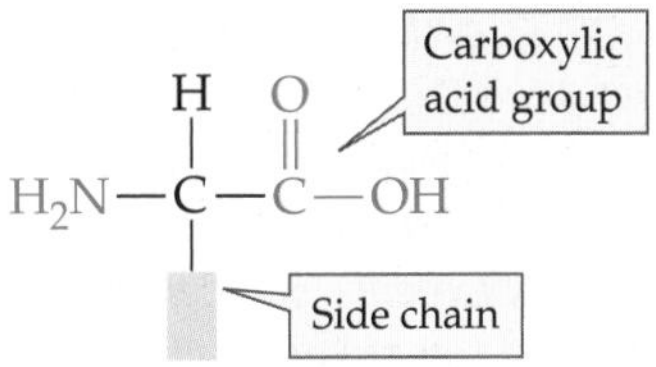

$$H_2N\text{—}\underset{\underset{\boxed{}}{|}}{\overset{\overset{H}{|}}{C}}\text{—}\overset{\overset{O}{\|}}{C}\text{—}OH$$

Carboxylic acid group

Side chain

An amino acid

⫷ Looking Ahead

All amino acids contain both the amino functional group, —NH_2, and the carboxylic acid functional group, —COOH (in addition to whatever functional groups are part of the side chain). The chemistry of the carboxylic acid group is discussed in Chapter 17. The amino acids and their combination to form proteins are covered in Chapter 18. ⫷

WORKED EXAMPLE   **15.1** Drawing and Classifying Amines from Their Names

Write the structure of *N,N*-diethylbutylamine and identify it as a primary, secondary, or tertiary amine.

ANALYSIS Look for terms within the name that provide clues about the parent compound and its substituents. For example, the word "butyl" immediately preceding the *-amine* suffix indicates that butylamine, the 4-carbon alkyl amine, is the parent compound. The *N,N* indicates that two other groups are bonded to the amino nitrogen, and the *diethyl* indicates they are ethyl groups.

SOLUTION
The structure shows that three alkyl groups are bonded to the N atom, so this must be a tertiary amine.

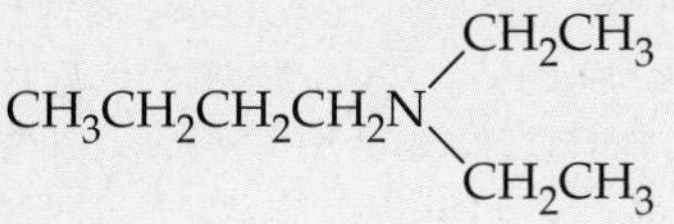

$$CH_3CH_2CH_2CH_2N\underset{\diagdown CH_2CH_3}{\overset{\diagup CH_2CH_3}{}}$$

APPLICATION ▶ Chemical Information

Look at the ingredient label on any package in your house or even that soda can in front of you. What if you wanted to know more about the chemicals contained in these ingredients. Where would you look for information?

Let us take, for example, caffeine. One place you might look for information on caffeine is a chemical handbook such as *Lange's Handbook of Chemistry*. Basic handbooks give the structure for the chemical and an entry like the one shown here. A typical entry includes some physical properties and possibly a reference to a more comprehensive handbook, such as Beilstein's *Handbook of Organic Chemistry*. Another standard chemical handbook is the *CRC Handbook of Chemistry and Physics* (CRC Press, Boca Raton, FL).

What if you want more information, perhaps something about a chemical's biological properties? Another generally available and reliable source of chemical information is *The Merck Index: An Encyclopedia of Chemicals and Drugs*, first published in 1889 as a list of products of the Merck pharmaceutical company. The index has since grown far beyond its original intended use and has evolved to become a standard reference work containing information on the preparation, properties (both chemical and biological), and uses of over 10,000 chemical compounds. It has a strong medical emphasis and, where appropriate, lists toxicity data, therapeutic uses in both human and veterinary medicine, and the physiological effects and cautions associated with hazardous chemicals. *The Merck Index* is now available on CD-ROM or online via the Internet.

The entry for caffeine in *The Merck Index* is reproduced on the facing page; it begins with the systematic name for caffeine followed by a list of alternative chemical names and a capitalized entry that is the name of a medication containing caffeine. One advantage of using *The Merck Index* is that all of the alternative names of every substance in the book

appear in the index, including registered drug names, so no matter what name you come across, you can discover exactly to what substance it refers.

Next, information about the molecular formula, structure, sources, patents, and scientific journal article references about caffeine (including an account of its first synthesis in 1895) are provided. After this, a paragraph describes the physical properties of caffeine and some of its important derivatives, with the final lines listing the therapeutic uses of caffeine, which is a central nervous system (CNS) stimulant.

If you know only the formula for an unknown molecule that happens to be caffeine ($C_8H_{10}N_4O_2$), you can also search in the molecular formula index of *The Merck Index*. In a current edition, you would find two molecules with that formula—caffeine and a bronchodilator with a structure similar to caffeine.

Having learned from *The Merck Index* that caffeine is a CNS stimulant, you can next turn to *The Physicians Desk Reference* (the *PDR*) to learn about the use of caffeine in drugs. Although intended primarily for physicians, the *PDR* is readily available in bookstores and libraries. It is also available on CD-ROM (but very expensive!). For each drug, it contains the full product labeling information provided by the manufacturers. By using the Generic and Chemical Name Index in both the *PDR* and the *PDR For Nonprescription Drugs and Diet Supplements*, you can also discover that caffeine is present in 31 other medications, ranging from nonprescription analgesics like Anacin and stimulants like Vivarin to a potent prescription drug for migraine headaches that combines caffeine with belladonna and ergotamine (both alkaloids, Section 15.6) and sodium pentobarbital. You can go directly to the entry for Vivarin via the product name index and learn more about the effects and cautions accompanying the intended use of this medication.

NAME	FORMULA	FORMULA WEIGHT	BEILSTEIN REFERENCE	DENSITY	REFRACTIVE INDEX	MELTING POINT	BOILING POINT	FLASH POINT	SOLUBILITY IN 100 PARTS SOLVENT
Caffeine		194.19	26,461	1.23^{18}_{4}		238	subl 178		2.1 aq; 1.5 alc; 18 chl; 0.19 eth; 1 bz

Caffeine information from Lange's Handbook of Chemistry. *Reproduced from* Dean/Lange's Handbook of Chemistry, *15th Ed. (1999), Entry C1, pp. 7–194, with permission of McGraw-Hill, Inc.*

You can, of course, also search the Internet for information about caffeine. A 2008 search for "caffeine" with "chemistry" together turned up about 380,000 hits using a well-known search engine, about 132,000 MORE than were found doing the same search in 2005! Since the Internet is not "peer-reviewed" (a sophisticated way of saying not checked for accuracy), starting with the reference books described here still remains the most efficient and trustworthy way to find the molecular structure and other fundamentals.

1635.   Caffeine. *3,7-Dihydro-1,3,7-trimethyl-1 H-purine-2,6-dione*; 1,3,7-trimethylxanthine; 1,3,7-trimethyl-2,6-dioxopurine; coffeine; thein; guaranine; methyltheobromine; No-Doz. $C_8H_{10}N_4O_2$; mol wt 194.19. C 49.48%, H 5.19%, N 28.85%, O 16.48%. Occurs in tea, coffee, maté leaves; also in guarana paste and cola nuts: Shuman, **U.S.** pat. **2,508,545** (1950 to General Foods). Obtained as a by-product from the manuf of caffeine-free coffee: Barch, **U.S.** pat. **2,817,588** (1957 to Standard Brands); Nutting, **U.S.** pat. **2,802,739** (1957 to Hill Bros. Coffee); Adler, Earle, U.S. pat. **2,933,395** (1960 to General Foods). Crystal structure: Sutor, *Acta Cryst.* **11**, 453 (1958). Synthesis: Fischer, Arch, *Ber.* **28**, 2473, 3135 (1895); Review of dietary sources: D. M. Graham, Nutr. Rev. 36, 97–102 (1978); of clinical pharmacology: N. L. Benowitz, Annu. Rev. Med. 41, 277–288 (1990); of CNS effects: A. Nehlig *et al.,* Brain Res. Rev. 17, 139–170 (1992); of therapeutic uses: J. Sawynok, Drugs 49, 37–50 (1996) . . .

Hexagonal prisms by sublimation, mp 238°. Odorless with bitter taste. Sublimes 178°. Fast sublimation is obtained at 160–165° under 1 mm press. at 5 mm distance. d_4^{18} 1.23. pH of 1% soln 6.9. Absorption spectrum: Hartley, *J. Chem. Soc.* **87**, 1802 (1905). One gram dissolves in 46 ml water, 5.5 ml water at 80°, 1.5 ml boiling water, 66 ml alcohol, 22 ml alcohol at 60°, 50 ml acetone, 5.5 ml chloroform, 530 ml ether, 100 ml benzene, 22 ml boiling benzene. Freely sol in pyrrole; in tetrahydrofuran contg about 4% water; also sol in ethyl acetate; slightly in petr ether. Soly in water is increased by alkali benzoates, cinnamates, citrates or salicylates. LD_{50} orally in mice, hamsters, rats, rabbits (mg/kg): 127, 230, 355, 246 (males); 137, 249, 247, 224 (females) (Palm).

Monohydrate, felted needles, contg 8.5% H_2O. Efflorescent in air; complete dehydration takes place at 80°. . . .

THERAP CAT: CNS stimulant; respiratory stimulant.

THERAP CAT (VET): Has been used as a cardiac and respiratory stimulant and as a diuretic.

Caffeine information from The Merck Index. *Reproduced from* The Merck Index, *15th Ed. (2005), S. Budavari, M.J. O'Neil, A. Smith, P.E. Heckelman, Eds., by permission of the copyright owner, Merck & Co., Inc., Rahway, N.J., U.S.A., © Merck & Co., Inc., 1989.*

Vivarin® Tablets & Caplets

Alertness Aid with Caffeine, Maximum Strength

Each tablet contains 200 mg. Caffeine, equal to about two cups of coffee. Take Vivarin for a safe, fast pick up anytime you feel drowsy and need to be alert. The caffeine in Vivarin is less irritating to your stomach than coffee, according to a government appointed panel of experts.

ACTIVE INGREDIENTS:
Caffeine, 200 mg

INACTIVE INGREDIENTS:
Tablet: colloidal Silicon Dioxide, D&C Yellow #10 Al. Lake, Dextrose, FD&C Yellow #6 Al. Lake, Magnesium Stearate, Microcrystalline Cellulose, Starch. . . .

DIRECTIONS:
Adults and children 12 years and over: One tablet (200 mg) not more often than every 3 to 4 hours.

WARNINGS:
The recommended dose of this product contains about as much caffeine as two cups of coffee. Limit the use of caffeine-containing medications, foods, or beverages while taking this product because too much caffeine may cause nervousness, irritability, sleeplessness, and occasionally, rapid heart beat. Not intended as a substitute for sleep. If fatigue or drowsiness persists or continues to recur, consult a doctor. Do not give to children under 12 years of age. As with any drug, if you are pregnant or nursing a baby, seek the advice of a health professional before using this product. **KEEP THIS AND ALL MEDICINES OUT OF THE REACH OF CHILDREN. IN CASE OF ACCIDENTAL OVERDOSE, SEEK PROFESSIONAL ASSISTANCE OR CONTACT A POISON CONTROL CENTER IMMEDIATELY.**

HOW SUPPLIED:
Tablets: consumer packages of 16, 40 and 80 tablets. . . .

Entry on Vivarin®+ from The Physicians Desk Reference For Nonprescription Drugs and Diet Supplements (26th Edition, 2005). *Reproduced from* The Physicians Desk Reference *with permission of Medical Economics Company, Inc., Oradell, N.J.*

See Additional Problem 15.43 at the end of the chapter.

WORKED EXAMPLE 15.2 Naming and Classifying an Amine from Its Structure

Name the following compound. Is it a primary, secondary, or tertiary amine?

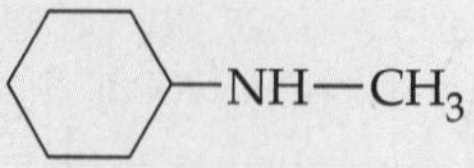

ANALYSIS Determine how many organic groups are attached to the nitrogen. We can see that two carbon groups are bonded to the nitrogen. Since the cyclohexyl group is the largest alkyl group bonded to N, the compound is named as a cyclohexylamine. One methyl group is bonded to the nitrogen; we indicate this with the prefix N.

SOLUTION
The name is N-methylcyclohexylamine. Because the compound has two groups bonded to N, it is a secondary amine.

WORKED EXAMPLE 15.3 Classifying a Cyclic Amine from Its Structure

The following heterocyclic amine is named octahydroindolizine. Is it a primary, secondary, or tertiary amine?

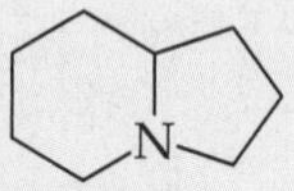

ANALYSIS Start by looking at the nitrogen; we can see that it is attached to three different carbons (as indicated by red, blue and black bond lines). Even when the nitrogen is part of a ring, an amine will be classified by the number of organic groups that are bonded to it.

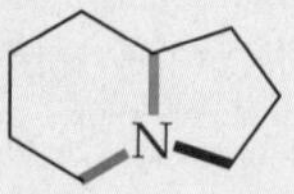

SOLUTION
In this molecule, three individual carbon groups are bound to N; it therefore is a tertiary amine.

PROBLEM 15.1

Identify the following compounds as primary, secondary, or tertiary amines.

(a) $CH_3(CH_2)_4CH_2NH_2$

(b) $CH_3CH_2CH_2NHCH(CH_3)_2$

(c)

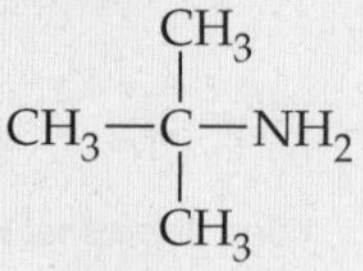

(d)

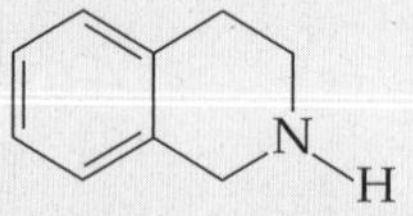

(e)

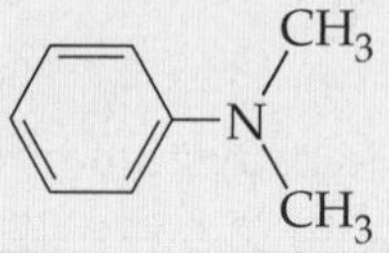

PROBLEM 15.2

What are the names of these amines?

(a) $(CH_3CH_2CH_2CH_2)_4N^+OH^-$

(b) H—N—CH₃
      |
      CH₃

(c) 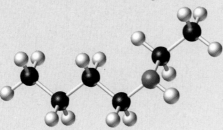 —NHCH₂CH₂CH₂CH₂CH₃

PROBLEM 15.3

Draw structures corresponding to the following names:

(a) Hexylamine   **(b)** *N*-Methylbutylamine

(c) *N*-Methylaniline   **(d)** 4-Amino-2-butanol

⚫ KEY CONCEPT PROBLEM 15.4

Draw the structure of tetramethylammonium ion, count its valence electrons, and explain why the ion has a positive charge.

⚫ KEY CONCEPT PROBLEM 15.5

Draw the condensed formula of the following molecule and name it.

15.2 Properties of Amines

The lone electron pair on the nitrogen in amines, like the lone electron pair in ammonia, causes amines to act as either weak Brønsted–Lowry bases or **Lewis bases** by forming a bond with an H^+ ion from an acid or water (Section 15.4).

> The lone pair on the nitrogen makes it a Lewis base.

Lewis base A compound containing an unshared pair of electrons (an amine, for example).

$$H-\overset{\overset{\displaystyle H}{|}}{\underset{\underset{\displaystyle H}{|}}{N}}:(aq) + H_2O\,(l) \;\rightleftharpoons\; H-\overset{\overset{\displaystyle H}{|}}{\underset{\underset{\displaystyle H}{|}}{N^+}}-H\,(aq) + OH^-(aq)$$

$$CH_3-\overset{\overset{\displaystyle H}{|}}{\underset{\underset{\displaystyle H}{|}}{N}}:(aq) + HCl\,(aq) \;\rightleftharpoons\; CH_3-\overset{\overset{\displaystyle H}{|}}{\underset{\underset{\displaystyle H}{|}}{N^+}}-H(aq) + Cl^-(aq)$$

In primary and secondary amines, hydrogen bonds can form between the lone pair on the very electronegative nitrogen atom and the slightly positive hydrogen atom on another primary or secondary amine (Figure 15.1). All amines (primary,

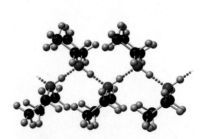

▲ **FIGURE 15.1 Hydrogen bonding of a secondary amine.** Hydrogen bonding between $(CH_3CH_2)_2NH$ molecules is shown by red dots.

secondary, and tertiary) can form hydrogen bonds with water.

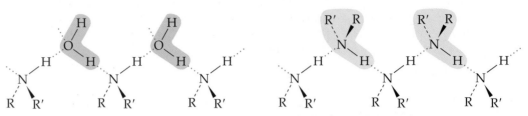

2° amine–H₂O hydrogen bonds   2° amine–2° amine hydrogen bonds

Because of the ability to engage in hydrogen bonding, primary and secondary amines have higher boiling points than alkanes of similar size (remember that, in the absence of hydrogen bonding, boiling points of molecules increase with increasing molecular mass). (⊂⊃, p. 385) Amines are, in general, lower boiling than alcohols of similar size. In fact, mono-, di-, and trimethylamine as well as ethylamine are all gases at room temperature. The boiling points of some simple amines can be found in Table 15.1.

TABLE 15.1 Boiling Points of Some Simple Amines

STRUCTURE	NAME	BOILING POINT (°C)	STRUCTURE	NAME	BOILING POINT (°C)
NH_3	Ammonia	−33.3	**Secondary amines**		
Primary amines			$(CH_3)_2NH$	Dimethylamine	7.4
			$(CH_3CH_2)_2NH$	Diethylamine	56.3
CH_3NH_2	Methylamine	−6.3	$((CH_3)_2CH)_2NH$	Diisopropylamine	84
$CH_3CH_2NH_2$	Ethylamine	16.6			
$(CH_3)_3CNH_2$	*tert*-Butylamine	44.4	**Tertiary amines**		
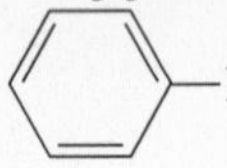			$(CH_3)_3N$	Trimethylamine	3
⬡—NH_2	Aniline	184.1	$(CH_3CH_2)_3N$	Triethylamine	89.3
			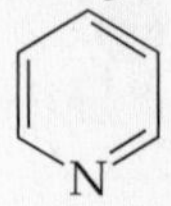 (pyridine)	Pyridine	115

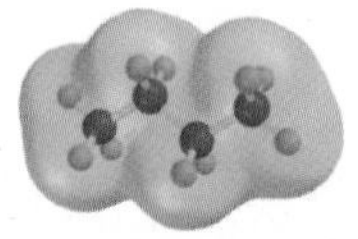

$CH_3CH_2CH_2CH_3$

Butane, bp 0 °C
MW = 58

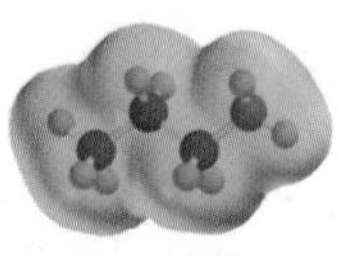

$CH_3CH_2CH_2NH_2$

Propylamine, bp 48 °C
MW = 59

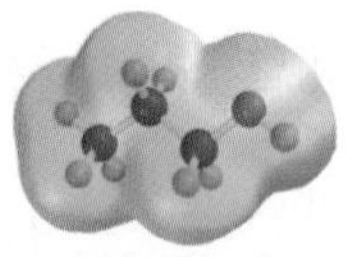

$CH_3CH_2CH_2OH$

Propanol, bp 97 °C
MW = 60

Tertiary amine molecules have no hydrogen atoms attached to nitrogen and therefore cannot hydrogen-bond with each other. As a result they are much lower boiling than alcohols or primary or secondary amines of similar molecular weight. Compare the boiling point of trimethylamine (3 °C) with those of propylamine (48 °C) and the other compounds shown in Table 15.1.

All amines, however, can hydrogen-bond to water molecules through the lone electron pair on their nitrogen atoms. As a result, amines with up to about 6 carbon atoms have appreciable solubility in water.

Many volatile amines have strong odors. Some smell like ammonia and others like stale fish or decaying meat. The protein in flesh contains amine groups, and the smaller, volatile amines produced during decay and protein breakdown are responsible for the odor of rotten meat. One such amine, 1,5-diaminopentane, is commonly known as cadaverine.

▲ Cadaver-sniffing dogs are used to detect the strong odor of the amines produced from decaying flesh.

Another significant property of amines is that many cause physiological responses. The simpler amines are irritating to the skin, eyes, and mucous membranes and are toxic by ingestion. Some of the more complex amines from plants (alkaloids, Section 15.6) can be very poisonous. All living things nevertheless contain a wide variety of amines, and many useful drugs are amines.

Properties of Amines

- Primary and secondary amines can hydrogen-bond and thus are higher boiling than alkanes but lower boiling than alcohols.
- Tertiary amines are lower boiling than secondary or primary amines because hydrogen bonding between tertiary amines is not possible.
- The simplest amines are gases; other common amines are liquids.
- Volatile amines have unpleasant odors.
- Simple amines are water-soluble because of hydrogen bonding with water.
- Amines are weak Brønsted–Lowry/Lewis bases (Section 15.4).
- Many amines are physiologically active, and many are toxic (see the Application "Toxicology," p. 482).

PROBLEM 15.6

Arrange the following compounds in order of increasing boiling point. Explain why you placed them in that order.

(a) CH$_3$—N(CH$_3$)—CH$_2$CH$_3$ **(b)** CH$_3$CH$_2$CH$_2$CH$_2$OH   **(c)** CH$_3$CH$_2$CH$_2$CH$_2$NH$_2$

PROBLEM 15.7

Draw the structures of (a) ethylamine and (b) trimethylamine. Use dashed lines to show how they would form hydrogen bonds to water molecules.

15.3 Heterocyclic Nitrogen Compounds

In many nitrogen-containing compounds, the nitrogen atom is in a ring with carbon atoms. Compounds that contain atoms other than carbon in the ring are known as **heterocycles**. Heterocyclic nitrogen compounds may be nonaromatic or aromatic. Piperidine, for example, is a saturated heterocyclic amine with a six-membered ring, and pyridine is an aromatic heterocyclic amine that, like other aromatic compounds, is often represented on paper as a ring with alternating double and single bonds.

Heterocycle A ring that contains nitrogen or some other atom in addition to carbon.

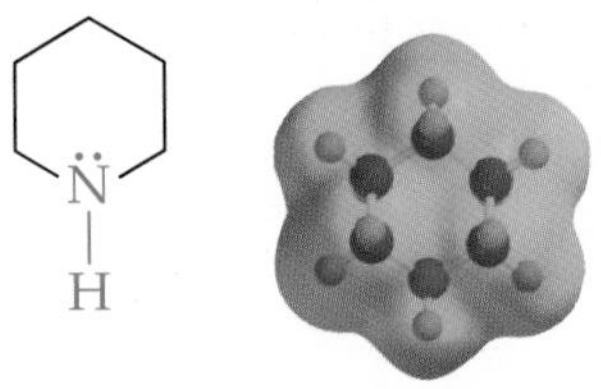

Piperidine
(a saturated cyclic amine)

Pyridine
(an aromatic amine)

The names and structures of several heterocyclic nitrogen compounds are given in Table 15.2; note how seemingly random the names are. You need not memorize these names and structures, but you should take note that such rings are very common in many natural compounds found in plants and animals. For example,

TABLE 15.2 Some Heterocyclic Nitrogen Compounds

 Pyrrolidine (in nicotine and other alkaloids)	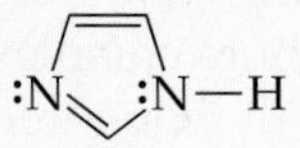 Imidazole (in histamine)	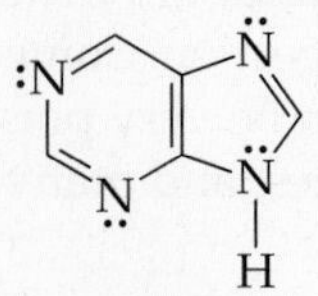 Purine (nitrogen ring system in DNA; present in anticancer drugs)	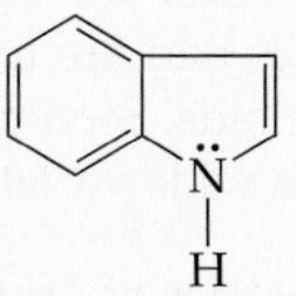 Indole (in many alkaloids and drugs)
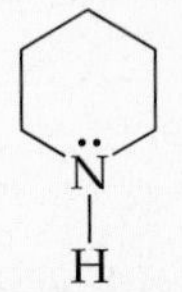 Piperidine (in many drugs)	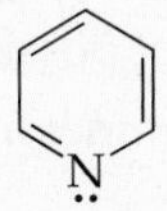 Pyridine (in several B vitamins)	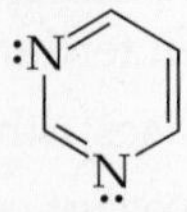 Pyrimidine (nitrogen ring present in some B vitamins)	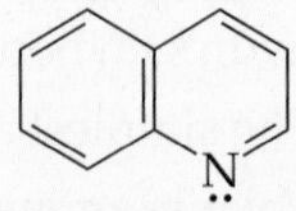 Quinoline (in antibacterial agents)

nicotine, from tobacco leaves, contains one pyridine ring and one pyrrolidine ring; quinine, an antimalarial drug isolated from the bark of the South American *Cinchona* tree, contains a quinoline ring system plus a nitrogen ring with a two-carbon bridge across it. The amino acid tryptophan contains an indole ring system in addition to its amino group.

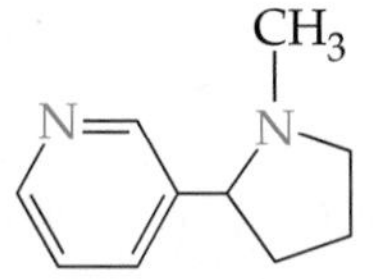

Nicotine from tobacco (an insecticide; an active ingredient in cigarette smoke)

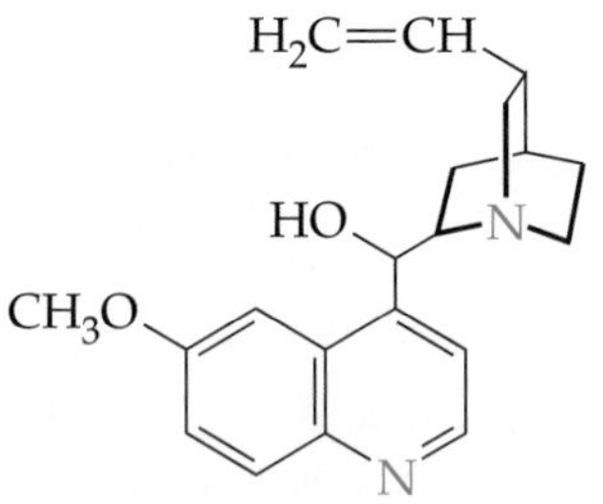

Quinine from the *Cinchona* Tree (an antimalarial drug)

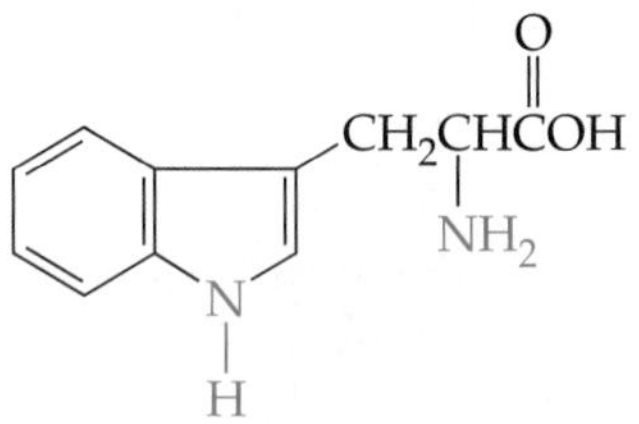

Tryptophan (an amino acid)

Adenine, a nitrogen-containing cyclic compound, is one of the four amines that compose the "bases" in DNA that code for genetic traits.

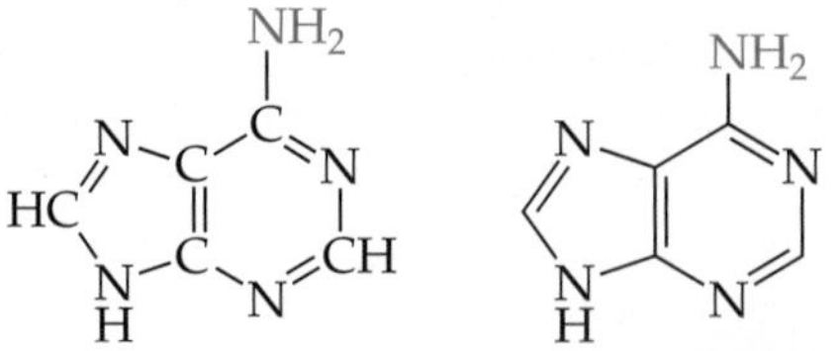

Adenine

Looking Ahead

Hydrogen bonding that occurs between hydrogen atoms on nitrogens and oxygens and the oxygen or nitrogen atoms of other groups within a molecule helps to determine the shape of many biomolecules. Such attractions contribute to the complex shapes into which large protein molecules are folded (Section 18.8). Hydrogen bonding of amine groups also plays a crucial role in the helical structure of the molecule that carries hereditary information—deoxyribonucleic acid, DNA (Section 26.4).

APPLICATION ▶ NO: A Small Molecule with Big Responsibilities

Imagine a molecule that lowers blood pressure, kills invading bacteria, enhances memory, and is a toxic air pollutant. What a major surprise it has been to discover that nitric oxide (NO) is just such a molecule. Even more surprising for a molecule with such a diversity of biochemical functions, pure nitric oxide is a gas at room and body temperatures.

A close look at NO shows that it has 11 valence electrons (5 from N and 6 from O). The atoms are joined by a double bond which gives O an electron octet and leaves N with 7 valence electrons. The unpaired electron on the nitrogen makes the molecule a *free radical* and therefore very reactive—it quickly grabs onto another electron in whatever way it can. In fact, the lifetime of free NO in the body is around 100 milliseconds.

Nitric oxide is synthesized in the linings of blood vessels and elsewhere from oxygen and the amino acid arginine. In blood vessels, NO activates reactions in smooth muscle cells that cause dilation and a resulting decrease in blood pressure. This discovery explains the action of drugs long used to treat angina, the pain experienced during exertion by individuals with partially blocked blood vessels. Nitroglycerin and other drugs of this type release NO.

$$CH_2CHCH_2$$
$$| \quad | \quad |$$
$$O \quad O \quad O$$
$$| \quad | \quad |$$
$$NO_2 \quad NO_2 \quad NO_2 \qquad :\dot{N}=\ddot{O}:$$
Nitroglycerin

Alfred Nobel, who became wealthy as the developer of dynamite, suffered from angina shortly before his death. He wrote to a friend, "Isn't it the irony of fate that I have been prescribed nitroglycerin, to be taken internally. They call it Trinitrin, so as not to scare the chemist and the public." Nobel refused to take Trinitrin. In a further extension of the irony, the Nobel Prize in Physiology or Medicine was awarded in 1998 to three individuals who studied the role of NO in physiology (Ferid Murad, Robert F. Furchgott, and Lous J. Ignarro). One of them discovered that it is the NO released by nitroglycerin that relieves the pain of angina.

Nitric oxide is now recognized as one of the modern-day heroes of human biology. This is a long way from the days when nitric oxide was best known as a smog-producing pollutant that comes out of automobile tailpipes.

As you might expect, the discoveries about the role of NO in human health and disease have opened the door to possible therapeutic applications. Current research suggests that increasing the production of nitric oxide (through exercise, for example) may help prevent heart disease. One of the

▲ Exercise increases levels of NO in humans, boosting levels of this amazing molecule in the body.

initial effects of arteriosclerosis is damage to the arterial lining, which exposes the vessels to harmful circulating cells. Nitric oxide released by the endothelium works to prevent red blood cells from sticking together and attaching to the vessel wall. NO also helps to control vascular tone, allowing the arteries to relax and stay clear. Compounds that release NO are being considered for inhibition of the growth of cancerous cells, protection of vessel walls after surgery, promotion of wound healing, and killing parasites. Inhibitors of NO are being investigated for their ability to treat chronic tension headaches. NO also appears to be important in cellular signaling in plants. As you can see, NO is an amazing molecule that has come a long way from once being known mainly for its role as a pollutant.

See Additional Problem 15.44 at the end of the chapter.

PROBLEM 15.8

Consult Table 15.1 and identify:

(a) Two amines that are gases at room temperature
(b) A heterocyclic amine
(c) A compound with an amine group on an aromatic ring

PROBLEM 15.9

Consult Table 15.2 and write the molecular formulas for piperidine and purine.

PROBLEM 15.10

Which of the following compounds are heterocyclic nitrogen compounds?

(a) 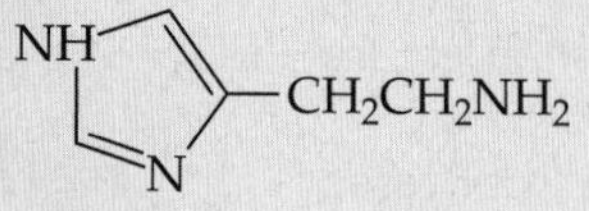 [structure: imidazole ring with NH and N, bearing $-CH_2CH_2NH_2$]
(b) 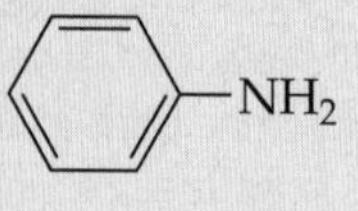 [structure: benzene ring bearing $-NH_2$]

(c) 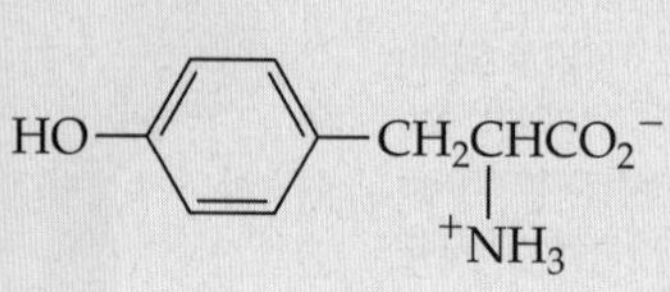 $HO-$[benzene ring]$-CH_2CHCO_2^-$ with $^+NH_3$
(d) 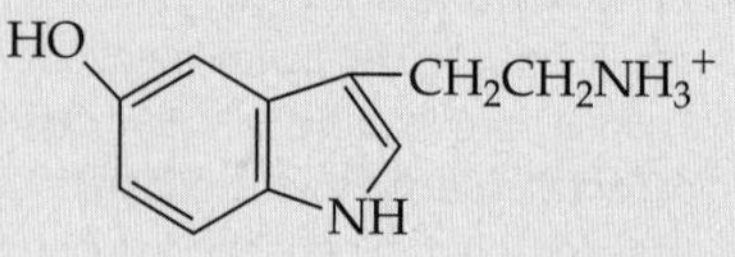 [structure: HO-substituted indole bearing $-CH_2CH_2NH_3^+$ and NH]

15.4 Basicity of Amines

Ammonium ion A positive ion formed by addition of hydrogen to ammonia or an amine (may be primary, secondary, or tertiary).

Just like ammonia, aqueous solutions of amines are weakly basic because of formation of the R_3NH^+ ion in water. Consider the following equilibria of the neutral amines and their **ammonium ions**:

$$CH_3CH_2NH_2 + H_2O \rightleftharpoons CH_3CH_2NH_3^+ + OH^-$$

$$(CH_3CH_2)_2NH + H_2O \rightleftharpoons (CH_3CH_2)_2NH_2^+ + OH^-$$

$$(CH_3CH_2)_3N + H_2O \rightleftharpoons (CH_3CH_2)_3NH^+ + OH^-$$

Notice that these are reversible reactions; ammonium ions can react as acids in the presence of bases to regenerate the amines. This equilibrium is found to exist in solutions with pH values as high as 8.

Ammonium ions are also formed in the reactions of amines with the hydronium ion in acidic solutions:

$$CH_3CH_2NH_2 + H_3O^+ \rightleftharpoons CH_3CH_2NH_3^+ + H_2O$$

$$(CH_3CH_2)_2NH + H_3O^+ \rightleftharpoons (CH_3CH_2)_2NH_2^+ + H_2O$$

$$(CH_3CH_2)_3N + H_3O^+ \rightleftharpoons (CH_3CH_2)_3NH^+ + H_2O$$

The positive ions formed by addition of H^+ to alkylamines are named by replacing the ending *-amine* with *-ammonium*. To name the ions of heterocyclic amines, the amine name is modified by replacing the *-e* with *-ium*. For example:

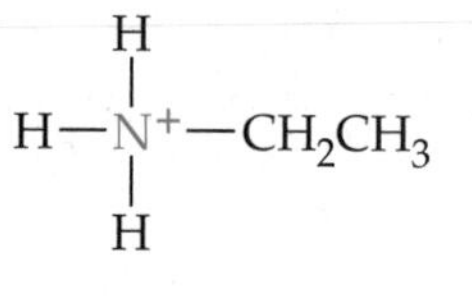

[structure: $H-N^+-CH_2CH_3$ with two H substituents]   [structure: $CH_3CH_2CH_2-N^+-CH_2CH_2CH_3$ with H substituents]   [structure: pyridine ring with N^+ and H]

Ethylammonium ion   Dipropylammonium ion   Pyridinium ion
(from ethylamine)   (from dipropylamine)   (from pyridine)

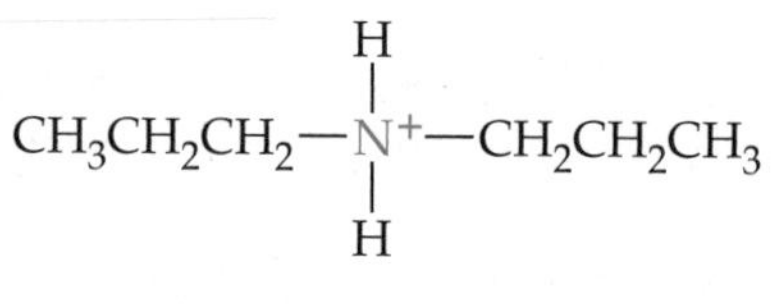

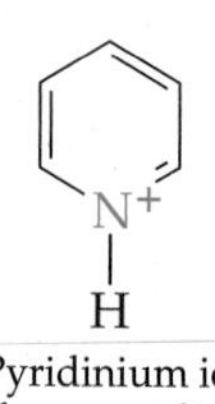

As a result of the equilibria shown above, amines exist as ammonium ions in the aqueous environment of blood and other body fluids, which have a typical pH value of 7.2; for this reason, they are written as ions in the context of biochemistry. For example, histamine and serotonin (both neurotransmitters; Sections 20.8, 20.9) are represented as follows:

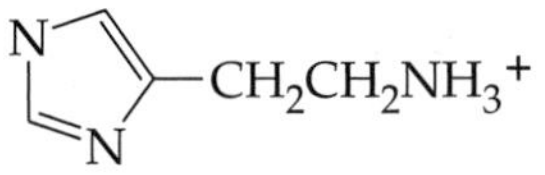

Histamine
(causes allergic reaction)

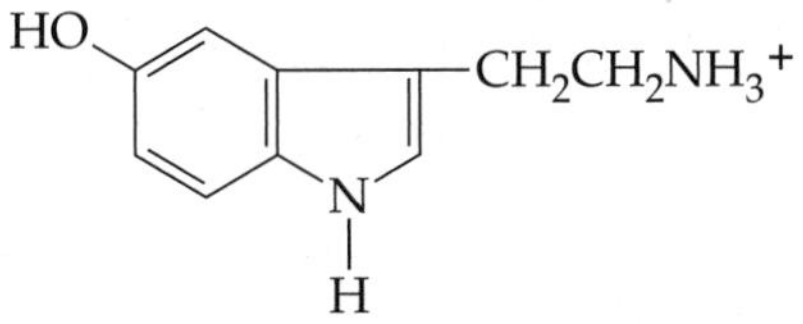

Serotonin
(a neurotransmitter active in the brain)

In general, nonaromatic amines (such as $CH_3CH_2NH_2$) are slightly stronger bases than ammonia, and aromatic amines (such as aniline or pyridine, Table 15.2) are weaker bases than ammonia:

Basicity: Nonaromatic amines > Ammonia > Aromatic amines

WORKED EXAMPLE **15.4** Amines as Bases in Water

Write balanced equations for the reaction of ammonia with water and for the reaction of ethylamine with water. Label each species in your equations as either an acid or a base.

ANALYSIS Determine which species is the base and which is the acid. Remember that the base will accept a hydrogen ion from the acid. Review the definitions for a Brønsted-Lowry base (Section 10.3) and a Lewis base (Section 15.2).

SOLUTION
Like ammonia, amines have a lone pair of electrons on the nitrogen atom. Because ammonia is a base that reacts with water to accept a hydrogen ion (which bonds to the lone pair), it is reasonable to expect that amines are bases that react in a similar manner.

$$NH_3 + H_2O \rightleftharpoons NH_4^+ + OH^-$$
Base — Acid — Acid — Base

$$CH_3CH_2NH_2 + H_2O \rightleftharpoons CH_3CH_2NH_3^+ + OH^-$$
Base — Acid — Acid — Base

Notice that in both cases water acts as an acid because it donates a hydrogen ion to the nitrogen.

PROBLEM 15.11

Write an equation for the acid–base equilibrium of aniline and water. Label each species in the equilibrium as either an acid or a base.

PROBLEM 15.12

Complete the following equations:

(a) $CH_3CH_2CHNH_2 + HBr(aq) \longrightarrow$? (b) 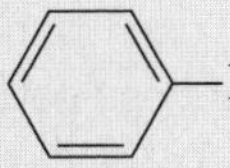 $-NH_2 + HCl(aq) \longrightarrow$?
　　　$\underset{\displaystyle CH_3}{|}$

(c) $CH_3CH_2NH_2 + CH_3COOH(aq) \longrightarrow$?
(d) $CH_3NH_3^+Cl^- + NaOH(aq) \longrightarrow$?

PROBLEM 15.13

Name the organic ions produced in reactions (a)–(c) in Problem 15.12.

PROBLEM 15.14

Which is the stronger base in each pair?

(a) Ammonia or ethylamine   **(b)** Triethylamine or pyridine

PROBLEM 15.15

Write the formulas for the ammonium ions formed by the following amines:

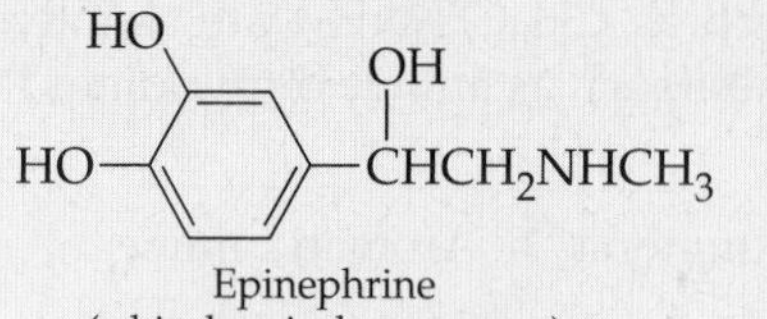

Epinephrine
(a biochemical messenger)

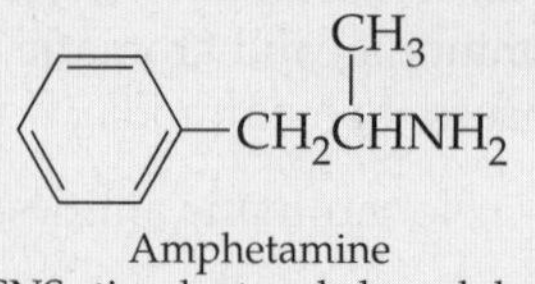

Amphetamine
(a CNS stimulant and abused drug)

APPLICATION ▶

Organic Compounds in Body Fluids and the "Solubility Switch"

The chemical reactions that keep us alive and functioning occur in the aqueous solutions known as *body fluids*—blood, digestive juices, and the fluid inside cells. Waste products from these metabolic reactions are excreted in urine. For organic compounds of all classes, water solubility decreases as the hydrocarbon-like portions of the molecules become larger and molecular weight increases. How does the body manage to carry out reactions in water, especially when large and complex biomolecules are involved?

Many biomolecules contain acidic and basic functional groups. At the pH of body fluids (for example, approximately 7.2 for blood), many of these groups are ionized and thus water-soluble, providing what is often called a *solubility switch*. (⊂⊃, p. 257) The most frequently seen ionized functional groups present in biomolecules are carboxylate (pronounced car-**box**-ill-late) groups (from carboxylic acids, —COOH, discussed in Section 17.3), phosphate groups (as well as diphosphates and triphosphates, discussed in Section 17.8), and ammonium groups.

For example, nicotinamide adenine dinucleotide (NAD$^+$, an important biochemical oxidizing agent; Section 21.7) has three charges as it exists in body fluids: two negative charges from a diphosphate and one positive charge from a quaternary amine.

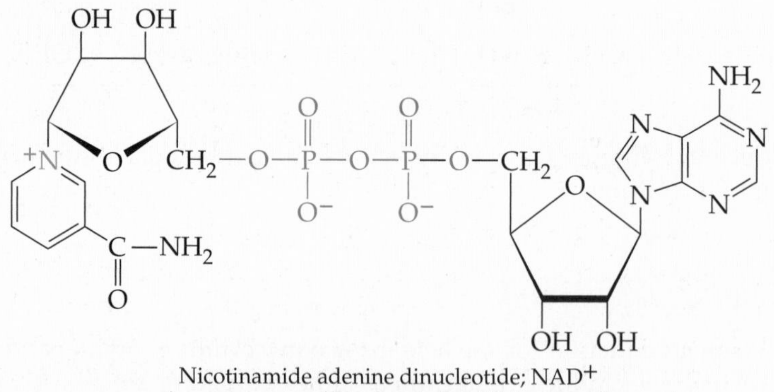

Nicotinamide adenine dinucleotide; NAD$^+$
(a coenzyme and biochemical oxidizing agent)

In the biochemistry that lies ahead, you will see that most of the major biochemical pathways occur in the aqueous medium of the cytosol found inside cells. It would be disastrous if these pathways could be shut down by diffusion of reactants out of the cells. Such diffusion would require passage of intermediates through the cell wall, which is a nonpolar medium. Diffusion does not occur because the intermediates are ionized

Ionic solubility switches

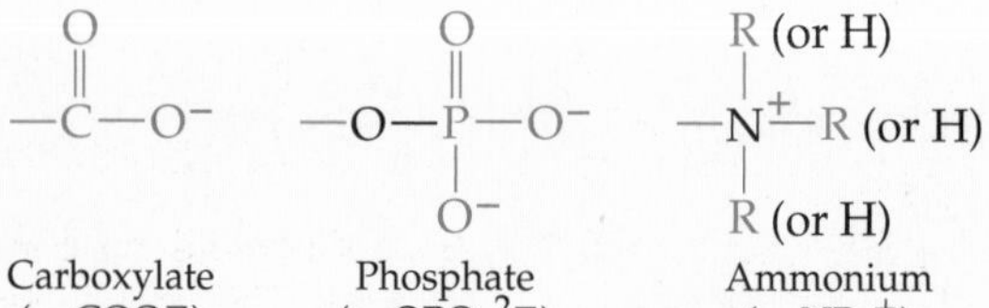

Carboxylate
(—COO$^-$)

Phosphate
(—OPO$_3{}^{2-}$)

Ammonium
(—NR$_3{}^+$)

within the cytosol and are therefore polar and cannot pass through the nonpolar cell wall.

Medications must be soluble in body fluids in order to be transported from their entry point in the body to their site of action. Many drugs are weak acids or bases and therefore are present as their ions in body fluids. Aspirin and amphetamine are two examples:

Aspirin
(an acid)

Amphetamine
(a base)

The extent of ionization of a drug helps determine how it is distributed in the body. Weak acids, such as aspirin,

are essentially un-ionized in the acidic environment in the stomach and are therefore readily absorbed there. Weak bases, however, are completely ionized in the stomach, and therefore no significant absorption occurs there. It is not until they reach the more basic environment of the small intestine that these weak bases revert to their neutral form and are absorbed.

Many pharmaceutical agents must be delivered to the body in their more water-soluble forms as salts. Converting amines such as phenylephrine (the decongestant in Neo-Synephrine) to ammonium hydrochlorides increases their solubility to the point where delivery in solution is possible.

Phenylephrine hydrochloride
(a decongestant)

See Additional Problems 15.45–15.47 at the end of the chapter.

15.5 Amine Salts

An **ammonium salt** (also known as an *amine salt*) is composed of a cation and an anion and is named by combining the ion names. For example, in methylammonium chloride ($CH_3NH_3^+Cl^-$), the methylammonium ion, $CH_3NH_3^+$, is the cation and the chloride ion is the anion.

Ammonium salts are generally odorless, white, crystalline solids that are much more water-soluble than neutral amines because they are ionic. For example:

> **Ammonium salt** An ionic compound composed of an ammonium cation and an anion; an amine salt.

$$CH_3CH_2CH_2CH_2-\underset{\underset{CH_2CH_2CH_2CH_3}{|}}{N}-CH_2CH_2CH_2CH_3 + HCl(aq) \rightleftharpoons CH_3CH_2CH_2CH_2-\overset{\overset{H}{|}}{\underset{\underset{CH_2CH_2CH_2CH_3}{|}}{N^+}}-CH_2CH_2CH_2CH_3 \ Cl^-(aq)$$

Tributylamine
(water-insoluble)

Hydrochloric
acid

Tributylammonium chloride
(water-soluble)

In medicinal chemistry, amine salt formulas are quite often written and named by combining the structures and names of the amine and the acid used to form its salt. By this system, methylammonium chloride is written $CH_3NH_2 \cdot HCl$ and named methylamine hydrochloride. You will often see this system used with drugs that are amine salts. For example, diphenhydramine is one of a family of antihistamines available in over-the-counter medications. Antihistamines of this type are oily liquids and difficult to formulate as such, so they are converted to amine salts for formulation into medications.

$$(C_6H_5)_2CHOCH_2CH_2N(CH_3)_2 \cdot HCl$$

or

$$(C_6H_5)_2CHOCH_2CH_2CH_2NH(CH_3)_2^+Cl^-$$

Diphenhydramine hydrochloride
(Benadryl), an antihistamine

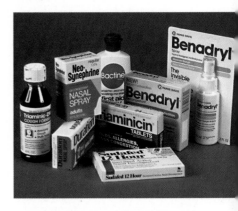

▲ **Over-the-counter ammonium salts.** The active ingredient in each of these over-the-counter medications is an ammonium salt.

If a free amine is needed, it is easily regenerated from an amine salt by treatment with a base:

$$CH_3NH_3{}^+Cl^-(aq) + NaOH(aq) \longrightarrow CH_3NH_2(aq) + NaCl(aq) + H_2O(l)$$

Quaternary ammonium ions have four organic groups bonded to the nitrogen atom, and this bonding give the nitrogen a positive charge. With no H atom that can be removed by a base and no lone pair on the nitrogen that can bond to H^+, ammonium ions are neither acidic nor basic, and their structures in solution are unaffected by changes in pH. Their salts are known as **quaternary ammonium salts**. One commonly encountered quaternary ammonium salt has the following structure, where R represents a range of C_8 to C_{18} alkyl groups:

Quaternary ammonium salt An ionic compound composed of a quaternary ammonium ion and an anion.

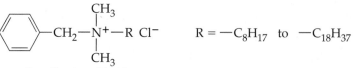

Benzalkonium chloride
(an antiseptic and disinfectant)

These benzalkonium chlorides have both antimicrobial and detergent properties. As dilute solutions, they are used in surgical scrubs and for sterile storage of instruments; concentrated solutions, however, are harmful to body tissues.

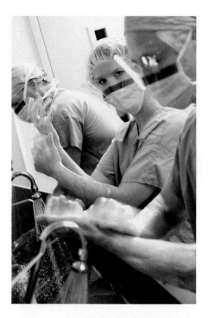

▲ **The preoperative surgical scrub.** Several quaternary ammonium compounds have germicidal properties that suit them for this use.

PROBLEM 15.16

Write the structures of the following compounds:

(a) Hexyldimethylammonium chloride

(b) Isopropylammonium bromide

PROBLEM 15.17

Identify each compound in Problem 15.16 as the salt of a primary, secondary, or tertiary amine.

PROBLEM 15.18

Write an equation for the formation of the free amine from butylammonium chloride by reaction with aqueous OH^-.

PROBLEM 15.19

Compare the structure of Benadryl (p. 479) with the general antihistamine structure:

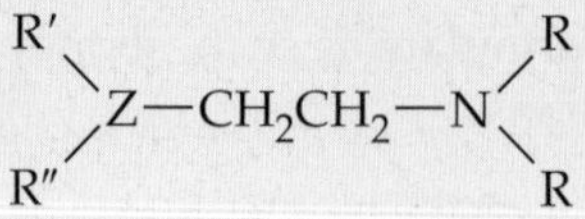

$$\begin{matrix} R' \\ \diagdown \\ \diagup \\ R'' \end{matrix} Z-CH_2CH_2-N \begin{matrix} R \\ \diagup \\ \diagdown \\ R \end{matrix}$$

$$Z = N, C, C-O$$

Does Benadryl have that general structure? Write a comparison of the structures.

PROBLEM 15.20

Write the structure of benzylamine hydrochloride in two different ways, and name the hydrochloride as an ammonium salt.

$$\text{⟨benzene ring⟩—CH}_2\text{NH}_2 \quad \text{Benzylamine}$$

15.6 Amines in Plants: Alkaloids

The roots, leaves, and fruits of flowering plants are a rich source of nitrogen compounds. These compounds, once called "vegetable alkali" because their water solutions are basic, are now referred to as **alkaloids**.

The molecular structures of many thousands of alkaloids have been determined. Most are bitter-tasting, physiologically active, structurally complex, and toxic to human beings and other animals in sufficiently high doses. Most people are familiar with the physiological activity of two alkaloids—caffeine (p. 469) and nicotine (p. 474), which are stimulants. Quinine (p. 474) is used as a standard for bitterness: Even a 1×10^{-6} M solution tastes bitter. For a long time, quinine was the only drug available for treating malaria (caused by a parasitic protozoan).

The bitterness and poisonous nature of alkaloids probably evolved to protect plants from being devoured by animals. The three poisonous compounds described here—coniine, atropine, and solanine—illustrate some more of the many types of alkaloid structures.

Alkaloid A naturally occurring, nitrogen-containing compound isolated from a plant; usually basic, bitter, and poisonous.

Examples of toxic alkaloids

Coniine

Atropine

Solanine
(X = a group of three sugar molecules)

- **Coniine** is extracted from poison hemlock (*Conium maculatum*). Socrates used this poison to end his life after being convicted of corrupting Greek youth with philosophical discussions.

- **Atropine** is the toxic substance in the herb known as *deadly nightshade* or *belladonna* (*Atropa belladonna*). In Meyerbeer's opera, *L'Africaine*, the heroine sings of the peaceful death this plant brings before committing suicide over her lost love. Like many other alkaloids, atropine acts on the central nervous system, a property sometimes applied in medications (in appropriately low dosage!) to reduce cramping of the digestive tract. Atropine is also used as an antidote against nerve gases, such as Sarin.

- **Solanine,** an even more potent poison than atropine, is found in potatoes and tomatoes, both of which belong to the same botanical family as the deadly nightshade (*Solanaceae*). The tiny amount of solanine in properly stored potatoes actually contributes to their characteristic flavor, but when potatoes are exposed to sunlight or stored under very cold or very warm conditions, the production of solanine is increased to levels that can be dangerous. The reason you are warned

▲ A potato that has turned green because of exposure to sunlight. Before it is eaten, this potato must be peeled to remove all of the green chlorophyll so that the poisonous alkaloid solanine is also removed.

▲ All parts of the poppy, including poppy seeds, contain morphine. Eating poppy seeds can introduce enough morphine into your body fluids to show up in a laboratory drug screen.

that you must peel green potatoes deeply is that alkaloids such as solanine are formed under the skin. Alkaloids are not destroyed during cooking but can be removed by peeling. Fortunately, sunlight also stimulates the formation of chlorophyll under the skin, and the green color of the chlorophyll provides a warning. By peeling away all of the green color, you most likely remove the excess solanine. Potato sprouts (the "eyes") also contain solanine and should be cut out before the potatoes are cooked.

Not all alkaloids are known for their poisonous nature; some are notable as pain relievers (*analgesics*), as sleep inducers, and for the euphoric states they can create. Raw opium, a paste derived from the opium poppy (*Papaver somniferum*), has been known for these properties since ancient times. About 20 alkaloids are present in the poppy, including morphine and codeine. The free alkaloids are oily liquids, and not very soluble in water. The medicinal use of morphine for pain was expanded in the sixteenth century when the German physician Paracelsus extracted opium into brandy to produce *laudanum*, essentially a solution of morphine in alcohol. A similar extract (10% opium by weight in alcohol) is still sometimes prescribed for diarrhea, as is *paregoric*, a more dilute solution of opium combined with anise oil, glycerin, benzoic acid, and camphor. Heroin does not occur naturally but is easily synthesized from its parent compound morphine. Within the body, removal of the $CH_3C{=}O$ groups converts heroin back to morphine.

Morphine

Heroin

Codeine

APPLICATION ▶ Toxicology

Toxicology is the science devoted to poisons—their identification, their effects, their modes of action, and methods of protecting against them or counteracting their effects. It is a science with many subspecialties. *Clinical* toxicology is concerned with the treatment of individuals harmed by toxic agents. *Forensic* toxicology deals with the effects of toxic agents as they relate to criminal cases, most notably in drug abuse or intentional poisonings. A third branch of toxicology—*environmental* toxicology—is concerned with toxic substances purposefully or accidentally introduced into our surroundings. Such environmental pollutants may be harmful to humans, to other animals, or to plants. Yet another important branch of toxicology focuses on the beneficial uses of poisons—for example, pesticides in agriculture or chemotherapeutic agents to kill cancer cells.

Some toxic substances have the general effect of harming any kind of tissue. Strong acids such as sulfuric acid and strong bases such as sodium hydroxide (lye) "burn" by destroying all cells they contact.

Many poisons have a molecular structure that allows them to interact with a specific biomolecule. Several of the

most poisonous substances known are *neurotoxins*, which bind to proteins that form ion channels in nerve cell membranes. Blocking these channels prevents transmission of nerve impulses, causing paralysis and death by suffocation. Tetrodotoxin, a poison of this type, is produced by the puffer fish. Despite this, puffer fish (also called Fugu) is considered a delicacy in Japan, where the risk is minimized by allowing only well-trained and certified chefs to remove the toxin-containing parts of the fish.

Tetrodotoxin

Chemical change in the body can be caused by poisons, medications, or natural chemical messengers (hormones or neurotransmitters, discussed in Chapter 20). Change is initiated when a messenger molecule encounters a structurally compatible receptor on the surface of a cell or an enzyme that catalyzes a necessary reaction.

chemical messages between nerve cells). For example, muscarine, the active ingredient in the highly poisonous mushroom *Amanita muscaria*, duplicates the action of acetylcholine, overstimulating certain of its receptors. In this case (though not always), you can easily see the structural similarity between the natural chemical messenger and its poisonous mimic:

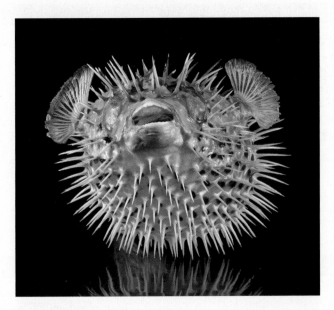

▲ A puffer fish.

A number of poisons act at receptors that normally bind acetylcholine, a neurotransmitter (a molecule that carries

Ideally, a thorough understanding of any poison includes knowing how it acts at the molecular level and knowing a molecular mechanism by which its effects can be reversed. Such knowledge is a major goal of toxicology.

See Additional Problem 15.48 at the end of the chapter.

SUMMARY: REVISITING THE CHAPTER GOALS

1. **What are the different types of amines?** *Amines* are classified as *primary, secondary,* or *tertiary,* depending on whether they have one, two, or three organic groups individually bonded to nitrogen. These amines can all add hydrogen to form *ammonium ions,* which have four bonds to the nitrogen, which bears a single positive charge. Ions with four organic groups bonded to nitrogen are known as *quaternary ammonium ions.* In *heterocyclic amines,* the nitrogen of the amine group is bonded to 2 carbon atoms that are part of a ring.

2. **How are amines named?** Primary amine names have *-amine* added to the alkyl group name, and secondary and tertiary amines with identical R groups have *di-* and *tri-* prefixes. When the R groups are different, amines are named as *N-substituted derivatives* of the amine with the largest R group. Ions derived from amines are named by replacing *-amine* in the name with *-ammonium.* The —NH_2 group as a substituent is called an *amino group.*

3. **What are the general properties of amines?** Amines have an unshared electron pair on nitrogen that is available to accept a proton or for hydrogen bonding. Primary and secondary amine molecules hydrogen-bond to each other, but tertiary amine molecules cannot do so. Thus, the general order of boiling points for molecules of comparable size is

   Hydrocarbons < Tertiary amines < Primary and secondary amines < Alcohols

All amines can, however, hydrogen-bond to other molecules containing OH and NH groups, and for this reason small amine molecules are water-soluble. Many amines are physiologically active. Volatile amines have strong, unpleasant odors.

4. **How do amines react with water and acids?** Amines are weak bases and establish equilibria with water by adding H^+ to form ammonium ions (RNH_3^+, $R_2NH_2^+$, R_3NH^+) and

KEY WORDS

Alkaloid, *p. 481*

Amine (primary, secondary, tertiary), *p. 465*

Amino group, *p. 467*

Ammonium ion, *p. 476*

Ammonium salt, *p. 479*

Heterocycle, *p. 473*

Lewis base, *p. 471*

Quaternary ammonium ion, *p. 466*

Quaternary ammonium salt, *p. 480*

hydroxide ions (OH^-). They react with acids to form ammonium salts. Ammonium ions react as acids (proton donors) in the presence of a base. *Quaternary ammonium ions* (R_4N^+) have no lone electron pair and are not bases, nor can they form hydrogen bonds.

5. **What are alkaloids?** *Alkaloids* are naturally occurring nitrogen compounds found in plants. They are all bases, most with a bitter taste. Like other amines, many are physiologically active, notably as poisons or analgesics.

SUMMARY OF REACTIONS

1. **Reactions of amines (Section 15.4)**

   (a) Acid–base reaction with water:

$$CH_3CH_2NH_2 + H_2O \rightleftharpoons CH_3CH_2NH_3^+ + OH^-$$

   (b) Acid–base reaction with a strong acid to yield an ammonium ion:

$$CH_3CH_2NH_2 + H_3O^+ \longrightarrow CH_3CH_2NH_3^+ + H_2O$$

2. **Reaction of ammonium ion (Section 15.4) or amine salt (Section 15.5)** Acid–base reaction of primary, secondary, or tertiary amine salt (or ion) with a base to regenerate the amine:

$$CH_3CH_2NH_3^+Cl^- + NaOH \longrightarrow CH_3CH_2NH_2 + NaCl + H_2O$$

UNDERSTANDING KEY CONCEPTS

15.21

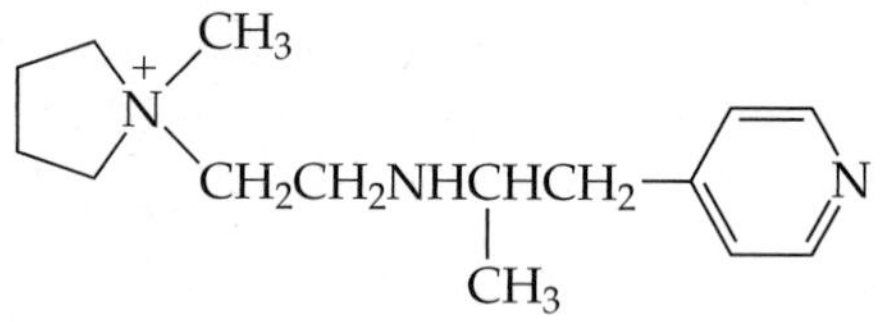

(a) For the compound above, identify each nitrogen as either a primary, secondary, tertiary, quaternary, or aromatic amine.

(b) Which amine group(s) would be able to provide a hydrogen bond? Which could accept a hydrogen bond?

15.22 The structure of the amino acid histidine (in its uncharged form) is shown below.

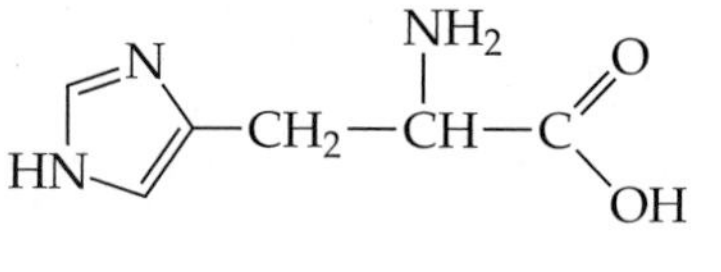

(a) Which amine groups would be able to participate in hydrogen bonding?

(b) Is histidine likely to be water-soluble? Explain.

15.23 Draw structures to illustrate hydrogen bonding (similar to those on p. 472) between the following compounds.

(a) $CH_3—NH—CH_2—CH_2—CH_3$ and $CH_3—NH—CH_2—CH_2—CH_3$

(b) 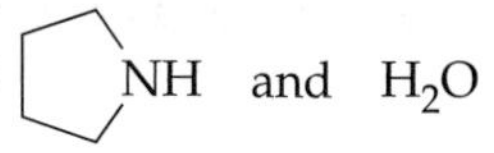 and H_2O

(c) $NH_3—CH_2—CH_2—CH_3$ and 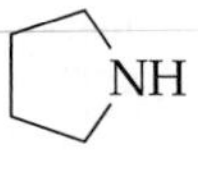

15.24 Explain what bonds must be made or broken and where the electrons go when the hydrogen-bonded water between the two amines shown at the top of page 472 reacts to form an amine, ammonium ion, and OH^-.

15.25 Arrange the following compounds in the order of decreasing base strength.

NH_3 $(CH_3)_2NH$ 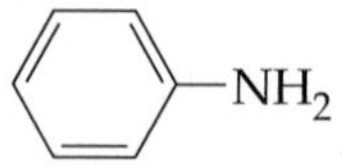

15.26 Complete the following equations:

(a) 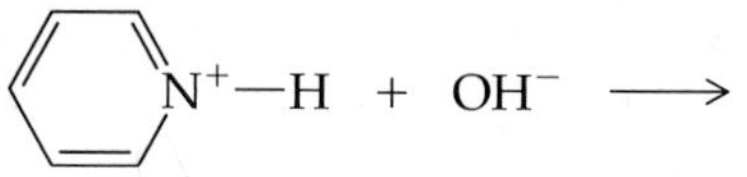

(b) $CH_3—\underset{H}{N}—\underset{CH_3}{CH}—CH_3 + H_2O \rightleftharpoons$

(c) $(CH_3)_4N^+ + OH^- \longrightarrow$

(d) $CH_3—\overset{CH_3}{\underset{CH_3}{C}}—NH_2 + HCl \longrightarrow$

ADDITIONAL PROBLEMS

AMINES AND AMMONIUM SALTS

15.27 Draw the structures corresponding to the following names:

(a) Cyclohexylamine
(b) Diisopropylamine
(c) *N,N*-Dimethylbutylamine

15.28 Draw the structures corresponding to the following names:

(a) *N*-Methylpentylamine
(b) *N*-Ethylcyclobutylamine
(c) *p*-Propylaniline

15.29 Name the following amines, and identify them as primary, secondary, or tertiary:

(a) 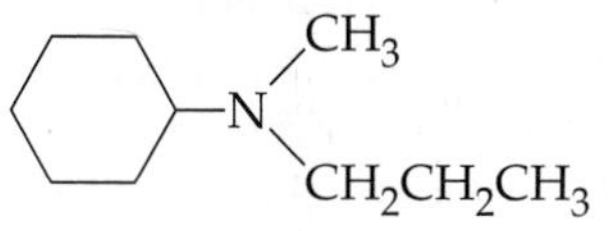 [cyclohexyl—N with CH₃ and CH₂CH₂CH₃ groups]

(b) 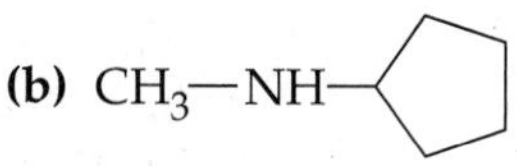 CH_3-NH- [cyclopentyl]

15.30 Name the following amines, and identify them as primary, secondary, or tertiary:

(a) 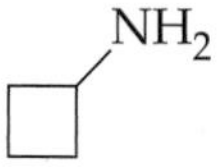 [cyclobutyl—NH₂]

(b) 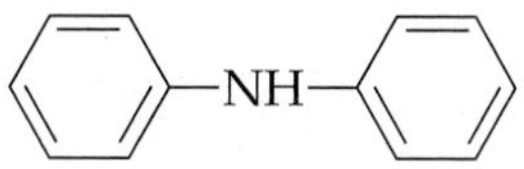 [phenyl—NH—phenyl]

15.31 Is water a weaker or stronger base than ammonia?

15.32 Which is a stronger base, diethyl ether or diethylamine?

15.33 Give names for structures, and structures for names, for the following ammonium salts. Indicate whether each is the ammonium salt of a primary, secondary, or tertiary amine.

(a) $CH_3NH_3^+ Cl^-$

(b) 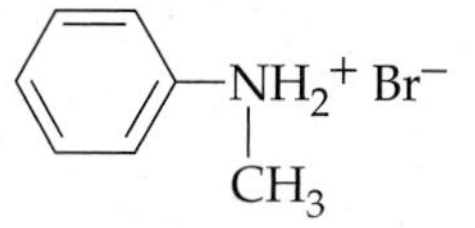 [phenyl—NH_2^+ Br⁻ with CH₃]

(c) *N*-Propylbutylammonium bromide

15.34 Give names for structures, and structures for names, for the following ammonium salts. Indicate whether each is the ammonium salt of a primary, secondary, or tertiary amine.

(a) 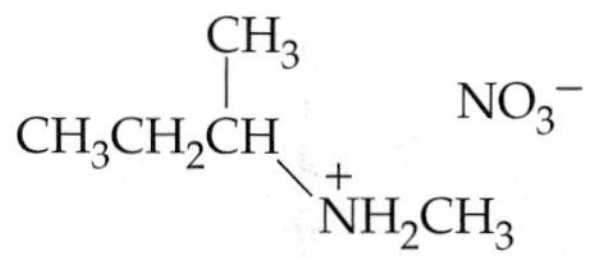 CH_3CH_2CH with CH_3 and $\overset{+}{N}H_2CH_3$, NO_3^-

(b) Pyridinium chloride
(c) *N*-Butyl-*N*-isopropylhexylammonium chloride

15.35 The compound Lidocaine is used medically as a local anesthetic. Identify the functional groups present in Lidocaine.

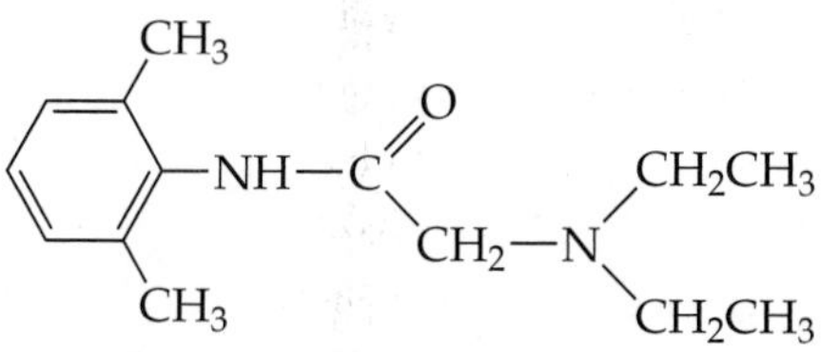

[structure of Lidocaine: dimethylphenyl—NH—C(=O)—CH₂—N(CH₂CH₃)(CH₂CH₃)]

Lidocaine

15.36 Identify the functional groups in cocaine.

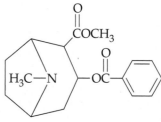

Cocaine

15.37 Most illicit cocaine is actually cocaine hydrochloride—the product of the reaction of cocaine (Problem 15.36) with HCl. Show the structure of cocaine hydrochloride.

15.38 When quinine (an antimalarial drug, p. 474) reacts with HCl, which nitrogen is first to form the ammonium salt? Show the structure of quinine hydrochloride.

REACTIONS OF AMINES

15.39 Complete the following equations:

(a) 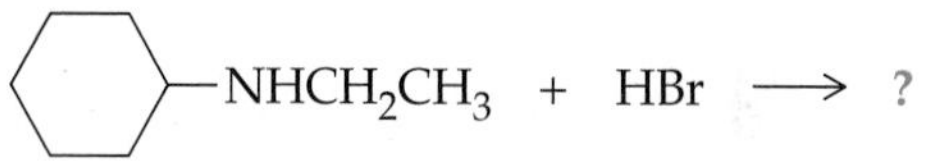 [cyclohexyl]$-NHCH_2CH_3$ + HBr $\longrightarrow$?

(b) 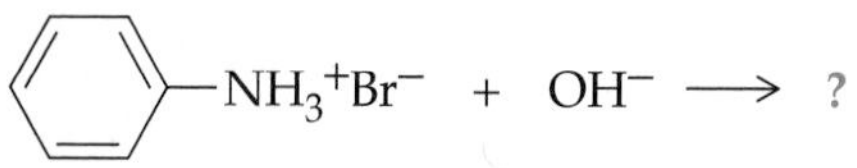 [phenyl]$-NH_3^+Br^-$ + OH^- $\longrightarrow$?

(c) 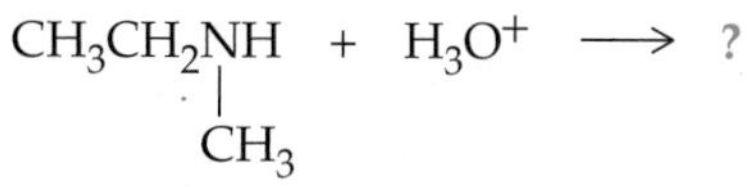 CH_3CH_2NH with CH_3 + H_3O^+ $\longrightarrow$?

15.40 Complete the following equations:

(a) 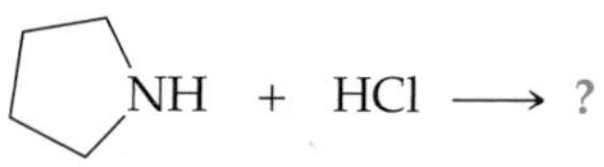 [cyclopentyl]NH + HCl $\longrightarrow$?

(b) 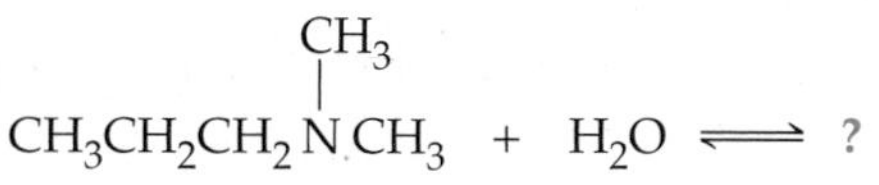 $CH_3CH_2CH_2\overset{\displaystyle CH_3}{N}CH_3$ + H_2O $\rightleftharpoons$?

(c) 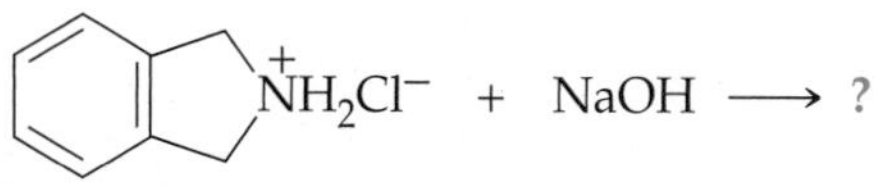 [bicyclic structure]$\overset{+}{N}H_2Cl^-$ + NaOH $\longrightarrow$?

15.41 Many hair conditioners contain an ammonium salt such as the following to help prevent "fly-away" hair. Will this salt react with acids or bases? Why or why not?

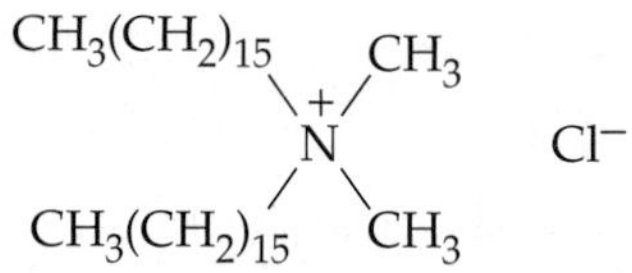

$CH_3(CH_2)_{15}$ and CH_3 on $\overset{+}{N}$ with $CH_3(CH_2)_{15}$ and CH_3, Cl^-

15.42 Choline has the following structure. Do you think that this substance reacts with aqueous hydrochloric acid? If so, what is the product? If not, why not?

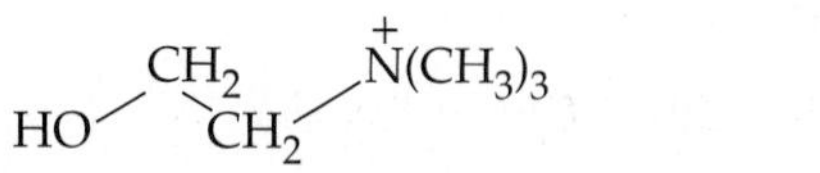

$HO\overset{CH_2}{\underset{CH_2}{}}\overset{+}{N}(CH_3)_3$

Applications

15.43 The *Handbook of Chemistry and Physics* indicates that adenine has a solubility of 0.09 g/100 mL of cold water, is slightly soluble in alcohol, and is insoluble in chloroform and ether. If you wished to extract caffeine, but not adenine, from ground coffee beans, what would be your solvent of choice? Why? [*Chemical Information, p. 468*]

15.44 In the last ten years or so, there has been a lot of interest in NO (nitric oxide). List five functions have been attributed to NO. [*NO: A Small Molecule with Big Responsibilities, p. 475*]

15.45 Which of the following drugs are more readily absorbed? [*Organic Compounds in Body Fluids and the "Solubility Switch," p. 478*]

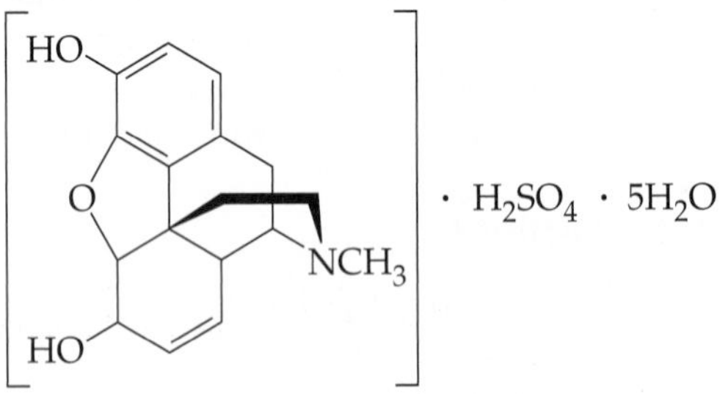

Morphine sulfate

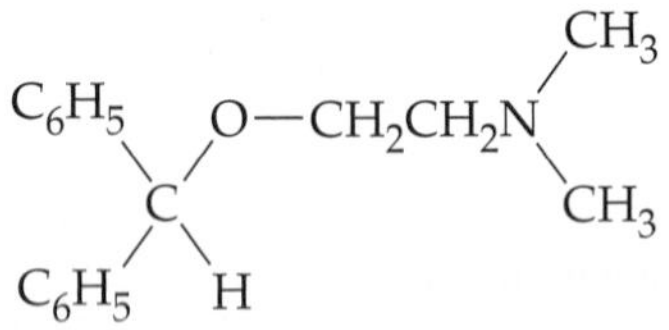

Benadryl®

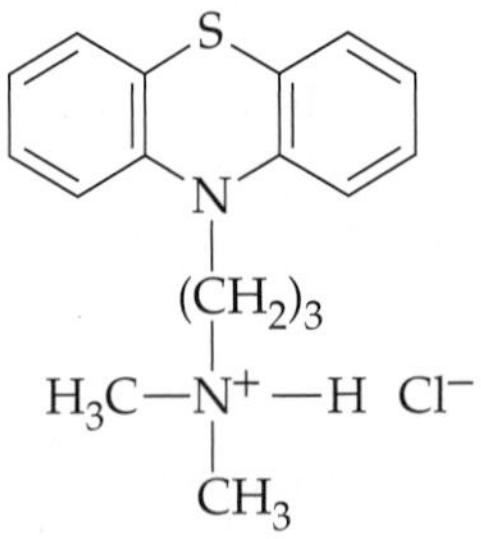

Thorazine hydrochloride

15.46 Promazine, a potent antipsychotic tranquilizer, is administered as the hydrochloride salt. Write the formula of the salt (there is only one HCl in the salt). [*Organic Compounds in Body Fluids and the "Solubility Switch," p. 478*]

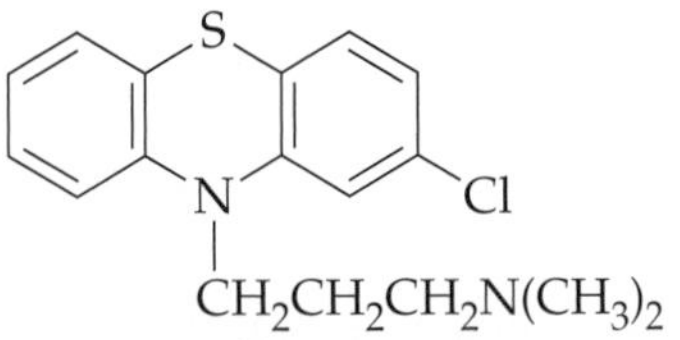

Promazine

15.47 Turn to the citric acid cycle (Figure 10.9, Section 10.8) and list the names of the intermediates in the cycle. Why are they not listed as acids? [*Organic Compounds in Body Fluids and the "Solubility Switch," p. 478*]

15.48 **(a)** What kind of work might a forensic toxicologist be called upon to do?
(b) As you study a new toxin, what three questions need to be answered so that you can better understand it, and hopefully develop an antidote for the toxin? [*Toxicology, p. 482*]

General Questions and Problems

15.49 1-Propylamine, 1-propanol, acetic acid, and butane have about the same molar masses. Which would you expect to have the (a) highest boiling point, (b) lowest boiling point, (c) least solubility in water, and (d) least chemical reactivity? Explain.

15.50 Explain why decylamine is much less soluble in water than ethylamine.

15.51 Propose structures for amines that fit these descriptions:
(a) A secondary amine with formula $C_5H_{13}N$
(b) A tertiary amine with formula $C_6H_{13}N$
(c) A cyclic quaternary amine that has the formula $C_6H_{14}N^+$

15.52 *para*-Aminobenzoic acid (PABA) is a common ingredient in sunscreens. Draw the structure of PABA.

15.53 PABA (Problem 15.52) is used by certain bacteria as a starting material from which folic acid (a necessary vitamin) is made. Sulfa drugs such as sodium sulfanilamide work because they resemble PABA. The bacteria try to metabolize the sulfa drug, fail to do so, and die due to lack of folic acid.

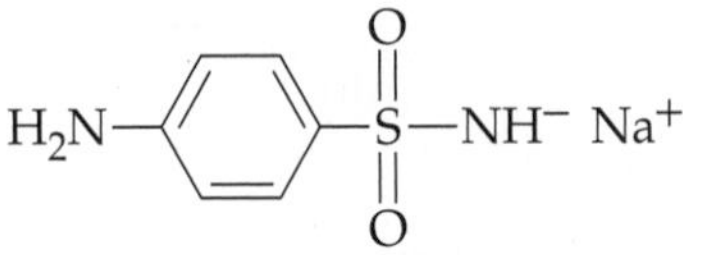

Sodium sulfanilamide

(a) Describe how this structure is similar to that of PABA.
(b) Why do you think the sodium salt, rather than the neutral compound, is used as the drug?

15.54 Acyclovir is an antiviral drug used to treat herpes infections. It has the following structure:

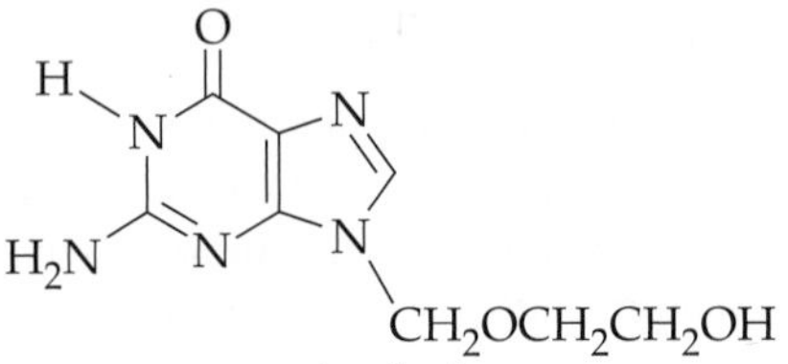

Acyclovir

(a) What heterocyclic base (Table 15.2) is the parent of this compound?
(b) Label the other functional groups present.

15.55 Which is the stronger base, trimethylamine or pyridine? In which direction will the following reaction proceed?

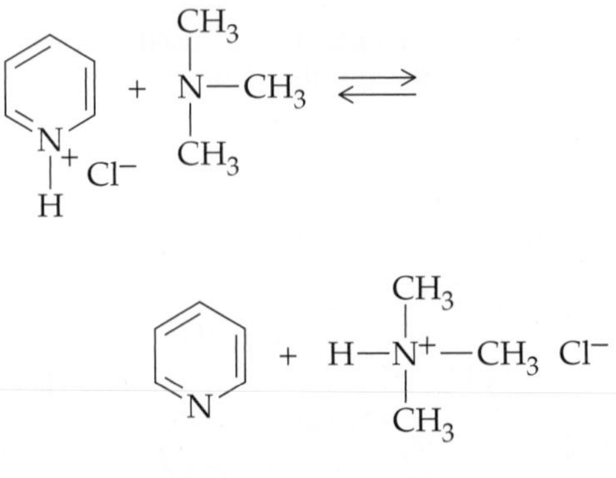

15.56 How do amines differ from analogous alcohols in (a) odor, (b) basicity, and (c) boiling point?

15.57 What two undesirable characteristics are often associated with alkaloids?

15.58 Name the following compounds:

(a) 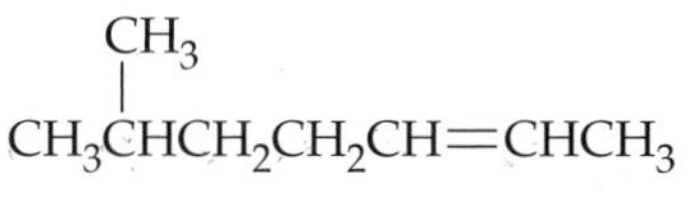
$$
\begin{array}{c}
CH_3 \\
| \\
CH_3CHCH_2CH_2CH=CHCH_3
\end{array}
$$

(b) 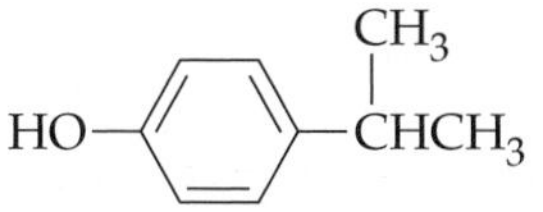
$$
HO-\!\!\!\bigcirc\!\!\!-\overset{\overset{\displaystyle CH_3}{|}}{CHCH_3}
$$

(c) $(CH_3CH_2CH_2CH_2)_2NH$

15.59 Complete the following equations:

(a) 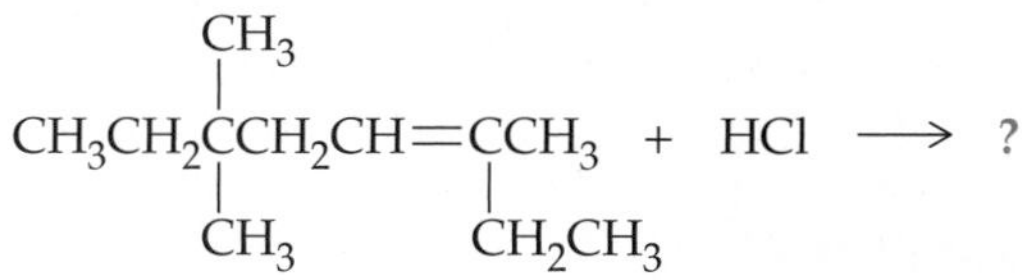
$$
\begin{array}{c}
\quad\ CH_3 \\
\quad\ | \\
CH_3CH_2CCH_2CH=CCH_3 \ +\ HCl \ \longrightarrow\ ? \\
\quad\ | \qquad\qquad\ | \\
\quad\ CH_3 \qquad CH_2CH_3
\end{array}
$$

(b) 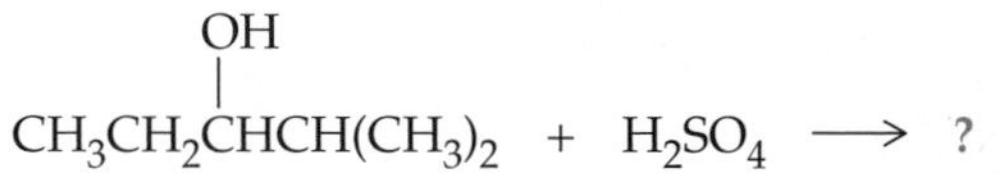
$$
\begin{array}{c}
\qquad\qquad OH \\
\qquad\qquad | \\
CH_3CH_2CHCH(CH_3)_2 \ +\ H_2SO_4 \ \longrightarrow\ ?
\end{array}
$$

(c) $2\ CH_3CH_2SH \ \xrightarrow{[O]}\ ?$

(d) 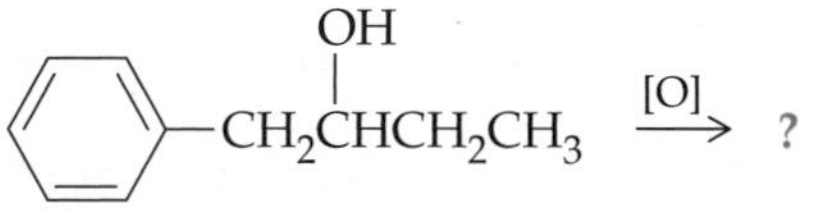
$$
\bigcirc\!\!-\overset{\overset{\displaystyle OH}{|}}{CH_2CHCH_2CH_3} \ \xrightarrow{[O]}\ ?
$$

(e) $(CH_3)_3N \ +\ H_2O \ \longrightarrow\ ?$

(f) $(CH_3)_3N \ +\ HCl \ \longrightarrow\ ?$

(g) $(CH_3)_3NH^+ \ +\ OH^- \ \longrightarrow\ ?$

15.60 Hexylamine and triethylamine have the same molar mass. The boiling point of hexylamine is 129 °C, whereas that of triethylamine is only 89 °C. Explain these observations.

15.61 Lemon juice, which contains citric acid, is traditionally recommended for removing the odor associated with cleaning fish. What functional group is responsible for a "fishy" odor, and why does lemon juice work to remove the odor?

15.62 Baeocystin is a hallucinogenic compound that is isolated from the mushroom Psilocybe baeocystis and has the structure shown below. What heterocyclic base (Table 15.2) is the parent of this compound?

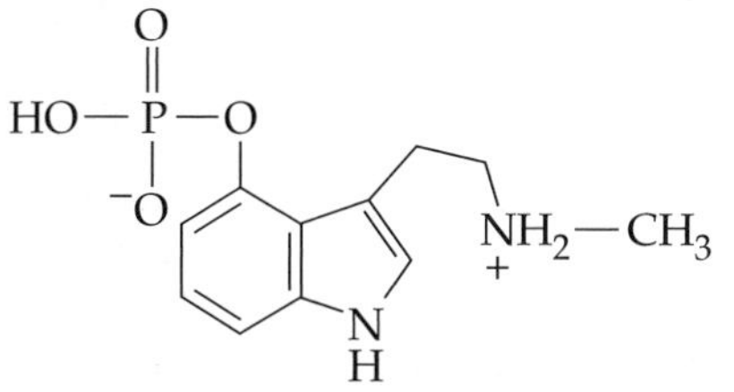

Baeocystin

15.63 Why is aniline not considered to be a heterocyclic nitrogen compound?

15.64 Benzene and pyridine are both single-ring, aromatic compounds. Benzene is a neutral compound that is insoluble in water. Pyridine, with a similar molar mass, is basic and completely miscible with water. Explain these phenomena.

15.65 Name the organic reactants in Problem 15.39.

CHAPTER 16

Aldehydes and Ketones

CONCEPTS TO REVIEW

Electronegativity and molecular polarity
(Sections 5.8, 5.9)

Hydrogen bonding
(Section 8.11)

Naming alkanes
(Section 12.6)

▲ The bombardier beetle defends itself by spraying boiling hot benzoquinone (a ketone) at a predator. The benzoquinone is produced in a fraction of a second by the oxidation of dihydroxybenzene.

CONTENTS

CHAPTER GOALS

In this chapter, we focus on the following questions:

1. What is the carbonyl group?

THE GOAL: Be able to recognize the carbonyl group and describe its polarity and shape.

2. How are ketones and aldehydes named?

THE GOAL: Be able to name the simple members of these families and write their structures, given the names.

3. What are the general properties of aldehydes and ketones?

THE GOAL: Be able to describe such properties as polarity, hydrogen bonding, and water solubility.

4. What are some of the significant occurrences and applications of aldehydes and ketones?

THE GOAL: Be able to specify where aldehydes and ketones are found, list their major applications, and discuss some important members of each family.

5. What are the results of the oxidation and reduction of aldehydes and ketones?

THE GOAL: Be able to describe and predict the products of the oxidation and reduction of aldehydes and ketones.

6. What are hemiacetals and acetals, how are they formed, and how do they react?

THE GOAL: Be able to recognize hemiacetals and acetals, describe the conditions under which they are formed, and predict the products of hemiacetal and acetal formation and acetal hydrolysis.

I n this and the next chapter, we will study several families of compounds that contain what is known as a *carbonyl group*. The carbonyl group has a carbon atom and an oxygen atom connected by a double bond, C=O. The two simplest families of carbonyl compounds are the subject of this chapter, the *aldehydes* and *ketones*. In aldehydes, the carbonyl group is bonded to at least one hydrogen atom, so that the —CHO group (the common abbreviation for the aldehyde functional group) falls at one end of a molecule. In ketones, the carbonyl group is bonded to 2 carbon atoms and thus can never be on the end of a molecule.

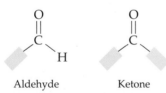

Aldehyde   Ketone

Where might you encounter aldehydes or ketones? The aromas of many flowers and plants derive largely from natural aldehydes. Cookies baking in the oven fill the air with the smell of cinnamon, vanilla, or almond—all natural flavors and aromas due to aldehydes. Have you ever burned a citronella candle to repel mosquitoes? Or enjoyed the smell of cherries? These too are the aromas of aldehydes. Among ketones, jasmone from the jasmine flower and muscone from the male musk deer are vital to the complex formulations of expensive perfumes. Aldehyde and ketone functional groups also play essential roles in the carbohydrates, biomolecules that will be our focus in Chapter 22.

16.1 The Carbonyl Group

Carbonyl compounds are distinguished by the presence of a **carbonyl group** (C=O) and are classified according to what is bonded to the carbonyl carbon, as illustrated in Table 16.1.

Since oxygen is more electronegative than carbon (Section 5.8), carbonyl groups are strongly polarized, with a partial positive charge on the carbon atom and a

Carbonyl compound Any compound that contains a carbonyl group C=O.

Carbonyl group A functional group that has a carbon atom joined to an oxygen atom by a double bond.

TABLE 16.1 Some Kinds of Carbonyl Compounds

FAMILY NAME	STRUCTURE	EXAMPLE	
Aldehyde	$R\overset{\overset{\displaystyle O}{\|\|}}{-C}-H$	$H_3C\overset{\overset{\displaystyle O}{\|\|}}{-C}-H$	Acetaldehyde
Ketone	$R\overset{\overset{\displaystyle O}{\|\|}}{-C}-R'$	$H_3C\overset{\overset{\displaystyle O}{\|\|}}{-C}-CH_3$	Acetone
Carboxylic acid	$R\overset{\overset{\displaystyle O}{\|\|}}{-C}-O-H$	$H_3C\overset{\overset{\displaystyle O}{\|\|}}{-C}-O-H$	Acetic acid
Ester	$R\overset{\overset{\displaystyle O}{\|\|}}{-C}-O-R'$	$H_3C\overset{\overset{\displaystyle O}{\|\|}}{-C}-O-CH_3$	Methyl acetate
Amide	$R\overset{\overset{\displaystyle O}{\|\|}}{-C}-N\diagup$	$H_3C\overset{\overset{\displaystyle O}{\|\|}}{-C}-NH_2$	Acetamide

partial negative charge on the oxygen atom. (⬭, p. 128) The polarity of the carbonyl group gives rise to its reactivity.

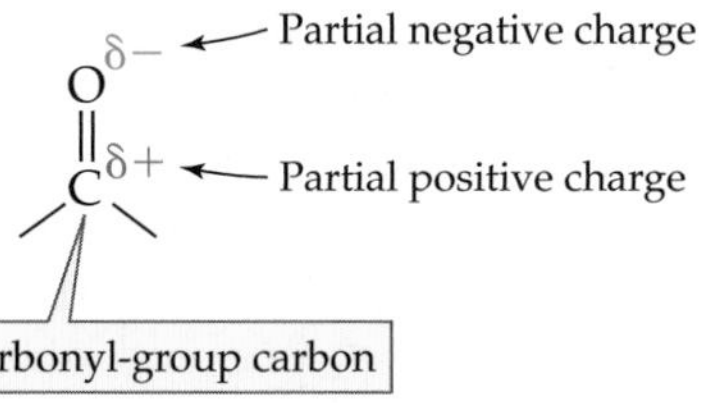

Partial negative charge

Partial positive charge

Carbonyl-group carbon

Another property common to all carbonyl groups is planarity. The bond angles between the three substituents on the carbonyl carbon atom are 120°, or close to it.

120° angles, in a planar triangle

Aldehyde A compound that has a carbonyl group bonded to at least one hydrogen, RCHO.

Ketone A compound that has a carbonyl group bonded to two carbons in organic groups that can be the same or different, $R_2C{=}O$, RCOR'.

Chemists find it useful to divide carbonyl compounds into two major classes based on their chemical properties. In one group are the **aldehydes** and **ketones**, which have similar properties because their carbonyl groups are bonded to atoms that do not attract electrons strongly—carbon and hydrogen. In the second group are *carboxylic acids, esters,* and *amides* (the *carboxyl* family). The carbonyl-group carbon in these compounds is bonded to an atom (other than carbon or hydrogen) that *does* attract electrons strongly, typically an oxygen or nitrogen atom. This second group of carbonyl-containing compounds is discussed in Chapter 17.

There are various ways of representing carbonyl compound structures on paper. Because of the trigonal planar arrangement of atoms around the carbonyl group, the bonds of the carbonyl carbon are often drawn at 120° angles to remind us that such angles are present in the molecules. Structures like those in Table 16.1, on the other hand, which emphasize the location of the double bond, do not fit well on a single line of type, so the simplified formulas shown below are often used for aldehydes and ketones:

Aldehydes

$R\overset{\overset{\displaystyle O}{\|\|}}{-C}-H$ RCHO

Ketones

$R\overset{\overset{\displaystyle O}{\|\|}}{-C}-R'$ RCOR' or $R_2C{=}O$

For example,

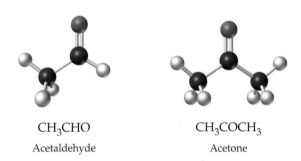

CH$_3$CHO
Acetaldehyde

CH$_3$COCH$_3$
Acetone

The aldehyde group, you will notice, can only be connected to one carbon atom and therefore is always at the end of a carbon chain. The ketone group, by contrast, must be connected to two carbon groups, and thus always occurs within a carbon chain.

PROBLEM 16.1

Which of the following molecules contain aldehyde or ketone functional groups? Copy the formulas and circle these functional groups.

(a)

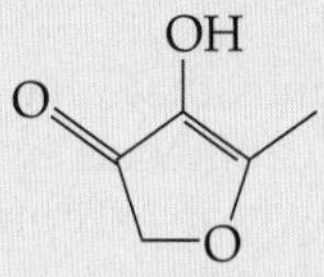

Component of beef flavoring

(b)

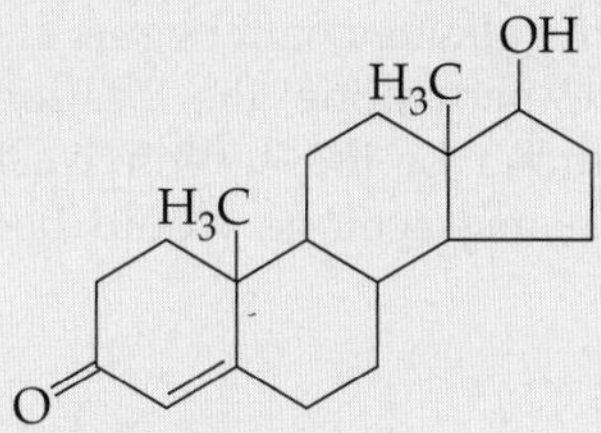

Testosterone
(a male hormone)

(c)

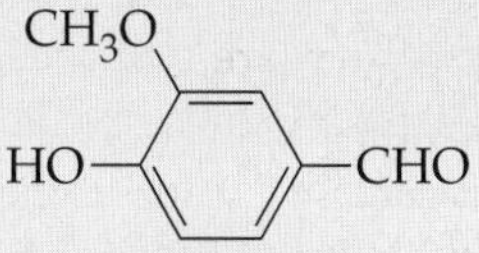

Vanillin
(a flavoring agent)

(d) C$_4$H$_9$COCH$_3$　　　(e) C$_4$H$_9$CHO　　　(f) C$_4$H$_9$COOCH$_3$

PROBLEM 16.2

Draw the structures of compounds (d) and (e) in Problem 16.1 to show all individual atoms and all covalent bonds.

⊂⊃ Looking Ahead

Aldehyde or ketone groups are present in biomolecules with a wide range of functions, from the steroid hormones that regulate sexual function (Section 20.5), to the bases that are essential to nucleic acids and the genetic code (Section 26.2). Most

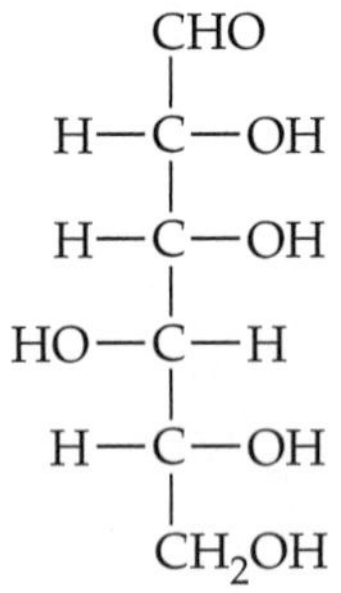

CHO
|
H—C—OH
|
H—C—OH
|
HO—C—H
|
H—C—OH
|
CH₂OH

▲ One of the most important of the aldehyde sugars is glucose, which is also known as dextrose.

distinctively, the structure and reactions of aldehydes and ketones are fundamental to the chemistry of carbohydrates, those in our diet and those that provide energy and structure to our bodies (Chapters 22 and 23).

16.2 Naming Aldehydes and Ketones

The simplest aldehydes are known by their common names, which end in *-aldehyde*, for example, formaldehyde, acetaldehyde, and benzaldehyde. To name aldehydes systematically in the IUPAC system, the final *-e* of the name of the alkane with the same number of carbons is replaced by *-al*. The 3-carbon aldehyde derived from propane is named systematically as propanal, the 4-carbon aldehyde as butanal, and so on. When substituents are present, the chain is numbered beginning with 1 for the carbonyl carbon , as illustrated below for 3-methylbutanal.

Aldehydes

| Formaldehyde | Acetaldehyde | Benzaldehyde | 3-Methylbutanal |

Most simple ketones are best known by common names that give the names of the two alkyl groups bonded to the carbonyl carbon followed by the word *ketone*—for example, methyl ethyl ketone, shown below. An exception to this common-name scheme is seen for the simplest ketone, acetone. Ketones are named systematically by replacing the final *-e* of the corresponding alkane name with *-one* (pronounced **own**). The numbering of the alkane chain begins at the end nearest the carbonyl group. As shown here for 2-pentanone, the location of the carbonyl group is indicated by placing the number of the carbonyl carbon in front of the name. Using this nomenclature scheme, acetone would be named 2-propanone.

Ketones

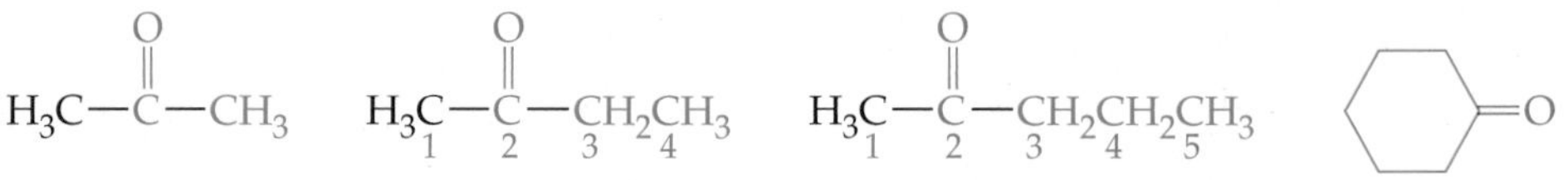

| Acetone (2-Propanone) | Methyl ethyl ketone (2-Butanone) | 2-Pentanone | Cyclohexanone |

WORKED EXAMPLE 16.1 Naming a Ketone Given Its Structure

Give both the systematic (IUPAC) name and the common name for the following compound:

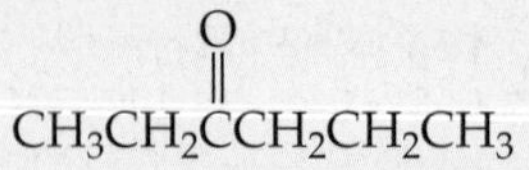

O
‖
CH₃CH₂CCH₂CH₂CH₃

ANALYSIS The compound is a ketone, as shown by the single carbonyl group bonded to two alkyl groups: an ethyl group on the left (CH_3CH_2-) and a propyl group on the right ($-CH_2CH_2CH_3$). The IUPAC system identifies and numbers carbon chains to indicate where the carbonyl group is located, counting in the direction that gives the carbonyl carbon the lowest number possible.

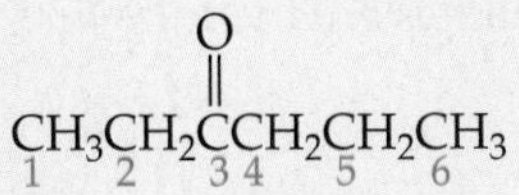

$$\underset{1}{CH_3}\underset{2}{CH_2}\underset{3}{\overset{\overset{\displaystyle O}{\|}}{C}}\underset{4}{CH_2}\underset{5}{CH_2}\underset{6}{CH_3}$$

The common name uses the names of the two alkyl groups.

SOLUTION

The IUPAC name is 3-hexanone. The common name is ethyl propyl ketone.

PROBLEM 16.3

Draw structures corresponding to the following names:

(a) Octanal   **(b)** Methyl phenyl ketone

(c) 4-Methylhexanal   **(d)** Methyl *tert*-butyl ketone

APPLICATION ▶   Chemical Warfare among the Insects

Life in the insect world is a jungle. Predators abound, just waiting to make a meal of any insect that happens along. To survive, insects have evolved extraordinarily effective means of chemical protection. Take the humble millipede *Apheloria corrugata*, for example. When attacked by ants, the millipede protects itself by discharging benzaldehyde cyanohydrin.

Cyanohydrins [RCH(OH)C≡N] are formed by addition of the toxic gas HCN (hydrogen cyanide) to ketones or aldehydes, not unlike the addition of HCl or H_2O to alkenes (Section 13.6). The reaction with HCN to yield a cyanohydrin is reversible, just like the reaction of a ketone or aldehyde with an alcohol to yield a hemiacetal (Section 16.7). Thus, the benzaldehyde cyanohydrin secreted by the millipede decomposes to yield benzaldehyde and HCN. This action protects the millipede because the decomposition reaction releases deadly hydrogen cyanide gas, a remarkably clever and very effective kind of chemical warfare.

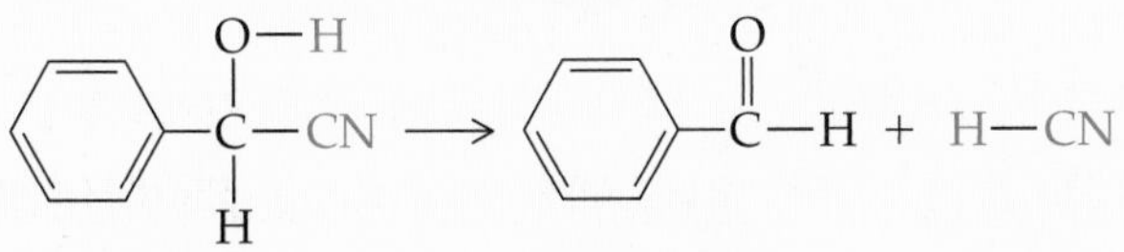

▲ The beautifully colored millipede *Apheloria corrugata* can produce as much as 0.6 mg of HCN to defend itself against attacks.

Benzaldehyde cyanohydrin   Benzaldehyde   Hydrogen cyanide

The potent chemical weapon of the bombardier beetle (featured at the opening of this chapter) is benzoquinone, the simplest member of a class of compounds that are cyclohexadienediones (cyclohexene rings with two double bonds and two carbonyl groups). When threatened, the bombardier beetle initiates the enzyme-catalyzed oxidation of dihydroxybenzene by hydrogen peroxide. A hot cloud (up to 100 °C) of irritating benzoquinone vapor shoots out of the beetle's defensive organ with such force that it sounds like a pistol shot.

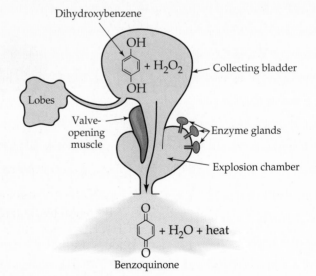

▲ Defensive organ of the bombardier beetle and the chemical warfare factory it contains.

See Additional Problems 16.48 and 16.49 at the end of the chapter.

PROBLEM 16.4

Give systematic, IUPAC names for the following compounds:

(a) $\underset{\displaystyle \text{O}}{\overset{\displaystyle \parallel}{}}$ CH$_3$CH$_2$CH$_2$CH$_2$CH (b) CH$_3$CH$_2$CCH$_2$CH$_3$

(c) $\underset{\displaystyle \text{CH}_3}{|}$ CH$_3$CH$_2$CHCH$_2$CH$_2$CH (d) Dipropyl ketone

KEY CONCEPT PROBLEM 16.5

Which of these two molecules is a ketone and which is an aldehyde? Write the condensed formulas for both of them.

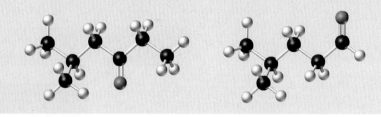

16.3 Properties of Aldehydes and Ketones

The polarity of the carbonyl group makes aldehydes and ketones moderately polar compounds. As a result, boil at a higher temperature than alkanes with similar molecular weights. Since they have no hydrogen atoms bonded to oxygen or nitrogen, individual molecules do not hydrogen-bond with each other, which makes aldehydes and ketones lower boiling than alcohols. In a series of compounds with similar molecular weights, the alkane is lowest boiling, the alcohol is highest boiling, and the aldehyde and ketone fall in between.

CH$_3$CH$_2$CH$_2$CH$_3$

Butane, bp 0 °C

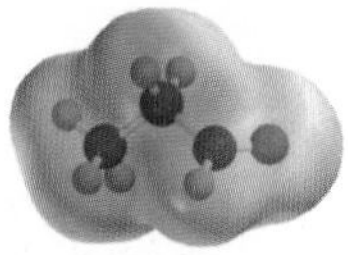

$\overset{\displaystyle \text{O}}{\overset{\displaystyle \parallel}{}}$
CH$_3$CH$_2$CH

Propanal, bp 50 °C

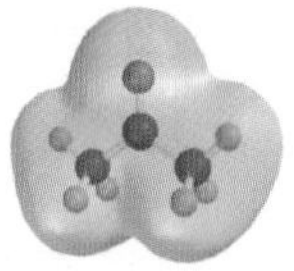

$\overset{\displaystyle \text{O}}{\overset{\displaystyle \parallel}{}}$
CH$_3$CCH$_3$

Acetone, bp 56 °C

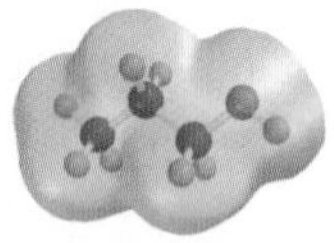

CH$_3$CH$_2$CH$_2$OH

Propanol, bp 97 °C

Formaldehyde (HCHO), the simplest aldehyde, is a gas; acetaldehyde (CH$_3$CHO) boils close to room temperature. The other simple aldehydes and ketones are liquids (Table 16.2), and those with more than 12 carbon atoms are solids.

Aldehydes and ketones are soluble in common organic solvents, and those with fewer than five or six carbon atoms are also soluble in water because they are able to hydrogen-bond with water molecules (Figure 16.1).

Simple ketones are excellent solvents because they dissolve both polar and nonpolar compounds. With increasing numbers of carbon atoms, aldehydes and ketones become more alkane-like and less water-soluble.

TABLE 16.2 Physical Properties of Some Simple Aldehydes and Ketones

STRUCTURE	NAME	BOILING POINT (°C)	WATER SOLUBILITY (g/100 mL H$_2$O)
HCHO	Formaldehyde	−21	55
CH$_3$CHO	Acetaldehyde	21	Soluble
CH$_3$CH$_2$CHO	Propanal	49	16
CH$_3$CH$_2$CH$_2$CHO	Butanal	76	7
CH$_3$CH$_2$CH$_2$CH$_2$CHO	Pentanal	103	1
⬡—CHO	Benzaldehyde	178	0.3
CH$_3$COCH$_3$	Acetone	56	Soluble
CH$_3$CH$_2$COCH$_3$	2-Butanone	80	26
CH$_3$CH$_2$CH$_2$COCH$_3$	2-Pentanone	102	6
⬡=O	Cyclohexanone	156	2

▲ A perfumer sits at a mixing table testing new combinations of fragrances, many of which are aldehydes and ketones.

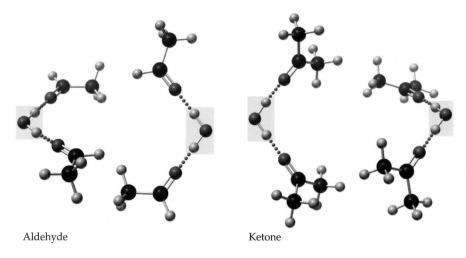

Aldehyde   Ketone

◀ **FIGURE 16.1 Hydrogen bonding with water (highlighted in blue) of an aldehyde (CH$_3$CHO) and a ketone.**

The structures of a few naturally occurring aldehydes and ketones with distinctive odors are shown below. Citronellal (used in citronella candles) is one of about a dozen compounds with similar structures that contribute to the aroma of oils extracted from geraniums, roses, citronella (a tropical plant), and lemon grass. All are used in soaps, cosmetics, and perfumes.

CH$_3$C=CHCH$_2$CH$_2$CHCH$_2$CHO
    | |
   CH$_3$ CH$_3$

Citronellal
(insect repellant, also used in perfumes; from citronella and lemon grass oils)

Cinnamaldehyde
(cinnamon flavor in foods, drugs; from cinnamon bark)

Camphor
(moth repellant from camphor tree)

Civetone
(musky odor in perfumes; from the scent gland of the civet cat)

The lower-boiling aldehydes and ketones are flammable and can form explosive mixtures with air. The simple ketones generally have low toxicity, but the simple aldehydes, especially formaldehyde, are toxic.

APPLICATION ▶ Vanilla: Which Kind Is Best?

You are standing at the flavoring shelf in the supermarket, thinking about the batch of sugar cookies you are going to bake. You need vanilla extract but are faced with a dilemma: which should you buy, the bottle of "artificial vanilla extract" or the bottle of "pure vanilla extract"? Chemical logic tells you that an "artificial" flavoring compound—that is, one made synthetically from other chemicals—*should* have exactly the same taste and aroma as the identical compound obtained from a plant. This is a correct assumption to make. The principal flavoring ingredient in vanilla extract is vanillin, and a vanillin molecule synthesized in the laboratory has the same structure and properties as a vanillin molecule extracted from vanilla beans harvested from nature.

▲ Vanilla beans. There is only about 1 g of vanillin in 400 g of beans.

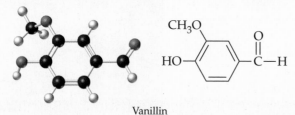

Vanillin

The other components of the two types of vanilla extract are not the same, however, and herein lies the reason why cookies made with pure vanilla extract taste slightly different from cookies made with artificial vanilla extract. The artificial extract contains no flavoring agents other than synthetic vanillin. The pure extract, however, contains more than 250 organic compounds in addition to vanillin, and it is this combination of so many ingredients that gives pure vanilla extract one of the most complex tastes in the world. The resulting richer flavor is, like that of all other natural flavorings and aromas, a blend of the contributions from many chemical compounds.

Pure vanilla extract may also contain added sugar, corn syrup, caramel, colors, and/or stabilizers. Just like fine wine, vanilla extracts continue to develop body and depth for about two years, at which time they reach their peak.

See Additional Problem 16.50 at the end of the chapter.

Properties of Aldehydes and Ketones

- Aldehydes and ketone molecules are polar due to the presence of the carbonyl group.
- Since aldehydes and ketones cannot hydrogen-bond with one another, they are lower boiling than alcohols but higher boiling than alkanes because of their polarity.
- Common aldehydes and ketones are typically liquids.
- Simple aldehydes and ketones are water-soluble due to hydrogen bonding with water molecules, and ketones are good solvents.
- Many aldehydes and ketones have distinctive odors.
- Simple ketones are less toxic than simple aldehydes.

◉ KEY CONCEPT PROBLEM 16.6

For the compound shown below, indicate which of each pair of properties is most applicable.

$$CH_3CH_2\overset{\overset{\displaystyle O}{\|}}{C}CH_2CH_3$$

(a) Polar or nonpolar
(b) Flammable or nonflammable
(c) Solid or liquid
(d) A boiling point of 250 °C or of 100 °C

KEY CONCEPT PROBLEM 16.7

Why do aldehydes and ketones have lower boiling points than alcohols with similar molecular weights? Why are their boiling points higher than those of alkanes with similar molecular weights?

16.4 Some Common Aldehydes and Ketones

Formaldehyde (HCHO): Toxic but Useful

At room temperature, pure formaldehyde is a colorless gas with a pungent, suffo-cating odor. Low concentrations in the air (0.1–1.1 ppm) can cause eye, throat, and bronchial irritation, and higher concentrations can trigger asthma attacks. Skin con-tact can produce dermatitis. Because formaldehyde is formed during incomplete combustion of hydrocarbon fuels, it is partly responsible for the irritation caused by smog-laden air. Formaldehyde is very toxic by ingestion, causing serious kidney damage, coma, and sometimes even death; it is a product of the biochemical break-down of ingested methanol (methyl alcohol, also known as wood alcohol) and is one of the reasons that drinking methanol is so toxic.

Formaldehyde is commonly sold as a 37% aqueous solution (weight/weight percent, meaning 37 g of formaldehyde in 100 g of solution) under the name *formalin*. It kills viruses, fungi, and bacteria by reaction with the —NH₂ groups in proteins, allowing for its use in disinfecting and sterilizing equipment. It is too harsh for use on the skin and was once commonly used as a preservative for bio-logical specimens. On standing, formaldehyde polymerizes into a solid known as *paraformaldehyde*. At one time, paraformaldehyde candles were burned to disinfect hospital rooms that had been occupied by patients with contagious diseases.

In the chemical industry, the major use of formaldehyde is found in the produc-tion of polymers with applications such as adhesives for binding plywood, foam insulation for buildings, textile finishes, and hard and durable manufactured objects such as telephone parts. The first completely synthetic and commercially successful plastic was a polymer of phenol and formaldehyde known as *Bakelite*, once widely used for such items as pot handles, fountain pens, and cameras. Urea–formaldehyde polymers are now more widely used than Bakelite. Once the final polymerization of such materials is finished, no further melting and reshaping is possible because of its cross-linked, three-dimensional structure. In the general formulas below, the red wavy lines indicate bonds to the rest of the polymer; the CH₂ groups are from the formaldehyde molecules.

Formaldehyde

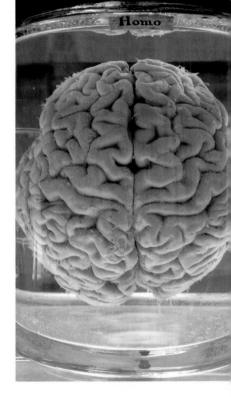

▲ Aqueous formaldehyde has long been used to preserve biological specimens, such as this brain.

Phenol–formaldehyde polymer

Urea–formaldehyde polymer

Formaldehyde polymers, especially when new, release formaldehyde into the air. Because of concern over the toxicity and possible carcinogenicity of formaldehyde from polymeric materials, their use in most household applications has been limited.

▲ Jewelery made from Bakelite.

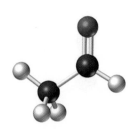

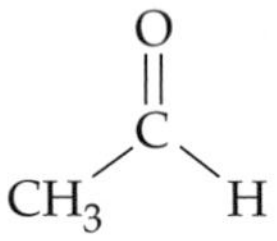

O
‖
C
CH₃   H
Acetaldehyde

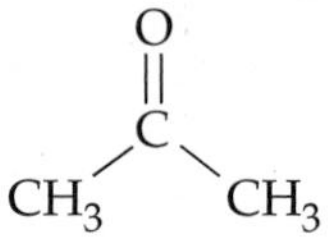

O
‖
C
CH₃   CH₃
Acetone

Acetaldehyde (CH₃CHO): Sweet Smelling but Narcotic

Acetaldehyde is a sweet-smelling, flammable liquid present in ripe fruits, notably apples. It is less toxic than formaldehyde, and small amounts are produced in the normal breakdown of carbohydrates. Acetaldehyde is, however, a general narcotic, and large doses can cause respiratory failure. Chronic exposure produces symptoms like those of alcoholism. (See Application "Ethyl Alcohol as a Drug and a Poison," Chapter 14, p. 447.) It is commonly used as an reagent for silvering mirrors.

Acetone (CH₃COCH₃): A Super Solvent

Acetone, a liquid at room temperature, is one of the most widely used of all organic solvents. It dissolves most organic compounds and is also miscible with water. In paint stores, acetone is sold for general-purpose cleanup work. Acetone is the solvent in many varnishes, lacquers, and nail polish removers.

Acetone is highly volatile and is a serious fire and explosion hazard when allowed to evaporate in a closed space. No chronic health risk has been associated with casual acetone exposure. Unfortunately, it is also one of a large group of readily available products that include volatile solvents (such as benzene, chloroform, model-airplane glue, and nail polish remover) that are abused by inhalation to produce alcohol-like intoxication, sometimes with serious outcomes.

When the biochemical breakdown of fats and carbohydrates to yield energy is out of balance (for example, in starvation or diabetes mellitus), acetone is produced in the liver, a condition known as ketosis that in severe cases leaves the odor of acetone on a patient's breath (Section 25.7).

⚭ Looking Ahead

In the biochemistry chapters that lie ahead, you will find that all of the simplest sugars—the monosaccharides (Section 22.4)—contain either an aldehyde group or a ketone group. Glucose, the 6-carbon sugar shown below, plays a major role in metabolism as the primary fuel molecule for energy generation (Section 23.2).

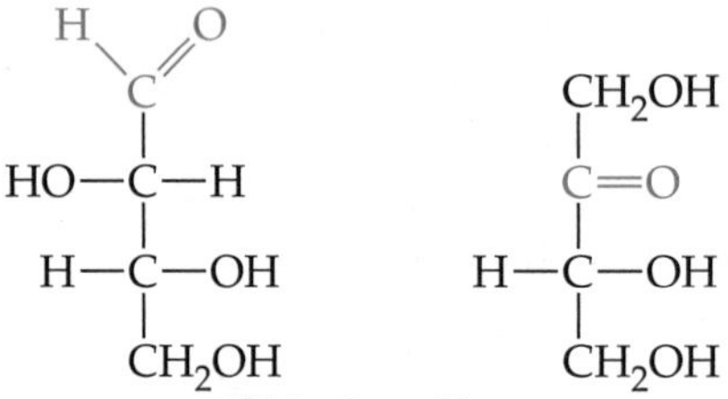

H   O
  \ ∥
   C
   |
HO—C—H
   |
H—C—OH
   |
CH₂OH

CH₂OH
   |
C=O
   |
H—C—OH
   |
CH₂OH

Aldehyde and ketone
four-carbon sugars

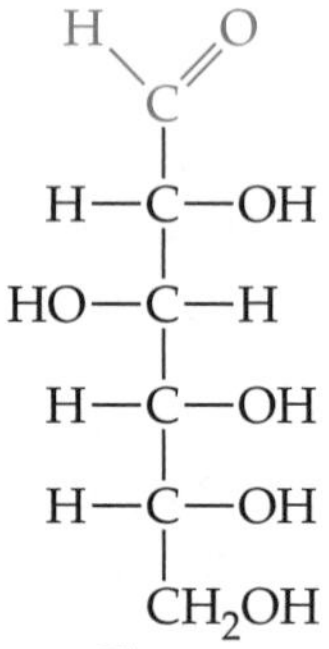

H   O
  \ ∥
   C
   |
H—C—OH
   |
HO—C—H
   |
H—C—OH
   |
H—C—OH
   |
CH₂OH

Glucose

⚬ KEY CONCEPT PROBLEM 16.8

Identify the functional groups in the following compounds.

(a) 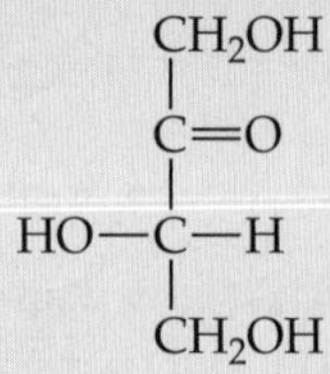
CH₂OH
   |
C=O
   |
HO—C—H
   |
CH₂OH

(b)
H   O
  \ ∥
   C
   |
HO—C—H
   |
CH₂OH

(c) ⬡—CH₂CHO

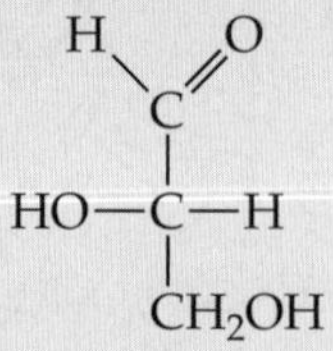

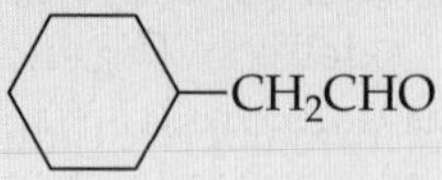

(d) H₂NCH₂CH₂COCH₃

16.5 Oxidation of Aldehydes

Alcohols can be oxidized to aldehydes or ketones (⬭, p. 444), and aldehydes can be further oxidized to carboxylic acids. In aldehyde oxidation, the hydrogen bonded to the carbonyl carbon is replaced by an —OH group. Ketones, because they do not have this hydrogen, do not react cleanly with oxidizing agents (except with those strong enough to destroy the molecule).

Oxidation of aldehydes and ketones

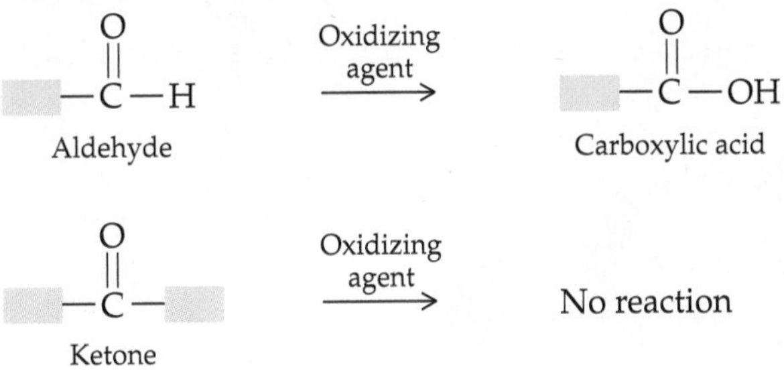

For example,

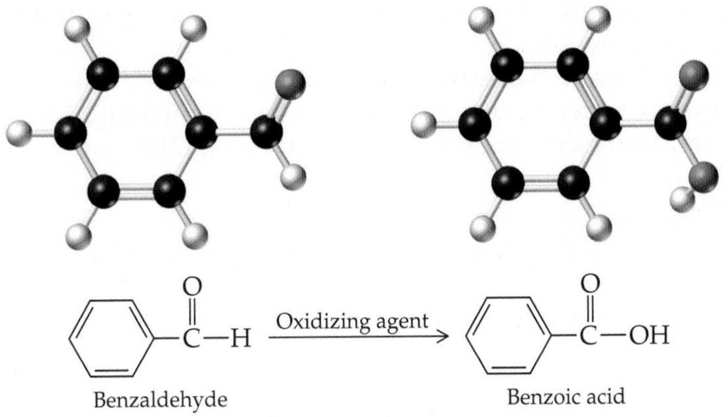

Benzaldehyde → Benzoic acid

Of the mild oxidizing agents that convert aldehydes to carboxylic acids, oxygen in the air is the simplest. Aldehydes typically have a musty odor; this is due to their partial oxidation to carboxylic acids, which generally have a strong, unpleasant odor. To prevent air oxidation, aldehydes are often stored under nitrogen gas.

Because ketones cannot be oxidized, treatment with a mild oxidizing agent is used as a test to distinguish between aldehydes and ketones. *Tollens' reagent,* which consists of a solution containing silver ion in aqueous ammonia, is the most visually appealing oxidizing agent for aldehydes. Treatment of an aldehyde with this reagent, in which the Ag^+ ion (present as $[Ag(NH_3)_2]^+$) is the oxidizing agent, rapidly yields the carboxylic acid anion and metallic silver. If the reaction is done in a clean glass container, metallic silver deposits on the inner walls, producing a beautiful shiny mirror (Figure 16.2a). Before modern instrumental methods were available, chemists had to rely on such visible chemical changes to identify chemical compounds.

Tollens' test

$$RCHO + Ag(NH_3)_2^+ \xrightarrow{NH_3, H_2O} RCOO^- + NH_4^+ + Ag$$

Tollens' reagent   Silver
(colorless)   mirror

A test with another mild oxidizing agent, known as *Benedict's reagent,* also relies on reduction of a metal to produce visible evidence of the presence of aldehydes. The reagent solution contains blue copper(II) ion, which is reduced to give a precipitate of red copper(I) oxide in the reaction with an aldehyde (Figure 16.2b). Benedict's reagent is also used to test for the presence of ketones that have an —OH on the carbon next to the carbonyl (*alpha* hydroxy ketones), a common grouping

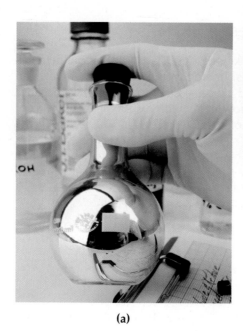

(a) **(b)**

► **FIGURE 16.2 The Tollens' and Benedict's tests for aldehydes.** (a) In the Tollens' test, colorless silver ion (Ag^+) is reduced to metallic silver. (b) In the Benedict's test for aldehyde-containing sugars, the blue copper(II) ion (Cu^{2+}) is reduced to copper(I) in brick-red copper(I) oxide (Cu_2O). In the photo, the Cu^{2+} containing reagent solution is on the left. A large quantity of glucose produces the brick-red precipitate on the right. In both tests, the aldehyde is oxidized to the carboxylic acid.

of atoms found in sugars. As with aldehydes, a red copper(I) precipitate is evidence of the presence of these ketones.

At one time, Benedict's reagent was extensively used as a test for sugars in the urine. Today, more specific, enzyme-based tests are preferred (see Application "Diagnosis and Monitoring of Diabetes," Chapter 23, p. 744).

Benedict's test

$$RCHO + Cu^{2+} \xrightarrow{\text{Buffer}} RCOO^- + Cu_2O$$

Blue
in solution
Brick-red
solid

PROBLEM 16.9

Draw structures of the products you would obtain by treating the following compounds with Tollens' reagent. If no reaction occurs, write "NR."

(a) $CH_3CHCH_2CH_2CH_2CHO$ with CH_3 above the first CH
    (b) 2,2-Dimethylpentanal
(c) 2-Methyl-3-pentanone

16.6 Reduction of Aldehydes and Ketones

The reduction of a carbonyl group occurs with the addition of hydrogen across the double bond to produce an —OH group, a reaction that is the reverse of the oxidation of an alcohol:

Aldehyde or ketone $\underset{\text{Oxidation}}{\overset{\text{Reduction}}{\rightleftharpoons}}$ Alcohol

Aldehydes are reduced to primary alcohols, and ketones are reduced to secondary alcohols:

Aldehyde $\xrightarrow{\text{Reducing agent}}$ Primary alcohol

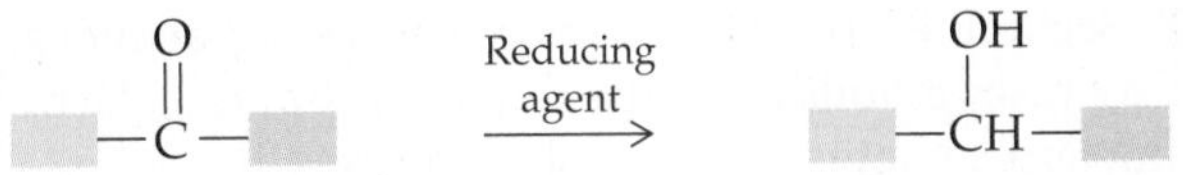

$$
\underset{\text{Ketone}}{\blacksquare\!-\!\overset{\displaystyle O}{\overset{\|}{C}}\!-\!\blacksquare}
\quad\xrightarrow[\text{agent}]{\text{Reducing}}\quad
\underset{\text{Secondary alcohol}}{\blacksquare\!-\!\overset{\displaystyle OH}{\overset{|}{CH}}\!-\!\blacksquare}
$$

These reductions occur by formation of a bond to the carbonyl carbon atom by a hydride ion (H^-) accompanied by bonding of a hydrogen ion (H^+) to the carbonyl oxygen atom. The reductions make good sense when you think about the polarity of the carbonyl group. The carbonyl-group carbon has a partial positive charge because electrons are drawn away by the electronegative oxygen atom, so the negatively charged hydride ion is drawn to this carbon atom. Because the oxygen atom has a partial negative charge, the positively charged hydrogen atom is attracted there.

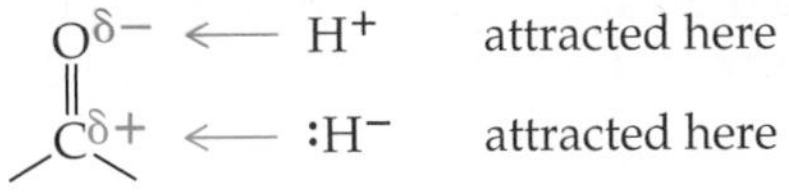

$$
\begin{array}{ll}
O^{\delta-} \longleftarrow H^+ & \text{attracted here}\\
\overset{\|}{C}{}^{\delta+} \longleftarrow :H^- & \text{attracted here}
\end{array}
$$

Note that a hydride ion (H^-) has a lone pair of valence electrons. Both electrons are used to form a covalent bond to the carbonyl carbon. This change leaves a negative charge on the carbonyl oxygen. Aqueous acid is then added, H^+ bonds to the oxygen, and a neutral alcohol results. Thus, the two new hydrogen atoms in the alcohol product come from different sources.

Reduction of an aldehyde

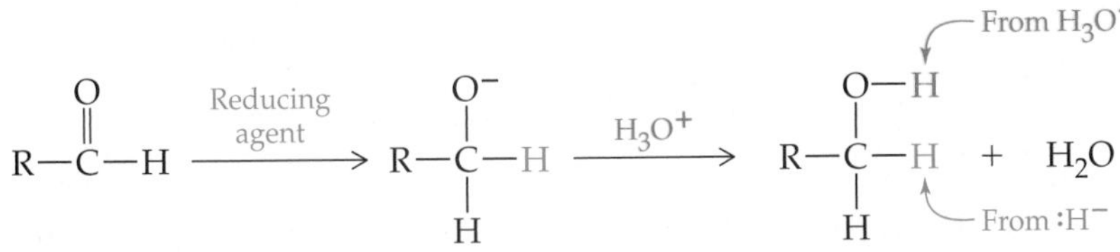

$$
\underset{\text{Aldehyde}}{R\!-\!\overset{\displaystyle O}{\overset{\|}{C}}\!-\!H}
\xrightarrow[\text{agent}]{\text{Reducing}}
R\!-\!\overset{\displaystyle O^-}{\underset{\displaystyle H}{\overset{|}{\underset{|}{C}}}}\!-\!H
\xrightarrow{H_3O^+}
\underset{\text{Primary alcohol}}{R\!-\!\overset{\displaystyle O\!-\!H}{\underset{\displaystyle H}{\overset{|}{\underset{|}{C}}}}\!-\!H}
\; + \; H_2O
$$

From H_3O^+
From $:H^-$

$$
\underset{\text{Propanal}}{CH_3CH_2\overset{\displaystyle O}{\overset{\|}{CH}}}
\xrightarrow[H_3O^+]{\underset{\text{agent}}{\text{Reducing}}}
\underset{\text{1-Propanol}}{CH_3CH_2CH_2OH}
$$

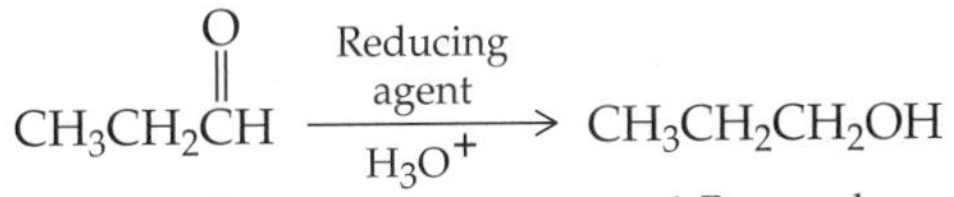

Reduction of a ketone

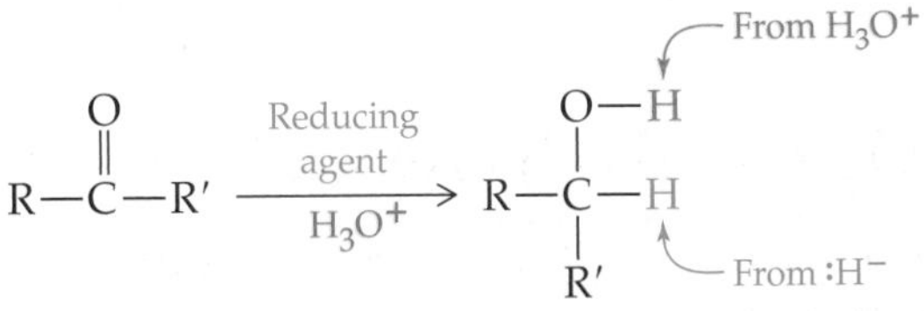

$$
\underset{\text{Ketone}}{R\!-\!\overset{\displaystyle O}{\overset{\|}{C}}\!-\!R'}
\xrightarrow[H_3O^+]{\underset{\text{agent}}{\text{Reducing}}}
\underset{\substack{\text{Secondary}\\\text{alcohol}}}{R\!-\!\overset{\displaystyle O\!-\!H}{\underset{\displaystyle R'}{\overset{|}{\underset{|}{C}}}}\!-\!H}
$$

From H_3O^+
From $:H^-$

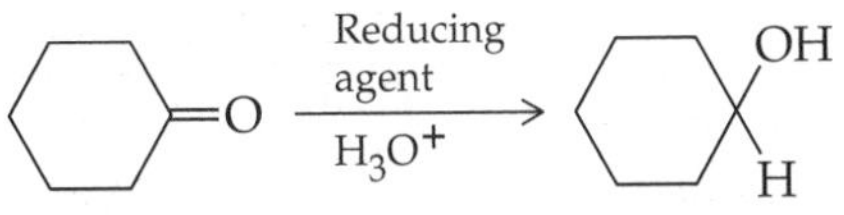

Cyclohexanone   →   Cyclohexanol

with *Reducing agent* / H_3O^+ over the arrow.

In biological systems, the reducing agent for a carbonyl group is often the coenzyme nicotinamide adenine dinucleotide (abbreviated as NAD), which cycles between reacting as a reducing agent (NADH) and an oxidizing agent (NAD^+) by the loss and gain of a hydride ion (H^-). The biochemical reduction of pyruvic acid, a ketone-containing acid that plays a pivotal role in energy production, utilizes

NADH (⚬⚬ , Section 21.4). The reaction occurs in active skeletal muscles. Vigorous exercise causes a buildup of the reduction product, lactic acid, leading to a tired, flat feeling and sometimes to muscle cramps.

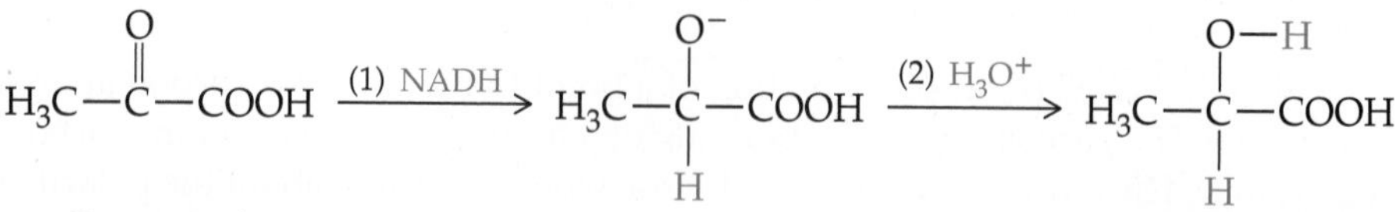

Pyruvic acid   Lactic acid

⚬⚬ **Looking Ahead**

The reduction of aldehydes and ketones to alcohols is an important reaction in living cells, and NADH is the common source of the hydride ion. It donates H^- to an aldehyde or ketone to yield an anion, which then picks up H^+ from surrounding aqueous fluids. The major role of NADH as a biochemical reducing agent is introduced in Section 21.7.   ⚬⚬

WORKED EXAMPLE **16.2** Writing the Products of a Carbonyl Reduction

What product would you obtain by reduction of benzaldehyde?

ANALYSIS First, draw the structure of the starting material, showing the double bond in the carbonyl group. Then rewrite the structure showing only a single bond between C and O, along with partial bonds to both C and O:

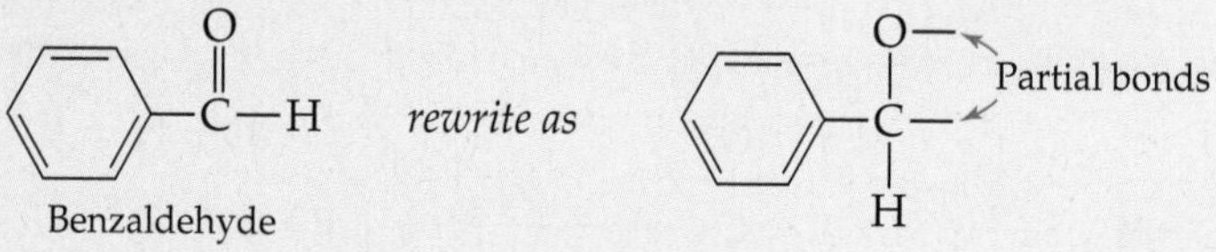

Benzaldehyde

Finally, attach hydrogen atoms to the two partial bonds and rewrite the product.

SOLUTION
The product obtained is benzyl alcohol

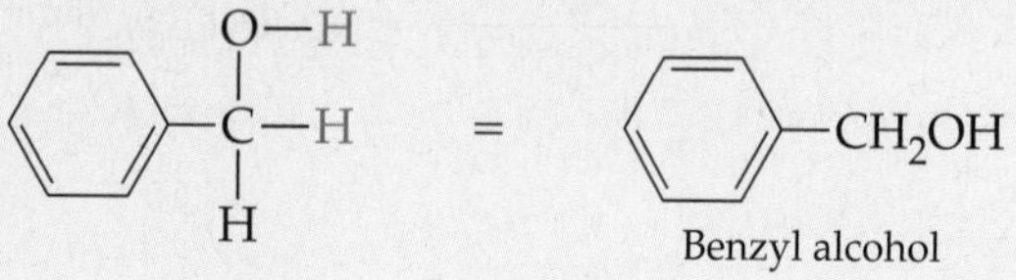

Benzyl alcohol

PROBLEM 16.10

What product would you obtain from reduction of the following ketones and aldehydes?

(a) $(CH_3)_2CH(C=O)CH_3$ (b) *p*-Hydroxybenzaldehyde
(c) 2-Methylcyclopentanone

PROBLEM 16.11

What ketones or aldehydes might be reduced to yield the following alcohols?

(a)   (b) $HOCH_2CH_2CH_2CHCH_3$
                                          |
                                        CH_3
(c) 2-Methyl-1-pentanol

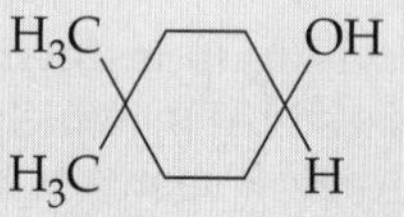

In its broadest meaning, the term *drug* refers to any chemical agent other than food that affects living organisms. More commonly, we use the term to mean substances that prevent or treat disease. By contrast, a *poison* or *toxic substance* is any chemical agent that harms living organisms. (See Application "Toxicology" in Chapter 15.) Strange as it may seem, the categories "drugs" and "poisons" are not mutually exclusive. Often a substance that in low concentrations cures disease or alleviates symptoms causes injury or death when taken in larger amounts. (This is precisely true, for instance, of almost all drugs used to treat cancer.) The sixteenth-century German physician Paracelsus understood that most substances cannot be absolutely categorized as being either safe or toxic. He expressed this insight in a famous phrase: "The dose makes the poison."

The term *dose* refers to the amount of substance that enters the body at one time. As just mentioned in the preceding paragraph, a small dose of a substance may be a life-saving medication, whereas a large dose of the same substance may be poisonous. Vitamin A, for example, is essential to our health, and our daily diet should include about 1 mg, but a single dose larger than 200 mg is classified as toxic because it can cause nausea, fatigue, and other unpleasant symptoms. Another example is botulinum toxin, a neurotoxin protein produced by the bacterium *Clostridium botulinum*. This protein is one of the most poisonous naturally occurring substances in the world and yet in minute doses is used both to treat painful muscle spasms and as a cosmetic treatment; you know it as Botox.

One standard method for reporting toxicity is the LD_{50}, or lethal dose, 50%, which is a measure of the toxicity of a single dose, known as *acute* toxicity. (Exposure over a long period of time to doses of a drug at sub-acute levels is known as *chronic* toxicity and is a much harder parameter to quantify.) The LD_{50} for a material is determined as follows: A substance is fed in varying doses to laboratory animals, frequently rats or mice, and the mortality rate of the animals recorded. The result of the test is reported as the LD_{50}, the dose that kills 50% of the animals in a uniform population, say all male rats with a similar genetic makeup being fed an identical diet. Because LD_{50} values are listed in standard chemical references, such as *The Merck Index* (see Application "Chemical Information," Chapter 15), they are the most easily found type of toxicity data.

By comparing LD_{50} values, relative toxicities of various laboratory chemicals can be evaluated. The LD_{50} values in the table, reported as the dose in grams per kilogram of body weight of the test animal, show that in this group of compounds formaldehyde is the most toxic and acetone the least toxic by ingestion.

	LD_{50}
Formaldehyde, 37% aqueous solution	800 mg/kg
Acetaldehyde	1.9 g/kg
Butyraldehyde	5.8 g/kg
Acetone	8.4 g/kg

▲ **Beautiful but deadly.** *Amanita muscaria*, a highly toxic mushroom. To pick wild mushrooms, one must know the difference between poisonous and nonpoisonous varieties very well.

To put these values in perspective, however, compare them with the LD_{50} of some naturally occurring toxins: 1 ng/kg for botulinum toxin or the LD_{50} of 0.23 mg/kg for muscarine, the poisonous chemical in *Amanita muscaria* mushrooms.

For therapeutic use in humans, a compound must show a comfortably wide margin between the dose that produces the desired effect and the dose that produces an acute toxic effect. Aspirin, for example, has an LD_{50} of 1.75 g/kg in rats. An average therapeutic dose for humans is two aspirin tablets, which contain only 650 mg, or 0.01 g/kg for an average 65 kg person—1/175 of the LD_{50}, if we assume that the results for rats can be extrapolated to human beings. This is a fairly comfortable safety margin, but not an extraordinarily generous one (overdoses of aspirin, accidental or deliberate, still cause a significant number of fatalities each year).

The LD_{50} test is controversial; it has many drawbacks and many advantages. Among the drawbacks are the need to sacrifice large numbers of animals; the wide variation of results with animal characteristics such as species, sex, age, and diet; and the difficulty of extrapolating results from test animals to humans. (Obviously we cannot conduct LD_{50} tests in humans.) The advantages include the relative speed of the test; the information it provides on the cause of toxicity (obtained via post mortem examination of the animals); and the value of the test as a first approximation of hazards to those exposed.

The alternative to animal testing of toxicity is *in vitro* testing: testing not carried out in living animals, but in laboratory glassware. (*In vitro* literally means "in glass.") *In vitro* tests rely on observing the effects of chemicals on cultured living cells; the results of these tests are also controversial and fraught with their own troubles. Clearly, much work remains to be done in the field of toxicology.

See Additional Problems 16.51–53 at the end of the chapter.

16.7 Addition of Alcohols: Hemiacetals and Acetals

Hemiacetal Formation

Addition reaction, aldehydes and ketones Addition of an alcohol or other compound to the carbon double bond to give a carbon–oxygen single bond.

Hemiacetal A compound with both an alcohol-like —OH group and an ether-like —OR group bonded to the same carbon atom.

Aldehydes and ketones undergo **addition reactions** in which an alcohol combines with the carbonyl carbon and oxygen. The carbonyl double bond is converted to a single bond in the addition reaction, which is similar to the addition of water to a carbon-carbon double bond (Section 13.6). The initial product of addition reactions with alcohols are known as *hemiacetals*. **Hemiacetals** have both an alcohol-like —OH group and an ether-like —OR group bonded to what was once the carbonyl carbon atom. The H from the alcohol bonds to the carbonyl-group oxygen, and the OR from the alcohol bonds to the carbonyl-group carbon.

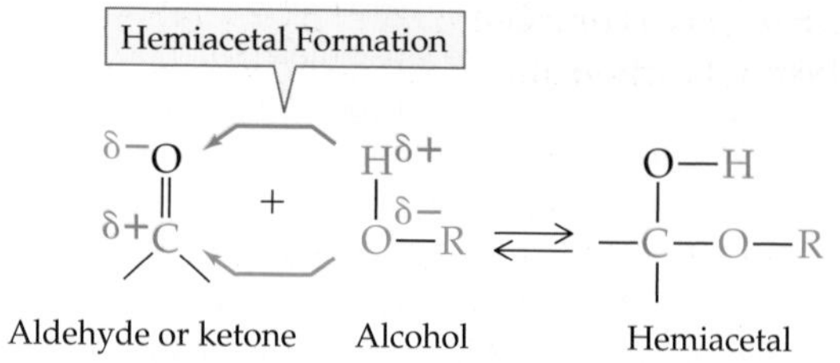

Hemiacetal Formation

Aldehyde or ketone · Alcohol · Hemiacetal

- **The negatively polarized alcohol oxygen atom adds to the positively polarized carbonyl carbon** (similar to what happens in reduction of the carbonyl group). Almost all carbonyl-group reactions follow this same polarity pattern.

- **The reaction is reversible.** Hemiacetals rapidly revert back to aldehydes or ketones by loss of alcohol and establish an equilibrium with the aldehyde or ketone.

Ethanol (CH_3CH_2OH) forms hemiacetals with acetaldehyde and acetone as follows:

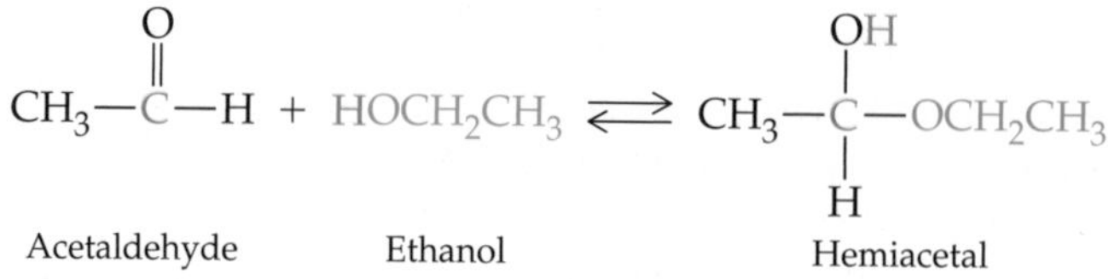

Acetaldehyde · Ethanol · Hemiacetal

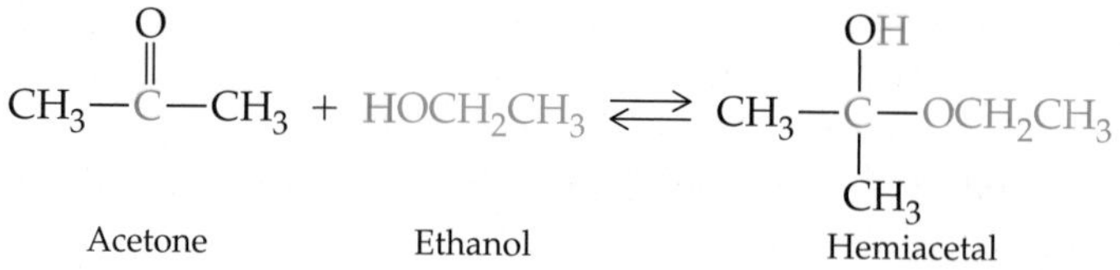

Acetone · Ethanol · Hemiacetal

In practice, hemiacetals are often too unstable to be isolated. When equilibrium is reached, very little hemiacetal is present. A major exception occurs when the —OH and —CHO functional groups that react are part of the *same* molecule. The resulting cyclic hemiacetal is more stable than a noncyclic hemiacetal. Because of their greater stability, most simple sugars exist mainly in the cyclic hemiacetal form, as shown below for glucose, rather than in the open-chain form shown on p. 492.

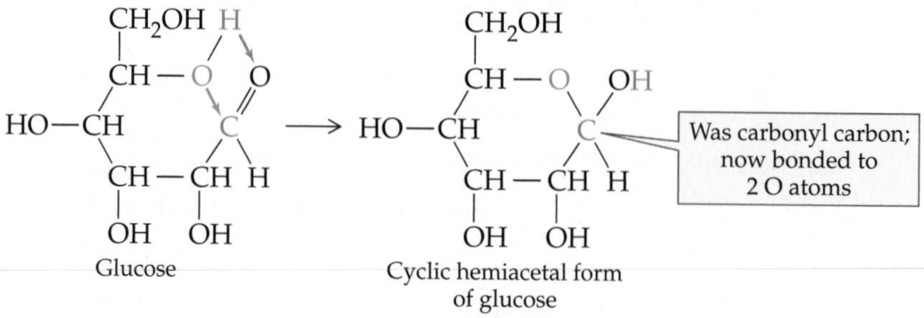

Glucose · Cyclic hemiacetal form of glucose

Was carbonyl carbon; now bonded to 2 O atoms

The cyclic form of glucose is customarily written as

$$CH_2OH$$

Acetal Formation

If a small amount of acid catalyst is added to the reaction of an alcohol with an aldehyde or ketone, the hemiacetal initially formed is converted into an *acetal* in a substitution reaction. An **acetal** is a compound that has *two* etherlike —OR groups bonded to what was the carbonyl carbon atom (the two —OR groups can be different):

Acetal A compound that has two ether-like —OR groups bonded to the same carbon atom.

Aldehyde or ketone   Hemiacetal   Acetal

For example,

Acetaldehyde   Ethanol   Acetal

Acetone   Ethanol   Acetal

(In old nomenclature, a distinction was made by calling the compounds from aldehydes *hemiacetals* and those from ketones *hemiketals*. Thus you may see the names *hemiketal* and *ketal* in some publications, but we do not make such a distinction in this text.)

The hemiacetal and acetal formed by reaction of acetaldehyde with ethanol are shown in Figure 16.3.

> **WORKED EXAMPLE** **16.3** Predicting the Products of Hemiacetal and Acetal Formation
>
> Write the structure of the intermediate hemiacetal and the acetal final product formed in the following reaction:
>
> $$CH_3CH_2CH + 2\,CH_3OH \xrightarrow{\text{Acid catalyst}} ?$$
>
> **ANALYSIS** First, rewrite the structure showing only a single bond between C and O, along with partial bonds to both C and O:
>
> $$CH_3CH_2-C-H \quad \text{is rewritten as} \quad CH_3CH_2-C-$$

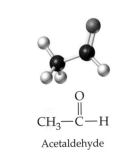

$$CH_3-C-H$$

Acetaldehyde

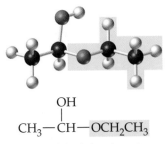

$$CH_3-CH-OCH_2CH_3$$

Acetaldehyde hemiacetal with ethanol

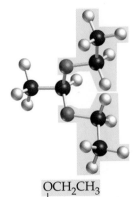

$$CH_3-CH-OCH_2CH_3$$

Acetaldehyde acetal with ethanol

▲ **FIGURE 16.3 Acetaldehyde and its hemiacetal and acetal.** This hemiacetal and acetal shown here are formed by reactions with ethanol (highlighted in blue).

Next, add one molecule of the alcohol (CH_3OH in this case) by attaching —H to the oxygen partial bond and —OCH_3 to the carbon partial bond. This yields the hemiacetal intermediate:

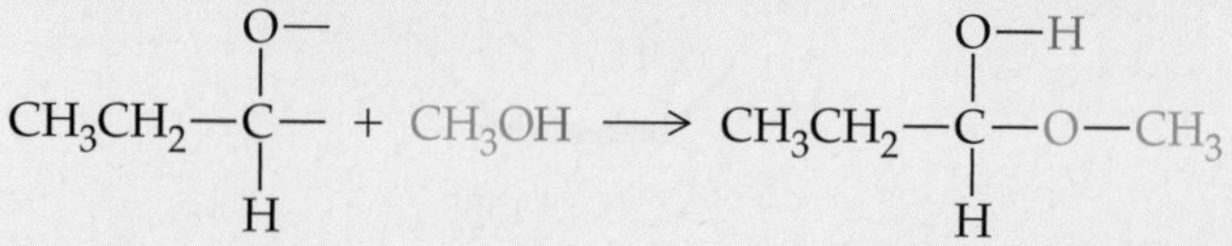

$$CH_3CH_2-\overset{\overset{O-}{|}}{\underset{\underset{H}{|}}{C}}- \ + \ CH_3OH \longrightarrow CH_3CH_2-\overset{\overset{O-H}{|}}{\underset{\underset{H}{|}}{C}}-O-CH_3$$

Hemiacetal

Finally, replace the —OH group of the hemiacetal with an —OCH_3 from a second molecule of alcohol.

SOLUTION

The reaction produces the acetal and water:

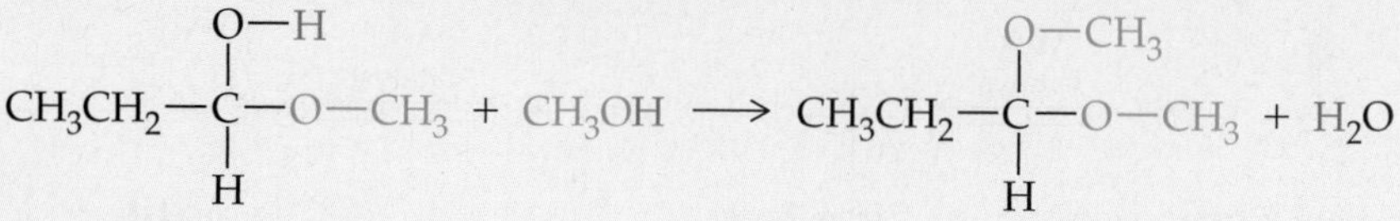

$$CH_3CH_2-\overset{\overset{O-H}{|}}{\underset{\underset{H}{|}}{C}}-O-CH_3 \ + \ CH_3OH \longrightarrow CH_3CH_2-\overset{\overset{O-CH_3}{|}}{\underset{\underset{H}{|}}{C}}-O-CH_3 \ + \ H_2O$$

Acetal

WORKED EXAMPLE **16.4** Identification of Hemiacetals

Which of the following compounds are hemiacetals?

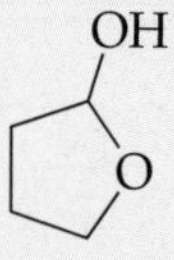

(a) 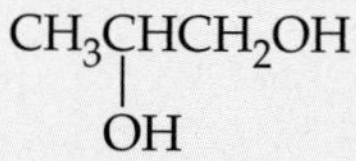 $CH_3\overset{\underset{\underset{OH}{|}}{|}}{C}HCH_2OH$

(b) [cyclic structure with OH and O]

(c) $CH_3-\overset{\overset{OH}{|}}{\underset{\underset{OCH_3}{|}}{C}}-CH_3$

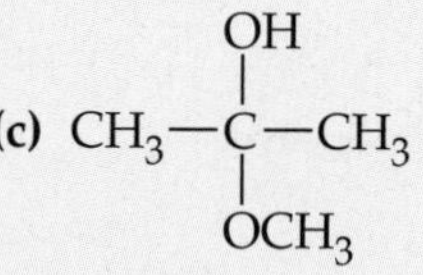

ANALYSIS To identify a hemiacetal, look for a carbon atom with single bonds to 2 oxygen atoms, one in an —OH group and one in an —OR group. Note that the O of the —OR group can be part of a ring.

SOLUTION

Compound (a) contains 2 O atoms, but they are bonded to *different* C atoms; it is not a hemiacetal. Compound (b) has 1 ring C atom bonded to 2 oxygen atoms, one in the substituent —OH group and one bonded to the rest of the ring, which is the R group; it is a cyclic hemiacetal. Compound (c) also contains a C atom bonded to one —OH group and one —OR group, so it too is a hemiacetal.

WORKED EXAMPLE **16.5** Identification of Acetals

Which of the following compounds are acetals?

(a) 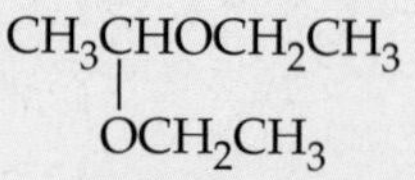 $CH_3\overset{\underset{\underset{OCH_2CH_3}{|}}{|}}{C}HOCH_2CH_3$

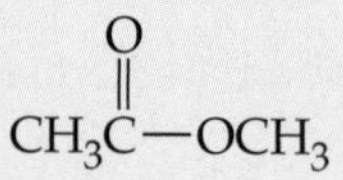

(b) $CH_3\overset{\overset{O}{\|}}{C}-OCH_3$

(c) [cyclic structure with O]—OCH_2CH_3

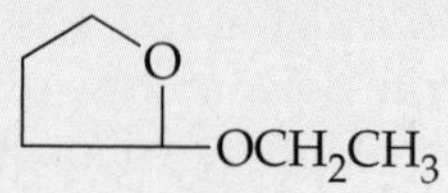

ANALYSIS As in identifying hemiacetals, look for a carbon atom that has single bonds to 2 oxygen atoms, but in this case both of them will be —OR groups. Note that the O of the —OR group can be part of a ring.

SOLUTION

In (a), the central carbon atom is bonded to —CH_3, —H, and two —OCH_2CH_3 groups, so the compound is an acetal. Compound (b) does have a carbon atom bonded to 2 oxygen atoms, but one of the bonds is a double bond rather than a single bond, so this is not an acetal. Compound (c) has an oxygen atom in a ring, making it also part of an —OR group, where R is the ring. Since one of the carbons connected to the O in the ring is also connected to an —OCH$_2$CH$_3$ group, compound (c) is an acetal.

PROBLEM 16.12

Which of the following compounds are hemiacetals?

(a) 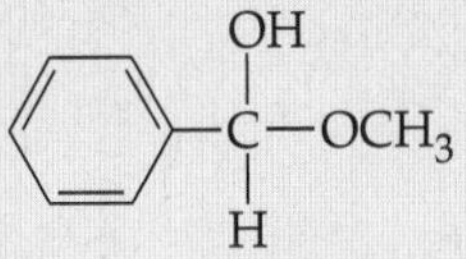 [structure: phenyl group bonded to C with OH above, H below, and —OCH$_3$]   **(b)** CH$_3$CHCHCH$_3$ with HO and OH below   **(c)** [cyclohexane ring with —OH and OCH$_3$ substituents]

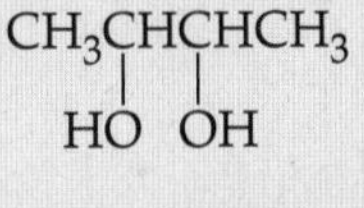

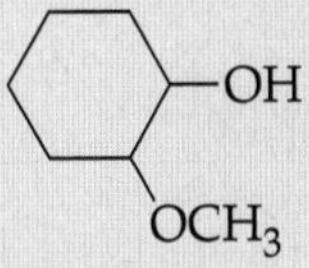

PROBLEM 16.13

Draw the structures of the hemiacetals formed in these reactions:

(a) CH$_3$CH$_2$CH$_2$CHO + CH$_3$CH$_2$OH $\longrightarrow$?

(b) CH$_3$CH$_2$CCH$_2$CH(CH$_3$)$_2$ + CH$_3$OH $\longrightarrow$?   [the second carbon bears a =O]

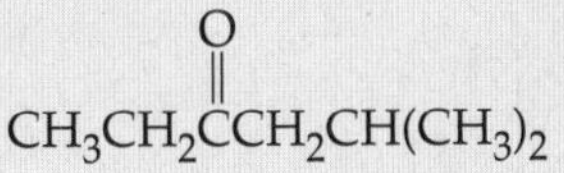

PROBLEM 16.14

Draw the structure of each acetal final product formed in the reactions shown in Problem 16.13.

PROBLEM 16.15

Which of the following compounds are hemiacetals or acetals?

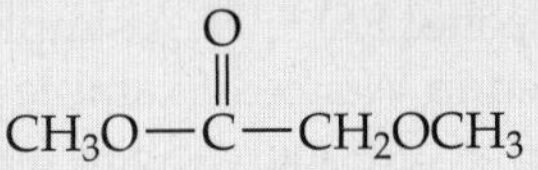

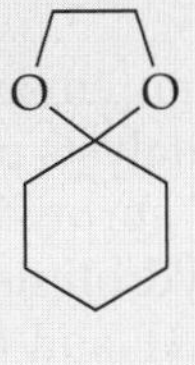

(a) 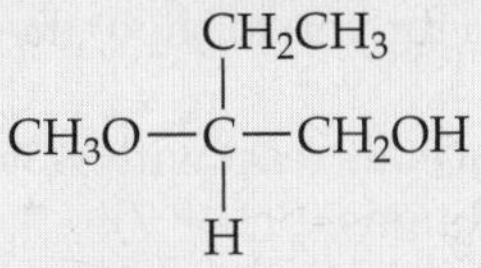 CH$_3$O—C—CH$_2$OH   [C bears CH$_2$CH$_3$ above and H below]   **(b)** CH$_3$O—C—CH$_2$OCH$_3$ [C bears =O above]

(c) [spiro cyclohexane with dioxolane ring, two O atoms]   **(d)** 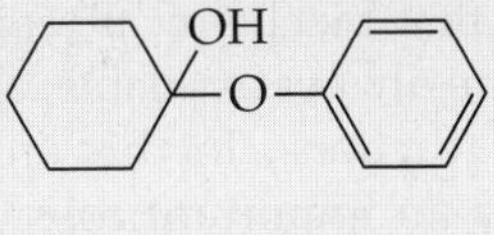 [cyclohexane ring bearing OH and —O—phenyl]

PROBLEM 16.16

Of the compounds in Problem 16.15 that are acetals or hemiacetals, which were formed from aldehydes and which were formed from ketones? Explain what indicates the difference.

Acetal Hydrolysis

Because acetal formation is an equilibrium reaction, the extent to which the reaction proceeds in either direction can be controlled by changing the reaction conditions. Therefore, the aldehyde or ketone from which an acetal is formed can be regenerated by reversing the reaction. Reversal requires an acid catalyst and a large quantity of water (a product of acetal formation) to drive the reaction back toward the aldehyde or ketone (Le Châtelier's principle) (⊂⊃, p. 203).

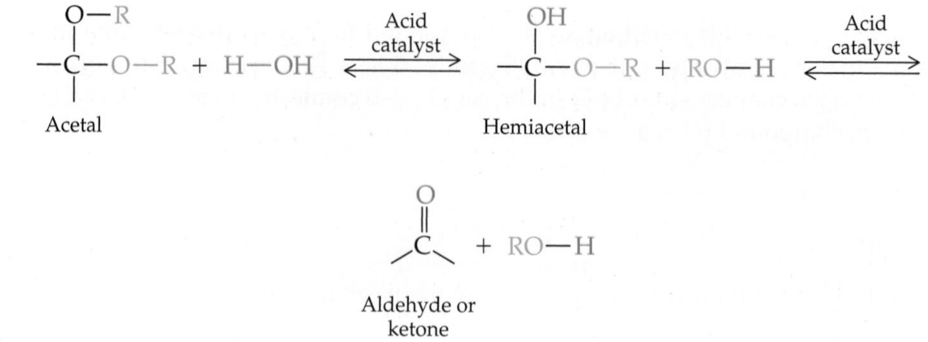

$$\underset{\text{Acetal}}{-\overset{\overset{\displaystyle O-R}{|}}{\underset{|}{C}}-O-R} + H-OH \underset{\text{catalyst}}{\overset{\text{Acid}}{\rightleftharpoons}} \underset{\text{Hemiacetal}}{-\overset{\overset{\displaystyle OH}{|}}{\underset{|}{C}}-O-R} + RO-H \underset{\text{catalyst}}{\overset{\text{Acid}}{\rightleftharpoons}}$$

$$\underset{\text{Aldehyde or}\atop\text{ketone}}{\overset{\displaystyle O}{\underset{\displaystyle \diagup C \diagdown}{\|}}} + RO-H$$

For example,

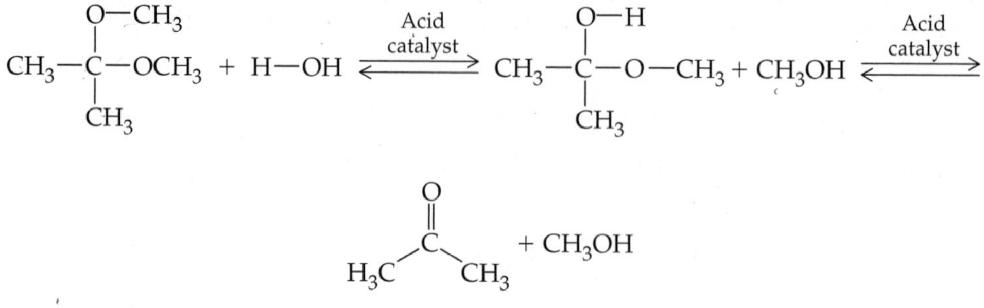

$$\underset{\displaystyle CH_3}{\overset{\displaystyle O-CH_3}{CH_3-\overset{|}{\underset{|}{C}}-OCH_3}} + H-OH \underset{\text{catalyst}}{\overset{\text{Acid}}{\rightleftharpoons}} \underset{\displaystyle CH_3}{\overset{\displaystyle O-H}{CH_3-\overset{|}{\underset{|}{C}}-O-CH_3}} + CH_3OH \underset{\text{catalyst}}{\overset{\text{Acid}}{\rightleftharpoons}}$$

$$\underset{H_3C \quad CH_3}{\overset{\displaystyle O}{\underset{\diagup C \diagdown}{\|}}} + CH_3OH$$

Hydrolysis A reaction in which a bond or bonds are broken and the —H and —OH of water add to the atoms of the broken bond or bonds.

The reaction of an acetal with water is an example of **hydrolysis** (*Latin:* "to split with water"), a reaction in which a bond or bonds are broken and the —H and —OH of water add to the atoms of the broken bond or bonds. With an acetal, the first step is formation of the hemiacetal as the water breaks one of the C—OR bonds and a C—OH bond is formed in its place. The carbonyl group is then formed as the bond to the H of the C—OH and the hemiacetal C—OR bond are broken. The result is the ketone or aldehyde from which the acetal was made plus two molecules of the alcohol RO—H. A simple way to for you to show the steps of this reaction is presented in Worked Example 16.6.

It should be noted that although acetals and hemiacetals react with water in the presence of acid, they are unreactive under basic conditions (pH > 7).

⊂◯⊃ Looking Ahead

Consider for a moment that biochemical reactions take place in an environment where water molecules are always available, along with enzyme catalysts precisely suited to the necessary reactions. In this environment, it is not surprising that hydrolysis reactions play an important role. During digestion, hydrolysis breaks bonds in carbohydrates (Section 23.1), triacylglycerols (Section 25.1), and proteins (Section 28.1). ⊂◯⊃

WORKED EXAMPLE **16.6** Writing the Products Obtained from Acetal Hydrolysis

Write the structure of the aldehyde or ketone that forms by hydrolysis of the following acetal:

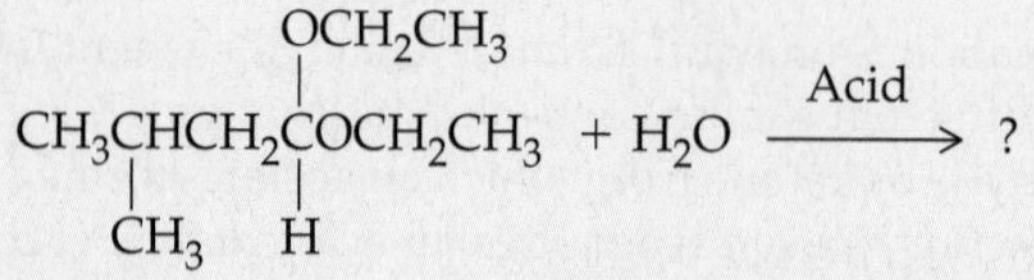

$$\underset{\underset{\displaystyle CH_3 \quad H}{|\qquad\quad|}}{CH_3CHCH_2\overset{\overset{\displaystyle OCH_2CH_3}{|}}{C}OCH_2CH_3} + H_2O \xrightarrow{\text{Acid}} ?$$

ANALYSIS The products are the aldehyde or ketone plus two molecules of the alcohol from which the acetal could have been formed. First, identify the two C—O acetal bonds, redrawing the structure if necessary:

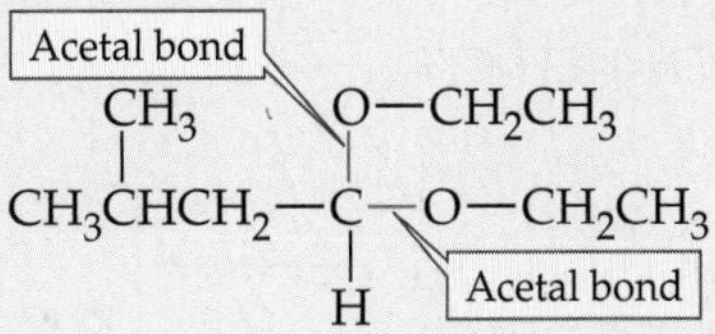

Next, break the H—OH bond and one of the acetal C—OR bonds (it does not matter which one); move the water OH to the acetal carbon to form the hemiacetal and the water H to the OR to form one molecule of HOR:

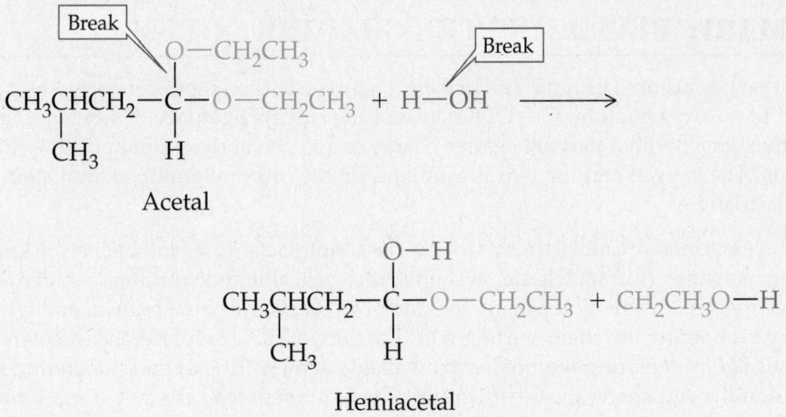

Remove the H and OR groups from the hemiacetal, and change the C—O single bond to a C=O double bond to give carbon the four bonds it must have. Combine the H and OR you removed from the second alcohol molecule.

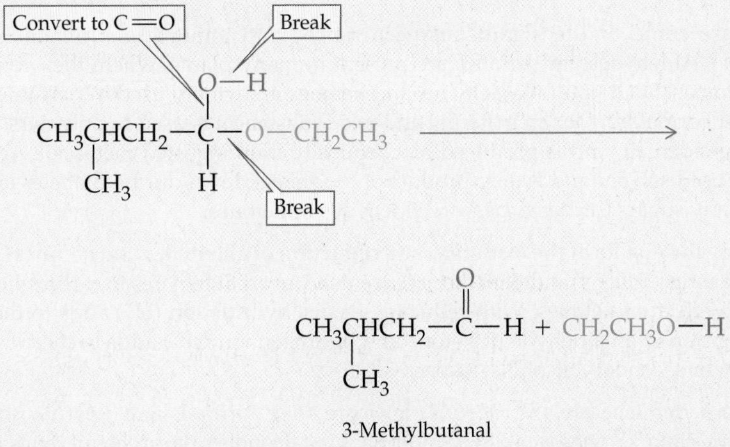

SOLUTION
In this example, the product is an aldehyde.

PROBLEM 16.17

What aldehydes or ketones result from the following acetal hydrolysis reactions? What alcohol is formed in each case?

(a) [benzene ring]—$CH_2C(OCH_3)_2CH_2CH_3$ $\xrightarrow{H_3O^+}$?

(b) $CH_3CH_2CH_2OCHOCH_2CH_2CH_3$ $\xrightarrow{H_3O^+}$?
$\qquad\qquad\qquad\quad\;|$
$\qquad\qquad\quad CH_2CH_3$

(c) $CH_3CH_2CH_2OCH_2OCH_2CH_2CH_3$ $\xrightarrow{H_3O^+}$?

SUMMARY: REVISITING THE CHAPTER GOALS

1. **What is the carbonyl group?** The *carbonyl group* is a carbon atom connected by a double bond to an oxygen atom, $C=O$. Because of the electronegativity of oxygen, the $C=O$ group is polar, with a partial negative charge on oxygen and a partial positive charge on carbon. The oxygen and the two substituents on the carbonyl-group carbon atom form a planar triangle.

2. **How are ketones and aldehydes named?** The simplest *aldehydes* and *ketones* are known by common names (formaldehyde, acetaldehyde, benzaldehyde, acetone). Aldehydes are named systematically by replacing the final *-e* in an alkane name with *-al* and when necessary numbering the chain starting with 1 at the —CHO group. Ketones are named systematically by replacing the final *-e* in an alkane name with *-one* and numbering starting with 1 at the end nearer the $C=O$ group. The location of the carbonyl group is indicated by placing the number of its carbon before the name. Some common names of ketones identify each alkyl group separately.

3. **What are the general properties of aldehydes and ketones?** Aldehyde and ketone molecules are moderately polar, do not hydrogen-bond with each other, but can hydrogen-bond with water molecules. The smaller ones are water-soluble, and the ketones are excellent solvents. In comparable series of compounds, aldehydes and ketones are higher boiling than alkanes but lower boiling than alcohols. Many aldehydes and ketones have distinctive, pleasant odors.

4. **What are some of the significant occurrences and applications of aldehydes and ketones?** Aldehydes and ketones are present in many plants, where they contribute to their aromas. Such natural aldehydes and ketones are widely used in perfumes and flavorings. Formaldehyde (an irritating and toxic substance) is used in polymers, is present in smog-laden air, and is produced biochemically from ingested methanol. Acetone is a widely used solvent and is a by-product of food breakdown during diabetes and starvation. Many sugars (*carbohydrates*) are aldehydes or ketones.

5. **What are the results of the oxidation and reduction of aldehydes and ketones?** Mild oxidizing agents (Tollens' and Benedict's reagents) convert aldehydes to carboxylic acids but have no effect on ketones. With reducing agents, hydride ion (H^-) adds to the C of the $C=O$ group in an aldehyde or ketone and hydrogen ion (H^+) adds to the O to produce primary or secondary alcohols, respectively.

6. **What are hemiacetals and acetals, how are they formed, and how do they react?** Aldehydes and ketones establish equilibria with alcohols to form hemiacetals or acetals. The relatively unstable *hemiacetals*, which have an —OH and an —OR on what was the carbonyl carbon, result from addition of one alcohol molecule to the $C=O$ bond. The more stable *acetals*, which have two —OR groups on what was the carbonyl carbon, form by addition of a second alcohol molecule to a hemiacetal. The aldehyde or ketone can be regenerated from an acetal by treatment with an acid catalyst and a large quantity of water, which is an example of a *hydrolysis* reaction.

SUMMARY OF REACTIONS

1. **Reactions of aldehydes**

   (a) Oxidation to yield a carboxylic acid (Section 16.5):

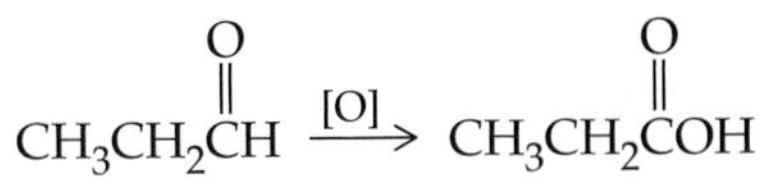

$$CH_3CH_2\overset{\displaystyle O}{\overset{\|}{C}}H \xrightarrow{[O]} CH_3CH_2\overset{\displaystyle O}{\overset{\|}{C}}OH$$

   (b) Reduction to yield a primary alcohol (Section 16.6):

$$CH_3CH_2\overset{\displaystyle O}{\overset{\|}{C}}H \xrightarrow{[H]} CH_3CH_2CH_2OH$$

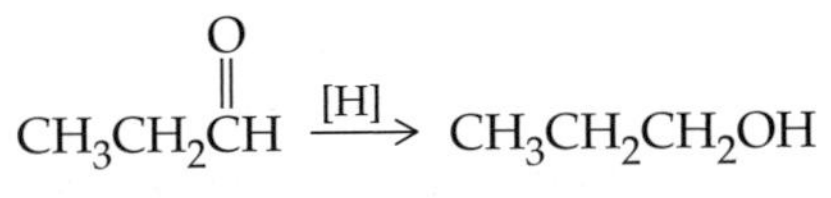

   (c) Addition of alcohol to yield a hemiacetal or acetal (Section 16.7):

$$CH_3\overset{\displaystyle O}{\overset{\|}{C}}H + CH_3CH_2OH \longrightarrow CH_3\overset{\displaystyle H}{\underset{\displaystyle OH}{\overset{\displaystyle |}{\underset{|}{C}}}}OCH_2CH_3$$

$$CH_3\overset{\displaystyle H}{\underset{\displaystyle OH}{\overset{\displaystyle |}{\underset{|}{C}}}}OCH_2CH_3 + CH_3CH_2OH \longrightarrow CH_3\overset{\displaystyle H}{\underset{\displaystyle OCH_2CH_3}{\overset{\displaystyle |}{\underset{|}{C}}}}OCH_2CH_3 + H_2O$$

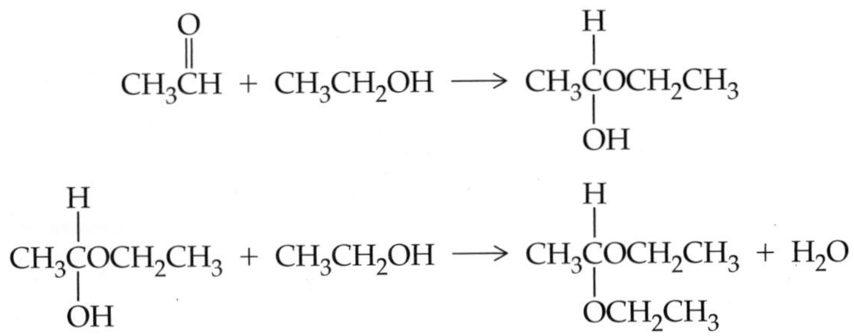

2. **Reactions of ketones**

   (a) Reduction to yield a secondary alcohol (Section 16.6):

$$CH_3\overset{\displaystyle O}{\overset{\|}{C}}CH_3 \xrightarrow{[H]} CH_3\underset{\displaystyle OH}{\underset{|}{C}}HCH_3$$

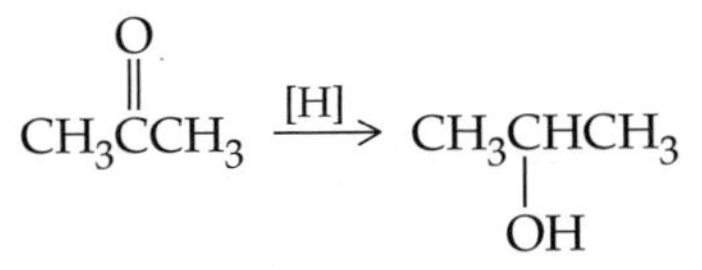

   (b) Addition of an alcohol to yield a hemiacetal or acetal (Section 16.7):

$$CH_3\overset{\displaystyle O}{\overset{\|}{C}}CH_3 + CH_3CH_2OH \longrightarrow CH_3\overset{\displaystyle CH_3}{\underset{\displaystyle OH}{\overset{\displaystyle |}{\underset{|}{C}}}}{-}OCH_2CH_3$$

$$CH_3\overset{\displaystyle CH_3}{\underset{\displaystyle OH}{\overset{\displaystyle |}{\underset{|}{C}}}}{-}OCH_2CH_3 + CH_3CH_2OH \longrightarrow CH_3\overset{\displaystyle CH_3}{\underset{\displaystyle OCH_2CH_3}{\overset{\displaystyle |}{\underset{|}{C}}}}{-}OCH_2CH_3 + H_2O$$

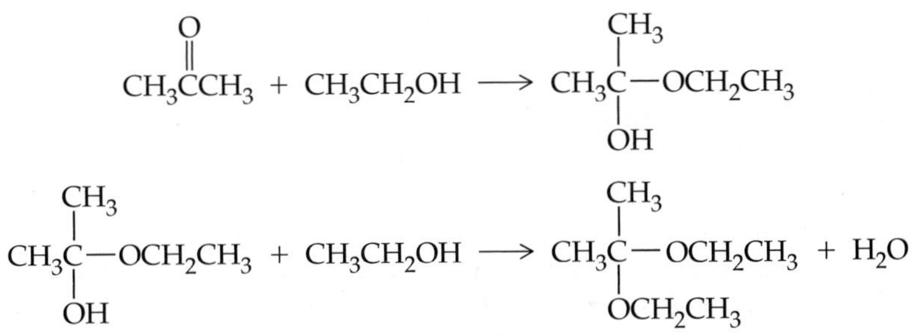

3. **Reaction of acetals**

   Hydrolysis to regenerate an aldehyde or ketone (Section 16.7):

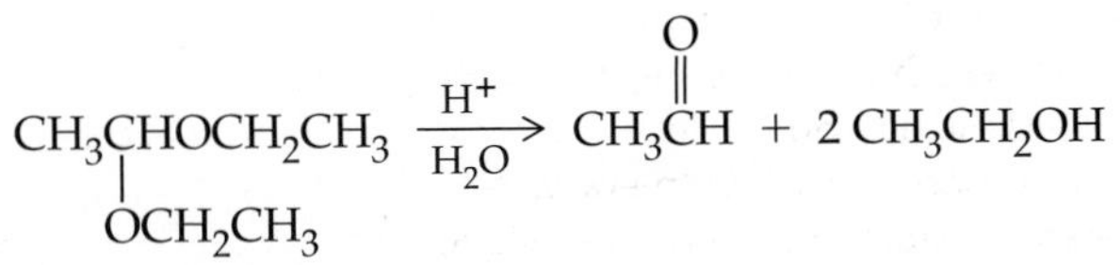

$$CH_3\underset{\displaystyle OCH_2CH_3}{\underset{|}{C}}HOCH_2CH_3 \xrightarrow[H_2O]{H^+} CH_3\overset{\displaystyle O}{\overset{\|}{C}}H + 2\ CH_3CH_2OH$$

16.18 The carbonyl group can be reduced by addition of a hydride ion (H^-) and a proton (H^+). Removal of H^- and H^+ from an alcohol results in a carbonyl group.

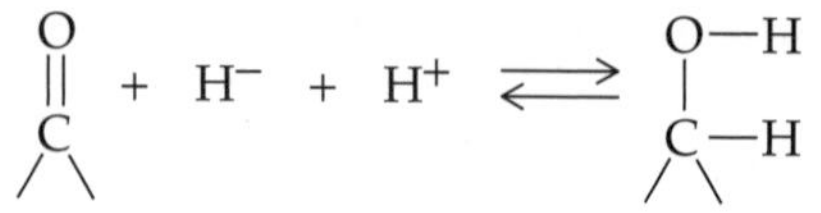

(a) To which atom of the carbonyl is the hydride ion added, and why?

(b) In the reaction above, indicate which reaction arrow represents reduction and which represents oxidation.

16.19 A fundamental difference between aldehydes and ketones is that one can be oxidized to carboxylic acids, but the other cannot. Which is which? Give an example of a test to differentiate aldehydes from ketones.

16.20 In the diagram below, indicate with dashed lines where hydrogen bonds would form. Explain why you chose these atoms to hydrogen-bond.

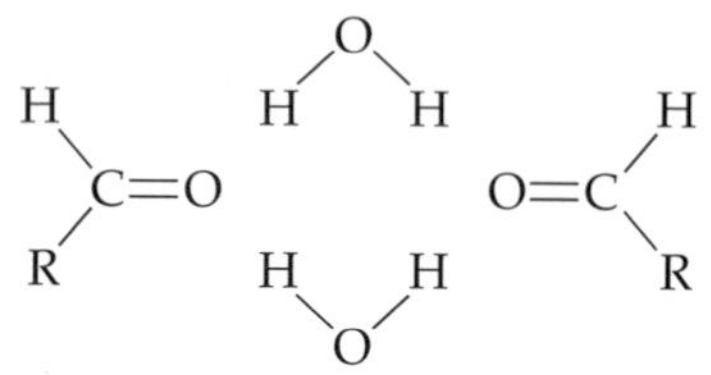

16.21 (a) Describe what happens in the reaction of an aldehyde with an alcohol. What is necessary for this reaction to occur?

(b) Copy the structures below and use lines to show where new bonds are formed. Cross out bonds that no longer exist as the aldehyde and alcohol react to form a hemiacetal.

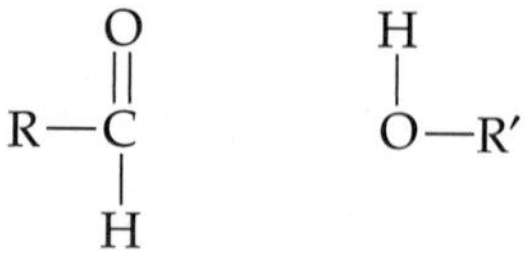

16.22 Glucose is the major sugar in mammalian blood. We often see it represented as either the "free aldehyde" or the cyclic hemiacetal forms shown here. Of the two forms of glucose, the cyclic hemiacetal is the preferred form found in blood. Can you suggest two reasons why?

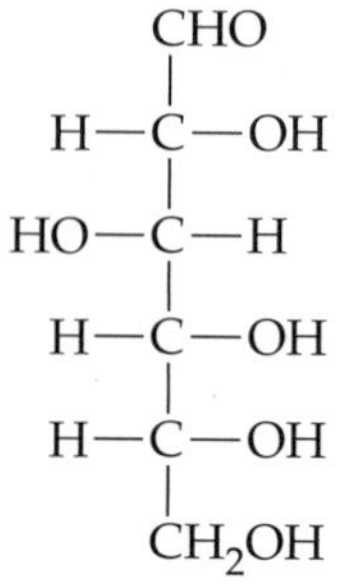

"Free aldehyde" 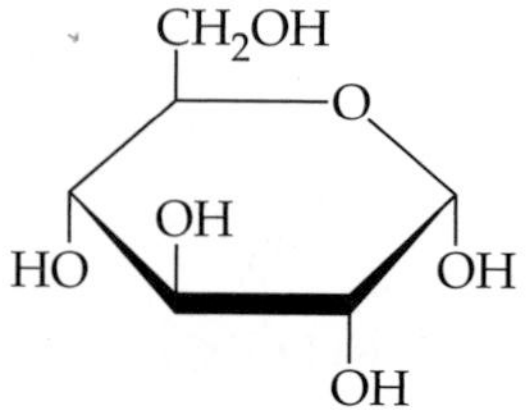   Cyclic hemiacetal

16.23 Describe the two types of addition reactions that aldehydes and ketones undergo with alcohols.

ADDITIONAL PROBLEMS

ALDEHYDES AND KETONES

16.24 Draw a structure for a compound that meets each description:

(a) A ketone, C_4H_8O

(b) An aldehyde with 5 carbons

(c) A ketoaldehyde, $C_5H_8O_2$

(d) A hydroxyketone, $C_4H_8O_2$

16.25 Draw a structure for a compound that meets each description:

(a) An aldehyde, C_5H_8O

(b) A ketone with 8 carbons

(c) A ketoaldehyde, $C_6H_{10}O_2$

(d) A cyclic hydroxyketone, $C_5H_8O_2$

16.26 Indicate which compounds contain aldehyde or ketone carbonyl groups.

(a) $CH_3CH_2\overset{\displaystyle H}{\underset{}{C}}=O$

(b) $O=CCH_2CH_2CHCH_3$ with NH_2 and CH_3 substituents

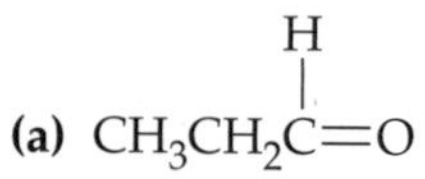

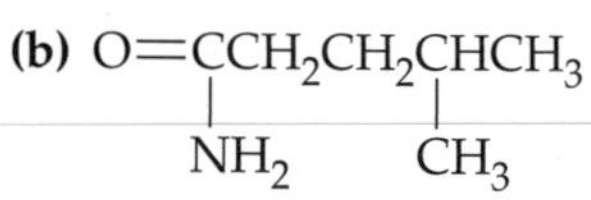

(c) $CH_3CH_2-O-CH=CH_2$

(d) $CH_3CH_2C(OCH_3)_3$

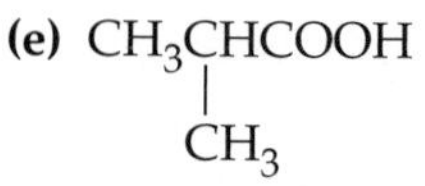

(e) $CH_3CHCOOH$ with CH_3

(f) $CH_3COCH_2CH_2OH$

16.27 Indicate which compounds contain aldehyde or ketone carbonyl groups.

(a) CH_3CH_2CHO

(b) $(CH_3)_2C(OH)CH_2CH_2CH_3$

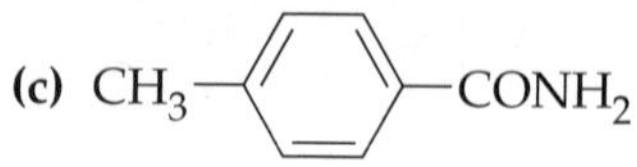

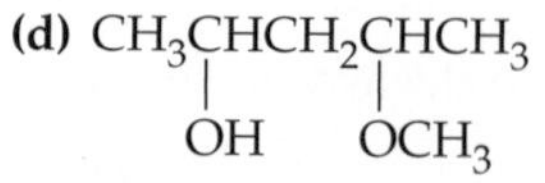

(c) $CH_3-\bigcirc-CONH_2$

(d) $CH_3CHCH_2CHCH_3$ with OH and OCH_3

(e) $CH_3CH_2COCH_2CH_3$

(f) CH_3COOCH_3

16.28 Draw structures corresponding to the following aldehyde and ketone names:

(a) 3-Methylbutanal

(b) 4-Chloro-2-hydroxypentanal

(c) *p*-Nitrobenzaldehyde

(d) 2-Octanone

(e) 2,4-Dimethyl-3-pentanone

(f) Phenyl methyl ketone (also known as acetophenone)

16.29 Draw structures corresponding to the following aldehyde and ketone names:

(a) 4-Hydroxy-2,2,4-trimethylheptanal

(b) 4-Ethyl-2-isopropylhexanal

(c) *p*-Bromobenzaldehyde

(d) 2,4-Dihydroxycyclohexanone

(e) 1,1,1-Trichloro-3-pentanone

(f) 2-Methyl-3-hexanone

16.30 Give systematic names for the following aldehydes and ketones:

(a) 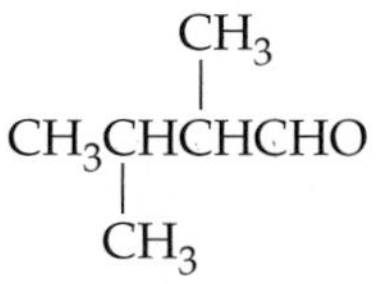
$$\underset{\underset{CH_3}{|}}{\overset{\overset{CH_3}{|}}{CH_3CHCHCHO}}$$

(b) 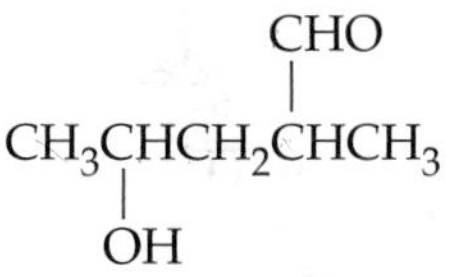
$$\underset{\underset{OH}{|}}{CH_3CHCH_2CHCH_3}\ \overset{CHO}{\overset{|}{}}$$

(c) $(CH_3)_3CCHO$

(d) 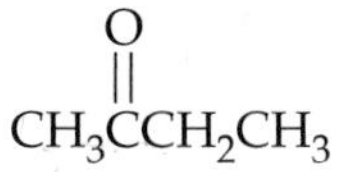
$$CH_3\overset{O}{\overset{\|}{C}}CH_2CH_3$$

(e) 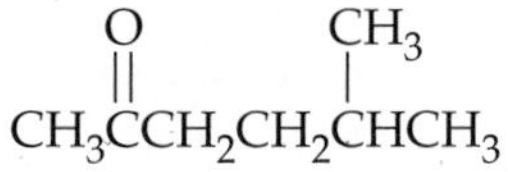
$$CH_3\overset{O}{\overset{\|}{C}}CH_2CH_2\overset{CH_3}{\overset{|}{C}H}CH_3$$

16.31 Give systematic names for the following aldehydes and ketones:

(a)

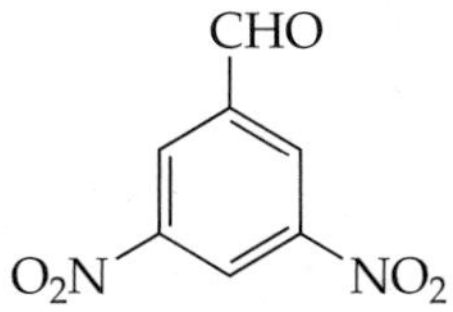

(b) 
$$CH_3CH_2\overset{O}{\overset{\|}{C}}C(CH_3)_3$$

(c)
$$CH_3-\underset{\underset{HO}{|}}{CH}-\underset{\underset{CH_3}{|}}{\overset{CH_3CH_2}{\overset{|}{C}}}-CH_2-\overset{O}{\overset{\|}{C}}-H$$

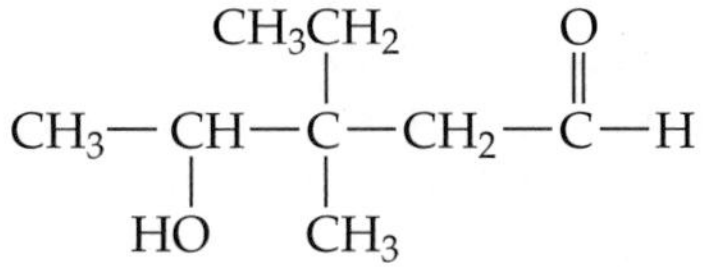

(d)

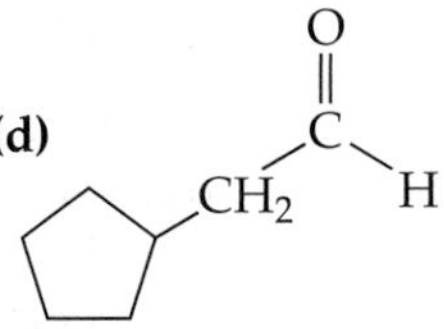

(e)

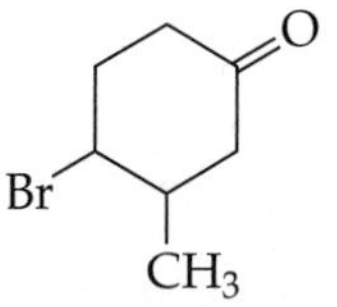

16.32 The following names are incorrect. What is wrong with each?

(a) 1-Pentanone

(b) 4-Methyl-3-pentanone

(c) 3-Butanone

16.33 The following names are incorrect. What is wrong with each?

(a) Cyclohexanal

(b) 2-Butanal

(c) 1-Methyl-1-pentanone

REACTIONS OF ALDEHYDES AND KETONES

16.34 What kind of compound is produced when an aldehyde reacts with an alcohol in a 1:1 ratio? Does this reaction also form a second product? Illustrate your answer using propanal and methanol.

16.35 What kind of compound is produced when an aldehyde reacts with an alcohol in a 1:2 ratio in the presence of an acid catalyst? Illustrate your answer using propanal and methanol.

16.36 Draw the structures of the products formed when the following compounds react with (1) Tollens' reagent and (2) a reducing agent.

(a) Cyclopentanone

(b) Hexanal

(c) 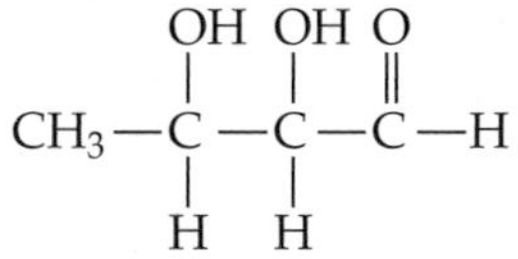
$$CH_3-\underset{\underset{H}{|}}{\overset{\overset{OH}{|}}{C}}-\underset{\underset{H}{|}}{\overset{\overset{OH}{|}}{C}}-\overset{O}{\overset{\|}{C}}-H$$

16.37 Draw the structures of the products formed when the following compounds react with (1) Tollens' reagent and (2) a reducing agent.

(a)

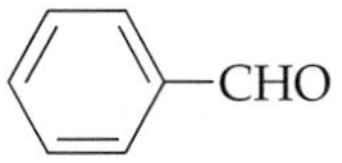

(b) 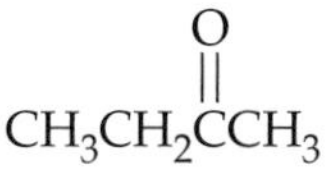
$$CH_3CH_2\overset{O}{\overset{\|}{C}}CH_3$$

(c)
$$Cl_2CH\overset{O}{\overset{\|}{C}}H$$

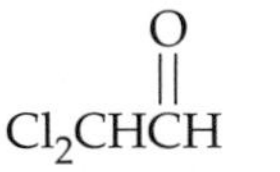

16.38 Draw the structures of the aldehydes and primary alcohols that might be oxidized to yield the following carboxylic acids:

(a)

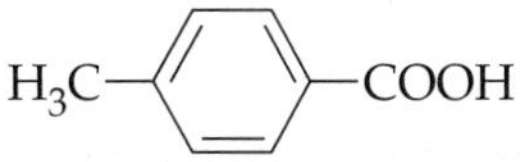

(b) 
$$\underset{}{CH_3CH_2CHCH_2CHCH_3}\ \overset{COOH\ CH_3}{\overset{|\qquad|}{}}$$

(c) $CH_3CH{=}CHCOOH$

16.39 Draw the structures of the aldehydes and primary alcohols that might be oxidized to yield the following carboxylic acids:

(a)

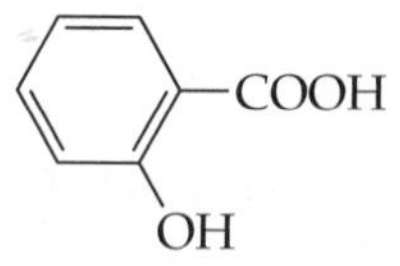

[benzene ring with —COOH and —OH substituents]

(b)

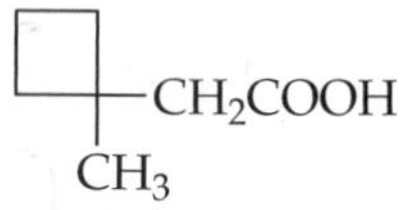

[cyclobutane ring with —CH₂COOH and CH₃ substituents]

(c) $CH_3CH{=}CHCH_2COOH$

16.40 Write the structures of the hemiacetals that result from reactions (a) and (b). Write the structures of the hydrolysis products of the compounds in (c) and (d).

(a) 2-Butanone + 1-Propanol ⟶ ?

(b) Butanal + Isopropyl alcohol ⟶ ?

(c) 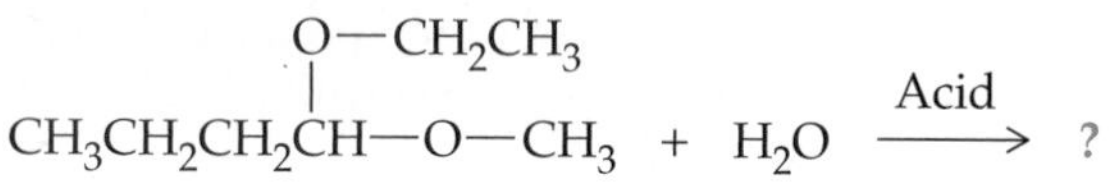 $CH_3CH_2CH_2\overset{\overset{\displaystyle O-CH_2CH_3}{|}}{CH}-O-CH_3$ + H_2O $\xrightarrow{\text{Acid}}$?

(d)

[structure] H_3C $O-CH_2$ / H_3C $O-CH_2$ + H_2O $\xrightarrow{\text{Acid}}$?

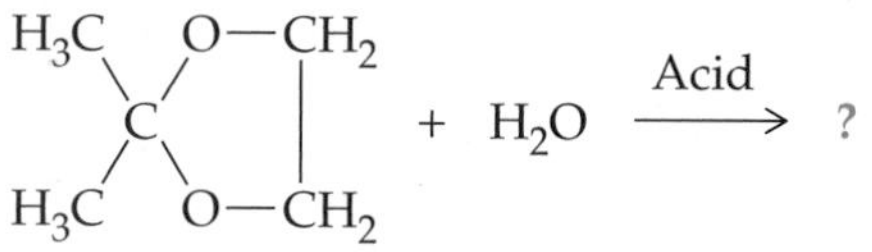

16.41 Write the structures of the hemiacetals that result from reactions (a) and (b). Write the structures of the hydrolysis products of the compounds in (c) and (d).

(a) Acetone + Ethanol ⟶ ?

(b) Hexanal + 2-Butanol ⟶ ?

(c)

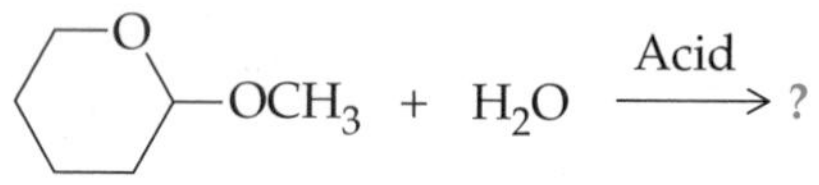

[ring structure with O and —OCH₃] + H_2O $\xrightarrow{\text{Acid}}$?

(d)

[benzodioxole ring with Br substituent] + H_2O $\xrightarrow{\text{Acid}}$?

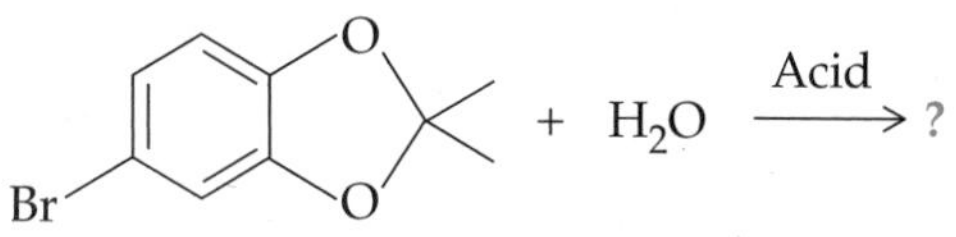

16.42 Cyclic hemiacetals commonly form if an alcohol group in one part of a molecule adds to a carbonyl group elsewhere in the same molecule, especially if a five- or six-membered ring results. What is the structure of the open-chain hydroxy ketone from which this hemiacetal might form?

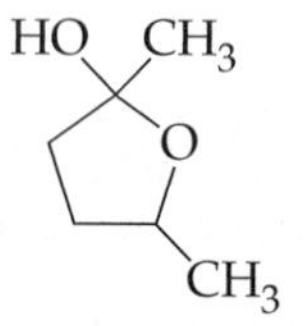

[cyclic hemiacetal structure with HO, CH₃, O, CH₃]

A cyclic hemiacetal

16.43 Glucosamine is found in the shells of lobsters; it exists largely in the cyclic hemiacetal form shown here. Draw

the structure of glucosamine in its open-chain hydroxy aldehyde form.

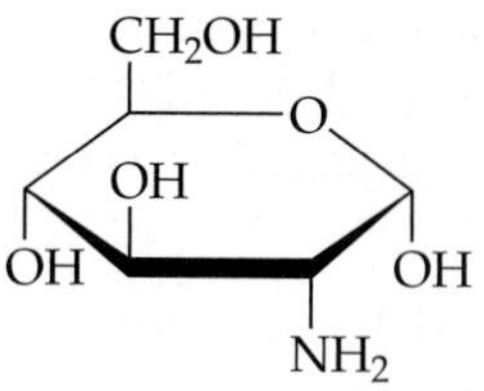

[glucosamine cyclic structure with CH₂OH, OH, OH, OH, NH₂]

16.44 What products result from hydrolysis of this cyclic acetal?

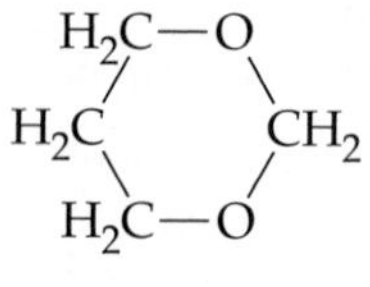

[cyclic acetal structure with H₂C—O, H₂C, CH₂, H₂C—O]

16.45 Acetals are usually made by reaction of an aldehyde or ketone with two molecules of a monoalcohol. If an aldehyde or ketone reacts with *one* molecule of a dialcohol, however, a cyclic acetal results. Draw the structure of the cyclic acetal formed in the reaction

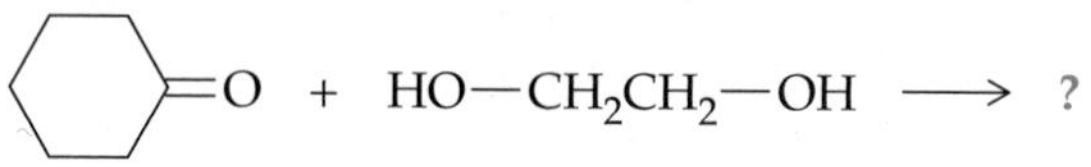

[cyclohexanone] $={=}O$ + $HO-CH_2CH_2-OH$ ⟶ ?

16.46 Aldosterone is a key steroid involved in controlling the sodium–potassium balance in the body. Identify the functional groups in aldosterone.

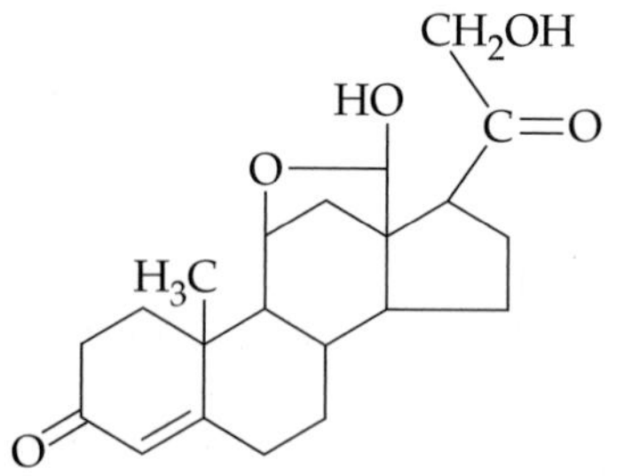

[aldosterone steroid structure with CH₂OH, HO, C=O, O, H₃C, O]

16.47 The compound carvone is responsible for the odor of spearmint. Identify the functional groups in ionone.

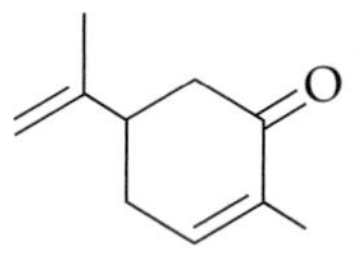

[carvone structure]

Carvone

Applications

16.48 In oxidation–reduction reactions, there must be a reduction associated with each oxidation and vice versa. When enzymes within the bombardier beetle's defensive organ catalyze oxidation of *p*-dihydroxybenzene, what is reduced in the reaction? Write the complete oxidation–reduction reaction and be sure to balance the equation. [*Chemical Warfare among the Insects*, p. 493]

16.49 HCN is quite toxic. How do you suppose the millipede uses this weapon without killing itself? [*Chemical Warfare among the Insects*, p. 493]

16.50 For both pure and imitation vanilla extract, the vanillin is extracted from the source with a solvent. Look at the structure of vanillin in the application; predict whether water, ethanol, or diethyl ether would be the best solvent for good extraction and explain your choice. Remember, the solvent also must be suitable for human consumption. Check a vanilla extract bottle to see what was used for the extraction. [*Vanilla: Which Kind Is Best?*, p. 496]

16.51 Both HCN and benzaldehyde have the aroma of almonds. How would you test a liquid with a faint almond odor to determine which of these two is present? [*How Toxic Is Toxic?*, p. 503]

16.52 **(a)** What is an advantage of *in vitro* acute toxicity testing?

(b) Why is considerable uncertainty still associated with *in vitro* acute toxicity testing? (*Hint*: Since you are testing only one certain cell type, what might be happening in other cell types?) [*How Toxic Is Toxic?*, p. 503]

16.53 The following list gives the LD$_{50}$ of three compounds. Which is the most toxic and which is the least toxic?

(a) 23 g/kg

(b) 23 mg/kg

(c) 18 mg/kg

(d) Tetrodotoxin, found in the puffer fish, has an oral LD$_{50}$ of 334 μg per kilogram of body weight. If an average adult male weighs 200 lb, how much tetrodotoxin is needed to kill him? [*How Toxic Is Toxic?*, p. 503]

General Questions and Problems

16.54 Name the following compound, which is used in the fragrance industry.

16.55 Can the alcohol (CH$_3$)$_3$COH be formed by the reduction of an aldehyde or ketone? Why or why not?

16.56 Many flavorings and perfumes are partially based on fragrant aldehydes and ketones. Why do you think the portion of the odor due to the ketone is more stable than that due to the aldehyde?

16.57 One problem with burning some plastics is the release of formaldehyde. What are some of the physiological effects of exposure to formaldehyde?

16.58 *Hydrates* are formed when water, rather than an alcohol, adds across the carbonyl carbon. Chloral hydrate, a potent sedative and component in "knockout" drops, is formed by reacting trichloroacetaldehyde with water in a reaction analogous to hemiacetal formation. Draw the formula of chloral hydrate.

16.59 Name the following compounds:

(a) CH$_3$CH$_2$CCH(CH$_3$)$_2$

(b) CH$_2$=CHCH$_2$CH$_2$CH=CH$_2$

(c)

(d) (CH$_3$)$_3$CCHCCH$_2$CH$_3$

16.60 Name the following compounds:

(a) —CH(CH$_3$)$_2$ **(b)** CH$_3$CH$_2$C≡CC(CH$_2$CH$_3$)$_3$

(c) —NH$_2$$^+$ Br$^-$ **(d)** (CH$_3$CH$_2$)$_2$N(CH$_2$)$_5$CH$_3$

16.61 Draw the structural formulas of the following compounds:

(a) 2,4-Dinitroacetophenone

(b) 2,4-Dihydroxycyclopentanone

(c) 2-Methoxy-2-methylpropane

(d) 2,3,4-Trimethyl-3-pentanol

16.62 Draw the structural formulas of the following compounds:

(a) 2,3,3-Triiodopentanal

(b) 1,1,3-Tribromoacetone

(c) 4-Amino-4-methyl-2-hexanone

16.63 Complete the following equations:

(a) CH$_3$CH=C(CH$_3$)$_2$ + H$_2$ $\xrightarrow{Pd}$?

(b) CH$_3$CH$_2$CCH$_2$CH$_3$ $\xrightarrow{[O]}$?

(c) CH$_3$CCH$_2$CH$_2$CH$_3$ $\xrightarrow[H_3O^+]{\text{Reducing agent}}$?

(d) CH$_2$CH + HOCH$_2$CH$_2$CH$_3$ ⟶ ? (Hemiacetal)

16.64 Complete the following equations:

(a) CH$_2$CH + 2 HOCH$_2$CH$_2$CH$_3$ ⟶ ? (Acetal)

(b) CH$_3$CH=CCH$_2$CH$_2$CH$_3$ + HCl ⟶ ?

(c) CH$_3$—⟨⟩—CH$_2$CH$_2$OH $\xrightarrow{H_2SO_4}$?

16.65 How can you differentiate between 3-hexanol and hexanal using a simple chemical test?

16.66 The liquids 1-butanol, 1-butylamine, and butanal have similar molar masses. Assign the observed boiling points of 78 °C, 75 °C, and 117 °C to these compounds and explain your choices.

16.67 2-Butanone has a solubility of 26 g/100 mL of H$_2$O, but 2-heptanone, which is found in clove and cinnamon bark oils, is only very slightly soluble in water. Explain the difference in solubility of these two ketones.

CHAPTER 17

Carboxylic Acids and Their Derivatives

CONCEPTS TO REVIEW

Acid–base chemistry
(Sections 10.1–10.6, 10.14)

Naming alkanes
(Section 12.6)

▲ A red wood ant colony reacts defensively by spraying the simplest of the carboxylic acids, formic acid, from a gland in the rear of their abdomens.

CONTENTS

CHAPTER GOALS

In this chapter, we will answer the following questions:

1. What are the general structures and properties of carboxylic acids and their derivatives?

THE GOAL: Be able to describe and compare the structures, reactions, hydrogen bonding, water solubility, boiling points, and acidity or basicity of carboxylic acids, esters, and amides.

2. How are carboxylic acids, esters, and amides named?

THE GOAL: Be able to name the simple members of these families and write their structures given the names.

3. What are some occurrences and applications of significant carboxylic acids, esters, and amides?

THE GOAL: Be able to identify the general occurrence and some important members of each family.

4. How are esters and amides synthesized from carboxylic acids and converted back to carboxylic acids?

THE GOAL: Be able to describe and predict the products of the ester- and amide-forming reactions of carboxylic acids and the hydrolysis of esters and amides.

5. What are the organic phosphoric acid derivatives?

THE GOAL: Be able to recognize and write the structures of phosphate esters and their ionized forms.

The simplest carbonyl-containing functional groups, the aldehydes and ketones, were described in Chapter 16. In this chapter, we move on to carbonyl compounds that are either *carboxylic acids* or *derivatives of carboxylic acids*—the *esters* and *amides*. Esters of phosphoric acid are introduced here also because of their major role in biochemistry and their chemical similarity to carboxylic acids and esters.

Carboxylic acid   Ester   Amide   Phosphoric acid ester

Carboxylic acids are easily converted to esters and amides, and the esters and amides are easily converted back to the carboxylic acids. These properties make molecules that contain carboxylic acids, amides, and esters important in biochemistry as well as in the chemical industry. The reactivity of carboxylic acids, for example, makes possible the formation of polymers—whether it be the synthetic kind (Section 17.7) or those that keep us alive: the proteins (which are polymers of amino acids, Chapter 18).

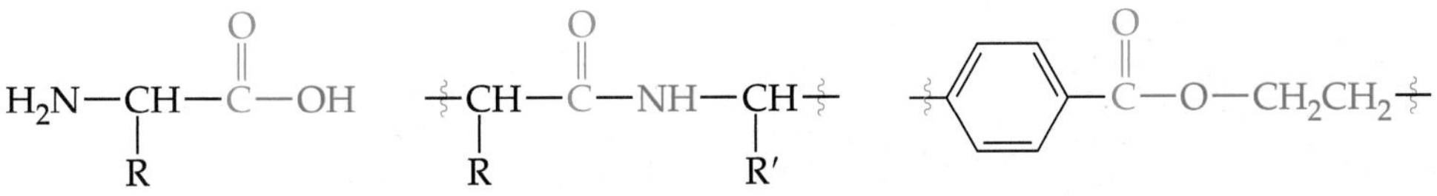

Amino acid   Amide group in a protein   Ester group in Dacron polyester

The study of these compounds will prepare you for Chapters 18 and 19, where we explore how proteins are put together and their roles in living things.

17.1 Carboxylic Acids and Their Derivatives: Properties and Names

Carboxylic acids have an —OH group bonded to the carbonyl carbon atom. In their derivatives, the —OH group is replaced by other groups. **Esters** have an —OR group bonded to the carbonyl carbon atom. **Amides** have an —NH₂, —NHR,

Carboxylic acid A compound that has a carbonyl group bonded to an —OH group, RCOOH.

Ester A compound that has a carbonyl group bonded to an —OR group, RCOOR'.

Amide A compound that has a carbonyl group bonded to a nitrogen atom group, RCONR'₂, where the R' groups may be alkyl groups or hydrogen atoms.

or —NR$_2$ group bonded to the carbonyl carbon atom. Finally, there are the acid anhydrides, formed when two carboxylic acids join together by eliminating a molecule of water. (These compounds are the subject of Worked Example 17.4.)

Note that carboxylic acids, esters, and amides have in common a carbonyl carbon atom bonded either to an oxygen or to a nitrogen. The resulting polarity of their functional groups and the structural similarities of carboxylic acids, esters, and amides account for many similarities in the properties of these compounds.

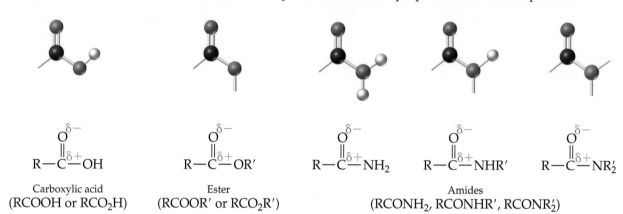

For example, since carboxylic acids and their derivatives all contain polar functional groups, all boil at a higher temperature than comparable alkanes. Carboxylic acids and amides can also take part in hydrogen bonding, which plays a prominent role in their chemical and physical properties.

Carbonyl-group substitution reaction A reaction in which a new group replaces (substitutes for) a group attached to a carbonyl-group carbon.

Carboxylic acids and their derivatives commonly undergo **carbonyl-group substitution reactions**, in which a group we represent as —Z replaces (substitutes for) the group bonded to the carbonyl carbon atom:

$$
\underset{}{R-\overset{O}{\underset{\|}{C}}-OH} + H-Z \rightleftarrows R-\overset{O}{\underset{\|}{C}}-Z + H-OH
$$

—OR′	—OR′
—NH$_2$	—NH$_2$
—NHR′	—NHR′
—NR$_2'$	—NR$_2'$

For example, esters are routinely made by such reactions:

$$
\underset{\substack{\text{Acetic acid}\\\text{(a carboxylic acid)}}}{CH_3-\overset{O}{\underset{\|}{C}}-OH} + \underset{\text{Ethanol}}{H-OCH_2CH_3} \rightleftarrows \underset{\substack{\text{Ethyl acetate}\\\text{(an ester)}}}{CH_3-\overset{O}{\underset{\|}{C}}-OCH_2CH_3} + \underset{\text{Water}}{H-OH}
$$

And esters can be converted back to carboxylic acids by reversing the reaction:

$$
\underset{\text{Ethyl acetate}}{CH_3-\overset{O}{\underset{\|}{C}}-OCH_2CH_3} + \underset{\text{Water}}{H-OH} \rightleftarrows \underset{\text{Acetic acid}}{CH_3-\overset{O}{\underset{\|}{C}}-OH} + \underset{\text{Ethanol}}{H-OCH_2CH_3}
$$

Acyl group An RC═O group.

The portion of the carboxylic acid that does not change during a carbonyl-group substitution reaction is known as an **acyl group**.

Acyl groups

$$
CH_3\overset{O}{\underset{\|}{C}}- \qquad CH_3CH_2\overset{O}{\underset{\|}{C}}- \qquad R-\overset{O}{\underset{\|}{C}}-
$$

In biochemistry, carbonyl-group substitution reactions are quite often called *acyl transfer reactions* and play an important role in the metabolism of a variety of biomolecules.

PROBLEM 17.1

Identify the following formulas as that of a carboxylic acid, an amide, an ester, or none of these.

(a) $CH_3\overset{\displaystyle O}{\overset{\displaystyle \|}{C}}NH_2$ (b) CH_3OCH_3 (c) CH_3COOH (d) $CH_3COOCH_2CH_3$

(e) CH_3COCH_3 (f) $CH_3CH_2CONHCH_3$ (g) $CH_3CH_2NH_2$ (h) $CH_3CH_2\overset{\displaystyle O}{\overset{\displaystyle \|}{C}}NH_2$

Carboxylic Acids

The most significant property of carboxylic acids is their behavior as weak acids. They surrender the hydrogen of the **carboxyl group**, COOH, to bases and establish acid–base equilibria in aqueous solution (a property further discussed in Section 17.3). The common carboxylic acids share the concentration-dependent corrosive properties of all acids but are not generally hazardous to human health.

Like alcohols, carboxylic acids form hydrogen bonds with each other so that even formic acid (HCOOH), the simplest carboxylic acid, is a liquid at room temperature with a boiling point of 101 °C.

Carboxyl group The —COOH functional group.

$$-\overset{\displaystyle O}{\overset{\displaystyle \|}{C}}-OH$$

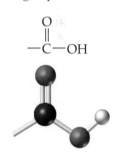

Carboxylic acids pair up by hydrogen bonding, as illustrated for formic acid.

$$H-C\overset{O\cdots H-O}{\underset{O-H\cdots O}{\Big\langle}}C-H$$

Acids with saturated straight-chain R groups of up to nine carbon atoms are volatile liquids with strong, pungent, and usually unpleasant odors; those with up to four carbons are water-soluble. Acids with larger saturated R groups are waxy, odorless solids. Their water solubility falls off as the size of the hydrophobic, alkane-like R group increases relative to the size of the water-soluble —COOH portion.

Carboxylic acids are named in the IUPAC system by replacing the final *-e* of the corresponding alkane name with *-oic acid*. The 3-carbon acid is propanoic acid; the straight-chain, 4-carbon acid is butanoic acid; and so on. If alkyl substituents or other functional groups are present, the chain is numbered beginning at the —COOH end, as in 3-methylbutanoic acid or in 2-hydroxypropanoic acid, which is better known as lactic acid, the acid present in sour milk.

$$CH_3CH_2-\overset{\displaystyle O}{\overset{\displaystyle \|}{C}}-OH \qquad \underset{4}{CH_3}\underset{3}{\overset{\displaystyle CH_3}{\overset{\displaystyle |}{C}H}}\underset{2}{CH_2}-\underset{1}{\overset{\displaystyle O}{\overset{\displaystyle \|}{C}}}-OH \qquad \underset{3}{CH_3}\underset{\underset{OH}{|}}{\overset{2}{C}H}-\underset{1}{\overset{\displaystyle O}{\overset{\displaystyle \|}{C}}}-OH$$

Propanoic acid   3-Methylbutanoic acid   Lactic acid
(2-Hydroxypropanoic acid)

Unlike other families of organic molecules, the common names of many of the carboxylic acids are used far more often than their IUPAC names, primarily because carboxylic acids were among the first organic compounds to be isolated and purified. Recognizing the common acid names given in Table 17.1 is important, as they provide the basis for the common names of derivatives of these acids. When using common names, the carbon atoms attached to the —COOH group are identified by Greek letters α, β, and so on, rather than numbers. In this system, for example,

TABLE 17.1 Physical Properties of Some Carboxylic Acids

STRUCTURE	COMMON NAME	MELTING POINT (°C)	BOILING POINT (°C)
HCOOH	Formic	8	101
CH₃COOH	Acetic	17	118
CH₃CH₂COOH	Propionic	−22	141
CH₃CH₂CH₂COOH	Butyric	−4	163
CH₃CH₂CH₂CH₂COOH	Valeric	−34	185
CH₃(CH₂)₁₆COOH	Stearic	70	383
HOOCCOOH	Oxalic	190	Decomposes
HOOCCH₂COOH	Malonic	135	Decomposes
HOOCCH₂CH₂COOH	Succinic	188	Decomposes
HOOCCH₂CH₂CH₂COOH	Glutaric	98	Decomposes
H₂C=CHCOOH	Acrylic	13	141
CH₃CH=CHCOOH	Crotonic	72	185
⬡—COOH	Benzoic	122	249
⬡(OH)—COOH	Salicylic	159	Decomposes

the 3-carbon acid is *propionic acid*, and the second C=O group (a *keto* group in common nomenclature) next to the —COOH group is an α-keto group.

$$CH_3\overset{\beta}{-}\underset{\overset{\parallel}{O}}{C}\overset{\alpha}{-}\underset{\overset{\parallel}{O}}{C}-OH \qquad \overset{\beta}{CH_3}-\overset{\alpha}{\underset{\underset{NH_2}{|}}{CH}}-\underset{\overset{\parallel}{O}}{C}-OH$$

α-Ketopropionic acid   α-Aminopropionic acid
(Pyruvic acid,   (Alanine)
a key biochemical
intermediate)

In alanine, as in all common amino acids, the –NH₂ group is on the alpha carbon atom (the C next to –COOH).

The acyl group that remains when a carboxylic acid loses its —OH is named by replacing the *-ic* at the end of the acid name with *-oyl*. One very important exception is the acyl group from acetic acid, which is traditionally called an **acetyl group**.

Acetyl group A CH₃C=O group.

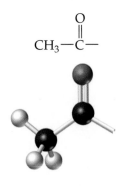

$$CH_3-\underset{\overset{\parallel}{O}}{C}- \qquad CH_3CH_2-\underset{\overset{\parallel}{O}}{C}- \qquad ⬡-\underset{\overset{\parallel}{O}}{C}-$$

Acetyl group   Propanoyl group   Benzoyl group

Dicarboxylic acids, which contain two —COOH groups, are named systematically by adding the ending *-dioic acid* to the alkane name (the *-e* is retained). Here again the simple dicarboxylic acids are usually referred to by their common names. Oxalic acid (IUPAC name: ethanedioic acid) is found in plants of the genus *Oxalis*, which includes rhubarb and spinach. You will encounter succinic acid, glutaric acid, and several other dicarboxylic acids when we come to the generation of biochemical energy and the citric acid cycle (Section 21.8).

$$HO-\underset{\overset{\parallel}{O}}{C}-\underset{\overset{\parallel}{O}}{C}-OH \qquad HO-\underset{\overset{\parallel}{O}}{C}-CH_2CH_2-\underset{\overset{\parallel}{O}}{C}-OH \qquad HO-\underset{\overset{\parallel}{O}}{C}-(CH_2)_3-\underset{\overset{\parallel}{O}}{C}-OH$$

Oxalic acid   Succinic acid   Glutaric acid
(Ethanedioic acid)   (Butanedioic acid)   (Pentanedioic acid)

Unsaturated acids are named systematically in the IUPAC system with the ending -*enoic*. For example, the simplest unsaturated acid, $H_2C=CHCOOH$, is named propenoic acid. It is, however, best known as acrylic acid, which is a raw material for acrylic polymers.

Looking Ahead

Biochemistry is dependent on the continual breakdown of food molecules. Frequently this process requires transfer of acetyl groups from one molecule to another. Acetyl-group transfer occurs, for example, at the beginning of the citric acid cycle, which is central to production of life-sustaining energy (Section 21.8).

WORKED EXAMPLE **17.1** Naming a Carboxylic Acid

Give the systematic and common names for this compound:

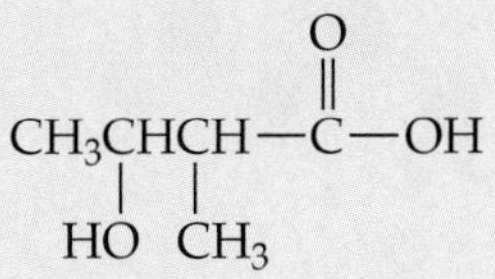

$$\begin{array}{c} \qquad\qquad\quad O \\ \qquad\qquad\quad \| \\ CH_3CHCH-C-OH \\ \quad\; | \quad\; | \\ \quad HO \quad CH_3 \end{array}$$

ANALYSIS For the systematic name of a carboxylic acid, first identify the longest chain containing the —COOH group and number it starting with the carboxyl-group carbon:

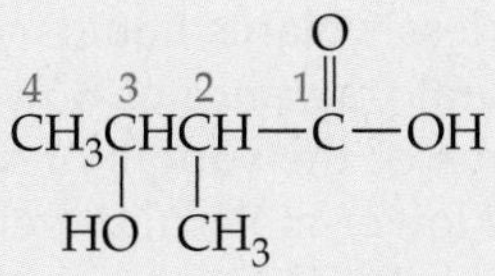

$$\begin{array}{c} \qquad\qquad\qquad\quad O \\ \; 4 \quad\; 3 \quad 2 \quad\; 1\| \\ CH_3CHCH-C-OH \\ \quad\;\; | \quad\; | \\ \quad\; HO \quad CH_3 \end{array}$$

The parent compound is the 4-carbon acid, butanoic acid. It has a methyl group on carbon 2 and a hydroxyl group on carbon 3. From Table 17.1 we see that the common name for the 4-carbon acid is butyric acid. In the common nomenclature scheme, substituents are located by Greek letters rather than numbers:

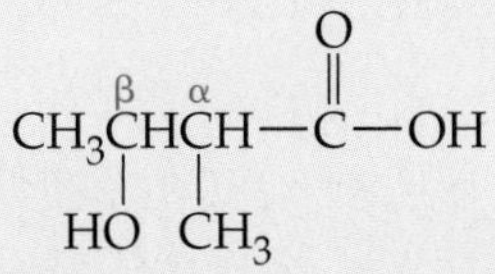

$$\begin{array}{c} \qquad\qquad\qquad\quad O \\ \quad\;\beta \quad\; \alpha \quad\;\; \| \\ CH_3CHCH-C-OH \\ \quad\;\; | \quad\; | \\ \quad\; HO \quad CH_3 \end{array}$$

SOLUTION
The IUPAC name of this molecule is 3-hydroxy-2-methylbutanoic acid; the common name of this acid is β-hydroxy-α-methylbutyric acid.

PROBLEM 17.2

Draw the structures of the following acids:

(a) 2-Ethyl-3-Hydroxyhexanoic acid

(b) *m*-Nitrobenzoic acid

PROBLEM 17.3

Write both the complete structural formula of succinic acid (refer to Table 17.1), showing all bonds, and the line-angle structural formula.

PROBLEM 17.4

Identify the acid that is formed by addition of Br_2 to the double bond in acrylic acid (refer to Table 17.1 and Section 13.6).

Esters

$-C(=O)-O-CH_3$

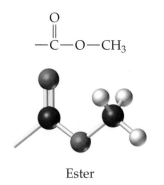

Ester

When the —OH of the carboxyl group is converted to the —OR of an ester group (—COOR), the ability of the molecules to hydrogen-bond with each other is lost. Simple esters are therefore lower boiling than the acids from which they are derived.

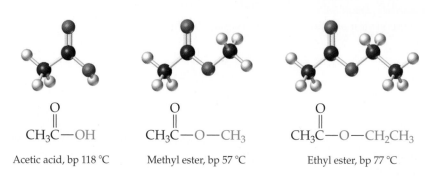

$CH_3C(=O)-OH$	$CH_3C(=O)-O-CH_3$	$CH_3C(=O)-O-CH_2CH_3$
Acetic acid, bp 118 °C	Methyl ester, bp 57 °C	Ethyl ester, bp 77 °C

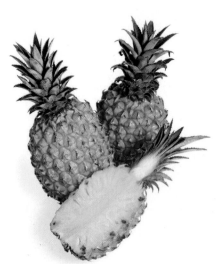

▲ Pineapples are among the many fruits with flavors derived from esters.

The simple esters are colorless, volatile liquids with pleasant odors, and many of them contribute to the natural fragrance of flowers and ripe fruits. Prolonged exposure to high concentrations of the vapors of volatile esters can be irritating and have a narcotic effect. The lower-molecular-weight esters are somewhat soluble in water and are quite flammable. Esters are neither acids nor bases in aqueous solution.

Ester names consist of two words. The first is the name of the alkyl group R in the ester group —COOR. The second is the name of the parent acid, with the family-name ending *-ic acid* replaced by *-ate.* (Note that the order of the two parts of the name is the reverse of the order in which ester condensed formulas are usually written.)

Naming an ester

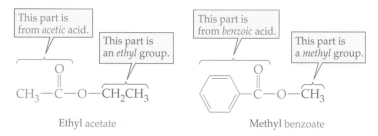

Ethyl acetate   Methyl benzoate

Both common and systematic names are derived in this manner. For example, an ester of a straight-chain, 4-carbon carboxylic acid is named systematically as a butanoate (from butanoic acid) or by its common name as a butyrate (from butyric acid):

$$CH_3CH_2CH_2C(=O)OCH_2CH_3$$

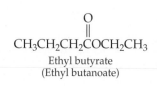

Ethyl butyrate
(Ethyl butanoate)

This ester is used as a food flavoring to give the taste and smell of pineapples.

WORKED EXAMPLE   **17.2** Writing the Structure of an Ester from Its Name

What is the structure of butyl acetate?

ANALYSIS The two-word name consisting of an alkyl group name followed by an acid name with an *-ate* ending shows that the compound is an ester. The name "acetate" shows that the RCO— part of the molecule is from acetic acid (CH_3COOH). The "butyl" part of the name indicates that a butyl group has replaced H in the carboxyl group.

SOLUTION
The structure of butyl acetate is

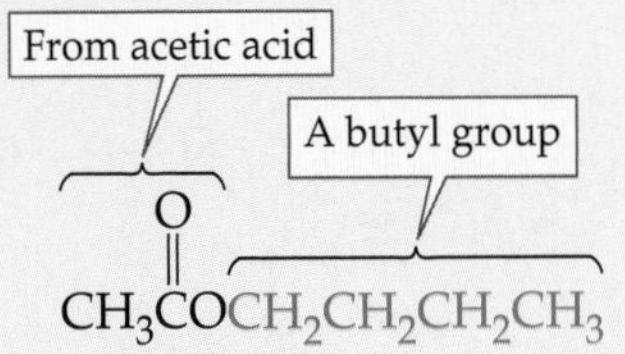

$$\underset{\text{A butyl group}}{\underset{\text{From acetic acid}}{CH_3CO CH_2CH_2CH_2CH_3}}$$

WORKED EXAMPLE   **17.3** Naming an Ester from Its Structure

What is the name of this compound?

$$CH_3(CH_2)_{16}\overset{O}{\overset{\|}{C}}OCH_2CH_2CH_3$$

ANALYSIS The compound has the general formula RCOOR', so it is an ester. The acyl part of the molecule (RCO—) is from stearic acid (see Table 17.1). The R' group has 3 C atoms and is therefore a propyl group.

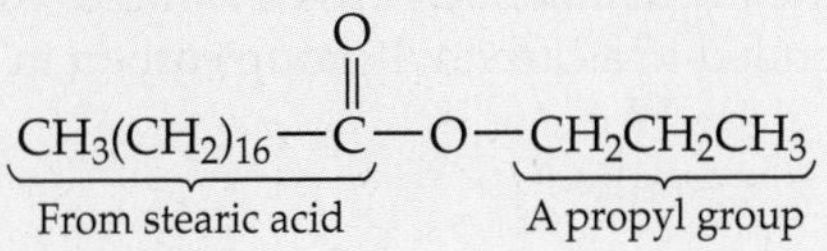

$$\underset{\text{From stearic acid}}{CH_3(CH_2)_{16}}-\overset{O}{\overset{\|}{C}}-O-\underset{\text{A propyl group}}{CH_2CH_2CH_3}$$

SOLUTION
The compound is propyl stearate.

PROBLEM 17.5

Draw the structures of:

(a) Isopropyl benzoate   **(b)** Ethyl valerate   **(c)** Isopropyl crotonate
(See Table 17.1.)

PROBLEM 17.6

Which of the following compounds has the highest boiling point and which has the lowest boiling point? Explain your answer.

(a) CH_3OCH_3 **(b)** CH_3COOH **(c)** $CH_3CH_2CH_3$

PROBLEM 17.7

In the following pairs of compounds, which would you expect to be more soluble in water? Why?

(a) $C_8H_{17}COOH$ or $CH_3CH_2CH_2COOH$

(b) $\underset{\underset{CH_3}{|}}{CH_3CHCOOH}$ or $\underset{\underset{CH_3}{|}}{CH_3CH_2COOCHCH_3}$

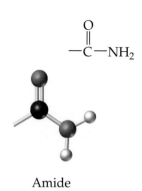

Amide

Amides

Compounds with a nitrogen directly attached to the carbonyl carbon atom are *amides*. The nitrogen of an amide may be an —NH$_2$ group or may have one or two R′ groups bonded to it. *Unsubstituted amides* (RCONH$_2$) can form multiple hydrogen bonds to other amide molecules and are thus higher melting and higher boiling than the acids from which they are derived.

Hydrogen bonding in RCNH$_2$

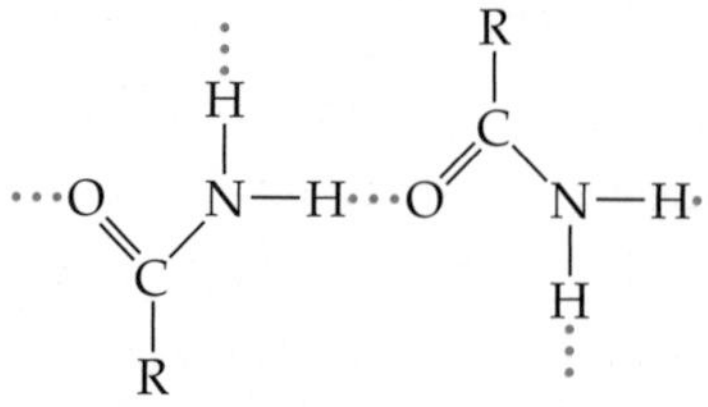

Low-molecular-weight unsubstituted amides are solids (except for the simplest amide (formamide, HCONH$_2$, a liquid)) that are soluble in both water (with which they form hydrogen bonds) and organic solvents. *Monosubstituted amides* (RCONHR′) can also form hydrogen bonds to each other, but *disubstituted amides* (RCONR′$_2$) cannot do so and are therefore lower boiling.

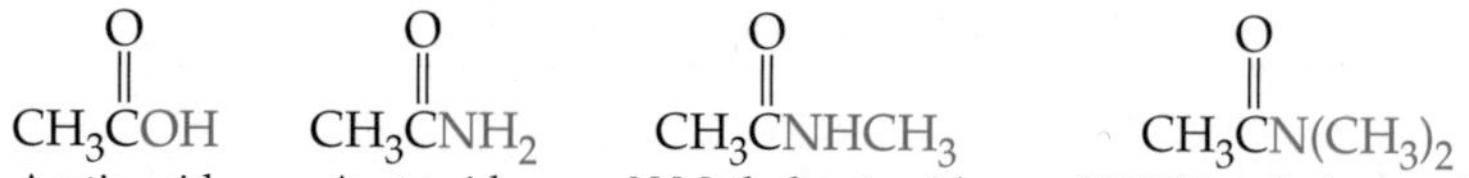

CH$_3$COH	CH$_3$CNH$_2$	CH$_3$CNHCH$_3$	CH$_3$CN(CH$_3$)$_2$
Acetic acid	Acetamide	*N*-Methylacetamide	*N,N*-Dimethylacetamide
(bp 118 °C)	(bp 222 °C)	(bp 206 °C)	(bp 165 °C)

It is important to note the distinction between amines (Chapter 15) and amides. The nitrogen atom is bonded to a carbonyl-group carbon in an amide, but *not* in an amine:

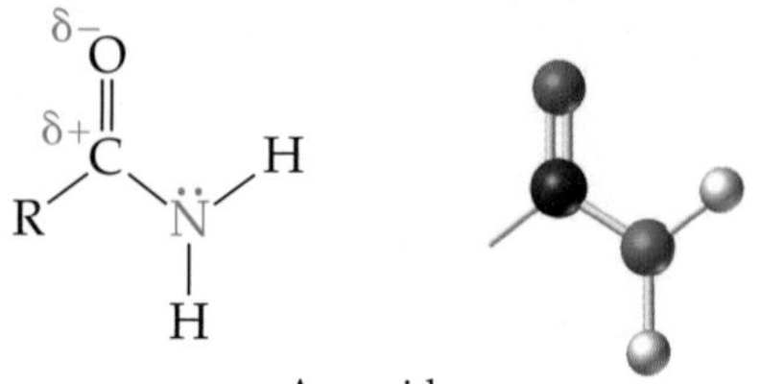

An amide
(RCONH$_2$)

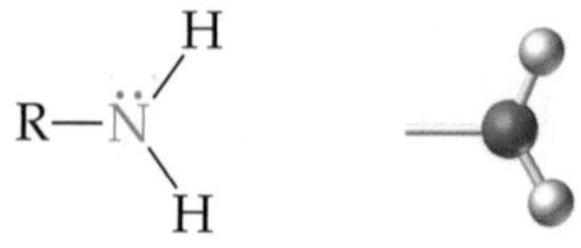

An amine
(RNH$_2$)

The positive end of the carbonyl group attracts the unshared pair of electrons on nitrogen strongly enough to prevent it from acting as a base by accepting a hydrogen atom. As a result, *amides are NOT basic like amines.*

Amides with an unsubstituted —NH$_2$ group are named by replacing the *-ic acid* or *-oic acid* of the corresponding carboxylic acid name with *-amide*. For example, the amide derived from acetic acid is called acetamide. If the nitrogen atom of the amide has alkyl substituents on it, the compound is named by first specifying the alkyl group and then identifying the amide name. The alkyl substituents are preceded by the letter *N* to identify them as being attached directly to nitrogen.

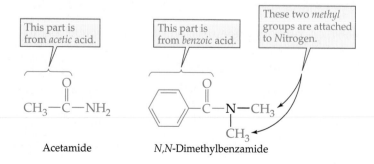

This part is from *acetic* acid.

This part is from *benzoic* acid.

These two *methyl* groups are attached to Nitrogen.

CH$_3$—C—NH$_2$

N,N-Dimethylbenzamide

Acetamide

To review, some derivatives of acetic acid are shown below:

Carbonyl derivatives of acetic acid

$$CH_3\overset{\displaystyle O}{\overset{\|}{C}}OCH_3 \quad CH_3\overset{\displaystyle O}{\overset{\|}{C}}NH_2 \quad CH_3\overset{\displaystyle O}{\overset{\|}{C}}NHCH_3 \quad CH_3\overset{\displaystyle O}{\overset{\|}{C}}N\overset{\displaystyle CH_3}{\underset{\displaystyle CH_3}{}} \quad CH_3\overset{\displaystyle O}{\overset{\|}{C}}N\overset{\displaystyle CH_3}{\underset{\displaystyle CH_2CH_3}{}}$$

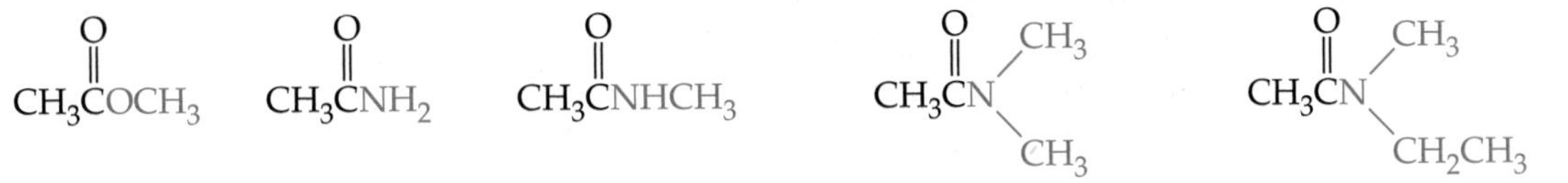

Methyl acetate   Acetamide   *N*-Methylacetamide   *N,N*-Dimethylacetamide   *N*-Ethyl-*N*-Methylacetamide

Properties of Carboxylic Acids, Esters, and Amides

- All undergo carbonyl-group substitution reactions.
- Esters and amides are made from carboxylic acids.
- Esters and amides can be converted back to carboxylic acids.
- Carboxylic acids and unsubstituted or monosubstituted amides exhibit strong hydrogen bonding to one another; disubstituted amides and esters do not hydrogen bond to one another.
- Simple acids and esters are liquids; all unsubstituted amides (except formamide) are solids.
- Carboxylic acids give acidic aqueous solutions. Esters and amides are neither acids nor bases (pH neutral).
- Small (low-molecular-weight) esters are somewhat water-soluble, and small amides are water-soluble.
- Volatile acids have strong, sharp odors; volatile esters have pleasant, fruity odors. Amides generally have no odors associated with them.

Looking Ahead

In later chapters, you will see that the fundamental bonding connections in proteins are amide bonds (Section 18.2) and those in oils and fats are ester bonds (Section 25.2).

WORKED EXAMPLE 17.4 Formation of Acid Anhydrides

An important but often overlooked class of carboxylic acid derivatives are the *acid anhydrides*. (*Anhydride* means "without water.") Acid anhydrides are formed when pairs of acid molecules react (dimerize) via a carbonyl-group substitution reaction to lose water. Relate the reactants below to the substitution reaction pattern (on p. 518) and complete the equation.

$$CH_3-\overset{\displaystyle O}{\overset{\|}{C}}-OH \ + \ HO-\overset{\displaystyle O}{\overset{\|}{C}}-CH_3 \longrightarrow$$

ANALYSIS The reaction here fits the substitution reaction pattern as follows:

$$R-\overset{\displaystyle O}{\overset{\|}{C}}-OH + HZ \longrightarrow R-\overset{\displaystyle O}{\overset{\|}{C}}-Z + H-OH$$

with Z equal to

$$-O-\overset{\displaystyle O}{\overset{\|}{C}}-CH_3$$

SOLUTION

The reaction is:

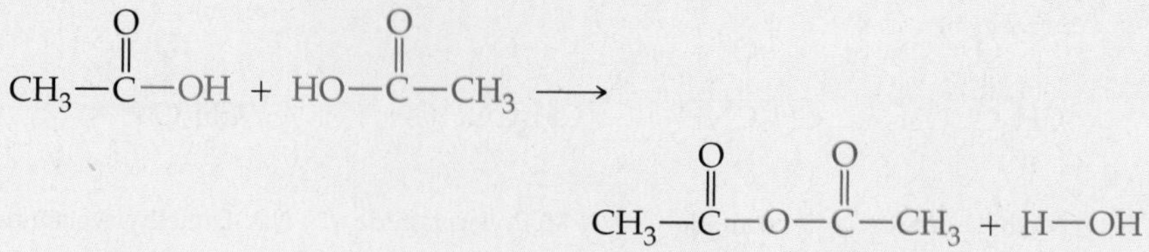

$$CH_3-\overset{\overset{\displaystyle O}{\|}}{C}-OH \; + \; HO-\overset{\overset{\displaystyle O}{\|}}{C}-CH_3 \longrightarrow$$

$$CH_3-\overset{\overset{\displaystyle O}{\|}}{C}-O-\overset{\overset{\displaystyle O}{\|}}{C}-CH_3 \; + \; H-OH$$
$$\text{Acetic anhydride}$$

(Carboxylic acid anhydrides, although important in an organic chemistry lab, are of little interest in biochemistry. They easily react with water to give back acids in the reverse of the reaction described here and are not found in biological systems. They are introduced here to prepare the way for the introduction of phosphoric acid anhydrides in Section 17.8)

PROBLEM 17.8

Write condensed structures for propionic acid and an ester and unsubstituted, N-monosubstituted, and N,N-disubstituted amides derived from propionic acid. Name the derivatives.

PROBLEM 17.9

What are the names of the following compounds?

(a) 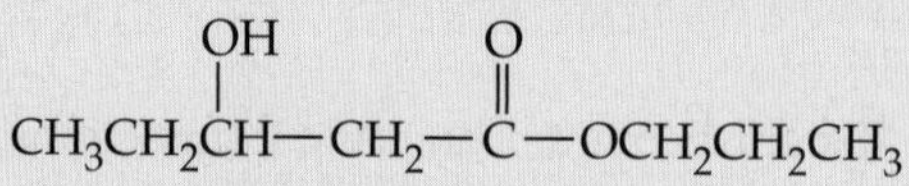 $CH_3CH_2\overset{\overset{\displaystyle OH}{|}}{CH}-CH_2-\overset{\overset{\displaystyle O}{\|}}{C}-OCH_2CH_2CH_3$

(b) 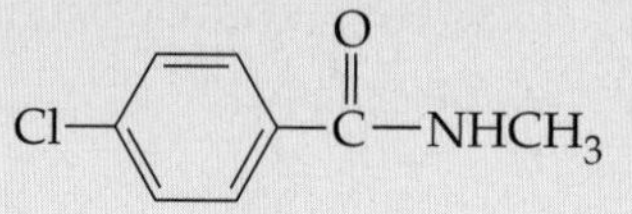 $Cl-\!\!\left\langle\bigcirc\right\rangle\!\!-\overset{\overset{\displaystyle O}{\|}}{C}-NHCH_3$

PROBLEM 17.10

Draw structures corresponding to these names:

(a) 4-Methylpentanamide
(b) N-Ethyl-N-methylpropanamide

PROBLEM 17.11

Match partial structures (a)–(d) with the following classes of compounds: (i) α-amino carboxylic acid, (ii) monosubstituted amide, (iii) methyl ester, (iv) carboxylic acid.

(a) 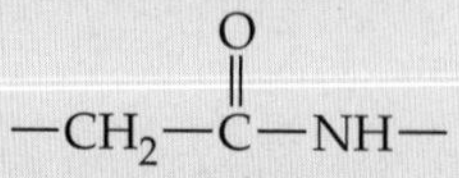 $-CH_2-\overset{\overset{\displaystyle O}{\|}}{C}-NH-$

(b) $-\overset{\overset{\displaystyle O}{\|}}{\underset{\underset{\displaystyle NH_2}{|}}{CH}}-\overset{\overset{\displaystyle O}{\|}}{C}-OH$

(c) $-CH_2-\overset{\overset{\displaystyle O}{\|}}{C}-OH$

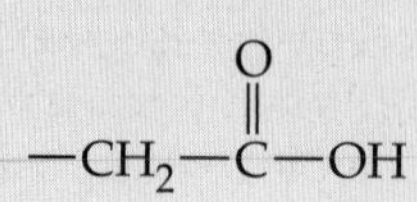

(d) 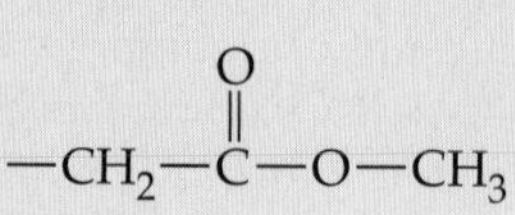 $-CH_2-\overset{\overset{\displaystyle O}{\|}}{C}-O-CH_3$

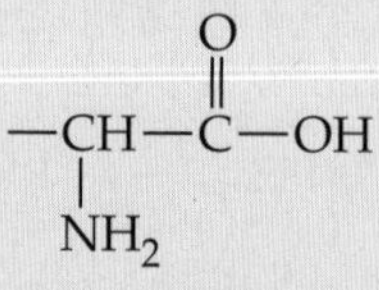

PROBLEM 17.12

Classify each compound (a)–(f) as one of the following: (i) amide, (ii) ester, (iii) carboxylic acid.

(a) CH₃COOCH₃

(b) RCONHR

(c) C₆H₅COOH

(d) CH₃CH₂C—N(CH₃)₂ (with =O above C)

(e) CH₃CH₂CH₂CONH₂

(f) HOOCCH₂—CH—CH₃ (with CH₃ below CH)

KEY CONCEPT PROBLEM 17.13

Identify the following molecules as an ester, carboxylic acid, or amide, and write the condensed molecular structural formula and the line angle structural formula for each.

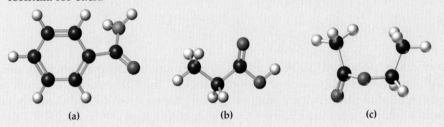

(a)   (b)   (c)

17.2 Some Common Carboxylic Acids

Carboxylic acids occur throughout the plant and animal kingdoms, and the common names of many come from the plants or animals in which they were first identified. Formic acid is known as the chemical that puts the sting in ant bites (from the Latin *formica*, "ant"). Acetic acid and lactic acid are named from the Latin *acetum*, meaning vinegar, and *lactis*, meaning milk. Butyric acid (from the Latin *butyrum*, "butter") is responsible for the terrible odor of rancid butter (in which it is formed by oxidation of butyraldehyde). Caproic acid (from the Latin *caper*, "goat") was first isolated from the skin of goats (which also have a distinctive odor). The esters of long-chain carboxylic acids such as stearic acid and oleic acid are components of all animal fats and vegetable oils.

Some naturally occurring carboxylic acids

CH₃CH₂CH₂—C—OH (with =O above C)
Butyric acid
(rancid butter)

CH₃CH₂CH₂CH₂CH₂—C—OH (with =O above C)
Caproic acid
(goats)

CH₃CH₂CH₂CH₂CH₂CH₂CH₂CH₂CH₂CH₂CH₂CH₂CH₂CH₂CH₂CH₂CH₂—C—OH (with =O above C)
Stearic acid (C₁₈H₃₆O₂, from animal fat)

Acetic Acid (CH₃COOH) : In Vinegar

Everyone recognizes the sour taste of the best known of the carboxylic acids, acetic acid, since it is the primary organic component of vinegar. Vinegar is a solution of 4–8% acetic acid in water (with various flavoring agents). When fermentation of grapes, apples, and other fruits proceeds in the presence of ample oxygen, oxidation goes beyond the formation of ethanol to the formation of acetic acid (Section 14.5).

CH₃C—OH (with =O above C)

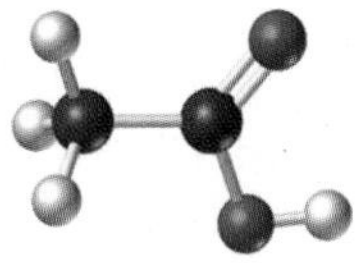

Acetic acid

The production of "boutique" vinegars from various wine varietals has become big business today; for example, it is common to find champagne vinegar in many cooking stores. Aqueous acetic acid solutions are also common laboratory reagents.

In concentrations over 50%, acetic acid is corrosive and can damage the skin, eyes, nose, and mouth. There is no pain when the concentrated acid is spilled on unbroken skin, but painful blisters form about 30 minutes later. Pure acetic acid is known as *glacial acetic acid* because, with just a slight amount of cooling below room temperature (to 17 °C), the liquid forms icy-looking crystals that resemble glaciers. Acetic acid is a reactant in many industrial processes and is sometimes used as a solvent. As a food additive, it is used to adjust acidity.

Citric Acid: In Citrus Fruits and Blood

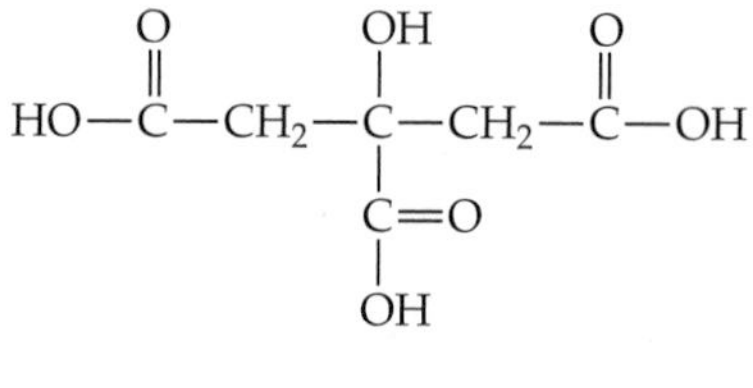

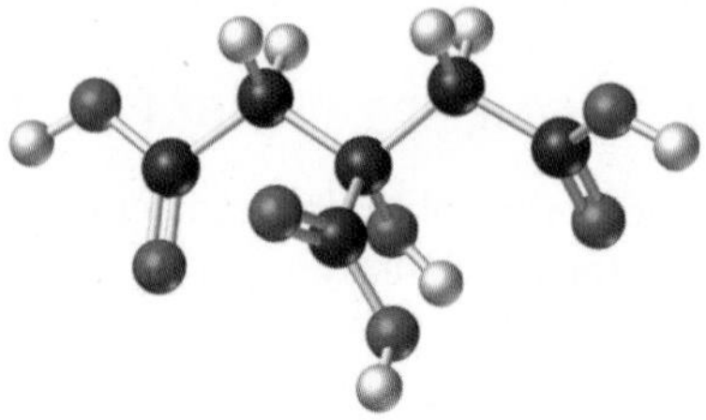

Citric acid

Citric acid is produced by almost all plants and animals during metabolism, and its normal concentration in human blood is about 2 mg/100 mL. Citrus fruits owe their tartness to citric acid; for example, lemon juice contains 4–8% and orange juice about 1% citric acid. Pure citric acid is a white, crystalline solid (mp 153 °C) that is very soluble in water. Citric acid and its salts are extensively used in pharmaceuticals, foods, and cosmetics. They buffer pH in shampoos and hair-setting lotions, add tartness to candies and soft drinks, and react with bicarbonate ion to produce the fizz in Alka-Seltzer.

⊂⊃ **Looking Ahead**

Citric acid lends its name to the *citric acid cycle*, part of the major biochemical pathway that leads directly to the generation of energy. Citric acid is the product of the first reaction of an eight-reaction cycle, which is presented in Section 21.8.   ⊂⊃

17.3 Acidity of Carboxylic Acids

Carboxylate anion The anion that results from ionization of a carboxylic acid, RCOO$^-$.

Carboxylic acids are weak acids that establish equilibria in aqueous solution with **carboxylate anions**, RCOO$^-$. The carboxylate anions are named by replacing the *-ic* ending in the carboxylic acid name with *-ate* (giving the same names and endings used in naming esters). At pH 7.4 in body fluids, carboxylic acids exist mainly as their carboxylate anions:

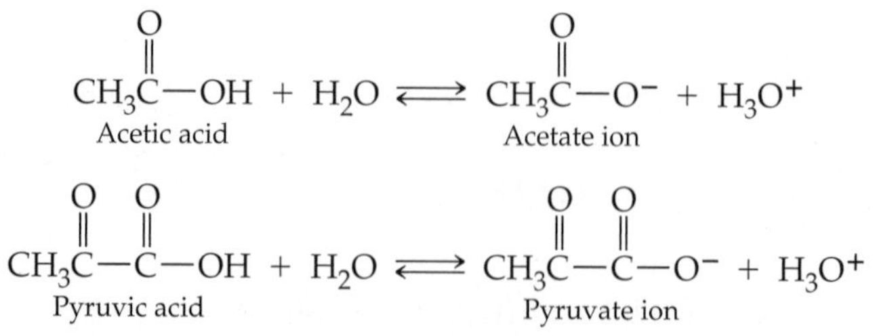

$$CH_3\overset{\overset{\displaystyle O}{\|}}{C}-OH + H_2O \rightleftharpoons CH_3\overset{\overset{\displaystyle O}{\|}}{C}-O^- + H_3O^+$$

Acetic acid — Acetate ion

$$CH_3\overset{\overset{\displaystyle O}{\|}}{C}-\overset{\overset{\displaystyle O}{\|}}{C}-OH + H_2O \rightleftharpoons CH_3\overset{\overset{\displaystyle O}{\|}}{C}-\overset{\overset{\displaystyle O}{\|}}{C}-O^- + H_3O^+$$

Pyruvic acid — Pyruvate ion

The comparative strength of an acid is measured by its acid dissociation constant (K_a); the smaller the value of K_a, the weaker the acid (Section 10.7). Many carboxylic acids have about the same acid strength as acetic acid, as shown by the acid ionization constants in Table 17.2. (⊂⊃, p. 529) There are some exceptions, though. Trichloroacetic acid, used to prepare microscope slides and to precipitate

TABLE 17.2 Carboxylic Acid Dissociation Constants*

NAME	STRUCTURE	ACID DISSOCIATION CONSTANT (K_a)
Trichloroacetic acid	Cl_3CCOOH	2.3×10^{-1}
Chloroacetic acid	$ClCH_2COOH$	1.4×10^{-3}
Formic acid	$HCOOH$	1.8×10^{-4}
Acetic acid	CH_3COOH	1.8×10^{-5}
Propanoic acid	CH_3CH_2COOH	1.3×10^{-5}
Hexanoic acid	$CH_3(CH_2)_4COOH$	1.3×10^{-5}
Benzoic acid	C_6H_5COOH	6.5×10^{-5}
Acrylic acid	$H_2C{=}CHCOOH$	5.6×10^{-5}
Oxalic acid	$HOOCCOOH$	5.4×10^{-2}
	$^-OOCCOOH$	5.2×10^{-5}
Glutaric acid	$HOOC(CH_2)_3COOH$	4.5×10^{-5}
	$^-OOC(CH_2)_3COOH$	3.8×10^{-6}

*The acid dissociation constant K_a is the equilibrium constant for the ionization of an acid; the smaller its value, the weaker the acid:

$$RCOOH + H_2O \rightleftharpoons RCOO^- + H_3O^+ \quad K_a = \frac{[RCOO^-][H_3O^+]}{[RCOOH]}$$

protein from body fluids, is a strong acid that must be handled with the same respect as sulfuric acid.

Carboxylic acids undergo neutralization reactions with bases in the same manner as other acids. With strong bases, such as sodium hydroxide, a carboxylic acid reacts to give water and a **carboxylic acid salt**, as shown below for the formation of sodium acetate. (Like all other such aqueous acid–base reactions, this reaction proceeds much more favorably in the forward direction than in the reverse direction, and is thus written with a single arrow.) As for all salts, a carboxylic acid salt is named with cation and anion names.

Carboxylic acid salt An ionic compound containing a cation and a carboxylic acid anion.

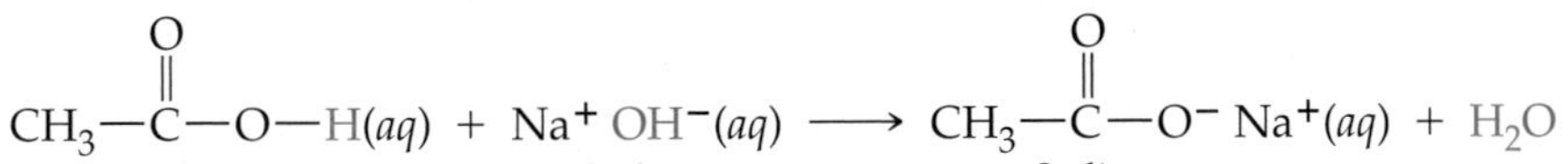

$$CH_3{-}\overset{\overset{\displaystyle O}{\|}}{C}{-}O{-}H(aq) + Na^+ OH^-(aq) \longrightarrow CH_3{-}\overset{\overset{\displaystyle O}{\|}}{C}{-}O^- Na^+(aq) + H_2O$$

Acetic acid   Sodium   Sodium acetate
(a weak acid)   hydroxide

The sodium and potassium salts of carboxylic acids are ionic solids that are usually far more soluble in water than the carboxylic acids themselves. For example, benzoic acid has a water solubility of only 3.4 g/L at 25 °C, whereas for sodium benzoate the water solubility is 550 g/L. The formation of carboxylic acid salts, like the formation of amine salts, is useful in creating water-soluble derivatives of drugs. (See Application "Organic Compounds in Body Fluids and the 'Solubility Switch'," Chapter 15, p. 478.)

WORKED EXAMPLE   **17.5** Effect of Structure on Carboxylic Acid Strength

Write the structural formulas of trichloroacetic acid and acetic acid and explain why trichloroacetic acid is the much stronger acid of the two.

ANALYSIS

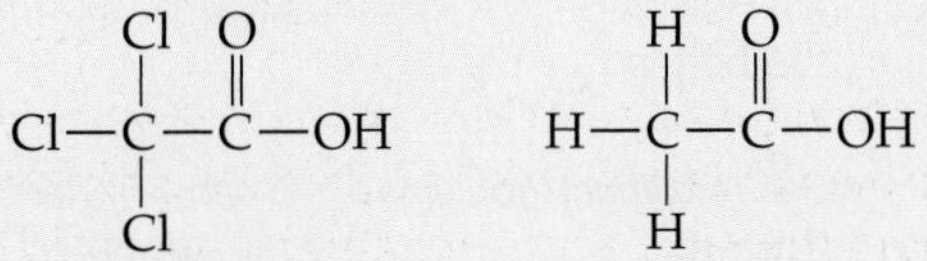

Trichloroacetic acid   Acetic acid

The structural difference is the replacement of 3 hydrogen atoms on the alpha carbon by 3 chlorine atoms. The chlorines are much more electronegative than the hydrogen and therefore draw electrons away from the rest of the molecule in trichloroacetic acid. (⊂⊃, p. 128) The result is that the hydrogen atom in the —COOH group in trichloroacetic acid is held less strongly and is much more easily removed than the corresponding hydrogen atom in acetic acid.

SOLUTION

Since the —COOH hydrogen atom in trichloroacetic acid is held less strongly, it is the stronger acid.

APPLICATION ▶ Acids for the Skin

A strongly acidic solution can damage the skin as seriously as a flame can. Nevertheless, physicians have found that such solutions are useful in the treatment of a variety of skin conditions. Weakly acidic solutions used in skin treatments fall at the borderline of the distinction between prescription drugs and cosmetics.

Trichloroacetic acid, a strong acid (Table 17.2), is used for chemical peeling of the skin (a treatment for eczema or psoriasis), for removal of acne scars, and sometimes for removing wrinkles. In what is the equivalent of a first- or second-degree burn, a surface layer of the skin is destroyed by reaction with the acid. The depth of removal (and the length of the healing period) varies with the strength of the acid and how long it is left on the skin. As the result of healing, old skin is replaced by a new, smoother skin surface.

A number of naturally occurring α-hydroxy acids (similar in acid strength to acetic acid) provide less aggressive skin treatments.

▲ Skin-care products relying on the benefits of an α-hydroxy acid.

Glycolic acid
(in sugar cane)

Lactic acid
(in sour milk)

Salicylic acid
(in willow trees)

Glycolic acid is used by physicians for a "mini" peel that can be done quickly and from which the skin returns to a normal appearance in just a few hours. In higher concentrations, glycolic acid is used for "spot" removal of precancerous lesions or unsightly brown, thickened skin (*keratoses*).

The effectiveness of α-hydroxy acids for conditions that produce dry, flaking, itchy skin has brought them into the cosmetics market. In recent years, "alpha-hydroxys" have been highly promoted for removing wrinkles, moisturizing skin, and generally giving us softer, smoother skin. The alpha-hydroxys have a further promotional advantage because they are all "nature's own" chemicals. For example, lactic acid, which at 12% concentration is a prescription drug, is present at lower concentration in many over-the-counter lotions and creams.

Although adult use of cosmetics that contain 10% or less of alpha-hydroxy acid is considered safe, one general caution has been noted. Industry-sponsored studies have suggested that alpha-hydroxys increase the sensitivity of the skin to ultraviolet radiation from the sun. Thus, an individual using alpha-hydroxys should be diligent about applying a good, strong sunscreen before going outside, the higher the SPF the better.

See Additional Problems 17.76 and 17.77 at the end of the chapter.

APPLICATION ▶ Acid Salts as Food Additives

Take a moment and examine the lists of ingredients on the labels of soft drinks, cookies, cakes, dried sauce mixes, or preserved meats. Chances are you will find the names of some acid salts.

Scanning the label on a package of strawberry jam-filled cookies, for example, turns up *sodium benzoate, sodium propionate, potassium sorbate,* and *sodium citrate.* What is the purpose of all these food additives? The first three are preservatives. Sodium benzoate prevents the growth of microorganisms, especially in acidic foods. Used in many soft drinks, it is one of the most common food additives. Sodium propionate, also very common, prevents the growth of mold in baked goods. Potassium sorbate, the salt of an unsaturated acid (sorbic acid, $CH_3CH=CHCH=CHCOOH$) that occurs naturally in many plants, is a good mold and fungus inhibitor. The fourth ingredient is the trisodium salt of citric acid. In cookies, it is combined with citric acid to buffer the acidity of the strawberry jam.

A package of dehydrated cream sauce with bacon bits and noodles includes a group of less familiar acid salts. Disodium guanylate, disodium inosinate, and monosodium glutamate are salts of acids that occur naturally in meats. As food additives, these salts serve as flavor enhancers by imparting a "meaty" flavor and are sometimes found in packaged foods that should taste "meaty" but do not contain much meat.

Many foods also contain two other additives that are salts, not of carboxylic acids, but of compounds with acidic hydroxyl groups: sodium ascorbate and sodium erythorbate. Ascorbic acid is vitamin C, and erythorbic acid is an isomer of vitamin C. Both salts act as antioxidants (Section 21.10) and help to maintain the color of cured meat such as bacon, though only the ascorbate has value as a vitamin.

All the additives mentioned are salts of acids that occur naturally in plants and animals, and all have been approved by the U.S. Food and Drug Administration (FDA). Some additives are essential—without preservatives certain foods would harbor disease-causing microorganisms. Other additives make convenience foods possible or make it more

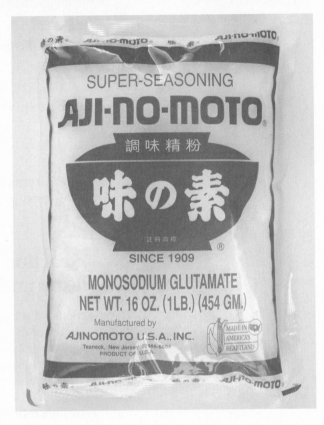

▲ Certain acid salts, such as monosodium glutamate (MSG) help to enhance the flavor of foods.

appealing by enhancing flavor, color, or consistency. It could be argued that such additives are not essential, although that does not necessarily mean they are harmful. Do we really need dry powders that turn into cream sauce when water, butter, and milk are added? Each of us must decide this for ourselves.

See Additional Problems 17.78 and 17.79 at the end of the chapter.

Common food additives

Sodium benzoate

Sodium ascorbate
(an antioxidant)

Monosodium glutamate
(flavor enhancer)

PROBLEM 17.14

Write the products of the following reactions:

(a) $CH_3CH(OH)COOH + NaOH \rightarrow ?$
(b) 2,2-Dimethylpentanoic acid $+ Ca(OH)_2 \rightarrow ?$

PROBLEM 17.15

Write the formulas of calcium salicylate and sodium acrylate (refer to Table 17.1).

PROBLEM 17.16

Suppose that potassium acetate and disodium glutarate are dissolved in water. Write the formulas of the ions present in the solution (refer to Table 17.1).

17.4 Reactions of Carboxylic Acids: Ester and Amide Formation

The reactions of alcohols and amines with carboxylic acids follow the same pattern—both result in substitution of other groups for the —OH of the acid and formation of water as a by-product. With alcohols, the —OH of the acid is replaced by the —OR of the alcohol. With amines, the —OH of the acid is replaced by the —NR$_2$ of the amine.

Ester formation

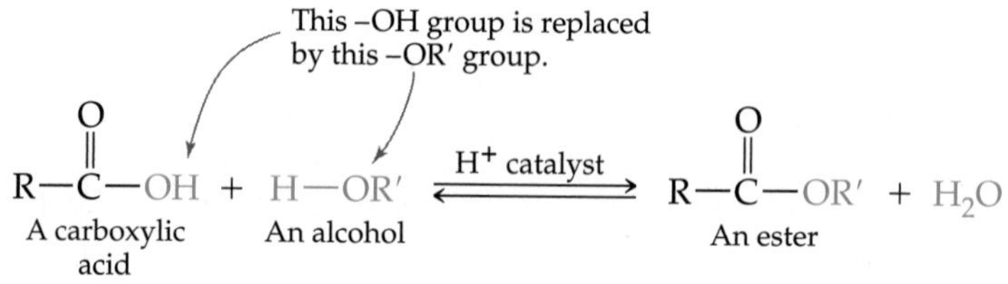

This –OH group is replaced by this –OR' group.

$$\underset{\substack{\text{A carboxylic} \\ \text{acid}}}{R-\overset{\overset{\displaystyle O}{\|}}{C}-OH} + \underset{\text{An alcohol}}{H-OR'} \underset{\longleftarrow}{\overset{H^+ \text{ catalyst}}{\longrightarrow}} \underset{\text{An ester}}{R-\overset{\overset{\displaystyle O}{\|}}{C}-OR'} + H_2O$$

Amide formation

This –OH group is replaced by this –NR'$_2$ group.

$$\underset{\substack{\text{A carboxylic} \\ \text{acid}}}{R-\overset{\overset{\displaystyle O}{\|}}{C}-OH} + \underset{\text{Amine}}{H-NR'_2} \overset{\text{heat}}{\longrightarrow} \underset{\text{Amide}}{R-\overset{\overset{\displaystyle O}{\|}}{C}-NR'_2} + H_2O$$

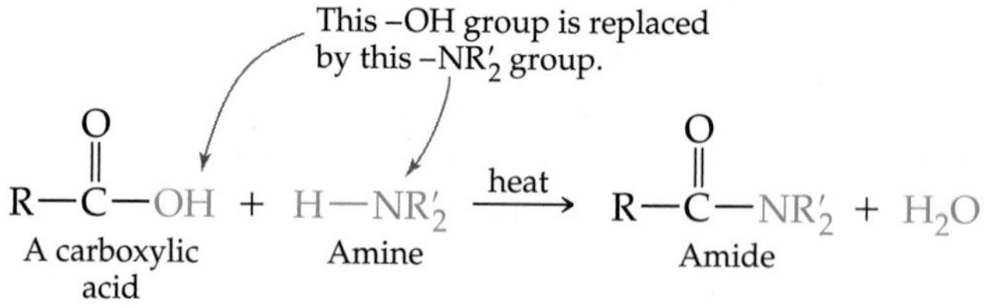

Esterification

Esterification The reaction between an alcohol and a carboxylic acid to yield an ester plus water.

In the laboratory, ester formation, known as **esterification**, is carried out by warming a carboxylic acid with an alcohol in the presence of a strong-acid catalyst such as sulfuric acid. For example,

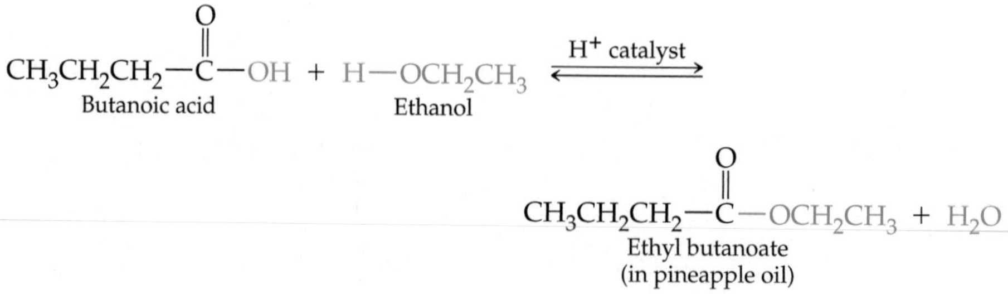

$$\underset{\text{Butanoic acid}}{CH_3CH_2CH_2-\overset{\overset{\displaystyle O}{\|}}{C}-OH} + \underset{\text{Ethanol}}{H-OCH_2CH_3} \underset{\longleftarrow}{\overset{H^+ \text{ catalyst}}{\longrightarrow}}$$

$$\underset{\substack{\text{Ethyl butanoate} \\ \text{(in pineapple oil)}}}{CH_3CH_2CH_2-\overset{\overset{\displaystyle O}{\|}}{C}-OCH_2CH_3} + H_2O$$

Esterification reactions are reversible and often reach equilibrium with approximately equal amounts of both reactants and products present. Ester formation is favored either by using a large excess of the alcohol or by continuously removing one of the products (for example, by distilling off a low-boiling ester, or removing water in a similar fashion). Both techniques are applications of Le Châtelier's principle. (, p. 203)

WORKED EXAMPLE 17.6 Writing the Products of an Esterification Reaction

The flavor ingredient in oil of wintergreen is an ester that is made by reaction of *o*-hydroxybenzoic acid (salicylic acid) with methanol. What is its structure?

$$\text{(benzene ring)}-\overset{\overset{\displaystyle O}{\|}}{C}-OH + CH_3OH \xrightarrow{H^+} ?$$

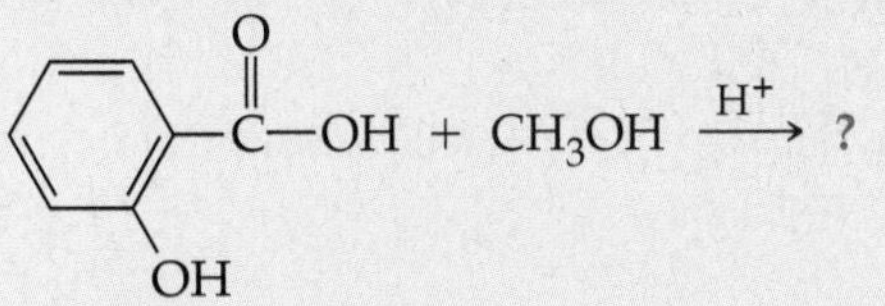

ANALYSIS First, write the two reaction partners so that the —COOH group of the acid and the —OH group of the alcohol face each other:

o-Hydroxybenzoic acid $\quad\text{(benzene ring)}-\overset{\overset{\displaystyle O}{\|}}{C}-\boxed{OH + H}-OCH_3 \quad$ Methanol
$\qquad\qquad\qquad\qquad\qquad\qquad\qquad\searrow H_2O$

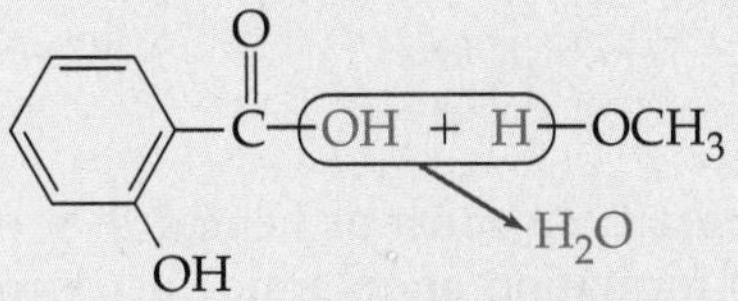

Next, remove —OH from the acid and —H from the alcohol to form water and then join the two resulting organic fragments with a single bond.

SOLUTION
The product is the ester:

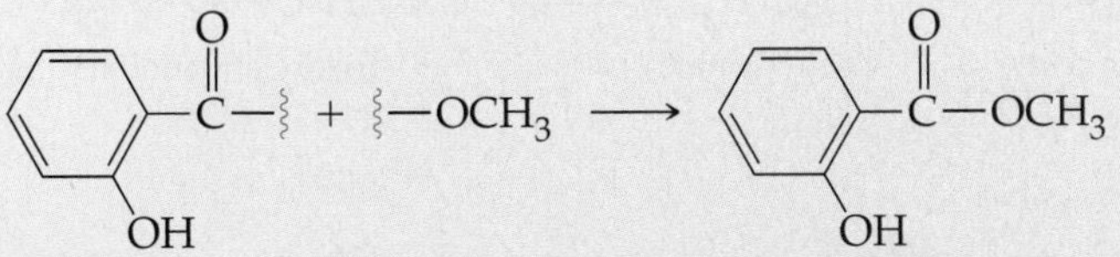

$$\text{(benzene ring)}-\overset{\overset{\displaystyle O}{\|}}{C}-\xi + \xi-OCH_3 \longrightarrow \text{(benzene ring)}-\overset{\overset{\displaystyle O}{\|}}{C}-OCH_3$$

Methyl *o*-hydroxybenzoate
(Methyl salicylate)

▲ The unique flavors and aromas of various beers are due in part to esters formed during fermentation.

PROBLEM 17.17

Raspberry oil contains an ester that is made by reaction of formic acid with 2-methyl-1-propanol. What is its structure?

$$HCOOH + (CH_3)_2CHCH_2OH \longrightarrow ?$$

PROBLEM 17.18

What carboxylic acid and alcohol are needed to make the following esters?

(a) 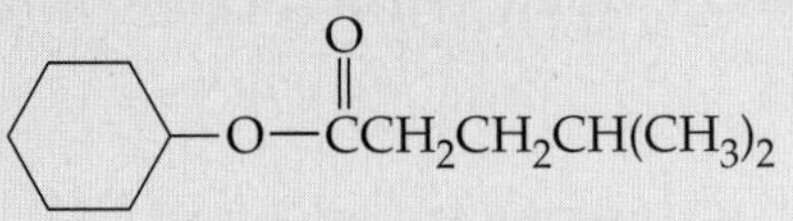 $\text{(cyclohexyl ring)}-O-\overset{\overset{\displaystyle O}{\|}}{C}CH_2CH_2CH(CH_3)_2$

(b) $CH_3CH_2CH_2CH_2\overset{\overset{\displaystyle O}{\|}}{C}-O-CH(CH_3)_2$

Amide Formation

Unsubstituted amides are formed by the reaction of carboxylic acids with ammonia:

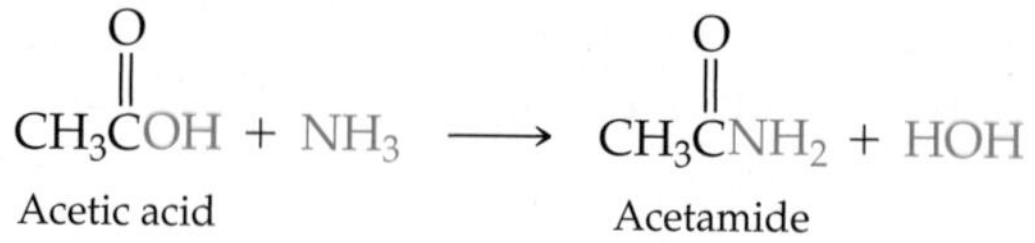

$$CH_3\overset{\overset{\displaystyle O}{\|}}{C}OH + NH_3 \longrightarrow CH_3\overset{\overset{\displaystyle O}{\|}}{C}NH_2 + HOH$$

Acetic acid   Acetamide

Substituted amides are produced in reactions between primary or secondary amines and carboxylic acids:

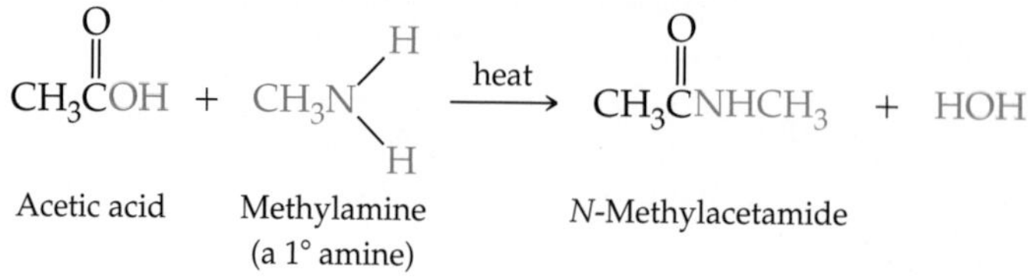

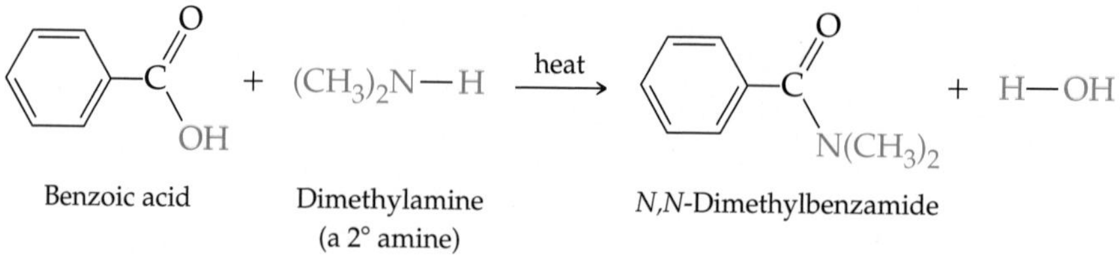

$$CH_3\overset{\overset{\displaystyle O}{\|}}{C}OH + CH_3\overset{\overset{\displaystyle H}{\diagup}}{\underset{\underset{\displaystyle H}{}}{N}} \xrightarrow{\text{heat}} CH_3\overset{\overset{\displaystyle O}{\|}}{C}NHCH_3 + HOH$$

Acetic acid   Methylamine   N-Methylacetamide
              (a 1° amine)

Benzoic acid   Dimethylamine   N,N-Dimethylbenzamide
             (a 2° amine)

The amide formation reactions must be heated to proceed as shown. In each case, the overall reaction is formation of an amide accompanied by formation of water by the —OH group of the acid and an —H atom from ammonia or an amine. Tertiary amines (such as triethylamine) do not form amides, generating ammonium salts instead:

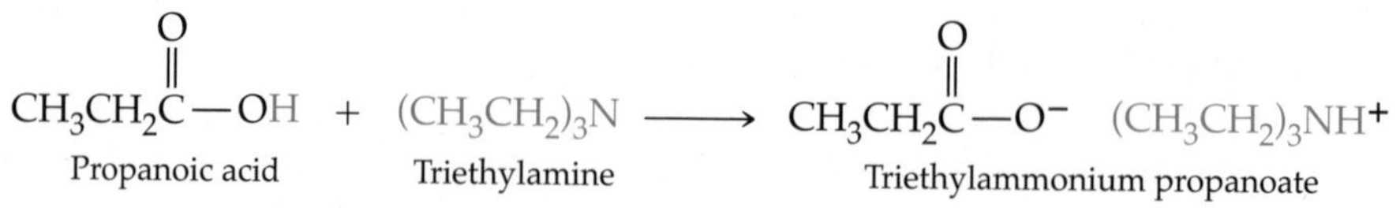

$$CH_3CH_2\overset{\overset{\displaystyle O}{\|}}{C}-OH + (CH_3CH_2)_3N \longrightarrow CH_3CH_2\overset{\overset{\displaystyle O}{\|}}{C}-O^- \;\; (CH_3CH_2)_3NH^+$$

Propanoic acid   Triethylamine   Triethylammonium propanoate

⊂▣⊃ Looking Ahead

Proteins are constructed of long chains of amino acids held together by amide bonds.

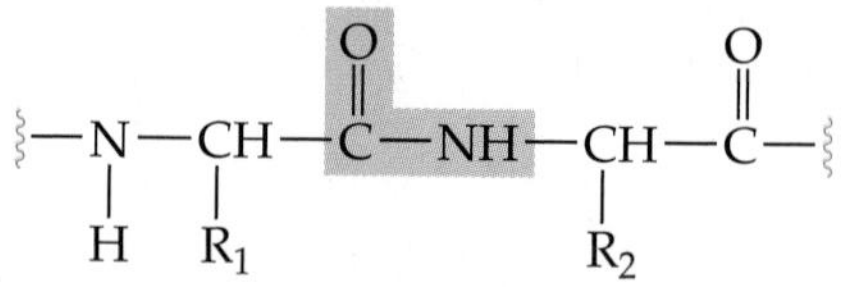

The biochemical synthesis of proteins, described in Section 26.10, is a strictly controlled process in which amino acids with different R groups must be assembled in an exact order that is determined by an organism's DNA sequence.   ⊂▣⊃

WORKED EXAMPLE   **17.7** Writing the Products of Amide Formation

The mosquito and tick repellent DEET (diethyltoluamide) is prepared by reaction of diethylamine with *m*-methylbenzoic acid (*m*-toluic acid). What is the structure of DEET?

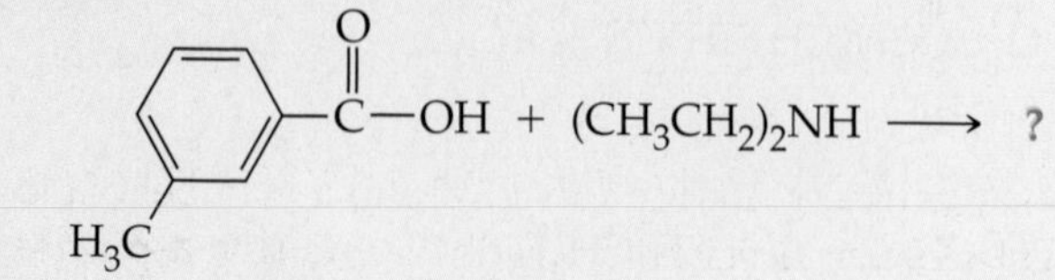

ANALYSIS First, rewrite the equation so that the —OH of the acid and the —H of the amine face one another:

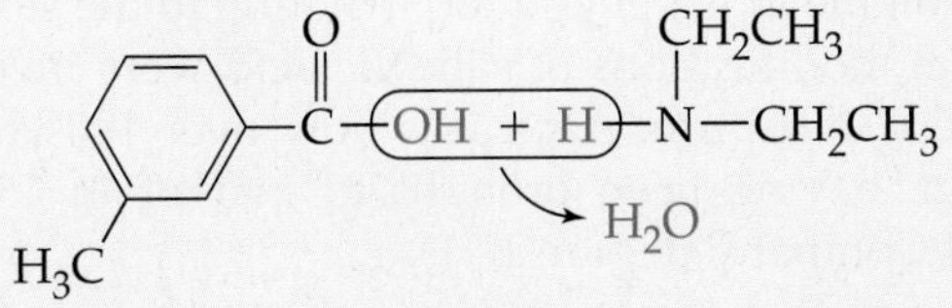

Next, remove the —OH from the acid and the —H from the nitrogen atom of the amine to form water. Then join the two resulting fragments together to form the amide product.

SOLUTION

The structure of DEET is

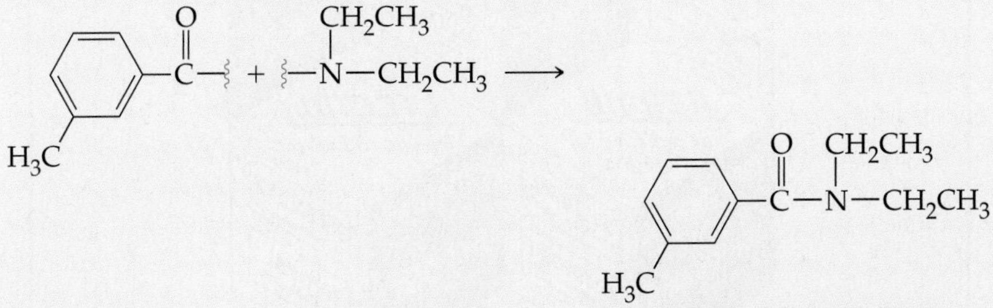

N,N-Diethyltoluamide (DEET)

PROBLEM 17.19

Draw structures of the amides that can be made from the following reactants:

(a) $CH_3NH_2 + (CH_3)_2CHCOOH \longrightarrow$?

(b) 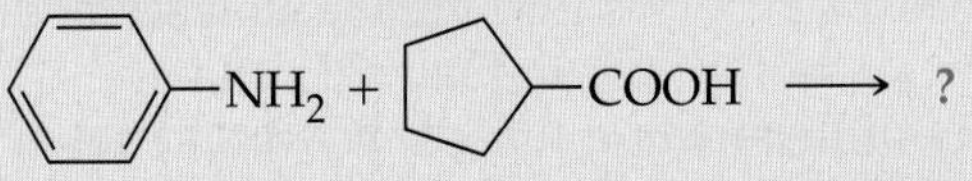 ⬡—NH_2 + ⬠—$COOH \longrightarrow$?

PROBLEM 17.20

What carboxylic acid and amine would you use if you wanted to prepare phenacetin, once used in headache remedies but now banned because of its potential for causing kidney damage?

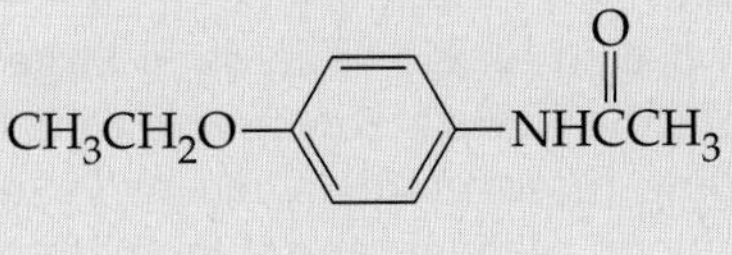

CH_3CH_2O—⬡—$NHCCH_3$ Phenacetin

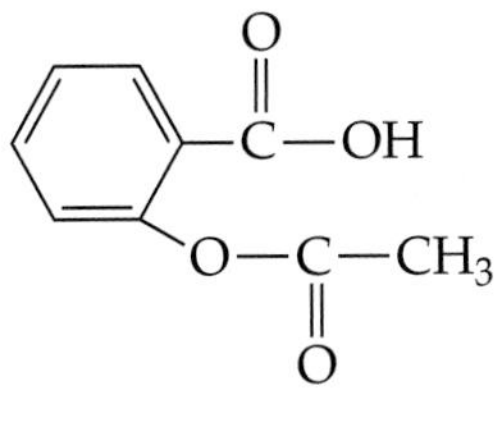

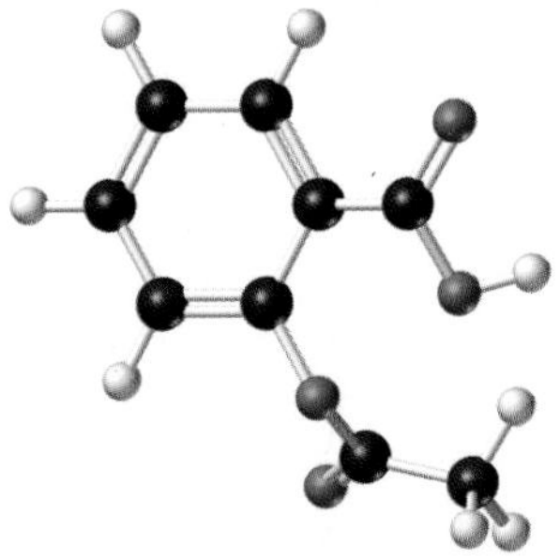

17.5 Aspirin and Other Over-the-Counter Carboxylic Acid Derivatives

Esters and amides have many uses in medicine, in industry, and in living systems. Their importance in the industrial production of polyesters and polyamides is explored in Section 17.7. Here we discuss a few of the most familiar over-the-counter medications that are carboxylic acid derivatives.

Aspirin and Other Salicylic Acid Derivatives

Aspirin is a white, crystalline solid (mp 135 °C) that is a member of the group of drugs known as salicylates: esters of salicyclic acid. Chemically, aspirin is acetylsalicylic acid, an ester formed between acetic acid and the —OH group of salicylic acid.

Aspirin

As early as the fifth century B.C., Hippocrates knew that chewing the bark of a willow tree relieved pain (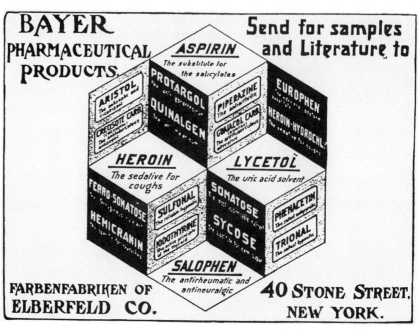 see Application: Aspirin—A Case Study, p. 9). By the 1800s, salicylic acid had been identified as the active ingredient in willow bark but was both too insoluble in water and too irritating to the stomach to be of widespread use. Chemical modifications of salicylic acid were investigated and led to the ultimate discovery of aspirin. Aspirin is now over 100 years old as a widely available medication, having been brought to market in 1899 in Germany by Frederich Bayer and Company (Figure 17.1).

► **FIGURE 17.1 Pharmaceutical advertisement from the 1900s.** The ad points out that aspirin was a substitute for salicylates. Substitutes have since been found for some of the other pharmaceuticals listed as well.

Aspirin is an example of a *prodrug*, which is a compound that is administered in an inactive (or significantly less active) form and then metabolized *in vivo* into the active drug form. Once taken, aspirin passes through the stomach and into the intestine, where it undergoes ester hydrolysis to give salicylic acid.

Aspirin has an amazing array of therapeutic actions. It is best known for providing pain relief (an *analgesic*), reducing fever (an *antipyretic*), and reducing inflammation (an *anti-inflammatory*). As an analgesic and anti-inflammatory, it is the first line of defense in several forms of arthritis. There is even a beef-flavored aspirin on the market for pain relief in arthritic dogs (Palaprin). (Because the risk of toxicity in dogs is greater, the canine aspirin is available only through veterinarians and must be used with careful supervision.)

For a drug that has been in use so long and in such large quantity (approximately 29 *billion* tablets per year in the United States), it is amazing that physiological effects of aspirin are still being discovered. In recent years, aspirin has been found to inhibit the clumping of blood platelets and thereby to protect against heart attacks caused by blood clots. Small regular doses of aspirin (for example, one baby aspirin, which is 85 mg, per day as compared to 325 mg in the usual tablet) are being recommended for some individuals at risk for heart attack. (See Section 24.9 for the chemical action of aspirin.) It is estimated that approximately one-fifth of all U.S. adults take one aspirin either daily or every other day.

In 1982 the relationship between *Reye's syndrome* (a rare children's liver disorder) and the use of aspirin to treat influenza or chickenpox was discovered. It is now strongly recommended that aspirin not be given to a child whose fever might be caused by these diseases.

The principal undesirable side effects of aspirin are gastric bleeding and gastrointestinal distress. Only slightly ionized in the acidic environment of the stomach, aspirin causes trouble once inside the stomach lining, where its dissociation produces ions that do not easily exit across cell membranes. Aspirin also inhibits the action of an enzyme necessary for coagulation of blood cells. After taking two aspirin tablets, the time it takes for bleeding to stop is doubled for several days;

accordingly, patients that are about to undergo surgical (or even dental) procedures are advised to stop taking aspirin for a few days prior. Regular use of aspirin as an anti-inflammatory can cause the loss of 3–8 mL of blood a day. The usual blood loss of a few milliliters per tablet is not harmful, but more extensive bleeding can occur in susceptible individuals.

Numerous efforts have been made to produce a modification of aspirin that retains all of its beneficial effects while eliminating the negative ones, but no salicylate has yet proved as effective. The sodium salt of aspirin causes less bleeding, but has no action against blood clotting. Conversion of the carboxylic acid group to an amide gives a medication with very little gastric irritation but no anti-inflammatory activity and unreliable benefits.

Methyl salicylate, or oil of wintergreen (Worked Example 17.6), is too poisonous to be of any value as an oral medication but does have value as a topical medication. Therapeutically it is useful as a *counterirritant*, a substance that relieves internal pain by stimulating nerve endings in the skin. Methyl salicylate is one of the active ingredients in liniments such as Bengay and Heet.

Acetaminophen

An alternative to aspirin for pain relief is acetaminophen (best known by the trade name Tylenol), an amide that also contains a hydroxyl group:

Acetaminophen

Like aspirin, acetaminophen reduces fever, but unlike aspirin it is not an anti-inflammatory agent. The major advantage of acetaminophen as compared to aspirin is that it does not induce internal bleeding. For this reason, it is the pain reliever of choice for individuals prone to bleeding or recovering from surgery or wounds. Overdoses of acetaminophen can cause kidney and liver damage, however.

Ibuprofen

Ibuprofen is a nonsteroidal anti-inflammatory drug (NSAID) marketed most notably as Advil and Motrin. It is used for relief of symptoms of arthritis, abdominal cramps, and fever and as an analgesic, especially where there is inflammation. Ibuprofen is also known to have an effect on blood-clotting times, though this effect is relatively mild and short-lived compared with that of aspirin. At low doses (those commonly prescribed), ibuprofen seems to have the lowest incidence of adverse gastrointestinal side effects of all general NSAIDs. It is a core medicine in the World Health Organization's "Essential Drugs List," which is a list of the most effective, safest, and cost-efficient medicines needed for a basic health care system

Ibuprofen

Benzocaine and Lidocaine

Benzocaine is a local anesthetic used in many over-the-counter *topical* preparations (those applied to the skin surface) for such conditions as cold sores, poison ivy, sore throats, and hemorrhoids. Like other local anesthetics, it works by blocking the transmission of impulses by sensory nerves. Benzocaine is one of a family of structurally related local anesthetics that includes lidocaine (Xylocaine), which is most commonly administered by injection to prevent pain during dental work. Because benzocaine is less soluble than lidocaine, it cannot be used in this manner.

Benzocaine

Lidocaine

▲ Lidocaine is commonly used as a dental pain killer.

PROBLEM 17.21

Salsalate is another salicylate used as an aspirin alternative for those who are hypersensitive to aspirin. Draw the structures of salicylic acid and salsalate, which is an ester formed by the reaction of two molecules of salicylic acid.

⊷● KEY CONCEPT PROBLEM 17.22

Examine the structures of aspirin, acetaminophen, benzocaine, and lidocaine. For each compound, indicate whether it is acidic, basic, or neither.

17.6 Hydrolysis of Esters and Amides

Recall that in hydrolysis a bond or bonds are broken and the H— and —OH of water add to the atoms that were part of the broken bond. Esters and amides undergo hydrolysis to give back carboxylic acids plus alcohols or amines in reactions that follow the carbonyl-group substitution pattern (⊂⊃, p. 532).

For esters, the net effect of hydrolysis is substitution of —OH for —OR':

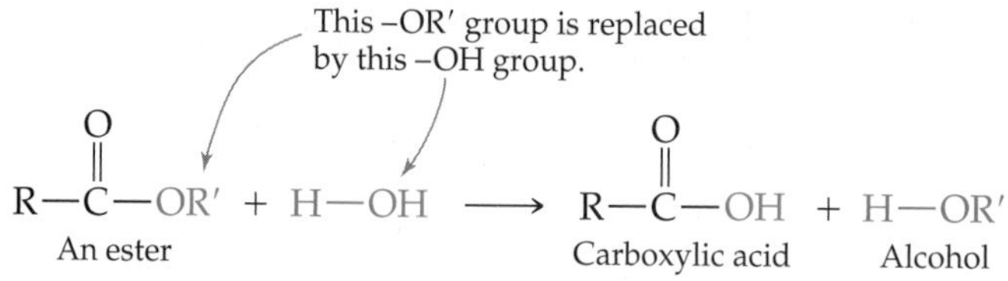

For amides, the net effect of hydrolysis is substitution of —OH for —NH₂ or the substituted amide nitrogen:

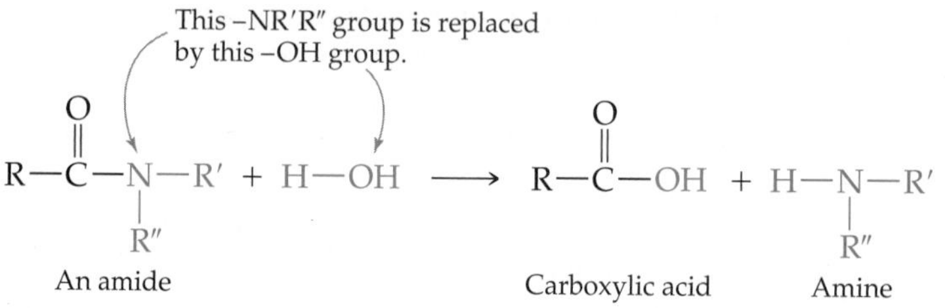

Ester Hydrolysis

Both acids and bases can cause ester hydrolysis. Acid-catalyzed hydrolysis is simply the reverse of the esterification. An ester is treated with water in the presence of a strong acid catalyst such as sulfuric acid, and hydrolysis takes place:

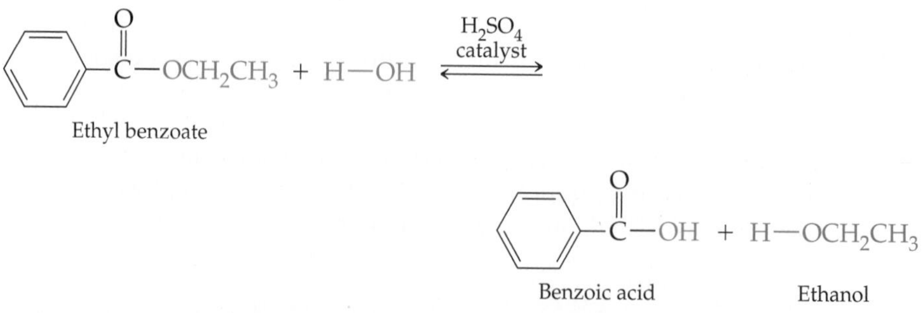

An excess of water pushes the equilibrium to the right.

Ester hydrolysis by reaction using a base such as NaOH or KOH is known as **saponification** (after the Latin word *sapo*, "soap"). The product of saponification

Saponification The reaction of an ester with aqueous hydroxide ion to yield an alcohol and the metal salt of a carboxylic acid.

is a carboxylate anion rather than a free carboxylic acid. (The use of saponification in making soap is discussed in Section 24.4.)

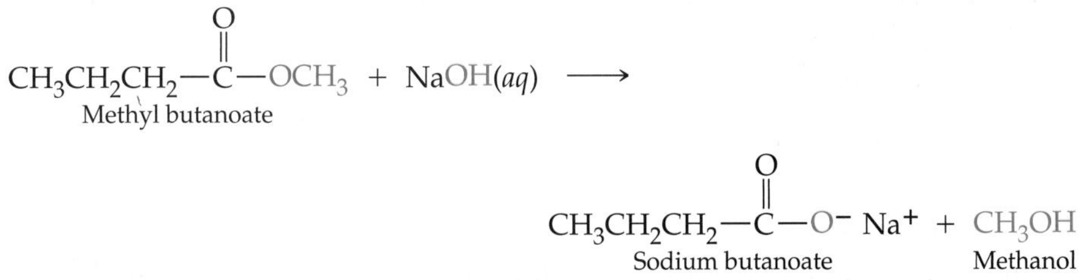

$$CH_3CH_2CH_2-\overset{\overset{\displaystyle O}{\|}}{C}-OCH_3 \ + \ NaOH(aq) \ \longrightarrow$$

Methyl butanoate

$$CH_3CH_2CH_2-\overset{\overset{\displaystyle O}{\|}}{C}-O^- \ Na^+ \ + \ CH_3OH$$

Sodium butanoate   Methanol

WORKED EXAMPLE   **17.8** Writing the Products of an Ester Hydrolysis

What product would you obtain from acid-catalyzed hydrolysis of ethyl formate, a flavor constituent of rum?

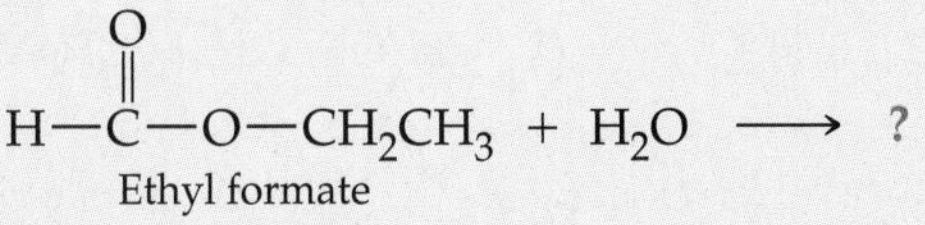

$$H-\overset{\overset{\displaystyle O}{\|}}{C}-O-CH_2CH_3 \ + \ H_2O \ \longrightarrow \ ?$$

Ethyl formate

ANALYSIS The name of an ester gives a good indication of the names of the two products. Thus, ethyl formate yields ethyl alcohol and formic acid. To find the product structures in a more systematic way, write the structure of the ester and locate the bond between the carbonyl-group carbon and the —OR′ group:

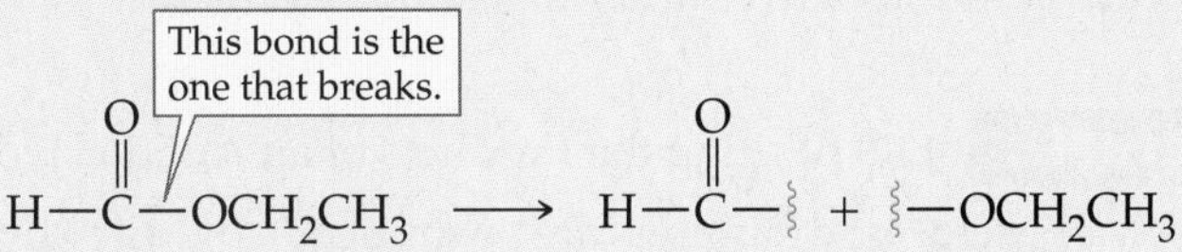

This bond is the one that breaks.

$$H-\overset{\overset{\displaystyle O}{\|}}{C}-OCH_2CH_3 \ \longrightarrow \ H-\overset{\overset{\displaystyle O}{\|}}{C}-\ + \ -OCH_2CH_3$$

SOLUTION
Carry out a hydrolysis reaction on paper. First form the carboxylic acid product by connecting an —OH to the carbonyl-group carbon. Then add an —H to the —OCH₂CH₃ group to form the alcohol product.

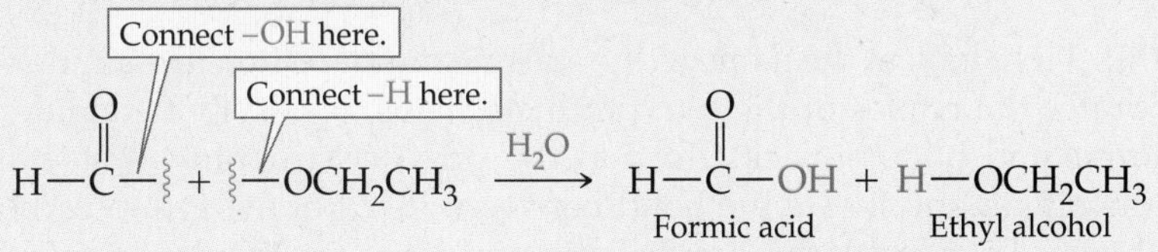

Connect –OH here.

Connect –H here.

$$H-\overset{\overset{\displaystyle O}{\|}}{C}-\ + \ -OCH_2CH_3 \ \overset{H_2O}{\longrightarrow} \ H-\overset{\overset{\displaystyle O}{\|}}{C}-OH \ + \ H-OCH_2CH_3$$

Formic acid   Ethyl alcohol

PROBLEM 17.23

If a bottle of aspirin tablets has the aroma of vinegar, it is time to discard those tablets. Explain why, and include a chemical equation in the explanation.

PROBLEM 17.24

What products would you obtain from acid-catalyzed hydrolysis of the following esters?

(a) Isopropyl *p*-nitrobenzoate

(b)

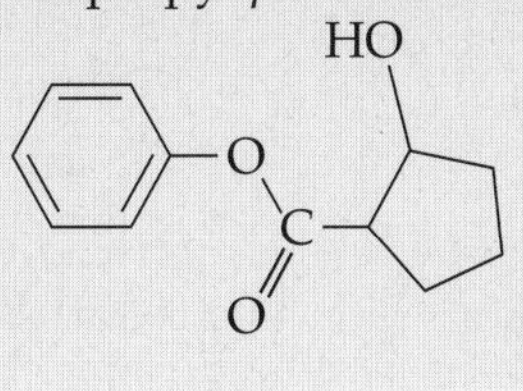

(c) $CH_2{=}CH-\overset{\overset{\displaystyle O}{\|}}{C}-O\,CH_2CH_3$

Amide Hydrolysis

Amides are extremely stable in water but undergo hydrolysis with heating in the presence of acids or bases. The products are the carboxylic acid and amine from which the amide was synthesized:

$$\underset{O}{\overset{\displaystyle O}{\overset{\displaystyle \|}{RC}}}-NHR + H-OH \longrightarrow \underset{}{\overset{\displaystyle O}{\overset{\displaystyle \|}{RC}}}-OH + RN\overset{H}{\underset{H}{\diagdown}}$$

In practice, the products obtained depend on whether the hydrolysis is done using acid or base. Under acidic conditions, the carboxylic acid and amine salt are obtained. Doing the hydrolysis using base produces the neutral amine and carboxylate anion. For example, in the hydrolysis of *N*-methylacetamide:

Hydrolysis products of **N-Methylacetamide**

$$CH_3\overset{\displaystyle O}{\overset{\displaystyle \|}{C}}-NHCH_3 + H_3O^+ \longrightarrow CH_3\overset{\displaystyle O}{\overset{\displaystyle \|}{C}}-OH + CH_3NH_3^+ \qquad \text{Acid hydrolysis}$$

$$CH_3\overset{\displaystyle O}{\overset{\displaystyle \|}{C}}-NHCH_3 + OH^- \longrightarrow CH_3\overset{\displaystyle O}{\overset{\displaystyle \|}{C}}-O^- + CH_3NH_2 \qquad \text{Base hydrolysis}$$

⊂▢⊃ Looking Ahead

In Chapter 28 you will see that the cleavage of amide bonds by hydrolysis is the key process that occurs in the stomach during digestion of proteins.   ⊂▢⊃

WORKED EXAMPLE   **17.9** Writing the Products of an Amide Hydrolysis

What carboxylic acid and amine are produced by the hydrolysis of *N*-ethylbutanamide?

$$CH_3CH_2CH_2\overset{\displaystyle O}{\overset{\displaystyle \|}{C}}-NHCH_2CH_3 + H_2O \longrightarrow ?$$

N-Ethylbutanamide

ANALYSIS First, look at the name of the starting amide. Often, the amide's name incorporates the names of the two products. Thus, *N*-ethylbutanamide yields ethylamine and butanoic acid. To find the product structures systematically, write the amide and locate the bond between the carbonyl-group carbon and the nitrogen. Then break this amide bond and write the two fragments:

This amide bond is the one that breaks.

$$CH_3CH_2CH_2\overset{\displaystyle O}{\overset{\displaystyle \|}{C}}-NHCH_2CH_3 \longrightarrow CH_3CH_2CH_2\overset{\displaystyle O}{\overset{\displaystyle \|}{C}}-\xi + \xi-NHCH_2CH_3$$

SOLUTION
Carry out a hydrolysis reaction on paper and form the products by connecting an —OH to the carbonyl-group carbon and an —H to the nitrogen:

Connect —OH here.

Connect —H here.

$$CH_3CH_2CH_2\overset{\displaystyle O}{\overset{\displaystyle \|}{C}}-\xi + \xi-NHCH_2CH_3 \xrightarrow{\text{H}_2\text{O}}$$

$$CH_3CH_2CH_2\overset{\displaystyle O}{\overset{\displaystyle \|}{C}}-OH + H-NHCH_2CH_3$$
Butanoic acid   Ethylamine

PROBLEM 17.25

What carboxylic acids and amines result from hydrolysis of the following amides?

(a) $CH_3CH{=}CHCNHCH_3$ (with C=O above)

(b) *N,N*-Dimethyl-*p*-nitrobenzamide

17.7 Polyamides and Polyesters

Imagine what would happen if a molecule with *two* carboxylic acid groups reacted with a molecule having *two* amino groups. Amide formation could join the two molecules together, but further reactions could then link more and more molecules together until a giant chain resulted. This is exactly what happens when certain kinds of synthetic polymers are made.

Nylons are *polyamides* produced by reaction of diamines with diacids. One such nylon, nylon 6,6 (pronounced "six-six"), is so named because of the structures of the two compounds that are used to produce it. Nylon 6,6 is made by heating adipic acid (hexanedioic acid, a six-carbon dicarboxylic acid) with hexamethylenediamine (1,6-hexanediamine, a six-carbon diamine) at 280 °C:

$$n \; HOOC-(CH_2)_4-COOH$$
Adipic acid

$$+$$

$$n \; H_2N-(CH_2)_6-NH_2$$
Hexamethylenediamine

$$\xrightarrow[-H_2O]{280°} \left[\overset{O}{\overset{\|}{C}}-(CH_2)_4-\overset{O}{\overset{\|}{C}}-NH-(CH_2)_6-NH \right]_n$$

Nylon 6,6, a polyamide
(repeating unit)

The polymer molecules are composed of thousands of the repeating unit, shown here enclosed in square brackets. (In the next chapter, you will see that proteins are also polyamides; unlike nylon, however, proteins do not normally have identical repeating units.)

The properties of nylon make it suitable for a wide range of applications. High impact strength, abrasion resistance, and a naturally slippery surface make nylon an excellent material for bearings and gears. It can be formed into very strong fibers, making it valuable for a range of applications from nylon stockings, to clothing, to mountaineering ropes and carpets. Sutures and replacement arteries are also fabricated from nylon, which is resistant to deterioration in body fluids.

Just as diacids and diamines react to yield polyamides, diacids and dialcohols react to yield *polyesters*. The most widely used polyester is made by the reaction of terephthalic acid (1,4-benzenedicarboxylic acid) with ethylene glycol:

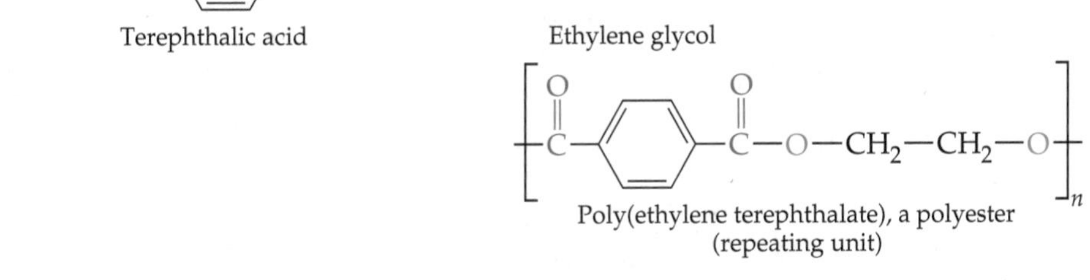

$$n \; HO-\overset{O}{\overset{\|}{C}}-\hspace{-0.3em}\raisebox{-0.3em}{\text{⬡}}\hspace{-0.3em}-\overset{O}{\overset{\|}{C}}-OH \; + \; n \; HO-CH_2-CH_2-OH \xrightarrow{-H_2O}$$

Terephthalic acid   Ethylene glycol

$$\left[\overset{O}{\overset{\|}{C}}-\hspace{-0.3em}\raisebox{-0.3em}{\text{⬡}}\hspace{-0.3em}-\overset{O}{\overset{\|}{C}}-O-CH_2-CH_2-O \right]_n$$

Poly(ethylene terephthalate), a polyester
(repeating unit)

▲ Nylon being pulled from the interface between adipic acid and hexamethylenediamine.

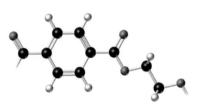

Repeating unit of polyester

We know this polyester best in clothing fiber, where it has the trade name Dacron. Under the name Mylar it is used in plastic film and recording tape. Its chemical name, poly(ethylene terephthalate) or PET, is usually applied when it is used in clear, flexible soft-drink bottles.

APPLICATION ▶ Kevlar: A Life-Saving Polymer

Imagine that you are a police officer about to take part in capturing a dangerous, well-armed criminal, or a member of the bomb disposal unit. Perhaps you are part of a military peacekeeping force going out on patrol in hostile environs. You know your temperament and training will serve you well. But there is another protection you are very grateful for—your bulletproof vest. It is filled with fibers made of Kevlar, an amazing polymer that protects police officers, soldiers, firefighters, bicycle riders, lumberjacks, and others engaged in hazardous activities.

Kevlar is a polyamide created at DuPont by chemist Stephanie L. Kwolek and introduced to commercial applications in the 1970s. Today there is an ever-expanding list of applications for Kevlar. In its various forms it is five times stronger than steel, almost half as dense as fiberglass, highly resistant to damage by chemicals, dimensionally stable, very difficult to cut or break, a poor electrical conductor, and flame-resistant (if ignited, it self-extinguishes). In fact, surgical gloves made partly of Kevlar are used by orthopedic surgeons to virtually eliminate the possibility of being cut by broken bone.

Like nylon, Kevlar is produced by the reaction of a dicarboxylic acid with a diamine, and because it contains aromatic rings, it is classified as a *polyaramide:*.

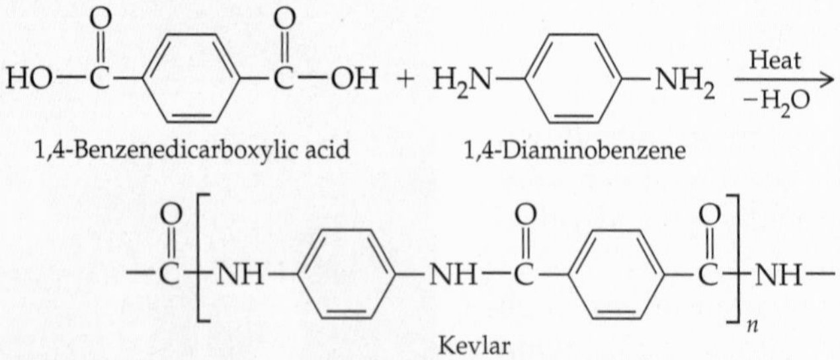

▲ A bulletproof vest made with Kevlar.

The great strength of Kevlar results from the way that uniformly arranged hydrogen bonding holds the polymer chains together, as indicated by the red dots in the structural diagram shown here.

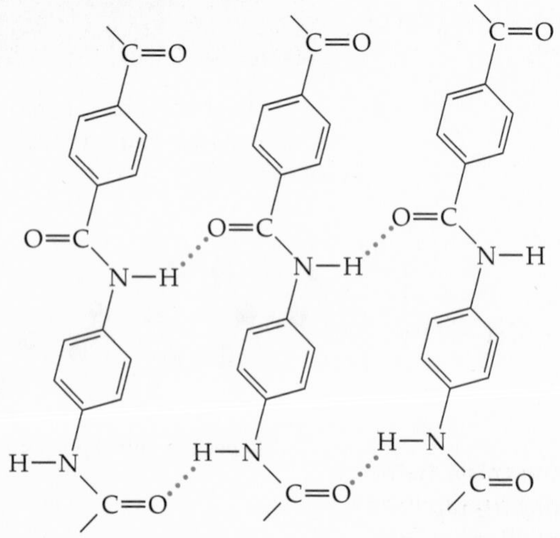

Hydrogen bonding in Kevlar

In 1995, Kwolek was inducted into the National Inventors Hall of Fame. Her achievements in designing a tough polymer are thus recognized alongside those of inventors such as Louis Pasteur, Alexander Graham Bell, Thomas Edison, and Henry Ford.

Some Uses of Kevlar

- Bulletproof vests
- Heat-protective apparel
- Cut-resistant gloves and other apparel
- Helmets for firefighters and bicycle riders
- Automotive and industrial hoses
- Structural composites for boats and aircraft
- Emergency tow lines for boats
- Brake linings and other friction-resistant applications

See Additional Problem 17.80 at the end of the chapter.

KEY CONCEPT PROBLEM 17.26

Give the structure of the repeating units in the polymers that are formed in the reactions of the following compounds.

(a) 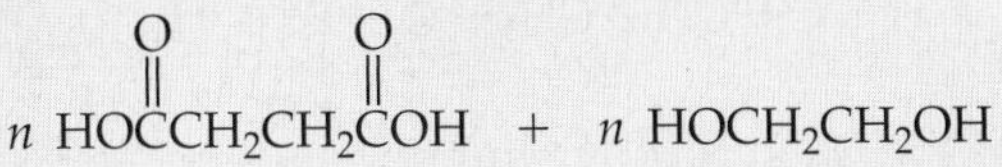 n HOCCH$_2$CH$_2$COH   + n HOCH$_2$CH$_2$OH

(b) 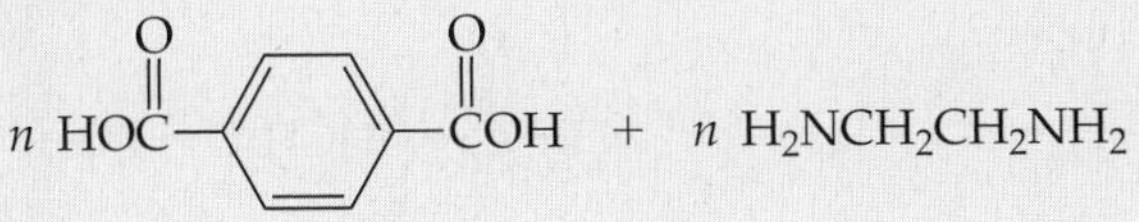 n HOC—⟨benzene ring⟩—COH   + n H$_2$NCH$_2$CH$_2$NH$_2$

17.8 Phosphoric Acid Derivatives

Phosphoric acid is an inorganic acid with 3 ionizable hydrogen atoms (red), allowing it to form three different anions:

| Phosphoric acid (H_3PO_4) | Dihydrogen phosphate ion ($H_2PO_4^-$) | Hydrogen phosphate ion (HPO_4^{2-}) | Phosphate ion (PO_4^{3-}) |

Notice the similarities between phosphoric acid and a carboxylic acid:

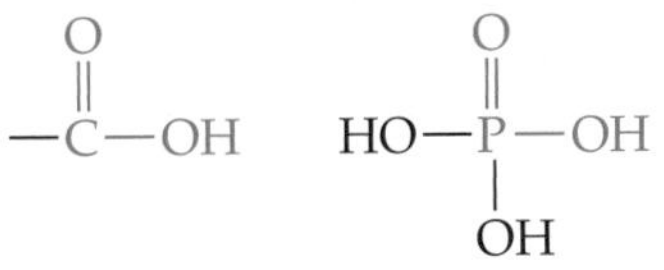

Just like a carboxylic acid, phosphoric acid reacts with alcohols to form **phosphate esters**. It may be esterified at one, two, or all three of its —OH groups by reaction with an alcohol. Reaction with one molecule of methanol gives the monoester:

$$HO-\overset{\overset{O}{\|}}{\underset{\underset{OH}{|}}{P}}-OH + CH_3OH \longrightarrow HO-\overset{\overset{O}{\|}}{\underset{\underset{OH}{|}}{P}}-OCH_3 + H-OH$$

Methyl phosphate
(a phosphate monoester)

Phosphate ester A compound formed by reaction of an alcohol with phosphoric acid; may be a monoester, $ROPO_3H_2$; a diester, $(RO)_2PO_3H$; or a triester, $(RO)_3PO$; also may be a di- or triphosphate.

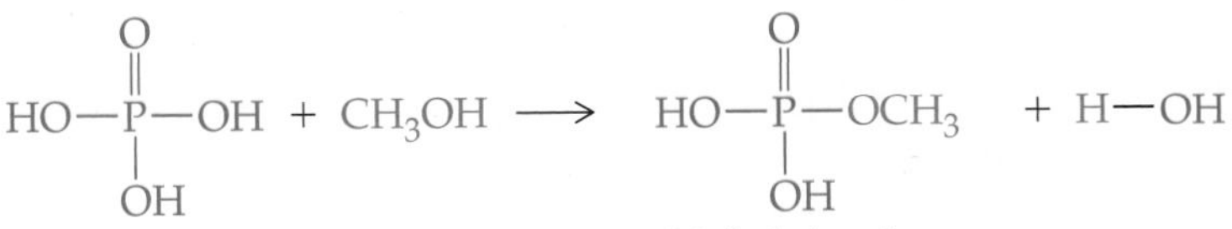

The corresponding diester and triester are

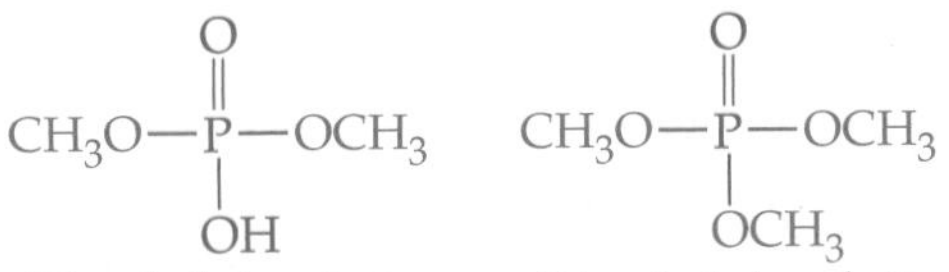

Dimethyl phosphate
(a phosphate diester)

Trimethyl phosphate
(a phosphate triester)

Phosphate monoesters and diesters are acidic because they still contain acidic hydrogen atoms. In neutral or alkaline solutions, including most body fluids, they are present as ions. In biochemical formulas and equations, the phosphate groups are usually written in their ionized forms. For example, you will most often see the formula for glyceraldehyde monophosphate, a key intermediate in the metabolism of glucose (Section 23.2) written as an ion in one of these two ways:

$$H-\overset{\overset{\displaystyle O}{\|}}{C}-\underset{\underset{\displaystyle OH}{|}}{C}HCH_2OH \qquad H-\overset{\overset{\displaystyle O}{\|}}{C}-\underset{\underset{\displaystyle OH}{|}}{C}H-CH_2-O-\overset{\overset{\displaystyle O}{\|}}{\underset{\underset{\displaystyle OH}{|}}{P}}-OH$$

Glyceraldehyde   Glyceraldehyde monophosphate

$$H-\overset{\overset{\displaystyle O}{\|}}{C}-\underset{\underset{\displaystyle OH}{|}}{C}HCH_2-O-\overset{\overset{\displaystyle O}{\|}}{\underset{\underset{\displaystyle O^-}{|}}{P}}-O^- \qquad or \qquad H-\overset{\overset{\displaystyle O}{\|}}{C}-\underset{\underset{\displaystyle OH}{|}}{C}HCH_2O-PO_3{}^{2-}$$

Phosphoryl group

Ionized glyceraldehyde monophosphate

Phosphoryl group The $-PO_3{}^{2-}$ group in organic phosphates.

The $-PO_3{}^{2-}$ group as part of a larger molecule is referred to as a **phosphoryl group** (pronounced fos-for-**eel**).

If two molecules of phosphoric acid combine to lose water, they form a phosphoric acid anhydride. (See Worked Example 17.4 for the analogous reaction involving carboxylic acids.) The resulting acid (*pyrophosphoric acid*, or *diphosphoric acid*) reacts with yet another phosphoric acid molecule to give *triphosphoric acid*.

$$HO-\overset{\overset{\displaystyle O}{\|}}{\underset{\underset{\displaystyle OH}{|}}{P}}-O-\overset{\overset{\displaystyle O}{\|}}{\underset{\underset{\displaystyle OH}{|}}{P}}-OH \qquad HO-\overset{\overset{\displaystyle O}{\|}}{\underset{\underset{\displaystyle OH}{|}}{P}}-O-\overset{\overset{\displaystyle O}{\|}}{\underset{\underset{\displaystyle OH}{|}}{P}}-O-\overset{\overset{\displaystyle O}{\|}}{\underset{\underset{\displaystyle OH}{|}}{P}}-OH$$

Pyrophosphoric acid   Triphosphoric acid

These acids can also form esters, which are known as diphosphates and triphosphates. In the following two methyl phosphate esters, written in their ionized forms, note the difference between the C—O—P ester linkage and the P—O—P phosphoric anhydride linkages:

$$H_3C-O-\overset{\overset{\displaystyle O}{\|}}{\underset{\underset{\displaystyle O^-}{|}}{P}}-O-\overset{\overset{\displaystyle O}{\|}}{\underset{\underset{\displaystyle O^-}{|}}{P}}-O^- \qquad H_3C-O-\overset{\overset{\displaystyle O}{\|}}{\underset{\underset{\displaystyle O^-}{|}}{P}}-O-\overset{\overset{\displaystyle O}{\|}}{\underset{\underset{\displaystyle O^-}{|}}{P}}-O-\overset{\overset{\displaystyle O}{\|}}{\underset{\underset{\displaystyle O^-}{|}}{P}}-O^-$$

Ester linkage   Anhydride linkage

Phosphorylation Transfer of a phosphoryl group, $-PO_3{}^{2-}$, between organic molecules.

Transfer of a phosphoryl group from one molecule to another is known as **phosphorylation**. In biochemical reactions, the phosphoryl groups are often provided by a triphosphate (adenosine triphosphate, ATP), which is converted to a diphosphate (adenosine diphosphate, ADP) in a reaction accompanied by the release of energy. The addition and removal of phosphoryl groups is a common mechanism for regulating the activity of biomolecules (Section 19.9).

$$Adenosine-O-\overset{\overset{\displaystyle O}{\|}}{\underset{\underset{\displaystyle O^-}{|}}{P}}-O-\overset{\overset{\displaystyle O}{\|}}{\underset{\underset{\displaystyle O^-}{|}}{P}}-O-\overset{\overset{\displaystyle O}{\|}}{\underset{\underset{\displaystyle O^-}{|}}{P}}-O^- + ROH \longrightarrow$$

ATP

$$Adenosine-O-\overset{\overset{\displaystyle O}{\|}}{\underset{\underset{\displaystyle O^-}{|}}{P}}-O-\overset{\overset{\displaystyle O}{\|}}{\underset{\underset{\displaystyle O^-}{|}}{P}}-O^- + RO-\overset{\overset{\displaystyle O}{\|}}{\underset{\underset{\displaystyle O^-}{|}}{P}}-O^- + Energy$$

ADP

Organic Phosphates

- Organic phosphates contain —C—O—P— linkages; those with one, two, or three R groups have the general formulas $ROPO_3H_2$, $(RO)_2PO_2H$, and $(RO)_3PO$.
- Organic phosphates with one or two R groups (monoesters, $ROPO_3^{2-}$, or diesters, $(RO)_2PO_2^{-}$) are acids and exist in ionized form in body fluids.
- The diphosphate and triphosphate groups, which are important in biomolecules, contain one or two P—O—P anhydride linkages, respectively.
- Phosphorylation is the transfer of a phosphoryl group ($—PO_3^{2-}$) from one molecule to another.

PROBLEM 17.27

Write the formula for the phosphate monoester formed from isopropyl alcohol in both its nonionized and ionized forms.

PROBLEM 17.28

Identify the functional group in the following compounds and give the structures of the products of hydrolysis for these compounds.

(a) $CH_3\overset{\overset{\displaystyle O}{\|}}{C}NH_2$ **(b)** $CH_3CH_2OPO_3^{2-}$ **(c)** $CH_3CH_2\overset{\overset{\displaystyle O}{\|}}{C}OCH_3$

PROBLEM 17.29

In the structure of acetyl coenzyme A drawn below, identify a phosphate monoester group, a phosphorus anhydride linkage, two amide groups, and the acetyl group.

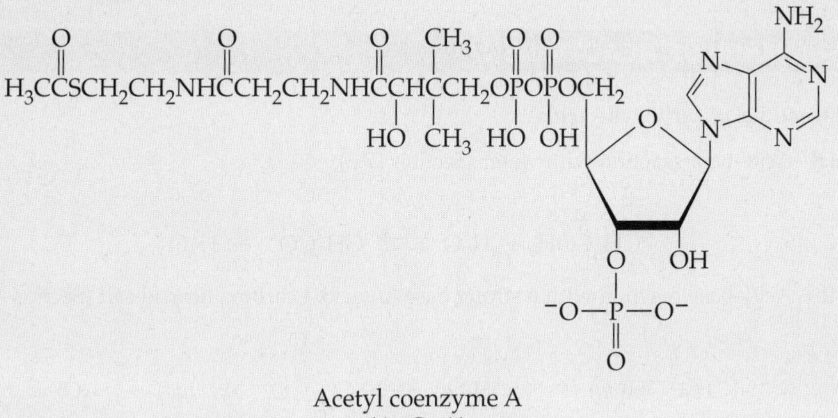

Acetyl coenzyme A
(AcCoA)

SUMMARY: REVISITING THE CHAPTER GOALS

1. What are the general structures and properties of carboxylic acids and their derivatives? *Carboxylic acids, amides,* and *esters* have the general structures

$$\underset{\text{Carboxylic acid}}{R-\overset{\overset{\displaystyle O}{\|}}{C}-OH} \qquad \underset{\text{Amide}}{R-\overset{\overset{\displaystyle O}{\|}}{C}-NH_2} \qquad \underset{\text{Ester}}{R-\overset{\overset{\displaystyle O}{\|}}{C}-OR'}$$

They undergo *carbonyl-group substitution reactions*. Most carboxylic acids are weak acids (a few are strong acids), but esters and amides are neither acids nor bases. Acids and unsubstituted or monosubstituted amides hydrogen-bond with each other, but ester and disubstituted amide molecules do not do so. Simple acids and esters are liquids; all amides

KEY WORDS

Acetyl group, *p. 520*

Acyl group, *p. 518*

Amide, *p. 517*

Carbonyl-group substitution reaction, *p. 518*

Carboxyl group, *p. 519*

Carboxylate anion, *p. 528*

(except formamide) are solids. The simpler compounds of all three classes are water-soluble or partially water-soluble.

2. **How are carboxylic acids, esters, and amides named?** Many carboxylic acids are best known by their common names (Table 17.1), and these names are the basis for the common names of esters and amides. Esters are named with two words: The first is the name of the alkyl group that has replaced the —H in —COOH, and the second is the name of the parent acid with -*ic acid* replaced by -*ate* (for example, methyl acetate). For amides, the ending -*amide* is used, and where there are R groups on the N, these are named first, preceded by *N* (as in *N*-methylacetamide).

3. **What are some occurrences and applications of significant carboxylic acids, esters, and amides?** Natural carboxylic acids and esters are common; the acids have bad odors, whereas esters contribute to the pleasant odors of fruits and flowers. Acetic acid and citric acid occur in vinegar and citrus fruits, respectively. Aspirin and other salicylates are esters; acetaminophen (Tylenol) is an amide; ibuprofen (Advil, Motrin) is a carboxylic acid; benzocaine is representative of a family of amides that are local anesthetics. Proteins and nylon are polymers containing amide bonds. Fats and oils are esters, as are polyesters such as Dacron.

4. **How are esters and amides synthesized from carboxylic acids and converted back to carboxylic acids?** In ester formation, the —OH of a carboxylic acid group is replaced by the —OR group of an alcohol. In amide formation, the —OH group of a carboxylic acid is replaced by —NH_2 from ammonia, or by —NHR or —NR_2 from an amine. Hydrolysis with acids or bases adds —H and —OH to the atoms from the broken bond to restore the carboxylic acid and the alcohol, ammonia, or amine.

5. **What are the organic phosphoric acid derivatives?** Phosphoric acid forms mono-, di-, and triesters: $ROPO_3H_2$, $(RO)_2PO_2H$, and $(RO)_3PO$. There are also esters that contain the diphosphate and triphosphate groups from pyrophosphoric acid and triphosphoric acid (p. 543). Esters that retain hydrogen atoms are ionized in body fluids—for example, $ROPO_3^{2-}$, $(RO)_2PO_2^-$. *Phosphorylation* is the transfer of a *phosphoryl group*, —PO_3^{2-}, from one molecule to another. In biochemical reactions, the phosphoryl group is often donated by a triphosphate (such as ATP) with release of energy.

SUMMARY OF REACTIONS

1. **Reactions of carboxylic acids**

   (a) Acid–base reaction with water (Section 17.3):

   $$\underset{O}{CH_3COH} + H_2O \rightleftharpoons \underset{O}{CH_3CO^-} + H_3O^+$$

   (b) Acid–base reaction with a strong base to yield a carboxylic acid salt (Section 17.3):

   $$CH_3COH(aq) + NaOH(aq) \longrightarrow CH_3CO^- \, Na^+(aq) + H_2O$$

   (c) Substitution with an alcohol to yield an ester (Section 17.4):

   $$CH_3COH + CH_3OH \xrightarrow{H^+} CH_3COCH_3 + H_2O$$

   (d) Substitution with an amine to yield an amide (Section 17.4):

   $$CH_3COH + CH_3NH_2 \xrightarrow{heat} CH_3CNHCH_3 + H_2O$$

2. **Reactions of esters**

   (a) Hydrolysis to yield an acid and an alcohol (Section 17.6):

   $$CH_3COCH_3 \xrightarrow[H_2O]{H^+} CH_3COH + CH_3OH$$

(b) Hydrolysis with a strong base to yield a carboxylate anion and an alcohol (saponification; Section 17.6):

$$CH_3CH_2CH_2CH_2CH_2\overset{\overset{\displaystyle O}{\|}}{C}OCH_3 + NaOH(aq) \xrightarrow{H_2O}$$

$$CH_3CH_2CH_2CH_2CH_2\overset{\overset{\displaystyle O}{\|}}{C}O^- \ Na^+ + CH_3OH$$

3. Reactions of amides

(a) Hydrolysis to yield an acid and an amine (Section 17.6):

$$CH_3\overset{\overset{\displaystyle O}{\|}}{C}NHCH_3 \xrightarrow[H_2O]{H^+ \text{ or } OH^-} CH_3\overset{\overset{\displaystyle O}{\|}}{C}OH + CH_3NH_2$$

4. Phosphate reactions (Section 17.8)

(a) Phosphate ester formation

$$HO-\overset{\overset{\displaystyle O}{\|}}{\underset{\underset{\displaystyle OH}{|}}{P}}-OH + CH_3OH \longrightarrow HO-\overset{\overset{\displaystyle O}{\|}}{\underset{\underset{\displaystyle OH}{|}}{P}}-OCH_3 + H_2O$$

(b) Phosphorylation

$$Adenosine-O-\overset{\overset{\displaystyle O}{\|}}{\underset{\underset{\displaystyle O^-}{|}}{P}}-O-\overset{\overset{\displaystyle O}{\|}}{\underset{\underset{\displaystyle O^-}{|}}{P}}-O-\overset{\overset{\displaystyle O}{\|}}{\underset{\underset{\displaystyle O^-}{|}}{P}}-O^- + ROH \longrightarrow$$

$$Adenosine-O-\overset{\overset{\displaystyle O}{\|}}{\underset{\underset{\displaystyle O^-}{|}}{P}}-O-\overset{\overset{\displaystyle O}{\|}}{\underset{\underset{\displaystyle O^-}{|}}{P}}-O^- + RO-\overset{\overset{\displaystyle O}{\|}}{\underset{\underset{\displaystyle O^-}{|}}{P}}-O^- + Energy$$

UNDERSTANDING KEY CONCEPTS

17.30 Muscle cells deficient in oxygen reduce pyruvate (an intermediate in metabolism) to lactate at a cellular pH of approximately 7.4:

$$CH_3-\overset{\overset{\displaystyle O}{\|}}{C}-COO^- \xrightarrow{[H]} CH_3-\overset{\overset{\displaystyle OH}{|}}{C}H-COO^-$$

Pyruvate   Lactate

(a) Why do we say pyruvate and lactate, rather than pyruvic acid and lactic acid?

(b) Alter the above structures to create pyruvic acid and lactic acid.

(c) Show hydrogen bonding of water to both pyruvate and lactate. Would you expect a difference in water solubility of lactate and pyruvate? Explain.

17.31 *N*-Acetylglucosamine (also known as NAG) is an important component on the surfaces of cells.

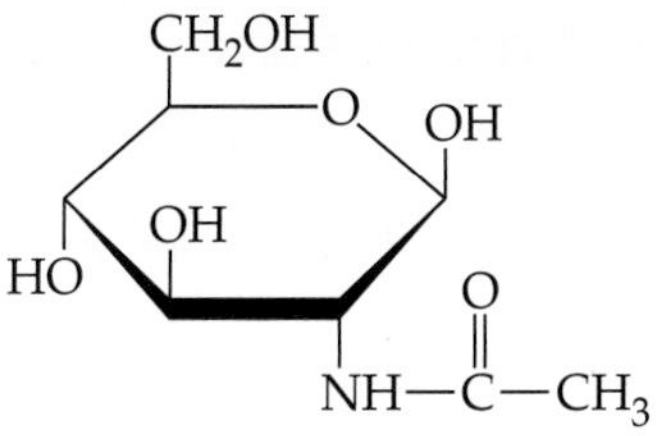

(a) Under what chemical conditions might the acetyl group be removed, changing the nature of the cell-surface components?

N-Acetylglucosamine

(b) Draw the structures of the products of acid hydrolysis.

17.32 One phosphorylated form of glycerate is 3-phosphoglycerate (a metabolic intermediate found in the glycolytic cycle, Section 23.3):

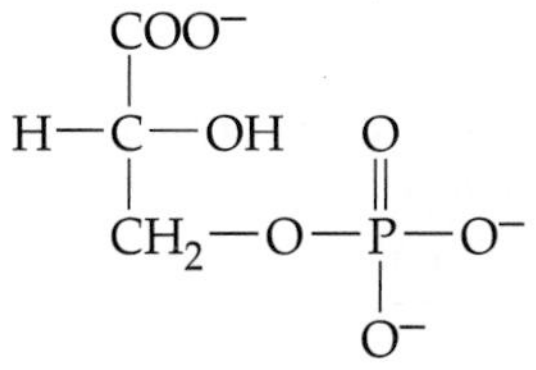

$$\begin{array}{c} COO^- \\ | \\ H-C-OH \quad\quad O \\ | \quad\quad\quad\quad \| \\ CH_2-O-P-O^- \\ | \\ O^- \end{array}$$

(a) Identify the type of linkage between glycerate and phosphate.

(b) 1,3-Bisphosphoglycerate (two phosphates on glycerate) has an anhydride linkage between the carbonyl at C1 of glycerate and phosphate. Draw the structure of 1,3-bisphosphoglycerate (another metabolic intermediate).

17.33 The names of the first four dicarboxylic acids can be remembered by using the first letter of each word of the saying "*Oh, my such good...*" to remind us of *oxalate, malonate, succinate,* and *glutarate* (the form in which these acids occur at physiological pH). Write the structures of these four dicarboxylate anions.

17.34 The amino acid homoserine has the following structure:

$$HOOC—\overset{\overset{\displaystyle NH_2}{\displaystyle |}}{CH}—CH_2—CH_2—OH$$

(a) If *two* molecules of homoserine react to form an ester, what is the structure of the ester product?

(b) If *two* molecules of homoserine react to form an amide, what is the structure of the amide product?

(c) Draw the cyclic ester resulting from the intramolecular reaction of the hydroxyl group of homoserine with its carboxyl group. Cyclic esters are called *lactones*. A common name for this lactone is homoserine lactone.

17.35 (a) Draw the structures of the following compounds and use dashed lines to indicate where they form hydrogen bonds to other molecules of the same kind: (i) acetic acid, (ii) methyl acetate, (iii) acetamide.

(b) Arrange these compounds in order of increasing boiling points and explain your rationale for the order.

17.36 Volicitin, in the "spit" from beet armyworms, causes corn plants to produce volatile compounds. Draw the three hydrolysis products that form from volicitin that match the common names given below.

(a) Glutamic acid (α-aminoglutaric acid)

(b) Ammonia

(c) 17-Hydroxylinolenic acid

Volicitin

17.37 For the following compounds, give the systematic name and indicate whether the compound is an acid, a base, or neither.

(a)

(b) $CH_3—CH_2CH_2CH_2—\overset{\overset{\displaystyle O}{\displaystyle \|}}{C}—OCH_2CH_3$

(c) $CH_3\overset{\overset{\displaystyle }{\displaystyle |}}{\underset{\underset{\displaystyle CH_3}{\displaystyle |}}{CH}}—CH_2\overset{\overset{\displaystyle }{\displaystyle |}}{\underset{\underset{\displaystyle OH}{\displaystyle |}}{CH}}—\overset{\overset{\displaystyle O}{\displaystyle \|}}{C}—OCH_3$

(d) $CH_3CH_2\overset{\overset{\displaystyle }{\displaystyle |}}{\underset{\underset{\displaystyle CH_2CH_3}{\displaystyle |}}{N}}—\overset{\overset{\displaystyle O}{\displaystyle \|}}{C}H$

ADDITIONAL PROBLEMS

CARBOXYLIC ACIDS

17.38 Write the equation for the ionization of benzoic acid in water.

17.39 Suppose you have a sample of propanoic acid dissolved in water at pH 7.

(a) Draw the structure of the major species present in the water solution.

(b) Now assume that aqueous HCl is added to the propanoic acid solution until pH 2 is reached. Draw the structure of the major species present.

(c) Finally, assume that aqueous NaOH is added to the propanoic acid solution until pH 12 is reached. Draw the structure of the major species present.

17.40 Draw and name all carboxylic acids with the formula $C_5H_{10}O_2$. (*Hint:* There are more than three.)

17.41 Draw and name three different carboxylic acids with the formula $C_7H_{14}O_2$.

17.42 Give systematic names for the following carboxylic acids:

(a) $CH_3CH_2—\overset{\overset{\displaystyle }{\displaystyle |}}{\underset{\underset{\displaystyle CH_3}{\displaystyle |}}{CH}}—\overset{\overset{\displaystyle }{\displaystyle |}}{\underset{\underset{\displaystyle CH_2CH_3}{\displaystyle |}}{CH}}—\overset{\overset{\displaystyle O}{\displaystyle \|}}{C}—OH$

(b) $CH_3(CH_2)_7—\overset{\overset{\displaystyle O}{\displaystyle \|}}{C}—OH$

(c)

(d)

17.43 Give systematic names for the following carboxylic acids:

(a) $BrCH_2CH_2\overset{\overset{\displaystyle O}{\|}}{C}HCOH$
$\quad\quad\quad\quad\underset{\displaystyle CH_3}{|}$

(b) 2-methylbenzoic acid structure: benzene ring with CH_3 and $COOH$

(c) $(CH_3CH_2)_3CCOOH$

(d) $CH_3(CH_2)_5COOH$

17.44 Give systematic names for the following carboxylic acid salts:

(a) $CH_3CH_2\overset{\displaystyle |}{\underset{\displaystyle CH_2CH_3}{C}}HCH_2\overset{\overset{\displaystyle O}{\|}}{C}O^-\ K^+$

(b) benzene ring $-\overset{\overset{\displaystyle O}{\|}}{C}O^-\ NH_4^+$

(c) $[CH_3CH_2\overset{\overset{\displaystyle O}{\|}}{C}O^-]_2\ Ca^{2+}$

17.45 Give systematic names and common names for the following carboxylic acid salts:

(a) $CH_3\overset{\overset{\displaystyle O}{\|}}{C}-O^-\ NH_4^+$

(b) $HOOC-\overset{\displaystyle |}{\underset{\displaystyle CH_2CH_3}{C}}H-(CH_2)_2-\overset{\overset{\displaystyle O}{\|}}{C}-O^-\ Na^+$

(c) $\overset{\overset{\displaystyle O}{\|}}{\underset{\underset{\displaystyle O}{\|}}{\overset{\displaystyle C}{\underset{\displaystyle C}{|}}}}\overset{\displaystyle -O^-}{\underset{\displaystyle -O^-}{}}\ Ca^{2+}$

17.46 Draw structures corresponding to these names:
(a) 3,4-Dimethylhexanoic acid
(b) Phenylacetic acid
(c) 3,4-Dinitrobenzoic acid
(d) Triethylammonium butanoate

17.47 Draw structures corresponding to these names:
(a) 2,2,3-Trifluorobutanoic acid
(b) 3-Hydroxybutanoic acid
(c) 3,3-Dimethyl-4-phenylpentanoic acid

17.48 Malic acid, a dicarboxylic acid found in apples, has the systematic name hydroxybutanedioic acid. Draw its structure.

17.49 Fumaric acid is a metabolic intermediate that has the systematic name *trans*-2-butenedioic acid. Draw its structure.

17.50 What is the formula for the diammonium salt of fumaric acid (see Problem 17.49)?

17.51 Aluminum acetate is used as an antiseptic ingredient in some skin-rash ointments. Draw its structure.

ESTERS AND AMIDES

17.52 Draw and name compounds that meet these descriptions:
(a) Three different amides with the formula $C_5H_{11}NO$
(b) Three different esters with the formula $C_6H_{12}O_2$

17.53 Draw and name compounds that meet these descriptions:
(a) Three different amides with the formula $C_6H_{13}NO$
(b) Three different esters with the formula $C_5H_{10}O_2$

17.54 Give systematic names for the following structures, and structures for the names:

(a) $CH_3\overset{\overset{\displaystyle O}{\|}}{C}OCH_2CH_2\overset{\displaystyle |}{\underset{\displaystyle }{C}}HCH_3$ with CH_3 on the CH

(b) $CH_3\overset{\displaystyle |}{\underset{\displaystyle CH_3}{C}}HCH_2CH_2\overset{\overset{\displaystyle O}{\|}}{C}OCH_3$

(c) Cyclohexyl acetate

(d) Phenyl *o*-hydroxybenzoate

17.55 Give systematic names for the following structures, and structures for the names:

(a) cyclopentyl $-O-\overset{\overset{\displaystyle O}{\|}}{C}-$ cyclohexyl

(b) Ethyl 2-hydroxypropanoate

(c) benzene ring $-\overset{\overset{\displaystyle O}{\|}}{C}-OCH_2CH_2CH_3$

(d) Butyl 3,3-dimethylhexanoate

(e) $(CH_3)_2CH\overset{\overset{\displaystyle O}{\|}}{C}OC(CH_3)_3$

17.56 Draw structures of the carboxylic acids and alcohols you would use to prepare each ester in Problem 17.54.

17.57 Draw structures of the carboxylic acids and alcohols you would use to prepare each ester in Problem 17.55.

17.58 Give systematic names for the following structures, and structures for the names:

(a) $CH_3CH_2\overset{\displaystyle |}{\underset{\displaystyle CH_2CH_3}{C}}H-\overset{\overset{\displaystyle O}{\|}}{C}-NH_2$

(b) benzene ring $-\overset{\overset{\displaystyle O}{\|}}{C}NH-$ benzene ring

(c) *N*-Ethyl-*N*-methylbenzamide

(d) 2,3-Dibromohexanamide

17.59 Give systematic names for the following structures, and structures for the names:

(a) 3-Methylpentanamide

(b) *N*-Phenylacetamide

(c) $\overset{\displaystyle O}{\overset{\displaystyle \|}{H C}} N(CH_3)_2$

(d) $CH_3CH_2\overset{\displaystyle O}{\overset{\displaystyle \|}{C}}NH\overset{\displaystyle CH_3}{\overset{\displaystyle |}{C}}H CH_3$

17.60 Show how you would prepare each amide in Problem 17.58 from the appropriate carboxylic acid and amine.

17.61 What compounds are produced from hydrolysis of each amide in Problem 17.59?

REACTIONS OF CARBOXYLIC ACIDS AND THEIR DERIVATIVES

17.62 Procaine, a local anesthetic whose hydrochloride is Novocain, has the following structure. Identify the functional groups present, and show the structures of the alcohol and carboxylic acids you would use to prepare procaine.

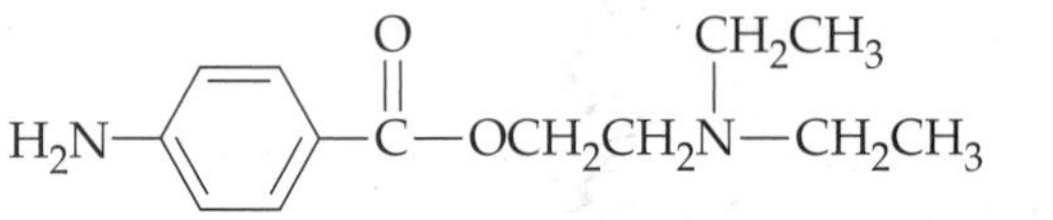

$$H_2N-\text{⟨benzene⟩}-\overset{O}{\overset{\|}{C}}-OCH_2CH_2\overset{CH_2CH_3}{\overset{|}{N}}-CH_2CH_3 \qquad \text{Procaine}$$

17.63 Lidocaine (Xylocaine) is a local anesthetic closely related to procaine. Identify the functional groups present in lidocaine, and show how you might prepare it from a carboxylic acid and an amine.

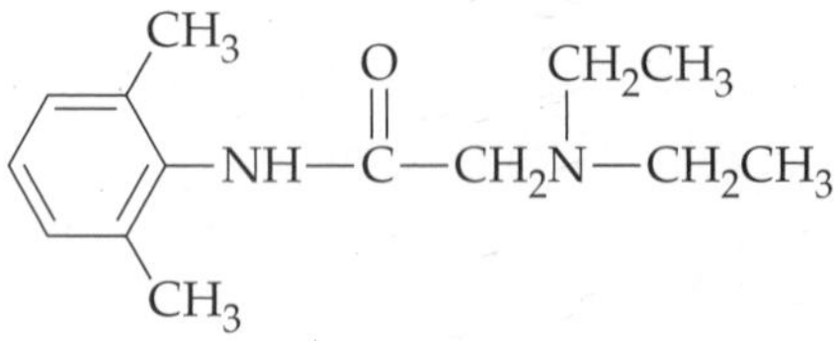

$$\text{⟨xylyl⟩}-NH-\overset{O}{\overset{\|}{C}}-CH_2\overset{CH_2CH_3}{\overset{|}{N}}-CH_2CH_3 \qquad \text{Lidocaine}$$

17.64 Lactones are cyclic esters in which the carboxylic acid part and the alcohol part are connected to form a ring. One of the most notorious lactones is gamma-butyrolactone (GBL), whose hydrolysis product is the date-rape drug GHB. Draw the structure of GHB

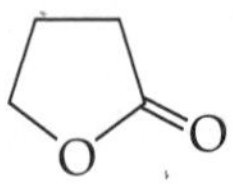

GBL

17.65 A *lactam* is a cyclic amide, where the amide group is part of the ring. The most well known of the lactams are the *beta lactams*, which are found in many antibiotics. Draw the structure of the product(s) obtained from hydrolysis of this beta lactam:

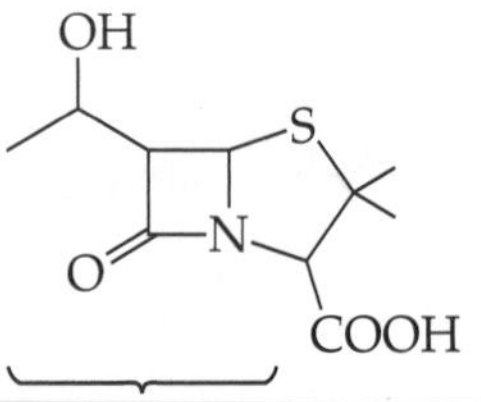

A beta lactam ring

17.66 LSD (lysergic acid diethylamide), a semisynthetic psychedelic drug of the ergoline family, has the structure shown here. Identify the functional groups present, and give the structures of the products you would obtain from hydrolysis of LSD.

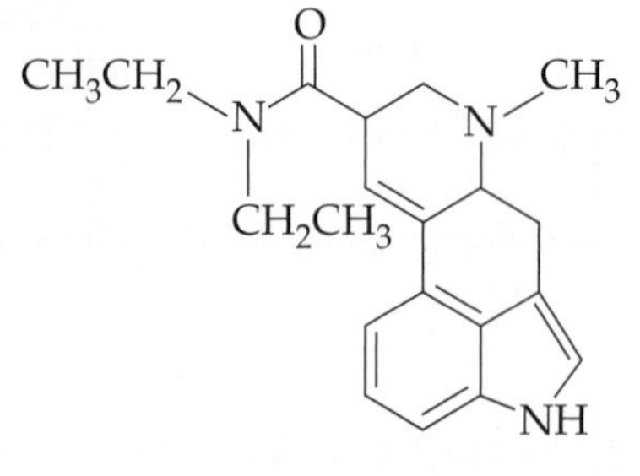

LSD

17.67 Household soap is a mixture of the sodium or potassium salts of long-chain carboxylic acids that arise from saponification of animal fat.

(a) Identify the functional groups present in the fat molecule shown in the reaction below.

(b) Draw the structures of the soap molecules produced in the following reaction:

$$\begin{array}{l} CH_2-O-CO(CH_2)_{14}CH_3 \\ CH-O-CO(CH_2)_7CH=CH(CH_2)_7CH_3 \\ CH_2-O-CO(CH_2)_{16}CH_3 \end{array} \xrightarrow{3\ KOH} ?$$

A fat

POLYESTERS AND POLYAMIDES

17.68 Baked-on paints used for automobiles and many appliances are often based on *alkyds*, such as can be made from terephthalic acid (p. 541) and glycerol (below). Sketch a section of the resultant polyester polymer. Note that the glycerol can be esterified at any of the three alcohol groups, providing *cross-linking* to form a very strong surface.

$$\begin{array}{l} CH_2OH \\ CHOH \\ CH_2OH \end{array}$$

Glycerol

17.69 A simple polyamide can be made from ethylenediamine and oxalic acid (p. 520). Draw the polymer formed when three units of ethylenediamine reacts with three units of oxalic acid.

$$NH_2-CH_2-CH_2-NH_2$$

Ethylenediamine

PHOSPHATE ESTERS AND ANHYDRIDES

17.70 The following phosphate ester is an important intermediate in carbohydrate metabolism. What products result from hydrolysis of this ester?

$$\begin{array}{l} CH_2OH \\ C=O \qquad\quad O \\ CH_2-O-\overset{\|}{P}-O^- \\ \qquad\qquad O^- \end{array}$$

17.71 In the compound

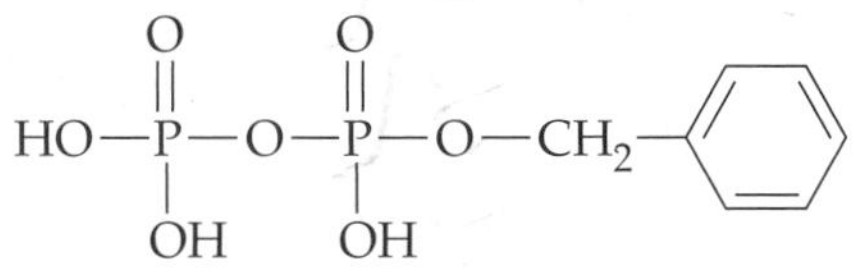

(a) Identify the ester linkage.
(b) Identify the anhydride linkage.
(c) Show the complete acid hydrolysis products.

17.72 The metabolic intermediate acetyl phosphate is a mixed anhydride formed from acetic acid and phosphoric acid. What is the structure of acetyl phosphate?

17.73 Acetyl phosphate (see Problem 17.72) has what is called "high phosphoryl-group transfer potential." Write a reaction in which there is phosphoryl-group transfer from acetyl phosphate to ethanol to make a phosphate ester.

17.74 Cyclic ribose nucleotide phosphates, which are important signaling agents in living cells, all have the general structure shown here. What kind of linkage holds the phosphate to the ribose (see arrows)?

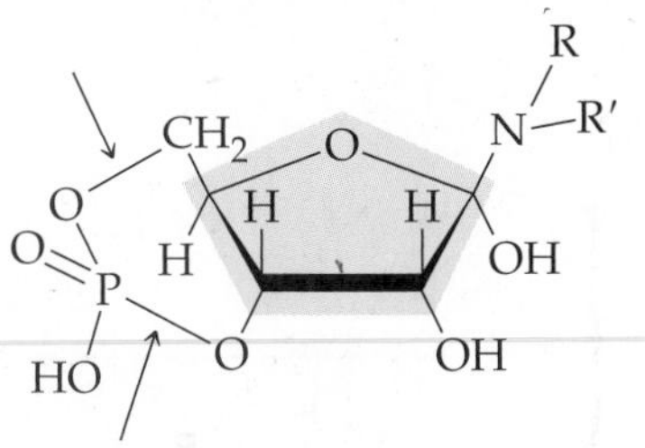

Ribose

17.75 Differentiate between a phosphate diester and a diphosphate. Give an example of each.

Applications

17.76 Name these two compounds, and explain what chemical properties account for how they are used in skin treatment. [*Acids for the Skin, p. 530*]

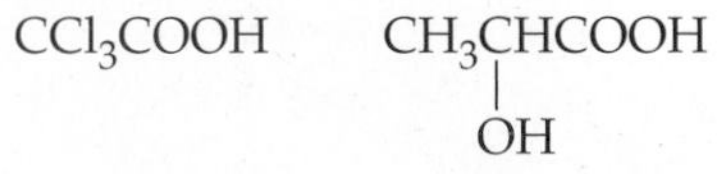

$$CCl_3COOH \qquad CH_3CHCOOH$$
$$\qquad\qquad\qquad OH$$

17.77 If you were working as a health care professional, what should you warn your clients about as far as using α-hydroxy acids during the summer beach season? [*Acids for the Skin, p. 530*]

17.78 Write the structure for the trisodium salt of citric acid and then show the reaction of trisodium citrate with hydrogen ions as it buffers against acidity from fruit juices. [*Acid Salts as Food Additives, p. 531*]

17.79 Against what type of microorganisms are sodium benzoate and potassium sorbate particularly effective? [*Acid Salts as Food Additives, p. 531*]

17.80 Kevlar appears to be nearly indestructible; however, there are still groups of chemicals to which it is not resistant. Based on the chemistry of Kevlar, name one group of chemicals that Kevlar is susceptible to, and explain why it is not resistant. [*Kevlar: A Life-Saving Polymer, p. 542*]

General Questions and Problems

17.81 Three amide isomers, *N,N*-dimethylformamide, *N*-methylacetamide, and propanamide, have respective boiling points of 153 °C, 202 °C, and 213 °C. Explain these boiling points in light of their structural formulas.

17.82 Salol, the phenyl ester of salicylic acid, is used as an intestinal antiseptic. Draw the structure of phenyl salicylate.

17.83 Propanamide and methyl acetate have about the same molar mass, both are quite soluble in water, and yet the boiling point of propanamide is 213 °C whereas that of methyl acetate is 57 °C. Explain.

17.84 Mention at least two simple chemical tests by which you can distinguish between benzaldehyde and benzoic acid.

17.85 Write the formula of the triester formed from glycerol and stearic acid (Table 17.1).

17.86 Name the following compounds.

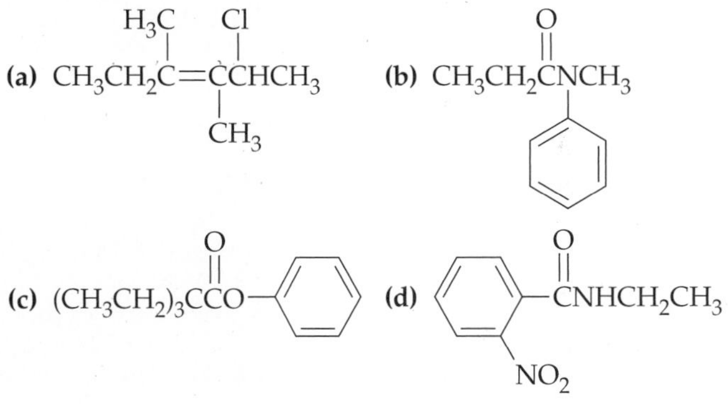

CHAPTER 18

Amino Acids and Proteins

CONCEPTS TO REVIEW

Acid–base properties
(Sections 6.10, 10.9, 17.3)

Hydrolysis reactions
(Section 17.6)

Intermolecular forces
(Section 8.11)

Polymers
(Applications, p. 127, p. 420;
Sections 13.8, 17.7)

▲ Meat, fish, dairy products, beans, and nuts are all high in protein content.

CONTENTS

CHAPTER GOALS

In this chapter, we will look at the following questions about amino acids and proteins:

1. What are the structural features of amino acids?

THE GOAL: Be able to describe and recognize amino acid structures and illustrate how they are connected in proteins.

2. What are the properties of amino acids?

THE GOAL: Be able to describe how the properties of amino acids depend on their side chains and how their ionic charges vary with pH.

3. Why do amino acids have "handedness"?

THE GOAL: Be able to explain what is responsible for handedness and recognize simple molecules that display this property.

4. What is the primary structure of a protein and what conventions are used for drawing and naming primary structures?

THE GOAL: Be able to define protein primary structure, explain how primary structures are represented, and draw and name a simple protein structure, given its amino acid sequence.

5. What types of interactions determine the overall shapes of proteins?

THE GOAL: Be able to describe and recognize disulfide bonds, hydrogen bonding along the protein backbone, and noncovalent interactions between amino acid side chains in proteins.

6. What are the secondary and tertiary structures of proteins?

THE GOAL: Be able to define these structures and the attractive forces that determine their nature, describe the α-helix and β-sheet, and distinguish between fibrous and globular proteins.

7. What is quaternary protein structure?

THE GOAL: Be able to define quaternary structure, identify the forces responsible for quaternary structure, and give examples of proteins with quaternary structure.

8. What chemical properties do proteins have?

THE GOAL: Be able to describe protein hydrolysis and denaturation, and give some examples of agents that cause denaturation.

The word *protein* is a familiar one. Taken from the Greek *proteios*, meaning "primary," "protein" is an apt description for the biological molecules that are of primary importance to all living organisms. Approximately 50% of your body's dry weight is protein. Some proteins, such as the collagen in connective tissue, serve a structural purpose. Others direct responses to internal and external conditions. And still other proteins defend the body against foreign invaders. Most importantly, as enzymes, proteins catalyze almost every chemical reaction that occurs in your body. Because of their importance and the role they play in all biochemical functions, we have chosen to discuss proteins, which are polymers of amino acids, in this first chapter devoted to biochemistry.

18.1 An Introduction to Biochemistry

Biochemistry, the study of molecules, and their reactions in living organisms, is based on the inorganic and organic chemical principles outlined in the first chapters of this book. Now we are ready to investigate the chemical basis of life. Physicians are faced with biochemistry every day because all diseases are associated with abnormalities in biochemistry. Nutritionists evaluate our dietary needs based on our biochemistry. And the pharmaceutical industry designs molecules that mimic or alter the action of biomolecules. The ultimate goal of biochemistry is to understand the structures of biomolecules and the relationship between their structures and functions.

Biochemistry is the common ground for the life sciences. Microbiology, botany, zoology, immunology, pathology, physiology, toxicology, neuroscience, cell biology—in all these fields, answers to fundamental questions are being found at the molecular level.

The principal classes of biomolecules are *proteins, carbohydrates, lipids*, and *nucleic acids*. Some biomolecules are small and have only a few functional

▲ **An electrophoretic gel.** Proteins have been separated by size and stained blue for viewing. Diseases can cause changes in the protein patterns seen.

groups. Others are huge and their biochemistry is governed by the interactions of large numbers of functional groups. Proteins, the subject of this chapter; nucleic acids (Chapters 26, 27); and large carbohydrates (Section 22.9) are all polymers, some containing hundreds, thousands, or even millions of repeating units.

Biochemical reactions must continuously break down food molecules, generate and store energy, build up new biomolecules, and eliminate waste. Each biomolecule has its own role to play in these processes. But despite the huge size of some biomolecules and the complexity of their interactions, their functional groups and chemical reactions are no different from those of simpler organic molecules. ***All the principles of chemistry introduced thus far apply to biochemistry.*** Of the functional groups introduced in previous chapters, those listed in Table 18.1 are of greatest importance in biomolecules.

TABLE 18.1 Functional Groups of Importance in Biochemical Molecules

FUNCTIONAL GROUP	STRUCTURE	TYPE OF BIOMOLECULE
Amino group	$-NH_3^+, \ -NH_2$	Amino acids and proteins (Sections 18.3, 18.7)
Hydroxyl group	$-OH$	Monosaccharides (carbohydrates) and glycerol: a component of triacylglycerols (lipids) (Sections 22.4, 24.2)
Carbonyl group	$\overset{\displaystyle O}{\overset{\|}{-C-}}$	Monosaccharides (carbohydrates); in acetyl group (CH_3CO) used to transfer carbon atoms during catabolism (Sections 22.4, 21.4, 21.8)
Carboxyl group	$\overset{\displaystyle O}{\overset{\|}{-C}}-OH, \ \overset{\displaystyle O}{\overset{\|}{-C}}-O^-$	Amino acids, proteins, and fatty acids (lipids) (Sections 18.3, 18.7, 24.2)
Amide group	$\overset{\displaystyle O}{\overset{\|}{-C}}-\overset{\displaystyle \ }{\underset{\|}{N}}-$	Links amino acids in proteins; formed by reaction of amino group and carboxyl group (Section 18.7)
Carboxylic acid ester	$\overset{\displaystyle O}{\overset{\|}{-C}}-O-R$	Triacylglycerols (and other lipids); formed by reaction of carboxyl group and hydroxyl group (Section 24.2)
Phosphates, mono-, di-, tri-	$-\overset{\|}{\underset{\|}{C}}-O-\overset{\displaystyle O}{\overset{\|}{\underset{\underset{O^-}{\|}}{P}}}-O^-$ $-\overset{\|}{\underset{\|}{C}}-O-\overset{\displaystyle O}{\overset{\|}{\underset{\underset{O^-}{\|}}{P}}}-O-\overset{\displaystyle O}{\overset{\|}{\underset{\underset{O^-}{\|}}{P}}}-O^-$ $-\overset{\|}{\underset{\|}{C}}-O-\overset{\displaystyle O}{\overset{\|}{\underset{\underset{O^-}{\|}}{P}}}-O-\overset{\displaystyle O}{\overset{\|}{\underset{\underset{O^-}{\|}}{P}}}-O-\overset{\displaystyle O}{\overset{\|}{\underset{\underset{O^-}{\|}}{P}}}-O^-$	ATP and many metabolism intermediates (Sections 17.8, 21.5, and throughout metabolism sections)
Hemiacetal group	$-\overset{\|}{\underset{\underset{OR}{\|}}{C}}-OH$	Cyclic forms of monosaccharides; formed by a reaction of carbonyl group with hydroxyl group (Sections 16.7, 22.4)
Acetal group	$-\overset{\|}{\underset{\underset{OR}{\|}}{C}}-OR$	Connects monosaccharides in disaccharides and larger carbohydrates; formed by reaction of carbonyl group with hydroxyl group (Sections 16.7, 22.7, 22.9)

⫘ **Looking Ahead**

The focus in the rest of this book is on human biochemistry and the essential structure–function relationships of biomolecules. In this and the next two chapters, we examine the structure of proteins and the roles of proteins and other molecules in controlling biochemical reactions. Next, we present an overview of metabolism and the production of energy (Chapter 21). Then we discuss the structure and function of carbohydrates (Chapters 22 and 23), the structure and function of lipids (Chapters 24 and 25), the role of nucleic acids in protein synthesis and heredity (Chapters 26 and 27), the metabolism of proteins (Chapter 28), and the chemistry of body fluids (Chapter 29). ⫘

18.2 Protein Structure and Function: An Overview

Proteins are polymers of **amino acids**. Every amino acid in a protein contains an amine group a carboxyl group (COOH), and an R group called a **side chain**, all bonded to a central carbon atom known as the alpha (α) carbon. The amino acids in proteins are **alpha-amino (α-amino) acids**—the amine group in each is connected to the carbon atom "*alpha* to" (next to) the carboxylic acid group. The R groups may be hydrocarbons, or they may contain a functional group:

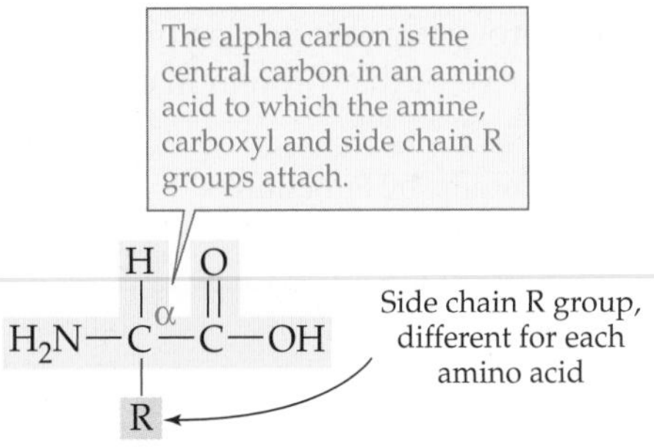

> The alpha carbon is the central carbon in an amino acid to which the amine, carboxyl and side chain R groups attach.

$$H_2N-\overset{\overset{\displaystyle H}{|}}{\underset{\underset{\displaystyle R}{|}}{C}}\overset{\alpha}{}-\overset{\overset{\displaystyle O}{\|}}{C}-OH$$

Side chain R group, different for each amino acid

An α-amino acid

Protein A large biological molecule made of many amino acids linked together through amide (peptide) bonds.

Amino acid A molecule that contains both an amino group and a carboxylic acid functional group.

Side chain (amino acid) The group bonded to the carbon next to the carboxyl group in an amino acid; different in different amino acids.

Alpha- (α−) amino acid An amino acid in which the amino group is bonded to the carbon atom next to the —COOH group.

Two or more amino acids can link together by forming amide bonds (⫘, Section 17.4), which are known as **peptide bonds** when they occur in proteins. A *dipeptide* results from the formation of a peptide bond between the —NH$_2$ group of one amino acid and the —COOH group of a second amino acid. For example, valine and cysteine are connected in a dipeptide as follows:

Peptide bond An amide bond that links two amino acids together.

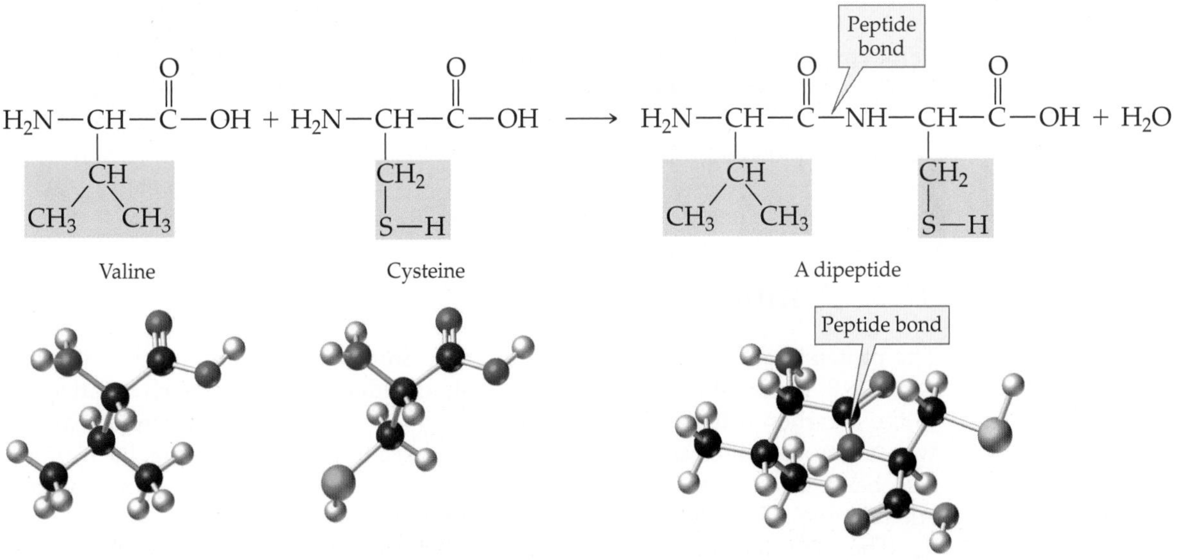

$$H_2N-CH-\overset{\overset{\displaystyle O}{\|}}{C}-OH + H_2N-CH-\overset{\overset{\displaystyle O}{\|}}{C}-OH \longrightarrow H_2N-CH-\overset{\overset{\displaystyle O}{\|}}{C}-NH-CH-\overset{\overset{\displaystyle O}{\|}}{C}-OH + H_2O$$

Valine ... Cysteine ... A dipeptide

Peptide bond

A *tripeptide* results from linkage of three amino acids via two peptide bonds, and so on. Any number of amino acids can link together to form a linear chainlike polymer—a *polypeptide*.

Proteins have four levels of structure, each of which is explored later in this chapter.

- *Primary structure* is the sequence of amino acids in a protein chain (Section 18.7).
- *Secondary structure* is the regular and repeating spatial organization of neighboring segments of single protein chains (Section 18.9).
- *Tertiary structure* is the overall shape of a protein molecule (Section 18.10) produced by regions of secondary structure combined with the overall bending and folding of the protein chain.
- *Quaternary structure* refers to the overall structure of proteins composed of more than one polypeptide chain (Section 18.11).

What roles do proteins play in living things? No doubt, you are aware that a hamburger is produced from muscle protein and that we depend on our own muscle proteins for every move we make. But this is only one of many essential roles of proteins. They provide *structure* and *support* to tissues and organs throughout our bodies. As *hormones* and *enzymes*, they control all aspects of metabolism. In body fluids, water-soluble proteins pick up other molecules for *storage* or *transport*. And the proteins of the immune system provide *protection* against invaders. To accomplish their biological functions, which are summarized in Table 18.2, some proteins must be tough and fibrous, whereas others must be globular and soluble in body fluids. The overall shape of a protein molecule, as you will see often in the following chapters, is essential to the role of that protein in our metabolism.

TABLE 18.2   Classification of Proteins by Function

TYPE	FUNCTION	EXAMPLE
Enzymes	Catalysts	*Amylase*—begins digestion of carbohydrates by hydrolysis
Hormones	Regulate body functions by carrying messages to receptors	*Insulin*—facilitates use of glucose for energy generation
Storage proteins	Make essential substances available when needed	*Myoglobin*—stores oxygen in muscles
Transport proteins	Carry substances through body fluids	*Serum albumin*—carries fatty acids in blood
Structural proteins	Provide mechanical shape and support	*Collagen*—provides structure to tendons and cartilage
Protective proteins	Defend the body against foreign matter	*Immunoglobulin*—aids in destruction of invading bacteria
Contractile proteins	Do mechanical work	*Myosin and actin*—govern muscle movement

18.3 Amino Acids

The proteins in all living organisms are built from the 20 α-amino acids listed in Table 18.3. For 19 of these amino acids, only the identity of the side chain attached to the α carbon differs. The remaining amino acid (proline) is a secondary amine whose nitrogen and α carbon atoms are joined in a five-membered ring. Each amino acid has a three-letter shorthand code that is included in the table—for example, Ala (alanine), Gly (glycine), and Pro (proline).

The 20 protein amino acids are classified as neutral, acidic, or basic, depending on the nature of their side chains. The 15 neutral amino acids are further divided into those with nonpolar side chains and those with polar functional groups such as amide or hydroxyl groups in their side chains. As we explore the structure and

TABLE 18.3 **The 20 Protein Amino Acids with Their Abbreviations and Isoelectric Points.** The structures are written here in their fully ionized forms. These ions and the isoelectric points given in parentheses are explained in Section 18.4.

Nonpolar Side Chains

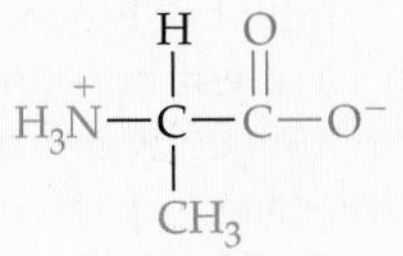

Alanine, Ala (6.0)

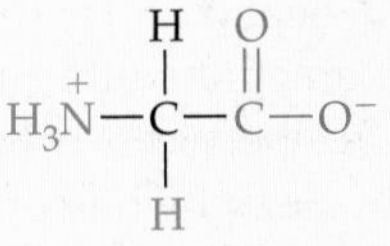

Glycine, Gly (6.0)

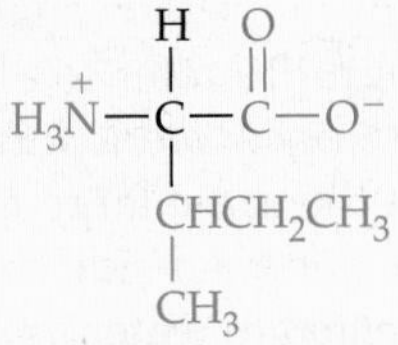

Isoleucine, Ile (6.0)

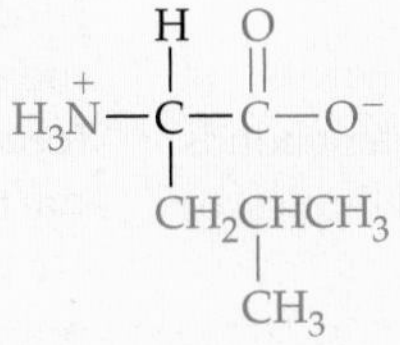

Leucine, Leu (6.0)

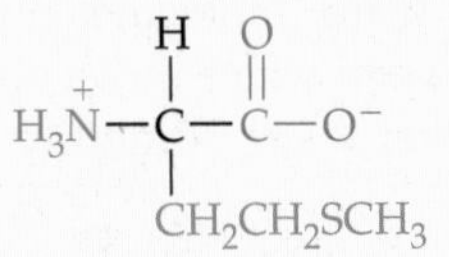

Methionine, Met (5.7)

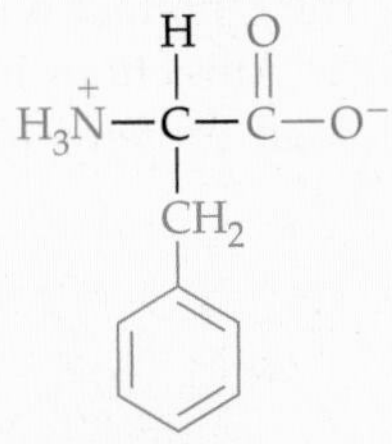

Phenylalanine, Phe (5.5)

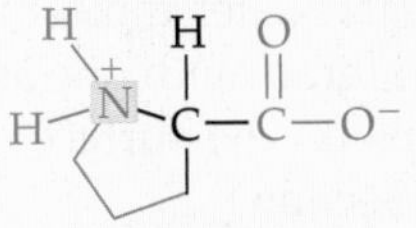

Proline, Pro (6.3)

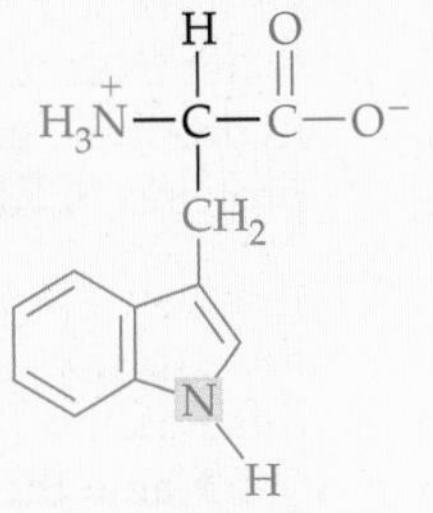

Tryptophan, Trp (5.9)

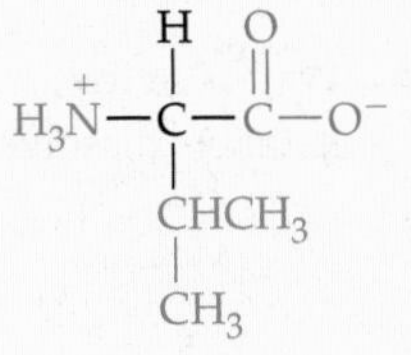

Valine, Val (6.0)

Polar, Neutral Side Chains

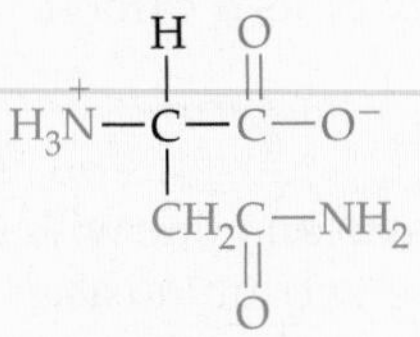

Asparagine, Asn (5.4)

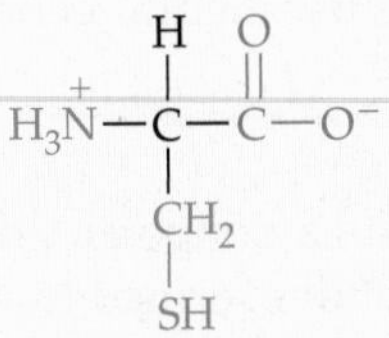

Cysteine, Cys (5.0)

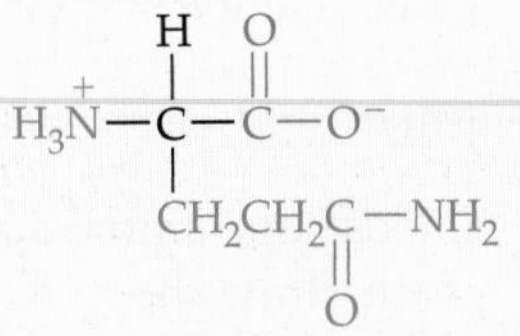

Glutamine, Gln (5.7)

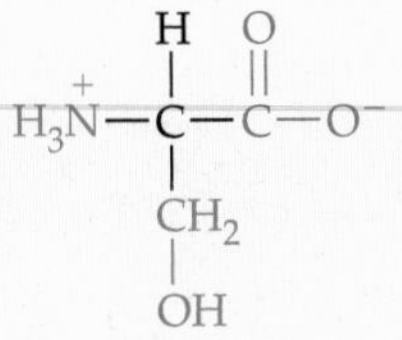

Serine, Ser (5.7)

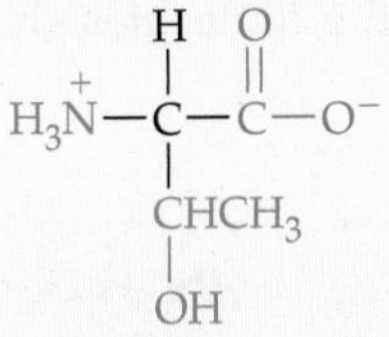

Threonine, Thr (5.6)

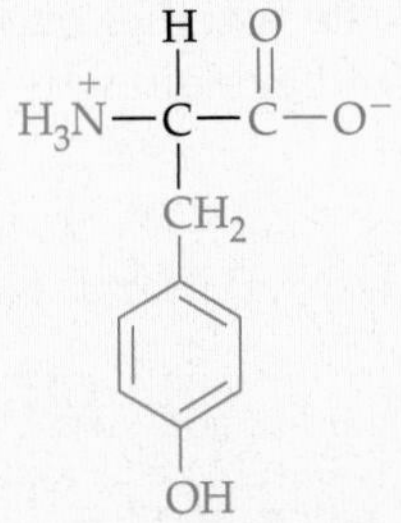

Tyrosine, Tyr (5.7)

Acidic Side Chains

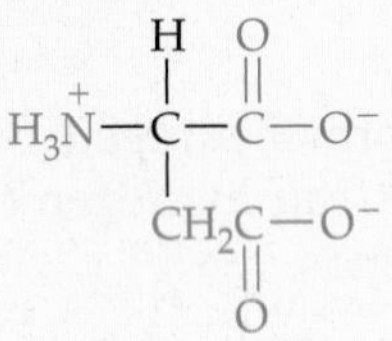

Aspartic acid, Asp (3.0)
(Aspartate)

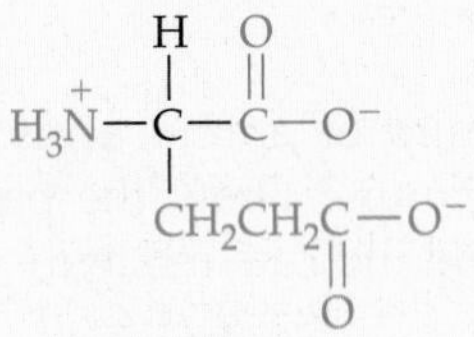

Glutamic acid, Glu (3.2)
(Glutamate)

Basic Side Chains

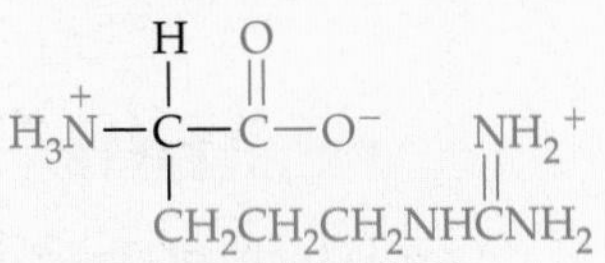

Arginine, Arg (10.8)

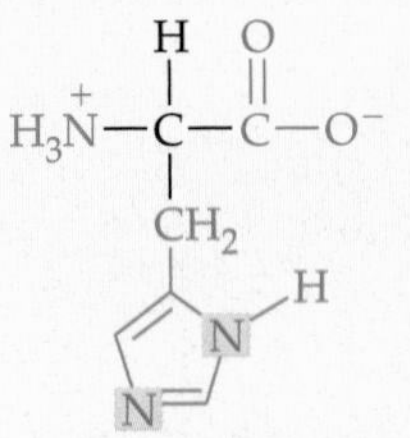

Histidine, His (7.6)

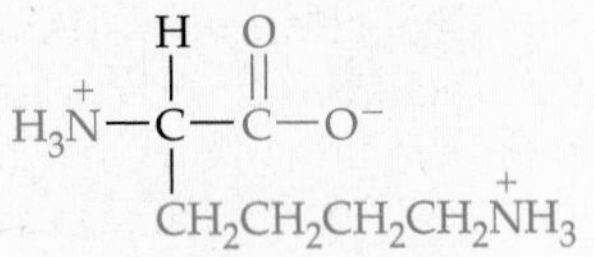

Lysine, Lys (9.7)

function of proteins, you will see that it is the sequence of amino acids in a protein and the chemical nature of their side chains that enable proteins to perform their varied functions.

Intermolecular forces are of central importance in determining the shapes and functions of proteins. (🔗, Section 8.11) In the context of biochemistry it is more meaningful to refer to all interactions other than covalent bonding as **noncovalent forces**. Noncovalent forces act between different molecules or between different parts of the same large molecule, which is often the case in proteins (Section 18.8).

The nonpolar side chains are described as **hydrophobic** (water-fearing)—they are not attracted to water molecules. To avoid aqueous body fluids, they gather into clusters that provide a water-free environment, often a pocket within a large protein molecule. The polar, acidic, and basic side chains are **hydrophilic** (water-loving)—they *are* attracted to polar water molecules. They interact with water molecules much as water molecules interact with one another. Attractions between water molecules and hydrophilic groups on the surface of folded proteins impart water solubility to the proteins. (🔗, Section 9.2)

Noncovalent forces Forces of attraction other than covalent bonds that can act between molecules or within molecules.

Hydrophobic Water-fearing; a hydrophobic substance does not dissolve in water.

Hydrophilic Water-loving; a hydrophilic substance dissolves in water.

PROBLEM 18.1

Name the common amino acids that contain an aromatic ring, contain sulfur, are alcohols, and have alkyl-group side chains.

PROBLEM 18.2

Draw alanine showing the tetrahedral geometry of its α carbon.

PROBLEM 18.3

Choose one amino acid with a nonpolar side chain and one with a polar side chain; draw the two dipeptides formed by these two amino acids.

PROBLEM 18.4

Indicate whether each of the molecules shown below is an α-amino acid or not and explain why.

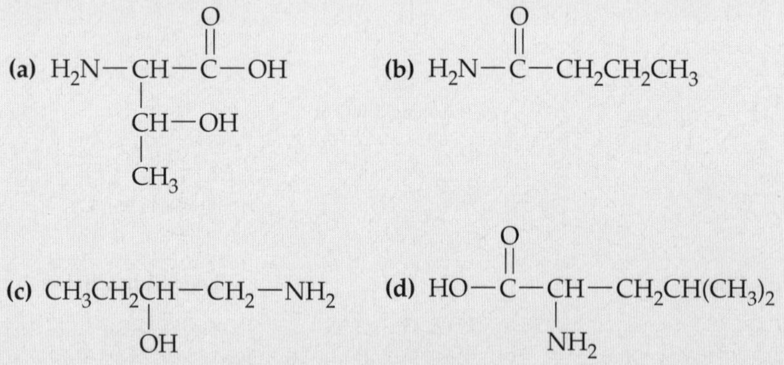

(a) $H_2N-CH-C(=O)-OH$, with $CH-OH$ below the CH, and CH_3 below that.

(b) $H_2N-C(=O)-CH_2CH_2CH_3$

(c) $CH_3CH_2CH-CH_2-NH_2$, with OH below the CH.

(d) $HO-C(=O)-CH-CH_2CH(CH_3)_2$, with NH_2 below the CH.

PROBLEM 18.5

Which of the following pairs of amino acids can form hydrogen bonds between their side-chain groups? Draw the pairs that can hydrogen-bond through their side chains and indicate the hydrogen bonds.

(a) Phe, Thr   (b) Asn, Ser   (c) Thr, Tyr   (d) Gly, Trp

PROBLEM 18.6

In the ball-and-stick model of valine near the beginning of Section 18.2, identify the carboxyl group, the amino group, and the R group.

18.4 Acid–Base Properties of Amino Acids

Amino acids contain both an acidic group, —COOH, and a basic group, —NH₂. As you might expect, these two groups can undergo an intramolecular acid–base reaction. The result is transfer of the hydrogen from the —COOH group to the —NH₂ group to form a *dipolar* ion, an ion that has one positive charge and one negative charge and is thus electrically neutral. Dipolar ions are known as **zwitterions** (from the German *zwitter*, "hybrid"). The zwitterion form of threonine is shown here, and those of the other amino acids commonly found in proteins in mammals are given in Table 18.3.

Zwitterion A neutral dipolar ion that has one + charge and one − charge.

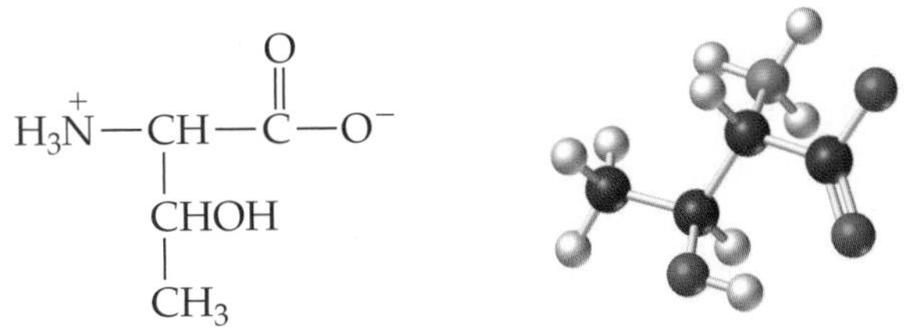

$$\overset{+}{H_3N}-CH-\overset{\overset{\displaystyle O}{\|}}{C}-O^-$$
$$|$$
$$CHOH$$
$$|$$
$$CH_3$$

Threonine—zwitterion

Because they are zwitterions, amino acids have many of the physical properties we associate with salts. (⊂⊃, Section 4.4) Pure amino acids can form crystals, have high melting points, and are soluble in water but not in hydrocarbon solvents.

In acidic solution (low pH), amino acid zwitterions accept protons on their basic —COO⁻ groups to leave only the positively charged —NH₃⁺ groups. In basic solution (high pH), amino acid zwitterions *lose* protons from their acidic —NH₃⁺ groups to leave only the negatively charged —COO⁻ groups:

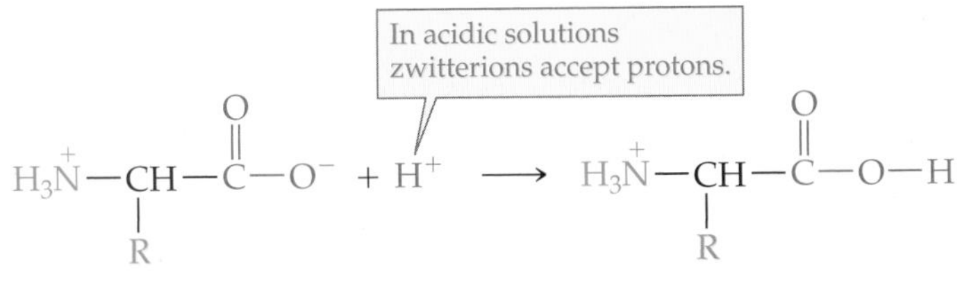

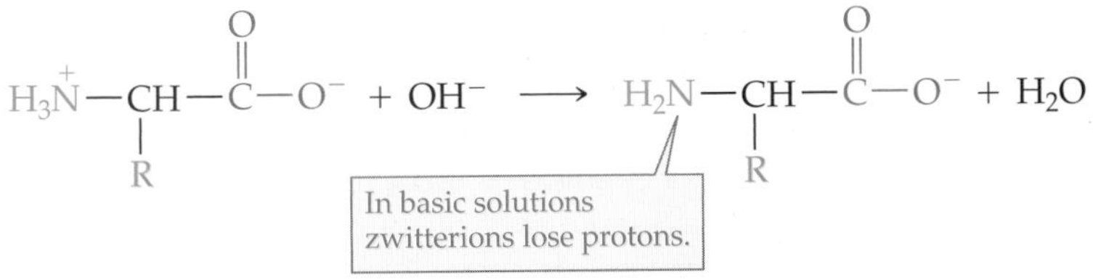

In acidic solutions zwitterions accept protons.

$$\overset{+}{H_3N}-CH-\overset{\overset{\displaystyle O}{\|}}{C}-O^- + H^+ \longrightarrow \overset{+}{H_3N}-CH-\overset{\overset{\displaystyle O}{\|}}{C}-O-H$$
$$\qquad | \qquad\qquad\qquad\qquad\qquad\quad |$$
$$\qquad R \qquad\qquad\qquad\qquad\qquad\quad R$$

$$\overset{+}{H_3N}-CH-\overset{\overset{\displaystyle O}{\|}}{C}-O^- + OH^- \longrightarrow H_2N-CH-\overset{\overset{\displaystyle O}{\|}}{C}-O^- + H_2O$$
$$\qquad | \qquad\qquad\qquad\qquad\qquad\qquad |$$
$$\qquad R \qquad\qquad\qquad\qquad\qquad\qquad R$$

In basic solutions zwitterions lose protons.

Amino acids are never present in the completely nonionized form in either the solid state or aqueous solution. The charge of an amino acid molecule at any given moment depends on the particular amino acid and the pH of the medium. The pH at which the net positive and negative charges are evenly balanced is the amino acid's **isoelectric point (pI)**. At this point, the net charge of all the amino acids in a sample is zero. (The mathematical relationship between isoelectric points and acid dissociation constants is discussed in Appendix C.)

Isoelectric point (pI) The pH at which a sample of an amino acid has equal numbers of + and − charges.

The two amino acids with acidic side chains, aspartic acid and glutamic acid, have isoelectric points at more acidic (lower) pH values than those with neutral side chains. Since the side-chain —COOH groups of these compounds are substantially ionized at physiological pH of 7.4, these amino acids are usually referred to as *aspartate* and *glutamate*, the names of the anions formed when the —COOH groups in the side chains are ionized. (Recall that the same convention is used, for example, for sulfate ion from sulfuric acid or nitrate ion from nitric acid.)

APPLICATION ▶ Nutrition in Health and Disease

To a professional, nutrition is more about the chemical components of what we eat than the flavor, form, and texture of our food. No matter what the recipe, once inside our bodies, it is only the quantity and fate of the proteins, carbohydrates, fats, minerals, and vitamins that matters.

A massive reevaluation of what we eat is occurring in the United States. Increasingly, fruits and vegetables are "in" and red meat is "out." The ongoing changes are in response to experimental evidence about the relationships between body chemistry and health. Fruits and vegetables contain a growing catalog of "phytochemicals" (which just means chemicals from plants) that appear to counteract cancer, heart disease, and a host of other conditions. And meat—especially the fat and cholesterol that naturally occur in meat—most certainly plays a role in the development of heart disease.

As a guide to a healthy diet, the U.S. Department of Agriculture, after lengthy study and consultation, remodeled the *Food Guide Pyramid* in 2005 into MyPyramid, Steps to a Healthier You. The dietary guidelines outlined in MyPyramid can be tailored to your own individual parameters such as weight, age, and activity level. MyPyramid emphasizes consumption of whole grains, fruits, and vegetables, represented by wide vertical stripes on the pyramid (see below), foods that should be eaten in larger amounts. Other foods, such as fats and oils, that should be used sparingly are represented by a narrow stripe. The principal sources of protein are in the milk and meat groups; these groups are represented by stripes of moderate widths. (In later chapters, we will return to the pyramid as we discuss the role of fats and carbohydrates in our diets.)

MyPyramid provides individualized guidelines for average healthy adults who eat a typical American diet. Guidelines for infants, the elderly, and persons with diseases must be tailored to specific needs. Designing a diet for such individuals depends on understanding body chemistry, food chemistry, and the distinctive characteristics of the individuals to be fed. In specifying diets for medical patients, nutritionists must know their patient's medical history, blood and urine levels of key chemicals, medications, physical dimensions (underweight or obese?), energy requirements, and allergies (if any). In the past decade, modified versions of the Food Guide Pyramid for various ethnic and cultural groups, for young children, for the elderly, for vegetarians, and individuals looking for "healthy weight" dietary guidelines have been developed. (One example of an alternative pyramid can be seen at www.usda.gov/kids/index.html; MyPyramid is found at www.mypyramid.gov.)

To give just one example of the implications of patient assessment, consider a person taking an amine antidepressant such as phenelzine (Nardil). Phenelzine acts by inhibiting the enzyme that removes amino groups from amino acids during their normal metabolism (Section 27.3). Some foods, especially those that are aged, fermented, or decayed, contain tyramine, an amine produced from the protein amino acid tyrosine. Notice the structural similarities among these compounds:

Phenelzine
(an antidepressant)

Tyramine
(a pressor)

Tyrosine
(α-amino acid)

Tyramine is a *pressor*—it constricts blood vessels and elevates blood pressure, which can cause irregular heartbeat, severe headache, and in serious cases, intracranial hemorrhage and cardiac failure. Ordinarily, tyramine is inactivated by removal of its amino group by the same enzymes that act on the protein amino acids. In the presence of phenelzine, however, this does not occur. A person taking phenelzine must avoid foods that contain high levels of tyramine. There is also some indication (though not proven) that tyramine in foods triggers migraine headaches.

Foods Containing Relatively High Levels of Tyramine

Cheese
Smoked fish
Chocolate
Pork products
Dry sausage
Sauerkraut
Peanuts
Beer, ale, and some wines
Bananas

Grains	Vegetables	Fruits	Milk	Meat & Beans
Eat 6 oz. every day	Eat 2.5 cups every day	Eat 2 cups every day	Get 3 cups every day; for kids aged 2 to 8, it's 2	Eat 5.5 oz. every day

▲ MyPyramid from the U.S. Department of Agriculture.

See Additional Problem 18.72 at the end of the chapter.

WORKED EXAMPLE **18.1** Determining Side-Chain Hydrophobicity/ Hydrophilicity

Consider the structures of phenylalanine and serine in Table 18.3. Which of these two amino acids has a hydrophobic side chain and which has a hydrophilic side chain?

ANALYSIS Identify the side chains. The side chain in phenylalanine is an alkane. The side chain in serine contains a hydroxyl group.

SOLUTION
The hydrocarbon side chain in phenylalanine is an alkane, which is nonpolar and hydrophobic. The hydroxyl group in the side chain of serine is polar and is therefore hydrophilic.

WORKED EXAMPLE **18.2** Drawing Zwitterion Forms

Look up the zwitterion structure of valine in Table 18.3. Draw valine as it would be found (a) at low pH and (b) at high pH.

ANALYSIS At low pH, which is acidic, basic groups may gain H^+ and at high pH, which is basic, acidic groups may lose H^+. In the zwitterion form of an amino acid, the $-COO^-$ group is basic and the $-NH_3^+$ is acidic.

SOLUTION
Valine has an alkyl-group side chain that is unaffected by pH. At low pH, which is acidic, valine adds a hydrogen ion to its carboxyl group to give the structure on the left below. At high pH, which is basic, valine loses a hydrogen ion from its acidic $-NH_3^+$ group to give the structure on the right below.

$$
\underset{\text{Low pH}}{\overset{\displaystyle O}{H_3\overset{+}{N}-CH-\overset{\|}{C}-OH}}
\qquad
\underset{\text{High pH}}{\overset{\displaystyle O}{H_2N-CH-\overset{\|}{C}-O^-}}
$$

(with side chains $CHCH_3$ and CH_3 below the central carbon in each structure)

PROBLEM 18.7

Draw the structure of glutamic acid at low pH and at high pH.

PROBLEM 18.8

Use the definitions of acids and bases as proton donors and proton acceptors to explain which functional group in the zwitterion form of an amino acid is an acid and which is a base. (See zwitterion of threonine, p. 559.)

18.5 Handedness

Are you right-handed or left-handed? Although you may not think about it very often, handedness affects almost everything you do. It also affects the biochemical activity of molecules.

Anyone who plays softball knows that the last available glove always fits the wrong hand. This happens because your hands are not identical. Rather, they are mirror images. When you hold your left hand up to a mirror, the image you see looks like your right hand (Figure 18.1). Try it.

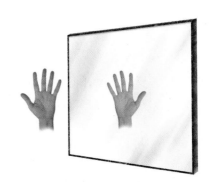

▲ **FIGURE 18.1 The meaning of mirror image.** If you hold your left hand up to a mirror, the image you see looks like your right hand.

Chiral Having right- or left-handedness with two *different* mirror-image forms.

Achiral The opposite of chiral; having superimposable mirror images and thus no right- or left-handedness.

▲ **FIGURE 18.2** **The meaning of superimposable.** It is easy to visualize the chair on top of its mirror image.

Additionally, note that the mirror images of your hand cannot be superimposed on each other; one does not completely fit on top of the other. Objects that have handedness in this manner are said to be **chiral** (pronounced **ky**-ral, from the Greek *cheir*, meaning "hand").

Not all objects are handed, of course. There is no such thing as a right-handed tennis ball or a left-handed coffee mug. When a tennis ball or a coffee mug is held up to a mirror, the image reflected is identical to the ball or mug itself. Objects like the coffee mug that lack handedness are said to be nonchiral, or **achiral**. Their mirror images are superimposable because they have a plane of symmetry. Take a minute to convince yourself of this by studying the chair in Figure 18.2.

PROBLEM 18.9

Which of the following objects are chiral?

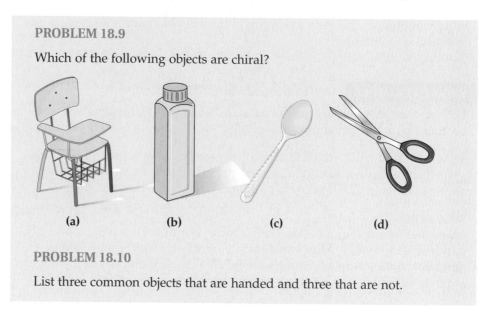

(a)　　　(b)　　　(c)　　　(d)

PROBLEM 18.10

List three common objects that are handed and three that are not.

18.6 Molecular Handedness and Amino Acids

Just as certain objects are chiral, certain molecules are also chiral. Alanine and propane provide a comparison between chiral and achiral molecules:

Alanine, a chiral molecule　　　　*Propane, an achiral molecule*

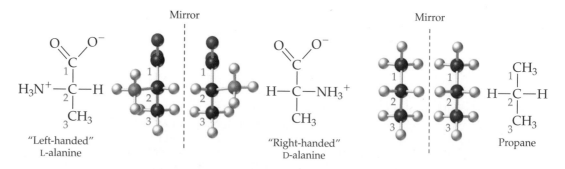

Alanine is a chiral molecule. Its mirror images cannot be superimposed. As a result, alanine exists in two forms that are mirror images of each other: a "right-handed" form known as D-alanine and a "left-handed" form known as L-alanine. (Note: The D and L designations are derived from the relationship of the structure of the amino acids to the structure of glyceraldehyde.) (⊂⊃, Section 22.2) Propane, by contrast, is an achiral molecule. The molecule and its mirror image are identical and it has no left- and right-handed isomers.

Why are some molecules chiral but others are not? Can we predict chirality from structural formulas? Recall from Section 5.7 that carbon forms four bonds oriented to the four corners of an imaginary tetrahedron. (⊂⊃, pp. 122–125) The formulas for alanine and propane are drawn below in a manner that emphasizes the four groups bonded to the central carbon atom. In alanine, this carbon is connected to *four different groups*: a $-COO^-$ group, an $-H$ atom, an $-NH_3^+$ group, and a $-CH_3$ group:

Alanine
(chiral)

1.$-COO^-$
2.$-H$
3.$-NH_3^+$
4.$-CH_3$ } Different

Propane
(achiral)

1.$-CH_3$
2.$-CH_3$ } Identical
3.$-H$
4.$-H$ } Identical

Such a carbon atom is referred to as a **chiral carbon atom**. The presence of one chiral carbon atom always produces a chiral molecule that exists in mirror-image forms. Thus, alanine is chiral. In propane the central carbon atom is bonded to two pairs of identical groups, and the two other carbon atoms are each bonded to three hydrogen atoms. The propane molecule has no chiral carbon atoms and is therefore achiral. (If a molecule has two or more chiral carbon atoms, it may or may not be chiral, depending on its overall shape.)

The two mirror-image forms of a chiral molecule like alanine are called either **enantiomers** (pronounced en-an-ti-o-mer) or **optical isomers** ("optical" because of their effect on polarized light; we will discuss this in Section 22.2). The mirror-image relationship of the enantiomers of a compound with four different groups on one carbon atom is illustrated in Figure 18.3.

Like other isomers, enantiomers have the same formula but different arrangements of their atoms. More specifically, enantiomers are one kind of **stereoisomer**, compounds that have the same formula and atoms with the same connections but different spatial arrangements. (Cis–trans isomers are stereoisomers, too. ⊂⊃, Section 13.3) Pairs of enantiomers have many of the same physical properties. Both enantiomers of alanine, for example, have the same melting point, the same solubility in water, the same isoelectric point, and the same density. But pairs of enantiomers always differ in their effect on polarized light and in how they react with other molecules that are also chiral. Most importantly, pairs of enantiomers often differ in their biological activity, odors, tastes, or activity as drugs. For example, the very different natural flavors of spearmint and caraway seeds are attributed to these two enantiomers:

L-carvone
(in spearmint)

Chiral
carbon

D-carvone
(in caraway)

What about the amino acids listed in Table 18.3? Are any of them chiral? Of the 20 common amino acids, 19 are chiral because they have four different groups bonded to their α carbons, $-H$, $-NH_2$, $-COOH$, and $-R$ (the side chain). Only glycine, H_2NCH_2COOH, is achiral; its α carbon is bonded to two hydrogen atoms. Even though the naturally occurring chiral α-amino acids have pairs of enantiomers, nature uses only a single isomer of each for making proteins. For historical reasons (as you will see in Section 22.2), the naturally occurring isomers are all classified as left-handed, or L-amino acids.

The artificial sweetener aspartame (sold as Equal or NutraSweet) provides another excellent illustration of the delicate nature of the structure–function relationship and its role in biochemistry. Aspartame is the methyl ester of a dipeptide

Chiral carbon atom A carbon atom bonded to four different groups.

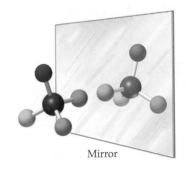

▲ **FIGURE 18.3** **A chiral molecule.** The central atom is bonded to four different groups; the molecule is therefore chiral.

Enantiomers (optical isomers) The two mirror-image forms of a chiral molecule.

Stereoisomers Isomers that have the same molecular and structural formulas but different spatial arrangements of their atoms.

▲ Spearmint leaves and caraway seeds. The very different flavors of these food seasonings are imparted by a pair of enantiomers, which interact in different ways with our taste buds.

made from aspartate and phenylalanine in which both amino acids have the naturally occurring "left-handed" or L chirality. By contrast, the methyl esters of aspartate and phenylalanine dipeptides that have two D isomers or one D isomer combined with one L isomer are bitter.

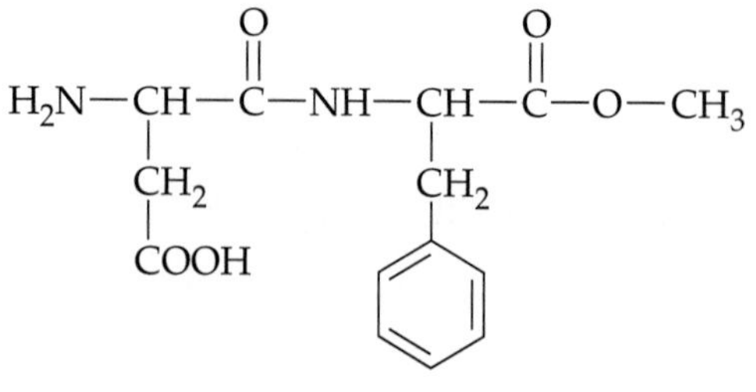

▲ This artificial sweetener would not taste sweet if either of the two amino acids in its molecular structure were D rather than L isomers.

Aspartame
(methyl ester of aspartylphenylalanine)

⫘ **Looking Ahead**

Amino acids, as you have seen, are chiral. Chirality is an important property of another major class of biomolecules. The individual sugar units in all carbohydrates are chiral, a topic addressed in Sections 22.2 and 22.3. ⫘

WORKED EXAMPLE | **18.3** Determining Chirality

Lactic acid can be isolated from sour milk. Is lactic acid chiral?

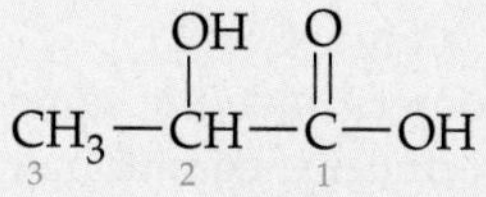

$$CH_3 \underset{3}{-} \overset{OH}{\underset{2}{CH}} \underset{}{-} \overset{O}{\underset{1}{C}} -OH$$

Lactic acid

ANALYSIS A molecule is chiral if it contains one C atom bonded to four different groups. Identify any C atoms that meet this condition.

SOLUTION
To find out if lactic acid is chiral, list the groups attached to each carbon atom:

Groups on carbon 1	Groups on carbon 2	Groups on carbon 3
1. —OH	1. —COOH	1. —CH(OH)COOH
2. ═O	2. —OH	2. —H
3. —CH(OH)CH₃	3. —H	3. —H
	4. —CH₃	4. —H

Next, look at the lists to see if any carbon atom is attached to four different groups. Of the three carbons, carbon 2 has four different groups, and lactic acid is therefore chiral.

PROBLEM 18.11

2-Aminopropane is an achiral molecule, but 2-aminobutane is chiral. Explain.

PROBLEM 18.12

Which of the following molecules are chiral?

(a) 3-Chloropentane   **(b)** 2-Chloropentane

(c) $CH_3CHCH_2CHCH_2CH_3$
        $\quad\quad |\quad\quad\quad |$
        $\quad\quad CH_3\quad CH_3$

PROBLEM 18.13

Two of the 20 common amino acids have two chiral carbon atoms in their structures. Identify these amino acids and their chiral carbon atoms.

KEY CONCEPT PROBLEM 18.14

Two isomers have the formula C_2H_4BrCl. Draw them and identify any chiral carbon atoms.

18.7 Primary Protein Structure

The **primary structure** of a protein is the sequence in which its amino acids are lined up and connected by peptide bonds. Along the *backbone* of the protein is a chain of alternating peptide bonds and α carbons. The amino acid side chains (R_1, R_2, ...) are substituents along the backbone, where they are bonded to the α carbons:

Primary protein structure The sequence in which amino acids are linked by peptide bonds in a protein.

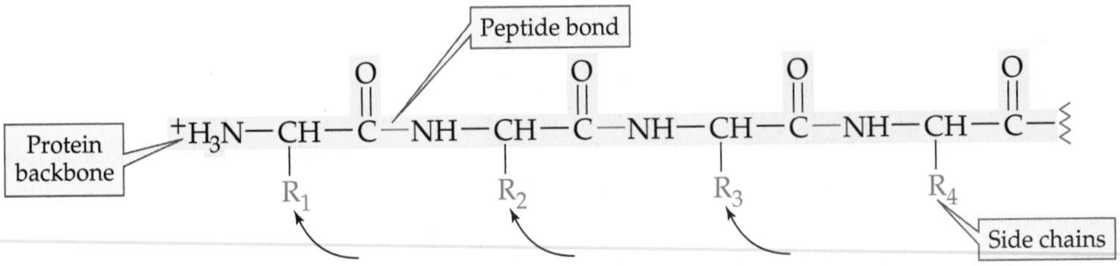

The carbon and nitrogen atoms along the backbone lie in a zigzag arrangement, with tetrahedral bonding around the α carbons. The electrons of each carbonyl-group double bond are shared to a considerable extent with the adjacent C—N bond. This sharing makes the C—N bond sufficiently like a double bond that there is no rotation around it. (⬤⬤, Section 13.3) The result is that the carbonyl group, the —NH group bonded to it, and the two adjacent α carbons form a rigid, planar unit. The side-chain groups extend to opposite sides of the chain. Whether in a dipeptide or a huge polymer chain, these peptide units are planar:

Planar units along a protein chain

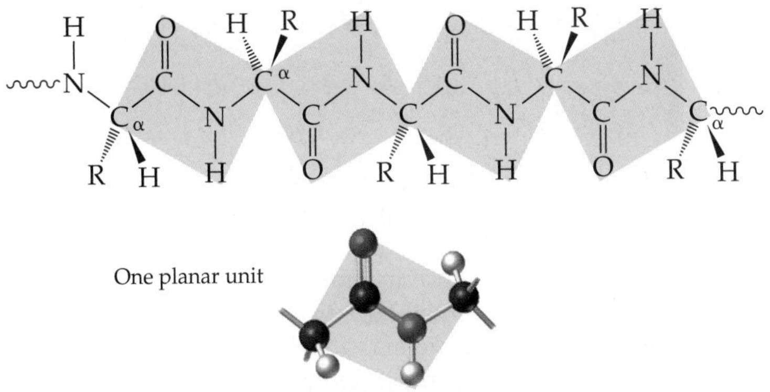

One planar unit

A pair of amino acids—for example, alanine and serine—can be combined to form two different dipeptides. The alanine —COO$^-$ can react with the serine —NH$_3{}^+$:

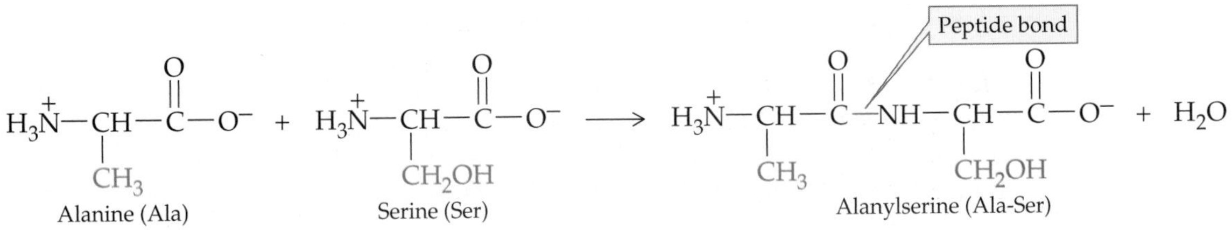

Or the serine —COO⁻ can react with the alanine —NH₃⁺:

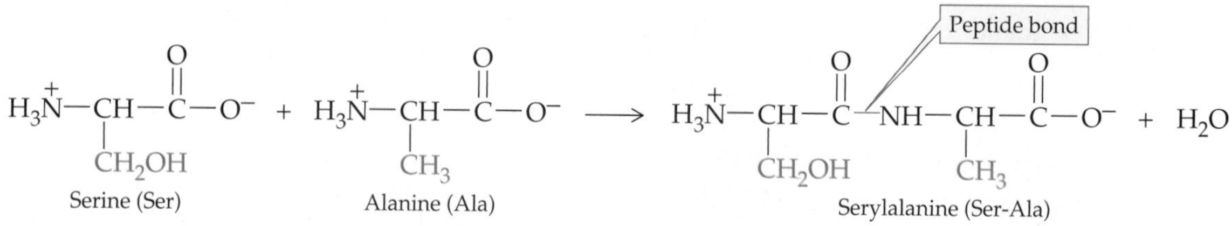

Serine (Ser) + Alanine (Ala) → Serylalanine (Ser-Ala) + H_2O

Amino-terminal (N-terminal) amino acid The amino acid with the free —NH₃⁺ group at the end of a protein.

Carboxyl-terminal (C-terminal) amino acid The amino acid with the free —COO⁻ group at the end of a protein.

Residue An amino acid unit in a polypeptide.

By convention, peptides and proteins are always written with the **amino-terminal amino acid** (also called N-terminal amino acid, the one with the free —NH₃⁺) on the left and the **carboxyl-terminal amino acid** (also called the C-terminal amino acid, the one with the free —COO⁻ group) on the right. The individual amino acids joined in the chain are referred to as **residues**.

A peptide is named by citing the amino acid residues in order, starting at the N-terminal acid and ending with the C-terminal acid. All residue names except the C-terminal one have the *-yl* ending instead of *-ine*, as in alanylserine (abbreviated Ala-Ser) or serylalanine (Ser-Ala).

The primary structure of a protein is the result of the amino acids being lined up one by one to form peptide bonds in precisely the correct order. Consider that there are six ways in which three different amino acids can be joined, more than 40,000 ways in which eight amino acids can be joined, and more than 360,000 ways in which ten amino acids can be joined. Despite the rapid increase in possible combinations with the number of amino acid residues present, only the one correct isomer can do the job. For example, human *angiotensin II* must have its eight amino acids arranged in exactly the correct order:

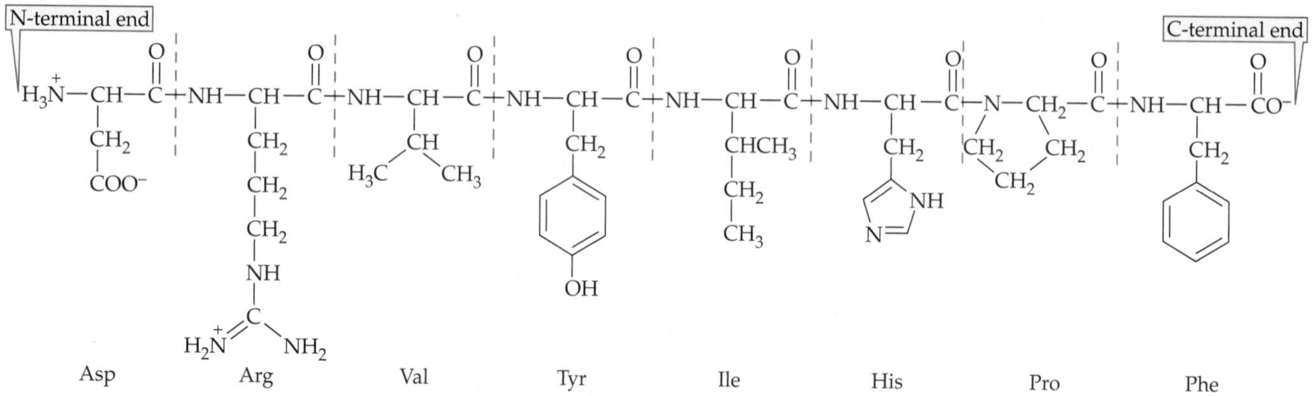

If not arranged properly, this hormone will not participate as it should in regulating blood pressure.

So crucial is primary structure to function—no matter how big the protein—that the change of only one amino acid can sometimes drastically alter a protein's biological properties. Sickle-cell anemia is the best-known example of the potentially devastating result of amino acid substitution. It is a hereditary disease caused by a genetic difference that replaces one amino acid (glutamate, Glu) in each of two polypeptide chains of the hemoglobin molecule with another (valine, Val).

Sickle-cell anemia is named for the "sickle" shape of affected red blood cells. (A sickle is a tool with a curved blade and short handle that is used to cut tall grass.) The sickling of the cells and the resultant painful, debilitating, and potentially fatal disease are entirely the result of the single amino acid substitution. The change replaces a hydrophilic, carboxylic acid–containing side chain (Glu) on hemoglobin with a hydrophobic, neutral hydrocarbon side chain (Val) and thus alters the shape of the hemoglobin molecule. (The effect of the change in charge on electrophoresis is illustrated in the Application, "Protein Analysis by Electrophoresis," p. 573.)

Hemoglobin, found solely inside red blood cells, is the molecule that carries oxygen in the blood and releases it where it is needed. Each red blood cell contains millions of hemoglobin molecules. Sickling takes place in red blood cells carrying the sickle-cell form of hemoglobin that has released oxygen. In this state, a hydrophobic pocket is exposed on the surface of the hemoglobin and the hydrophobic valine side chain on another hemoglobin molecule is drawn into this pocket. As this combining takes place in more and more hemoglobin molecules in a red blood cell, insoluble fibrous chains are formed. The stiff fibers force the cell into the sickled shape. Normal hemoglobin molecules that have released oxygen do not form such fibers because the —COO⁻ side chain in glutamate is too hydrophilic to enter the hydrophobic pocket on another hemoglobin molecule. Thus, each individual molecule of normal hemoglobin does not form part of any chain and no deformation of a normal red blood cell occurs. Furthermore, the hydrophobic pocket is not available in any oxygen-carrying hemoglobin molecule because of a change in shape that occurs when the molecule picks up oxygen.

Sickled red blood cells are fragile, and because they are inflexible, they tend to collect and block capillaries, causing inflammation and pain, and possibly blocking blood flow in a manner that damages major organs. Also, they have a short lifespan, which causes afflicted individuals to become severely anemic.

The percentage of individuals carrying the genetic trait for sickle-cell anemia is highest among people in ethnic groups with origins in tropical regions where malaria is prevalent. The ancestors of these individuals survived because if they were infected with malaria it was not fatal. Malaria-causing parasites enter red blood cells and reproduce there. In a person with the sickle-cell trait, the cells respond by sickling and the parasites cannot multiply. As a result, the genetic trait for sickle-cell anemia is carried forward in the surviving population.

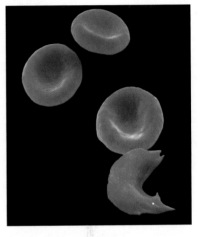

▲ Three normal (convex) and one sickled red blood cells. Because of their shape, sickled cells tend to clog vessels.

⊂◯⊃ Looking Ahead

More than any other kind of biomolecule, proteins are in control of our biochemistry. Are you wondering how each of our thousands of proteins is produced with all their amino acids lined up in the correct order? The information necessary to do this is stored in DNA, and the remarkable machinery that does the job resides in the nuclei of our cells. Chapter 26 provides the details of how protein synthesis is accomplished. ⊂◯⊃

WORKED EXAMPLE **18.4** Drawing Dipeptides

Draw the structure of the dipeptide Ala-Gly.

ANALYSIS You need the names and structures of the two amino acids. Since alanine is named first, it is amino-terminal and glycine is carboxyl-terminal. Ala-Gly must have a peptide bond between the alanine —COO⁻ and the glycine —NH₃⁺.

SOLUTION
The structures of alanine and glycine, and the structure of the Ala-Gly dipeptide are

$$H_3\overset{+}{N}-CH-\overset{\overset{\textstyle O}{\|}}{C}-O^- \qquad H_3\overset{+}{N}-CH_2-\overset{\overset{\textstyle O}{\|}}{C}-O^-$$
$$\underset{CH_3}{|}$$

Alanine (Ala)   Glycine (Gly)

and

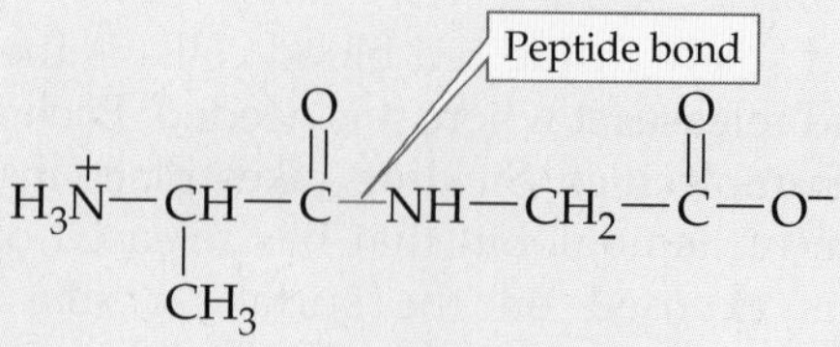

Ala-Gly

PROBLEM 18.15

(a) Use the three-letter shorthand notations to name all the isomeric tripeptides that can be made from serine, tyrosine, and glycine.

(b) Draw the complete structure of the tripeptides that have glycine as the amino-terminal amino acid.

PROBLEM 18.16

Using three-letter abbreviations, show the six tripeptides that contain leucine, tryptophan, and serine.

PROBLEM 18.17

Identify the amino acids in the following dipeptide and tripeptide, and write the abbreviated forms of the peptide names.

(a)

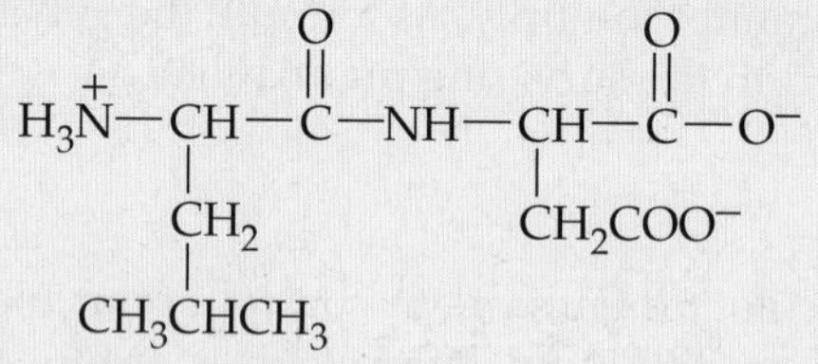

(b)

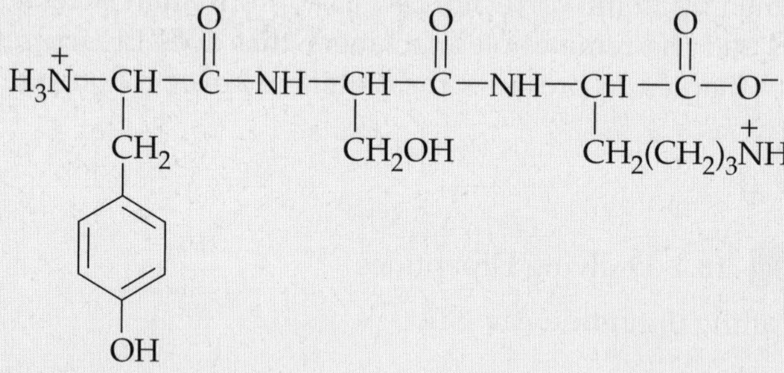

PROBLEM 18.18

Copy the structure of the tripeptide in Problem 18.17b and circle the two planar regions along the backbone.

KEY CONCEPT PROBLEM 18.19

Endoproteases are enzymes that hydrolyze proteins at specific points within their sequences. Chymotrypsin is an endoprotease that cuts on the C-terminal side of aromatic amino acids. In Table 18.3 identify the three amino acids that have aromatic side chains. Now determine the number of fragments that result when chymotrypsin reacts with vasopressin, which has the structure

Asp-Tyr-Phe-Glu-Asn-Cys-Pro-Lys-Gly

and then write out the sequences of these fragments using the standard three-letter designator for each amino acid.

APPLICATION ▶ Proteins in the Diet

From a biochemical viewpoint, what are our protein requirements? Proteins are a necessary part of the daily diet because our bodies do not store proteins like they do carbohydrates and fats. Children need large amounts of protein for proper growth, and adults need protein to replace what is lost each day by normal biochemical reactions. Furthermore, 9 of the 20 amino acids are not synthesized by adult humans and must be obtained in the diet. These are known as the *essential amino acids* (histidine, isoleucine, leucine, lysine, methionine, phenylalanine, threonine, tryptophan, valine).

The total recommended daily amount of protein for an adult, which is the *minimum* required for good health, is 0.8 g per kilogram of body weight. For a 70 kg (154 lb) male, this is 56 g, and for a 55 kg (121 lb) female, it is 44 g. For reference, a McDonald's Big Mac contains 25 g of protein (along with 29 g of fat). The average protein intake in the United States is about 110 g/day, well above what most of us need.

Not all foods are equally good sources of protein. A *complete* protein source provides each of the nine essential amino acids in sufficient amounts to meet our minimum daily needs. Most meat and dairy products meet this requirement, but many vegetable sources such as wheat and corn do not.

Vegetarians must be careful to adopt a diet that includes all of the essential amino acids, which means consuming a variety of foods. In some regions, food combinations that automatically provide *complementary* proteins (proteins that together supply all of the essential amino acids) are traditional, for example, rice and lentils in India, corn tortillas and beans in Mexico, and rice and black-eyed peas in the southern United States. The grains are low in lysine and threonine, whereas the legumes (lentils, beans, and peas) supply these amino acids, but are low in methionine and tryptophan, which are present in grains.

Protein is the major source of nitrogen in the diet. A healthy adult is normally at nitrogen equilibrium, meaning that the amount of nitrogen taken in each day is equal to the amount excreted. Infants and children, pregnant women, those recovering from starvation, and those with healing wounds are usually in *positive nitrogen balance*—they are excreting less nitrogen than they consume, a condition to be expected when new tissue is growing. The reverse condition, *negative nitrogen balance*, occurs when more nitrogen is excreted than consumed. This happens when protein intake is inadequate, during starvation, and in a number of pathologic

▲ This typical Mexican meal contains a complementary protein food combination: beans and rice.

conditions including malignancies, malabsorption syndromes, and kidney disease.

Health and nutrition professionals group all disorders caused by inadequate protein intake as *protein-energy malnutrition* (PEM). Children, because of their higher protein needs, suffer most from this kind of malnutrition. The problem is rampant where meat and milk are in short supply and where the dietary staples are vegetables or grains. An individual is malnourished to some degree if *any* of the essential amino acids are deficient in their diet.

Protein deficiency alone is rare, however, and its symptoms are usually accompanied by those of vitamin deficiencies, infectious diseases, and starvation. At one end of the spectrum is *kwashiorkor*, in which protein is deficient although caloric intake may be adequate. Children with kwashiorkor have edema (swelling due to water retention) and an enlarged liver, and are underdeveloped. The word "kwashiorkor" is from the language of Ghana and translates as "the sickness the older child gets when the next child is born." The onset of kwashiorkor comes when weaning from mother's milk results in conversion to a high-carbohydrate, low-protein diet consisting primarily of corn or cassava gruel. At the other end of the spectrum is *marasmus*, which is the result of starvation. As distinguished from kwashiorkor, marasmus in children is identified with severe muscle wasting, below-normal stature, and poor response to treatment.

See Additional Problems 18.73 through 18.76 and 18.84 at the end of the chapter.

18.8 Shape-Determining Interactions in Proteins

Without interactions between atoms in amino acid side chains or along the backbone, protein chains would twist about randomly in body fluids like spaghetti strands in boiling water. The essential structure–function relationship for each protein depends on the polypeptide chain being held in its necessary shape by these interactions. Before we look at the secondary, tertiary, and quaternary structures of proteins, it will be helpful to understand the kinds of interactions that determine the shapes of protein molecules.

Hydrogen Bonds along the Backbone

Hydrogen bonds form when a hydrogen atom bonded to a highly electronegative atom is attracted to another highly electronegative atom that has an unshared electron pair. The hydrogens in the —NH— groups and the oxygens in the —C═O groups along protein backbones meet these conditions:

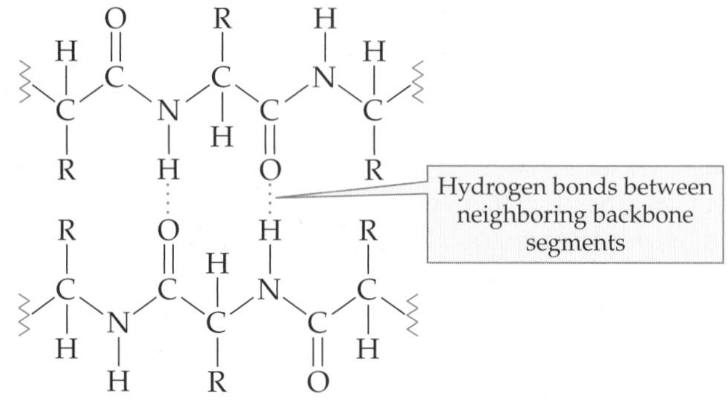

Hydrogen bonds between neighboring backbone segments

This type of hydrogen bonding creates pleated sheet and helical secondary structures, as described in Section 18.9 and as illustrated in the imaginary protein in Figure 18.4.

Hydrogen Bonds of R Groups with Each Other or with Backbone Atoms

Some amino acid side chains contain atoms that can form hydrogen bonds. Side-chain hydrogen bonds can connect different parts of a protein molecule, sometimes nearby and sometimes far apart along the chain. In the protein in Figure 18.4, hydrogen bonds between side chains have created folds in two places. Often hydrogen-bonding side chains are present on the surface of a folded protein, where they can hydrogen-bond with surrounding water molecules.

Ionic Attractions between R Groups (Salt Bridges)

Where there are ionized acidic and basic side chains, the attraction between their positive and negative charges creates what are sometimes known as *salt bridges*. A basic lysine side chain and an acidic aspartate side chain have formed a salt bridge in the middle of the protein shown in Figure 18.4.

Hydrophobic Interactions between R Groups

Hydrocarbon side chains are attracted to each other by the dispersion forces caused by the momentary uneven distribution of electrons. (⬤▭⬤, Section 8.11) The result is that these groups cluster together in the same way that oil molecules cluster on the surface of water (⬤▭⬤, Section 9.2), so that these interactions are often referred to as *hydrophobic*. By clustering in this manner, the hydrophobic groups shown in Figure 18.4 create a water-free pocket in the protein chain. Although the individual

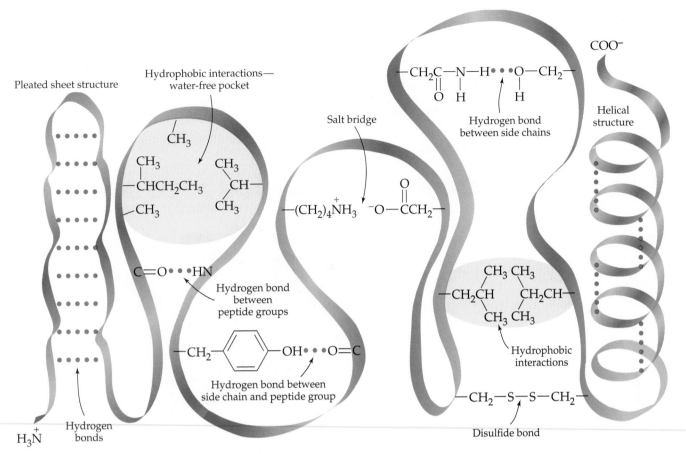

▲ **FIGURE 18.4** **Interactions that determine protein shape.** The regular pleated sheet (*left*) and helical structure (*right*) are created by hydrogen bonding between neighboring backbone atoms; the other interactions involve side-chain groups that can be nearby or quite far apart in the protein chain.

attractions are weak, their large number in proteins plays a major role in stabilizing the folded structures.

Covalent Sulfur–Sulfur Bonds

In addition to the noncovalent interactions, one type of covalent bond plays a role in determining protein shape. Cysteine amino acid residues have side chains containing thiol functional groups (—SH) that can react to form sulfur–sulfur bonds (—S—S—) (⬤⬤, Section 14.9):

If the cysteines are in different protein chains, the otherwise separate chains are linked together. If the cysteines are in the same chain, a loop is formed in the chain. Insulin provides a good example. It consists of two polypeptide chains connected by **disulfide bonds** in two places. One of the chains also has a loop caused by a third disulfide bond.

Disulfide bond An S—S bond formed between two cysteine side chains; can join two peptide chains together or cause a loop in a peptide chain.

Structure of insulin

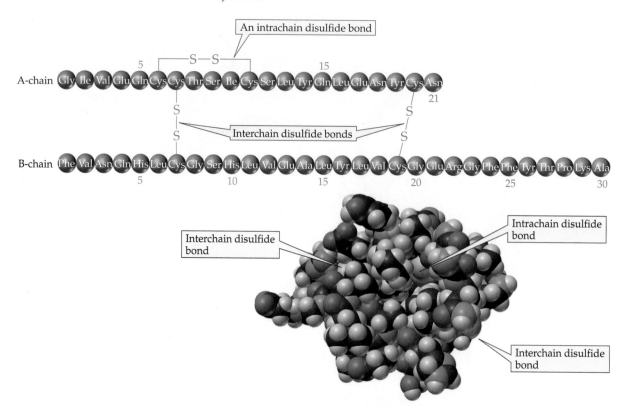

The structure and function of insulin are of intense interest because of its role in glucose metabolism and the need for supplementary insulin by individuals with diabetes (discussed in Section 23.9). Undoubtedly because of this need, studies of insulin have led the way in our still-developing ability to determine the structure of a biomolecule and prepare it synthetically.

In a historically important accomplishment, the amino acid sequence of insulin was determined in 1951—it was the *first* protein for which this was done. It took 15 years before the cross-linking and complete structure of the molecule were determined and a successful laboratory synthesis was carried out. With the advent of biotechnology in the 1980s, once again insulin was first. Until then, diabetic individuals relied on insulin extracted from the pancreas of cows, and because of differences in three amino acids from human insulin, allergic reactions occasionally resulted. In 1982, human insulin became the first commercial product of genetic engineering to be licensed by the U.S. government for clinical use.

⊂⊃ Looking Ahead

Insulin and angiotensin II (p. 566) are representative of a class of small polypeptides that are hormones—they are released when a chemical message must be carried from one place to another. Hormones are discussed in Sections 20.2–20.5.   ⊂⊃

WORKED EXAMPLE   **18.5** Drawing Side-Chain Interactions

What type of noncovalent interaction occurs between the threonine and glutamine side chains? Draw the structures of these amino acids to show the interaction.

ANALYSIS The side chains of threonine and glutamine contain an amide group and a hydroxyl group, respectively. These groups do not form salt bridges because they do not ionize. They are polar and therefore not hydrophobic. They form a hydrogen bond between the oxygen of the amide carbonyl group and the hydrogen of the hydroxyl group.

SOLUTION

The noncovalent, hydrogen bond interaction between threonine and glutamine is as follows:

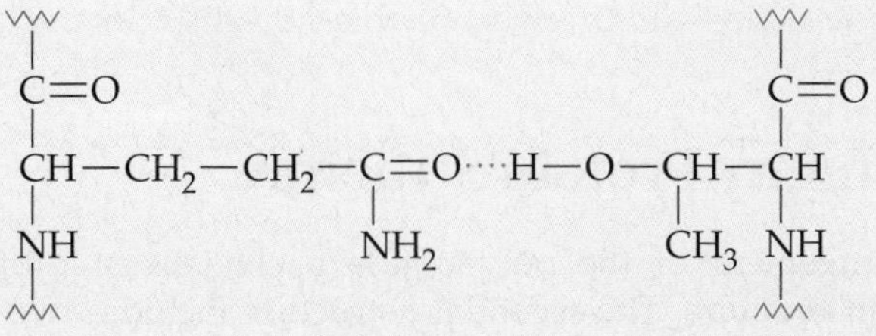

$$
\begin{array}{ccc}
\overset{\text{\tiny wwww}}{|} & & \overset{\text{\tiny wwww}}{|} \\
\text{C=O} & & \text{C=O} \\
| & & | \\
\text{CH—CH}_2\text{—CH}_2\text{—C=O}\cdots\text{H—O—CH—CH} \\
| \qquad\qquad\qquad | \qquad\quad\ \ | \quad\ | \\
\text{NH} \qquad\qquad\ \ \text{NH}_2 \qquad\quad \text{CH}_3 \ \ \text{NH} \\
\overset{\text{\tiny wwww}}{} & & \overset{\text{\tiny wwww}}{}
\end{array}
$$

PROBLEM 18.20

Look at Table 18.3 and identify the type of noncovalent interaction expected between the side chains of the following pairs of amino acids:

(a) Glutamine and tyrosine   (b) Leucine and proline

(c) Aspartate and arginine   (d) Isoleucine and phenylalanine

---

APPLICATION ▶ Protein Analysis by Electrophoresis

Protein molecules in solution can be separated from each other by taking advantage of their net charges. In the electric field between two electrodes, a positively charged particle moves toward the negative electrode and a negatively charged particle moves toward the positive electrode. This movement, known as *electrophoresis*, varies with the strength of the electric field, the charge of the particle, the size and shape of the particle, and the nature of the medium in which the protein is moving.

The net charge on a protein is determined by how many of the acidic or basic side-chain functional groups in the protein are ionized, and this, like the charge of an amino acid (Section 18.4), depends on the pH. Thus, the mobility of a protein during electrophoresis depends on the pH of the medium. If the medium is at a pH equal to the isoelectric point of the protein, the protein does not move.

By varying the nature of the medium between the electrodes and other conditions, proteins can be separated in a variety of ways, including by their molecular weight. Once the separation is complete, the various proteins are made visible by the addition of a dye.

Electrophoresis is routinely used in the clinical laboratory for determining which proteins are present in a blood sample. One application is in the diagnosis of sickle-cell anemia (p. 566). Normal adult hemoglobin (HbA) and hemoglobin showing the inherited sickle-cell trait (HbS) differ in their charges. Therefore, HbA and HbS move different distances during electrophoresis in a medium with constant pH. The accompanying diagram compares the results of electrophoresis of the hemoglobin extracted from red blood cells for a normal individual, one with sickle-cell anemia (two inherited sickle-cell genes), and one with sickle-cell trait (one normal and one inherited sickle-cell gene). With sickle-cell trait, an individual is likely to suffer symptoms of the disease only under conditions of severe oxygen deprivation.

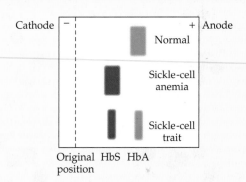

▲ **Gel electrophoresis of hemoglobin.** Hemoglobin in samples placed at the original position have moved left to right as shown. The normal individual has only HbA. The individual with sickle-cell anemia has no HbA, and the individual with sickle-cell trait has roughly equal amounts of HbA and HbS. HbA and HbS have negative charges of different magnitudes because HbS has two fewer Glu residues than HbA.

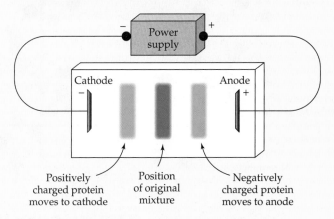

▲ Movement of charged molecules in electrophoresis.

See Additional Problems 18.77–18.78 and 18.83 at the end of the chapter.

PROBLEM 18.21

In Figure 18.4, identify the amino acids that have formed (a) hydrogen bonds from their side chains and (b) hydrophobic side-chain interactions.

18.9 Secondary Protein Structure

Secondary protein structure
Regular and repeating structural patterns (for example, α-helix, β-sheet) created by hydrogen bonding between backbone atoms in neighboring segments of protein chains.

Alpha- (α-) helix Secondary protein structure in which a protein chain forms a right-handed coil stabilized by hydrogen bonds between peptide groups along its backbone.

The spatial arrangement of the polypeptide backbones of proteins constitutes **secondary protein structure**. The secondary structure includes two kinds of repeating patterns known as the **alpha-helix** (α-helix), and the **beta-sheet** (β-sheet). In both, hydrogen bonding between *backbone* atoms holds the polypeptide chain in place. The hydrogen bonding connects the carbonyl oxygen atom of one peptide unit with the amide hydrogen atom of another peptide unit ($-C{=}O\cdots H{-}N{-}$). In large protein molecules, regions of α-helix and β-sheet structure are connected either by randomly arranged loops or coils that are a third type of secondary structure.

α-Helix

A single protein chain coiled in a spiral with a right-handed (clockwise) twist is known as an **alpha-helix (α-helix)** (Figure 18.5a). The helix, which resembles a coiled telephone cord, is stabilized by hydrogen bonds between each backbone carbonyl

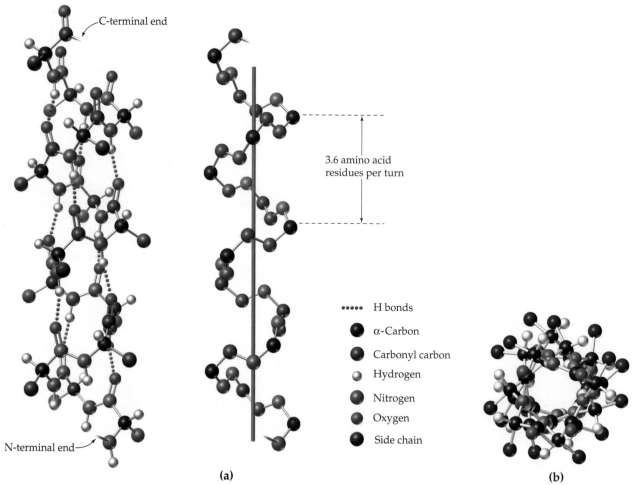

3.6 amino acid
residues per turn

····· H bonds
● α-Carbon
● Carbonyl carbon
○ Hydrogen
● Nitrogen
● Oxygen
● Side chain

C-terminal end

N-terminal end

(a)

(b)

▲ **FIGURE 18.5 Alpha-helix secondary structure.** (a) The coil is held in place by hydrogen bonds (dotted red lines) between each carbonyl oxygen and the amide hydrogen four amino acid residues above it. The chain is a right-handed coil (shown separately on the right), and the hydrogen bonds lie parallel to the vertical axis. (b) Viewed from the top into the center of the helix, the side chains point to the exterior of the helix.

oxygen and an amide hydrogen four amino acid residues farther along the backbone. The hydrogen bonds lie vertically along the helix, and the amino acid R groups extend to the outside of the coil. Although the strength of each individual hydrogen bond is small, the large number of bonds in the helix results in an extremely stable secondary structure. A view of the helix from the top (Figure 18.5b) clearly shows the side chains on the amino acids oriented to the exterior of the helix.

β-Sheet

In the **beta-sheet (β-sheet)** structure, the polypeptide chains are held in place by hydrogen bonds between pairs of peptide units along neighboring backbone segments. The protein chains, which are extended to their full length, bend at each α carbon so that the sheet has a pleated contour, with the R groups extending above and below the sheet (Figure 18.6).

Beta- (β-) sheet Secondary protein structure in which adjacent protein chains either in the same molecule or in different molecules are held in place by hydrogen bonds along the backbones.

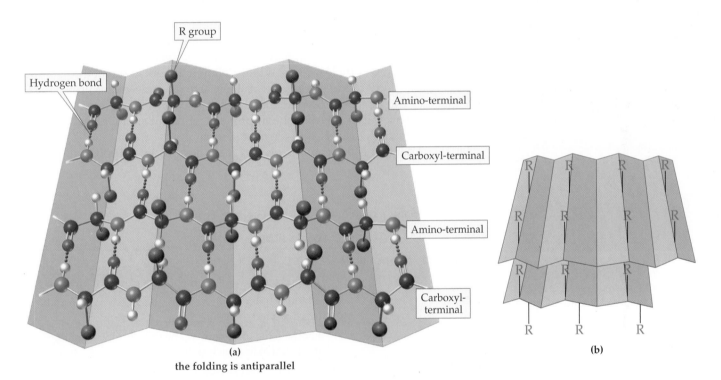

(a)
the folding is antiparallel

(b)

▲ **FIGURE 18.6 Beta-sheet secondary structure.** (a) The hydrogen bonds between neighboring protein chains. The protein chains usually lie side by side so that alternating chains run from the N-terminal end to the C-terminal end and from the C-terminal end to the N-terminal end (known as the *antiparallel* arrangement). (b) A pair of stacked pleated sheets illustrate how the R groups point above and below the sheets.

PROBLEM 18.22

Examine the α-helix in Figure 18.5 and determine how many backbone C and N atoms are included in the loop between an amide hydrogen atom and the carbonyl oxygen to which it is hydrogen-bonded.

Secondary Structure in Fibrous and Globular Proteins

Proteins are classified in several ways, one of which is to identify them as either *fibrous proteins* or *globular proteins*. In an example of the integration of molecular structure and function that is central to biochemistry, fibrous and globular proteins each have functions made possible by their distinctive structures.

Fibrous protein A tough, insoluble protein whose protein chains form fibers or sheets.

Globular protein A water-soluble protein whose chain is folded in a compact shape with hydrophilic groups on the outside.

▲ The proteins found in eggs, milk, and cheese are examples of globular proteins. A spider web is made from fibrous protein.

Secondary structure is primarily responsible for the nature of **fibrous proteins**—tough, insoluble proteins in which the chains form long fibers or sheets. Wool, hair, and fingernails are made of fibrous proteins known as *α-keratins*, which are composed almost completely of α-helixes. In α-keratins, pairs of α-helixes are twisted together into small fibrils that are in turn twisted into larger and larger bundles. The hardness, flexibility, and stretchiness of the material varies with the number of disulfide bonds. In fingernails, for example, large numbers of disulfide bonds hold the bundles in place.

Natural silk and spider webs are made of *fibroin*, a fibrous protein almost entirely composed of stacks of β-sheets. For such close stacking, the R groups must be relatively small (see Figure 18.6). Fibroin contains regions of alternating glycine (—H on α carbon) and alanine (—CH_3 on α carbon). The sheets stack so that sides with the smaller glycine hydrogens face each other and sides with the larger alanine methyl groups face each other.

Globular proteins are water-soluble proteins whose chains are folded into compact, globe-like shapes. Their structures, which vary widely with their functions, are not regular like those of fibrous proteins. Where the protein chain folds back on itself, sections of α-helix and β-sheet are usually present, as illustrated in Figure 18.4. The presence of hydrophilic side chains on the outer surfaces of globular proteins accounts for their water solubility, allowing them to travel through the blood and other body fluids to sites where their activity is needed. Furthermore, many globular proteins are enzymes that are dissolved in the intercellular fluids inside cells. The overall shapes of globular proteins represent another level of structure, tertiary structure, discussed in the next section.

Table 18.4 compares the occurrences and functions of some fibrous and globular proteins.

TABLE 18.4   Some Common Fibrous and Globular Proteins

NAME	OCCURRENCE AND FUNCTION
Fibrous proteins (insoluble)	
Keratins	Found in skin, wool, feathers, hooves, silk, fingernails
Collagens	Found in animal hide, tendons, bone, eye cornea, and other connective tissue
Elastins	Found in blood vessels and ligaments, where ability of the tissue to stretch is important
Myosins	Found in muscle tissue
Fibrin	Found in blood clots
Globular proteins (soluble)	
Insulin	Regulatory hormone for controlling glucose metabolism
Ribonuclease	Enzyme that catalyzes RNA hydrolysis
Immunoglobulins	Proteins involved in immune response
Hemoglobin	Protein involved in oxygen transport
Albumins	Proteins that perform many transport functions in blood; protein in egg white

18.10 Tertiary Protein Structure

Tertiary protein structure The way in which an entire protein chain is coiled and folded into its specific three-dimensional shape.

The overall three-dimensional shape that results from the folding of a protein chain is the protein's **tertiary structure**. In contrast to secondary structure, which depends mainly on attraction between backbone atoms, tertiary structure depends mainly on interactions of amino acid side chains that are far apart along the same backbone.

Although the bends and twists of the protein chain in a globular protein may appear irregular and the three-dimensional structure may appear random, this is not the case. Each protein molecule folds in a distinctive manner that is determined by its primary structure and results in its maximum stability. A protein with the shape in which it functions in living systems is known as a **native protein**.

The noncovalent interactions and disulfide covalent bonds described in Section 18.8 govern tertiary structure. The enzyme *ribonuclease*, shown here as an example, is drawn in a style that shows the combination of α-helix and β-sheet regions, the loops connecting them, and four disulfide bonds:

Native protein A protein with the shape (secondary, tertiary, and quaternary structure) in which it exists naturally in living organisms.

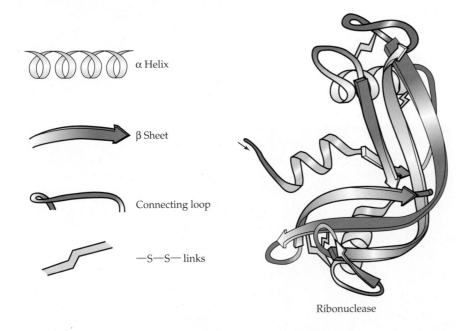

α Helix

β Sheet

Connecting loop

—S—S— links

Ribonuclease

The structure of ribonuclease is representative of the tertiary structure of globular, water-soluble proteins. The hydrophobic, nonpolar side chains congregate in a hydrocarbon-like interior, and the hydrophilic side chains, which provide water solubility, congregate on the outside. Ribonuclease is classified as a **simple protein** because it is composed only of amino acid residues (124 of them). The drawing shows ribonuclease in a style that clearly represents the combination of secondary structures in the overall tertiary structure of a globular protein.

Myoglobin is another example of a small globular protein. A relative of hemoglobin (described in the next section), myoglobin stores oxygen in skeletal muscles for use when there is an immediate need for energy. Structurally, the 153 amino acid residues of myoglobin are arranged in eight α-helical segments connected by short segments looped so that hydrophilic amino acid residues are on the exterior of the compact, spherical tertiary structure. Like many proteins, myoglobin is not a simple protein, but is a **conjugated protein**—a protein that is aided in its function by an associated non–amino acid nonprotein group. The oxygen-carrying portion of myoglobin has a heme group embedded within the polypeptide chain. In Figure 18.7 the myoglobin molecule is drawn in four different ways, each often used to illustrate the shapes of protein molecules. Some examples of other kinds of conjugated proteins are listed in Table 18.5.

Simple protein A protein composed of only amino acid residues.

Conjugated protein A protein that incorporates one or more non–amino acid units in its structure.

⊸○ KEY CONCEPT PROBLEM 18.23

Hydrogen bonds are important in stabilizing both the secondary and tertiary structures of proteins. How do the groups that form hydrogen bonds in the secondary and tertiary structures differ?

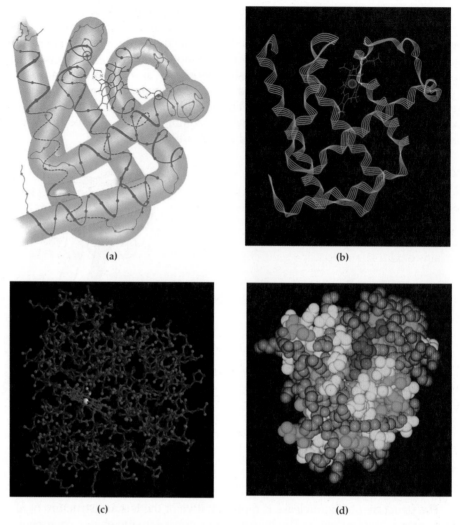

▲ **FIGURE 18.7** **Myoglobin, drawn in four styles.** In each panel the red structure embedded in the protein is a molecule of heme, to which O_2 binds. (a) The sausage-like shape is often used alone to represent the helical portions of a globular protein. (b) A protein *ribbon model* shows the helical portions as a ribbon. (c) A ball-and-stick molecular model of myoglobin. (d) A space-filling model of myoglobin in which the hydrophobic residues are blue and the hydrophilic residues are purple.

TABLE 18.5 Some Examples of Conjugated Proteins

CLASS OF PROTEIN	NONPROTEIN PART	EXAMPLES
Glycoproteins	Carbohydrates	Glycoproteins in cell membranes (Section 24.7)
Lipoproteins	Lipids	High- and low-density lipoproteins that transport cholesterol and other lipids through the body (Section 25.2)
Metalloproteins	Metal ions	The enzyme cytochrome oxidase, necessary for biological energy production, and many other enzymes
Phosphoproteins	Phosphate groups	Milk casein, which provides essential nutrients to infants
Hemoproteins	Heme	Hemoglobin (transplants oxygen) and myoglobin (stores oxygen)
Nucleoproteins	RNA (ribonucleic acid)	Found in cell ribosomes, where they take part in protein synthesis

18.11 Quaternary Protein Structure

The fourth and final level of protein structure, and the most complex, is **quaternary protein structure**—the way in which two or more polypeptide subunits associate to form a single three-dimensional protein unit. The individual polypeptides are held together by the same noncovalent forces responsible for tertiary structure. In some cases, there are also covalent bonds and the protein may incorporate a non–amino acid portion. *Hemoglobin* and *collagen* are both well-understood examples of proteins with quaternary structure essential to their function.

Quaternary protein structure The way in which two or more protein chains aggregate to form large, ordered structures.

Hemoglobin

Hemoglobin (Figure 18.8a) is a conjugated quaternary protein composed of four polypeptide chains (two α chains and two β chains) held together primarily by the interaction of hydrophobic groups, and four heme groups. The polypeptides are similar in composition and tertiary structure to myoglobin (Figure 18.7). The α chains have 141 amino acids, and the β chains have 146 amino acids.

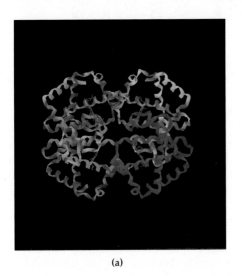

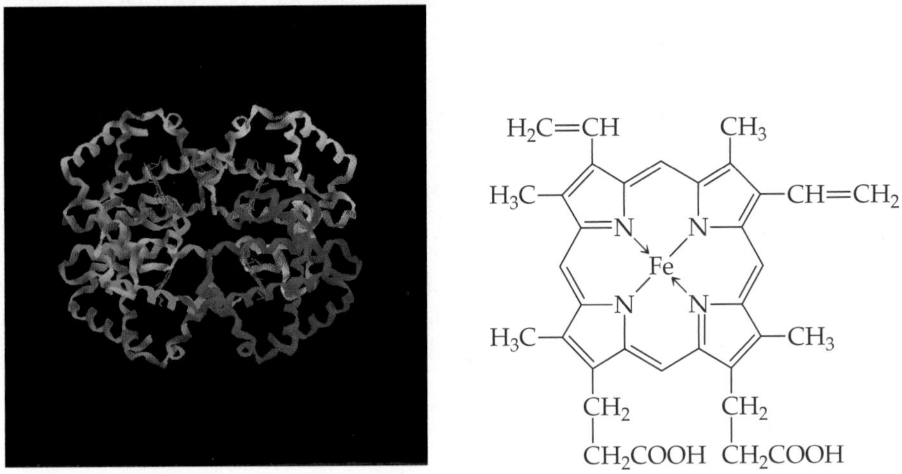

(a)

$$H_2C=CH \qquad CH_3$$
$$H_3C- \qquad -CH=CH_2$$
$$N \quad N$$
$$Fe$$
$$N \quad N$$
$$H_3C- \qquad -CH_3$$
$$CH_2 \qquad CH_2$$
$$CH_2COOH \quad CH_2COOH$$

(b)

◄ **FIGURE 18.8 Heme and hemoglobin, a protein with quaternary structure.** (a) The polypeptides are shown in purple, green, blue, and yellow, with their heme units in red. Each polypeptide resembles myoglobin in structure. (b) A heme unit is present in each of the four polypeptides in hemoglobin.

The hemes (Figure 18.8b), one in each of the four polypeptides, each contain an iron atom that is essential to their function. Hemoglobin is the oxygen carrier in red blood cells. In the lungs, O_2 binds to the Fe^{2+}, so that each hemoglobin can carry a maximum of four O_2 molecules. In tissues in need of oxygen, the O_2 is released, and CO_2 (the product of respiration) is picked up and carried back to the lungs. (Oxygen transport is discussed further in Section 29.6.)

Collagen

Collagen is the most abundant of all proteins in mammals, making up 30% or more of the total. A fibrous protein, collagen is the major constituent of skin, tendons, bones, blood vessels, and other connective tissues. The basic structural unit of collagen (*tropocollagen*) consists of three intertwined chains of about 1000 amino acids each. Each stiff, rod-like chain is loosely coiled in a left-handed (counterclockwise) direction (Figure 18.9a). Three of these coiled chains wrap around one another (in a clockwise direction) to form a stiff, rod-like triple helix (Figure 18.9b) in which the chains are held together by hydrogen bonds.

The various kinds of collagen have in common a glycine residue at every third position. Only glycine residues (with —H as the side chain on the α carbon) can fit in the center of the tightly coiled triple helix. The larger side chains face the exterior of the helix. Proline is incorporated into the originally synthesized collagen molecules. A hydroxyl group is then added to some proline residues in a reaction that requires vitamin C (Section 19.10). Herein lies the explanation for the symptoms of scurvy, the disease that results from vitamin C deficiency. When vitamin C is in short supply, collagen is deficient in hydroxylated proline residues and, as a result,

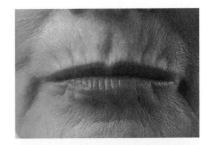

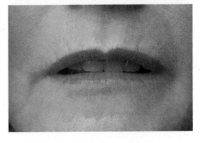

▲ Collagen injections are used to smooth away facial wrinkles, as seen in these before and after photos.

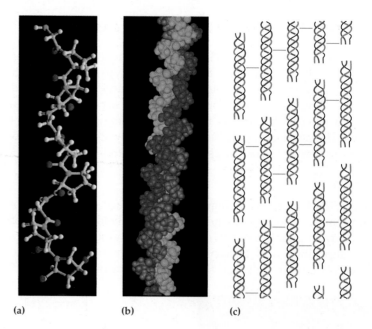

► **FIGURE 18.9 Collagen.** (a) A single collagen helix (carbon, green; hydrogen, light blue; nitrogen, dark blue; oxygen, red). (b) The triple helix of tropocollagen. (c) The quaternary structure of a cross-linked collagen, showing the assemblage of tropocollagen molecules.

(a)   (b)   (c)

forms fibers poorly. The results are the skin lesions and fragile blood vessels that accompany scurvy.

The tropocollagen triple helixes are assembled into collagen in a quaternary structure formed by a great many strands overlapping lengthwise (Figure 18.9c). Depending on the exact purpose collagen serves in the body, further structural modifications occur. In connective tissue like tendons, covalent bonds between strands give collagen fibers a rigid, cross-linked structure. In teeth and bones, calcium hydroxyapatite $[Ca_5(PO_4)_3OH]$ deposits in the gaps between chains to further harden the overall assembly.

Protein Structure Summary

- **Primary structure**—the sequence of amino acids connected by peptide bonds in the polypeptide chain; for example, Asp-Arg-Val-Tyr.

- **Secondary structure**—the arrangement in space of the polypeptide chain, which includes the regular patterns of the α-helix and the β-sheet (held together by hydrogen bonds between backbone carbonyl and amino groups in amino acid residues along adjacent chains segments) plus the loops and coils that connect these segments.

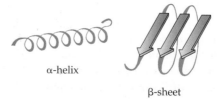

α-helix

β-sheet

- **Tertiary structure**—the folding of a protein molecule into a specific three-dimensional shape held together by noncovalent interactions primarily between amino acid side chains that can be quite far apart along the backbone and, in some cases, by disulfide bonds between side-chain thiol groups.

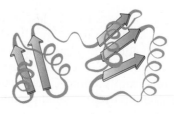

- **Quaternary structure**—two or more protein chains assembled in a larger three-dimensional structure held together by noncovalent interactions.

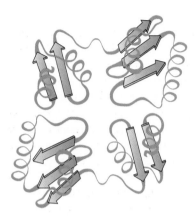

- **Classes of proteins**
- *Fibrous proteins* are tough, insoluble, and composed of fibers and sheets; *globular proteins* are water-soluble and have chains folded into compact shapes.
- *Simple proteins* contain only amino acid residues; *conjugated proteins* include one or more non–amino acid units.

KEY CONCEPT PROBLEM 18.24

Identify the following statements as descriptive of the secondary, tertiary, or quaternary structure of a protein. What type(s) of interaction(s) stabilize each type of structure?

(a) The polypeptide chain has a number of bends and twists resulting in a compact structure.

(b) The polypeptide backbone forms a right-handed coil.

(c) The four polypeptide chains are arranged in a spherical shape.

APPLICATION ▶ Collagen—A Tale of Two Diseases

Case 1: Aboard a British naval ship around 1740. Approximately half of the crew, able-bodied seamen and officers alike, cannot work due to illness. Symptoms include joint pain and swelling, blackened bruises on the skin, and swollen, bleeding gums accompanied by tooth loss. Several sailors have died after spontaneous bleeding from nasal mucous membranes.

Case 2: A hospital emergency room in 2008. A 4-month-old infant girl is brought to the emergency room by her parents. She is in pain and her left arm does not appear "normal." An X-ray establishes that she has a broken arm.

Both of these cases are the result of defects in collagen synthesis. The curable disease scurvy is responsible for the first; the infant in Case 2 suffers from osteogenesis imperfecta, also known as brittle bone disease, an incurable genetic disease.

The symptoms of scurvy have been recognized for centuries due to its prevalence; it is experienced whenever fresh fruits and vegetables are not available for long periods of time. Armies, navies, and medieval people in northern regions in late winter were particularly susceptible to scurvy due to lack of fresh produce. In modern times scurvy is rarely seen except in the elderly, those on highly unusual diets and sometimes in infants. It may also be seen on college campuses

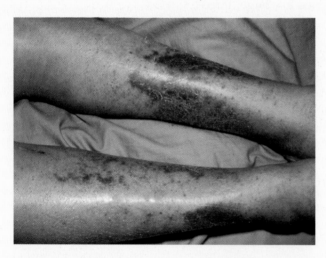

▲ Visible signs of scurvy. Note the swollen ankles and the bruising and lesions on the lower legs that are characteristic of scurvy.

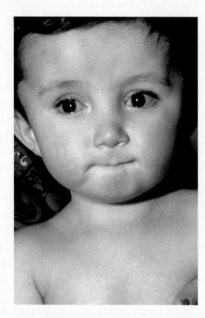

▲ Child with osteogenesis imperfecta. One characteristic of osteogenesis imperfecta is the blue color of the sclera (whites) of the eyes.

where students away from home forget why they should eat fresh fruits and vegetables.

We now know that scurvy results from the synthesis of defective collagen. Collagen is synthesized continuously by the body as old cells and tissues are remodeled and the process is dependent upon vitamin C. Humans neither produce nor store vitamin C in their bodies; it must be obtained daily from the diet. Vitamin C is necessary for the addition of hydroxyl groups to proline by the enzyme prolyhydroxylase after the synthesis of a collagen chain. Proline makes up

about one-third of the amino acids in collagen. About half of the proline residues are hydroxylated, resulting in stabilization of the helical structure of each individual collagen chain. This stabilization further enhances the formation of the triple helix of tropocollagen (Figure 18.9). Lack of vitamin C leads to the formation of weak collagen and tropocollagen. Since tropocollagen is part of capillary walls, it is not surprising that weak collagen leads to the spontaneous bruising, bleeding, and soft tissue swelling that are characteristic of scurvy.

Scurvy is both preventable and curable. Prior to the discovery of vitamin C and the recognition of its role at the molecular level in the prevention of scurvy, fresh food, which inevitably included fruits and vegetables, was recognized as a cure for the disease. The British naval surgeon James Lind is generally attributed with the definitive experiment (in 1747) showing that citrus fruit, in particular lemons and limes, cured scurvy developed during long sea voyages. Subsequently, the British Navy supplied its sailors with lemons and limes during voyages as a preventative treatment for scurvy. This simple move both improved the health and survival rate of the sailors and also resulted in British sailors being known as "limeys."

The disease of the infant in Case 2, osteogenesis imperfecta, is also a collagen disease. Unlike scurvy, which is a dietary disease, osteogenesis imperfecta is inherited. The genetic defect is dominant, meaning it can be inherited from only one parent. The primary symptom of this disease is spontaneous broken bones.

Collagen forms the scaffold for bone. Collagen fibers are the bone matrix, which is filled in with calcium containing crystals of hydroxyapatite ($Ca_5[PO_4]_3OH$). The combination of collagen and hydroxyapaptite makes strong bone tissue. In osteogenesis imperfecta, incorrectly synthesized collagen leads to weaker bone structures. Type I osteogenesis imperfecta is most common and fractures occur primarily before puberty. Other symptoms are grayness of the sclera (white part) of the eye, loose joints, low muscle tone, and brittle teeth. In more severe types of osteogenesis imperfecta, the child may have numerous, frequent fractures, even before birth, small stature, and respiratory problems. Treatment is supportive, aimed at preventing fractures and strengthening muscles. There is no cure for osteogenesis imperfecta, although current research is directed at understanding the underlying biochemical defect in hopes of designing better treatment.

It is difficult to distinguish osteogenesis imperfecta from child abuse. However, the types of spontaneous bone fractures seen in osteogenesis imperfecta are not the typical fractures seen in child abuse cases. A definitive diagnosis of osteogenesis imperfecta requires genetic testing of tissue from the child. Only a small amount of skin tissue is needed. (⬤▭, See Chapter 27: Genomics for DNA testing.)

See Additional Problems 18.79 and 18.80 at the end of the chapter.

18.12 Chemical Properties of Proteins

Protein Hydrolysis

Just as a simple amide can be hydrolyzed to yield an amine and a carboxylic acid, a protein can be hydrolyzed. (⬭ , Section 17.6) In protein hydrolysis, the reverse of protein formation, peptide bonds are hydrolyzed to yield amino acids. In fact, digestion of proteins in the diet involves nothing more than hydrolyzing peptide bonds. For example,

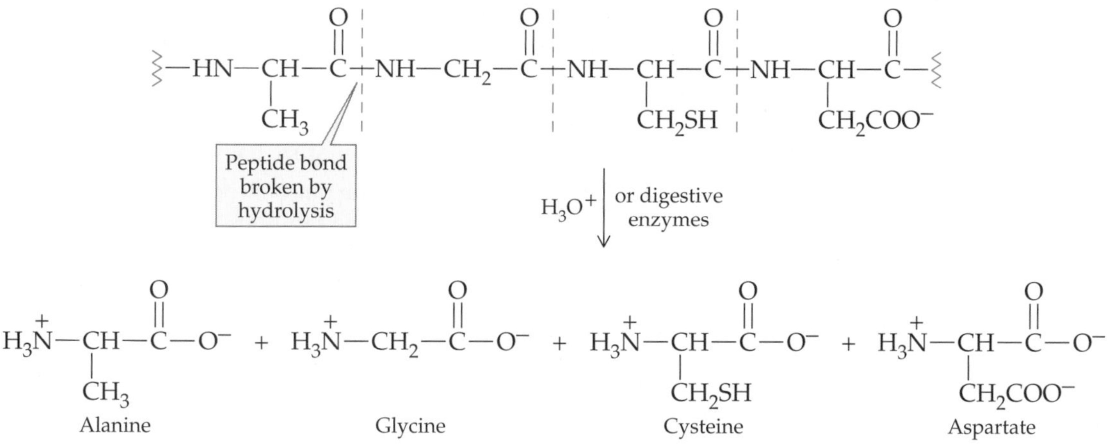

$$—HN—CH—\overset{\displaystyle O}{\overset{\|}{C}}\!+\!NH—CH_2—\overset{\displaystyle O}{\overset{\|}{C}}\!+\!NH—CH—\overset{\displaystyle O}{\overset{\|}{C}}\!+\!NH—CH—\overset{\displaystyle O}{\overset{\|}{C}}—$$

with side chains CH_3 , CH_2SH , and CH_2COO^-

Peptide bond broken by hydrolysis

H_3O^+ | or digestive enzymes ↓

$$\overset{+}{H_3N}—CH—\overset{\displaystyle O}{\overset{\|}{C}}—O^- \ + \ \overset{+}{H_3N}—CH_2—\overset{\displaystyle O}{\overset{\|}{C}}—O^- \ + \ \overset{+}{H_3N}—CH—\overset{\displaystyle O}{\overset{\|}{C}}—O^- \ + \ \overset{+}{H_3N}—CH—\overset{\displaystyle O}{\overset{\|}{C}}—O^-$$

with side chains CH_3 (Alanine), (Glycine), CH_2SH (Cysteine), CH_2COO^- (Aspartate)

A chemist in the laboratory might choose to hydrolyze a protein by heating it with a solution of hydrochloric acid. Most digestion of proteins in the body takes place in the stomach and small intestine, where the process is catalyzed by enzymes (Section 28.1). Once formed, individual amino acids are absorbed through the wall of the intestine and transported in the bloodstream to wherever they are needed.

Protein Denaturation

Since the overall shape of a protein is determined by a delicate balance of noncovalent forces, it is not surprising that a change in protein shape often results when the balance is disturbed. Such a disruption in shape that does not affect the protein's primary structure is known as **denaturation**. When denaturation of a globular protein occurs, for example, the structure unfolds from a well-defined globular shape to a randomly looped chain:

Denaturation The loss of secondary, tertiary, or quaternary protein structure due to disruption of noncovalent interactions and/or disulfide bonds that leaves peptide bonds and primary structure intact.

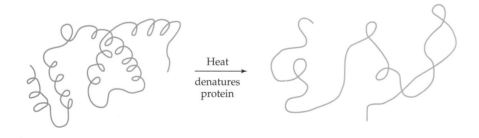

Heat denatures protein →

Denaturation is accompanied by changes in physical, chemical, and biological properties. Solubility is often decreased by denaturation, as occurs when egg white is cooked and the albumins coagulate into an insoluble white mass. Enzymes lose their catalytic activity and other proteins are no longer able to carry out their biological functions when their shapes are altered by denaturation.

▲ Protein denaturation in action: The egg white denatures as the egg fries.

Agents that cause denaturation include heat, mechanical agitation, detergents, organic solvents, extremely acidic or basic pH, and inorganic salts.

- **Heat** The weak side-chain attractions in globular proteins are easily disrupted by heating, in many cases only to temperatures above 50 °C. Cooking meat converts some of the insoluble collagen into soluble gelatin, which can be used in glue and for thickening sauces.

- **Mechanical agitation** The most familiar example of denaturation by agitation is the foam produced by beating egg whites. Denaturation of proteins at the surface of the air bubbles stiffens the protein and causes the bubbles to be held in place.

- **Detergents** Even very low concentrations of detergents can cause denaturation by disrupting the association of hydrophobic side chains.

- **Organic compounds** Polar solvents such as acetone and ethanol interfere with hydrogen bonding by competing for bonding sites. The disinfectant action of ethanol, for example, results from its ability to denature bacterial protein.

- **pH change** Excess H^+ or OH^- ions react with the basic or acidic side chains in amino acid residues and disrupt salt bridges. One familiar example of denaturation by pH change is the protein coagulation that occurs when milk turns sour because it has become acidic.

- **Inorganic salts** Sufficiently high concentrations of ions can disturb salt bridges.

Most denaturation is irreversible: Hard-boiled eggs do not soften when their temperature is lowered. Many cases are known, however, in which unfolded proteins spontaneously undergo *renaturation*—a return to their native state when placed in a nondenaturing medium. Renaturation is accompanied by recovery of biological activity, indicating that the protein has completely refolded to its stable secondary and tertiary structure. By spontaneously refolding into their native shapes, proteins demonstrate that all the information needed to determine these shapes is present in the primary structure.

APPLICATION ▶ Prions: Proteins that Cause Disease

No one believed it at first. The existence of a protein that duplicates itself and causes disease seemed impossible. Only bacteria, viruses, and other microorganisms that have DNA are able to reproduce and cause disease. That had been believed for a long time.

Even more unbelievable was the proposal that a form of protein could be responsible for disease that can be inherited, that can be transmitted between individuals, and that can arise spontaneously in the absence of inheritance or transmission. Stanley B. Prusiner, a neurologist, received the Nobel Prize for Physiology and Medicine in 1997 for his research demonstrating that indeed such proteins exist and do cause disease in all of these ways.

Prusiner named these proteins *prions* (pronounced **pree**-ons), for "proteinaceous infectious particles." Dr. Prusiner began his prion research in 1974. Some are still skeptical and there is much to be learned, but accumulating evidence indicates that all of these unbelievable premises are correct.

Prion-caused disease leaped into worldwide notice in the 1990s when individuals in Great Britain began to die from Creutzfeldt–Jakob disease (CJD) and it became apparent that the cause was eating beef from cows infected with mad-cow disease, known technically as bovine spongiform encephalopathy (BSE). The BSE name summarizes a major symptom of these and other prion diseases in which open spaces develop in the brain tissue, thereby becoming sponge-like. Other diseases in the spongiform encephalopathy family include scrapie in sheep; a chronic wasting disease in elk and mule deer; and in humans, inherited CJD and kuru, which occurred among natives in New Guinea who honored their dead relatives by eating their brains but disappeared when the cannibalism ceased. There is no therapy for the spongiform encephalopathies. All types are fatal and are characterized by loss of muscular control and symptoms of dementia.

The following statements summarize some well-supported facts known about prion proteins:

- Humans and all animals tested thus far have a gene for making a normal prion protein that resides in the brain and does not cause disease.

- An inherited genetic defect can cause one amino acid (proline) in normal prion protein to be replaced by another (leucine), resulting in a prion that is responsible for one of the inherited human prion diseases.

- The difference between normal and disease-causing prions lies in their secondary structure. Alpha helixes in the normal prion are replaced by β-sheets, resulting in a prion with a different shape.

- A misfolded, disease-causing prion can induce a normal prion to flip from the normal shape to the disease-causing shape. It does not matter whether the disease-causing prion arises from a genetic defect, enters the body in food or in some other manner, or is formed randomly and spontaneously. Exactly how the change in shape occurs is not yet understood, but the result is an accelerating spread of the disease-causing form.

- The infectious nature of prions is not affected by either heat or UV radiation treatment, both of which destroy bacteria and viruses. Therefore, cooked beef from infected cows can contain infectious prions.

- Synthetic prions (made using recombinant DNA technology) cause neurological disease in mice similar to mad cow disease or CJD.

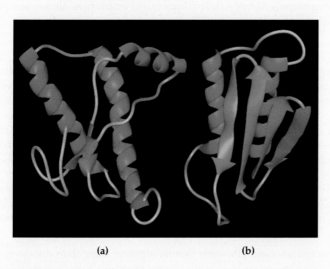

(a) (b)

▲ (a) A normal prion. (b) Proposed conformation for a disease-causing prion. (To interpret these drawings, see the picture of ribonuclease, p. 577.)

Many unanswered questions lie behind the ongoing efforts to understand prions and their disease-causing potential. Are some individuals more susceptible to prion-caused diseases than others? To what extent can the disease move from one species to another? Can a diagnostic test be developed to identify those who are susceptible or already exposed to abnormal prions? Looming over all is a most intriguing, major question: Could it be that other neurodegenerative diseases that create abnormal structures in the brain, including Alzheimer's disease and Parkinson's disease, are also prion diseases? Some of these questions may be answered soon because Dr. Prusiner and his colleagues reported the synthesis of infectious prions in bacteria in 2004. This ground-breaking work is leading to new approaches in studying diseases associated with amyloid plaque formation and/or brain structure degeneration.

Meanwhile, the epidemic of mad-cow disease in Great Britain was devastating. Millions of cattle with BSE were slaughtered, and more than 120 people in Great Britain have died from CJD since 1995.

An initial ban on the export of beef from Great Britain to other European countries was lifted in 1998, but their beef exports remain just a fraction of what they were and bans on import continue in some countries. BSE has, however, been identified in cattle in several other European countries and fears are growing that it can be transmitted by sheep and that it might spread outside of Europe.

In November, 2001, the U.S. Department of Agriculture released the results of a major scientific study that assessed the risk of mad cow disease in the United States. The principal conclusions of the study were that early actions prevented the disease from entering the United States, but that vigilance and testing must continue. Unfortunately, in December, 2003, a dairy cow in Washington State tested positive for mad cow disease after slaughter, resulting in temporary bans (since removed) on the importation of U.S. beef by several countries. This particular cow was born in Canada, and no other positive cases have been found in the United States. It is unlikely that humans will develop CJD from eating beef in the United States because all nerve and brain tissue as well as intestinal tissue are excluded from the food chain—both human and animal.

See Additional Problems 18.81 and 18.82 at the end of the chapter.

SUMMARY: REVISITING THE CHAPTER GOALS

1. **What are the structural features of amino acids?** *Amino acids* in body fluids have an ionized carboxylic acid group ($-COO^-$), an ionized amino group ($-NH_3^+$), and a side-chain R group bonded to a central carbon atom (the α carbon). Twenty different amino acids occur in *proteins* (Table 18.3), connected by *peptide bonds* (amide bonds) formed between the carboxyl group of one amino acid and the amino group of the next.

KEY WORDS

Achiral, *p. 562*

Alpha- (α-) amino acid, *p. 555*

Alpha- (α-) helix, *p. 574*

2. **What are the properties of amino acids?** Amino acid side chains have acidic or basic functional groups or neutral groups that are either polar or nonpolar. In glycine, the "side chain" is a hydrogen atom. The dipolar ion in which the amino and carboxylic acid groups are both ionized is known as a *zwitterion*. For each amino acid, there is a distinctive *isoelectric point*—the pH at which the numbers of positive and negative charges in a solution are equal. At more acidic pH, some carboxylic acid groups are not ionized; at more basic pH, some amino groups are not ionized.

3. **Why do amino acids have "handedness"?** An object, including a molecule, has "handedness"—is *chiral*—when it has no plane of symmetry and thus has mirror images that cannot be superimposed on each other. A simple molecule can be identified as chiral if it contains a carbon atom bonded to four different groups. All α-amino acids except glycine meet this condition by having four different groups bonded to the α carbon.

4. **What is the primary structure of a protein and what conventions are used for drawing and naming primary structures?** Proteins are polymers of amino acids (*polypeptides*). Their *primary structure* is the linear sequence in which the amino acids are connected by peptide bonds. Using formulas or amino acid abbreviations, the primary structures are written with the amino-terminal end on the left (^+H_3N—) and the carboxyl-terminal end on the right (—COO^-). To name a peptide, the names of the amino acids are combined, starting at the amino-terminal end, with the endings of all but the carboxyl-terminal amino acid changed to -*yl*. Primary structures are often represented by combining three-letter abbreviations for the amino acids.

5. **What types of interactions determine the overall shapes of proteins?** Protein chains are drawn into their distinctive and biochemically active shapes by attractions between atoms along their backbones and between atoms in side-chain groups. Hydrogen bonding can occur between the backbone carbonyl groups and amide hydrogens of adjacent protein chains. *Noncovalent interactions* between side chains include ionic bonding between acidic and basic groups (*salt bridges*), and *hydrophobic interactions* among nonpolar groups. Covalent sulfur–sulfur bonds (*disulfide bonds*) can form bridges between the side chains in cysteine.

6. **What are the secondary and tertiary structures of proteins?** *Secondary structures* include the regular, repeating three-dimensional structures held in place by hydrogen bonding between backbone atoms within a chain or in adjacent chains. The *α-helix* is a coil with hydrogen bonding between carbonyl oxygens and amide hydrogens four amino acid residues farther along the same chain. The *β-sheet* is a pleated sheet with adjacent protein-chain segments connected by hydrogen bonding between peptide groups. The adjacent chains in the β-sheet may be parts of the same protein chain or different protein chains. Secondary structure mainly determines the properties of *fibrous proteins*, which are tough and insoluble. *Tertiary structure* is the overall three-dimensional shape of a folded protein chain. Tertiary structure determines the properties of *globular proteins*, which are water-soluble, with hydrophilic groups on the outside and hydrophobic groups on the inside. Globular proteins often contain regions of α-helix and/or β-sheet secondary structures.

7. **What is quaternary protein structure?** Proteins that incorporate more than one peptide chain are said to have *quaternary structure*. In a quaternary structure, two or more folded protein subunits are united in a single structure by noncovalent interactions. Hemoglobin, for example, consists of two pairs of subunits, with a nonprotein heme molecule in each of the four subunits. Collagen is a fibrous protein composed of protein chains twisted together in triple helixes.

8. **What chemical properties do proteins have?** The peptide bonds are broken by *hydrolysis*, which may occur in acidic solution or during enzyme-catalyzed digestion of proteins in food. The end result of hydrolysis is production of the individual amino acids from the protein. *Denaturation* is the loss of overall structure by a protein while retaining its primary structure. Among the agents that cause denaturation are heat, mechanical agitation, pH change, and exposure to a variety of chemical agents, including detergents.

UNDERSTANDING KEY CONCEPTS

18.25 Draw the structure of the following tripeptides at low pH and high pH. At each pH, assume that all functional groups that might do so are ionized.

(a) Val-Gly-Leu
(b) Arg-Lys-His
(c) Tyr-Pro-Ser
(d) Glu-Asp-Phe
(e) Gln-Ala-Asn
(f) Met-Trp-Cys

18.26 Interactions of amino acids on the interior of proteins are key to the shapes of proteins. In group (a) below, which pairs of amino acids form hydrophobic interactions? In group (b), which pairs form ionic interactions? Which pairs in group (c) form hydrogen bonds?

(a) 1. Pro...Phe
  2. Lys...Ser
  3. Thr...Leu
  4. Ala...Gly
(b) 1. Val...Leu
  2. Glu...Lys
  3. Met...Cys
  4. Asp...His
(c) 1. Cys...Cys
  2. Asp...Ser
  3. Val...Gly
  4. Met...Cys

18.27 Draw the hexapeptide Asp-Gly-Phe-Leu-Glu-Ala in linear form showing all of the atoms, and show (using dotted lines) the hydrogen bonding that stabilizes this structure if it is part of an α-helix.

18.28 Compare and contrast the characteristics of fibrous and globular proteins. Consider biological function, water solubility, amino acid composition, secondary structure, and tertiary structure. Give examples of three fibrous and three globular proteins. (Hint: Make a table.)

18.29 Cell membranes are studded with proteins. Some of these proteins, involved in the transport of molecules across the membrane into the cell, span the entire membrane and are called trans-membrane proteins. The interior of the cell membrane is hydrophobic and nonpolar, whereas both the extracellular and intracellular fluid are water-based.

(a) List three amino acids you would expect to find in the trans-membrane protein in the part that lies within the cell membrane.
(b) List three amino acids you would expect to find in the trans-membrane protein in the part that lies outside the cell.
(c) List three amino acids you would expect to find in the trans-membrane protein in the part that lies inside the cell.

ADDITIONAL PROBLEMS

AMINO ACIDS

18.30 The amino acids in most biological systems are said to be α-L-acids. What does the prefix "α" mean?

18.31 What does the prefix "L" in α-L-acid mean?

18.32 What amino acids do the following abbreviations stand for? Draw the structure of each.

(a) Ala
(b) Cys
(c) Asp

18.33 What amino acids do the following abbreviations stand for? Draw the structure of each.

(a) Leu
(b) Tyr
(c) Asn

18.34 Name and draw the structures of the amino acids that fit these descriptions:

(a) Contains a thiol group
(b) Contains a phenol group

18.35 Name and draw the structures of the amino acids that fit these descriptions:

(a) Contains an isopropyl group
(b) Contains a secondary alcohol group

18.36 At neutral pH, which of the following amino acids has a net positive charge, which has a net negative charge, and which is neutral?

(a) Glutamine
(b) Histidine
(c) Methionine

18.37 At neutral pH, which of the following amino acids has a net positive charge, which has a net negative charge, and which is neutral?

(a) Glutamic acid
(b) Arginine
(c) Leucine

18.38 Which of the following forms of aspartic acid would you expect to predominate at low pH, neutral pH, and high pH?

(a) $HO\overset{O}{\overset{\|}{C}}-CH_2CH-\overset{O}{\overset{\|}{C}}O^-$
    $\quad\quad\quad\quad\overset{|}{{}^+NH_3}$

(b) $^-O\overset{O}{\overset{\|}{C}}-CH_2CH-\overset{O}{\overset{\|}{C}}O^-$
    $\quad\quad\quad\quad\overset{|}{NH_2}$

(c) $HO\overset{O}{\overset{\|}{C}}-CH_2CH-\overset{O}{\overset{\|}{C}}OH$
    $\quad\quad\quad\quad\overset{|}{{}^+NH_3}$

18.39 Which of the following forms of lysine would you expect to predominate at low pH, neutral pH, and high pH?

(a) $^+NH_3-\overset{H}{\underset{|}{\overset{|}{C}}}-\overset{O}{\overset{\|}{C}}-O^-$
    $\quad\quad\quad\overset{|}{(CH_2)_4}$
    $\quad\quad\quad\overset{|}{NH_3{}^+}$

(b)

$$NH_3^+-\overset{\displaystyle \overset{H}{|}}{C}-\overset{\displaystyle \overset{O}{\parallel}}{C}-OH$$
$$\underset{\displaystyle \underset{NH_3^+}{|}}{\overset{\displaystyle \overset{|}{(CH_2)_4}}{}}$$

(c)

$$NH_3^+-\overset{\displaystyle \overset{H}{|}}{C}-\overset{\displaystyle \overset{O}{\parallel}}{C}-O^-$$
$$\underset{\displaystyle \underset{NH_2}{|}}{\overset{\displaystyle \overset{|}{(CH_2)_4}}{}}$$

HANDEDNESS IN MOLECULES

18.40 What does the term *chiral* mean? Give two examples.

18.41 What does the term *achiral* mean? Give two examples.

18.42 Which of the following objects are chiral?

(a) A mayonnaise jar
(b) A rocking chair
(c) A coin

18.43 Which of the following objects are achiral?

(a) A pair of scissors
(b) A comb
(c) A vase

18.44 Draw the structures of the following compounds. Which of them is chiral? Mark each chiral carbon with an asterisk.

(a) 2-Bromo-2-chloropropane
(b) 2-Bromo-2-chlorobutane
(c) 2-Bromo-2-chloro-3-methylbutane

18.45 Draw the structures of the following compounds. Which of them is chiral? Mark each chiral carbon with an asterisk.

(a) 2-Chloropentane
(b) Cyclopentane
(c) 2-Methylpropanol

18.46 Which of the carbon atoms marked with arrows in the following compound are chiral?

$$CH_3CHCH_2CH_3$$
$$\underset{F}{|}$$

18.47 Which of the carbon atoms marked with arrows in the following compound are chiral?

PEPTIDES AND PROTEINS

18.48 What is the difference between a simple protein and a conjugated protein?

18.49 What kinds of molecules are found in the following classes of conjugated proteins in addition to the protein part?

(a) Metalloproteins
(b) Hemoproteins
(c) Lipoproteins
(d) Nucleoproteins

18.50 Name four biological functions of proteins in the human body, and give an example of a protein for each function.

18.51 What is meant by the following terms as they apply to proteins, and what primary interactions stabilize the structure?

(a) Primary structure
(b) Secondary structure
(c) Tertiary structure
(d) Quaternary structure

18.52 Why is cysteine such an important amino acid for defining the tertiary structure of some proteins?

18.53 What conditions are required for disulfide bonds to form between cysteine residues in a protein?

18.54 How do the following noncovalent interactions help to stabilize the tertiary and quaternary structure of a protein? Give an example of a pair of amino acids that could give rise to each interaction.

(a) Hydrophobic interactions
(b) Salt bridges (ionic interactions)

18.55 How do the following interactions help to stabilize the tertiary and quaternary structure of a protein? Give an example of a pair of amino acids that could give rise to each interaction.

(a) Side-chain hydrogen bonding
(b) Disulfide bonds

18.56 What kinds of changes take place in a protein when it is denatured?

18.57 Explain how a protein is denatured by the following:

(a) Heat
(b) Strong acids
(c) Organic solvents

18.58 Use the three-letter abbreviations to name all tripeptides that contain valine, methionine, and leucine.

18.59 Write structural formulas for the two dipeptides that contain leucine and aspartate.

18.60 Which of the following amino acids are most likely to be found on the outside of a globular protein, and which of them are more likely to be found on the inside? Explain each answer. (Hint: Consider the effect of the amino acid side chain in each case.)

(a) Valine
(b) Aspartate
(c) Histidine
(d) Alanine

18.61 Which of the following amino acids are most likely to be found on the outside of a globular protein? Which are more likely to be found on the inside? Explain each answer. (Hint: Consider the effect of the amino acid side chain in each case.)

(a) Leucine   (b) Glutamate
(c) Phenylalanine   (d) Glutamine

18.62 Why do you suppose diabetics must receive insulin subcutaneously by injection rather than orally?

18.63 Individuals with phenylketonuria (PKU) are sensitive to phenylalanine in their diet. Why is a warning on foods containing aspartame (L-aspartyL-L-phenylalanine methyl ester) of concern to PKU individuals?

18.64 The *endorphins* are a group of naturally occurring neurotransmitters that act in a manner similar to morphine to control pain. Research has shown that the biologically active parts of the endorphin molecules are simple pentapeptides called *enkephalins*. Draw the structure of the methionine enkephalin with the sequence Tyr-Gly-Gly-Phe-Met. Identify the N-terminal and C-terminal amino acids.

18.65 Refer to Problem 18.64. Draw the structure of the leucine enkephalin with the sequence Tyr-Gly-Gly-Phe-Leu. Identify the N-terminal and C-terminal amino acids.

PROPERTIES AND REACTIONS OF AMINO ACIDS AND PROTEINS

18.66 Much of the chemistry of amino acids is the familiar chemistry of carboxylic acids and amine functional groups. What products would you expect to obtain from the following reactions of glycine?

(a) $H_3\overset{+}{N}-CH_2-\overset{O}{\overset{||}{C}}O^- + HCl \longrightarrow$?

(b) $H_3\overset{+}{N}-CH_2-\overset{O}{\overset{||}{C}}OH + CH_3OH \xrightarrow{H^+ \text{ catalyst}}$?

18.67 A scientist tried to prepare the simple dipeptide glycylglycine by the following reaction:

$$2\ H_3\overset{+}{N}CH_2\overset{O}{\overset{||}{C}}O^- \longrightarrow H_3\overset{+}{N}CH_2\overset{O}{\overset{||}{C}}NHCH_2\overset{O}{\overset{||}{C}}O^-$$

An unexpected product formed during the reaction. This product is found to have the molecular formula $C_4H_6N_2O_2$ and to contain two peptide bonds. What happened?

18.68 (a) Identify the amino acids present in the peptide shown below.
(b) Identify the N-terminal and C-terminal amino acids of the peptide.
(c) Show the structures of the products that are obtained on digestion of the peptide at physiological pH.

$$H_3\overset{+}{N}-\underset{\underset{CH_3CHCH_3}{|}}{CH}-\overset{O}{\overset{||}{C}}-\underset{H}{N}-\underset{\underset{H}{|}}{CH}-\overset{O}{\overset{||}{C}}-\underset{H}{N}-\underset{\underset{CH_2OH}{|}}{CH}-\overset{O}{\overset{||}{C}}-\underset{H}{N}-\underset{\underset{CH_3}{|}}{CH}-\overset{O}{\overset{||}{C}}-\underset{H}{N}-\underset{\underset{CH_2COO^-}{|}}{CH}CO^-$$

18.69 (a) Identify the amino acids present in the peptide shown below.
(b) Identify the N-terminal and C-terminal amino acids of the peptide.
(c) Show the structures of the products that are obtained on digestion of the peptide at physiological pH.

$$H_3\overset{+}{N}-\underset{\underset{SH}{\overset{|}{\underset{CH_2}{|}}}}{CH}-\overset{O}{\overset{||}{C}}-\underset{H}{N}-\underset{\underset{NH_3}{\overset{+}{\underset{(CH_2)_4}{|}}}}{CH}-\overset{O}{\overset{||}{C}}-\underset{H}{N}-\underset{\underset{\underset{O}{\overset{||}{C}}-O^-}{\overset{|}{\underset{(CH_2)_2}{|}}}}{CH}-\overset{O}{\overset{||}{C}}-N-\underset{\underset{CH_2}{|}\ \underset{CH_2}{|}}{CH}-COO^-$$

18.70 Which would you expect to be more soluble in water, a peptide rich in alanine and leucine, or a peptide rich in lysine and aspartate? Explain.

18.71 Proteins are usually less soluble in water at their isoelectric points. Explain.

Applications

18.72 (a) Tyramine is said to be the "decarboxylation" product of the amino acid tyrosine. Write the reaction and explain what is meant by "decarboxylation."
(b) Phenelzine inhibits the "deamination" of tyramine. What key characteristic of phenelzine makes it similar enough to tyramine that it is able to block deamination of tyramine? [*Nutrition in Health and Disease, p. 560*]

18.73 Why is it more important to have a daily source of protein than a daily source of fat or carbohydrates? [*Proteins in the Diet, p. 569*]

18.74 What is an incomplete protein? [*Proteins in the Diet, p. 569*]

18.75 In general, which is more likely to contain a complete (balanced) protein for human use—food from plant sources or food from animal sources? Explain. [*Proteins in the Diet, p. 569*]

18.76 Two of the most complete (balanced) proteins (that is, proteins that have the best ratio of the amino acids for humans) are cow's milk protein (casein) and egg white protein. Explain why (not surprisingly) these are very balanced proteins for human growth and development. [*Proteins in the Diet, p. 569*]

18.77 The proteins collagen, bovine insulin, and human hemoglobin have isoelectric points of 6.6, 5.4, and 7.1, respectively. Suppose a sample containing these proteins is applied to an electrophoresis strip in a buffer at pH 6.6. Describe the motion of each with respect to the positive and negative electrodes in the electrophoresis apparatus. [*Protein Analysis by Electrophoresis, p. 573*]

18.78 Three dipeptides are separated by electrophoresis at pH 5.8. If the dipeptides are Arg-Trp, Asp-Thr, and Val-Met, describe the motion of each with respect to the positive and negative electrodes in the electrophoresis apparatus. [*Protein Analysis by Electrophoresis, p. 573*]

18.79 In the middle ages both citizens of besieged cities and members of the armies laying siege would die of scurvy, a non-contagious disease, during long sieges. Explain why this would occur. [*Collagen—A Tale of Two Diseases, p. 581*]

18.80 Describe the cause and biochemical defect that results in osteogenesis imperfecta. [*Collagen—A Tale of Two Diseases, p. 581*]

18.81 The change from a normal to a disease-causing prion results in a change from α-helices to β-sheets. How might this change alter the overall structure and intermolecular forces in the prion? [*Prions: Proteins that Cause Disease, p. 584*]

18.82 List the properties of disease-causing prions that made their existence difficult to accept. [*Prions: Proteins that Cause Disease, p. 584*]

18.83 A family visits their physician with their sick child. The four-month-old baby is pale, has obvious episodes of pain, and is not thriving. The doctor orders a series of blood tests, including a test for hemoglobin types. The results show that the infant is not only anemic but that the anemia is due to sickle-cell anemia. The family wants to know if their other two children have sickle-cell anemia, sickle-cell trait, or no sickle-cell gene at all.

   (a) What test will be used?

   (b) Sketch the expected results if samples for each child are tested at the same time.

   (c) What is the difference between sickle-cell anemia and sickle-cell trait?

   [*Protein Analysis by Electrophoresis, p. 573*]

18.84 What could you prepare for dinner for a strict vegan that provides all of the essential amino acids in appropriate amounts? (Remember, strict vegans do not eat meat, eggs, milk, or products that contain those animal products.) [*Proteins in the Diet, p. 569*]

General Questions and Problems

18.85 What is the difference between protein digestion and protein denaturation? Both occur after a meal.

18.86 Fresh pineapple cannot be used in gelatin desserts because it contains an enzyme that hydrolyzes the proteins in gelatin, destroying the gelling action. Canned pineapple can be added to gelatin with no problem. Why?

18.87 Both α-keratin and tropocollagen have helical secondary structure. How do they differ?

18.88 Bradykinin, a peptide that helps to regulate blood pressure, has the primary structure Arg-Pro-Pro-Gly-Phe-Ser-Pro-Phe-Arg.

   (a) Draw the complete structural formula of bradykinin.

   (b) Bradykinin has a very kinked secondary structure. Why?

18.89 For each amino acid listed, tell whether its influence on tertiary structure is largely through hydrophobic interactions, hydrogen bonding, formation of salt bridges, covalent bonding, or some combination of these effects.

   (a) Tyrosine

   (b) Cysteine

   (c) Asparagine

   (d) Lysine

   (e) Tryptophan

   (f) Alanine

   (g) Leucine

   (h) Methionine

18.90 When subjected to oxidation, the chiral carbon in 2-pentanol becomes achiral. Why does this happen?

18.91 Why is hydrolysis of a protein not considered to be denaturation?

18.92 Oxytocin is a small peptide that is used to induce labor by causing contractions in uterine walls. It has the primary structure Cys-Tyr-Ile-Gln-Asn-Cys-Pro-Leu-Gln. This peptide is held in a cyclic configuration by a disulfide bridge. Draw a diagram of oxytocin, showing the disulfide bridge.

18.93 Methionine has a sulfur atom in its formula. Explain why methionine does not form disulfide bridges.

18.94 List the amino acids that are capable of hydrogen bonding if included in a peptide chain. Draw an example of two of these amino acids hydrogen-bonding to one another. For each one, draw a hydrogen bond to water in a separate sketch. Refer to Section 8.11 for help with drawing hydrogen bonds.

18.95 Four of the most abundant amino acids in proteins are leucine, alanine, glycine, and valine. What do these amino acids have in common? Would you expect these amino acids to be found on the interior or on the exterior of the protein?

18.96 Globular proteins are water-soluble, whereas fibrous proteins are insoluble in water. Indicate whether you expect the following amino acids to be on the surface of a globular protein or on the surface of a fibrous protein.

   (a) Ala

   (b) Glu

   (c) Leu

   (d) Phe

   (e) Ser

   (f) Val

18.97 In Figure 18.7, notice the small purple segment in each rendering of the molecule. These purple segments connecting adjacent regions of secondary structure are often referred to either as "reverse turns" or as "bends." The two most common amino acids in reverse turns are glycine and proline. Use your knowledge of the structures of these two amino acids to speculate on why they might be found in reverse turns.

18.98 During sickle-cell anemia research to determine the modification involved in sickling, sequencing of the affected person's hemoglobin β-subunit reveals that the sixth amino acid is valine rather than glutamate; thus, the replacement of glutamate by valine severely alters the three-dimensional structure of hemoglobin. Which amino acid, if it replaced the Glu, would cause the least disruption in hemoglobin structure? Why?

18.99 As a chef, you prepare a wide variety of foods daily. The following dishes all contain protein. What method (if any) has been used to denature the protein present in each food?

(a) Charcoal grilled steak
(b) Pickled pigs' feet
(c) Meringue
(d) Steak tartare (raw chopped beef)
(e) Salt pork

CHAPTER 19

Enzymes and Vitamins

CONCEPTS TO REVIEW

Coordinate covalent bonds
(Section 5.4)

Reaction rates
(Section 7.5)

pH
(Section 10.9)

**Effects of conditions
on reaction rates**
(Section 7.6)

Tertiary protein structure
(Section 18.10)

▲ Luciferinase, an enzyme found in firefly tails, yields light as a reaction product.

CONTENTS

CHAPTER GOALS

Here, we will address the following questions:

1. What are enzymes?

THE GOAL: Be able to describe the chemical nature of enzymes and their function in biochemical reactions.

2. How do enzymes work, and why are they so specific?

THE GOAL: Be able to provide an overview of what happens as one or more substrates and an enzyme come together so that the catalyzed reaction can occur, and be able to list the properties of enzymes that make their specificity possible.

3. What effects do temperature, pH, enzyme concentration, and substrate concentration have on enzyme activity?

THE GOAL: Be able to describe the changes in enzyme activity that result when temperature, pH, enzyme concentration, or substrate concentration change.

4. How is enzyme activity regulated?

THE GOAL: Be able to define and identify feedback control, allosteric control, reversible and irreversible inhibition, inhibition by covalent modification, and genetic control of enzymes.

5. What are vitamins?

THE GOAL: Be able to describe the two major classes of vitamins, the reasons vitamins are necessary in our diets, and the general results of excesses or deficiencies.

Think of your body as a living chemical laboratory. Although the analogy is not perfect, there is a good deal of truth to it. In your body, as in the laboratory, chemical reactions are the major activity. In a laboratory, however, chemical reactions are carried out singly by mixing pure chemicals in individual containers. In your body, many thousands of reactions take place simultaneously in the same cells or body fluids.

Another difference between chemistry in a laboratory and chemistry in a living organism is control. In a laboratory, the speed of a reaction is controlled by adjusting experimental conditions such as temperature, solvent, and pH. In an organism, these conditions cannot be adjusted. The human body must maintain a temperature of 37 °C, the solvent must be water, and the pH must be close to 7.4 in most body fluids.

Animals and plants are composed of numerous cells organized into different functional types. Among the many thousands of protein molecules in each cell there are more than 2000 different specialized proteins, called enzymes, each one used in a different reaction. Although enzymes—powerful and highly selective biological catalysts—carry out the chemical reactions in cells, how do cells organize so many different reactions so that all occur to the proper extent? The answer is that all enzyme reactions in living organisms are under tight regulation by a variety of mechanisms.

In this chapter, the focus will be on enzymes and the regulation of enzymatic reactions. We will also look at *vitamins*, because they are essential to the function of certain enzymes. Chapter 20 is devoted to the role of *hormones* and *neurotransmitters* in keeping our biochemistry under control.

19.1 Catalysis by Enzymes

Catalysts accelerate the rates of chemical reactions but at the end of the reaction have undergone no change. **Enzymes**, the catalysts for biochemical reactions, fit this definition. Like all catalysts, an enzyme does not affect the equilibrium point of a reaction and cannot bring about a reaction that is energetically unfavorable. What an enzyme does is decrease the time it takes to reach equilibrium by lowering the activation energy. (⊂⊃, p. 194)

Enzymes, with few exceptions, are water-soluble globular proteins. (⊂⊃, p. 576) As proteins, they are far larger and more complex molecules than simple inorganic catalysts. Because of their size and complexity, enzymes have more ways

Enzyme A protein or other molecule that acts as a catalyst for a biological reaction.

Active site A pocket in an enzyme with the specific shape and chemical makeup necessary to bind a substrate.

Substrate A reactant in an enzyme-catalyzed reaction.

Specificity (enzyme) The limitation of the activity of an enzyme to a specific substrate, specific reaction, or specific type of reaction.

available in which to connect with reactants, speed up reactions, and be controlled by other molecules.

Within the folds of an enzyme's protein chain is the **active site**—the region where the reaction takes place. The active site has the specific shape and chemical reactivity needed to catalyze the reaction. One or more **substrates**, the reactants in an enzyme-catalyzed reaction, are held in place by attractions to groups that line the active site.

The extent to which an enzyme's activity is limited to a certain substrate and a certain type of reaction is referred to as the **specificity** of the enzyme. Enzymes differ greatly in their specificity. *Catalase*, for example, is almost completely specific for one reaction—the decomposition of hydrogen peroxide (Figure 19.1), a reaction needed to destroy the peroxide before it damages essential biomolecules by oxidizing them:

$$2 \, H_2O_2 \xrightleftharpoons[]{\text{Catalase}} 2 \, H_2O + O_2 \, (g)$$

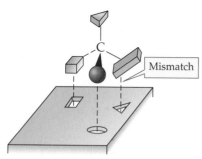

▶ **FIGURE 19.1 Catalase in action.** As ground beef liver is added to hydrogen peroxide solution in the beaker, catalase in the liver catalyzes rapid decomposition of the hydrogen peroxide, as shown by the formation of oxygen-filled bubbles.

Thrombin is specific for catalyzing hydrolysis of a peptide bond following an arginine and does so primarily in a protein essential to blood clotting. When this bond breaks, the product (fibrin) proceeds to polymerize into a blood clot (Section 29.5). *Carboxypeptidase A* is less limited—it removes many different C-terminal amino acid residues from protein chains during digestion. And the enzyme *papain* from papaya fruit catalyzes the hydrolysis of peptide bonds in many locations. (⊂⊃, p. 583) It is this ability to break down proteins that accounts for the use of papain in meat tenderizers, in contact-lens cleaners, and in cleansing dead or infected tissue from wounds (*debridement*).

Since the amino acids in enzymes are all L-amino acids, it should come as no surprise that enzymes are also specific with respect to stereochemistry. If a substrate is chiral, an enzyme usually catalyzes the reaction of only one of the pair of enantiomers because only one fits the active site in such a way that the reaction can occur. The enzyme lactate dehydrogenase, for example, catalyzes the removal of hydrogen from L-lactate but not from D-lactate:

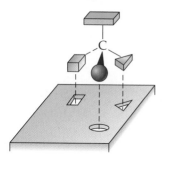

▲ **FIGURE 19.2 A chiral reactant and a chiral reaction site.** The enantiomer at the top fits the reaction site like a hand in a glove, but the enantiomer at the bottom does not fit and therefore cannot be a substrate for this enzyme.

This is another example of the importance of molecular shape in biochemistry. The specificity of an enzyme for one of two enantiomers is a matter of fit. A left-handed enzyme cannot fit with a right-handed substrate any more than a left-handed glove fits on a right hand (Figure 19.2).

The catalytic activity of an enzyme is measured by its **turnover number**, the maximum number of substrate molecules acted upon by one molecule of enzyme per unit time (Table 19.1). Most enzymes turn over 10–1000 molecules per second, but some are much faster. Catalase, with its essential role in protecting against molecular damage, is one of the fastest–it can turn over 10 million molecules per second. This is the fastest reaction rate attainable in the body because it is the rate at which molecules collide.

Turnover number The maximum number of substrate molecules acted upon by one molecule of enzyme per unit time.

TABLE 19.1   Turnover Numbers for Some Enzymes

ENZYME	REACTION CATALYZED	TURNOVER NUMBER (MAXIMUM NUMBER OF CATALYTIC EVENTS PER SECOND)
Papain	Hydrolysis of peptide bonds	10
Ribonuclease	Hydrolysis of phosphate ester link in RNA	10^2
Kinases	Transfer of phosphoryl group between substrates	10^3
Acetylcholinesterase	Deactivation of the neurotransmitter acetylcholine	10^4
Carbonic anhydrase	Converts CO_2 to HCO_3^-	10^6
Catalase	Decomposition of H_2O_2 to $H_2O + O_2$	10^7

PROBLEM 19.1

Which of the enzymes listed in Table 19.1 catalyzes a maximum of 1000 reactions per second?

PROBLEM 19.2

Why is it essential to rinse contact lenses in saline solution after they have been cleaned by an enzyme solution and before they are placed in the eyes?

19.2 Enzyme Cofactors

Many enzymes are conjugated proteins that require nonprotein portions known as **cofactors** as part of their functional structure. Some cofactors are metal ions, and others are nonprotein organic molecules called **coenzymes**. To be active, an enzyme may require a metal ion, a coenzyme, or both. Some enzyme cofactors are tightly held by noncovalent attractions or are covalently bound to their enzymes; others are more loosely bound so that they can enter and leave the active site as needed.

Why are cofactors necessary? The functional groups in proteins are limited to those of the amino acid side chains. By combining with cofactors, enzymes acquire chemically reactive groups not available in side chains. For example, the NAD^+ reactant in the equation for the dehydrogenation of L-lactate shown on p. 594, is a coenzyme and is the oxidizing agent that makes the reaction possible. (Vitamins that function as cofactors are discussed in Section 19.10.)

The requirement that many enzymes have for metal ion cofactors explains our dietary need for trace minerals. The ions of iron, zinc, copper, manganese, molybdenum, cobalt, nickel, vanadium, and selenium all function as enzyme cofactors. These ions are able to form coordinate covalent bonds by accepting lone-pair electrons present on nitrogen or oxygen atoms in enzymes or substrates. (⟨▭⟩, Section 5.4)

Cofactor A nonprotein part of an enzyme that is essential to the enzyme's catalytic activity; a metal ion or a coenzyme.

Coenzyme An organic molecule that acts as an enzyme cofactor.

This bonding may anchor a substrate in the active site and may also allow the metal ion to participate in the catalyzed reaction. For example, every molecule of the digestive enzyme carboxypeptidase A contains one Zn^{2+} ion that is essential for its catalytic action. We say that the zinc ion is "coordinated" to two nitrogens in histidine side chains and one oxygen in a glutamate side chain. In this way the ion is held in place in the active site of the enzyme.

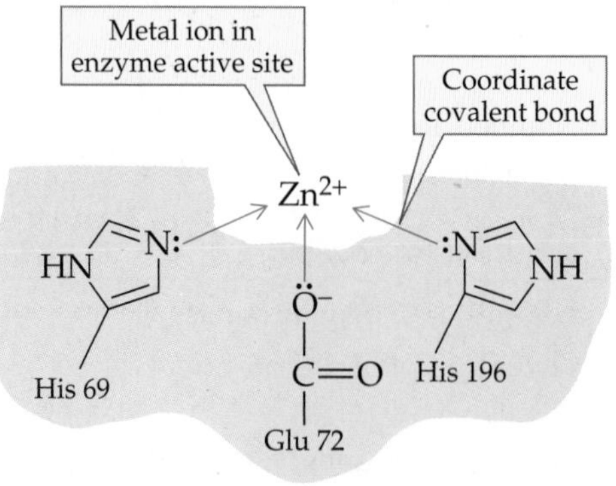

Like the trace minerals, certain vitamins are a dietary necessity for humans because they function as building blocks for coenzymes and we cannot synthesize them.

PROBLEM 19.3

Check the label on a bottle of multivitamin/multimineral tablets and identify the metal ion cofactors listed above that are included in the supplement.

⚷ KEY CONCEPT PROBLEM 19.4

The cofactors NAD^+, Cu^{2+}, Zn^{2+}, coenzyme A, FAD, and Ni^{2+} are all needed by your body for enzymatic reactions.

(a) Which coenzymes have vitamins as part of their structure? (Hint: look ahead in this chapter to the structures of the vitamins.)

(b) What is the primary difference between the coenzymes and the remaining cofactors in the list?

19.3 Enzyme Classification

Enzymes are divided into six main classes according to the general kind of reaction they catalyze, and each main class is further subdivided based on substrate specificity (Table 19.2). Most of the names of the main classes are self-explanatory; they are listed below with examples:

- *Oxidoreductases* catalyze oxidation–reduction reactions of substrate molecules, most commonly addition or removal of oxygen or hydrogen. Because oxidation and reduction must occur together, these enzymes require coenzymes that are reduced or oxidized as the substrate is oxidized or reduced.

$$A(\text{Reduced}) + B(\text{Oxidized}) \longrightarrow A'(\text{Oxidized}) + B'(\text{Reduced})$$

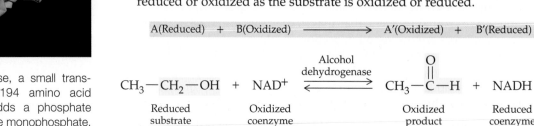

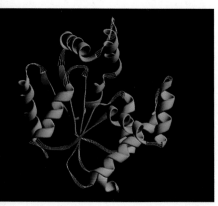

▲ Adenylate kinase, a small transferase enzyme (194 amino acid residues), that adds a phosphate group to adenosine monophosphate.

TABLE 19.2   Classification of Enzymes

MAIN CLASS AND SUBCLASS	EXAMPLES OF REACTION TYPES CATALYZED
Oxidoreductases	**Oxidation–reduction reactions**
Oxidases	Addition of O_2 to a substrate
Reductases	Reduction of a substrate
Dehydrogenases	Removal of 2 Hs to form a double bond
Transferases	**Transfer of functional groups**
Transaminases	Transfer of amino group between substrates
Kinases	Transfer of a phosphoryl group between substrates
Hydrolases	**Hydrolysis reactions**
Lipases	Hydrolysis of ester groups in lipids
Proteases	Hydrolysis of peptide bonds in proteins
Nucleases	Hydrolysis of phosphate ester bonds in nucleic acids
Isomerases	**Isomerization of a substrate**
Lyases	**Group elimination to form double bond or addition to a double bond**
Dehydrases	Removal of H_2O from substrate to give double bond
Decarboxylases	Replacement of a carboxyl group by a hydrogen
Synthases	Addition of small molecule to a double bond
Ligases	**Bond formation coupled with ATP hydrolysis to provide energy**
Synthetases	Formation of bond between two substrates
Carboxylases	Formation of bond between substrate and CO_2 to add a carboxyl group ($-COO^-$)

- **Transferases** catalyze transfer of a group from one molecule to another. Kinases, for example, transfer a phosphate group from adenosine triphosphate (ATP) to give adenosine diphosphate (ADP) and a phosphorylated product.

$$A + B\!-\!C \rightleftharpoons A\!-\!B + C$$

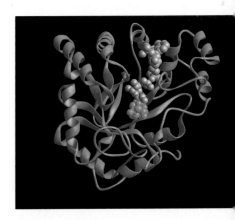

▲ Aldose reductase, an oxidoreductase enzyme that reduces a $C{=}O$ group in a sugar molecule to a $-C-OH$ group with the aid of the coenzyme NADH. The sugar glucose (orange) and the NADH (gray) are shown in the active site of the enzyme.

Fructose 6-phosphate + Adenosine triphosphate ⇌ (Phosphofructokinase) Fructose 1,6-*bis*phosphate + Adenosine diphosphate

- **Hydrolases** catalyze the hydrolysis of substrates—the breaking of bonds with addition of water. The digestion of carbohydrates and proteins by hydrolysis requires these enzymes.

$$A\!-\!B + H_2O \longrightarrow A\!-\!OH + B\!-\!H$$

Polypeptide + H_2O → (A protease) → Shortened polypeptide + Amino acid

- *Isomerases* catalyze the isomerization (rearrangement of atoms) of a substrate in reactions that have but one substrate and one product.

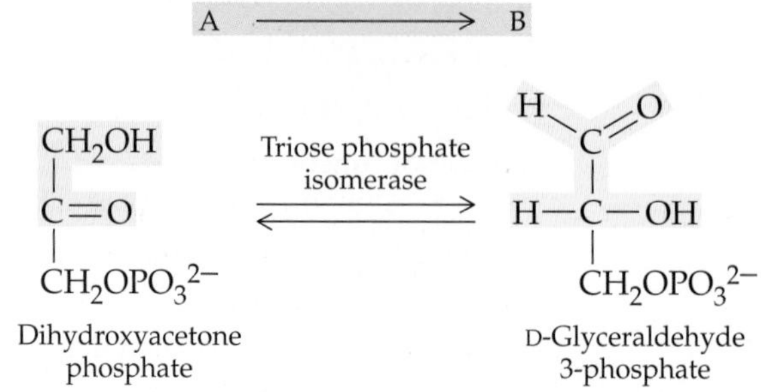

$$A \longrightarrow B$$

$$\begin{array}{c} CH_2OH \\ | \\ C{=}O \\ | \\ CH_2OPO_3{}^{2-} \end{array} \quad \underset{\rightleftharpoons}{\overset{\text{Triose phosphate}}{\underset{\text{isomerase}}{\longrightarrow}}} \quad \begin{array}{c} H{-}C{\overset{O}{\diagup}} \\ | \\ H{-}C{-}OH \\ | \\ CH_2OPO_3{}^{2-} \end{array}$$

Dihydroxyacetone phosphate   D-Glyceraldehyde 3-phosphate

- *Lyases* (from the Greek *lein,* meaning "to break") catalyze the addition of a molecule such as H_2O, CO_2, or NH_3 to a double bond or the reverse reaction in which a molecule is eliminated to leave a double bond.

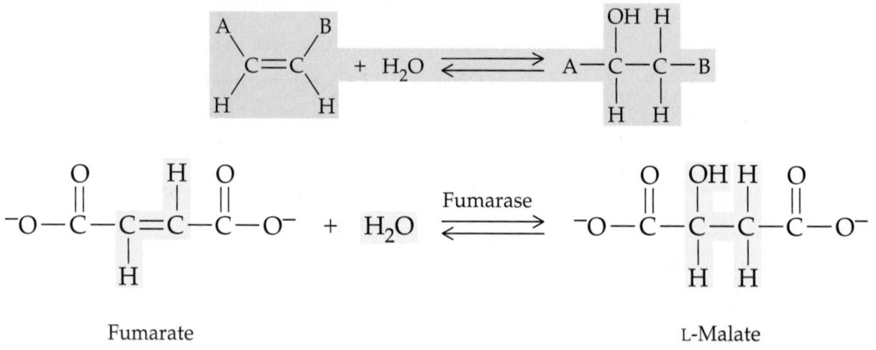

$$\begin{array}{c} A \\ {\diagdown} \\ C{=}C \\ H \end{array} \begin{array}{c} B \\ {\diagup} \\ H \end{array} + H_2O \rightleftharpoons \begin{array}{c} OH\ H \\ | \ \ | \\ A{-}C{-}C{-}B \\ | \ \ | \\ H\ \ H \end{array}$$

$$\begin{array}{c} O \ \ \ H \ O \\ \| \ \ \ | \ \| \\ {}^-O{-}C{-}C{=}C{-}C{-}O^- \\ | \\ H \end{array} + H_2O \underset{\rightleftharpoons}{\overset{\text{Fumarase}}{\longrightarrow}} \begin{array}{c} O \ OH\ H\ O \\ \| \ \ | \ \ | \ \| \\ {}^-O{-}C{-}C{-}C{-}C{-}O^- \\ | \ \ | \\ H\ \ H \end{array}$$

Fumarate   L-Malate

- *Ligases* (from the Latin *ligare,* meaning "to tie together") catalyze the bonding together of two substrate molecules. Because such reactions are generally not favorable, they require the simultaneous release of energy by a hydrolysis reaction, usually by the conversion of ATP to ADP (such energy release is discussed in Section 21.5).

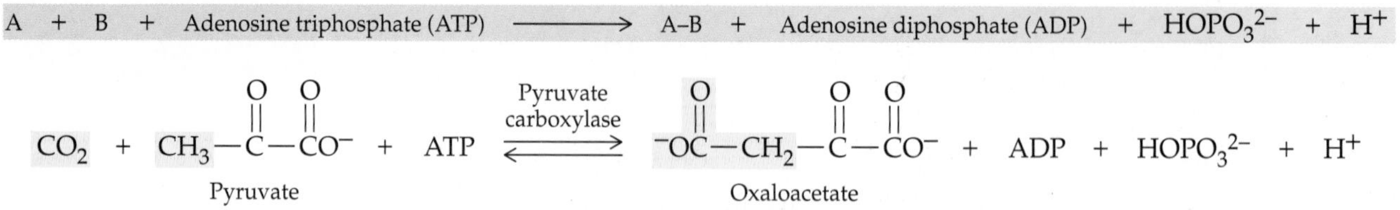

$$A + B + \text{Adenosine triphosphate (ATP)} \longrightarrow A{-}B + \text{Adenosine diphosphate (ADP)} + HOPO_3{}^{2-} + H^+$$

$$CO_2 + \begin{array}{c} O \ \ O \\ \| \ \ \| \\ CH_3{-}C{-}CO^- \end{array} + ATP \underset{\rightleftharpoons}{\overset{\text{Pyruvate}}{\underset{\text{carboxylase}}{\longrightarrow}}} \begin{array}{c} O \ \ \ \ \ O \ \ O \\ \| \ \ \ \ \ \| \ \ \| \\ {}^-OC{-}CH_2{-}C{-}CO^- \end{array} + ADP + HOPO_3{}^{2-} + H^+$$

Pyruvate   Oxaloacetate

Note in the preceding examples that the enzymes have the family-name ending *-ase.* Exceptions to this rule occur for enzymes such as papain and trypsin, which are still referred to by older common names. The more informative modern systematic names always have two parts: The first identifies the substrate on which the enzyme operates, and the second part is an enzyme subclass name like those shown in Table 19.2. For example, *pyruvate carboxylase* is a ligase that acts on the substrate *pyruvate* to add a *carboxyl group.* Note also that some enzymes are capable of catalyzing both forward and reverse reactions, and where both directions are of significance, the equations are often written with both arrows.

WORKED EXAMPLE   **19.1** Classifying Enzymes

To what class does the enzyme that catalyzes the following reaction belong?

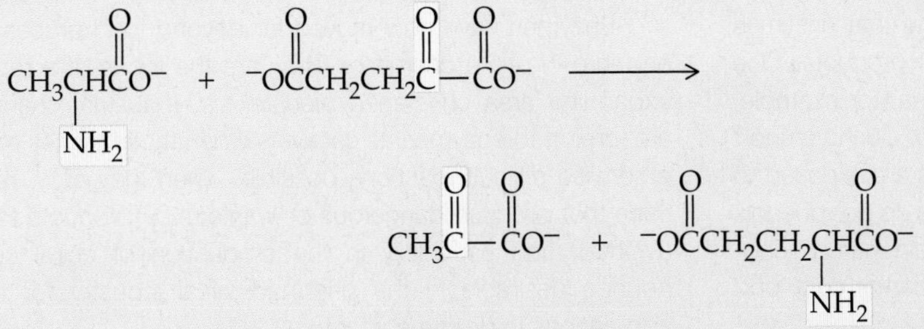

$$CH_3\overset{\overset{\displaystyle O}{\|}}{C}\underset{\underset{\displaystyle NH_2}{|}}{H}CO^- \ + \ {}^-O\overset{\overset{\displaystyle O}{\|}}{C}CH_2CH_2\overset{\overset{\displaystyle O}{\|}}{C}-\overset{\overset{\displaystyle O}{\|}}{C}O^- \longrightarrow$$

$$CH_3\overset{\overset{\displaystyle O}{\|}}{C}-\overset{\overset{\displaystyle O}{\|}}{C}O^- \ + \ {}^-O\overset{\overset{\displaystyle O}{\|}}{C}CH_2CH_2\overset{\overset{\displaystyle O}{\|}}{C}\underset{\underset{\displaystyle NH_2}{|}}{H}CO^-$$

ANALYSIS First, identify the type of reaction that has occurred. An amino group and a carbonyl group have changed places. Then, determine what class of enzyme catalyzes this type of reaction.

SOLUTION
The reaction is a transfer of an amino functional group, meaning that the enzyme is a transferase.

PROBLEM 19.5

Describe the reactions that you would expect these enzymes to catalyze.

(a) Glutamate dehydrogenase   (b) Alanine aminotransferase
(c) Carbamoyl phosphate synthetase I   (d) Triose phosphate isomerase

PROBLEM 19.6

Name the enzyme whose substrate is

(a) Arginine   (b) Maltose

PROBLEM 19.7

To what class of enzymes does glucose 6-phosphate isomerase belong? Describe in general the reaction it catalyzes.

PROBLEM 19.8

Identify and describe the chemical change in the lyase-catalyzed reaction on p. 598. Identify the substrate(s) and product(s).

PROBLEM 19.9

Which of the following reactions can be catalyzed by a decarboxylase?

(a)

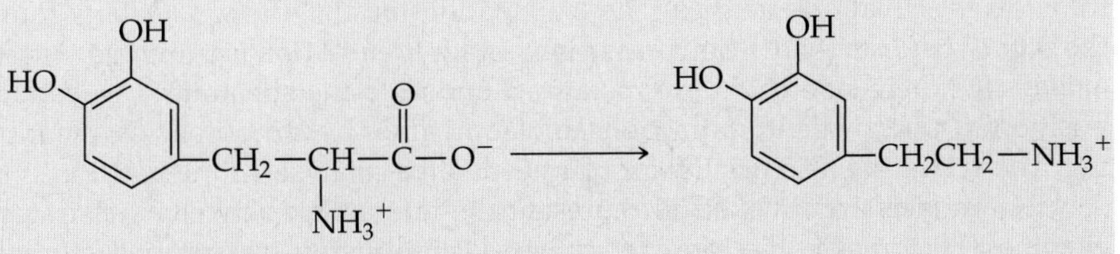

(b)

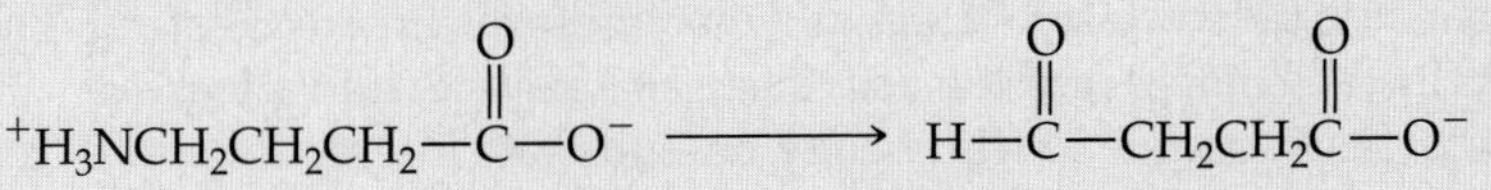

$${}^+H_3NCH_2CH_2CH_2-\overset{\overset{\displaystyle O}{\|}}{C}-O^- \longrightarrow H-\overset{\overset{\displaystyle O}{\|}}{C}-CH_2CH_2\overset{\overset{\displaystyle O}{\|}}{C}-O^-$$

APPLICATION ▶ Biocatalysis: Industrial Biochemistry

We have known for a very long time that enzymes create desirable changes in food products. One step in the preparation of black tea, for example, relies on an enzyme present in the tea leaves. During a short period in which the dried, crushed tea leaves are exposed to the oxygen in air, a polyphenoloxidase breaks up polyphenols into tannins, smaller segments that also contain many phenolic hydroxyl groups. The tannins impart the darker color and characteristic flavors to black tea.

And then there is corn syrup, which is created by the action of enzymes on starch granules from corn. Starch (as you will see in Section 22.9) is initially a polymer of sugars. Digestion of the starch by yeast enzymes yields a mixture of individual sugar molecules that make corn syrup sweet and larger carbohydrate molecules that have no sweetness but thicken the syrup and make it sticky.

Creative application of an enzyme is also responsible for chocolate candies with liquid centers. When these candies are produced, the center is a moist fudge-like mixture of flavoring, sucrose (which is table sugar), and an enzyme made by yeast that breaks sucrose into its two smaller sugar components. Because these sugars are more water-soluble than sucrose, the center gradually liquefies.

Enzymes have now moved far beyond food processing and into big-time chemistry. They are the focus of a rapidly expanding area christened *biocatalysis*. Industrial chemists are turning to enzymes as catalysts when the enzymes make otherwise difficult reactions possible, when they allow reactions that produce dangerous or wasteful by-products to be avoided, and especially in the production of chiral compounds desirable in the pharmaceutical industry for their applications in drug manufacture.

A simple example of biocatalysis is the hydration of acrylonitrile to produce acrylamide, an important industrial chemical used in permanent press fabrics, dyes, adhesives, paper, and many other products:

$$CH_2\!\!=\!\!CHC\!\!\equiv\!\!N \xrightarrow[\text{hydratase}]{\text{Nitrile}} CH_2\!\!=\!\!CH\!\!-\!\!\overset{\displaystyle O}{\overset{\|}{C}}\!\!-\!\!NH_2$$

Acrylonitrile — Acrylamide

Biocatalysis has often been exploited in the synthesis of intermediates—chemicals that then become reactants in the synthesis of other substances. It is used, for example, in the production of *p*-hydroxyphenylglycine, which is needed in the synthesis of the antibiotic amoxicillin.

p-Hydroxyphenylglycine

Amoxicillin

The production of agricultural chemicals and synthetic vitamins are other major areas of chemistry in which biocatalysis is an active area of study. Already on the horizon is biocatalysis by microorganisms that have been genetically altered to produce a desired chemical (Section 27.5).

See Additional Problems 19.74 and 19.75 at the end of the chapter.

19.4 How Enzymes Work

Any theory of how enzymes work must explain why they are so specific and how they lower activation energies. The explanation for enzyme *specificity* is found in the active site. Exactly the right environment for the reaction is provided within the active site. There side-chain groups attract and hold the substrate or substrates in position by noncovalent interactions and sometimes by temporary covalent bonding. The active site also has the groups needed for catalysis of the reaction.

Two models are invoked to represent the interaction between substrates and enzymes. Historically, the *lock-and-key model* came first; it was proposed when the need for a spatial fit between substrates and enzymes was first recognized. The substrate is described as fitting into the active site as a key fits into a lock.

When it became possible to study enzyme–substrate interaction more closely, a new model was needed. Our modern understanding of molecular structure makes it clear that enzyme molecules are not totally rigid, like locks. The **induced-fit model** accounts for changes in the shape of the enzyme active site that accommodate the substrate and facilitate the reaction. As an enzyme and substrate come together, their interaction *induces* exactly the right fit for catalysis of the reaction.

Induced-fit model A model of enzyme action in which the enzyme has a flexible active site that changes shape to best fit the substrate and catalyze the reaction.

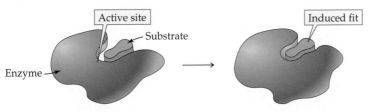

A well-studied example of induced fit, the interaction between glucose and hexokinase, is illustrated in Figure 19.3. The reaction, a common one, is a phosphorylation—the addition of a phosphoryl group to an —OH group, catalyzed by a kinase. The reaction is the first step in glucose (a hexose) metabolism (Section 23.2). Notice in Figure 19.3 how the enzyme closes in once the glucose molecule has entered the active site—this is the induced fit.

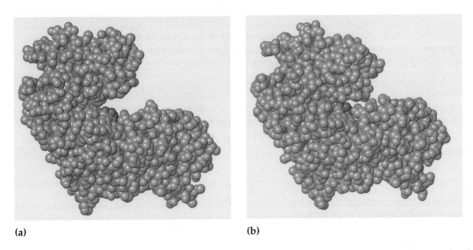

(a) (b)

▲ **FIGURE 19.3 The induced fit of hexokinase (blue) and its substrate, glucose (red).** (a) The active site is a groove in the hexokinase molecule. (b) When glucose enters the active site, the enzyme changes shape, wrapping itself more snugly around the substrate.

Enzyme-catalyzed reactions begin with migration of the substrate or substrates into the active site to form an *enzyme–substrate complex*. The substrate is first drawn into position by the same kinds of noncovalent forces that govern the shapes of protein molecules. (⬤⬤⬤, Figure 18.4, p. 571)

Before complex formation, the substrate molecule is in its most stable, lowest-energy shape. Within the complex, the molecule is forced into a less stable shape, and bonding electrons may be drawn away from some bonds in preparation for breaking them and forming new bonds. The result is to *lower the activation energy barrier* between substrate and product.

Within the enzyme–substrate complex, atoms that will form new bonds must connect with each other. The new bonds might be with a second substrate or temporary bonds with atoms in the enzyme. Also, groups needed for catalysis must be close to the necessary locations in the substrate. Many organic reactions, for example, require acidic, basic, or metal ion catalysts. An enzyme's active site can provide acidic and basic groups without disrupting the constant-pH environment in body fluids, while the necessary metal ions are present as cofactors. Once the chemical reaction is completed, enzyme and product molecules separate from each other and the enzyme, restored to its original condition, becomes available for another substrate.

The hydrolysis of a peptide bond by chymotrypsin, shown in Figure 19.4, illustrates how an enzyme functions. Chymotrypsin is one of the enzymes that participates in the digestion of proteins by breaking them down to smaller molecules.

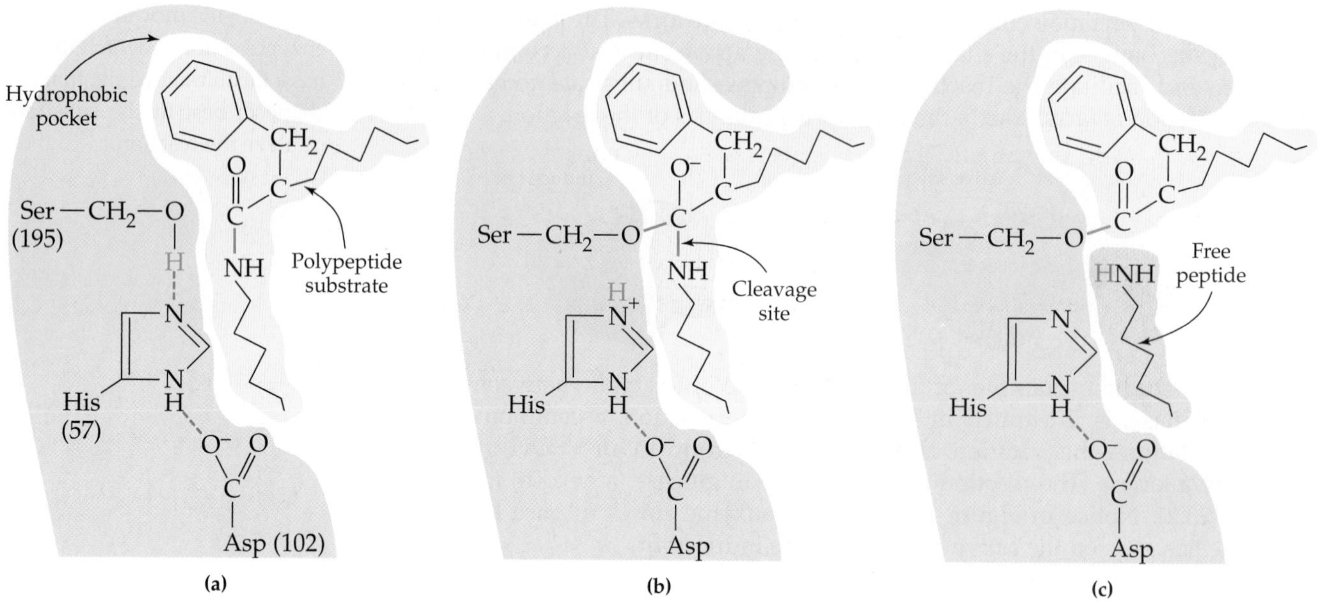

▲ **FIGURE 19.4 Hydrolysis of a peptide bond by chymotrypsin.** (a) The polypeptide enters the active site with its hydrophobic side chain in the hydrophobic pocket and the peptide bond to be broken (red) opposite serine and histidine residues. (b) Hydrogen transfer from serine to histidine allows formation of a strained intermediate in which the serine side chain bonds to the peptide bond carbon (green). (c) The peptide bond is broken and the segment with the new terminal —NH₂ group leaves the active site. In subsequent steps (not shown) a water molecule enters the active site; its H atom restores the serine side chain and its —OH bonds to the other piece of the substrate protein to give a new terminal —COOH group so that this piece can leave the active site.

It cleaves polypeptide chains by breaking the peptide bond on the carbonyl side of amino acid residues that include an aromatic ring:

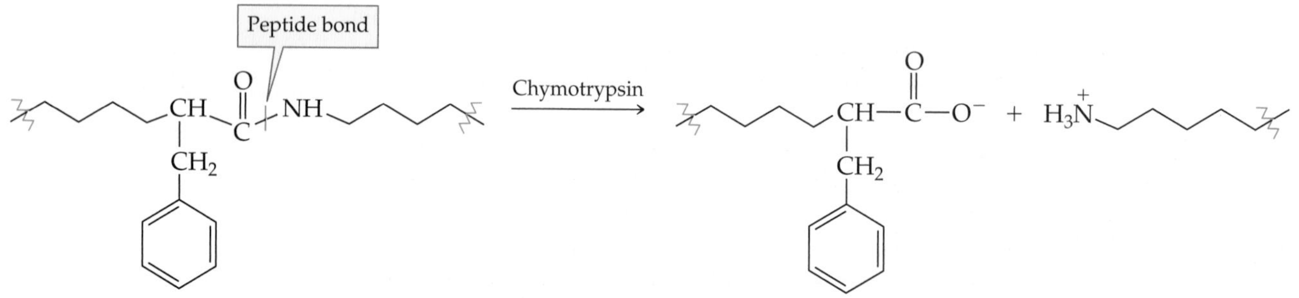

The enzyme–substrate complex forms (Figure 19.4a and b) by attraction of a hydrophobic side chain of the substrate into a hydrophobic pocket in the active site and the subsequent formation of a covalent bond (green) to the substrate. The result is to position the substrate with the peptide bond to be broken (red) next to the amino acid side chains that function as catalysts. The enzyme has not only joined up with the substrate (the *proximity effect*), but has done so in such a way as to bring the groups that must connect close to each other (the *orientation effect*). Aspartate, histidine, and serine provide functional groups needed for catalysis within the active site (the *catalytic effect*). As an illustration of the critical nature of protein folding, note that in the 241-amino-acid primary structure of chymotrypsin, aspartate is number 102, histidine is number 57, and serine is number 195. These amino acids are distant from each other along the backbone but are brought close together so that their side chains are in exactly the positions needed in the active site.

With the peptide bond carbon temporarily bonded to serine in the active site, it is easier for the peptide bond to break because the activation energy barrier has been lowered (the *energy effect*). As the bond breaks, the nitrogen picks up a hydrogen (blue) from histidine to form the new terminal amino group and this portion of the substrate is set free (Figure 19.4c). Reaction with a water molecule then restores the hydrogen to serine and supplies an OH group to form the new terminal carboxyl group. This part of the substrate is set free and the enzyme is restored to its original state.

In summary, enzymes act as catalysts because of their ability to

* Bring substrate(s) and catalytic sites together (*proximity effect*)

* Hold substrate(s) at the exact distance and in the exact orientation necessary for reaction (*orientation effect*)

* Provide acidic, basic, or other types of groups required for catalysis (*catalytic effect*)

* Lower the energy barrier by inducing strain in bonds in the substrate molecule (*energy effect*)

🔑 KEY CONCEPT PROBLEM 19.10

The active sites of enzymes usually contain amino acids with acidic, basic, and polar side chains. Some enzymes also have amino acids with nonpolar side chains in their active sites. Which types of side chains would you expect to participate in holding the substrate in the active site? Which types would you expect to be involved in the catalytic activity of the enzyme?

19.5 Effect of Concentration on Enzyme Activity

For a reaction to occur, the enzyme and substrate molecules must come together and form the enzyme–substrate complex. Therefore, variation in the reaction rate can be expected if the enzyme or substrate concentration changes.

Substrate Concentration

Consider the common situation in which the substrate concentration varies while the enzyme concentration remains unchanged. If the substrate concentration is low relative to that of the enzyme, not all the enzyme molecules are in use. The rate therefore increases with the concentration of substrate because more of the enzyme molecules are put to work. In this situation, shown at the far left of the curve in Figure 19.5, the rate increases as the available substrate increases. If the substrate concentration doubles, the rate doubles (a directly proportional relationship).

As the substrate concentration continues to increase, however, the increase in the rate begins to level off as more and more of the active sites are occupied. (Think of people waiting on line to take their seats in a theater. The line moves more slowly as more seats fill and it becomes more difficult to find an empty one.) Eventually, the substrate concentration reaches a point at which none of the available active sites are free. Since the reaction rate is now determined by how fast the enzyme–substrate complex is converted to product, the reaction rate becomes constant—the enzyme is saturated.

Once the enzyme is saturated, increasing substrate concentration has no effect on the rate. In the absence of a change in the concentration of the enzyme, the rate when the enzyme is saturated is determined by the efficiency of the enzyme, the pH, and the temperature.

Under most conditions, an enzyme is not likely to be saturated. Therefore, at a given pH and temperature, the reaction rate is controlled by the amount of substrate and the overall efficiency of the enzyme. If the enzyme–substrate complex is rapidly converted to product, the rate at which enzyme and substrate combine to form the complex becomes the limiting factor. Calculations show an upper limit

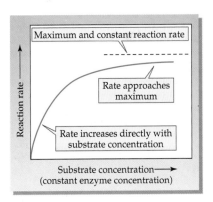

▲ **FIGURE 19.5 Change of reaction rate with substrate concentration when enzyme concentration is constant.** At low substrate concentration (at left), the reaction rate is directly proportional to the substrate concentration (at constant pH and temperature). With increasing substrate concentration, the rate drops off as more of the active sites are occupied. Eventually, with all active sites occupied, the rate reaches a maximum and constant rate.

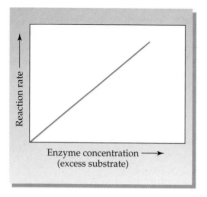

▲ **FIGURE 19.6 Change of reaction rate with enzyme concentration in the presence of excess substrate.**

to this rate: Enzyme and substrate molecules moving at random in solution can collide with each other no more often than about 10^8 collisions per mole per liter per second. Remarkably, a few enzymes actually operate with close to this efficiency–every one of the collisions results in the formation of product! One example of such an efficient enzyme is catalase, the enzyme that breaks down hydrogen peroxide at the rate of 10^7 catalytic events per second (Table 19.1, p. 595).

Enzyme Concentration

It is possible for the concentration of an active enzyme to vary according to our metabolic needs. So long as the concentration of substrate does not become a limitation, the reaction rate varies directly with the enzyme concentration (Figure 19.6). If the enzyme concentration doubles, the rate doubles; if the enzyme concentration triples, the rate triples; and so on.

> **KEY CONCEPT PROBLEM 19.11**
>
> What do we mean when we say an enzyme is saturated with substrate? When an enzyme is saturated with substrate, how does adding more (a) substrate and (b) enzyme affect the rate of the reaction?

19.6 Effect of Temperature and pH on Enzyme Activity

Enzymes have been finely tuned through evolution so that their maximum catalytic activity is highly dependent on pH and temperature. As you might expect, optimum conditions vary slightly for each enzyme but are generally near normal body temperature and the pH of the body fluid in which the enzyme functions.

Effect of Temperature on Enzyme Activity

An increase in temperature increases the rate of most chemical reactions, and enzyme-catalyzed reactions are no exception. Unlike many simple reactions, however, the rates of enzyme-catalyzed reactions do not increase continuously with rising temperature. Instead, the rates reach a maximum and then begin to decrease, as shown in Figure 19.7a. This falloff in rate occurs because enzymes begin to denature when heated too strongly. (⬤⬤, p. 583) The noncovalent

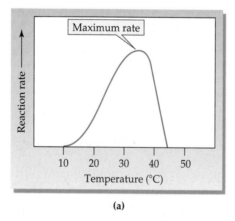

(a)

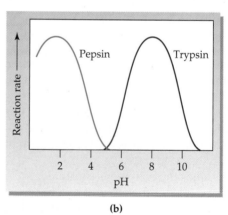

(b)

▲ **FIGURE 19.7 Effect of temperature (a) and pH (b) on reaction rate.** (a) The rate increases with increasing temperature until a temperature is reached at which the protein begins to denature; then the rate decreases rapidly. (b) The optimum activity for an enzyme occurs at the pH where it acts, as illustrated for two protein hydrolysis enzymes—pepsin, which acts in the highly acidic environment of the stomach, and trypsin, which acts in the small intestine, an alkaline environment.

attractions between protein side chains are disrupted, the delicately maintained three-dimensional shape of the enzyme begins to come apart, and as a result the active site needed for catalytic activity is destroyed.

Most enzymes denature and lose their catalytic activity above 50–60 °C, a fact that explains why medical instruments and laboratory glassware can be sterilized by heating with steam in an autoclave. The high temperature of the steam permanently denatures the enzymes of any bacteria present, thereby killing them.

A severe drop in body temperature creates the potentially fatal condition of hypothermia, which is accompanied by a slowdown in metabolic reactions. This effect is used to advantage by cooling the body during cardiac surgery. Upon gentle warming, enzymatic reaction rates return to normal because cooling does not denature proteins.

Effect of pH on Enzyme Activity

The catalytic activity of many enzymes depends on pH and usually has a well-defined optimum point at the normal, buffered pH of the enzyme's environment. For example, pepsin, which initiates protein digestion in the highly acidic environment in the stomach, has its optimum activity at pH 2 (Figure 19.7b). By contrast, trypsin—like chymotrypsin, an enzyme that aids digestion of proteins in the small intestine—has optimum activity at pH 8. Most enzymes have their maximum activity between the pH values of 5 to 9. Eventually, extremes of pH will denature a protein.

WORKED EXAMPLE   **19.2** Enzymatic Activity: Determining Optimum pH

Enzymatic activity is shown for three different enzymes as a function of pH in the graph below. What is the optimum pH for pepsin (curve A), for urease (curve B), and for alanine dehydrogenase (curve C)?

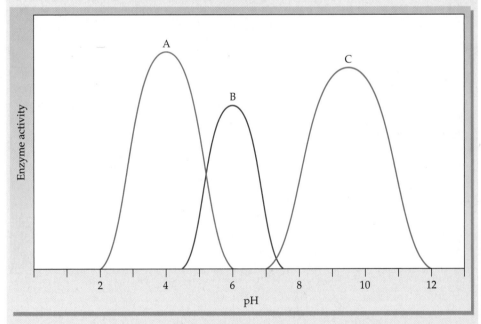

ANALYSIS Recall that the optimum pH is the pH at which the enzyme shows the highest activity; therefore, the highest point on the curve, representing maximum activity, is the optimum pH for the enzyme.

SOLUTION
Find the correct curve for each enzyme, then find the apex of the activity curve. Drop a vertical line to the pH axis and read the optimum pH directly from the axis scale. The optimum pH for pepsin is approximately 4.0, that for urease approximately 6.0, and that for alanine dehydrogenase approximately 9.5.

WORKED EXAMPLE **19.3** Enzymatic Activity: Determining Optimum Temperature

Consider the temperature activity curve below. Enzymatic activity is shown for muscle lactate dehydrogenase from 0 °C to 60 °C. Suppose you wish to test a sample for lactate dehydrogenase activity; what is the best temperature for the test?

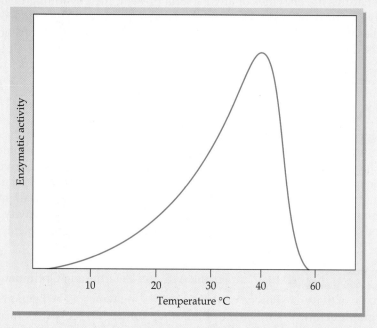

ANALYSIS An enzyme shows its highest catalytic activity at a certain temperature, with less activity at temperatures below and above the optimum temperature. Look at the curve of activity versus temperature and find the highest point on the curve—that point represents the optimum activity.

SOLUTION
From the highest point on the preceding curve of activity versus temperature, drop a vertical line down to the *x*-axis (the one that reads "Temperature") to find the optimum temperature. The temperature optimum for lactate dehydrogenase is 40 °C.

PROBLEM 19.12

Will the reaction catalyzed by the enzyme represented in Figure 19.7a have a higher rate of reaction at 20 °C or at 30 °C? Will it have a higher rate of reaction at 30 °C or at 40 °C?

PROBLEM 19.13

How will the rates of the reaction catalyzed by trypsin (Figure 19.7b) compare at pH 6 and pH 8?

19.7 Enzyme Regulation: Feedback and Allosteric Control

The control of biochemical reactions by enzymes is only part of the enzyme story. In the body, the concentrations of thousands of different compounds must vary continuously to meet changing conditions as we eat, sleep, exercise, or fall ill. Thus,

APPLICATION ▶ Extremozymes—Enzymes from the Edge

What do your laundry detergent, dishwasher detergent, oil drilling slurry, and the production of "stone washed" fabric have in common? One thing—the presence of enzymes that are active under unusual conditions. Most mammalian enzymes display optimum activity around 40 °C near pH 7.0 at 1 atmosphere of pressure. *Extremozymes* are enzymes from extremophiles, organisms that live in conditions hostile to mammalian cells. The conditions under which extremophiles live would inactivate mammalian enzymes as well as those of many other organisms. Of greatest interest are bacteria that live under conditions of extreme heat or cold, in highly acid or alkaline environments or in very salty environments. Useful enzymes have been developed from bacteria that have optimum growth temperatures as high as 106 °C and from those with an optimum as low as 4 °C. Other enzymes have been isolated from various bacteria species that grow in pH as low as 0.7 and as high as 10.

▲ Extremophiles such as the bacteria that inhabit hot springs in Yellowstone National Park live in conditions that would denature most mammalian enzymes.

Enzymes from thermophiles (heat lovers) include ones that break down starch and cellulose. These enzymes find wide use in the food, feed, textile ("stone washing"), and paper industries. One of the widest uses in the food industry is the hydrolysis of starch to glucose. Biobleaching of paper pulp using heat-stable xylanases lowers the use of halogen-containing bleaching products, decreasing halogen release into the environment. Forensic science and basic molecular biology research both take advantage of the thermostable enzyme Taq polymerase, first identified in a microorganism isolated from a hot spring in Yellowstone National Park. For a description of its use see the Application: "Serendipity and the Polymerase Chain Reaction" (p. 850).

Enzymes from cold-environment microorganisms (psychrophiles) are used in products such as cold-water-wash laundry detergents. Typical enzymes used in these products are lipases to break down greasy deposits and proteases to hydrolyze proteins deposited on clothing. Other applications for these enzymes exist in the cold-food industry where meat is tenderized and fruit juice is clarified using pectinases prior to sale. If you wear contact lenses, your cleaning solution contains room temperature stable, active proteases.

How do extremozymes withstand conditions that denature most enzymes? The strategy appears to vary depending on the type of extreme condition under which the organism lives. Thermophiles synthesize special proteins called *chaperonins*. Chaperonins recognize and bind to heat-denatured proteins, refolding them into their active forms. (⟐⟐⟐, Section 18.8) Also, proteins from thermophiles have tightly folded, highly nonpolar cores that resist denaturing forces. This sticky core is reinforced by the absence of amino acids that provide flexibility and by the presence of numerous ionic bonds on the surface of the protein. The result is a rigid protein that resists heat denaturation. In contrast, psychrophiles have more polar, flexible proteins than thermophiles. This structure is necessary to maintain activity at low temperatures.

So why do oil drillers use enzymes? Often oil and gas is trapped in rock. A thick, gooey mixture of enzymes, guar gum, sand, and water is forced into the well hole. The drillers then set off an explosion, simultaneously cracking the rock and forcing the enzyme-laced mixture into the cracks. The thermophilic enzymes hydrolyze the guar gum turning the viscous mixture into a thin solution that allows the oil and gas to flow out of the rock and up the well. Because of the tremendous heat in the drill hole, only heat-stable, thermophilic enzymes work. Also, the use of enzymes with optimum activity at high temperatures ensures the guar gum will not be hydrolyzed prematurely.

See Additional Problems 19.76 and 19.77 at the end of the chapter.

---

enzymes do more than merely speed up reactions. At a moment's notice, they turn some reactions off, slow some down a bit, or quickly accelerate others to their maximum possible rate. Clearly, then, the enzymes themselves must be regulated. How is this regulation achieved?

A variety of strategies are utilized to adjust the rates of enzyme-catalyzed reactions. Any process that starts or increases the action of an enzyme is an **activation**. Any process that slows or stops the action of an enzyme is an **inhibition**. Although

Activation (of an enzyme) Any process that initiates or increases the action of an enzyme.

Inhibition (of an enzyme) Any process that slows or stops the action of an enzyme.

APPLICATION ▶ Enzymes in Medical Diagnosis

In a healthy person, certain enzymes, such as those responsible for forming and dissolving blood clots, are normally present in high concentration in blood serum. Other enzymes function mainly within cells and are normally in low concentration in blood serum, which they enter only during normal degeneration of healthy cells. When tissue is injured, however, large quantities of enzymes are released into the blood from dying cells, with the distribution of enzymes dependent on the identity of the injured cells. Measurement of blood levels of enzymes is therefore a valuable diagnostic tool. For example, higher-than-normal activities of the enzymes included in a routine blood analysis indicate the following conditions:

Aspartate transaminase (AST)	Damage to heart or liver
Alanine transaminase (ALT)	Damage to heart or liver
Lactate dehydrogenase (LH)	Damage to heart, liver, or red blood cells
Alkaline phosphatase (ALP)	Damage to bone and liver cells
γ-Glutamyl transferase (GGT)	Damage to liver cells; alcoholism

Enzyme analysis measures the activity of an enzyme rather than its concentration. Because activity is influenced by pH, temperature, and substrate concentration, it is measured in international units at standard conditions. One international unit (U) is defined as the amount of an enzyme that converts 1 μmol of its substrate to product per minute under defined standard conditions of pH, temperature, and substrate concentration. The analytical results are reported in units per liter (U/L).

Among the most useful enzyme assays are those done to diagnose heart attacks (*myocardial infarctions, MI*), which involves measuring activities of three enzymes: creatine phosphokinase (CPK), aspartate transaminase (AST, formerly referred to as GOT), and lactate dehydrogenase (LDH). The CPK level rises almost immediately following an MI, reaching a six-fold increase over normal values after about 30 hours; the AST level triples after about 40 hours; and the LDH level doubles after about 4 days.

More specific verification of a heart attack is gained analyzing the *isoenzymes* of LDH. Isoenzymes are structural variations of the same enzyme that catalyze the same reaction but in different tissues. Lactate dehydrogenase is a mixture of five isoenzymes, denoted LDH_1, LDH_2, LDH_3, LDH_4 and LDH_5. Of the five, heart tissue contains primarily LDH_1. A characteristic change following a heart attack is an elevated level of LDH_1 above LDH_2. Many other changes in the LDH isoenzyme levels are indicative of different conditions. For example, elevation of LDH_4 and LDH_5 indicates acute hepatitis.

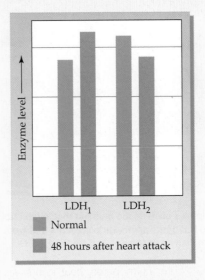

▲ Blood levels of lactate dehydrogenase (LDH) isoenzymes in a normal person (blue) and in a heart attack victim after 48 hours (red). Note the flip of LDH_1 and LDH_2 levels following a heart attack.

See Additional Problems 19.78 and 19.79 at the end of the chapter.

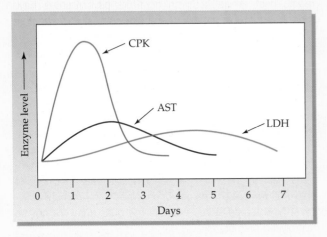

▲ Blood levels of creatine phosphokinase (CPK), aspartate transaminase (AST), and lactate dehydrogenase (LDH) in the days following a heart attack.

we will describe the strategies of enzyme control one by one, keep in mind that several strategies usually operate together. Considering that a cell contains thousands of proteins—many molecules of some proteins and only a few molecules of other proteins—and hundreds of other kinds of biomolecules, all in concentrations required to maintain constant conditions, the achievement of enzyme control is awe-inspiring.

Feedback Control

As you will see in subsequent chapters, biochemical reaction pathways are dependent on series of consecutive reactions in which the product of one reaction is the reactant for the next. Such pathways are subject to *feedback control*, which occurs when the result of a process feeds information back to affect the beginning of the process. Any device that maintains a constant temperature, such as an oven, is regulated by feedback control. Ovens have sensors that detect temperature and feed back that information to turn heating elements on or off.

Consider a biochemical pathway in which A is converted to B, then B is converted to C, and so on, with each reaction catalyzed by its own enzyme:

$$A \xrightarrow{\text{Enzyme 1}} B \xrightarrow{\text{Enzyme 2}} C \xrightarrow{\text{Enzyme 3}} D$$

What happens if product D inhibits enzyme 1? This inhibition causes the amount of A converted to B to decrease, so the synthesis of B and C decreases in turn. The effect of this **feedback control** mechanism is to control the concentration of D. When more D is present than is needed for other biochemical pathways, enzyme 1 is inhibited and its reaction is slowed or stopped. No energy is then wasted making the unneeded intermediates B and C. When all of the available D is used up in other reactions, none is available for feedback control. As a result, enzyme 1 is no longer inhibited and the production of D accelerates. Often, biochemists refer to feedback control and feedback inhibition interchangeably; these terms have the same meaning.

Suppose that the conversion of A to B, the reaction catalyzed by enzyme 1 in the pathway above, is at a point where control is critical. Perhaps the amount of product D must vary over a wide range, or perhaps it is needed only in an emergency. Feedback control is a common strategy at such key points in biochemical pathways.

WORKED EXAMPLE **19.4** Determining Feedback Control Points

Look at the three-step pathway for the conversion of 3-phosphoglycerate to serine:

$$\text{3-phosphoglycerate} \xrightarrow{1} \text{3-phosphohydroxypyruvate} \xrightarrow{2}$$
$$\text{3-phosphoserine} \xrightarrow{3} \text{serine}$$

When the cell has plenty of serine available, which enzyme in the pathway, 1, 2 or 3, is most likely to be inhibited?

ANALYSIS This is a simple, linear pathway (with no branching). The pathway is most likely controlled by feedback control of the final product.

SOLUTION
Assuming that feedback control is the simplest control mechanism for this linear pathway, serine, the product of the pathway, will inhibit the first enzyme in the pathway when sufficient serine is available in the cell.

Allosteric Control

Feedback control is commonly exercised by what is known as *allosteric* control (from the Greek *allos*, meaning "other" and *steros*, meaning "space"). In **allosteric control** the binding of a molecule (an *allosteric regulator* or *effector*) at one site on a protein affects the binding of another molecule at a different site. Most **allosteric enzymes** have more than one protein chain and two kinds of binding sites—those for substrate and those for regulators (Figure 19.8). Binding of a regulator, usually by noncovalent interactions, changes the shape of the enzyme. This change alters the shape of the active site, thereby affecting the ability of the enzyme to bind

Feedback control Regulation of an enzyme's activity by the product of a reaction later in a pathway.

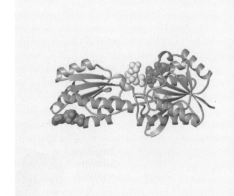

(a)

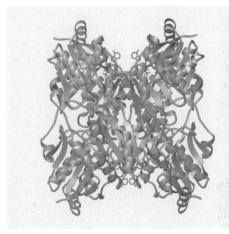

(b)

▲ **FIGURE 19.8 An allosteric enzyme.** (a) One of the four identical subunits in a phosphofructokinase, an enzyme that catalyzes transfer of a phosphoryl group from ATP to fructose 6-phosphate (see transferase reaction on p. 597). The subunit is shown with the diphosphate (yellow) and ADP (green) in the active site and the allosteric activator (red, also ADP) in the regulatory site. (b) The four subunits of the enzyme, two blue and two in purple.

Allosteric control An interaction in which the binding of a regulator at one site on a protein affects the protein's ability to bind another molecule at a different site.

Allosteric enzyme An enzyme whose activity is controlled by the binding of an activator or inhibitor at a location other than the active site.

its substrate and catalyze its reaction. One advantage of allosteric enzyme control is that the regulators need not be structurally similar to the substrate because they do not bind to the active site.

Allosteric control can be either positive or negative. Binding a positive regulator changes the active sites so that the enzyme becomes a better catalyst and the rate accelerates. Binding a negative regulator changes the active sites so that the enzyme is a less effective catalyst and the rate slows down. Because allosteric enzymes can have several substrate binding sites and several regulator binding sites and because there may be interaction among them, very fine control is achieved.

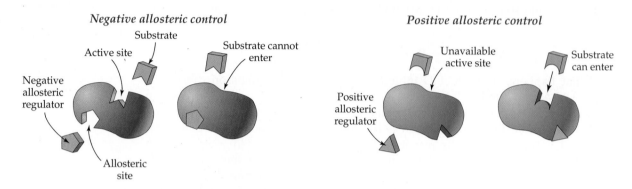

19.8 Enzyme Regulation: Inhibition

The inhibition of an enzyme can be *reversible* or *irreversible*. In reversible inhibition, the inhibitor can leave, restoring the enzyme to its uninhibited level of activity. In irreversible inhibition, the inhibitor remains permanently bound and the enzyme is permanently inhibited. The inhibition can also be *competitive* or *noncompetitive*, depending on whether the inhibitor binds to the active site or elsewhere.

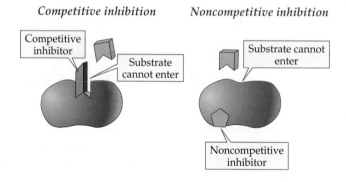

Reversible Noncompetitive Inhibition

Noncompetitive (enzyme) inhibition Enzyme regulation in which an inhibitor binds to an enzyme elsewhere than at the active site, thereby changing the shape of the enzyme's active site and reducing its efficiency.

In **noncompetitive inhibition**, the inhibitor does not compete with the substrate for the active site. It binds to the enzyme in a different place. A noncompetitive inhibitor exerts allosteric control by changing the enzyme's shape so that the active site is less accessible or the reaction occurs less efficiently.

The rates with and without a noncompetitive inhibitor are compared in the bottom and top curves in Figure 19.9. With the inhibitor, the reaction rate increases with increasing substrate concentration more gradually than when no inhibitor is present. The maximum rate is lowered and once that rate is reached, no amount of substrate can increase it further. So long as the inhibitor is connected to the enzyme, this upper limit does not change.

Reversible Competitive Inhibition

What happens if an enzyme encounters a molecule very much like its normal substrate in shape, size, and functional groups? The impostor molecule enters the

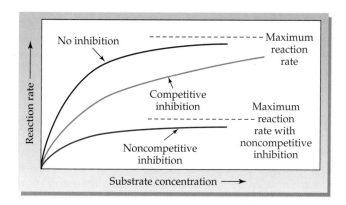

◀ FIGURE 19.9 Enzyme inhibition. The top curve and dashed line show the reaction rate and maximum rate with no inhibitor. With a competitive inhibitor (middle curve), the maximum rate is unchanged, but a higher substrate concentration is required to reach it. With a noncompetitive inhibitor (bottom curve) the maximum rate (bottom dashed line) is lowered.

Competitive (enzyme) inhibition
Enzyme regulation in which an inhibitor competes with a substrate for binding to the enzyme active site.

enzyme's active site, binds to it, and thereby prevents the usual substrate molecule from binding to the same site. Consequently, the enzyme is tied up, making it unavailable as a catalyst. This situation is **competitive inhibition**—the inhibitor *competes* with substrate for binding to the active site. A competitive inhibitor binds reversibly to an active site through noncovalent interactions, but undergoes no reaction. While it is there, it prevents the substrate from entering the active site. Whether the substrate or the inhibitor occupies the active site depends on their relative concentrations.

$$\text{Substrate} + \text{Enzyme} \rightleftharpoons \text{Substrate–enzyme complex}$$

$$\text{Inhibitor} + \text{Enzyme} \rightleftharpoons \text{Inhibitor–enzyme complex}$$

A substrate in relatively high concentration occupies more of the active sites, so the reaction is less inhibited. An inhibitor in relatively high concentration occupies more of the active sites, so the reaction is more inhibited.

The middle curve in Figure 19.9 shows that in the presence of a competitive inhibitor, the reaction rate increases more gradually with increasing substrate concentration than in its absence. Unlike noncompetitive inhibition, however, the maximum reaction rate is unchanged. Eventually all of an enzyme's active sites can be occupied by substrate, but a higher substrate concentration is required to reach that condition.

The product of a reaction may be a competitive inhibitor for the enzyme that catalyzes that reaction. For example, glucose 6-phosphate is an allosteric competitive inhibitor for hexokinase, which catalyzes formation of this phosphorylated form of glucose. Thus, when supplies of glucose 6-phosphate are ample, the glucose is available for other reactions.

A competitive inhibitor is sometimes put to use in treating an unhealthy condition because the inhibitor mimics the structure of the substrate and fits into the enzyme's active site. For example, competitive inhibition is used to good advantage in the treatment of methanol poisoning. Although not harmful itself, methanol (wood alcohol) is oxidized in the body to formaldehyde, which is highly toxic ($CH_3OH \longrightarrow H_2C{=}O$). Because of its molecular similarity to methanol, ethanol acts as a competitive inhibitor of the methanol dehydrogenase enzyme. With the oxidation of methanol blocked by ethanol, the methanol is excreted without causing harm. Thus, the medical treatment of methanol poisoning includes administering ethanol.

Another example of reversible inhibition involves lead poisoning. Lead can poison animals, including humans, in two ways. One way is by displacing an essential metal cofactor from the active site of an enzyme. When lead displaces zinc in an enzyme essential to the synthesis of heme, which is the oxygen-carrying part of hemoglobin, anemia can result. Physicians can treat this sort of lead poisoning with chelation therapy. Ethylenediaminetetraacetic acid (EDTA) forms coordinate covalent bonds preferentially with lead in the body, and the lead is then excreted in the urine as a chelated compound.

The second way lead can poison involved the process knows as irreversible inhibition, the topic we look at next.

Irreversible Inhibition

Irreversible (enzyme) inhibition
Enzyme deactivation in which an inhibitor forms covalent bonds to the active site, permanently blocking it.

If an inhibitor forms a bond that is not easily broken with a group in an active site, the result is **irreversible inhibition**. The enzyme's reaction cannot occur because the substrate cannot connect appropriately with the active site. Many irreversible inhibitors are poisons as a result of their ability to completely shut down the active site. Heavy metal ions, such as mercury (Hg^{2+}) and lead (Pb^{2+}), are irreversible inhibitors that form covalent bonds to the sulfur atoms in the —SH groups of cysteine residues. (⬤⬤, p. 571)

Targets for this effect include enzymes that function in the nervous system. At low levels, lead can thus cause decreased attention span and mental difficulties. These symptoms are noticed in children who develop the habit of ingesting flakes of lead-containing paint, which have a sweet taste. Primarily for this reason, lead-containing paint has not been used since the 1950s. It is, however, still present in older homes. Small amounts of mercury in the diet cause similar problems. For this reason, children and pregnant women are advised to severely limit their intake of fish, particularly deep-sea fish such as tuna. Tuna bioaccumulate mercury in their tissues; the mercury is absorbed from our digestive system and remains in our bodies.

Organophosphorus insecticides, such as parathion and malathion, and nerve gases, such as Sarin, are irreversible inhibitors of the enzyme acetylcholinesterase, which breaks down a chemical messenger (*acetylcholine*) that transmits nerve impulses (Section 20.7) The acetylcholinesterase inhibitors bond covalently to a serine residue in the enzyme's active site:

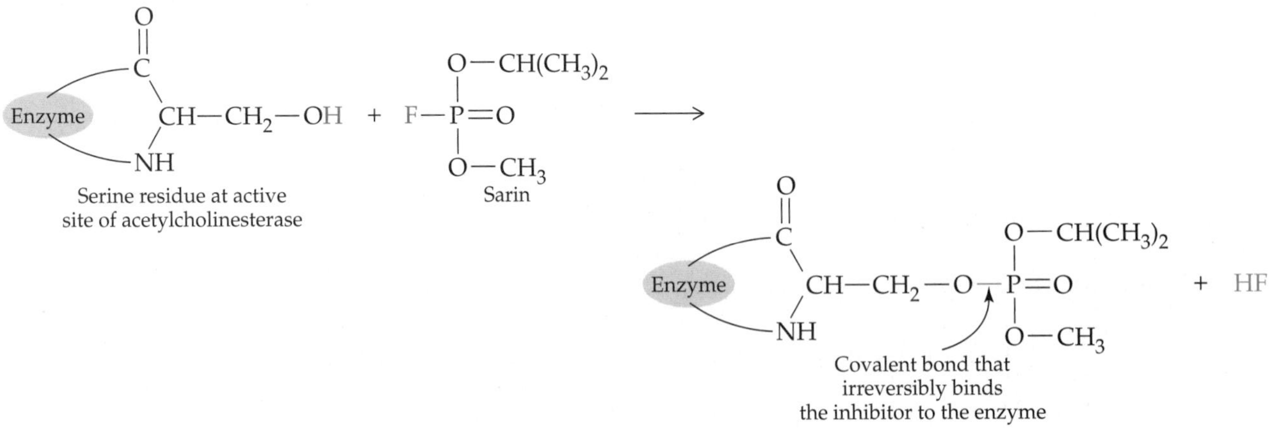

Serine residue at active
site of acetylcholinesterase

Sarin

Covalent bond that
irreversibly binds
the inhibitor to the enzyme

Normally, acetylcholinesterase breaks down acetylcholine immediately after that molecule transmits a nerve impulse. Without this enzyme activity, acetylcholine blocks transmission of further nerve impulses, resulting in paralysis of muscle fibers and death from respiratory failure. Sarin was the poison released by terrorists in the Tokyo subway system in 1995. The attack resulted in 12 deaths and varying degrees of injury to more than 5000 people. There is no effective treatment to counteract this irreversible inhibitor of acetylcholinesterase.

PROBLEM 19.14

Could either of the molecules shown below be a competitive inhibitor for the enzyme that has *p*-aminobenzoate as its substrate? If so, why?

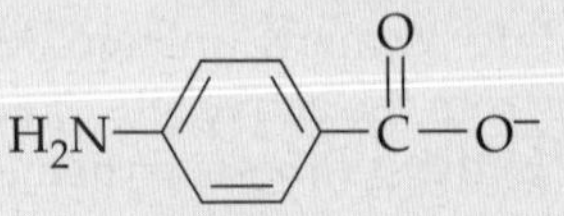

p-Aminobenzoate, the substrate

(a) $H_2NCH_2CH_3$ (b)

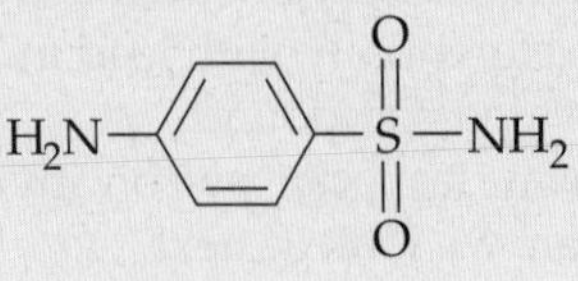

PROBLEM 19.15

What kind of reaction product might be a competitive inhibitor for the enzyme that catalyzes its formation?

APPLICATION ▶ Enzyme Inhibitors as Drugs

Consider a situation in which the chemical structures of a substrate and the active site to which it binds are known. A drug designer can than create a molecule sufficiently similar to the substrate that it binds to the active site and acts as an inhibitor. Inhibiting a particular enzyme can be desirable for many reasons.

The family of drugs known as ACE inhibitors is a good example of enzyme inhibitors as drugs. Angiotensin II, the octapeptide illustrated earlier (⬤▭⬤, p. 566), is a potent *pressor*—it elevates blood pressure, in part by causing contraction of blood vessels. Angiotensin I, a decapeptide, is an inactive precursor of angiotensin II. To become active, two amino acid residues—His and Leu—must be cut off the end of angiotensin I, a reaction catalyzed by angiotensin-converting enzyme (ACE).

$$\text{Asp-Arg-Val-Tyr-Ile-His-Pro-Phe-His-Leu} \xrightarrow{\substack{\text{Angiotensin-}\\\text{converting}\\\text{enzyme}}}$$

Angiotensin I

$$\text{Asp-Arg-Val-Tyr-Ile-His-Pro-Phe} + \text{His-Leu}$$

Angiotensin II

This reaction is part of a normal pathway for blood pressure control and is accelerated when blood pressure drops because of bleeding or dehydration.

The development of inhibitors for ACE was aided by knowing that a zinc(II) ion is present in the ACE active site.

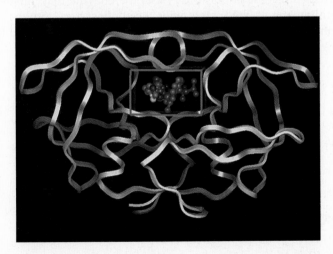

▲ Ritonovir, an enzyme inhibitor, in the active site of HIV protease.

Knowing that the extract of venom from a South American pit viper is a mild ACE inhibitor and that this extract contains a pentapeptide with a proline residue at the carboxyl-terminal end was also helpful. This information led to a search for a proline-containing molecule that would bind to the zinc(II) ion.

The first ACE inhibitor on the market, *captopril*, was developed by experimenting with modifications of the proline structure. Success was achieved by introducing an —SH group that binds to the zinc ion in the active site:

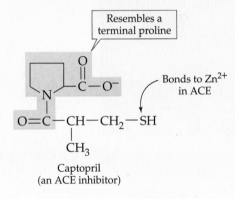

Captopril
(an ACE inhibitor)

Several other ACE inhibitors have subsequently been developed, and they are now common medications for patients with high blood pressure.

The development of enzyme inhibitors also plays a continuing, major role in the battle against *acquired immunodeficiency disease*, AIDS. The battle is far from won, but two important AIDS-fighting drugs are enzyme inhibitors. The first, known as AZT (*azidothymidine*, also called *zidovudine*), resembles in structure a molecule essential to reproduction of the AIDS-causing *human immunodeficiency virus (HIV)*. Because AZT is accepted by an HIV enzyme as a substrate, it prevents the virus from producing duplicate copies of itself.

The most successful AIDS drug thus far inhibits a *protease*, an enzyme that cuts a long protein chain into smaller pieces needed by the HIV. *Protease inhibitors*, such as Ritonovir, cause dramatic decreases in the virus population and AIDS symptoms. The success is only achieved, however, by taking a "cocktail" of several drugs including AZT. The cocktail is expensive and requires precise adherence to a schedule of taking 20 pills a day. These conditions make it unavailable or too difficult for many individuals to use.

See Additional Problems 19.80 and 19.81 at the end of the chapter.

19.9 Enzyme Regulation: Covalent Modification and Genetic Control

Covalent Modification

Zymogen A compound that becomes an active enzyme after undergoing a chemical change.

There are two general modes of enzyme regulation by covalent modification—removal of a covalently bonded portion of an enzyme, or addition of a group. Some enzymes are synthesized in inactive forms that differ from the active forms in composition. Activation of such enzymes, known as **zymogens** or *proenzymes*, requires a chemical reaction that splits off part of the molecule. Blood clotting, for example, is initiated by activation of zymogens.

Three of the enzymes that digest proteins in the small intestine, for example, are produced in the pancreas as the zymogens *trypsinogen*, *chymotrypsinogen*, and *proelastase*. These enzymes must be inactive when they are synthesized so that they do not immediately digest the pancreas. Each zymogen has a polypeptide segment at one end that is not present in the active enzymes. The extra segments are snipped off to produce trypsin, chymotrypsin, and elastase, the active enzymes, when the zymogens reach the small intestine, where protein digestion occurs.

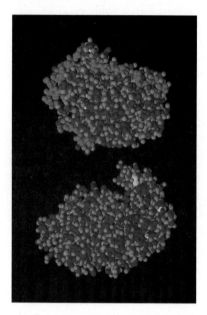

▲ Pepsinogen (a zymogen) at top and the active enzyme pepsin at bottom.

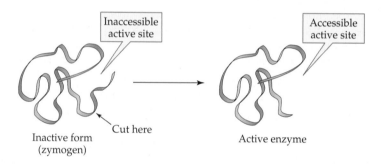

One danger of traumatic injury to the pancreas or the duct that leads to the small intestine is premature activation of these zymogens, resulting in acute pancreatitis, a painful and potentially fatal condition in which the activated enzymes attack the pancreas.

The reversible addition of phosphoryl groups ($-PO_3^{2-}$) to a serine, tyrosine or threonine residue is another mode of covalent modification. *Kinase* enzymes catalyze the addition of a phosphoryl group supplied by ATP (*phosphorylation*). *Phosphatase enzymes* catalyze the removal of the phosphoryl group (*dephosphorylation*). This control strategy swings into action, for example, when glycogen stored in muscles must be hydrolyzed to glucose that is needed for quick energy. Two serine residues in glycogen phosphorylase, the enzyme that initiates glycogen breakdown, are phosphorylated. Only with these phosphoryl groups in place is glycogen phosphorylase active. The groups are removed once the need to break down glycogen for quick energy has passed.

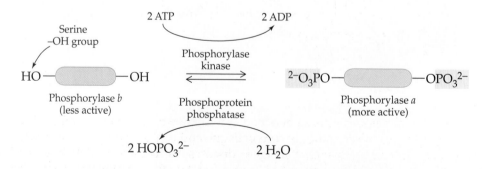

The curved arrows shown above are used frequently in biochemical equations in later chapters. Their focus is on changes in the major biomolecule reactant. The participation of other reactants needed to accomplish the chemical change is shown by the curved arrows adjacent to the main reaction arrow. Coenzymes and energy-providing molecules like ATP are often included in this manner. Here, the top curved arrow shows that the reaction in the forward direction requires ATP to supply the phosphoryl groups and produces ADP. The bottom curved arrow shows that water is needed for the reverse reaction, the hydrolysis that removes the phosphoryl groups as hydrogen phosphate anions.

Genetic Control

Yet another enzyme control strategy affects the supply of the enzyme itself. The synthesis of enzymes, like that of all proteins, is regulated by genes (Chapter 26). The **genetic control** strategy is especially useful for enzymes needed only at certain stages of development. Mechanisms controlled by hormones (Section 20.2) can accelerate or decelerate enzyme synthesis.

In summary, we have described the most important strategies that control the activity of enzymes. In any given biochemical pathway in a healthy individual, several of these strategies are likely in use at once.

Genetic (enzyme) control Regulation of enzyme activity by control of the synthesis of enzymes.

Mechanisms of Enzyme Control

- *Feedback control*, which is exerted on an earlier reactant by a later product in a reaction pathway and is made possible by *allosteric control*, which is bonding with the feedback molecule in a way that alters the shape and therefore the efficiency of the enzyme.

- *Inhibition*, which is *reversible* and can occur away from the active site (*noncompetitive inhibition*) or at the active site (*competitive inhibition*) by molecules that often mimic substrate structure, or which is *irreversible* because of covalent bonding of the inhibitor to the enzyme. Competitive inhibition is a strategy often utilized in medications, and irreversible inhibition is a mode of action of poisons.

- *Production of inactive enzymes (zymogens)*, which must be activated by cleaving a portion of the molecule.

- *Covalent modification of an enzyme by addition and removal of a phosphoryl group*, with the phosphoryl group supplied by ATP.

- *Genetic control*, whereby the quantity of enzyme is regulated by limiting the amount that is synthesized.

PROBLEM 19.16

Which type of enzyme regulation is best for the following situations?

(a) An enzyme that becomes overactive during a disease

(b) An enzyme needed only when there is low blood glucose

(c) An enzyme that springs into action when a traumatic injury occurs

(d) An enzyme needed only during adolescence

19.10 Vitamins

Long before the reasons were understood, it was known that lime and other citrus juices cure scurvy, meat and milk cure pellagra, and cod-liver oil prevents rickets. Eventually, it was discovered that these diseases are caused by deficiencies of **vitamins**—organic molecules required in only trace amounts that must be obtained through the diet. Vitamins are a dietary necessity for humans because our bodies do not have the ability to synthesize them.

▲ Vitamins in capsule and tablet forms.

Vitamin An organic molecule, essential in trace amounts that must be obtained in the diet because it is not synthesized in the body.

The symptoms of scurvy—muscle weakness; swollen, bleeding gums; and the easy bruising that result from defective collagen synthesis—are cured by consuming foods that contain vitamin C, the coenzyme necessary for collagen synthesis. (⚭, p. 579) Pellagra, with such varied symptoms as weight loss, dermatitis, depression, and dementia, results from a deficiency of niacin. And rickets, the occurrence of soft bones in children because of inadequate availability of calcium and phosphate, is due to a deficiency of vitamin D, which is essential to the incorporation of calcium in bones.

Water-Soluble Vitamins

Vitamins are grouped by solubility into two classes: water-soluble and fat-soluble. The water-soluble vitamins, listed in Table 19.3, are found in the aqueous environment inside cells, where most of them are needed as components of coenzymes. Over time, an assortment of names, letters, and numbers for designating vitamins have accumulated. (One reason is that what was originally known as vitamin B turned out to be several different vitamins.) Among the water-soluble vitamins,

TABLE 19.3   The Water-Soluble Vitamins*

VITAMIN	SIGNIFICANCE	SOURCES	REFERENCE DAILY INTAKE**	EFFECTS OF DEFICIENCY	EFFECTS OF EXCESS
Thiamine (B_1)	In coenzyme for decarboxylation reactions	Milk, meat, bread, legumes	1.5 mg	Muscle weakness, and cardiovascular problems including heart disease; causes beriberi	Low blood pressure
Riboflavin (B_2)	In coenzymes FMN and FAD	Milk, meat	1.7 mg	Skin and mucous membrane deterioration	Itching, tingling sensations
Niacin (nicotinic acid, nicotinamide, B_3)	In coenzyme NAD^+	Meat, bread, potatoes	2.0 mg	Nervous system, gastrointestinal, skin, and mucous membrane deterioration; causes pellagra	Itching, burning sensations, blood vessel dilation, death after large dose
B_6 (pyridoxine)	In coenzyme for amino acid and lipid metabolism	Meat, legumes	2.0 mg	Retarded growth, anemia, convulsions, epithelial changes	Central nervous system alterations, perhaps fatal
Folic acid	In coenzyme for amino acid and nucleic acid metabolism	Vegetables, cereal, bread	0.4 mg	Retarded growth, anemia, gastrointestinal disorders; neural tube defects	Few noted except at massive doses
B_{12} (cobalamin)	In coenzyme for nucleic acid metabolism	Milk, meat	6 μg	Pernicious anemia	Excess red blood cells
Biotin	In coenzyme for carboxylation reactions	Eggs, meat, vegetables	0.3 mg	Fatigue, muscular pain, nausea, dermatitis	None reported
Pantothenic acid (B_5)	In coenzyme A	Milk, meat	10 mg	Retarded growth, central nervous system disturbances	None reported
C (ascorbic acid)	Coenzyme; delivers hydride ions; antioxidant	Citrus fruits, broccoli; greens	60 mg	Epithelial and mucosal deterioration, causing scurvy	Kidney stones

*Adapted in part from Frederic H. Martini, Fundamentals of Anatomy and Physiology, 4th edition (Prentice Hall, 1998).

**RDI values are the basis for information on the Nutrition Facts Label included on most packaged foods. The values are based on the Recommended Dietary Allowances of 1968.

three remain best known by letters rather than names—vitamins C, B$_6$, and B$_{12}$. Structurally, the water-soluble vitamins have in common the presence of —OH, —COOH, or other polar groups that impart their water solubility, but otherwise they range from simple molecules like vitamin C to quite large and complex structures like vitamin B$_{12}$.

Vitamin C is biologically active without any change in structure from the molecules present in foods. It is increasingly shown to be a valuable *antioxidant*, as described later in this section. *Biotin* is connected to enzymes by an amide bond at its carboxyl group but otherwise undergoes no structural change from dietary biotin.

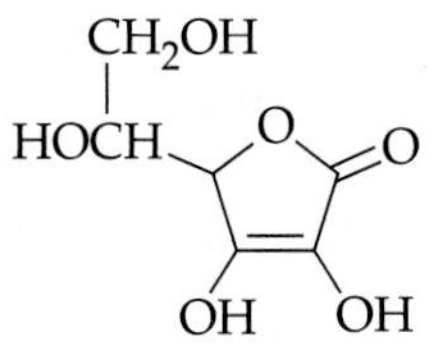

Vitamin C
(Ascorbic acid)

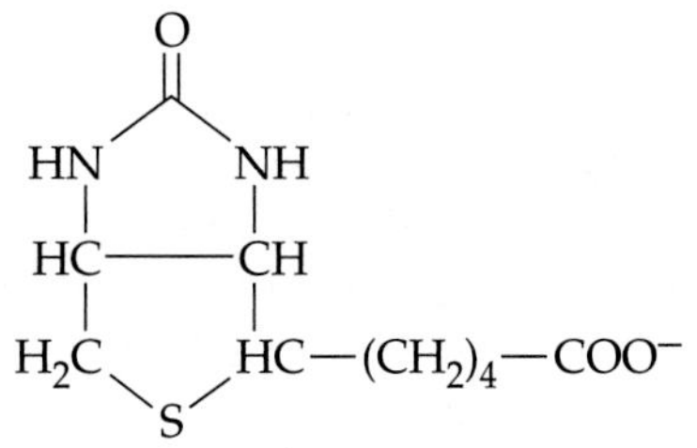

Biotin

The other water-soluble vitamins are incorporated into coenzymes. The vitamin-derived portions of two of the most important coenzymes, NAD$^+$ and coenzyme A, are illustrated in Figure 19.10. The functions, deficiency symptoms, and major dietary sources of the water-soluble vitamins are included in Table 19.3.

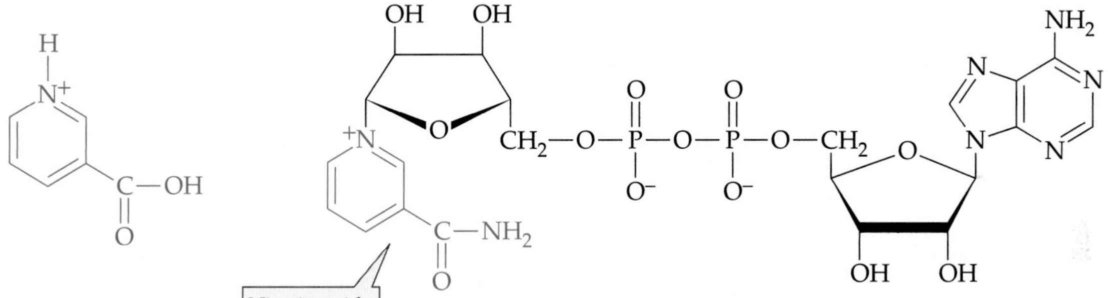

Niacin
(Nicotinic acid)

Nicotinamide

Nicotinamide adenine dinucleotide (NAD$^+$), a coenzyme

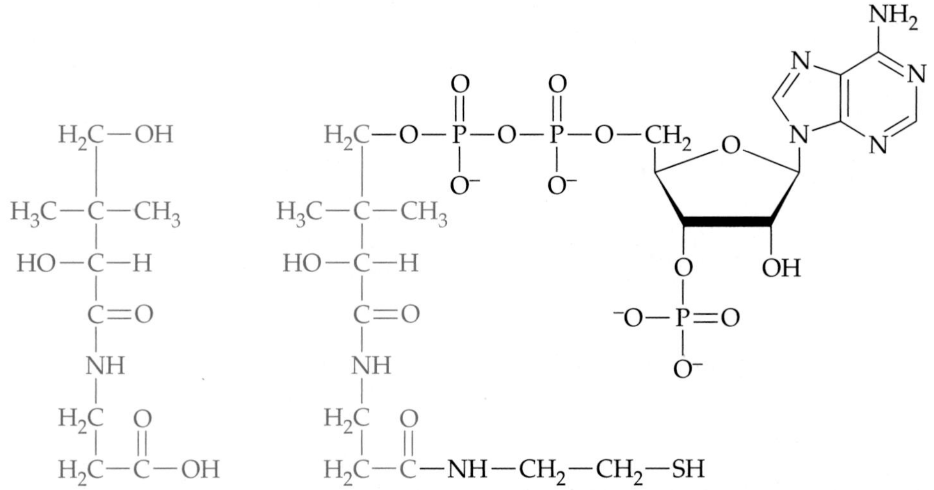

Pantothenic acid

Coenzyme A

▲ **FIGURE 19.10** **The vitamin-derived portions of NAD$^+$ and coenzyme A.**

WORKED EXAMPLE   19.4 Identifying Coenzymes

Identify the substrate, product, and coenzyme in the reaction shown below. The reaction is catalyzed by the enzyme alcohol dehydrogenase.

$$\text{Ethanol} + \text{NAD}^+ \longrightarrow \text{Acetaldehyde} + \text{NADH} + \text{H}^+$$

ANALYSIS Identify which molecules have been changed and how, starting from the left side of the arrow (the beginning of the reaction) to the right side of the arrow (the end of the reaction). In this case, ethanol is oxidized to acetaldehyde and NAD^+ is reduced to NADH/H^+. Recognize that nicotinamide adenine dinucleotide (NAD^+) is a coenzyme involved in oxidation/reduction reactions.

SOLUTION
Since NAD^+ is a coenzyme involved in oxidation/reduction reactions, ethanol (the other molecule on the left side of the equation) is the substrate and acetaldehyde (on the right side of the arrow) is the product of the reaction. $\text{NADH} + \text{H}^+$ is the reduced form of nicotinamide adenine dinucleotide and is considered to be reduced coenzyme only—not a product of the reaction.

Fat-Soluble Vitamins

The fat-soluble vitamins A, D, E, and K are stored in the body's fat deposits. Although the clinical effects of deficiencies of these vitamins are well documented, the molecular mechanisms by which they act are not nearly as well understood as those of the water-soluble vitamins. None has been identified as a coenzyme. Their functions, sources, and deficiency symptoms are summarized in Table 19.4. The hazards of overdosing on fat-soluble vitamins are greater than the hazards of overdosing on water-soluble vitamins because the fat-soluble vitamins accumulate

TABLE 19.4 The Fat-Soluble Vitamins*

VITAMIN	SIGNIFICANCE	SOURCES	REFERENCE DAILY INTAKE**	EFFECTS OF DEFICIENCY	EFFECTS OF EXCESS
A	Maintains epithelia; required for synthesis of visual pigments; antioxidant	Leafy green and yellow vegetables	1000 μg	Retarded growth, night blindness, deterioration of epithelial membranes	Liver damage, skin peeling, central nervous system effects (nausea, anorexia)
D	Required for normal bone growth, calcium and phosphorus absorption at gut, and retention at kidneys	Synthesized in skin exposed to sunlight	10 μg	Rickets, skeletal deterioration	Calcium deposits in many tissues, disrupting functions
E	Prevents breakdown of vitamin A and fatty acids; antioxidant	Meat, milk, vegetables	10 mg	Anemia; other problems suspected	None reported
K	Essential for liver synthesis of prothrombin and other clotting factors	Vegetables; production by intestinal bacteria	80 μg	Bleeding disorders	Liver dysfunction, jaundice

*Adapted in part from Frederic H. Martini, Fundamentals of Anatomy and Physiology, 4th edition (Prentice Hall, 1998).

**RDI values are the basis for information on the Nutrition Facts Label included on most packaged foods. The values are based on the Recommended Dietary Allowances of 1968. RDIs for fat-soluble vitamins are often reported in International Units (IU), which are defined differently for each vitamin. The values given here are approximate equivalents in mass units.

in body fats. Excesses of the water-soluble vitamins are more likely to be excreted in the urine.

Vitamin A, which is essential for night vision, healthy eyes, and normal development of epithelial tissue, has three active forms: retinol, retinal, and retinoic acid. It is produced in the body by cleavage of β-carotene, the molecule that gives an orange color to carrots and other vegetables. (⬭, p.406)

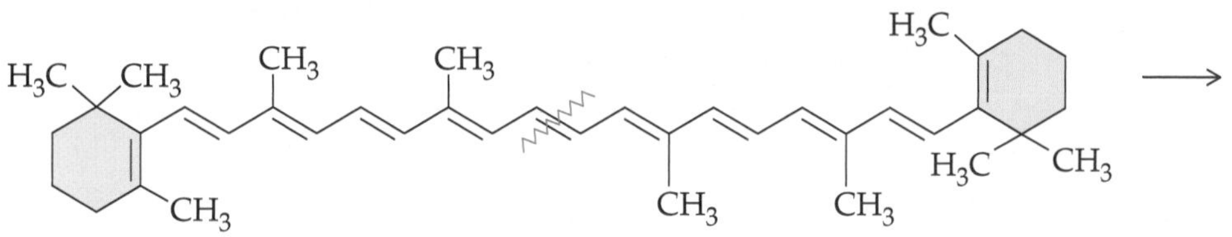

β-Carotene

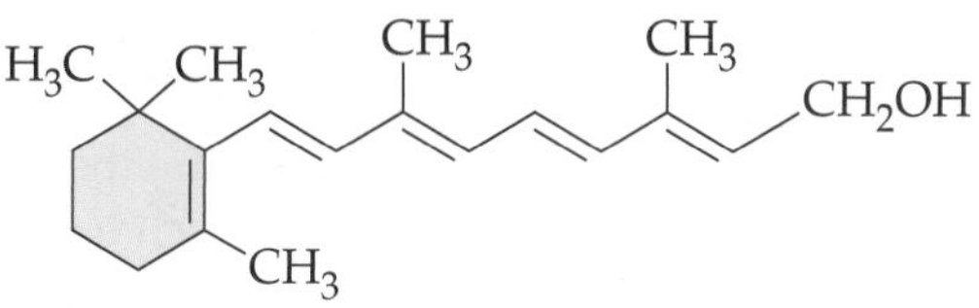

Vitamin A
(Retinol)

Vitamin D, which is related in structure to cholesterol, is synthesized when ultraviolet light from the sun strikes a cholesterol derivative in the skin. In the kidney, vitamin D is converted to a hormone that regulates calcium absorption and bone formation. Vitamin D deficiencies are most likely to occur in malnourished individuals living where there is little sunlight. (It is interesting to note that a sunscreen of SPF factor 6–8 completely blocks vitamin D synthesis.) For many people, as little as 15 minutes of sunshine daily is sufficient for maintenance of adequate vitamin D levels.

▲ Deeply pigmented vegetables and fruits contain vitamins.

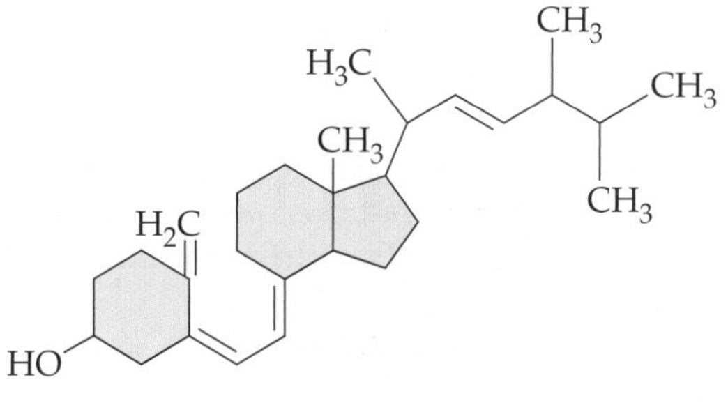

Vitamin D

Vitamin E comprises a group of structurally similar compounds called tocopherols, the most active of which is α-tocopherol. Like vitamin C, it is an antioxidant: It prevents the breakdown by oxidation of vitamin A and polyunsaturated fats. Vitamin E apparently is not toxic in overdosage as are the other fat-soluble vitamins. (Nevertheless, it is best to avoid excessively large doses of vitamin E.)

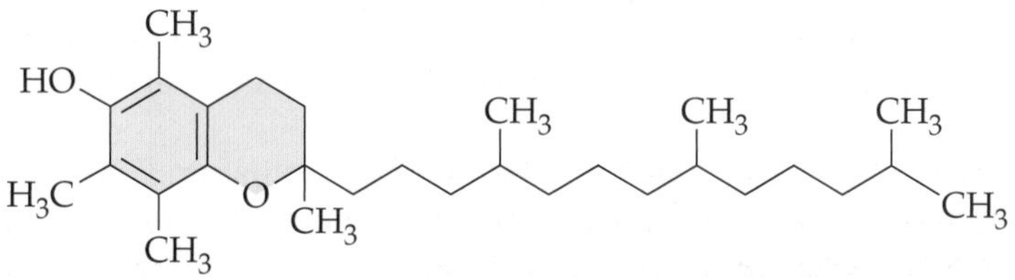

Vitamin E

Vitamin K also includes a number of structurally related compounds, in this case with hydrocarbon side chains of varying length. This vitamin is essential to the synthesis of several blood-clotting factors. It is produced by intestinal bacteria, so deficiencies are rare.

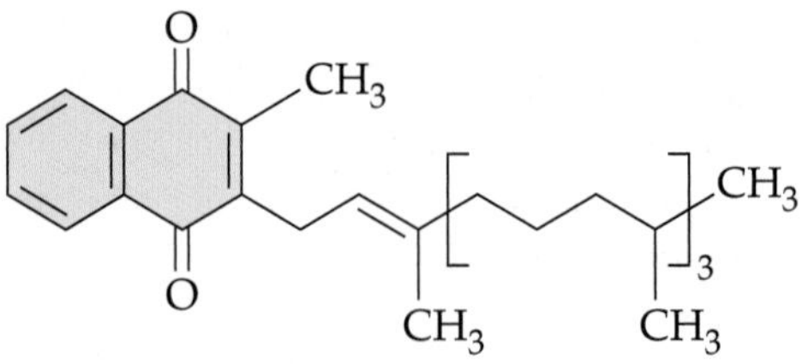

Vitamin K

PROBLEM 19.17

Compare the structures of vitamin A and vitamin C. What structural features does each have that make one water-soluble and the other fat-soluble?

PROBLEM 19.18

Based on the structure shown above for retinol (vitamin A) and the names of the two related forms of vitamin A, retinal and retinoic acid, what do you expect to be the structural differences among these three compounds?

Antioxidants

Antioxidant A substance that prevents oxidation by reacting with an oxidizing agent.

An **antioxidant** is a substance that prevents oxidation. Earlier we noted that antioxidants are used to combat air oxidation of unsaturated fats that cause deterioration of baked goods. (, p. 450) In the body, we need similar protection against active oxidizing agents that are by-products of normal metabolism.

Our principal dietary antioxidants are vitamin C, vitamin E, β-carotene, and the mineral selenium. They work together to defuse the potentially harmful action of **free radicals**, highly reactive molecular fragments with unpaired electrons (for example, superoxide ion, $\cdot O_2^-$). Free radicals quickly gain stability by picking up electrons from nearby molecules, which are thereby damaged. (We will have more to say about this in Section 21.10.)

Free radical An atom or molecule with an unpaired electron.

Vitamin E is unique in having antioxidant activity as its principal biochemical role. It acts by giving up the hydrogen from its —OH group to oxygen-containing free radicals. The hydrogen is then restored by reaction with vitamin C. Selenium joins the list of important antioxidants because it is a cofactor in an enzyme that converts hydrogen peroxide (H_2O_2) to water before the peroxide can go on to produce free radicals.

Evidence for the benefits of antioxidants in disease prevention, especially in the prevention of cancer and heart disease, is accumulating. Laboratory experiments have demonstrated anticancer activity for vitamin C, vitamin E, β-carotene, and selenium, and the results are supported by a variety of studies in defined human populations. For example, vitamin E appears to reduce the risk of cancer among smokers, though not among nonsmokers. Low levels of serum selenium and vitamin E have been associated with a greater risk of breast cancer in a group of Finnish women.

In a study of over 100,000 people, those who took vitamin E supplements had fewer heart attacks. Vitamin C may slow the development of blocked arteries and may also prevent cancer by inhibiting the formation of carcinogens in the gastrointestinal tract.

APPLICATION ▶ Vitamins, Minerals, and Food Labels

It is not uncommon to encounter incomplete or incorrect information about vitamins and minerals. We have been frightened by the possibility that aluminum causes Alzheimer's disease and tantalized by the possibility that vitamin C defeats the common cold. Sorting out fact from fiction or distinguishing preliminary research results from scientifically proven relationships is especially difficult in this area of nutrition. Much is yet to be learned about the functions of vitamins and minerals in the body, and new information is continuously being reported. It is tempting for health-conscious individuals to look for guaranteed routes to better health by taking vitamins or minerals, and taking advantage of this motivation is just as tempting to profit-making organizations.

One consistent source of information on nutrition is the Food and Nutrition Board of the National Academy of Sciences-National Research Council. They periodically survey the latest nutritional information and publish Recommended Dietary Allowances (RDAs) that are "designed for the maintenance of good nutrition of the majority of healthy persons in the United States." Another source is the U.S. Food and Drug Administration (FDA), which has among its many responsibilities setting the rules for food labeling.

Since 1994, as mandated by the FDA, most packaged food products carry standardized *Nutrition Facts* labels. The nutritional value of a food serving of a specified size is reported as *% Daily Value*. For vitamins and minerals, these percentages are calculated from *Reference Daily Intake* values (RDIs). The RDIs are mostly derived from the 1968 RDAs and were designed to avoid deficiencies. The RDIs are averages for adults and children over 4 years of age. The values for vitamins are included in Tables 19.3 and 19.4. For minerals, they are listed in the accompanying Table.

In choosing which vitamins and minerals *must* be listed on the new labels, the government has focused on those of greatest importance in maintaining good health. The choices reflect a new emphasis on preventing disease rather than preventing deficiencies. The *mandatory* listings are for vitamin A, vitamin C, calcium, and iron. These recommendations are based on evidence for the benefits of high dietary levels of the antioxidants vitamin A (or the related compound, β-carotene) and vitamin C. Calcium deficiencies are related to osteoporosis, and iron deficiencies are a special concern for women because of their menstrual blood loss. Thiamin, riboflavin, and niacin listings are no longer mandatory because deficiencies of these vitamins are no longer a public health problem in the United States.

Reference Daily Intake Values* for Minerals

MINERAL	RDI	MINERAL	RDI
Calcium	1.0 g	Selenium	70 μg
Iron	18 mg	Manganese	2 mg
Phosphorus	1.0 g	Fluoride	2.5 mg
Iodine	150 μg	Chromium	120 μg
Magnesium	400 mg	Molybdenum	75 μg
Zinc	15 mg	Chloride	3.4 g
Copper	2 mg		

On Nutrition Facts Labels, calcium and iron must be listed; phosphorus, iodine, magnesium, zinc, and copper listings are optional; the others cannot be listed.

Nutrition Facts
Serving Size 55 pieces (30g/1.1oz)
Servings Per Container About 6

Amount Per Serving

Calories 140   Calories from Fat 45

	% Daily Value*
Total Fat 5g	**8%**
Saturated Fat 1g	**5%**
Trans Fat 0g	
Polyunsaturated Fat 1.5g	
Monounsaturated Fat 2.5g	
Cholesterol Less than 5mg	**1%**
Sodium 250mg	**10%**
Total Carbohydrate 19g	**6%**
Dietary Fiber 2g	**7%**
Sugars Less than 1g	
Protein 4g	

Vitamin A	0%	Vitamin C	0%
Calcuim	4%	Iron	6%

*Percent Daily Values are based on a 2,000 calorie diet. Your daily values may be higher or lower depending on your caloric needs:

	Calories:	2,000	2,500
Total Fat	Less than	65g	80g
Sat. Fat	Less than	20g	25g
Cholesterol	Less than	300mg	300mg
Sodium	Less than	2,400mg	2,400mg
Total Carbohydrate		300g	375g
Dietary Fiber		25g	30g

See Additional Problems 19.82–19.85 at the end of the chapter.

◀● KEY CONCEPT PROBLEM 19.19

Vitamins are a diverse group of compounds that must be present in the diet. List four functions of vitamins in the body.

SUMMARY: REVISITING THE CHAPTER GOALS

1. **What are enzymes?** *Enzymes* are the catalysts for biochemical reactions. They are mostly water-soluble, globular proteins, and many incorporate *cofactors*, which are either metal ions or the nonprotein organic molecules known as *coenzymes*. One or more *substrate* molecules (the reactants) enter an *active site* lined by those protein side chains and cofactors necessary for catalyzing the reaction. Six major classes and many subclasses of reactions are catalyzed by enzymes (Table 19.2).

2. **How do enzymes work, and why are they so specific?** A substrate is drawn into the active site by noncovalent interactions. As the substrate enters the active site, the enzyme shape adjusts to best accommodate the substrate and catalyze the reaction (the *induced fit*). Within the *enzyme–substrate complex*, the substrate is held in the best orientation for reaction and in a strained condition that allows the activation energy to be lowered. When the reaction is complete, the product is released and the enzyme returns to its original condition. The *specificity* of each enzyme is determined by the presence within the active site of catalytically active groups, hydrophobic pockets, and ionic or polar groups that exactly fit the chemical makeup of the substrate.

3. **What effects do temperature, pH, enzyme concentration, and substrate concentration have on enzyme activity?** With increasing temperature, reaction rate increases to a maximum and then decreases as the enzyme protein denatures. Reaction rate is maximal at a pH that reflects the pH of the enzyme's site of action in the body. In the presence of excess substrate, reaction rate is directly proportional to enzyme concentration. With fixed enzyme concentration, reaction rate first increases with increasing substrate concentration and then approaches a fixed maximum at which all active sites are occupied. (See Figures 19.5, 19.6, and 19.7.)

4. **How is enzyme activity regulated?** The effectiveness of enzymes is controlled by a variety of *activation* and *inhibition* strategies. A product of a later reaction can exercise *feedback control* over an enzyme for an earlier reaction in a pathway. Feedback control acts through *allosteric control* of enzymes that have regulatory sites separate from their active sites. Binding a regulator induces a change of shape in the active site, increasing or decreasing the efficiency of the enzyme. *Allosteric regulators* are *noncompetitive inhibitors* because they act away from the active site; they lower the maximum reaction rate. *Competitive inhibitors* typically resemble the substrate and reversibly block the active site; they slow the reaction rate but do not change the maximum rate. *Irreversible inhibitors* form covalent bonds to an enzyme that permanently inactivate it; most are poisons. Enzyme activity is also regulated by *reversible* phosphorylation and dephosphorylation, and by synthesis of inactive *zymogens* that are later activated by removal of part of the molecule. *Genetic control* is exercised by regulation of the synthesis of enzymes.

5. **What are vitamins?** *Vitamins* are organic molecules required in small amounts in the diet because our bodies cannot synthesize them. The water-soluble vitamins (Table 19.3) are coenzymes or parts of coenzymes. The fat-soluble vitamins (Table 19.4) have diverse and less well understood functions. In general, excesses of water-soluble vitamins are excreted and excesses of fat-soluble vitamins are stored in body fat, making excesses of the fat-soluble vitamins potentially more harmful. Vitamin C, β-carotene (a precursor of vitamin A), vitamin E, and selenium work together as *antioxidants* to protect biomolecules from damage by free radicals.

UNDERSTANDING KEY CONCEPTS

19.20 On the diagram shown below, indicate with dotted lines the bonding between the enzyme (a dipeptidase; select amino acid residues in black) and the substrate (in blue) that might occur to form the enzyme–substrate complex. What are the two types of bonding likely to occur?

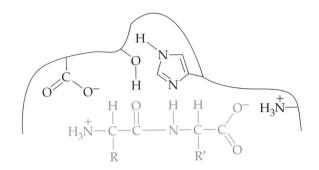

19.21 Answer questions (a)–(e) concerning the following reaction:

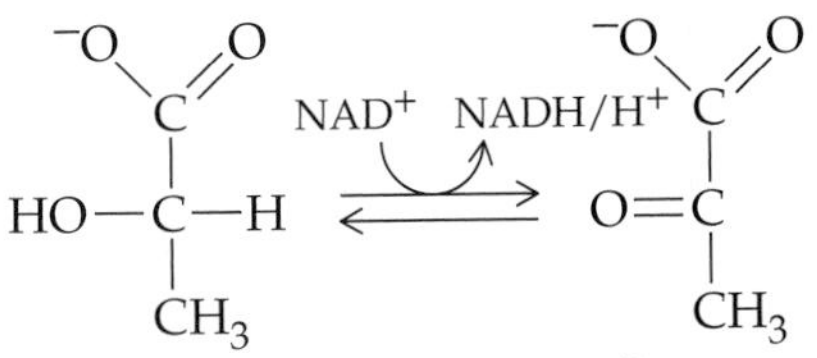

$$\text{L-Lactate} \xrightleftharpoons[\text{NAD}^+ \;\; \text{NADH/H}^+]{} \text{Pyruvate}$$

(a) The enzyme involved in this reaction belongs to what class of enzymes?

(b) Since hydrogens are removed, the enzyme belongs to what subclass of the enzyme class from part (a)?

(c) What is the substrate for the reaction as written?

(d) What is the product for the reaction as written?

(e) The enzyme name is derived from the substrate name and the subclass of the enzyme, and ends in the family-name ending for an enzyme. Name the enzyme.

19.22 In the reaction shown in Problem 19.21, will the enzyme likely use D-lactate as a substrate? Explain your answer. If D-lactate binds to the enzyme, how is it likely to affect the enzyme?

19.23 In the reaction shown in Problem 19.21, identify the coenzyme required for catalytic activity. Is the coenzyme an oxidizing agent or a reducing agent? Coenzymes that are modified during the reaction are called *stoichiometric*

coenzymes, since they are needed in stoichiometric proportions to the substrate. What vitamin is a part of the coenzyme for this reaction?

19.24 Explain how the following changes affect the rate of an enzyme-catalyzed reaction in the presence of a noncompetitive inhibitor: (a) increasing the substrate concentration at a constant inhibitor concentration, (b) decreasing the inhibitor concentration at a constant substrate concentration.

19.25 Explain how the following mechanisms regulate enzyme activity.

(a) Covalent modification
(b) Genetic control
(c) Allosteric regulation
(d) Inhibition

19.26 What type of enzyme inhibition occurs in the following situations?

(a) Buildup of the product of the pathway that converts glucose to pyruvate stops at the first enzyme in the multistep process.

(b) Sarin, a nerve gas, covalently binds to acetylcholinesterase, stopping nerve signal transmission.

(c) Lactase is not produced in the adult.

(d) Conversion of isocitrate to α-ketoglutarate is inhibited by high levels of ATP. (Hint: ATP is neither a product nor a substrate in this reaction.)

19.27 Acidic and basic groups are often found in the active sites of enzymes. Identify the acidic and basic amino acids in the active site in the adjacent diagram.

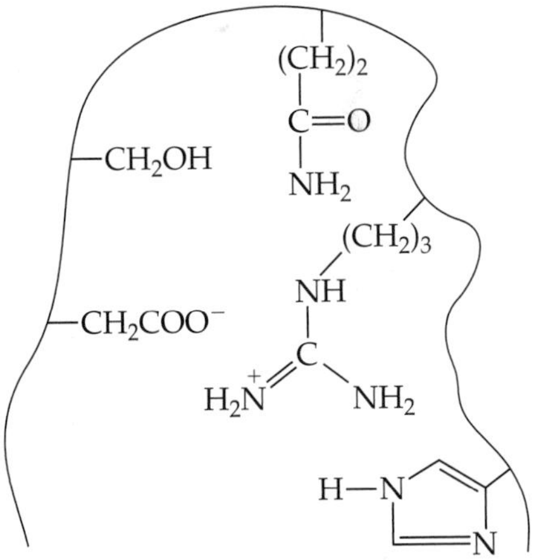

ADDITIONAL PROBLEMS

STRUCTURE AND CLASSIFICATION OF ENZYMES

19.28 What general kinds of reactions do the following enzymes catalyze?

(a) Oxidoreductases
(b) Lyases
(c) Transferases

19.29 What general kinds of reactions do the following enzymes catalyze?

(a) Hydrolases
(b) Isomerases
(c) Ligases

19.30 Name the enzyme that acts on each molecule.

(a) Sucrose
(b) Fumarate
(c) RNA

19.31 Name the enzyme that acts on each molecule.

(a) Galactose
(b) Urea
(c) DNA

19.32 What features of enzymes makes them so specific in their action?

19.33 Describe in general terms how enzymes act as catalysts.

19.34 What classes of enzymes would you expect to catalyze the following reactions?

(a) $H_2NCHCNHCHCOH + H_2O \longrightarrow$

$H_2NCHCOH + H_2NCHCOH$

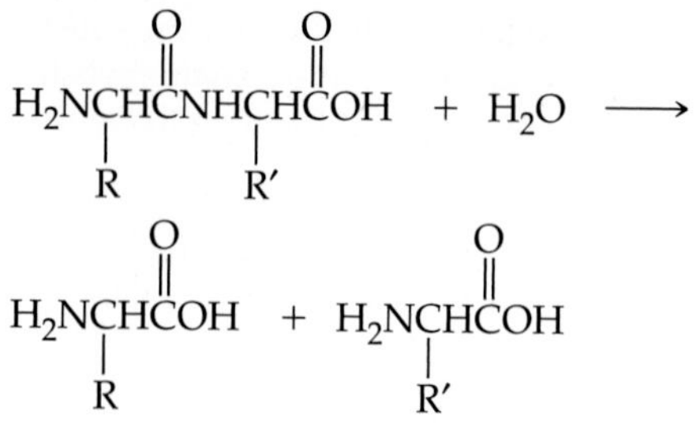

(b) $HOOC-CH_2-C-COOH \longrightarrow$

$CH_3-C-COOH + CO_2$

(c) $HOCCH_2CH_2COH \longrightarrow HOCCH=CHCOH$

19.35 What classes of enzymes would you expect to catalyze the following reactions?

(a) Pyruvate + L-Aspartate $\xrightarrow{\text{Vitamin B}_6}$ L-Alanine + Oxaloacetate

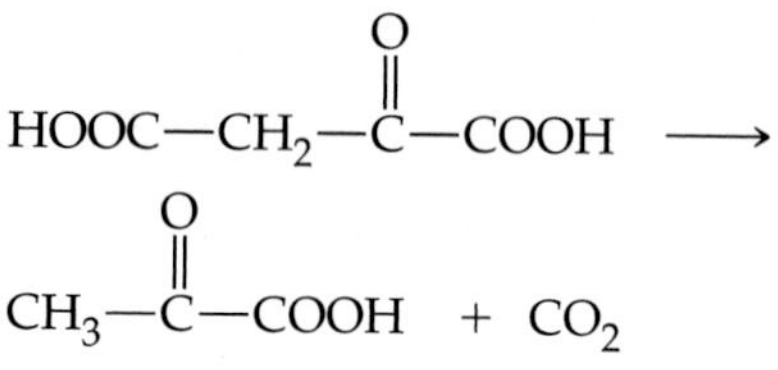

(b) 3-Phosphoglyceraldehyde $\rightleftharpoons$ Dihydroxyacetone phosphate

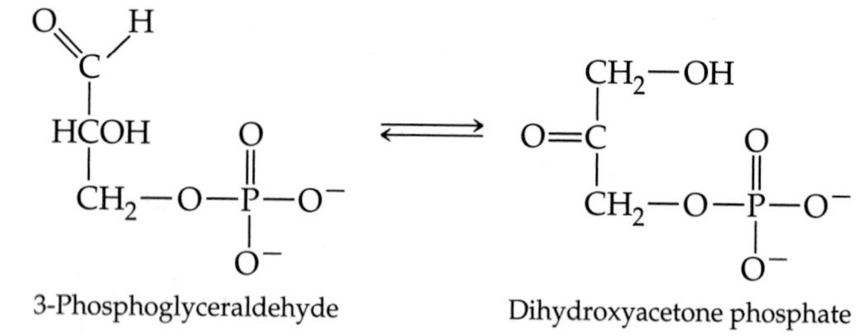

(c) Pyruvate + CO_2 $\xrightarrow{\text{ATP} \quad \text{ADP}}$ Oxaloacetate

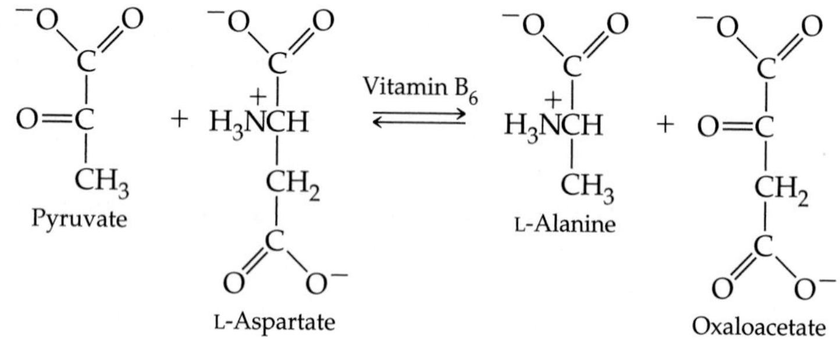

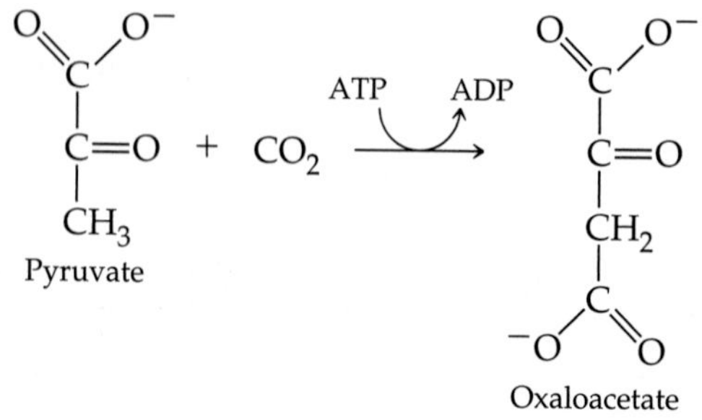

19.36 What kind of reaction does each of these enzymes catalyze?

(a) A decarboxylase
(b) A transmethylase
(c) A dehydrogenase

19.37 What kind of reaction does each of these enzymes catalyze?

(a) A kinase   (b) A ligase   (c) A peptidase

19.38 The following reaction is catalyzed by the enzyme urease. To what class of enzymes does urease belong?

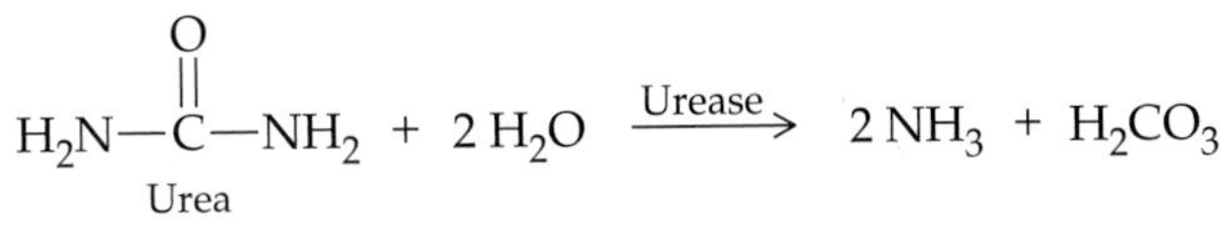

$$H_2N-\overset{\overset{\displaystyle O}{\|}}{C}-NH_2 \ + \ 2\,H_2O \ \xrightarrow{\text{Urease}} \ 2\,NH_3 \ + \ H_2CO_3$$

Urea

19.39 Alcohol dehydrogenase (ADH) catalyzes the following reaction. To what class of enzymes does ADH belong?

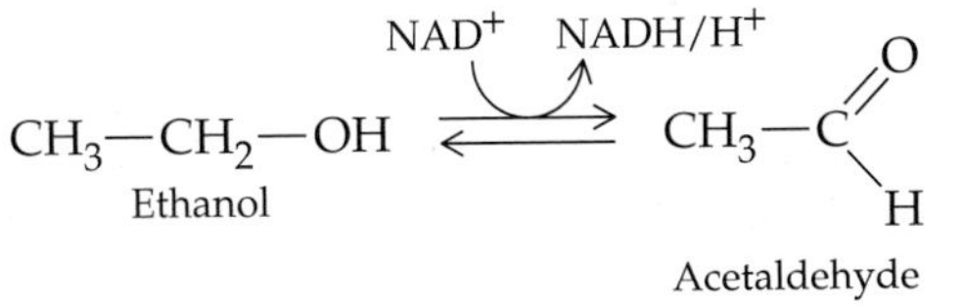

$$CH_3-CH_2-OH \ \underset{}{\overset{NAD^+ \quad NADH/H^+}{\rightleftharpoons}} \ CH_3-\overset{\overset{\displaystyle O}{\|}}{C}_{\diagdown H}$$

Ethanol   Acetaldehyde

19.40 Name the vitamin to which each of these coenzymes is related.

(a) FAD   (b) Coenzyme A
(c) Pyridoxal phosphate

19.41 Which of the following is a cofactor and which is a coenzyme?

(a) Cu^{2+} (b) Tetrahydrofolate
(c) NAD   (d) Mg^{2+}

ENZYME FUNCTION AND REGULATION

19.42 What is the difference between the lock-and-key model of enzyme action and the induced-fit model?

19.43 Why is the induced-fit model a more likely model than the lock-and-key model?

19.44 Must the amino acid residues in the active site be near each other along the polypeptide chain? Explain.

19.45 The active site of an enzyme is a small portion of the enzyme molecule. What is the function of the rest of the huge molecule?

19.46 How do you explain the observation that pepsin, a digestive enzyme found in the stomach, has a high catalytic activity at pH 1.5, while trypsin, an enzyme of the small intestine, has no activity at pH 1.5?

19.47 Amino acid side chains in the active sites of enzymes can act as acids or bases during catalysis. List the amino acid side chains that can accept H^+ and those that can donate H^+ during enzyme-catalyzed reactions.

19.48 Draw an energy diagram for the exothermic enzyme-catalyzed hydrolysis of urea (Problem 19.38). Label the energy levels of reactants and products, the activation energy, and the overall energy difference of the reaction.

19.49 Draw an energy diagram for the exothermic enzyme-catalyzed oxidation of ethanol (Problem 19.39). Label the energy levels of reactants and products, the activation energy, and the overall energy difference of the reaction.

19.50 Draw a graph showing how the rate of reaction changes as enzyme concentration is increased. Explain your diagram.

19.51 Draw a graph showing how the rate of reaction changes as substrate concentration is increased. Explain your diagram.

19.52 Do increases in concentration of substrate or enzyme change reaction rates in exactly the same way? Explain.

19.53 Do decreases in concentration of substrate or enzyme change reaction rates in exactly the same way? Explain.

19.54 What general effects would you expect the following changes to have on the rate of an enzyme-catalyzed reaction for an enzyme that has its maximum activity at body temperature (about 37 °C)?

(a) Raising the temperature from 37 °C to 60 °C
(b) Lowering the pH from 8 to 3
(c) Adding an organic solvent, such as methanol

19.55 What general effects would you expect the following changes to have on the rate of an enzyme-catalyzed reaction for an enzyme that has its maximum activity at body temperature (about 37 °C)?

(a) Lowering the reaction temperature from 85 °C to 40 °C
(b) Adding a drop of a dilute $PbCl_2$ solution
(c) Adding an oxidizing agent, such as hydrogen peroxide

19.56 The text discusses three forms of enzyme inhibition: noncompetitive inhibition, competitive inhibition, and irreversible inhibition.

(a) Describe how an enzyme inhibitor of each type works.
(b) What kinds of bonds are formed between an enzyme and each of these three kinds of inhibitors?

19.57 Refer to Problem 19.56. Poisoning by which of the three kinds of inhibitors is the most difficult to treat medically? Why?

19.58 Eco R1, an enzyme that hydrolyzes DNA strands, requires Mg^{2+} as a cofactor for activity. Ethylenediaminetetraacetic acid (EDTA) chelates divalent metal ions in solution. In the graphs shown here, the arrow indicates the point at which EDTA is added to a reaction mediated by Eco R1. Which graph represents the activity curve you would expect to see? (Activity is shown as total product from the reaction as time increases.)

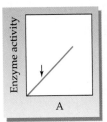
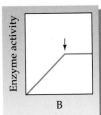
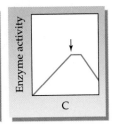

19.59 The enzyme lactic dehydrogenase converts lactic acid to pyruvate with the aid of the coenzyme NAD^+. In the graphs of Problem 19.58, the arrow indicates the point at which ethylenediaminetetraacetic acid (EDTA) is added to a reaction mixture of lactic dehydrogenase and lactic acid. Which graph represents the activity curve you would you expect to see? (Activity is shown as total product from the reaction as time increases.)

19.60 Lead exerts its poisonous effect on enzymes by two mechanisms. Which mechanism is irreversible and why?

19.61 One mechanism by which lead exerts its poisonous effect on enzymes can be stopped by chelation therapy with EDTA. Describe this type of lead poisoning and explain why it is reversible.

19.62 The meat tenderizer used in cooking is primarily papain, a protease enzyme isolated from the papaya tree. Why do you suppose papain is so effective at tenderizing meat?

19.63 Refer to Problem 19.62. Papain is also used to help relieve the pain of bee stings. Why do you suppose it works?

19.64 Why do allosteric enzymes have two types of binding sites?

19.65 Discuss the purpose of positive and negative regulation.

19.66 What is feedback inhibition?

19.67 What would be a reason to have "feed-forward activation"?

19.68 What is a zymogen? Why must some enzymes be secreted as zymogens?

19.69 Activation of a zymogen is by covalent modification. How might phosphorylation or dephosphorylation (also covalent modification) modify an enzyme to make it more active (or more inactive)?

VITAMINS

19.70 What criteria make a compound a vitamin?

19.71 What is the relationship between vitamins and enzymes?

19.72 Why is daily ingestion of vitamin C more critical than daily ingestion of vitamin A?

19.73 List the four fat-soluble vitamins.. Why is excess consumption of three of these vitamins of concern?

Applications

19.74 Give two reasons why an industrial chemist might choose to use enzymes rather than nonbiological catalysts for a particular reaction. [*Biocatalysis: Food and Chemicals, p. 600*]

19.75 A pharmaceutical chemist needs to synthesize *p*-hydroxyphenylglycine, whose structure is shown below. What feature of enzymes makes biocatalysis particularly useful in this reaction? [*Biocatalysis: Food and Chemicals, p. 600*]

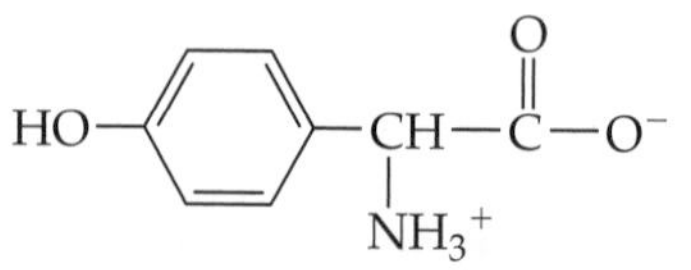

p-Hydroxyphenylglycine

19.76 Taq polymerase, an enzyme that joins nucleotides in DNA synthesis, is used in the polymerase chain reaction—a repeated sequence of automated steps that involve heating to 90 °C at one point. Taq polymerase is found in thermophilic bacteria. Explain why this enzyme retains its activity through repeated heating and cooling steps. [*Extremozymes: Enzymes from the Edge, p. 607*]

19.77 When energy companies drill for oil, are enzymes from thermophiles or psychrophiles used? Why? [*Extremozymes: Enzymes from the Edge, p. 607*]

19.78 What three enzymes show marked concentration increases in the blood after a heart attack? Why? [*Enzymes in Medical Diagnosis, p. 608*]

19.79 Why must enzyme activity be monitored under standard conditions? [*Enzymes in Medical Diagnosis, p. 608*]

19.80 The primary structure of angiotensin II has . . . Pro-Phe at the C-terminal end of the octapeptide. The angiotensin-converting enzyme (ACE) inhibitor from the South American pit viper is a pentapeptide with a C-terminal proline and is a mild ACE inhibitor. Captopril has a modified proline structure and is also a mild ACE inhibitor. Why do you suppose that a mild ACE inhibitor is more valuable for the treatment of high blood pressure than a very potent ACE inhibitor? What structural modifications to the pit viper peptide might make it a more powerful ACE inhibitor? [*Enzyme Inhibitors as Drugs, p. 613*]

19.81 AZT (zivovudine) inhibits the synthesis of the HIV virus RNA because AZT resembles substrate molecules. Which kind of inhibition is most likely taking place in this reaction? [*Enzyme Inhibitors as Drugs, p. 613*]

19.82 Why are overdoses less common with water-soluble vitamins than with fat-soluble vitamins? [*Vitamins, Minerals, and Food Labels, p. 621*]

19.83 Read the labels on foods that you eat for a day, or look the foods up in a nutrition table and determine what percent of your daily dosage of vitamins and minerals you get from each. Are you getting the recommended amounts from the food you eat, or should you be taking a vitamin or mineral supplement? [*Vitamins, Minerals, and Food Labels, p. 621*]

19.84 For what reasons are listings for vitamin A, vitamin C, iron, and calcium mandatory on food labels? [*Vitamins, Minerals, and Food Labels, p. 621*]

19.85 In addition to the four nutrients named in Problem 19.84, what other nutrients may be listed on food labels? (Hint, look at all the ingredients that have amounts listed on the label shown in the application box.) [*Vitamins, Minerals, and Food Labels, p. 621*]

General Questions and Problems

19.86 Look up the structures of vitamin C and vitamin E in Section 19.10, and identify the functional groups in these vitamins.

19.87 The adult recommended daily allowance (RDA) of riboflavin is 1.6 mg. If one glass (100 mL) of apple juice contains 0.014 mg of riboflavin, how much apple juice would an adult have to consume to obtain the RDA?

19.88 Many vegetables are "blanched" (dropped into boiling water) for a few minutes before being frozen. Why is blanching necessary?

19.89 How can you distinguish between a competitive inhibitor and a noncompetitive inhibitor experimentally?

19.90 Trypsin is an enzyme that cleaves on the C-terminal side (that is, to the right of) all basic residues. Consider the peptide shown below. Predict the fragments that would be formed by treatment of this peptide with trypsin:

N terminal end-Leu-Gly-Arg-Ile-Met-His-Tyr-Trp-Ala

19.91 The ability to change a selected amino acid residue to another amino acid is referred to as "point mutation" by biochemists. Referring to the reaction for peptide bond hydrolysis in Figure 19.4, speculate on the effects that the following point mutations might have on the chymotrypsin mechanism shown in Figure 19.4: serine to valine; aspartate to glutamate.

19.92 Ingestion of methanol is a medical emergency. It is often treated by the administration of ethanol, which prevents the dangerous effects of methanol metabolism. In the body, methanol is oxidized to formaldehyde, a toxic molecule that cannot be further oxidized and that damages proteins. Ethanol is oxidized, by the same enzyme (alcohol dehydrogenase), to acetaldehyde, a metabolite. If ethanol and methanol are substrates for the same enzyme, how does ethanol prevent the oxidation of methanol?

CHAPTER 20

Chemical Messengers: Hormones, Neurotransmitters, and Drugs

CONCEPTS TO REVIEW

Amino acids
(Sections 18.3, 18.4)

Shapes of proteins
(Section 18.8)

Tertiary and quaternary protein structure
(Sections 18.10, 18.11)

How enzymes work
(Section 19.4)

▲ A hard-played moment in the 2006 World Cup final. Floods of chemical messengers help these soccer players keep up the physical effort needed to finish the game.

CONTENTS

628

CHAPTER GOALS

In this chapter, we will address the following questions about hormones, neurotransmitters, and the drugs that affect their actions:

1. What are hormones, and how do they function?

THE GOAL: Be able to describe in general the origins, pathways, and actions of hormones.

2. What is the chemical nature of hormones?

THE GOAL: Be able to list, with examples, the different chemical types of hormones.

3. How does the hormone epinephrine deliver its message, and what is its major mode of action?

THE GOAL: Be able to outline the sequence of events in epinephrine's action as a hormone.

4. What are neurotransmitters, and how do they function?

THE GOAL: Be able to describe the general origins, pathways, and actions of neurotransmitters.

5. How does acetylcholine deliver its message, and how do drugs alter its function?

THE GOAL: Be able to outline the sequence of events in acetylcholine's action as a neurotransmitter and give examples of its agonists and antagonists.

6. Which neurotransmitters and what kinds of drugs play roles in allergies, mental depression, drug addiction, and pain?

THE GOAL: Be able to identify neurotransmitters and drugs active in these conditions.

7. What are some of the methods used in drug discovery and design?

THE GOAL: Be able to explain the general roles of ethnobotany, chemical synthesis, combinatorial chemistry, and computer-aided design in the development of new drugs.

At this point, you have seen a few of the many kinds of enzyme-catalyzed reactions that take place in cells. How are the individual reactions tied together? Clearly, the many thousands of reactions taking place in the billions of individual cells of our bodies do not occur randomly. There must be overall control mechanisms that coordinate these reactions, keeping us in chemical balance.

Two systems share major responsibility for regulating body chemistry—the *endocrine system* and the *nervous system*. The endocrine system depends on *hormones*, chemical messengers that circulate in the bloodstream. The nervous system relies primarily on a much faster means of communication—electrical impulses in nerve cells, triggered by its own chemical messengers, the *neurotransmitters*. Neurotransmitters carry signals from one nerve cell to another and also from nerve cells to their targets, the ultimate recipients of the messages.

Given the crucial role of hormones and neurotransmitters in the proper functioning of our bodies, it should not be surprising to find that many drugs act by mimicking, modifying, or opposing the action of chemical messengers.

Receptor A molecule or portion of a molecule with which a hormone, neurotransmitter, or other biochemically active molecule interacts to initiate a response in a target cell.

Hormone A chemical messenger secreted by cells of the endocrine system and transported through the bloodstream to target cells with appropriate receptors where it elicits a response.

20.1 Messenger Molecules

Coordination and control of your body's vital functions are accomplished by chemical messengers. Whether the messengers are hormones that arrive via the bloodstream or neurotransmitters released by nerve cells, such messengers ultimately connect with a *target*. The message is delivered by interaction between the chemical messenger and a **receptor** at the target. The receptor then acts like a light switch, causing some biochemical response to occur—the contraction of a muscle, for example, or the secretion of another biomolecule.

Non-covalent attractions draw messengers and receptors together, much as a substrate is drawn into the active site of an enzyme. (⬡⬡⬡, Sections 18.8, 19.4) These attractions hold the messenger and receptor together long enough for the message to be delivered, but without any permanent chemical change to the messenger or the receptor. The results of the interaction are chemical changes within the target cell.

The chemical messengers of the endocrine system are the **hormones**. These molecules are produced by endocrine glands and tissues in various parts of the body, quite often at distances far away from their ultimate site of action. Because of this, hormones must travel through the bloodstream to their targets and the responses

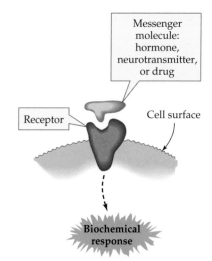

▲ A general representation of the interaction between a messenger molecule and a cellular receptor.

they produce can require anywhere from seconds to hours to begin. The action or actions they elicit, however, may last a long time and can be wide-ranging. A single hormone will often affect many different tissues and organs—any cell with the appropriate receptors is a target. Insulin, for example, is a hormone secreted by the pancreas in response to elevated blood glucose levels. At target cells throughout the body, insulin accelerates uptake and utilization of glucose; in muscles it accelerates formation of glycogen, a glucose polymer that is metabolized when muscles need quick energy; and in fatty tissue it stimulates storage of fat molecules (lipids).

The chemical messengers of the nervous system are a set of molecules referred to as **neurotransmitters**. The electrical signals of the nervous system travel along nerve fibers, taking only a fraction of a second to reach their highly specific destinations. Most nerve cells, however, do not make direct contact with the cells they stimulate. A neurotransmitter must carry the message across the tiny gap separating the nerve cell from its target. Because neurotransmitters are released in very short bursts and are quickly broken down or reabsorbed by the nerve cell, their effects are short-lived. The nervous system is organized so that nearly all of its vital switching, integrative, and information-processing functions depend on neurotransmitters. Neurotransmitters are typically synthesized and released very close to their site of action.

In this chapter, we first discuss hormones and the endocrine system and include a detailed description of how one hormone—epinephrine (also known as adrenalin)—performs its functions. Then we will cover neurotransmitters, using the action of acetylcholine to illustrate how neurotransmitters act. It is essential to recognize that hormones and neurotransmitters play a fundamental role in maintaining your health by their influence on metabolic processes. (See the Application "Homeostasis.") Finally, we will look briefly at the discovery and design of drugs as chemical messengers.

Neurotransmitter A chemical messenger that travels between a neuron and a neighboring neuron or other target cell to transmit a nerve impulse.

20.2 Hormones and the Endocrine System

The **endocrine system** includes all cells that secrete hormones into the bloodstream. Some of these cells are found in organs that also have non-endocrine functions (for example, the pancreas, which also produces digestive enzymes); others occur in glands devoted solely to hormonal control (for example, the thyroid gland). Hormones, however, do NOT carry out chemical reactions. Hormones are simply messengers that alter the biochemistry of a cell by signaling the inhibition or activation of an existing enzyme, by initiating or altering the rate of synthesis of a specific protein, or in other ways.

Endocrine system A system of specialized cells, tissues, and ductless glands that excretes hormones and shares with the nervous system the responsibility for maintaining constant internal body conditions and responding to changes in the environment.

The major endocrine glands are the thyroid gland, the adrenal glands, the ovaries and testes, and the pituitary gland (found in the brain). The hypothalamus, a section of the brain just above the pituitary gland, is in charge of the endocrine system. It communicates with other tissues in three ways:

- **Direct neural control** A nervous system message from the hypothalamus initiates release of hormones by the adrenal gland. For example,

$$\text{Hypothalamus} \xrightarrow{\text{Nerve message}} \text{Adrenal gland} \longrightarrow \text{Epinephrine}$$

Epinephrine is targeted to many cells; it increases heart rate, blood pressure, and glucose availability.

- **Direct release of hormones** Hormones move from the hypothalamus to the posterior pituitary gland, where they are stored until needed. For example,

$$\text{Hypothalamus} \longrightarrow \text{Antidiuretic hormone}$$

Antidiuretic hormone, which is stored in the posterior pituitary gland, targets the kidneys and causes retention of water and elevation of blood pressure.

- **Indirect control through release of regulatory hormones** In the most common control mechanism, *regulatory hormones* from the hypothalamus stimulate or

APPLICATION ▶   Homeostasis

*H*omeostasis—the maintenance of a constant internal environment in the body—is as important to the study of living things as atomic structure is to the study of chemistry. The phrase "internal environment" is a general way to describe all the conditions within cells, organs, and body systems. Conditions such as body temperature, the availability of chemical compounds that supply energy, and the disposal of waste products must remain within specific limits for an organism to function properly. Throughout our bodies, sensors track the internal environment and send signals to restore proper balance if the environment changes. If oxygen is in short supply, for example, a signal is sent that makes us breathe harder. When we are cold, a signal is sent to constrict surface blood vessels and prevent further loss of heat.

At the chemical level, homeostasis requires that the concentrations of ions and many different organic compounds stay near normal levels. The predictability of the concentrations of such substances is the basis for *clinical chemistry*—the chemical analysis of body tissues and fluids. In the clinical lab, various tests measure concentrations of significant ions and compounds in blood, urine, feces, spinal fluid, or other samples from a patient's body. Comparing the lab results with "norms" (average concentration ranges in a population of healthy individuals) shows which body systems are struggling, or possibly failing, to maintain homeostasis. To give just one example, urate (commonly known as uric acid) is an anion that helps to carry waste nitrogen from the body. A uric acid concentration higher than the normal range of about 2.5–7.7 mg/dL in blood can indicate the onset of gout or signal possible kidney malfunction.

A copy of a clinical lab report for a routine blood analysis is shown in the Figure. (Fortunately, this individual has no significant variations from normal.) The metal names in the report refer to the various cations, and the heading "Phosphorus" refers to the phosphate anion.

See Additional Problems 20.78 and 20.79 at the end of the chapter.

TEST	RESULT	NORMAL RANGE	TEST	RESULT	NORMAL RANGE
Albumin	4.3 g/dL	3.5–5.3 g/dL	SGOT*	23 U/L	0–28 U/L
Alk. Phos.*	33 U/L	25–90 U/L	Total protein	5.9 g/dL	6.2–8.5 g/dL
BUN*	8 mg/dL	8–23 mg/dL	Triglycerides	75 mg/dL	36–165 mg/dL
Bilirubin T.*	0.1 mg/dL	0.2–1.6 mg/dL	Uric Acid	4.1 mg/dL	2.5–7.7 mg/dL
Calcium	8.6 mg/dL	8.5–10.5 mg/dL	GGT*	23 U/L	0–45 U/L
Cholesterol	227 mg/dL	120–250 mg/dL	Magnesium	1.7 mEq/L	1.3–2.5 mEq/L
Chol., HDL*	75 mg/dL	30–75 mg/dL	Phosphorus	2.6 mg/dL	2.5–4.8 mg/dL
Creatinine	0.6 mg/dL	0.7–1.5 mg/dL	SGPT*	13 U/L	0–26 U/L
Glucose	86 mg/dL	65–110 mg/dL	Sodium	137.7 mEq/L	135–155 mEq/L
Iron	101 mg/dL	35–140 mg/dL	Potassium	3.8 mEq/L	3.5–5.5 mEq/L
LDH*	48 U/L	50–166 U/L			

▲ A clinical lab report for routine blood analysis. The abbreviations marked with asterisks are for the following tests (alternative standard abbreviations are in parentheses): Alk. Phos., alkaline phosphatase (ALP); BUN, blood urea nitrogen; Bilirubin T., total bilirubin; Chol., HDL, cholesterol, high-density lipoproteins; LDH, lactate dehydrogenase; SGOT, serum glutamic oxaloacetic transaminase (AST); GGT, γ-glutamyl transferase; SGPT, serum glutamic pyruvic transaminase (ALT).

inhibit the release of hormones by the anterior pituitary gland. Many of these pituitary hormones in turn stimulate release of still other hormones by their own target tissues. For example,

Hypothalamus $\xrightarrow{\text{Releasing factor}}$ Pituitary gland $\longrightarrow$

Thyrotropin (a regulatory hormone) $\longrightarrow$

Thyroid gland $\longrightarrow$ Thyroid hormones

Thyroid hormones are targeted to cells throughout the body; they affect oxygen availability, blood pressure, and other endocrine tissues.

Chemically, hormones are of three major types: (1) amino acid derivatives, small molecules containing amino groups; (2) polypeptides, which range from just a few amino acids to several hundred amino acids; and (3) steroids, which are lipids with a distinctive molecular structure based on four connected rings.

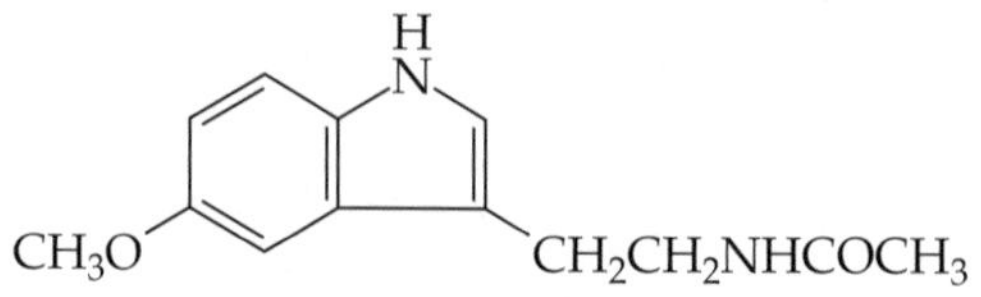

Melatonin, an amino acid derivative
(regulates day–night cycle)

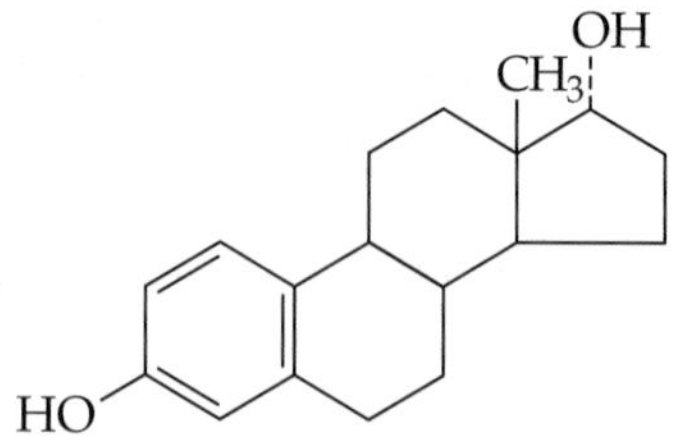

Estradiol, a steroid
(an estrogen that acts in ovulation)

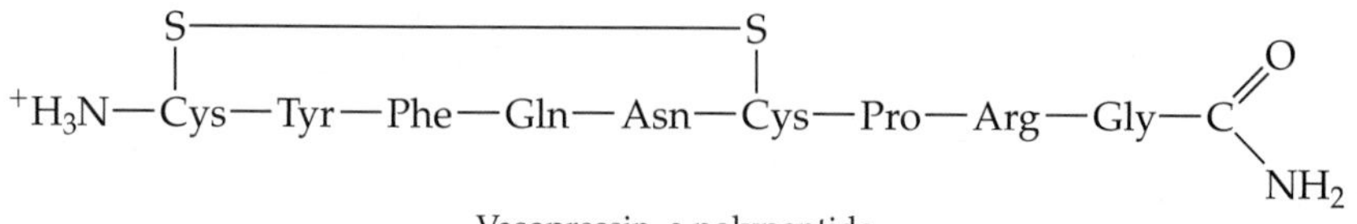

^+H_3N—Cys—Tyr—Phe—Gln—Asn—Cys—Pro—Arg—Gly—C

Vasopressin, a polypeptide
(controls urine volume)

Examples of the targets and actions of some hormones of each type are given in Table 20.1.

Upon arrival at its target cell, a hormone must deliver its signal to create a chemical response inside the cell. The signal enters the cell in ways determined by

TABLE 20.1 Examples of Each Chemical Class of Hormones

CHEMICAL CLASS	HORMONE EXAMPLES	SOURCE	TARGET	MAJOR ACTION
Amino acid derivatives	Epinephrine and norepinephrine	Adrenal medulla	Most cells	Release glucose from storage; increase heart rate and blood pressure
	Thyroxine	Thyroid gland	Most cells	Influence energy use, oxygen consumption, growth, and development
Polypeptides (regulatory hormones)	Adrenocorticotropic hormone	Anterior pituitary	Adrenal cortex	Stimulate release of glucocorticoids (steroids), which control glucose metabolism
	Growth hormone	Anterior pituitary	Peripheral tissues	Stimulate growth of muscle and skeleton
	Follicle-stimulating hormone, luteinizing hormone	Anterior pituitary	Ovaries and testes	Stimulate release of steroid hormones
	Vasopressin	Posterior pituitary	Kidneys	Cause retention of water, elevation of blood volume and blood pressure
	Thyrotropin	Anterior pituitary	Thyroid gland	Stimulates release of thyroid hormones
Steroids	Cortisone and cortisol (glucocorticoids)	Adrenal cortex	Most cells	Counteract inflammation; control metabolism when glucose must be conserved
	Testosterone; estrogen, progesterone	Testes; ovaries	Most cells	Control development of secondary sexual characteristics, maturation of sperm and eggs

the chemical nature of the hormone (Figure 20.1). Because the cell is surrounded by a membrane composed of hydrophobic molecules, only nonpolar, hydrophobic molecules can move across it on their own. The steroid hormones are nonpolar, so they can enter the cell directly; this is one of the ways a hormone delivers its message. Once within the cell's cytoplasm, a steroid hormone encounters a receptor that carries it to its target, DNA in the nucleus of the cell. The result is some variation in production of a protein governed by a particular gene.

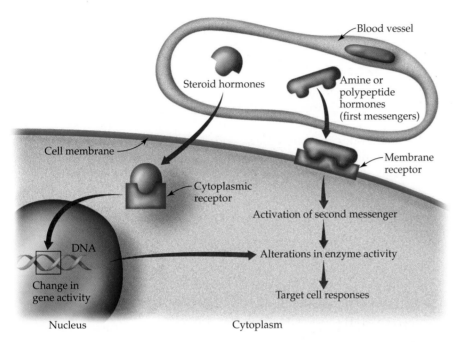

▲ **FIGURE 20.1   Interaction of hormones and receptors at the cellular level.**   Steroid hormones are hydrophobic and can cross the cell membrane to find receptors inside the cell. Amine and polypeptide hormones are hydrophilic and, because they cannot cross the cell membrane, act via second messengers.

By contrast, the polypeptide and amine hormones are water-soluble molecules and as such cannot cross the hydrophobic cell membranes. Rather than entering the cell, they deliver their message by bonding noncovalently with receptors on cell surfaces. The result is release of a **second messenger** within the cell. There are several different second messengers, and the specific sequence of events varies. In general, three membrane-bound proteins participate in release of the second messenger: the receptor (1) and a *G protein* (a member of the guanine nucleotide-binding protein family) (2) that transfer the message to an enzyme (3). First, interaction of the hormone with its receptor causes a change in the receptor (much like the effect of an allosteric regulator on an enzyme; Section 19.7). Next, the G protein activates an enzyme that participates in release of the second messenger. The action of epinephrine by way of a second messenger is described in Section 20.3. Further examples of amino acid, polypeptide, and steroid hormones are given in Sections 20.4 and, 20.5.

Second messenger Chemical messenger released inside a cell when a hydrophilic hormone or neurotransmitter interacts with a receptor on the cell surface.

⫯⫰⫰ Looking Ahead

The cell membrane is hydrophobic because it is composed primarily of *lipids*—molecules that are not water-soluble and cling together like oil floating on the surface of a pond. Even though steroids are classified as lipids because they are not water-soluble, they are discussed here because they function as hormones. The chemical nature of lipids other than steroids and of cell membranes is explored in Chapter 24.

> **WORKED EXAMPLE** **20.1** Classifying Hormones Based on Structure
>
> Classify the following hormones as an amino acid derivative, a polypeptide, or a steroid.
>
> (a) 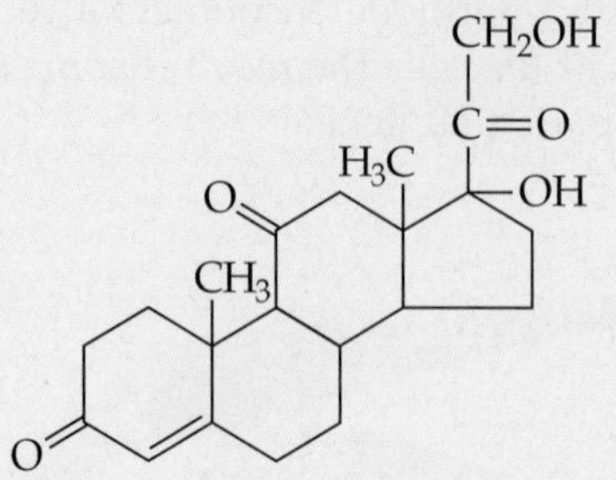
>
> (b) 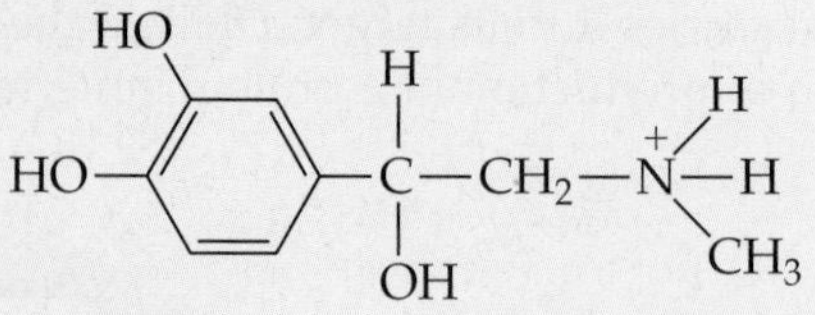
>
> (c) ^+H_3N—His—Ser—Glu— $\cdots$ Thr—COO$^-$
>
> **ANALYSIS** Hormones that are amino acid derivatives are recognized by the presence of amino groups. Those that are polypeptides are composed of amino acids. Steroids are recognizable by their distinctive four-ring structures.
>
> **SOLUTION**
> Compound (a) is a steroid, (b) is an amino acid derivative, and (c) is a polypeptide.

20.3 How Hormones Work: Epinephrine and Fight-or-Flight

Epinephrine (pronounced ep-pin-**eff**-rin), also known as *adrenaline*, is often called the *fight-or-flight hormone* because it is released when we need an instant response to danger.

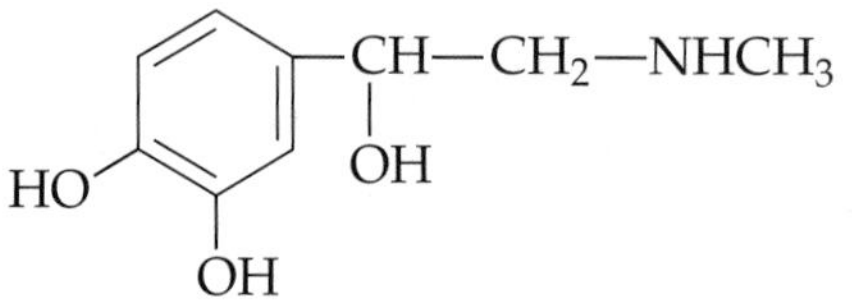

Epinephrine
(Adrenaline)

We have all felt the rush of epinephrine that accompanies a near-miss accident or a sudden loud noise. The main function of epinephrine in a "startle" reaction is a dramatic increase in the availability of glucose as a source of energy to deal with whatever stress is at hand. The time elapsed from initial stimulus to glucose release into the bloodstream is only a few seconds.

Epinephrine acts via *cyclic adenosine monophosphate* (*cyclic AMP*, or *cAMP*), an important second messenger. The sequence of events in this action, shown in Figure 20.2 and described below, illustrates one type of biochemical response to a change in an individual's external or internal environment.

- Epinephrine, a hormone carried in the bloodstream, binds to a receptor on the surface of a cell.

- The hormone–receptor complex activates a nearby G protein embedded in the interior surface of the cell membrane.

- GDP (guanosine diphosphate) associated with the G protein is converted to GTP (guanosine triphosphate) by addition of a phosphate group.

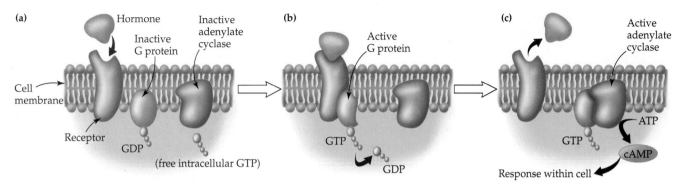

▲ **FIGURE 20.2 Activation of cyclic AMP as a second messenger.** (a) The hormone receptor, inactive G protein, and inactive adenylate cyclase enzyme reside in the cell membrane. (b) On formation of the hormone–receptor complex an allosteric change occurs and the guanosine diphosphate (GDP) of the G protein is replaced by a free intracellular guanosine triphosphate (GTP). (c) The active G protein–GTP complex activates adenylate cyclase, causing production of cyclic AMP inside the cell, where it initiates the action called for by the hormone.

- The G protein–GTP complex activates *adenylate cyclase*, an enzyme that also is embedded in the interior surface of the cell membrane.
- Adenylate cyclase catalyzes production within the cell of the second messenger—*cyclic AMP*—from ATP, as shown in Figure 20.3.
- Cyclic AMP initiates reactions that activate glycogen phosphorylase, the enzyme responsible for release of glucose from storage. (Interaction of other hormones with their specific receptors results in initiation by cyclic AMP of other reactions.)
- When the emergency has passed, cyclic AMP is converted back to ATP.

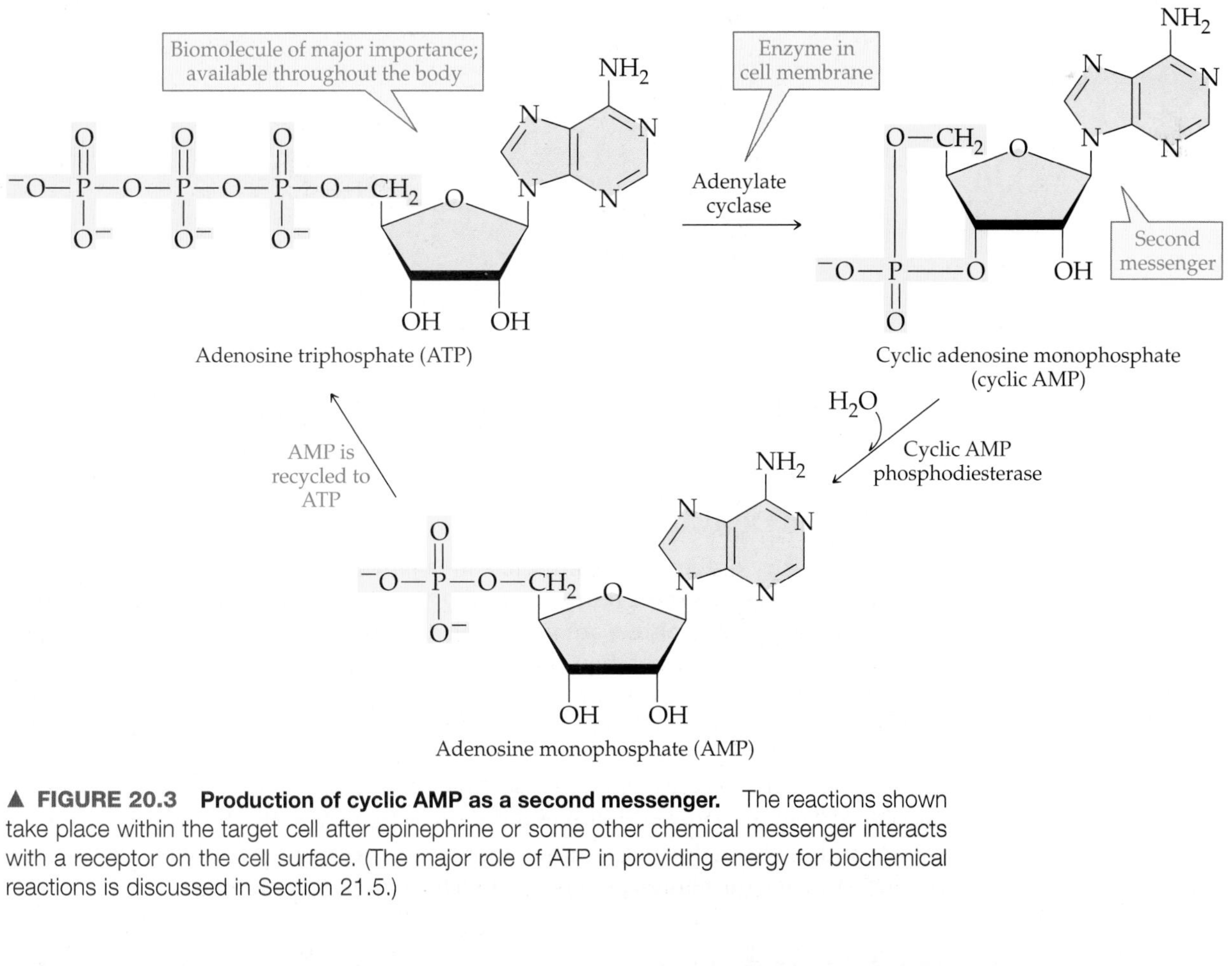

▲ **FIGURE 20.3 Production of cyclic AMP as a second messenger.** The reactions shown take place within the target cell after epinephrine or some other chemical messenger interacts with a receptor on the cell surface. (The major role of ATP in providing energy for biochemical reactions is discussed in Section 21.5.)

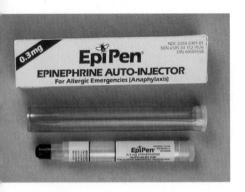

▲ An epinephrine autoinjection pen. Such devices are carried by individuals at risk of an anaphylactic reaction to an allergen.

In addition to making glucose available, epinephrine reacts with other receptors that increase blood pressure, heart rate, and respiratory rate; decrease blood flow to the digestive system (digestion is not important during an emergency); and counteract spasms in the respiratory system. The resulting combined and rapid effects make epinephrine the most crucial drug for treatment of *anaphylactic shock*. Anaphylactic shock is the result of a severe allergic reaction, perhaps to a bee or wasp sting or to a drug or even to something seemingly as benign as peanuts; it is an extremely serious medical emergency. The major symptoms include a severe drop in blood pressure due to blood vessel dilation and difficulty breathing due to bronchial constriction. Epinephrine directly counters these symptoms. Individuals who know they are susceptible to these life-threatening allergic responses carry epinephrine with them at all times (typically in the form of an autoinjector known as an "EpiPen").

PROBLEM 20.1

A phosphorus-containing anion is removed from ATP in its conversion to cyclic AMP, as shown in Figure 20.3. The anion is often abbreviated as PP_i. Which of the following anions is represented by PP_i?

(a) $H_2PO_4^-$ (b) PO_4^{3-} (c) $P_2O_7^{4-}$ (d) $P_3O_{10}^{5-}$

KEY CONCEPT PROBLEM 20.2

Caffeine and theobromine (from chocolate) act as stimulants. They work by altering the cAMP signal. Refer to Figure 20.3 and decide how these molecules might interact with an enzyme in the cAMP pathway to enhance the effect of cAMP.

Caffeine   Theobromine

20.4 Amino Acid Derivatives and Polypeptides as Hormones

Amino Acid Derivatives

The biochemistry of the brain is an active area of research. As our understanding of chemical messages in the brain grows, the traditional distinctions between hormones and neurotransmitters are vanishing. Several amino acid derivatives classified as hormones because of their roles in the endocrine system are also synthesized in neurons and function as neurotransmitters in the brain. (Because a barrier—the *blood–brain barrier*—limits entry into the brain of chemicals traveling in the bloodstream, the brain cannot rely on a supply of chemical messengers synthesized elsewhere; see Section 29.4.) Epinephrine, the fight-or-flight hormone described in Section 20.3, is one of the amino acid derivatives that are both a hormone and a neurotransmitter. The pathway for the synthesis of epinephrine is shown in Figure 20.4; several other chemical messengers are also formed in this pathway.

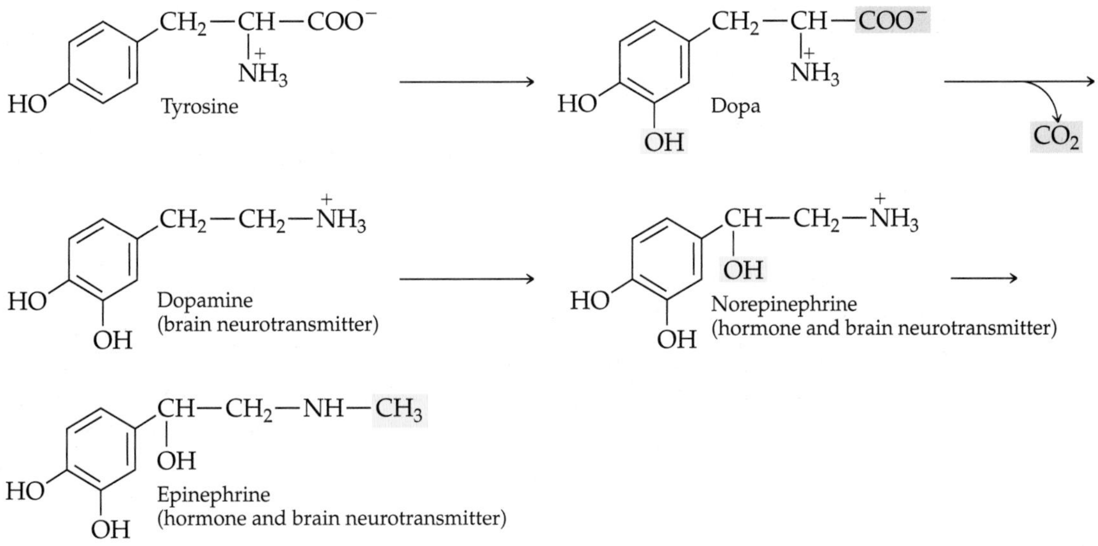

Tyrosine → Dopa (with CO_2 loss) → Dopamine (brain neurotransmitter) → Norepinephrine (hormone and brain neurotransmitter) → Epinephrine (hormone and brain neurotransmitter)

▲ **FIGURE 20.4 Synthesis of chemical messengers from tyrosine.** The changes in each step are highlighted in gold for additions and in green for losses.

Thyroxine, another amino acid derivative, is also a hormone. It is one of two iodine-containing hormones produced by the thyroid gland, and our need for dietary iodine is due to these hormones. Unlike other hormones derived from amino acids, thyroxine is a nonpolar compound that can cross cell membranes and enter cells, where it activates the synthesis of various enzymes. When dietary iodine is insufficient, the thyroid gland compensates by enlarging in order to produce more thyroxine. Thus, a greatly enlarged thyroid gland (a goiter) is a symptom of iodine deficiency. In developed countries where iodine is added to table salt, goiter is uncommon. In some regions of the world, however, iodine deficiency is a common and serious problem that results not only in goiter but also in severe mental retardation in infants (*cretinism*).

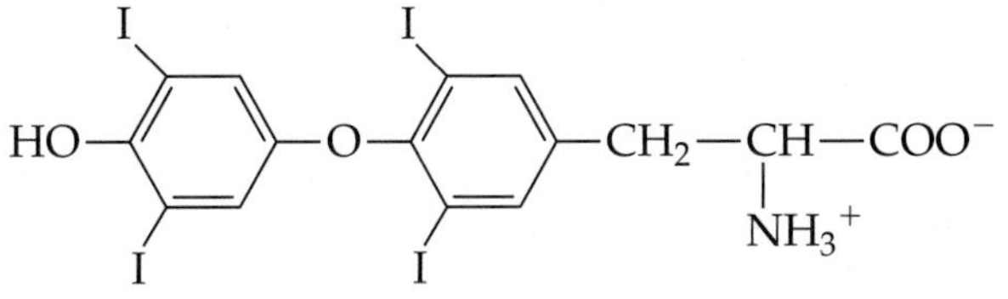

Thyroxine

Polypeptides

Polypeptides are the largest class of hormones. They range widely in molecular size and complexity as illustrated by two hormones that control the thyroid gland, *thyrotropin-releasing hormone (TRH)* and *thyroid-stimulating hormone (TSH)*. TRH, a modified tripeptide, is a regulatory hormone released by the hypothalamus. At the pituitary gland, TRH activates release of TSH, a protein that has 208 amino acid residues in two chains. The TSH in turn triggers release of amino acid derivative hormones from the thyroid gland.

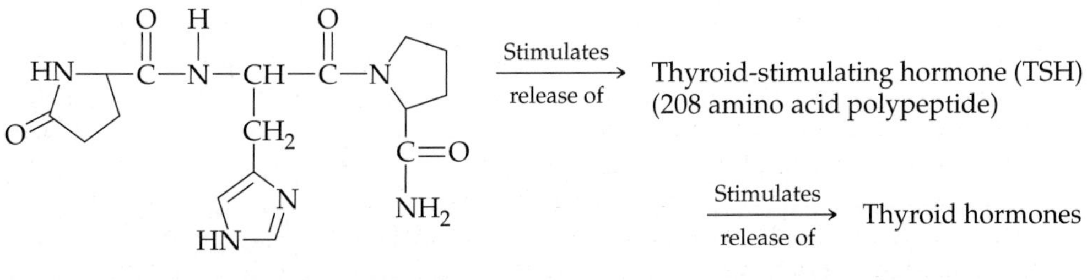

Thyrotropin-releasing hormone (TRH) — Stimulates release of → Thyroid-stimulating hormone (TSH) (208 amino acid polypeptide) — Stimulates release of → Thyroid hormones

Insulin, a protein containing 51 amino acids (⬤▭, p. 572), is released by the pancreas in response to high concentrations of glucose in the blood. It stimulates cells to take up glucose and either put it to use or store it.

⬤▭ **Looking Ahead**

Because of its importance in glucose metabolism and diabetes mellitus, the function of insulin as a hormone is described in Section 23.9. ⬤▭

PROBLEM 20.3

Examine the TRH structure and identify the three amino acids from which it is derived. (The N-terminal amino acid has undergone ring formation, and the carboxyl group at the C-terminal end has been converted to an amide.)

◆ KEY CONCEPT PROBLEM 20.4

Look at the structure of thyroxine above. Is thyroxine, an amino acid derivative, hydrophobic or hydrophilic? Explain.

20.5 Steroid Hormones

Steroid A lipid whose structure is based on the following tetracyclic (four-ring) carbon skeleton:

The steroid nucleus

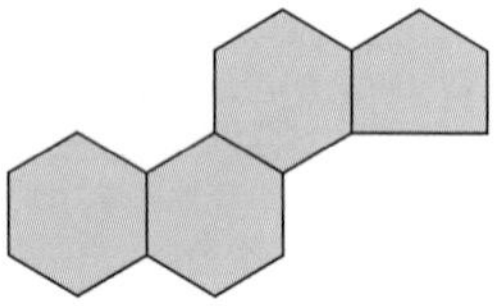

Steroids have in common a central structure composed of the four connected rings shown in the margin. Because they are soluble in hydrophobic solvents and not in water, steroids are classified as lipids. The steroid hormones are divided according to function into three types:

- *Mineralocorticoids*, such as *aldosterone*, regulate the delicate cellular fluid balance between Na^+ and K^+ ions (hence the "mineral" in their name).

- *Glucocorticoids*, such as cortisol (also known as *hydrocortisone*) and its close relative cortisone, help to regulate glucose metabolism and inflammation. You have probably used an anti-inflammatory ointment containing hydrocortisone to reduce the swelling and itching of poison ivy or some other skin irritation.

- *Sex hormones.* The two most important male sex hormones, or *androgens*, are *testosterone* and *androsterone*. They are responsible for the development of male secondary sex characteristics during puberty and for promoting tissue and muscle growth.

Male sex hormones (androgens)

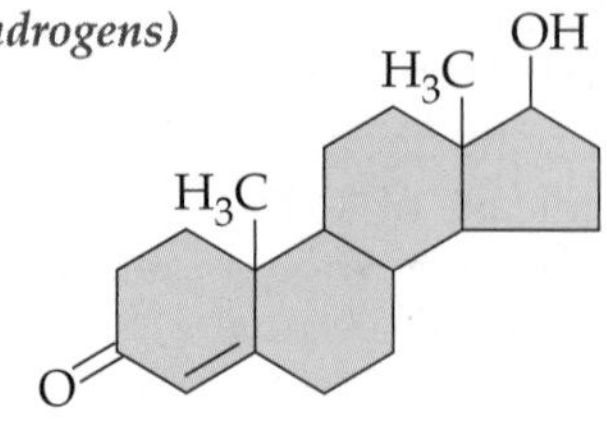

Testosterone

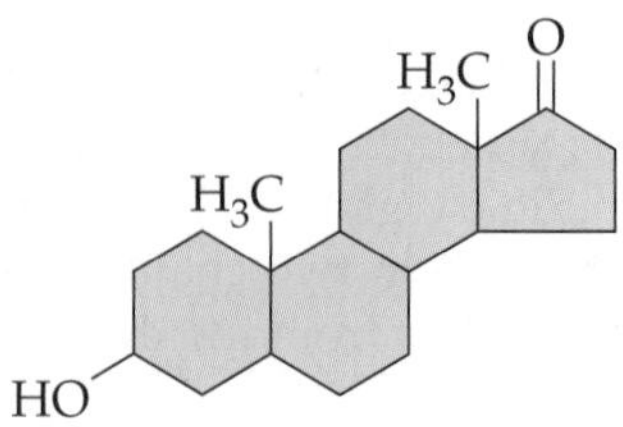

Androsterone

Estrone and *estradiol*, the female hormones known as *estrogens*, are synthesized from testosterone, primarily in the ovaries but also to a small extent in the adrenal cortex. Estrogens govern development of female secondary sex characteristics and participate in regulation of the menstrual cycle. The *progestins*, principally *progesterone*, are released by the ovaries during the second half of the menstrual cycle and prepare the uterus for implantation of a fertilized ovum should conception occur.

Female sex hormones

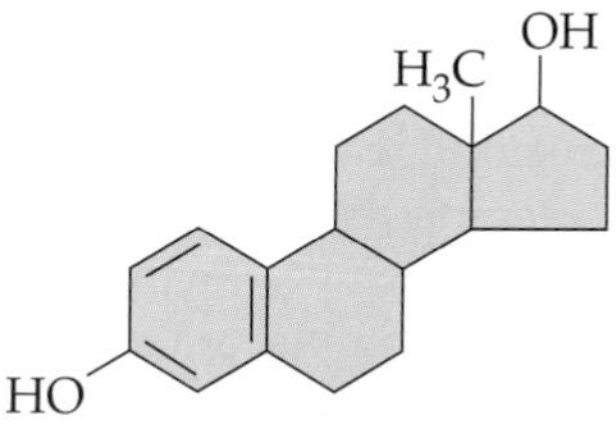

Estradiol (an estrogen)	Estrone (an estrogen)	Progesterone (a progestin)

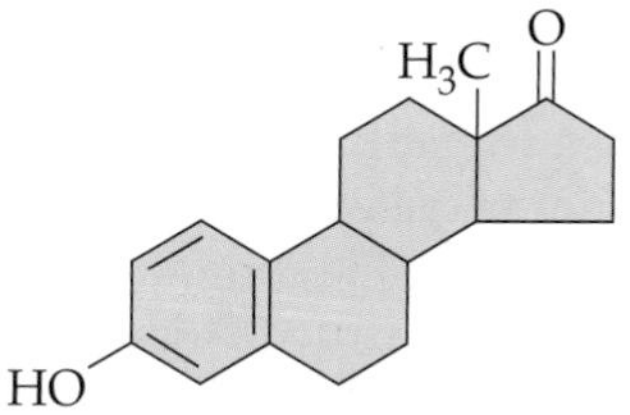

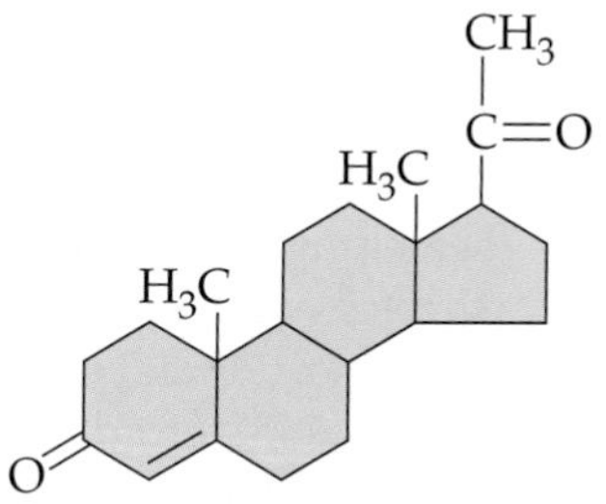

In addition to the several hundred known steroids isolated from plants and animals, a great many more have been synthesized in the laboratory in the search for new drugs. Most birth control pills are a mixture of the synthetic estrogen *ethynyl estradiol* and the synthetic progestin *norethindrone*. These steroids function by tricking the body into a false pregnant state, making it temporarily infertile. The compound known as *RU-486*, or *mifepristone*, is effective as a "morning-after" pill. It prevents pregnancy by binding strongly to the progesterone receptor, thereby blocking implantation in the uterus of a fertilized egg cell. This "morning after" pill is available in the United States, but only by prescription.

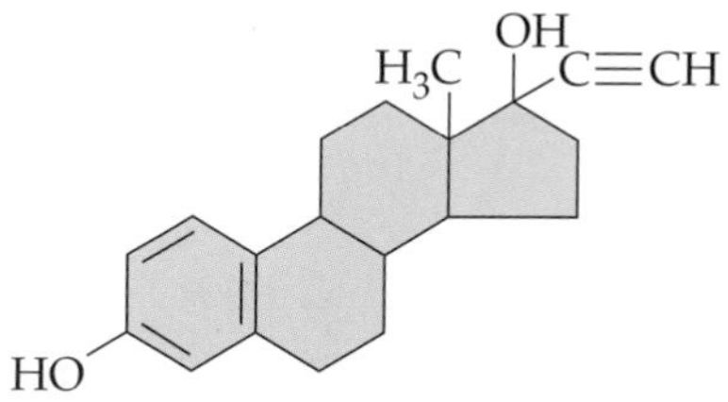

Ethynyl estradiol (a synthetic estrogen)	Norethindrone (a synthetic progestin)	RU-486 (Mifepristone)

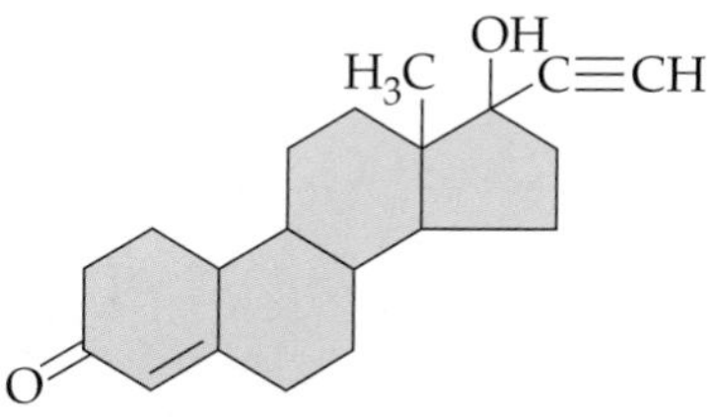

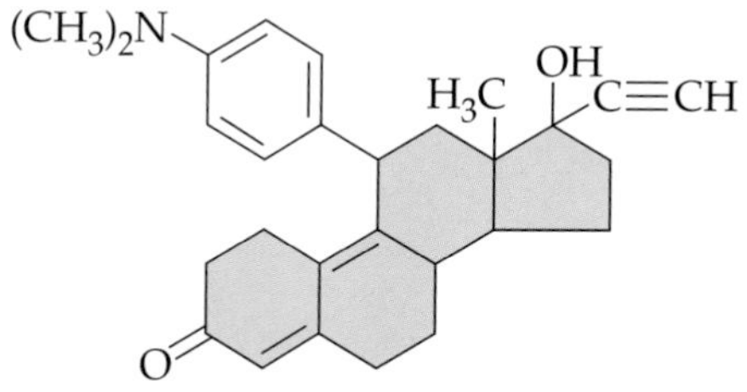

Anabolic steroids, which have the ability to increase muscle mass and consequently strength, are drugs that resemble androgenic (male) hormones, such as testosterone. These steroids have been used by body-builders for decades to change their body shape to a more muscular, bulky form, and some professional and semi-professional athletes (both men and women) have used them in the hope of gaining weight, strength, power, speed, endurance, and aggressiveness. Unfortunately, many serious side effects can arise from abuse of anabolic steroids. Stunted bone growth in adolescents; cancer of the liver, prostate, and kidney; high blood pressure; aggressive behavior; liver damage; irregular heart beat; and nosebleeds (arising out of blood coagulation disorders) are but a few of the short- and long-term side effects of these agents. Today, most organized amateur and professional sports have banned the use of these and other "performance enhancing" drugs.

Despite bans, the use of "roids" is widespread in sports. For example, so many baseball players have apparently used anabolic steroids that a Congressional hearing was held in 2007. Several trainers and players from baseball and other sports testified, and as a result some individual records are now in question. One of the first athletes outside of baseball investigated was Marion Jones, a high-profile track star. She pled guilty in 2007 to using steroids while training and during her Olympic medal events. She has been striped of her medals, as have her relay teammates. The list of anabolic steroid users is long and includes cyclists, shot putters, and sprinters. Bulgaria's weightlifters did not participate in the 2008 Summer Olympic Games because routine testing revealed the presence of a banned steroid in every team member.

Did you know that it is legal to use anabolic steroids on race horses in nearly every state? However, this use may be curtailed, as it has been in humans, by the events of the 2008 Kentucky Derby, when the horse Eight Belles collapsed at the finish line and had to be destroyed. The consensus of racing officials and horse owners is that because Eight Belles was given anabolic steroids, her muscle growth far out-paced bone growth. The combined stress of disproportionately high muscle mass and running the race caused both front ankles to break at the finish line. Currently, ana-bolic steroids are also legally administered to greyhounds and other racing dogs.

To enforce the ban on anabolic steroids, athletes are subjected to random drug screening, but some athletes attempt to get around the screenings by using *designer steroids*—steroids that cannot be detected with current screening methods because identification depends on knowing the compound's structure. However, analysis of a synthetic steroid to determine its structure is easily done. For example, in October 2003 chemists announced that they had identified a new performance-enhancing (and previously undetectable) synthetic steroid. The illegal use of this new com-pound, tetrahydrogestrinone (THG), was discovered when an anonymous coach sent a spent syringe to U.S. antidoping officials because he was concerned that ath-letes might be using a mysterious performance-enhancing drug. THG is similar in chemical structure to two other previously banned synthetic anabolic steroids, tren-bolone (used by cattle ranchers to increase the size of cattle) and gestrinone (used to treat endometriosis in women). A test for THG was quickly developed and is now used routinely. Despite the apparent victory over THG, chemists know that it is just a matter of time before the next designer steroid becomes available; new tests and procedures for detection are constantly being developed.

Designer Anabolic Steroids

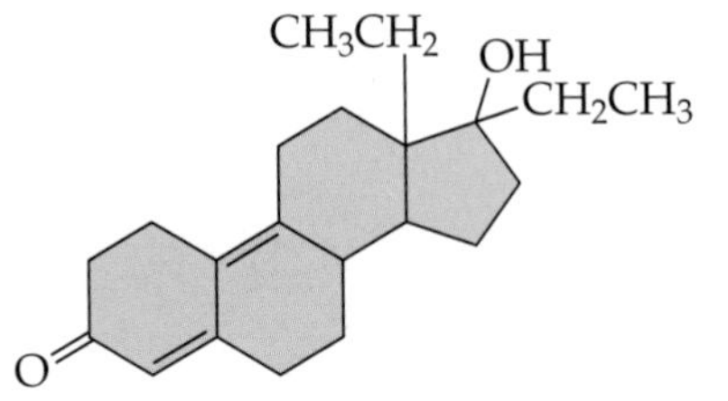

Tetrahydrogestrinone
(THG)

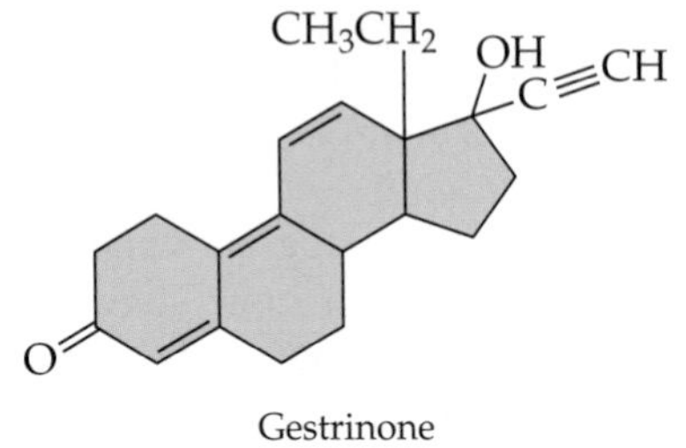

Gestrinone

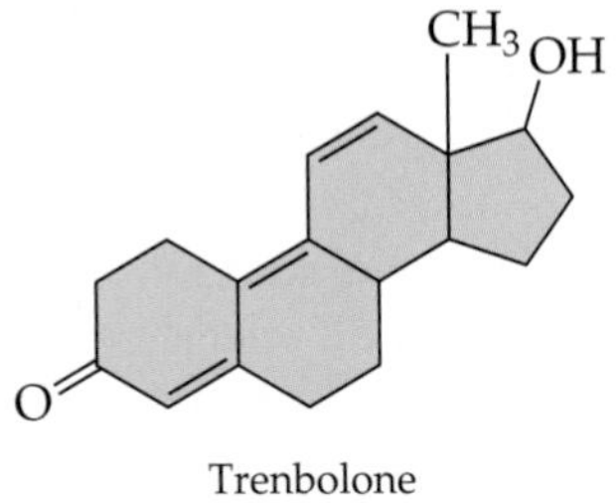

Trenbolone

PROBLEM 20.5

Nandrolone is an anabolic, or tissue-building, steroid sometimes taken by ath-letes seeking to build muscle mass (it is banned by the International Olympic Committee as well as other athletic organizations). Among its effects is a high level of androgenic activity. Which of the androgens shown on p. 638 does it most closely resemble? How does it differ from that androgen?

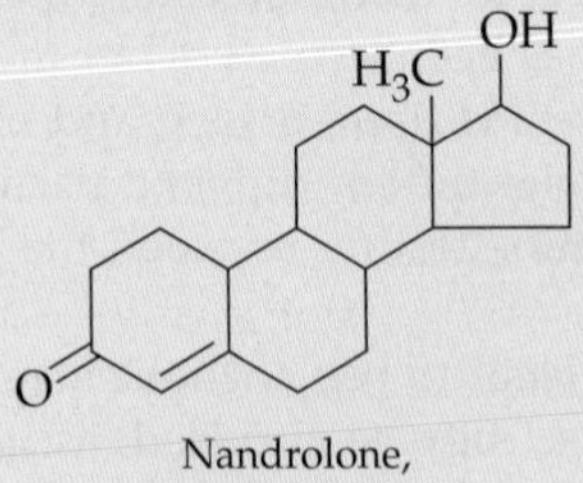

Nandrolone,
an anabolic steroid

APPLICATION ▶ Plant Hormones

Would you believe that plants have hormones? Actually, plants do, and these hormones are just as important to the health and development of plants as human hormones are to us. But since plants are not animals, there are some differences in how hormones work in them. Plants do not have endocrine systems, nor do they have fluids like blood that continuously circulate so that chemicals can be picked up where they are created and distributed to wherever they are needed.

Unlike animal hormones, which must be transported to their targets, plant hormones (known as *phytohormones*) affect the cells in which they are synthesized. They may also reach nearby cells by diffusion or travel upward with water from the roots or downward with sugars made by photosynthesis in the leaves. A very simple alkene, ethylene, $CH_2{=}CH_2$, a gas, functions as a hormone in plants. At one time, citrus growers ripened oranges that they picked green in rooms heated with kerosene stoves. Mysteriously, when the stoves were replaced with other means of heating, the oranges no longer ripened. It turned out that ripening is hastened by the ethylene released by burning kerosene, not by the warmth. Plants produce ethylene when it is time for fruit to ripen. Today, bananas, tomatoes, and other fruits are picked hard and unripe to make them easier to ship; they are thereafter ripened by exposure to ethylene. You can try this at home. Enclose some less than perfectly ripe pears or peaches in a brown paper bag along with, for example, a very ripe banana. Be careful, though. "One rotten apple can spoil the barrel." With too much exposure to ethylene, ripening can be overdone. The artificial ripening produced commercially by ethylene exposure is, however, by and large superficial; true ripening is much more complex. The majority of ethylene production occurs in the final stages of the ripening process.

Just by watching our houseplants, we know that plants turn toward the sun, a phenomenon known as *phototropism*.

Charles Darwin was one of the first to wonder why this happens. He observed that covering the growing tips of the plants prevented phototropism. The explanation lies in the formation in the tip of an *auxin*, a hormone that travels downward and stimulates elongation of the stem. When light distribution is uneven, auxin concentrates on the shady side of the stem, causing it to grow faster so that the stem bends toward the sun. Auxin is produced in seed embryos, young leaves, and growing tips of plants. Interestingly, plants synthesize auxin from tryptophan, the starting compound in the synthesis of several mammalian chemical messengers. As is the case with ethylene, the effects of auxin can also be overdone. An excessive concentration of auxin kills plants by overaccelerating their growth. The most familiar synthetic auxin, 2,4-D, an herbicide, is widely used to kill broad-leaved weeds in this manner.

▲ Phototropism in plants is caused by auxin production.

See Additional Problems 20.80 and 20.81 at the end of the chapter.

20.6 Neurotransmitters

Neurotransmitters are the chemical messengers of the nervous system. They are released by nerve cells (*neurons*) and transmit signals to neighboring target cells, such as other nerve cells, muscle cells, or endocrine cells. Structurally, nerve cells that rely on neurotransmitters typically have a bulb-like body connected to a long, thin stem called an *axon* (Figure 20.5). Short, tentacle-like appendages, the *dendrites*, protrude from the bulbous end of the neuron, and numerous filaments protrude from the axon at the opposite end. The filaments lie close to the target cell, separated only by a narrow gap—the **synapse**.

A nerve impulse is transmitted along a nerve cell by variations in electrical potential caused by the exchange of positive and negative ions across the cell membrane. Chemical transmission of the impulse between a nerve cell and its target occurs when neurotransmitter molecules are released from a *presynaptic neuron*, cross the synapse, and bind to receptors on the target cell. When the target is

Synapse The place where the tip of a neuron and its target cell lie adjacent to each other.

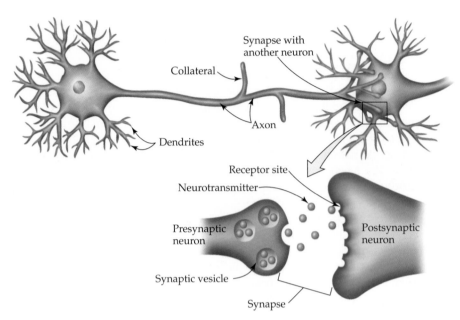

▲ FIGURE 20.5 A nerve cell and transmission of a nerve signal by neurotransmitters. Transmission occurs between neurons when a neurotransmitter is released by the presynaptic neuron, crosses the synapse, and fits into a receptor on the postsynaptic neuron.

another nerve cell, the receptors lie on the dendrites of the next, *postsynaptic* neuron, as shown in Figure 20.5. Once neurotransmitter–receptor binding has occurred, the message has been delivered. The postsynaptic neuron then transmits the nerve impulse down its own axon until a neurotransmitter delivers the message to the next neuron or other target cell.

Neurotransmitter molecules are synthesized in the presynaptic neurons and stored there in small pockets, known as *vesicles*, from which they are released as needed. After a neurotransmitter has done its job, it must be *rapidly* removed so that the postsynaptic neuron is ready to receive another impulse. This occurs in one of two ways. Either a chemical change catalyzed by an enzyme available in the synaptic cleft inactivates the neurotransmitter, or alternatively, the neurotransmitter is returned to the presynaptic neuron and placed in storage until it is needed again.

Most neurotransmitters are amines and are synthesized from amino acids. The synthesis of dopamine, norepinephrine, and epinephrine from tyrosine is shown in Figure 20.4. The synthesis of serotonin and melatonin from tryptophan is shown in Figure 20.6. Some neurotransmitters act directly by causing changes in adjacent cells as soon as they connect with their receptors. Others rely on second messengers, often cyclic AMP, the same second messenger utilized by hormones. Individual neurotransmitters have been associated with emotions, drug addiction, pain relief, and other brain functions, as we shall see in the following sections.

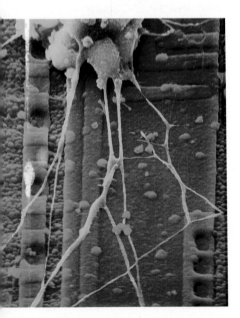

▲ Two ways to send messages. This photograph shows a human neuron growing on the surface of a silicon computer chip.

PROBLEM 20.6

Which of the following transformations of amines in Figure 20.6 is (1) an acetylation, (2) a methylation, (3) a decarboxylation?

(a) 5-Hydroxytryptophan to serotonin

(b) Serotonin to *N*-acetylserotonin

(c) *N*-Acetylserotonin to melatonin

Tryptophan

5-Hydroxytryptophan

CO_2

Serotonin
(brain neurotransmitter)

N-Acetylserotonin

Melatonin (hormone)

▲ **FIGURE 20.6 Synthesis of chemical messengers from tryptophan.** The changes in each step are highlighted in gold for additions and in green for losses.

20.7 How Neurotransmitters Work: Acetylcholine, Its Agonists and Antagonists

Acetylcholine in Action

Acetylcholine (ACh) is a neurotransmitter responsible for the control of skeletal muscles. It is also widely distributed in the brain, where it may play a role in the sleep–wake cycle, learning, memory, and mood. Nerves that rely on ACh as their neurotransmitter are classified as *cholinergic* nerves.

ACh is synthesized in presynaptic neurons and stored in their vesicles. The rapid sequence of events in the action of ACh in communicating between nerve cells, illustrated in Figure 20.7, is as follows:

Acetylcholine

- A nerve impulse arrives at the presynaptic neuron.
- The vesicles move to the cell membrane, fuse with it, and release their ACh molecules (several thousand molecules from each vesicle).
- ACh crosses the synapse and binds to receptors on the postsynaptic neuron, causing a change in membrane permeability to ions.
- The resulting change in the permeability to ions of the postsynaptic neuron initiates the nerve impulse in that neuron.
- With the message delivered, acetylcholinesterase present in the synaptic cleft catalyzes the decomposition of acetylcholine:

$$CH_3-\overset{\overset{\displaystyle O}{\|}}{C}-O-CH_2-CH_2-\overset{+}{N}(CH_3)_3 \xrightarrow[\text{H}_2\text{O}]{\text{Acetylcholinesterase}} CH_3COO^- + HO-CH_2-CH_2-\overset{+}{N}(CH_3)_3$$

Acetylcholine (ACh)   Acetate   Choline

- Choline is absorbed back into the presynaptic neuron where new ACh is synthesized.

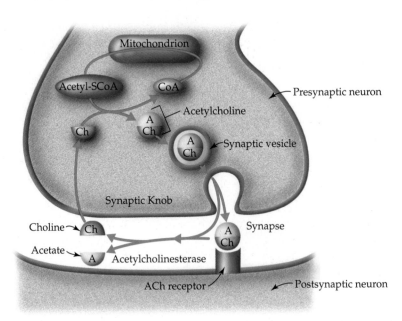

► **FIGURE 20.7 Acetylcholine (ACh) release and re-uptake.** ACh is stored in vesicles in the presynaptic neuron. After it is released into the synapse and connects with its receptor, it is broken down by hydrolysis into acetate and choline in the reaction catalyzed by acetylcholinesterase. The choline is taken back into the synaptic knob and used to synthesize ACh, which is then stored in the vesicles until needed.

PROBLEM 20.7

Propranolol (trade name, Inderal) is an antagonist for certain epinephrine receptors and is a member of the class of drugs known as beta blockers (because they block what are known as beta receptors). Circle the functional groups in propranolol and name them. Compare the structure of propranolol with the structure of epinephrine and describe the differences.

Propranolol
(Inderal®)

Epinephrine
(adrenaline)

Drugs and Acetylcholine

Drug Any substance that alters body function when it is introduced from an external source.

Many drugs act at acetylcholine synapses (Section 20.6), the places where the tip of a neuron that releases acetylcholine and its target cell lie adjacent to each other. A **drug** is any molecule that alters normal functions when it enters the body from an external source. The action is at the molecular level, and it can be either therapeutic or poisonous. To have an effect, many drugs must connect with a receptor just as a substrate must bind to an enzyme or as a hormone or neurotransmitter must bind to a receptor. In fact, many drugs are designed to mimic a given hormone or neurotransmitter and in so doing elicit either an enhanced or attenuated effect.

Agonist A substance that interacts with a receptor to cause or prolong the receptor's normal biochemical response.

Antagonist A substance that blocks or inhibits the normal biochemical response of a receptor.

Pharmacologists classify some drugs as **agonists**—substances that act to produce or prolong the normal biochemical response of a receptor. Other drugs are classified as **antagonists**—substances that block or inhibit the normal response of a receptor. Many agonists and antagonists compete with normal signaling molecules for interaction with the receptor, just as inhibitor molecules compete with substrate for the active site in an enzyme (Chapter 19). To illustrate the ways in which drugs can affect our biochemical activity, we next describe the action of a group of drugs. These drugs are all members of the same drug family in the sense that their biochemical activity occurs at acetylcholine synapses in the central nervous system. The locations of their actions can be seen in Figure 20.7.

- *Botulinus toxin (an antagonist), blocks acetylcholine release and causes botulism.* The toxin, which is produced by bacterial growth in improperly canned food, binds irreversibly to the presynaptic neuron, where acetylcholine would be released. It prevents this release, frequently causing death due to muscle paralysis. Commercially, this toxin (marketed as Botox) has found use in cosmetic surgery, where carefully controlled doses of it are used to temporarily tighten up wrinkled skin without the need for invasive surgical procedures. Recent experiments have shown that Botox does not stay at the injection site but migrates along neurons to the brain, far from the original injection.

- *Black widow spider venom (an agonist), releases excess acetylcholine.* In the opposite reaction from that of botulism, the synapse is flooded with acetylcholine, resulting in muscle cramps and spasms.

- *Organophosphorus insecticides (antagonists), inhibit acetylcholinesterase.* All of the organophosphorus insecticides (a few examples are shown below) prevent the cholinesterase enzyme from breaking down acetylcholine within the synapse. As a result, the nerves are overstimulated, causing a variety of symptoms including muscle contraction and weakness, lack of coordination, and at high doses, convulsions. Recently, the death of thousands of honeybees was attributed to clothianidin dust released from the protective coating on corn seeds during planting. Many organophosphorus compounds are also used as nerve gasses. This is why agriculturists receive special training before using these insecticides on crops.

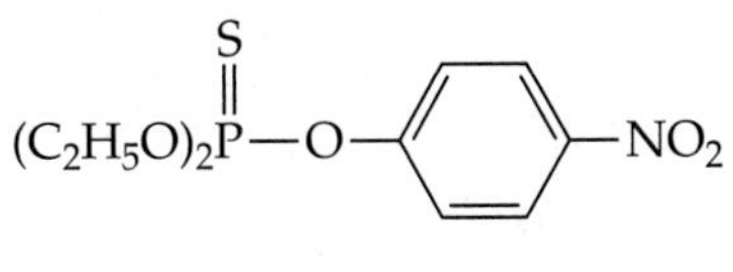

Parathion 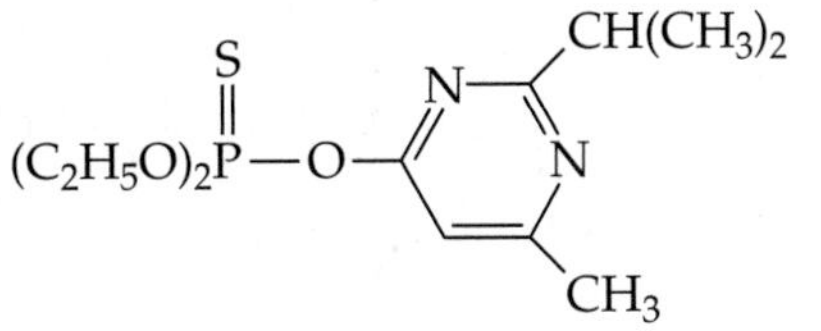   Diazinon 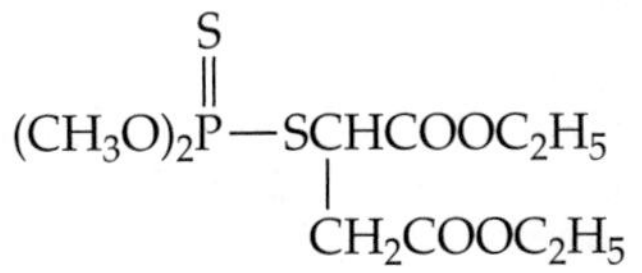   Malathion

- *Nicotine binds to acetylcholine receptors.* Nicotine at low doses is a stimulant (an agonist) because it activates acetylcholine receptors. The sense of alertness and well-being produced by inhaling tobacco smoke is a result of this effect. At high doses, nicotine is an antagonist. It irreversibly blocks the acetylcholine receptors and can cause their degeneration. Nicotine has no therapeutic use in humans other than in overcoming addiction to smoking. Nicotine transdermal patches, which release a small, controlled dose of nicotine through the skin, and chewing gum containing nicotine are used to help smokers overcome nicotine addiction. Nicotine (along with its sulfate salt) is one of the most toxic botanical insecticides known and has been used since the 1600s as a contact poison.

- *Atropine (an antagonist), competes with acetylcholine at receptors.* Atropine, found naturally in the plant named deadly nightshade, is an alkaloid that is a poison at high doses. At controlled doses, its therapeutic uses include acceleration of abnormally slow heart rate, paralysis of eye muscles during surgery, and relaxation of intestinal muscles in gastrointestinal disorders. Most importantly, it is a specific antidote for cholinesterase poisons such as organophosphorus insecticides. By blocking activation of the receptors, it counteracts the excess acetylcholine created by cholinesterase inhibitors.

- *Tubocurarine (an antagonist), competes with acetylcholine at receptors.* Purified from curare, a mixture of chemicals extracted from a plant found in South America, the alkaloid tubocurarine is used to paralyze patients in conjunction with anesthesia drugs prior to surgery. In the last 30 years, researchers have developed safer, more easily purified synthetic derivatives that have the same mode of action. These molecules have nearly replaced use of tubocurarine in medical procedures.

PROBLEM 20.8

The LD_{50} values (lethal dose in mg/kg, for rats; see p. 503) for the three organophosphorus insecticides listed above are parathion, 3–13 mg/kg; diazinon, 250–285 mg/kg; and malathion, 1000–1375 mg/kg. Which would you choose for use in your garden and why?

KEY CONCEPT PROBLEM 20.9

Some drugs are classified as agonists whereas others are classified as antagonists. Explain the difference between agonists and antagonists and give an example of each.

20.8 Histamine and Antihistamines

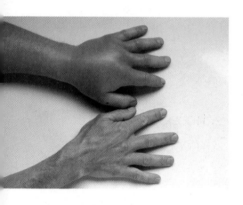

▲ The left hand is swelling due to a histamine response.

Histamine is the neurotransmitter responsible for the symptoms of the allergic reaction so familiar to hay fever sufferers or those who are allergic to animals. It is also the chemical that causes an itchy bump when an insect bites you. In the body, histamine is produced by decarboxylation of the amino acid histidine:

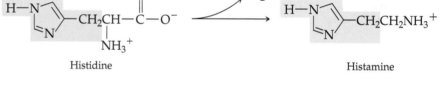

Histidine   Histamine

The *antihistamines* are a family of drugs that counteract the effect of histamine because they are histamine receptor antagonists. They competitively block the attachment of histamine to its receptors. Members of this family have in common a disubstituted ethylamine side chain, usually with two *N*-methyl groups. As illustrated by the examples below, the R′ and R″ groups at the other end of the molecule tend to be bulky and aromatic.

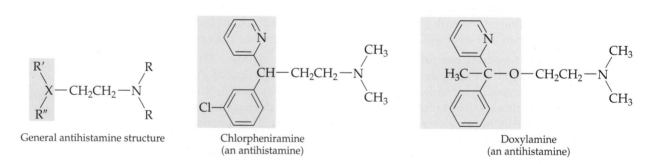

General antihistamine structure

Chlorpheniramine
(an antihistamine)

Doxylamine
(an antihistamine)

Histamine also activates secretion of acid in the stomach. Development of an antagonist for this function of histamine was accomplished by what is today commonly known as *rational drug design*. After synthesis of about 200 different compounds with systematic variations on the histamine structure, the goal of attaining a histamine antagonist was achieved. The result was *cimetidine*, widely publicized as a treatment for heartburn under its trade name of Tagamet. Today many other histamine antagonists exist, including ranitidine, sold under its trade name of Zantac.

APPLICATION ▶   And from This Little Frog . . .

Pain management is an important topic in medicine; from relief of chronic back pain to help for those suffering with cancer, effective treatment of pain is an ongoing challenge. Chemists have risen to this challenge and identified a chemical compound that may fill a major gap in our arsenal of painkillers. The story of this compound is still unfolding, as the compound works its way through tests for safety and efficacy. But the story is worth telling because it illustrates a not uncommon sequence of events in drug development.

The story began in a tropical rain forest in Ecuador more than 30 years ago. Dr. John W. Daly of the National Institutes of Health traveled often to Central and South America to collect new species of frogs. In 1978, Daly returned to Ecuador to collect more of a poison frog (*Epipedobates tricolor*) in order to follow up on an exciting lead. Something in these frogs acted like morphine, but appeared to do so by a completely different mechanism.

Back home, he set out to find the structure of this compound by analyzing 60 mg of a mixture of frog skin extract. Step by step, with the skill needed to handle such tiny amounts, he isolated the pure compound, determined its molecular weight, broke it into two fragments, and identified the atoms in each fragment (one fragment with 6 C atoms, 10 H atoms, and 1 N atom; the other with 5 C atoms, 3 H atoms, 1 N atom, and 1 Cl atom). At this point he had only 0.5 mg of the compound left. Meanwhile, he had also discovered that frogs grown in the laboratory do not produce the compound of interest. And no more frogs could be collected, because they had been put on the endangered species list.

Afraid to destroy his tiny sample without finishing the identification, Daly put the sample away and waited 13 years. By then, modern instrumentation had developed to the point where the analysis could be finished, and he succeeded in determining the structure of the compound he named *epibatidine* (in honor of the frog in which it was discovered). Unfortunately, the remaining sample of the compound was too small for studies of its physiological effects. The next step was for someone to devise a way to synthesize the compound in the laboratory. Other chemists accomplished this in 1996.

With an adequate supply, epibatidine's effect on living things could be studied. The bad news was that it is too toxic for use as a drug. The exciting news was that it does indeed act as a painkiller (approximately 50 times more potent than morphine), but by a completely different mechanism than morphine and other opioids. Instead of acting at an opioid receptor, the compound acts at a class of acetylcholine receptors in the central nervous system known as the *neuronal nicotinic acetylcholine receptors* (nAChR or NNR). Nicotine, which is similar in structure to epibatidine, acts at the same receptor but is only a weak painkiller.

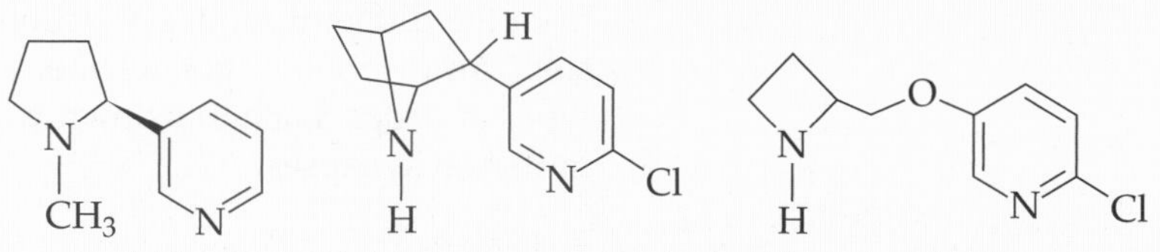

Nicotine   Epibatidine   ABT-594

These discoveries revealed the possibility that a compound structurally similar to nicotine and epibatidine, but less poisonous, might be an effective painkiller but without the addictive properties and negative side effects of morphine and other opioids. The hunt was on. In 1998, after screening more than 500 similar compounds, the chemists at Abbott Laboratories zeroed in on a compound they labeled ABT-594 with the structure shown here as the best candidate for the long-sought completely new type of painkiller. ABT-594 is about 200 times more effective than morphine at pain management but appears to be nonaddictive. Phase II safety trials on ABT-594 have been completed and look promising; however, nausea and gastrointestinal side effects have led Abbott Laboratories in 2004 to test what they call the next generation NNR: ABT-894. This new compound (the structure of which is proprietary) is much better tolerated than ABT-594 and extends the domain that once belonged solely to a small poison frog.

▲ *Epipedobates tricolor*. Identification of a chemical from the skin of this rare poison arrow frog may lead to a new class of painkilling drugs.

See Additional Problems 20.82 and 20.83 at the end of the chapter.

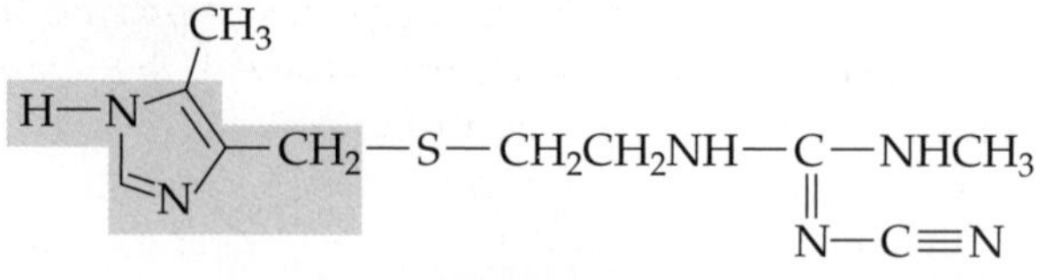

Cimetidine
(Tagamet®)

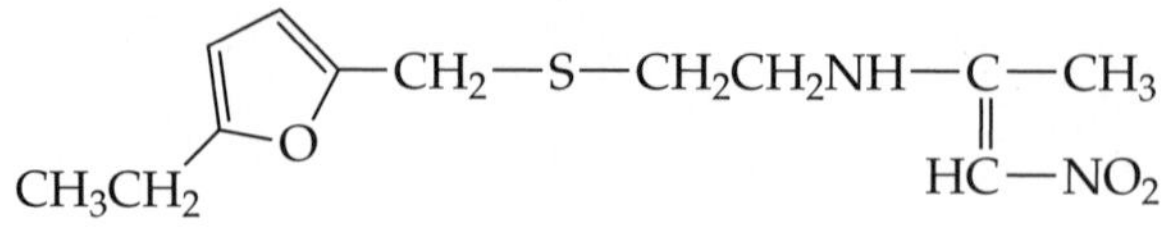

Ranitidine
(Zantac®)

20.9 Serotonin, Norepinephrine, and Dopamine

The Monoamines and Therapeutic Drugs

Serotonin, norepinephrine, and dopamine could be called the "big three" of neuro-transmitters. They regularly make news as discoveries about them accumulate. Collectively, serotonin, norepinephrine, and dopamine are known as *monoamines*. (Their biochemical syntheses are shown in Figures 20.4 and 20.6.) All are active in the brain and all have been identified in various ways with mood, the experiences of fear and pleasure, mental illness, and drug addiction. Needless to say, chemistry plays a central role in mental illness—that has become an inescapable conclusion.

One well-established relationship is the connection between major depression and a deficiency of serotonin, norepinephrine, and dopamine. The evidence comes from the different modes of action of three families of drugs used to treat depression. Amitriptyline, phenelzine, and fluoxetine are representative of these three types of drugs. Each in its own way increases the concentration of the neurotransmitters at synapses.

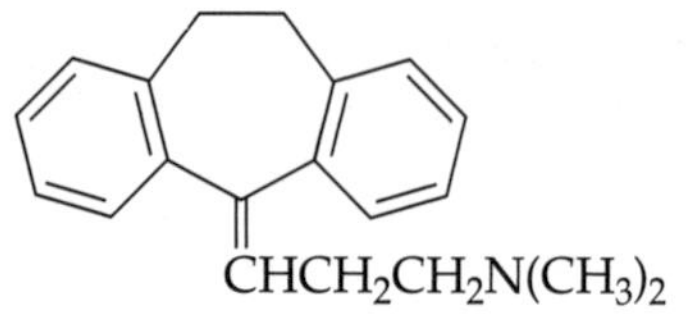

Amitriptyline, a tricyclic antidepressant
(Elavil®)

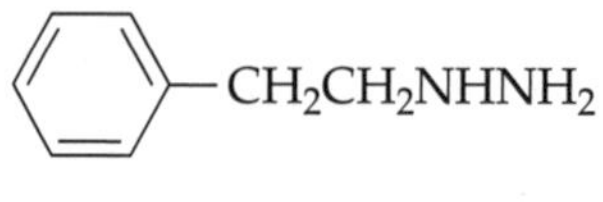

Phenelzine, an MAO inhibitor
(Nardil®)

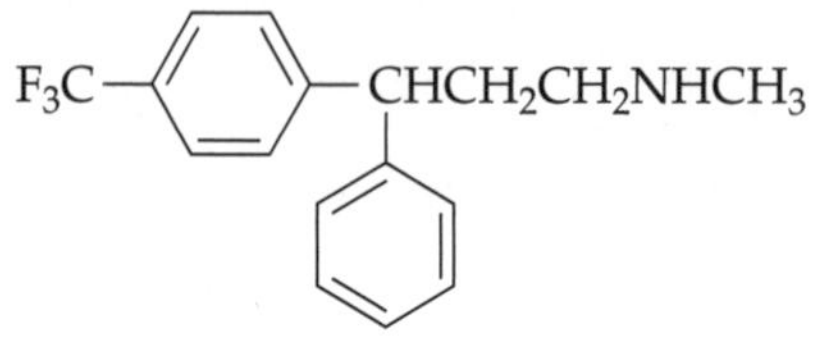

Fluoxetine, an SSRI
(Prozac®)

- Amitriptyline is representative of the *tricyclic antidepressants*, which were the first generation of these drugs. The tricyclics prevent the re-uptake of serotonin and norepinephrine from within the synapse. Serotonin is important in mood-control pathways and functions more slowly than other neurotransmitters; slowing its re-uptake often improves mood in depressed patients.

- Phenelzine is a *monoamine oxidase (MAO) inhibitor*, one of a group of medications that inhibit the enzyme that breaks down monoamine neurotransmitters. This inhibition of MAO allows the concentrations of monoamines at synapses to increase.

- Fluoxetine represents the newest class of antidepressants, the *selective serotonin re-uptake inhibitors (SSRI)*. They are more selective than the tricyclics because they inhibit only the re-uptake of serotonin. Fluoxetine (Prozac) has rapidly become the most widely prescribed drug for all but the most severe forms of depression. Most antidepressants cause unpleasant side effects; fluoxetine does not, a major benefit.

It is important to note that the relief of the symptoms of depression by these drugs is not evidence that either the chemical basis of depression is fully understood or that increasing neurotransmitter concentration is the only action of these drugs. The brain still holds many secrets. As one of the pharmacologists who developed fluoxetine put it, "If the human brain were simple enough for us to understand, we would be too simple to understand it."

The complex and not yet fully understood relationships between neurotransmitter activity and behavior are illustrated by the use of fluoxetine for conditions other than depression. It is used to treat obsessive compulsive disorder, bulimia, obesity, panic disorder, body dysmorphic disorder, teen depression, and premenstrual dysphoric disorder (formerly known as PMS). New uses for this class of drugs are constantly being explored.

Dopamine and Drug Addiction

Dopamine plays a role in the brain in processes that control movement, emotional responses, and the experiences of pleasure and pain. It interacts with five different kinds of receptors in different parts of the brain. An oversupply of dopamine is associated with schizophrenia, and an undersupply results in the loss of fine motor control in Parkinson's disease (see the Application "The Blood–Brain Barrier," Section 29.4). Dopamine also plays an important role in the brain's reward system. An ample supply of brain dopamine produces the pleasantly satisfied feeling that results from a rewarding experience—a "natural high." Herein lies the role of dopamine in drug addiction: The more the dopamine receptors are stimulated, the greater the high.

Experiments show that cocaine blocks re-uptake of dopamine from the synapse, and amphetamines accelerate release of dopamine. Studies have linked increased brain levels of dopamine to alcohol and nicotine addiction as well. The higher-than-normal stimulation of dopamine receptors by drugs results in tolerance. In the drive to maintain constant conditions (see the Application "Homeostasis"), the number of dopamine receptors decreases and the sensitivity of those that remain decreases. Consequently, brain cells require more and more of a drug for the same result, a condition that contributes to addiction.

Marijuana also creates an increase in dopamine levels, in the same brain areas where dopamine levels increase after administration of heroin and cocaine. The most active ingredient in marijuana is tetrahydrocannabinol (THC). The use of marijuana medically for chronic pain relief has become a controversial topic in recent years, as questions about its benefits and drawbacks are debated.

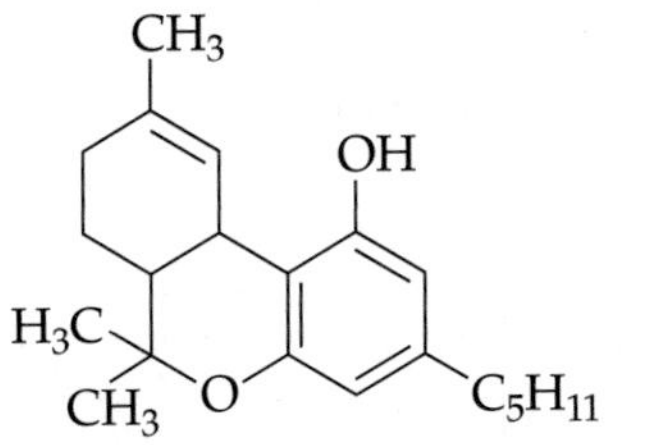

Tetrahydrocannabinol (THC)

●○ KEY CONCEPT PROBLEM 20.10

Identify the functional groups present in THC. Is the molecule likely to be hydrophilic or hydrophobic? Would you expect THC to build up in fatty tissues in the body, or would it be readily eliminated in the bloodstream?

WORKED EXAMPLE **20.2** Predicting Biological Activity Based on Structure

The relationship between the structure of a molecule and its biochemical function is an essential area of study in biochemistry and the design of drugs. Terfenadine (Seldane) was one of the first of the new generation of "nondrowsy" antihistamines (it was removed from the market due to potential cardiotoxicity). Based solely on what you have learned so far, suggest which of its structural features make it an antihistamine.

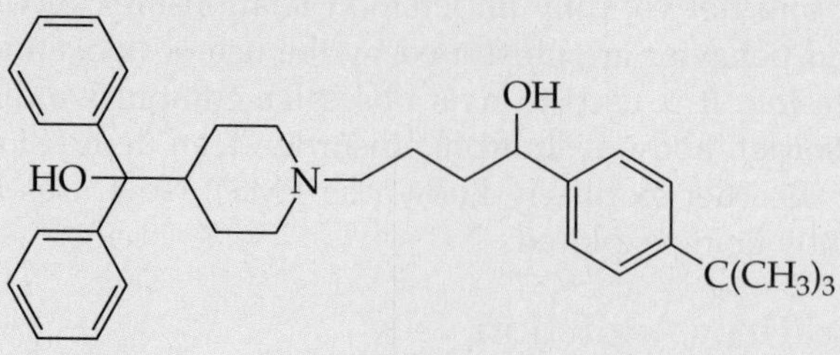

Terfenadine

ANALYSIS From Section 20.8, we see that members of the antihistamine family have in common the general structure shown here: an X group (usually a CH) to which two aromatic groups (noted as *aryl* in the drawing) are attached. The X is also attached to a disubstituted nitrogen by a carbon chain:

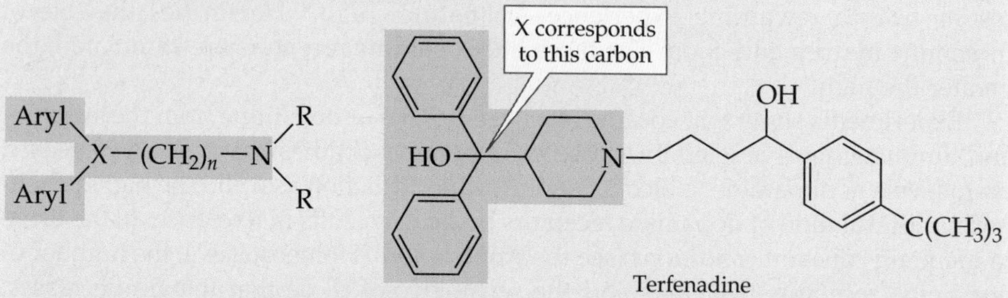

Terfenadine

SOLUTION
Since terfenadine contains the same basic structure as a general antihistamine, its biological function should be similar.

◖● KEY CONCEPT PROBLEM 20.11

Predict which of the following compounds is an antihistamine and which is an antidepressant.

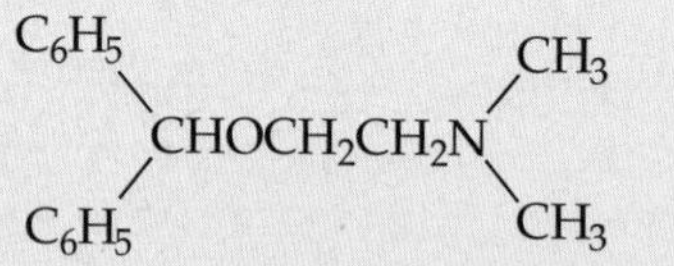

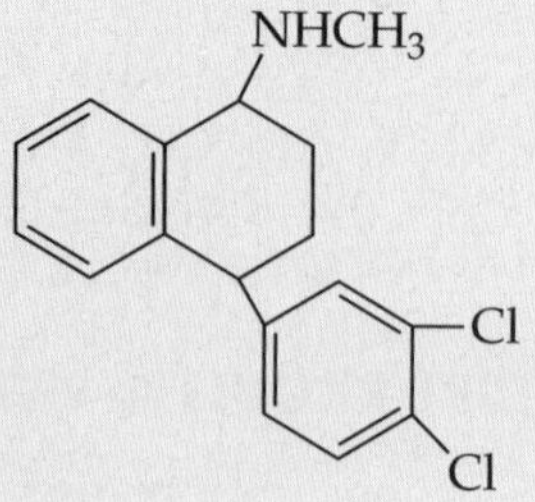

20.10 Neuropeptides and Pain Relief

Studies of morphine and other opium derivatives in the 1970s revealed that these addictive but effective pain-killing substances act via their own specific brain receptors. This raised some interesting questions: Why are there brain receptors for chemicals from a plant? Could it be that there are animal neurotransmitters that act at the same receptors?

The two pentapeptides *Met-enkephalin* and *Leu-enkephalin* (Met and Leu stand for the carboxy terminal amino acids, Section 18.3) were discovered in the effort to answer these questions.

Met-enkephalin: Tyr-Gly-Gly-Phe-Met

Leu-enkephalin: Tyr-Gly-Gly-Phe-Leu

Both exert morphine-like suppression of pain when injected into the brains of experimental animals. The structural similarity between Met-enkephalin and morphine, highlighted below, supports the concept that both interact with the same receptors, which are located in regions of the brain and spinal cord that act in the perception of pain.

Met-enkephalin   Morphine

▲ Endorphin Rush Hot Sauce.

Subsequently, about a dozen natural pain-killing polypeptides that act via the opiate receptors have been found. They are classified as *endorphins*. A 31-amino-acid polypeptide that ends with the same 5-amino-acid sequence as Met-enkephalin is a more potent pain suppressor than morphine. Disappointingly, though, none of these compounds is the long-sought ideal, nonaddicting painkiller—all are addictive.

Although there is much to be learned, enkephalins have been implicated in the runner's "high," the regulation of complex behavior states such as anger and sexual excitement, and the suppression of pain by acupuncture or during extreme stress—for example, in the competitive athlete who continues to play though injured. The term *endorphin* has entered the popular idiom to the extent that there is an endurance trial known as the Endorphin Fix Adventure Race, endorphin label running shoes, and even an Endorphin Rush hot sauce.

20.11 Drug Discovery and Drug Design

In a tropical rain forest, a botanist trudges after a native healer, taking notes about the plants the healer chooses. In a pristine laboratory, scientists monitor an army of robots and computer screens. In yet another laboratory, researchers stare at computer-drawn pictures of candidate molecules connecting with receptors. Any of these activities can start a new drug on its path to medical success.

Plants were our first source for drugs. By trial and error, primitive peoples learned which plants dulled pain, caused "visions," and cured diseases. This was the beginning of drug discovery. From generation to generation, the knowledge was added to and passed along. Eventually, chemists learned how to identify the structures of the active molecules and sometimes to improve upon them. This is how we got codeine for pain (from opium poppies), quinine for malaria (from fever tree), vinblastine for Hodgkin's disease (from rosy periwinkle), scopolamine for motion sickness (from Jimson weed), and others. Estimates are that 25% of the prescriptions written each year in North America are for plant-derived drugs.

Today *ethnobotanists* work in remote regions of the world to learn what indigenous people have discovered about the healing powers of plants. The botanists are pursuing drug discovery in a race against time, both because forests and jungles are disappearing with the pressures of population expansion and because the aging healers who learned their skills years ago as apprentices to their elders are not finding new apprentices to teach.

Probably the first *synthetic* chemicals used in medicine were diethyl ether and chloroform as anesthetics:

$$CH_3CH_2OCH_2CH_3 \qquad CHCl_3$$

Diethyl ether   Chloroform

The technique of modifying a known structure to improve its biochemical activity was developed after cocaine was first used as a local anesthetic in 1884. The actual structure of cocaine was not known, but its hydrolysis products could be identified and showed that cocaine might be an ester of benzoic acid. Experiments with other benzoic acid esters in the early 1900s yielded benzocaine and procaine (novocaine), both still in use:

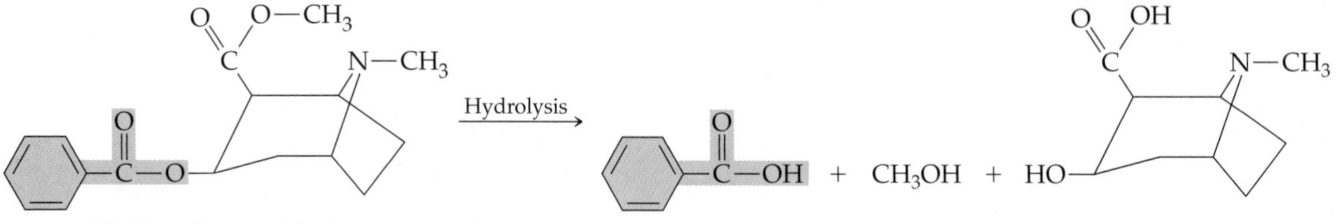

Cocaine

Hydrolysis →

Benzoic acid

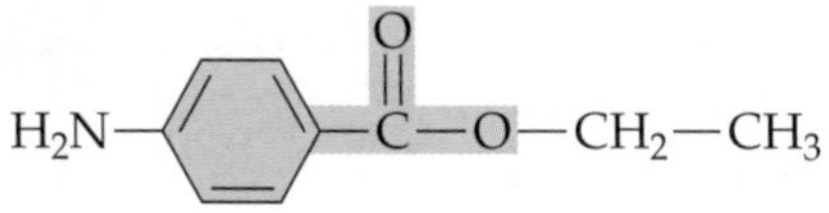

Benzocaine

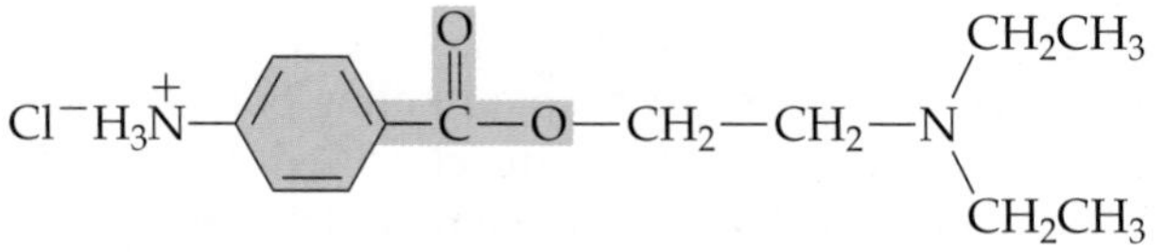

Procaine hydrochloride
(Novocaine)

Also in the late 1800s, phenacetin was introduced as an analgesic. The use of acetanilide, preceded it; however, acetanilide was soon withdrawn because of its toxicity. Derived as it was from the results of animal experiments with aniline, phenacetin was one of the first drugs designed with some knowledge of biochemistry. It remained on the market for many years until it was eventually withdrawn as evidence for its toxicity and possible carcinogenicity accumulated. Acetaminophen, introduced in 1893, is widely used today under such familiar trade names as Tylenol. We now know that it is produced in the body during metabolism of acetanilide and phenacetin. Other analgesics, such as ibuprofen (Motrin) and naproxen (Aleve), also share this general structure.

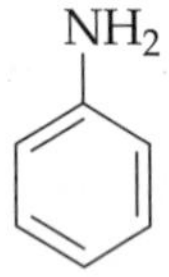

Aniline

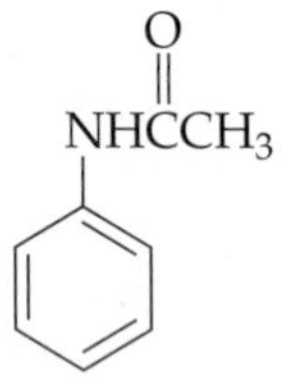

Acetanilide

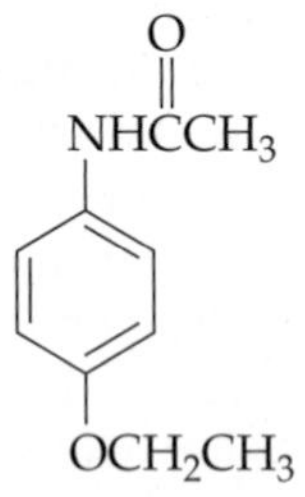

Phenacetin

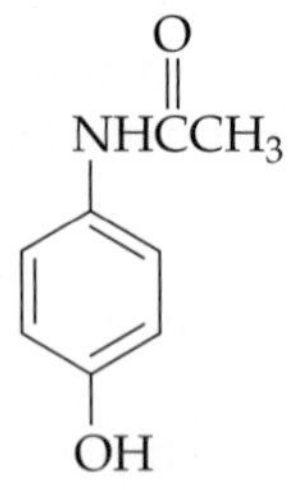

Acetaminophen
(Tylenol®)

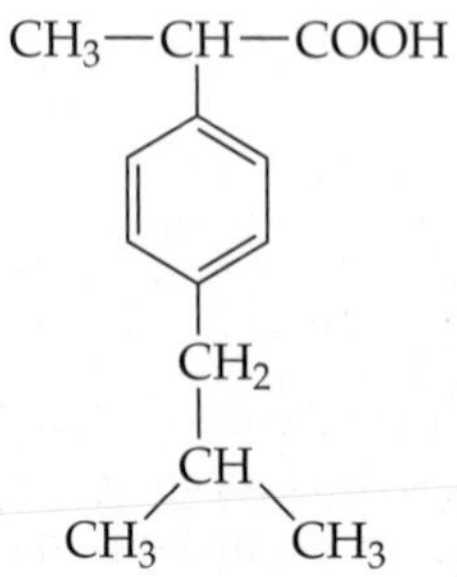

Ibuprofen
(Motrin®)

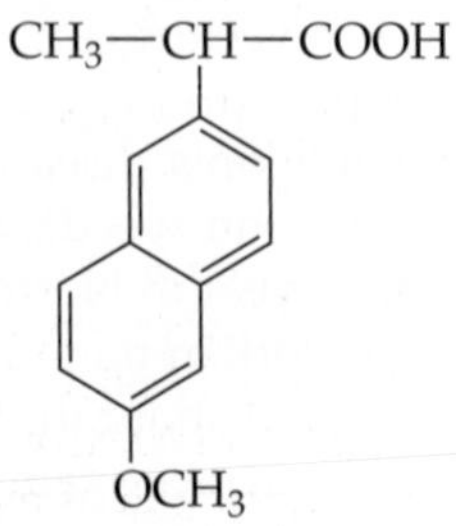

Naproxen
(Aleve®)

Interestingly, the mode of action of these well-known pain relievers is still unclear. It is believed that they inhibit the formation of the prostaglandins (Section 24.9). Meanwhile, expanding knowledge of the structure of biochemically active molecules combined with advancing technology have opened the door to a new era. Drug *discovery* is merging with drug *design*.

One new technology, *combinatorial chemistry*, arrived on the scene in 1991 and since 2005 has become a routine and powerful tool in drug discovery and design. It involves mass production at the molecular level. Suppose it is believed that some combination of a defined set of molecular building blocks will yield an effective drug. The techniques of combinatorial chemistry allow the building blocks to react in every possible combination, not one reaction at a time but hundreds at a time. Reactions are carried out on a microgram scale in tiny tubes or with molecules held down on solid supports. By combining reactants, dividing up the products, adding other reactants, and continuing this process, millions of related compounds can be synthesized. Robots help with the mixing of chemicals. Computers track the combinations and screen the products for some type of activity. Since, on average, only 5 in 5000 compounds prepared in the lab ever make it to human testing and only one of these five will ever be approved for general clinical use, hope runs high that this combinatorial approach will lead to a significant decrease in the average of 12 years and $802 million (in 2005 dollars) needed for the initial discovery and ultimate development of a new drug.

In another rapidly developing technology, supercomputers and molecular graphics now allow an approach right to the heart of drug action—the drug–receptor connection. The ability to find the structure of proteins, once a tedious and lengthy activity, is advancing every day. Let us say that the complete tertiary structure of an enzyme has been found, the active site identified, and a search for an inhibitor for this enzyme is underway. The computer can consult a database of quantitative information about drug–receptor interactions and other important properties such as hydrophobic versus hydrophilic solubilities. Once potential inhibitors are identified, pictures of such molecules entering the active site can be created on the computer screen. The pictures can be rotated and the fit examined from many angles. In this way, it is increasingly possible to design a molecule with just the right chemical and physical properties needed to connect with a biomolecule and produce a desired result. For those students who would like to see the results of computer modeling first hand, a free web-browser plug-in called Chime (from Elsevier MDL) exists. After installation on your personal computer, Chime allows you to view thousands of drugs that have been modeled. A quick Google search of the internet using the combination search terms *chime + drugs* gave 272,000 hits in 2008. Indeed, many of the molecules discussed in this chapter are easily found on the web as models that are viewable using Chime.

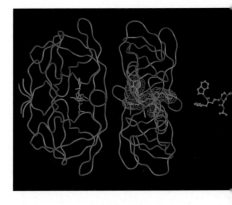

▲ Computer modeling of HIV-1 proteinase has led to the discovery of new drugs for the treatment of AIDS.

SUMMARY: REVISITING THE CHAPTER GOALS

1. **What are hormones, and how do they function?** *Hormones* are the chemical messengers of the *endocrine system*. Under control of the hypothalamus they are released from various locations, many in response to intermediate, regulatory hormones. Hormones travel in the bloodstream to target cells, where they connect with receptors that initiate chemical changes within cells.

2. **What is the chemical nature of hormones?** Hormones are *polypeptides, steroids,* or *amino acid derivatives*. Many are polypeptides, which range widely in size and include small molecules such as vasopressin and oxytocin, larger ones like insulin, and all of the regulatory hormones. Steroids have a distinctive four-ring structure and are classified as lipids because they are hydrophobic. All of the sex hormones are steroids. Hormones that are amino acid derivatives are synthesized from amino acids (Figures 20.5 and 20.6). Epinephrine and norepinephrine act as hormones throughout the body and also act as neurotransmitters in the brain.

KEY WORDS

Agonist, *p. 644*

Antagonist, *p. 644*

Drug, *p. 644*

Endocrine system, *p. 630*

Hormone, *p. 629*

Neurotransmitter, *p. 630*

Receptor, *p. 629*

Second messenger, *p. 633*

Steroid, *p. 638*

Synapse, *p. 641*

3. **How does the hormone epinephrine deliver its message, and what is its mode of action?** Epinephrine, the fight-or-flight hormone, acts via a cell-surface receptor and a G protein that connects with an enzyme, both of which are embedded in the cell membrane. The enzyme adenylate cyclase transfers the message to a *second messenger*, a cyclic adenosine mono-phosphate (cyclic AMP), which acts within the target cell.

4. **What are neurotransmitters, and how do they function?** *Neurotransmitters* are synthesized in presynaptic neurons and stored there in vesicles from which they are released when needed. They travel across a *synaptic cleft* to *receptors* on adjacent target cells. Some act directly via their receptors; others utilize cyclic AMP or other second messengers. After their message is delivered, neurotransmitters must be quickly broken down or taken back into the presynaptic neuron so that the receptor is free to receive further messages.

5. **How does acetylcholine deliver its message, and how do drugs alter its function?** Acetylcholine is released from the vesicles of a presynaptic neuron and connects with receptors that initiate continuation of a nerve impulse in the postsynaptic neuron. It is then broken down in the synaptic cleft by acetylcholinesterase to form choline that is returned to the presynaptic neuron where it is converted back to acetylcholine. *Agonists* such as nicotine at low doses activate acetylcholine receptors and are stimulants. *Antagonists* such as tubocurarine or atropine, which block activation of the receptors, are toxic in high doses, but at low doses are useful as muscle relaxants.

6. **Which neurotransmitters and what kinds of drugs play roles in allergies, mental depression, drug addiction, and pain?** *Histamine*, an amino acid derivative, causes allergic symptoms. *Antihistamines* are antagonists with a general structure that resembles histamines, but with bulky groups at one end. Monoamines (serotonin, norepinephrine, and dopamine) are brain neurotransmitters; a deficiency of any of these molecules is associated with mental depression. *Drugs* that increase their activity include *tricyclic antidepressants* (for example, amitriptyline), *monoamine oxidase (MAO) inhibitors* (for example, phenelzine), and *selective serotonin re-uptake inhibitors (SSRI)* (for example, fluoxetine). An increase of dopamine activity in the brain is associated with the effects of most addictive substances. A group of neuropeptides acts at opiate receptors to counteract pain; all may be addictive.

7. **What are some of the methods used in drug discovery and design?** *Ethnobotanists* work to identify the medicinal products of plants known to native peoples. *Chemical synthesis* is used to improve on the medicinal properties of known compounds by creating similar structures. *Combinatorial chemistry* produces many related molecules for drug screening. *Computer design* is used to select the precise molecular structure to fit a given receptor.

UNDERSTANDING KEY CONCEPTS

20.12 In many species of animals, at the onset of pregnancy, luteinizing hormone is released; it promotes the synthesis of progesterone—a major hormone in maintaining the pregnancy.

(a) Where is LH produced, and to what class of hormones does it belong?

(b) Where is progesterone produced, and to what class of hormones does it belong?

(c) Do progesterone-producing cells have LH receptors on their surface, or does LH enter the cell to carry out its function?

(d) Does progesterone bind to a cell-surface receptor, or does it enter the cell to carry out its function? Explain.

20.13 The "rush" of epinephrine in response to danger causes the release of glucose in muscle cells so that those muscles can carry out either "fight or flight." Very small amounts of the hormone produced in the adrenal gland cause a powerful response. To get such a response, the original signal (epinephrine) must be amplified many times. At what step in the sequence of events (Section 20.3) would you predict that the signal is amplified? Explain. How might that amplification take place?

20.14 Diabetes occurs when there is a malfunction in the uptake of glucose from the bloodstream into the cells. Your friend's little brother was just diagnosed with type I diabetes, and she has asked you the following questions. How would you answer them?

(a) What hormone is involved, and what class is it?

(b) Where is the hormone released?

(c) How is this hormone transported to the cells that need it to allow glucose to enter?

(d) Would you expect the hormone to enter the cell to carry out its function? Explain.

20.15 Give two mechanisms by which neurotransmitters exert their effects.

20.16 When an impulse arrives at the synapse, the synaptic vesicles open and release neurotransmitters into the cleft within a thousandth of a second. Within another ten-thousandth of a second, these molecules have diffused across the cleft and bound to receptor sites in the effector cell. In what two ways is transmission across a synapse terminated so that the neuron's signal is concluded?

20.17 What is the significance of dopamine in the addictive effects of cocaine, amphetamines, and alcohol?

ADDITIONAL PROBLEMS

CHEMICAL MESSENGERS

20.18 What is a hormone? What is the function of a hormone? How is the presence of a hormone detected by its target?

20.19 What is the difference between a hormone and a vitamin?

20.20 What is the difference between a hormone and a neurotransmitter?

20.21 Is a hormone changed as a result of binding to a receptor? Is the receptor changed as a result of binding the hormone? What are the binding forces between hormone and receptor?

20.22 How is hormone binding to its receptor more like an allosteric regulator binding to an enzyme than a substrate binding to an enzyme?

20.23 Describe what is meant by the terms *chemical messenger*, *target tissue*, and *hormone receptor*.

HORMONES AND THE ENDOCRINE SYSTEM

20.24 What is the purpose of the body's endocrine system?

20.25 Name as many endocrine glands as you can.

20.26 List the three major classes of hormones.

20.27 Give two examples of each of the three major classes of hormones.

20.28 What is the structural difference between an enzyme and a hormone?

20.29 What is the relationship between enzyme specificity and tissue specificity for a hormone?

20.30 Describe in general terms how a peptide hormone works.

20.31 Describe in general terms how a steroid hormone works.

HOW HORMONES WORK: EPINEPHRINE

20.32 In what gland is epinephrine produced and released?

20.33 Under what circumstances is epinephrine released?

20.34 How does epinephrine reach its target tissues?

20.35 What is the main function of epinephrine at its target tissues?

20.36 In order of their involvement, name the three membrane-bound proteins involved in transmitting the epinephrine message across the cell membrane.

20.37 What is the "second messenger" inside the cell that results from the epinephrine message? Is the ratio of epinephrine molecules to second messenger less than 1:1, 1:1, or greater than 1:1? Explain.

20.38 What role does the second messenger play in a cell stimulated by epinephrine?

20.39 What enzyme catalyzes hydrolysis of the second messenger to terminate the message? What is the product called?

20.40 Epinephrine is used clinically in the treatment of what life-threatening allergic response?

20.41 People susceptible to anaphylactic shock due to insect stings or certain food allergies must be prepared to treat themselves in case of exposure. How are they prepared and what must they do?

HORMONES

20.42 Give an example of a polypeptide hormone. How many amino acids are in the hormone? Where is the hormone released? Where does the hormone function? What is the result of the hormone message?

20.43 Give an example of a steroid hormone. What is the structure of the hormone? Where is the hormone released? Where does the hormone function? What is the result of the hormone message?

20.44 What are the three major classes of steroid hormones? Give an example of each class.

20.45 Name the two primary male sex hormones.

20.46 Name the three principal female sex hormones.

20.47 What characteristics in their mechanism of action does thyroxine share with the steroid hormones?

20.48 List two hormones that also function as neurotransmitters.

20.49 Explain why epinephrine can act as both a neurotransmitter and a hormone without "crossover" between the two functions.

20.50 Identify the class to which each of these hormones belongs:

(a) HO—⟨ ⟩—$CH_2CH_2NH_2$ (with HO at two positions)

(b) Insulin

(c)

20.51 Identify the class to which each of these hormones belongs:

(a) Glucagon

(b)

Thyroxine

(c)

Estradiol

NEUROTRANSMITTERS

20.52 What is a synapse, and what role does it play in nerve transmission?

20.53 What is an axon, and what role does it play in nerve transmission?

20.54 List three cell types that might receive a message transmitted by a neurotransmitter.

20.55 What kinds of cellular or organ actions would you expect to be influenced by neurotransmitters?

20.56 Describe in general terms how a nerve impulse is passed from one neuron to another.

20.57 What are the two methods for removing the neurotransmitter once its job is done?

20.58 List the three steps in chemical transmission of the impulse between a nerve cell and its target.

20.59 Write an equation for the reaction that is catalyzed by acetylcholinesterase.

20.60 Why are enkephalins sometimes called *neurohormones*?

20.61 Outline the six steps in cholinergic nerve transmission.

CHEMICAL MESSENGERS AND DRUGS

20.62 Describe the difference between drugs that are agonists and those that are antagonists.

20.63 Give an example of a drug that acts as an agonist for acetylcholine receptors and one that acts as an antagonist for these receptors.

20.64 Give examples of two histamine antagonists that have very different tissue specificities and functions.

20.65 Name three families of drugs used to treat depression.

20.66 Give an example of a drug from each family in Problem 20.65.

20.67 Name the "big three" monoamine neurotransmitters.

20.68 What is the impact and mode of action of cocaine on dopamine levels in the brain?

20.69 What is the impact and mode of action of amphetamines on dopamine levels in the brain?

20.70 How is the tetrahydrocannabinol of marijuana similar in action to heroin and cocaine?

20.71 Why do we have brain receptors that respond to morphine and other opium derivatives from plants?

20.72 What are endorphins? Where in the body are they found?

20.73 "Runner's high," sexual excitement, and other complex behaviors are believed to involve which neuropeptides?

20.74 What does an ethnobotanist do?

20.75 Combinatorial chemistry has added hundreds of drugs to the pharmaceutical market in recent years. What is the basis of the combinatorial approach to drug design? What advantages might the combinatorial approach have for the pharmaceutical industry?

20.76 In what ways are studies of the exact size and shape of biomolecules (such as enzymes, receptors, signal transducers, and so on) leading to the development of new drugs to treat disease?

20.77 How are computers used in the development of new drugs to treat disease?

Applications

20.78 One of the responsibilities of the endocrine system is maintenance of homeostasis in the body. Briefly explain what is meant by the term *homeostasis*. [*Homeostasis, p. 631*]

20.79 What is the goal of the measurements of clinical chemistry? [*Homeostasis, p. 631*]

20.80 In animals, hormones are produced by the endocrine glands and tissues in various parts of the body. Why is it necessary for plants to synthesize the hormones in the cells where they are needed rather than in specialized cells? [*Plant Hormones, p. 641*]

20.81 How does 2,4-D, a weed killer, take advantage of the function of a plant hormone? [*Plant Hormones, p. 641*]

20.82 What is believed to be the mode of action of epibatidine? [*And from This Little Frog . . . p. 647*]

20.83 Suggest a chemical modification to epibatidine that might be synthesized and tested as a painkiller. What is your reasoning for the suggested chemical modification? [*And from This Little Frog . . ., p. 647*]

General Questions and Problems

20.84 List and describe the functions of the three types of proteins involved in transmission of a hormone signal.

20.85 The cyclic AMP (second messenger) of signal transmission is very reactive and breaks down rapidly after synthesis. Why is this important to the signal transmission process?

20.86 We say that there is signal amplification in the transmission process. Explain how signal amplification occurs and what it means for transmission of the signal to the sites of cellular activity.

20.87 The phosphodiesterase that catalyzes hydrolysis of cyclic AMP is inhibited by caffeine. What overall effect would caffeine have on a signal that is mediated by cAMP?

20.88 Compare the structures of the sex hormones testosterone and progesterone. What portions of the structures are the same? Where do they differ?

20.89 When you compare the structures of ethynyl estradiol to norethindrone, where do they differ? Where is ethynyl estradiol similar to estradiol? Where is norethindrone similar to progesterone?

20.90 Anandamides have been isolated from brain tissues and appear to be the natural ligand for the receptor that also binds tetrahydrocannabinol. Anandamides have also been discovered in chocolate and cocoa powder. How might the craving for chocolate be explained?

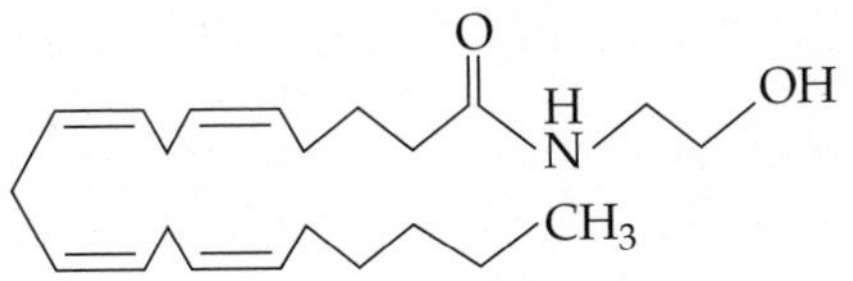

An anandamide structure

20.91 Identify the structural changes that occur in the first two steps in the conversion of tyrosine to epinephrine (Figure 20.4). To what main classes and subclasses of enzymes do the enzymes that catalyze these reactions belong?

20.92 Look at the structures of the two male sex hormones shown on p. 638. Identify the type of functional-group change that interconverts testosterone and androsterone. To which class of chemical reactions does this change belong?

20.93 Look at the structures of the three female sex hormones shown on p. 639. Identify the type of functional-group change that interconverts estradiol and estrone. To which class of chemical reactions does this change belong?

CHAPTER 21

The Generation of Biochemical Energy

CONCEPTS TO REVIEW

Oxidation–reduction reactions
(Sections 6.11, 6.12)

Energy in chemical reactions
(Sections 7.2, 7.4)

Enzymes
(Sections 19.1, 19.4)

▲ A running cheetah generates and uses energy at a rapid rate.

CONTENTS

CHAPTER GOALS

The major questions to be answered include the following:

1. **What is the source of our energy, and what is its fate in the body?**

THE GOAL: Be able to provide an overview of the sources of our energy and how we use it, identify the cellular location of energy generation, and explain the significance of exergonic and endergonic reactions in metabolism.

2. **How are the reactions that break down food molecules organized?**

THE GOAL: Be able to list the stages in catabolism and describe the role of each.

3. **What are the major strategies of metabolism?**

THE GOAL: Be able to explain and give examples of the roles of ATP, coupled reactions, and oxidized and reduced coenzymes in metabolic pathways.

4. **What is the citric acid cycle?**

THE GOAL: Be able to describe what happens in the citric acid cycle and explain its role in energy production.

5. **How is ATP generated in the final stage of catabolism?**

THE GOAL: Be able to describe in general the electron-transport chain, oxidative phosphorylation, and how they are coupled.

6. **What are the harmful by-products produced from oxygen, and what protects against them?**

THE GOAL: Be able to identify the highly reactive oxygen-containing products formed during metabolism and the enzymes and vitamins that counteract them.

All organisms obtain energy from their surroundings to stay alive. In animals, the energy comes from food and is released in the exquisitely interconnected reaction pathways of metabolism. We are powered by the oxidation of biomolecules made mainly of carbon, hydrogen, and oxygen. The end products are carbon dioxide, water, and energy:

$$C, H, O \text{ (Food molecules)} + O_2 \longrightarrow CO_2 + H_2O + \text{Energy}$$

The principal food molecules—lipids, proteins, and carbohydrates—differ in structure and are broken down by individual pathways that are examined in later chapters. The product of these individual pathways, usually acetyl coenzyme A, enters the central, final pathways to yield usable energy. For the present, we are going to concentrate on these final, common pathways by which energy is released from all types of food molecules.

21.1 Energy and Life

Living things must do mechanical work—microorganisms engulf food, plants bend toward the sun, humans walk about. All organisms must also do the chemical work of synthesizing biomolecules needed for energy storage, growth, repair, and replacement. In addition, cells need energy for the work of moving molecules and ions across cell membranes. In humans, it is the energy released from food that allows these various kinds of work to be done.

Energy can be converted from one form to another but can be neither created nor destroyed. (⬤, Section 7.2) Ultimately, the energy used by all but a very few living things on earth comes from the sun (Figure 21.1). Plants convert sunlight to potential energy stored mainly in the chemical bonds of carbohydrates.

Plant-eating animals then utilize this energy, some of it for immediate needs and the rest to be stored for future needs, mainly in the chemical bonds of fats. Other animals, including humans, are able to eat plants or animals and use the chemical energy these organisms have stored.

Our bodies do not produce energy by burning up a steak all at once, however, because the large quantity of heat released would be harmful to us. Furthermore, it is difficult to capture energy for storage once it has been converted to heat. What we need is energy that can be stored and then released in the right amounts when and where it is needed, whether we are running away from an angry dog, studying for

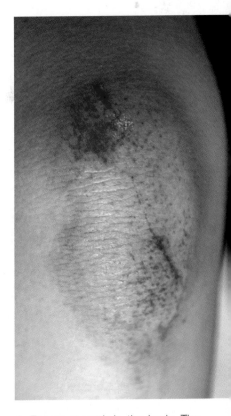

▲ Energy at work in the body. The biomolecules needed to heal this wound will be synthesized using energy from the catabolism of food molecules.

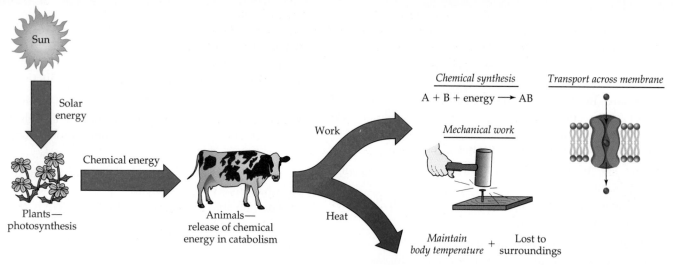

▲ **FIGURE 21.1** **The flow of energy through the biosphere.** Energy from the sun is ultimately stored in chemical bonds, used for work, used to maintain body temperature, or lost as heat.

an exam, or sleeping peacefully. We therefore have some specific requirements for energy:

- Energy must be released from food gradually.
- Energy must be stored in readily accessible forms.
- Release of energy from storage must be finely controlled so that it is available exactly when and where it is needed.
- Just enough energy must be released as heat to maintain constant body temperature.
- Energy in a form other than heat must be available to drive chemical reactions that are not favorable at body temperatures.

This chapter looks at some of the ways these requirements for energy management are met. We begin by reviewing basic concepts about energy. Then we take an overview of *metabolism* and the strategies on which it relies. Next, we look at the *citric acid cycle* and *oxidative phosphorylation*, which together form the common pathway for the production of energy from all sources and for all needs.

21.2 Energy and Biochemical Reactions

Chemical reactions either release energy as they proceed or absorb energy in order to proceed. For a reaction to be favorable and proceed on its own is dependent on either the release or absorption of energy as heat (the change in enthalpy, ΔH (Section 7.2)), together with the increase or decrease in disorder (ΔS, the entropy change (Section 7.4)) caused by the reaction. The net effect of these changes is given by the free-energy change of a reaction: $\Delta G = \Delta H - T\Delta S$. (⟨▭▭⟩, p. 190)

Reactions in living organisms are no different from reactions in a chemistry laboratory. Both follow the same laws, and both have the same energy requirements. Spontaneous reactions—that is, those that are *favorable* in the forward direction—release free energy and the energy released is available to do work. Such reactions, described as *exergonic*, are the source of our biochemical energy. Remember the difference between the terms *exothermic* and *exergonic*. "Exergonic" applies to the release of free energy, represented by a negative ΔG. "Exothermic" applies only to the release of heat, represented by a negative value for the heat of reaction, ΔH. (⟨▭▭⟩, p. 185)

As shown by the energy diagram in Figure 21.2a, the products of a favorable, exergonic reaction are farther *downhill* on the energy scale than the reactants. That is, the products are more stable than the reactants, and as a result the free-energy change (ΔG) has a negative value. Oxidation reactions, for example, are usually downhill reactions that release energy. Oxidation of glucose, the principal source of energy for animals, produces 686 kcal of free energy per mole of glucose:

$$C_6H_{12}O_6 + 6\ O_2 \longrightarrow 6\ CO_2 + 6\ H_2O \qquad \Delta G = -686\ \text{kcal/mol}$$

The greater the amount of free energy released, the further a reaction proceeds toward product formation before reaching equilibrium.

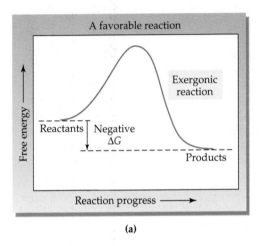
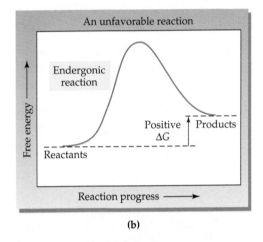

(a)   (b)

▲ **FIGURE 21.2** **Energy diagrams for favorable and unfavorable reactions.** (a) In a favorable reaction, the products have less energy than the reactants. (b) In an unfavorable reaction, the products have more energy than the reactants.

Reactions in which the products are higher in energy than the reactants can also take place, but such *unfavorable* reactions cannot occur without the input of energy from an external source. In other words, energy has to be added to the reactants for an energetically *uphill* change to occur (Figure 21.2b). You might think of these as reactions that have to be *pushed* up the hill. Such reactions are described as *endergonic*. The larger the positive free-energy change, the greater the amount of energy that must be added to convert the reactants to products. Remember the difference here also: "Endergonic" applies to reactions that require an input of free energy and have a positive value of ΔG. "Endothermic" refers to reactions that absorb heat from their surroundings and have a positive value for the heat of reaction, ΔH.

Like the heat of reaction, the free-energy change switches sign for the reverse of a reaction, but the value does not change. Photosynthesis, the process whereby plants convert CO_2 and H_2O to glucose plus O_2, is the reverse of the oxidation of glucose. Its ΔG is therefore positive and of the same numerical value as that for the oxidation of glucose (see the Application "Plants and Photosynthesis," p. 686). The sun provides the necessary external energy for photosynthesis (686 kcal per mole of glucose formed):

▲ Our source of energy for photosynthesis.

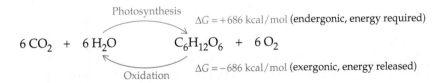

Living systems make constant use of this principle in the series of chemical reactions we know as the biochemical *pathways*. Energy is stored in the products of

Endergonic   A nonspontaneous reaction or process that absorbs free energy and has a positive ΔG.

Exergonic   A spontaneous reaction or process that releases free energy and has a negative ΔG.

an overall **endergonic** reaction pathway. This stored energy is released as needed in an overall **exergonic** reaction pathway that regenerates the original reactants. It is not necessary that every reaction in the pathways between the reactants and products be the same, so long as the pathways connect the same reactants and products.

WORKED EXAMPLE   **21.1** Determining Reaction Energy

Are the following reactions exergonic or endergonic?

(a) Glucose 6-phosphate $\rightarrow$ Fructose 6-phosphate $\Delta G = 0.5$ kcal/mol

(b) Fructose 6-phosphate + ATP $\rightarrow$ Fructose 1,6-bisphosphate + ADP
$\Delta G = -3.4$ kcal/mol

ANALYSIS Exergonic reactions release free energy, and ΔG is negative. Endergonic reactions gain free energy, and so ΔG is positive.

SOLUTION
Reaction (a), the conversion of glucose 6-phosphate to fructose 6-phosphate has a positive ΔG; therefore it is endergonic. Reaction (b), the conversion of fructose 6-phosphate to fructose 1,6-bisphosphate has a negative ΔG; therefore it is exergonic.

WORKED EXAMPLE   **21.2** Determining Reaction Energy
for Reverse Reactions

Write the reverse reaction for each reaction in Worked Example 21.1. For each reverse reaction, determine ΔG and characterize the reaction as either exergonic or endergonic.

ANALYSIS First, remember that reactions are written left to right with the reaction arrow pointing to the right. Second, remember that the compounds that are products in the original reaction are reactants in the reverse reaction and the compounds that are reactants in the original reaction are products in the reverse reaction. (We are assuming the reaction is directly reversible; this is not always true inside cells.) Third, remember that if ΔG for the forward reaction is positive, ΔG for the reverse reaction has the same numeric value but is negative. If ΔG for the forward reaction is negative, ΔG for the reverse reaction has the same number value but is positive. Negative ΔG values indicate exergonic reactions, and positive ΔG values indicate endergonic reactions.

SOLUTION

(a) Fructose 6-phosphate $\rightarrow$ Glucose 6-phosphate $\Delta G = -0.5$ kcal/mol
This reaction is exergonic.

(b) Fructose 1,6-bisphosphate + ADP $\rightarrow$ Fructose 6-phosphate + ATP
$\Delta G = 3.4$ kcal/mol
This reaction is endergonic.

PROBLEM 21.1

The following reactions occur in the citric acid cycle, an energy-producing sequence of reactions that we will discuss later in this chapter. Which of the reactions listed is (are) exergonic? Which is (are) endergonic? Which will release the most energy? Write the complete equation for the reverse of reaction (c). (Recall that organic acids are usually referred to in biochemistry with the *-ate* ending because they exist as anions in body fluids.)

(a) Acetyl coenzyme A + Oxaloacetate + $H_2O \longrightarrow$ Citrate + Coenzyme A
$\Delta G = -9$ kcal/mol

(b) Citrate $\longrightarrow$ Isocitrate $\Delta G = +3$ kcal/mol

(c) Fumarate + $H_2O \longrightarrow$ L-Malate $\Delta G = -0.9$ kcal/mol

APPLICATION ▶ Life without Sunlight

Before we had the equipment to descend deep into the ocean, no one imagined that life existed there. What could provide the food and energy? Textbooks firmly stated that all life depends on sunlight.

Not true. In 1977, hydrothermal vents—openings spewing water heated to 400 °C deep within the earth—were found on the ocean floor. The hydrothermal vents were dubbed "black smokers" because the water was black with mineral sulfides precipitating from the hot, acidic water as it exited the vents. At 2200 m below the ocean surface, there is no chance for the penetration of energy from sunlight. Therefore, the discovery of thriving clusters of tube worms, giant clams, mussels, and other creatures surrounding the black smokers was a great surprise.

Distinctive types of bacteria form the basis for the web of life in these locations. What replaces sunlight as their source of energy? The hot water is rich in dissolved inorganic substances that are reducing agents and therefore electron donors. Life-supporting energy is set free by their oxidation. Hydrogen sulfide, for example, is abundant in the hot seawater, which has passed through sulfur-bearing mineral deposits on its way to the surface. This is the same gas produced during anaerobic decomposition of organic matter in a swamp; it is also the gas that gives the awful odor to rotten eggs. As the hydrogen sulfide is converted to sulfate ions, the electrons set free in the oxidation move through an electron-transport chain that makes ATP formation possible.

Carbon dioxide dissolved in the seawater is the raw material used by the bacteria to make their own essential carbon-containing biomolecules. Experiments have shown that the tube worms, giant clams, and other creatures surrounding the black smokers do not eat the bacteria. Rather, the bacteria colonize their gastrointestinal tracts, where the waste products and dead bodies of the bacteria are the carbon source for biosynthesis by their hosts.

An opportunity to observe the colonization of a hot deep-ocean environment came in 1991 when scientists discovered a volcano erupting underneath the ocean. Initially, all life in the

▲ **Tube worms at a hydrothermal vent in the Galapagos rift.**

vicinity was wiped out, yet soon afterward, the area was thriving with bacteria. This discovery and others have raised some intriguing questions. The same black smoker bacteria have been found in the vicinity of the Mount St. Helens volcanic eruption, and hydrothermal vents with their communities of living things have been found in the fresh waters of the deepest lake on earth, Lake Baikal in Russia. Could it be that a thriving population of bacteria has been living in the hot interior of the earth ever since it formed? Were these anaerobic bacteria earth's first inhabitants, and could they exist beneath the surface of other planets?

See Additional Problems 21.77 and 21.78 at the end of the chapter.

●○ KEY CONCEPT PROBLEM 21.2

In a cell, sugar can be oxidized via metabolic pathways. Alternatively, you could burn sugar in the laboratory. Which of these methods consumes or produces more energy?

●○ KEY CONCEPT PROBLEM 21.3

The overall equation in this section,

$$6\,CO_2 + 6\,H_2O \underset{\text{oxidation}}{\overset{\text{photosynthesis}}{\rightleftharpoons}} C_6H_{12}O_6 + 6\,O_2,$$

shows the cycle between photosynthesis and oxidation. Pathways operating in opposite directions cannot be exergonic in both directions.

(a) Which of the two pathways in this cycle is exergonic and which is endergonic?

(b) Where does the energy for the endergonic pathway come from?

21.3 Cells and Their Structure

Before we proceed with our overview of metabolism, it is important to see where the energy-generating reactions take place within the cells of living organisms. There are two main categories of cells: *prokaryotic cells*, usually found in single-celled organisms including bacteria and blue-green algae, and *eukaryotic cells*, found in some single-celled organisms and all plants and animals.

Eukaryotic cells are about 1000 times larger than bacterial cells, have a membrane-enclosed nucleus that contains their DNA, and include several other kinds of internal structures known as *organelles*—small, functional units that perform specialized tasks. A generalized eukaryotic cell is shown in Figure 21.3; the accompanying Table describes the functions of its major parts. Everything between the cell membrane and the nuclear membrane in a eukaryotic cell, including the various organelles, is referred to as the **cytoplasm**. The organelles are surrounded by the fluid part of the cytoplasm, the **cytosol**, which contains electrolytes, nutrients, and many enzymes, all in aqueous solution.

The **mitochondria** (singular, **mitochondrion**), often called the cell's "power plants," are the most important of the organelles for energy production. It is in the mitochondria that about 90% of the body's energy-carrying molecule, ATP, is produced.

A mitochondrion is a roughly egg-shaped structure composed of a smooth outer membrane and a folded inner membrane (Figure 21.4). The space enclosed by the inner membrane is the **mitochondrial matrix**. It is within the matrix that the citric acid cycle (Section 21.8) and the production of most of the body's **adenosine triphosphate (ATP)** take place. The coenzymes and proteins that manage the transfer of energy to the chemical bonds of ATP (Section 21.9) are embedded in the inner membrane of the mitochondrion.

It is believed that millions of years ago mitochondria were free-living bacteria that became trapped within single-celled plants and animals. As evidence for this, consider that mitochondria contain their own DNA, can synthesize some of their own proteins, and can multiply without outside assistance. The relationship of mitochondria to their host cells became a symbiotic one—the mitochondria produced energy needed by the eukaryotic cells, and the cells provided the mitochondria with nutrients. (This kind of relationship is known as endosymbiosis.) Thus, the mitochondria remained within the cells throughout evolution. The number of mitochondria is greatest in eye, brain, heart, and muscle cells, where the need for

Cytoplasm The region between the cell membrane and the nuclear membrane in a eukaryotic cell.

Cytosol The fluid part of the cytoplasm surrounding the organelles within a cell.

Mitochondrion (plural, **mitochondria**) An egg-shaped organelle where small molecules are broken down to provide the energy for an organism.

Mitochondrial matrix The space surrounded by the inner membrane of a mitochondrion.

Adenosine triphosphate (ATP) The principal energy-carrying molecule; removal of a phosphoryl group to give ADP releases free energy.

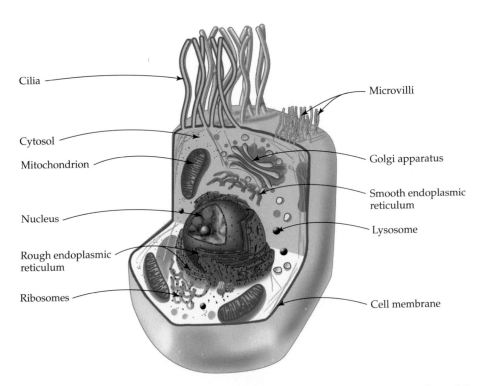

Cilia

Microvilli

Cytosol

Golgi apparatus

Mitochondrion

Smooth endoplasmic
reticulum

Nucleus

Lysosome

Rough endoplasmic
reticulum

Ribosomes

Cell membrane

▲ **FIGURE 21.3   A generalized eukaryotic cell.**   The table below lists the functions of the cell components most important for metabolism.

CELL COMPONENT	PRINCIPAL FUNCTION
Cilia	Movement of materials; for example, mucus in lungs (not present in all cells)
Golgi apparatus	Synthesis and packaging of secretions and cell membrane
Mitochondrion	Synthesis of ATP from ADP
Rough endoplasmic reticulum	Protein synthesis and transport
Nucleus	Replication of DNA, which carries genetic information and governs protein synthesis
Ribosome	Protein synthesis
Microvilli	Absorption of extracellular substances; for example, in digestive tract (not present in all cells)
Cytosol	Intracellular fluid; contains dissolved proteins and nutrients
Lysosome	Removal of pathogens or damaged organelles
Smooth endoplasmic reticulum	Lipid and carbohydrate synthesis
Cell membrane	Composed of lipids plus proteins that govern entry and exit from cell and deliver signals to interior of cell

energy is greatest. The ability of mitochondria to reproduce is called upon in athletes who put heavy energy demands on their bodies—they develop an increased number of mitochondria to aid in energy production.

Interestingly, all mitochondria in our bodies develop from those in the egg that was fertilized, meaning that only our mothers contribute our inherited mitochondrial DNA. This fact is useful in anthropological and archeological studies. For instance, the migration patterns of prehistoric peoples are currently being studied

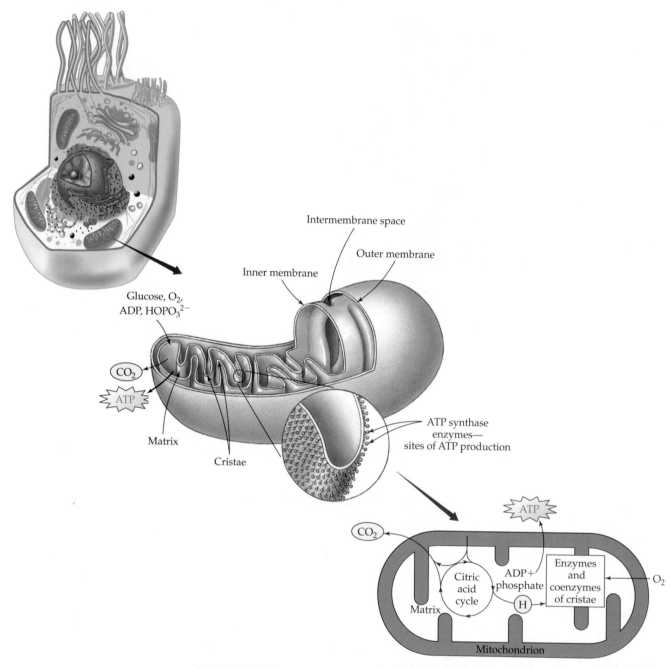

▲ **FIGURE 21.4 The mitochondrion.** Cells have many mitochondria. The citric acid cycle takes place in the matrix. Electron transport and ATP production, the final stage in biochemical energy generation (described in Section 21.9), takes place at the inner surface of the inner membrane. The numerous folds in the inner membrane—known as *cristae*—increase the surface area over which these pathways can take place.

with techniques that compare variations in mitochondrial DNA between populations over time and over location. The Genographic Study sponsored by The National Geographic Society is one such study.

21.4 An Overview of Metabolism and Energy Production

Together, all of the chemical reactions that take place in an organism constitute its metabolism. Most of these reactions occur in the reaction sequences of *metabolic pathways*. Such pathways may be linear (that is, the product of one reaction serves as the starting material for the next); cyclic (a series of reactions regenerates one of the first reactants); or spiral (the same set of enzymes progressively builds up or breaks down a molecule):

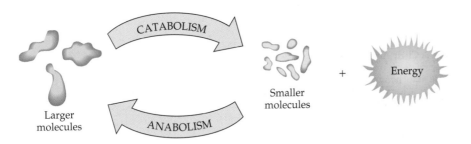

Those pathways that break molecules apart are known collectively as **catabolism**, whereas those that put building blocks back together to assemble larger molecules are known collectively as **anabolism**. The purpose of catabolism is to release energy from food, and the purpose of anabolism is to synthesize new biomolecules, including those that store energy.

Catabolism Metabolic reaction pathways that break down food molecules and release biochemical energy.

Anabolism Metabolic reactions that build larger biological molecules from smaller pieces.

The overall picture of digestion, catabolism, and energy production is simple: Eating provides fuel, breathing provides oxygen, and our bodies oxidize the fuel to extract energy. The process can be roughly divided into the four stages described below and shown in Figure 21.5.

Stage 1: Digestion   Enzymes in saliva, the stomach, and the small intestine convert the large molecules of lipids, carbohydrates, and proteins to smaller molecules. Carbohydrates are broken down to glucose and other sugars, proteins are broken down to amino acids, and triacylglycerols, the lipids commonly known as fats and oils, are broken down to glycerol plus long-chain carboxylic acids, termed fatty acids. These smaller molecules are transferred into the blood for transport to cells throughout the body.

Stage 2: Acetyl-S-coenzyme A production   The small molecules from digestion follow separate pathways that move their carbon atoms into two-carbon acetyl groups. The acetyl groups are attached to coenzyme A (, Figure 19.10, p. 617) by a bond between the sulfur atom of the thiol (—SH) group at the end of the coenzyme A molecule and the carbonyl carbon atom of the acetyl group:

Attachment of acetyl group to coenzyme A

$$\underset{\text{Acetyl group}}{\boxed{CH_3-}}\overset{\overset{\displaystyle O}{\|}}{C}-S-[\text{Coenzyme A}]$$

The resultant compound, **acetyl-S-coenzyme A**, which we abbreviate **acetyl-SCoA**, is an intermediate in the breakdown of *all* classes of food molecules. It carries the acetyl groups into the common pathways of catabolism—Stage 3, the citric acid cycle (Section 21.8) and Stage 4, electron transport and ATP production (Section 21.9).

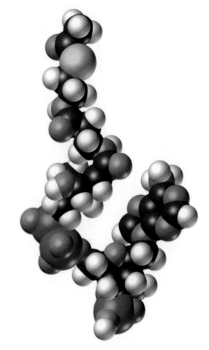

Acetyl-S-coenzyme A

Acetyl-S-coenzyme A (acetyl-SCoA) Acetyl-substituted coenzyme A—the common intermediate that carries acetyl groups into the citric acid cycle.

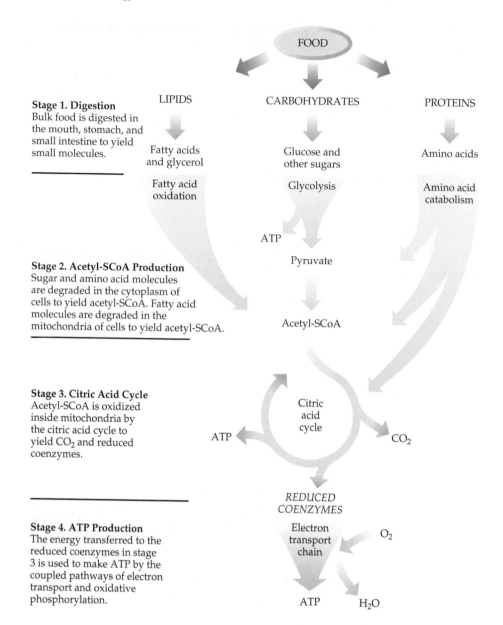

Stage 1. Digestion
Bulk food is digested in the mouth, stomach, and small intestine to yield small molecules.

Stage 2. Acetyl-SCoA Production
Sugar and amino acid molecules are degraded in the cytoplasm of cells to yield acetyl-SCoA. Fatty acid molecules are degraded in the mitochondria of cells to yield acetyl-SCoA.

Stage 3. Citric Acid Cycle
Acetyl-SCoA is oxidized inside mitochondria by the citric acid cycle to yield CO_2 and reduced coenzymes.

Stage 4. ATP Production
The energy transferred to the reduced coenzymes in stage 3 is used to make ATP by the coupled pathways of electron transport and oxidative phosphorylation.

► **FIGURE 21.5 Pathways for the digestion of food and the production of biochemical energy.** This diagram summarizes pathways covered in this chapter (the citric acid cycle and electron transport), and also the pathways discussed in Chapter 23 for carbohydrate metabolism, in Chapter 25 for lipid metabolism, and in Chapter 28 for protein metabolism.

Stage 3: Citric acid cycle Within mitochondria, the acetyl-group carbon atoms are oxidized to the carbon dioxide that we exhale. Most of the energy released in the oxidation leaves the citric acid cycle in the chemical bonds of reduced coenzymes (NADH, $FADH_2$). Some energy also leaves the cycle stored in the chemical bonds of adenosine triphosphate (ATP) or a related triphosphate.

Stage 4: ATP production Electrons from the reduced coenzymes are passed from molecule to molecule down an electron-transport chain. Along the way, their energy is harnessed to produce more ATP. At the end of the process, these electrons—along with hydrogen ions from the reduced coenzymes—combine with oxygen we breathe to produce water. Thus, the reduced coenzymes are in effect oxidized by atmospheric oxygen, and the energy that they carried is stored in the chemical bonds of ATP molecules.

⊂⊃ Looking Ahead

Digestion and conversion of food molecules to acetyl-SCoA, Stages 1 and 2 in Figure 21.5, occur by different metabolic pathways for carbohydrates, lipids, and proteins. Each of these pathways is discussed separately in later chapters: carbohydrate metabolism in Chapter 23, lipid metabolism in Chapter 25, and protein metabolism in Chapter 28.

WORKED EXAMPLE **21.3** Identifying Metabolic Pathways That Convert Basic Molecules to Energy

(a) Identify in Figure 21.5 the stages in the catabolic pathway in which lipids ultimately yield ATP.

(b) Identify in Figure 21.5 the place at which the products of lipid catabolism can join the common metabolism pathway.

ANALYSIS Look at Figure 21.5 and find the pathway for lipids. Follow the arrows to trace the flow of energy. Note that Stage 3 is the point at which the products of lipid, carbohydrate, and protein catabolism all feed into a central, common metabolic pathway, the citric acid cycle. The lipid molecules that feed into Stage 3 do so via acetyl-SCoA (Stage 2). Note also that most products of Stage 3 catabolism feed into Stage 4 catabolism to produce ATP.

SOLUTION
The lipids in food are broken down in Stage 1 (digestion) to fatty acids and glycerol. Stage 2 (acetyl-SCoA production) results in fatty acid oxidation to acetyl-SCoA. In Stage 3 (Citric Acid Cycle) acetyl-SCoA enters the citric acid cycle, which produces ATP, reduced coenzymes, and CO_2. In Stage 4 (ATP production) the energy stored in the reduced coenzymes is converted to ATP energy.

PROBLEM 21.4

(a) Identify in Figure 21.5 the stages in the pathway for the conversion of the energy from carbohydrates to energy stored in ATP molecules.

(b) Identify in Figure 21.5 the three places at which the products of amino acid catabolism can join the central metabolism pathway.

21.5 Strategies of Metabolism: ATP and Energy Transfer

We have described ATP as the body's energy-transporting molecule. What exactly does that mean? Consider that the molecule has three $-PO_3^-$ groups:

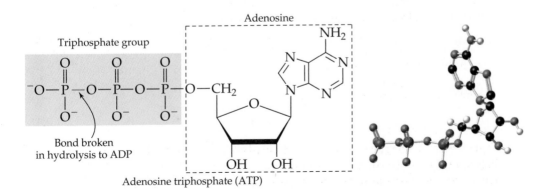

Adenosine triphosphate (ATP)

Removal of one of the $-PO_3^{2-}$ groups from ATP by hydrolysis gives adenosine diphosphate (ADP). The ATP → ADP reaction is exergonic; it releases chemical energy that was held in the bond to the $-PO_3^{2-}$ group:

$$ATP + H_2O \longrightarrow ADP + HOPO_3^{2-} + H^+ \quad \Delta G = -7.3 \text{ kcal/mol}$$

The reverse of ATP hydrolysis—a phosphorylation reaction—is, of course, endergonic (Section 21.2):

$$ADP + HOPO_3^{2-} + H^+ \longrightarrow ATP + H_2O \quad \Delta G = +7.3 \text{ kcal/mol}$$

(In equations for biochemical reactions, we represent ATP and other energy-carrying molecules in red and their lower-energy equivalent molecules in blue.)

ATP is an energy transporter because its production from ADP requires an input of energy that is then released wherever the reverse reaction occurs. Biochemical energy is gathered from exergonic reactions that produce ATP. The ATP then travels to where energy is needed inside the cell, and ATP hydrolysis releases the energy for whatever energy-requiring work must take place. *Biochemical energy production, transport, and use all depend upon the ATP $\rightleftharpoons$ ADP interconversion.*

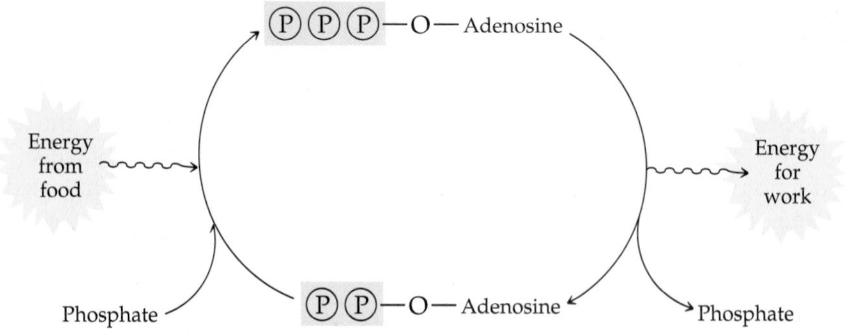

The hydrolysis of ATP to give ADP and its reverse, the phosphorylation of ADP, are reactions perfectly suited to their role in metabolism for two major reasons. One reason is the slow rate of ATP hydrolysis in the absence of a catalyst, so that the stored energy is released only in the presence of the appropriate enzymes.

The second reason is the intermediate value of the free energy of hydrolysis of ATP, as illustrated in Table 21.1. Since the primary metabolic function of ATP is to

TABLE 21.1 Free Energies of Hydrolysis of Some Phosphates

$$R-O-\overset{\displaystyle O}{\underset{\displaystyle O^-}{\overset{\displaystyle \|}{P}}}-O^- + H_2O \rightleftharpoons ROH + HO-\overset{\displaystyle O}{\underset{\displaystyle O^-}{\overset{\displaystyle \|}{P}}}-O^-$$

Compound Name	Function	ΔG (kcal/mol)
Phosphoenol pyruvate	Final intermediate in conversion of glucose to pyruvate (glycolysis)—stage 2, Figure 21.5	−14.8
1, 3-Bisphosphoglycerate	Another intermediate in glycolysis	−11.8
Creatine phosphate	Energy storage in muscle cells	−10.3
ATP (⟶ADP)	Principal energy carrier	−7.3
Glucose 1-phosphate	First intermediate in breakdown of carbohydrates stored as starch or glycogen	−5.0
Glucose 6-phosphate	First intermediate in glycolysis	−3.3
Fructose 6-phosphate	Second intermediate in glycolysis	−3.3

transport energy, it is often referred to as a "high-energy" molecule or as containing "high-energy" phosphorus–oxygen bonds. These terms are misleading because they promote the idea that ATP is somehow different from other compounds. The terms mean only that ATP is reactive and that a useful amount of energy is released when a phosphoryl group is removed from it by hydrolysis.

In fact, if removal of a phosphoryl group from ATP released *unusually* large amounts of energy, other reactions would not be able to provide enough energy to convert ADP back to ATP. ATP is a convenient energy carrier in metabolism *because* its free energy of hydrolysis has an intermediate value. For this reason, the phosphorylation of ADP can be driven by coupling this reaction with a more exergonic reaction, as illustrated in the next section.

PROBLEM 21.5

Acetyl phosphate, whose structure is given here, is another compound with a relatively high free energy of hydrolysis.

$$CH_3-\overset{\overset{\displaystyle O}{\|}}{C}-O-\overset{\overset{\displaystyle O}{\|}}{\underset{\underset{\displaystyle O^-}{|}}{P}}-O^-$$

Using structural formulas, write the equation for the hydrolysis of this phosphate.

PROBLEM 21.6

A common metabolic strategy is the lack of reactivity—that is, the slowness to react—of compounds whose breakdown is exergonic. For example, hydrolysis of ATP to ADP or AMP is exergonic but does not take place without an appropriate enzyme present. Why would the cell use this metabolic strategy?

21.6 Strategies of Metabolism: Metabolic Pathways and Coupled Reactions

Now that you are acquainted with ATP, we will explore how stored chemical energy is gradually released and how it can be used to drive endergonic (uphill) reactions. We have noted before that our bodies cannot burn up the energy obtained from consuming a steak all at once. As shown in Figure 21.2a, however, the energy difference between a reactant (the steak) and the ultimate products of its catabolism (mainly carbon dioxide and water) is a fixed quantity. The same amount of energy is released no matter what pathway is taken between reactants and products. The metabolic pathways of catabolism take advantage of this fact by releasing energy bit by bit in a series of reactions, somewhat like the stepwise release of potential energy as water flows down an elaborate waterfall (Figure 21.6).

The overall reaction and the overall free-energy change for any series of reactions can be found by summing up the equations and the free-energy changes for the individual steps. For example, glucose is converted to pyruvate via the 10 reactions of the glycolysis pathway (part of Stage 2, Figure 21.5). The overall free-energy change for glycolysis is about -8 kcal, showing that the pathway is exergonic—that is, downhill and favorable. The reactions of all metabolic pathways *add up* to favorable processes with negative free-energy changes.

▲ **FIGURE 21.6 Stepwise release of potential energy.** No matter what the pathway from the top to the bottom of this waterfall, the amount of potential energy released as the water falls from the top to the very bottom is the same.

Unlike the waterfall, however, not every individual step in every metabolic pathway is downhill. The metabolic strategy for dealing with what would be an energetically unfavorable reaction is to *couple* it with an energetically favorable reaction so that the overall energy change for the two reactions is favorable. For example, consider the reaction of glucose with hydrogen phosphate ion ($HOPO_3^{2-}$) to yield glucose 6-phosphate plus water, for which $\Delta G = +3.3$ kcal/mol. The reaction is unfavorable because the two products are 3.3 kcal higher in energy than the starting materials. This phosphorylation of glucose is, however, the essential first step toward all metabolic use of glucose. To accomplish this reaction, it is coupled with the exergonic hydrolysis of ATP to give ADP:

(*Unfavorable*)	Glucose + $HOPO_3^{2-} \longrightarrow$ Glucose 6-phosphate + H_2O	$\Delta G = +3.3$ kcal/mol
(*Favorable*)	ATP + $H_2O \longrightarrow$ ADP + $HOPO_3^{2-}$ + H^+	$\Delta G = -7.3$ kcal/mol
(*Favorable*)	Glucose + ATP $\longrightarrow$ Glucose 6-phosphate + ADP	$\Delta G = -4.0$ kcal/mol

The net energy change for these two coupled reactions is favorable: 4.0 kcal of free energy is released for each mole of glucose that is phosphorylated. Only by such coupling can the energy stored in one chemical compound be transferred to other compounds. Any excess energy is released as heat and contributes to maintaining body temperature (Figure 21.7).

Although we have written these reactions separately to show how their energies combine, coupled reactions do not take place separately. The net change occurs all at once as represented by the overall equation. The phosphoryl group is transferred directly from ATP to glucose without the intermediate formation of $HOPO_3^{2-}$. (Also, under physiological conditions, a reaction may be more or less exergonic than in the examples given here. We have stated the free-energy values at standard conditions.)

What about the endergonic synthesis of ATP from ADP, which has $\Delta G = +7.3$ kcal/mol? The same principle of coupling is put to use. For this endergonic reaction to occur, it must be coupled with a reaction that releases *more* than 7.3 kcal/mol. In a different step of glycolysis, for example, the formation of ATP is coupled with the hydrolysis of phosphoenolpyruvate, a phosphate of higher energy than ATP (Table 21.1). Here, the overall reaction is transfer of a phosphoryl group from phosphoenolpyruvate to ADP:

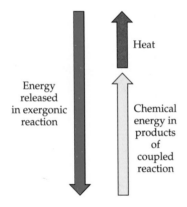

▲ **FIGURE 21.7 Energy exchange in coupled reactions.** The energy provided by an exergonic reaction is either released as heat or stored as chemical potential energy in the bonds of products of the coupled endergonic reaction.

$$\underset{\text{Phosphoenolpyruvate}}{\overset{\overset{\displaystyle O-PO_3^{2-}}{|}}{H_2C=C-COO^-}} + H_2O \longrightarrow \underset{\text{Pyruvate}}{\overset{\overset{\displaystyle O}{||}}{CH_3-C-COO^-}} + HOPO_3^{2-} \qquad \Delta G = -14.8 \text{ kcal/mol}$$

$$ADP + HOPO_3^{2-} + H^+ \longrightarrow ATP + H_2O \qquad \Delta G = +7.3 \text{ kcal/mol}$$

$$\underset{}{\overset{\overset{\displaystyle O-PO_3^{2-}}{|}}{H_2C=C-COO^-}} + ADP \longrightarrow \overset{\overset{\displaystyle O}{||}}{CH_3C-COO^-} + ATP \qquad \Delta G = -7.5 \text{ kcal/mol}$$

Remember that in equations representing coupled reactions, a curved arrow often connects the reactants and products in one of the two chemical changes. (⊂⊃, p. 614) For example, the reaction of phosphoenolpyruvate illustrated above can be written

$$\overset{\overset{\displaystyle O-PO_3^{2-}}{|}}{H_2C=C-COO^-} \xrightarrow{\quad\overset{\displaystyle ADP \quad ATP}{\curvearrowright}\quad} \overset{\overset{\displaystyle O}{||}}{CH_3-C-COO^-}$$

PROBLEM 21.7

One of the steps in lipid metabolism is the reaction of glycerol (1,2,3-propanetriol, $HOCH_2CH(OH)CH_2OH$), with ATP to yield glycerol 1-phosphate. Write the equation for this reaction using the curved arrow symbolism.

PROBLEM 21.8

Why must a metabolic pathway that synthesizes a given molecule occur by a different series of reactions than a pathway that breaks down the same molecule?

APPLICATION ▶ Basal Metabolism

The minimum amount of energy expenditure required per unit of time to stay alive—to breathe, maintain body temperature, circulate blood, and keep all body systems functioning—is referred to as the *basal metabolic rate*. Ideally, it is measured in a person who is awake, is lying down at a comfortable temperature, has fasted and avoided strenuous exercise for 12 hours, and is not under the influence of any medications. The basal metabolic rate can be measured by monitoring respiration and finding the rate of oxygen consumption, which is proportional to the energy used.

An *average* basal metabolic rate is 70 kcal/hr, or about 1700 kcal/day. The rate varies with many factors, including sex, age, weight, and physical condition. A rule of thumb used by nutritionists to estimate basal energy needs per day is the requirement for 1 kcal/hr per kilogram of body weight by a male and 0.95 kcal/hr per kilogram of body weight by a female. For example, a 50 kg (110 lb) female has an estimated basal metabolic rate of (50 kg) (0.95 kcal/kg hr) = (48 kcal/hr) giving a daily requirement of approximately 1200 kcal.

The total calories a person needs each day is determined by his or her basal requirements plus the energy used in additional physical activities. The caloric consumption rates associated with some activities are listed in the accompanying Table. A relatively inactive person requires about 30% above basal requirements per day, a lightly active person requires about 50% above basal, and a very active person such as an athlete or construction worker can use 100% above basal requirements in a day. Each day that you consume food with more calories than you use, the excess calories are stored as potential energy in the chemical bonds of fats in your body and your weight rises. Each day that you consume food with fewer calories than you burn, some chemical energy in your body is taken out of storage to make up the deficit. Fat is metabolized to CO_2 and H_2O, which the body gets rid of, and your weight drops.

Calories Used in Various Activities

ACTIVITY	KILOCALORIES (NUTRITION CALORIES) USED PER MINUTE
Sleeping	1.2
Reading	1.3
Listening to lecture	1.7
Weeding garden	5.6
Walking, 3.5 mph	5.6
Pick-and-shovel work	6.7
Recreational tennis	7.0
Soccer, basketball	9.0
Walking up stairs	10.0–18.0
Running, 12 min/mi (5 mph)	10.0
Running, 5 min/mi (12 mph)	25.0

▲ The cola drink contains 160 Cal (kcal) and the hamburger contains 500 Cal. How long would you have to jog at 5 mph to burn off these calories?

See Additional Problems 21.79 through 21.82 at the end of the chapter.

> **WORKED EXAMPLE**   **21.4** Determining if a Coupled Reaction Is Favorable
>
> The hydrolysis of succinyl-SCoA is coupled with the production of GTP. The equations for the reactions are given below. Combine the equations appropriately and determine if the coupled reaction is favorable.
>
> $$\text{Succinyl-SCoA} \longrightarrow \text{Succinate} + \text{SCoA} \qquad \Delta G = -9.4 \text{ kcal/mol}$$
> $$\text{GDP} + \text{HOPO}_3{}^{2-} + \text{H}^+ \longrightarrow \text{GTP} + \text{H}_2\text{O} \qquad \Delta G = +7.3 \text{ kcal/mol}$$
>
> **ANALYSIS** Add the two equations together to produce the equation for the coupled reaction. Also add the ΔG values together, paying close attention to the signs. If the ΔG is positive, the reaction is not favorable and will not occur; if the ΔG is negative, the reaction is favorable and will occur.
>
> **SOLUTION**
>
> $$\text{Succinyl-SCoA} + \text{GDP} + \text{HOPO}_3{}^{2-} + \text{H}^+ \longrightarrow \text{Succinate} + \text{GTP} + \text{H}_2$$
> $$\Delta G = -2.1 \text{ kcal/mol}$$
>
> Since ΔG is negative, the coupled reaction will occur as written.

> **PROBLEM 21.9**
>
> The hydrolysis of acetyl phosphate to give acetate and hydrogen phosphate ion has $\Delta G = -10.3$ kcal/mol. Combine the equations and ΔG values to determine whether coupling of this reaction with phosphorylation of ADP to produce ATP is favorable. (You need give only compound names or abbreviations in the equations.)

21.7 Strategies of Metabolism: Oxidized and Reduced Coenzymes

The net result of catabolism is the oxidation of food molecules to release energy. Many metabolic reactions are therefore oxidation–reduction reactions, which means that a steady supply of oxidizing and reducing agents must be available. To deal with this requirement, a few coenzymes cycle continuously between their oxidized and reduced forms, just as adenosine cycles continuously between its triphosphate and diphosphate forms:

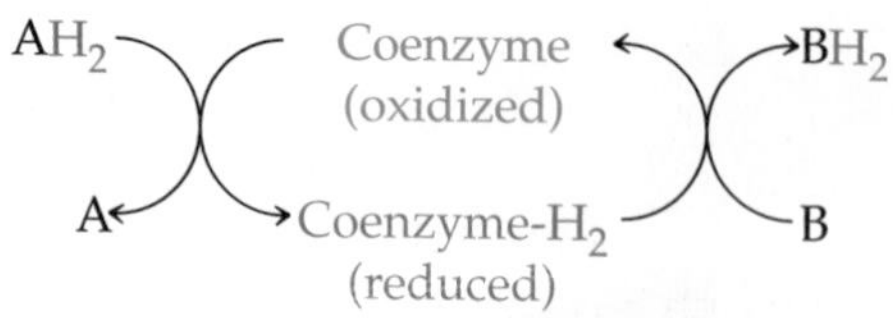

COENZYME	AS OXIDIZING AGENT	AS REDUCING AGENT
Nicotinamide adenine dinucleotide	NAD^+	NADH/H^+
Nicotinamide adenine dinucleotide phosphate	NADP^+	NADPH/H^+
Flavin adenine dinucleotide	FAD	FADH_2
Flavin mononucleotide	FMN	FMNH_2

To review briefly, keep in mind these important points about oxidation and reduction:

- Oxidation can be loss of electrons, loss of hydrogen, or addition of oxygen.
- Reduction can be gain of electrons, gain of hydrogen, or loss of oxygen.
- Oxidation and reduction always occur together.

Each increase in the number of carbon–oxygen bonds is an oxidation, and each decrease in the number of carbon–hydrogen bonds is a reduction, as shown in Table 21.2.

Nicotinamide adenine dinucleotide and its phosphate are widespread, independent coenzymes that enter and leave enzyme active sites in which they are required for redox reactions. As oxidizing agents (NAD^+ and $NADP^+$) they remove hydrogen from a substrate, and as reducing agents (NADH and NADPH) they provide hydrogen that adds to a substrate. The complete structure of NAD^+ is shown below with the change that converts it to NADH. The only difference between the structures of NAD^+/NADH and $NADP^+$/NADPH is that the color-shaded —OH group here is instead a —OPO_3^{2-} group in $NADP^+$ and NADPH:

TABLE 21.2 Oxidation of Carbon by Increased Bonding to Oxygen

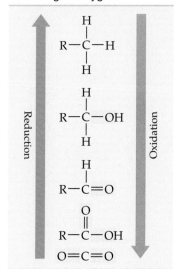

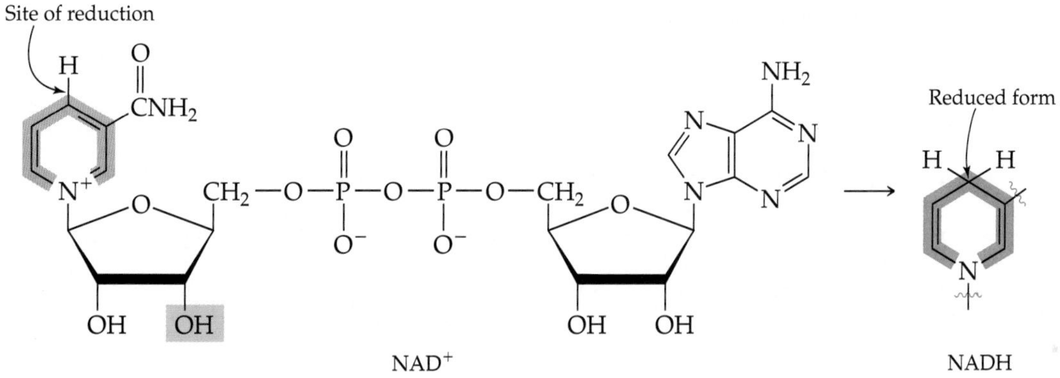

Site of reduction

NAD^+

Reduced form

NADH

As an example, consider a reaction in the citric acid cycle (Step 8 in Figure 21.9, Section 21.8) from the oxidation–reduction, or redox, point of view:

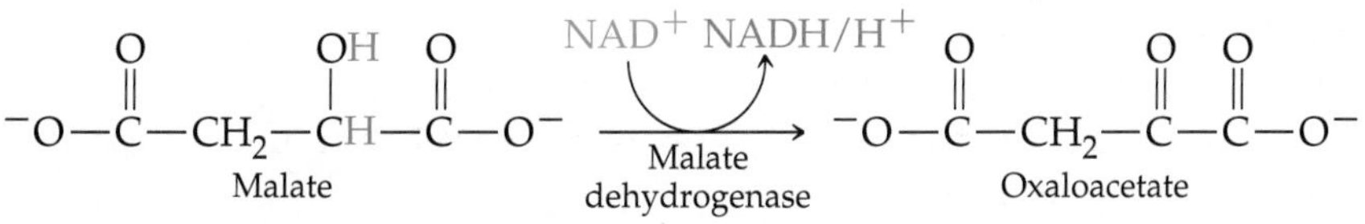

Malate

NAD^+ NADH/H^+

Malate dehydrogenase

Oxaloacetate

The oxidation of malate to oxaloacetate requires the removal of two hydrogen atoms to convert a secondary alcohol to a ketone. (⊂⊃, Section 14.5) The oxidizing agent, which will be reduced, is NAD^+, a *coenzyme*, in this case for malate dehydrogenase. (Sometimes NAD^+ is written as a reactant or product to emphasize its role in a reaction. Keep in mind that although it is free to enter and leave the active site, it always functions as a coenzyme with the appropriate enzyme for the reaction.)

When considering enzyme-catalyzed redox reactions, it is important to recognize that a hydrogen atom is equivalent to a hydrogen *ion*, H^+, plus an electron, e^-. Thus, for the two hydrogen atoms removed in the oxidation of malate,

$$2 \text{ H atoms} = 2 \text{ H}^+ + 2 \text{ e}^-$$

When NAD^+ is reduced, both electrons accompany one of the hydrogens to give a hydride ion,

$$\text{H}^+ + 2 \text{ e}^- = \ :\text{H}^-$$

The reduction of NAD occurs by addition of H^- to the ring in the nicotinamide part of the structure, where the two electrons of H^- form a covalent bond:

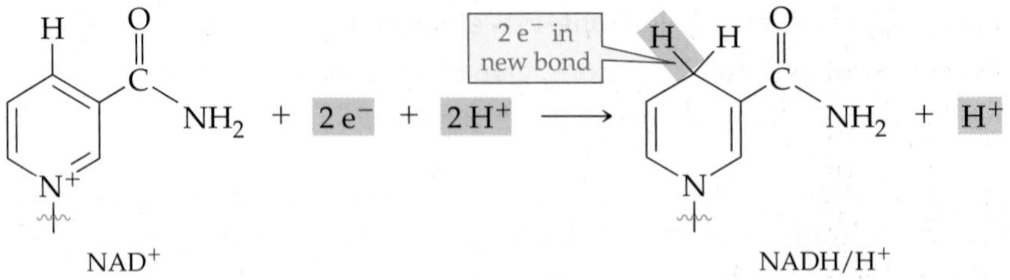

NAD$^+$ NADH/H$^+$

The second hydrogen removed from the oxidized substrate enters the surrounding aqueous solution as a hydrogen ion, H^+. The product of NAD$^+$ reduction is therefore often represented as NADH/H$^+$ to show that two hydrogen atoms have been removed from the reactant, one of which has bonded to NAD$^+$ and the other of which is a hydrogen ion in solution. (NADP$^+$ is reduced in the same way to form NADPH/H$^+$.)

Flavin adenine dinucleotide (FAD), another common oxidizing agent in catabolic reactions, is reduced by the formation of covalent bonds to two hydrogen atoms to give FADH$_2$. It participates in several reactions of the citric acid cycle, which is described in the next section.

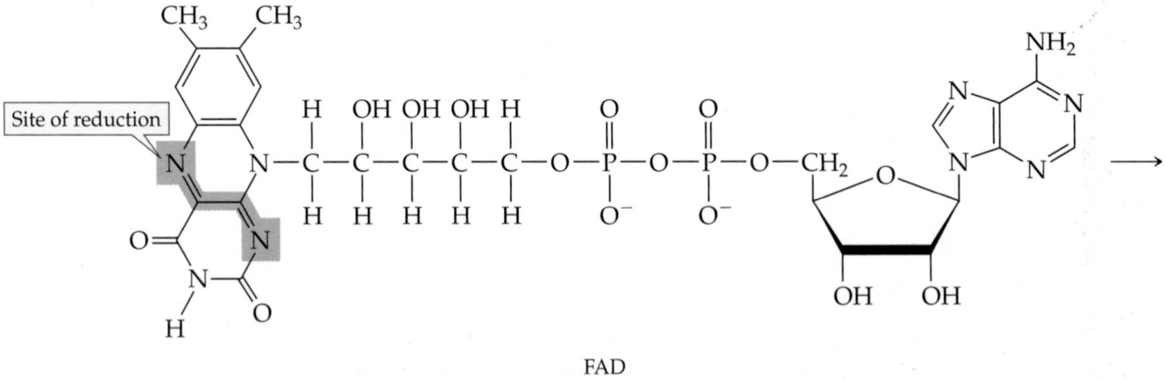

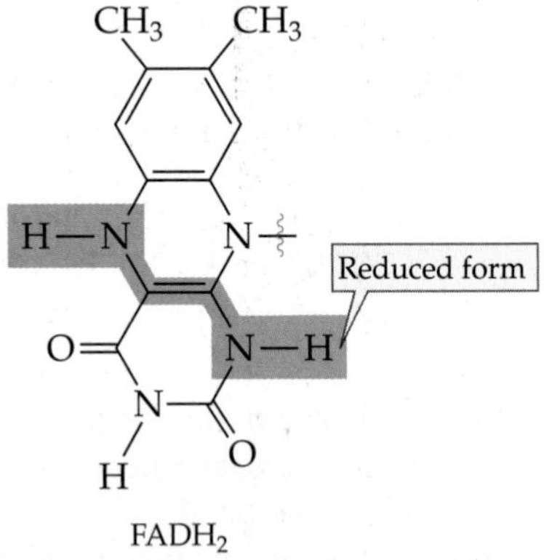

FAD

FADH$_2$

Because the reduced coenzymes, NADH and FADH$_2$, have picked up electrons (in their bonds to hydrogen) that are passed along in subsequent reactions, they are often referred to as *electron carriers*. As these coenzymes cycle through their oxidized and reduced forms, they also carry energy along from reaction to reaction. Ultimately, this energy is passed on to the bonds in ATP, as described in Section 21.9.

PROBLEM 21.10

Is adenosine diphosphate one of the building blocks of the FAD coenzyme?

> **PROBLEM 21.11**
>
> Look ahead to Figure 21.9 for the citric acid cycle. Draw the structures of the reactants in Steps 3, 6, and 8, and indicate which hydrogen atoms are removed in these reactions.

21.8 The Citric Acid Cycle

The carbon atoms from the first two stages of catabolism are carried into the third stage as acetyl groups bonded to coenzyme A. Like the phosphoryl groups in ATP molecules, the acetyl groups in acetyl-SCoA molecules are readily removed in an energy-releasing hydrolysis reaction:

Citric acid cycle The series of biochemical reactions that breaks down acetyl groups to produce energy carried by reduced coenzymes and carbon dioxide.

$$CH_3-\overset{\overset{\textstyle O}{\|}}{C}-SCoA \ + \ H_2O \ \longrightarrow \ CH_3-\overset{\overset{\textstyle O}{\|}}{C}-O^- \ + \ H-SCoA \ + \ H^+ \qquad \Delta G = -7.5 \ kcal/mol$$

Acetyl-SCoA   Coenzyme A

Oxidation of 2 carbons to give CO_2 and transfer of energy to reduced coenzymes occurs in the **citric acid cycle**, also known as the *tricarboxylic acid cycle (TCA)* or *Krebs cycle* (after Sir Hans Krebs, who unraveled its complexities in 1937). As its name implies, the citric acid *cycle* is a closed loop of reactions in which the product of the final step (Step 8), oxaloacetate, a 4-carbon molecule, is the reactant in the first step. The pathway of carbon atoms through the cycle and the significant products formed are summarized in Figure 21.8 and shown in greater detail in Figure 21.9. The 2 carbon atoms of the acetyl group add to the 4 carbon atoms of oxaloacetate in Step 1, and 2 carbon atoms are set free as carbon dioxide in Steps 3 and 4. The cycle continues as 4-carbon intermediates progress toward regeneration of oxaloacetate and production of additional reduced coenzymes.

A brief description of the eight steps of the citric acid cycle is given in Figure 21.9. The enzymes involved in each step are listed in the accompanying Table. The cycle takes place in mitochondria, where seven of the enzymes are dissolved in the matrix and one (for Step 6) is embedded in the inner mitochondrial membrane (Section 21.3).

The cycle operates as long as (1) acetyl groups are available from acetyl-SCoA and (2) the oxidizing agent coenzymes NAD^+ and FAD are available. To meet condition 2, the reduced coenzymes NADH and $FADH_2$ must be reoxidized via the electron-transport chain in Stage 4 of catabolism (described in the next section). Because Stage 4 relies on oxygen as the final electron acceptor, the cycle is also dependent on the availability of oxygen.

The steps of the citric acid cycle are summarized below, with an emphasis on what each step accomplishes.

STEPS 1 and 2: The first two steps set the stage for oxidation. Acetyl groups enter the cycle at Step 1 by addition to 4-carbon oxaloacetate to give citrate, a 6-carbon intermediate. Citrate is a tertiary alcohol and cannot be oxidized; it is converted in Step 2 to its isomer, isocitrate, a secondary alcohol that can be oxidized to a ketone in Step 3. The two steps of the isomerization are catalyzed by the same enzyme, aconitase. Water is first removed and then added back to the intermediate, which remains in the active site, so that the —OH is on a different carbon atom:

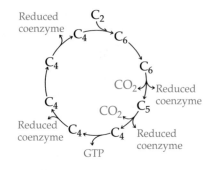

▲ **FIGURE 21.8 Significant outcomes of the citric acid cycle.** The eight steps of the cycle produce two molecules of carbon dioxide, four molecules of reduced coenzymes, and one energy-rich phosphate (GTP). The final step regenerates the reactant for Step 1 of the next turn of the cycle.

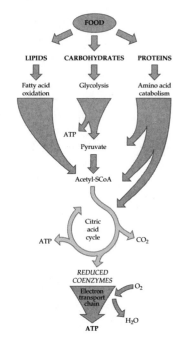

Citrate		Aconitate		Isocitrate
$\begin{array}{c} COO^- \\ | \\ CH_2 \\ | \\ HO-C-COO^- \\ | \\ CH_2 \\ | \\ COO^- \end{array}$	$\xrightarrow[\text{Aconitase}]{+H_2O}$	$\begin{array}{c} COO^- \\ | \\ CH_2 \\ | \\ C-COO^- \\ \| \\ CH \\ | \\ COO^- \end{array}$	$\xrightarrow[\text{Aconitase}]{+H_2O}$	$\begin{array}{c} COO^- \\ | \\ CH_2 \\ | \\ H-C-COO^- \\ | \\ HO-CH \\ | \\ COO^- \end{array}$

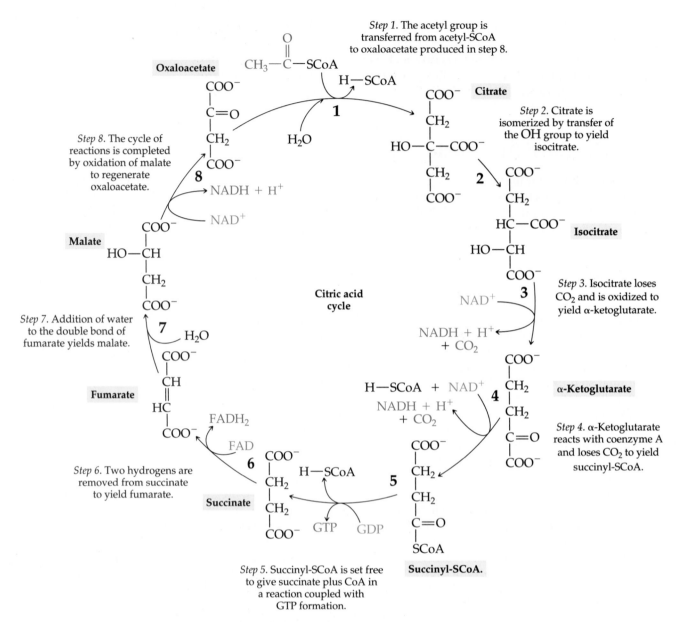

Step 1. The acetyl group is transferred from acetyl-SCoA to oxaloacetate produced in step 8.

Step 2. Citrate is isomerized by transfer of the OH group to yield isocitrate.

Step 3. Isocitrate loses CO_2 and is oxidized to yield α-ketoglutarate.

Step 4. α-Ketoglutarate reacts with coenzyme A and loses CO_2 to yield succinyl-SCoA.

Step 5. Succinyl-SCoA is set free to give succinate plus CoA in a reaction coupled with GTP formation.

Step 6. Two hydrogens are removed from succinate to yield fumarate.

Step 7. Addition of water to the double bond of fumarate yields malate.

Step 8. The cycle of reactions is completed by oxidation of malate to regenerate oxaloacetate.

Citric acid cycle

Enzymes of the Citric Acid Cycle

STEP NO.	ENZYME NAME	REACTION PRODUCT
1	Citrate synthase	Citrate
2	Aconitase	Isocitrate
3	Isocitrate dehydrogenase complex	α-Ketoglutarate
4	α-Ketoglutarate dehydrogenase complex	Succinyl-SCoA
5	Succinyl CoA synthetase	Succinate
6	Succinate dehydrogenase	Fumarate
7	Fumarase	Malate
8	Malate dehydrogenase	Oxaloacetate

▲ **FIGURE 21.9 The citric acid cycle.** The net effect of this eight-step cycle of reactions is the metabolic breakdown of acetyl groups (from acetyl-SCoA) into two molecules of carbon dioxide and energy carried by reduced coenzymes. Here and throughout this and the following chapters, energy-rich reactants or products (ATP, reduced coenzymes) are shown in red and their lower-energy counterparts (ADP, oxidized coenzymes) are shown in blue.

STEPS 3 and 4: Both steps are oxidations that rely on NAD^+ as the oxidizing agent. One CO_2 leaves at Step 3 as the —OH group of isocitrate is simultaneously oxidized to a keto group. A second CO_2 leaves at Step 4, and the resulting succinyl group is added to coenzyme A. In both steps, electrons and energy are transferred in the reduction of NAD^+. Succinyl-SCoA carries four carbon atoms along to the next step.

STEP 5: With 2 carbon atoms now removed as carbon dioxide (though not the original two from the acetyl group), the 4-carbon molecule oxaloacetate must be restored for Step 1 of the next cycle. In Step 5, the exergonic conversion of succinyl-SCoA to succinate is coupled with phosphorylation of **guanosine diphosphate (GDP)** to give **guanosine triphosphate (GTP)**. GTP is similar in structure to ATP and, like ATP, carries energy that can be released during transfer of one of its phosphoryl groups. In many cells, GTP is directly converted to ATP. Step 5 is the only step in the cycle that generates an energy-rich triphosphate.

STEP 6: Next, succinate from Step 5 is oxidized by removal of 2 hydrogen atoms to give fumarate. The enzyme for this reaction, succinate dehydrogenase, is part of the inner mitochondrial membrane. The reaction also requires the coenzyme FAD, which is covalently bound to its enzyme rather than free to come and go. Succinate dehydrogenase and FAD participate in Stage 4 of catabolism by passing electrons directly into electron transport.

STEPS 7 and 8: The citric acid cycle is completed by regeneration of oxaloacetate, a reactant for Step 1. Water is added across the double bond of fumarate to give malate (Step 7) and oxidation of malate, a secondary alcohol, gives oxaloacetate (Step 8).

Guanosine diphosphate (GDP)
An energy-carrying molecule that can gain or lose a phosphoryl group to transfer energy.

Guanosine triphosphate (GTP)
An energy-carrying molecule similar to ATP; removal of a phosphoryl group to give GDP releases free energy.

Net result of citric acid cycle

$$\text{Acetyl-SCoA} + 3\ NAD^+ + FAD + ADP + HOPO_3^{2-} + H_2O \longrightarrow$$

$$HSCoA + 3\ NADH + 3\ H^+ + FADH_2 + ATP + 2\ CO_2$$

- Production of four reduced coenzyme molecules (3 NADH, 1 $FADH_2$)
- Conversion of an acetyl group to two CO_2 molecules
- Production of one energy-rich molecule (GTP, converted immediately to ATP)

The rate of the citric acid cycle is controlled by the body's cellular need for ATP and reduced coenzymes, and for the energy derived from them. For example, when energy is being used at a high rate, ADP accumulates and acts as an allosteric activator (positive regulator, ⊂⊃, p. 609) for isocitrate dehydrogenase, the enzyme for Step 3. When the body's supply of energy is abundant, NADH is present in excess and acts as an inhibitor of isocitrate dehydrogenase. By such feedback mechanisms, as well as by variations in the concentrations of necessary reactants, the cycle is activated when energy is needed and inhibited when energy is in good supply.

| **WORKED EXAMPLE** | **21.5** Identifying Reactants and Products in the Citric Acid Cycle |

What substance or substances are the substrate(s) for the citric acid cycle? What are the products of the citric acid cycle?

ANALYSIS Study Figure 21.9. Note that acetyl-SCoA feeds into the cycle, but does not come out anywhere. Can you see that all of the other reaction substrates are integral to the cycle and are always present, being continuously synthesized and degraded? Note also that the coenzymes NAD^+ and FAD are reduced and the reduced versions are considered energy-carrying products of the cycle. Also, CO_2 is produced at two different steps in the cycle. Finally, GDP is converted to GTP in Step 5 of the cycle.

SOLUTION

Acetyl-SCoA is the substrate for the cycle. Along with GDP and CoA, the oxidized coenzymes NAD^+ and FAD might also be considered substrates despite their status as coenzymes because these substances cycle between the reduced and oxidized states. The products of the cycle are CO_2 and the energy-rich reduced coenzymes $NADH/H^+$ and $FADH_2$ as well as GTP.

PROBLEM 21.12

Which substances in the citric acid cycle are tricarboxylic acids (thus giving the cycle its alternative name)?

PROBLEM 21.13

In Figure 21.9, identify the steps at which reduced coenzymes are produced.

PROBLEM 21.14

Describe the reaction in the citric acid cycle that is catalyzed by succinate dehydrogenase.

PROBLEM 21.15

Identify the participants in the citric acid cycle that are α-keto acids.

PROBLEM 21.16

Which of the reactants in the citric acid cycle have 2 chiral carbon atoms?

KEY CONCEPT PROBLEM 21.17

The citric acid cycle can be divided into two stages. In one stage, carbon atoms are added and removed, and in the second stage, oxaloacetate is regenerated. Which steps of the citric acid cycle correspond to each stage?

21.9 The Electron-Transport Chain and ATP Production

Keep in mind that in some ways catabolism is just like burning petroleum or natural gas. In both cases, the goal is to produce useful energy and the reaction products are water and carbon dioxide. The difference is that in catabolism the products are not released all at once and not all of the energy is released as heat.

At the conclusion of the citric acid cycle, the reduced coenzymes formed in the cycle are ready to donate their energy to making additional ATP. The energy is released in a series of oxidation–reduction reactions that move electrons from one electron carrier to the next as each carrier is reduced (gains an electron from the preceding carrier) and then oxidized (loses an electron by passing it along to the next carrier). Each reaction in the series is favorable; that is, it is exergonic. You can think of each reaction as a step along the way down our waterfall. The sequence of reactions that move the electrons along is known as the **electron-transport chain** (also the *respiratory chain*). The enzymes and coenzymes of the chain and ATP synthesis are embedded in the inner membrane of the mitochondrion (Figure 21.10).

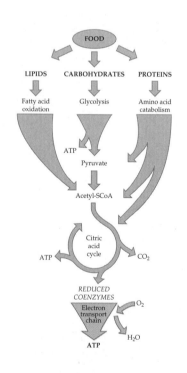

Electron-transport chain The series of biochemical reactions that passes electrons from reduced coenzymes to oxygen and is coupled to ATP formation.

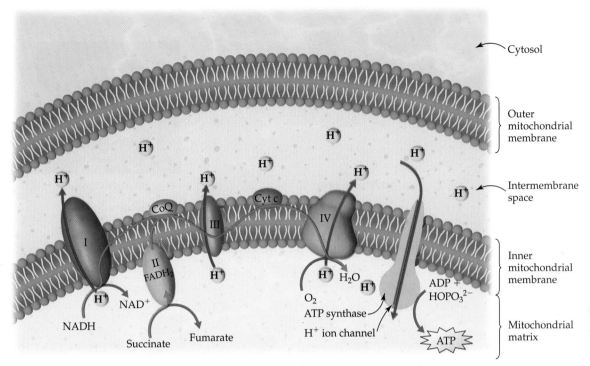

▲ **FIGURE 21.10** **The mitochondrial electron-transport chain and ATP synthase.** The red line shows the path of electrons, and the green lines show the paths of hydrogen ions. The movement of hydrogen ions across the inner membrane at complexes I, III, and IV creates a higher concentration on the intermembrane side of the inner membrane than on the matrix side. The energy released by hydrogen ions returning to the matrix through ATP synthase provides the energy needed for ATP synthesis.

In the last step of the chain, the electrons combine with the oxygen that we breathe and with hydrogen ions from their surroundings to produce water:

$$O_2 + 4\,e^- + 4\,H^+ \longrightarrow 2\,H_2O$$

This reaction is fundamentally the combination of hydrogen and oxygen gases. Carried out all at once with the gases themselves, the reaction is explosive. What happens to all that energy during electron transport?

As electrons move down the electron-transport pathway, the energy released is used to move hydrogen ions out of the mitochondrial matrix and into the inter-membrane space. Because the inner membrane is otherwise impermeable to the H^+ ion, the result is a higher H^+ concentration in the intermembrane space than in the mitochondrial matrix. Moving ions from a region of lower concentration to one of higher concentration opposes the natural tendency for random motion to equalize concentrations throughout a mixture and therefore requires energy to make it happen. This energy is recaptured for use in ATP synthesis.

Electron Transport

Electron transport proceeds in four enzyme complexes held in fixed positions within the inner membrane of mitochondria and two electron carriers that move from one complex to another. The complexes and mobile electron carriers are organized in the sequence of their ability to pick up electrons, as illustrated in Figure 21.10. The four fixed complexes are very large assemblages of polypeptides and electron acceptors. The most important electron acceptors are of three types: (1) various cytochromes, which are proteins that contain heme groups (Figure 21.11a) in which the iron cycles between Fe^{2+} and Fe^{3+}; (2) proteins with iron–sulfur groups in which the iron also cycles between Fe^{2+} and Fe^{3+}; and (3) coenzyme Q (CoQ), often

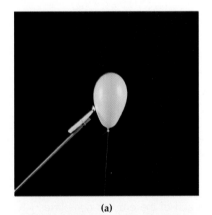

(a)

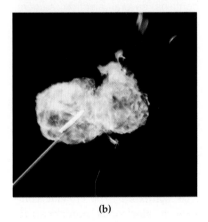

(b)

▲ Explosion of a balloon filled with a hydrogen–oxygen mixture. The amount of energy released is apparent.

▶ **FIGURE 21.11 A heme group and a cytochrome** (a) Heme groups, in which the substituents at the bonds marked in red vary, are iron-containing coenzymes in the cytochromes of the electron-transport chain. They are also the oxygen carriers in hemoglobin in red blood cells. (b) In the cytochrome shown here, the coiled olive green ribbon is the amino acid chain and the heme group is in red with its central iron atom in gray.

(a) A heme group

(b) A cytochrome

known as *ubiquinone* because of its ubiquitous (widespread) occurrence and because its ring structure with the two ketone groups is a *quinone*:

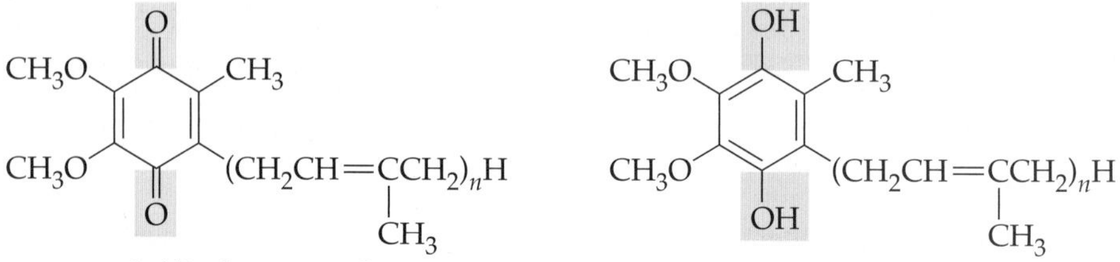

Oxidized coenzyme Q

Reduced coenzyme Q

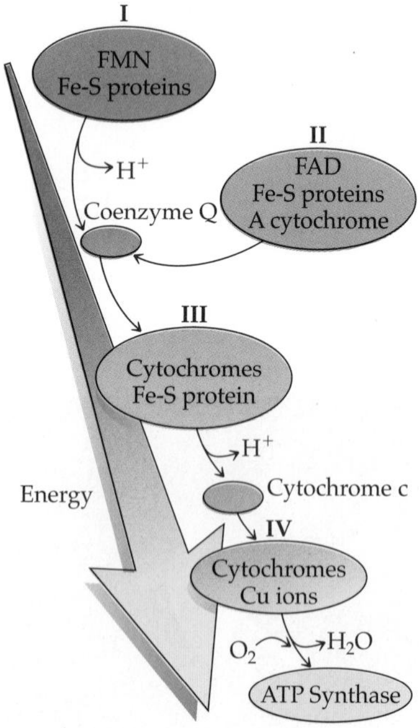

▲ **FIGURE 21.12 Pathway of electrons in electron transport.** Each of the enzyme complexes I–IV contains several electron carriers. (FMN in complex I is similar in structure to FAD.)

The details of the reactions that move electrons in the electron-transport chain are not important to us here. We need only focus on the following essential features of the pathway (Figure 21.12; refer also to Figure 21.10).

- Hydrogen and electrons from NADH and $FADH_2$ enter the electron-transport chain at enzyme complexes I and II, respectively. (In this case, the complexes function independently and not necessarily in numerical order.) The enzyme for Step 6 of the citric acid cycle is part of complex II, so $FADH_2$ is produced when that step of the cycle occurs. $FADH_2$ does not leave complex II. It is immediately oxidized there by reaction with coenzyme Q. Following formation of the mobile coenzyme Q, hydrogen ions no longer participate directly in the reductions. Instead, electrons are transferred one by one.

- Electrons are passed from weaker to increasingly stronger oxidizing agents, with energy released at each transfer.

- Hydrogen ions are released for transport through the inner membrane at complexes I, III, and IV. Some of these ions come from the reduced coenzymes and some from the matrix—exactly how the hydrogen ions are transported to the intermembrane space is not yet fully understood.

- The H^+ concentration difference creates a potential energy difference across the two sides of the inner membrane (like the energy difference between water at the top and bottom of the waterfall). The maintenance of this concentration gradient across the membrane is *crucial*—it is the mechanism by which energy for ATP formation is made available.

ATP Synthesis

Oxidative phosphorylation The synthesis of ATP from ADP using energy released in the electron-transport chain.

ATP synthase The enzyme complex in the inner mitochondrial membrane at which hydrogen ions cross the membrane and ATP is synthesized from ADP.

The reactions of the electron-transport chain are tightly coupled to **oxidative phosphorylation**, the conversion of ADP to ATP by a reaction that is both an oxidation and a phosphorylation. Hydrogen ions can return to the matrix only by passing through a channel that is part of the **ATP synthase** enzyme complex (green pathway at the right in Figure 21.10). In doing so, they release the potential energy they gained as they were moved against the concentration gradient at the enzyme complexes of the electron-transport chain. The energy they release drives the phosphorylation of ADP by reaction with hydrogen phosphate ion ($HOPO_3^{2-}$):

$$ADP + HOPO_3^{2-} \longrightarrow ATP + H_2O$$

ATP synthase has knob-tipped stalks that protrude into the matrix and are clearly visible in electron micrographs, as seen in the accompanying photograph. ADP and $HOPO_3^{2-}$ are attracted into the knob portion. As hydrogen ions flow through the complex, ATP is produced and released back into the matrix. The reaction is facilitated by changes in the shape of the enzyme complex that are induced by the flow of hydrogen ions.

How much ATP energy is produced from a molecule of NADH or a molecule of $FADH_2$ by oxidative phosphorylation? The electrons from molecules of NADH enter the electron-transport chain at complex I, while those from $FADH_2$ enter at complex II. These different entry points into the electron-transport chain result in different yields of ATP molecules. Recent research suggests that each NADH molecule yields about 2.5 molecules of ATP and that each $FADH_2$ molecule yields approximately 1.5 molecules of ATP. In this book, we round these numbers up and use the older yields of 3 ATP molecules generated for every NADH molecule and 2 ATP molecules generated from every $FADH_2$ molecule during oxidative phosphorylation.

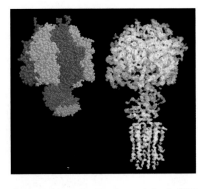

▲ A color enhanced model of ATP synthase. The structure on the left is the catalytic site; the structure on the right also shows the stalk.

PROBLEM 21.18

Within the mitochondrion, is the pH higher in the intermembrane space or in the mitochondrial matrix?

◀○ KEY CONCEPT PROBLEM 21.19

The reduced coenzymes NADH and $FADH_2$ are oxidized in the electron-transport system. What is the final electron acceptor of the electron-transport system? What is the function of the H^+ ion in ATP synthesis?

APPLICATION ▶ Energy Undone: Blockers and Uncouplers of Oxidative Phosphorylation

Cyanide and barbiturates such as amytal have long been known to be so dangerous—even fatal—that mystery writers often use these substances as the murder weapon. What makes them so dangerous? They are among a group of substances that block respiration (oxidative phosphorylation) at one of the electron transfer stages, resulting in blockage of electron flow through the electron-transport system and no ATP production. Continuous production of ATP at tightly regulated levels is crucial to the organism's survival. ATP is the energy link between the oxidation of fuels and energy-requiring processes. Without continuous ATP production the organism will die.

The blockers act at the cytochromes in the electron-transport chain, with different blockers acting on different cytochromes. Rotenone, derived from plants and used to kill both fish and insects, and barbiturates like amytal inhibit complex I proteins in the electron-transport system, so electrons are not transferred to coenzyme Q from complex I. (See Figure 21.12.) The antibiotic Antimycin A inhibits some of the cytochromes and proteins of complex III from transferring electrons to cytochrome c. The third blockage point in the electron-transport system occurs between complex IV and oxygen. Here cyanide and hydrogen sulfide bind tightly to the iron and copper in the enzymes involved, preventing the conversion of oxygen to water. So, if you take barbiturates (prescription only) or use rotenone, follow the directions!

Carbon monoxide, an odorless, colorless gas, acts on the electron-transport system in much the same way as cyanide—by binding the central iron and copper enzymes involved in the conversion of oxygen to water. Similarly, carbon monoxide binds strongly to the iron in the heme group in hemoglobin, the component in blood cells that transports oxygen, replacing oxygen. Herein lies its danger. As you may recall (⊂⊃, Section 12.8), carbon monoxide is a by-product

H—C≡N

Hydrogen cyanide

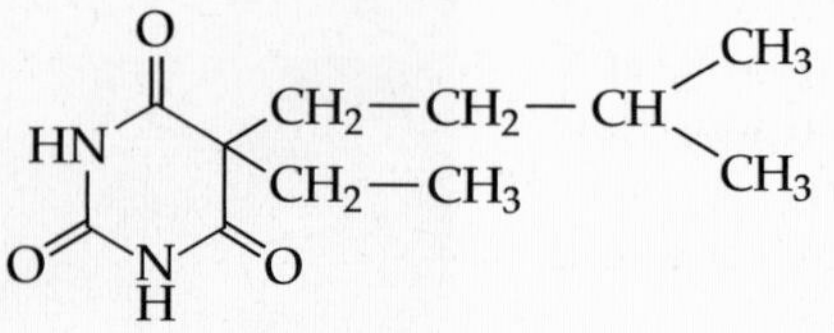

Amytal

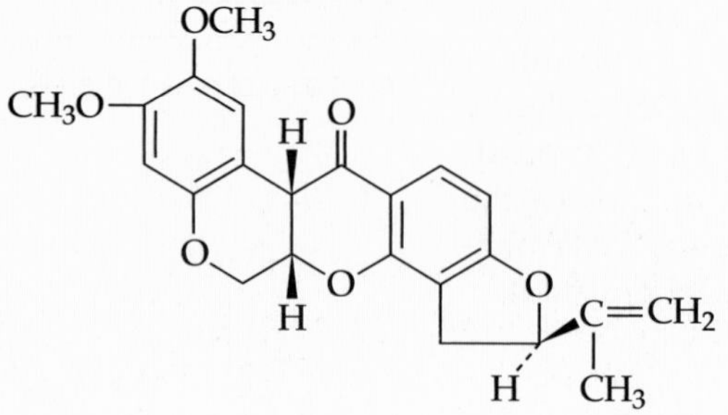

Rotenone

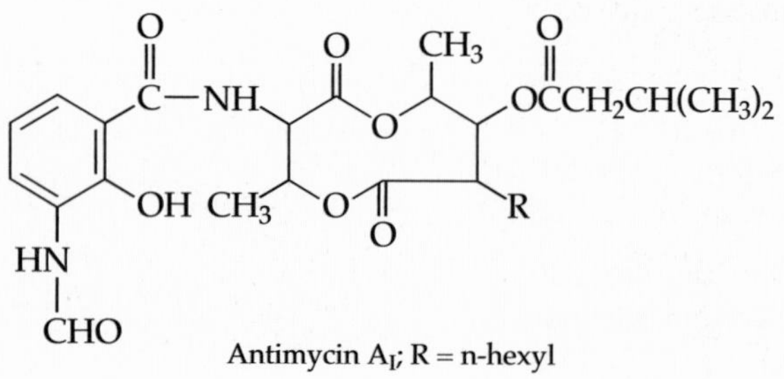

Antimycin A$_I$; R = n-hexyl

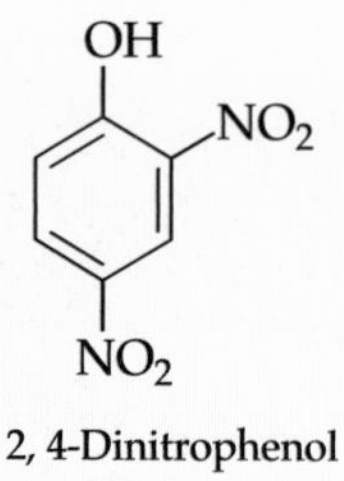

2, 4-Dinitrophenol

▲ The structures of several inhibitors and one uncoupler of oxidative phosphorylation.

of the incomplete combustion of fuels. It is ubiquitous in modern life: It is produced when natural gas is burned in home heating, in automobile exhaust, and when using charcoal for grilling. Proper ventilation will ensure that it does not poison us; well-ventilated, outdoor areas lower the carbon monoxide concentration near cars and gas grills to levels that are not toxic, although some carbon monoxide will bind to hemoglobin in the blood. Safety experts recommend homes with gas appliances have carbon monoxide detectors installed. One type employs a sensor that has hemoglobin inside it; when the hemoglobin absorbs a preset amount of carbon monoxide, the detector's alarm sounds.

Just as the production of ATP can be blocked, some substances, such as dicumarol, an anticoagulant, allow electron transport to occur but prevent the conversion of ADP to ATP by ATP synthase. If this happens, the rate of oxygen use increases as the proton gradient between the mitochondrial matrix and the intermembrane space dissipates with the simultaneous formation of water. When ATP production is thus severed from energy use, it is said that ATP production is *uncoupled* from the energy of the proton gradient. One chemical that has this effect, once used as a weight-reducing agent, is 2,4-dinitrophenol. Uncoupling electron transport does result in weight loss; however, the toxic dose is too close to the therapeutic dose, so 2,4-dinitrophenol is no longer used as a reducing agent.

The body does have a tissue with the capacity to uncouple oxidative phosphorylation intentionally. It comes into play when the environment is cold. Brown fat (brown due to a high concentration of mitochondria in the fat storage cells) can uncouple oxidative phosphorylation in order to generate heat through dissipation of the proton gradient. This is accomplished by the presence of a special protein, thermogenin (uncoupling protein), in the inner membrane of the mitochondrion in these cells. Human infants and other newborn mammals have deposits of brown fat in order to keep warm. This type of fat disappears in most humans unless they have an occupation that routinely puts them in a cold environment for an extended period. Pearl divers, who spend several hours daily diving in the ocean, tend to have deposits of brown fat as do mammals that hibernate.

See Additional Problems 21.83 through 21.86 at the end of the chapter.

WORKED EXAMPLE **21.6** Determining Phosphorylation Type

Consider the two reactions shown below. Which one represents oxidative phosphorylation and which one represents substrate-level phosphorylation?

(a) Succinyl-SCoA + P$_i$ + GDP ⇌ Succinate + GTP + CoA

(b) ADP + HOPO$_3$$^{2-}$ $\xrightarrow{\text{ATP synthase}}$ ATP + H$_2$O

ANALYSIS Both substrate-level phosphorylation and oxidative phosphorylation involve the transfer of a phosphate group and its energy to another molecule.

Both may result in the production of ATP. The key difference is that substrate-level phosphorylation involves the transfer of a phosphate group from one molecule to another, whereas oxidative phosphorlyation adds a phosphate ion directly to ADP with the aid of ATP synthase.

SOLUTION
Reaction (a), involving the formation of GTP coupled with the conversion of succinyl-SCoA to succinate, is an example of substrate-level phosphorylation. Reaction (b), involving the direct addition of phosphate to ADP by ATP synthase, is oxidative phosphorylation.

21.10 Harmful Oxygen By-Products and Antioxidant Vitamins

More than 90% of the oxygen we breathe is used in the coupled electron-transport–ATP synthesis reactions. In these and other oxygen-consuming redox reactions, the product may not be water, but one or more of three highly reactive species. Two are free radicals, which contain unpaired electrons (represented by single dots in the formulas). Like all free radicals, these two oxygen-containing species, the superoxide ion ($\cdot O_2^-$) and the hydroxyl free radical ($\cdot OH^-$) react as soon as possible to get rid of the unpaired electron. Often, they do this by grabbing an electron from a bond in another molecule, which results in breaking that bond. The third oxygen by-product is hydrogen peroxide, H_2O_2, which is a relatively strong oxidizing agent. The conditions that can enhance production of these three reactive oxygen species are represented in the drawing below. Some causes are environmental, such as exposure to smog or radiation. Others are physiological changes, including aging and inflammation. (Reperfusion is the return of blood to an area that had little blood circulation.)

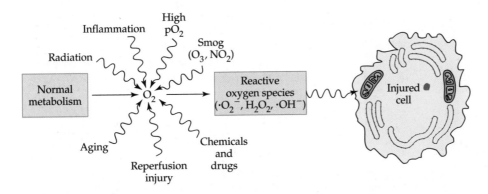

The reactivity of the superoxide free radical is beneficial in destroying infectious microorganisms. In what is known as a respiratory burst, the *phagocytes* (cells that engulf bacteria) produce superoxide ions that react destructively with bacteria:

$$2\,O_2 + NADPH \longrightarrow 2\cdot O_2^- + NADP^+ + H^+$$

Reactive oxygen species are also dangerous, however. They can break covalent bonds in enzymes and other proteins, DNA, and the lipids in cell membranes. Among the possible outcomes of such destruction are cancer, liver damage, rheumatoid arthritis, heart disease, immune system damage, and, according to some theories, the deterioration that normally accompanies aging. One hazard of breathing polluted air and cigarette smoke is breathing free radicals.

Potentially harmful oxygen species are constantly being generated in the body. Our protection against them is provided by superoxide dismutase and

APPLICATION ▶ Plants and Photosynthesis

The principal biochemical difference between ourselves and plants is that plants can derive energy directly from sunlight and we cannot. In the process of *photosynthesis*, plants use solar energy to synthesize oxygen and energy-rich carbohydrates from energy-poor reactants: CO_2 and water. Our own metabolism breaks down energy-rich reactants to extract the useful energy and produce energy-poor products: CO_2 and water. Is it surprising to discover that despite this difference in the direction of their reactions, plants rely on biochemical pathways very much like our own?

The energy-capturing phase of photosynthesis takes place mainly in green leaves. Plant cells contain *chloroplasts*, which, though larger and more complex in structure, resemble mitochondria. Embedded in membranes within the chloroplasts are large groups of *chlorophyll* molecules and the enzymes of an electron-transport chain. Chlorophyll is similar in structure to heme (Figure 21.12), but contains magnesium ions (Mg^{2+}) instead of iron ions (Fe^{2+}).

As solar energy is absorbed, chlorophyll molecules pass it along to specialized reaction centers, where it is used to boost the energy of electrons. The excited electrons then give up their extra energy as they pass down a pair of electron-transport chains.

Some of this energy is used to oxidize water, splitting it into oxygen, hydrogen ions, and electrons (which replace those entering the electron-transport chain). At the end of the chain, the hydrogen ions, together with the electrons, are used to reduce $NADP^+$ to NADPH. Along the way, part of the

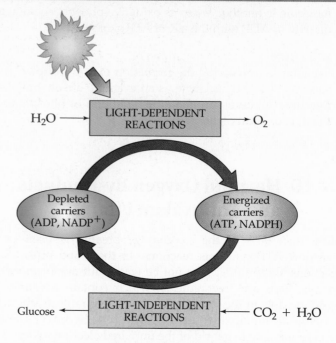

▲ **The coupled reactions of photosynthesis.**

▲ These flowers are converting the potential energy of the sun into chemical potential energy stored in the bonds of carbohydrates.

energy of the electrons is used to pump hydrogen ions across a membrane to create a concentration gradient. As in mitochondria, the hydrogen ions can only return across the membrane at enzyme complexes that convert ADP to ATP. Water needed for these *light-dependent reactions* enters the plant through the roots and leaves, and the oxygen that is formed is released through openings in the leaves.

The energy-carrying ATP and NADPH enter the fluid interior of the chloroplasts. Here their energy is used to drive the synthesis of carbohydrate molecules. So long as ATP and NADH are available, this part of photosynthesis is *light-independent*—it can proceed in the absence of sunlight.

Plants have mitochondria as well as chloroplasts, so they can also carry out the release of energy from stored carbohydrates. Because the breakdown of carbohydrates continues in many harvested fruits and vegetables, the goal in storage is to slow it down. Refrigeration is one measure that is taken, since (like most chemical reactions) the rate of respiration decreases at lower temperatures. Another is replacement of air over stored fruits and vegetables with carbon dioxide or nitrogen.

See Additional Problems 21.87 through 21.90 at the end of the chapter.

catalase, which, as we have mentioned earlier, are among the fastest-acting enzymes. (⬡, Section 19.5)

$$2 \cdot O_2^- + 2\,H^+ \xrightarrow{\text{Superoxide dismutase}} H_2O_2 + O_2$$

Superoxide ion   Hydrogen peroxide

$$2\,H_2O_2 \xrightarrow{\text{Catalase}} 2\,H_2O + O_2$$

These and other enzymes are active inside cells where oxygen by-products are generated. Nevertheless, it is estimated that 1 in 50 of the harmful oxygen species escapes destruction inside a cell.

Protection is also provided by the antioxidant vitamins E, C, and A (or its precursor β-carotene), all of which disarm free radicals by bonding with them. (⬡, Section 19.10) Vitamin E is fat-soluble, and its major function is to protect cell membranes from potential damage initiated when a cell membrane lipid (RH) is converted to an oxygen-containing free radical ROO·. Vitamin E gives up electrons and hydrogen atoms to convert the lipid free radical to ROOH, a peroxide that is then enzymatically converted to ROH.

Vitamin E   Product of vitamin E
reacting as an antioxidant

Because Vitamin C is water-soluble, it is a free-radical scavenger in the blood. There are also many other natural antioxidants among the chemical compounds distributed in fruits and vegetables. We still have much to learn about them.

SUMMARY: REVISITING THE CHAPTER GOALS

1. **What is the source of our energy, and what is its fate in the body?** *Endergonic* reactions are unfavorable and require an external source of free energy to occur. *Exergonic* reactions are favorable, proceed spontaneously, and release free energy. We derive energy by oxidation of food molecules that contain energy captured by plants from sunlight. The energy is released gradually in exergonic reactions and is available to do work, to drive endergonic reactions, to provide heat, or to be stored until needed. Energy generation in eukaryotic cells takes place in *mitochondria*.

2. **How are the reactions that break down food molecules organized?** Food molecules undergo *catabolism* (are broken down) to provide energy in four stages (Figure 21.5): (1) digestion to form smaller molecules that can be absorbed into cells; (2) decomposition (by separate pathways for lipids, carbohydrates, and proteins) into 2-carbon acetyl groups that are bonded to coenzyme A in *acetyl coenzyme A*; (3) reaction of the acetyl groups via the *citric acid cycle* to generate energy-rich reduced coenzymes and liberate carbon dioxide; and (4) *electron transport* and transfer of the energy of the reduced coenzymes from the citric acid cycle to our principal energy transporter, ATP.

3. **What are the major strategies of metabolism?** Using the energy from exergonic reactions, ADP is *phosphorylated* to give ATP. Where energy must be expended, it is released by removal of a phosphoryl group from ATP to give back ADP. An otherwise "uphill" reaction in a metabolic pathway is driven by coupling with an exergonic, "downhill" reaction that provides enough energy so that their combined outcome is exergonic and favorable. The oxidizing and reducing agents needed by the many redox reactions of

KEY WORDS

Acetyl-S-coenzyme A (acetyl-SCoA), *p. 667*

Adenosine triphosphate (ATP), *p. 664*

Anabolism, *p. 667*

ATP synthase, *p. 682*

Catabolism, *p. 667*

Citric acid cycle, *p. 677*

Cytoplasm, *p. 664*

Cytosol, *p. 664*

Electron-transport chain, *p. 680*

Endergonic, *p. 662*

Exergonic, *p. 662*

Guanosine diphosphate (GDP), *p. 679*

metabolism are coenzymes that constantly cycle between their oxidized and reduced forms.

4. **What is the citric acid cycle?** The *citric acid cycle* (Figure 21.9) is a cyclic pathway of eight reactions, in which the product of the final reaction is the substrate for the first reaction. The reactions of the citric acid cycle (1) set the stage for oxidation of the acetyl group (Steps 1 and 2); (2) remove two carboxyl groups as CO_2 molecules (oxidative decarboxylation) from the tricarboxylic acid isocitrate (Steps 3 and 4); and (3) oxidize the 4-carbon dicarboxylic acid succinate and regenerate oxaloacetate so that the cycle can start again (Steps 5–8). Along the way, four reduced coenzyme molecules and one molecule of ATP are produced for each acetyl group oxidized. The reduced coenzymes carry energy for the subsequent production of additional ATP. The cycle is activated when energy is in short supply and inhibited when energy is in good supply.

5. **How is ATP generated in the final stage of catabolism?** ATP generation is accomplished by a series of enzyme complexes in the inner membranes of mitochondria (Figure 21.10). Electrons and hydrogen ions enter the first two complexes of the electron-transport chain from succinate (in the citric acid cycle) and NADH, where they are transferred to *coenzyme Q*. Then, the electrons and hydrogen ions proceed independently, the electrons gradually giving up their energy to the transport of hydrogen ions across the inner mitochondrial membrane to maintain different concentrations on opposite sides of the membrane. The hydrogen ions return to the matrix by passing through *ATP synthase*, where the energy they release is used to convert ADP to ATP.

6. **What are the harmful by-products produced from oxygen, and what protects against them?** Harmful by-products of oxygen-consuming reactions are the hydroxyl free radical, superoxide ion (also a free radical), and hydrogen peroxide. These reactive species damage other molecules by breaking bonds. Superoxide dismutase and catalase are enzymes that disarm these oxygen by-products. Vitamins E, C, and A (or its precursor β-carotene) are also antioxidants.

UNDERSTANDING KEY CONCEPTS

21.20 The following coupled reaction is the result of an exergonic reaction and an endergonic reaction:

Succinyl phosphate + ADP ⟶ Succinate + ATP

(a) Write the exergonic portion of the reaction.
(b) Write the endergonic portion of the reaction.

21.21 Each of these reactions is involved in one of the four stages of metabolism shown in Figure 21.5. Identify the stage in which each reaction occurs.

(a) Hydrolysis of starch to produce glucose
(b) Oxidation of NADH coupled with synthesis of ATP
(c) Conversion of glucose to acetyl-SCoA
(d) Oxidation of acetyl-SCoA in a series of reactions where NAD^+ is reduced and CO_2 is produced

21.22 For the first step in fatty acid catabolism, we say that ATP is used to "drive" the reaction that links the fatty acid with coenzyme-A. Without the ATP hydrolysis, would you predict that the linking of fatty acid to coenzyme-A would be exergonic or endergonic? In the fatty acid SCoA synthesis, the hydrolysis of the ATP portion is based on what major strategy of metabolism?

21.23 Since no molecular oxygen participates in the citric acid cycle, the steps in which acetyl groups are oxidized to CO_2 involve removal of hydride ions and hydrogen ions. What is the acceptor of hydride ions? What is the acceptor of hydrogen ions?

21.24 The reaction at the top of the next page is catalyzed by isocitrate dehydrogenase and occurs in two steps, the first of which (Step A) is formation of an unstable intermediate (shown in brackets).

(a) In which step is a coenzyme needed? Identify the coenzyme.
(b) In which step is CO_2 evolved and a hydrogen ion added?
(c) Which of the structures shown can be described as a β-keto acid?

$$
\begin{array}{c}
\text{COO}^- \\
| \\
\text{CH}_2 \\
| \\
\text{H---C---COO}^- \\
| \\
\text{HO---C---H} \\
| \\
\text{COO}^-
\end{array}
\xrightarrow{\text{A}}
$$

Isocitrate

$$
\left[
\begin{array}{c}
\text{COO}^- \\
| \\
\text{CH}_2 \\
| \\
\text{H---C---COO}^- \\
| \\
\text{C=O} \\
| \\
\text{COO}^-
\end{array}
\right]
\xrightarrow{\text{B}}
\begin{array}{c}
\text{COO}^- \\
| \\
\text{CH}_2 \\
| \\
\text{CH}_2 \\
| \\
\text{C=O} \\
| \\
\text{COO}^-
\end{array}
$$

α-Ketoglutarate

ADDITIONAL PROBLEMS

FREE ENERGY AND BIOCHEMICAL REACTIONS

21.27 What energy requirements must be met in order for a reaction to be favorable?

21.28 What is the difference between an endergonic process and an exergonic process?

21.29 Why is ΔG a useful quantity for predicting the favorability of biochemical reactions?

21.30 Many biochemical reactions are catalyzed by enzymes. Do enzymes have an influence on the magnitude or sign of ΔG? Why or why not?

21.31 The following reactions occur during the catabolism of acetyl-SCoA. Which is exergonic? Which is endergonic? Which reaction produces a phosphate that later yields energy by giving up a phosphate group?

(a) Succinyl-SCoA + GDP + Phosphate (P_i) →
Succinate + CoA-SH + GTP + H_2O
$\Delta G = -0.4$ kcal/mol

(b) Acetyl-SCoA + Oxaloacetate → Citrate + CoA-SH
$\Delta G = -8$ kcal/mol

(c) L-Malate + NAD$^+$ → Oxaloacetate + NADH + H$^+$
$\Delta G = +7$ kcal/mol

21.32 The following reactions occur during the catabolism of glucose. Which is exergonic? Which is endergonic? Which proceeds farthest toward products at equilibrium?

(a) 1,3-Bisphosphoglycerate + H_2O →
3-Phosphoglycerate + P_i
$\Delta G = -11.8$ kcal/mol

(b) Phosphoenol pyruvate + H_2O →
Pyruvate + Phosphate (P_i)
$\Delta G = -14.8$ kcal/mol

(c) Glucose + P_i → Glucose 6-phosphate + H_2O
$\Delta G = +3.3$ kcal/mol

CELLS AND THEIR STRUCTURE

21.33 What kinds of organisms have prokaryotic cells, and what kinds have eukaryotic cells?

21.25 For each reaction in the citric acid cycle, give the type of reaction occurring, name the enzyme involved, and indicate which of the six classes of enzymes it belongs to. (Some may have more than one kind of enzyme activity.)

21.26 The electron-transport chain uses several different metal ions, especially iron, copper, zinc, and manganese. Why are metals used frequently in these two pathways? What can metals do better than organic biomolecules?

21.34 List five differences between prokaryotic cells and eukaryotic cells.

21.35 Sketch a eukaryotic cell and label the major components. What is the function of each labeled component?

21.36 What is the difference between the cytoplasm and the cytosol?

21.37 What is an organelle?

21.38 Describe in general terms the structural makeup of a mitochondrion.

21.39 What is the function of cristae in the mitochondrion?

21.40 Why are mitochondria called the "power plants" of the cell?

METABOLISM

21.41 What is the difference between catabolism and anabolism?

21.42 What is the difference between digestion and metabolism?

21.43 Arrange the following events in the order in which they occur in a catabolic process: electron transport, digestion, oxidative phosphorylation, citric acid cycle.

21.44 What key metabolic intermediate is formed from the catabolism of all three major classes of foods: carbohydrates, lipids, and proteins?

STRATEGIES OF METABOLISM

21.45 What is the full name of the substance formed during catabolism that is used to store chemical energy?

21.46 Why is ATP sometimes called a high-energy molecule?

21.47 What general kind of chemical reaction does ATP carry out?

21.48 What is the chemical difference between ATP and ADP?

21.49 What does it mean when we say that two reactions are coupled?

21.50 Show why coupling the reaction for the hydrolysis of 1,3-bisphosphoglycerate to the phosphorylation of ADP is energetically favorable. Combine the equations and calculate ΔG for the coupled process. You need only give names or abbreviations in your equations, not chemical structures.

21.51 Is the hydrolysis of fructose 6-phosphate favorable for phosphorylating ADP? Why or why not? (Refer to Table 21.1.)

21.52 Write the reaction for the hydrolysis of 1,3-bisphosphoglycerate coupled to the phosphorylation of ADP using the curved arrow symbolism.

21.53 FAD is a coenzyme for dehydrogenation.

(a) When a molecule is dehydrogenated, is FAD oxidized or reduced?
(b) Is FAD an oxidizing agent or a reducing agent?
(c) What type of substrate is FAD associated with, and what is the type of product molecule after dehydrogenation?
(d) What is the form of FAD after dehydrogenation?
(e) Use the curved arrow symbolism to write a general equation for a reaction involving FAD.

21.54 NAD^+ is a coenzyme for dehydrogenation.

(a) When a molecule is dehydrogenated, is NAD^+ oxidized or reduced?
(b) Is NAD^+ an oxidizing agent or a reducing agent?
(c) What type of substrate is NAD^+ associated with, and what type of product molecule is formed after dehydrogenation?
(d) What is the form of NAD^+ after dehydrogenation?
(e) Use the curved arrow symbolism to write a general equation for a reaction involving NAD^+.

THE CITRIC ACID CYCLE

21.55 By what other names is the citric acid cycle known?

21.56 Where in the cell does the citric acid cycle take place?

21.57 What substance acts as the starting point of the citric acid cycle, reacting with acetyl-SCoA in the first step and being regenerated in the last step? Draw its structure.

21.58 What is the final fate of the carbons in acetyl-SCoA after several turns of the citric acid cycle?

21.59 Look at the eight steps of the citric acid cycle (Figure 21.9) and answer these questions:

(a) Which steps involve oxidation reactions?
(b) Which steps involve decarboxylation (loss of CO_2)?
(c) Which step or steps involve a hydration reaction?

21.60 How many NADH and how many $FADH_2$ molecules are formed in the citric acid cycle?

21.61 Which reaction(s) of the citric acid cycle store energy as $FADH_2$?

21.62 Which reaction(s) of the citric acid cycle store energy as NADH?

THE ELECTRON-TRANSPORT CHAIN; OXIDATIVE PHOSPHORYLATION

21.63 What are the two primary functions of the electron-transport chain?

21.64 How are the processes of the citric acid cycle and the electron-transport chain interrelated?

21.65 What two coenzymes are involved with initial events of the electron-transport chain?

21.66 What are the ultimate products of the electron-transport chain?

21.67 Where are the following found in the cell?

(a) FAD
(b) CoQ
(c) $NADH/H^+$
(d) Cyt c

21.68 What do the following abbreviations stand for?

(a) FAD
(b) CoQ
(c) $NADH/H^+$
(d) Cyt c

21.69 What atom in the cytochromes undergoes oxidation and reduction in the electron-transport chain? What atoms in coenzyme Q undergo oxidation and reduction in the electron-transport chain?

21.70 Put the following substances in the correct order of their action in the electron-transport chain: cytochrome c, coenzyme Q, NADH.

21.71 Fill in the missing substances in these coupled reactions:

$$FAD \longleftarrow \quad \longrightarrow CoQH_2 \quad \longrightarrow ?$$
$$? \quad \longleftarrow ? \quad \longrightarrow 2\ Fe^{2+}$$

21.72 What would happen to the citric acid cycle if NADH and $FADH_2$ were not reoxidized?

21.73 Across what membrane is there a pH differential caused by the release of H^+ ions? On which side of this membrane are there more H^+ ions?

21.74 What does the term "oxidative phosphorylation" mean?

21.75 In oxidative phosphorylation, what is oxidized and what is phosphorylated?

21.76 Oxidative phosphorylation has three reaction products.

(a) Name the energy-carrying product.
(b) Name the other two products.

Applications

21.77 Photosynthetic plants use a sunlight-driven electron-transport system to remove electrons from H_2O to produce O_2 and generate ATP and NADPH. In the bacteria surrounding black smokers, H_2S is the electron donor and corresponds to what component of the electron-transport system in photosynthetic plants? [*Life without Sunlight*, p. 663]

21.78 Why do you suppose bacteria found around deep-sea hydrothermal vents and near the Mount St. Helens volcanic eruption use H_2S as a source of electrons in energy-generating reactions? [*Life without Sunlight*, p. 663]

21.79 How is basal metabolic rate defined? [*Basal Metabolism*, p. 673]

21.80 Estimate your basal metabolic rate using the guidelines in the application. [*Basal Metabolism*, p. 673]

21.81 Calculate the total calories needed in a day for an 80 kg lightly active male. Use the "Kilocalories used per minutes" values given in the table in the application. [*Basal Metabolism, p. 673*]

21.82 Why do activities such as walking raise a body's needs above the basal metabolic rate? [*Basal Metabolism, p. 673*]

21.83 The antibiotic piericidin, a nonpolar molecule, is structurally similar to ubiquinone (coenzyme Q) and can cross the mitochondrial membrane. What effect might the presence of piericidin have on oxidative phosphorylation? [*Energy Undone: Blockers and Uncouplers of Oxidative Phosphorylation, p. 683*]

21.84 When oxidative phosphorylation is uncoupled, does oxygen consumption decrease, increase, or stay the same? Explain. [*Energy Undone: Blockers and Uncouplers of Oxidative Phosphorylation, p. 683*]

21.85 Why is 2,4-dinitrophenol no longer used as a dieting aid? [*Energy Undone: Blockers and Uncouplers of Oxidative Phosphorylation, p. 683*]

21.86 Which animal would you expect to have more brown fat, a seal or a domestic cat? Explain. [*Energy Undone: Blockers and Uncouplers of Oxidative Phosphorylation, p. 683*]

21.87 Chlorophyll is similar in structure to heme in red blood cells but does not have an iron atom. What metal ion is present in chlorophyll? [*Plants and Photosynthesis, p. 686*]

21.88 Photosynthesis consists of both light-dependent and light-independent reactions. What is the purpose of each type of reaction? [*Plants and Photosynthesis, p. 686*]

21.89 One step of the cycle that incorporates CO_2 into glyceraldehyde in plants is the production of two 3-phosphoglycerates. This reaction has $\Delta G = -0.84$ kcal/mol. Is this process endergonic or exergonic? [*Plants and Photosynthesis, p. 686*]

21.90 What general process does refrigeration of harvested fruits and vegetables slow? What cellular processes are slowed by refrigeration? [*Plants and Photosynthesis, p. 686*]

General Questions and Problems

21.91 Why must the breakdown of molecules for energy in the body occur in several steps, rather than in one step?

21.92 The first step in the citric acid cycle involves the reaction of acetyl-SCoA and oxaloacetate. Show the product of this reaction before hydrolysis to yield citrate.

21.93 The fumarate produced in Step 6 of the citric acid cycle must have a trans double bond to continue on in the cycle. Suggest a reason why the corresponding cis double-bond isomer cannot continue in the cycle.

21.94 With what class of enzymes are the coenzymes NAD^+ and FAD associated?

21.95 We talk of burning food in a combustion process, producing CO_2 and H_2O from food and O_2. Explain how O_2 is involved in the process although no O_2 is directly involved in the citric acid cycle.

21.96 One of the steps that occurs when lipids are metabolized is shown below. Does this process require FAD or NAD^+ as the coenzyme? What is the general class of enzyme that catalyzes this process?

21.97 Solutions of hydrogen peroxide can be kept for months in a brown closed bottle with only moderate decomposition. When used on a cut as an antiseptic, hydrogen peroxide begins to bubble rapidly. Give a possible explanation for this observation.

21.98 Identify the chiral intermediates in the TCA cycle.

21.99 If you use a flame to burn a pile of glucose completely to give carbon dioxide and water, the overall reaction is identical to the metabolic oxidation of glucose. Explain the differences in the fate of the energy released in each case.

21.100 What highly reactive oxygen species are by-products of oxygen-consuming reactions in the body? Which enzymes and vitamins are used to destroy these reactive species?

21.101 After running a mile, you stop and breathe heavily for a short period due to oxygen debt. Why do you need to breathe so heavily? (Hint: Think about the metabolic pathway that uses oxygen.)

21.102 Put in order, from lowest to highest number of mitochondria per cell, the following tissues: adipose tissue (regular), heart muscle, skin cells, skeletal muscle.

CHAPTER 22

Carbohydrates

CONCEPTS TO REVIEW

Molecular shape
(Section 5.7)

Chirality
(Section 18.5)

Oxidation–reduction reactions
(Sections 6.11, 6.12, 14.5, 16.5, 16.6)

▲ Corn stores glucose in the polysaccharides starch and cellulose.

CONTENTS

CHAPTER GOALS

In this chapter, we will answer the following questions about carbohydrates:

1. What are the different kinds of carbohydrates?

THE GOAL: Be able to define monosaccharides, disaccharides, and polysaccharides, and recognize examples.

2. Why are monosaccharides chiral, and how does this influence the numbers and types of their isomers?

THE GOAL: Be able to identify the chiral carbon atoms in monosaccharides, predict the number of isomers for different monosaccharides, and identify pairs of enantiomers.

3. What are the structures of monosaccharides, and how are they represented in written formulas?

THE GOAL: Be able to explain relationships among open-chain and cyclic monosaccharide structures, describe the isomers of monosaccharides, and show how they are represented by Fischer projections and cyclic structural formulas.

4. How do monosaccharides react with oxidizing agents and alcohols?

THE GOAL: Be able to identify reducing sugars and the products of their oxidation, recognize acetals of monosaccharides, and describe glycosidic linkages in disaccharides.

5. What are the structures of some important disaccharides?

THE GOAL: Be able to identify the monosaccharides combined in maltose, lactose, and sucrose, and describe the types of linkages between the monosaccharides.

6. What are the functions of some important carbohydrates that contain modified monosaccharide structures?

THE GOAL: Be able to identify the functions of chitin, connective-tissue polysaccharides, heparin, and glycoproteins.

7. What are the structures and functions of cellulose, starch, and glycogen?

THE GOAL: Be able to describe the monosaccharides and linkages in these polysaccharides, their uses and fates in metabolism.

The word *carbohydrate* originally described glucose, the simplest and most readily available sugar. Because glucose has the formula $C_6H_{12}O_6$, it was once thought to be a "hydrate of carbon," $C_6(H_2O)_6$. Although this view has been abandoned, the name "carbohydrate" persisted, and we now use it to refer to a large class of biomolecules with similar structures. Carbohydrates have in common many hydroxyl groups on adjacent carbons together with either an aldehyde or ketone group. Glucose, for example, has five —OH groups and one —CHO group:

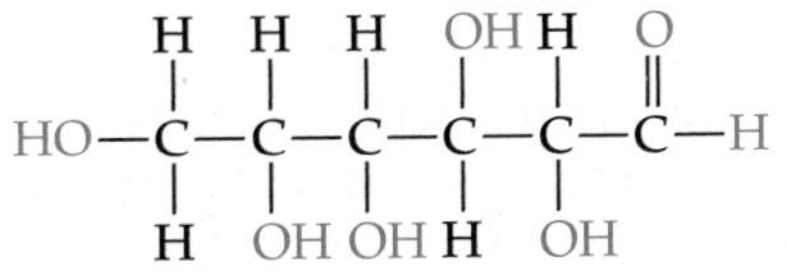

$$
\begin{array}{c}
\text{H}\ \ \text{H}\ \ \text{H}\ \ \text{OH}\ \text{H}\ \ \text{O} \\
|\ \ \ |\ \ \ |\ \ \ |\ \ \ |\ \ \ \| \\
\text{HO}-\text{C}-\text{C}-\text{C}-\text{C}-\text{C}-\text{C}-\text{H} \\
|\ \ \ |\ \ \ |\ \ \ |\ \ \ | \\
\text{H}\ \ \text{OH}\ \text{OH}\ \text{H}\ \ \text{OH}
\end{array}
$$

Glucose

Carbohydrates are synthesized by plants and stored as starch, a polymer of glucose. When starch is eaten and digested, the freed glucose becomes a major source of the energy required by living organisms. Thus, carbohydrates are intermediaries by which energy from the sun is made available to animals.

22.1 An Introduction to Carbohydrates

Carbohydrates are a large class of naturally occurring polyhydroxy aldehydes and ketones. **Monosaccharides**, sometimes known as **simple sugars**, are the simplest carbohydrates. They have from three to seven carbon atoms, and each contains one aldehyde or one ketone functional group. If the sugar has an aldehyde group, it is classified as an **aldose**. If it has a ketone group, the sugar is classified as a **ketose**. The aldehyde group is always at the end of the carbon chain, and the ketone group is always on the second carbon of the chain. In either case, there is a —CH₂OH group at the other end of the chain.

Carbohydrate A member of a large class of naturally occurring polyhydroxy ketones and aldehydes.

Monosaccharide (simple sugar) A carbohydrate with three to seven carbon atoms.

Aldose A monosaccharide that contains an aldehyde carbonyl group.

Ketose A monosaccharide that contains a ketone carbonyl group.

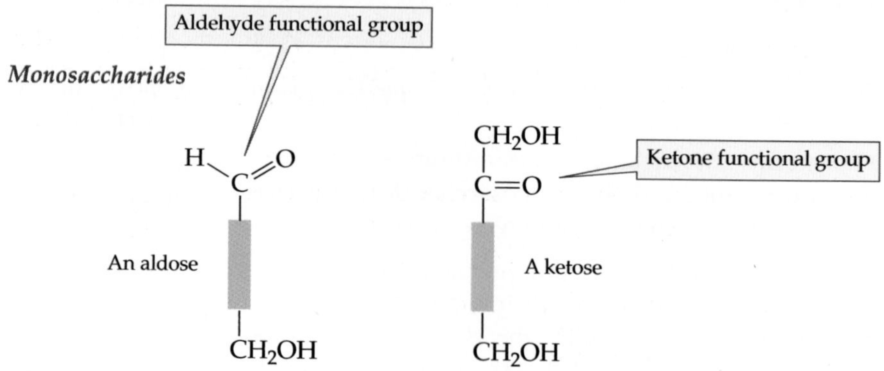

Monosaccharides

Aldehyde functional group

Ketone functional group

An aldose

A ketose

There are hydroxyl groups on all the carbon atoms between the carbonyl carbon atom and the —CH_2OH at the other end, and also on the end carbon next to a ketone group, as illustrated in the following three structures. The family-name ending *-ose* indicates a carbohydrate, and simple sugars are known by common names like *glucose, ribose,* and *fructose* rather than systematic names.

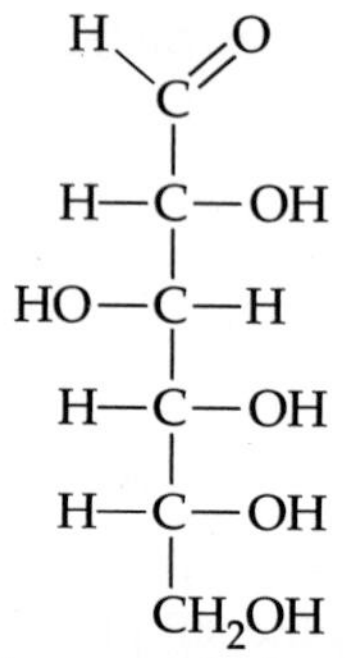

Glucose, an aldohexose
(monomer for starch and cellulose;
major source of energy)

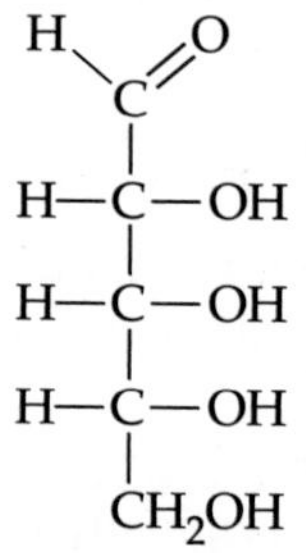

Ribose, an aldopentose
(a component of ATP,
coenzymes, and RNA)

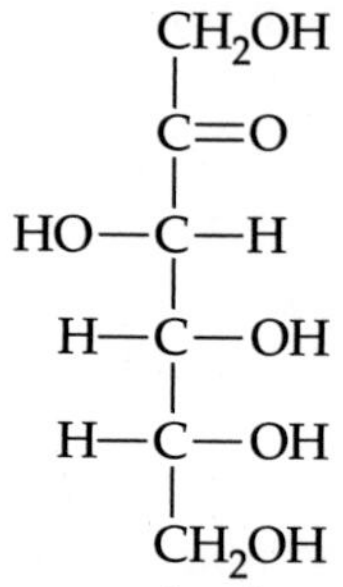

Fructose, a ketohexose
(present in corn syrup
and fruit)

The number of carbon atoms in an aldose or ketose is specified by the prefixes *tri-, tetr-, pent-, hex-,* or *hept-*. Thus, glucose is an aldo*hex*ose (*aldo-* = aldehyde, *-hex* = 6 carbons; *-ose* = sugar); fructose is a keto*hex*ose (a 6-carbon ketone sugar); and ribose is an aldo*pent*ose (a 5-carbon aldehyde sugar). Most naturally occurring simple sugars are aldehydes with either 5 or 6 carbons.

Because of their many functional groups, monosaccharides undergo a variety of structural changes and chemical reactions. They react with each other to form **disaccharides** and **polysaccharides** (also known as **complex carbohydrates**), which are polymers of monosaccharides. Their functional groups are involved in reactions with alcohols, lipids or proteins to form biomolecules with specialized functions. These and other carbohydrates are introduced in later sections of this chapter. First, we are going to discuss two important aspects of carbohydrate structure:

Disaccharide A carbohydrate composed of two monosaccharides.

Polysaccharide (complex carbohydrate) A carbohydrate that is a polymer of monosaccharides.

- Monosaccharides are chiral molecules (Sections 22.2, 22.3).
- Monosaccharides exist mainly in cyclic forms rather than the straight-chain forms shown above (Section 22.4).

WORKED EXAMPLE 22.1 Classifying Monosaccharides

Classify the monosaccharide shown as an aldose or a ketose, and name it according to its number of carbon atoms.

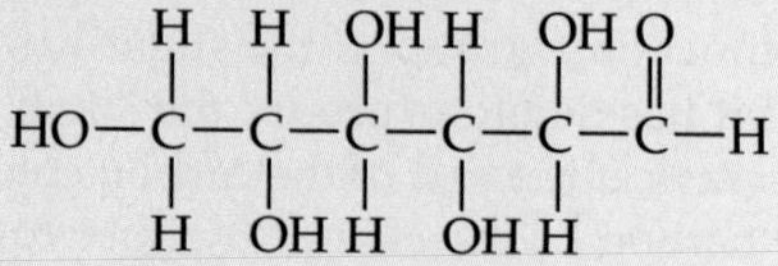

ANALYSIS First determine if the monosaccharide is an aldose or a ketose. Then determine the number of carbon atoms present. This monosaccharide is an aldose because an aldehyde group is present. It contains 6 carbon atoms.

SOLUTION
The monosaccharide is a 6-carbon aldose, so we refer to it as an aldohexose.

PROBLEM 22.1

Classify the following monosaccharides as an aldose or a ketose, and name each according its number of carbon atoms.

(a) 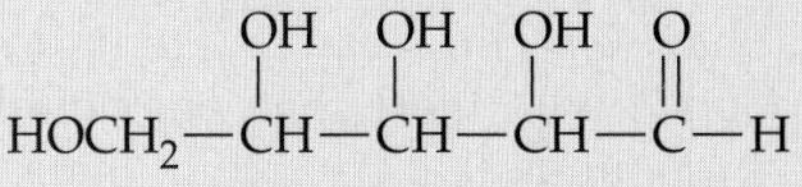
$$\underset{}{HOCH_2}-\overset{\overset{OH}{|}}{CH}-\overset{\overset{OH}{|}}{CH}-\overset{\overset{OH}{|}}{CH}-\overset{\overset{O}{\|}}{C}-H$$

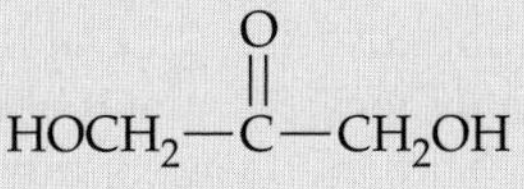

(b)
$$HOCH_2-\overset{\overset{O}{\|}}{C}-CH_2OH$$

(c)
$$HOCH_2-\overset{\overset{OH}{|}}{CH}-\overset{\overset{OH}{|}}{CH}-\overset{\overset{O}{\|}}{C}-H$$

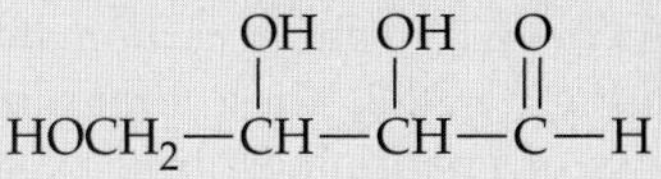

PROBLEM 22.2

Draw the structures of an aldopentose and a ketohexose.

22.2 Handedness of Carbohydrates

You have seen that amino acids are chiral (that is, not superimposable on their mirror images) because they contain carbon atoms bonded to four different groups. Glyceraldehyde, an aldotriose and the simplest naturally occurring carbohydrate, has the structure shown below. Because four different groups are bonded to the number 2 carbon atom ($-CHO$, $-H$, $-OH$, and $-CH_2OH$), this molecule is also chiral. (⬭, Section 18.5)

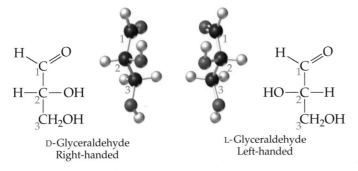

D-Glyceraldehyde
Right-handed

L-Glyceraldehyde
Left-handed

Chiral compounds lack a plane of symmetry and exist as a pair of enantiomers in either a "right-handed" D form or a "left-handed" L form. Like all enantiomers, the two forms of glyceraldehyde have the same physical properties except for the way in which they affect polarized light.

Light as we usually see it consists of electromagnetic waves oscillating in all planes at right angles to the direction of travel of the light beam. (⬭, p. 72) When ordinary light is passed through a polarizer, only waves in one plane get through, producing what is known as *plane-polarized light*. (Polaroid sunglasses work on a similar principle.) Solutions of *optically active* chemical compounds change the plane in which the light is polarized. The angle by which the plane is rotated can be measured in an instrument known as a *polarimeter*, which works on the principle diagrammed in Figure 22.1. Each enantiomer of a pair rotates the plane of the light

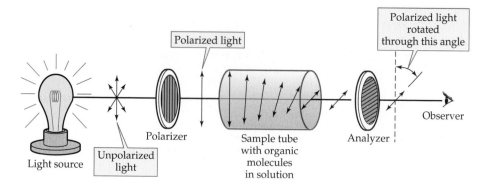

▶ **FIGURE 22.1 Principle of a polarimeter, used to determine optical activity.** A solution of an optically active isomer rotates the plane of the polarized light by a characteristic amount.

by the same amount, but the directions of rotation are *opposite*. If one enantiomer rotates the plane of the light to the left, the other rotates it to the right.

Compounds like glyceraldehyde that have *one* chiral carbon atom can exist as two enantiomers. But what about compounds with more than one chiral carbon atom? How many isomers are there of compounds that have two, three, four, or more chiral carbons? Aldotetroses, for example, have two chiral carbon atoms and can exist in the four isomeric forms shown in Figure 22.2. These four aldotetrose stereoisomers consist of two mirror-image pairs of enantiomers, one pair named *erythrose* and one pair named *threose*. Because erythrose and threose are stereoisomers but not mirror images of each other, they are described as **diastereomers**.

Diastereomers Stereoisomers that are not mirror images of each other.

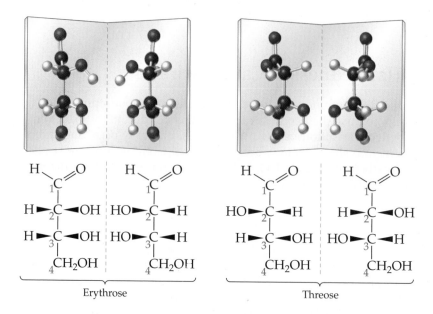

▶ **FIGURE 22.2 Two pairs of enantiomers: The four isomeric aldotetroses (2,3,4-trihydroxybutanals).** Carbon atoms 2 and 3 are chiral. Their —H atoms and —OH groups are written here to show their mirror-image relationship. Erythrose and threose exist as enantiomeric pairs.

In general, a compound with *n* chiral carbon atoms has a maximum of 2^n possible stereoisomers and half that many pairs of enantiomers. The aldotetroses, for example, have $n = 2$ so that $2^n = 2^2 = 4$, meaning that four stereoisomers are possible. Glucose, an aldohexose, has four chiral carbon atoms and a total of $2^4 = 16$ possible stereoisomers (8 pairs of isomers). All 16 stereoisomers of glucose are known. (In some cases, fewer than the maximum predicted number of stereoisomers exist because some of the molecules have symmetry planes that make them identical to their mirror images.)

PROBLEM 22.3

Aldopentoses have three chiral carbon atoms. What is the maximum possible number of aldopentose stereoisomers?

PROBLEM 22.4

From monosaccharides (a)–(d) in Problem 22.5, choose the one that is the enantiomer of the unlabeled monosaccharide shown.

PROBLEM 22.5

Notice in structures (a)–(d) below that the bottom carbon and its substituents are written as CH_2OH in every case. How does the C in this group differ in each case from the C atoms above it? Why must the locations of the H atoms and —OH groups attached to the carbons between this one and the carbonyl group be shown?

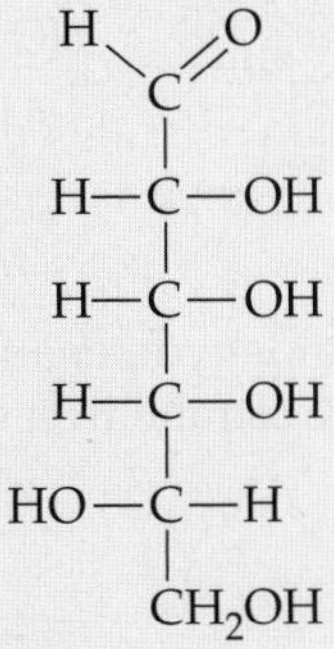

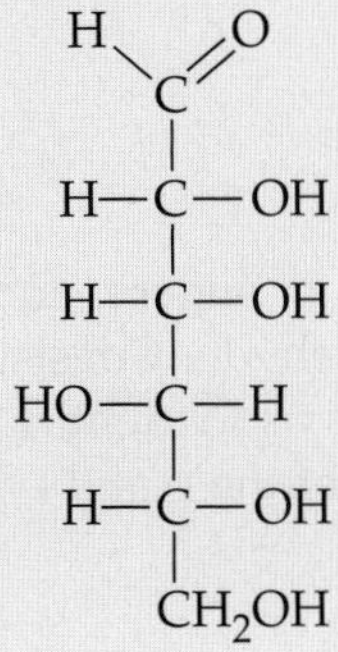

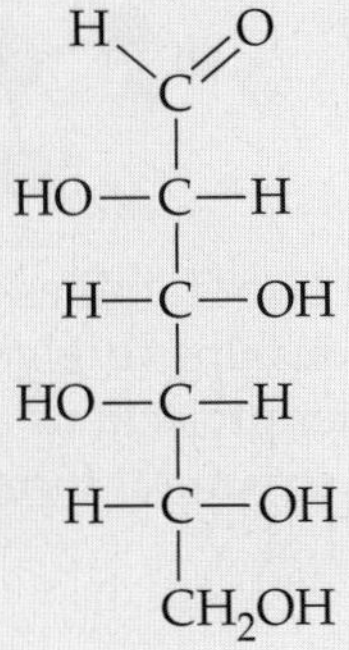

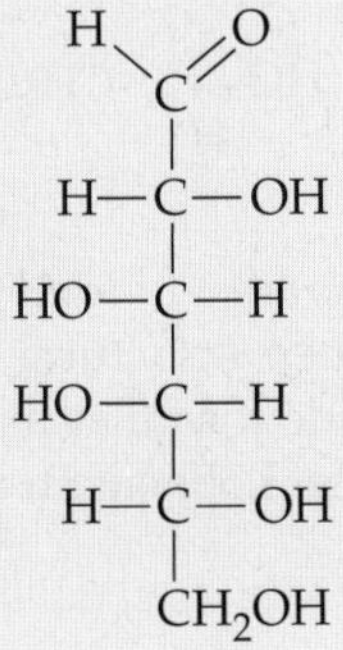

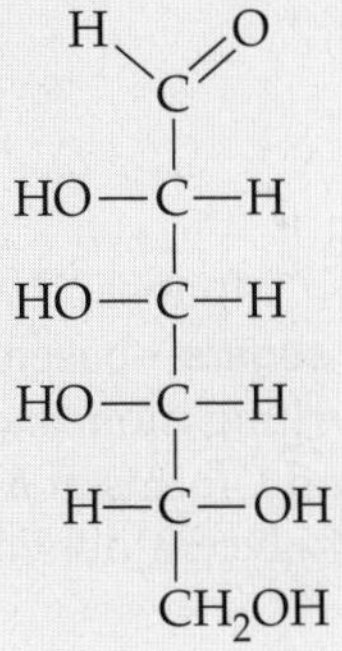

22.3 The D and L Families of Sugars: Drawing Sugar Molecules

A standard method of representation called a **Fischer projection** has been adopted for drawing stereoisomers on a flat page so that we can tell one from another. A chiral carbon atom is represented in a Fischer projection as the intersection of two crossed lines, and this carbon atom is considered to be on the printed page. Bonds that point up and out of the page are shown as horizontal lines, and bonds that point behind the page are shown as vertical lines. Until now, we have used solid wedges and dashed lines to represent bonds above and behind the printed page, respectively, with ordinary solid lines for bonds in the plane of the page. (⊂⊃, p. 124) The relationship between such a structure and a Fischer projection is as follows:

Fischer projection Structure that represents chiral carbon atoms as the intersections of two lines, with the horizontal lines representing bonds pointing out of the page and the vertical lines representing bonds pointing behind the page. For sugars, the aldehyde or ketone is at the top.

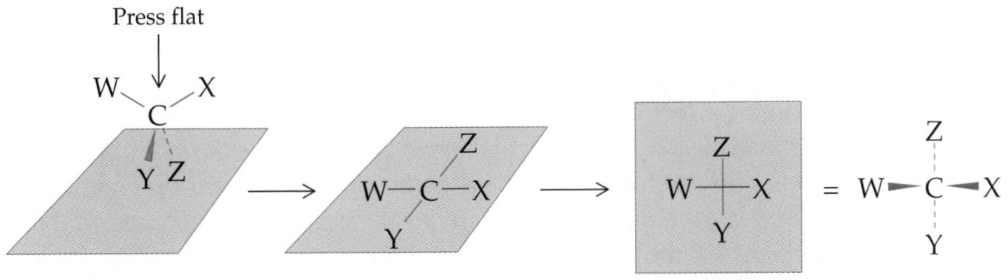

In a Fischer projection, the aldehyde or ketone carbonyl group of a monosaccharide is always placed at the top. The result is that —H and —OH groups projecting above the page are on the left and right of the chiral carbons, and groups projecting behind the page are above and below the chiral carbons. The Fischer projection of one of the enantiomers of glyceraldehyde is therefore interpreted as follows:

Fischer projection of a glyceraldehyde enantiomer

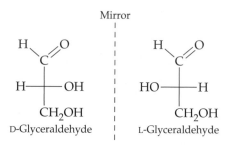

For comparison, the same glyceraldehyde enantiomer is represented below in the conventional manner, showing the tetrahedron of bonds to the chiral carbon.

D Sugar Monosaccharide with the —OH group on the chiral carbon atom farthest from the carbonyl group pointing to the right in a Fischer projection.

L Sugar Monosaccharide with the —OH group on the chiral carbon atom farthest from the carbonyl group pointing to the left in a Fischer projection.

Monosaccharides are divided into two families—the **D sugars** and the **L sugars**—based on their structural relationships to glyceraldehyde. Consistently writing monosaccharide formulas as Fischer projections allows us to identify the D and L forms at a glance. Look again at the structural formulas of the D and L forms of glyceraldehyde.

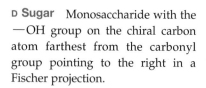

In the D form, the —OH group on carbon 2 comes out of the plane of the paper and points to the *right*; in the L form, the —OH group at carbon 2 comes out of the plane of the paper and points to the *left*. If you mentally place a mirror plane between these Fischer projections, you can see that they are mirror images.

Nature has a strong preference for one type of handedness in carbohydrates, just as it does in amino acids and in snail shells. It happens, however, that carbohydrates and amino acids have opposite handedness. Most naturally occurring α-amino acids belong to the L family, but most carbohydrates belong to the D family.

The designations D and L derive from the Latin *dextro* for "right" and *levo* for "left." *In all Fischer projections, the D form of a monosaccharide has the hydroxyl group on the chiral carbon atom farthest from the carbonyl group pointing toward the right, whereas the mirror-image L form has the hydroxyl group on this same carbon pointing toward the left.*

Fischer projections of molecules with more than one chiral carbon atom are written with the chiral carbons one above the other in a vertical line. To simplify visualizing the structures, we often include the C's for the chiral carbons in the

▲ Nature's preference. Snail shells have a preferred handedness, as do many molecules.

APPLICATION ▶ Chirality and Drugs

Nature is better at synthesizing single optical isomers than chemists are. Laboratory synthesis schemes that yield a mixture of optical isomers (a *racemic mixture*) are easy to develop. It is more difficult to devise synthetic schemes that yield a single isomer, but the goal of modern drug development is often creation of a drug molecule that binds with a specific hormone, enzyme, or cellular receptor. Because most biomolecules are chiral, a chiral drug molecule (single isomer) is likely to meet the need most effectively by fitting most closely with the target. Whether a pharmaceutical company decides to produce a racemic mixture of a particular drug molecule or a single isomer of the molecule is an issue of scientific, medical, and commercial importance in the drug industry.

The route to a chiral drug molecule might start with natural chiral reactants, with a natural enzyme, or with synthesis of a pair of enantiomers that are then separated from each other. It is often important to have the separate enantiomers available during the elaborate testing needed to prove a drug effective. Sometimes, marketing the mixture of isomers is the wrong thing to do. Naproxen, for example, the active ingredient in the pain killer and anti-inflammatory Aleve, is sold as a single enantiomer. The other enantiomer causes liver damage.

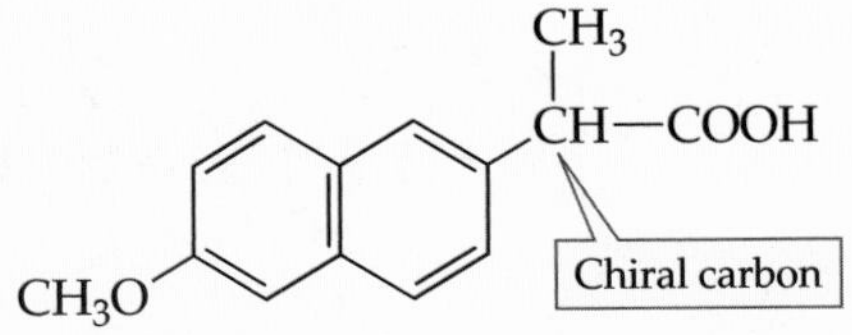

The size of the single-enantiomer drug market has increased every year since its inception about 15 years ago. With expanding research efforts, the ability to produce such drugs is growing easier. The top five classes of single-enantiomer drugs are cardiovascular drugs, antibiotics, hormones, cancer drugs, and those for central-nervous-system disorders.

See Additional Problems 22.77 and 22.78 at the end of the chapter.

---

plane of the page. Otherwise, the structures are interpreted like Fischer projections. Two pairs of aldohexose enantiomers are represented below in this manner. Given the Fischer projection of one enantiomer, you can draw the other by reversing the substituents on the left and right of each chiral atom. Note that each pair of enantiomers has a different name.

Two pairs of aldohexose enantiomers

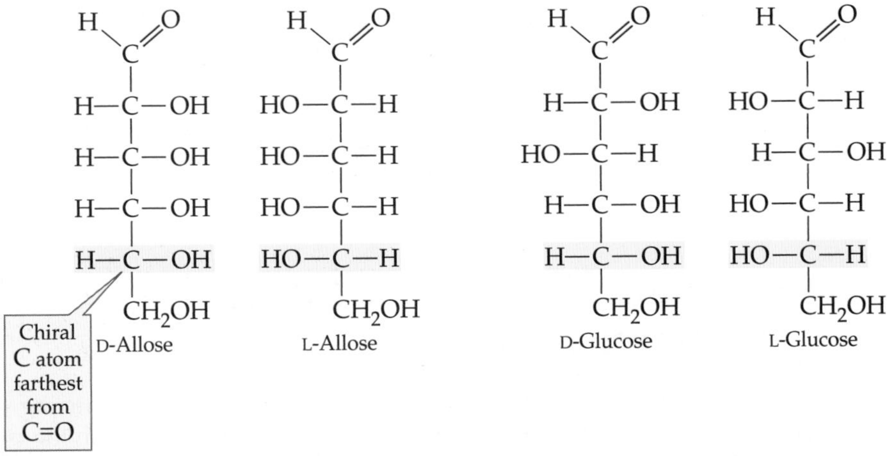

It would be easy to assume that the use of D and L, because they stand for *dextro* and *levo*, carries some meaning about the direction of rotation of plane-polarized light. Logical as it seems, this is not the case. The D and L relate directly only to the position of that —OH group on the chiral carbon farthest from the carbonyl carbon in a Fischer projection. And the D and L isomers do indeed rotate plane-polarized light in opposite directions. But—and here is the point to remember—*the direction of rotation cannot be predicted*. There are D isomers that rotate polarized light to the left and L isomers that rotate it to the right.

WORKED EXAMPLE 22.2 Identifying D and L Isomers

Identify the following monosaccharides as (a) D-ribose or L-ribose, (b) D-mannose or L-mannose.

(a)

$$
\begin{array}{c}
H-C=O \\
H-C-OH \\
H-C-OH \\
H-C-OH \\
CH_2OH
\end{array}
$$

(b)

$$
\begin{array}{c}
H-C=O \\
H-C-OH \\
H-C-OH \\
HO-C-H \\
HO-C-H \\
CH_2OH
\end{array}
$$

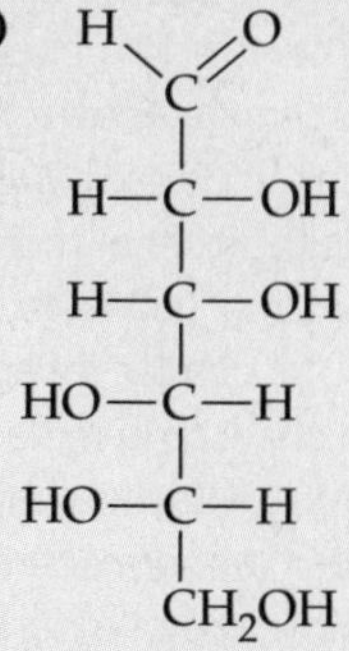

ANALYSIS To identify D or L isomers, you must check the location of the —OH group on the chiral carbon atom farthest from the carbonyl group. In a Fischer projection, this is the carbon atom above the bottom one. The —OH group points left in an L enantiomer and right in a D enantiomer.

SOLUTION

In (a) the —OH group on the chiral carbon above the bottom of the structure points to the right, so this is D-ribose. In (b) this —OH group points to the left, so this is L-mannose.

PROBLEM 22.6

Draw the enantiomer of the following monosaccharides, and in each pair identify the D sugar and the L sugar.

(a)

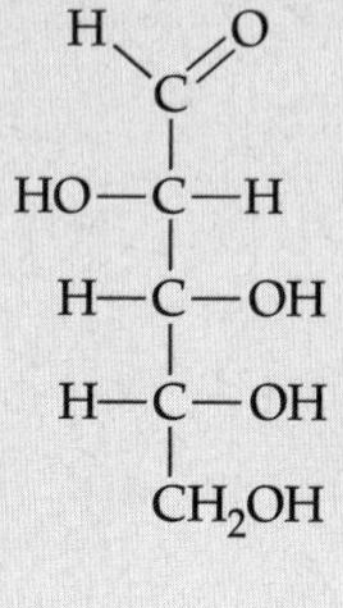

$$
\begin{array}{c}
H-C=O \\
HO-C-H \\
H-C-OH \\
H-C-OH \\
CH_2OH
\end{array}
$$

(b)

$$
\begin{array}{c}
CH_2OH \\
C=O \\
H-C-OH \\
HO-C-H \\
HO-C-H \\
CH_2OH
\end{array}
$$

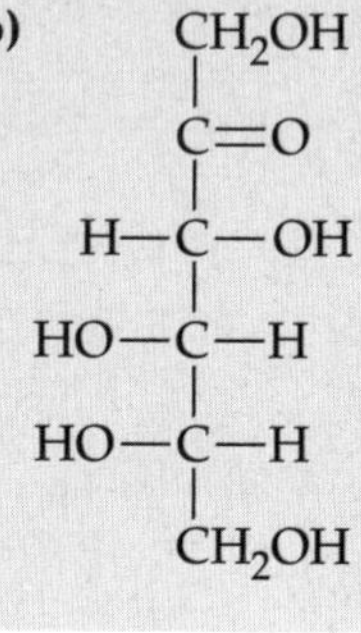

22.4 Structure of Glucose and Other Monosaccharides

D-Glucose, also called *dextrose* or *blood sugar,* is the most widely occurring of all monosaccharides and has the most important function. In nearly all living organisms, D-glucose serves as a source of energy to fuel biochemical reactions. It is stored as starch in plants and glycogen in animals (Section 22.9). Our discussion here of the structure of D-glucose illustrates a major point about the structure of monosaccharides: Although they can be written with the carbon atoms in a straight chain, monosaccharides with five or six carbon atoms exist primarily in their cyclic forms.

Look at the Fischer projection of D-glucose at the top left-hand corner of Figure 22.3 (p. 701) and notice the locations of the aldehyde group and the hydroxyl groups. You have seen that aldehydes and ketones react reversibly with alcohols to

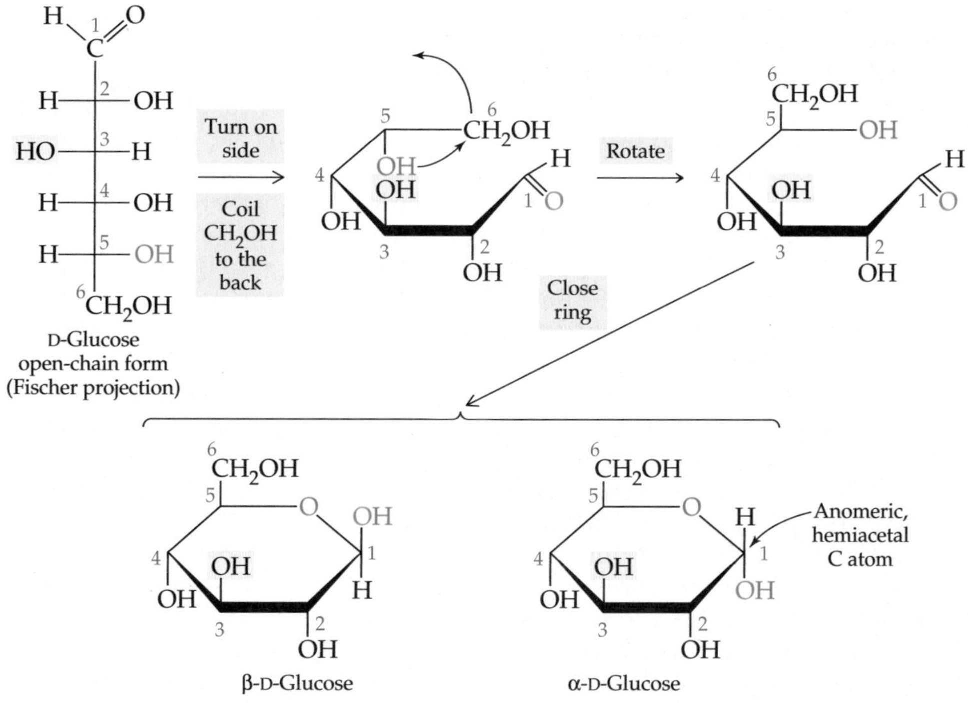

▲ **FIGURE 22.3 The structure of D-glucose.** D-Glucose can exist as an open-chain hydroxy aldehyde or as a pair of cyclic hemiacetals. The cyclic forms differ only at C1, where the —OH group is either on the opposite side of the six-membered ring from the CH_2OH (α) or on the same side (β). To convert the Fischer projection into the six-membered ring formula, the Fischer projection is laid down with C1 to the right and the other end curled around at the back. Then the single bond between C4 and C5 is rotated so that the —CH_2OH group is vertical. Finally, the hemiacetal O—R bond is formed by connecting oxygen from the —OH group on C5 to C1, and the hemiacetal O—H group is placed on C1. (H's on carbons 2–5 are omitted here for clarity.)

yield hemiacetals as shown below. (⬭, Section 16.7) (Remember that the key to recognizing the hemiacetal is a carbon atom bonded to both an —OH and an —OR group.)

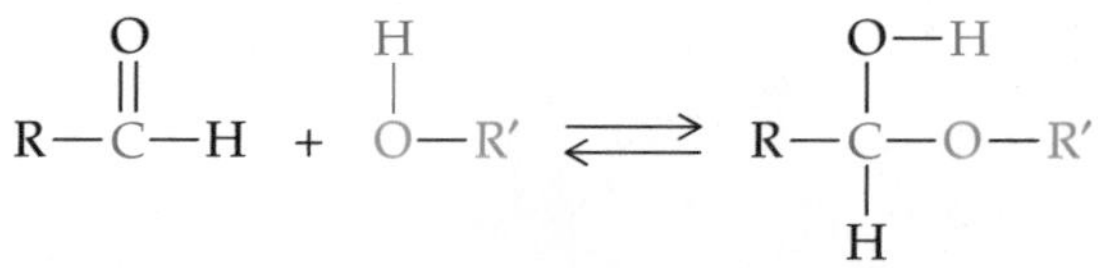

Since glucose has alcohol hydroxyl groups and an aldehyde carbonyl group in the same molecule, *internal* hemiacetal formation is possible. The aldehyde carbonyl group at carbon 1 (C1) and the hydroxyl group at carbon 5 (C5) in glucose react to form a six-membered ring that is a hemiacetal. Monosaccharides with five or six carbon atoms form rings in this manner.

The three structures at the top in Figure 22.3 show how to picture the 5-hydroxyl and the aldehyde group approaching each other for hemiacetal formation. When visualized in this manner, Fischer projections are converted to cyclic structures that (like the Fischer projections) can be interpreted consistently because the same relative arrangements of the groups on the chiral carbon atoms are maintained.

In the cyclic structures at the bottom of Figure 22.3, note how the —OH group on carbon 3, which is on the left in the Fischer projection, points *up* in the cyclic structure, and —OH groups that were on the right on carbons 2 and 4 point *down*.

When cyclic structures (called *Haworth projections*) are drawn as shown in Figure 22.3, such relationships are always maintained. Note also that the —CH₂OH group in D sugars is always *above* the plane of the ring.

The hemiacetal carbon atom (C1) in the cyclic structures, like that in other hemiacetals, is bonded to two oxygen atoms (one in —OH and one in the ring). This carbon is chiral. As a result, there are two cyclic forms of glucose, known as the α and β forms. To see the difference, compare the locations of the hemiacetal —OH groups on C1 in the two bottom structures in Figure 22.3. In the β form, the hydroxyl at C1 points *up* and is on the same side of the ring as the —CH₂OH group at C5. In the α form, the hydroxyl at C1 points *down* and is on the opposite side of the ring from the —CH₂OH group. This relationship is maintained in cyclic monosaccharide structures drawn like those in Figure 22.3.

Cyclic monosaccharides that differ only in the positions of substituents at carbon 1 are known as **anomers**, and carbon 1 is said to be an **anomeric carbon atom**. It was the carbonyl carbon atom (C1 in an aldose and C2 in a hexose) that is now bonded to two O atoms. Note that the α and β anomers of a given sugar are not optical isomers because they are not mirror images.

Although the structural difference between anomers appears small, it has enormous biological consequences. For example, this one small change in structure accounts for the vast difference between the digestibility of starch, which we can digest, and that of cellulose, which we cannot digest (Section 22.9).

Ordinary crystalline glucose is entirely in the cyclic α form. Once dissolved in water, however, equilibrium is established among the open-chain form and the two anomers. The optical rotation of a freshly made solution of α-D-glucose gradually changes from its original value until it reaches a constant value that represents the optical activity of the equilibrium mixture. A solution of β-D-glucose or a mixture of the α and β forms also undergoes this gradual change in rotation, known as **mutarotation**, until the ring opening and closing reactions come to the following equilibrium:

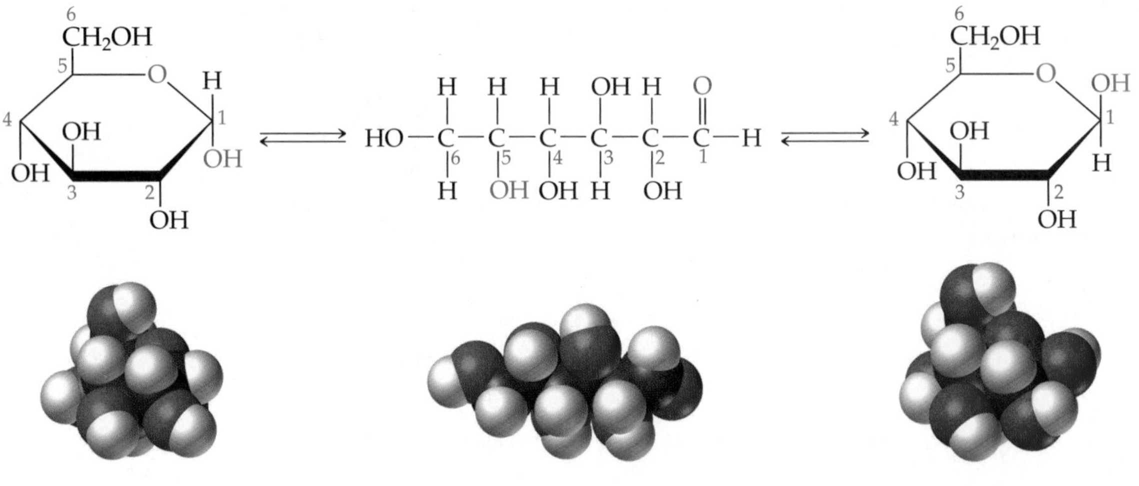

Anomers Cyclic sugars that differ only in positions of substituents at the hemiacetal carbon (the anomeric carbon); the α form has the —OH on the opposite side from the —CH₂OH; the β form has the —OH on the same side as the —CH₂OH.

Anomeric carbon atom The hemiacetal C atom in a cyclic sugar; the C atom bonded to an —OH group and an O in the ring.

Mutarotation Change in rotation of plane-polarized light resulting from the equilibrium between cyclic anomers and the open-chain form of a sugar.

α-D-Glucose (36%) ⇌ Open-chain D-Glucose (0.02%) ⇌ β-D-Glucose (64%)

All monosaccharides with five or six carbon atoms establish similar equilibria, but with different percentages of the different forms present.

Monosaccharide Structures—Summary

- Monosaccharides are polyhydroxy aldehydes or ketones.
- Monosaccharides have three to seven carbon atoms, and a maximum of 2^n possible stereoisomers, where n is the number of chiral carbon atoms.
- D and L enantiomers differ in the orientation of the —OH group on the chiral carbon atom farthest from the carbonyl. In Fischer projections, D sugars have this —OH on the right and L sugars have this —OH on the left.

- D-Glucose (and other 6-carbon aldoses) forms cyclic hemiacetals conventionally represented (as in Figure 22.3) so that —OH groups on chiral carbons on the left in Fischer projections point up and those on the right in Fischer projections point down.

- In glucose, the hemiacetal carbon (*the anomeric carbon*) is chiral, and α and β anomers differ in the orientation of the —OH groups on this carbon. The α anomer has the —OH on the opposite side from the —CH₂OH, and the β anomer has the —OH on the same side as the —CH₂OH.

WORKED EXAMPLE **22.3** Converting Fisher Projections to Cyclic Hemiacetals

The open-chain form of D-altrose, an aldohexose isomer of glucose, has the following structure. Draw D-altrose in its cyclic hemiacetal form:

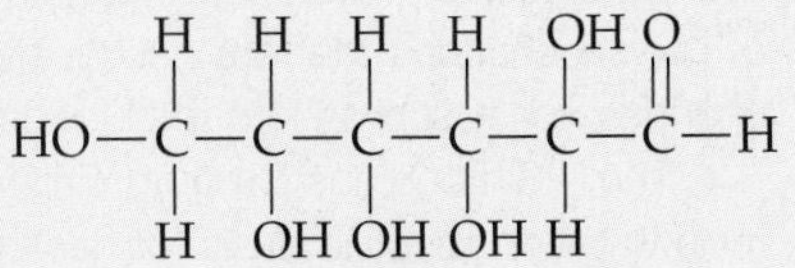

D-Altrose

SOLUTION

First, coil D-altrose into a circular shape by mentally grasping the end farthest from the carbonyl group and bending it backward into the plane of the paper:

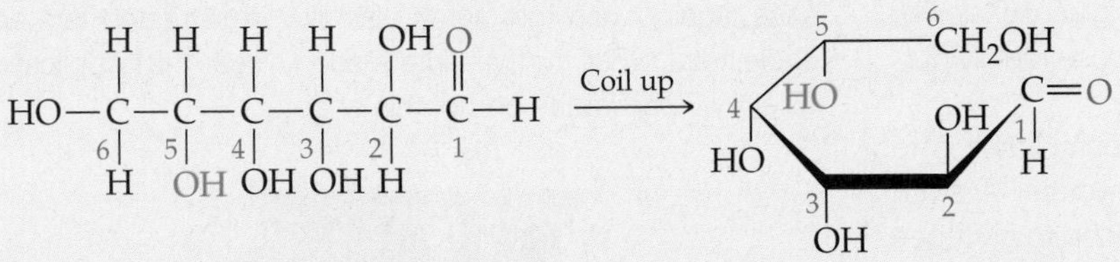

Next, rotate the bottom of the structure around the single bond between C4 and C5 so that the —CH₂OH group at the end of the chain points up and the —OH group on C5 points toward the aldehyde carbonyl group on the right:

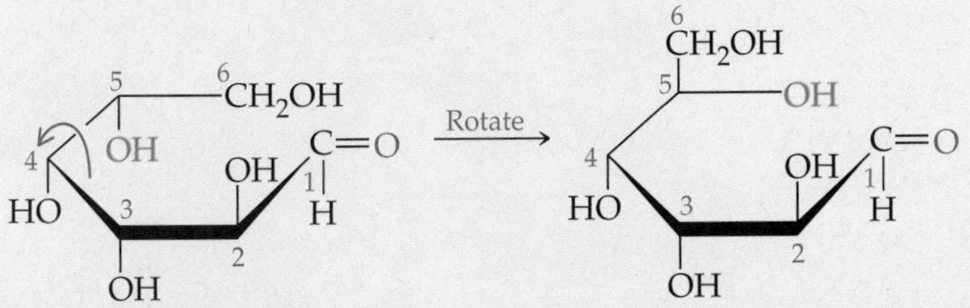

Finally, add the —OH group at C5 to the carbonyl C=O to form a hemiacetal ring. The new —OH group formed on C1 can be either up (β) or down (α):

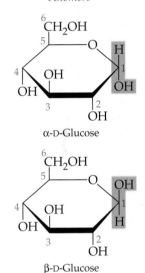

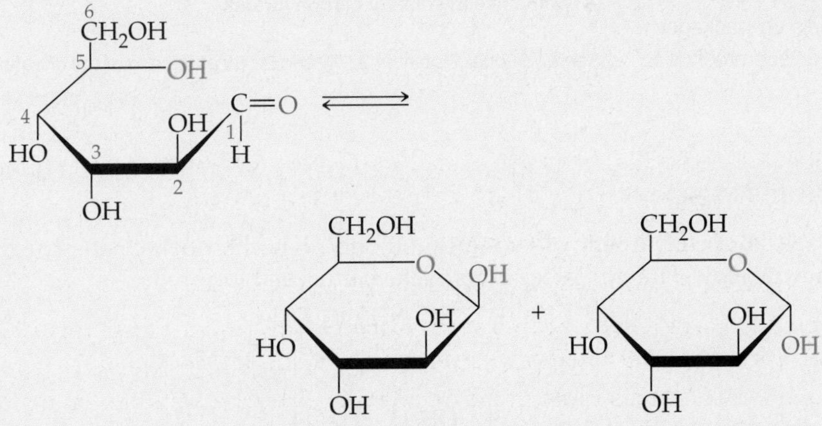

Enantioners

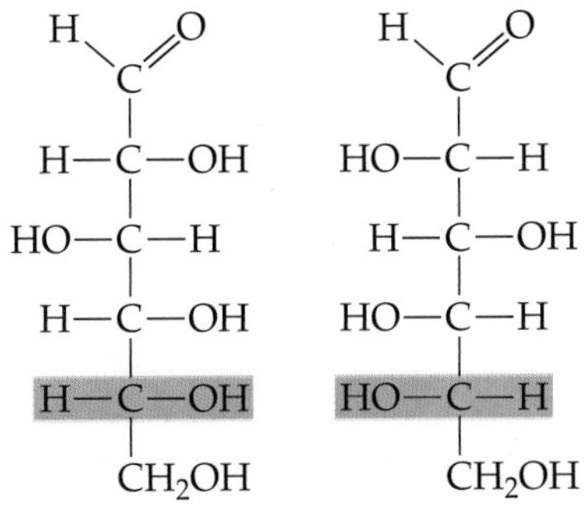

D-Glucose   L-Glucose

Anomers

α-D-Glucose

β-D-Glucose

APPLICATION ▶ Carbohydrates in the Diet

The major monosaccharides in our diets are fructose and glucose from fruits and honey. The major disaccharides are sucrose, commonly called table sugar and refined from both sugar cane and sugar beets, and lactose from milk. In addition, our diets contain large amounts of the digestible polysaccharide starch, present in grains such as wheat and rice, root vegetables such as potatoes, and legumes such as beans and peas. Nutritionists often refer to polysaccharides as *complex carbohydrates*.

How easily and how rapidly complex carbohydrates are digested and absorbed affects blood sugar levels. Consumption of the easily digested carbohydrates found in refined foods, such as white bread and rice, or in potatoes results in rapid elevation of blood glucose levels followed by lower-than-desired levels a few hours later. Carbohydrates that are digested and absorbed more slowly are associated with healthier blood sugar responses. The glycemic index is a scale that compares the blood sugar response to eating a complex carbohydrate with the response evoked by glucose. Foods with a low glycemic index release glucose slowly; foods with a high glycemic index release glucose rapidly during digestion and should be limited in the diet.

The body's major use of digestible carbohydrates is to provide energy, 4 kcal per gram of carbohydrate. A small amount of any excess carbohydrate is converted to glycogen for storage in the liver and muscles, but most dietary carbohydrate in excess of our immediate needs for energy is converted into fat.

The MyPyramid meal-planning tool (p. 560) reflects the emphasis on decreasing the amounts of meat and increasing the amounts of nonmeat products in our diet. The widest stripe of the pyramid is the recommendation for six 8-ounce equivalent servings per day of bread, cereal, rice, and pasta, all foods high in complex carbohydrates. Half of these servings should consist of whole grains. If your reaction is, I cannot possibly eat that much, you should know that a 1-ounce equivalent serving, as defined for the pyramid, is quite small: 1 slice of bread; 1/2 cup of cooked rice, pasta, or cereal; or 1 cup of cold cereal.

In terms of *total* carbohydrate, which includes both simple sugars and fiber, the *Nutrition Facts* labels on packaged foods (p. 621) give percentages based on a recommended 300 g per day of total carbohydrate and 25 g per day of dietary fiber (the nondigestible carbohydrates). These quantities provide 2000 Cal a day, with 60% of the calories from carbohydrates. The label also gives the total grams of sugars in the food, without a percentage because there is no recommended daily quantity of sugars. For purposes of the label, "sugars" are defined as all monosaccharides and disaccharides, whether naturally present or added.

As an option, the label may also include grams of *soluble fiber* and *insoluble fiber*. Taken together, these are the types of polysaccharides that are neither hydrolyzed to monosaccharides nor absorbed into the bloodstream.

The U.S. Food and Drug Administration is responsible for reviewing the scientific basis for health claims for foods. Two allowed claims relate to carbohydrates. The first states that a diet high in fiber may lower the risk of cancer and heart disease if the diet is also low in saturated fats and cholesterol. The second states that foods high in the soluble fiber from whole oats (oat bran) may also reduce the risk of heart disease, again when the diet is also low in saturated fats and cholesterol. (For further information, see the Application "Dietary Fiber" at the end of this chapter.)

▲ Some healthy dietary carbohydrates.

See Additional Problems 22.79 and 22.80 at the end of the chapter.

PROBLEM 22.7

D-Talose, a constituent of certain antibiotics, has the open-chain structure shown below. Draw D-talose in its cyclic hemiacetal form.

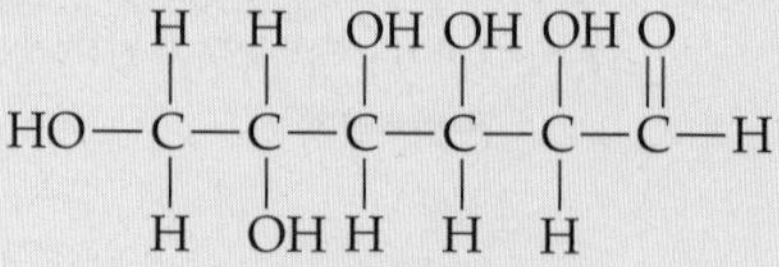

$$\underset{\text{D-Talose}}{HO-\overset{\overset{\displaystyle H}{|}}{\underset{\underset{\displaystyle H}{|}}{C}}-\overset{\overset{\displaystyle H}{|}}{\underset{\underset{\displaystyle OH}{|}}{C}}-\overset{\overset{\displaystyle OH}{|}}{\underset{\underset{\displaystyle H}{|}}{C}}-\overset{\overset{\displaystyle OH}{|}}{\underset{\underset{\displaystyle H}{|}}{C}}-\overset{\overset{\displaystyle OH}{|}}{\underset{\underset{\displaystyle H}{|}}{C}}-\overset{\overset{\displaystyle O}{\|}}{C}-H}$$

PROBLEM 22.8

The cyclic structure of D-idose, an aldohexose, is shown below. Convert this to the straight-chain Fischer projection structure.

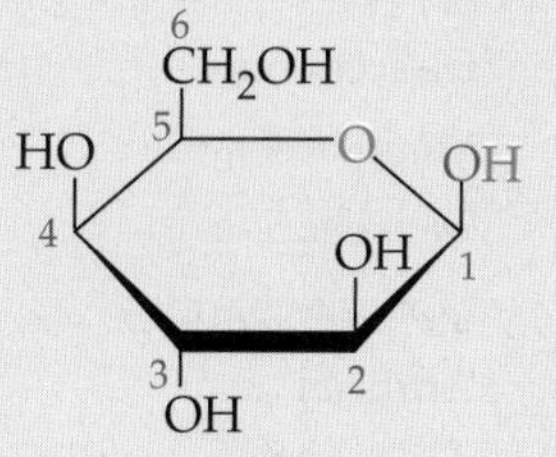

D-Idose

🔑 **KEY CONCEPT PROBLEM 22.9**

Ouabain is a potent poison derived from a plant and used as a dart poison.

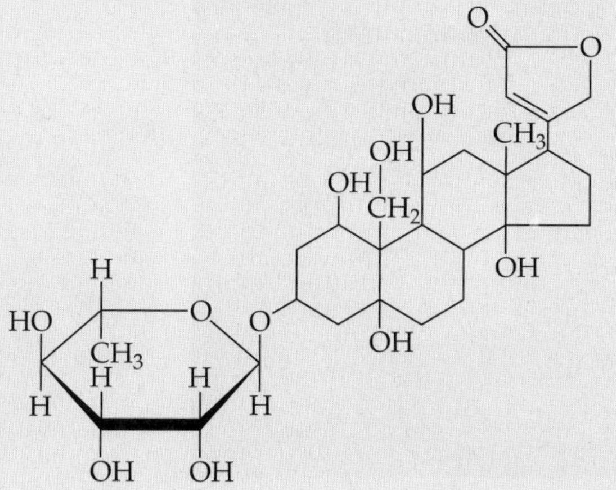

Ouabain

The structure can be roughly divided into three sections: a monosaccharide ring, a four-ring system, and an oxygen-containing ring known as a lactone. (a) Identify the monosaccharide ring according to the number of carbons in the ring. Based on the location of the linkage between the monosaccharide ring and the larger ring system, is the monosaccharide the α or the β form? (b) The large ring system is similar to that in a class of molecules that you encountered in an earlier chapter. Identify this class of molecule. (c) Look closely at the "lactone" ring. A lactone is a cyclic version of what common organic functional group?

22.5 Some Important Monosaccharides

The monosaccharides, with their many opportunities for hydrogen bonding through their hydroxyl groups, are generally high-melting, white, crystalline solids that are soluble in water and insoluble in nonpolar solvents. Most monosaccharides and disaccharides are sweet-tasting (Table 22.1), digestible, and nontoxic (Figure 22.4). Except for glyceraldehyde (an aldotriose) and fructose (a ketohexose), the carbohydrates of interest in human biochemistry are all aldohexoses or aldopentoses. Most are in the D family.

Glucose

Glucose is the most important simple carbohydrate in human metabolism. It is the final product of carbohydrate digestion and provides acetyl groups for entry into

TABLE 22.1   Relative Sweetness of Some Sugars and Sugar Substitutes

NAME	TYPE	SWEETNESS
Lactose	Disaccharide	16
Galactose	Monosaccharide	30
Maltose	Disaccharide	33
Glucose	Monosaccharide	75
Sucrose	Disaccharide	100
Fructose	Monosaccharide	175
Cyclamate	Artificial	3000
Aspartame	Artificial	15,000
Saccharin	Artificial	35,000
Sucralose	Artificial	60,000

(a)

(b)

(c)

▲ **FIGURE 22.4   Common sugars.**   (a) Sucrose (glucose + fructose) is found in sugar cane and sugar beets. (b) Jam, with galactose in the pectin that stiffens it. (c) Honey, which is high in fructose.

the citric acid cycle as acetyl-SCoA. (⬤⬤⬤, Section 21.8) Maintenance of an appropriate blood glucose level is essential to human health. The hormones insulin and glucagon regulate blood glucose concentration. Because glucose is metabolized without further digestion, glucose solutions can be supplied intravenously to restore blood glucose levels.

⬤⬤⬤ Looking Ahead

In Chapter 23 we will describe the metabolic pathway (glycolysis) by which glucose is converted to pyruvate and then to acetyl-SCoA for entry into the citric acid cycle. The role of insulin in controlling blood glucose concentrations and the way in which those concentrations are affected by diabetes mellitus are also examined there.   ⬤⬤⬤

Galactose

D-Galactose is widely distributed in plant gums and pectins, the sticky polysaccharides present in plant cells. It is also a component of the disaccharide lactose (milk sugar) and is produced from lactose during digestion. Like glucose, galactose is an aldohexose; it differs from glucose only in the spatial orientation of the —OH group at carbon 4. In the body, galactose is converted to glucose to provide energy and is synthesized from glucose to produce lactose for milk and compounds needed in brain tissue.

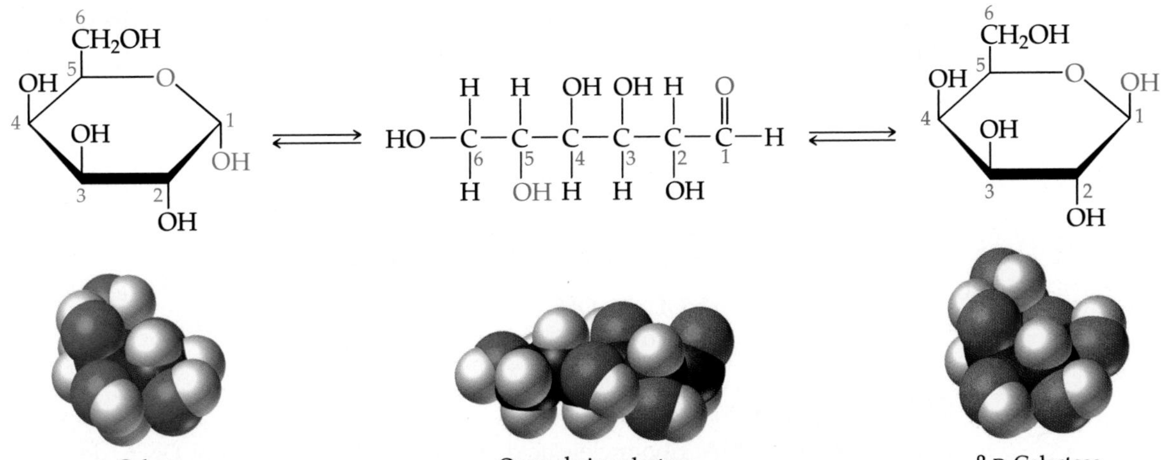

α-D-Galactose   Open-chain galactose   β-D-Galactose

A group of genetic disorders known as *galactosemias* result from an inherited deficiency of any of several enzymes needed to metabolize galactose. The result is a buildup of galactose or galactose 1-phosphate in blood and tissues. Early symptoms in infants include vomiting, an enlarged liver, and general failure to thrive. Other possible outcomes are liver failure, mental retardation, and development of cataracts when galactose in the eye is reduced to a polyhydroxy alcohol that accumulates. Treatment of galactosemia consists of a galactose-free diet for life.

Fructose

D-Fructose, often called *levulose* or *fruit sugar*, occurs in honey and many fruits. It is one of the two monosaccharides combined in the disaccharide sucrose. Fructose is produced commercially in large quantities by hydrolysis of cornstarch to make high fructose corn syrup. Like glucose and galactose, fructose is a 6-carbon sugar. However, it is a ketohexose rather than an aldohexose. In solution, fructose forms five-membered rings:

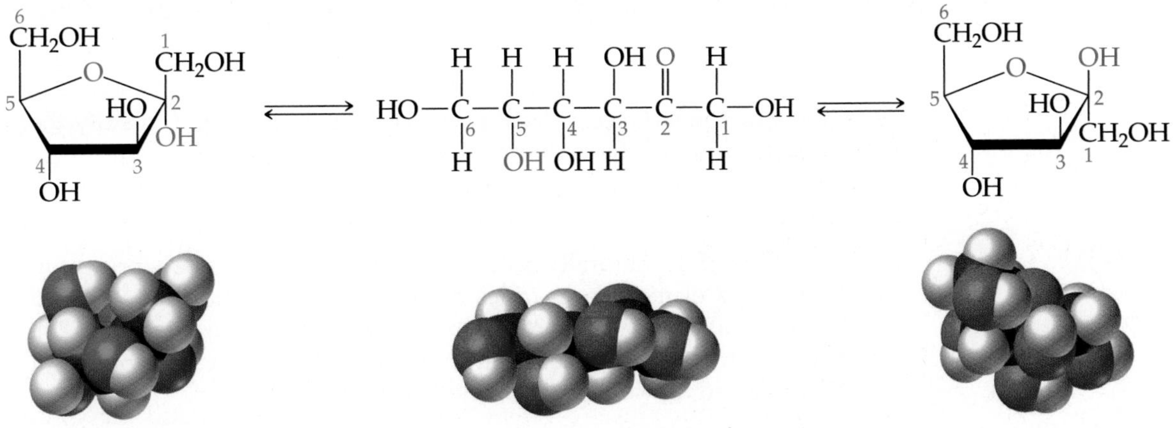

α-D-Fructose   Open-chain D-Fructose   β-D-Fructose

Fructose is sweeter than sucrose and is an ingredient in many sweetened beverages and prepared foods. As a phosphate, it is an intermediate in glucose metabolism.

Ribose and 2-Deoxyribose

Ribose and its relative 2-deoxyribose are both 5-carbon aldehyde sugars. These two sugars are most important as parts of larger biomolecules. You have already seen ribose as a constituent of coenzyme A (Figure 19.10), in ATP and the second messenger cyclic AMP (Figure 20.3) and in oxidizing and reducing agent coenzymes (p. 675).

As its name indicates, *2-deoxy*ribose differs from ribose by the absence of one oxygen atom, that in the —OH group at C2. Both ribose and 2-deoxyribose exist in the usual mixture of open-chain and cyclic hemiacetal forms.

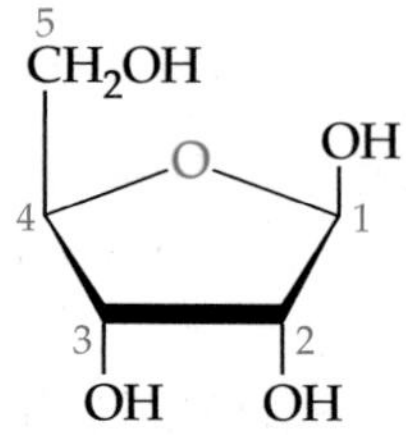

β-D-Ribose 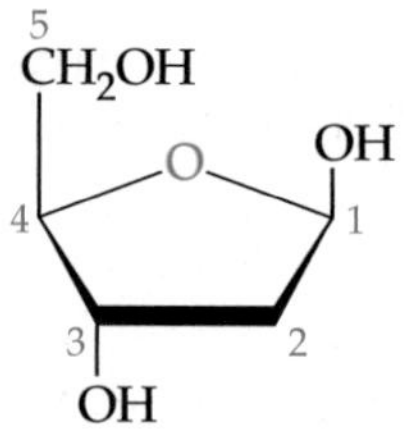   β-D-2-Deoxyribose

Looking Ahead

Ribose is part of RNA, ribonucleic acid, and deoxyribose is part of DNA, deoxyribonucleic acid. Chapter 26 is devoted to the roles of DNA in protein synthesis and heredity.

PROBLEM 22.10

In the following monosaccharide hemiacetal, identify the anomeric carbon atom, number all the carbon atoms, and identify it as the α or β anomer.

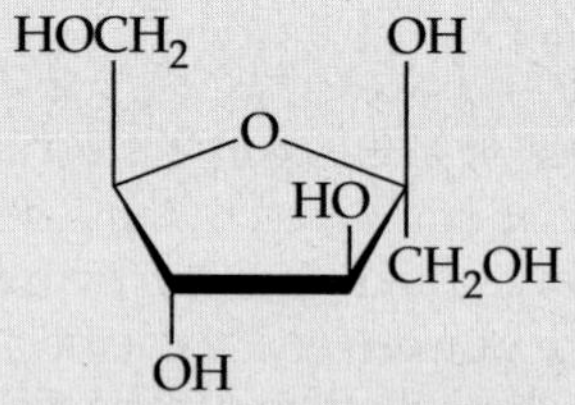

PROBLEM 22.11

Identify the chiral carbons in α-D-fructose, α-D-ribose, and β-D-2-deoxyribose.

PROBLEM 22.12

Draw the structures of cyclic AMP and ATP (Figure 20.3), and identify the portion of the molecule from ribose.

PROBLEM 22.13

L-Fucose is one of the naturally occurring L monosaccharides. It is present in the short chains of monosaccharides by which blood groups are classified (see the Application "Cell-Surface Carbohydrates and Blood Type," p. 720). Compare the structure of L-fucose given below with the structures of α- and β-D-galactose and answer the following questions.

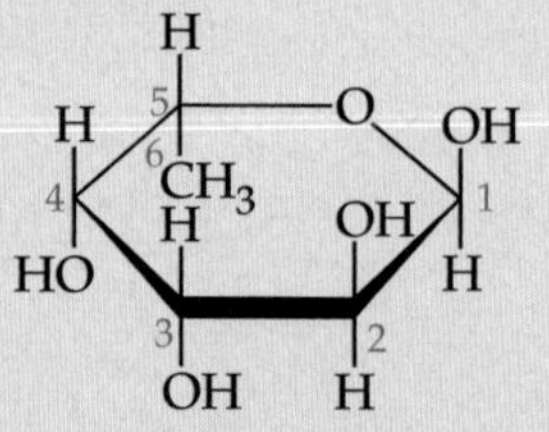

L-Fucose

(a) Is L-fucose an α or β anomer?

(b) Compared with galactose, on which carbon is L-fucose missing an oxygen?

(c) How do the positions of the —OH groups above and below the plane of the ring on carbons 2, 3, and 4 compare in D-galactose and L-fucose?

(d) "Fucose" is a common name. Is 6-deoxy-L-galactose a correct name for fucose?

22.6 Reactions of Monosaccharides

Reaction with Oxidizing Agents: Reducing Sugars

Aldehydes can be oxidized to carboxylic acids (RCHO → RCOOH), a reaction that applies to the open-chain form of aldose monosaccharides. (⊂⊃, Section 16.5) As the open-chain aldehyde is oxidized, its equilibrium with the cyclic form is displaced, in accord with Le Châtelier's principle, so that the open-chain form continues to be produced. (⊂⊃, p. 203) As a result, the aldehyde group of the monosaccharide is ultimately oxidized to a carboxylic acid group. For glucose, the reaction is

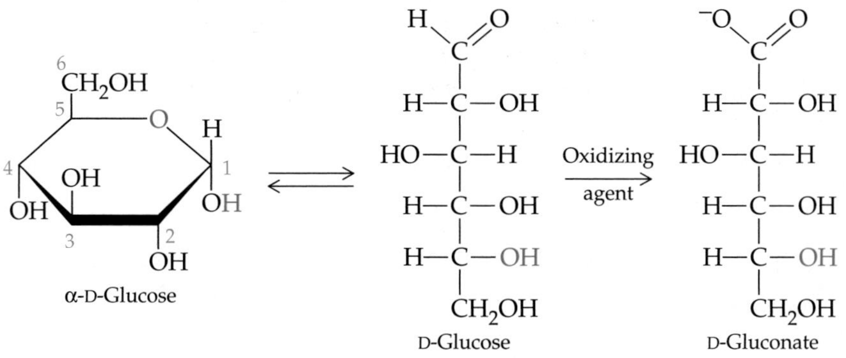

α-D-Glucose   D-Glucose   D-Gluconate

Carbohydrates that react with mild oxidizing agents are classified as **reducing sugars** (they reduce the oxidizing agent).

You probably would not predict it, but in basic solution, ketoses are also reducing sugars. The explanation is that, under these conditions, a ketone that has a hydrogen atom on the carbon adjacent to the carbonyl carbon undergoes a rearrangement. This hydrogen moves over to the carbonyl oxygen. The product is an *enediol*, "ene" for the double bond and "diol" for the two hydroxyl groups. The enediol rearranges to give an aldose, which is susceptible to oxidation.

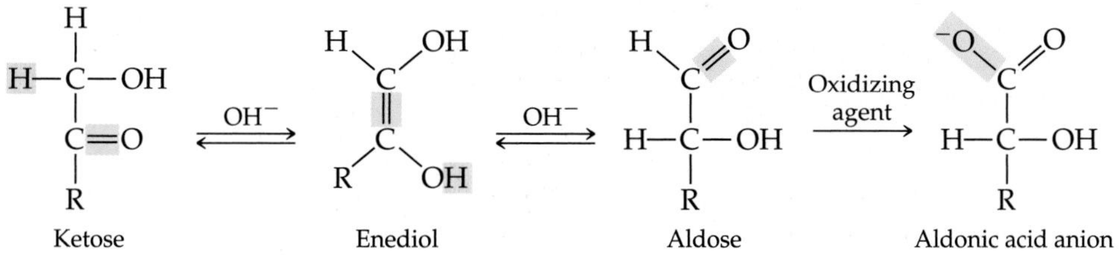

Reducing sugar A carbohydrate that reacts in basic solution with a mild oxidizing agent.

Ketose   Enediol   Aldose   Aldonic acid anion

Here also, oxidation of the aldehyde to an acid drives the equilibria toward the right, and complete oxidation of the ketose occurs. Thus, *in basic solution, all monosaccharides, whether aldoses or ketoses, are reducing sugars*. This ability to act as reducing agents is the basis for most laboratory tests for the presence of monosaccharides.

The first equilibrium above—between the ketose and the enediol—is an example of *keto–enol tautomerism*, an equilibrium that results from a shift in position of a hydrogen atom and a double bond. Keto–enol tautomerism is possible whenever there is a hydrogen atom on a carbon adjacent to a carbonyl carbon.

Reaction with Alcohols: Glycoside and Disaccharide Formation

Hemiacetals react with alcohols with the loss of water to yield acetals, compounds with two —OR groups bonded to the same carbon. (⬭⬭, Section 16.7)

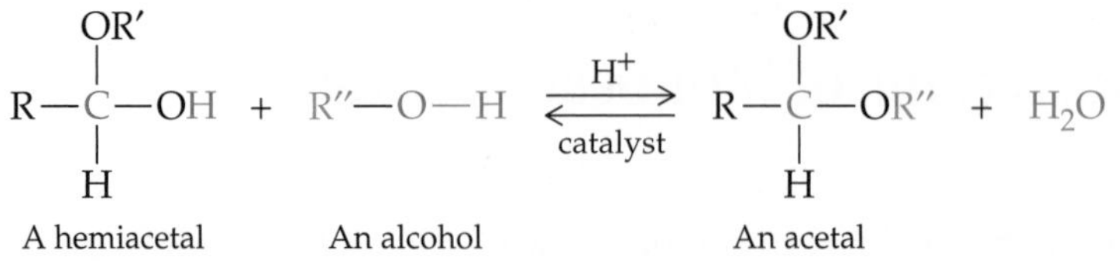

$$\underset{\text{A hemiacetal}}{R-\underset{H}{\overset{OR'}{C}}-OH} + \underset{\text{An alcohol}}{R''-O-H} \underset{\text{catalyst}}{\overset{H^+}{\rightleftharpoons}} \underset{\text{An acetal}}{R-\underset{H}{\overset{OR'}{C}}-OR''} + H_2O$$

Glycoside A cyclic acetal formed by reaction of a monosaccharide with an alcohol, accompanied by loss of H_2O.

Because glucose and other monosaccharides are cyclic hemiacetals, they also react with alcohols to form acetals, which are called **glycosides**. In a glycoside, the —OH group on the anomeric carbon atom is replaced by an —OR group. For example, glucose reacts with methanol to produce methyl glucoside. (Note that a *gluc*oside is a cyclic acetal formed by glucose. A cyclic acetal derived from *any* sugar is a *glyc*oside.)

Formation of a glycoside

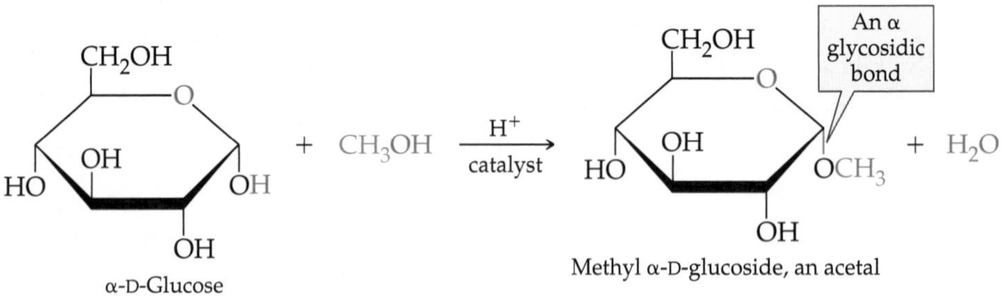

α-D-Glucose → Methyl α-D-glucoside, an acetal

Glycosidic bond Bond between the anomeric carbon atom of a monosaccharide and an —OR group.

The bond between the anomeric carbon atom of the monosaccharide and the oxygen atom of the —OR group is called a **glycosidic bond**. Since glycosides like the one shown above do not contain hemiacetal groups that establish equilibria with open-chain forms, they are *not* reducing sugars.

In larger molecules, including disaccharides and polysaccharides, monosaccharides are connected to each other by glycosidic bonds. For example, a disaccharide forms by reaction of the anomeric carbon of one monosaccharide with an —OH group of a second monosaccharide.

Formation of a glycosidic bond between two monosaccharides

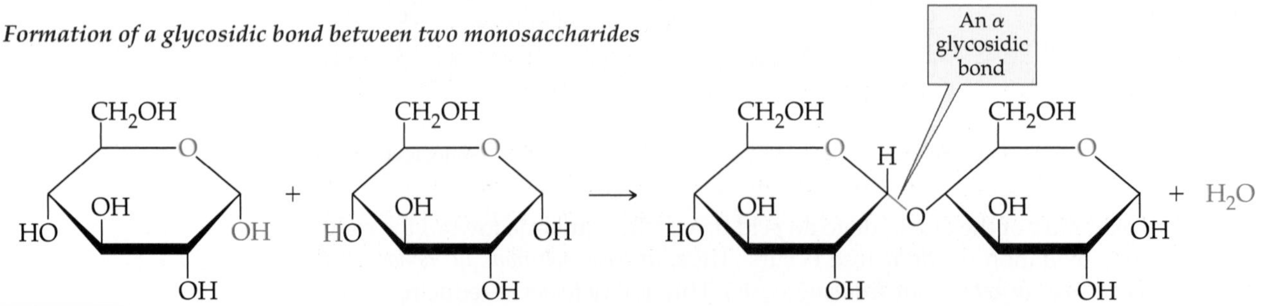

The reverse of this reaction is a *hydrolysis* and is the reaction that takes place during digestion of all carbohydrates.

Hydrolysis of a disaccharide

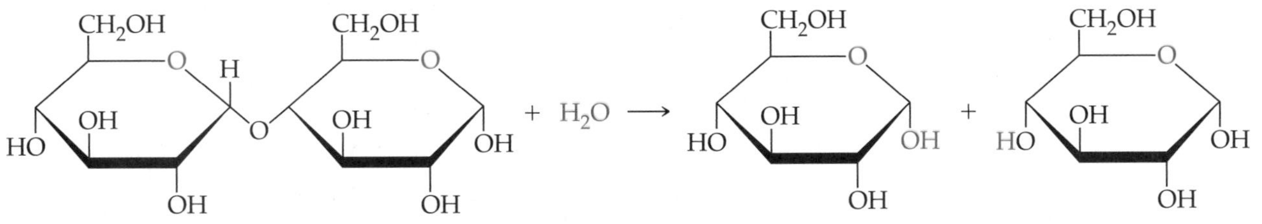

PROBLEM 22.14

Draw the structure of the α and β anomers that result from the reaction of methanol and ribose (see p. 710). Are these compounds acetals or hemiacetals?

Formation of Phosphate Esters of Alcohols

Phosphate esters of alcohols contain a $-PO_3^{2-}$ group bonded to the oxygen atom of an $-OH$ group. The $-OH$ groups of sugars can add $-PO_3^{2-}$ groups to form phosphate esters in the same manner. The resulting phosphate esters of monosaccharides appear as reactants and products throughout the metabolism of carbohydrates. Glucose phosphate is the first to be formed and sets the stage for subsequent reactions. It is produced by the transfer of a $-PO_3^{2-}$ group from ATP to glucose in the first step of glycolysis, the metabolic pathway followed by glucose and other sugars, which is described in Chapter 23. Glycolysis leads to the ultimate conversion of glucose to the acetyl groups that are carried into the citric acid cycle.

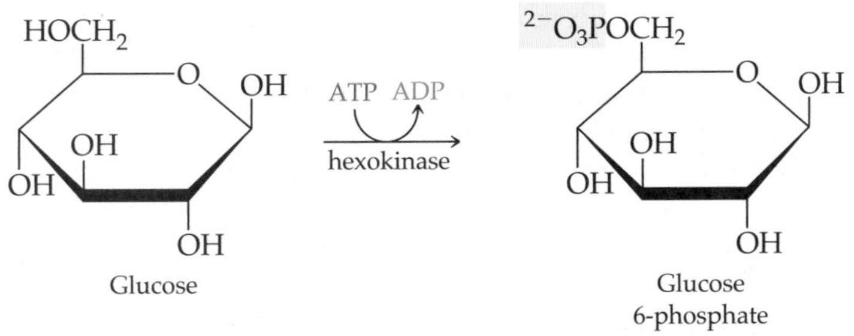

22.7 Disaccharides

Every day, you eat a disaccharide—sucrose, common table sugar. Sucrose is made of two monosaccharides, one glucose and one fructose, covalently bonded to each other. Sucrose is present in modest amounts, along with other mono- and disaccharides, in most fresh fruits and many fresh vegetables. But most sucrose in our diets has been added to something. Perhaps you add it to your coffee or tea. Or it is there in a ready-to-eat food product that you buy, maybe breakfast cereal, ice cream, or a "super-sized" soda, or even bread. Excessive consumption of high-sucrose foods has been blamed for everything from criminal behavior to heart disease to hyperactivity in children, but without any widely accepted scientific proof. A proven connection with heart disease does exist, of course, but by way of the contribution of excess sugar calories to obesity.

Disaccharide Structure

The two monosaccharides in a disaccharide are connected by a glycosidic bond. The bond may be α or β as in cyclic monosaccharides: α points below the ring and β points above the ring (see Figure 22.3). The structures include glycosidic bonds that create a **1,4 link**, that is, a link between C1 of one monosaccharide and C4 of the second monosaccharide:

1,4 Link A glycosidic link between the hemiacetal hydroxyl group at C1 of one sugar and the hydroxyl group at C4 of another sugar.

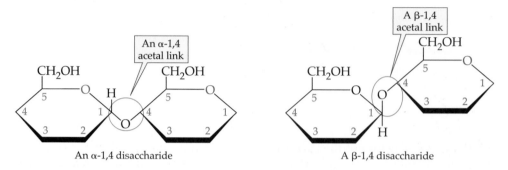

An α-1,4 disaccharide   A β-1,4 disaccharide

The three naturally occurring disaccharides discussed in the following sections are the most common ones. They illustrate the three different ways monosaccharides are linked: by a glycosidic bond in the α orientation (maltose), a glycosidic bond in the β orientation (lactose), or a bond that connects two anomeric carbon atoms (sucrose).

Maltose

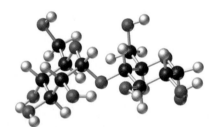

Maltose, often called malt sugar, is present in fermenting grains and can be prepared by enzyme-catalyzed degradation of starch. It is used in prepared foods as a sweetener. In the body, it is produced during starch digestion by α-amylase in the small intestine and then hydrolyzed to glucose by a second enzyme, maltase.

Two α-D-glucose molecules are joined in maltose by an α-1,4 link. A careful look at maltose shows that it is both an acetal (at C1 in the glucose on the left below) and a hemiacetal (at C1 in the glucose on the right below). Since the acetal ring on the left does not open and close spontaneously, it cannot react with an oxidizing agent. The hemiacetal group on the right, however, establishes equilibrium with the aldehyde, making maltose a reducing sugar.

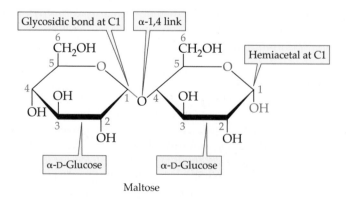

Maltose

Lactose

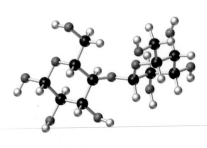

Lactose, or milk sugar, is the major carbohydrate in mammalian milk. Human milk, for example, is about 7% lactose. Structurally, lactose is a disaccharide composed of β-D-galactose and β-D-glucose. The two monosaccharides are connected by

a β-1,4 link. Like maltose, lactose is a reducing sugar because the glucose ring (on the right in the following structure) is a hemiacetal at C1.

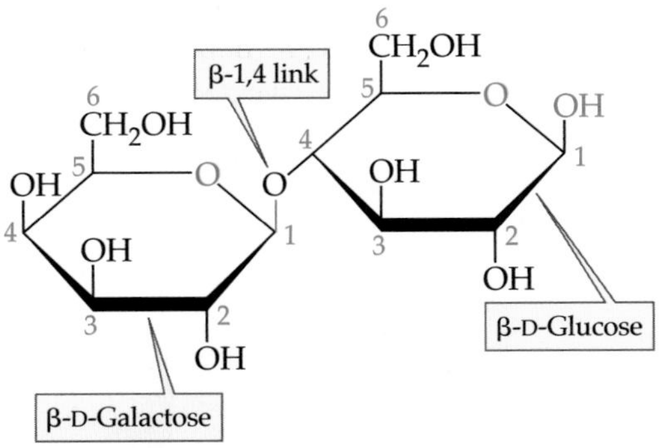

▲ Milk for lactose-intolerant individuals. The lactose content of the milk has been decreased by treating it with lactase.

Lactose

Lactose intolerance in adults is an unpleasant, though not life-threatening, condition that is prevalent in all populations. In fact, it has been suggested that the *absence* of this condition in adults rather than its presence is the deviation from the norm. The activity of lactase, the enzyme that allows lactose digestion by infants, apparently gradually diminishes over the years. Because lactose remains in the intestines rather than being absorbed, it raises the osmolarity there, which draws in excess water. (⬤◼︎, Section 9.12) Bacteria in the intestine also ferment the lactose to produce lactate, carbon dioxide, hydrogen, and methane. The result is bloating, cramps, flatulence, and diarrhea. The condition may be treated by a lactose-free diet, which extends to limitations on taking the many medications and artificial sweeteners in which lactose is an inactive ingredient. An alternative is the use of commercial enzyme preparations taken before milk products are consumed and Lactaid, milk that has been treated with lactase to reduce its lactose content.

Sucrose

Sucrose—plain table sugar—is probably the most common highly purified organic chemical used in the world. Sugar beets and sugarcane are the most common sources of sucrose. Hydrolysis of sucrose yields one molecule of D-glucose and one molecule of D-fructose. The 50:50 mixture of glucose and fructose that results, often referred to as *invert sugar,* is commonly used as a food additive because it is sweeter than sucrose.

Sucrose differs from maltose and lactose in that it has no hemiacetal group because a 1,2 link joins *both* anomeric carbon atoms. The absence of a hemiacetal group means that sucrose is not a reducing sugar. Sucrose is the only common disaccharide that is not a reducing sugar.

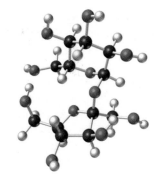

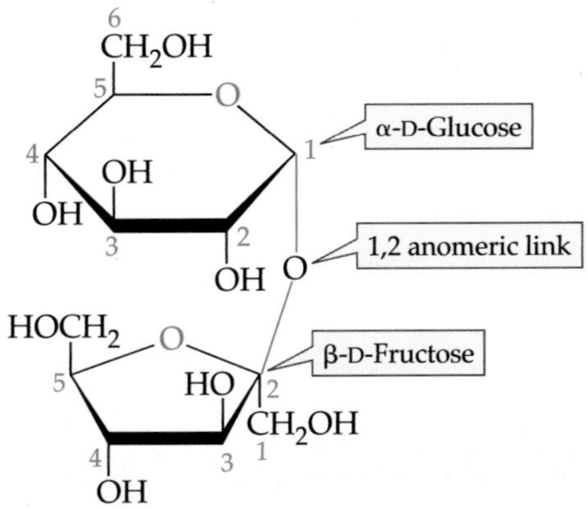

Sucrose

APPLICATION ▶ Cell Walls: Rigid Defense Systems

All cells are defined by the presence of a plasma membrane, which confines the cell's contents inside a lipid bilayer studded with proteins. (●◯◯, Section 21.3) Bacteria and higher plants surround the plasma membrane with a rigid cell wall. Cell walls differ markedly in composition, but not in primary function between higher plants and bacteria. The primary function of a cell wall is to make the cell rigid. The rigidity of the wall prevents the cell from bursting due to osmotic pressure, because the dissolved metabolites and ions inside the cell are at a greater concentration than outside. In addition to its rigidity, the cell wall also gives shape to the cell and protects it from pathogens.

Plant cell walls are composed of fibrils of cellulose in a polymer matrix of pectins, lignin, and hemicellulose. Although you might think each cell is isolated from the others, plant cell walls contain small perforations that permit contact between adjacent cells. This allows for the transfer of nutrients and signals. Cellulose chains range from about 6000 to 16,000 glucose units in length. Neighboring chains of cellulose form hydrogen bonds between them, thereby strengthening the cell wall.

In addition to providing strength and shape, bacterial cell walls provide a rigid platform for the attachment of flagella and pilli. Furthermore, the composition of the cell wall provides attachment sites for bacteriophages (viruses that infect bacteria). Although bacterial cell walls do contain modified sugar polymers, they do not contain cellulose. Cell wall composition varies among bacterial species and is an important factor in distinguishing between some groups of bacteria. A majority of bacterial cell walls are composed of a polymer of *peptidoglycan*, an alternating sequence of the modified sugars *N*-acetylglucosamine (NAG) and *N*-acetylmuraminic acid (NAMA). Peptidoglycan strands are cross-linked to one another by short peptide bridges; these bridges are unique in that both D-alanine and L-alanine are present. The interlocked strands form a porous, multilayered grid over the bacterial plasma membrane.

Fortunately, animals have developed natural defenses that can control many bacteria. For example, lysozyme—an enzyme found naturally in tears, saliva, and egg white—hydrolyzes the peptidoglycan cell wall of pathogenic bacteria, thereby killing them. In the middle of the twentieth century the antibiotic penicillin was developed. The penicillin family members all contain a beta-lactam ring that allows these compounds to act as "suicide inhibitors" of the enzymes that synthesize the peptidoglycan cross-linking peptide chain. Penicillin and its relatives target only reproducing bacteria. Mammals do not contain the enzyme pathway that synthesizes peptidoglycans, and this is what allows us to kill the bacteria without harming ourselves.

Today we take the availability and effectiveness of antibiotics for granted. When penicillin was discovered, it was hailed as a "magic bullet" because it could cure bacterial infections that were often fatal. Unfortunately, bacteria have developed resistance to penicillin and its relatives; resistant bacteria have developed enzymes that destroy the beta-lactam

ring, thereby destroying the effectiveness of penicillin. Other antibiotics have since been developed, but the spread of antibiotic-resistant bacterial strains is a public health concern due to the "bullet-proof vest" nature of the bacterial cell wall.

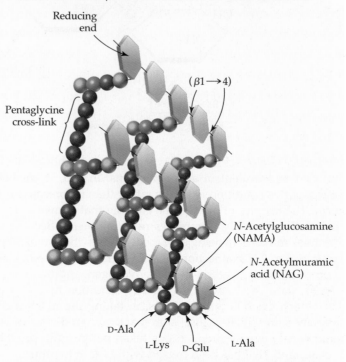

▲ **Peptidoglycan structure:** Strands of alternating NAG and NAMA connected by peptides form a mesh covering the bacterial cell membrane.

Penicillin

where R = 〈benzene〉—CH₂—

or 〈benzene〉—O—CH₂—

or 〈benzene with O—CH₃ substituents〉

See Additional Problems 22.81 through 22.84 at the end of the chapter.

WORKED EXAMPLE   **22.4** Identifying Reducing Sugars

The disaccharide cellobiose can be obtained by enzyme-catalyzed hydrolysis of cellulose. Do you expect cellobiose to be a reducing or a nonreducing sugar?

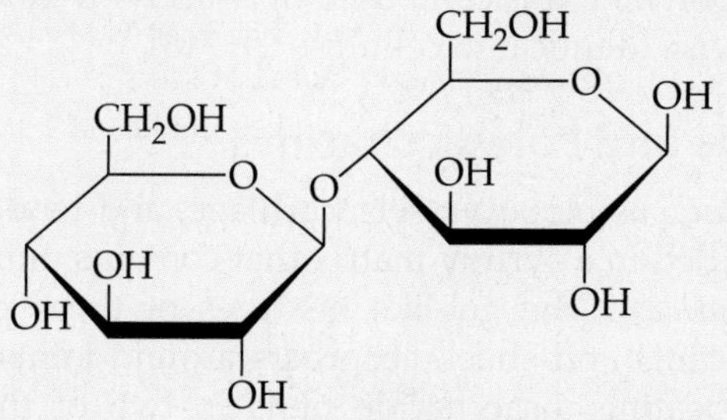

Cellobiose

ANALYSIS To be a reducing sugar, a disaccharide must contain a hemiacetal group, that is, a carbon bonded to one —OH group and one —OR group. The ring at the right in the structure above has such a group.

SOLUTION
Cellobiose is a reducing sugar.

PROBLEM 22.15

Refer to the cellobiose structure in Worked Example 22.4. How would you classify the link between the monosaccharides in cellobiose?

PROBLEM 22.16

Refer to the cellobiose structure in Worked Example 22.4. Show the structures of the two monosaccharides that are formed on hydrolysis of cellobiose. What are their names?

KEY CONCEPT PROBLEM 22.17

Identify the following disaccharides. (a) The disaccharide contains two glucose units joined by an α-glycosidic linkage. (b) The disaccharide contains fructose and glucose. (c) The disaccharide contains galactose and glucose.

22.8 Variations on the Carbohydrate Theme

Monosaccharides with modified functional groups are components of a wide variety of biomolecules. Also, short chains of monosaccharides (known as *oligosaccharides*) enhance the functions of proteins and lipids to which they are bonded.

In this section we mention a few of the more interesting and important variations on the carbohydrate theme, several of which incorporate the modified glucose molecules shown here. Their distinctive functional groups are highlighted in yellow:

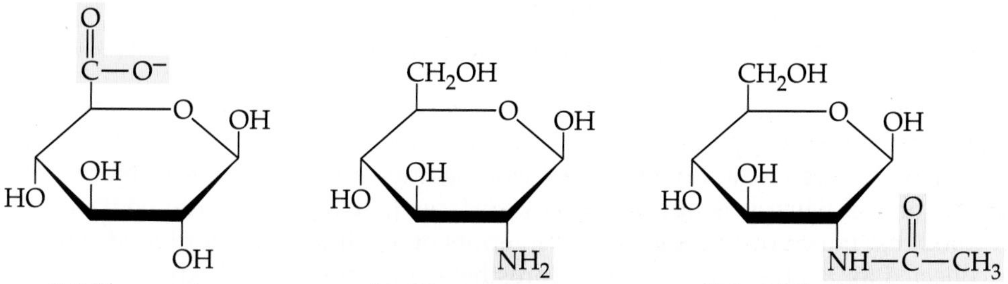

β-D-Glucuronate   β-D-Glucosamine   N-Acetyl-β-D-Glucosamine

▲ A fungus beetle from the Amazon rainforest in its purple-spotted exoskeleton made of chitin.

Chitin

The shells of lobsters, beetles, and spiders are made of chitin, the second most abundant polysaccharide in the natural world. (Cellulose is the most abundant.) Chitin is a hard, structural polymer. It is composed of N-acetyl-D-glucosamine rather than glucose but is otherwise identical to cellulose (p. 718).

Connective Tissue and Polysaccharides

Connective tissues such as blood vessels, cartilage, and tendons are composed of protein fibers embedded in a syrupy matrix that contains unbranched polysaccharides (*mucopolysaccharides*). The gel-like mixtures of these polysaccharides with water serve as lubricants and shock absorbers around joints and in extracellular spaces. Note the repeating disaccharide units in two of these polysaccharides, hyaluronate and chondroitin:

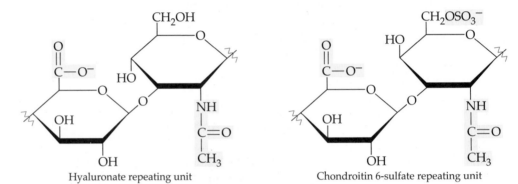

Hyaluronate repeating unit   Chondroitin 6-sulfate repeating unit

Hyaluronate molecules contain up to 25,000 disaccharide units and form a quite rigid, very viscous mixture with water molecules attracted to its negative charges. This mixture is the *synovial fluid* that lubricates joints. It is also present within the eye. *Chondroitin 6-sulfate* (also the 4-sulfate) is present in tendons and cartilage, where it is linked to proteins. It has been used in artificial skin. Chondroitin sulfates and glucosamine sulfate are available as dietary supplements in health food stores and are promoted as cures for osteoarthritis, in which cartilage at joints deteriorates. They are prescribed by veterinarians for arthritic dogs, and there is anecdotal evidence for benefits in humans.

Heparin

Another of the polysaccharides associated with connective tissue, heparin is valuable medically as an *anticoagulant* (an agent that prevents or retards the clotting of blood). Heparin is composed of a variety of different monosaccharides, many of them containing sulfate groups.

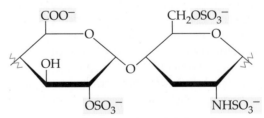

Example of repeating unit in heparin

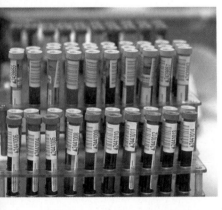

▲ Blood on its way to the clinical lab. The vials with the lavender tops contain heparin and are used for blood destined for a routine hematology screen.

Notice the large number of negative charges in this heparin repeating unit. Heparin binds strongly to a blood-clotting factor and in this way prevents clot formation. It is used clinically to prevent clotting after surgery or serious injury. Also, a coating of heparin is applied to any surfaces that will come into contact with blood that must not clot, such as the interiors of test tubes used for blood samples collected for analysis or materials in prosthetic implants for the body.

Glycoproteins

Proteins that contain short carbohydrate chains (*oligosaccharide* chains) are known as **glycoproteins**. (The prefix *glyco-* always refers to carbohydrates.) The carbohydrate is connected to the protein by a glycosidic bond between an anomeric carbon and a side chain of the protein. The bond is either a C—N glycosidic bond or a C—O glycosidic bond:

Glycoprotein A protein that contains a short carbohydrate chain.

OUTSIDE OF CELL

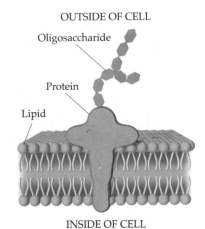

INSIDE OF CELL

Glycoproteins have important functions on the surfaces of all cells. (You might say our cells are sugar-coated.) The protein portion of the molecule lies within the cell membrane, and the hydrophilic carbohydrate portion extends into the surrounding fluid. There, the oligosaccharide chains function as receptors for molecular messengers, other cells, pathogenic microorganisms, or drugs. They are also responsible for the familiar A, B, O system of typing blood. (See the Application "Cell-Surface Carbohydrates and Blood Type," p. 720.)

One unusual glycoprotein, found in the bloodstream and other body fluids of some Antarctic fish species, functions as an antifreeze. This glycoprotein is a polymer with a repeating tripeptide (alanine-alanine-threonine) unit that has a disaccharide (galactosyl-*N*-acetylgalactosamine) bonded to every threonine. The polymer varies in length from 17 to 50 units. It does not protect against freezing by lowering the freezing point, as does the antifreeze used in car radiators; instead the polar groups on the glycoprotein bind with water molecules at the surface of tiny ice crystals, slowing the growth of the crystals. As the blood circulates through the liver, it warms enough for the ice crystals to melt before they harm the organism.

▲ Atlantic codfish, which can survive in frigid water.

⬤⬤ Looking Ahead

The basic components of cell membranes are lipid molecules. The wonderfully complex structure and function of the membrane are explored in Sections 24.5 and 24.6. Glycolipids—carbohydrates bonded to lipids—are, like glycoproteins, essential in cell membranes and are also discussed in Section 24.5. ⬤⬤

PROBLEM 22.18

In *N*-linked glycoproteins, the sugar is usually attached to the protein by a bond to the N atom in a side-chain amide. Which amino acids can form such a bond?

PROBLEM 22.19

Identify the type of glycosidic linkage in the repeating unit of heparin illustrated in this section.

22.9 Some Important Polysaccharides

Polysaccharides are polymers of tens, hundreds, or even many thousands of monosaccharides linked together through glycosidic bonds of the same type as in maltose and lactose. Three of the most important polysaccharides are *cellulose, starch,* and *glycogen*. The repeating units making up cellulose and starch are compared in the following structures:

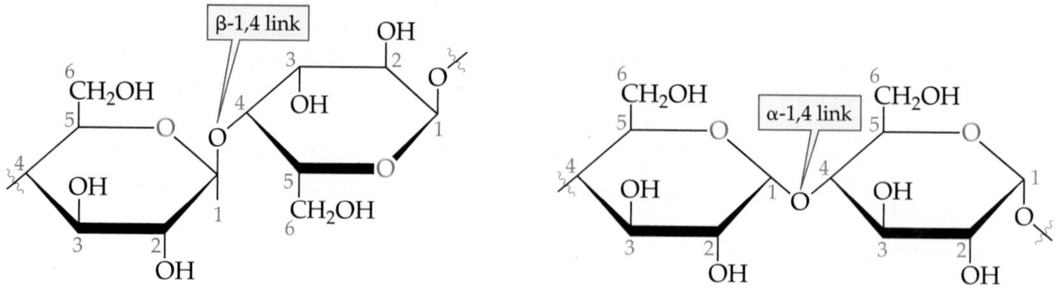

Cellulose repeating unit

Starch and glycogen repeating unit

Cellulose

Cellulose is the fibrous substance that provides structure in plants. Each huge cellulose molecule consists of several thousand β-D-glucose units joined in a long, straight chain by β-1,4 links. The bonding in cellulose is illustrated above by the flat hexagons we have used so far for monosaccharides. In reality, because of the tetrahedral bonding at each carbon atom, the carbohydrate rings are not flat but are bent up at one end and down at the other in what is known as the *chair conformation*:

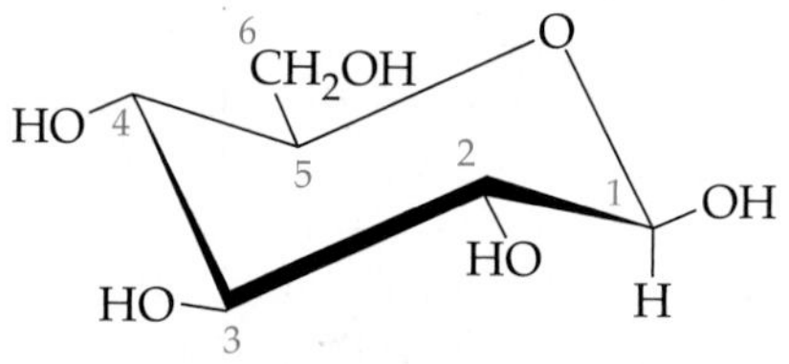

Chair conformation of β-D-glucose

Inspection of the chair conformation shows that the bulkier hydroxyl groups point toward the sides of the ring as does the —CH$_2$OH. This *equatorial* position minimizes interactions between these bulky substituents on the ring. The smaller substituents on the ring (the H atoms) extend either above or below the ring in the *axial* position. This axial/equatorial arrangement resulting from the chair conformation is the most energetically stable form of a six-membered carbohydrate ring.

When the cellulose structure is drawn with all the rings in the chair conformation it is much easier to see how each glucose ring is reversed relative to the next by comparing the locations of the ring O atoms:

Cellulose

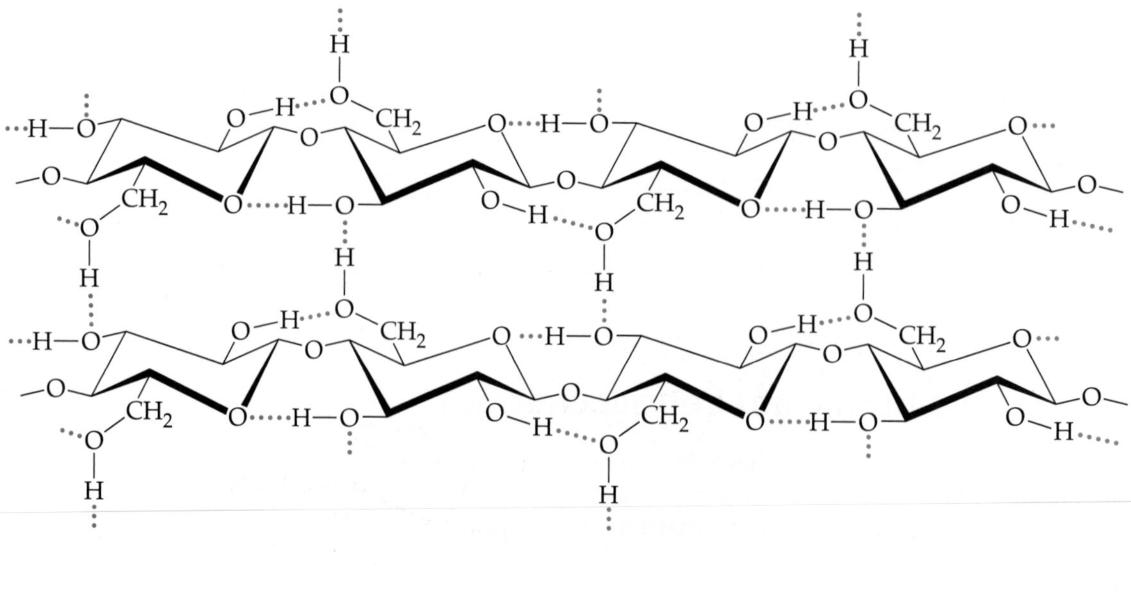

Note in this drawing how the ring O in the top left ring is at the bottom of the ring, the ring O in the ring to the right is at the top of the ring, and so on. The hydrogen bonds within chains and between chains (shown in red) contribute to the rigidity and toughness of cellulose fibers.

Earlier we noted that the seemingly minor distinction between the α and β forms of cyclic sugars accounts for a vast difference between cellulose and starch. Cows and other grazing animals, termites, and moths are able to digest cellulose because microorganisms colonizing their digestive tracts produce enzymes that hydrolyze its β glycosidic bonds. Humans neither produce such enzymes nor harbor such organisms, and therefore cannot hydrolyze cellulose, although some is broken down by bacteria in the large intestine. Cellulose is what grandma used to call "roughage," and we need it in our diets in addition to starch.

Starch

Starch, like cellulose, is a polymer of glucose. In starch, individual glucose units are joined by α-1,4 links rather than by the β-1,4 links of cellulose. Starch is fully digestible and is an essential part of the human diet. It is present only in plant material; our major sources are beans, the grains wheat and rice, and potatoes.

Unlike cellulose, which has only one form, there are two kinds of starch—amylose and amylopectin. *Amylose*, which accounts for about 20% of starch, is somewhat soluble in hot water and consists of several hundred to a thousand α-D-glucose units linked in long chains by the α-1,4 glycosidic bonds. Instead of lying side by side and flat as in cellulose, amylose tends to coil into helices (Figure 22.5). Dissolved amylose makes the cooking water cloudy when you boil potatoes.

▲ **FIGURE 22.5 Helical structure of amylose.**

Amylose

Amylopectin, which accounts for about 80% of starch, is similar to amylose but has much larger molecules (up to 100,000 glucose units per molecule) and has α-1,6 branches approximately every 25 units along its chain. A glucose molecule at one of these branch points (shaded below) is linked to three other sugars. Amylopectin is not water-soluble.

Branch point in amylopectin (also glycogen)

APPLICATION ▶ Cell-Surface Carbohydrates and Blood Type

Nearly 100 years ago, scientists discovered that human blood can be classified into four blood group types, called A, B, AB, and O. This classification results from the presence on red blood cell surfaces of three different oligosaccharide units, designated A, B, and O (see the diagram). Individuals with type AB blood have both A and B oligosaccharides displayed on the same cells.

Selecting a matching blood type is vitally important in choosing blood for transfusions because a major component of the body's immune system (, Chapter 28) is a collection of proteins called *antibodies* that recognize and attack foreign substances, such as viruses, bacteria, potentially harmful macromolecules and foreign blood cells. Among the targets of these antibodies are cell-surface molecules that are not present on the individual's own cells and are thus "foreign blood cells." For example, if you have type A blood, your plasma (the liquid portion of the blood) contains antibodies to the type B oligosaccharide. Thus, if type B blood enters your body, its red blood cells will be recognized as foreign and your immune system will launch an attack on them. The result is clumping of the cells (agglutination), blockage of capillaries, and possibly death.

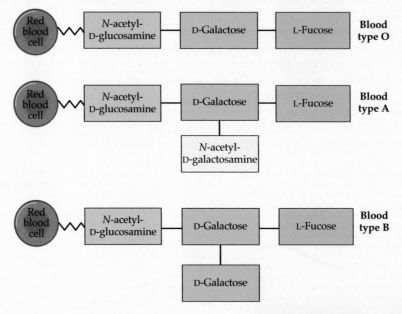

Because of the danger of such interactions, both the blood types that individuals can receive and the blood types of recipients to whom they can donate blood are limited, as indicated in the accompanying table. A few features of the table deserve special mention:

- Note in the diagram that type O cell-surface oligosaccharides are similar in composition to those of types A and B. Consequently, people with blood types A, B, and AB all lack antibodies to type O cells. Individuals with type O blood are therefore known as "universal donors"—in an emergency, their blood can safely be given to individuals of all blood types.

- Similarly, type AB individuals are known as "universal recipients." Because people with type AB blood have both A and B molecules on their red cells, their blood contains no antibodies to A, B, or O, and they can, if necessary, receive blood of all types.

- In theory, antibodies in the plasma of donated blood could also attack the red cells of the recipient. In practice, such reactions are unlikely to cause significant harm. Unless very large quantities of whole blood or

▲ A unit of packed blood cells stored in blood bank refrigeration until needed for transfusion.

plasma (the fluid portion of the blood) are transfused, the donor's blood is quickly diluted by mixing with the much larger volume of the recipient's blood. Moreover, many transfusions today consist of packed red cells, with a minimum of the antibody-containing plasma. Nevertheless, exact matching of blood types is preferred whenever possible.

See Additional Problems 22.85 and 22.86 at the end of the chapter.

INDIVIDUALS WITH BLOOD TYPE . . .	. . . HAVE ANTIBODIES TO TYPE . . .,	. . . CAN RECEIVE FROM TYPE . . .,	. . . AND CAN DONATE TO TYPE
O	A and B	O	O, A, and B*
A	B	O and A	A and AB
B	A	O and B	B and AB
AB	None	O, A and B*	AB

*Red blood cells only

Starch molecules are digested mainly in the small intestine by α-amylase, which catalyzes hydrolysis of the α-1,4 links. As is usually the case in enzyme-catalyzed reactions, α-amylase is highly specific in its action. It hydrolyzes only α acetal links between glucose units (found in starch) and leaves β acetal links (found in cellulose) untouched.

PROBLEM 22.20

An individual starch molecule contains thousands of glucose units but has only a single hemiacetal group at the end of the long polymer chain. Would you expect starch to be a reducing carbohydrate? Explain.

Glycogen

Glycogen, sometimes called *animal starch*, serves the same energy storage role in animals that starch serves in plants. Some of the glucose from starches in our diet is used immediately as fuel, and some is stored as glycogen for later use. The largest amounts of glycogen are stored in the liver and muscles. In the liver, glycogen is a source of glucose, which is formed there when hormones signal a need for glucose in the blood. In muscles, glycogen is converted to glucose 6-phosphate for the synthesis of ATP.

Structurally, glycogen is similar to amylopectin in being a long polymer of α-D-glucose with the same type of branch points in its chain. Glycogen has many more branches than amylopectin, however, and is much larger—up to one million glucose units per molecule.

Comparison of branching in amylopectin and glycogen

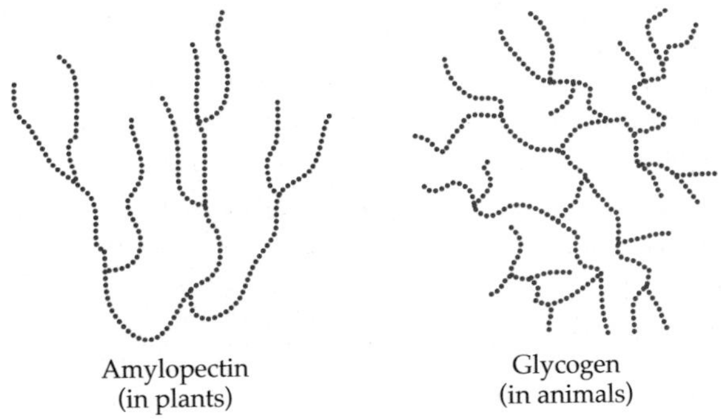

Amylopectin
(in plants)

Glycogen
(in animals)

APPLICATION ▶   Dietary Fiber

Dietary fiber includes cellulose and all other indigestible polysaccharides in vegetables, both soluble and insoluble. The major categories of noncellulose fiber are hemicellulose, pectins and gums, and lignins.

Hemicellulose is a collective term for insoluble plant polysaccharides other than cellulose. These polysaccharides are composed of xylose, mannose, galactose, and modifications of these monosaccharides.

Pectins and vegetables gums, which contain galactose modified by the addition of carboxylic acid and *N*-acetyl groups, comprise the "soluble" portion of dietary fiber. Their outstanding characteristic is solubility in water or the formation of sticky or gelatinous dispersions with water. Pectins, which are present in fruits, are responsible for the "gel" in jelly. Because this texture of their dispersions in water is a desirable characteristic, pectins are often added to prepared foods to retain moisture, thicken sauces, or give a creamier texture.

Lignin, which like cellulose provides rigid structure in plants and especially in trees, is an insoluble dietary fiber. It is not a polysaccharide, however, but a polymer of complex structure that contains phenyl groups connected by carbon–carbon and carbon–oxygen bonds.

Foods high in insoluble fiber include wheat, bran cereals, and brown rice. Beans, peas, and other legumes contain both soluble and insoluble fiber and are high in small polysaccharides that contain galactose bonded to glucose residues. These small polysaccharides are digested by bacteria in the gut, with the production of lactate, short-chain fatty acids, and gaseous by-products including hydrogen, carbon dioxide, and methane.

Fiber functions in the body to soften and add bulk to solid waste. Studies have shown that increased fiber in the diet may reduce the risk of colon and rectal cancer, hemorrhoids, diverticulosis, and cardiovascular disease. A reduction in the risk of developing colon and rectal cancer may also occur because potentially carcinogenic substances are absorbed on fiber surfaces and eliminated before doing any harm. Pectin may also absorb and carry away bile acids, causing an increase in their synthesis from cholesterol in the liver and a resulting decrease in blood cholesterol levels.

▲ Beano, a product that contains α-galactosidase. Beano promises to diminish the production of gas in the large intestine.

The U.S. Food and Drug Administration periodically reviews the MyPyramid meal-planning tool (p. 560) and the related Dietary Guidelines for Americans. In the sixth edition, released in 2005, the guidelines were described as providing "science-based advice to promote health and to reduce risk for chronic diseases through diet and physical activity." A significant change in the new edition is the recommendation "Choose fiber-rich fruits, vegetables and whole grains often." To accomplish this, you should choose foods with ingredients such as these whole grains listed *first* on the ingredients label: brown rice, oatmeal, graham flour, pearl barley, whole oats, whole wheat, or whole rye. (Note that ingredients listed as "wheat flour" and "enriched flour" are not whole grains.) Unlike refined grains, all of the recommended whole grains have a low glycemic index. The new guidelines emphasize eating two cups each of fiber-rich fruits and vegetables daily.

See Additional Problems 22.87 and 22.88 at the end of the chapter.

KEY WORDS

1,4 Link, *p. 712*

Aldose, *p. 693*

Anomeric carbon atom, *p. 702*

Anomers, *p. 702*

Carbohydrate, *p. 693*

D Sugar, *p. 698*

Diastereomers, *p. 696*

Disaccharide, *p. 694*

SUMMARY: REVISITING THE CHAPTER GOALS

1. **What are the different kinds of carbohydrates?** *Monosaccharides* are compounds with three to seven carbons, an aldehyde group on carbon 1 (an *aldose*) or a ketone group on carbon 2 (a *ketose*), and hydroxyl groups on all other carbons. *Disaccharides* consist of two monosaccharides; *polysaccharides* are polymers composed of up to thousands of monosaccharides.

2. **Why are monosaccharides chiral, and how does this influence the numbers and types of their isomers?** Monosaccharides can contain several chiral carbon atoms, each bonded to one —H, one —OH, and two other carbon atoms in the carbon chain. A monosaccharide with n chiral carbon atoms may have 2^n stereoisomers and half that number of pairs of enantiomers. The members of different enantiomeric pairs are *diastereomers*—they are *not* mirror images of each other.

3. What are the structures of monosaccharides, and how are they represented in written formulas? *Fischer projection formulas* represent the open-chain structures of monosaccharides. They are interpreted as shown below, with the D and L enantiomers in a pair identified by having the —OH group on the chiral carbon farthest from the carbonyl group on the right (the D isomer) or the left (the L isomer).

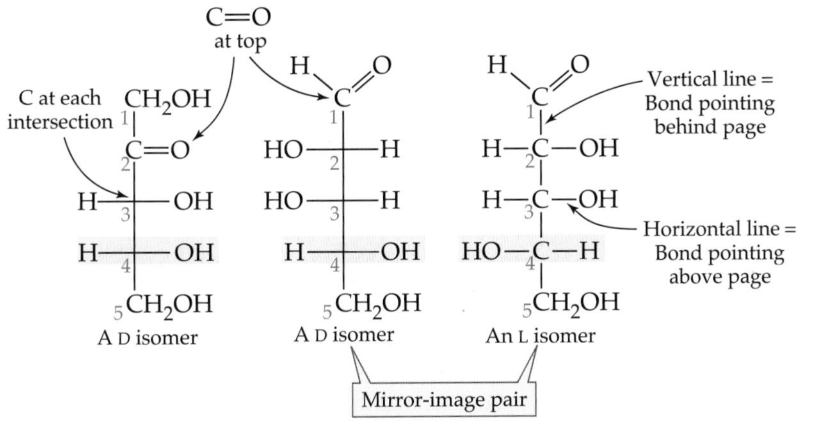
C=O
at top
C at each intersection
Vertical line = Bond pointing behind page
Horizontal line = Bond pointing above page
A D isomer
A D isomer
An L isomer
Mirror-image pair

In solution, open-chain monosaccharides with five or six carbons establish equilibria with cyclic forms that are hemiacetals. The hemiacetal carbon (bonded to two O atoms) is referred to as the *anomeric carbon*, and this carbon is chiral. Two isomers of the cyclic form of a D or L monosaccharide, known as *anomers*, are possible because the —OH on the anomeric carbon may lie above or below the plane of the ring.

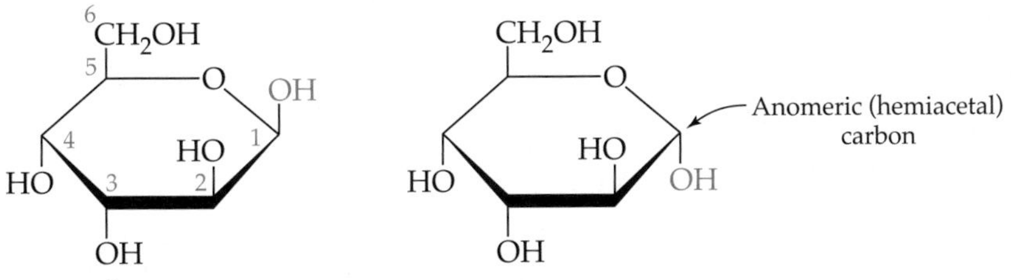
Anomeric (hemiacetal) carbon

β anomer
-OH on C1 on same side as $-CH_2OH$ on C5

α anomer
-OH on C1 on opposite side from $-CH_2OH$ on C5

4. How do monosaccharides react with oxidizing agents and alcohols? Oxidation of a monosaccharide can result in a carboxyl group on the first carbon atom (C1 in the Fischer projection). Ketoses, as well as aldoses, are *reducing sugars* because the ketose is in equilibrium with an aldose form (via an enediol) that can be oxidized.

Reaction of a hemiacetal with an alcohol produces an acetal. For a cyclic monosaccharide, reaction with an alcohol converts the —OH group on the anomeric carbon to an —OR group. The bond to the —OR group, known as a *glycosidic bond*, is α or β to the ring as was the —OH group. Disaccharides result from glycosidic bond formation between two monosaccharides.

5. What are the structures of some important disaccharides? In *maltose*, two D-glucose molecules are joined by an α-glycosidic bond that connects C1 (the anomeric carbon) of one molecule to C4 of the other—an *α-1,4 link*. In *lactose*, D-galactose and D-glucose are joined by a β-1,4 link. In *sucrose*, D-fructose and D-glucose are joined by a glycosidic bond between the two anomeric carbons, a 1,2 link. Unlike maltose and lactose, sucrose is not a *reducing sugar* because it has no hemiacetal that can establish equilibrium with an aldehyde.

6. What are the functions of some important carbohydrates that contain modified monosaccharide structures? *Chitin* is a hard structural polysaccharide found in the shells of lobsters and insects. Joints and intracellular spaces are lubricated by polysaccharides like *hyaluronate* and *chondroitin 6-sulfate*, which have ionic functional groups and form gel-like mixtures with water. *Heparin*, a polysaccharide with many ionized sulfate groups, binds to a clotting factor and thus acts as an anticoagulant. *Glycoproteins* have short carbohydrate chains bonded to proteins; the carbohydrate segments (*oligosaccharides*) function as receptors at cell surfaces.

Fischer projection, *p. 697*

Glycoprotein, *p. 717*

Glycoside, *p. 710*

Glycosidic bond, *p. 710*

Ketose, *p. 693*

L Sugar, *p. 698*

Monosaccharide (simple sugar), *p. 693*

Mutarotation, *p. 702*

Polysaccharide (complex carbohydrate), *p. 694*

Reducing sugar, *p. 709*

7. What are the structures and functions of cellulose, starch, and glycogen? *Cellulose* is a straight-chain polymer of β-D-glucose with β-1,4 links; it provides structure in plants. Cellulose is not digestible by humans, but is digestible by animals whose digestive tract contains bacteria that provide enzymes to hydrolyze the β-glycosidic bonds. *Starch* is a polymer of α-D-glucose connected by α-1,4 links in straight-chain (*amylose*) and branched-chain (*amylopectin*) forms. Starch is a storage form of glucose for plants and is digestible by humans. *Glycogen* is a storage form of glucose for animals, including humans. It is structurally similar to amylopectin, but is more highly branched. Glycogen from meat in the diet is also digestible.

UNDERSTANDING KEY CONCEPTS

22.21 During the digestion of starch from potatoes, the enzyme α-amylase catalyzes the hydrolysis of starch into maltose. Subsequently, the enzyme maltase catalyzes the hydrolysis of maltose into two glucose units. Write a word equation for the enzymatic conversion of starch to glucose. Classify each of the carbohydrates in the equation as a disaccharide, monosaccharide, or polysaccharide.

22.22 Identify the following as diastereomers, enantiomers, and/or anomers. (a) α-D-fructose and β-D-fructose (b) D-galactose and L-galactose (c) L-allose and D-glucose (both aldohexoses)

22.23 Consider the trisaccharide A, B, C at the bottom of this page.

(a) Identify the hemiacetal and acetal linkages.

(b) Identify the anomeric carbons, and indicate whether each is α or β.

(c) State the numbers of the carbons that form glycosidic linkages between monosaccharide A and monosaccharide B.

(d) State the numbers of the carbons that form glycosidic linkages between monosaccharide B and monosaccharide C.

22.24 Hydrolysis of both glycosidic bonds in the trisaccharide A, B, C at the bottom of this page yields three monosaccharides.

(a) Are any two of these monosaccharides the same?

(b) Are any two of these monosaccharides enantiomers?

(c) Draw the Fischer projections for the three monosaccharides.

(d) Assign a name to each monosaccharide.

22.25 The trisaccharide shown with Problem 22.24 has a specific sequence of monosaccharides. To determine this sequence, we could react the trisaccharide with an oxidizing agent. Since one of the monosaccharides in the trisaccharide is a reducing sugar, it would be oxidized from an aldehyde to a carboxylate. Which of the monosaccharides (A, B, or C) is oxidized? Write the structure of the oxidized monosaccharide that results after hydrolysis of the trisaccharide. How does this reaction assist in identifying the sequence of the trisaccharide?

22.26 Are one or more of the disaccharides maltose, lactose, cellobiose, and sucrose part of the trisaccharide in Problem 22.24? If so, identify which disaccharide and its location. (Hint: Look for an α-1,4 link, β-1,4 link, or 1,2 link, and then determine if the correct monosaccharides are present.)

22.27 Cellulose, amylose, amylopectin, and glycogen are the polysaccharides of glucose that we examined in this chapter. The major criteria that distinguish these four polysaccharides include α-glycosidic links or β-glycosidic links, 1,4 links or both 1,4 and 1,6 links, and the degree of branching. Describe each polysaccharide using these five criteria. (Hint: Make a table.)

22.28 In solution, glucose exists predominantly in the cyclic hemiacetal form, which does not contain an aldehyde group. How is it possible for mild oxidizing agents to oxidize glucose?

A   B   C

ADDITIONAL PROBLEMS

CLASSIFICATION AND STRUCTURE OF CARBOHYDRATES

22.29 What is a carbohydrate?

22.30 What is the family-name ending for a sugar?

22.31 What is the structural difference between an aldose and a ketose?

22.32 Classify the following carbohydrates by indicating the nature of the carbonyl group and the number of carbon atoms present. For example, glucose is an aldohexose.

(a)

H   O
 \ //
  C
  |
HO—C—H
  |
H—C—OH
  |
 CH_2OH
Threose

(b)

 CH_2OH
  |
  C=O
  |
H—C—OH
  |
H—C—OH
  |
 CH_2OH
Ribulose

(c)

H   O
 \ //
  C
  |
H—C—OH
  |
HO—C—H
  |
H—C—OH
  |
 CH_2OH
Xylose

(d)

 CH_2OH
  |
  C=O
  |
HO—C—H
  |
HO—C—H
  |
H—C—OH
  |
 CH_2OH
Tagatose

22.33 How many chiral carbon atoms are present in each of the molecules shown in Problem 22.32?

22.34 How many chiral carbon atoms are there in each of the two parts of the repeating unit in heparin (p. 716)? What is the total number of chiral carbon atoms in the repeating unit?

22.35 Draw the open-chain structure of a ketoheptose.

22.36 Draw the open-chain structure of a 4-carbon deoxy sugar.

22.37 Name four important monosaccharides and tell where each occurs in nature.

22.38 Name a common use for each monosaccharide listed in Problem 22.37.

HANDEDNESS IN CARBOHYDRATES

22.39 How are enantiomers related to each other?

22.40 What is the structural relationship between L-glucose and D-glucose?

22.41 Only three stereoisomers are possible for 2,3-dibromo-2,3-dichlorobutane. Draw them, indicating which pair are enantiomers (optical isomers). Why does the other isomer not have an enantiomer?

22.42 In Section 16.6 you saw that aldehydes react with reducing agents to yield primary alcohols (RCH=O → RCH_2OH). The structures of two D-aldotetroses are shown. One of them can be reduced to yield a chiral product, but the other yields an achiral product. Explain.

H   O
 \ //
  C
  |
H—C—OH
  |
H—C—OH
  |
 CH_2OH
D-Erythrose

H   O
 \ //
  C
  |
HO—C—H
  |
H—C—OH
  |
 CH_2OH
D-Threose

22.43 What is the definition of an optically active compound?

22.44 What does a polarimeter measure?

22.45 Sucrose and D-glucose rotate plane-polarized light to the right; D-fructose rotates light to the left. When sucrose is hydrolyzed, the glucose–fructose mixture rotates light to the left.

  (a) What does this indicate about the relative degrees of rotation of light of glucose and fructose?

  (b) Why do you think the mixture is called "invert sugar"?

22.46 What generalization can you make about the direction and degree of rotation of light by enantiomers?

REACTIONS OF CARBOHYDRATES

22.47 What does the term *reducing sugar* mean?

22.48 What structural property makes a sugar a reducing sugar?

22.49 What is mutarotation? Do all chiral molecules do this?

22.50 What are anomers, and how do the anomers of a given sugar differ from each other?

22.51 What is the structural difference between the α hemiacetal form of a carbohydrate and the β form?

22.52 D-Gulose, an aldohexose isomer of glucose, has the cyclic structure shown here. Which is shown, the α form or the β form?

D-Gulose

22.53 In its open-chain form, D-mannose, an aldohexose found in orange peels, has the structure shown here. Coil mannose around and draw it in the cyclic hemiacetal α and β forms.

H   H   H   OH OH O
| | | | | ‖
HO—C—C—C—C—C—C—H
| | | | |
H   OH OH H   H
D-Mannose

22.54 In its open-chain form, D-altrose has the structure shown here. Coil altrose around and draw it in the cyclic hemiacetal α and β forms.

D-Altrose

22.55 Draw D-gulose (Problem 22.52) in its open-chain aldehyde form, both coiled and uncoiled.

22.56 D-Ribulose, a ketopentose related to ribose, has the following structure in open-chain form. Coil ribulose around and draw it in its five-membered cyclic β hemiacetal form.

D-Ribulose

22.57 D-Allose, an aldohexose, is identical to D-glucose except that the hydroxyl group at C3 points down rather than up in the cyclic hemiacetal form. Draw the β form of this cyclic form of D-allose.

22.58 Draw D-allose (Problem 22.57) in its open-chain form.

22.59 Treatment of D-glucose with a reducing agent yields sorbitol, a substance used as a sugar substitute by diabetics. Draw the structure of sorbitol.

22.60 Reduction of D-fructose with a reducing agent yields a mixture of D-sorbitol along with a second, isomeric product. What is the structure of the second product?

22.61 Treatment of an aldose with an oxidizing agent such as Tollens' reagent (Section 16.5) yields a carboxylic acid. Gluconic acid, the product of glucose oxidation, is used as its magnesium salt for the treatment of magnesium deficiency. Draw the structure of gluconic acid.

22.62 Oxidation of the aldehyde group of ribose yields a carboxylic acid. Draw the structure of ribonic acid.

22.63 What is the structural difference between a hemiacetal and an acetal?

22.64 What are glycosides, and how can they be formed?

22.65 Look at the structure of D-mannose (Problem 22.53) and draw the two glycosidic products that you expect to obtain by reacting D-mannose with methanol.

22.66 Draw a disaccharide of two mannose molecules attached by an α-1,4 glycosidic linkage. Explain why the glycosidic products in Problem 22.65 are *not* reducing sugars, but the product in this problem *is* a reducing sugar.

DISACCHARIDES AND POLYSACCHARIDES

22.67 Give the names of three important disaccharides. Tell where each occurs in nature. From which two monosaccharides is each made?

22.68 Lactose and maltose are reducing disaccharides, but sucrose is a nonreducing disaccharide. Explain.

22.69 Amylose (a form of starch) and cellulose are both polymers of glucose. What is the main structural difference between them? What roles do these two polymers have in nature?

22.70 How are amylose and amylopectin similar to each other, and how are they different from each other?

22.71 *Gentiobiose*, a rare disaccharide found in saffron, has the following structure. What simple sugars do you obtain on hydrolysis of gentiobiose?

Gentiobiose

22.72 Does gentiobiose (Problem 22.71) have an acetal grouping? A hemiacetal grouping? Do you expect gentiobiose to be a reducing or nonreducing sugar? How would you classify the linkage (α or β and carbon numbers) between the two monosaccharides?

22.73 *Trehalose*, a disaccharide found in the blood of insects, has the following structure. What simple sugars would you obtain on hydrolysis of trehalose?

Trehalose

22.74 Does trehalose (Problem 22.73) have an acetal grouping? A hemiacetal grouping? Do you expect trehalose to be a reducing or nonreducing sugar? Classify the linkage between the two monosaccharides.

22.75 Amylopectin (a form of starch) and glycogen are both α-linked polymers of glucose. What is the structural difference between them? What roles do they serve in nature?

22.76 *Amygdalin* (Laetrile) is a glycoside isolated in 1830 from almond and apricot seeds. It is called a cyanogenic glycoside because hydrolysis with aqueous acid liberates hydrogen cyanide (HCN) along with benzaldehyde and two molecules of glucose. Structurally, amygdalin is a glycoside composed of gentiobiose (Problem 22.71) and an alcohol, mandelonitrile. Draw the structure of amygdalin.

Mandelonitrile

Applications

22.77 Give some advantages and disadvantages of synthesizing and marketing a single-enantiomer drug. [*Chirality and Drugs, p. 699*]

22.78 What is the advantage of using enzymes in the synthesis of single-enantiomer drugs? [*Chirality and Drugs, p. 699*]

22.79 Carbohydrates provide 4 kcal per gram. If a person eats 200 g per day of digestible carbohydrates, what percentage of a 2000 kcal daily diet would be digestible carbohydrate? [*Carbohydrates in the Diet, p. 704*]

22.80 Give an example of a complex carbohydrate in the diet and a simple carbohydrate in the diet. Are soluble fiber and insoluble fiber complex or simple carbohydrates? [*Carbohydrates in the Diet, p. 704*]

22.81 List three functions of all cell walls. [*Cell Walls: Rigid Defense Systems, p. 714*]

22.82 Name the monomeric unit and the polymer that makes up most of a plant cell wall. [*Cell Walls: Rigid Defense Systems, p. 714*]

22.83 Name the individual units and the crosslink for the polymer that makes up most of a bacterial cell wall. [*Cell Walls: Rigid Defense Systems, p. 714*]

22.84 When you take the antibiotic penicillin when you are ill, why does the penicillin kill a bacterial cell but not your liver cells? [*Cell Walls: Rigid Defense Systems, p. 714*]

22.85 Look at the structures of the blood group determinants. What groups do all blood types have in common? [*Cell-Surface Carbohydrates and Blood Type, p. 720*]

22.86 People with type O blood can donate blood to anyone, but they cannot receive blood from everyone. From whom can they not receive blood? People with type AB blood can receive blood from anyone, but they cannot give blood to everyone. To whom can they give blood? Why? [*Cell-Surface Carbohydrates and Blood Type, p. 720*]

22.87 Our bodies do not have the enzymes required to digest cellulose, yet it is a necessary addition to a healthy diet. Why? [*Dietary Fiber, p. 722*]

22.88 Name two types of soluble fiber and their sources. [*Dietary Fiber, p. 722*]

General Questions and Problems

22.89 What is the relationship between D-ribose (p. 700) and D-xylose (Problem 22.32c)? What generalizations can you make about D-ribose and D-xylose with respect to the following?

(a) Melting point
(b) Rotation of plane-polarized light
(c) Density
(d) Solubility in water
(e) Chemical reactivity

22.90 What is the relationship between D-ribose and L-ribose? What generalizations can you make about D-ribose and L-ribose with respect to the following?

(a) Melting point
(b) Rotation of plane-polarized light
(c) Density
(d) Solubility in water
(e) Chemical reactivity

22.91 Are the α and β forms of monosaccharides enantiomers of each other? Why or why not?

22.92 Are the α and β forms of the disaccharide lactose enantiomers of each other? Why or why not?

22.93 L-Sorbose, which is used in the commercial production of vitamin C, differs from D-fructose only at carbon 5. Draw the open-chain structure of D-sorbose and the five-membered ring in the β form.

22.94 D-Fructose can form a six-membered cyclic hemiacetal as well as the more prevalent five-membered cyclic form. Draw the α isomer of D-fructose in the six-membered ring.

22.95 *Raffinose*, found in sugar beets, is the most prevalent trisaccharide. It is formed by an α-1,6 linkage of D-galactose to the glucose portion of sucrose. Draw the structure of raffinose.

22.96 Does raffinose (Problem 22.95) have a hemiacetal grouping? An acetal grouping? Is raffinose a reducing sugar?

22.97 When you chew a cracker for several minutes, it begins to taste sweet. What do you think the saliva in your mouth does to the starch in the cracker?

22.98 Write the open-chain structure of the only ketotriose. Name this compound and explain why it has no optical isomers.

22.99 Write the open-chain structure of the only ketotetrose. Name this compound. Does it have an optical isomer?

22.100 What is lactose intolerance, and what are its symptoms?

22.101 Many people who are lactose intolerant can eat yogurt, which is prepared from milk curdled by bacteria, with no problems. Give a reason why this is possible.

22.102 What is the group of disorders that result when the body lacks an enzyme necessary to digest galactose? What are the symptoms?

22.103 When a person cannot digest galactose, its reduced form, called dulcitol, often accumulates in the blood and tissues. Write the structure of the open-chain form of dulcitol. Does dulcitol have an enantiomer? Why or why not?

22.104 L-Fucose is also known as 6-deoxy-α-L-galactose. How many chiral carbons are present in L-fucose?

CHAPTER 23

Carbohydrate Metabolism

CONCEPTS TO REVIEW

Phosphorylation
(Section 17.8)

Function of ATP
(Sections 21.5, 21.9)

Oxidized and reduced
coenzymes
(Section 21.7)

Carbohydrate structure
(Chapter 22)

▲ The complex carbohydrates in this meal provide fuel for metabolism.

CONTENTS

CHAPTER GOALS

In this chapter, we will answer the following questions about carbohydrate metabolism:

1. What happens during digestion of carbohydrates?

THE GOAL: Be able to describe carbohydrate digestion, its location, the enzymes involved, and name the major products of this process.

2. What are the major pathways in the metabolism of glucose?

THE GOAL: Be able to identify the pathways by which glucose is (1) synthesized and (2) broken down, and describe their interrelationships.

3. What is glycolysis?

THE GOAL: Be able to give an overview of the glycolysis pathway and its products, and to identify where the major monosaccharides enter the pathway.

4. What happens to pyruvate once it is formed?

THE GOAL: Be able to describe the pathways involving pyruvate and their respective outcomes.

5. How is glucose metabolism regulated, and what are the influences of starvation and diabetes mellitus?

THE GOAL: Be able to identify the hormones that influence glucose metabolism and describe the changes in metabolism during starvation and diabetes mellitus.

6. What are glycogenesis and glycogenolysis?

THE GOAL: Be able to define these pathways and their purpose.

7. What is the role of gluconeogenesis in metabolism?

THE GOAL: Be able to identify the functions, substrates, and products of this pathway.

The story of carbohydrate metabolism is essentially the story of glucose: how it is converted to acetyl-SCoA for entrance into the citric acid cycle, how it is stored and then released, and how it is synthesized when carbohydrates are in short supply. Because of the importance of glucose, the body has several alternative strategies for maintaining an even glucose concentration in blood and providing glucose to cells that depend on it.

Digestion A general term for the breakdown of food into small molecules.

23.1 Digestion of Carbohydrates

The first stage in catabolism is **digestion**, the breakdown of food into small molecules. Digestion entails the physical grinding, softening, and mixing of food, as well as the enzyme-catalyzed hydrolysis of carbohydrates, proteins, and fats. Digestion begins in the mouth, continues in the stomach, and concludes in the small intestine.

The products of digestion are mostly small molecules that are absorbed from the intestinal tract. The absorption happens through millions of tiny projections (the *villi*) that provide a total surface area as big as a football field. Once in the bloodstream, the small molecules are transported into target cells where many are further broken down for the purpose of releasing energy as their carbon atoms are converted to carbon dioxide. Others are excreted, and some are used as building blocks to synthesize new biomolecules.

The digestion of carbohydrates is summarized in Figure 23.1. α-Amylase present in saliva catalyzes hydrolysis of the α glycosidic bonds in the carbohydrates amylose and amylopectin—starch (⚭, p. 721). Starch from plants, but not cellulose, and glycogen from meat are hydrolyzed to give smaller polysaccharides and the disaccharide maltose. Salivary α-amylase continues to act on dietary polysaccharides in the stomach until it is inactivated by stomach acid. No further carbohydrate digestion takes place in the stomach.

α-Amylase is also secreted by the pancreas and enters the small intestine, where conversion of polysaccharides to maltose continues. Other enzymes secreted from the mucous lining of the small intestine hydrolyze maltose and the dietary disaccharides sucrose and lactose to the monosaccharides glucose, fructose, and galactose, which are then transported across the intestinal wall into the bloodstream. Our focus in this chapter is on the metabolism of glucose; both fructose and

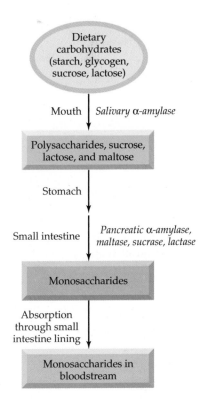

▲ **FIGURE 23.1** The digestion of carbohydrates.

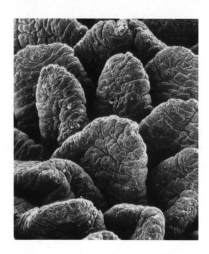

▲ A micrograph showing *villi*, the projections that line the small intestine. Each villus is covered with microvilli, where the digested food molecules are absorbed into the bloodstream.

galactose can be converted to intermediates that enter the same metabolic pathway followed by glucose.

23.2 Glucose Metabolism: An Overview

Glucose is the major fuel for your body. It yields the energy carried by ATP. The initial metabolic fate of glucose is conversion into pyruvate and then usually to acetyl-SCoA, the common intermediate in the catabolism of all foods. The acetyl-SCoA proceeds down the central pathway of metabolism to the ultimate formation of ATP.

In this chapter, we examine the central position of glucose in metabolism, as summarized in Figure 23.2. Glycolysis is the major catabolic pathway leading to ATP synthesis. The reverse of glycolysis (gluconeogenesis) and the pathways leading to and from glycogen are discussed in the sections of this chapter noted in the figure. As you read this chapter, you will find it helpful to use Figure 23.2 and its accompanying Table to sort out the pathways that have such similar names.

When glucose enters a cell from the bloodstream, it is immediately converted to glucose 6-phosphate. Once this phosphate is formed, glucose is trapped within the cell because phosphorylated molecules cannot cross the cell membrane. Like the first step in many metabolic pathways, the formation of glucose-6-phosphate is highly exergonic and not reversible in the glycolytic pathway, thereby committing the initial substrate to subsequent reactions.

Several pathways are available to the glucose 6-phosphate:

- When energy is needed, glucose 6-phosphate moves down the central catabolic pathway shown in light brown in Figure 23.2, proceeding via the reactions of *glycolysis* to pyruvate and then to acetyl-S-coenzyme A, which enters the citric acid cycle. (⬤⬤, Section 21.8)

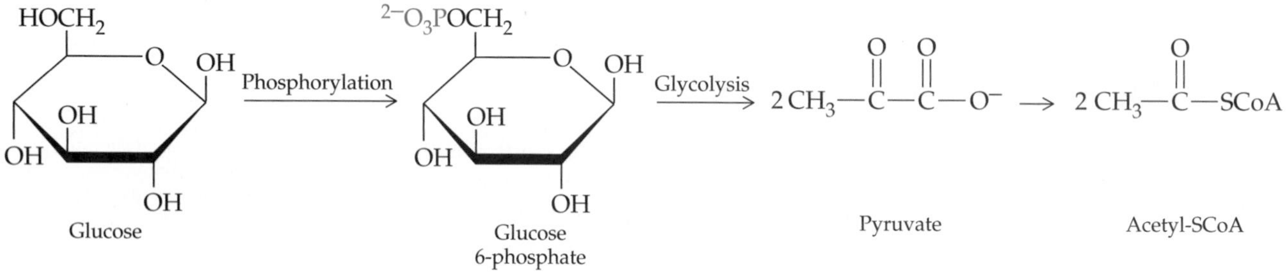

Glucose $\xrightarrow{\text{Phosphorylation}}$ Glucose 6-phosphate $\xrightarrow{\text{Glycolysis}}$ $2\ CH_3-\overset{O}{\underset{}{C}}-\overset{O}{\underset{}{C}}-O^-$ $\rightarrow$ $2\ CH_3-\overset{O}{\underset{}{C}}-SCoA$

Pyruvate   Acetyl-SCoA

- When cells are already well supplied with glucose, the excess glucose is converted to other forms for storage: to glycogen, the glucose storage polymer, by the *glycogenesis* pathway, or to fatty acids by entrance of acetyl-SCoA into the pathways of lipid metabolism (Chapter 25) rather than the citric acid cycle.

Pentose phosphate pathway The biochemical pathway that produces ribose (a pentose), NADPH, and other sugar phosphates from glucose; an alternative to glycolysis.

- Glucose-6-phosphate can also enter the **pentose phosphate pathway**. This multistep pathway yields two products of importance to our metabolism. One is a supply of the coenzyme NADPH, a reducing agent that is essential for various biochemical reactions. The other is ribose 5-phosphate, which is necessary for the synthesis of nucleic acids (DNA and RNA). Glucose-6-phosphate enters the pentose phosphate pathway when a cell's need for NADPH or ribose-5-phosphate exceeds its need for ATP.

PROBLEM 23.1

Name the following pathways:

(a) Pathway for synthesis of glycogen

(b) Pathway for release of glucose from glycogen

(c) Pathway for synthesis of glucose from lactate

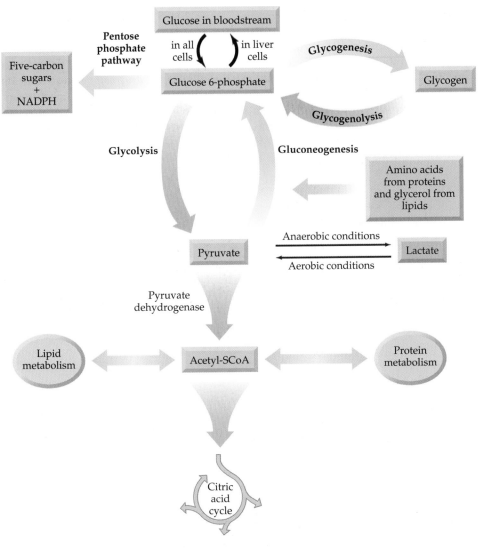

Metabolic Pathways of Glucose

NAME	DERIVATION OF NAME	FUNCTION
Glycolysis (Section 23.3)	*glyco*-, glucose (from Greek, meaning "sweet") *-lysis*, decomposition	Conversion of glucose to pyruvate
Gluconeogenesis (Section 23.11)	*gluco*-, glucose *-neo*-, new *-genesis*, creation	Synthesis of glucose from amino acids, pyruvate, and other noncarbohydrates
Glycogenesis (Section 23.10)	*glyco(gen)*-, glycogen *-genesis*, creation	Synthesis of glycogen from glucose
Glycogenolysis (Section 23.10)	*glycogen*-, glycogen *-lysis*, decomposition	Breakdown of glycogen to glucose
Pentose phosphate pathway (Section 23.2)	*pentose*-, a five-carbon sugar *phosphate*	Conversion of glucose to five-carbon sugar phosphates

◀ **FIGURE 23.2 Glucose metabolism.** Synthetic pathways (anabolism) are shown in blue, pathways that break down biomolecules (catabolism) are shown in light brown, and connections to lipid and protein metabolism are shown in green.

PROBLEM 23.2

Name the synthetic pathways that have glucose 6-phosphate as their first reactant.

23.3 Glycolysis

Glycolysis The biochemical pathway that breaks down a molecule of glucose into two molecules of pyruvate plus energy.

Glycolysis is a series of 10 enzyme-catalyzed reactions that breaks down each glucose molecule into two pyruvate molecules, and in the process yields two ATPs and two NADHs. The steps of glycolysis (also called the *Embden–Meyerhoff pathway* after its co-discoverers) are summarized in Figure 23.3, where the reactions and structures of intermediates should be noted as you read the following paragraphs. Almost all organisms carry out glycolysis; in humans it occurs in the cytosol of all cells.

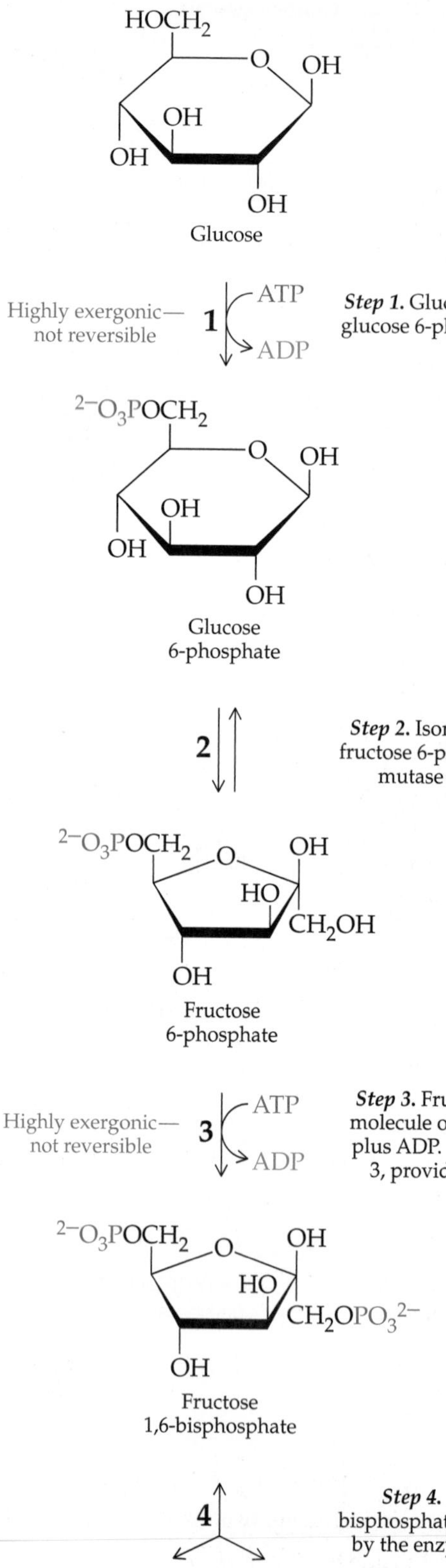

HOCH$_2$

OH

OH

O OH

OH

Glucose

Highly exergonic—
not reversible

1

ATP

ADP

Step 1. Glucose undergoes reaction with ATP to yield glucose 6-phosphate plus ADP in a reaction catalyzed by *hexokinase*.

$^{2-}$O$_3$POCH$_2$

OH

OH

O OH

OH

Glucose
6-phosphate

2

Step 2. Isomerization of glucose 6-phosphate yields fructose 6-phosphate. The reaction is catalyzed by the mutase enzyme, *glucose 6-phosphate isomerase*.

$^{2-}$O$_3$POCH$_2$

O OH

HO

CH$_2$OH

OH

Fructose
6-phosphate

Highly exergonic—
not reversible

3

ATP

ADP

Step 3. Fructose 6-phosphate reacts with a second molecule of ATP to yield fructose 1,6-bisphosphate plus ADP. *Phosphofructokinase*, the enzyme for step 3, provides a major control point in glycolysis.

$^{2-}$O$_3$POCH$_2$

O OH

HO

CH$_2$OPO$_3$$^{2-}$

OH

Fructose
1,6-bisphosphate

▶ **FIGURE 23.3 The glycolysis pathway for converting glucose to pyruvate.**

4

Step 4. The six-carbon chain of fructose 1,6-bisphosphate is cleaved into two three-carbon pieces by the enzyme *aldolase*. (Continued on next page.)

STEPS 1–3 of Glycolysis: Phosphorylation Glucose is carried in blood to cells where it is transported across the cell membrane. As soon as it enters the cell, glucose is phosphorylated in *Step 1* of glycolysis, which requires an energy investment from ATP. (Phosphorylation is addition of a —PO_3^{2-} group. (⊂⊃, Section 17.8) This is the first of three highly exergonic and irreversible steps in glycolysis. From here on, all intermediates are sugar phosphates and are trapped within the cells because phosphates cannot cross cell membranes.

The product of Step 1, glucose 6-phosphate, is an allosteric inhibitor for the enzyme for this step (*hexokinase*), and therefore plays an important role in the elaborate and delicate control of glucose metabolism. (⊂⊃, Figure 19.3, p. 601)

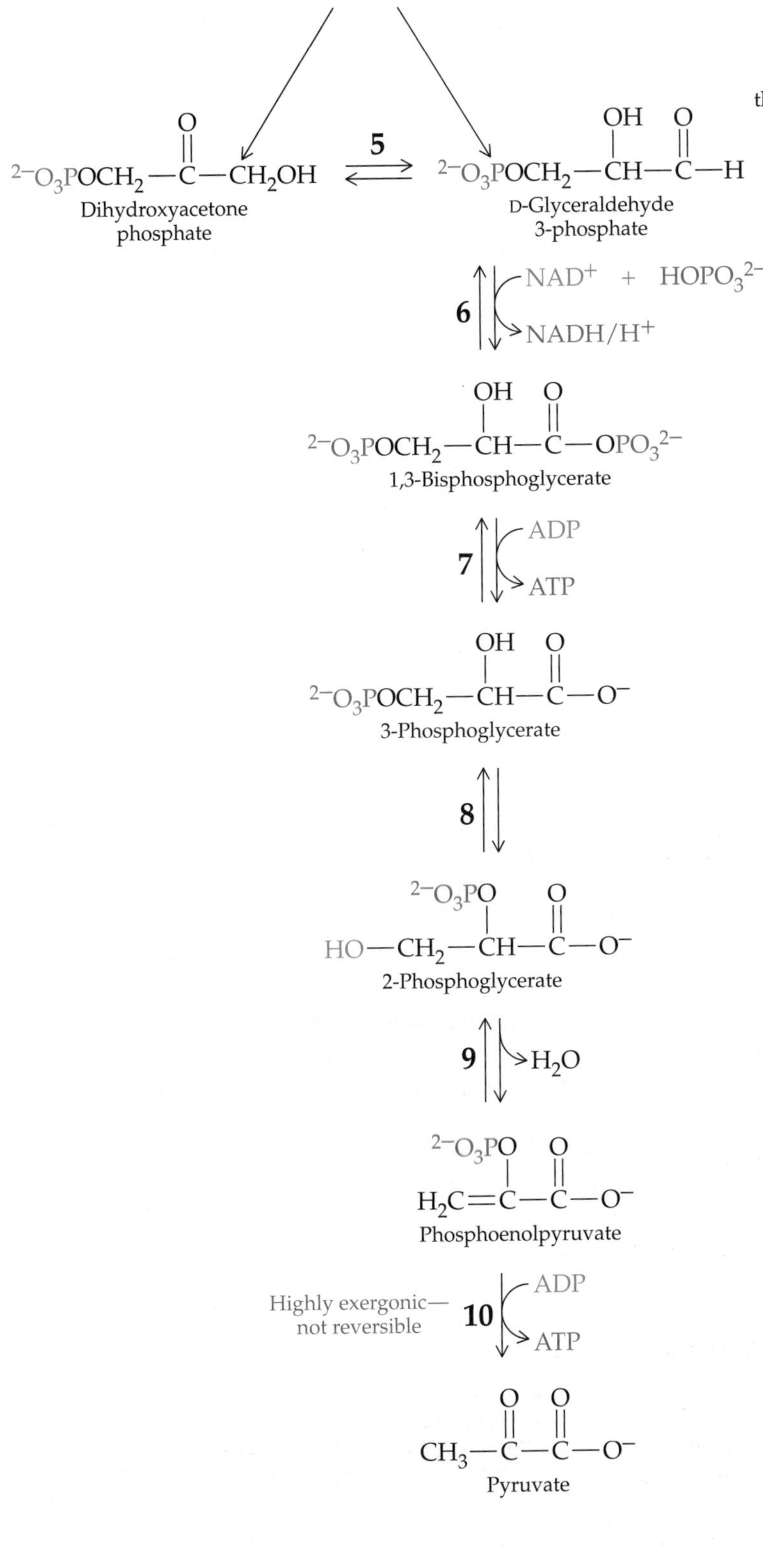

$$^{2-}O_3POCH_2-\overset{\overset{\displaystyle O}{\|}}{C}-CH_2OH \quad \underset{\longleftarrow}{\overset{5}{\longrightarrow}} \quad ^{2-}O_3POCH_2-\overset{\overset{\displaystyle OH}{|}}{CH}-\overset{\overset{\displaystyle O}{\|}}{C}-H$$

Dihydroxyacetone phosphate   D-Glyceraldehyde 3-phosphate

Step 5. The two products of step 4 are both three-carbon sugars, but only glyceraldehyde 3-phosphate can continue in the glycolysis pathway. Dihydroxyacetone phosphate must first be isomerized by the enzyme *triose phosphate isomerase*.

6 NAD^+ + $HOPO_3^{2-}$ → $NADH/H^+$

Step 6. Two reactions occur as glyceraldehyde 3-phosphate is first oxidized to a carboxylic acid and then phosphorylated by the enzyme *glyceraldehyde 3-phosphate dehydrogenase*. The coenzyme nicotinamide adenine dinucleotide (NAD^+) and inorganic phosphate ion ($HOPO_3^{2-}$) are required.

$$^{2-}O_3POCH_2-\overset{\overset{\displaystyle OH}{|}}{CH}-\overset{\overset{\displaystyle O}{\|}}{C}-OPO_3^{2-}$$

1,3-Bisphosphoglycerate

7   ADP → ATP

Step 7. A phosphate group from 1,3-bisphosphoglycerate is transferred to ADP, resulting in synthesis of ATP, and catalyzed by *phosphoglycerate kinase*.

$$^{2-}O_3POCH_2-\overset{\overset{\displaystyle OH}{|}}{CH}-\overset{\overset{\displaystyle O}{\|}}{C}-O^-$$

3-Phosphoglycerate

8

Step 8. A phosphate group is next transferred from carbon 3 to carbon 2 of phosphoglycerate in a step catalyzed by the enzyme *phosphoglycerate mutase*.

$$HO-CH_2-\overset{\overset{\displaystyle ^{2-}O_3PO}{|}}{CH}-\overset{\overset{\displaystyle O}{\|}}{C}-O^-$$

2-Phosphoglycerate

9 → H_2O

Step 9. Loss of water from 2-phosphoglycerate produces phosphoenolpyruvate (PEP). The dehydration is catalyzed by the enzyme *enolase*.

$$H_2C=\overset{\overset{\displaystyle ^{2-}O_3PO}{|}}{C}-\overset{\overset{\displaystyle O}{\|}}{C}-O^-$$

Phosphoenolpyruvate

Highly exergonic— not reversible   **10**   ADP → ATP

Step 10. Transfer of the phosphate group from phosphoenolpyruvate to ADP yields pyruvate and generates ATP, catalyzed by *pyruvate kinase*.

$$CH_3-\overset{\overset{\displaystyle O}{\|}}{C}-\overset{\overset{\displaystyle O}{\|}}{C}-O^-$$

Pyruvate

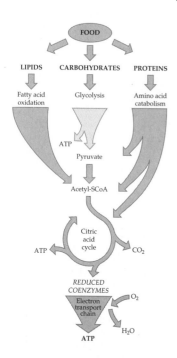

STEP 2 **is the isomerization of glucose 6-phosphate to fructose 6-phosphate.** The enzyme (*glucose 6-phosphate isomerase*) acts by converting glucose 6-phosphate (an aldohexose) to fructose 6-phosphate (a ketohexose). The result is conversion of the six-membered glucose ring to a five-membered ring with a —CH$_2$OH group, which prepares the molecule for addition of another phosphoryl group in the next step.

STEP 3 **makes a second energy investment as fructose 6-phosphate is converted to fructose 1,6-bisphosphate by reaction with ATP in another exergonic reaction.** (*Bis-* means "two"—that is, fructose with two phosphate groups. The "bis" prefix is used to distinguish between a molecule containing two phosphate groups in different locations and a "diphosphate"—a compound that contains a single diphosphate group, —OP$_2$O$_6^{4-}$.) (▭, Section 17.8) Step 3 is another major control point for glycolysis. When the cell is short of energy, ADP and AMP (adenosine monophosphate) concentrations build up and activate the Step 3 enzyme, *phosphofructokinase*. When energy is in good supply, ATP and citrate build up and allosterically inhibit the enzyme. The outcome of Steps 1–3 is the formation of a molecule ready to be split into the 2-carbon intermediates that will ultimately become two molecules of pyruvate.

STEPS 4 and 5 of Glycolysis: Cleavage and Isomerization *Step 4* converts the 6-carbon bisphosphate from Step 3 into two 3-carbon monophosphates, one an aldose phosphate and one a ketose phosphate. Aldolase catalyzes the breakage of the bond between carbons 3 and 4 in fructose 1,6-bisphosphate, and a C=O group is formed.

$$
\begin{array}{ccc}
\overset{1}{\text{C}}\text{H}_2\text{OPO}_3{}^{2-} & & \overset{}{\text{C}}\text{H}_2\text{OPO}_3{}^{2-} \\
| & & | \\
\overset{2}{\text{C}}{=}\text{O} & & \text{C}{=}\text{O} \\
| & & | \\
\text{HO}{-}\overset{3}{\text{C}}{-}\text{H} & \xrightarrow{\text{Aldolase}} & \text{CH}_2\text{OH} \\
| & \rightleftharpoons & \\
\text{H}{-}\overset{4}{\text{C}}{-}\text{OH} & & + \\
| & & \\
\text{H}{-}\overset{5}{\text{C}}{-}\text{OH} & & \\
| & & \\
\overset{6}{\text{C}}\text{H}_2\text{OPO}_3{}^{2-} & & \\
\end{array}
$$

Dihydroxyacetone phosphate

$$
\begin{array}{c}
\text{H}{-}\overset{}{\text{C}}{\nwarrow}{}^{\text{O}} \\
| \\
\text{H}{-}\text{C}{-}\text{OH} \\
| \\
\text{CH}_2\text{OPO}_3{}^{2-}
\end{array}
$$

D-Glyceraldehyde 3-phosphate

Fructose 1,6-bisphosphate

The two 3-carbon sugar phosphates produced in Step 4 are isomers that are interconvertible in an aldose–ketose equilibrium (*Step 5* in Figure 23.3) catalyzed by triose phosphate isomerase. Only glyceraldehyde 3-phosphate can continue on the glycolysis pathway, however. As the glyceraldehyde 3-phosphate reacts in Step 6, the equilibrium of Step 5 shifts to the right. The overall result of Steps 4 and 5 is therefore the production of *two* molecules of glyceraldehyde 3-phosphate.

Steps 1–5 are referred to as the *energy investment* part of glycolysis. So far, two ATPs have been invested and no income earned, but the stage is now set for a small energy profit. Note that since one glucose molecule gives two glyceraldehyde 3-phosphates that pass separately down the rest of the pathway, Steps 6–10 of glycolysis each take place twice for every glucose molecule that enters at Step 1.

STEPS 6–10 of Glycolysis: Energy Generation The second half of glycolysis is devoted to generating molecules with phosphate groups that can be transferred to ATP.

Step 6 is the oxidation of glyceraldehyde 3-phosphate to 1,3-bisphosphoglycerate by glyceraldehyde 3-phosphate dehydrogenase. The enzyme cofactor NAD$^+$ is the oxidizing agent. Some of the energy from the exergonic oxidation is captured in NADH, and some is devoted to forming the phosphate. This is the first energy-generating step of glycolysis.

Step 7 generates the first ATP of glycolysis by transferring a phosphate group from 1,3-bisphosphoglycerate to ADP; the enzyme phosphoglycerate kinase

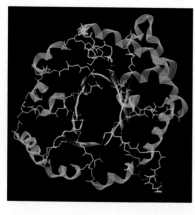

▲ **Triosephosphate isomerase, enzyme for Step 5 of glycolysis.** β-Sheets are shown in blue, α-helixes in green, and random coil regions in yellow.

accomplishes this. The other product of the reaction is 3-phosphoglycerate. Because this step occurs twice for each glucose molecule, the ATP energy balance sheet in glycolysis is even after Step 7. Two ATPs were spent in Steps 1–5, and now they have been replaced.

Steps 8 and 9—an isomerization of 3-phosphoglycerate to 2-phosphoglycerate, catalyzed by phosphoglycerate mutase followed by dehydration of 2-phosphoglycerate by enolase—generate phosphoenolpyruvate, the second energy-providing phosphate of glycolysis. Step 8 sets the stage for Step 9 by rearranging the position of the phosphate group in the molecule. Then, Step 9 generates a double bond by way of water loss.

Step 10 is a highly exergonic, irreversible transfer of a phosphate group to ADP catalyzed by pyruvate kinase. The production of ATP by transfer of a phosphate group to ADP from another molecule is called *substrate level phosphorylation*. Once the phosphate group leaves phosphoenolpyruvate, the less stable enol form of pyruvate spontaneously rearranges into the more stable keto form of pyruvate. The large amount of free energy released is accounted for by this rearrangement. (⬤▭, Section 22.6)

The two ATPs formed by the reactions in Step 10 are pure profit, and the overall results of glycolysis are as follows:

Net result of glycolysis

$$C_6H_{12}O_6 + 2\,NAD^+ + 2\,HOPO_3^{2-} + 2\,ADP \longrightarrow 2\,CH_3{-}\overset{\overset{\displaystyle O}{\|}}{C}{-}\overset{\overset{\displaystyle O}{\|}}{C}{-}O^- + 2\,NADH + 2\,ATP + 2\,H_2O + 2\,H^+$$

Glucose   Pyruvate

- Conversion of glucose to two pyruvates
- Production of two ATPs
- Production of two molecules of reduced coenzyme NADH from NAD^+

WORKED EXAMPLE **23.1** Relating Enzyme Names with Reaction Steps of Glycolysis

How do the names of the enzymes involved in the first two steps of glycolysis relate to the reactions involved?

ANALYSIS Look at the names of the enzymes and the reactions. Also recall the enzyme classification scheme from Chapter 19 (⬤▭, p. 597).

SOLUTION
In the first reaction, a phosphoryl group is added to glucose. The enzyme name is hexokinase; *kinase* because kinases transfer phosphoryl groups, and *hexo-* for a hexose sugar as the substrate. In the second reaction glucose 6-phosphate is rearranged to fructose 6-phosphate by phosphoglucose isomerase. This enzyme belongs to the enzyme class isomerases, enzymes that rearrange molecules to an isomer of the original molecule. The phosphoglucose part of the name tells us that a phosphorylated glucose molecule will be rearranged; inspection of the reaction shows this is true.

PROBLEM 23.3

Identify the two pairs of steps in glycolysis in which phosphate intermediates are synthesized and their energy harvested as ATP.

PROBLEM 23.4

Identify each step in glycolysis that is an isomerization.

PROBLEM 23.5

Verify the isomerization that occurs in Step 2 of glycolysis by drawing the open-chain forms of glucose 6-phosphate and fructose 6-phosphate.

KEY CONCEPT PROBLEM 23.6

In Figure 23.3 compare the starting compound (glucose) and the final product (pyruvate).

(a) Which is oxidized to a greater extent?

(b) Are there any steps in the glycolytic pathway in which an oxidation or reduction occurs? Identify the oxidizing or reducing agents that are involved in these steps.

23.4 Entry of Other Sugars into Glycolysis

The major monosaccharides from digestion other than glucose also eventually join the glycolysis pathway.

Fructose, from fruits or hydrolysis of the disaccharide sucrose, is converted to glycolysis intermediates in two ways: In muscle, it is phosphorylated to fructose 6-phosphate, and in the liver, it is converted to glyceraldehyde 3-phosphate. Fructose 6-phosphate is the substrate for Step 3 of glycolysis; glyceraldehyde 3-phosphate is the substrate for Step 6.

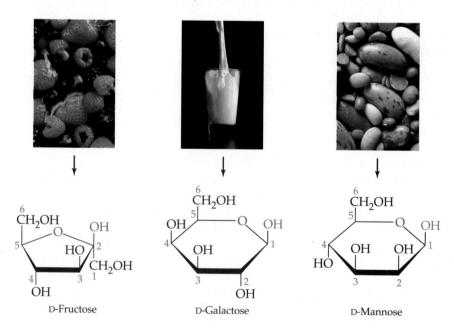

▶ Major dietary monosaccharides other than glucose.

D-Fructose   D-Galactose   D-Mannose

Galactose from hydrolysis of the disaccharide lactose is converted to glucose 6-phosphate, the substrate for Step 2 of glycolysis, by a five-step pathway. A hereditary defect affecting any enzyme in this pathway can cause galactosemia. (⊂⊃, p. 707)

Mannose is a product of the hydrolysis of plant polysaccharides other than starch. It is converted by hexokinase to a 6-phosphate, which then undergoes a multistep, enzyme-catalyzed rearrangement and enters glycolysis as fructose 6-phosphate as the substrate for Step 3.

PROBLEM 23.7

Use curved arrows (like those in Figure 23.3) to write an equation for the conversion of fructose to fructose 6-phosphate by ATP. At what step does fructose 6-phosphate enter glycolysis?

PROBLEM 23.8

Compare glucose and galactose (see Section 22.5), and explain how their structures differ.

APPLICATION ▶ Tooth Decay

Tooth decay is a complex interaction between food, bacteria, and a host organism. The clinical term for tooth decay is *dental caries*. It is recognized as an infectious microbial disease that results in the destruction of the calcified structures of the teeth.

The mouth is home to many different species of bacteria. A variety of habitats are provided by the diverse surfaces of the teeth, tongue, gums, and cheeks, and there are nutrients specific to each. Two permanent bacterial residents of the oral cavity, *Streptococcus sanguis* and *Streptococcus mutans*, compete for the same habitat on the biting surfaces of the teeth.

Dental plaque is defined as bacterial aggregations on the teeth that cannot be removed by the mechanical action of a strong water spray. Immediately after plaque has been removed by scrubbing with an abrasive paste, a coating of organic material composed of glycoproteins from the saliva begins to form. It completely covers the teeth within two hours after a visit to the dentist's office.

Bacteria then quickly colonize this newly formed film. They secrete a sticky matrix of an insoluble polysaccharide known as *dextran*, a branched polymer of the glucose that the bacteria have produced by hydrolysis of sucrose from food. The dextran allows the bacteria to stick firmly to the teeth so that the bacteria cannot be washed away by the saliva and swallowed. Bacteria that live comfortably in the mouth do not take well to the acidic environment of the stomach, so staying on the teeth is essential for their survival. The mass of bacteria, their sticky matrix, and the glycoprotein film together comprise dental plaque. Plaque is therefore not simply adherent food debris, but rather a community of microorganisms (known as a *biofilm*) that forms through an orderly sequence of events.

The bacteria resident in plaque release products consisting of proteins and carbohydrates. Some polysaccharides form intracellular granules that serve as energy storage depots for periods of low nutrient availability (between meals for the host). Other products are toxic to the gums and can promote periodontal disease. Carbohydrates, including structural components of the bacteria themselves, the storage granules, and the sticky matrix, constitute 20% of the dry weight of plaque.

What our dentists and parents told us—that eating candy would create cavities—is true! A diet high in sucrose favors the growth of *S. mutans* over that of *S. sanguis*. Although both bacteria can cause tooth decay, *S. mutans* attacks teeth much more vigorously. It has an enzyme (a glucosyltransferase) that transfers glucose units from sucrose to the dextran polymer. The enzyme is specific to sucrose, and does not act on free glucose or the glucose from other carbohydrates. The mature plaque community then metabolizes fructose from the sucrose to lactate, and this acid causes the local pH in the area of the tooth to drop dramatically. If the pH stays low enough for a long enough time, the minerals in the teeth are dissolved away and the tooth begins to decay.

Cleaning teeth by brushing and flossing disrupts the bacterial plaque, removing many of the bacteria. However, enough bacteria always remain so that the colonization process can begin anew almost immediately. The disruption of plaque via oral hygiene and a diet low in sucrose, however, favors the growth of *S. sanguis* over *S. mutans*. To control the decay process, it is necessary to limit both the amount of sucrose in the diet and the frequency with which it is ingested.

The third factor in tooth decay is the host—ourselves. Many variables prevent or promote tooth decay, including the composition of saliva, the shape of the teeth, and exposure to fluoride. As individuals, however, we have little control over these elements, leaving us reliant on proper oral hygiene habits, low sucrose diets, and preventive maintenance by dental professionals. Modern preventative maintenance includes the use of sealants on the chewing surfaces of both deciduous and permanent teeth as they develop in children. The result is many more young adults with healthy, unblemished teeth.

See Additional Problems 23.69 through 23.72 at the end of the chapter.

23.5 The Fate of Pyruvate

Aerobic In the presence of oxygen.

Anaerobic In the absence of oxygen.

The conversion of glucose to pyruvate is a central metabolic pathway in most living systems. The further reactions of pyruvate, however, depend on metabolic conditions and on the nature of the organism. Under normal oxygen-rich (**aerobic**) conditions, pyruvate is converted to acetyl-SCoA. This pathway, however, is short-circuited in some tissues, especially when there is not enough oxygen present (**anaerobic** conditions). Under anaerobic conditions, pyruvate is instead reduced to lactate. When sufficient oxygen again becomes available, the lactate is recycled to pyruvate. A third pathway for pyruvate is conversion back to glucose by *gluconeogenesis* (Section 23.11). This pathway is essential when the body is starved for glucose. The pyruvate for gluconeogenesis may come not only from glycolysis, but also from amino acids or glycerol from lipids. Use of protein and lipid for glucose synthesis occurs when calories needed exceed calorie intake as in starvation, certain diseases, and some carbohydrate-restricted diets.

$$CH_3CH_2OH \leftarrow CH_3-\overset{\overset{O}{\|}}{C}-\overset{\overset{O}{\|}}{C}-O^- \rightarrow CH_3-\overset{\overset{OH}{|}}{CH}-\overset{\overset{O}{\|}}{C}-O^-$$

Ethyl alcohol ← Pyruvate → Lactate

Acetyl-SCoA

$$CH_3-\overset{\overset{O}{\|}}{C}-SCoA$$

▲ The biochemical transformations of pyruvate.

Yeast is an organism with a different pathway for pyruvate, one that we put to use in a variety of ways. Yeast converts pyruvate to ethanol under anaerobic conditions.

Aerobic Oxidation of Pyruvate to Acetyl-SCoA

For aerobic oxidation to proceed, pyruvate first diffuses across the outer mitochondrial membrane from the cytosol where it was produced. Then it must be carried by a transporter protein across the otherwise impenetrable inner mitochondrial membrane. Once within the mitochondrial matrix, pyruvate encounters the *pyruvate dehydrogenase complex* (Figure 23.4), a large multienzyme complex that catalyzes the conversion of pyruvate to acetyl-SCoA.

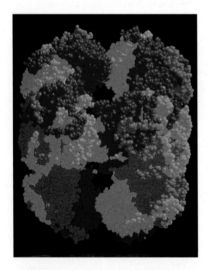

▲ **FIGURE 23.4 The central core of the pyruvate dehydrogenase complex.** The core is composed of 24 proteins arranged in three kinds of subunits shown in three different colors. The entire huge complex contains 60 subunits of three different enzymes. It requires NAD^+, CoA, FAD, and two other coenzymes. The enzyme subunits, which are adjacent to each other, swing into position one after the other as pyruvate loses CO_2 and is converted to an acetyl group that is then transferred to coenzyme A.

$$CH_3-\overset{\overset{O}{\|}}{C}-\overset{\overset{O}{\|}}{C}-O^- + HS-CoA \xrightarrow[\substack{\text{Pyruvate} \\ \text{dehydrogenase} \\ \text{complex}}]{NAD^+ \quad NADH/H^+} CH_3\overset{\overset{O}{\|}}{C}-SCoA + CO_2$$

Pyruvate   Acetyl-SCoA

Anaerobic Reduction to Lactate

Why does pyruvate take an alternative pathway when oxygen is in short supply? Since oxygen has not been needed in glucose catabolism thus far, what is the connection? The problem lies with the NADH formed in Step 6 of glycolysis (Figure 23.3).

Under aerobic conditions, NADH is continually reoxidized to NAD^+ during electron transport (⬤, Section 21.9), so NAD^+ is in constant supply. If electron transport slows down because of insufficient oxygen, however, NADH concentration increases, decreasing the supply of NAD^+, and glycolysis cannot continue. An alternative way to reoxidize NADH is therefore essential because glycolysis *must* continue—it is the only available source of fresh ATP.

The reduction of pyruvate to lactate solves the problem. NADH serves as the reducing agent and is reoxidized to NAD^+, which is then available in the cytosol for glycolysis. Lactate formation serves no purpose other than NAD^+ production, and the lactate is reoxidized to pyruvate when oxygen is available.

$$CH_3-\overset{\overset{\displaystyle O}{\|}}{C}-\overset{\overset{\displaystyle O}{\|}}{C}-O^- \xrightleftharpoons[\substack{\text{Aerobic} \\ \text{conditions}}]{\substack{NADH/H^+ \quad NAD^+ \\ \text{Anaerobic} \\ \text{conditions}}} CH_3-\overset{\overset{\displaystyle OH}{|}}{CH}-\overset{\overset{\displaystyle O}{\|}}{C}-O^-$$

Pyruvate   Lactate

Red blood cells have no mitochondria and therefore always form lactate as the end product of glycolysis. Tissues where oxygen is in short supply also rely on the anaerobic production of ATP by glycolysis. Examples are the cornea of the eye, where there is little blood circulation, and muscles during intense activity. The resulting buildup of lactate in working muscles causes fatigue and discomfort (see the Application "The Biochemistry of Running," p. 748).

Alcoholic Fermentation

Microorganisms often must survive in the absence of oxygen and have evolved numerous anaerobic strategies for energy production, generally known as **fermentation**. When pyruvate undergoes fermentation by yeast, it is converted into ethanol plus carbon dioxide. This process, known as **alcoholic fermentation**, is used to produce beer, wine, and other alcoholic beverages and also to make bread. The carbon dioxide causes the bread to rise, and the alcohol evaporates during baking. The first leavened, or raised, bread was probably made by accident when airborne yeasts got into the dough. The tempting aroma of baking bread includes the aroma of alcohol vapors. Beer can be made by exposing the mash to outside air, where airborne yeasts drift in from the surroundings.

Fermentation The production of energy under anaerobic conditions.

Alcoholic fermentation The anaerobic breakdown of glucose to ethanol plus carbon dioxide by the action of yeast enzymes.

WORKED EXAMPLE **23.2** Identifying Catabolic Stages

Complete oxidation of glucose produces six molecules of carbon dioxide. Describe the stage of catabolism at which each one is formed.

ANALYSIS Look at each stage of catabolism for the complete oxidation of glucose to carbon dioxide. Notice how many molecules of carbon dioxide are produced and by which step. Pathways to consider (in order) are glycolysis, conversion of pyruvate to acetyl-SCoA, and the citric acid cycle. (There is no need to consider oxidative phosporylation since glucose is completely oxidized at the end of the citric acid cycle.)

SOLUTION
No molecules of carbon dioxide are produced during glycolysis. Conversion of one molecule of pyruvate to one molecule of acetyl-SCoA yields one molecule of carbon dioxide. In the citric acid cycle, two molecules of carbon dioxide are released for each molecule of acetyl-SCoA oxidized. One is released in Step 3 when isocitrate is converted to α-ketoglutarate and the other when α-ketoglutarate is converted to succinyl-SCoA in Step 4. Since each glucose molecule produces two pyruvate molecules, the total is three molecules twice, or six molecules of carbon dioxide.

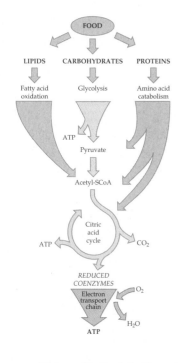

KEY CONCEPT PROBLEM 23.9

In alcoholic fermentation, each mole of pyruvate is converted to one mole of carbon dioxide and one mole of ethanol. In the process, about 50 kcal/mol of pyruvate is produced. Under the most favorable conditions, more than half of this energy is stored as ATP.

(a) What happens to the remaining energy produced in alcoholic fermentation?

(b) Give two reasons why it would be nearly impossible to reverse the reaction that converts pyruvate to ethanol and carbon dioxide.

23.6 Energy Output in Complete Catabolism of Glucose

The total energy output from oxidation of glucose is the combined result of (a) glycolysis, (b) conversion of pyruvate to acetyl-SCoA, (c) conversion of two acetyl groups to four molecules of CO_2 in the citric acid cycle, and, finally, (d) the passage of reduced coenzymes from each of these pathways through electron transport and the production of ATP by oxidative phosphorylation.

To determine the total number of ATP molecules generated from one glucose molecule, we first sum the net equations for each pathway that precedes oxidative phosphorylation. Since each glucose yields two pyruvates and two acetyl-SCoAs, the net equations for pyruvate oxidation and the citric acid cycle are multiplied by 2:

Net result of catabolism of one glucose molecule

Glycolysis (Section 23.3)

$$\text{Glucose} + 2\,\text{NAD}^+ + 2\,\text{HOPO}_3^{2-} + 2\,\text{ADP} \longrightarrow 2\,\text{Pyruvate} + 2\,\text{NADH} + 2\,\text{ATP} + 2\,\text{H}_2\text{O} + 2\,\text{H}^+$$

Pyruvate oxidation (Section 23.5)

$$2\,\text{Pyruvate} + 2\,\text{NAD}^+ + 2\,\text{HSCoA} \longrightarrow 2\,\text{Acetyl-SCoA} + 2\,\text{CO}_2 + 2\,\text{NADH} + 2\,\text{H}^+$$

Citric acid cycle (Section 21.8)

$$2\,\text{Acetyl-SCoA} + 6\,\text{NAD}^+ + 2\,\text{FAD} + 2\,\text{ADP} + 2\,\text{HOPO}_3^{2-} + 4\,\text{H}_2\text{O} \longrightarrow$$
$$2\,\text{HSCoA} + 6\,\text{NADH} + 6\,\text{H}^+ + 2\,\text{FADH}_2 + 2\,\text{ATP} + 4\,\text{CO}_2$$

$$\text{Glucose} + 10\,\text{NAD}^+ + 2\,\text{FAD} + 2\,\text{H}_2\text{O} + 4\,\text{ADP} + 4\,\text{HOPO}_3^{2-} \longrightarrow$$
$$10\,\text{NADH} + 10\,\text{H}^+ + 2\,\text{FADH}_2 + 4\,\text{ATP} + 6\,\text{CO}_2$$

The summation above shows a total of 4 ATPs per glucose molecule. The remainder of our ATP is generated via electron transport and oxidative phosphorylation. Thus, the total number of ATPs per glucose molecule is the 4 ATPs from glucose catabolism plus the number of ATPs produced for each reduced coenzyme that enters electron transport.

Based on an energy yield assumption of 3 ATPs per NADH and 2 ATPs per $FADH_2$, the maximum yield for the complete catabolism of one molecule of glucose is 38 ATPs, as calculated below:

$$10\,\text{NADH}\left(\frac{3\,\text{ATP}}{\text{NADH}}\right) + 2\,\text{FADH}_2\left(\frac{2\,\text{ATP}}{\text{FADH}_2}\right) + 4\,\text{ATP} = 38\,\text{ATP}$$

Our ever-expanding understanding of biochemical pathways has led to a revision in the potential number of ATPs per reduced coenzyme. The 38 ATPs per glucose

APPLICATION ▶

Microbial Fermentations: Ancient and Modern

Archaeological evidence shows that several ancient civilizations used fermentation in the preparation of food and drink. Residues found in pottery in both China and Egypt point to wine making as long as 9000 years ago. Analysis of lees (yeast residue) from jars dating to the sixth century B.C. found ribosomal DNA sequences from *Saccharomyces cerevisiae*, the same organism used today in wine making, brewing beer, and leavening bread. Other microorganisms also perform fermentations and are important in producing cheese, sauerkraut, soy sauce, and other foods.

The fermentations most widely exploited by man involve converting grain and other vegetable matter into ethanol through "alcohol fermentation," as described by Louis Pasteur, who studied wine making and developed the pasteurization process. Under anaerobic conditions in yeast and some other microorganisms, pyruvate, the product of glycolysis, is converted first to acetaldehyde and carbon dioxide by pyruvate decarboxylase and then to ethanol by alcohol dehydrogenase:

$$CH_3COCOOH \xrightleftharpoons{\text{pyruvate decarboxylase}} CH_3CHO +$$

$$CO_2 \xrightleftharpoons{\text{alcohol dehydrogenase}} CH_3CH_2OH$$

Any fruit can be fermented but many societies have focused on fermenting grapes. Fermentation of fruit is a way to preserve some of the food value without spoilage. Grapes take up little space when grown with other crops and are high in fermentable sugars. An added bonus is the natural coating of several varieties of yeast on the grapes, although during fermentation *Saccharomyces cerevisiae* rapidly replaces the others. Many wines contain about 12% ethanol; when ethanol in the fermenting mixture reaches that level, the yeast die and fermentation ceases. It was known for many years that if the fermentation vessel was tightly sealed, the gas (CO_2) produced made the wine "bubbly" if the seal held against the increased pressure inside the container; unfortunately, most containers were not strong enough to yield a reliably fizzy product. In the seventeenth century, glass bottles and corks were introduced for wine storage, replacing wooden barrels, and the production of champagne became possible.

Similarly, beer making from the fermentation of grain is nearly as old as wine production. In South America beer was fermented from boiled maize or manioc root with some of the cooked starch chewed and added, along with salivary enzymes, to the mix. The presence of salivary enzymes hydrolyzed the starch to maltose for fermentation. In much of Asia, millet, sorghum, and rice were made into beer; in Europe and Africa, malted (sprouted) grains like millet, wheat, rye, and barley were used. The enzymes present in malted grains hydrolyze the starch in the grain to maltose. The maltose is fermented to produce ethanol and carbon dioxide by *Saccharomyces cerevisiae* and related yeasts.

When curdled milk is fermented, cheese is produced. Humans have been making cheese for over 5000 years. The characteristic flavors and aromas of the different varieties of cheeses depend partly on the bacteria used in the fermentation and the molecules that are changed. For example, the characteristic flavor and appearance of Swiss cheese are due to the conversion of lactic acid to propionic acid, acetic acid and carbon dioxide gas, which produces the holes in the cheese. Fresh milk can also be fermented; we know these products as yogurt, sour cream, and buttermilk.

Not all bacterial fermentations are friendly. *Clostridium* species are responsible for gas gangrene, a condition that causes the death of living tissue and can cause the death of the infected person. The bacteria invade the body, generally through a puncture wound, and multiply in an anaerobic area. Fermentation products of these bacteria are butyric acid, butanol, acetone, ethanol, and carbon dioxide. CO_2 infiltrates surrounding tissues, maintains an anaerobic environment by excluding O_2 from the cells, and causes necrosis (tissue death). The presence of the organic acids, CO_2 and toxins secreted by the bacteria leads to the spread of gangrene and make it difficult to treat. Treatment usually involves surgical removal of necrotic tissue and sometimes hyperbaric (at a pressure above atmospheric pressure) oxygen treatment since O_2 is toxic to these bacteria.

See Additional Problems 23.73 and 23.74 at the end of the chapter.

molecule is viewed as a maximum yield of ATP, most likely possible in bacteria and other prokaryotes. In humans and other mammals the yield is possibly lower; the maximum is most likely 30–32 ATPs per glucose molecule.

23.7 Regulation of Glucose Metabolism and Energy Production

Normal blood glucose concentration a few hours after a meal ranges roughly from 65 to 110 mg/dL. When departures from normal occur, we are in trouble (Figure 23.5). Low blood glucose (**hypoglycemia**) causes weakness, sweating, and rapid heartbeat,

Hypoglycemia Lower-than-normal blood glucose concentration.

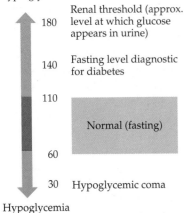

Glucose concentration (mg/dL)

Hyperglycemia

180 — Renal threshold (approx. level at which glucose appears in urine)

140 — Fasting level diagnostic for diabetes

110

Normal (fasting)

60

30 — Hypoglycemic coma

Hypoglycemia

▲ **FIGURE 23.5 Blood glucose.** The ranges for low blood glucose (in green; hypoglycemia), normal blood glucose (in purple), and high blood glucose (in orange; hyperglycemia) are indicated.

and in severe cases, low glucose in brain cells causes mental confusion, convulsions, coma, and eventually death. The brain can use *only* glucose as a source of energy. At a blood glucose level of 30 mg/dL, consciousness is impaired or lost, and prolonged hypoglycemia can cause permanent dementia. High blood glucose (**hyperglycemia**) causes increased urine flow as the normal osmolarity balance of fluids within the kidney is disturbed. Prolonged hyperglycemia can cause low blood pressure, coma, and death.

Two hormones from the pancreas have the major responsibility for blood glucose regulation. The first, insulin, is released when blood glucose concentration rises (Figure 23.6). Its role is to decrease blood glucose concentrations by accelerating the uptake of glucose by cells where it is used for energy production, and by stimulating synthesis of glycogen, proteins, and lipids.

The second hormone, glucagon, is released when blood glucose concentration drops. In a reversal of insulin's effects, glucagon stimulates the breakdown of glycogen in the liver and release of glucose. Proteins and lipids are also broken down so that amino acids from proteins and glycerol from lipids can be converted to glucose in the liver by the gluconeogenesis pathways (see Section 23.11). Epinephrine (the "fight-or-flight" hormone) also accelerates the breakdown of glycogen, but primarily in muscle tissue, where glucose is used to generate energy needed for quick action. (⊂□⊃, Section 20.3)

Rising blood glucose concentration

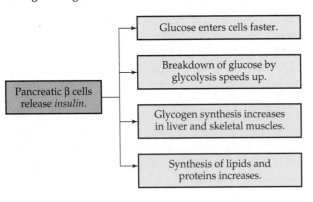

Falling blood glucose concentration

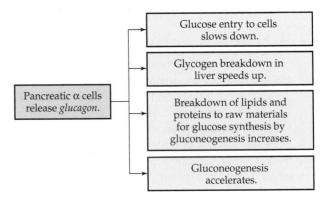

▲ **FIGURE 23.6** Regulation of glucose concentration by insulin and glucagon from the pancreas.

Hyperglycemia Higher-than-normal blood glucose concentration.

23.8 Metabolism in Fasting and Starvation

Imagine that you are lost in the woods. You have had no carbohydrates and very little else to eat for hours, and you are exhausted. The glycogen stored in your liver and muscles will soon be used up, but your brain relies on glucose to keep functioning. What happens next? Fortunately, your body is not ready to give up yet. It has mechanisms to assure that the limited glucose supplies will be delivered preferentially to your brain. In the liver, the gluconeogenesis pathway (Section 23.11) can make glucose from proteins. And if you are lost for a long time, there is a further backup system that extracts energy from compounds other than glucose.

The metabolic changes in the absence of food begin with a gradual decline in blood glucose concentration accompanied by an increased release of glucose from glycogen (Figure 23.7) (glycogenolysis, Section 23.10). All cells contain glycogen, but most is stored in liver cells (about 90 g in a 70 kg man) and muscle cells (about 350 g in a 70 kg man). Free glucose and glycogen represent less than 1% of our energy reserves and are used up in 15–20 hours of normal activity (three hours in a marathon race).

Fats are our largest energy reserve, but adjusting to dependence on them for energy takes time as there is no direct pathway to glucose for the fatty acids from

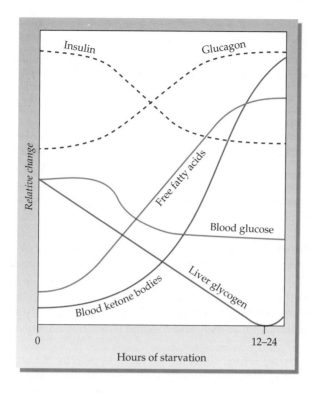

◄ **FIGURE 23.7** Relative changes during early stages of starvation.

fats (as shown in Figure 23.2). Energy from fatty tissue must be generated by catabolism of fatty acids to acetyl-SCoA, oxidation of acetyl-SCoA via the citric acid cycle, and production of ATP energy from electron transport.

As glucose and glycogen reserves are exhausted, metabolism turns first to breakdown of proteins and glucose production from amino acids via gluconeogenesis in the liver. During the first few days of starvation, protein is used up at a rate as high as 75 g/day. Meanwhile, lipid catabolism is mobilized, and acetyl-SCoA molecules derived from breakdown of lipids accumulate. Eventually, the citric acid cycle is overloaded and cannot degrade acetyl-SCoA as rapidly as it is produced. Acetyl-SCoA therefore builds up inside cells and begins to be removed by a new series of metabolic reactions that transform it into a group of compounds known as *ketone bodies*.

Ketone bodies

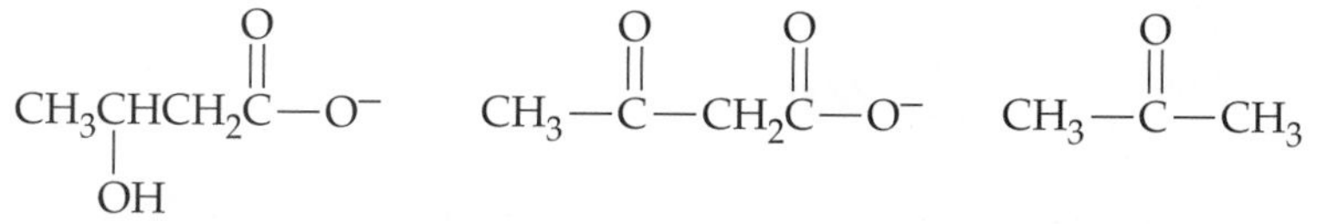

| 3-Hydroxybutyrate | Acetoacetate | Acetone |

These ketone bodies enter the bloodstream and, as starvation continues, the brain and other tissues are able to switch over to producing up to 50% of their ATP from catabolizing ketone bodies instead of glucose. By the fortieth day of starvation, metabolism has stabilized at the use of about 25 g of protein and 180 g of fat each day, a condition that conserves glucose and protein as much as possible. So long as adequate water is available, an average person can survive in this state for several months; those with more fat can survive longer.

Looking Ahead

The breakdown of triacylglycerols from the fatty tissue produces not only ketone bodies but also glycerol, one of the compounds that can be converted to glucose by gluconeogenesis. The production of glycerol and ketone bodies from triacylglycerols is described in Chapter 25, which is devoted to lipid metabolism.

PROBLEM 23.10

Refer to Figure 23.7 and summarize the changes in each substance shown during the starvation period represented in the figure.

23.9 Metabolism in Diabetes Mellitus

Diabetes mellitus A chronic condition due to either insufficient insulin or failure of insulin to activate crossing of cell membranes, by glucose.

Diabetes mellitus is one of the most common metabolic diseases. It is not a single disease but is classified into two major types, insulin-dependent and non-insulin-dependent; recently identified is a pre-diabetic condition called metabolic syndrome. The insulin-dependent disease, also called Type I or juvenile-onset diabetes (because it often appears in childhood) is caused by failure of the pancreatic cells to produce enough insulin. By contrast, in non-insulin-dependent diabetes, also called Type II or adult-onset diabetes (because it usually occurs in individuals over about 40 years of age), insulin is in good supply but fails to promote the passage of glucose across cell membranes. An estimated 6.6% of the U.S. population has Type II diabetes. Another 24% of the U.S. population is estimated to have metabolic syndrome. Although often thought of only as a disease of glucose metabolism, diabetes affects protein and fat metabolism as well, and in some ways the metabolic response resembles starvation.

The symptoms by which diabetes (Type I) is usually detected are excessive thirst accompanied by frequent urination, abnormally high glucose concentrations in urine and blood, and wasting of the body despite a good diet. These symptoms result when available glucose does not enter cells where it is needed. Glucose

APPLICATION ▶ Diagnosis and Monitoring of Diabetes

Glucose measurements are essential in the diagnosis of diabetes mellitus and in the management of diabetic patients, both in a clinical setting and on a day-to-day basis by patients themselves. The glucose tolerance test is among the clinical laboratory tests usually done to pin down a diagnosis of diabetes mellitus. The patient must fast for 10–16 hours, avoid a diet high in carbohydrates prior to the fast, and refrain from taking any of a long list of drugs that can interfere with the test.

First, a blood sample is drawn to determine the fasting glucose concentration. The average normal fasting glucose level is 65–100 mg/dL. Then, the patient drinks a solution containing 75 g of glucose, and additional blood samples are taken at regular intervals thereafter. The accompanying Figure compares the changes in diabetic, prediabetic, and normal individuals. The diabetic patient has a higher fasting blood glucose level than the nondiabetic individual. In both, blood glucose concentration rises in the first hour. A difference becomes apparent after two hours, when the concentration in a normal individual has dropped to close to the fasting level but that in a diabetic individual remains high. The metabolic syndrome, pre-diabetic patient has an intermediate response; the fasting glucose level is greater than 100 mg/dL and the challenge response is intermediate between that of the diabetic and non-diabetic patient.

As listed below, a fasting blood glucose concentration of 140 mg/dL or higher and/or a glucose tolerance test concentration that remains above 200 mg/dL beyond one hour are considered diagnostic criteria for diabetes. For a firm diagnosis, the glucose tolerance test is usually given more than once.

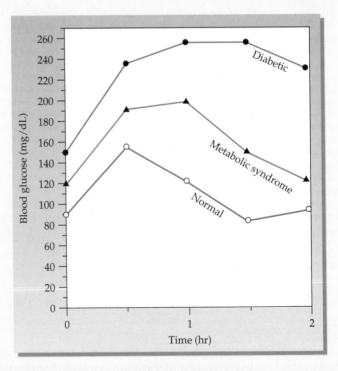

▲ Blood glucose concentration in glucose tolerance test for normal and diabetic individuals.

Key Diagnostic Features of Diabetes Mellitus

- Classic symptoms such as frequent urination, excessive thirst, rapid weight loss (Type I only)
- Random blood glucose concentration (without fasting) greater than 200 mg/dL
- Fasting blood glucose greater than 140 mg/dL
- Sustained blood glucose concentration greater than 200 mg/dL after glucose challenge in glucose tolerance test

Diabetics must monitor their blood glucose levels at home daily, often several times a day. Most tests for glucose in urine or blood rely on detecting a color change that accompanies the oxidation of glucose. Because glucose and its oxidation product, gluconate, are colorless, the oxidation must be tied chemically to the color change of a suitable indicator. Modern methods for glucose detection rely on the action of an enzyme specific for glucose. The most commonly used enzyme is glucose oxidase, and the products of the oxidation are gluconate and hydrogen peroxide (H_2O_2). A second enzyme in the reaction mixture, a peroxidase, catalyzes the reaction of hydrogen peroxide with a dye that gives a detectable color change.

$$\text{Glucose} + O_2 \xrightarrow{\text{Glucose oxidase}} \text{Gluconate} + H_2O_2$$

$$\underset{\text{(Colorless)}}{H_2O_2 + \text{Reduced dye}} \xrightarrow{\text{Peroxidase}} \underset{\text{(Colored)}}{H_2O_2 + \text{Oxidized dye}}$$

The glucose oxidase test is available for urine and blood. Many diabetic individuals monitor their glucose levels in blood rather than urine, using an instrument that reads the color change electronically and can store several weeks' data at a time. The enzymes needed for the reactions are embedded

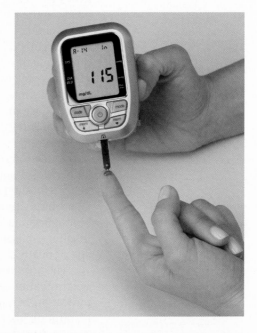

▲ Glucose blood test. A tiny drop of blood is absorbed on the test strip in the blood glucose monitor. The results of the test are read in less than 10 seconds by most modern monitors and displayed on an LCD screen.

in the test strip itself and only a miniscule drop of blood is needed. The blood test is desirable because it is more specific and it detects rising glucose levels earlier than the urine test. It is used to achieve tighter control of blood glucose levels to help diabetics live longer, healthier lives.

See Additional Problems 23.75 through 23.78 at the end of the chapter.

builds up in the blood, causing the symptoms of hyperglycemia and spilling over into the urine (glucosuria). In untreated diabetes, metabolism responds to the glucose shortage within cells by proceeding through the same stages as in starvation, from depletion of glycogen stores to breakdown of proteins and fats.

Type II diabetes is thought to result when cell membrane receptors fail to recognize insulin. This state is sometimes referred to as insulin resistance. Drugs that increase either insulin or insulin receptor levels are an effective treatment because more of the undamaged receptors are put to work. Treatment also includes diet modification and exercise.

Metabolic syndrome resembles a pre-diabetic state with slightly elevated fasting blood glucose levels and an impaired glucose response. Long-term population studies show that metabolic syndrome is a strong predictor for the development of diabetes as the population ages, as well as coronary heart disease and stroke. Metabolic syndrome is characterized by abdominal obesity, elevated blood pressure and impaired glucose metabolism. Current treatment recommends lifestyle changes involving diet and exercise.

Type I diabetes is classified as an autoimmune disease, a condition in which the body misidentifies some part of itself as an invader (Section 29.4). Gradually, the immune system wrongly identifies pancreatic beta cells as foreign matter, develops antibodies to them, and destroys them. To treat Type I diabetes, the missing insulin must be supplied by injection. Commercially available human insulin is now produced by bacteria modified by recombinant DNA techniques (Section 27.4) and was the first product of genetic engineering approved for use in humans.

Diabetic individuals are subject to several serious conditions that result from elevated blood glucose levels. One reasonably well understood outcome is blindness due to cataracts. Increased glucose levels within the eye increase the quantity of glucose converted to sorbitol (in which the —CHO group of glucose is converted to —CH_2OH). Because sorbitol is not transported out of the cell, as is glucose, its rising concentration increases the osmolarity of fluid in the eye, causing increased pressure and cataracts. Elevated sorbitol is also associated with blood vessel lesions and gangrene in the legs, conditions that can accompany long-term diabetes.

An insulin-dependent diabetic is at risk for two types of medical emergencies: ketoacidosis and hypoglycemia. Ketoacidosis results from the buildup of acidic ketones in the late stages of uncontrolled diabetes. It can lead to coma and diminished brain function but can also be reversed by timely insulin administration. Hypoglycemia, or "insulin shock," by contrast, may be due to an overdose of insulin or failure to eat. If untreated, diabetic hypoglycemia can cause nerve damage or death.

The arrival at the emergency room of a diabetic patient in a coma requires quick determination of whether the condition is due to ketoacidosis or excess insulin. One indication of ketoacidosis is the aroma of acetone on the breath. Another is rapid respiration driven by the need to diminish acid concentration by eliminating carbon dioxide:

$$H^+ + HCO_3^- \longrightarrow H_2CO_3 \longrightarrow H_2O + CO_2 \text{ (Exhaled)}$$

An overdose of insulin does not cause rapid respiration.

Observations are backed up by bedside tests for glucose and ketones in blood and urine. A patient in insulin shock will, for example, have a very low blood glucose concentration.

PROBLEM 23.11

Sorbitol is the alcohol that accumulates in the eye and can cause cataracts. Draw the open-chain structure of sorbitol, which is identical to that of D-glucose except that the aldehyde group has been reduced to an alcohol group. Can sorbitol form a five- or six-membered cyclic hemiacetal? Explain why or why not.

KEY CONCEPT PROBLEM 23.12

Ketoacidosis is relieved by rapid breathing, which converts bicarbonate ions and hydrogen ions in the blood to gaseous carbon dioxide and water, as shown in the equation on p. 746.

(a) Assuming that these reactions can go in either direction, how does a state of acidosis help to increase the generation of carbon dioxide?

(b) What principle describes the effect of added reactants and products on an equilibrium?

23.10 Glycogen Metabolism: Glycogenesis and Glycogenolysis

Glycogenesis The biochemical pathway for synthesis of glycogen.

Glycogen, the storage form of glucose in animals, is a branched polymer of glucose. (⚬⚬, p. 721) Glycogen synthesis, known as **glycogenesis**, occurs when glucose concentrations are high. It begins with glucose 6-phosphate and occurs via the three steps shown on the right in Figure 23.8.

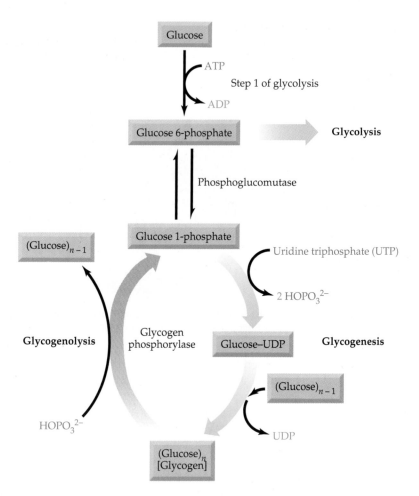

◀ **FIGURE 23.8 Glycogenolysis and glycogenesis.** Reading from the top down shows the pathway for glycogen synthesis from glucose (glycogenesis). Reading from the bottom up shows the pathway for release of glucose from glycogen (glycogenolysis).

Glucose 6-phosphate is first isomerized to glucose 1-phosphate by phosphoglucomutase. The glucose residue is then attached to uridine diphosphate (UDP) in a reaction catalyzed by UDP-glucose pyrophosphorylase and driven by the release of inorganic pyrophosphate:

Glucose-UDP, the activated carrier of glucose in glycogen synthesis

The resulting glucose-UDP transfers glucose to a growing glycogen chain in an exergonic reaction catalyzed by glycogen synthase.

As is usually true in metabolism, synthesis and breakdown are not accomplished by exactly reverse pathways. **Glycogenolysis** occurs in the two steps on the left in Figure 23.8. The first step is formation of glucose 1-phosphate by the action of glycogen phosphorylase on a terminal glucose residue in glycogen. Glucose 1-phosphate is then converted to glucose 6-phosphate by phosphoglucomutase in the reverse of the reaction by which it is formed. These reactions occur inside liver and muscle cells and are different from the reactions involved in the hydrolysis of glycogen during the digestion of muscle that you have eaten, perhaps in a hamburger.

In muscle cells, glycogenolysis occurs when there is an immediate need for energy. The glucose 6-phosphate produced from glycogen goes directly into glycolysis. In the liver, glycogenolysis occurs when blood glucose is low (for example, during starvation; see Section 23.8). Because glucose 6-phosphate cannot cross cell membranes, liver cells contain glucose 6-phosphatase, an enzyme that hydrolyzes glucose 6-phosphate to free glucose, which can then be released into the bloodstream.

Glycogenolysis The biochemical pathway for breakdown of glycogen to free glucose.

APPLICATION ▶ The Biochemistry of Running

A runner is poised tense and expectant, waiting for the sound of the starting gun. Long hours of training have prepared heart, lungs, and red blood cells to deliver the maximum amount of oxygen to the muscles, which have been conditioned to use it as efficiently as possible. In the moments before the race, mounting levels of epinephrine have readied the body for action. Now, everything depends on biochemistry: Chemical reactions in muscle cells will provide the energy to see the race through. How will that energy be produced?

The first source is the supply of immediately available ATP, but this is used up very quickly—probably within a matter of seconds. Additional ATP is then provided by the reaction of ADP with creatine phosphate, an amino acid phosphate in muscle cells that maintains the following equilibrium:

$$\text{ADP} + \text{Creatine phosphate} \rightleftharpoons \text{ATP} + \text{Creatine}$$

After about 30 seconds to a minute, stores of creatine phosphate are depleted, and glucose from glycogenolysis becomes the chief energy source. During maximum muscle exertion, oxygen cannot enter muscle cells fast enough to keep the citric acid cycle and oxidative phosphorylation going. Under these anaerobic conditions, the pyruvate from glycolysis is converted to lactate rather than entering the citric acid cycle.

In a 100 m sprint, all the energy comes from available ATP, creatine phosphate (CP in the figure), and glycolysis of glucose from muscle glycogen. Anaerobic glycolysis suffices for only a minute or two of maximum exertion, because a buildup of lactate causes muscle fatigue.

▲ Glycogen stores helped provide the energy needed for Dire Tune of Ethiopia to win the Women's 2008 Boston Marathon.

Beyond this, other pathways must come into action. As breathing and heart rate speed up and oxygen-carrying blood flows more quickly to muscles, the aerobic pathway is activated and ATP is once again generated by oxidative phosphorylation. The trick to avoiding muscle exhaustion in a long race is to run at a speed just under the "anaerobic threshold"—the rate of exertion at which oxygen is in short supply, ATP is supplied only by glycolysis, and lactate is produced.

Now the question is, which fuel will metabolism rely on during a long race—carbohydrate or fat? Burning fatty acids from fats is more efficient. More than twice as many calories are generated by burning a gram of fat than by burning a gram of carbohydrate. When we are sitting quietly, in fact, our muscle cells are burning mostly fat, and the fat in storage could support the exertion of marathon running for several days. By contrast, glycogen alone can provide enough glucose to fuel only two to three hours of such running under aerobic conditions.

The difficulty is that fatty acids cannot be delivered to muscle cells fast enough to maintain the ATP level needed for running, so metabolism compromises and the glycogen stored in muscles remains the limiting factor for the marathon runner. Once glycogen is gone, extreme exhaustion and mental confusion set in—the condition known as "hitting the wall." Running speed is then limited to that sustainable by fats only. To delay this point as long as possible, a runner encourages glycogen synthesis by a diet high in carbohydrates prior to and during a race. In the hours just before the race, however, carbohydrates are avoided. Their effect of triggering insulin release is undesirable at this point because the resulting faster use of glucose will hasten depletion of glycogen.

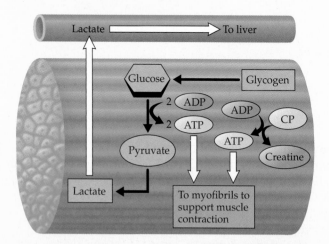

▲ At peak activity, ATP formation relies on creatine phosphate (CP) and glucose from muscle glycogen. Pyruvate is converted to lactate, which enters the bloodstream for transport to the liver, where it is recycled to pyruvate.

See Additional Problems 23.79 through 23.81 at the end of the chapter.

WORKED EXAMPLE **23.3** Analyzing a Reaction

The following overall reaction is a good example of the coupling of endergonic and exergonic reactions:

$$\text{UTP} + \text{Glucose 1-phosphate} + H_2O \longrightarrow \text{Glucose-UDP} + 2 \, HOPO_3^{2-}$$

The coupled reactions are

$$\text{UTP} + \text{Glucose 1-phosphate} \longrightarrow$$
$$\text{Glucose-UDP} + OP_2O_6^{4-} \qquad \Delta G = 1.1 \text{ kcal/mol}$$
$$OP_2O_6^{4-} + H_2O \longrightarrow 2 \, HOPO_3^{2-} \qquad \Delta G = -8.0 \text{ kcal/mol}$$

What is the common intermediate in these coupled reactions? What is ΔG for the coupled reactions? Based on these ΔG values, is the change favorable or unfavorable?

ANALYSIS To find the common intermediate, look for a moiety that appears on the left-hand side of one of the equations and on the right-hand side of the other. In this case $OP_2O_6^{4-}$ appears in each equation, but on opposite sides. It must be the common intermediate. To determine the ΔG for the coupled reactions, simply add the ΔGs. If the sign of the sum is negative, the reaction is exergonic and favored; if the sign of the sum is positive, the reaction is endergonic and unfavored.

SOLUTION
Inspection of the coupled reactions shows $OP_2O_6^{4-}$ to be the common intermediate because it appears on opposite sides of the arrow in the two reactions. The sum of the ΔGs is -6.9 kcal/mol. The reaction is exergonic and is favored.

23.11 Gluconeogenesis: Glucose from Noncarbohydrates

Gluconeogenesis, which occurs mainly in the liver, is the pathway for making glucose from noncarbohydrate molecules—lactate, amino acids, and glycerol. This pathway becomes critical during fasting and the early stages of starvation. Failure of gluconeogenesis is usually fatal.

Lactate is a normal product of glycolysis in red blood cells and of vigorous muscle activity. The body deals with the conditions during and after vigorous muscle activity—that is, high concentrations of lactate and depleted muscle glycogen—via the pathway shown in Figure 23.9. This pathway allows the liver to assume the burden of producing glucose by recycling it from lactate. Lactate absorbed from the

Gluconeogenesis The biochemical pathway for the synthesis of glucose from noncarbohydrates, such as lactate, amino acids, or glycerol.

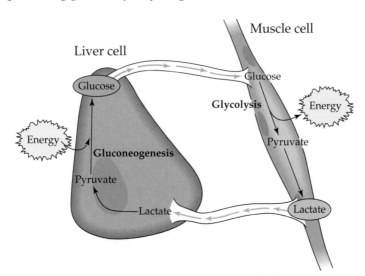

◀ **FIGURE 23.9 Glucose production during exercise (the Cori cycle).** L-Lactate produced in muscles under anaerobic conditions during exercise is sent to the liver, where it is converted back to glucose. The glucose can then return via the bloodstream to the muscles, to be stored as glycogen or used for energy production. Gluconeogenesis requires energy, so shifting this pathway to the liver frees the muscles from the burden of having to produce even more energy.

blood is converted to pyruvate, the reactant for the first step of gluconeogenesis. The glucose synthesized in the liver is then returned to the muscles, where it can provide energy once more or be placed back into storage as muscle glycogen. This cycle, referred to as the Cori cycle, is named for the husband-wife scientist team Carl and Gerti Cori who elucidated its intricacies and won the Nobel Prize in Medicine in 1947 for their work on glycogen metabolism.

We noted earlier that for metabolic pathways to be favorable, they must be exergonic. As a result, most are not reversible, because the amount of energy required by the reverse, endergonic pathway would be too large to be supplied by cellular metabolism. Glycolysis and gluconeogenesis provide another good example of this relationship and of the way around it.

Three reactions in glycolysis—Steps 1, 3, and 10 in Figure 23.3—are too exergonic to be directly reversed. Gluconeogenesis therefore uses alternative reactions catalyzed by different enzymes than those used in glycolysis that effectively reverse these steps. For example, gluconeogenesis begins in the mitochondria with conversion of pyruvate to phosphoenolpyruvate, the reverse of the highly exergonic Step 10 of glycolysis. Pyruvate is transported from the cytosol into the mitochondria or is produced there from amino acids before gluconeogenesis can begin. In gluconeogenesis, two steps are required, utilizing two enzymes and the energy provided by two triphosphates, ATP and GTP. (Guanosine triphosphate, GTP, is similar in structure to ATP, and its phosphate groups can be removed in exergonic reactions. GDP is guanosine diphosphate.) In Step 1, pyruvate is converted to oxaloacetate by pyruvate carboxylase, which adds a carboxyl group to pyruvate in an energy expensive reaction that bypasses the last, irreversible step in glycolysis. This is the first of three reactions that bypass irreversible steps in glycolysis. Next, oxaloacetate is reduced to malate and transported out of the mitochondrion into the cytosol where malate is immediately reconverted to oxaloacetate. Then, in Step 2, phosphoenolpyruvate carboxykinase adds a phosphate group and rearranges oxaloacetate to produce phosphoenolpyruvate.

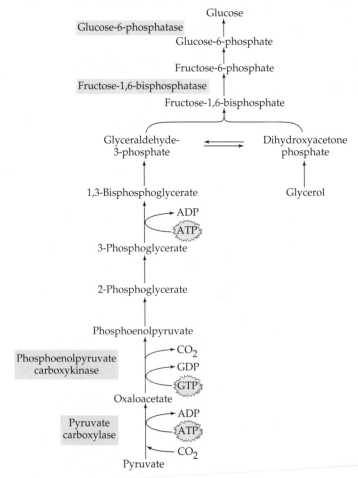

APPLICATION ▶ Polysaccharides—What Are They Good For?

For a moment, forget about the metabolism of carbohydrates and think of them as raw materials—sources of connected carbon, hydrogen, and oxygen atoms. For years, we have taken advantage of the fiber-forming properties of cellulose. Cotton clothing is about 90% cellulose, and we make rayon (cellulose acetate) by converting the hydroxyl groups in cellulose to acetate ester groups. Many chemists are now hard at work exploring new ways to make useful substances from carbohydrates. This research is part of a movement to expand the use of *biomass*—raw materials from plants and animals. Biomass can be derived from fast-growing trees and grasses; agricultural leftovers such as wheat straw; and wood waste, which includes sawdust, tree prunings, waste paper, and yard waste.

One possible way to use cellulose is in production of a nonpolluting, energy-efficient fuel—hydrogen. This would be environmentally friendly. The only by-product of burning hydrogen is water, so replacement of fossil fuels with hydrogen is a tantalizing concept. Burning hydrogen from biomass would also be economical. It has been calculated that the cellulose from the huge quantity of U.S. waste newspaper alone could provide enough hydrogen to meet the energy needs of one million city residents. And as another benefit, biomass is renewable—we can keep growing it, so that it holds promise for the future as supplies of fossil fuels dwindle and become more difficult to recover.

The first step would be production of glucose by cooking cellulose with strong acid. Then, a bacterial enzyme (a glucose dehydrogenase) could be used to oxidize the glucose while converting $NADP^+$ to NADPH. Finally, another enzyme could be used to produce hydrogen from NADPH. Will we ever see this process in practical use? Right now, no one knows.

We are quite familiar with another use of renewable biomass—the production of ethyl alcohol in beverages by yeast fermentation of the glucose in grapes, corn, rye, and other plant materials. Did you know that, with some modification, automobile engines can run on straight ethyl alcohol? A 10% alcohol/90% gasoline mixture that can be used in existing engines is already commercially available. Several states and countries require that all gasoline sold contain ethanol. Currently, most ethanol for fuel is produced from corn; however, the search is on for a combination of chemical and enzymatic treatments that would permit large-scale production of fuel-grade ethyl alcohol from materials such as agricultural waste (straw, corn cobs, hulls, and so on), hardwood sawdust, and waste from pulp and paper production. Switchgrass, a rapid growing, dry-land native grass is under consideration as a major source of ethanol and biodiesel; switchgrass grows readily on marginal cropland. Another

promising source is commercial production of algae that contain a high amount of oil as well as fermentable carbohydrate for cellulosic ethanol production.

A variety of carbohydrate derivatives are widely used in commercial food products. Their major advantage is that they provide a texture and mouth feel like that of fats, but accompanied by zero calories or fewer calories than fat. The Calorie Control Council, a nonprofit association that represents manufacturers of low-calorie foods and their ingredients, lists 12 carbohydrate-based fat replacers. They include zero-calorie food additives from cellulose and plant fibers. A finely ground cellulose, for example, can replace some of the fat in dairy products, frozen desserts, and salad dressings. A product from the insoluble fibers in oat, soybean, and other plant hulls is heat stable and can thus be used in baked goods, hot dogs, and hamburgers. There are soluble gums (guar gum, xanthan gum from bacteria, carrageenan from seaweed, and pectin, for example) that incorporate water molecules within their highly branched structures to produce thick, creamy textures. The gums, which also have just about zero calories, are used in salad dressings, desserts, and processed meats. Fat replacers with a few calories per gram include modified starches, dextrins (found in tapioca), and a group of polyols that can serve as both sweeteners with fewer calories than sugar and fat replacers.

You might try reading the ingredients list of prepared foods to see how many fat replacers you are consuming every day.

▲ **This biomass could be put to practical use.**

See Additional Problems 23.82 and 23.83 at the end of the chapter.

The next six reactions of the 10-step glycolysis pathway are reversible because they operate at near-equilibrium conditions. They occur in reverse during gluconeogenesis, catalyzed by the same enzymes as in glycolysis. The second bypass of an irreversible step occurs in the conversion of fructose 1,6 bisphosphate to fructose 6-phosphate (Step 9). This step does not produce ATP, as an exact reversal would; instead fructose 1,6-bisphosphatase removes a phosphate group by hydrolysis in an energetically favorable reaction.

The next to last reaction in gluconeogenesis is the reverse of the glycolysis reaction, resulting in the conversion of fructose 6-phosphate to glucose 6-phosphate. The final reaction in the pathway is the third bypass of a glycolysis pathway reaction. Because it is energetically unfavorable to reverse hexokinase, regenerating ATP from ADP in the process, a different enzyme, glucose 6-phosphatase hydrolyzes glucose 6-phosphate producing glucose and inorganic phosphate. The glucose produced crosses the cell membrane and is then transported to cells that use only glucose for energy generation, such as red blood cells or brain tissue.

Glycerol from triacylglycerol catabolism (Section 25.3) is converted to dihydroxyacetone phosphate and enters the gluconeogenesis pathway at Step 5 (Figure 23.3). The carbon atoms from certain amino acids (the glucogenic amino acids, Section 28.5) enter gluconeogenesis as either pyruvate or oxaloacetate.

PROBLEM 23.13

What two types of reactions convert glycerol to dihydroxyacetone phosphate?

$$
\begin{array}{ccc}
CH_2OH & & CHOPO_3^{2-} \\
| & & | \\
HO-C-H & \xrightarrow{?} & C=O \\
| & & | \\
CH_2OH & & CH_2OH \\
\text{Glycerol} & & \text{Dihydroxyacetone} \\
& & \text{phosphate}
\end{array}
$$

SUMMARY: REVISITING THE CHAPTER GOALS

1. **What happens during digestion of carbohydrates?** Carbohydrate *digestion,* the hydrolysis of disaccharides and polysaccharides, begins in the mouth and continues in the stomach and small intestine. The products that enter the bloodstream from the small intestine are monosaccharides—mainly glucose, fructose, and galactose.

2. **What are the major pathways in the metabolism of glucose?** The major pathway for glucose, once inside a cell and converted to glucose 6-phosphate, is *glycolysis.* Pyruvate, the end product of glycolysis, is then fed into the citric acid cycle via acetyl-SCoA. One alternative pathway for glucose is *glycogenesis,* the synthesis of glycogen, which is stored mainly in the liver and muscles. Another alternative is the *pentose phosphate pathway,* which provides the 5-carbon sugars and NADPH needed for biosynthesis (see Figure 23.2).

3. **What is glycolysis?** Glycolysis (Figure 23.3) is a 10-step pathway that produces two molecules of pyruvate, two reduced coenzymes (NADH), and two ATPs for each molecule of glucose metabolized. Glycolysis begins with phosphorylation (Steps 1–3) to form fructose 1,6-bisphosphate, followed by cleavage and isomerization that produce two molecules of glyceraldehyde 3-phosphate (Steps 4–5). Each glyceraldehyde 3-phosphate then proceeds through the energy-generating steps (Steps 6–10) in which phosphates are alternately created and then donate their phosphate groups to ADP. Dietary monosaccharides other than glucose enter glycolysis at various points–fructose as fructose 6-phosphate or glyceraldehyde 3-phosphate, galactose as glucose 6-phosphate, and mannose as fructose 6-phosphate.

4. **What happens to pyruvate once it is formed?** When oxygen is in good supply, pyruvate is transported into mitochondria and converted to acetyl-SCoA for energy generation via the citric acid cycle and oxidative phosphorylation. When there is insufficient oxygen,

pyruvate is reduced to L-lactate, with the production of the oxidized coenzyme NAD^+ that is essential to the continuation of glycolysis. This production of NAD^+ compensates for the shortage of NAD^+ created by the slowdown of electron transport when oxygen is in short supply. The lactate is transported in the bloodstream to the liver, where oxygen is available, and is oxidized back to pyruvate. In the presence of yeast, pyruvate undergoes *anaerobic fermentation* to yield ethyl alcohol.

5. **How is glucose metabolism regulated, and what are the influences of starvation and diabetes mellitus?** *Insulin*, produced when blood glucose concentration rises, accelerates glycolysis and glycogen synthesis. *Glucagon*, produced when blood glucose concentration drops, accelerates production of glucose in the liver from stored glycogen and from other precursors via the *gluconeogenesis* pathway. Adaptation to starvation begins with the effects of glucagon and energy production from protein, and then proceeds to reliance on ketone bodies from fatty tissue for energy generation. *Diabetes mellitus* may be insulin-dependent (the pancreas fails to produce insulin) or non-insulin-dependent (insulin receptors fail to recognize insulin). Among the serious outcomes of uncontrolled diabetes are cataracts, blood vessel lesions, ketoacidosis, and *hypoglycemia*.

6. **What are glycogenesis and glycogenolysis?** *Glycogenesis* (Figure 23.8), the synthesis of the polysaccharide glycogen, puts excess glucose into storage, mainly in muscles and the liver. *Glycogenolysis* is the release of stored glucose from glycogen. Glycogenolysis occurs in muscles when there is an immediate need for energy, producing glucose 6-phosphate for intracellular glycolysis. It occurs in the liver when blood glucose concentration is low and must be elevated. The liver has an enzyme to convert glucose 6-phosphate to glucose, which is released to the bloodstream.

7. **What is the role of gluconeogenesis in metabolism?** *Gluconeogenesis* (Figure 23.9) maintains glucose levels by synthesizing it from lactate, from certain amino acids derived from proteins, and from glycerol derived from fatty tissue; it is part of normal metabolism and is critical during fasting and starvation. The gluconeogenesis pathway uses alternate enzymes for the reverse of the three highly exergonic steps of glycolysis, but otherwise utilizes the same enzymes for reactions that run in reverse of their direction in glycolysis.

UNDERSTANDING KEY CONCEPTS

23.14 What class of enzymes catalyzes the majority of the reactions involved in carbohydrate digestion?

23.15 Glucose 6-phosphate is in a pivotal position in metabolism. Depending on conditions, glucose 6-phosphate follows one of several pathways. Under what conditions do the following occur?

(a) Glycolysis
(b) Hydrolysis to free glucose
(c) Pentose phosphate pathway
(d) Glycogenesis

23.16 What "chemical investments" are made to get glycolysis started, and why are they made? What happens in the middle of the pathway to generate two 3-carbon compounds? What are the outcomes of the reactions of these 3-carbon compounds?

23.17 Outline the condition(s) that directs pyruvate toward:

(a) Entry into the citric acid cycle
(b) Conversion to ethanol and CO_2
(c) Conversion to lactate
(d) Glucose synthesis (gluconeogenesis)

In what tissues or organisms is each pathway present?

23.18 Classify each enzyme of glycolysis into one of the six classes of enzymes. What class of enzymes has the most representatives in glycolysis? Why is this consistent with the goals of glycolysis? Why are ligases *not* represented in glycolysis?

23.19 When blood glucose levels rise following a meal, the events listed below occur. Arrange these events in the appropriate sequence.

(a) Glucagon is secreted.
(b) Glycolysis replenishes ATP supplies.
(c) Glucose is absorbed by cells.
(d) The liver releases glucose into the bloodstream.
(e) Glycogen synthesis (glycogenesis) occurs with excess glucose.
(f) Blood levels pass through normal to below normal (hypoglycemic).
(g) Insulin levels rise.

23.20 What are the sources of molecules for gluconeogenesis? When and in what kind of cell is each molecule the predominant one used?

23.21 Fatty acids from stored triacylglycerols (fat) are *not* available for gluconeogenesis. Speculate why we do not have the enzymes to convert fatty acids into glucose. Plants (especially seeds) *do* have enzymes to convert fatty acids into carbohydrates. Why are they so lucky?

23.22 The pathway that converts glucose to acetyl-SCoA is often referred to as an "aerobic oxidation pathway." (a) Is molecular oxygen involved in any of the steps of glycolysis? (b) Thinking back to Chapter 21, where does molecular oxygen enter the picture?

ADDITIONAL PROBLEMS

DIGESTION AND METABOLISM

23.23 Where in the body does digestion occur, and what kinds of chemical reactions does it involve?

23.24 Complete the following word equation:

$$\text{Lactose} + H_2O \longrightarrow ? + ?$$

Where in the digestive system does this process occur?

23.25 What are the major monosaccharide products produced by digestion of carbohydrates?

23.26 What are the products of digestion of proteins, triacylglycerols, maltose, sucrose, lactose, and starch?

23.27 What do the words *aerobic* and *anaerobic* mean?

23.28 What three products are formed from pyruvate under aerobic, anaerobic, and fermentation conditions?

23.29 Differentiate between glycolysis and gluconeogenesis.

23.30 Differentiate between glycogenolysis and glycogenesis.

23.31 Differentiate between gluconeogenesis and glycogenesis.

23.32 Differentiate between glycolysis and glycogenolysis.

23.33 What is the major purpose of the pentose phosphate pathway? What cofactor (coenzyme) is used?

23.34 Depending on the body's needs, into what type of compounds is glucose converted in the pentose phosphate pathway?

GLYCOLYSIS

23.35 Where in a eukaryotic cell do the following pathways occur?

(a) Glycolysis
(b) Gluconeogenesis
(c) Glycogenesis
(d) Glycogenolysis

23.36 Which cell types, liver, muscle, or brain, uses the following pathways?

(a) Glycolysis
(b) Gluconeogenesis
(c) Glycogenesis
(d) Glycogenolysis

23.37 What is the name of the final product of glycolysis? Draw it. Is there also a changed coenzyme product? If so, what is it?

23.38 Glycolysis can occur under anaerobic conditions. What two possible pathways handle the NADH generated in this case? What are the products of the two pathways?

23.39 Which glycolysis reactions are catalyzed by the following enzymes?

(a) Pyruvate kinase
(b) Glyceraldehyde 3-phosphate dehydrogenase
(c) Hexokinase
(d) Phosphoglycerate mutase
(e) Aldolase

23.40 Review the 10 steps in glycolysis (Figure 23.3) and then answer the following questions:

(a) Which steps involve phosphorylation?
(b) Which step is an oxidation?
(c) Which step is a dehydration?

23.41 How many moles of ATP are produced by phosphorylation in the following?

(a) Glycolysis of 1 mol of glucose
(b) Aerobic conversion of 1 mol of pyruvate to 1 mol of acetyl-SCoA
(c) Catabolism of 1 mol of acetyl-SCoA in the citric acid cycle

23.42 For each reaction in problem 23.42 tell if the ATP formed is produced by oxidative phosphorylation or substrate level phosphorylation. What is the difference in the two types of ATP formation?

23.43 Why is pyruvate converted to lactate under anaerobic conditions?

23.44 Lactate can be converted into pyruvate by the enzyme lactate dehydrogenase and the coenzyme NAD^+. Write the reaction in the standard biochemical format, using a curved arrow to show the involvement of NAD^+.

23.45 How many moles of CO_2 are produced by the complete catabolism of 1 mol of sucrose?

23.46 How many moles of acetyl-SCoA are produced by the complete catabolism of 1 mol of sucrose?

REGULATION OF GLUCOSE METABOLISM/METABOLISM IN DIABETES MELLITUS

23.47 Differentiate between the effect of insulin and glucagon on blood sugar concentration.

23.48 Differentiate between blood sugar levels and resulting symptoms in hyperglycemia and hypoglycemia.

23.49 What molecules are used initially during starvation or fasting to produce glucose?

23.50 As starvation continues, acetyl-SCoA is converted to _____ to prevent its buildup in the cells. (Fill in the blank.)

23.51 What are the symptoms of Type I diabetes?

23.52 What are the characteristics of Type II diabetes?

23.53 Explain the relationship between metabolic syndrome and diabetes.

23.54 Many diabetics suffer blindness due to cataracts or undergo amputation of limbs. Why are these conditions associated with this disease?

23.55 Why is Type I diabetes considered to be an autoimmune disease?

23.56 What is the difference between Type I and Type II diabetes?

23.57 Where is most of the glycogen in the body stored?

23.58 What major site of glycogen storage is not able to release glucose to the bloodstream?

23.59 How is UTP used in the formation of glycogen from glucose?

23.60 Glycogenolysis is not the exact reverse of the process of glycogenesis. Explain why.

GLUCOSE ANABOLISM

23.61 Name the anabolic pathway for making glucose.

23.62 Name the two molecules that serve as starting materials for glucose synthesis.

23.63 Pyruvate is initially converted to _____ in the anabolism of glucose. That molecule in turn is converted to _____. (Fill in the blanks.)

23.64 Explain why pyruvate cannot be converted to glucose in an exact reverse of the glycolysis pathway.

23.65 Explain how the energy releasing steps of glycolysis are reversed in gluconeogenesis.

23.66 How many steps in gluconeogenesis are not the exact reversal of the steps in glycolysis? What kind of conversion of substrate to product does each involve? What is the common theme in each of these reactions?

23.67 What is the Cori cycle?

23.68 Explain why the Cori cycle is necessary and when an organism would use this cycle.

Applications

23.69 What is the function of the insoluble polysaccharide known as dextran in formation of dental plaque? [*Tooth Decay, p. 737*]

23.70 Name four of the major components of dental plaque. [*Tooth Decay, p. 737*]

23.71 How is dental plaque associated with periodontal disease? [*Tooth Decay, p. 737*]

23.72 Explain the process that leads to cavities after dental plaque has formed. [*Tooth Decay, p. 737*]

23.73 Compare alcohol fermentation in yeast to lactate production in muscle. [*Microbial Fermentations: Ancient and Modern, p. 741*]

23.74 What common fermentation products can be found in a typical grocery store? [*Microbial Fermentations: Ancient and Modern, p. 741*]

23.75 Briefly describe the enzymatic process for determination of glucose. [*Diagnosis and Monitoring of Diabetes, p. 744*]

23.76 How do fasting levels of glucose in a diabetic person compare to those in a nondiabetic person? [*Diagnosis and Monitoring of Diabetes, p. 744*]

23.77 Discuss the differences in the response of a diabetic person compared to those of a nondiabetic person after drinking a glucose solution. [*Diagnosis and Monitoring of Diabetes, p. 744*]

23.78 Describe the response curve of a pre-diabetic person to a glucose challenge. How does this differ from the response of a normal person and that of a diabetic? [*Diagnosis and Monitoring of Diabetes, p. 744*]

23.79 Why is creatine phosphate a better source of quick energy than glucose? [*The Biochemistry of Running, p. 748*]

23.80 Why is it not possible for a person to sprint for miles? [*The Biochemistry of Running, p. 748*]

23.81 Order the following sources of energy (from first used to last used) when muscles are called upon to do extensive work:

(a) Fatty acids from triacylglycerols
(b) ATP
(c) Glycogen
(d) Creatine phosphate
(e) Glucose

[*The Biochemistry of Running, p. 748*]

23.82 Cellulose is the most abundant biomolecule in nature. Describe three products on the market that are derived from cellulose. What derivatives of cellulose are used to make these products? [*Polysaccharides—What Are They Good For? p. 751*]

23.83 Ethanol is fermented for use as a gasoline replacement from which major food crop? [*Polysaccharides—What Are They Good For? p. 751*]

General Questions and Problems

23.84 Why can pyruvate cross the mitochondrial membrane but no other molecule after Step 1 in glycolysis can?

23.85 Look at the glycolysis pathway (Figure 23.3). With what type of process are kinase enzymes usually associated?

23.86 Is the same net production of ATP observed in the complete oxidation of fructose as is observed in the oxidation of glucose? Why or why not?

23.87 Explain why one more ATP is produced when glucose is obtained from glycogen rather than used directly from the blood.

23.88 Why is it important that glycolysis be tightly controlled by the cell?

23.89 It is important to avoid air when making wine, so a novice winemaker added yeast to fresh grape juice and placed it in a sealed bottle to avoid air. Several days later, the lid exploded off the bottle. Explain the biochemistry responsible for the exploding lid.

Lipids

▲ An abundant supply of fish is converted to energy stored as triacylglycerols, a lipid, in fat tissue to supply energy during hibernation.

CONTENTS

CHAPTER GOALS

In this chapter, we will answer the following questions about lipids:

1. What are the major classes of fatty acids and lipids?

THE GOAL: Be able to describe the chemical structures and general properties of fatty acids, waxes, fats, and oils.

2. What reactions do triacylglycerols undergo?

THE GOAL: Be able to describe the results of hydrogenation and hydrolysis of triacylglycerols, and, given the reactants, predict the products.

3. What are the membrane lipids?

THE GOAL: Be able to identify the membrane lipids, describe their structures, and roles.

4. What is the nature of a cell membrane?

THE GOAL: Be able to describe the general structure of a cell membrane and its chemical composition.

5. How do substances cross cell membranes?

THE GOAL: Be able to distinguish between passive transport and active transport and between simple diffusion and facilitated diffusion.

6. What are eicosanoids?

THE GOAL: Be able to describe the general structure of prostaglandins and leukotrienes, and some of their functions.

L ipids are less well known than carbohydrates and proteins, yet lipids are just as essential to our diet and well-being. They have three major roles in human biochemistry: (1) Within fat cells (*adipocytes*), they store energy from metabolism of food. (2) As part of all cell membranes, they keep separate the different chemical environments inside and outside the cells. (3) In the endocrine system and elsewhere, they serve as chemical messengers.

Chemically, lipids are defined as naturally occurring organic molecules that are nonpolar and therefore dissolve in nonpolar organic solvents but not in water. For example, if a sample of plant or animal tissue is placed in a kitchen blender, finely ground, and then extracted with ether, any molecule that dissolves in the ether is a lipid and any molecule that does not dissolve in the ether (including carbohydrates, proteins, and inorganic salts) is not a lipid.

24.1 Structure and Classification of Lipids

Since **lipids** are defined by solubility in nonpolar solvents (a physical property) rather than by chemical structure, it should not surprise you that there are a great many different kinds and that they serve a variety of functions in the body. In the following examples of lipid structures, note that the molecules contain large hydrocarbon portions and not many polar groups, which accounts for their solubility behavior. Many lipids have hydrocarbon or modified hydrocarbon structure, properties, and behavior. This similarity to hydrocarbons and their derivatives unifies a set of highly diverse molecules into one class.

Lipid A naturally occurring molecule from a plant or animal that is soluble in nonpolar organic solvents.

$$CH_3(CH_2)_{28}CH_2-O-\overset{\overset{\displaystyle O}{\|}}{C}-(CH_2)_{14}CH_3$$

A wax

$$CH_2-O-\overset{\overset{\displaystyle O}{\|}}{C}-(CH_2)_{14}CH_3$$
$$CH-O-\overset{\overset{\displaystyle O}{\|}}{C}-(CH_2)_7CH=CH(CH_2)_7CH_3$$
$$CH_2-O-\overset{\overset{\displaystyle O}{\|}}{C}-(CH_2)_{16}CH_3$$

A triacylglycerol

Cholesterol, a steroid

A prostaglandin

757

Fatty acid A long-chain carboxylic acid; those in animal fats and vegetable oils often have 12–22 carbon atoms.

Figure 24.1 organizes the classes of lipids discussed in this chapter according to their chemical structures. Many lipids are esters or amides of carboxylic acids with long, unbranched hydrocarbon chains, which are known as **fatty acids**. The fatty acids that contain unbranched hydrocarbon chains are loosely referred to as *straight-chain fatty acids*.

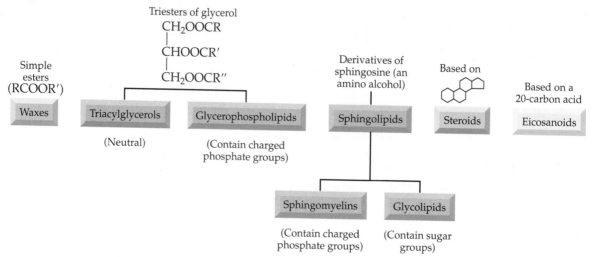

▲ **FIGURE 24.1 Families of lipids.**

Lipids that are esters or amides of fatty acids

- *Waxes* are carboxylic acid esters (RCOOR′) with long, straight hydrocarbon chains in both R groups; they are secreted by sebaceous glands in the skin of animals and perform mostly external protective functions (Section 24.2).

- *Triacylglycerols* are carboxylic acid triesters of glycerol, a three-carbon trialcohol. Triacylglycerols (Sections 24.2–24.3) are found as the fats stored in our bodies and in most dietary fats and oils. They are a major source of biochemical energy, a function described in Chapter 25.

- *Glycerophospholipids* (Section 24.5) are triesters of glycerol that contain charged phosphate diester groups and are abundant in cell membranes. Together with other lipids, they help to control the flow of molecules into and out of cells.

- *Sphingomyelins*, amides derived from an amino alcohol (*sphingosine*), also contain charged phosphate diester groups; they are essential to the structure of cell membranes (Section 24.5) and are especially abundant in nerve cell membranes.

- *Glycolipids*, different amides derived from *sphingosine*, contain polar carbohydrate groups; on cell surfaces the carbohydrate portion is recognized and interact with intercellular messengers (Section 24.5).

There are also two groups of lipids that are not esters or amides: the *steroids* and the *eicosanoids*.

Other types of lipids

- *Steroids*. You have already seen the role of steroids as hormones. (⊂⊃, Section 20.5) In this chapter, we focus on cholesterol (Section 24.6), which contributes to the structure of cell membranes.

- *Eicosanoids*. The eicosanoids (Section 24.9) are carboxylic acids that are a special type of intercellular chemical messenger.

WORKED EXAMPLE 24.1 Identifying Lipid Families

Use Figure 24.1 to identify the family of lipids to which each of these molecules belongs.

(a)

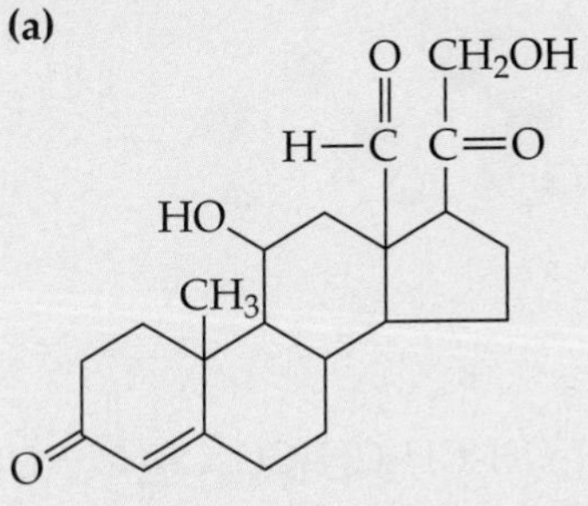

(b)

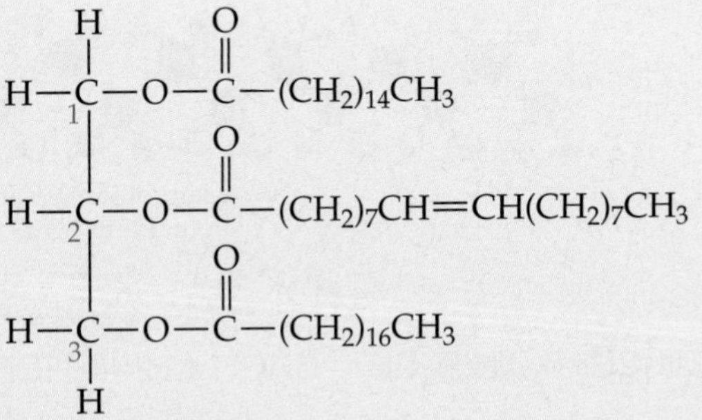

ANALYSIS Inspect the molecules and note their distinguishing characteristics. Molecule (a) has a four-member fused ring system. Only steroids have this structure. Molecule (b) has three fatty acids esterified to a single backbone molecule—glycerol. Thus, (b) must be a member of the triacylglycerol family.

SOLUTION
Molecule (a) is a steroid (sterol), and molecule (b) is a triacylglycerol.

PROBLEM 24.1

Use Figure 24.1 to identify the family of lipids to which each of these molecules belongs.

(a)

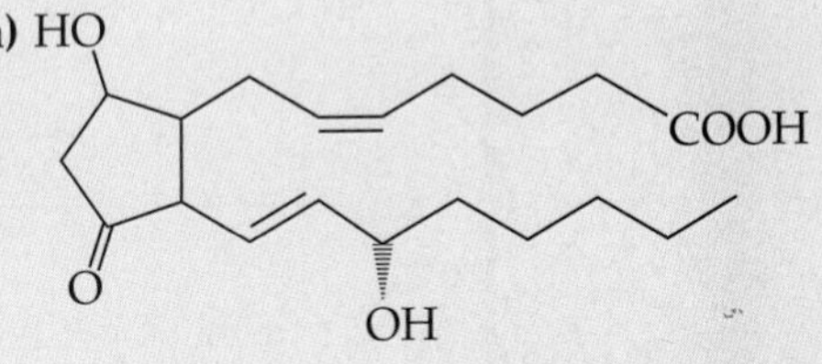

(b)

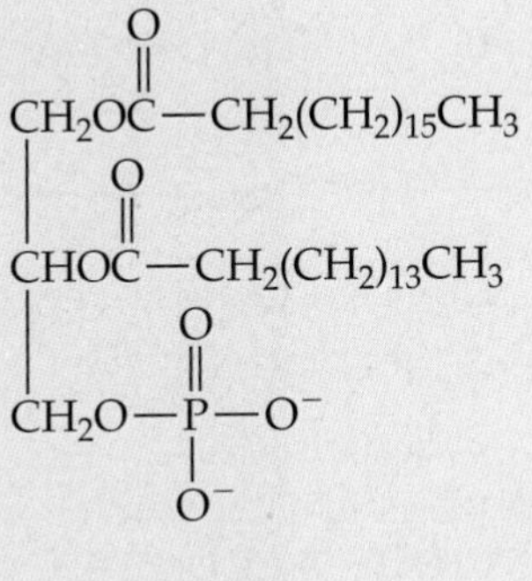

(c) $CH_3(CH_2)_{16}\overset{\displaystyle O}{\overset{\|}{C}}-O-CH_2(CH_2)_6CH=CH(CH_2)_6CH_3$

24.2 Fatty Acids and Their Esters

Naturally occurring fats and oils are triesters formed between glycerol and fatty acids. (Recall that an ester, RCOOR′, is formed from a carboxylic acid and an alcohol. (⬭, Section 17.4) Fatty acids are long, unbranched hydrocarbon chains with a carboxylic acid group at one end. Most have even numbers of carbon atoms. Fatty acids may or may not contain carbon–carbon double bonds. Those without double bonds are known as **saturated fatty acids**; those with double bonds are known as **unsaturated fatty acids**. If double bonds are present in naturally occurring fats and oils, the double bonds are usually cis rather than trans. (⬭, Section 13.3)

Saturated fatty acid A long-chain carboxylic acid containing only carbon–carbon single bonds.

Unsaturated fatty acid A long-chain carboxylic acid containing one or more carbon–carbon double bonds.

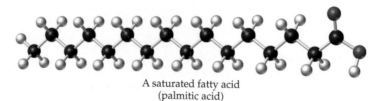

A saturated fatty acid
(palmitic acid)

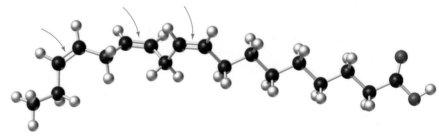

A cis unsaturated fatty acid
(linolenic acid)

Polyunsaturated fatty acid A long-chain carboxylic acid that has two or more carbon–carbon double bonds.

Some of the common fatty acids are listed in Table 24.1. Palmitic acid (16 carbons) and stearic acid (18 carbons) are the most common saturated acids; oleic and linoleic acids (both with 18 carbons) are the most common unsaturated ones. Oleic acid is *monounsaturated*, that is, it has only one carbon–carbon double bond. The **polyunsaturated fatty acids** have more than one carbon–carbon double bond.

TABLE 24.1   Structures of Some Common Fatty Acids

NAME	TYPICAL SOURCE	NUMBER OF CARBONS	NUMBER OF DOUBLE BONDS	CONDENSED FORMULA	MELTING POINT (°C)
Saturated					
Lauric	Coconut oil	12	0	$CH_3(CH_2)_{10}COOH$	44
Myristic	Butter fat	14	0	$CH_3(CH_2)_{12}COOH$	58
Palmitic	Most fats and oils	16	0	$CH_3(CH_2)_{14}COOH$	63
Stearic	Most fats and oils	18	0	$CH_3(CH_2)_{16}COOH$	70
Unsaturated					
Oleic	Olive oil	18	1	$CH_3(CH_2)_7CH{=}CH(CH_2)_7COOH(cis)$	4
Linoleic	Vegetable oils	18	2	$CH_3(CH_2)_4CH{=}CHCH_2CH{=}CH(CH_2)_7COOH(all\ cis)$	−5
Linolenic	Soybean and canola oils	18	3	$CH_3CH_2CH{=}CHCH_2CH{=}CHCH_2CH{=}CH(CH_2)_7COOH(all\ cis)$	−11
Arachidonic	Lard	20	4	$CH_3(CH_2)_4(CH{=}CHCH_2)_4CH_2CH_2COOH(all\ cis)$	−50

Two of the polyunsaturated fatty acids, linoleic and linolenic, are essential in the human diet because the body does not synthesize them even though they are needed for the synthesis of other lipids. Infants grow poorly and develop severe skin lesions if fed a diet lacking these acids. Adults usually have sufficient reserves in body fat to avoid such problems. A deficiency in adults can arise, however, after long-term intravenous feeding that contains inadequate essential fatty acids or among those surviving on limited and inadequate diets. Malnutrition in the developed world also results from many other causes; two common ones are unusual slimming diets and anorexia.

Waxes

The simplest fatty acid esters in nature are waxes. A **wax** is a mixture of fatty acid—long-chain alcohol esters. The acids usually have an even number from 16 to 36 carbons, whereas the alcohols have an even number from 24 to 36 carbons. For example, a major component in beeswax is the ester formed from a 30-carbon alcohol (triacontanol) and a 16-carbon acid (palmitic acid). The waxy protective coatings on most fruits, berries, leaves, and animal furs have similar structures. Aquatic birds have a water-repellent waxy coating on their feathers. When caught in an oil spill, the waxy coating dissolves in the oil and the birds lose their buoyancy.

Wax A mixture of monoesters of long-chain carboxylic acids with long-chain alcohols.

▲ This grebe is coated with oil spilled by a tanker that sank off Brittany on the northwest coast of France. If the oil is not removed from its feathers, the bird will perish.

Example of a wax

| Long-chain alcohol | Long-chain acid |

$$CH_3(CH_2)_{28}CH_2 - O - \overset{\overset{\displaystyle O}{\|}}{C}(CH_2)_{14}CH_3$$

Triacontanyl hexadecanoate (from beeswax)

Triacylglycerols

Animal fats and vegetable oils are the most plentiful lipids in nature. Although they appear different—animal fats like butter and lard are solid, whereas vegetable oils like corn, olive, soybean, and peanut oil are liquid—their structures are closely related. All fats and oils are composed of triesters of glycerol (1,2,3-propanetriol, also known as glycerine) with three fatty acids. They are named chemically as **triacylglycerols**, but are often called **triglycerides**.

Triacylglycerol (triglyceride) A triester of glycerol with three fatty acids.

Triacylglycerols

$$
\begin{array}{l}
CH_2OH \\
| \\
CHOH \\
| \\
CH_2OH
\end{array}
\;+\;
\begin{array}{l}
R\overset{\overset{\displaystyle O}{\|}}{C}-OH \\
R'\overset{\overset{\displaystyle O}{\|}}{C}-OH \\
R''\overset{\overset{\displaystyle O}{\|}}{C}-OH
\end{array}
\;\longrightarrow\;
\begin{array}{l}
CH_2-O-\overset{\overset{\displaystyle O}{\|}}{C}-R \\
| \\
CH-O-\overset{\overset{\displaystyle O}{\|}}{C}-R' \\
| \\
CH_2-O-\overset{\overset{\displaystyle O}{\|}}{C}-R''
\end{array}
$$

Glycerol   Fatty acids

	Fatty acid
Glycerol	Fatty acid
	Fatty acid

The three fatty acids of any specific triacylglycerol are not necessarily the same, as is the case in the molecule below.

Example of a triacylglycerol

$$CH_2-O-\overset{\overset{\displaystyle O}{\|}}{C}-CH_2CH_2CH_2CH_2CH_2CH_2CH_2CH_2CH_2CH_2CH_2CH_2CH_2CH_2CH_3 \quad \text{Palmitic acid}$$

$$CH-O-\overset{\overset{\displaystyle O}{\|}}{C}-CH_2CH_2CH_2CH_2CH_2CH_2CH_2CH=CHCH_2CH_2CH_2CH_2CH_2CH_2CH_2CH_3 \quad \text{Oleic acid}$$

$$CH_2-O-\overset{\overset{\displaystyle O}{\|}}{C}-CH_2CH_2CH_2CH_2CH_2CH_2CH_2CH=CHCH_2CH=CHCH_2CH_2CH_2CH_3 \quad \text{Linoleic acid}$$

Furthermore, the fat or oil from a given natural source is a complex mixture of many different triacylglycerols. Table 24.2 lists the average composition of fats and oils from several different sources. Note particularly that vegetable oils consist almost entirely of unsaturated fatty acids, whereas animal fats contain a much larger percentage of saturated fatty acids. This difference in composition is the primary reason for the different melting points of fats and oils, as explained in the next section.

TABLE 24.2 Approximate Composition of Some Common Fats and Oils*

SOURCE	SATURATED FATTY ACIDS (%)				UNSATURATED FATTY ACIDS (%)	
	C_{12} LAURIC	C_{14} MYRISTIC	C_{16} PALMITIC	C_{18} STEARIC	C_{18} OLEIC	C_{18} LINOLEIC
Animal Fat						
Lard	—	1	25	15	50	6
Butter	2	10	25	10	25	5
Human fat	1	3	25	8	46	10
Whale blubber	—	8	12	3	35	10
Vegetable Oil						
Corn	—	1	8	4	46	42
Olive	—	1	5	5	83	7
Peanut	—	—	7	5	60	20
Soybean	—	—	7	4	34	53

Where totals are less than 100%, small quantities of several other acids are present, with cholesterol also present in animal fats.

PROBLEM 24.2

One of the constituents of the carnauba wax used in floor and furniture polish is an ester of a C_{32} straight-chain alcohol with a C_{20} straight-chain carboxylic acid. Draw the structure of this ester. (Use subscripts to show the numbers of connected CH_2 groups, as in Table 24.1.)

PROBLEM 24.3

Draw the structure of a triacylglycerol whose components are glycerol and three oleic acid acyl groups.

KEY CONCEPT PROBLEM 24.4

(a) Which animal fat has the largest percentage of saturated fatty acids?

(b) Which vegetable oil has the largest percentage of polyunsaturated fatty acids?

(c) Which fat or oil has the largest percentage of the essential fatty acid linoleic acid?

24.3 Properties of Fats and Oils

The melting points listed in Table 24.1 show that the more double bonds a fatty acid has, the lower its melting point. For example, the saturated 18-carbon acid (stearic) melts at 70 °C, the monounsaturated 18-carbon acid (oleic) melts at 4 °C, and the

APPLICATION ▶ Lipids in the Diet

The major recognizable sources of fats and oils in our diet are butter and margarine, vegetable oils, the visible fat in meat, and chicken skin. In addition, triacylglycerols in meat, poultry, fish, dairy products, and eggs add saturated fats to our diet, along with small quantities of cholesterol. Vegetable oils, such as those in nuts and seeds, whole-grain cereals, and prepared foods, have a higher unsaturated fatty acid content and no cholesterol. They never contain cholesterol because plants do not synthesize cholesterol. Thus, salad oil labels proclaiming, "No cholesterol" are stating a fact, not announcing something special the manufacturer has done for us.

Fats and oils are a popular component of our diet: They taste good, give a pleasant texture to food, and, because they are digested slowly, give a feeling of satisfaction after a meal. The percentage of calories from fats and oils in the average U.S. diet in the early 1980s was 40–45%, considerably more than needed. Current research suggests that the average person in the United States now obtains about 35% of her or his daily calories from fats and oils, a number approaching the recommended 30%. Excess energy from dietary fats and oils is mostly stored as fat in adipose tissue.

Concern for the relationships among saturated fats, cholesterol levels, and various diseases—most notably heart disease and cancer (see the Application "Lipids and Atherosclerosis," Chapter 25)—caused a modest decrease to 37% of the average calories from fats and oils in the U.S. diet by 1990. We significantly decreased our consumption of butter, eggs, beef, and whole milk (all containing relatively high proportions of saturated fat and cholesterol). This change did not, however, coincide with a reduction in obesity, which is a weight 20% over the desirable weight for a person's height, sex, and activity level or a body mass index (BMI) of 30 or greater. In fact, concern has been accelerating in recent years over a rise in obesity in the U.S. population and its inevitable association with heart disease. Recent surveys of the U.S. population report that 66% of the adult population are overweight or obese and 33% are obese. Even more concerning is the fact that 17% of children and young adults between the ages of 2 and 18 are overweight.

Several organizations, including the U.S. Food and Drug Administration, recommend a diet with not more than 30% of its calories from fats and oils. In a daily diet of 2200 Cal, which is about right for teenage girls, active women, and sedentary men, 30% from fats and oils is approximately 73 g, the amount in about 6 tablespoons of butter. Teenage boys and men, or women with very active lifestyles, require more daily calories, and therefore can include proportionately more fats in their diets.

▲ A selection of appealing but high-fat foods.

The Nutrition Facts labels (illustrated in Application "Vitamins, Minerals, and Food Labels," Chapter 19) list the calories from fat, grams of fat (which includes all triacylglycerols), and grams of saturated fat in a single serving of a commercially prepared food. To check up on what you are eating, remember that you have to check the number of servings in the container. The label on a package of hot dogs, for example, says 12 g of fat. But checking shows that one hot dog = a single serving. If you eat two or three hot dogs, you will be getting 24 g or 36 g of fat.

The FDA further recommends that not more than 10% of daily calories come from saturated fat and not more than 300 mg of cholesterol be included in the daily diet. For those with the goal of 2200 Cal/day, the limit is 24 g of saturated fats. (Those hot dogs contain 5 g of saturated fat per hot dog.) To reduce saturated fats requires choosing the low-fat varieties of foods when possible. For example, 1% milk provides 1.6 g of saturated fat per cup compared with 5.1 g per cup in whole milk. The foods highest in cholesterol are high-fat dairy products, liver, and egg yolks (see Section 24.6, Cell Membrane Lipids: Cholesterol, p. 774).

See Additional Problems 24.81 and 24.82 at the end of the chapter.

Oil A mixture of triacylglycerols that is liquid because it contains a high proportion of unsaturated fatty acids.

Fat A mixture of triacylglycerols that is solid because it contains a high proportion of saturated fatty acids.

diunsaturated 18-carbon acid (linoleic) melts at −5 °C. The same trend also holds true for triacylglycerols: The more highly unsaturated the acyl groups in a triacylglycerol, the lower its melting point. The difference in melting points between fats and oils is a consequence of this difference. Vegetable **oils** are lower melting because oils generally have a higher proportion of unsaturated fatty acids than animal **fats**.

How do the double bonds make such a significant difference in the melting point? Compare the shapes of a saturated and an unsaturated fatty acid molecule:

A saturated fat has only single C-C bonds and appears straight

Unsaturated fats bend due to cis double bonds

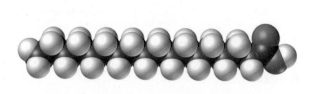

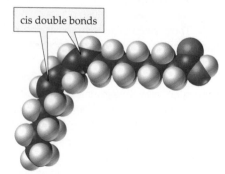

Stearic acid, an 18-carbon saturated fatty acid

Linoleic acid, an 18-carbon unsaturated fatty acid

The hydrocarbon chains in saturated acids are uniform in shape with identical angles at each carbon atom, and the chains are flexible, allowing them to nestle together. By contrast, the carbon chains in unsaturated acids have rigid kinks wherever they contain cis double bonds. The kinks make it difficult for such chains to fit next to each other in the orderly fashion necessary to form a solid. The more double bonds there are in a triacylglycerol, the harder it is for it to solidify. The shapes of the molecular models in Figure 24.2 further illustrate this concept.

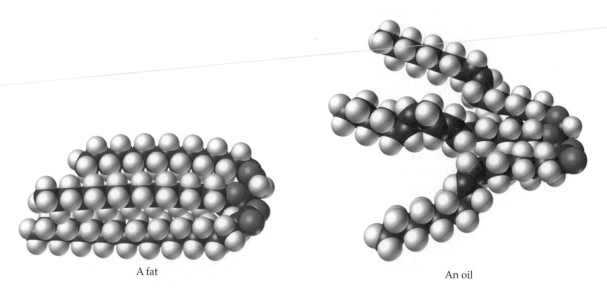

A fat

An oil

▲ **FIGURE 24.2** **Triacylglycerols from a fat and an oil.**

Triacylglycerols are uncharged, nonpolar, hydrophobic molecules. When stored in fatty tissue they coalesce, and the interior of an adipocyte is occupied by one large fat droplet with the cell's nucleus pushed to one side. The primary function of triacylglycerols is long-term storage of energy for the organism. In addition, adipose tissue serves to provide thermal insulation and protective padding. Most fatty tissue is located under the skin or in the abdominal cavity, where it cushions the organs.

We are accustomed to the characteristic yellow color and flavors of cooking oils, but these are contributed by natural materials carried along during production of the oils from plants; pure oils are colorless and odorless. Overheating, or exposure to air or oxidizing agents, causes decomposition to products with unpleasant odors or flavors, creating what we call a *rancid oil*. Antioxidants are added to prepared foods to prevent oxidation of their oils. (See the Application, "Phenols as Antioxidants," p. 450.)

Properties of the Triacylglycerols in Natural Fats and Oils

- Nonpolar and hydrophobic
- No ionic charges
- Solid triacylglycerols (fats)—high proportion of saturated fatty acid chains
- Liquid triacylglycerols (oils)—high proportion of unsaturated fatty acid chains

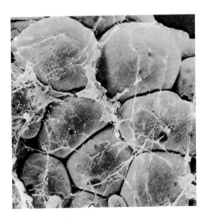

▲ Where the fat goes. Each of these adipose tissue cells holds a globule of fat. (Magnified more than 500 times.)

 **Looking Ahead**

Triacylglycerols from plants and animals are a major component of our diet. In our bodies, they are the depots for energy storage. Therefore, in considering the metabolism of lipids, it is the metabolism of triacylglycerols that is of greatest interest. This topic is discussed in Chapter 25.

PROBLEM 24.5

Draw the complete structural formula of arachidonic acid (Table 24.1) in a way that shows the cis stereochemistry of its four double bonds.

WORKED EXAMPLE **24.2** Comparing Melting Points

Which of these two fatty acids has the higher melting point?

(a) $CH_3(CH_2)_4CH{=}CHCH_2CH{=}CHCH_2(CH_2)_6\overset{\overset{\displaystyle O}{\|}}{C}{-}OH$

(b) $CH_3(CH_2)_5CH{=}CHCH_2(CH_2)_6\overset{\overset{\displaystyle O}{\|}}{C}{-}OH$

ANALYSIS First determine the chain length (number of carbon atoms) and the number of unsaturated bonds present. In general, the more carbon atoms present in a molecule, the higher the melting point. However, the higher the number of unsaturated bonds, the lower the melting point. The **degree of unsaturation** (defined as the number of double bonds) is more important than the number of carbon atoms when the number of carbon atoms is identical or similar.

Degree of unsaturation The number of carbon–carbon double bonds in a molecule.

SOLUTION
Molecule (a) has 18 carbon atoms and 2 unsaturated bonds. Molecule (b) has 16 carbon atoms and 1 unsaturated bond. Although molecule (a) is slightly larger than molecule (b) and would be expected to have a higher melting point, molecule (a) has two double bonds, whereas molecule (b) has only one double bond. Since the degree of unsaturation is more important in these similarly sized molecules, molecule (b) has a higher melting point.

PROBLEM 24.6

Can there be any chiral carbon atoms in triacylglycerols? If so, which one(s) can be chiral and what determines their chirality?

KEY CONCEPT PROBLEM 24.7

What noncovalent interactions (covered in Section 8.11) bind lipid molecules together? Are these forces generally weak or strong? Why do lipids not mix readily with water?

24.4 Chemical Reactions of Triacylglycerols

Hydrogenation

The carbon–carbon double bonds in vegetable oils can be hydrogenated to yield saturated fats in the same way that any alkene can react with hydrogen to yield an alkane. (⊂⊃ , Section 13.6) Margarine and solid cooking fats (shortenings) are produced commercially by hydrogenation of vegetable oils to give a product chemically similar to that found in animal fats:

Partial structure of an unsaturated vegetable oil

$$-O-\overset{\overset{\displaystyle O}{\|}}{C}-CH_2CH_2CH_2CH_2CH_2CH_2CH_2CH = CHCH_2CH = CHCH_2CH_2CH_2CH_2CH_3$$

$$2\,H_2 \;|\; \text{Pd catalyst}$$

Partial structure of hydrogenated oil

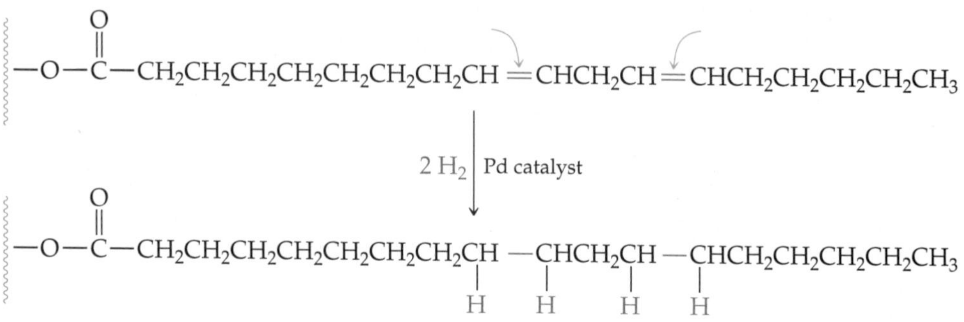

$$-O-\overset{\overset{\displaystyle O}{\|}}{C}-CH_2CH_2CH_2CH_2CH_2CH_2CH_2CH-\underset{H}{\overset{|}{C}}HCH_2CH-\underset{H}{\overset{|}{C}}HCH_2CH_2CH_2CH_2CH_3$$

The extent of hydrogenation varies with the number of double bonds in the unsaturated acids and their locations. In general, the number of double bonds is reduced in stepwise fashion from three to two to one. By controlling the extent of hydrogenation and monitoring the composition of the product, it is possible to control consistency. In margarine, for example, only about two-thirds of the double bonds present in the starting vegetable oil are hydrogenated. The remaining double bonds, which vary in their locations, are left unhydrogenated so that the margarine has exactly the right consistency to remain soft in the refrigerator and melt on warm toast. (See the Application "Butter and Its Substitutes," p. 774.)

PROBLEM 24.8

Write an equation for the complete hydrogenation of triolein, the triacylglycerol with three oleic acid acyl groups for which you drew the structure in Problem 24.3. Name the fatty acid from which the resulting acyl groups are derived.

Hydrolysis of Triacylglycerols; Soap

Triacylglycerols, like all esters, can be hydrolyzed—that is, they can react with water to form their carboxylic acids and alcohols. In the body, this hydrolysis is catalyzed by enzymes (hydrolases) and is the first reaction in the digestion of dietary fats and oils (Section 25.1).

Soap making is an ancient art involving the base-catalyzed hydrolysis of triacylglycerols. This process was probably discovered accidentally. Throughout the centuries soap has been made both at home and in "factories" by much the same process; the principal variation was in the source of fat. Northern societies, like the English, used solid animal fats but the most abundant fat in southern Italy was olive oil. Indeed, any mixture of triacylglycerols, whether from animal or plant sources can be

Soap The mixture of salts of fatty acids formed by saponification of animal fat.

APPLICATION ▶ Detergents

Strictly speaking, anything that washes away dirt is a *detergent*. The term is usually applied, however, to synthetic materials made from petroleum chemicals. In the 1950s, synthetic detergents began to replace natural soaps. The goal was to overcome the problems caused by soap used in hard water, which contains metal ions (mostly Ca^{2+} and Mg^{2+}) that have dissolved into it. When the metal ions in solution encounter fatty acid anions, they form what we call soap scum—precipitates of salts (for example, $[CH_3(CH_2)_{14}COO^-]_2Ca^{2+}$). The results are that soap is wasted, residues are left in washed clothing, and the hard-to-remove scum is left behind in bathtubs and washing machines.

Like soaps, synthetic detergent molecules have hydrophobic hydrocarbon tails and hydrophilic heads, and they cleanse by the same mechanism as soap—forming micelles around greasy dirt. All substances that function in this manner are described as surface-active agents, or *surfactants*. The hydrophilic heads may be anionic, cationic, or non-ionic. Anionic surfactants are commonly used in home laundry products; cationic surfactants are used in fabric softeners and in disinfectant soaps; and non-ionic surfactants are low-sudsing and are effective at low temperatures.

Note that the hydrocarbon chains in the representative surfactants shown below are unbranched. Some of the first detergents contained branched-chain hydrocarbons, but this was soon discovered to be a mistake. The bacteria in natural waters and sewage treatment plants are slow to consume branched-chain hydrocarbons, so detergents containing them were not decomposed and produced suds in streams and lakes.

See Additional Problems 24.83 through 24.85 at the end of the chapter.

$$Na^+ \; {}^-O-\overset{\displaystyle O}{\underset{\displaystyle O}{\overset{\|}{\underset{\|}{S}}}}-\langle benzene \rangle-CH_2CH_2CH_2CH_2CH_2CH_2CH_2CH_2CH_2CH_2CH_2CH_3$$

Sodium dodecylbenzenesulfonate
(An ionic detergent)

$$CH_3(CH_2)_{10}CH_2OCH_2CH_2(OCH_2CH_2)_7OH$$

A polyether
(A non-ionic detergent)

$$\langle benzene \rangle-CH_2-\overset{\displaystyle CH_3}{\underset{\displaystyle CH_3}{\overset{|}{\underset{|}{N^+}}}}-R \; Cl^-$$

A benzalkonium chloride; R=C_8H_{17} to $C_{18}H_{37}$
(A cationic detergent)

used. American colonials made soap at home, generally once a year when animal fat was abundant. The second ingredient needed was lye or a potash solution. This was obtained by soaking wood ashes in rainwater and slowly straining out the ashes. Saved and fresh fat was rendered and cleaned—freed from meat, skin and any other material that was not fat. Then water, cleaned fat, and lye were mixed and boiled for several hours outdoors over an open fire. This dangerous process was complete when the soap foamed up well in the pot. After cooling, the soft soap was poured into storage barrels, to be dipped out for use. This process made soft soap because wood ashes contain KOH; potassium fatty acid salts dissolved readily in water and the soap retained some water. Hard soap was made by adding salt (NaCl) at the last step. Na^+ replaces K^+ in the fatty acid salt formed and these salts are solids. The smooth soap that precipitates is then dried, perfumed, and pressed into bars for household use. When a new method of producing NaOH became available in the middle of the nineteenth century, factory soap production increased and home production became less important. Today's soaps contain less residual lye, more additives, and are kinder to our skin and clothes than the soap produced at home by the colonists. Although soap making was a standard home process on the frontier and in small factories in towns, the chemistry was not understood until the twentieth century.

In the laboratory and in commercial production of soap, hydrolysis of fats and oils is usually carried out by strong aqueous bases (NaOH or KOH) and is called *saponification* (pronounced sae-**pon**-if-i-**ka**-tion, from the Latin *sapon*, soap). (⬤▭⬤, Section 17.6) The initial products of saponification of a fat or oil molecule are one

▲ Where are the lipids in this picture?

molecule of glycerol and three molecules of fatty acid carboxylate salts:

Saponification

Strong aqueous base catalyzes fat hydrolysis

$$
\begin{array}{c}
CH_2-O-\overset{\displaystyle O}{\overset{\displaystyle \|}{C}}-R \\[2ex]
CH-O-\overset{\displaystyle O}{\overset{\displaystyle \|}{C}}-R' \\[2ex]
CH_2-O-\overset{\displaystyle O}{\overset{\displaystyle \|}{C}}-R''
\end{array}
\quad \xrightarrow[H_2O]{NaOH} \quad
\begin{array}{c}
CH_2-OH \\[2ex]
CH-OH \\[2ex]
CH_2-OH
\end{array}
\; + \;
\begin{array}{c}
R-\overset{\displaystyle O}{\overset{\displaystyle \|}{C}}-O^-\,Na^+ \\
+ \\
R'-\overset{\displaystyle O}{\overset{\displaystyle \|}{C}}-O^-\,Na^+ \\
+ \\
R''-\overset{\displaystyle O}{\overset{\displaystyle \|}{C}}-O^-\,Na^+
\end{array}
$$

A fat or oil   Glycerol   Fatty acid salts (soap)

How does soap do its job? Soaps work as cleaning agents because the two ends of a soap molecule are so different. The sodium salt end is ionic and therefore hydrophilic (water-loving); it tends to dissolve in water. The long hydrocarbon chain portion of the molecule, however, is nonpolar and therefore hydrophobic (water-fearing). Like an alkane, it tends to avoid water and to dissolve in nonpolar substances such as grease, fat, and oil. Because of these opposing tendencies, soap molecules are attracted to both grease and water.

When soap is dispersed in water, the big organic anions cluster together so that their long, hydrophobic hydrocarbon tails are in contact. By doing so, they avoid disrupting the strong hydrogen bond interactions of water and instead create a nonpolar microenvironment. At the same time, their hydrophilic ionic heads on the surface of the cluster stick out into the water. The resulting spherical clusters are called **micelles** (Figure 24.3). Grease and dirt are suspended in water because they are coated by the nonpolar tails of the soap molecules and trapped in the center of the micelles. Once suspended within micelles, the grease and dirt can be rinsed away.

Micelle A spherical cluster formed by the aggregation of soap or detergent molecules so that their hydrophobic ends are in the center and their hydrophilic ends are on the surface.

▶ **FIGURE 24.3 Soap or detergent molecules in water.** The hydrophilic ionic ends remain in the water. At the surface of the water, a film forms with the hydrocarbon chains on the surface. Within the solution, the hydrocarbon chains cluster together at the centers of micelles. Greasy dirt is dissolved in the oily center and carried away. Lipids are transported in the bloodstream in similar micelles, as described in Section 25.2.

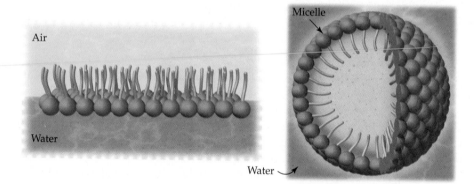

PROBLEM 24.9

Write the complete equation for the hydrolysis of a triacylglycerol in which the fatty acids are two molecules of stearic acid and one of oleic acid (see Table 24.1).

24.5 Cell Membrane Lipids: Phospholipids and Glycolipids

Cell membranes separate the aqueous interior of cells from the aqueous environment surrounding the cells. To accomplish this, the membranes establish a hydrophobic barrier between the two watery environments. Lipids are ideal for this function.

The three major kinds of cell membrane lipids in animals are phospholipids, glycolipids, and cholesterol. The **phospholipids** contain a phosphate ester link. They are built up from either glycerol (to give *glycerophospholipids*) or from the alcohol sphingosine (to give *sphingomyelins*).

Phospholipid A lipid that has an ester link between phosphoric acid and an alcohol (either glycerol or sphingosine).

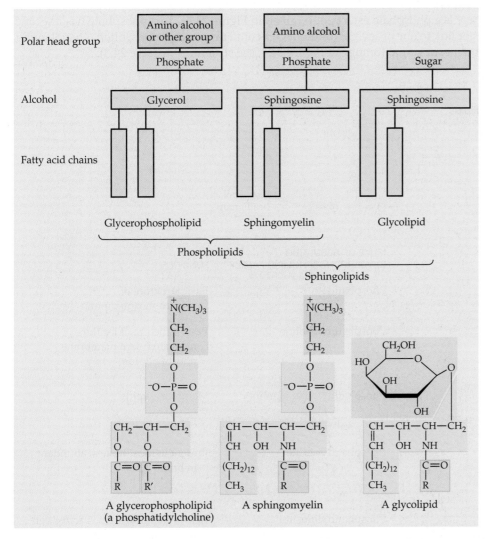

Glycerol 3-phosphate
(alcohol in glycerophospholipids)

Sphingosine
(alcohol in sphingolipids)

The **glycolipids** are also derived from sphingosine. They contain no phosphate group, but have an attached carbohydrate that is a monosaccharide or a short chain of monosaccharides. The general structures of these lipids and the relationships of their classification are shown at the top in Figure 24.4. Note the overlapping classes

Glycolipid A lipid with a fatty acid bonded to the C2—NH$_2$ group and a sugar bonded to the C1—OH group of sphingosine.

▲ **FIGURE 24.4** **Membrane lipids.** All have two hydrocarbon tails and polar, hydrophilic head groups. In the sphingolipids (sphingomyelins and glycolipids), one of the two hydrocarbon tails is part of the alcohol sphingosine (blue).

Sphingolipid A lipid derived from the amino alcohol sphingosine.

Glycerophospholipid (phosphoglyceride) A lipid in which glycerol is linked by ester bonds to two fatty acids and one phosphate, which is in turn linked by another ester bond to an amino alcohol (or other alcohol).

▲ Lecithin is the emulsifying agent in most chocolates.

of membrane lipids. Glycolipids and sphingomyelins both contain sphingosine and are therefore classified as **sphingolipids**, whereas glycerophospholipids and sphingomyelins both contain phosphate groups and are therefore classified as phospholipids.

Cholesterol is a steroid, a class of biomolecules that are characterized by a system of four fused rings. We have discussed steroids in the context of their most significant role as hormones. (⊂⊃, Section 20.5) The presence of cholesterol in cell membranes is explored in Section 24.6, and its connection to heart disease is discussed in the Application "Lipids and Atherosclerosis," in Chapter 25 (p. 791). Cholesterol is modified in liver cells to produce bile acids, essential in the digestion of dietary fats.

Phospholipids

Because phospholipids have ionized phosphate groups at one end, they are similar to soap and detergent molecules in having ionic, hydrophilic heads and hydrophobic tails (see Figure 24.3). They differ, however, in having *two* tails instead of one.

Glycerophospholipids (also known as **phosphoglycerides**) are triesters of glycerol 3-phosphate, and are the most abundant membrane lipids. Two of the ester bonds are with fatty acids, which provide the two hydrophobic tails (pink in the general glycerophospholipid structure in Figure 24.4). The fatty acids may be any of the fatty acids normally present in fats or oils. The fatty acid acyl group (R—C=O) bonded to C1 of glycerol is usually saturated, whereas the fatty acyl group at C2 is usually unsaturated. At the third position in glycerophospholipids there is a phosphate ester group (green in Figure 24.4). This phosphate has a second ester link to one of several different OH-containing compounds, often ethanolamine, choline, or serine (orange in Figure 24.4; see structures in Table 24.3).

TABLE 24.3 Some Glycerophospholipids

PRECURSOR OF X (HO–X)	FORMULA OF X	NAME OF RESULTING GLYCEROPHOSPHOLIPID FAMILY	FUNCTION
Water	—H	Phosphatidate	Basic structure of glycerophospholipids
Choline	—CH$_2$CH$_2$N$^+$(CH$_3$)$_3$	Phosphatidylcholine	Basic structure of lecithins; most abundant membrane phospholipids
Ethanolamine	—CH$_2$CH$_2$NH$_3^+$	Phosphatidylethanolamine	Membrane lipids
Serine	—CH$_2$—CH with NH$_3^+$ and COO$^-$	Phosphatidylserine	Present in most tissues; abundant in brain
myo-Inositol	(inositol ring structure, bond site)	Phosphatidylinositol	Relays chemical signals across cell membranes

The glycerophospholipids are named as derivatives of phosphatidic acids. In the molecule below on the right, for example, the phosphate ester link to the right of the P is to the amino alcohol choline, $HOCH_2CH_2N^+(CH_3)_3$. Lipids of this type are known as either *phosphatidylcholines*, or *lecithins*. (A substance referred to in the singular as either lecithin or phosphatidylserine, or any of the other classes of phospholipids, is usually a mixture of molecules with different R and R' tails.) Examples of some other classes of glycerophospholipids are included in Table 24.3.

A phosphatidate

A phosphatidatidylcholine
(a glycerophospholipid that is a lecithin)

Because of their combination of hydrophobic tails and hydrophilic head groups, the glycerophospholipids are *emulsifying agents*—substances that surround droplets of nonpolar liquids and hold them in suspension in water (see micelle in Figure 24.3). You will find lecithin, usually obtained from soybean oil, listed as an ingredient in chocolate bars and other foods where it is added to keep oils from separating out. It is the lecithin in egg yolk that emulsifies the oil droplets in mayonnaise.

In sphingolipids, the amino alcohol sphingosine provides one of the two hydrophobic hydrocarbon tails (blue here and in Figure 24.4). The second hydrocarbon tail is from a fatty acid acyl group connected by an amide link to the $-NH_2$ group in sphingosine (red here; pink in Figure 24.4).

A sphingomyelin (a sphingolipid)

Sphingomyelins are sphingosine derivatives with a phosphate ester group at C1 of sphingosine. The sphingomyelins are major components of the coating around nerve fibers (the *myelin sheath*) and are present in large quantities in brain tissue. A diminished amount of sphingomyelins and phospholipids in brain myelin has been associated with multiple sclerosis. The orientation of the hydrophilic and hydrophobic regions of a sphingomyelin is shown in Figure 24.5, together with a general representation of this and other types of cell membrane lipids used in drawing cell membranes.

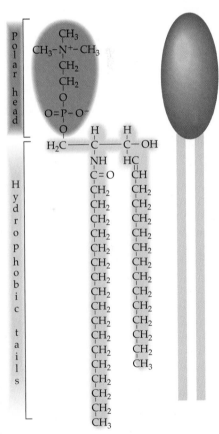

A sphingomyelin

▲ **FIGURE 24.5 A sphingomyelin, showing its polar, hydrophilic head group and its two hydrophobic tails.** The drawing on the right is the representation of phospholipids used in picturing cell membranes. It shows the relative positions of the hydrophilic head and the hydrophobic tails.

PROBLEM 24.10

For (a) the phosphatidylcholine on p. 770 and (b) the sphingomyelin in Figure 24.5, identify the products formed by complete hydrolysis of all ester bonds.

Glycolipids

A glycolipid

Glycolipids, like sphingomyelins, are derived from sphingosine. They differ in having a carbohydrate group at C1 (orange in the glycolipid in Figure 24.4) instead of a phosphate bonded to choline.

Glycolipids reside in cell membranes with their carbohydrate segments extending into the fluid surrounding the cells, just as do the carbohydrate segments of glycoproteins. (⟨◯▭⟩, p. 717) In this location they function as receptors that, as you saw in Chapter 20, are essential for recognizing chemical messengers, other cells, pathogens, and drugs.

The glycolipid molecule is classified as a *cerebroside*. Cerebrosides, which contain a single monosaccharide, are particularly abundant in nerve cell membranes in the brain, where the monosaccharide is D-galactose. They are also found in other cell membranes, where the sugar unit is D-glucose.

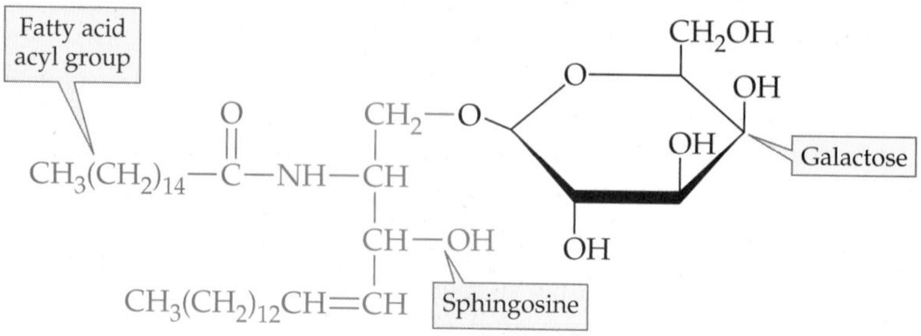

A glycolipid
(A cerebroside)

Gangliosides are glycolipids in which the carbohydrate is a small polysaccharide (an oligosaccharide) rather than a monosaccharide. Over 60 gangliosides are known. The oligosaccharides responsible for blood types are part of ganglioside molecules (see the Application "Cell-Surface Carbohydrates and Blood Type" in Chapter 22).

Tay-Sachs disease, a genetic disorder found mainly in persons of Eastern European descent, is the result of a defect in lipid metabolism, a deficiency in the enzyme β-hexosaminidase A, that causes a greatly elevated concentration of a particular ganglioside in the brain. An infant born with this defect suffers mental retardation and liver enlargement, and usually dies by age three. Tay-Sachs is one of a group of sphingolipid storage diseases. Another well-known, fatal disease in this group is Niemann-Pick disease, in which sphingomyelin accumulates due to a deficiency in the enzyme sphingomyelinase. These metabolic diseases result from deficiencies in the supply of enzymes that break down sphingolipids.

Currently there is no known therapy for Tay-Sachs disease or for Niemann-Pick disease. The harmful consequences result from the *storage* of the excess sphingolipids. A more promising outcome may be available for those with Gaucher's disease, the most common lipid storage disease. In Gaucher's patients, fats accumulate in many organs (liver, lungs, and brain) due to a deficiency in the enzyme glucocerebrosidase. Enzyme replacement therapy allows many of these patients to avoid some of the non-neurological effects of Gaucher's disease, although the treatment must be given frequently.

WORKED EXAMPLE 24.3 Identifying Complex Lipid Components

A class of membrane lipids known as *plasmalogens* has the general structure shown here. Identify the component parts of this lipid and choose the terms that apply to it: phospholipid, glycerophospholipid, sphingolipid, glycolipid.

Is it most similar to a phosphatidylethanolamine, a phosphatidylcholine, a cerebroside, or a ganglioside?

$$
\begin{array}{l}
\text{R—CH=CH—O—CH}_2 \\
\qquad\qquad\qquad\quad | \\
\qquad\text{O}\qquad\quad | \\
\qquad || \qquad\quad | \\
\text{R'—C—O—CH} \\
\qquad\qquad\quad | \\
\qquad\qquad\quad\quad \text{O} \\
\qquad\qquad\quad\quad || \\
\qquad\text{CH}_2\text{—O—P—O—CH}_2\text{CH}_2\overset{+}{\text{NH}}_3 \\
\qquad\qquad\qquad\quad | \\
\qquad\qquad\qquad\quad \text{O}^-
\end{array}
$$

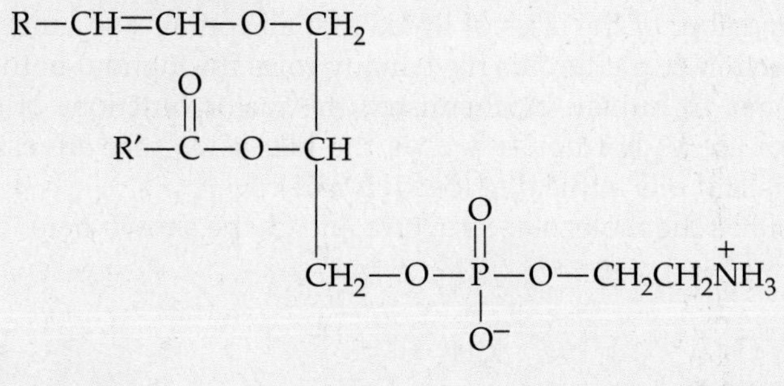

ANALYSIS Compare each part of the molecule with the basic components found in complex lipids and decide which lipid component the part resembles most. The molecule contains a phosphate group and thus is a phospholipid. The glycerol backbone of three carbon atoms bonded to three oxygen atoms is also present, so the compound is a glycerophospholipid, but one in which there is an ether linkage ($-\text{CH}_2-\text{O}-\text{CH}=\text{CHR}$) in place of one of the ester linkages. The phosphate group is bonded to ethanolamine ($\text{HOCH}_2\text{CH}_2\text{NH}_2$). The compound is not a sphingolipid or a glycolipid because it is not derived from sphingosine; for the same reason it is not a cerebroside or a ganglioside. Except for the ether group in place of an ester group, the compound has the same structure as a phosphatidylethanolamine.

SOLUTION
The terms that apply to this plasmalogen are *phospholipids* and *glycerophospholipid*. It has a structure nearly identical to phosphatidylethanolamine, so it is most similar to phosphatidylethanolamine.

PROBLEM 24.11

Draw the structure of the sphingomyelin that contains a myristic acid acyl group. Identify the hydrophilic head group and the hydrophobic tails in this molecule.

PROBLEM 24.12

Draw the structure of the glycerophospholipid that contains a stearic acid acyl group, an oleic acid acyl group, and a phosphate bonded to ethanolamine.

PROBLEM 24.13

Which of the following terms apply to the compound shown below?

(a) A phospholipid   **(b)** A steroid

(c) A sphingolipid   **(d)** A glycerophospholipid

(e) A lipid   **(f)** A phosphate ester

(g) A ketone

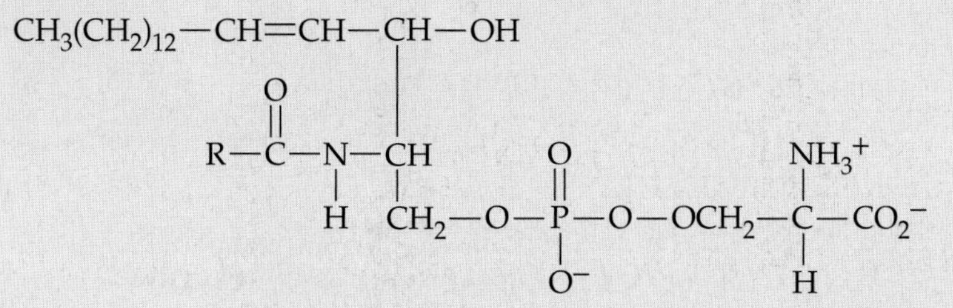

$$
\begin{array}{l}
\text{CH}_3(\text{CH}_2)_{12}\text{—CH=CH—CH—OH} \\
\qquad\qquad\qquad\qquad\qquad | \\
\qquad\text{O}\qquad\qquad\qquad | \\
\qquad ||\qquad\qquad\qquad | \\
\text{R—C—N—CH}\qquad\quad\text{O}\qquad\qquad\overset{+}{\text{NH}}_3 \\
\qquad\quad |\quad |\qquad\qquad || \qquad\qquad\quad | \\
\qquad\quad\text{H}\quad\text{CH}_2\text{—O—P—O—OCH}_2\text{—C—CO}_2^- \\
\qquad\qquad\qquad\qquad\quad |\qquad\qquad\qquad | \\
\qquad\qquad\qquad\qquad\quad \text{O}^-\qquad\qquad\qquad \text{H}
\end{array}
$$

24.6 Cell Membrane Lipids: Cholesterol

Animal cell membranes also contain significant amounts of cholesterol. Cholesterol is a steroid, a member of the class of lipids that all contain the same four-ring system. (🔗, Section 20.5) Steroids have many roles throughout both the plant and animal kingdoms. In human biochemistry, the major functions of steroids other than cholesterol are as hormones and as the bile acids that are essential for the digestion of fats and oils in the diet (Section 25.1).

Cholesterol has the molecular structure and shape shown here:

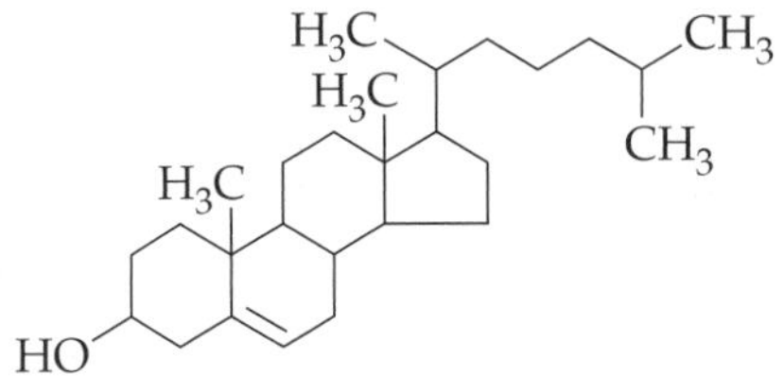

It is the most abundant animal steroid. The body of a 60 kg person contains about 175 g of cholesterol that serves two important functions: as a component of cell membranes and as the starting material for the synthesis of all other steroids.

APPLICATION ▶ | Butter and Its Substitutes

Sometimes the more scientific evidence we accumulate about the relationship between diet and health, the more difficult it is to choose what to eat. The choice between butter and margarine provides an excellent example.

It has become medically accepted that butter can contribute to elevated blood cholesterol, which is to be avoided because of cholesterol's role in heart disease. (We will have more to say about the cholesterol–heart disease connection in the Application "Lipids and Atherosclerosis," Chapter 25.) In response to this information, many individuals switched from butter to margarine. Margarine, which is made from vegetable oils, contains no cholesterol and much less saturated fat than butter.

Gradually, however, information has accumulated that margarine contains what *might* be an even more unhealthful ingredient—*trans fatty acids*. Oils are catalytically hydrogenated to give them a firmer consistency and also to lessen their tendency to become rancid as oxidation breaks the double bonds. (🔗, Section 13.6) During the partial hydrogenation, some of the cis double bonds are inevitably converted to trans double bonds.

Numerous studies have linked the quantity of trans fatty acids in a person's diet to a greater risk for heart disease and cancer. One such study, widely reported in 1997, came from following the diet and health of 80,000 nurses for 14 years. Those with higher quantities of hydrogenated oils in their diets had a significantly higher risk of heart disease. One suggested explanation is that trans fats alter the metabolism of polyunsaturated fats, which are protective against heart disease.

Meat and dairy products contain a very small amount of trans fatty acids (about 0.2% in butter), but the quantities in foods containing hydrogenated oils are much higher—up to 40% in the stiffer margarines. If you choose to be serious about avoiding trans fats, much more is involved than choosing a softer margarine, however. By reading the lists of ingredients on food labels, you will discover that almost all commercial baked goods, cookies, and crackers, as well as many other packaged food products, contain partially hydrogenated oils. Food labels must now list the quantity of trans fats present per serving.

▲ Is butter better?

See Additional Problems 24.86 and 24.87 at the end of the chapter.

"Cholesterol" has become a household word because of its presence in the arterial plaque that contributes to heart disease (see the Application "Lipids and Atherosclerosis," Chapter 25). Some cholesterol is obtained from the diet, but cholesterol is also synthesized in the liver. Even on a strict no-cholesterol diet, the body of an adult can manufacture approximately 800 mg of cholesterol per day.

The molecular model of cholesterol reveals the nearly flat shape of the molecule. Except for its —OH group, cholesterol is hydrophobic. Within a cell membrane, cholesterol molecules are distributed among the hydrophobic tails of the phospholipids. Because they are more rigid than the hydrophobic tails, the cholesterol molecules help to maintain the structural rigidity of the membrane. Approximately 25% of liver cell membrane lipid is cholesterol.

24.7 Structure of Cell Membranes

Phospholipids provide the basic structure of cell membranes, where they aggregate in a closed, sheet-like, *double leaflet* structure—the **lipid bilayer** (Figure 24.6). The bilayer is formed by two parallel layers of lipids oriented so that the ionic head groups are exposed to the aqueous environments on either side of the bilayer. The nonpolar tails cluster together in the middle of the bilayer, where they interact and avoid water. Each half of the bilayer is termed a *leaflet*.

The bilayer is a favorable arrangement for phospholipids—it is highly ordered and stable, but still flexible. When phospholipids are shaken vigorously with water, they spontaneously form **liposomes**—small spherical vesicles with a lipid bilayer surrounding an aqueous center, as shown in Figure 24.6. Water-soluble substances can be trapped in the center of liposomes, and lipid-soluble substances can be incorporated into the bilayer. Liposomes are potentially useful as carriers for drug delivery because they can fuse with cell membranes and empty their contents into the cell (see the Application, "Liposomes for Health and Beauty").

Lipid bilayer The basic structural unit of cell membranes; composed of two parallel sheets of membrane lipid molecules arranged tail to tail.

Liposome A spherical structure in which a lipid bilayer surrounds a water droplet.

APPLICATION ▶ Liposomes for Health and Beauty

Because they can fuse with a cell's membrane and then empty their contents directly into the cell, liposomes have an exciting future in health care. Imagine, for instance, a delivery system for chemotherapy drugs that can deliver the drugs only to cancer cells in a patient's body and leave healthy cells alone. Research is under way throughout the industrial and academic communities to discover just how this potential can be put to work.

A major difficulty with cancer chemotherapy is that the drugs are toxic in varying degrees to both cancer cells and healthy cells. A properly designed liposome will travel intact to the location of a tumor, carrying its water-soluble, tumor-destroying drug safely in its interior. Once it arrives at the tumor site, the liposome will fuse with the membrane of a cancerous cell and deliver the drug directly to where it can act most effectively.

Another possibility is the use of liposomes in gene therapy. It has already been shown that DNA (deoxyribonucleic acid, Section 26.1), the carrier of genes, can be incorporated into liposomes. The hope is that DNA carrying a normal gene can be inserted from the liposome into cells where it can replace the flawed gene.

Liposome drug delivery has already met the requirements of the Food and Drug Administration in an intravenous drug that targets systemic fungal infections. Individuals with compromised immune systems due to AIDS are especially susceptible to this kind of infection. The liposomes carry amphotericin B, an antibiotic that attacks the fungal cell membrane. By delivering amphotericin to the fungal cells, the liposomal drug diminishes the serious side effects of attack by this antibiotic on kidney cells and cells in other healthy organs.

Meanwhile, a broad array of liposome-based products has reached the marketplace as cosmetics. These products are expensive, but they promise what is obviously desirable to many individuals. The creams and lotions are described as creating healthier and more beautiful skin by "energizing" cells, preventing oxidative damage, and erasing wrinkles. Because they can merge with cell membranes, the liposomes deliver their moisturizers, perfumes, or vitamins directly into the skin rather than just onto the surface as a conventional lotion does.

See Additional Problems 24.88 and 24.89 at the end of the chapter.

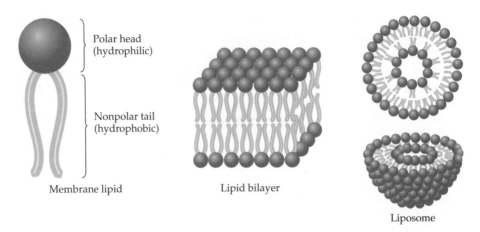

Polar head
(hydrophilic)

Nonpolar tail
(hydrophobic)

Membrane lipid   Lipid bilayer

Liposome

▲ **FIGURE 24.6 Aggregation of membrane lipids.** The lipid bilayer provides the basic structure of a cell membrane.

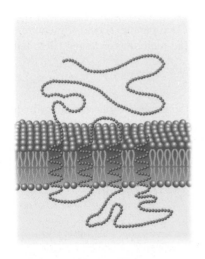

▲ **General structure for an integral membrane protein.** The green circles represent the amino acids. Many membrane proteins pass in and out of the membrane numerous times.

The overall structure of cell membranes is represented by the *fluid-mosaic model*. The membrane is described as *fluid* because it is not rigid and molecules can move around within it, and as a *mosaic* because it contains many kinds of molecules. The components of the cell membrane are shown in Figure 24.7.

Glycolipids and cholesterol are present in cell membranes, and 20% or more of the weight of a membrane consists of protein molecules, many of them glycoproteins (⬤▭, p. 717). *Peripheral proteins* are associated with just one face of the bilayer (that is, with one leaflet) and are held within the membrane by noncovalent interactions with the hydrophobic lipid tails or the hydrophilic head groups. *Integral proteins* extend completely through the cell membrane and are anchored by hydrophobic regions that extend through the bilayer. In some cases, the hydrophobic amino acid chain may twist in and out of the membrane many times before ending on the exterior of the membrane with a hydrophilic sugar group. The carbohydrate parts of glycoproteins and glycolipids mediate the interactions of the cell with outside agents. Some integral proteins form channels to allow specific molecules or ions to enter or leave the cell.

Because the bilayer membrane is fluid rather than rigid, it is not easily ruptured. The lipids in the bilayer simply flow back together to repair any small hole or puncture. The effect is similar to what is observed in cooking when a thin film of oil or melted butter lies on top of water in a cooking pot. The film can be punctured and broken, but it immediately flows together when left alone.

One consequence of membrane fluidity is the movement of proteins within the membrane. For example, low-density lipoprotein receptors, which are glycoproteins that interact with lipoproteins in the extracellular fluid [⬤▭, Section 25.2], move sideways within the membrane to form clusters of receptors on the cell surface. The glycoproteins move sideways in the membrane layers continuously, not unlike floating on a pond; this is an energetically neutral motion. However, phospholipids and other membrane components do not flip from the inside leaflet of the membrane to the outside leaflet, or vice versa. That is an energetically unfavored action because it would force polar and nonpolar interactions between membrane components.

Two other consequences of bilayer fluidity are that small *nonpolar* molecules can easily enter the cell through the membrane and that some individual lipid or protein molecules can diffuse rapidly from place to place within the membrane.

The fluidity of the membrane varies with the relative amounts of saturated and unsaturated fatty acids in the glycerophospholipids. Such variation is put to use in the adaptation of organisms to their environment. In reindeer, for example, the membranes of cells near the hooves contain a higher proportion of unsaturated fatty acid chains than in other cells. These chains do not pack tightly together. The result is a membrane that remains fluid while the animals stand in snow.

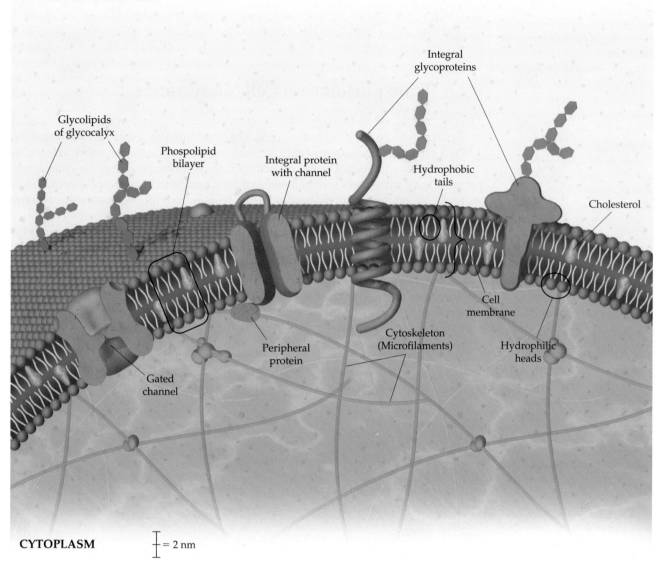

EXTRACELLULAR FLUID

CYTOPLASM ⌐ = 2 nm

▲ **FIGURE 24.7** **The cell membrane.** Cholesterol forms part of the membrane, proteins are embedded in the lipid bilayer, and the carbohydrate chains of glycoproteins and glycolipids extend into the extracellular space, where they act as receptors. Integral proteins form channels to the outside of the cell and also participate in transporting large molecules across the membrane.

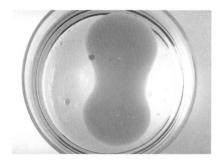

▲ **Oil floating on water, an analogy for the fluid-mosaic cell membrane model.** When the oil layer is disturbed, it soon flows back together.

━○ KEY CONCEPT PROBLEM 24.14

Integral membrane proteins are not water-soluble. Why? How must these proteins differ from globular proteins?

24.8 Transport Across Cell Membranes

The cell membrane must accommodate opposing needs in allowing the passage of molecules and ions into and out of a cell. On one hand, the membrane surrounding a living cell cannot be impermeable, because nutrients must enter and waste products must leave the cell. On the other hand, the membrane cannot be completely permeable, or substances would just move back and forth until their concentrations were equal on both sides—hardly what is required for homeostasis (see the Application "Homeostasis," p. 631).

The problem is solved by two modes of passage across the membrane (Figure 24.8). In **passive transport**, substances move across the membrane freely by diffusion from regions of higher concentration to regions of lower concentration. In **active transport**, substances can cross the membrane only when energy is supplied because they must go in the reverse direction—from lower to higher concentration regions.

Passive transport Movement of a substance across a cell membrane without the use of energy, from a region of higher concentration to a region of lower concentration.

Active transport Movement of substances across a cell membrane with the assistance of energy (for example, from ATP).

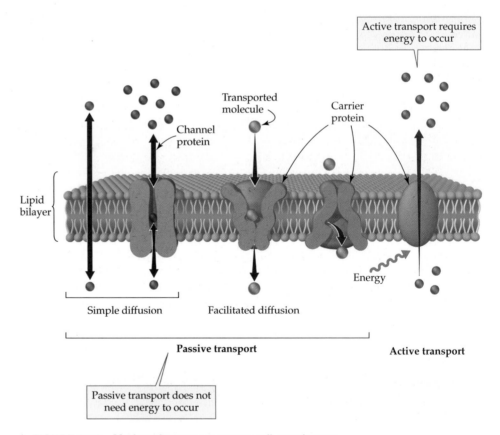

▲ **FIGURE 24.8** **Modes of transport across cell membranes.**

Passive Transport by Simple Diffusion

Simple diffusion Passive transport by the random motion of diffusion through the cell membrane.

Some solutes enter and leave cells by **simple diffusion**—they simply wander by normal molecular motion into areas of lower concentration. Small, nonpolar molecules, such as CO_2 and O_2, and lipid-soluble substances, including steroid hormones, move through the hydrophobic lipid bilayer in this way. Hydrophilic substances similarly pass through the aqueous solutions inside channels formed by integral proteins. What can pass through the protein channels is limited by the size of the molecules relative to the size of the openings. The lipid bilayer is essentially impermeable to ions and larger polar molecules, which are not soluble in the non-polar hydrocarbon region.

Passive Transport by Facilitated Diffusion

Like simple diffusion, **facilitated diffusion** is passive transport and requires no energy input. The difference is that in facilitated diffusion solutes are helped across the membrane by proteins. The interaction is similar to that between enzymes and sub-strates. The molecule to be transported binds to a membrane protein, which changes shape so that the transported molecule is released on the other side of the membrane. Glucose is transported into many cells in this fashion.

Facilitated diffusion Passive transport across a cell membrane with the assistance of a protein that changes shape.

PROBLEM 24.15

Does an NO molecule cross a lipid bilayer by simple diffusion? Explain.

PROBLEM 24.16

As noted earlier (Section 23.3), the first step in glycolysis, which occurs within cells, is phosphorylation of glucose to glucose 6-phosphate. Why does this step prevent passive diffusion of glucose back out of the cell?

Active Transport

It is essential to life that the concentrations of some solutes be different inside and outside cells. Such differences are contrary to the natural tendency of solutes to move about until the concentration equalizes. Therefore, maintaining **concentration gradients** (differences in concentration within the same system) requires the expen-diture of energy. An important example of active transport is the continuous move-ment of sodium and potassium ions across cell membranes. Only by this means is it possible to maintain homeostasis, which requires low Na^+ concentrations within cells and higher Na^+ concentrations in extracellular fluids, with the opposite concentration ratio for K^+. Energy from the conversion of ATP to ADP is used to change the shape of an integral membrane protein (an ATPase referred to as the sodium/potassium pump), simultaneously bringing two K^+ ions into the cell and moving three Na^+ ions out of the cell (Figure 24.9).

Here are some important points to remember about cell membranes:

Concentration gradient A differ-ence in concentration within the same system.

Properties of cell membranes

*   Cell membranes are composed of a fluid-like phospholipid bilayer.
*   The bilayer incorporates cholesterol, proteins (including glycoproteins), and glycolipids.
*   Small nonpolar molecules cross by simple diffusion through the lipid bilayer.
*   Small ions and polar molecules diffuse across the membrane via protein pores (simple diffusion) .
*   Glucose and certain other substances (including amino acids) cross with the aid of proteins and without energy input (*facilitated diffusion*).
* Na^+, K^+, and other substances that maintain concentration gradients across the cell membrane cross with expenditure of energy and the aid of proteins (*active transport*).

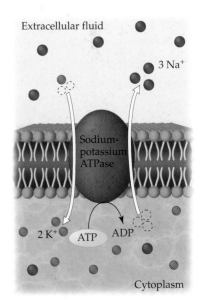

▲ **FIGURE 24.9 An example of active transport.** A protein known as sodium–potassium ATPase uses energy from ATP to move Na^+ and K^+ ions across cell membranes against their concentration gradients.

KEY CONCEPT PROBLEM 24.17

The compositions of the inner and outer surfaces of the lipid bilayer are different. Why do these differences exist and how might they be of use to a living cell?

24.9 Eicosanoids: Prostaglandins and Leukotrienes

Eicosanoid A lipid derived from a 20-carbon unsaturated carboxylic acid.

The **eicosanoids** are a group of compounds derived from 20-carbon unsaturated fatty acids (*eicosanoic acids*) and synthesized throughout the body. They function as short-lived chemical messengers that act near their points of synthesis ("local hormones").

The *prostaglandins* (named for their discovery in prostate cells) and the *leukotrienes* (named for their discovery in leukocytes) are two classes of eicosanoids that differ somewhat in their structure. The prostaglandins all contain a five-membered ring, which the leukotrienes lack.

Prostaglandins and leukotrienes are synthesized in the body from the 20-carbon unsaturated fatty acid arachidonic acid. Arachidonic acid, in turn, is synthesized from linolenic acid, helping to explain why linolenic is one of the two essential fatty acids. The relationships among arachidonic acid, the prostaglandins, and the leukotrienes are illustrated here; arachidonic acid and the leukotriene are drawn "bent" into a shape similar to the shape they take within the cell:

20 19 18 17 16 15 14 13 12 11 10 9 8 7 6 5 4 3 2 1

$$CH_3CH_2CH_2CH_2CH_2CH{=}CHCH_2CH{=}CHCH_2CH{=}CHCH_2CH{=}CHCH_2CH_2CH_2COOH$$

Arachidonic acid

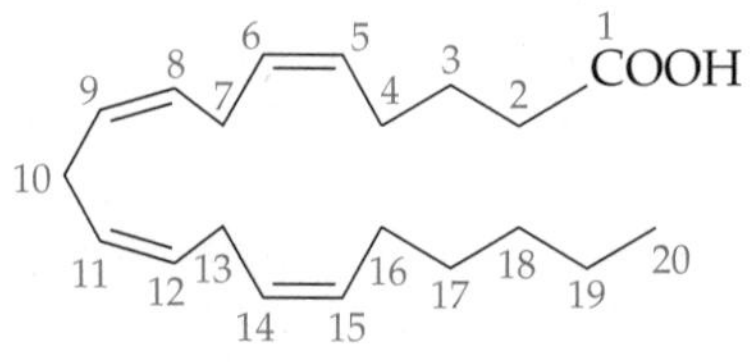

Arachidonic acid (bent)

Multistep enzyme-catalyzed synthesis

Multistep enzyme-catalyzed synthesis

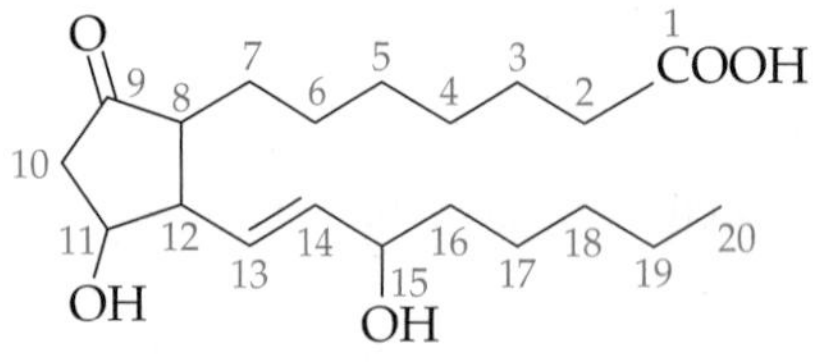

PGE$_1$, a prostaglandin

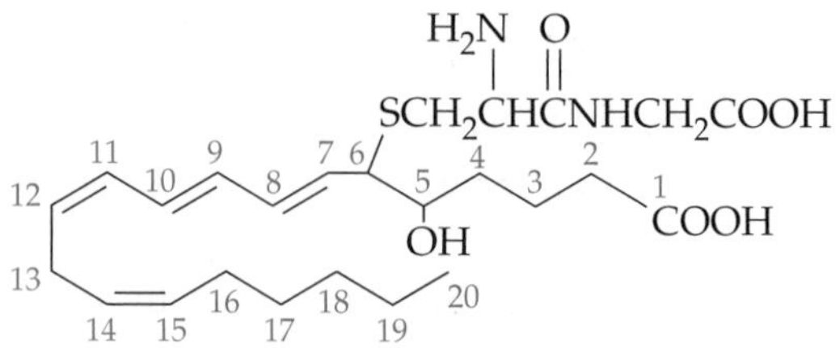

Leukotriene D$_4$

Research targets the myriad of potential therapeutic applications prostaglandins promise. The several dozen known prostaglandins have an extraordinary range of biological effects. They can lower blood pressure, influence platelet aggregation during blood clotting, stimulate uterine contractions, and lower the extent of gastric secretions. In addition, they are responsible for some of the pain and swelling that accompany inflammation. The first approved clinical uses of prostaglandins include stimulation of uterine contractions in therapeutic abortion (for example, of a dead fetus) and to halt persistent bleeding after delivery of a baby (postpartum bleeding).

In 1971, it was discovered that the anti-inflammatory and fever-reducing (*antipyretic*) action of aspirin results in part from its inhibition of prostaglandin synthesis. Aspirin transfers its acetyl group to a serine side chain in cyclooxygenase (COX), the enzyme that catalyzes the first step in conversion of arachidonic acid to prostaglandins, irreversibly inhibiting the enzyme. This inhibition is also thought to explain the effect of aspirin on combating heart attacks. Cyclooxygenase is present in two forms in cells, referred to as COX-1 and COX-2. Drugs have been designed to inhibit either one or the other of these enzymes. Of great interest are drugs that block the activity of COX-2, the enzyme responsible for prostaglandins that are involved in inflammation and pain responses in diseases such as arthritis. Celebrex® and Vioxx®, two drugs introduced in the 1990s, both inhibit COX-2. Unfortunately, Vioxx® is no longer available because of unexpected potentially lethal side effects, and Celebrex® is prescribed with a strong warning of side effects on the label. While basic research for alternate COX-2 inhibitors continues, current medical practice prescribes these drugs sparingly and depends on older, better-understood analgesics such as aspirin and acetaminophen to lessen pain and fever.

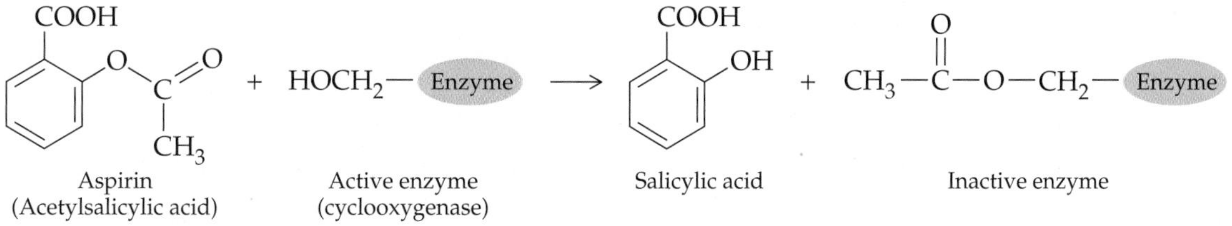

Aspirin (Acetylsalicylic acid) + Active enzyme (cyclooxygenase) → Salicylic acid + Inactive enzyme

There is also great interest in the leukotrienes. Leukotriene release has been found to trigger the asthmatic response, severe allergic reactions, and inflammation. Asthma treatment with drugs that inhibit leukotriene synthesis is being studied.

⚷ KEY CONCEPT PROBLEM 24.18

In the eicosanoid shown here, identify all the functional groups. Which groups are capable of hydrogen bonding? Which are most acidic? Is this molecule primarily nonpolar, polar, or something in between?

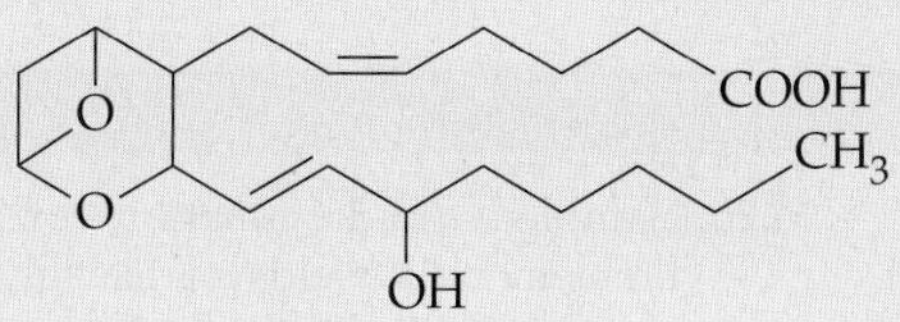

SUMMARY: REVISITING THE CHAPTER GOALS

KEY WORDS

Active transport, *p. 778*
Concentration gradient, *p. 779*

1. **What are the major classes of fatty acids and lipids?** *Fatty acids* are carboxylic acids with long, straight (unbranched) hydrocarbon chains; they may be saturated or unsaturated. *Waxes* are esters of fatty acids and alcohols with long, straight hydrocarbon chains. *Fats* and *oils* are *triacylglycerols*—triesters of glycerol with fatty acids. In fats, the fatty acid

chains are mostly saturated; in oils, the proportions of unsaturated fatty acid chains vary. Fats are solid because the saturated hydrocarbon chains pack together neatly; oils are liquids because the kinks at the cis double bonds prevent such packing.

2. What reactions do triacylglycerols undergo? The principal reactions of triacylglycerols are catalytic *hydrogenation* and *hydrolysis*. Hydrogen adds to the double bonds of unsaturated hydrocarbon chains in oils, thereby thickening the consistency of the oils and raising their melting points. Treatment of a fat or oil with a strong base such as NaOH hydrolyzes the triacylglycerols to give glycerol and salts of fatty acids. Such *saponification* reactions produce soap, a mixture of fatty acid salts.

3. What are the membrane lipids? The membrane lipids include *phospholipids* and *glycolipids* (which have hydrophilic, polar head groups and a pair of hydrophobic tails) and cholesterol (a steroid). *Phospholipids*, which are either *glycerophospholipids* (derived from glycerol) or *sphingomyelins* (derived from the amino alcohol sphingosine), have charged phosphate diester groups in their hydrophilic heads. *Sphingolipids*, which are either sphingomyelins or *glycolipids*, are sphingosine derivatives. The glycolipids have carbohydrate head groups.

4. What is the nature of a cell membrane? The basic structure of cell membranes is a *bilayer of lipids*, with their hydrophilic heads in the aqueous environment outside and inside the cells, and their hydrophobic tails clustered together in the center of the bilayer. *Cholesterol* molecules fit between the hydrophobic tails and help maintain membrane structure and rigidity. The membrane also contains *glycoproteins* and *glycolipids* (with their carbohydrate segments at the cell surface, where they serve as receptors), as well as *proteins*. Some of the proteins extend through the membrane (*integral proteins*), and others are only partially embedded at one surface (*peripheral proteins*).

5. How do substances cross cell membranes? Small molecules and lipid-soluble substances can cross the lipid bilayer by simply diffusing through it. Ions and hydrophilic substances can move through aqueous fluid-filled channels in membrane proteins. Some substances cross the membrane by binding to an integral protein, which then releases them inside the cell. These modes of crossing are all *passive transport*—they do not require energy because the substances move from regions of higher concentration to regions of lower concentration. Passive transport takes the form of *simple diffusion*, crossing the membrane by passing through it unimpeded, or *facilitated diffusion*, crossing the membrane with the aid of a protein embedded in the membrane. *Active transport*, which requires energy and is carried out by certain integral membrane proteins, moves substances against their *concentration gradients*.

6. What are eicosanoids? The *eicosanoids* are a group of compounds derived from 20-carbon unsaturated fatty acids. They are *local hormones*—that is, they act near their point of origin and are short-lived. *Prostaglandins*, which contain a five-membered ring, have a wide range of actions (such as stimulating uterine contractions and causing inflammation). *Leukotrienes*, which do not contain a five-membered ring, trigger the asthmatic response, severe allergic reactions, and inflammation.

UNDERSTANDING KEY CONCEPTS

24.19 The fatty acid composition of three triacylglycerols (A, B, and C) is reported below. Predict which one has the highest melting point. Which one do you expect to be liquid (oil) at room temperature? Explain.

	Palmitic acid	Stearic acid	Oleic acid	Linoleic acid
A	21.4%	27.8%	35.6%	11.9%
B	12.2%	16.7%	48.2%	22.6%
C	11.2%	8.3%	28.2%	48.6%

24.20 Complete hydrogenation of triacylglycerol C in Problem 24.19 yields a triacylglycerol of what fatty acid composition? Will the hydrogenation product of triacylglycerol C be more like the hydrogenation product of triacylglycerol A or B? Explain.

24.21 A membrane lipid was isolated and completely hydrolyzed. The following products were detected: ethanolamine, phosphate, glycerol, palmitic acid, and oleic acid. Propose a structure for this membrane lipid, and name the family (Table 24.3) to which it belongs.

24.22 According to the fluid-mosaic model (Figure 24.7), the cell membrane is held together mostly by hydrophobic interactions. Considering the forces applied, why does the cell membrane not rupture as you move, press against objects, etc.?

24.23 Dipalmitoyl phosphatidylcholine (DPPC) is a surfactant on the surface of the alveoli in the lungs. What is the nature of its fatty acid groups? In what arrangement is it likely to exist at the lung surfaces?

ADDITIONAL PROBLEMS

WAXES, FATS, AND OILS

24.24 What is a lipid?

24.25 Name the general classes of lipids.

24.26 Draw an 18-carbon saturated fatty acid. Is this a "straight chain" molecule or a "bent" molecule?

24.27 Draw an 18-carbon unsaturated fatty acid that contains two carbon-carbon double bonds, one on carbon 6 and one on carbon 9 (count starting with the carboxyl carbon). Is this a "straight chain" molecule or a "bent" molecule?

24.28 Differentiate between saturated, monounsaturated, and polyunsaturated fatty acids.

24.29 Are the carbon-carbon double bonds in naturally occurring fatty acids primarily cis or trans?

24.30 What does it mean when we say that a fatty acid is an essential fatty acid?

24.31 Name two essential fatty acids. What are good sources of these fatty acids?

24.32 Why does the presence of double bonds lower the melting temperature of a fatty acid?

24.33 Do cis fatty acids have a higher or lower melting temperature than the corresponding trans fatty acids? Why?

24.34 What does it mean to say that fats and oils are triacylglycerols?

24.35 Is a triacylglycerol composed of linoleic, oleic, and palmitic acids a liquid or a solid at room temperature?

24.36 How do fats differ from oils in terms of physical properties?

24.37 List typical food sources for oils and fats. Are there similarities or differences in the sources for each?

24.38 Draw the structure of glyceryl trilaurate, which is made from glycerol and three lauric acid molecules.

24.39 There are two isomeric triacylglycerol molecules whose components are glycerol, one palmitic acid unit, and two stearic acid units. Draw the structures of both, and explain how they differ.

24.40 What function does a wax serve in a plant or animal?

24.41 What functions do fats serve in an animal?

24.42 *Spermaceti*, a fragrant substance isolated from sperm whales, was commonly used in cosmetics until it was banned in 1976 to protect the whales from extinction. Chemically, spermaceti is cetyl palmitate, the ester of palmitic acid with cetyl alcohol (the straight-chain 16-carbon alcohol). Draw the structure of spermaceti.

24.43 What kind of lipid is spermaceti—a fat, a wax, or a steroid?

24.44 A major ingredient in peanut butter cup candy is soy lecithin. Draw the structure of lecithin.

24.45 Which kind of lipid is lecithin?

CHEMICAL REACTIONS OF LIPIDS

24.46 Name the process that converts unsaturated fatty acids to saturated fatty acids.

24.47 Of what commercial use is the process that converts unsaturated fatty acids to saturated fatty acids?

24.48 How would you convert a vegetable oil like corn oil into a soft margarine?

24.49 How would you convert a vegetable oil like soybean oil into a solid cooking fat?

24.50 When a vegetable oil is converted to a soft margarine, a non-natural product is synthesized. What is this product?

24.51 Is the reaction shown here esterification, hydrogenation, hydrolysis, saponification, or substitution?

$$\begin{array}{c}
\text{H} \quad\quad \text{O} \\
| \quad\quad\quad || \\
\text{H}-\text{C}-\text{O}-\text{C}(\text{CH}_2)_{14}\text{CH}_3 \\
| \quad\quad\quad\quad \text{O} \\
| \quad\quad\quad\quad || \\
\text{H}-\text{C}-\text{O}-\text{C}(\text{CH}_2)_{14}\text{CH}_3 + 3\text{KOH} \xrightarrow{\Delta} 3\text{CH}_3(\text{CH}_2)_{14}\overset{\text{O}}{\overset{||}{\text{C}}}\text{O}^-\text{K}^+ + \begin{array}{c} \text{H} \\ | \\ \text{H}-\text{C}-\text{OH} \\ | \\ \text{H}-\text{C}-\text{OH} \\ | \\ \text{H}-\text{C}-\text{OH} \\ | \\ \text{H} \end{array} \\
| \quad\quad\quad\quad \text{O} \\
| \quad\quad\quad\quad || \\
\text{H}-\text{C}-\text{O}-\text{C}(\text{CH}_2)_{14}\text{CH}_3
\end{array}$$

24.52 Draw the structures of all products you would obtain by saponification of the following lipid with aqueous KOH. What are the names of the products?

$$\begin{array}{c}
\text{O} \\
|| \\
\text{CH}_2\text{OC}(\text{CH}_2)_{16}\text{CH}_3 \\
| \quad\quad \text{O} \quad\quad \text{H} \quad \text{H} \\
| \quad\quad || \quad\quad | \quad\quad | \\
\text{CHOC}(\text{CH}_2)_7\text{C}=\text{C}(\text{CH}_2)_7\text{CH}_3 \\
| \quad\quad \text{O} \quad\quad \text{H} \quad \text{H} \quad\quad \text{H} \quad \text{H} \quad\quad \text{H} \quad \text{H} \\
| \quad\quad || \quad\quad | \quad\quad | \quad\quad | \quad\quad | \quad\quad | \quad\quad | \\
\text{CH}_2\text{OC}(\text{CH}_2)_7\text{C}=\text{CCH}_2\text{C}=\text{CCH}_2\text{C}=\text{CCH}_2\text{CH}_3
\end{array}$$

24.53 Draw the structure of the product you would obtain on complete hydrogenation of the triacylglycerol in Problem 24.52. What is its name? Does it have a higher or lower melting temperature than the original triacylglycerol?

24.54 Estimate how many different products you would obtain on incomplete hydrogenation of the triacylglycerol in Problem 24.52.

PHOSPHOLIPIDS, GLYCOLIPIDS, AND CELL MEMBRANES

24.55 Describe the difference between a triacylglycerol and a phospholipid.

24.56 Why are glycerophospholipids, rather than triacylglycerols, found in cell membranes?

24.57 How do sphingomyelins and cerebrosides differ structurally?

24.58 Name the two different kinds of sphingosine-based lipids.

24.59 Why are glycerophospholipids more soluble in water than triacylglycerols?

24.60 What are the functions of glycerophospholipids in the human body? Of triacylglycerides in the human body?

24.61 Explain how a soap micelle differs from a membrane bilayer.

24.62 Describe the similarities and differences between a liposome and a micelle.

24.63 What constituents besides phospholipids are present in a cell membrane?

24.64 What would happen if cell membranes were freely permeable to all molecules?

24.65 Show the structure of a cerebroside made up of D-galactose, sphingosine, and myristic acid.

24.66 Draw the structure of a sphingomyelin that contains a stearic acid unit.

24.67 Draw the structure of a glycerophospholipid that contains palmitic acid, oleic acid, and the phosphate bonded to propanolamine.

24.68 *Cardiolipin*, a compound found in heart muscle, has the following structure. What products are formed if all ester bonds in the molecule are saponified by treatment with aqueous NaOH?

Cardiolipin

24.69 Which process requires energy—passive or active transport? Why is energy sometimes required to move solute across the cell membrane?

24.70 How does facilitated diffusion differ from simple diffusion?

24.71 Based on the information in Section 24.8, how would you expect each of these common metabolites to cross the cell membrane?

   (a) NO (nitrous oxide)
   (b) Fructose
   (c) Ca^{2+}

24.72 Based on the information in Section 24.8, how would you expect each of these common metabolites to cross the cell membrane?

   (a) Galactose
   (b) CO
   (c) Mg^{2+}

EICOSANOIDS

24.73 Why are the eicosanoids often called "local hormones"?

24.74 Give an example of an eicosanoid serving as a local hormone.

24.75 Arachidonic acid is used to produce prostaglandins and leukotrienes. From what common fatty acid is arachidonic acid synthesized?

24.76 *Thromboxane A_2* is a lipid involved in the blood-clotting process. To what category of lipids does thromboxane A_2 belong?

Thromboxane A_2

What fatty acid do you think serves as a biological precursor of thromboxane A_2?

24.77 Why is it desirable to inhibit the production of leukotrienes?

24.78 After a fall hike through fields of goldenrod and ragweed, you develop severe hay-fever followed by an asthmatic attack. What class of molecule is most likely responsible for the asthmatic attack?

24.79 How does aspirin function to inhibit the formation of prostaglandins from arachidonic acid?

24.80 Typically the pain and swelling from a scrape or cut is confined to the area around the injury. What class of molecules is responsible for this effect?

Applications

24.81 Fats and oils are major sources of triacylglycerols. List some other foods that are associated with high lipid content. [*Lipids in the Diet, p. 763*]

24.82 According to the FDA, what is the maximum percentage of your daily calories that should come from fats and oils? [*Lipids in the Diet, p. 763*]

24.83 Describe the mechanism by which soaps and detergents provide cleaning action. [*Detergents, p. 767*]

24.84 Cationic detergents are rarely used for cleaning. For what purposes are they used? [*Detergents, p. 767*]

24.85 Why are branched-chain hydrocarbons no longer used for detergents? [*Detergents, p. 767*]

24.86 Butter and an equally solid margarine both contain an abundance of saturated fatty acids. What lipid that has been identified as a health hazard is not present in margarine but is present in butter? [*Butter and Its Substitutes, p. 774*]

24.87 Recently it has been suggested that using oils with more monounsaturated fatty acids (for example, oleic acid) is better for our health than those with polyunsaturated fatty acids or saturated fatty acids. What are good sources of oils with predominantly monounsaturated fatty acids? (Hint: See Table 24.2.) [*Butter and Its Substitutes, p. 774*]

24.88 A liposome designed to deliver a chemotherapy drug should be able to distinguish between the tumor cell and noncancerous cells. One way to accomplish this is to attach a molecule that recognizes specific groups on the tumor cells to the surface of the liposome. What kind(s) of groups might be recognized? [*Liposomes for Health and Beauty, p. 775*]

24.89 Why does a liposome deliver a moisturizer to skin cells more efficiently than applying the moisturizer directly to the skin? [*Liposomes for Health and Beauty, p. 775*]

General Questions and Problems

24.90 Which of the following are saponifiable lipids?
(a) Progesterone   **(b)** Glyceryl trioleate
(c) A sphingomyelin   **(d)** Prostaglandin E_1
(e) A cerebroside   **(f)** A lecithin

24.91 Identify the component parts of each saponifiable lipid listed in Problem 24.90.

24.92 Draw the structure of a triacylglycerol made from two molecules of myristic acid and one molecule of linolenic acid.

24.93 Would the triacylglycerol described in Problem 24.92 have a higher or lower melting temperature than the triacylglycerol made from one molecule each of linolenic, myristic, and stearic acids? Why?

24.94 Common names for some triacylglycerols depend on their source. Identify the source. Choices are plant oils (soybean, canola, corn, sunflower, and so on), beef fat, pork fat.
(a) tallow
(b) cooking oil
(c) lard

24.95 Explain why cholesterol is not saponifiable.

24.96 Draw cholesterol acetate. Is this molecule saponifiable? Explain.

24.97 Jojoba wax, used in candles and cosmetics, is partially composed of the ester of stearic acid and a straight-chain C_{22} alcohol. Draw the structure of this wax component. Compare this structure with the structure drawn for spermaceti in Problem 24.42. Do you think jojoba wax could replace spermaceti in the cosmetic industry?

24.98 Which three types of lipids are particularly abundant in brain tissue?

24.99 Name two major roles of cholesterol in the body.

24.100 List some of the functions prostaglandins serve in the body.

24.101 Lecithins are often used as food additives to provide emulsification. How do they accomplish this purpose?

24.102 If the average molar mass of a sample of soybean oil is 1500 amu, how many grams of NaOH are needed to saponify 5.0 g of the oil?

24.103 The concentration of cholesterol in the blood serum of a normal adult is approximately 200 mg/dL. How many grams of cholesterol does a person with a blood volume of 5.75 L have circulating in his or her blood?

Lipid Metabolism

▲ These Emperor penguins will survive for several months on the energy supplied by lipid metabolism.

CONTENTS

CHAPTER GOALS

In this chapter, we will address the following questions about lipid metabolism:

1. **What happens during the digestion of triacylglycerols?**

   THE GOAL: Be able to list the sequence of events in the digestion of dietary triacylglycerols and their transport into the bloodstream.

2. **What are the various roles of lipoproteins in lipid transport?**

   THE GOAL: Be able to name the major classes of lipoproteins, specify the nature and function of the lipids they transport, and identify their destinations.

3. **What are the major pathways in the metabolism of triacylglycerols?**

   THE GOAL: Be able to name the major pathways for the synthesis and breakdown of triacylglycerols and fatty acids, and identify their connections to other metabolic pathways.

4. **How are triacylglycerols moved into and out of storage in adipose tissue?**

   THE GOAL: Be able to explain the reactions by which triacylglycerols are stored and mobilized, and how these reactions are regulated.

5. **How are fatty acids oxidized, and how much energy is produced by their oxidation?**

   THE GOAL: Be able to explain what happens to a fatty acid from its entry into a cell until its conversion to acetyl-SCoA.

6. **What is the function of ketogenesis?**

   THE GOAL: Be able to identify ketone bodies, describe their properties and synthesis, and explain their role in metabolism.

7. **How are fatty acids synthesized?**

   THE GOAL: Be able to compare the pathways for fatty acid synthesis and oxidation, and describe the reactions of the synthesis pathway.

C arbohydrate metabolism (discussed in Chapter 23) is one of our two major sources of energy. Lipid metabolism, the topic of this chapter, is the other. Of the various classes of lipids you saw in Chapter 24, the majority of the lipids in our diet are triacylglycerols. Therefore, our focus here is on the metabolism of triacylglycerols, which are stored in fatty tissue and constitute our chief energy reserve.

25.1 Digestion of Triacylglycerols

When food containing triacylglycerols (TAGs) is eaten, the triacylglycerols pass through the mouth unchanged and enter the stomach (Figure 25.1). (Recall that an *acyl* group is the R—C=O portion of an ester. The acyl groups from fatty acids have relatively long-chain R groups.) The heat and churning action of the stomach break the triacylglycerols into smaller droplets, a process that takes longer than the physical breakdown and digestion of other foods in the stomach. To be sure that there is time for this breakdown, the presence of triacylglycerols in consumed food slows down the rate at which the mixture of partially digested foods leaves the stomach. (One reason foods containing lipids are a pleasing part of the diet is that the stomach feels full for a longer time after a fatty meal.)

The pathway of dietary triacylglycerols from the mouth to their ultimate biochemical fate in the body is not as straightforward as that of carbohydrates. Complications arise because triacylglycerols are not water-soluble but nevertheless must enter an aqueous environment. To be moved around within the body by the blood and lymph systems, they must therefore be dispersed and surrounded by a water-soluble coating, a process that must happen more than once as the triacylglycerols travel along their metabolic pathways. During these travels, they are packaged in various kinds of **lipoproteins**, which consist of droplets of hydrophobic lipids surrounded by phospholipids (⊂▭⊃, Section 24.5) and other molecules with their hydrophilic ends to the outside (Figure 25.2).

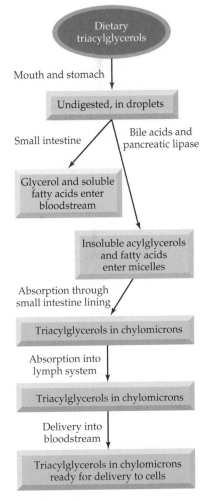

▲ **FIGURE 25.1 Digestion of triacylglycerols.**

Lipoprotein A lipid–protein complex that transports lipids.

787

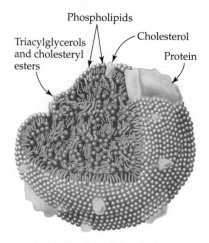

▲ **FIGURE 25.2 A lipoprotein.** A lipoprotein contains a core of neutral lipids, including triacylglycerols and cholesteryl esters. Surrounding the core is a layer of phospholipids in which varying proportions of proteins and cholesterol are embedded.

Bile Fluid secreted by the liver and released into the small intestine from the gallbladder during digestion; contains bile acids, bicarbonate ion, and other electrolytes.

Bile acids Steroid acids derived from cholesterol that are secreted in bile.

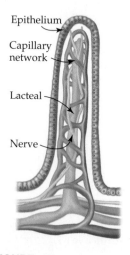

▲ **FIGURE 25.3 A villus, site of absorption in the intestinal lining.** A huge number of villi provide the surface at which lipids and other nutrients are absorbed. Small molecules enter the capillary network, and larger lipids enter the lacteals.

When partially digested food leaves the stomach, it enters the upper end of the small intestine (the *duodenum*), where its arrival triggers the release of *pancreatic lipases*—enzymes for the hydrolysis of lipids. The gallbladder simultaneously releases **bile**, a mixture that is manufactured in the liver and stored in the gallbladder until needed. Among other components, bile contains **bile acids** (derived from cholesterol) and cholesterol, which are steroids, and phospholipids.

By the time dietary triacylglycerols enter the small intestine, they are dispersed as small, greasy, insoluble droplets, and for this reason enzymes in the small intestine cannot attack them. It is the job of the bile acids and phospholipids to emulsify the triacylglycerols by forming micelles much like soap micelles. (⟨▭▭⟩, p. 768) The major bile acid is cholic acid, and you can see from the structure of its anion that it resembles soaps and detergents because it contains both hydrophilic and hydrophobic regions that allow it to act as an emulsifying agent:

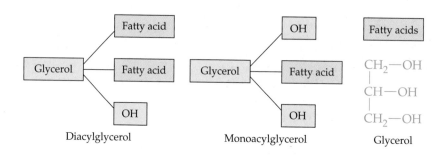

Cholate

Pancreatic lipase partially hydrolyzes the emulsified triacylglycerols, producing mono- and diacylglycerols plus fatty acids and a small amount of glycerol:

Small fatty acids and glycerol are water-soluble and are absorbed directly through the surface of the villi that line the small intestine. Once they are in the villi (Figure 25.3), these molecules enter the bloodstream through capillaries and are carried by the blood to the liver (via the hepatic portal vein).

The water-insoluble acylglycerols and larger fatty acids within the intestine are once again emulsified. Then, at the intestinal lining they are released from the micelles and absorbed. Because these lipids, and also cholesterol and partially hydrolyzed phospholipids, must next enter the aqueous bloodstream for transport, they are again packaged into water-soluble units—in this case, the lipoproteins known as *chylomicrons*.

Chylomicrons are too large to enter the bloodstream through capillary walls. Instead, they are absorbed into the lymphatic system through lacteals within the villi (see Figure 25.3). Then, chylomicrons are carried to the thoracic duct (just below the collarbone), where the lymphatic system empties into the bloodstream. At this point, these lipids are ready to be used for energy generation or put into storage. The pathways of lipids through the villi and into the transport systems of the bloodstream and the lymphatic system are summarized in Figure 25.4.

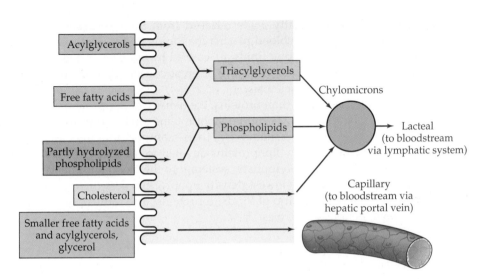

◀ **FIGURE 25.4 Pathways of lipids through the villi.**

◄**O KEY CONCEPT PROBLEM 25.1**

Cholesterol (see structure in margin) and cholate (a bile acid anion, structure shown on p. 788) are steroids with very similar structures. However, the roles they play in the body are different: Cholate is an emulsifier, whereas cholesterol plays an important role in membrane structure. Identify the small differences in their structures that make them well suited to their jobs in the body. Given their similar structures, can the roles of these molecules be reversed?

Cholesterol

25.2 Lipoproteins for Lipid Transport

The lipids used in the body's metabolic pathways have three sources. They enter the pathways (1) from the digestive tract as food is broken down, (2) from adipose tissue, where excess lipids have been stored, and (3) from the liver, where lipids are synthesized. Whatever their source, these lipids must eventually be transported in blood, an aqueous medium, as summarized in Figure 25.5.

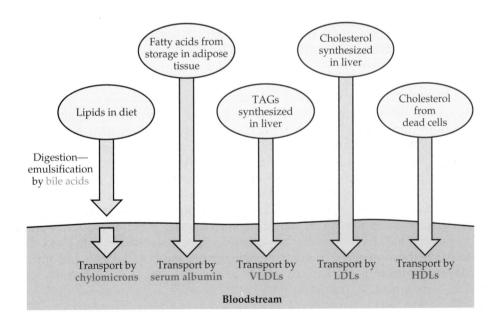

◀ **FIGURE 25.5 Transport of lipids.** Fatty acids released from storage are carried by albumin, which is a large protein. All of the other lipids are carried packaged in various lipoproteins.

To become water-soluble, fatty acids released from adipose tissue associate with albumin, a protein found in blood plasma that binds up to 10 fatty acid molecules per protein molecule. All other lipids are carried by lipoproteins. (The role of lipoproteins in heart disease, where they are of great concern, is discussed in the Application "Lipids and Atherosclerosis.")

Because lipids are less dense than proteins, the density of lipoproteins depends on their ratio of lipids to proteins. Therefore, lipoproteins can be arbitrarily divided into five major types distinguishable by their composition and densities. Chylomicrons, which are the only lipoproteins devoted to transport of lipids from the diet, carry TAGs through the lymphatic system into the blood and thence to the liver for processing. These are the lowest-density lipoproteins (less than 0.95 g/cm^3) because they carry the highest ratio of lipids to proteins. The four denser lipoprotein fractions have the following roles:

- **Very-low-density lipoproteins (VLDLs)** ($0.96-1.006 \text{ g/cm}^3$) carry TAGs from the liver (where they are synthesized) to peripheral tissues for storage or energy generation.

- **Intermediate-density lipoproteins (IDLs)** ($1.006-1.019 \text{ g/cm}^3$) carry remnants of the VLDLs from peripheral tissues back to the liver for use in synthesis.

- **Low-density lipoproteins (LDLs)** ($1.019-1.063 \text{ g/cm}^3$) transport cholesterol from the liver to peripheral tissues, where it is used in cell membranes or for steroid synthesis (and is also available for formation of arterial plaque).

- **High-density lipoproteins (HDLs)** ($1.063-1.210 \text{ g/cm}^3$) transport cholesterol *from* dead or dying cells back to the liver, where it is converted to bile acids. The bile acids are then available for use in digestion or are excreted via the digestive tract when in excess.

WORKED EXAMPLE **25.1** Digesting and Transporting Fats

Describe how the fat in an ice cream cone gets from the ice cream to a liver cell.

ANALYSIS Dietary fat from animal sources (such as the whole milk often found in ice cream) is primarily triacylglycerides with a small amount of cholesterol present. Fat-digesting enzymes are secreted by the pancreas and delivered via the common duct to the small intestine, along with bile acids (, Section 24.6). As discussed above, only free fatty acids, mono- and diacylglycerides can cross the intestinal cell wall before being passed on to the blood stream in special packaging called lipoproteins.

SOLUTION
As the ice cream cone is eaten, it passes through the mouth to the stomach where mixing occurs. This mixing action promotes the formation of triacylglycerols into small droplets. No enzymatic digestion of lipids occurs in the stomach. When the stomach contents move to the small intestine, bile acids and pancreatic lipases are secreted into the mixture. The bile acids help to emulsify the fat droplets into micelles. Once micelles have formed, lipases hydrolyze the triacylglycerides to mono- and diacylglycerides; the hydrolysis also produces fatty acids. These three hydrolysis products cross into the cells lining the small intestine, are rearranged, and secreted into the bloodstream in the form of chylomicrons. Chylomicrons travel to the liver and enter cells for processing.

APPLICATION ▶ Lipids and Atherosclerosis

According to the U.S. Food and Drug Administration (FDA), and in agreement with many other authorities, there is "strong, convincing, and consistent evidence" for the connection between heart disease and diets high in saturated fats and cholesterol. (The FDA has also found evidence that high dietary fat is one risk factor for certain types of cancer.)

Several points are clear:

- A diet rich in saturated animal fats leads to an increase in blood-serum cholesterol.

- A diet lower in saturated fat and higher in unsaturated fat can lower the serum cholesterol level.

- High levels of cholesterol are correlated with *atherosclerosis*, a condition in which yellowish deposits (*arterial plaque*) composed of cholesterol and other lipid-containing materials form within the larger arteries. The result of atherosclerosis is an increased risk of coronary artery disease and heart attack brought on by blockage of blood flow to heart muscles, or an increased risk of stroke due to blockage of blood flow to the brain.

Factors considered in an overall evaluation of an individual's risk of heart disease are the following:

Risk factors for heart disease

High blood levels of cholesterol and low levels of high-density lipoproteins (HDLs)

Cigarette smoking

High blood pressure

Diabetes

Obesity

Low level of physical activity

Family history of early heart disease

As discussed in Section 25.2, lipoproteins are complex assemblages of lipids and proteins that transport lipids throughout the body. If LDL (the so-called "bad" cholesterol) delivers more cholesterol than is needed to peripheral tissues, and if not enough HDL (the so-called "good" cholesterol) is present to remove it, the excess cholesterol is deposited in cells and arteries. Thus, the higher the HDL level, the less the likelihood of deposits and the lower risk of heart disease. There is some evidence that a low HDL level (less than 35 mg/dL) may be the single best predictor of heart attack potential. Also, LDL has the negative potential to trigger inflammation and the buildup of plaque in artery walls. (Remember it this way—**low L**DL is good; **high H**DL is good.)

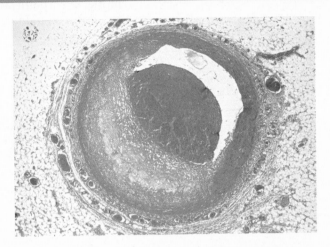

▲ **Plaque.** Deposits of cholesterol and associated lipids (known collectively as plaque) have partially blocked the flow of blood in this artery.

Many groups recommend that individuals strive for the following blood levels:

Total cholesterol	200 mg/dL or lower
LDL	160 mg/dL or lower
HDL	60 mg/dL or higher

To further assess the risk level represented by an individual's cholesterol and HDL values, the total cholesterol/HDL ratio is calculated. The ideal ratio is considered to be 3.5. A ratio of 4.5 indicates an average risk, and a ratio of 5 or higher shows a high and potentially dangerous risk. The ratio overcomes the difficulty in evaluating the significance of, for example, a negatively high cholesterol level of 290 mg/dL combined with a positively high HDL value of 75 mg/dL. The resultant ratio of 3.9 indicates a low risk level.

Decreasing saturated fats and cholesterol in the diet, adopting an exercise program, and ceasing to smoke constitute the first line of defense for those at risk. For those at high risk or for whom the first-line defenses are inadequate, drugs are available that prevent or slow the progress of coronary artery disease by lowering serum cholesterol levels. Among the drugs are indigestible resins (*cholestyramine* and *colestipol*) that bind bile acids and accelerate their excretion, causing the liver to use up more cholesterol in bile acid synthesis. Another class of effective drugs are the statins (for example, lovastatin), which inhibit an enzyme crucial to the synthesis of cholesterol.

See Additional Problems 25.52 through 25.55 at the end of the chapter.

25.3 Triacylglycerol Metabolism: An Overview

The metabolic pathways for triacylglycerols are summarized in Figure 25.6 and further explained in following sections of this chapter.

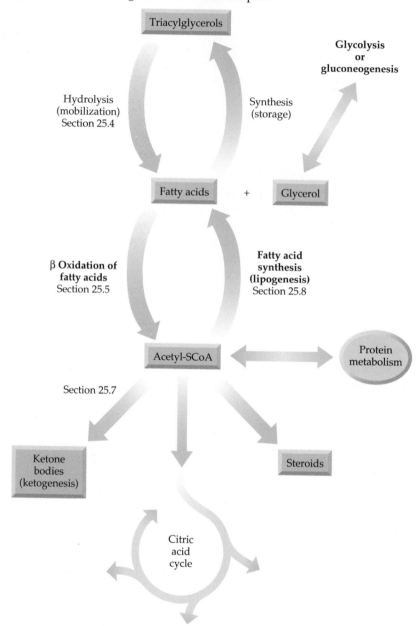

▶ **FIGURE 25.6 Metabolism of triacylglycerols.** Pathways that break down molecules (catabolism) are shown in light brown, and synthetic pathways (anabolism) are shown in blue. Connections to other pathways or intermediates of metabolism are shown in green.

Dietary Triacylglycerols

For TAGs from the diet, hydrolysis occurs when chylomicrons in the bloodstream encounter lipoprotein lipase anchored in capillary walls. The resulting fatty acids then have two possible fates: (1) When energy is in good supply, they are converted back to TAGs for storage in adipose tissue. (2) When cells need energy, the fatty acid carbon atoms are activated by conversion to fatty acyl-SCoA and then oxidized as acetyl-SCoA, shortening the fatty acyl-SCoA molecule by two carbon atoms for each oxidation.

The primary metabolic fate of acetyl-SCoA is the generation of energy via the citric acid cycle and oxidative phosphorylation. (, Figure 21.5, p. 668) Acetyl-SCoA has several important roles in lipid metabolism as well. In resting muscle, fatty acids are the major energy source. Acetyl-SCoA serves as the starting material for the biosynthesis of fatty acids (*lipogenesis*) in the liver (Section 25.8). In addition, it enters the *ketogenesis* pathway for production of ketone bodies, a source of energy called on when glucose is in short supply (Section 25.7). Acetyl-SCoA is also the starting material for the synthesis of cholesterol, from which all other steroids are made.

Triacylglycerols from Adipocytes

When stored TAGs are needed as an energy source, lipases within fat cells are activated by hormones (insulin and glucagon, Section 23.7). The stored TAGs are hydrolyzed to fatty acids, and the free fatty acids and glycerol are then released into the bloodstream. These fatty acids travel in association with *albumins* (blood plasma proteins) to cells (primarily muscle and liver cells), where they are converted to acetyl-SCoA for energy generation.

Glycerol from Triacylglycerols

The glycerol produced from TAG hydrolysis is carried in the bloodstream to the liver or kidneys, where it is converted in a series of reactions to glyceraldehyde 3-phosphate and dihydroxyacetone phosphate (DHAP):

$$
\underset{\text{Glycerol}}{
\begin{array}{c}
CH_2OH \\
| \\
HO-C-H \\
| \\
CH_2OH
\end{array}}
\;\;\xrightarrow{\text{ATP} \;\; \text{ADP}}\;\;
\underset{\substack{\text{Glycerol} \\ \text{3-phosphate}}}{
\begin{array}{c}
CH_2OH \\
| \\
HO-C-H \\
| \\
CH_2-O-PO_3{}^{2-}
\end{array}}
\;\;\xrightarrow{\text{NAD}^+ \;\; \text{NADH/H}^+}\;\;
\underset{\substack{\text{Dihydroxyacetone} \\ \text{phosphate (DHAP)}}}{
\begin{array}{c}
CH_2OH \\
| \\
C=O \\
| \\
CH_2-O-PO_3{}^{2-}
\end{array}}
$$

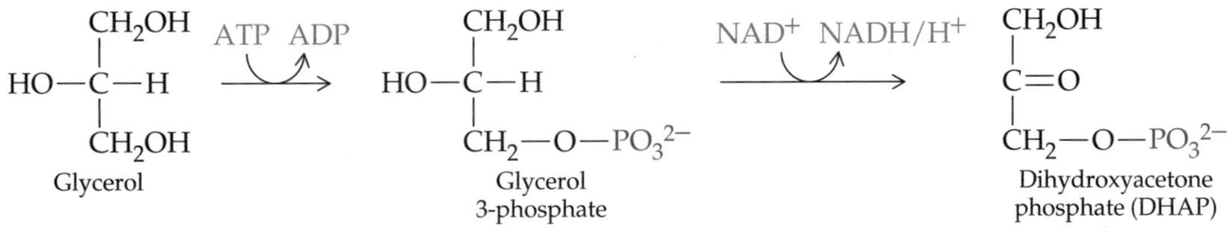

DHAP, which is a reactant in the synthesis of triacylglycerols (Section 25.4), enters the glycolysis pathway (at Step 5, Figure 23.3, p. 732) and thus is a link between carbohydrate metabolism and lipid metabolism.

The varied possible metabolic destinations of the fatty acids, glycerol, and acetyl-SCoA from dietary TAGs are summarized as follows:

Fate of Dietary Triacylglycerols

- *Triacylglycerols* undergo hydrolysis to fatty acids and glycerol.
- *Fatty acids* undergo
  - Resynthesis of triacylglycerols for storage
  - Conversion to acetyl-SCoA
- *Glycerol* is converted to glyceraldehyde 3-phosphate and DHAP, which participate in
  - *Glycolysis*—energy generation (⬭, Section 23.3)
  - *Gluconeogenesis*—glucose formation (⬭, Section 23.11)
  - *Triacylglycerol synthesis*—energy storage (Section 25.4)
- *Acetyl-SCoA* participates in
  - *Triacylglycerol synthesis* (Section 25.4)
  - Ketone body synthesis (*ketogenesis*, Section 25.7)
  - Synthesis of steroids and other lipids
  - *Citric acid cycle and oxidative phosphorylation* (⬭, Sections 21.8, 21.9)

PROBLEM 25.2

Examine Figure 23.3 (p. 732) and explain how dihydroxyacetone phosphate can enter the glycolysis pathway and be converted to pyruvate.

APPLICATION ▶ Fat Storage: A Good Thing or Not?

Bears do it, ground hogs do it, even humans do it. Do what? Store fat. Mammals store excess dietary calories as triacylglycerols in adipocytes (which are fat cells, that is, the cells that make up adipose tissue). Some mammals, like bears and ground hogs, eat to store energy for use during hibernation; others, humans among them, seem simply to eat more calories than necessary when given the opportunity. The body can do several things with extra calories. It can burn fuel through exercise, use it to create heat, or store it for future use. Our bodies are very efficient at storing the extra calories against future need.

Excessive storage of triacylglycerols is a predictor of serious health problems, and has been associated with increased risk of developing Type II diabetes and colon cancer, as well as an increased risk of having a heart attack or stroke. It is a predictor of serious health problems. Health professionals have developed charts based on Body Mass Index (⬭ Application "Obesity and Body Fat," p. 41) to estimate obesity. Although these charts do not allow for individual variation in percent body fat, the general trend is clear. For example, those with a Body Mass Index of 30 (defined as obese) or greater develop Type II diabetes at a higher rate than those with a normal Body Mass Index. The problem is even more acute in obese children. Not only do they risk developing serious health problems at an earlier age than those who become obese as adults, but they also actually have more fat cells than adults. Clinical research shows that adipocytes in an obese child can divide, making more fat cells and allowing for storage of even more triacylglycerols; this process does not occur in adults.

Why do people eat too much? Nature or nurture? Many factors are intertwined; environmental factors (including availability of food and cultural attitudes towards food choices and exercise) are involved as are "natural" factors such as variations in metabolic rates and hormone levels. Some of the latter variations may be due to genetic differences. It has recently been discovered that a deficiency in *leptin* can lead to overeating. Leptin, a peptide hormone, is synthesized in adipocytes and acts on the brain to stop eating—it suppresses appetite. *Grehlin*, another peptide hormone, influences appetite in a different way. Made in cells lining the stomach, grehlin stimulates intense sensations of hunger. Other hormones, including insulin, are also apparently involved in appetite and satiety regulation. Research shows that genetics may be involved in extreme cases of obesity. Mice with defective leptin genes apparently have little control over the amount of food eaten and can grow to three times the size of mice with normal leptin levels. Although this discovery initiated promising new treatments for obese individuals found to be leptin deficient (leptin treatment can help those individuals lose weight), the majority of obese people are not leptin deficient.

Weight control is a complex process. The bodies of mammals are seemingly "programmed" to conserve all extra calories as fat against a time when calories might be scarce. Remember that metabolic pathways exist to convert carbohydrate and protein into fat for storage; it is not only dietary fat that is stored. Scientists do not yet understand all the hormonal, metabolic connections in the storage process. It is known, however, that a sensible diet combined with regular exercise habits can sustain a stable weight without excess fat accumulation. Today's supersized meals, fatty snacks, and sedentary habits are undoubtedly tied to increasing rates of energy storage as fat. Remember, our basal metabolism rates have not changed, only our diets and activity levels. (⬭ Application "Basal Metabolism," p. 673)

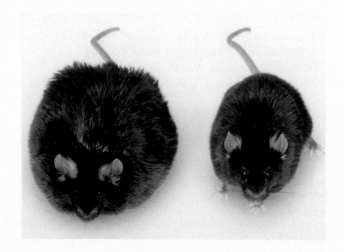

▲ Both mice have a mutation that causes increased fat storage. The mouse on the right is normal-sized due to treatment with leptin, while the untreated mouse on the left is obese.

See Additional Problems 25.56 through 25.59 at the end of the chapter.

25.4 Storage and Mobilization of Triacylglycerols

We have noted that adipose tissue is the storage depot for TAGs and that TAGs are our primary energy storage form. TAGs do not just sit unused until needed, however. The passage of fatty acids in and out of storage in adipose tissue is a continuous process essential to maintaining homeostasis (see the Application "Homeostasis," p. 631).

TAG Synthesis

To see how our bodies regulate the storage and **mobilization** of TAGs, review Figure 23.6 (, p. 742), which shows the effects of the hormones insulin and glucagon on metabolism. After a meal, blood glucose levels increase rapidly, insulin levels rise, and glucagon levels drop. Glucose enters cells, and the rate of glycolysis increases. Under these conditions, insulin activates the synthesis of TAGs for storage.

The reactants in TAG synthesis are glycerol 3-phosphate and fatty acid acyl groups carried by coenzyme A. TAG synthesis proceeds by transfer of first one and then another fatty acid acyl group from coenzyme A to glycerol 3-phosphate. The reaction is catalyzed by acyl transferase, and the product is phosphatidic acid:

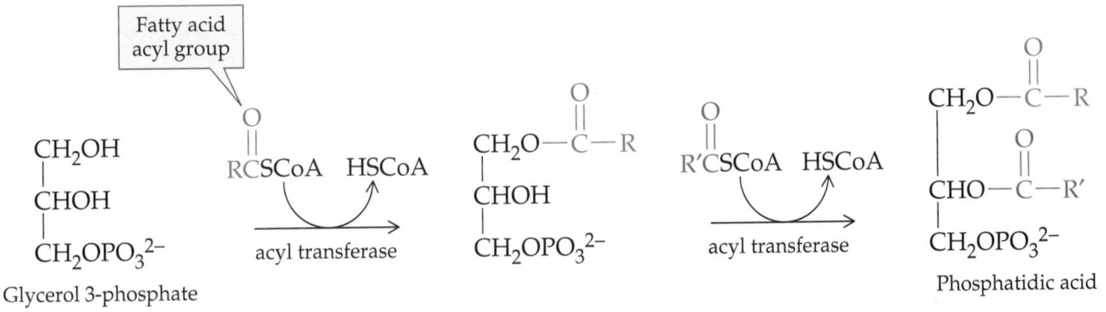

> **Mobilization (of triacylglycerols)** Hydrolysis of triacylglycerols in adipose tissue and release of fatty acids into the bloodstream.

Next, the phosphate group is removed from phosphatidic acid by phosphatidic acid phosphatase to produce 1,2-diacylglycerol. In the presence of acyl transferase, the third fatty acid group is then added to give a triacylglycerol:

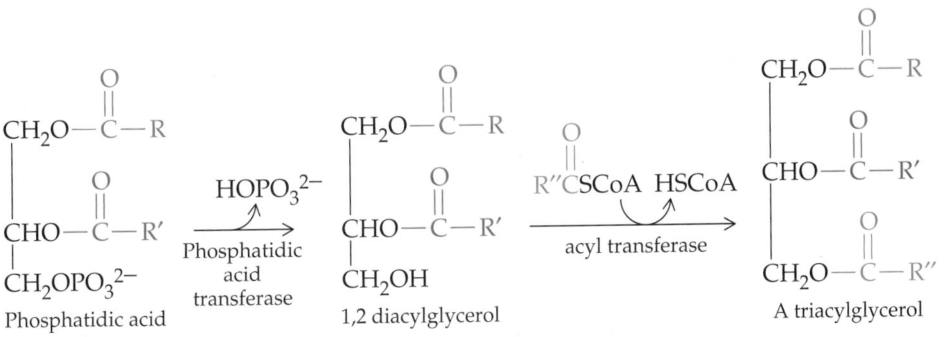

As the reaction on p. 793 shows, glycerol is one source of glycerol-3-phosphate. Since adipocytes do not have the kinase needed to convert glycerol to glycerol 3-phosphate, they cannot synthesize triacylglycerols unless dihydroxyacetone phosphate (DHAP) is available from some other pathway. In adipocytes, this pathway is called *glyceroneogenesis,* and it supplies the DHAP necessary to become glycerol 3-phosphate. Glyceroneogenesis is an abbreviated form of gluconeogenesis (p. 749), ending with the conversion of DHAP to glycerol 3-phosphate followed by TAG synthesis.

TAG Mobilization

When digestion of a meal is finished, blood glucose levels return to normal; consequently insulin levels drop and glucagon levels rise. The lower insulin level and higher glucagon level together activate *triacylglycerol lipase,* the enzyme within adipocytes that controls hydrolysis of stored TAGs. When glycerol 3-phosphate is in short supply—an indication that glycolysis is not producing sufficient energy—the fatty acids and glycerol produced by hydrolysis of the stored TAGs are released to the bloodstream for transport to energy-generating cells. Otherwise, the fatty acids and glycerol are cycled back into new TAGs for storage. Dieters on special low-carbohydrate diets are trying to produce this metabolic state in order to "burn fat." An undesirable side effect of these diets is ketosis and the production of ketone bodies (Section 25.7).

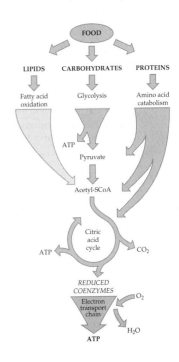

25.5 Oxidation of Fatty Acids

Once a fatty acid enters the cytosol of a cell that needs energy, three successive processes must occur.

1. **Activation** The fatty acid must be activated by conversion to fatty acyl-SCoA. This activation, which occurs in the cytosol, serves the same purpose as the first few steps in oxidation of glucose by glycolysis. Initially some energy from ATP must be invested in converting the fatty acid to fatty acyl-SCoA, a form that breaks down more easily.

$$R-\overset{\overset{O}{\|}}{C}-O^- + HSCoA + ATP \longrightarrow R-\overset{\overset{O}{\|}}{C}-SCoA + AMP + P_2O_7^{4-}$$

Fatty acid   Fatty acyl-SCoA

2. **Transport** The fatty acyl-SCoA, which cannot cross the mitochondrial membrane by diffusion, must be transported from the cytosol into the mitochondrial matrix, where energy generation will occur. Carnitine, an amino-oxy acid, undergoes a transesterification reaction with the fatty acyl-SCoA, resulting in a fatty acyl-carnitine ester that moves across the membrane into the mitochondria by facilitated diffusion. There, another transesterification reaction regenerates the fatty acyl-SCoA and carnitine.

3. **Oxidation** The fatty acyl-SCoA must be oxidized by enzymes in the mitochondrial matrix to produce acetyl-SCoA plus the reduced coenzymes to be used in ATP generation. The oxidation occurs by repeating the series of four reactions shown in Figure 25.7, which make up the **β-oxidation pathway**. Each repetition of these reactions cleaves a 2-carbon acetyl group from the end of a fatty acid acyl group and produces one acetyl-SCoA. This pathway is a spiral rather than a cycle because the long-chain fatty acyl group must continue to return to the pathway until each pair of carbon atoms is removed.

β-Oxidation pathway A repetitive series of biochemical reactions that degrades fatty acids to acetyl-SCoA by removing carbon atoms two at a time.

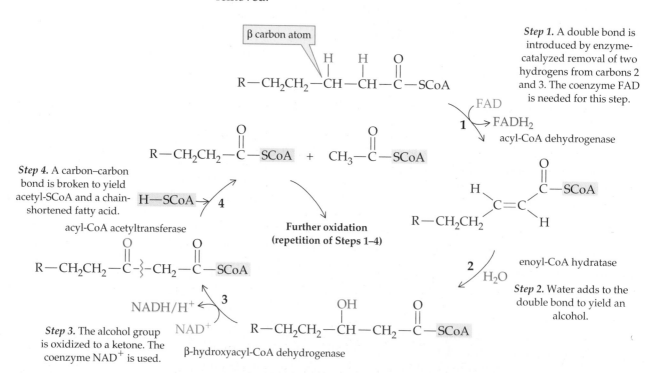

▲ FIGURE 25.7 β Oxidation of fatty acids. Passage of an acyl-SCoA through these four steps cleaves one acetyl group from the end of the fatty acid chain. In this manner, carbon atoms are removed from a fatty acid two at a time.

The β-Oxidation Pathway

The name *β oxidation* refers to the oxidation of the carbon atom *β* to the thioester linkage in two steps of the pathway.

R—CH₂CH₂—CH—CH—C—SCoA

A fatty acyl-SCoA

STEP 1: The first β oxidation Acyl-CoA dehydrogenase and its coenzyme FAD remove hydrogen atoms from the carbon atoms *α* and *β* to the C=O group in the fatty acyl-SCoA, forming a carbon–carbon double bond. These hydrogen atoms and their electrons are passed directly from $FADH_2$ to coenzyme Q so that the electrons can enter the electron transport chain. (⬭ , Section 21.9)

STEP 2: Hydration Enoyl-CoA hydratase adds a water molecule across the newly created double bond to give an alcohol with the —OH group on the *β* carbon.

STEP 3: The second β oxidation The coenzyme NAD^+ is the oxidizing agent for conversion of the *β*—OH group to a carbonyl group by *β*-hydroxyacyl-CoA dehydrogenase.

STEP 4: Cleavage to remove an acetyl group An acetyl group is split off by thiolase (acyl-CoA acetyltransferase) and attached to a new coenzyme A molecule, leaving behind an acyl-SCoA that is two carbon atoms shorter.

For a fatty acid with an even number of carbon atoms, all of the carbons are transferred to acetyl-SCoA molecules by an appropriate number of trips through the *β*-oxidation spiral. Additional steps are required to oxidize fatty acids with odd numbers of carbon atoms and those with double bonds. Ultimately, all fatty acid carbons are released for further oxidation in the citric acid cycle.

◉ KEY CONCEPT PROBLEM 25.3

In *β* oxidation, (a) identify the steps that are oxidations and describe the changes that occur; (b) identify the oxidizing agents; (c) identify the reaction that is an addition; (d) identify the reaction that is a substitution.

25.6 Energy from Fatty Acid Oxidation

The total energy output from fatty acid catabolism, like that from glucose catabolism, is measured by the total number of ATPs produced. In the case of fatty acids, this is the total number of ATPs from the passage of acetyl-SCoA through the citric acid cycle including those produced from the reduced coenzymes NADH and $FADH_2$ during oxidative phosphorylation, plus those produced by the reduced coenzymes (NADH and $FADH_2$) during fatty acid oxidation.

To compute the ATPs gained in the process, we first need to know the number of molecules of acetyl-SCoA from the fatty acids. Two carbon atoms are removed with each acetyl-SCoA, so the total number of acetyl-SCoA molecules for a fatty acid is its number of carbon atoms divided by 2. These acetyl-SCoA molecules proceed to the citric acid cycle, where each one yields 1 ATP and a total of 3 NADH molecules and 1 $FADH_2$ molecule. (⬭ , Section 21.8) Using the estimates of 3 ATPs for each NADH and 2 ATPs for each $FADH_2$ (⬭ , Section 23.6), each acetyl-SCoA generates 11 ATPs from reduced coenzymes, giving a total of 12 ATPs per acetyl-SCoA.

In addition, we must take into account the number of ATPs derived from the two reduced coenzymes (1 $FADH_2$ and 1 NADH) from each repetition of *β* oxidation. With 2 ATPs per $FADH_2$ and 3 ATPs per NADH, the total is 5 ATPs for each *β* oxidation. Note that the number of repetitions is always one fewer than the

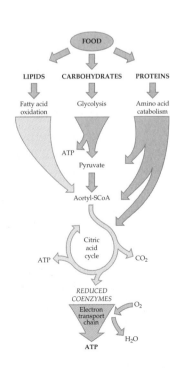

number of acetyl-SCoA molecules produced because the last β oxidation cleaves a 4-carbon chain to give 2 acetyl-SCoA molecules. Also, we must subtract the equivalent of 2 ATPs spent in activation of a fatty acid, the step that precedes the β oxidations.

As an example, the 12-carbon fatty acid, lauric acid $(CH_3(CH_2)_{10}COOH)$, yields 95 ATPs:

From the citric acid cycle:

$$12 \text{ C atoms}/2 = 6 \text{ acetyl-SCoAs}$$

$$\frac{12 \text{ ATPs}}{\text{acetyl-SCoA}} \times 6 \text{ acetyl-SCoAs} = 72 \text{ ATPs}$$

Activation of the fatty acid: $= -2$ ATPs

From the 5 β oxidations:

$$\frac{5 \text{ ATPs}}{\beta \text{ oxidation}} \times 5 \beta \text{ oxidations} = 25 \text{ ATPs}$$

$$Total = 95 \text{ ATPs}$$

Comparing the amount of ATP produced by fatty acid catabolism with the amount produced by glucose catabolism illustrates why our bodies use triacylglycerols rather than carbohydrates for long-term energy storage. We use lauric acid as our example because it has a molar mass close to that of glucose. Our best estimates show that 1 mol of glucose (180 g) generates 38 mol of ATP (, Section 23.6), whereas 1 mol of lauric acid (200 g) generates 95 mol of ATP. Thus, fatty acids yield nearly three times as much energy per gram as carbohydrates. In terms of nutritional Calories (that is, kilocalories), carbohydrates yield 4 Cal/g, whereas fats and oils yield 9 Cal/g.

In addition, stored fats have a greater "energy density" than stored carbohydrates. Because glycogen—the storage form of carbohydrates—is hydrophilic, about 2 g of water are held with each gram of glycogen. The hydrophobic fats do not hold water in this manner.

▲ **Fat as a source of water.** A camel's hump is almost entirely fat, which serves as a source of energy and also water. As reduced coenzymes from fatty acid oxidation pass through electron transport to generate ATP, large amounts of water are formed (about one water molecule for each C atom in a fatty acid). This water sustains camels during long periods when no drinking water is available.

WORKED EXAMPLE **25.2** Spiraling through β Oxidation

How many times does stearic acid $(CH_3(CH_2)_{16}COOH)$ spiral through the β-oxidation pathway to produce acetyl-SCoA?

ANALYSIS Each turn of the β-oxidation spiral pathway produces 1 acetyl-SCoA. To determine the number of turns, divide the total number of carbon atoms in the fatty acid, 18 in this case, by 2 since an acetyl group contains two carbon atoms and they come from the fatty acid. Subtract 1 turn, since the last turn produces two acetyl-SCoA molecules.

SOLUTION
Stearic acid contains 18 carbon atoms; the acetyl group contains 2 carbon atoms. Therefore 8 β oxidations occur, and 9 molecules of acetyl-SCoA are produced.

PROBLEM 25.4

How many molecules of acetyl-SCoA are produced by catabolism of the following fatty acids, and how many β oxidations are needed?

(a) Lauric acid, $CH_3(CH_2)_{10}COOH$

(b) Myrstic acid, $CH_3(CH_2)_{12}COOH$

PROBLEM 25.5

Look back at the reactions of the citric acid cycle (Figure 21.9, p. 678) and identify the three reactions in that cycle that are similar to the first three reactions of the β oxidation of a fatty acid.

We have now examined the metabolism of carbohydrates and lipids, which together provide most of our energy. Can proteins be used for energy production? Protein catabolism is utilized for energy production primarily when the carbohydrate and lipid supply is inadequate. Proteins are too important for their essential roles in providing structure and regulating function to be routinely used for energy production.

25.7 Ketone Bodies and Ketoacidosis

What happens if lipid catabolism (or any other condition) produces more acetyl-SCoA than the citric acid cycle can handle? The energy is preserved by conversion of acetyl-SCoA in liver mitochondria to 3-hydroxybutyrate and acetoacetate. Because it is a β-keto acid and therefore somewhat unstable, acetoacetate undergoes spontaneous, nonenzymatic decomposition to acetone:

Ketone bodies

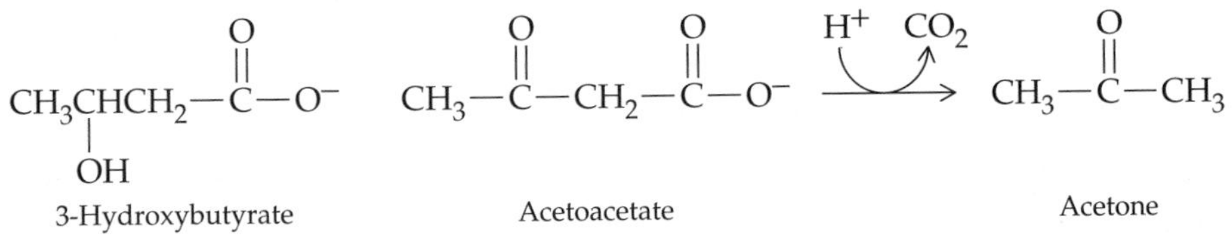

3-Hydroxybutyrate   Acetoacetate   Acetone

These compounds are traditionally known as **ketone bodies**, although one of them, 3-hydroxybutyrate, contains no ketone functional group. Because they are water-soluble, ketone bodies do not need protein carriers to travel in the bloodstream. Once formed, they become available to all tissues.

The formation of the three ketone bodies, a process known as **ketogenesis**, occurs in four enzyme-catalyzed steps plus the spontaneous decomposition of acetoacetate.

Ketone bodies Compounds produced in the liver that can be used as fuel by muscle and brain tissue; 3-hydroxybutyrate, acetoacetate, and acetone.

Ketogenesis The synthesis of ketone bodies from acetyl-SCoA.

Ketogenesis

STEPS 1 and 2: Assembly of 6-Carbon Intermediate

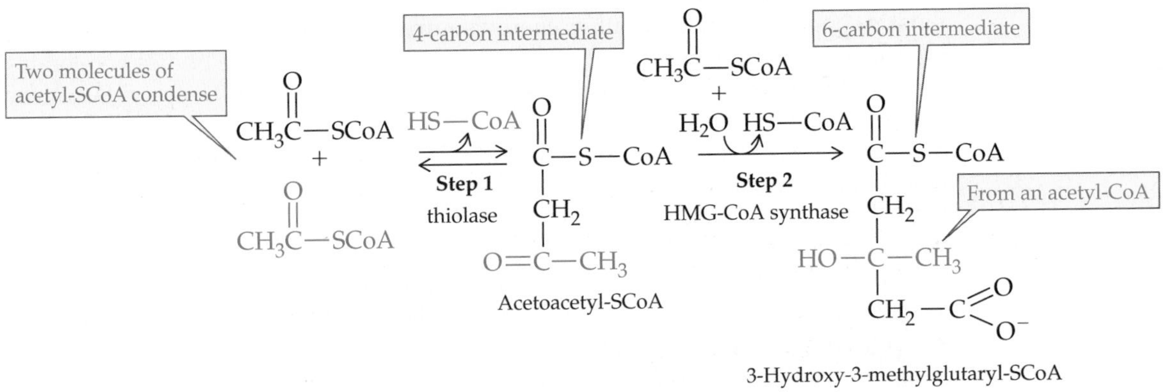

3-Hydroxy-3-methylglutaryl-SCoA

In Step 1, which is the reverse of the final step of β oxidation (Step 4 in Figure 25.7), two acetyl-SCoA molecules combine in a reaction catalyzed by thiolase to produce acetoacetyl-SCoA. Then, in Step 2, a third acetyl-SCoA and a water molecule react with acetoacetyl-SCoA to give 3-hydroxy-3-methylglutaryl-SCoA (HMG-CoA). The enzyme for this step, HMG-CoA synthase, is found only in mitochondria and is specific only for the D isomer of the substrate. The enzyme for the β-oxidation pathway, also found in mitochondria, has the same name but is specific for the L form of 3-hydroxy-3-methylglutaryl-SCoA. The pathways are separated by the specificity of the enzymes for their respective substrates.

STEPS 3 and 4 OF KETOGENESIS: Formation of the Ketone Bodies

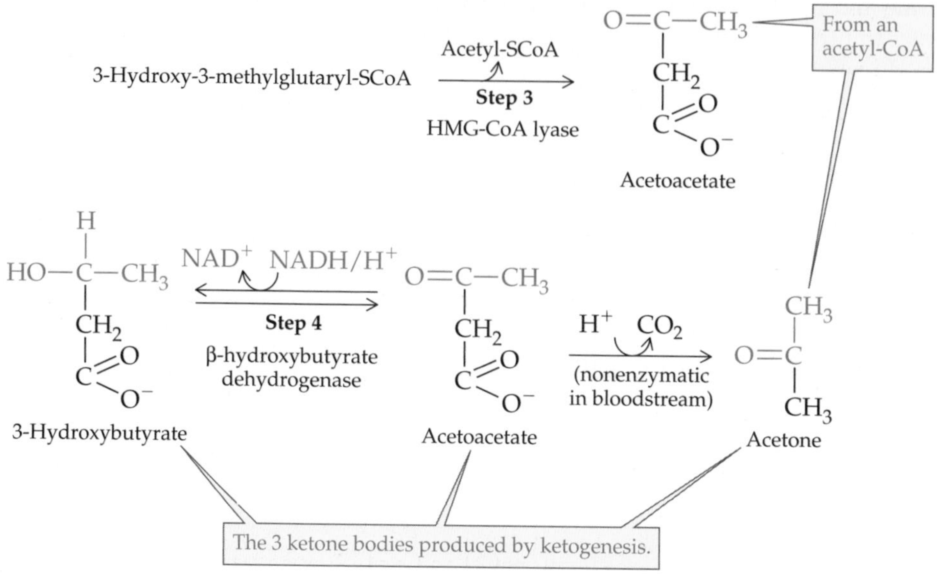

The 3 ketone bodies produced by ketogenesis.

In Step 3, removal of acetyl-SCoA from the product of Step 2 by HMG-CoA lyase produces the first of the ketone bodies, *acetoacetate*. Acetoacetate is the precursor of the other two ketone bodies produced by ketogenesis, 3-hydroxybutyrate and acetone. In Step 4, the acetoacetate produced in Step 3 is reduced to 3-hydroxybutyrate by 3-hydroxybutyrate dehydrogenase. (Note in the equation for Step 4 that 3-hydroxybutyrate and acetoacetate are connected by a reversible reaction. In tissues that need energy, acetoacetate is produced by different enzymes than those used for ketogenesis. Acetyl-SCoA can then be produced from the acetoacetate.) As acetoacetate and 3-hydroxybutyrate are synthesized by ketogenesis in liver mitochondria, they are released to the bloodstream. Acetone is then formed in the bloodstream by the decomposition of acetoacetate and is excreted primarily by exhalation.

In a person who is well fed and healthy, skeletal muscles derive a small portion of their daily energy needs from acetoacetate, and heart muscles use it in preference to glucose when fatty acids are in short supply. But consider the situation when energy production from glucose is inadequate due to starvation or because glucose is not being metabolized normally due to diabetes. (⬤, Section 23.9) The body must respond by providing other energy sources in what can become a precarious balancing act. Under these conditions, the production of ketone bodies accelerates because acetoacetate and 3-hydroxybutyrate can be converted to acetyl-SCoA for oxidation in the citric acid cycle.

During the early stages of starvation, heart and muscle tissues burn larger quantities of acetoacetate, thereby preserving glucose for use in the brain. In prolonged starvation, even the brain can switch to ketone bodies to meet up to 75% of its energy needs.

The condition in which ketone bodies are produced faster than they are utilized (*ketosis*) occurs in diabetes. It is indicated by the odor of acetone (a highly volatile ketone) on the patient's breath and the presence of ketone bodies in the urine (*ketonuria*) and the blood (*ketonemia*).

APPLICATION ▶ The Liver, Clearinghouse for Metabolism

The liver is the largest reservoir of blood in the body and also the largest internal organ, making up about 2.5% of the body's mass. Blood carrying the end products of digestion (glucose, other sugars, amino acids, and so forth) enters the liver through the hepatic portal vein before going into general circulation, so the liver is therefore ideally situated to regulate the concentrations of nutrients and other substances in the blood. The liver is important as the gateway for entry of drugs into the circulation and also contains the enzymes needed to inactivate toxic substances as well.

Various functions of the liver have been described in scattered sections of this book, but it is only by taking an overview that the central role of the liver in metabolism can be appreciated. Among its many functions, the liver synthesizes glycogen from glucose, glucose from noncarbohydrates, triacylglycerols from mono- and diacylglycerols, and fatty acids from acetyl-SCoA. It is also the site of synthesis of cholesterol, bile acids, plasma proteins, and blood clotting factors. In addition, liver cells can catabolize glucose, fatty acids, and amino acids to yield carbon dioxide and energy stored in ATP. The *urea cycle*, by which nitrogen from amino acids is converted to urea for excretion, takes place in the liver (Section 28.4).

The liver stores reserves of glycogen, certain lipids and amino acids, iron, and fat-soluble vitamins, and releases them as needed to maintain homeostasis. In addition, only liver cells have the enzyme needed to convert glucose 6-phosphate from glycogenolysis and gluconeogenesis to glucose that can enter the bloodstream.

Given its central role in metabolism, the liver is subject to a number of pathologic conditions based on excessive accumulation of various metabolites. One example is *cirrhosis*, the development of fibrous tissue that is preceded by excessive triacylglycerol buildup. Cirrhosis occurs in alcoholism, uncontrolled diabetes, and metabolic conditions in which the synthesis of lipoproteins from triacylglycerols is blocked. Another

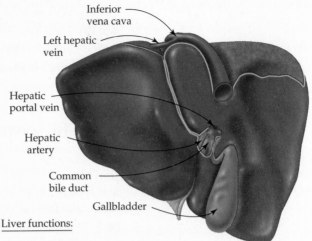

Liver functions:

- Balances level of circulating glucose
- Balances levels of circulating triacylglycerol, fatty acid and cholesterol
- Removes excess amino acids from circulation; converts their nitrogen to urea for excretion
- Stores reserves of fat-soluble vitamins and iron
- Removes drugs from circulation and breaks them down

▲ **Anatomy of the liver.** Blood carrying metabolites from the digestive system enters the liver through the hepatic portal vein. The gallbladder is the site for storage of bile.

example is *Wilson's disease*, a genetic defect in copper metabolism. In Wilson's disease, copper accumulates in the liver rather than being excreted or recycled for use in coenzymes. Chronic liver disease, as well as brain damage and anemia, are symptoms of Wilson's disease. The disease is treated by a low-copper diet and drugs that enhance the excretion of copper.

See Additional Problems 25.60 and 25.61 at the end of the chapter.

Because two of the ketone bodies are carboxylic acids, continued ketosis such as might occur in untreated diabetes leads to the potentially serious condition known as **ketoacidosis**—acidosis resulting from increased concentrations of ketone bodies in the blood. The blood's buffers are overwhelmed and blood pH drops. An individual experiences dehydration due to increased urine flow, labored breathing because acidic blood is a poor oxygen carrier, and depression. Ultimately, if untreated, the condition leads to coma and death.

Ketoacidosis Lowered blood pH due to accumulation of ketone bodies.

PROBLEM 25.6

Which of the following classifications apply to the formation of 3-hydroxybutyrate from acetoacetate?

(a) Condensation   **(b)** Hydrolysis

(c) Oxidation   **(d)** Reduction

> **PROBLEM 25.7**
>
> Consider the reactions of ketogenesis.
>
> **(a)** What role does acetyl-SCoA play?
>
> **(b)** How many acetyl-SCoA molecules are used in the production of the ketone bodies?
>
> **(c)** What is the essential role of ketone bodies during prolonged starvation?

25.8 Biosynthesis of Fatty Acids

Lipogenesis The biochemical pathway for synthesis of fatty acids from acetyl-SCoA.

The biosynthesis of fatty acids from acetyl-SCoA, a process known as **lipogenesis**, provides a link between carbohydrate, lipid, and protein metabolism. Because acetyl-SCoA is an end product of carbohydrate and amino acid catabolism, using it to make fatty acids allows the body to divert the energy of excess carbohydrates and amino acids into storage as triacylglycerols.

Fatty acid synthesis and catabolism are similar in that they both proceed two carbon atoms at a time and in that they are both recursive, spiral pathways. But as is usually the case, the biochemical pathway in one direction is not the exact reverse of the pathway in the other direction because the reverse of an energetically favorable pathway is energetically unfavorable. This principle applies to β oxidation of fatty acids and its reverse, lipogenesis. Furthermore, catabolism of fatty acids occurs in the mitochondria and anabolism is located in the cytoplasm. The two pathways are compared in Table 25.1.

TABLE 25.1 Comparison of Fatty Acid Oxidation and Synthesis

OXIDATION	SYNTHESIS
Occurs in mitochondria	Occurs in cytosol
Enzymes different from synthesis	Enzymes different from oxidation
Intermediates carried by coenzyme A	Intermediates carried by acyl carrier protein
Coenzymes: FAD, NAD$^+$	Coenzyme: NADPH
Carbon atoms removed two at a time	Carbon atoms added two at a time

The stage is set for lipogenesis by two reactions: (1) transfer of an acetyl group from acetyl-SCoA to a carrier enzyme in the fatty acid synthase complex (S-enzyme 1) and (2) conversion of acetyl-SCoA to malonyl-SCoA in a reaction that requires investment of energy from ATP:

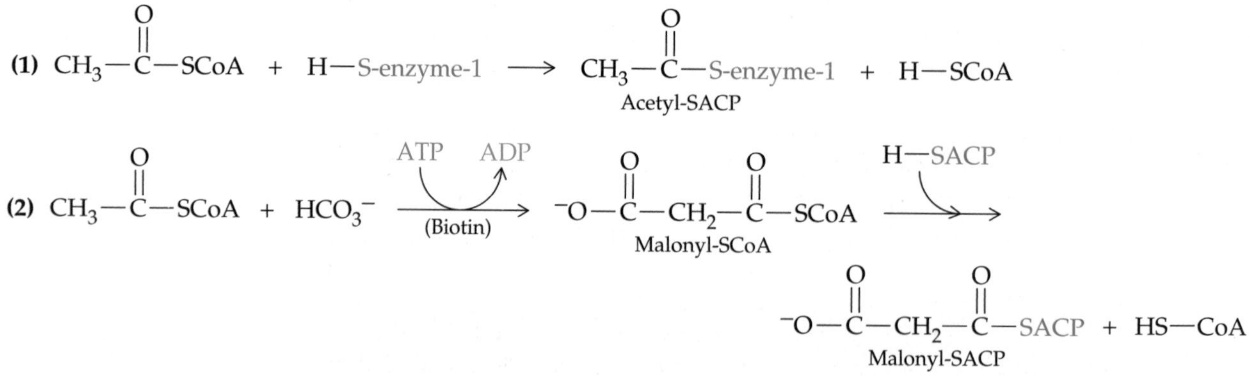

(1) $\underset{\text{Acetyl-SACP}}{CH_3-\overset{\overset{\textstyle O}{\|}}{C}-S\text{-enzyme-1}}$ reaction:

(1) $CH_3-\overset{\overset{\textstyle O}{\|}}{C}-SCoA$ + H—S-enzyme-1 ⟶ $CH_3-\overset{\overset{\textstyle O}{\|}}{C}-S\text{-enzyme-1}$ + H—SCoA
Acetyl-SACP

(2) $CH_3-\overset{\overset{\textstyle O}{\|}}{C}-SCoA$ + HCO_3^- $\xrightarrow[\text{(Biotin)}]{\text{ATP} \quad \text{ADP}}$ $^-O-\overset{\overset{\textstyle O}{\|}}{C}-CH_2-\overset{\overset{\textstyle O}{\|}}{C}-SCoA$ $\xrightarrow{\text{H—SACP}}$
Malonyl-SCoA

$^-O-\overset{\overset{\textstyle O}{\|}}{C}-CH_2-\overset{\overset{\textstyle O}{\|}}{C}-SACP$ + HS—CoA
Malonyl-SACP

(*Fatty acid synthase* is a multienzyme complex that contains all six of the enzymes needed for lipogenesis, with a protein called *acyl carrier protein* (ACP) anchored in the center of the complex.) The malonyl group of reaction (2) carries the carbon atoms that will be incorporated two at a time into the fatty acid. The malonyl-SCoA is then transferred to the ACP part of fatty acid synthase forming malonyl-SACP.

Once malonyl-SACP and the acetyl group on S-enzyme 1 have been readied, a series of four reactions lengthens the growing fatty acid chain by two carbon atoms with each repetition (Figure 25.8). Fatty acids with up to 16 carbon atoms (palmitic acid) are produced by this route.

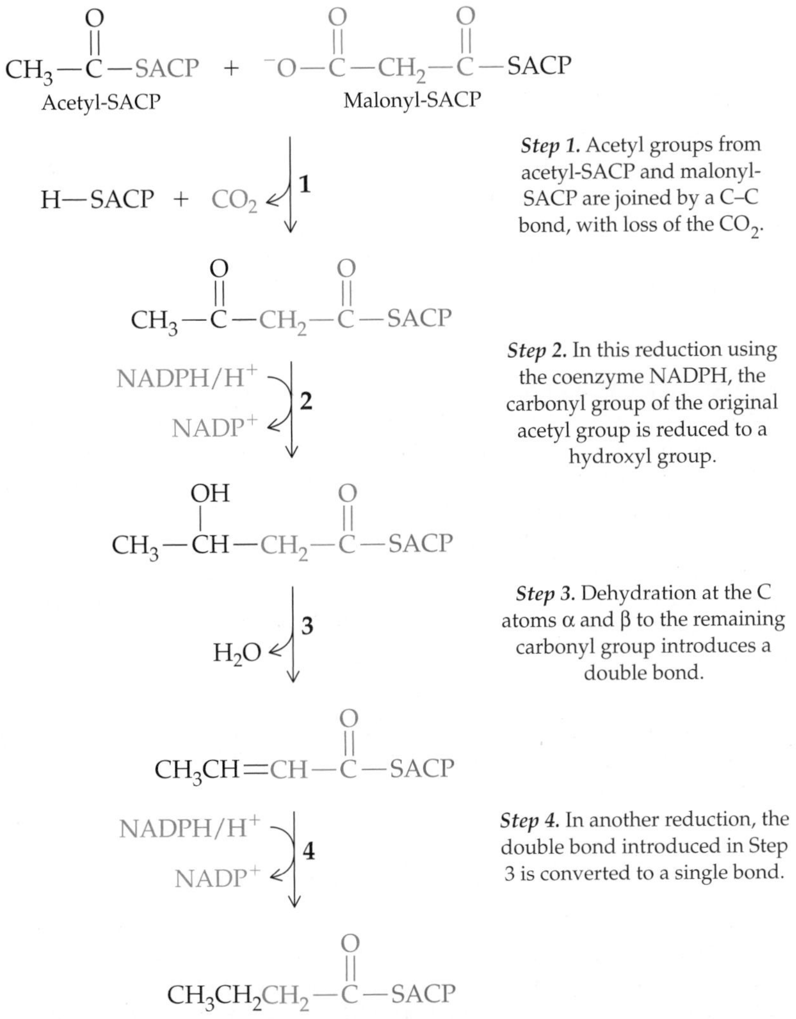

Step 1. Acetyl groups from acetyl-SACP and malonyl-SACP are joined by a C–C bond, with loss of the CO_2.

Step 2. In this reduction using the coenzyme NADPH, the carbonyl group of the original acetyl group is reduced to a hydroxyl group.

Step 3. Dehydration at the C atoms α and β to the remaining carbonyl group introduces a double bond.

Step 4. In another reduction, the double bond introduced in Step 3 is converted to a single bond.

▲ **FIGURE 25.8 Chain elongation in the biosynthesis of fatty acids.** The steps shown begin with acetyl acyl carrier protein (acetyl-SACP), the reactant in the first spiral of palmitic acid synthesis. Each new pair of carbon atoms is carried into the next spiral by a new malonyl-SACP. The growing chain remains attached to the carrier protein from the original acetyl-SACP.

Chain Elongation of Fatty Acid

STEP 1: Condensation The malonyl group from malonyl-SACP transfers to acetyl-SACP with the loss of CO_2. The loss of the CO_2 that was added in the endothermic, ATP-driven formation of malonyl-SACP releases energy to drive the reaction.

STEPS 2–4: Reduction, Dehydration, and Reduction These three reactions accomplish the reverse of Steps 3, 2, and 1 in the β oxidation of fatty acids shown in Figure 25.7, p. 796. The carbonyl group is reduced to an —OH group, dehydration yields a carbon–carbon double bond, and the double bond is reduced by addition of hydrogen. Both reductions involve the coenzyme nicotinamide adenine dinucleotide phosphate, in the reduced form, NADPH.

The result of the first cycle in fatty acid synthesis is the addition of two carbon atoms to an acetyl group to give a 4-carbon acyl group still attached to the carrier protein in fatty acid synthase. The next cycle then adds two more carbon atoms to give a 6-carbon acyl group by repeating the four steps of chain elongation shown here:

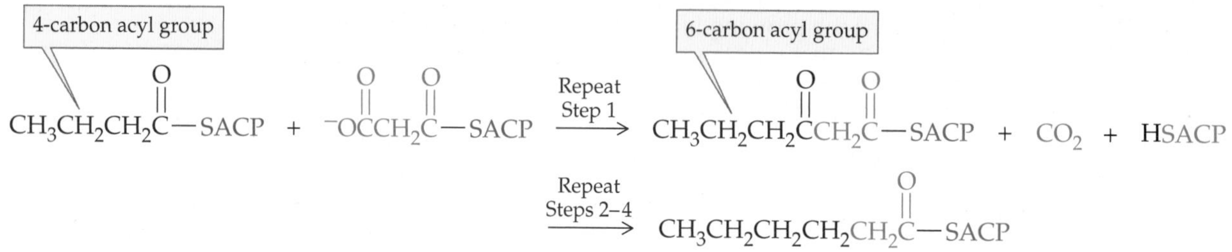

After seven trips through the elongation spiral, a 16-carbon palmitoyl group is produced and released from the fatty acid synthase. Larger fatty acids are synthesized from palmitoyl-SCoA with the aid of specific enzymes in the endoplasmic reticulum.

KEY WORDS

β-oxidation pathway, *p. 796*

Bile, *p. 788*

Bile acids, *p. 788*

Ketoacidosis, *p. 801*

Ketogenesis, *p. 799*

Ketone bodies, *p. 799*

Lipogenesis, *p. 802*

Lipoprotein, *p. 787*

**Mobilization
(of triacylglycerols),** *p. 795*

SUMMARY: REVISITING THE CHAPTER GOALS

1. **What happens during the digestion of triacylglycerols?** *Triacylglycerols (TAGs)* from the diet are broken into droplets in the stomach and enter the small intestine, where they are emulsified by *bile acids* and form micelles. Pancreatic lipases partially hydrolyze the TAGs in the micelles. Small fatty acids and glycerol from TAG hydrolysis are absorbed directly into the bloodstream at the intestinal surface. Insoluble hydrolysis products are carried to the lining in micelles, where they are absorbed and reassembled into TAGs. These TAGs are then assembled into *chylomicrons* (which are *lipoproteins*) and absorbed into the lymph system for transport to the bloodstream.

2. **What are the various roles of lipoproteins in lipid transport?** In addition to chylomicrons, which carry TAGs from the diet into the bloodstream, there are VLDLs (*very-low-density lipoproteins*), which carry TAGs synthesized in the liver to peripheral tissues for energy generation or storage; LDLs (*low-density lipoproteins*), which transport cholesterol from the liver to peripheral tissues for cell membranes or steroid synthesis; and HDLs (*high-density lipoproteins*), which transport cholesterol from peripheral tissues back to the liver for conversion to bile acids that are used in digestion or excreted.

3. **What are the major pathways in the metabolism of triacylglycerols?** Dietary TAGs carried by chylomicrons in the bloodstream undergo hydrolysis to fatty acids and glycerol by enzymes in capillary walls. TAGs in storage are similarly hydrolyzed within adipocytes. The fatty acids from either source undergo *β oxidation* to acetyl-SCoA or resynthesis into TAGs for storage. Acetyl-SCoA can participate in resynthesis of fatty acids (*lipogenesis*), formation of *ketone bodies* (*ketogenesis*), steroid synthesis, or energy generation via the citric acid cycle and oxidative phosphorylation. Glycerol can participate in glycolysis, gluconeogenesis, or TAG synthesis.

4. **How are triacylglycerols moved into and out of storage in adipose tissue?** Synthesis of TAGs for storage is activated by insulin when glucose levels are high. The synthesis requires dihydroxyacetone phosphate (from glycolysis or glycerol) for conversion to glycerol 3-phosphate, to which fatty acyl groups are added one at a time to yield TAGs. Hydrolysis of TAGs stored in adipocytes is activated by glucagon when glucose levels drop and also by epinephrine.

5. **How are fatty acids oxidized, and how much energy is produced by their oxidation?** Fatty acids are activated (in the cytosol) by conversion to fatty acyl coenzyme A, a reaction that requires the equivalent of two ATPs in the conversion of ATP to AMP. The fatty acyl-SCoA molecules are transported into the mitochondrial matrix and are then oxidized two carbon atoms at a time to acetyl-SCoA by repeated trips through the β-oxidation spiral.

6. **What is the function of ketogenesis?** The ketone bodies are 3-hydroxybutyrate, acetoacetate, and acetone. They are produced from two acetyl-SCoA molecules. Their production

is increased when energy generation from the citric acid cycle cannot keep pace with the quantity of acetyl-SCoA available. This occurs during the early stages of starvation and in unregulated diabetes. The ketone bodies are water-soluble and can travel unassisted in the bloodstream to tissues where acetyl-SCoA is produced from acetoacetate and 3-hydroxybutyrate. In this way, acetyl-SCoA is made available for energy generation when glucose is in short supply.

7. **How are fatty acids synthesized?** Fatty acid synthesis (lipogenesis), like β oxidation, proceeds two carbon atoms at a time in a four-step pathway. The pathways utilize different enzymes and coenzymes. In synthesis, the initial four carbons are transferred from acetyl-SCoA to the malonyl carrier protein. Each additional pair of carbons is then added to the growing chain bonded to the carrier protein, with the final three steps of the four-step synthesis sequence the reverse of the first three steps in β oxidation.

UNDERSTANDING KEY CONCEPTS

25.8 Oxygen is not a reactant in the β oxidation of fatty acids. Can β oxidation occur under anaerobic conditions? Explain.

25.9 Identify each lipoprotein described below as either chylomicron, HDL, LDL, or VLDL.

(a) Which lipoprotein has the lowest density? Why?

(b) Which lipoprotein carries TAGs from the diet?

(c) Which lipoprotein removes cholesterol from circulation?

(d) Which lipoprotein contains "bad cholesterol" from a vascular disease risk standpoint?

(e) Which lipoprotein has the highest ratio of protein to lipid?

(f) Which lipoprotein carries TAGs from the liver to peripheral tissues? How are TAGs used?

(g) Which lipoprotein transports cholesterol from the liver to peripheral tissues?

25.10 Lipid metabolism, especially TAG anabolism and catabolism, is closely associated with carbohydrate (glucose) metabolism. Insulin and glucagon levels in blood are regulated by the glucose levels in blood. Draw lines from the appropriate phrases in column A to appropriate phrases in columns B and C.

A	B	C
High blood glucose	High glucagon/ low insulin	Fatty acid and TAGs synthesis
Low blood glucose	High insulin/ low glucagon	TAG hydrolysis; fatty acid oxidation

25.11 One strategy used in many different biochemical pathways is an initial investment of energy early on, and a large payoff in energy at the end of the pathway. How is this strategy utilized in the catabolism of fats?

25.12 When oxaloacetate in liver tissue is being used for gluconeogenesis, what impact does this have on the citric acid cycle? Explain.

25.13 Why is it more efficient to store energy as TAGs rather than as glycogen?

25.14 Explain the rationale for the production of ketone bodies during starvation.

25.15 Compare the differences between β oxidation and fatty acid synthesis (lipogenesis). Are these pathways the reverse of each other?

ADDITIONAL PROBLEMS

DIGESTION AND CATABOLISM OF LIPIDS

25.16 Why do lipids make you feel full for a long time after a meal?

25.17 Where does digestion of lipids occur?

25.18 What is the purpose of bile acids in lipid digestion?

25.19 Where are bile acids synthesized, and what is the starting molecule?

25.20 Write the equation for the hydrolysis of a triacylglycerol composed of stearic acid, oleic acid, and linoleic acid by pancreatic lipase.

25.21 Lipases break down triacylglycerols by catalyzing hydrolysis. What are the products of this hydrolysis?

25.22 What are chylomicrons, and how are they involved in lipid metabolism?

25.23 What is the origin of the triacylglycerols transported by very-low-density lipoproteins?

25.24 How are the fatty acids from adipose tissue transported?

25.25 How is cholesterol transported around the body? When it leaves the liver, what is its destination and use?

25.26 The glycerol derived from lipolysis of triacylglycerols is converted into glyceraldehyde 3-phosphate, which then enters into Step 6 of the glycolysis pathway. What further transformations are necessary to convert glyceraldehyde 3-phosphate into pyruvate?

25.27 If the conversion of glycerol to glyceraldehyde 3-phosphate releases one molecule of ATP, how many molecules of ATP are released during the conversion of glycerol to pyruvate?

25.28 How many molecules of ATP are released in the overall catabolism of glycerol to acetyl-SCoA? How many molecules of ATP are released in the complete catabolism of glycerol to CO_2 and H_2O?

25.29 How many molecules of acetyl-SCoA result from catabolism of one molecule of glyceryl trimyristate?

25.30 What is an adipocyte?

25.31 What is the primary function of adipose tissues, and where in the body are they located?

25.32 Which tissues carry out fatty acid oxidation as their primary source of energy?

25.33 Where in the cell does β oxidation take place?

25.34 What initial chemical transformation takes place on a fatty acid to activate it for catabolism?

25.35 What must take place before an activated fatty acid undergoes β oxidation?

25.36 Why is the stepwise oxidation of fatty acids called β oxidation?

25.37 Why is the sequence of reactions that catabolize fatty acids a *spiral* rather than a *cycle*?

25.38 How many moles of ATP are produced by one cycle of β oxidation?

25.39 How many moles of ATP are produced by the complete oxidation of 1 mol of lauric acid?

25.40 Arrange these four molecules in order of their biological energy content (per mole):
  (a) Sucrose
  (b) Myristic acid, $CH_3(CH_2)_{12}COOH$
  (c) Glucose
  (d) Capric acid, $CH_3(CH_2)_8COOH$

25.41 Arrange these four molecules in order of their biological energy content per mole:
  (a) Mannose
  (b) Stearic acid, $CH_3(CH_2)_{16}COOH$
  (c) Fructose
  (d) Palmitic acid, $CH_3(CH_2)_{14}COOH$

25.42 Show the products of each step in the fatty acid oxidation of hexanoic acid:

(a) $CH_3(CH_2)_4\overset{\displaystyle O}{\overset{\displaystyle \|}{C}}SCoA \xrightarrow[\text{Acetyl-SCoA}]{\text{FAD} \quad \text{FADH}_2} \underset{\text{dehydrogenase}}{} ?$

(b) Product of (a) + H_2O $\xrightarrow[\text{hydratase}]{\text{Enoyl-SCoA}}$?

(c) Product of (b) $\xrightarrow[\text{$\beta$-Hydroxyacyl-SCoA}]{\text{NAD}^+ \quad \text{NADH/H}^+}$?
     dehydrogenase

(d) Product of (c) + HSCoA $\xrightarrow[\text{transferase}]{\text{Acetyl-SCoA}}$?

25.43 Write the equation for the final step in the catabolism of any fatty acid with an even number of carbons.

25.44 How many molecules of acetyl-SCoA result from complete catabolism of the following compounds?
  (a) Myristic acid, $CH_3(CH_2)_{12}COOH$
  (b) Caprylic acid, $CH_3(CH_2)_6COOH$

25.45 How many cycles of β oxidation are necessary to completely catabolize myristic and caprylic acids?

FATTY ACID ANABOLISM

25.46 Name the anabolic pathway that synthesizes fatty acids.

25.47 Explain why β oxidation cannot proceed backward to produce triacylglycerols.

25.48 Name the starting material for fatty acid synthesis.

25.49 Why are fatty acids generally composed of an even number of carbons?

25.50 How many rounds of the lipogenesis cycle are needed to synthesize palmitic acid, $C_{15}H_{31}COOH$?

25.51 How many molecules of NADPH are needed to synthesize palmitic acid, $C_{15}H_{31}COOH$?

Applications

25.52 What are desirable goals for total cholesterol, HDL, and LDL values? [*Lipids and Atherosclerosis, p. 791*]

25.53 What is atherosclerosis? [*Lipids and Atherosclerosis, p. 791*]

25.54 What is the difference in the roles of LDL and HDL? [*Lipids and Atherosclerosis, p. 791*]

25.55 Explain the significance of cholesterol/HDL ratios of 3.5, 4.5, and 5.5. [*Lipids and Atherosclerosis, p. 791*]

25.56 What diseases are obese people at high risk of developing? [*Fat Storage: A Good Thing or Not?, p. 794*]

25.57 What role is leptin thought to play in fat storage? [*Fat Storage: A Good Thing or Not?, p. 794*]

25.58 What factors contribute to storage of excess energy as triacylglycerols? [*Fat Storage: A Good Thing or Not?, p. 794*]

25.59 Propose a reason for why excess weight, once gained and stored as fat, is so difficult to lose. [*Fat Storage: A Good Thing or Not?, p. 794*]

25.60 Give some reasons why the liver is so vital to proper metabolic function. [*The Liver, Clearinghouse for Metabolism, p. 801*]

25.61 What is cirrhosis of the liver, and what can trigger it? [*The Liver, Clearinghouse for Metabolism, p. 801*]

General Questions and Problems

25.62 Consuming too many carbohydrates causes deposition of fats in adipose tissue. How can this happen?

25.63 Why are extra calories consumed as carbohydrates stored as fat and not as glycogen?

25.64 Are any of the intermediates in the β-oxidation pathway chiral? Explain.

25.65 What three compounds are classified as ketone bodies? Why are they so designated? What process in the body produces them? Why do they form?

25.66 What is ketosis? What condition results from prolonged ketosis? Why is it dangerous?

25.67 What causes acetone to be present in the breath of someone with uncontrolled diabetes?

25.68 Individuals suffering from ketoacidosis have acidic urine. What effect do you expect ketones to have on pH? Why is pH lowered when ketone bodies are present?

25.69 Compare fats and carbohydrates as energy sources in terms of the amount of energy released per mole, and account for the observed energy difference.

25.70 Lipoproteins that transport lipids from the diet are described as exogenous. Those that transport lipids produced in metabolic pathways are described as endogenous. Which of the following lipoproteins is exogenous and which is endogenous?

(a) High-density lipoprotein (HDL)
(b) Chylomicrons

25.71 High blood cholesterol levels are dangerous because of their correlation with atherosclerosis and consequent heart attacks and strokes. Is it possible to eliminate all cholesterol from the bloodstream by having a diet that includes no cholesterol? Is it desirable to have no cholesterol at all in your body?

25.72 In the synthesis of cholesterol, acetyl-SCoA is converted to 2-methyl-1,3-butadiene. Molecules of 2-methyl-1,3-butadiene are then joined to give the carbon skeleton of cholesterol. Draw the condensed structure of 2-methyl-1,3-butadiene. How many carbon atoms does cholesterol contain? What minimum number of 2-methyl-1,3-butadiene molecules is required to make one molecule of cholesterol?

25.73 A low-fat diet of pasta, bread, beer, and soda can easily lead to an increase in weight. The increase is stored triacylglycerols in adipocytes. Explain the weight increase and why the excess carbohydrate is stored as fat.

CHAPTER 26

Nucleic Acids and Protein Synthesis

CONCEPTS TO REVIEW

Hydrogen bonding
(Section 8.11)

Phosphoric acid derivatives
(Section 17.8)

Protein structure
(Sections 18.7, 18.8)

▲ The symmetrical beauty of the DNA molecule has inspired more than just scientific research, as demonstrated by this sculpture of the double helix in Valencia, Spain.

CONTENTS

CHAPTER GOALS

In this chapter, we will answer the following questions about nucleic acids:

1. What is the composition of the nucleic acids, DNA and RNA?

THE GOAL: Be able to describe and identify the components of nucleosides, nucleotides, DNA, and RNA.

2. What is the structure of DNA?

THE GOAL: Be able to describe the double helix and base pairing in DNA.

3. How is DNA reproduced?

THE GOAL: Be able to explain the process of DNA replication.

4. What are the functions of RNA?

THE GOAL: Be able to list the types of RNA, their locations in the cell, and their functions.

5. How do organisms synthesize messenger RNA?

THE GOAL: Be able to explain the process of transcription.

6. How does RNA participate in protein synthesis?

THE GOAL: Be able to explain the genetic code, and describe the initiation, elongation, and termination steps of translation.

How does a seed know what kind of plant to become? How does a fertilized egg know how to grow into a human being? And how does a cell in that fertilized egg know to become a finger, a liver, or a heart? The answers to these and a multitude of other fundamental questions about all living organisms reside in the biological molecules known as *nucleic acids*.

Nucleic acids are the chemical carriers of an organism's genetic information. Coded in an organism's *deoxyribonucleic acid (DNA)* is all the information that determines the nature of the organism, be it a dandelion, goldfish, or human being. DNA contains the blueprint for every protein in the body. As you saw in Chapters 18 and 19, proteins have a wide array of structures and functions, and nearly all reactions in the body are catalyzed by enzymes, which are proteins. An organism's proteins determine the nature of the organism.

26.1 DNA, Chromosomes, and Genes

The terms *chromosome* and *gene* were coined long before the chemical nature of these cell components was understood. A *chromosome* (which means "colored body") was a structure in the cell nucleus thought to be the carrier of genetic information—all the information needed by an organism to duplicate itself. A *gene* was presumed to be the portion of a chromosome that controlled a specific inheritable trait such as brown eyes or red hair.

Our knowledge of cell biology and biochemistry has increased remarkably since these terms were introduced. We now understand the molecular structure of chromosomes and genes. We can describe the sequence of events in which genetic information is reproduced when a cell divides. The relationship of genes to the synthesis of proteins is no longer a mystery. These concepts are described in this chapter. The story continues in the next chapter with an introduction to molecular biology, the study of biology and biochemistry at the DNA level. The achievements of molecular biology have led to a revolution in science and technology, with the biggest development of all being that a complete map of the genetic information passed along during cell division is now available for numerous organisms, including humans.

When a cell is not actively dividing, its nucleus is occupied by *chromatin*, which is a compact, orderly tangle of **DNA (deoxyribonucleic acid)**, the carrier of genetic information, twisted around organizing proteins known as *histones*. During cell division, chromatin organizes itself into **chromosomes**. Each chromosome contains a different DNA molecule, and the DNA is duplicated so that each new cell receives a complete copy.

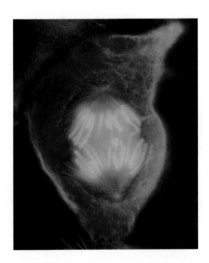

▲ Chromosomes during cell division.

DNA (deoxyribonucleic acid) The nucleic acid that stores genetic information; a polymer of deoxyribonucleotides.

Chromosome A complex of proteins and DNA; visible during cell division.

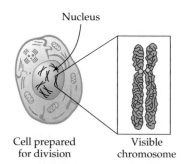

Nondividing cell — Chromatin in nucleus — Cell prepared for division — Visible chromosome

Gene Segment of DNA that directs the synthesis of a single polypeptide.

Each DNA molecule, in turn, is composed of many **genes**—individual segments of DNA containing the instructions that direct the synthesis of a single polypeptide. Interestingly, not all genes coded for by an organism's DNA are expressed as protein (Chapter 27).

Organisms differ widely in their numbers of chromosomes. A horse, for example, has 64 chromosomes (32 pairs), a cat has 38 chromosomes (19 pairs), a mosquito has 6 chromosomes (3 pairs), and a corn plant has 20 chromosomes (10 pairs). A human has 46 chromosomes (23 pairs).

26.2 Composition of Nucleic Acids

Nucleic acid A polymer of nucleotides.

Nucleotide A 5-carbon sugar bonded to a cyclic amine base and a phosphate group; monomer for nucleic acids.

Like proteins and carbohydrates, nucleic acids are polymers. Proteins are polypeptides, carbohydrates are polysaccharides, and **nucleic acids** are *polynucleotides*. Each **nucleotide** has three parts: a five-membered cyclic monosaccharide, a nitrogen-containing cyclic compound known as a *nitrogenous base*, and a phosphate group ($-OPO_3^{2-}$).

A nucleotide

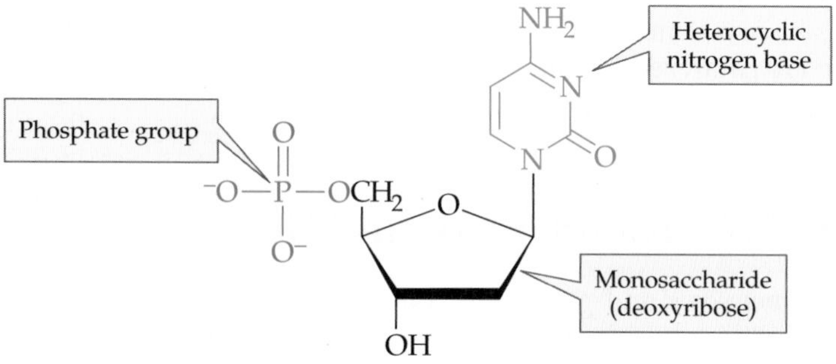

Phosphate group — Heterocyclic nitrogen base — Monosaccharide (deoxyribose)

RNA (ribonucleic acid) Nucleic acids responsible for putting the genetic information to use in protein synthesis; polymer of ribonucleotides. Includes messenger, transfer, and ribosomal RNA.

There are two types of nucleic acids, DNA and **RNA (ribonucleic acid)**, with RNA coming in different types. The function of one type of RNA is to put the information stored in DNA to use. Other types of RNA assist in the conversion of the message a specific RNA carries into protein. Before we discuss how the nucleic acids fulfill their functions, we need to understand how their component parts are joined together and how DNA and RNA differ from each other.

The Sugars

One difference between DNA and RNA is found in the sugar portion of the molecules. In RNA, the sugar is D-ribose (Sections 22.4 and 22.5; hereafter simply referred to as ribose), as indicated by the name *ribonucleic acid*. In DNA, the sugar is 2-*deoxy*ribose, giving *deoxyribonucleic acid*. (The prefix 2-*deoxy*- means that an oxygen atom is missing from the C2 position of ribose.)

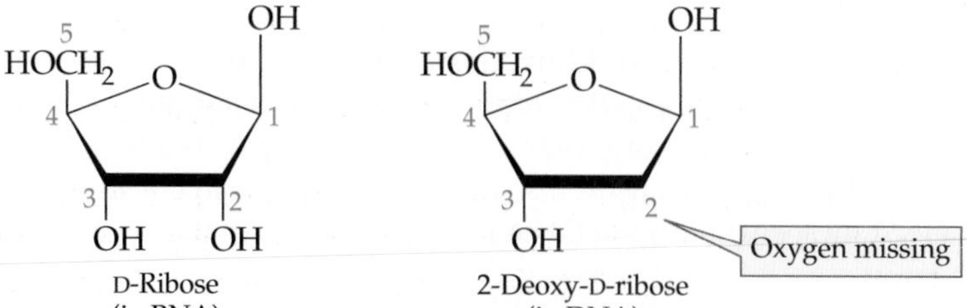

D-Ribose (in RNA)

2-Deoxy-D-ribose (in DNA) — Oxygen missing

The Bases

The five nitrogenous bases found in DNA and RNA are shown highlighted in tan in Table 26.1, along with the two parent bases, purine and pyrimidine. The nitrogenous bases that are purine derivatives, adenine and guanine, contain two fused nitrogen-containing rings. The bases that are pyrimidine derivatives—cytosine, thymine, and uracil—contain only one nitrogen-containing ring.

TABLE 26.1 Bases in DNA and RNA

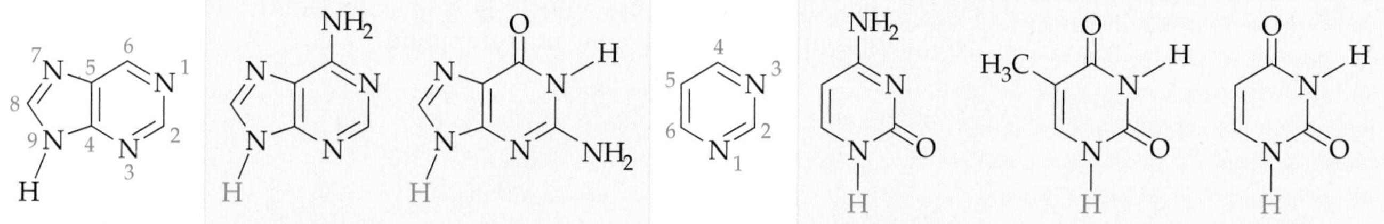

| PURINE BASES IN NUCLEIC ACIDS | | | | PYRIMIDINE BASES IN NUCLEIC ACIDS | | |

Purine (Parent) • Adenine (DNA, RNA) • Guanine (DNA, RNA) • Pyrimidine (Parent) • Cytosine (DNA, RNA) • Thymine* (DNA) • Uracil (RNA)

Thymine occurs in a few cases of RNA.

In addition to differing in the sugars they contain, RNA and DNA differ in their bases. As Table 26.1 notes,

- Thymine is present only in DNA molecules (with rare exceptions).
- Uracil is present only in RNA molecules.
- Adenine, guanine, and cytosine are present in both DNA and RNA.

Sugar + Base = Nucleoside

A molecule composed of either ribose or deoxyribose and one of the five nitrogenous bases found in DNA and/or RNA is called a **nucleoside**. The combination of ribose and adenine, for example, gives the nucleoside known as adenosine, which you should recognize as the parent molecule of adenosine triphosphate (ATP). (⊂⊃, p. 669):

Nucleoside A 5-carbon sugar bonded to a cyclic amine base; like a nucleotide but with no phosphate group.

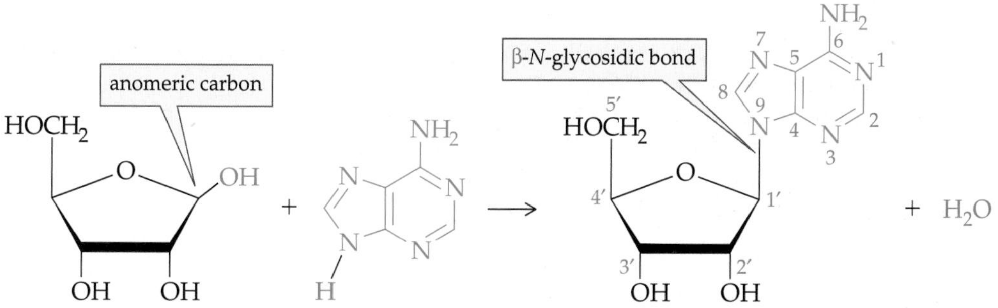

Ribose + Adenine → Adenosine (a nucleoside) + H_2O

The sugar and base are connected by a bond between one of the nitrogen atoms in the base and the anomeric carbon atom (the one bonded to two O atoms) of the sugar. This bond is a β-N-glycosidic bond. (⊂⊃, p. 717) Notice that this linkage (the 1′ position of the sugar to the 9 position nitrogen of the adenine) is closely related to an acetal (Section 16.7).

In each of the nucleic acid bases in Table 26.1, the hydrogen atom lost in nucleoside formation is shown in red.

Nucleoside names are the nitrogenous base name modified by the suffix *-osine* for the purine bases (as we just saw for adenosine) and the suffix *-idine* for the pyrimidine bases. No prefix is used for nucleosides containing ribose, but the prefix *deoxy-* is added for those that contain deoxyribose. Therefore the four nucleosides found in RNA are named adenosine, guanosine, cytidine, and uridine, and the four found in DNA are named deoxyadenosine, deoxyguanosine, deoxycytidine, and deoxythymidine.

To distinguish between atoms in the sugar ring of a nucleoside and atoms in the base ring (or rings), numbers without primes are used for atoms in the base ring (or rings), and numbers with primes are used for atoms in the sugar ring.

WORKED EXAMPLE | **26.1** Naming a Nucleic Acid Component from Its Structure

Is the compound shown here a nucleoside or a nucleotide? Identify its sugar and base components, and name the compound.

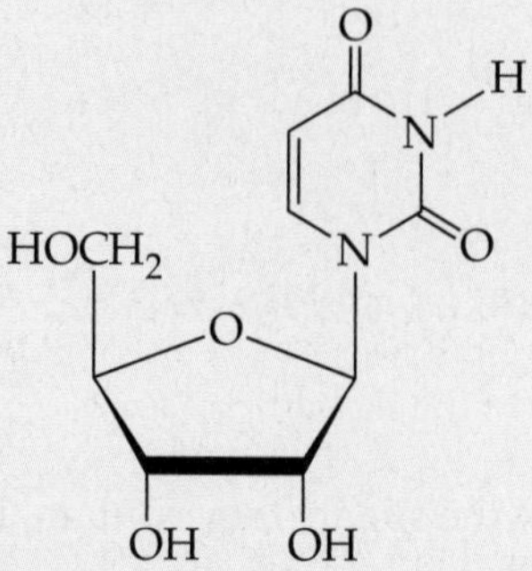

ANALYSIS The compound contains a sugar, recognizable by the O atom in the ring and the —OH groups. It also contains a nitrogenous base, recognizable by the nitrogen-containing ring. The sugar has an —OH in the 2′ position and is therefore ribose (if it were missing the —OH in the 2′ position, it would be a *deoxy*ribose). Checking the base structures in Table 26.1 shows that this is uracil, a pyrimidine base, requiring its name to end in *-idine*.

SOLUTION
The compound is nucleoside, and its name is uridine.

KEY CONCEPT PROBLEM 26.1

Name the nucleoside shown here. Copy the structure, and number the C and N atoms (refer to Table 26.1).

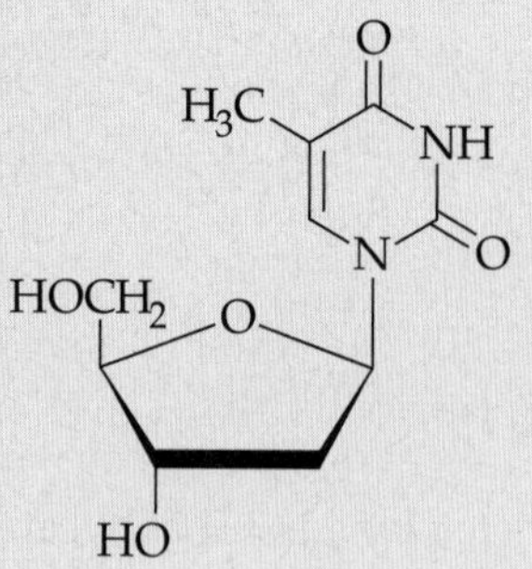

PROBLEM 26.2

Write the molecular formulas for the sugars D-ribose and 2-deoxy-D-ribose. Exactly how do they differ in composition? Can you think of one chemical property that might differ slightly between the two?

Nucleoside + Phosphate = Nucleotide

Nucleotides are the building blocks of nucleic acids; they are the monomers of the DNA and RNA polymers. Each nucleotide is a 5'-monophosphate ester of a nucleoside:

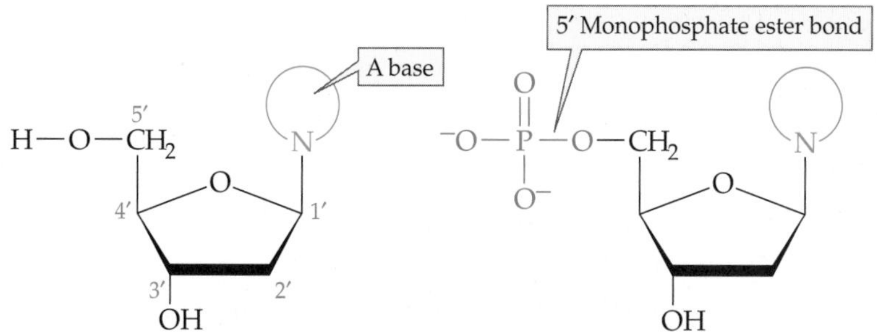

A deoxyribonucleoside

A deoxyribonucleotide

Nucleotides are named by adding 5'-*monophosphate* at the end of the name of the nucleoside. The nucleotides corresponding, for example, to adenosine and deoxycytidine are thus adenosine 5'-monophosphate (AMP) and deoxycytidine 5'-monophosphate (dCMP). Nucleotides that contain ribose are classified as **ribonucleotides** and those that contain 2-deoxy-D-ribose are known as **deoxyribonucleotides** (and are designated by leading their abbreviations with a lower case "d"). For example

Ribonucleotide A nucleotide that contains D-ribose.

Deoxyribonucleotide A nucleotide that contains 2-deoxy-D-ribose.

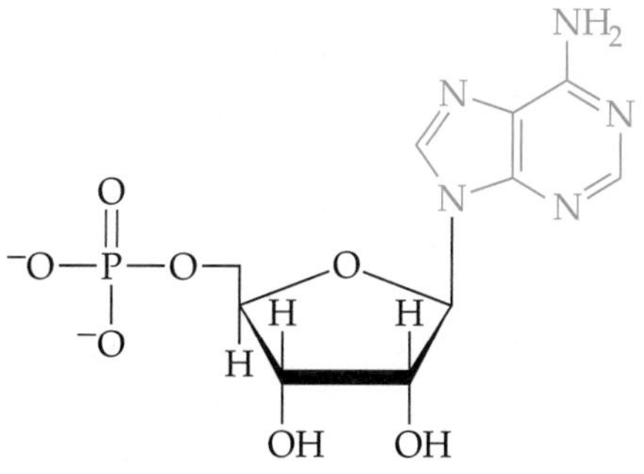

Adenosine 5'-monophosphate (AMP)
(a ribonucleotide)

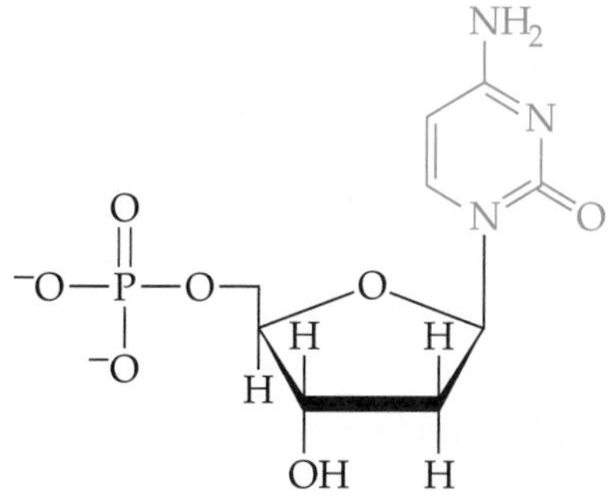

Deoxycytidine 5'-monophosphate (dCMP)
(a deoxyribonucleotide)

Phosphate groups can be added to any of the nucleotides to form diphosphate or triphosphate esters. As illustrated by *adenosine triphosphate*, these esters are named with the nucleoside name plus *diphosphate* or *triphosphate*. In preceding chapters, you have seen that adenosine triphosphate (ATP) plays an essential role as a source of biochemical energy, which is released during its conversion to adenosine diphosphate (ADP).

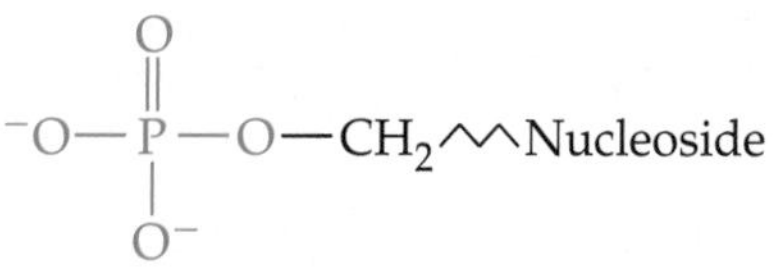

Nucleoside monophosphate

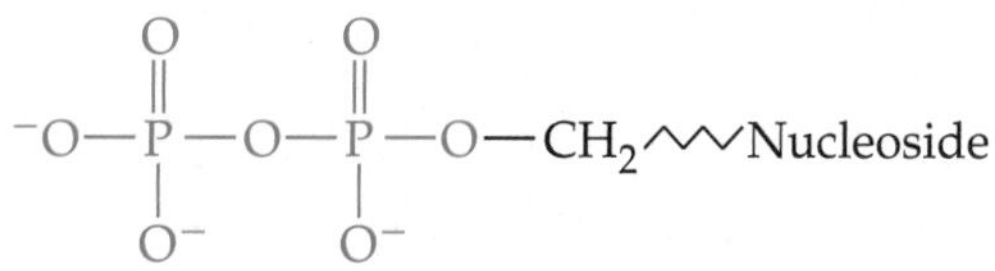

Nucleoside diphosphate

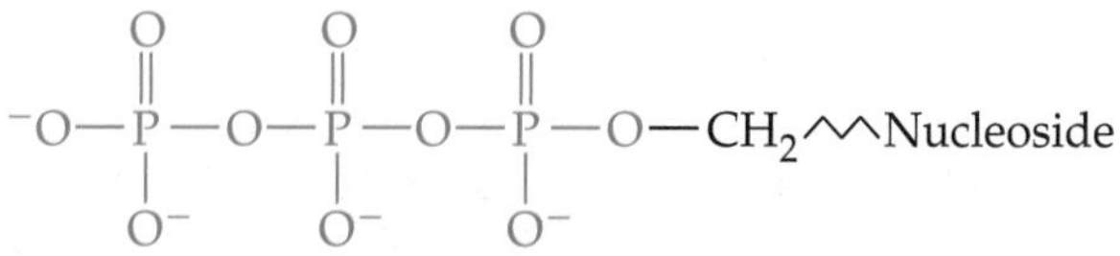

Nucleoside triphosphate

The names of the bases, nucleosides, and nucleotides are summarized in Table 26.2 together with their abbreviations, which are commonly used in writing about biochemistry.

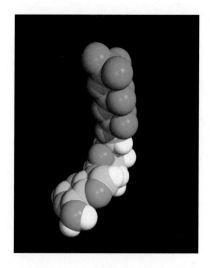

▲ ATP is the triphosphate of the adenosine nucleotide (Section 21.5).

TABLE 26.2 Names of Bases, Nucleosides, and Nucleotides in DNA and RNA

BASES	NUCLEOSIDES	NUCLEOTIDES*
DNA		
	Deoxyribonucleosides	**Deoxyribonucleotides**
Adenine (A)	Deoxyadenosine	Deoxyadenosine 5'-monophosphate (dAMP)
Guanine (G)	Deoxyguanosine	Deoxyguanosine 5'-monophosphate (dGMP)
Cytosine (C)	Deoxycytidine	Deoxycytidine 5'-monophosphate (dCMP)
Thymine (T)	Deoxythymidine	Deoxythymidine 5'-monophosphate (dTMP)
RNA		
	Ribonucleosides	**Ribonucleotides**
Adenine (A)	Adenosine	Adenosine 5'-monophosphate (AMP)
Guanine (G)	Guanosine	Guanosine 5'-monophosphate (GMP)
Cytosine (C)	Cytidine	Cytidine 5'-monophosphate (CMP)
Uracil (U)	Uridine	Uridine 5'-monophosphate (UMP)

The nucleotides are also named as, for example, deoxyadenylate and adenylate.

Summary—Nucleoside, Nucleotide, and Nucleic Acid Composition

Nucleoside
- A sugar and a base

Nucleotide
- A sugar, a base, and a phosphate group ($-OPO_3^{2-}$)

DNA (deoxyribonucleic acid)
- A polymer of deoxyribonucleotides
- The sugar is 2-deoxy-D-ribose
- The bases are adenine, guanine, cytosine, and *thymine*

RNA (ribonucleic acid)
- A polymer of ribonucleotides
- The sugar is D-ribose
- The bases are adenine, guanine, cytosine, and *uracil*

WORKED EXAMPLE **26.2** Drawing a Nucleic Acid Component from Its Name

Draw the structure of the nucleotide represented by dTMP.

ANALYSIS From Table 26.2 we see that dTMP is deoxythymidine 5'-monophosphate. Therefore, the nitrogen base in this nucleotide is thymine, whose structure is shown in Table 26.1. This base must be bonded (by replacing the H that is red in Table 26.1) to deoxyribose, and there must be a phosphate group in the 5' position of the deoxyribose.

SOLUTION
The structure is

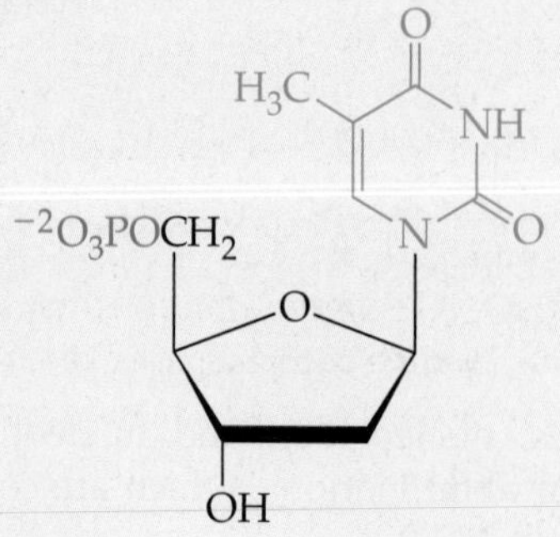

KEY CONCEPT PROBLEM 26.3

Draw the structure of 2′-deoxyadenosine 5′-monophosphate, and use the primed-unprimed format to number all the atoms in the rings.

PROBLEM 26.4

Draw the structure of the triphosphate of guanosine, a triphosphate that, like ATP, provides energy for certain reactions.

PROBLEM 26.5

Write the full names of dCMP, CMP, UDP, AMP, and ATP.

26.3 The Structure of Nucleic Acid Chains

Keep in mind that nucleic acids are polymers of nucleotides. The nucleotides in DNA and RNA are connected by phosphate diester linkages between the —OH group on C3′ of the sugar ring of one nucleotide and the phosphate group on C5′ of the next nucleotide:

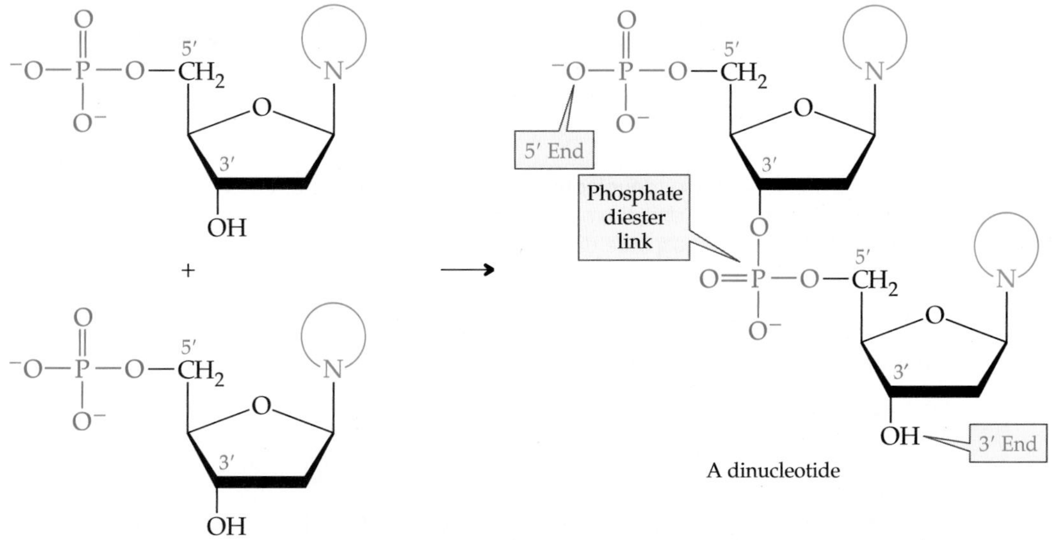

A dinucleotide

A nucleotide chain commonly has a free phosphate group on a 5′ carbon at one end (known as the 5′ *end*) and a free —OH group on a 3′ carbon at the other end (the 3′ *end*), as illustrated in the dinucleotide just above and in the trinucleotide in Figure 26.1. Additional nucleotides join by forming additional phosphate diester linkages between these groups until the polynucleotide chain of a DNA molecule is formed.

Just as the structure and function of a protein depend on the sequence in which the amino acids are connected (⊂⊙⊃, Section 18.7), the structure and function of a nucleic acid depend on the sequence in which the nucleotides are connected. With a nucleic acid, however, we have a second detail to consider: structure and function both depend on the direction in which the nucleic acid is read by enzymes involved in making gene products. Like proteins, nucleic acids have backbones that do not vary in composition. The differences between different proteins and between different nucleic acids result from the *order* of the groups bonded to the backbone—amino acid side chains in proteins and bases in nucleic acids.

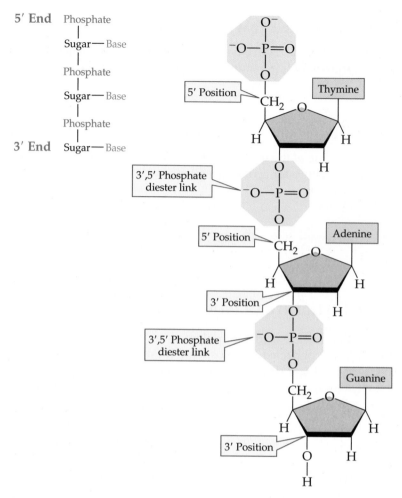

▲ **FIGURE 26.1 A deoxytrinucleotide.** In all polynucleotides, as shown here, there is a phosphate group at the 5′ end; there is a sugar —OH group at the 3′ end; and the nucleotides are connected by 3′,5′-phosphate diester links.

Comparison of protein and nucleic acid backbones and side chains

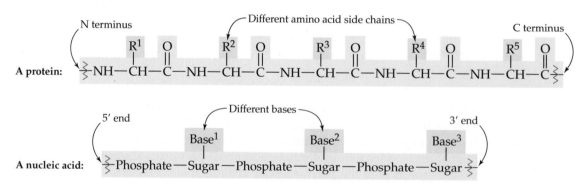

The sequence of nucleotides in a nucleic acid chain is read by starting at the 5′ end and identifying the bases in order of occurrence. Rather than write the full name of each nucleotide or each base, one-letter abbreviations of the bases are commonly used to designate the order in which they are attached to the sugar–phosphate backbone: A for adenine, G for guanine, C for cytosine, T for thymine, and U for uracil in RNA. The trinucleotide in Figure 26.1, for example, would be represented by T-A-G or TAG.

PROBLEM 26.6

Identify the bases in their order in the C-G-A-U-A pentanucleotide. Does this come from RNA or DNA? Explain.

PROBLEM 26.7

Draw the full structure of the DNA dinucleotide G-T. Identify the 5′ and 3′ ends of this dinucleotide.

26.4 Base Pairing in DNA: The Watson–Crick Model

DNA samples from different cells of the same species have the same relative proportions of nucleotides in each pair of nitrogenous bases (A:T and C:G), but the proportions vary from one species to another. For example, human DNA contains 30% each of adenine and thymine, and 20% each of guanine and cytosine, whereas the bacterium *Escherichia coli* contains 24% each of adenine and thymine, and 26% each of guanine and cytosine. Note that in both cases, A and T are present in equal amounts and G and C are present in equal amounts. This observation, known as Chargaff's rule, suggests that the bases occur in discrete pairs. Why should this be?

In 1953, James Watson and Francis Crick proposed a structure for DNA that not only accounts for the pairing of bases but also accounts for the storage and transfer of genetic information. According to the Watson–Crick model, a DNA molecule consists of *two* polynucleotide strands coiled around each other in a helical, screw-like fashion. The sugar–phosphate backbone is on the *outside* of this right-handed **double helix**, and the heterocyclic bases are on the *inside*, so that a base on one strand points directly toward a base on the second strand. The double helix resembles a twisted ladder, with the sugar–phosphate backbone making up the sides and the hydrogen-bonded base pairs, the rungs:

Double helix Two strands coiled around each other in a screw-like fashion; in most organisms the two polynucleotides of DNA form a double helix.

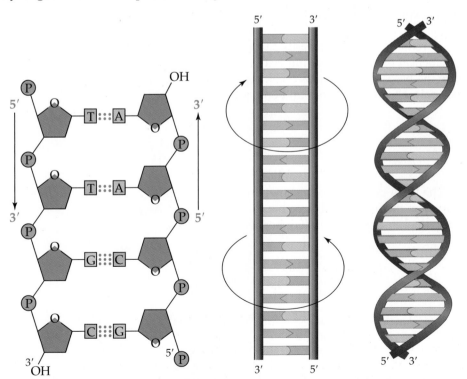

The two strands of the DNA double helix run in opposite directions—one in the 5′ to 3′ direction, the other in the 3′ to 5′ direction (the strands are said to be *antiparallel to each other*). The stacking of the hydrophobic bases in the interior and the alignment of the hydrophilic sugars and phosphate groups on the exterior provide stability to the structure. Hydrogen bonding also enhances DNA stability. Each pair of bases in the center of the double helix is connected by hydrogen bonding. As shown in Figure 26.2, adenine and thymine (A-T) form two hydrogen bonds to each other, and cytosine and guanine (C-G) form three hydrogen bonds to each other. Although individual hydrogen bonds are not especially strong, the several thousand along a DNA chain collectively contribute to stability.

Thymine–Adenine

Cytosine–Guanine

▶ **FIGURE 26.2 Base pairing in DNA.** Hydrogen bonds of similar lengths connect the pairs of bases; thymine with adenine, cytosine with guanine.

Base pairing The pairing of bases connected by hydrogen bonding (G-C and A-T), as in the DNA double helix.

The pairing of the bases strung like beads along the two polynucleotide strands of the DNA double helix is described as *complementary*. Wherever a thymine occurs in one strand, an adenine falls opposite it in the other strand; wherever a cytosine occurs in one strand, a guanine falls opposite it on the other strand. This **base pairing** explains why A and T occur in equal amounts in double-stranded DNA, as do C and G.

To remember how the bases pair up, note that if the symbols are arranged in alphabetical order, the first and last ones pair, and the two middle ones pair. (Or, since the bases are written as block capital letters remember that the two straight ones pair, AT, and the two round ones pair, CG.)

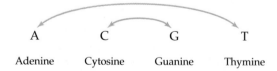

| A | C | G | T |
| Adenine | Cytosine | Guanine | Thymine |

The DNA double helix is shown in Figure 26.3. Both its strength and its shape depend on the fit and hydrogen bonding of the bases. As you will see, base pairing is also the key to understanding how DNA functions.

▶ **FIGURE 26.3 A segment of DNA.** (a) In this model, notice that the base pairs are nearly perpendicular to the sugar–phosphate backbones. (b) A space-filling model of the same DNA segment. (c) An abstract representation of the DNA double helix and base pairing.

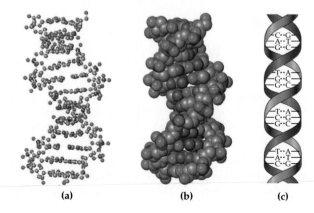

(a)   (b)   (c)

WORKED EXAMPLE **26.3** Writing Complementary Nucleic Acid Sequences

What sequence of bases of one strand of DNA (reading in the 3′ to 5′ direction) is complementary to the sequence 5′ T-A-T-G-C-A-G 3′ on the other strand?

ANALYSIS Remembering that A always bonds to T and C always bonds to G, go through the original 5′ to 3′ sequence, replacing each A by T, each T by A, each C by G, and each G by C. Keep in mind that when a 5′ to 3′ strand is matched in this manner to its complementary strand, the complementary strand will read from left to right in the 3′ to 5′ direction. (Where the direction in which a base sequence is written is *not* specified, you can assume it follows the customary 5′ to 3′ direction.)

SOLUTION

Original strand   5′ T-A-T-G-C-A-G 3′
Complementary strand   3′ A-T-A-C-G-T-C 5′

PROBLEM 26.8

What sequence of bases on one DNA strand is complementary to the following sequences on another strand?

(a) 5′ G-C-C-T-A-G-T 3′   **(b)** 5′ A-A-T-G-G-C-T-C-A 3′

PROBLEM 26.9

Draw the structures of adenine and uracil (which replaces thymine in RNA), and show the hydrogen bonding that occurs between them.

PROBLEM 26.10

Is a DNA molecule neutral, negatively charged, or positively charged? Explain.

KEY CONCEPT PROBLEM 26.11

(a) DNA and RNA, like proteins, can be denatured to produce unfolded or uncoiled strands. Heating DNA to what is referred to as its "melting temperature" denatures it. Why does a longer strand of DNA have a higher melting temperature than a shorter one? (b) The DNA melting temperature also varies with base composition. Would you expect a DNA with a high percentage of G:C base pairs to have a higher or lower melting point than one with a high percentage of A:T base pairs? How do you account for your choice?

26.5 Nucleic Acids and Heredity

Your heredity is determined by the DNA in the fertilized egg from which you grew. A sperm cell carrying DNA from your father united with an egg cell carrying DNA from your mother. Their combination produced the full complement of chromosomes and genes that you carry through life. Each of your 23 pairs of chromosomes contains one DNA molecule copied from that of your father and one DNA molecule copied from that of your mother. Most cells in your body contain copies of these originals. (The exceptions are red blood cells, which have no nuclei and no DNA, and egg or sperm cells, which have 23 single DNA molecules, rather than pairs.)

APPLICATION ▶ Viruses and AIDS

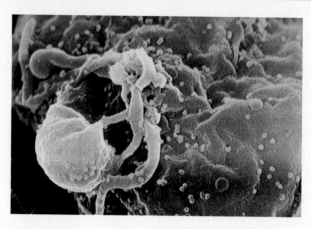

▲ **HIV viruses emerging from an infected lymphocyte.**

Viruses are submicroscopic infectious agents consisting of a piece of nucleic acid, typically wrapped in a protective coat of protein. The viral nucleic acid may be either DNA or RNA, it may be either single-stranded or double-stranded, and it may consist of either a single piece or several pieces. Many hundreds of viruses are known, each of which can infect a particular plant or animal cell.

Viruses occupy the gray area between living and nonliving. By itself, a virus has none of the cellular machinery necessary for replication. But once it enters a living cell, a virus takes over the host cell and forces it to produce virus copies. Some of the infected cells may eventually die, but others continue to produce copies of the virus, which then leave the host cell and spread the infection to other cells.

Viral infection begins when a virus enters a host cell, loses its protein coat, and releases its nucleic acid. What happens next depends on whether the virus is based on DNA or RNA. If it is a DNA virus, the host cell first replicates the viral DNA, producing many copies, and then decodes it in the normal way. The viral DNA is transcribed to produce RNA, and the RNA is translated to synthesize viral coat proteins. Copies of the viral DNA are packaged in newly synthesized protein envelopes, producing new virus particles that are then released from the cell.

If the infectious agent is an RNA virus, however, replication is not so straightforward. Either the cell must transcribe and produce proteins directly from the viral RNA template, or else it must first produce DNA from the viral RNA by *reverse transcription*. Reverse transcription is the process by which viral RNA that has entered the host cell is transcribed into a complementary DNA sequence (this is the reverse of the normal DNA to RNA process, hence the name). The enzyme that accomplishes this is called *reverse transcriptase* and is provided by the virus itself. Without reverse transcriptase, the viral genome could not become incorporated into the host cell, and the virus could not reproduce. Viruses that follow the reverse transcription route are known as *retroviruses*.

The *human immunodeficiency virus (HIV-1)* responsible for most cases of AIDS is a retrovirus. As shown in the diagram, *reverse transcriptase* first produces single DNA strands complementary to the HIV RNA and then produces double-stranded DNA that enters a host cell chromosome. There, the cell's normal replication and translation produce RNA and proteins that are assembled into new virus particles.

The HIV virus attacks mainly *T lymphocytes*, which are part of the immune system that defends the body against foreign invaders (Section 29.4). As T lymphocytes die off, the body is open to attack. Most AIDS victims succumb to infection by bacteria, other viruses, or fungi that thrive in the absence of defense by T lymphocytes.

Viral infections, whether HIV infections or the common cold, are difficult to treat with chemical agents. The challenge is to design a drug that can act on viruses within cells without damaging the cells and their genetic machinery. Development of drugs for the treatment of AIDS is especially challenging because HIV has the highest mutation rate of any known virus. Drugs active against one strain of HIV may soon encounter a mutant they cannot combat.

The best success with AIDS drugs thus far has been with a three-drug therapy, the first therapy that allows HIV-infected individuals to survive. Two of the drugs (AZT, azidothymidine; 3TC, lamivudine) are false nucleosides.

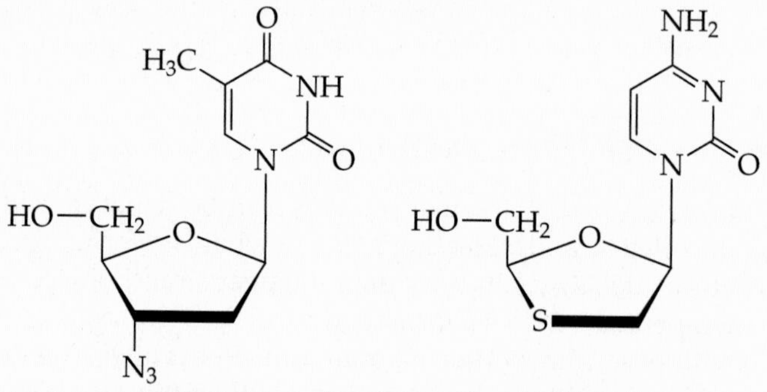

AZT

3TC
(Lamivudine)

The viral reverse transcriptase incorporates them into the viral DNA, and they then slow down production of new viral RNA. The third drug (saquinavir) inhibits an enzyme (protease) that is necessary for production of proteins coded for by the viral genes. Taken together, these three drugs can reduce the amount of HIV in a patient's body to undetectable levels.

The success of the three-drug therapy is not without problems. It requires the patient to follow carefully a challenging

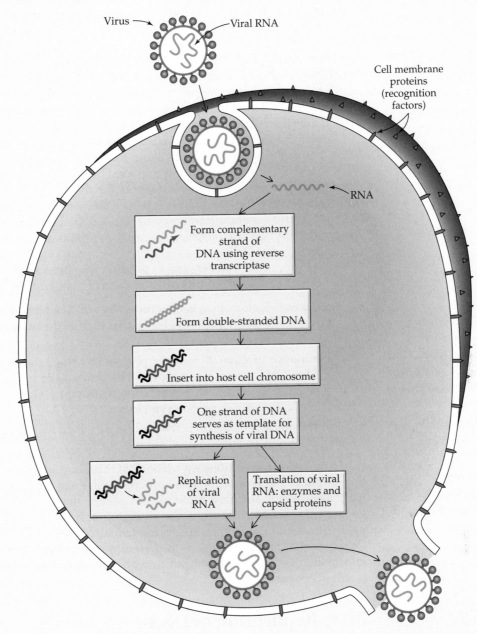

schedule for taking a large number of pills daily. It is not effective for everyone. It is very expensive—so expensive that it is beyond the reach of people in developing countries, where AIDS is rampant. It is feared that the virus will eventually develop resistance to the protease inhibitor that is greatly responsible for the success of this therapy. Thus, the search for other kinds of therapies continues. One unattained goal is development of a vaccine that prevents the disease. Many researchers have turned their attention toward this goal, including the late Dr. Jonas Salk, inventor of the polio vaccine.

A major problem in developing an AIDS vaccine lies in the extreme variability of the virus; for example, currently there are nine subtypes of the HIV virus known. Given that HIV has the capacity to mutate within a single infected organism, it is unclear if a single vaccine is even a realistic goal. Medical research, however, has shown the capacity to overcome obstacles in the past, and new and novel approaches to reaching the goal of an AIDS vaccine are currently under development. Although a small trial of an AIDS vaccine conducted in Italy in 2005 showed promise, larger trials of similar vaccines in 2007 were halted. There were indications the newer vaccine was at best ineffective and at worst may have made some recipients more susceptible to the AIDS virus. Strategies to produce an effective vaccine treatment are under review by health organizations and AIDS researchers.

See Additional Problems 26.66 through 26.68 at the end of the chapter.

Cell division is an ongoing process—no single cell has a lifespan equal to that of the organism in which it is found. Therefore, every time a cell divides, its DNA must be copied. The double helix of DNA and complementary base pairing make this duplication possible. Because of how bases pair, each strand of the double helix is a blueprint for the other strand. The copying process is an awesome aspect of how DNA functions—a process that we discuss in Section 26.6.

But first, we need to answer two related questions: How do nucleic acids carry the information that determines our inherited traits? And how is that stored information interpreted and put into action?

Genetic information is conveyed not just in the numbers and kinds of bases in DNA, but in the *sequence* of bases along the DNA strands (Section 26.9); any mistakes in either copying or reading a given DNA sequence can lead to mutations, which may have disastrous consequences for the resulting daughter cells. Every time a cell divides, the information is passed along to the daughter cells, which ultimately pass this genetic information to their daughter cells. Within cells, the genetic information encoded in the DNA directs the synthesis of proteins, a process known as the *expression* of DNA.

The duplication, transfer, and expression of genetic information occur as the result of three fundamental processes: *replication, transcription,* and *translation.*

Replication The process by which copies of DNA are made when a cell divides.

Transcription The process by which the information in DNA is read and used to synthesize RNA.

Translation The process by which RNA directs protein synthesis.

- **Replication** (Section 26.6) is the process by which a replica, or identical copy, of DNA is made when a cell divides, so that each of the two daughter cells has the same DNA (Figure 26.4).

- **Transcription** (Section 26.8) is the process by which the genetic messages contained in DNA are read and copied. The products of transcription are individual ribonucleic acids, which carry the instructions stored in DNA out of the nucleus to the sites of protein synthesis.

- **Translation** (Section 26.10) is the process by which the genetic messages carried by RNA are decoded and used to build proteins.

In the following sections we will look at these important processes. Replication, transcription, and translation must proceed with great accuracy and require participation by many auxiliary molecules to ensure the integrity (or fidelity) of the genetic information. Many enzymes working in harmony with one another, coupled with energy-supplying nucleoside triphosphates (NTPs) play essential roles. Our next goal in this chapter is to present a simple overview of how the genetic information is duplicated and put to work, as the full elucidation of this process is still in progress.

26.6 Replication of DNA

The Watson–Crick double-helix model of DNA does more than explain base pairing. In the short publication that announced their model, Watson and Crick made the following simple statement, suggesting the significance of what they had just discovered:

It has not escaped our notice that the specific pairing we have postulated immediately suggests a possible copying mechanism for the genetic material.

DNA replication begins in the nucleus with partial unwinding of the double helix; this process involves enzymes known as *helicases.* The unwinding occurs simultaneously in many specific locations known as *origins of replication* (Figure 26.4). The DNA strands separate, exposing the bases. These branch points, called replication forks, provide a "bubble" in which the replication process can begin. Enzymes called DNA polymerase move into position on the separated strands—their function is to facilitate transcription of the exposed single-stranded DNA. Nucleoside triphosphates carrying each of the four bases are available in the vicinity. One by one, the triphosphates move into place by forming hydrogen bonds

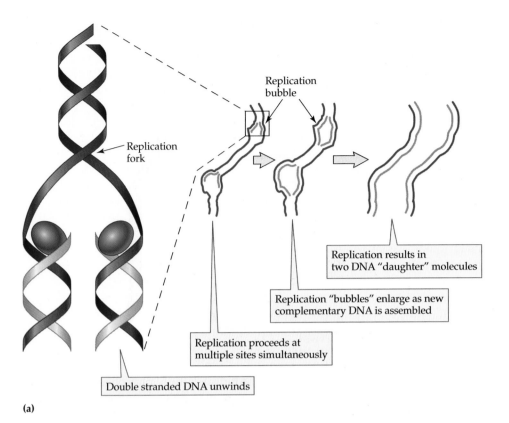

(a)

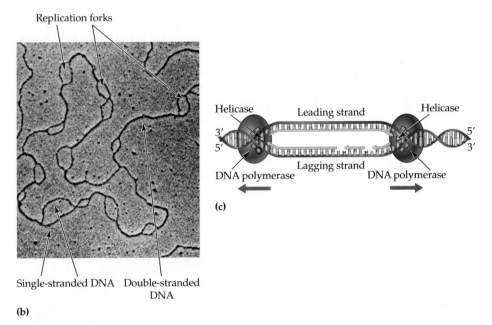

(b)

(c)

▲ **FIGURE 26.4 DNA replication sites.** (a) Replication takes place when DNA unwinds, exposing single strands. This occurs in multiple sites simultaneously. (b) Single-stranded DNA is exposed at numerous replication forks as DNA unwinds. (c) DNA polymerase enzymes facilitate copying of the single-stranded DNA in the 5′ to 3′ direction on both exposed single strands. One single strand, the leading strand, is copied continuously; the other single strand, called the lagging strand, is copied in segments.

to the bases exposed on the DNA template strand. The hydrogen bond formation requires that A pair only with T, and G pair only with C. DNA polymerase then catalyzes bond formation between the 5′ phosphate group of the arriving nucleoside triphosphate and the 3′—OH of the growing polynucleotide strand, as the two extra phosphate groups are removed:

Bond formation in DNA replication

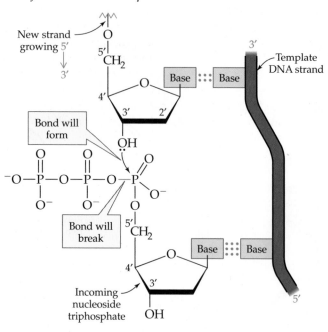

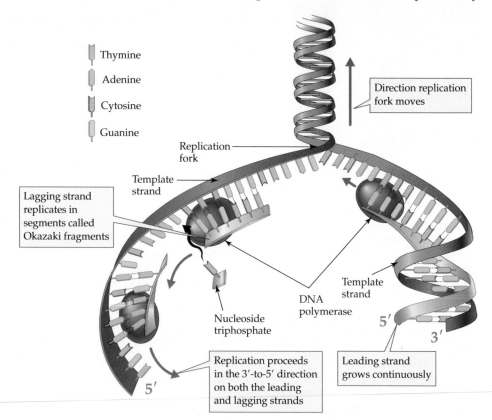

▲ Semiconservative replication produces a pair of DNA double helixes in which one strand (dark green) is the original strand and the other (light green) is the strand that has been copied from the original.

DNA polymerase catalyzes the reaction between the 5′ phosphate on an incoming nucleotide and the free 3′—OH on the growing polynucleotide. Therefore, the template strand can only be read in the 3′ to 5′ direction, and the new DNA strand can grow only in the 5′ to 3′ direction.

Since each new strand is complementary to its template strand, two identical copies of the DNA double helix are produced during replication. In each new double helix, one strand is the template and the other is the newly synthesized strand. We describe the result as *semiconservative* replication (one of the two strands is conserved).

Note in Figure 26.5 that the incoming nucleoside triphosphate is added to the 3′ end of the new strand. Because the original DNA strands are complementary,

▶ **FIGURE 26.5 DNA replication.** Because the polynucleotide must grow in the 5′ to 3′ direction, the leading strand (shown at the right, in light green) grows continuously toward the replication fork while the lagging strand (at the left in light green) grows in segments as the fork moves. The segments are later joined by a DNA ligase enzyme.

only one new strand, known as the *leading strand*, grows continuously in this manner as the point of replication (the *replication fork*) moves along (review Figure 26.4c). If the other strand were to grow continuously, it would have to occur by addition of nucleotides at the 5′ end, but since the DNA polymerase enzyme (which is bound to BOTH strands at the same location) does not carry out that reaction, it must be accomplished by another mechanism. Since DNA polymerase performs replication only in the 3′ to 5′ direction, this other strand, called the *lagging strand*, is replicated in short 3′ to 5′ segments called *Okazaki fragments* (after the Japanese scientist who discovered them). The directions of growth are shown in Figure 26.5, where the leading strand is the continuously growing strand of the new DNA and the lagging strand is the one composed of the short Okazaki fragments. To form the lagging strand from the Okazaki fragments, these short 3′ to 5′ sections are joined by the action of an enzyme known as *DNA ligase*.

Consider the magnitude of the job in replication. The total number of base pairs in a human cell—the human **genome**—is 3 *billion* base pairs. Yet the base sequences of these huge DNA molecules are faithfully copied during replication, and a random error occurs only about once in each 10 billion to 100 billion bases. The complete copying process in human cells takes several hours. To replicate a huge molecule such as human DNA at this speed requires not one, but many replication forks, producing many segments of DNA strands that are ultimately joined to produce a faithful copy of the original.

Genome All of the genetic material in the chromosomes of an organism; its size is given as the number of base pairs.

26.7 Structure and Function of RNA

RNA is similar to DNA—both are sugar–phosphate polymers and both have nitrogen-containing bases attached—but there are important differences (Table 26.3). We have already seen that RNA and DNA differ in composition (Section 26.2): The sugar in RNA is ribose rather than deoxyribose, and the base uracil in RNA pairs up with adenine rather than with thymine. RNA and DNA also differ in size and structure—RNA strands are smaller (that is, they have lower total molar masses) than DNA molecules. The RNAs are almost always single-stranded molecules (as distinct from DNA, which is almost always double-stranded); RNA molecules also often have complex folds, sometimes folding back on themselves to form double helixes in some regions.

TABLE 26.3   Comparison of DNA and RNA

	SUGAR	BASES	SHAPE AND SIZE	FUNCTION
DNA	Deoxyribose	Adenine Guanine Cytosine Thymine	Paired strands in double helix; 50 million or more nucleotides per strand	Stores genetic information
RNA	Ribose	Adenine Guanine Cytosine Uracil	Single-stranded with folded regions; <100 to about 50,000 nucleotides per RNA	**mRNA**—Encodes copy of genetic information ("blueprints" for protein synthesis) **tRNA**—Carries amino acids for incorporation into protein **rRNA**—Component of ribosomes (sites of protein synthesis)

There are also different kinds of RNA, each type with its own unique function in the flow of genetic information, whereas DNA has only one function—storing genetic information. Working together, the three types of RNA make it possible for the encoded information carried by DNA to be put to use in the synthesis of proteins:

- **Ribosomal RNAs**   Outside the nucleus but within the cytoplasm of a cell are the **ribosomes**—small granular organelles where protein synthesis takes place.

Ribosome The structure in the cell where protein synthesis occurs; composed of protein and rRNA.

APPLICATION ▶ It's a Ribozyme!

What is involved in protein synthesis but is not a protein enzyme, not a tRNA, and not exactly mRNA? Why, it is a **ribozyme**, which is RNA acting as an enzyme by catalyzing a chemical reaction.

For example, the reaction that removes unneeded sections from newly synthesized mRNA is accomplished by the mRNA itself acting as the reaction catalyst. Any RNA that acts as an enzyme (a term used for protein catalysts) is called a ribozyme. The mRNA loops back on itself and rearranges the ester bonds (a transesterification reaction) between nucleotides, neatly excising the unneeded sections. The slicing out of unneeded bases (introns) and splicing together the rest of the mRNA is termed spliceosome activity and was originally attributed to a protein that researchers were unable to isolate. In the 1970s Thomas Cech discovered that spliceosome activity was due to mRNA acting on itself; Sidney Altman later confirmed the observation with his work on tRNA and an enzyme named RNaseP. In 1989, Cech and Altman were awarded the Nobel Prize in Chemistry for their discovery of the catalytic (ribozymal) function of RNA.

Since then more than 500 ribozymes in different organisms have been identified. One class, the viroids, are small, circular RNA molecules that infect plants. Viroids are self-splicing and direct the plant host cell to synthesize new viroids in long, repeated strands that then splice themselves out of the strand into individual infective particles. Viroids cause diseases in economically important plants, such as flowers.

Perhaps the largest, best-known ribozyme is the 23S RNA molecule, which is one part of the structure of ribosomes. Structural evidence shows that during protein synthesis in a ribosome, 23S RNA catalyzes the formation of the peptide bond between amino acids in the growing protein molecule. The associated proteins, all 31 of them, appear to provide the necessary structure to maintain the correct three-dimensional relationship among the molecules involved in protein synthesis.

Basic research into the nature and functions of ribozymes continues. However, applied research is directed toward their use in medical treatments. One such undertaking, intended as a treatment for cancer, targets the mRNA responsible for the cell surface receptor *vascular endothelial growth factor*. The hope is that breaking up the mRNA for this receptor will allow fewer blood vessels to grow to supply a tumor with nutrients, thereby slowing tumor growth.

See Additional Problems 26.69 and 26.70 at the end of the chapter.

Ribozyme RNA that acts as an enzyme.

Ribosomal RNA (rRNA) The RNA that is complexed with proteins in ribosomes.

Messenger RNA (mRNA) The RNA that carries the code transcribed from DNA and directs protein synthesis.

Transfer RNA (tRNA) The RNA that transports amino acids into position for protein synthesis.

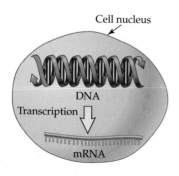

Cell nucleus

DNA

Transcription

mRNA

(Their location in the cell is shown in Figure 21.3, p. 665.) Each ribosome is a complex consisting of about 60% **ribosomal RNA (rRNA)** and 40% protein, with a total molecular mass of approximately 5,000,000 amu.

- **Messenger RNAs** The **messenger RNAs (mRNA)** carry information transcribed from DNA. They are formed in the cell nucleus and transported out to the ribosomes, where proteins will be synthesized. They are polynucleotides that carry the same code for proteins as does the DNA.

- **Transfer RNAs** The **transfer RNAs (tRNA)** are smaller RNAs that deliver amino acids one by one to protein chains growing at ribosomes. Each tRNA carries only one amino acid.

26.8 Transcription: RNA Synthesis

Ribonucleic acids are synthesized in the cell nucleus. Before leaving the nucleus, all types of RNA molecules are modified in various ways that enable them to perform their different functions. We focus here on messenger RNA (mRNA; in eukaryotes) because its synthesis (transcription) is the first step in transferring the information carried by DNA into protein synthesis.

In transcription, as in replication, a small section of the DNA double helix unwinds, the bases on the two strands are exposed, and one by one the complementary nucleotides are attached. rRNA, tRNA, and mRNA are all synthesized in essentially the same manner. Only one of the two DNA strands is transcribed during RNA synthesis. The DNA strand that is transcribed is the *template strand*; its complement in the original helix is the *informational strand*. The mRNA molecule is complementary to the template strand, which makes it an exact RNA-duplicate of the DNA informational strand, with the exception that a U replaces each T in the

DNA strand. The relationships are illustrated by the following short DNA and mRNA segments:

DNA informational strand	5′ ATG	CCA	GTA	GGC	CAC	TTG	TCA	3′
DNA template strand	3′ TAC	GGT	CAT	CCG	GTG	AAC	AGT	5′
mRNA	5′ AUG	CCA	GUA	GGC	CAC	UUG	UCA	3′

The transcription process, shown in Figure 26.6, begins when RNA polymerase recognizes a control segment in DNA that precedes the nucleotides to be transcribed. *The genetic code*, which we will discuss in Section 26.9, consists of triplets of consecutive bases known as *codons*. The nucleotide triplets carried by mRNA code for amino acids to be assembled into proteins (Section 26.10). The sequence of nucleic acid code that corresponds to a complete protein is known as a *gene*. RNA polymerase moves down the DNA segment to be transcribed, adding complementary nucleotides one by one to the growing RNA strand as it goes. Transcription ends when the RNA polymerase reaches a codon triplet that signals the end of the sequence to be copied.

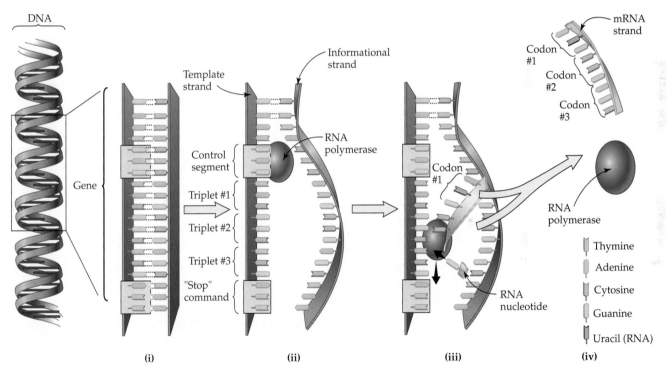

▲ **FIGURE 26.6 Transcription of DNA to produce mRNA.** The transcription shown here produces a hypothetical three-codon mRNA. From left to right, (i) the DNA unwinds; (ii) the RNA polymerase connects with the control, or start, segment on the template strand; (iii) the mRNA is assembled as the polymerase moves along the template strand; and (iv) transcription ends when the polymerase reaches the stop command, releasing both the new mRNA strand and the RNA polymerase.

At the end of transcription, the mRNA molecule contains a matching base for every base that was on the informational DNA strand, from the start to the stop codon. Some of these bases, however, do not code for genes. It turns out that genes occupy only about 10% of the base pairs in DNA. The code for a gene is contained in one or more small sections of DNA called an **exon** (exons carry code that is *ex*pressed). The cellular production of the protein encoded by a particular gene is called *expression*. The process includes transcription of DNA, processing of the resulting mRNA product, and its translation into an active protein. The code for a given gene may be interrupted by a sequence of bases called an **intron** (a section that

Exon A nucleotide sequence in a gene that codes for part of a protein.

Intron A nucleotide sequence in mRNA that does not code for part of a protein; removed before mRNA proceeds to protein synthesis.

*in*tervenes or *in*terrupts), and then resumed farther down the chain in another exon. Introns are sections of DNA that do not code for any part of the protein to be synthesized. The mRNA strand (the "primary transcript"), like the DNA from which it was synthesized, contains both exons and introns, and is known as **heterogeneous nuclear RNA (or hnRNA)**. Further steps are necessary before the mRNA can direct protein synthesis. In the final mRNA molecule released from the nucleus, the intron sections have been cut out and the remaining pieces (consisting of the exons) are spliced together through the action of a structure known as a *spliceosome* (a protein–RNA complex that removes introns from nuclear RNA).

Heterogeneous nuclear RNA (hnRNA) The initially synthesized mRNA strand containing both introns and exons.

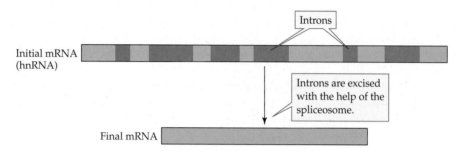

The possible functions of introns, which are noncoding DNA, are the subject of great speculation and study. Thus far, no clear-cut function has been assigned to introns. One intriguing possibility is that they are leftovers from an earlier stage in evolution. An interesting observation is that the exon portions of different genes often code for protein segments of similar structure and function. It has been suggested that introns might separate exons that can be mixed and matched to create new genes, or that the intron portions of one gene may actually be the exon portion of another gene. It has also been postulated that in DNA introns may act as regulatory sites for other genes or somehow help to stabilize the genome; in any case, introns undoubtedly play an important, not yet fully understood role in an organism's genetic construction.

WORKED EXAMPLE **26.4** Writing Complementary DNA and RNA Strands from Informational DNA Strands

The nucleotide sequence in a segment of a DNA informational strand is given below. What is the nucleotide sequence in the complementary DNA template strand? What is the sequence transcribed from the template strand into mRNA?

5′ AAC GTT CCA ACT GTC 3′

ANALYSIS Recall:

1. In the informational and template strands of DNA the base pairs are A-T and C-G.
2. Matching base pairs along the informational strand gives the template strand written in the 3′ to 5′ direction.
3. The mRNA strand is identical to the DNA informational strand except that it has a U wherever the informational strand has a T.
4. Matching base pairs along the template strand gives the mRNA strand written in the 5′ to 3′ direction.

SOLUTION
Applying these principles gives

DNA informational strand	5′ AAC GTT CAA ACT GTC 3′
DNA template strand	3′ TTG CAA GTT TGA CAG 5′
mRNA	5′ AAC GUU CAA ACU GUC 3′

PROBLEM 26.12

What mRNA base sequences are complementary to the following DNA template sequences? Be sure to label the 5' and 3' ends of the complementary sequences.

(a) 5' CAG ACT GTA CAC 3'   **(b)** 3' TAG TAT CGA GCG 5'

26.9 The Genetic Code

The ribonucleotide sequence in an mRNA chain is like a coded sentence that spells out the order in which amino acid residues should be joined to form a protein. Each "word" consists of a triplet of ribonucleotides, or **codon**, in the mRNA sentence, which in turn corresponds to a specific amino acid. That is, a series of codons spells out a sequence of amino acids. For example, the series uracil-uracil-guanine (UUG) on an mRNA transcript is a codon directing incorporation of the amino acid leucine into a growing protein chain. Similarly, the sequence guanine-adenine-uracil (GAU) codes for aspartate.

Of the 64 possible three-base combinations in RNA, 61 code for specific amino acids and 3 code for chain termination (the *stop codons*). The "meaning" of each codon—the **genetic code** universal to all but a few living organisms—is given in Table 26.4. Note that most amino acids are specified by more than one codon and that codons are always written in the 5' to 3' direction.

Codon A sequence of three ribonucleotides in the messenger RNA chain that codes for a specific amino acid; also a three-nucleotide sequence that is a stop codon and stops translation.

Genetic code The sequence of nucleotides, coded in triplets (codons) in mRNA, that determines the sequence of amino acids in protein synthesis.

TABLE 26.4   Codon Assignments of Base Triplets in mRNA

FIRST BASE (5' END)	SECOND BASE	THIRD BASE (3' END)			
		U	C	A	G
U	U	Phe	Phe	Leu	Leu
	C	Ser	Ser	Ser	Ser
	A	Tyr	Tyr	Stop	Stop
	G	Cys	Cys	Stop	Trp
C	U	Leu	Leu	Leu	Leu
	C	Pro	Pro	Pro	Pro
	A	His	His	Gln	Gln
	G	Arg	Arg	Arg	Arg
A	U	Ile	Ile	Ile	Met
	C	Thr	Thr	Thr	Thr
	A	Asn	Asn	Lys	Lys
	G	Ser	Ser	Arg	Arg
G	U	Val	Val	Val	Val
	C	Ala	Ala	Ala	Ala
	A	Asp	Asp	Glu	Glu
	G	Gly	Gly	Gly	Gly

The relationship between the DNA informational and template strand segments illustrated earlier is repeated here, along with the protein segment for which they code:

DNA informational strand	5' ATG CCA GTA GGC CAC TTG TCA 3'
DNA template strand	3' TAC GGT CAT CCG GTG AAC AGT 5'
mRNA	5' AUG CCA GUA GGC CAC UUG UCA 3'
Protein	Met Pro Val Gly His Leu Ser

Notice that 5′ end of the mRNA strand codes for the N-terminal amino acid, whereas the 3′ end of the mRNA strand codes for the C-terminal amino acid.

WORKED EXAMPLE **26.5** Translating RNA into Protein

In Worked Example 26.4, we derived the mRNA sequence of nucleotides shown below. What is the sequence of amino acids coded for by the mRNA sequence?

5′ AAC GUU CAA ACU GUC   3′

ANALYSIS The codons must be identified by consulting Table 26.4. They are

5′ AAC GUU CAA ACU GUC   3′
    Asn Val Gln Thr Val

SOLUTION
Written out in full, the protein sequence is

asparagine-valine-glutamine-threonine-valine

PROBLEM 26.13

List possible codon sequences for the following amino acids.

(a) Ala   **(b)** Pro   **(c)** Ser

(d) Lys   **(e)** Tyr

PROBLEM 26.14

Name the base represented by the codon GUG and identify the amino acid for which this codon codes.

PROBLEM 26.15

What amino acids do the following sequences code for?

(a) AUA   **(b)** GCC   **(c)** CGU   **(d)** AAA

PROBLEM 26.16

A hypothetical tripeptide Leu-Leu-Leu could be synthesized by the cell. What three different base triplets in mRNA could be combined to code for this tripeptide?

26.10 Translation: Transfer RNA and Protein Synthesis

How are the messages carried by mRNA translated, and how does the translation process result in the synthesis of proteins? Protein synthesis occurs at ribosomes, which are located outside the nucleus in the cytoplasm of cells. First, mRNA binds to the ribosome; then, amino acids, which are available in the cytosol, are delivered one by one by transfer RNA (tRNA) molecules to be joined into a specific protein by the ribosomal "machinery." All of the RNA molecules required for translation were synthesized from DNA by transcription in the nucleus and moved to the cytosol for translation.

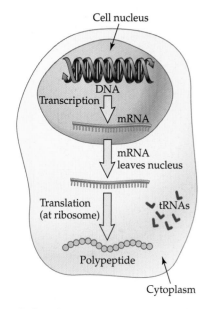

▲ **Overview of protein synthesis.** The codons of mature mRNA are translated in the ribosomes, where tRNAs deliver amino acids to be assembled into proteins (polypeptides).

First, let us examine the structure of the tRNAs. Every cell contains more than 20 different tRNAs, each designed to carry a specific amino acid, even though they are similar in overall structure. A tRNA molecule is a single polynucleotide chain held together by regions of base pairing in a partially helical structure something like a cloverleaf (Figure 26.7a). In three dimensions, a tRNA molecule is L-shaped, as shown in Figures 26.7b and c.

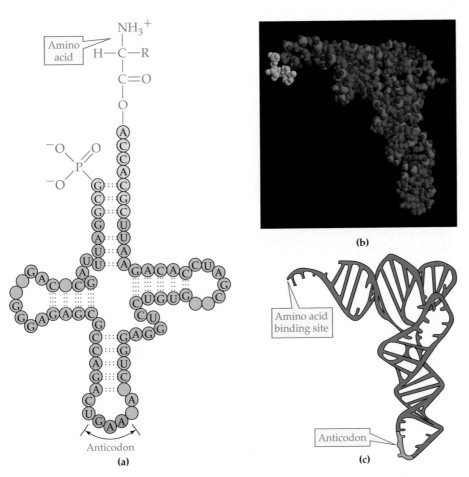

▲ **FIGURE 26.7** **Structure of tRNA.** (a) Schematic, flattened tRNA molecule. The cloverleaf-shaped tRNA contains an anticodon triplet on one "leaf" and a covalently bonded amino acid at its 3′ end. (The example shown is a yeast tRNA that codes for phenylalanine. All tRNAs have similar structures. The nucleotides not identified (blank circles) are slightly altered analogs of the four normal ribonucleotides.) (b) A computer-generated model of the serine tRNA molecule. The serine binding site is shown in yellow and the anticodon in red. (c) The three-dimensional shape (the tertiary structure) of a tRNA molecule. Note how the anticodon is at one end and the amino acid is at the other end.

Each amino acid is bonded to its specific tRNA by an ester linkage between the —COOH of the amino acid and an —OH group on a ribose at the 3′ end of the tRNA chain (which is at one end of the "L"). Individual synthetase enzymes are responsible for connecting each amino acid with its partner tRNA in an energy-requiring reaction. This reaction is referred to as *charging* the tRNA. Once charged, the tRNA is ready to be used in the synthesis of new protein.

At the other end of the L-shaped tRNA molecule is a sequence of three nucleotides called an **anticodon** (Figure 26.7). The anticodon of each tRNA is complementary to an mRNA codon—*always the one designating the particular amino acid that the tRNA carries.* For example, the tRNA carrying the amino acid leucine, which is coded for by 5′ CUG 3′ in mRNA, has the complementary sequence 3′ GAC 5′ as its anticodon. This is how the genetic message of nucleotide triplets, the codons,

Anticodon A sequence of three ribonucleotides on tRNA that recognizes the complementary sequence (the codon) on mRNA.

is translated into the sequence of amino acids in a protein. When the tRNA codon pairs off with its complementary mRNA codon, leucine is delivered to its proper place in the growing protein chain.

The three stages in protein synthesis are *initiation, elongation,* and *termination.* As you read the descriptions, follow along in the diagram of translation in Figure 26.8.

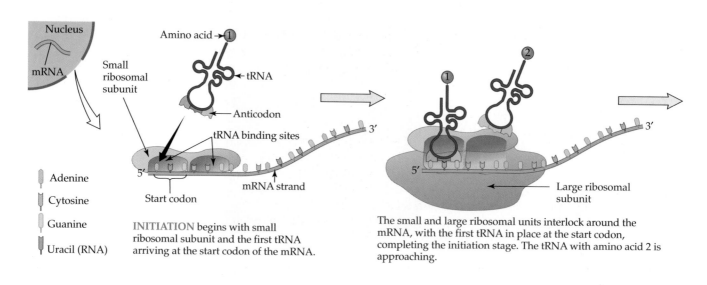

INITIATION begins with small ribosomal subunit and the first tRNA arriving at the start codon of the mRNA.

The small and large ribosomal units interlock around the mRNA, with the first tRNA in place at the start codon, completing the initiation stage. The tRNA with amino acid 2 is approaching.

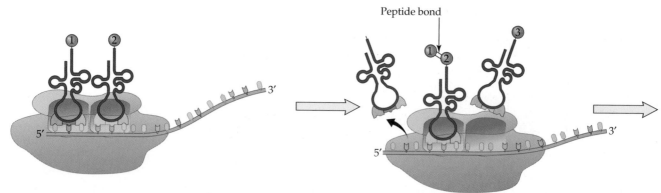

ELONGATION begins as the tRNA with amino acid 2 binds to its codon at the second site within the ribosome.

A peptide bond forms between amino acid 1 and 2, the first tRNA is released, the ribosome moves one codon to the right, and the tRNA with amino acid 3 is arriving.

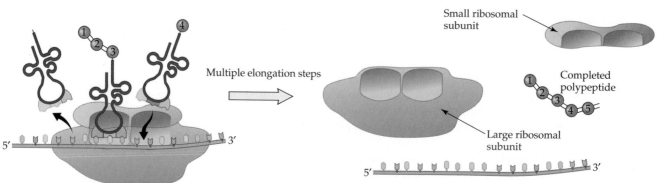

Elongation continues with three amino acids in the growing chain and the fourth one arriving with its tRNA.

TERMINATION occurs after the elongation steps have been repeated until the stop codon is reached. The ribosomal units, the mRNA, and the polypeptide separate.

▲ **FIGURE 26.8** Translation: The initiation, elongation, and termination stages in protein synthesis.

(1) Initiation

Each ribosome in a cell is made up of two subunits of markedly different sizes, called, logically enough, the *small subunit* and the *large subunit*. Each subunit contains protein enzymes and ribosomal RNA (rRNA). Protein synthesis begins with the binding of an mRNA to the small subunit of a ribosome, joined by the first tRNA. The first codon on the 5′ end of mRNA, an AUG, acts as a "start" signal for the translation machinery and codes for a methionine-carrying tRNA. Initiation is completed when the large ribosomal subunit joins the small one and the methionine-bearing tRNA occupying one of the two binding sites on the united ribosome. (Not all proteins have methionine at one end. If it is not needed, the methionine from chain initiation is removed by *post-translational modification* before the new protein goes to work.)

(2) Elongation

Next to the first binding site on the ribosome is a second binding site where the next codon on mRNA is exposed and the tRNA carrying the next amino acid will be attached. All available tRNA molecules can approach and try to fit, but only one with the appropriate anticodon sequence can bind. Once the tRNA with amino acid 2 arrives, an enzyme in the large subunit catalyzes formation of the new peptide bond and breaks the bond linking amino acid 1 to its tRNA. These energy-requiring steps are fueled by the hydrolysis of GTP to GDP. The first tRNA then leaves the ribosome, and the entire ribosome shifts three positions (one codon) along the mRNA chain. As a result, the second binding site is opened up to accept the tRNA carrying the next amino acid.

The three elongation steps now repeat:

- The next appropriate tRNA binds to the ribosome.
- Peptide bond formation attaches the newly arrived amino acid to the growing chain and the tRNA carrying it is released.
- Ribosome position shifts to free the second binding site for the next tRNA.

A single mRNA can be "read" simultaneously by many ribosomes. The growing polypeptides increase in length as the ribosomes move down the mRNA strand.

(3) Termination

When synthesis of the protein is completed, a "stop" codon signals the end of translation. An enzyme called a *releasing factor* then catalyzes cleavage of the polypeptide chain from the last tRNA, the tRNA and mRNA molecules are released from the ribosome, and the two ribosome subunits separate. This step also requires energy from GTP. Overall, to add one amino acid to the growing polypeptide chain requires four molecules of GTP, excluding the energy needed to charge the tRNA.

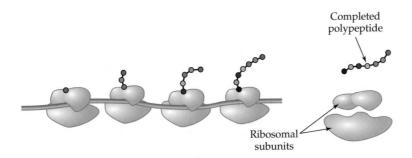

Completed polypeptide

Ribosomal subunits

In our discussion in this and the preceding sections, we have left many questions about replication, transcription, and translation unanswered. What tells a cell when to start replication? Are there mechanisms to repair damaged DNA or correct random errors made during replication? (There are.) Keep in mind that synthesis of

APPLICATION ▶   "Bird Flu": The Next Epidemic?

When we talk of "the flu," we take for granted that it is solely a human condition. Flu is caused by the influenza virus, of which there are three major types—A, B, and C—with many subtypes of each type. Influenza A and influenza B viruses cause human flu epidemics almost every winter. In the United States, these winter epidemics can cause illness in 10% to 20% of the human population and are associated with an average of 36,000 deaths and 114,000 hospitalizations per year.

Getting a flu shot can prevent illness from types A and B influenza. Influenza type C infections cause a mild respiratory illness and are not thought to cause epidemics. Flu shots do not protect against type C influenza.

Can animals get the "flu"? The answer is yes. Many subtypes of influenza A viruses are also found in a variety of animals, including ducks, chickens, pigs, whales, horses, and seals. Certain strains (subtypes) of influenza A virus are specific only to certain species. Unlike other animals, however, birds are susceptible to all known subtypes of the influenza A virus.

Influenza viruses that infect birds are called *avian influenza viruses*. These viruses occur naturally among birds. Wild birds, most notably migratory waterfowl such as wild ducks, carry the viruses in their intestines; these birds are also the most resistant to infection by the viruses they carry. However, avian influenza is very contagious among birds if infection does occur. Domesticated birds, such as chickens, ducks, and turkeys, are particularly susceptible to infection, which either makes them very sick or kills them. Direct or indirect contact of domestic flocks with wild migratory waterfowl has been implicated as a frequent cause of these epidemics. Avian influenza, first identified in Italy more than 100 years ago, occurs worldwide. Humans also are susceptible to influenza A viruses, but avian influenza viruses do not usually infect humans, as there are substantial genetic differences between the subtypes that typically infect both people and birds.

The story would normally end here, but alarmingly several cases of human infection with avian influenza viruses have occurred since 1997. These viruses may be transmitted to humans directly from birds, from an environment contaminated by avian virus, or through an intermediate host, such as a pig. Because pigs are susceptible to infection by both avian and human viruses, they can serve as a "mixing vessel" for the scrambling of genetic material from human and avian viruses, resulting in the emergence of a novel viral subtype. For example, if a pig is infected with a human influenza virus and an avian influenza virus at the same time, the viruses can re-assort and produce a new virus that has most of the genes from the human virus, but surface proteins from the avian virus. This process is known as an *antigenic shift*. Antigenic shift results when a new influenza A subtype to which most people have little or no immune protection infects humans.

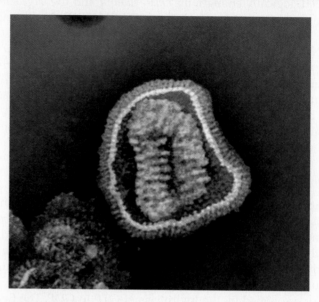

▲ A tranmission electron micrograph of negatively stained influenza A virus particles.

This is how a new virus is born, a virus against which humans will have little or no immunity and that could result in sustained human-to-human transmission and ultimately an influenza epidemic. Conditions favorable for the emergence of antigenic shift have long been thought to involve humans living in close proximity to domestic poultry and pigs. However, recent events suggest humans themselves can serve as the "mixing vessel." This scenario has frightening consequences; so frightening that the Center for Disease Control and Prevention (CDC) considers the control of avian influenza to be a top priority.

In 1918, a strain of influenza that became known as *Spanish flu* killed an estimated 20–50 million people worldwide. Most of those who died were not in the highly fragile groups of very young and very old but rather were healthy adults between 20 and 40 years old. Analysis and reconstruction of the 1918 influenza virus showed it was Type A, variant H1N1. The avian influenza virus we are seeing today is also a type A virus, but a different variant than the virus that caused the 1918 pandemic. However, research has shown that the hemagglutinin genes and gene products (protein) of the avian flu virus are closely similar to those of the 1918 flu virus, suggesting an explanation for why the avian flu is able to infect humans.

The first documented infection of humans with an avian influenza virus occurred in Hong Kong in 1997, when the avian influenza strain H5N1 caused severe respiratory disease in 18 humans, 6 of whom died. Symptoms of bird flu in humans range from typical flu-like symptoms (fever, cough, sore throat, and muscle aches) to eye infections, pneumonia, severe respiratory diseases (such as acute respiratory distress), and other severe and life-threatening complications. The infection of humans coincided with an epidemic of a highly pathogenic avian influenza, caused by H5N1, in Hong Kong's poultry population. Studies at the genetic level determined that the virus jumped directly from birds to humans, raising concern that a pandemic may be in the making. A disease that quickly and severely affects a large number of people and then subsides is an *epidemic*. A *pandemic* is a widespread epidemic that may affect entire continents or even the world. To avert a pandemic, Hong Kong's entire poultry population (approximately 1.5 million birds) was destroyed within three days. This drastic measure reduced opportunities for further direct transmission to humans. This event, marking the first time that an avian influenza virus was transmitted directly to humans and which caused severe illness with high mortality, put public health authorities on high alert. Alarms sounded again in February 2003, when an outbreak of H5N1 in Hong Kong caused two cases and one death in members of a family who had recently traveled to southern China. The World Health Organization reports fewer than 100 cases of human avian flu per year worldwide; however, the mortality rate is approximately 60%. As more humans become infected over time, the likelihood increases that humans, if concurrently infected with human and avian influenza strains, could serve as the mixing vessel for the emergence of an even more novel and dangerous virus strain

easily transmitted from person to person. Such an event would mark the start of an influenza pandemic. Studies suggest that only two of the four prescription medicines approved for human flu viruses should work in preventing bird flu infection in humans, but due to their extreme variability, viruses can quickly become resistant to these drugs.

▲ Thai workers remove chickens to be destroyed at a farm in the central province of Suphanburi, Thailand.

See Additional Problems 26.71 and 26.72 at the end of the chapter.

mRNA is the beginning of synthesis of a protein. How are proteins modified, and where does this modification occur? Since each cell contains the entire genome and since cells differ widely in their function, what keeps genes for unneeded proteins turned off? What determines just when a particular gene in a particular cell is transcribed? What indicates the spot on DNA where transcription should begin? How is hnRNA converted into mRNA? How are transcription and the resulting protein synthesis regulated? You have seen that steroid hormones function by directly entering the nucleus to activate enzyme synthesis. (⊂◯⊃, Section 20.5) This is just one mode of gene regulation. It can also occur by modification of DNA during transcription and during translation. Much is known beyond what we have covered (and if you are curious, take a genetics course!). There are also many questions that we cannot yet fully answer.

PROBLEM 26.17

What amino acid sequence is coded for by the mRNA base sequence CAG-AUG-CCU-UGG-CCC-UUA?

PROBLEM 26.18

What anticodon sequences of tRNAs match the mRNA codons in Problem 26.17?

SUMMARY: REVISITING THE CHAPTER GOALS

1. **What is the composition of the nucleic acids, DNA and RNA?** *Nucleic acids* are polymers of *nucleotides*. Each nucleotide contains a sugar, a base, and a phosphate group. The sugar is D-ribose in *ribonucleic acids (RNAs)* and 2-deoxy-D-ribose in *deoxyribonucleic acids (DNAs)*. The C5—OH of the sugar is bonded to the phosphate group, and the anomeric carbon of the sugar is connected by an *N*-glycosidic bond to one of five heterocyclic nitrogen bases (Table 26.1). A *nucleoside* contains a sugar and a base, but not the phosphate group. In DNA and RNA, the nucleotides are connected by phosphate diester linkages between the 3′—OH group of one nucleotide and the 5′ phosphate group of the next nucleotide. DNA and RNA both contain adenine, guanine, and cytosine; thymine occurs in DNA and uracil occurs in RNA.

2. **What is the structure of DNA?** The DNA in each *chromosome* consists of two polynucleotide strands twisted together in a *double helix*. The sugar–phosphate backbones are on the outside, and the bases are in the center of the helix. The bases on the two strands are complementary—opposite every thymine is an adenine, opposite every guanine is a cytosine. The base pairs are connected by hydrogen bonds (two between T and A; three between G and C). Because of the *base pairing*, the DNA strands are *antiparallel*: One DNA strand runs in the 5′ to 3′ direction and its complementary partner in the 3′ to 5′ direction.

3. **How is DNA reproduced?** *Replication* (Figure 26.5) requires DNA polymerase enzymes and deoxyribonucleoside triphosphates. The DNA helix partially unwinds and the enzymes move along incorporating nucleotides with bases complementary to those on the unwound DNA strand being copied. The enzymes copy only in the 3′ to 5′ direction of the template strand, so that one strand is copied continuously and the other strand is copied in segments as the replication fork moves along. In each resulting double helix, one strand is the original template strand and the other is the new copy.

4. **What are the functions of RNA?** *Messenger RNA (mRNA)* carries the genetic information out of the nucleus to the *ribosomes* in the cytosol, where protein synthesis occurs. *Transfer RNAs (tRNAs)* circulate in the cytosol, where they bond to amino acids that they then deliver to protein synthesis. *Ribosomal RNAs (rRNAs)* are incorporated into ribosomes.

5. **How do organisms synthesize messenger RNA?** In *transcription* (Figure 26.6), one DNA strand serves as the template and the other, the informational strand, is not copied. Nucleotides carrying bases complementary to the template bases between a control segment and a stop *codon* are connected one by one to form mRNA. The primary transcript mRNA (or hnRNA) is identical to the matching segment of the informational strand, but with uracil replacing thymine. *Introns*, which are base sequences that do not code for amino acids in the protein, are cut out before the final transcript mRNA leaves the nucleus.

6. **How does RNA participate in protein synthesis?** The genetic information is read as a sequence of codons—triplets of bases in DNA that give the sequence of amino acids in a protein. Of the 64 possible codons (Table 26.4), 61 specify amino acids and 3 are stop codons. Each tRNA has at one end an *anticodon* consisting of three bases complementary to those of the codon that specifies the amino acid it carries. Initiation of *translation* (Figure 26.8) is the coming together of the large and small subunits of the ribosome, an mRNA, and the first amino acid–bearing tRNA connected at the first of the two binding sites in the ribosome. *Elongation* proceeds as the next tRNA arrives at the second binding site, its amino acid is bonded to the first one, the first tRNA leaves, and the ribosome moves along so that once again there is a vacant second site. These steps repeat until the stop codon is reached. The termination step consists of separation of the two ribosome subunits, the mRNA, and the protein.

UNDERSTANDING KEY CONCEPTS

26.19 Combine the structures below to create a ribonucleotide. Show where water is removed to form an *N*-glycosidic linkage and where water is removed to form a phosphate ester. Draw the resulting ribonucleotide structure, and name it.

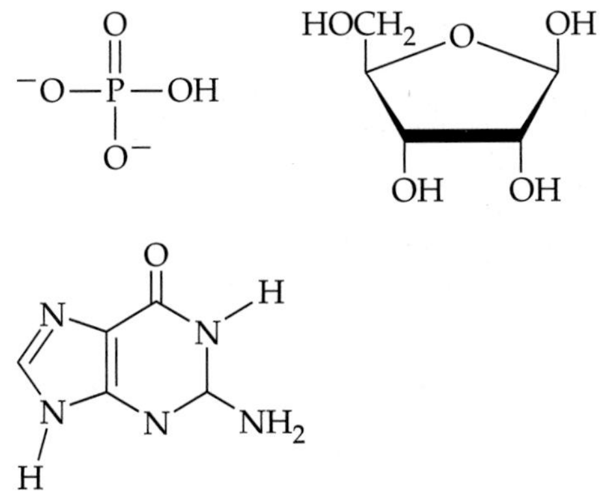

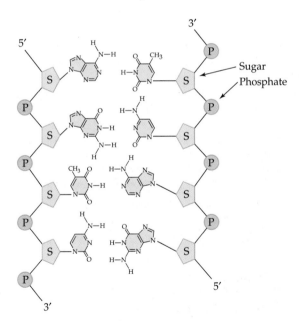

26.20 Copy the diagram shown below and use dotted lines to indicate where hydrogen bonding occurs between the complementary strands of DNA. What is the sequence of each strand of DNA drawn (remember that the sequence is written from 5′ to 3′ end)?

26.21 Copy this simplified drawing of a DNA replication fork:

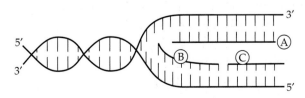

(a) On the drawing, indicate the direction of synthesis of the new strand labeled A and the location of DNA polymerase on the strand.

(b) On the drawing, indicate the direction of synthesis of the new strand labeled B and the location of DNA polymerase on the strand.

(c) How will strand C and strand B be connected?

26.22 What groups are found on the exterior of the DNA double helix? In the nucleus, DNA strands are wrapped around proteins called histones. Would you expect histones to be neutral, positively charged, or negatively charged? Based on your answer, which amino acids do you expect to be abundant in histones, and why?

26.23 In addition to RNA polymerase, transcription of DNA for the synthesis of RNA requires (a) a control segment of DNA (also called an initiation sequence), (b) an informational strand, (c) a template strand of DNA, and (d) an end of the sequence (termination sequence). Determine the direction of RNA synthesis on the RNA strand in the following diagram. Draw in the locations of elements (a)–(d).

26.24 Gln-His-Pro-Gly is the sequence of a molecule known as progenitor thyrotropin-releasing hormone (pro-TRH). If we were searching for pro-TRH genes, we would need to know what sequence of bases in DNA we should be looking for. Use the boxes below to indicate answers to parts (a)–(d).

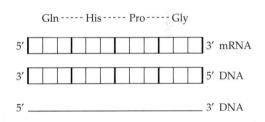

(a) What RNA sequence could code for these four amino acids?

(b) What double-stranded DNA sequence (gene) could code for these amino acids?

(c) Which strand of DNA is the template strand, and which is the informational strand?

(d) How many possible DNA sequences are there?

ADDITIONAL PROBLEMS

STRUCTURE AND FUNCTION OF NUCLEIC ACIDS

26.25 For the following molecule:

(a) Label the three nucleic acid building blocks it contains.

(b) Draw a box around the nucleoside portion of the molecule.

(c) Draw a circle around the nucleotide portion of the molecule.

26.26 What are the sugars in DNA and RNA, and how do they differ?

26.27 (a) What are the four major heterocyclic bases in DNA?
(b) What are the four major heterocyclic bases in RNA?
(c) Structurally, how do the heterocyclic bases in RNA differ from those in DNA? (See Table 26.1.)

26.28 What are the two structural types of bases in DNA and RNA? Which bases correspond to each type?

26.29 What are the three main kinds of RNA, and what are their functions?

26.30 Rank the following in order of size: tRNA, DNA, mRNA.

26.31 (a) What is meant by the term *base pairing*?
(b) Which bases pair with which other bases?
(c) How many hydrogen bonds does each base pair have?

26.32 How are replication, transcription, and translation similar? How are they different?

26.33 What is the difference between a gene and a chromosome?

26.34 What are the two major components of chromatin?

26.35 What genetic information does a single gene contain?

26.36 How many chromosomes are present in a human cell?

26.37 What kind of intermolecular attraction holds the DNA double helix together?

26.38 What does it mean to speak of bases as being *complementary*?

26.39 The DNA from sea urchins contains about 32% A and about 18% G. What percentages of T and C would you expect in sea urchin DNA? Explain.

26.40 If a double-stranded DNA molecule is 19% G, what is the percentage of A, T, and C? Explain.

26.41 What is the difference between the 3′ end and the 5′ end of a polynucleotide?

26.42 Are polynucleotides written 3′ to 5′, or 5′ to 3′?

26.43 Draw structures to show how the phosphate and sugar components of a nucleic acid are joined. What kind of linkage forms between the sugar and the phosphate?

26.44 Draw structures to show how the sugar and heterocyclic base components of a nucleic acid are joined. What small molecule is formed?

26.45 Draw the complete structure of uridine 5′-phosphate, one of the four major ribonucleotides.

26.46 Draw the complete structure of the RNA dinucleotide U-C. Identify the 5′ and 3′ ends of the dinucleotide.

26.47 The segment of DNA that encompasses a gene typically contains *introns* and *exons*. Define each of these terms. What are some possible roles introns might play?

NUCLEIC ACIDS AND HEREDITY

26.48 Transcribed RNA is complementary to which strand of DNA?

26.49 Why is more than one replication fork needed when human DNA is duplicated?

26.50 Why do we say that DNA replication is semiconservative?

26.51 What is a codon, and on what kind of nucleic acid is it found?

26.52 What is an anticodon, and on what kind of nucleic acid is it found?

26.53 What is the general shape and structure of a tRNA molecule?

26.54 There are different tRNAs for each amino acid. What is one major way to differentiate among the tRNAs for each amino acid?

26.55 Which amino acid(s) have the most codons? Which amino acid(s) have the fewest codons? Can you think of a reason why multiple codons code for certain amino acids but other amino acids are coded for by very few codons?

26.56 Look at Table 26.4 and find codons for the following amino acids:
(a) Ala
(b) His
(c) Pro

26.57 What amino acids are specified by the following codons?
(a) C-C-U
(b) G-C-A
(c) A-U-U

26.58 What anticodon sequences are complementary to the codons listed in Problem 26.57? (Remember that the anticodons are opposite in direction to the codons, so label the 3′ and 5′ ends!)

26.59 What anticodon sequences are complementary to the codons for the amino acids given in Problem 26.56? (Remember that the anticodons are opposite in direction to the codons, so label the 3′ and 5′ ends!)

26.60 If the sequence A-A-C-G-G-A appears on the informational strand of DNA, what sequence appears opposite it on the template strand? Label your answer with 3′ and 5′ ends.

26.61 Refer to Problem 26.60. What sequence appears on the mRNA molecule transcribed from the DNA sequence A-A-C-G-G-A? Label your answer with 3′ and 5′ ends.

26.62 Refer to Problems 26.60 and 26.61. What dipeptide is synthesized from the informational DNA sequence A-A-C-G-G-A?

26.63 What tetrapeptide is synthesized from the informational DNA sequence G-C-T-C-A-G-C-C-G-A-A-T?

26.64 Metenkephalin is a small peptide found in animal brains that has morphine-like properties. Give an mRNA sequence that could code for the synthesis of metenkephalin: Tyr-Gly-Gly-Phe-Met. Label your answer with 3′ and 5′ ends.

26.65 Refer to Problem 26.64. Give a double-stranded DNA sequence that could code for metenkephalin. Label your answer with 3′ and 5′ ends.

Applications

26.66 How do viruses differ from living organisms? [*Viruses and AIDS, p. 820*]

26.67 Explain the process of reverse transcription. What is the name given to viruses that use this process? [*Viruses and AIDS, p. 820*]

26.68 How do vaccines work? Why is it so difficult to design drugs effective against AIDS? [*Viruses and AIDS, p. 820*]

26.69 What is a ribozyme? What is unusual about the function of a ribozyme? [*It's a Ribozyme!, p. 826*]

26.70 Some scientists think that a "RNA world" preceded our current cellular reactions dominated by enzymes. What would lead them to assert this? [*It's a Ribozyme!, p. 826*]

26.71 Describe how the avian influenza virus is transmitted to humans. [*"Bird Flu," The Next Epidemic?, p. 834*]

26.72 The influenza virus H1N1 can infect both humans and other animals. Use the Internet to collect information that allows you to describe some of the similarities and some of differences between the H1N1 virus and the virus responsible for avian influenza. [*"Bird Flu," The Next Epidemic?, p. 834*]

General Questions and Problems

26.73 A normal hemoglobin protein has a glutamic acid at position 6; in sickle-cell hemoglobin, this glutamic acid has been replaced by a valine. List all the possible mRNA codons that could be present for each type of hemoglobin. Can a single base change result in a change from Glu to Val in hemoglobin?

26.74 Insulin is synthesized as preproinsulin, which has 81 amino acids. How many heterocyclic bases must be present in the informational DNA strand to code for preproinsulin (assuming no introns are present)?

26.75 Human and horse insulin are both composed of two polypeptide chains with one chain containing 21 amino acids and the other containing 30 amino acids. Human and horse insulin differ at two amino acids: position 9 in one chain (human has serine and horse has glycine) and position 30 on the other chain (human has threonine and horse has alanine). How must the DNA differ to account for this? Identify the 5′ and 3′ ends of the four trinucleotide complementary DNA sequences.

26.76 If the initiation codon for proteins is AUG, how do you account for the case of a protein that does not include methionine as its first amino acid?

26.77 Suppose that 22% of the nucleotides of a DNA molecule are deoxyadenosine, and during replication the relative amounts of available deoxynucleoside triphosphates are 22% dATP, 22% dCTP, 28% dGTP, and 28% dTTP. What deoxynucleoside triphosphate is limiting to the replication? Explain.

CHAPTER 27

Genomics

CONCEPTS TO REVIEW

Structure, synthesis, and
function of DNA

(Chapter 26)

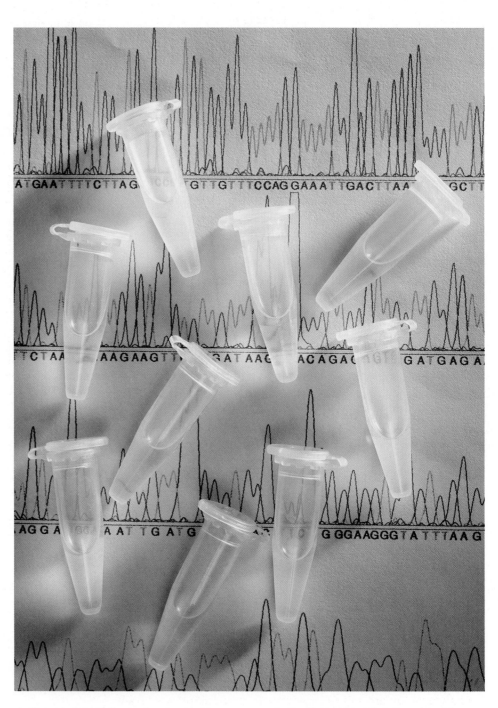

▲ The vials hold DNA samples dissolved in an aqueous buffer. Behind the vials is the printout from an automatic DNA sequencing machine

CONTENTS

CHAPTER GOALS

In this chapter, we will address the following questions about genomics:

1. **What is the working draft of the human genome and the circumstances of its creation?**

   THE GOAL: Be able to describe the genome mapping projects and the major accomplishments of their working drafts.

2. **What are the various segments along the length of the DNA in a chromosome?**

   THE GOAL: Be able to describe the nature of telomeres, centromeres, exons and genes, and noncoding DNA.

3. **What are mutations?**

   THE GOAL: Be able to define mutations, identify what can cause them, and also identify their possible results.

4. **What are polymorphisms and single nucleotide polymorphisms (SNPs) and how can identifying them be useful?**

   THE GOAL: Be able to define polymorphisms and SNPs, and explain the significance of knowing the locations of SNPs.

5. **What is recombinant DNA?**

   THE GOAL: Be able to define recombinant DNA and explain how it is used for production of proteins by bacteria.

6. **What does the future hold for uses of genomic information?**

   THE GOAL: Be able to provide an overview of the current and possible future applications of the human genome map.

In Chapter 26 we described the fundamentals of DNA structure and function, the biosynthesis of DNA, and its role in making proteins from amino acids. Out of all this work on DNA has come a scientific revolution: the sequencing of the entire genome for many organisms. The crowning achievement of this endeavor is that an almost complete and accurate map of human DNA is now freely available. Creation of this map has been compared to such landmark achievements as harnessing nuclear power and flight into outer space. In significance for individual human beings, there has never been anything like it.

This chapter begins with a description of how this genomic map was obtained. We then explore variations in the content of the DNA in each chromosome and a technique for manipulating DNA. Finally, we look briefly at ways in which genomic information can be put to use.

27.1 Mapping the Human Genome

How to Map a Genome

Many people tend to think of mapping the genome of a given organism like reading a novel: you start at the first page and continue until you reach the end. Unfortunately, mapping an organism's genome, whether the organism be a bacterium or a human, cannot be done by starting at one end of the DNA in each chromosome and proceeding base by base. For example, when you consider that the nucleotides that code for proteins in humans (the *exons*) (, Section 26.8) are interrupted by noncoding nucleotides (the *introns*), it should be clear what mapping challenges exist for any organism whose genome contains only a few dozen genes. These challenges are greatly magnified for the human genome, which contains between twenty and twenty-five *thousand* genes! Consider also that there is neither spacing between "words" in the genetic code nor any "punctuation." Using the English language as an analogy, try to find a meaningful phrase in this:

sfdggmaddrydkdkdkrrrsjfljhadxccctmctmaqqqoumlittgklejagkjghjoailambrsslj

The phrase is "mary had a little lamb":

sfdgg**maddry**dkdkdkrrrsjflj**had**xccctmctm**a**qqqoum**little**gkl**eja**gkjghjoail**ambr**sslj

Now consider how hard finding meaning would be if the phrase you were looking for were in a language you were not familiar with! It has been estimated that the

▲ A sample of DNA ready for analysis.

string of C's, G's, T's, and A's that make up the human genome would fill 75,490 pages of standard-size type in a newspaper like the *New York Times*.

Two organizations led the effort to map the human genome: the Human Genome Project (a collection of 20 groups at not-for-profit institutes and universities) and Celera Genomics (a commercial biotechnology company). These two groups used different approaches to taking DNA apart, analyzing its base sequences, and reassembling the information. The Human Genome Project proceeded through a series of maps of finer and finer resolution (think of a satellite map program such as Google Earth, where you can progress from a satellite photo of the United States to a map of your state to a map of the city where you live to the street you live on and ultimately to a picture of the house you live in). Celera followed a seemingly random approach in which they fragmented DNA and then relied on instrumental and computer-driven techniques to establish the sequence. It was believed that data obtained via the combination of these two approaches should speed up the enormous task of sequencing the human genome.

In 2001 the stunning announcement was made that 90% of the human genome sequence had been mapped in 15 months instead of the originally anticipated four years. By October 2004, an analysis of the Human Genome Project reported that 99% of the gene-containing parts of the genome were sequenced and declared to be 99.999% accurate. Additionally, the mapped sequence reportedly correctly identifies almost all known genes (99.74% of them, to be exact). At a practical level, this "gold-standard" sequence data allows researchers to rely on highly accurate sequence information, priming new biomedical research.

The strategy utilized by the Human Genome Project for generating the complete map is shown in Figure 27.1. Pictured at the top is a type of chromosome drawing known as an *ideogram* for human chromosome 21. The light and dark blue shadings represent the location of banding visible in electron micrographs. Chromosome 21 is the smallest human chromosome, with 37 million base pairs (abbreviated 37 Mb) and was the second chromosome to be mapped (chromosome 22 was the first).

Chromosome 21 (37 Mb)

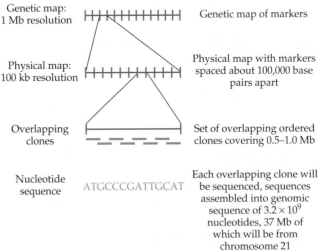

▶ **FIGURE 27.1 Human Genome Project mapping strategy.**

The first step is to generate a *genetic map*, which shows the physical location of *markers*, identifiable physical locations on either introns or exons that are known to be inherited. The markers are an average of one million nucleotides apart. This is known as a genetic map because the order and locations of the markers are established by genetic studies of inheritance in related individuals.

The next map, the *physical map*, refines the distance between markers to about 100,000 base pairs. The physical map includes markers identified by a variety of experimental methods.

To proceed to a map of finer resolution, a chromosome is cut into large segments and multiple copies of the segments are produced. The segments are **clones**, a term that refers to identical copies of organisms, cells, or in this case, DNA segments.

The overlapping clones, which cover the entire length of the chromosome, are organized in sequence to produce the next level of map (see Figure 27.1). The clones are fragmented into 500 base pieces, each of which ends in a C, G, T, or A and in which each C, G, T, or A fluoresces in a different color. These fragments are separated by electrophoresis (⬭, p. 573). In the final step, the sequences are assembled into a completed nucleotide map of the chromosome.

Clones Identical copies of organisms, cells, or DNA segments from a single ancestor.

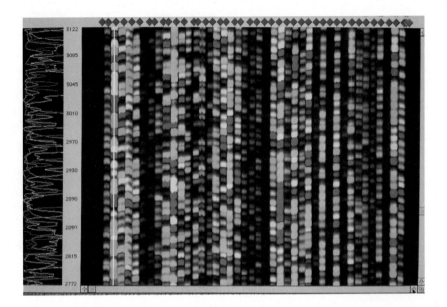

◀ Computer readout showing a DNA sequencing gel and computer-assisted analysis of it.

The approach taken by Celera Genomics was much bolder. In what has come to be known as their *shotgun approach*, Celera broke the human genome into fragments without identifying the origin of any given fragment. The fragments were copied many times and modified; ultimately they were cut into 500 base pieces containing fluorescently labeled bases that could be sequenced by high-speed machines. The resulting sequences were reassembled by identifying overlapping ends. At Celera, this monumental reassembly task was carried out using the world's largest nongovernmental supercomputing center.

PROBLEM 27.1

Decode the following sequence of letters to find an English phrase. (Hint: First look for a word you recognize, then work forward and backward from there.)
uuiioouoppagfdtttrroncetrtrnnnaedigopuponsldjflsjfxxxbvgfaqqqeutimeabrrx

27.2 A Trip Along a Chromosome

In this section we take a trip along the major regions and structural variations in the DNA folded into each chromosome. Understanding how DNA is structured should give you some insight into the biotechnology revolution ushered in by the Human Genome Project.

Telomeres and Centromeres

The DNA in a chromosome begins with the **telomeres** that are found at both ends of every chromosome. The telomeres in human DNA are long, noncoding series of

Telomeres The ends of chromosomes; in humans contain long series of repeating groups of nucleotides.

APPLICATION ▶ Whose Genome Is It?

One might wonder whose genome provided the standard against which those of all other human beings will be evaluated. What individual symbolizes all of us? A star athlete? A brilliant scientist? A truly average person?

It does not take more than a moment's thought to realize that using the DNA of a single individual to map the human genome is a bad idea. What if the person chosen had a genetic aberration? How does a project of this importance deal with ethnic differences? Since no two individuals, other than identical twins, have exactly the same base sequences in their genomes, some sort of normalized average is needed.

The obvious choice, recognized by both genome mapping groups, was to employ DNA from a group of anonymous individuals. In the Human Genome Project, researchers collected blood (female) or sperm (male) samples from a large number of donors of diverse backgrounds. After removing all identifying labels, only a randomly selected few of the many collected samples were used for sequencing, so that neither donors nor scientists could know the origins of the DNA being sequenced. The ultimate map comes from a composite of these random samples.

The Celera project relied on anonymous donors as well. DNA from five individuals (the leader of the Celera team, Craig Venter, has since acknowledged being one of these individuals) was collected, mixed, and processed for sequencing. The anonymous donors were of European, African, American

(North, Central, South), and Asian ancestry. As a result, one of the most frequently asked questions about the human genome, "Whose DNA was sequenced?" can never truly be answered because the DNA sequenced is a composite of the DNA of many anonymous individuals.

▲ From the four nucleotides that compose DNA comes the incredibly diverse population that makes up the human race.

See Additional Problems 27.48 and 27.49 at the end of the chapter.

Centromeres The central regions of chromosomes.

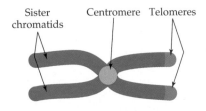

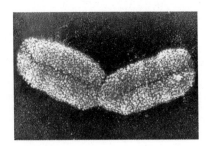

▲ **Top:** A duplicated chromosome immediately prior to cell division showing location of telomeres and centromere. **Bottom:** Color-enhanced electron microscope image showing the constriction of the centromere during metaphase.

a repeating group of nucleotides, (TTAGGG)$_n$. They act as "endcaps," or "covers," protecting the ends of the chromosome from accidental changes that might alter the more important DNA coding sequences. Telomeres also prevent the DNA from bonding to the DNA in other chromosomes or DNA fragments. Telomeres were not sequenced in the mapping projects described in Section 27.1, nor were the **centromeres** of chromosomes, the constrictions that appear in duplicated chromosomes during the process of cell division. The duplicated chromosomes bound at the centromere are known as *sister chromatids*. The centromere, like the telomeres, contain large repetitive base sequences that do not code for proteins.

Each new cell starts life with a long stretch of telomeric DNA on each of its chromosomes, 1000 or more sets of the repeating group. Some of this repeating sequence is lost with each cell division, so that as the cell ages, the telomere gets shorter and shorter. A very short telomere is associated with the stage at which a cell stops dividing (known as *senescence*). Continuation of shortening beyond this stage is associated with DNA instability and cell death.

Telomerase is the enzyme responsible for adding telomeres to DNA. Under normal, healthy conditions, telomerase is active in young cells destined to become specialized. It is not active in other cells (the *somatic* cells in adults). There is widespread speculation that telomere shortening plays a role in the natural progression of human aging. Some support for this concept comes from experiments with mice whose telomerase activity has been destroyed ("knocked out" in genetic research vernacular). These mice age prematurely, and if they become pregnant, their embryos do not survive.

What would happen if telomerase remains active in a cell rather than declining in activity with age? With the length of its telomeres constantly being replenished by telomerase, the cell would not age and instead would continue to divide.

Consider that such continuing division is one characteristic of cancer cells. In fact, the majority of cancer cells are known to contain active telomerase, which is thought to confer immortality on these tumor cells. Where this activity stands in relation to the presence of cancer-causing genes and environmental factors is not yet understood because neither amplification nor mutation of the telomerase gene has been identified in tumors. As a result, a causal role for telomerase in tumorigenesis has yet to be established. Current research suggests that it is likely that regulators of its expression are where control is exerted and that these regulators are located away from the telomerase gene. As you might suspect, there are ongoing experiments on the consequences of telomerase inactivation on cancer cells.

Noncoding DNA

In addition to the noncoding telomeres, centromeres, and introns along a chromosome, there are noncoding promoter sequences, which are regulatory regions of DNA that determine which of its genes are turned on. Remember that all of your cells (except red blood cells) contain all of your genes, but only the genes needed by the individual cells will be activated. At the onset of the Human Genome Project, most scientists believed that humans had about 100,000 genes; as of October 2007 researchers have confirmed the existence of approximately 20,000 protein-coding genes in the human genome and identified another 2000 DNA segments that are predicted to be protein-coding genes. This current data suggests that only about 2% of all DNA in the human genome actually codes for protein. It is interesting to note that the human genome has much more noncoding DNA than do the genomes known for other organisms. This evidence raises the question of the role played by the vast amount of noncoding DNA present in our genome The question arises out of the observation that genome size does not correlate with organismal complexity, with one example being that many plants have larger genomes than humans. Some scientists have suggested that the segments of repetitive but noncoding DNA are needed to accommodate the folding of DNA within the nucleus, others think these segments may have played a role in evolution, while still others argue that the segments are functional but the functions are not yet understood. The function of noncoding DNA remains to be discovered; meanwhile the debate over its role continues.

Genes

In learning about transcription (⬭⬮, Section 26.8), you saw that the nucleotides of a single gene are not consecutive along a stretch of DNA, having coding segments (the *exons*) punctuated by noncoding segments (the *introns*, p. 827). As an example of what must be dealt with in mapping the human genome, consider a "small," 2900-nucleotide sequence found in a much simpler organism (corn) that codes for the enzyme triose phosphate isomerase:

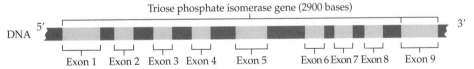

This gene consists of 9 exons that account for 759 of the 2900 nucleotides (26%), with the 8 introns accounting for the remaining bases. Now imagine the human genome, which is much more complex; it is estimated that somewhere between 1–3% of our genetic material is coding sequence. For example, chromosome 22 is one of the smaller human chromosomes to have had all of its nonrepetitive DNA mapped. The chromosome map identified 34 million bases attributable to 545 genes, with an average of 7 exons and 6 introns per gene. Chromosome 22 is of medical interest because it carries genes known to be associated with the immune system, congenital heart disease, schizophrenia, leukemia, various cancers, and many other genetically related conditions. The map also revealed several hundred previously unknown genes. With the signal (exon) to noise (intron) ratio being so low, it will be challenging to completely identify all the coding sequences present in the human genome.

▲ An error in nucleic acid composition that occurs once in 3–4 million lobsters is responsible for the beautiful color of this crustacean.

Mutation An error in base sequence that is carried along in DNA replication.

Mutagen A substance that causes mutations.

27.3 Mutations and Polymorphisms

The base-pairing mechanism of DNA replication and RNA transcription provides an extremely efficient and accurate method for preserving and using genetic information, but it is not perfect. Occasionally an error occurs, resulting in the incorporation of an incorrect base at some point.

An occasional error during the transcription of a messenger RNA molecule may not create a serious problem. After all, large numbers of mRNA molecules are continually being produced and an error that occurs perhaps one out of a million times would hardly be noticed in the presence of many correct mRNAs. If an error occurs during the replication of a DNA molecule, however, the consequences can be far more damaging. Each chromosome in a cell contains only *one* kind of DNA, and if it is miscopied during replication, then the error is passed on when the cell divides.

An error in base sequence that is carried along during DNA replication is called a **mutation**. Mutation commonly refers to variations in DNA sequence found in a very small number of individuals of a species. Some mutations result from spontaneous and random events. Others are induced by exposure to a **mutagen**—an external agent that can cause a mutation. Viruses, chemicals, and ionizing radiation can all be mutagenic. The most common types of mutations are listed in Table 27.1.

TABLE 27.1 Types of Mutations

TYPE	DESCRIPTION
Point mutations	A single base change
Silent	A change that specifies the same amino acid; for example, GUU → GUC, gives Val → Val
Missense	A change that specifies a different amino acid; for example, GUU → GCU gives Val → Ala
Nonsense	A change that produces a stop codon, for example, CGA → UGA gives Arg → Stop
Frameshift	The number of inserted or deleted bases is not a multiple of 3, so that all triplets following the mutation are read differently
Insertion	Addition of one or more bases
Deletion	Loss of one or more bases

The biological effects of incorporating an incorrect amino acid into a protein range from negligible to catastrophic, depending on both the nature and location of the change. The effect might result in a hereditary disease like sickle-cell anemia or a birth defect. There are thousands of known human hereditary diseases. Some of the more common ones are listed in Table 27.2. Mutations, sometimes the combination

TABLE 27.2 Some Common Hereditary Diseases, Their Causes, and Their Prevalence

NAME	NATURE AND CAUSE OF DEFECT	PREVALENCE IN POPULATION
Phenylketonuria (PKU)	Brain damage in infants caused by the defective enzyme phenylalanine hydroxylase	1 in 40,000
Albinism	Absence of skin pigment caused by the defective enzyme tyrosinase	1 in 20,000
Tay-Sachs disease	Mental retardation caused by a defect in production of the enzyme hexosaminidase A	1 in 6000 (Ashkenazi Jews) 1 in 100,000 (General population)
Cystic fibrosis	Bronchopulmonary, liver, and pancreatic obstructions by thickened mucus; defective gene and protein identified	1 in 3000
Sickle-cell anemia	Anemia and obstruction of blood flow caused by a defect in hemoglobin	1 in 185 (African-Americans)

of several mutations, can also produce vulnerability to certain diseases, which may or may not develop in an individual.

Polymorphisms are also variations in the nucleotide sequence of DNA, but here the reference is to variations that are common within a given population. The location of polymorphisms responsible for some inherited human diseases are shown in Figure 27.2.

Polymorphism A variation in DNA sequence within a population.

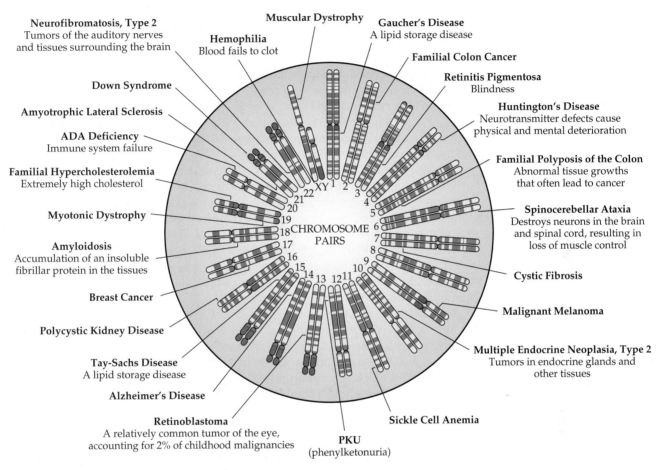

Neurofibromatosis, Type 2
Tumors of the auditory nerves and tissues surrounding the brain

Hemophilia
Blood fails to clot

Muscular Dystrophy

Gaucher's Disease
A lipid storage disease

Familial Colon Cancer

Down Syndrome

Retinitis Pigmentosa
Blindness

Amyotrophic Lateral Sclerosis

Huntington's Disease
Neurotransmitter defects cause physical and mental deterioration

ADA Deficiency
Immune system failure

Familial Polyposis of the Colon
Abnormal tissue growths that often lead to cancer

Familial Hypercholesterolemia
Extremely high cholesterol

Spinocerebellar Ataxia
Destroys neurons in the brain and spinal cord, resulting in loss of muscle control

Myotonic Dystrophy

Amyloidosis
Accumulation of an insoluble fibrillar protein in the tissues

Cystic Fibrosis

Breast Cancer

Malignant Melanoma

Polycystic Kidney Disease

Multiple Endocrine Neoplasia, Type 2
Tumors in endocrine glands and other tissues

Tay-Sachs Disease
A lipid storage disease

Alzheimer's Disease

Sickle Cell Anemia

Retinoblastoma
A relatively common tumor of the eye, accounting for 2% of childhood malignancies

PKU
(phenylketonuria)

CHROMOSOME PAIRS

▲ **FIGURE 27.2 A human chromosome map.** Regions on each chromosome that have been identified as responsible for inherited diseases are indicated.

The replacement of one nucleotide by another in the same location along the DNA sequence is known as a **single-nucleotide polymorphism** (**SNP**, pronounced "snip"). In other words, two different nucleotides at the same position along two defined stretches of DNA are SNPs. A SNP is expected to occur in at least 1% of a specific population and therefore provides a link to a genetic characteristic of that population.

Single-nucleotide polymorphism (SNP) Common single-base-pair variation in DNA.

The biological effects of SNPs are wide ranging, from being negligible, to normal variations such as those in eye or hair color, to genetic diseases. *SNPs are the most common source of variations between individual human beings.* Most genes carry one or more SNPs, and in different individuals most SNPs occur in the same location.

Imagine that the sequence A-T-G on the informational strand of DNA is miscopied to give messenger RNA with the codon sequence A-C-G rather than the correct sequence A-U-G. Because A-C-G codes for threonine, whereas A-U-G codes for methionine, threonine will be inserted into the corresponding protein during translation. Furthermore, every copy of the protein will have the same variation.

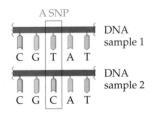

A SNP

DNA sample 1
C G T A T

DNA sample 2
C G C A T

The seriousness of the outcome depends on the function of the protein and the effect of the amino acid change on its structure and activity.

In addition to producing a change in the identity of an amino acid, a SNP might specify the same amino acid (for example, changing GUU to GUC, both of which code for valine), or it might terminate protein synthesis by introducing a stop codon (like changing CGA to UGA).

Concurrent with the work of the Human Genome Project, an international team of industrial and academic scientists is compiling a catalog of SNPs. As of 2007, the exact locations in the human genome of over 3.1 million SNPs had been recorded. Their frequency is roughly one SNP for about every 2000-5000 bases, with many of them in coding regions. These are considered but a fraction of the total that will eventually be identified.

We have described the single amino acid change that results in sickle-cell anemia (, p. 566). It took years of research to identify the SNP responsible for that disease. With a computerized catalog of SNPs, it might have been found in a few hours. Another known SNP is associated with the risk of developing Alzheimer's disease. Not all SNPs create susceptibility to diseases; for example, there is also one that imparts a resistance to HIV and AIDS.

The SNP catalog, although far from complete, has been of value from the start. Early in its development it was used to locate SNPs responsible for 30 abnormal conditions, including total color blindness, one type of epilepsy, and susceptibility to the development of breast cancer. For example, the catalog has been used to find 15 SNPs in a gene that affects testosterone levels. Examination of the DNA from prostate cancer patients showed that these SNPs occur in four combinations in these people. The next step is to hunt down the role of each of those four genetic variations in the disease. It is hoped that this information will inspire the development of new treatments for diseases.

The search for disease-related SNPs is not easy, although as the SNP catalog expands and the related proteins are understood, this is expected to change. One research group set out to identify SNPs associated with diabetes. Eventually they sequenced a 66,000-nucleotide sequence in a stretch of DNA implicated in diabetes. It contained 180 SNPs. Comparing these sequences for 100 diabetics and 100 non-diabetic controls turned up a combination of three SNPs that signal a susceptibility to diabetes.

The cataloging of SNPs has ushered in the era of genetic medicine. Ultimately, the SNP catalog may allow physicians to predict for an individual the potential age at which inherited diseases will become active, their severity, and their reactions to various types of treatment. The therapeutic course will be designed to meet the distinctive genomic profile of the person.

WORKED EXAMPLE **27.1** Determining the Effect of Changes in DNA on Proteins

The result of a mutation that changes a single amino acid in a DNA sequence depends on the type of amino acid replaced and the nature of the new amino acid. (a) What kind of change would have little effect on the protein containing the alternative amino acid? (b) What kind of change could have a major effect on the protein that contains the alternative amino acid? Give an example of each type of mutation.

ANALYSIS The result of exchanging one amino acid for another depends on the change in the nature of the amino acid side chains. To speculate on the result of such a change requires us to think about the structure of the side chains, which are shown in Table 18.3. The question to consider is whether the mutation introduces an amino acid with such a different side chain character that it is likely to alter the function of the resulting protein.

SOLUTION

(a) Exchange of an amino acid with a small nonpolar side chain for another with the same type of side chain (for example, glycine for alanine) or exchange of amino acids with very similar side chains (say, serine for threonine) might have little effect.

(b) Conversion of an amino acid with a nonpolar side chain to one with a polar, acidic, or basic side chain could have a major effect because the side chain interactions that affect protein folding may change (see Figure 18.4). Some examples of this type include exchanging threonine, glutamate, or lysine for isoleucine.

🔑 KEY CONCEPT PROBLEM 27.2

Consider that a SNP alters the base sequence in an mRNA codon by changing UGU to UGG. Speculate on the significance of this change.

27.4 Recombinant DNA

In this section, we describe a technique that makes it possible to manipulate, alter, and reproduce pieces of DNA. The technique requires the creation of **recombinant DNA**—DNA that contains two or more DNA segments not found together in nature. Progress in all aspects of genomics has built upon information gained in the application of recombinant DNA. The two other techniques that play major roles in DNA studies are the polymerase chain reaction (PCR) and electrophoresis. PCR is a method by which large quantities of identical pieces of DNA can be synthesized (see the Application "Serendipity and the Polymerase Chain Reaction"). Electrophoresis, which can be carried out simultaneously on large numbers of samples, separates proteins or DNA fragments according to their size (⬤▭◯, Application p. 573).

Using recombinant DNA technology, it is possible to cut a gene out of one organism and splice it into (*recombine* it with) the DNA of a second organism. Bacteria provide excellent hosts for recombinant DNA. Bacterial cells, unlike the cells of higher organisms, contain part of their DNA in circular pieces called *plasmids*, each of which carries just a few genes. Plasmids are extremely easy to isolate, several copies of each plasmid may be present in a cell, and each plasmid replicates through the normal base-pairing pathway. The ease of isolating and manipulating plasmids plus the rapid replication of bacteria create ideal conditions for production of recombinant DNA and the proteins whose synthesis it directs in bacteria.

To prepare a plasmid for insertion of a foreign gene, the plasmid is cut open with a bacterial enzyme, known as a *restriction endonuclease*, that recognizes a specific sequence in a DNA molecule and cleaves between the same two nucleotides in that sequence. For example, the restriction endonuclease *Eco*RI cuts a plasmid between G and A in the sequence G-A-A-T-T-C. This restriction enzyme makes its cut at the same spot in the sequence of both strands of the double-stranded DNA read in the same 5′ to 3′ direction. As a result, the cut is offset so that both DNA strands are left with a few unpaired bases on each end. These groups of unpaired bases are known as *sticky ends* because they are available to match up with complementary base sequences.

Recombinant DNA DNA that contains two or more DNA segments not found together in nature.

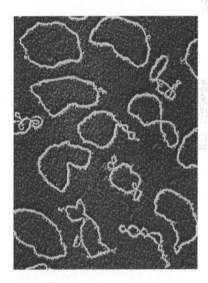

▲ Plasmids from the bacterium *Escherichia coli*, hosts for recombinant DNA.

APPLICATION ▶ Serendipity and the Polymerase Chain Reaction

Before the 1980s, studying DNA involved the frustration of working with very small, hard-to-obtain samples. Everyone wished there were a way to copy DNA, to make millions of copies of a sample. Their wish was granted one evening by Kary B. Mullis, a young biochemist who was mentally attacking a different problem—how to identify the sequence of nucleotides in DNA. In his own words, here is what happened:

> Sometimes a good idea comes to you when you are not looking for it. Through an improbable combination of coincidences, naiveté, and lucky mistakes, such a revelation came to me one Friday night in April 1983 as I gripped the steering wheel of my car and snaked along a moonlit road into northern California's redwood country. . . . I liked night driving; every weekend I went north to my cabin and sat still for three hours in the car, my hands occupied, my mind free. On that particular night I was thinking about my proposed DNA sequencing experiment.*

Mullis drove along, trying out in his mind and then rejecting various ways to approach the experiment, which required combining DNA polymerase, natural DNA (the target DNA), nucleoside triphosphates, and short synthetic nucleotide chains (*oligonucleotides*) in just the right way. Then,

> I was suddenly jolted by a realization: The strands of DNA in the target and the extended oligonucleotides would have the same base sequences. In effect, the mock reaction would have doubled the number of DNA targets in the sample! . . . Excited, I started running powers of two in my head: 2, 4, 8, 16, 32— I remembered vaguely that 2 to the 10th power was about 1000 and that therefore 2 to the 20th was around a million.*

The outcome of this night drive was development of the *polymerase chain reaction (PCR)*, now carried out automatically by instruments in every molecular biology lab. In 1993 Mullis shared the Nobel Prize in chemistry for this work. Today, PCR is so common and so simple a technique that it is routinely taught and carried out in undergraduate lab courses.

The goal of the PCR is to produce a large quantity of a specific segment of DNA. The DNA might be part of a genome study, it might be from a crime scene or a fossil, or it might be from a specimen preserved as a medical record. The raw materials for the reaction are the DNA that contains the nucleotide sequence to be amplified, *primers* (synthetic oligonucleotides with bases complementary to the base sequences on either side of the sequence of interest), the deoxyribonucleoside triphosphates that carry the four DNA bases, and DNA polymerase that will copy the DNA between the primers.

The reaction is carried out in three steps:

STEP 1: Heating of a DNA solution to cause the helix to unravel into single strands:

STEP 2: Addition of primers complementary to the DNA on either side of the single-stranded DNA sequence to be amplified. Because DNA polymerase copies DNA at the point where the double helix is unwinding, it is necessary to create double-stranded DNA at the point where copying is to start. The primers indicate this starting point:

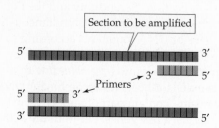

STEP 3: Extension of the primers by DNA polymerase to create double-stranded DNA identical to the original. The DNA polymerase starts copying at the ends of the primers so that the new DNA segment incorporates the primers:

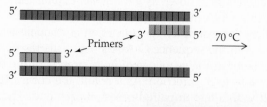

The reactants are combined in a closed container and the temperature cycled from about 90 °C for Step 1, to about

50 °C for Step 2, and to about 70 °C for Step 3. The temperature cycle requires only a few minutes and can be repeated over and over again for the same mixture. The first cycle produces two molecules of DNA; the second produces four molecules; and so on, with doubling at each cycle. Just 25 amplification cycles yield over 30 million copies of the original DNA segment.

Automation of the PCR was made possible by the discovery of a heat-stable polymerase (*Taq polymerase*) isolated from a bacterium that lives in hot springs. Because the enzyme survives the temperature needed for separating the DNA strands, it is not necessary to add fresh enzyme for each three-step cycle.

The quotations above are from "The Unusual Origin of the Polymerase Chain Reaction," Kary B. Mullis, Scientific American, April 1990, p. 56, which gives an extended account of the thought processes that led to the discovery.

See Additional Problems 27.50 and 27.51 at the end of the chapter.

Recombinant DNA is produced by cutting the two DNA segments to be combined with the same restriction endonuclease. The result is DNA fragments with sticky ends that are complementary to each other.

Consider a gene fragment that has been cut from human DNA and is to be inserted into a plasmid. The gene and the plasmid are both cut with the same enzyme, one that produces sticky ends. The sticky ends on the gene fragment are complementary to the sticky ends on the opened plasmid. The two are mixed in the presence of a DNA ligase enzyme that joins them together by re-forming their phosphodiester bonds and reconstitutes the now-altered plasmid.

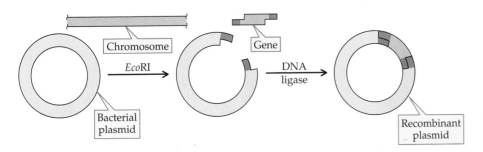

Once the altered plasmid is made, it is inserted back into a bacterial cell where the normal processes of transcription and translation take place to synthesize the protein encoded by the inserted gene. Since bacteria multiply rapidly, there are soon a large number of them, all containing the recombinant DNA and all manufacturing the protein encoded by the recombinant DNA. Huge numbers of the bacteria can be put to work as a protein factory.

As ideal as this strategy sounds, there are tremendous technical hurdles that have to be overcome before a protein manufactured in this way can be used commercially. One hurdle is getting the plasmid into a bacterium. Another is finding a host organism that does not post-translationally modify the protein you are trying to make; for example, yeast cells are known to attach carbohydrates to various amino acids in a protein, rendering the protein inactive. The most serious hurdle of all is isolation of the protein of interest from unwanted endotoxins. *Endotoxins* are potentially toxic natural compounds (usually structural components released when bacteria are lysed) found inside the host organism. Because the presence of even small amounts of endotoxins can lead to serious inflammatory responses, rigorous purification and screening protocols are necessary before the protein can be used in humans.

Despite the aforementioned obstacles, proteins manufactured in this manner have already reached the marketplace, and many more are on the way. Human insulin was the first such protein to become available. Others now include human growth hormone used for children who would otherwise be abnormally small, and blood clotting factors for hemophiliacs. A major advantage of this technology is that large amounts of these proteins can be made thus allowing their practical therapeutic use.

APPLICATION ▶ DNA Fingerprinting

A crime scene does not always yield fingerprints. It may, instead, yield samples of blood or semen or bits of hair. DNA analysis of such samples provides a new kind of "fingerprinting" for identifying criminals or proving suspects innocent.

DNA fingerprinting relies on finding variations between two or more DNA samples; for example, DNA isolated from a crime scene can be examined to determine if its variations match those of a suspect or victim. The naturally occurring variability of the base sequence in DNA is like a fingerprint. It is the same in all cells from a given individual and is sufficiently different from that of other individuals that it can be used for identification.

Throughout a DNA sample there are regions of noncoding DNA that contain repeating nucleotide sequences. The repetitive patterns used in DNA fingerprinting are known as *variable number tandem repeats (VNTRs)*. They are from 15 to 100 base pairs long, lie within or between genes, and consist of repetitive base pair patterns. For statistical significance, lab technicians examine the known VNTRs on several genes to create a DNA fingerprint. The probability of a fingerprint match with someone other than the correct individual is estimated at 1 in 1.5 billion.

Fingerprinting relies on use of a restriction endonuclease (an enzyme used to cut DNA) that recognizes a given VNTR

sequence and then cuts the DNA at either end of the repeating sequence. The general procedure is as follows:

- Digest the DNA sample with the restriction endonuclease.
- Separate the resulting DNA fragments according to their size by gel electrophoresis. (p. 573)
- Transfer the fragments to a nylon membrane (a *blotting* technique).
- Treat the blot with a radioactive DNA probe complementary to the sticky ends of the DNA fragments, so that the probe binds to its matching DNA fragments.
- Identify the locations of the now-radioactive fragments by exposing an X-ray film to the blot. (The film result of this procedure is known as an *autoradiogram*.)

An autoradiogram resembles a bar code, with dark bands arrayed in order of increasing molecular size of the DNA fragments. To compare the DNA of different individuals, the DNA samples are run in parallel columns on the same electrophoresis gel. In this way, the comparison is validated by having been run under identical conditions. The illustration below shows hypothetical DNA fingerprint patterns of six members of a family, where three of the children share the same mother and father and the fourth has been adopted. As you can see, even individuals in the same family will have distinguishably different DNA fingerprints; only identical twins have identical DNA fingerprints. There are always some similarities in the DNA patterns of offspring and their parents, making such fingerprints valuable in proving or disproving paternity.

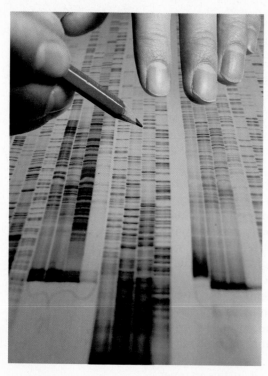

▲ Examination of an autoradiogram

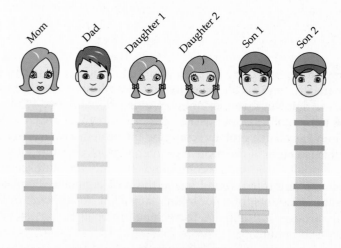

See Additional Problems 27.52 and 27.53 at the end of the chapter.

PROBLEM 27.3

A restriction enzyme known as *Bgl*II cuts DNA in the place marked below.

5′-A-//-G-A-T-C-T -3′

Draw the complementary 3′ to 5′ strand and show where it is cut by the same enzyme.

PROBLEM 27.4

Are the following base sequences "sticky" or not?

(a) A-C-G-G-A and T-G-C-C-T

(b) G-T-G-A-C and C-A-T-G-G

(c) G-T-A-T-A and A-C-G-C-G

27.5 Genomics: Using What We Know

Genomics has a simple and straightforward definition. It is the study of whole sets of genes and their functions. Genomics is inspiring studies that reach into all aspects of plant and animal life. For example, the study of bacterial genomics has been instrumental in linking the three domains of life—Archaea (formerly archeabacteria), Bacteria, and Eukarya—to one another from an evolutionary standpoint. The study of bacterial genomics is not only giving us a better understanding of how bacteria cause disease, it is also helping in the development of new therapies. Plant genomics is enhancing the value and utility of agricultural crops. The genomic study of farm animals is improving their health and availability. Humans will benefit from all of these studies, as well as those that contribute to their own health.

To glimpse where genomics is headed, we have provided descriptions of some of its applications in Table 27.3. These descriptions are not quite definitions. Many of these fields of study are so new that their territory is viewed differently by different individuals. At the opening of this chapter, we noted that we stand at the beginning of a revolution. You may well encounter some of the endeavors listed in Table 27.3 as the revolution proceeds.

Genomics The study of whole sets of genes and their functions.

Genetically Modified Plants and Animals

The development of new varieties of plants and animals has been proceeding for centuries as the result of natural accidents and occasional success in the hybridization of known varieties. The techniques for mapping, studying, and modifying human genes apply equally as well to plants and animals. Their application can greatly accelerate our ability to generate crop plants and farm animals with desirable characteristics and lacking undesirable ones.

Some genetically modified crops have already been planted in large quantities in the United States. Each year millions of tons of corn are destroyed by a caterpillar (the European corn borer) that does its damage deep inside the corn stalk and out of reach of pesticides. To solve this problem, a bacterial gene (from *Bacillus thuringiensis*, Bt) has been transplanted into corn. The gene causes the corn to produce a toxin that kills the caterpillars. In 2000, one-quarter of all corn planted in the United States was Bt corn. Soybeans genetically modified to withstand herbicides are also widely grown. The soybean crop remains unharmed when the surrounding weeds are killed by the herbicide.

Tests are under way with genetically modified coffee beans that are caffeine-free, potatoes that absorb less fat when they are fried, and "Golden Rice," a yellow

TABLE 27.3   Genomics-Related Fields of Study

Biotechnology

A collective term for the application of biological and biochemical research to the development of products that improve the health of humans, other animals, and plants.

Bioinformatics

The use of computers to manage and interpret genomic information, and to make predictions about biological systems. Applications of bioinformatics include studies of individual genes and their functions, drug design, and drug development.

Functional genomics

Use of genome sequences to solve biological problems.

Comparative genomics

Comparison of the genome sequences of different organisms to discover regions with similar functions and perhaps similar evolutionary origins.

Proteomics

Study of the complete set of proteins coded for by a genome or synthesized within a given type of cell, including the quest for an understanding of the role of each protein in healthy or diseased conditions. This understanding has potential application in drug design and is being pursued by more than one commercial organization.

Pharmacogenomics

The genetic basis of responses to drug treatment. Goals include the design of more effective drugs and an understanding of why certain drugs work in some patients but not in others.

Pharmacogenetics

The matching of drugs to individuals based on the content of their personal genome in order to avoid administration of drugs that are ineffective or toxic and focus on drugs that are most effective for that individual.

Toxicogenomics

A newly developing application that combines genomics and bioinformatics in studying how toxic agents affect genes and in screening possibly harmful agents.

Genetic engineering

Alteration of the genetic material of a cell or an organism. The goals may be to make the organism produce new substances or perform new functions. Examples are introduction of a gene that causes bacteria to produce a desired protein or allows a crop plant to withstand the effects of a pesticide that repels harmful insects.

Gene therapy

Alteration of an individual's genetic makeup with the goal of curing or preventing a disease.

Bioethics

The ethical implications of how knowledge of the human genome is used.

▲ "Golden Rice" has been genetically modified to provide vitamin A.

rice that provides the vitamin A desperately needed in poor populations where insufficient vitamin A causes death and blindness.

Fish farming is an expanding industry as natural populations of fish diminish. There are genetically engineered salmon that can grow to 7–10 pounds, a marketable size, in up to one-half the time of their unmodified cousins. Similar genetic modifications are anticipated for other varieties of fish. And there is the prospect of cloning leaner pigs.

Will genetically modified plants and animals intermingle with natural varieties and cause harm to them? Should food labels state whether the food contains genetically modified ingredients? Might unrecognized harmful substances enter the food supply? These are hotly debated questions and have lead to the establishment of the Non-GMO Project, where the GMO stands for genetically modified organism. The goal of this project is to offer consumers a non-GMO choice for organic and natural products that are produced without genetic engineering or recombinant DNA technologies. Many foods found in stores are labeled "Non-GMO."

But genetic modifications can also be used to produce previously unseen beauty. Consider the blue rose shown on the cover of this text. Suntory Limited, in a joint venture with Florigene, has recently been able to successfully implant into roses the gene that leads to the synthesis of blue pigments. It is expected that the first blue roses will be sold in Japan in early 2009. Even more exciting is the expectation that the introduction of blue pigments into roses will lead to an explosion in the variety of possible rose colors available to the average consumer.

Gene Therapy

Gene therapy is based on the premise that replacement of a disease-causing gene with the healthy gene will cure or prevent a disease. The most clear-cut expectations for gene therapy lie in treating monogenic diseases, those that result from flawed DNA in a single gene.

The focus has been on using viruses as *vectors,* the agents that deliver therapeutic quantities of DNA directly into cell nuclei. The expectation was that this method could result in lifelong elimination of an inherited disease, and many studies have been undertaken. Unfortunately, expectations remain greater than achievements thus far. Investigations into the direct injection of "naked DNA" have begun, with one early report of success in encouraging blood vessel growth in patients with inadequate blood supply to their hearts. The Food and Drug Administration (FDA) has, as of August 2008, not yet approved any human gene therapy product for sale. Current gene therapy is still experimental and has not proven to be widely successful in clinical trials. Although little progress has been made since the first gene therapy clinical trial began in 1990, vigorous research into this area continues as new approaches continue to be examined.

A Personal Genomic Survey

Suppose that during the examination prior to diagnosis and treatment for a health problem that a patient's entire genome could be surveyed. What benefits might result? One possibility is that the choice of drugs could be directed toward those that are most effective for that individual. It is no secret that not everyone reacts in the same manner to a given medication. Perhaps the patient lacks an enzyme needed for a drug's metabolism. It is known, for example, that codeine is ineffective as a pain killer in people who lack the enzyme that converts codeine into morphine, which is the active analgesic (, p. 481). Perhaps the patient has a *monogenic defect,* a flaw in a single gene that is the direct cause of the disease. Such a patient might, at some time in the future, be a candidate for gene therapy.

In cancer therapy there may be advantages in understanding the genetic differences between normal cells and tumor cells. Such knowledge could assist in chemotherapy, where the goal is use of an agent that kills the tumor cells but does the least possible amount of harm to noncancerous cells.

Another application of human genetic information may arise from genetic screening of infants. The immediate use of gene therapy might eliminate the threat of a monogenically based disease. Or perhaps a lifestyle adjustment would be in order for an individual with one or more SNPs that predict a susceptibility to heart disease, diabetes, or some other disease that results from combinations of genetic and environmental influences. And consider that once done, an individual's genetic map would be available for the rest of their life. Perhaps someday we may even carry a wallet card encoded with our genetic information. With this knowledge, however, also come ethical dilemmas that have made this use of genomics a hotly debated topic.

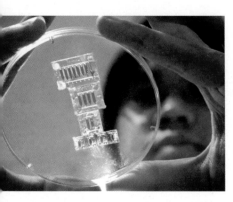

▲ A DNA chip used for genetic screening.

Snips and Chips

Our understanding of SNPs is already at work in screening implemented by DNA chips. A chip can be used, for example, to screen for the polymorphism that wipes out the analgesic effect of codeine. Or consider that a gene with several polymorphisms codes for an enzyme that metabolizes cardiovascular agents, antipsychotics, and many other drugs. Genomic screening can determine whether an individual with polymorphisms in that gene will get no effect from a drug, the expected effect, or perhaps have a greater-than-normal response to the drug. Eventually, screening tests for polymorphisms of this enzyme will be a diagnostic test carried out by a DNA chip in a doctor's office. The results will aid in choosing the right drug and dosage.

DNA chip screening has already revealed the genetic variations responsible for two types of pediatric leukemia, a distinction that could not be made by examining diseased cells under the microscope. Because the two leukemias require quite different therapies, use of the chip to identify the types is a valuable development.

A DNA chip is a solid support bearing large numbers of short, single-stranded bits of DNA of known composition. The DNA is organized on the chip in whatever manner is best for a particular type of screening, for example, to identify the presence or absence of polymorphisms. A sample to be screened is labeled with a fluorescent tag and applied to the chip. During an incubation period, the sample will bond to the DNA segments with complementary nucleic acid sequences. Then, the fluorescence is read to discover where the bonding has occurred and what DNA variations are present.

Bioethics

We can mention only briefly an area of major concern that arises from the revolution in genomics. This concern is not chemical, nor is it directly related to curing and preventing disease. The existence of this concern is recognized in the ELSI program of the National Human Genome Research Institute. ELSI deals with the Ethical, Legal, and Social Implications of human genetic research. The scope of ELSI is broad and thought-provoking. It deals with many questions such as the following:

- Who should have access to personal genetic information and how will it be used?
- Who should own and control genetic information?
- Should genetic testing be performed when no treatment is available?
- Are disabilities diseases? Do they need to be cured or prevented?
- Preliminary attempts at gene therapy are exorbitantly expensive. Who will have access to these therapies? Who will pay for their use?
- Should we re-engineer the genes we pass on to our children?

If you are interested in the ELSI program, their web page is an excellent resource (http://www.genome.gov/10001618).

> **PROBLEM 27.5**
>
> Classify the following activities according to the fields of study listed in Table 27.3.
>
> **(a)** Identification of genes that perform identical functions in mice and humans
>
> **(b)** Creation of a variety of wheat that will not be harmed by an herbicide that kills weeds that threaten wheat crops

(c) Screening of an individual's genome to choose the most appropriate pain-killing medication for that person

(d) Computer analysis of base sequence information from groups of people with and without a given disease to discover where the disease-causing polymorphism lies

SUMMARY: REVISITING THE CHAPTER GOALS

1. **What is the working draft of the human genome and the circumstances of its creation?** The Human Genome Project, an international consortium of not-for-profit institutions, and Celera Genomics, a for-profit company, have both announced completion of working drafts of the human genome. With the exception of large areas of repetitive DNA, the DNA base sequences of all chromosomes have been examined. The Human Genome Project utilized successive maps of DNA segments of known location. Celera began by fragmenting all of the DNA. In both groups the fragments were cloned, labeled, ordered, and the individual maps assembled by computers. The results of the two projects are generally supportive of each other. There are about three billion base pairs and 20,000–25,000 genes, each able to direct the synthesis of more than one protein. The bulk of the genome consists of noncoding, repetitive sequences. About 200 of the human genes are identical to those in bacteria.

2. **What are the various segments along the length of the DNA in a chromosome?** Telomeres, which fall at the ends of chromosomes, are regions of noncoding, repetitive DNA that protect the ends from accidental changes. At each cell division, the telomeres are shortened, with significant shortening associated with senescence and death of the cell. In cancer cells, telomerase remains active in telomere synthesis. Centromeres are the constricted regions of chromosomes that form during cell division and also carry noncoding DNA. Exons are the protein coding regions of DNA and together make up the genes that direct protein synthesis. The repetitive noncoding segments of DNA are of either no function or unknown function.

3. **What are mutations?** A *mutation* is an error in the base sequence of DNA that is passed along during replication. Mutations arise by random error during replication but may also be caused by ionizing radiation, viruses, or chemical agents (*mutagens*). Mutations cause inherited diseases and the tendency to acquire others.

4. **What are polymorphisms and single nucleotide polymorphisms (SNPs) and how can identifying them be useful?** A polymorphism is a variation in DNA that is linked to a trait within a population. A SNP is the replacement of one nucleotide by another. The result might be the replacement of one amino acid by another in a protein, no change because the new codon specifies the same amino acid, or the introduction of a stop codon. Many inherited diseases are known to be caused by SNPs; they can also be beneficial. Understanding the location and effect of SNPs is expected to lead to new therapies.

5. **What is recombinant DNA?** Recombinant DNA is produced by combining DNA segments that do not normally occur together. A gene from one organism is inserted into the DNA of another organism. Recombinant DNA is conveniently created in plasmids (circular DNA) from bacteria. Bacteria carrying these plasmids then serve as factories for the synthesis of large quantities of the encoded protein.

6. **What does the future hold for uses of genomic information?** Mapping the human genome holds major promise for applications in health and medicine. Drugs can be precisely chosen based on a patient's own DNA, thereby avoiding drugs that are ineffective or toxic for that individual. Perhaps one day inherited diseases will be prevented or cured by gene therapy. By genetic modification of crop plants and farm animals, the productivity, marketability, and health benefits of these products can be enhanced. Progress in each of these areas is bound to be accompanied by controversy and ethical dilemmas.

KEY WORDS

Clones, *p. 843*

Centromeres, *p. 844*

Genomics, *p. 853*

Mutagen, *p. 846*

Mutation, *p. 846*

Polymorphisms, *p. 847*

Recombinant DNA, *p. 849*

Single-nucleotide polymorphism (SNP), *p. 847*

Telomeres, *p. 843*

UNDERSTANDING KEY CONCEPTS

27.6 What steps are necessary in the mapping of the human genome, as outlined by the Human Genome Project?

27.7 Clearly, all humans have variations in their DNA sequences. How is it possible to sequence the human genome if every individual is unique? How was the diversity of the human genome addressed?

27.8 List the types of noncoding DNA. Give the function of each, if it is known.

27.9 What are the similarities and differences between mutations and polymorphisms?

27.10 What is recombinant DNA? How can it be used to produce human proteins in bacteria?

27.11 Identify some major potential benefits of the applications of genomics and some major negative outcomes.

ADDITIONAL PROBLEMS

THE HUMAN GENOME MAP

27.12 How did the private corporation Celera Genomics approach the sequencing of the human genome? What was the advantage of this approach?

27.13 How did the competition that developed between the groups developing the human genome map benefit the Human Genome Project?

27.14 Approximately what portion of the human genome is composed of repeat sequences?

27.15 Approximately how many base pairs were identified in the human genome working drafts?

27.16 Among results of the genome working drafts, (a) were any human genes found to be identical to genes in bacteria and (b) what was learned about the number of proteins produced by a given gene?

27.17 What is the most surprising result found thus far in the human genome studies?

27.18 You may have heard of Dolly, the cloned sheep grown from an embryo created in a laboratory. In the context of DNA mapping, what are clones and what essential role do they play?

CHROMOSOMES, MUTATIONS, AND POLYMORPHISMS

27.19 What is thought to be the primary purpose of telomeres?

27.20 How is the age of a cell predicted by its telomeric sequences?

27.21 What is the role of the enzyme telomerase? In what kind of cell is it normally most active and most inactive?

27.22 What is the centromere?

27.23 What is a mutagen?

27.24 What is a silent mutation?

27.25 Why is a mutation of a base in a DNA sequence much more serious than a mutation in a transcribed mRNA sequence?

27.26 What are the two general and common ways that mutations occur in a DNA sequence?

27.27 What is a SNP?

27.28 How are SNPs linked to traits in individual human beings?

27.29 List some potential biological effects of SNPs.

27.30 What would be a medical advantage of having a catalog of SNPs?

27.31 Does a single base pair substitution in a strand of DNA always result in a new amino acid in the protein coded for by that gene? Why or why not?

27.32 What determines the significance of a change in the identity of an amino acid in a protein?

27.33 Compare the severity of DNA mutations that produce the following changes in mRNA codons:
  (a) UUU to UCU
  (b) CUU to CUG

27.34 Compare the severity of DNA mutations that produce the following changes in mRNA codons:
  (a) GUC to GCC
  (b) CCC to CAC

RECOMBINANT DNA

27.35 Why are bacterial plasmids the preferred host for DNA sequences that are to be cloned?

27.36 What is an advantage of using recombinant DNA to make proteins such as insulin, human growth hormone, or blood clotting factors?

27.37 What are some of the hurdles that have to be overcome to make recombinant gene products commercially available?

27.38 In the formation of recombinant DNA, a restriction endonuclease cuts a bacterial plasmid to give sticky ends. The DNA segments that are to be added to the plasmid are cleaved with the same restriction endonuclease. What are sticky ends and why is it important that the target DNA and the plasmid it will be incorporated in have complementary sticky ends?

27.39 Give the sequence of unpaired bases that would be sticky with the following sequences:
  (a) CCATG
  (b) TGGGT
  (c) ACACA

27.40 Are the following base sequences sticky or not sticky?
  (a) TTAGC and AAACG
  (b) CGTACG and GTACGT

GENOMICS

27.41 How might the work of a person who practices pharmacogenomics differ from that of a pharmacogeneticist?

27.42 What is proteomics and how might it benefit health care?

27.43 Genetic engineering and gene therapy are similar fields within genomics. What do they have in common and what distinguishes them?

27.44 Provide two examples of genetically engineered crops that are improvements over their predecessors.

27.45 Imagine that you become a parent in an age when a full genetic workup is available for every baby. What advantages and disadvantages might there be to having this information?

27.46 What type of technology might be used to diagnose inherited diseases in the doctor's office of the future?

27.47 What is a DNA chip?

Applications

27.48 Whose DNA was used by scientists working on the two projects that sequenced the human genome? Why was this important? [*Whose Genome Is It?, p. 844*]

27.49 Explain why using either the DNA of a single individual or the DNA of a group of individuals from the same ethnic group would have been a bad choice in mapping the human genome. [*Whose Genome Is It?, p. 844*]

27.50 What is the purpose of a polymerase chain reaction? [*Serendipity and the Polymerase Chain Reaction, p. 850*]

27.51 Briefly describe how a polymerase chain reaction works. [*Serendipity and the Polymerase Chain Reaction, p. 850*]

27.52 State the five basic steps of DNA fingerprinting. Why is it important that DNA fingerprinting techniques be standardized? [*DNA Fingerprinting, p. 852*]

27.53 Why is it possible in DNA fingerprinting to compare segments of DNA from different tissues (for example, semen and blood samples)? [*DNA Fingerprinting, p. 852*]

General Questions and Problems

27.54 What is a monogenic disease?

27.55 What is the role of a vector in gene therapy?

27.56 Write the base sequence that would be sticky with the sequence T-A-T-G-A-C-T.

27.57 If the DNA sequence A-T-T-G-G-C-C-T-A on an informational strand mutated and became A-C-T-G-G-C-C-T-A, what effect would the mutation have on the sequence of the protein produced?

27.58 In general terms, what is the cause of hereditary diseases?

27.59 In the DNA of what kind of cell must a mutation occur for the genetic change to be passed down to future generations?

Protein and Amino Acid Metabolism

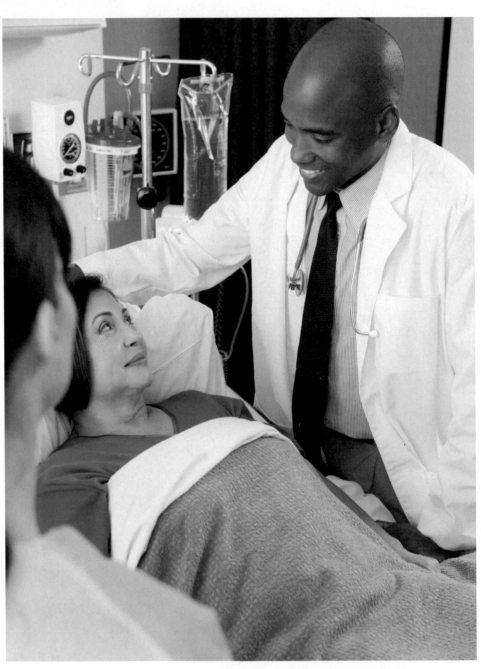

▲ Understanding the metabolic basis of disease can lead to new and more effective treatments in medicine.

CONTENTS

In this chapter, we will answer the following questions about the metabolism of proteins and amino acids:

1. **What happens during the digestion of proteins, and what are the fates of the amino acids?**

   THE GOAL: Be able to list the sequence of events in the digestion of proteins, and describe the nature of the amino acid pool.

2. **What are the major strategies in the catabolism of amino acids?**

   THE GOAL: Be able to identify the major reactions and products of amino acid catabolism and the fate of the products.

3. **What is the urea cycle?**

   THE GOAL: Be able to list the major reactants and products of the urea cycle.

4. **What are the essential and nonessential amino acids, and how, in general, are amino acids synthesized?**

   THE GOAL: Be able to define essential and nonessential amino acids, and describe the general strategy of amino acid biosynthesis.

B efore discussing protein and amino acid metabolism, a review of what we have already covered on proteins and amino acids is in order. We discussed the structures of the amino acids and the various levels of protein structure in Chapter 18; the essential function of proteins as enzymes in Chapter 19; and their function within cell membranes in Section 24.8. The roles of amino acids and proteins as hormones or precursors to neurotransmitter synthesis were examined in Chapter 20, and the biosynthesis of proteins as directed by the genetic code in Chapter 26. The remaining major aspect of protein biochemistry, to which we now turn, is the metabolic fate of proteins and ultimately amino acids. Although we have the biochemical machinery necessary to make them, the hydrolysis of dietary protein is our major source for amino acids.

28.1 Digestion of Protein

The end result of protein digestion is simple—the hydrolysis of all peptide bonds to produce a collection of amino acids:

Hydrolysis of peptide bonds

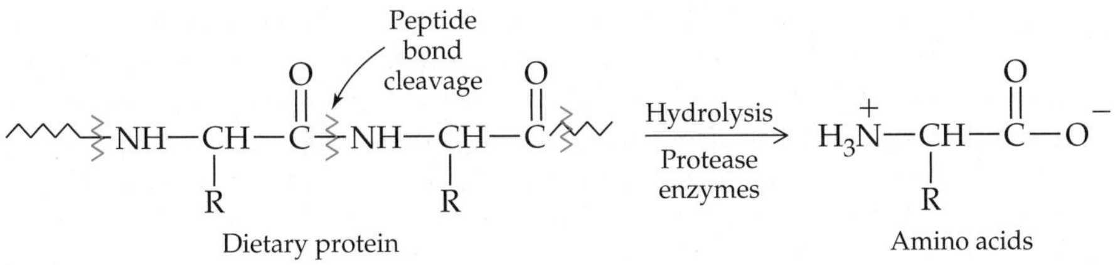

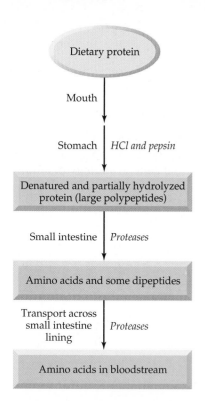

▲ **FIGURE 28.1 Digestion of proteins.**

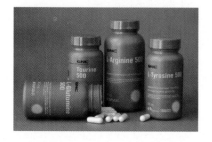

▲ Individual amino acids are promoted for a variety of unproven health benefits. Because amino acids are classified as foods, they need not undergo the stringent testing for purity, safety, and efficacy required for FDA approval.

Amino acid pool The entire collection of free amino acids in the body.

For example:

Alanine   Serine   Phenylalanine

Figure 28.1 summarizes the digestive processes involved in the conversion of protein to amino acids. The breakdown of protein begins in the mouth, where large pieces of food are converted into smaller, more digestible portions. Although no chemical digestion of the protein has begun, this step is necessary to increase the surface area of the food to be digested. The chemical digestion of dietary proteins begins with their denaturation in the strongly acidic environment of the stomach (pH 1–2), where the tertiary and secondary structures of consumed proteins begin to unfold. In addition to hydrochloric acid, gastric secretions include pepsinogen, a zymogen that is activated by acid to give the enzyme pepsin. Unlike most proteins, pepsin is stable and active at pH 1–2. Protein hydrolysis begins as pepsin breaks some of the peptide bonds in the denatured proteins, producing polypeptides.

The polypeptides produced by pepsin then enter the small intestine, where the pH is about 7–8. Pepsin is inactivated in the less acidic environment, and a group of pancreatic zymogens is secreted. The activated enzymes (proteases such as trypsin, chymotrypsin, and carboxypeptidase) then take over further hydrolysis of peptide bonds in the partially digested proteins.

The combined action of the pancreatic proteases in the small intestine and other proteases in the cells of the intestinal lining completes the conversion of dietary proteins into free amino acids. After active transport across cell membranes lining the intestine, the amino acids are absorbed directly into the bloodstream.

The active transport of amino acids into cells is managed by several transport systems devoted to different groups of amino acids. For this reason, an excess of one amino acid in the diet can dominate the transport and produce a deficiency of others. This condition usually arises only in individuals taking large quantities of a single amino acid dietary supplement, such as those often sold in health food stores.

28.2 Amino Acid Metabolism: An Overview

The entire collection of free amino acids throughout the body—the **amino acid pool**—occupies a central position in protein and amino acid metabolism (Figure 28.2). All tissues and biomolecules in the body are constantly being degraded, repaired, and replaced. Cells throughout the body have the enzymes for hydrolysis

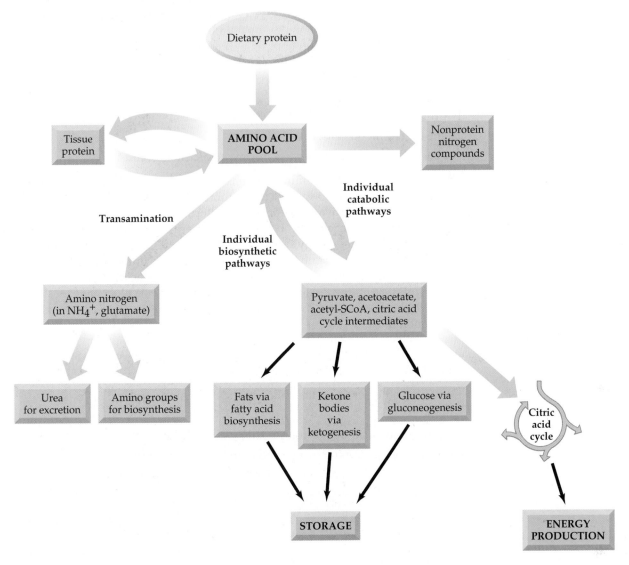

▲ **FIGURE 28.2** **Protein and amino acid metabolism.** Amino acids move in and out of the amino acid pool, as shown. Amino acids and their catabolic products are used for synthesis of other biomolecules as well as for production of energy.

of waste protein, and a healthy adult turns over about 300 g of protein every day. Thus, amino acids are continuously entering the pool, not only from digestion but also from the breakdown of old protein, and are continuously being withdrawn for synthesis of new nitrogen-containing biomolecules.

Each of the 20 amino acids is degraded via its own unique pathway. The important point to remember is that the general scheme is the same for each one.

General scheme for amino acid catabolism:

- Removal of the amino group
- Use of nitrogen in synthesis of new nitrogen compounds
- Passage of nitrogen into the urea cycle
- Incorporation of the carbon atoms into compounds that can enter the citric acid cycle

Our bodies do not store nitrogen-containing compounds and ammonia is toxic to cells. Therefore, the amino nitrogen from dietary protein has just two possible fates. Amino nitrogen must either be incorporated into urea and excreted, or be used in the synthesis of new nitrogen-containing compounds, including the following types of biomolecules:

Nitric oxide (NO, a chemical messenger, p. 475)

Hormones (Section 20.4)

Neurotransmitters (Section 20.6)

Nicotinamide (in NAD^+ and $NADP^+$; Section 21.7)

Heme (in red blood cells; p. 682)

Purine and pyrimidine bases (for nucleic acids; Section 26.2)

The carbon portion of the amino acid has a much more varied fate. The carbon atoms of amino acids are converted to compounds that can enter the citric acid cycle; from there they are available for several alternative pathways. They continue through the citric acid cycle (the body's main energy-generating pathway; Section 21.8) to give CO_2 and energy stored in ATP. About 10–20% of our energy is normally produced in this way from amino acids. If not needed immediately for energy, the carbon-carrying intermediates produced from amino acids enter storage as triacylglycerols (via lipogenesis, Section 25.8) or glycogen (via gluconeogenesis and glycogen synthesis, Sections 23.10, 23.11). They can also be converted to ketone bodies (Section 25.7).

PROBLEM 28.1

Decide whether each of the following statements is true or false. If false, explain why.

(a) The amino acid pool is found mainly in the liver.

(b) Nitrogen-containing compounds can be stored in fatty tissue.

(c) Some hormones and neurotransmitters are synthesized from amino acids.

(d) Amino groups can be stored in fatty tissue.

(e) Glycine is an essential amino acid because it is present in every protein.

KEY CONCEPT PROBLEM 28.2

Serotonin is a monoamine neurotransmitter. It is formed in the body from the amino acid tryptophan (Figure 20.6, p. 643). What class of enzyme catalyzes each of the two steps that converts tryptophan to serotonin?

28.3 Amino Acid Catabolism: The Amino Group

Transamination The interchange of the amino group of an amino acid and the keto group of an α-keto acid.

The first step in amino acid catabolism is removal of the amino group. In this process, known as **transamination**, the amino group of the amino acid and the keto group of an α-keto acid change places:

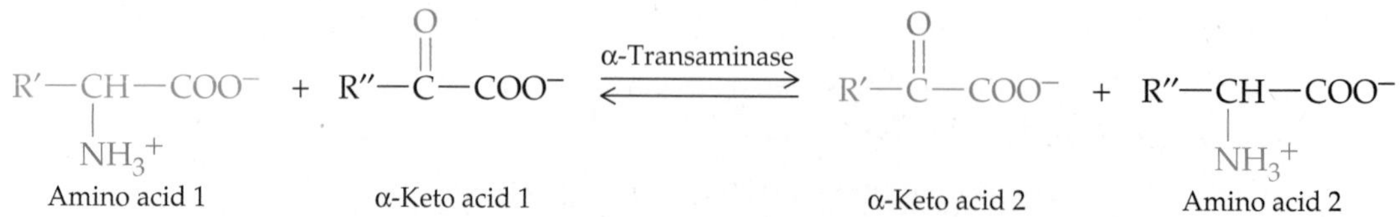

$$\underset{\substack{\text{Amino acid 1}}}{R'\!-\!\underset{\substack{| \\ NH_3^+}}{CH}\!-\!COO^-} + \underset{\substack{\text{α-Keto acid 1}}}{R''\!-\!\overset{\substack{O \\ ||}}{C}\!-\!COO^-} \underset{}{\overset{\text{α-Transaminase}}{\rightleftharpoons}} \underset{\substack{\text{α-Keto acid 2}}}{R'\!-\!\overset{\substack{O \\ ||}}{C}\!-\!COO^-} + \underset{\substack{\text{Amino acid 2}}}{R''\!-\!\underset{\substack{| \\ NH_3^+}}{CH}\!-\!COO^-}$$

A number of transaminase enzymes are responsible for "transporting" (hence the prefix *trans-*) an amino group from one molecule to another. Most are specific for α-ketoglutarate as the amino group acceptor and work with several amino acids.

The α-ketoglutarate is converted to glutamate, and the amino acid is converted to an α-keto acid. For example, alanine is converted to pyruvate by transamination:

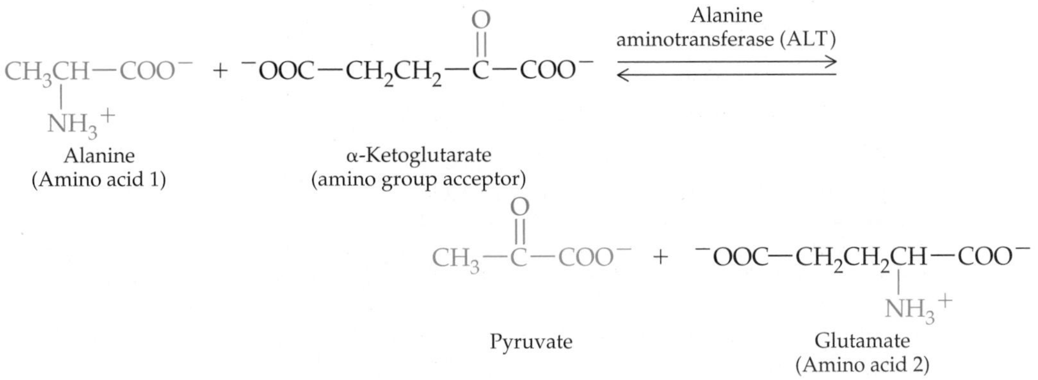

$$CH_3CH-COO^- \quad + \quad ^-OOC-CH_2CH_2-\overset{\overset{\displaystyle O}{\|}}{C}-COO^- \quad \underset{\underset{\displaystyle \text{aminotransferase (ALT)}}{\longleftarrow}}{\overset{\overset{\displaystyle \text{Alanine}}{\longrightarrow}}{\rule{0pt}{0pt}}}$$

$$\underset{\text{NH}_3^+}{\overset{|}{}}$$

Alanine $\qquad\qquad\qquad$ α-Ketoglutarate
(Amino acid 1) $\qquad\qquad$ (amino group acceptor)

$$CH_3-\overset{\overset{\displaystyle O}{\|}}{C}-COO^- \quad + \quad ^-OOC-CH_2CH_2CH-COO^-$$

$$\underset{\text{NH}_3^+}{\overset{|}{}}$$

Pyruvate $\qquad\qquad\qquad$ Glutamate
$\qquad\qquad\qquad\qquad\qquad$ (Amino acid 2)

The enzyme for this conversion, alanine aminotransferase (ALT), is especially abundant in the liver, and above-normal ALT concentrations in the blood are taken as an indication of liver damage that has allowed ALT to leak into the bloodstream.

Transamination is a key reaction in many biochemical pathways, where it interconverts amino acid amino groups and carbonyl groups as necessary. The transamination reactions are reversible and go easily in either direction, depending on the concentrations of the reactants. In this way, amino acid concentrations are regulated by keeping synthesis and breakdown in balance. For example, the reaction of pyruvate with glutamate (the reverse of the preceding reaction) is the main synthetic route for alanine.

The glutamate from transamination serves as an amino group carrier. Glutamate can be used to provide amino groups for the synthesis of new amino acids, but most of the glutamate formed in this way is recycled to regenerate α-ketoglutarate. This process, known as **oxidative deamination**, oxidatively removes the glutamate amino group as ammonium ion to give back α-ketoglutarate:

Oxidative deamination Conversion of an amino acid $-NH_2$ group to an α-keto group, with removal of NH_4^+.

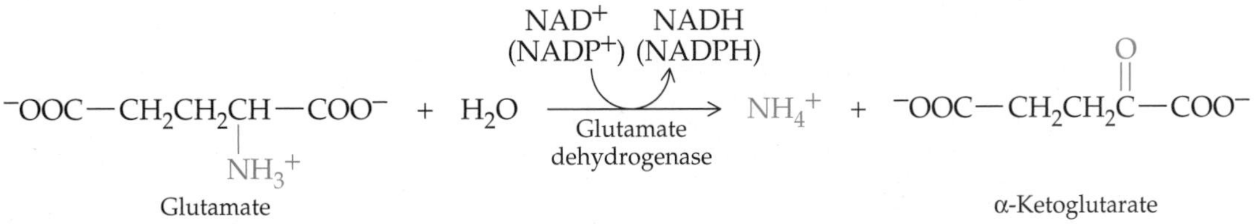

$$^-OOC-CH_2CH_2CH-COO^- \quad + \quad H_2O \quad \xrightarrow[\substack{\text{Glutamate} \\ \text{dehydrogenase}}]{\substack{\text{NAD}^+ \quad \text{NADH} \\ \text{(NADP}^+\text{)} \quad \text{(NADPH)}}} \quad NH_4^+ \quad + \quad ^-OOC-CH_2CH_2\overset{\overset{\displaystyle O}{\|}}{C}-COO^-$$

$$\underset{\text{NH}_3^+}{\overset{|}{}}$$

Glutamate $\qquad\qquad\qquad\qquad\qquad\qquad\qquad\qquad\qquad\qquad\qquad\qquad$ α-Ketoglutarate

The ammonium ion formed in this reaction proceeds to the urea cycle where it is eliminated in the urine as urea. The pathway of nitrogen from an amino acid to urea is summarized in Figure 28.3.

WORKED EXAMPLE **28.1** Predicting Transamination Products

The blood serum concentration of the transaminase from heart muscle, aspartate aminotransferase (AST), is used in the diagnosis of heart disease because the enzyme escapes into the serum from damaged heart cells. AST catalyzes transamination of aspartate with α-ketoglutarate. What are the products of this reaction?

ANALYSIS The reaction is the interchange of an amino group from aspartate with the keto group from α-ketoglutarate. We know that α-ketoglutarate always gives glutamate in transamination, so one product is glutamate. The product from the amino acid will have a keto group instead of the amino group; we need to consider various amino acid structures to identify a candidate.

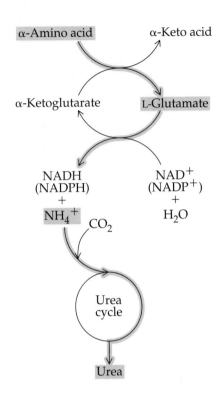

▲ **FIGURE 28.3 Pathway of nitrogen from an amino acid to urea.** The nitrogen-bearing compounds and their pathway are highlighted in orange.

Consulting Table 18.3 (which lists the 20 protein amino acid structures), we see that the structure of aspartate (aspartic acid) is

$$^-OOC-CH_2CH-COO^-$$

with α labeled on the α carbon and NH_3^+ circled below.

Aspartate

Removing the $-NH_3^+$ and $-H$ groups bonded to the α carbon and replacing them by a $=O$ gives the desired α-keto acid, which in this case happens to be oxaloacetate:

$$^-OOC-CH_2-\overset{\overset{O}{\|}}{C}-COO^-$$

Oxaloacetate

SOLUTION
The overall reaction is therefore

Aspartate + α-Ketoglutarate → Oxaloacetate + Glutamate

PROBLEM 28.3

What are the structure and the IUPAC name of the α-keto acid formed by transamination of the amino acid threonine? (Refer to Table 18.3.)

PROBLEM 28.4

What is the product of the reaction

$$CH_3-S-CH_2CH_2CH-COO^- \xrightarrow{\substack{\text{α-Ketoglutarate} \quad \text{Glutamate}}} \; ?$$
with NH_3^+ on the α carbon

PROBLEM 28.5

Explain how the conversion of alanine to pyruvic acid (pyruvate) can be identified as an oxidation reaction.

PROBLEM 28.6

Unlike most amino acids, branched-chain amino acids are broken down in tissues other than the liver. Identify the three amino acids with branched-chain R groups (⊂⊃, Table 18.3, p. 557). For any one of these amino acids, write the equation for its transamination.

28.4 The Urea Cycle

Ammonia is highly toxic to living things and must be eliminated in a way that does no harm. Fish are able to excrete ammonia through their gills directly into their watery surroundings where it is immediately diluted and its toxic effects effectively neutralized. Since mammals do not live in an environment where this immediate dilution is feasible, they must find other ways to get rid of ammonia. Excretion of ammonia in urine is not feasible for mammals, because the volume of water needed to accomplish this safely would cause dehydration. Mammals must first convert ammonia, in solution as ammonium ion, to nontoxic urea via the **urea cycle**.

The conversion of ammonium ion to urea takes place in the liver. From there, the urea is transported to the kidneys and transferred to urine for excretion. Like many other biochemical pathways, urea formation begins with an energy investment. Ammonium ion (from oxidative deamination of amino acids), bicarbonate ion (from carbon dioxide from the citric acid cycle), and ATP combine to form carbamoyl phosphate. This reaction, like the citric acid cycle, takes place in the mitochondrial matrix. Two ATPs are invested and one phosphate is transferred to form the carbamoyl phosphate (an energy-rich phosphate-like ATP):

Urea cycle The cyclic biochemical pathway that produces urea for excretion.

▲ Fish do not need to convert ammonia to urea for elimination because it is quickly diluted in the surrounding water; this is why the water in fish tanks must be constantly monitored to ensure that ammonia levels do not reach toxic levels.

$$NH_4^+ + HCO_3^- \xrightarrow[\text{phosphate synthetase I}]{\text{Carbamoyl}} \underset{\text{Carbamoyl phosphate}}{H_3\overset{+}{N}-\overset{\overset{O}{\|}}{C}-O-PO_3^{2-}} + HOPO_3^{2-} + H_2O$$

(over the arrow: 2 ATP, 2 ADP)

Carbamoyl phosphate next reacts in the first step of the four-step urea cycle, shown in Figure 28.4.

STEPS 1 AND 2 OF THE UREA CYCLE: Building Up a Reactive Intermediate The first step of the urea cycle transfers the carbamoyl group, $H_2NC=O$, from carbamoyl phosphate to ornithine, an amino acid not found in proteins, to give citrulline, another nonprotein amino acid. This exergonic reaction introduces the first urea nitrogen into the urea cycle.

Next, a molecule of aspartate combines with citrulline in a reaction driven by conversion of ATP to AMP and pyrophosphate ($P_2O_7^{4-}$), followed by the additional exergonic hydrolysis of pyrophosphate. Both nitrogen atoms destined for elimination as urea are now bonded to the same carbon atom in argininosuccinate (red C atom in Figure 28.4).

STEPS 3 AND 4 OF THE UREA CYCLE: Cleavage and Hydrolysis of the Step 2 Product Step 3 cleaves argininosuccinate into two pieces: arginine, an amino acid, and fumarate, which you may recall is an intermediate in the citric acid cycle (Figure 21.9, p. 678). Now all that remains, in Step 4, is hydrolysis of arginine to give urea and regenerate the reactant in Step 1 of the cycle, ornithine.

Net Result of the Urea Cycle

$$HCO_3^- + NH_4^+ + 3\,ATP + \underset{\text{Aspartate}}{{}^-OOC-CH_2-\underset{\underset{NH_3^+}{|}}{CH}-COO^-} + 2\,H_2O \longrightarrow$$

$$\underset{\text{Urea}}{H_2N-\overset{\overset{O}{\|}}{C}-NH_2} + 2\,ADP + AMP + 4\,HOPO_3^{2-} + \underset{\text{Fumarate}}{{}^-OOC-CH=CH-COO^-}$$

We can summarize the results of the urea cycle as follows:

- Formation of urea from the carbon of CO_2, NH_4^+, and one nitrogen from the amino acid aspartate, followed by biological elimination through urine

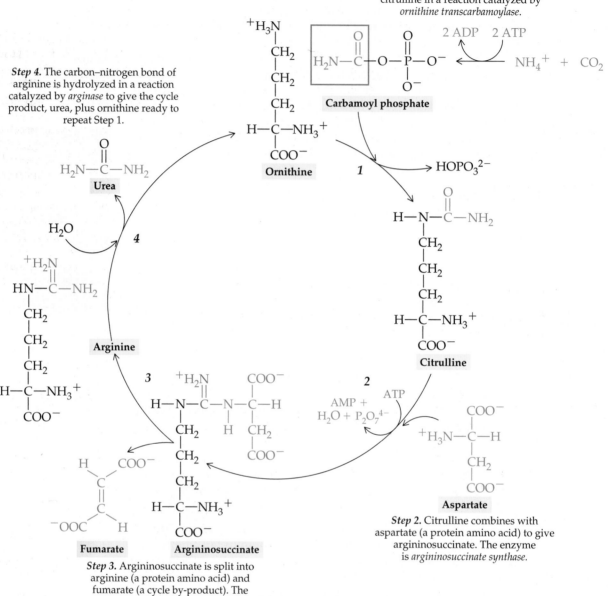

Step 1. Carbamoyl phosphate transfers its H₂NC=O group to ornithine (a nonprotein amino acid) to give citrulline in a reaction catalyzed by *ornithine transcarbamoylase.*

Step 4. The carbon–nitrogen bond of arginine is hydrolyzed in a reaction catalyzed by *arginase* to give the cycle product, urea, plus ornithine ready to repeat Step 1.

Step 2. Citrulline combines with aspartate (a protein amino acid) to give argininosuccinate. The enzyme is *argininosuccinate synthase.*

Step 3. Argininosuccinate is split into arginine (a protein amino acid) and fumarate (a cycle by-product). The enzyme is *argininosuccinase.*

▲ **FIGURE 28.4** **The urea cycle.** The formation of carbamoyl phosphate and Step 1, the formation of citrulline, take place in the mitochondrial matrix. Steps 2–4 take place in the cytosol. The carbamoyl group is shown boxed in red at the top of the figure.

- Breaking of four high-energy phosphate bonds to provide energy
- Production of the citric acid cycle intermediate, fumarate

Hereditary diseases associated with defects in the enzymes for each step in the urea cycle have been identified. The resulting abnormally high levels of ammonia in the blood (*hyperammonemia*) cause vomiting in infancy, lethargy, irregular muscle coordination (*ataxia*), and mental retardation. Immediate treatment consists of transfusions, blood dialysis (*hemodialysis*), and use of chemical agents to remove ammonia. Long-term treatment requires a low-protein diet and frequent small meals to avoid protein overload.

APPLICATION ▶   Gout: When Biochemistry Goes Awry

Gout is a severely painful condition caused by the precipitation of sodium urate crystals in joints. (⊂⊃, p. 165) A small amount of our waste nitrogen is excreted in urine and feces as urate rather than urea. Because the urate salt is highly insoluble, any excess of the urate anion causes precipitation of sodium urate. The pain of gout results from a cascade of inflammatory responses to these crystals in the affected tissue.

Even though it has been known for a very long time that the symptoms of gout are caused by urate crystals, understanding the many possible causes of the crystal formation is far from complete, even with modern medicine and all its sophisticated technology. Looking at a few of the pathways to gout illustrates some of the many ways that the delicate balance of our biochemistry can be disrupted.

Uric acid is an end product of the breakdown of purine nucleosides, and loss of its acidic H (in red) gives urate ion. Adenosine, for example, undergoes a number of enzymatic steps to produce xanthine, which is eventually converted to uric acid:

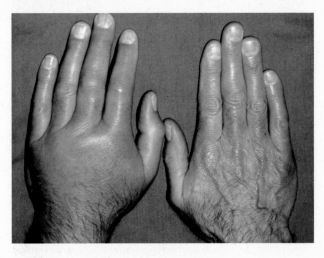

▲ The hand on the left shows the effect of gout while the one on the right is normal.

Adenosine   Xanthine

Uric acid   Urate ion

Anything that increases the production of uric acid or inhibits its excretion in the urine is a possible cause of gout. For example, several known hereditary enzyme defects increase the quantity of purines and therefore of uric acid. Sometimes, gouty attacks follow injury or severe muscle exertion. Complicating matters is the observation that the presence of crystals in a joint is not always accompanied by inflammation and pain.

One significant cause of increased uric acid production is accelerated breakdown of ATP, ADP, or AMP. For example,

alcohol abuse generates acetaldehyde that must be metabolized in the kidney by a pathway that requires ATP and produces excess AMP. Inherited fructose intolerance, glycogen storage diseases, and circulation of poorly oxygenated blood also accelerate uric acid production by this route. With low oxygen, ATP is not efficiently regenerated from ADP in mitochondria, leaving the ADP to be disposed of.

Conditions that diminish excretion of uric acid include kidney disease, dehydration, hypertension, lead poisoning, and competition for excretion from anions produced by ketoacidosis.

One treatment for gout relies on allopurinol, a structural analog of hypoxanthine, which is a precursor of xanthine in the formation of urate. Allopurinol inhibits the enzyme for conversion of hypoxanthine and xanthine to urate. Since hypoxanthine and xanthine are more soluble than sodium urate, they are more easily eliminated.

Hypoxanthine   Allopurinol

See Additional Problems 28.41 and 28.42 at the end of the chapter.

APPLICATION ▶ The Importance of Essential Amino Acids and Effects of Deficiencies

Remember when your mother told you to "eat all your vegetables if you want to grow up big and strong"? Although you may not have appreciated it at the time, she was doing nothing more than ensuring that you got your daily intake of essential vitamins and minerals. Regardless of the amazing numbers of biomolecules our bodies can synthesize, there are some nutrients we need but cannot make. These molecules, called "essential" nutrients, must be harvested daily from the foods we eat. Although there are no known essential carbohydrates, there are essential fatty acids and essential amino acids. We have discussed the two essential fatty acids, linoleic and linolenic, previously (Section 24.2).

In Chapter 18 you learned there are nine essential amino acids: histidine, isoleucine, leucine, lysine, methionine, phenylalanine, threonine, tryptophan, and valine. Histidine's classification as essential is argued by some biochemists; although it is essential in growing children, it is considered a nonessential amino acid for adults, as healthy adults are capable of synthesizing enough to meet their biochemical requirements except under physiological requirements imposed by certain stress or disease situations. Some nutritionists also place arginine on the essential list, not because there is no biochemical pathway for its synthesis but because it is synthesized by cells at an insufficient rate to meet the growth needs of developing mammals. Since the majority of arginine is synthesized as part of the urea cycle (see Figure 28.4), and as such is cleaved to form urea and ornithine, a certain amount must still be taken in daily, especially for men of reproduction age, since 80% of the amino acid composition of male seminal fluid is made of arginine. Both arginine and histidine are sometimes called *conditional amino acids*, as they are truly essential only under certain conditions.

But what would happen if you do not heed your mother's advice? What if your diet is such that one or more of the essential amino acids are missing? Among other topics of research, nutritional biochemists seek to understand what physiological fate will befall someone whose diet is deficient in one of the essential amino acids. This challenging research requires tight controls over the food given to the animal models under study. For example, to determine the effect of a valine deficiency on a mouse, researchers must ensure that its diet contains very little (if any) valine, while not lacking in any other nutrient. Studies like this, as well as observations in humans, have given rise to a number of conclusions concerning the functions of individual essential amino acids and the effects a deficiency may cause:

Histidine: An essential amino acid during the period of growth and conditionally essential in adults, this amino acid may also be required in the diet during old age and in those suffering from degenerative diseases. A deficiency of histidine

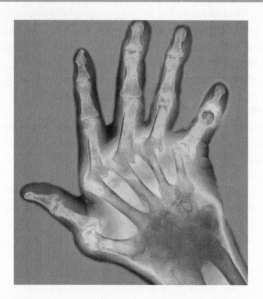

▲ Colored X ray of the deformed hand of a patient suffering from rheumatoid arthritis. Joint damage (shown in red) has caused the fingers to bend abnormally.

can cause pain in bony joints and may have a link to rheumatoid arthritis.

Isoleucine, Leucine, and Valine: These three hydrophobic amino acids are essential for the production and maintenance of body proteins; in addition, leucine and valine have been reported to increase mental alertness. It is difficult to tell exactly what are the true effects of the deficiency of any of these three amino acids. Since leucine is especially important in controlling the net synthesis of protein, its deficiency may severely limit regeneration of protein, which, for example, could affect healing after surgery. Valine deficiency has been reported to cause sensitivity to touch and sound.

Lysine: This extremely important amino acid plays a role in absorption of calcium; formation of collagen for bones, cartilage, and connective tissues; and the production of antibodies, hormones, and enzymes. Lysine deficiency can lead to a poor appetite, reduction in body weight, anemia, and a reduced ability to concentrate. Lysine deficiency in the body has also been associated with pneumonia, kidney disease (nephritis), and acidosis, as well as with malnutrition and rickets in children.

Methionine: The most important role of this amino acid is to act as a primary metabolic source of sulfur; it is really only necessary when cysteine intake is limited. Methionine seems to play a role in lowering cholesterol and reducing liver fat, protecting kidneys, and promoting hair growth. Methionine deficiency may ultimately lead to chronic rheumatic fever in children, hardening of the liver (cirrhosis), and nephritis.

Phenylalanine: As a starting material for the synthesis of tyrosine, phenylalanine is the primary source of the aromatic rings needed for a whole array of biomolecules, most notably the neurotransmitters (Section 20.4). Deficiency of phenylalanine can lead to behavioral changes such as psychotic and schizophrenic behavior (presumably due to being needed for the synthesis of tyrosine, dopamine, and epinephrine).

Threonine: This amino acid is key in the formation of collagen, elastin, and tooth enamel. Its deficiency can result in irritability in children and has been suggested by some as being essential in the prevention and treatment of mental illness.

Tryptophan: Considered to be a natural relaxant, it has been used to help relieve insomnia. Tryptophan has also been recommended for the treatment of migraines and mild depression (as it is the metabolic starting material for serotonin) and as such has been called "nature's Prozac." A deficiency of tryptophan can lead to serotonin deficiency syndrome, which in turn can lead to a broad array of emotional and behavioral problems such as depression, PMS, anxiety, alcoholism, insomnia, violence, aggression, and suicide.

See Additional Problems 28.43 and 28.44 at the end of the chapter.

KEY CONCEPT PROBLEM 28.7

Fumarate from Step 3 of the urea cycle may be recycled into aspartate for use in Step 2 of the cycle. The sequence of reactions for this process is

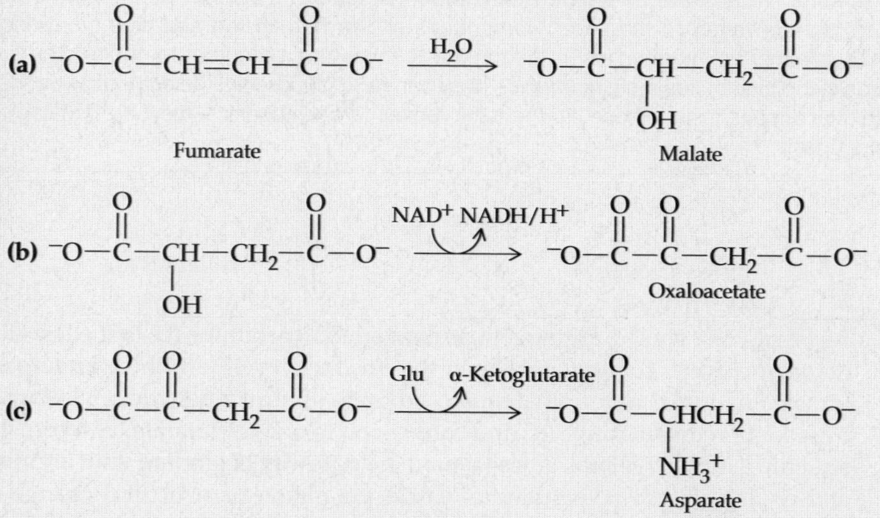

Classify each reaction as one of the following:

(1) Oxidation (2) Reduction (3) Transamination
(4) Elimination (5) Addition

28.5 Amino Acid Catabolism: The Carbon Atoms

The carbon atoms of each protein amino acid arrive by distinctive pathways at pyruvate, acetyl-SCoA, or one of the citric acid cycle intermediates shown in blue type in Figure 28.5. Eventually, all of the amino acid carbon skeletons can be used to generate energy, either by passing through the citric acid cycle and into the gluconeogenesis pathway to form glucose or by entering the ketogenesis pathway to form ketone bodies.

Those amino acids that are converted to acetoacetyl-SCoA or acetyl-SCoA (pink boxes in Figure 28.5) then enter the ketogenesis pathway, and are called *ketogenic amino acids.* (⊂⊃, Section 25.7)

Those amino acids that proceed by way of oxaloacetate to the gluconeogenesis pathway (⊂⊃, Section 23.11) are known as *glucogenic amino acids* (Table 28.1; blue boxes in Figure 28.5). Both ketogenic and glucogenic amino acids are able to enter fatty acid biosynthesis via acetyl-SCoA. (⊂⊃, Section 25.8)

TABLE 28.1 Glucogenic and Ketogenic Amino Acids

Glucogenic	
Alanine	Glycine
Arginine	Histidine
Asparagine	Methionine
Aspartate	Proline
Cysteine	Serine
Glutamate	Threonine
Glutamine	Valine

Both glucogenic and ketogenic
Isoleucine
Lysine
Phenylalanine
Tryptophan
Tyrosine

Ketogenic
Leucine

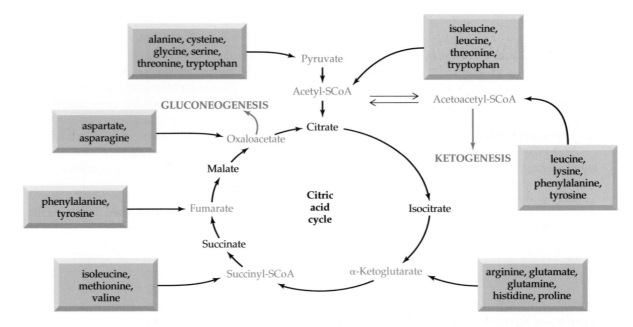

▲ **FIGURE 28.5 Fate of amino acid carbon atoms.** The carbon atoms of the amino acids are converted to the seven compounds shown here in red and blue type, each of which is either an intermediate in the citric acid cycle or a precursor to citrate. The amino acids in the blue boxes are glucogenic—they can form glucose via the entry of oxaloacetate into gluconeogenesis. Those in the pink boxes are ketogenic—they are available for ketogenesis.

PROBLEM 28.8

As Figure 28.4 shows, arginine (a) is converted to ornithine (b) in the last step of the urea cycle. To ultimately enter the citric acid cycle, ornithine undergoes transamination at its terminal amino group to give an aldehyde (c), followed by oxidation to glutamate (d) and conversion to α-ketoglutarate (e). Write the structures of the five molecules (a–e) in the pathway beginning with arginine and ending with α-ketoglutarate. Circle the region of structural change in each.

28.6 Biosynthesis of Nonessential Amino Acids

Nonessential amino acid One of 11 amino acids that are synthesized in the body and are therefore not necessary in the diet.

Essential amino acid An amino acid that cannot be synthesized by the body and thus must be obtained in the diet.

Humans are able to synthesize about half of the 20 amino acids found in proteins. These are known as the **nonessential amino acids** because they do not have to be supplied by our diet. The remaining amino acids—the **essential amino acids** (Table 28.2)—are synthesized only by plants and microorganisms. Humans must obtain the essential amino acids from food (see the Application "Proteins in the Diet," p. 569). Meats contain all of the essential amino acids. The foods that do not have all of them are described as having *incomplete amino acids*, and dietary deficiencies of the essential amino acids can lead to a number of health problems (see the Application "The Importance of Essential Amino Acids and Effects of Deficiencies," p. 870). Food combinations that together contain all of the amino acids are *complementary* sources of protein. It is interesting to note that we synthesize the nonessential amino acids in pathways containing only one to three steps, whereas synthesis of the essential amino acids by other organisms is much more complicated, requiring many more steps and a substantial energy investment.

TABLE 28.2   Essential Amino Acids

Amino acids essential for adults

Histidine	Lysine	Threonine
Isoleucine	Methionine	Tryptophan
Leucine	Phenylalanine	Valine

Some foods with incomplete amino acids

Grains, nuts, and seeds: High in methionine, low in lysine

Legumes: High in lysine, low in methionine

Corn: High in methionine, low in lysine and tryptophan

Some examples of complementary sources of protein

Peanut butter on bread	Nuts and soybeans
Rice and beans	Black-eyed peas and corn bread
Beans and corn	

All of the nonessential amino acids derive their amino groups from glutamate. As you have previously seen, this is the molecule that picks up ammonia in amino acid catabolism and carries it into the urea cycle. Glutamate can also be made from NH_4^+ and α-ketoglutarate by **reductive deamination**, the reverse of oxidative deamination (Section 28.3). The same glutamate dehydrogenate enzyme carries out the reaction:

Reductive deamination Conversion of an α-keto acid to an amino acid by reaction with NH_4^+.

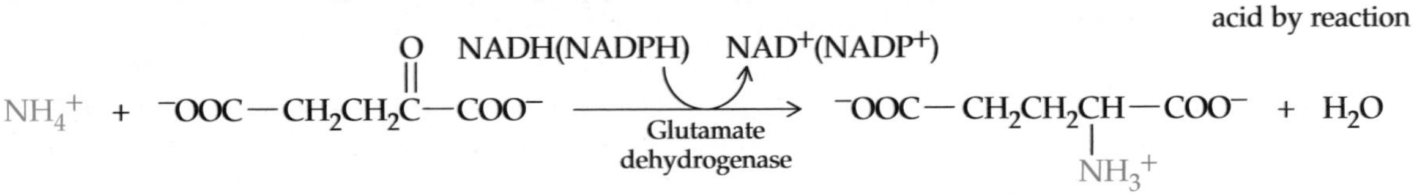

$$NH_4^+ + {}^-OOC-CH_2CH_2C(=O)-COO^- \xrightarrow[\text{Glutamate dehydrogenase}]{\text{NADH(NADPH)} \quad \text{NAD}^+(\text{NADP}^+)} {}^-OOC-CH_2CH_2CH(NH_3^+)-COO^- + H_2O$$

α-Ketoglutarate → Glutamate

Glutamate also provides nitrogen for the synthesis of other nitrogen-containing compounds, including the purines and pyrimidines that are part of DNA. (⊂⊃, Section 26.2)

The following four common metabolic intermediates, which you have seen play many roles, are the precursors for synthesis of the nonessential amino acids:

Precursors in synthesis of nonessential amino acids

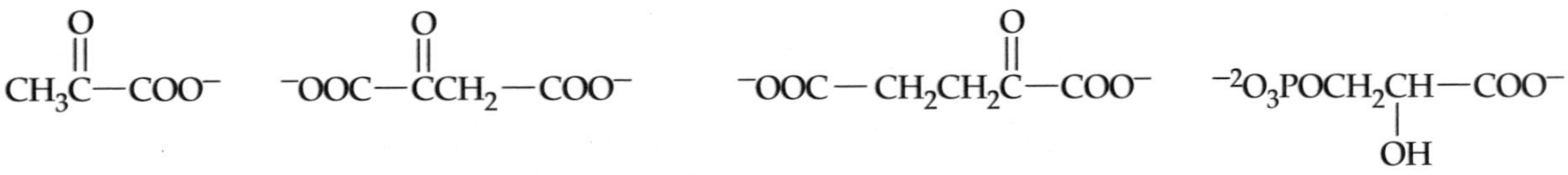

$CH_3C(=O)-COO^-$	${}^-OOC-C(=O)CH_2-COO^-$	${}^-OOC-CH_2CH_2C(=O)-COO^-$	${}^{-2}O_3POCH_2CH(OH)-COO^-$
Pyruvate	Oxaloacetate	α-Ketoglutarate	3-Phosphoglycerate

Glutamine is made from glutamate, and asparagine is made by reaction of glutamine with aspartate:

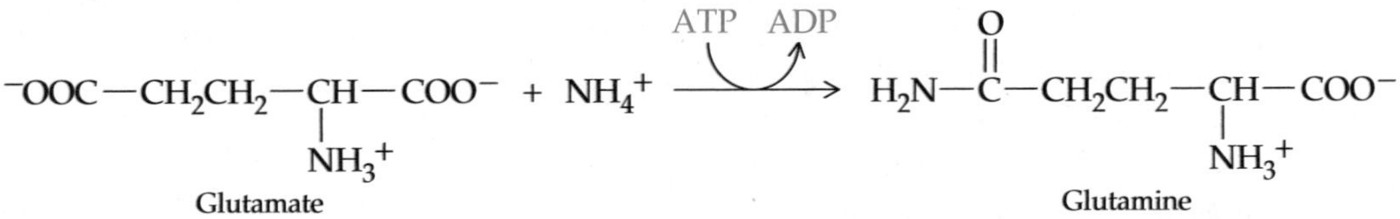

$${}^-OOC-CH_2CH_2-CH(NH_3^+)-COO^- + NH_4^+ \xrightarrow[]{\text{ATP} \quad \text{ADP}} H_2N-C(=O)-CH_2CH_2-CH(NH_3^+)-COO^-$$

Glutamate → Glutamine

$${}^-OOC-CH_2-CH(NH_3^+)-COO^- \underset{\text{ATP} \quad \text{AMP}}{\overset{\text{Glutamine} \quad \text{Glutamate}}{\rightleftarrows}} H_2N-C(=O)-CH_2-CH(NH_3^+)-COO^-$$

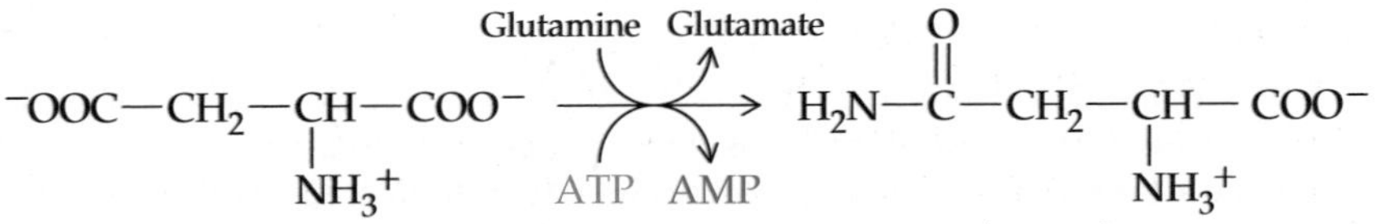

Aspartate → Asparagine

The amino acid tyrosine is classified as nonessential because we can synthesize it from phenylalanine, an essential amino acid:

$$\underset{\text{Phenylalanine}}{\text{CH}_2\text{CH}-\text{COO}^- \underset{\text{NH}_3^+}{|}} \longrightarrow \underset{\text{Tyrosine}}{\text{HO}-\text{CH}_2\text{CH}-\text{COO}^- \underset{\text{NH}_3^+}{|}}$$

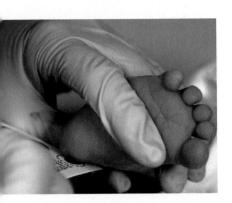

▲ Drawing blood from a newborn infant for the PKU test.

Whatever the classification, we have a high nutritional requirement for phenylalanine, and several metabolic diseases are associated with defects in the enzymes needed to convert it to tyrosine and other metabolites. The best known of these diseases is phenylketonuria (PKU), the first inborn error of metabolism for which the biochemical cause was recognized. In 1947 it was found that failure to convert phenylalanine to tyrosine causes PKU.

PKU results in elevated blood serum and urine concentrations of phenylalanine, phenylpyruvate, and several other metabolites produced when the body diverts phenylalanine to metabolism by other pathways. Undetected PKU causes mental retardation by the second month of life. Estimates are that, prior to the 1960s, 1% of those institutionalized for mental retardation were PKU victims. Widespread screening of newborn infants is the only defense against PKU and similar treatable metabolic disorders that take their toll early in life. In the 1960s a test for PKU was introduced, and virtually all hospitals in the United States now routinely screen for it. Treatment consists of a diet low in phenylalanine, which is maintained in infants with special formulas and in older individuals by eliminating meat and using low-protein grain products. Individuals with PKU must be on alert for foods sweetened with aspartame (Nutrasweet, for example), which is a derivative of phenylalanine.

KEY CONCEPT PROBLEM 28.9

In the pathway for synthesis of serine,

$$\underset{\substack{|\\\text{OH}\\\text{3-Phosphoglycerate}}}{^-\text{OOCCHCH}_2\text{OPO}_3^{2-}} \longrightarrow \underset{\text{3-Phosphohydroxypyruvate}}{^-\text{OOCCCH}_2\text{OPO}_3^{2-}} \longrightarrow$$

$$\underset{\substack{|\\\text{NH}_3^+\\\text{3-Phosphoserine}}}{^-\text{OOCCHCH}_2\text{OPO}_3^{2-}} \longrightarrow \underset{\substack{|\\\text{NH}_3^+\\\text{Serine}}}{^-\text{OOCCHCH}_2\text{OH}}$$

identify which step of the reaction is

(a) A transamination   **(b)** A hydrolysis   **(c)** An oxidation

SUMMARY: REVISITING THE CHAPTER GOALS

KEY WORDS

Amino acid pool, *p. 862*

Essential amino acids, *p. 872*

1. *What happens during the digestion of proteins, and what is the fate of the amino acids?* Protein digestion begins in the stomach and continues in the small intestine. The result is virtually complete hydrolysis to yield free amino acids. The amino acids enter the bloodstream after active transport into cells lining the intestine. The body does not store nitrogen compounds, but amino acids are constantly entering the amino acid pool from dietary protein or broken down body protein and being withdrawn from the pool for biosynthesis or further catabolism.

2. **What are the major strategies in the catabolism of amino acids?** Each amino acid is catabolized by a distinctive pathway, but in most of them the amino group is removed by *transamination* (the transfer of an amino group from an amino acid to a keto acid), usually to form glutamate. Then, the amino group of glutamate is removed as ammonium ion by *oxidative deamination*. The ammonium ion is destined for the *urea cycle*. The carbon atoms from amino acids are incorporated into compounds that can enter the *citric acid cycle*. These carbon compounds are also available for conversion to fatty acids or glycogen for storage, or for synthesis of ketone bodies.

3. **What is the urea cycle?** Ammonium ion (from amino acid catabolism) and bicarbonate ion (from carbon dioxide) react to produce carbamoyl phosphate, which enters the urea cycle. The first two steps of the urea cycle produce a reactive intermediate in which both of the nitrogens that will be part of the urea end product are bonded to the same carbon atom. Then arginine is formed and split by hydrolysis to yield urea, which will be excreted. The net result of the urea cycle is reaction of ammonium ion with aspartate to give urea and fumarate.

4. **What are the essential and nonessential amino acids, and how, in general, are amino acids synthesized?** *Essential amino acids* must be obtained in the diet because our bodies do not synthesize them. They are made only by plants and microorganisms, and their synthetic pathways are complex. Our bodies do synthesize the so-called *nonessential amino acids*. Their synthetic pathways are quite simple and generally begin with pyruvate, oxaloacetate, α-ketoglutarate, or 3-phosphoglycerate. The nitrogen is commonly supplied by glutamate.

Nonessential amino acid, *p. 872*

Oxidative deamination, *p. 865*

Reductive deamination, *p. 873*

Transamination, *p. 864*

Urea cycle, *p. 867*

UNDERSTANDING KEY CONCEPTS

28.10 In the diagram shown here, fill in the sources for the amino acid pool.

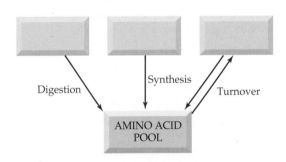

28.11 What are the fates of the carbon and nitrogen atoms in a catabolized amino acid?

28.12 A treatment for hyperammonemia (excess NH_4^+ in the blood) is to administer pyruvate. What two enzymes are necessary to detoxify the ammonium ion in the presence of pyruvate? What is the product?

28.13 Three metabolites that can result from the breakdown of the carbon skeleton of amino acids are ketone bodies, acetyl-SCoA, and glucose. Briefly describe how each of these metabolites can be produced from amino acid catabolism.

28.14 Define what an "essential" nutrient is and explain how it differs from a "nonessential" nutrient.

28.15 In the liver, the relative activity of ornithine transcarbamylase is high, that of argininosuccinate synthetase is low, and that of arginase is high. Why is it important that ornithine transcarbamylase activity be high? What might be the consequence if arginase activity is low or defective?

ADDITIONAL PROBLEMS

AMINO ACID POOL

28.16 Where is the body's amino acid pool?

28.17 In what part of the digestive tract does the digestion of proteins begin?

28.18 What glycolytic intermediates are precursors to amino acids?

28.19 What citric acid cycle intermediates are precursors to amino acids?

AMINO ACID CATABOLISM

28.20 What is meant by transamination?

28.21 Pyruvate and oxaloacetate can be acceptors for the amino group in transamination. Write the structures for the products formed from transamination of these two compounds.

28.22 What is the structure of the α-keto acid formed from transamination of the following amino acids?

 (a) Histidine
 (b) Cysteine

28.23 What is the structure of the α-keto acid formed from transamination of the following amino acids?

 (a) Tryptophan
 (b) Serine

28.24 What is meant by an oxidative deamination?

28.25 What coenzymes are associated with oxidative deamination?

28.26 Write the structure of the α-keto acid produced by oxidative deamination of the following amino acids:

 (a) Phenylalanine
 (b) Tyrosine

28.27 What other product is formed in oxidative deamination besides an α-keto acid?

28.28 What is a ketogenic amino acid? Give three examples.

28.29 What is a glucogenic amino acid? Give three examples.

UREA CYCLE

28.30 Why does the body convert NH_4^+ to urea for excretion?

28.31 What is the source of carbon in the formation of urea?

28.32 What are the sources of the two nitrogens in the formation of urea?

28.33 Why might the urea cycle be called the arginine cycle?

AMINO ACID BIOSYNTHESIS

28.34 How do essential and nonessential amino acids differ from each other in the number of steps required for their synthesis in organisms that synthesize both?

28.35 Which amino acid serves as the source of nitrogen for synthesis of the other amino acids?

28.36 What is the process by which amino acids are made from common non-nitrogen metabolites? This is the reverse of what process from amino acid catabolism?

28.37 How is tyrosine biosynthesized in the body? What disease prevents this biosynthesis, thereby making tyrosine an essential amino acid for those who have this condition?

28.38 PKU is an abbreviation for what disorder? What are the symptoms of PKU? How can PKU be treated for a nearly normal life?

28.39 Diet soft drinks that are sweetened with aspartame carry a warning label for phenylketonurics. Why?

28.40 Which of the following biomolecules contain nitrogen from an amino acid?

 (a) Glycogen
 (b) Nitric oxide
 (c) Collagen
 (d) Epinephrine
 (e) Stearic acid
 (f) Fructose

Applications

28.41 Your grandfather complains of pain in his swollen and inflamed big toe, and the doctor indicates that it is caused by gout. [*Gout: When Biochemistry Goes Awry, p. 869*]

 (a) How would you explain to him what gout is and its biochemical cause?
 (b) What can you suggest to him to prevent these gouty attacks?

28.42 Allopurinol is a drug often used to assist in the control of gout. At which step(s) in the catabolism of purines is allopurinol effective? What is its effect? Compare the structure of allopurinol with the structures of hypoxanthine and xanthine. Where does allopurinol differ in structure from hypoxanthine? Is this the site on the molecule that corresponds to the site where hypoxanthine or xanthine is oxidized? [*Gout: When Biochemistry Goes Awry, p. 869*]

28.43 Why is arginine sometimes considered an essential amino acid, even though our cells synthesize it? [*The Importance of Essential Amino Acids and Effects of Deficiencies, p. 870*]

28.44 What essential amino acid has been called "nature's Prozac"? What are some of the symptoms seen if deficiencies of it occur? [*The Importance of Essential Amino Acids and Effects of Deficiencies, p. 870*]

General Questions and Problems

28.45 What energy source is used in the formation of urea?

28.46 Write the equation for the transamination reaction that occurs between isoleucine and pyruvate.

28.47 Name the four products (carbon skeletons) of amino acid catabolism that can enter the citric acid cycle, and show where in the cycle they enter.

28.48 Can an amino acid be both glucogenic and ketogenic? Explain why or why not.

28.49 Briefly explain how the carbons from amino acids can end up in fatty acids located in adipose tissues.

28.50 Consider all of the metabolic processes we have studied. Why do we say that tissue biochemistry is dynamic? Describe some examples of these dynamic relationships.

28.51 Two major differences between the amino acid pool and the fat and carbohydrate pools in the body center on storage and on energy. Discuss these major differences.

28.52 When some of the carbons of glutamate are converted to glycogen, what is the order of the following compounds in that pathway?

(a) Glucose

(b) Glutamate

(c) Glycogen

(d) Oxaloacetate

(e) α-Ketoglutarate

(f) Phosphoenolpyruvate

28.53 The pancreatic proteases are synthesized and stored as zymogens. They are activated after the pancreatic juices enter the small intestine. Why is it essential that these enzymes be synthesized and stored in their inactive forms?

28.54 Why might an excess of a particular amino acid in the diet cause a deficiency of other amino acids?

28.55 The net reaction for the urea cycle shows that 3 ATPs are hydrolyzed; however, the total energy "cost" is 4 ATPs. Explain why this is true.

28.56 What are the two possible fates of the amino nitrogen from dietary proteins in animals?

CHAPTER 29

Body Fluids

CONCEPTS TO REVIEW

Solutions
(Sections 9.1, 9.2, 9.10)

pH
(Sections 10.8, 10.9)

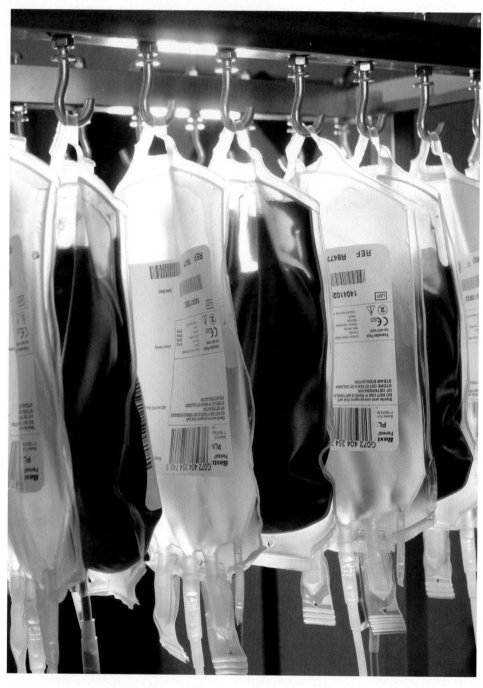

▲ Blood and other body fluids help maintain the delicate balance between life and death often faced during medical emergencies.

CONTENTS

CHAPTER GOALS

In this chapter, we will answer the following questions about the central role of chemistry in understanding physiology:

1. How are body fluids classified?

THE GOAL: Be able to describe the major categories of body fluids, their general composition, and the exchange of solutes between them.

2. What are the roles of blood in maintaining homeostasis?

THE GOAL: Be able to explain the composition and functions of blood.

3. How do blood components participate in the body's defense mechanisms?

THE GOAL: Be able to identify and describe the roles of blood components that participate in inflammation, the immune response, and blood clotting.

4. How do red blood cells participate in the transport of blood gases?

THE GOAL: Be able to explain the relationships among O_2 and CO_2 transport, and acid–base balance.

5. How is the composition of urine controlled?

THE GOAL: Be able to describe the transfer of water and solutes during urine formation, and give an overview of the composition of urine.

W e have chosen to put this chapter as the last one in your text because just about every aspect of chemistry you have studied so far applies to the subject of this chapter—body fluids. Electrolytes, nutrients and waste products, metabolic intermediates, and chemical messengers flow through your body in blood, in lymph fluid, and exit as waste in the urine and feces. The chemical compositions of blood and urine mirror chemical reactions throughout the body. Fortunately, samples of these fluids are easily collected and studied. Many advances in understanding biological chemistry have been based on information obtained from analysis of blood and urine. As a result, studies of blood and urine chemistry provide information essential for the diagnosis and treatment of disease.

29.1 Body Water and Its Solutes

All body fluids have water as the solvent; in fact the water content of the human body averages about 60% (by weight). Physiologists describe body water as occupying two different "compartments"—the *intracellular* and the *extracellular* compartments. We have looked primarily at the chemical reactions occurring in the **intracellular fluid** (the fluid inside cells), which includes about two-thirds of all body water (Figure 29.1). We now turn our attention to the remaining one-third of body water, the **extracellular fluid**, which includes mainly **blood plasma** (the fluid portion of blood) and **interstitial fluid** (the fluid that fills the spaces between cells).

To be soluble in water, a substance must be an ion, a gas, a small polar molecule, or a large molecule having many polar, hydrophilic (water-loving) or ionic groups on its surface. All four types of solutes are present in body fluids. The majority are inorganic ions and ionized biomolecules (mainly proteins), as shown in the comparison of blood plasma, interstitial fluid, and intracellular fluid in Figure 29.2. Although these fluids have different compositions, their **osmolarities** are the same; that is, they have the same number of moles of dissolved solute particles (ions or molecules) per liter. The osmolarity is kept in balance by the passage of water across cell membranes by osmosis, which occurs in response to osmolarity differences. (In osmosis, water moves across a membrane from the more dilute solution to the more concentrated solution.)(⬜⬜, Section 9.12)

Inorganic ions, known collectively as *electrolytes* (Section 9.9), are major contributors to the osmolarity of body fluids, and move about as necessary to maintain charge balance. Water-soluble proteins make up a large proportion of the solutes in blood plasma and intracellular fluid; 100 mL of blood contains about 7 g of protein. Blood proteins are used to transport lipids and other molecules and they play

Intracellular fluid Fluid inside cells.

Extracellular fluid Fluid outside cells.

Blood plasma Liquid portion of the blood: an extracellular fluid.

Interstitial fluid Fluid surrounding cells: an extracellular fluid.

Osmolarity Amount of dissolved solute per volume of solution.

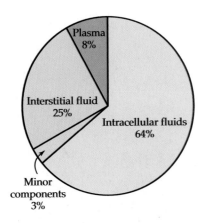

▲ **FIGURE 29.1 Distribution of body water.** About two-thirds of body water is intracellular—within cells. The extracellular fluids include blood plasma, fluids surrounding cells (interstitial), and such minor components as lymph, cerebrospinal fluid, and the fluid that lubricates joints (synovial fluid).

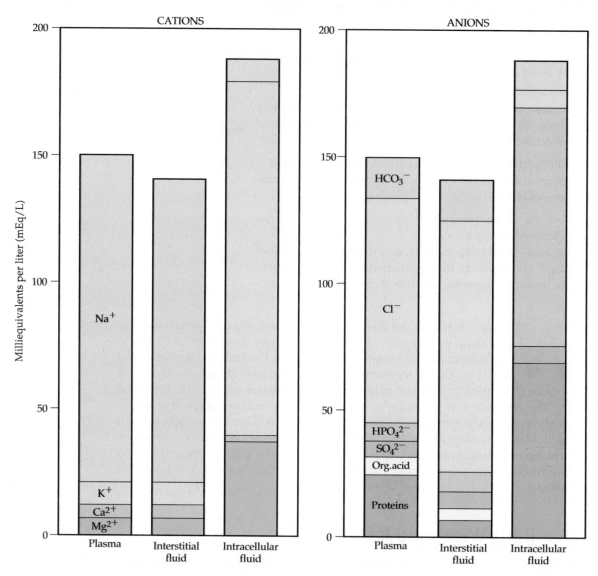

▲ **FIGURE 29.2** **The distribution of cations and anions in body fluids.** Outside cells, Na^+ is the major cation and Cl^- is the major anion. Inside cells, K^+ is the major cation and HPO_4^{2-} is the major anion. Note that at physiological pH, proteins are negatively charged.

essential roles in blood clotting (Section 29.5) and the immune response (Section 29.4). The blood gases (oxygen and carbon dioxide), along with glucose, amino acids, and the nitrogen-containing by-products of protein catabolism, are the major small molecules in body fluids.

Blood travels through peripheral tissue in a network of tiny, hair-like capillaries that connect the arterial and venous parts of the circulatory system (Figure 29.3). Capillaries are where nutrients and end products of metabolism are exchanged between blood and interstitial fluid. Capillary walls consist of a single layer of loosely spaced cells. Water and many small solutes move freely across the capillary walls in response to differences in fluid pressure and concentration (Figure 29.3).

Solutes that can cross membranes freely (passive diffusion) move from regions of high solute concentration to regions of low solute concentration. On the arterial ends of capillaries, blood pressure is higher than interstitial fluid pressure and solutes and water are pushed into interstitial fluid. On the venous ends of the capillaries, blood pressure is lower, and water and solutes from the surrounding tissues are able to reenter the blood plasma. The combined result of water and solute exchange at capillaries is that blood plasma and interstitial fluid are similar in composition (except for protein content; see Figure 29.2).

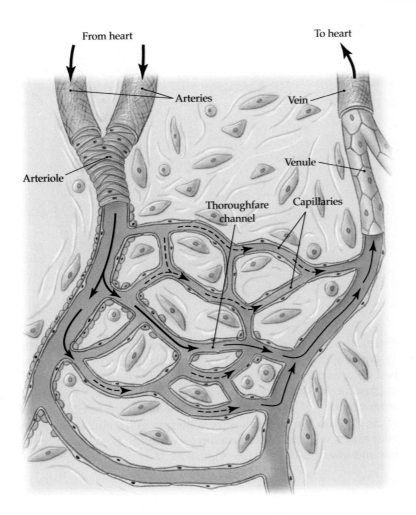

◀ **FIGURE 29.3** **The capillary network.** Solute exchange between blood and interstitial fluid occurs across capillary walls.

In addition to blood capillaries, peripheral tissue is networked with lymph capillaries (Figure 29.4). The lymphatic system collects excess interstitial fluid, debris from cellular breakdown, and proteins and lipid droplets too large to pass through capillary walls. Interstitial fluid and the substances that accompany it into the lymphatic system are referred to as *lymph*, and the walls of lymph capillaries are constructed so that lymph cannot return to the surrounding tissue. Ultimately lymph enters the bloodstream at the thoracic duct.

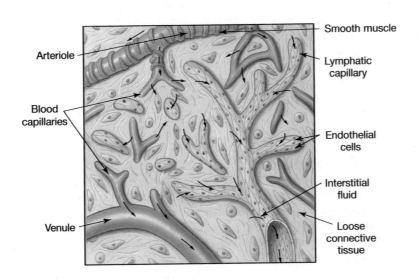

◀ **FIGURE 29.4** **Blood and lymph capillaries.** The thicker black arrows show the flow of fluids.

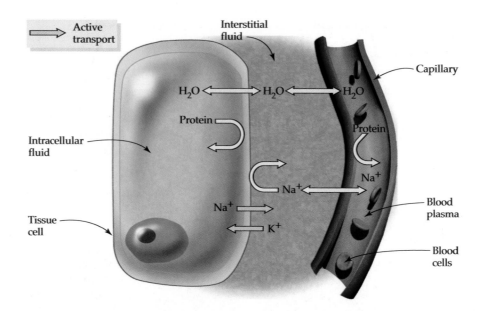

▶ **FIGURE 29.5 Exchange among body fluids.** Water exchanges freely in most tissues, with the result that the osmolarities of blood plasma, interstitial fluid, and intracellular fluid are the same. Large proteins cross neither capillary walls nor cell membranes, leaving the interstitial fluid protein concentration low. Concentration differences between interstitial fluid and intracellular fluid are maintained by active transport of Na^+ and K^+.

Exchange of solutes between the interstitial fluid and the intracellular fluid occurs by crossing cell membranes. Here, major differences in concentration are maintained by active transport (transport requiring energy) *against* concentration gradients (*from* regions of *low* concentration *to* regions of *high* concentration) and by the impermeability of cell membranes to certain solutes, notably the sodium ion (Figure 29.5). Sodium ion concentration is high in extracellular fluids and low in intracellular fluids, whereas potassium ion concentrations are just the reverse: high inside cells and low outside cells (see Figure 29.2).

KEY CONCEPT PROBLEM 29.1

The drug cisplatin is used to treat various forms of cancer in humans. As with many other drugs, the difficult part in designing the cisplatin molecule was to have a structure that ensures transport into the cell. The equilibrium reaction that takes place in the body when cisplatin is administered is:

$$\begin{bmatrix} Cl & NH_3 \\ & Pt & \\ Cl & NH_3 \end{bmatrix}(aq) + H_2O(l) \rightleftharpoons \begin{bmatrix} Cl & NH_3 \\ & Pt & \\ H_2O & NH_3 \end{bmatrix}^+ (aq) + Cl^-(aq)$$

Cisplatin   Monoaquacisplatin

(This is an example of *ligand exchange.*) Which form of cisplatin would you expect to exist inside the cell (where chloride concentrations are small)? Which form of cisplatin would you expect to exist outside the cell (where chloride concentrations are high)? Which form—cisplatin or monoaquacisplatin—enters the cell most readily? Why?

29.2 Fluid Balance

As you might imagine, preserving fluid balance is crucial in maintaining physiological homeostasis. One way we accomplish this is by ensuring that our daily intake of water is roughly enough to equal our daily output of water. Consider the following average intake/output water data for an adult human under normal environmental conditions:

WATER INTAKE (mL/DAY)		WATER OUTPUT (mL/DAY)	
Drinking water	1200	Urine	1400
Water from food	1000	Skin	400
Water from metabolic oxidation of food	300	Lungs	400
		Sweat	100
		Feces	200
Total	2500		2500

What are the physiological effects if this delicate balance is not maintained? This question is especially important to endurance athletes such as marathon runners and cyclists. During the course of a typical endurance event, especially when performed in the heat, much fluid loss occurs with minimal fluid intake to counter it. This typically results in a loss of body mass during the event and makes it easy to monitor performance versus fluid loss. This has been studied and the results can be summarized as follows:

% LOSS OF BODY MASS	SYMPTOMS AND PERFORMANCE
0%	Normal heat regulation and performance
1%	Thirst is stimulated, heat regulation during exercise is altered, performance begins to decline
2–3%	Further decrease in heat regulation, increased thirst, worsening performance
4%	Exercise performance cut by 20–30%
5%	Headache, irritability, "spaced-out" feeling, fatigue
6%	Weakness, severe loss of thermoregulation
7%	Collapse is likely unless exercise is stopped

Exercise physiologists consider 4% body mass loss and above to be the "danger zone." In fact, the sports drink Gatorade was developed in 1965 for just this reason. Doctors at the University of Florida developed the original formula to solve a serious problem for the school's football team: dehydration. So successful was this formula that by 1968 Gatorade had become the official sports drink of the National Football League and today commands an 80% share of the sports drink market, with gross sales of over two billion dollars per year. One can see why research into hydration strategies has led to the plethora of "sports drinks" that are now available on the market. (See Application, Electrolytes, Fluid Replacement, and Sports Drinks, on p. 279.)

Physiologically, the intake of water and electrolytes is regulated, but not closely regulated; however, the output of these substances *is* very closely controlled. Both the intake and output of water are controlled by hormones. Receptors in the hypothalamus monitor the concentration of solutes in blood plasma, and as little as a 2% change in osmolarity can cause an adjustment in hormone secretion (see Table 20.1). For example, when a rise in blood osmolarity indicates an increased concentration of solutes and therefore a shortage of water, secretion of *antidiuretic hormone* (ADH; also known as *vasopressin*) increases. One key role of the kidneys is to keep water and electrolytes in balance by increasing or decreasing the amounts eliminated. In the kidney, antidiuretic hormone causes a decrease in the water content of the urine. At the same time, osmoreceptors in the hypothalamus and baroreceptors in the heart and blood vessels activate the thirst mechanism, triggering increased water intake.

Antidiuretic hormone (ADH) is so tightly regulated that both oversecretion and undersecretion of this hormone can lead to serious disease states. Excess secretion can lead to what physicians refer to as the *syndrome of inappropriate antidiuretic hormone secretion (SIADH)*. Two of the many causes of SIADH are regional low blood volume arising from decreased blood return to the heart (caused by, for example, asthma, pneumonia, pulmonary obstruction, or heart failure) and misinterpretation by the hypothalamus of osmolarity (due, for example, to central-nervous-system disorders, barbiturates, or morphine). When ADH secretion is too high, the kidney excretes too little water, the water content of body compartments increases, and serum concentrations of electrolytes drop to dangerously low levels.

The reverse problem, inadequate secretion of antidiuretic hormone, is often a result of injury to the hypothalamus, and causes *diabetes insipidus*. In this condition (unrelated to diabetes mellitus), up to 15 L of dilute urine is excreted each day. Administration of synthetic hormone can control the problem.

29.3 Blood

Blood flows through the body in the circulatory system, which in the absence of trauma or disease, is an essentially closed system. About 55% of blood is plasma, which contains the proteins and other solutes shown in Figure 29.6; the remaining 45% is a mixture of red blood cells (**erythrocytes**), platelets, and white blood cells.

Erythrocytes Red blood cells; transporters of blood gases.

Whole blood Blood plasma plus blood cells.

The plasma and cells together make up **whole blood**, which is what is usually collected for clinical laboratory analysis. The whole blood sample is collected directly into evacuated tubes that contain an anticoagulant to prevent clotting (which normally occurs within 20–60 minutes at room temperature). Typical anticoagulants include heparin (which interferes with the action of enzymes needed for clotting) and citrate or oxalate ion (either of which form precipitates with calcium ion, which is also needed for blood clotting, thereby removing it from solution). Plasma is separated from blood cells by spinning the sample in a centrifuge, which causes the blood cells to clump together at the bottom of the tube, leaving the plasma at the top.

Blood serum Fluid portion of blood remaining after clotting has occurred.

Many laboratory analyses are performed on **blood serum**, the fluid remaining after blood has completely clotted. When a serum sample is desired, whole blood is collected in the presence of an agent that hastens clotting. Thrombin, a natural component of the clotting system (Section 29.5), is often used for this purpose. Centrifugation separates the clot and cells to leave behind the serum.

▲ **Result of separating blood serum from blood clot.** Analysis of the serum is an essential part of medical diagnosis.

Major Components of Blood

- **Whole blood**

   Blood plasma—fluid part of blood containing water-soluble solutes
   Blood cells—red blood cells (carry gases)
   —white blood cells (part of immune system)
- **Blood serum**—fluid portion of plasma left after blood has clotted

The functions of the major protein and cellular components of blood are summarized in Table 29.1. These functions fall into three categories:

- **Transport** The circulatory system is the body's equivalent of the interstate highway network, transporting materials from where they enter the system to where they are used or disposed of. Oxygen and carbon dioxide are carried to and from by red blood cells. Nutrients are carried from the intestine to the sites of their catabolism. Waste products of metabolism are carried to the kidneys. Hormones from endocrine glands are delivered to their target tissues.

- **Regulation** Blood redistributes body heat as it flows along, thereby participating in the regulation of body temperature. It also picks up or delivers water and electrolytes as they are needed. In addition, blood buffers are essential to the maintenance of acid–base balance.

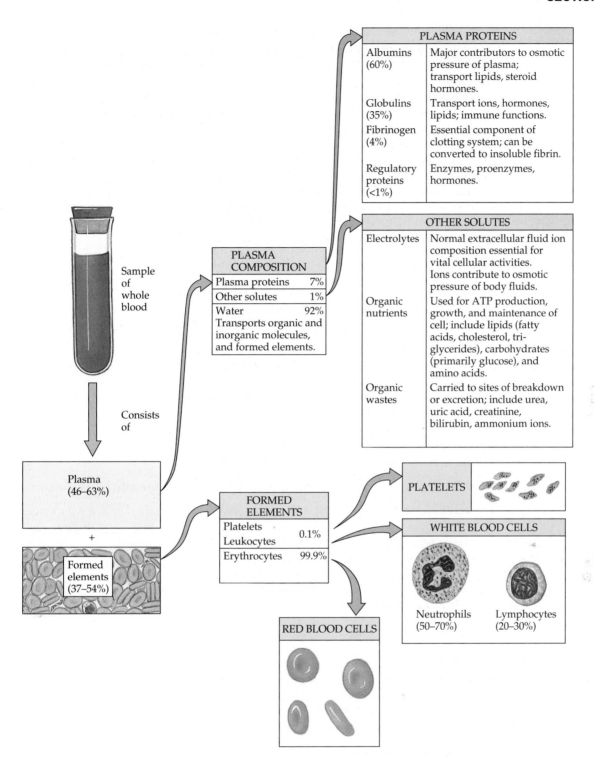

PLASMA PROTEINS

Albumins (60%)	Major contributors to osmotic pressure of plasma; transport lipids, steroid hormones.
Globulins (35%)	Transport ions, hormones, lipids; immune functions.
Fibrinogen (4%)	Essential component of clotting system; can be converted to insoluble fibrin.
Regulatory proteins (<1%)	Enzymes, proenzymes, hormones.

OTHER SOLUTES

Electrolytes	Normal extracellular fluid ion composition essential for vital cellular activities. Ions contribute to osmotic pressure of body fluids.
Organic nutrients	Used for ATP production, growth, and maintenance of cell; include lipids (fatty acids, cholesterol, tri-glycerides), carbohydrates (primarily glucose), and amino acids.
Organic wastes	Carried to sites of breakdown or excretion; include urea, uric acid, creatinine, bilirubin, ammonium ions.

Sample of whole blood

Consists of

PLASMA COMPOSITION

Plasma proteins	7%
Other solutes	1%
Water	92%

Transports organic and inorganic molecules, and formed elements.

Plasma (46–63%)

+

Formed elements (37–54%)

FORMED ELEMENTS

Platelets Leukocytes	0.1%
Erythrocytes	99.9%

PLATELETS

WHITE BLOOD CELLS

Neutrophils (50–70%)   Lymphocytes (20–30%)

RED BLOOD CELLS

▲ **FIGURE 29.6** **The composition of whole blood.**

- **Defense** Blood carries the molecules and cells needed for two major defense mechanisms: (1) the immune response, which destroys foreign invaders; and (2) blood clotting, which prevents loss of blood and begins the healing of wounds.

We will take a closer look at the defense functions of blood—the immune response and blood clotting (Sections 29.4 and 29.5) and then finish our discussion by examining the transport of blood gases (Section 29.6). (Lipid transport was discussed in Chapter 25.)

TABLE 29.1   Protein and Cellular Components of Blood

BLOOD COMPONENT	FUNCTION
Proteins	
Albumins	Transport lipids, hormones, drugs; major contributor to plasma osmolarity
Globulins	
Immunoglobulins (γ-globulins, antibodies)	Identify antigens (microorganisms and other foreign invaders) and initiate their destruction
Transport globulins	Transport lipids and metal ions
Fibrinogen	Forms fibrin, the basis of blood clots
Blood cells	
Red blood cells (erythrocytes)	Transport O_2, CO_2, H^+
White blood cells	
Lymphocytes (T cells and B cells)	Defend against specific pathogens and foreign substances
Neutrophils, eosinophils, and monocytes	Carry out phagocytosis—engulf foreign invaders
Basophils	Release histamine during inflammatory response of injured tissue
Platelets	Help to initiate blood clotting

PROBLEM 29.2

Match each term in the **(a)–(e)** group with its definition from the **(i)–(v)** group:

(a) Interstitial fluid
(b) Whole blood
(c) Blood serum
(d) Intracellular fluid
(e) Blood plasma

(i) Fluid that remains when blood cells are removed
(ii) Fluid, solutes, and cells that together flow through veins and arteries
(iii) Fluid that fills spaces between cells
(iv) Fluid that remains when blood clotting agents are removed from plasma
(v) Fluid within cells

29.4 Plasma Proteins, White Blood Cells, and Immunity

Antigen A substance foreign to the body that triggers the immune response.

Inflammatory response A nonspecific defense mechanism triggered by antigens or tissue damage.

Immune response Defense mechanism of the immune system dependent on the recognition of specific antigens, including viruses, bacteria, toxic substances, and infected cells; either cell-mediated or antibody-mediated.

An **antigen** is any molecule or portion of a molecule recognized by the body as a foreign invader. An antigen might be a molecule never seen before by the body or a molecular segment recognized as an invader (on the surface of a microorganism, for example). Antigens can also be small molecules, known as *haptens*, that are only recognized as antigens after they have bonded to carrier proteins. Haptens include some antibiotics, environmental pollutants, and allergens from plants and animals.

The recognition of an antigen can initiate three different responses. The first, the **inflammatory response**, is a localized response that is not specific to a given antigen. The two remaining types of **immune response** (cell-mediated response and antibody-mediated response) do depend on recognition of *specific* invaders (such as viruses, bacteria, toxic substances, or infected cells; Figure 29.7). At the molecular level, the invading antigen is detected by an interaction very much like that between an enzyme and its substrate. Noncovalent attraction allows a spatial fit between the antigen and a defender that is specific to that antigen. The *cell-mediated immune response* depends

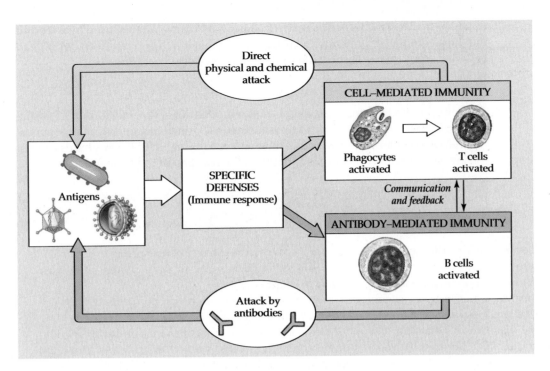

▲ **FIGURE 29.7 The immune response.** The attack on antigens occurs by cell-mediated and antibody-mediated immune responses.

on white blood cells known as *T cells*. The *antibody-mediated immune response* depends on **antibodies** (or **immunoglobulins**) produced by the white blood cells known as *B cells*.

Both inflammation and the immune responses require normal numbers of white blood cells to be effective (5 to 10 million white blood cells per milliliter). If the white blood cell count falls below 1000 per milliliter of blood, any infection can be life-threatening. The devastating results of white blood cell destruction in AIDS is an example of this condition (see the Application "Viruses and AIDS," p. 820).

Antibody (immunoglobulin) Glycoprotein molecule that identifies antigens.

Inflammatory Response

Cell damage due to infection or injury initiates **inflammation**, a nonspecific defense mechanism that produces swelling, redness, warmth, and pain. For example, the swollen, painful, red bump that develops around a splinter in your finger is an inflammation (this is generally known as a *wheal-and-flare reaction*). Chemical messengers released at the site of the injury direct the inflammatory response. One such messenger is histamine, which is synthesized from the amino acid histidine and is stored in cells throughout the body. Histamine release is also triggered by an allergic response.

Inflammation Result of the inflammatory response; includes swelling, redness, warmth, and pain.

$$\underset{\text{Histidine}}{\text{Imidazole—CH}_2\text{CH—NH}_3^+} \xrightarrow[\text{decarboxylase}]{\text{Histidine}} \underset{\text{Histamine}}{\text{Imidazole—CH}_2\text{CH}_2\text{—NH}_3^+} + CO_2$$

Histamine sets off dilation of capillaries and increases the permeability of capillary walls. The resulting increased blood flow into the damaged area reddens and warms the skin, and swelling occurs as plasma carrying blood-clotting factors and defensive proteins enters the intercellular space. At the same time, white blood cells cross capillary walls to attack invaders.

Bacteria or other antigens at the site of inflammation are destroyed by white blood cells known as *phagocytes*, which engulf invading cells and destroy them by enzyme-catalyzed hydrolysis reactions. Phagocytes also emit chemical messengers that help to direct the inflammatory response. An inflammation caused by a wound will heal completely only after all infectious agents have been removed, with dead cells and other debris absorbed into the lymph system.

Cell-Mediated Immune Response

The cell-mediated immune response is under the control of several kinds of *T lympho-cytes*, or *T cells*. The cell-mediated immune response principally guards against abnormal cells and bacteria or viruses that have entered normal cells; it also guards against the invasion of some cancer cells and causes the rejection of transplanted organs.

APPLICATION ▶ The Blood–Brain Barrier

Nowhere in human beings is the maintenance of a constant internal environment more important than in the brain. If the brain were exposed to the fluctuations in concentrations of hormones, amino acids, neurotransmitters, and potassium that occur elsewhere in the body, inappropriate nervous activity would result. Therefore, the brain must be rigorously isolated from variations in blood composition.

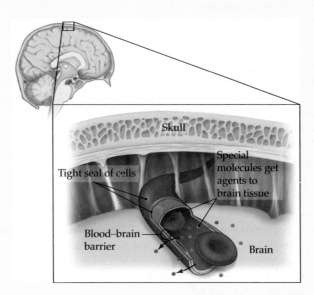

▲ The blood–brain barrier.

How can the brain receive nutrients from the blood in capillaries and yet be protected? The answer lies in the unique structure of the *endothelial cells* that form the walls of brain capillaries. Unlike the cells in most other capillaries, those in brain capillaries form a series of continuous tight junctions so that nothing can pass between them. To reach the brain, therefore, a substance must cross this blood–brain barrier (BBB) by crossing the endothelial cell membranes. The BBB serves as internal protection for the brain just as the skull serves as the brain's external protection.

The brain, of course, cannot be completely isolated or it will die from lack of nourishment. Glucose, the main source of energy for brain cells, and certain amino acids the brain cannot manufacture are recognized and brought across the cell membranes by transport mechanisms specific to each nutrient. Similar specific transporters move surplus substances out of the brain.

An asymmetric (one-way) transport system exists for glycine, a small amino acid that is a potent neurotransmitter. Glycine inhibits rather than activates transmission of nerve signals, and its concentration must be held at a lower level in the brain than in the blood. To accomplish this, there is a glycine transport system in the cell membrane closest to the brain, but no matching transport system on the other side. Thus, glycine can be transported out of the brain but not into it.

The brain is also protected by the "metabolic" blood–brain barrier. In this case, a compound that gets into an endothelial cell is converted within the cell to a metabolite that is unable to enter the brain. A striking demonstration of the metabolic brain barrier is provided by *dopamine*, a neurotransmitter, and L-*dopa*, a metabolic precursor of dopamine.

L-Dopa can both enter and leave the brain because it is recognized by an amino acid transporter. However, the brain is protected from an entering excess of L-dopa by its conversion to dopamine within the endothelial cells. Like glycine, dopamine, which is also produced from L-dopa within the brain, can leave the brain but cannot enter it. The dopamine deficiency that occurs in Parkinson's disease is therefore treated by administration of L-dopa.

L-Dopa

Dopamine

Since crossing the endothelial cell membrane is the route into the brain, substances soluble in the membrane lipids readily breach the blood–brain barrier. Among such substances are nicotine, caffeine, codeine, diazepam (Valium, an antidepressant), and heroin. Heroin differs from morphine in having two nonpolar acetyl groups where the morphine has polar hydroxyl groups. The resulting difference in lipid solubility allows heroin to enter the brain much more efficiently than morphine. Once heroin is inside the brain, enzymes remove the acetyl groups to give morphine, in essence trapping it in the brain, a general strategy many medicinal chemists try to capitalize upon. Finding ways to breach the blood–brain barrier is of major concern to medicinal chemists. For example, brain tumors are currently treated with either radiation or surgery, as the chemical agents used to typically treat cancer cannot cross the BBB. Researchers at the St. Louis University School of Medicine have been studying a cancer-killing compound (JV-1-36) that can sneak past the barrier, a discovery that could help doctors better treat a range of invasive brain malignancies. "The bottom line is, if you can get drugs into the brain, you can cure brain cancer," states Dr. William Banks, a member of the St. Louis research team. Researchers at UCLA have begun to examine *chimeric therapeutics*, materials that are half drug (which do not cross the BBB) and half "molecular Trojan horse" (genetically engineered proteins which do cross the BBB). As our understanding of this crucial barrier unfolds, we can expect many advances in the treatment of diseases of the brain that thus far have been treatable by only the most invasive of techniques.

See Additional Problems 29.53 through 29.56 at the end of the chapter.

A complex series of events begins when a T cell recognizes an antigenic cell. The result of these events is production of *cytotoxic*, or *killer*, T cells that can destroy the invader (for example, by releasing a toxic protein that kills by perforating cell membranes) and *helper* T cells that enhance the body's defenses against the invader. Thousands of *memory* T cells are also produced; they remain on guard and will immediately generate the appropriate killer T cells if the same pathogen reappears.

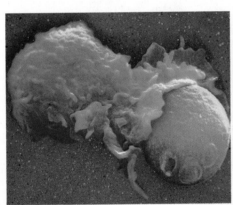 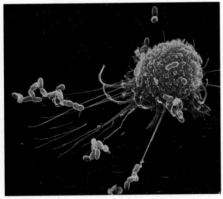

◀ **White blood cells.** (left) A lymphocyte phagocytizing a yeast cell. (right) A lymphocyte reaches out to snare several *E. coli* bacteria.

Antibody-Mediated Immune Response

The white blood cells known as *B lymphocytes* or *B cells*, with the assistance of T cells, are responsible for the antibody-mediated immune response. Unlike T cells, which identify only antigenic cells, B cells identify antigens adrift in body fluids. A B cell is activated when it first bonds to an antigen and then encounters a helper T cell that recognizes the same antigen. This activation can take place anywhere in the body but often occurs in lymph nodes, tonsils, or the spleen, which have large concentrations of lymphocytes.

Once activated, B cells divide to form plasma cells that secrete antibodies specific to the antigen. The antibodies are immunoglobulins. The body contains up to 10,000 different immunoglobulins at any given time, and we have the capacity to make more than 100 million others. The immunoglobulins are glycoproteins composed of two "heavy" polypeptide chains and two "light" polypeptide chains joined by disulfide bonds, as shown in Figure 29.8(a). The variable regions are sequences of amino acids that will bind a specific antigen. Once synthesized, antibodies spread out to find their antigens.

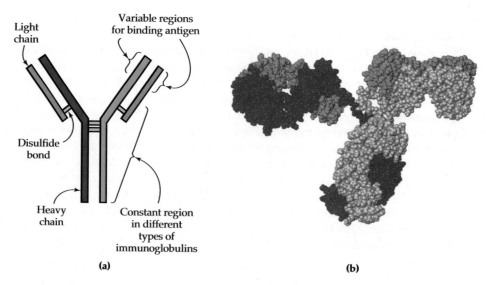

► **FIGURE 29.8 Structure of an immunoglobulin, which is an antibody.** (a) The regions of an immunoglobulin. The disulfide bridges that hold the chains together are shown in orange. (b) Molecular model of an immunoglobulin; the heavy chains are gray and blue and both light chains are red.

Formation of an antigen–antibody complex (Figure 29.9) inactivates the antigen by one of several methods. The complex may, for example, attract phagocytes, or it may block the mechanism by which the invader connects with a target cell.

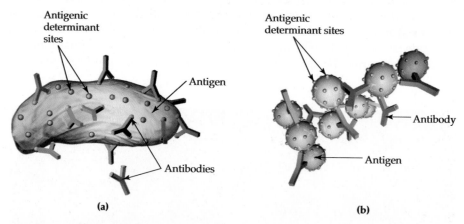

► **FIGURE 29.9 Antigen–antibody complexes.** (a) Antigens bond to antigenic-determinant sites on the surface of, for example, a bacterium. (b) Because each antibody has two binding sites, the interaction of many antigens and antibodies creates a large immune complex.

Activated B-cell division also yields memory cells that remain on guard and quickly produce more plasma cells if the same antigen reappears. The long-lived B and T memory cells are responsible for long-term immunity to diseases after the first illness or after a vaccination.

Several kinds of immunoglobulins have been identified. *Immunoglobulin G antibodies* (known as *gamma globulins*), for example, protect against viruses and bacteria. Allergies and asthma are caused by an oversupply of *immunoglobulin E*. Numerous disorders result from the mistaken identification of normal body constituents as foreign and the overproduction of antibodies to combat them. These **autoimmune diseases** include attack on connective tissue at joints in rheumatoid arthritis, attack on pancreatic islet cells in some forms of diabetes mellitus, and a generalized attack on nucleic acids and blood components in systemic lupus erythematosus.

Autoimmune disease Disorder in which the immune system identifies normal body components as antigens and produces antibodies to them.

Fibrin Insoluble protein that forms the fiber framework of a blood clot.

29.5 Blood Clotting

A blood clot consists of blood cells trapped in a mesh of the insoluble fibrous protein known as **fibrin**. Clot formation is a multiple-step process requiring participation of 12 clotting factors. Calcium ion is one of the clotting factors. Others, most of which are glycoproteins, are synthesized in the liver by pathways that require vitamin K as a coenzyme. Therefore, a deficiency of vitamin K, the presence of a competitive inhibitor of vitamin K, or a deficiency of a clotting factor can cause excessive bleeding, sometimes from even minor tissue damage. Hemophilia is a disorder caused by an inherited genetic defect that results in the absence of one or

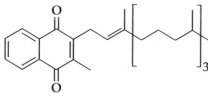

Vitamin K
(Phylloquinone)

more of the clotting factors. Hemophilia occurs in 1 in 10,000 individuals, with 80–90% of hemophiliacs being male.

The body's mechanism for halting blood loss from even the tiniest capillary is referred to as **hemostasis**. The first events in hemostasis are (1) constriction of surrounding blood vessels and (2) formation of a plug composed of the blood cells known as *platelets* at the site of tissue damage.

Next, a **blood clot** is formed in a process that is triggered by two pathways: (1) The *intrinsic pathway* begins when blood makes contact with the negatively charged surface of the fibrous protein collagen, which is exposed at the site of tissue damage. Clotting is activated in exactly the same manner when blood is placed in a glass tube because glass is also negatively charged. (2) The *extrinsic pathway* begins when damaged tissue releases an integral membrane glycoprotein known as *tissue factor*.

The result of either pathway is a cascade of reactions that is initiated when an inactive clotting factor (a zymogen, Section 19.9) is converted to its active form by cleavage of specific polypeptide sequences on its surface. Commonly, the newly activated enzyme then catalyzes the activation of the next factor in the cascade. The two pathways merge and, in the final step of the common pathway, the enzyme *thrombin* catalyzes cleavage of small polypeptides from the soluble plasma protein fibrinogen. Negatively charged groups in these polypeptides make fibrinogen soluble and keep the molecules apart. Once these polypeptides are removed, the resulting insoluble fibrin molecules immediately associate with each other by noncovalent interactions. Then they are bound into fibers by formation of amide cross-links between lysine and glutamine side chains in a reaction catalyzed by another of the clotting factors:

Hemostasis The stopping of bleeding.

Blood clot A network of fibrin fibers and trapped blood cells that forms at the site of blood loss.

▲ Colorized electron micrograph of a blood clot. Red blood cells can be seen enmeshed in the network of fibrin threads.

Gln—CH_2CH_2—$\overset{\displaystyle O}{\overset{\displaystyle \|}{C}}$—$NH_2$ + $\overset{+}{H_3N}CH_2CH_2CH_2CH_2$—Lys $\longrightarrow$

Protein chain

Gln—CH_2CH_2—$\overset{\displaystyle O}{\overset{\displaystyle \|}{C}}$—$NHCH_2CH_2CH_2CH_2$—Lys + NH_4^+

Cross-link between protein chains

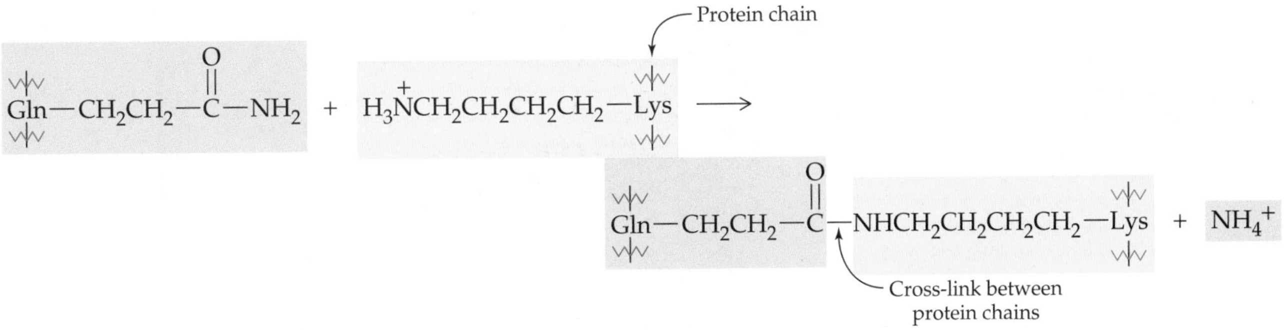

Once the clot has done its job of preventing blood loss and binding together damaged surfaces as they heal, the clot is broken down by hydrolysis of its peptide bonds.

29.6 Red Blood Cells and Blood Gases

Red blood cells, or erythrocytes, have one major purpose: to transport blood gases. Erythrocytes in mammals have no nuclei or ribosomes and cannot replicate themselves. In addition, they have no mitochondria or glycogen and must obtain glucose from the surrounding plasma. Their enormous number—about 250 million in a single drop of blood—and their large surface area provide for rapid exchange of gases throughout the body. Because they are small and flexible, erythrocytes can squeeze through the tiniest capillaries one at a time.

Of the protein in an erythrocyte, 95% is hemoglobin, the transporter of oxygen and carbon dioxide. Hemoglobin (Hb) is composed of four protein chains with the quaternary structure shown earlier in Section 18.11. Each protein chain has a central heme molecule in a crevice in its nonpolar interior, and each of the four hemes can combine with one O_2 molecule.

Oxygen Transport

The iron(II) ion, Fe^{2+}, sits in the center of each heme molecule and is the site to which O_2 binds through one of oxygen's unshared electron pairs. In contrast to the

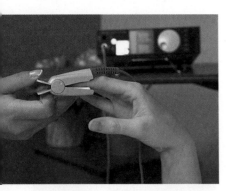

▲ **FIGURE 29.10 A pulse oxime-try sensor for continuous monitor-ing of blood oxygen.** One side of the sensor contains two light-emitting diodes (LEDs), one that emits in the visible red range (better absorbed by dark-red deoxygenated blood) and one that emits in the infrared range (better absorbed by oxygenated blood which is bright red). On the opposite side of the sensor, a pho-todetector measures the light that passes through and sends the signal to an instrument that computes the percent oxygen saturation of the blood and also records the pulse. Normal oxygen saturation is 95–100%. Below 85% tissues are at risk, and below 70% is typically life-threatening.

cytochromes of the respiratory chain, where iron cycles between Fe^{2+} and Fe^{3+}, heme iron must remain in the reduced Fe^{2+} state to maintain its oxygen-carrying ability. Hemoglobin (Hb) carrying four oxygens (oxyhemoglobin) is bright red. Hemoglobin that has lost one or more oxygens (deoxyhemoglobin) is dark red-purple, which accounts for the darker color of venous blood. Dried blood is brown, because exposure to atmospheric oxygen has oxidized the iron (think of rust). The color of arterial blood carrying oxygen is used in a clinically valuable method for monitoring oxygenation (known as *pulse oximetry*, Figure 29.10).

At normal physiological conditions, the percentage of heme molecules that carry oxygen, known as the *percent saturation*, is dependent on the partial pressure of oxy-gen in surrounding tissues (Figure 29.11). The shape of the curve indicates that bind-ing of oxygen to heme is allosteric in nature. (⬤⬤, Section 19.7) Each O_2 that binds causes changes in the hemoglobin quaternary structure that enhance binding of the next O_2, and releasing each oxygen enhances release of the next. As a result, oxygen is more readily released to tissue where the partial pressure of oxygen is low. The aver-age oxygen partial pressure in peripheral tissue is 40 mmHg, a pressure at which Hb remains 75% saturated by oxygen, leaving a large amount of O_2 in reserve for emer-gencies. Note, however, the rapid drop in the curve between 40 mmHg and 20 mmHg, which is the oxygen pressure in tissue where metabolism is occurring rapidly.

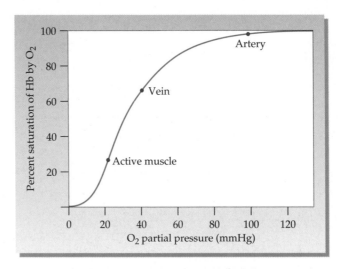

▲ **FIGURE 29.11 Oxygen saturation of hemoglobin at normal physiological conditions.** Oxygen pressure is about 100 mmHg in arteries and 20 mmHg in active muscles. Note the large release of oxygen as the partial pressure drops from 40 mmHg to 20 mmHg.

Carbon Dioxide Transport, Acidosis, and Alkalosis

Oxygen and carbon dioxide are the "blood gases" transported by erythrocytes. By way of the bicarbonate ion/carbon dioxide buffer, the intimate relationships among H^+ and HCO_3^- concentrations and O_2 and CO_2 partial pressures are essential to maintaining electrolyte and acid–base balance:

$$CO_2(aq) + H_2O(l) \rightleftharpoons H_2CO_3(aq) \rightleftharpoons HCO_3^-(aq) + H^+(aq)$$
Controlled by the lungs   Controlled by the kidneys

In a clinical setting, monitoring "blood gases" usually refers to measuring the pH of blood as well as the gas concentrations. Carbon dioxide from metabolism in periph-eral cells diffuses into interstitial fluid and then into capillaries, where it is trans-ported in blood three ways: (1) as dissolved $CO_2(aq)$, (2) bonded to Hb, or (3) as HCO_3^- in solution. About 7% of the CO_2 produced dissolves in blood plasma. The rest enters erythrocytes, where some of it bonds to the protein portion of hemoglo-bin by reaction with the nonionized amino acid $—NH_2$ groups present:

$$Hb—NH_2 + CO_2 \rightleftharpoons Hb—NHCOO^- + H^+$$

Most of the CO_2 is rapidly converted to bicarbonate ion within erythrocytes, which contain a large concentration of carbonic anhydrase. The resulting water-soluble HCO_3^- ion can leave the erythrocyte and travel in the blood to the lungs, where it will be converted back to CO_2 for exhalation. To maintain electrolyte balance, a Cl^- ion enters the erythrocyte for every HCO_3^- ion that leaves, and the process is reversed when the blood reaches the lungs:

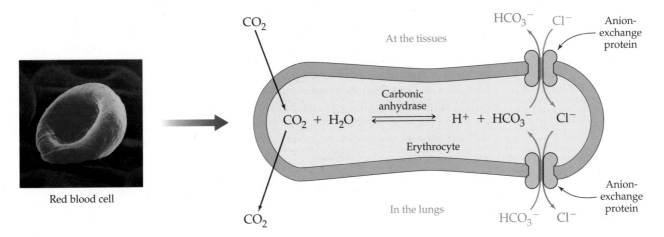

A cell-membrane protein controls this ion exchange. The exchange is passive, as the ions move from higher to lower concentrations.

Without some compensating change, the result of hemoglobin reacting with CO_2 and the action of carbonic anhydrase would be an unacceptably large increase in acidity. To cope with this, hemoglobin responds by reversibly binding hydrogen ions:

$$Hb \cdot 4\,O_2 + 2\,H^+ \rightleftharpoons Hb \cdot 2\,H^+ + 4\,O_2$$

The release of oxygen is enhanced by allosteric effects when the hydrogen ion concentration increases, and oxygen is held more firmly when the hydrogen ion concentration decreases.

The changes in the oxygen saturation curve with CO_2 and H^+ concentrations and with temperature are shown in Figure 29.12. The curve shifts to the right, indicating decreased affinity of Hb for O_2, when the H^+ and CO_2 concentrations increase and when the temperature increases. These are exactly the conditions in muscles that are working hard and need more oxygen. The curve shifts to the left,

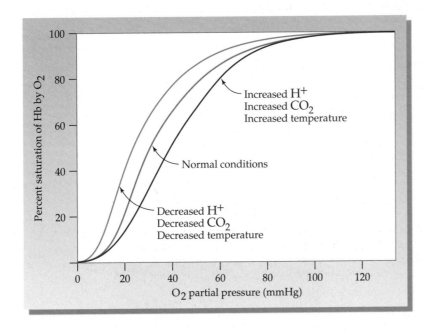

◀ **FIGURE 29.12 Changes in oxygen affinity of hemoglobin with changing conditions.** The normal curve of Figure 29.11 is shown in red here.

Acidosis The abnormal condition associated with a blood plasma pH below 7.35; may be respiratory or metabolic.

Alkalosis The abnormal condition associated with a blood plasma pH above 7.45; may be respiratory or metabolic.

indicating increased affinity of Hb for oxygen, under the opposite conditions of decreased H^+ and CO_2 concentrations and lower temperature.

Homeostasis requires a blood pH between 7.35 and 7.45. A pH outside this range creates either **acidosis** or **alkalosis**.

Acidosis	Normal	Alkalosis
Blood pH Below 7.35	Blood pH 7.35–7.45	Blood pH Above 7.45

The wide variety of conditions that cause acidosis or alkalosis are divided between respiratory malfunctions and metabolic malfunctions. Examples of each are given in Table 29.2. *Respiratory* disruption of acid–base balance can result when carbon dioxide generation by metabolism and carbon dioxide removal at the lungs are out of balance. *Metabolic* disruption of acid–base balance can result from abnormally high acid generation or failure of buffer systems and kidney function to regulate bicarbonate concentration.

TABLE 29.2   Causes of Acidosis and Alkalosis

TYPE OF IMBALANCE	CAUSES
Respiratory acidosis	CO_2 buildup due to:
	Decreased respiratory activity (hypoventilation)
	Cardiac insufficiency (for example, congestive failure, cardiac arrest)
	Deterioration of pulmonary function (for example, asthma, emphysema, pulmonary obstruction, pneumonia)
Respiratory alkalosis	Loss of CO_2 due to:
	Excessive respiratory activity (hyperventilation, due, for example, to high fever, nervous condition)
Metabolic acidosis	Increased production of metabolic acids due to:
	Fasting or starvation
	Diabetes
	Excessive exercise
	Decreased acid excretion in urine due to:
	Poisoning
	Renal failure
	Decreased plasma bicarbonate concentration due to:
	Diarrhea
Metabolic alkalosis	Elevated plasma bicarbonate concentration due to:
	Vomiting
	Diuretics
	Antacid overdose

➾ KEY CONCEPT PROBLEM 29.3

Carbon dioxide dissolved in body fluids has a pronounced effect on pH.

(a) Does pH go up or down when carbon dioxide dissolves in these fluids? Does this change indicate higher or lower acidity?

(b) What does a blood gas analysis measure?

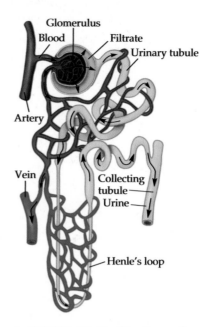

PROBLEM 29.4

Classify the following conditions as a cause of respiratory or metabolic acidosis or alkalosis (consult Table 29.2).

(a) Emphysema   **(b)** Kidney failure

(c) Overdose of an antacid   **(d)** Severe panic attack

(e) Congestive heart failure

29.7 The Kidney and Urine Formation

The kidneys bear the major responsibility for maintaining a constant internal environment in the body. By managing the elimination of appropriate amounts of water, electrolytes, hydrogen ions, and nitrogen-containing wastes, the kidneys respond to changes in health, diet, and physical activity.

About 25% of the blood pumped from the heart goes directly to the kidneys, where the functional units are the *nephrons* (Figure 29.13). Each kidney contains over a million of them. Blood enters a nephron at a *glomerulus* (at the top in Figure 29.13), a tangle of capillaries surrounded by a fluid-filled space. **Filtration**, the first of three essential kidney functions, occurs here. The pressure of blood pumped into the glomerulus directly from the heart is high enough to push plasma and all its solutes except large proteins across the capillary membrane into the surrounding fluid, the **glomerular filtrate**. The filtrate flows from the capsule into the tubule that makes up the rest of the nephron, and the blood enters the network of capillaries intertwined with the tubule.

About 125 mL of filtrate per minute enters the kidneys, and they produce 180 L of filtrate per day. This filtrate contains not only waste products but also many solutes the body cannot afford to lose, such as glucose and electrolytes. Since we excrete only about 1.4 L of urine each day you can see that another important function of the kidneys is **reabsorption**—the recapture of water and essential solutes by moving them out of the tubule.

Reabsorption alone, however, is not sufficient to provide the kind of control over urine composition that is needed. More of certain solutes must be excreted than are present in the filtrate. This situation is dealt with by **secretion**—the transfer of solutes *into* the kidney tubule.

Reabsorption and secretion require the transfer of solutes and water among the filtrate, the interstitial fluid surrounding the tubule, and blood in the capillaries. Some of the substances reabsorbed and secreted are listed in Table 29.3. Solutes cross the tubule and capillary membranes by passive diffusion in response to concentration or ionic charge differences, or by active transport. Water moves in response to differences in the osmolarity of the fluids on the two sides of the membranes. Solute and water movement is also controlled by hormone-directed variations in the permeability of the tubule membrane.

29.8 Urine Composition and Function

Urine contains the products of glomerular filtration, minus the substances reabsorbed in the tubules, plus the substances secreted in the tubules. The actual concentrations of these substances in urine at any time are determined by the amount of water being excreted, which can vary significantly with water intake, exercise, temperature, and state of health. (For identical quantities of solutes, concentration *decreases* when the quantity of solvent water *increases*, and concentration *increases* when the quantity of water *decreases*.)

About 50 g of solids in solution are excreted every day—about 20 g of electrolytes and 30 g of nitrogen-containing wastes (urea and ammonia from amino acid catabolism, creatinine from breakdown of creatine phosphate in muscles, and

▲ **FIGURE 29.13 Structure of a nephron.** Water moves out of the urinary tubule and the collecting tubule. The concentration of solutes in urine is established as they move both in and out along the tubules.

Filtration (kidney) Filtration of blood plasma through a glomerulus and into a kidney nephron.

Glomerular filtrate Fluid that enters the nephron from the glomerulus; filtered blood plasma.

Reabsorption (kidney) Movement of solutes out of filtrate in a kidney tubule.

Secretion (kidney) Movement of solutes into filtrate in a kidney tubule.

TABLE 29.3 Reabsorption and Secretion in Kidney Tubules

REABSORBED
Ions
$Na^+, Cl^-, K^+, Ca^{2+}, Mg^{2+}, PO_4^{3-}, SO_4^{2-}, HCO_3^-$
Metabolites
Glucose
Amino acids
Proteins
Vitamins
SECRETED
Ions
K^+, H^+, Ca^{2+}
Wastes
Creatinine
Ammonia
Organic acids and bases
Miscellaneous
Neurotransmitters
Histamine
Drugs (penicillin, atropine, morphine, numerous others)

uric acid from purine catabolism). Normal urine composition is usually reported as the quantity of each solute excreted per day, and laboratory urinalysis often requires collection of all urine excreted during a 24-hour period.

The following paragraphs briefly describe a few of the mechanisms that control the composition of urine.

Acid–Base Balance

Respiration, buffers, and excretion of hydrogen ions in urine combine to maintain acid–base balance. Metabolism normally produces an excess of hydrogen ions, a portion of which must be excreted each day to prevent acidosis. Very little free hydrogen ion exists in blood plasma, and therefore very little enters the glomerular filtrate. Instead, the H^+ to be eliminated is produced by the reaction of CO_2 with water in the cells lining the tubules of the nephrons:

$$CO_2 + H_2O \xrightarrow{\text{Carbonic anhydrase}} H^+ + HCO_3^-$$

To bloodstream
To filtrate

The HCO_3^- ions return to the bloodstream, and the H^+ ions enter the filtrate. Thus, the more hydrogen ions there are to be excreted, the more bicarbonate ions are returned to the bloodstream.

The urine must carry away the necessary quantity of H^+ without becoming excessively acid. To accomplish this, the H^+ is tied up by reaction with HPO_4^{2-} absorbed at the glomerulus, or by reaction with NH_3 produced in the tubule cells by deamination of glutamate:

$$H^+ + HPO_4^{2-} \longrightarrow H_2PO_4^-$$
$$H^+ + NH_3 \longrightarrow NH_4^+$$

APPLICATION ▶ Automated Clinical Laboratory Analysis

What happens when a physician orders chemical tests of blood, urine, or spinal fluid? The sample goes to a clinical chemistry laboratory, often in a hospital, where most tests are done by automated clinical chemistry analyzers. There are basically two types of chemical analysis, one for the quantity of a chemical (a natural biochemical, a drug, or a toxic substance) and the other for the quantity of an enzyme with a specific metabolic activity.

The quantity of a given chemical in the blood is determined either directly or indirectly. Many chemical components are measured directly by mixing a reagent with the sample—the *analyte*—and noting the quantity of a colored product formed by using a photometer, an instrument that measures the absorption of light of a wavelength specific to the product. For each test specified, a portion of the sample is mixed with the appropriate reagent and the photometer is adjusted to the exact wavelength necessary.

When it is not possible to utilize this direct technique, other indirect methods that produce a detectable product have been devised. Many analytes are also substrates for enzyme-catalyzed reactions. Analysis of the substrate concentration is therefore often made possible by treating the analyte with appropriate enzymes. Glucose is determined in this manner by utilizing a pair of enzyme-catalyzed reactions: The glucose is converted to glucose 6-phosphate using its hexokinase-catalyzed reaction with ATP; the glucose 6-phosphate is then oxidized by $NADP^+$; and the quantity of NADPH produced is measured photometrically.

The second type of analysis, determination of the quantity of a specific enzyme or the ratio of two or more enzymes, is invaluable in detecting organ damage that allows enzymes to leak into body fluids. For example, elevation of both ALT (alanine aminotransferase) and AST (aspartate aminotransferase) with an AST/ALT ratio greater than 1.0 is characteristic of liver disease. If, however, the AST is greatly elevated and the AST/ALT ratio is higher than 1.5, a myocardial infarction (heart attack) is likely to have occurred. When the substance being analyzed is an enzyme, its presence is detected, monitored, and quantified with an assay that employs a substrate of the enzyme in question; levels of the enzyme are measured by monitoring the substrate's appearance or disappearance. ALT, for example, is determined by photometrically monitoring the disappearance of NADH in the following pair of coupled reactions (where LD = lactate dehydrogenase):

$$\text{L-Alanine} + \alpha\text{-Ketoglutarate} \xrightarrow{\text{ALT}} \text{Pyruvate} + \text{L-Glutamate}$$

$$\text{Pyruvate} + \text{NADH/H}^+ \xrightarrow{\text{LD}} \text{Lactate} + \text{NAD}^+$$

As ALT causes pyruvate to form, LD causes the pyruvate to react with NADH to form lactate and NAD^+. By knowing how fast this reaction will occur with a given amount of LD, and by knowing how fast a given amount of ALT carries out the first reaction, the amount of ALT can be directly quantified in the sample being examined.

Automated analyzers rely on premixed reagents and automatic division of a fluid sample into small portions for each test. A low-volume analyzer that provides rapid results for a few tests accepts a bar-coded serum or plasma sample cartridge followed by bar-coded reagent cartridges. The instrument software reads the bar codes and directs an automatic pipette (which removes small samples of precisely measured volumes) to transfer the appropriate volume of sample to each test cartridge. The instrument then moves the test cartridge along as the sample and reagents are mixed, the reaction takes place for a measured amount of time, and the photometer reading is taken and converted to the test result.

A high-volume analyzer with more complex software randomly accesses 40 or more tests and runs over 400 tests per hour at a cost of less than 10 cents per test. The end result is a printed report on each sample listing the types of tests, the sample values, and a normal range for each test.

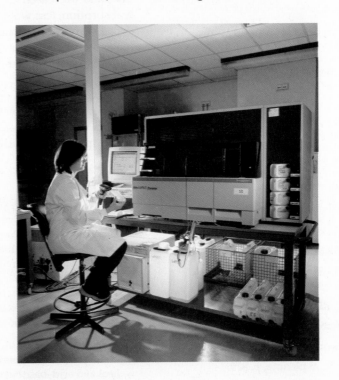

See Additional Problems 29.57 through 29.59 at the end of the chapter.

When acidosis occurs, the kidney responds by synthesizing more ammonia, thereby increasing the quantity of H^+ eliminated.

A further outcome of H^+ production in tubule cells is the net reabsorption of the HCO_3^- that entered the filtrate at the glomerulus. The body cannot afford to lose its primary buffering ion, HCO_3^-. If HCO_3^- were to be lost, the body would have to produce more; the result would be production of additional acid from carbon dioxide by reaction with water. Instead, H^+ secreted into the filtrate combines with HCO_3^- in the filtrate to produce CO_2 and water:

$$H^+ + HCO_3^- \longrightarrow CO_2 + H_2O$$

In the filtrate → To bloodstream

Upon returning to the bloodstream, the CO_2 is reconverted to HCO_3^-.

In summary, acid–base reactions in the kidneys have the following results:

- Secreted H^+ is eliminated in the urine as NH_4^+ or $H_2PO_4^-$.
- Secreted H^+ combines with filtered HCO_3^-, producing CO_2 that returns to the bloodstream and again is converted to HCO_3^-.

Fluid and Na^+ Balance

The amount of water reabsorbed is dependent on the osmolarity of the fluid passing through the kidneys, the antidiuretic hormone–controlled permeability of the collecting duct membrane, and the amount of Na^+ actively reabsorbed. Increased sodium reabsorption means higher interstitial osmolarity, greater water reabsorption, and decreased urine volume. In the opposite condition of decreased sodium reabsorption, less water is reabsorbed and urine volume increases. "Loop diuretic" drugs such as furosemide (trademarked as Lasix), which is used in treating hypertension and congestive heart failure, act by inhibiting the active transport of Na^+ out of the region of the urinary tubule called Henle's loop. Caffeine acts as a diuretic in a similar way.

The reabsorption of Na^+ is normally under the control of the steroid hormone aldosterone. The arrival of chemical messengers signaling a decrease in total blood plasma volume accelerates the secretion of aldosterone. The result is increased Na^+ reabsorption in the kidney tubules accompanied by increased water reabsorption.

KEY WORDS

Acidosis, p. 894

Alkalosis, p. 894

Antibody (immunoglobulin), p. 887

Antigen, p. 886

Autoimmune disease, p. 890

Blood clot, p. 891

Blood plasma, p. 879

Blood serum, p. 884

Erythrocytes, p. 884

Extracellular fluid, p. 879

Fibrin, p. 890

Filtration (kidney), p. 895

SUMMARY: REVISITING THE CHAPTER GOALS

1. **How are body fluids classified?** Body fluids are either intracellular or extracellular. *Extracellular fluid* includes *blood plasma* (the fluid part of blood) and *interstitial fluid*. *Blood serum* is the fluid remaining after blood has clotted. Solutes in body fluids include blood gases, electrolytes, metabolites, and proteins. Solutes are carried throughout the body in blood and lymph. Exchange of solutes between blood and interstitial fluid occurs at the network of capillaries in peripheral tissues. Exchange of solutes between interstitial fluid and intracellular fluid occurs by passage across cell membranes.

2. **What are the roles of blood in maintaining homeostasis?** The principal functions of blood are (1) transport of solutes and blood gases, (2) regulation, including regulation of heat and acid–base balance, and (3) defense, which includes the *immune response* and *blood clotting*. In addition to plasma and proteins, blood is composed of red blood cells (*erythrocytes*), which transport oxygen; white blood cells (for defense functions); and *platelets*, which participate in blood clotting (Table 29.1).

3. **How do blood components participate in the body's defense mechanisms?** The presence of an *antigen* (a substance foreign to the body) initiates (1) the inflammatory response, (2) the cell-mediated immune response, and (3) the antibody-mediated immune response. The *inflammatory response* is initiated by histamine and accompanied by the destruction of

invaders by *phagocytes*. The *cell-mediated response* is effected by *T cells* that can, for example, release a toxic protein that kills invaders. The *antibody-mediated response* is effected by *B cells*, which generates *antibodies (immunoglobulins)*, proteins that complex with antigens and destroy them. Blood clotting occurs in a cascade of reactions in which a series of zymogens are activated, ultimately resulting in the formation of a clot composed of *fibrin* and platelets.

4. **How do red blood cells participate in the transport of blood gases?** Oxygen is transported bonded to Fe^{2+} ions in hemoglobin. The percent saturation of hemoglobin with oxygen (Figure 29.12) is governed by the partial pressure of oxygen in surrounding tissues and allosteric variations in hemoglobin structure. Carbon dioxide is transported in blood as a solute, bonded to hemoglobin, or in solution as bicarbonate ion. In peripheral tissues, carbon dioxide diffuses into red blood cells, where it is converted to bicarbonate ion. Acid–base balance is controlled as hydrogen ions generated by bicarbonate formation are bound by hemoglobin. At the lungs, oxygen enters the cells, and bicarbonate and hydrogen ions leave. A blood pH outside the normal range of 7.35–7.45 can be caused by respiratory or metabolic imbalance, resulting in the potentially serious conditions of *acidosis* or *alkalosis*.

5. **How is the composition of urine controlled?** The first essential kidney function is *filtration*, in which plasma and most of its solute cross capillary membranes and enter the *glomerular filtrate*. Water and essential solutes are then reabsorbed, whereas additional solutes for elimination are secreted into the filtrate. Urine is thus composed of the products of filtration, minus the substances reabsorbed, plus the secreted substances. It is composed of water, nitrogen-containing wastes, and electrolytes (including $H_2PO_4^-$ and NH_4^+) that are excreted to help to maintain acid–base balance. The balance between water and Na^+ excreted or absorbed is governed by the osmolarity of fluid in the kidney, the hormone aldosterone, and various chemical messengers.

Glomerular filtrate, *p. 895*
Hemostasis, *p. 891*
Immune response, *p. 886*
Inflammation, *p. 887*
Inflammatory response, *p. 886*
Interstitial fluid, *p. 879*
Intracellular fluid, *p. 879*
Osmolarity, *p. 879*
Reabsorption (kidney), *p. 895*
Secretion (kidney), *p. 895*
Whole blood, *p. 884*

UNDERSTANDING KEY CONCEPTS

29.5 Body fluids occupy two different compartments, either inside the cells or outside the cells.

 (a) What are body fluids found inside the cell called?
 (b) What are body fluids found outside the cell called?
 (c) What are the two major subclasses of fluids found outside the cells?
 (d) What major electrolytes are found inside the cells?
 (e) What major electrolytes are found outside the cells?

29.6 In the diagram shown here, fill in the blanks with the names of the principal components of whole blood:

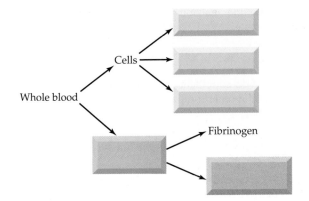

29.7 Fill in the blanks to identify some of the major functions of blood:

 (a) Blood carries _____ from lungs to tissues.
 (b) Blood carries _____ from the tissues to lungs.
 (c) Blood transports _____ from the digestive system to the tissues.
 (d) Blood carries _____ from the tissues to the site of excretion.
 (e) Blood transports _____ from the endocrine glands to their site of binding.
 (f) Blood transports defensive agents such as _____ to destroy foreign material and _____ to prevent blood loss.

29.8 List four symptoms of inflammation.

29.9 Explain how the chemical messenger histamine is biosynthesized, and how it elicits each symptom of inflammation.

29.10 Differentiate between cell-mediated immune response and antibody-mediated immune response.

29.11 How does the composition of urine help to maintain a healthy physiological acid–base balance?

BODY FLUIDS

29.12 What are the three principal body fluids and the approximate percentage of total body water accounted for by each?

29.13 What characteristics are needed for a substance to be soluble in body fluids?

29.14 Give an example of a substance that is not soluble in blood. How are components that are not normally soluble in blood transported?

29.15 What effects do the differences in pressure between arterial capillaries, interstitial fluids, and venous capillaries have on solutes crossing cell membranes?

29.16 How does blood pressure compare with the interstitial fluid pressure in arterial capillaries? With the interstitial fluid pressure in venous capillaries?

29.17 What is the purpose of the lymphatic system?

29.18 Where in the body does the lymph enter the bloodstream?

29.19 What is vasopresssin?

29.20 What is the purpose of antidiuretic hormone?

29.21 What is the difference between blood plasma and blood serum?

29.22 At what percent of body mass loss is collapse very likely to occur?

29.23 What are the three main types of blood cells?

29.24 What is the major function of the three types of blood cells?

29.25 What solutes in body fluids are referred to as electrolytes?

29.26 What are the major electrolytes inside cells and outside cells?

29.27 What is an antigen, and what are the three types of responses the body makes upon exposure to an antigen?

29.28 Antihistamines are often prescribed to counteract the effects of allergies. Explain how these drugs work. (Hint: See also Section 9.9.)

29.29 How are specific immune responses similar to the enzyme–substrate interaction?

29.30 What class of plasma proteins is involved in the antibody-mediated immune response?

29.31 What kinds of cells are associated with the antibody-directed immune response, and how do they work?

29.32 State the three major functions of T lymphocytes or T cells.

29.33 T cells are often discussed in conjunction with the disease AIDS, in which a virus destroys these cells. How do T cells work to combat disease?

29.34 What are memory cells, and what is their role in the immune response?

29.35 What is a blood clot? What is it composed of?

29.36 What vitamin and what mineral are specifically associated with the clotting process?

29.37 What two pathways trigger blood clotting?

29.38 Why do you suppose that many of the enzymes involved in blood clotting are secreted by the body as zymogens?

29.39 How many O_2 molecules can be bound by each hemoglobin tetramer?

29.40 What must be the charge of the iron in hemoglobin for it to perform its function?

29.41 How do deoxyhemoglobin and oxyhemoglobin differ in color?

29.42 How does the degree of saturation of hemoglobin vary with the partial pressure of O_2 in the tissues?

29.43 Oxygen has an allosteric interaction with hemoglobin. What are the results of this interaction as oxygen is bonded and as it is released?

29.44 What are the three ways of transporting CO_2 in the body?

29.45 Use Figure 29.11 to estimate the partial pressure of O_2 at which hemoglobin is 50% saturated with oxygen under normal conditions. Dry air at sea level is about 21% oxygen. What is the percentage saturation of your hemoglobin under these conditions?

29.46 When an actively metabolizing tissue produces CO_2, the H^+ concentration of blood increases. Explain how this happens using a chemical equation.

29.47 Do the following conditions cause hemoglobin to release more O_2 to the tissues or to absorb more O_2?

(a) Raising the temperature
(b) Production of CO_2
(c) Increasing the H^+ concentration

29.48 What is considered to be the normal range for blood serum pH? What is the metabolic condition called if the pH is lower than this range? If the pH is higher?

29.49 Ketoacidosis is a condition that can arise in an individual with diabetes due to excessive production of ketone bodies. Is this condition classified as metabolic acidosis or respiratory acidosis? Explain.

29.50 Is the acidosis caused by cardiac insufficiency due to congestive heart failure classified as metabolic acidosis or respiratory acidosis? Explain.

29.51 Kidneys are often referred to as filters that purify the blood. What other two essential functions do the kidneys perform to help maintain homeostasis?

29.52 Write the reactions by which HPO_4^{2-} and HCO_3^- absorb excess H^+ from the urine before elimination.

Applications

29.53 How do endothelial cells in brain capillaries differ from those in other capillary systems? [*The Blood–Brain Barrier, p. 888*]

29.54 What is meant by an asymmetric transport system? Give one specific example of such a system. [*The Blood–Brain Barrier, p. 888*]

29.55 What type of substance is likely to breach the blood–brain barrier? Would ethanol be likely to cross this barrier? Why or why not? [*The Blood–Brain Barrier, p. 888*]

29.56 Give a reason why a medicinal chemist would want to find a way to breach the blood–brain barrier and a disease this strategy could be used to treat? [*The Blood–Brain Barrier, p. 888*]

29.57 How are photometers used in automated analysis? [*Automated Clinical Laboratory Analysis, p. 897*]

29.58 Why is automated analysis useful to test for enzyme levels in body fluids? [*Automated Clinical Laboratory Analysis, p. 897*]

29.59 In analyzing body fluids for medical diagnoses, what are some advantages of using automated analyzers rather than technicians? [*Automated Clinical Laboratory Analysis, p. 897*]

General Questions and Problems

29.60 Why is ethanol soluble in blood?

29.61 Nursing mothers are able to impart some immunity to their infants. Why do you think this is so?

29.62 Many people find they retain water after eating salty food, evidenced by swollen fingers and ankles. Explain this phenomenon in terms of how the kidneys operate.

29.63 How does active transport differ from osmosis?

29.64 When is active transport necessary to move substances through cell membranes?

29.65 Discuss the importance of the CO_2/HCO_3^- equilibrium in blood and in urine.

29.66 We have discussed homeostasis throughout this text. But what is *hemostasis*? Is it related to homeostasis?

29.67 When people panic, cry, or have a high fever, they often begin to hyperventilate. Hyperventilation is abnormally fast or deep respiration, which results in the loss of carbon dioxide from the blood. Explain how hyperventilation changes the blood chemistry. Why can breathing into a paper bag alleviate hyperventilation?

APPENDIX A
Scientific Notation

What Is Scientific Notation?

The numbers that you encounter in chemistry are often either very large or very small. For example, there are about 33,000,000,000,000,000,000,000 H_2O molecules in 1.0 mL of water, and the distance between the H and O atoms in an H_2O molecule is 0.000 000 000 095 7 m. These quantities are more conveniently written in *scientific notation* as 3.3×10^{22} molecules and 9.57×10^{-11} m, respectively. In scientific notation (also known as *exponential notation*), a quantity is represented as a number between 1 and 10 multiplied by a power of 10. In this kind of expression, the small raised number to the right of the 10 is the exponent.

NUMBER	EXPONENTIAL FORM	EXPONENT
1,000,000	1×10^6	6
100,000	1×10^5	5
10,000	1×10^4	4
1,000	1×10^3	3
100	1×10^2	2
10	1×10^1	1
1		
0.1	1×10^{-1}	−1
0.01	1×10^{-2}	−2
0.001	1×10^{-3}	−3
0.000 1	1×10^{-4}	−4
0.000 01	1×10^{-5}	−5
0.000 001	1×10^{-6}	−6
0.000 000 1	1×10^{-7}	−7

Numbers greater than 1 have *positive* exponents, which tell how many times a number must be *multiplied* by 10 to obtain the correct value. For example, the expression 5.2×10^3 means that 5.2 must be multiplied by 10 three times:

$$5.2 \times 10^3 = 5.2 \times 10 \times 10 \times 10 = 5.2 \times 1000 = 5200$$

Note that doing this means moving the decimal point three places to the right:

5200.
123

The value of a positive exponent indicates *how many places to the right the decimal point must be moved* to give the correct number in ordinary decimal notation.

Numbers less than 1 have *negative* exponents, which tell how many times a number must be *divided* by 10 (or multiplied by one-tenth) to obtain the correct value. Thus, the expression 3.7×10^{-2} means that 3.7 must be divided by 10 two times:

$$3.7 \times 10^{-2} = \frac{3.7}{10 \times 10} = \frac{3.7}{100} = 0.037$$

Note that doing this means moving the decimal point two places to the left:

$$0.037$$
$$2\,1$$

The value of a negative exponent indicates *how may places to the left the decimal point must be moved* to give the correct number in ordinary decimal notation.

Representing Numbers in Scientific Notation

How do you convert a number from ordinary notation to scientific notation? If the number is greater than or equal to 10, shift the decimal point to the *left* by n places until you obtain a number between 1 and 10. Then, multiply the result by 10^n. For example, the number 8137.6 is written in scientific notation as 8.1376×10^3:

Number of places decimal point was shifted to the left

$$8137.6 = 8.1376 \times 10^3$$

Shift decimal point to the left by 3 places to get a number between 1 and 10

When you shift the decimal point to the left by three places, you are in effect dividing the number by $10 \times 10 \times 10 = 1000 = 10^3$. Therefore, you must multiply the result by 10^3 so that the value of the number is unchanged.

To convert a number less than 1 to scientific notation, shift the decimal point to the *right* by n places until you obtain a number between 1 and 10. Then, multiply the result by 10^{-n}. For example, the number 0.012 is written in scientific notation as 1.2×10^{-2}:

Number of places decimal point was shifted to the right

$$0.012 = 1.2 \times 10^{-2}$$

Shift decimal point to the right by 2 places to get a number between 1 and 10

When you shift the decimal point to the right by two places, you are in effect multiplying the number by $10 \times 10 = 100 = 10^2$. Therefore, you must multiply the result by 10^{-2} so that the value of the number is unchanged. ($10^2 \times 10^{-2} = 10^0 = 1$.)

The following table gives some additional examples. To convert from scientific notation to ordinary notation, simply reverse the preceding process. Thus, to write the number 5.84×10^4 in ordinary notation, drop the factor of 10^4 and move the decimal point 4 places to the *right* ($5.84 \times 10^4 = 58,400$). To write the number 3.5×10^{-1} in ordinary notation, drop the factor of 10^{-1} and move the decimal point 1 place to the *left* ($3.5 \times 10^{-1} = 0.35$). Note that you don't need scientific notation for numbers between 1 and 10 because $10^0 = 1$.

NUMBER	SCIENTIFIC NOTATION
58,400	5.84×10^4
0.35	3.5×10^{-1}
7.296	$7.296 \times 10^0 = 7.296 \times 1$

Mathematical Operations with Scientific Notation

Addition and Subtraction in Scientific Notation

To add or subtract two numbers expressed in scientific notation, both numbers must have the same exponent. Thus, to add 7.16×10^3 and 1.32×10^2, first write the latter number as 0.132×10^3 and then add:

$$\begin{array}{r} 7.16 \ \times 10^3 \\ +0.132 \times 10^3 \\ \hline 7.29 \ \times 10^3 \end{array}$$

The answer has three significant figures. (Significant figures are discussed in Section 2.4.) Alternatively, you can write the first number as 71.6×10^2 and then add:

$$\begin{array}{r} 71.6 \ \times 10^2 \\ + \ 1.32 \times 10^2 \\ \hline 72.9 \ \times 10^2 = 7.29 \times 10^3 \end{array}$$

Subtraction of these two numbers is carried out in the same manner.

$$\begin{array}{r} 7.16 \ \times 10^3 \\ -0.132 \times 10^3 \\ \hline 7.03 \ \times 10^3 \end{array} \quad \text{or} \quad \begin{array}{r} 71.6 \ \times 10^2 \\ - \ 1.32 \times 10^2 \\ \hline 70.3 \ \times 10^2 = 7.03 \times 10^3 \end{array}$$

Multiplication in Scientific Notation

To multiply two numbers expressed in scientific notation, multiply the factors in front of the powers of 10 and then add the exponents. For example,

$$(2.5 \times 10^4)(4.7 \times 10^7) = (2.5)(4.7) \times 10^{4+7} = 12 \times 10^{11} = 1.2 \times 10^{12}$$

$$(3.46 \times 10^5)(2.2 \times 10^{-2}) = (3.46)(2.2) \times 10^{5+(-2)} = 7.6 \times 10^3$$

Both answers have two significant figures.

Division in Scientific Notation

To divide two numbers expressed in scientific notation, divide the factors in front of the powers of 10 and then subtract the exponent in the denominator from the exponent in the numerator. For example,

$$\frac{3 \times 10^6}{7.2 \times 10^2} = \frac{3}{7.2} \times 10^{6-2} = 0.4 \times 10^4 = 4 \times 10^3 \quad \text{(1 significant figure)}$$

$$\frac{7.50 \times 10^{-5}}{2.5 \times 10^{-7}} = \frac{7.50}{2.5} \times 10^{-5-(-7)} = 3.0 \times 10^2 \quad \text{(2 significant figures)}$$

Scientific Notation and Electronic Calculators

With a scientific calculator you can carry out calculations in scientific notation. You should consult the instruction manual for your particular calculator to learn how to enter and manipulate numbers expressed in an exponential format. On most calculators, you enter the number $A \times 10^n$ by (i) entering the number A, (ii) pressing a key labeled EXP or EE, and (iii) entering the exponent n. If the exponent is negative, you press a key labeled $+/-$ before entering the value of n. (Note that you do not enter the number 10.) The calculator displays the number $A \times 10^n$ with the number A on the left followed by some space and then the exponent n. For example,

$$4.625 \times 10^2 \quad \text{is displayed as} \quad 4.625 \ 02$$

To add, subtract, multiply, or divide exponential numbers, use the same sequence of keystrokes as you would in working with ordinary numbers. When you add or subtract on a calculator, the numbers need not have the same exponent; the calculator automatically takes account of the different exponents. Remember, though, that the calculator often gives more digits in the answer than the allowed number of significant figures. It's sometimes helpful to outline the calculation on paper, as in the preceding examples, to keep track of the number of significant figures.

PROBLEM A.1

Perform the following calculations, expressing the results in scientific notation with the correct number of significant figures. (You don't need a calculator for these.)

(a) $(1.50 \times 10^4) + (5.04 \times 10^3)$

(b) $(2.5 \times 10^{-2}) - (5.0 \times 10^{-3})$

(c) $(6.3 \times 10^{15}) \times (10.1 \times 10^3)$

(d) $(2.5 \times 10^{-3}) \times (3.2 \times 10^{-4})$

(e) $(8.4 \times 10^4) \div (3.0 \times 10^6)$

(f) $(5.530 \times 10^{-2}) \div (2.5 \times 10^{-5})$

ANSWERS

(a) 2.00×10^4 (b) 2.0×10^{-2} (c) 6.4×10^{19}

(d) 8.0×10^{-7} (e) 2.8×10^{-2} (f) 2.2×10^3

PROBLEM A.2

Perform the following calculations, expressing the results in scientific notation with the correct number of significant figures. (Use a calculator for these.)

(a) $(9.72 \times 10^{-1}) + (3.4823 \times 10^2)$

(b) $(3.772 \times 10^3) - (2.891 \times 10^4)$

(c) $(1.956 \times 10^3) \div (6.02 \times 10^{23})$

(d) $3.2811 \times (9.45 \times 10^{21})$

(e) $(1.0015 \times 10^3) \div (5.202 \times 10^{-9})$

(f) $(6.56 \times 10^{-6}) \times (9.238 \times 10^{-4})$

ANSWERS

(a) 3.4920×10^2 (b) -2.514×10^4 (c) 3.25×10^{-21}

(d) 3.10×10^{22} (e) 1.925×10^{11} (f) 6.06×10^{-9}

APPENDIX B
Conversion Factors

Length   SI Unit: Meter (m)

1 meter = 0.001 kilometer (km)

= 100 centimeters (cm)

= 1.0936 yards (yd)

1 centimeter = 10 millimeters (mm)

= 0.3937 inch (in.)

1 nanometer = 1×10^{-9} meter

1 Angstrom (Å) = 1×10^{-10} meter

1 inch = 2.54 centimeters

1 mile = 1.6094 kilometers

Volume   SI Unit: Cubic meter (m^3)

1 cubic meter = 1000 liters (L)

1 liter = 1000 cubic centimeters (cm^3)

= 1000 milliliters (mL)

= 1.056710 quarts (qt)

1 cubic inch = 16.4 cubic centimeters

Temperature   SI Unit: Kelvin (K)

0 K = $-273.15\ ^{\circ}C$

= $-459.67\ ^{\circ}F$

$^{\circ}F = (9/5)\ ^{\circ}C + 32^{\circ};\ ^{\circ}F = (1.8 \times\ ^{\circ}C) + 32^{\circ}$

$^{\circ}C = (5/9)(^{\circ}F - 32^{\circ});\ ^{\circ}C = \dfrac{(^{\circ}F - 32^{\circ})}{1.8}$

$K =\ ^{\circ}C + 273.15^{\circ}$

Mass   SI Unit: Kilogram (kg)

1 kilogram = 1000 grams (g)

= 2.205 pounds (lb)

1 gram = 1000 milligrams (mg)

= 0.03527 ounce (oz)

1 pound = 453.6 grams

1 atomic mass unit = 1.66054×10^{-24} gram

Pressure   SI Unit: Pascal (Pa)

1 pascal = 9.869×10^{-6} atmosphere

1 atmosphere = 101,325 pascals

= 760 mmHg (Torr)

= 14.70 lb/in^2

Energy   SI Unit: Joule (J)

1 joule = 0.23901 calorie (cal)

1 calorie = 4.184 joules

1 Calorie (nutritional unit) = 1000 calories

= 1 kcal

Glossary

1,4 Link   A glycosidic link between the hemiacetal hydroxyl group at C1 of one sugar and the hydroxyl group at C4 of another sugar.

Acetal   A compound that has two ether-like —OR groups bonded to the same carbon atom.

Acetyl coenzyme A (acetyl-SCoA)   Acetyl-substituted coenzyme A—the common intermediate that carries acetyl groups into the citric acid cycle.

Acetyl group   A $CH_3C{=}O$ group.

Achiral   The opposite of chiral; having no right- or left-handedness and no nonsuperimposable mirror images.

Acid   A substance that provides H^+ ions in water.

Acid dissociation constant (K_a)   The equilibrium constant for the dissociation of an acid (HA), equal to $[H^+][A^-]/[HA]$

Acidosis   The abnormal condition associated with a blood plasma pH below 7.35; may be respiratory or metabolic.

Acid–base indicator   A dye that changes color depending on the pH of a solution.

Activation (of an enzyme)   Any process that initiates or increases the action of an enzyme.

Activation energy (E_{act})   The amount of energy necessary for reactants to surmount the energy barrier to reaction; affects reaction rate.

Active site   A pocket in an enzyme with the specific shape and chemical makeup necessary to bind a substrate.

Active transport   Movement of substances across a cell membrane with the assistance of energy (for example, from ATP).

Actual Yield   The amount of product actually formed in a reaction.

Acyl group   An $RC{=}O$ group.

Addition reaction   A general reaction type in which a substance X—Y adds to the multiple bond of an unsaturated reactant to yield a saturated product that has only single bonds.

Addition reaction, aldehydes and ketones   Addition of an alcohol or other compound to the carbon–oxygen double bond to give a carbon–oxygen single bond.

Adenosine triphosphate (ATP)   The principal energy-carrying molecule; removal of a phosphoryl group to give ADP releases free energy.

Aerobic   In the presence of oxygen.

Agonist   A substance that interacts with a receptor to cause or prolong the receptor's normal biochemical response.

Alcohol   A compound that has an —OH group bonded to a saturated, alkane-like carbon atom, R—OH.

Alcoholic fermentation   The anaerobic breakdown of glucose to ethanol plus carbon dioxide by the action of yeast enzymes.

Aldehyde   A compound that has a carbonyl group bonded to one carbon and one hydrogen, RCHO.

Aldose   A monosaccharide that contains an aldehyde carbonyl group.

Alkali metal   An element in group 1A of the periodic table.

Alkaline earth metal   An element in group 2A of the periodic table.

Alkaloid   A naturally occurring nitrogen-containing compound isolated from a plant; usually basic, bitter, and poisonous.

Alkalosis   The abnormal condition associated with a blood plasma pH above 7.45; may be respiratory or metabolic.

Alkane   A hydrocarbon that has only single bonds.

Alkene   A hydrocarbon that contains a carbon–carbon double bond.

Alkoxide ion   The anion resulting from deprotonation of an alcohol, RO^-.

Alkoxy group   An —OR group.

Alkyl group   The part of an alkane that remains when a hydrogen atom is removed.

Alkyl halide   A compound that has an alkyl group bonded to a halogen atom, R—X.

Alkyne   A hydrocarbon that contains a carbon–carbon triple bond.

Allosteric control   An interaction in which the binding of a regulator at one site on a protein affects the protein's ability to bind another molecule at a different site.

Allosteric enzyme   An enzyme whose activity is controlled by the binding of an activator or inhibitor at a location other than the active site.

Alpha (α) particle   A helium nucleus (He^{2+}), emitted as α-radiation.

Alpha- (α-) amino acid   An amino acid in which the amino group is bonded to the carbon atom next to the —COOH group.

Alpha- (α-) helix   Secondary protein structure in which a protein chain forms a right-handed coil stabilized by hydrogen bonds between peptide groups along its backbone.

Amide   A compound that has a carbonyl group bonded to a carbon atom and a nitrogen atom group, $RCONR'_2$, where the R' groups may be alkyl groups or hydrogen atoms.

Amine   A compound that has one or more organic groups bonded to nitrogen; primary, RNH_2; secondary, R_2NH; or tertiary, R_3N.

Amino acid   A molecule that contains both an amino group and a carboxylic acid functional group.

Amino acid pool   The entire collection of free amino acids in the body.

Amino group   The —NH_2 functional group.

Amino-terminal (N-terminal) amino acid   The amino acid with the free —NH_3^+ group at the end of a protein.

Ammonium ion   A positive ion formed by addition of hydrogen to ammonia or an amine (may be primary, secondary, or tertiary).

Ammonium salt   An ionic compound composed of an ammonium cation and an anion; an amine salt.

Amorphous solid   A solid whose particles do not have an orderly arrangement.

Amphoteric   Describing a substance that can react as either an acid or a base.

Anabolism   Metabolic reactions that build larger biological molecules from smaller pieces.

Anaerobic   In the absence of oxygen.

Anion   A negatively charged ion.

Anomeric carbon atom   The hemiacetal C atom in a cyclic sugar; the C atom bonded to an —OH group and an O in the ring.

Anomers   Cyclic sugars that differ only in positions of substituents at the hemiacetal carbon (the anomeric carbon); the α form has the —OH on the opposite side from the —CH_2OH; the β form has the —OH on the same side as the —CH_2OH.

Antagonist   A substance that blocks or inhibits the normal biochemical response of a receptor.

Antibody (immunoglobulin)   Glycoprotein molecule that identifies antigens.

Anticodon   A sequence of three ribonucleotides on tRNA that recognizes the complementary sequence (the codon) on mRNA.

Antigen   A substance foreign to the body that triggers the immune response.

Antioxidant   A substance that prevents oxidation by reacting with an oxidizing agent.

Aqueous solution   A solution in which water is the solvent.

Aromatic   The class of compounds containing benzene-like rings.

Artificial transmutation   The change of one atom into another brought about by a nuclear bombardment reaction.

Atom   The smallest and simplest particle of an element.

Atomic mass unit (amu)   A convenient unit for describing the mass of an atom; 1 amu = 1/12 the mass of a carbon-12 atom.

Atomic number (Z)   The number of protons in an atom.

Atomic theory   A set of assumptions proposed by English scientist John Dalton to explain the chemical behavior of matter.

Atomic weight   The weighted average mass of an element's atoms.

ATP synthase   The enzyme complex in the inner mitochondrial membrane at which hydrogen ions cross the membrane and ATP is synthesized from ADP.

Autoimmune disease   Disorder in which the immune system identifies normal body components as antigens and produces antibodies to them.

Avogadro's law   Equal volumes of gases at the same temperature and pressure contain equal numbers of molecules (V/n = constant, or $V_1/n_1 = V_2/n_2$).

Avogadro's number (N_A) The number of units in 1 mole of anything; 6.02×10^{23}.

Balanced equation Describing a chemical equation in which the numbers and kinds of atoms are the same on both sides of the reaction arrow.

Base A substance that provides OH^- ions in water.

Base pairing The pairing of bases connected by hydrogen bonding (G-C and A-T), as in the DNA double helix.

Beta- (β-) Oxidation pathway A repetitive series of biochemical reactions that degrades fatty acids to acetyl-SCoA by removing carbon atoms two at a time.

Beta (β) particle An electron (e^-), emitted as β radiation.

Beta- (β-) Sheet Secondary protein structure in which adjacent protein chains either in the same molecule or in different molecules are held in place by hydrogen bonds along the backbones.

Bile acids Steroid acids derived from cholesterol that are secreted in bile.

Bile Fluid secreted by the liver and released into the small intestine from the gallbladder during digestion; contains bile acids, bicarbonate ion, and other electrolytes.

Binary compound A compound formed by combination of two different elements.

Blood clot A network of fibrin fibers and trapped blood cells that forms at the site of blood loss.

Blood plasma Liquid portion of the blood: an extracellular fluid.

Blood serum Fluid portion of blood remaining after clotting has occurred.

Boiling point (bp) The temperature at which liquid and gas are in equilibrium.

Bond angle The angle formed by three adjacent atoms in a molecule.

Bond dissociation energy The amount of energy that must be supplied to break a bond and separate the atoms in an isolated gaseous molecule.

Bond length The optimum distance between nuclei in a covalent bond.

Boyle's law The pressure of a gas at constant temperature is inversely proportional to its volume ($PV = $ constant, or $P_1V_1 = P_2V_2$).

Branched-chain alkane An alkane that has a branching connection of carbons.

Brønsted–Lowry acid A substance that can donate a hydrogen ion, H^+, to another molecule or ion.

Brønsted–Lowry base A substance that can accept H^+ from an acid.

Buffer A combination of substances that act together to prevent a drastic change in pH; usually a weak acid and its conjugate base.

Carbohydrate A member of a large class of naturally occurring polyhydroxy ketones and aldehydes.

Carbonyl compound Any compound that contains a carbonyl group $C=O$.

Carbonyl group A functional group that has a carbon atom joined to an oxygen atom by a double bond, $C=O$.

Carbonyl-group substitution reaction A reaction in which a new group replaces (substitutes for) a group attached to a carbonyl-group carbon in an acyl group.

Carboxyl group The $-COOH$ functional group.

Carboxyl-terminal (C-terminal) amino acid The amino acid with the free $-COO^-$ group at the end of a protein.

Carboxylate anion The anion that results from ionization of a carboxylic acid, $RCOO^-$.

Carboxylic acid A compound that has a carbonyl group bonded to a carbon atom and an $-OH$ group, RCOOH.

Carboxylic acid salt An ionic compound containing a carboxylic anion and a cation.

Catabolism Metabolic reaction pathways that break down food molecules and release biochemical energy.

Catalyst A substance that speeds up the rate of a chemical reaction but is itself unchanged.

Cation A positively charged ion.

Centromeres The central regions of chromosomes.

Chain reaction A reaction that, once started, is self-sustaining.

Change of state The conversion of a substance from one state to another—for example, from a liquid to a gas.

Charles's law The volume of a gas at constant pressure is directly proportional to its Kelvin temperature ($V/T = $ constant, or $V_1/T_1 = V_2/T_2$).

Chemical change A change in the chemical makeup of a substance.

Chemical compound A pure substance that can be broken down into simpler substances by chemical reactions.

Chemical equation An expression in which symbols and formulas are used to represent a chemical reaction.

Chemical equilibrium A state in which the rates of forward and reverse reactions are the same.

Chemical formula A notation for a chemical compound using element symbols and subscripts to show how many atoms of each element are present.

Chemical reaction A process in which the identity and composition of one or more substances are changed.

Chemistry The study of the nature, properties, and transformations of matter.

Chiral carbon atom (chirality center) A carbon atom bonded to four different groups.

Chiral Having right- or left-handedness; able to have two different mirror-image forms.

Chromosome A complex of proteins and DNA; visible during cell division.

Cis-trans isomers Alkenes that have the same connections between atoms but differ in their three-dimensional structures because of the way that groups are attached to different sides of the double bond. The cis isomer has hydrogen atoms on the same side of the double bond; the trans isomer has them on opposite sides.

Citric acid cycle The series of biochemical reactions that breaks down acetyl groups to produce energy carried by reduced coenzymes and carbon dioxide.

Clones Identical copies of organisms, cells, or DNA segments from a single ancestor.

Codon A sequence of three ribonucleotides in the messenger RNA chain that codes for a specific amino acid; also the three nucleotide sequence (a stop codon) that stops translation.

Coefficient A number placed in front of a formula to balance a chemical equation.

Coenzyme An organic molecule that acts as an enzyme cofactor.

Cofactor A nonprotein part of an enzyme that is essential to the enzyme's catalytic activity; a metal ion or a coenzyme.

Colligative property A property of a solution that depends only on the number of dissolved particles, not on their chemical identity.

Colloid A homogeneous mixture that contains particles that range in diameter from 2 to 500 nm.

Combined gas law The product of the pressure and volume of a gas is proportional to its temperature ($PV/T = $ constant, or $P_1V_1/T_1 = P_2V_2/T_2$).

Combustion A chemical reaction that produces a flame, usually because of burning with oxygen.

Competititve (enzyme) inhibition Enzyme regulation in which an inhibitor competes with a substrate for binding to the enzyme active site.

Concentration A measure of the amount of a given substance in a mixture.

Concentration gradient A difference in concentration within the same system.

Condensed structure A shorthand way of drawing structures in which $C-C$ and $C-H$ bonds are understood rather than shown.

Conformation The specific three-dimensional arrangement of atoms in a molecule at a given instant.

Conformers Molecular structures having identical connections between atoms.

Conjugate acid The substance formed by addition of H^+ to a base.

Conjugate acid–base pair Two substances whose formulas differ by only a hydrogen ion, H^+.

Conjugate base The substance formed by loss of H^+ from an acid.

Conjugated protein A protein that incorporates one or more non-amino acid units in its structure.

Constitutional isomers Compounds with the same molecular formula but different connections among their atoms.

Conversion factor An expression of the relationship between two units.

Coordinate covalent bond The covalent bond that forms when both electrons are donated by the same atom.

Cosmic rays A mixture of high-energy particles—primarily of protons and various atomic nuclei—that shower the earth from outer space.

Covalent bond A bond formed by sharing electrons between atoms.

Critical mass The minimum amount of radioactive material needed to sustain a nuclear chain reaction.

Crystalline solid A solid whose atoms, molecules, or ions are rigidly held in an ordered arrangement.

Cycloalkane An alkane that contains a ring of carbon atoms.

Cycloalkene A cyclic hydrocarbon that contains a double bond.

Cytoplasm The region between the cell membrane and the nuclear membrane in a eukaryotic cell.

Cytosol The fluid part of the cytoplasm surrounding the organelles within a cell.

d-Block element A transition metal element that results from the filling of d orbitals.

D-Sugar Monosaccharide with the $-OH$ group on the chiral carbon atom farthest from the carbonyl group pointing to the right in a Fischer projection.

Dalton's law The total pressure exerted by a mixture of gases is equal to the sum of the partial pressures exerted by each individual gas.

Decay series A sequential series of nuclear disintegrations leading from a heavy radioisotope to a nonradioactive product.

Degree of unsaturation The number of carbon–carbon double bonds in a molecule.

Dehydration The loss of water from an alcohol to yield an alkene.

Denaturation The loss of secondary, tertiary, or quaternary protein structure due to disruption of noncovalent interactions and/or disulfide bonds that leaves peptide bond and primary structure intact.

Density The physical property that relates the mass of an object to its volume; mass per unit volume.

Deoxyribonucleotide A nucleotide containing 2-deoxy-D-ribose.

Diabetes mellitus A chronic condition due to either insufficient insulin or failure of insulin to activate crossing of cell membranes by glucose.

Diastereomers Stereoisomers that are not mirror images of each other.

Digestion A general term for the breakdown of food into small molecules.

Dilution factor The ratio of the initial and final solution volumes (V_1/V_2).

Dipole–dipole force The attractive force between positive and negative ends of polar molecules.

Disaccharide A carbohydrate composed of two monosaccharides.

Dissociation The splitting apart of an acid in water to give H^+ and an anion.

Disulfide A compound that contains a sulfur–sulfur bond, RS–SR.

Disulfide bond (in protein) An S–S bond formed between two cysteine side chains; can join two peptide chains together or cause a loop in a peptide chain.

DNA (deoxyribonucleic acid) The nucleic acid that stores genetic information; a polymer of deoxyribonucleotides.

Double bond A covalent bond formed by sharing two electron pairs.

Double helix Two strands coiled around each other in a screwlike fashion; in most organisms the two polynucleotides of DNA form a double helix.

Drug Any substance that alters body function when it is introduced from an external source.

Eicosanoid A lipid derived from a 20-carbon unsaturated carboxylic acid.

Electrolyte A substance that produces ions and therefore conducts electricity when dissolved in water.

Electron A negatively charged subatomic particle.

Electron affinity The energy released on adding an electron to a single atom in the gaseous state.

Electron capture A process in which the nucleus captures an inner-shell electron from the surrounding electron cloud, thereby converting a proton into a neutron.

Electron configuration The specific arrangement of electrons in an atom's shells and subshells.

Electron shell A grouping of electrons in an atom according to energy.

Electron subshell A grouping of electrons in a shell according to the shape of the region of space they occupy.

Electron-dot symbol An atomic symbol with dots placed around it to indicate the number of valence electrons.

Electron-transport chain The series of biochemical reactions that passes electrons from reduced coenzymes to oxygen and is coupled to ATP formation.

Electronegativity The ability of an atom to attract electrons in a covalent bond.

Element A fundamental substance that can't be broken down chemically into any simpler substance.

Elimination reaction A general reaction type in which a saturated reactant yields an unsaturated product by losing groups from two adjacent carbon atoms.

Enantiomers, optical isomers The two mirror-image forms of a chiral molecule.

Endergonic A nonspontaneous reaction or process that absorbs free energy and has a positive ΔG.

Endocrine system A system of specialized cells, tissues, and ductless glands that excretes hormones and shares with the nervous system the responsibility for maintaining constant internal body conditions and responding to changes in the environment.

Endothermic A process or reaction that absorbs heat and has a positive ΔH.

Energy The capacity to do work or supply heat.

Enthalpy A measure of the amount of energy associated with substances involved in a reaction.

Enthalpy change (ΔH) An alternative name for heat of reaction.

Entropy (S) The amount of disorder in a system.

Enzyme A protein or other molecule that acts as a catalyst for a biological reaction.

Equilibrium constant (K) Value of the equilibrium constant expression for a given reaction.

Equivalent For ions, the amount equal to 1 mol of charge.

Equivalent of acid Amount of an acid that contains 1 mole of H^+ ions.

Equivalent of base Amount of base that contains 1 mole of OH^- ions.

Erythrocytes Red blood cells; transporters of blood gases.

Essential amino acid An amino acid that cannot be synthesized by the body and thus must be obtained in the diet.

Ester A compound that has a carbonyl group bonded to a carbon atom and an —OR′ group, RCOOR′.

Esterification The reaction between an alcohol and a carboxylic acid to yield an ester plus water.

Ether A compound that has an oxygen atom bonded to two organic groups, R—O—R.

Ethyl group The —CH_2CH_3 alkyl group.

Exergonic A spontaneous reaction or process that releases free energy and has a negative ΔG.

Exon A nucleotide sequence in DNA that is part of a gene and codes for part of a protein.

Exothermic A process or reaction that releases heat and has a negative ΔH.

Extracellular fluid Fluid outside cells.

ƒ-Block element An inner transition metal element that results from the filling of ƒ orbitals.

Facilitated diffusion Passive transport across a cell membrane with the assistance of a protein that changes shape.

Factor–label method A problem-solving procedure in which equations are set up so that unwanted units cancel and only the desired units remain.

Fat A mixture of triacylglycerols that is solid because it contains a high proportion of saturated fatty acids.

Fatty acid A long-chain carboxylic acid; those in animal fats and vegetable oils often have 12–22 carbon atoms.

Feedback control Regulation of an enzyme's activity by the product of a reaction later in a pathway.

Fermentation The production of energy under anaerobic conditions.

Fibrin Insoluble protein that forms the fiber framework of a blood clot.

Fibrous protein A tough, insoluble protein whose protein chains form fibers or sheets.

Filtration (kidney) Filtration of blood plasma through a glomerulus and into a kidney nephron.

Fischer projection Structure that represents chiral carbon atoms as the intersections of two lines, with the horizontal lines representing bonds pointing out of the page and the vertical lines representing bonds pointing behind the page. For sugars, the aldehyde or ketone is at the top.

Formula unit The formula that identifies the smallest neutral unit of a compound.

Formula weight The sum of the atomic weights of the atoms in one formula unit of any compound.

Free radical An atom or molecule with an unpaired electron.

Free-energy change (ΔG) The criterion for spontaneous change (negative ΔG; $\Delta G = \Delta H - T\,\Delta S$).

Functional group An atom or group of atoms within a molecule that has a characteristic structure and chemical behavior.

Gamma (γ) radiation Radioactivity consisting of high-energy light waves.

Gas A substance that has neither a definite volume nor a definite shape.

Gas constant (R) The constant R in the ideal gas law, $PV = nRT$.

Gas laws A series of laws that predict the influence of pressure (P), volume (V), and temperature (T) on any gas or mixture of gases.

Gay-Lussac's law For a fixed amount of gas at a constant voume, pressure is directly proportional to the Kelvin temperature (P/T = constant, or $P_1/T_1 = P_2/T_2$).

Gene Segment of DNA that directs the synthesis of a single polypeptide.

Genetic (enzyme) control Regulation of enzyme activity by control of the synthesis of enzymes.

Genetic code The sequence of nucleotides, coded in triplets (codons) in mRNA, that determines the sequence of amino acids in protein synthesis.

Genome All of the genetic material in the chromosomes of an organism; its size is given as the number of base pairs.

Genomics The study of whole sets of genes and their functions.

Globular protein A water–soluble protein whose chain is folded in a compact shape with hydrophilic groups on the outside.

Glomerular filtrate Fluid that enters the nephron from the glomerulus; filtered blood plasma.

Gluconeogenesis The biochemical pathway for the synthesis of glucose from noncarbohydrates, such as lactate, amino acids, or glycerol.

Glycerophospholipid (phosphoglyceride) A lipid in which glycerol is linked by ester bonds to two fatty acids and one phosphate, which is in turn linked by another ester bond to an amino alcohol (or other alcohol).

Glycogenesis The biochemical pathway for synthesis of glycogen.

Glycogenolysis The biochemical pathway for breakdown of glycogen to free glucose.

Glycol A dialcohol, or diol having the two –OH groups on adjacent carbons.

Glycolipid A lipid with a fatty acid bonded to the $C2-NH_2$ and a sugar bonded to the $C1-OH$ group of sphingosine.

Glycolysis The biochemical pathway that breaks down a molecule of glucose into two molecules of pyruvate plus energy.

Glycoprotein A protein that contains a short carbohydrate chain.

Glycoside A cyclic acetal formed by reaction of a monosaccharide with an alcohol, accompanied by loss of H_2O.

Glycosidic bond Bond between the anomeric carbon atom of a monosaccharide and an $-OR$ group.

Gram-equivalent For ions, the molar mass of the ion divided by the ionic charge.

Group One of the 18 vertical columns of elements in the periodic table.

Guanosine diphosphate (GDP) An energy-carrying molecule that can gain or lose a phosphoryl group to transfer energy.

Guanosine triphosphate (GTP) An energy-carrying molecule similar to ATP; removal of a phosphoryl group to give GDP releases free energy.

Half-life ($t_{1/2}$) The amount of time required for one-half of a radioactive sample to decay.

Halogen An element in group 7A of the periodic table.

Halogenation (alkene) The addition of Cl_2 or Br_2 to a multiple bond to give a 1,2-dihalide product.

Halogenation (aromatic) The substitution of a halogen group $(-X)$ for a hydrogen on an aromatic ring.

Heat The kinetic energy transferred from a hotter object to a colder object when the two are in contact.

Heat of fusion The quantity of heat required to completely melt a substance once it has reached its melting point.

Heat of reaction (ΔH) The amount of heat absorbed or released in a reaction.

Heat of vaporization The quantity of heat needed to completely vaporize a liquid once it has reached its boiling point.

Hemiacetal A compound with both an alcohol-like $-OH$ group and an ether-like $-OR$ group bonded to the same carbon atom.

Hemostasis The stopping of bleeding.

Henderson-Hasselbalch equation The logarithmic form of the K_a equation for a weak acid, used in applications involving buffer solutions.

Henry's law The solubility of a gas in a liquid is directly proportional to its partial pressure over the liquid at constant temperature.

Heterocycle A ring that contains nitrogen or some other atom in addition to carbon.

Heterogeneous mixture A nonuniform mixture that has regions of different composition.

Heterogeneous nuclear RNA The initially synthesized mRNA strand containing both introns and exons.

Homogeneous mixture A uniform mixture that has the same composition throughout.

Hormone A chemical messenger secreted by cells of the endocrine system and transported through the bloodstream to target cells with appropriate receptors where it elicits a response.

Hydration The addition of water to a multiple bond to give an alcohol product.

Hydrocarbon An organic compound that contains only carbon and hydrogen.

Hydrogen bond The attraction between a hydrogen atom bonded to an electronegative O, N, or F atom and another nearby electronegative O, N, or F atom.

Hydrogenation The addition of H_2 to a multiple bond to give a saturated product.

Hydrohalogenation The addition of HCl or HBr to a multiple bond to give an alkyl halide product.

Hydrolysis A reaction in which a bond or bonds are broken and the $H-$ and $-OH$ of water add to the atoms of the broken bond or bonds.

Hydronium ion The H_3O^+ ion, formed when an acid reacts with water.

Hydrophilic Water-loving; a hydrophilic substance dissolves in water.

Hydrophobic Water-fearing; a hydrophobic substance does not dissolve in water.

Hygroscopic Having the ability to pull water molecules from the surrounding atmosphere.

Hyperglycemia Higher-than-normal blood glucose concentration.

Hypertonic Having an osmolarity greater than the surrounding blood plasma or cells.

Hypoglycemia Lower-than-normal blood glucose concentration.

Hypotonic Having an osmolarity less than the surrounding blood plasma or cells.

Ideal gas A gas that obeys all the assumptions of the kinetic–molecular theory.

Ideal gas law A general expression relating pressure, volume, temperature, and amount for an ideal gas: $PV = nRT$.

Immune response Defense mechanism of the immune system dependent on the recognition of specific antigens, including viruses, bacteria, toxic substances, and infected cells; either cell-mediated or antibody-mediated.

Induced-fit model A model of enzyme action in which the enzyme has a flexible active site that changes shape to best fit the substrate and catalyze the reaction.

Inflammation Result of the inflammatory response: includes swelling, redness, warmth, and pain.

Inflammatory response A nonspecific defense mechanism triggered by antigens or tissue damage.

Inhibition (of an enzyme) Any process that slows or stops the action of an enzyme.

Inner transition metal element An element in one of the 14 groups shown separately at the bottom of the periodic table.

Intermolecular force A force that acts between molecules and holds molecules close to one another in liquids and solids.

Interstitial fluid Fluid surrounding cells: an extracellular fluid.

Intracellular fluid Fluid inside cells.

Intron A portion of DNA between coding regions of a gene (exons); is transcribed and then removed from final messenger RNA.

Ion An electrically charged atom or group of atoms.

Ion-product constant for water (K_w) The product of the H_3O^+ and OH^- molar concentrations in water or any aqueous solution ($K_w = [H_3O^+][OH^-] = 1.00 \times 10^{-14}$).

Ionic bond The electrical attractions between ions of opposite charge in a crystal.

Ionic compound A compound that contains ionic bonds.

Ionic equation An equation in which ions are explicitly shown.

Ionic solid A crystalline solid held together by ionic bonds.

Ionization energy The energy required to remove one electron from a single atom in the gaseous state.

Ionizing radiation A general name for high-energy radiation of all kinds.

Irreversible (enzyme) inhibition Enzyme deactivation in which an inhibitor forms covalent bonds to the active site, permanently blocking it.

Isoelectric point (pI) The pH at which a sample of an amino acid has equal number of + and – charges.

Isomers Compounds with the same molecular formula but different structures.

Isopropyl group The branched-chain alkyl group $-CH(CH_3)_2$.

Isotonic Having the same osmolarity.

Isotopes Atoms with identical atomic numbers but different mass numbers.

Ketoacidosis Lowered blood pH due to accumulation of ketone bodies.

Ketogenesis The synthesis of ketone bodies from acetyl-SCoA.

Ketone A compound that has a carbonyl group bonded to two carbons in organic groups that can be the same or different, $R_2C=O$, RCOR'.

Ketone bodies Compounds produced in the liver that can be used as fuel by muscle and brain tissue; 3-hydroxybutyrate, acetoacetate, and acetone.

Ketose A monosaccharide that contains a ketone carbonyl group.

Kinetic energy The energy of an object in motion.

Kinetic–molecular theory (KMT) of gases A group of assumptions that explain the behavior of gases.

L-Sugar Monosaccharide with the $-OH$ group on the chiral carbon atom farthest from the carbonyl group pointing to the left in a Fischer projection.

Law of conservation of energy Energy can be neither created nor destroyed in any physical or chemical change.

Law of conservation of mass Matter can be neither created nor destroyed in any physical or chemical change.

Le Châtelier's principle When a stress is applied to a system in equilibrium, the equilibrium shifts to relieve the stress.

Lewis base A compound containing an unshared pair of electrons.

Lewis structure A molecular representation that shows both the connections among atoms and the locations of lone-pair valence electrons.

Limiting reagent The reactant that runs out first in a chemical reaction.

Line structure A shorthand way of drawing structures in which atoms aren't shown; instead, a carbon atom is understood to be at every intersection of lines, and hydrogens are filled in mentally.

Lipid A naturally occurring molecule from a plant or animal that is soluble in nonpolar organic solvents.

Lipid bilayer The basic structural unit of cell membranes; composed of two parallel sheets of membrane lipid molecules arranged tail to tail.

Lipogenesis The biochemical pathway for synthesis of fatty acids from acetyl-SCoA.

Lipoprotein A lipid–protein complex that transports lipids.

Liposome A spherical structure in which a lipid bilayer surrounds a water droplet.

Liquid A substance that has a definite volume but that changes shape to fit its container.

London dispersion force The short-lived attractive force due to the constant motion of electrons within molecules.

Lone pair A pair of electrons that is not used for bonding.

Main group element An element in one of the two groups on the left or the six groups on the right of the periodic table.

Markovnikov's rule In the addition of HX to an alkene, the H becomes attached to the carbon that already has the most H's, and the X becomes attached to the carbon that has fewer H's.

Mass A measure of the amount of matter in an object.

Mass number (A) The total number of protons and neutrons in an atom.

Matter The physical material that makes up the universe; anything that has mass and occupies space.

Melting point (mp) The temperature at which solid and liquid are in equilibrium.

Messenger RNA (mRNA) The RNA that carries the code transcribed from DNA and directs protein synthesis.

Metal A malleable element with a lustrous appearance that is a good conductor of heat and electricity.

Metalloid An element whose properties are intermediate between those of a metal and a nonmetal.

Methyl group The —CH_3 alkyl group.

Methylene Another name for a CH_2 unit.

Micelle A spherical cluster formed by the aggregation of soap or detergent molecules so that their hydrophobic ends are in the center and their hydrophilic ends are on the surface.

Miscible Mutually soluble in all proportions.

Mitochondrial matrix The space surrounded by the inner membrane of a mitochondrion.

Mitochondrion (plural, mitochondria) An egg-shaped organelle where small molecules are broken down to provide the energy for an organism.

Mixture A blend of two or more substances, each of which retains its chemical identity.

Mobilization (of triacylglycerols) Hydrolysis of triacylglycerols in adipose tissue and release of fatty acids into the bloodstream.

Molar mass The mass in grams of one mole of a substance, numerically equal to the molecular weight.

Molarity (M) Concentration expressed as the number of moles of solute per liter of solution.

Mole The amount of a substance corresponding to 6.02×10^{23} units.

Molecular compound A compound that consists of molecules rather than ions.

Molecular formula A formula that shows the numbers and kinds of atoms in one molecule of a compound.

Molecular weight The sum of the atomic weights of the atoms in a molecule.

Molecule A group of atoms held together by covalent bonds.

Monomer A small molecule that is used to prepare a polymer.

Monosaccharide (simple sugar) A carbohydrate with 3–7 carbon atoms.

Mutagen A substance that causes mutations.

Mutarotation Change in rotation of plane-polarized light resulting from the equilibrium between cyclic anomers and the open-chain form of a sugar.

Mutation An error in base sequence that is carried along in DNA replication.

***n*-propyl group** The straight-chain alkyl group —$CH_2CH_2CH_3$.

Native protein A protein with the shape (secondary, tertiary, and quaternary structure) in which it exists naturally in living organisms.

Net ionic equation An equation that does not include spectator ions.

Neurotransmitter A chemical messenger that travels between a neuron and a neighboring neuron or other target cell to transmit a nerve impulse.

Neutralization reaction The reaction of an acid with a base.

Neutron An electrically neutral subatomic particle.

Nitration The substitution of a nitro group (—NO_2) for a hydrogen on an aromatic ring.

Noble gas An element in group 8A of the periodic table.

Noncompetitive (enzyme) inhibition Enzyme regulation in which an inhibitor binds to an enzyme elsewhere than at the active site, thereby changing the shape of the enzyme's active site and reducing its efficiency.

Noncovalent forces Forces of attraction other than covalent bonds that can act between molecules or within molecules.

Nonelectrolyte A substance that does not produce ions when dissolved in water.

Nonessential amino acid One of eleven amino acids that are synthesized in the body and are therefore not necessary in the diet.

Nonmetal An element that is a poor conductor of heat and electricity.

Normal boiling point The boiling point at a pressure of exactly 1 atmosphere.

Normality (N) A measure of acid (or base) concentration expressed as the number of acid (or base) equivalents per liter of solution.

Nuclear decay The spontaneous emission of a particle from an unstable nucleus.

Nuclear fission The fragmenting of heavy nuclei.

Nuclear fusion The joining together of light nuclei.

Nuclear reaction A reaction that changes an atomic nucleus, usually causing the change of one element into another.

Nucleic acid A polymer of nucleotides.

Nucleon A general term for both protons and neutrons.

Nucleoside A five-carbon sugar bonded to a cyclic amine base; like a nucleotide but missing the phosphate group.

Nucleotide A five-carbon sugar bonded to a cyclic amine base and one phosphate group (a nucleoside monophosphate); monomer for nucleic acids.

Nucleus The dense, central core of an atom that contains protons and neutrons.

Nuclide The nucleus of a specific isotope of an element.

Octet rule Main-group elements tend to undergo reactions that leave them with 8 valence electrons.

Oil A mixture of triacylglycerols that is liquid because it contains a high proportion of unsaturated fatty acids.

Orbital A region of space within an atom where an electron in a given subshell can be found.

Organic chemistry The study of carbon compounds.

Osmolarity (osmol) The sum of the molarities of all dissolved particles in a solution.

Osmosis The passage of solvent through a semipermeable membrane separating two solutions of different concentration.

Osmotic pressure The amount of external pressure applied to the more concentrated solution to halt the passage of solvent molecules across a semipermeable membrane.

Oxidation The loss of one or more electrons by an atom.

Oxidation number A number that indicates whether an atom is neutral, electron-rich, or electron-poor.

Oxidation–Reduction, or Redox, reaction A reaction in which electrons are transferred from one atom to another.

Oxidative deamination Conversion of an amino acid —NH_2 group to an α-keto group, with removal of NH_4^+.

Oxidative phosphorylation The synthesis of ATP from ADP using energy released in the electron-transport chain.

Oxidizing agent A reactant that causes an oxidation by taking electrons from or increasing the oxidation number of another reactant.

***p*-Block element** A main group element that results from the filling of p orbitals.

Partial pressure The pressure exerted by a gas in a mixture.

Parts per billion (ppb) Number of parts per one billion (10^9) parts.

Parts per million (ppm) Number of parts per one million (10^6) parts.

Passive transport Movement of a substance across a cell membrane without the use of energy, from a region of higher concentration to a region of lower concentration.

Pentose phosphate pathway The biochemical pathway that produces ribose (a pentose), NADPH, and other sugar phosphates from glucose; an alternative to glycolysis.

Peptide bond An amide bond that links two amino acids together.

Percent yield The percent of the theoretical yield actually obtained from a chemical reaction.

Period One of the 7 horizontal rows of elements in the periodic table.

Periodic table A table of the elements in order of increasing atomic number and grouped according to their chemical similarities.

pH A measure of the acid strength of a solution; the negative common logarithm of the H_3O^+ concentration.

Phenol A compound that has an —OH group bonded directly to an aromatic, benzene-like ring, Ar—OH.

Phenyl The C_6H_5— group.

Phosphate ester A compound formed by reaction of an alcohol with phosphoric acid; may be a monoester, $ROPO_3H_2$; a diester, $(RO)_2PO_3H$; or a triester, $(RO)_3PO$; also may be a di- or triphosphate.

Phospholipid A lipid that has an ester link between phosphoric acid and an alcohol (glycerol or sphingosine).

Phosphoryl group The $—PO_3^{2-}$ group in organic phosphates.

Phosphorylation Transfer of a phosphoryl group, $—PO_3^{2-}$, between organic molecules.

Physical change A change that does not affect the chemical makeup of a substance or object.

Physical quantity A physical property that can be measured.

Polar covalent bond A bond in which the electrons are attracted more strongly by one atom than by the other.

Polyatomic ion An ion that is composed of more than one atom.

Polymer A large molecule formed by the repetitive bonding together of many smaller molecules.

Polymorphism A variation in DNA sequence within a population.

Polysaccharide (complex carbohydrate) A carbohydrate that is a polymer of monosaccharides.

Polyunsaturated fatty acid A long-chain fatty acid that has two or more carbon–carbon double bonds.

Positron A "positive electron," which has the same mass as an electron but a positive charge.

Potential energy Energy that is stored because of position, composition, or shape.

Precipitate An insoluble solid that forms in solution during a chemical reaction.

Pressure The force per unit area pushing against a surface.

Primary carbon atom A carbon atom with one other carbon attached to it.

Primary protein structure The sequence in which amino acids are linked by peptide bonds in a protein.

Product A substance that is formed in a chemical reaction and is written on the right side of the reaction arrow in a chemical equation.

Property A characteristic useful for identifying a substance or object.

Protein A large biological molecule made of many amino acids linked together through amide (peptide) bonds.

Proton A positively charged subatomic particle.

Pure substance A substance that has uniform chemical composition throughout.

Quaternary ammonium ion A positive ion with four organic groups bonded to the nitrogen atom.

Quaternary ammonium salt An ionic compound composed of a quaternary ammonium ion and an anion.

Quaternary carbon atom A carbon atom with four other carbons attached to it.

Quaternary protein structure The way in which two or more protein chains aggregate to form large, ordered structures.

Radioactivity The spontaneous emission of radiation from a nucleus.

Radioisotope A radioactive isotope.

Radionuclide The nucleus of a radioactive isotope.

Reabsorption (kidney) Movement of solutes out of filtrate in a kidney tubule.

Reactant A substance that undergoes change in a chemical reaction and is written on the left side of the reaction arrow in a chemical equation.

Reaction mechanism A description of the individual steps by which old bonds are broken and new bonds are formed in a reaction.

Reaction rate A measure of how rapidly a reaction occurs.

Rearrangement reaction A general reaction type in which a molecule undergoes bond reorganization to yield an isomer.

Receptor A molecule or portion of a molecule with which a hormone, neurotransmitter, or other biochemically active molecule interacts to initiate a response in a target cell.

Recombinant DNA DNA that contains segments from two different species.

Reducing agent A reactant that causes a reduction by giving up electrons or increasing the oxidation number of another reactant.

Reducing sugar A carbohydrate that reacts in basic solution with a mild oxidizing agent.

Reduction The gain of one or more electrons by an atom.

Reductive deamination Conversion of an α-keto acid to an amino acid by reaction with NH_4^+.

Regular tetrahedron A geometric figure with four identical triangular faces.

Replication The process by which copies of DNA are made when a cell divides.

Residue (amino acid) An amino acid unit in a polypeptide.

Resonance The phenomenon where the true structure of a molecule is an average among two or more conventional structures.

Reversible reaction A reaction that can go in either the forward direction or the reverse direction, from products to reactants or reactants to products.

Ribonucleotide A nucleotide containing D-ribose.

Ribosomal RNA (rRNA) The RNA that is complexed with proteins in ribosomes.

Ribosome The structure in the cell where protein synthesis occurs; composed of protein and rRNA.

RNA (ribonucleic acids) The nucleic acids (messenger, transfer, and ribosomal) responsible for putting the genetic information to use in protein synthesis; polymers of ribonucleotides.

Rounding off A procedure used for deleting nonsignificant figures.

s-Block element A main group element that results from the filling of an s orbital.

Salt An ionic compound formed from reaction of an acid with a base.

Saponification The reaction of an ester with aqueous hydroxide ion to yield an alcohol and the metal salt of a carboxylic acid.

Saturated A molecule whose carbon atoms bond to the maximum number of hydrogen atoms.

Saturated fatty acid A long-chain carboxylic acid containing only carbon–carbon single bonds.

Saturated solution A solution that contains the maximum amount of dissolved solute at equilibrium.

Scientific Method Systematic process of observation, hypothesis, and experimentation to expand and refine a body of knowledge.

Scientific notation A number expressed as the product of a number between 1 and 10, times the number 10 raised to a power.

Second messenger Chemical messenger released inside a cell when a hydrophilic hormone or neurotransmitter interacts with a receptor on the cell surface.

Secondary carbon atom A carbon atom with two other carbons attached to it.

Secondary protein structure Regular and repeating structural patterns (for example, α-helix, β-sheet) created by hydrogen bonding between backbone atoms in neighboring segments of protein chains.

Secretion (kidney) Movement of solutes into filtrate in a kidney tubule.

SI units Units of measurement defined by the International System of Units.

Side chain (amino acid) The group bonded to the carbon next to the carboxyl group in an amino acid; different in different amino acids.

Significant figures The number of meaningful digits used to express a value.

Simple diffusion Passive transport by the random motion of diffusion through the cell membrane.

Simple protein A protein composed of only amino acid residues.

Single bond A covalent bond formed by sharing one electron pair.

Single nucleotide polymorphism Common single-base-pair variation in DNA.

Soap The mixture of salts of fatty acids formed on saponification of animal fat.

Solid A substance that has a definite shape and volume.

Solubility The maximum amount of a substance that will dissolve in a given amount of solvent at a specified temperature.

Solute A substance dissolved in a liquid.

Solution A homogeneous mixture that contains particles the size of a typical ion or small molecule.

Solvation The clustering of solvent molecules around a dissolved solute molecule or ion.

Solvent The liquid in which another substance is dissolved.

Specific gravity The density of a substance divided by the density of water at the same temperature.

Specific heat The amount of heat that will raise the temperature of 1 g of a substance by 1 °C.

Specificity (enzyme) The limitation of the activity of an enzyme to a specific substrate, specific reaction, or specific type of reaction.

Spectator ion An ion that appears unchanged on both sides of a reaction arrow.

Sphingolipid A lipid derived from the amino alcohol sphingosine.

Spontaneous process A process or reaction that, once started, proceeds on its own without any external influence.

Standard molar volume The volume of one mole of a gas at standard temperature and pressure (22.4 L).

Standard temperature and pressure (STP) Standard conditions for a gas, defined as 0 °C (273 K) and 1 atm (760 mmHg) pressure.

State of matter The physical state of a substance as a solid, a liquid, or a gas.

Stereoisomers Isomers that have the same molecular and structural formulas, but different spatial arrangements of their atoms.

Steroid A lipid whose structure is based on the following tetracyclic (four-ring) carbon skeleton:

Straight-chain alkane An alkane that has all its carbons connected in a row.

Strong acid An acid that gives up H^+ easily and is essentially 100% dissociated in water.

Strong base A base that has a high affinity for H^+ and holds it tightly.

Strong electrolyte A substance that ionizes completely when dissolved in water.

Structural formula A molecular representation that shows the connections among atoms by using lines to represent covalent bonds.

Subatomic particles Three kinds of fundamental particles from which atoms are made: protons, neutrons, and electrons.

Substituent An atom or group of atoms attached to a parent compound.

Substitution reaction A general reaction type in which an atom or group of atoms in a molecule is replaced by another atom or group of atoms.

Substrate A reactant in an enzyme catalyzed reaction.

Sulfonation The substitution of a sulfonic acid group ($—SO_3H$) for a hydrogen on an aromatic ring.

Supersaturated solution A solution that contains more than the maximum amount of dissolved solute; a nonequilibrium situation.

Synapse The place where the tip of a neuron and its target cell lie adjacent to each other.

Telomeres The ends of chromosomes; in humans, contain long series of repeating groups of nucleotides.

Temperature The measure of how hot or cold an object is.

Tertiary carbon atom A carbon atom with three other carbons attached to it.

Tertiary protein structure The way in which an entire protein chain is coiled and folded into its specific three-dimensional shape.

Theoretical yield The amount of product formed assuming complete reaction of the limiting reagent.

Thiol A compound that contains an —SH group, R—SH.

Titration A procedure for determining the total acid or base concentration of a solution.

Transamination The interchange of the amino group of an amino acid and the keto group of an α-keto acid.

Transcription The process by which the information in DNA is read and used to synthesize RNA.

Transfer RNA (tRNA) The RNA that transports amino acids into position for protein synthesis.

Transition metal element An element in one of the 10 smaller groups near the middle of the periodic table.

Translation The process by which RNA directs protein synthesis.

Transmutation The change of one element into another.

Triacylglycerol (triglyceride) A triester of glycerol with three fatty acids.

Triple bond A covalent bond formed by sharing three electron pairs.

Turnover number The maximum number of substrate molecules acted upon by one molecule of enzyme per unit time.

Unit A defined quantity used as a standard of measurement.

Unsaturated A molecule that contains a carbon–carbon multiple bond, to which more hydrogen atoms can be added.

Unsaturated fatty acid A long-chain carboxylic acid containing one or more carbon–carbon double bonds.

Urea cycle The cyclic biochemical pathway that produces urea for excretion.

Valence electron An electron in the outermost, or valence, shell of an atom.

Valence shell The outermost electron shell of an atom.

Valence-shell electron-pair repulsion (VSEPR) model A method for predicting molecular shape by noting how many electron charge clouds surround atoms and assuming that the clouds orient as far away from one another as possible.

Vapor The gas molecules in equilibrium with a liquid.

Vapor pressure The partial pressure of gas molecules in equilibrium with a liquid.

Vitamin An organic molecule, essential in trace amounts that must be obtained in the diet because it is not synthesized in the body.

Volume/volume percent concentration [(v/v)%] Concentration expressed as the number of milliliters of solute dissolved in 100 mL of solution.

Wax A mixture of esters of long-chain carboxylic acids with long-chain alcohols.

Weak acid An acid that gives up H^+ with difficulty and is less than 100% dissociated in water.

Weak base A base that has only a slight affinity for H^+ and holds it weakly.

Weak electrolyte A substance that is only partly ionized in water.

Weight A measure of the gravitational force that the earth or other large body exerts on an object.

Weight/volume percent concentration [(w/v)%] Concentration expressed as the number of grams of solute per 100 mL of solution.

Whole blood Blood plasma plus blood cells.

X rays Electromagnetic radiation with an energy somewhat less than that of γ rays.

Zwitterion A neutral dipolar ion that has one + charge and one − charge.

Zymogen A compound that becomes an active enzyme after undergoing a chemical change.

Answers to Selected Problems

Short answers are given for in-chapter problems, *Understanding Key Concepts* problems, and even-numbered end-of-chapter problems. Explanations and full answers for all problems are provided in the accompanying *Study Guide and Solutions Manual*.

Chapter 1

1.1 all; natural: (a), (d); synthetic: (b), (c) **1.2** physical: (a), (d); chemical: (b), (c) **1.3** solid **1.4** mixture: (a), (d); pure: (b), (c) **1.5** physical: (a), (c); chemical: (b) **1.6** chemical change **1.7** (a) Na (b) Ti (c) Sr (d) Y (e) F (f) H **1.8** (a) uranium (b) calcium (c) neodymium (d) potassium (e) tungsten (f) tin **1.9** (a) 1 nitrogen atom, 3 hydrogen atoms (b) 1 sodium atom, 1 hydrogen atom, 1 carbon atom, 3 oxygen atoms (c) 8 carbon atoms, 18 hydrogen atoms (d) 6 carbon atoms, 8 hydrogen atoms, 6 oxygen atoms **1.10** (a) chromium (24) (b) potassium (19) (c) sulfur (16) (d) radon (86) **1.11** Metalloids are at the boundary between metals and nonmetals. **1.12** helium (He), neon (Ne), argon (Ar), krypton (Kr), xenon (Xe), radon (Rn) **1.13** copper (Cu), silver (Ag), gold (Au) **1.14** red: vanadium, metal; green: boron, metalloid; blue: bromine, nonmetal **1.15** Americium, a metal **1.16** Chemistry is the study of matter. **1.18** physical: (a), (c), (e); chemical (b), (d) **1.20** A gas has no definite shape or volume; a liquid has no definite shape but has a definite volume; a solid has a definite volume and a definite shape. **1.22** gas **1.24** mixture: (a), (b), (d), (f); pure: (c), (e) **1.26** element: (a); compound: (b), (c); mixture: (d), (e), (f) **1.28** (a) reactant: hydrogen peroxide; products: water, oxygen (b) compounds: hydrogen peroxide, water; element: oxygen **1.30** 117 elements; 90 occur naturally **1.32** Metals: lustrous, malleable, conductors of heat and electricity; nonmetals: gases or brittle solids, nonconductors; metalloids: properties intermediate between metals and nonmetals. **1.34** (a) Gd (b) Ge (c) Tc (d) As (e) Cd **1.36** (a) nitrogen (b) potassium (c) chlorine (d) calcium (e) phosphorus (f) manganese **1.38** Only the first letter of a chemical symbol is capitalized. **1.40** (a) Br (b) Mn (c) C (d) K **1.42** (a) A mixture doesn't have a chemical formula. (b) The symbol for nitrogen is N. **1.44** (a) magnesium, sulfur, oxygen (b) iron, bromine (c) cobalt, phosphorus (d) arsenic, hydrogen (e) calcium, chromium, oxygen **1.46** Carbon, hydrogen, nitrogen, oxygen; ten atoms **1.48** $C_{13}H_{18}O_2$ **1.50** (a) metal (b) nonmetal **1.52** 9 carbons, 8 hydrogens, 4 oxygens; 21 atoms **1.54** (a) A physical change doesn't alter chemical makeup; a chemical change alters a substance's chemical makeup. (b) Melting point is the temperature at which a change of state from solid to liquid occurs; boiling point is the temperature at which a change of state from liquid to gas occurs. (c) A reactant is a substance that undergoes change in a chemical reaction; a product is a substance formed as a result of a chemical reaction. (d) A metal is a lustrous malleable element that is a good conductor of heat and electricity; a nonmetal is an element that is a poor conductor. **1.56** compounds: (a), (c), (e); elements: (b), (d) **1.58** mixture **1.60** (a) Fe (b) Cu (c) Co (d) Mo (e) Cr (f) F (g) S **1.62** Elements 115,119: metals Element 117: metal or metalloid

Chapter 2

2.1 (a) deciliter (b) milligram (c) nanosecond (d) kilometer (e) microgram **2.2** (a) L (b) kg (c) nm (d) Mm **2.3** (a) 0.000 000 001 m (b) 0.1 g (c) 1000 m (d) 0.000 001 s (e) 0.000 000 001 g **2.4** (a) 3 (b) 4 (c) 5 (d) exact **2.5** 32.3 °C; three significant figures **2.6** (a) 5.8×10^{-2} g (b) 4.6792×10^4 m (c) 6.072×10^{-3} cm (d) 3.453×10^2 kg **2.7** (a) 48,850 mg (b) 0.000 008 3 m (c) 0.0400 m **2.8** (a) 6.3000×10^5 (b) 1.30×10^3 (c) 7.942×10^{11} **2.9** 2.78×10^{-10} m $= 2.78 \times 10^2$ pm **2.10** (a) 2.30 g (b) 188.38 mL (c) 0.009 L (d) 1.000 kg **2.11** (a) 50.9 mL (b) 0.078 g (c) 11.9 m (d) 51 mg (e) 103 **2.12** (a) 1 L = 1000 mL; 1 mL = 0.001 L (b) 1 g = 0.03527 oz; 1 oz = 28.35 g (c) 1 L = 1.057 qt; 1 qt = 0.9464 L **2.13** (a) 454 g (b) 2.5 L (c) 105 qt **2.14** 795 mL **2.15** 7.36 m/s **2.16** (a) 3.4 kg (b) 120 mL **2.17** (a) 10.6 mg/kg (b) 36 mg/kg **2.18** 57.8 °C **2.19** −38.0 °F; 234.3 K **2.20** 7,700 cal

2.21 0.21 cal/g °C **2.22** float: ice, human fat, cork, balsa wood; sink: gold, table sugar, earth **2.23** 8.392 mL **2.24** 2.21 g/cm³ **2.25** more dense **2.26** (a) 34 mL (b) 2.7 cm; two significant figures **2.27** (no answer) **2.28** (a) 0.977 (b) three (c) less dense **2.29** The smaller cylinder is more precise because the gradations are smaller. **2.30** 3 1/8 in.; 8.0 cm **2.31** start: 0.11 mL stop: 0.25 mL volume: 0.14 mL **2.32** (a) Both are equally precise. (b) 355 mL **2.33** higher in chloroform **2.34** A physical quantity consists of a number and a unit. **2.36** mass (kg); volume (m³); length (m); temperature (K) **2.38** They are the same. **2.40** (a) centiliter (b) decimeter (c) millimeter (d) nanoliter (e) milligram (f) cubic meter (g) cubic centimeter **2.42** 10^9 pg, 3.5×10^4 pg **2.44** (a) 9.457×10^3 (b) 7×10^{-5} (c) 2.000×10^{10} (d) 1.2345×10^{-2} (e) 6.5238×10^2 **2.46** (a) 6 (b) 3 (c) 3 (d) 4 (e) 1–5 (f) 2–3 **2.48** (a) 7,926 mi, 7,900 mi, 7,926.38 mi (b) 7.926 381 × 10³ mi **2.50** (a) 12.1 g (b) 96.19 cm (c) 263 mL (d) 20.9 mg **2.52** (a) 0.3614 cg (b) 0.0120 ML (c) 0.0144 mm (d) 60.3 ng (e) 1.745 dL (f) 1.5×10^3 cm **2.54** (a) 97.8 kg (b) 0.133 mL (c) 0.46 ng (d) 2.99 Mm **2.56** (a) 62.1 mi/hr (b) 91.1 ft/s **2.58** 4×10^3 cells/in. **2.60** 10 g **2.62** 6×10^{10} cells **2.64** 37.0 °C, 310.2 K **2.66** 537 cal = 0.537 kcal **2.68** 0.092 cal/g·°C **2.70** Hg: 76 °C; Fe: 40.7 °C **2.72** 0.179 cm³ **2.74** 11.4 g/cm³ **2.76** 0.7856 g/mL; sp gr = 0.7856 **2.78** (a) 2.0×10^{-6} cm (b) 1.3×10^6 cells/in (c) 1 mL = 0.269 fluidram (d) 1 mL = 16 minim (e) 1 minim = 2.08×10^{-3} fluid ounce **2.80** −2 °C; 271 K **2.82** (a) BMI = 29 (b) BMI = 23.7 (c) BMI = 24.4; individual (a) **2.84** 3.12 in; 7.92 cm Discrepancies are due to rounding errors and changes in significant figures. **2.86** 177 °C **2.88** 3.9×10^{-2} g/dL iron, 8.3×10^{-3} g/dL calcium, 2.24×10^{-1} g/dL cholesterol **2.90** 7.8×10^6 mL/day **2.92** 0.13 g **2.94** 4.4 g; 0.0097 lb **2.96** 2200 mL **2.98** 2.2 tablespoons **2.100** iron **2.102** 4.99×10^{10} L **2.104** (a) 2×10^3 mg/L (b) 2×10^3 μg/mL (c) 2 g/L (d) 2×10^3 ng/μL **2.106** 34.1 °C **2.108** float

Chapter 3

3.1 14.0 amu **3.2** 4.99×10^{-8} g **3.3** 6.02×10^{23} atoms in all cases **3.4** When the mass in grams is numerically equal to the mass in amu, there are 6.02×10^{23} atoms. **3.5** (a) Re (b) Ca (c) Te **3.6** 27 protons, 27 electrons, 33 neutrons **3.7** technetium **3.8** The answers agree **3.9** $^{79}_{35}\text{Br}$, $^{81}_{35}\text{Br}$ **3.10** (a) $^{11}_{5}\text{B}$ (b) $^{56}_{26}\text{Fe}$ (c) $^{37}_{17}\text{Cl}$ **3.11** group 3A, period 3 **3.12** silver, calcium **3.13** Nitrogen (1), phosphorus (2), arsenic (3), antimony (4), bismuth (5) **3.14** Metals: titanium, scandium; nonmetals: selenium, argon; metalloids: tellurium, astatine **3.15** (a) nonmetal, main group, noble gas (b) metal, main group (c) nonmetal, main group (d) metal, transition element **3.16** red = zinc, metal, period 4, group 2B; blue = oxygen, period 6, group 6A **3.17** 6, 2, 6 **3.18** 10, neon **3.19** 12, magnesium **3.20** (a) $1s^2\,2s^2\,2p^2$ (b) $1s^2\,2s^2\,2p^6\,3s^2\,3p^3$ (c) $1s^2\,2s^2\,2p^6\,3s^2\,3p^5$ (d) $1s^2\,2s^2\,2p^6\,3s^2\,3p^6\,4s^1$ **3.21** $1s^2\,2s^2\,2p^6\,3s^2\,3p^2$; $1s^2\,2s^2\,2p^6\,3s^2\,3p^6\,4s^2\,3d^{10}\,4p^6$ **3.22** $4p^3$, all are unpaired **3.23** gallium **3.24** (a) $1s^2\,2s^2\,2p^5$; [He] $2s^2\,2p^5$ (b) $1s^2\,2s^2\,2p^6\,3s^2\,3p^1$; [Ne] $3s^2\,3p^1$ (c) $1s^2\,2s^2\,2p^6\,3s^2\,3p^6\,4s^2\,3d^{10}\,4p^3$; [Ne] $4s^2\,3d^{10}\,4p^6$ **3.25** group 2A **3.26** group 7A, $1s^2\,2s^2\,2p^6\,3s^2\,3p^5$ **3.27** group 6A, $ns^2\,np^4$ **3.28** $\cdot\dot{\text{X}}\cdot$ **3.29** $:\!\ddot{\text{Rn}}\!:$ $\cdot\text{Pb}\cdot$ $:\!\ddot{\text{Xe}}\!:$ $\cdot\text{Ra}\cdot$ **3.30** (a) p orbital (b) s orbital

3.31

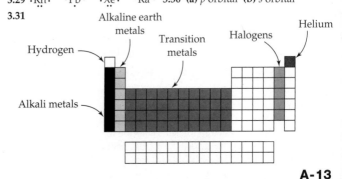

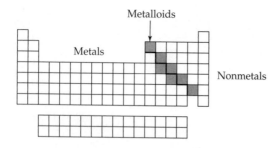

3.32 red: gas (fluorine); blue: atomic number 79 (gold); green: (calcium); beryllium, magnesium, strontium, barium, and radium are similar.
3.33

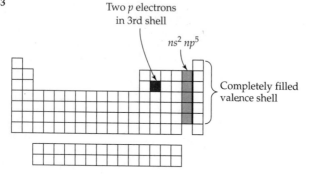

3.34 selenium **3.35** $1s^2 2s^2 2p^6 3s^2 3p^6 4s^2 3d^{10} 4p^3$ **3.36** Matter is composed of atoms. Atoms of different elements differ. Compounds consist of different atoms combined in specific proportions. Atoms do not change in chemical reactions. **3.38 (a)** 3.4702×10^{-22} g **(b)** 2.1801×10^{-22} g **(c)** 6.6465×10^{-24} g **3.40** 14.01 g **3.42** 6.022×10^{23} atoms **3.44** protons (+ charge, 1 amu); neutrons (no charge, 1 amu); electrons (− charge, 0.0005 amu). **3.46 (a)** potassium **(b)** tin **(c)** zinc **3.48** 18, 20, 22 **3.50 (a)** and **(c)** **3.52 (a)** F-19 **(b)** Ne-19 **(c)** F-21 **(d)** Mg-21 **3.54 (a)** $^{14}_{6}\text{C}$ **(b)** $^{39}_{19}\text{K}$ **(c)** $^{20}_{10}\text{Ne}$ **3.56** $^{12}_{6}\text{C}$—six neutrons $^{13}_{6}\text{C}$—seven neutrons $^{14}_{6}\text{C}$—eight neutrons **3.58** 63.55 amu **3.60** Eight electrons are needed to fill the 3s and 3p subshells. **3.62** Am, metal **3.64 (a, b)** transition metals **(c)** 3d **3.66 (a)** Rb: (i), (v), (vii) **(b)** W: (i), (iv) **(c)** Ge: (iii), (v) **(d)** Kr: (ii), (v), (vi) **3.68** selenium **3.70** sodium, potassium, rubidium, cesium, francium **3.72** 2 **3.74** 2, 8, 18 **3.76** 3, 4, 4 **3.78** 10, neon **3.80 (a)** two paired, two unpaired **(b)** four paired, one unpaired **(c)** two unpaired **3.82** 2, 1, 2, 0, 3, 4 **3.84** 2 **3.86** beryllium, 2s; arsenic, 4p **3.88 (a)** 8 **(b)** 4 **(c)** 2 **(d)** 1 **(e)** 3 **(f)** 7 **3.90** A scanning tunneling microscope has much higher resolution. **3.92** H, He **3.94 (a)** ultraviolet **(b)** gamma waves **(c)** X rays **3.96** He, Ne, Ar, Kr, Xe, Rn **3.98** Tellurium atoms have more neutrons than iodine atoms. **3.100** 1 (2 e), 2 (8 e), 3 (18 e), 4 (32 e), 5 (18 e), 6 (4 e) **3.102** 79.90 amu **3.104** carbon weighs more **3.106** 12 g **3.108** Sr, metal, group 2A, period 5, 38 protons **3.110** 2, 8, 18, 10, 2; metal **3.112 (a)** The 4s subshell fills before 3d. **(b)** The 2s subshell fills before 2p. **(c)** Silicon has 14 electrons: $1s^2 2s^2 2p^6 3s^2 3p^2$. **(d)** The 3s electrons have opposite spins. **3.114** An electron will fill or half-fill a d subshell instead of filling an s subshell of a higher shell.

Chapter 4
4.1 Mg^{2+} is a cation **4.2** S^{2-} is an anion **4.3** O^{2-} is an anion. **4.4** less than Kr, but higher than most other elements **4.5 (a)** B **(b)** Ca **(c)** Sc **4.6 (a)** H **(b)** S **(c)** Cr **4.7** Potassium ($1s^2 2s^2 2p^6 3s^2 3p^6 4s^1$) can gain the argon configuration by losing 1 electron. **4.8** Aluminum can lose 3 electrons to form Al^{3+}. **4.9** X: would be a 2A metal and would lose electrons; ·Ÿ· would be a 6A nonmetal and would gain electrons.

$$X\colon + \cdot\ddot{Y}\cdot \longrightarrow X^{2+} + \colon\ddot{Y}\colon^{2-}$$ **4.10** cation **4.11 (b)**

4.12 (a) $Se + 2 e^- \rightarrow Se^{2-}$ **(b)** $Ba \rightarrow Ba^{2+} + 2 e^-$ **(c)** $Br + e^- \rightarrow Br^-$
4.13 (a) copper(II) ion **(b)** fluoride ion **(c)** magnesium ion **(d)** sulfide ion **4.14 (a)** Ag^+ **(b)** Fe^{2+} **(c)** Cu^+ **(d)** Te^{2-} **4.15** Na^+, sodium ion; K^+, potassium ion; Ca^{2+}, calcium ion; Cl^-, chloride ion **4.16 (a)** nitrate ion **(b)** cyanide ion **(c)** hydroxide ion **(d)** hydrogen phosphate ion **4.17 (a)** HCO_3^- **(b)** NH_4^+ **(c)** PO_4^{3-} **(d)** MnO_4^- **4.18 (a)** AgI **(b)** Ag_2O

(c) Ag_3PO_4 **4.19 (a)** Na_2SO_4 **(b)** $FeSO_4$ **(c)** $Cr_2(SO_4)_3$ **4.20** $(NH_4)_2CO_3$ **4.21** $Al_2(SO_4)_3$, $Al(CH_3CO_2)_3$ **4.22** blue: K_2S; red: $BaBr_2$ green: Al_2O_3 **4.23** Calcium ion = Ca^{2+}; nitride ion = N^{3-}; calcium nitride formula = Ca_3N_2 **4.24** $BaSO_4$ **4.25** silver(I) sulfide **4.26 (a)** tin(IV) oxide **(b)** calcium cyanide **(c)** sodium carbonate **(d)** copper(I) sulfate **(e)** barium hydroxide **(f)** iron(II) nitrate **4.27 (a)** Li_3PO_4 **(b)** $CuCO_3$ **(c)** $Al_2(SO_3)_3$ **(d)** CuF **(e)** $Fe_2(SO_4)_3$ **(f)** NH_4Cl **4.28** Cr_2O_3 chromium (III) oxide **4.29** Acids: **(a)**, **(d)**; bases **(b)**, **(c)** **4.30 (a)** HCl **(b)** H_2SO_4
4.31

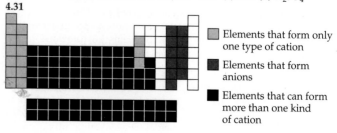

☐ Elements that form only one type of cation

■ Elements that form anions

■ Elements that can form more than one kind of cation

All of the other elements form neither anions nor cations readily.
4.32

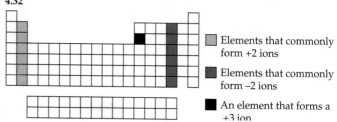

☐ Elements that commonly form +2 ions

■ Elements that commonly form −2 ions

■ An element that forms a +3 ion

4.33 (a) O^{2-} **(b)** Na^+ **(c)** Ca **4.34 (a)** sodium atom (larger) **(b)** Na^+ ion (smaller) **4.35 (a)** chlorine atom (smaller) **(b)** Cl^- anion (larger) **4.36** red: MgO; blue: LiCl; green: $AlBr_3$ **4.37 (a)** ZnS **(b)** $PbBr_2$ **(c)** CrF_3 **(d)** Al_2O_3 **4.38 (a)** ·Be· **(b)** :N̈e: **(c)** ·Sr· **(d)** ·Al· **4.40 (a)** $Ca \rightarrow Ca^{2+} + 2 e^-$ **(b)** $Au \rightarrow Au^+ + e^-$ **(c)** $F + e^- \rightarrow F^-$ **(d)** $Cr \rightarrow Cr^{3+} + 3 e^-$ **4.42** true: **(d)**; false: **(a)**, **(b)**, **(c)** **4.44** Main group atoms undergo reactions that leave them with a noble gas electron configuration. **4.46** −2 **4.48 (a)** Sr **(b)** Br **4.50 (a)** $1s^2 2s^2 2p^6 3s^2 3p^6 4s^2 3d^{10} 4p^6$ **(b)** $1s^2 2s^2 2p^6 3s^2 3p^6 4s^2 3d^{10} 4p^6$ **(c)** $1s^2 2s^2 2p^6 3s^2 3p^6$ **(d)** $1s^2 2s^2 2p^6 3s^2 3p^6 4s^2 3d^{10} 4p^6 5s^2 4d^{10} 5p^6$ **(e)** $1s^2 2s^2 2p^6$ **4.52 (a)** O **(b)** Li **(c)** Zn **(d)** N **4.54** none **4.56** Cr^{2+}: $1s^2 2s^2 2p^6 3s^2 3p^6 3d^4$; Cr^{3+}: $1s^2 2s^2 2p^6 3s^2 3p^6 3d^3$ **4.58** greater **4.60 (a)** sulfide ion **(b)** tin(II) ion **(c)** strontium ion **(d)** magnesium ion **(e)** gold(I) ion **4.62 (a)** Se^{2-} **(b)** O^{2-} **(c)** Ag^+ **4.64 (a)** OH^- **(b)** HSO_4^- **(c)** $CH_3CO_2^-$ **(d)** MnO_4^- **(e)** OCl^- **(f)** NO_3^- **(g)** CO_3^{2-} **(h)** $Cr_2CO_7^{2-}$ **4.66 (a)** $Al_2(SO_4)_3$ **(b)** Ag_2SO_4 **(c)** $ZnSO_4$ **(d)** $BaSO_4$ **4.68 (a)** $Al_2(SO_4)_3$ **(b)** Ag_2SO_4 **(c)** $ZnSO_4$ **(d)** $BaSO_4$
4.70

	S^{2-}	Cl^-	PO_4^{3-}	CO_3^{2-}
copper(II)	CuS	$CuCl_2$	$Cu_3(PO_4)_2$	$CuCO_3$
Ca^{2+}	CaS	$CaCl_2$	$Ca_3(PO_4)_2$	$CaCO_3$
NH_4^+	$(NH_4)_2S$	NH_4Cl	$(NH_4)_3PO_4$	$(NH_4)_2CO_3$
ferric ion	Fe_2S_3	$FeCl_3$	$FePO_4$	$Fe_2(CO_3)_3$

4.72 copper(II) sulfide, copper(II) chloride, copper(II) phosphate, copper(II) carbonate; calcium sulfide, calcium chloride, calcium phosphate, calcium carbonate; ammonium sulfide, ammonium chloride, ammonium phosphate, ammonium carbonate; iron(III) sulfide, iron(III) chloride, iron(III) phosphate, iron(III) carbonate **4.74 (a)** magnesium carbonate **(b)** calcium acetate **(c)** silver(I) cyanide **(d)** sodium dichromate **4.76** $Ca_3(PO_4)_2$ **4.78** An acid gives H^+ ions in water; a base gives OH^- ions. **4.80 (a)** $H_2CO_3 \rightarrow 2 H^+ + CO_3^{2-}$ **(b)** $HCN \rightarrow H^+ + CN^-$ **(c)** $Mg(OH)_2 \rightarrow Mg^{2+} + 2 OH^-$ **(d)** $KOH \rightarrow K^+ + OH^-$ **4.82** To a geologist, a mineral is a naturally occurring crystalline compound. To a nutritionist, a mineral is a metal ion essential for human health. **4.84** slight **4.86** Sodium protects against fluid loss and is necessary for muscle contraction and transmission of nerve impulses.

4.88 10 Ca^{2+}, 6 PO_4^{3-}, 2 OH^- **4.90** H^- has the helium configuration, $1s^2$ **4.92 (a)** CrO_3 **(b)** VCl_5 **(c)** MnO_2 **(d)** MoS_2 **4.94 (a)** -1 **(b)** 3 gluconate ions per iron(III)   **4.96 (a)** $Co(CN)_2$ **(b)** UO_3 **(c)** $SnSO_4$ **(d)** MnO_2 **(e)** K_3PO_4 **(f)** Ca_3P_2 **(g)** $LiHSO_4$ **(h)** $Al(OH)_3$ **4.98 (a)** metal **(b)** nonmetal **(c)** X_2Y_3 **(d)** X: group 3A; Y: group 6A

Chapter 5

5.1 $:\ddot{I}:\ddot{I}:$; xenon   **5.2 (a)** P 3, H 1 **(b)** Se 2, H 1 **(c)** H 1, Cl 1 **(d)** Si 4, F 1

5.3 $PbCl_4$ **5.4 (a)** CH_2Cl_2 **(b)** BH_3 **(c)** NI_3 **(d)** $SiCl_4$

5.5

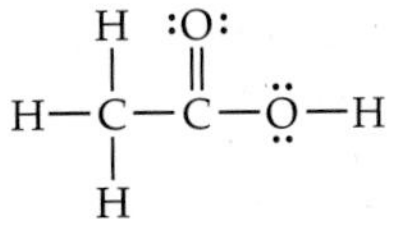

5.6

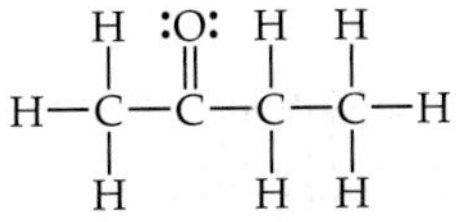

5.7

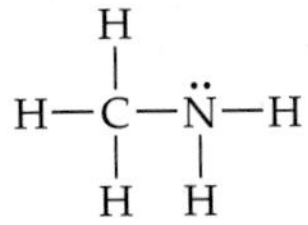

5.8 (a) 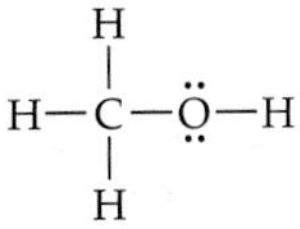   **(b)** 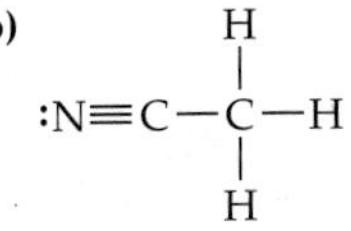   **(c)**

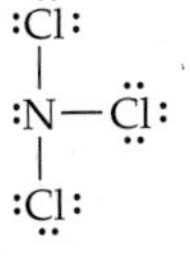

5.9 (a) 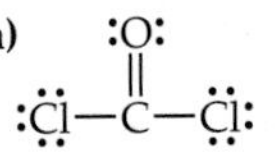   **(b)** 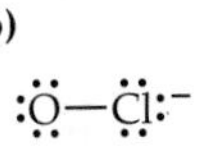   **(c)** 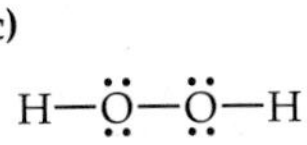   **(d)**

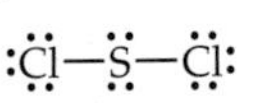

5.10

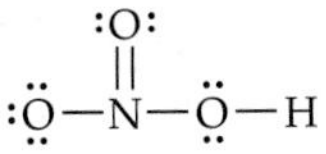

5.11 (a) $C_6H_{10}O_2$ **(b)**

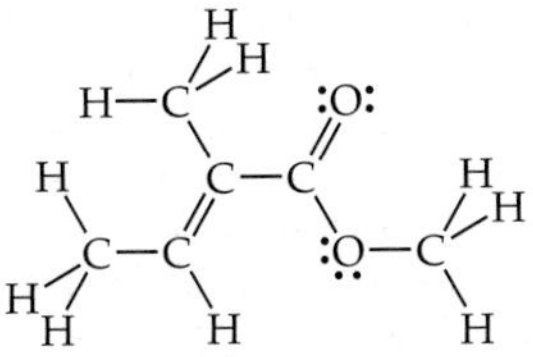

5.12 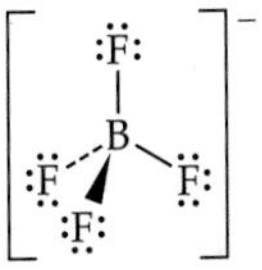 tetrahedral

5.13 chloroform: tetrahedral; dichloroethylene: planar   **5.14** Both are tetrahedral.   **5.15** Both are bent.

5.16 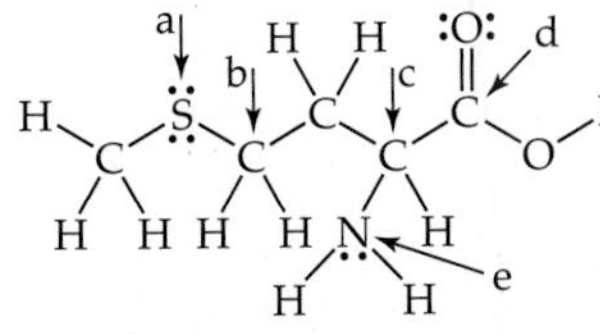
(a) bent
(b) tetrahedral
(c) tetrahedral
(d) planar triangular
(e) pyramidal

5.17 H < P < S < N < O   **5.18 (a)** polar covalent **(b)** ionic **(c)** covalent **(d)** polar covalent   **5.19 (a)** $\overset{\delta+}{S}-\overset{\delta-}{F}$ **(b)** $\overset{\delta+}{P}-\overset{\delta-}{O}$ **(c)** $\overset{\delta+}{As}-\overset{\delta-}{Cl}$
5.20

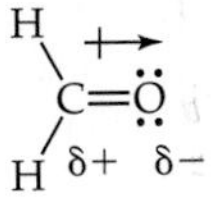

5.21 The carbons are tetrahedral; the oxygen is bent.

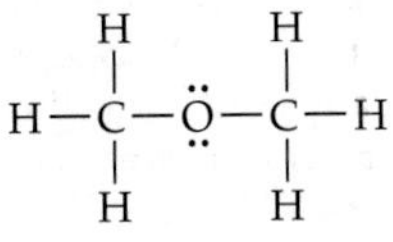

5.22

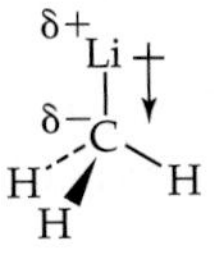

5.23 (a) disulfur dichloride **(b)** iodine chloride **(c)** iodine trichloride

5.24 (a) SeF_4 **(b)** P_2O_5 **(c)** BrF_3 **5.25** molecular solid   **5.26** $AlCl_3$ is a covalent compound, and Al_2O_3 is ionic   **5.27 (a)** ionic **(b)** covalent   **5.28 (a)**   **5.29 (a)** tetrahedral **(b)** pyramidal **(c)** planar triangular   **5.30 (c)** is square planar   **5.31 (a)** $C_8H_9NO_2$ **(b)**

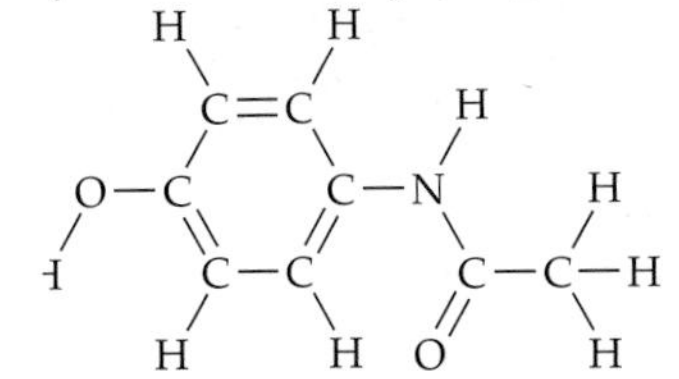

(c) All carbons are trigonal planar except the —CH_3 carbon. Nitrogen is pyramidal.   **5.32**

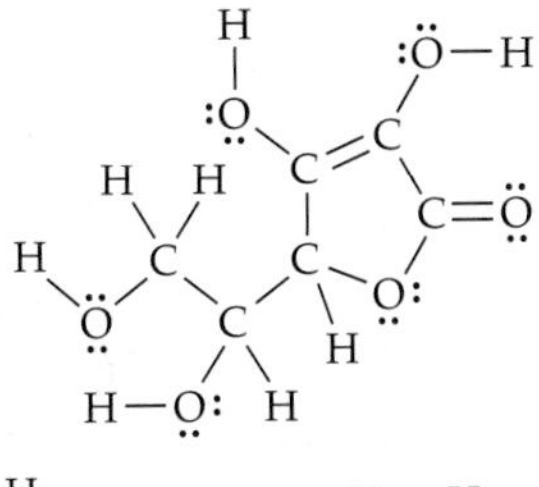

5.33 $C_{13}H_{10}N_2O_4$

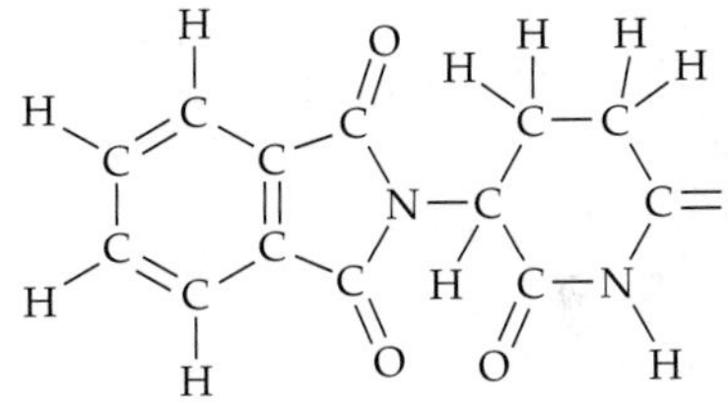

5.34 $:\ddot{O}: \leftarrow$ electron-rich

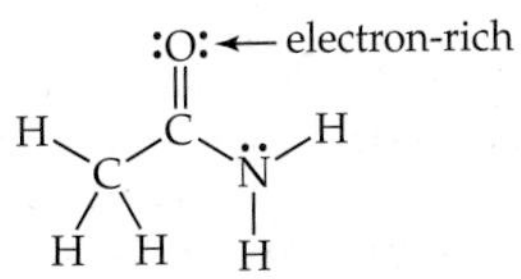

5.36 In a coordinate covalent bond, both electrons in the bond come from the same atom.   **5.38** covalent bonds: **(a) (b)**; ionic bonds: **(c) (d) (e)**   **5.40** two covalent bonds   **5.42 (b), (c)**   **5.44** $SnCl_4$ **5.46** the N–O bond   **5.48 (a)** A molecular formula shows the numbers and kinds of atoms; a structural formula shows how the atoms are bonded to one another. **(b)** A structural formula shows the bonds between atoms; a condensed structure shows atoms but not bonds.   **(c)** A lone pair of valence electrons is not shared in a bond; a shared pair of electrons is shared between two atoms.   **5.50 (a)** 10 **(b)** 10 **(c)** 24 **(d)** 20   **5.52** Too many hydrogens   **5.54 (a)**   **(b)** 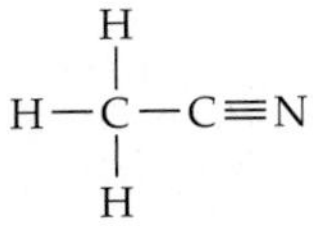   **(c)**

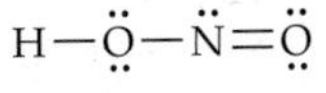

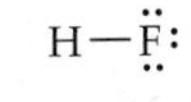

5.56 (a) $CH_3CH_2CH_3$ **(b)** $H_2C=CHCH_3$ **(c)** CH_3CH_2Cl
5.58 CH_3CO_2H
5.60
(a)   **(b)** 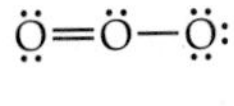   **(c)**

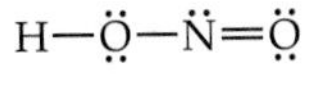

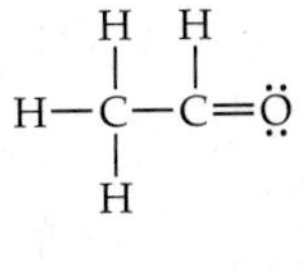

5.62 $H-\overset{\,}{C}-\ddot{O}-\overset{\,}{C}-H$ Dimethyl ether

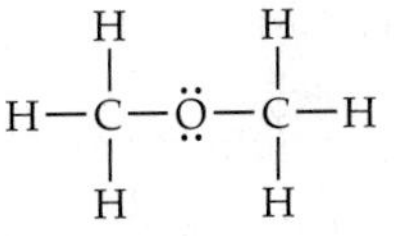

5.64   Tetrachloroethylene contains a double bond

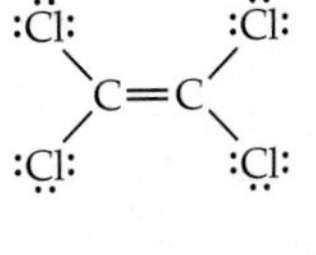

5.66

$$H-\overset{..}{N}-\overset{..}{O}-H$$
$$\quad\ \ |$$
$$\quad\ \ H$$

5.68 (a)

$$\left[\ H-\overset{O}{\underset{\overset{..}{O}:}{\overset{\|}{C}}}-\overset{..}{O}:\ \right]^{-}$$

(b)

$$\left[\ :\overset{:O:}{\underset{..}{\overset{\|}{C}}}-\overset{..}{O}:\ \right]^{2-}$$

(c)

$$\left[\ :\overset{:\overset{..}{O}:}{\underset{..}{\overset{|}{S}}}-\overset{..}{O}:\ \right]^{2-}$$

(d)

$$\left[\ :\overset{..}{S}-C\equiv N:\ \right]^{-}$$

(e)

$$\left[\ :\overset{:\overset{..}{O}:}{\underset{:\overset{..}{O}:}{\overset{|}{P}}}-\overset{..}{O}:\ \right]^{3-}$$

(f)

$$\left[\ :\overset{..}{O}-\overset{..}{\underset{..}{Cl}}-\overset{..}{O}:\ \right]^{-}$$

5.70 tetrahedral; pyramidal; bent   **5.72** (a), (b) tetrahedral (c), (d) planar triangular (e) pyramidal   **5.74** All are planar triangular, except for the $-CH_3$ carbon, which is tetrahedral.   **5.76** It should have low electronegativity, like other alkali metals.   **5.78** Cl > C > Cu > Ca > Cs

5.80
(a) $\overset{\delta-}{O}-\overset{\delta+}{Br}$ (b) $\overset{\delta-}{N}-\overset{\delta+}{H}$ (c) $\overset{\delta+}{P}-\overset{\delta-}{O}$

(d) nonpolar   (e) $\overset{\delta-}{C}-\overset{\delta+}{Li}$

5.82 $PH_3 < HCl < H_2O < CF_4$

5.84 (a)

$$H\overset{\longrightarrow}{-}Cl$$

polar

(b)

$$\overset{\ \ \overset{\uparrow}{(\cdot\cdot)}}{\underset{H}{\underset{|}{P}}}\overset{+}{\diagdown}_{H}^{H}$$

polar

(c)

$$\overset{(\cdot\cdot)}{\underset{H}{\underset{|}{O}}}\overset{+}{\diagdown}_{H}$$

polar

(d) nonpolar

5.86 S—H bonds are less polar because S is less electronegative than O.   **5.88** (a) silicon tetrachloride (b) sodium hydride (c) antimony pentafluoride (d) osmium tetroxide   **5.90** (a) SeO_2 (b) XeO_4 (c) N_2S_5 (d) P_3Se_4 **5.92** It relaxes arterial walls.   **5.94** Carbohydrates, DNA, and proteins are all polymers that occur in nature.   **5.96** no

5.98 (a)

$$H-\overset{H}{\underset{H}{\overset{|}{C}}}-\overset{:\overset{..}{O}:}{\overset{|}{C}}-\overset{H}{\underset{H}{\overset{|}{C}}}-H \qquad H-\overset{H}{\underset{H}{\overset{|}{C}}}-\overset{H}{\underset{H}{\overset{|}{C}}}-\overset{:\overset{..}{O}:}{\overset{\|}{C}}-H$$

(b) The C=O carbons are planar triangular; the other carbons are tetrahedral.   (c) The C=O bonds are polar.   **5.100** (a) C forms 4 bonds (b) N forms 3 bonds (c) S forms 2 bonds (d) could be correct
5.102 (b) tetrahedral (c) contains a coordinate covalent bond (d) has 19 p and 18 e⁻

$$\left[\ H-\overset{H}{\underset{H}{\overset{|}{P}}}-H\ \right]^{+}$$

5.104 (a) calcium chloride (b) tellurium dichloride (c) boron trifluoride (d) magnesium sulfate (e) potassium oxide (f) iron(III) fluoride (g) phosphorus trifluoride

5.106

$$:\overset{..}{\underset{:\overset{..}{Cl}:}{Cl}}-\overset{:\overset{..}{Cl}:}{\underset{:\overset{..}{Cl}:}{\overset{|}{C}}}-\overset{H}{\underset{H}{\overset{|}{C}}}-\overset{..}{O}-H$$

5.108

$$H-\overset{..}{O}-\overset{:O:}{\underset{..}{\overset{\|}{C}}}-\overset{:O:}{\underset{..}{\overset{\|}{C}}}-\overset{..}{O}-H$$

5.110 (a)

$$:\overset{..}{\underset{..}{Cl}}-\overset{:O:}{\overset{\|}{C}}-\overset{..}{O}-\overset{H}{\underset{H}{\overset{|}{C}}}-H$$

(b)

$$H-\overset{H}{\underset{H}{\overset{|}{C}}}-C\equiv C-H$$

Chapter 6

6.1 (a) Solid cobalt(II) chloride plus gaseous hydrogen fluoride gives solid cobalt(II) fluoride plus gaseous hydrogen chloride.   (b) Aqueous

lead(II) nitrate plus aqueous potassium iodide gives solid lead(II) iodide plus aqueous potassium nitrate.   **6.2** balanced: (a), (c)   **6.3** $3 O_2 \rightarrow 2 O_3$ **6.4** (a) $Ca(OH)_2 + 2 HCl \rightarrow CaCl_2 + 2 H_2O$ (b) $4 Al + 3 O_2 \rightarrow 2 Al_2O_3$ (c) $2 CH_3CH_3 + 7 O_2 \rightarrow 4 CO_2 + 6 H_2O$ (d) $2 AgNO_3 + MgCl_2 \rightarrow 2 AgCl + Mg(NO_3)_2$ **6.5** $2 A + B_2 \rightarrow A_2B_2$ **6.6** (a) 206.0 amu (b) 232.0 amu   **6.7** 1.71×10^{21} molecules   **6.8** 0.15 g   **6.9** 111.0 amu   **6.10** 0.217 mol; 4.6 g   **6.11** 5.00 g weighs more   **6.12** (a) $Ni + 2 HCl \rightarrow NiCl_2 + H_2$; 4.90 mol (b) 6.00 mol   **6.13** $6 CO_2 + 6 H_2O \rightarrow C_6H_{12}O_6 + 6 O_2$; 90.0 mol CO_2 **6.14** (a) 39.6 mol (b) 13.8 g   **6.15** 6.31 g WO_3; 0.165 g H_2 **6.16** 44.7 g; 57.0%   **6.17** 49.3 g   **6.18** A_2 **6.19** (a) precipitation (b) redox (c) acid–base neutralization   **6.20** Soluble: (b), (d); insoluble: (a), (c), (e)   **6.21** precipitation: (a), (b)
6.22 (a) $2 CsOH(aq) + H_2SO_4(aq) \rightarrow Cs_2SO_4(aq) + 2 H_2O(l)$
(b) $Ca(OH)_2(aq) + 2 CH_3CO_2H(aq) \rightarrow Ca(CH_3CO_2)_2(aq) + 2 H_2O(l)$
(c) $NaHCO_3(aq) + HBr(aq) \rightarrow NaBr(aq) + CO_2(g) + H_2O(l)$
6.23 (a) oxidizing: Cu; reducing: Fe (b) oxidizing: Cl; reducing: Mg (c) oxidizing: Cr_2O_3; reducing: Al   **6.24** $2 K + Br_2 \rightarrow 2 KBr$; oxidizing: Br_2; reducing: K   **6.25** (a) V(III) (b) Sn(IV) (c) Cr(VI) (d) Cu(II) (e) Ni(II)
6.26 (a) not redox (b) Na oxidized from 0 to +1; H reduced from +1 to 0 (c) C oxidized from 0 to +4; O reduced from 0 to −2 (d) not redox (e) S oxidized from +4 to +6; Mn reduced from +7 to +2
6.27 (a) $Zn(s) + Pb^{2+}(aq) \rightarrow Zn^{2+}(aq) + Pb(s)$
(b) $OH^-(aq) + H^+(aq) \rightarrow H_2O(l)$
(c) $2 Fe^{3+}(aq) + Sn^{2+}(aq) \rightarrow 2 Fe^{2+}(aq) + Sn^{4+}(aq)$
6.28 (d)   **6.29** (c)   **6.30** reactants: (d); products: (c)
6.31 $C_5H_{11}NO_2S$; MW = 149.1 amu   **6.32** (a) $A_2 + 3 B \rightarrow 2 AB_3$ (b) 2 mol AB_3; 0.67 mol AB_3 **6.33** (a) box 1 (b) box 2 (c) box 3
6.34 Ag_2CO_3 and $Ag_2Cr_2O_4$ are possible   **6.35** 22 g, 31 g   **6.36** In a balanced equation, the numbers and kinds of atoms are the same on both sides of the reaction arrow.   **6.38** (a) $SO_2(g) + H_2O(g) \rightarrow H_2SO_3(aq)$
(b) $2 K(s) + Br_2(l) \rightarrow 2 KBr(s)$
(c) $C_3H_8(g) + 5 O_2(g) \rightarrow 3 CO_2(g) + 4 H_2O(l)$
6.40 (a) $2 C_2H_6(g) + 7 O_2(g) \rightarrow 4 CO_2(g) + 6 H_2O(g)$
(b) balanced   (c) $2 Mg(s) + O_2(g) \rightarrow 2 MgO(s)$
(d) $2 K(s) + 2 H_2O(l) \rightarrow 2 KOH(aq) + H_2(g)$
6.42 (a) $Hg(NO_3)_2(aq) + 2 LiI(aq) \rightarrow 2 LiNO_3(aq) + HgI_2(s)$
(b) $I_2(s) + 5 Cl_2(g) \rightarrow 2 ICl_5(s)$ (c) $4 Al(s) + 3 O_2(g) \rightarrow 2 Al_2O_3(s)$
(d) $CuSO_4(aq) + 2 AgNO_3(aq) \rightarrow Ag_2SO_4(s) + Cu(NO_3)_2(aq)$
(e) $2 Mn(NO_3)_3(aq) + 3 Na_2S(aq) \rightarrow Mn_2S_3(s) + 6 NaNO_3(aq)$
(f) $4 NO_2(g) + O_2(g) \rightarrow 2 N_2O_5(g)$
(g) $P_4O_{10}(s) + 6 H_2O(l) \rightarrow 4 H_3PO_4(aq)$
6.44 (a) $2 C_4H_{10}(g) + 13 O_2(g) \rightarrow 8 CO_2(g) + 10 H_2O(l)$
(b) $C_2H_6O(g) + 3 O_2(g) \rightarrow 2 CO_2(g) + 3 H_2O(l)$
(c) $2 C_8H_{18}(g) + 25 O_2(g) \rightarrow 16 CO_2(g) + 18 H_2O(l)$

6.46 molecular weight = sum of the weights of individual atoms in a molecule; formula weight = sum of weights of individual atoms in a formula unit; molar mass = mass in grams of 6.022×10^{23} molecules or formula units of any substance   **6.48** 5.25 mol ions   **6.50** 10.6 g uranium
6.52 (a) 1 mol (b) 1 mol (c) 2 mol   **6.54** 6.44×10^{-4} mol   **6.56** 284.5 g
6.58 (a) 0.0132 mol (b) 0.0536 mol (c) 0.0608 mol (d) 0.0129 mol
6.60 0.27 g; 9.0×10^{20} molecules   **6.62** 1.4×10^{-3} mol; 0.18 g
6.64 (a) $C_4H_8O_2(l) + 2 H_2(g) \rightarrow 2 C_2H_6O(l)$ (b) 3.0 mol (c) 138 g (d) 12.5 g (e) 0.55 g   **6.66** (a) $N_2(g) + 3 H_2(g) \rightarrow 2 NH_3(g)$ (b) 0.471 mol (c) 16.1 g
6.68 (a) $Fe_2O_3(s) + 3 CO(g) \rightarrow 2 Fe(s) + 3 CO_2(g)$ (b) 1.59 g (c) 141 g
6.70 158 kg   **6.72** (a) CO is limiting (b) 11.4 g (c) 83.8%
6.74 (a) $CH_4(g) + 2 Cl_2(g) \rightarrow CH_2Cl_2(l) + 2 HCl(g)$ (b) 444 g (c) 202 g
6.76 (a) redox (b) neutralization (c) precipitation (d) neutralization
6.78 (a) $Ba^{2+}(aq) + SO_4^{2-}(aq) \rightarrow BaSO_4(s)$
(b) $Zn(s) + 2 H^+(aq) \rightarrow Zn^{2+}(aq) + H_2(g)$ **6.80** $Ba(NO_3)_2$
6.82 (a) $2 NaBr(aq) + Hg_2(NO_3)_2(aq) \rightarrow Hg_2Br_2(s) + 2 NaNO_3(aq)$
(d) $(NH_4)_2CO_3(aq) + CaCl_2(aq) \rightarrow CaCO_3(s) + 2 NH_4Cl(aq)$
(e) $2 KOH(aq) + MnBr_2(aq) \rightarrow Mn(OH)_2(s) + 2 KBr(aq)$
(f) $3 Na_2S(aq) + 2 Al(NO_3)_3(aq) \rightarrow Al_2S_3(s) + 6 NaNO_3(aq)$
6.84 (a) $2 Au^{3+}(aq) + 3 Sn(s) \rightarrow 3 Sn^{2+}(aq) + 2 Au(s)$
(b) $2 I^-(aq) + Br_2(l) \rightarrow 2 Br^-(aq) + I_2(s)$
(c) $2 Ag^+(aq) + Fe(s) \rightarrow Fe^{2+}(aq) + 2 Ag(s)$
6.86 Most easily oxidized: metals on left side; most easily reduced: groups 6A and 7A   **6.88** oxidation number increases: (b), (c) oxidation

number decreases **(a)**, **(d)** **6.90 (a)** Co: +3 **(b)** Fe: +2 **(c)** U: +6 **(d)** Cu: +2 **(e)** Ti: +4 **(f)** Sn: +2 **6.92 (a)** oxidized: S; reduced: O **(b)** oxidized: Na; reduced: Cl **(c)** oxidized: Zn; reduced: Cu **(d)** oxidized: Cl; reduced: F **6.94** $FeSO_4$; 151.9 g/mol; 91.8 mg Fe **6.96** Zn is the reducing agent, and Mn^{2+} is the oxidizing agent. **6.98** 132 kg Li_2O; not a redox reaction **6.100** 6×10^{13} molecules
6.102 (a) $C_{12}H_{22}O_{11}(s) \rightarrow 12\ C(s) + 11\ H_2O(l)$ **(b)** 25.3 g C **(c)** 8.94 g H_2O
6.104 (a) 6.40 g **(b)** 104 g
6.106 (a) $Al(OH)_3(aq) + 3\ HNO_3(aq) \rightarrow Al(NO_3)_3(aq) + 3\ H_2O(l)$
(b) $3\ AgNO_3(aq) + FeCl_3(aq) \rightarrow 3\ AgCl(s) + Fe(NO_3)_3(aq)$
(c) $(NH_4)_2Cr_2O_7(s) \rightarrow Cr_2O_3(s) + 4\ H_2O(g) + N_2(g)$
(d) $Mn_2(CO_3)_3(s) \rightarrow Mn_2O_3(s) + 3\ CO_2(g)$
6.108 (a) $2\ SO_2(g) + O_2(g) \rightarrow 2\ SO_3(g)$
(b) $SO_3(g) + H_2O(g) \rightarrow H_2SO_4(l)$ **(c)** SO_2: +4; SO_3, H_2SO_4: +6

Chapter 7
7.1 (a) endothermic **(b)** $\Delta H = +678$ kcal
(c) $C_6H_{12}O_6(aq) + 6\ O_2(g) \rightarrow 6\ CO_2(g) + 6\ H_2O(l) + 678$ kcal
7.2 (a) endothermic **(b)** 200 kcal **(c)** 74.2 kcal **7.3** 91 kcal
7.4 (a) increase **(b)** increase **(c)** decrease **7.5 (a)** no **(b)** increases **(c)** yes
7.6 (a) +0.06 kcal/mol; nonspontaneous **(b)** 0.00 kcal/mol; equilibrium **(c)** −0.05 kcal/mol; spontaneous **7.7 (a)** positive **(b)** spontaneous at all temperatures
7.8   **7.9**

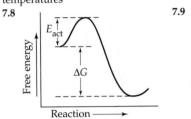

7.10 (a) rate increases **(b)** rate decreases **(c)** rate decreases

7.11 (a) $K = \dfrac{[NO_2]^2}{[N_2O_4]}$ **(b)** $K = \dfrac{[CH_3Cl][HCl]}{[CH_4][Cl_2]}$

(c) $K = \dfrac{[Br_2][F_2]^5}{[BrF_5]^2}$ **7.12 (a)** products favored **(b)** reactants favored

(c) products favored **7.13** $K = 29.0$ **7.14** The reaction forming CD has larger K. **7.15** reaction favored by high pressure and low temperature
7.16 (a) favors reactants **(b)** favors product **(c)** favors product **7.17** ΔH is positive; ΔS is positive; ΔG is negative **7.18** ΔH is negative; ΔS is negative; ΔG is negative **7.19 (a)** $2\ A_2 + B_2 \rightarrow 2\ A_2B$ **(b)** ΔH is negative; ΔS is negative; ΔG is negative **7.20 (a)** blue curve represents faster reaction **(b)** red curve is spontaneous **7.21** red curve represents catalyzed reaction
7.22

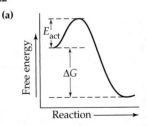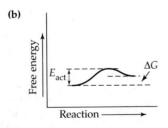

7.23 (a) positive **(b)** nonspontaneous at low temperature; spontaneous at high temperature **7.24** lower enthalpy for reactants
7.26 (a) positive **(b)** 43 kcal **(c)** 3.8 kcal
7.28 (a) $C_6H_{12}O_6 + 6\ O_2 \rightarrow 6\ CO_2 + 6\ H_2O$ **(b)** 1.0×10^3 kcal **(c)** 57 kcal
7.30 increased disorder: **(a)**; decreased disorder: **(b)**, **(c)**
7.32 A spontaneous process, once started, continues without external influence. **7.34** release or absorption of heat, and increase or decrease

in entropy **7.36** ΔH is usually larger than $T\Delta S$ **7.38 (a)** endothermic **(b)** increases **(c)** $T\Delta S$ is larger than ΔH
7.40 (a) $H_2(g) + Br_2(l) \rightarrow 2\ HBr(g)$ **(b)** increases **(c)** yes, because ΔH is negative and ΔS is positive **(d)** $\Delta G = -25.6$ kcal/mol
7.42 the amount of energy needed for reactants to surmount the barrier to reaction
7.44

(a)   **(b)**

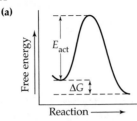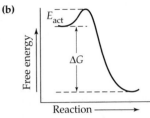

7.46 Collisions increase in frequency and occur with more energy.
7.48 A catalyst lowers the activation energy. **7.50 (a)** yes **(b)** reaction rate is slow **7.52** At equilibrium, the rates of forward and reverse reactions are equal. Amounts of reactants and products need not be equal.

7.54 (a) $K = \dfrac{[CO_2]^2}{[CO]^2[O_2]}$ **(b)** $K = \dfrac{[HCl]^2[C_2H_4Cl_2]}{[Cl_2]^2[C_2H_6]}$

(c) $K = \dfrac{[H_3O^+][F^-]}{[HF][H_2O]}$ **(d)** $K = \dfrac{[O_3]^2}{[O_2]^3}$ **7.56** $K = 7.2 \times 10^{-3}$

7.58 (a) 0.087 mol/L **(b)** 0.023 mol/L **7.60** more reactant
7.62 (a) endothermic **(b)** reactants are favored **(c)** (1) favors ozone; (2) favors ozone; (3) favors O_2; (4) no effect; (5) favors ozone
7.64 (a) decrease **(b)** no effect **(c)** increase **7.66** increase
7.68 (a) increase **(b)** decrease **(c)** no effect **(d)** decrease **7.70** fat
7.72 thyroid, hypothalamus **7.74** conversion of N_2 into chemically useful compounds; microorganisms and lightning
7.76 (a) $2\ C_2H_3OH(l) + 3\ O_2(g) \rightarrow 2\ CO_2(g) + 2\ H_2O(g)$
(b) negative **(c)** 35.5 kcal **(d)** 5.63 g **(e)** 5.60 kcal/mL
7.78 (a) $Fe_3O_4(s) + 4\ H_2(g) \rightarrow 3\ Fe(s) + 4\ H_2O(g)$ $\Delta H = +36$ kcal/mol
(b) 12 kcal **(c)** 3.6 g H_2 **(d)** reactants **7.80** 450 g
7.82 (a) $4\ NH_3(g) + 5\ O_2(g) \rightarrow 4\ NO(g) + 6\ H_2O(g)$ + heat

(b) $K = \dfrac{[NO]^4[H_2O]^6}{[NH_3]^4[O_2]^5}$ **(c)** (1) favors reactants (2) favors reactants

(3) favors reactants (4) favors products **7.84 (a)** exergonic
(b) $\Delta G = -10$ kcal/mol **7.86** 1.91 kcal; exothermic

Chapter 8
8.1 (a) disfavored by ΔH; favored by ΔS **(b)** +0.02 kcal/mol
(c) $\Delta H = -9.72$ kcal/mol; $\Delta S = -26.1$ cal/(mol·K) **8.2** 0.29 atm; 4.3 psi; 29,000 Pa **8.3** 1000 mmHg **8.4** 450 L **8.5** 1.3 atm, 64 atm
8.6 0.27 L; 0.88 L **8.7** 33 psi **8.8** 352 L **8.9** balloon **(a)**
8.10 4460 mol; 7.14×10^4 g CH_4; 1.96×10^5 g CO_2
8.11 5.0 atm **8.12** 1,100 mol; 4,400 g
8.13 (a)   **(b)**

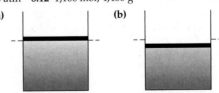

8.14 9.3 atm He; 0.19 atm O_2 **8.15** 75.4% N_2, 13.2% O_2, 5.3% CO_2, 6.2% H_2O **8.16** 35.0 mmHg **8.17 (c)** **8.18 (a)** decrease **(b)** increase **8.19 (a)**, **(c)** **8.20 (a)** London forces **(b)** hydrogen bonds, dipole–dipole forces, London forces **(c)** dipole–dipole forces, London forces
8.21 1.93 kcal, 14.3 kcal

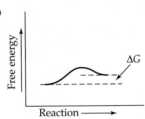

8.22

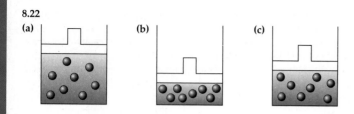

(a) volume increases by 50% (b) volume decreases by 50% (c) volume unchanged **8.23** (c) **8.24** (c)

8.25

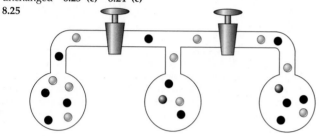

8.26

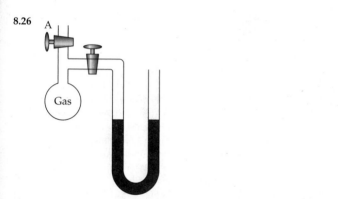

8.27 (a) 10 °C (b) 75 °C (c) 1.3 kcal/mol (d) 7.5 kcal/mol

8.28

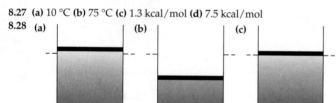

8.29 red = 300 mmHg; yellow = 100 mmHg; green = 200 mmHg
8.30 One atmosphere is equal to exactly 760 mmHg. **8.32** (1) A gas consists of tiny particles moving at random with no forces between them. (2) The amount of space occupied by the gas particles is small. (3) The average kinetic energy of the gas particles is proportional to the Kelvin temperature. (4) Collisions between particles are elastic.
8.34 (a) 760 mmHg (b) 1310 mmHg (c) 5.7×10^3 mmHg (d) 711 mmHg (e) 0.314 mmHg **8.36** 930 mmHg **8.38** V varies inversely with P when n and T are constant. **8.40** 101 mL **8.42** 1.75 L **8.44** V varies directly with T when n and P are constant. **8.46** 364 K = 91 °C
8.48 220 mL **8.50** P varies directly with T when n and V are constant. **8.52** 1.2 atm **8.54** 493 K = 220 °C **8.56** 68.4 mL **8.58** (a) P increases by factor of 4 (b) P decreases by factor of 4 **8.60** 484 mL **8.62** Because gas particles are so far apart and have no interactions, their chemical identity does not matter. **8.64** 22.4 L **8.66** the same number of molecules; the O_2 sample weighs more **8.68** 11.8 g **8.70** 15 kg **8.72** $PV = nRT$ **8.74** Cl_2 has fewer molecules but weighs more **8.76** 370 atm (3.7×10^2 atm) **8.78** 2.2×10^4 mmHg **8.80** 22.3 L **8.82** the pressure contribution of one component in a mixture of gases **8.84** 93 mmHg **8.86** the partial pressure of the vapor above the liquid **8.88** Increased pressure raises a liquid's boiling point; decreased pressure lowers it. **8.90** (a) all molecules (b) molecules with polar covalent bonds (c) molecules with —OH or —NH bonds **8.92** Ethanol forms hydrogen bonds. **8.94** (a) 29.2 kcal (b) 173 kcal **8.96** Atoms in a crystalline solid have a regular, orderly

arrangement. **8.98** 4.82 kcal **8.100** Systolic pressure is the maximum pressure just after contraction; diastolic pressure is the minimum pressure at the end of the heart cycle. **8.102** Increase in atmospheric [CO_2]; increase in global temperatures **8.104** The supercritical state is intermediate in properties between liquid and gas. **8.106** As temperature increases, molecular collisions become more violent. **8.108** 0.13 mol; 4.0 L **8.110** 590 g/day **8.112** 0.92 g/L; less dense than air at STP **8.114** (a) 0.714 g/L (b) 1.96 g/L (c) 1.43 g/L
8.116 (a)

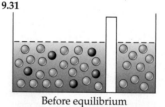

(b)

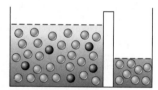

(c) Ethylene glycol forms hydrogen bonds.
8.118 (a) 492 °R (b) $R = 0.0455$ (L·atm)/(mol·°R)

Chapter 9
9.1 (a) heterogeneous mixture (b) homogeneous solution (c) homogeneous colloid (d) homogeneous solution **9.2** (c), (d)
9.3 $Na_2SO_4 \cdot 10H_2O$ **9.4** 322 g **9.5** 80 g/100 mL **9.6** 5.6 g/100 mL **9.7** 6.8×10^{-5} g/100 mL **9.8** 0.925 M **9.9** (a) 0.061 mol (b) 0.67 mol **9.10** 0.48 g **9.11** 0.39 g **9.12** 0.0086% (w/v) **9.13** 6.6% (w/v)
9.14 (a) 20 g (b) 1.2 g **9.15** Place 38 mL acetic acid in flask and dilute to 500.0 mL. **9.16** (a) 22 mL (b) 18 mL **9.17** 1.6 ppm **9.18** Pb: 0.015 ppm, 0.0015 mg; Cu: 1.3 ppm, 0.13 mg **9.19** 2.40 M **9.20** 39.1 mL
9.21 750 L **9.22** (a) 39.1 g; 39.1 mg (b) 79.9 g; 79.9 mg (c) 12.2 g; 12.2 mg (d) 48.0 g; 48.0 mg (e) 9.0 g; 9.0 mg (f) 31.7 g; 31.7 mg **9.23** 9.0 mg
9.24 102.0 °C **9.25** weak electrolyte **9.26** (a) red curve is pure solvent; green curve is solution (b) solvent bp = 62 °C; solution bp = 69 °C (c) 2 M **9.27** –1.9 °C **9.28** 3 ions/mol **9.29** (a) 0.70 osmol (b) 0.30 osmol **9.30** 0.33 osmol
9.31

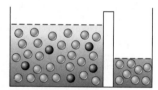

Before equilibrium   At equilibrium

9.32 HCl completely dissociates into ions; acetic acid dissociates only slightly. **9.33** HBr is completely dissociated; HF is not. **9.34** Upper red line: liquid; lower green line: gas **9.35** (a) **9.36** (d) **9.37** green curve is solution; red curve is solvent **9.38** homogeneous: mixing is uniform; heterogeneous: mixing is nonuniform **9.40** polarity **9.42** (b), (d) **9.44** 15.3 g/100 mL **9.46** Concentrated solutions can be saturated or not; saturated solutions can be concentrated or not. **9.48** Molarity is the number of moles of solute per liter of solution. **9.50** Dissolve 45.0 mL of ethyl alcohol in water and dilute to 750.0 mL **9.52** Dissolve 1.5 g NaCl in water to a final volume of 250 mL. **9.54** (a) 7.7% (w/v) (b) 3.9% (w/v) **9.56** (a) 4.0 g (b) 15 g **9.58** 230 mL, 1600 mL **9.60** 10 ppm **9.62** (a) 0.425 M (b) 1.53 M (c) 1.03 M **9.64** 5.3 mL **9.66** 37 g **9.68** 400 mL **9.70** 0.53 L **9.72** 600 mL **9.74** a substance that conducts electricity when dissolved in water **9.76** Ca^{2+} concentration is 0.0015 M **9.78** 40 mEq **9.80** 0.28 L **9.82** $Ba(OH)_2$ **9.84** 861 g **9.86** The inside of the cell has higher osmolarity than water, so water passes in and increases pressure. **9.88** (a) 0.20 M Na_2SO_4 (b) 3% (w/v) NaOH **9.90** 2.4 osmol **9.92** The body manufactures more hemoglobin. **9.94** Sports drinks contain electrolytes, carbohydrate, and vitamins. **9.96** (a) 6.84 mmHg (b) 1.9 g/100 mL **9.98** (a) 0.0067% (w/v) (b) 67 ppm (c) 0.000 40 M **9.100** 9.4 mL **9.102** NaCl: 0.147 M; KCl: 0.0040 M; $CaCl_2$: 0.0030 M **9.104** 0.00020 % (w/v) **9.106** 4.0 mL **9.108** (a) $CoCl_2 + 6 H_2O \rightarrow CoCl_2 \cdot 6H_2O$ (b) 1.13 g **9.110** (a) 1.36 mol particles (b) 2.53 °C

Chapter 10
10.1 (a), (b) **10.2** (a), (c) **10.3** (a) H_2S (b) HPO_4^{2-} (c) HCO_3^- (d) NH_3 **10.4** acids: HF, H_2S; bases: HS^-, F^-; conjugate acid–base pairs: H_2S and

HS^-, HF and F^- **10.5 (a)** base **(b)** acid **(c)** base **10.6 (a)** NH_4^+ **(b)** H_2SO_4 **(c)** H_2CO_3 **10.7 (a)** F^- **(b)** OH^- **10.8** HPO_4^{2-} + OH^- $\rightleftharpoons$ PO_4^{3-} + H_2O; favored in forward direction **10.9** The $-NH_3^+$ hydrogens are most acidic **10.10** citric acid **10.11 (a)** acidic, $[OH^-]$ = 3.1×10^{-10} M **(b)** basic, $[OH^-]$ = 3.2×10^{-3} M **10.12** pH = 5 has higher $[H^+]$; pH 9 has higher $[OH^-]$ **10.13 (a)** 5 **(b)** 5 **10.14 (a)** 1×10^{-13} M **(b)** 1×10^{-3} M **(c)** 1×10^{-8} M; **(b)** is most acidic; **(a)** is most basic **10.15** pH = 4 **10.16 (a)** acidic, 3×10^{-7} M **(b)** most basic, 1×10^{-8} M **(c)** acidic 2×10^{-4} M **(d)** most acidic, 3×10^{-4} M **10.17 (a)** 8.28 **(b)** 5.05 **10.18** 2.60 **10.19** 3.38 **10.20** 9.45 **10.21** 9.13 **10.22 (a)** 0.079 Eq **(b)** 0.338 Eq **(c)** 0.14 Eq **10.23 (a)** 0.26 N **(b)** 1.13 N **(c)** 0.47 N

10.24 $Al(OH)_3$ + 3 HCl → $AlCl_3$ + 3 H_2O;
$Mg(OH)_2$ + 2 HCl → $MgCl_2$ + 2 H_2O
10.25 (a) 2 $KHCO_3$ + H_2SO_4 → 2 H_2O + 2 CO_2 + K_2SO_4
(b) $MgCO_3$ + 2 HNO_3 → H_2O + CO_2 + $Mg(NO_3)_2$
10.26 H_2SO_4 + 2 NH_3 → $(NH_4)_2SO_4$
10.27 $CH_3CH_2NH_2$ + HCl → $CH_3CH_2NH_3^+$ Cl^-
10.28 0.730 M **10.29** 133 mL **10.30** 0.225 M **10.31 (a)** neutral **(b)** basic **(c)** basic **(d)** acidic

10.32 (a)

HCO_3^- + H_2O ⟶ CO_3^{2-} + H_3O^+
Acid   Base   Base   Acid

(b)

HCO_3^- + HF ⟶ H_2CO_3 + F^-
Base   Acid   Acid   Base

10.33 (a) box 2 **(b)** box 3 **(c)** box 1 **10.34** The O—H hydrogen in each is most acidic; acetic acid **10.35 (a)** box 1 **(b)** box 2 **(c)** box 1 **10.36 (a)** box 3 **(b)** box 1 **10.37** 0.67 M **10.38** HBr dissociates into ions **10.40** KOH dissociates into ions **10.42** A monoprotic acid can donate one proton; a diprotic acid can donate two. **10.44 (a)**, **(e)** **10.46 (a)** acid **(b)** base **(c)** neither **(d)** acid **(e)** neither **(f)** acid **10.48 (a)** CH_2ClCO_2H **(b)** $C_5H_5NH^+$ **(c)** $HSeO_4^-$ **(d)** $(CH_3)_3NH^+$ **10.50 (a)** HCO_3^- + HCl → H_2O + CO_2 + Cl^-; HCO_3^- + NaOH → H_2O + Na^+ + CO_3^{2-} **(b)** $H_2PO_4^-$ + HCl → H_3PO_4 + Cl^-; $H_2PO_4^-$ + NaOH → H_2O + Na^+ + HPO_4^{2-} **10.52** 2 HCl + $CaCO_3$ → H_2O + CO_2 + $CaCl_2$

10.54 $K_a = \dfrac{[H_3O^+][A^-]}{[HA]}$

10.56 $K_w = [H_3O^+][OH^-] = 1.0 \times 10^{-14}$
10.58 CH_3CO_2H is a weak acid and is only partially dissociated.

10.60
$K_a = \dfrac{[H_2PO_4^{2-}][H_3O^+]}{[H_3PO_4]}$ $K_a = \dfrac{[HPO_4^{2-}][H_3O^+]}{[H_2PO_4^-]}$ $K_a = \dfrac{[PO_4^{3-}][H_3O^+]}{[HPO_4^{2-}]}$

10.62 (a) OH^- **(b)** NO_2^- **(c)** OH^- **(d)** CN^- **(e)** HPO_4^{2-} **10.64** basic; 1.3×10^{-8} **10.66** 1×10^{-2} M **10.68** 1.0; 13.0 **10.70 (a)** 7.60 **(b)** 3.30 **(c)** 11.64 **10.72 (a)** 1×10^{-4} M; 1×10^{-10} M; **(b)** 1×10^{-11} M; 1×10^{-3} M **(c)** 1 M; 1×10^{-14} M **(d)** 4.2×10^{-2} M; 2.4×10^{-13} M **(e)** 1.1×10^{-8} M; 9.1×10^{-7} M **10.74** A buffer contains a weak acid and its anion. The acid neutralizes any added base, and the anion neutralizes any added acid.

10.76 (a) pH = pK_a + $\log\dfrac{[CH_3CO_2^-]}{[CH_3CO_2H]}$ = 4.74 + $\log\dfrac{[0.100]}{[0.100]}$ = 4.74

(b) $CH_3CO_2^-$ Na^+ + H^+ → CH_3CO_2H + Na^+; CH_3CO_2H + OH^- → $CH_3CO_2^-$ + H_2O **10.78** 9.19 **10.80** 9.07 **10.82** An equivalent is the formula weight in grams divided by the number of H_3O^+ or OH^- ions produced. **10.84** 63.0 g; 32.7 g; 56.1 g; 29.3 g **10.86** 25 mL; 25 mL **10.88 (a)** 0.50 Eq **(b)** 0.084 Eq **(c)** 0.25 Eq **10.90** 0.13 M; 0.26 N **10.92** 0.22 M **10.94** 0.075 M **10.96 (a)** pH = 2 to 3 **(b)** $NaHCO_3$ + HCl → CO_2 + H_2O + $NaCl^+$ **(c)** 20 mg

10.98 Intravenous bicarbonate neutralizes the hydrogen ions in the blood and restores pH. **10.100** 2×10^{-6} M **10.102** Citric acid reacts with sodium bicarbonate to release CO_2. **10.104** Both have the same amount of acid; HCl has higher $[H_3O^+]$ and lower pH. **10.106** 0.35 M **10.108 (a)** NH_4^+, acid; OH^-, base; NH_3, conjugate base; H_2O, conjugate acid **(b)** 5.56 g **10.110 (a)** $Na_2O(aq)$ + $H_2O(l)$ → 2 NaOH(aq) **(b)** 13.0 **(c)** 5000 mL

Chapter 11

11.1 $^{218}_{84}Po$ **11.2** $^{226}_{88}Ra$ **11.3** $^{14}_{6}C$ → $^{0}_{-1}e$ + $^{14}_{7}N$ **11.4 (a)** $^{3}_{1}H$ → $^{0}_{-1}e$ + $^{3}_{2}He$ **(b)** $^{210}_{82}Pb$ → $^{0}_{-1}e$ + $^{210}_{83}Bi$ **(c)** $^{20}_{9}F$ → $^{0}_{-1}e$ + $^{20}_{10}Ne$ **11.5 (a)** $^{38}_{20}Ca$ → $^{0}_{1}e$ + $^{38}_{19}K$ **(b)** $^{118}_{54}Xe$ → $^{0}_{1}e$ + $^{118}_{53}I$ **(c)** $^{79}_{37}Rb$ → $^{0}_{1}e$ + $^{79}_{36}Kr$ **11.6 (a)** $^{62}_{30}Zn$ → $^{0}_{-1}e$ + $^{62}_{29}Cu$ **(b)** $^{110}_{50}Sn$ → $^{0}_{-1}e$ + $^{110}_{49}In$ **(c)** $^{81}_{36}Kr$ + $^{0}_{-1}e$ → $^{81}_{35}Br$ **11.7** $^{120}_{49}In$ → $^{0}_{-1}e$ + $^{120}_{50}Sn$ **11.8** 12% **11.9** 3 days **11.10** 13 m **11.11** 2% **11.12** 4.0 mL **11.13** $^{237}_{93}Np$ **11.14** $^{241}_{95}Am$ + $^{4}_{2}He$ → 2 $^{1}_{0}n$ + $^{243}_{97}Bk$ **11.15** $^{40}_{18}Ar$ + $^{1}_{1}H$ → $^{1}_{0}n$ + $^{40}_{19}K$ **11.16** $^{235}_{92}U$ + $^{1}_{0}n$ → 2 $^{1}_{0}n$ + $^{137}_{52}Te$ + $^{97}_{40}Zr$ **11.17** 2 half-lives **11.18** $^{28}_{12}Mg$ → $^{0}_{-1}e$ + $^{28}_{13}Al$

11.19

11.20 $^{14}_{6}C$ **11.21** The shorter arrows represent β emission; longer arrows represent α emission. **11.22** $^{241}_{94}Pu$ → $^{241}_{95}Am$ → $^{237}_{93}Np$ → $^{233}_{91}Pa$ → $^{233}_{92}U$ **11.23** $^{148}_{69}Tm$ → $^{0}_{1}e$ + $^{148}_{68}Er$ or $^{148}_{69}Tm$ + $^{0}_{-1}e$ → $^{148}_{68}Er$ **11.24** 3.5 years **11.25** The curve doesn't represent nuclear decay. **11.26** It emits radiation. **11.28** A nuclear reaction changes the identity of the atoms, is unaffected by temperature or catalysts, and often releases a large amount of energy. A chemical reaction does not change the identity of the atoms, is affected by temperature and catalysts, and involves relatively small energy changes. **11.30** $^{4}_{2}He$ **11.32** Gamma is highest and alpha is lowest. **11.34** by breaking bonds in DNA **11.36** A neutron decays to a proton and an electron. **11.38** The number of nucleons and the number of charges is the same on both sides. **11.40** α emission: Z decreases by 2 and A decreases by 4; β emission: Z increases by 1 and A is unchanged **11.42** In fission, a nucleus fragments to smaller pieces. **11.44 (a)** $^{35}_{17}Cl$ **(b)** $^{24}_{11}Na$ **(c)** $^{90}_{39}Y$ **11.46 (a)** $^{109}_{47}Ag$ **(b)** $^{10}_{5}B$ **11.48 (a)** 4$^{1}_{0}n$ **(b)** $^{146}_{57}La$ **11.50** $^{198}_{80}Hg$ + $^{1}_{0}n$ → $^{198}_{79}Au$ + $^{1}_{1}H$; a proton **11.52** $^{228}_{90}Th$ **11.54** Half of a sample decays in that time. **11.56** 0.006 g **11.58** 1 ng; 2×10^{-3} ng **11.60** The inside walls of a Geiger counter tube are negatively charged, and a wire in the center is positively charged. Radiation ionizes argon gas inside the tube, which creates a conducting path for current between the wall and the wire. **11.62** In a scintillation counter, a phosphor emits a flash of light when struck by radiation, and the flashes are counted. **11.64** more than 25 rems **11.66** 1.9 mL **11.68** 1.9 rem **11.70** *in vivo* procedures, therapeutic procedures, boron neutron capture **11.72** Irradiation kills harmful microorganisms by destroying their DNA. **11.74** They yield more data, including three-dimensional images. **11.76** Only living organisms incorporate C-14. After death, the ratio of C-14/C-12 decreases, and the ratio can be measured to determine age. **11.78** Nuclear decay is an intrinsic property of a nucleus and is not affected by external conditions. **11.80 (a)** β emission **(b)** Mo-98 **11.82 (a)** $^{238}_{94}Pu$ → $^{4}_{2}He$ + $^{234}_{92}U$ **(b)** for radiation shielding **11.84** Their cells divide rapidly. **11.86** advantages: few harmful byproducts, fuel is inexpensive; disadvantage: needs a high temperature **11.88 (a)** $^{253}_{99}Es$ + $^{4}_{2}He$ → $^{256}_{101}Md$ + $^{1}_{0}n$ **(b)** $^{250}_{98}Cf$ + $^{11}_{5}B$ → $^{257}_{103}Lr$ + 4$^{1}_{0}n$ **11.90** $^{10}_{5}B$ + $^{1}_{0}n$ → $^{7}_{3}Li$ + $^{4}_{2}He$ **11.92** $^{238}_{92}U$ + 3$^{4}_{2}He$ → $^{246}_{98}Cf$ + 4$^{1}_{0}n$

Chapter 12

12.1 (a) alcohol, carboxylic acid **(b)** double bond, ester **(c)** aromatic ring, amine, carboxylic acid **12.2 (a)** CH_3CHO **(b)** $CH_3CH_2CO_2H$ **12.3** $CH_3CH_2CH_2CH_2CH_2CH_2CH_3$

12.4

$$CH_3CH_2CH_2CH_2\overset{\underset{\displaystyle CH_3}{|}}{C}HCH_3 \qquad CH_3CH_2CH_2\overset{\underset{\displaystyle CH_3}{|}}{C}HCH_2CH_3$$

$$CH_3CH_2CH_2\overset{\overset{\displaystyle CH_3}{|}}{\underset{\underset{\displaystyle CH_3}{|}}{C}}CH_3 \qquad CH_3CH_2\overset{\overset{\displaystyle CH_3}{|}}{\underset{\underset{\displaystyle CH_3}{|}}{C}}HCHCH_3 \qquad CH_3CH_2\overset{\overset{\displaystyle CH_3}{|}}{\underset{\underset{\displaystyle CH_3}{|}}{C}}CH_2CH_3$$

$$CH_3\overset{\underset{\displaystyle CH_3}{|}}{C}HCH_2\overset{\underset{\displaystyle CH_3}{|}}{C}HCH_3 \qquad CH_3CH_2\overset{\underset{\displaystyle CH_2CH_3}{|}}{C}HCH_2CH_3 \qquad \overset{\underset{\displaystyle H_3C}{|}}{CH_3C}-\overset{\underset{\displaystyle }{}}{C}HCH_3$$
(with H$_3$C and CH$_3$ substituents)

12.5 **(a)**

$$CH_3CH_2CH_2CH_2CH_3$$
Pentane

(b)
$$CH_3\overset{\underset{\displaystyle CH_3}{|}}{C}HCH_2CH_3$$
2-Methylbutane

(c)
$$CH_3\overset{\overset{\displaystyle CH_3}{|}}{\underset{\underset{\displaystyle CH_3}{|}}{C}}CH_3$$
2,2-Dimethylpropane

12.6 (a) **(b)** **(c)**

12.7 (a)
$$CH_3CH_2\overset{\overset{\displaystyle H_3C}{|}}{\underset{\underset{\displaystyle CH_2CHCH_3}{|}}{C}}-\overset{\underset{\displaystyle Cl}{|}}{C}HCH_2CH_3$$
with CH$_2$CHCH$_3$ and CH$_3$ below

(b)
$$CH_3\overset{\overset{\displaystyle H_3C\ \ CH_3}{\diagdown\diagup}}{C}-\overset{\overset{\displaystyle H_3C\ \ CH_3}{\diagdown\diagup}}{C}H-\overset{\underset{\displaystyle CH_2CH_3}{|}}{C}CH_3$$

12.8 Structures **(a)** and **(c)** are identical, and are isomers of **(b)**.

12.9

$$CH_3CH_2CH_2CH_2CH_2CH_2CH_2CH_3 \qquad CH_3CH_2CH_2CH_2CH_2\overset{\underset{\displaystyle CH_3}{|}}{C}HCH_3 \qquad CH_3CH_2CH_2CH_2\overset{\underset{\displaystyle CH_3}{|}}{C}HCH_2CH_3 \qquad CH_3CH_2CH_2\overset{\underset{\displaystyle CH_3}{|}}{C}HCH_2CH_2CH_3$$

$$CH_3CH_2CH_2CH_2\overset{\overset{\displaystyle CH_3}{|}}{\underset{\underset{\displaystyle CH_3}{|}}{C}}CH_3 \qquad CH_3CH_2CH_2\overset{\underset{\displaystyle CH_3}{|}}{C}HCHCH_3 \qquad CH_3CH_2\overset{\underset{\displaystyle CH_3}{|}}{C}HCH_2\overset{\underset{\displaystyle CH_3}{|}}{C}HCH_3 \qquad CH_3\overset{\underset{\displaystyle CH_3}{|}}{C}HCH_2CH_2\overset{\underset{\displaystyle CH_3}{|}}{C}HCH_3$$

$$CH_3CH_2CH_2\overset{\overset{\displaystyle CH_3}{|}}{\underset{\underset{\displaystyle CH_3}{|}}{C}}CH_2CH_3 \qquad CH_3\overset{\overset{\displaystyle CH_3}{|}}{\underset{\underset{\displaystyle CH_3}{|}}{C}}-\overset{\underset{\displaystyle CH_3}{|}}{C}CH_3 \qquad CH_3CH_2\overset{\underset{\displaystyle CH_3}{|}}{C}HCHCH_2CH_3 \qquad CH_3CH_2CH_2\overset{\underset{\displaystyle CH_2CH_3}{|}}{C}HCH_2CH_3 \qquad CH_3CH_2\overset{\underset{\displaystyle CH_3}{|}}{C}HCCH_3$$
(last one with H$_3$C and CH$_3$)

$$CH_3\overset{\underset{\displaystyle CH_3}{|}}{C}HCH_2\overset{\underset{\displaystyle CH_3}{|}}{C}CH_3 \qquad CH_3CH_2\overset{\overset{\displaystyle CH_3}{|}}{\underset{\underset{\displaystyle CH_3}{|}}{C}}-\overset{\underset{\displaystyle CH_3}{|}}{C}HCH_3 \qquad CH_3\overset{\underset{\displaystyle CH_3}{|}}{C}HCHCHCH_3 \qquad CH_3CH_2\overset{\underset{\displaystyle CH_2CH_3}{|}}{C}HCHCH_3 \qquad CH_3CH_2\overset{\underset{\displaystyle CH_2CH_3}{|}}{C}CH_2CH_3$$
(with various CH$_3$ substituents)

12.10 (a) 2,6-dimethyloctane **(b)** 3,3-diethylheptane

12.11 (a)
$$\underset{p}{CH_3}\underset{s}{CH_2}\underset{s}{CH_2}\overset{\overset{\displaystyle P\ CH_3}{|}}{\underset{t}{C}}H\underset{s}{CH_2}\underset{p}{CH_3}$$

(b)
$$\underset{p}{CH_3}\underset{s}{CH_2}\underset{s}{CH_2}\underset{s}{CH_2}\overset{\overset{\displaystyle P\ CH_3}{\underset{t}{|}}}{\underset{t}{C}}H\underset{s}{CH}\underset{s}{CH_2}\underset{p}{CH_3}$$
with P CH$_3$ below

(c)
$$\underset{p}{CH_3}\overset{\overset{\displaystyle P\ CH_3}{|}}{\underset{t}{C}}H\underset{s}{CH_2}\overset{\overset{\displaystyle P\ CH_3}{|}}{\underset{}{C}}CH_3$$
with q P CH$_3$

12.12 There are many possible answers; for example:

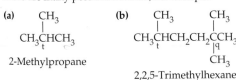

(a) 2-Methylpropane

(b) 2,2,5-Trimethylhexane

12.13 **(a)** 2,2-dimethylpentane **(b)** 2,3,3-trimethylpentane
12.14 $2 C_2H_6 + 7 O_2 \rightarrow 4 CO_2 + 6 H_2O$
12.15

$CH_3CH_2CH_2Cl + CH_3CHClCH_3 + CH_3CCl_2CH_3 +$

$CH_3CH_2CHCl_2 + CH_3CHClCH_2Cl + ClCH_2CH_2CH_2Cl$

12.16 **(a)** 1-ethyl-4-methylcyclohexane **(b)** 1-ethyl-3-isopropylcyclopentane
12.17 **(a)**   **(b)**

12.18 propylcyclohexane
12.19 **(a)** 12 hydrogens **(b)** 10 hydrogens **(c)** 8 hydrogens
12.20 **(a)**   **(b)**

12.21 **(a)**   **(b)**

12.22 **(a)** double bond, ketone, ether **(b)** double bond, amine, carboxylic acid   **12.23** **(a)** 2,3-dimethylpentane **(b)** 2,5-dimethylhexane
12.24 **(a)** 1,1-dimethylcyclopentane **(b)** isopropylcyclobutane
12.25 The methyl groups are on the same side of the ring in one structure and on opposite sides in the other.   **12.26** Carbon can form four strong bonds to itself and to many other elements.   **12.28** Organic compounds are nonpolar.   **12.30** A polar covalent bond is a bond in which electrons are shared unequally.   **12.32** **(a)** amine, aromatic ring, ether, amide, sulfonamide double bond **(b)** aromatic ring, carboxylic acid, ester
12.34 **(a)**   **(b)**

Ketone   Ester

(c)

Amine
carboxylic acid

12.36 They must have the same formula but different structures.
12.38 A primary carbon is bonded to one other carbon; a secondary carbon is bonded to two other carbons; a tertiary carbon is bonded to three other carbons; and a quaternary carbon is bonded to four other carbons.
12.40 **(a)** 2,3-dimethylbutane **(b)** cyclopentane
12.42

$CH_3CH_2CH_2OH$ $CH_3CHOHCH_3$ $CH_3CH_2-O-CH_3$

12.44
(a)

$CH_3CH_2CH_2CH_2OH$ $CH_3CH_2CHOHCH_3$ CH_3CHCH_2OH

$CH_3COH(CH_3)CH_3$

(b)

$CH_3CH_2CH_2NH_2$ $CH_3CHNH_2CH_3$ $CH_3CH_2NHCH_3$ $CH_3NCH_3(CH_3)$

(c)

$CH_3CH_2CH_2CCH_3$ $CH_3CH_2CCH_2CH_3$ CH_3CHCCH_3

12.46 identical: **(a)**; isomers: **(b)**, **(d)**, **(e)**; unrelated: **(c)**   **12.48** All have a carbon with five bonds.   **12.50** **(a)** 4-ethyl-3-methyloctane **(b)** 5-isopropyl-3-methyloctane   **(c)** 2,2,6-trimethylheptane **(d)** 4-isopropyl-4-methyloctane   **(e)** 2,2,4,4-tetramethylpentane **(f)** 4,4-diethyl-2-methylhexane   **(g)** 2,2-dimethyldecane
12.52

(a)   **(b)**

(c)

(d)

(e)   **(f)**

12.54 **(a)** 1-ethyl-3-methylcyclobutane **(b)** 1,1,3,3-tetramethylcyclopentane **(c)** 1-ethyl-3-propylcyclohexane **(d)** 4-butyl-1,1,2,2-tetramethylcyclopentane
12.56 **(a)** 2,2-dimethylpentane **(b)** 2,4-dimethylpentane **(c)** isobutylcyclobutane   **12.58** heptane, 2-methylhexane, 3-methylhexane, 2,2-dimethylpentane, 2,3-dimethylpentane, 2,4-dimethylpentane, 3,3-dimethylpentane, 3-ethylpentane, 2,2,3-trimethylbutane
12.60 $C_3H_8 + 5 O_2 \rightarrow 3 CO_2 + 4 H_2O$
12.62

$CH_3CH_2CCH_2Cl(CH_3) + CH_3CHCHClCH_3(CH_3) + ClCH_2CH_2CCH_3(CH_3)$

12.64 Minor differences in shape cause differences in behavior.
12.66 Branched-chain hydrocarbons   **12.68** **(a)** ketone, alcohol, double bond **(b)** carboxylic acid, amine, amide, ester, aromatic ring   **12.70** nonpolar solvents dissolve nonpolar substances   **12.72** pentane; more London forces

Chapter 13

13.1 (a) 2-methyl-3-heptene (b) 2-methyl-1,5-hexadiene
(c) 3-methyl-3-hexene

13.2

(a)
$$CH_3CH_2CH_2CH_2\overset{\overset{\displaystyle CH_3}{|}}{CH}CH=CH_2$$

(b)
$$H_3C-\overset{\overset{\displaystyle CH_3}{|}}{\underset{\underset{\displaystyle CH_3}{|}}{C}}-C\equiv C-CH_3$$

(c)
$$CH_3CH_2CH_2CH=\overset{\overset{\displaystyle CH_3}{|}}{CH}CHCH_3$$

(d)
$$CH_3CH_2CH=\overset{\overset{\displaystyle CH_3}{|}}{C}-\overset{\overset{\displaystyle CH_3}{|}}{\underset{\underset{\displaystyle CH_3}{|}}{C}}-CH_3$$

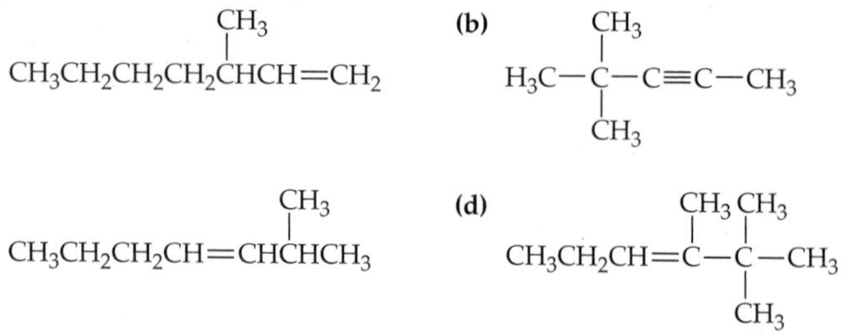

13.3 (a) 2,3-dimethyl-1-pentene (b) 2,3-dimethyl-2-hexene **13.4** (a), (c)

13.5

$$\underset{\text{cis-3,4-Dimethyl-3-hexene}}{\overset{\displaystyle \underset{H_3C}{\overset{CH_3CH_2}{}}C=C\overset{CH_2CH_3}{\underset{CH_3}{}}}{}}$$ $$\underset{\text{trans-3,4-Dimethyl-3-hexene}}{\overset{\displaystyle \underset{H_3C}{\overset{CH_3CH_2}{}}C=C\overset{CH_3}{\underset{CH_2CH_3}{}}}{}}$$

13.6 (a) cis-4-methyl-2-hexene (b) trans-5,6-dimethyl-3-heptene
13.7 (a) substitution (b) addition (c) elimination **13.8** (a), (b),
(c) $CH_3CH_2CH_2CH_3$ (d) methylcyclohexane **13.9** (a) 1,2-dibromo-
2-methylpropane (b) 1,2-dichloropentane (c) 4,5-dichloro-
2,4-dimethylheptane **13.10** (a) 1-chloro-1-methylcyclopentane
(b) 2-bromobutane (c) 2-chloro-2-methylbutane **13.11** (a) 3-ethyl-
2-pentene (b) 2,3-dimethyl-1-butene or 2,3-dimethyl-2-butene
13.12 2-bromo-2,4-dimethylhexane
13.13 (a), (b) Same product is obtained.

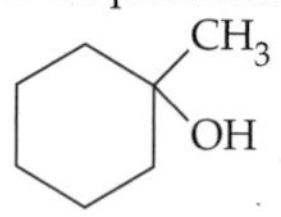

(c)
$$\overset{\overset{\displaystyle OH}{|}}{CH_3CHCH_2CH_2CH_3} \ + \ CH_3CH_2\overset{\overset{\displaystyle OH}{|}}{CH}CH_2CH_3$$

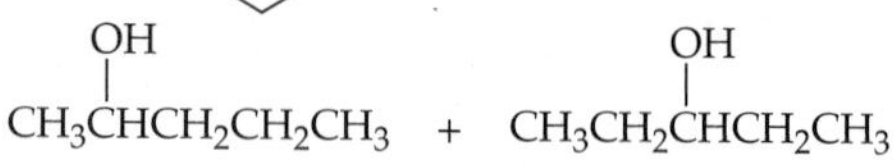

13.14 2-ethyl-1-butene or 3-methyl-2-pentene **13.15** $(CH_3)_3C^+$

13.16

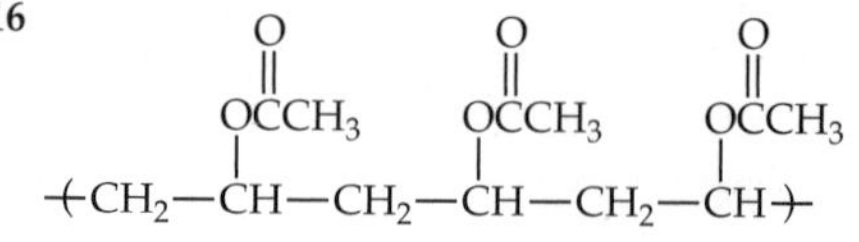

13.17 (a)
$$\overset{\overset{\displaystyle CN}{|}}{H_2C=CCl}$$
(b)
$$\overset{\overset{\displaystyle CO_2CH_3}{|}}{H_2C=CH}$$

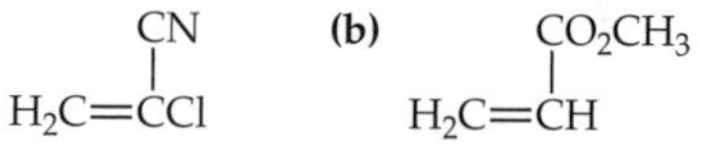

13.18 (a) m-ethylphenol (b) p-bromoaniline (c) 2-methyl-2-phenylbutane

13.19

(a)   (b)   (c)

(d)

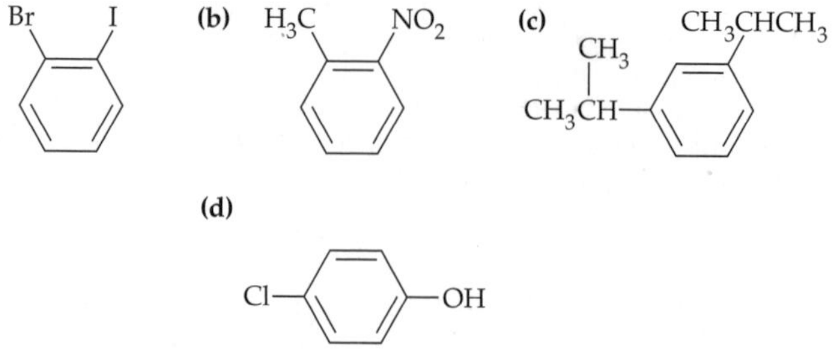

13.20 (a) o-isopropylphenol (b) p-bromoaniline

13.21 (a)   (b)

(c)

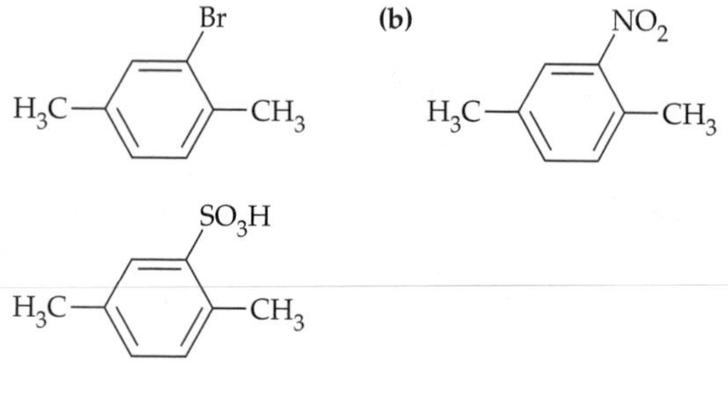

13.22 o-, m-, and p-bromotoluene

13.23

(a) 2,5-Dimethyl-2-heptene

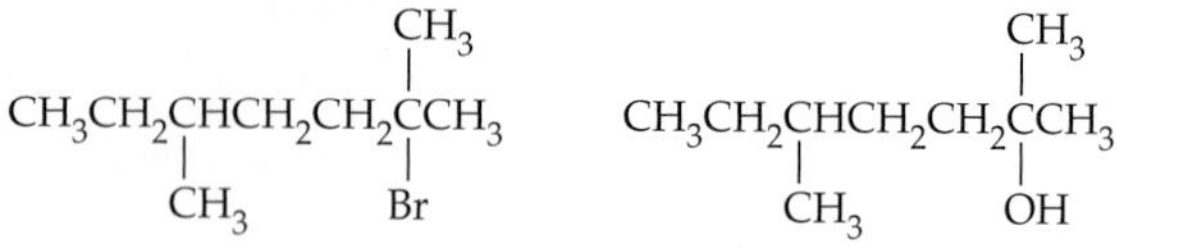

(b) 3,3-Dimethylcyclopentene

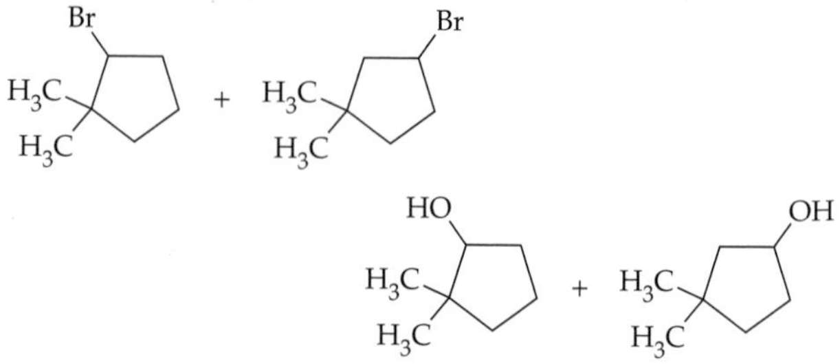

13.24 (a) 4,4-dimethyl-1-hexyne (b) 2,7-dimethyl-4-octyne
13.25 (a) m-isopropylphenol (b) o-bromobenzoic acid
13.26 (a)

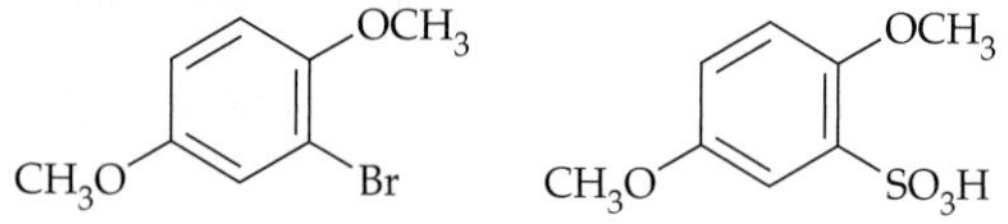

(b)

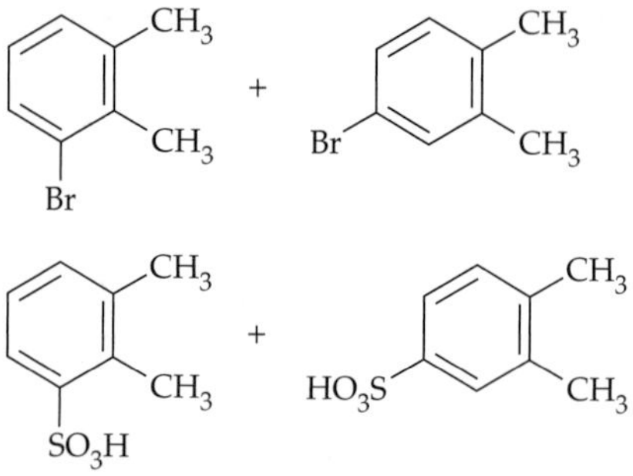

13.27

(a)   (b)

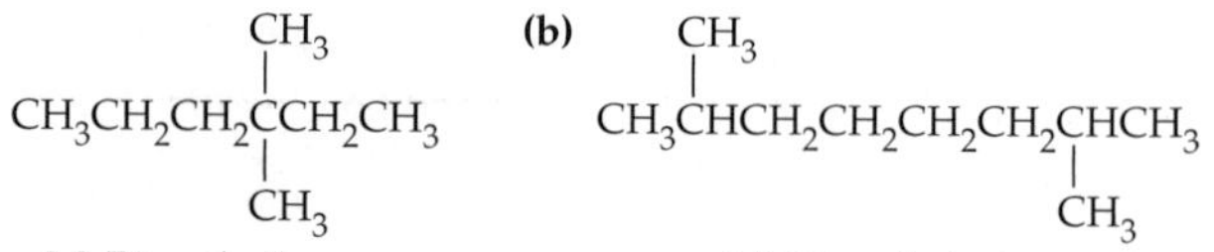

3,3-Dimethylhexane   2,7-Dimethyloctane

13.28

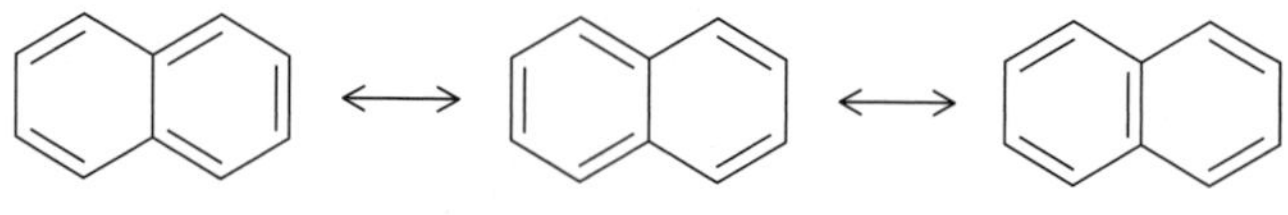

13.29 2-methyl-2-pentene, 2-methyl-1-pentene **13.30** They have C—C
multiple bonds and can add hydrogen. **13.32** alkene: –ene; alkyne: –yne;
aromatic: –benzene

13.34

(a)
$$CH_3CH_2CH_2CH_2CH_2CH=CH_2$$
(b)
$$CH_3CH_2CH_2C\equiv CH$$

(c)

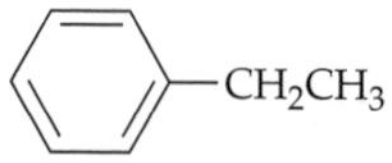

13.36 (a) 1-pentene (b) 2,5-dimethyl-3-hexyne (c) 2,3-dimethyl-2-butene
(d) 2-ethyl-3-methyl-1,3-pentadiene (e) 2-ethyl-1,3-dimethylcyclohexene
(f) 3-ethylisopropylcyclobutene

13.38 (a)

(b)

$$CH_3CH_2CH=CHCHCH_3 \quad (CH_3)$$

(c)

(d)

(e)

(f)

(g) $CH_3CH_2CH_2$ $CH_2CH_2CH_3$

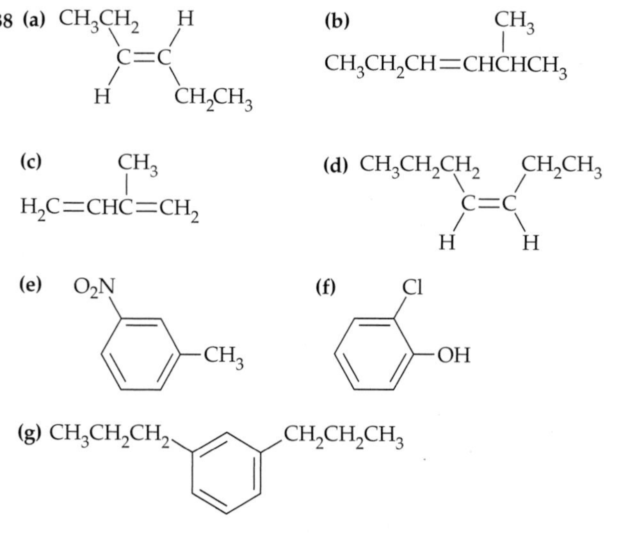

13.40 1-hexyne, 2-hexyne, 3-hexyne, 3-methyl-1-pentyne, 4-methyl-1-pentyne, 4-methyl-2-pentyne, 3,3-dimethyl-1-butyne **13.42** 1-pentene, 2-pentene, 2-methyl-1-butene, 3-methyl-1-butene, 2-methyl-2-butene **13.44** Each double bond carbon must be bonded to two different groups. **13.46** 2-pentene
13.48

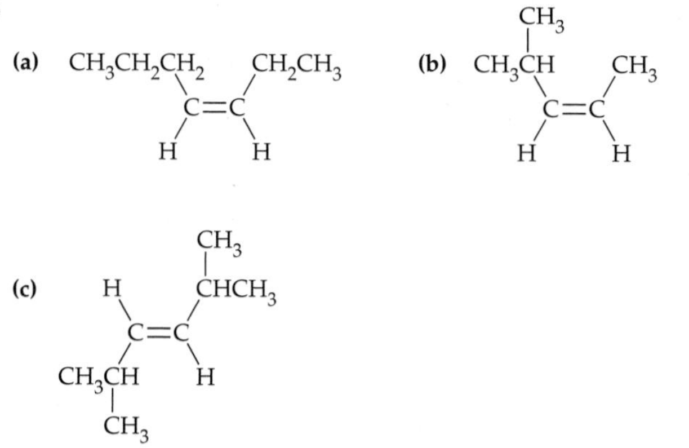

(a)

(b)

(c)

13.50 (a) identical **(b)** identical **13.52** substitution: two reactants exchange parts to give two products; addition: two reactants add to give one product **13.54** rearrangement **13.56 (a)** substitution **(b)** rearrangement
13.58

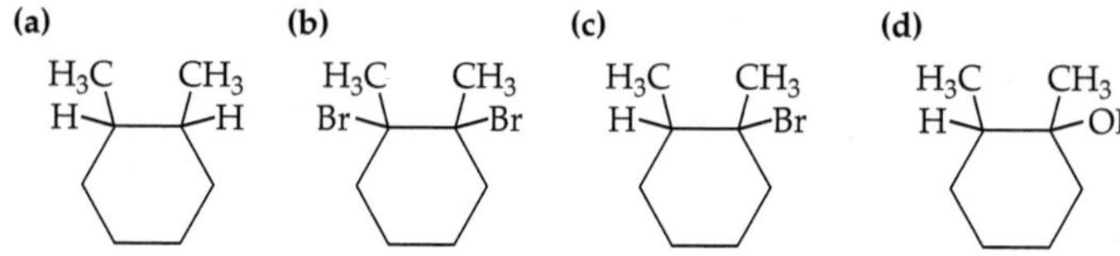

(a) **(b)** **(c)** **(d)**

13.60

(a) $CH_3CH=CHCCH_3 \ (CH_3)(CH_3) \ + \ Cl_2$ **(b)** $CH_3CH=CH_2 \ + \ H_2$

(c) $CH_3CH=CHCH_3$ or $H_2C=CHCH_2CH_3 \ + \ HBr$

(d)

$+ \ HCl$

(e)

$=CH_2 \ + \ Cl_2$

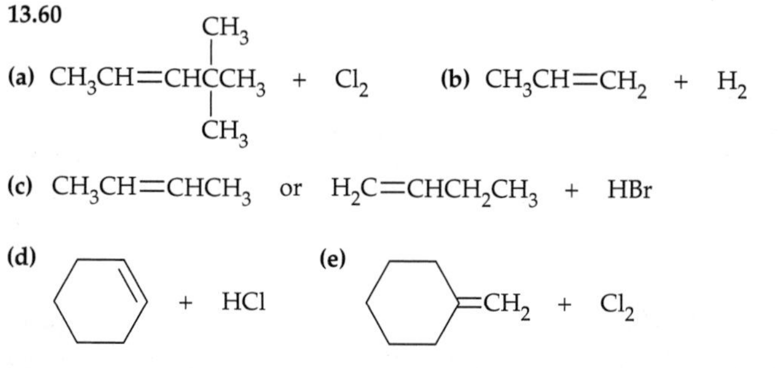

13.62

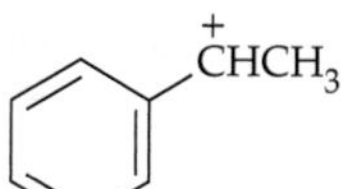

13.64

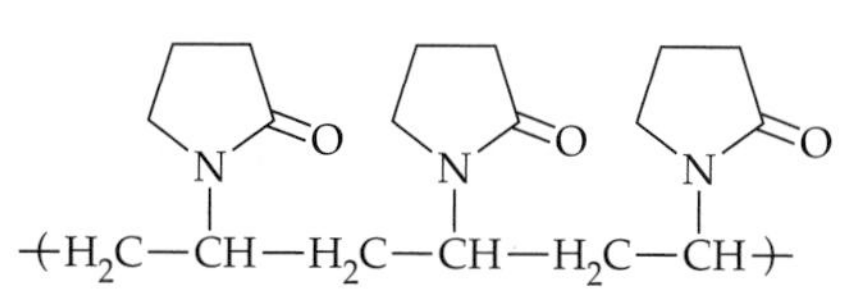

13.66 (b) benzene $+ \ Br_2 \rightarrow$ bromobenzene $+ \ HBr$ **13.68** methylcyclohexane **13.70** Rod cells are responsible for vision in dim light; cone cells are responsible for color vision.
13.72

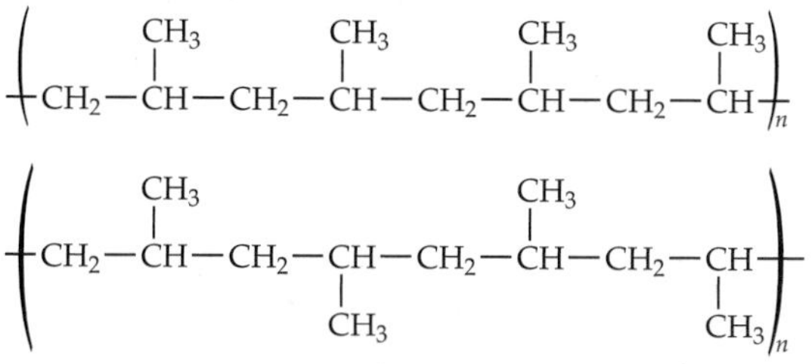

13.74 The body converts it to a water-soluble diol epoxide that can react with DNA and lead to cancer. **13.76** yellow
13.78

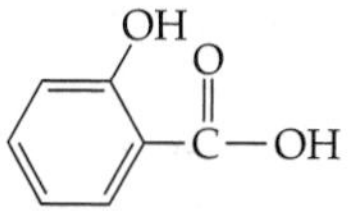

Salicylic acid

13.80 (a) 5-methyl-2-hexene **(b)** 4-methyl-2-heptyne **(c)** 2,3-dimethyl-1-butene **(d)** 1,2,4-trinitrobenzene **(e)** 3,4-dimethylcyclohexene **(f)** 3-methyl-1,3-pentadiene **13.82** Cyclohexene reacts with Br_2; benzene doesn't.
13.84

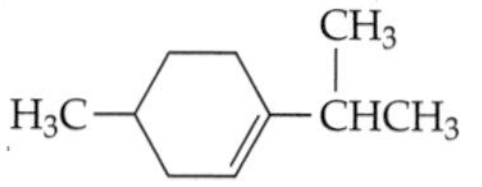

Menthene

13.86

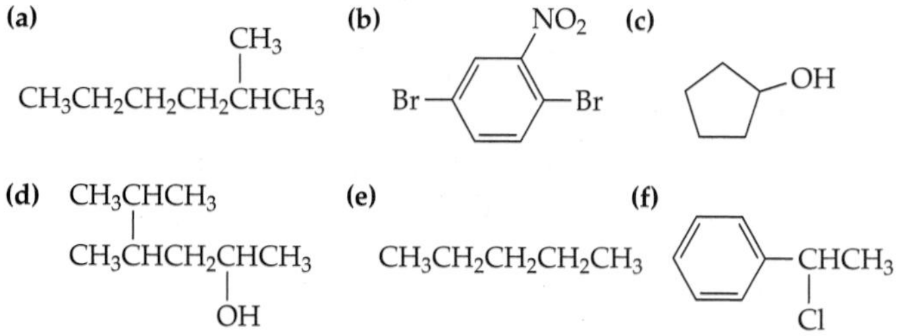

(a)

$CH_3CH_2CH_2CH_2CHCH_3 \ (CH_3)$

(b)

(c)

(d) CH_3CHCH_3 $|$ $CH_3CHCH_2CHCH_3$ (OH)

(e)

$CH_3CH_2CH_2CH_2CH_3$

(f)

13.88 Naphthalene has a greater molar mass.
13.90

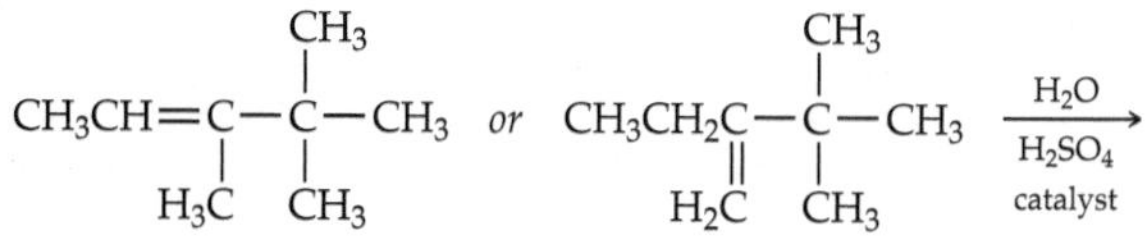

Chapter 14
14.1 (a) alcohol **(b)** alcohol **(c)** phenol **(d)** alcohol **(e)** ether **(f)** ether
14.2 A hydroxyl group is a part of a larger molecule.
14.3 (a)

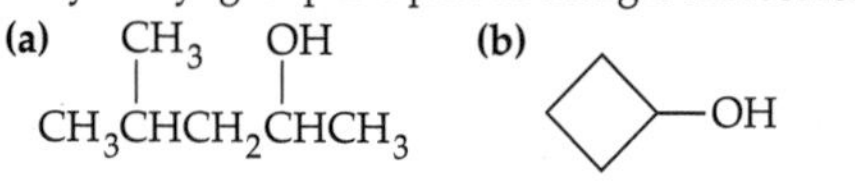

$CH_3CHCH_2CHCH_3$ $(CH_3)(OH)$

secondary alcohol

(b)

OH

secondary alcohol

(c)

$CH_3CH_2CH_2CH_2CHCH_2CHCH_3$ $(OH)(CH_3)$

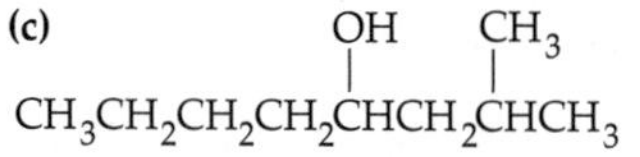

secondary alcohol

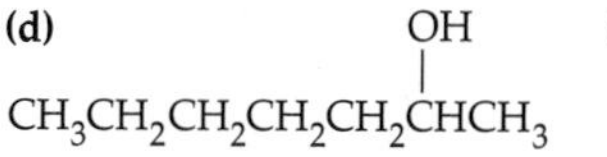

(d)

$CH_3CH_2CH_2CH_2CH_2CHCH_3$ (OH)

secondary alcohol

(e)

$ClCH_2CHCH_2OH$ (Cl)

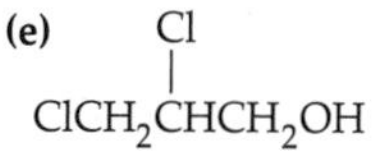

primary alcohol

14.4 (a) 3-pentanol, secondary **(b)** 2-ethyl-1-pentanol, primary **(c)** 5,6-dichloro-2-ethyl-1-hexanol, primary **(d)** 2-isopropyl-4-methylcyclohexanol, secondary **14.5** See 14.3 and 14.4 **14.6 (a)** **14.7 (b)**

14.8 (a) propene **(b)** cyclohexene **(c)** 4-methyl-1-pentene (minor) and 4-methyl-2-pentene (major) **14.9 (a)** 2,3-dimethyl-2-butanol **(b)** 1-butanol or 2-butanol

14.10

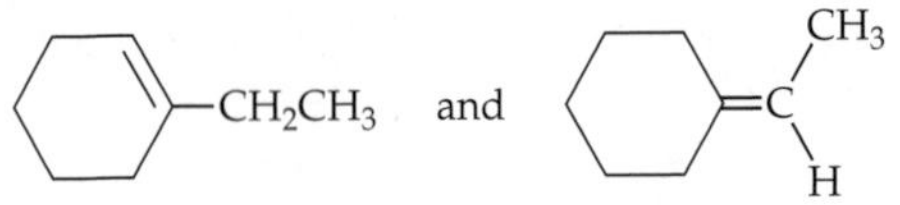

14.11 (a)

(b)

(c)

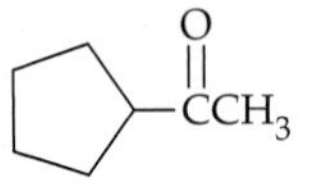

14.12 (a) 2-propanol **(b)** cycloheptanol **(c)** 3-methyl-1-butanol

14.13 (a)

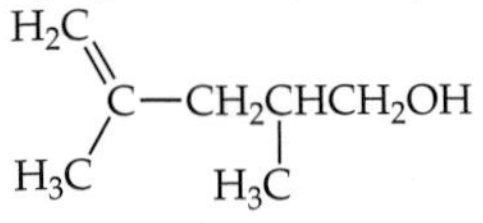

(b)

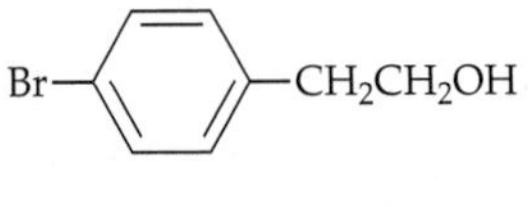

14.14 (a)

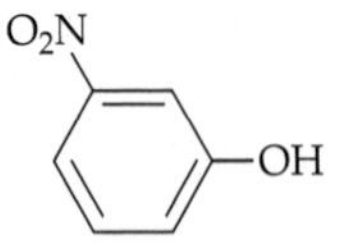

(b)

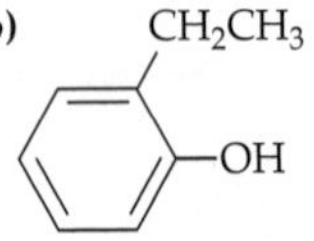

14.15 (a) *p*-chlorophenol **(b)** 4-bromo-2-methylphenol **14.16 (a)** methyl propyl ether **(b)** diisopropyl ether **(c)** methyl phenyl ether

14.17 (a) $CH_3CH_2CH_2S-SCH_2CH_2CH_3$

(b) $(CH_3)_2CHCH_2CH_2S-SCH_2CH_2CH(CH_3)_2$

14.18 (a) 1-chloro-1-ethylcyclopentane **(b)** 3-bromo-5-methylheptane

14.19 (a) 5-methyl-3-hexanol **(b)** *m*-methoxytoluene **(c)** 3-methylcyclohexanol

14.20

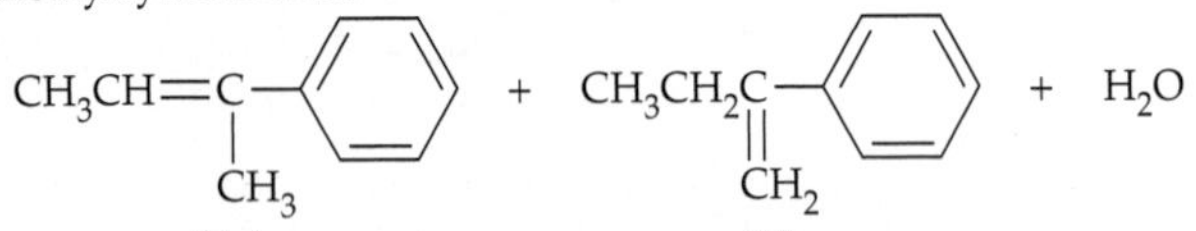

Major   Minor

14.21 $(CH_3)_2CHCH_2CH_2CHO$, $(CH_3)_2CHCH_2CH_2CO_2H$

14.22

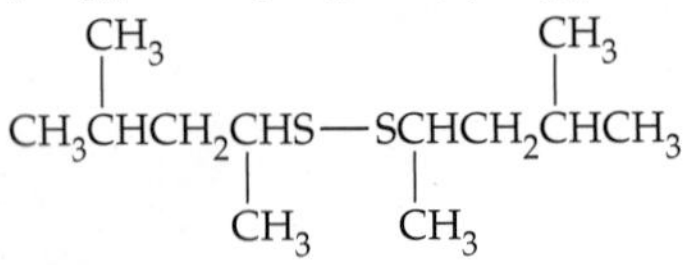

14.23 (a)

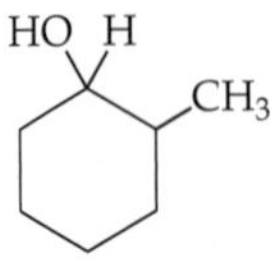

(b)

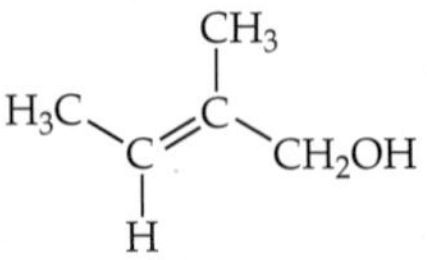

(c)

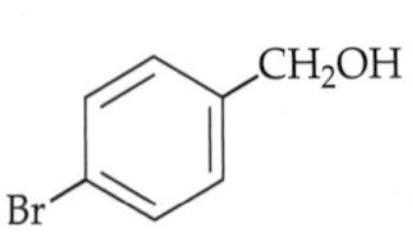

14.24 Alcohols have an —OH group bonded to an alkane-like carbon atom; ethers have an oxygen atom bonded to two carbon atoms; and phenols have an —OH group bonded to a carbon of an aromatic ring.
14.26 Alcohols form hydrogen bonds. **14.28** ketone, carbon–carbon double bond, alcohol **14.30 (a)** 2-ethyl-1-pentanol **(b)** 3-methyl-1-butanol **(c)** 1,2,4-butanetriol **(d)** 2-methyl-2-phenyl-1-propanol **(e)** 2-ethyl-3-methylcyclohexanol **(f)** 3,3-dimethyl-2-hexanol

14.32 (a)

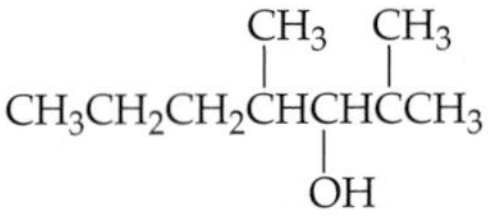

(b)

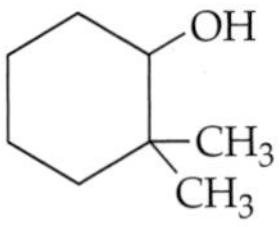

(c)

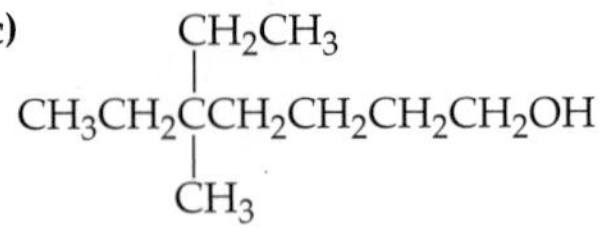

(d)

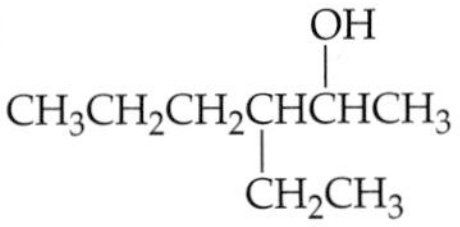

(e)

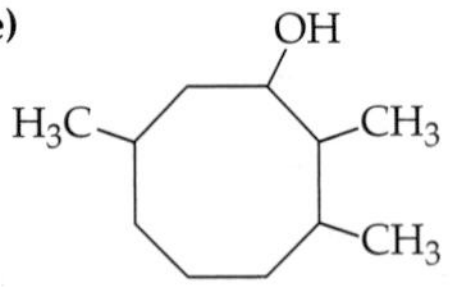

(f)

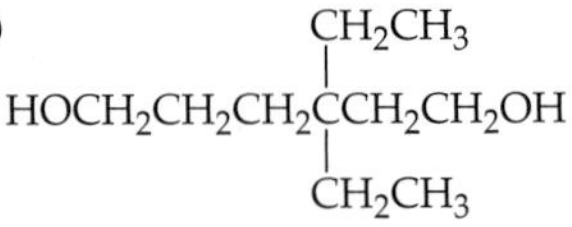

14.34 (a) primary **(b)** primary **(c)** primary, secondary **(d)** primary **(e)** secondary **(f)** secondary **14.36 (a)** < **(c)** < **(b)** **14.38** a ketone **14.40** aldehyde or carboxylic acid **14.42** Phenols dissolve in aqueous NaOH; alcohols don't.

14.44

(a)

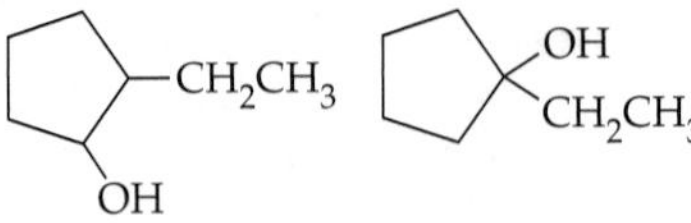

(b)

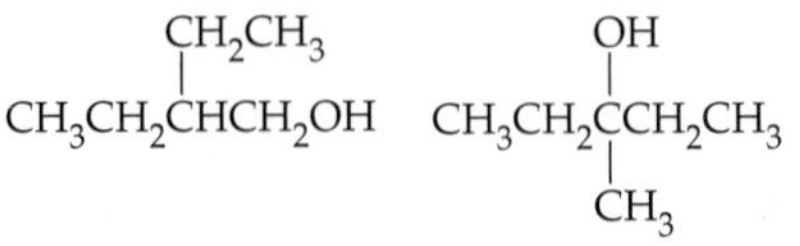

(c)

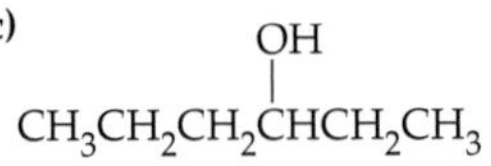

(d)

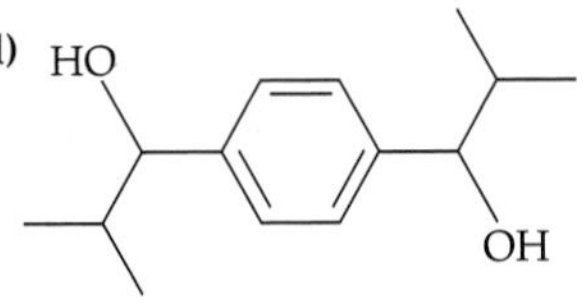

(e)

$HOCH_2CH_2CH_2CH_2CH_2OH$

(f)

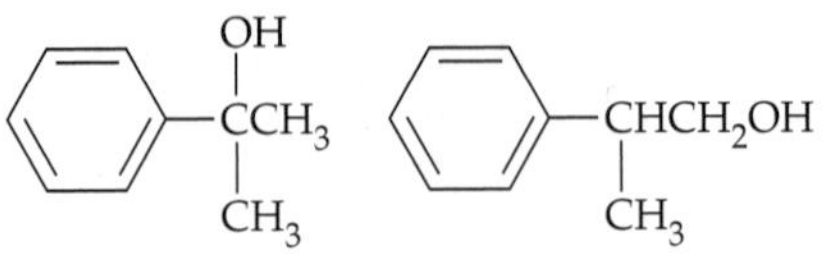

14.46

(a)

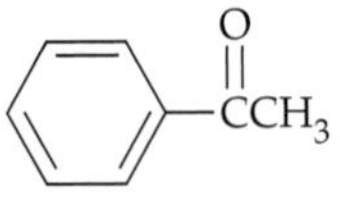

(b)

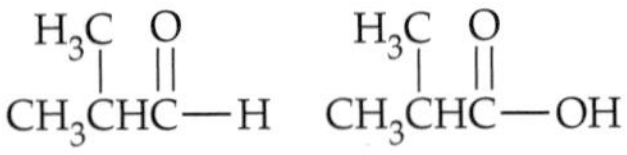

(c) NR **(d)**

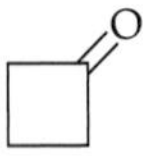

(e) NR **(f)**

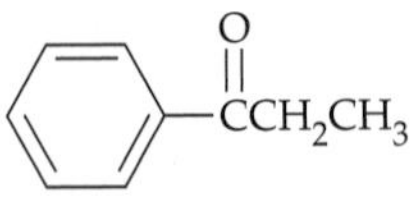

14.48 odor

14.50

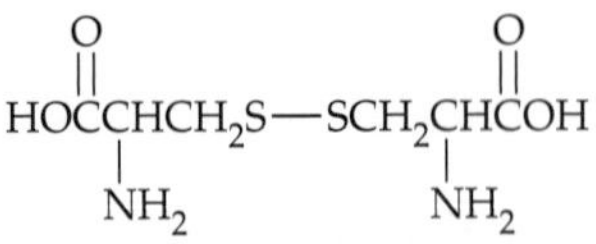

14.52 Alcohols can form hydrogen bonds; thiols and alkyl chlorides cannot. **14.54** depressant **14.56** The liver is the site of alcohol metabolism. **14.58** a reactive species that contains an unpaired electron **14.60** diethyl ether **14.62** The ozone layer shields the earth from intense solar radiation. **14.64** alcohols: 1-butanol, 2-butanol, 2-methyl-1-propanol, 2-methyl-2-propanol; ethers: diethyl ether, methyl propyl ether, isopropyl methyl ether **14.66** Alcohols become less soluble as their

nonpolar part becomes larger. **14.68** An antiseptic kills microorganisms on living tissue. **14.70** (a) *p*-dibromobenzene (b) 1,2-dibromo-1-butene (c) *m*-propylanisole (d) 1,1-dibromocyclopentane (e) 6-chloro-2,4-dimethyl-2,4-hexanediol (f) 4-methyl-2,4,5-heptanetriol (g) 4-bromo-6,6-dimethyl-2-heptyne (h) 1-bromo-2-iodoocyclobutane
14.72 3,7-Dimethyl-2,6-octadiene-1-ol

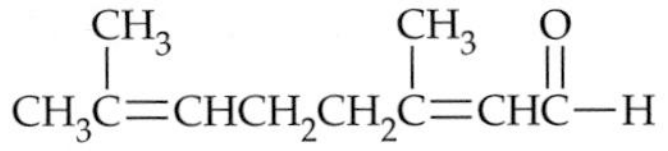

$$CH_3\overset{\underset{\mid}{CH_3}}{C}=CHCH_2CH_2\overset{\underset{\mid}{CH_3}}{C}=CH\overset{\overset{O}{\|}}{C}-H$$

14.74 $C_2H_6O + 3 O_2 \rightarrow 2 CO_2 + 3 H_2O$

Chapter 15
15.1 (a) primary (b) secondary (c) primary (d) secondary (e) tertiary
15.2 (a) Tetrabutylammonium hydroxide (b) Dimethylamine (c) *N*-Pentylaniline
15.3 (a)

$$CH_3CH_2CH_2CH_2CH_2CH_2NH_2$$

(b)

$$CH_3CH_2CH_2CH_2\overset{\underset{\mid}{CH_3}}{NH}$$

(c)

(d)

$$\overset{\underset{\mid}{NH_2}}{CH_2}\overset{\underset{\mid}{OH}}{CH_2}CHCH_3$$

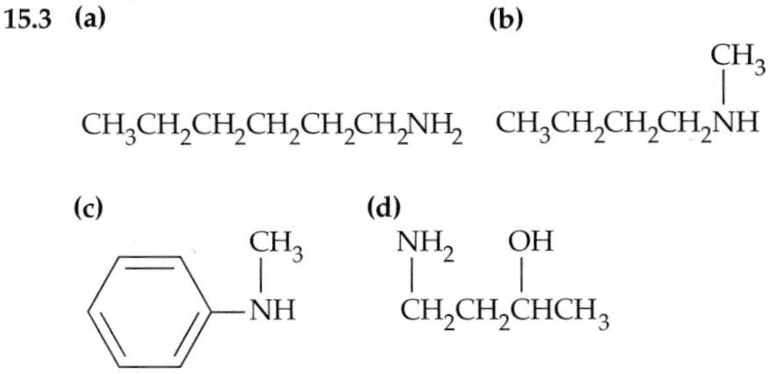

15.4 The ion has one less electron than the neutral atoms.

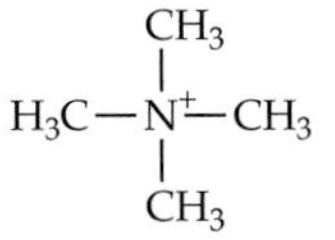

$$H_3C-\overset{\overset{\displaystyle CH_3}{\mid}}{\underset{\underset{\displaystyle CH_3}{\mid}}{N^+}}-CH_3$$

15.5 $CH_3CH_2CH_2CH_2NHCH_2CH_3$ *N*-Ethylbutylamine
15.6 Compound (a) is lowest boiling; (b) is highest boiling (strongest hydrogen bonds).
15.7 (a) (b)

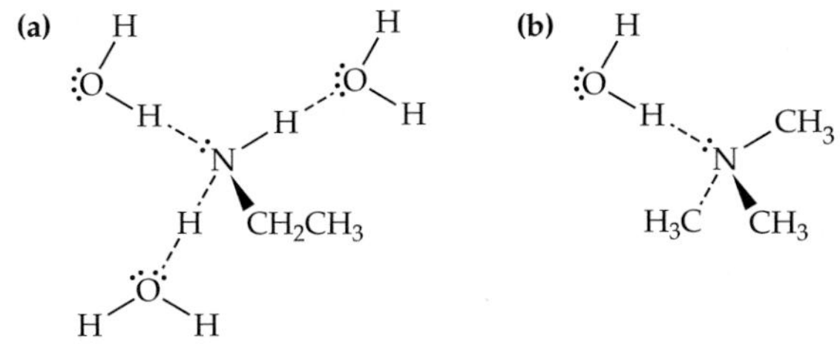

15.8 (a) Methylamine, Ethylamine, Dimethylamine, Trimethylamine (b) Pyridine (c) Aniline
15.9 (a) Piperidine: $C_5H_{11}N$ (b) Purine: $C_5H_4N_4$
15.10 (a) and (d)
15.11

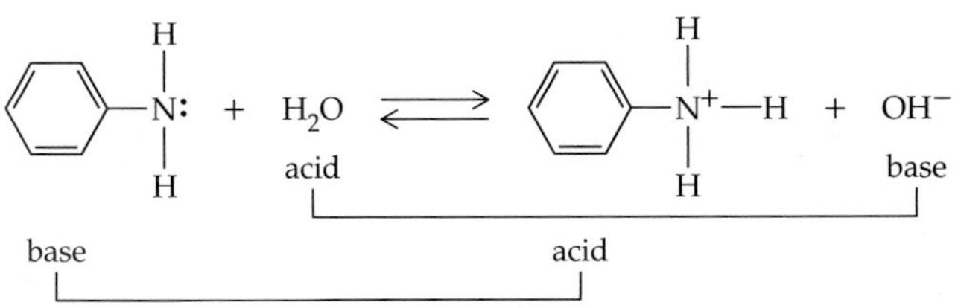

15.12

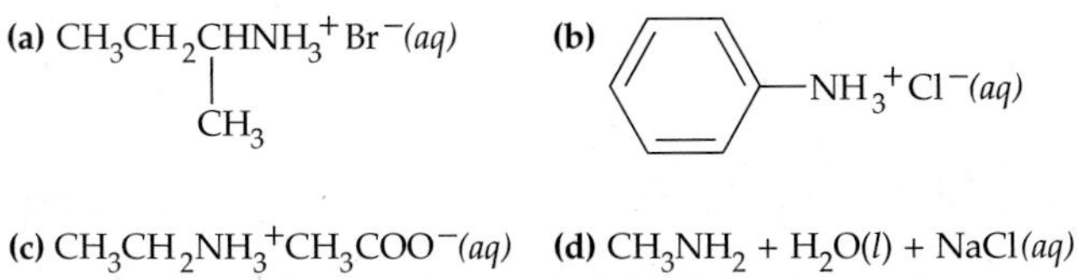

(a) $CH_3CH_2\overset{\underset{\mid}{CH_3}}{CH}NH_3^+ Br^-(aq)$ (b) $\text{—}NH_3^+ Cl^-(aq)$

(c) $CH_3CH_2NH_3^+CH_3COO^-(aq)$ (d) $CH_3NH_2 + H_2O(l) + NaCl(aq)$

15.13 (a) *sec*-Butylammonium bromide (b) Anilinium chloride (c) Ethylammonium acetate (d) Methylamine **15.14** (a) Ethylamine (b) Triethylamine

15.15
(a) HO

$$HO\text{—}\overset{\overset{OH}{\mid}}{\text{—}CHCH_2\overset{+}{N}H_2CH_3}$$

(b)

$$\text{—}CH_2\overset{\underset{\mid}{CH_3}}{CH}NH_3^+$$

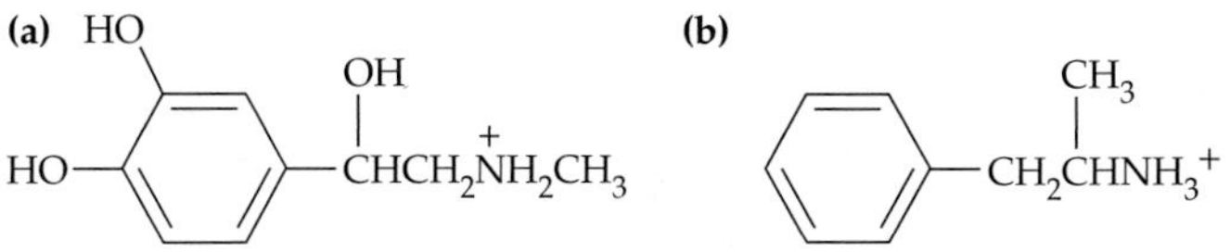

15.16–15.17 (a)

$$CH_3CH_2CH_2CH_2CH_2CH_2\overset{\overset{\displaystyle CH_3}{\mid}}{\underset{\underset{\displaystyle CH_3}{\mid}}{N}}H^+Cl^-$$

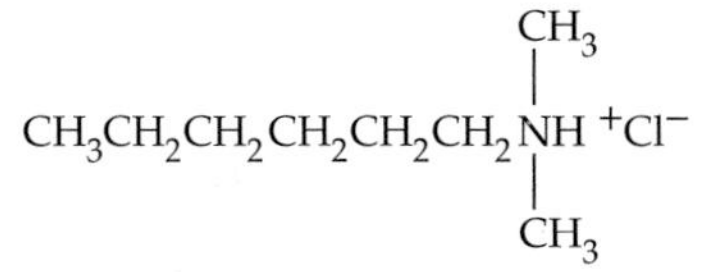

Hexyldimethylammonium chloride
or N, N-Dimethylhexylammonium chloride
Salt of a tertiary amine

(b)

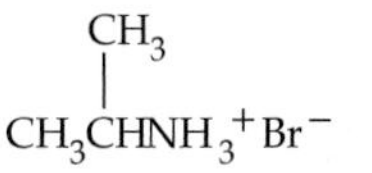

$$CH_3\overset{\underset{\mid}{CH_3}}{CH}NH_3^+Br^-$$

Isopropylammonium bromide
Salt of a primary amine

15.18 $CH_3CH_2CH_2CH_2NH_3^+ Cl^-(aq) + NaOH(aq) \rightarrow CH_3CH_2CH_2NH_2 + H_2O(l) + NaCl(aq)$
15.19 Benadryl has the general structure. In Benadryl, R = $-CH_3$, and R′ = R″ = C_6H_5-.
15.20

$$\text{—}CH_2NH_3^+ Cl^- \qquad \text{—}CH_2NH_2 \cdot HCl$$

Benzylammonium chloride

15.21

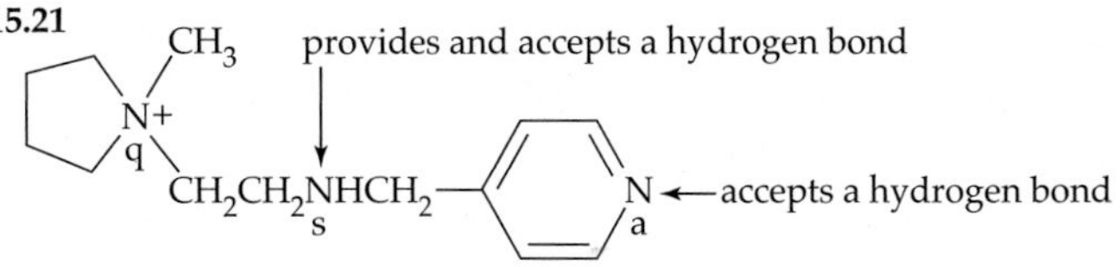

$$\begin{array}{l} \text{CH}_3 \text{ provides and accepts a hydrogen bond} \\ N+ \\ \text{CH}_2\text{CH}_2\text{NHCH}_2\text{—}N\leftarrow\text{accepts a hydrogen bond} \end{array}$$

15.22 (a) All amine groups can participate in hydrogen bonding. (b) Histidine is water-soluble because it can form hydrogen bonds with water.
15.23
(a)

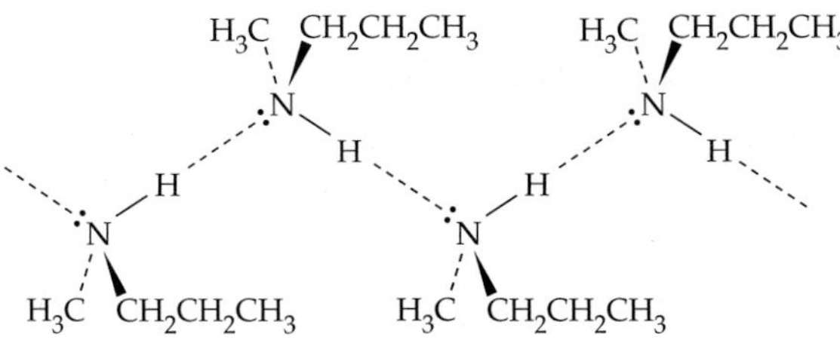

(b)

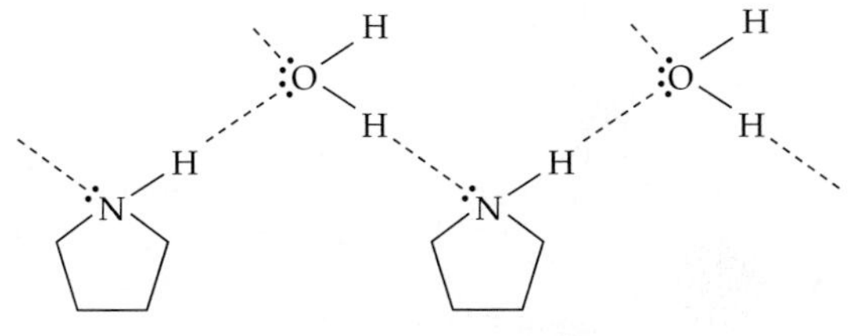

(c)

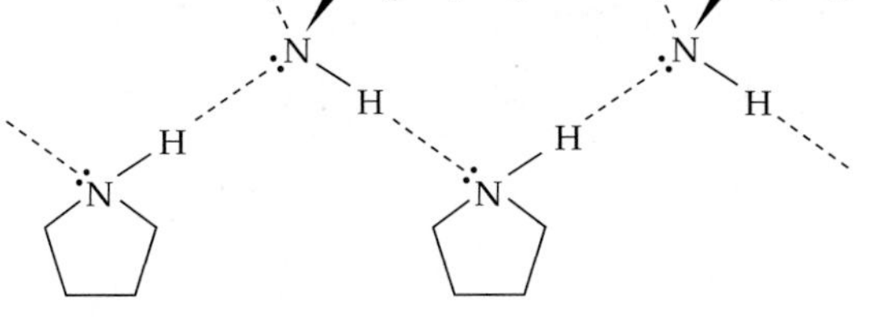

15.24

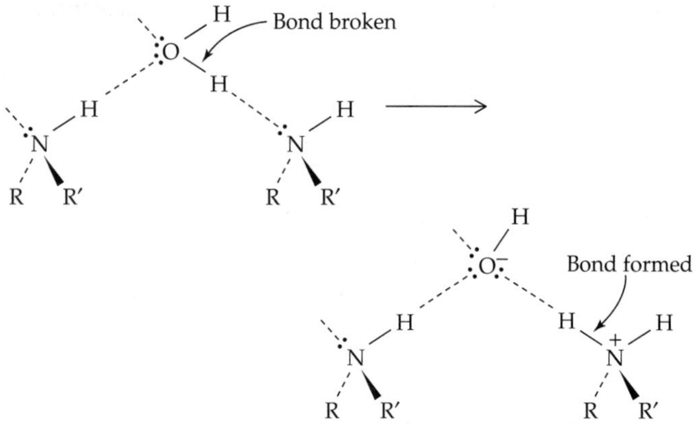

15.25 most basic: $(CH_3)_2NH$ least basic: $C_6H_5NH_2$

15.26

(a)

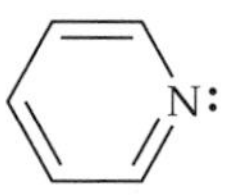

$+ H_2O$

(b) $CH_3\overset{+}{N}H_2CH(CH_3)_2 + OH^-$

(c) No reaction

(d) $(CH_3)_3CNH_3{}^+Cl^-$

15.28 **(a)**

$$CH_3CH_2CH_2CH_2CH_2\overset{H}{\underset{|}{N}}CH_3$$

(b)

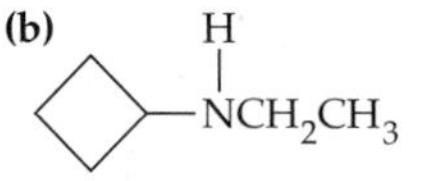

(c) $CH_3CH_2CH_2\!-\!$ $-NH_2$

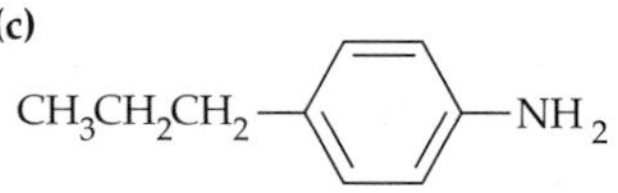

15.30 **(a)** Cyclobutylamine (primary) **(b)** Diphenylamine (secondary)
15.32 Diethylamine   **15.34** **(a)** N-Methyl-2-butylammonium nitrate
(salt of a secondary amine).

(b)

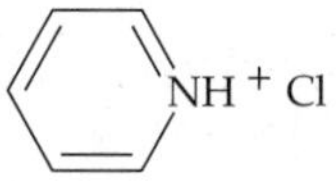

(salt of a tertiary amine)

(c)

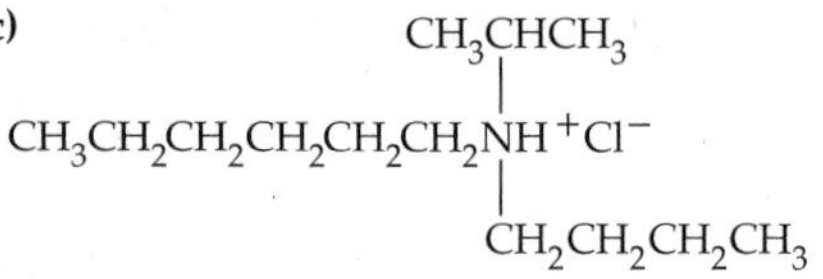

$$CH_3CH_2CH_2CH_2CH_2CH_2\overset{CH_3CHCH_3}{\underset{|}{\underset{CH_2CH_2CH_2CH_3}{N}H}}{}^+Cl^-$$

(salt of a tertiary amine)

15.36

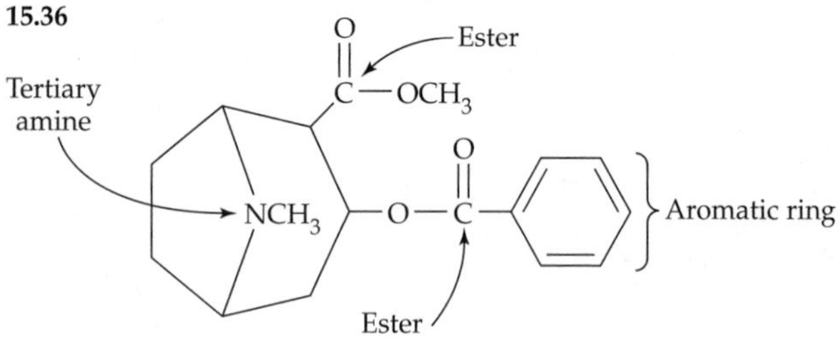

Cocaine

15.38

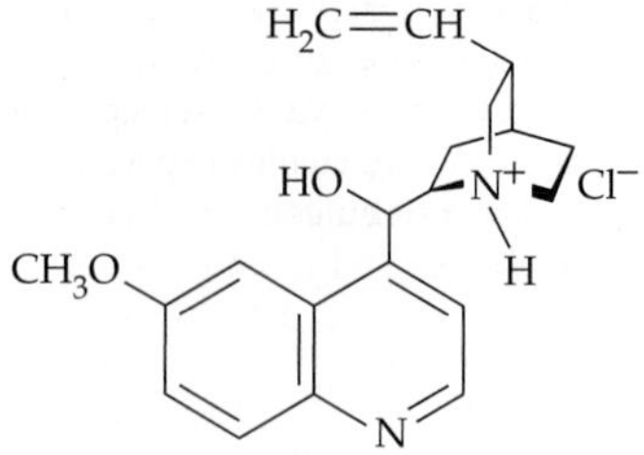

Quinine hydrochloride

15.40

(a)

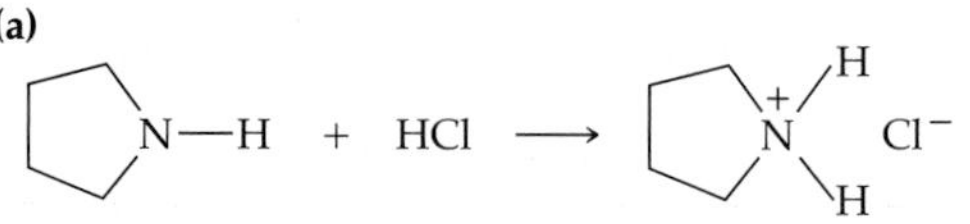

(b)

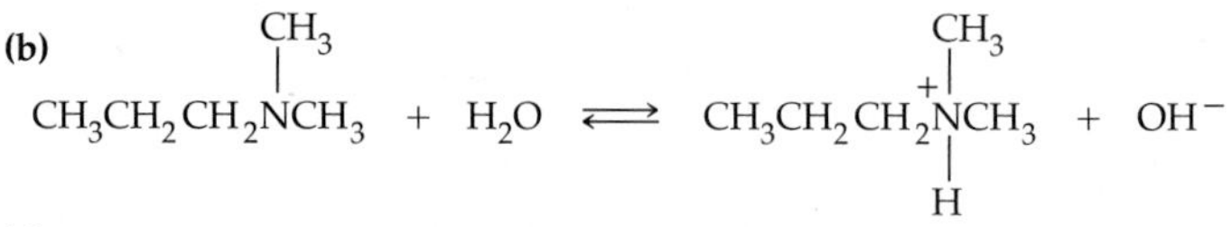

$$CH_3CH_2CH_2\overset{CH_3}{\underset{|}{N}}CH_3 + H_2O \rightleftharpoons CH_3CH_2CH_2\overset{CH_3}{\underset{\underset{H}{|}}{\overset{+}{N}}}CH_3 + OH^-$$

(c)

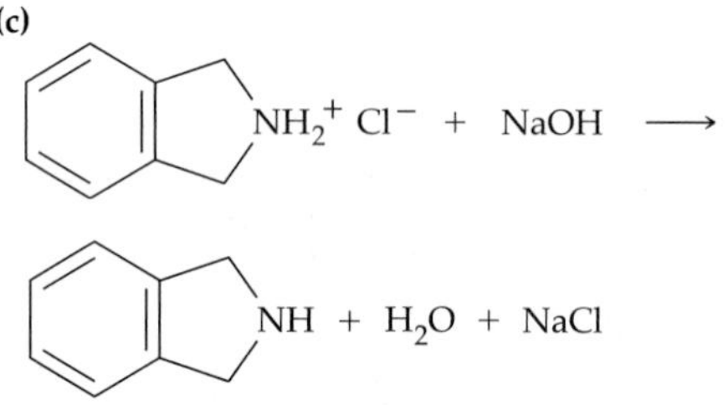

15.42 Choline doesn't react with HCl because its nitrogen isn't basic.
15.44 (1) lowering of blood pressure (2) memory enhancement (3) reduction of sickling in sickle-cell anemia (4) destruction of malaria parasites (5) destruction of the tuberculosis bacterium.

15.46

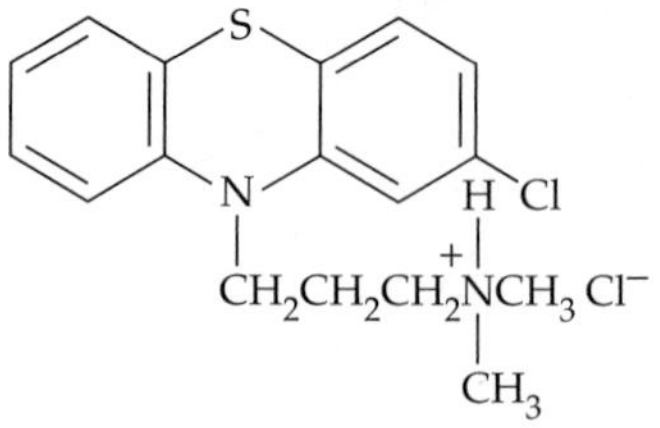

15.48 **(a)** A forensic toxicologist deals with criminal cases involving drug abuse and poisoning **(b)** the structure of the toxin, its mode of action, a mechanism to reverse its effects
15.50 Its large hydrocarbon region is water-insoluble.
15.52

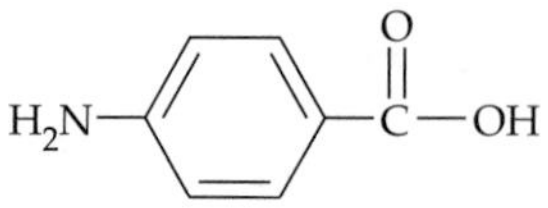

PABA

15.54

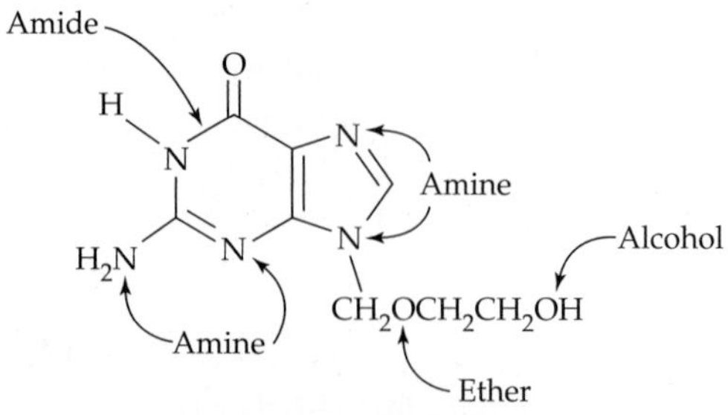

Acyclovir—related to purine

15.56 Amines: foul-smelling, somewhat basic, lower boiling (weaker hydrogen bonds); Alcohols: pleasant-smelling, not basic, higher boiling (stronger hydrogen bonds) **15.58 (a)** 6-Methyl-2-heptene **(b)** *p*-Isopropylphenol **(c)** Dibutylamine **15.60** Molecules of hexylamine can form hydrogen bonds to each other, but molecules of triethylamine can't. **15.62** Baeocystin is related to indole. **15.64** Pyridine forms H-bonds with water; benzene doesn't form H-bonds.

Chapter 16

16.1

(a)
Ketone

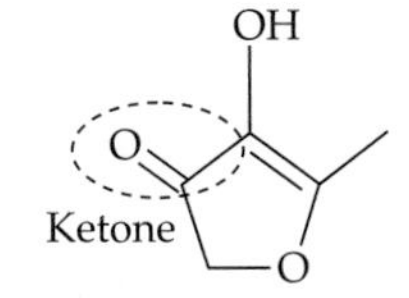

(b) 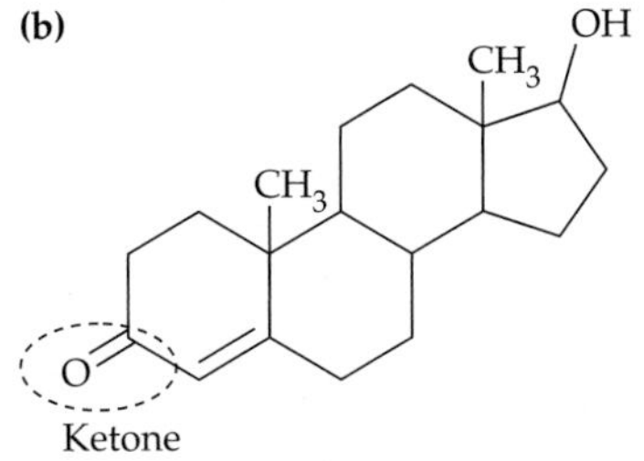
Ketone
Testosterone

(c) 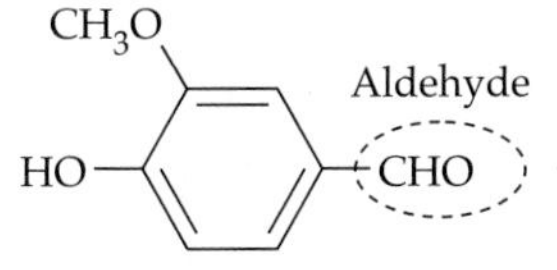
Aldehyde
Vanillin

(d) C₄H₉COCH₃
Ketone

(e) C₄H₉CHO
Aldehyde

(f) C₄H₉COOCH₃
Ester

16.2 (d)

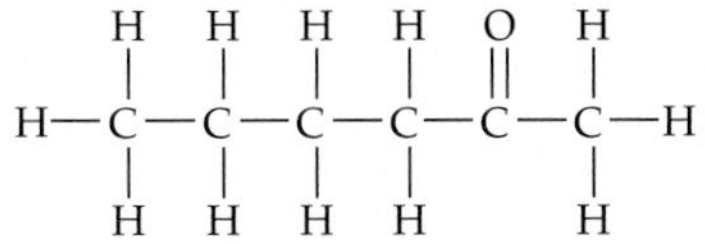

(e)

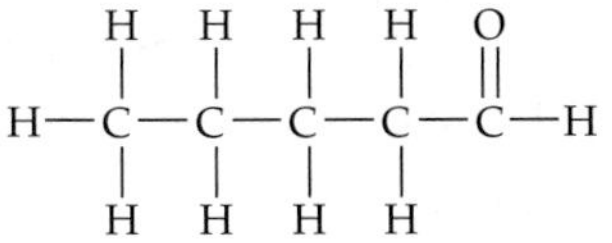

16.3 (a) CH₃CH₂CH₂CH₂CH₂CH₂CH₂CH

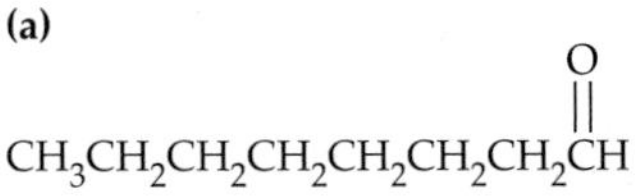

(b)

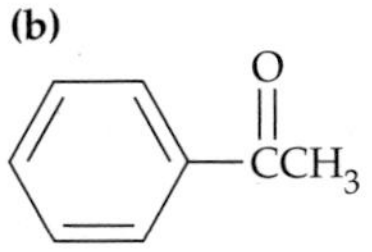

(c)

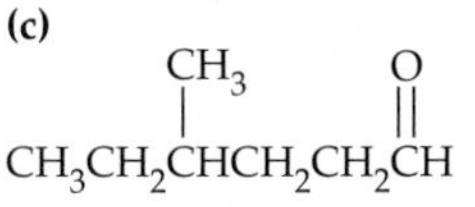

(d)

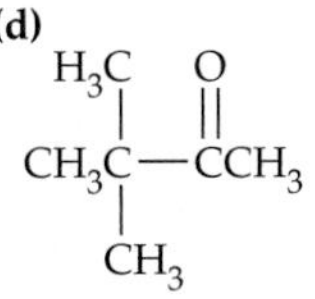

16.4 (a) Pentanal **(b)** 3-Pentanone **(c)** 4-Methylhexanal **(d)** 4-Heptanone

16.5 (a)
C₇H₁₄O
5-Methyl-3-hexanone
A ketone

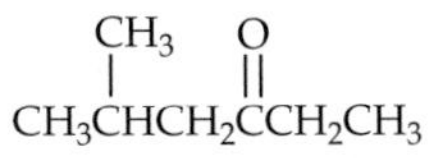

(b)
C₆H₁₂O
4-Methylpentanal
An aldehyde

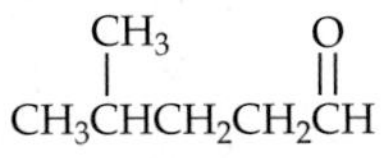

16.6 (a) polar **(b)** flammable **(c)** liquid **(d)** bp of 100 °C

16.7 Alcohols form hydrogen bonds, which raise their boiling points. Aldehydes and ketones have higher boiling points than alkanes because they are polar.

16.8 (a)

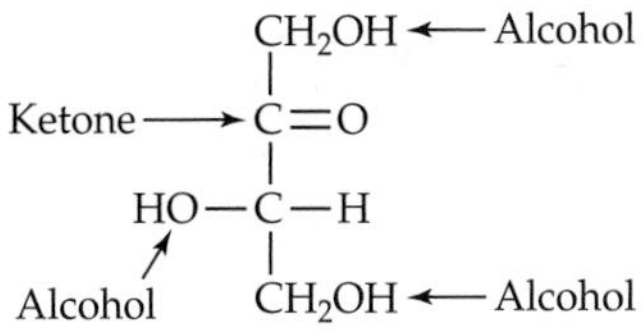

(b)

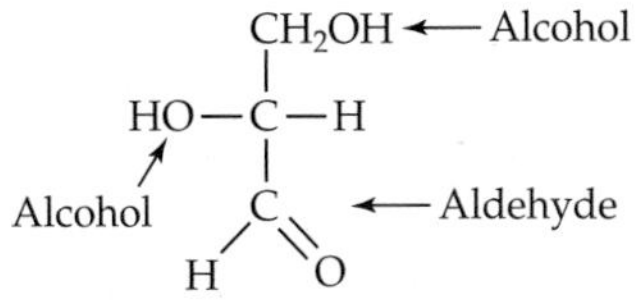

(c)
Aldehyde 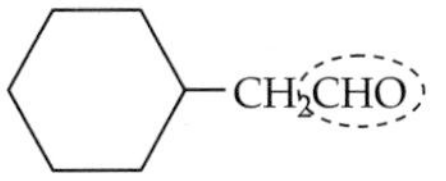

(d) H₂NCH₂CH₂COCH₃
Amine   Ketone

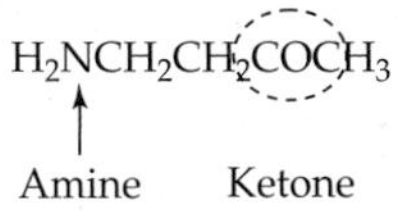

16.9

(a)

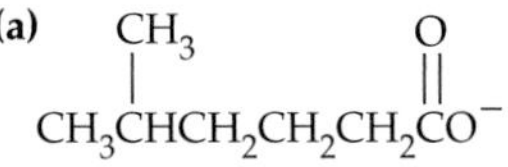

(b)

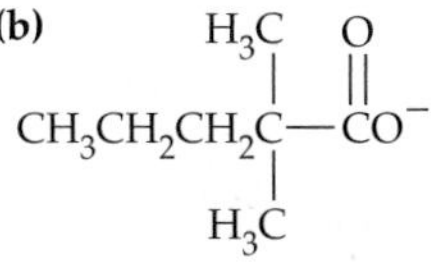

(c) NR

16.10

(a)

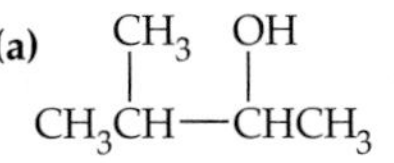

(b)

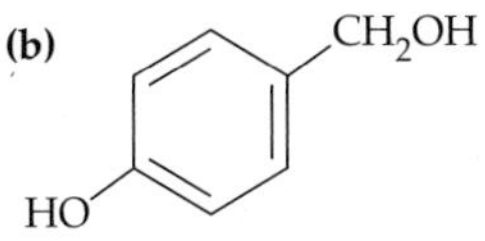

(c)

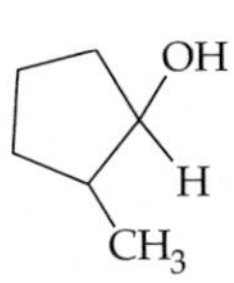

16.11 (a)

(b)

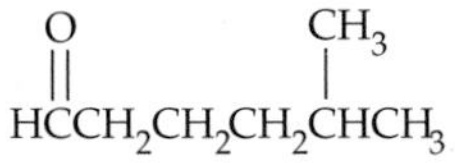

(c)

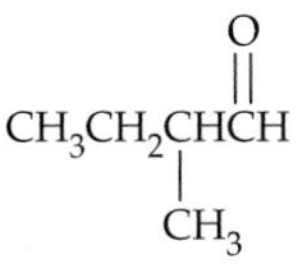

16.12 Compound **(a)**

16.13

(a)

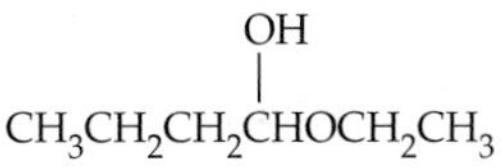

(b)

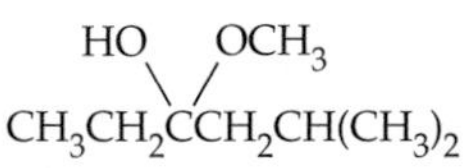

16.14

(a)

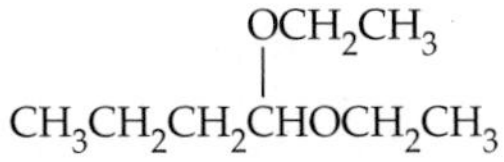

(b)

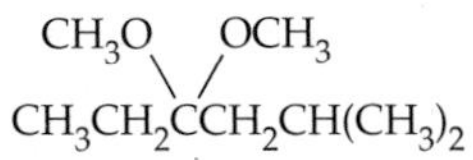

16.15 (a) neither **(b)** neither **(c)** acetal **(d)** hemiacetal **16.16** The acetal and the hemiacetal were both formed from ketones.

16.17 (a)

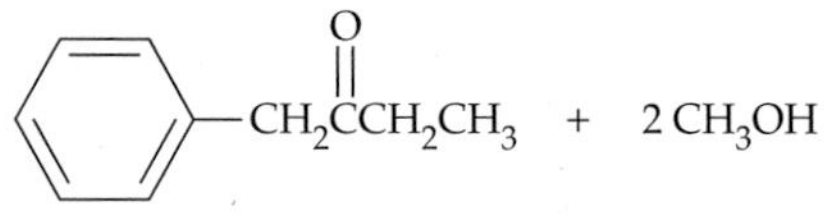

$$-CH_2CCH_2CH_3 + 2 CH_3OH$$

(b)

$$CH_3CH_2CHO + 2 CH_3CH_2CH_2OH$$

(c)

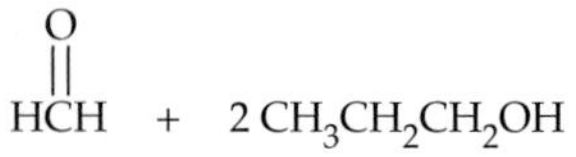

$$HCH + 2 CH_3CH_2CH_2OH$$

16.18 (a) Hydride adds to the carbonyl carbon. **(b)** The arrow to the right represents reduction, and the arrow to the left represents oxidation.
16.19 Aldehydes can be oxidized to carboxylic acids. Tollens' reagent differentiates an aldehyde from a ketone.
16.20

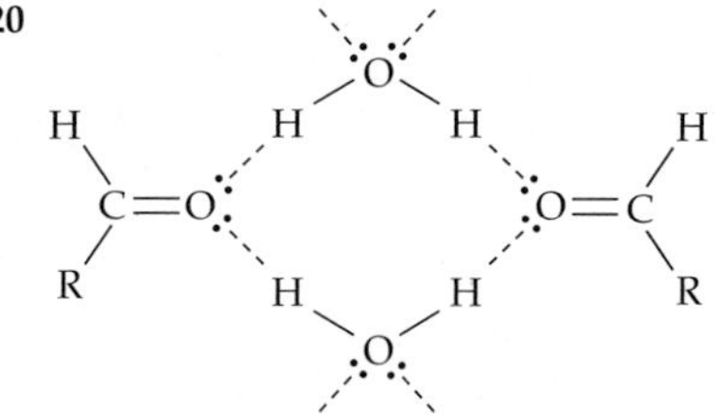

16.21 (a) Under acidic conditions, an alcohol adds to the carbonyl group of an aldehyde to form a hemiacetal, which is unstable and further reacts to form an acetal.
(b)

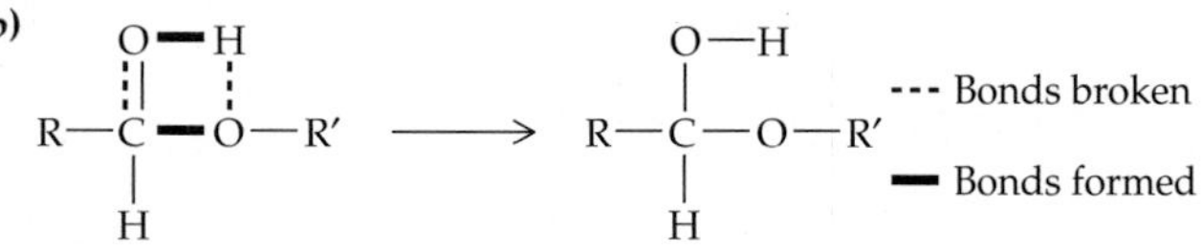

--- Bonds broken
— Bonds formed

16.22 In solution, glucose exists as a cyclic hemiacetal because this structure is more stable. **16.23** One equivalent of an alcohol adds to the carbonyl group of an aldehyde or ketone to form a hemiacetal. Two equivalents of alcohol add to the carbonyl group to yield an acetal.
16.24 (a) **(b)** **(c)**

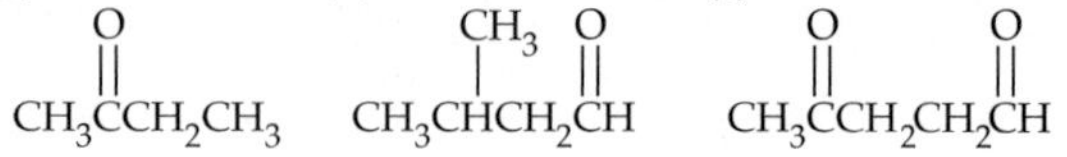

$$CH_3CCH_2CH_3 \quad CH_3CHCH_2CH \quad CH_3CCH_2CH_2CH$$

(d)

$$HOCH_2CH_2CCH_3$$

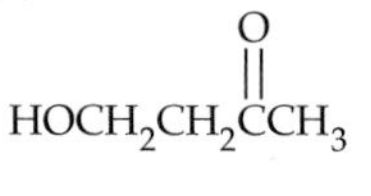

16.26 Structure **(a)** is an aldehyde, and structure **(f)** is a ketone.
16.28

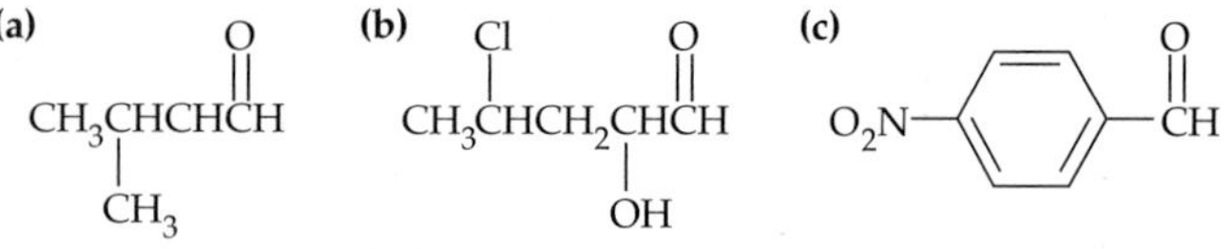

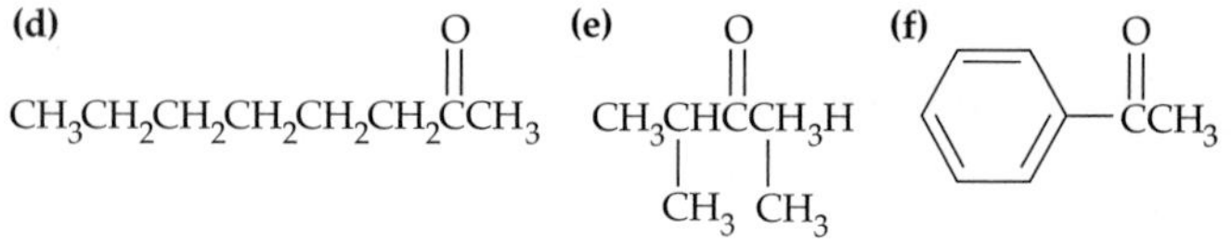

(a)
$$CH_3CHCHCH$$
$$\quad CH_3$$

(b)
$$CH_3CHCH_2CHCH$$
$$\quad OH$$

(c)
$$O_2N- \bigcirc -CH$$

(d)
$$CH_3CH_2CH_2CH_2CH_2CH_2CCH_3$$

(e)
$$CH_3CHCCH_3H$$
$$\quad CH_3 \ CH_3$$

(f)
$$\bigcirc -CCH_3$$

16.30 (a) 2,3-Dimethylbutanal **(b)** 4-Hydroxy-2-methylpentanal **(c)** 2,2-Dimethylpropanal **(d)** 2-Butanone **(e)** 5-Methyl-2-hexanone **16.32** For **(a)**, a ketone can't occur at the end of a carbon chain. For **(b)**, the methyl group receives the lowest possible number. For **(c)**, numbering must start at the end of the carbon chain closer to the carbonyl group.
16.34 A hemiacetal is produced.
16.36 (a) NR; cyclopentanol

(b)

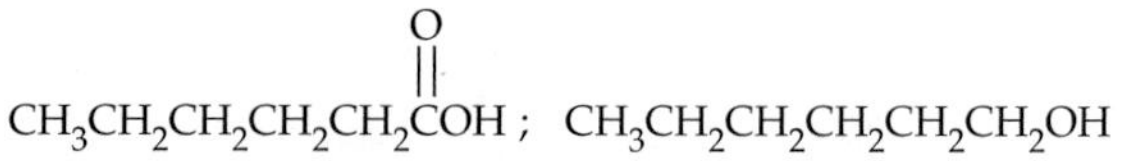

$$CH_3CH_2CH_2CH_2CH_2COH \ ; \ CH_3CH_2CH_2CH_2CH_2CH_2OH$$

(c)

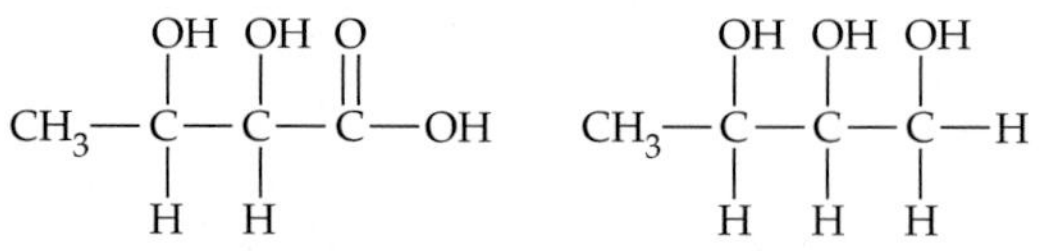

$$CH_3-\overset{OH}{\underset{H}{C}}-\overset{OH}{\underset{H}{C}}-\overset{O}{C}-OH \quad CH_3-\overset{OH}{\underset{H}{C}}-\overset{OH}{\underset{H}{C}}-\overset{OH}{\underset{H}{C}}-H$$

16.38 (a)

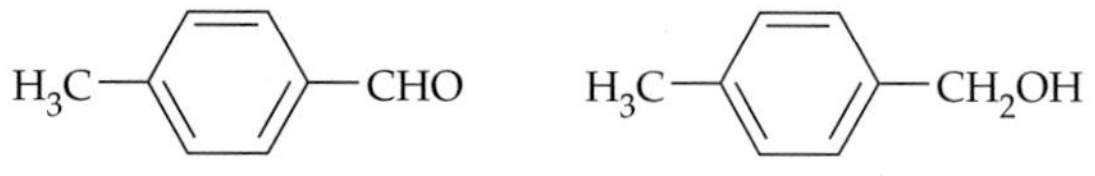

$$H_3C-\bigcirc -CHO \qquad H_3C-\bigcirc -CH_2OH$$

(b)

$$\overset{CHO}{\underset{}{|}} \qquad \overset{CH_2OH}{\underset{}{|}}$$
$$CH_3CH_2CHCH_2CHCH_3 \qquad CH_3CH_2CHCH_2CHCH_3$$
$$\qquad\qquad CH_3 \qquad\qquad\qquad CH_3$$

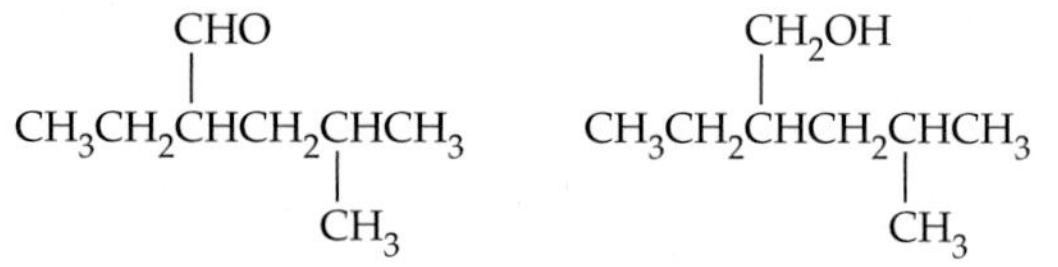

(c) $CH_3CH=CHCHO \qquad CH_3CH=CHCH_2OH$

16.40

(a)
$$\overset{OH}{\underset{}{|}}$$
$$CH_3CH_2COCH_2CH_2CH_3$$
$$\overset{}{\underset{CH_3}{|}}$$

(b)
$$\overset{OH}{\underset{}{|}}$$
$$CH_3CH_2CH_2COCH(CH_3)_2$$
$$\overset{}{\underset{H}{|}}$$

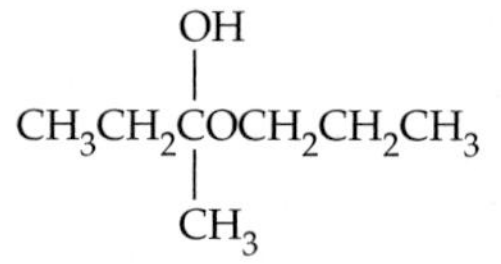

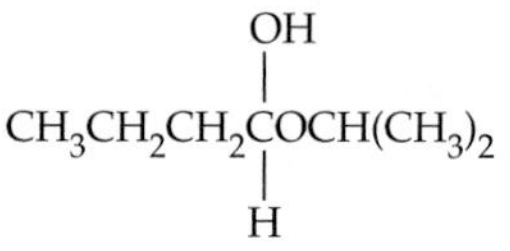

(c)
$$\overset{O}{\overset{||}{}}$$
$$CH_3CH_2CH_2CH + CH_3CH_2OH + CH_3OH$$

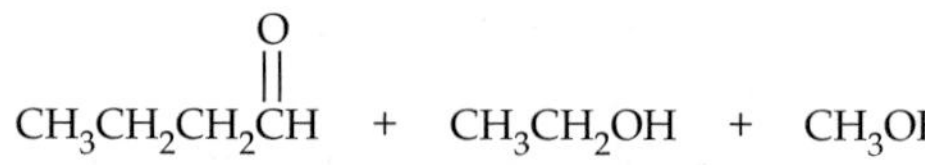

(d) H_3C
$$\overset{H_3C}{\underset{H_3C}{>}}C=O + HOCH_2CH_2OH$$

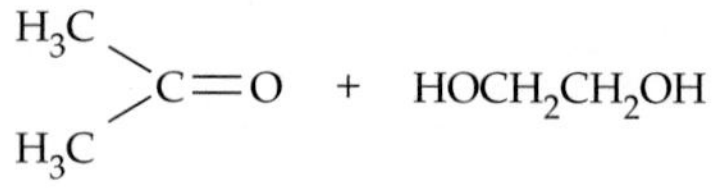

16.42

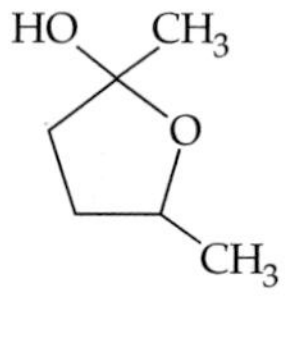

16.44 $HOCH_2CH_2CH_2OH$ and CH_2O (formaldehyde).
16.46

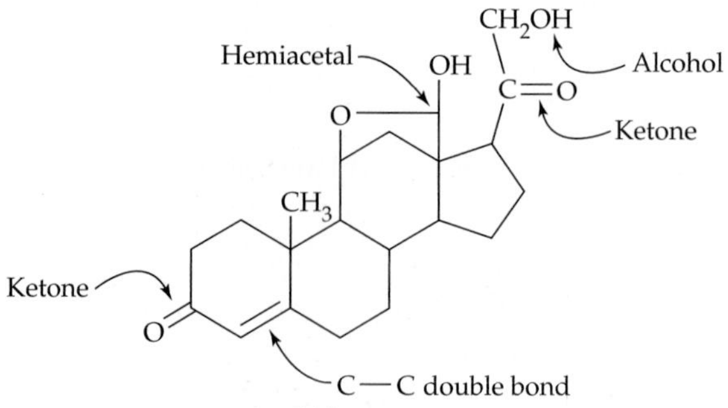

Hemiacetal
OH
Alcohol
Ketone
CH_3
Ketone
O
C—C double bond
Aldosterone

16.48

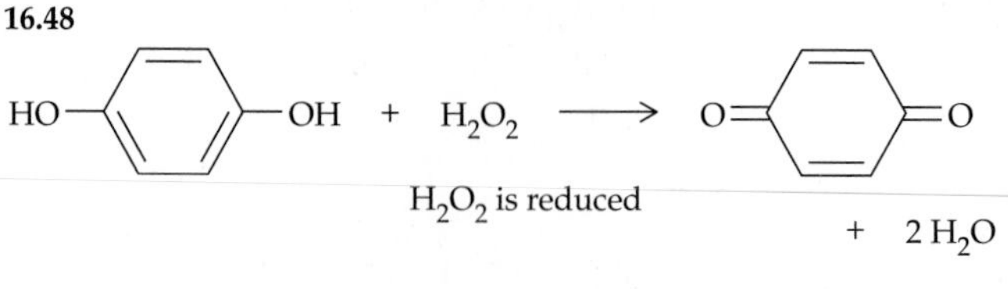

$$HO-\bigcirc -OH + H_2O_2 \longrightarrow O=\bigcirc =O$$

H_2O_2 is reduced

$$+ \ 2 \ H_2O$$

16.50 Ethanol **16.52** (a) Advantages: inexpensive, no need to sacrifice animals. (b) Disadvantage: results of tests on cultured cells may not be reliable for more complex organisms. **16.54** *p*-Methoxybenzaldehyde **16.56** Aldehydes are easily oxidized.

16.58

$$Cl_3CC\overset{\overset{\displaystyle OH}{|}}{\underset{\underset{\displaystyle H}{|}}{C}}-OH$$

16.60 (a) *o*-Isopropylmethoxybenzene (b) 5,5-Diethyl-3-heptyne (c) *N*-Ethylcyclopentylammonium bromide (d) *N*,*N*-Diethylhexylamine

16.62

(a)

$$CH_3CH_2\overset{\overset{\displaystyle I}{|}}{\underset{\underset{\displaystyle I}{|}}{C}}\overset{\overset{\displaystyle O}{||}}{C}H\overset{\overset{}{\underset{\underset{\displaystyle I}{|}}{C}}}{H}CH$$

(b)

$$BrCH_2\overset{\overset{\displaystyle O}{||}}{C}CHBr_2$$

(c)

$$CH_3\overset{\overset{\displaystyle NH_2}{|}}{C}CH_2CH_2\overset{\overset{\displaystyle O}{||}}{C}CH_3\\ \underset{\displaystyle CH_3}{|}$$

16.64 (a)

$$\text{(phenyl)}CH_2\overset{\overset{\displaystyle OCH_2CH_2CH_3}{|}}{C}H-OCH_2CH_2CH_3$$

(b)

$$CH_3CH_2\overset{\overset{\displaystyle CH_2CH_3}{|}}{C}CH_2CH_2CH_3\\ \underset{\displaystyle Cl}{|}$$

(c)

$$H_3C-\text{(benzene)}-CH=CH_2$$

16.66 Highest boiling = 1-butanol (strongest hydrogen bonds)

Chapter 17
17.1 carboxylic acid: (c); amides: (a) (f) (h); ester: (d); none: (b) (e) (g)

17.2 (a)

$$\underset{6}{C}H_3\underset{5}{C}H_2\underset{4}{C}H_2\underset{3}{C}\overset{\overset{\displaystyle OH}{|}}{H}\underset{2}{C}H\underset{1}{C}\overset{\overset{\displaystyle O}{||}}{O}OH\\ \qquad\qquad\underset{\displaystyle CH_2CH_3}{|}$$

(b)

$$O_2N-\text{(benzene)}-\overset{\overset{\displaystyle O}{||}}{C}OH$$

17.3

$$H-O-\overset{\overset{\displaystyle O}{||}}{C}-\overset{\overset{\displaystyle H}{|}}{\underset{\underset{\displaystyle H}{|}}{C}}-\overset{\overset{\displaystyle H}{|}}{\underset{\underset{\displaystyle H}{|}}{C}}-\overset{\overset{\displaystyle O}{||}}{C}-O-H$$

$$HO\overset{\overset{\displaystyle O}{||}}{C}CH_2CH_2\overset{\overset{\displaystyle O}{||}}{C}OH$$

17.4 2,3-Dibromopropanoic acid

17.5 (a)

$$\text{(benzene)}-\overset{\overset{\displaystyle O}{||}}{C}OCH(CH_3)_2$$

(b)

$$CH_3CH_2CH_2CH_2\overset{\overset{\displaystyle O}{||}}{C}OCH_2CH_3$$

(c)

$$CH_3CH=CH\overset{\overset{\displaystyle O}{||}}{C}OCH(CH_3)_2$$

17.6 CH$_3$COOH is highest boiling (most H-bonding). CH$_3$CH$_2$CH$_3$ is lowest boiling (nonpolar). **17.7** (a) C$_3$H$_7$COOH is more soluble (smaller —R group). (b) (CH$_3$)$_2$CHCOOH is more soluble (carboxylic acid).

17.8

$$CH_3CH_2\overset{\overset{\displaystyle O}{||}}{C}OH$$
Propanoic acid

$$CH_3CH_2\overset{\overset{\displaystyle O}{||}}{C}OCH_3$$
Methyl propanoate

$$CH_3CH_2\overset{\overset{\displaystyle O}{||}}{C}NH_2$$
Propanamide

$$CH_3CH_2\overset{\overset{\displaystyle O}{||}}{C}NHCH_3$$
N-Methylpropanamide

$$CH_3CH_2\overset{\overset{\displaystyle O}{||}}{C}NCH_3\\ \qquad\qquad\underset{\displaystyle CH_3}{|}$$
N,*N*-Dimethylpropanamide

17.9 (a) Propyl 3-hydroxypentanoate (b) *N*-Methyl-*p*-chlorobenzamide

17.10 (a)

$$CH_3\overset{\overset{\displaystyle CH_3}{|}}{C}HCH_2CH_2\overset{\overset{\displaystyle O}{||}}{C}NH_2$$

(b)

$$CH_3CH_2\overset{\overset{\displaystyle O}{||}}{C}NCH_2CH_3\\ \qquad\underset{\displaystyle CH_3}{|}$$

17.11 (a) (ii) (b) (i) (c) (iv) (d) (iii)   **17.12** (a) (ii) (b) (i) (c) (iii) (d) (i) (e) (i) (f) (iii)

17.13 (a)

$$C_6H_5\overset{\overset{\displaystyle O}{||}}{C}NH_2$$
Amide (C$_7$H$_7$NO)

(b)

$$CH_3CH_2\overset{\overset{\displaystyle O}{||}}{C}OH$$
Carboxylic acid (C$_3$H$_6$O$_2$)

(c)

$$CH_3\overset{\overset{\displaystyle O}{||}}{C}OCH_2CH_3$$
Ester (C$_3$H$_6$O$_2$)

17.14 (a)

$$CH_3\overset{\overset{\displaystyle HO}{|}}{C}H\overset{\overset{\displaystyle O}{||}}{C}-O^-\,Na^+\;+\;H_2O$$

(b)

$$\left[CH_3CH_2CH_2\overset{\overset{\displaystyle H_3C\quad O}{\diagdown\;||}}{C}-C-O^-\\ \qquad\qquad\quad\underset{\displaystyle CH_3}{|}\right]_2 Ca^{2+}$$

17.15 (a)

$$\left[\text{(benzene)}\underset{\diagdown OH}{\diagup COO^-}\right]_2 Ca^{2+}$$

(b)

$$H_2C=CHCOO^-\,Ca^+$$

17.16 CH$_3$COO$^-$ $^-$OOCCH$_2$CH$_2$CH$_2$COO$^-$ Na$^+$ K$^+$
17.17 HCOOCH$_2$CH(CH$_3$)$_{2,,}$

17.18 (a)

$$\text{(cyclohexane)}-OH\;+\;HO\overset{\overset{\displaystyle O}{||}}{C}CH_2CH_2CH(CH_3)_2$$

(b)

$$CH_3CH_2CH_2CH_2\overset{\overset{\displaystyle O}{||}}{C}OH\;+\;HOCH(CH_3)_2$$

17.19

(a)

$$CH_3\overset{\overset{\displaystyle O}{||}}{C}H\overset{}{C}-NHCH_3\\ \underset{\displaystyle CH_3}{|}$$

(b)

$$\text{(cyclopentane)}-\overset{\overset{\displaystyle O}{||}}{C}-NH-\text{(benzene)}$$

17.20

$$CH_3CH_2O-\text{(benzene)}-NH_2\;+\;HOOCCH_3$$

17.21

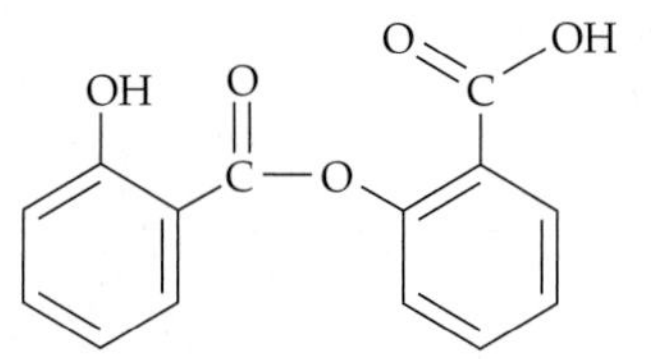

17.22 Aspirin is acidic (—COOH), lidocaine is basic (amine), benzocaine is weakly basic (aromatic amine), acetaminophen is weakly acidic (phenol).
17.23 Moisture in the air hydrolyzes the ester bond.

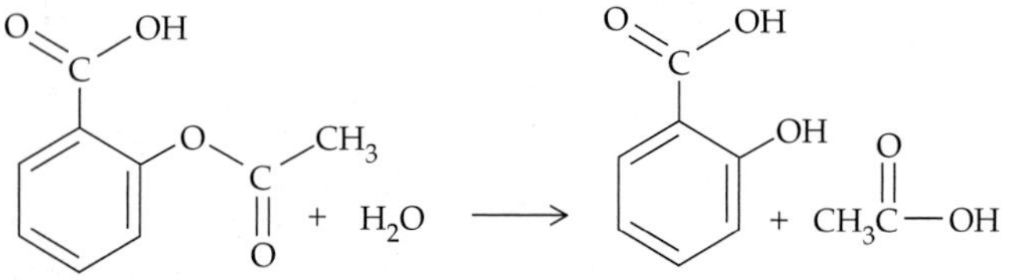

$$+ H_2O \longrightarrow + CH_3\overset{O}{\overset{\|}{C}}{-}OH$$

17.24 (a) *p*-Nitrobenzoic acid + 2-Propanol (b) Phenol + 2-Hydroxycyclopentanecarboxylic acid (c) Acrylic acid + Ethanol
17.25 (a) 2-Butenoic acid + Methylamine (b) *p*-Nitrobenzoic acid + Dimethylamine
17.26 (a)

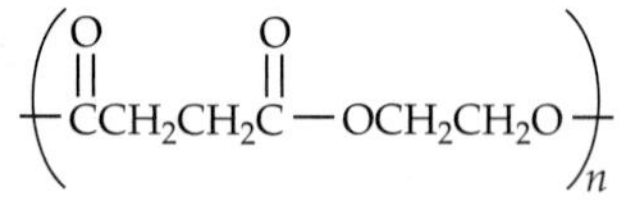

(b)

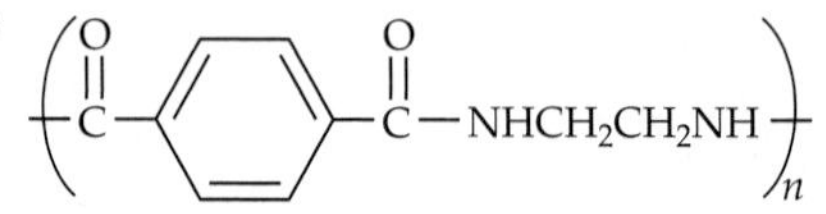

17.27

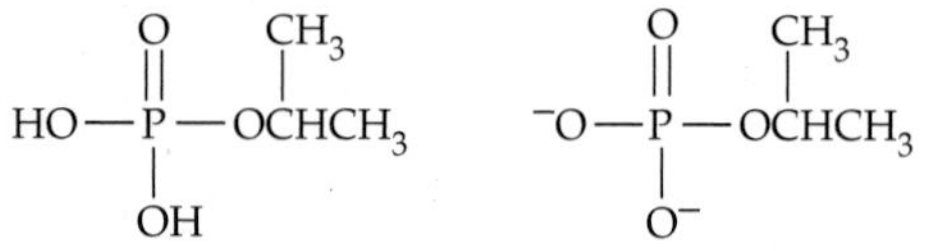

17.28 (a) amide + $H_2O \rightarrow CH_3COOH + NH_3$
(b) phosphate monoester + $H_2O \rightarrow CH_3CH_2OH + HOPO_3^{2-}$
(c) carboxylic acid ester + $H_2O \rightarrow CH_3CH_2COOH + HOCH_3$
17.29

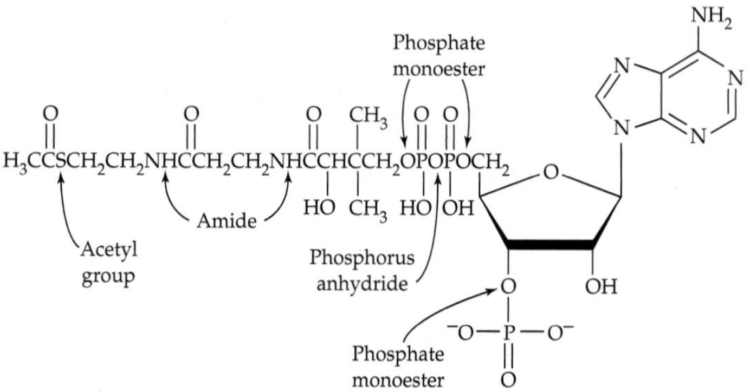

17.30 (a) At pH = 7.4, pyruvate and lactate are anions.
(b)

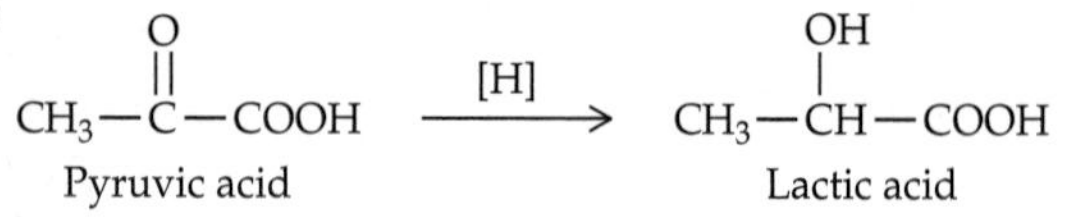

$$CH_3-\overset{O}{\overset{\|}{C}}-COOH \xrightarrow{[H]} CH_3-\overset{OH}{\overset{|}{CH}}-COOH$$

Pyruvic acid   Lactic acid

(c) Pyruvate and lactate have similar solubilities in water.

(d)

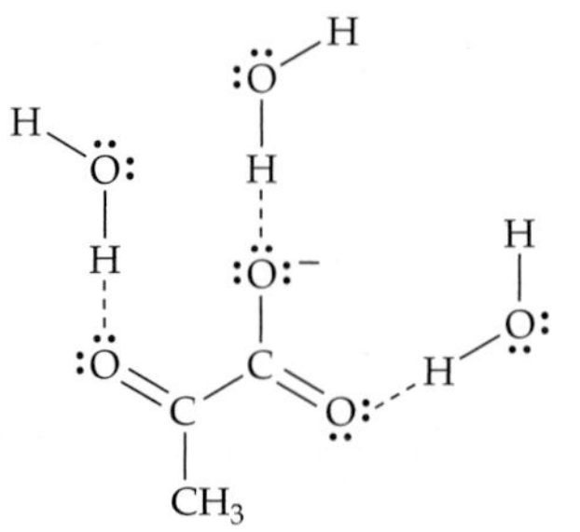

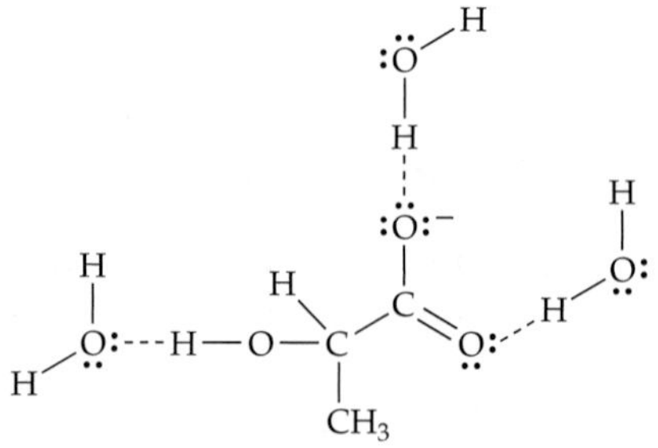

17.31 (a) H_2O + acid or base
(b)

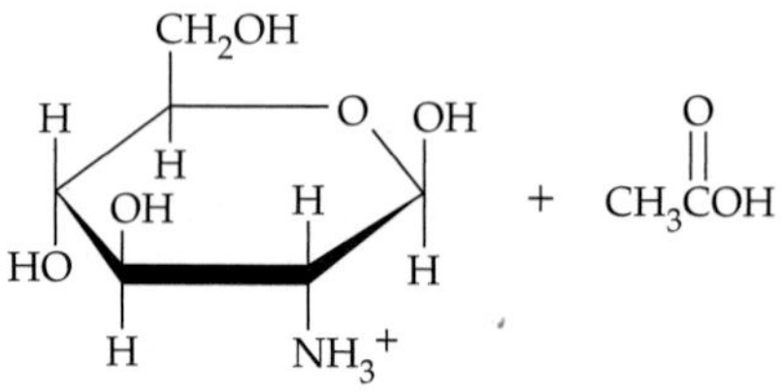

$$+ CH_3\overset{O}{\overset{\|}{C}}OH$$

17.32 (a) a phosphate ester linkage
(b)

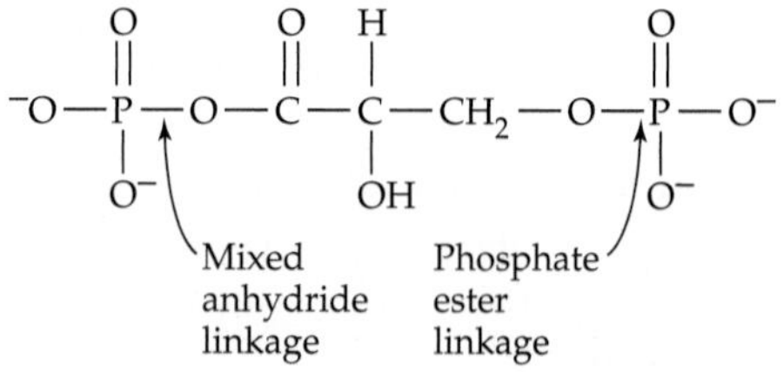

Mixed anhydride linkage   Phosphate ester linkage

17.33

$^-OOCCOO^-$ $^-OOCCH_2COO^-$
Oxalate   Malonate

$^-OOCCH_2CH_2COO^-$ $^-OOCCH_2CH_2CH_2COO^-$
Succinate   Glutarate

17.34
(a)

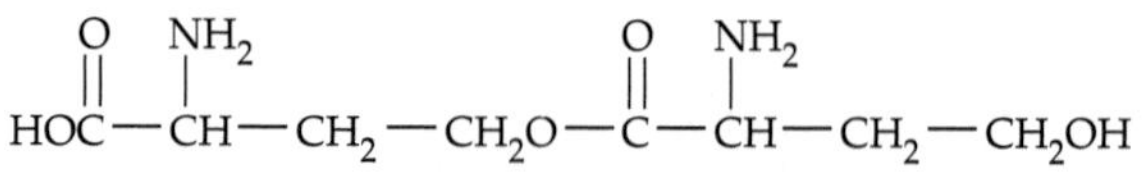

(b)

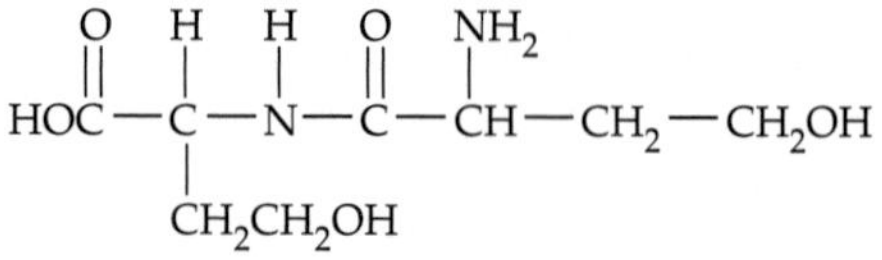

(c)

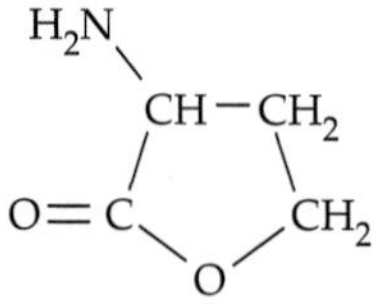

17.35

(a) (i)

(ii)

(iii)

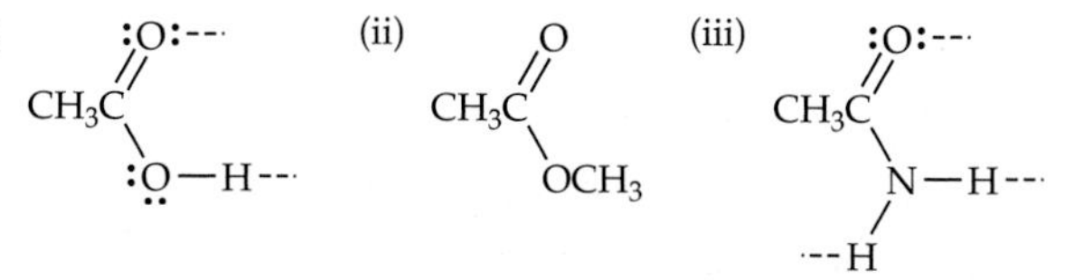

Acetic acid   Methyl acetate   Acetamide

(b) Methyl acetate is lowest boiling (no hydrogen bonds); acetamide is highest boiling.

17.36

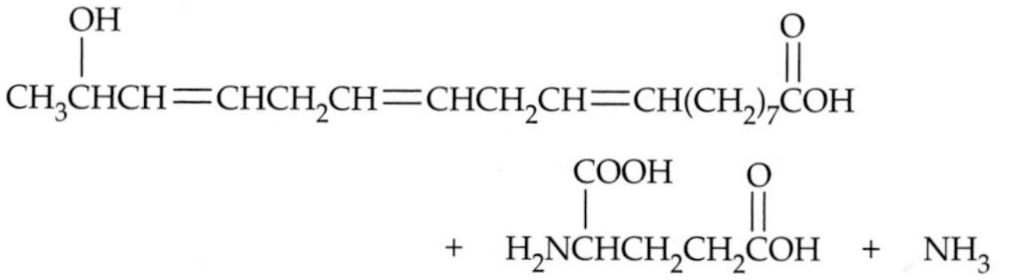

$$CH_3CHCH=CHCH_2CH=CHCH_2CH=CH(CH_2)_7COOH$$

$$+ \quad H_2NCHCH_2CH_2COOH \quad + \quad NH_3$$

17.37 **(a)** *N*-Ethyl benzamide **(b)** Ethyl pentanoate **(c)** Methyl isobutyl-malonate **(d)** *N,N*-dimethylformamide. Compound **(c)** is acidic; all other compounds are neutral.

17.38

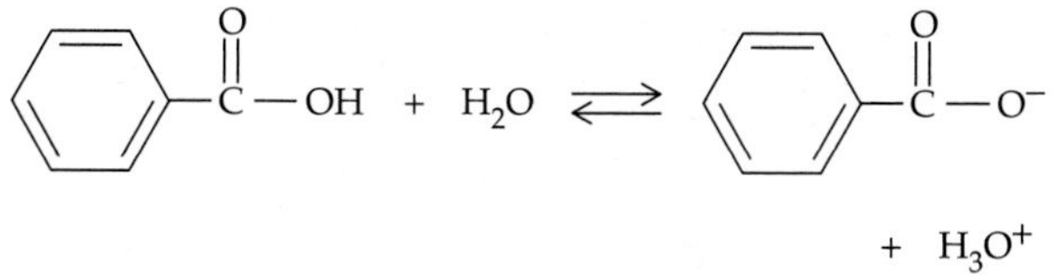

17.40

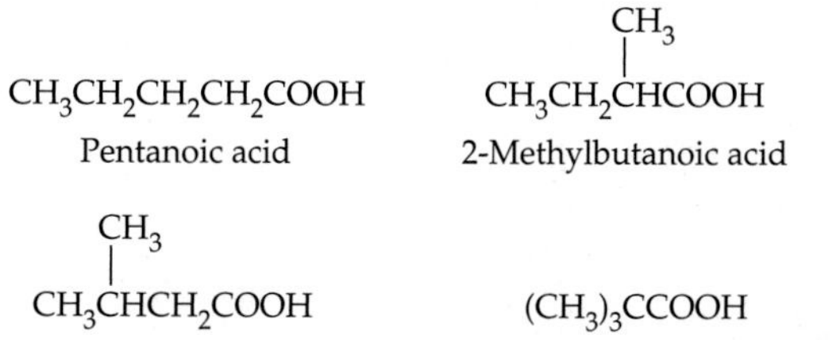

$$CH_3CH_2CH_2CH_2COOH$$
Pentanoic acid

$$CH_3CH_2CHCOOH$$ (with CH$_3$)
2-Methylbutanoic acid

$$CH_3CHCH_2COOH$$ (with CH$_3$)
3-Methylbutanoic acid

$$(CH_3)_3CCOOH$$
2,2-Dimethylpropanoic acid

17.42 **(a)** 2-Ethyl-3-methylpentanoic acid **(b)** Nonanoic acid **(c)** Cyclo-hexanecarboxylic acid **(d)** *p*-Methylbenzoic acid

17.44 **(a)** Potassium 3-ethylpentanoate **(b)** Ammonium benzoate **(c)** Calcium propanoate

17.46

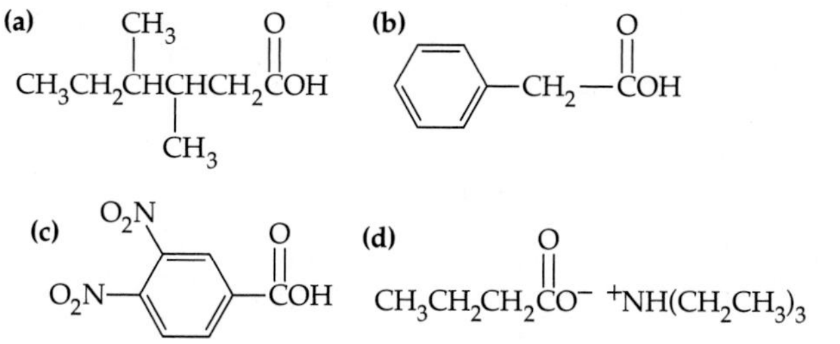

(a)

(b)

(c)

(d)

$$CH_3CH_2CH_2CO^- \; {}^+NH(CH_2CH_3)_3$$

17.48

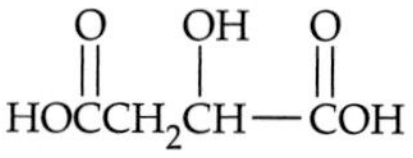

17.50

$$NH_4^+ \; {}^-OCCH=CHCO^- NH_4^+$$

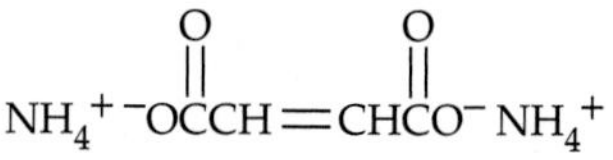

17.52 **(a)** $CH_3CH_2CH_2CH_2CONH_2$ $CH_3CH_2CONHCH_2CH_3$
      Pentanamide   *N*-Ethylpropanamide

$$HCON(CH_2CH_3)_2$$
N,N-Diethylformamide

(b) $CH_3CH_2CH_2CH_2COOCH_3$ $CH_3CH_2COOCH_2CH_2CH_3$
      Methyl pentanoate   Propyl propanoate

$$HCOOCH_2CH_2CH_2CH_2CH_3$$
Pentyl formate

17.54 **(a)** 3-Methylbutyl acetate **(b)** Methyl 4-methylpentanoate

(c)

(d)

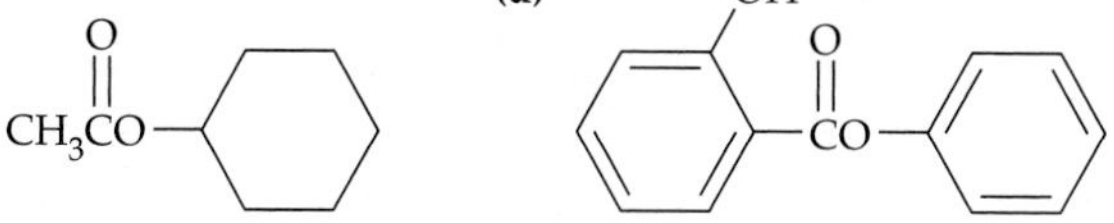

17.56 **(a)** $CH_3COOH + HOCH_2CH_2CH(CH_3)_2$
(b) $(CH_3)_2CHCH_2CH_2COOH + HOCH_3$

(c)

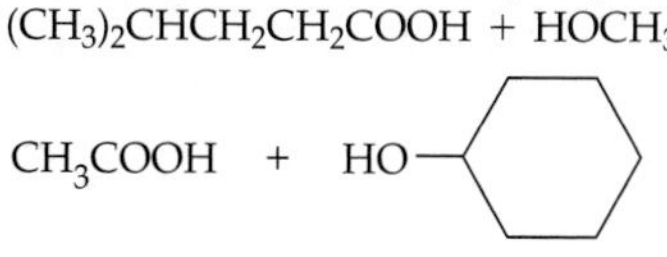

(d)

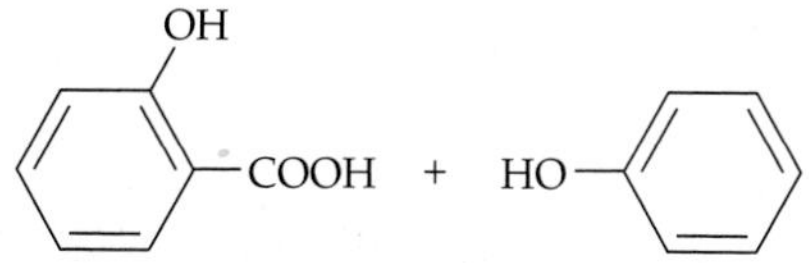

17.58 **(a)** 2-Ethylbutanamide **(b)** *N*-Phenylbenzamide

(c)

(d)

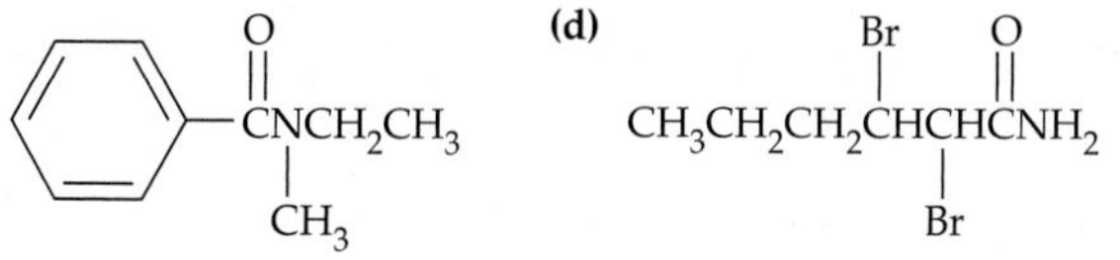

17.60 **(a)** 2-Ethylbutanoic acid + Ammonia **(b)** Benzoic acid + Aniline **(c)** Benzoic acid + *N*-Methylethylamine **(d)** 2,3-Dibromohexanoic acid + Ammonia

17.62

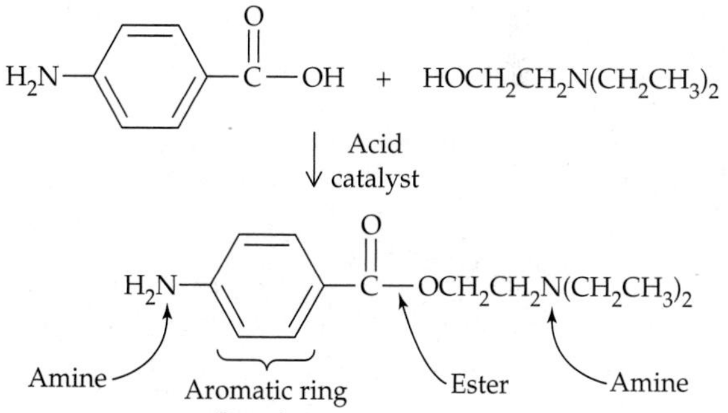

17.64 $HOCH_2CH_2CH_2COOH$

17.66

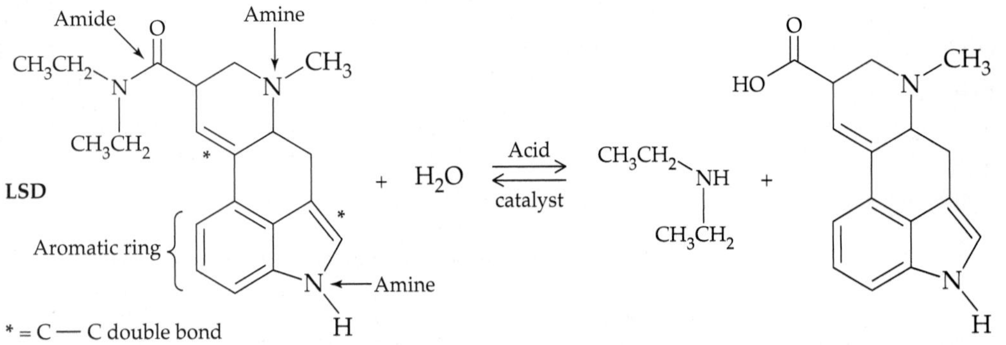

Amide · Amine

LSD

Aromatic ring

* = C — C double bond

$+ H_2O$ ⇌ (Acid catalyst)

CH_3CH_2—NH—CH_2CH_3 +

17.68

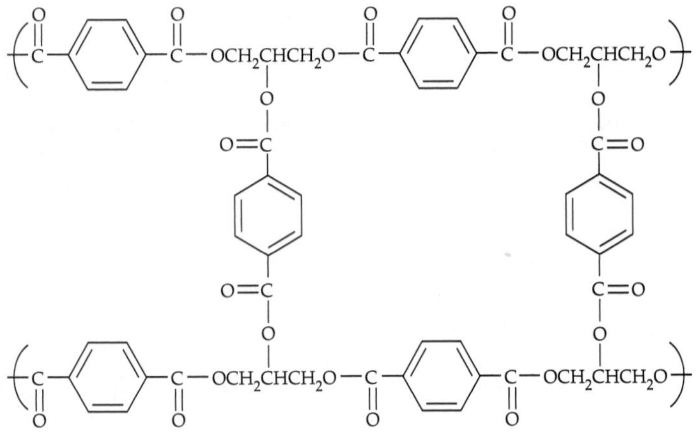

17.70 Dihydroxyacetone and hydrogen phosphate anion.

17.72

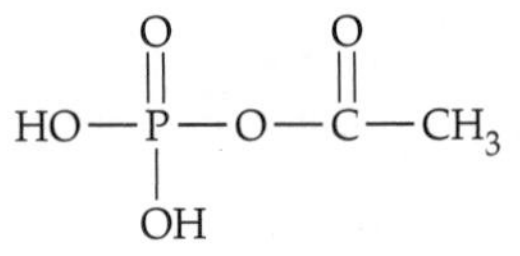

HO—P(=O)(OH)—O—C(=O)—CH₃

17.74 A cyclic phosphate diester is formed when a phosphate group forms an ester with two hydroxyl groups in the same molecule.

17.76 Trichloroacetic acid: used for chemical peeling of the skin. Lactic acid: used for wrinkle removal and moisturizing.

17.78

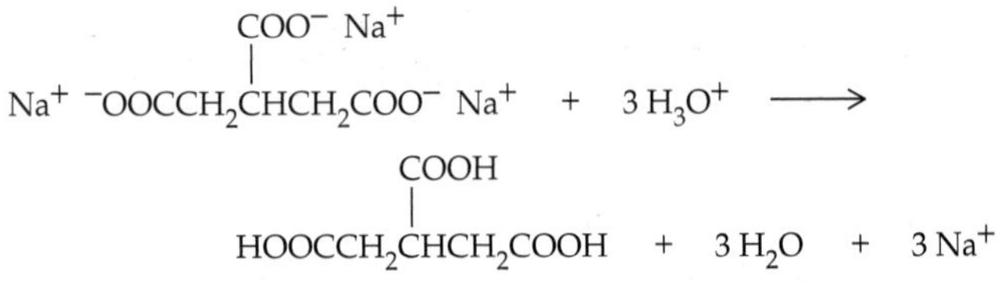

Na^+ $^-OOCCH_2CHCH_2COO^-$ Na^+ + 3 H_3O^+ ⟶ (with COO^- Na^+ side group)

$HOOCCH_2CHCH_2COOH$ + 3 H_2O + 3 Na^+ (with COOH side group)

17.80 strong acids and bases.

17.82

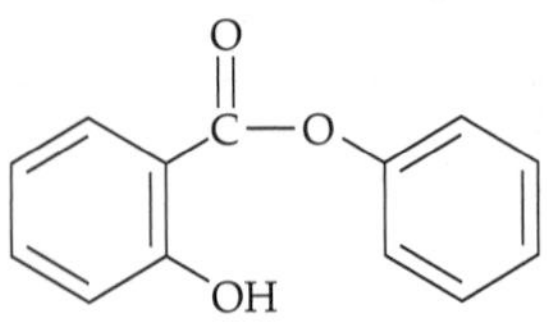

17.84 pH; Tollens' test   **17.86** (a) 2-Chloro-3,4-dimethyl-3-hexene (b) N-Methyl-N-phenylpropanamide (c) Phenyl 2,2-diethylbutanoate (d) N-Ethyl-o-nitrobenzamide

Chapter 18
18.1 Aromatic ring: phenylalanine, tyrosine, tryptophan
Contain sulfur: cysteine, methionine
Alcohols: serine, threonine, tyrosine (phenol)
Alkyl side chain: alanine, valine, leucine, isoleucine

18.2

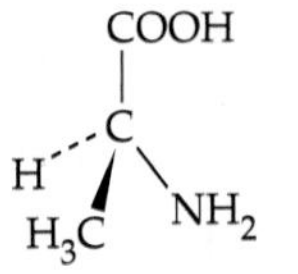

18.3

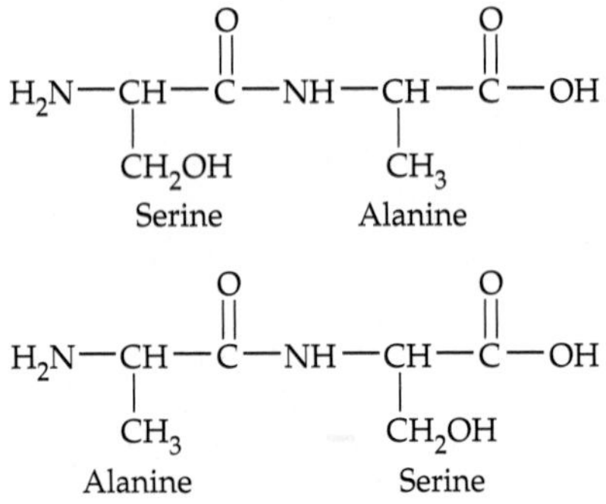

H_2N—CH—C(=O)—NH—CH—C(=O)—OH
with CH₂OH (Serine) and CH₃ (Alanine)

H_2N—CH—C(=O)—NH—CH—C(=O)—OH
with CH₃ (Alanine) and CH₂OH (Serine)

18.4 α–amino acids: **(a)**, **(d)**
18.5 **(b)** Asn, Ser **(c)** Thr, Tyr

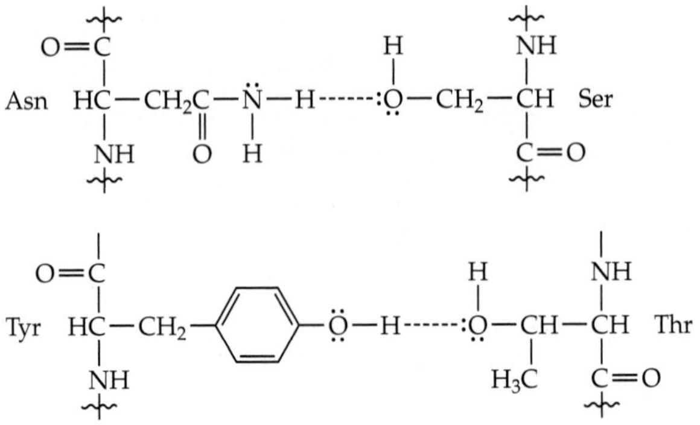

18.6

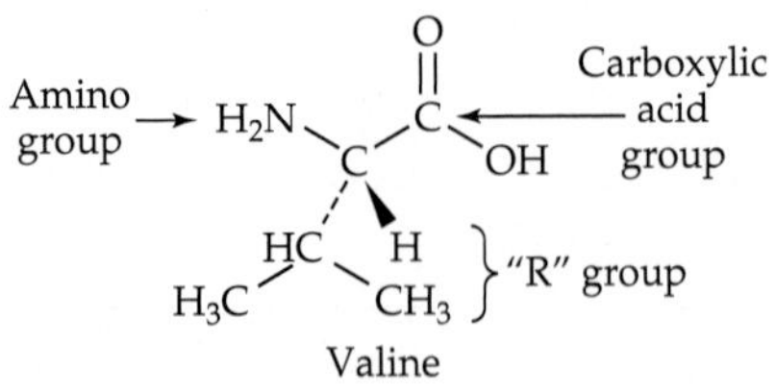

Amino group → H_2N
Carboxylic acid group → C(=O)OH
"R" group → HC(CH₃)(CH₃)
Valine

18.7

$H_3\overset{+}{N}$—CH—C(=O)—OH with CH₂—CH₂—C(=O)OH   at low pH

H_2N—CH—C(=O)—O^- with CH₂—CH₂—C(=O)O^- at high pH

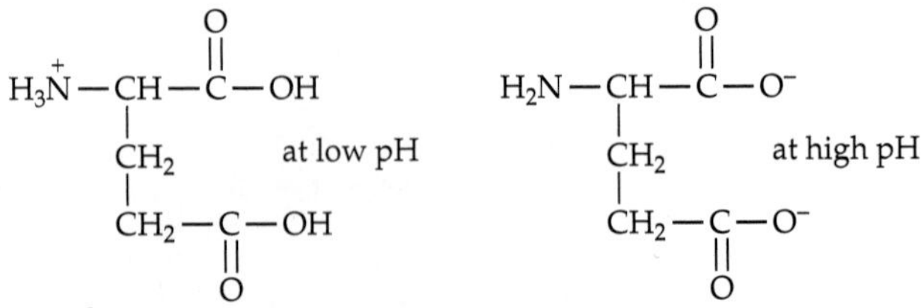

18.8 In the zwitterionic form of an amino acid, the $-NH_3^+$ group is an acid, and the $-COO^-$ group is a base.   **18.9** Chiral: **(a)**, **(b)**, and **(d)**
18.10 Handed: wrench, corkscrew, jar lid
Not handed: thumbtack, pencil, straw

18.11 2-Aminobutane has a carbon with 4 different groups bonded to it.

18.12 chiral: **(b)**, **(c)**

18.13

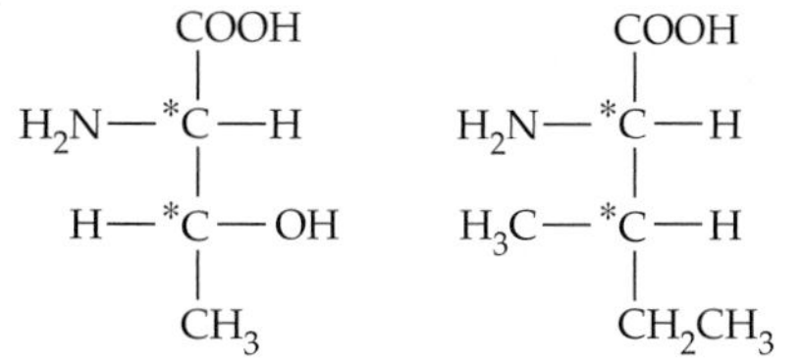

Threonine   Isoleucine

18.14

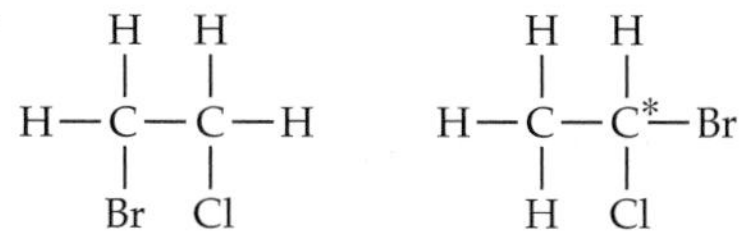

18.15 (a) Gly–Ser–Tyr   Tyr–Ser–Gly   Ser–Tyr–Gly
Gly–Tyr–Ser   Tyr–Gly–Ser   Ser–Gly–Tyr

(b)

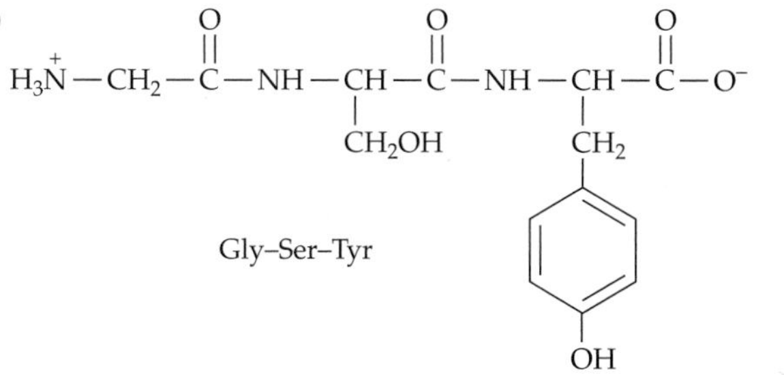

Gly–Ser–Tyr

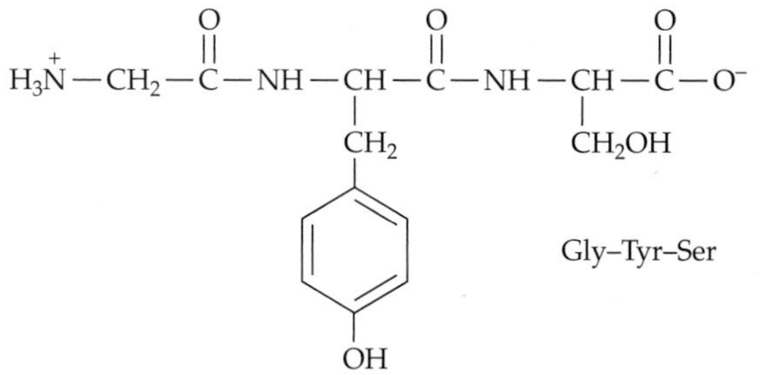

Gly–Tyr–Ser

18.16   Leu–Trp–Ser   Trp–Leu–Ser   Ser–Trp–Leu
Leu–Ser–Trp   Trp–Ser–Leu   Ser–Leu–Trp

18.17 (a) Leu–Asp **(b)** Tyr–Ser–Lys

18.18

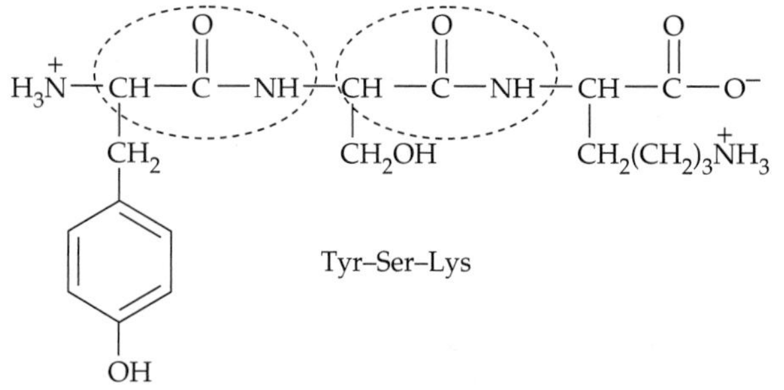

Tyr–Ser–Lys

18.19 Asp−Tyr + Phe + Glu−Asn−Cys−Pro−Lys−Gly

18.20 (a) hydrogen bond **(b)** hydrophobic interaction **(c)** salt bridge **(d)** hydrophobic interaction   **18.21 (a)** Tyr, Asp, Ser **(b)** Ala, Ile, Val, Leu

18.22 eleven backbone atoms   **18.23** Secondary structure: stabilized by hydrogen bonds between amide nitrogens and carbonyl oxygens of polypeptide backbone. Tertiary structure: stabilized by hydrogen bonds between amino acid side-chain groups.   **18.24 (a)** tertiary; **(b)** secondary; **(c)** quaternary   **18.25** At low pH, the groups at the end of the polypeptide chain exist as $-NH_3^+$ and $-COOH$. At high pH, they exist as $-NH_2$ and $-COO^-$. In addition, side chain functional groups may be ionized as follows: **(a)** no change **(b)** Lys, His, Arg positively charged at low pH; Lys, His neutral at high pH: **(c)** Tyr neutral at low pH, negatively charged at high pH: **(d)** Glu, Asp neutral at low pH, negatively charged at high pH: **(e)** no change: **(f)** Cys neutral at low pH, negatively charged at high pH.   **18.26 (a)** 1, 4 **(b)** 2, 4 **(c)** 2

18.27 *See below for answer.*

18.28 *Fibrous Proteins*: structural proteins, water-insoluble, contain many Gly and Pro residues, contain large regions of α-helix or β-sheet, few side-chain interactions. Examples: Collagen, α-Keratin, Fibroin. *Globular Proteins*: enzymes and hormones, usually water-soluble, contain most amino acids, contain smaller regions of α-helix and β-sheet, complex tertiary structure. Examples: Ribonuclease, hemoglobin, insulin.

18.29 (a) Leu, Phe, Ala or any other amino acid with a nonpolar side chain. **(b)** and **(c)** Asp, Lys, Thr or any other amino acid with a polar side chain.   **18.30** The prefix "α" means that $-NH_2$ and $-COOH$ are bonded to the same carbon, the carbon atom in the alpha position (next to) the carbonyl carbon atom in the carboxyl group.

18.32 (a)

(b)

(c)

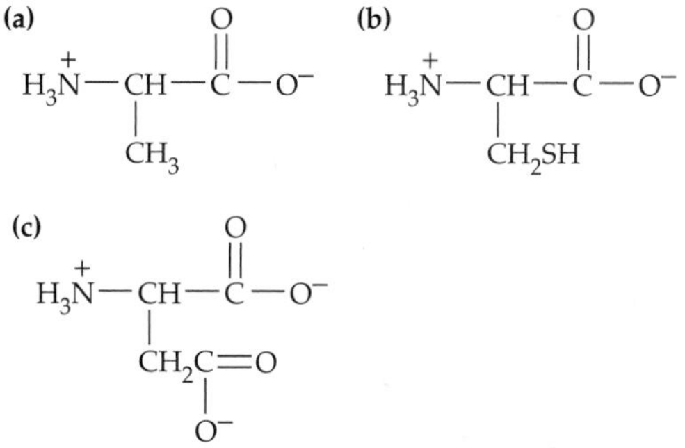

18.34 (a)

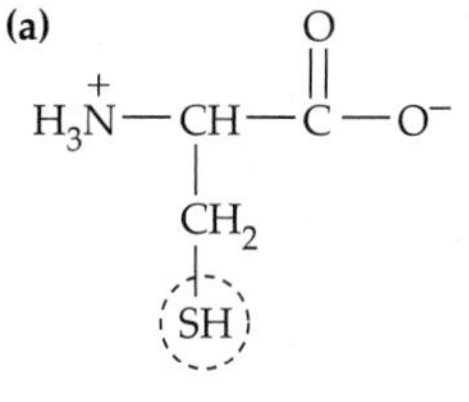

Cysteine (Cys)

(b)

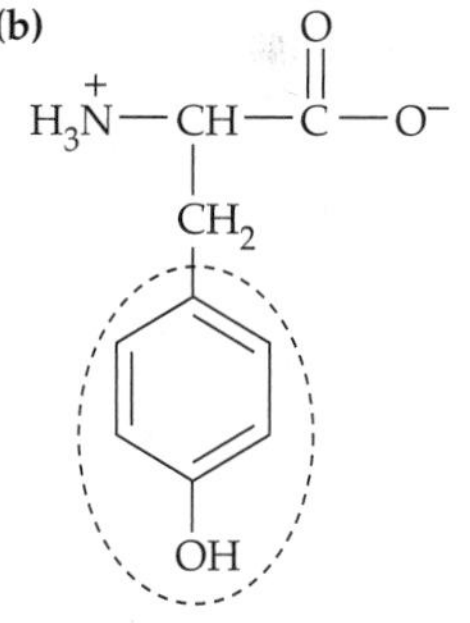

Tyrosine (Tyr)

18.27

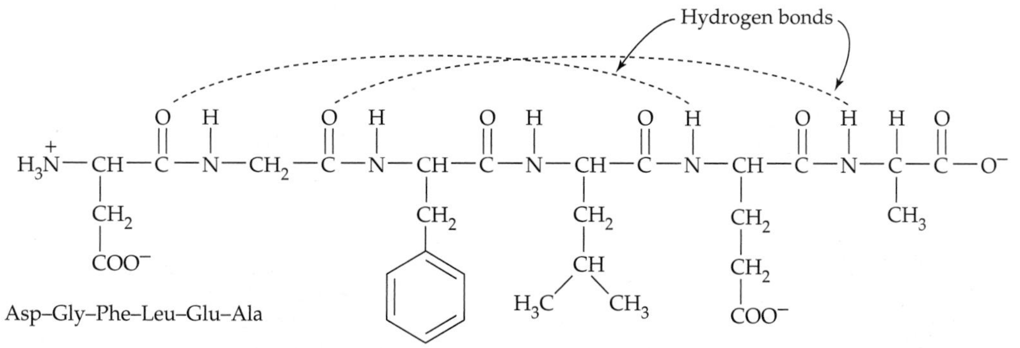

Hydrogen bonds

Asp–Gly–Phe–Leu–Glu–Ala

18.36 neutral: **(a) (c)** positive charge: **(b)** **18.38 (a)**, **(c)** low pH **(b)** high pH **18.40** A chiral object is handed. Examples: glove, car. **18.42 (a)**

18.44

(a)

$$CH_3\overset{\underset{\Large|}{Cl}}{\underset{\underset{\Large Br}{|}}{C}}CH_3$$

Achiral

(b)

$$CH_3CH_2\overset{\underset{\Large|}{Cl}}{\overset{*}{\underset{\underset{\Large Br}{|}}{C}}}CH_3$$

Chiral

(c)

$$(CH_3)_2CH\overset{\underset{\Large|}{Cl}}{\overset{*}{\underset{\underset{\Large Br}{|}}{C}}}CH_3$$

Chiral

18.46 Chiral ⟋ ⟍ Achiral

$$CH_3\underset{\underset{\Large F}{|}}{CH}CH_2CH_3$$

↙ Achiral ⟍ Achiral

18.48 A simple protein is composed only of amino acids. A conjugated protein consists of a protein associated with one or more nonprotein molecules.

18.50

'TYPE OF PROTEIN	FUNCTION	EXAMPLE
Enzymes:	Catalyze biochemical reactions	Ribonuclease
Hormones:	Regulate body functions	Insulin
Storage proteins:	Store essential substances	Myoglobin
Transport proteins:	Transport substances through body fluids	Serum albumin
Structural proteins:	Provide shape and support	Collagen
Protective proteins:	Defend the body against foreign matter	Immunoglobulins
Contractile proteins:	Do mechanical work	Myosin and actin

18.52 Disulfide bonds stabilize tertiary structure. **18.54** In *hydrophobic interactions,* hydrocarbon side chains cluster in the center of proteins and make proteins spherical. Examples: Phe, Ile. *Salt bridges* bring together distant parts of a polypeptide chain. Examples: Lys, Asp. **18.56** When a protein is denatured, its nonprimary structure is disrupted, and it can no longer catalyze reactions. **18.58** Val–Met–Leu, Met–Val–Leu, Leu–Met–Val, Val–Leu–Met, Met–Leu–Val, Leu–Val–Met.
18.60 *Outside*: Asp, His (They can form H-bonds.) *Inside*: Val, Ala (They have hydrophobic interactions.) **18.62** Digestive enzymes would hydrolyze insulin if it were swallowed.

18.64

N-terminal   C-terminal

18.66 **(a)** $H_3{}^+NCH_2COOH$ **(b)** $H_3{}^+NCH_2COOCH_3$
18.68 N-terminal: Val–Gly–Ser–Ala–Asp C-terminal
18.70 A peptide rich in Asp and Lys is more soluble, because its side chains are more polar and can form hydrogen bonds with water.
18.72

(a)

Decarboxylation is the loss of CO_2. **(b)** Phenelzine resembles tyramine (it has a phenyl group and an amino group) and blocks the enzyme that catalyzes deamination of tyramine. **18.74** An incomplete protein lacks one or more essential amino acids. **18.76** They must provide complete nutrition to developing organisms. **18.78** Arg–Trp moves to negative electrode, Val–Met doesn't move, Asp–Thr moves to positive electrode. **18.80** Osteogenesis imperfecta is a dominant genetic defect in which collagen is synthesized incorrectly, resulting in weak bones. **18.82** It was hard to accept that a protein might duplicate itself, cause disease, be responsible for inherited disease, be transmitted, and might arise spontaneously. **18.84** A combination of grains, legumes, and nuts in each meal provides all of the essential amino acids. **18.86** Canned pineapple has been heated to inactivate enzymes

18.88 (a)

Arg———Pro———Pro———Gly———Phe———Ser———Pro———Phe———Arg

(b) Proline rings introduce kinks and bends and prevent hydrogen bonds from forming.

18.90 Carbon is no longer bonded to four different groups.

18.92

Asn—Cys—Pro—Leu—Gln

Gln | | S

Ile | | S   Oxytocin

Tyr—Cys

18.94 Arg, Asp, Asn, Glu, Gln, His, Lys, Ser, Thr, Tyr

$$O=C$$
Asn HC—CH$_2$C—N̈—H------:Ö—CH$_2$—CH Ser
       | || | |
      NH   O H   C=O

$$O=C$$
Asn HC—CH$_2$C—N—H------:Ö: :Ö:-----H—Ö—CH$_2$—CH Ser
       | || | | | |
      NH   O H   H H   C=O
                                        NH

18.96 On the outside of a globular protein: Glu, Ser. On the outside of a fibrous protein: Ala, Val. On the outside of neither: Leu, Phe.

18.98 Asp is similar in size and function to Glu.

Chapter 19

19.1 Kinases **19.2** (1) The enzyme might catalyze reactions within the eye. (2) Saline is sterile and isotonic. **19.3** iron, copper, manganese, molybdenum, vanadium, cobalt, nickel, chromium **19.4 (a)** NAD$^+$, coenzyme A, FAD; **(b)** They are minerals. **19.5 (a)** catalyzes the removal of two —H from glutamate. **(b)** catalyzes the transfer of an amino group from alanine to a second substrate. **(c)** catalyzes the formation of a bond between carbamoyl phosphate and another substrate. **(d)** catalyzes the isomerization of triose phosphate. **19.6 (a)** arginase **(b)** maltase **19.7** isomerase. It catalyzes the isomerization of glucose 6-phosphate. **19.8** Water adds to fumarate (substrate) to give L-malate (product). **19.9** Reaction **(a)** **19.10** Acidic, basic, and polar side chains take part in catalytic activity. All types of side chains hold the enzyme in the active site. **19.11** Substrate molecules are bound to all of the active sites. **(a)** no effect; **(b)** increases the rate. **19.12** higher at 30°; somewhat higher at 40°. **19.13** The rate is much greater at pH = 8. **19.14** molecule **(b)**, because it resembles the substrate. **19.15** a product that resembles the substrate. **19.16 (a)** competitive inhibition **(b)** covalent modification or feedback control **(c)** covalent modification **(d)** genetic control **19.17** Vitamin A—long hydrocarbon chain. Vitamin C—polar hydroxyl groups. **19.18** Retinal—aldehyde. Retinoic acid—carboxylic acid. Retinol—alcohol. Same functional group modified in each molecule. **19.19** enzyme cofactors; antioxidants; aid in absorption of calcium and phosphate ions; aid in synthesis of visual pigments and blood clotting factors.

19.20

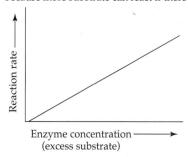

---- Hydrogen bonds

⸺ Salt bridges

19.21 (a) oxidoreductase **(b)** dehydrogenase **(c)** L-lactate **(d)** pyruvate **(e)** L-lactate dehydrogenase

19.22 No. An enzyme usually catalyzes the reaction of only one isomer. D-Lactate might be a competitive inhibitor. **19.23** NAD$^+$ is an oxidizing agent and includes the vitamin niacin. **19.24 (a)** Rate increases when [substrate] is low, but max. rate is soon reached; max. rate is always lower than max. rate of uninhibited reaction. **(b)** Rate increases. **19.25 (a)** Addition or removal of a covalently bonded group changes the activity of an enzyme. **(b)** Hormones control the synthesis of enzymes. **(c)** Binding of the regulator at a site away from the catalytic site changes the shape of the enzyme. **(d)** Noncompetitive inhibition is a type of allosteric regulation (see part **(c)**). Competitive inhibition occurs when an inhibitor reversibly occupies an enzyme's active site. Irreversible inhibition results when an inhibitor covalently binds to an enzyme and destroys its ability to catalyze a reaction. **19.26 (a)** feedback inhibition **(b)** irreversible inhibition **(c)** genetic control **(d)** noncompetitive inhibition **19.27** From left to right: aspartate (acidic), serine, glutamine, arginine (basic), histidine (basic). **19.28 (a)** oxidation or reduction of a substrate; **(b)** addition of a small molecule to a double bond, or removal of a small molecule to form a double bond; **(c)** transfer of a functional group from one substrate to another **19.30 (a)** sucrase **(b)** fumarase **(c)** RNAse **19.32** An enzyme is a large three-dimensional molecule with a catalytic site into which a substrate can fit. Enzymes are specific in their action because only one or a few molecules have the appropriate shape and functional groups to fit into the catalytic site. **19.34 (a)** hydrolase **(b)** lyase **(c)** oxidoreductase **19.36 (a)** loss of CO_2 from a substrate; **(b)** transfer of a methyl group between substrates; **(c)** removal of two —H to form a double bond **19.38** hydrolase **19.40 (a)** riboflavin (B$_2$) **(b)** pantothenic acid (B$_5$) **(c)** pyridoxine (B$_6$) **19.42** Lock-and-key: An enzyme is rigid (lock) and only one specific substrate (key) can fit in the active site. Induced fit: An enzyme can change its shape to accommodate the substrate and to catalyze the reaction. **19.44** No. Protein folding can bring the residues close to each other. **19.46** In the stomach, an enzyme must be active at an acidic pH. In the intestine, an enzyme needs to be active at a higher pH and need not be active at pH = 1.5.

19.48

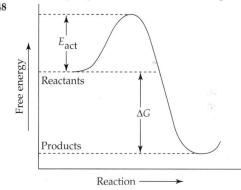

19.50 At a high substrate concentration relative to enzyme concentration, the rate of reaction increases as the enzyme concentration increases because more substrate can react if there are more active sites.

Reaction rate ↑

Enzyme concentration → (excess substrate)

19.52 Increasing enzyme concentration increases the rate; increasing substrate concentration increases the rate until all active sites are filled, when the rate levels off. **19.54 (a) (b)** lowers rate; **(c)** denatures the enzyme and stops reaction **19.56** *Noncompetitive inhibition:* Inhibitor binds reversibly and noncovalently away from the active site and changes the shape of the site to make it difficult for the enzyme to

catalyze reactions. *Competitive inhibition*: Inhibitor binds reversibly and noncovalently at the active site and keeps the substrate from entering. *Irreversible inhibition*: Inhibitor irreversibly forms a covalent bond at the active site and destroys the catalytic ability of the enzyme. **19.58** diagram **(c)** **19.60** (1) displacing an essential metal from an active site; (2) bonding to a cysteine residue (irreversible) **19.62** Papain catalyzes the hydrolysis of peptide bonds and partially digests the proteins in meat. **19.64** One site is for catalysis, and one site is for regulation. **19.66** The end product of a reaction series is an inhibitor for an earlier step. **19.68** A zymogen is an enzyme synthesized in a form different from its active form because it might otherwise harm the organism. **19.70** Vitamins are small essential organic molecules that must be obtained from food. **19.72** Vitamin C is excreted, but vitamin A is stored in fatty tissue. **19.74** (1) Enzymes produce only one enantiomer of a product. (2) Enzymes allow difficult reactions to take place (3) Use of enzymes avoids reactions with hazardous byproducts. **19.76** Two strategies keep the enzyme active: (1) Use of chaperonins, enzymes that return a protein to its active form; (2) The protein itself is rigid and resists heat denaturation. **19.78** CPK, ALT, and LDH1 leak from damaged heart vessels. **19.80** A mild inhibitor allows closer control of blood pressure. A modification to pit viper protein might be to introduce a –SH residue near proline. **19.82** The body excretes excess water-soluble vitamins, but fat-soluble vitamins are stored in tissue. **19.84** They are the most important for maintaining good health.
19.86 *See below for answer.*
19.88 Blanching destroys enzymes that would cause deterioration in food quality. **19.90** Leu−Gly−Arg, Ile−Met−His, Tyr−Trp−Ala **19.92** Excess ethanol displaces methanol from the active site.

Chapter 20

20.1 (c) 20.2 The molecules resemble the heterocyclic part of cAMP, and they might act as inhibitors to the enzyme that inactivates cAMP. **20.3** Glu−His−Pro **20.4** The hydrophobic part of the structure is larger than the polar, hydrophilic part. **20.5** Testosterone has a —CH$_3$ group between the first two rings; nandrolone doesn't. Otherwise, their structures are identical. **20.6 (a)** 3 **(b)** 1 **(c)** 2 **20.7** Similarities: both structures have aromatic rings, secondary amine groups, alcohol groups. Differences: propranolol has an ether group and a naphthalene ring system; epinephrine has two phenol hydroxyl groups; the compounds have different side chain carbon skeletons. **20.8** Malathion: it's the least toxic. **20.9** An agonist (black widow spider venom) prolongs the biochemical response of a receptor. An antagonist (botulinus toxin) blocks or inhibits the normal response of a receptor. **20.10** phenol hydroxyl group, ether, carbon-carbon double bond, aromatic ring. THC is hydrophobic and is likely to accumulate in fatty tissue. **20.11 (a)** antihistamine **(b)** antidepressant **20.12 (a)** polypeptide hormone (produced in the anterior pituitary gland) **(b)** steroid hormone (produced in ovaries) **(c)** Progesterone-producing cells have LH receptors. **(d)** Progesterone is lipid-soluble and can enter cells. **20.13** (1) Adenylate cyclase can produce a great many molecules of cAMP. (2) A large amount of glucose is released when glycogen phosphorylase is activated. **20.14 (a)** insulin (polypeptide hormone) **(b)** pancreas **(c)** in the bloodstream **(d)** Insulin doesn't enter cells directly because it can't pass through cell membranes. Instead, it binds with a cell surface receptor. **20.15** Neurotransmitters can act either by binding to receptors or by activating second messengers. **20.16** Enzymatic inactivation; reuptake of neurotransmitter molecules by presynaptic neuron. **20.17** These substances increase dopamine levels in the brain. The brain responds by decreasing the number and sensitivity of dopamine receptors. Thus more of the substance is needed to elevate dopamine levels, leading to addiction. **20.18** A hormone is a molecule that travels through the bloodstream to its target tissue, where it binds to a receptor and regulates biochemical reactions. **20.20** A hormone transmits a chemical message from an endocrine gland to a target tissue. A neurotransmitter carries an impulse between neighboring nerve cells. **20.22** Hormone binding is noncovalent and regulates a reaction, rather than catalyzing it. **20.24** The endocrine system manufactures and secretes hormones. **20.26** polypeptide hormones, steroid hormones, amino acid derivatives **20.28** Enzymes are proteins; hormones may be polypeptides, proteins, steroids, or amino acid derivatives. **20.30** Polypeptide hormones travel through the bloodstream and bind to cell receptors, which are on the outside of a cell. The receptors cause production within cells of "second messengers" that activate enzymes. **20.32** the adrenal medulla **20.34** through the bloodstream **20.36** In order of involvement; the hormone receptor, G protein, and adenylate cyclase. **20.38** It initiates reactions that release glucose from storage. Termination occurs when phosphodiesterase converts cAMP to AMP. **20.40** anaphylaxis **20.42** Insulin contains 51 amino acids, is released from the pancreas, and acts at cells, causing them to take up glucose. **20.44** mineralocorticoids (aldosterone), glucocorticoids (cortisone) sex hormones (testosterone, estrone). **20.46** estrone, estradiol, progesterone **20.48** epinephrine, norepinephrine, dopamine **20.50 (a)** amino acid derivative **(b)** polypeptide hormone **(c)** steroid hormone **20.52** A synapse is the gap between two nerve cells that neurotransmitters cross to transmit their message. **20.54** nerve cell, muscle cell, endocrine cell **20.56** A nerve impulse arrives at the presynaptic end of a neuron. The nerve impulse stimulates the movement of a vesicle, containing neurotransmitter molecules, to the cell membrane. The vesicle fuses with the cell membrane and releases the neurotransmitter, which crosses the synaptic cleft to a receptor site on the postsynaptic end of a second neuron. After reception, the cell transmits an electrical signal down its axon and passes on the impulse. Enzymes then deactivate the neurotransmitter so that the neuron can receive the next impulse. Alternatively, the neurotransmitter may be returned to the presynaptic neuron. **20.58** (1) Neurotransmitter

19.86

Vitamin E

molecules are released from a presynaptic neuron. (2) Neurotransmitter molecules bind to receptors on the target cell. (3) The neurotransmitter is deactivated **20.60** They are secreted in the central nervous system and have receptors in brain tissue. **20.62** Agonists prolong the response of a receptor. Antagonists block the response of a receptor. **20.64** Antihistamines such as doxylamine counteract allergic responses caused by histamine by blocking histamine receptors in mucous membranes. Antihistamines such as cimetidine block receptors for histamine that stimulate production of stomach acid. **20.66** Tricyclic antidepressant: Elavil MAO inhibitor: Nardil SSRI: Prozac **20.68** Cocaine increases dopamine levels by blocking reuptake. **20.70** Tetrahydrocannabinol (THC) increases dopamine levels in the same brain areas where dopamine levels increase after administration of heroin and cocaine. **20.72** Endorphins are polypeptides with morphine-like activity. They are produced by the pituitary gland and have receptors in the brain. **20.74** An ethnobotanist discovers what indigenous people have learned about the healing power of plants. **20.76** Scientists who know about the exact size and shape of enzymes and receptors can design drugs that interact with the active sites of these biomolecules. **20.78** Homeostasis is the maintenance of a constant internal environment. **20.80** Plants don't have endocrine systems or a circulatory fluid like blood. **20.82** Epibatidine acts as a painkiller at acetylcholine receptors in the central nervous system. **20.84** The *hormone receptor* recognizes the hormone, sets in motion the stimulation process, and interacts with *G-protein*. This protein binds GTP and activates *adenylate cyclase*, which catalyzes the formation of cAMP. **20.86** Signal amplification is the process by which a small signal induces a response much larger than the original signal. In the case of hormones, a small amount of hormone can bring about a very large response. **20.88** Testosterone has an —OH group in the 5-membered ring; progesterone has an acetyl group in that position. Otherwise, the two molecules are identical. **20.90** Chocolate acts at dopamine receptors and produces feelings of satisfaction. **20.92** Testosterone is converted to androsterone by reductions (C=O, C=C) in the first ring, and an oxidation of —OH in the five-membered ring.

Chapter 21

21.1 exergonic: **(a)**, **(c)**; endergonic: **(b)**; releases the most energy: **(a)**
21.2 Both pathways produce the same amount of energy.
21.3 (a) exergonic: oxidation of glucose; endergonic: photosynthesis **(b)** sunlight
21.4 (a)

Carbohydrates $\xrightarrow{\text{Digestion}}$ Glucose, sugars $\xrightarrow{\text{Glycolysis}}$

Pyruvate $\longrightarrow$ Acetyl-SCoA $\xrightarrow[\text{cycle}]{\text{Citric acid}}$

Reduced coenzymes $\xrightarrow[\text{transport}]{\text{Electron}}$ ATP

(b) pyruvate, acetyl-SCoA, citric acid cycle intermediates.
21.5

$$H_3C-\overset{\overset{O}{\|}}{C}-O-\overset{\overset{O}{\|}}{\underset{\underset{O^-}{|}}{P}}-O^- + H_2O \longrightarrow$$

$$H_3C-\overset{\overset{O}{\|}}{C}-O^- + \ ^-O-\overset{\overset{O}{\|}}{\underset{\underset{OH}{|}}{P}}-O^- + H^+$$

21.6 Energy is produced only when it is needed.
21.7

$$HOCH_2CHCH_2OH \ \underset{\overset{\displaystyle\frown}{\text{ATP} \quad \text{ADP}}}{\xrightarrow{\hspace{2cm}}} \ HOCH_2CHCH_2O-\overset{\overset{O}{\|}}{\underset{\underset{O^-}{|}}{P}}-O^-$$

with OH on the middle carbon on both sides.

21.8 If a process is exergonic, its exact reverse is endergonic and can't occur unless it is coupled with an exergonic reaction in a different pathway. **21.9** favorable ($\Delta G = -3.0$ kcal/mol). **21.10** yes

21.11

21.12 Citric acid, isocitric acid. **21.13** steps 3, 4, 6, 8. **21.14** Succinic dehydrogenase catalyzes the removal of two hydrogens from succinate to yield fumarate. **21.15** α-Ketoglutarate, oxaloacetate. **21.16** isocitrate **21.17** Steps 1–4 correspond to the first stage, and steps 5–8 correspond to the second stage. **21.18** Mitochondrial matrix **21.19** O_2. Movement of H^+ from a region of high $[H^+]$ to a region of low $[H^+]$ releases energy that is used in ATP synthesis.
21.20 (a) Succinyl phosphate + $H_2O \rightarrow$ Succinate + $HOPO_3^{2-}$ + H^+ **(b)** ADP + $HOPO_3^{2-}$ + $H^+ \rightarrow$ ATP + H_2O $\quad \Delta G = +7.3$ kcal/mol
21.21 (a) Stage 1 (digestion) **(b)** Stage 4 (ATP synthesis) **(c)** Stage 2 (glycolysis) **(d)** Stage 3 (citric acid cycle). **21.22** Endergonic; coupled reactions **21.23** NAD^+ accepts hydride ions; hydrogen ions are released to the mitochondrial matrix, and ultimately combine with reduced O_2 to form H_2O. **21.24 (a)** Step A (NAD^+) **(b)** Step B **(c)** product of A (in brackets) **21.25** Step 1: lyase Step 2: isomerase Step 3: oxidoreductase Step 4: oxidoreductase, lyase Step 5: ligase Step 6: oxidoreductase Step 7: lyase Step 8: oxidoreductase **21.26** Metals are better oxidizing and reducing agents. Also, they can accept and donate electrons in one-electron increments. **21.28** An endergonic reaction requires energy, and an exergonic reaction releases energy. **21.30** Enzymes affect only the rate of a reaction, not the size or sign of ΔG. **21.32** Exergonic: **(a)**, **(b)**; endergonic: **(c)**. Reaction **(a)** proceeds farthest toward products.
21.34

PROKARYOTIC CELLS	EUKARYOTIC CELLS
Quite small	Relatively large
No nucleus	Nucleus
Dispersed DNA	DNA in nucleus
No organelles	Organelles
Occur in single-celled organisms	Occur in single-celled and higher organisms

21.36 The cytoplasm consists of everything between the cell membrane and the nuclear membrane. The cytosol is the medium that fills the interior of the cell and contains electrolytes, nutrients, and many enzymes, in aqueous solution. **21.38** A mitochondrion is egg-shaped and consists of a smooth, outer membrane and a folded inner membrane. The intermembrane space lies between the outer and inner membranes, and the space enclosed by the inner membrane is called the *mitochondrial matrix*. **21.40** 90% of the body's ATP is synthesized in mitochondria.
21.42 Metabolism refers to all reactions that take place inside cells. Digestion is a part of metabolism in which food is broken down into small organic molecules. **21.44** acetyl—ScoA **21.46** Energy is released when ATP transfers a phosphoryl group. **21.48** ATP has a triphosphate group bonded to C5 of ribose, and ADP has a diphosphate group in that position. **21.50** $\Delta G = -4.5$ kcal/mol.
21.52

$$\text{1,3-Bisphosphoglycerate} \ \underset{\overset{\displaystyle\frown}{\text{ADP} \quad \text{ATP}}}{\xrightarrow{\hspace{2cm}}} \ \text{3-Phosphoglycerate}$$

21.54 (a) NAD^+ is reduced. **(b)** NAD^+ is an oxidizing agent. **(c)** NAD^+ oxidizes secondary alcohol to a ketone. **(d)** $NADH/H^+$
(e)

$$H-\overset{|}{\underset{|}{C}}-OH \ \underset{\overset{\displaystyle\frown}{NAD^+ \quad NADH/H^+}}{\xrightarrow{\hspace{2cm}}} \ C=O$$

21.56 cellular mitochondria **21.58** Both carbons are oxidized to CO_2. **21.60** 3 NADH, one $FADH_2$. **21.62** Step 3 (isocitrate $\rightarrow$ α-ketoglutarate),

Step 4 (α-ketoglutarate $\rightarrow$ succinyl-SCoA) and Step 8 (malate $\rightarrow$ oxaloacetate) store energy as NADH. **21.64** One complete citric acid cycle produces four reduced coenzymes, which enter the electron transfer chain and ultimately generate ATP. **21.66** H_2O, ATP, oxidized coenzymes **21.68 (a)** FAD = flavin adenine dinucleotide; **(b)** CoQ = coenzyme Q; **(c)** NADH/H^+ = reduced nicotinamide adenine dinucleotide, plus hydrogen ion; **(d)** Cyt c = Cytochrome c **21.70** NADH, coenzyme Q, cytochrome c **21.72** The citric acid cycle would stop. **21.74** formation of ATP from the reactions of reduced coenzymes in the electron transport system **21.76 (a)** ATP **(b)** H_2O, oxidized coenzymes **20.78** Bacteria use H_2S because no light is available for the usual light-dependent reaction of H_2O that provides O_2 and electrons. **21.80** Answer based on reader's own weight and activity level **21.82** Daily activities such as walking use energy, and thus the body requires a larger caloric intake than that needed to maintain basal metabolism. **21.84** Oxygen consumption increases because the proton gradient from ATP production dissipates. **21.86** A seal has more brown fat because it needs to keep warm. **21.88** The light reaction produces O_2, NADPH, and ATP. The dark reaction produces carbohydrates from water and CO_2. **21.90** Refrigeration slows the breakdown of carbohydrates by decreasing the rate of respiration. Enzyme reactions are slower at low temperatures. **21.92**

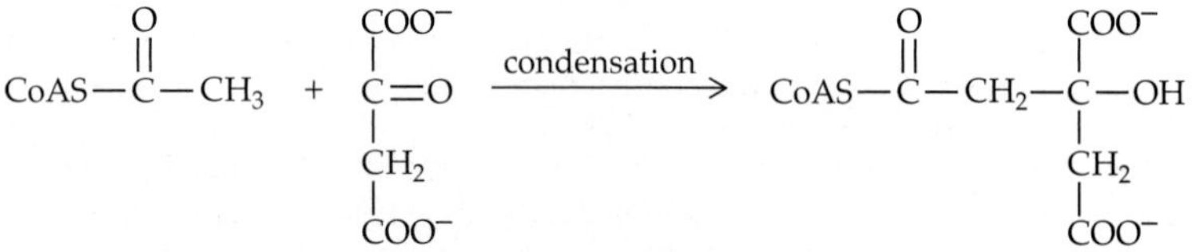

21.94 oxidoreductases **21.96** FAD; oxidoreductases **21.98** isocitrate; malate **21.100** $O_2^-\cdot$, $OH^-\cdot$, and H_2O_2; superoxide dismutase, catalase, vitamins E, C, and A. **21.102** adipose tissue, skin cells, skeletal muscle, heart muscle.

Chapter 22

22.1 (a) aldopentose **(b)** ketotriose **(c)** aldotetrose
22.2

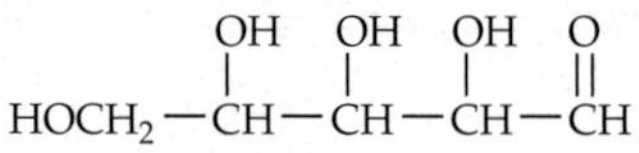

An aldopentose

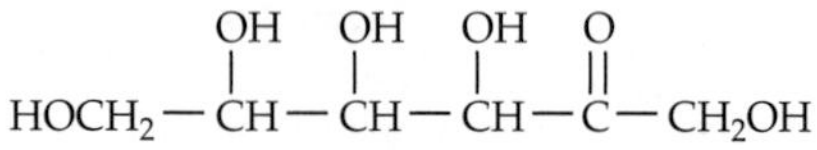

A ketohexose

22.3 eight stereoisomers **22.4 (d)** **22.5** The bottom carbon is not chiral. The orientations of the hydroxyl groups bonded to the chiral carbons must be shown in order to indicate which stereoisomer is pictured.

22.6 (a)

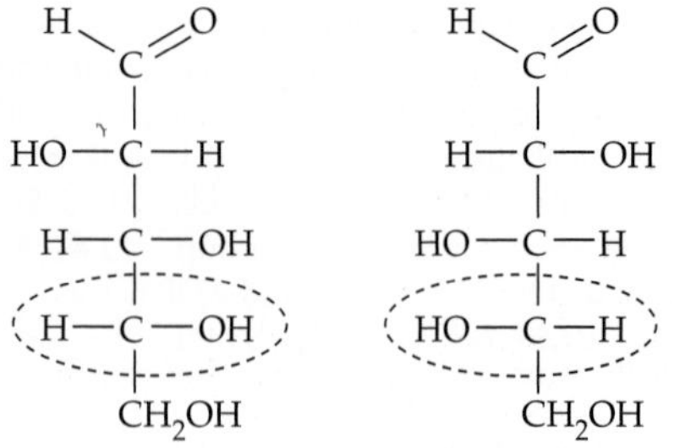

A D-aldopentose   An L-aldopentose

(b)

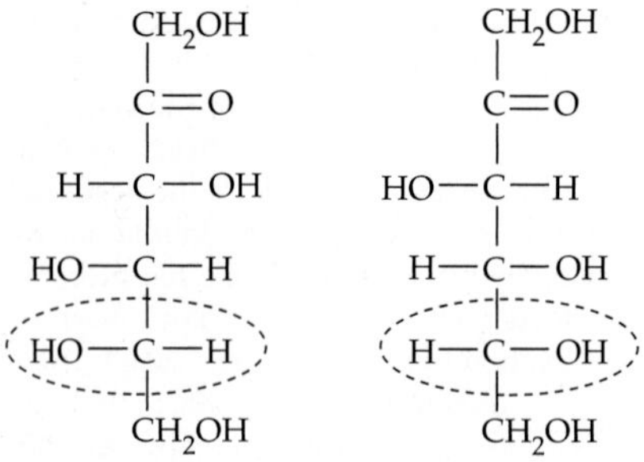

An L-ketohexose   A D-ketohexose

22.7

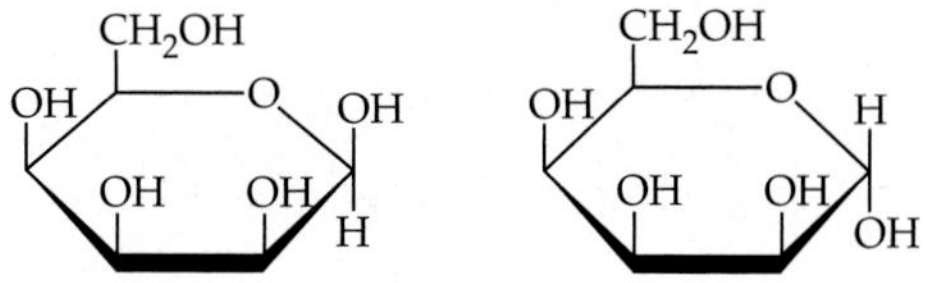

β-anomer α-anomer

22.8

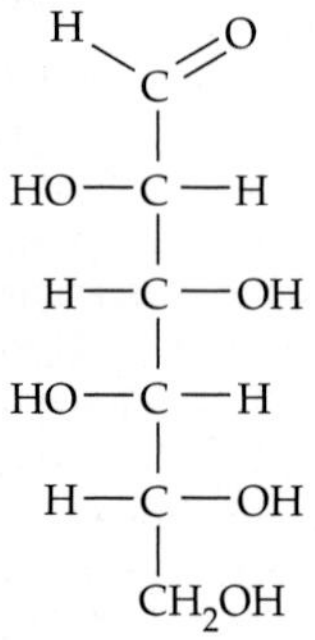

D-Idose

22.9 (a) an α-hexose **(b)** a steroid **(c)** an ester
22.10

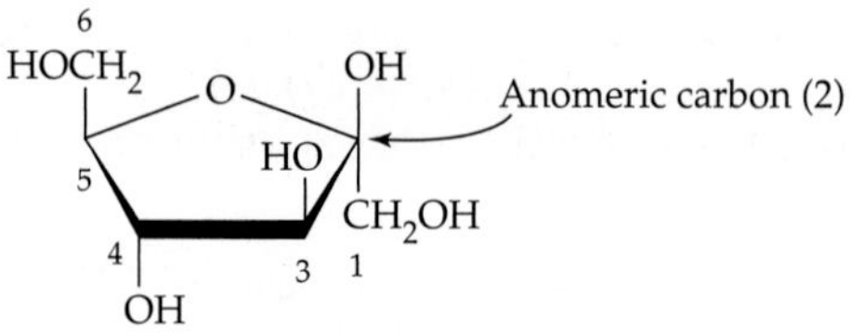

β-anomer

22.11

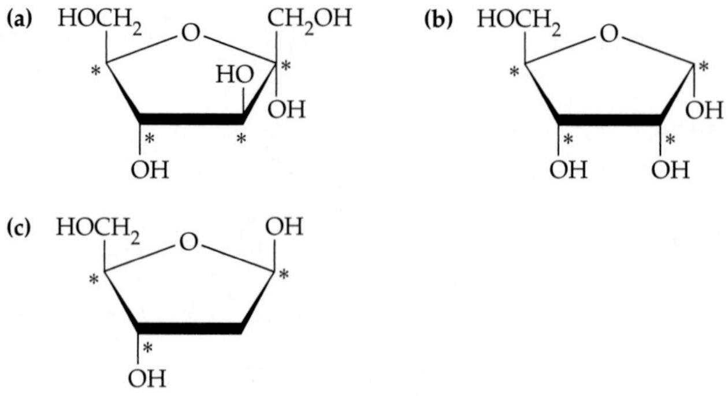

(a)

(b)

(c)

22.12

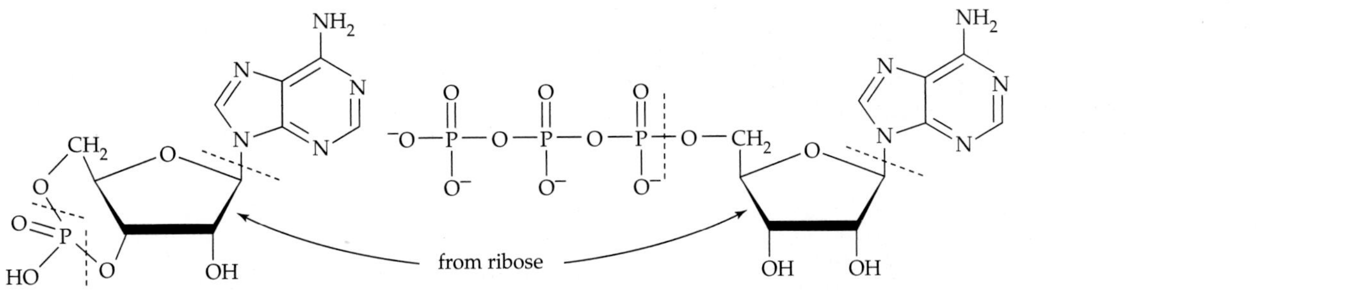

Cyclic AMP

from ribose

ATP

22.13 (a) an α anomer (b) carbon 6 (c) Groups that are below the plane of the ring in D-galactose are above the plane of the ring in L-fucose. Groups that are above the plane of the ring in D-galactose are below the plane of the ring in L-fucose. (d) yes

22.14

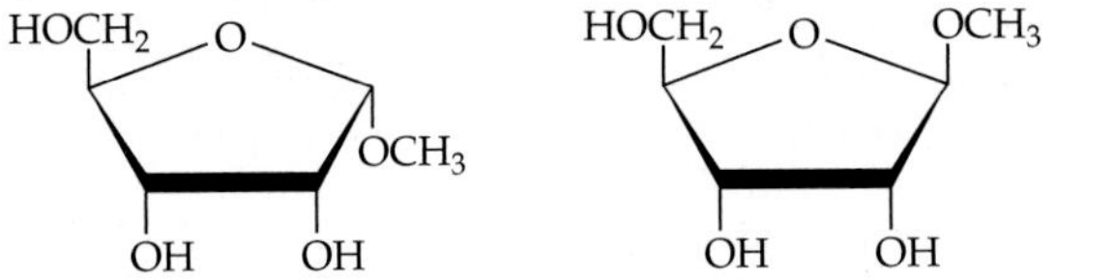

Methyl α-D-riboside   Methyl β-D-riboside

22.15 a β-1,4 glycosidic link   **22.16** β-D-Glucose + β-D-Glucose
22.17 (a) maltose (b) sucrose (c) lactose   **22.18** glutamine, asparagine
22.19 an α-1,4 glycosidic link   **22.20** No. There are too few hemiacetal units to give a detectable result.

22.21 Starch $\xrightarrow{\text{Amylase}}$ Maltose $\xrightarrow{\text{Maltase}}$ Glucose
polysaccharide   disaccharide   monosaccharide

22.22 (a) diastereomers, anomers (b) enantiomers (c) diastereomers
22.23 (a) (b)

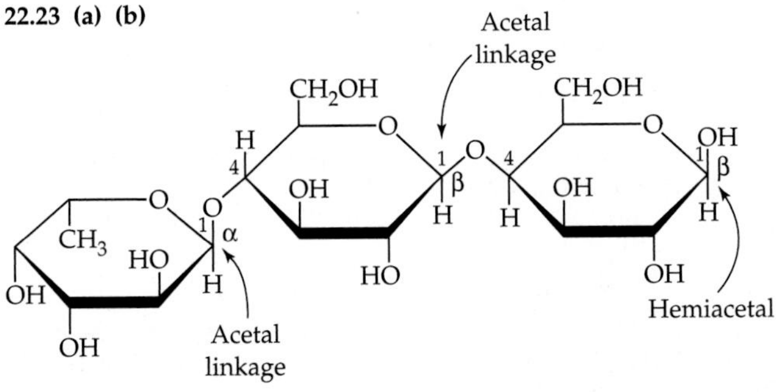

A   B   C
α-anomer β-anomer β-anomer

(c) α-1,4 linkage between C4 of B and C1 of A (d) β-1,4 linkage between C4 of C and C1 of B   **22.24** (a) (b) No monosaccharides are identical, and none are enantiomers.
(c) (d)

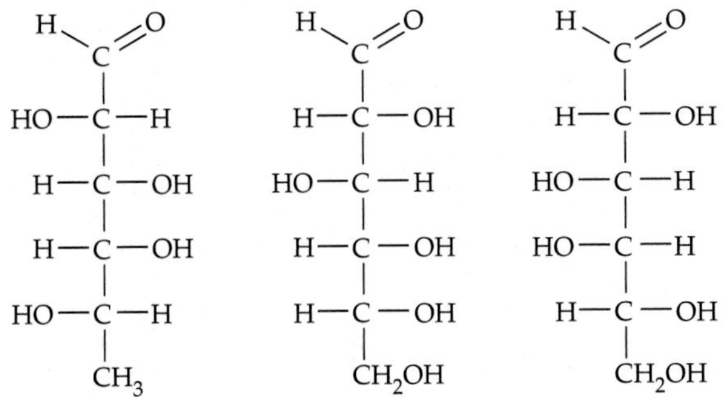

L-Fucose   D-Glucose   D-Galactose

22.25 Monosaccharide C is oxidized. Identification of the carboxylic acid also identifies the terminal monosaccharide.

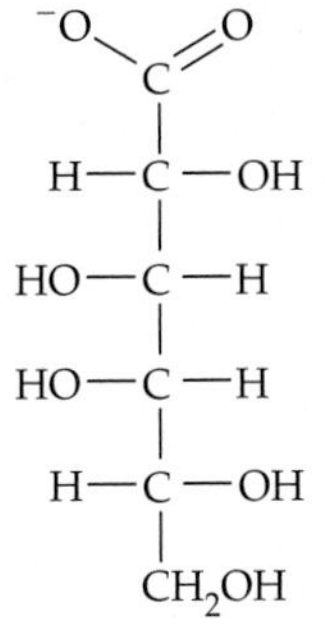

22.26 No
22.27

POLYSACCHARIDE	LINKAGE	BRANCHING?
Cellulose	β-1,4	no
Amylose	α-1,4	no
Amylopectin	α-1,4	yes: α-1,6 branches occur ~ every 25 units
Glycogen	α-1,4	yes: even more α-1,6 branches than in amylopectin

22.28 Glucose is in equilibrium with its open-chain aldehyde form, which reacts with an oxidizing agent.   **22.30** -ose   **22.32** (a) aldotetrose (b) ketopentose (c) aldopentose (d) ketohexose
22.34 right part = 4 chiral carbons; left part = 5 chiral carbons; total = 9 chiral carbons
22.36

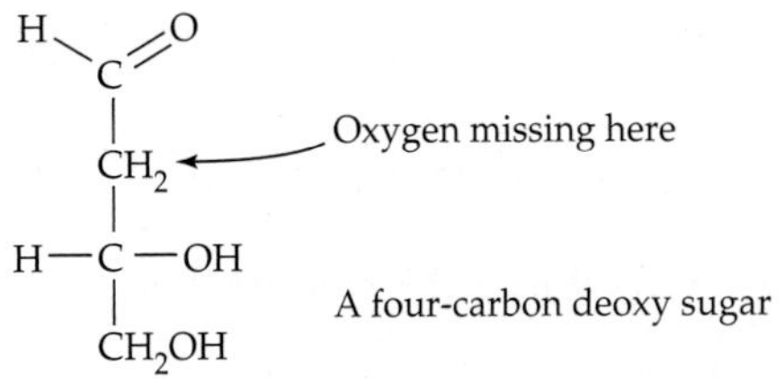

Oxygen missing here

A four-carbon deoxy sugar

22.38 glucose – food, all living organisms; galactose – brain tissue, milk; fructose – fruit; ribose – nucleic acids   **22.40** They are mirror images.
22.42 The reduction product of D-erythrose is achiral.   **22.44** A polarimeter measures the degree of rotation of plane-polarized light by a solution of an optically active compound.   **22.46** Equimolar solutions of enantiomers rotate light to the same degree but in opposite directions.
22.48 A reducing sugar contains an aldehyde or ketone group.
22.50 An anomer is one of a pair of hemiacetal stereoisomers formed when an open-chain sugar cyclizes. Anomers differ in the orientation of the hydroxyl group at the anomeric carbon.   **22.52** the α form

22.54

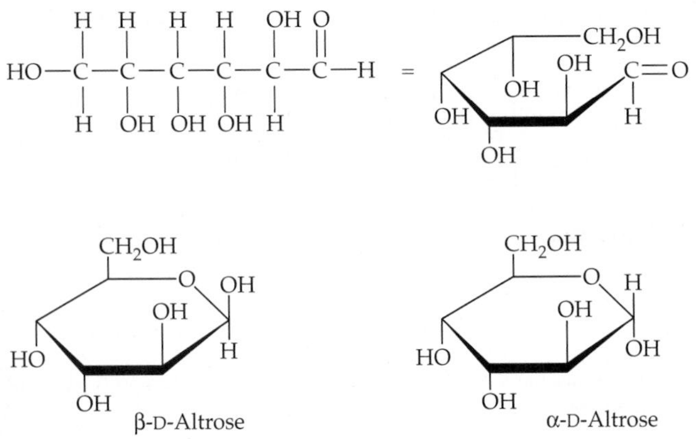

β-D-Altrose   α-D-Altrose

22.56

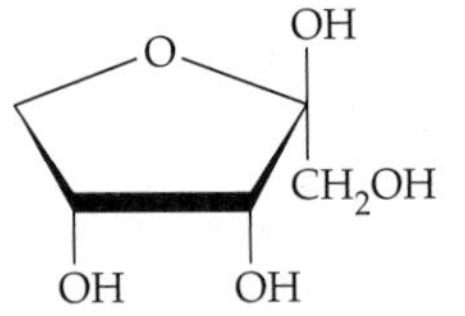

22.58

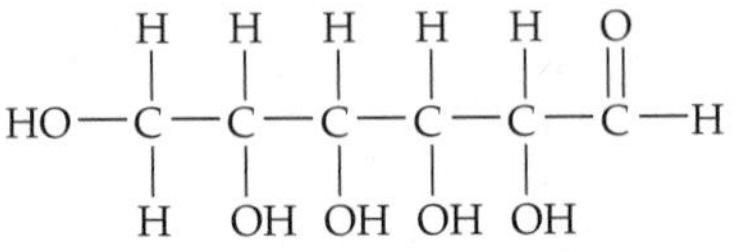

22.60

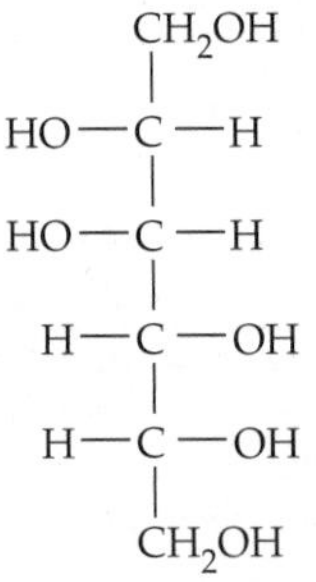

22.62

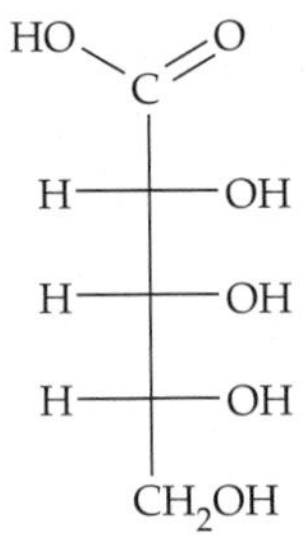

22.64 A glycoside is an acetal that is formed when the hemiacetal —OH group of a carbohydrate reacts with an alcohol.

22.66

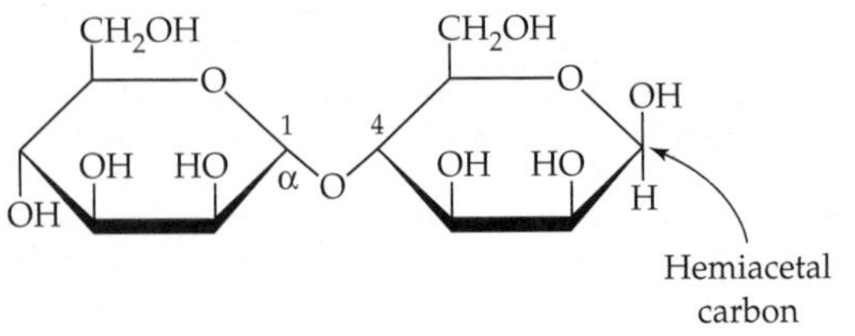

The hemiacetal carbon in this problem is in equilibrium with an open-chain aldehyde that is a reducing sugar. **22.68** Sucrose has no hemiacetal group. **22.70** Amylose and amylopectin are both components of starch and both consist of long polymers of α-D-glucose linked by α-1,4 glycosidic bonds. Amylopectin is much larger and has α-1,6 branches every 25 units or so along the chain. **22.72** Gentiobiose contains both an acetal grouping and a hemiacetal grouping. Gentiobiose is a reducing sugar. A β-1,6 linkage connects the two monosaccharides. **22.74** Trehalose is a nonreducing sugar because it contains no hemiacetal linkages. The two D-glucose monosaccharides are connected by an α-1,1 acetal link.

22.76

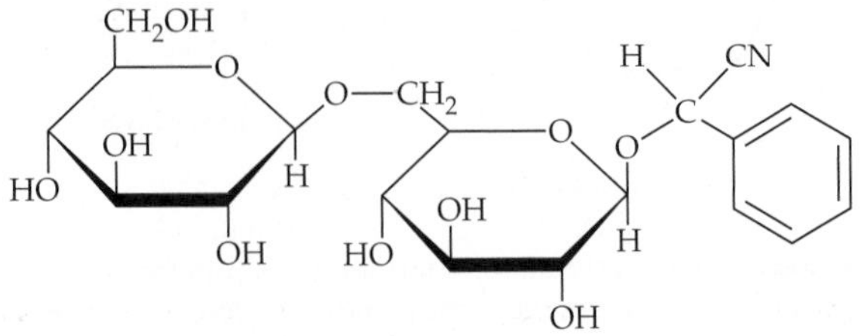

22.78 Enzyme-catalyzed reactions usually produce only one enantiomer. **22.80** Starch is a complex carbohydrate. Glucose is a simple carbohydrate. Soluble and insoluble fiber are complex carbohydrates. **22.82** glucose, cellulose **22.84** Penicillin inhibits the enzyme that synthesizes bacterial cell walls. Mammals do not have this synthetic pathway. **22.86** People with type O blood can receive blood only from other donors that have type O blood. People with type AB blood can give blood only to other people with type AB blood. **22.88** pectin and vegetable gum: found in fruits, barley, oats, and beans. **22.90** D-Ribose and L-ribose are enantiomers that are identical in all properties (melting point, density, solubility, and chemical reactivity) except for the direction that they rotate plane-polarized light (b). **22.92** No, because they are not mirror images.

22.94

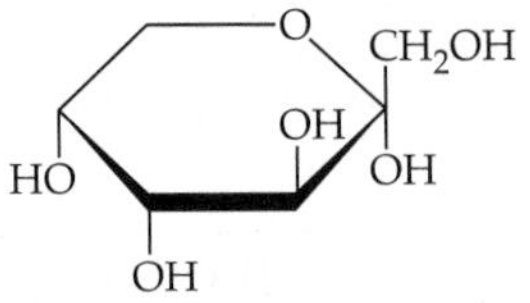

22.96 Raffinose is not a reducing sugar because it has no hemiacetal group. (It has acetal groups.)

22.98

HOCH$_2$CCH$_2$OH

1,3-Dihydroxyacetone has no optical isomers because it has no chiral carbons. **22.100** Lactose intolerance is an inability to digest lactose. Symptoms include bloating, cramps, and diarrhea. **22.102** Symptoms of galactosemia: vomiting, liver failure, mental retardation, cataracts. **22.104** 4 chiral carbons.

Chapter 23

23.1 (a) glycogenesis **(b)** glycogenolysis **(c)** gluconeogenesis **23.2** glycogenesis, pentose phosphate pathway, glycolysis **23.3 (a)** steps 6 and 7 **(b)** steps 9 and 10 **23.4** Isomerizations: steps 2, 5, 8

23.5

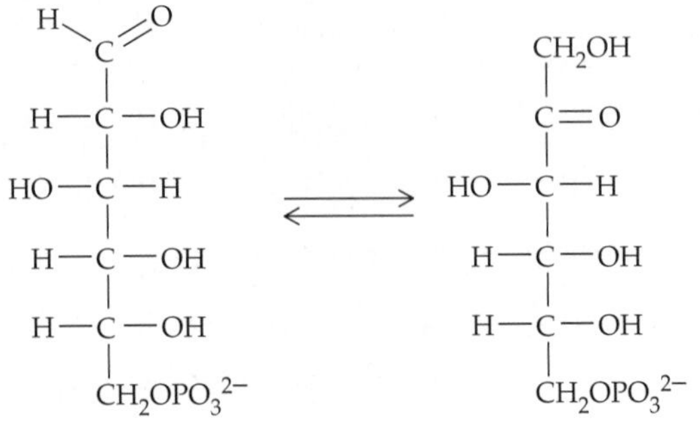

23.6 (a) pyruvate **(b)** Step 6: glyceraldehyde 3-phosphate is oxidized; NAD$^+$ is the oxidizing agent

23.7 Fructose 6-phosphate enters glycolysis at step 3.

23.8 Glucose and galactose differ in configuration at C4. **23.9** **(a)** The energy is lost as heat **(b)** The reverse of fermentation is very endothermic; loss of CO_2 drives the reaction to completion in the forward direction. **23.10** Insulin decreases; blood glucose decreases, the level of glucagon increases. Glucagon causes the breakdown of liver glycogen and the release of glucose. As glycogen is used up, the level of free fatty acids and ketone bodies increases.
23.11 Sorbitol can't form a cyclic acetal because it doesn't have a carbonyl group.

Sorbitol

23.12 **(a)** The increase in $[H^+]$ drives the equilibrium shown in Section 23.9 to the right, causing the production of CO_2. **(b)** Le Châtelier's Principle. **23.13** phosphorylation, oxidation **23.14** hydrolases **23.15** **(a)** when the supply of glucose is adequate and the body needs energy. **(b)** when the body needs free glucose. **(c)** when ribose 5-phosphate or NADPH are needed. **(d)** when glucose supply is adequate, and the body does not need to use glucose for energy production. **23.16** Phosphorylations of glucose and fructose 6-phosphate produce important intermediates that repay the initial energy investment. Fructose 1,6-bisphosphate is cleaved into two three-carbon compounds, which are converted to pyruvate. **23.17** **(a)** when the body needs energy, in mitochondria; **(b)** under anaerobic conditions, in yeast; **(c)** under anaerobic conditions, in muscle, red blood cells; **(d)** when the body needs free glucose, in the liver **23.18** Step 1: transferase Step 2: isomerase Step 3: transferase Step 4: lyase Step 5: isomerase Step 6: oxidoreductase, transferase Step 7: transferase Step 8: isomerase Step 9: lyase Step 10: transferase; transferases (because many reactions involve phosphate transfers). Ligases are associated with reactions that synthesize molecules, not with reactions that break down molecules. **23.19** (g), (c), (b), (e), (f), (a), (d) **23.20** Sources of compounds for gluconeogenesis: pyruvate, lactate, citric acid cycle intermediates, many amino acids. **23.21** Germinating seeds need to synthesize carbohydrates from fats; humans obtain carbohydrates from food. **23.22** **(a)** No **(b)** Molecular oxygen appears in the last step of the electron transport chain, where it combines with H^+ and electrons (from electron transport) to form H_2O. **23.24** glucose + galactose; in the lining of the small intestine

23.26

TYPE OF FOOD MOLECULES	PRODUCTS OF DIGESTION
Proteins	Amino acids
Triacylglycerols	Glycerol and fatty acids
Sucrose	Glucose and fructose
Lactose	Glucose and galactose
Starch	Glucose

23.28 acetyl-SCoA; lactate; ethanol + CO_2 **23.30** glycogenolysis: breakdown of glycogen to form glucose glycogenesis: synthesis of glycogen from glucose **23.32** glycolysis: catabolism of glucose to pyruvate glycogenolysis: breakdown of glycogen to form glucose **23.34** ribose 5-phosphate, glycolysis intermediates **23.36** **(a)** all organs; **(b)** liver; **(c)** **(d)** muscle, liver **23.38** **(a)** pyruvate $\rightarrow$ lactate; **(b)** pyruvate $\rightarrow$ ethanol + CO_2 **23.40** **(a)** Steps 1, 3, 4, 7, 10; **(b)** Step 6; **(c)** Step 9 **23.42** Direct (substrate level) phosphorylation: **(a)** 2 mol ATP **(b)** 0 **(c)** 1 mol ATP Oxidative phosphorylation (ideal): **(a)** 6 ATP **(b)** 3 ATP **(c)** 11 ATP Most of the ATP in the citric acid cycle is produced from reduced coenzymes in the electron transport chain.

23.44

23.46 4 mol acetyl-SCoA **23.48** Hypoglycemia: low blood sugar; weakness, sweating, rapid heartbeat, confusion, coma, death; Hyperglycemia: high blood sugar; increased urine flow, low blood pressure, coma, death **23.50** ketone bodies **23.52** In Type 2 diabetes, insulin is in good supply, but cell membrane receptors fail to recognize insulin. Individuals are often overweight. **23.54** Excess glucose is converted to sorbitol, which can't be transported out of cells. This buildup changes osmolarity and causes cataracts and blindness. Excess glucose also causes neuropathy and poor circulation leading to limb amputation due to tissue death. **23.56** *Type 1 diabetes* is caused by insufficient production of insulin in the pancreas. *Type 2 diabetes* is caused by the failure of cell membrane receptors to recognize insulin. **23.58** muscle cells **23.60** The exact reverse of an energetically favorable pathway must occur by an alternate route in order to be favorable. **23.62** pyruvate, lactate **23.64** Several steps in the reverse of glycolysis are energetically unfavorable. **23.66** Steps 1, 3, 10 of glycolysis; all involve phosphate transfers and require energy. **23.68** when muscle glucose is depleted and oxygen is in short supply **23.70** glycoproteins, bacteria, dextran, polysaccharide storage granules **23.72** In an environment rich in sucrose, bacteria secrete an enzyme that transfers glucose units from digested sucrose to the dextran polymer. The residual fructose is metabolized to lactate, which lowers pH. The resulting acidic environment in the mouth dissolves minerals in teeth, leading to cavities. **23.74** beer, wine, cheese, yogurt, sour cream, and buttermilk **23.76** 140 g/dL (diabetic) vs 90 g/dL (normal) **23.78** between the curve for a diabetic and a nondiabetic **23.80** Creatine phosphate and glycogen are quickly used up. **23.82** Cotton fabric, paper, and rayon. Cotton fabric and paper are made from unmodified cellulose. In rayon, the hydroxyl groups of cellulose are converted to acetate groups. **23.84** Pyruvate is not phosphorylated. **23.86** Yes. Fructose 6-phosphate enters glycolysis as a glycolysis intermediate. **23.88** The body must avoid extreme fluctuations in glucose concentration.

Chapter 24

24.1 (a) eicosanoid **(b)** glycerophospholipid **(c)** wax

24.2

$$CH_3(CH_2)_{18}\overset{O}{\overset{\|}{C}}-OCH_2(CH_2)_{30}CH_3$$

24.3

$$CH_2O\overset{O}{\overset{\|}{C}}(CH_2)_7CH=CH(CH_2)_7CH_3$$
$$CHO\overset{O}{\overset{\|}{C}}(CH_2)_7CH=CH(CH_2)_7CH_3$$
$$CH_2O\overset{O}{\overset{\|}{C}}(CH_2)_7CH=CH(CH_2)_7CH_3$$

24.4 (a) butter **(b)** soybean oil **(c)** soybean oil   **24.5** *See below for answer.*
24.6 When two different fatty acids are bonded to C1 and C3 of glycerol, C2 is chiral.   **24.7** London forces; weak; hydrogen bonds between water molecules are stronger than London forces.
24.8 The acyl groups are from stearic acid.

$$CH_2O\overset{O}{\overset{\|}{C}}(CH_2)_7CH=CH(CH_2)_7CH_3$$
$$CHO\overset{O}{\overset{\|}{C}}(CH_2)_7CH=CH(CH_2)_7CH_3 \quad \xrightarrow{3\ H_2}$$
$$CH_2O\overset{O}{\overset{\|}{C}}(CH_2)_7CH=CH(CH_2)_7CH_3$$

$$CH_2O\overset{O}{\overset{\|}{C}}(CH_2)_{16}CH_3$$
$$CHO\overset{O}{\overset{\|}{C}}(CH_2)_{16}CH_3$$
$$CH_2O\overset{O}{\overset{\|}{C}}(CH_2)_{16}CH_3$$

24.9

$$CH_2O\overset{O}{\overset{\|}{C}}(CH_2)_{16}CH_3$$
$$CHO\overset{O}{\overset{\|}{C}}(CH_2)_{16}CH_3 \quad \xrightarrow{NaOH,\ H_2O}$$
$$CH_2O\overset{O}{\overset{\|}{C}}(CH_2)_7CH=CH(CH_2)_7CH_3$$
or the isomer

$$\begin{array}{l} CH_2OH \\ | \\ CHOH \quad + \\ | \\ CH_2OH \end{array}$$

$$2\ CH_3(CH_2)_{16}COO^-\ Na^+$$
$$CH_3(CH_2)_7CH=CH(CH_2)_7COO^-\ Na^+$$

24.10 (a) glycerol, phosphate ion, choline, $RCOO^-Na^+$, $R'COO^-Na^+$
(b) sphingosine, phosphate ion, choline, sodium palmitate

24.11

$$(CH_3)_3\overset{+}{N}CH_2CH_2O-\overset{O}{\overset{\|}{P}}-O-CH_2$$

with O^- below the P, and to the right:

Myristic acid
$$CHNH-\overset{O}{\overset{\|}{C}}(CH_2)_{12}CH_3$$
$$CHOH \quad \text{Hydrophobic tail}$$
$$CH=CH(CH_2)_{12}CH_3$$

Choline   Phosphate

Hydrophilic head   Hydrophobic tail

24.12

$$CH_2-O-\overset{O}{\overset{\|}{C}}-(CH_2)_{16}CH_3 \quad \text{Stearic acid acyl group}$$
$$CH-O-\overset{O}{\overset{\|}{C}}-(CH_2)_7CH=CH(CH_2)_7CH_3 \quad \text{Oleic acid acyl group}$$
$$CH_2-O-\overset{O}{\overset{\|}{P}}-OCH_2CH_2NH_3^+$$
$$O^-$$

Phosphate   Ethanolamine

24.13 (a), (c), (e), (f)   24.14 They must be hydrophobic, contain many amino acids with nonpolar side chains, and must be folded so that the hydrophilic regions face outward.   **24.15** yes; gasses diffuse through the cell membrane due to small size and no charge   **24.16** Glucose 6-phosphate has a charged phosphate group and can't pass through the hydrophobic lipid bilayer.   **24.17** The surfaces are in different environments and serve different functions.   **24.18** carboxylic acid (most acidic), alcohol, C—C double bonds, ethers. The molecule has both polar and nonpolar regions. Form hydrogen bonds: —COOH, —OH.
24.19 A has the highest melting point. B and C are probably liquids at room temperature due to the high percentage of unsaturated fatty acids present.   **24.20** B is 12.2% palmitic acid, 87.5% stearic acid after hydrogenation; C is 11.2% palmitic acid and 85.1% stearic acid after hydrogenation. These are very similar.
24.21

$$CH_2O\overset{O}{\overset{\|}{C}}(CH_2)_{14}CH_3$$
$$CHO\overset{O}{\overset{\|}{C}}(CH_2)_7CH=CH(CH_2)_7CH_3$$
$$CH_2O-\overset{O}{\overset{\|}{P}}-OCH_2CH_2$$
$$O^- \qquad NH_3^+$$

A glycerophospholipid

24.22 Because the membrane is fluid, it can flow together after an injury.   **24.23** C_{16} saturated fatty acids. The polar head lies in lung tissue, and

24.5

$$H_3C \overset{CH_2}{\diagdown} \overset{CH_2}{\diagdown} \overset{CH_2}{\diagup} C=C \overset{CH_2}{\diagup} C=C \overset{CH_2}{\diagup} C=C \overset{CH_2}{\diagup} C=C \overset{CH_2}{\diagdown} \overset{CH_2}{\diagup} \overset{OH}{\underset{\|}{C}} $$
with H's on the double bond carbons and O double-bonded at the end.

the hydrocarbon tails protrude into the alveoli. **24.24** a naturally-occurring molecule that dissolves in nonpolar solvents **24.26** $CH_3(CH_2)_{16}COOH$; straight chain **24.28** *Saturated fatty acids* are long-chain carboxylic acids that contain no carbon–carbon double bonds. *Monounsaturated fatty acids* contain one carbon–carbon double bond. *Polyunsaturated fatty acids* contain two or more carbon–carbon double bonds. **24.30** An essential fatty acid can't be synthesized by the human body and must be part of the diet. **24.32** The double bonds in unsaturated fatty acids make it harder for them to be arranged in a crystal. **24.34** a triester of glycerol and 3 fatty acids **24.36** Fats: composed of TAGs containing saturated and unsaturated fatty acids, solids; Oils: composed of TAGs containing mostly unsaturated fatty acids, liquids.

24.38

$$CH_2-O-\overset{\displaystyle O}{\overset{\|}{C}}-CH_2(CH_2)_9CH_3$$
$$CH-O-\overset{\displaystyle O}{\overset{\|}{C}}-CH_2(CH_2)_9CH_3$$
$$CH_2-O-\overset{\displaystyle O}{\overset{\|}{C}}-CH_2(CH_2)_9CH_3$$

24.40 a protective coating

24.42

$$CH_3(CH_2)_{13}CH_2\overset{\displaystyle O}{\overset{\|}{C}}-OCH_2(CH_2)_{14}CH_3$$

24.44

$$CH_2-O-\overset{\displaystyle O}{\overset{\|}{C}}-(CH_2)_nCH_3$$
$$CH-O-\overset{\displaystyle O}{\overset{\|}{C}}-(CH_2)_nCH=CH(CH_2)_nCH_3$$
$$CH_2-O-\overset{\displaystyle O}{\overset{\|}{P}}-OCH_2CH_2CH_2\overset{+}{N}(CH_3)_3$$
$$\underset{\displaystyle O^-}{}$$

24.46 hydrogenation **24.48** Hydrogenate some of the double bonds. **24.50** a product with cis and trans double bonds; "trans fatty acids" **24.52** glycerol, K^+ stearate, K^+ oleate, K^+ linolenate **24.54** The products have one or more of the double bonds hydrogenated. There could be up to 12 different products. **24.56** Glycerophospholipids have polar heads (point outward) and nonpolar tails that cluster to form the membrane. Triacylglycerols don't have polar heads. **24.58** sphingomyelins, glycolipids **24.60** Glycerophospholipids are components of cell membranes. Stored fats in the body are triacylglycerols. **24.62** Both liposomes and micelles are spherical clusters of lipids. A liposome resembles a spherical lipid bilayer, in which polar heads cluster both inside and outside of the sphere. A micelle has a single layer of lipid molecules. **24.64** Concentrations of all substances would be the same on both sides of the membrane.

24.66

$$CH_2-O-\overset{\displaystyle O}{\overset{\|}{P}}-OCH_2CH_2\overset{+}{N}(CH_3)_3$$
$$\underset{\displaystyle O^-}{}$$
$$CH-NH-\overset{\displaystyle O}{\overset{\|}{C}}-(CH_2)_{16}CH_3$$
$$CHOH$$
$$CH=CH(CH_2)_{12}CH_3$$

24.68 3 glycerols, $RCOO^-\ Na^+$, $R'COO^-\ Na^+$, $R'COO^-\ Na^+$ $R'COO^-\ Na^+$, 2 phosphates **24.70** Facilitated diffusion requires carrier proteins. **24.72 (a)** facilitated diffusion **(b)** simple diffusion **(c)** active transport **24.74** A prostaglandin that stimulates uterine contractions.

24.76 an eicosanoid; Arachidonic acid is a precursor. **24.78** leukotrienes **24.80** prostaglandins **24.82** no more than 30% **24.84** fabric softeners, disinfecting soaps **24.86** cholesterol **24.88** glycolipids **24.90 (b) (c) (e) (f)**

24.92

$$CH_2-O-\overset{\displaystyle O}{\overset{\|}{C}}-(CH_2)_{12}CH_3$$
$$CH-O-\overset{\displaystyle O}{\overset{\|}{C}}-(CH_2)_7CH=CHCH_2CH=CHCH_2CH_3$$
$$CH_2-O-\overset{\displaystyle O}{\overset{\|}{C}}-(CH_2)_{12}CH_3$$

or

$$CH_2-O-\overset{\displaystyle O}{\overset{\|}{C}}-(CH_2)_{12}CH_3$$
$$CH-O-\overset{\displaystyle O}{\overset{\|}{C}}-(CH_2)_{12}CH_3$$
$$CH_2-O-\overset{\displaystyle O}{\overset{\|}{C}}-(CH_2)_7CH=CHCH_2CH=CHCH_2CH_3$$

24.94 (a) beef fat **(b)** plant oil **(c)** pork fat **24.96** It is saponifiable. **24.98** sphingomyelins, cerebrosides, gangliosides **24.100** lower blood pressure, assist in blood clotting, stimulate uterine contractions, lower gastric secretions, cause swelling **24.102** 0.4g NaOH

Chapter 25

25.1 Cholate has 4 polar groups on its hydrophilic side that allow it to interact with an aqueous environment; its hydrophobic side interacts with TAGs. Cholate and cholesterol can't change roles. **25.2** Dihydroxyacetone phosphate is isomerized to glyceraldehyde 3-phosphate, which enters glycolysis. **25.3 (a)**, **(b)** *Step 1*; a $C=C$ double bond is introduced; FAD is the oxidizing agent. *Step 3*; an alcohol is oxidized to a ketone; NAD^+ is the oxidizing agent. **(c)** *Step 2*; water is added to a carbon-carbon double bond. **(d)** *Step 4*; HSCoA displaces acetyl-SCoA, producing a chain-shortened acyl-SCoA fatty acid **25.4 (a)** 6 acetyl-SCoA, 5 β oxidations **(b)** 7 acetyl-SCoA, 6 β oxidations **25.5** Step 6, Step 7, Step 8 **25.6 (d)** **25.7 (a)** Acetyl-SCoA provides the acetyl groups used in synthesis of ketone bodies. **(b)** 3 **(c)** The body uses ketone bodies as an energy source during starvation. **25.8** Oxygen is needed to reoxidize reduced coenzymes, formed in β oxidation, that enter the electron transport chain. **25.9 (a)** chylomicrons; because they have the greatest ratio of lipid to protein **(b)** chylomicrons **(c)** HDL **(d)** LDL **(e)** HDL **(f)** VLDL; used for storage or energy production **(g)** LDL **25.10** high blood glucose → high insulin/low glucagon → fatty acid and TAG synthesis: low blood glucose → low insulin/high glucagon → TAG hydrolysis; fatty acid oxidation **25.11** Formation of a fatty acyl-SCoA is coupled with conversion of ATP to AMP and pyrophosphate. This energy expenditure is recaptured in β oxidation. **25.12** Less acetyl-SCoA can be catabolized in the citric acid cycle, and acetyl-SCoA is diverted to ketogenesis. **25.13** Catabolism of fat provides more calories/gram than does catabolism of glycogen, and, thus, fats are a more efficient way to store calories. **25.14** Ketone bodies can be metabolized to form acetyl-SCoA, which provides energy. **25.15** No. Although both these processes add or remove two carbon units, one is not the reverse of the other. The two processes involve different enzymes, coenzymes, and activation steps. **25.16** They slow the rate of movement of food through the stomach. **25.18** Bile emulsifies lipid droplets. **25.20** mono- and diacylglycerols, stearic acid, oleic acid, linoleic acid, glycerol **25.22** Acylglycerols, fatty acids, and protein are combined to form *chylomicrons*, which are lipoproteins used to transport lipids from the diet into the bloodstream. **25.24** by albumins **25.26** Steps 6–10 of the glycolysis pathway. **25.28** 9 molecules ATP;

21 molecules ATP **25.30** An adipocyte is a cell, almost entirely filled with fat globules, in which TAGs are stored and mobilized. **25.32** heart, liver, resting muscle cells **25.34** A fatty acid is converted to its fatty acyl-SCoA in order to activate it for catabolism. **25.36** The carbon β to the thioester group (two carbons away from the thioester group) is oxidized in the process. **25.38** 17 ATP **25.40** *Least* glucose, sucrose, capric acid, myristic acid *Most*
25.42

(a)

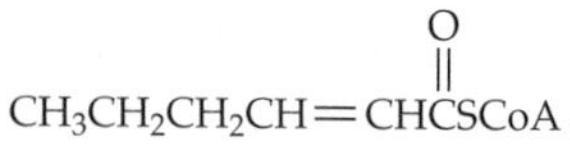

$$CH_3CH_2CH_2CH=CHCSCoA$$ (with C=O above CSCoA)

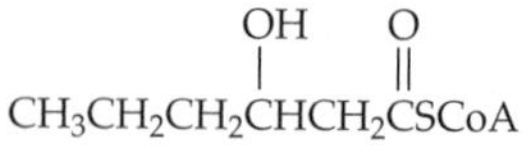

(b)

$$CH_3CH_2CH_2CHCH_2CSCoA$$ (with OH and C=O groups)

(c)

$$CH_3CH_2CH_2CCH_2CSCoA$$ (with two C=O groups)

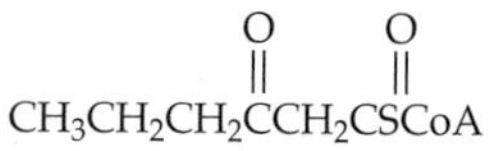

(d)

$$CH_3CH_2CH_2CSCoA \;+\; CH_3CSCoA$$ (with C=O groups)

25.44 (a) 7 acetyl-SCoA, 6 cycles **(b)** 4 acetyl-SCoA, 3 cycles **25.46** lipogenesis **25.48** acetyl-SCoA **25.50** 7 cycles **25.52** Total cholesterol: 200 mg/dL or lower. LDL: 160 mg/dL or lower. HDL: 60 mg/dL or higher. **25.54** LDL carries cholesterol from the liver to tissues; HDL carries cholesterol from tissues to the liver, where it is converted to bile and excreted. **25.56** Type II diabetes, colon cancer, heart attacks, stroke **25.58** calorie-dense food, lack of exercise **25.60** The liver synthesizes many important biomolecules, it catabolizes glucose, fatty acids and amino acids, it stores many substances, and it inactivates toxic substances. **25.62** The excess acetyl-SCoA from catabolism of carbohydrates is stored as fat. The body can't resynthesize carbohydrate from acetyl-SCoA. **25.64** The alcohol intermediate is chiral. **25.66** Ketosis is a condition in which ketone bodies accumulate in the blood faster than they can be metabolized. Since two of the ketone bodies are carboxylic acids, they lower the pH of the blood, producing the condition known as ketoacidosis. Symptoms of ketoacidosis include dehydration, labored breathing, and depression; prolonged ketoacidosis may lead to coma and death. **25.68** Ketones have little effect on pH, but the two other ketone bodies are acidic, and they lower the pH of urine. **25.70 (a)** endogenous **(b)** exogenous **25.72** $H_2C=CHC(CH_3)=CH_2$. Since cholesterol has 27 carbons, at least 6 2-methyl-1,3-butadiene molecules are needed.

Chapter 26
26.1

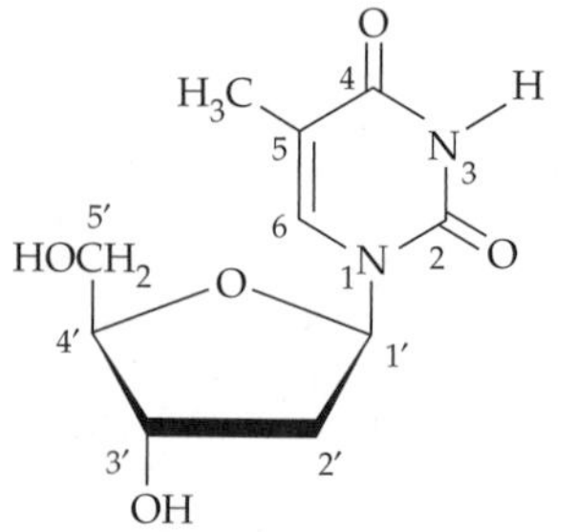

2'-Deoxythymidine

26.2 D-Ribose ($C_5H_{10}O_5$) has one more oxygen atom than 2-deoxy-D-ribose ($C_5H_{10}O_4$), and thus can form more hydrogen bonds.
26.3

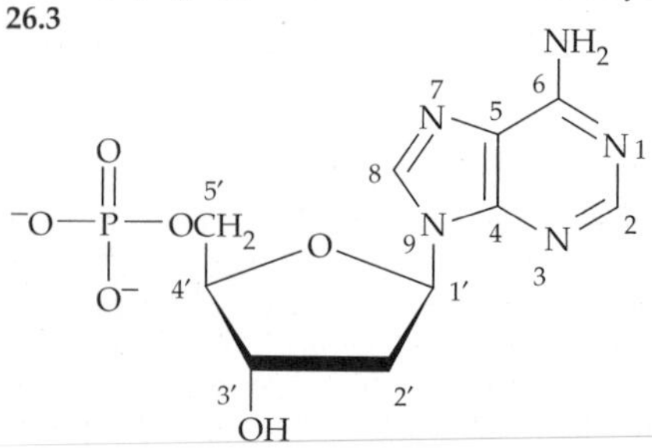

2'-Deoxyadenosine 5'-monophosphate

26.4

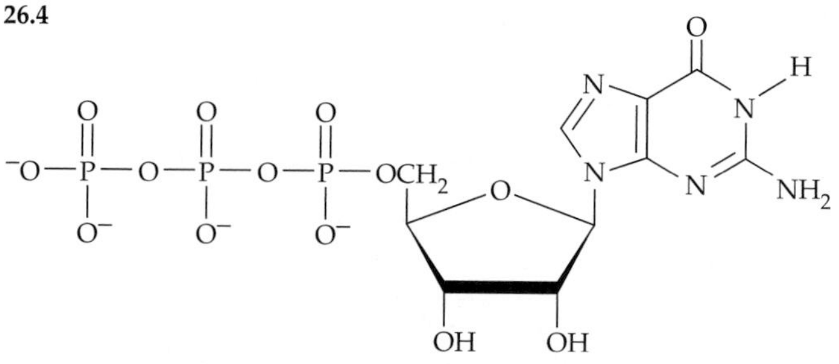

Guanosine 5'-triphosphate (GTP)

26.5 dCMP—2'-Deoxycytidine 5'-monophosphate; CMP—Cytidine 5'-monophosphate; UDP—Uridine 5'-diphosphate; AMP—Adenosine 5'-monophosphate; ATP—Adenosine 5'-triphosphate **26.6** cytosine–guanine–adenine–uracil–adenine. The pentanucleotide comes from RNA because uracil is present.
26.7

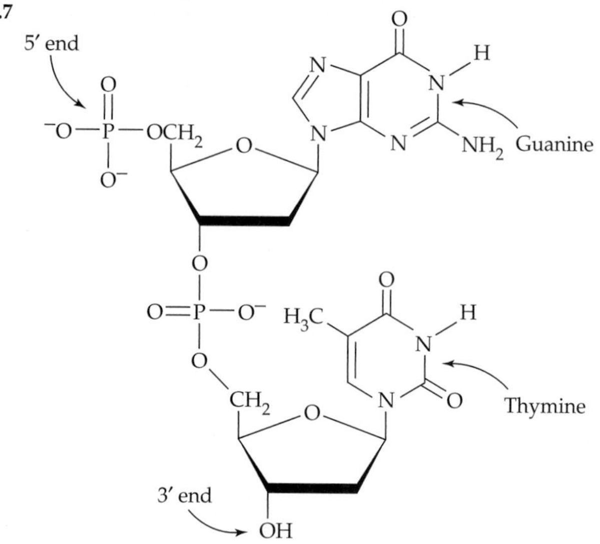

26.8 (a) 3′ C-G-G-A-T-C-A 5′ **(b)** 3′ T-T-A-C-C-G-A-G-T 5′
26.9

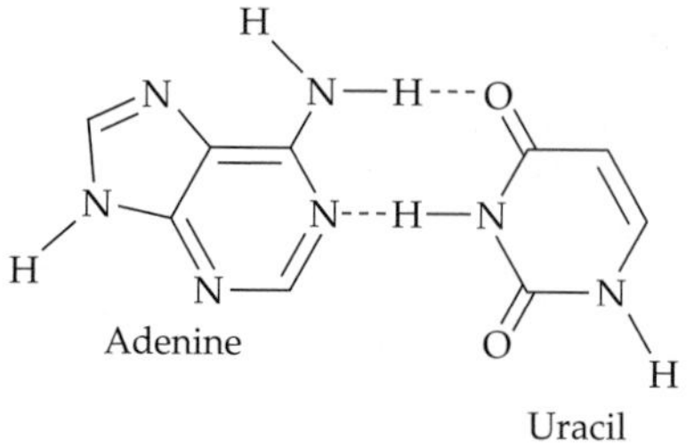

26.10 negatively charged (because of the phosphate groups)
26.11 (a) A longer strand has more hydrogen bonds. **(b)** A chain with a higher percent of G/C pairs has a higher melting point, because it has more hydrogen bonds. **26.12 (a)** 3′ G-U-C-U-G-A-C-A-U-G-U-G 5′ **(b)** 5′ A-U-C-A-U-A-C-G-U-C-G-C 3′ **26.13 (a)** GCU GCC GCA GCG **(b)** CCU CCC CCA CCG **(c)** UCU UCC UCA UCG AGU AGC **(d)** AAA AAG **(e)** UAU UAC **26.14** The sequence guanine-uracil-guanine codes for valine. **26.15 (a)** Ile **(b)** Ala **(c)** Arg **(d)** Lys **26.16** Six mRNA triplets can code for Leu: CUU, CUC, CUA, CUG, UUA, and UUG Among the many possible combinations for Leu-Leu-Leu are:

5′ UUAUUGCUU 3′   5′ UUAUUGCUC 3′   5′ UUAUUGCUA 3′
5′ UUAUUGCUG 3′   5′ UUACUUCUC 3′   5′ UUACUUCUA 3′

26.17–26.18

mRNA sequence:	5′ CAG—AUG—CCU—UGG—CCC—UUA 3′				
Amino-acid sequence:	Gln—Met—Pro—Trp—Pro—Leu				
tRNA anticodons:	3′ GUC—UAC—GGA—ACC—GGG—AAU 5′				

26.19

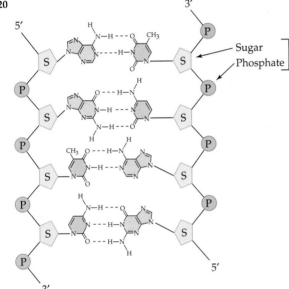

26.20

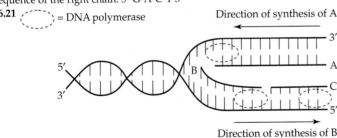

Sequence of the left chain: 5′ A-G-T-C 3′
Sequence of the right chain: 5′ G-A-C-T 3′

26.21 ⟨ ⟩ = DNA polymerase

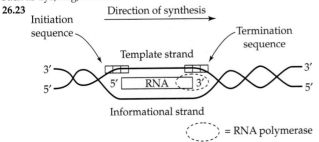

Direction of synthesis of A
Direction of synthesis of B

Segments B and C are joined by the action of a DNA ligase.
26.22 The sugar-phosphate backbone is found on the outside of the DNA double helix. Histones are positively charged; they contain groups such as Lys, Arg, and His.

26.23

Direction of synthesis
Initiation sequence
Termination sequence
Template strand
RNA
Informational strand
⟨ ⟩ = RNA polymerase

26.24 More than one codon can code for each amino acid. Only one possibility is shown.

(a) 5′ | C | A | A | C | A | C | C | C | C | G | G | G | 3′ mRNA

(b) 3′ | G | T | T | G | T | G | G | G | G | C | C | C | 5′ DNA template strand

(c) 5′ | C | A | A | C | A | C | C | C | C | G | G | G | 3′ DNA informational strand

(d) 64 possible sequences

26.26 2-deoxyribose (DNA); ribose (RNA). 2-Deoxyribose is missing an —OH group at C2. **26.28** The purine bases (two fused heterocyclic rings) are adenine and guanine. The pyrimidine bases (one heterocyclic ring) are cytosine, thymine (in DNA), and uracil (in RNA). **26.30** DNA is largest; tRNA is smallest. **26.32** *Similarities*: All are polymerizations; all use a nucleic acid as a template; all use hydrogen bonding to bring the subunits into position. *Differences*: In replication, DNA makes a copy of itself. In transcription, DNA is used as a template for the synthesis of mRNA. In translation, mRNA is used as a template for the synthesis of proteins. Replication and transcription take place in the nucleus of cells, and translation takes place in ribosomes. **26.34** DNA, protein **26.36** 46 chromosomes (23 pairs) **26.38** They always occur in pairs: they always H-bond with each other. **26.40** 19% G, 19% C, 31% A, 31% T. (%G = %C; %A = %T: %T + %A + %C + %G = 100%) **26.42** 5′ to 3′

26.44

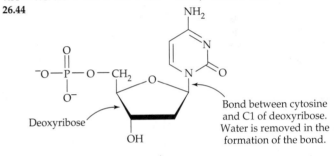

Bond between cytosine and C1 of deoxyribose. Water is removed in the formation of the bond.
Deoxyribose

26.46

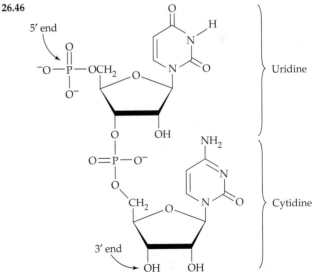

5′ end
Uridine
Cytidine
3′ end

26.48 the template strand **26.50** Each new DNA has one newly synthesized strand and one old strand. **26.52** An anticodon is a 3 nucleotide tRNA sequence that is complementary to an mRNA codon for a specific amino acid. **26.54** tRNAs for each amino acid differ in their anticodon sequences. **26.56 (a)** GCU GCC GCA GCG **(b)** CAU CAC **(c)** CCU CCC CCA CCG **26.58** (3′ → 5′) **(a)** GGA **(b)** CGU **(c)** UAA **26.60** (3′ → 5′) TTGCCT **26.62** Asn-Gly **26.64** (5′ → 3′) UAU–GGU–GGU–UUU–AUG–UAA Other sequences are possible. **26.66** Viruses consist of a strand of nucleic acid wrapped in a protein coat; viruses can't replicate or manufacture protein independent of a host cell. **26.68** To be effective, a drug must be powerful enough to act on viruses within cells without damaging the cells and their genetic material. Vaccines prime the body to recognize viruses as foreign and to destroy them. **26.70** Ribozyme activity is common among the simplest and most primitive life forms, such as viroids, leading scientists to speculate that ribozyme catalysis might have preceded enzyme catalysis. **26.72** Influenza A viruses are described by a code that describes the hemagglutinins (H) and the neuraminidases (N) in the virus. The H1N1 virus was responsible for the 1918 influenza pandemic, and the H5N1 virus is present in avian flu. Since these viruses can undergo antigenic drift in host animals, there is concern when infected birds and animals harbor influenza viruses. **26.74** 249 bases **26.76** Met is removed after synthesis is complete.

Chapter 27

27.1 "once upon a time" **27.2** As a result of the SNP, the base sequence codes for Trp, instead of Cys. This change would probably affect the functioning of the protein. **27.3** $3'-T-C-T-A-G-//-A-5'$ **27.4 (a)** sticky **(b) (c)** not sticky **27.5 (a)** comparative genomics **(b)** genetic engineering **(c)** pharmacogenetics **(d)** bioinformatics **27.6** (1) A genetic map, which shows the location of markers one million nucleotides apart, is created. (2) Next comes a physical map, which refines the distance between markers to 100,000 base pairs. (3) The chromosome is cleaved into large segments of overlapping clones. (4) The clones are fragmented into 500 base pieces, which are sequenced. **27.7** The variations are only a small part of the genome; the rest is identical among humans. A diverse group of individuals contributed DNA to the project. **27.8** *telomeres* (protect the chromosome from damage, involved with aging), *centromeres* (involved with cell division), *promoter sequences* (determine which genes will be replicated), *introns* (function unknown) **27.9** Similarities: both are variations in base sequences. Differences: A mutation is an error that is transferred during replication and affects only a few people; a polymorphism is a variation in sequence that is common within a population. **27.10** Recombinant DNA contains two or more DNA segments that do not occur together in nature. The DNA that codes for a specific human protein can be incorporated into a bacterial plasmid using recombinant DNA technology. The plasmid is then reinserted into a bacterial cell, where its protein-synthesizing machinery makes the desired protein. **27.11** Major benefits of genomics: creation of disease-resistant and nutrient-rich crops, gene therapy, genetic screening. Major negative outcomes: misuse of an individual's genetic information, prediction of a genetic disease for which there is no cure. **27.12** Celera broke the genome into many unidentified fragments. The fragments were multiplied and cut into 500 base pieces, which were sequenced. A supercomputer was used to determine the order of the bases. **27.14** 50% **27.16 (a)** Approx. 200 genes are shared between bacteria and humans. **(b)** A single gene may produce several proteins. **27.18** The clones used in DNA mapping are identical copies of DNA segments from a single individual. In mapping, it is essential to have a sample large enough for experimental manipulation. **27.20** The youngest cells have long telomeres, and the oldest cells have short telomeres. **27.22** It is the constriction that determines the shape of a chromosome during cell division. **27.24** A silent mutation is a single base change that specifies the same amino acid. **27.26** random and spontaneous events, exposure to a mutagen **27.28** A SNP can result in the change in identity of an amino acid inserted into a protein a particular location in a polypeptide chain. The effect of a SNP depends on the function of the protein and the nature of the SNP. **27.30** A physician could predict the age at which inherited diseases might become active, their severity, and the response to various types of treatment. **27.32** a change in the type of side chain **27.34 (a)** Substitution of Ala for Val may have minor effects **(b)** Substitution of His for Pro is more serious because the amino acids have very different side chains. **27.36** Proteins can be produced in large quantities. **27.38** Sticky ends are unpaired bases at the end of a DNA fragment. Recombinant DNA is formed when the sticky ends of the DNA of interest and of the DNA of the plasmid have complementary base pairs and can be joined by a DNA ligase. **27.40 (a) (b)** not sticky **27.42** Proteomics, the study of the complete set

of proteins coded for by a genome or synthesized by a given type of cell, might provide information about the role of a protein in both healthy and diseased cells. **27.44** corn, soybeans **27.46** a DNA chip **27.48** a group of anonymous individuals **27.50** production of a large quantity of a specific segment of DNA **27.52** (1) digestion with a restriction endonuclease (2) separation of fragments by electrophoresis (3) fragments transferred to a nylon membrane (4) treatment of the blot with a radioactive DNA probe (5) identification of fragments by exposure to X-ray film. All samples must be analyzed under the same conditions in order to be compared. **27.54** A monogenic disease is caused by the variation in just one gene. **27.56** ATACTGA **27.58** A hereditary disease is caused by a mutation in the DNA of a germ cell and is passed from parent to offspring. The mutation affects the amino-acid sequence of an important protein and causes a change in the biological activity of the protein.

Chapter 28

28.1 (a) false **(b)** true **(c)** true **(d)** false **(e)** false **28.2** oxidoreductase; lyase

28.3

$$
\underset{\text{4-Hydroxy-}\alpha\text{-ketopentanoate}}{CH_3\overset{OH}{\underset{|}{CH}}CH_2\overset{O}{\underset{||}{C}}COO^-}
$$

28.4

$$
CH_3SCH_2CH_2\overset{O}{\underset{||}{C}}HCOO^-
$$

28.5 by the loss of two hydrogens to either NAD^+ or $NADP^+$
28.6 valine, leucine, isoleucine

$$
\underset{\text{Valine}}{CH_3\overset{H_3C}{\underset{|}{CH}}\overset{NH_3^+}{\underset{|}{CH}}COO^-} + \underset{\alpha\text{-Ketoglutarate}}{^-OOCCH_2CH_2\overset{O}{\underset{||}{C}}COO^-}
$$

$$
\downarrow
$$

$$
\underset{\alpha\text{-Keto-3-methylbutanoate}}{CH_3\overset{H_3C}{\underset{|}{CH}}\overset{O}{\underset{||}{C}}COO^-} + \underset{\text{Glutamate}}{^-OOCCH_2CH_2\overset{NH_3^+}{\underset{|}{CH}}COO^-}
$$

28.7 (a) 5 **(b)** 1 **(c)** 3

28.8

28.9 3-Phosphoglycerate → 3-Phosphohydroxypyruvate (oxidation)
3-Phosphohydroxypyruvate → 3-Phosphoserine (transamination)
3-Phosphoserine → Serine (hydrolysis)
28.10

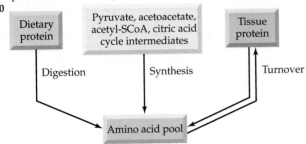

28.11 (1) Catabolism of an amino acid begins with a transamination reaction that removes the amino nitrogen (2) The resulting α-keto acid, which contains the carbon atoms, is converted to a common metabolic intermediate. (3) The amino group of glutamate (from the amino acid) is removed by oxidative deamination. (4) The amino nitrogen is transformed to urea in the urea cycle and is excreted. **28.12** glutamate dehydrogenase; alanine aminotransferase. Alanine is the product. **28.13** The carbon atoms from ketogenic amino acids can be converted to ketone bodies or to acetyl-SCoA. The carbon atoms from glucogenic amino acids can be converted to compounds that can enter gluconeogenesis and can form glucose, which can enter glycolysis and also yield acetyl-SCoA. **28.14** All amino acids are necessary for protein synthesis. The body can synthesize only some of them; the others must be provided by food and are thus essential in the diet. **28.15** to quickly remove ammonia from the body; buildup of urea and shortage of ornithine **28.16** throughout the body **28.18** pyruvate, 3-phosphoglycerate **28.20** In transamination, a keto group of an α-keto acid and an amino group of an amino acid change places.
28.22 (a)   **(b)**

$$N\!\!\diagdown\!\!-CH_2-\overset{O}{\overset{\|}{C}}-COO^- \qquad HSCH_2-\overset{O}{\overset{\|}{C}}-COO^-$$

28.24 An —NH_3^+ group of an amino acid is replaced by a carbonyl group, and ammonium ion is eliminated.
28.26
(a)   **(b)**

$$\bigcirc\!\!-CH_2-\overset{O}{\overset{\|}{C}}-COO^- \qquad HO-\bigcirc\!\!-CH_2-\overset{O}{\overset{\|}{C}}-COO^-$$

28.28 A ketogenic amino acid is catabolized to acetoacetyl-SCoA or acetyl-SCoA. Examples: leucine, isoleucine, lysine **28.30** Ammonia is toxic. **28.32** One nitrogen comes from carbamoyl phosphate, which is synthesized from ammonium ion by oxidative deamination. The other nitrogen comes from aspartate. **28.34** Nonessential amino acids are synthesized in humans in 1–3 steps. Essential amino acids are synthesized in microorganisms in 7–10 steps. **28.36** reductive amination; the reverse of oxidative deamination **28.38** phenylketonuria; mental retardation; restriction of phenylalanine in the diet **28.40** (b) (c) (d) **28.42** Oxidized allopurinol inhibits the enzyme that converts xanthine to uric acid. The more soluble intermediates are excreted. The nitrogen at position 7 of hypoxanthine is at position 8 in allopurinol, where it blocks oxidation of xanthine. **28.44** tryptophan; emotional and behavioral problems **28.46** isoleucine + pyruvate → α-keto-3-methylpentanoate + alanine **28.48** Yes. Some amino acids yield two kinds of products—those that can enter the citric acid cycle and those that are intermediates of fatty acid metabolism. **28.50** Tissue is dynamic because its components are constantly being broken down and reformed.

28.52 (b) → (e) → (d) → (f) → (a) → (c) **28.54** An excess of one amino acid might overwhelm a transport system that other amino acids use, resulting in a deficiency of those amino acids. **28.56** The nitrogen may be converted to urea and excreted in the urine, or it may be used in the synthesis of a new nitrogen-containing compound.

Chapter 29
29.1 In the cell: the charged form. Outside the cell: the uncharged form. The uncharged form enters the cell more readily. **29.2 (a)** iii **(b)** ii **(c)** iv **(d)** v **(e)** i **29.3 (a)** pH goes down **(b)** [O_2], [CO_2], pH **29.4 (a)** respiratory acidosis **(b)** metabolic acidosis **(c)** metabolic alkalosis **(d)** respiratory alkalosis **(e)** respiratory acidosis **29.5 (a)** intracellular fluid **(b)** extracellular fluid **(c)** blood plasma, interstitial fluid **(d)** K^+, Mg^{2+}, HPO_4^{2-} **(e)** Na^+, Cl^-
29.6

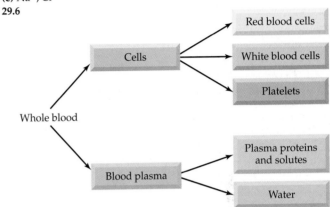

29.7 (a) O_2 **(b)** CO_2 **(c)** nutrients **(d)** waste products **(e)** hormones **(f)** white blood cells, platelets **29.8** swelling, redness, warmth, pain **29.9** enzymatic decarboxylation of histidine; Histamine dilates capillaries, increasing blood flow that reddens and warms the skin. Blood-clotting factors and defensive proteins cause pain and swelling. **29.10** *Cell-mediated immune response*: under control of T cells; arises when abnormal cells, bacteria, or viruses enter cells; invaders killed by T cells. *Antibody-mediated immune response*: under control of B cells, assisted by T cells; occurs when antigens enter cells; B cells divide to produce plasma cells, which form antibodies; an antibody-antigen complex inactivates the antigen. **29.11** Excess hydrogen ions are excreted by reaction with NH_3 or HPO_4^{2-}. H^+ ions also combine with bicarbonate, producing CO_2 that returns to the bloodstream. **29.12** intracellular fluid (64%), interstitial fluid (25%), plasma (8%) **29.14** Substances not soluble in blood, such as lipids, are transported by blood proteins. **29.16** Blood pressure in arterial capillaries is higher than interstitial fluid pressure, and blood pressure in venous capillaries is lower than interstitial fluid pressure. **29.18** the thoracic duct **29.20** it causes a decrease in the water content of the urine **29.22** seven percent **29.24** *Red blood cells* transport blood gases. *White blood cells* protect the body from foreign substances. *Platelets* assist in blood clotting. **29.26** *Inside cells:* K^+, Mg^{2+}, HPO_4^{2-}; *Outside cells:* Na^+, Cl^- **29.28** Antihistamines block attachment of the neurotransmitter histamine to its receptors. **29.30** immunoglobulins **29.32** Killer T cells destroy the invader; helper T cells enhance defenses; memory T cells can produce new killer T cells if needed. **29.34** Memory cells "remember" an antigen and are capable of producing antibodies to it for a long time. **29.36** Vitamin K, Ca^{2+} **29.38** They are released as zymogens in order to avoid undesirable clotting in noninjured tissues. **29.40** +2 **29.42** If pO_2 is below 10 mmHg, hemoglobin is unsaturated. If pO_2 is greater than 100 mmHg, hemoglobin is completely saturated. Between these pressures, hemoglobin is partially saturated. **29.44** a dissolved gas, bound to hemoglobin, bicarbonate ion
29.46

$$CO_2 + H_2O \overset{\text{Carbonic anhydrase}}{\rightleftharpoons} HCO_3^- + H^+$$

29.48 7.35–7.45 below 7.35 = acidosis; above 7.35 = alkalosis
29.50 respiratory acidosis
29.52

$$H^+ + HCO_3^- \rightleftharpoons CO_2 + H_2O$$

$$H^+ + HPO_4^{2-} \rightleftharpoons H_2PO_4^-$$

29.54 Substances can either be transported into a cell or be transported out of a cell, but not both.   **29.56** The chemist might need to deliver a medication to brain cells of someone suffering from Parkinson's disease.

29.58 Automated analysis can reproducibly detect changes in enzyme levels that might indicate organ damage.   **29.60** It is a small, polar molecule.   **29.62** When $[Na^+]$ is high, secretion of ADH increases and causes the amount of water retained by the body to increase, causing swelling.   **29.64** Active transport is necessary when a cell needs a substance that has a higher concentration inside the cell than outside, or when a cell needs to secrete a substance that has a higher concentration outside the cell than inside.   **29.66** Hemostasis is the body's mechanism for preventing blood loss, and might be considered to be a part of homeostasis.

Photo Credits

Frontmatter: vii, Shutterstock; **viii,** Escher, M.C. (1898-1972), The M. C. Escher Company; **ix,** Corbis; **x,** iStockphoto.com; **xi,** Phil Degginger/Alamy; **xii,** Luca Bruno/AP; **xiii,** Shutterstock; **xiv,** Andrew Brookes/CORBIS.

Chapter 1: Opener, Bjanka Kadic/Alamy; **4** Richard Megna/Fundamental Photographs, NYC; **Fig. 1.2,** Eric Schrader/Pearson; **6,** Shutterstock; **7,** Richard Megna/Fundamental Photographs, NYC; **8,** iStockphoto; **9 (top),** Richard Megna/Fundamental Photographs, NYC; **9 (bottom),** Richard Megna/Fundamental Photographs, NYC; **12(a) (top left),** iStockphoto; **12(b) (top middle),** Wikipedia.org; **12(c) (top right),** Shutterstock; **12(a) (bottom left),** Shutterstock; **12(b) (bottom middle),** iStockphoto.com; **12(c) (bottom right),** Wikipedia.org; **13 (bottom left),** Lester V. Bergman/Corbis/Bettmann; **13 (bottom right),** Photo Courtesty of Texas Instruments Incorporated; **14,** Wikipedia.org.

Chapter 2: Opener, Pearson Asset Library/Pearson; **20,** CDC/C. Goldsmith, P. Feorino, E.L. Palmer, W.R. McManus; **Fig. 2.1(a),** Ohaus Corporation; **Fig 2.1(b),** Ohaus Corporation; **22 (bottom),** McCracken Photographers/Pearson Education/PH College; **24,** McCracken Photographers/Pearson Education/PH College; **25,** Corbis Royalty Free; **27,** Pearson Asset Library/Eric Schrader; **28(a) (left),** CDC/Dr. Thomas F. Sellers/Emory University; **28(b) (middle),** Eye of Science/Photo Researchers, Inc.; **28(c) (right),** Pearson Asset Library/Pearson; **29,** Frank LaBua/Pearson Education/PH College; **31,** Comstock Complete; **33,** Dennis Kunkel Phototake, NYC; **34,** Michal Heron/Pearson Education/PH College; **35,** iStockphoto.com; **36,** stockbyte/Photolibrary; **38,** Pearson Asset Library/Pearson; **39,** Richard Megna/Fundamental Photographs, NYC; **40,** Shutterstock; **41,** iStockphoto.com; **43,** Michael Wright; **44,** Eric Schrader/Pearson.

Chapter 3: Opener, Shutterstock; **49,** Paul Silverman/Fundamental Photographs, NYC; **50 (top),** Shutterstock; **50 (bottom),** AP Wide World Photos; **52,** Pearson Asset Library/Pearson; **57,** Richard Megna/Fundamental Photographs, NYC; **58,** Richard Megna/Fundamental Photographs, NYC; **60 (top),** Phillipe Plaily/Photo Researcher, Inc.; **60 (middle),** Pearson Asset Library/Pearson; **60 (bottom),** iStockphoto.com; **61,** NASA; **62,** iStockphoto.com; **73,** iStockphoto.com.

Chapter 4: Opener, Shutterstock; **79,** Richard Megna/Fundamental Photographs, NYC; **Fig. 4.2 (left and right),** Pearson Asset Library/Pearson; **Fig. 4.3 (bottom),** Ed Degginer/Color-Pic, Inc.; **84,** Richard Megna/Fundamental Photographs, NYC; **85(a) (left),** Chip Clark; **85(b) (middle)** and **85(c) (right),** Jeffrey A. Scovil; **89,** iStockphoto.com; **91,** Jeffrey A. Scovil; **97,** CNRI/Science Photo Library/Photo Researchers Inc.; **100,** Prof. P. Motta, Department of Anatomy, University "La Sapienza," Rome/Science Photo Library/Photo Researchers, Inc.

Chapter 5: Opener, The M.C. Escher Company BV; **108,** Comstock Complete; **114,** Comstock Complete; **116,** iStockphoto.com; **117,** Farmland Industries, Inc.; **124,** Shutterstock; **127,** Courtesy of DuPont Nomex©; **135,** Shutterstock.

Chapter 6: Opener, Charles O'Rear/CORBIS/CORBIS-NY; **147,** David R. Frazier/Photo Researchers, Inc.; **148,** Richard Megna/Fundamental Photographs, NYC; **Fig. 6.1(a),** Phil Degginger/Color-Pic, Inc.; **149,** Tom Bochsler/Pearson Education/PH College; **153 (left),** Library of Congress; **153 (right),** Science Photo Library/Photo Researchers, Inc.; **158,** Richard Megna/Fundamental Photographs, NYC; **162 (top),** iStockphoto.com; **162 (bottom),** Richard Megna/Fundamental Photographs, NYC; **164,** Wikipedia.org; **167,** McCracken Photographers/Pearson Education/PH College; **168,** McCracken Photographers/Pearson Education/PH College; **171 (top),** iStockphoto.com; **171 (bottom),** Tony Freeman/PhotoEdit Inc.; **172,** Richard Megna/Fundamental Photographs, NYC.

Chapter 7: Opener, Getty Images, Inc.; **185,** Pearson Asset Library/Pearson; **186,** Shutterstock; **188,** Getty Images, Inc.—PhotoDisc; **189,** iStockphoto.com; **Fig. 7.1,** Tom Bochsler/Pearson Education/PH College; **193 (bottom),** AC/General Motors/Peter Arnold, Inc.; **198,** Reuters; **199,** IndexOpen; **201,** Richard Megna/Fundamental Photographs, NYC.

Chapter 8: Opener, Don King/Pacific Stock; **Fig. 8.3,** NASA Headquarters; **223,** iStockphoto.com; **224,** iStockphoto.com; **226,** iStockphoto.com; **227,** Pearson Asset Library/Pearson; **234,** ©John Van Hesselt/CORBIS All Rights Reserved; **241 (top),** Richard Megna/Fundamental Photographs, NYC; **241 (bottom),** iStockphoto.com; **242 (top),** iStockphoto.com; **Fig. 8.21 (middle),** Pearson Asset Library/Pearson; **242 (bottom),** AGE Fotostock America, Inc.; **243 (bottom left),** Jeffrey A. Scovil; **243 (bottom right),** iStickphoto.com; **244,** Shutterstock; **246,** Pearson Asset Library/Pearson; **248,** Shutterstock.

Chapter 9: Opener, Phil Schermeiser/National Geographic Imag Collection; **256 (left and middle),** iStockphoto.com; **256 (right),** Shutterstock; **258 (top),** Shutterstock; **258 (bottom)** Pearson Asset Library/Pearson; **259,** Jonathan Blair/Corbis/Bettmann; **262,** Richard Megna/Fundamental Photographs, NYC; **263,** Shutterstock; **Fig. 9.6,** Richard Megna/Fundamental Photographs, NYC; **266,** AP Wide World Photos; **273,** Pearson Asset Library/Tony Freeman/Photo Edit; **Fig.9.8,** Richard Megna/Fundamental Photographs, NYC; **279,** Gero Breloe/NewsCom; **280,** NewsCom; **Fig. 9.12 (a),** Pearson Asset Library/Dennis Kunkel/Phototake, **(b)** and **(c)** Dennis Kunkel/Phototake; **286 (left),** Martin Dohrn/Science Photo Library/Photo Researchers, Inc.; **286 (right),** Amet Jean Pierre/Corbis/Sygma.

Chapter 10: Opener, iStockphoto.com; **294,** Richard Megna/Fundamental Photographs, NYC; **295,** Eric Schrader/Pearson; **302,** ISM/PhototakeUSA.com; **306,** iStockphoto.com; **Fig. 10.4 (left),** Richard Megna/Fundamental Photographs, NYC; **Fig. 10.4 (right),** Tom Bochsler/Pearson Education/PH College; **Fig. 10.5,** Tom Bochsler/Pearson Education/PH College; **320,** Pearson Asset Library/Pearson; **321,** iStockphoto.com; **Fig. 10.8,** Ed Degginger/Color-Pic, Inc.; **324 (top and bottom),** National Atmospheric Deposition Program/National Trends Network.

Chapter 11: Opener, Corbis/Reuters America LLC; **342 (top),** Simon Fraser/Medical Physics, Royal Victoria Infirmar, Newcastle-upon-Tyne, England/Science Photo Library/Photo Researchers, Inc.; **342 (bottom),** Martin Dohrn/Science Photo Library/Photo Researchers, Inc.; **344,** Pearson Asset Library/Pearson; **347 (top),** Pearson Asset Library/Pearson; **347 (bottom),** Pearson Asset Library/Pearson; **Fig. 11.6 (top),** iStockphoto.com; **Fig. 11.6 (bottom),** Rennie Van Munchow/Phototake, NYC; **349** International Atomic Energy Agency; **350,** ©Shirley Clive/Greenpeace International; **351,** iStockphoto.com; **352,** Roger Tully/Getty Images Inc.—Stone Allstock; **353,** iStockphoto.com; **355 (top),** Colorfoto Hans Hinz, Basel, Switzerland; **355 (bottom),** Courtesy of General Atomics.

Chapter 12: Opener, Corbis Royalty Free; **363,** Andy Levin/Photo Researchers, Inc; **379,** John McMurry; **386,** iStockphoto.com; **388,** iStockphoto.com.

Chapter 13: Opener, iStockphoto.com; **397,** iStockphoto.com; **406,** Omikron/Photo Researchers, Inc.; **Fig. 13.1,** Richard Megna/Fundamental Photographs, NYC; **411 (left and right),** Richard Megna/Fundamental Photographs, NYC; **418 (left and right),** Eric Schrader/Pearson; **419,** Michal Heron/Pearson; **420 (top),** Museum of Applied Arts, Helsinki, Finland; **420 (bottom),** Mitch Kezar, Getty Images Inc.—Stone Allstock; **421,** iStockphoto.com; **423,** Shutterstock; **427,** Michael Holford/Michael Holford Photographs.

Chapter 14: Opener, iStockphoto.com; **437,** Pearson Asset Library/Pearson; **438,** Pearson Asset Library/Pearson; **447,** Advanced Safety Devices, Inc.; **448,** iStockphoto.com; **450,** Pearson Asset Library/Pearson; **452,** Rod Planck/Photo Researchers, Inc.; **453,** iStockphoto.com; **454,** Corbis/Bettmann; **456,** Photo Researchers, Inc.; **457,** NASA/Goddard Space Flight Center.

Chapter 15: Opener, iStockphoto.com; **472,** Olivier Matthys/Landov LLC; **475,** iStockphoto.com; **479,** Donald Clegg and Roxy Wilson/Pearson; **480,** SuperStock, Inc.; **481,** Pearson Asset Library/Pearson; **482,** Allan Rosenberg/Getty Images, Inc.—PhotoDisc; **483,** iStockphoto.com.

Chapter 16: Opener (main), Charles S. Lewallen; **Opener (inset),** Thomas Eisner and Daniel Aneshansley, Cornell University; **493 (photo),** Paul Marek, Department of Biology, East Carolina University; **493 (drawing),** Adapted from *Introduction to Ecoloical Biochemistry* 2/e by J.B. Harbone with permission of Academic Press, Inc., San Diego; **495,** ©Gail Mooney/CORBIS All Rights Reserved; **496,** iStockphoto.com; **497 (top),** iStockphoto.com; **497 (bottom),** iStockphoto.com; **Fig. 16.2 (a)** and **(b),** Richard Megna/Fundamental Photographs; **495,** iStockphoto.com; **503,** iStockphoto.com.

Chapter 17: Opener, Konrad Wothe/Minden Pictures; **522,** iStockphoto.com; **530,** Eric Schrader/Pearson; **531,** Pearson Asset Library/Pearson; **533,** Alan Levenson/Getty Images Inc.—Stone Allstock; **Fig. 17.1,** The Granger Collection; **537,** iStockPhoto.com; **541,** Pearson Asset Library/Pearson; **542,** Michael Temchine/NewsCom.

Chapter 18: Opener, Pixtal/AGE Fotostock; **553,** Peter Ginter/Science Faction; **553 (top),** iStockphoto.com; **553 (bottom),** iStockphoto.com; **564,** Pearson Asset Library/Pearson; **567,** Visuals Unlimited; **569,** iStockphoto.com; **576 (top),** Shutterstock; **576 (bottom),** Shutterstock; **Fig. 18.7(b),** Kim M. Gernert/Pearson Education/PH College; **Fig. 18.7(c),** Ken Eward/Photo

Researchers, Inc.; **Fig. 18.7(d)**, Kim M. Gernert/Pearson Education/PH College; **579 (top)** and **(bottom)**, Phototake NYC; **Fig. 18.9(a)**, Pearson Asset Library/Pearson; **Fig. 18.9(b)**, Pearson Asset Library/Pearson; **582 (top)**, St. Mary's Hospital Medical School/Photo Researchers; **582 (bottom)**, NMSB/Custom Medical Stock Photo; **583**, iStockphoto.com.

Chapter 19: Opener, Phil Degginger/Alamy; **Fig. 19.1**, Richard Megna/Fundamental Photographs, NYC; **596**, Manuel C. Peitsch/Corbis/Bettmann; **597**, Manuel C. Peitsch/Corbis/Bettmann; **607**, iStockphoto; **613**, Abbott Laboratories; **614**, Ken Eward/Science Source/Photo Researchers; **615**, iStockphoto.com; **619**, iStockphoto.com.

Chapter 20: Opener, Luca Bruno/AP; **636**, Michal Heron/Pearson Education/PH College; **641**, Martin Shields/Alamy; **642**, Don W. Fawcett/Science Source/Photo Researchers, Inc.; **646**, Alan Sirulnikoff/Photo Researchers, Inc.; **647**, Shutterstock; **651**, Garden Raw Foods; **653**, Prasanna, M.D., Vondrasek, J., Wlodawer, A., Bhat, T.N., Application of InChI to curae, index and query 3-D structures. *PROTEINS. Structure, Function and Bioinformatics* 60, 1-4 (2005). (http://xpdb.nist.gov/hivsdb/hivsdb.html)

Chapter 21: Opener, Martin Harvey/Peter Arnold, Inc.; **659**, iStockphoto.com; **661**, iStockphoto.com; **663**, Al Giddings/Al Giddings Images, Inc.; **671**, iStockphoto.com; **673**, iStockphoto.com; **681 (a)** and **(b)**, Donald Clegg/Pearson Education/PH College; **Fig. 21.11(b)**, Manuel C. Peitsch/Corbis/Bettmann; **683**, Left-hand image: Clyde Gibbons, Martin C. Montgomery, Andrew G.W. Leslie & John E. Walker, The structure of the central stalk in bovine F1-ATPase at 2.4 × resolution in *Nature Structural Biology 7*, 1055-1061 (2000) Fig. 1a on page 1055. Right-hand image: Daniela Stock, Andrew G.W. Leslie, John E. Walker, Molecular Architecture of the Rotary Motor in AT Synthase, *Science* 26 November 1999: Vol. 286, no. 5445, pp. 1700-705 Fig. 2a (left) page 1702; **686**, iStockphoto.com.

Chapter 22: Opener, iStockphoto.com; **698**, iStockphoto.com; **704**, Shutterstock; **Fig. 22.4(a)**, Corbis Premium RF/Alamy; **Fig. 22.4(b)**, Eric Schrader/Pearson; **Fig. 22.4(c)**, iStockphoto.com; **713**, Eric Schrader/Pearson; **716 (top)**, Michael & Patricia Fogden/Minden Pictures; **716 (bottom)**, David Polack/Corbiss/Stock Market; **717**, Doug Allan/Oxford Scientific Films/Animals Animals/Earth Scenes; **720**, Larry Mulvehill/Science Source/Photo Researchers, Inc.; **722**, GlaxoSmithKline plc.

Chapter 23: Opener, Batista Moon/Shutterstock; **730**, Photo Lennart Nilsson/Albert Bonniers Forlag; **734**, Coordinates by T. Alber, G.A. Petsko

and E. Lolis; image by Molecular Graphics and Modelling, Duke University. Simon & Schuster/PH College; **736 (left, middle, and right)**, iStockphoto.com; **Fig. 23.4**, Andrea Mattevi and Wim G.J. Hol/Pearson Education/PH College; **745**, iStockphoto.com; **748**, Winslow Townson/AP; **751**, Benelux/Photo Researchers, Inc.

Chapter 24: Opener, iStockphoto.com; **761**, Marcel Mochet/NewsCom; **763**, Eric Schrader/Pearson; **765**, Frank Lane Picture Agency/Corbis/Bettmann; **767**, iStockphoto.com; **759**, C Squared Studios/Getty Images, Inc.—Photodisc; **762**, Royalty Free/CORBIS All Rights Reserved; **777**, Kristen Brochmann/Fundamental Photographs, NYC.

Chapter 25: Opener, Art Wolfe/Getty Images Inc.—Stone Allstock; **791**, SPL/Photo Researchers, Inc.; **794**, John Sholtis/Amgen Inc.; **798**, iStockphoto.com.

Chapter 26: Opener, Javier Larea/SuperStock, Inc.; **809**, Micrograph by Conly L. Rieder, Division of Molecular Medicine, Wadsworth Center, Albany, New York 12201-0509; **813**, Prof. K. Seddon & Dr. T. Evans, Queen's University, Belfast/Photo Researchers, Inc.; **820**, Centers for Disease Control; **Fig. 26.4(b)**, reproduced by permission from H.J. Kreigstein and D.S. Hogness, *Proceedings of the National Academy of Sciences* 71:136 (1974), page 137, Fig. 2; **Fig. 26.7(b)**, Ken Eward/Science Source/Photo Researchers, Inc.; **834**; Centers for Disease Control; **822**, AP Wide World Photos.

Chapter 27: Opener, James King-Holmes/Photo Researchers, Inc.; **841**, BSIP/Ermakorr/Photo Researchers, Inc.; **843**, SPL/Photo Researchers, Inc.; **844 (right)**, Shutterstock; **844 (left, top)**, Dinodia/The Image Works; **844 (left, bottom)**, Biophoto Associates/Photo Researchers, Inc.; **834**, NewsCom; **849**, Dr. Gopal Murti/Photo Researchers, Inc.; **852**, Sinclair Stammers/Photo Researchers, Inc.; **854**, Courtesy Syngenta; **856**, Wong Maye-e/AP.

Chapter 28: Opener, Shutterstock; **862**, Eric Schrader/Pearson; **867**, iStockphoto.com; **869**, Dr. P. Marazzi/Photo Researchers, Inc.; **870**, Photo Researchers, Inc.; **874**, Staff Sgt. Eric T. Sheler/U.S. Air Force.

Chapter 29: Opener, ©Andrew Brookes/CORBIS All Rights Reserved; **884 (top)**, Bryan F. Peterson/Corbis/Stock Market; **884 (bottom)**, Mark Burnett/Photo Researchers, Inc.; **889 (left)**, Biology Media/Science Source/Photo Researchers, Inc.; **889 (right)**, Photo Lennart Nilsson/Albert Bonniers Forlag; **891**, Volker Steger/Peter Arnold, Inc.; **Fig. 29.10**, Michal Heron/Pearson Education/PH College; **893**, Bill Longcore/Photo Researchers, Inc.; **897**, Colin Cuthbert/Photo Researchers, Inc.

Index

Functional Groups of Importance in Biochemical Molecules

Functional Group	Structure	Type of Biomolecule
Amino group	$-NH_3^+$, $-NH_2$	Alkaloids and neurotransmitters; amino acids and proteins (Sections 15.1, 15.3, 15.6, 18.3, 18.7, 20.6)
Hydroxyl group	$-OH$	Monosaccharides (carbohydrates) and glycerol: a component of triacylglycerols (lipids) (Sections 17.4, 22.4, 24.2)
Carbonyl group	$\overset{\displaystyle O}{\underset{\displaystyle \|}{-C-}}$	Monosaccharides (carbohydrates); in acetyl group (CH_3CO) used to transfer carbon atoms during catabolism (Sections 16.1, 17.4, 21.4, 21.8, 22.4)
Carboxyl group	$-\overset{O}{\overset{\|}{C}}-OH$, $-\overset{O}{\overset{\|}{C}}-O^-$	Amino acids, proteins, and fatty acids (lipids) (Sections 17.1, 18.3, 18.7, 24.2)
Amide group	$-\overset{O}{\overset{\|}{C}}-N-$	Links amino acids in proteins; formed by reaction of amino group and carboxyl group (Sections 17.1, 17.4, 18.7)
Carboxylic acid ester	$-\overset{O}{\overset{\|}{C}}-O-R$	Triacylglycerols (and other lipids); formed by reaction of carboxyl group and hydroxyl group (Sections 17.1, 17.4, 24.2)
Phosphates: mono-, di-, tri-	$-\overset{\|}{\underset{\|}{C}}-O-\overset{O}{\overset{\|}{\underset{\underset{O^-}{\|}}{P}}}-O^-$ $-\overset{\|}{\underset{\|}{C}}-O-\overset{O}{\overset{\|}{\underset{\underset{O^-}{\|}}{P}}}-O-\overset{O}{\overset{\|}{\underset{\underset{O^-}{\|}}{P}}}-O^-$ $-\overset{\|}{\underset{\|}{C}}-O-\overset{O}{\overset{\|}{\underset{\underset{O^-}{\|}}{P}}}-O-\overset{O}{\overset{\|}{\underset{\underset{O^-}{\|}}{P}}}-O-\overset{O}{\overset{\|}{\underset{\underset{O^-}{\|}}{P}}}-O^-$	ATP and many metabolism intermediates (Sections 17.8, 21.5, and throughout metabolism sections)
Hemiacetal group	$-\overset{\|}{\underset{\underset{OR}{\|}}{C}}-OH$	Cyclic forms of monosaccharides; formed by a reaction of carbonyl group with hydroxyl group (Sections 16.7, 22.4)
Acetal group	$-\overset{\|}{\underset{\underset{OR}{\|}}{C}}-OR$	Connects monosaccharides in disaccharides and larger carbohydrates; formed by reaction of carbonyl group with hydroxyl group (Sections 16.7, 22.7, 22.9)